BOOKS | À LA CARTE EDITION

D1072343

Blitzer

College Algebra
An Early Functions Approach
3e

YOUR TEXTBOOK—IN A BINDER-READY EDITION!

This unbound, three-hole punched version of your textbook
lets you take only what you need to class and incorporate your
own notes—all at an affordable price!

ISBN-13: 978-0-321-83786-8
ISBN-10: 0-321-83786-X

EAN

9 780321 837868

ALWAYS LEARNING

PEARSON

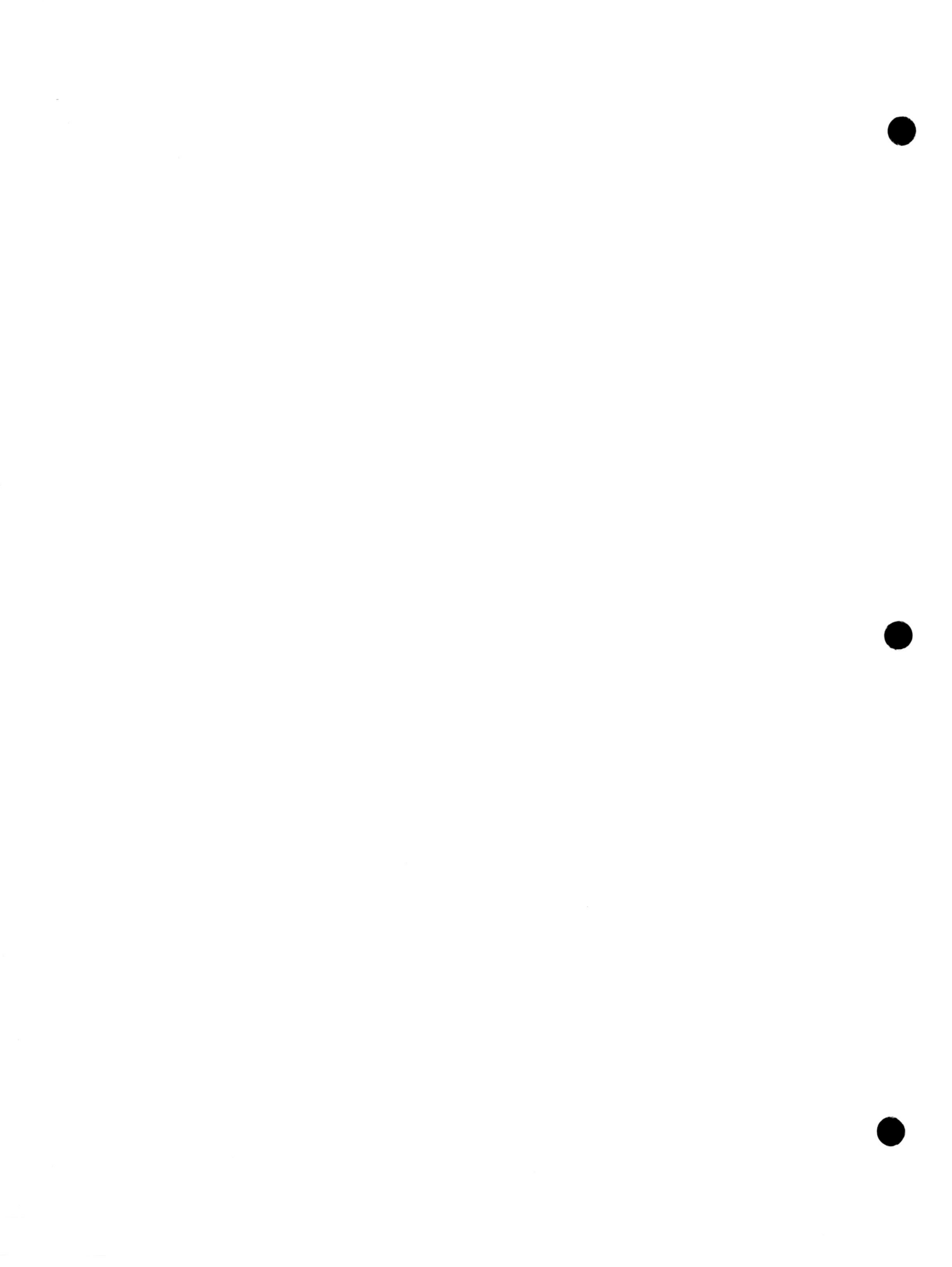

A Brief Guide to **Getting the Most** from this Book

1 Read the Book

Feature	Description	Benefit
Section-Opening Scenarios	Every section opens with a scenario presenting a unique application of algebra in your life outside the classroom.	Realizing that algebra is everywhere will help motivate your learning. **(See page 238.)**
Detailed Worked-Out Examples	Examples are clearly written and provide step-by-step solutions. No steps are omitted, and each step is thoroughly explained to the right of the mathematics.	The blue annotations will help you understand the solutions by providing the reason why every algebraic step is true. **(See page 240.)**
Applications Using Real-World Data	Interesting applications from nearly every discipline, supported by up-to-date real-world data, are included in every section.	Ever wondered how you'll use algebra? This feature will show you how algebra can solve real problems. **(See page 161.)**
Great Question!	NEW to this Edition. Answers to students' questions offer suggestions for problem solving, point out common errors to avoid, and provide informal hints and suggestions.	By seeing common mistakes, you'll be able to avoid them. This feature should help you not to feel anxious or threatened when asking questions in class. **(See page 26.)**
Explanatory Voice Balloons	Voice balloons help to demystify algebra. They translate algebraic language into plain English, clarify problem-solving procedures, and present alternative ways of understanding.	Does math ever look foreign to you? This feature often translates math into everyday English. **(See page 325.)**
Learning Objectives	Every section begins with a list of objectives. Each objective is restated in the margin where the objective is covered.	The objectives focus your reading by emphasizing what is most important and where to find it. **(See page 340.)**
Technology	The screens displayed in the technology boxes show how graphing utilities verify and visualize algebraic results.	Even if you are not using a graphing utility in the course, this feature will help you understand different approaches to problem solving. **(See page 363.)**

2 Work the Problems

Feature	Description	Benefit
Check Point Examples	Each example is followed by a similar matched problem, called a Check Point, that offers you the opportunity to work a similar exercise. The answers to the Check Points are provided in the answer section.	You learn best by doing. You'll solidify your understanding of worked examples if you try a similar problem right away to be sure you understand what you've just read. **(See page 184.)**
Concept and Vocabulary Checks	NEW to this Edition. These short-answer questions, mainly fill-in-the-blank and true/false items, assess your understanding of the definitions and concepts presented in each section.	It is difficult to learn algebra without knowing its special language. These exercises test your understanding of the vocabulary and concepts. **(See page 131.)**
Extensive and Varied Exercise Sets	An abundant collection of exercises is included in an Exercise Set at the end of each section. Exercises are organized within several categories. Your instructor will usually provide guidance on which exercises to work. The exercises in the first category, Practice Exercises, follow the same order as the section's worked examples.	The parallel order of the Practice Exercises lets you refer to the worked examples and use them as models for solving these problems. **(See page 370.)**
Practice Plus Problems	This category of exercises contains more challenging problems that often require you to combine several skills or concepts.	It is important to dig in and develop your problem-solving skills. Practice Plus Exercises provide you with ample opportunity to do so. **(See page 397.)**
Preview Problems	Each Exercise Set concludes with three problems to help you prepare for the next section.	These exercises let you review previously covered material that you'll need to be successful for the forthcoming section. Some of these problems will get you thinking about concepts you'll soon encounter. **(See page 208.)**

3 Review for Quizzes and Tests

Feature	Description	Benefit
Mid-Chapter Check Points	At approximately the midway point in the chapter, an integrated set of review exercises allows you to review the skills and concepts y*f*ou learned separately over several sections.	By combining exercises from the first half of the chapter, the Mid-Chapter Check Points give a comprehensive review before you move on to the material in the remainder of the chapter. **(See page 177.)**
Chapter Review Grids	Each chapter contains a review chart that summarizes the definitions and concepts in every section of the chapter. Examples that illustrate these key concepts are also referenced in the chart.	Review this chart and you'll know the most important material in the chapter! **(See page 593.)**
Chapter Review Exercises	A comprehensive collection of review exercises for each of the chapter's sections follows the grid.	Practice makes perfect. These exercises contain the most significant problems for each of the chapter's sections. **(See page 332.)**
Chapter Tests	Each chapter contains a practice test with approximately 25 problems that cover the important concepts in the chapter. Take the practice test, check your answers, and then watch the Chapter Test Prep Videos to see worked-out solutions for any exercises you miss.	You can use the chapter test to determine whether you have mastered the material covered in the chapter. **(See page 449.)**
Chapter Test Prep Videos	These videos contain worked-out solutions to every exercise in each chapter test and can be found in MyMathLab and on YouTube.	The videos let you review any exercises you miss on the chapter test.
Cumulative Review Exercises	Beginning with Chapter 2, each chapter concludes with a comprehensive collection of mixed cumulative review exercises. These exercises combine problems from previous chapters and the present chapter, providing an ongoing cumulative review.	Ever forget what you've learned? These exercises ensure that you are not forgetting anything as you move forward. **(See page 450.)**

COLLEGE ALGEBRA: AN EARLY FUNCTIONS APPROACH

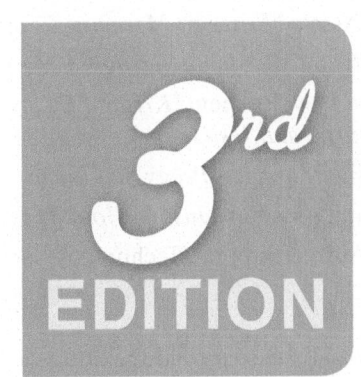

3rd EDITION

Robert Blitzer
Miami Dade College

PEARSON

Boston • Columbus • Indianapolis • New York • San Francisco • Upper Saddle River
Amsterdam • Cape Town • Dubai • London • Madrid • Milan • Munich • Paris • Montréal • Toronto
Delhi • Mexico City • São Paulo • Sydney • Hong Kong • Seoul • Singapore • Taipei • Tokyo

Editorial Director: *Christine Hoag*
Acquisitions Editor: *Katie O'Connor*
Senior Sponsoring Editor: *Dawn Murrin*
Assistant Editor: *Joseph Colella*
Senior Managing Editor: *Karen Wernholm*
Senior Production Project Manager: *Kathleen A. Manley*
Digital Assets Manager: *Marianne Groth*
Media Producer: *Shana Siegmund*
Software Development: *Kristina Evans, MathXL; and Mary Durnwald, TestGen*
Marketing Manager: *Peggy Sue Lucas*
Marketing Assistant: *Justine Goulart*
Senior Author Support/Technology Specialist: *Joe Vetere*

Rights and Permissions Advisor: *Cheryl Besenjak*
Image Manager: *Rachel Youdelman*
Procurement Manager: *Evelyn Beaton*
Procurement Specialist: *Debbie Rossi*
Senior Media Buyer: *Ginny Michaud*
Associate Director of Design: *Andrea Nix*
Art Direction and Cover Design: *Beth Paquin*
Text Design: *Lisa Kuhn, Curio Press, and Ellen Pettengell Design*
Production Coordination: *Rebecca Dunn/cMPreparé*
Composition: *cMPreparé*
Illustrations: *Scientific Illustrators/Laserwords*

For permission to use copyrighted material, grateful acknowledgment is made to the copyright holders on page C1, which is hereby made part of this copyright page.

Many of the designations used by manufacturers and sellers to distinguish their products are claimed as trademarks. Where those designations appear in this book, and Pearson Education was aware of a trademark claim, the designations have been printed in initial caps or all caps.

Library of Congress Cataloging-in-Publication Data

Blitzer, Robert.
College algebra: an early functions approach / Robert Blitzer. —3rd ed.
 p. cm.
Includes index.
ISBN 978-0-321-72964-4
1. Functions—Textbooks. 2. Algebra—Textbooks. I. Title.

QA331.3.B55 2014
512—dc23

2012023477

Copyright © 2014, 2010, 2007 Pearson Education, Inc.

All rights reserved. No part of this publication may be reproduced, stored in a retrieval system, or transmitted, in any form or by any means, electronic, mechanical, photocopying, recording, or otherwise, without the prior written permission of the publisher. Printed in the United States of America. For information on obtaining permission for use of material in this work, please submit a written request to Pearson Education, Inc., Rights and Contracts Department, 501 Boylston Street, Suite 900, Boston, MA 02116, fax your request to 617-671-3447, or e-mail at http://www.pearsoned.com/legal/permissions.htm.

6 7 8 9 10–DHB–16 15 14

www.pearsonhighered.com

ISBN 10: 0-321-72964-1
ISBN 13: 978-0-321-72964-4

DEDICATION

For Jerid (1985–2012)
And for those who have loved their pets
and have been loved by them

CONTENTS

P Prerequisites: Fundamental Concepts of Algebra 1

1 Functions and Graphs 105

8 Sequences, Induction, and Probability 715

PREFACE

I've written *College Algebra: An Early Functions Approach,* **Third Edition**, to help diverse students, with different backgrounds and future goals, to succeed. The book has three fundamental goals:

1. To help students acquire a solid foundation in algebra, preparing them for other courses such as calculus, business calculus, and finite mathematics.
2. To show students how algebra can model and solve authentic real-world problems.
3. To enable students to develop problem-solving skills, while fostering critical thinking, within an interesting setting.

One major obstacle in the way of achieving these goals is the fact that very few students actually read their textbook. This has been a regular source of frustration for me and for my colleagues in the classroom. Anecdotal evidence gathered over years highlights two basic reasons that students do not take advantage of their textbook:

- "I'll never use this information."
- "I can't follow the explanations."

I've written every page of the Third Edition with the intent of eliminating these two objections. The ideas and tools I've used to do so are described for the student in "A Brief Guide to Getting the Most from This Book," which appears at the front of the book.

How Does College Algebra: An Early Functions Approach Differ from College Algebra, Sixth Edition?

Functions, the central idea of the course, are introduced in Chapter 1. All prerequisite topics necessary for this early introduction of functions are reviewed in Chapter P, including interval notation, intersections and unions of intervals, using properties of equality to solve (basic) linear equations, solving a formula for one of its variables, solving equations using the square root property, and completing the square of a binomial. Equation-solving in Chapter 2 emphasizes how equations are related to functions and graphs. There are frequent connections among the real solutions of equations, an associated function's real zeros, and the *x*-intercepts of its graph. Chapters 3 through 8 are identical in *An Early Functions Approach* and the *Sixth Edition*, with the theme of functions integrated into each chapter's topics.

What's New in the Third Edition?

New Applications and Real-World Data. I'm on a constant search for data that can be used to illustrate unique algebraic applications. I researched hundreds of books, magazines, newspapers, almanacs, and online sites to prepare the Third Edition. Among the 108 worked-out examples and exercises based on new data sets, you'll find applications involving modeling blood-alcohol concentration (Section P.1), starting salaries for college graduates (Section 2.2), the world's vanishing tiger population (Section 3.2), and the year humans become immortal (Section 4.1).

Concept and Vocabulary Checks. The Third Edition contains 470 new short-answer exercises, mainly fill-in-the-blank and true/false items, that assess students' understanding of the definitions and concepts presented in each section. The Concept and Vocabulary Checks appear as separate features preceding the Exercise Sets.

Great Question! This feature takes the content of each Study Tip in the Second Edition and presents it in the context of a student question. Answers to questions offer suggestions for problem solving, point out common errors to avoid, and provide informal hints and suggestions. 'Great Question!' should draw students' attention and curiosity more than the 'Study Tips.' As a secondary benefit, this new feature should help students not to feel anxious or threatened when asking questions in class.

New Chapter-Opening and Section-Opening Scenarios. Every chapter and every section open with a scenario based on an application, many of which are unique to the Third Edition. These scenarios are revisited in the course of the chapter or section in one of the book's new examples, exercises, or discussions. The often humorous tone of these openers is intended to help fearful and reluctant students overcome their negative perceptions about math.

New Blitzer Bonuses. The Third Edition contains a variety of new but optional enrichment essays. Examples include "Using Algebra to Measure Blood-Alcohol Concentration" (Section P.1), "Seven Ways to Spend $1 Trillion" (Section P.3), "Addressing Stress Parabolically" (Section 3.1), "Five Things Scientists Learned from the Hubble Space Telescope" (Section 7.3), and "Ponzi Schemes and Geometric Sequences" (Section 8.3).

Sample Homework Assignments. Within each Exercise Set, I have chosen odd-numbered problems, primarily from the Practice Exercises, that can serve as sample homework assignments. These are indicated by a <u>blue underline</u> in the Annotated Instructor's Edition. Based on the goals and objectives of your course, you may wish to enrich each sample homework assignment with additional exercises from the other categories in the Exercise Set.

New Interactive Figures. These new figures bring mathematical concepts to life and are included in MyMathLab. Used as a lecture tool, the figures help engage students more fully and save the time spent drawing figures by hand. Questions pertaining to each figure are assignable in MyMathLab and reinforce active learning and critical thinking. Each figure has an accompanying Exploratory Exercise that encourages further study and can be used as a presentation tool or as an open-ended learning assignment.

What Content and Organizational Changes Have Been Made to the Third Edition?

- **Section P.1 (Algebraic Expressions, Mathematical Models, and Real Numbers)** contains a new essay, now called a Blitzer Bonus, on using algebra to measure blood-alcohol concentration. This Blitzer Bonus should set the stage for the book's engaging collection of unique applications.

- **Section P.7 (Rational Expressions)** presents a new example on excluding numbers from a rational expression with a trinomial denominator.

- **Section 1.6 (Transformations of Functions)** has a more thoroughly developed discussion of how stretching or shrinking changes a graph's shape.

- **Section 1.7 (Combinations of Functions; Composite Functions)** has a new example that ties in with the section opener (number of births and deaths in the United States) and illustrates an application of the algebra of functions.

- **Section 2.1 (Linear Equations and Rational Equations)** contains examples involving both inconsistent equations and identities.

- **Section 3.2 (Polynomial Functions and Their Graphs)** contains a new example on graphing $f(x) = -2(x - 1)^2(x + 2)$, a polynomial function whose equation is given in factored form.

- **Section 3.5 (Rational Functions and Their Graphs)** has a variety of exercises where students must factor to find vertical asymptotes or holes.

- **Section 3.6 (Polynomial and Rational Inequalities)** contains a new example on solving a polynomial inequality with irrational boundary points that requires the use of the quadratic formula.

- **Section 4.1 (Exponential Functions)** presents an intriguing new Blitzer Bonus on the year humans become immortal. The section also contains a new table clarifying interest plans in which interest is paid more than once a year.

- **Section 4.4 (Exponential and Logarithmic Equations)** has a new discussion (within the context of the Great Question! feature) on whether a negative number can belong to the solution set of a logarithmic equation.

- **Section 5.3 (Partial Fractions)** uses the Great Question! feature to include a discussion on speeding up the process of finding partial fraction decompositions.

What Familiar Features Have Been Retained in the Third Edition?

- **Detailed Worked-Out Examples.** Each worked example is titled, making clear the purpose of the example. Examples are clearly written and provide students with detailed step-by-step solutions. No steps are omitted and key steps are thoroughly explained to the right of the mathematics.

- **Explanatory Voice Balloons.** Voice balloons are used in a variety of ways to demystify mathematics. They translate algebraic ideas into everyday English, help clarify problem-solving procedures, present alternative ways of understanding concepts, and connect problem solving to concepts students have already learned.

- **Check Point Examples.** Each example is followed by a similar matched problem, called a Check Point, offering students the opportunity to test their understanding of the example by working a similar exercise. The answers to the Check Points are provided in the answer section.

- **Extensive and Varied Exercise Sets.** An abundant collection of exercises is included in an Exercise Set at the end of each section. Exercises are organized within eight category types: Practice Exercises, Practice Plus Exercises, Application Exercises, Writing in Mathematics, Technology Exercises, Critical Thinking Exercises, Group Exercises, and Preview Exercises. This format makes it easy to create well-rounded homework assignments. The order of the Practice Exercises is exactly the same as the order of the section's worked

examples. This parallel order enables students to refer to the titled examples and their detailed explanations to achieve success working the Practice Exercises.

- **Practice Plus Problems.** This category of exercises contains more challenging practice problems that often require students to combine several skills or concepts. With an average of ten Practice Plus problems per Exercise Set, instructors are provided with the option of creating assignments that take Practice Exercises to a more challenging level.

- **Mid-Chapter Check Points.** At approximately the midway point in each chapter, an integrated set of Review Exercises allows students to review and assimilate the skills and concepts they learned separately over several sections.

- **Graphing and Functions.** Graphing and functions are introduced in Chapter 1, with an integrated graphing functional approach emphasized throughout the book. Graphs and functions that model data appear in nearly every section and Exercise Set. Examples and exercises use graphs of functions to explore relationships between data and to provide ways of visualizing a problem's solution. Because functions are the core of this course, students are repeatedly shown how functions relate to equations and graphs.

- **Section Objectives.** Learning objectives are clearly stated at the beginning of each section. These objectives help students recognize and focus on the section's most important ideas. The objectives are restated in the margin at their point of use.

- **Integration of Technology Using Graphic and Numerical Approaches to Problems.** Side-by-side features in the Technology boxes connect algebraic solutions to graphic and numerical approaches to problems. Although the use of graphing utilities is optional, students can use the explanatory voice balloons to understand different approaches to problems even if they are not using a graphing utility in the course.

- **Chapter Summaries.** Each chapter contains a review chart that summarizes the definitions and concepts in every section of the chapter. Examples that illustrate these key concepts are also referenced in the chart.

- **End-of-Chapter Materials.** A comprehensive collection of Review Exercises for each of the chapter's sections follows the Summary. This is followed by a Chapter Test that enables students to test their understanding of the material covered in the chapter. Beginning with Chapter 2, each chapter concludes with a comprehensive collection of mixed Cumulative Review Exercises.

- **Discovery.** Discovery boxes, found throughout the text, encourage students to further explore algebraic concepts. These explorations are optional and their omission does not interfere with the continuity of the topic at hand.

I hope that my passion for teaching, as well as my respect for the diversity of students I have taught and learned from over the years, is apparent throughout this new edition. By connecting algebra to the whole spectrum of learning, it is my intent to show students that their world is profoundly mathematical, and indeed, π is in the sky.

Robert Blitzer

Acknowledgments

An enormous benefit of authoring a successful series is the broad-based feedback I receive from the students, dedicated users, and reviewers. Every change to this edition is the result of their thoughtful comments and suggestions. I would like to express my appreciation to all the reviewers, whose collective insights form the backbone of this revision. In particular, I would like to thank the following people for reviewing *College Algebra, Algebra and Trigonometry, Precalculus*, and *Trigonometry*.

Barnhill, Kayoko Yates, Clark College
Beaver, Timothy, Isothermal Community College
Becan, Jaromir, University of Texas-San Antonio
Benjelloun, Imad, Delaware Valley College
Best, Lloyd, Pacific Union College
Brown, Denise, Collin College, Spring Creek Campus
Burgin, Bill, Gaston College
Cabaniss, Jennifer, Central Texas College
Chang, Jimmy, St. Petersburg College
Colt, Diana, University of Minnesota-Duluth
Cornell, Shannon, Amarillo College
Davidson, Wendy, Georgia Perimeter College-Newton
Densmore, Donna, Bossier Parish Community College
Enegren, Disa, Rose State College
Erickson, Keith A., Georgia Gwinnett College
Fisher, Nancy, University of Alabama
Gerken, Donna, Miami Dade College
Glickman, Cynthia, Community College of Southern Nevada
Goel, Sudhir Kumar, Valdosta State University
Gordon, Donald, Manatee Community College
Gross, David L., University of Connecticut
Groves, Jason W., South Plains College
Haack, Joel K., University of Northern Iowa
Haefner, Jeremy, University of Colorado
Hague, Joyce, University of Wisconsin at River Falls
Hall, Mike, University of Mississippi
Hassani, Mahshid, Hillsborough Community College
Hayes, Tom, Montana State University
Hay-Jahans, Christopher N., University of South Dakota
Hernandez, Celeste, Richland College
Horney, Jo Beth, South Plains College
Howard, Heidi, Florida State College at Jacksonville, South Campus
Ihlow, Winfield A., SUNY College at Oswego
Johnson, Nancy Raye, Manatee Community College
Larue, Dennine, Fairmont State University
Leesburg, Mary, Manatee Community College
Lehman, Christine Heinecke, Purdue University North Central
Levichev, Alexander, Boston University
Lin, Zongzhu, Kansas State University
Marlin, Benjamin, Northwestern Oklahoma State University
Massey, Marilyn, Collin County Community College
McCarthy-Germaine, Yvelyne, University of New Orleans

Mertens, Owen, Missouri State University-Springfield
Miller, James, West Virginia University
Nega, Martha, Georgia Perimeter College-Decatur
Peterman, Shahla, University of Missouri St. Louis
Pharo, Debra A., Northwestern Michigan College
Phoenix, Gloria, North Carolina Agricultural and Technical State University
Platt, David, Front Range Community College
Pohjanpelto, Juha, Oregon State University
Quinlan, Brooke, Hillsborough Community College
Rech, Janice, University of Nebraska at Omaha
Rody, Joseph W., Arizona State University
Rouhani, Behnaz, Georgia Perimeter College-Dunwoody
Salmon, Judith, Fitchburg State University
Schramm, Michael, Indian River State College
Schultz, Cynthia, Illinois Valley Community College
Shelton, Pat, North Carolina Agricultural and Technical State University
Spillman, Caroline, Georgia Perimeter College-Clarkston
Stadler, Jonathan, Capital University
Stallworth, Franotis R., Gwinnett Technical College
Stump, Chris, Bethel College
Sykes, Scott, University of West Georgia
Townsend, Richard, North Carolina Central University
Trim, Pamela, Southwest Tennessee Community College
Turner, Chris, Arkansas State University
Van Lommel, Richard E., California State University-Sacramento
Van Peursem, Dan, University of South Dakota
Van Veldhuizen, Philip, University of Nevada at Reno
Weaver, Jeffrey, Baton Rouge Community College
Wheeler, Amanda, Amarillo College
White, David, The Victoria College
Wienckowski, Tracy, University of Buffalo

Additional acknowledgments are extended to Dan Miller and Kelly Barber for preparing the solutions manuals; Brad Davis for preparing the answer section, serving as accuracy checker, and writing the new learning guide; cMPreparé formatting team for the book's brilliant paging; Brian Morris and Kevin Morris at Scientific Illustrators for superbly illustrating the book; Sheila Norman, photo researcher, for obtaining the book's new photographs; Rebecca Dunn, project manager, and Kathleen Manley, production editor, whose collective talents kept every aspect of this complex project moving through its many stages.

I would like to thank my editors at Pearson, Katie O'Connor and Dawn Murrin, who, with the assistance of Joseph Colella, guided and coordinated the book from manuscript through production. Thanks to Beth Paquin for the beautiful covers and interior design. Finally, thanks to Peggy Lucas for her innovative marketing efforts and to the entire Pearson sales force for their confidence and enthusiasm about the book.

Robert Blitzer

DYNAMIC RESOURCES

MyMathLab® Online Course (access code required)

MyMathLab delivers **proven results** in helping individual students succeed. It provides **engaging experiences** that personalize, stimulate, and measure learning for each student. And it comes from a **trusted partner** with educational expertise and an eye on the future.

To learn more about how MyMathLab combines proven learning applications with powerful assessment, visit **www.mymathlab.com** or contact your Pearson representative.

In Blitzer's **MyMathLab®** course, you have access to the most cutting-edge, innovative study solutions proven to increase student success. Noteworthy features include the following:

Ready to Go Courses.

These new courses provide students with all the same great MyMathLab features that you're used to but make it easier for instructors to get started. Each course includes author-chosen, preassigned homework, integrated review questions, quizzes, and cumulative review exercises to make creating your course even simpler.

Interactive Figures.

These *Mathematica*-based figures make the figures from the text come alive. Used during a lecture, interactive figures engage students more fully and save time that would otherwise be spent drawing them by hand. Exercises pertaining to each interactive figure are assignable in MyMathLab to reinforce active learning, critical thinking, and conceptual reasoning.

Integrated Review.

Skill review quizzes are assignable throughout the course, testing students on prerequisite knowledge. From these quizzes, each student receives a personalized, just-in-time review assignment, allowing them to refresh forgotten concepts.

MathTalk Videos.

Engaging videos connect mathematics to real-life events and interesting applications. These fun, instructional videos show students that math is relevant to their daily lives and are assignable in MyMathLab.

Video Assessment.

Assignable MXL exercises are available for MathTalk videos to help students retain valuable information presented in the videos.

Section-Lecture Videos.

These videos provide lectures for each section of the text to help students review important concepts and procedures 24/7.

Concept and Vocabulary Check.

New and assignable in MyMathLab, these short-answer and fill-in-the blank exercises provide a quick check for understanding of concepts. These questions also test for reading comprehension before the student moves on to the exercises.

www.youtube.com/BlitzerColAlgEFA3e

Chapter Test Prep Videos.

Students can watch instructors work through step-by-step solutions to all the Chapter Test exercises from the textbook. These are available in MyMathLab and on YouTube.

Instructor Resources

Additional resources can be downloaded from www.pearsonhighered.com or hardcopy resources can be ordered from your sales representative.

TestGen.

Enables instructors to build, edit, print, and administer tests using a computerized bank of algorithmic questions developed to cover all the objectives of the text.

PowerPoint Lecture Slides.

Fully editable lecture slides that correlate to the textbook.

Instructor's Solutions Manual.

Fully worked solutions to all textbook exercises.

Mini Lecture Notes.

Additional examples and helpful teaching tips for each section.

Annotated Instructor's Edition.

Shorter answers are on the page beside the exercises. Longer answers are in the back of the text.

Student Resources

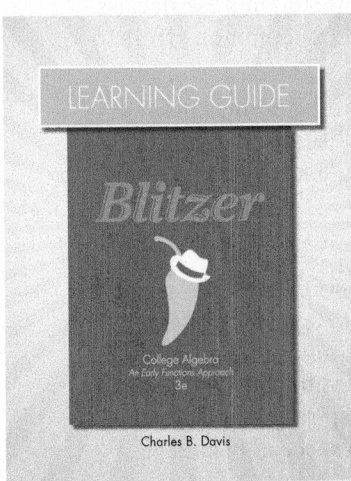

Additional resources to help student success are available to be packaged with the Blitzer textbook and MyMathLab access code.

Student's Solutions Manual.

Fully worked solutions to odd-numbered exercises and available to be packaged with the textbook.

Learning Guide.

This note-taking guide is organized by objective and begins each chapter with an engaging application, providing additional examples and exercises for students to work through for greater conceptual understanding and mastery of mathematical topics. The Learning Guide is available as PDFs and customizable Word files in MyMathLab. They can also be packaged with the textbook and MyMathLab access code.

TO THE STUDENT

The bar graph shows some of the qualities that students say make a great teacher. It was my goal to incorporate each of these qualities throughout the pages of this book.

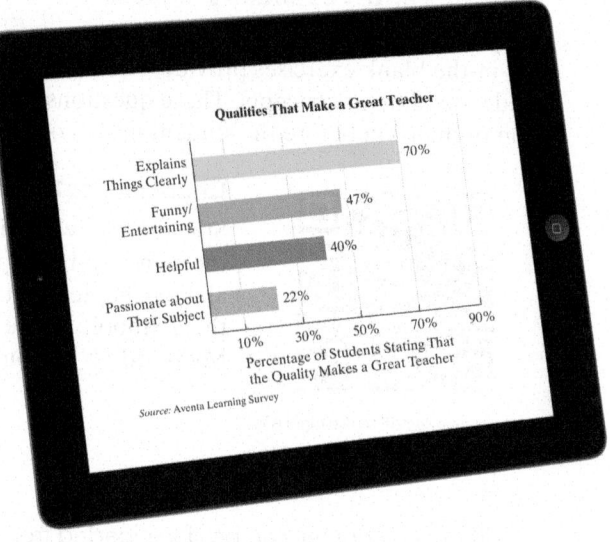

Explains Things Clearly

I understand that your primary purpose in reading *College Algebra: An Early Functions Approach* is to acquire a solid understanding of the required topics in your algebra course. In order to achieve this goal, I've carefully explained each topic. Important definitions and procedures are set off in boxes, and worked-out examples that present solutions in a step-by-step manner appear in every section. Each example is followed by a similar matched problem, called a Check Point, for you to try so that you can actively participate in the learning process as you read the book. (Answers to all Check Points appear in the back of the book.)

Funny & Entertaining

Who says that an algebra textbook can't be entertaining? From our quirky cover to the photos in the chapter and section openers, prepare to expect the unexpected. I hope some of the book's enrichment essays, called Blitzer Bonuses, will put a smile on your face from time to time.

Helpful

I designed the book's features to help you acquire knowledge of college algebra, as well as to show you how algebra can solve authentic problems that apply to your life. These helpful features include

- **Explanatory Voice Balloons:** Voice balloons are used in a variety of ways to make math less intimidating. They translate algebraic language into everyday English, help clarify problem-solving procedures, present alternative ways of understanding concepts, and connect new concepts to concepts you have already learned.

- **Great Question!:** The book's Great Question! boxes are based on questions students ask in class. The answers to these questions give suggestions for problem solving, point out common errors to avoid, and provide informal hints and suggestions.

- **Chapter Summaries:** Each chapter contains a review chart that summarizes the definitions and concepts in every section of the chapter. Examples from the chapter that illustrate these key concepts are also referenced in the chart. Review these summaries and you'll know the most important material in the chapter!

Passionate about the Subject

I passionately believe that no other discipline comes close to math in offering a more extensive set of tools for application and development of your mind. I wrote the book in Point Reyes National Seashore, 40 miles north of San Francisco. The park consists of 75,000 acres with miles of pristine surf-washed beaches, forested ridges, and bays bordered by white cliffs. It was my hope to convey the beauty and excitement of mathematics using nature's unspoiled beauty as a source of inspiration and creativity. Enjoy the pages that follow as you empower yourself with the algebra needed to succeed in college, your career, and your life.

Regards,

Bob

Robert Blitzer

ABOUT THE AUTHOR

Bob Blitzer is a native of Manhattan and received a Bachelor of Arts degree with dual majors in mathematics and psychology (minor: English literature) from the City College of New York. His unusual combination of academic interests led him toward a Master of Arts in mathematics from the University of Miami and a doctorate in behavioral sciences from Nova University. Bob's love for teaching mathematics was nourished for nearly 30 years at Miami Dade College, where he received numerous teaching awards, including Innovator of the Year from the League for Innovations in the Community College and an endowed chair based on excellence in the classroom. In addition to *College Algebra: An Early Functions Approach*, Bob has written textbooks covering introductory algebra, intermediate algebra, trigonometry, algebra and trigonometry, precalculus, and liberal arts mathematics, all published by Pearson. When not secluded in his Northern California writer's cabin, Bob can be found hiking the beaches and trails of Point Reyes National Seashore and tending to the chores required by his beloved entourage of horses, chickens, and irritable roosters.

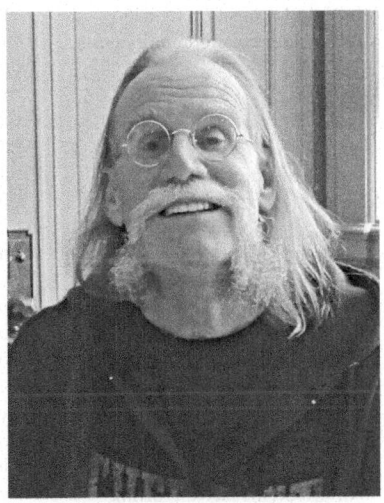

APPLICATIONS INDEX

PREREQUISITES:
FUNDAMENTAL CONCEPTS OF ALGEBRA

What can algebra possibly have to tell me about

- the skyrocketing cost of a college education?
- my workouts?
- the effects of alcohol?
- the meaning of the national debt that exceeds $15 trillion?
- time dilation on a futuristic high-speed journey to a nearby star?
- ethnic diversity in the United States?
- the widening imbalance between numbers of women and men on college campuses?

This chapter reviews fundamental concepts of algebra that are prerequisites for the study of college algebra. Throughout the chapter, you will see how the special language of algebra describes your world.

HERE'S WHERE YOU'LL FIND THESE APPLICATIONS:

College costs: Section P.1, Example 2; Exercise Set P.1, Exercises 129–130

Workouts: Exercise Set P.1, Exercises 127–128

The effects of alcohol: Blitzer Bonus on page 15

The national debt: Section P.3, Example 12

Time dilation: Blitzer Bonus on page 58

U.S. ethnic diversity: Chapter P Review, Exercise 35

College gender imbalance: Chapter P Test, Exercise 33.

Algebraic Expressions, Mathematical Models, and Real Numbers

Objectives

1 Evaluate algebraic expressions.

2 Use mathematical models.

3 Find the intersection of two sets.

4 Find the union of two sets.

5 Recognize subsets of the real numbers.

6 Use inequality symbols.

7 Use interval notation.

8 Find intersections and unions of intervals.

9 Evaluate absolute value.

10 Use absolute value to express distance.

How would your lifestyle change if a gallon of gas cost $9.15? Or if the price of a staple such as milk was $15? That's how much those products would cost if their prices had increased at the same rate college tuition has increased since 1980. (*Source:* Center for College Affordability and Productivity) In this section, you will learn how the special language of algebra describes your world, including the skyrocketing cost of a college education.

Algebraic Expressions

Algebra uses letters, such as x and y, to represent numbers. If a letter is used to represent various numbers, it is called a **variable**. For example, imagine that you are basking in the sun on the beach. We can let x represent the number of minutes that you can stay in the sun without burning with no sunscreen. With a number 6 sunscreen, exposure time without burning is six times as long, or 6 times x. This can be written $6 \cdot x$, but it is usually expressed as $6x$. Placing a number and a letter next to one another indicates multiplication.

Notice that $6x$ combines the number 6 and the variable x using the operation of multiplication. A combination of variables and numbers using the operations of addition, subtraction, multiplication, or division, as well as powers or roots, is called an **algebraic expression**. Here are some examples of algebraic expressions:

$$x + 6, \quad x - 6, \quad 6x, \quad \frac{x}{6}, \quad 3x + 5, \quad x^2 - 3, \quad \sqrt{x} + 7.$$

Many algebraic expressions involve *exponents*. For example, the algebraic expression

$$4x^2 + 341x + 3194$$

approximates the average cost of tuition and fees at public U.S. colleges for the school year ending x years after 2000. The expression x^2 means $x \cdot x$ and is read "x to the second power" or "x squared." The exponent, 2, indicates that the base, x, appears as a factor two times.

Exponential Notation

If n is a counting number (1, 2, 3, and so on),

$$\overset{\text{Exponent or Power}}{b^n} = \underbrace{b \cdot b \cdot b \cdots \cdots b.}_{\substack{b \text{ appears as a} \\ \text{factor } n \text{ times.}}}$$

b^n is read "the nth power of b" or "b to the nth power." Thus, the nth power of b is defined as the product of n factors of b. The expression b^n is called an **exponential expression**. Furthermore, $b^1 = b$.

For example,

$$8^2 = 8 \cdot 8 = 64, \quad 5^3 = 5 \cdot 5 \cdot 5 = 125, \quad \text{and} \quad 2^4 = 2 \cdot 2 \cdot 2 \cdot 2 = 16.$$

① Evaluate algebraic expressions.

Evaluating Algebraic Expressions

Evaluating an algebraic expression means to find the value of the expression for a given value of the variable.

Many algebraic expressions involve more than one operation. Evaluating an algebraic expression without a calculator involves carefully applying the following order of operations agreement:

The Order of Operations Agreement

1. Perform operations within the innermost parentheses and work outward. If the algebraic expression involves a fraction, treat the numerator and the denominator as if they were each enclosed in parentheses.
2. Evaluate all exponential expressions.
3. Perform multiplications and divisions **as they occur**, working **from left to right**.
4. Perform additions and subtractions **as they occur**, working **from left to right**.

EXAMPLE 1 Evaluating an Algebraic Expression

Evaluate $7 + 5(x - 4)^3$ for $x = 6$.

SOLUTION

$$7 + 5(x - 4)^3 = 7 + 5(6 - 4)^3 \quad \text{Replace x with 6.}$$
$$= 7 + 5(2)^3 \quad \text{First work inside parentheses: } 6 - 4 = 2.$$
$$= 7 + 5(8) \quad \text{Evaluate the exponential expression:}$$
$$\quad 2^3 = 2 \cdot 2 \cdot 2 = 8.$$
$$= 7 + 40 \quad \text{Multiply: } 5(8) = 40.$$
$$= 47 \quad \text{Add.} \qquad \bullet\bullet\bullet$$

✓ **Check Point 1** Evaluate $8 + 6(x - 3)^2$ for $x = 13$.

② Use mathematical models.

Formulas and Mathematical Models

An **equation** is formed when an equal sign is placed between two algebraic expressions. One aim of algebra is to provide a compact, symbolic description of the world. These descriptions involve the use of *formulas*. A **formula** is an equation that uses variables to express a relationship between two or more quantities.

Here are two examples of formulas related to heart rate and exercise.

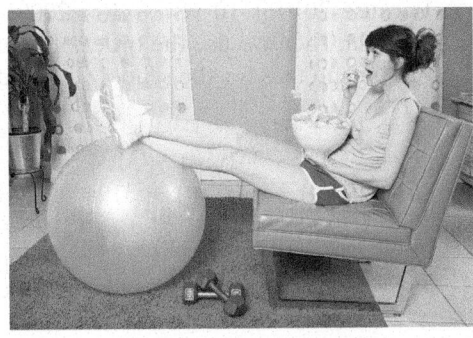

Couch-Potato Exercise

$$H = \tfrac{1}{5}(220 - a)$$

| Heart rate, in beats per minute, | is | $\frac{1}{5}$ of | the difference between 220 and your age. |

Working It

$$H = \tfrac{9}{10}(220 - a)$$

| Heart rate, in beats per minute, | is | $\frac{9}{10}$ of | the difference between 220 and your age. |

The process of finding formulas to describe real-world phenomena is called **mathematical modeling**. Such formulas, together with the meaning assigned to the variables, are called **mathematical models**. We often say that these formulas model, or describe, the relationships among the variables.

EXAMPLE 2 Modeling the Cost of Attending a Public College

The bar graph in **Figure P.1** shows the average cost of tuition and fees for public four-year colleges, adjusted for inflation. The formula

$$T = 4x^2 + 341x + 3194$$

models the average cost of tuition and fees, T, for public U.S. colleges for the school year ending x years after 2000.

 a. Use the formula to find the average cost of tuition and fees at public U.S. colleges for the school year ending in 2010.

 b. By how much does the formula underestimate or overestimate the actual cost shown in **Figure P.1**?

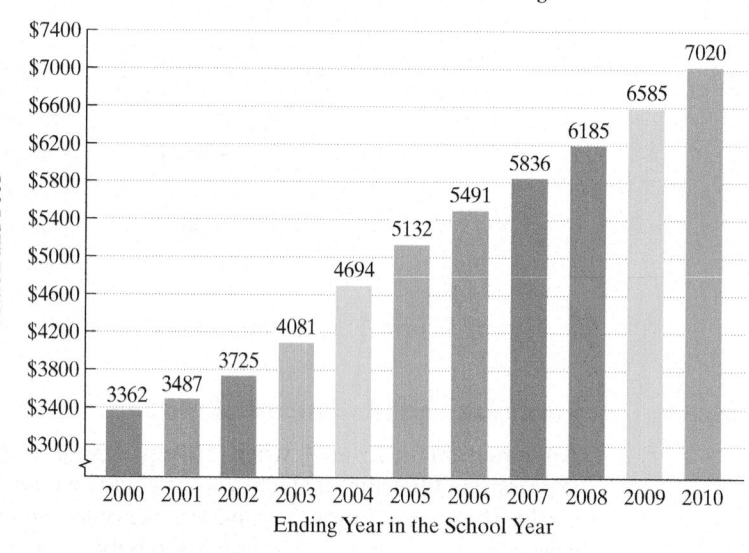

Average Cost of Tuition and Fees at Public Four-Year United States Colleges

Tuition and Fees: 2000: 3362; 2001: 3487; 2002: 3725; 2003: 4081; 2004: 4694; 2005: 5132; 2006: 5491; 2007: 5836; 2008: 6185; 2009: 6585; 2010: 7020

Ending Year in the School Year

FIGURE P.1
Source: The College Board

SOLUTION

 a. Because 2010 is 10 years after 2000, we substitute 10 for x in the given formula. Then we use the order of operations to find T, the average cost of tuition and fees for the school year ending in 2010.

$T = 4x^2 + 341x + 3194$	This is the given mathematical model.
$T = 4(10)^2 + 341(10) + 3194$	Replace each occurrence of x with 10.
$T = 4(100) + 341(10) + 3194$	Evaluate the exponential expression: $10^2 = 10 \cdot 10 = 100$.
$T = 400 + 3410 + 3194$	Multiply from left to right: 4(100) = 400 and 341(10) = 3410.
$T = 7004$	Add.

The formula indicates that for the school year ending in 2010, the average cost of tuition and fees at public U.S. colleges was $7004.

 b. **Figure P.1** shows that the average cost of tuition and fees for the school year ending in 2010 was $7020.

The cost obtained from the formula, $7004, underestimates the actual data value by $7020 − $7004, or by $16.

• • •

 Check Point **2** Assuming trends indicated by the data in **Figure P.1** continue, use the formula $T = 4x^2 + 341x + 3194$, described in Example 2, to project the average cost of tuition and fees at public U.S. colleges for the school year ending in 2015.

Sometimes a mathematical model gives an estimate that is not a good approximation or is extended to include values of the variable that do not make sense. In these cases, we say that **model breakdown** has occurred. For example, it is not likely that the formula in Example 2 would give a good estimate of tuition and fees in 2050 because it is too far in the future. Thus, model breakdown would occur.

Sets

Before we describe the set of real numbers, let's be sure you are familiar with some basic ideas about sets. A **set** is a collection of objects whose contents can be clearly determined. The objects in a set are called the **elements** of the set. For example, the set of numbers used for counting can be represented by

$$\{1, 2, 3, 4, 5, \dots\}.$$

The braces, { }, indicate that we are representing a set. This form of representation, called the **roster method**, uses commas to separate the elements of the set. The symbol consisting of three dots after the 5, called an *ellipsis*, indicates that there is no final element and that the listing goes on forever.

A set can also be written in **set-builder notation**. In this notation, the elements of the set are described but not listed. Here is an example:

$$\{x \mid x \text{ is a counting number less than 6}\}.$$

The set of all x such that x is a counting number less than 6.

The same set written using the roster method is

$$\{1, 2, 3, 4, 5\}.$$

If A and B are sets, we can form a new set consisting of all elements that are in both A and B. This set is called the *intersection* of the two sets.

Definition of the Intersection of Sets

The **intersection** of sets A and B, written $A \cap B$, is the set of elements common to both set A **and** set B. This definition can be expressed in set-builder notation as follows:

$$A \cap B = \{x \mid x \text{ is an element of } A \text{ AND } x \text{ is an element of } B\}.$$

Figure P.2 shows a useful way of picturing the intersection of sets A and B. The figure indicates that $A \cap B$ contains those elements that belong to both A and B at the same time.

EXAMPLE 3 Finding the Intersection of Two Sets

Find the intersection: $\{7, 8, 9, 10, 11\} \cap \{6, 8, 10, 12\}$.

SOLUTION

The elements common to $\{7, 8, 9, 10, 11\}$ and $\{6, 8, 10, 12\}$ are 8 and 10. Thus,

$$\{7, 8, 9, 10, 11\} \cap \{6, 8, 10, 12\} = \{8, 10\}.$$ • • •

 Check Point **3** Find the intersection: $\{3, 4, 5, 6, 7\} \cap \{3, 7, 8, 9\}$.

GREAT QUESTION!

Can I use symbols other than braces when writing sets using the roster method?

No. Grouping symbols such as parentheses, (), and square brackets, [], are not used to represent sets in the roster method. Furthermore, only commas are used to separate the elements of a set. Separators such as colons or semicolons are not used.

③ Find the intersection of two sets.

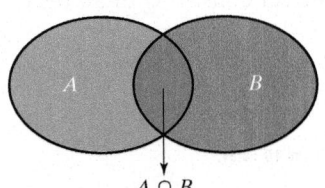

$A \cap B$

FIGURE P.2 Picturing the intersection of two sets

If a set has no elements, it is called the **empty set**, or the **null set**, and is represented by the symbol \varnothing (the Greek letter phi). Here is an example that shows how the empty set can result when finding the intersection of two sets:

$$\{2, 4, 6\} \cap \{3, 5, 7\} = \varnothing.$$

These sets have no common elements.

Their intersection has no elements and is the empty set.

④ Find the union of two sets.

Another set that we can form from sets A and B consists of elements that are in A or B or in both sets. This set is called the *union* of the two sets.

Definition of the Union of Sets

The **union** of sets A and B, written $A \cup B$, is the set of elements that are members of set A **or** of set B or of both sets. This definition can be expressed in set-builder notation as follows:

$$A \cup B = \{x \mid x \text{ is an element of } A \text{ OR } x \text{ is an element of } B\}.$$

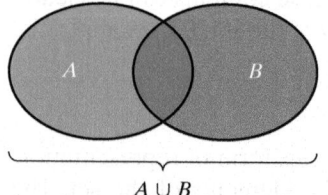

$A \cup B$

FIGURE P.3 Picturing the union of two sets

Figure P.3 shows a useful way of picturing the union of sets A and B. The figure indicates that $A \cup B$ is formed by joining the sets together.

We can find the union of set A and set B by listing the elements of set A. Then we include any elements of set B that have not already been listed. Enclose all elements that are listed with braces. This shows that the union of two sets is also a set.

GREAT QUESTION!

How can I use the words *union* and *intersection* to help me distinguish between these two operations?

Union, as in a marriage union, suggests joining things, or uniting them. Intersection, as in the intersection of two crossing streets, brings to mind the area common to both, suggesting things that overlap.

EXAMPLE 4 Finding the Union of Two Sets

Find the union: $\{7, 8, 9, 10, 11\} \cup \{6, 8, 10, 12\}$.

SOLUTION

To find $\{7, 8, 9, 10, 11\} \cup \{6, 8, 10, 12\}$, start by listing all the elements from the first set, namely, 7, 8, 9, 10, and 11. Now list all the elements from the second set that are not in the first set, namely, 6 and 12. The union is the set consisting of all these elements. Thus,

$$\{7, 8, 9, 10, 11\} \cup \{6, 8, 10, 12\} = \{6, 7, 8, 9, 10, 11, 12\}.$$

Although 8 and 10 appear in both sets, do not list 8 and 10 twice.

● ● ●

✓ Check Point **4** Find the union: $\{3, 4, 5, 6, 7\} \cup \{3, 7, 8, 9\}$.

⑤ Recognize subsets of the real numbers.

The Set of Real Numbers

The sets that make up the real numbers are summarized in **Table P.1** at the top of the next page. We refer to these sets as **subsets** of the real numbers, meaning that all elements in each subset are also elements in the set of real numbers.

Notice the use of the symbol \approx in the examples of irrational numbers. The symbol means "is approximately equal to." Thus,

$$\sqrt{2} \approx 1.414214.$$

TECHNOLOGY

A calculator with a square root key gives a decimal approximation for $\sqrt{2}$, not the exact value.

We can verify that this is only an approximation by multiplying 1.414214 by itself. The product is very close to, but not exactly, 2:

$$1.414214 \times 1.414214 = 2.000001237796.$$

Table P.1 Important Subsets of the Real Numbers

Name/Symbol	Description	Examples	
Natural numbers \mathbb{N}	$\{1, 2, 3, 4, 5, \ldots\}$ These are the numbers that we use for counting.	$2, 3, 5, 17$	
Whole numbers \mathbb{W}	$\{0, 1, 2, 3, 4, 5, \ldots\}$ The set of whole numbers includes 0 and the natural numbers.	$0, 2, 3, 5, 17$	
Integers \mathbb{Z}	$\{\ldots, -5, -4, -3, -2, -1, 0, 1, 2, 3, 4, 5, \ldots\}$ The set of integers includes the negatives of the natural numbers and the whole numbers.	$-17, -5, -3, -2, 0, 2, 3, 5, 17$	
Rational numbers \mathbb{Q}	$\left\{ \dfrac{a}{b} \,\middle	\, a \text{ and } b \text{ are integers and } b \neq 0 \right\}$ This means that b is not equal to zero. The set of rational numbers is the set of all numbers that can be expressed as a quotient of two integers, with the denominator not 0. Rational numbers can be expressed as terminating or repeating decimals.	$-17 = \frac{-17}{1}, -5 = \frac{-5}{1}, -3, -2,$ $0, 2, 3, 5, 17,$ $\frac{2}{5} = 0.4,$ $\frac{-2}{3} = -0.6666\ldots = -0.\overline{6}$
Irrational numbers \mathbb{I}	The set of irrational numbers is the set of all numbers whose decimal representations are neither terminating nor repeating. Irrational numbers cannot be expressed as a quotient of integers.	$\sqrt{2} \approx 1.414214$ $-\sqrt{3} \approx -1.73205$ $\pi \approx 3.142$ $-\frac{\pi}{2} \approx -1.571$	

FIGURE P.4 Every real number is either rational or irrational.

Not all square roots are irrational. For example, $\sqrt{25} = 5$ because $5^2 = 5 \cdot 5 = 25$. Thus, $\sqrt{25}$ is a natural number, a whole number, an integer, and a rational number $\left(\sqrt{25} = \frac{5}{1}\right)$.

The set of *real numbers* is formed by taking the union of the sets of rational numbers and irrational numbers. Thus, every real number is either rational or irrational, as shown in **Figure P.4**.

Real Numbers

The set of **real numbers** is the set of numbers that are either rational or irrational:

$$\{x \,|\, x \text{ is rational or } x \text{ is irrational}\}.$$

The symbol \mathbb{R} is used to represent the set of real numbers. Thus,

$$\mathbb{R} = \{x \,|\, x \text{ is rational}\} \cup \{x \,|\, x \text{ is irrational}\}.$$

EXAMPLE 5 Recognizing Subsets of the Real Numbers

Consider the following set of numbers:

$$\left\{ -7, -\frac{3}{4}, 0, 0.\overline{6}, \sqrt{5}, \pi, 7.3, \sqrt{81} \right\}.$$

List the numbers in the set that are

a. natural numbers. **b.** whole numbers. **c.** integers.

d. rational numbers. **e.** irrational numbers. **f.** real numbers.

SOLUTION

a. Natural numbers: The natural numbers are the numbers used for counting. The only natural number in the set $\left\{ -7, -\frac{3}{4}, 0, 0.\overline{6}, \sqrt{5}, \pi, 7.3, \sqrt{81} \right\}$ is $\sqrt{81}$ because $\sqrt{81} = 9$. (9 multiplied by itself, or 9^2, is 81.)

b. Whole numbers: The whole numbers consist of the natural numbers and 0. The elements of the set $\left\{-7, -\frac{3}{4}, 0, 0.\overline{6}, \sqrt{5}, \pi, 7.3, \sqrt{81}\right\}$ that are whole numbers are 0 and $\sqrt{81}$.

c. Integers: The integers consist of the natural numbers, 0, and the negatives of the natural numbers. The elements of the set $\left\{-7, -\frac{3}{4}, 0, 0.\overline{6}, \sqrt{5}, \pi, 7.3, \sqrt{81}\right\}$ that are integers are $\sqrt{81}$, 0, and -7.

d. Rational numbers: All numbers in the set $\left\{-7, -\frac{3}{4}, 0, 0.\overline{6}, \sqrt{5}, \pi, 7.3, \sqrt{81}\right\}$ that can be expressed as the quotient of integers are rational numbers. These include $-7\left(-7 = \frac{-7}{1}\right), -\frac{3}{4}, 0\left(0 = \frac{0}{1}\right)$, and $\sqrt{81}\left(\sqrt{81} = \frac{9}{1}\right)$. Furthermore, all numbers in the set that are terminating or repeating decimals are also rational numbers. These include $0.\overline{6}$ and 7.3.

e. Irrational numbers: The irrational numbers in the set $\left\{-7, -\frac{3}{4}, 0, 0.\overline{6}, \sqrt{5}, \pi, 7.3, \sqrt{81}\right\}$ are $\sqrt{5}\left(\sqrt{5} \approx 2.236\right)$ and $\pi(\pi \approx 3.14)$. Both $\sqrt{5}$ and π are only approximately equal to 2.236 and 3.14, respectively. In decimal form, $\sqrt{5}$ and π neither terminate nor have blocks of repeating digits.

f. Real numbers: All the numbers in the given set $\left\{-7, -\frac{3}{4}, 0, 0.\overline{6}, \sqrt{5}, \pi, 7.3, \sqrt{81}\right\}$ are real numbers. • • •

☑ Check Point **5** Consider the following set of numbers:

$$\left\{-9, -1.3, 0, 0.\overline{3}, \frac{\pi}{2}, \sqrt{9}, \sqrt{10}\right\}.$$

List the numbers in the set that are

a. natural numbers. **b.** whole numbers. **c.** integers.

d. rational numbers. **e.** irrational numbers. **f.** real numbers.

The Real Number Line

The **real number line** is a graph used to represent the set of real numbers. An arbitrary point, called the **origin**, is labeled 0. Select a point to the right of 0 and label it 1. The distance from 0 to 1 is called the **unit distance**. Numbers to the right of the origin are **positive** and numbers to the left of the origin are **negative**. The real number line is shown in **Figure P.5**.

FIGURE P.5 The real number line

GREAT QUESTION!

How did you locate $\sqrt{2}$ as a precise point on the number line in Figure P.6?

We used a right triangle with two legs of length 1. The remaining side has a length measuring $\sqrt{2}$.

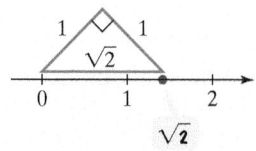

We'll have lots more to say about right triangles later in the book.

Real numbers are **graphed** on a number line by placing a dot at the correct location for each number. The integers are easiest to locate. In **Figure P.6**, we've graphed six rational numbers and three irrational numbers on a real number line.

FIGURE P.6 Graphing numbers on a real number line

Every real number corresponds to a point on the number line and every point on the number line corresponds to a real number. We say that there is a **one-to-one correspondence** between all the real numbers and all points on a real number line.

⑥ Use inequality symbols.

Ordering the Real Numbers

On the real number line, the real numbers increase from left to right. The lesser of two real numbers is the one farther to the left on a number line. The greater of two real numbers is the one farther to the right on a number line.

Look at the number line in **Figure P.7**. The integers −4 and −1 are graphed.

$$-5 \quad -4 \quad -3 \quad -2 \quad -1 \quad 0 \quad 1 \quad 2 \quad 3 \quad 4 \quad 5$$ FIGURE P.7

Observe that −4 is to the left of −1 on the number line. This means that −4 is less than −1.

$$-4 < -1$$ −4 is less than −1 because −4 is to the **left** of −1 on the number line.

In **Figure P.7**, we can also observe that −1 is to the right of −4 on the number line. This means that −1 is greater than −4.

$$-1 > -4$$ −1 is greater than −4 because −1 is to the **right** of −4 on the number line.

The symbols $<$ and $>$ are called **inequality symbols**. These symbols always point to the lesser of the two real numbers when the inequality statement is true.

−4 is less than −1. $-4 < -1$ The symbol points to −4, the lesser number.

−1 is greater than −4. $-1 > -4$ The symbol still points to −4, the lesser number.

The symbols $<$ and $>$ may be combined with an equal sign, as shown in the following table:

	Symbols	Meaning	Examples	Explanation
This inequality is true if either the < part or the = part is true.	$a \leq b$	a is less than or equal to b.	$2 \leq 9$ $9 \leq 9$	Because $2 < 9$ Because $9 = 9$
This inequality is true if either the > part or the = part is true.	$b \geq a$	b is greater than or equal to a.	$9 \geq 2$ $2 \geq 2$	Because $9 > 2$ Because $2 = 2$

⑦ Use interval notation.

Interval Notation

Some sets of real numbers can be represented using **interval notation**. Suppose that a and b are two real numbers such that $a < b$.

Interval Notation	Graph
The **open interval** (a, b) represents the set of real numbers between, but not including, a and b. $$(a, b) = \{x \mid a < x < b\}$$ *x is greater than a (a < x) and x is less than b (x < b).*	 *The parentheses in the graph and in interval notation indicate that a and b, the endpoints, are excluded from the interval.*
The **closed interval** $[a, b]$ represents the set of real numbers between, and including, a and b. $$[a, b] = \{x \mid a \le x \le b\}$$ *x is greater than or equal to a (a ≤ x) and x is less than or equal to b (x ≤ b).*	 *The square brackets in the graph and in interval notation indicate that a and b, the endpoints, are included in the interval.*
The **infinite interval** (a, ∞) represents the set of real numbers that are greater than a. $$(a, \infty) = \{x \mid x > a\}$$ *The infinity symbol does not represent a real number. It indicates that the interval extends indefinitely to the right.*	 *The parenthesis indicates that a is excluded from the interval.*
The **infinite interval** $(-\infty, b]$ represents the set of real numbers that are less than or equal to b. $$(-\infty, b] = \{x \mid x \le b\}$$ *The negative infinity symbol indicates that the interval extends indefinitely to the left.*	 *The square bracket indicates that b is included in the interval.*

Parentheses and Brackets in Interval Notation

Parentheses indicate endpoints that are not included in an interval. Square brackets indicate endpoints that are included in an interval. Parentheses are always used with ∞ or $-\infty$.

Table P.2 lists nine possible types of intervals used to describe subsets of real numbers.

Table P.2 Intervals on the Real Number Line

Let a and b be real numbers such that $a < b$.		
Interval Notation	**Set-Builder Notation**	**Graph**
(a, b)	$\{x \mid a < x < b\}$	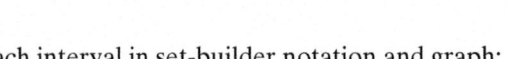
$[a, b]$	$\{x \mid a \le x \le b\}$	
$[a, b)$	$\{x \mid a \le x < b\}$	
$(a, b]$	$\{x \mid a < x \le b\}$	
(a, ∞)	$\{x \mid x > a\}$	
$[a, \infty)$	$\{x \mid x \ge a\}$	
$(-\infty, b)$	$\{x \mid x < b\}$	
$(-\infty, b]$	$\{x \mid x \le b\}$	
$(-\infty, \infty)$	$\{x \mid x$ is a real number$\}$ or \mathbb{R} (set of real numbers)	

EXAMPLE 6 Using Interval Notation

Express each interval in set-builder notation and graph:

 a. $(-1, 4]$ **b.** $[2.5, 4]$ **c.** $(-4, \infty)$.

SOLUTION

 a. $(-1, 4] = \{x \mid -1 < x \le 4\}$

 b. $[2.5, 4] = \{x \mid 2.5 \le x \le 4\}$

 c. $(-4, \infty) = \{x \mid x > -4\}$

• • •

✓ **Check Point 6** Express each interval in set-builder notation and graph:

 a. $[-2, 5)$ **b.** $[1, 3.5]$ **c.** $(-\infty, -1)$.

⑧ Find intersections and unions of intervals.

Intersections and Unions of Intervals

Graphs are helpful in finding intersections and unions of intervals.

> **Finding Intersections and Unions of Two Intervals**
>
> **1.** Graph each interval on a number line.
> **2. a.** To find the intersection, take the portion of the number line that the two graphs have in common.
> **b.** To find the union, take the portion of the number line representing the total collection of numbers in the two graphs.

EXAMPLE 7 Finding Intersections and Unions of Intervals

Use graphs to find each set:

 a. $(1, 4) \cap [2, 8]$ **b.** $(1, 4) \cup [2, 8]$.

SOLUTION

a. $(1, 4) \cap [2, 8]$, the intersection of the intervals $(1, 4)$ and $[2, 8]$, consists of the numbers that are in both intervals.

Graph of $(1, 4)$: $\{x \mid 1 < x < 4\}$

Graph of $[2, 8]$: $\{x \mid 2 \le x \le 8\}$

To find $(1, 4) \cap [2, 8]$, take the portion of the number line that the two graphs have in common.

Numbers in both
$(1, 4)$ and $[2, 8]$: $\{x \mid 1 < x < 4$ and $2 \le x \le 8\}$

> The numbers common to both intervals are those that are greater than or equal to 2 and less than 4: [2, 4).

Thus, $(1, 4) \cap [2, 8] = [2, 4)$.

b. $(1, 4) \cup [2, 8]$, the union of the intervals $(1, 4)$ and $[2, 8]$, consists of the numbers that are in either one interval or the other (or both).

Graph of $(1, 4)$: $\{x \mid 1 < x < 4\}$

Graph of $[2, 8]$: $\{x \mid 2 \le x \le 8\}$

To find $(1, 4) \cup [2, 8]$, take the portion of the number line representing the total collection of numbers in the two graphs.

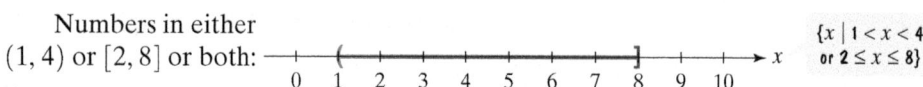

Numbers in either
$(1, 4)$ or $[2, 8]$ or both: $\{x \mid 1 < x < 4$ or $2 \le x \le 8\}$

> The numbers in either one interval or the other (or both) are those that are greater than 1 and less than or equal to 8: (1, 8].

Thus, $(1, 4) \cup [2, 8] = (1, 8]$.　　• • •

✓ Check Point **7** Use graphs to find each set:

　　a. $[1, 3] \cap (2, 6)$　　　　**b.** $[1, 3] \cup (2, 6)$.

⑨ Evaluate absolute value.

Absolute Value

The **absolute value** of a real number a, denoted by $|a|$, is the distance from 0 to a on the number line. This distance is always taken to be nonnegative. For example, the real number line in **Figure P.8** shows that

$$|-3| = 3 \quad \text{and} \quad |5| = 5.$$

The absolute value of -3 is 3 because -3 is 3 units from 0 on the number line. The absolute value of 5 is 5 because 5 is 5 units from 0 on the number line. The absolute value of a positive real number or 0 is the number itself. The absolute value of a negative real number, such as -3, is the number without the negative sign.

FIGURE P.8 Absolute value as the distance from 0

We can define the absolute value of the real number x without referring to a number line. The algebraic definition of the absolute value of x is given as follows:

Definition of Absolute Value

$$|x| = \begin{cases} x & \text{if } x \geq 0 \\ -x & \text{if } x < 0 \end{cases}$$

If x is nonnegative (that is, $x \geq 0$), the absolute value of x is the number itself. For example,

$$|5| = 5 \qquad |\pi| = \pi \qquad \left|\frac{1}{3}\right| = \frac{1}{3} \qquad |0| = 0.$$

Zero is the only number whose absolute value is 0.

If x is a negative number (that is, $x < 0$), the absolute value of x is the opposite of x. This makes the absolute value positive. For example,

$$|-3| = -(-3) = 3 \qquad |-\pi| = -(-\pi) = \pi \qquad \left|-\frac{1}{3}\right| = -\left(-\frac{1}{3}\right) = \frac{1}{3}.$$

This middle step is usually omitted.

EXAMPLE 8 Evaluating Absolute Value

Rewrite each expression without absolute value bars:

 a. $|\sqrt{3} - 1|$ **b.** $|2 - \pi|$ **c.** $\dfrac{|x|}{x}$ if $x < 0$.

SOLUTION

a. Because $\sqrt{3} \approx 1.7$, the number inside the absolute value bars, $\sqrt{3} - 1$, is positive. The absolute value of a positive number is the number itself. Thus,

$$|\sqrt{3} - 1| = \sqrt{3} - 1.$$

b. Because $\pi \approx 3.14$, the number inside the absolute value bars, $2 - \pi$, is negative. The absolute value of x when $x < 0$ is $-x$. Thus,

$$|2 - \pi| = -(2 - \pi) = \pi - 2.$$

c. If $x < 0$, then $|x| = -x$. Thus,

$$\frac{|x|}{x} = \frac{-x}{x} = -1.$$

 • • •

⊘ **Check Point 8** Rewrite each expression without absolute value bars:

 a. $|1 - \sqrt{2}|$ **b.** $|\pi - 3|$ **c.** $\dfrac{|x|}{x}$ if $x > 0$.

DISCOVERY

Verify the triangle inequality if $a = 4$ and $b = 5$. Verify the triangle inequality if $a = 4$ and $b = -5$.

When does equality occur in the triangle inequality and when does inequality occur? Verify your observation with additional number pairs.

Listed below are several basic properties of absolute value. Each of these properties can be derived from the definition of absolute value.

> **Properties of Absolute Value**
>
> For all real numbers a and b,
>
> **1.** $|a| \geq 0$ **2.** $|-a| = |a|$ **3.** $a \leq |a|$
>
> **4.** $|ab| = |a||b|$ **5.** $\left|\dfrac{a}{b}\right| = \dfrac{|a|}{|b|}, \quad b \neq 0$
>
> **6.** $|a + b| \leq |a| + |b|$ (called the triangle inequality)

🔟 Use absolute value to express distance.

Distance between Points on a Real Number Line

Absolute value is used to find the distance between two points on a real number line. If a and b are any real numbers, the **distance between a and b** is the absolute value of their difference. For example, the distance between 4 and 10 is 6. Using absolute value, we find this distance in one of two ways:

$$|10 - 4| = |6| = 6 \qquad \text{or} \qquad |4 - 10| = |-6| = 6.$$

> The distance between 4 and 10 on the real number line is 6.

Notice that we obtain the same distance regardless of the order in which we subtract.

> **Distance between Two Points on the Real Number Line**
>
> If a and b are any two points on a real number line, then the distance between a and b is given by
>
> $$|a - b| \quad \text{or} \quad |b - a|.$$

EXAMPLE 9 Distance between Two Points on a Number Line

Find the distance between -5 and 3 on the real number line.

SOLUTION

Because the distance between a and b is given by $|a - b|$, the distance between -5 and 3 is

$$|-5 - 3| = |-8| = 8.$$

> $a = -5 \qquad b = 3$

FIGURE P.9 The distance between -5 and 3 is 8.

Figure P.9 verifies that there are 8 units between -5 and 3 on the real number line. We obtain the same distance if we reverse the order of the subtraction:

$$|3 - (-5)| = |8| = 8.$$

• • •

✓ Check Point **9** Find the distance between -4 and 5 on the real number line.

Blitzer Bonus ‖ Using Algebra to Measure Blood-Alcohol Concentration

The amount of alcohol in a person's blood is known as blood-alcohol concentration (BAC), measured in grams of alcohol per deciliter of blood. A BAC of 0.08, meaning 0.08%, indicates that a person has 8 parts alcohol per 10,000 parts blood. In every state in the United States, it is illegal to drive with a BAC of 0.08 or higher.

How Do I Measure My Blood-Alcohol Concentration?

Here's a formula that models BAC for a person who weighs w pounds and who has n drinks* per hour.

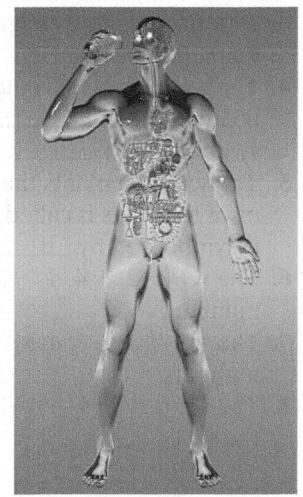

$$\text{BAC} = \frac{600n}{w(0.6n + 169)}$$

Blood-alcohol concentration

Body weight, in pounds

Number of drinks consumed in an hour

*A drink can be a 12-ounce can of beer, a 5-ounce glass of wine, or a 1.5-ounce shot of liquor. Each contains approximately 14 grams, or $\frac{1}{2}$ ounce, of alcohol.

Blood-alcohol concentration can be used to quantify the meaning of "tipsy."

BAC	Effects on Behavior
0.05	Feeling of well-being; mild release of inhibitions; absence of observable effects
0.08	Feeling of relaxation; mild sedation; exaggeration of emotions and behavior; slight impairment of motor skills; increase in reaction time
0.12	Muscle control and speech impaired; difficulty performing motor skills; uncoordinated behavior
0.15	Euphoria; major impairment of physical and mental functions; irresponsible behavior; some difficulty standing, walking, and talking
0.35	Surgical anesthesia; lethal dosage for a small percentage of people
0.40	Lethal dosage for 50% of people; severe circulatory and respiratory depression; alcohol poisoning/overdose

Source: National Clearinghouse for Alcohol and Drug Information

Keeping in mind the meaning of "tipsy," we can use our model to compare blood-alcohol concentrations of a 120-pound person and a 200-pound person for various numbers of drinks.

We determined each BAC using a calculator, rounding to three decimal places.

Blood-Alcohol Concentrations of a 120-Pound Person

$$\text{BAC} = \frac{600n}{120(0.6n + 169)}$$

n (number of drinks per hour)	1	2	3	4	5	6	7	8	9	10
BAC (blood-alcohol concentration)	0.029	0.059	0.088	0.117	0.145	0.174	0.202	0.230	0.258	0.286

Illegal to drive

Blood-Alcohol Concentrations of a 200-Pound Person

$$\text{BAC} = \frac{600n}{200(0.6n + 169)}$$

n (number of drinks per hour)	1	2	3	4	5	6	7	8	9	10
BAC (blood-alcohol concentration)	0.018	0.035	0.053	0.070	0.087	0.104	0.121	0.138	0.155	0.171

Illegal to drive

Like all mathematical models, the formula for BAC gives approximate rather than exact values. There are other variables that influence blood-alcohol concentration that are not contained in the model. These include the rate at which an individual's body processes alcohol, how quickly one drinks, sex, age, physical condition, and the amount of food eaten prior to drinking.

CONCEPT AND VOCABULARY CHECK

Fill in each blank so that the resulting statement is true.

1. A combination of numbers, variables, and operation symbols is called an algebraic _____.

2. If n is a counting number, b^n, read _____, indicates that there are n factors of b. The number b is called the _____ and the number n is called the _____.

3. An equation that expresses a relationship between two or more variables, such as $H = \frac{9}{10}(220 - a)$, is called a/an _____. The process of finding such equations to describe real-world phenomena is called mathematical _____. Such equations, together with the meaning assigned to the variables, are called mathematical _____.

4. The set of elements common to both set A and set B is called the _____ of sets A and B, and is symbolized by _____.

5. The set of elements that are members of set A or set B or of both sets is called the _____ of sets A or B and is symbolized by _____.

6. The set $\{1, 2, 3, 4, 5, \ldots\}$ is called the set of _____ numbers.

7. The set $\{0, 1, 2, 3, 4, 5, \ldots\}$ is called the set of _____ numbers.

8. The set $\{\ldots, -4, -3, -2, -1, 0, 1, 2, 3, 4, \ldots\}$ is called the set of _____.

9. The set of numbers in the form $\frac{a}{b}$, where a and b belong to the set in Exercise 8 and $b \neq 0$, is called the set of _____ numbers.

10. The set of numbers whose decimal representations are neither terminating nor repeating is called the set of _____ numbers.

11. Every real number is either a/an _____ number or a/an _____ number.

12. In interval notation, $[2, 5)$ represents the set of real numbers between ____ and ____, including ____ but not including ____.

13. In interval notation, $(-2, \infty)$ represents the set of real numbers _____ -2.

14. In interval notation, $(-\infty, -1]$ represents the set of real numbers _____ -1.

15. The set of elements common to both $(-\infty, 9)$ and $(-\infty, 12)$ is _____. This represents the _____ of these intervals.

16. The set of elements in $(-\infty, 9)$ or $(-\infty, 12)$ or in both sets is _____. This represents the _____ of these intervals.

17. The notation $|x|$ is read the _____ of x. If $x \geq 0$, then $|x| =$ ____. If $x < 0$, then $|x| =$ ____.

EXERCISE SET P.1

Practice Exercises

In Exercises 1–16, evaluate each algebraic expression for the given value or values of the variable(s).

1. $7 + 5x$, for $x = 10$

2. $8 + 6x$, for $x = 5$

3. $6x - y$, for $x = 3$ and $y = 8$

4. $8x - y$, for $x = 3$ and $y = 4$

5. $x^2 + 3x$, for $x = 8$

6. $x^2 + 5x$, for $x = 6$

7. $x^2 - 6x + 3$, for $x = 7$

8. $x^2 - 7x + 4$, for $x = 8$

9. $4 + 5(x - 7)^3$, for $x = 9$

10. $6 + 5(x - 6)^3$, for $x = 8$

11. $x^2 - 3(x - y)$, for $x = 8$ and $y = 2$

12. $x^2 - 4(x - y)$, for $x = 8$ and $y = 3$

13. $\dfrac{5(x + 2)}{2x - 14}$, for $x = 10$

14. $\dfrac{7(x - 3)}{2x - 16}$, for $x = 9$

15. $\dfrac{2x + 3y}{x + 1}$, for $x = -2$ and $y = 4$

16. $\dfrac{2x + y}{xy - 2x}$, for $x = -2$ and $y = 4$

The formula

$$C = \frac{5}{9}(F - 32)$$

expresses the relationship between Fahrenheit temperature, F, and Celsius temperature, C. In Exercises 17–18, use the formula to convert the given Fahrenheit temperature to its equivalent temperature on the Celsius scale.

17. $50°F$ **18.** $86°F$

A football was kicked vertically upward from a height of 4 feet with an initial speed of 60 feet per second. The formula

$$h = 4 + 60t - 16t^2$$

describes the ball's height above the ground, h, in feet, t seconds after it was kicked. Use this formula to solve Exercises 19–20.

19. What was the ball's height 2 seconds after it was kicked?

20. What was the ball's height 3 seconds after it was kicked?

In Exercises 21–28, find the intersection of the sets.

21. $\{1, 2, 3, 4\} \cap \{2, 4, 5\}$ **22.** $\{1, 3, 7\} \cap \{2, 3, 8\}$

23. $\{s, e, t\} \cap \{t, e, s\}$ **24.** $\{r, e, a, l\} \cap \{l, e, a, r\}$

25. $\{1, 3, 5, 7\} \cap \{2, 4, 6, 8, 10\}$

26. $\{0, 1, 3, 5\} \cap \{-5, -3, -1\}$

27. $\{a, b, c, d\} \cap \varnothing$ **28.** $\{w, y, z\} \cap \varnothing$

In Exercises 29–34, find the union of the sets.

29. $\{1, 2, 3, 4\} \cup \{2, 4, 5\}$

30. $\{1, 3, 7, 8\} \cup \{2, 3, 8\}$

31. $\{1, 3, 5, 7\} \cup \{2, 4, 6, 8, 10\}$

32. $\{0, 1, 3, 5\} \cup \{2, 4, 6\}$

33. $\{a, e, i, o, u\} \cup \varnothing$

34. $\{e, m, p, t, y\} \cup \varnothing$

In Exercises 35–38, list all numbers from the given set that are **a.** *natural numbers,* **b.** *whole numbers,* **c.** *integers,* **d.** *rational numbers,* **e.** *irrational numbers,* **f.** *real numbers.*

35. $\left\{-9, -\frac{4}{5}, 0, 0.25, \sqrt{3}, 9.2, \sqrt{100}\right\}$

36. $\left\{-7, -0.\overline{6}, 0, \sqrt{49}, \sqrt{50}\right\}$

37. $\left\{-11, -\frac{5}{6}, 0, 0.75, \sqrt{5}, \pi, \sqrt{64}\right\}$

38. $\left\{-5, -0.\overline{3}, 0, \sqrt{2}, \sqrt{4}\right\}$

39. Give an example of a whole number that is not a natural number.

40. Give an example of a rational number that is not an integer.

41. Give an example of a number that is an integer, a whole number, and a natural number.

42. Give an example of a number that is a rational number, an integer, and a real number.

Determine whether each statement in Exercises 43–50 is true or false.

43. $-13 \le -2$

44. $-6 > 2$

45. $4 \ge -7$

46. $-13 < -5$

47. $-\pi \ge -\pi$

48. $-3 > -13$

49. $0 \ge -6$

50. $0 \ge -13$

In Exercises 51–58, use the roster method to list the elements in each set.

51. $\{x \mid x$ is a natural number and $x < 5\}$

52. $\{x \mid x$ is a natural number and $x < 4\}$

53. $\{x \mid x$ is an integer and $-8 < x < -3\}$

54. $\{x \mid x$ is an integer and $-7 < x < -2\}$

55. $\{x \mid x$ is a natural number and $x \ge 7\}$

56. $\{x \mid x$ is a natural number and $x \ge 9\}$

57. $\{x \mid x$ is an odd whole number and $x < 11\}$

58. $\{x \mid x$ is an odd whole number and $x < 9\}$

In Exercises 59–72, express each interval in set-builder notation and graph the interval on a number line.

59. $(1, 6]$

60. $(-2, 4]$

61. $[-5, 2)$

62. $[-4, 3)$

63. $[-3, 1]$

64. $[-2, 5]$

65. $(2, \infty)$

66. $(3, \infty)$

67. $[-3, \infty)$

68. $[-5, \infty)$

69. $(-\infty, 3)$

70. $(-\infty, 2)$

71. $(-\infty, 5.5)$

72. $(-\infty, 3.5]$

In Exercises 73–84, use graphs to find each set.

73. $(-3, 0) \cap [-1, 2]$

74. $(-4, 0) \cap [-2, 1]$

75. $(-3, 0) \cup [-1, 2]$

76. $(-4, 0) \cup [-2, 1]$

77. $(-\infty, 5) \cap [1, 8)$

78. $(-\infty, 6) \cap [2, 9)$

79. $(-\infty, 5) \cup [1, 8)$

80. $(-\infty, 6) \cup [2, 9)$

81. $[3, \infty) \cap (6, \infty)$

82. $[2, \infty) \cap (4, \infty)$

83. $[3, \infty) \cup (6, \infty)$

84. $[2, \infty) \cup (4, \infty)$

In Exercises 85–94, rewrite each expression without absolute value bars.

85. $|300|$

86. $|-203|$

87. $|12 - \pi|$

88. $|7 - \pi|$

89. $|\sqrt{2} - 5|$

90. $|\sqrt{5} - 13|$

91. $\dfrac{-3}{|-3|}$

92. $\dfrac{-7}{|-7|}$

93. $\||-3| - |-7|\|$

94. $\||-5| - |-13|\|$

In Exercises 95–100, evaluate each algebraic expression for $x = 2$ and $y = -5$.

95. $|x + y|$

96. $|x - y|$

97. $|x| + |y|$

98. $|x| - |y|$

99. $\dfrac{y}{|y|}$

100. $\dfrac{|x|}{x} + \dfrac{|y|}{y}$

In Exercises 101–108, express the distance between the given numbers using absolute value. Then find the distance by evaluating the absolute value expression.

101. 2 and 17

102. 4 and 15

103. -2 and 5

104. -6 and 8

105. -19 and -4

106. -26 and -3

107. -3.6 and -1.4

108. -5.4 and -1.2

Practice Plus

In Exercises 109–116, insert either $<$, $>$, *or* $=$ *in the shaded area to make a true statement.*

109. $|-6|$ ▨ $|-3|$

110. $|-20|$ ▨ $|-50|$

111. $\left|\dfrac{3}{5}\right|$ ▨ $|-0.6|$

112. $\left|\dfrac{5}{2}\right|$ ▨ $|-2.5|$

113. $\dfrac{30}{40} - \dfrac{3}{4}$ ▨ $\dfrac{14}{15} \cdot \dfrac{15}{14}$

114. $\dfrac{17}{18} \cdot \dfrac{18}{17}$ ▨ $\dfrac{50}{60} - \dfrac{5}{6}$

115. $\dfrac{8}{13} \div \dfrac{8}{13}$ ▨ $|-1|$

116. $|-2|$ ▨ $\dfrac{4}{17} \div \dfrac{4}{17}$

In Exercises 117–126, use the order of operations to simplify each expression.

117. $8^2 - 16 \div 2^2 \cdot 4 - 3$

118. $10^2 - 100 \div 5^2 \cdot 2 - 3$

119. $\dfrac{5 \cdot 2 - 3^2}{[3^2 - (-2)]^2}$

120. $\dfrac{10 \div 2 + 3 \cdot 4}{(12 - 3 \cdot 2)^2}$

121. $8 - 3[-2(2 - 5) - 4(8 - 6)]$

122. $8 - 3[-2(5 - 7) - 5(4 - 2)]$

123. $\dfrac{2(-2) - 4(-3)}{5 - 8}$

124. $\dfrac{6(-4) - 5(-3)}{9 - 10}$

125. $\dfrac{(5 - 6)^2 - 2|3 - 7|}{89 - 3 \cdot 5^2}$

126. $\dfrac{12 \div 3 \cdot 5|2^2 + 3^2|}{7 + 3 - 6^2}$

Application Exercises

The maximum heart rate, in beats per minute, that you should achieve during exercise is 220 minus your age:

$$220 - a.$$

This algebraic expression gives maximum heart rate in terms of age, a.

The following bar graph shows the target heart rate ranges for four types of exercise goals. The lower and upper limits of these ranges are fractions of the maximum heart rate, $220 - a$. Exercises 127–128 are based on the information in the graph.

Target Heart Rate Ranges for Exercise Goals

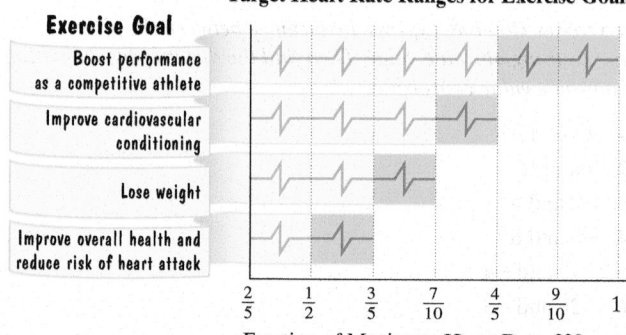

Fraction of Maximum Heart Rate, $220 - a$

127. If your exercise goal is to improve cardiovascular conditioning, the graph shows the following range for target heart rate, H, in beats per minute:

Lower limit of range $H = \dfrac{7}{10}(220 - a)$

Upper limit of range $H = \dfrac{4}{5}(220 - a)$

a. What is the lower limit of the heart range, in beats per minute, for a 20-year-old with this exercise goal?

b. What is the upper limit of the heart range, in beats per minute, for a 20-year-old with this exercise goal?

128. If your exercise goal is to improve overall health, the graph shows the following range for target heart rate, H, in beats per minute:

Lower limit of range $H = \dfrac{1}{2}(220 - a)$

Upper limit of range $H = \dfrac{3}{5}(220 - a)$

a. What is the lower limit of the heart range, in beats per minute, for a 30-year-old with this exercise goal?

b. What is the upper limit of the heart range, in beats per minute, for a 30-year-old with this exercise goal?

The bar graph shows the average cost of tuition and fees at private four-year colleges in the United States.

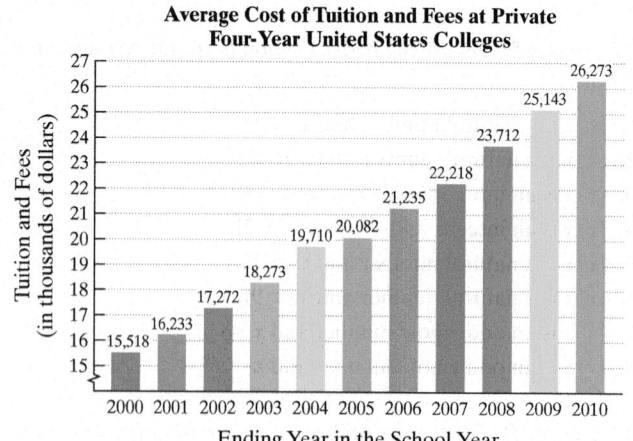

Average Cost of Tuition and Fees at Private Four-Year United States Colleges

Source: The College Board

The formula

$$T = 26x^2 + 819x + 15{,}527$$

models the average cost of tuition and fees, T, at private U.S. colleges for the school year ending x years after 2000. Use this information to solve Exercises 129–130.

129. a. Use the formula to find the average cost of tuition and fees at private U.S. colleges for the school year ending in 2010.

b. By how much does the formula underestimate or overestimate the actual cost shown by the graph for the school year ending in 2010?

c. Use the formula to project the average cost of tuition and fees at private U.S. colleges for the school year ending in 2013.

130. a. Use the formula to find the average cost of tuition and fees at private U.S. colleges for the school year ending in 2009.

b. By how much does the formula underestimate or overestimate the actual cost shown by the graph for the school year ending in 2009?

c. Use the formula to project the average cost of tuition and fees at private U.S. colleges for the school year ending in 2012.

131. Read the Blitzer Bonus on page 15. Use the formula

$$\text{BAC} = \frac{600n}{w(0.6n + 169)}$$

and replace w with your body weight. Using this formula and a calculator, compute your BAC for integers from $n = 1$ to $n = 10$. Round to three decimal places. According to this model, how many drinks can you consume in an hour without exceeding the legal measure of drunk driving?

Writing in Mathematics

Writing about mathematics will help you learn mathematics. For all writing exercises in this book, use complete sentences to respond to the question. Some writing exercises can be answered in a sentence; others require a paragraph or two. You can decide how much you need to write as long as your writing clearly and directly answers the question in the exercise. Standard references such as a dictionary and a thesaurus should be helpful.

132. What is an algebraic expression? Give an example with your explanation.

133. If n is a natural number, what does b^n mean? Give an example with your explanation.

134. What does it mean when we say that a formula models real-world phenomena?

135. What is the intersection of sets A and B?

136. What is the union of sets A and B?

137. How do the whole numbers differ from the natural numbers?

138. Can a real number be both rational and irrational? Explain your answer.

139. If you are given two real numbers, explain how to determine which is the lesser.

140. If $a < b$, describe each of the following sets: (a, b); $[a, b]$; (a, ∞); $(-\infty, b)$.

141. Explain how to use a number line to find the intersection of intervals.

142. Explain how to use a number line to find the union of intervals.

143. How can $\dfrac{|x|}{x}$ be equal to 1 or -1?

Critical Thinking Exercises

Make Sense? *In Exercises 144–147, determine whether each statement makes sense or does not make sense, and explain your reasoning.*

144. My mathematical model describes the data for tuition and fees at public four-year colleges for the past ten years extremely well, so it will serve as an accurate prediction for the cost of public colleges in 2050.

145. A model that describes the average cost of tuition and fees at private U.S. colleges for the school year ending x years after 2000 cannot be used to estimate the cost of private education for the school year ending in 2000.

146. The humor in this cartoon is based on the fact that the football will never be hiked.

Foxtrot copyright © 2003, 2009 by Bill Amend/Distributed by Universal Uclick.

147. I used a parenthesis to indicate that an endpoint is included in an interval.

In Exercises 148–151, determine whether each statement is true or false. If the statement is false, make the necessary change(s) to produce a true statement.

148. Every rational number is an integer.

149. Some whole numbers are not integers.

150. Some rational numbers are not positive.

151. Irrational numbers cannot be negative.

In Exercises 152–154, insert either $<$ or $>$ in the shaded area between the numbers to make the statement true.

152. $\sqrt{2} \quad\quad 1.5$

153. $-\pi \quad\quad -3.5$

154. $-\dfrac{3.14}{2} \quad\quad -\dfrac{\pi}{2}$

Preview Exercises

Exercises 155–157 will help you prepare for the material covered in the next section.

155. a. Simplify: $3(5 + 7)$.

b. Simplify: $3 \cdot 5 + 3 \cdot 7$.

c. Based on your simplifications in parts (a) and (b), what can you conclude?

156. If 7 is substituted for x in $2x + 3 = 17$, is the resulting statement true or false?

157. The formula $B = 14 + 0.2x$ models the percentage of college students majoring in business, B, x years after 1970. Use the formula to project the percentage of college students who will major in business in 2015.

SECTION P.2 — Basic Rules of Algebra

Objectives

1. Identify properties of the real numbers.
2. Simplify algebraic expressions.
3. Use properties of equality to solve basic equations.
4. Solve a formula for a variable.
5. Solve applied problems using mathematical models.

The bar graph in **Figure P.10** shows the percentage of college students in 1970 and 2010 for the five most popular majors in 2010.

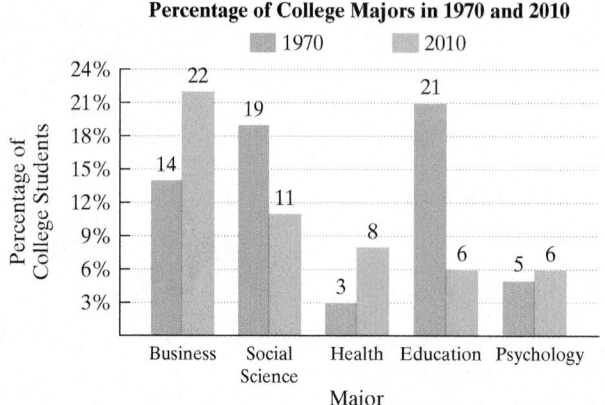

Percentage of College Majors in 1970 and 2010

FIGURE P.10
Source: *Newsweek*, August 20, 2012

The graph indicates that in 1970, 14% of college students majored in business. For the period from 1970 through 2010, this percentage increased by approximately 0.2 each year. The percentage of business majors x years after 1970 can be modeled by the formula

$$B = 14 + 0.2x.$$

| Percentage of business majors | The 1970 percentage | Increased by | 0.2 each year for x years |

So, if the trend from 1970 through 2010 continues, when will 24% of college students major in business? Substitute 24 for B in the formula $B = 14 + 0.2x$:

$$24 = 14 + 0.2x.$$

Our goal is to determine the value of x, the number of years after 1970 when 24% of college students will major in business.

In this section, we review basic algebraic properties, using these properties to determine values of x in equations such as $24 = 14 + 0.2x$. Armed with these properties, you will be able to use mathematical models to project what might occur in the future.

Properties of Real Numbers and Algebraic Expressions

① Identify properties of the real numbers.

When you use your calculator to add two real numbers, you can enter them in any order. The fact that two real numbers can be added in any order is called the **commutative property of addition**. You probably use this property, as well as other properties of real numbers listed in **Table P.3**, without giving it much thought. The

properties of the real numbers are especially useful when working with algebraic expressions. For each property listed in **Table P.3**, a, b, and c represent real numbers, variables, or algebraic expressions.

Blitzer Bonus

The Associative Property and the English Language

In the English language, phrases can take on different meanings depending on the way the words are associated with commas.

Here are three examples.

- Woman, without her man, is nothing.
 Woman, without her, man is nothing.

- What's the latest dope?
 What's the latest, dope?

- Population of Amsterdam broken down by age and sex
 Population of Amsterdam, broken down by age and sex

Table P.3 Properties of the Real Numbers

Name	Meaning	Examples
Commutative Property of Addition	Changing order when adding does not affect the sum. $a + b = b + a$	• $13 + 7 = 7 + 13$ • $13x + 7 = 7 + 13x$
Commutative Property of Multiplication	Changing order when multiplying does not affect the product. $ab = ba$	• $\sqrt{2} \cdot \sqrt{5} = \sqrt{5} \cdot \sqrt{2}$ • $x \cdot 6 = 6x$
Associative Property of Addition	Changing grouping when adding does not affect the sum. $(a + b) + c = a + (b + c)$	• $3 + (8 + x) = (3 + 8) + x$ $\quad = 11 + x$
Associative Property of Multiplication	Changing grouping when multiplying does not affect the product. $(ab)c = a(bc)$	• $-2(3x) = (-2 \cdot 3)x = -6x$
Distributive Property of Multiplication over Addition	Multiplication distributes over addition. $a \cdot (b + c) = a \cdot b + a \cdot c$	• $7(4 + \sqrt{3}) = 7 \cdot 4 + 7 \cdot \sqrt{3}$ $\quad = 28 + 7\sqrt{3}$ • $5(3x + 7) = 5 \cdot 3x + 5 \cdot 7$ $\quad = 15x + 35$
Identity Property of Addition	Zero can be deleted from a sum. $a + 0 = a$ $0 + a = a$	• $\sqrt{3} + 0 = \sqrt{3}$ • $0 + 6x = 6x$
Identity Property of Multiplication	One can be deleted from a product. $a \cdot 1 = a$ $1 \cdot a = a$	• $1 \cdot \pi = \pi$ • $13x \cdot 1 = 13x$
Inverse Property of Addition	The sum of a real number and its additive inverse gives 0, the additive identity. $a + (-a) = 0$ $(-a) + a = 0$	• $\sqrt{5} + (-\sqrt{5}) = 0$ • $-\pi + \pi = 0$ • $6x + (-6x) = 0$ • $(-4y) + 4y = 0$
Inverse Property of Multiplication	The product of a nonzero real number and its multiplicative inverse gives 1, the multiplicative identity. $a \cdot \dfrac{1}{a} = 1, \quad a \neq 0$ $\dfrac{1}{a} \cdot a = 1, \quad a \neq 0$	• $7 \cdot \dfrac{1}{7} = 1$ • $\left(\dfrac{1}{x - 3}\right)(x - 3) = 1, \quad x \neq 3$

The properties of the real numbers in **Table P.3** apply to the operations of addition and multiplication. Subtraction and division are defined in terms of addition and multiplication.

Definitions of Subtraction and Division

Let a and b represent real numbers.

Subtraction: $a - b = a + (-b)$

We call $-b$ the **additive inverse** or **opposite** of b.

Division: $a \div b = a \cdot \frac{1}{b}$, where $b \neq 0$

We call $\frac{1}{b}$ the **multiplicative inverse** or **reciprocal** of b. The quotient of a and b, $a \div b$, can be written in the form $\frac{a}{b}$, where a is the **numerator** and b the **denominator** of the fraction.

Because subtraction is defined in terms of adding an inverse, the distributive property can be applied to subtraction:

$$a(b - c) = ab - ac$$
$$(b - c)a = ba - ca.$$

For example,

$$4(2x - 5) = 4 \cdot 2x - 4 \cdot 5 = 8x - 20.$$

② Simplify algebraic expressions.

Simplifying Algebraic Expressions

The **terms** of an algebraic expression are those parts that are separated by addition. For example, consider the algebraic expression

$$7x - 9y + z - 3,$$

which can be expressed as

$$7x + (-9y) + z + (-3).$$

This expression contains four terms, namely, $7x$, $-9y$, z, and -3.

The numerical part of a term is called its **coefficient**. In the term $7x$, the 7 is the coefficient. If a term containing one or more variables is written without a coefficient, the coefficient is understood to be 1. Thus, z means $1z$. If a term is a constant, its coefficient is that constant. Thus, the coefficient of the constant term -3 is -3.

$$7x + (-9y) + z + (-3)$$

| Coefficient is 7. | Coefficient is −9. | Coefficient is 1; z means 1z. | Coefficient is −3. |

The parts of each term that are multiplied are called the **factors** of the term. The factors of the term $7x$ are 7 and x.

Like terms are terms that have exactly the same variable factors. For example, $3x$ and $7x$ are like terms. The distributive property in the form

$$ba + ca = (b + c)a$$

enables us to add or subtract like terms. For example,

$$3x + 7x = (3 + 7)x = 10x$$
$$7y^2 - y^2 = 7y^2 - 1y^2 = (7 - 1)y^2 = 6y^2.$$

This process is called **combining like terms**.

GREAT QUESTION!

What is the bottom line for combining like terms?

To combine like terms mentally, add or subtract the coefficients of the terms. Use this result as the coefficient of the terms' variable factor(s).

An algebraic expression is **simplified** when parentheses have been removed and like terms have been combined.

EXAMPLE 1 Simplifying an Algebraic Expression

Simplify: $6(2x^2 + 4x) + 10(4x^2 + 3x)$.

SOLUTION

$$6(2x^2 + 4x) + 10(4x^2 + 3x)$$

$52x^2$ and $54x$ are not like terms. They contain different variable factors, x^2 and x, and cannot be combined.

$$= 6 \cdot 2x^2 + 6 \cdot 4x + 10 \cdot 4x^2 + 10 \cdot 3x \quad \text{Use the distributive property to}$$
remove the parentheses.
$$= 12x^2 + 24x + 40x^2 + 30x \quad \text{Multiply.}$$
$$= (12x^2 + 40x^2) + (24x + 30x) \quad \text{Group like terms.}$$
$$= 52x^2 + 54x \quad \text{Combine like terms.} \quad \bullet\bullet\bullet$$

✓ Check Point **1** Simplify: $7(4x^2 + 3x) + 2(5x^2 + x)$.

Properties of Negatives

The distributive property can be extended to cover more than two terms within parentheses. For example,

This sign represents subtraction.

This sign tells us that the number is negative.

$$-3(4x - 2y + 6) = -3 \cdot 4x - (-3) \cdot 2y - 3 \cdot 6$$
$$= -12x - (-6y) - 18$$
$$= -12x + 6y - 18.$$

The voice balloons illustrate that negative signs can appear side by side. They can represent the operation of subtraction or the fact that a real number is negative. Here is a list of properties of negatives and how they are applied to algebraic expressions:

Properties of Negatives

Let a and b represent real numbers, variables, or algebraic expressions.

Property	**Examples**
1. $(-1)a = -a$	$(-1)4xy = -4xy$
2. $-(-a) = a$	$-(-6y) = 6y$
3. $(-a)b = -ab$	$(-7)4xy = -7 \cdot 4xy = -28xy$
4. $a(-b) = -ab$	$5x(-3y) = -5x \cdot 3y = -15xy$
5. $-(a + b) = -a - b$	$-(7x + 6y) = -7x - 6y$
6. $-(a - b) = -a + b$	$-(3x - 7y) = -3x + 7y$
$\qquad\quad = b - a$	$\qquad\qquad\quad = 7y - 3x$

It is not uncommon to see algebraic expressions with parentheses preceded by a negative sign or subtraction. Properties 5 and 6 in the box, $-(a + b) = -a - b$ and $-(a - b) = -a + b$, are related to this situation. An expression of the form $-(a + b)$ can be simplified as follows:

$$-(a + b) = -1(a + b) = (-1)a + (-1)b = -a + (-b) = -a - b.$$

Do you see a fast way to obtain the simplified expression on the right in $-(a + b) = -a - b$? **If a negative sign or a subtraction symbol appears outside parentheses, drop the parentheses and change the sign of every term within the parentheses.** For example,

$$-(3x^2 - 7x - 4) = -3x^2 + 7x + 4.$$

EXAMPLE 2 Simplifying an Algebraic Expression

Simplify: $8x + 2[5 - (x - 3)]$.

SOLUTION

$$8x + 2[5 - (x - 3)]$$

$= 8x + 2[5 - x + 3]$	Drop parentheses and change the sign of each term in parentheses: $-(x - 3) = -x + 3$.
$= 8x + 2[8 - x]$	Simplify inside brackets: $5 + 3 = 8$.
$= 8x + 16 - 2x$	Apply the distributive property: $2[8 - x] = 2 \cdot 8 - 2x = 16 - 2x$.
$= (8x - 2x) + 16$	Group like terms.
$= (8 - 2)x + 16$	Apply the distributive property.
$= 6x + 16$	Simplify. • • •

✓ **Check Point** **2** Simplify: $6 + 4[7 - (x - 2)]$.

③ Use properties of equality to solve basic equations.

Properties of Equality

Properties of equality are used to *solve equations*. For example, consider the equation

$$4x + 12 = 0.$$

Solving an equation in x involves determining all values of x that result in a true statement when substituted into the equation. Such values are **solutions** or **roots** of the equation. For example, substitute -3 for x in $4x + 12 = 0$. We obtain

$$4(-3) + 12 = 0, \quad \text{or} \quad -12 + 12 = 0.$$

This simplifies to the true statement $0 = 0$. Thus, -3 is a solution of the equation $4x + 12 = 0$. We also say that -3 **satisfies** the equation $4x + 12 = 0$, because when we substitute -3 for x, a true statement results. The set of all such solutions is called the equation's **solution set**. For example, the solution set of the equation $4x + 12 = 0$ is $\{-3\}$.

Two or more equations that have the same solution set are called **equivalent equations**. For example, the equations

$$4x + 12 = 0 \quad \text{and} \quad 4x = -12 \quad \text{and} \quad x = -3$$

are equivalent equations because the solution set for each is $\{-3\}$. To solve an equation in x, we transform the equation into an equivalent equation one or more times. Our final equivalent equation should be of the form

$$x = \text{a number.}$$

The solution set of this equation is the set consisting of the number.

To generate equivalent equations, we will use the following properties:

The Addition and Multiplication Properties of Equality

The Addition Property of Equality

The same real number or algebraic expression may be added to both sides of an equation without changing the equation's solution set.

$$a = b \text{ and } a + c = b + c \text{ are equivalent equations.}$$

The Multiplication Property of Equality

The same nonzero real number may multiply both sides of an equation without changing the equation's solution set.

$$a = b \text{ and } ac = bc \text{ are equivalent equations as long as } c \neq 0.$$

Because subtraction is defined in terms of addition, the addition property also lets us subtract the same number from both sides of an equation without changing the equation's solution set. Similarly, because division is defined in terms of multiplication, the multiplication property of equality can be used to divide both sides of an equation by the same nonzero number to obtain an equivalent equation.

Table P.4 illustrates how these properties are used to isolate x to obtain an equation of the form $x =$ a number.

Table P.4 Using Properties of Equality to Solve Equations

	Equation	How to Isolate x	Solving the Equation	The Equation's Solution Set
These equations are solved using the Addition Property of Equality.	$x - 3 = 8$	Add 3 to both sides.	$x - 3 + 3 = 8 + 3$ $x = 11$	$\{11\}$
	$x + 7 = -15$	Subtract 7 from both sides.	$x + 7 - 7 = -15 - 7$ $x = -22$	$\{-22\}$
These equations are solved using the Multiplication Property of Equality.	$6x = 30$	Divide both sides by 6 (or multiply both sides by $\frac{1}{6}$).	$\dfrac{6x}{6} = \dfrac{30}{6}$ $x = 5$	$\{5\}$
	$\dfrac{x}{5} = 9$	Multiply both sides by 5.	$5 \cdot \dfrac{x}{5} = 5 \cdot 9$ $x = 45$	$\{45\}$

EXAMPLE 3 Using Properties of Equality to Solve an Equation

Solve and check: $2x + 3 = 17$.

SOLUTION

Our goal is to obtain an equivalent equation with x isolated on one side and a number on the other side.

$$2x + 3 = 17 \qquad \text{This is the given equation.}$$
$$2x + 3 - 3 = 17 - 3 \qquad \text{Subtract 3 from both sides.}$$
$$2x = 14 \qquad \text{Simplify.}$$
$$\frac{2x}{2} = \frac{14}{2} \qquad \text{Divide both sides by 2.}$$
$$x = 7 \qquad \text{Simplify.}$$

Now we check the proposed solution, 7, by replacing x with 7 in the original equation.

$$2x + 3 = 17 \quad \text{This is the original equation.}$$

$$2 \cdot 7 + 3 \overset{?}{=} 17 \quad \text{Substitute 7 for x. The question mark indicates}$$
$$\text{that we do not yet know if the two sides are equal.}$$

$$14 + 3 \overset{?}{=} 17 \quad \text{Multiply: } 2 \cdot 7 = 14.$$

This statement is true.

$$17 = 17 \quad \text{Add: } 14 + 3 = 17.$$

Because the check results in a true statement, we conclude that the solution set of the given equation is {7}. ● ● ●

 Check Point **3** Solve and check: $4x + 5 = 29$.

GREAT QUESTION!

What are the differences between what I'm supposed to do with algebraic expressions and algebraic equations?

We simplify algebraic expressions. We solve algebraic equations. Notice the differences between the procedures:

Simplifying an Algebraic Expression	**Solving an Algebraic Equation**
Simplify: $3(x - 7) - (5x - 11)$.	Solve: $3(x - 7) - (5x - 11) = 14$.

This is not an equation. There is no equal sign.

This is an equation. There is an equal sign.

Solution $3(x - 7) - (5x - 11)$
$= 3x - 21 - 5x + 11$
$= (3x - 5x) + (-21 + 11)$
$= -2x + (-10)$
$= -2x - 10$

Stop! Further simplification is not possible. Avoid the common error of setting $-2x - 10$ equal to 0.

Solution $3(x - 7) - (5x - 11) = 14$
$3x - 21 - 5x + 11 = 14$
$-2x - 10 = 14$

Add 10 to both sides. $-2x - 10 + 10 = 14 + 10$
$-2x = 24$

Divide both sides by -2. $\dfrac{-2x}{-2} = \dfrac{24}{-2}$
$x = -12$

The solution set is $\{-12\}$.

④ Solve a formula for a variable.

Solving a Formula for One of Its Variables

We know that solving an equation is the process of finding the number (or numbers) that make the equation a true statement. The equation we solved in Example 3, $2x + 3 = 17$, contained only one letter, x.

By contrast, basic formulas contain two or more letters, representing two or more variables. An example is the formula for the perimeter of a rectangle:

$$P = 2l + 2w.$$

The perimeter, P, of a rectangle is the sum of twice the length, l, and twice the width, w.

We say that this formula is solved for the variable P because P is alone on one side of the equation and the other side does not contain a P.

Solving a formula for a variable means rewriting the formula using the addition and multiplication properties of equality so that the variable is isolated on one side of the equation. It does not mean obtaining a numerical value for that variable.

To solve a formula for one of its variables, treat that variable as if it were the only variable in the equation. Think of the other variables as if they were numbers. Isolate all terms with the specified variable on one side of the equation and all terms without the specified variable on the other side. Then divide both sides by the same nonzero quantity to get the specified variable alone. The next two examples show how to do this.

EXAMPLE 4 Solving a Formula for a Variable

Solve the formula $P = 2l + 2w$ for l.

SOLUTION

First, isolate $2l$ on the right by subtracting $2w$ from both sides. Then solve for l by dividing both sides by 2.

We need to isolate l.

$$P = 2l + 2w \qquad \text{This is the given formula.}$$

$$P - 2w = 2l + 2w - 2w \qquad \text{Isolate } 2l \text{ by subtracting } 2w \text{ from both sides.}$$

$$P - 2w = 2l \qquad \text{Simplify.}$$

$$\frac{P - 2w}{2} = \frac{2l}{2} \qquad \text{Solve for } l \text{ by dividing both sides by 2.}$$

$$\frac{P - 2w}{2} = l \qquad \text{Simplify.}$$

Equivalently,

$$l = \frac{P - 2w}{2}.$$

$\bullet\bullet\bullet$

⬛ Check Point **4** Solve the formula $P = 2l + 2w$ for w.

EXAMPLE 5 Solving a Formula for a Variable That Occurs Twice

The formula

$$A = P + Prt$$

describes the amount, A, that a principal of P dollars is worth after t years when invested at a simple annual interest rate, r. Solve this formula for P.

SOLUTION

Notice that all the terms with P already occur on the right side of the formula.

We need to isolate P.

$$A = P + Prt$$

We can use the distributive property on the right to convert the two occurrences of P into one.

$$A = P + Prt \qquad \text{This is the given formula.}$$

$$A = P(1 + rt) \qquad \text{Use the distributive property to obtain a single occurrence of } P.$$

$$\frac{A}{1 + rt} = \frac{P(1 + rt)}{1 + rt} \qquad \text{Divide both sides by } 1 + rt.$$

$$\frac{A}{1 + rt} = P \qquad \text{Simplify: } \frac{P(1 + rt)}{1(1 + rt)} = \frac{P}{1} = P.$$

GREAT QUESTION!

Can I solve $A = P + Prt$ for P by subtracting Prt from both sides and writing

$$A - Prt = P?$$

No. When a formula is solved for a specified variable, that variable must be isolated on one side. The variable P occurs on both sides of

$$A - Prt = P.$$

Equivalently,

$$P = \frac{A}{1 + rt}.$$

 Check Point **5** Solve the formula $P = C + MC$ for C.

⑤ Solve applied problems using mathematical models.

Applications

Our next example shows how properties of equality can be used to find the value of a variable in a mathematical model.

EXAMPLE 6 Modeling College Majors

The formula

$$B = 14 + 0.2x$$

models the percentage of college students majoring in business, B, x years after 1970. When will 24% of college students major in business?

SOLUTION

We are interested in the year when 24% of college students will major in business, so substitute 24 for B in the formula and solve for x, the number of years after 1970.

$B = 14 + 0.2x$	This is the given formula.
$24 = 14 + 0.2x$	Replace B with 24.
$24 - 14 = 14 + 0.2x - 14$	Subtract 14 from both sides.
$10 = 0.2x$	Simplify.
$\dfrac{10}{0.2} = \dfrac{0.2x}{0.2}$	Divide both sides by 0.2.
$50 = x$	Simplify.

The model indicates that 50 years after 1970, or in 2020, 24% of college students will major in business. • • •

 Check Point **6** The formula $S = 19 - 0.2x$ models the percentage of college students majoring in social science, S, x years after 1970. When will 8% of college students major in social science?

CONCEPT AND VOCABULARY CHECK

Fill in each blank so that the resulting statement is true.

1. The commutative properties state that $a + b =$ _____ and $ab =$ ___.
2. The associative properties state that $(a + b) + c =$ _____ and _____ $= a(bc)$.
3. The distributive property states that $a(b + c) =$ _____.
4. $a + (-a) =$ ___ : The sum of a real number and its additive _____ is ___, the additive _____.
5. $a \cdot \frac{1}{a} = 1, a \neq 0$: The product of a nonzero real number and its multiplicative _____ is ___, the multiplicative _____.
6. An algebraic expression is _____ when parentheses have been removed and like terms have been combined.

7. $-(-a) =$ ___.
8. Two or more equations that have the same solution set are called _____ equations.
9. The first step in solving $4x - 5 = 15$ is to _____.
10. The first step in solving $-3x = 21$ is to _____.
11. Solving a formula for a variable means rewriting the formula so that the variable is _____.
12. The first step in solving $IR + Ir = E$ for I is to obtain a single occurrence of I by applying the _____ property on the left side.

EXERCISE SET P.2

Practice Exercises

In Exercises 1–10, state the name of the property illustrated.

1. $6 + (-4) = (-4) + 6$
2. $11 \cdot (7 + 4) = 11 \cdot 7 + 11 \cdot 4$
3. $6 + (2 + 7) = (6 + 2) + 7$
4. $6 \cdot (2 \cdot 3) = 6 \cdot (3 \cdot 2)$
5. $(2 + 3) + (4 + 5) = (4 + 5) + (2 + 3)$
6. $7 \cdot (11 \cdot 8) = (11 \cdot 8) \cdot 7$
7. $2(-8 + 6) = -16 + 12$
8. $-8(3 + 11) = -24 + (-88)$
9. $\dfrac{1}{(x + 3)} (x + 3) = 1, x \neq -3$
10. $(x + 4) + [-(x + 4)] = 0$

In Exercises 11–22, simplify each algebraic expression.

11. $5(3x + 4) - 4$
12. $2(5x + 4) - 3$
13. $5(3x - 2) + 12x$
14. $2(5x - 1) + 14x$
15. $7(3y - 5) + 2(4y + 3)$
16. $4(2y - 6) + 3(5y + 10)$
17. $5(3y - 2) - (7y + 2)$
18. $4(5y - 3) - (6y + 3)$
19. $7 - 4[3 - (4y - 5)]$
20. $6 - 5[8 - (2y - 4)]$
21. $18x^2 + 4 - [6(x^2 - 2) + 5]$
22. $14x^2 + 5 - [7(x^2 - 2) + 4]$

In Exercises 23–28, write each algebraic expression without parentheses.

23. $-(-14x)$
24. $-(-17y)$
25. $-(2x - 3y - 6)$
26. $-(5x - 13y - 1)$
27. $\frac{1}{3}(3x) + [(4y) + (-4y)]$
28. $\frac{1}{2}(2y) + [(-7x) + 7x]$

In Exercises 29–38, solve and check each equation.

29. $5x + 3 = 18$
30. $3x + 8 = 50$
31. $6x - 3 = 63$
32. $5x - 8 = 72$
33. $14 - 5x = -41$
34. $25 - 6x = -83$
35. $11x - (6x - 5) = 40$
36. $5x - (2x - 8) = 35$
37. $16 = 3(x - 1) - (x - 7)$
38. $7 = 2(x + 5) - 3(x - 2)$

In Exercises 39–48, simplify or solve, whichever is appropriate.

39. $3(x - 2) - 2(x + 5)$
40. $x - 3(x + 1) - 2(x - 1)$
41. $3(x - 2) - 2(x + 5) = -7$
42. $x - 3(x + 1) - 2(x - 1) = 3$
43. $4[x - (3 - x)] - 7x = 7$
44. $2[3x - (4x - 6)] - 5x = -30$
45. $\frac{1}{2}(4y + 8) + \frac{2}{3}(9y - 12)$
46. $\frac{3}{4}(24 - 8y) + \frac{2}{3}(6y - 9)$
47. $\frac{1}{2}(4y + 8) + \frac{2}{3}(9y - 12) = 16$
48. $\frac{3}{4}(24 - 8y) + \frac{2}{3}(6y - 9) = 16$

In Exercises 49–68, solve each formula for the specified variable. Do you recognize the formula? If so, what does it describe?

49. $d = rt$ for r
50. $d = rt$ for t
51. $I = Prt$ for P
52. $I = Prt$ for r
53. $C = 2\pi r$ for r
54. $C = \pi d$ for d
55. $E = mc^2$ for m
56. $V = \pi r^2 h$ for h
57. $y = mx + b$ for m
58. $y = mx + b$ for x
59. $T = D + pm$ for p
60. $P = C + MC$ for M
61. $Ax + By + C = 0$ for x
62. $Ax + By + C = 0$ for y
63. $y - y_1 = m(x - x_1)$ for m
64. $y_2 - y_1 = m(x_2 - x_1)$ for m
65. $L = a + (n - 1)d$ for n
66. $L = a + (n - 1)d$ for d
67. $A = 2lw + 2lh + 2wh$ for l
68. $A = 2lw + 2lh + 2wh$ for h

Practice Plus

In Exercises 69–76, write each English phrase as an algebraic expression. Then simplify the expression. Let x represent the number.

69. A number decreased by the sum of the number and four
70. A number decreased by the difference between eight and the number
71. Six times the product of negative five and a number
72. Ten times the product of negative four and a number
73. The difference between the product of five and a number and twice the number
74. The difference between the product of six and a number and negative two times the number
75. The difference between eight times a number and six more than three times the number
76. Eight decreased by three times the sum of a number and six

Application Exercises

77. You had $10,000 to invest. You put x dollars in a safe, government-insured certificate of deposit paying 5% per year. You invested the remainder of the money in noninsured corporate bonds paying 12% per year. Your total interest earned at the end of the year is given by the algebraic expression
$$0.05x + 0.12(10,000 - x).$$
 a. Simplify the algebraic expression.
 b. Use each form of the algebraic expression to determine your total interest earned at the end of the year if you invested $6000 in the safe, government-insured certificate of deposit.

78. It takes you 50 minutes to get to campus. You spend t minutes walking to the bus stop and the rest of the time riding the bus. Your walking rate is 0.06 mile per minute and the bus travels at a rate of 0.5 mile per minute. The total distance walking and traveling by bus is given by the algebraic expression
$$0.06t + 0.5(50 - t).$$
 a. Simplify the algebraic expression.
 b. Use each form of the algebraic expression to determine the total distance that you travel if you spend 20 minutes walking to the bus stop.

79. The formula $E = 21 - 0.375x$ models the percentage of college students majoring in education, E, x years after 1970. Use the formula to project when only 3% of college students will major in education.

80. The formula $H = 3 + 0.125x$ models the percentage of college students majoring in health, H, x years after 1970. Use the formula to project when 9% of college students will major in health.

Writing in Mathematics

81. Describe the difference between the commutative and the associative properties of addition.

82. Why is $3(x + 7) - 4x$ not simplified? What must be done to simplify the expression?

83. Describe how the addition and multiplication properties of equality are used to solve the equation $2x - 3 = 7$.

84. Is it possible to solve $3(x - 2) - (x + 7)$? When variables appear, describe how to determine whether to simplify or to solve.

85. Describe what it means to solve a formula for a variable.

Critical Thinking Exercises

Make Sense? *In Exercises 86–89, determine whether each statement makes sense or does not make sense, and explain your reasoning.*

86. Just as the commutative properties change groupings, the associative properties change order.

87. I simplified $-(3x^2 - 5x - 11)$ without showing any steps.

88. I used properties of equality to solve $3(x - 7) - (5x - 11)$.

89. I solved $P = C + MC$ for C by subtracting MC from both sides.

In Exercises 90–93, determine whether each statement is true or false. If the statement is false, make the necessary change(s) to produce a true statement.

90. The term x has no coefficient.

91. $5 + 3(x - 4) = 8(x - 4) = 8x - 32$

92. $-x - x = -x + (-x) = 0$

93. $x - 0.02(x + 200) = 0.98x - 4$

94. The height, h, in feet, of water in a fountain is described by the formula

$$h = -16t^2 + 64t$$

and the velocity, v, in feet per second, of water in the fountain is described by $v = -32t + 64$. Find the time when the water's velocity is 16 feet per second and then find the water's height at that time.

Preview Exercises

Exercises 95–97 will help you prepare for the material covered in the next section.

95. In parts (a) and (b), complete each statement.

 a. $b^4 \cdot b^3 = (b \cdot b \cdot b \cdot b)(b \cdot b \cdot b) = b^?$

 b. $b^5 \cdot b^5 = (b \cdot b \cdot b \cdot b \cdot b)(b \cdot b \cdot b \cdot b \cdot b) = b^?$

 c. Generalizing from parts (a) and (b), what should be done with the exponents when multiplying exponential expressions with the same base?

96. In parts (a) and (b), complete each statement.

 a. $\dfrac{b^7}{b^3} = \dfrac{b \cdot b \cdot b \cdot b \cdot b \cdot b \cdot b}{b \cdot b \cdot b} = b^?$

 b. $\dfrac{b^8}{b^2} = \dfrac{b \cdot b \cdot b \cdot b \cdot b \cdot b \cdot b \cdot b}{b \cdot b} = b^?$

 c. Generalizing from parts (a) and (b), what should be done with the exponents when dividing exponential expressions with the same base?

97. If 6.2 is multiplied by 10^3, what does this multiplication do to the decimal point in 6.2?

SECTION P.3 Exponents and Scientific Notation

Objectives

1. Use the product rule.
2. Use the quotient rule.
3. Use the zero-exponent rule.
4. Use the negative-exponent rule.
5. Use the power rule.
6. Find the power of a product.
7. Find the power of a quotient.
8. Simplify exponential expressions.
9. Use scientific notation.

Bigger than the biggest thing ever and then some. Much bigger than that in fact, really amazingly immense, a totally stunning size, real 'wow, that's big', time ... Gigantic multiplied by colossal multiplied by staggeringly huge is the sort of concept we're trying to get across here.

Douglas Adams, *The Restaurant at the End of the Universe*

Although Adams's description may not quite apply to this $15.2 trillion national debt, exponents can be used to explore the meaning of this "staggeringly huge" number. In this section, you will learn to use exponents to provide a way of putting large and small numbers in perspective.

The Product and Quotient Rules

We have seen that exponents are used to indicate repeated multiplication. Now consider the multiplication of two exponential expressions, such as $b^4 \cdot b^3$. We are multiplying 4 factors of b and 3 factors of b. We have a total of 7 factors of b:

4 factors 3 factors
of b of b

$$b^4 \cdot b^3 = (b \cdot b \cdot b \cdot b)(b \cdot b \cdot b) = b^7.$$

Total: 7 factors of b

The product is exactly the same if we add the exponents:

$$b^4 \cdot b^3 = b^{4+3} = b^7.$$

The fact that $b^4 \cdot b^3 = b^7$ suggests the following rule:

① Use the product rule.

The Product Rule

$$b^m \cdot b^n = b^{m+n}$$

When multiplying exponential expressions with the same base, add the exponents. Use this sum as the exponent of the common base.

EXAMPLE 1 Using the Product Rule

Multiply each expression using the product rule:

a. $2^2 \cdot 2^3$ **b.** $(6x^4y^3)(5x^2y^7)$.

SOLUTION

a. $2^2 \cdot 2^3 = 2^{2+3} = 2^5$ or 32 $\boxed{2^5 = 2 \cdot 2 \cdot 2 \cdot 2 \cdot 2 = 32}$

b. $(6x^4y^3)(5x^2y^7)$

$\qquad = 6 \cdot 5 \cdot x^4 \cdot x^2 \cdot y^3 \cdot y^7$ Use the associative and commutative properties. This step can be done mentally.

$\qquad = 30x^{4+2}y^{3+7}$

$\qquad = 30x^6y^{10}$ • • •

✓ Check Point **1** Multiply each expression using the product rule:
a. $3^3 \cdot 3^2$ **b.** $(4x^3y^4)(10x^2y^6)$.

② Use the quotient rule.

Now, consider the division of two exponential expressions, such as the quotient of b^7 and b^3. We are dividing 7 factors of b by 3 factors of b.

$$\frac{b^7}{b^3} = \frac{b \cdot b \cdot b \cdot b \cdot b \cdot b \cdot b}{b \cdot b \cdot b} = \boxed{\frac{b \cdot b \cdot b}{b \cdot b \cdot b}} \cdot b \cdot b \cdot b \cdot b = 1 \cdot b \cdot b \cdot b \cdot b = b^4$$

This factor is equal to 1.

The quotient is exactly the same if we subtract the exponents:

$$\frac{b^7}{b^3} = b^{7-3} = b^4.$$

This suggests the following rule:

The Quotient Rule

$$\frac{b^m}{b^n} = b^{m-n}, \quad b \neq 0$$

When dividing exponential expressions with the same nonzero base, subtract the exponent in the denominator from the exponent in the numerator. Use this difference as the exponent of the common base.

EXAMPLE 2 Using the Quotient Rule

Divide each expression using the quotient rule:

a. $\dfrac{(-2)^7}{(-2)^4}$ **b.** $\dfrac{30x^{12}y^9}{5x^3y^7}$.

SOLUTION

a. $\dfrac{(-2)^7}{(-2)^4} = (-2)^{7-4} = (-2)^3$ or -8 $(-2)^3 = (-2)(-2)(-2) = -8$

b. $\dfrac{30x^{12}y^9}{5x^3y^7} = \dfrac{30}{5} \cdot \dfrac{x^{12}}{x^3} \cdot \dfrac{y^9}{y^7} = 6x^{12-3}y^{9-7} = 6x^9y^2$

$\bullet\bullet\bullet$

☑ **Check Point 2** Divide each expression using the quotient rule:

a. $\dfrac{(-3)^6}{(-3)^3}$ **b.** $\dfrac{27x^{14}y^8}{3x^3y^5}$.

③ Use the zero-exponent rule.

Zero as an Exponent

A nonzero base can be raised to the 0 power. The quotient rule can be used to help determine what zero as an exponent should mean. Consider the quotient of b^4 and b^4, where b is not zero. We can determine this quotient in two ways.

$$\frac{b^4}{b^4} = 1 \qquad\qquad \frac{b^4}{b^4} = b^{4-4} = b^0$$

Any nonzero expression divided by itself is 1. Use the quotient rule and subtract exponents.

This means that b^0 must equal 1.

The Zero-Exponent Rule

If b is any real number other than 0,

$$b^0 = 1.$$

Here are examples involving simplification using the zero-exponent rule:

$$8^0 = 1, \quad (-6)^0 = 1, \quad -6^0 = -1, \quad (5x)^0 = 1, \quad 5x^0 = 5.$$

Because there are no parentheses, only 6 is raised to the 0 power: $-6^0 = -(6^0) = -1$. Because there are no parentheses, only x is raised to the 0 power: $5x^0 = 5 \cdot 1 = 5$.

④ Use the negative-exponent rule.

Negative Integers as Exponents

A nonzero base can be raised to a negative power. The quotient rule can be used to help determine what a negative integer as an exponent should mean. Consider the quotient of b^3 and b^5, where b is not zero. We can determine this quotient in two ways.

$$\frac{b^3}{b^5} = \frac{\cancel{b} \cdot \cancel{b} \cdot \cancel{b}}{\cancel{b} \cdot \cancel{b} \cdot \cancel{b} \cdot b \cdot b} = \frac{1}{b^2} \qquad\qquad \frac{b^3}{b^5} = b^{3-5} = b^{-2}$$

After dividing common factors, we have two factors of b in the denominator.

Use the quotient rule and subtract exponents.

Notice that $\dfrac{b^3}{b^5}$ equals both b^{-2} and $\dfrac{1}{b^2}$. This means that b^{-2} must equal $\dfrac{1}{b^2}$. This example is a special case of the **negative-exponent rule**.

The Negative-Exponent Rule

If b is any real number other than 0 and n is a natural number, then

$$b^{-n} = \frac{1}{b^n}.$$

EXAMPLE 3 Using the Negative-Exponent Rule

Use the negative-exponent rule to write each expression with a positive exponent. Simplify, if possible:

a. 9^{-2} **b.** $(-2)^{-5}$ **c.** $\dfrac{1}{6^{-2}}$ **d.** $7x^{-5}y^2$.

SOLUTION

a. $9^{-2} = \dfrac{1}{9^2} = \dfrac{1}{81}$

b. $(-2)^{-5} = \dfrac{1}{(-2)^5} = \dfrac{1}{(-2)(-2)(-2)(-2)(-2)} = \dfrac{1}{-32} = -\dfrac{1}{32}$

Only the sign of the exponent, −5, changes. The base, −2, does not change sign.

c. $\dfrac{1}{6^{-2}} = \dfrac{1}{\dfrac{1}{6^2}} = 1 \cdot \dfrac{6^2}{1} = 6^2 = 36$

d. $7x^{-5}y^2 = 7 \cdot \dfrac{1}{x^5} \cdot y^2 = \dfrac{7y^2}{x^5}$ • • •

⊘ Check Point 3 Use the negative-exponent rule to write each expression with a positive exponent. Simplify, if possible:

a. 5^{-2} **b.** $(-3)^{-3}$ **c.** $\dfrac{1}{4^{-2}}$ **d.** $3x^{-6}y^4$.

In Example 3 and Check Point 3, did you notice that

$$\frac{1}{6^{-2}} = 6^2 \qquad \text{and} \qquad \frac{1}{4^{-2}} = 4^2?$$

In general, if a negative exponent appears in a denominator, an expression can be written with a positive exponent using

$$\frac{1}{b^{-n}} = b^n.$$

> **Negative Exponents in Numerators and Denominators**
>
> If b is any real number other than 0 and n is a natural number, then
>
> $$b^{-n} = \frac{1}{b^n} \quad \text{and} \quad \frac{1}{b^{-n}} = b^n.$$
>
> When a negative number appears as an exponent, switch the position of the base (from numerator to denominator or from denominator to numerator) and make the exponent positive. The sign of the base does not change.

⑤ Use the power rule.

The Power Rule for Exponents (Powers to Powers)

The next property of exponents applies when an exponential expression is raised to a power. Here is an example:

$$(b^2)^4.$$

> The exponential expression b^2
> is raised to the fourth power.

There are four factors of b^2. Thus,

$$(b^2)^4 = b^2 \cdot b^2 \cdot b^2 \cdot b^2 = b^{2+2+2+2} = b^8.$$

> Add exponents when multiplying
> with the same base.

We can obtain the answer, b^8, by multiplying the original exponents:

$$(b^2)^4 = b^{2 \cdot 4} = b^8.$$

This suggests the following rule:

> **The Power Rule (Powers to Powers)**
>
> $$(b^m)^n = b^{mn}$$
>
> When an exponential expression is raised to a power, multiply the exponents. Place the product of the exponents on the base and remove the parentheses.

EXAMPLE 4 Using the Power Rule (Powers to Powers)

Simplify each expression using the power rule:

a. $(2^2)^3$ **b.** $(y^5)^{-3}$ **c.** $(b^{-4})^{-2}.$

SOLUTION

a. $(2^2)^3 = 2^{2 \cdot 3} = 2^6$ or 64 **b.** $(y^5)^{-3} = y^{5(-3)} = y^{-15} = \dfrac{1}{y^{15}}$

c. $(b^{-4})^{-2} = b^{(-4)(-2)} = b^8$

• • •

> ✓ Check Point **4** Simplify each expression using the power rule:
>
> **a.** $(3^3)^2$ **b.** $(y^7)^{-2}$ **c.** $(b^{-3})^{-4}.$

⑥ Find the power of a product.

The Products-to-Powers Rule for Exponents

The next property of exponents applies when we are raising a product to a power. Here is an example:

$$(2x)^4.$$

The product $2x$ is raised to the fourth power.

There are four factors of $2x$. Thus,

$$(2x)^4 = 2x \cdot 2x \cdot 2x \cdot 2x = 2 \cdot 2 \cdot 2 \cdot 2 \cdot x \cdot x \cdot x \cdot x = 2^4 x^4.$$

We can obtain the answer, $2^4 x^4$, by raising each factor within the parentheses to the fourth power:

$$(2x)^4 = 2^4 x^4.$$

The fact that $(2x)^4 = 2^4 x^4$ suggests the following rule:

Products to Powers

$$(ab)^n = a^n b^n$$

When a product is raised to a power, raise each factor to that power.

EXAMPLE 5　Raising a Product to a Power

Simplify: $(-2y^2)^4$.

SOLUTION

$$(-2y^2)^4 = (-2)^4 (y^2)^4 \quad \text{Raise each factor to the fourth power.}$$
$$= (-2)^4 y^{2 \cdot 4} \quad \text{To raise an exponential expression to a power, multiply exponents: } (b^m)^n = b^{mn}.$$
$$= 16y^8 \quad \text{Simplify: } (-2)^4 = (-2)(-2)(-2)(-2) = 16. \quad \bullet\bullet\bullet$$

✓ Check Point **5** Simplify: $(-4x)^3$.

The rule for raising a product to a power can be extended to cover three or more factors. For example,

$$(-2xy)^3 = (-2)^3 x^3 y^3 = -8x^3 y^3.$$

⑦ Find the power of a quotient.

The Quotients-to-Powers Rule for Exponents

The following rule is used to raise a quotient to a power:

Quotients to Powers

If b is a nonzero real number, then

$$\left(\frac{a}{b}\right)^n = \frac{a^n}{b^n}.$$

When a quotient is raised to a power, raise the numerator to that power and divide by the denominator to that power.

EXAMPLE 6　Raising Quotients to Powers

Simplify by raising each quotient to the given power:

a. $\left(-\dfrac{3}{x}\right)^4$ 　　　**b.** $\left(\dfrac{x^2}{4}\right)^3.$

SOLUTION

a. $\left(-\dfrac{3}{x}\right)^4 = \dfrac{(-3)^4}{x^4} = \dfrac{(-3)(-3)(-3)(-3)}{x^4} = \dfrac{81}{x^4}$

b. $\left(\dfrac{x^2}{4}\right)^3 = \dfrac{(x^2)^3}{4^3} = \dfrac{x^{2 \cdot 3}}{4 \cdot 4 \cdot 4} = \dfrac{x^6}{64}$

⦸ Check Point **6** Simplify:

a. $\left(-\dfrac{2}{y}\right)^5$ b. $\left(\dfrac{x^5}{3}\right)^3$.

8 Simplify exponential expressions.

Simplifying Exponential Expressions

Properties of exponents are used to simplify exponential expressions. An exponential expression is **simplified** when

- No parentheses appear.
- No powers are raised to powers.
- Each base occurs only once.
- No negative or zero exponents appear.

Simplifying Exponential Expressions

	Example
1. If necessary, remove parentheses by using $$(ab)^n = a^n b^n \quad \text{or} \quad \left(\dfrac{a}{b}\right)^n = \dfrac{a^n}{b^n}.$$	$(xy)^3 = x^3 y^3$
2. If necessary, simplify powers to powers by using $$(b^m)^n = b^{mn}.$$	$(x^4)^3 = x^{4 \cdot 3} = x^{12}$
3. If necessary, be sure that each base appears only once by using $$b^m \cdot b^n = b^{m+n} \quad \text{or} \quad \dfrac{b^m}{b^n} = b^{m-n}.$$	$x^4 \cdot x^3 = x^{4+3} = x^7$
4. If necessary, rewrite exponential expressions with zero powers as 1 ($b^0 = 1$). Furthermore, write the answer with positive exponents by using $$b^{-n} = \dfrac{1}{b^n} \quad \text{or} \quad \dfrac{1}{b^{-n}} = b^n.$$	$\dfrac{x^5}{x^8} = x^{5-8} = x^{-3} = \dfrac{1}{x^3}$

The following example shows how to simplify exponential expressions. Throughout the example, assume that no variable in a denominator is equal to zero.

EXAMPLE 7 Simplifying Exponential Expressions

Simplify:

a. $(-3x^4y^5)^3$ b. $(-7xy^4)(-2x^5y^6)$ c. $\dfrac{-35x^2y^4}{5x^6y^{-8}}$ d. $\left(\dfrac{4x^2}{y}\right)^{-3}$.

SOLUTION

a. $(-3x^4y^5)^3 = (-3)^3(x^4)^3(y^5)^3$ Raise each factor inside the parentheses to the third power.

$$= (-3)^3 x^{4 \cdot 3} y^{5 \cdot 3}$$ Multiply the exponents when raising powers to powers.

$$= -27x^{12}y^{15}$$ $(-3)^3 = (-3)(-3)(-3) = -27$

b. $(-7xy^4)(-2x^5y^6) = (-7)(-2)xx^5y^4y^6$ Group factors with the same base.

$$= 14x^{1+5}y^{4+6}$$ When multiplying expressions with the same base, add the exponents.

$$= 14x^6y^{10}$$ Simplify.

c. $\dfrac{-35x^2y^4}{5x^6y^{-8}} = \left(\dfrac{-35}{5}\right)\left(\dfrac{x^2}{x^6}\right)\left(\dfrac{y^4}{y^{-8}}\right)$ Group factors with the same base.

$$= -7x^{2-6}y^{4-(-8)}$$ When dividing expressions with the same base, subtract the exponents.

$$= -7x^{-4}y^{12}$$ Simplify. Notice that $4 - (-8) = 4 + 8 = 12$.

$$= \dfrac{-7y^{12}}{x^4}$$ Write as a fraction and move the base with the negative exponent, x^{-4}, to the other side of the fraction bar and make the negative exponent positive.

d. $\left(\dfrac{4x^2}{y}\right)^{-3} = \dfrac{(4x^2)^{-3}}{y^{-3}}$ Raise the numerator and the denominator to the -3 power.

$$= \dfrac{4^{-3}(x^2)^{-3}}{y^{-3}}$$ Raise each factor inside the parentheses to the -3 power.

$$= \dfrac{4^{-3}x^{-6}}{y^{-3}}$$ Multiply the exponents when raising a power to a power: $(x^2)^{-3} = x^{2(-3)} = x^{-6}$.

$$= \dfrac{y^3}{4^3 x^6}$$ Move each base with a negative exponent to the other side of the fraction bar and make each negative exponent positive.

$$= \dfrac{y^3}{64x^6}$$ $4^3 = 4 \cdot 4 \cdot 4 = 64$ • • •

✓ Check Point **7** Simplify:

a. $(2x^3y^6)^4$

b. $(-6x^2y^5)(3xy^3)$

c. $\dfrac{100x^{12}y^2}{20x^{16}y^{-4}}$

d. $\left(\dfrac{5x}{y^4}\right)^{-2}$.

GREAT QUESTION!

Simplifying exponential expressions seems to involve lots of steps. Are there common errors I can avoid along the way?

Yes. Here's a list. The first column has the correct simplification. The second column contains common errors you should try to avoid.

Correct	Incorrect	Description of Error
$b^3 \cdot b^4 = b^7$	$b^3 \cdot b^4 = b^{12}$	The exponents should be added, not multiplied.
$3^2 \cdot 3^4 = 3^6$	$3^2 \cdot 3^4 = 9^6$	The common base should be retained, not multiplied.
$\dfrac{5^{16}}{5^4} = 5^{12}$	$\dfrac{5^{16}}{5^4} = 5^4$	The exponents should be subtracted, not divided.
$(4a)^3 = 64a^3$	$(4a)^3 = 4a^3$	Both factors should be cubed.
$b^{-n} = \dfrac{1}{b^n}$	$b^{-n} = -\dfrac{1}{b^n}$	Only the exponent should change sign.
$(a+b)^{-1} = \dfrac{1}{a+b}$	$(a+b)^{-1} = \dfrac{1}{a} + \dfrac{1}{b}$	The exponent applies to the entire expression $a + b$.

⑨ Use scientific notation.

Table P.5 Names of Large Numbers

10^2	hundred
10^3	thousand
10^6	million
10^9	billion
10^{12}	trillion
10^{15}	quadrillion
10^{18}	quintillion
10^{21}	sextillion
10^{24}	septillion
10^{27}	octillion
10^{30}	nonillion
10^{100}	googol
10^{googol}	googolplex

Scientific Notation

As of December 2011, the national debt of the United States was about $15.2 trillion. This is the amount of money the government has had to borrow over the years, mostly by selling bonds, because it has spent more than it has collected in taxes. A stack of $1 bills equaling the national debt would measure more than 950,000 miles. That's more than two round trips from Earth to the moon. Because a trillion is 10^{12} (see **Table P.5**), the national debt can be expressed as

$$15.2 \times 10^{12}.$$

Because $15.2 = 1.52 \times 10$, the national debt can be expressed as

$$15.2 \times 10^{12} = (1.52 \times 10) \times 10^{12} = 1.52 \times (10 \times 10^{12})$$
$$= 1.52 \times 10^{1+12} = 1.52 \times 10^{13}$$

The number 1.52×10^{13} is written in a form called *scientific notation.*

Scientific Notation

A number is written in **scientific notation** when it is expressed in the form

$$a \times 10^n,$$

where the absolute value of a is greater than or equal to 1 and less than 10 $(1 \le |a| < 10)$, and n is an integer.

It is customary to use the multiplication symbol, \times, rather than a dot, when writing a number in scientific notation.

Converting from Scientific to Decimal Notation

Here are two examples of numbers in scientific notation:

$$6.4 \times 10^5 \quad \text{means} \quad 640{,}000.$$
$$2.17 \times 10^{-3} \quad \text{means} \quad 0.00217.$$

Do you see that the number with the positive exponent is relatively large and the number with the negative exponent is relatively small?

We can use n, the exponent on the 10 in $a \times 10^n$, to change a number in scientific notation to decimal notation. If n is **positive**, move the decimal point in a to the **right** n places. If n is **negative**, move the decimal point in a to the **left** $|n|$ places.

EXAMPLE 8 Converting from Scientific to Decimal Notation

Write each number in decimal notation:

 a. 6.2×10^7 **b.** -6.2×10^7 **c.** 2.019×10^{-3} **d.** -2.019×10^{-3}.

SOLUTION

In each case, we use the exponent on the 10 to determine how far to move the decimal point and in which direction. In parts (a) and (b), the exponent is positive, so we move the decimal point to the right. In parts (c) and (d), the exponent is negative, so we move the decimal point to the left.

a. $6.2 \times 10^7 = 62,000,000$

 $n = 7$ Move the decimal point 7 places to the right.

b. $-6.2 \times 10^7 = -62,000,000$

 $n = 7$ Move the decimal point 7 places to the right.

c. $2.019 \times 10^{-3} = 0.002019$

 $n = -3$ Move the decimal point $|-3|$ places, or 3 places, to the left.

d. $-2.019 \times 10^{-3} = -0.002019$

 $n = -3$ Move the decimal point $|-3|$ places, or 3 places, to the left.

 • • •

✓ Check Point **8** Write each number in decimal notation:

 a. -2.6×10^9 **b.** 3.017×10^{-6}.

Converting from Decimal to Scientific Notation

To convert from decimal notation to scientific notation, we reverse the procedure of Example 8.

> **Converting from Decimal to Scientific Notation**
>
> Write the number in the form $a \times 10^n$.
>
> • Determine a, the numerical factor. Move the decimal point in the given number to obtain a number whose absolute value is between 1 and 10, including 1.
> • Determine n, the exponent on 10^n. The absolute value of n is the number of places the decimal point was moved. The exponent n is positive if the decimal point was moved to the left, negative if the decimal point was moved to the right, and 0 if the decimal point was not moved.

EXAMPLE 9 Converting from Decimal Notation to Scientific Notation

Write each number in scientific notation:

 a. 34,970,000,000,000 **b.** $-34,970,000,000,000$

 c. 0.0000000000802 **d.** -0.0000000000802.

SOLUTION

a. $34,970,000,000,000 = 3.497 \times 10^{13}$

 Move the decimal point to get a number whose absolute value is between 1 and 10. The decimal point was moved 13 places to the left, so $n = 13$.

b. $-34,970,000,000,000 = -3.497 \times 10^{13}$

TECHNOLOGY

You can use your calculator's EE (enter exponent) or EXP key to convert from decimal to scientific notation. Here is how it's done for 0.0000000000802.

Many Scientific Calculators

Keystrokes

.0000000000802 EE =

Display

8.02 − 11

Many Graphing Calculators

Use the mode setting for scientific notation.

Keystrokes

.0000000000802 ENTER

Display

8.02E − 11

c. $0.0000000000802 = 8.02 \times 10^{-11}$

> Move the decimal point to get a number whose absolute value is between 1 and 10.

> The decimal point was moved 11 places to the right, so $n = -11$.

d. $-0.0000000000802 = -8.02 \times 10^{-11}$

⊘ Check Point **9** Write each number in scientific notation:

a. 5,210,000,000 **b.** −0.00000006893.

GREAT QUESTION!

In scientific notation, which numbers have positive exponents and which have negative exponents?

If the absolute value of a number is greater than 10, it will have a positive exponent in scientific notation. If the absolute value of a number is less than 1, it will have a negative exponent in scientific notation.

EXAMPLE 10 Expressing the U.S. Population in Scientific Notation

As of December 2011, the population of the United States was approximately 312 million. Express the population in scientific notation.

SOLUTION

Because a million is 10^6, the 2011 population can be expressed as

$$312 \times 10^6.$$

> This factor is not between 1 and 10, so the number is not in scientific notation.

The voice balloon indicates that we need to convert 312 to scientific notation.

$$312 \times 10^6 = (3.12 \times 10^2) \times 10^6 = 3.12 \times 10^{2+6} = 3.12 \times 10^8$$

> $312 = 3.12 \times 10^2$

In scientific notation, the population is 3.12×10^8. •••

⊘ Check Point **10** Express 410×10^7 in scientific notation.

Computations with Scientific Notation

Properties of exponents are used to perform computations with numbers that are expressed in scientific notation.

TECHNOLOGY

$(6.1 \times 10^5)(4 \times 10^{-9})$
On a Calculator:

Many Scientific Calculators

6.1 EE 5 × 4 EE 9 +/− =

Display

2.44 − 03

Many Graphing Calculators

6.1 EE 5 × 4 EE (−) 9 ENTER

Display (in scientific notation mode)

2.44E − 3

EXAMPLE 11 Computations with Scientific Notation

Perform the indicated computations, writing the answers in scientific notation:

a. $(6.1 \times 10^5)(4 \times 10^{-9})$ **b.** $\dfrac{1.8 \times 10^4}{3 \times 10^{-2}}$.

SOLUTION

a. $(6.1 \times 10^5)(4 \times 10^{-9})$

$= (6.1 \times 4) \times (10^5 \times 10^{-9})$ Regroup factors.

$= 24.4 \times 10^{5+(-9)}$ Add the exponents on 10 and multiply the other parts.

$= 24.4 \times 10^{-4}$ Simplify.

$= (2.44 \times 10^1) \times 10^{-4}$ Convert 24.4 to scientific notation: $24.4 = 2.44 \times 10^1$.

$= 2.44 \times 10^{-3}$ $10^1 \times 10^{-4} = 10^{1+(-4)} = 10^{-3}$

b. $\dfrac{1.8 \times 10^4}{3 \times 10^{-2}} = \left(\dfrac{1.8}{3}\right) \times \left(\dfrac{10^4}{10^{-2}}\right)$ Regroup factors.

$= 0.6 \times 10^{4-(-2)}$ Subtract the exponents on 10 and divide the other parts.

$= 0.6 \times 10^6$ Simplify: $4 - (-2) = 4 + 2 = 6$.

$= (6 \times 10^{-1}) \times 10^6$ Convert 0.6 to scientific notation: $0.6 = 6 \times 10^{-1}$.

$= 6 \times 10^5$ $10^{-1} \times 10^6 = 10^{-1+6} = 10^5$ •••

🗸 Check Point **11** Perform the indicated computations, writing the answers in scientific notation:

a. $(7.1 \times 10^5)(5 \times 10^{-7})$ **b.** $\dfrac{1.2 \times 10^6}{3 \times 10^{-3}}$.

Applications: Putting Numbers in Perspective

Due to tax cuts and spending increases, the United States began accumulating large deficits in the 1980s. To finance the deficit, the government had borrowed $15.2 trillion as of December 2011. The graph in **Figure P.11** shows the national debt increasing over time.

The National Debt

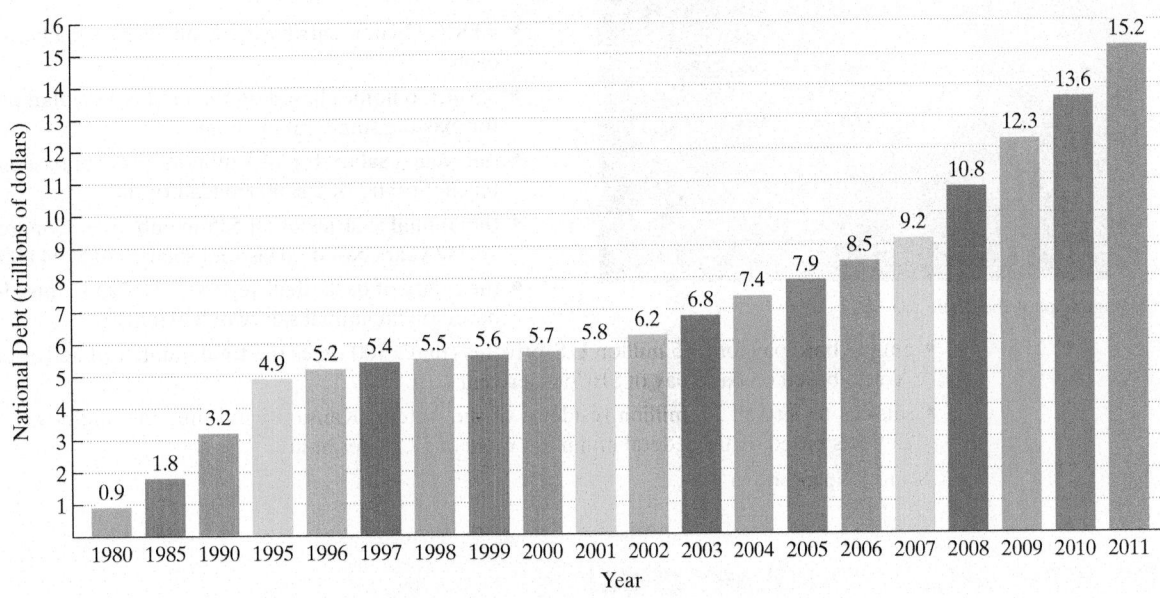

FIGURE P.11
Source: Office of Management and Budget

Example 12 shows how we can use scientific notation to comprehend the meaning of a number such as 15.2 trillion.

EXAMPLE 12 The National Debt

As of December 2011, the national debt was $15.2 trillion, or 15.2×10^{12} dollars. At that time, the U.S. population was approximately 312,000,000 (312 million), or 3.12×10^8. If the national debt was evenly divided among every individual in the United States, how much would each citizen have to pay?

SOLUTION

The amount each citizen must pay is the total debt, 15.2×10^{12} dollars, divided by the number of citizens, 3.12×10^{8}.

$$\frac{15.2 \times 10^{12}}{3.12 \times 10^{8}} = \left(\frac{15.2}{3.12}\right) \times \left(\frac{10^{12}}{10^{8}}\right)$$

$$\approx 4.87 \times 10^{12-8}$$

$$= 4.87 \times 10^{4}$$

$$= 48{,}700$$

Every U.S. citizen would have to pay approximately $48,700 to the federal government to pay off the national debt. •••

☑ Check Point **12** As of December 2011, the United States had spent $2.6 trillion for the wars in Iraq and Afghanistan. (Source: costsofwar.org) At that time, the U.S. population was approximately 312 million (3.12×10^{8}). If the cost of these wars was evenly divided among every individual in the United States, how much, to the nearest hundred dollars, would each citizen have to pay?

Blitzer Bonus ‖ Seven Ways to Spend $1 Trillion

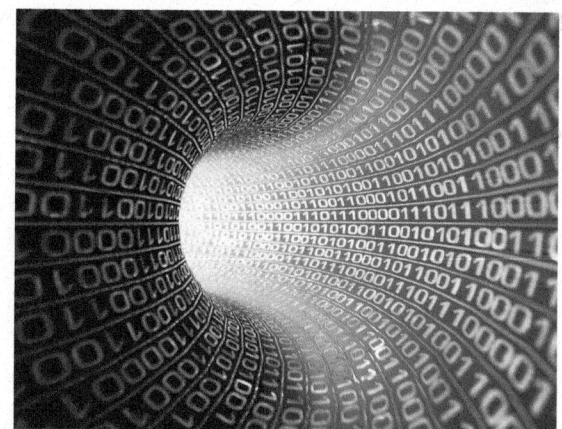

Image © photobank.kiev.ua, 2009

Confronting a national debt of $15.2 trillion starts with grasping just how colossal $1 trillion ($1 \times 10^{12}$) actually is. To help you wrap your head around this mind-boggling number, and to put the national debt in further perspective, consider what $1 trillion will buy:

- 40,816,326 new cars based on an average sticker price of $24,500 each
- 5,574,136 homes based on the national median price of $179,400 for existing single-family homes
- one year's salary for 14.7 million teachers based on the average teacher salary of $68,000 in California
- the annual salaries of all 535 members of Congress for the next 10,742 years based on current salaries of $174,000 per year
- the salary of basketball superstar LeBron James for 50,000 years based on an annual salary of $20 million
- annual base pay for 59.5 million U.S. privates (that's 100 times the total number of active-duty soldiers in the Army) based on basic pay of $16,794 per year
- salaries to hire all 2.8 million residents of the state of Kansas in full-time minimum-wage jobs for the next 23 years based on the federal minimum wage of $7.25 per hour

Source: Kiplinger.com

CONCEPT AND VOCABULARY CHECK

Fill in each blank so that the resulting statement is true.

1. The product rule for exponents states that $b^{m} \cdot b^{n} =$ _____. When multiplying exponential expressions with the same base, _____ the exponents.

2. The quotient rule for exponents states that $\dfrac{b^{m}}{b^{n}} =$
 _____, $b \neq 0$. When dividing exponential expressions with the same nonzero base, _____ the exponents.

3. If $b \neq 0$, then $b^{0} =$ ____.

4. The negative-exponent rule states that $b^{-n} =$ ____, $b \neq 0$.

5. True or false: $5^{-2} = -5^{2}$ _____

6. Negative exponents in denominators can be evaluated using $\dfrac{1}{b^{-n}} =$ ____, $b \neq 0$.

7. True or false: $\dfrac{1}{8^{-2}} = 8^{2}$ _____

8. A positive number is written in scientific notation when it is expressed in the form $a \times 10^{n}$, where a is _____ and n is a/an _____.

9. True or false: 4×10^{3} is written in scientific notation. _____

10. True or false: 40×10^{2} is written in scientific notation. _____

EXERCISE SET P.3

Practice Exercises

Evaluate each exponential expression in Exercises 1–22.

1. $5^2 \cdot 2$ **2.** $6^2 \cdot 2$

3. $(-2)^6$ **4.** $(-2)^4$

5. -2^6 **6.** -2^4

7. $(-3)^0$ **8.** $(-9)^0$

9. -3^0 **10.** -9^0

11. 4^{-3} **12.** 2^{-6}

13. $2^2 \cdot 2^3$ **14.** $3^3 \cdot 3^2$

15. $(2^2)^3$ **16.** $(3^3)^2$

17. $\dfrac{2^8}{2^4}$ **18.** $\dfrac{3^8}{3^4}$

19. $3^{-3} \cdot 3$ **20.** $2^{-3} \cdot 2$

21. $\dfrac{2^3}{2^7}$ **22.** $\dfrac{3^4}{3^7}$

Simplify each exponential expression in Exercises 23–64.

23. $x^{-2}y$ **24.** xy^{-3}

25. $x^0 y^5$ **26.** $x^7 y^0$

27. $x^3 \cdot x^7$ **28.** $x^{11} \cdot x^5$

29. $x^{-5} \cdot x^{10}$ **30.** $x^{-6} \cdot x^{12}$

31. $(x^3)^7$ **32.** $(x^{11})^5$

33. $(x^{-5})^3$ **34.** $(x^{-6})^4$

35. $\dfrac{x^{14}}{x^7}$ **36.** $\dfrac{x^{30}}{x^{10}}$

37. $\dfrac{x^{14}}{x^{-7}}$ **38.** $\dfrac{x^{30}}{x^{-10}}$

39. $(8x^3)^2$ **40.** $(6x^4)^2$

41. $\left(-\dfrac{4}{x}\right)^3$ **42.** $\left(-\dfrac{6}{y}\right)^3$

43. $(-3x^2y^5)^2$ **44.** $(-3x^4y^6)^3$

45. $(3x^4)(2x^7)$ **46.** $(11x^5)(9x^{12})$

47. $(-9x^3y)(-2x^6y^4)$ **48.** $(-5x^4y)(-6x^7y^{11})$

49. $\dfrac{8x^{20}}{2x^4}$ **50.** $\dfrac{20x^{24}}{10x^6}$

51. $\dfrac{25a^{13}b^4}{-5a^2b^3}$ **52.** $\dfrac{35a^{14}b^6}{-7a^7b^3}$

53. $\dfrac{14b^7}{7b^{14}}$ **54.** $\dfrac{20b^{10}}{10b^{20}}$

55. $(4x^3)^{-2}$ **56.** $(10x^2)^{-3}$

57. $\dfrac{24x^3y^5}{32x^7y^{-9}}$ **58.** $\dfrac{10x^4y^9}{30x^{12}y^{-3}}$

59. $\left(\dfrac{5x^3}{y}\right)^{-2}$ **60.** $\left(\dfrac{3x^4}{y}\right)^{-3}$

61. $\left(\dfrac{-15a^4b^2}{5a^{10}b^{-3}}\right)^3$ **62.** $\left(\dfrac{-30a^{14}b^8}{10a^{17}b^{-2}}\right)^3$

63. $\left(\dfrac{3a^{-5}b^2}{12a^3b^{-4}}\right)^0$ **64.** $\left(\dfrac{4a^{-5}b^3}{12a^3b^{-5}}\right)^0$

In Exercises 65–76, write each number in decimal notation without the use of exponents.

65. 3.8×10^2 **66.** 9.2×10^2

67. 6×10^{-4} **68.** 7×10^{-5}

69. -7.16×10^6 **70.** -8.17×10^6

71. 7.9×10^{-1} **72.** 6.8×10^{-1}

73. -4.15×10^{-3} **74.** -3.14×10^{-3}

75. -6.00001×10^{10} **76.** -7.00001×10^{10}

In Exercises 77–86, write each number in scientific notation.

77. 32,000 **78.** 64,000

79. 638,000,000,000,000,000 **80.** 579,000,000,000,000,000

81. -5716 **82.** -3829

83. 0.0027 **84.** 0.0083

85. -0.00000000504 **86.** -0.00000000405

In Exercises 87–106, perform the indicated computations. Write the answers in scientific notation. If necessary, round the decimal factor in your scientific notation answer to two decimal places.

87. $(3 \times 10^4)(2.1 \times 10^3)$ **88.** $(2 \times 10^4)(4.1 \times 10^3)$

89. $(1.6 \times 10^{15})(4 \times 10^{-11})$ **90.** $(1.4 \times 10^{15})(3 \times 10^{-11})$

91. $(6.1 \times 10^{-8})(2 \times 10^{-4})$ **92.** $(5.1 \times 10^{-8})(3 \times 10^{-4})$

93. $(4.3 \times 10^8)(6.2 \times 10^4)$ **94.** $(8.2 \times 10^8)(4.6 \times 10^4)$

95. $\dfrac{8.4 \times 10^8}{4 \times 10^5}$ **96.** $\dfrac{6.9 \times 10^8}{3 \times 10^5}$

97. $\dfrac{3.6 \times 10^4}{9 \times 10^{-2}}$ **98.** $\dfrac{1.2 \times 10^4}{2 \times 10^{-2}}$

99. $\dfrac{4.8 \times 10^{-2}}{2.4 \times 10^6}$ **100.** $\dfrac{7.5 \times 10^{-2}}{2.5 \times 10^6}$

101. $\dfrac{2.4 \times 10^{-2}}{4.8 \times 10^{-6}}$ **102.** $\dfrac{1.5 \times 10^{-2}}{3 \times 10^{-6}}$

103. $\dfrac{480,000,000,000}{0.00012}$ **104.** $\dfrac{282,000,000,000}{0.00141}$

105. $\dfrac{0.00072 \times 0.003}{0.00024}$ **106.** $\dfrac{66,000 \times 0.001}{0.003 \times 0.002}$

Practice Plus

In Exercises 107–114, simplify each exponential expression. Assume that variables represent nonzero real numbers.

107. $\dfrac{(x^{-2}y)^{-3}}{(x^2y^{-1})^3}$ **108.** $\dfrac{(xy^{-2})^{-2}}{(x^{-2}y)^{-3}}$

109. $(2x^{-3}yz^{-6})(2x)^{-5}$ **110.** $(3x^{-4}yz^{-7})(3x)^{-3}$

111. $\left(\dfrac{x^3y^4z^5}{x^{-3}y^{-4}z^{-5}}\right)^{-2}$ **112.** $\left(\dfrac{x^4y^5z^6}{x^{-4}y^{-5}z^{-6}}\right)^{-4}$

113. $\dfrac{(2^{-1}x^{-2}y^{-1})^{-2}(2x^{-4}y^3)^{-2}(16x^{-3}y^3)^0}{(2x^{-3}y^{-5})^2}$

114. $\dfrac{(2^{-1}x^{-3}y^{-1})^{-2}(2x^{-6}y^4)^{-2}(9x^3y^{-3})^0}{(2x^{-4}y^{-6})^2}$

Application Exercises

The bar graph shows the total amount Americans paid in federal taxes, in trillions of dollars, and the U.S. population, in millions, from 2007 through 2010. Exercises 115–116 are based on the numbers displayed by the graph.

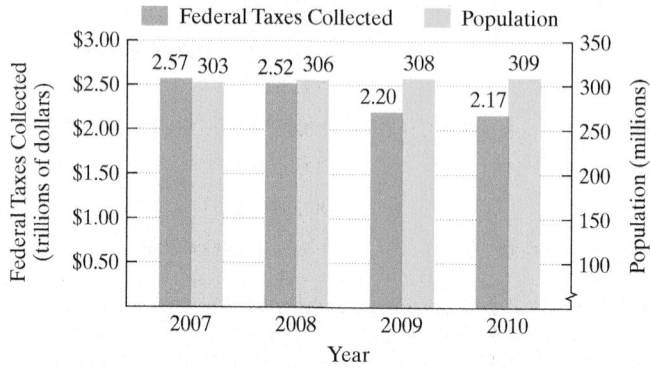

Federal Taxes and the United States Population

Sources: Internal Revenue Service and U.S. Census Bureau

115. a. In 2010, the United States government collected $2.17 trillion in taxes. Express this number in scientific notation.

 b. In 2010, the population of the United States was approximately 309 million. Express this number in scientific notation.

 c. Use your scientific notation answers from parts (a) and (b) to answer this question: If the total 2010 tax collections were evenly divided among all Americans, how much would each citizen pay? Express the answer in decimal notation, rounded to the nearest dollar.

116. a. In 2009, the United States government collected $2.20 trillion in taxes. Express this number in scientific notation.

 b. In 2009, the population of the United States was approximately 308 million. Express this number in scientific notation.

 c. Use your scientific notation answers from parts (a) and (b) to answer this question: If the total 2009 tax collections were evenly divided among all Americans, how much would each citizen pay? Express the answer in decimal notation, rounded to the nearest dollar.

In the dramatic arts, ours is the era of the movies. As individuals and as a nation, we've grown up with them. Our images of love, war, family, country—even of things that terrify us—owe much to what we've seen on screen. The bar graph at the top of the next column quantifies our love for movies by showing the number of tickets sold, in millions, and the average price per ticket for five selected years. Exercises 117–118 are based on the numbers displayed by the graph.

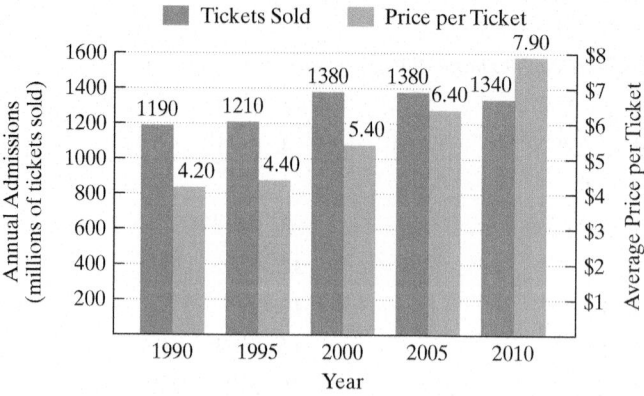

United States Film Admissions and Admission Charges

Source: Motion Picture Association of America

117. Use scientific notation to compute the amount of money that the motion picture industry made from box-office receipts in 2010. Express the answer in scientific notation.

118. Use scientific notation to compute the amount of money that the motion picture industry made from box office receipts in 2005. Express the answer in scientific notation.

119. The mass of one oxygen molecule is 5.3×10^{-23} gram. Find the mass of 20,000 molecules of oxygen. Express the answer in scientific notation.

120. The mass of one hydrogen atom is 1.67×10^{-24} gram. Find the mass of 80,000 hydrogen atoms. Express the answer in scientific notation.

121. There are approximately 3.2×10^7 seconds in a year. According to the United States Department of Agriculture, Americans consume 127 chickens per second. How many chickens are eaten per year in the United States? Express the answer in scientific notation.

122. Convert 365 days (one year) to hours, to minutes, and, finally, to seconds, to determine how many seconds there are in a year. Express the answer in scientific notation.

Writing in Mathematics

123. Describe what it means to raise a number to a power. In your description, include a discussion of the difference between -5^2 and $(-5)^2$.

124. Explain the product rule for exponents. Use $2^3 \cdot 2^5$ in your explanation.

125. Explain the power rule for exponents. Use $(3^2)^4$ in your explanation.

126. Explain the quotient rule for exponents. Use $\dfrac{5^8}{5^2}$ in your explanation.

127. Why is $(-3x^2)(2x^{-5})$ not simplified? What must be done to simplify the expression?

128. How do you know if a number is written in scientific notation?

129. Explain how to convert from scientific to decimal notation and give an example.

130. Explain how to convert from decimal to scientific notation and give an example.

131. Refer to the Blitzer Bonus on page 42. Use scientific notation to verify any three of the bulleted items on ways to spend $1 trillion.

Critical Thinking Exercises

Make Sense? *In Exercises 132–135, determine whether each statement makes sense or does not make sense, and explain your reasoning.*

132. There are many exponential expressions that are equal to $36x^{12}$, such as $(6x^6)^2$, $(6x^3)(6x^9)$, $36(x^3)^9$, and $6^2(x^2)^6$.

133. If 5^{-2} is raised to the third power, the result is a number between 0 and 1.

134. The population of Colorado is approximately 4.6×10^{12}.

135. I just finished reading a book that contained approximately 1.04×10^5 words.

In Exercises 136–143, determine whether each statement is true or false. If the statement is false, make the necessary change(s) to produce a true statement.

136. $4^{-2} < 4^{-3}$

137. $5^{-2} > 2^{-5}$

138. $(-2)^4 = 2^{-4}$

139. $5^2 \cdot 5^{-2} > 2^5 \cdot 2^{-5}$

140. $534.7 = 5.347 \times 10^3$

141. $\dfrac{8 \times 10^{30}}{4 \times 10^{-5}} = 2 \times 10^{25}$

142. $(7 \times 10^5) + (2 \times 10^{-3}) = 9 \times 10^2$

143. $(4 \times 10^3) + (3 \times 10^2) = 4.3 \times 10^3$

144. The mad Dr. Frankenstein has gathered enough bits and pieces (so to speak) for $2^{-1} + 2^{-2}$ of his creature-to-be. Write a fraction that represents the amount of his creature that must still be obtained.

145. If $b^A = MN$, $b^C = M$, and $b^D = N$, what is the relationship among A, C, and D?

146. Our hearts beat approximately 70 times per minute. Express in scientific notation how many times the heart beats over a lifetime of 80 years. Round the decimal factor in your scientific notation answer to two decimal places.

Group Exercise

147. Putting Numbers into Perspective. A large number can be put into perspective by comparing it with another number. For example, we put the \$15.2 trillion national debt (Example 12) and the \$2.17 trillion the government collected in taxes (Exercise 115) by comparing these numbers to the number of U.S. citizens.

For this project, each group member should consult an almanac, a newspaper, or the Internet to find a number greater than one million. Explain to other members of the group the context in which the large number is used. Express the number in scientific notation. Then put the number into perspective by comparing it with another number.

Preview Exercises

Exercises 148–150 will help you prepare for the material covered in the next section.

148. a. Find $\sqrt{16} \cdot \sqrt{4}$.

 b. Find $\sqrt{16 \cdot 4}$.

 c. Based on your answers to parts (a) and (b), what can you conclude?

149. a. Use a calculator to approximate $\sqrt{300}$ to two decimal places.

 b. Use a calculator to approximate $10\sqrt{3}$ to two decimal places.

 c. Based on your answers to parts (a) and (b), what can you conclude?

150. a. Simplify: $21x + 10x$.

 b. Simplify: $21\sqrt{2} + 10\sqrt{2}$.

SECTION P.4 Radicals and Rational Exponents

Objectives

1. Evaluate square roots.
2. Solve equations using the square root property.
3. Simplify expressions of the form $\sqrt{a^2}$.
4. Use the product rule to simplify square roots.
5. Use the quotient rule to simplify square roots.
6. Add and subtract square roots.
7. Rationalize denominators.
8. Evaluate and perform operations with higher roots.
9. Understand and use rational exponents.

This photograph shows mathematical models used by Albert Einstein at a lecture on relativity. Notice the radicals that appear in many of the formulas. Among these models, there is one describing how an astronaut in a moving spaceship ages more slowly than friends who remain on Earth. No description of your world can be complete without roots and radicals. In this section, in addition to reviewing the basics of radical expressions and the use of rational exponents to indicate radicals, you will see how radicals model time dilation for a futuristic high-speed trip to a nearby star.

① Evaluate square roots.

Square Roots

From our earlier work with exponents, we are aware that the square of both 5 and -5 is 25:

$$5^2 = 25 \quad \text{and} \quad (-5)^2 = 25.$$

The reverse operation of squaring a number is finding the *square root* of the number. For example,

- One square root of 25 is 5 because $5^2 = 25$.
- Another square root of 25 is -5 because $(-5)^2 = 25$.

In general, **if $b^2 = a$, then b is a square root of a.**

The symbol $\sqrt{\ }$ is used to denote the *nonnegative* or *principal square root* of a number. For example,

- $\sqrt{25} = 5$ because $5^2 = 25$ and 5 is positive.
- $\sqrt{100} = 10$ because $10^2 = 100$ and 10 is positive.

The symbol $\sqrt{\ }$ that we use to denote the principal square root is called a **radical sign**. The number under the radical sign is called the **radicand**. Together we refer to the radical sign and its radicand as a **radical expression**.

Radical sign $\quad \sqrt{a} \quad$ Radicand

Radical expression

Definition of the Principal Square Root

If a is a nonnegative real number, the nonnegative number b such that $b^2 = a$, denoted by $b = \sqrt{a}$, is the **principal square root** of a.

The symbol $-\sqrt{\ }$ is used to denote the negative square root of a number. For example,

- $-\sqrt{25} = -5$ because $(-5)^2 = 25$ and -5 is negative.
- $-\sqrt{100} = -10$ because $(-10)^2 = 100$ and -10 is negative.

EXAMPLE 1 Evaluating Square Roots

Evaluate:

a. $\sqrt{64}$ b. $-\sqrt{49}$ c. $\sqrt{\dfrac{1}{4}}$ d. $\sqrt{9 + 16}$ e. $\sqrt{9} + \sqrt{16}$.

SOLUTION

a. $\sqrt{64} = 8$ The principal square root of 64 is 8. Check: $8^2 = 64$.

b. $-\sqrt{49} = -7$ The negative square root of 49 is -7. Check: $(-7)^2 = 49$.

c. $\sqrt{\dfrac{1}{4}} = \dfrac{1}{2}$ The principal square root of $\frac{1}{4}$ is $\frac{1}{2}$. Check: $\left(\frac{1}{2}\right)^2 = \frac{1}{4}$.

d. $\sqrt{9 + 16} = \sqrt{25}$ First simplify the expression under the radical sign.

 $= 5$ Then take the principal square root of 25, which is 5.

e. $\sqrt{9} + \sqrt{16} = 3 + 4$ $\sqrt{9} = 3$ because $3^2 = 9$. $\sqrt{16} = 4$ because $4^2 = 16$.

 $= 7$ • • •

GREAT QUESTION!

Is $\sqrt{a + b}$ equal to $\sqrt{a} + \sqrt{b}$?

No. In Example 1, parts (d) and (e), observe that $\sqrt{9 + 16}$ is not equal to $\sqrt{9} + \sqrt{16}$. In general,

$$\sqrt{a + b} \neq \sqrt{a} + \sqrt{b}$$

and

$$\sqrt{a - b} \neq \sqrt{a} - \sqrt{b}.$$

✓ **Check Point 1** Evaluate:

a. $\sqrt{81}$ b. $-\sqrt{9}$ c. $\sqrt{\dfrac{1}{25}}$

d. $\sqrt{36 + 64}$ e. $\sqrt{36} + \sqrt{64}$.

A number that is the square of a rational number is called a **perfect square**. All the radicands in Example 1 and Check Point 1 are perfect squares.

- 64 is a perfect square because $64 = 8^2$. Thus, $\sqrt{64} = 8$.
- $\frac{1}{4}$ is a perfect square because $\frac{1}{4} = \left(\frac{1}{2}\right)^2$. Thus, $\sqrt{\frac{1}{4}} = \frac{1}{2}$.

Let's see what happens to the radical expression \sqrt{x} if x is a negative number. Is the square root of a negative number a real number? For example, consider $\sqrt{-25}$. Is there a real number whose square is -25? No. Thus, $\sqrt{-25}$ is not a real number. In general, **a square root of a negative number is not a real number**.

If a number a is nonnegative ($a \geq 0$), then $(\sqrt{a})^2 = a$. For example,

$$(\sqrt{2})^2 = 2, \quad (\sqrt{3})^2 = 3, \quad (\sqrt{4})^2 = 4, \quad \text{and} \quad (\sqrt{5})^2 = 5.$$

② Solve equations using the square root property.

The Square Root Property

Now that we know that the reverse operation of squaring a number is finding the square root of the number, let's apply this idea to a relatively simple equation:

$$x^2 = 9.$$

The value of x must be a number whose square is 9. There are two numbers whose square is 9:

$$x = \sqrt{9} = 3 \quad \text{or} \quad x = -\sqrt{9} = -3.$$

Thus, the solutions of $x^2 = 9$ are 3 and -3. This is an example of the **square root property**.

The Square Root Property

If u is an algebraic expression and d is a positive real number, then $u^2 = d$ has exactly two solutions:

$$\text{If } u^2 = d, \quad \text{then } u = \sqrt{d} \text{ or } u = -\sqrt{d}.$$

Equivalently,

$$\text{If } u^2 = d, \quad \text{then } u = \pm\sqrt{d}.$$

Notice that $u = \pm\sqrt{d}$ is a shorthand notation to indicate that $u = \sqrt{d}$ or $u = -\sqrt{d}$. Although we usually read $u = \pm\sqrt{d}$ as "u equals plus or minus the square root of d," we actually mean that u is the positive square root of d or the negative square root of d.

EXAMPLE 2 Solving Equations by the Square Root Property

Solve by the square root property:

a. $x^2 = 49$ **b.** $4x^2 = 20$.

SOLUTION

a. $x^2 = 49$ *This is the original equation.*

$x = \sqrt{49}$ or $x = -\sqrt{49}$ *Apply the square root property. You can also write $x = \pm\sqrt{49}$.*

$x = 7$ or $x = -7$ *In abbreviated notation, $x = \pm 7$.*

Substitute both values into the original equation and confirm that the solutions are 7 and -7. The solution set is $\{-7, 7\}$.

b. To apply the square root property, we need a squared expression by itself on one side of the equation.

$$4x^2 = 20$$

We want x^2 by itself.

We can get x^2 by itself if we divide both sides of $4x^2 = 20$ by 4.

$$4x^2 = 20 \qquad \text{This is the original equation.}$$

$$\frac{4x^2}{4} = \frac{20}{4} \qquad \text{Divide both sides by 4.}$$

$$x^2 = 5 \qquad \text{Simplify.}$$

$$x = \sqrt{5} \quad \text{or} \quad x = -\sqrt{5} \qquad \text{Apply the square root property.}$$

Now let's check these proposed solutions in the original equation.

Check $\sqrt{5}$:

$$4x^2 = 20$$
$$4(\sqrt{5})^2 \overset{?}{=} 20$$
$$4 \cdot 5 \overset{?}{=} 20$$
$$20 = 20, \quad \text{true}$$

Check $-\sqrt{5}$:

$$4x^2 = 20$$
$$4(-\sqrt{5})^2 \overset{?}{=} 20$$
$$4 \cdot 5 \overset{?}{=} 20$$
$$20 = 20, \quad \text{true}$$

The solutions are $\sqrt{5}$ and $-\sqrt{5}$. The solution set is $\{-\sqrt{5}, \sqrt{5}\}$ or $\{\pm\sqrt{5}\}$. ● ● ●

☑ **Check Point 2** Solve by the square root property:
 a. $x^2 = 36$ **b.** $5x^2 = 15$.

③ Simplify expressions of the form $\sqrt{a^2}$.

Simplifying Expressions of the Form $\sqrt{a^2}$

You may think that $\sqrt{a^2} = a$. However, this is not necessarily true. Consider the following examples:

$$\sqrt{4^2} = \sqrt{16} = 4$$
$$\sqrt{(-4)^2} = \sqrt{16} = 4.$$

The result is not −4, but rather the absolute value of −4, or 4.

Here is a rule for simplifying expressions of the form $\sqrt{a^2}$:

Simplifying $\sqrt{a^2}$

For any real number a,

$$\sqrt{a^2} = |a|.$$

In words, the principal square root of a^2 is the absolute value of a.

For example, $\sqrt{6^2} = |6| = 6$ and $\sqrt{(-6)^2} = |-6| = 6$.

④ Use the product rule to simplify square roots.

The Product Rule for Square Roots

A rule for multiplying square roots can be generalized by comparing $\sqrt{25} \cdot \sqrt{4}$ and $\sqrt{25 \cdot 4}$. Notice that

$$\sqrt{25} \cdot \sqrt{4} = 5 \cdot 2 = 10 \quad \text{and} \quad \sqrt{25 \cdot 4} = \sqrt{100} = 10.$$

Because we obtain 10 in both situations, the original radical expressions must be equal. That is,

$$\sqrt{25} \cdot \sqrt{4} = \sqrt{25 \cdot 4}.$$

This result is a special case of the **product rule for square roots** that can be generalized as follows:

The Product Rule for Square Roots

If a and b represent nonnegative real numbers, then

$$\sqrt{ab} = \sqrt{a} \cdot \sqrt{b} \quad \text{and} \quad \sqrt{a} \cdot \sqrt{b} = \sqrt{ab}.$$

The square root of a product is the product of the square roots.

The product of two square roots is the square root of the product of the radicands.

A square root is **simplified** when its radicand has no factors other than 1 that are perfect squares. For example, $\sqrt{500}$ is not simplified because it can be expressed as $\sqrt{100 \cdot 5}$ and 100 is a perfect square. Example 3 shows how the product rule is used to remove from the square root any perfect squares that occur as factors.

EXAMPLE 3 Using the Product Rule to Simplify Square Roots

Simplify:

 a. $\sqrt{500}$ **b.** $\sqrt{6x} \cdot \sqrt{3x}$.

SOLUTION

a.
$$\begin{aligned}
\sqrt{500} &= \sqrt{100 \cdot 5} && \text{Factor 500. 100 is the greatest perfect square factor.} \\
&= \sqrt{100}\sqrt{5} && \text{Use the product rule: } \sqrt{ab} = \sqrt{a} \cdot \sqrt{b}. \\
&= 10\sqrt{5} && \text{Write } \sqrt{100} \text{ as 10. We read } 10\sqrt{5} \text{ as "ten times the square} \\
& && \text{root of 5."}
\end{aligned}$$

b. We can simplify $\sqrt{6x} \cdot \sqrt{3x}$ using the product rule only if $6x$ and $3x$ represent nonnegative real numbers. Thus, $x \geq 0$.

$$\begin{aligned}
\sqrt{6x} \cdot \sqrt{3x} &= \sqrt{6x \cdot 3x} && \text{Use the product rule: } \sqrt{a}\sqrt{b} = \sqrt{ab}. \\
&= \sqrt{18x^2} && \text{Multiply in the radicand.} \\
&= \sqrt{9x^2 \cdot 2} && \text{Factor 18. 9 is the greatest perfect square factor.} \\
&= \sqrt{9x^2}\sqrt{2} && \text{Use the product rule: } \sqrt{ab} = \sqrt{a} \cdot \sqrt{b}. \\
&= \sqrt{9}\sqrt{x^2}\sqrt{2} && \text{Use the product rule to write } \sqrt{9x^2} \text{ as the} \\
& && \text{product of two square roots.} \\
&= 3x\sqrt{2} && \sqrt{x^2} = |x| = x \text{ because } x \geq 0.
\end{aligned}$$

• • •

GREAT QUESTION!

When simplifying square roots, what happens if I use a perfect square factor that isn't the greatest perfect square factor possible?

You'll need to simplify even further. For example, consider the following factorization:

$$\sqrt{500} = \sqrt{25 \cdot 20} = \sqrt{25}\sqrt{20} = 5\sqrt{20}.$$

25 is a perfect square factor of 500, but not the greatest perfect square factor.

Because 20 contains a perfect square factor, 4, the simplification is not complete.

$$5\sqrt{20} = 5\sqrt{4 \cdot 5} = 5\sqrt{4}\sqrt{5} = 5 \cdot 2\sqrt{5} = 10\sqrt{5}$$

Although the result checks with our simplification using $\sqrt{500} = \sqrt{100 \cdot 5}$, **more work is required when the greatest perfect square factor is not used.**

⊘ Check Point **3** Simplify:

 a. $\sqrt{75}$ **b.** $\sqrt{5x} \cdot \sqrt{10x}$.

⑤ Use the quotient rule to simplify square roots.

The Quotient Rule for Square Roots

Another property for square roots involves division.

The Quotient Rule for Square Roots

If a and b represent nonnegative real numbers and $b \neq 0$, then

$$\sqrt{\frac{a}{b}} = \frac{\sqrt{a}}{\sqrt{b}} \quad \text{and} \quad \frac{\sqrt{a}}{\sqrt{b}} = \sqrt{\frac{a}{b}}.$$

The square root of a quotient is the quotient of the square roots.

The quotient of two square roots is the square root of the quotient of the radicands.

EXAMPLE 4 Using the Quotient Rule to Simplify Square Roots

Simplify:

a. $\sqrt{\dfrac{100}{9}}$ **b.** $\dfrac{\sqrt{48x^3}}{\sqrt{6x}}$.

SOLUTION

a. $\sqrt{\dfrac{100}{9}} = \dfrac{\sqrt{100}}{\sqrt{9}} = \dfrac{10}{3}$

b. We can simplify the quotient of $\sqrt{48x^3}$ and $\sqrt{6x}$ using the quotient rule only if $48x^3$ and $6x$ represent nonnegative real numbers and $6x \neq 0$. Thus, $x > 0$.

$$\dfrac{\sqrt{48x^3}}{\sqrt{6x}} = \sqrt{\dfrac{48x^3}{6x}} = \sqrt{8x^2} = \sqrt{4x^2}\sqrt{2} = \sqrt{4}\sqrt{x^2}\sqrt{2} = 2x\sqrt{2}$$

$\boxed{\sqrt{x^2} = |x| = x \text{ because } x > 0.}$ •••

✓ Check Point **4** Simplify:

a. $\sqrt{\dfrac{25}{16}}$ **b.** $\dfrac{\sqrt{150x^3}}{\sqrt{2x}}$.

⑥ Add and subtract square roots.

GREAT QUESTION!

Should like radicals remind me of like terms?

Yes. Adding or subtracting like radicals is similar to adding or subtracting like terms:

$$7x + 6x = 13x$$

and

$$7\sqrt{11} + 6\sqrt{11} = 13\sqrt{11}.$$

Adding and Subtracting Square Roots

Two or more square roots can be combined using the distributive property provided that they have the same radicand. Such radicals are called **like radicals**. For example,

$$7\sqrt{11} + 6\sqrt{11} = (7 + 6)\sqrt{11} = 13\sqrt{11}.$$

$\boxed{\text{7 square roots of 11 plus 6 square roots of 11 result in 13 square roots of 11.}}$

EXAMPLE 5 Adding and Subtracting Like Radicals

Add or subtract as indicated:

a. $7\sqrt{2} + 5\sqrt{2}$ **b.** $\sqrt{5x} - 7\sqrt{5x}$.

SOLUTION

a. $7\sqrt{2} + 5\sqrt{2} = (7 + 5)\sqrt{2}$ Apply the distributive property.

$= 12\sqrt{2}$ Simplify.

b. $\sqrt{5x} - 7\sqrt{5x} = 1\sqrt{5x} - 7\sqrt{5x}$ Write $\sqrt{5x}$ as $1\sqrt{5x}$.

$= (1 - 7)\sqrt{5x}$ Apply the distributive property.

$= -6\sqrt{5x}$ Simplify. •••

✓ Check Point **5** Add or subtract as indicated:

a. $8\sqrt{13} + 9\sqrt{13}$ **b.** $\sqrt{17x} - 20\sqrt{17x}$.

In some cases, radicals can be combined once they have been simplified. For example, to add $\sqrt{2}$ and $\sqrt{8}$, we can write $\sqrt{8}$ as $\sqrt{4 \cdot 2}$ because 4 is a perfect square factor of 8.

$$\sqrt{2} + \sqrt{8} = \sqrt{2} + \sqrt{4 \cdot 2} = 1\sqrt{2} + 2\sqrt{2} = (1 + 2)\sqrt{2} = 3\sqrt{2}$$

EXAMPLE 6 Combining Radicals That First Require Simplification

Add or subtract as indicated:

 a. $7\sqrt{3} + \sqrt{12}$ **b.** $4\sqrt{50x} - 6\sqrt{32x}.$

SOLUTION

a. $7\sqrt{3} + \sqrt{12}$

 $= 7\sqrt{3} + \sqrt{4 \cdot 3}$ Split 12 into two factors such that one is a perfect square.

 $= 7\sqrt{3} + 2\sqrt{3}$ $\sqrt{4 \cdot 3} = \sqrt{4}\sqrt{3} = 2\sqrt{3}$

 $= (7 + 2)\sqrt{3}$ Apply the distributive property. You will find that this step is usually done mentally.

 $= 9\sqrt{3}$ Simplify.

b. $4\sqrt{50x} - 6\sqrt{32x}$

 $= 4\sqrt{25 \cdot 2x} - 6\sqrt{16 \cdot 2x}$ 25 is the greatest perfect square factor of 50x and 16 is the greatest perfect square factor of 32x.

 $= 4 \cdot 5\sqrt{2x} - 6 \cdot 4\sqrt{2x}$ $\sqrt{25 \cdot 2x} = \sqrt{25}\sqrt{2x} = 5\sqrt{2x}$ and $\sqrt{16 \cdot 2x} = \sqrt{16}\sqrt{2x} = 4\sqrt{2x}.$

 $= 20\sqrt{2x} - 24\sqrt{2x}$ Multiply: $4 \cdot 5 = 20$ and $6 \cdot 4 = 24.$

 $= (20 - 24)\sqrt{2x}$ Apply the distributive property.

 $= -4\sqrt{2x}$ Simplify. • • •

⌀ **Check Point 6** Add or subtract as indicated:

 a. $5\sqrt{27} + \sqrt{12}$ **b.** $6\sqrt{18x} - 4\sqrt{8x}.$

⑦ Rationalize denominators.

Rationalizing Denominators

The calculator screen in **Figure P.12** shows approximate values for $\dfrac{1}{\sqrt{3}}$ and $\dfrac{\sqrt{3}}{3}$. The two approximations are the same. This is not a coincidence:

$$\frac{1}{\sqrt{3}} = \frac{1}{\sqrt{3}} \cdot \boxed{\frac{\sqrt{3}}{\sqrt{3}}} = \frac{\sqrt{3}}{\sqrt{9}} = \frac{\sqrt{3}}{3}.$$

Any number divided by itself is 1. Multiplication by 1 does not change the value of $\frac{1}{\sqrt{3}}$.

```
1/√(3)
           .5773502692
√(3)/3
           .5773502692
```

FIGURE P.12 The calculator screen shows approximate values for $\dfrac{1}{\sqrt{3}}$ and $\dfrac{\sqrt{3}}{3}$.

This process involves rewriting a radical expression as an equivalent expression in which the denominator no longer contains any radicals. The process is called **rationalizing the denominator**. If the denominator consists of the square root of a natural number that is not a perfect square, **multiply the numerator and the denominator by the smallest number that produces the square root of a perfect square in the denominator.**

EXAMPLE 7 Rationalizing Denominators

Rationalize the denominator:

 a. $\dfrac{15}{\sqrt{6}}$ **b.** $\dfrac{12}{\sqrt{8}}.$

GREAT QUESTION!

What exactly does rationalizing a denominator do to an irrational number in the denominator?

Rationalizing a numerical denominator makes that denominator a rational number.

SOLUTION

a. If we multiply the numerator and the denominator of $\dfrac{15}{\sqrt{6}}$ by $\sqrt{6}$, the denominator becomes $\sqrt{6} \cdot \sqrt{6} = \sqrt{36} = 6$. Therefore, we multiply by 1, choosing $\dfrac{\sqrt{6}}{\sqrt{6}}$ for 1.

$$\frac{15}{\sqrt{6}} = \frac{15}{\sqrt{6}} \cdot \frac{\sqrt{6}}{\sqrt{6}} = \frac{15\sqrt{6}}{\sqrt{36}} = \frac{15\sqrt{6}}{6} = \frac{5\sqrt{6}}{2}$$

Multiply by 1. Simplify: $\frac{15}{6} = \frac{15 \div 3}{6 \div 3} = \frac{5}{2}$.

b. The *smallest* number that will produce the square root of a perfect square in the denominator of $\dfrac{12}{\sqrt{8}}$ is $\sqrt{2}$, because $\sqrt{8} \cdot \sqrt{2} = \sqrt{16} = 4$. We multiply by 1, choosing $\dfrac{\sqrt{2}}{\sqrt{2}}$ for 1.

$$\frac{12}{\sqrt{8}} = \frac{12}{\sqrt{8}} \cdot \frac{\sqrt{2}}{\sqrt{2}} = \frac{12\sqrt{2}}{\sqrt{16}} = \frac{12\sqrt{2}}{4} = 3\sqrt{2}$$

• • •

⌀ Check Point **7** Rationalize the denominator:

a. $\dfrac{5}{\sqrt{3}}$ **b.** $\dfrac{6}{\sqrt{12}}$.

Radical expressions that involve the sum and difference of the same two terms are called **conjugates**. Thus,

$$\sqrt{a} + \sqrt{b} \quad \text{and} \quad \sqrt{a} - \sqrt{b}$$

are conjugates. Conjugates are used to rationalize denominators because the product of such a pair contains no radicals:

Multiply each term of $\sqrt{a} - \sqrt{b}$
by each term of $\sqrt{a} + \sqrt{b}$.

$$\left(\sqrt{a} + \sqrt{b}\right)\left(\sqrt{a} - \sqrt{b}\right)$$
$$= \sqrt{a}\left(\sqrt{a} - \sqrt{b}\right) + \sqrt{b}\left(\sqrt{a} - \sqrt{b}\right)$$

Distribute \sqrt{a} Distribute \sqrt{b}
over $\sqrt{a} - \sqrt{b}$. over $\sqrt{a} - \sqrt{b}$.

$$= \sqrt{a} \cdot \sqrt{a} - \sqrt{a} \cdot \sqrt{b} + \sqrt{b} \cdot \sqrt{a} - \sqrt{b} \cdot \sqrt{b}$$
$$= \left(\sqrt{a}\right)^2 - \sqrt{ab} + \sqrt{ab} - \left(\sqrt{b}\right)^2$$

$-\sqrt{ab} + \sqrt{ab} = 0$

$$= \left(\sqrt{a}\right)^2 - \left(\sqrt{b}\right)^2$$
$$= a - b.$$

Multiplying Conjugates

$$\left(\sqrt{a} + \sqrt{b}\right)\left(\sqrt{a} - \sqrt{b}\right) = \left(\sqrt{a}\right)^2 - \left(\sqrt{b}\right)^2 = a - b$$

How can we rationalize a denominator if the denominator contains two terms with one or more square roots? **Multiply the numerator and the denominator by the conjugate of the denominator.** Here are three examples of such expressions:

• $\dfrac{7}{5 + \sqrt{3}}$ • $\dfrac{8}{3\sqrt{2} - 4}$ • $\dfrac{h}{\sqrt{x + h} - \sqrt{x}}$

The conjugate of the denominator is $5 - \sqrt{3}$. The conjugate of the denominator is $3\sqrt{2} + 4$. The conjugate of the denominator is $\sqrt{x + h} + \sqrt{x}$.

The product of the denominator and its conjugate is found using the formula

$$\left(\sqrt{a} + \sqrt{b}\right)\left(\sqrt{a} - \sqrt{b}\right) = \left(\sqrt{a}\right)^2 - \left(\sqrt{b}\right)^2 = a - b.$$

The simplified product will not contain a radical.

EXAMPLE 8 Rationalizing a Denominator Containing Two Terms

Rationalize the denominator: $\dfrac{7}{5 + \sqrt{3}}$.

SOLUTION

The conjugate of the denominator is $5 - \sqrt{3}$. If we multiply the numerator and denominator by $5 - \sqrt{3}$, the simplified denominator will not contain a radical. Therefore, we multiply by 1, choosing $\dfrac{5 - \sqrt{3}}{5 - \sqrt{3}}$ for 1.

$$\frac{7}{5 + \sqrt{3}} = \frac{7}{5 + \sqrt{3}} \cdot \frac{5 - \sqrt{3}}{5 - \sqrt{3}} = \frac{7(5 - \sqrt{3})}{5^2 - (\sqrt{3})^2} = \frac{7(5 - \sqrt{3})}{25 - 3}$$

Multiply by 1. $(\sqrt{a} + \sqrt{b})(\sqrt{a} - \sqrt{b})$
$= (\sqrt{a})^2 - (\sqrt{b})^2$

$$= \frac{7(5 - \sqrt{3})}{22} \quad \text{or} \quad \frac{35 - 7\sqrt{3}}{22}$$

In either form of the answer, there is no radical in the denominator.

• • •

✓ **Check Point 8** Rationalize the denominator: $\dfrac{8}{4 + \sqrt{5}}$.

⑧ Evaluate and perform operations with higher roots.

Other Kinds of Roots

We define the **principal nth root** of a real number a, symbolized by $\sqrt[n]{a}$, as follows:

Definition of the Principal nth Root of a Real Number

$$\sqrt[n]{a} = b \text{ means that } b^n = a.$$

If n, the **index**, is even, then a is nonnegative ($a \geq 0$) and b is also nonnegative ($b \geq 0$). If n is odd, a and b can be any real numbers.

For example,

$$\sqrt[3]{64} = 4 \text{ because } 4^3 = 64 \quad \text{and} \quad \sqrt[5]{-32} = -2 \text{ because } (-2)^5 = -32.$$

The same vocabulary that we learned for square roots applies to nth roots. The symbol $\sqrt[n]{\ }$ is called a **radical** and the expression under the radical is called the **radicand**.

GREAT QUESTION!

Should I know the higher roots of certain numbers by heart?

Some higher roots occur so frequently that you might want to memorize them.

Cube Roots	
$\sqrt[3]{1} = 1$	$\sqrt[3]{125} = 5$
$\sqrt[3]{8} = 2$	$\sqrt[3]{216} = 6$
$\sqrt[3]{27} = 3$	$\sqrt[3]{1000} = 10$
$\sqrt[3]{64} = 4$	

Fourth Roots	Fifth Roots
$\sqrt[4]{1} = 1$	$\sqrt[5]{1} = 1$
$\sqrt[4]{16} = 2$	$\sqrt[5]{32} = 2$
$\sqrt[4]{81} = 3$	$\sqrt[5]{243} = 3$
$\sqrt[4]{256} = 4$	
$\sqrt[4]{625} = 5$	

A number that is the nth power of a rational number is called a **perfect nth power**. For example, 8 is a perfect third power, or perfect cube, because $8 = 2^3$. Thus, $\sqrt[3]{8} = \sqrt[3]{2^3} = 2$. In general, one of the following rules can be used to find the nth root of a perfect nth power:

Finding nth Roots of Perfect nth Powers

If n is odd, $\sqrt[n]{a^n} = a$.
If n is even, $\sqrt[n]{a^n} = |a|$.

For example,

$$\sqrt[3]{(-2)^3} = -2 \qquad \text{and} \qquad \sqrt[4]{(-2)^4} = |-2| = 2.$$

> Absolute value is not needed with odd roots, but is necessary with even roots.

The Product and Quotient Rules for Other Roots

The product and quotient rules apply to cube roots, fourth roots, and all higher roots.

The Product and Quotient Rules for nth Roots

For all real numbers a and b, where the indicated roots represent real numbers,

$$\sqrt[n]{ab} = \sqrt[n]{a} \cdot \sqrt[n]{b} \qquad \text{and} \qquad \sqrt[n]{a} \cdot \sqrt[n]{b} = \sqrt[n]{ab}$$

> The nth root of a product is the product of the nth roots.

> The product of two nth roots is the nth root of the product of the radicands.

$$\sqrt[n]{\frac{a}{b}} = \frac{\sqrt[n]{a}}{\sqrt[n]{b}}, b \neq 0 \qquad \text{and} \qquad \frac{\sqrt[n]{a}}{\sqrt[n]{b}} = \sqrt[n]{\frac{a}{b}}, b \neq 0.$$

> The nth root of a quotient is the quotient of the nth roots.

> The quotient of two nth roots is the nth root of the quotient of the radicands.

EXAMPLE 9 Simplifying, Multiplying, and Dividing Higher Roots

Simplify: **a.** $\sqrt[3]{24}$ **b.** $\sqrt[4]{8} \cdot \sqrt[4]{4}$ **c.** $\sqrt[4]{\dfrac{81}{16}}$.

SOLUTION

a. $\sqrt[3]{24} = \sqrt[3]{8 \cdot 3}$ Find the greatest perfect cube that is a factor of 24. $2^3 = 8$, so 8 is a perfect cube and is the greatest perfect cube factor of 24.

$\qquad = \sqrt[3]{8} \cdot \sqrt[3]{3}$ $\sqrt[n]{ab} = \sqrt[n]{a} \cdot \sqrt[n]{b}$

$\qquad = 2\sqrt[3]{3}$ $\sqrt[3]{8} = 2$

b. $\sqrt[4]{8} \cdot \sqrt[4]{4} = \sqrt[4]{8 \cdot 4}$ $\sqrt[n]{a} \cdot \sqrt[n]{b} = \sqrt[n]{ab}$

$\qquad = \sqrt[4]{32}$ Find the greatest perfect fourth power that is a factor of 32.

$\qquad = \sqrt[4]{16 \cdot 2}$ $2^4 = 16$, so 16 is a perfect fourth power and is the greatest perfect fourth power that is a factor of 32.

$\qquad = \sqrt[4]{16} \cdot \sqrt[4]{2}$ $\sqrt[n]{ab} = \sqrt[n]{a} \cdot \sqrt[n]{b}$

$\qquad = 2\sqrt[4]{2}$ $\sqrt[4]{16} = 2$

c. $\sqrt[4]{\dfrac{81}{16}} = \dfrac{\sqrt[4]{81}}{\sqrt[4]{16}}$ $\sqrt[n]{\dfrac{a}{b}} = \dfrac{\sqrt[n]{a}}{\sqrt[n]{b}}$

$= \dfrac{3}{2}$ $\sqrt[4]{81} = 3$ because $3^4 = 81$ and $\sqrt[4]{16} = 2$ because $2^4 = 16.$ •••

⊘ Check Point **9** Simplify:

a. $\sqrt[3]{40}$ b. $\sqrt[5]{8} \cdot \sqrt[5]{8}$ c. $\sqrt[3]{\dfrac{125}{27}}.$

We have seen that adding and subtracting square roots often involves simplifying terms. The same idea applies to adding and subtracting higher roots.

EXAMPLE 10 Combining Cube Roots

Subtract: $5\sqrt[3]{16} - 11\sqrt[3]{2}.$

SOLUTION

$$5\sqrt[3]{16} - 11\sqrt[3]{2}$$

$= 5\sqrt[3]{8 \cdot 2} - 11\sqrt[3]{2}$ Factor 16. 8 is the greatest perfect cube factor: $2^3 = 8$ and $\sqrt[3]{8} = 2$.

$= 5 \cdot 2\sqrt[3]{2} - 11\sqrt[3]{2}$ $\sqrt[3]{8 \cdot 2} = \sqrt[3]{8}\sqrt[3]{2} = 2\sqrt[3]{2}$

$= 10\sqrt[3]{2} - 11\sqrt[3]{2}$ Multiply: $5 \cdot 2 = 10$.

$= (10 - 11)\sqrt[3]{2}$ Apply the distributive property.

$= -1\sqrt[3]{2}$ or $-\sqrt[3]{2}$ Simplify. •••

⊘ Check Point **10** Subtract: $3\sqrt[3]{81} - 4\sqrt[3]{3}.$

⑨ Understand and use rational exponents.

Rational Exponents

We define rational exponents so that their properties are the same as the properties for integer exponents. For example, we know that exponents are multiplied when an exponential expression is raised to a power. For this to be true,

$$\left(7^{\frac{1}{2}}\right)^2 = 7^{\frac{1}{2} \cdot 2} = 7^1 = 7.$$

We also know that

$$\left(\sqrt{7}\right)^2 = \sqrt{7} \cdot \sqrt{7} = \sqrt{49} = 7.$$

Can you see that the square of both $7^{\frac{1}{2}}$ and $\sqrt{7}$ is 7? It is reasonable to conclude that

$$7^{\frac{1}{2}} \quad \text{means} \quad \sqrt{7}.$$

We can generalize the fact that $7^{\frac{1}{2}}$ means $\sqrt{7}$ with the following definition:

The Definition of $a^{\frac{1}{n}}$

If $\sqrt[n]{a}$ represents a real number, where $n \geq 2$ is an integer, then

$$a^{\frac{1}{n}} = \sqrt[n]{a}.$$

The denominator of the rational exponent is the radical's index.

Furthermore,

$$a^{-\frac{1}{n}} = \dfrac{1}{a^{\frac{1}{n}}} = \dfrac{1}{\sqrt[n]{a}}, \quad a \neq 0.$$

EXAMPLE 11 Using the Definition of $a^{\frac{1}{n}}$

Simplify:

a. $64^{\frac{1}{2}}$ **b.** $125^{\frac{1}{3}}$ **c.** $-16^{\frac{1}{4}}$ **d.** $(-27)^{\frac{1}{3}}$ **e.** $64^{-\frac{1}{3}}$.

SOLUTION

a. $64^{\frac{1}{2}} = \sqrt{64} = 8$

b. $125^{\frac{1}{3}} = \sqrt[3]{125} = 5$

> The denominator is the index.

c. $-16^{\frac{1}{4}} = -(\sqrt[4]{16}) = -2$

> The base is 16 and the negative sign is not affected by the exponent.

d. $(-27)^{\frac{1}{3}} = \sqrt[3]{-27} = -3$

> Parentheses show that the base is −27 and that the negative sign is affected by the exponent.

e. $64^{-\frac{1}{3}} = \dfrac{1}{64^{\frac{1}{3}}} = \dfrac{1}{\sqrt[3]{64}} = \dfrac{1}{4}$

• • •

○ Check Point **11** Simplify:

a. $25^{\frac{1}{2}}$ **b.** $8^{\frac{1}{3}}$ **c.** $-81^{\frac{1}{4}}$ **d.** $(-8)^{\frac{1}{3}}$ **e.** $27^{-\frac{1}{3}}$.

In Example 11 and Check Point 11, each rational exponent had a numerator of 1. If the numerator is some other integer, we still want to multiply exponents when raising a power to a power. For this reason,

$$a^{\frac{2}{3}} = (a^{\frac{1}{3}})^2 \quad \text{and} \quad a^{\frac{2}{3}} = (a^2)^{\frac{1}{3}}.$$

> This means $(\sqrt[3]{a})^2$. This means $\sqrt[3]{a^2}$.

Thus,

$$a^{\frac{2}{3}} = (\sqrt[3]{a})^2 = \sqrt[3]{a^2}.$$

Do you see that the denominator, 3, of the rational exponent is the same as the index of the radical? The numerator, 2, of the rational exponent serves as an exponent in each of the two radical forms. We generalize these ideas with the following definition:

The Definition of $a^{\frac{m}{n}}$

If $\sqrt[n]{a}$ represents a real number and $\dfrac{m}{n}$ is a positive rational number, $n \geq 2$, then

$$a^{\frac{m}{n}} = (\sqrt[n]{a})^m.$$

Also,

$$a^{\frac{m}{n}} = \sqrt[n]{a^m}.$$

Furthermore, if $a^{-\frac{m}{n}}$ is a nonzero real number, then

$$a^{-\frac{m}{n}} = \dfrac{1}{a^{\frac{m}{n}}}.$$

The first form of the definition of $a^{\frac{m}{n}}$, shown again below, involves taking the root first. This form is often preferable because smaller numbers are involved. Notice that the rational exponent consists of two parts, indicated by the following voice balloons:

The numerator is the exponent.

$$a^{\frac{m}{n}} = (\sqrt[n]{a})^m.$$

The denominator is the radical's index.

EXAMPLE 12 Using the Definition of $a^{\frac{m}{n}}$

Simplify:

 a. $27^{\frac{2}{3}}$ **b.** $9^{\frac{3}{2}}$ **c.** $81^{-\frac{3}{4}}$.

SOLUTION

a. $27^{\frac{2}{3}} = (\sqrt[3]{27})^2 = 3^2 = 9$

b. $9^{\frac{3}{2}} = (\sqrt{9})^3 = 3^3 = 27$

c. $81^{-\frac{3}{4}} = \dfrac{1}{81^{\frac{3}{4}}} = \dfrac{1}{(\sqrt[4]{81})^3} = \dfrac{1}{3^3} = \dfrac{1}{27}$

 • • •

☑ Check Point **12** Simplify:

 a. $27^{\frac{4}{3}}$ **b.** $4^{\frac{3}{2}}$ **c.** $32^{-\frac{2}{5}}$.

TECHNOLOGY

Here are the calculator keystroke sequences for $81^{-\frac{3}{4}}$:

Many Scientific Calculators

81 $\boxed{y^x}$ $\boxed{(}$ 3 $\boxed{+/_}$ $\boxed{\div}$ 4 $\boxed{)}$ $\boxed{=}$

Many Graphing Calculators

81 $\boxed{\wedge}$ $\boxed{(}$ $\boxed{(-)}$ 3 $\boxed{\div}$ 4 $\boxed{)}$ \boxed{ENTER}.

Properties of exponents can be applied to expressions containing rational exponents.

EXAMPLE 13 Simplifying Expressions with Rational Exponents

Simplify using properties of exponents:

 a. $\left(5x^{\frac{1}{2}}\right)\left(7x^{\frac{3}{4}}\right)$ **b.** $\dfrac{32x^{\frac{5}{3}}}{16x^{\frac{3}{4}}}$.

SOLUTION

a. $\left(5x^{\frac{1}{2}}\right)\left(7x^{\frac{3}{4}}\right) = 5 \cdot 7x^{\frac{1}{2}} \cdot x^{\frac{3}{4}}$ *Group numerical factors and group variable factors with the same base.*

 $= 35x^{\frac{1}{2}+\frac{3}{4}}$ *When multiplying expressions with the same base, add the exponents.*

 $= 35x^{\frac{5}{4}}$ $\frac{1}{2} + \frac{3}{4} = \frac{2}{4} + \frac{3}{4} = \frac{5}{4}$

b. $\dfrac{32x^{\frac{5}{3}}}{16x^{\frac{3}{4}}} = \left(\dfrac{32}{16}\right)\left(\dfrac{x^{\frac{5}{3}}}{x^{\frac{3}{4}}}\right)$ *Group numerical factors and group variable factors with the same base.*

 $= 2x^{\frac{5}{3}-\frac{3}{4}}$ *When dividing expressions with the same base, subtract the exponents.*

 $= 2x^{\frac{11}{12}}$ $\frac{5}{3} - \frac{3}{4} = \frac{20}{12} - \frac{9}{12} = \frac{11}{12}$

 • • •

☑ Check Point **13** Simplify using properties of exponents:

 a. $\left(2x^{\frac{4}{3}}\right)\left(5x^{\frac{8}{3}}\right)$ **b.** $\dfrac{20x^4}{5x^{\frac{3}{2}}}$.

Rational exponents are sometimes useful for simplifying radicals by reducing the index.

EXAMPLE 14 Reducing the Index of a Radical

Simplify: $\sqrt[9]{x^3}$.

SOLUTION

$$\sqrt[9]{x^3} = x^{\frac{3}{9}} = x^{\frac{1}{3}} = \sqrt[3]{x}$$

• • •

✓ Check Point **14** Simplify: $\sqrt[6]{x^3}$.

Blitzer Bonus ‖ A Radical Idea: Time Is Relative

The Persistence of Memory (1931), Salvador Dali:. © 2011 MOMA/ARS.

What does travel in space have to do with radicals? Imagine that in the future we will be able to travel at velocities approaching the speed of light (approximately 186,000 miles per second). According to Einstein's theory of special relativity, time would pass more quickly on Earth than it would in the moving spaceship. The special-relativity equation

$$R_a = R_f\sqrt{1 - \left(\frac{v}{c}\right)^2}$$

gives the aging rate of an astronaut, R_a, relative to the aging rate of a friend, R_f, on Earth. In this formula, v is the astronaut's speed and c is the speed of light. As the astronaut's speed approaches the speed of light, we can substitute c for v.

$$R_a = R_f\sqrt{1 - \left(\frac{v}{c}\right)^2} \qquad \text{Einstein's equation gives the aging rate of an astronaut, } R_a, \text{ relative to the aging rate of a friend, } R_f, \text{ on Earth.}$$

$$R_a = R_f\sqrt{1 - \left(\frac{c}{c}\right)^2} \qquad \text{The velocity, } v, \text{ is approaching the speed of light, } c, \text{ so let } v = c.$$

$$= R_f\sqrt{1 - 1} \qquad \left(\frac{c}{c}\right)^2 = 1^2 = 1 \cdot 1 = 1$$

$$= R_f\sqrt{0} \qquad \text{Simplify the radicand: } 1 - 1 = 0.$$

$$= R_f \cdot 0 \qquad \sqrt{0} = 0$$

$$= 0 \qquad \text{Multiply: } R_f \cdot 0 = 0.$$

Close to the speed of light, the astronaut's aging rate, R_a, relative to a friend, R_f, on Earth is nearly 0. What does this mean? As we age here on Earth, the space traveler would barely get older. The space traveler would return to an unknown futuristic world in which friends and loved ones would be long gone.

CONCEPT AND VOCABULARY CHECK

Fill in each blank so that the resulting statement is true.

1. The symbol $\sqrt{\ }$ is used to denote the nonnegative, or _____, square root of a number.
2. $\sqrt{64} = 8$ because ___ = 64.
3. The square root property states that if $u^2 = d$, then $u =$ _____.
4. If $x^2 = 17$, then $x =$ _____.
5. $\sqrt{a^2} =$ ___
6. The product rule for square roots states that if a and b are nonnegative, then $\sqrt{ab} =$ _____.
7. The quotient rule for square roots states that if a and b are nonnegative and $b \neq 0$, then $\sqrt{\dfrac{a}{b}} =$ _____.
8. $8\sqrt{3} + 10\sqrt{3} =$ _____

9. $\sqrt{3} + \sqrt{75} = \sqrt{3} + \sqrt{25 \cdot 3} = \sqrt{3} + \underline{\ \ }\sqrt{3} =$ _____
10. The conjugate of $7 + \sqrt{3}$ is _____.
11. We rationalize the denominator of $\dfrac{5}{\sqrt{10} - \sqrt{2}}$ by multiplying the numerator and denominator by _____.
12. In the expression $\sqrt[3]{64}$, the number 3 is called the _____ and the number 64 is called the _____.
13. $\sqrt[5]{-32} = -2$ because _____ = -32.
14. If n is odd, $\sqrt[n]{a^n} =$ ___. If n is even, $\sqrt[n]{a^n} =$ ___.
15. $a^{\frac{1}{n}} =$ _____
16. $16^{\frac{3}{4}} = \left(\sqrt[4]{16}\right)^3 = (\underline{\ \ })^3 =$ ___

EXERCISE SET P.4

Practice Exercises

Evaluate each expression in Exercises 1–12, or indicate that the root is not a real number.

1. $\sqrt{36}$
2. $\sqrt{25}$
3. $-\sqrt{36}$
4. $-\sqrt{25}$
5. $\sqrt{-36}$
6. $\sqrt{-25}$
7. $\sqrt{25 - 16}$
8. $\sqrt{144 + 25}$
9. $\sqrt{25} - \sqrt{16}$
10. $\sqrt{144} + \sqrt{25}$
11. $\sqrt{(-13)^2}$
12. $\sqrt{(-17)^2}$

In Exercises 13–24, solve each equation by the square root property.

13. $x^2 = 1$
14. $x^2 = 25$
15. $x^2 = 7$
16. $x^2 = 13$
17. $5x^2 = 20$
18. $3x^2 = 27$
19. $2y^2 + 1 = 51$
20. $3y^2 - 1 = 47$
21. $4y^2 = 49$
22. $16y^2 = 25$
23. $3z^2 - 6 = 24$
24. $4z^2 - 17 = 3$

Use the product rule to simplify the expressions in Exercises 25–34. In Exercises 29–34, assume that variables represent nonnegative real numbers.

25. $\sqrt{50}$
26. $\sqrt{27}$
27. $\sqrt{45x^2}$
28. $\sqrt{125x^2}$
29. $\sqrt{2x} \cdot \sqrt{6x}$
30. $\sqrt{10x} \cdot \sqrt{8x}$
31. $\sqrt{x^3}$
32. $\sqrt{y^3}$
33. $\sqrt{2x^2} \cdot \sqrt{6x}$
34. $\sqrt{6x} \cdot \sqrt{3x^2}$

Use the quotient rule to simplify the expressions in Exercises 35–44. Assume that $x > 0$.

35. $\sqrt{\dfrac{1}{81}}$
36. $\sqrt{\dfrac{1}{49}}$
37. $\sqrt{\dfrac{49}{16}}$
38. $\sqrt{\dfrac{121}{9}}$
39. $\dfrac{\sqrt{48x^3}}{\sqrt{3x}}$
40. $\dfrac{\sqrt{72x^3}}{\sqrt{8x}}$
41. $\dfrac{\sqrt{150x^4}}{\sqrt{3x}}$
42. $\dfrac{\sqrt{24x^4}}{\sqrt{3x}}$

43. $\dfrac{\sqrt{200x^3}}{\sqrt{10x^{-1}}}$
44. $\dfrac{\sqrt{500x^3}}{\sqrt{10x^{-1}}}$

In Exercises 45–56, add or subtract terms whenever possible.

45. $7\sqrt{3} + 6\sqrt{3}$
46. $8\sqrt{5} + 11\sqrt{5}$
47. $6\sqrt{17x} - 8\sqrt{17x}$
48. $4\sqrt{13x} - 6\sqrt{13x}$
49. $\sqrt{8} + 3\sqrt{2}$
50. $\sqrt{20} + 6\sqrt{5}$
51. $\sqrt{50x} - \sqrt{8x}$
52. $\sqrt{63x} - \sqrt{28x}$
53. $3\sqrt{18} + 5\sqrt{50}$
54. $4\sqrt{12} - 2\sqrt{75}$
55. $3\sqrt{8} - \sqrt{32} + 3\sqrt{72} - \sqrt{75}$
56. $3\sqrt{54} - 2\sqrt{24} - \sqrt{96} + 4\sqrt{63}$

In Exercises 57–66, rationalize the denominator.

57. $\dfrac{1}{\sqrt{7}}$
58. $\dfrac{2}{\sqrt{10}}$
59. $\dfrac{\sqrt{2}}{\sqrt{5}}$
60. $\dfrac{\sqrt{7}}{\sqrt{3}}$
61. $\dfrac{13}{3 + \sqrt{11}}$
62. $\dfrac{3}{3 + \sqrt{7}}$
63. $\dfrac{7}{\sqrt{5} - 2}$
64. $\dfrac{5}{\sqrt{3} - 1}$
65. $\dfrac{6}{\sqrt{5} + \sqrt{3}}$
66. $\dfrac{11}{\sqrt{7} - \sqrt{3}}$

Evaluate each expression in Exercises 67–78, or indicate that the root is not a real number.

67. $\sqrt[3]{125}$
68. $\sqrt[3]{8}$
69. $\sqrt[3]{-8}$
70. $\sqrt[3]{-125}$
71. $\sqrt[4]{-16}$
72. $\sqrt[4]{-81}$
73. $\sqrt[4]{(-3)^4}$
74. $\sqrt[4]{(-2)^4}$
75. $\sqrt[5]{(-3)^5}$
76. $\sqrt[5]{(-2)^5}$
77. $\sqrt[5]{-\frac{1}{32}}$
78. $\sqrt[6]{\frac{1}{64}}$

Simplify the radical expressions in Exercises 79–86 if possible.

79. $\sqrt[3]{32}$

80. $\sqrt[3]{150}$

81. $\sqrt[3]{x^4}$

82. $\sqrt[3]{x^5}$

83. $\sqrt[3]{9} \cdot \sqrt[3]{6}$

84. $\sqrt[3]{12} \cdot \sqrt[3]{4}$

85. $\dfrac{\sqrt[5]{64x^6}}{\sqrt[5]{2x}}$

86. $\dfrac{\sqrt[4]{162x^5}}{\sqrt[4]{2x}}$

In Exercises 87–94, add or subtract terms whenever possible.

87. $4\sqrt[5]{2} + 3\sqrt[5]{2}$

88. $6\sqrt[5]{3} + 2\sqrt[5]{3}$

89. $5\sqrt[3]{16} + \sqrt[3]{54}$

90. $3\sqrt[3]{24} + \sqrt[3]{81}$

91. $\sqrt[3]{54xy^3} - y\sqrt[3]{128x}$

92. $\sqrt[3]{24xy^3} - y\sqrt[3]{81x}$

93. $\sqrt{2} + \sqrt[3]{8}$

94. $\sqrt{3} + \sqrt[3]{15}$

In Exercises 95–102, evaluate each expression without using a calculator.

95. $36^{\frac{1}{2}}$

96. $121^{\frac{1}{2}}$

97. $8^{\frac{1}{3}}$

98. $27^{\frac{1}{3}}$

99. $125^{\frac{2}{3}}$

100. $8^{\frac{2}{3}}$

101. $32^{-\frac{4}{5}}$

102. $16^{-\frac{5}{2}}$

In Exercises 103–112, simplify using properties of exponents.

103. $\left(7x^{\frac{1}{3}}\right)\left(2x^{\frac{1}{4}}\right)$

104. $\left(3x^{\frac{2}{3}}\right)\left(4x^{\frac{3}{4}}\right)$

105. $\dfrac{20x^{\frac{1}{2}}}{5x^{\frac{1}{4}}}$

106. $\dfrac{72x^{\frac{3}{4}}}{9x^{\frac{1}{3}}}$

107. $\left(x^{\frac{2}{3}}\right)^3$

108. $\left(x^{\frac{4}{5}}\right)^5$

109. $(25x^4y^6)^{\frac{1}{2}}$

110. $(125x^9y^6)^{\frac{1}{3}}$

111. $\dfrac{\left(3y^{\frac{1}{4}}\right)^3}{y^{\frac{1}{12}}}$

112. $\dfrac{\left(2y^{\frac{1}{5}}\right)^4}{y^{\frac{3}{10}}}$

In Exercises 113–120, simplify by reducing the index of the radical.

113. $\sqrt[4]{5^2}$

114. $\sqrt[4]{7^2}$

115. $\sqrt[3]{x^6}$

116. $\sqrt[4]{x^{12}}$

117. $\sqrt[6]{x^4}$

118. $\sqrt[9]{x^6}$

119. $\sqrt[9]{x^6y^3}$

120. $\sqrt[12]{x^4y^8}$

Practice Plus

In Exercises 121–122, evaluate each expression.

121. $\sqrt[3]{\sqrt[4]{16} + \sqrt{625}}$

122. $\sqrt[3]{\sqrt{\sqrt{169} + \sqrt{9}} + \sqrt{\sqrt[3]{1000} + \sqrt[3]{216}}}$

In Exercises 123–126, simplify each expression. Assume that all variables represent positive numbers.

123. $(49x^{-2}y^4)^{-\frac{1}{2}}\left(xy^{\frac{1}{2}}\right)$

124. $(8x^{-6}y^3)^{\frac{1}{3}}\left(x^{\frac{5}{6}}y^{-\frac{1}{3}}\right)^6$

125. $\left(\dfrac{x^{-\frac{5}{4}}y^{\frac{1}{3}}}{x^{-\frac{3}{4}}}\right)^{-6}$

126. $\left(\dfrac{x^{\frac{1}{2}}y^{-\frac{7}{4}}}{y^{-\frac{5}{4}}}\right)^{-4}$

Application Exercises

127. The popular comic strip *FoxTrot* follows the off-the-wall lives of the Fox family. Youngest son Jason is forever obsessed by his love of math. In the math-themed strip shown at the top of the next column, Jason shares his opinion in a coded message about the mathematical abilities of his sister Paige.

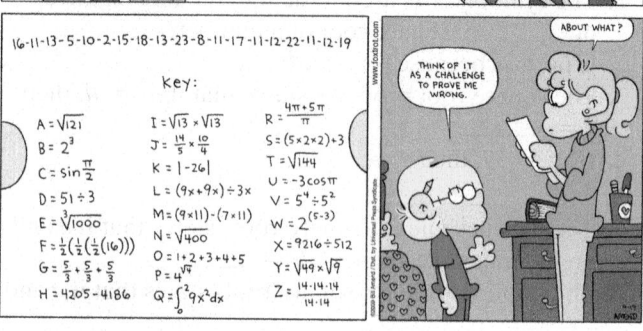

Foxtrot © 2003, 2009 by Bill Amend/Used by permission of Universal Uclick. All rights reserved.

Solve problems A through Z in the left panel. Then decode Jason Fox's message involving his opinion about the mathematical abilities of his sister Paige shown on the first line.

Hints: Here is the solution for problem C and partial solutions for problems Q and U.

These are from trigonometry.

$$C = \sin\frac{\pi}{2} = \sin 90° = 1$$

$$Q = \int_0^2 9x^2\,dx = 3x^3\Big|_0^2 = 3 \cdot 2^3 - 3 \cdot 0^3 = \underline{\quad}$$

This is from calculus.

$$U = -3\cos\pi = -3\cos 180° = -3(-1) = \underline{\quad}$$

Note: The comic strip *FoxTrot* is now printed in more than one thousand newspapers. What made cartoonist Bill Amend, a college physics major, put math in the comic? "I always try to use math in the strip to make the joke accessible to anyone," he said. "But if you understand math, hopefully you'll like it that much more!" We highly recommend the math humor in Amend's *FoxTrot* collection *Math, Science, and Unix Underpants* (Andrews McMeel Publishing, 2009).

128. America is getting older. The graph shows the projected elderly U.S. population for ages 65–84 and for ages 85 and older.

Projected Elderly United States Population

Source: U.S. Census Bureau

The formula $E = 5\sqrt{x} + 34.1$ models the projected number of elderly Americans ages 65–84, E, in millions, x years after 2010.

a. Use the formula to find the projected increase in the number of Americans ages 65–84, in millions, from 2020 to 2050. Express this difference in simplified radical form.

b. Use a calculator and write your answer in part (a) to the nearest tenth. Does this rounded decimal overestimate or underestimate the difference in the projected data shown by the bar graph on the previous page? By how much?

129. The early Greeks believed that the most pleasing of all rectangles were **golden rectangles**, whose ratio of width to height is

$$\frac{w}{h} = \frac{2}{\sqrt{5} - 1}.$$

The Parthenon at Athens fits into a golden rectangle once the triangular pediment is reconstructed.

Rationalize the denominator of the golden ratio. Then use a calculator and find the ratio of width to height, correct to the nearest hundredth, in golden rectangles.

130. Use Einstein's special-relativity equation

$$R_a = R_f\sqrt{1 - \left(\frac{v}{c}\right)^2},$$

described in the Blitzer Bonus on page 58, to solve this exercise. You are moving at 90% of the speed of light. Substitute $0.9c$ for v, your velocity, in the equation. What is your aging rate, correct to two decimal places, relative to a friend on Earth? If you are gone for 44 weeks, approximately how many weeks have passed for your friend?

The perimeter, P, of a rectangle with length l and width w is given by the formula $P = 2l + 2w$. The area, A, is given by the formula $A = lw$. In Exercises 131–132, use these formulas to find the perimeter and area of each rectangle. Express answers in simplified radical form. Remember that perimeter is measured in linear units, such as feet or meters, and area is measured in square units, such as square feet, ft^2, or square meters, m^2.

131. **132.**

Writing in Mathematics

133. What is the square root property?
134. Explain how to simplify $\sqrt{10} \cdot \sqrt{5}$.
135. Explain how to add $\sqrt{3} + \sqrt{12}$.

136. Describe what it means to rationalize a denominator. Use both $\dfrac{1}{\sqrt{5}}$ and $\dfrac{1}{5 + \sqrt{5}}$ in your explanation.

137. What difference is there in simplifying $\sqrt[3]{(-5)^3}$ and $\sqrt[4]{(-5)^4}$?

138. What does $a^{\frac{m}{n}}$ mean?

139. Describe the kinds of numbers that have rational fifth roots.

140. Why must a and b represent nonnegative numbers when we write $\sqrt{a} \cdot \sqrt{b} = \sqrt{ab}$? Is it necessary to use this restriction in the case of $\sqrt[3]{a} \cdot \sqrt[3]{b} = \sqrt[3]{ab}$? Explain.

141. Read the Blitzer Bonus on page 58. The future is now: You have the opportunity to explore the cosmos in a starship traveling near the speed of light. The experience will enable you to understand the mysteries of the universe in deeply personal ways, transporting you to unimagined levels of knowing and being. The downside: You return from your two-year journey to a futuristic world in which friends and loved ones are long gone. Do you explore space or stay here on Earth? What are the reasons for your choice?

Critical Thinking Exercises

Make Sense? *In Exercises 142–145, determine whether each statement makes sense or does not make sense, and explain your reasoning.*

142. The joke in this Peanuts cartoon would be more effective if Woodstock had rationalized the denominator correctly in the last frame.

Peanuts © 1978 Peanuts Worldwide LLC. Used by permission of Universal Uclick. All rights reserved.

143. Using my calculator, I determined that $6^7 = 279,936$, so 6 must be a seventh root of 279,936.

144. I simplified the terms of $2\sqrt{20} + 4\sqrt{75}$, and then I was able to add the like radicals.

145. When I use the definition for $a^{\frac{m}{n}}$, I usually prefer to first raise a to the m power because smaller numbers are involved.

In Exercises 146–149, determine whether each statement is true or false. If the statement is false, make the necessary change(s) to produce a true statement.

146. $7^{\frac{1}{2}} \cdot 7^{\frac{1}{2}} = 49$ **147.** $8^{-\frac{1}{3}} = -2$

148. The cube root of -8 is not a real number.

149. $\dfrac{\sqrt{20}}{8} = \dfrac{\sqrt{10}}{4}$

In Exercises 150–151, fill in each box to make the statement true.

150. $(5 + \sqrt{\boxed{}})(5 - \sqrt{\boxed{}}) = 22$

151. $\sqrt[\boxed{}]{x^{\boxed{}}} = 5x^7$

152. Find the exact value of $\sqrt{13 + \sqrt{2} + \dfrac{7}{3 + \sqrt{2}}}$ without the use of a calculator.

153. Place the correct symbol, $>$ or $<$, in the shaded area between the given numbers. *Do not use a calculator.* Then check your result with a calculator.

a. $3^{\frac{1}{2}}$ $3^{\frac{1}{3}}$ **b.** $\sqrt{7} + \sqrt{18}$ ▨ $\sqrt{7 + 18}$

154. a. A mathematics professor recently purchased a birthday cake for her son with the inscription

$$\text{Happy}\left(2^{\frac{5}{2}} \cdot 2^{\frac{3}{4}} \div 2^{\frac{1}{4}}\right)\text{th Birthday.}$$

How old is the son?

b. The birthday boy, excited by the inscription on the cake, tried to wolf down the whole thing. Professor Mom, concerned about the possible metamorphosis of her son into a blimp, exclaimed, "Hold on! It is your birthday, so why not take $\dfrac{8^{-\frac{4}{3}} + 2^{-2}}{16^{-\frac{3}{4}} + 2^{-1}}$ of the cake? I'll eat half of what's left over." How much of the cake did the professor eat?

Preview Exercises

Exercises 155–157 will help you prepare for the material covered in the next section.

155. Multiply: $(2x^3y^2)(5x^4y^7)$.

156. Use the distributive property to multiply:
$$2x^4(8x^4 + 3x).$$

157. Simplify and express the answer in descending powers of x:
$$2x(x^2 + 4x + 5) + 3(x^2 + 4x + 5).$$

SECTION P.5 Polynomials

Objectives

1. Understand the vocabulary of polynomials.
2. Add and subtract polynomials.
3. Multiply polynomials.
4. Use FOIL in polynomial multiplication.
5. Use special products in polynomial multiplication.
6. Perform operations with polynomials in several variables.

Can that be Axl, your author's yellow lab, sharing a special moment with a baby chick? And if it is (it is), what possible relevance can this have to polynomials? An answer is promised before you reach the Exercise Set. For now, we open the section by defining and describing polynomials.

Old Dog…New Chicks

How We Define Polynomials

More education results in a higher income. The mathematical models

$$M = 0.6x^3 + 285x^2 - 2256x + 15{,}112$$

and $$W = -1.2x^3 + 367x^2 - 4900x + 26{,}561$$

describe the median, or middlemost, annual income for men, M, and women, W, who have completed x years of education. We'll be working with these models and the data upon which they are based in the Exercise Set.

The algebraic expressions that appear on the right sides of the models are examples of *polynomials*. A **polynomial** is a single term or the sum of two or more terms containing variables with whole-number exponents. The polynomials above each contain four terms. Equations containing polynomials are used in such diverse areas as science, business, medicine, psychology, and sociology. In this section, we review basic ideas about polynomials and their operations.

How We Describe Polynomials

Consider the polynomial

1. Understand the vocabulary of polynomials.

$$7x^3 - 9x^2 + 13x - 6.$$

We can express this polynomial as

$$7x^3 + (-9x^2) + 13x + (-6).$$

The polynomial contains four terms. It is customary to write the terms in the order of descending powers of the variable. This is the **standard form** of a polynomial.

Some polynomials contain only one variable. Each term of such a polynomial in x is of the form ax^n. If $a \neq 0$, the **degree** of ax^n is n. For example, the degree of the term $7x^3$ is 3.

GREAT QUESTION!

Why doesn't the constant 0 have a degree?

We can express 0 in many ways, including $0x$, $0x^2$, and $0x^3$. It is impossible to assign a single exponent on the variable. This is why 0 has no defined degree.

The Degree of ax^n

If $a \neq 0$, the degree of ax^n is n. The degree of a nonzero constant is 0. The constant 0 has no defined degree.

Here is an example of a polynomial and the degree of each of its four terms:

$$6x^4 - 3x^3 + 2x - 5.$$

degree 4 degree 3 degree 1 degree of nonzero constant: 0

Notice that the exponent on x for the term $2x$, meaning $2x^1$, is understood to be 1. For this reason, the degree of $2x$ is 1. You can think of -5 as $-5x^0$; thus, its degree is 0.

A polynomial is simplified when it contains no grouping symbols and no like terms. A simplified polynomial that has exactly one term is called a **monomial**. A **binomial** is a simplified polynomial that has two terms. A **trinomial** is a simplified polynomial with three terms. Simplified polynomials with four or more terms have no special names.

The **degree of a polynomial** is the greatest degree of all the terms of the polynomial. For example, $4x^2 + 3x$ is a binomial of degree 2 because the degree of the first term is 2, and the degree of the other term is less than 2. Also, $7x^5 - 2x^2 + 4$ is a trinomial of degree 5 because the degree of the first term is 5, and the degrees of the other terms are less than 5.

Up to now, we have used x to represent the variable in a polynomial. However, any letter can be used. For example,

- $7x^5 - 3x^3 + 8$ is a polynomial (in x) of degree 5. Because there are three terms, the polynomial is a trinomial.
- $6y^3 + 4y^2 - y + 3$ is a polynomial (in y) of degree 3. Because there are four terms, the polynomial has no special name.
- $z^7 + \sqrt{2}$ is a polynomial (in z) of degree 7. Because there are two terms, the polynomial is a binomial.

We can tie together the threads of our discussion with the formal definition of a polynomial in one variable. In this definition, the coefficients of the terms are represented by a_n (read "a sub n"), a_{n-1} (read "a sub n minus 1"), a_{n-2}, and so on. The small letters to the lower right of each a are called **subscripts** and are *not exponents*. Subscripts are used to distinguish one constant from another when a large and undetermined number of such constants are needed.

Definition of a Polynomial in x

A **polynomial in x** is an algebraic expression of the form

$$a_n x^n + a_{n-1} x^{n-1} + a_{n-2} x^{n-2} + \cdots + a_1 x + a_0,$$

where $a_n, a_{n-1}, a_{n-2}, \ldots, a_1$, and a_0 are real numbers, $a_n \neq 0$, and n is a nonnegative integer. The polynomial is of **degree n**, a_n is the **leading coefficient**, and a_0 is the **constant term**.

2 Add and subtract polynomials.

Adding and Subtracting Polynomials

Polynomials are added and subtracted by combining like terms. For example, we can combine the monomials $-9x^3$ and $13x^3$ using addition as follows:

$$-9x^3 + 13x^3 = (-9 + 13)x^3 = 4x^3.$$

| These like terms both contain x to the third power. | Add coefficients and keep the same variable factor, x^3. |

EXAMPLE 1 Adding and Subtracting Polynomials

Perform the indicated operations and simplify:

a. $(-9x^3 + 7x^2 - 5x + 3) + (13x^3 + 2x^2 - 8x - 6)$
b. $(7x^3 - 8x^2 + 9x - 6) - (2x^3 - 6x^2 - 3x + 9).$

SOLUTION

a. $(-9x^3 + 7x^2 - 5x + 3) + (13x^3 + 2x^2 - 8x - 6)$

$= (-9x^3 + 13x^3) + (7x^2 + 2x^2) + (-5x - 8x) + (3 - 6)$ Group like terms.

$= 4x^3 + 9x^2 + (-13x) + (-3)$ Combine like terms.

$= 4x^3 + 9x^2 - 13x - 3$ Simplify.

b. $(7x^3 - 8x^2 + 9x - 6) - (2x^3 - 6x^2 - 3x + 9)$

Rewrite subtraction as addition of the additive inverse.

Change the sign of each coefficient.

$= (7x^3 - 8x^2 + 9x - 6) + (-2x^3 + 6x^2 + 3x - 9)$

$= (7x^3 - 2x^3) + (-8x^2 + 6x^2) + (9x + 3x) + (-6 - 9)$ Group like terms.

$= 5x^3 + (-2x^2) + 12x + (-15)$ Combine like terms.

$= 5x^3 - 2x^2 + 12x - 15$ Simplify. • • •

⌀ Check Point **1** Perform the indicated operations and simplify:

a. $(-17x^3 + 4x^2 - 11x - 5) + (16x^3 - 3x^2 + 3x - 15)$
b. $(13x^3 - 9x^2 - 7x + 1) - (-7x^3 + 2x^2 - 5x + 9).$

Multiplying Polynomials

The product of two monomials is obtained by using properties of exponents. For example,

$$(-8x^6)(5x^3) = -8 \cdot 5x^{6+3} = -40x^9.$$

Multiply coefficients and add exponents.

Furthermore, we can use the distributive property to multiply a monomial and a polynomial that is not a monomial. For example,

$$3x^4(2x^3 - 7x + 3) = 3x^4 \cdot 2x^3 - 3x^4 \cdot 7x + 3x^4 \cdot 3 = 6x^7 - 21x^5 + 9x^4.$$

Monomial

Trinomial

How do we multiply two polynomials if neither is a monomial? For example, consider

$$(2x + 3)(x^2 + 4x + 5).$$

Binomial

Trinomial

GREAT QUESTION!

Can I use a vertical format to add and subtract polynomials?

Yes. Arrange like terms in columns and combine vertically:

$$\begin{array}{r} 7x^3 - 8x^2 + 9x - 6 \\ -2x^3 + 6x^2 + 3x - 9 \\ \hline 5x^3 - 2x^2 + 12x - 15 \end{array}$$

The like terms can be combined by adding their coefficients and keeping the same variable factor.

3 Multiply polynomials.

GREAT QUESTION!

Because monomials with the same base and different exponents can be multiplied, can they also be added?

No. Don't confuse adding and multiplying monomials.

Addition:

$$5x^4 + 6x^4 = 11x^4$$

Multiplication:

$$(5x^4)(6x^4) = (5 \cdot 6)(x^4 \cdot x^4)$$
$$= 30x^{4+4}$$
$$= 30x^8$$

Only like terms can be added or subtracted, but unlike terms may be multiplied.

Addition:

$5x^4 + 3x^2$ cannot be simplified.

Multiplication:

$$(5x^4)(3x^2) = (5 \cdot 3)(x^4 \cdot x^2)$$
$$= 15x^{4+2}$$
$$= 15x^6$$

One way to perform $(2x + 3)(x^2 + 4x + 5)$ is to distribute $2x$ throughout the trinomial

$$2x(x^2 + 4x + 5)$$

and 3 throughout the trinomial

$$3(x^2 + 4x + 5).$$

Then combine the like terms that result.

Multiplying Polynomials When Neither Is a Monomial

Multiply each term of one polynomial by each term of the other polynomial. Then combine like terms.

EXAMPLE 2 Multiplying a Binomial and a Trinomial

Multiply: $(2x + 3)(x^2 + 4x + 5)$.

SOLUTION

$(2x + 3)(x^2 + 4x + 5)$
$= 2x(x^2 + 4x + 5) + 3(x^2 + 4x + 5)$ Multiply the trinomial by each term of the binomial.

$= 2x \cdot x^2 + 2x \cdot 4x + 2x \cdot 5 + 3x^2 + 3 \cdot 4x + 3 \cdot 5$ Use the distributive property.
$= 2x^3 + 8x^2 + 10x + 3x^2 + 12x + 15$ Multiply monomials: Multiply coefficients and add exponents.

$= 2x^3 + 11x^2 + 22x + 15$ Combine like terms: $8x^2 + 3x^2 = 11x^2$ and $10x + 12x = 22x$. •••

Another method for performing the multiplication is to use a vertical format similar to that used for multiplying whole numbers.

$$
\begin{array}{r}
x^2 + 4x + 5 \\
2x + 3 \\
\hline
3x^2 + 12x + 15 \\
2x^3 + 8x^2 + 10x \\
\hline
2x^3 + 11x^2 + 22x + 15
\end{array}
$$

$3(x^2 + 4x + 5)$

$2x(x^2 + 4x + 5)$

Write like terms in the same column.
Combine like terms.

✓ **Check Point 2** Multiply: $(5x - 2)(3x^2 - 5x + 4)$.

④ Use FOIL in polynomial multiplication.

The Product of Two Binomials: FOIL

Frequently, we need to find the product of two binomials. One way to perform this multiplication is to distribute each term in the first binomial through the second binomial. For example, we can find the product of the binomials $3x + 2$ and $4x + 5$ as follows:

$$(3x + 2)(4x + 5) = 3x(4x + 5) + 2(4x + 5)$$
$$= 3x(4x) + 3x(5) + 2(4x) + 2(5)$$
$$= 12x^2 + 15x + 8x + 10.$$

Distribute $3x$ over $4x + 5$. Distribute 2 over $4x + 5$.

We'll combine these like terms later. For now, our interest is in how to obtain *each of these four terms.*

We can also find the product of $3x + 2$ and $4x + 5$ using a method called FOIL, which is based on our preceding work. Any two binomials can be quickly multiplied

by using the FOIL method, in which **F** represents the product of the **first** terms in each binomial, **O** represents the product of the **outside** terms, **I** represents the product of the **inside** terms, and **L** represents the product of the **last**, or second, terms in each binomial. For example, we can use the FOIL method to find the product of the binomials $3x + 2$ and $4x + 5$ as follows:

$$(3x + 2)(4x + 5) = 12x^2 + 15x + 8x + 10$$

| Product of First terms | Product of Outside terms | Product of Inside terms | Product of Last terms |

$$= 12x^2 + 23x + 10 \qquad \text{Combine like terms.}$$

In general, here's how to use the FOIL method to find the product of $ax + b$ and $cx + d$:

Using the FOIL Method to Multiply Binomials

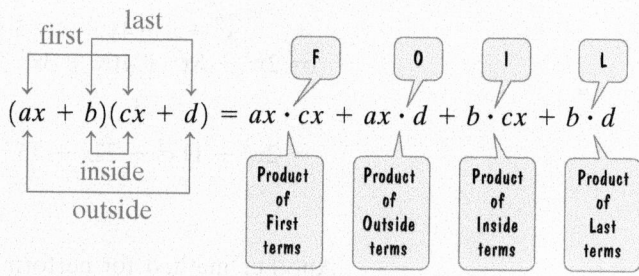

$$(ax + b)(cx + d) = ax \cdot cx + ax \cdot d + b \cdot cx + b \cdot d$$

| Product of First terms | Product of Outside terms | Product of Inside terms | Product of Last terms |

EXAMPLE 3 Using the FOIL Method

Multiply: $(3x + 4)(5x - 3)$.

SOLUTION

$$
\begin{aligned}
(3x + 4)(5x - 3) &= 3x \cdot 5x + 3x(-3) + 4 \cdot 5x + 4(-3) \\
&= 15x^2 - 9x + 20x - 12 \\
&= 15x^2 + 11x - 12 \qquad \text{Combine like terms.} \quad \bullet\bullet\bullet
\end{aligned}
$$

⌀ Check Point **3** Multiply: $(7x - 5)(4x - 3)$.

5 Use special products in polynomial multiplication.

Multiplying the Sum and Difference of Two Terms

We can use the FOIL method to multiply $A + B$ and $A - B$ as follows:

$$(A + B)(A - B) = A^2 - AB + AB - B^2 = A^2 - B^2.$$

Notice that the outside and inside products have a sum of 0 and the terms cancel. The FOIL multiplication provides us with a quick rule for multiplying the sum and difference of two terms, referred to as a *special-product formula*.

The Product of the Sum and Difference of Two Terms

$$(A + B)(A - B) = A^2 - B^2$$

| The product of the sum and the difference of the same two terms | is | the square of the first term minus the square of the second term. |

EXAMPLE 4 Finding the Product of the Sum and Difference of Two Terms

Multiply:

 a. $(4y + 3)(4y - 3)$ **b.** $(5a^4 + 6)(5a^4 - 6)$.

SOLUTION

Use the special-product formula shown.

$$(A + B)(A - B) \quad = \quad A^2 \quad - \quad B^2$$

| | First term squared | − | Second term squared | = | Product |

 a. $(4y + 3)(4y - 3) \quad = \quad (4y)^2 \quad - \quad 3^2 \quad = \quad 16y^2 - 9$

 b. $(5a^4 + 6)(5a^4 - 6) \quad = \quad (5a^4)^2 \quad - \quad 6^2 \quad = \quad 25a^8 - 36$ •••

⊘ Check Point **4** Multiply:

 a. $(7x + 8)(7x - 8)$ **b.** $(2y^3 - 5)(2y^3 + 5)$.

The Square of a Binomial

Let us find $(A + B)^2$, the square of a binomial sum. To do so, we begin with the FOIL method and look for a general rule.

 F O I L

$$(A + B)^2 = (A + B)(A + B) = A \cdot A + A \cdot B + A \cdot B + B \cdot B$$
$$= A^2 + 2AB + B^2$$

This result implies the following rule, which is another example of a special-product formula:

The Square of a Binomial Sum

$$(A + B)^2 \quad = \quad A^2 \quad + \quad 2AB \quad + \quad B^2$$

| The square of a binomial sum | is | first term squared | plus | 2 times the product of the terms | plus | last term squared. |

GREAT QUESTION!

When finding $(x + 3)^2$, why can't I just write $x^2 + 3^2$, or $x^2 + 9$?

Caution! The square of a sum is *not* the sum of the squares.

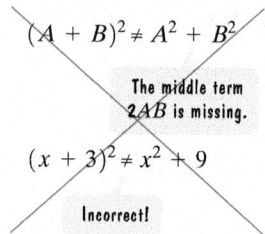

$(A + B)^2 \neq A^2 + B^2$

The middle term $2AB$ is missing.

$(x + 3)^2 \neq x^2 + 9$

Incorrect!

Show that $(x + 3)^2$ and $x^2 + 9$ are not equal by substituting 5 for x in each expression and simplifying.

EXAMPLE 5 Finding the Square of a Binomial Sum

Multiply:

a. $(x + 3)^2$

b. $(3x + 7)^2$.

SOLUTION

Use the special-product formula shown.

$$(A + B)^2 = \quad A^2 \quad + \quad 2AB \quad + \quad B^2$$

	(First Term)2	+	2 · Product of the Terms	+	(Last Term)2	= Product
a. $(x + 3)^2 =$	x^2	+	$2 \cdot x \cdot 3$	+	3^2	$= x^2 + 6x + 9$
b. $(3x + 7)^2 =$	$(3x)^2$	+	$2(3x)(7)$	+	7^2	$= 9x^2 + 42x + 49$

• • •

✓ Check Point **5** Multiply:

a. $(x + 10)^2$

b. $(5x + 4)^2$.

A similar pattern occurs for $(A - B)^2$, the square of a binomial difference. Using the FOIL method on $(A - B)^2$, we obtain the following rule:

The Square of a Binomial Difference

$$(A - B)^2 = A^2 - 2AB + B^2$$

| The square of a binomial difference | is | first term squared | minus | 2 times the product of the terms | plus | last term squared. |

EXAMPLE 6 Finding the Square of a Binomial Difference

Multiply:

a. $(x - 4)^2$

b. $(5y - 6)^2$.

SOLUTION

Use the special-product formula shown.

$$(A - B)^2 = \quad A^2 \quad - \quad 2AB \quad + \quad B^2$$

	(First Term)2	−	2 · Product of the Terms	+	(Last Term)2	= Product
a. $(x - 4)^2 =$	x^2	−	$2 \cdot x \cdot 4$	+	4^2	$= x^2 - 8x + 16$
b. $(5y - 6)^2 =$	$(5y)^2$	−	$2(5y)(6)$	+	6^2	$= 25y^2 - 60y + 36$

• • •

✓ Check Point **6** Multiply:

a. $(x - 9)^2$

b. $(7x - 3)^2$.

Special Products

There are several products that occur so frequently that it's convenient to memorize the form, or pattern, of these formulas.

GREAT QUESTION!

Do I have to memorize the special products shown in the table on the right?

Not necessarily. Although it's convenient to memorize these forms, the FOIL method can be used on all five examples in the box. To cube $x + 4$, you can first square $x + 4$ using FOIL and then multiply this result by $x + 4$. We suggest memorizing these special forms because they let you multiply far more rapidly than using the FOIL method.

Special Products

Let A and B represent real numbers, variables, or algebraic expressions.

Special Product	Example
Sum and Difference of Two Terms	
$(A + B)(A - B) = A^2 - B^2$	$(2x + 3)(2x - 3) = (2x)^2 - 3^2$ $= 4x^2 - 9$
Squaring a Binomial	
$(A + B)^2 = A^2 + 2AB + B^2$	$(y + 5)^2 = y^2 + 2 \cdot y \cdot 5 + 5^2$ $= y^2 + 10y + 25$
$(A - B)^2 = A^2 - 2AB + B^2$	$(3x - 4)^2$ $= (3x)^2 - 2 \cdot 3x \cdot 4 + 4^2$ $= 9x^2 - 24x + 16$
Cubing a Binomial	
$(A + B)^3 = A^3 + 3A^2B + 3AB^2 + B^3$	$(x + 4)^3$ $= x^3 + 3x^2(4) + 3x(4)^2 + 4^3$ $= x^3 + 12x^2 + 48x + 64$
$(A - B)^3 = A^3 - 3A^2B + 3AB^2 - B^3$	$(x - 2)^3$ $= x^3 - 3x^2(2) + 3x(2)^2 - 2^3$ $= x^3 - 6x^2 + 12x - 8$

⑥ Perform operations with polynomials in several variables.

Polynomials in Several Variables

A **polynomial in two variables**, x and y, contains the sum of one or more monomials in the form $ax^n y^m$. The constant, a, is the **coefficient**. The exponents, n and m, represent whole numbers. The **degree** of the monomial $ax^n y^m$ is $n + m$.

Here is an example of a polynomial in two variables:

The coefficients are 7, −17, 1, −6, and 9.

$$7x^2y^3 \quad - \quad 17x^4y^2 \quad + \quad xy \quad - \quad 6y^2 \quad + \quad 9.$$

Degree of monomial: $2 + 3 = 5$ Degree of monomial: $4 + 2 = 6$ Degree of monomial $(1x^1y^1)$: $1 + 1 = 2$ Degree of monomial $(-6x^0y^2)$: $0 + 2 = 2$ Degree of monomial $(9x^0y^0)$: $0 + 0 = 0$

The **degree of a polynomial in two variables** is the highest degree of all its terms. For the preceding polynomial, the degree is 6.

Polynomials containing two or more variables can be added, subtracted, and multiplied just like polynomials that contain only one variable. For example, we can add the monomials $-7xy^2$ and $13xy^2$ as follows:

$$-7xy^2 + 13xy^2 = (-7 + 13)xy^2 = 6xy^2.$$

These like terms both contain the variable factors x and y^2. Add coefficients and keep the same variable factors, xy^2.

EXAMPLE 7 Subtracting Polynomials in Two Variables

Subtract:

$$(5x^3 - 9x^2y + 3xy^2 - 4) - (3x^3 - 6x^2y - 2xy^2 + 3).$$

SOLUTION

$(5x^3 - 9x^2y + 3xy^2 - 4) - (3x^3 - 6x^2y - 2xy^2 + 3)$

Change the sign of each coefficient.

$= (5x^3 - 9x^2y + 3xy^2 - 4) + (-3x^3 + 6x^2y + 2xy^2 - 3)$ Add the opposite of the polynomial being subtracted.

$= (5x^3 - 3x^3) + (-9x^2y + 6x^2y) + (3xy^2 + 2xy^2) + (-4 - 3)$ Group like terms.

$= 2x^3 - 3x^2y + 5xy^2 - 7$ Combine like terms by adding coefficients and keeping the same variable factors.

• • •

⊘ Check Point **7** Subtract: $(x^3 - 4x^2y + 5xy^2 - y^3) - (x^3 - 6x^2y + y^3).$

EXAMPLE 8 Multiplying Polynomials in Two Variables

Multiply: **a.** $(x + 4y)(3x - 5y)$ **b.** $(5x + 3y)^2.$

SOLUTION

We will perform the multiplication in part (a) using the FOIL method. We will multiply in part (b) using the formula for the square of a binomial sum, $(A + B)^2.$

a. $(x + 4y)(3x - 5y)$ Multiply these binomials using the FOIL method.

 F O I L

$= (x)(3x) + (x)(-5y) + (4y)(3x) + (4y)(-5y)$

$= 3x^2 - 5xy + 12xy - 20y^2$

$= 3x^2 + 7xy - 20y^2$ Combine like terms.

$(A + B)^2 \ = \ A^2 \ + \ 2 \cdot A \cdot B \ + \ B^2$

b. $(5x + 3y)^2 = (5x)^2 + 2(5x)(3y) + (3y)^2$

 $= 25x^2 + 30xy + 9y^2$

• • •

⊘ Check Point **8** Multiply:

 a. $(7x - 6y)(3x - y)$ **b.** $(2x + 4y)^2.$

Blitzer Bonus || Labrador Retrievers and Polynomial Multiplication

The color of a Labrador retriever is determined by its pair of genes. A single gene is inherited at random from each parent. The black-fur gene, B, is dominant. The yellow-fur gene, Y, is recessive. This means that labs with at least one black-fur gene (BB or BY) have black coats. Only labs with two yellow-fur genes (YY) have yellow coats.

Axl, your author's yellow lab, inherited his genetic makeup from two black BY parents.

Second BY parent, a black lab
with a recessive yellow-fur gene

	B	Y
B	BB	BY
Y	BY	YY

First BY parent, a black lab
with a recessive yellow-fur gene

The table shows the four possible combinations of color genes that BY parents can pass to their offspring.

Because YY is one of four possible outcomes, the probability that a yellow lab like Axl will be the offspring of these black parents is $\frac{1}{4}$.

The probabilities suggested by the table can be modeled by the expression $\left(\frac{1}{2}B + \frac{1}{2}Y\right)^2$.

$$\left(\frac{1}{2}B + \frac{1}{2}Y\right)^2 = \left(\frac{1}{2}B\right)^2 + 2\left(\frac{1}{2}B\right)\left(\frac{1}{2}Y\right) + \left(\frac{1}{2}Y\right)^2$$

$$= \frac{1}{4}BB + \frac{1}{2}BY + \frac{1}{4}YY$$

The probability of a black lab with two dominant black genes is $\frac{1}{4}$.

The probability of a black lab with a recessive yellow gene is $\frac{1}{2}$.

The probability of a yellow lab with two recessive yellow genes is $\frac{1}{4}$.

CONCEPT AND VOCABULARY CHECK

Fill in each blank so that the resulting statement is true.

1. A polynomial is a single term or the sum of two or more terms containing variables with exponents that are _____ numbers.
2. It is customary to write the terms of a polynomial in the order of descending powers of the variable. This is called the _____ form of a polynomial.
3. A simplified polynomial that has exactly one term is called a/an _____.
4. A simplified polynomial that has two terms is called a/an _____.
5. A simplified polynomial that has three terms is called a/an _____.
6. If $a \neq 0$, the degree of ax^n is _____.
7. Polynomials are added by combining _____ terms.
8. To multiply $7x^3(4x^5 - 8x^2 + 6)$, use the _____ property to multiply each term of the trinomial _____ by the monomial _____.
9. To multiply $(5x + 3)(x^2 + 8x + 7)$, begin by multiplying each term of $x^2 + 8x + 7$ by _____. Then multiply each term of $x^2 + 8x + 7$ by _____. Then combine _____ terms.

10. When using the FOIL method to find $(x + 7)(3x + 5)$, the product of the first terms is _____, the product of the outside terms is _____, the product of the inside terms is _____, and the product of the last terms is _____.
11. $(A + B)(A - B) = $ _____. The product of the sum and difference of the same two terms is the square of the first term _____ the square of the second term.
12. $(A + B)^2 = $ _____. The square of a binomial sum is the first term _____ plus 2 times the _____ plus the last term _____.
13. $(A - B)^2 = $ _____. The square of a binomial difference is the first term squared _____ 2 times the _____ _____ the last term squared.

 plus or minus?

14. If $a \neq 0$, the degree of $ax^n y^m$ is _____.

EXERCISE SET P.5

Practice Exercises

In Exercises 1–4, is the algebraic expression a polynomial? If it is, write the polynomial in standard form.

1. $2x + 3x^2 - 5$
2. $2x + 3x^{-1} - 5$
3. $\dfrac{2x + 3}{x}$
4. $x^2 - x^3 + x^4 - 5$

In Exercises 5–8, find the degree of the polynomial.

5. $3x^2 - 5x + 4$
6. $-4x^3 + 7x^2 - 11$
7. $x^2 - 4x^3 + 9x - 12x^4 + 63$
8. $x^2 - 8x^3 + 15x^4 + 91$

In Exercises 9–14, perform the indicated operations. Write the resulting polynomial in standard form and indicate its degree.

9. $(-6x^3 + 5x^2 - 8x + 9) + (17x^3 + 2x^2 - 4x - 13)$
10. $(-7x^3 + 6x^2 - 11x + 13) + (19x^3 - 11x^2 + 7x - 17)$
11. $(17x^3 - 5x^2 + 4x - 3) - (5x^3 - 9x^2 - 8x + 11)$
12. $(18x^4 - 2x^3 - 7x + 8) - (9x^4 - 6x^3 - 5x + 7)$
13. $(5x^2 - 7x - 8) + (2x^2 - 3x + 7) - (x^2 - 4x - 3)$
14. $(8x^2 + 7x - 5) - (3x^2 - 4x) - (-6x^3 - 5x^2 + 3)$

In Exercises 15–58, find each product.

15. $(x + 1)(x^2 - x + 1)$
16. $(x + 5)(x^2 - 5x + 25)$
17. $(2x - 3)(x^2 - 3x + 5)$
18. $(2x - 1)(x^2 - 4x + 3)$
19. $(x + 7)(x + 3)$
20. $(x + 8)(x + 5)$
21. $(x - 5)(x + 3)$
22. $(x - 1)(x + 2)$
23. $(3x + 5)(2x + 1)$
24. $(7x + 4)(3x + 1)$
25. $(2x - 3)(5x + 3)$
26. $(2x - 5)(7x + 2)$
27. $(5x^2 - 4)(3x^2 - 7)$
28. $(7x^2 - 2)(3x^2 - 5)$
29. $(8x^3 + 3)(x^2 - 5)$
30. $(7x^3 + 5)(x^2 - 2)$
31. $(x + 3)(x - 3)$
32. $(x + 5)(x - 5)$
33. $(3x + 2)(3x - 2)$
34. $(2x + 5)(2x - 5)$
35. $(5 - 7x)(5 + 7x)$
36. $(4 - 3x)(4 + 3x)$
37. $(4x^2 + 5x)(4x^2 - 5x)$
38. $(3x^2 + 4x)(3x^2 - 4x)$
39. $(1 - y^5)(1 + y^5)$
40. $(2 - y^5)(2 + y^5)$
41. $(x + 2)^2$
42. $(x + 5)^2$
43. $(2x + 3)^2$
44. $(3x + 2)^2$
45. $(x - 3)^2$
46. $(x - 4)^2$
47. $(4x^2 - 1)^2$
48. $(5x^2 - 3)^2$
49. $(7 - 2x)^2$
50. $(9 - 5x)^2$
51. $(x + 1)^3$
52. $(x + 2)^3$
53. $(2x + 3)^3$
54. $(3x + 4)^3$
55. $(x - 3)^3$
56. $(x - 1)^3$
57. $(3x - 4)^3$
58. $(2x - 3)^3$

In Exercises 59–66, perform the indicated operations. Indicate the degree of the resulting polynomial.

59. $(5x^2y - 3xy) + (2x^2y - xy)$
60. $(-2x^2y + xy) + (4x^2y + 7xy)$
61. $(4x^2y + 8xy + 11) + (-2x^2y + 5xy + 2)$
62. $(7x^4y^2 - 5x^2y^2 + 3xy) + (-18x^4y^2 - 6x^2y^2 - xy)$
63. $(x^3 + 7xy - 5y^2) - (6x^3 - xy + 4y^2)$
64. $(x^4 - 7xy - 5y^3) - (6x^4 - 3xy + 4y^3)$
65. $(3x^4y^2 + 5x^3y - 3y) - (2x^4y^2 - 3x^3y - 4y + 6x)$
66. $(5x^4y^2 + 6x^3y - 7y) - (3x^4y^2 - 5x^3y - 6y + 8x)$

In Exercises 67–82, find each product.

67. $(x + 5y)(7x + 3y)$
68. $(x + 9y)(6x + 7y)$
69. $(x - 3y)(2x + 7y)$
70. $(3x - y)(2x + 5y)$
71. $(3xy - 1)(5xy + 2)$
72. $(7x^2y + 1)(2x^2y - 3)$
73. $(7x + 5y)^2$
74. $(9x + 7y)^2$
75. $(x^2y^2 - 3)^2$
76. $(x^2y^2 - 5)^2$
77. $(x - y)(x^2 + xy + y^2)$
78. $(x + y)(x^2 - xy + y^2)$
79. $(3x + 5y)(3x - 5y)$
80. $(7x + 3y)(7x - 3y)$
81. $(7xy^2 - 10y)(7xy^2 + 10y)$
82. $(3xy^2 - 4y)(3xy^2 + 4y)$

Practice Plus

In Exercises 83–90, perform the indicated operation or operations.

83. $(3x + 4y)^2 - (3x - 4y)^2$
84. $(5x + 2y)^2 - (5x - 2y)^2$
85. $(5x - 7)(3x - 2) - (4x - 5)(6x - 1)$
86. $(3x + 5)(2x - 9) - (7x - 2)(x - 1)$
87. $(2x + 5)(2x - 5)(4x^2 + 25)$
88. $(3x + 4)(3x - 4)(9x^2 + 16)$
89. $\dfrac{(2x - 7)^5}{(2x - 7)^3}$
90. $\dfrac{(5x - 3)^6}{(5x - 3)^4}$

Application Exercises

As you complete more years of education, you can count on a greater income. The bar graph shows the median, or middlemost, annual income for Americans, by level of education, in 2009.

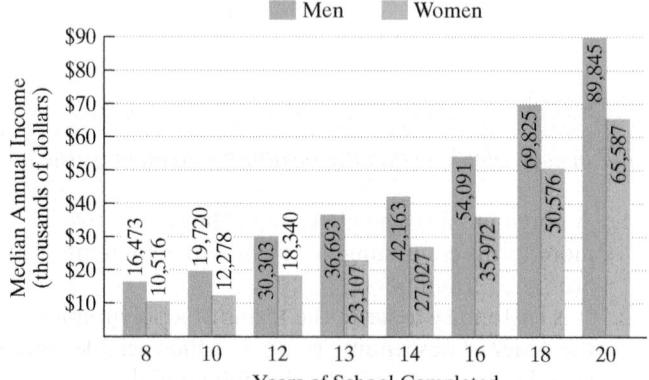

Median Annual Income, by Level of Education, 2009

Source: Bureau of the Census

Here are polynomial models that describe the median annual income for men, M, and for women, W, who have completed x years of education:

$$M = 312x^2 - 2615x + 16,615$$
$$W = 316x^2 - 4224x + 23,730$$
$$M = 0.6x^3 + 285x^2 - 2256x + 15,112$$
$$W = -1.2x^3 + 367x^2 - 4900x + 26,561$$

Exercises 91–92 are based on these models and the data displayed by the graph.

91. a. Use the equation defined by a polynomial of degree 2 to find the median annual income for a man with 16 years of education. Does this underestimate or overestimate the median income shown by the bar graph? By how much?

b. Use the equations defined by polynomials of degree 3 to find a mathematical model for $M - W$.

c. According to the model in part (b), what is the difference, rounded to the nearest dollar, in the median annual income between men and women with 14 years of education?

d. According to the data displayed by the graph, what is the actual difference in the median annual income between men and women with 14 years of education? Did the result of part (c) underestimate or overestimate this difference? By how much?

92. a. Use the equation defined by a polynomial of degree 2 to find the median annual income for a woman with 18 years of education. Does this underestimate or overestimate the median income shown by the bar graph? By how much?

b. Use the equations defined by polynomials of degree 3 to find a mathematical model for $M - W$.

c. According to the model in part (b), what is the difference, rounded to the nearest dollar, in the median annual income between men and women with 16 years of education?

d. According to the data displayed by the graph, what is the actual difference in the median annual income between men and women with 16 years of education? Did the result of part (c) underestimate or overestimate this difference? By how much?

The volume, V, of a rectangular solid with length l, width w, and height h is given by the formula V = lwh. In Exercises 93–94, use this formula to write a polynomial in standard form that models, or represents, the volume of the open box.

93.

94.

In Exercises 95–96, write a polynomial in standard form that models, or represents, the area of the shaded region.

95.

96.

Writing in Mathematics

97. What is a polynomial in x?

98. Explain how to subtract polynomials.

99. Explain how to multiply two binomials using the FOIL method. Give an example with your explanation.

100. Explain how to find the product of the sum and difference of two terms. Give an example with your explanation.

101. Explain how to square a binomial difference. Give an example with your explanation.

102. Explain how to find the degree of a polynomial in two variables.

Critical Thinking Exercises

Make Sense? *In Exercises 103–106, determine whether each statement makes sense or does not make sense, and explain your reasoning.*

103. Knowing the difference between factors and terms is important: In $(3x^2y)^2$, I can distribute the exponent 2 on each factor, but in $(3x^2 + y)^2$, I cannot do the same thing on each term.

104. I used the FOIL method to find the product of $x + 5$ and $x^2 + 2x + 1$.

105. Many English words have prefixes with meanings similar to those used to describe polynomials, such as *monologue, binocular,* and *tricuspid.*

106. Special-product formulas have patterns that make their multiplications quicker than using the FOIL method.

In Exercises 107–110, determine whether each statement is true or false. If the statement is false, make the necessary change(s) to produce a true statement.

107. $(3x^3 + 2)(3x^3 - 2) = 9x^9 - 4$

108. $(x - 5)^2 = x^2 - 5x + 25$

109. $(x + 1)^2 = x^2 + 1$

110. Suppose a square garden has an area represented by $9x^2$ square feet. If one side is made 7 feet longer and the other side is made 2 feet shorter, then the trinomial that models the area of the larger garden is $9x^2 + 15x - 14$ square feet.

In Exercises 111–113, perform the indicated operations.

111. $[(7x + 5) + 4y][(7x + 5) - 4y]$

112. $[(3x + y) + 1]^2$

113. $(x^n + 2)(x^n - 2) - (x^n - 3)^2$

114. Express the area of the plane figure shown as a polynomial in standard form.

Preview Exercises

Exercises 115–117 will help you prepare for the material covered in the next section. In each exercise, replace the boxed question mark with an integer that results in the given product. Some trial and error may be necessary.

115. $(x + 3)(x + \boxed{?}) = x^2 + 7x + 12$

116. $(x - \boxed{?})(x - 12) = x^2 - 14x + 24$

117. $(4x + 1)(2x - \boxed{?}) = 8x^2 - 10x - 3$

CHAPTER P Mid-Chapter Check Point

WHAT YOU KNOW: We defined the real numbers [{$x|x$ is rational} \cup {$x|x$ is irrational}] and represented subsets of real numbers using interval notation. We reviewed the basic rules of algebra, using these properties to both simplify algebraic expressions and solve algebraic equations. We expanded our knowledge of exponents to include exponents other than natural numbers:

$$b^0 = 1; \quad b^{-n} = \frac{1}{b^n}; \quad \frac{1}{b^{-n}} = b^n; \quad b^{\frac{1}{n}} = \sqrt[n]{b};$$

$$b^{\frac{m}{n}} = (\sqrt[n]{b})^m = \sqrt[n]{b^m}; \quad b^{-\frac{m}{n}} = \frac{1}{b^{\frac{m}{n}}}.$$

We used properties of exponents to simplify exponential expressions and properties of radicals to simplify radical expressions. We applied the square root property to solve equations of the form $u^2 = d$, where $d > 0$, obtaining two solutions: $u = \sqrt{d}$ or $u = -\sqrt{d}$. Finally, we performed operations with polynomials. We used a number of fast methods for finding products of polynomials, including the FOIL method for multiplying binomials, a special-product formula for the product of the sum and difference of two terms $[(A + B)(A - B) = A^2 - B^2]$, and special-product formulas for squaring binomials $[(A + B)^2 = A^2 + 2AB + B^2; (A - B)^2 = A^2 - 2AB + B^2]$.

In Exercises 1–30, simplify the given expression, perform the indicated operation (and simplify, if possible), or solve the equation, whichever is appropriate.

1. $(3x + 5)(4x - 7)$

2. $(3x + 5) - (4x - 7)$

3. $5x + 3 = 48$

4. $6x^2 - 5 = 49$

5. $\sqrt{6} + 9\sqrt{6}$

6. $3\sqrt{12} - \sqrt{27}$

7. $7x + 3[9 - (2x - 6)]$

8. $(8x - 3)^2$

9. $\left(x^{\frac{1}{3}}y^{-\frac{1}{2}}\right)^6$

10. $\left(\frac{2}{7}\right)^0 - 32^{-\frac{2}{5}}$

11. $(2x - 5) - (x^2 - 3x + 1)$

12. $(2x - 5)(x^2 - 3x + 1)$

13. $x^3 + x^3 - x^3 \cdot x^3$

14. $(9a - 10b)(2a + b)$

15. $\{a, c, d, e\} \cup \{c, d, f, h\}$

16. $\{a, c, d, e\} \cap \{c, d, f, h\}$

17. $(2, 5) \cap [3, 9]$

18. $(2, 5) \cup [3, 9]$

19. $(3x^2y^3 - xy + 4y^2) - (-2x^2y^3 - 3xy + 5y^2)$

20. $\dfrac{24x^2y^{13}}{-2x^5y^{-2}}$

21. $\left(\frac{1}{3}x^{-5}y^4\right)(18x^{-2}y^{-1})$

22. $\sqrt[12]{x^4}$

23. $-5 + 3(x + 5) = 28$

24. $\dfrac{24 \times 10^3}{2 \times 10^6}$ (Express the answer in scientific notation.)

25. $\dfrac{\sqrt[3]{32}}{\sqrt[3]{2}}$

26. $(x^3 + 2)(x^3 - 2)$

27. $(x^2 + 2)^2$

28. $\sqrt{50} \cdot \sqrt{6}$

29. $\dfrac{11}{7 - \sqrt{3}}$

30. $\dfrac{11}{\sqrt{3}}$

31. List all the rational numbers in this set:

$$\left\{-11, -\frac{3}{7}, 0, 0.45, \sqrt{23}, \sqrt{25}\right\}.$$

In Exercises 32–33, express each interval in set-builder notation and graph the interval on a number line.

32. $[-2, 6)$

33. $(-\infty, 6]$

In Exercises 34–35, rewrite each expression without absolute value bars.

34. $|2 - \sqrt{13}|$

35. $x^2|x|$ if $x < 0$

36. Solve for x: $Ax - By + C = 0$.

37. If the population of the United States is approximately 3.0×10^8 and each person produces about 4.6 pounds of garbage per day, express the total number of pounds of garbage produced in the United States in one day in scientific notation.

38. A human brain contains 3×10^{10} neurons and a gorilla brain contains 7.5×10^9 neurons. How many times as many neurons are in the brain of a human as in the brain of a gorilla?

39. TVs keep getting fancier and bigger, but prices do not. The bar graph shows the average price of a TV in the United States from 2007 through 2012.

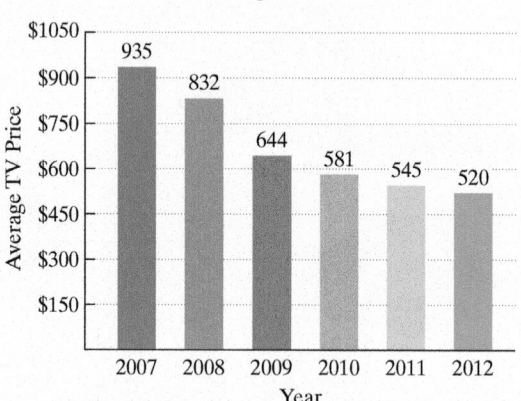

Average Price of a TV

Source: Consumer Electronics Association

Here are two mathematical models for the data shown by the graph. In each formula, P represents the average price of a TV x years after 2007.

Model 1 $P = -86x + 890$

Model 2 $P = 18x^2 - 175x + 950$

a. Which model better describes the data for 2007?

b. Does the polynomial model of degree 2 underestimate or overestimate the average TV price for 2012? By how much?

c. According to the polynomial model of degree 1, when will the average price of a TV be $202?

SECTION P.6 Factoring Polynomials

Objectives

1. Factor out the greatest common factor of a polynomial.
2. Factor by grouping.
3. Factor trinomials.
4. Factor the difference of squares.
5. Factor perfect square trinomials.
6. Factor the sum or difference of two cubes.
7. Use a general strategy for factoring polynomials.
8. Factor algebraic expressions containing fractional and negative exponents.
9. Complete the square of a binomial.

A two-year-old boy is asked, "Do you have a brother?" He answers, "Yes." "What is your brother's name?" "Tom." Asked if Tom has a brother, the two-year-old replies, "No." The child can go in the direction from self to brother, but he cannot reverse this direction and move from brother back to self.

As our intellects develop, we learn to reverse the direction of our thinking. Reversibility of thought is found throughout algebra. For example, we can multiply polynomials and show that

$$5x(2x + 3) = 10x^2 + 15x.$$

We can also reverse this process and express the resulting polynomial as

$$10x^2 + 15x = 5x(2x + 3).$$

Factoring a polynomial expressed as the sum of monomials means finding an equivalent expression that is a product.

Factoring $10x^2 + 15x$

Sum of monomials Equivalent expression that is a product

$$10x^2 + 15x = 5x(2x + 3)$$

The factors of $10x^2 + 15x$ are $5x$ and $2x + 3$.

In this section, we will be **factoring over the set of integers**, meaning that the coefficients in the factors are integers. Polynomials that cannot be factored using integer coefficients are called **irreducible over the integers**, or **prime**.

The goal in factoring a polynomial is to use one or more factoring techniques until each of the polynomial's factors, except possibly for a monomial factor, is prime or irreducible. In this situation, the polynomial is said to be **factored completely**.

We will now discuss basic techniques for factoring polynomials.

1. Factor out the greatest common factor of a polynomial.

Common Factors

In any factoring problem, the first step is to look for the *greatest common factor*. The **greatest common factor**, abbreviated GCF, is an expression of the highest degree that divides each term of the polynomial. The distributive property in the reverse direction

$$ab + ac = a(b + c)$$

can be used to factor out the greatest common factor.

EXAMPLE 1 Factoring Out the Greatest Common Factor

Factor:

a. $18x^3 + 27x^2$ **b.** $x^2(x + 3) + 5(x + 3).$

GREAT QUESTION!

Is there a rule that can help me determine the greatest common factor?

Yes. The variable part of the greatest common factor always contains the *smallest* power of a variable or algebraic expression that appears in all terms of the polynomial.

SOLUTION

a. To factor $18x^3 + 27x^2$, first determine the greatest common factor.

9 is the greatest integer that divides 18 and 27.

$$18x^3 + 27x^2$$

x^2 is the greatest expression that divides x^3 and x^2.

The GCF of the two terms of the polynomial is $9x^2$.

$$18x^3 + 27x^2$$
$$= 9x^2(2x) + 9x^2(3) \quad \text{Express each term as the product of the GCF and its other factor.}$$
$$= 9x^2(2x + 3) \quad \text{Factor out the GCF.}$$

b. To factor $x^2(x + 3) + 5(x + 3)$, the greatest common factor is the common binomial factor $(x + 3)$. We factor out this common factor as follows:

$$x^2(x + 3) + 5(x + 3) = (x + 3)(x^2 + 5). \quad \text{Factor out the common binomial factor.}$$

● ● ●

⊘ Check Point **1** Factor:
 a. $10x^3 - 4x^2$ **b.** $2x(x - 7) + 3(x - 7)$.

② Factor by grouping.

Factoring by Grouping

Some polynomials have only a greatest common factor of 1. However, by a suitable grouping of the terms, it still may be possible to factor. This process, called **factoring by grouping**, is illustrated in Example 2.

EXAMPLE 2 Factoring by Grouping

Factor: $x^3 + 4x^2 + 3x + 12$.

SOLUTION

There is no factor other than 1 common to all terms. However, we can group terms that have a common factor:

$$\boxed{x^3 + 4x^2} \quad + \quad \boxed{3x + 12}.$$

Common factor is x^2. Common factor is 3.

We now factor the given polynomial as follows:

$$x^3 + 4x^2 + 3x + 12$$
$$= (x^3 + 4x^2) + (3x + 12) \quad \text{Group terms with common factors.}$$
$$= x^2(x + 4) + 3(x + 4) \quad \text{Factor out the greatest common factor from the grouped terms. The remaining two terms have } x + 4 \text{ as a common binomial factor.}$$
$$= (x + 4)(x^2 + 3). \quad \text{Factor out the GCF, } x + 4.$$

DISCOVERY

In Example 2, group the terms as follows:

$$(x^3 + 3x) + (4x^2 + 12).$$

Factor out the greatest common factor from each group and complete the factoring process. Describe what happens. What can you conclude?

Thus, $x^3 + 4x^2 + 3x + 12 = (x + 4)(x^2 + 3)$. Check the factorization by multiplying the right side of the equation using the FOIL method. Because the factorization is correct, you should obtain the original polynomial.

● ● ●

⊘ Check Point **2** Factor: $x^3 + 5x^2 - 2x - 10$.

③ Factor trinomials.

Factoring Trinomials

To factor a trinomial of the form $ax^2 + bx + c$, a little trial and error may be necessary.

A Strategy for Factoring $ax^2 + bx + c$

Assume, for the moment, that there is no greatest common factor.

1. Find two **First** terms whose product is ax^2:

$$(\square x + \quad)(\square x + \quad) = ax^2 + bx + c.$$

2. Find two **Last** terms whose product is c:

$$(\square x + \square)(\square x + \square) = ax^2 + bx + c.$$

3. By trial and error, perform steps 1 and 2 until the sum of the **O**utside product and **I**nside product is bx:

$$(\square x + \square)(\square x + \square) = ax^2 + bx + c.$$

Sum of O + I

If no such combination exists, the polynomial is prime.

GREAT QUESTION!

Should I feel discouraged if it takes me a while to get the correct factorization?

The *error* part of the factoring strategy plays an important role in the process. If you do not get the correct factorization the first time, this is not a bad thing. This error is often helpful in leading you to the correct factorization.

EXAMPLE 3 Factoring a Trinomial Whose Leading Coefficient Is 1

Factor: $x^2 + 6x + 8$.

SOLUTION

Step 1 Find two First terms whose product is x^2.
$$x^2 + 6x + 8 = (x \quad)(x \quad)$$

Step 2 Find two Last terms whose product is 8.

| Factors of 8 | 8, 1 | 4, 2 | $-8, -1$ | $-4, -2$ |

Step 3 Try various combinations of these factors. The correct factorization of $x^2 + 6x + 8$ is the one in which the sum of the **Outside** and **Inside** products is equal to $6x$. Here is a list of the possible factorizations:

Possible Factorizations of $x^2 + 6x + 8$	Sum of Outside and Inside Products (Should Equal $6x$)
$(x + 8)(x + 1)$	$x + 8x = 9x$
$(x + 4)(x + 2)$	$2x + 4x = 6x$ — This is the required middle term.
$(x - 8)(x - 1)$	$-x - 8x = -9x$
$(x - 4)(x - 2)$	$-2x - 4x = -6x$

Thus, $x^2 + 6x + 8 = (x + 4)(x + 2)$ or $(x + 2)(x + 4)$. • • •

In factoring a trinomial of the form $x^2 + bx + c$, you can speed things up by listing the factors of c and then finding their sums. We are interested in a sum of b. For example, in factoring $x^2 + 6x + 8$, we are interested in the factors of 8 whose sum is 6.

| Factors of 8 | 8, 1 | 4, 2 | $-8, -1$ | $-4, -2$ |
| Sum of Factors | 9 | 6 | -9 | -6 |

This is the desired sum.

Thus, $x^2 + 6x + 8 = (x + 4)(x + 2)$.

⟡ Check Point **3** Factor: $x^2 + 13x + 40$.

EXAMPLE 4 Factoring a Trinomial Whose Leading Coefficient Is 1

Factor: $x^2 + 3x - 18$.

SOLUTION

Step 1 Find two First terms whose product is x^2.

$$x^2 + 3x - 18 = (x\qquad)(x\qquad)$$

To find the second term of each factor, we must find two integers whose product is -18 and whose sum is 3.

Step 2 Find two Last terms whose product is -18.

| Factors of -18 | $18, -1$ | $-18, 1$ | $9, -2$ | $-9, 2$ | $6, -3$ | $-6, 3$ |

Step 3 Try various combinations of these factors. We are looking for the pair of factors whose sum is 3.

Factors of -18	$18, -1$	$-18, 1$	$9, -2$	$-9, 2$	$6, -3$	$-6, 3$
Sum of Factors	17	-17	7	-7	3	-3

> This is the desired sum.

Thus, $x^2 + 3x - 18 = (x + 6)(x - 3)$ or $(x - 3)(x + 6)$. • • •

GREAT QUESTION!

Is there a way to eliminate some of the combinations of factors for a trinomial whose leading coefficient is 1?

Yes. To factor $x^2 + bx + c$ when c is positive, find two numbers with the same sign as the middle term.

$$x^2 + 6x + 8 = (x + 4)(x + 2)\qquad x^2 - 5x + 6 = (x - 3)(x - 2)$$

Same signs Same signs

To factor $x^2 + bx + c$ when c is negative, find two numbers with opposite signs whose sum is the coefficient of the middle term.

$$x^2 + 3x - 18 = (x + 6)(x - 3)\qquad x^2 - 2x - 99 = (x + 9)(x - 11)$$

Negative Opposite signs Negative Opposite signs

⟡ Check Point **4** Factor: $x^2 - 5x - 14$.

EXAMPLE 5 Factoring a Trinomial Whose Leading Coefficient Is Not 1

Factor: $8x^2 - 10x - 3$.

GREAT QUESTION!

When factoring trinomials, must I list every possible factorization before getting the correct one?

With practice, you will find that it is not necessary to list every possible factorization of the trinomial. As you practice factoring, you will be able to narrow down the list of possible factors to just a few. When it comes to factoring, practice makes perfect.

SOLUTION

Step 1 Find two First terms whose product is $8x^2$.

$$8x^2 - 10x - 3 \overset{?}{=} (8x \quad)(x \quad)$$
$$8x^2 - 10x - 3 \overset{?}{=} (4x \quad)(2x \quad)$$

Step 2 Find two Last terms whose product is -3. The possible factorizations are $1(-3)$ and $-1(3)$.

Step 3 Try various combinations of these factors. The correct factorization of $8x^2 - 10x - 3$ is the one in which the sum of the **O**utside and **I**nside products is equal to $-10x$. Here is a list of the possible factorizations:

These four factorizations use $(8x \quad)(x \quad)$ with $1(-3)$ and $-1(3)$ as factorizations of -3.

These four factorizations use $(4x \quad)(2x \quad)$ with $1(-3)$ and $-1(3)$ as factorizations of -3.

Possible Factorizations of $8x^2 - 10x - 3$	Sum of Outside and Inside Products (Should Equal $-10x$)
$(8x + 1)(x - 3)$	$-24x + x = -23x$
$(8x - 3)(x + 1)$	$8x - 3x = 5x$
$(8x - 1)(x + 3)$	$24x - x = 23x$
$(8x + 3)(x - 1)$	$-8x + 3x = -5x$
$(4x + 1)(2x - 3)$	$-12x + 2x = -10x$
$(4x - 3)(2x + 1)$	$4x - 6x = -2x$
$(4x - 1)(2x + 3)$	$12x - 2x = 10x$
$(4x + 3)(2x - 1)$	$-4x + 6x = 2x$

This is the required middle term.

Thus, $8x^2 - 10x - 3 = (4x + 1)(2x - 3)$ or $(2x - 3)(4x + 1)$.

Use FOIL multiplication to check either of these factorizations. • • •

✓ **Check Point 5** Factor: $6x^2 + 19x - 7$.

EXAMPLE 6 Factoring a Trinomial in Two Variables

Factor: $2x^2 - 7xy + 3y^2$.

SOLUTION

Step 1 Find two First terms whose product is $2x^2$.

$$2x^2 - 7xy + 3y^2 = (2x \quad)(x \quad)$$

Step 2 Find two Last terms whose product is $3y^2$. The possible factorizations are $(y)(3y)$ and $(-y)(-3y)$.

Step 3 Try various combinations of these factors. The correct factorization of $2x^2 - 7xy + 3y^2$ is the one in which the sum of the **O**utside and **I**nside products is equal to $-7xy$. Here is a list of possible factorizations:

Possible Factorizations of $2x^2 - 7xy + 3y^2$	Sum of Outside and Inside Products (Should Equal $-7xy$)
$(2x + 3y)(x + y)$	$2xy + 3xy = 5xy$
$(2x + y)(x + 3y)$	$6xy + xy = 7xy$
$(2x - 3y)(x - y)$	$-2xy - 3xy = -5xy$
$(2x - y)(x - 3y)$	$-6xy - xy = -7xy$

This is the required middle term.

GREAT QUESTION!

I zone out reading your long lists of possible factorizations. Are there any rules for shortening these lists?

Here are some suggestions for reducing the list of possible factorizations for $ax^2 + bx + c$:

1. If b is relatively small, avoid the larger factors of a.

2. If c is positive, the signs in both binomial factors must match the sign of b.

3. If the trinomial has no common factor, no binomial factor can have a common factor.

4. Reversing the signs in the binomial factors changes the sign of bx, the middle term.

Thus,

$$2x^2 - 7xy + 3y^2 = (2x - y)(x - 3y) \quad \text{or} \quad (x - 3y)(2x - y).$$

Use FOIL multiplication to check either of these factorizations.

\oslash **Check Point 6** Factor: $3x^2 - 13xy + 4y^2$.

④ Factor the difference of squares.

Factoring the Difference of Two Squares

A method for factoring the difference of two squares is obtained by reversing the special product for the sum and difference of two terms.

> #### The Difference of Two Squares
>
> If A and B are real numbers, variables, or algebraic expressions, then
> $$A^2 - B^2 = (A + B)(A - B).$$
>
> In words: The difference of the squares of two terms factors as the product of a sum and a difference of those terms.

EXAMPLE 7 Factoring the Difference of Two Squares

Factor: **a.** $x^2 - 4$ **b.** $81x^2 - 49$.

SOLUTION

We must express each term as the square of some monomial. Then we use the formula for factoring $A^2 - B^2$.

a. $x^2 - 4 = x^2 - 2^2 = (x + 2)(x - 2)$

$$A^2 - B^2 = (A + B)(A - B)$$

b. $81x^2 - 49 = (9x)^2 - 7^2 = (9x + 7)(9x - 7)$

\oslash **Check Point 7** Factor:
 a. $x^2 - 81$ **b.** $36x^2 - 25$.

We have seen that a polynomial is factored completely when it is written as the product of prime polynomials. To be sure that you have factored completely, check to see whether any factors with more than one term in the factored polynomial can be factored further. If so, continue factoring.

GREAT QUESTION!

Why isn't factoring $x^4 - 81$ as $(x^2 + 9)(x^2 - 9)$ a complete factorization?

The second factor, $x^2 - 9$, is itself a difference of two squares and can be factored.

EXAMPLE 8 A Repeated Factorization

Factor completely: $x^4 - 81$.

SOLUTION

$$\begin{aligned}
x^4 - 81 &= (x^2)^2 - 9^2 \\
&= (x^2 + 9)(x^2 - 9) \\
&= (x^2 + 9)(x^2 - 3^2) \\
&= (x^2 + 9)(x + 3)(x - 3)
\end{aligned}$$

Express as the difference of two squares.

The factors are the sum and the difference of the expressions being squared.

The factor $x^2 - 9$ is the difference of two squares and can be factored.

The factors of $x^2 - 9$ are the sum and the difference of the expressions being squared.

⊘ Check Point **8** Factor completely: $81x^4 - 16$.

⑤ Factor perfect square trinomials.

Factoring Perfect Square Trinomials

Our next factoring technique is obtained by reversing the special products for squaring binomials. The trinomials that are factored using this technique are called **perfect square trinomials**.

Factoring Perfect Square Trinomials

Let A and B be real numbers, variables, or algebraic expressions.

1. $A^2 + 2AB + B^2 = (A + B)^2$ **2.** $A^2 - 2AB + B^2 = (A - B)^2$

 Same sign Same sign

The two items in the box show that perfect square trinomials, $A^2 + 2AB + B^2$ and $A^2 - 2AB + B^2$, come in two forms: one in which the coefficient of the middle term is positive and one in which the coefficient of the middle term is negative. Here's how to recognize a perfect square trinomial:

1. The first and last terms are squares of monomials or integers.

2. The middle term is twice the product of the expressions being squared in the first and last terms.

EXAMPLE 9 Factoring Perfect Square Trinomials

Factor: **a.** $x^2 + 6x + 9$ **b.** $25x^2 - 60x + 36$.

SOLUTION

a. $x^2 + 6x + 9 = x^2 + 2 \cdot x \cdot 3 + 3^2 = (x + 3)^2$ The middle term has a positive sign.

$A^2 + 2AB + B^2 = (A + B)^2$

b. We suspect that $25x^2 - 60x + 36$ is a perfect square trinomial because $25x^2 = (5x)^2$ and $36 = 6^2$. The middle term can be expressed as twice the product of $5x$ and 6.

$$25x^2 - 60x + 36 = (5x)^2 - 2 \cdot 5x \cdot 6 + 6^2 = (5x - 6)^2$$

$A^2 - 2AB + B^2 = (A - B)^2$ • • •

⊘ Check Point **9** Factor:

 a. $x^2 + 14x + 49$ **b.** $16x^2 - 56x + 49$.

6 Factor the sum or difference of two cubes.

Factoring the Sum or Difference of Two Cubes

We can use the following formulas to factor the sum or the difference of two cubes:

Factoring the Sum or Difference of Two Cubes

1. Factoring the Sum of Two Cubes

$$A^3 + B^3 = (A + B)(A^2 - AB + B^2)$$

Same signs Opposite signs Always positive

2. Factoring the Difference of Two Cubes

$$A^3 - B^3 = (A - B)(A^2 + AB + B^2)$$

Same signs Opposite signs Always positive

GREAT QUESTION!

A Cube of SOAP

The formulas for factoring $A^3 + B^3$ and $A^3 - B^3$ are difficult to remember and easy to confuse. Can you help me out?

When factoring sums or differences of cubes, observe the sign patterns shown by the voice balloons in the box. The word *SOAP* is a way to remember these patterns:

S O A P.

| Same signs | Opposite signs | Always Positive |

EXAMPLE 10 Factoring Sums and Differences of Two Cubes

Factor: **a.** $x^3 + 8$ **b.** $64x^3 - 125$.

SOLUTION

a. To factor $x^3 + 8$, we must express each term as the cube of some monomial. Then we use the formula for factoring $A^3 + B^3$.

$$x^3 + 8 = x^3 + 2^3 = (x + 2)(x^2 - x \cdot 2 + 2^2) = (x + 2)(x^2 - 2x + 4)$$

$$A^3 + B^3 = (A + B)(A^2 - AB + B^2)$$

b. To factor $64x^3 - 125$, we must express each term as the cube of some monomial. Then we use the formula for factoring $A^3 - B^3$.

$$64x^3 - 125 = (4x)^3 - 5^3 = (4x - 5)[(4x)^2 + (4x)(5) + 5^2]$$

$$A^3 - B^3 = (A - B)(A^2 + AB + B^2)$$

$$= (4x - 5)(16x^2 + 20x + 25)$$ • • •

✓ Check Point **10** Factor:

a. $x^3 + 1$ **b.** $125x^3 - 8$.

7 Use a general strategy for factoring polynomials.

A Strategy for Factoring Polynomials

It is important to practice factoring a wide variety of polynomials so that you can quickly select the appropriate technique. The polynomial is factored completely when all its polynomial factors, except possibly for monomial factors, are prime. Because of the commutative property, the order of the factors does not matter.

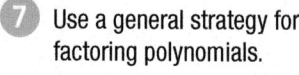

A Strategy for Factoring a Polynomial

1. If there is a common factor, factor out the GCF.

2. Determine the number of terms in the polynomial and try factoring as follows:

 a. If there are two terms, can the binomial be factored by using one of the following special forms?

 Difference of two squares: $A^2 - B^2 = (A + B)(A - B)$

 Sum of two cubes: $A^3 + B^3 = (A + B)(A^2 - AB + B^2)$

 Difference of two cubes: $A^3 - B^3 = (A - B)(A^2 + AB + B^2)$

 b. If there are three terms, is the trinomial a perfect square trinomial? If so, factor by using one of the following special forms:

$$A^2 + 2AB + B^2 = (A + B)^2$$
$$A^2 - 2AB + B^2 = (A - B)^2.$$

 If the trinomial is not a perfect square trinomial, try factoring by trial and error.

 c. If there are four or more terms, try factoring by grouping.

3. Check to see if any factors with more than one term in the factored polynomial can be factored further. If so, factor completely.

EXAMPLE 11 Factoring a Polynomial

Factor: $2x^3 + 8x^2 + 8x$.

SOLUTION

Step 1 If there is a common factor, factor out the GCF. Because $2x$ is common to all terms, we factor it out.

$$2x^3 + 8x^2 + 8x = 2x(x^2 + 4x + 4) \quad \text{Factor out the GCF.}$$

Step 2 Determine the number of terms and factor accordingly. The factor $x^2 + 4x + 4$ has three terms and is a perfect square trinomial. We factor using $A^2 + 2AB + B^2 = (A + B)^2$.

$$2x^3 + 8x^2 + 8x = 2x(x^2 + 4x + 4)$$

$$= 2x(x^2 + 2 \cdot x \cdot 2 + 2^2)$$

$$\underset{A^2 \quad + \quad 2AB \quad + \quad B^2}{}$$

$$= 2x(x + 2)^2 \qquad\qquad A^2 + 2AB + B^2 = (A + B)^2$$

Step 3 Check to see if factors can be factored further. In this problem, they cannot. Thus,

$$2x^3 + 8x^2 + 8x = 2x(x + 2)^2. \qquad \bullet\bullet\bullet$$

 ⊘ Check Point **11** Factor: $3x^3 - 30x^2 + 75x$.

EXAMPLE 12 Factoring a Polynomial

Factor: $x^2 - 25a^2 + 8x + 16$.

SOLUTION

Step 1 If there is a common factor, factor out the GCF. Other than 1 or −1, there is no common factor.

Step 2 Determine the number of terms and factor accordingly. There are four terms. We try factoring by grouping. It can be shown that grouping into two groups

of two terms does not result in a common binomial factor. Let's try grouping as a difference of squares.

$$x^2 - 25a^2 + 8x + 16$$
$$= (x^2 + 8x + 16) - 25a^2$$ Rearrange terms and group as a perfect square trinomial minus $25a^2$ to obtain a difference of squares.

$$= (x + 4)^2 - (5a)^2$$ Factor the perfect square trinomial.

$$= (x + 4 + 5a)(x + 4 - 5a)$$ Factor the difference of squares. The factors are the sum and difference of the expressions being squared.

Step 3 Check to see if factors can be factored further. In this case, they cannot, so we have factored completely. • • •

⟠ Check Point **12** Factor: $x^2 - 36a^2 + 20x + 100$.

⑧ Factor algebraic expressions containing fractional and negative exponents.

Factoring Algebraic Expressions Containing Fractional and Negative Exponents

Although expressions containing fractional and negative exponents are not polynomials, they can be simplified using factoring techniques.

EXAMPLE 13 Factoring Involving Fractional and Negative Exponents

Factor and simplify: $x(x + 1)^{-\frac{3}{4}} + (x + 1)^{\frac{1}{4}}$.

SOLUTION

The greatest common factor of $x(x + 1)^{-\frac{3}{4}} + (x + 1)^{\frac{1}{4}}$ is $x + 1$ with the *smaller exponent* in the two terms. Thus, the greatest common factor is $(x + 1)^{-\frac{3}{4}}$.

$$x(x + 1)^{-\frac{3}{4}} + (x + 1)^{\frac{1}{4}}$$

$$= (x + 1)^{-\frac{3}{4}}x + (x + 1)^{-\frac{3}{4}}(x + 1)^{\frac{4}{4}}$$ Express each term as the product of the greatest common factor and its other factor. Note that $(x + 1)^{-\frac{3}{4}}(x + 1)^{\frac{4}{4}} = (x + 1)^{-\frac{3}{4}+\frac{4}{4}} = (x + 1)^{\frac{1}{4}}$.

$$= (x + 1)^{-\frac{3}{4}}x + (x + 1)^{-\frac{3}{4}}(x + 1)$$ Simplify: $(x + 1)^{\frac{4}{4}} = (x + 1)$.

$$= (x + 1)^{-\frac{3}{4}}[x + (x + 1)]$$ Factor out the greatest common factor.

$$= \frac{2x + 1}{(x + 1)^{\frac{3}{4}}}$$ $b^{-n} = \dfrac{1}{b^n}$ • • •

⟠ Check Point **13** Factor and simplify: $x(x - 1)^{-\frac{1}{2}} + (x - 1)^{\frac{1}{2}}$

⑨ Complete the square of a binomial.

Completing the Square

Throughout algebra, it will frequently be convenient to create perfect square trinomials from binomials.

Completing the Square

If $x^2 + bx$ is a binomial, then by adding $\left(\dfrac{b}{2}\right)^2$, which is the square of half the coefficient of x, a perfect square trinomial will result. That is,

$$x^2 + bx + \left(\frac{b}{2}\right)^2 = \left(x + \frac{b}{2}\right)^2.$$

Area of blue region:
$$x^2 + 2\left(\frac{b}{2}\right)x = x^2 + bx$$

x $\frac{b}{2}$

x

$\frac{b}{2}$

Adding this red square of area $\left(\frac{b}{2}\right)^2$ completes the larger square.

FIGURE P.13 Visualizing completing the square

Squares and rectangles make it possible to visualize completing the square, as shown in **Figure P.13**.

EXAMPLE 14 Completing the Square

What term should be added to each binomial so that it becomes a perfect square trinomial? Write and factor the trinomial.

a. $x^2 + 8x$ **b.** $x^2 - 7x$ **c.** $x^2 + \frac{3}{5}x$

SOLUTION

To complete the square, we must add a term to each binomial. The term that should be added is the square of half the coefficient of x.

$$x^2 + 8x \qquad x^2 - 7x \qquad x^2 + \frac{3}{5}x$$

Add $\left(\frac{8}{2}\right)^2 = 4^2$.
Add 16 to complete the square.

Add $\left(\frac{-7}{2}\right)^2$, or $\frac{49}{4}$, to complete the square.

Add $\left(\frac{1}{2} \cdot \frac{3}{5}\right)^2 = \left(\frac{3}{10}\right)^2$.
Add $\frac{9}{100}$ to complete the square.

a. The coefficient of the x-term in $x^2 + 8x$ is 8. Half of 8 is 4, and $4^2 = 16$. Add 16. The result is a perfect square trinomial.

$$x^2 + 8x + 16 = (x + 4)^2$$

b. The coefficient of the x-term in $x^2 - 7x$ is -7. Half of -7 is $-\frac{7}{2}$, and $\left(-\frac{7}{2}\right)^2 = \frac{49}{4}$. Add $\frac{49}{4}$. The result is a perfect square trinomial.

$$x^2 - 7x + \frac{49}{4} = \left(x - \frac{7}{2}\right)^2$$

c. The coefficient of the x-term in $x^2 + \frac{3}{5}x$ is $\frac{3}{5}$. Half of $\frac{3}{5}$ is $\frac{1}{2} \cdot \frac{3}{5}$, or $\frac{3}{10}$, and $\left(\frac{3}{10}\right)^2 = \frac{9}{100}$. Add $\frac{9}{100}$. The result is a perfect square trinomial.

$$x^2 + \frac{3}{5}x + \frac{9}{100} = \left(x + \frac{3}{10}\right)^2$$ • • •

GREAT QUESTION!

I'm not accustomed to factoring perfect square trinomials in which fractions are involved. Is there a rule or an observation that can make the factoring easier?

Yes. The constant in the factorization is always half the coefficient of x.

$$x^2 - 7x + \frac{49}{4} = \left(x - \frac{7}{2}\right)^2 \qquad x^2 + \frac{3}{5}x + \frac{9}{100} = \left(x + \frac{3}{10}\right)^2$$

Half the coefficient of x, -7, is $-\frac{7}{2}$.

Half the coefficient of x, $\frac{3}{5}$, is $\frac{3}{10}$.

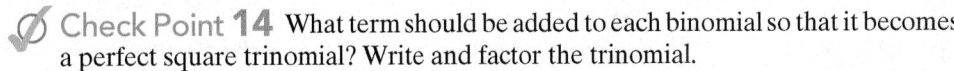 Check Point **14** What term should be added to each binomial so that it becomes a perfect square trinomial? Write and factor the trinomial.

a. $x^2 + 6x$ **b.** $x^2 - 5x$ **c.** $x^2 + \frac{2}{3}x$

CONCEPT AND VOCABULARY CHECK

Here is a list of the factoring techniques that we have discussed.

a. Factoring out the GCF

b. Factoring by grouping

c. Factoring trinomials by trial and error

d. Factoring the difference of two squares
$$A^2 - B^2 = (A + B)(A - B)$$

e. Factoring perfect square trinomials
$$A^2 + 2AB + B^2 = (A + B)^2$$
$$A^2 - 2AB + B^2 = (A - B)^2$$

f. Factoring the sum of two cubes
$$A^3 + B^3 = (A + B)(A^2 - AB + B^2)$$

g. Factoring the difference of two cubes
$$A^3 - B^3 = (A - B)(A^2 + AB + B^2)$$

Fill in each blank by writing the letter of the technique (a through g) for factoring the polynomial.

1. $16x^2 - 25$ _____

2. $27x^3 - 1$ _____

3. $x^2 + 7x + xy + 7y$ _____

4. $4x^2 + 8x + 3$ _____

5. $9x^2 + 24x + 16$ _____

6. $5x^2 + 10x$ _____

7. $x^3 + 1000$ _____

8. The algebraic expression $(x + 1)^{\frac{1}{2}} - \frac{1}{3}(x + 1)^{\frac{3}{2}}$ can be factored using _____ as the greatest common factor.

9. To complete the square on $x^2 - 3x$, add ____.

10. To complete the square on $x^2 - \frac{4}{5}x$, add ____.

EXERCISE SET P.6

Practice Exercises

In Exercises 1–10, factor out the greatest common factor.

1. $18x + 27$ **2.** $16x - 24$

3. $3x^2 + 6x$ **4.** $4x^2 - 8x$

5. $9x^4 - 18x^3 + 27x^2$ **6.** $6x^4 - 18x^3 + 12x^2$

7. $x(x + 5) + 3(x + 5)$ **8.** $x(2x + 1) + 4(2x + 1)$

9. $x^2(x - 3) + 12(x - 3)$ **10.** $x^2(2x + 5) + 17(2x + 5)$

In Exercises 11–16, factor by grouping.

11. $x^3 - 2x^2 + 5x - 10$ **12.** $x^3 - 3x^2 + 4x - 12$

13. $x^3 - x^2 + 2x - 2$ **14.** $x^3 + 6x^2 - 2x - 12$

15. $3x^3 - 2x^2 - 6x + 4$ **16.** $x^3 - x^2 - 5x + 5$

In Exercises 17–38, factor each trinomial, or state that the trinomial is prime.

17. $x^2 + 5x + 6$ **18.** $x^2 + 8x + 15$

19. $x^2 - 2x - 15$ **20.** $x^2 - 4x - 5$

21. $x^2 - 8x + 15$ **22.** $x^2 - 14x + 45$

23. $3x^2 - x - 2$ **24.** $2x^2 + 5x - 3$

25. $3x^2 - 25x - 28$ **26.** $3x^2 - 2x - 5$

27. $6x^2 - 11x + 4$ **28.** $6x^2 - 17x + 12$

29. $4x^2 + 16x + 15$ **30.** $8x^2 + 33x + 4$

31. $9x^2 - 9x + 2$ **32.** $9x^2 + 5x - 4$

33. $20x^2 + 27x - 8$ **34.** $15x^2 - 19x + 6$

35. $2x^2 + 3xy + y^2$ **36.** $3x^2 + 4xy + y^2$

37. $6x^2 - 5xy - 6y^2$ **38.** $6x^2 - 7xy - 5y^2$

In Exercises 39–48, factor the difference of two squares.

39. $x^2 - 100$ **40.** $x^2 - 144$

41. $36x^2 - 49$ **42.** $64x^2 - 81$

43. $9x^2 - 25y^2$ **44.** $36x^2 - 49y^2$

45. $x^4 - 16$ **46.** $x^4 - 1$

47. $16x^4 - 81$ **48.** $81x^4 - 1$

In Exercises 49–56, factor each perfect square trinomial.

49. $x^2 + 2x + 1$ **50.** $x^2 + 4x + 4$

51. $x^2 - 14x + 49$ **52.** $x^2 - 10x + 25$

53. $4x^2 + 4x + 1$ **54.** $25x^2 + 10x + 1$

55. $9x^2 - 6x + 1$ **56.** $64x^2 - 16x + 1$

In Exercises 57–64, factor using the formula for the sum or difference of two cubes.

57. $x^3 + 27$ **58.** $x^3 + 64$

59. $x^3 - 64$ **60.** $x^3 - 27$

61. $8x^3 - 1$ **62.** $27x^3 - 1$

63. $64x^3 + 27$ **64.** $8x^3 + 125$

In Exercises 65–92, factor completely, or state that the polynomial is prime.

65. $3x^3 - 3x$ **66.** $5x^3 - 45x$

67. $4x^2 - 4x - 24$ **68.** $6x^2 - 18x - 60$

69. $2x^4 - 162$ **70.** $7x^4 - 7$

71. $x^3 + 2x^2 - 9x - 18$ **72.** $x^3 + 3x^2 - 25x - 75$

73. $2x^2 - 2x - 112$ **74.** $6x^2 - 6x - 12$

75. $x^3 - 4x$ **76.** $9x^3 - 9x$

77. $x^2 + 64$ **78.** $x^2 + 36$

79. $x^3 + 2x^2 - 4x - 8$ **80.** $x^3 + 2x^2 - x - 2$

81. $y^5 - 81y$ **82.** $y^5 - 16y$

83. $20y^4 - 45y^2$ **84.** $48y^4 - 3y^2$

85. $x^2 - 12x + 36 - 49y^2$ **86.** $x^2 - 10x + 25 - 36y^2$

87. $9b^2x - 16y - 16x + 9b^2y$

88. $16a^2x - 25y - 25x + 16a^2y$

89. $x^2y - 16y + 32 - 2x^2$ **90.** $12x^2y - 27y - 4x^2 + 9$

91. $2x^3 - 8a^2x + 24x^2 + 72x$

92. $2x^3 - 98a^2x + 28x^2 + 98x$

In Exercises 93–102, factor and simplify each algebraic expression.

93. $x^{\frac{3}{2}} - x^{\frac{1}{2}}$ **94.** $x^{\frac{3}{4}} - x^{\frac{1}{4}}$

95. $4x^{-\frac{2}{3}} + 8x^{\frac{1}{3}}$ **96.** $12x^{-\frac{3}{4}} + 6x^{\frac{1}{4}}$

97. $(x + 3)^{\frac{1}{2}} - (x + 3)^{\frac{3}{2}}$ **98.** $(x^2 + 4)^{\frac{3}{2}} + (x^2 + 4)^{\frac{7}{2}}$

99. $(x + 5)^{-\frac{1}{2}} - (x + 5)^{-\frac{3}{2}}$

100. $(x^2 + 3)^{-\frac{2}{3}} + (x^2 + 3)^{-\frac{5}{3}}$

101. $(4x - 1)^{\frac{1}{2}} - \frac{1}{3}(4x - 1)^{\frac{3}{2}}$

102. $-8(4x + 3)^{-2} + 10(5x + 1)(4x + 3)^{-1}$

In Exercises 103–112, determine the constant that should be added to the binomial so that it becomes a perfect square trinomial. Then write and factor the trinomial.

103. $x^2 + 2x$ **104.** $x^2 + 4x$

105. $x^2 - 14x$ **106.** $x^2 - 10x$

107. $x^2 + 7x$ **108.** $x^2 + 9x$

109. $x^2 - \frac{1}{2}x$ **110.** $x^2 - \frac{1}{3}x$

111. $x^2 + \frac{4}{3}x$ **112.** $x^2 + \frac{4}{5}x$

Practice Plus

In Exercises 113–124, factor completely.

113. $10x^2(x + 1) - 7x(x + 1) - 6(x + 1)$

114. $12x^2(x - 1) - 4x(x - 1) - 5(x - 1)$

115. $6x^4 + 35x^2 - 6$ **116.** $7x^4 + 34x^2 - 5$

117. $y^7 + y$ **118.** $(y + 1)^3 + 1$

119. $x^4 - 5x^2y^2 + 4y^4$ **120.** $x^4 - 10x^2y^2 + 9y^4$

121. $(x - y)^4 - 4(x - y)^2$ **122.** $(x + y)^4 - 100(x + y)^2$

123. $2x^2 - 7xy^2 + 3y^4$ **124.** $3x^2 + 5xy^2 + 2y^4$

Application Exercises

125. Your computer store is having an incredible sale. The price on one model is reduced by 40%. Then the sale price is reduced by another 40%. If x is the computer's original price, the sale price can be modeled by

$$(x - 0.4x) - 0.4(x - 0.4x).$$

 a. Factor out $(x - 0.4x)$ from each term. Then simplify the resulting expression.

 b. Use the simplified expression from part (a) to answer these questions. With a 40% reduction followed by a 40% reduction, is the computer selling at 20% of its original price? If not, at what percentage of the original price is it selling?

126. Your local electronics store is having an end-of-the-year sale. The price on a plasma television had been reduced by 30%. Now the sale price is reduced by another 30%. If x is the television's original price, the sale price can be modeled by

$$(x - 0.3x) - 0.3(x - 0.3x).$$

 a. Factor out $(x - 0.3x)$ from each term. Then simplify the resulting expression.

 b. Use the simplified expression from part (a) to answer these questions. With a 30% reduction followed by a 30% reduction, is the television selling at 40% of its original price? If not, at what percentage of the original price is it selling?

In Exercises 127–130,

 a. *Write an expression for the area of the shaded region.*

 b. *Write the expression in factored form.*

127. **128.**

129. **130.**

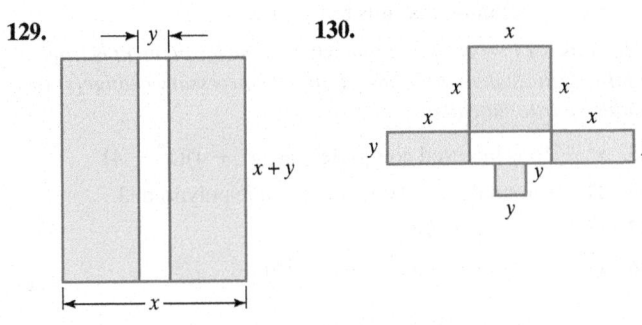

In Exercises 131–132, find the formula for the volume of the region outside the smaller rectangular solid and inside the larger rectangular solid. Then express the volume in factored form.

131.

132.

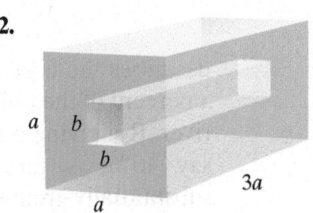

Writing in Mathematics

133. Using an example, explain how to factor out the greatest common factor of a polynomial.

134. Suppose that a polynomial contains four terms. Explain how to use factoring by grouping to factor the polynomial.

135. Explain how to factor $3x^2 + 10x + 8$.

136. Explain how to factor the difference of two squares. Provide an example with your explanation.

137. What is a perfect square trinomial and how is it factored?

138. Explain how to factor $x^3 + 1$.

139. What does it mean to factor completely?

140. Explain how to complete the square for a binomial. Use $x^2 + 6x$ to illustrate your explanation.

Critical Thinking Exercises

Make Sense? *In Exercises 141–144, determine whether each statement makes sense or does not make sense, and explain your reasoning.*

141. Although $20x^3$ appears in both $20x^3 + 8x^2$ and $20x^3 + 10x$, I'll need to factor $20x^3$ in different ways to obtain each polynomial's factorization.

142. You grouped the polynomial's terms using different groupings than I did, yet we both obtained the same factorization.

143. I factored $4x^2 - 100$ completely and obtained $(2x + 10)(2x - 10)$.

144. First factoring out the greatest common factor makes it easier for me to determine how to factor the remaining factor, assuming that it is not prime.

In Exercises 145–148, determine whether each statement is true or false. If the statement is false, make the necessary change(s) to produce a true statement.

145. $x^4 - 16$ is factored completely as $(x^2 + 4)(x^2 - 4)$.

146. The trinomial $x^2 - 4x - 4$ is a prime polynomial.

147. $x^2 + 36 = (x + 6)^2$

148. $x^3 - 64 = (x + 4)(x^2 + 4x - 16)$

In Exercises 149–152, factor completely.

149. $x^{2n} + 6x^n + 8$

150. $-x^2 - 4x + 5$

151. $x^4 - y^4 - 2x^3y + 2xy^3$

152. $(x - 5)^{-\frac{1}{2}}(x + 5)^{-\frac{1}{2}} - (x + 5)^{\frac{1}{2}}(x - 5)^{-\frac{3}{2}}$

In Exercises 153–154, find all integers b so that the trinomial can be factored.

153. $x^2 + bx + 15$

154. $x^2 + 4x + b$

Preview Exercises

Exercises 155–157 will help you prepare for the material covered in the next section.

155. Factor the numerator and the denominator. Then simplify by dividing out the common factor in the numerator and the denominator.

$$\frac{x^2 + 6x + 5}{x^2 - 25}$$

In Exercises 156–157, perform the indicated operation. Where possible, reduce the answer to its lowest terms.

156. $\dfrac{5}{4} \cdot \dfrac{8}{15}$

157. $\dfrac{1}{2} + \dfrac{2}{3}$

SECTION P.7 Rational Expressions

Objectives

1. Specify numbers that must be excluded from the domain of a rational expression.
2. Simplify rational expressions.
3. Multiply rational expressions.
4. Divide rational expressions.
5. Add and subtract rational expressions.
6. Simplify complex rational expressions.

DISCOVERY

What happens if you try substituting 100 for x in

$$\frac{250x}{100 - x}?$$

What does this tell you about the cost of cleaning up all of the river's pollutants?

How do we describe the costs of reducing environmental pollution? We often use algebraic expressions involving quotients of polynomials. For example, the algebraic expression

$$\frac{250x}{100 - x}$$

describes the cost, in millions of dollars, to remove x percent of the pollutants that are discharged into a river. Removing a modest percentage of pollutants, say 40%, is far less costly than removing a substantially greater percentage, such as 95%. We see this by evaluating the algebraic expression for $x = 40$ and $x = 95$.

Evaluating $\dfrac{250x}{100 - x}$ for

$x = 40$:

Cost is $\dfrac{250(40)}{100 - 40} \approx 167.$

$x = 95$:

Cost is $\dfrac{250(95)}{100 - 95} = 4750.$

The cost increases from approximately \$167 million to a possibly prohibitive \$4750 million, or \$4.75 billion. Costs spiral upward as the percentage of removed pollutants increases.

Many algebraic expressions that describe costs of environmental projects are examples of *rational expressions*. First we will define rational expressions. Then we will review how to perform operations with such expressions.

① Specify numbers that must be excluded from the domain of a rational expression.

Rational Expressions

A **rational expression** is the quotient of two polynomials. Some examples are

$$\frac{x-2}{4}, \quad \frac{4}{x-2}, \quad \frac{x}{x^2-1}, \quad \text{and} \quad \frac{x^2+1}{x^2+2x-3}.$$

The set of real numbers for which an algebraic expression is defined is the **domain** of the expression. Because rational expressions indicate division and division by zero is undefined, **we must exclude numbers from a rational expression's domain that make the denominator zero.**

EXAMPLE 1 Excluding Numbers from the Domain

Find all the numbers that must be excluded from the domain of each rational expression:

a. $\dfrac{4}{x-2}$ **b.** $\dfrac{x}{x^2-1}$ **c.** $\dfrac{9x}{x^2+3x-18}.$

SOLUTION

To determine the numbers that must be excluded from each domain, examine the denominators.

a. $\dfrac{4}{x-2}$ **b.** $\dfrac{x}{x^2-1} = \dfrac{x}{(x+1)(x-1)}$ **c.** $\dfrac{9x}{x^2+3x-18} = \dfrac{9x}{(x+6)(x-3)}$

| This denominator would equal zero if $x = 2$. | This factor would equal zero if $x = -1$. | This factor would equal zero if $x = 1$. | This factor would equal zero if $x = -6$. | This factor would equal zero if $x = 3$. |

For the rational expression in part (a), we must exclude 2 from the domain. For the rational expression in part (b), we must exclude both −1 and 1 from the domain. For the rational expression in part (c), we must exclude both −6 and 3 from the domain. These excluded numbers are often written to the right of a rational expression:

$$\frac{4}{x-2}, x \neq 2 \qquad \frac{x}{x^2-1}, x \neq -1, x \neq 1 \qquad \frac{9x}{x^2+3x-18}, x \neq -6, x \neq 3. \quad \bullet\bullet\bullet$$

✓ Check Point **1** Find all the numbers that must be excluded from the domain of each rational expression:

a. $\dfrac{7}{x+5}$ **b.** $\dfrac{x}{x^2-36}$ **c.** $\dfrac{7x}{x^2-5x-14}.$

② Simplify rational expressions.

Simplifying Rational Expressions

A rational expression is **simplified** if its numerator and denominator have no common factors other than 1 or −1. The following procedure can be used to simplify rational expressions:

Simplifying Rational Expressions

1. Factor the numerator and the denominator completely.
2. Divide both the numerator and the denominator by any common factors.

EXAMPLE 2 Simplifying Rational Expressions

Simplify:

a. $\dfrac{x^3 + x^2}{x + 1}$ b. $\dfrac{x^2 + 6x + 5}{x^2 - 25}$.

SOLUTION

a. $\dfrac{x^3 + x^2}{x + 1} = \dfrac{x^2(x + 1)}{x + 1}$ Factor the numerator. Because the denominator is x + 1, x ≠ −1.

$= \dfrac{x^2(\overset{1}{\cancel{x + 1}})}{\underset{1}{\cancel{x + 1}}}$ Divide out the common factor, x + 1.

$= x^2, x \neq -1$ Denominators of 1 need not be written because $\frac{a}{1} = a$.

b. $\dfrac{x^2 + 6x + 5}{x^2 - 25} = \dfrac{(x + 5)(x + 1)}{(x + 5)(x - 5)}$ Factor the numerator and denominator. Because the denominator is (x + 5)(x − 5), x ≠ −5 and x ≠ 5.

$= \dfrac{(\overset{1}{\cancel{x + 5}})(x + 1)}{(\underset{1}{\cancel{x + 5}})(x - 5)}$ Divide out the common factor, x + 5.

$= \dfrac{x + 1}{x - 5}, \quad x \neq -5, \quad x \neq 5$

● ● ●

✓ **Check Point 2** Simplify:

a. $\dfrac{x^3 + 3x^2}{x + 3}$ b. $\dfrac{x^2 - 1}{x^2 + 2x + 1}$.

③ Multiply rational expressions.

Multiplying Rational Expressions

The product of two rational expressions is the product of their numerators divided by the product of their denominators. Here is a step-by-step procedure for multiplying rational expressions:

Multiplying Rational Expressions

1. Factor all numerators and denominators completely.
2. Divide numerators and denominators by common factors.
3. Multiply the remaining factors in the numerators and multiply the remaining factors in the denominators.

EXAMPLE 3 Multiplying Rational Expressions

Multiply: $\dfrac{x - 7}{x - 1} \cdot \dfrac{x^2 - 1}{3x - 21}$.

SOLUTION

$$\frac{x-7}{x-1} \cdot \frac{x^2-1}{3x-21}$$

This is the given multiplication problem.

$$= \frac{x-7}{x-1} \cdot \frac{(x+1)(x-1)}{3(x-7)}$$

Factor as many numerators and denominators as possible. Because the denominators have factors of $x-1$ and $x-7$, $x \neq 1$ and $x \neq 7$.

$$= \frac{\overset{1}{\cancel{x-7}}}{\underset{1}{\cancel{x-1}}} \cdot \frac{(x+1)(\overset{1}{\cancel{x-1}})}{3(\underset{1}{\cancel{x-7}})}$$

Divide numerators and denominators by common factors.

$$= \frac{x+1}{3}, x \neq 1, x \neq 7$$

Multiply the remaining factors in the numerators and denominators.

These excluded numbers from the domain must also be excluded from the simplified expression's domain.

• • •

✓ **Check Point 3** Multiply:

$$\frac{x+3}{x^2-4} \cdot \frac{x^2-x-6}{x^2+6x+9}.$$

④ Divide rational expressions.

Dividing Rational Expressions

The quotient of two rational expressions is the product of the first expression and the multiplicative inverse, or reciprocal, of the second expression. The reciprocal is found by interchanging the numerator and the denominator. Thus, **we find the quotient of two rational expressions by inverting the divisor and multiplying.**

EXAMPLE 4 Dividing Rational Expressions

Divide: $\dfrac{x^2-2x-8}{x^2-9} \div \dfrac{x-4}{x+3}.$

SOLUTION

$$\frac{x^2-2x-8}{x^2-9} \div \frac{x-4}{x+3}$$

This is the given division problem.

$$= \frac{x^2-2x-8}{x^2-9} \cdot \frac{x+3}{x-4}$$

Invert the divisor and multiply.

$$= \frac{(x-4)(x+2)}{(x+3)(x-3)} \cdot \frac{x+3}{x-4}$$

Factor as many numerators and denominators as possible. For nonzero denominators, $x \neq -3$, $x \neq 3$, and $x \neq 4$.

$$= \frac{(\overset{1}{\cancel{x-4}})(x+2)}{(\cancel{x+3})(x-3)} \cdot \frac{(\overset{1}{\cancel{x+3}})}{(\underset{1}{\cancel{x-4}})}$$

Divide numerators and denominators by common factors.

$$= \frac{x+2}{x-3}, x \neq -3, x \neq 3, x \neq 4$$

Multiply the remaining factors in the numerators and in the denominators.

• • •

⊘ Check Point **4** Divide:

$$\frac{x^2 - 2x + 1}{x^3 + x} \div \frac{x^2 + x - 2}{3x^2 + 3}.$$

5 Add and subtract rational expressions.

Adding and Subtracting Rational Expressions with the Same Denominator

We add or subtract rational expressions with the same denominator by (1) adding or subtracting the numerators, (2) placing this result over the common denominator, and (3) simplifying, if possible.

EXAMPLE 5 Subtracting Rational Expressions with the Same Denominator

Subtract: $\dfrac{5x + 1}{x^2 - 9} - \dfrac{4x - 2}{x^2 - 9}.$

SOLUTION

GREAT QUESTION!

When subtracting a numerator containing more than one term, do I have to insert parentheses like you did in Example 5?

Yes. When a numerator is being subtracted, we must **subtract every term in that expression**. Parentheses indicate that every term inside is being subtracted.

Don't forget the parentheses.

$$\frac{5x + 1}{x^2 - 9} - \frac{4x - 2}{x^2 - 9} = \frac{5x + 1 - (4x - 2)}{x^2 - 9}$$

Subtract numerators and include parentheses to indicate that both terms are subtracted. Place this difference over the common denominator.

$$= \frac{5x + 1 - 4x + 2}{x^2 - 9}$$

Remove parentheses and then change the sign of each term in parentheses.

$$= \frac{x + 3}{x^2 - 9}$$

Combine like terms.

$$= \frac{\overset{1}{x + 3}}{\underset{1}{(x + 3)}(x - 3)}$$

Factor and simplify ($x \neq -3$ and $x \neq 3$).

$$= \frac{1}{x - 3}, x \neq -3, x \neq 3$$

• • •

⊘ Check Point **5** Subtract: $\dfrac{x}{x + 1} - \dfrac{3x + 2}{x + 1}.$

Adding and Subtracting Rational Expressions with Different Denominators

Rational expressions that have no common factors in their denominators can be added or subtracted using one of the following properties:

$$\frac{a}{b} + \frac{c}{d} = \frac{ad + bc}{bd} \qquad \frac{a}{b} - \frac{c}{d} = \frac{ad - bc}{bd}, b \neq 0, d \neq 0.$$

The denominator, bd, is the product of the factors in the two denominators. Because we are considering rational expressions that have no common factors in their denominators, the product bd gives the least common denominator.

EXAMPLE 6 Subtracting Rational Expressions Having No Common Factors in Their Denominators

Subtract: $\dfrac{x+2}{2x-3} - \dfrac{4}{x+3}$.

SOLUTION

We need to find the least common denominator. This is the product of the distinct factors in each denominator, namely $(2x-3)(x+3)$. We can therefore use the subtraction property given previously as follows:

$$\dfrac{a}{b} - \dfrac{c}{d} = \dfrac{ad-bc}{bd}$$

$$\dfrac{x+2}{2x-3} - \dfrac{4}{x+3} = \dfrac{(x+2)(x+3) - (2x-3)4}{(2x-3)(x+3)}$$

Observe that
$a = x+2,\ b = 2x-3,\ c = 4,$
and $d = x+3$.

$$= \dfrac{x^2 + 5x + 6 - (8x - 12)}{(2x-3)(x+3)}$$

Multiply in the numerator.

$$= \dfrac{x^2 + 5x + 6 - 8x + 12}{(2x-3)(x+3)}$$

Remove parentheses and then change the sign of each term in parentheses.

$$= \dfrac{x^2 - 3x + 18}{(2x-3)(x+3)},\ x \neq \dfrac{3}{2},\ x \neq -3$$

Combine like terms in the numerator. • • •

✓ **Check Point 6** Add: $\dfrac{3}{x+1} + \dfrac{5}{x-1}$.

The **least common denominator**, or LCD, of several rational expressions is a polynomial consisting of the product of all prime factors in the denominators, with each factor raised to the greatest power of its occurrence in any denominator. When adding and subtracting rational expressions that have different denominators with one or more common factors in the denominators, it is efficient to find the least common denominator first.

Finding the Least Common Denominator

1. Factor each denominator completely.
2. List the factors of the first denominator.
3. Add to the list in step 2 any factors of the second denominator that do not appear in the list.
4. Form the product of the factors from the list in step 3. This product is the least common denominator.

EXAMPLE 7 Finding the Least Common Denominator

Find the least common denominator of

$$\dfrac{7}{5x^2 + 15x} \quad \text{and} \quad \dfrac{9}{x^2 + 6x + 9}.$$

SOLUTION

Step 1 Factor each denominator completely.

$$5x^2 + 15x = 5x(x + 3)$$
$$x^2 + 6x + 9 = (x + 3)^2 \quad \text{or} \quad (x + 3)(x + 3)$$

Factors are 5, x, and $x + 3$. $\dfrac{7}{5x^2 + 15x}$ $\dfrac{9}{x^2 + 6x + 9}$ Factors are $x + 3$ and $x + 3$.

Step 2 List the factors of the first denominator.

$$5, x, x + 3$$

Step 3 Add any unlisted factors from the second denominator. One factor of $x^2 + 6x + 9$ is already in our list. That factor is $x + 3$. However, the other factor of $x + 3$ is not listed in step 2. We add a second factor of $x + 3$ to the list. We have

$$5, x, x + 3, x + 3.$$

Step 4 The least common denominator is the product of all factors in the final list. Thus,

$$5x(x + 3)(x + 3) \quad \text{or} \quad 5x(x + 3)^2$$

is the least common denominator. • • •

⌀ Check Point **7** Find the least common denominator of

$$\frac{3}{x^2 - 6x + 9} \quad \text{and} \quad \frac{7}{x^2 - 9}.$$

Finding the least common denominator for two (or more) rational expressions is the first step needed to add or subtract the expressions.

Adding and Subtracting Rational Expressions That Have Different Denominators

1. Find the LCD of the rational expressions.
2. Rewrite each rational expression as an equivalent expression whose denominator is the LCD. To do so, multiply the numerator and the denominator of each rational expression by any factor(s) needed to convert the denominator into the LCD.
3. Add or subtract numerators, placing the resulting expression over the LCD.
4. If possible, simplify the resulting rational expression.

EXAMPLE 8 **Adding Rational Expressions with Different Denominators**

Add: $\dfrac{x + 3}{x^2 + x - 2} + \dfrac{2}{x^2 - 1}.$

SOLUTION

Step 1 Find the least common denominator. Start by factoring the denominators.

$$x^2 + x - 2 = (x + 2)(x - 1)$$
$$x^2 - 1 = (x + 1)(x - 1)$$

The factors of the first denominator are $x + 2$ and $x - 1$. The only factor from the second denominator that is not listed is $x + 1$. Thus, the least common denominator is

$$(x + 2)(x - 1)(x + 1).$$

Step 2 Write equivalent expressions with the LCD as denominators. We must rewrite each rational expression with a denominator of $(x + 2)(x - 1)(x + 1)$. We do so by multiplying both the numerator and the denominator of each rational expression by any factor(s) needed to convert the expression's denominator into the LCD.

$$\frac{x + 3}{(x + 2)(x - 1)} \cdot \frac{x + 1}{x + 1} = \frac{(x + 3)(x + 1)}{(x + 2)(x - 1)(x + 1)} \qquad \frac{2}{(x + 1)(x - 1)} \cdot \frac{x + 2}{x + 2} = \frac{2(x + 2)}{(x + 2)(x - 1)(x + 1)}$$

Multiply the numerator and denominator by $x + 1$ to get $(x + 2)(x - 1)(x + 1)$, the LCD.

Multiply the numerator and denominator by $x + 2$ to get $(x + 2)(x - 1)(x + 1)$, the LCD.

Because $\dfrac{x + 1}{x + 1} = 1$ and $\dfrac{x + 2}{x + 2} = 1$, we are not changing the value of either rational expression, only its appearance.

Now we are ready to perform the indicated addition.

$$\frac{x + 3}{x^2 + x - 2} + \frac{2}{x^2 - 1}$$ This is the given problem.

$$= \frac{x + 3}{(x + 2)(x - 1)} + \frac{2}{(x + 1)(x - 1)}$$ Factor the denominators. The LCD is $(x + 2)(x - 1)(x + 1)$.

$$= \frac{(x + 3)(x + 1)}{(x + 2)(x - 1)(x + 1)} + \frac{2(x + 2)}{(x + 2)(x - 1)(x + 1)}$$ Rewrite equivalent expressions with the LCD.

Step 3 Add numerators, putting this sum over the LCD.

$$= \frac{(x + 3)(x + 1) + 2(x + 2)}{(x + 2)(x - 1)(x + 1)}$$

$$= \frac{x^2 + 4x + 3 + 2x + 4}{(x + 2)(x - 1)(x + 1)}$$ Perform the multiplications in the numerator.

$$= \frac{x^2 + 6x + 7}{(x + 2)(x - 1)(x + 1)}, x \neq -2, x \neq 1, x \neq -1$$ Combine like terms in the numerator: $4x + 2x = 6x$ and $3 + 4 = 7$.

Step 4 If necessary, simplify. Because the numerator is prime, no further simplification is possible. • • •

✓ Check Point **8** Subtract: $\dfrac{x}{x^2 - 10x + 25} - \dfrac{x - 4}{2x - 10}$.

⑥ Simplify complex rational expressions.

Complex Rational Expressions

Complex rational expressions, also called **complex fractions,** have numerators or denominators containing one or more rational expressions. Here are two examples of such expressions:

$$\frac{1 + \dfrac{1}{x}}{1 - \dfrac{1}{x}}$$ Separate rational expressions occur in the numerator and the denominator.

$$\frac{\dfrac{1}{x + h} - \dfrac{1}{x}}{h}$$ Separate rational expressions occur in the numerator.

One method for simplifying a complex rational expression is to combine its numerator into a single expression and combine its denominator into a single expression. Then perform the division by inverting the denominator and multiplying.

EXAMPLE 9 Simplifying a Complex Rational Expression

Simplify: $\dfrac{1 + \dfrac{1}{x}}{1 - \dfrac{1}{x}}$.

SOLUTION

Step 1 Add to get a single rational expression in the numerator.

$$1 + \frac{1}{x} = \frac{1}{1} + \frac{1}{x} = \frac{1 \cdot x}{1 \cdot x} + \frac{1}{x} = \frac{x}{x} + \frac{1}{x} = \frac{x + 1}{x}$$

> The LCD is $1 \cdot x$, or x.

Step 2 Subtract to get a single rational expression in the denominator.

$$1 - \frac{1}{x} = \frac{1}{1} - \frac{1}{x} = \frac{1 \cdot x}{1 \cdot x} - \frac{1}{x} = \frac{x}{x} - \frac{1}{x} = \frac{x - 1}{x}$$

> The LCD is $1 \cdot x$, or x.

Step 3 Perform the division indicated by the main fraction bar: Invert and multiply. If possible, simplify.

$$\frac{1 + \dfrac{1}{x}}{1 - \dfrac{1}{x}} = \frac{\dfrac{x+1}{x}}{\dfrac{x-1}{x}} = \frac{x+1}{x} \cdot \frac{x}{x-1} = \frac{x+1}{\cancel{x}} \cdot \frac{\cancel{x}}{x-1} = \frac{x+1}{x-1}$$

> Invert and multiply.

• • •

⊘ **Check Point 9** Simplify: $\dfrac{\dfrac{1}{x} - \dfrac{3}{2}}{\dfrac{1}{x} + \dfrac{3}{4}}$.

A second method for simplifying a complex rational expression is to find the least common denominator of all the rational expressions in its numerator and denominator. Then multiply each term in its numerator and denominator by this least common denominator. Because we are multiplying by a form of 1, we will obtain an equivalent expression that does not contain fractions in its numerator or denominator. Here we use this method to simplify the complex rational expression in Example 9.

$$\frac{1 + \dfrac{1}{x}}{1 - \dfrac{1}{x}} = \frac{\left(1 + \dfrac{1}{x}\right)}{\left(1 - \dfrac{1}{x}\right)} \cdot \frac{x}{x}$$

> The least common denominator of all the rational expressions is x. Multiply the numerator and denominator by x. Because $\dfrac{x}{x} = 1$, we are not changing the complex fraction ($x \neq 0$).

$$= \frac{1 \cdot x + \dfrac{1}{x} \cdot x}{1 \cdot x - \dfrac{1}{x} \cdot x}$$

> Use the distributive property. Be sure to distribute x to every term.

$$= \frac{x + 1}{x - 1}, x \neq 0, x \neq 1$$

> Multiply. The complex rational expression is now simplified.

EXAMPLE 10 Simplifying a Complex Rational Expression

Simplify: $\dfrac{\dfrac{1}{x+h}-\dfrac{1}{x}}{h}$.

SOLUTION

We will use the method of multiplying each of the three terms, $\dfrac{1}{x+h}, \dfrac{1}{x}$, and h, by the least common denominator. The least common denominator is $x(x+h)$.

$$\dfrac{\dfrac{1}{x+h}-\dfrac{1}{x}}{h}$$

$$=\dfrac{\left(\dfrac{1}{x+h}-\dfrac{1}{x}\right)x(x+h)}{hx(x+h)}$$ Multiply the numerator and denominator by $x(x+h), h \neq 0, x \neq 0, x \neq -h$.

$$=\dfrac{\dfrac{1}{x+h}\cdot x(x+h)-\dfrac{1}{x}\cdot x(x+h)}{hx(x+h)}$$ Use the distributive property in the numerator.

$$=\dfrac{x-(x+h)}{hx(x+h)}$$ Simplify: $\dfrac{1}{x+h}\cdot x(x+h) = x$ and $\dfrac{1}{x}\cdot x(x+h) = x+h$.

$$=\dfrac{x-x-h}{hx(x+h)}$$ Subtract in the numerator. Remove parentheses and change the sign of each term in parentheses.

$$=\dfrac{\overset{1}{-h}}{\underset{1}{h}x(x+h)}$$ Simplify: $x - x - h = -h$.

$$=-\dfrac{1}{x(x+h)}, h \neq 0, x \neq 0, x \neq -h$$ Divide the numerator and denominator by h.

● ● ●

⬿ **Check Point 10** Simplify: $\dfrac{\dfrac{1}{x+7}-\dfrac{1}{x}}{7}$.

CONCEPT AND VOCABULARY CHECK

Fill in each blank so that the resulting statement is true.

1. A rational expression is the quotient of two _____.
2. The set of real numbers for which a rational expression is defined is the _____ of the expression. We must exclude all numbers from this set that make the denominator of the rational expression __.
3. We simplify a rational expression by _____ the numerator and the denominator completely. Then we divide the numerator and the denominator by any _____.
4. $\dfrac{x}{5}\cdot\dfrac{x}{3} =$ ___

5. $\dfrac{x}{5}\div\dfrac{x}{3} =$ ___, $x \neq 0$
6. $\dfrac{x^2}{3}-\dfrac{x-4}{3} =$ _____
7. Consider the following subtraction problem:
$$\dfrac{x-1}{x^2+x-6}-\dfrac{x-2}{x^2+4x+3}.$$
The factors of the first denominator are _____.
The factors of the second denominator are _____.
The LCD is _____.

8. An equivalent expression for $\dfrac{3x + 2}{x - 5}$ with a denominator of $(3x + 4)(x - 5)$ can be obtained by multiplying the numerator and denominator by _____.

9. A rational expression whose numerator or denominator or both contain rational expressions is called a/an _____ rational expression or a/an _____ fraction.

10. $\dfrac{\dfrac{1}{x + 3} - \dfrac{1}{x}}{3} = \dfrac{x(x + 3)}{x(x + 3)} \cdot \dfrac{\left(\dfrac{1}{x + 3} - \dfrac{1}{x}\right)}{3} = \dfrac{-- - (\underline{\quad})}{3x(x + 3)}$

$= \dfrac{\overline{\underline{\quad\quad}}}{3x(x + 3)}$

$= \underline{\quad\quad}$

EXERCISE SET P.7

Practice Exercises

In Exercises 1–6, find all numbers that must be excluded from the domain of each rational expression.

1. $\dfrac{7}{x - 3}$

2. $\dfrac{13}{x + 9}$

3. $\dfrac{x + 5}{x^2 - 25}$

4. $\dfrac{x + 7}{x^2 - 49}$

5. $\dfrac{x - 1}{x^2 + 11x + 10}$

6. $\dfrac{x - 3}{x^2 + 4x - 45}$

In Exercises 7–14, simplify each rational expression. Find all numbers that must be excluded from the domain of the simplified rational expression.

7. $\dfrac{3x - 9}{x^2 - 6x + 9}$

8. $\dfrac{4x - 8}{x^2 - 4x + 4}$

9. $\dfrac{x^2 - 12x + 36}{4x - 24}$

10. $\dfrac{x^2 - 8x + 16}{3x - 12}$

11. $\dfrac{y^2 + 7y - 18}{y^2 - 3y + 2}$

12. $\dfrac{y^2 - 4y - 5}{y^2 + 5y + 4}$

13. $\dfrac{x^2 + 12x + 36}{x^2 - 36}$

14. $\dfrac{x^2 - 14x + 49}{x^2 - 49}$

In Exercises 15–32, multiply or divide as indicated.

15. $\dfrac{x - 2}{3x + 9} \cdot \dfrac{2x + 6}{2x - 4}$

16. $\dfrac{6x + 9}{3x - 15} \cdot \dfrac{x - 5}{4x + 6}$

17. $\dfrac{x^2 - 9}{x^2} \cdot \dfrac{x^2 - 3x}{x^2 + x - 12}$

18. $\dfrac{x^2 - 4}{x^2 - 4x + 4} \cdot \dfrac{2x - 4}{x + 2}$

19. $\dfrac{x^2 - 5x + 6}{x^2 - 2x - 3} \cdot \dfrac{x^2 - 1}{x^2 - 4}$

20. $\dfrac{x^2 + 5x + 6}{x^2 + x - 6} \cdot \dfrac{x^2 - 9}{x^2 - x - 6}$

21. $\dfrac{x^3 - 8}{x^2 - 4} \cdot \dfrac{x + 2}{3x}$

22. $\dfrac{x^2 + 6x + 9}{x^3 + 27} \cdot \dfrac{1}{x + 3}$

23. $\dfrac{x + 1}{3} \div \dfrac{3x + 3}{7}$

24. $\dfrac{x + 5}{7} \div \dfrac{4x + 20}{9}$

25. $\dfrac{x^2 - 4}{x} \div \dfrac{x + 2}{x - 2}$

26. $\dfrac{x^2 - 4}{x - 2} \div \dfrac{x + 2}{4x - 8}$

27. $\dfrac{4x^2 + 10}{x - 3} \div \dfrac{6x^2 + 15}{x^2 - 9}$

28. $\dfrac{x^2 + x}{x^2 - 4} \div \dfrac{x^2 - 1}{x^2 + 5x + 6}$

29. $\dfrac{x^2 - 25}{2x - 2} \div \dfrac{x^2 + 10x + 25}{x^2 + 4x - 5}$

30. $\dfrac{x^2 - 4}{x^2 + 3x - 10} \div \dfrac{x^2 + 5x + 6}{x^2 + 8x + 15}$

31. $\dfrac{x^2 + x - 12}{x^2 + x - 30} \cdot \dfrac{x^2 + 5x + 6}{x^2 - 2x - 3} \div \dfrac{x + 3}{x^2 + 7x + 6}$

32. $\dfrac{x^3 - 25x}{4x^2} \cdot \dfrac{2x^2 - 2}{x^2 - 6x + 5} \div \dfrac{x^2 + 5x}{7x + 7}$

In Exercises 33–58, add or subtract as indicated.

33. $\dfrac{4x + 1}{6x + 5} + \dfrac{8x + 9}{6x + 5}$

34. $\dfrac{3x + 2}{3x + 4} + \dfrac{3x + 6}{3x + 4}$

35. $\dfrac{x^2 - 2x}{x^2 + 3x} + \dfrac{x^2 + x}{x^2 + 3x}$

36. $\dfrac{x^2 - 4x}{x^2 - x - 6} + \dfrac{4x - 4}{x^2 - x - 6}$

37. $\dfrac{4x - 10}{x - 2} - \dfrac{x - 4}{x - 2}$

38. $\dfrac{2x + 3}{3x - 6} - \dfrac{3 - x}{3x - 6}$

39. $\dfrac{x^2 + 3x}{x^2 + x - 12} - \dfrac{x^2 - 12}{x^2 + x - 12}$

40. $\dfrac{x^2 - 4x}{x^2 - x - 6} - \dfrac{x - 6}{x^2 - x - 6}$

41. $\dfrac{3}{x + 4} + \dfrac{6}{x + 5}$

42. $\dfrac{8}{x - 2} + \dfrac{2}{x - 3}$

43. $\dfrac{3}{x + 1} - \dfrac{3}{x}$

44. $\dfrac{4}{x} - \dfrac{3}{x + 3}$

45. $\dfrac{2x}{x + 2} + \dfrac{x + 2}{x - 2}$

46. $\dfrac{3x}{x - 3} - \dfrac{x + 4}{x + 2}$

47. $\dfrac{x + 5}{x - 5} + \dfrac{x - 5}{x + 5}$

48. $\dfrac{x + 3}{x - 3} + \dfrac{x - 3}{x + 3}$

49. $\dfrac{3}{2x + 4} + \dfrac{2}{3x + 6}$

50. $\dfrac{5}{2x + 8} + \dfrac{7}{3x + 12}$

51. $\dfrac{4}{x^2 + 6x + 9} + \dfrac{4}{x + 3}$

52. $\dfrac{3}{5x + 2} + \dfrac{5x}{25x^2 - 4}$

53. $\dfrac{3x}{x^2 + 3x - 10} - \dfrac{2x}{x^2 + x - 6}$

54. $\dfrac{x}{x^2 - 2x - 24} - \dfrac{x}{x^2 - 7x + 6}$

55. $\dfrac{x + 3}{x^2 - 1} - \dfrac{x + 2}{x - 1}$

56. $\dfrac{x + 5}{x^2 - 4} - \dfrac{x + 1}{x - 2}$

57. $\dfrac{4x^2 + x - 6}{x^2 + 3x + 2} - \dfrac{3x}{x + 1} + \dfrac{5}{x + 2}$

58. $\dfrac{6x^2 + 17x - 40}{x^2 + x - 20} + \dfrac{3}{x - 4} - \dfrac{5x}{x + 5}$

In Exercises 59–72, simplify each complex rational expression.

59. $\dfrac{\dfrac{x}{3} - 1}{x - 3}$

60. $\dfrac{\dfrac{x}{4} - 1}{x - 4}$

61. $\dfrac{1 + \dfrac{1}{x}}{3 - \dfrac{1}{x}}$

62. $\dfrac{8 + \dfrac{1}{x}}{4 - \dfrac{1}{x}}$

63. $\dfrac{\dfrac{1}{x} + \dfrac{1}{y}}{x + y}$

64. $\dfrac{1 - \dfrac{1}{x}}{xy}$

65. $\dfrac{x - \dfrac{x}{x + 3}}{x + 2}$

66. $\dfrac{x - 3}{x - \dfrac{3}{x - 2}}$

67. $\dfrac{\dfrac{3}{x - 2} - \dfrac{4}{x + 2}}{\dfrac{7}{x^2 - 4}}$

68. $\dfrac{\dfrac{x}{x - 2} + 1}{\dfrac{3}{x^2 - 4} + 1}$

69. $\dfrac{\dfrac{1}{x + 1}}{\dfrac{1}{x^2 - 2x - 3} + \dfrac{1}{x - 3}}$

70. $\dfrac{\dfrac{6}{x^2 + 2x - 15} - \dfrac{1}{x - 3}}{\dfrac{1}{x + 5} + 1}$

71. $\dfrac{\dfrac{1}{(x + h)^2} - \dfrac{1}{x^2}}{h}$

72. $\dfrac{\dfrac{x + h}{x + h + 1} - \dfrac{x}{x + 1}}{h}$

Practice Plus

In Exercises 73–80, perform the indicated operations. Simplify the result, if possible.

73. $\left(\dfrac{2x + 3}{x + 1} \cdot \dfrac{x^2 + 4x - 5}{2x^2 + x - 3}\right) - \dfrac{2}{x + 2}$

74. $\dfrac{1}{x^2 - 2x - 8} \div \left(\dfrac{1}{x - 4} - \dfrac{1}{x + 2}\right)$

75. $\left(2 - \dfrac{6}{x + 1}\right)\left(1 + \dfrac{3}{x - 2}\right)$

76. $\left(4 - \dfrac{3}{x + 2}\right)\left(1 + \dfrac{5}{x - 1}\right)$

77. $\dfrac{y^{-1} - (y + 5)^{-1}}{5}$

78. $\dfrac{y^{-1} - (y + 2)^{-1}}{2}$

79. $\left(\dfrac{1}{a^3 - b^3} \cdot \dfrac{ac + ad - bc - bd}{1}\right) - \dfrac{c - d}{a^2 + ab + b^2}$

80. $\dfrac{ab}{a^2 + ab + b^2} + \left(\dfrac{ac - ad - bc + bd}{ac - ad + bc - bd} \div \dfrac{a^3 - b^3}{a^3 + b^3}\right)$

Application Exercises

81. The rational expression

$$\dfrac{130x}{100 - x}$$

describes the cost, in millions of dollars, to inoculate x percent of the population against a particular strain of flu.

 a. Evaluate the expression for $x = 40$, $x = 80$, and $x = 90$. Describe the meaning of each evaluation in terms of percentage inoculated and cost.

 b. For what value of x is the expression undefined?

 c. What happens to the cost as x approaches 100%? How can you interpret this observation?

82. The average rate on a round-trip commute having a one-way distance d is given by the complex rational expression

$$\dfrac{2d}{\dfrac{d}{r_1} + \dfrac{d}{r_2}},$$

in which r_1 and r_2 are the average rates on the outgoing and return trips, respectively. Simplify the expression. Then find your average rate if you drive to campus averaging 40 miles per hour and return home on the same route averaging 30 miles per hour. Explain why the answer is not 35 miles per hour.

83. The bar graph shows the estimated number of calories per day needed to maintain energy balance for various gender and age groups for moderately active lifestyles. (Moderately active means a lifestyle that includes physical activity equivalent to walking 1.5 to 3 miles per day at 3 to 4 miles per hour, in addition to the light physical activity associated with typical day-to-day life.)

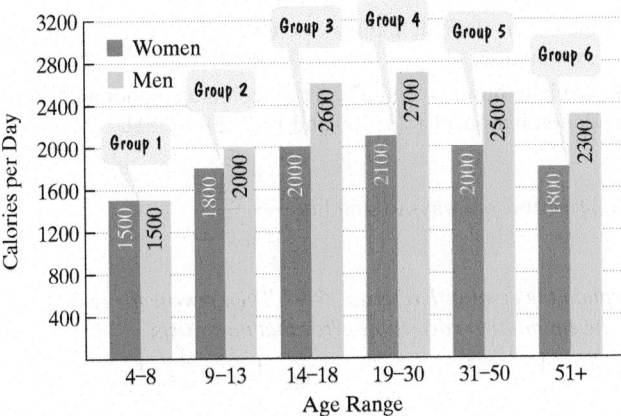

Calories Needed to Maintain Energy Balance for Moderately Active Lifestyles

Source: U.S.D.A.

(Be sure to refer to the graph at the bottom of the previous page.)

a. The mathematical model
$$W = -66x^2 + 526x + 1030$$
describes the number of calories needed per day, W, by women in age group x with moderately active lifestyles. According to the model, how many calories per day are needed by women between the ages of 19 and 30, inclusive, with this lifestyle? Does this underestimate or overestimate the number shown by the graph? By how much?

b. The mathematical model
$$M = -120x^2 + 998x + 590$$
describes the number of calories needed per day, M, by men in age group x with moderately active lifestyles. According to the model, how many calories per day are needed by men between the ages of 19 and 30, inclusive, with this lifestyle? Does this underestimate or overestimate the number shown by the graph? By how much?

c. Write a simplified rational expression that describes the ratio of the number of calories needed per day by women in age group x to the number of calories needed per day by men in age group x for people with moderately active lifestyles.

In Exercises 84–85, express the perimeter of each rectangle as a single rational expression.

84.

85.

Writing in Mathematics

86. What is a rational expression?

87. Explain how to determine which numbers must be excluded from the domain of a rational expression.

88. Explain how to simplify a rational expression.

89. Explain how to multiply rational expressions.

90. Explain how to divide rational expressions.

91. Explain how to add or subtract rational expressions with the same denominators.

92. Explain how to add rational expressions having no common factors in their denominators. Use $\dfrac{3}{x+5} + \dfrac{7}{x+2}$ in your explanation.

93. Explain how to find the least common denominator for denominators of $x^2 - 100$ and $x^2 - 20x + 100$.

94. Describe two ways to simplify $\dfrac{\dfrac{3}{x} + \dfrac{2}{x^2}}{\dfrac{1}{x^2} + \dfrac{2}{x}}$.

Explain the error in Exercises 95–97. Then rewrite the right side of the equation to correct the error that now exists.

95. $\dfrac{1}{a} + \dfrac{1}{b} = \dfrac{1}{a+b}$

96. $\dfrac{1}{x} + 7 = \dfrac{1}{x+7}$

97. $\dfrac{a}{x} + \dfrac{a}{b} = \dfrac{a}{x+b}$

Critical Thinking Exercises

Make Sense? *In Exercises 98–101, determine whether each statement makes sense or does not make sense, and explain your reasoning.*

98. I evaluated $\dfrac{3x-3}{4x(x-1)}$ for $x = 1$ and obtained 0.

99. The rational expressions
$$\frac{7}{14x} \quad \text{and} \quad \frac{7}{14+x}$$
can both be simplified by dividing each numerator and each denominator by 7.

100. When performing the division
$$\frac{7x}{x+3} \div \frac{(x+3)^2}{x-5},$$
I began by dividing the numerator and the denominator by the common factor, $x + 3$.

101. I subtracted $\dfrac{3x-5}{x-1}$ from $\dfrac{x-3}{x-1}$ and obtained a constant.

In Exercises 102–105, determine whether each statement is true or false. If the statement is false, make the necessary change(s) to produce a true statement.

102. $\dfrac{x^2 - 25}{x - 5} = x - 5$

103. The expression $\dfrac{-3y-6}{y+2}$ simplifies to the consecutive integer that follows -4.

104. $\dfrac{2x-1}{x-7} + \dfrac{3x-1}{x-7} - \dfrac{5x-2}{x-7} = 0$

105. $6 + \dfrac{1}{x} = \dfrac{7}{x}$

In Exercises 106–108, perform the indicated operations.

106. $\dfrac{1}{x^n - 1} - \dfrac{1}{x^n + 1} - \dfrac{1}{x^{2n} - 1}$

107. $\left(1 - \dfrac{1}{x}\right)\left(1 - \dfrac{1}{x+1}\right)\left(1 - \dfrac{1}{x+2}\right)\left(1 - \dfrac{1}{x+3}\right)$

108. $(x-y)^{-1} + (x-y)^{-2}$

109. In one short sentence, five words or less, explain what
$$\frac{\dfrac{1}{x} + \dfrac{1}{x^2} + \dfrac{1}{x^3}}{\dfrac{1}{x^4} + \dfrac{1}{x^5} + \dfrac{1}{x^6}}$$
does to each number x.

Preview Exercises

Exercises 110–112 will help you prepare for the material covered in the first section of the next chapter.

110. If $y = 4 - x^2$, find the value of y that corresponds to values of x for each integer starting with -3 and ending with 3.

111. If $y = 1 - x^2$, find the value of y that corresponds to values of x for each integer starting with -3 and ending with 3.

112. If $y = |x + 1|$, find the value of y that corresponds to values of x for each integer starting with -4 and ending with 2.

CHAPTER P Summary, Review, and Test

SUMMARY: BASIC FORMULAS

Definition of Absolute Value

$$|x| = \begin{cases} x & \text{if } x \geq 0 \\ -x & \text{if } x < 0 \end{cases}$$

Distance between Points a and b on a Number Line

$$|a - b| \quad \text{or} \quad |b - a|$$

Properties of Real Numbers

Commutative	$a + b = b + a$
	$ab = ba$
Associative	$(a + b) + c = a + (b + c)$
	$(ab)c = a(bc)$
Distributive	$a(b + c) = ab + ac$
Identity	$a + 0 = a$
	$a \cdot 1 = a$
Inverse	$a + (-a) = 0$
	$a \cdot \dfrac{1}{a} = 1, a \neq 0$

Properties for Solving Equations

The Addition Property of Equality
 $a = b$ and $a + c = b + c$ are equivalent equations.
The Multiplication Property of Equality
 $a = b$ and $ac = bc$ are equivalent equations, $c \neq 0$.
The Square Root Property
 If $u^2 = d$, then $u = \pm \sqrt{d}$, $d > 0$.

Properties of Exponents

$$b^{-n} = \frac{1}{b^n}, \quad b^0 = 1, \quad b^m \cdot b^n = b^{m+n},$$

$$(b^m)^n = b^{mn}, \quad \frac{b^m}{b^n} = b^{m-n}, \quad (ab)^n = a^n b^n, \quad \left(\frac{a}{b}\right)^n = \frac{a^n}{b^n}$$

Product and Quotient Rules for nth Roots

$$\sqrt[n]{ab} = \sqrt[n]{a} \cdot \sqrt[n]{b}, \qquad \sqrt[n]{\frac{a}{b}} = \frac{\sqrt[n]{a}}{\sqrt[n]{b}}$$

Rational Exponents

$$a^{\frac{1}{n}} = \sqrt[n]{a}, \quad a^{-\frac{1}{n}} = \frac{1}{a^{\frac{1}{n}}} = \frac{1}{\sqrt[n]{a}},$$

$$a^{\frac{m}{n}} = (\sqrt[n]{a})^m = \sqrt[n]{a^m}, \quad a^{-\frac{m}{n}} = \frac{1}{a^{\frac{m}{n}}}$$

Special Products

$$(A + B)(A - B) = A^2 - B^2$$
$$(A + B)^2 = A^2 + 2AB + B^2$$
$$(A - B)^2 = A^2 - 2AB + B^2$$
$$(A + B)^3 = A^3 + 3A^2B + 3AB^2 + B^3$$
$$(A - B)^3 = A^3 - 3A^2B + 3AB^2 - B^3$$

Factoring Formulas

$$A^2 - B^2 = (A + B)(A - B)$$
$$A^2 + 2AB + B^2 = (A + B)^2$$
$$A^2 - 2AB + B^2 = (A - B)^2$$
$$A^3 + B^3 = (A + B)(A^2 - AB + B^2)$$
$$A^3 - B^3 = (A - B)(A^2 + AB + B^2)$$

Completing the Square

If $x^2 + bx$ is a binomial, adding $\left(\dfrac{b}{2}\right)^2$, the square of half the coefficient of x, results in a perfect square trinomial.

$$x^2 + bx + \left(\frac{b}{2}\right)^2 = \left(x + \frac{b}{2}\right)^2$$

REVIEW EXERCISES

You can use these review exercises, like the review exercises at the end of each chapter, to test your understanding of the chapter's topics. However, you can also use these exercises as a prerequisite test to check your mastery of the fundamental algebra skills needed in this book.

P.1

In Exercises 1–2, evaluate each algebraic expression for the given value or values of the variable(s).

1. $3 + 6(x - 2)^3$ for $x = 4$

2. $x^2 - 5(x - y)$ for $x = 6$ and $y = 2$

3. You are riding along an expressway traveling x miles per hour. The formula

$$S = 0.015x^2 + x + 10$$

models the recommended safe distance, S, in feet, between your car and other cars on the expressway. What is the recommended safe distance when your speed is 60 miles per hour?

In Exercises 4–7, let $A = \{a, b, c\}, B = \{a, c, d, e\}$, and $C = \{a, d, f, g\}$. Find the indicated set.

4. $A \cap B$ **5.** $A \cup B$

6. $A \cup C$ **7.** $C \cap A$

8. Consider the set:

$$\left\{-17, -\tfrac{9}{13}, 0, 0.75, \sqrt{2}, \pi, \sqrt{81}\right\}.$$

List all numbers from the set that are **a.** natural numbers, **b.** whole numbers, **c.** integers, **d.** rational numbers, **e.** irrational numbers, **f.** real numbers.

In Exercises 9–11, express each interval in set-builder notation and graph the interval on a number line.

9. $[-3, 5)$ **10.** $(-2, \infty)$ **11.** $(-\infty, 0]$

In Exercises 12–15, use graphs to find each set.

12. $(-2, 1] \cap [-1, 3)$ **13.** $(-2, 1] \cup [-1, 3)$

14. $[1, 3) \cap (0, 4)$ **15.** $[1, 3) \cup (0, 4)$

In Exercises 16–19, rewrite each expression without absolute value bars.

16. $|-103|$ **17.** $|\sqrt{2} - 1|$ **18.** $|3 - \sqrt{17}|$

19. Express the distance between the numbers -17 and 4 using absolute value. Then evaluate the absolute value.

P.2

In Exercises 20–25, state the name of the property illustrated.

20. $3 + 17 = 17 + 3$ **21.** $(6 \cdot 3) \cdot 9 = 6 \cdot (3 \cdot 9)$

22. $\sqrt{3}(\sqrt{5} + \sqrt{3}) = \sqrt{15} + 3$

23. $(6 \cdot 9) \cdot 2 = 2 \cdot (6 \cdot 9)$

24. $\sqrt{3}(\sqrt{5} + \sqrt{3}) = (\sqrt{5} + \sqrt{3})\sqrt{3}$

25. $(3 \cdot 7) + (4 \cdot 7) = (4 \cdot 7) + (3 \cdot 7)$

In Exercises 26–32, simplify or solve, whichever is appropriate.

26. $5(2x - 3) + 7x$

27. $\dfrac{1}{5}(5x) + [(3y) + (-3y)] - (-x)$

28. $2x - 5 = 7$ **29.** $7(x - 4) - x = 2$

30. $3(4y - 5) - (7y + 2)$ **31.** $8 - 2[3 - (5x - 1)]$

32. $2(x - 4) + 3(x + 5) - 2x = -2$

33. Solve for g: $vt + gt^2 = s$.

34. Solve for g: $T = gr + gvt$.

35. The diversity index, from 0 (no diversity) to 100, measures the chance that two randomly selected people are a different race or ethnicity. The diversity index in the United States varies widely from region to region, from as high as 81 in Hawaii to as low as 11 in Vermont. The bar graph shows the national diversity index for the United States for four years in the period from 1980 through 2010.

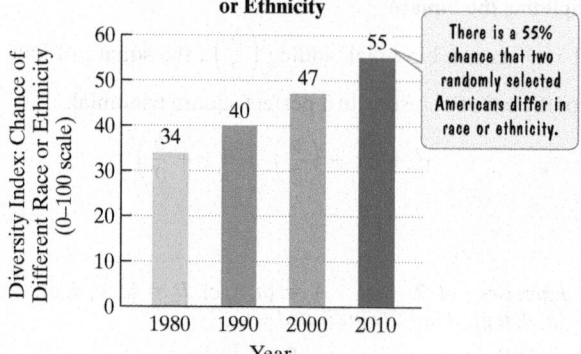

Chance That Two Randomly Selected Americans Are a Different Race or Ethnicity

There is a 55% chance that two randomly selected Americans differ in race or ethnicity.

Source: *USA Today*

The data in the graph can be modeled by the formula
$$D = 0.005x^2 + 0.55x + 34,$$
where D is the national diversity index in the United States x years after 1980. According to the formula, what was the U.S. diversity index in 2010? How does this compare with the index displayed by the bar graph?

P.3

Evaluate each exponential expression in Exercises 36–39.

36. $(-3)^3(-2)^2$ **37.** $2^{-4} + 4^{-1}$

38. $5^{-3} \cdot 5$ **39.** $\dfrac{3^3}{3^6}$

Simplify each exponential expression in Exercises 40–43.

40. $(-2x^4y^3)^3$ **41.** $(-5x^3y^2)(-2x^{-11}y^{-2})$

42. $(2x^3)^{-4}$ **43.** $\dfrac{7x^5y^6}{28x^{15}y^{-2}}$

In Exercises 44–45, write each number in decimal notation.

44. 3.74×10^4 **45.** 7.45×10^{-5}

In Exercises 46–47, write each number in scientific notation.

46. $3,590,000$ **47.** 0.00725

In Exercises 48–49, perform the indicated operation and write the answer in decimal notation.

48. $(3 \times 10^3)(1.3 \times 10^2)$ **49.** $\dfrac{6.9 \times 10^3}{3 \times 10^5}$

In 2009, the United States government spent more than it had collected in taxes, resulting in a budget deficit of $1.35 trillion. In Exercises 50–52, you will use scientific notation to put a number like 1.35 trillion in perspective. Use 10^{12} for 1 trillion.

50. Express 1.35 trillion in scientific notation.

51. There are approximately 32,000,000 seconds in a year. Express this number in scientific notation.

52. Use your scientific notation answers from Exercises 50 and 51 to answer this question: How many years is 1.35 trillion seconds? Round to the nearest year. (*Note:* 1.35 trillion seconds would take us back in time to a period when Neanderthals were using stones to make tools.)

P.4

In Exercises 53–54, solve each equation by the square root property.

53. $x^2 = 19$ **54.** $3x^2 + 5 = 17$

Use the product rule to simplify the expressions in Exercises 55–58. In Exercises 57–58, assume that variables represent nonnegative real numbers.

55. $\sqrt{300}$ **56.** $\sqrt{12x^2}$

57. $\sqrt{10x} \cdot \sqrt{2x}$ **58.** $\sqrt{r^3}$

Use the quotient rule to simplify the expressions in Exercises 59–60.

59. $\sqrt{\dfrac{121}{4}}$ **60.** $\dfrac{\sqrt{96x^3}}{\sqrt{2x}}$ (Assume that $x > 0$.)

In Exercises 61–63, add or subtract terms whenever possible.

61. $7\sqrt{5} + 13\sqrt{5}$ **62.** $2\sqrt{50} + 3\sqrt{8}$

63. $4\sqrt{72} - 2\sqrt{48}$

In Exercises 64–67, rationalize the denominator.

64. $\dfrac{30}{\sqrt{5}}$ **65.** $\dfrac{\sqrt{2}}{\sqrt{3}}$

66. $\dfrac{5}{6 + \sqrt{3}}$ **67.** $\dfrac{14}{\sqrt{7} - \sqrt{5}}$

Evaluate each expression in Exercises 68–71 or indicate that the root is not a real number.

68. $\sqrt[3]{125}$ **69.** $\sqrt[5]{-32}$

70. $\sqrt[4]{-125}$ **71.** $\sqrt[4]{(-5)^4}$

Simplify the radical expressions in Exercises 72–76.

72. $\sqrt[3]{81}$ **73.** $\sqrt[3]{y^5}$

74. $\sqrt[4]{8} \cdot \sqrt[4]{10}$ **75.** $4\sqrt[3]{16} + 5\sqrt[3]{2}$

76. $\dfrac{\sqrt[4]{32x^5}}{\sqrt[4]{16x}}$ (Assume that $x > 0$.)

In Exercises 77–82, evaluate each expression.

77. $16^{\frac{1}{2}}$ **78.** $25^{-\frac{1}{2}}$ **79.** $125^{\frac{1}{3}}$

80. $27^{-\frac{1}{3}}$ **81.** $64^{\frac{2}{3}}$ **82.** $27^{-\frac{4}{3}}$

In Exercises 83–85, simplify using properties of exponents.

83. $\left(5x^{\frac{2}{3}}\right)\left(4x^{\frac{1}{4}}\right)$ **84.** $\dfrac{15x^{\frac{3}{4}}}{5x^{\frac{1}{2}}}$

85. $(125x^6)^{\frac{2}{3}}$

86. Simplify by reducing the index of the radical: $\sqrt[6]{y^3}$.

P.5

In Exercises 87–88, perform the indicated operations. Write the resulting polynomial in standard form and indicate its degree.

87. $(-6x^3 + 7x^2 - 9x + 3) + (14x^3 + 3x^2 - 11x - 7)$

88. $(13x^4 - 8x^3 + 2x^2) - (5x^4 - 3x^3 + 2x^2 - 6)$

In Exercises 89–95, find each product.

89. $(3x - 2)(4x^2 + 3x - 5)$ **90.** $(3x - 5)(2x + 1)$

91. $(4x + 5)(4x - 5)$ **92.** $(2x + 5)^2$

93. $(3x - 4)^2$ **94.** $(2x + 1)^3$

95. $(5x - 2)^3$

In Exercises 96–97, perform the indicated operations. Indicate the degree of the resulting polynomial.

96. $(7x^2 - 8xy + y^2) + (-8x^2 - 9xy - 4y^2)$

97. $(13x^3y^2 - 5x^2y - 9x^2) - (-11x^3y^2 - 6x^2y + 3x^2 - 4)$

In Exercises 98–102, find each product.

98. $(x + 7y)(3x - 5y)$ **99.** $(3x - 5y)^2$

100. $(3x^2 + 2y)^2$ **101.** $(7x + 4y)(7x - 4y)$ *

102. $(a - b)(a^2 + ab + b^2)$

P.6

In Exercises 103–119, factor completely, or state that the polynomial is prime.

103. $15x^3 + 3x^2$ **104.** $x^2 - 11x + 28$

105. $15x^2 - x - 2$ **106.** $64 - x^2$

107. $x^2 + 16$ **108.** $3x^4 - 9x^3 - 30x^2$

109. $20x^7 - 36x^3$ **110.** $x^3 - 3x^2 - 9x + 27$

111. $16x^2 - 40x + 25$ **112.** $x^4 - 16$

113. $y^3 - 8$ **114.** $x^3 + 64$

115. $3x^4 - 12x^2$ **116.** $27x^3 - 125$

117. $x^5 - x$ **118.** $x^3 + 5x^2 - 2x - 10$

119. $x^2 + 18x + 81 - y^2$

In Exercises 120–122, factor and simplify each algebraic expression.

120. $16x^{-\frac{3}{4}} + 32x^{\frac{1}{4}}$

121. $(x^2 - 4)(x^2 + 3)^{\frac{1}{2}} - (x^2 - 4)^2(x^2 + 3)^{\frac{3}{2}}$

122. $12x^{-\frac{1}{2}} + 6x^{-\frac{3}{2}}$

In Exercises 123–124, determine the constant that should be added to the binomial so that it becomes a perfect square trinomial. Then write and factor the trinomial.

123. $x^2 - 8x$ **124.** $x^2 + \dfrac{1}{2}x$

P.7

In Exercises 125–127, simplify each rational expression. Also, list all numbers that must be excluded from the domain.

125. $\dfrac{x^3 + 2x^2}{x + 2}$ **126.** $\dfrac{x^2 + 3x - 18}{x^2 - 36}$ **127.** $\dfrac{x^2 + 2x}{x^2 + 4x + 4}$

In Exercises 128–130, multiply or divide as indicated.

128. $\dfrac{x^2 + 6x + 9}{x^2 - 4} \cdot \dfrac{x + 3}{x - 2}$ **129.** $\dfrac{6x + 2}{x^2 - 1} \div \dfrac{3x^2 + x}{x - 1}$

130. $\dfrac{x^2 - 5x - 24}{x^2 - x - 12} \div \dfrac{x^2 - 10x + 16}{x^2 + x - 6}$

In Exercises 131–134, add or subtract as indicated.

131. $\dfrac{2x - 7}{x^2 - 9} - \dfrac{x - 10}{x^2 - 9}$ **132.** $\dfrac{3x}{x + 2} + \dfrac{x}{x - 2}$

133. $\dfrac{x}{x^2 - 9} + \dfrac{x - 1}{x^2 - 5x + 6}$

134. $\dfrac{4x - 1}{2x^2 + 5x - 3} - \dfrac{x + 3}{6x^2 + x - 2}$

In Exercises 135–137, simplify each complex rational expression.

135. $\dfrac{\frac{1}{x} - \frac{1}{2}}{\frac{1}{3} - \frac{x}{6}}$ **136.** $\dfrac{3 + \frac{12}{x}}{1 - \frac{16}{x^2}}$ **137.** $\dfrac{3 - \frac{1}{x + 3}}{3 + \frac{1}{x + 3}}$

CHAPTER P TEST

In Exercises 1–18, simplify the given expression, perform the indicated operation (and simplify, if possible), or solve the equation, whichever is appropriate.

1. $5(2x^2 - 6x) - (4x^2 - 3x)$

2. $3(2x - 4) + 3(x + 1) = 9$

3. $2x^2 - 1 = 49$

4. $[-1, 2) \cup (0, 5]$

5. $[-1, 2) \cap (0, 5]$

6. $\dfrac{30x^3y^4}{6x^9y^{-4}}$

7. $\sqrt{6r}\sqrt{3r}$ (Assume that $r \geq 0$.)

8. $4\sqrt{50} - 3\sqrt{18}$ **9.** $\dfrac{3}{5 + \sqrt{2}}$

10. $\sqrt[3]{16x^4}$ **11.** $\dfrac{x^2 + 2x - 3}{x^2 - 3x + 2}$

12. $\dfrac{5 \times 10^{-6}}{20 \times 10^{-8}}$ (Express the answer in scientific notation.)

13. $(2x - 5)(x^2 - 4x + 3)$ **14.** $(5x + 3y)^2$

15. $\dfrac{2x + 8}{x - 3} \div \dfrac{x^2 + 5x + 4}{x^2 - 9}$ **16.** $\dfrac{x}{x + 3} + \dfrac{5}{x - 3}$

17. $\dfrac{2x + 3}{x^2 - 7x + 12} - \dfrac{2}{x - 3}$ **18.** $\dfrac{\frac{1}{x} - \frac{1}{3}}{\frac{1}{x}}$

In Exercises 19–24, factor completely, or state that the polynomial is prime.

19. $x^2 - 9x + 18$

20. $x^3 + 2x^2 + 3x + 6$

21. $25x^2 - 9$

22. $36x^2 - 84x + 49$

23. $y^3 - 125$

24. $x^2 + 10x + 25 - 9y^2$

25. Factor and simplify:

$$x(x + 3)^{-\frac{3}{5}} + (x + 3)^{\frac{2}{5}}.$$

26. List all the rational numbers in this set:

$$\left\{ -7, -\tfrac{4}{5}, 0, 0.25, \sqrt{3}, \sqrt{4}, \tfrac{22}{7}, \pi \right\}.$$

In Exercises 27–28, state the name of the property illustrated.

27. $3(2 + 5) = 3(5 + 2)$

28. $6(7 + 4) = 6 \cdot 7 + 6 \cdot 4$

29. Express in scientific notation: 0.00076.

30. Evaluate: $27^{-\frac{5}{3}}$.

31. Solve for y: $Ax + By = C$.

32. In 2007, world population was approximately 6.6×10^9. By some projections, world population will double by 2040. Express the population at that time in scientific notation.

33. Big (Lack of) Men on Campus In 2007, 135 women received bachelor's degrees for every 100 men. According to the U.S. Department of Education, that gender imbalance will widen in the coming years, as shown by the bar graph.

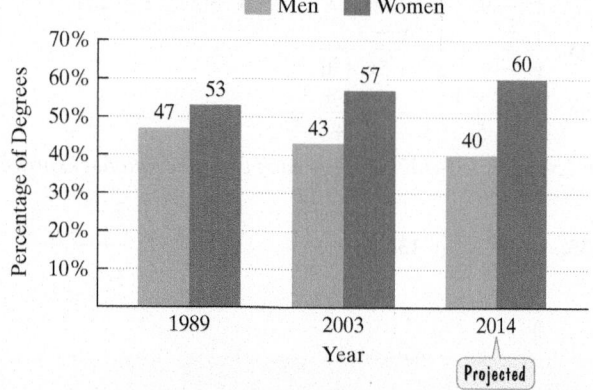

Percentage of Bachelor's Degrees Awarded to United States Men and Women

Source: U.S. Department of Education

The data for bachelor's degrees can be described by the following mathematical models:

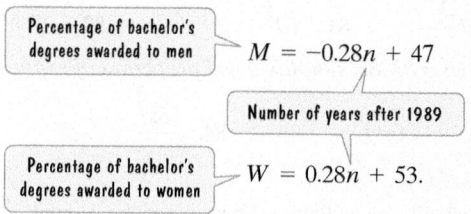

Percentage of bachelor's degrees awarded to men — $M = -0.28n + 47$

Number of years after 1989

Percentage of bachelor's degrees awarded to women — $W = 0.28n + 53$.

a. According to the first formula, what percentage of bachelor's degrees were awarded to men in 2003? Does this underestimate or overestimate the actual percent shown by the bar graph? By how much?

b. Use the given formulas to write a new formula with a rational expression that models the ratio of the percentage of bachelor's degrees received by men to the percentage received by women n years after 1989. Name this new mathematical model R, for ratio.

c. Use the formula for R to find the projected ratio of bachelor's degrees received by men to degrees received by women in 2014. According to the model, how many women will receive bachelor's degrees for every two men in 2014? How well does this describe the projections shown by the graph?

FUNCTIONS AND GRAPHS

HERE'S WHERE YOU'LL FIND THESE APPLICATIONS:

A mathematical model involving global warming is developed in Example 9 in Section 1.4.

Using mathematics in a way that is similar to making a movie is discussed in the Blitzer Bonus on page 174.

A vast expanse of open water at the top of our world was once covered with ice. The melting of the Arctic ice caps has forced polar bears to swim as far as 40 miles, causing them to drown in significant numbers. Such deaths were rare in the past.

There is strong scientific consensus that human activities are changing the Earth's climate. Scientists now believe that there is a striking correlation between atmospheric carbon dioxide concentration and global temperature. As both of these variables increase at significant rates, there are warnings of a planetary emergency that threatens to condemn coming generations to a catastrophically diminished future.*

In this chapter, you'll learn to approach our climate crisis mathematically by creating formulas, called functions, that model data for average global temperature and carbon dioxide concentration over time. Understanding the concept of a function will give you a new perspective on many situations, ranging from global warming to using mathematics in a way that is similar to making a movie.

*Sources: Al Gore, *An Inconvenient Truth*, Rodale, 2006; *Time*, April 3, 2006

Graphs and Graphing Utilities

Objectives

① Plot points in the rectangular coordinate system.

② Graph equations in the rectangular coordinate system.

③ Interpret information about a graphing utility's viewing rectangle or table.

④ Use a graph to determine intercepts.

⑤ Interpret information given by graphs.

The beginning of the seventeenth century was a time of innovative ideas and enormous intellectual progress in Europe. English theatergoers enjoyed a succession of exciting new plays by Shakespeare. William Harvey proposed the radical notion that the heart was a pump for blood rather than the center of emotion. Galileo, with his newfangled invention called the telescope, supported the theory of Polish astronomer Copernicus that the sun, not the Earth, was the center of the solar system. Monteverdi was writing the world's first grand operas. French mathematicians Pascal and Fermat invented a new field of mathematics called probability theory.

Into this arena of intellectual electricity stepped French aristocrat René Descartes (1596–1650). Descartes (pronounced "day cart"), propelled by the creativity surrounding him, developed a new branch of mathematics that brought together algebra and geometry in a unified way—a way that visualized numbers as points on a graph, equations as geometric figures, and geometric figures as equations. This new branch of mathematics, called *analytic geometry*, established Descartes as one of the founders of modern thought and among the most original mathematicians and philosophers of any age. We begin this section by looking at Descartes's deceptively simple idea, called the **rectangular coordinate system** or (in his honor) the **Cartesian coordinate system**.

① Plot points in the rectangular coordinate system.

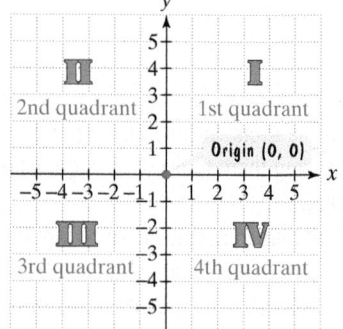

FIGURE 1.1 The rectangular coordinate system

GREAT QUESTION!

What's the significance of the word "ordered" when describing a pair of real numbers?

The phrase *ordered pair* is used because order is important. The order in which coordinates appear makes a difference in a point's location. This is illustrated in **Figure 1.2**.

Points and Ordered Pairs

Descartes used two number lines that intersect at right angles at their zero points, as shown in **Figure 1.1**. The horizontal number line is the **x-axis**. The vertical number line is the **y-axis**. The point of intersection of these axes is their zero points, called the **origin**. Positive numbers are shown to the right and above the origin. Negative numbers are shown to the left and below the origin. The axes divide the plane into four quarters, called **quadrants**. The points located on the axes are not in any quadrant.

Each point in the rectangular coordinate system corresponds to an **ordered pair** of real numbers, (x, y). Examples of such pairs are $(-5, 3)$ and $(3, -5)$. The first number in each pair, called the **x-coordinate**, denotes the distance and direction from the origin along the x-axis. The second number in each pair, called the **y-coordinate**, denotes vertical distance and direction along a line parallel to the y-axis or along the y-axis itself.

Figure 1.2 shows how we **plot**, or locate, the points corresponding to the ordered pairs $(-5, 3)$ and $(3, -5)$. We plot $(-5, 3)$ by going 5 units from 0 to the left along the x-axis. Then we go 3 units up parallel to the y-axis. We plot $(3, -5)$ by going 3 units from 0 to the right along the x-axis and 5 units down parallel to the y-axis. The phrase "the points corresponding to the ordered pairs $(-5, 3)$ and $(3, -5)$" is often abbreviated as "the points $(-5, 3)$ and $(3, -5)$."

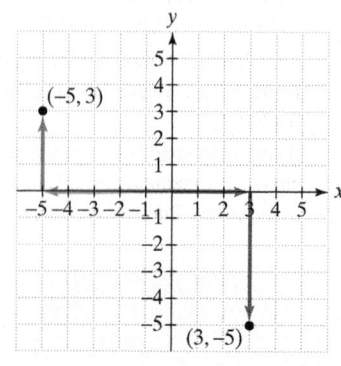

FIGURE 1.2 Plotting $(-5, 3)$ and $(3, -5)$

EXAMPLE 1 Plotting Points in the Rectangular Coordinate System

Plot the points: $A(-3, 5)$, $B(2, -4)$, $C(5, 0)$, $D(-5, -3)$, $E(0, 4)$, and $F(0, 0)$.

SOLUTION

See **Figure 1.3**. We move from the origin and plot the points in the following way:

$A(-3, 5)$: 3 units left, 5 units up

$B(2, -4)$: 2 units right, 4 units down

$C(5, 0)$: 5 units right, 0 units up or down

$D(-5, -3)$: 5 units left, 3 units down

$E(0, 4)$: 0 units right or left, 4 units up

$F(0, 0)$: 0 units right or left, 0 units up or down

Notice that the origin is represented by $(0, 0)$.

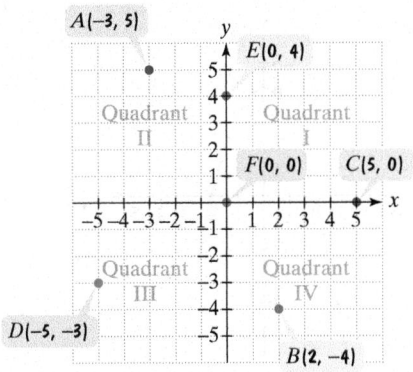

FIGURE 1.3 Plotting points

• • •

GREAT QUESTION!

Why is it so important to work each of the book's Check Points?

You learn best by doing. Do not simply look at the worked examples and conclude that you know how to solve them. To be sure you understand the worked examples, try each Check Point. Check your answer in the answer section before continuing your reading. Expect to read this book with pencil and paper handy to work the Check Points.

② Graph equations in the rectangular coordinate system.

☑ Check Point **1** Plot the points: $A(-2, 4)$, $B(4, -2)$, $C(-3, 0)$, and $D(0, -3)$.

Graphs of Equations

A relationship between two quantities can be expressed as an **equation in two variables**, such as

$$y = 4 - x^2.$$

A **solution of an equation in two variables**, x and y, is an ordered pair of real numbers with the following property: When the x-coordinate is substituted for x and the y-coordinate is substituted for y in the equation, we obtain a true statement. For example, consider the equation $y = 4 - x^2$ and the ordered pair $(3, -5)$. When 3 is substituted for x and -5 is substituted for y, we obtain the statement $-5 = 4 - 3^2$, or $-5 = 4 - 9$, or $-5 = -5$. Because this statement is true, the ordered pair $(3, -5)$ is a solution of the equation $y = 4 - x^2$. We also say that $(3, -5)$ **satisfies** the equation.

We can generate as many ordered-pair solutions as desired to $y = 4 - x^2$ by substituting numbers for x and then finding the corresponding values for y. For example, suppose we let $x = 3$:

Start with x.	Compute y.	Form the ordered pair (x, y).
x	$y = 4 - x^2$	Ordered Pair (x, y)
3	$y = 4 - 3^2 = 4 - 9 = -5$	$(3, -5)$

Let $x = 3$. $(3, -5)$ is a solution of $y = 4 - x^2$.

The **graph of an equation in two variables** is the set of all points whose coordinates satisfy the equation. One method for graphing such equations is the **point-plotting method**. First, we find several ordered pairs that are solutions of the equation. Next, we plot these ordered pairs as points in the rectangular coordinate system. Finally, we connect the points with a smooth curve or line. This often gives us a picture of all ordered pairs that satisfy the equation.

EXAMPLE 2 Graphing an Equation Using the Point-Plotting Method

Graph $y = 4 - x^2$. Select integers for x, starting with -3 and ending with 3.

SOLUTION

For each value of x, we find the corresponding value for y.

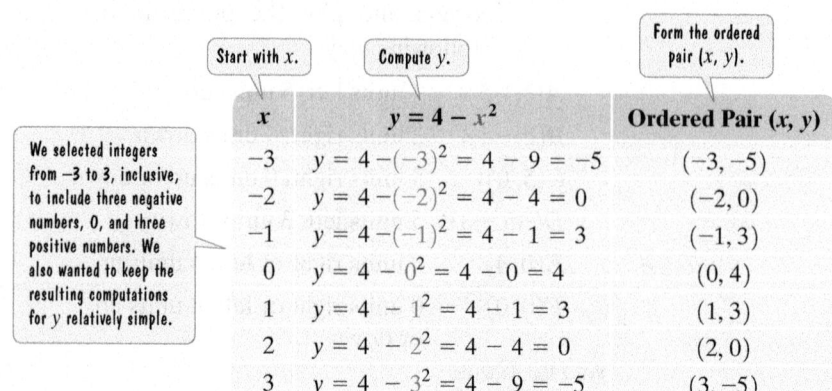

We selected integers from -3 to 3, inclusive, to include three negative numbers, 0, and three positive numbers. We also wanted to keep the resulting computations for y relatively simple.

x	$y = 4 - x^2$	Ordered Pair (x, y)
-3	$y = 4 - (-3)^2 = 4 - 9 = -5$	$(-3, -5)$
-2	$y = 4 - (-2)^2 = 4 - 4 = 0$	$(-2, 0)$
-1	$y = 4 - (-1)^2 = 4 - 1 = 3$	$(-1, 3)$
0	$y = 4 - 0^2 = 4 - 0 = 4$	$(0, 4)$
1	$y = 4 - 1^2 = 4 - 1 = 3$	$(1, 3)$
2	$y = 4 - 2^2 = 4 - 4 = 0$	$(2, 0)$
3	$y = 4 - 3^2 = 4 - 9 = -5$	$(3, -5)$

FIGURE 1.4 The graph of $y = 4 - x^2$

Now we plot the seven points and join them with a smooth curve, as shown in **Figure 1.4**. The graph of $y = 4 - x^2$ is a curve where the part of the graph to the right of the y-axis is a reflection of the part to the left of it and vice versa. The arrows on the left and the right of the curve indicate that it extends indefinitely in both directions. •••

✓ **Check Point 2** Graph $y = 4 - x$. Select integers for x, starting with -3 and ending with 3.

EXAMPLE 3 Graphing an Equation Using the Point-Plotting Method

Graph $y = |x|$. Select integers for x, starting with -3 and ending with 3.

SOLUTION

For each value of x, we find the corresponding value for y.

| x | $y = |x|$ | Ordered Pair (x, y) |
|---|---|---|
| -3 | $y = |-3| = 3$ | $(-3, 3)$ |
| -2 | $y = |-2| = 2$ | $(-2, 2)$ |
| -1 | $y = |-1| = 1$ | $(-1, 1)$ |
| 0 | $y = |0| = 0$ | $(0, 0)$ |
| 1 | $y = |1| = 1$ | $(1, 1)$ |
| 2 | $y = |2| = 2$ | $(2, 2)$ |
| 3 | $y = |3| = 3$ | $(3, 3)$ |

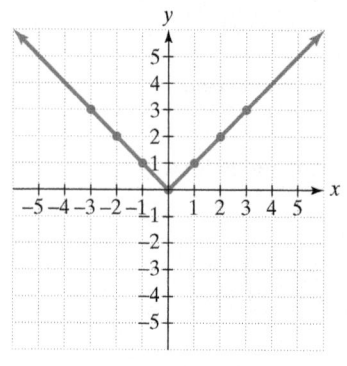

FIGURE 1.5 The graph of $y = |x|$

We plot the points and connect them, resulting in the graph shown in **Figure 1.5**. The graph is V-shaped and centered at the origin. For every point (x, y) on the graph, the point $(-x, y)$ is also on the graph. This shows that the absolute value of a positive number is the same as the absolute value of its opposite. •••

✓ **Check Point 3** Graph $y = |x + 1|$. Select integers for x, starting with -4 and ending with 2.

③ Interpret information about a graphing utility's viewing rectangle or table.

Graphing Equations and Creating Tables Using a Graphing Utility

Graphing calculators and graphing software packages for computers are referred to as **graphing utilities** or graphers. A graphing utility is a powerful tool that quickly generates the graph of an equation in two variables. **Figures 1.6(a)** and **1.6(b)** show two such graphs for the equations in Examples 2 and 3.

FIGURE 1.6(a) The graph of $y = 4 - x^2$

FIGURE 1.6(b) The graph of $y = |x|$

GREAT QUESTION!

I'm not using a graphing calculator, so should I skip this part of the section?

Even if you are not using a graphing utility in the course, read this part of the section. Knowing about viewing rectangles will enable you to understand the graphs that we display in the technology boxes throughout the book.

What differences do you notice between these graphs and the graphs that we drew by hand? They do seem a bit "jittery." Arrows do not appear on the left and right ends of the graphs. Furthermore, numbers are not given along the axes. For both graphs in **Figure 1.6**, the x-axis extends from -10 to 10 and the y-axis also extends from -10 to 10. The distance represented by each consecutive tick mark is one unit. We say that the **viewing rectangle**, or the **viewing window**, is $[-10, 10, 1]$ by $[-10, 10, 1]$.

Xmin	Xmax	Xscl		Ymin	Ymax	Yscl
$[-10,$	$10,$	$1]$	by	$[-10,$	$10,$	$1]$
The minimum x-value along the x-axis is -10.	The maximum x-value along the x-axis is 10.	Distance between consecutive tick marks on the x-axis is one unit.		The minimum y-value along the y-axis is -10.	The maximum y-value along the y-axis is 10.	Distance between consecutive tick marks on the y-axis is one unit.

To graph an equation in x and y using a graphing utility, enter the equation and specify the size of the viewing rectangle. The size of the viewing rectangle sets minimum and maximum values for both the x- and y-axes. Enter these values, as well as the values representing the distances between consecutive tick marks, on the respective axes. The $[-10, 10, 1]$ by $[-10, 10, 1]$ viewing rectangle used in **Figure 1.6** is called the **standard viewing rectangle**.

EXAMPLE 4 Understanding the Viewing Rectangle

What is the meaning of a $[-2, 3, 0.5]$ by $[-10, 20, 5]$ viewing rectangle?

SOLUTION

We begin with $[-2, 3, 0.5]$, which describes the x-axis. The minimum x-value is -2 and the maximum x-value is 3. The distance between consecutive tick marks is 0.5.

Next, consider $[-10, 20, 5]$, which describes the y-axis. The minimum y-value is -10 and the maximum y-value is 20. The distance between consecutive tick marks is 5.

Figure 1.7 illustrates a $[-2, 3, 0.5]$ by $[-10, 20, 5]$ viewing rectangle. To make things clearer, we've placed numbers by each tick mark. These numbers do not appear on the axes when you use a graphing utility to graph an equation. • • •

FIGURE 1.7 A $[-2, 3, 0.5]$ by $[-10, 20, 5]$ viewing rectangle

⊘ Check Point **4** What is the meaning of a $[-100, 100, 50]$ by $[-100, 100, 10]$ viewing rectangle? Create a figure like the one in **Figure 1.7** that illustrates this viewing rectangle.

On most graphing utilities, the display screen is two-thirds as high as it is wide. By using a square setting, you can equally space the x and y tick marks. (This does not occur in the standard viewing rectangle.) Graphing utilities can also *zoom in* and *zoom out*.

When you zoom in, you see a smaller portion of the graph, but you do so in greater detail. When you zoom out, you see a larger portion of the graph. Thus, zooming out may help you to develop a better understanding of the overall character of the graph. With practice, you will become more comfortable with graphing equations in two variables using your graphing utility. You will also develop a better sense of the size of the viewing rectangle that will reveal needed information about a particular graph.

Graphing utilities can also be used to create tables showing solutions of equations in two variables. Use the Table Setup function to choose the starting value of x and to input the increment, or change, between the consecutive x-values. The corresponding y-values are calculated based on the equation(s) in two variables in the $\boxed{Y=}$ screen. In **Figure 1.8**, we used a TI-84 Plus to create a table for $y = 4 - x^2$ and $y = |x|$, the equations in Examples 2 and 3.

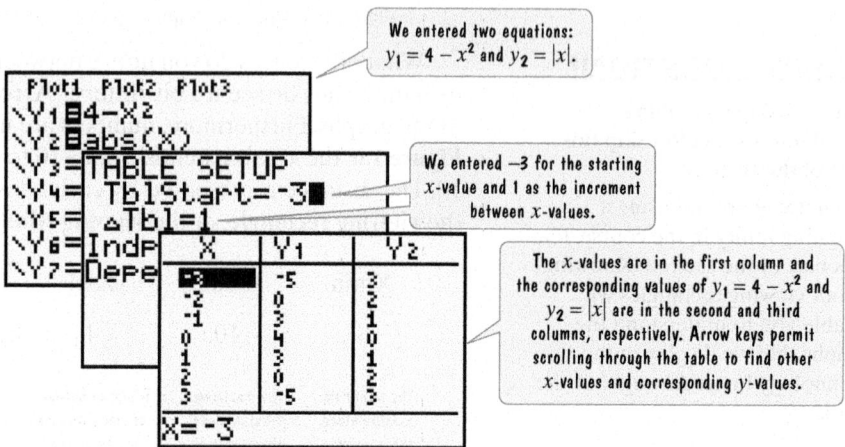

We entered two equations: $y_1 = 4 - x^2$ and $y_2 = |x|$.

We entered −3 for the starting x-value and 1 as the increment between x-values.

The x-values are in the first column and the corresponding values of $y_1 = 4 - x^2$ and $y_2 = |x|$ are in the second and third columns, respectively. Arrow keys permit scrolling through the table to find other x-values and corresponding y-values.

FIGURE 1.8 Creating a table for $y_1 = 4 - x^2$ and $y_2 = |x|$

④ Use a graph to determine intercepts.

Intercepts

An **x-intercept** of a graph is the x-coordinate of a point where the graph intersects the x-axis. For example, look at the graph of $y = 4 - x^2$ in **Figure 1.9**. The graph crosses the x-axis at $(-2, 0)$ and $(2, 0)$. Thus, the x-intercepts are -2 and 2. **The y-coordinate corresponding to an x-intercept is always zero.**

A **y-intercept** of a graph is the y-coordinate of a point where the graph intersects the y-axis. The graph of $y = 4 - x^2$ in **Figure 1.9** shows that the graph crosses the y-axis at $(0, 4)$. Thus, the y-intercept is 4. **The x-coordinate corresponding to a y-intercept is always zero.**

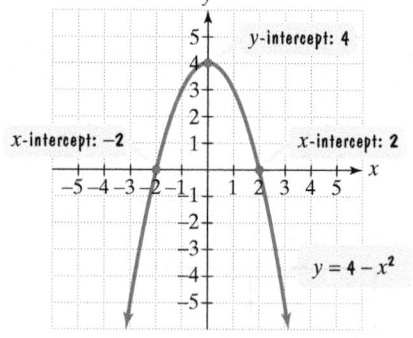

FIGURE 1.9 Intercepts of $y = 4 - x^2$

GREAT QUESTION!

Are single numbers the only way to represent intercepts? Can ordered pairs also be used?

Mathematicians tend to use two ways to describe intercepts. Did you notice that we are using single numbers? If a is an x-intercept of a graph, then the graph passes through the point $(a, 0)$. If b is a y-intercept of a graph, then the graph passes through the point $(0, b)$.

Some books state that the x-intercept is the *point* $(a, 0)$ and the x-intercept is *at a* on the x-axis. Similarly, the y-intercept is the *point* $(0, b)$ and the y-intercept is *at b* on the y-axis. In these descriptions, the intercepts are the actual points where the graph intersects the axes.

Although we'll describe intercepts as single numbers, we'll immediately state the point on the x- or y-axis that the graph passes through. Here's the important thing to keep in mind:

x-intercept: The corresponding value of y is 0.

y-intercept: The corresponding value of x is 0.

EXAMPLE 5 Identifying Intercepts

Identify the *x*- and *y*-intercepts.

a.

b.

c.
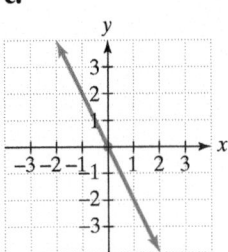

SOLUTION

a. The graph crosses the *x*-axis at $(-1, 0)$. Thus, the *x*-intercept is -1. The graph crosses the *y*-axis at $(0, 2)$. Thus, the *y*-intercept is 2.

b. The graph crosses the *x*-axis at $(3, 0)$, so the *x*-intercept is 3. This vertical line does not cross the *y*-axis. Thus, there is no *y*-intercept.

c. This graph crosses the *x*- and *y*-axes at the same point, the origin. Because the graph crosses both axes at $(0, 0)$, the *x*-intercept is 0 and the *y*-intercept is 0.

● ● ●

✓ **Check Point 5** Identify the *x*- and *y*-intercepts.

a.

b.

c.
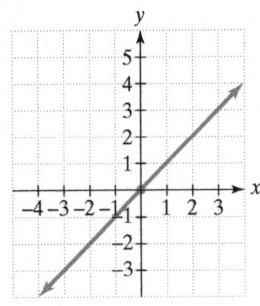

Figure 1.10 illustrates that a graph may have no intercepts or several intercepts.

No *x*-intercept One *y*-intercept	One *x*-intercept No *y*-intercept	No intercepts	One *x*-intercept Three *y*-intercepts	The same *x*-intercept and *y*-intercept

FIGURE 1.10

⑤ Interpret information given by graphs.

Interpreting Information Given by Graphs

Line graphs are often used to illustrate trends over time. Some measure of time, such as months or years, frequently appears on the horizontal axis. Amounts are generally listed on the vertical axis. Points are drawn to represent the given information. The graph is formed by connecting the points with line segments.

A line graph displays information in the first quadrant of a rectangular coordinate system. By identifying points on line graphs and their coordinates, you can interpret specific information given by the graph.

EXAMPLE 6 Age at Marriage and the Probability of Divorce

Divorce rates are considerably higher for couples who marry in their teens. The line graphs in **Figure 1.11** show the percentages of marriages ending in divorce based on the wife's age at marriage.

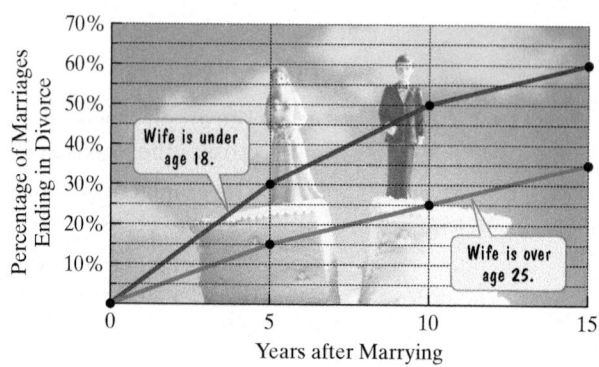

Probability of Divorce, by Wife's Age at Marriage

FIGURE 1.11
Source: B. E. Pruitt et al., *Human Sexuality*, Prentice Hall, 2007.

Here are two mathematical models that approximate the data displayed by the line graphs:

Wife is under 18 at time of marriage.

Wife is over 25 at time of marriage.

$$d = 4n + 5 \qquad d = 2.3n + 1.5.$$

In each model, the variable n is the number of years after marriage and the variable d is the percentage of marriages ending in divorce.

 a. Use the appropriate formula to determine the percentage of marriages ending in divorce after 10 years when the wife is over 25 at the time of marriage.

 b. Use the appropriate line graph in **Figure 1.11** to determine the percentage of marriages ending in divorce after 10 years when the wife is over 25 at the time of marriage.

 c. Does the value given by the mathematical model underestimate or overestimate the actual percentage of marriages ending in divorce after 10 years as shown by the graph? By how much?

SOLUTION

 a. Because the wife is over 25 at the time of marriage, we use the formula on the right, $d = 2.3n + 1.5$. To find the percentage of marriages ending in divorce after 10 years, we substitute 10 for n and evaluate the formula.

$$d = 2.3n + 1.5 \qquad \text{This is one of the two given mathematical models.}$$
$$d = 2.3(10) + 1.5 \qquad \text{Replace } n \text{ with 10.}$$
$$d = 23 + 1.5 \qquad \text{Multiply: } 2.3(10) = 23.$$
$$d = 24.5 \qquad \text{Add.}$$

The model indicates that 24.5% of marriages end in divorce after 10 years when the wife is over 25 at the time of marriage.

b. Now let's use the line graph that shows the percentage of marriages ending in divorce when the wife is over 25 at the time of marriage. The graph is shown again in **Figure 1.12**. To find the percentage of marriages ending in divorce after 10 years:

- Locate 10 on the horizontal axis and locate the point above 10.
- Read across to the corresponding percent on the vertical axis.

FIGURE 1.12

The actual data displayed by the graph indicate that 25% of these marriages end in divorce after 10 years.

c. The value obtained by evaluating the mathematical model, 24.5%, is close to, but slightly less than, the actual percentage of divorces, 25.0%. The difference between these percents is 25.0% − 24.5%, or 0.5%. The value given by the mathematical model, 24.5%, underestimates the actual percent, 25%, by only 0.5, providing a fairly accurate description of the data. • • •

Check Point 6

a. Use the appropriate formula from Example 6 to determine the percentage of marriages ending in divorce after 15 years when the wife is under 18 at the time of marriage.

b. Use the appropriate line graph in **Figure 1.11** to determine the percentage of marriages ending in divorce after 15 years when the wife is under 18 at the time of marriage.

c. Does the value given by the mathematical model underestimate or overestimate the actual percentage of marriages ending in divorce after 15 years as shown by the graph? By how much?

GREAT QUESTION!

What am I supposed to do with the exercises in the Concept and Vocabulary Check?

It is impossible to learn algebra without knowing its special language. The exercises in the Concept and Vocabulary Check, mainly fill-in-the-blank and true/false items, test your understanding of the definitions and concepts presented in each section. **Work all of the exercises in the Concept and Vocabulary Check** regardless of which exercises your professor assigns in the Exercise Set that follows.

CONCEPT AND VOCABULARY CHECK

Fill in each blank so that the resulting statement is true.

1. In the rectangular coordinate system, the horizontal number line is called the _____.

2. In the rectangular coordinate system, the vertical number line is called the _____.

3. In the rectangular coordinate system, the point of intersection of the horizontal axis and the vertical axis is called the _____.

4. The axes of the rectangular coordinate system divide the plane into regions, called _____. There are _____ of these regions.

5. The first number in an ordered pair such as (8, 3) is called the _____. The second number in such an ordered pair is called the _____.

6. The ordered pair $(4, 19)$ is a/an _____ of the equation $y = x^2 + 3$ because when 4 is substituted for x and 19 is substituted for y, we obtain a true statement. We also say that $(4, 19)$ _____ the equation.

7. The x-coordinate of a point where a graph crosses the x-axis is called a/an _____. The y-coordinate of such a point is always _____.

8. The y-coordinate of a point where a graph crosses the y-axis is called a/an _____. The x-coordinate of such a point is always _____.

EXERCISE SET 1.1

Practice Exercises

In Exercises 1–12, plot the given point in a rectangular coordinate system.

1. $(1, 4)$ **2.** $(2, 5)$ **3.** $(-2, 3)$
4. $(-1, 4)$ **5.** $(-3, -5)$ **6.** $(-4, -2)$
7. $(4, -1)$ **8.** $(3, -2)$ **9.** $(-4, 0)$
10. $(0, -3)$ **11.** $\left(\frac{7}{2}, -\frac{3}{2}\right)$ **12.** $\left(-\frac{5}{2}, \frac{3}{2}\right)$

Graph each equation in Exercises 13–28. Let $x = -3, -2, -1, 0, 1, 2,$ and 3.

13. $y = x^2 - 2$ **14.** $y = x^2 + 2$ **15.** $y = x - 2$
16. $y = x + 2$ **17.** $y = 2x + 1$ **18.** $y = 2x - 4$
19. $y = -\frac{1}{2}x$ **20.** $y = -\frac{1}{2}x + 2$ **21.** $y = 2|x|$
22. $y = -2|x|$ **23.** $y = |x| + 1$ **24.** $y = |x| - 1$
25. $y = 9 - x^2$ **26.** $y = -x^2$ **27.** $y = x^3$
28. $y = x^3 - 1$

In Exercises 29–32, match the viewing rectangle with the correct figure. Then label the tick marks in the figure to illustrate this viewing rectangle.

29. $[-5, 5, 1]$ by $[-5, 5, 1]$ **30.** $[-10, 10, 2]$ by $[-4, 4, 2]$
31. $[-20, 80, 10]$ by $[-30, 70, 10]$
32. $[-40, 40, 20]$ by $[-1000, 1000, 100]$

a. **b.**

c. **d.**

The table of values was generated by a graphing utility with a TABLE feature. Use the table to solve Exercises 33–40.

X	Y1	Y2
-3	9	5
-2	4	4
-1	1	3
0	0	2
1	1	1
2	4	0
3	9	-1

X=-3

33. Which equation corresponds to Y_2 in the table?
a. $y_2 = x + 8$ **b.** $y_2 = x - 2$
c. $y_2 = 2 - x$ **d.** $y_2 = 1 - 2x$

34. Which equation corresponds to Y_1 in the table?
a. $y_1 = -3x$ **b.** $y_1 = x^2$
c. $y_1 = -x^2$ **d.** $y_1 = 2 - x$

35. Does the graph of Y_2 pass through the origin?

36. Does the graph of Y_1 pass through the origin?

37. At which point does the graph of Y_2 cross the x-axis?

38. At which point does the graph of Y_2 cross the y-axis?

39. At which points do the graphs of Y_1 and Y_2 intersect?

40. For which values of x is $Y_1 = Y_2$?

*In Exercises 41–46, use the graph to **a.** determine the x-intercepts, if any; **b.** determine the y-intercepts, if any. For each graph, tick marks along the axes represent one unit each.*

41. **42.**

43. **44.**

45. **46.**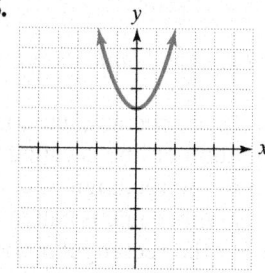

Practice Plus

In Exercises 47–50, write each English sentence as an equation in two variables. Then graph the equation.

47. The *y*-value is four more than twice the *x*-value.

48. The *y*-value is the difference between four and twice the *x*-value.

49. The *y*-value is three decreased by the square of the *x*-value.

50. The *y*-value is two more than the square of the *x*-value.

In Exercises 51–54, graph each equation.

51. $y = 5$ (Let $x = -3, -2, -1, 0, 1, 2,$ and 3.)

52. $y = -1$ (Let $x = -3, -2, -1, 0, 1, 2,$ and 3.)

53. $y = \dfrac{1}{x}$ (Let $x = -2, -1, -\dfrac{1}{2}, -\dfrac{1}{3}, \dfrac{1}{3}, \dfrac{1}{2}, 1,$ and 2.)

54. $y = -\dfrac{1}{x}$ (Let $x = -2, -1, -\dfrac{1}{2}, -\dfrac{1}{3}, \dfrac{1}{3}, \dfrac{1}{2}, 1,$ and 2.)

Application Exercises

The graphs show the percentage of high school seniors who used alcohol or marijuana during the 30 days prior to being surveyed for the University of Michigan's Monitoring the Future *study.*

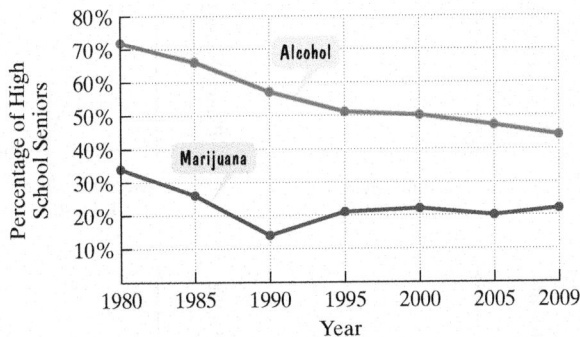

Alcohol and Marijuana Use by United States High School Seniors

Source: U.S. Department of Health and Human Services

The data can be described by the following mathematical models:

Percentage of seniors using alcohol $A = -0.9n + 69$

Number of years after 1980

Percentage of seniors using marijuana $M = -0.3n + 27$

Use this information to solve Exercises 55–56.

55. a. Use the appropriate line graph to determine the percentage of seniors who used marijuana in 2005.

b. Use the appropriate formula to determine the percentage of seniors who used marijuana in 2005. Does the formula underestimate or overestimate the actual percentage displayed by the graph? By how much?

c. Use the appropriate line graph to estimate the percentage of seniors who used alcohol in 2005.

d. Use the appropriate formula to determine the percentage of seniors who used alcohol in 2005. How does this compare with your estimate in part (c)?

e. For the period from 1980 through 2009, in which year was marijuana use by seniors at a minimum? Estimate the percentage of seniors who used marijuana in that year.

56. a. Use the appropriate line graph to determine the percentage of seniors who used alcohol in 2000.

b. Use the appropriate formula to determine the percentage of seniors who used alcohol in 2000. Does the formula underestimate or overestimate the actual percentage displayed by the graph? By how much?

c. Use the appropriate line graph to estimate the percentage of seniors who used marijuana in 2000.

d. Use the appropriate formula to determine the percentage of seniors who used marijuana in 2000. How does this compare with your estimate in part (c)?

e. For the period from 1980 through 2009, in which year was alcohol use by seniors at a maximum? Estimate the percentage of seniors who used alcohol in that year.

Contrary to popular belief, older people do not need less sleep than younger adults. However, the line graphs show that they awaken more often during the night. The numerous awakenings are one reason why some elderly individuals report that sleep is less restful than it had been in the past. Use the line graphs to solve Exercises 57–60.

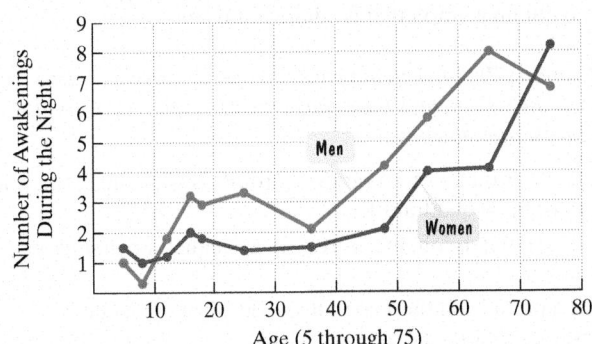

Average Number of Awakenings During the Night, by Age and Gender

Source: Stephen Davis and Joseph Palladino, *Psychology*, 5th Edition, Prentice Hall, 2007.

57. At which age, estimated to the nearest year, do women have the least number of awakenings during the night? What is the average number of awakenings at that age?

58. At which age do men have the greatest number of awakenings during the night? What is the average number of awakenings at that age?

59. Estimate, to the nearest tenth, the difference between the average number of awakenings during the night for 25-year-old men and 25-year-old women.

60. Estimate, to the nearest tenth, the difference between the average number of awakenings during the night for 18-year-old men and 18-year-old women.

Writing in Mathematics

61. What is the rectangular coordinate system?

62. Explain how to plot a point in the rectangular coordinate system. Give an example with your explanation.

63. Explain why $(5, -2)$ and $(-2, 5)$ do not represent the same point.

64. Explain how to graph an equation in the rectangular coordinate system.

65. What does a $[-20, 2, 1]$ by $[-4, 5, 0.5]$ viewing rectangle mean?

Technology Exercise

66. Use a graphing utility to verify each of your hand-drawn graphs in Exercises 13–28. Experiment with the settings for the viewing rectangle to make the graph displayed by the graphing utility resemble your hand-drawn graph as much as possible.

Critical Thinking Exercises

Make Sense? *In Exercises 67–70, determine whether each statement makes sense or does not make sense, and explain your reasoning.*

67. The rectangular coordinate system provides a geometric picture of what an equation in two variables looks like.

68. There is something wrong with my graphing utility because it is not displaying numbers along the *x*- and *y*-axes.

69. I used the ordered pairs $(-2, 2)$, $(0, 0)$, and $(2, 2)$ to graph a straight line.

70. I used the ordered pairs

(time of day, calories that I burned)

to obtain a graph that is a horizontal line.

In Exercises 71–74, determine whether each statement is true or false. If the statement is false, make the necessary change(s) to produce a true statement.

71. If the product of a point's coordinates is positive, the point must be in quadrant I.

72. If a point is on the *x*-axis, it is neither up nor down, so $x = 0$.

73. If a point is on the *y*-axis, its *x*-coordinate must be 0.

74. The ordered pair $(2, 5)$ satisfies $3y - 2x = -4$.

In Exercises 75–78, list the quadrant or quadrants satisfying each condition.

75. $xy > 0$

76. $\dfrac{y}{x} < 0$

77. $x^3 > 0$ and $y^3 < 0$

78. $x^3 < 0$ and $y^3 > 0$

In Exercises 79–82, match the story with the correct figure. The figures are labeled (a), (b), (c), and (d).

79. As the blizzard got worse, the snow fell harder and harder.

80. The snow fell more and more softly.

81. It snowed hard, but then it stopped. After a short time, the snow started falling softly.

82. It snowed softly, and then it stopped. After a short time, the snow started falling hard.

a.

b.

c.

d.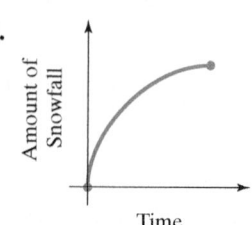

In Exercises 83–86, select the graph that best illustrates each story.

83. An airplane flew from Miami to San Francisco.

a.

b.

c.

d.

84. At noon, you begin to breathe in.

a.

b.

c.

d.

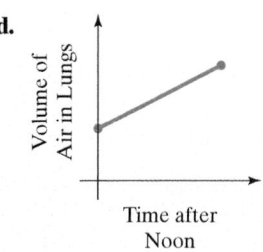

85. Measurements are taken of a person's height from birth to age 100.

a.

b.

c.

d.

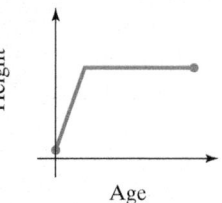

86. You begin your bike ride by riding down a hill. Then you ride up another hill. Finally, you ride along a level surface before coming to a stop.

a.

b.

c.

d.

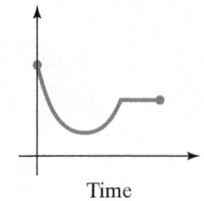

Preview Exercises

Exercises 87–89 will help you prepare for the material covered in the next section.

87. Here are two sets of ordered pairs:

$$\text{set 1: } \{(1, 5), (2, 5)\}$$
$$\text{set 2: } \{(5, 1), (5, 2)\}.$$

In which set is each x-coordinate paired with only one y-coordinate?

88. Graph $y = 2x$ and $y = 2x + 4$ in the same rectangular coordinate system. Select integers for x, starting with -2 and ending with 2.

89. Use the following graph to solve this exercise.

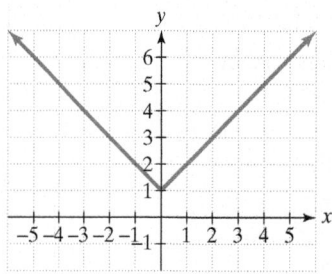

a. What is the y-coordinate when the x-coordinate is 2?

b. What are the x-coordinates when the y-coordinate is 4?

c. Describe the x-coordinates of all points on the graph.

d. Describe the y-coordinates of all points on the graph.

Basics of Functions and Their Graphs

Objectives

1. Find the domain and range of a relation.
2. Determine whether a relation is a function.
3. Determine whether an equation represents a function.
4. Evaluate a function.
5. Graph functions by plotting points.
6. Use the vertical line test to identify functions.
7. Obtain information about a function from its graph.
8. Identify the domain and range of a function from its graph.
9. Identify intercepts from a function's graph.

Magnified 6000 times, this color-scanned image shows a T-lymphocyte blood cell (green) infected with the HIV virus (red). Depletion of the number of T cells causes destruction of the immune system.

The average number of T cells in a person with HIV is a *function* of time after infection. In this section, you will be introduced to the basics of functions and their graphs. We will analyze the graph of a function using an example that illustrates the progression of HIV and T cell count. Much of our work in this course will be devoted to the important topic of functions and how they model your world.

Relations

Forbes magazine published a list of the highest-paid TV celebrities between June 2010 and June 2011. The results are shown in **Figure 1.13**.

1 Find the domain and range of a relation.

Highest Paid TV Celebrities between June 2010 and June 2011

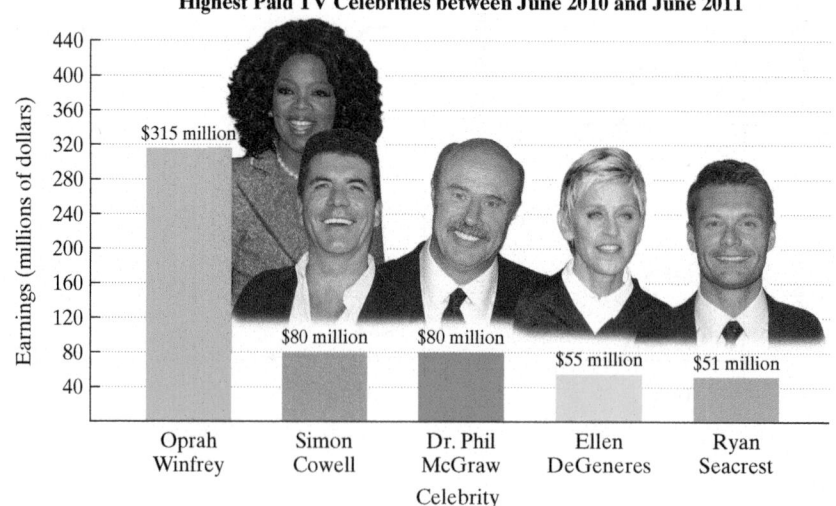

FIGURE 1.13
Source: *Forbes*

The graph indicates a correspondence between a TV celebrity and that person's earnings, in millions of dollars. We can write this correspondence using a set of ordered pairs:

{(Winfrey, 315), (Cowell, 80), (McGraw, 80), (DeGeneres, 55), (Seacrest, 51)}.

These braces indicate we are representing a set.

The mathematical term for a set of ordered pairs is a *relation*.

Definition of a Relation

A **relation** is any set of ordered pairs. The set of all first components of the ordered pairs is called the **domain** of the relation and the set of all second components is called the **range** of the relation.

EXAMPLE 1 Finding the Domain and Range of a Relation

Find the domain and range of the relation:

{(Winfrey, 315), (Cowell, 80), (McGraw, 80), (DeGeneres, 55), (Seacrest, 51)}.

SOLUTION

The domain is the set of all first components. Thus, the domain is

{Winfrey, Cowell, McGraw, DeGeneres, Seacrest}.

The range is the set of all second components. Thus, the range is

{315, 80, 55, 51}.

> Although Cowell and McGraw both earned $80 million, it is not necessary to list 80 twice.

• • •

☑ Check Point **1** Find the domain and range of the relation:
{(0, 9.1), (10, 6.7), (20, 10.7), (30, 13.2), (40, 21.2)}.

As you worked Check Point 1, did you wonder if there was a rule that assigned the "inputs" in the domain to the "outputs" in the range? For example, for the ordered pair (30, 13.2), how does the output 13.2 depend on the input 30? The ordered pair is based on the data in **Figure 1.14(a)**, which shows the percentage of first-year U.S. college students claiming no religious affiliation.

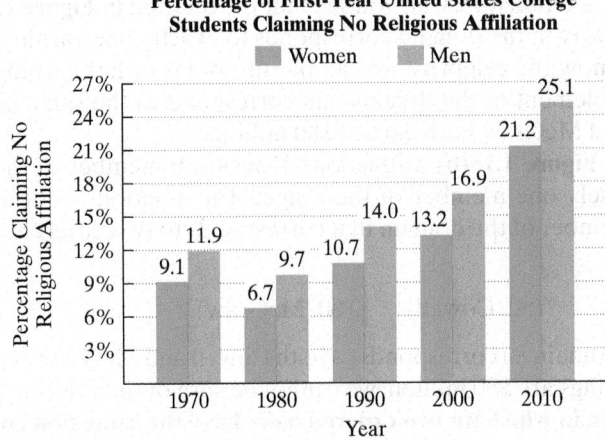

FIGURE 1.14(a) Data for women and men
Source: John Macionis, *Sociology,* Fourteenth Edition, Pearson, 2012.

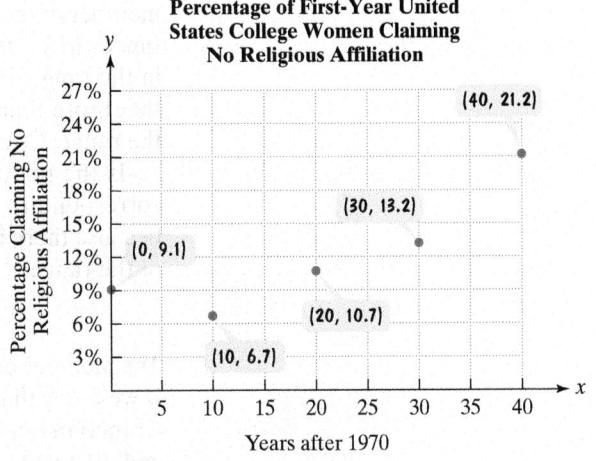

FIGURE 1.14(b) Visually representing the relation for the women's data

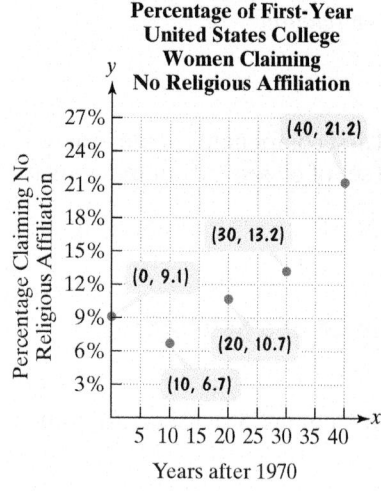

Percentage of First-Year United States College Women Claiming No Religious Affiliation

FIGURE 1.14(b) (repeated) Visually representing the relation for the women's data

In **Figure 1.14(b)**, we used the data for college women to create the following ordered pairs:

$$\left(\text{years after 1970,} \quad \begin{array}{c} \text{percentage of first-year college} \\ \text{women claiming no religious} \\ \text{affiliation} \end{array} \right).$$

Consider, for example, the ordered pair $(30, 13.2)$.

$$(30, 13.2)$$

30 years after 1970, or in 2000, 13.2% of first-year college women claimed no religious affiliation.

The five points in **Figure 1.14(b)** visually represent the relation formed from the women's data. Another way to visually represent the relation is as follows:

Domain Range

② Determine whether a relation is a function.

Functions

Table 1.1, based on our earlier discussion, shows the highest-paid TV celebrities and their earnings between June 2010 and June 2011, in millions of dollars. We've used this information to define two relations.

Figure 1.15(a) shows a correspondence between celebrities and their earnings. **Figure 1.15(b)** shows a correspondence between earnings and celebrities.

Table 1.1 Highest-Paid TV Celebrities

Celebrity	Earnings (millions of dollars)
Winfrey	315
Cowell	80
McGraw	80
DeGeneres	55
Seacrest	51

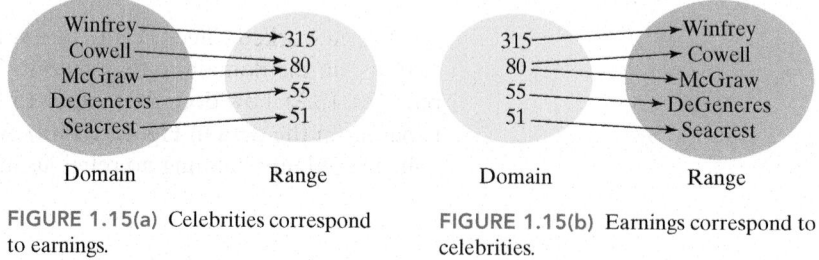

FIGURE 1.15(a) Celebrities correspond to earnings.

FIGURE 1.15(b) Earnings correspond to celebrities.

A relation in which each member of the domain corresponds to exactly one member of the range is a **function**. Can you see that the relation in **Figure 1.15(a)** is a function? Each celebrity in the domain corresponds to exactly one earnings amount in the range: If we know the celebrity, we can be sure of his or her earnings. Notice that more than one element in the domain can correspond to the same element in the range: Cowell and McGraw both earned $80 million.

Is the relation in **Figure 1.15(b)** a function? Does each member of the domain correspond to precisely one member of the range? This relation is not a function because there is a member of the domain that corresponds to two different members of the range:

$$(80, \text{Cowell}) \qquad (80, \text{McGraw}).$$

The member of the domain 80 corresponds to both Cowell and McGraw in the range. If we know that earnings are $80 million, we cannot be sure of the celebrity. Because **a function is a relation in which no two ordered pairs have the same first component and different second components**, the ordered pairs (80, Cowell) and (80, McGraw) are not ordered pairs of a function.

Same first component

(80, Cowell) (80, McGraw)

Different second components

Definition of a Function

A **function** is a correspondence from a first set, called the **domain**, to a second set, called the **range**, such that each element in the domain corresponds to *exactly one* element in the range.

In Check Point 1, we considered a relation that gave a correspondence between years after 1970 and the percentage of first-year college women claiming no religious affiliation. Can you see that this relation is a function?

Each element in the domain

$$\{(0, 9.1), (10, 6.7), (20, 10.7), (30, 13.2), (40, 21.2)\}$$

corresponds to exactly one element in the range.

However, Example 2 illustrates that not every correspondence between sets is a function.

EXAMPLE 2 Determining Whether a Relation Is a Function

Determine whether each relation is a function:

 a. $\{(1, 6), (2, 6), (3, 8), (4, 9)\}$ **b.** $\{(6, 1), (6, 2), (8, 3), (9, 4)\}$.

SOLUTION

We begin by making a figure for each relation that shows the domain and the range (**Figure 1.16**).

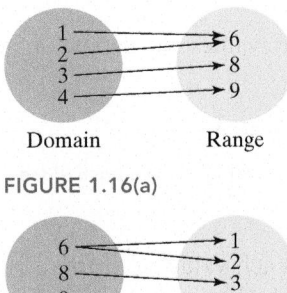

Domain Range

FIGURE 1.16(a)

Domain Range

FIGURE 1.16(b)

 a. **Figure 1.16(a)** shows that every element in the domain corresponds to exactly one element in the range. The element 1 in the domain corresponds to the element 6 in the range. Furthermore, 2 corresponds to 6, 3 corresponds to 8, and 4 corresponds to 9. No two ordered pairs in the given relation have the same first component and different second components. Thus, the relation is a function.

 b. **Figure 1.16(b)** shows that 6 corresponds to both 1 and 2. If any element in the domain corresponds to more than one element in the range, the relation is not a function. This relation is not a function; two ordered pairs have the same first component and different second components.

Same first component

(6, 1) (6, 2)

Different second components

 ● ● ●

 Look at **Figure 1.16(a)** again. The fact that 1 and 2 in the domain correspond to the same number, 6, in the range does not violate the definition of a function. **A function can have two different first components with the same second component.** By contrast, a relation is not a function when two different ordered pairs have the same first component and different second components. Thus, the relation in **Figure 1.16(b)** is not a function.

GREAT QUESTION!

If I reverse a function's components, will this new relation be a function?

If a relation is a function, reversing the components in each of its ordered pairs may result in a relation that is not a function.

⊘ Check Point **2** Determine whether each relation is a function:

 a. $\{(1, 2), (3, 4), (5, 6), (5, 8)\}$ **b.** $\{(1, 2), (3, 4), (6, 5), (8, 5)\}$.

3 Determine whether an equation represents a function.

Functions as Equations

Functions are usually given in terms of equations rather than as sets of ordered pairs. For example, here is an equation that models the percentage of first-year college women claiming no religious affiliation as a function of time:

$$y = 0.014x^2 - 0.24x + 8.8.$$

The variable x represents the number of years after 1970. The variable y represents the percentage of first-year college women claiming no religious affiliation. The variable y is a function of the variable x. For each value of x, there is one and only one value of y. The variable x is called the **independent variable** because it can be assigned any value from the domain. Thus, x can be assigned any nonnegative integer representing the number of years after 1970. The variable y is called the **dependent variable** because its value depends on x. The percentage claiming no religious affiliation depends on the number of years after 1970. The value of the dependent variable, y, is calculated after selecting a value for the independent variable, x.

We have seen that not every set of ordered pairs defines a function. Similarly, not all equations with the variables x and y define functions. If an equation is solved for y and more than one value of y can be obtained for a given x, then the equation does not define y as a function of x.

EXAMPLE 3 Determining Whether an Equation Represents a Function

Determine whether each equation defines y as a function of x:

 a. $x^2 + y = 4$ **b.** $x^2 + y^2 = 4$.

SOLUTION

Solve each equation for y in terms of x. If two or more values of y can be obtained for a given x, the equation is not a function.

 a. $x^2 + y = 4$ This is the given equation.

 $x^2 + y - x^2 = 4 - x^2$ Solve for y by subtracting x^2 from both sides.

 $y = 4 - x^2$ Simplify.

From this last equation we can see that for each value of x, there is one and only one value of y. For example, if $x = 1$, then $y = 4 - 1^2 = 3$. The equation defines y as a function of x.

 b. $x^2 + y^2 = 4$ This is the given equation.

 $x^2 + y^2 - x^2 = 4 - x^2$ Isolate y^2 by subtracting x^2 from both sides.

 $y^2 = 4 - x^2$ Simplify.

 $y = \pm\sqrt{4 - x^2}$ Apply the square root property: If $u^2 = d$, then $u = \pm\sqrt{d}$.

The \pm in this last equation shows that for certain values of x (all values between -2 and 2), there are two values of y. For example, if $x = 1$, then $y = \pm\sqrt{4 - 1^2} = \pm\sqrt{3}$. For this reason, the equation does not define y as a function of x. • • •

⊘ Check Point **3** Solve each equation for *y* and then determine whether the equation defines *y* as a function of *x*:

a. $2x + y = 6$ b. $x^2 + y^2 = 1$.

④ Evaluate a function.

Input *x*

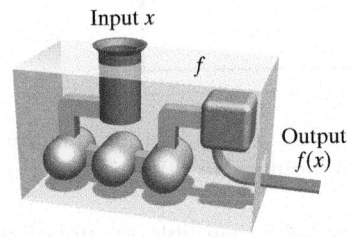

FIGURE 1.17 A "function machine" with inputs and outputs

Output $f(x)$

Function Notation

If an equation in *x* and *y* gives one and only one value of *y* for each value of *x*, then the variable *y* is a function of the variable *x*. When an equation represents a function, the function is often named by a letter such as *f, g, h, F, G,* or *H*. Any letter can be used to name a function. Suppose that *f* names a function. Think of the domain as the set of the function's inputs and the range as the set of the function's outputs. As shown in **Figure 1.17**, input is represented by *x* and the output by $f(x)$. The special notation $f(x)$, read "*f* of *x*" or "*f* at *x*," represents the **value of the function at the number *x***.

Let's make this clearer by considering a specific example. We know that the equation

$$y = 0.014x^2 - 0.24x + 8.8$$

defines *y* as a function of *x*. We'll name the function *f*. Now, we can apply our new function notation.

We read this equation as "*f* of *x* equals $0.014x^2 - 0.24x + 8.8$."

Input	Output	Equation
x	$f(x)$	$f(x) = 0.014x^2 - 0.24x + 8.8$

GREAT QUESTION!

Doesn't $f(x)$ indicate that I need to multiply *f* and *x*?

The notation $f(x)$ does *not* mean "*f* times *x*." The notation describes the value of the function at *x*.

Suppose we are interested in finding $f(30)$, the function's output when the input is 30. To find the value of the function at 30, we substitute 30 for *x*. We are **evaluating the function** at 30.

$f(x) = 0.014x^2 - 0.24x + 8.8$	This is the given function.
$f(30) = 0.014(30)^2 - 0.24(30) + 8.8$	Replace each occurrence of x with 30.
$= 0.014(900) - 0.24(30) + 8.8$	Evaluate the exponential expression: $30^2 = 30 \cdot 30 = 900$.
$= 12.6 - 7.2 + 8.8$	Perform the multiplications.
$f(30) = 14.2$	Subtract and add from left to right.

The statement $f(30) = 14.2$, read "*f* of 30 equals 14.2," tells us that the value of the function at 30 is 14.2. When the function's input is 30, its output is 14.2. **Figure 1.18** illustrates the input and output in terms of a function machine.

$f(x) = 0.014x^2 - 0.24x + 8.8$

Input $x = 30$

$0.014(30)^2 - 0.24(30) + 8.8$

Output $f(30) = 14.2$

FIGURE 1.18 A function machine at work

$$f(30) = 14.2$$

| 30 years after 1970, or in 2000, | 14.2% of first-year college women claimed no religious affiliation. |

We have seen that in 2000, 13.2% actually claimed nonaffiliation, so our function that models the data slightly overestimates the percent for 2000.

TECHNOLOGY

Graphing utilities can be used to evaluate functions. The screens below show the evaluation of

$$f(x) = 0.014x^2 - 0.24x + 8.8$$

at 30 on a TI-84 Plus graphing calculator. The function f is named Y_1.

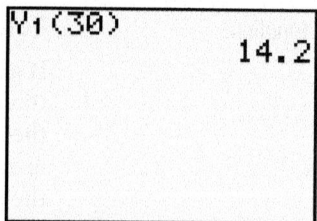

We used $f(x) = 0.014x^2 - 0.24x + 8.8$ to find $f(30)$. To find other function values, such as $f(40)$ or $f(55)$, substitute the specified input value, 40 or 55, for x in the function's equation.

If a function is named f and x represents the independent variable, the notation $f(x)$ corresponds to the y-value for a given x. Thus,

$$f(x) = 0.014x^2 - 0.24x + 8.8 \quad \text{and} \quad y = 0.014x^2 - 0.24x + 8.8$$

define the same function. This function may be written as

$$y = f(x) = 0.014x^2 - 0.24x + 8.8.$$

EXAMPLE 4 Evaluating a Function

If $f(x) = x^2 + 3x + 5$, evaluate each of the following:

 a. $f(2)$ **b.** $f(x + 3)$ **c.** $f(-x)$.

SOLUTION

We substitute 2, $x + 3$, and $-x$ for x in the equation for f. When replacing x with a variable or an algebraic expression, you might find it helpful to think of the function's equation as

$$f(\boxed{}) = \boxed{}^2 + 3\boxed{} + 5.$$

a. We find $f(2)$ by substituting 2 for x in the equation.

$$f(2) = 2^2 + 3 \cdot 2 + 5 = 4 + 6 + 5 = 15$$

Thus, $f(2) = 15$.

b. We find $f(x + 3)$ by substituting $x + 3$ for x in the equation.

$$f(x + 3) = (x + 3)^2 + 3(x + 3) + 5$$

Equivalently,

$$f(x + 3) = (x + 3)^2 + 3(x + 3) + 5$$

$$= x^2 + 6x + 9 + 3x + 9 + 5 \quad \text{Square } x + 3 \text{ using}$$
$$(A + B)^2 = A^2 + 2AB + B^2.$$
$$\text{Distribute 3 throughout the parentheses.}$$

$$= x^2 + 9x + 23. \quad \text{Combine like terms.}$$

c. We find $f(-x)$ by substituting $-x$ for x in the equation.

$$f(-x) = (-x)^2 + 3(-x) + 5$$

Equivalently,

$$f(-x) = (-x)^2 + 3(-x) + 5$$
$$= x^2 - 3x + 5.$$

DISCOVERY

Using $f(x) = x^2 + 3x + 5$ and the answers in parts (b) and (c):

1. Is $f(x + 3)$ equal to $f(x) + f(3)$?

2. Is $f(-x)$ equal to $-f(x)$?

• • •

✓ Check Point **4** If $f(x) = x^2 - 2x + 7$, evaluate each of the following:

a. $f(-5)$ **b.** $f(x + 4)$ **c.** $f(-x)$.

⑤ Graph functions by plotting points.

Graphs of Functions

The **graph of a function** is the graph of its ordered pairs. For example, the graph of $f(x) = 2x$ is the set of points (x, y) in the rectangular coordinate system satisfying $y = 2x$. Similarly, the graph of $g(x) = 2x + 4$ is the set of points (x, y) in the rectangular coordinate system satisfying the equation $y = 2x + 4$. In the next example, we graph both of these functions in the same rectangular coordinate system.

EXAMPLE 5 Graphing Functions

Graph the functions $f(x) = 2x$ and $g(x) = 2x + 4$ in the same rectangular coordinate system. Select integers for x, starting with -2 and ending with 2.

SOLUTION

We begin by setting up a partial table of coordinates for each function. Then we plot the five points in each table and connect them, as shown in **Figure 1.19**. The graph of each function is a straight line. Do you see a relationship between the two graphs? The graph of g is the graph of f shifted vertically up by 4 units.

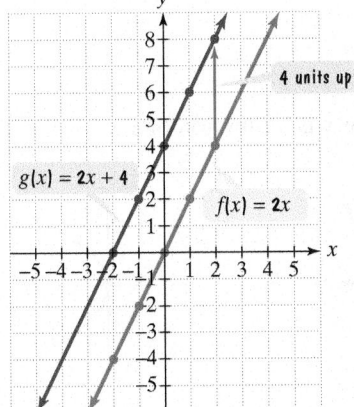

FIGURE 1.19

x	$f(x) = 2x$	(x, y) or $(x, f(x))$	x	$g(x) = 2x + 4$	(x, y) or $(x, g(x))$
-2	$f(-2) = 2(-2) = -4$	$(-2, -4)$	-2	$g(-2) = 2(-2) + 4 = 0$	$(-2, 0)$
-1	$f(-1) = 2(-1) = -2$	$(-1, -2)$	-1	$g(-1) = 2(-1) + 4 = 2$	$(-1, 2)$
0	$f(0) = 2 \cdot 0 = 0$	$(0, 0)$	0	$g(0) = 2 \cdot 0 + 4 = 4$	$(0, 4)$
1	$f(1) = 2 \cdot 1 = 2$	$(1, 2)$	1	$g(1) = 2 \cdot 1 + 4 = 6$	$(1, 6)$
2	$f(2) = 2 \cdot 2 = 4$	$(2, 4)$	2	$g(2) = 2 \cdot 2 + 4 = 8$	$(2, 8)$

Choose x. Compute $f(x)$ by evaluating f at x. Form the ordered pair. Choose x. Compute $g(x)$ by evaluating g at x. Form the ordered pair.

• • •

The graphs in Example 5 are straight lines. All functions with equations of the form $f(x) = mx + b$ graph as straight lines. Such functions, called **linear functions**, will be discussed in detail in Section 1.4.

TECHNOLOGY

We can use a graphing utility to check the tables and the graphs in Example 5 for the functions

$$f(x) = 2x \quad \text{and} \quad g(x) = 2x + 4.$$

Enter $y_1 = 2x$ in the $\boxed{y=}$ screen. Enter $y_2 = 2x + 4$ in the $\boxed{y=}$ screen.

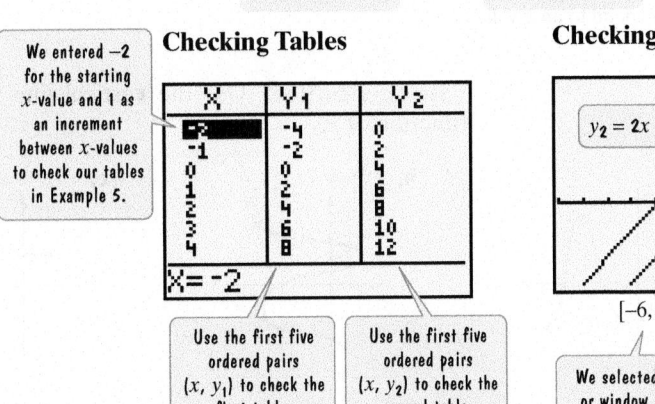

We entered -2 for the starting x-value and 1 as an increment between x-values to check our tables in Example 5.

Checking Tables

Checking Graphs

$[-6, 6, 1]$ by $[-6, 9, 1]$

Use the first five ordered pairs (x, y_1) to check the first table. Use the first five ordered pairs (x, y_2) to check the second table.

We selected this viewing rectangle, or window, to match **Figure 1.19**.

⊘ Check Point **5** Graph the functions $f(x) = 2x$ and $g(x) = 2x - 3$ in the same rectangular coordinate system. Select integers for x, starting with -2 and ending with 2. How is the graph of g related to the graph of f?

6 Use the vertical line test to identify functions.

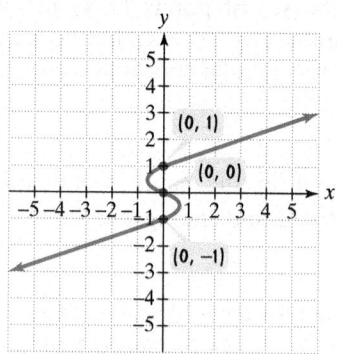

FIGURE 1.20 y is not a function of x because 0 is paired with three values of y, namely, 1, 0, and -1.

The Vertical Line Test

Not every graph in the rectangular coordinate system is the graph of a function. The definition of a function specifies that no value of x can be paired with two or more different values of y. Consequently, if a graph contains two or more different points with the same first coordinate, the graph cannot represent a function. This is illustrated in **Figure 1.20**. Observe that points sharing a common first coordinate are vertically above or below each other.

This observation is the basis of a useful test for determining whether a graph defines y as a function of x. The test is called the **vertical line test**.

> ### The Vertical Line Test for Functions
>
> If any vertical line intersects a graph in more than one point, the graph does not define y as a function of x.

EXAMPLE 6 Using the Vertical Line Test

Use the vertical line test to identify graphs in which y is a function of x.

a. b. c. d.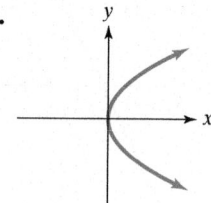

SOLUTION

y is a function of x for the graphs in (b) and (c).

a. b. c. d.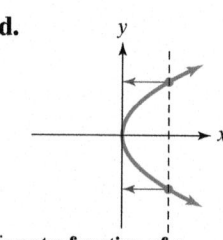

y is **not a function** of x.
Two values of y
correspond to an x-value.

y is **a function** of x.

y is **a function** of x.

y is **not a function** of x.
Two values of y
correspond to an x-value.

• • •

⊘ Check Point **6** Use the vertical line test to identify graphs in which y is a function of x.

a. b. c. d.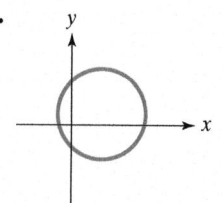

7 Obtain information about a function from its graph.

Obtaining Information from Graphs

You can obtain information about a function from its graph. At the right or left of a graph, you will find closed dots, open dots, or arrows.

- A closed dot indicates that the graph does not extend beyond this point and the point belongs to the graph.
- An open dot indicates that the graph does not extend beyond this point and the point does not belong to the graph.
- An arrow indicates that the graph extends indefinitely in the direction in which the arrow points.

EXAMPLE 7 Analyzing the Graph of a Function

The human immunodeficiency virus, or HIV, infects and kills helper T cells. Because T cells stimulate the immune system to produce antibodies, their destruction disables the body's defenses against other pathogens. By counting the number of T cells that remain active in the body, the progression of HIV can be monitored. The fewer helper T cells, the more advanced the disease. **Figure 1.21** shows a graph that is used to monitor the average progression of the disease. The average number of T cells, $f(x)$, is a function of time after infection, x.

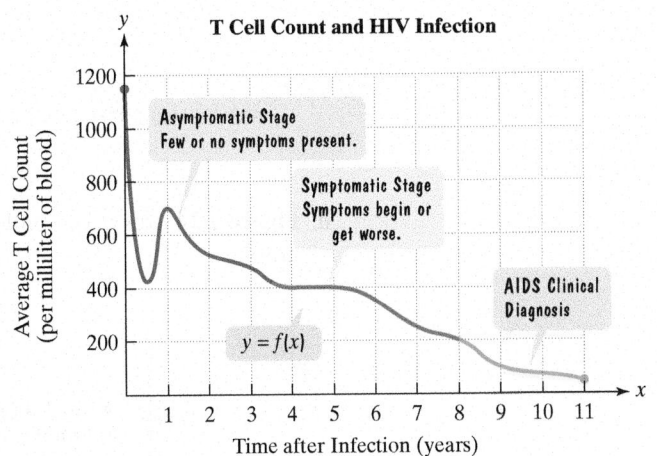

FIGURE 1.21
Source: B. E. Pruitt et al., *Human Sexuality,* Prentice Hall, 2007.

a. Explain why f represents the graph of a function.
b. Use the graph to find $f(8)$.
c. For what value of x is $f(x) = 350$?
d. Describe the general trend shown by the graph.

SOLUTION

a. No vertical line can be drawn that intersects the graph of f more than once. By the vertical line test, f represents the graph of a function.

b. To find $f(8)$, or f of 8, we locate 8 on the x-axis. **Figure 1.22** shows the point on the graph of f for which 8 is the first coordinate. From this point, we look to the y-axis to find the corresponding y-coordinate. We see that the y-coordinate is 200. Thus,

$$f(8) = 200.$$

When the time after infection is 8 years, the average T cell count is 200 cells per milliliter of blood. (AIDS clinical diagnosis is given at a T cell count of 200 or below.)

FIGURE 1.22 Finding $f(8)$

c. To find the value of x for which $f(x) = 350$, we find the approximate location of 350 on the y-axis. **Figure 1.23** shows that there is one point on the graph of f for which 350 is the second coordinate. From this point, we look to the x-axis to find the corresponding x-coordinate. We see that the x-coordinate is 6. Thus,

$$f(x) = 350 \text{ for } x = 6.$$

An average T cell count of 350 occurs 6 years after infection.

FIGURE 1.23 Finding x for which $f(x) = 350$

d. Figure 1.24 uses voice balloons to describe the general trend shown by the graph.

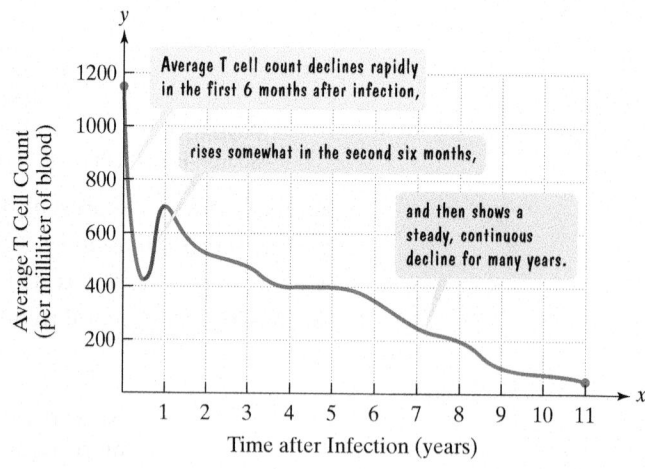

FIGURE 1.24 Describing changing T cell count over time in a person infected with HIV

● ● ●

✐ Check Point 7

a. Use the graph of f in **Figure 1.21** on page 127 to find $f(5)$.

b. For what value of x is $f(x) = 100$?

c. Estimate the minimum average T cell count during the asymptomatic stage.

⑧ Identify the domain and range of a function from its graph.

Identifying Domain and Range from a Function's Graph

GREAT QUESTION!

I peeked below and saw that you are using interval notation. What should I already know about this notation?

Recall that square brackets indicate endpoints that are included in an interval. Parentheses indicate endpoints that are not included in an interval. Parentheses are always used with ∞ or −∞. For more detail on interval notation, see Section P.1, pages 10–12.

Figure 1.25 illustrates how the graph of a function is used to determine the function's domain and its range.

Domain: set of inputs

Found on the *x*-axis

Range: set of outputs

Found on the *y*-axis

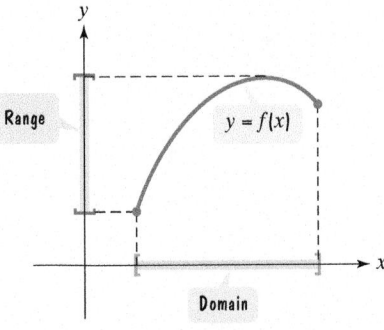

FIGURE 1.25 Domain and range of *f*

Let's apply these ideas to the graph of the function shown in **Figure 1.26**. To find the domain, look for all the inputs on the *x*-axis that correspond to points on the graph. Can you see that they extend from −4 to 2, inclusive? The function's domain can be represented as follows:

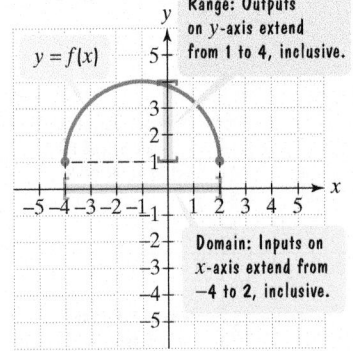

FIGURE 1.26 Domain and range of *f*

Using Set-Builder Notation

$$\{ x \mid -4 \leq x \leq 2 \}$$

The set of all *x* such that *x* is greater than or equal to −4 and less than or equal to 2.

Using Interval Notation

$$[-4, 2].$$

The square brackets indicate −4 and 2 are included. Note the square brackets on the *x*-axis in **Figure 1.26.**

To find the range, look for all the outputs on the *y*-axis that correspond to points on the graph. They extend from 1 to 4, inclusive. The function's range can be represented as follows:

Using Set-Builder Notation

$$\{ y \mid 1 \leq y \leq 4 \}$$

The set of all *y* such that *y* is greater than or equal to 1 and less than or equal to 4.

Using Interval Notation

$$[1, 4].$$

The square brackets indicate 1 and 4 are included. Note the square brackets on the *y*-axis in **Figure 1.26.**

EXAMPLE 8 Identifying the Domain and Range of a Function from Its Graph

Use the graph of each function to identify its domain and its range.

a.

b.

c.

d.

e.

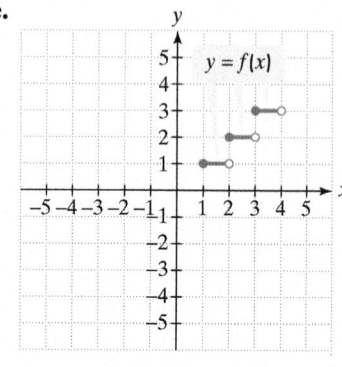

SOLUTION

For the graph of each function, the domain is highlighted in purple on the *x*-axis and the range is highlighted in green on the *y*-axis.

a.

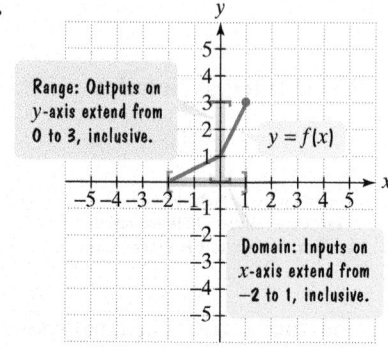

Range: Outputs on *y*-axis extend from 0 to 3, inclusive.

Domain: Inputs on *x*-axis extend from −2 to 1, inclusive.

Domain = {*x*|−2 ≤ *x* ≤ 1} or [−2, 1]
Range = {*y*|0 ≤ *y* ≤ 3} or [0, 3]

b.

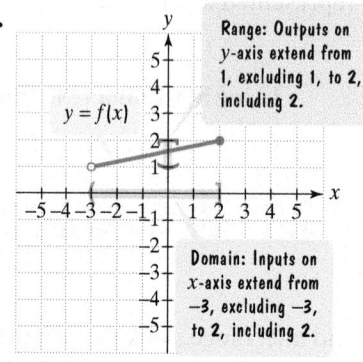

Range: Outputs on *y*-axis extend from 1, excluding 1, to 2, including 2.

Domain: Inputs on *x*-axis extend from −3, excluding −3, to 2, including 2.

Domain = {*x*|−3 < *x* ≤ 2} or (−3, 2]
Range = {*y*|1 < *y* ≤ 2} or (1, 2]

c.

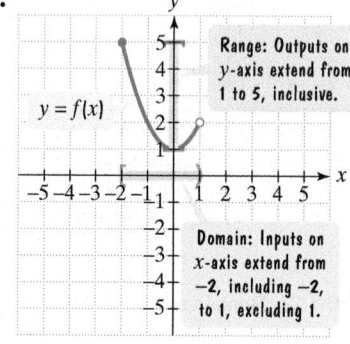

Range: Outputs on *y*-axis extend from 1 to 5, inclusive.

Domain: Inputs on *x*-axis extend from −2, including −2, to 1, excluding 1.

Domain = {*x*|−2 ≤ *x* < 1} or [−2, 1)
Range = {*y*|1 ≤ *y* ≤ 5} or [1, 5]

GREAT QUESTION!

The range in Example 8(e) was identified as {*y*|*y* = 1, 2, 3}. Why didn't you also use interval notation like you did in the other parts of Example 8?

Interval notation is not appropriate for describing a set of distinct numbers such as {1, 2, 3}. Interval notation, [1, 3], would mean that numbers such as 1.5 and 2.99 are in the range, but they are not. That's why we only used set-builder notation.

d.

Range: Outputs on *y*-axis include real numbers greater than or equal to 0.

Domain: Inputs on *x*-axis include real numbers less than or equal to 4.

Domain = {*x*|*x* ≤ 4} or (−∞, 4]
Range = {*y*|*y* ≥ 0} or [0, ∞)

e.

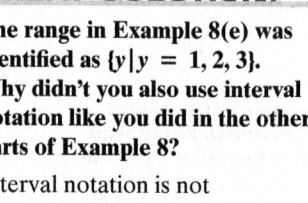

Range: Outputs on *y*-axis "step" from 1 to 2 to 3.

Domain: Inputs on *x*-axis extend from 1, including 1, to 4, excluding 4.

Domain = {*x*|1 ≤ *x* < 4} or [1, 4)
Range = {*y*|*y* = 1, 2, 3}

• • •

✓ Check Point **8** Use the graph of each function to identify its domain and its range.

a.

b.

c.

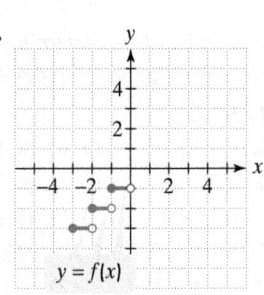

⑨ Identify intercepts from a function's graph.

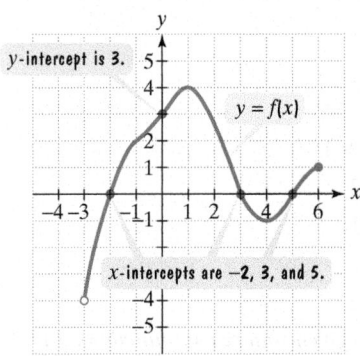

FIGURE 1.27 Identifying intercepts

Identifying Intercepts from a Function's Graph

Figure 1.27 illustrates how we can identify intercepts from a function's graph. To find the x-intercepts, look for the points at which the graph crosses the x-axis. There are three such points: $(-2, 0)$, $(3, 0)$, and $(5, 0)$. Thus, the x-intercepts are $-2, 3$, and 5. We express this in function notation by writing $f(-2) = 0, f(3) = 0$, and $f(5) = 0$. We say that $-2, 3$, and 5 are the *zeros of the function*. The **zeros of a function** f are the x-values for which $f(x) = 0$. Thus, the real zeros are the x-intercepts.

To find the y-intercept, look for the point at which the graph crosses the y-axis. This occurs at $(0, 3)$. Thus, the y-intercept is 3. We express this in function notation by writing $f(0) = 3$.

By the definition of a function, for each value of x we can have at most one value for y. What does this mean in terms of intercepts? **A function can have more than one x-intercept but at most one y-intercept.**

CONCEPT AND VOCABULARY CHECK

Fill in each blank so that the resulting statement is true.

1. Any set of ordered pairs is called a/an _____. The set of all first components of the ordered pairs is called the _____. The set of all second components of the ordered pairs is called the _____.

2. A set of ordered pairs in which each member of the set of first components corresponds to exactly one member of the set of second components is called a/an _____.

3. The notation $f(x)$ describes the value of _____ at _____.

4. True or false: $y = x^2 - 1$ defines y as a function of x. _____

5. True or false : $y = \pm\sqrt{x^2 - 1}$ defines y as a function of x. _____

6. If $f(x) = x^2 - 5x + 4$, we can find $f(x + 6)$ by replacing each occurrence of _____ by _____.

7. The graph of a function is the graph of its _____.

8. If any vertical line intersects a graph _____, the graph does not define y as a/an _____ of x.

9. The shaded set of numbers shown on the x-axis can be expressed in interval notation as _____. This set represents the function's _____.

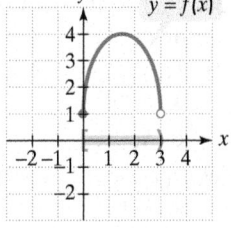

10. The shaded set of numbers shown on the y-axis can be expressed in interval notation as _____. This set represents the function's _____.

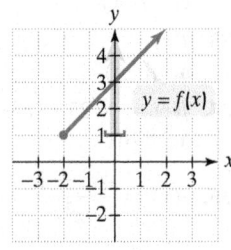

11. If the x-intercepts of a function are -1 and 3, then $f(-1) =$ ____ and $f(3) =$ ____. The x-intercepts, -1 and 3, are called the _____ of the function.

12. True or false: A function can have more than one y-intercept. _____

EXERCISE SET 1.2

Practice Exercises

In Exercises 1–10, determine whether each relation is a function. Give the domain and range for each relation.

1. $\{(1, 2), (3, 4), (5, 5)\}$ **2.** $\{(4, 5), (6, 7), (8, 8)\}$
3. $\{(3, 4), (3, 5), (4, 4), (4, 5)\}$
4. $\{(5, 6), (5, 7), (6, 6), (6, 7)\}$
5. $\{(3, -2), (5, -2), (7, 1), (4, 9)\}$
6. $\{(10, 4), (-2, 4), (-1, 1), (5, 6)\}$
7. $\{(-3, -3), (-2, -2), (-1, -1), (0, 0)\}$
8. $\{(-7, -7), (-5, -5), (-3, -3), (0, 0)\}$
9. $\{(1, 4), (1, 5), (1, 6)\}$
10. $\{(4, 1), (5, 1), (6, 1)\}$

In Exercises 11–26, determine whether each equation defines y as a function of x.

11. $x + y = 16$ **12.** $x + y = 25$
13. $x^2 + y = 16$ **14.** $x^2 + y = 25$
15. $x^2 + y^2 = 16$ **16.** $x^2 + y^2 = 25$
17. $x = y^2$ **18.** $4x = y^2$
19. $y = \sqrt{x + 4}$ **20.** $y = -\sqrt{x + 4}$
21. $x + y^3 = 8$ **22.** $x + y^3 = 27$
23. $xy + 2y = 1$ **24.** $xy - 5y = 1$
25. $|x| - y = 2$ **26.** $|x| - y = 5$

In Exercises 27–38, evaluate each function at the given values of the independent variable and simplify.

27. $f(x) = 4x + 5$
 a. $f(6)$ **b.** $f(x + 1)$ **c.** $f(-x)$
28. $f(x) = 3x + 7$
 a. $f(4)$ **b.** $f(x + 1)$ **c.** $f(-x)$
29. $g(x) = x^2 + 2x + 3$
 a. $g(-1)$ **b.** $g(x + 5)$ **c.** $g(-x)$
30. $g(x) = x^2 - 10x - 3$
 a. $g(-1)$ **b.** $g(x + 2)$ **c.** $g(-x)$
31. $h(x) = x^4 - x^2 + 1$
 a. $h(2)$ **b.** $h(-1)$
 c. $h(-x)$ **d.** $h(3a)$
32. $h(x) = x^3 - x + 1$
 a. $h(3)$ **b.** $h(-2)$
 c. $h(-x)$ **d.** $h(3a)$
33. $f(r) = \sqrt{r + 6} + 3$
 a. $f(-6)$ **b.** $f(10)$ **c.** $f(x - 6)$
34. $f(r) = \sqrt{25 - r} - 6$
 a. $f(16)$ **b.** $f(-24)$ **c.** $f(25 - 2x)$
35. $f(x) = \dfrac{4x^2 - 1}{x^2}$

 a. $f(2)$ **b.** $f(-2)$ **c.** $f(-x)$

36. $f(x) = \dfrac{4x^3 + 1}{x^3}$

 a. $f(2)$ **b.** $f(-2)$ **c.** $f(-x)$

37. $f(x) = \dfrac{x}{|x|}$
 a. $f(6)$ **b.** $f(-6)$ **c.** $f(r^2)$

38. $f(x) = \dfrac{|x + 3|}{x + 3}$
 a. $f(5)$ **b.** $f(-5)$ **c.** $f(-9 - x)$

In Exercises 39–50, graph the given functions, f and g, in the same rectangular coordinate system. Select integers for x, starting with −2 and ending with 2. Once you have obtained your graphs, describe how the graph of g is related to the graph of f.

39. $f(x) = x, g(x) = x + 3$
40. $f(x) = x, g(x) = x - 4$
41. $f(x) = -2x, g(x) = -2x - 1$
42. $f(x) = -2x, g(x) = -2x + 3$
43. $f(x) = x^2, g(x) = x^2 + 1$
44. $f(x) = x^2, g(x) = x^2 - 2$
45. $f(x) = |x|, g(x) = |x| - 2$
46. $f(x) = |x|, g(x) = |x| + 1$
47. $f(x) = x^3, g(x) = x^3 + 2$
48. $f(x) = x^3, g(x) = x^3 - 1$
49. $f(x) = 3, g(x) = 5$
50. $f(x) = -1, g(x) = 4$

In Exercises 51–54, graph the given square root functions, f and g, in the same rectangular coordinate system. Use the integer values of x given to the right of each function to obtain ordered pairs. Because only nonnegative numbers have square roots that are real numbers, be sure that each graph appears only for values of x that cause the expression under the radical sign to be greater than or equal to zero. Once you have obtained your graphs, describe how the graph of g is related to the graph of f.

51. $f(x) = \sqrt{x}$ $(x = 0, 1, 4, 9)$ and
 $g(x) = \sqrt{x} - 1$ $(x = 0, 1, 4, 9)$
52. $f(x) = \sqrt{x}$ $(x = 0, 1, 4, 9)$ and
 $g(x) = \sqrt{x} + 2$ $(x = 0, 1, 4, 9)$
53. $f(x) = \sqrt{x}$ $(x = 0, 1, 4, 9)$ and
 $g(x) = \sqrt{x - 1}$ $(x = 1, 2, 5, 10)$
54. $f(x) = \sqrt{x}$ $(x = 0, 1, 4, 9)$ and
 $g(x) = \sqrt{x + 2}$ $(x = -2, -1, 2, 7)$

In Exercises 55–64, use the vertical line test to identify graphs in which y is a function of x.

55.

56.

57.

58.

59.

60.

61.

62.

63.

64.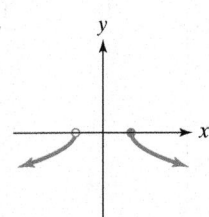

In Exercises 65–70, use the graph of f to find each indicated function value.

65. $f(-2)$

66. $f(2)$

67. $f(4)$

68. $f(-4)$

69. $f(-3)$

70. $f(-1)$

Use the graph of g to solve Exercises 71–76.

71. Find $g(-4)$.

72. Find $g(2)$.

73. Find $g(-10)$.

74. Find $g(10)$.

75. For what value of x is $g(x) = 1$?

76. For what value of x is $g(x) = -1$?

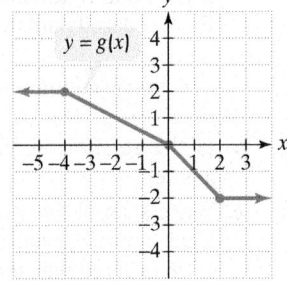

In Exercises 77–92, use the graph to determine **a.** *the function's domain;* **b.** *the function's range;* **c.** *the x-intercepts, if any;* **d.** *the y-intercept, if any; and* **e.** *the missing function values, indicated by question marks, below each graph.*

77.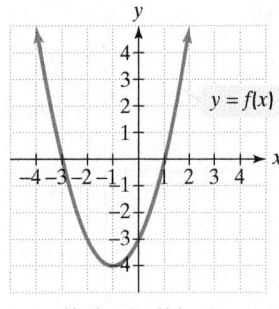

$f(-2) = ?$ $f(2) = ?$

78.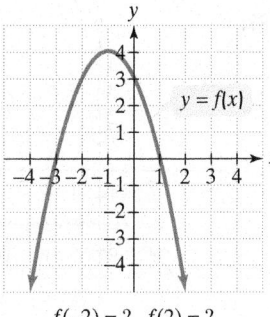

$f(-2) = ?$ $f(2) = ?$

79.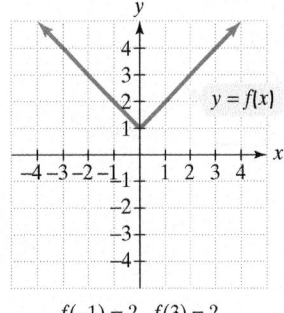

$f(-1) = ?$ $f(3) = ?$

80.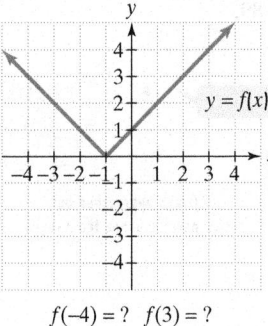

$f(-4) = ?$ $f(3) = ?$

81.

$f(3) = ?$

82.

$f(-5) = ?$

83.

$f(4) = ?$

84.

$f(3) = ?$

85.

$f(-1) = ?$

86.

$f(-2) = ?$

87.

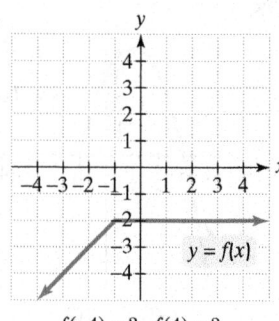

$f(-4) = ?$ $f(4) = ?$

88.

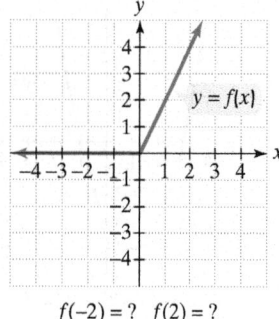

$f(-2) = ?$ $f(2) = ?$

89.

$f(4) = ?$

90.

$f(2) = ?$

91.

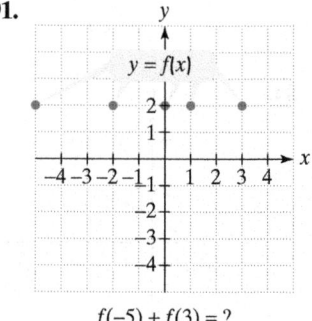

$f(-5) + f(3) = ?$

92.

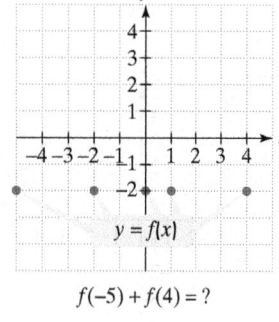

$f(-5) + f(4) = ?$

Practice Plus

In Exercises 93–94, let $f(x) = x^2 - x + 4$ and $g(x) = 3x - 5$.

93. Find $g(1)$ and $f(g(1))$. **94.** Find $g(-1)$ and $f(g(-1))$.

In Exercises 95–96, let f and g be defined by the following table:

x	f(x)	g(x)
−2	6	0
−1	3	4
0	−1	1
1	−4	−3
2	0	−6

95. Find $\sqrt{f(-1) - f(0)} - [g(2)]^2 + f(-2) \div g(2) \cdot g(-1)$.

96. Find $|f(1) - f(0)| - [g(1)]^2 + g(1) \div f(-1) \cdot g(2)$.

In Exercises 97–98, find $f(-x) - f(x)$ for the given function f. Then simplify the expression.

97. $f(x) = x^3 + x - 5$

98. $f(x) = x^2 - 3x + 7$

Application Exercises

The bar graph shows minimum legal ages for sex and marriage in five selected countries. Use this information to solve Exercises 99–100. (We did not include data for the United States because the legal age of sexual consent varies according to state law. Furthermore, women are allowed to marry younger than men: 16 for women and 18 for men.)

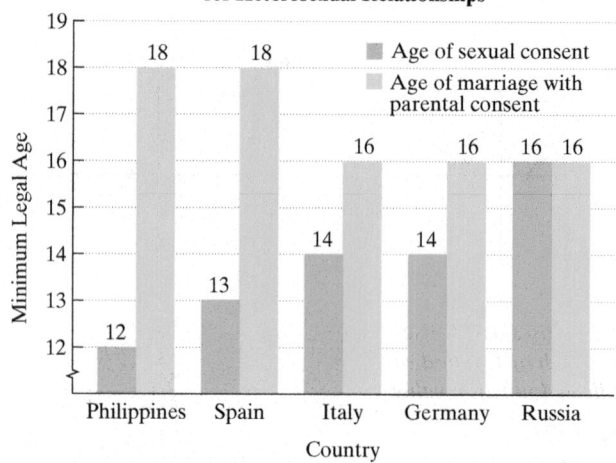

Minimum Legal Ages for Sex and Marriage, for Heterosexual Relationships

Source: Mitchell Beazley, *Snapshot: The Visual Almanac for Our World Today.* Octopus Publishing, 2009

99. a. Write a set of five ordered pairs in which countries correspond to the minimum legal age of sexual consent. Each ordered pair should be in the form

(country, minimum legal age of sexual consent).

b. Is the relation in part (a) a function? Explain your answer.

c. Write a set of five ordered pairs in which the minimum legal age of sexual consent corresponds to a country. Each ordered pair should be in the form

(minimum legal age of sexual consent, country).

d. Is the relation in part (c) a function?

100. a. Write a set of five ordered pairs in which countries correspond to the minimum legal age of marriage with parental consent. Each ordered pair should be in the form

(country, minimum legal age of marriage with parental consent).

b. Is the relation in part (a) a function? Explain your answer.

c. Write a set of five ordered pairs in which the minimum legal age of marriage with parental consent corresponds to a country. Each ordered pair should be in the form

(minimum legal age of marriage with parental consent, country).

d. Is the relation in part (c) a function?

The bar graph shows your chances of surviving to various ages once you reach 60.

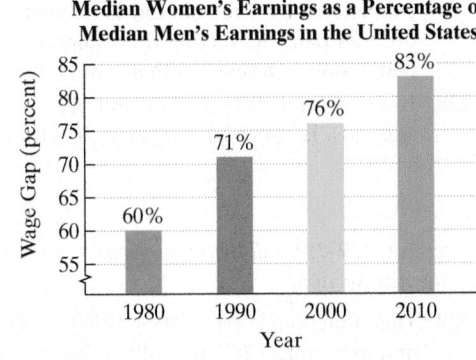

Chances of 60-Year-Olds Surviving to Various Ages

Source: National Center for Health Statistics

The functions

$$f(x) = -2.9x + 286$$

$$\text{and} \quad g(x) = 0.01x^2 - 4.9x + 370$$

model the chance, as a percent, that a 60-year-old will survive to age x. Use this information to solve Exercises 101–102.

101. a. Find and interpret $f(70)$.

b. Find and interpret $g(70)$.

c. Which function serves as a better model for the chance of surviving to age 70?

102. a. Find and interpret $f(90)$.

b. Find and interpret $g(90)$.

c. Which function serves as a better model for the chance of surviving to age 90?

The wage gap is used to compare the status of women's earnings relative to men's. The wage gap is expressed as a percent and is calculated by dividing the median, or middlemost, annual earnings for women by the median annual earnings for men. The bar graph shows the wage gap for selected years from 1980 through 2010.

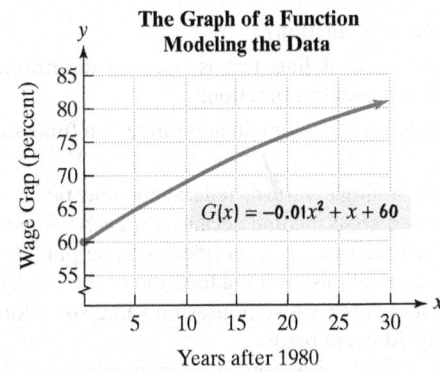

Median Women's Earnings as a Percentage of Median Men's Earnings in the United States

Source: Bureau of Labor Statistics

The Graph of a Function Modeling the Data

$G(x) = -0.01x^2 + x + 60$

The function $G(x) = -0.01x^2 + x + 60$ models the wage gap, as a percent, x years after 1980. The graph of function G is shown to the right of the actual data. Use this information to solve Exercises 103–104.

103. a. Find and interpret $G(30)$. Identify this information as a point on the graph of the function.

b. Does $G(30)$ overestimate or underestimate the actual data shown by the bar graph? By how much?

104. a. Find and interpret $G(10)$. Identify this information as a point on the graph of the function.

b. Does $G(10)$ overestimate or underestimate the actual data shown by the bar graph? By how much?

In Exercises 105–108, you will be developing functions that model given conditions.

105. A company that manufactures bicycles has a fixed cost of $100,000. It costs $100 to produce each bicycle. The total cost for the company is the sum of its fixed cost and variable costs. Write the total cost, C, as a function of the number of bicycles produced, x. Then find and interpret $C(90)$.

106. A car was purchased for $22,500. The value of the car decreased by $3200 per year for the first six years. Write a function that describes the value of the car, V, after x years, where $0 \le x \le 6$. Then find and interpret $V(3)$.

107. You commute to work a distance of 40 miles and return on the same route at the end of the day. Your average rate on the return trip is 30 miles per hour faster than your average rate on the outgoing trip. Write the total time, T, in hours, devoted to your outgoing and return trips as a function of your rate on the outgoing trip, x. Then find and interpret $T(30)$. Hint:

$$\text{Time traveled} = \frac{\text{Distance traveled}}{\text{Rate of travel}}.$$

108. A chemist working on a flu vaccine needs to mix a 10% sodium-iodine solution with a 60% sodium-iodine solution to obtain a 50-milliliter mixture. Write the amount of sodium iodine in the mixture, S, in milliliters, as a function of the number of milliliters of the 10% solution used, x. Then find and interpret $S(30)$.

Writing in Mathematics

109. What is a relation? Describe what is meant by its domain and its range.

110. Explain how to determine whether a relation is a function. What is a function?

111. How do you determine if an equation in x and y defines y as a function of x?

112. Does $f(x)$ mean f times x when referring to a function f? If not, what does $f(x)$ mean? Provide an example with your explanation.

113. What is the graph of a function?

114. Explain how the vertical line test is used to determine whether a graph represents a function.

115. Explain how to identify the domain and range of a function from its graph.

116. For people filing a single return, federal income tax is a function of adjusted gross income because for each value of adjusted gross income there is a specific tax to be paid. By contrast, the price of a house is not a function of the lot size on which the house sits because houses on same-sized lots can sell for many different prices.
 a. Describe an everyday situation between variables that is a function.
 b. Describe an everyday situation between variables that is not a function.

Technology Exercise

117. Use a graphing utility to verify any five pairs of graphs that you drew by hand in Exercises 39–54.

Critical Thinking Exercises

Make Sense? *In Exercises 118–121, determine whether each statement makes sense or does not make sense, and explain your reasoning.*

118. My body temperature is a function of the time of day.

119. Using $f(x) = 3x + 2$, I found $f(50)$ by applying the distributive property to $(3x + 2)50$.

120. I graphed a function showing how paid vacation days depend on the number of years a person works for a company. The domain was the number of paid vacation days.

121. I graphed a function showing how the average number of annual physician visits depends on a person's age. The domain was the average number of annual physician visits.

Use the graph of f to determine whether each statement in Exercises 122–125 is true or false.

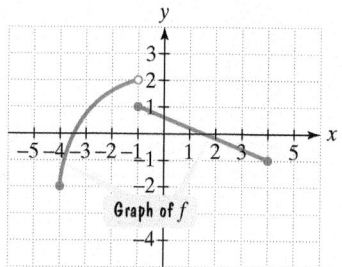

Graph of f

122. The domain of f is $[-4, -1) \cup (-1, 4]$.

123. The range of f is $[-2, 2]$.

124. $f(-1) - f(4) = 2$

125. $f(0) = 2.1$

126. If $f(x) = 3x + 7$, find $\dfrac{f(a + h) - f(a)}{h}$.

127. Give an example of a relation with the following characteristics: The relation is a function containing two ordered pairs. Reversing the components in each ordered pair results in a relation that is not a function.

128. If $f(x + y) = f(x) + f(y)$ and $f(1) = 3$, find $f(2), f(3)$, and $f(4)$. Is $f(x + y) = f(x) + f(y)$ for all functions?

Preview Exercises

Exercises 129–131 will help you prepare for the material covered in the next section.

129. The function $C(t) = 20 + 0.40(t - 60)$ describes the monthly cost, $C(t)$, in dollars, for a cellphone plan for t calling minutes, where $t > 60$. Find and interpret $C(100)$.

130. Use point plotting to graph $f(x) = x + 2$ if $x \le 1$.

131. Simplify: $2(x + h)^2 + 3(x + h) + 5 - (2x^2 + 3x + 5)$.

SECTION 1.3

More on Functions and Their Graphs

Objectives

1. Identify intervals on which a function increases, decreases, or is constant.
2. Use graphs to locate relative maxima or minima.
3. Identify even or odd functions and recognize their symmetries.
4. Understand and use piecewise functions.
5. Find and simplify a function's difference quotient.

It's hard to believe that this gas-guzzler, with its huge fins and overstated design, was available in 1957 for approximately $1800. The line graph in **Figure 1.28** shows the average fuel efficiency, in miles per gallon, of new U.S. passenger cars for selected years from 1955 through 2009. Based on the averages shown by the graph, it's unlikely that this classic 1957 Chevy got more than 15 miles per gallon.

FIGURE 1.28
Source: U.S. Department of Transportation

You are probably familiar with the words used to describe the graph in **Figure 1.28**:

<div align="center">

decreased increased minimum maximum

</div>

In this section, you will enhance your intuitive understanding of ways of describing graphs by viewing these descriptions from the perspective of functions.

Increasing and Decreasing Functions

1. Identify intervals on which a function increases, decreases, or is constant.

Too late for that flu shot now! It's only 8 A.M. and you're feeling lousy. Your temperature is 101°F. Fascinated by the way that algebra models the world (your author is projecting a bit here), you decide to construct graphs showing your body temperature as a function of the time of day. You decide to let x represent the number of hours after 8 A.M. and $f(x)$ your temperature at time x.

At 8 A.M. your temperature is 101°F and you are not feeling well. However, your temperature starts to decrease. It reaches normal (98.6°F) by 11 A.M. Feeling energized, you construct the graph shown on the right, indicating decreasing temperature for $\{x \mid 0 < x < 3\}$, or on the interval $(0, 3)$.

Temperature decreases on $(0, 3)$, reaching 98.6° by 11 A.M.

Did creating that first graph drain you of your energy? Your temperature starts to rise after 11 A.M. By 1 P.M., 5 hours after 8 A.M., your temperature reaches 100°F. However, you keep plotting points on your graph. At the right, we can see that your temperature increases for $\{x \mid 3 < x < 5\}$, or on the interval $(3, 5)$.

The graph of f is decreasing to the left of $x = 3$ and increasing to the right of $x = 3$. Thus, your temperature 3 hours after 8 A.M. was at its lowest point. Your relative minimum temperature was 98.6°.

Temperature increases on $(3, 5)$.

By 3 P.M., your temperature is no worse than it was at 1 P.M.: It is still 100°F. (Of course, it's no better, either.) Your temperature remained the same, or constant, for $\{x \mid 5 < x < 7\}$, or on the interval $(5, 7)$.

The time-temperature flu scenario illustrates that a function f is increasing when its graph rises from left to right, decreasing when its graph falls from left to right, and remains constant when it neither rises nor falls. Let's now provide a more precise algebraic description for these intuitive concepts.

Temperature remains constant at 100° on $(5, 7)$.

Increasing, Decreasing, and Constant Functions

1. A function is **increasing** on an open interval, I, if $f(x_1) < f(x_2)$ whenever $x_1 < x_2$ for any x_1 and x_2 in the interval.
2. A function is **decreasing** on an open interval, I, if $f(x_1) > f(x_2)$ whenever $x_1 < x_2$ for any x_1 and x_2 in the interval.
3. A function is **constant** on an open interval, I, if $f(x_1) = f(x_2)$ for any x_1 and x_2 in the interval.

Increasing	Decreasing	Constant

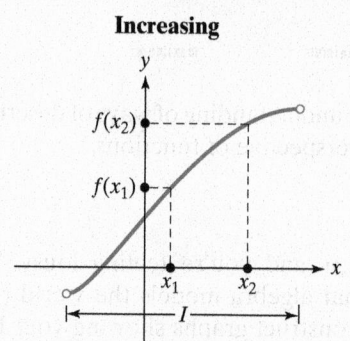

(1) For $x_1 < x_2$ in I, $f(x_1) < f(x_2)$; f is increasing on I.

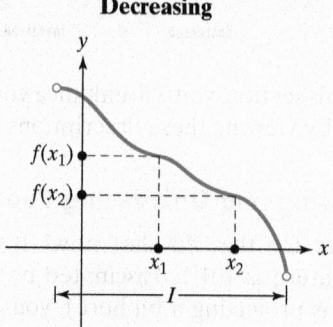

(2) For $x_1 < x_2$ in I, $f(x_1) > f(x_2)$; f is decreasing on I.

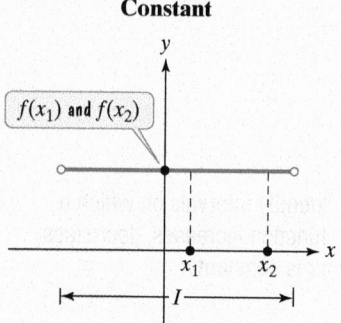

(3) For x_1 and x_2 in I, $f(x_1) = f(x_2)$; f is constant on I.

GREAT QUESTION!

GREAT QUESTION!

Do you use x-coordinates or y-coordinates when describing where functions increase, decrease, or are constant?

The open intervals describing where functions increase, decrease, or are constant use x-coordinates and not y-coordinates. Furthermore, points are not used in these descriptions.

EXAMPLE 1 Intervals on Which a Function Increases, Decreases, or Is Constant

State the intervals on which each given function is increasing, decreasing, or constant.

a.

b.

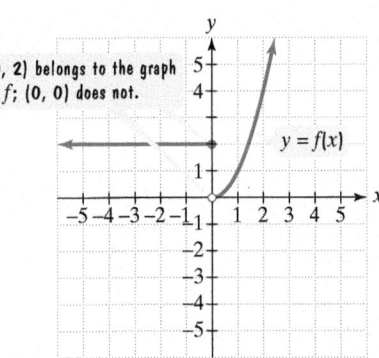

SOLUTION

a. The function is decreasing on the interval $(-\infty, 0)$, increasing on the interval $(0, 2)$, and decreasing on the interval $(2, \infty)$.

b. Although the function's equations are not given, the graph indicates that the function is defined in two pieces. The part of the graph to the left of the y-axis shows that the function is constant on the interval $(-\infty, 0)$. The part to the right of the y-axis shows that the function is increasing on the interval $(0, \infty)$. • • •

⊘ Check Point **1** State the intervals on which the given function is increasing, decreasing, or constant.

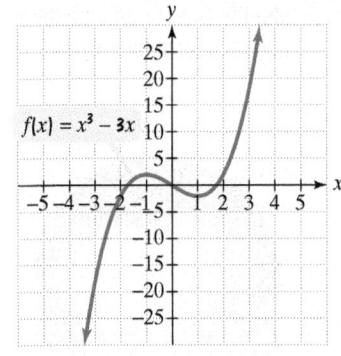

② Use graphs to locate relative maxima or minima.

Relative Maxima and Relative Minima

The points at which a function changes its increasing or decreasing behavior can be used to find any *relative maximum* or *relative minimum* values of the function.

Definitions of Relative Maximum and Relative Minimum

1. A function value $f(a)$ is a **relative maximum** of f if there exists an open interval containing a such that $f(a) > f(x)$ for all $x \neq a$ in the open interval.

2. A function value $f(b)$ is a **relative minimum** of f if there exists an open interval containing b such that $f(b) < f(x)$ for all $x \neq b$ in the open interval.

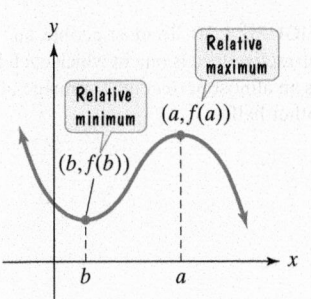

The word *local* is sometimes used instead of *relative* when describing maxima or minima.

GREAT QUESTION!

What's the difference between where a function has a relative maximum or minimum and the relative maximum or minimum?

- If $f(a)$ is greater than all other values of f near a, then f has a relative maximum at the input value, a. The relative maximum is the output value, $f(a)$. Equivalently, $f(a)$ is a relative maximum *value* of the function.
- If $f(b)$ is less than all other values of f near b, then f has a relative minimum at the input value, b. The relative minimum is the output value, $f(b)$. Equivalently, $f(b)$ is a relative minimum *value* of the function.

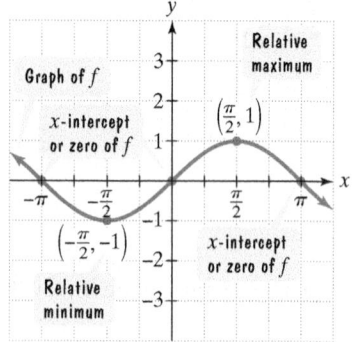

FIGURE 1.29 Using a graph to locate where a function has a relative maximum or minimum

If the graph of a function is given, we can often visually locate the number(s) at which the function has a relative maximum or a relative minimum. For example, the graph of f in **Figure 1.29** shows that

- f has a relative maximum at $\dfrac{\pi}{2}$.

 The relative maximum is $f\left(\dfrac{\pi}{2}\right) = 1$.

- f has a relative minimum at $-\dfrac{\pi}{2}$.

 The relative minimum is $f\left(-\dfrac{\pi}{2}\right) = -1$.

Notice that f does not have a relative maximum or minimum at $-\pi, 0$, and π, the x-intercepts, or zeros, of the function.

③ Identify even or odd functions and recognize their symmetries.

Even and Odd Functions and Symmetry

Is beauty in the eye of the beholder? Or are there certain objects (or people) that are so well balanced and proportioned that they are universally pleasing to the eye? What constitutes an attractive human face? In **Figure 1.30**, we've drawn lines between paired features and marked the midpoints. Notice how the features line up almost perfectly. Each half of the face is a mirror image of the other half through the white vertical line.

Did you know that graphs of some equations exhibit exactly the kind of symmetry shown by the attractive face in **Figure 1.30**? The word *symmetry* comes from the Greek *symmetria*, meaning "the same measure." We can identify a function whose graph has symmetry by looking at the function's equation and determining if the function is *even* or *odd*.

FIGURE 1.30 To most people, an attractive face is one in which each half is an almost perfect mirror image of the other half.

Definitions of Even and Odd Functions

The function f is an **even function** if

$$f(-x) = f(x) \quad \text{for all } x \text{ in the domain of } f.$$

The right side of the equation of an even function does not change if x is replaced with $-x$.

The function f is an **odd function** if

$$f(-x) = -f(x) \quad \text{for all } x \text{ in the domain of } f.$$

Every term on the right side of the equation of an odd function changes sign if x is replaced with $-x$.

EXAMPLE 2 Identifying Even or Odd Functions

Determine whether each of the following functions is even, odd, or neither:

 a. $f(x) = x^3 - 6x$ **b.** $g(x) = x^4 - 2x^2$ **c.** $h(x) = x^2 + 2x + 1.$

SOLUTION

In each case, replace x with $-x$ and simplify. If the right side of the equation stays the same, the function is even. If every term on the right side changes sign, the function is odd.

a. We use the given function's equation, $f(x) = x^3 - 6x$, to find $f(-x)$.

Use $f(x) = x^3 - 6x$.

Replace x with $-x$.

$$f(-x) = (-x)^3 - 6(-x) = (-x)(-x)(-x) - 6(-x) = -x^3 + 6x$$

There are two terms on the right side of the given equation, $f(x) = x^3 - 6x$, and each term changed sign when we replaced x with $-x$. Because $f(-x) = -f(x)$, f is an odd function.

b. We use the given function's equation, $g(x) = x^4 - 2x^2$, to find $g(-x)$.

Use $g(x) = x^4 - 2x^2$.

Replace x with $-x$.

$$g(-x) = (-x)^4 - 2(-x)^2 = (-x)(-x)(-x)(-x) - 2(-x)(-x)$$
$$= x^4 - 2x^2$$

The right side of the equation of the given function, $g(x) = x^4 - 2x^2$, did not change when we replaced x with $-x$. Because $g(-x) = g(x)$, g is an even function.

c. We use the given function's equation, $h(x) = x^2 + 2x + 1$, to find $h(-x)$.

Use $h(x) = x^2 + 2x + 1$.

Replace x with $-x$.

$$h(-x) = (-x)^2 + 2(-x) + 1 = x^2 - 2x + 1$$

The right side of the equation of the given function, $h(x) = x^2 + 2x + 1$, changed when we replaced x with $-x$. Thus, $h(-x) \neq h(x)$, so h is not an even function. The sign of *each* of the three terms in the equation for $h(x)$ did not change when we replaced x with $-x$. Only the second term changed signs. Thus, $h(-x) \neq -h(x)$, so h is not an odd function. We conclude that h is neither an even nor an odd function. ● ● ●

 Check Point 2 Determine whether each of the following functions is even, odd, or neither:

a. $f(x) = x^2 + 6$ **b.** $g(x) = 7x^3 - x$ **c.** $h(x) = x^5 + 1$.

Now, let's see what even and odd functions tell us about a function's graph. Begin with the even function $f(x) = x^2 - 4$, shown in **Figure 1.31**. The function is even because

$$f(-x) = (-x)^2 - 4 = x^2 - 4 = f(x).$$

Examine the pairs of points shown, such as $(3, 5)$ and $(-3, 5)$. Notice that we obtain the same y-coordinate whenever we evaluate the function at a value of x and the value of its opposite, $-x$. Like the attractive face, each half of the graph is a mirror image of the other half through the y-axis. If we were to fold the paper along the y-axis, the two halves of the graph would coincide. This is what it means for the graph to be *symmetric with respect to the y-axis*. A graph is **symmetric with respect to the y-axis** if, for every point (x, y) on the graph, the point $(-x, y)$ is also on the graph. All even functions have graphs with this kind of symmetry.

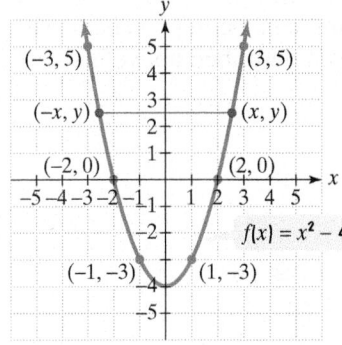

FIGURE 1.31 y-axis symmetry with $f(-x) = f(x)$

Even Functions and y-Axis Symmetry

The graph of an even function in which $f(-x) = f(x)$ is symmetric with respect to the y-axis.

Now, consider the graph of the function $f(x) = x^3$, shown in **Figure 1.32**. The function is odd because

$$f(-x) = (-x)^3 = (-x)(-x)(-x) = -x^3 = -f(x).$$

Although the graph in **Figure 1.32** is not symmetric with respect to the y-axis, it is symmetric in another way. Look at the pairs of points, such as $(2, 8)$ and $(-2, -8)$. For each point (x, y) on the graph, the point $(-x, -y)$ is also on the graph. The points $(2, 8)$ and $(-2, -8)$ are reflections of one another through the origin. This means that the origin is the midpoint of the line segment connecting the points.

A graph is **symmetric with respect to the origin** if, for every point (x, y) on the graph, the point $(-x, -y)$ is also on the graph. Observe that the first- and third-quadrant portions of $f(x) = x^3$ are reflections of one another with respect to the origin. Notice that $f(x)$ and $f(-x)$ have opposite signs, so that $f(-x) = -f(x)$. All odd functions have graphs with origin symmetry.

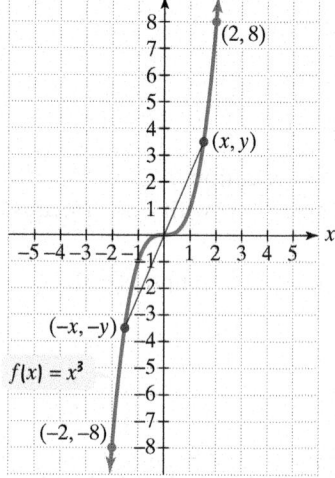

FIGURE 1.32 Origin symmetry with $f(-x) = -f(x)$

Odd Functions and Origin Symmetry

The graph of an odd function in which $f(-x) = -f(x)$ is symmetric with respect to the origin.

④ Understand and use piecewise functions.

Piecewise Functions

A cellphone company offers the following plan:

- \$20 per month buys 60 minutes.
- Additional time costs \$0.40 per minute.

We can represent this plan mathematically by writing the total monthly cost, C, as a function of the number of calling minutes, t.

$$C(t) = \begin{cases} 20 & \text{if } 0 \le t \le 60 \\ 20 + 0.40(t - 60) & \text{if } t > 60 \end{cases}$$

The cost is \$20 for up to and including 60 calling minutes.

The cost is \$20 plus \$0.40 per minute for additional time for more than 60 calling minutes.

\$20 for first 60 minutes | \$0.40 per minute | times the number of calling minutes exceeding 60

A function that is defined by two (or more) equations over a specified domain is called a **piecewise function**. Many cellphone plans can be represented with piecewise functions. The graph of the piecewise function described above is shown in **Figure 1.33**.

FIGURE 1.33

EXAMPLE 3 Evaluating a Piecewise Function

Use the function that describes the cellphone plan

$$C(t) = \begin{cases} 20 & \text{if } 0 \le t \le 60 \\ 20 + 0.40(t - 60) & \text{if } t > 60 \end{cases}$$

to find and interpret each of the following:

 a. $C(30)$ **b.** $C(100)$.

SOLUTION

a. To find $C(30)$, we let $t = 30$. Because 30 lies between 0 and 60, we use the first line of the piecewise function.

$$C(t) = 20 \qquad \text{This is the function's equation for } 0 \le t \le 60.$$
$$C(30) = 20 \qquad \text{Replace } t \text{ with 30. Regardless of this function's} \\ \text{input, the constant output is 20.}$$

This means that with 30 calling minutes, the monthly cost is $20. This can be visually represented by the point $(30, 20)$ on the first piece of the graph in **Figure 1.33**.

b. To find $C(100)$, we let $t = 100$. Because 100 is greater than 60, we use the second line of the piecewise function.

$$C(t) = 20 + 0.40(t - 60) \qquad \text{This is the function's equation for } t > 60.$$
$$C(100) = 20 + 0.40(100 - 60) \qquad \text{Replace } t \text{ with 100.}$$
$$= 20 + 0.40(40) \qquad \text{Subtract within parentheses: } 100 - 60 = 40.$$
$$= 20 + 16 \qquad \text{Multiply: } 0.40(40) = 16.$$
$$= 36 \qquad \text{Add: } 20 + 16 = 36.$$

Thus, $C(100) = 36$. This means that with 100 calling minutes, the monthly cost is $36. This can be visually represented by the point $(100, 36)$ on the second piece of the graph in **Figure 1.33**. •••

⌀ Check Point **3** Use the function in Example 3 to find and interpret each of the following:

a. $C(40)$ b. $C(80)$.

Identify your solutions by points on the graph in **Figure 1.33**.

EXAMPLE 4 Graphing a Piecewise Function

Graph the piecewise function defined by

$$f(x) = \begin{cases} x + 2 & \text{if } x \le 1 \\ 4 & \text{if } x > 1. \end{cases}$$

GREAT QUESTION!

When I graphed the function in Example 4, here's what I got:

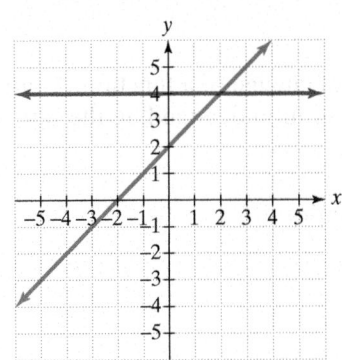

Is my graph ok?

No. You incorrectly ignored the domain for each piece of the function and graphed each equation as if its domain was $(-\infty, \infty)$.

SOLUTION

We graph f in two parts, using a partial table of coordinates to create each piece. The tables of coordinates and the completed graph are shown in **Figure 1.34**.

Graph $f(x) = x + 2$ for $x \le 1$.
Domain $= (-\infty, 1]$

x $(x \le 1)$	$f(x) = x + 2$	$(x, f(x))$
1	$f(1) = 1 + 2 = 3$	$(1, 3)$
0	$f(0) = 0 + 2 = 2$	$(0, 2)$
−1	$f(-1) = -1 + 2 = 1$	$(-1, 1)$
−2	$f(-2) = -2 + 2 = 0$	$(-2, 0)$
−3	$f(-3) = -3 + 2 = -1$	$(-3, -1)$
−4	$f(-4) = -4 + 2 = -2$	$(-4, -2)$

Graph $f(x) = 4$ for $x > 1$.
Domain $= (1, \infty)$

x $(x > 1)$	$f(x) = 4$	$(x, f(x))$
1.5	$f(1.5) = 4$	$(1.5, 4)$
2	$f(2) = 4$	$(2, 4)$
3	$f(3) = 4$	$(3, 4)$
4	$f(4) = 4$	$(4, 4)$
5	$f(5) = 4$	$(5, 4)$

FIGURE 1.34 Graphing a piecewise function •••

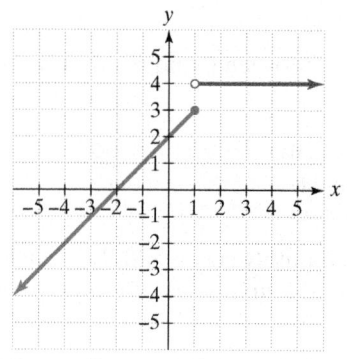

FIGURE 1.34 (repeated)

We can use the graph of the piecewise function in **Figure 1.34** to find the range of f. The range of the blue piece on the left is $\{y \mid y \leq 3\}$. The range of the red horizontal piece on the right is $\{y \mid y = 4\}$. Thus, the range of f is

$$\{y \mid y \leq 3\} \cup \{y \mid y = 4\}, \text{ or } (-\infty, 3] \cup \{4\}.$$

⊘ **Check Point 4** Graph the piecewise function defined by

$$f(x) = \begin{cases} 3 & \text{if } x \leq -1 \\ x - 2 & \text{if } x > -1. \end{cases}$$

Some piecewise functions are called **step functions** because their graphs form discontinuous steps. One such function is called the **greatest integer function**, symbolized by $\text{int}(x)$ or $[\![x]\!]$, where

$$\text{int}(x) = \text{the greatest integer that is less than or equal to } x.$$

For example,

$$\text{int}(1) = 1, \quad \text{int}(1.3) = 1, \quad \text{int}(1.5) = 1, \quad \text{int}(1.9) = 1.$$

> 1 is the greatest integer that is less than or equal to 1, 1.3, 1.5, and 1.9.

Here are some additional examples:

$$\text{int}(2) = 2, \quad \text{int}(2.3) = 2, \quad \text{int}(2.5) = 2, \quad \text{int}(2.9) = 2.$$

> 2 is the greatest integer that is less than or equal to 2, 2.3, 2.5, and 2.9.

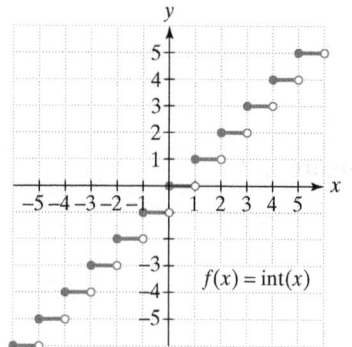

FIGURE 1.35 The graph of the greatest integer function

Notice how we jumped from 1 to 2 in the function values for $\text{int}(x)$. In particular,

$$\text{If } 1 \leq x < 2, \quad \text{then} \quad \text{int}(x) = 1.$$
$$\text{If } 2 \leq x < 3, \quad \text{then} \quad \text{int}(x) = 2.$$

The graph of $f(x) = \text{int}(x)$ is shown in **Figure 1.35**. The graph of the greatest integer function jumps vertically one unit at each integer. However, the graph is constant between each pair of consecutive integers. The rightmost horizontal step shown in the graph illustrates that

$$\text{If } 5 \leq x < 6, \quad \text{then} \quad \text{int}(x) = 5.$$

In general,

$$\text{If } n \leq x < n + 1, \text{ where } n \text{ is an integer}, \quad \text{then} \quad \text{int}(x) = n.$$

⑤ Find and simplify a function's difference quotient.

Functions and Difference Quotients

In Section 1.5, we will be studying the average rate of change of a function. A ratio, called the *difference quotient*, plays an important role in understanding the rate at which functions change.

> Definition of the Difference Quotient of a Function
>
> The expression
>
> $$\frac{f(x + h) - f(x)}{h}$$
>
> for $h \neq 0$ is called the **difference quotient** of the function f.

EXAMPLE 5 Evaluating and Simplifying a Difference Quotient

If $f(x) = 2x^2 - x + 3$, find and simplify each expression:

a. $f(x + h)$ **b.** $\dfrac{f(x + h) - f(x)}{h}, h \neq 0.$

SOLUTION

a. We find $f(x + h)$ by replacing x with $x + h$ each time that x appears in the equation.

$$f(x) = 2x^2 \quad - \quad x \quad + \quad 3$$

| Replace x with $x + h$. | Replace x with $x + h$. | Replace x with $x + h$. | Copy the 3. There is no x in this term. |

$$f(x + h) = 2(x + h)^2 - (x + h) \quad + \quad 3$$
$$= 2(x^2 + 2xh + h^2) - x - h + 3$$
$$= 2x^2 + 4xh + 2h^2 - x - h + 3$$

b. Using our result from part (a), we obtain the following:

This is $f(x + h)$ from part (a).

This is $f(x)$ from the given equation.

$$\frac{f(x + h) - f(x)}{h} = \frac{\boxed{2x^2 + 4xh + 2h^2 - x - h + 3} - (2x^2 - x + 3)}{h}$$

$$= \frac{2x^2 + 4xh + 2h^2 - x - h + 3 - 2x^2 + x - 3}{h} \qquad \text{Remove parentheses and change the sign of each term in the parentheses.}$$

$$= \frac{(2x^2 - 2x^2) + (-x + x) + (3 - 3) + 4xh + 2h^2 - h}{h} \qquad \text{Group like terms.}$$

$$= \frac{4xh + 2h^2 - 1h}{h} \qquad \text{Simplify.}$$

We wrote $-h$ as $-1h$ to avoid possible errors in the next factoring step.

$$= \frac{h(4x + 2h - 1)}{h} \qquad \text{Factor h from the numerator.}$$

$$= 4x + 2h - 1 \qquad \text{Divide out identical factors of h in the numerator and denominator.} \qquad \bullet\bullet\bullet$$

⌀ **Check Point 5** If $f(x) = -2x^2 + x + 5$, find and simplify each expression:

a. $f(x + h)$ **b.** $\dfrac{f(x + h) - f(x)}{h}, h \neq 0.$

CONCEPT AND VOCABULARY CHECK

Fill in each blank so that the resulting statement is true.

1. Assume that f is a function defined on an open interval I and x_1 and x_2 are any elements in the interval I.
 f is increasing on I if $f(x_1)$ _____ when $x_1 < x_2$.
 f is decreasing on I if $f(x_1)$ _____ when $x_1 < x_2$.
 f is constant on I if $f(x_1)$ _____.

2. If $f(a) > f(x)$ in an open interval containing $a, x \neq a$, then the function value $f(a)$ is a relative _____ of f. If $f(b) < f(x)$ in an open interval containing $b, x \neq b$, then the function value $f(b)$ is a relative _____ of f.

3. If f is an even function, then $f(-x) =$ _____. Such a function is symmetric with respect to the _____.

4. If f is an odd function, then $f(-x) =$ _____. The graph of such a function is symmetric with respect to the _____.

5. A function defined by two or more equations over a specified domain is called a/an _____ function.

6. The greatest integer function is defined by $\text{int}(x) =$ the greatest integer that is _____. For example, $\text{int}(2.5) =$ ____, $\text{int}(-2.5) =$ ____, and $\text{int}(0.5) =$ ____.

7. The expression
 $$\frac{f(x + h) - f(x)}{h}, h \neq 0,$$
 is called the _____ of the function f. We find this expression by replacing x with _____ each time x appears in the function's equation. Then we subtract _____. After simplifying, we factor ____ from the numerator and divide out identical factors of ____ in the numerator and denominator.

8. True or false: $f(x + h) = f(x) + h$ _____

9. True or false: $f(x + h) = f(x) + f(h)$ _____

EXERCISE SET 1.3

Practice Exercises

In Exercises 1–12, use the graph to determine

 a. *intervals on which the function is increasing, if any.*
 b. *intervals on which the function is decreasing, if any.*
 c. *intervals on which the function is constant, if any.*

1.

2.

3.

4.

5.

6.

7.

8.

9.

10.

11.

12.
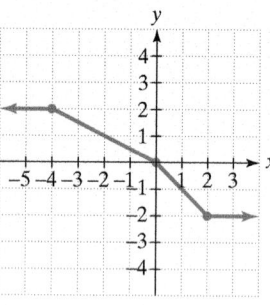

In Exercises 13–16, the graph of a function f is given. Use the graph to find each of the following:

 a. *The numbers, if any, at which f has a relative maximum. What are these relative maxima?*

 b. *The numbers, if any, at which f has a relative minimum. What are these relative minima?*

13.

14.

15.

$f(x) = 2x^3 + 3x^2 - 12x + 1$

$[-4, 4, 1]$ by $[-15, 25, 5]$

16.

$f(x) = 2x^3 - 15x^2 + 24x + 19$

$[-2, 6, 1]$ by $[-15, 35, 5]$

In Exercises 17–28, determine whether each function is even, odd, or neither.

17. $f(x) = x^3 + x$

18. $f(x) = x^3 - x$

19. $g(x) = x^2 + x$

20. $g(x) = x^2 - x$

21. $h(x) = x^2 - x^4$

22. $h(x) = 2x^2 + x^4$

23. $f(x) = x^2 - x^4 + 1$

24. $f(x) = 2x^2 + x^4 + 1$

25. $f(x) = \frac{1}{5}x^6 - 3x^2$

26. $f(x) = 2x^3 - 6x^5$

27. $f(x) = x\sqrt{1 - x^2}$

28. $f(x) = x^2\sqrt{1 - x^2}$

In Exercises 29–32, use possible symmetry to determine whether each graph is the graph of an even function, an odd function, or a function that is neither even nor odd.

29.

30.

31.

32.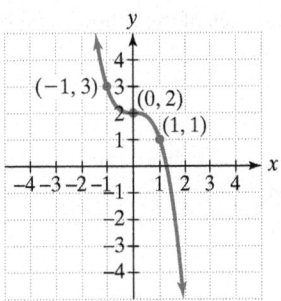

33. Use the graph of f to determine each of the following. Where applicable, use interval notation.

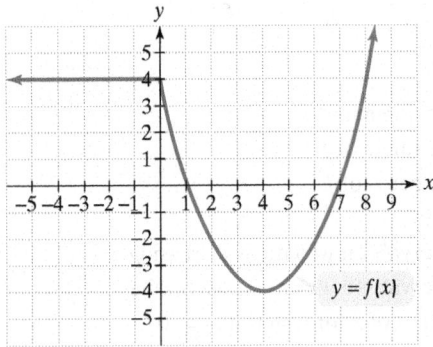

$y = f(x)$

a. the domain of f

b. the range of f

c. the x-intercepts

d. the y-intercept

e. intervals on which f is increasing

f. intervals on which f is decreasing

g. intervals on which f is constant

h. the number at which f has a relative minimum

i. the relative minimum of f

j. $f(-3)$

k. the values of x for which $f(x) = -2$

l. Is f even, odd, or neither?

34. Use the graph of f to determine each of the following. Where applicable, use interval notation.

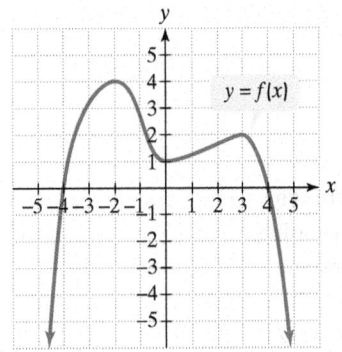

$y = f(x)$

a. the domain of f

b. the range of f

c. the x-intercepts

d. the y-intercept

e. intervals on which f is increasing

f. intervals on which f is decreasing

g. values of x for which $f(x) \leq 0$

h. the numbers at which f has a relative maximum

i. the relative maxima of f

j. $f(-2)$

k. the values of x for which $f(x) = 0$

l. Is f even, odd, or neither?

35. Use the graph of f to determine each of the following. Where applicable, use interval notation.

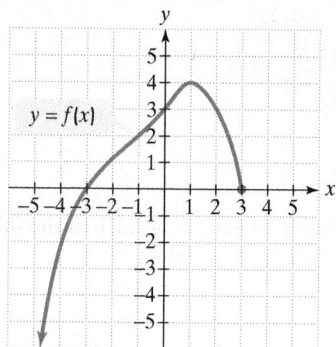

a. the domain of f
b. the range of f
c. the zeros of f
d. $f(0)$
e. intervals on which f is increasing
f. intervals on which f is decreasing
g. values of x for which $f(x) \le 0$
h. any relative maxima and the numbers at which they occur
i. the value of x for which $f(x) = 4$
j. Is $f(-1)$ positive or negative?

36. Use the graph of f to determine each of the following. Where applicable, use interval notation.

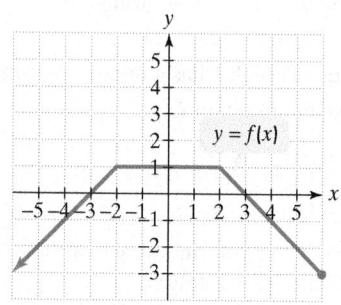

a. the domain of f
b. the range of f
c. the zeros of f
d. $f(0)$
e. intervals on which f is increasing
f. intervals on which f is decreasing
g. intervals on which f is constant
h. values of x for which $f(x) > 0$
i. values of x for which $f(x) = -2$
j. Is $f(4)$ positive or negative?
k. Is f even, odd, or neither?
l. Is $f(2)$ a relative maximum?

In Exercises 37–42, evaluate each piecewise function at the given values of the independent variable.

37. $f(x) = \begin{cases} 3x + 5 & \text{if } x < 0 \\ 4x + 7 & \text{if } x \ge 0 \end{cases}$

 a. $f(-2)$ b. $f(0)$ c. $f(3)$

38. $f(x) = \begin{cases} 6x - 1 & \text{if } x < 0 \\ 7x + 3 & \text{if } x \ge 0 \end{cases}$

 a. $f(-3)$ b. $f(0)$ c. $f(4)$

39. $g(x) = \begin{cases} x + 3 & \text{if } x \ge -3 \\ -(x + 3) & \text{if } x < -3 \end{cases}$

 a. $g(0)$ b. $g(-6)$ c. $g(-3)$

40. $g(x) = \begin{cases} x + 5 & \text{if } x \ge -5 \\ -(x + 5) & \text{if } x < -5 \end{cases}$

 a. $g(0)$ b. $g(-6)$ c. $g(-5)$

41. $h(x) = \begin{cases} \dfrac{x^2 - 9}{x - 3} & \text{if } x \ne 3 \\ 6 & \text{if } x = 3 \end{cases}$

 a. $h(5)$ b. $h(0)$ c. $h(3)$

42. $h(x) = \begin{cases} \dfrac{x^2 - 25}{x - 5} & \text{if } x \ne 5 \\ 10 & \text{if } x = 5 \end{cases}$

 a. $h(7)$ b. $h(0)$ c. $h(5)$

In Exercises 43–54, the domain of each piecewise function is $(-\infty, \infty)$.

 a. *Graph each function.*
 b. *Use your graph to determine the function's range.*

43. $f(x) = \begin{cases} -x & \text{if } x < 0 \\ x & \text{if } x \ge 0 \end{cases}$

44. $f(x) = \begin{cases} x & \text{if } x < 0 \\ -x & \text{if } x \ge 0 \end{cases}$

45. $f(x) = \begin{cases} 2x & \text{if } x \le 0 \\ 2 & \text{if } x > 0 \end{cases}$

46. $f(x) = \begin{cases} \frac{1}{2}x & \text{if } x \le 0 \\ 3 & \text{if } x > 0 \end{cases}$

47. $f(x) = \begin{cases} x + 3 & \text{if } x < -2 \\ x - 3 & \text{if } x \ge -2 \end{cases}$

48. $f(x) = \begin{cases} x + 2 & \text{if } x < -3 \\ x - 2 & \text{if } x \ge -3 \end{cases}$

49. $f(x) = \begin{cases} 3 & \text{if } x \le -1 \\ -3 & \text{if } x > -1 \end{cases}$

50. $f(x) = \begin{cases} 4 & \text{if } x \le -1 \\ -4 & \text{if } x > -1 \end{cases}$

51. $f(x) = \begin{cases} \frac{1}{2}x^2 & \text{if } x < 1 \\ 2x - 1 & \text{if } x \ge 1 \end{cases}$

52. $f(x) = \begin{cases} -\frac{1}{2}x^2 & \text{if } x < 1 \\ 2x + 1 & \text{if } x \ge 1 \end{cases}$

53. $f(x) = \begin{cases} 0 & \text{if} \quad x < -4 \\ -x & \text{if} \quad -4 \le x < 0 \\ x^2 & \text{if} \quad x \ge 0 \end{cases}$

54. $f(x) = \begin{cases} 0 & \text{if} \quad x < -3 \\ -x & \text{if} \quad -3 \le x < 0 \\ x^2 - 1 & \text{if} \quad x \ge 0 \end{cases}$

In Exercises 55–76, find and simplify the difference quotient

$$\frac{f(x + h) - f(x)}{h}, h \ne 0$$

for the given function.

55. $f(x) = 4x$

56. $f(x) = 7x$

57. $f(x) = 3x + 7$

58. $f(x) = 6x + 1$

59. $f(x) = x^2$

60. $f(x) = 2x^2$

61. $f(x) = x^2 - 4x + 3$

62. $f(x) = x^2 - 5x + 8$

63. $f(x) = 2x^2 + x - 1$

64. $f(x) = 3x^2 + x + 5$

65. $f(x) = -x^2 + 2x + 4$

66. $f(x) = -x^2 - 3x + 1$

67. $f(x) = -2x^2 + 5x + 7$

68. $f(x) = -3x^2 + 2x - 1$

69. $f(x) = -2x^2 - x + 3$

70. $f(x) = -3x^2 + x - 1$

71. $f(x) = 6$

72. $f(x) = 7$

73. $f(x) = \dfrac{1}{x}$

74. $f(x) = \dfrac{1}{2x}$

75. $f(x) = \sqrt{x}$

76. $f(x) = \sqrt{x - 1}$

Practice Plus

In Exercises 77–78, let f be defined by the following graph:

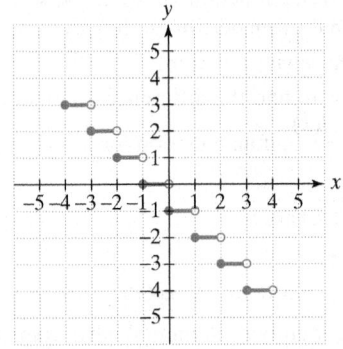

77. Find

$$\sqrt{f(-1.5) + f(-0.9)} - [f(\pi)]^2 + f(-3) \div f(1) \cdot f(-\pi).$$

78. Find

$$\sqrt{f(-2.5) - f(1.9)} - [f(-\pi)]^2 + f(-3) \div f(1) \cdot f(\pi).$$

A cellphone company offers the following plans. Also given are the piecewise functions that model these plans. Use this information to solve Exercises 79–80.

Plan A

- *$30 per month buys 120 minutes.*
- *Additional time costs $0.30 per minute.*

$$C(t) = \begin{cases} 30 & \text{if} \quad 0 \le t \le 120 \\ 30 + 0.30(t - 120) & \text{if} \quad t > 120 \end{cases}$$

Plan B

- *$40 per month buys 200 minutes.*
- *Additional time costs $0.30 per minute.*

$$C(t) = \begin{cases} 40 & \text{if} \quad 0 \le t \le 200 \\ 40 + 0.30(t - 200) & \text{if} \quad t > 200 \end{cases}$$

79. Simplify the algebraic expression in the second line of the piecewise function for plan A. Then use point-plotting to graph the function.

80. Simplify the algebraic expression in the second line of the piecewise function for plan B. Then use point-plotting to graph the function.

In Exercises 81–82, write a piecewise function that models each cellphone billing plan. Then graph the function.

81. $50 per month buys 400 minutes. Additional time costs $0.30 per minute.

82. $60 per month buys 450 minutes. Additional time costs $0.35 per minute.

Application Exercises

With aging, body fat increases and muscle mass declines. The line graphs show the percent body fat in adult women and men as they age from 25 to 75 years. Use the graphs to solve Exercises 83–90.

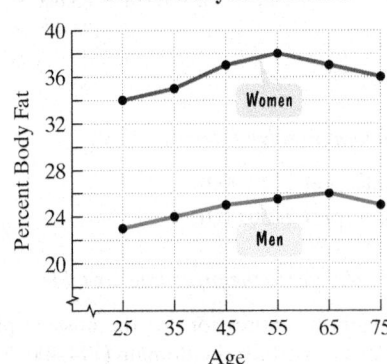

Percent Body Fat in Adults

Source: Thompson et al., *The Science of Nutrition,* Benjamin Cummings, 2008

83. State the intervals on which the graph giving the percent body fat in women is increasing and decreasing.

84. State the intervals on which the graph giving the percent body fat in men is increasing and decreasing.

85. For what age does the percent body fat in women reach a maximum? What is the percent body fat for that age?

86. At what age does the percent body fat in men reach a maximum? What is the percent body fat for that age?

87. Use interval notation to give the domain and the range for the graph of the function for women.

88. Use interval notation to give the domain and the range for the graph of the function for men.

89. The function $p(x) = -0.002x^2 + 0.15x + 22.86$ models percent body fat, $p(x)$, where x is the number of years a person's age exceeds 25. Use the graphs to determine whether this model describes percent body fat in women or in men.

90. The function $p(x) = -0.004x^2 + 0.25x + 33.64$ models percent body fat, $p(x)$, where x is the number of years a person's age exceeds 25. Use the graphs to determine whether this model describes percent body fat in women or in men.

Here is the 2011 Federal Tax Rate Schedule X that specifies the tax owed by a single taxpayer.

If Your Taxable Income Is Over	But Not Over	The Tax You Owe Is	Of the Amount Over
$ 0	$ 8500	10%	$ 0
$ 8500	$ 34,500	$ 850.00 + 15%	$ 8500
$ 34,500	$ 83,600	$ 4750.00 + 25%	$ 34,500
$ 83,600	$174,400	$ 17,025.00 + 28%	$ 83,600
$174,400	$379,150	$ 42,449.00 + 33%	$174,400
$379,150	–	$110,016.50 + 35%	$379,150

The preceding tax table can be modeled by a piecewise function, where x represents the taxable income of a single taxpayer and T(x) is the tax owed:

$$T(x) = \begin{cases} 0.10x & \text{if} & 0 < x \le 8500 \\ 850.00 + 0.15(x - 8500) & \text{if} & 8500 < x \le 34,500 \\ 4750.00 + 0.25(x - 34,500) & \text{if} & 34,500 < x \le 83,600 \\ 17,025.00 + 0.28(x - 83,600) & \text{if} & 83,600 < x \le 174,400 \\ \underline{\quad ? \quad} & \text{if} & 174,400 < x \le 379,150 \\ \underline{\quad ? \quad} & \text{if} & x > 379,150. \end{cases}$$

Use this information to solve Exercises 91–94.

91. Find and interpret $T(20,000)$.

92. Find and interpret $T(50,000)$.

In Exercises 93–94, refer to the preceding tax table.

93. Find the algebraic expression for the missing piece of $T(x)$ that models tax owed for the domain $(174,400, 379,150]$.

94. Find the algebraic expression for the missing piece of $T(x)$ that models tax owed for the domain $(379,150, \infty)$.

The figure shows the cost of mailing a first-class letter, f(x), as a function of its weight, x, in ounces. Use the graph to solve Exercises 95–98.

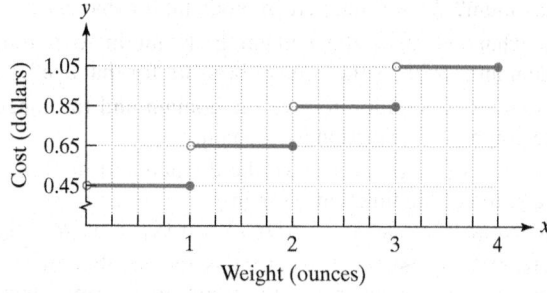

Source: Lynn E. Baring, Postmaster, Inverness, CA

95. Find $f(3)$. What does this mean in terms of the variables in this situation?

96. Find $f(3.5)$. What does this mean in terms of the variables in this situation?

97. What is the cost of mailing a letter that weighs 1.5 ounces?

98. What is the cost of mailing a letter that weighs 1.8 ounces?

99. Furry Finances A pet insurance policy has a monthly rate that is a function of the age of the insured dog or cat. For pets whose age does not exceed 4, the monthly cost is $20. The cost then increases by $2 for each successive year of the pet's age.

Age Not Exceeding	Monthly Cost
4	$20
5	$22
6	$24

The cost schedule continues in this manner for ages not exceeding 10. The cost for pets whose ages exceed 10 is $40. Use this information to create a graph that shows the monthly cost of the insurance, $f(x)$, for a pet of age x, where the function's domain is $[0, 14]$.

Writing in Mathematics

100. What does it mean if a function f is increasing on an interval?

101. Suppose that a function f whose graph contains no breaks or gaps on (a, c) is increasing on (a, b), decreasing on (b, c), and defined at b. Describe what occurs at $x = b$. What does the function value $f(b)$ represent?

102. If you are given a function's equation, how do you determine if the function is even, odd, or neither?

103. If you are given a function's graph, how do you determine if the function is even, odd, or neither?

104. What is a piecewise function?

105. Explain how to find the difference quotient of a function f, $\dfrac{f(x + h) - f(x)}{h}$, if an equation for f is given.

Technology Exercises

106. The function

$$f(x) = -0.00002x^3 + 0.008x^2 - 0.3x + 6.95$$

models the number of annual physician visits, $f(x)$, by a person of age x. Graph the function in a $[0, 100, 5]$ by $[0, 40, 2]$ viewing rectangle. What does the shape of the graph indicate about the relationship between one's age and the number of annual physician visits? Use the $\boxed{\text{TABLE}}$ or minimum function capability to find the coordinates of the minimum point on the graph of the function. What does this mean?

In Exercises 107–112, use a graphing utility to graph each function. Use a $[-5, 5, 1]$ by $[-5, 5, 1]$ viewing rectangle. Then find the intervals on which the function is increasing, decreasing, or constant.

107. $f(x) = x^3 - 6x^2 + 9x + 1$

108. $g(x) = |4 - x^2|$

109. $h(x) = |x - 2| + |x + 2|$

110. $f(x) = x^{\frac{1}{3}}(x - 4)$

111. $g(x) = x^{\frac{2}{3}}$

112. $h(x) = 2 - x^{\frac{2}{5}}$

113. **a.** Graph the functions $f(x) = x^n$ for $n = 2, 4$, and 6 in a $[-2, 2, 1]$ by $[-1, 3, 1]$ viewing rectangle.

b. Graph the functions $f(x) = x^n$ for $n = 1, 3$, and 5 in a $[-2, 2, 1]$ by $[-2, 2, 1]$ viewing rectangle.

c. If n is positive and even, where is the graph of $f(x) = x^n$ increasing and where is it decreasing?

d. If n is positive and odd, what can you conclude about the graph of $f(x) = x^n$ in terms of increasing or decreasing behavior?

e. Graph all six functions in a $[-1, 3, 1]$ by $[-1, 3, 1]$ viewing rectangle. What do you observe about the graphs in terms of how flat or how steep they are?

Critical Thinking Exercises

Make Sense? *In Exercises 114–117, determine whether each statement makes sense or does not make sense, and explain your reasoning.*

114. My graph is decreasing on $(-\infty, a)$ and increasing on (a, ∞), so $f(a)$ must be a relative maximum.

115. This work by artist Scott Kim (1955–) has the same kind of symmetry as an even function.

"DYSLEXIA," 1981

116. I graphed
$$f(x) = \begin{cases} 2 & \text{if} \quad x \neq 4 \\ 3 & \text{if} \quad x = 4 \end{cases}$$
and one piece of my graph is a single point.

117. I noticed that the difference quotient is always zero if $f(x) = c$, where c is any constant.

118. Sketch the graph of f using the following properties. (More than one correct graph is possible.) f is a piecewise function that is decreasing on $(-\infty, 2)$, $f(2) = 0$, f is increasing on $(2, \infty)$, and the range of f is $[0, \infty)$.

119. Define a piecewise function on the intervals $(-\infty, 2]$, $(2, 5)$, and $[5, \infty)$ that does not "jump" at 2 or 5 such that one piece is a constant function, another piece is an increasing function, and the third piece is a decreasing function.

120. Suppose that $h(x) = \dfrac{f(x)}{g(x)}$. The function f can be even, odd, or neither. The same is true for the function g.

a. Under what conditions is h definitely an even function?

b. Under what conditions is h definitely an odd function?

Group Exercise

121. (For assistance with this exercise, refer to the discussion of piecewise functions beginning on page 142, as well as to Exercises 79–80.) Group members who have cellphone plans should describe the total monthly cost of the plan as follows:

$____$ per month buys $____$ minutes. Additional time costs $\$____$ per minute.

(For simplicity, ignore other charges.) The group should select any three plans, from "basic" to "premier." For each plan selected, write a piecewise function that describes the plan and graph the function. Graph the three functions in the same rectangular coordinate system. Now examine the graphs. For any given number of calling minutes, the best plan is the one whose graph is lowest at that point. Compare the three calling plans. Is one plan always a better deal than the other two? If not, determine the interval of calling minutes for which each plan is the best deal. (You can check out cellphone plans by visiting www.point.com.)

Preview Exercises

Exercises 122–124 will help you prepare for the material covered in the next section.

122. If $(x_1, y_1) = (-3, 1)$ and $(x_2, y_2) = (-2, 4)$, find $\dfrac{y_2 - y_1}{x_2 - x_1}$.

123. Find the ordered pairs $(____, 0)$ and $(0, ____)$ satisfying $4x - 3y - 6 = 0$.

124. Solve for y: $3x + 2y - 4 = 0$.

●

SECTION 1.4 Linear Functions and Slope

Objectives

① Calculate a line's slope.

② Write the point-slope form of the equation of a line.

③ Write and graph the slope-intercept form of the equation of a line.

④ Graph horizontal or vertical lines.

⑤ Recognize and use the general form of a line's equation.

⑥ Use intercepts to graph the general form of a line's equation.

⑦ Model data with linear functions and make predictions.

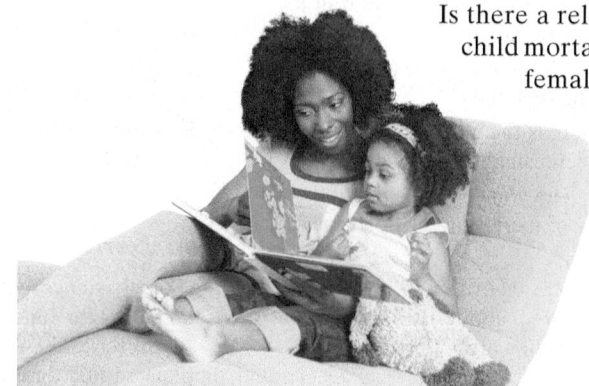

Is there a relationship between literacy and child mortality? As the percentage of adult females who are literate increases, does the mortality of children under five decrease? **Figure 1.36** indicates that this is, indeed, the case. Each point in the figure represents one country.

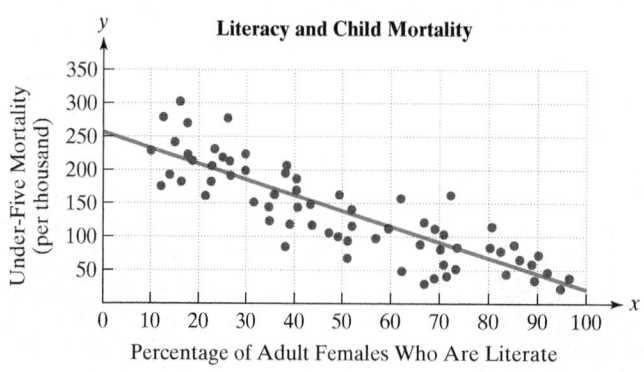

FIGURE 1.36
Source: United Nations

Data presented in a visual form as a set of points is called a **scatter plot**. Also shown in **Figure 1.36** is a line that passes through or near the points. A line that best fits the data points in a scatter plot is called a **regression line**. By writing the equation of this line, we can obtain a model for the data and make predictions about child mortality based on the percentage of literate adult females in a country.

Data often fall on or near a line. In this section, we will use functions to model such data and make predictions. We begin with a discussion of a line's steepness.

① Calculate a line's slope.

The Slope of a Line

Mathematicians have developed a useful measure of the steepness of a line, called the *slope* of the line. Slope compares the vertical change (the **rise**) to the horizontal change (the **run**) when moving from one fixed point to another along the line. To calculate the slope of a line, we use a ratio that compares the change in y (the rise) to the corresponding change in x (the run).

Definition of Slope

The **slope** of the line through the distinct points (x_1, y_1) and (x_2, y_2) is

$$\frac{\text{Change in } y}{\text{Change in } x} = \frac{\text{Rise}}{\text{Run}} \quad \begin{array}{l}\text{Vertical change}\\ \text{Horizontal change}\end{array}$$

$$= \frac{y_2 - y_1}{x_2 - x_1},$$

where $x_2 - x_1 \neq 0$.

It is common notation to let the letter m represent the slope of a line. The letter m is used because it is the first letter of the French verb *monter*, meaning "to rise" or "to ascend."

EXAMPLE 1 Using the Definition of Slope

Find the slope of the line passing through each pair of points:

a. $(-3, -1)$ and $(-2, 4)$ **b.** $(-3, 4)$ and $(2, -2)$.

SOLUTION

a. Let $(x_1, y_1) = (-3, -1)$ and $(x_2, y_2) = (-2, 4)$. We obtain the slope as follows:

$$m = \frac{\text{Change in } y}{\text{Change in } x} = \frac{y_2 - y_1}{x_2 - x_1} = \frac{4 - (-1)}{-2 - (-3)} = \frac{4 + 1}{-2 + 3} = \frac{5}{1} = 5.$$

The situation is illustrated in **Figure 1.37**. The slope of the line is 5. For every vertical change, or rise, of 5 units, there is a corresponding horizontal change, or run, of 1 unit. The slope is positive and the line rises from left to right.

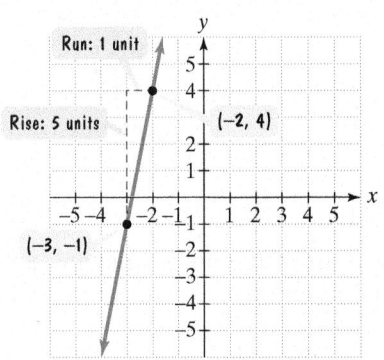

Run: 1 unit
Rise: 5 units
$(-2, 4)$
$(-3, -1)$

FIGURE 1.37 Visualizing a slope of 5

GREAT QUESTION!

When using the definition of slope, how do I know which point to call (x_1, y_1) and which point to call (x_2, y_2)?

When computing slope, it makes no difference which point you call (x_1, y_1) and which point you call (x_2, y_2). If we let $(x_1, y_1) = (-2, 4)$ and $(x_2, y_2) = (-3, -1)$, the slope is still 5:

$$m = \frac{\text{Change in } y}{\text{Change in } x} = \frac{y_2 - y_1}{x_2 - x_1} = \frac{-1 - 4}{-3 - (-2)} = \frac{-5}{-1} = 5.$$

However, you should not subtract in one order in the numerator $(y_2 - y_1)$ and then in a different order in the denominator $(x_1 - x_2)$.

$$\frac{-1 - 4}{-2 - (-3)} = \frac{-5}{1} = -5. \quad \text{Incorrect! The slope is not } -5.$$

b. To find the slope of the line passing through $(-3, 4)$ and $(2, -2)$, we can let $(x_1, y_1) = (-3, 4)$ and $(x_2, y_2) = (2, -2)$. The slope of the line is computed as follows:

$$m = \frac{\text{Change in } y}{\text{Change in } x} = \frac{y_2 - y_1}{x_2 - x_1} = \frac{-2 - 4}{2 - (-3)} = \frac{-6}{5} = -\frac{6}{5}.$$

The situation is illustrated in **Figure 1.38**. The slope of the line is $-\frac{6}{5}$. For every vertical change of -6 units (6 units down), there is a corresponding horizontal change of 5 units. The slope is negative and the line falls from left to right. ● ● ●

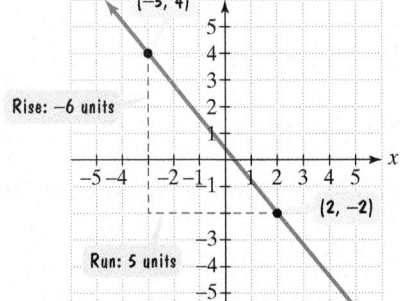

$(-3, 4)$
Rise: -6 units
Run: 5 units
$(2, -2)$

FIGURE 1.38 Visualizing a slope of $-\frac{6}{5}$

GREAT QUESTION!

Is it OK to say that a vertical line has no slope?

Always be clear in the way you use language, especially in mathematics. For example, it's not a good idea to say that a line has "no slope." This could mean that the slope is zero or that the slope is undefined.

✓ Check Point **1** Find the slope of the line passing through each pair of points:
a. $(-3, 4)$ and $(-4, -2)$ **b.** $(4, -2)$ and $(-1, 5)$.

Example 1 illustrates that a line with a positive slope is increasing and a line with a negative slope is decreasing. By contrast, a horizontal line is a constant function and has a slope of zero. A vertical line has no horizontal change, so $x_2 - x_1 = 0$ in the formula for slope. Because we cannot divide by zero, the slope of a vertical line is undefined. This discussion is summarized in **Table 1.2** at the top of the next page.

Table 1.2 Possibilities for a Line's Slope

Positive Slope	Negative Slope	Zero Slope	Undefined Slope
$m > 0$	$m < 0$	$m = 0$	m is undefined.
Line rises from left to right.	Line falls from left to right.	Line is horizontal.	Line is vertical.

② Write the point-slope form of the equation of a line.

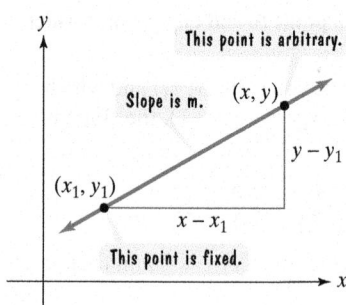

FIGURE 1.39 A line passing through (x_1, y_1) with slope m

The Point-Slope Form of the Equation of a Line

We can use the slope of a line to obtain various forms of the line's equation. For example, consider a nonvertical line that has slope m and that contains the point (x_1, y_1).

The line in **Figure 1.39** has slope m and contains the point (x_1, y_1). Let (x, y) represent any other point on the line.

Regardless of where the point (x, y) is located, the steepness of the line in **Figure 1.39** remains the same. Thus, the ratio for the slope stays a constant m. This means that for all points (x, y) along the line

$$m = \frac{\text{Change in } y}{\text{Change in } x} = \frac{y - y_1}{x - x_1}.$$

We used (x, y) instead of (x_2, y_2) because (x_2, y_2) represents a fixed point and (x, y) is not a fixed point on the line.

We can clear the fraction by multiplying both sides by $x - x_1$, the least common denominator.

$$m = \frac{y - y_1}{x - x_1} \qquad \text{This is the slope of the line in Figure 1.39.}$$

$$m(x - x_1) = \frac{y - y_1}{x - x_1} \cdot (x - x_1) \qquad \text{Multiply both sides by } x - x_1.$$

$$m(x - x_1) = y - y_1 \qquad \text{Simplify: } \frac{y - y_1}{x - x_1} \cdot (x - x_1) = y - y_1.$$

Now, if we reverse the two sides, we obtain the *point-slope form* of the equation of a line.

GREAT QUESTION!

When using $y - y_1 = m(x - x_1)$, for which variables do I substitute numbers?

When writing the point-slope form of a line's equation, you will never substitute numbers for x and y. You will substitute values for $x_1, y_1,$ and m.

Point-Slope Form of the Equation of a Line

The **point-slope form of the equation** of a nonvertical line with slope m that passes through the point (x_1, y_1) is

$$y - y_1 = m(x - x_1).$$

For example, the point-slope form of the equation of the line passing through $(1, 5)$ with slope 2 $(m = 2)$ is

$$y - 5 = 2(x - 1).$$

We will soon be expressing the equation of a nonvertical line in function notation. To do so, we need to solve the point-slope form of a line's equation for y. Example 2 illustrates how to isolate y on one side of the equal sign.

EXAMPLE 2 Writing an Equation for a Line in Point-Slope Form

Write an equation in point-slope form for the line with slope 4 that passes through the point $(-1, 3)$. Then solve the equation for y.

SOLUTION

We use the point-slope form of the equation of a line with $m = 4$, $x_1 = -1$, and $y_1 = 3$.

$$y - y_1 = m(x - x_1) \quad \text{This is the point-slope form of the equation.}$$

$$y - 3 = 4[x - (-1)] \quad \text{Substitute the given values: } m = 4 \text{ and } (x_1, y_1) = (-1, 3).$$

$$y - 3 = 4(x + 1) \quad \text{We now have an equation in point-slope form for the given line.}$$

Now we solve this equation for y.

We need to isolate y.

$$y - 3 = 4(x + 1) \quad \text{This is the point-slope form of the equation.}$$

$$y - 3 = 4x + 4 \quad \text{Use the distributive property.}$$

$$y = 4x + 7 \quad \text{Add 3 to both sides.} \qquad \bullet\bullet\bullet$$

✓ Check Point **2** Write an equation in point-slope form for the line with slope 6 that passes through the point $(2, -5)$. Then solve the equation for y.

EXAMPLE 3 Writing an Equation for a Line in Point-Slope Form

Write an equation in point-slope form for the line passing through the points $(4, -3)$ and $(-2, 6)$. (See **Figure 1.40**.) Then solve the equation for y.

SOLUTION

To use the point-slope form, we need to find the slope. The slope is the change in the y-coordinates divided by the corresponding change in the x-coordinates.

$$m = \frac{6 - (-3)}{-2 - 4} = \frac{9}{-6} = -\frac{3}{2} \quad \begin{array}{l}\text{This is the definition of slope using} \\ (4, -3) \text{ and } (-2, 6).\end{array}$$

We can take either point on the line to be (x_1, y_1). Let's use $(x_1, y_1) = (4, -3)$. Now, we are ready to write the point-slope form of the equation.

$$y - y_1 = m(x - x_1) \quad \text{This is the point-slope form of the equation.}$$

$$y - (-3) = -\tfrac{3}{2}(x - 4) \quad \text{Substitute: } (x_1, y_1) = (4, -3) \text{ and } m = -\tfrac{3}{2}.$$

$$y + 3 = -\tfrac{3}{2}(x - 4) \quad \text{Simplify.}$$

We now have an equation in point-slope form for the line shown in **Figure 1.40**. Now, we solve this equation for y.

We need to isolate y.

$$y + 3 = -\tfrac{3}{2}(x - 4) \quad \text{This is the point-slope form of the equation.}$$

$$y + 3 = -\tfrac{3}{2}x + 6 \quad \text{Use the distributive property.}$$

$$y = -\tfrac{3}{2}x + 3 \quad \text{Subtract 3 from both sides.} \qquad \bullet\bullet\bullet$$

✓ Check Point **3** Write an equation in point-slope form for the line passing through the points $(-2, -1)$ and $(-1, -6)$. Then solve the equation for y.

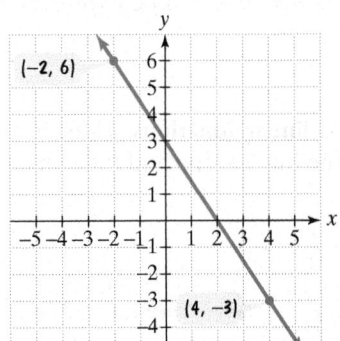

FIGURE 1.40 Write an equation in point-slope form for this line.

DISCOVERY

You can use either point for (x_1, y_1) when you write a point-slope equation for a line. Rework Example 3 using $(-2, 6)$ for (x_1, y_1). Once you solve for y, you should still obtain

$$y = -\tfrac{3}{2}x + 3.$$

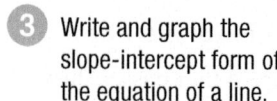

③ Write and graph the slope-intercept form of the equation of a line.

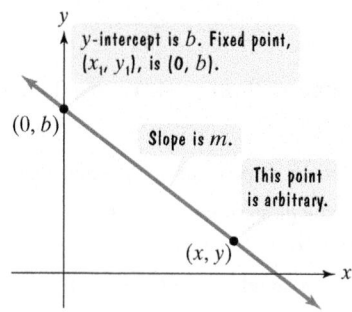

FIGURE 1.41 A line with slope m and y-intercept b

The Slope-Intercept Form of the Equation of a Line

Let's write the point-slope form of the equation of a nonvertical line with slope m and y-intercept b. The line is shown in **Figure 1.41**. Because the y-intercept is b, the line passes through $(0, b)$. We use the point-slope form with $x_1 = 0$ and $y_1 = b$.

$$y - y_1 = m(x - x_1)$$

Let $y_1 = b$. Let $x_1 = 0$.

We obtain

$$y - b = m(x - 0).$$

Simplifying on the right side gives us

$$y - b = mx.$$

Finally, we solve for y by adding b to both sides.

$$y = mx + b$$

Thus, if a line's equation is written with y isolated on one side, the coefficient of x is the line's slope and the constant term is the y-intercept. This form of a line's equation is called the *slope-intercept form* of the line.

Slope-Intercept Form of the Equation of a Line

The **slope-intercept form of the equation** of a nonvertical line with slope m and y-intercept b is

$$y = mx + b.$$

The slope-intercept form of a line's equation, $y = mx + b$, can be expressed in function notation by replacing y with $f(x)$:

$$f(x) = mx + b.$$

We have seen that functions in this form are called **linear functions**. Thus, in the equation of a linear function, the coefficient of x is the line's slope and the constant term is the y-intercept. Here are two examples:

$$y = 2x - 4 \qquad\qquad f(x) = \tfrac{1}{2}x + 2.$$

The slope is 2. The y-intercept is −4. The slope is $\tfrac{1}{2}$. The y-intercept is 2.

If a linear function's equation is in slope-intercept form, we can use the y-intercept and the slope to obtain its graph.

GREAT QUESTION!

If the slope is an integer, such as 2, why should I express it as $\frac{2}{1}$ for graphing purposes?

Writing the slope, m, as a fraction allows you to identify the rise (the fraction's numerator) and the run (the fraction's denominator).

Graphing $y = mx + b$ Using the Slope and y-Intercept

1. Plot the point containing the y-intercept on the y-axis. This is the point $(0, b)$.
2. Obtain a second point using the slope, m. Write m as a fraction, and use rise over run, starting at the point containing the y-intercept, to plot this point.
3. Use a straightedge to draw a line through the two points. Draw arrowheads at the ends of the line to show that the line continues indefinitely in both directions.

EXAMPLE 4 Graphing Using the Slope and y-Intercept

Graph the linear function: $f(x) = -\dfrac{3}{2}x + 2.$

SOLUTION

The equation of the line is in the form $f(x) = mx + b$. We can find the slope, m, by identifying the coefficient of x. We can find the y-intercept, b, by identifying the constant term.

$$f(x) = -\tfrac{3}{2}x + 2$$

The slope is $-\tfrac{3}{2}$. The y-intercept is 2.

Now that we have identified the slope, $-\tfrac{3}{2}$, and the y-intercept, 2, we use the three-step procedure to graph the equation.

Step 1 Plot the point containing the y-intercept on the y-axis. The y-intercept is 2. We plot $(0, 2)$, shown in **Figure 1.42**.

Step 2 Obtain a second point using the slope, m. Write m as a fraction, and use rise over run, starting at the point containing the y-intercept, to plot this point. The slope, $-\tfrac{3}{2}$, is already written as a fraction.

$$m = -\frac{3}{2} = \frac{-3}{2} = \frac{\text{Rise}}{\text{Run}}$$

We plot the second point on the line by starting at $(0, 2)$, the first point. Based on the slope, we move 3 units *down* (the rise) and 2 units to the *right* (the run). This puts us at a second point on the line, $(2, -1)$, shown in **Figure 1.42**.

Step 3 Use a straightedge to draw a line through the two points. The graph of the linear function $f(x) = -\tfrac{3}{2}x + 2$ is shown as a blue line in **Figure 1.42**. •••

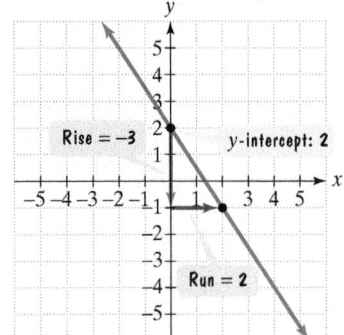

FIGURE 1.42 The graph of $f(x) = -\tfrac{3}{2}x + 2$

⊘ Check Point **4** Graph the linear function: $f(x) = \tfrac{3}{5}x + 1$.

④ Graph horizontal or vertical lines.

Equations of Horizontal and Vertical Lines

If a line is horizontal, its slope is zero: $m = 0$. Thus, the equation $y = mx + b$ becomes $y = b$, where b is the y-intercept. All horizontal lines have equations of the form $y = b$.

EXAMPLE 5 Graphing a Horizontal Line

Graph $y = -4$ in the rectangular coordinate system.

SOLUTION

All ordered pairs that are solutions of $y = -4$ have a value of y that is always -4. Any value can be used for x. In the table on the right, we have selected three of the possible values for x: $-2, 0$, and 3. The table shows that three ordered pairs that are solutions of $y = -4$ are $(-2, -4)$, $(0, -4)$, and $(3, -4)$. Drawing a line that passes through the three points gives the horizontal line shown in **Figure 1.43**. •••

x	$y = -4$	(x, y)
-2	-4	$(-2, -4)$
0	-4	$(0, -4)$
3	-4	$(3, -4)$

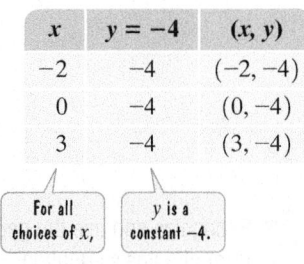

For all choices of x, y is a constant -4.

FIGURE 1.43 The graph of $y = -4$ or $f(x) = -4$

⊘ Check Point **5** Graph $y = 3$ in the rectangular coordinate system.

Equation of a Horizontal Line

A horizontal line is given by an equation of the form

$$y = b,$$

where b is the y-intercept of the line. The slope of a horizontal line is zero.

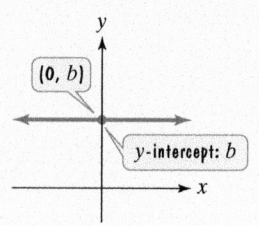

Because any vertical line can intersect the graph of a horizontal line $y = b$ only once, a horizontal line is the graph of a function. Thus, we can express the equation $y = b$ as $f(x) = b$. This linear function is often called a **constant function**.

Next, let's see what we can discover about the graph of an equation of the form $x = a$ by looking at an example.

EXAMPLE 6 Graphing a Vertical Line

Graph the linear equation: $x = 2$.

SOLUTION

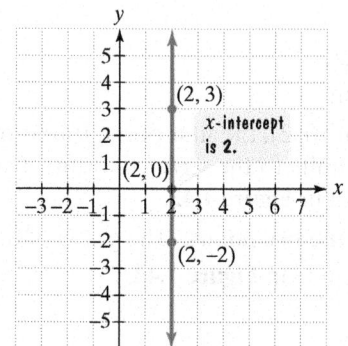

FIGURE 1.44 The graph of $x = 2$

All ordered pairs that are solutions of $x = 2$ have a value of x that is always 2. Any value can be used for y. In the table on the right, we have selected three of the possible values for y: -2, 0, and 3. The table shows that three ordered pairs that are solutions of $x = 2$ are $(2, -2)$, $(2, 0)$, and $(2, 3)$. Drawing a line that passes through the three points gives the vertical line shown in **Figure 1.44**.

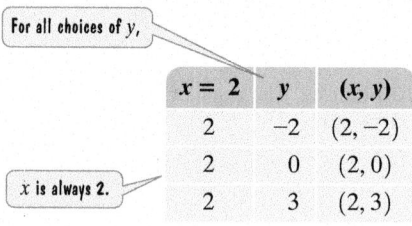

$x = 2$	y	(x, y)
2	-2	$(2, -2)$
2	0	$(2, 0)$
2	3	$(2, 3)$

• • •

Equation of a Vertical Line

A vertical line is given by an equation of the form

$$x = a,$$

where a is the x-intercept of the line. The slope of a vertical line is undefined.

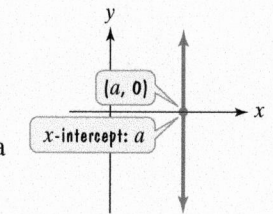

Does a vertical line represent the graph of a linear function? No. Look at the graph of $x = 2$ in **Figure 1.44**. A vertical line drawn through $(2, 0)$ intersects the graph infinitely many times. This shows that infinitely many outputs are associated with the input 2. **No vertical line represents a linear function.**

✓ Check Point 6 Graph the linear equation: $x = -3$.

⑤ Recognize and use the general form of a line's equation.

The General Form of the Equation of a Line

The vertical line whose equation is $x = 5$ cannot be written in slope-intercept form, $y = mx + b$, because its slope is undefined. However, every line has an equation that can be expressed in the form $Ax + By + C = 0$. For example, $x = 5$ can be expressed as $1x + 0y - 5 = 0$, or $x - 5 = 0$. The equation $Ax + By + C = 0$ is called the *general form* of the equation of a line.

General Form of the Equation of a Line

Every line has an equation that can be written in the **general form**

$$Ax + By + C = 0,$$

where A, B, and C are real numbers, and A and B are not both zero.

GREAT QUESTION!

In the general form
$Ax + By + C = 0$, **can I immediately determine that the slope is A and the y-intercept is B?**

No. Avoid this common error. You need to solve $Ax + By + C = 0$ for y before finding the slope and the y-intercept.

If the equation of a nonvertical line is given in general form, it is possible to find the slope, m, and the y-intercept, b, for the line. We solve the equation for y, transforming it into the slope-intercept form $y = mx + b$. In this form, the coefficient of x is the slope of the line and the constant term is its y-intercept.

EXAMPLE 7 Finding the Slope and the y-Intercept

Find the slope and the y-intercept of the line whose equation is $3x + 2y - 4 = 0$.

SOLUTION

The equation is given in general form. We begin by rewriting it in the form $y = mx + b$. We need to solve for y.

> Our goal is to isolate y.

$3x + 2y - 4 = 0$	This is the given equation.
$2y = -3x + 4$	Isolate the term containing y by adding $-3x + 4$ to both sides.
$\dfrac{2y}{2} = \dfrac{-3x + 4}{2}$	Divide both sides by 2.
$y = -\dfrac{3}{2}x + 2$	On the right, divide each term in the numerator by 2 to obtain slope-intercept form.

slope y-intercept

The coefficient of x, $-\frac{3}{2}$, is the slope and the constant term, 2, is the y-intercept. This is the form of the equation that we graphed in **Figure 1.42** on page 157. • • •

⌀ Check Point **7** Find the slope and the y-intercept of the line whose equation is $3x + 6y - 12 = 0$. Then use the y-intercept and the slope to graph the equation.

⑥ Use intercepts to graph the general form of a line's equation.

Using Intercepts to Graph $Ax + By + C = 0$

Example 7 and Check Point 7 illustrate that one way to graph the general form of a line's equation is to convert to slope-intercept form, $y = mx + b$. Then use the slope and the y-intercept to obtain the graph.

A second method for graphing $Ax + By + C = 0$ uses intercepts. This method does not require rewriting the general form in a different form.

Using Intercepts to Graph $Ax + By + C = 0$

1. Find the x-intercept. Let $y = 0$ and solve for x. Plot the point containing the x-intercept on the x-axis.

2. Find the y-intercept. Let $x = 0$ and solve for y. Plot the point containing the y-intercept on the y-axis.

3. Use a straightedge to draw a line through the two points containing the intercepts. Draw arrowheads at the ends of the line to show that the line continues indefinitely in both directions.

EXAMPLE 8 Using Intercepts to Graph a Linear Equation

Graph using intercepts: $4x - 3y - 6 = 0$.

SOLUTION

FIGURE 1.45 The graph of $4x - 3y - 6 = 0$

Step 1 Find the x-intercept. Let $y = 0$ and solve for x.

$$4x - 3 \cdot 0 - 6 = 0 \qquad \text{Replace y with 0 in } 4x - 3y - 6 = 0.$$
$$4x - 6 = 0 \qquad \text{Simplify.}$$
$$4x = 6 \qquad \text{Add 6 to both sides.}$$
$$x = \frac{6}{4} = \frac{3}{2} \qquad \text{Divide both sides by 4.}$$

The x-intercept is $\frac{3}{2}$, so the line passes through $\left(\frac{3}{2}, 0\right)$ or $(1.5, 0)$, as shown in **Figure 1.45**.

Step 2 Find the y-intercept. Let $x = 0$ and solve for y.

$$4 \cdot 0 - 3y - 6 = 0 \qquad \text{Replace x with 0 in } 4x - 3y - 6 = 0.$$
$$-3y - 6 = 0 \qquad \text{Simplify.}$$
$$-3y = 6 \qquad \text{Add 6 to both sides.}$$
$$y = -2 \qquad \text{Divide both sides by } -3.$$

The y-intercept is -2, so the line passes through $(0, -2)$, as shown in **Figure 1.45**.

Step 3 Graph the equation by drawing a line through the two points containing the intercepts. The graph of $4x - 3y - 6 = 0$ is shown in **Figure 1.45**. • • •

☑ **Check Point 8** Graph using intercepts: $3x - 2y - 6 = 0$.

We've covered a lot of territory. Let's take a moment to summarize the various forms for equations of lines.

Equations of Lines

1. Point-slope form	$y - y_1 = m(x - x_1)$
2. Slope-intercept form	$y = mx + b$ or $f(x) = mx + b$
3. Horizontal line	$y = b$
4. Vertical line	$x = a$
5. General form	$Ax + By + C = 0$

⑦ Model data with linear functions and make predictions.

Applications

Linear functions are useful for modeling data that fall on or near a line.

EXAMPLE 9 Modeling Global Warming

The amount of carbon dioxide in the atmosphere, measured in parts per million, has been increasing as a result of the burning of oil and coal. The buildup of gases and particles traps heat and raises the planet's temperature. The bar graph in **Figure 1.46(a)** at the top of the next page gives the average atmospheric concentration of carbon dioxide and the average global temperature for six selected years. The data are displayed as a set of six points in a rectangular coordinate system in **Figure 1.46(b)**.

FIGURE 1.46(a)

FIGURE 1.46(b)

Source: National Oceanic and Atmospheric Administration

a. Shown on the scatter plot in **Figure 1.46(b)** is a line that passes through or near the six points. Write the slope-intercept form of this equation using function notation.

b. The preindustrial concentration of atmospheric carbon dioxide was 280 parts per million. The United Nations' Intergovernmental Panel on Climate Change predicts global temperatures will rise between 2°F and 5°F if carbon dioxide concentration doubles from the preindustrial level. Compared to the average global temperature of 57.99°F for 2009, how well does the function from part (a) model this prediction?

SOLUTION

a. The line in **Figure 1.46(b)** passes through (326, 57.06) and (385, 57.99). We start by finding its slope.

$$m = \frac{\text{Change in } y}{\text{Change in } x} = \frac{57.99 - 57.06}{385 - 326} = \frac{0.93}{59} \approx 0.02$$

The slope indicates that for each increase of one part per million in carbon dioxide concentration, the average global temperature is increasing by approximately 0.02°F.

Now we write the line's equation in slope-intercept form.

$y - y_1 = m(x - x_1)$ Begin with the point-slope form.

$y - 57.06 = 0.02(x - 326)$ Either ordered pair can be (x_1, y_1). Let $(x_1, y_1) = (326, 57.06)$. From above, $m \approx 0.02$.

$y - 57.06 = 0.02x - 6.52$ Apply the distributive property: $0.02(326) = 6.52$.

$y = 0.02x + 50.54$ Add 57.06 to both sides and solve for y.

A linear function that models average global temperature, $f(x)$, for an atmospheric carbon dioxide concentration of x parts per million is

$$f(x) = 0.02x + 50.54.$$

TECHNOLOGY

You can use a graphing utility to obtain a model for a scatter plot in which the data points fall on or near a straight line. After entering the data in **Figure 1.46(a)** on the previous page, a graphing utility displays a scatter plot of the data and the regression line, that is, the line that best fits the data.

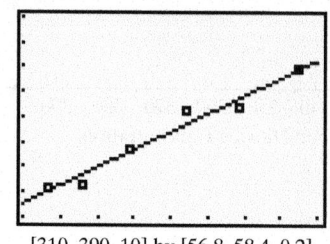

[310, 390, 10] by [56.8, 58.4, 0.2]

Also displayed is the regression line's equation.

```
LinReg
y=ax+b
a=.0142119834
b=52.50782579
```

b. If carbon dioxide concentration doubles from its preindustrial level of 280 parts per million, which many experts deem very likely, the concentration will reach 280×2, or 560 parts per million. We use the linear function to predict average global temperature at this concentration.

$$f(x) = 0.02x + 50.54 \qquad \text{Use the function from part (a).}$$

$$f(560) = 0.02(560) + 50.54 \qquad \text{Substitute 560 for } x.$$

$$= 11.2 + 50.54 = 61.74$$

Our model projects an average global temperature of 61.74°F for a carbon dioxide concentration of 560 parts per million. Compared to the average global temperature of 57.99° for 2009 shown in **Figure 1.46(a)** on the previous page, this is an increase of

$$61.74°F - 57.99°F = 3.75°F.$$

This is consistent with a rise between 2°F and 5°F as predicted by the Intergovernmental Panel on Climate Change. •••

 Check Point **9** Use the data points (317, 57.04) and (354, 57.64), shown, but not labeled, in **Figure 1.46(b)** on the previous page to obtain a linear function that models average global temperature, $f(x)$, for an atmospheric carbon dioxide concentration of x parts per million. Round m to three decimal places and b to one decimal place. Then use the function to project average global temperature at a concentration of 600 parts per million.

CONCEPT AND VOCABULARY CHECK

Fill in each blank so that the resulting statement is true.

1. Data presented in a visual form as a set of points is called a/an _____. A line that best fits this set of points is called a/an _____ line.

2. The slope, m, of a line through the distinct points (x_1, y_1) and (x_2, y_2) is given by the formula $m = $ _____.

3. If a line rises from left to right, the line has _____ slope.

4. If a line falls from left to right, the line has _____ slope.

5. The slope of a horizontal line is _____.

6. The slope of a vertical line is _____.

7. The point-slope form of the equation of a nonvertical line with slope m that passes through the point (x_1, y_1) is _____.

8. The slope-intercept form of the equation of a line is _____, where m represents the _____ and b represents the _____.

9. In order to graph the line whose equation is

$y = \dfrac{2}{5}x + 3$, begin by plotting the point _____.

From this point, we move _____ units up (the rise) and _____ units to the right (the run).

10. The graph of the equation $y = 3$ is a/an _____ line.

11. The graph of the equation $x = -2$ is a/an _____ line.

12. The equation $Ax + By + C = 0$, where A and B are not both zero, is called the _____ form of the equation of a line.

EXERCISE SET 1.4

Practice Exercises

In Exercises 1–10, find the slope of the line passing through each pair of points or state that the slope is undefined. Then indicate whether the line through the points rises, falls, is horizontal, or is vertical.

1. $(4, 7)$ and $(8, 10)$
2. $(2, 1)$ and $(3, 4)$
3. $(-2, 1)$ and $(2, 2)$
4. $(-1, 3)$ and $(2, 4)$
5. $(4, -2)$ and $(3, -2)$
6. $(4, -1)$ and $(3, -1)$
7. $(-2, 4)$ and $(-1, -1)$
8. $(6, -4)$ and $(4, -2)$
9. $(5, 3)$ and $(5, -2)$
10. $(3, -4)$ and $(3, 5)$

In Exercises 11–38, use the given conditions to write an equation for each line in point-slope form and slope-intercept form.

11. Slope $= 2$, passing through $(3, 5)$
12. Slope $= 4$, passing through $(1, 3)$
13. Slope $= 6$, passing through $(-2, 5)$
14. Slope $= 8$, passing through $(4, -1)$
15. Slope $= -3$, passing through $(-2, -3)$
16. Slope $= -5$, passing through $(-4, -2)$
17. Slope $= -4$, passing through $(-4, 0)$
18. Slope $= -2$, passing through $(0, -3)$
19. Slope $= -1$, passing through $\left(-\frac{1}{2}, -2\right)$
20. Slope $= -1$, passing through $\left(-4, -\frac{1}{4}\right)$
21. Slope $= \frac{1}{2}$, passing through the origin
22. Slope $= \frac{1}{3}$, passing through the origin
23. Slope $= -\frac{2}{3}$, passing through $(6, -2)$
24. Slope $= -\frac{3}{5}$, passing through $(10, -4)$
25. Passing through $(1, 2)$ and $(5, 10)$
26. Passing through $(3, 5)$ and $(8, 15)$
27. Passing through $(-3, 0)$ and $(0, 3)$
28. Passing through $(-2, 0)$ and $(0, 2)$
29. Passing through $(-3, -1)$ and $(2, 4)$
30. Passing through $(-2, -4)$ and $(1, -1)$
31. Passing through $(-3, -2)$ and $(3, 6)$
32. Passing through $(-3, 6)$ and $(3, -2)$
33. Passing through $(-3, -1)$ and $(4, -1)$
34. Passing through $(-2, -5)$ and $(6, -5)$
35. Passing through $(2, 4)$ with x-intercept $= -2$
36. Passing through $(1, -3)$ with x-intercept $= -1$
37. x-intercept $= -\frac{1}{2}$ and y-intercept $= 4$
38. x-intercept $= 4$ and y-intercept $= -2$

In Exercises 39–48, give the slope and y-intercept of each line whose equation is given. Then graph the linear function.

39. $y = 2x + 1$
40. $y = 3x + 2$
41. $f(x) = -2x + 1$
42. $f(x) = -3x + 2$
43. $f(x) = \dfrac{3}{4}x - 2$
44. $f(x) = \dfrac{3}{4}x - 3$
45. $y = -\dfrac{3}{5}x + 7$
46. $y = -\dfrac{2}{5}x + 6$
47. $g(x) = -\dfrac{1}{2}x$
48. $g(x) = -\dfrac{1}{3}x$

In Exercises 49–58, graph each equation in a rectangular coordinate system.

49. $y = -2$
50. $y = 4$
51. $x = -3$
52. $x = 5$
53. $y = 0$
54. $x = 0$
55. $f(x) = 1$
56. $f(x) = 3$
57. $3x - 18 = 0$
58. $3x + 12 = 0$

In Exercises 59–66,

 a. *Rewrite the given equation in slope-intercept form.*
 b. *Give the slope and y-intercept.*
 c. *Use the slope and y-intercept to graph the linear function.*

59. $3x + y - 5 = 0$
60. $4x + y - 6 = 0$
61. $2x + 3y - 18 = 0$
62. $4x + 6y + 12 = 0$
63. $8x - 4y - 12 = 0$
64. $6x - 5y - 20 = 0$
65. $3y - 9 = 0$
66. $4y + 28 = 0$

In Exercises 67–72, use intercepts to graph each equation.

67. $6x - 2y - 12 = 0$
68. $6x - 9y - 18 = 0$
69. $2x + 3y + 6 = 0$
70. $3x + 5y + 15 = 0$
71. $8x - 2y + 12 = 0$
72. $6x - 3y + 15 = 0$

Practice Plus

In Exercises 73–76, find the slope of the line passing through each pair of points or state that the slope is undefined. Assume that all variables represent positive real numbers. Then indicate whether the line through the points rises, falls, is horizontal, or is vertical.

73. $(0, a)$ and $(b, 0)$
74. $(-a, 0)$ and $(0, -b)$
75. (a, b) and $(a, b + c)$
76. $(a - b, c)$ and $(a, a + c)$

In Exercises 77–78, give the slope and y-intercept of each line whose equation is given. Assume that $B \neq 0$.

77. $Ax + By = C$
78. $Ax = By - C$

In Exercises 79–80, find the value of y if the line through the two given points is to have the indicated slope.

79. $(3, y)$ and $(1, 4)$, $m = -3$
80. $(-2, y)$ and $(4, -4)$, $m = \frac{1}{3}$

In Exercises 81–82, graph each linear function.

81. $3x - 4f(x) - 6 = 0$
82. $6x - 5f(x) - 20 = 0$

83. If one point on a line is $(3, -1)$ and the line's slope is -2, find the y-intercept.

84. If one point on a line is $(2, -6)$ and the line's slope is $-\frac{3}{2}$, find the y-intercept.

Use the figure to make the lists in Exercises 85–86.

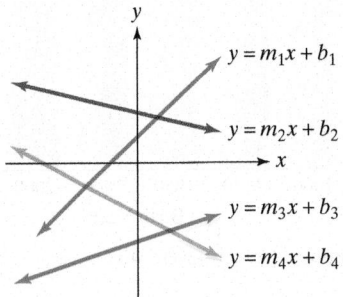

85. List the slopes m_1, m_2, m_3, and m_4 in order of decreasing size.
86. List the y-intercepts b_1, b_2, b_3, and b_4 in order of decreasing size.

Application Exercises

Americans are getting married later in life or not getting married at all. In 2008, nearly half of Americans ages 25 through 29 were unmarried. The following bar graph shows the percentage of never-married men and women in this age group. The data are displayed as two sets of four points each, one scatter plot for the percentage of never-married American men and one for the percentage of never-married American women. Also shown for each scatter plot is a line that passes through or near the four points. Use these lines to solve Exercises 87–88.

Percentage of United States Population Never Married, Ages 25–29

Source: U.S. Census Bureau

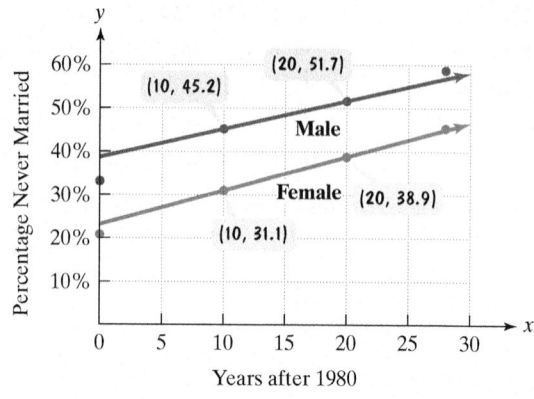

87. In this exercise, you will use the blue line for the women shown on the scatter plot to develop a model for the percentage of never-married American females ages 25–29.

 a. Use the two points whose coordinates are shown by the voice balloons to find the point-slope form of the equation of the line that models the percentage of never-married American females ages 25–29, y, x years after 1980.

 b. Write the equation from part (a) in slope-intercept form. Use function notation.

 c. Use the linear function to predict the percentage of never-married American females, ages 25–29, in 2020.

The bar graph gives the life expectancy for American men and women born in six selected years. In Exercises 89–90, you will use the data to obtain models for life expectancy and make predictions about how long American men and women will live in the future.

Life Expectancy in the United States, by Year of Birth

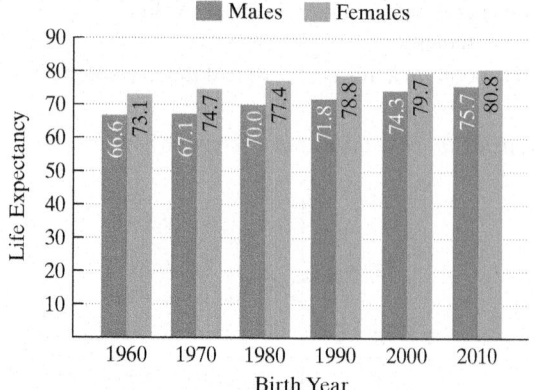

Source: National Center for Health Statistics

88. In this exercise, you will use the red line for the men shown on the scatter plot to develop a model for the percentage of never-married American males ages 25–29.

 a. Use the two points whose coordinates are shown by the voice balloons to find the point-slope form of the equation of the line that models the percentage of never-married American males ages 25–29, y, x years after 1980.

 b. Write the equation from part (a) in slope-intercept form. Use function notation.

 c. Use the linear function to predict the percentage of never-married American males, ages 25–29, in 2015.

89. Use the data for males shown in the bar graph at the bottom of the previous column to solve this exercise.

 a. Let x represent the number of birth years after 1960 and let y represent male life expectancy. Create a scatter plot that displays the data as a set of six points in a rectangular coordinate system.

 b. Draw a line through the two points that show male life expectancies for 1980 and 2000. Use the coordinates of these points to write a linear function that models life expectancy, $E(x)$, for American men born x years after 1960.

 c. Use the function from part (b) to project the life expectancy of American men born in 2020.

90. Use the data for females shown in the bar graph at the bottom of the previous column to solve this exercise.

 a. Let x represent the number of birth years after 1960 and let y represent female life expectancy. Create a scatter plot that displays the data as a set of six points in a rectangular coordinate system.

 b. Draw a line through the two points that show female life expectancies for 1970 and 2000. Use the coordinates of these points to write a linear function that models life expectancy, $E(x)$, for American women born x years after 1960. Round the slope to two decimal places.

 c. Use the function from part (b) to project the life expectancy of American women born in 2020.

91. Shown, again, is the scatter plot that indicates a relationship between the percentage of adult females in a country who are literate and the mortality of children under five. Also shown is a line that passes through or near the points. Find a linear function that models the data by finding the slope-intercept form of the line's equation. Use the function to make a prediction about child mortality based on the percentage of adult females in a country who are literate.

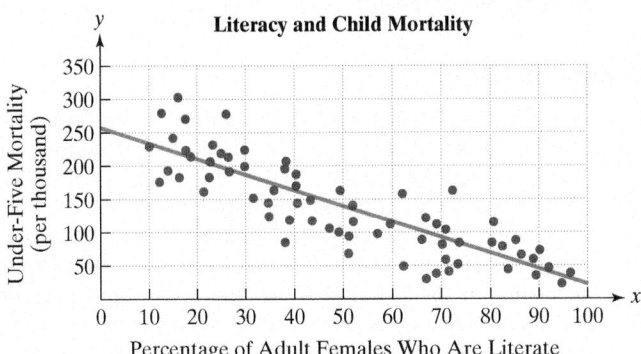

Literacy and Child Mortality

Source: United Nations

92. Just as money doesn't buy happiness for individuals, the two don't necessarily go together for countries either. However, the scatter plot does show a relationship between a country's annual per capita income and the percentage of people in that country who call themselves "happy."

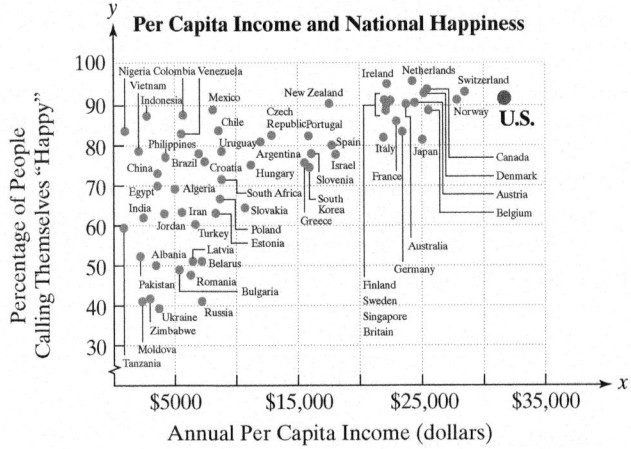

Per Capita Income and National Happiness

Source: Richard Layard, *Happiness: Lessons from a New Science*, Penguin, 2005

Draw a line that fits the data so that the spread of the data points around the line is as small as possible. Use the coordinates of two points along your line to write the slope-intercept form of its equation. Express the equation in function notation and use the linear function to make a prediction about national happiness based on per capita income.

Writing in Mathematics

93. What is the slope of a line and how is it found?

94. Describe how to write the equation of a line if the coordinates of two points along the line are known.

95. Explain how to derive the slope-intercept form of a line's equation, $y = mx + b$, from the point-slope form
$$y - y_1 = m(x - x_1).$$

96. Explain how to graph the equation $x = 2$. Can this equation be expressed in slope-intercept form? Explain.

97. Explain how to use the general form of a line's equation to find the line's slope and y-intercept.

98. Explain how to use intercepts to graph the general form of a line's equation.

99. Take another look at the scatter plot in Exercise 91. Although there is a relationship between literacy and child mortality, we cannot conclude that increased literacy causes child mortality to decrease. Offer two or more possible explanations for the data in the scatter plot.

Technology Exercises

Use a graphing utility to graph each equation in Exercises 100–103. Then use the [TRACE] *feature to trace along the line and find the coordinates of two points. Use these points to compute the line's slope. Check your result by using the coefficient of x in the line's equation.*

100. $y = 2x + 4$ **101.** $y = -3x + 6$

102. $y = -\frac{1}{2}x - 5$ **103.** $y = \frac{3}{4}x - 2$

104. Is there a relationship between wine consumption and deaths from heart disease? The table gives data from 19 developed countries.

Country	A	B	C	D	E	F	G France
Liters of alcohol from drinking wine, per person per year (x)	2.5	3.9	2.9	2.4	2.9	0.8	9.1
Deaths from heart disease, per 100,000 people per year (y)	211	167	131	191	220	297	71

Country	H	I	J	K	L	M	N	O	P	Q	R U.S.	S
(x)	0.8	0.7	7.9	1.8	1.9	0.8	6.5	1.6	5.8	1.3	1.2	2.7
(y)	211	300	107	167	266	227	86	207	115	285	199	172

Source: *New York Times*

a. Use the statistical menu of your graphing utility to enter the 19 ordered pairs of data items shown in the table.

b. Use the scatter plot capability to draw a scatter plot of the data.

c. Select the linear regression option. Use your utility to obtain values for a and b for the equation of the regression line, $y = ax + b$. You may also be given a **correlation coefficient**, r. Values of r close to 1 indicate that the points can be described by a linear relationship and the regression line has a positive slope. Values of r close to -1 indicate that the points can be described by a linear relationship and the regression line has a negative slope. Values of r close to 0 indicate no linear relationship between the variables. In this case, a linear model does not accurately describe the data.

d. Use the appropriate sequence (consult your manual) to graph the regression equation on top of the points in the scatter plot.

Critical Thinking Exercises

Make Sense? *In Exercises 105–108, determine whether each statement makes sense or does not make sense, and explain your reasoning.*

105. The graph of my linear function at first increased, reached a maximum point, and then decreased.

106. A linear function that models tuition and fees at public four-year colleges from 2000 through 2012 has negative slope.

107. Because the variable m does not appear in $Ax + By + C = 0$, equations in this form make it impossible to determine the line's slope.

108. The federal minimum wage was \$5.15 per hour from 1997 through 2006, so $f(x) = 5.15$ models the minimum wage, $f(x)$, in dollars, for the domain $\{1997, 1998, 1999, \ldots, 2006\}$.

In Exercises 109–112, determine whether each statement is true or false. If the statement is false, make the necessary change(s) to produce a true statement.

109. The equation $y = mx + b$ shows that no line can have a y-intercept that is numerically equal to its slope.

110. Every line in the rectangular coordinate system has an equation that can be expressed in slope-intercept form.

111. The graph of the linear function $5x + 6y - 30 = 0$ is a line passing through the point $(6, 0)$ with slope $-\frac{5}{6}$.

112. The graph of $x = 7$ in the rectangular coordinate system is the single point $(7, 0)$.

In Exercises 113–114, find the coefficients that must be placed in each shaded area so that the function's graph will be a line satisfying the specified conditions.

113. $\quad x + \quad y - 12 = 0$; x-intercept $= -2$; y-intercept $= 4$

114. $\quad x + \quad y - 12 = 0$; y-intercept $= -6$; slope $= \dfrac{1}{2}$

115. Prove that the equation of a line passing through $(a, 0)$ and $(0, b)(a \neq 0, b \neq 0)$ can be written in the form $\dfrac{x}{a} + \dfrac{y}{b} = 1$. Why is this called the *intercept form* of a line?

116. Excited about the success of celebrity stamps, post office officials were rumored to have put forth a plan to institute two new types of thermometers. On these new scales, $°E$ represents degrees Elvis and $°M$ represents degrees Madonna. If it is known that $40°E = 25°M, 280°E = 125°M$, and degrees Elvis is linearly related to degrees Madonna, write an equation expressing E in terms of M.

Group Exercise

117. In Exercises 87–88, we used the data in a bar graph to develop linear functions that modeled the percentage of never-married American females and males, ages 25–29. For this group exercise, you might find it helpful to pattern your work after Exercises 87 and 88. Group members should begin by consulting an almanac, newspaper, magazine, or the Internet to find data that appear to lie approximately on or near a line. Working by hand or using a graphing utility, group members should construct scatter plots for the data that were assembled. If working by hand, draw a line that approximately fits the data in each scatter plot and then write its equation as a function in slope-intercept form. If using a graphing utility, obtain the equation of each regression line. Then use each linear function's equation to make predictions about what might occur in the future. Are there circumstances that might affect the accuracy of the prediction? List some of these circumstances.

Preview Exercises

Exercises 118–120 will help you prepare for the material covered in the next section.

118. Write the slope-intercept form of the equation of the line passing through $(-3, 1)$ whose slope is the same as the line whose equation is $y = 2x + 1$.

119. Write an equation in general form of the line passing through $(3, -5)$ whose slope is the negative reciprocal (the reciprocal with the opposite sign) of $-\frac{1}{4}$.

120. If $f(x) = x^2$, find

$$\frac{f(x_2) - f(x_1)}{x_2 - x_1},$$

where $x_1 = 1$ and $x_2 = 4$.

SECTION 1.5 More on Slope

Objectives

1. Find slopes and equations of parallel and perpendicular lines.
2. Interpret slope as rate of change.
3. Find a function's average rate of change.

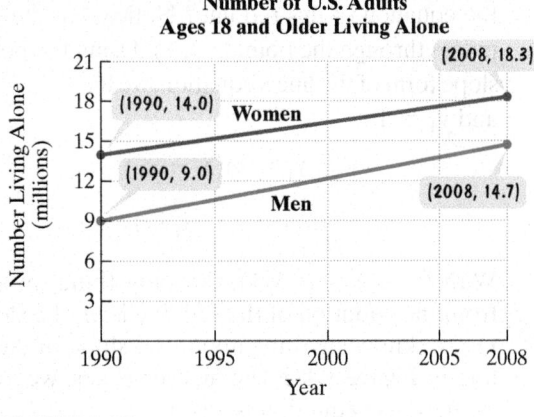

Number of U.S. Adults Ages 18 and Older Living Alone

(1990, 14.0) Women
(2008, 18.3)
(1990, 9.0)
Men
(2008, 14.7)

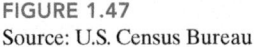

FIGURE 1.47
Source: U.S. Census Bureau

A best guess at the future of our nation indicates that the numbers of men and women living alone will increase each year. **Figure 1.47** shows that in 2008, 14.7 million men and 18.3 million women lived alone, an increase over the numbers displayed in the graph for 1990.

Take a second look at **Figure 1.47**. Can you tell that the green graph representing men has a greater slope than the red graph representing women? This indicates a greater rate of change in the number of men living alone than in the number of women living alone over the period from 1990 through 2008. In this section, you will learn to interpret slope as a rate of change. You will also explore the relationships between slopes of parallel and perpendicular lines.

① Find slopes and equations of parallel and perpendicular lines.

Parallel and Perpendicular Lines

Two nonintersecting lines that lie in the same plane are **parallel**. If two lines do not intersect, the ratio of the vertical change to the horizontal change is the same for both lines. Because two parallel lines have the same "steepness," they must have the same slope.

Slope and Parallel Lines

1. If two nonvertical lines are parallel, then they have the same slope.
2. If two distinct nonvertical lines have the same slope, then they are parallel.
3. Two distinct vertical lines, both with undefined slopes, are parallel.

EXAMPLE 1 Writing Equations of a Line Parallel to a Given Line

Write an equation of the line passing through $(-3, 1)$ and parallel to the line whose equation is $y = 2x + 1$. Express the equation in point-slope form and slope-intercept form.

SOLUTION

The situation is illustrated in **Figure 1.48**. We are looking for the equation of the red line passing through $(-3, 1)$ and parallel to the blue line whose equation is $y = 2x + 1$. How do we obtain the equation of this red line? Notice that the line passes through the point $(-3, 1)$. Using the point-slope form of the line's equation, we have $x_1 = -3$ and $y_1 = 1$.

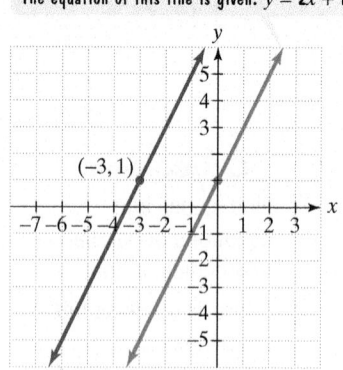

The equation of this line is given: $y = 2x + 1$.

$$y - y_1 = m(x - x_1)$$

$$y_1 = 1 \qquad x_1 = -3$$

With $(x_1, y_1) = (-3, 1)$, the only thing missing from the equation of the red line is m, the slope. Do we know anything about the slope of either line in **Figure 1.48**? The answer is yes; we know the slope of the blue line on the right, whose equation is given.

We must write the equation of this line.

FIGURE 1.48

$$y = 2x + 1$$

The slope of the blue line on the right in **Figure 1.48** is 2.

Parallel lines have the same slope. Because the slope of the blue line is 2, the slope of the red line, the line whose equation we must write, is also 2: $m = 2$. We now have values for x_1, y_1, and m for the red line.

$$y - y_1 = m(x - x_1)$$

$$y_1 = 1 \qquad m = 2 \qquad x_1 = -3$$

The point-slope form of the red line's equation is

$$y - 1 = 2[x - (-3)] \text{ or}$$
$$y - 1 = 2(x + 3).$$

Solving for y, we obtain the slope-intercept form of the equation.

$$y - 1 = 2x + 6 \quad \text{Apply the distributive property.}$$
$$y = 2x + 7 \quad \text{Add 1 to both sides. This is the slope-intercept form, } y = mx + b, \text{ of the equation. Using function notation, the equation is } f(x) = 2x + 7. \quad \bullet\bullet\bullet$$

⊘ **Check Point 1** Write an equation of the line passing through $(-2, 5)$ and parallel to the line whose equation is $y = 3x + 1$. Express the equation in point-slope form and slope-intercept form.

Two lines that intersect at a right angle (90°) are said to be **perpendicular**, shown in **Figure 1.49**. The relationship between the slopes of perpendicular lines is not as obvious as the relationship between parallel lines. **Figure 1.49** shows line AB, with slope $\frac{c}{d}$. Rotate line AB counterclockwise 90° to the left to obtain line $A'B'$, perpendicular to line AB. The figure indicates that the rise and the run of the new line are reversed from the original line, but the former rise, the new run, is now negative. This means that the slope of the new line is $-\frac{d}{c}$. Notice that the product of the slopes of the two perpendicular lines is -1:

$$\left(\frac{c}{d}\right)\left(-\frac{d}{c}\right) = -1.$$

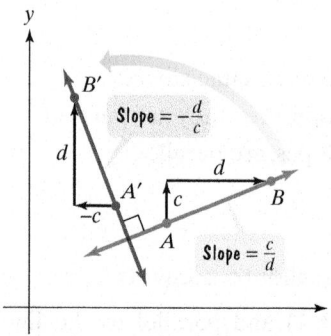

FIGURE 1.49 Slopes of perpendicular lines

This relationship holds for all perpendicular lines and is summarized in the box at the top of the next page.

Slope and Perpendicular Lines

1. If two nonvertical lines are perpendicular, then the product of their slopes is -1.
2. If the product of the slopes of two lines is -1, then the lines are perpendicular.
3. A horizontal line having zero slope is perpendicular to a vertical line having undefined slope.

An equivalent way of stating this relationship is to say that **one line is perpendicular to another line if its slope is the _negative reciprocal_ of the slope of the other line.** For example, if a line has slope 5, any line having slope $-\frac{1}{5}$ is perpendicular to it. Similarly, if a line has slope $-\frac{3}{4}$, any line having slope $\frac{4}{3}$ is perpendicular to it.

EXAMPLE 2 Writing Equations of a Line Perpendicular to a Given Line

a. Find the slope of any line that is perpendicular to the line whose equation is $x + 4y - 8 = 0$.

b. Write the equation of the line passing through $(3, -5)$ and perpendicular to the line whose equation is $x + 4y - 8 = 0$. Express the equation in general form.

SOLUTION

a. We begin by writing the equation of the given line, $x + 4y - 8 = 0$, in slope-intercept form. Solve for y.

$x + 4y - 8 = 0$ This is the given equation.

$4y = -x + 8$ To isolate the y-term, subtract x and add 8 on both sides.

$y = -\frac{1}{4}x + 2$ Divide both sides by 4.

Slope is $-\frac{1}{4}$.

The given line has slope $-\frac{1}{4}$. Any line perpendicular to this line has a slope that is the negative reciprocal of $-\frac{1}{4}$. Thus, the slope of any perpendicular line is 4.

b. Let's begin by writing the point-slope form of the perpendicular line's equation. Because the line passes through the point $(3, -5)$, we have $x_1 = 3$ and $y_1 = -5$. In part (a), we determined that the slope of any line perpendicular to $x + 4y - 8 = 0$ is 4, so the slope of this particular perpendicular line must also be 4: $m = 4$.

$$y - y_1 = m(x - x_1)$$

$y_1 = -5$ $m = 4$ $x_1 = 3$

The point-slope form of the perpendicular line's equation is

$$y - (-5) = 4(x - 3) \text{ or}$$

$$y + 5 = 4(x - 3).$$

How can we express this equation, $y + 5 = 4(x - 3)$, in general form $(Ax + By + C = 0)$? We need to obtain zero on one side of the equation. Let's do this and keep A, the coefficient of x, positive.

$$y + 5 = 4(x - 3)$$ This is the point-slope form of the line's equation.

$$y + 5 = 4x - 12$$ Apply the distributive property.

$$y - y + 5 - 5 = 4x - y - 12 - 5$$ To obtain 0 on the left, subtract y and subtract 5 on both sides.

$$0 = 4x - y - 17$$ Simplify.

In general form, the equation of the perpendicular line is $4x - y - 17 = 0$.

● ● ●

☑ Check Point 2

a. Find the slope of any line that is perpendicular to the line whose equation is $x + 3y - 12 = 0$.

b. Write the equation of the line passing through $(-2, -6)$ and perpendicular to the line whose equation is $x + 3y - 12 = 0$. Express the equation in general form.

② Interpret slope as rate of change.

Slope as Rate of Change

Slope is defined as the ratio of a change in y to a corresponding change in x. It describes how fast y is changing with respect to x. For a linear function, slope may be interpreted as the rate of change of the dependent variable per unit change in the independent variable.

Our next example shows how slope can be interpreted as a rate of change in an applied situation. When calculating slope in applied problems, keep track of the units in the numerator and the denominator.

EXAMPLE 3 Slope as a Rate of Change

The line graphs for the number of women and men living alone are shown again in **Figure 1.50**. Find the slope of the line segment for the women. Describe what this slope represents.

SOLUTION

We let x represent a year and y the number of women living alone in that year. The two points shown on the line segment for women have the following coordinates:

$$(1990, 14.0) \qquad (2008, 18.3).$$

In 1990, 14 million U.S. women lived alone. In 2008, 18.3 million U.S. women lived alone.

Now we compute the slope:

$$m = \frac{\text{Change in } y}{\text{Change in } x} = \frac{18.3 - 14.0}{2008 - 1990}$$

The unit in the numerator is *million women*.

The unit in the denominator is *year*.

$$= \frac{4.3}{18} \approx \frac{0.24 \text{ million women}}{\text{year}}.$$

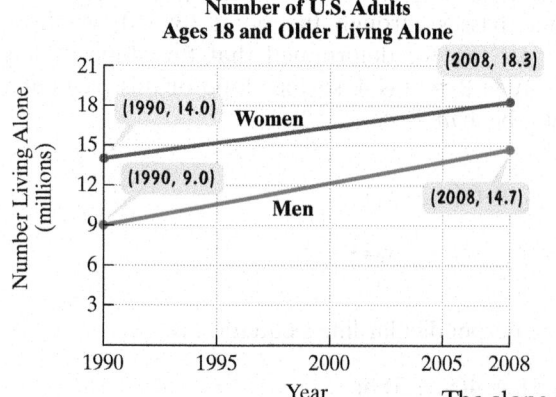

Number of U.S. Adults Ages 18 and Older Living Alone

FIGURE 1.50

The slope indicates that the number of American women living alone increased at a rate of approximately 0.24 million each year for the period from 1990 through 2008. The rate of change is 0.24 million women per year.

● ● ●

✓ Check Point **3** Use the ordered pairs in **Figure 1.50** to find the slope of the green line segment for the men. Express the slope correct to two decimal places and describe what it represents.

In Check Point 3, did you find that the slope of the line segment for the men is different from that of the women? The rate of change for the number of men living alone is greater than the rate of change for the number of women living alone. The green line segment representing men in **Figure 1.50** is steeper than the red line segment representing women. If you extend the line segments far enough, the resulting lines will intersect. They are not parallel.

3 Find a function's average rate of change.

The Average Rate of Change of a Function

If the graph of a function is not a straight line, the **average rate of change** between any two points is the slope of the line containing the two points. This line is called a **secant line**. For example, **Figure 1.51** shows the graph of a particular man's height, in inches, as a function of his age, in years. Two points on the graph are labeled: $(13, 57)$ and $(18, 76)$. At age 13, this man was 57 inches tall and at age 18, he was 76 inches tall.

FIGURE 1.51 Height as a function of age

The man's average growth rate between ages 13 and 18 is the slope of the secant line containing $(13, 57)$ and $(18, 76)$:

$$m = \frac{\text{Change in } y}{\text{Change in } x} = \frac{76 - 57}{18 - 13} = \frac{19}{5} = 3\frac{4}{5}.$$

This man's average rate of change, or average growth rate, from age 13 to age 18 was $3\frac{4}{5}$, or 3.8, inches per year.

The Average Rate of Change of a Function

Let $(x_1, f(x_1))$ and $(x_2, f(x_2))$ be distinct points on the graph of a function f. (See **Figure 1.52**.) The **average rate of change of f** from x_1 to x_2 is

$$\frac{f(x_2) - f(x_1)}{x_2 - x_1}.$$

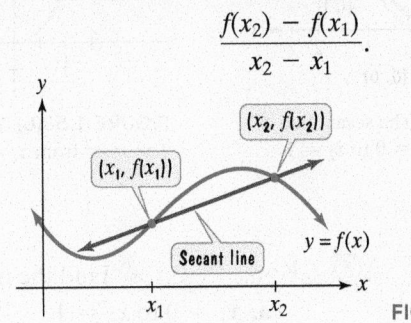

FIGURE 1.52

EXAMPLE 4 Finding the Average Rate of Change

Find the average rate of change of $f(x) = x^2$ from

a. $x_1 = 0$ to $x_2 = 1$ **b.** $x_1 = 1$ to $x_2 = 2$ **c.** $x_1 = -2$ to $x_2 = 0$.

SOLUTION

a. The average rate of change of $f(x) = x^2$ from $x_1 = 0$ to $x_2 = 1$ is

$$\frac{f(x_2) - f(x_1)}{x_2 - x_1} = \frac{f(1) - f(0)}{1 - 0} = \frac{1^2 - 0^2}{1} = 1.$$

Figure 1.53(a) shows the secant line of $f(x) = x^2$ from $x_1 = 0$ to $x_2 = 1$. The average rate of change is positive and the function is increasing on the interval $(0, 1)$.

b. The average rate of change of $f(x) = x^2$ from $x_1 = 1$ to $x_2 = 2$ is

$$\frac{f(x_2) - f(x_1)}{x_2 - x_1} = \frac{f(2) - f(1)}{2 - 1} = \frac{2^2 - 1^2}{1} = 3.$$

Figure 1.53(b) shows the secant line of $f(x) = x^2$ from $x_1 = 1$ to $x_2 = 2$. The average rate of change is positive and the function is increasing on the interval $(1, 2)$. Can you see that the graph rises more steeply on the interval $(1, 2)$ than on $(0, 1)$? This is because the average rate of change from $x_1 = 1$ to $x_2 = 2$ is greater than the average rate of change from $x_1 = 0$ to $x_2 = 1$.

c. The average rate of change of $f(x) = x^2$ from $x_1 = -2$ to $x_2 = 0$ is

$$\frac{f(x_2) - f(x_1)}{x_2 - x_1} = \frac{f(0) - f(-2)}{0 - (-2)} = \frac{0^2 - (-2)^2}{2} = \frac{-4}{2} = -2.$$

Figure 1.53(c) shows the secant line of $f(x) = x^2$ from $x_1 = -2$ to $x_2 = 0$. The average rate of change is negative and the function is decreasing on the interval $(-2, 0)$.

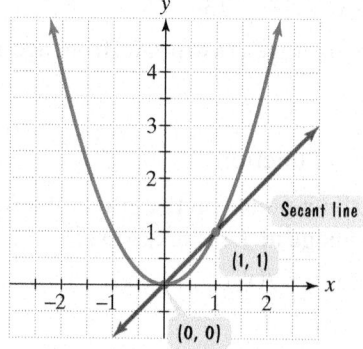

FIGURE 1.53(a) The secant line of $f(x) = x^2$ from $x_1 = 0$ to $x_2 = 1$

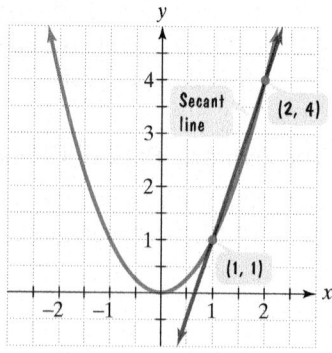

FIGURE 1.53(b) The secant line of $f(x) = x^2$ from $x_1 = 1$ to $x_2 = 2$

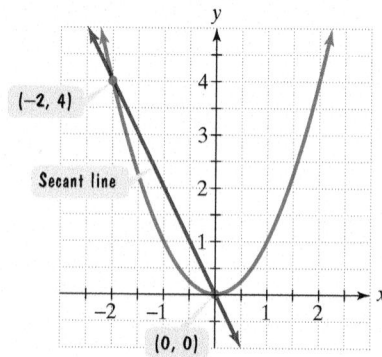

FIGURE 1.53(c) The secant line of $f(x) = x^2$ from $x_1 = -2$ to $x_2 = 0$ • • •

Check Point **4** Find the average rate of change of $f(x) = x^3$ from
a. $x_1 = 0$ to $x_2 = 1$ **b.** $x_1 = 1$ to $x_2 = 2$ **c.** $x_1 = -2$ to $x_2 = 0$.

Suppose we are interested in the average rate of change of f from $x_1 = x$ to $x_2 = x + h$. In this case, the average rate of change is

$$\frac{f(x_2) - f(x_1)}{x_2 - x_1} = \frac{f(x + h) - f(x)}{x + h - x} = \frac{f(x + h) - f(x)}{h}.$$

Do you recognize the last expression? It is the difference quotient that you used in Section 1.3. Thus, the difference quotient gives the average rate of change of a function from x to $x + h$. In the difference quotient, h is thought of as a number very close to 0. In this way, the average rate of change can be found for a very short interval.

EXAMPLE 5 Finding the Average Rate of Change

When a person receives a drug injected into a muscle, the concentration of the drug in the body, measured in milligrams per 100 milliliters, is a function of the time elapsed after the injection, measured in hours. **Figure 1.54** shows the graph of such a function, where x represents hours after the injection and $f(x)$ is the drug's concentration at time x. Find the average rate of change in the drug's concentration between 3 and 7 hours.

SOLUTION

At 3 hours, the drug's concentration is 0.05 and at 7 hours, the concentration is 0.02. The average rate of change in its concentration between 3 and 7 hours is

$$\frac{f(x_2) - f(x_1)}{x_2 - x_1} = \frac{f(7) - f(3)}{7 - 3} = \frac{0.02 - 0.05}{7 - 3} = \frac{-0.03}{4} = -0.0075.$$

The average rate of change is -0.0075. This means that the drug's concentration is decreasing at an average rate of 0.0075 milligram per 100 milliliters per hour. •••

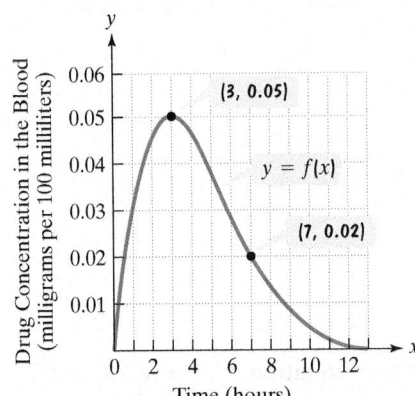

FIGURE 1.54 Concentration of a drug as a function of time

GREAT QUESTION!

Can you clarify how you determine the units that you use when describing slope as a rate of change?

Units used to describe x and y tend to "pile up" when expressing the rate of change of y with respect to x. The unit used to express the rate of change of y with respect to x is

<div align="center">

the unit used to describe y **per** the unit used to describe x.

In **Figure 1.54**, y, or drug concentration, is described in milligrams per 100 milliliters. In **Figure 1.54**, x, or time, is described in hours.

</div>

In Example 5, the rate of change is described in terms of milligrams per 100 milliliters per hour.

✓ Check Point **5** Use **Figure 1.54** to find the average rate of change in the drug's concentration between 1 hour and 3 hours.

Blitzer Bonus ‖ How Calculus Studies Change

Take a rapid sequence of still photographs of a moving scene and project them onto a screen at thirty shots a second or faster. Our eyes see the results as continuous motion. The small difference between one frame and the next cannot be detected by the human visual system. The idea of calculus likewise regards continuous motion as made up of a sequence of still configurations. Calculus masters the mystery of movement by "freezing the frame" of a continuous changing process, instant by instant. For example, **Figure 1.55** shows a male's changing height over intervals of time. Over the period of time from P to D, his average rate of growth is his change in height—that is, his height at time D minus his height at time P—divided by the change in time from P to D. This is the slope of secant line PD.

The secant lines PD, PC, PB, and PA shown in **Figure 1.55** have slopes that show average growth rates for successively shorter periods of time. Calculus makes these time frames so small that they approach a single point—that is, a single instant in time. This point is shown as point P in **Figure 1.55**. The slope of the line that touches the graph at P gives the male's growth rate at one instant in time, P.

FIGURE 1.55 Analyzing continuous growth over intervals of time and at an instant in time

CONCEPT AND VOCABULARY CHECK

Fill in each blank so that the resulting statement is true.

1. If two nonvertical lines are parallel, then they have _____ slope.
2. If two nonvertical lines are perpendicular, then the product of their slopes is ____.
3. Consider the line whose equation is $y = -\frac{1}{3}x + 5$. The slope of any line that is parallel to this line is ____. The slope of any line that is perpendicular to this line is ____.

4. Consider the line whose equation is $2x + y - 6 = 0$. The slope of any line that is parallel to this line is ____. The slope of any line that is perpendicular to this line is ____.
5. The slope of the line through the distinct points (x_1, y_1) and (x_2, y_2) can be interpreted as the rate of change in ____ with respect to ____.
6. If $(x_1, f(x_1))$ and $(x_2, f(x_2))$ are distinct points on the graph of a function f, the average rate of change of f from x_1 to x_2 is _____.

EXERCISE SET 1.5

Practice Exercises

In Exercises 1–4, write an equation for line L in point-slope form and slope-intercept form.

1.

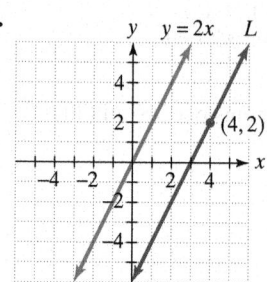

L is parallel to $y = 2x$.

2.

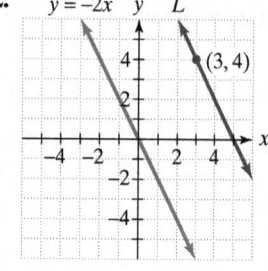

L is parallel to $y = -2x$.

3.

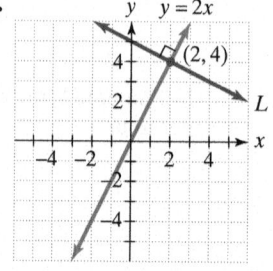

L is perpendicular to $y = 2x$.

4.

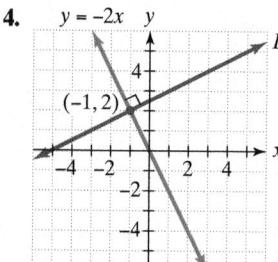

L is perpendicular to $y = -2x$.

In Exercises 5–8, use the given conditions to write an equation for each line in point-slope form and slope-intercept form.

5. Passing through $(-8, -10)$ and parallel to the line whose equation is $y = -4x + 3$

6. Passing through $(-2, -7)$ and parallel to the line whose equation is $y = -5x + 4$

7. Passing through $(2, -3)$ and perpendicular to the line whose equation is $y = \frac{1}{5}x + 6$

8. Passing through $(-4, 2)$ and perpendicular to the line whose equation is $y = \frac{1}{3}x + 7$

In Exercises 9–12, use the given conditions to write an equation for each line in point-slope form and general form.

9. Passing through $(-2, 2)$ and parallel to the line whose equation is $2x - 3y - 7 = 0$

10. Passing through $(-1, 3)$ and parallel to the line whose equation is $3x - 2y - 5 = 0$

11. Passing through $(4, -7)$ and perpendicular to the line whose equation is $x - 2y - 3 = 0$

12. Passing through $(5, -9)$ and perpendicular to the line whose equation is $x + 7y - 12 = 0$

In Exercises 13–18, find the average rate of change of the function from x_1 to x_2.

13. $f(x) = 3x$ from $x_1 = 0$ to $x_2 = 5$

14. $f(x) = 6x$ from $x_1 = 0$ to $x_2 = 4$

15. $f(x) = x^2 + 2x$ from $x_1 = 3$ to $x_2 = 5$

16. $f(x) = x^2 - 2x$ from $x_1 = 3$ to $x_2 = 6$

17. $f(x) = \sqrt{x}$ from $x_1 = 4$ to $x_2 = 9$

18. $f(x) = \sqrt{x}$ from $x_1 = 9$ to $x_2 = 16$

Practice Plus

In Exercises 19–24, write an equation in slope-intercept form of a linear function f whose graph satisfies the given conditions.

19. The graph of f passes through $(-1, 5)$ and is perpendicular to the line whose equation is $x = 6$.

20. The graph of f passes through $(-2, 6)$ and is perpendicular to the line whose equation is $x = -4$.

21. The graph of f passes through $(-6, 4)$ and is perpendicular to the line that has an x-intercept of 2 and a y-intercept of -4.

22. The graph of f passes through $(-5, 6)$ and is perpendicular to the line that has an x-intercept of 3 and a y-intercept of -9.

23. The graph of f is perpendicular to the line whose equation is $3x - 2y - 4 = 0$ and has the same y-intercept as this line.

24. The graph of f is perpendicular to the line whose equation is $4x - y - 6 = 0$ and has the same y-intercept as this line.

Application Exercises

The bar graph shows that as costs changed over the decades, Americans devoted less of their budget to groceries and more to health care.

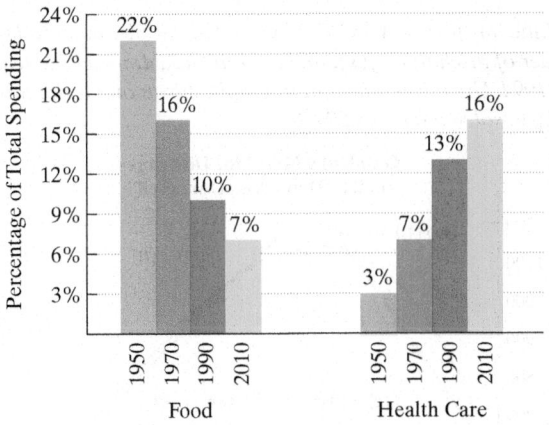

Percentage of Total Spending in the United States on Food and Health Care

Source: *Time*, October 10, 2011

In Exercises 25–26, find a linear function in slope-intercept form that models the given description. Each function should model the percentage of total spending, $p(x)$, by Americans x years after 1950.

25. In 1950, Americans spent 22% of their budget on food. This has decreased at an average rate of approximately 0.25% per year since then.

26. In 1950, Americans spent 3% of their budget on health care. This has increased at an average rate of approximately 0.22% per year since then.

The stated intent of the 1994 "don't ask, don't tell" policy was to reduce the number of discharges of gay men and lesbians from the military. Nearly 14,000 active-duty gay servicemembers were dismissed under the policy, which officially ended in 2011, after 18 years. The line graph shows the number of discharges under "don't ask, don't tell" from 1994 through 2010. Use the data displayed by the graph to solve Exercises 27–28.

Number of Active-Duty Gay Servicemembers Discharged from the Military for Homosexuality

Source: General Accountability Office

(In Exercises 27–28, be sure to refer to the graph at the bottom of the previous page.)

27. Find the average rate of change, rounded to the nearest whole number, from 1994 through 1998. Describe what this means.

28. Find the average rate of change, rounded to the nearest whole number, from 2001 through 2006. Describe what this means.

The function $f(x) = 1.1x^3 - 35x^2 + 264x + 557$ models the number of discharges, $f(x)$, under "don't ask, don't tell" x years after 1994. Use this model and its graph, shown on the domain [0, 12], to solve Exercises 29–30.

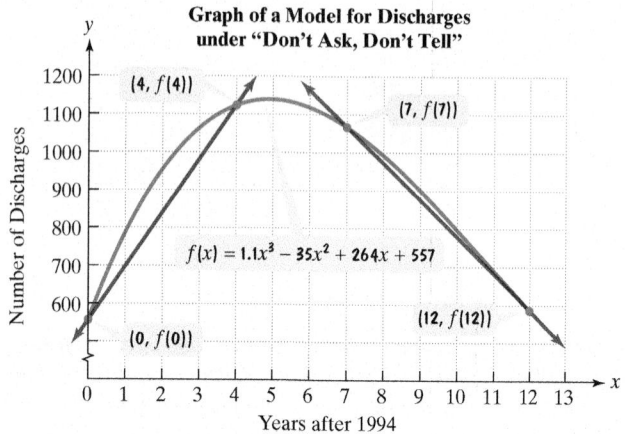

Graph of a Model for Discharges under "Don't Ask, Don't Tell"

$f(x) = 1.1x^3 - 35x^2 + 264x + 557$

Years after 1994

29. a. Find the slope of the secant line, rounded to the nearest whole number, from $x_1 = 0$ to $x_2 = 4$.

b. Does the slope from part (a) underestimate or overestimate the average yearly increase that you determined in Exercise 27? By how much?

30. a. Find the slope of the secant line, rounded to the nearest whole number, from $x_1 = 7$ to $x_2 = 12$.

b. Does the slope from part (b) underestimate or overestimate the average yearly decrease that you determined in Exercise 28? By how much?

Writing in Mathematics

31. If two lines are parallel, describe the relationship between their slopes.

32. If two lines are perpendicular, describe the relationship between their slopes.

33. If you know a point on a line and you know the equation of a line perpendicular to this line, explain how to write the line's equation.

34. A formula in the form $y = mx + b$ models the average retail price, y, of a new car x years after 2000. Would you expect m to be positive, negative, or zero? Explain your answer.

35. What is a secant line?

36. What is the average rate of change of a function?

Technology Exercise

37. a. Why are the lines whose equations are $y = \frac{1}{3}x + 1$ and $y = -3x - 2$ perpendicular?

b. Use a graphing utility to graph the equations in a [−10, 10, 1] by [−10, 10, 1] viewing rectangle. Do the lines appear to be perpendicular?

c. Now use the zoom square feature of your utility. Describe what happens to the graphs. Explain why this is so.

Critical Thinking Exercises

Make Sense? *In Exercises 38–41, determine whether each statement makes sense or does not make sense, and explain your reasoning.*

38. Some of the steel girders in this photo of the Eiffel Tower appear to have slopes that are negative reciprocals of each other.

39. I have linear functions that model changes for men and women over the same time period. The functions have the same slope, so their graphs are parallel lines, indicating that the rate of change for men is the same as the rate of change for women.

40. The graph of my function is not a straight line, so I cannot use slope to analyze its rates of change.

41. According to the Blitzer Bonus on page 174, calculus studies change by analyzing slopes of secant lines over successively shorter intervals.

42. What is the slope of a line that is perpendicular to the line whose equation is $Ax + By + C = 0, A \neq 0$ and $B \neq 0$?

43. Determine the value of A so that the line whose equation is $Ax + y - 2 = 0$ is perpendicular to the line containing the points $(1, -3)$ and $(-2, 4)$.

Preview Exercises

Exercises 44–46 will help you prepare for the material covered in the next section. In each exercise, graph the functions in parts (a) and (b) in the same rectangular coordinate system.

44. a. Graph $f(x) = |x|$ using the ordered pairs $(-3, f(-3))$, $(-2, f(-2)), (-1, f(-1)), (0, f(0)), (1, f(1)), (2, f(2))$, and $(3, f(3))$.

b. Subtract 4 from each y-coordinate of the ordered pairs in part (a). Then graph the ordered pairs and connect them with two linear pieces.

c. Describe the relationship between the graph in part (b) and the graph in part (a).

45. a. Graph $f(x) = x^2$ using the ordered pairs $(-3, f(-3))$, $(-2, f(-2)), (-1, f(-1)), (0, f(0)), (1, f(1)), (2, f(2))$, and $(3, f(3))$.

b. Add 2 to each x-coordinate of the ordered pairs in part (a). Then graph the ordered pairs and connect them with a smooth curve.

c. Describe the relationship between the graph in part (b) and the graph in part (a).

46. a. Graph $f(x) = x^3$ using the ordered pairs $(-2, f(-2))$, $(-1, f(-1))$, $(0, f(0))$, $(1, f(1))$, and $(2, f(2))$.

 b. Replace each x-coordinate of the ordered pairs in part (a) with its opposite, or additive inverse. Then graph the ordered pairs and connect them with a smooth curve.

c. Describe the relationship between the graph in part (b) and the graph in part (a).

CHAPTER 1 Mid-Chapter Check Point

WHAT YOU KNOW: We learned that a function is a relation in which no two ordered pairs have the same first component and different second components. We represented functions as equations and used function notation. We graphed functions and applied the vertical line test to identify graphs of functions. We determined the domain and range of a function from its graph, using inputs on the x-axis for the domain and outputs on the y-axis for the range. We used graphs to identify intervals on which functions increase, decrease, or are constant, as well as to locate relative maxima or minima. We identified even functions [$f(-x) = f(x)$: y-axis symmetry] and odd functions [$f(-x) = -f(x)$: origin symmetry]. Finally, we studied linear functions and slope, using slope (change in y divided by change in x) to develop various forms for equations of lines:

Point-slope form	Slope-intercept form	Horizontal line	Vertical line	General form
$y - y_1 = m(x - x_1)$	$y = f(x) = mx + b$	$y = f(x) = b$	$x = a$	$Ax + By + C = 0.$

We saw that parallel lines have the same slope and that perpendicular lines have slopes that are negative reciprocals. For linear functions, slope was interpreted as the rate of change of the dependent variable per unit change in the independent variable. For nonlinear functions, the slope of the secant line between $(x_1, f(x_1))$ and $(x_2, f(x_2))$ described the average rate of change of f from x_1 to x_2: $\dfrac{f(x_2) - f(x_1)}{x_2 - x_1}$.

In Exercises 1–6, determine whether each relation is a function. Give the domain and range for each relation.

1. $\{(2, 6), (1, 4), (2, -6)\}$ **2.** $\{(0, 1), (2, 1), (3, 4)\}$

3.

4.

5.

6.
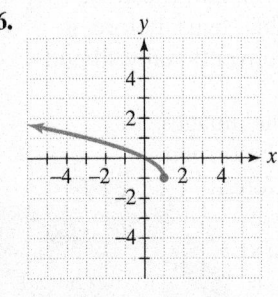

In Exercises 7–8, determine whether each equation defines y as a function of x.

7. $x^2 + y = 5$ **8.** $x + y^2 = 5$

Use the graph of f to solve Exercises 9–24. Where applicable, use interval notation.

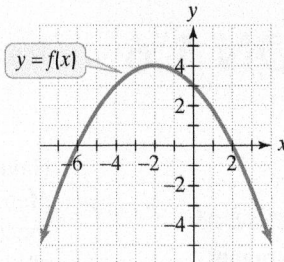

9. Explain why f represents the graph of a function.

10. Find the domain of f.

11. Find the range of f.

12. Find the x-intercept(s).

13. Find the y-intercept.

14. Find the interval(s) on which f is increasing.

15. Find the interval(s) on which f is decreasing.

16. At what number does f have a relative maximum?

17. What is the relative maximum of f?

18. Find $f(-4)$.

19. For what value or values of x is $f(x) = -2$?

20. For what value or values of x is $f(x) = 0$?

21. For what values of x is $f(x) > 0$?

22. Is $f(100)$ positive or negative?

23. Is f even, odd, or neither?

24. Find the average rate of change of f from $x_1 = -4$ to $x_2 = 4$.

In Exercises 25–36, graph each equation in a rectangular coordinate system.

25. $y = -2x$ **26.** $y = -2$ **27.** $x + y = -2$

28. $y = \frac{1}{3}x - 2$ **29.** $x = 3.5$ **30.** $4x - 2y = 8$

31. $f(x) = x^2 - 4$ **32.** $f(x) = x - 4$ **33.** $f(x) = |x| - 4$

34. $5y = -3x$ **35.** $5y = 20$

36. $f(x) = \begin{cases} -1 & \text{if} \quad x \le 0 \\ 2x + 1 & \text{if} \quad x > 0 \end{cases}$

37. Let $f(x) = -2x^2 + x - 5$.

 a. Find $f(-x)$. Is f even, odd, or neither?

 b. Find $\dfrac{f(x + h) - f(x)}{h}$, $h \ne 0$.

38. Let $C(x) = \begin{cases} 30 & \text{if} \quad 0 \le x \le 200 \\ 30 + 0.40(x - 200) & \text{if} \quad x > 200 \end{cases}$.

 a. Find $C(150)$. **b.** Find $C(250)$.

In Exercises 39–42, write a function in slope-intercept form whose graph satisfies the given conditions.

39. Slope $= -2$, passing through $(-4, 3)$

40. Passing through $(-1, -5)$ and $(2, 1)$

41. Passing through $(3, -4)$ and parallel to the line whose equation is $3x - y - 5 = 0$

42. Passing through $(-4, -3)$ and perpendicular to the line whose equation is $2x - 5y - 10 = 0$

43. Determine whether the line through $(2, -4)$ and $(7, 0)$ is parallel to a second line through $(-4, 2)$ and $(1, 6)$.

44. Exercise is useful not only in preventing depression, but also as a treatment. The following graphs show the percentage of patients with depression in remission when exercise (brisk walking) was used as a treatment. (The control group that engaged in no exercise had 11% of the patients in remission.)

 a. Find the slope of the line passing through the two points shown by the voice balloons. Express the slope as a decimal.

Exercise and Percentage of Patients with Depression in Remission

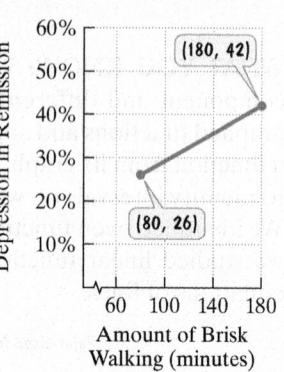

Source: *Newsweek*, March 26, 2007

 b. Use your answer from part (a) to complete this statement: For each minute of brisk walking, the percentage of patients with depression in remission increased by _____%. The rate of change is _____% per _____.

45. Find the average rate of change of $f(x) = 3x^2 - x$ from $x_1 = -1$ to $x_2 = 2$.

SECTION 1.6 Transformations of Functions

Objectives

1. Recognize graphs of common functions.
2. Use vertical shifts to graph functions.
3. Use horizontal shifts to graph functions.
4. Use reflections to graph functions.
5. Use vertical stretching and shrinking to graph functions.
6. Use horizontal stretching and shrinking to graph functions.
7. Graph functions involving a sequence of transformations.

Have you seen *Terminator 2, The Mask*, or *The Matrix?* These were among the first films to use spectacular effects in which a character or object having one shape was transformed in a fluid fashion into a quite different shape. The name for such a transformation is **morphing**. The effect allows a real actor to be seamlessly transformed into a computer-generated animation. The animation can be made to perform impossible feats before it is morphed back to the conventionally filmed image.

Like transformed movie images, the graph of one function can be turned into the graph of a different function. To do this, we need to rely on a function's equation. Knowing that a graph is a transformation of a familiar graph makes graphing easier.

① Recognize graphs of common functions.

Graphs of Common Functions

Table 1.3 gives names to seven frequently encountered functions in algebra. The table shows each function's graph and lists characteristics of the function. Study the shape of each graph and take a few minutes to verify the function's characteristics from its graph. Knowing these graphs is essential for analyzing their transformations into more complicated graphs.

Table 1.3 Algebra's Common Graphs

Constant Function	Identity Function	Absolute Value Function
$f(x) = c$	$f(x) = x$	$f(x) = \lvert x \rvert$
• Domain: $(-\infty, \infty)$	• Domain: $(-\infty, \infty)$	• Domain: $(-\infty, \infty)$
• Range: the single number c	• Range: $(-\infty, \infty)$	• Range: $[0, \infty)$
• Constant on $(-\infty, \infty)$	• Increasing on $(-\infty, \infty)$	• Decreasing on $(-\infty, 0)$ and increasing on $(0, \infty)$
• Even function	• Odd function	• Even function

Standard Quadratic Function	Square Root Function	Standard Cubic Function	Cube Root Function
$f(x) = x^2$	$f(x) = \sqrt{x}$	$f(x) = x^3$	$f(x) = \sqrt[3]{x}$
• Domain: $(-\infty, \infty)$	• Domain: $[0, \infty)$	• Domain: $(-\infty, \infty)$	• Domain: $(-\infty, \infty)$
• Range: $[0, \infty)$	• Range: $[0, \infty)$	• Range: $(-\infty, \infty)$	• Range: $(-\infty, \infty)$
• Decreasing on $(-\infty, 0)$ and increasing on $(0, \infty)$	• Increasing on $(0, \infty)$	• Increasing on $(-\infty, \infty)$	• Increasing on $(-\infty, \infty)$
• Even function	• Neither even nor odd	• Odd function	• Odd function

DISCOVERY

The study of how changing a function's equation can affect its graph can be explored with a graphing utility. Use your graphing utility to verify the hand-drawn graphs as you read this section.

② Use vertical shifts to graph functions.

Vertical Shifts

Let's begin by looking at three graphs whose shapes are the same. **Figure 1.56** on the next page shows the graphs. The black graph in the middle is the standard quadratic function, $f(x) = x^2$. Now, look at the blue graph on the top. The equation of this graph, $g(x) = x^2 + 2$, adds 2 to the right side of $f(x) = x^2$. The y-coordinate of each

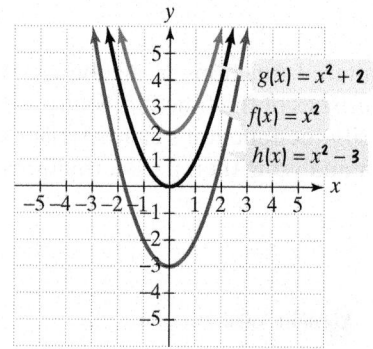

FIGURE 1.56 Vertical shifts

point of g is 2 more than the corresponding y-coordinate of each point of f. What effect does this have on the graph of f? It shifts the graph vertically up by 2 units.

$$g(x) = x^2 + 2 = f(x) + 2$$

The graph of g shifts the graph of f up 2 units.

Finally, look at the red graph on the bottom in **Figure 1.56**. The equation of this graph, $h(x) = x^2 - 3$, subtracts 3 from the right side of $f(x) = x^2$. The y-coordinate of each point of h is 3 less than the corresponding y-coordinate of each point of f. What effect does this have on the graph of f? It shifts the graph vertically down by 3 units.

$$h(x) = x^2 - 3 = f(x) - 3$$

The graph of h shifts the graph of f down 3 units.

In general, if c is positive, $y = f(x) + c$ shifts the graph of f upward c units and $y = f(x) - c$ shifts the graph of f downward c units. These are called **vertical shifts** of the graph of f.

Vertical Shifts

Let f be a function and c a positive real number.

- The graph of $y = f(x) + c$ is the graph of $y = f(x)$ shifted c units vertically upward.
- The graph of $y = f(x) - c$ is the graph of $y = f(x)$ shifted c units vertically downward.

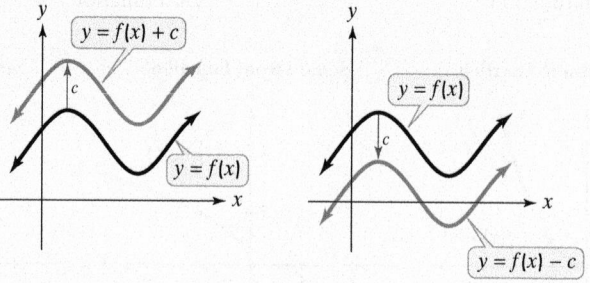

GREAT QUESTION!

If I'm using the graph of a familiar function, how do I actually obtain the graph of a transformation?

To keep track of transformations and obtain their graphs, identify a number of points on the given function's graph. Then analyze what happens to the coordinates of these points with each transformation.

EXAMPLE 1 Vertical Shift Downward

Use the graph of $f(x) = |x|$ to obtain the graph of $g(x) = |x| - 4$.

SOLUTION

The graph of $g(x) = |x| - 4$ has the same shape as the graph of $f(x) = |x|$. However, it is shifted down vertically 4 units.

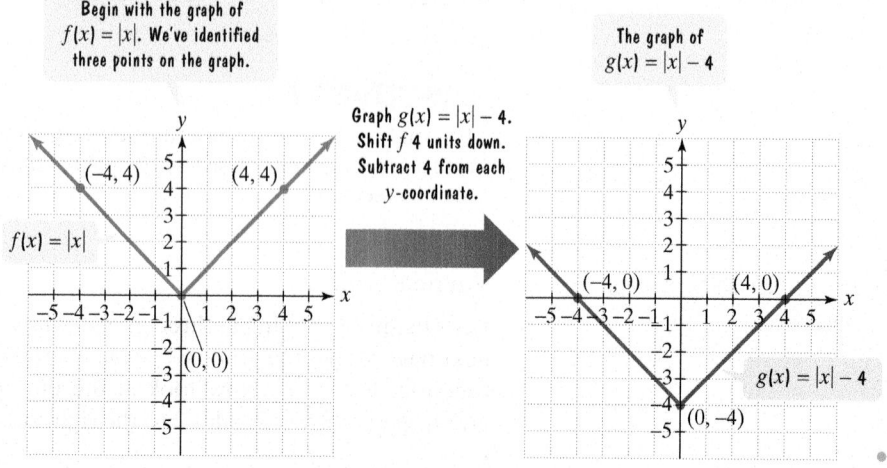

 • • •

✓ Check Point **1** Use the graph of $f(x) = |x|$ to obtain the graph of $g(x) = |x| + 3$.

③ Use horizontal shifts to graph functions.

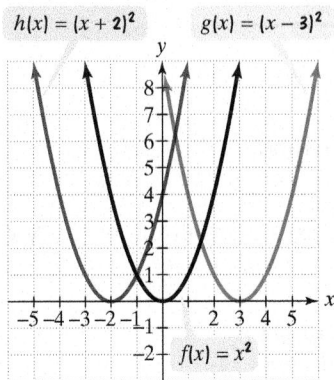

$h(x) = (x + 2)^2$ $g(x) = (x - 3)^2$

FIGURE 1.57 Horizontal shifts

Horizontal Shifts

We return to the graph of $f(x) = x^2$, the standard quadratic function. In **Figure 1.57**, the graph of function f is in the middle of the three graphs. By contrast to the vertical shift situation, this time there are graphs to the left and to the right of the graph of f. Look at the blue graph on the right. The equation of this graph, $g(x) = (x - 3)^2$, subtracts 3 from each value of x before squaring it. What effect does this have on the graph of $f(x) = x^2$? It shifts the graph horizontally to the right by 3 units.

$$g(x) = (x - 3)^2 = f(x - 3)$$

The graph of g shifts the graph of f 3 units to the right.

Does it seem strange that *subtracting* 3 in the domain causes a shift of 3 units to the *right*? Perhaps a partial table of coordinates for each function will numerically convince you of this shift.

x	$f(x) = x^2$
-2	$(-2)^2 = 4$
-1	$(-1)^2 = 1$
0	$0^2 = 0$
1	$1^2 = 1$
2	$2^2 = 4$

x	$g(x) = (x - 3)^2$
1	$(1 - 3)^2 = (-2)^2 = 4$
2	$(2 - 3)^2 = (-1)^2 = 1$
3	$(3 - 3)^2 = \quad 0^2 = 0$
4	$(4 - 3)^2 = \quad 1^2 = 1$
5	$(5 - 3)^2 = \quad 2^2 = 4$

Notice that for the values of $f(x)$ and $g(x)$ to be the same, the values of x used in graphing g must each be 3 units greater than those used to graph f. For this reason, the graph of g is the graph of f shifted 3 units to the right.

Now, look at the red graph on the left in **Figure 1.57**. The equation of this graph, $h(x) = (x + 2)^2$, adds 2 to each value of x before squaring it. What effect does this have on the graph of $f(x) = x^2$? It shifts the graph horizontally to the left by 2 units.

$$h(x) = (x + 2)^2 = f(x + 2)$$

The graph of h shifts the graph of f **2** units to the left.

In general, if c is positive, $y = f(x + c)$ shifts the graph of f to the left c units and $y = f(x - c)$ shifts the graph of f to the right c units. These are called **horizontal shifts** of the graph of f.

GREAT QUESTION!

Using my intuition, it seems that $f(x + c)$ should cause a shift to the right and $f(x - c)$ should cause a shift to the left. Is my intuition on target when it comes to these horizontal shifts?

No. On a number line, if x represents a number and c is positive, then $x + c$ lies c units to the right of x and $x - c$ lies c units to the left of x. This orientation does not apply to horizontal shifts: $f(x + c)$ causes a shift of c units to the left and $f(x - c)$ causes a shift of c units to the right.

Horizontal Shifts

Let f be a function and c a positive real number.

- The graph of $y = f(x + c)$ is the graph of $y = f(x)$ shifted to the left c units.
- The graph of $y = f(x - c)$ is the graph of $y = f(x)$ shifted to the right c units.

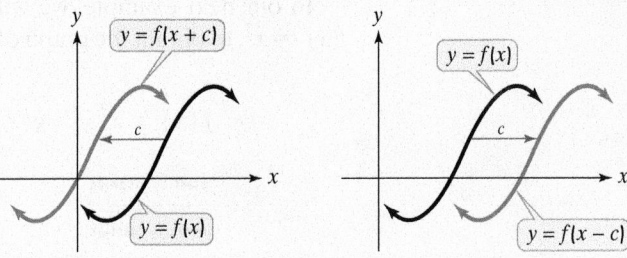

EXAMPLE 2 Horizontal Shift to the Left

Use the graph of $f(x) = \sqrt{x}$ to obtain the graph of $g(x) = \sqrt{x + 5}$.

SOLUTION

Compare the equations for $f(x) = \sqrt{x}$ and $g(x) = \sqrt{x + 5}$. The equation for g adds 5 to each value of x before taking the square root.

$$y = g(x) = \sqrt{x + 5} = f(x + 5)$$

| The graph of g | shifts the graph of f 5 units to the left. |

The graph of $g(x) = \sqrt{x + 5}$ has the same shape as the graph of $f(x) = \sqrt{x}$. However, it is shifted horizontally to the left 5 units.

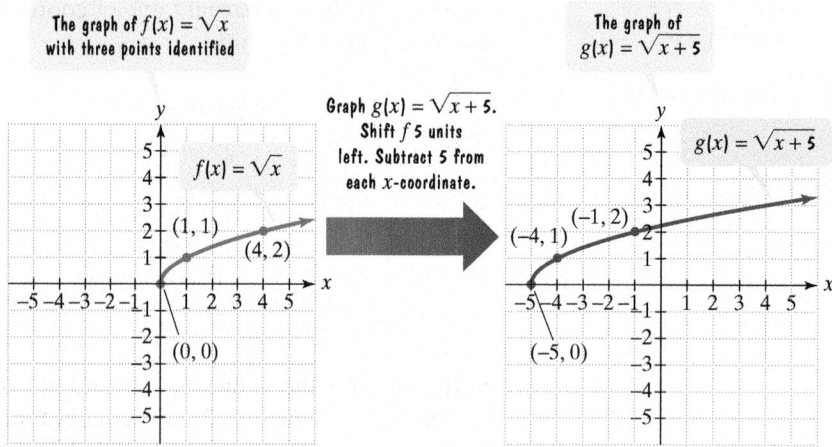

GREAT QUESTION!

What's the difference between $f(x) + c$ and $f(x + c)$?

- $y = f(x) + c$ shifts the graph of $y = f(x)$ c units vertically upward.
- $y = f(x + c)$ shifts the graph of $y = f(x)$ c units horizontally to the left.

There are analogous differences between $f(x) - c$ and $f(x - c)$.

⌀ Check Point **2** Use the graph of $f(x) = \sqrt{x}$ to obtain the graph of $g(x) = \sqrt{x - 4}$.

Some functions can be graphed by combining horizontal and vertical shifts. These functions will be variations of a function whose equation you know how to graph, such as the standard quadratic function, the standard cubic function, the square root function, the cube root function, or the absolute value function.

In our next example, we will use the graph of the standard quadratic function, $f(x) = x^2$, to obtain the graph of $h(x) = (x + 1)^2 - 3$. We will graph three functions:

$$f(x) = x^2 \qquad g(x) = (x + 1)^2 \qquad h(x) = (x + 1)^2 - 3.$$

| Start by graphing the standard quadratic function. | Shift the graph of f horizontally one unit to the left. | Shift the graph of g vertically down 3 units. |

EXAMPLE 3 Combining Horizontal and Vertical Shifts

Use the graph of $f(x) = x^2$ to obtain the graph of $h(x) = (x + 1)^2 - 3$.

SOLUTION

The graph of $f(x) = x^2$ with three points identified

Graph $g(x) = (x + 1)^2$.
Shift f horizontally 1 unit left. Subtract 1 from each x-coordinate.

The graph of $g(x) = (x + 1)^2$

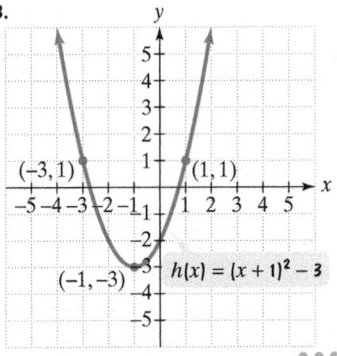

Graph $h(x) = (x + 1)^2 - 3$.
Shift g vertically down 3 units. Subtract 3 from each y-coordinate.

The graph of $h(x) = (x + 1)^2 - 3$

DISCOVERY

Work Example 3 by first shifting the graph of $f(x) = x^2$ three units down, graphing $g(x) = x^2 - 3$. Now, shift this graph one unit left to graph $h(x) = (x + 1)^2 - 3$. Did you obtain the last graph shown in the solution of Example 3? What can you conclude?

✓ Check Point **3** Use the graph of $f(x) = \sqrt{x}$ to obtain the graph of $h(x) = \sqrt{x - 1} - 2$.

④ Use reflections to graph functions.

Reflections of Graphs

This photograph shows a reflection of an old bridge in a Maryland river. This perfect reflection occurs because the surface of the water is absolutely still. A mild breeze rippling the water's surface would distort the reflection.

Is it possible for graphs to have mirror-like qualities? Yes. **Figure 1.58** shows the graphs of $f(x) = x^2$ and $g(x) = -x^2$. The graph of g is a **reflection about the x-axis** of the graph of f. For corresponding values of x, the y-coordinates of g are the opposites of the y-coordinates of f. In general, the graph of $y = -f(x)$ reflects the graph of f about the x-axis. Thus, the graph of g is a reflection of the graph of f about the x-axis because

$$g(x) = -x^2 = -f(x).$$

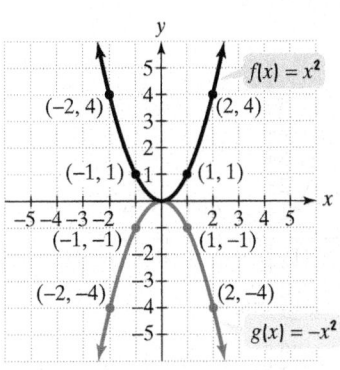

FIGURE 1.58 Reflection about the x-axis

Reflection about the x-Axis

The graph of $y = -f(x)$ is the graph of $y = f(x)$ reflected about the x-axis.

EXAMPLE 4 Reflection about the x-Axis

Use the graph of $f(x) = \sqrt[3]{x}$ to obtain the graph of $g(x) = -\sqrt[3]{x}$.

SOLUTION

Compare the equations for $f(x) = \sqrt[3]{x}$ and $g(x) = -\sqrt[3]{x}$. The graph of g is a reflection about the x-axis of the graph of f because

$$g(x) = -\sqrt[3]{x} = -f(x).$$

The graph of $f(x) = \sqrt[3]{x}$ with three points identified

The graph of $g(x) = -\sqrt[3]{x}$

Graph $g(x) = -\sqrt[3]{x}$.
Reflect f about the x-axis. Replace each y-coordinate with its opposite.

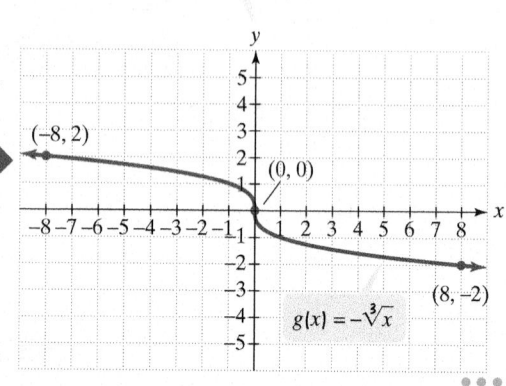

✓ Check Point **4** Use the graph of $f(x) = |x|$ to obtain the graph of $g(x) = -|x|$.

It is also possible to reflect graphs about the y-axis.

Reflection about the y-Axis

The graph of $y = f(-x)$ is the graph of $y = f(x)$ reflected about the y-axis.

For each point (x, y) on the graph of $y = f(x)$, the point $(-x, y)$ is on the graph of $y = f(-x)$.

EXAMPLE 5 Reflection about the y-Axis

Use the graph of $f(x) = \sqrt{x}$ to obtain the graph of $h(x) = \sqrt{-x}$.

SOLUTION

Compare the equations for $f(x) = \sqrt{x}$ and $h(x) = \sqrt{-x}$. The graph of h is a reflection about the y-axis of the graph of f because

$$h(x) = \sqrt{-x} = f(-x).$$

The graph of $f(x) = \sqrt{x}$ with three points identified

The graph of $h(x) = \sqrt{-x}$

Graph $h(x) = \sqrt{-x}$.
Reflect f about the y-axis. Replace each x-coordinate with its opposite.

✓ Check Point **5** Use the graph of $f(x) = \sqrt[3]{x}$ to obtain the graph of $h(x) = \sqrt[3]{-x}$.

⑤ Use vertical stretching and shrinking to graph functions.

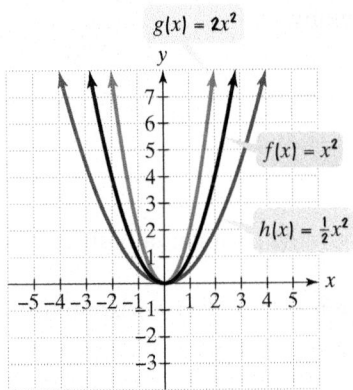

FIGURE 1.59 Vertically stretching and shrinking $f(x) = x^2$

GREAT QUESTION!

Does vertical stretching or shrinking change a graph's shape?

Yes. A vertical stretch moves a function's graph away from the x-axis. A vertical shrink compresses a function's graph toward the x-axis. The other transformations we have discussed (vertical shifts, horizontal shifts, and reflections) change only the position of a function's graph without changing the shape of the basic graph.

Vertical Stretching and Shrinking

Morphing does much more than move an image horizontally, vertically, or about an axis. An object having one shape is transformed into a different shape. Horizontal shifts, vertical shifts, and reflections do not change the basic shape of a graph. Graphs remain rigid and proportionally the same when they undergo these transformations. How can we shrink and stretch graphs, thereby altering their basic shapes?

Look at the three graphs in **Figure 1.59**. The black graph in the middle is the graph of the standard quadratic function, $f(x) = x^2$. Now, look at the blue graph on the top. The equation of this graph is $g(x) = 2x^2$, or $g(x) = 2f(x)$. Thus, for each x, the y-coordinate of g is two times as large as the corresponding y-coordinate on the graph of f. The result is a narrower graph because the values of y are rising faster. We say that the graph of g is obtained by vertically *stretching* the graph of f. Now, look at the red graph on the bottom. The equation of this graph is $h(x) = \frac{1}{2}x^2$, or $h(x) = \frac{1}{2}f(x)$. Thus, for each x, the y-coordinate of h is one-half as large as the corresponding y-coordinate on the graph of f. The result is a wider graph because the values of y are rising more slowly. We say that the graph of h is obtained by vertically *shrinking* the graph of f.

These observations can be summarized as follows:

Vertically Stretching and Shrinking Graphs

Let f be a function and c a positive real number.

- If $c > 1$, the graph of $y = cf(x)$ is the graph of $y = f(x)$ vertically stretched by multiplying each of its y-coordinates by c.
- If $0 < c < 1$, the graph of $y = cf(x)$ is the graph of $y = f(x)$ vertically shrunk by multiplying each of its y-coordinates by c.

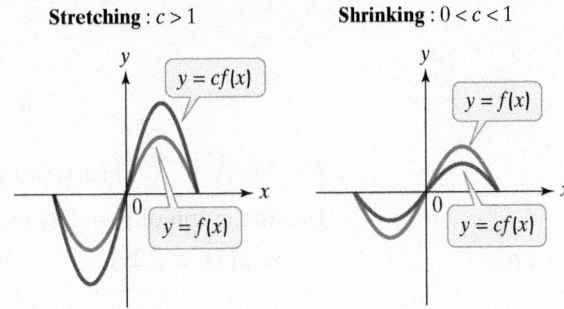

EXAMPLE 6 Vertically Shrinking a Graph

Use the graph of $f(x) = x^3$ to obtain the graph of $h(x) = \frac{1}{2}x^3$.

SOLUTION

The graph of $h(x) = \frac{1}{2}x^3$ is obtained by vertically shrinking the graph of $f(x) = x^3$.

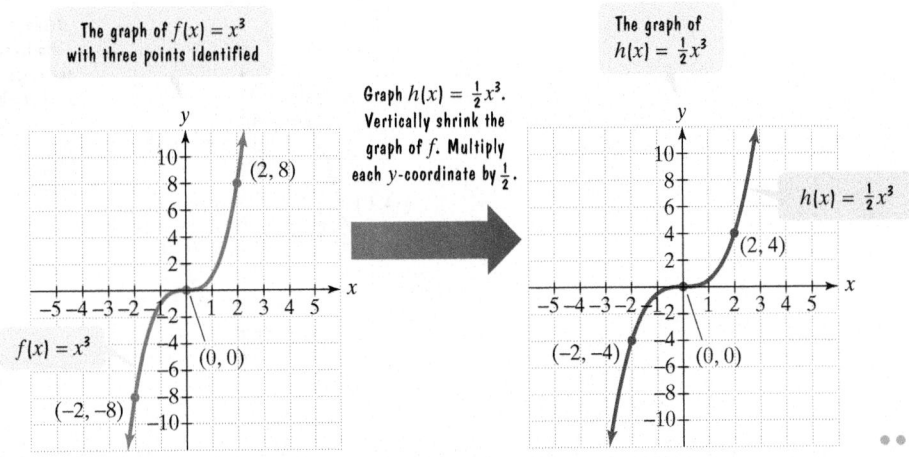

● ● ●

⊘ Check Point **6** Use the graph of $f(x) = |x|$ to obtain the graph of $g(x) = 2|x|$.

6 Use horizontal stretching and shrinking to graph functions.

Horizontal Stretching and Shrinking

It is also possible to stretch and shrink graphs horizontally.

GREAT QUESTION!

How does horizontal shrinking or stretching change a graph's shape?

A horizontal shrink compresses a function's graph toward the y-axis. A horizontal stretch moves a function's graph away from the y-axis.

Horizontally Stretching and Shrinking Graphs

Let f be a function and c a positive real number.

• If $c > 1$, the graph of $y = f(cx)$ is the graph of $y = f(x)$ horizontally shrunk by dividing each of its x-coordinates by c.

• If $0 < c < 1$, the graph of $y = f(cx)$ is the graph of $y = f(x)$ horizontally stretched by dividing each of its x-coordinates by c.

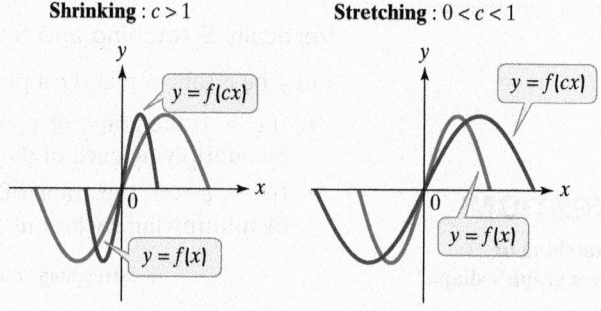

EXAMPLE 7 Horizontally Stretching and Shrinking a Graph

Use the graph of $y = f(x)$ in **Figure 1.60** to obtain each of the following graphs:

a. $g(x) = f(2x)$ **b.** $h(x) = f\left(\frac{1}{2}x\right)$.

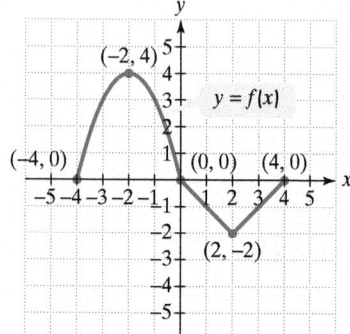

FIGURE 1.60

SOLUTION

a. The graph of $g(x) = f(2x)$ is obtained by horizontally shrinking the graph of $y = f(x)$.

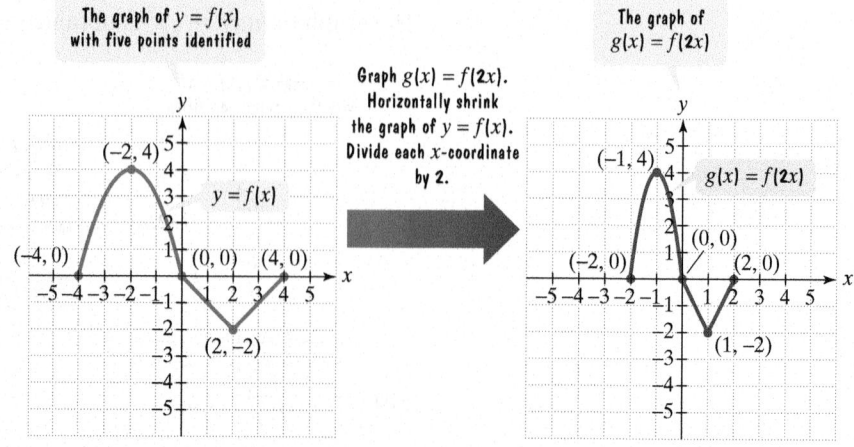

b. The graph of $h(x) = f\left(\frac{1}{2}x\right)$ is obtained by horizontally stretching the graph of $y = f(x)$.

FIGURE 1.61

⊘ **Check Point 7** Use the graph of $y = f(x)$ in **Figure 1.61** to obtain each of the following graphs:

 a. $g(x) = f(2x)$ **b.** $h(x) = f\left(\frac{1}{2}x\right)$.

⑦ Graph functions involving a sequence of transformations.

Sequences of Transformations

Table 1.4 summarizes the procedures for transforming the graph of $y = f(x)$.

Table 1.4 Summary of Transformations

In each case, c represents a positive real number.

To Graph:	Draw the Graph of f and:	Changes in the Equation of $y = f(x)$
Vertical shifts $y = f(x) + c$ $y = f(x) - c$	Raise the graph of f by c units. Lower the graph of f by c units.	c is added to $f(x)$. c is subtracted from $f(x)$.
Horizontal shifts $y = f(x + c)$ $y = f(x - c)$	Shift the graph of f to the left c units. Shift the graph of f to the right c units.	x is replaced with $x + c$. x is replaced with $x - c$.
Reflection about the x-axis $y = -f(x)$	Reflect the graph of f about the x-axis.	$f(x)$ is multiplied by -1.
Reflection about the y-axis $y = f(-x)$	Reflect the graph of f about the y-axis.	x is replaced with $-x$.
Vertical stretching or shrinking $y = cf(x), c > 1$ $y = cf(x), 0 < c < 1$	Multiply each y-coordinate of $y = f(x)$ by c, vertically stretching the graph of f. Multiply each y-coordinate of $y = f(x)$ by c, vertically shrinking the graph of f.	$f(x)$ is multiplied by $c, c > 1$. $f(x)$ is multiplied by $c, 0 < c < 1$.
Horizontal stretching or shrinking $y = f(cx), c > 1$ $y = f(cx), 0 < c < 1$	Divide each x-coordinate of $y = f(x)$ by c, horizontally shrinking the graph of f. Divide each x-coordinate of $y = f(x)$ by c, horizontally stretching the graph of f.	x is replaced with $cx, c > 1$. x is replaced with $cx, 0 < c < 1$.

A function involving more than one transformation can be graphed by performing transformations in the following order:

1. Horizontal shifting
2. Stretching or shrinking
3. Reflecting
4. Vertical shifting

EXAMPLE 8 Graphing Using a Sequence of Transformations

Use the graph of $y = f(x)$ given in **Figure 1.60** of Example 7 on page 186, and repeated below, to graph $y = -\frac{1}{2}f(x - 1) + 3$.

SOLUTION

Our graphs will evolve in the following order:

1. Horizontal shifting: Graph $y = f(x - 1)$ by shifting the graph of $y = f(x)$ 1 unit to the right.
2. Shrinking: Graph $y = \frac{1}{2}f(x - 1)$ by shrinking the previous graph by a factor of $\frac{1}{2}$.
3. Reflecting: Graph $y = -\frac{1}{2}f(x - 1)$ by reflecting the previous graph about the x-axis.
4. Vertical shifting: Graph $y = -\frac{1}{2}f(x - 1) + 3$ by shifting the previous graph up 3 units.

The graph of $y = f(x)$ with five points identified

Graph $y = f(x - 1)$. Shift 1 unit to the right. Add 1 to each x-coordinate.

The graph of $y = f(x - 1)$

Graph $y = \frac{1}{2}f(x - 1)$. Shrink vertically by a factor of $\frac{1}{2}$. Multiply each y-coordinate by $\frac{1}{2}$.

The graph of $y = \frac{1}{2}f(x - 1)$

Graph $y = -\frac{1}{2}f(x - 1)$. Reflect about the x-axis. Replace each y-coordinate with its opposite.

Graph $y = -\frac{1}{2}f(x - 1) + 3$. Shift up 3 units. Add 3 to each y-coordinate.

The graph of $y = -\frac{1}{2}f(x - 1)$

The graph of $y = -\frac{1}{2}f(x - 1) + 3$

● ● ●

✓ Check Point **8** Use the graph of $y = f(x)$ given in **Figure 1.61** of Check Point 7 on page 187 to graph $y = -\frac{1}{3}f(x + 1) - 2$.

EXAMPLE 9 Graphing Using a Sequence of Transformations

Use the graph of $f(x) = x^2$ to graph $g(x) = 2(x + 3)^2 - 1$.

SOLUTION

Our graphs will evolve in the following order:

1. **Horizontal shifting:** Graph $y = (x + 3)^2$ by shifting the graph of $f(x) = x^2$ three units to the left.

2. **Stretching:** Graph $y = 2(x + 3)^2$ by stretching the previous graph by a factor of 2.

3. **Vertical shifting:** Graph $g(x) = 2(x + 3)^2 - 1$ by shifting the previous graph down 1 unit.

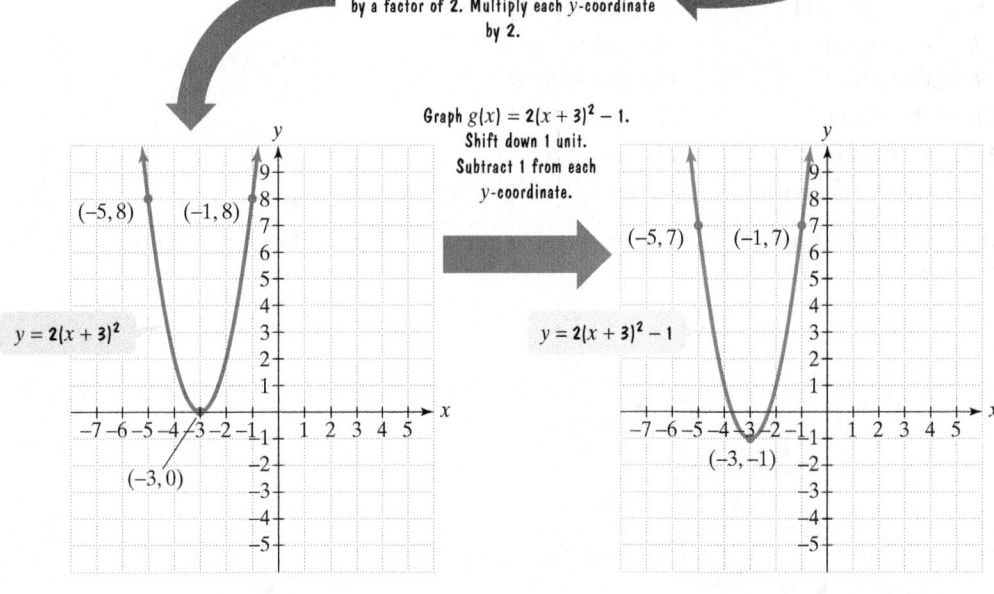

The graph of $y = 2(x + 3)^2$ The graph of $g(x) = 2(x + 3)^2 - 1$

● ● ●

✓ Check Point **9** Use the graph of $f(x) = x^2$ to graph $g(x) = 2(x - 1)^2 + 3$.

CONCEPT AND VOCABULARY CHECK

Fill in each blank so that the resulting statement is true.

1. The graph of $y = f(x) - 5$ is obtained by a/an _____ shift of the graph of $y = f(x)$ _____ a distance of 5 units.

2. The graph of $y = f(x - 5)$ is obtained by a/an _____ shift of the graph of $y = f(x)$ _____ a distance of 5 units.

3. The graph of $y = -f(x)$ is the graph of $y = f(x)$ reflected about the _____.

4. The graph of $y = f(-x)$ is the graph of $y = f(x)$ reflected about the _____.

5. The graph of $y = 5f(x)$ is obtained by a/an _____ stretch of the graph of $y = f(x)$ by multiplying each of its ____-coordinates by 5.

6. The graph of $y = f\left(\frac{1}{5}x\right)$ is obtained by a/an _____ stretch of the graph of $y = f(x)$ by multiplying each of its ____-coordinates by 5.

7. True or false: The graph of $g(x) = \sqrt{x + 4}$ is the graph of $f(x) = \sqrt{x}$ shifted horizontally to the right by 4 units. _____

EXERCISE SET 1.6

Practice Exercises

In Exercises 1–16, use the graph of $y = f(x)$ to graph each function g.

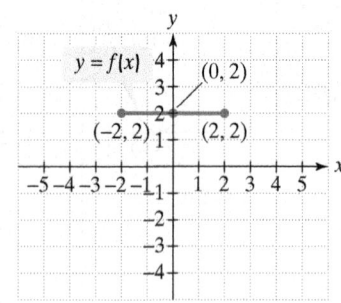

1. $g(x) = f(x) + 1$
2. $g(x) = f(x) - 1$
3. $g(x) = f(x + 1)$
4. $g(x) = f(x - 1)$
5. $g(x) = f(x - 1) - 2$
6. $g(x) = f(x + 1) + 2$
7. $g(x) = f(-x)$
8. $g(x) = -f(x)$
9. $g(x) = -f(x) + 3$
10. $g(x) = f(-x) + 3$
11. $g(x) = \frac{1}{2}f(x)$
12. $g(x) = 2f(x)$
13. $g(x) = f\left(\frac{1}{2}x\right)$
14. $g(x) = f(2x)$
15. $g(x) = -f\left(\frac{1}{2}x\right) + 1$
16. $g(x) = -f(2x) - 1$

In Exercises 17–32, use the graph of $y = f(x)$ to graph each function g.

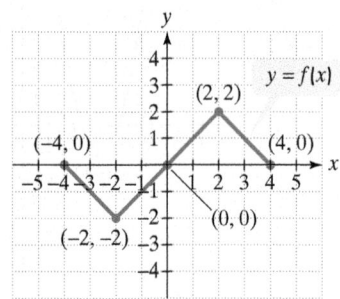

17. $g(x) = f(x) - 1$
18. $g(x) = f(x) + 1$
19. $g(x) = f(x - 1)$
20. $g(x) = f(x + 1)$
21. $g(x) = f(x - 1) + 2$
22. $g(x) = f(x + 1) - 2$
23. $g(x) = -f(x)$
24. $g(x) = f(-x)$

25. $g(x) = f(-x) + 1$
26. $g(x) = -f(x) + 1$
27. $g(x) = 2f(x)$
28. $g(x) = \frac{1}{2}f(x)$
29. $g(x) = f(2x)$
30. $g(x) = f\left(\frac{1}{2}x\right)$
31. $g(x) = 2f(x + 2) + 1$
32. $g(x) = 2f(x + 2) - 1$

In Exercises 33–44, use the graph of $y = f(x)$ to graph each function g.

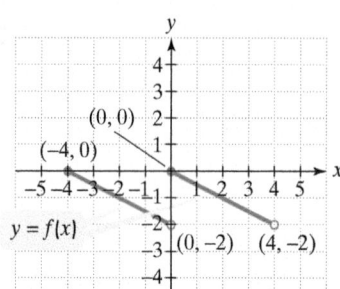

33. $g(x) = f(x) + 2$
34. $g(x) = f(x) - 2$
35. $g(x) = f(x + 2)$
36. $g(x) = f(x - 2)$
37. $g(x) = -f(x + 2)$
38. $g(x) = -f(x - 2)$
39. $g(x) = -\frac{1}{2}f(x + 2)$
40. $g(x) = -\frac{1}{2}f(x - 2)$
41. $g(x) = -\frac{1}{2}f(x + 2) - 2$
42. $g(x) = -\frac{1}{2}f(x - 2) + 2$
43. $g(x) = \frac{1}{2}f(2x)$
44. $g(x) = 2f\left(\frac{1}{2}x\right)$

In Exercises 45–52, use the graph of $y = f(x)$ to graph each function g.

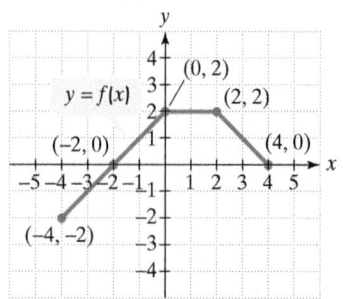

45. $g(x) = f(x - 1) - 1$
46. $g(x) = f(x + 1) + 1$
47. $g(x) = -f(x - 1) + 1$
48. $g(x) = -f(x + 1) - 1$

49. $g(x) = 2f\left(\frac{1}{2}x\right)$

50. $g(x) = \frac{1}{2}f(2x)$

51. $g(x) = \frac{1}{2}f(x + 1)$

52. $g(x) = 2f(x - 1)$

In Exercises 53–66, begin by graphing the standard quadratic function, $f(x) = x^2$. Then use transformations of this graph to graph the given function.

53. $g(x) = x^2 - 2$

54. $g(x) = x^2 - 1$

55. $g(x) = (x - 2)^2$

56. $g(x) = (x - 1)^2$

57. $h(x) = -(x - 2)^2$

58. $h(x) = -(x - 1)^2$

59. $h(x) = (x - 2)^2 + 1$

60. $h(x) = (x - 1)^2 + 2$

61. $g(x) = 2(x - 2)^2$

62. $g(x) = \frac{1}{2}(x - 1)^2$

63. $h(x) = 2(x - 2)^2 - 1$

64. $h(x) = \frac{1}{2}(x - 1)^2 - 1$

65. $h(x) = -2(x + 1)^2 + 1$

66. $h(x) = -2(x + 2)^2 + 1$

In Exercises 67–80, begin by graphing the square root function, $f(x) = \sqrt{x}$. Then use transformations of this graph to graph the given function.

67. $g(x) = \sqrt{x} + 2$

68. $g(x) = \sqrt{x} + 1$

69. $g(x) = \sqrt{x + 2}$

70. $g(x) = \sqrt{x + 1}$

71. $h(x) = -\sqrt{x + 2}$

72. $h(x) = -\sqrt{x + 1}$

73. $h(x) = \sqrt{-x + 2}$

74. $h(x) = \sqrt{-x + 1}$

75. $g(x) = \frac{1}{2}\sqrt{x + 2}$

76. $g(x) = 2\sqrt{x + 1}$

77. $h(x) = \sqrt{x + 2} - 2$

78. $h(x) = \sqrt{x + 1} - 1$

79. $g(x) = 2\sqrt{x + 2} - 2$

80. $g(x) = 2\sqrt{x + 1} - 1$

In Exercises 81–94, begin by graphing the absolute value function, $f(x) = |x|$. Then use transformations of this graph to graph the given function.

81. $g(x) = |x| + 4$

82. $g(x) = |x| + 3$

83. $g(x) = |x + 4|$

84. $g(x) = |x + 3|$

85. $h(x) = |x + 4| - 2$

86. $h(x) = |x + 3| - 2$

87. $h(x) = -|x + 4|$

88. $h(x) = -|x + 3|$

89. $g(x) = -|x + 4| + 1$

90. $g(x) = -|x + 4| + 2$

91. $h(x) = 2|x + 4|$

92. $h(x) = 2|x + 3|$

93. $g(x) = -2|x + 4| + 1$

94. $g(x) = -2|x + 3| + 2$

In Exercises 95–106, begin by graphing the standard cubic function, $f(x) = x^3$. Then use transformations of this graph to graph the given function.

95. $g(x) = x^3 - 3$

96. $g(x) = x^3 - 2$

97. $g(x) = (x - 3)^3$

98. $g(x) = (x - 2)^3$

99. $h(x) = -x^3$

100. $h(x) = -(x - 2)^3$

101. $h(x) = \frac{1}{2}x^3$

102. $h(x) = \frac{1}{4}x^3$

103. $r(x) = (x - 3)^3 + 2$

104. $r(x) = (x - 2)^3 + 1$

105. $h(x) = \frac{1}{2}(x - 3)^3 - 2$

106. $h(x) = \frac{1}{2}(x - 2)^3 - 1$

In Exercises 107–118, begin by graphing the cube root function, $f(x) = \sqrt[3]{x}$. Then use transformations of this graph to graph the given function.

107. $g(x) = \sqrt[3]{x} + 2$

108. $g(x) = \sqrt[3]{x} - 2$

109. $g(x) = \sqrt[3]{x + 2}$

110. $g(x) = \sqrt[3]{x - 2}$

111. $h(x) = \frac{1}{2}\sqrt[3]{x + 2}$

112. $h(x) = \frac{1}{2}\sqrt[3]{x - 2}$

113. $r(x) = \frac{1}{2}\sqrt[3]{x + 2} - 2$

114. $r(x) = \frac{1}{2}\sqrt[3]{x - 2} + 2$

115. $h(x) = -\sqrt[3]{x + 2}$

116. $h(x) = -\sqrt[3]{x - 2}$

117. $g(x) = \sqrt[3]{-x - 2}$

118. $g(x) = \sqrt[3]{-x + 2}$

Practice Plus

*In Exercises 119–122, use transformations of the graph of the greatest integer function, $f(x) = \text{int}(x)$, to graph each function. (The graph of $f(x) = \text{int}(x)$ is shown in **Figure 1.35** on page 144.)*

119. $g(x) = 2 \text{ int } (x + 1)$

120. $g(x) = 3 \text{ int } (x - 1)$

121. $h(x) = \text{int}(-x) + 1$

122. $h(x) = \text{int}(-x) - 1$

In Exercises 123–126, write a possible equation for the function whose graph is shown. Each graph shows a transformation of a common function.

123.

$[-2, 8, 1]$ by $[-1, 4, 1]$

124.

$[-3, 3, 1]$ by $[-6, 6, 1]$

125.

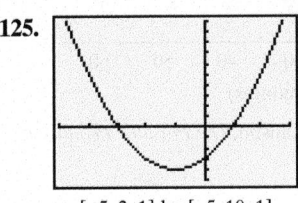

$[-5, 3, 1]$ by $[-5, 10, 1]$

126.

$[-1, 9, 1]$ by $[-1, 5, 1]$

Application Exercises

127. The function $f(x) = 2.9\sqrt{x} + 20.1$ models the median height, $f(x)$, in inches, of boys who are x months of age. The graph of f is shown.

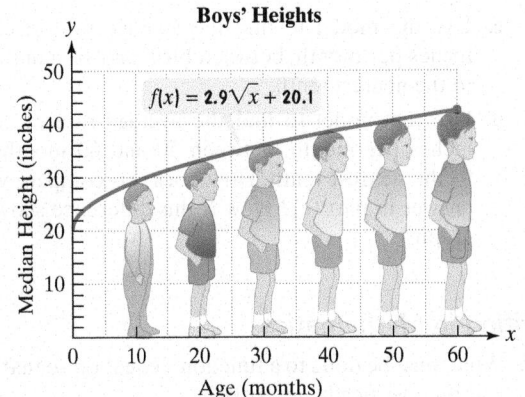

Source: Laura Walther Nathanson, *The Portable Pediatrician for Parents*

a. Describe how the graph can be obtained using transformations of the square root function $f(x) = \sqrt{x}$.

b. According to the model, what is the median height of boys who are 48 months, or four years, old? Use a calculator and round to the nearest tenth of an inch. The actual median height for boys at 48 months is 40.8 inches. How well does the model describe the actual height?

(This exercise continues on the next page.)

c. Use the model to find the average rate of change, in inches per month, between birth and 10 months. Round to the nearest tenth.

d. Use the model to find the average rate of change, in inches per month, between 50 and 60 months. Round to the nearest tenth. How does this compare with your answer in part (c)? How is this difference shown by the graph?

128. The function $f(x) = 3.1\sqrt{x} + 19$ models the median height, $f(x)$, in inches, of girls who are x months of age. The graph of f is shown.

Girls' Heights

Source: Laura Walther Nathanson, *The Portable Pediatrician for Parents*

a. Describe how the graph can be obtained using transformations of the square root function $f(x) = \sqrt{x}$.

b. According to the model, what is the median height of girls who are 48 months, or 4 years, old? Use a calculator and round to the nearest tenth of an inch. The actual median height for girls at 48 months is 40.2 inches. How well does the model describe the actual height?

c. Use the model to find the average rate of change, in inches per month, between birth and 10 months. Round to the nearest tenth.

d. Use the model to find the average rate of change, in inches per month, between 50 and 60 months. Round to the nearest tenth. How does this compare with your answer in part (c)? How is this difference shown by the graph?

Writing in Mathematics

129. What must be done to a function's equation so that its graph is shifted vertically upward?

130. What must be done to a function's equation so that its graph is shifted horizontally to the right?

131. What must be done to a function's equation so that its graph is reflected about the x-axis?

132. What must be done to a function's equation so that its graph is reflected about the y-axis?

133. What must be done to a function's equation so that its graph is stretched vertically?

134. What must be done to a function's equation so that its graph is shrunk horizontally?

Technology Exercises

135. a. Use a graphing utility to graph $f(x) = x^2 + 1$.

b. Graph $f(x) = x^2 + 1$, $g(x) = f(2x)$, $h(x) = f(3x)$, and $k(x) = f(4x)$ in the same viewing rectangle.

c. Describe the relationship among the graphs of $f, g, h,$ and k, with emphasis on different values of x for points on all four graphs that give the same y-coordinate.

d. Generalize by describing the relationship between the graph of f and the graph of g, where $g(x) = f(cx)$ for $c > 1$.

e. Try out your generalization by sketching the graphs of $f(cx)$ for $c = 1, c = 2, c = 3$, and $c = 4$ for a function of your choice.

136. a. Use a graphing utility to graph $f(x) = x^2 + 1$.

b. Graph $f(x) = x^2 + 1$, $g(x) = f(\frac{1}{2}x)$, and $h(x) = f(\frac{1}{4}x)$ in the same viewing rectangle.

c. Describe the relationship among the graphs of $f, g,$ and h, with emphasis on different values of x for points on all three graphs that give the same y-coordinate.

d. Generalize by describing the relationship between the graph of f and the graph of g, where $g(x) = f(cx)$ for $0 < c < 1$.

e. Try out your generalization by sketching the graphs of $f(cx)$ for $c = 1$, and $c = \frac{1}{2}$, and $c = \frac{1}{4}$ for a function of your choice.

Critical Thinking Exercises

Make Sense? *During the winter, you program your home thermostat so that at midnight, the temperature is 55°. This temperature is maintained until 6 A.M. Then the house begins to warm up so that by 9 A.M. the temperature is 65°. At 6 P.M. the house begins to cool. By 9 P.M., the temperature is again 55°. The graph illustrates home temperature, $f(t)$, as a function of hours after midnight, t.*

In Exercises 137–140, determine whether each statement makes sense or does not make sense, and explain your reasoning. If the statement makes sense, graph the new function on the domain $[0, 24]$. If the statement does not make sense, correct the function in the statement and graph the corrected function on the domain $[0, 24]$.

137. I decided to keep the house 5° warmer than before, so I reprogrammed the thermostat to $y = f(t) + 5$.

138. I decided to keep the house 5° cooler than before, so I reprogrammed the thermostat to $y = f(t) - 5$.

139. I decided to change the heating schedule to start one hour earlier than before, so I reprogrammed the thermostat to $y = f(t - 1)$.

140. I decided to change the heating schedule to start one hour later than before, so I reprogrammed the thermostat to $y = f(t + 1)$.

In Exercises 141–144, determine whether each statement is true or false. If the statement is false, make the necessary change(s) to produce a true statement.

141. If $f(x) = |x|$ and $g(x) = |x + 3| + 3$, then the graph of g is a translation of the graph of f three units to the right and three units upward.

142. If $f(x) = -\sqrt{x}$ and $g(x) = \sqrt{-x}$, then f and g have identical graphs.

143. If $f(x) = x^2$ and $g(x) = 5(x^2 - 2)$, then the graph of g can be obtained from the graph of f by stretching f five units followed by a downward shift of two units.

144. If $f(x) = x^3$ and $g(x) = -(x - 3)^3 - 4$, then the graph of g can be obtained from the graph of f by moving f three units to the right, reflecting about the x-axis, and then moving the resulting graph down four units.

In Exercises 145–148, functions f and g are graphed in the same rectangular coordinate system. If g is obtained from f through a sequence of transformations, find an equation for g.

145.

146.

147.

148.

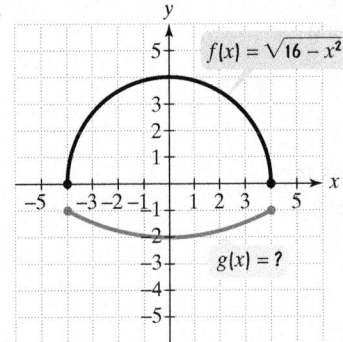

For Exercises 149–152, assume that (a, b) is a point on the graph of f. What is the corresponding point on the graph of each of the following functions?

149. $y = f(-x)$

150. $y = 2f(x)$

151. $y = f(x - 3)$

152. $y = f(x) - 3$

Preview Exercises

Exercises 153–155 will help you prepare for the material covered in the next section.

In Exercises 153–154, perform the indicated operation or operations.

153. $(2x - 1)(x^2 + x - 2)$

154. $(f(x))^2 - 2f(x) + 6$, where $f(x) = 3x - 4$

155. Simplify: $\dfrac{2}{\dfrac{3}{x} - 1}$.

Combinations of Functions; Composite Functions

Objectives

1. Find the domain of a function.
2. Combine functions using the algebra of functions, specifying domains.
3. Form composite functions.
4. Determine domains for composite functions.
5. Write functions as compositions.

We're born. We die. **Figure 1.62** quantifies these statements by showing the number of births and deaths in the United States from 2000 through 2009.

Number of Births and Deaths in the United States

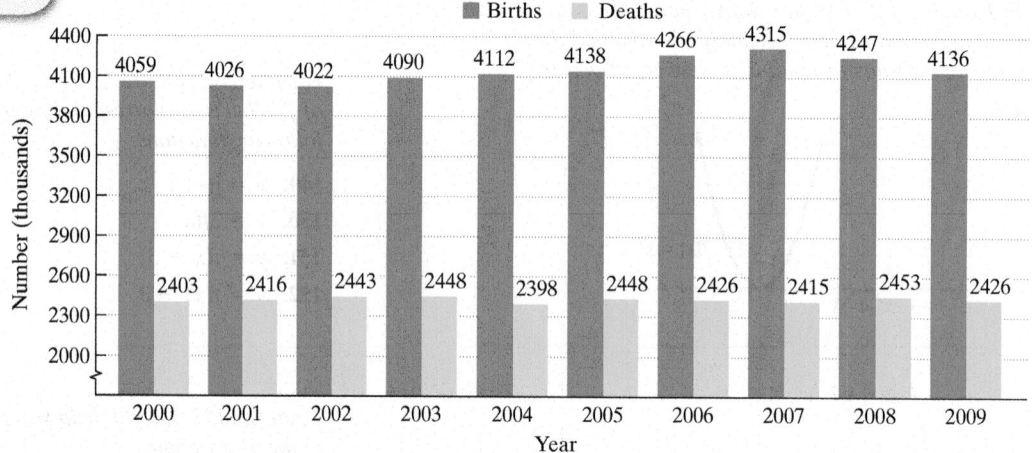

FIGURE 1.62
Source: U.S. Department of Health and Human Services

In this section, we look at these data from the perspective of functions. By considering the yearly change in the U.S. population, you will see that functions can be subtracted using procedures that will remind you of combining algebraic expressions.

1. Find the domain of a function.

The Domain of a Function

We begin with two functions that model the data in **Figure 1.62**.

$$B(x) = -2.6x^2 + 49x + 3994 \qquad D(x) = -0.6x^2 + 7x + 2412$$

Number of births, $B(x)$, in thousands, x years after **2000**

Number of deaths, $D(x)$, in thousands, x years after **2000**

The years in **Figure 1.62** extend from 2000 through 2009. Because x represents the number of years after 2000,

$$\text{Domain of } B = \{0, 1, 2, 3, \ldots, 9\}$$

and

$$\text{Domain of } D = \{0, 1, 2, 3, \ldots, 9\}.$$

Functions that model data often have their domains explicitly given with the function's equation. However, for most functions, only an equation is given and the domain is not specified. In cases like this, the domain of a function f is the largest set of real numbers for which the value of $f(x)$ is a real number. For example, consider the function

$$f(x) = \frac{1}{x - 3}.$$

Because division by 0 is undefined, the denominator, $x - 3$, cannot be 0. Thus, x cannot equal 3. The domain of the function consists of all real numbers other than 3, represented by

Domain of $f = \{x \mid x \text{ is a real number and } x \neq 3\}$.

Using interval notation,

Domain of $f = (-\infty, 3) \cup (3, \infty)$.

| All real numbers less than 3 | or | All real numbers greater than 3 |

Now consider a function involving a square root:

$$g(x) = \sqrt{x - 3}.$$

Because only nonnegative numbers have square roots that are real numbers, the expression under the square root sign, $x - 3$, must be nonnegative. We can use inspection to see that $x - 3 \geq 0$ if $x \geq 3$. The domain of g consists of all real numbers that are greater than or equal to 3:

Domain of $g = \{x \mid x \geq 3\}$ or $[3, \infty)$.

GREAT QUESTION!

Despite the voice balloons, the notation for the domain of f on the right is a mouthful! Will you be using set operations with interval notation in this section? What should I already know?

Yes, you'll be seeing the intersection and the union of sets that are expressed in interval notation. Recall that the intersection of sets A and B, written $A \cap B$, is the set of elements common to both set A and set B. When sets A and B are in interval notation, to find the intersection, graph each interval and take the portion of the number line that the two graphs have in common. We will also be using notation involving the union of sets A and B, $A \cup B$, meaning the set of elements in A or in B or in both. For more detail, see Section P.1, pages 5–6 and pages 11–12.

Finding a Function's Domain

If a function f does not model data or verbal conditions, its domain is the largest set of real numbers for which the value of $f(x)$ is a real number. Exclude from a function's domain real numbers that cause division by zero and real numbers that result in an even root, such as a square root, of a negative number.

EXAMPLE 1 Finding the Domain of a Function

Find the domain of each function:

a. $f(x) = 3x + 2$ **b.** $g(x) = \dfrac{3x + 2}{x + 1}$ **c.** $h(x) = \sqrt{x + 3}$.

SOLUTION

a. The function $f(x) = 3x + 2$ contains neither division nor a square root. For every real number, x, the algebraic expression $3x + 2$ is a real number. Thus, the domain of f is the set of all real numbers.

Domain of $f = (-\infty, \infty)$

b. The function $g(x) = \dfrac{3x + 2}{x + 1}$ contains division. Because division by 0 is undefined, we must exclude from the domain the values of x that cause $x + 1$ to be 0. Thus, x cannot equal -1.

Domain of $g = (-\infty, -1) \cup (-1, \infty)$

c. The function $h(x) = \sqrt{x + 3}$ contains a square root. Because the square root of a negative number is not a real number, $x + 3$ must be nonnegative. Thus, the domain of h is all x such that $x + 3 \geq 0$. Using inspection, x must be greater than or equal to -3: $x \geq -3$.

$$\text{Domain of } h = [-3, \infty)$$

\oslash Check Point **1** Find the domain of each function:

a. $f(x) = \frac{1}{2}x + 3$ **b.** $g(x) = \dfrac{7x + 4}{x + 5}$ **c.** $h(x) = \sqrt{x - 2}$.

2 Combine functions using the algebra of functions, specifying domains.

The Algebra of Functions

We can combine functions using addition, subtraction, multiplication, and division by performing operations with the algebraic expressions that appear on the right side of the equations. For example, the functions $f(x) = 2x$ and $g(x) = x - 1$ can be combined to form the sum, difference, product, and quotient of f and g. Here's how it's done:

For each function, $f(x) = 2x$ and $g(x) = x - 1$.

Sum: $f + g$

$$(f + g)(x) = f(x) + g(x)$$
$$= 2x + (x - 1) = 3x - 1$$

Difference: $f - g$

$$(f - g)(x) = f(x) - g(x)$$
$$= 2x - (x - 1) = 2x - x + 1 = x + 1$$

Product: fg

$$(fg)(x) = f(x) \cdot g(x)$$
$$= 2x(x - 1) = 2x^2 - 2x$$

Quotient: $\dfrac{f}{g}$

$$\left(\frac{f}{g}\right)(x) = \frac{f(x)}{g(x)} = \frac{2x}{x - 1}, x \neq 1.$$

The domain for each of these functions consists of all real numbers that are common to the domains of f and g. Using D_f to represent the domain of f and D_g to represent the domain of g, the domain for each function is $D_f \cap D_g$. In the case of the quotient function $\dfrac{f(x)}{g(x)}$, we must remember not to divide by 0, so we add the further restriction that $g(x) \neq 0$.

The Algebra of Functions: Sum, Difference, Product, and Quotient of Functions

Let f and g be two functions. The **sum** $f + g$, the **difference** $f - g$, the **product** fg, and the **quotient** $\dfrac{f}{g}$ are functions whose domains are the set of all real numbers common to the domains of f and g ($D_f \cap D_g$), defined as follows:

1. Sum: $(f + g)(x) = f(x) + g(x)$
2. Difference: $(f - g)(x) = f(x) - g(x)$
3. Product: $(fg)(x) = f(x) \cdot g(x)$
4. Quotient: $\left(\dfrac{f}{g}\right)(x) = \dfrac{f(x)}{g(x)}$, provided $g(x) \neq 0$.

EXAMPLE 2 Combining Functions

Let $f(x) = 2x - 1$ and $g(x) = x^2 + x - 2$. Find each of the following functions:

a. $(f + g)(x)$ **b.** $(f - g)(x)$ **c.** $(fg)(x)$ **d.** $\left(\dfrac{f}{g}\right)(x)$.

Determine the domain for each function.

SOLUTION

a. $(f + g)(x) = f(x) + g(x)$ *This is the definition of the sum f + g.*

 $= (2x - 1) + (x^2 + x - 2)$ *Substitute the given functions.*

 $= x^2 + 3x - 3$ *Remove parentheses and combine like terms.*

b. $(f - g)(x) = f(x) - g(x)$ *This is the definition of the difference f − g.*

 $= (2x - 1) - (x^2 + x - 2)$ *Substitute the given functions.*

 $= 2x - 1 - x^2 - x + 2$ *Remove parentheses and change the sign of each term in the second set of parentheses.*

 $= -x^2 + x + 1$ *Combine like terms and arrange terms in descending powers of x.*

c. $(fg)(x) = f(x) \cdot g(x)$ *This is the definition of the product fg.*

 $= (2x - 1)(x^2 + x - 2)$ *Substitute the given functions.*

 $= 2x(x^2 + x - 2) - 1(x^2 + x - 2)$ *Multiply each term in the second factor by 2x and −1, respectively.*

 $= 2x^3 + 2x^2 - 4x - x^2 - x + 2$ *Use the distributive property.*

 $= 2x^3 + (2x^2 - x^2) + (-4x - x) + 2$ *Rearrange terms so that like terms are adjacent.*

 $= 2x^3 + x^2 - 5x + 2$ *Combine like terms.*

d. $\left(\dfrac{f}{g}\right)(x) = \dfrac{f(x)}{g(x)}$ *This is the definition of the quotient $\dfrac{f}{g}$.*

 $= \dfrac{2x - 1}{x^2 + x - 2}$ *Substitute the given functions. This rational expression cannot be simplified.*

GREAT QUESTION!

Should I simplify a quotient function before finding its domain?

No. If the function $\dfrac{f}{g}$ can be simplified, determine the domain *before* simplifying. All values of x for which $g(x) = 0$ must be excluded from the domain.

Example:

$$f(x) = x^2 - 4 \text{ and}$$
$$g(x) = x - 2$$

$$\left(\frac{f}{g}\right)(x) = \frac{x^2 - 4}{x - 2}$$

$x \neq 2$. The domain of $\dfrac{f}{g}$ is $(-\infty, 2) \cup (2, \infty)$.

$$= \frac{\overset{1}{(x + 2)(x - 2)}}{\underset{1}{(x - 2)}} = x + 2$$

Because the equations for f and g do not involve division or contain even roots, the domain of both f and g is the set of all real numbers. Thus, the domain of $f + g, f - g$, and fg is the set of all real numbers, $(-\infty, \infty)$.

The function $\dfrac{f}{g}$ contains division. We must exclude from its domain values of x that cause the denominator, $x^2 + x - 2$, to be 0. Let's identify these values by factoring the denominator as follows:

$$\left(\frac{f}{g}\right)(x) = \frac{2x - 1}{x^2 + x - 2} = \frac{2x - 1}{(x + 2)(x - 1)}.$$

Because $x + 2 \neq 0$, Because $x - 1 \neq 0$,
$x \neq -2$. $x \neq 1$.

We must exclude -2 and 1 from the domain of $\dfrac{f}{g}$.

$$\text{Domain of } \frac{f}{g} = (-\infty, -2) \cup (-2, 1) \cup (1, \infty) \qquad \bullet\bullet\bullet$$

✓ **Check Point 2** Let $f(x) = x - 5$ and $g(x) = x^2 - 1$. Find each of the following functions:

 a. $(f + g)(x)$ **b.** $(f - g)(x)$ **c.** $(fg)(x)$ **d.** $\left(\dfrac{f}{g}\right)(x)$.

Determine the domain for each function.

EXAMPLE 3 Adding Functions and Determining the Domain

Let $f(x) = \sqrt{x + 3}$ and $g(x) = \sqrt{x - 2}$. Find each of the following:

 a. $(f + g)(x)$ **b.** the domain of $f + g$.

SOLUTION

 a. $(f + g)(x) = f(x) + g(x) = \sqrt{x + 3} + \sqrt{x - 2}$

 b. The domain of $f + g$ is the set of all real numbers that are common to the domain of f and the domain of g. Thus, we must find the domains of f and g before finding their intersection.

• $f(x) = \sqrt{x + 3}$ • $g(x) = \sqrt{x - 2}$

> $x + 3$ must be nonnegative: $x - 2$ must be nonnegative:
> $x + 3 \geq 0.\ D_f = [-3, \infty)$ $x - 2 \geq 0.\ D_g = [2, \infty)$

FIGURE 1.63 Finding the domain of the sum $f + g$

Now, we can use a number line to determine $D_f \cap D_g$, the domain of $f + g$. **Figure 1.63** shows the domain of f in blue and the domain of g in red. Can you see that all real numbers greater than or equal to 2 are common to both domains? This is shown in purple on the number line. Thus, the domain of $f + g$ is $[2, \infty)$. $\bullet\bullet\bullet$

TECHNOLOGY

Graphic Connections

Domain: $[2, \infty)$

The graph on the left is the graph of

$$y = \sqrt{x + 3} + \sqrt{x - 2}$$

in a $[-3, 10, 1]$ by $[0, 8, 1]$ viewing rectangle. The graph reveals what we discovered algebraically in Example 3(b). The domain of this function is $[2, \infty)$.

⊘ Check Point **3** Let $f(x) = \sqrt{x - 3}$ and $g(x) = \sqrt{x + 1}$. Find each of the following:

a. $(f + g)(x)$ **b.** the domain of $f + g$.

EXAMPLE 4 Applying the Algebra of Functions

We opened the section with functions that model the number of births and deaths in the United States from 2000 through 2009:

$$B(x) = -2.6x^2 + 49x + 3994 \qquad D(x) = -0.6x^2 + 7x + 2412.$$

Number of births, $B(x)$, in thousands, x years after 2000

Number of deaths, $D(x)$, in thousands, x years after 2000

a. Write a function that models the change in U.S. population for the years from 2000 through 2009.

b. Use the function from part (a) to find the change in U.S. population in 2008.

c. Does the result in part (b) overestimate or underestimate the actual population change in 2008 obtained from the data in **Figure 1.62** on page 194? By how much?

SOLUTION

a. The change in population is the number of births minus the number of deaths. Thus, we will find the difference function, $B - D$.

$(B - D)(x)$

$= B(x) - D(x)$

$= (-2.6x^2 + 49x + 3994) - (-0.6x^2 + 7x + 2412)$ Substitute the given functions.

$= -2.6x^2 + 49x + 3994 + 0.6x^2 - 7x - 2412$ Remove parentheses and change the sign of each term in the second set of parentheses.

$= (-2.6x^2 + 0.6x^2) + (49x - 7x) + (3994 - 2412)$ Group like terms.

$= -2x^2 + 42x + 1582$ Combine like terms.

The function

$$(B - D)(x) = -2x^2 + 42x + 1582$$

models the change in U.S. population, in thousands, x years after 2000.

b. Because 2008 is 8 years after 2000, we substitute 8 for x in the difference function $(B - D)(x)$.

$$(B - D)(x) = -2x^2 + 42x + 1582 \quad \text{Use the difference function } B - D.$$

$$(B - D)(8) = -2(8)^2 + 42(8) + 1582 \quad \text{Substitute 8 for x.}$$

$$= -2(64) + 42(8) + 1582 \quad \text{Evaluate the exponential expression: } 8^2 = 64.$$

$$= -128 + 336 + 1582 \quad \text{Perform the multiplications.}$$

$$= 1790 \quad \text{Add from left to right.}$$

We see that $(B - D)(8) = 1790$. The model indicates that there was a population increase of 1790 thousand, or approximately 1,790,000 people, in 2008.

c. The data for 2008 in **Figure 1.62** on page 194 show 4247 thousand births and 2453 thousand deaths.

$$\text{population change} = \text{births} - \text{deaths}$$

$$= 4247 - 2453 = 1794$$

The actual population increase was 1794 thousand, or 1,794,000. Our model gave us an increase of 1790 thousand. Thus, the model underestimates the actual increase by $1794 - 1790$, or 4 thousand people.　● ● ●

⊘ Check Point **4** Use the birth and death models from Example 4.

a. Write a function that models the total number of births and deaths in the United States for the years from 2000 through 2009.

b. Use the function from part (a) to find the total number of births and deaths in the United States in 2003.

c. Does the result in part (b) overestimate or underestimate the actual number of total births and deaths in 2003 obtained from the data in **Figure 1.62** on page 194? By how much?

❸ Form composite functions.

Composite Functions

There is another way of combining two functions. To help understand this new combination, suppose that your local computer store is having a sale. The models that are on sale cost either $300 less than the regular price or 85% of the regular price. If x represents the computer's regular price, the discounts can be modeled with the following functions:

$$f(x) = x - 300 \quad g(x) = 0.85x.$$

The computer is on sale for $300 less than its regular price.　　The computer is on sale for 85% of its regular price.

At the store, you bargain with the salesperson. Eventually, she makes an offer you can't refuse. The sale price will be 85% of the regular price followed by a $300 reduction:

$$0.85x - 300.$$

85% of the regular price　　followed by a $300 reduction

In terms of the functions f and g, this offer can be obtained by taking the output of $g(x) = 0.85x$, namely, $0.85x$, and using it as the input of f:

$$f(x) = x - 300$$

Replace x with $0.85x$, the output of $g(x) = 0.85x$.

$$f(0.85x) = 0.85x - 300.$$

Because $0.85x$ is $g(x)$, we can write this last equation as

$$f(g(x)) = 0.85x - 300.$$

We read this equation as "f of g of x is equal to $0.85x - 300$." We call $f(g(x))$ the **composition of the function f with g**, or a **composite function**. This composite function is written $f \circ g$. Thus,

$$(f \circ g)(x) = f(g(x)) = 0.85x - 300.$$

This can be read "f of g of x" or "f composed with g of x."

Like all functions, we can evaluate $f \circ g$ for a specified value of x in the function's domain. For example, here's how to find the value of the composite function describing the offer you cannot refuse at 1400:

$$(f \circ g)(x) = 0.85x - 300$$

Replace x with 1400.

$$(f \circ g)(1400) = 0.85(1400) - 300 = 1190 - 300 = 890.$$

This means that a computer that regularly sells for \$1400 is on sale for \$890 subject to both discounts. We can use a partial table of coordinates for each of the discount functions, g and f, to verify this result numerically.

Computer's regular price	85% of the regular price		85% of the regular price	\$300 reduction
x	$g(x) = 0.85x$		x	$f(x) = x - 300$
1200	1020		1020	720
1300	1105		1105	805
1400	1190		1190	890

Using these tables, we can find $(f \circ g)(1400)$:

$$(f \circ g)(1400) = f(g(1400)) = f(1190) = 890.$$

The table for g shows that $g(1400) = 1190$.　　The table for f shows that $f(1190) = 890$.

This verifies that a computer that regularly sells for \$1400 is on sale for \$890 subject to both discounts.

Before you run out to buy a computer, let's generalize our discussion of the computer's double discount and define the composition of any two functions.

The Composition of Functions

The **composition of the function f with g** is denoted by $f \circ g$ and is defined by the equation

$$(f \circ g)(x) = f(g(x)).$$

The **domain of the composite function $f \circ g$** is the set of all x such that

1. x is in the domain of g and

2. $g(x)$ is in the domain of f.

The composition of f with g, $f \circ g$, is illustrated in **Figure 1.64**.

Step 1 Input x into g. **Step 2** Input $g(x)$ into f.

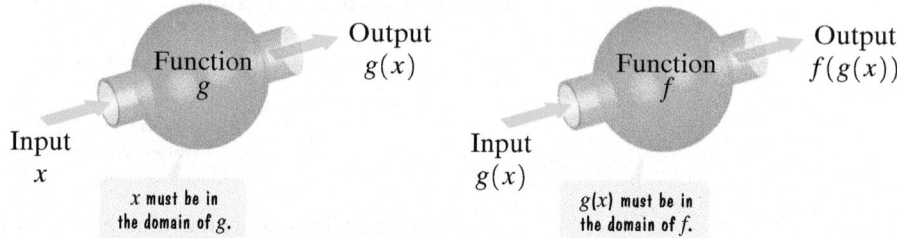

FIGURE 1.64

The figure reinforces the fact that the inside function g in $f(g(x))$ is done first.

EXAMPLE 5 Forming Composite Functions

Given $f(x) = 3x - 4$ and $g(x) = x^2 - 2x + 6$, find each of the following:

a. $(f \circ g)(x)$ **b.** $(g \circ f)(x)$ **c.** $(g \circ f)(1)$.

SOLUTION

a. We begin with $(f \circ g)(x)$, the composition of f with g. Because $(f \circ g)(x)$ means $f(g(x))$, we must replace each occurrence of x in the equation for f with $g(x)$.

$$f(x) = 3x - 4 \qquad \text{This is the given equation for f.}$$

Replace x with $g(x)$.

$$
\begin{aligned}
(f \circ g)(x) = f(g(x)) &= 3g(x) - 4 \\
&= 3(x^2 - 2x + 6) - 4 && \text{Because } g(x) = x^2 - 2x + 6, \\
&&& \text{replace } g(x) \text{ with } x^2 - 2x + 6. \\
&= 3x^2 - 6x + 18 - 4 && \text{Use the distributive property.} \\
&= 3x^2 - 6x + 14 && \text{Simplify.}
\end{aligned}
$$

Thus, $(f \circ g)(x) = 3x^2 - 6x + 14$.

b. Next, we find $(g \circ f)(x)$, the composition of g with f. Because $(g \circ f)(x)$ means $g(f(x))$, we must replace each occurrence of x in the equation for g with $f(x)$.

$$g(x) = x^2 - 2x + 6 \qquad \text{This is the equation for g.}$$

Replace x with $f(x)$.

$$
\begin{aligned}
(g \circ f)(x) = g(f(x)) &= (f(x))^2 - 2f(x) + 6 \\
&= (3x - 4)^2 - 2(3x - 4) + 6 && \text{Because } f(x) = 3x - 4, \\
&&& \text{replace } f(x) \text{ with } 3x - 4.
\end{aligned}
$$

$$= 9x^2 - 24x + 16 - 6x + 8 + 6$$ Use $(A - B)^2 =$ $A^2 - 2AB + B^2$ to square $3x - 4$.

$$= 9x^2 - 30x + 30$$ Simplify: $-24x - 6x = -30x$ and $16 + 8 + 6 = 30$.

Thus, $(g \circ f)(x) = 9x^2 - 30x + 30$. **Notice that $(f \circ g)(x)$ is not the same function as $(g \circ f)(x)$.**

c. We can use $(g \circ f)(x)$ to find $(g \circ f)(1)$.

$$(g \circ f)(x) = 9x^2 - 30x + 30$$

Replace x with 1.

$$(g \circ f)(1) = 9 \cdot 1^2 - 30 \cdot 1 + 30 = 9 - 30 + 30 = 9$$

It is also possible to find $(g \circ f)(1)$ without determining $(g \circ f)(x)$.

$$(g \circ f)(1) = g(f(1)) = g(-1) = 9$$

First find $f(1)$.
$f(x) = 3x - 4$, so
$f(1) = 3 \cdot 1 - 4 = -1$.

Next find $g(-1)$.
$g(x) = x^2 - 2x + 6$, so
$g(-1) = (-1)^2 - 2(-1) + 6$
$= 1 + 2 + 6 = 9$.

• • •

⊘ **Check Point 5** Given $f(x) = 5x + 6$ and $g(x) = 2x^2 - x - 1$, find each of the following:

a. $(f \circ g)(x)$ **b.** $(g \circ f)(x)$ **c.** $(f \circ g)(-1)$.

④ Determine domains for composite functions.

We need to be careful in determining the domain for a composite function.

Excluding Values from the Domain of $(f \circ g)(x) = f(g(x))$

The following values must be excluded from the input x:

• If x is not in the domain of g, it must not be in the domain of $f \circ g$.
• Any x for which $g(x)$ is not in the domain of f must not be in the domain of $f \circ g$.

EXAMPLE 6 Forming a Composite Function and Finding Its Domain

GREAT QUESTION!

Can you remind me of how you simplified the complex fraction on the right?

One method for simplifying complex fractions is to find the least common denominator of all the rational expressions in the numerator and the denominator. Then multiply each term in the numerator and denominator by this least common denominator. The procedures for simplifying complex fractions can be found in Section P.7, pages 95–97.

Given $f(x) = \dfrac{2}{x - 1}$ and $g(x) = \dfrac{3}{x}$, find each of the following:

a. $(f \circ g)(x)$ **b.** the domain of $f \circ g$.

SOLUTION

a. Because $(f \circ g)(x)$ means $f(g(x))$, we must replace x in $f(x) = \dfrac{2}{x - 1}$ with $g(x)$.

$$(f \circ g)(x) = f(g(x)) = \frac{2}{g(x) - 1} = \frac{2}{\frac{3}{x} - 1} = \frac{2}{\frac{3}{x} - 1} \cdot \frac{x}{x} = \frac{2x}{3 - x}$$

$g(x) = \frac{3}{x}$

Simplify the complex fraction by multiplying by $\frac{x}{x}$, or 1.

Thus, $(f \circ g)(x) = \dfrac{2x}{3 - x}$.

b. We determine values to exclude from the domain of $(f \circ g)(x)$ in two steps.

Rules for Excluding Numbers from the Domain of $(f \circ g)(x) = f(g(x))$	Applying the Rules to $f(x) = \dfrac{2}{x-1}$ and $g(x) = \dfrac{3}{x}$
If x is not in the domain of g, it must not be in the domain of $f \circ g$.	Because $g(x) = \dfrac{3}{x}$, 0 is not in the domain of g. Thus, 0 must be excluded from the domain of $f \circ g$.
Any x for which $g(x)$ is not in the domain of f must not be in the domain of $f \circ g$.	Because $f(g(x)) = \dfrac{2}{g(x) - 1}$, we must exclude from the domain of $f \circ g$ any x for which $g(x) = 1$. $\dfrac{3}{x} = 1$ Set $g(x)$ equal to 1. $3 = x$ Multiply both sides by x. 3 must be excluded from the domain of $f \circ g$.

We see that 0 and 3 must be excluded from the domain of $f \circ g$. The domain of $f \circ g$ is

$$(-\infty, 0) \cup (0, 3) \cup (3, \infty).$$ • • •

 Check Point **6** Given $f(x) = \dfrac{4}{x+2}$ and $g(x) = \dfrac{1}{x}$, find each of the following:

a. $(f \circ g)(x)$ **b.** the domain of $f \circ g$.

⑤ Write functions as compositions.

Decomposing Functions

When you form a composite function, you "compose" two functions to form a new function. It is also possible to reverse this process. That is, you can "decompose" a given function and express it as a composition of two functions. Although there is more than one way to do this, there is often a "natural" selection that comes to mind first. For example, consider the function h defined by

$$h(x) = (3x^2 - 4x + 1)^5.$$

GREAT QUESTION!

Is there a procedure I can use to write a function as a composition?

Yes. Suppose the form of function h is $h(x) = (\text{algebraic expression})^{\text{power}}$. Function h can be expressed as a composition, $f \circ g$, using

$f(x) = x^{\text{power}}$

$g(x) = \text{algebraic expression}.$

The function h takes $3x^2 - 4x + 1$ and raises it to the power 5. A natural way to write h as a composition of two functions is to raise the function $g(x) = 3x^2 - 4x + 1$ to the power 5. Thus, if we let

$$f(x) = x^5 \text{ and } g(x) = 3x^2 - 4x + 1, \text{ then}$$

$$(f \circ g)(x) = f(g(x)) = f(3x^2 - 4x + 1) = (3x^2 - 4x + 1)^5.$$

EXAMPLE 7 Writing a Function as a Composition

Express $h(x)$ as a composition of two functions:

$$h(x) = \sqrt[3]{x^2 + 1}.$$

SOLUTION

The function h takes $x^2 + 1$ and takes its cube root. A natural way to write h as a composition of two functions is to take the cube root of the function $g(x) = x^2 + 1$. Thus, we let

$$f(x) = \sqrt[3]{x} \text{ and } g(x) = x^2 + 1.$$

We can check this composition by finding $(f \circ g)(x)$. This should give the original function, namely, $h(x) = \sqrt[3]{x^2 + 1}$.

$$(f \circ g)(x) = f(g(x)) = f(x^2 + 1) = \sqrt[3]{x^2 + 1} = h(x)$$ • • •

⌀ Check Point **7** Express $h(x)$ as a composition of two functions:
$$h(x) = \sqrt{x^2 + 5}.$$

CONCEPT AND VOCABULARY CHECK

Fill in each blank so that the resulting statement is true.

1. We exclude from a function's domain real numbers that cause division by _____.
2. We exclude from a function's domain real numbers that result in a square root of a/an _____ number.
3. $(f + g)(x) = $ _____
4. $(f - g)(x) = $ _____
5. $(fg)(x) = $ _____
6. $\dfrac{f}{g}(x) = $ _____, provided _____ $\neq 0$
7. The domain of $f(x) = 5x + 7$ consists of all real numbers, represented in interval notation as _____.
8. The domain of $g(x) = \dfrac{3}{x - 2}$ consists of all real numbers except 2, represented in interval notations as $(-\infty, 2) \cup$ _____.

9. The domain of $h(x) = \dfrac{1}{x} + \dfrac{7}{x - 3}$ consists of all real numbers except 0 and 3, represented in interval notation as $(-\infty, 0) \cup$ _____ \cup _____.
10. The notation $f \circ g$, called the _____ of the function f with g, is defined by $(f \circ g)(x) = $ _____.
11. I find $(f \circ g)(x)$ by replacing each occurrence of x in the equation for ____ with _____.
12. The notation $g \circ f$, called the _____ of the function g with f, is defined by $(g \circ f)(x) = $ _____.
13. I find $(g \circ f)(x)$ by replacing each occurrence of x in the equation for ____ with _____.
14. True or false: $f \circ g$ is the same function as $g \circ f$. _____
15. True or false: $f(g(x)) = f(x) \cdot g(x)$ _____
16. If $f(g(x)) = \dfrac{3}{g(x) - 4}$ and $g(x) = \dfrac{8}{x}$, then 0 and ____ must be excluded from the domain of $f \circ g$.

EXERCISE SET 1.7

Practice Exercises

In Exercises 1–20, find the domain of each function.

1. $f(x) = 3x + 5$
2. $f(x) = 4x + 7$
3. $g(x) = \dfrac{1}{x + 4}$
4. $g(x) = \dfrac{1}{x + 5}$
5. $f(x) = \dfrac{2x}{x - 3}$
6. $f(x) = \dfrac{4x}{x - 2}$
7. $f(x) = \dfrac{1}{x + 7} + \dfrac{3}{x - 9}$
8. $f(x) = \dfrac{1}{x + 8} + \dfrac{3}{x - 10}$
9. $h(x) = \dfrac{4}{\dfrac{3}{x} - 1}$
10. $h(x) = \dfrac{5}{\dfrac{4}{x} - 1}$
11. $g(x) = \dfrac{1}{\dfrac{4}{x - 1} - 2}$
12. $g(x) = \dfrac{1}{\dfrac{4}{x - 2} - 3}$
13. $f(x) = \sqrt{x + 4}$
14. $f(x) = \sqrt{x + 5}$

15. $f(x) = \dfrac{1}{\sqrt{x - 4}}$
16. $f(x) = \dfrac{1}{\sqrt{x - 5}}$
17. $g(x) = \sqrt{4 - x}$
18. $g(x) = \sqrt{5 - x}$
19. $h(x) = \sqrt{x - 2} + \sqrt{x + 3}$
20. $h(x) = \sqrt{x - 3} + \sqrt{x + 4}$

In Exercises 21–38, find $f + g, f - g, fg,$ and $\frac{f}{g}$. Determine the domain for each function.

21. $f(x) = 2x + 3, g(x) = x - 1$
22. $f(x) = 3x - 4, g(x) = x + 2$
23. $f(x) = x - 5, g(x) = 3x^2$
24. $f(x) = x - 6, g(x) = 5x^2$
25. $f(x) = 2x^2 - x - 3, g(x) = x + 1$
26. $f(x) = 6x^2 - x - 1, g(x) = x - 1$

27. $f(x) = 3 - x^2, g(x) = x^2 + 2x - 15$

28. $f(x) = 5 - x^2, g(x) = x^2 + 4x - 12$

29. $f(x) = \sqrt{x}, g(x) = x - 4$

30. $f(x) = \sqrt{x}, g(x) = x - 5$

31. $f(x) = 2 + \dfrac{1}{x}, g(x) = \dfrac{1}{x}$

32. $f(x) = 6 - \dfrac{1}{x}, g(x) = \dfrac{1}{x}$

33. $f(x) = \dfrac{5x + 1}{x^2 - 9}, g(x) = \dfrac{4x - 2}{x^2 - 9}$

34. $f(x) = \dfrac{3x + 1}{x^2 - 25}, g(x) = \dfrac{2x - 4}{x^2 - 25}$

35. $f(x) = \sqrt{x + 4}, g(x) = \sqrt{x - 1}$

36. $f(x) = \sqrt{x + 6}, g(x) = \sqrt{x - 3}$

37. $f(x) = \sqrt{x - 2}, g(x) = \sqrt{2 - x}$

38. $f(x) = \sqrt{x - 5}, g(x) = \sqrt{5 - x}$

In Exercises 39–54, find

 a. $(f \circ g)(x)$ **b.** $(g \circ f)(x)$ **c.** $(f \circ g)(2)$ **d.** $(g \circ f)(2)$.

39. $f(x) = 2x, g(x) = x + 7$

40. $f(x) = 3x, g(x) = x - 5$

41. $f(x) = x + 4, g(x) = 2x + 1$

42. $f(x) = 5x + 2, g(x) = 3x - 4$

43. $f(x) = 4x - 3, g(x) = 5x^2 - 2$

44. $f(x) = 7x + 1, g(x) = 2x^2 - 9$

45. $f(x) = x^2 + 2, g(x) = x^2 - 2$

46. $f(x) = x^2 + 1, g(x) = x^2 - 3$

47. $f(x) = 4 - x, g(x) = 2x^2 + x + 5$

48. $f(x) = 5x - 2, g(x) = -x^2 + 4x - 1$

49. $f(x) = \sqrt{x}, g(x) = x - 1$

50. $f(x) = \sqrt{x}, g(x) = x + 2$

51. $f(x) = 2x - 3, g(x) = \dfrac{x + 3}{2}$

52. $f(x) = 6x - 3, g(x) = \dfrac{x + 3}{6}$

53. $f(x) = \dfrac{1}{x}, g(x) = \dfrac{1}{x}$

54. $f(x) = \dfrac{2}{x}, g(x) = \dfrac{2}{x}$

In Exercises 55–62, find

 a. $(f \circ g)(x)$ **b.** *the domain of* $f \circ g$.

55. $f(x) = \dfrac{2}{x + 3}, g(x) = \dfrac{1}{x}$

56. $f(x) = \dfrac{5}{x + 4}, g(x) = \dfrac{1}{x}$

57. $f(x) = \dfrac{x}{x + 1}, g(x) = \dfrac{4}{x}$

58. $f(x) = \dfrac{x}{x + 5}, g(x) = \dfrac{6}{x}$

59. $f(x) = \sqrt{x}, g(x) = x - 2$

60. $f(x) = \sqrt{x}, g(x) = x - 3$

61. $f(x) = x^2 + 4, g(x) = \sqrt{1 - x}$

62. $f(x) = x^2 + 1, g(x) = \sqrt{2 - x}$

In Exercises 63–70, express the given function h as a composition of two functions f and g so that $h(x) = (f \circ g)(x)$.

63. $h(x) = (3x - 1)^4$

64. $h(x) = (2x - 5)^3$

65. $h(x) = \sqrt[3]{x^2 - 9}$

66. $h(x) = \sqrt{5x^2 + 3}$

67. $h(x) = |2x - 5|$

68. $h(x) = |3x - 4|$

69. $h(x) = \dfrac{1}{2x - 3}$

70. $h(x) = \dfrac{1}{4x + 5}$

Practice Plus

Use the graphs of f and g to solve Exercises 71–78.

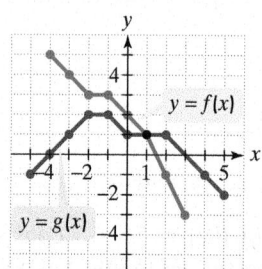

71. Find $(f + g)(-3)$.

72. Find $(g - f)(-2)$.

73. Find $(fg)(2)$.

74. Find $\left(\dfrac{g}{f}\right)(3)$.

75. Find the domain of $f + g$.

76. Find the domain of $\dfrac{f}{g}$.

77. Graph $f + g$.

78. Graph $f - g$.

In Exercises 79–82, use the graphs of f and g to evaluate each composite function.

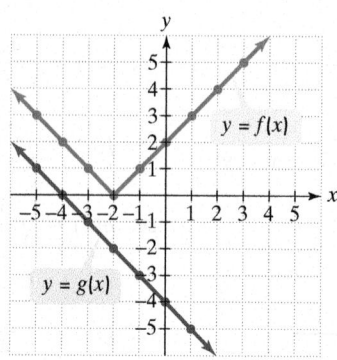

79. $(f \circ g)(-1)$ **80.** $(f \circ g)(1)$

81. $(g \circ f)(0)$ **82.** $(g \circ f)(-1)$

Application Exercises

The bar graph shows the population of the United States, in millions, for six selected years.

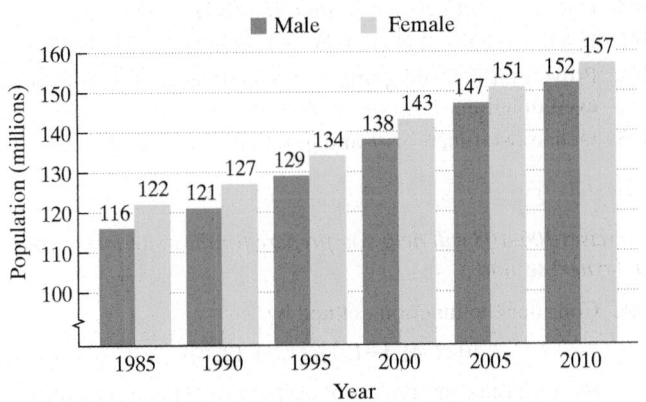

Population of the United States

Source: U.S. Census Bureau

Here are two functions that model the data:

$$M(x) = 1.53x + 114.8$$

> Male U.S. population,
> $M(x)$, in millions,
> x years after 1985

$$F(x) = 1.46x + 120.7.$$

> Female U.S. population,
> $F(x)$, in millions,
> x years after 1985

Use the functions to solve Exercises 83–84.

83. a. Write a function that models the total U.S. population for the years shown in the bar graph.
 b. Use the function from part (a) to find the total U.S. population in 2005.
 c. Does the result in part (b) overestimate or underestimate the actual total U.S. population in 2005 shown by the bar graph? By how much?

84. a. Write a function that models the difference between the female U.S. population and the male U.S. population for the years shown in the bar graph.
 b. Use the function from part (a) to find how many more women than men there were in the U.S. population in 2005.
 c. Does the result in part (b) overestimate or underestimate the actual difference between the female and male population in 2005 shown by the bar graph? By how much?

85. A company that sells radios has yearly fixed costs of $600,000. It costs the company $45 to produce each radio. Each radio will sell for $65. The company's costs and revenue are modeled by the following functions, where x represents the number of radios produced and sold:

$$C(x) = 600,000 + 45x \quad \text{This function models the company's costs.}$$

$$R(x) = 65x. \quad \text{This function models the company's revenue.}$$

Find and interpret $(R - C)(20,000), (R - C)(30,000)$, and $(R - C)(40,000)$.

86. A department store has two locations in a city. From 2008 through 2012, the profits for each of the store's two branches are modeled by the functions $f(x) = -0.44x + 13.62$ and $g(x) = 0.51x + 11.14$. In each model, x represents the number of years after 2008, and f and g represent the profit, in millions of dollars.
 a. What is the slope of f? Describe what this means.
 b. What is the slope of g? Describe what this means.
 c. Find $f + g$. What is the slope of this function? What does this mean?

87. The regular price of a computer is x dollars. Let $f(x) = x - 400$ and $g(x) = 0.75x$.
 a. Describe what the functions f and g model in terms of the price of the computer.
 b. Find $(f \circ g)(x)$ and describe what this models in terms of the price of the computer.
 c. Repeat part (b) for $(g \circ f)(x)$.
 d. Which composite function models the greater discount on the computer, $f \circ g$ or $g \circ f$? Explain.

88. The regular price of a pair of jeans is x dollars. Let $f(x) = x - 5$ and $g(x) = 0.6x$.
 a. Describe what functions f and g model in terms of the price of the jeans.
 b. Find $(f \circ g)(x)$ and describe what this models in terms of the price of the jeans.
 c. Repeat part (b) for $(g \circ f)(x)$.
 d. Which composite function models the greater discount on the jeans, $f \circ g$ or $g \circ f$? Explain.

Writing in Mathematics

89. If a function is defined by an equation, explain how to find its domain.

90. If equations for f and g are given, explain how to find $f - g$.

91. If equations for two functions are given, explain how to obtain the quotient function and its domain.

92. Describe a procedure for finding $(f \circ g)(x)$. What is the name of this function?

93. Describe the values of x that must be excluded from the domain of $(f \circ g)(x)$.

Technology Exercises

94. Graph $y_1 = x^2 - 2x, y_2 = x$, and $y_3 = y_1 \div y_2$ in the same $[-10, 10, 1]$ by $[-10, 10, 1]$ viewing rectangle. Then use the $\boxed{\text{TRACE}}$ feature to trace along y_3. What happens at $x = 0$? Explain why this occurs.

95. Graph $y_1 = \sqrt{2 - x}, y_2 = \sqrt{x}$, and $y_3 = \sqrt{2 - y_2}$ in the same $[-4, 4, 1]$ by $[0, 2, 1]$ viewing rectangle. If y_1 represents f and y_2 represents g, use the graph of y_3 to find the domain of $f \circ g$. Then verify your observation algebraically.

Critical Thinking Exercises

Make Sense? *In Exercises 96–99, determine whether each statement makes sense or does not make sense, and explain your reasoning.*

96. I used a function to model data from 1980 through 2005. The independent variable in my model represented the number of years after 1980, so the function's domain was $\{x \mid x = 0, 1, 2, 3, \ldots, 25\}$.

97. I have two functions. Function f models total world population x years after 2000 and function g models population of the world's more-developed regions x years after 2000. I can use $f - g$ to determine the population of the world's less-developed regions for the years in both function's domains.

98. I must have made a mistake in finding the composite functions $f \circ g$ and $g \circ f$, because I notice that $f \circ g$ is not the same function as $g \circ f$.

99. This diagram illustrates that $f(g(x)) = x^2 + 4$.

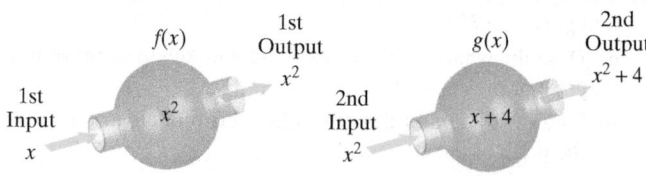

In Exercises 100–103, determine whether each statement is true or false. If the statement is false, make the necessary change(s) to produce a true statement.

100. If $f(x) = x^2 - 4$ and $g(x) = \sqrt{x^2 - 4}$, then $(f \circ g)(x) = -x^2$ and $(f \circ g)(5) = -25$.

101. There can never be two functions f and g, where $f \neq g$, for which $(f \circ g)(x) = (g \circ f)(x)$.

102. If $f(7) = 5$ and $g(4) = 7$, then $(f \circ g)(4) = 35$.

103. If $f(x) = \sqrt{x}$ and $g(x) = 2x - 1$, then $(f \circ g)(5) = g(2)$.

104. Prove that if f and g are even functions, then fg is also an even function.

105. Define two functions f and g so that $f \circ g = g \circ f$.

Preview Exercises

Exercises 106–108 will help you prepare for the material covered in the next section.

106. Consider the function defined by

$$\{(-2, 4), (-1, 1), (1, 1), (2, 4)\}.$$

Reverse the components of each ordered pair and write the resulting relation. Is this relation a function?

107. Solve for y: $x = \dfrac{5}{y} + 4$.

108. Solve by the square root property: $4x^2 - 15 = 5$.

SECTION 1.8 Inverse Functions

Objectives

1. Verify inverse functions.
2. Find the inverse of a function.
3. Use the horizontal line test to determine if a function has an inverse function.
4. Use the graph of a one-to-one function to graph its inverse function.
5. Find the inverse of a function and graph both functions on the same axes.

Based on Shakespeare's *Romeo and Juliet*, the film *West Side Story* swept the 1961 Academy Awards with ten Oscars. The top four movies to win the most Oscars are shown in **Table 1.5**.

Table 1.5 Films Winning the Most Oscars

Movie	Year	Number of Academy Awards
Ben-Hur	1960	11
Titanic	1998	11
The Lord of the Rings: The Return of the King	2003	11
West Side Story	1961	10

Source: Russell Ash, *The Top 10 of Everything, 2011*

We can use the information in **Table 1.5** to define a function. Let the domain of the function be the set of four movies shown in the table. Let the range be the number of Academy Awards for each of the respective films. The function can be written as follows:

f: {(*Ben-Hur*, 11), (*Titanic*, 11), (*The Lord of the Rings*, 11), (*West Side Story*, 10)}.

Now let's "undo" f by interchanging the first and second components in each of the ordered pairs. Switching the inputs and outputs of f, we obtain the following relation:

Same first component

Undoing f: {(11, *Ben-Hur*), (11, *Titanic*), (11, *The Lord of the Rings*), (10, *West Side Story*)}.

Different second components

Can you see that this relation is not a function? Three of its ordered pairs have the same first component and different second components. This violates the definition of a function.

If a function f is a set of ordered pairs, (x, y), then the changes produced by f can be "undone" by reversing the components of all the ordered pairs. The resulting relation, (y, x), may or may not be a function. In this section, we will develop these ideas by studying functions whose compositions have a special "undoing" relationship.

Inverse Functions

Here are two functions that describe situations related to the price of a computer, x:

$$f(x) = x - 300 \qquad g(x) = x + 300.$$

Function f subtracts $300 from the computer's price and function g adds $300 to the computer's price. Let's see what $f(g(x))$ does. Put $g(x)$ into f:

$$f(x) = x - 300 \qquad \text{This is the given equation for } f.$$

Replace x with $g(x)$.

$$f(g(x)) = g(x) - 300$$
$$= x + 300 - 300 \qquad \text{Because } g(x) = x + 300,$$
$$= x. \qquad \text{replace } g(x) \text{ with } x + 300.$$

This is the computer's original price.

Using $f(x) = x - 300$ and $g(x) = x + 300$, we see that $f(g(x)) = x$. By putting $g(x)$ into f and finding $f(g(x))$, the computer's price, x, went through two changes: the first, an increase; the second, a decrease:

$$x + 300 - 300.$$

The final price of the computer, x, is identical to its starting price, x.

In general, if the changes made to x by a function g are undone by the changes made by a function f, then

$$f(g(x)) = x.$$

Assume, also, that this "undoing" takes place in the other direction:

$$g(f(x)) = x.$$

Under these conditions, we say that each function is the *inverse function* of the other. The fact that g is the inverse of f is expressed by renaming g as f^{-1}, read "f-inverse." For example, the inverse functions

$$f(x) = x - 300 \qquad g(x) = x + 300$$

are usually named as follows:

$$f(x) = x - 300 \qquad f^{-1}(x) = x + 300.$$

We can use partial tables of coordinates for f and f^{-1} to gain numerical insight into the relationship between a function and its inverse function.

Computer's regular price	$300 reduction		Price with $300 reduction	$300 price increase
x	$f(x) = x - 300$		x	$f^{-1}(x) = x + 300$
1200	900		900	1200
1300	1000		1000	1300
1400	1100		1100	1400

Ordered pairs for f:
(1200, 900), (1300, 1000), (1400, 1100)

Ordered pairs for f^{-1}:
(900, 1200), (1000, 1300), (1100, 1400)

The tables illustrate that if a function f is the set of ordered pairs (x, y), then its inverse, f^{-1}, is the set of ordered pairs (y, x). Using these tables, we can see how one function's changes to x are undone by the other function:

$$(f^{-1} \circ f)(1300) = f^{-1}(f(1300)) = f^{-1}(1000) = 1300.$$

The table for f shows that $f(1300) = 1000$.

The table for f^{-1} shows that $f^{-1}(1000) = 1300$.

The final price of the computer, $1300, is identical to its starting price, $1300.

With these ideas in mind, we present the formal definition of the inverse of a function:

Definition of the Inverse of a Function

Let f and g be two functions such that

$$f(g(x)) = x \qquad \text{for every } x \text{ in the domain of } g$$

and

$$g(f(x)) = x \qquad \text{for every } x \text{ in the domain of } f.$$

The function g is the **inverse of the function f** and is denoted by f^{-1} (read "f-inverse"). Thus, $f(f^{-1}(x)) = x$ and $f^{-1}(f(x)) = x$. The domain of f is equal to the range of f^{-1}, and vice versa.

GREAT QUESTION!

Is the -1 in f^{-1} an exponent?

The notation f^{-1} represents the inverse function of f. The -1 is *not* an exponent. The notation f^{-1} does *not* mean $\dfrac{1}{f}$:

$$f^{-1} \neq \frac{1}{f}.$$

① Verify inverse functions.

EXAMPLE 1 Verifying Inverse Functions

Show that each function is the inverse of the other:

$$f(x) = 3x + 2 \quad \text{and} \quad g(x) = \frac{x - 2}{3}.$$

SOLUTION

To show that f and g are inverses of each other, we must show that $f(g(x)) = x$ and $g(f(x)) = x$. We begin with $f(g(x))$.

$$f(x) = 3x + 2 \quad \text{This is the equation for } f.$$

Replace x with $g(x)$.

$$f(g(x)) = 3g(x) + 2 = 3\left(\frac{x-2}{3}\right) + 2 = x - 2 + 2 = x$$

$$g(x) = \frac{x-2}{3}$$

Next, we find $g(f(x))$.

$$g(x) = \frac{x-2}{3} \quad \text{This is the equation for } g.$$

Replace x with $f(x)$.

$$g(f(x)) = \frac{f(x) - 2}{3} = \frac{(3x+2) - 2}{3} = \frac{3x}{3} = x$$

$$f(x) = 3x + 2$$

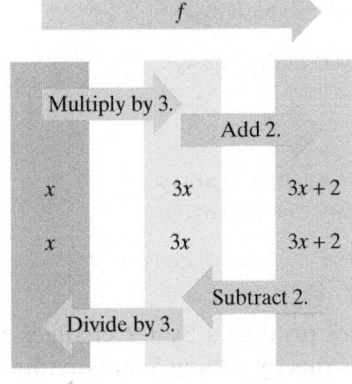

FIGURE 1.65 f^{-1} undoes the changes produced by f.

Because g is the inverse of f (and vice versa), we can use inverse notation and write

$$f(x) = 3x + 2 \quad \text{and} \quad f^{-1}(x) = \frac{x-2}{3}.$$

Notice how f^{-1} undoes the changes produced by f: f changes x by *multiplying* by 3 and *adding* 2, and f^{-1} undoes this by *subtracting* 2 and *dividing* by 3. This "undoing" process is illustrated in **Figure 1.65**. • • •

⊘ Check Point **1** Show that each function is the inverse of the other:

$$f(x) = 4x - 7 \quad \text{and} \quad g(x) = \frac{x+7}{4}.$$

② Find the inverse of a function.

Finding the Inverse of a Function

The definition of the inverse of a function tells us that the domain of f is equal to the range of f^{-1}, and vice versa. This means that if the function f is the set of ordered pairs (x, y), then the inverse of f is the set of ordered pairs (y, x). If a function is defined by an equation, we can obtain the equation for f^{-1}, the inverse of f, by interchanging the role of x and y in the equation for the function f.

Finding the Inverse of a Function

The equation for the inverse of a function f can be found as follows:

1. Replace $f(x)$ with y in the equation for $f(x)$.
2. Interchange x and y.
3. Solve for y. If this equation does not define y as a function of x, the function f does not have an inverse function and this procedure ends. If this equation does define y as a function of x, the function f has an inverse function.
4. If f has an inverse function, replace y in step 3 by $f^{-1}(x)$. We can verify our result by showing that $f(f^{-1}(x)) = x$ and $f^{-1}(f(x)) = x$.

The procedure for finding a function's inverse uses a *switch-and-solve* strategy. Switch x and y, and then solve for y.

EXAMPLE 2 Finding the Inverse of a Function

Find the inverse of $f(x) = 7x - 5$.

SOLUTION

Step 1 Replace $f(x)$ with y:

$$y = 7x - 5.$$

Step 2 Interchange x and y:

$$x = 7y - 5. \quad \text{This is the inverse function.}$$

Step 3 Solve for y:

$$x + 5 = 7y \qquad \text{Add 5 to both sides.}$$
$$\frac{x + 5}{7} = y. \qquad \text{Divide both sides by 7.}$$

DISCOVERY

In Example 2, we found that if $f(x) = 7x - 5$, then

$$f^{-1}(x) = \frac{x + 5}{7}.$$

Verify this result by showing that

$$f(f^{-1}(x)) = x$$

and

$$f^{-1}(f(x)) = x.$$

Step 4 Replace y with $f^{-1}(x)$:

$$f^{-1}(x) = \frac{x + 5}{7}. \quad \begin{array}{l}\text{The equation is written}\\\text{with } f^{-1} \text{ on the left.}\end{array}$$

Thus, the inverse of $f(x) = 7x - 5$ is $f^{-1}(x) = \dfrac{x + 5}{7}$.

The inverse function, f^{-1}, undoes the changes produced by f. f changes x by multiplying by 7 and subtracting 5. f^{-1} undoes this by adding 5 and dividing by 7. • • •

✓ Check Point **2** Find the inverse of $f(x) = 2x + 7$.

EXAMPLE 3 Finding the Inverse of a Function

Find the inverse of $f(x) = x^3 + 1$.

SOLUTION

Step 1 Replace $f(x)$ with y: $y = x^3 + 1$.
Step 2 Interchange x and y: $x = y^3 + 1$.
Step 3 Solve for y:

> Our goal is to isolate y.
> Because $\sqrt[3]{y^3} = y$, we will take the cube root of both sides of the equation.

$$x - 1 = y^3 \qquad \text{Subtract 1 from both sides.}$$
$$\sqrt[3]{x - 1} = \sqrt[3]{y^3} \qquad \text{Take the cube root on both sides.}$$
$$\sqrt[3]{x - 1} = y. \qquad \text{Simplify.}$$

Step 4 Replace y with $f^{-1}(x)$: $f^{-1}(x) = \sqrt[3]{x - 1}$.

Thus, the inverse of $f(x) = x^3 + 1$ is $f^{-1}(x) = \sqrt[3]{x - 1}$. • • •

✓ Check Point **3** Find the inverse of $f(x) = 4x^3 - 1$.

EXAMPLE 4 Finding the Inverse of a Function

Find the inverse of $f(x) = \dfrac{5}{x} + 4$.

SOLUTION

Step 1 Replace $f(x)$ with y:

$$y = \frac{5}{x} + 4.$$

Step 2 Interchange x and y:

$$x = \frac{5}{y} + 4.$$

> Our goal is to isolate y. To get y out of the denominator, we will multiply both sides of the equation by y, $y \neq 0$.

Step 3 Solve for y:

$$x = \frac{5}{y} + 4 \qquad \text{This is the equation from step 2.}$$

$$xy = \left(\frac{5}{y} + 4\right)y \qquad \text{Multiply both sides by } y, y \neq 0.$$

$$xy = \frac{5}{y} \cdot y + 4y \qquad \text{Use the distributive property.}$$

$$xy = 5 + 4y \qquad \text{Simplify: } \frac{5}{y} \cdot y = 5.$$

$$xy - 4y = 5 \qquad \text{Subtract } 4y \text{ from both sides.}$$

$$y(x - 4) = 5 \qquad \text{Factor out } y \text{ from } xy - 4y \text{ to obtain a single occurrence of } y.$$

$$\frac{y(x - 4)}{x - 4} = \frac{5}{x - 4} \qquad \text{Divide both sides by } x - 4, x \neq 4.$$

$$y = \frac{5}{x - 4}. \qquad \text{Simplify.}$$

Step 4 Replace y with $f^{-1}(x)$:

$$f^{-1}(x) = \frac{5}{x - 4}.$$

Thus, the inverse of $f(x) = \dfrac{5}{x} + 4$ is $f^{-1}(x) = \dfrac{5}{x - 4}$. • • •

✓ **Check Point 4** Find the inverse of $f(x) = \dfrac{3}{x} - 1$.

③ Use the horizontal line test to determine if a function has an inverse function.

The Horizontal Line Test and One-to-One Functions

Let's see what happens if we try to find the inverse of the standard quadratic function, $f(x) = x^2$.

Step 1 Replace $f(x)$ with y: $y = x^2$.

Step 2 Interchange x and y: $x = y^2$.

Step 3 Solve for y: We apply the square root property to solve $y^2 = x$ for y. We obtain

$$y = \pm\sqrt{x}.$$

The \pm in $y = \pm\sqrt{x}$ shows that for certain values of x (all positive real numbers), there are two values of y. Because this equation does not represent y as a function of x, the standard quadratic function $f(x) = x^2$ does not have an inverse function.

We can use a few of the solutions of $y = x^2$ to illustrate numerically that this function does not have an inverse:

Four solutions of $y = x^2$

$(-2, 4), \quad (-1, 1), \quad (1, 1), \quad (2, 4)$

Interchange x and y in each ordered pair.

$(4, -2), \quad (1, -1), \quad (1, 1), \quad (4, 2).$

The input 1 is associated with two outputs, −1 and 1.

The input 4 is associated with two outputs, −2 and 2.

A function provides exactly one output for each input. Thus, the ordered pairs in the bottom row do not define a function.

Can we look at the graph of a function and tell if it represents a function with an inverse? Yes. The graph of the standard quadratic function $f(x) = x^2$ is shown in **Figure 1.66**. Four units above the x-axis, a horizontal line is drawn. This line intersects the graph at two of its points, $(-2, 4)$ and $(2, 4)$. Inverse functions have ordered pairs with the coordinates reversed. We just saw what happened when we interchanged x and y. We obtained $(4, -2)$ and $(4, 2)$, and these ordered pairs do not define a function.

If any horizontal line, such as the one in **Figure 1.66**, intersects a graph at two or more points, the set of these points will not define a function when their coordinates are reversed. This suggests the **horizontal line test** for inverse functions.

The Horizontal Line Test for Inverse Functions

A function f has an inverse that is a function, f^{-1}, if there is no horizontal line that intersects the graph of the function f at more than one point.

EXAMPLE 5 Applying the Horizontal Line Test

Which of the following graphs represent functions that have inverse functions?

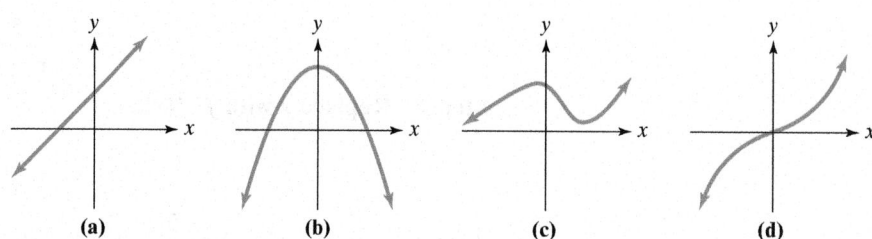

(a) (b) (c) (d)

SOLUTION

Notice that horizontal lines can be drawn in graphs **(b)** and **(c)** that intersect the graphs more than once. These graphs do not pass the horizontal line test. These are not the graphs of functions with inverse functions. By contrast, no horizontal line can be drawn in graphs **(a)** and **(d)** that intersects the graphs more than once. These graphs pass the horizontal line test. Thus, the graphs in parts **(a)** and **(d)** represent functions that have inverse functions.

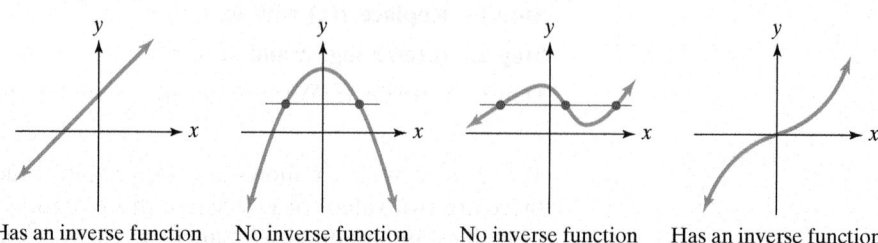

Has an inverse function No inverse function No inverse function Has an inverse function
(a) (b) (c) (d)

• • •

(left margin)

$f(x) = x^2$

$(-2, 4)$ $(2, 4)$

FIGURE 1.66 The horizontal line intersects the graph twice.

DISCOVERY

How might you restrict the domain of $f(x) = x^2$, graphed in **Figure 1.66**, so that the remaining portion of the graph passes the horizontal line test?

Check Point **5** Which of the following graphs represent functions that have inverse functions?

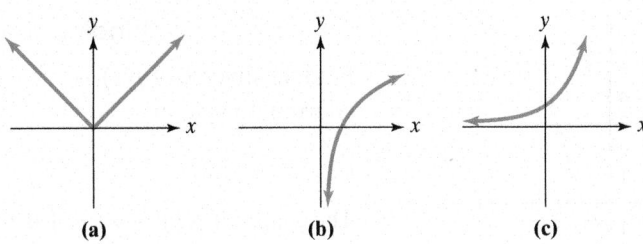

(a) (b) (c)

A function passes the horizontal line test when no two different ordered pairs have the same second component. This means that if $x_1 \neq x_2$, then $f(x_1) \neq f(x_2)$. Such a function is called a **one-to-one function**. Thus, **a one-to-one function is a function in which no two different ordered pairs have the same second component. Only one-to-one functions have inverse functions.** Any function that passes the horizontal line test is a one-to-one function. Any one-to-one function has a graph that passes the horizontal line test.

④ Use the graph of a one-to-one function to graph its inverse function.

Graphs of f and f^{-1}

There is a relationship between the graph of a one-to-one function, f, and its inverse, f^{-1}. Because inverse functions have ordered pairs with the coordinates interchanged, if the point (a, b) is on the graph of f, then the point (b, a) is on the graph of f^{-1}. The points (a, b) and (b, a) are symmetric with respect to the line $y = x$. Thus, **the graph of f^{-1} is a reflection of the graph of f about the line $y = x$.** This is illustrated in **Figure 1.67**.

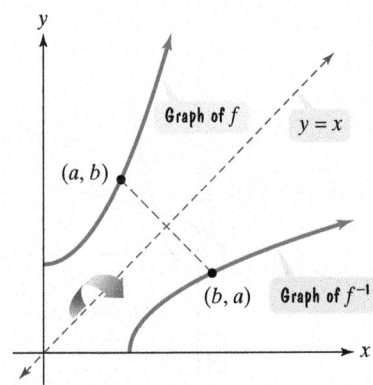

FIGURE 1.67 The graph of f^{-1} is a reflection of the graph of f about $y = x$.

EXAMPLE 6 Graphing the Inverse Function

Use the graph of f in **Figure 1.68** to draw the graph of its inverse function.

FIGURE 1.68

SOLUTION

We begin by noting that no horizontal line intersects the graph of f at more than one point, so f does have an inverse function. Because the points $(-3, -2), (-1, 0)$, and $(4, 2)$ are on the graph of f, the graph of the inverse function, f^{-1}, has points with these ordered pairs reversed. Thus, $(-2, -3), (0, -1)$, and $(2, 4)$ are on the graph of f^{-1}. We can use these points to graph f^{-1}. The graph of f^{-1} is shown in green in **Figure 1.69**. Note that the green graph of f^{-1} is the reflection of the blue graph of f about the line $y = x$. ●●●

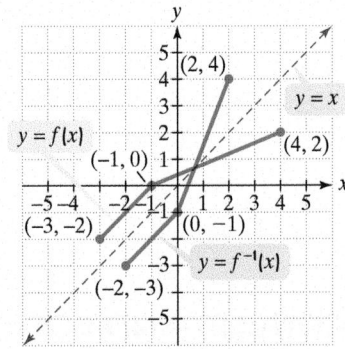

FIGURE 1.69 The graphs of f and f^{-1}

Check Point **6** The graph of function f consists of two line segments, one segment from $(-2, -2)$ to $(-1, 0)$ and a second segment from $(-1, 0)$ to $(1, 2)$. Graph f and use the graph to draw the graph of its inverse function.

⑤ Find the inverse of a function and graph both functions on the same axes.

FIGURE 1.70

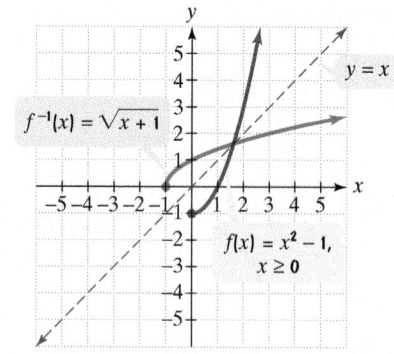

FIGURE 1.71

In our final example, we will first find f^{-1}. Then we will graph f and f^{-1} in the same rectangular coordinate system.

EXAMPLE 7 Finding the Inverse of a Domain-Restricted Function

Find the inverse of $f(x) = x^2 - 1$ if $x \geq 0$. Graph f and f^{-1} in the same rectangular coordinate system.

SOLUTION

The graph of $f(x) = x^2 - 1$ is the graph of the standard quadratic function shifted vertically down 1 unit. **Figure 1.70** shows the function's graph. This graph fails the horizontal line test, so the function $f(x) = x^2 - 1$ does not have an inverse function. By restricting the domain to $x \geq 0$, as given, we obtain a new function whose graph is shown in red in **Figure 1.70**. This red portion of the graph is increasing on the interval $(0, \infty)$ and passes the horizontal line test. This tells us that $f(x) = x^2 - 1$ has an inverse function if we restrict its domain to $x \geq 0$. We use our four-step procedure to find this inverse function. Begin with $f(x) = x^2 - 1, x \geq 0$.

Step 1 Replace $f(x)$ with y: $y = x^2 - 1, x \geq 0$.

Step 2 Interchange x and y: $x = y^2 - 1, y \geq 0$.

Step 3 Solve for y:

$$x = y^2 - 1, y \geq 0 \quad \text{This is the equation from step 2.}$$

$$x + 1 = y^2 \quad \text{Add 1 to both sides.}$$

$$\sqrt{x + 1} = y \quad \text{Apply the square root property.}$$

> Because $y \geq 0$, take only the principal square root and not the negative square root.

Step 4 Replace y with $f^{-1}(x)$: $f^{-1}(x) = \sqrt{x + 1}$.

Thus, the inverse of $f(x) = x^2 - 1, x \geq 0$, is $f^{-1}(x) = \sqrt{x + 1}$. The graphs of f and f^{-1} are shown in **Figure 1.71**. We obtained the graph of $f^{-1}(x) = \sqrt{x + 1}$ by shifting the graph of the square root function, $y = \sqrt{x}$, horizontally to the left 1 unit. Note that the green graph of f^{-1} is the reflection of the red graph of f about the line $y = x$. ● ● ●

✓ Check Point **7** Find the inverse of $f(x) = x^2 + 1$ if $x \geq 0$. Graph f and f^{-1} in the same rectangular coordinate system.

CONCEPT AND VOCABULARY CHECK

Fill in each blank so that the resulting statement is true.

1. The notation f^{-1} means the _____ of the function f.

2. If the function g is the inverse of the function f, then $f(g(x)) = $ _____ and $g(f(x)) = $ _____.

3. A function f has an inverse that is a function if there is no _____ line that intersects the graph of f at more than one point. Such a function is called a/an _____ function.

4. The graph of f^{-1} is a reflection of the graph of f about the line whose equation is _____.

EXERCISE SET 1.8

Practice Exercises

In Exercises 1–10, find f(g(x)) and g(f(x)) and determine whether each pair of functions f and g are inverses of each other.

1. $f(x) = 4x$ and $g(x) = \dfrac{x}{4}$

2. $f(x) = 6x$ and $g(x) = \dfrac{x}{6}$

3. $f(x) = 3x + 8$ and $g(x) = \dfrac{x - 8}{3}$

4. $f(x) = 4x + 9$ and $g(x) = \dfrac{x - 9}{4}$

5. $f(x) = 5x - 9$ and $g(x) = \dfrac{x + 5}{9}$

6. $f(x) = 3x - 7$ and $g(x) = \dfrac{x + 3}{7}$

7. $f(x) = \dfrac{3}{x - 4}$ and $g(x) = \dfrac{3}{x} + 4$

8. $f(x) = \dfrac{2}{x - 5}$ and $g(x) = \dfrac{2}{x} + 5$

9. $f(x) = -x$ and $g(x) = -x$

10. $f(x) = \sqrt[3]{x} - 4$ and $g(x) = x^3 + 4$

The functions in Exercises 11–28 are all one-to-one. For each function,

 a. *Find an equation for $f^{-1}(x)$, the inverse function.*

 b. *Verify that your equation is correct by showing that $f(f^{-1}(x)) = x$ and $f^{-1}(f(x)) = x$.*

11. $f(x) = x + 3$ **12.** $f(x) = x + 5$

13. $f(x) = 2x$ **14.** $f(x) = 4x$

15. $f(x) = 2x + 3$ **16.** $f(x) = 3x - 1$

17. $f(x) = x^3 + 2$ **18.** $f(x) = x^3 - 1$

19. $f(x) = (x + 2)^3$ **20.** $f(x) = (x - 1)^3$

21. $f(x) = \dfrac{1}{x}$ **22.** $f(x) = \dfrac{2}{x}$

23. $f(x) = \sqrt{x}$ **24.** $f(x) = \sqrt[3]{x}$

25. $f(x) = \dfrac{7}{x} - 3$ **26.** $f(x) = \dfrac{4}{x} + 9$

27. $f(x) = \dfrac{2x + 1}{x - 3}$ **28.** $f(x) = \dfrac{2x - 3}{x + 1}$

Which graphs in Exercises 29–34 represent functions that have inverse functions?

29.

30.

31.

32.

33.

34.

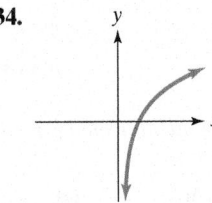

In Exercises 35–38, use the graph of f to draw the graph of its inverse function.

35.

36.

37.

38.

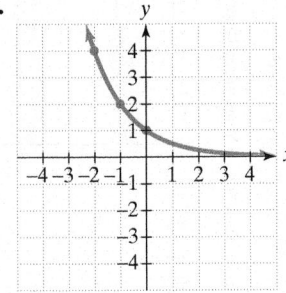

In Exercises 39–52,

 a. *Find an equation for $f^{-1}(x)$.*

 b. *Graph f and f^{-1} in the same rectangular coordinate system.*

 c. *Use interval notation to give the domain and the range of f and f^{-1}.*

39. $f(x) = 2x - 1$ **40.** $f(x) = 2x - 3$

41. $f(x) = x^2 - 4, x \geq 0$ **42.** $f(x) = x^2 - 1, x \leq 0$

43. $f(x) = (x - 1)^2, x \leq 1$ **44.** $f(x) = (x - 1)^2, x \geq 1$

45. $f(x) = x^3 - 1$ **46.** $f(x) = x^3 + 1$

47. $f(x) = (x + 2)^3$ **48.** $f(x) = (x - 2)^3$

(Hint for Exercises 49–52: To solve for a variable involving an nth root, raise both sides of the equation to the nth power: $(\sqrt[n]{y})^n = y$.)

49. $f(x) = \sqrt{x - 1}$ **50.** $f(x) = \sqrt{x} + 2$

51. $f(x) = \sqrt[3]{x} + 1$ **52.** $f(x) = \sqrt[3]{x - 1}$

Practice Plus

In Exercises 53–58, f and g are defined by the following tables. Use the tables to evaluate each composite function.

x	$f(x)$
-1	1
0	4
1	5
2	-1

x	$g(x)$
-1	0
1	1
4	2
10	-1

53. $f(g(1))$

54. $f(g(4))$

55. $(g \circ f)(-1)$

56. $(g \circ f)(0)$

57. $f^{-1}(g(10))$

58. $f^{-1}(g(1))$

In Exercises 59–64, let

$$f(x) = 2x - 5$$
$$g(x) = 4x - 1$$
$$h(x) = x^2 + x + 2.$$

Evaluate the indicated function without finding an equation for the function.

59. $(f \circ g)(0)$ **60.** $(g \circ f)(0)$ **61.** $f^{-1}(1)$
62. $g^{-1}(7)$ **63.** $g(f[h(1)])$ **64.** $f(g[h(1)])$

Application Exercises

In most societies, women say they prefer to marry men who are older than themselves, whereas men say they prefer women who are younger. The bar graph shows the preferred age difference in a mate in five selected countries. Use this information to solve Exercises 65–66.

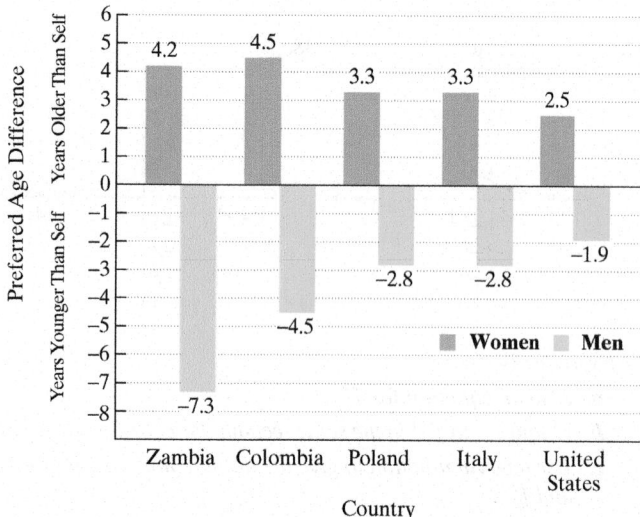

Preferred Age Difference in a Mate

Source: Carole Wade and Carol Tavris, *Psychology*, 6th Edition, Prentice Hall, 2000.

65. a. Consider a function, *f*, whose domain is the set of the five countries shown in the graph. Let the range be the set of the average number of years women in each of the respective countries prefer men who are older than themselves. Write the function *f* as a set of ordered pairs.

 b. Write the relation that is the inverse of *f* as a set of ordered pairs. Based on these ordered pairs, is *f* a one-to-one function? Explain your answer.

66. a. Consider a function, *g*, whose domain is the set of the five countries shown in the graph. Let the range be the set of negative numbers representing the average number of years men in each of the respective countries prefer women who are younger than themselves. Write the function *g* as a set of ordered pairs.

 b. Write the relation that is the inverse of *g* as a set of ordered pairs. Based on these ordered pairs, is *g* a one-to-one function? Explain your answer.

67. The graph represents the probability of two people in the same room sharing a birthday as a function of the number of people in the room. Call the function *f*.

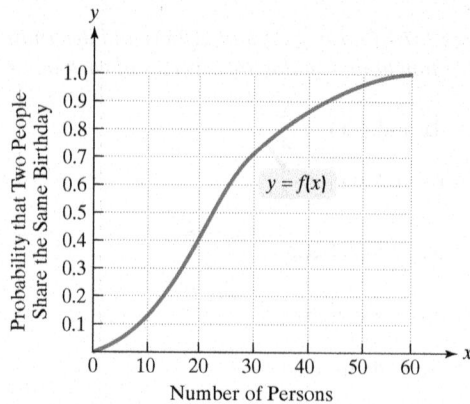

 a. Explain why *f* has an inverse that is a function.

 b. Describe in practical terms the meaning of $f^{-1}(0.25), f^{-1}(0.5)$, and $f^{-1}(0.7)$.

68. A study of 900 working women in Texas showed that their feelings changed throughout the day. As the graph indicates, the women felt better as time passed, except for a blip (that's slang for relative maximum) at lunchtime.

Average Level of Happiness at Different Times of Day

Source: D. Kahneman et al., "A Survey Method for Characterizing Daily Life Experience," *Science*.

 a. Does the graph have an inverse that is a function? Explain your answer.

 b. Identify two or more times of day when the average happiness level is 3. Express your answers as ordered pairs.

 c. Do the ordered pairs in part (b) indicate that the graph represents a one-to-one function? Explain your answer.

69. The formula

$$y = f(x) = \frac{9}{5}x + 32$$

is used to convert from *x* degrees Celsius to *y* degrees Fahrenheit. The formula

$$y = g(x) = \frac{5}{9}(x - 32)$$

is used to convert from *x* degrees Fahrenheit to *y* degrees Celsius. Show that *f* and *g* are inverse functions.

Writing in Mathematics

70. Explain how to determine if two functions are inverses of each other.

71. Describe how to find the inverse of a one-to-one function.

72. What is the horizontal line test and what does it indicate?

73. Describe how to use the graph of a one-to-one function to draw the graph of its inverse function.

74. How can a graphing utility be used to visually determine if two functions are inverses of each other?

75. What explanations can you offer for the trends shown by the graph in Exercise 68?

Technology Exercises

In Exercises 76–83, use a graphing utility to graph the function. Use the graph to determine whether the function has an inverse that is a function (that is, whether the function is one-to-one).

76. $f(x) = x^2 - 1$

77. $f(x) = \sqrt[3]{2 - x}$

78. $f(x) = \dfrac{x^3}{2}$

79. $f(x) = \dfrac{x^4}{4}$

80. $f(x) = \text{int}(x - 2)$

81. $f(x) = |x - 2|$

82. $f(x) = (x - 1)^3$

83. $f(x) = -\sqrt{16 - x^2}$

In Exercises 84–86, use a graphing utility to graph f and g in the same viewing rectangle. In addition, graph the line $y = x$ and visually determine if f and g are inverses.

84. $f(x) = 4x + 4, g(x) = 0.25x - 1$

85. $f(x) = \dfrac{1}{x} + 2, g(x) = \dfrac{1}{x - 2}$

86. $f(x) = \sqrt[3]{x} - 2, g(x) = (x + 2)^3$

Critical Thinking Exercises

Make Sense? *In Exercises 87–90, determine whether each statement makes sense or does not make sense, and explain your reasoning.*

87. I found the inverse of $f(x) = 5x - 4$ in my head: The reverse of multiplying by 5 and subtracting 4 is adding 4 and dividing by 5, so $f^{-1}(x) = \dfrac{x + 4}{5}$.

88. I'm working with the linear function $f(x) = 3x + 5$ and I do not need to find f^{-1} in order to determine the value of $(f \circ f^{-1})(17)$.

Exercises 89–90 are based on the following cartoon.

© John L. Hart FLP. All rights reserved.

89. Assuming that there is no such thing as metric crickets, I modeled the information in the first frame of the cartoon with the function

$$T(n) = \dfrac{n}{4} + 40,$$

where $T(n)$ is the temperature, in degrees Fahrenheit, and n is the number of cricket chirps per minute.

90. I used the function in Exercise 89 and found an equation for T^{-1}, which expresses the number of cricket chirps per minute as a function of Fahrenheit temperature.

In Exercises 91–94, determine whether each statement is true or false. If the statement is false, make the necessary change(s) to produce a true statement.

91. The inverse of $\{(1, 4), (2, 7)\}$ is $\{(2, 7), (1, 4)\}$.

92. The function $f(x) = 5$ is one-to-one.

93. If $f(x) = 3x$, then $f^{-1}(x) = \dfrac{1}{3x}$.

94. The domain of f is the same as the range of f^{-1}.

95. If $f(x) = 3x$ and $g(x) = x + 5$, find $(f \circ g)^{-1}(x)$ and $(g^{-1} \circ f^{-1})(x)$.

96. Show that

$$f(x) = \dfrac{3x - 2}{5x - 3}$$

is its own inverse.

97. *Freedom 7* was the spacecraft that carried the first American into space in 1961. Total flight time was 15 minutes and the spacecraft reached a maximum height of 116 miles. Consider a function, s, that expresses *Freedom 7*'s height, $s(t)$, in miles, after t minutes. Is s a one-to-one function? Explain your answer.

98. If $f(2) = 6$, and f is one-to-one, find x satisfying $8 + f^{-1}(x - 1) = 10$.

Group Exercise

99. In Tom Stoppard's play *Arcadia*, the characters dream and talk about mathematics, including ideas involving graphing, composite functions, symmetry, and lack of symmetry in things that are tangled, mysterious, and unpredictable. Group members should read the play. Present a report on the ideas discussed by the characters that are related to concepts that we studied in this chapter. Bring in a copy of the play and read appropriate excerpts.

Preview Exercises

Exercises 100–102 will help you prepare for the material covered in the next section.

100. Let $(x_1, y_1) = (7, 2)$ and $(x_2, y_2) = (1, -1)$. Find $\sqrt{(x_2 - x_1)^2 + (y_2 - y_1)^2}$. Express the answer in simplified radical form.

101. Use a rectangular coordinate system to graph the circle with center $(1, -1)$ and radius 1.

102. What term should be added to each binomial so that it becomes a perfect square trinomial? Write and factor the trinomial.
 a. $x^2 + 4x$
 b. $y^2 - 6y$

Objectives

1. Find the distance between two points.
2. Find the midpoint of a line segment.
3. Write the standard form of a circle's equation.
4. Give the center and radius of a circle whose equation is in standard form.
5. Convert the general form of a circle's equation to standard form.

It's a good idea to know your way around a circle. Clocks, angles, maps, and compasses are based on circles. Circles occur everywhere in nature: in ripples on water, patterns on a butterfly's wings, and cross sections of trees. Some consider the circle to be the most pleasing of all shapes.

The rectangular coordinate system gives us a unique way of knowing a circle. It enables us to translate a circle's geometric definition into an algebraic equation. To do this, we must first develop a formula for the distance between any two points in rectangular coordinates.

1 Find the distance between two points.

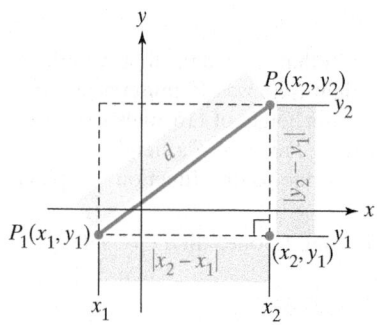

FIGURE 1.72

The Distance Formula

Using the Pythagorean Theorem, we can find the distance between the two points $P_1(x_1, y_1)$ and $P_2(x_2, y_2)$ in the rectangular coordinate system. The two points are illustrated in **Figure 1.72**.

The distance that we need to find is represented by d and shown in blue. Notice that the distance between the two points on the dashed horizontal line is the absolute value of the difference between the x-coordinates of the points. This distance, $|x_2 - x_1|$, is shown in pink. Similarly, the distance between the two points on the dashed vertical line is the absolute value of the difference between the y-coordinates of the points. This distance, $|y_2 - y_1|$, is also shown in pink.

Because the dashed lines are horizontal and vertical, a right triangle is formed. Thus, we can use the Pythagorean Theorem to find the distance d. Squaring the lengths of the triangle's sides results in positive numbers, so absolute value notation is not necessary.

$$d^2 = (x_2 - x_1)^2 + (y_2 - y_1)^2$$ Apply the Pythagorean Theorem to the right triangle in Figure 1.72.

$$d = \pm\sqrt{(x_2 - x_1)^2 + (y_2 - y_1)^2}$$ Apply the square root property.

$$d = \sqrt{(x_2 - x_1)^2 + (y_2 - y_1)^2}$$ Because distance is nonnegative, write only the principal square root.

This result is called the **distance formula**.

The Distance Formula

The distance, d, between the points (x_1, y_1) and (x_2, y_2) in the rectangular coordinate system is

$$d = \sqrt{(x_2 - x_1)^2 + (y_2 - y_1)^2}.$$

To compute the distance between two points, find the square of the difference between the x-coordinates plus the square of the difference between the y-coordinates. The principal square root of this sum is the distance.

When using the distance formula, it does not matter which point you call (x_1, y_1) and which you call (x_2, y_2).

EXAMPLE 1 Using the Distance Formula

Find the distance between $(-1, 4)$ and $(3, -2)$. Express the answer in simplified radical form and then round to two decimal places.

SOLUTION

We will let $(x_1, y_1) = (-1, 4)$ and $(x_2, y_2) = (3, -2)$.

$$d = \sqrt{(x_2 - x_1)^2 + (y_2 - y_1)^2} \qquad \text{Use the distance formula.}$$

$$= \sqrt{[3 - (-1)]^2 + (-2 - 4)^2} \qquad \text{Substitute the given values.}$$

$$= \sqrt{4^2 + (-6)^2} \qquad \text{Perform operations inside grouping symbols:}$$
$$\qquad\qquad\qquad\qquad 3 - (-1) = 3 + 1 = 4 \text{ and } -2 - 4 = -6.$$

$$= \sqrt{16 + 36} \qquad \boxed{\begin{array}{c}\text{Caution: This}\\ \text{does not equal}\\ \sqrt{16} + \sqrt{36}.\end{array}} \qquad \text{Square 4 and } -6.$$

$$= \sqrt{52} \qquad \text{Add.}$$

$$= \sqrt{4 \cdot 13} = 2\sqrt{13} \approx 7.21 \qquad \sqrt{52} = \sqrt{4 \cdot 13} = \sqrt{4}\sqrt{13} = 2\sqrt{13}$$

The distance between the given points is $2\sqrt{13}$ units, or approximately 7.21 units. The situation is illustrated in **Figure 1.73**. • • •

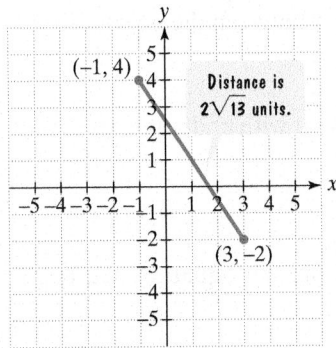

FIGURE 1.73 Finding the distance between two points

☑ Check Point **1** Find the distance between $(-1, -3)$ and $(2, 3)$. Express the answer in simplified radical form and then round to two decimal places.

② Find the midpoint of a line segment.

The Midpoint Formula

The distance formula can be used to derive a formula for finding the midpoint of a line segment between two given points. The formula is given as follows:

> ### The Midpoint Formula
>
> Consider a line segment whose endpoints are (x_1, y_1) and (x_2, y_2). The coordinates of the segment's midpoint are
>
> $$\left(\frac{x_1 + x_2}{2}, \frac{y_1 + y_2}{2}\right).$$
>
> To find the midpoint, take the average of the two x-coordinates and the average of the two y-coordinates.

GREAT QUESTION!

Do I add or subtract coordinates when applying the midpoint and distance formulas?

The midpoint formula requires finding the *sum* of coordinates. By contrast, the distance formula requires finding the *difference* of coordinates:

$$\underbrace{\left(\frac{x_1 + x_2}{2}, \frac{y_1 + y_2}{2}\right)}_{\text{Midpoint: Sum of coordinates}} \qquad \underbrace{\sqrt{(x_2 - x_1)^2 + (y_2 - y_1)^2}}_{\text{Distance: Difference of coordinates}}$$

It's easy to confuse the two formulas. Be sure to use addition, not subtraction, when applying the midpoint formula.

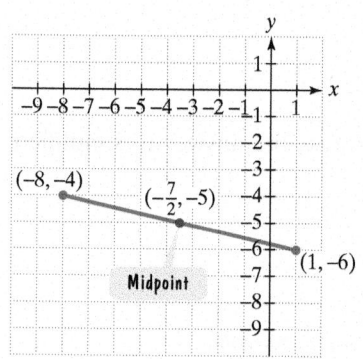

FIGURE 1.74 Finding a line segment's midpoint

EXAMPLE 2 Using the Midpoint Formula

Find the midpoint of the line segment with endpoints $(1, -6)$ and $(-8, -4)$.

SOLUTION

To find the coordinates of the midpoint, we average the coordinates of the endpoints.

$$\text{Midpoint} = \left(\frac{1 + (-8)}{2}, \frac{-6 + (-4)}{2} \right) = \left(\frac{-7}{2}, \frac{-10}{2} \right) = \left(-\frac{7}{2}, -5 \right)$$

Average the
x-coordinates.

Average the
y-coordinates.

Figure 1.74 illustrates that the point $\left(-\frac{7}{2}, -5 \right)$ is midway between the points $(1, -6)$ and $(-8, -4)$.

• • •

✓ Check Point **2** Find the midpoint of the line segment with endpoints $(1, 2)$ and $(7, -3)$.

Circles

Our goal is to translate a circle's geometric definition into an equation. We begin with this geometric definition.

Definition of a Circle

A **circle** is the set of all points in a plane that are equidistant from a fixed point, called the **center**. The fixed distance from the circle's center to any point on the circle is called the **radius**.

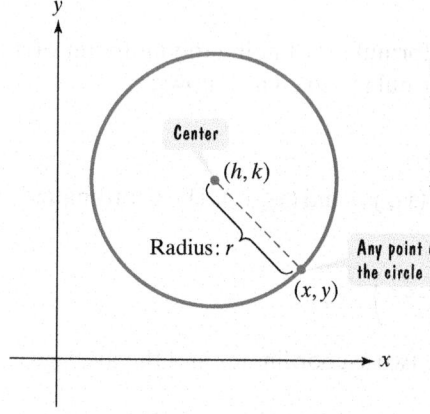

FIGURE 1.75 A circle centered at (h, k) with radius r

Figure 1.75 is our starting point for obtaining a circle's equation. We've placed the circle into a rectangular coordinate system. The circle's center is (h, k) and its radius is r. We let (x, y) represent the coordinates of any point on the circle.

What does the geometric definition of a circle tell us about the point (x, y) in **Figure 1.75**? The point is on the circle if and only if its distance from the center is r. We can use the distance formula to express this idea algebraically:

The distance between
(x, y) and (h, k) is always r.

$$\sqrt{(x - h)^2 + (y - k)^2} = r.$$

Squaring both sides of $\sqrt{(x - h)^2 + (y - k)^2} = r$ yields the *standard form of the equation of a circle.*

③ Write the standard form of a circle's equation.

The Standard Form of the Equation of a Circle

The **standard form of the equation of a circle** with center (h, k) and radius r is

$$(x - h)^2 + (y - k)^2 = r^2.$$

EXAMPLE 3 Finding the Standard Form of a Circle's Equation

Write the standard form of the equation of the circle with center $(0, 0)$ and radius 2. Graph the circle.

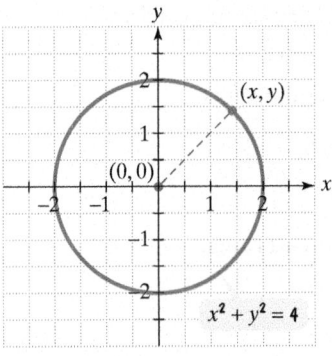

FIGURE 1.76 The graph of $x^2 + y^2 = 4$

SOLUTION

The center is $(0, 0)$. Because the center is represented as (h, k) in the standard form of the equation, $h = 0$ and $k = 0$. The radius is 2, so we will let $r = 2$ in the equation.

$(x - h)^2 + (y - k)^2 = r^2$ This is the standard form of a circle's equation.

$(x - 0)^2 + (y - 0)^2 = 2^2$ Substitute 0 for h, 0 for k, and 2 for r.

$x^2 + y^2 = 4$ Simplify.

The standard form of the equation of the circle is $x^2 + y^2 = 4$. **Figure 1.76** shows the graph. • • •

🗸 Check Point **3** Write the standard form of the equation of the circle with center $(0, 0)$ and radius 4.

TECHNOLOGY

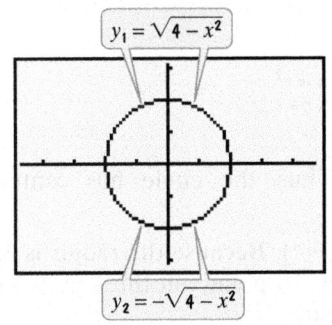

To graph a circle with a graphing utility, first solve the equation for y.

$$x^2 + y^2 = 4$$

$$y^2 = 4 - x^2$$

$$y = \pm\sqrt{4 - x^2}$$ y is not a function of x.

Graph the two equations

$$y_1 = \sqrt{4 - x^2} \quad \text{and} \quad y_2 = -\sqrt{4 - x^2}$$

in the same viewing rectangle. The graph of $y_1 = \sqrt{4 - x^2}$ is the top semicircle because y is always positive. The graph of $y_2 = -\sqrt{4 - x^2}$ is the bottom semicircle because y is always negative. Use a ZOOM SQUARE setting so that the circle looks like a circle. (Many graphing utilities have problems connecting the two semicircles because the segments directly to the left and to the right of the center become nearly vertical.)

Example 3 and Check Point 3 involved circles centered at the origin. The standard form of the equation of all such circles is $x^2 + y^2 = r^2$, where r is the circle's radius. Now, let's consider a circle whose center is not at the origin.

EXAMPLE 4 Finding the Standard Form of a Circle's Equation

Write the standard form of the equation of the circle with center $(-2, 3)$ and radius 4.

SOLUTION

The center is $(-2, 3)$. Because the center is represented as (h, k) in the standard form of the equation, $h = -2$ and $k = 3$. The radius is 4, so we will let $r = 4$ in the equation.

$(x - h)^2 + (y - k)^2 = r^2$ This is the standard form of a circle's equation.

$[x - (-2)]^2 + (y - 3)^2 = 4^2$ Substitute −2 for h, 3 for k, and 4 for r.

$(x + 2)^2 + (y - 3)^2 = 16$ Simplify.

The standard form of the equation of the circle is $(x + 2)^2 + (y - 3)^2 = 16$. • • •

🗸 Check Point **4** Write the standard form of the equation of the circle with center $(0, -6)$ and radius 10.

④ Give the center and radius of a circle whose equation is in standard form.

EXAMPLE 5 Using the Standard Form of a Circle's Equation to Graph the Circle

a. Find the center and radius of the circle whose equation is

$$(x - 2)^2 + (y + 4)^2 = 9.$$

b. Graph the equation.

c. Use the graph to identify the relation's domain and range.

SOLUTION

a. We begin by finding the circle's center, (h, k), and its radius, r. We can find the values for h, k, and r by comparing the given equation to the standard form of the equation of a circle, $(x - h)^2 + (y - k)^2 = r^2$.

$$(x - 2)^2 + (y + 4)^2 = 9$$
$$(x - 2)^2 + (y - (-4))^2 = 3^2$$

This is $(x - h)^2$, with $h = 2$. This is $(y - k)^2$, with $k = -4$. This is r^2, with $r = 3$.

We see that $h = 2$, $k = -4$, and $r = 3$. Thus, the circle has center $(h, k) = (2, -4)$ and a radius of 3 units.

b. To graph this circle, first locate the center $(2, -4)$. Because the radius is 3, you can locate at least four points on the circle by going out three units to the right, to the left, up, and down from the center.

The points three units to the right and to the left of $(2, -4)$ are $(5, -4)$ and $(-1, -4)$, respectively. The points three units up and down from $(2, -4)$ are $(2, -1)$ and $(2, -7)$, respectively.

Using these points, we obtain the graph in **Figure 1.77**.

c. The four points that we located on the circle can be used to determine the relation's domain and range. The points $(-1, -4)$ and $(5, -4)$ show that values of x extend from -1 to 5, inclusive:

$$\text{Domain} = [-1, 5].$$

The points $(2, -7)$ and $(2, -1)$ show that values of y extend from -7 to -1, inclusive:

$$\text{Range} = [-7, -1]. \qquad \bullet \bullet \bullet$$

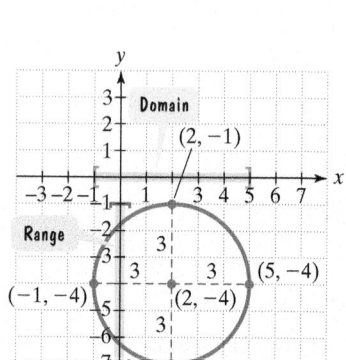

FIGURE 1.77 The graph of $(x - 2)^2 + (y + 4)^2 = 9$

GREAT QUESTION!

What's the bottom line when identifying h and k in the standard form of a circle's equation?

It's easy to make sign errors when finding h and k, the coordinates of a circle's center, (h, k). Keep in mind that h and k are the numbers that *follow the subtraction signs* in a circle's equation:

$$(x - 2)^2 + (y + 4)^2 = 9$$
$$(x - 2)^2 + (y - (-4))^2 = 9.$$

The number *after* the subtraction is 2: $h = 2$. The number *after* the subtraction is -4: $k = -4$.

☑ Check Point **5**

a. Find the center and radius of the circle whose equation is

$$(x + 3)^2 + (y - 1)^2 = 4.$$

b. Graph the equation.

c. Use the graph to identify the relation's domain and range.

If we square $x - 2$ and $y + 4$ in the standard form of the equation in Example 5, we obtain another form for the circle's equation.

$$(x - 2)^2 + (y + 4)^2 = 9 \qquad \text{This is the standard form of the equation in Example 5.}$$

$$x^2 - 4x + 4 + y^2 + 8y + 16 = 9 \qquad \text{Square } x - 2 \text{ and } y + 4.$$

$$x^2 + y^2 - 4x + 8y + 20 = 9 \qquad \text{Combine constants and rearrange terms.}$$

$$x^2 + y^2 - 4x + 8y + 11 = 0 \qquad \text{Subtract 9 from both sides.}$$

This result suggests that an equation in the form $x^2 + y^2 + Dx + Ey + F = 0$ can represent a circle. This is called the *general form of the equation of a circle.*

> ### The General Form of the Equation of a Circle
>
> The **general form of the equation of a circle** is
>
> $$x^2 + y^2 + Dx + Ey + F = 0,$$
>
> where D, E, and F are real numbers.

⑤ Convert the general form of a circle's equation to standard form.

We can convert the general form of the equation of a circle to the standard form $(x - h)^2 + (y - k)^2 = r^2$. We do so by completing the square on x and y. Let's see how this is done.

GREAT QUESTION!

Egad! I forgot how to complete the square. Any suggestions?

To complete the square, on $x^2 + bx$, take half the coefficient of x. Then square this number. By adding the square of half the coefficient of x, a perfect square trinomial will result. For more detail, see Section P.6, pages 84–85.

EXAMPLE 6 Converting the General Form of a Circle's Equation to Standard Form and Graphing the Circle

Write in standard form and graph: $x^2 + y^2 + 4x - 6y - 23 = 0.$

SOLUTION

Because we plan to complete the square on both x and y, let's rearrange the terms so that x-terms are arranged in descending order, y-terms are arranged in descending order, and the constant term appears on the right.

$$x^2 + y^2 + 4x - 6y - 23 = 0 \qquad \text{This is the given equation.}$$

$$(x^2 + 4x \quad) + (y^2 - 6y \quad) = 23 \qquad \text{Rewrite in anticipation of completing the square.}$$

$$(x^2 + 4x + 4) + (y^2 - 6y + 9) = 23 + 4 + 9 \qquad \text{Complete the square on x: } \tfrac{1}{2} \cdot 4 = 2 \text{ and } 2^2 = 4, \text{ so add 4 to both sides. Complete the square on y: } \tfrac{1}{2}(-6) = -3 \text{ and } (-3)^2 = 9, \text{ so add 9 to both sides.}$$

> Remember that numbers added on the left side must also be added on the right side.

$$(x + 2)^2 + (y - 3)^2 = 36 \qquad \text{Factor on the left and add on the right.}$$

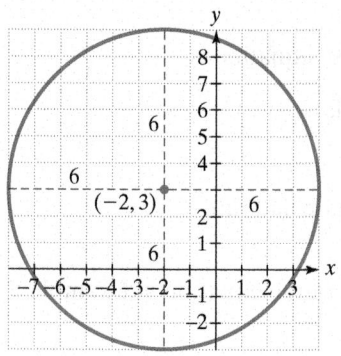

FIGURE 1.78 The graph of $(x + 2)^2 + (y - 3)^2 = 36$

This last equation, $(x + 2)^2 + (y - 3)^2 = 36$, is in standard form. We can identify the circle's center and radius by comparing this equation to the standard form of the equation of a circle, $(x - h)^2 + (y - k)^2 = r^2$.

$$(x + 2)^2 + (y - 3)^2 = 36$$
$$(x - (-2))^2 + (y - 3)^2 = 6^2$$

This is $(x - h)^2$, with $h = -2$.　　This is $(y - k)^2$, with $k = 3$.　　This is r^2, with $r = 6$.

We use the center, $(h, k) = (-2, 3)$, and the radius, $r = 6$, to graph the circle. The graph is shown in **Figure 1.78**.　　•••

⌀ Check Point **6** Write in standard form and graph:

$$x^2 + y^2 + 4x - 4y - 1 = 0.$$

CONCEPT AND VOCABULARY CHECK

Fill in each blank so that the resulting statement is true.

1. The distance, d, between the points (x_1, y_1) and (x_2, y_2) in the rectangular coordinate system is
$d =$ _____.

2. The midpoint of a line segment whose endpoints are

(x_1, y_1) and (x_2, y_2) is (_____, _____).

3. The set of all points in a plane that are equidistant from a fixed point is a/an _____. The fixed point is called the _____. The distance from this fixed point to any point on the geometric figure is called the _____.

4. The standard form of the equation of a circle with center (h, k) and radius r is _____.

5. The equation $x^2 + y^2 + Dx + Ey + F = 0$ is called the _____ form of the equation of a circle.

6. In the equation $(x^2 + 4x \quad) + (y^2 - 8y \quad) = 5$, we complete the square on x by adding ____ to both sides. We complete the square on y by adding ____ to both sides.

EXERCISE SET 1.9

Practice Exercises

In Exercises 1–18, find the distance between each pair of points. If necessary, round answers to two decimals places.

1. $(2, 3)$ and $(14, 8)$
2. $(5, 1)$ and $(8, 5)$
3. $(4, -1)$ and $(-6, 3)$
4. $(2, -3)$ and $(-1, 5)$
5. $(0, 0)$ and $(-3, 4)$
6. $(0, 0)$ and $(3, -4)$
7. $(-2, -6)$ and $(3, -4)$
8. $(-4, -1)$ and $(2, -3)$
9. $(0, -3)$ and $(4, 1)$
10. $(0, -2)$ and $(4, 3)$
11. $(3.5, 8.2)$ and $(-0.5, 6.2)$
12. $(2.6, 1.3)$ and $(1.6, -5.7)$
13. $\left(0, -\sqrt{3}\right)$ and $\left(\sqrt{5}, 0\right)$
14. $\left(0, -\sqrt{2}\right)$ and $\left(\sqrt{7}, 0\right)$
15. $\left(3\sqrt{3}, \sqrt{5}\right)$ and $\left(-\sqrt{3}, 4\sqrt{5}\right)$
16. $\left(2\sqrt{3}, \sqrt{6}\right)$ and $\left(-\sqrt{3}, 5\sqrt{6}\right)$
17. $\left(\frac{7}{3}, \frac{1}{5}\right)$ and $\left(\frac{1}{3}, \frac{6}{5}\right)$
18. $\left(-\frac{1}{4}, -\frac{1}{7}\right)$ and $\left(\frac{3}{4}, \frac{6}{7}\right)$

In Exercises 19–30, find the midpoint of each line segment with the given endpoints.

19. $(6, 8)$ and $(2, 4)$
20. $(10, 4)$ and $(2, 6)$
21. $(-2, -8)$ and $(-6, -2)$
22. $(-4, -7)$ and $(-1, -3)$
23. $(-3, -4)$ and $(6, -8)$
24. $(-2, -1)$ and $(-8, 6)$
25. $\left(-\frac{7}{2}, \frac{3}{2}\right)$ and $\left(-\frac{5}{2}, -\frac{11}{2}\right)$
26. $\left(-\frac{2}{5}, \frac{7}{15}\right)$ and $\left(-\frac{2}{5}, -\frac{4}{15}\right)$
27. $\left(8, 3\sqrt{5}\right)$ and $\left(-6, 7\sqrt{5}\right)$
28. $\left(7\sqrt{3}, -6\right)$ and $\left(3\sqrt{3}, -2\right)$
29. $\left(\sqrt{18}, -4\right)$ and $\left(\sqrt{2}, 4\right)$
30. $\left(\sqrt{50}, -6\right)$ and $\left(\sqrt{2}, 6\right)$

In Exercises 31–40, write the standard form of the equation of the circle with the given center and radius.

31. Center $(0,0)$, $r = 7$

32. Center $(0,0)$, $r = 8$

33. Center $(3,2)$, $r = 5$

34. Center $(2,-1)$, $r = 4$

35. Center $(-1,4)$, $r = 2$

36. Center $(-3,5)$, $r = 3$

37. Center $(-3,-1)$, $r = \sqrt{3}$

38. Center $(-5,-3)$, $r = \sqrt{5}$

39. Center $(-4,0)$, $r = 10$

40. Center $(-2,0)$, $r = 6$

In Exercises 41–52, give the center and radius of the circle described by the equation and graph each equation. Use the graph to identify the relation's domain and range.

41. $x^2 + y^2 = 16$

42. $x^2 + y^2 = 49$

43. $(x-3)^2 + (y-1)^2 = 36$

44. $(x-2)^2 + (y-3)^2 = 16$

45. $(x+3)^2 + (y-2)^2 = 4$

46. $(x+1)^2 + (y-4)^2 = 25$

47. $(x+2)^2 + (y+2)^2 = 4$

48. $(x+4)^2 + (y+5)^2 = 36$

49. $x^2 + (y-1)^2 = 1$

50. $x^2 + (y-2)^2 = 4$

51. $(x+1)^2 + y^2 = 25$

52. $(x+2)^2 + y^2 = 16$

In Exercises 53–64, complete the square and write the equation in standard form. Then give the center and radius of each circle and graph the equation.

53. $x^2 + y^2 + 6x + 2y + 6 = 0$

54. $x^2 + y^2 + 8x + 4y + 16 = 0$

55. $x^2 + y^2 - 10x - 6y - 30 = 0$

56. $x^2 + y^2 - 4x - 12y - 9 = 0$

57. $x^2 + y^2 + 8x - 2y - 8 = 0$

58. $x^2 + y^2 + 12x - 6y - 4 = 0$

59. $x^2 - 2x + y^2 - 15 = 0$

60. $x^2 + y^2 - 6y - 7 = 0$

61. $x^2 + y^2 - x + 2y + 1 = 0$

62. $x^2 + y^2 + x + y - \frac{1}{2} = 0$

63. $x^2 + y^2 + 3x - 2y - 1 = 0$

64. $x^2 + y^2 + 3x + 5y + \frac{9}{4} = 0$

Practice Plus

In Exercises 65–66, a line segment through the center of each circle intersects the circle at the points shown.

a. *Find the coordinates of the circle's center.*

b. *Find the radius of the circle.*

c. *Use your answers from parts (a) and (b) to write the standard form of the circle's equation.*

65.

66.

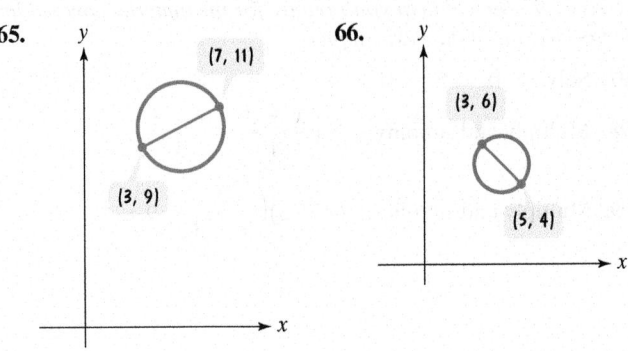

In Exercises 67–70, graph both equations in the same rectangular coordinate system and find all points of intersection. Then show that these ordered pairs satisfy the equations.

67. $x^2 + y^2 = 16$
$x - y = 4$

68. $x^2 + y^2 = 9$
$x - y = 3$

69. $(x-2)^2 + (y+3)^2 = 4$
$y = x - 3$

70. $(x-3)^2 + (y+1)^2 = 9$
$y = x - 1$

Application Exercises

The cellphone screen shows coordinates of six cities from a rectangular coordinate system placed on North America by long-distance telephone companies. Each unit in this system represents $\sqrt{0.1}$ mile.

In Exercises 71–72, use this information to find the distance, to the nearest mile, between each pair of cities.

Source: Peter H. Dana

71. Boston and San Francisco

72. New Orleans and Houston

73. A rectangular coordinate system with coordinates in miles is placed with the origin at the center of Los Angeles. The figure indicates that the University of Southern California is located 2.4 miles west and 2.7 miles south of central Los Angeles. A seismograph on the campus shows that a small earthquake occurred. The quake's epicenter is estimated to be approximately 30 miles from the university. Write the standard form of the equation for the set of points that could be the epicenter of the quake.

74. The Ferris wheel in the figure has a radius of 68 feet. The clearance between the wheel and the ground is 14 feet. The rectangular coordinate system shown has its origin on the ground directly below the center of the wheel. Use the coordinate system to write the equation of the circular wheel.

Writing in Mathematics

75. In your own words, describe how to find the distance between two points in the rectangular coordinate system.

76. In your own words, describe how to find the midpoint of a line segment if its endpoints are known.

77. What is a circle? Without using variables, describe how the definition of a circle can be used to obtain a form of its equation.

78. Give an example of a circle's equation in standard form. Describe how to find the center and radius for this circle.

79. How is the standard form of a circle's equation obtained from its general form?

80. Does $(x - 3)^2 + (y - 5)^2 = 0$ represent the equation of a circle? If not, describe the graph of this equation.

81. Does $(x - 3)^2 + (y - 5)^2 = -25$ represent the equation of a circle? What sort of set is the graph of this equation?

82. Write and solve a problem about the flying time between a pair of cities shown on the cellphone screen for Exercises 71–72. Do not use the pairs in Exercise 71 or Exercise 72. Begin by determining a reasonable average speed, in miles per hour, for a jet flying between the cities.

Technology Exercises

In Exercises 83–84, use a graphing utility to graph each circle whose equation is given.

83. $x^2 + y^2 = 25$

84. $(y + 1)^2 = 36 - (x - 3)^2$

Critical Thinking Exercises

Make Sense? *In Exercises 85–88, determine whether each statement makes sense or does not make sense, and explain your reasoning.*

85. I've noticed that in mathematics, one topic often leads logically to a new topic:

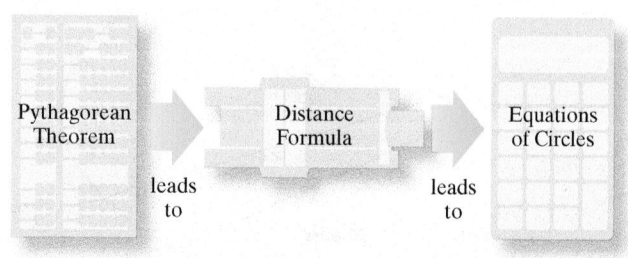

86. To avoid sign errors when finding h and k, I place parentheses around the numbers that follow the subtraction signs in a circle's equation.

87. I used the equation $(x + 1)^2 + (y - 5)^2 = -4$ to identify the circle's center and radius.

88. My graph of $(x - 2)^2 + (y + 1)^2 = 16$ is my graph of $x^2 + y^2 = 16$ translated two units right and one unit down.

In Exercises 89–92, determine whether each statement is true or false. If the statement is false, make the necessary change(s) to produce a true statement.

89. The equation of the circle whose center is at the origin with radius 16 is $x^2 + y^2 = 16$.

90. The graph of $(x - 3)^2 + (y + 5)^2 = 36$ is a circle with radius 6 centered at $(-3, 5)$.

91. The graph of $(x - 4) + (y + 6) = 25$ is a circle with radius 5 centered at $(4, -6)$.

92. The graph of $(x - 3)^2 + (y + 5)^2 = -36$ is a circle with radius 6 centered at $(3, -5)$.

93. Show that the points $A(1, 1 + d), B(3, 3 + d)$, and $C(6, 6 + d)$ are collinear (lie along a straight line) by showing that the distance from A to B plus the distance from B to C equals the distance from A to C.

94. Prove the midpoint formula by using the following procedure.
 a. Show that the distance between (x_1, y_1) and $\left(\dfrac{x_1 + x_2}{2}, \dfrac{y_1 + y_2}{2}\right)$ is equal to the distance between (x_2, y_2) and $\left(\dfrac{x_1 + x_2}{2}, \dfrac{y_1 + y_2}{2}\right)$.
 b. Use the procedure from Exercise 93 and the distances from part (a) to show that the points (x_1, y_1), $\left(\dfrac{x_1 + x_2}{2}, \dfrac{y_1 + y_2}{2}\right)$, and (x_2, y_2) are collinear.

95. Find the area of the donut-shaped region bounded by the graphs of $(x - 2)^2 + (y + 3)^2 = 25$ and $(x - 2)^2 + (y + 3)^2 = 36$.

96. A **tangent line** to a circle is a line that intersects the circle at exactly one point. The tangent line is perpendicular to the radius of the circle at this point of contact. Write an equation in point-slope form for the line tangent to the circle whose equation is $x^2 + y^2 = 25$ at the point $(3, -4)$.

Preview Exercises

Exercises 97–99 will help you prepare for the material covered in the first section of the next chapter.

97. Solve: $4x - 5 = 31$.

98. Multiply and simplify: $12\left(\dfrac{x + 2}{4} - \dfrac{x - 1}{3}\right)$.

99. Multiply and simplify: $(x - 3)\left(\dfrac{3}{x - 3} + 9\right)$.

CHAPTER 1 Summary, Review, and Test

SUMMARY

DEFINITIONS AND CONCEPTS	EXAMPLES

1.1 Graphs and Graphing Utilities

a. The rectangular coordinate system consists of a horizontal number line, the x-axis, and a vertical number line, the y-axis, intersecting at their zero points, the origin. Each point in the system corresponds to an ordered pair of real numbers (x, y). The first number in the pair is the x-coordinate; the second number is the y-coordinate. See Figure 1.1 on page 106. — Ex. 1, p. 107

b. An ordered pair is a solution of an equation in two variables if replacing the variables by the corresponding coordinates results in a true statement. The ordered pair is said to satisfy the equation. The graph of the equation is the set of all points whose coordinates satisfy the equation. One method for graphing an equation is to plot ordered-pair solutions and connect them with a smooth curve or line. — Ex. 2, p. 108; Ex. 3, p. 108

c. An x-intercept of a graph is the x-coordinate of a point where the graph intersects the x-axis. The y-coordinate corresponding to an x-intercept is always zero.
A y-intercept of a graph is the y-coordinate of a point where the graph intersects the y-axis. The x-coordinate corresponding to a y-intercept is always zero. — Ex. 5, p. 111

1.2 Basics of Functions and Their Graphs

a. A relation is any set of ordered pairs. The set of first components is the domain of the relation and the set of second components is the range. — Ex. 1, p. 119

b. A function is a correspondence from a first set, called the domain, to a second set, called the range, such that each element in the domain corresponds to exactly one element in the range. If any element in a relation's domain corresponds to more than one element in the range, the relation is not a function. — Ex. 2, p. 121

c. Functions are usually given in terms of equations involving x and y, in which x is the independent variable and y is the dependent variable. If an equation is solved for y and more than one value of y can be obtained for a given x, then the equation does not define y as a function of x. If an equation defines a function, the value of the function at x, $f(x)$, often replaces y. — Ex. 3, p. 122; Ex. 4, p. 124

d. The graph of a function is the graph of its ordered pairs. — Ex. 5, p. 125

e. The vertical line test for functions: If any vertical line intersects a graph in more than one point, the graph does not define y as a function of x. — Ex. 6, p. 126

f. The graph of a function can be used to determine the function's domain and its range. To find the domain, look for all the inputs on the x-axis that correspond to points on the graph. To find the range, look for all the outputs on the y-axis that correspond to points on the graph. — Ex. 8, p. 129

g. The zeros of a function, f, are the values of x for which $f(x) = 0$. At the real zeros, the graph of f has x-intercepts. The graph of a function can have more than one x-intercept but at most one y-intercept. — Figure 1.27, p. 131

1.3 More on Functions and Their Graphs

a. A function is increasing on intervals where its graph rises, decreasing on intervals where it falls, and constant on intervals where it neither rises nor falls. Precise definitions are given in the box on page 138. — Ex. 1, p. 139

b. If the graph of a function is given, we can often visually locate the number(s) at which the function has a relative maximum or relative minimum. Precise definitions are given in the box on page 139. — Figure 1.29, p. 140

c. The graph of an even function in which $f(-x) = f(x)$ is symmetric with respect to the y-axis. The graph of an odd function in which $f(-x) = -f(x)$ is symmetric with respect to the origin. — Ex. 2, p. 140

d. Piecewise functions are defined by two (or more) equations over a specified domain. Some piecewise functions are called step functions because their graphs form discontinuous steps. An example is $f(x) = \text{int}(x)$, where $\text{int}(x)$ is the greatest integer that is less than or equal to x. — Ex. 3, p. 142; Ex. 4, p. 143; Figure 1.35, p. 144

e. The difference quotient of a function f is
$$\frac{f(x + h) - f(x)}{h}, h \neq 0.$$
— Ex. 5, p. 144

DEFINITIONS AND CONCEPTS	**EXAMPLES**

1.4 Linear Functions and Slope

a. The slope, m, of the line through (x_1, y_1) and (x_2, y_2) is $m = \dfrac{y_2 - y_1}{x_2 - x_1}$.

Ex. 1, p. 153

b. Equations of lines include point-slope form, $y - y_1 = m(x - x_1)$, slope-intercept form, $y = mx + b$, and general form, $Ax + By + C = 0$. The equation of a horizontal line is $y = b$; a vertical line is $x = a$. A vertical line is not a linear function.

Ex. 2, p. 155;
Ex. 3, p. 155;
Ex. 5, p. 157;
Ex. 6, p. 158

c. Linear functions in the form $f(x) = mx + b$ can be graphed using the slope, m, and the y-intercept, b. (See the box on page 156.) Linear equations in the general form $Ax + By + C = 0$ can be solved for y and graphed using the slope and the y-intercept. Intercepts can also be used to graph $Ax + By + C = 0$. (See the box on page 159.)

Ex. 4, p. 156;
Ex. 7, p. 159;
Ex. 8, p. 160

1.5 More on Slope

a. Parallel lines have equal slopes. Perpendicular lines have slopes that are negative reciprocals.

Ex. 1, p. 167;
Ex. 2, p. 169

b. The slope of a linear function is the rate of change of the dependent variable per unit change of the independent variable.

Ex. 3, p. 170

c. The average rate of change of f from x_1 to x_2 is

$$\frac{f(x_2) - f(x_1)}{x_2 - x_1}.$$

Ex. 4, p. 172;
Ex. 5, p. 173

1.6 Transformations of Functions

a. Table 1.3 on page 179 shows the graphs of the constant function, $f(x) = c$, the identity function, $f(x) = x$, the absolute value function, $f(x) = |x|$, the standard quadratic function, $f(x) = x^2$, the square root function, $f(x) = \sqrt{x}$, the standard cubic function, $f(x) = x^3$, and the cube root function, $f(x) = \sqrt[3]{x}$. The table also lists characteristics of each function.

b. Table 1.4 on page 187 summarizes how to graph a function using vertical shifts, $y = f(x) \pm c$, horizontal shifts, $y = f(x \pm c)$, reflections about the x-axis, $y = -f(x)$, reflections about the y-axis, $y = f(-x)$, vertical stretching, $y = cf(x), c > 1$, vertical shrinking, $y = cf(x), 0 < c < 1$, horizontal shrinking, $y = f(cx), c > 1$, and horizontal stretching, $y = f(cx), 0 < c < 1$.

Ex. 1, p. 180;
Ex. 2, p. 182;
Ex. 3, p. 183;
Ex. 4, p. 184;
Ex. 5, p. 184;
Ex. 6, p. 185;
Ex. 7, p. 186

c. A function involving more than one transformation can be graphed in the following order: (1) horizontal shifting; (2) stretching or shrinking; (3) reflecting; (4) vertical shifting.

Ex. 8, p. 188;
Ex. 9, p. 189

1.7 Combinations of Functions; Composite Functions

a. If a function f does not model data or verbal conditions, its domain is the largest set of real numbers for which the value of $f(x)$ is a real number. Exclude from a function's domain real numbers that cause division by zero and real numbers that result in an even root, such as a square root, of a negative number.

Ex. 1, p. 195

b. When functions are given as equations, they can be added, subtracted, multiplied, or divided by performing operations with the algebraic expressions that appear on the right side of the equations. Definitions for the sum $f + g$, the difference $f - g$, the product fg, and the quotient $\frac{f}{g}$ functions, with domains $D_f \cap D_g$, and $g(x) \neq 0$ for the quotient function, are given in the box on page 196.

Ex. 2, p. 197;
Ex. 3, p. 198;
Ex. 4, p. 199

c. The composition of functions f and g, $f \circ g$, is defined by $(f \circ g)(x) = f(g(x))$. The domain of the composite function $f \circ g$ is given in the box on page 202. This composite function is obtained by replacing each occurrence of x in the equation for f with $g(x)$.

Ex. 5, p. 202;
Ex. 6, p. 203

d. Expressing a given function as a composition of two functions is called decomposing the given function.

Ex. 7, p. 204

1.8 Inverse Functions

a. If $f(g(x)) = x$ and $g(f(x)) = x$, function g is the inverse of function f, denoted f^{-1} and read "f-inverse." Thus, to show that f and g are inverses of each other, one must show that $f(g(x)) = x$ and $g(f(x)) = x$.

Ex. 1, p. 210

b. The procedure for finding a function's inverse uses a switch-and-solve strategy. Switch x and y, and then solve for y. The procedure is given in the box on page 211.

Ex. 2, p. 212;
Ex. 3, p. 212;
Ex. 4, p. 213

c. The horizontal line test for inverse functions: A function f has an inverse that is a function, f^{-1}, if there is no horizontal line that intersects the graph of the function f at more than one point.

Ex. 5, p. 214

DEFINITIONS AND CONCEPTS ## EXAMPLES

d. A one-to-one function is one in which no two different ordered pairs have the same second component. Only one-to-one functions have inverse functions.

e. If the point (a, b) is on the graph of f, then the point (b, a) is on the graph of f^{-1}. The graph of f^{-1} is a reflection of the graph of f about the line $y = x$.

Ex. 6, p. 215;
Ex. 7, p. 216

1.9 Distance and Midpoint Formulas; Circles

a. The distance, d, between the points (x_1, y_1) and (x_2, y_2) is given by $d = \sqrt{(x_2 - x_1)^2 + (y_2 - y_1)^2}$.

Ex. 1, p. 221

b. The midpoint of the line segment whose endpoints are (x_1, y_1) and (x_2, y_2) is the point with coordinates $\left(\dfrac{x_1 + x_2}{2}, \dfrac{y_1 + y_2}{2}\right)$.

Ex. 2, p. 222

c. The standard form of the equation of a circle with center (h, k) and radius r is $(x - h)^2 + (y - k)^2 = r^2$.

Ex. 3, p. 222;
Ex. 4, p. 223;
Ex. 5, p. 224

d. The general form of the equation of a circle is $x^2 + y^2 + Dx + Ey + F = 0$.

e. To convert from the general form to the standard form of a circle's equation, complete the square on x and y.

Ex. 6, p. 225

REVIEW EXERCISES

1.1

Graph each equation in Exercises 1–4. Let $x = -3, -2, -1, 0, 1, 2,$ and 3.

1. $y = 2x - 2$ **2.** $y = x^2 - 3$

3. $y = x$ **4.** $y = |x| - 2$

5. What does a $[-20, 40, 10]$ by $[-5, 5, 1]$ viewing rectangle mean? Draw axes with tick marks and label the tick marks to illustrate this viewing rectangle.

In Exercises 6–8, use the graph and determine the x-intercepts if any, and the y-intercepts if any. For each graph, tick marks along the axes represent one unit each.

6.

7.

8.

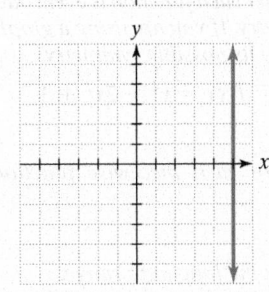

Salary after College. In 2010, MonsterCollege surveyed 1250 U.S. college students expecting to graduate in the next several years. Respondents were asked the following question:

> *What do you think your starting salary will be at your first job after college?*

The line graph shows the percentage of college students who anticipated various starting salaries. Use the graph to solve Exercises 9–14.

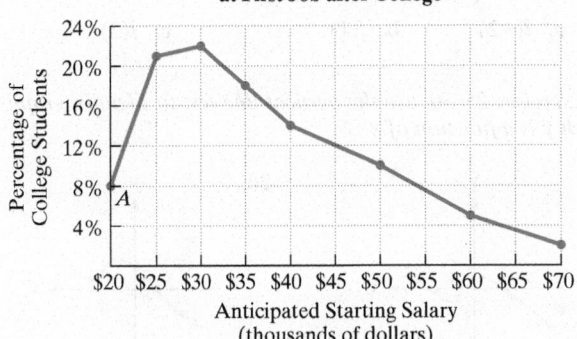

Source: MonsterCollege™

9. What are the coordinates of point A? What does this mean in terms of the information given by the graph?

10. What starting salary was anticipated by the greatest percentage of college students? Estimate the percentage of students who anticipated this salary.

11. What starting salary was anticipated by the least percentage of college students? Estimate the percentage of students who anticipated this salary.

12. What starting salaries were anticipated by more than 20% of college students?

13. Estimate the percentage of students who anticipated a starting salary of $40 thousand.

14. The mathematical model

$$p = -0.01s^2 + 0.8s + 3.7$$

describes the percentage of college students, p, who anticipated a starting salary s, in thousands of dollars. Use this formula to find the percentage of students who anticipated a starting salary of $40 thousand. How does this compare with your estimate in Exercise 13?

1.2 and 1.3

In Exercises 15–17, determine whether each relation is a function. Give the domain and range for each relation.

15. $\{(2, 7), (3, 7), (5, 7)\}$ **16.** $\{(1, 10), (2, 500), (13, \pi)\}$

17. $\{(12, 13), (14, 15), (12, 19)\}$

In Exercises 18–20, determine whether each equation defines y as a function of x.

18. $2x + y = 8$ **19.** $3x^2 + y = 14$

20. $2x + y^2 = 6$

In Exercises 21–24, evaluate each function at the given values of the independent variable and simplify.

21. $f(x) = 5 - 7x$
 a. $f(4)$ **b.** $f(x + 3)$ **c.** $f(-x)$

22. $g(x) = 3x^2 - 5x + 2$
 a. $g(0)$ **b.** $g(-2)$
 c. $g(x - 1)$ **d.** $g(-x)$

23. $g(x) = \begin{cases} \sqrt{x - 4} & \text{if } x \geq 4 \\ 4 - x & \text{if } x < 4 \end{cases}$
 a. $g(13)$ **b.** $g(0)$ **c.** $g(-3)$

24. $f(x) = \begin{cases} \dfrac{x^2 - 1}{x - 1} & \text{if } x \neq 1 \\ 12 & \text{if } x = 1 \end{cases}$
 a. $f(-2)$ **b.** $f(1)$ **c.** $f(2)$

In Exercises 25–30, use the vertical line test to identify graphs in which y is a function of x.

25. **26.**

27. **28.**

29. **30.**

*In Exercises 31–33, use the graph to determine **a.** the function's domain; **b.** the function's range; **c.** the x-intercepts if any; **d.** the y-intercept. if there is one; **e.** intervals on which the function is increasing, decreasing, or constant; and **f.** the missing function values, indicated by question marks, below each graph.*

31.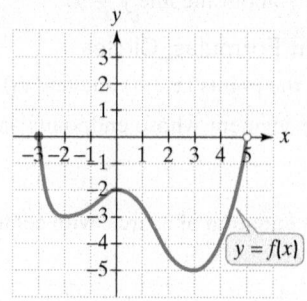

$f(-2) = ?\quad f(3) = ?$

32.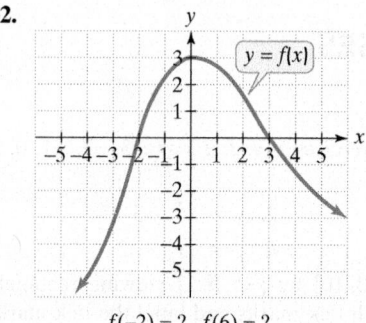

$f(-2) = ?\quad f(6) = ?$

33.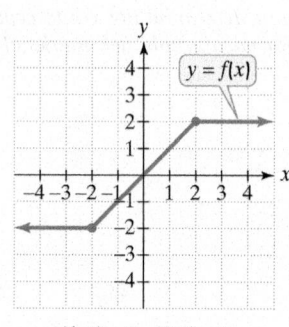

$f(-9) = ?\quad f(14) = ?$

In Exercises 34–35, find each of the following:

 a. The numbers, if any, at which f has a relative maximum. What are these relative maxima?

 b. The numbers, if any, at which f has a relative minimum. What are these relative minima?

34. Use the graph in Exercise 31.

35. Use the graph in Exercise 32.

In Exercises 36–38, determine whether each function is even, odd, or neither. State each function's symmetry. If you are using a graphing utility, graph the function and verify its possible symmetry.

36. $f(x) = x^3 - 5x$ **37.** $f(x) = x^4 - 2x^2 + 1$

38. $f(x) = 2x\sqrt{1 - x^2}$

In Exercises 39–40, the domain of each piecewise function is $(-\infty, \infty)$.

 a. Graph each function.

 b. Use the graph to determine the function's range.

39. $f(x) = \begin{cases} 5 & \text{if } x \le -1 \\ -3 & \text{if } x > -1 \end{cases}$

40. $f(x) = \begin{cases} 2x & \text{if } x < 0 \\ -x & \text{if } x \ge 0 \end{cases}$

In Exercises 41–42, find and simplify the difference quotient

$$\frac{f(x + h) - f(x)}{h}, \quad h \ne 0$$

for the given function.

41. $f(x) = 8x - 11$

42. $f(x) = -2x^2 + x + 10$

43. The graph shows the height, in meters, of an eagle in terms of its time, in seconds, in flight.

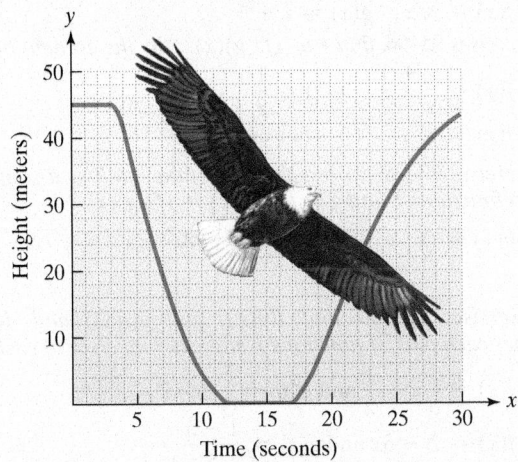

Time (seconds)

a. Is the eagle's height a function of time? Use the graph to explain why or why not.

b. On which interval is the function decreasing? Describe what this means in practical terms.

c. On which intervals is the function constant? What does this mean for each of these intervals?

d. On which interval is the function increasing? What does this mean?

44. A cargo service charges a flat fee of $5 plus $1.50 for each pound or fraction of a pound. Graph shipping cost, $C(x)$, in dollars, as a function of weight, x, in pounds, for $0 < x \le 5$.

1.4 and 1.5

In Exercises 45–48, find the slope of the line passing through each pair of points or state that the slope is undefined. Then indicate whether the line through the points rises, falls, is horizontal, or is vertical.

45. $(3, 2)$ and $(5, 1)$

46. $(-1, -2)$ and $(-3, -4)$

47. $\left(-3, \frac{1}{4}\right)$ and $\left(6, \frac{1}{4}\right)$

48. $(-2, 5)$ and $(-2, 10)$

In Exercises 49–52, use the given conditions to write an equation for each line in point-slope form and slope-intercept form.

49. Passing through $(-3, 2)$ with slope -6

50. Passing through $(1, 6)$ and $(-1, 2)$

51. Passing through $(4, -7)$ and parallel to the line whose equation is $3x + y - 9 = 0$

52. Passing through $(-3, 6)$ and perpendicular to the line whose equation is $y = \frac{1}{3}x + 4$

53. Write an equation in general form for the line passing through $(-12, -1)$ and perpendicular to the line whose equation is $6x - y - 4 = 0$.

In Exercises 54–57, give the slope and y-intercept of each line whose equation is given. Then graph the line.

54. $y = \frac{2}{5}x - 1$

55. $f(x) = -4x + 5$

56. $2x + 3y + 6 = 0$

57. $2y - 8 = 0$

58. Graph using intercepts: $2x - 5y - 10 = 0$.

59. Graph: $2x - 10 = 0$.

60. The points in the scatter plot show the number of firearms per 100 persons and the number of deaths per 100,000 persons for industrialized countries with the highest death rates.

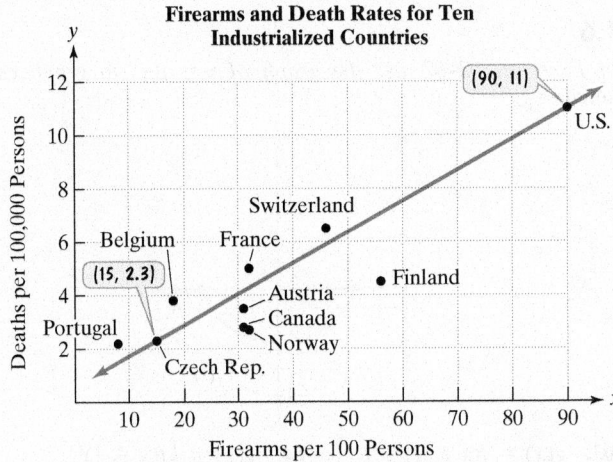

Source: International Action Network on Small Arms

a. Use the two points whose coordinates are shown by the voice balloons to find an equation in point-slope form for the line that models deaths per 100,000 persons, y, as a function of firearms per 100 persons, x.

b. Write the equation in part (a) in slope-intercept form. Use function notation.

c. France has 32 firearms per 100 persons. Use the appropriate point in the scatter plot to estimate that country's deaths per 100,000 persons.

d. Use the function from part (b) to find the number of deaths per 100,000 persons for France. Round to one decimal place. Does the function underestimate or overestimate the deaths per 100,000 persons that you estimated in part (c)? How is this shown by the line in the scatter plot?

61. In a 2010 survey of more than 200,000 freshmen at 279 colleges, only 52% rated their emotional health high or above average, a drop from 64% in 1985.

Source: UCLA Higher Education Research Institute

a. Find the slope of the line passing through the two points shown by the voice ballons at the bottom of the previous page.

b. Use your answer from part (a) to complete this statement: For each year from 1985 through 2010, the percentage of U.S. college freshmen rating their emotional health high or above average decreased by___. The rate of change was _____ per ___.

62. Find the average rate of change of $f(x) = x^2 - 4x$ from $x_1 = 5$ to $x_2 = 9$.

1.6

In Exercises 63–67, use the graph of $y = f(x)$ to graph each function g.

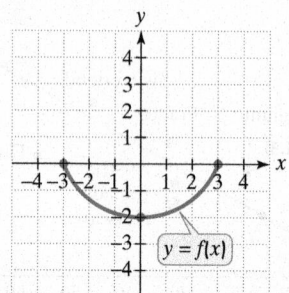

63. $g(x) = f(x + 2) + 3$
64. $g(x) = \frac{1}{2}f(x - 1)$
65. $g(x) = -f(2x)$
66. $g(x) = 2f(\frac{1}{2}x)$
67. $g(x) = -f(-x) - 1$

In Exercises 68–71, begin by graphing the standard quadratic function, $f(x) = x^2$. Then use transformations of this graph to graph the given function.

68. $g(x) = x^2 + 2$
69. $h(x) = (x + 2)^2$
70. $r(x) = -(x + 1)^2$
71. $y(x) = \frac{1}{2}(x - 1)^2 + 1$

In Exercises 72–74, begin by graphing the square root function, $f(x) = \sqrt{x}$. Then use transformations of this graph to graph the given function.

72. $g(x) = \sqrt{x + 3}$
73. $h(x) = \sqrt{3 - x}$
74. $r(x) = 2\sqrt{x + 2}$

In Exercises 75–77, begin by graphing the absolute value function, $f(x) = |x|$. Then use transformations of this graph to graph the given function.

75. $g(x) = |x + 2| - 3$
76. $h(x) = -|x - 1| + 1$
77. $r(x) = \frac{1}{2}|x + 2|$

In Exercises 78–80, begin by graphing the standard cubic function, $f(x) = x^3$. Then use transformations of this graph to graph the given function.

78. $g(x) = \frac{1}{2}(x - 1)^3$
79. $h(x) = -(x + 1)^3$
80. $r(x) = \frac{1}{4}x^3 - 1$

In Exercises 81–83, begin by graphing the cube root function, $f(x) = \sqrt[3]{x}$. Then use transformations of this graph to graph the given function.

81. $g(x) = \sqrt[3]{x + 2} - 1$
82. $h(x) = -\sqrt[3]{2x}$
83. $r(x) = -2\sqrt[3]{-x}$

1.7

In Exercises 84–89, find the domain of each function.

84. $f(x) = x^2 + 6x - 3$
85. $g(x) = \frac{4}{x - 7}$

86. $f(x) = \sqrt{4 - x}$
87. $f(x) = \dfrac{x}{x^2 + 4x - 21}$
88. $g(x) = \dfrac{\sqrt{x - 2}}{x - 5}$
89. $f(x) = \sqrt{x - 1} + \sqrt{x + 5}$

In Exercises 90–92, find $f + g, f - g, fg,$ and $\frac{f}{g}$. Determine the domain for each function.

90. $f(x) = 3x - 1, \quad g(x) = x - 5$
91. $f(x) = x^2 + x + 1, \quad g(x) = x^2 - 1$
92. $f(x) = \sqrt{x + 7}, \quad g(x) = \sqrt{x - 2}$

*In Exercises 93–94, find **a.** $(f \circ g)(x)$; **b.** $(g \circ f)(x)$; **c.** $(f \circ g)(3)$.*

93. $f(x) = x^2 + 3, \quad g(x) = 4x - 1$
94. $f(x) = \sqrt{x}, \quad g(x) = x + 1$

*In Exercises 95–96, find **a.** $(f \circ g)(x)$; **b.** the domain of $(f \circ g)$.*

95. $f(x) = \dfrac{x + 1}{x - 2}, \quad g(x) = \dfrac{1}{x}$
96. $f(x) = \sqrt{x - 1}, \quad g(x) = x + 3$

In Exercises 97–98, express the given function h as a composition of two functions f and g so that $h(x) = (f \circ g)(x)$.

97. $h(x) = (x^2 + 2x - 1)^4$
98. $h(x) = \sqrt[3]{7x + 4}$

1.8

In Exercises 99–100, find $f(g(x))$ and $g(f(x))$ and determine whether each pair of functions f and g are inverses of each other.

99. $f(x) = \dfrac{3}{5}x + \dfrac{1}{2}$ and $g(x) = \dfrac{5}{3}x - 2$

100. $f(x) = 2 - 5x$ and $g(x) = \dfrac{2 - x}{5}$

The functions in Exercises 101–103 are all one-to-one. For each function,

 a. Find an equation for $f^{-1}(x)$, the inverse function.

 b. Verify that your equation is correct by showing that $f(f^{-1}(x)) = x$ and $f^{-1}(f(x)) = x$.

101. $f(x) = 4x - 3$

102. $f(x) = 8x^3 + 1$

103. $f(x) = \dfrac{2}{x} + 5$

Which graphs in Exercises 104–107 represent functions that have inverse functions?

104.

105.

106.

107.

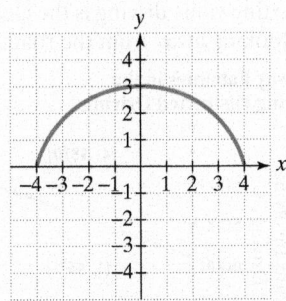

108. Use the graph of f in the figure shown to draw the graph of its inverse function.

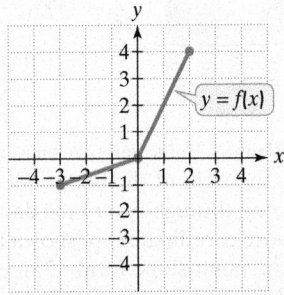

In Exercises 109–110, find an equation for $f^{-1}(x)$. Then graph f and f^{-1} in the same rectangular coordinate system.

109. $f(x) = 1 - x^2, x \geq 0$
110. $f(x) = \sqrt{x} + 1$

1.9

In Exercises 111–112, find the distance between each pair of points. If necessary, round answers to two decimal places.

111. $(-2, 3)$ and $(3, -9)$
112. $(-4, 3)$ and $(-2, 5)$

In Exercises 113–114, find the midpoint of each line segment with the given endpoints.

113. $(2, 6)$ and $(-12, 4)$
114. $(4, -6)$ and $(-15, 2)$

In Exercises 115–116, write the standard form of the equation of the circle with the given center and radius.

115. Center $(0, 0)$, $r = 3$
116. Center $(-2, 4)$, $r = 6$

In Exercises 117–119, give the center and radius of each circle and graph its equation. Use the graph to identify the relation's domain and range.

117. $x^2 + y^2 = 1$
118. $(x + 2)^2 + (y - 3)^2 = 9$
119. $x^2 + y^2 - 4x + 2y - 4 = 0$

CHAPTER 1 TEST

1. List by letter all relations that are not functions.
 a. $\{(7, 5), (8, 5), (9, 5)\}$
 b. $\{(5, 7), (5, 8), (5, 9)\}$
 c.

 d. $x^2 + y^2 = 100$
 e.

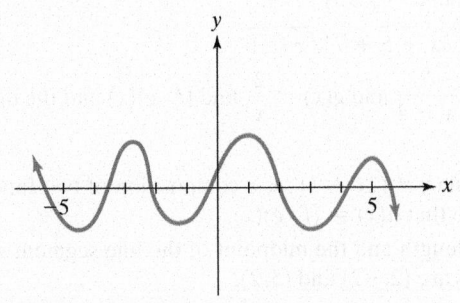

2. Use the graph of $y = f(x)$ to solve this exercise.

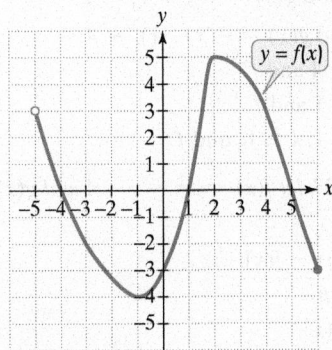

 a. What is $f(4) - f(-3)$?
 b. What is the domain of f?
 c. What is the range of f?
 d. On which interval or intervals is f increasing?
 e. On which interval or intervals is f decreasing?
 f. For what number does f have a relative maximum? What is the relative maximum?
 g. For what number does f have a relative minimum? What is the relative minimum?
 h. What are the x-intercepts?
 i. What is the y-intercept?

3. Use the graph of $y = f(x)$ to solve this exercise.

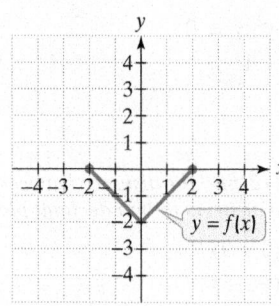

a. What are the zeros of f?

b. Find the value(s) of x for which $f(x) = -1$.

c. Find the value(s) of x for which $f(x) = -2$.

d. Is f even, odd, or neither?

e. Does f have an inverse function?

f. Is $f(0)$ a relative maximum, a relative minimum, or neither?

g. Graph $g(x) = f(x + 1) - 1$.

h. Graph $h(x) = \frac{1}{2}f(\frac{1}{2}x)$.

i. Graph $r(x) = -f(-x) + 1$.

j. Find the average rate of change of f from $x_1 = -2$ to $x_2 = 1$.

In Exercises 4–15, graph each equation in a rectangular coordinate system. If two functions are indicated, graph both in the same system. Then use your graphs to identify each relation's domain and range.

4. $x + y = 4$ **5.** $x^2 + y^2 = 4$

6. $f(x) = 4$ **7.** $f(x) = -\frac{1}{3}x + 2$

8. $(x + 2)^2 + (y - 1)^2 = 9$ **9.** $f(x) = \begin{cases} 2 & \text{if } x \le 0 \\ -1 & \text{if } x > 0 \end{cases}$

10. $x^2 + y^2 + 4x - 6y - 3 = 0$

11. $f(x) = |x|$ and $g(x) = \frac{1}{2}|x + 1| - 2$

12. $f(x) = x^2$ and $g(x) = -(x - 1)^2 + 4$

13. $f(x) = 2x - 4$ and f^{-1}

14. $f(x) = x^3 - 1$ and f^{-1}

15. $f(x) = x^2 - 1, x \ge 0,$ and f^{-1}

In Exercises 16–23, let $f(x) = x^2 - x - 4$ and $g(x) = 2x - 6$.

16. Find $f(x - 1)$.

17. Find $\dfrac{f(x + h) - f(x)}{h}$.

18. Find $(g - f)(x)$.

19. Find $\left(\dfrac{f}{g}\right)(x)$ and its domain.

20. Find $(f \circ g)(x)$.

21. Find $(g \circ f)(x)$.

22. Find $g(f(-1))$.

23. Find $f(-x)$. Is f even, odd, or neither?

In Exercises 24–25, use the given conditions to write an equation for each line in point-slope form and slope-intercept form.

24. Passing through $(2, 1)$ and $(-1, -8)$

25. Passing through $(-4, 6)$ and perpendicular to the line whose equation is $y = -\frac{1}{4}x + 5$

26. Write an equation in general form for the line passing through $(-7, -10)$ and parallel to the line whose equation is $4x + 2y - 5 = 0$.

27. Studies show that texting while driving is as risky as driving with a 0.08 blood alcohol level, the standard for drunk driving. The bar graph shows the number of fatalities in the United States involving distracted driving from 2004 through 2008. Although the distracted category involves such activities as talking on cellphones, conversing with passengers, and eating, experts at the National Highway Traffic Safety Administration claim that texting while driving is the clearest menace because it requires looking away from the road.

Number of Highway Fatalities in the United States Involving Distracted Driving

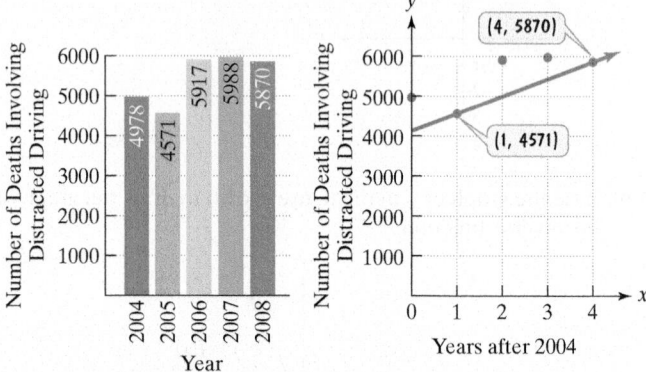

Source: National Highway Traffic Safety Administration

a. Shown to the right of the bar graph is a scatter plot with a line passing through two of the data points. Use the two points whose coordinates are shown by the voice balloons to write the point-slope form of an equation that models the number of highway fatalities involving distracted driving, y, in the United States x years after 2004.

b. Write the equation from part (a) in slope-intercept form. Use function notation.

c. In 2010, surveys showed overwhelming public support to ban texting while driving, although at that time only 19 states and Washington, D.C., outlawed the practice. Without additional laws that penalize texting drivers, use the linear function you obtained from part (b) to project the number of fatalities in the United States in 2014 involving distracted driving.

28. Find the average rate of change of $f(x) = 3x^2 - 5$ from $x_1 = 6$ to $x_2 = 10$.

29. If $g(x) = \begin{cases} \sqrt{x - 3} & \text{if } x \ge 3 \\ 3 - x & \text{if } x < 3 \end{cases}$, find $g(-1)$ and $g(7)$.

In Exercises 30–31, find the domain of each function.

30. $f(x) = \dfrac{3}{x + 5} + \dfrac{7}{x - 1}$

31. $f(x) = 3\sqrt{x + 5} + 7\sqrt{x - 1}$

32. If $f(x) = \dfrac{7}{x - 4}$ and $g(x) = \dfrac{2}{x}$, find $(f \circ g)(x)$ and the domain of $f \circ g$.

33. Express $h(x) = (2x + 3)^7$ as a composition of two functions f and g so that $h(x) = (f \circ g)(x)$.

34. Find the length and the midpoint of the line segment whose endpoints are $(2, -2)$ and $(5, 2)$.

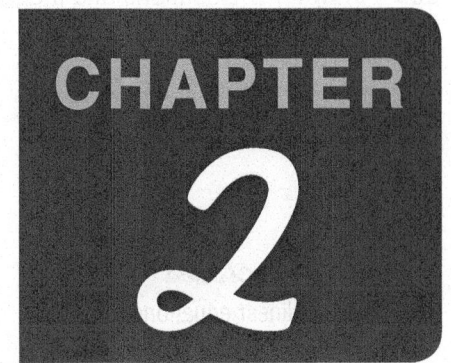

EQUATIONS AND INEQUALITIES

CHAPTER 2

The belief that humor and laughter can have positive benefits on our lives is not new. The Bible tells us, "A merry heart doeth good like a medicine, but a broken spirit drieth the bones." (Proverbs 17:22)

Some random humor factoids:

- The average adult laughs 15 times each day. (Newhouse News Service)
- Forty-six percent of people who are telling a joke laugh more than the people they are telling it to. (*U.S. News and World Report*)
- Eighty percent of adult laughter does not occur in response to jokes or funny situations. (*Independent*)
- Algebra can be used to model the influence that humor plays in our responses to negative life events. (Bob Blitzer, *College Algebra*)

That last tidbit that your author threw into the list is true. Based on our sense of humor, there is actually a formula that predicts how we will respond to difficult life events.

Functions can be used to explain what is happening in the present and to make predictions about what might occur in the future. In this chapter, you will learn to use functions in new ways that will help you to recognize patterns, logic, and order in a world that can appear chaotic to the untrained eye.

HERE'S WHERE YOU'LL FIND THESE APPLICATIONS:

A function that includes sense of humor as a variable is developed in Example 8 of Section 2.1.

SECTION 2.1 Linear Equations and Rational Equations

Objectives

① Solve linear equations in one variable.

② Solve linear equations containing fractions.

③ Solve rational equations with variables in the denominators.

④ Recognize identities, conditional equations, and inconsistent equations.

⑤ Solve applied problems using linear functions.

Sense of Humor and Depression

FIGURE 2.1

Source: Stephen Davis and Joseph Palladino, *Psychology,* 5th Edition, Prentice Hall, 2007.

The belief that humor and laughter can have positive benefits on our lives is not new. The graphs in **Figure 2.1** indicate that persons with a low sense of humor have higher levels of depression in response to negative life events than those with a high sense of humor. These graphs can be modeled by the following functions:

Low-Humor Group

$$D(x) = \frac{10}{9}x + \frac{53}{9}$$

High-Humor Group

$$D(x) = \frac{1}{9}x + \frac{26}{9}.$$

In each function, x represents the intensity of a negative life event (from 1, low, to 10, high) and $D(x)$ is the average level of depression in response to that event. The slope of the low-humor model, $\frac{10}{9}$, is greater than the slope of the high-humor model, $\frac{1}{9}$. This indicates a greater rate of increase for the low-humor group in response to negative life events per unit change in the intensity of that event.

Suppose that the low-humor group averages a level of depression of 10 in response to a negative life event. We can determine the intensity of that event by substituting 10 for $D(x)$ in the low-humor model and solving for x:

$$D(x) = \frac{10}{9}x + \frac{53}{9} = 10.$$

Setting the linear function equal to a constant results in a *linear equation*. In this section, we will study how to solve linear equations.

① Solve linear equations in one variable.

Solving Linear Equations in One Variable

We know that a linear function in x can be expressed as $f(x) = mx + b$. A **linear equation in x** can be written in the form $mx + b = 0$, with $m \neq 0$.

A linear equation $mx + b = 0$, with $m \neq 0$, has exactly one solution, or root. Recall that the zeros of a function, f, are the x-values for which $f(x) = 0$. For this reason, the solution, or root, of a linear equation $mx + b = 0$ is also called the zero of the linear function $f(x) = mx + b$.

Two or more linear equations with the same solution are called **equivalent equations**. To generate equivalent equations, we will use the following principles:

Generating Equivalent Equations

An equation can be transformed into an equivalent equation by one or more of the following operations:

Operation	Example
1. Simplify an expression by removing grouping symbols and combining like terms.	• $3(x - 6) = 6x - x$ $3x - 18 = 5x$
2. Add (or subtract) the same real number or variable expression on *both* sides of the equation.	• $3x - 18 = 5x$ $3x - 18 - 3x = 5x - 3x$ Subtract $3x$ from both sides of the equation. $-18 = 2x$
3. Multiply (or divide) by the same *nonzero* quantity on *both* sides of the equation.	• $-18 = 2x$ $\dfrac{-18}{2} = \dfrac{2x}{2}$ Divide both sides of the equation by 2. $-9 = x$
4. Interchange the two sides of the equation.	• $-9 = x$ $x = -9$

If you look closely at the equations in the box, you will notice that we have solved the equation $3(x - 6) = 6x - x$. The final equation, $x = -9$, with x isolated on the left side, shows that $\{-9\}$ is the solution set. The idea in solving a linear equation is to get the variable by itself on one side of the equal sign and a number by itself on the other side.

Questions about linear functions often lead to linear equations. For example, consider the linear functions

$$f(x) = 2(x - 3) - 17 \text{ and } g(x) = 13 - 3(x + 2).$$

Suppose we are interested in finding all values of x for which $f(x) = g(x)$. One way to do this is to set the expressions that define f and g equal to each other:

$$2(x - 3) - 17 = 13 - 3(x + 2).$$

In Example 1, we will solve this linear equation. Can you sense that a number of steps are necessary to determine the solution?

Here is a step-by-step procedure for solving a linear equation in one variable. Not all of these steps are necessary to solve every equation.

Solving a Linear Equation

1. Simplify the algebraic expression on each side by removing grouping symbols and combining like terms.
2. Collect all the variable terms on one side and all the numbers, or constant terms, on the other side.
3. Isolate the variable and solve.
4. Check the proposed solution in the original equation.

EXAMPLE 1 Solving a Linear Equation

Solve and check: $2(x - 3) - 17 = 13 - 3(x + 2)$.

SOLUTION

Step 1 Simplify the algebraic expression on each side.

> Do not begin with 13 − 3. Multiplication (the distributive property) is applied before subtraction.

$$2(x - 3) - 17 = 13 - 3(x + 2)$$ This is the given equation.
$$2x - 6 - 17 = 13 - 3x - 6$$ Use the distributive property.
$$2x - 23 = -3x + 7$$ Combine like terms.

Step 2 Collect variable terms on one side and constant terms on the other side. We will collect variable terms on the left by adding $3x$ to both sides. We will collect the numbers on the right by adding 23 to both sides.

$$2x - 23 + 3x = -3x + 7 + 3x$$ Add 3x to both sides.
$$5x - 23 = 7$$ Simplify: 2x + 3x = 5x.
$$5x - 23 + 23 = 7 + 23$$ Add 23 to both sides.
$$5x = 30$$ Simplify.

Step 3 Isolate the variable and solve. We isolate the variable, x, by dividing both sides of $5x = 30$ by 5.

$$\frac{5x}{5} = \frac{30}{5}$$ Divide both sides by 5.

$$x = 6$$ Simplify.

Step 4 Check the proposed solution in the original equation. Substitute 6 for x in the original equation.

$$2(x - 3) - 17 = 13 - 3(x + 2)$$ This is the original equation.
$$2(6 - 3) - 17 \stackrel{?}{=} 13 - 3(6 + 2)$$ Substitute 6 for x.
$$2(3) - 17 \stackrel{?}{=} 13 - 3(8)$$ Simplify inside parentheses.
$$6 - 17 \stackrel{?}{=} 13 - 24$$ Multiply.
$$-11 = -11$$ Subtract.

The true statement $-11 = -11$ verifies that the solution set is $\{6\}$. • • •

DISCOVERY

Solve the equation in Example 1 by collecting terms with the variable on the right and numerical terms on the left. What do you observe?

TECHNOLOGY

Numeric and Graphic Connections

In many algebraic situations, technology provides numeric and visual insights into problem solving. For example, you can use a graphing utility to check the solution of a linear equation, giving numeric and geometric meaning to the solution. Enter the function defined by each side of the equation separately under y_1 and y_2. Then use the table or the graphs to locate the x-value for which the function values, or y-values, are the same. This x-value is the solution.

Let's verify our work in Example 1 and show that 6 is the solution of

$$2(x - 3) - 17 = 13 - 3(x + 2).$$

Enter $y_1 = 2(x - 3) - 17$ Enter $y_2 = 13 - 3(x + 2)$
in the $\boxed{y=}$ screen. in the $\boxed{y=}$ screen.

Numeric Check

Display a table for y_1 and y_2.

When $x = 6$, y_1 and y_2 have the same value, namely -11. This verifies that 6 is the solution of $2(x - 3) - 17 = 13 - 3(x + 2)$.

Graphic Check

Display graphs for y_1 and y_2 and use the intersection feature. The solution is the x-coordinate of the intersection point.

$[-2, 16, 1]$ by $[-14, 8, 1]$

Choose a large enough viewing rectangle so that you can see the intersection point.

⊘ **Check Point 1** Solve and check: $4(2x + 1) = 29 + 3(2x - 5)$.

2 Solve linear equations containing fractions.

Linear Equations with Fractions

Equations are easier to solve when they do not contain fractions. How do we remove fractions from an equation? We begin by multiplying both sides of the equation by the least common denominator of any fractions in the equation. The least common denominator is the smallest number that all denominators will divide into. Multiplying every term on both sides of the equation by the least common denominator will eliminate the fractions in the equation. Example 2 shows how we "clear an equation of fractions."

EXAMPLE 2 Solving a Linear Equation Involving Fractions

Solve and check: $\dfrac{x + 2}{4} - \dfrac{x - 1}{3} = 2$.

SOLUTION

The fractional terms have denominators of 4 and 3. The smallest number that is divisible by 4 and 3 is 12. We begin by multiplying both sides of the equation by 12, the least common denominator.

$$\frac{x + 2}{4} - \frac{x - 1}{3} = 2 \qquad \text{This is the given equation.}$$

$$12\left(\frac{x + 2}{4} - \frac{x - 1}{3}\right) = 12 \cdot 2 \qquad \text{Multiply both sides by 12.}$$

$$12\left(\frac{x + 2}{4}\right) - 12\left(\frac{x - 1}{3}\right) = 24 \qquad \begin{array}{l}\text{Use the distributive property and}\\ \text{multiply each term on the left by 12.}\end{array}$$

$$\overset{3}{\cancel{12}}\left(\frac{x + 2}{\underset{1}{\cancel{4}}}\right) - \overset{4}{\cancel{12}}\left(\frac{x - 1}{\underset{1}{\cancel{3}}}\right) = 24 \qquad \begin{array}{l}\text{Divide out common factors in each}\\ \text{multiplication on the left.}\end{array}$$

$$3(x + 2) - 4(x - 1) = 24 \qquad \text{The fractions are now cleared.}$$

$$3x + 6 - 4x + 4 = 24 \qquad \text{Use the distributive property.}$$

$$-x + 10 = 24 \qquad \begin{array}{l}\text{Combine like terms: } 3x - 4x = -x\\ \text{and } 6 + 4 = 10.\end{array}$$

$$-x + 10 - 10 = 24 - 10 \qquad \text{Subtract 10 from both sides.}$$

$$-x = 14 \qquad \text{Simplify.}$$

We're not finished. A negative sign should not precede the variable.

Isolate x by multiplying or dividing both sides of $-x = 14$ by -1.

$$\frac{-x}{-1} = \frac{14}{-1} \qquad \text{Divide both sides by } -1.$$

$$x = -14 \qquad \text{Simplify.}$$

Check the proposed solution. Substitute -14 for x in the original equation. You should obtain $2 = 2$. This true statement verifies that the solution set is $\{-14\}$. • • •

☑ Check Point **2** Solve and check: $\dfrac{x - 3}{4} = \dfrac{5}{14} - \dfrac{x + 5}{7}$.

③ Solve rational equations with variables in the denominators.

Rational Equations

A **rational equation** is an equation containing one or more rational expressions. In Example 2, we solved a rational equation with constants in the denominators. This rational equation was a linear equation. Now, let's consider a rational equation such as

$$\frac{1}{x} = \frac{1}{5} + \frac{3}{2x}.$$

Can you see how this rational equation differs from the rational equation that we solved earlier? The variable, x, appears in two of the denominators. Although this rational equation is not a linear equation, the solution procedure still involves multiplying each side by the least common denominator. However, we must avoid any values of the variable that make a denominator zero. For example, examine the denominators in the equation

$$\frac{1}{x} = \frac{1}{5} + \frac{3}{2x}.$$

This denominator would equal zero if $x = 0$.　　This denominator would equal zero if $x = 0$.

We see that x cannot equal zero. With this in mind, let's solve the equation.

EXAMPLE 3　Solving a Rational Equation

Solve: $\dfrac{1}{x} = \dfrac{1}{5} + \dfrac{3}{2x}$.

SOLUTION

The denominators are x, 5, and $2x$. The least common denominator is $10x$. We begin by multiplying both sides of the equation by $10x$. We will also write the restriction that x cannot equal zero to the right of the equation.

$$\frac{1}{x} = \frac{1}{5} + \frac{3}{2x}, \quad x \neq 0 \qquad \text{This is the given equation.}$$

$$10x \cdot \frac{1}{x} = 10x\left(\frac{1}{5} + \frac{3}{2x}\right) \qquad \text{Multiply both sides by 10x.}$$

$$10x \cdot \frac{1}{x} = 10x \cdot \frac{1}{5} + 10x \cdot \frac{3}{2x} \qquad \text{Use the distributive property. Be sure to multiply each term by 10x.}$$

$$10 \overset{}{\cancel{x}} \cdot \frac{1}{\cancel{x}} = \overset{2}{\cancel{10}} \, x \cdot \frac{1}{\cancel{5}} + \overset{5}{\cancel{10}} \, \cancel{x} \cdot \frac{3}{\cancel{2} \, \cancel{x}} \qquad \text{Divide out common factors in the multiplications.}$$

$$10 = 2x + 15 \qquad \text{Complete the multiplications.}$$

Observe that the resulting equation,

$$10 = 2x + 15,$$

is now cleared of fractions. With the variable term, $2x$, already on the right, we will collect constant terms on the left by subtracting 15 from both sides.

$$10 - 15 = 2x + 15 - 15 \qquad \text{Subtract 15 from both sides.}$$
$$-5 = 2x \qquad \text{Simplify.}$$

Finally, we isolate the variable, x, in $-5 = 2x$ by dividing both sides by 2.

$$\frac{-5}{2} = \frac{2x}{2} \qquad \text{Divide both sides by 2.}$$

$$-\frac{5}{2} = x \qquad \text{Simplify.}$$

We check our solution by substituting $-\frac{5}{2}$ into the original equation or by using a calculator. With a calculator, evaluate each side of the equation for $x = -\frac{5}{2}$, or for $x = -2.5$. Note that the original restriction that $x \neq 0$ is met. The solution set is $\left\{-\frac{5}{2}\right\}$. • • •

⊘ **Check Point 3** Solve: $\dfrac{5}{2x} = \dfrac{17}{18} - \dfrac{1}{3x}$.

EXAMPLE 4 Solving a Rational Equation

Solve: $\dfrac{x}{x-3} = \dfrac{3}{x-3} + 9$.

SOLUTION

We must avoid any values of the variable x that make a denominator zero.

$$\frac{x}{x-3} = \frac{3}{x-3} + 9$$

These denominators are zero if $x = 3$.

We see that x cannot equal 3. With denominators of $x - 3, x - 3$, and 1, the least common denominator is $x - 3$. We multiply both sides of the equation by $x - 3$. We also write the restriction that x cannot equal 3 to the right of the equation.

$$\frac{x}{x-3} = \frac{3}{x-3} + 9, \quad x \neq 3 \qquad \text{This is the given equation.}$$

$$(x-3) \cdot \frac{x}{x-3} = (x-3)\left(\frac{3}{x-3} + 9\right) \qquad \text{Multiply both sides by } x-3.$$

$$(x-3) \cdot \frac{x}{x-3} = (x-3) \cdot \frac{3}{x-3} + (x-3) \cdot 9 \qquad \text{Use the distributive property.}$$

$$\cancel{(x-3)} \cdot \frac{x}{\cancel{x-3}} = \cancel{(x-3)} \cdot \frac{3}{\cancel{x-3}} + 9(x-3) \qquad \text{Divide out common factors in two of the multiplications.}$$

$$x = 3 + 9(x-3) \qquad \text{Simplify.}$$

The resulting equation, $x = 3 + 9(x - 3)$, is cleared of fractions. We now solve for x.

$$x = 3 + 9x - 27 \qquad \text{Use the distributive property.}$$
$$x = 9x - 24 \qquad \text{Combine numerical terms.}$$
$$x - 9x = 9x - 24 - 9x \qquad \text{Subtract } 9x \text{ from both sides.}$$
$$-8x = -24 \qquad \text{Simplify.}$$
$$\frac{-8x}{-8} = \frac{-24}{-8} \qquad \text{Divide both sides by } -8.$$
$$x = 3 \qquad \text{Simplify.}$$

The proposed solution, 3, is *not* a solution because of the restriction that $x \neq 3$. There is *no solution to this equation.* The solution set for this equation contains no elements. The solution set is \varnothing, the empty set. • • •

GREAT QUESTION!

Cut to the chase: When do I get rid of proposed solutions in rational equations?

Reject any proposed solution that causes any denominator in a rational equation to equal 0.

⊘ Check Point **4** Solve: $\dfrac{x}{x-2} = \dfrac{2}{x-2} - \dfrac{2}{3}$.

EXAMPLE 5 Solving a Rational Equation to Answer a Question about Functions

Consider the functions

$$f(x) = \frac{3}{x+6} + \frac{1}{x-2} \quad \text{and} \quad g(x) = \frac{4}{x^2 + 4x - 12}.$$

Find all values of x for which $f(x) = g(x)$.

SOLUTION

Because we are interested in one or more inputs, x, that cause $f(x)$ and $g(x)$ to be equal, we set the expressions that define f and g equal to each other:

$$\frac{3}{x+6} + \frac{1}{x-2} = \frac{4}{x^2 + 4x - 12}.$$

To identify values of x that make denominators zero, let's factor $x^2 + 4x - 12$, the denominator on the right. This factorization, $x^2 + 4x - 12 = (x + 6)(x - 2)$, is also necessary in identifying the least common denominator.

$$\frac{3}{x+6} + \frac{1}{x-2} = \frac{4}{(x+6)(x-2)}$$

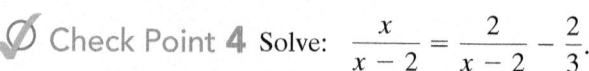

| This denominator | This denominator | This denominator is zero |
| is zero if $x = -6$. | is zero if $x = 2$. | if $x = -6$ or $x = 2$. |

We see that x cannot equal -6 or 2. The least common denominator is $(x + 6)(x - 2)$.

$$\frac{3}{x+6} + \frac{1}{x-2} = \frac{4}{(x+6)(x-2)}, \quad x \neq -6, \quad x \neq 2 \qquad \text{This is the given equation with a denominator factored.}$$

$$(x+6)(x-2)\left(\frac{3}{x+6} + \frac{1}{x-2}\right) = (x+6)(x-2) \cdot \frac{4}{(x+6)(x-2)} \qquad \text{Multiply both sides by } (x+6)(x-2), \text{ the LCD.}$$

$$(x+6)(x-2) \cdot \frac{3}{x+6} + (x+6)(x-2) \cdot \frac{1}{x-2} = (x+6)(x-2) \cdot \frac{4}{(x+6)(x-2)} \qquad \text{Use the distributive property and divide out common factors.}$$

$$3(x-2) + 1(x+6) = 4 \qquad \text{Simplify. This equation is cleared of fractions.}$$

$$3x - 6 + x + 6 = 4$$

Use the distributive property.

$$4x = 4$$

Combine like terms.

$$\frac{4x}{4} = \frac{4}{4}$$

Divide both sides by 4.

$$x = 1$$

Simplify. This is not part of the restriction that $x \neq -6$ and $x \neq 2$.

The value of x for which $f(x) = g(x)$ is 1.

Check

Is $f(x) = g(x)$ when $x = 1$? We use the given functions

$$f(x) = \frac{3}{x+6} + \frac{1}{x-2} \quad \text{and} \quad g(x) = \frac{4}{x^2 + 4x - 12}$$

to answer the question.

Using $f(x) = \dfrac{3}{x+6} + \dfrac{1}{x-2}$ and $g(x) = \dfrac{4}{x^2 + 4x - 12}$, we can check 1 by hand or with a graphing utility.

Checking by Hand

Substitute 1 for x in $f(x)$ and $g(x)$.

$$f(1) = \frac{3}{1+6} + \frac{1}{1-2} = \frac{3}{7} + \frac{1}{-1}$$

$$= \frac{3}{7} - 1 = \frac{3}{7} - \frac{7}{7} = -\frac{4}{7}$$

$$g(1) = \frac{4}{1^2 + 4 \cdot 1 - 12} = \frac{4}{1 + 4 - 12}$$

$$= \frac{4}{-7} = -\frac{4}{7}$$

When $x = 1$, f and g have the same value, namely, $-\dfrac{4}{7}$.

Checking with a Graphing Utility

Display a table showing function values for f and g. Enter the functions as y_1 and y_2, and be careful with parentheses.

$$\boxed{y_1 = 3 \div (x + 6) + 1 \div (x - 2)} \quad \boxed{y_2 = 4 \div (x^2 + 4x - 12)}$$

X	Y1	Y2
-3	.8	-.2667
-2	.5	-.25
-1	.26667	-.2667
0	0	-.3333
1	-.5714	-.5714
2	ERROR	ERROR
3	1.3333	.44444

X=1

No matter how far up or down you scroll, $y_1 = y_2$ only when $x = 1$.

• • •

⊘ **Check Point 5** Consider the functions

$$f(x) = \frac{1}{x+4} + \frac{1}{x-4} \quad \text{and} \quad g(x) = \frac{22}{x^2 - 16}.$$

Find all values of x for which $f(x) = g(x)$ and check.

④ Recognize identities, conditional equations, and inconsistent equations.

Types of Equations

Equations can be placed into categories that depend on their solution sets.

An equation that is true for all real numbers for which both sides are defined is called an **identity**. An example of an identity is

$$x + 3 = x + 2 + 1.$$

Every number plus 3 is equal to that number plus 2 plus 1. Therefore, the solution set to this equation is the set of all real numbers, expressed as $\{x \mid x \text{ is a real number}\}$. Another example of an identity is

$$\frac{2x}{x} = 2.$$

GREAT QUESTION!

What's the bottom line on all the vocabulary associated with types of equations?

If you are concerned by the vocabulary of equation types, keep in mind that there are three possible situations. We can state these situations informally as follows:

1. x = a real number

 conditional equation

2. x = all real numbers

 identity

3. x = no real numbers.

 inconsistent equation

Because division by 0 is undefined, the equation $\dfrac{2x}{x} = 2$ is true for all real number values of x except 0. The solution set is the set of nonzero real numbers, expressed as $\{x \mid x \text{ is a real number and } x \neq 0\}$.

An equation that is not an identity, but that is true for at least one real number, is called a **conditional equation**. The equation $2x + 3 = 17$ is an example of a conditional equation. The equation is not an identity and is true only if x is 7.

An **inconsistent equation** is an equation that is not true for even one real number. An example of an inconsistent equation is

$$x = x + 7.$$

There is no number that is equal to itself plus 7. The equation's solution set is \varnothing, the empty set. Some inconsistent equations are less obvious than this. Consider the equation in Example 4,

$$\frac{x}{x - 3} = \frac{3}{x - 3} + 9.$$

This equation is not true for any real number and has no solution. Thus, it is inconsistent.

If you eliminate the variable while solving an equation, a true statement such as $6 = 6$ or a false statement such as $2 = 3$ will be the result. **If a true statement results, the equation is an identity that is true for all real numbers. If a false statement results, the equation is an inconsistent equation with no solution.**

EXAMPLE 6 Categorizing an Equation

Solve and determine whether the equation

$$2(x + 1) = 2x + 3$$

is an identity, a conditional equation, or an inconsistent equation.

SOLUTION

Begin by applying the distributive property on the left side. We obtain

$$2x + 2 = 2x + 3.$$

Does something look strange? Can doubling a number and increasing the product by 2 give the same result as doubling the same number and increasing the product by 3? No. Let's continue solving the equation by subtracting $2x$ from both sides.

$$2x - 2x + 2 = 2x - 2x + 3$$

Keep reading. **2 = 3** is not the solution. $2 = 3$

The original equation, $2(x + 1) = 2x + 3$, is equivalent to the statement $2 = 3$, which is false for every value of x. The equation is inconsistent and has no solution. The solution set is \varnothing, the empty set. • • •

TECHNOLOGY

Graphic Connections

How can technology visually reinforce the fact that the equation

$$2(x + 1) = 2x + 3$$

has no solution? Enter the linear functions $y_1 = 2(x + 1)$ and $y_2 = 2x + 3$. The lines have the same slope, 2, but different y-intercepts, 2 and 3. Thus, the graphs of y_1 and y_2 are parallel lines with no intersection point. This supports our conclusion that $2(x + 1) = 2x + 3$ is an inconsistent equation with no solution.

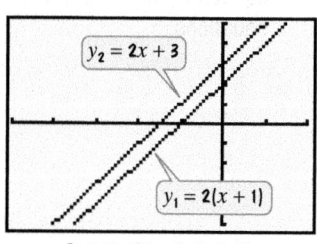

$[-5, 2, 1]$ by $[-5, 5, 1]$

⊘ Check Point **6** Solve and determine whether the equation
$$4x - 7 = 4(x - 1) + 3$$
is an identity, a conditional equation, or an inconsistent equation.

EXAMPLE 7 Categorizing an Equation

Solve and determine whether the equation
$$4x + 6 = 6(x + 1) - 2x$$
is an identity, a conditional equation, or an inconsistent equation.

SOLUTION

$$4x + 6 = 6(x + 1) - 2x$$ This is the given equation.

$$4x + 6 = 6x + 6 - 2x$$ Apply the distributive property on the right side.

$$4x + 6 = 4x + 6$$ Combine like terms on the right side: $6x - 2x = 4x$.

Can you see that the equation $4x + 6 = 4x + 6$ is true for every value of x? Let's continue solving the equation by subtracting $4x$ from both sides.

$$4x - 4x + 6 = 4x - 4x + 6$$

Keep reading. $6 = 6$ is not the solution. $$6 = 6$$

The original equation is equivalent to the statement $6 = 6$, which is true for every value of x. The equation is an identity, and all real numbers are solutions. You can express this by using the following notation:

$$\{x | x \text{ is a real number}\}.$$ • • •

TECHNOLOGY

Numeric Connections

A graphing utility's ⎢TABLE⎢ feature can be used to numerically verify that the solution set of
$$4x + 6 = 6(x + 1) - 2x$$
is the set of all real numbers.

No matter how far up or down you scroll, y_1 and y_2 have the same corresponding value for each x-value. The expressions $4x + 6$ and $6(x + 1) - 2x$ are always equal.

⊘ Check Point **7** Solve and determine whether the equation
$$7x + 9 = 9(x + 1) - 2x$$
is an identity, a conditional equation, or an inconsistent equation.

⑤ Solve applied problems using linear functions.

Applications

Our next example shows how the procedure for solving linear equations can be used to find the value of a variable in a mathematical model.

Sense of Humor and Depression

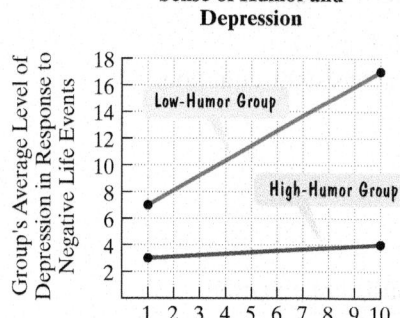

FIGURE 2.1 (repeated)

EXAMPLE 8 An Application: Responding to Negative Life Events

In the section opener, we introduced line graphs, repeated in **Figure 2.1**, indicating that persons with a low sense of humor have higher levels of depression in response to negative life events than those with a high sense of humor. These graphs can be modeled by the following linear functions:

Low-Humor Group

$$D(x) = \frac{10}{9}x + \frac{53}{9}$$

High-Humor Group

$$D(x) = \frac{1}{9}x + \frac{26}{9}.$$

In each function, x represents the intensity of a negative life event (from 1, low, to 10, high) and $D(x)$ is the average level of depression in response to that event. If the high-humor group averages a level of depression of 3.5, or $\frac{7}{2}$, in response to a negative life event, what is the intensity of that event? How is the solution shown on the red line graph in **Figure 2.1**?

SOLUTION

We are interested in the intensity of a negative life event with an average level of depression of $\frac{7}{2}$ for the high-humor group. We substitute $\frac{7}{2}$ for $D(x)$ in the high-humor model and solve for x, the intensity of the negative life event.

$$D(x) = \frac{1}{9}x + \frac{26}{9}$$

This is the given function for the high-humor group.

$$\frac{7}{2} = \frac{1}{9}x + \frac{26}{9}$$

Replace $D(x)$ with $\frac{7}{2}$.

$$18 \cdot \frac{7}{2} = 18\left(\frac{1}{9}x + \frac{26}{9}\right)$$

Multiply both sides by 18, the least common denominator.

$$18 \cdot \frac{7}{2} = 18 \cdot \frac{1}{9}x + 18 \cdot \frac{26}{9}$$

Use the distributive property.

$$\overset{9}{\cancel{18}} \cdot \frac{7}{\underset{1}{\cancel{2}}} = \overset{2}{\cancel{18}} \cdot \frac{1}{\underset{1}{\cancel{9}}}x + \overset{2}{\cancel{18}} \cdot \frac{26}{\underset{1}{\cancel{9}}}$$

Divide out common factors in the multiplications.

$$63 = 2x + 52$$

Complete the multiplications. The fractions are now cleared.

$$63 - 52 = 2x + 52 - 52$$

Subtract 52 from both sides to get constants on the left.

$$11 = 2x$$

Simplify.

$$\frac{11}{2} = \frac{2x}{2}$$

Divide both sides by 2.

$$\frac{11}{2} = x$$

Simplify.

The function indicates that if the high-humor group averages a level of depression of 3.5 in response to a negative life event, the intensity of that event is $\frac{11}{2}$, or 5.5. This is illustrated by the point $(5.5, 3.5)$ on the line graph for the high-humor group in **Figure 2.2**.

FIGURE 2.2

Check Point **8** Use the model for the low-humor group given in Example 8 to solve this problem. If the low-humor group averages a level of depression of 10 in response to a negative life event, what is the intensity of that event? How is the solution shown on the blue line graph in **Figure 2.1**?

CONCEPT AND VOCABULARY CHECK

Fill in each blank so that the resulting statement is true.

1. An equation in the form $mx + b = 0, m \neq 0$, such as $3x + 17 = 0$, is called a/an _____ equation in one variable.
2. Two or more equations that have the same solution set are called _____ equations.
3. The first step in solving $7 + 3(x - 2) = 2x + 10$ is to _____.
4. The fractions in the equation

$$\frac{x}{4} = 2 + \frac{x - 3}{3}$$

 can be eliminated by multiplying both sides by the _____ of $\frac{x}{4}$ and $\frac{x-3}{3}$, which is _____.
5. We reject any proposed solution of a rational equation that causes a denominator to equal _____.
6. The first step in solving

$$\frac{4}{x} + \frac{1}{2} = \frac{5}{x}$$

 is to multiply both sides by _____.
7. The first step in solving

$$\frac{x - 6}{x + 5} = \frac{x - 3}{x + 1}$$

 is to multiply both sides by _____.

8. The restrictions on the variable in the rational equation

$$\frac{1}{x - 2} - \frac{2}{x + 4} = \frac{2x - 1}{x^2 + 2x - 8}$$

 are _____ and _____.
9. $\dfrac{5}{x + 4} + \dfrac{3}{x + 3} = \dfrac{12x + 9}{(x + 4)(x + 3)}$

$$(x + 4)(x + 3)\left(\frac{5}{x + 4} + \frac{3}{x + 3}\right)$$
$$= (x + 4)(x + 3)\left(\frac{12x + 9}{(x + 4)(x + 3)}\right)$$

 The resulting equation cleared of fractions is _____.

10. An equation that is true for all real numbers for which both sides are defined is called a/an _____.
11. An equation that is not true for even one real number is called a/an _____ equation.

EXERCISE SET 2.1

Practice Exercises

In Exercises 1–16, solve and check each linear equation.

1. $7x - 5 = 72$
2. $6x - 3 = 63$
3. $11x - (6x - 5) = 40$
4. $5x - (2x - 10) = 35$
5. $2x - 7 = 6 + x$
6. $3x + 5 = 2x + 13$
7. $7x + 4 = x + 16$
8. $13x + 14 = 12x - 5$
9. $3(x - 2) + 7 = 2(x + 5)$
10. $2(x - 1) + 3 = x - 3(x + 1)$
11. $3(x - 4) - 4(x - 3) = x + 3 - (x - 2)$
12. $2 - (7x + 5) = 13 - 3x$
13. $16 = 3(x - 1) - (x - 7)$
14. $5x - (2x + 2) = x + (3x - 5)$
15. $25 - [2 + 5y - 3(y + 2)] =$
 $-3(2y - 5) - [5(y - 1) - 3y + 3]$
16. $45 - [4 - 2y - 4(y + 7)] =$
 $-4(1 + 3y) - [4 - 3(y + 2) - 2(2y - 5)]$

Exercises 17–30 contain linear equations with constants in denominators. Solve each equation.

17. $\dfrac{x}{3} = \dfrac{x}{2} - 2$
18. $\dfrac{x}{5} = \dfrac{x}{6} + 1$
19. $20 - \dfrac{x}{3} = \dfrac{x}{2}$
20. $\dfrac{x}{5} - \dfrac{1}{2} = \dfrac{x}{6}$
21. $\dfrac{3x}{5} = \dfrac{2x}{3} + 1$
22. $\dfrac{x}{2} = \dfrac{3x}{4} + 5$
23. $\dfrac{3x}{5} - x = \dfrac{x}{10} - \dfrac{5}{2}$
24. $2x - \dfrac{2x}{7} = \dfrac{x}{2} + \dfrac{17}{2}$

25. $\dfrac{x + 3}{6} = \dfrac{3}{8} + \dfrac{x - 5}{4}$
26. $\dfrac{x + 1}{4} = \dfrac{1}{6} + \dfrac{2 - x}{3}$
27. $\dfrac{x}{4} = 2 + \dfrac{x - 3}{3}$
28. $5 + \dfrac{x - 2}{3} = \dfrac{x + 3}{8}$
29. $\dfrac{x + 1}{3} = 5 - \dfrac{x + 2}{7}$
30. $\dfrac{3x}{5} - \dfrac{x - 3}{2} = \dfrac{x + 2}{3}$

*Exercises 31–50 contain rational equations with variables in denominators. For each equation, **a.** write the value or values of the variable that make a denominator zero. These are the restrictions on the variable. **b.** Keeping the restrictions in mind, solve the equation.*

31. $\dfrac{4}{x} = \dfrac{5}{2x} + 3$
32. $\dfrac{5}{x} = \dfrac{10}{3x} + 4$
33. $\dfrac{2}{x} + 3 = \dfrac{5}{2x} + \dfrac{13}{4}$
34. $\dfrac{7}{2x} - \dfrac{5}{3x} = \dfrac{22}{3}$
35. $\dfrac{2}{3x} + \dfrac{1}{4} = \dfrac{11}{6x} - \dfrac{1}{3}$
36. $\dfrac{5}{2x} - \dfrac{8}{9} = \dfrac{1}{18} - \dfrac{1}{3x}$
37. $\dfrac{x - 2}{2x} + 1 = \dfrac{x + 1}{x}$
38. $\dfrac{4}{x} = \dfrac{9}{5} - \dfrac{7x - 4}{5x}$
39. $\dfrac{1}{x - 1} + 5 = \dfrac{11}{x - 1}$
40. $\dfrac{3}{x + 4} - 7 = \dfrac{-4}{x + 4}$
41. $\dfrac{8x}{x + 1} = 4 - \dfrac{8}{x + 1}$
42. $\dfrac{2}{x - 2} = \dfrac{x}{x - 2} - 2$
43. $\dfrac{3}{2x - 2} + \dfrac{1}{2} = \dfrac{2}{x - 1}$

44. $\dfrac{3}{x + 3} = \dfrac{5}{2x + 6} + \dfrac{1}{x - 2}$

45. $\dfrac{3}{x + 2} + \dfrac{2}{x - 2} = \dfrac{8}{(x + 2)(x - 2)}$

46. $\dfrac{5}{x + 2} + \dfrac{3}{x - 2} = \dfrac{12}{(x + 2)(x - 2)}$

47. $\dfrac{2}{x + 1} - \dfrac{1}{x - 1} = \dfrac{2x}{x^2 - 1}$

48. $\dfrac{4}{x + 5} + \dfrac{2}{x - 5} = \dfrac{32}{x^2 - 25}$

49. $\dfrac{1}{x - 4} - \dfrac{5}{x + 2} = \dfrac{6}{x^2 - 2x - 8}$

50. $\dfrac{6}{x + 3} - \dfrac{5}{x - 2} = \dfrac{-20}{x^2 + x - 6}$

In Exercises 51–56, find all values of x satisfying the given conditions.

51. $f(x) = 5(2x - 8) - 2, g(x) = 5(x - 3) + 3$, and $f(x) = g(x)$.

52. $f(x) = 7(3x - 2) + 5, g(x) = 6(2x - 1) + 24$, and $f(x) = g(x)$.

53. $f(x) = \dfrac{x - 3}{5}, g(x) = \dfrac{x - 5}{4}$, and $(f - g)(x) = 1$.

54. $f(x) = \dfrac{x + 1}{4}, g(x) = \dfrac{x - 2}{3}$, and $(f - g)(x) = -4$.

55. $f(x) = \dfrac{5}{x + 4}, g(x) = \dfrac{3}{x + 3}, h(x) = \dfrac{12x + 19}{x^2 + 7x + 12}$, and $(f + g)(x) = h(x)$.

56. $f(x) = \dfrac{2x - 1}{x^2 + 2x - 8}, g(x) = \dfrac{2}{x + 4}, h(x) = \dfrac{1}{x - 2}$, and $(f + g)(x) = h(x)$.

In Exercises 57–60, find the zeros of each function.

57. $f(x) = 4[x - (3 - x)] - 7(x + 1)$

58. $f(x) = 2[3x - (4x - 6)] - 5(x - 6)$

59. $g(x) = \dfrac{x + 6}{3x - 12} - \dfrac{5}{x - 4} - \dfrac{2}{3}$

60. $g(x) = \dfrac{1}{5x + 5} - \dfrac{3}{x + 1} + \dfrac{7}{5}$

In Exercises 61–68, solve each equation. Then determine whether the equation is an identity, a conditional equation, or an inconsistent equation.

61. $5x + 9 = 9(x + 1) - 4x$ **62.** $4x + 7 = 7(x + 1) - 3x$

63. $3(x + 2) = 7 + 3x$ **64.** $4(x + 5) = 21 + 4x$

65. $10x + 3 = 8x + 3$ **66.** $5x + 7 = 2x + 7$

67. $\dfrac{2x}{x - 3} = \dfrac{6}{x - 3} + 4$ **68.** $\dfrac{3}{x - 3} = \dfrac{x}{x - 3} + 3$

The equations in Exercises 69–80 combine the types of equations we have discussed in this section. Solve each equation. Then state whether the equation is an identity, a conditional equation, or an inconsistent equation.

69. $\dfrac{x + 5}{2} - 4 = \dfrac{2x - 1}{3}$ **70.** $\dfrac{x + 2}{7} = 5 - \dfrac{x + 1}{3}$

71. $\dfrac{2}{x - 2} = 3 + \dfrac{x}{x - 2}$ **72.** $\dfrac{6}{x + 3} + 2 = \dfrac{-2x}{x + 3}$

73. $8x - (3x + 2) + 10 = 3x$

74. $2(x + 2) + 2x = 4(x + 1)$

75. $\dfrac{2}{x} + \dfrac{1}{2} = \dfrac{3}{4}$ **76.** $\dfrac{3}{x} - \dfrac{1}{6} = \dfrac{1}{3}$

77. $\dfrac{4}{x - 2} + \dfrac{3}{x + 5} = \dfrac{7}{(x + 5)(x - 2)}$

78. $\dfrac{1}{x - 1} = \dfrac{1}{(2x + 3)(x - 1)} + \dfrac{4}{2x + 3}$

79. $\dfrac{4x}{x + 3} - \dfrac{12}{x - 3} = \dfrac{4x^2 + 36}{x^2 - 9}$

80. $\dfrac{4}{x^2 + 3x - 10} - \dfrac{1}{x^2 + x - 6} = \dfrac{3}{x^2 - x - 12}$

In Exercises 81–84, use the $\boxed{Y=}$ *screen to write the equation being solved. Then use the table to solve the equation.*

81.

82.

83.

84.

Practice Plus

85. If $f(x) = x^2 - x$, find $f(c)$, where c satisfies
$4(c - 2) + 2 = 4c - 2(2 - c)$.

86. If $f(x) = x^2 - x$, find $f(c)$, where c satisfies
$2(c - 6) = 3c + 2(2c - 1)$.

In Exercises 87–90, find all values of x satisfying the given conditions.

87. $f(x) = 3x - 2$, $g(x) = 2x + 5$, and $(f \circ g)(x) = -11$.

88. $f(x) = 2x + 3$, $g(x) = 4x - 1$, and $(f \circ g)(x) = -15$.

89. $f(x) = 2x - 3$ and $f^{-1}(x) = 5$.

90. $f(x) = 4x - 5$ and $f^{-1}(x) = 3$.

Application Exercises

In the years after warning labels were put on cigarette packs, the number of smokers dropped from approximately two in five adults to one in five. The bar graph shows the percentage of American adults who smoked cigarettes for selected years from 1970 through 2010.

Percentage of American Adults Who Smoke Cigarettes

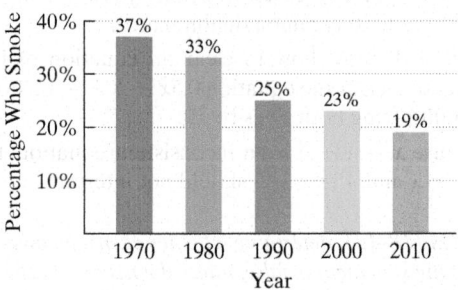

Source: Centers for Disease Control and Prevention

The mathematical model

$$p + \frac{x}{2} = 37$$

describes the percentage of Americans who smoked cigarettes, p, x years after 1970. Use this model to solve Exercises 91–92.

91. a. Does the mathematical model underestimate or overestimate the percentage of American adults who smoked cigarettes in 2010? By how much?

b. Use the mathematical model to project the year when only 7% of American adults will smoke cigarettes.

92. a. Does the mathematical model underestimate or overestimate the percentage of American adults who smoked cigarettes in 2000? By how much?

b. Use the mathematical model to project the year when only 2% of American adults will smoke cigarettes.

The line graph shows the cost of inflation. What cost $10,000 in 1967 would cost the amount shown by the graph in subsequent years. (The symbol ⌁ on the horizontal axis shows that there is a break in values between 0 and 80. Thus, the first tick mark on the horizontal axis represents 1980.)

The Cost of Inflation

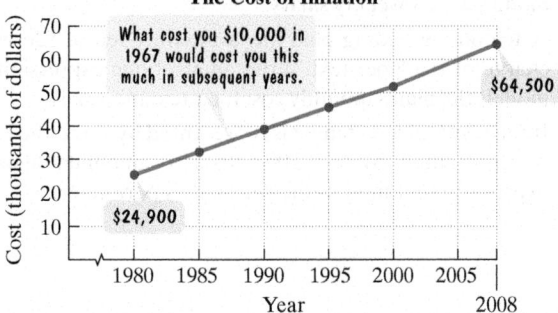

Source: U.S. Bureau of Labor Statistics

Here are two functions for the data shown by the graph. In each function, C(x) represents the cost x years after 1980 of what cost $10,000 in 1967.

Function 1 $\quad C(x) = 1388x + 24{,}963$

Function 2 $\quad C(x) = 3x^2 + 1308x + 25{,}268$

Use these models to solve Exercises 93–96.

93. a. Use the graph to estimate the cost in 2000, to the nearest thousand dollars, of what cost $10,000 in 1967.

b. Use the linear function to determine the cost in 2000. How well does this describe your estimate from part (a)?

c. Use the quadratic function to determine the cost in 2000. How well does this describe your estimate from part (a)?

94. a. Use the graph to estimate the cost in 1990, to the nearest thousand dollars, of what cost $10,000 in 1967.

b. Use the linear function to determine the cost in 1990. How well does this describe your estimate from part (a)?

c. Use the quadratic function to determine the cost in 1990. How well does this describe your estimate from part (a)?

95. Use the linear function to determine in which year the cost will be $77,707 for what cost $10,000 in 1967.

96. Use the linear function to determine in which year the cost will be $80,483 for what cost $10,000 in 1967.

Rational functions are often used to model learning. Many of these functions model the proportion of correct responses in terms of the number of trials of a particular task. One such model, called a learning curve, is

$$P(x) = \frac{0.9x - 0.4}{0.9x + 0.1},$$

where P(x) is the proportion of correct responses after x trials. If P(x) = 0, there are no correct responses. If P(x) = 1, all responses are correct. The graph of the rational function is shown. Use the function to solve Exercises 97–98.

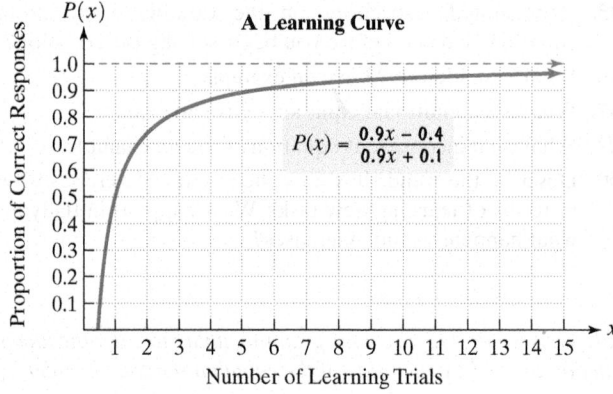

97. How many learning trials are necessary for 0.95 of the responses to be correct? Identify your solution as a point on the graph.

98. How many learning trials are necessary for 0.5 of the responses to be correct? Identify your solution as a point on the graph.

99. A company wants to increase the 10% peroxide content of its product by adding pure peroxide (100% peroxide). If x liters of pure peroxide are added to 500 liters of its 10% solution, the concentration, $C(x)$, of the new mixture is given by

$$C(x) = \frac{x + 0.1(500)}{x + 500}.$$

How many liters of pure peroxide should be added to produce a new product that is 28% peroxide?

100. Suppose that x liters of pure acid are added to 200 liters of a 35% acid solution.

 a. Write a function that gives the concentration, $C(x)$, of the new mixture. (*Hint:* See Exercise 99.)

 b. How many liters of pure acid should be added to produce a new mixture that is 74% acid?

Writing in Mathematics

101. What is a linear equation in one variable? Give an example of this type of equation.

102. Suppose that you solve $\dfrac{x}{5} - \dfrac{x}{2} = 1$ by multiplying both sides by 20 rather than the least common denominator (namely, 10). Describe what happens. If you get the correct solution, why do you think we clear the equation of fractions by multiplying by the *least* common denominator?

103. Suppose you are an algebra teacher grading the following solution on an examination:

$$-3(x - 6) = 2 - x$$
$$-3x - 18 = 2 - x$$
$$-2x - 18 = 2$$
$$-2x = -16$$
$$x = 8.$$

You should note that 8 checks, so the solution set is {8}. The student who worked the problem therefore wants full credit. Can you find any errors in the solution? If full credit is 10 points, how many points should you give the student? Justify your position.

104. Explain how to find restrictions on the variable in a rational equation.

105. Why should restrictions on the variable in a rational equation be listed before you begin solving the equation?

106. What is an identity? Give an example.

107. What is a conditional equation? Give an example.

108. What is an inconsistent equation? Give an example.

109. Describe the trend shown by the graph in Exercises 97–98 in terms of learning new tasks. What happens initially and what happens as time increases?

Technology Exercises

In Exercises 110–113, use your graphing utility to enter the linear function defined by each side of the equation separately under y_1 and y_2. Then use the utility's TABLE *or* GRAPH *feature to solve the equation.*

110. $5x + 2(x - 1) = 3x + 10$

111. $2x + 3(x - 4) = 4x - 7$

112. $\dfrac{x - 3}{5} - 1 = \dfrac{x - 5}{4}$

113. $\dfrac{2x - 1}{3} - \dfrac{x - 5}{6} = \dfrac{x - 3}{4}$

Critical Thinking Exercises

Make Sense? *In Exercises 114–117, determine whether each statement makes sense or does not make sense, and explain your reasoning.*

114. The function $P(x) = -0.18n + 2.1$ describes the number of pay phones, $P(x)$, in millions, n years after 2000, so I have to solve a linear equation to determine the number of pay phones in 2010.

115. Although I can solve $3x + \frac{1}{5} = \frac{1}{4}$ by first subtracting $\frac{1}{5}$ from both sides, I find it easier to begin by multiplying both sides by 20, the least common denominator.

116. Because I know how to clear an equation of fractions, I decided to clear the equation $0.5x + 8.3 = 12.4$ of decimals by multiplying both sides by 10.

117. Because $x = x + 5$ is an inconsistent equation, the graphs of $y = x$ and $y = x + 5$ should not intersect.

In Exercises 118–121, determine whether each statement is true or false. If the statement is false, make the necessary change(s) to produce a true statement.

118. The equation $-7x = x$ has no solution.

119. The equations $\dfrac{x}{x - 4} = \dfrac{4}{x - 4}$ and $x = 4$ are equivalent.

120. The equations $3y - 1 = 11$ and $3y - 7 = 5$ are equivalent.

121. If m and b are any real numbers, then $mx + b = 0$ always has one number in its solution set.

122. If x represents a number, write an English sentence about the number that results in an inconsistent equation.

123. Find b such that $\dfrac{7x + 4}{b} + 13 = x$ has a solution set given by {−6}.

124. Find b such that $\dfrac{4x - b}{x - 5} = 3$ has a solution set given by \varnothing.

Preview Exercises

Exercises 125–127 will help you prepare for the material covered in the next section.

125. Jane's salary exceeds Jim's by $150 per week. If x represents Jim's weekly salary, write an algebraic expression that models Jane's weekly salary.

126. A telephone texting plan has a monthly fee of $20 with a charge of $0.05 per text. Write an algebraic expression that models the plan's monthly cost for x text messages.

127. If the width of a rectangle is represented by x and the length is represented by $x + 200$, write a simplified algebraic expression that models the rectangle's perimeter.

SECTION 2.2 Models and Applications

How Long It Takes to Earn $1000

Howard Stern
Radio host
24 sec.

Dr. Phil McGraw
Television host
2 min. 24 sec.

Brad Pitt
Actor
4 min. 48 sec.

Kobe Bryant
Basketball player
5 min. 30 sec.

Chief executive
U.S. average
2 hr. 55 min.

Doctor, G.P.
U.S. average
13 hr. 5 min.

High school teacher
U.S. average
43 hours

Janitor
U.S. average
103 hours

Source: *Time*

Objective

① Use linear equations to solve problems.

In this section, you'll see examples and exercises focused on how much money Americans earn. These situations illustrate a step-by-step strategy for solving problems. As you become familiar with this strategy, you will learn to solve a wide variety of problems.

① Use linear equations to solve problems.

Problem Solving with Linear Equations

We have seen that a model is a mathematical representation of a real-world situation. In Chapter 1, we used real-world data to obtain models that were linear functions. In this section, we will be solving problems that are presented in English. This means that we must obtain models by translating from the ordinary language of English into the language of algebraic equations. To translate, however, we must understand the English prose and be familiar with the forms of algebraic language. Following are some general steps we will use in solving word problems:

GREAT QUESTION!

Why are word problems important?

There is great value in reasoning through the steps for solving a word problem. This value comes from the problem-solving skills that you will attain and is often more important than the specific problem or its solution.

Strategy for Solving Word Problems

Step 1 Read the problem carefully several times until you can state in your own words what is given and what the problem is looking for. Let x (or any variable) represent one of the unknown quantities in the problem.

Step 2 If necessary, write expressions for any other unknown quantities in the problem in terms of x.

Step 3 Write an equation in x that models the verbal conditions of the problem.

Step 4 Solve the equation and answer the problem's question.

Step 5 Check the solution *in the original wording* of the problem, not in the equation obtained from the words.

EXAMPLE 1 Starting Salaries for College Graduates
with Undergraduate Degrees

The bar graph in **Figure 2.3** shows the ten most popular college majors with median, or middlemost, starting salaries for recent college graduates.

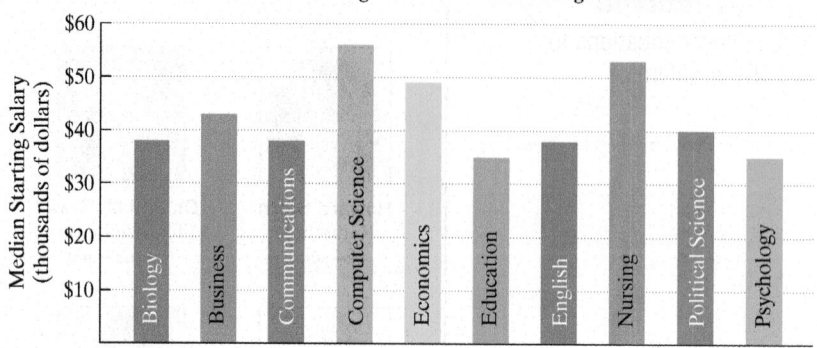

Median Starting Salaries for U.S. College Graduates

FIGURE 2.3
Source: PayScale (2010 data)

The median starting salary of a business major exceeds that of a psychology major by $8 thousand. The median starting salary of an English major exceeds that of a psychology major by $3 thousand. Combined, their median starting salaries are $116 thousand. Determine the median starting salaries of psychology majors, business majors, and English majors with bachelor's degrees.

SOLUTION

Step 1 Let x represent one of the unknown quantities. We know something about the median starting salaries of business majors and English majors: Business majors earn $8 thousand more than psychology majors and English majors earn $3 thousand more than psychology majors. We will let

x = the median starting salary, in thousands of dollars, of psychology majors.

Step 2 Represent other unknown quantities in terms of x. Because business majors earn $8 thousand more than psychology majors, let

Salary exceeds
a psychology
major, x, by
$8 thousand.

$x + 8$ = the median starting salary, in thousands of dollars, of business majors.

Because English majors earn $3 thousand more than psychology majors, let

Salary exceeds
a psychology
major, x, by
$3 thousand.

$x + 3$ = the median starting salary, in thousands of dollars, of English majors.

Step 3 Write an equation in x that models the conditions. Combined, the median starting salaries for psychology, business, and English majors are $116 thousand.

The median starting salary for psychology majors	plus	the median starting salary for business majors	plus	the median starting salary for English majors	is	$116 thousand.
x	$+$	$(x + 8)$	$+$	$(x + 3)$	$=$	116

Step 4 Solve the equation and answer the question.

$$x + (x + 8) + (x + 3) = 116$$ This is the equation that models the problem's conditions.

$$3x + 11 = 116$$ Remove parentheses, regroup, and combine like terms.

$$3x = 105$$ Subtract 11 from both sides.

$$x = 35$$ Divide both sides by 3.

Thus,

$$\text{starting salary of psychology majors} = x = 35$$
$$\text{starting salary of business majors} = x + 8 = 35 + 8 = 43$$
$$\text{starting salary of English majors} = x + 3 = 35 + 3 = 38.$$

The median starting salary of psychology majors is $35 thousand, the median starting salary of business majors is $43 thousand, and the median starting salary of English majors is $38 thousand.

Step 5 Check the proposed solution in the original wording of the problem. The problem states that combined, the median starting salaries are $116 thousand. Using the median salaries we determined in step 4, the sum is

$$\$35 \text{ thousand} + \$43 \text{ thousand} + \$38 \text{ thousand,}$$

or $116 thousand, which verifies the problem's conditions. • • •

GREAT QUESTION!

Example 1 involves using the word *exceeds* to represent two of the unknown quantities. Can you help me to write algebraic expressions for quantities described using *exceeds*?

Modeling with the word *exceeds* can be a bit tricky. It's helpful to identify the smaller quantity. Then add to this quantity to represent the larger quantity. For example, suppose that Tim's height exceeds Tom's height by a inches. Tom is the shorter person. If Tom's height is represented by x, then Tim's height is represented by $x + a$.

⊘ Check Point **1** Three of the bars in **Figure 2.3** on page 254 represent median starting salaries of education, computer science, and economics majors. The median starting salary of a computer science major exceeds that of an education major by $21 thousand. The median starting salary of an economics major exceeds that of an education major by $14 thousand. Combined, their median starting salaries are $140 thousand. Determine the median starting salaries of education majors, computer science majors, and economics majors with bachelor's degrees.

EXAMPLE 2 Modeling Attitudes of College Freshmen

Researchers have surveyed college freshmen every year since 1969. **Figure 2.4** at the top of the next page shows that attitudes about some life goals have changed dramatically over the years. In particular, the freshman class of 2009 was more interested in making money than the freshmen of 1969 had been. In 1969, 42% of first-year college students considered "being well-off financially" essential or very important. For the period from 1969 through 2009, this percentage increased by approximately 0.9 each year. If this trend continues, by which year will all college freshmen consider "being well-off financially" essential or very important?

Your author teaching math in 1969

Life Objectives of College Freshmen, 1969–2009

FIGURE 2.4
Source: Higher Education
Research Institute

SOLUTION

Step 1 Let x represent one of the unknown quantities. We are interested in the year when all college freshmen, or 100% of the freshmen, will consider this life objective essential or very important. Let

x = the number of years after 1969 when all freshmen will consider "being well-off financially" essential or very important.

Step 2 Represent other unknown quantities in terms of x. There are no other unknown quantities to find, so we can skip this step.

Step 3 Write an equation in x that models the conditions.

The 1969 percentage	increased by	0.9 each year for x years	equals	100% of the freshmen.
42	+	$0.9x$	=	100

Step 4 Solve the equation and answer the question.

$$42 + 0.9x = 100 \qquad \text{This is the equation that models the problem's conditions.}$$

$$42 - 42 + 0.9x = 100 - 42 \qquad \text{Subtract 42 from both sides.}$$

$$0.9x = 58 \qquad \text{Simplify.}$$

$$\frac{0.9x}{0.9} = \frac{58}{0.9} \qquad \text{Divide both sides by 0.9.}$$

$$x = 64.\overline{4} \approx 64 \qquad \text{Simplify and round to the nearest whole number.}$$

Using current trends, by approximately 64 years after 1969, or in 2033, all freshmen will consider "being well-off financially" essential or very important.

Step 5 Check the proposed solution in the original wording of the problem. The problem states that all freshmen (100%, represented by 100 using the model) will consider the objective essential or very important. Does this approximately occur if we increase the 1969 percentage, 42%, by 0.9 each year for 64 years, our proposed solution?

$$42 + 0.9(64) = 42 + 57.6 = 99.6 \approx 100$$

This verifies that using trends shown in **Figure 2.4**, all first-year college students will consider the objective essential or very important approximately 64 years after 1969. • • •

 ⊘ Check Point **2** **Figure 2.4** above shows that the freshman class of 2009 was less interested in developing a philosophy of life than the freshmen of 1969 had been. In 1969, 85% of the freshmen considered this objective essential or very important.

Since then, this percentage has decreased by approximately 0.9 each year. If this trend continues, by which year will only 25% of college freshmen consider "developing a meaningful philosophy of life" essential or very important?

EXAMPLE 3 Selecting a Monthly Text Message Plan

You are choosing between two texting plans. Plan A has a monthly fee of $20 with a charge of $0.05 per text. Plan B has a monthly fee of $5 with a charge of $0.10 per text. Both plans include photo and video texts. For how many text messages will the costs for the two plans be the same?

SOLUTION

Step 1 Let x represent one of the unknown quantities. Let

x = the number of text messages for which the two plans cost the same.

Step 2 Represent other unknown quantities in terms of x. There are no other unknown quantities, so we can skip this step.

Step 3 Write an equation in x that models the conditions. The monthly cost for plan A is the monthly fee, $20, plus the per-text charge, $0.05, times the number of text messages, x. The monthly cost for plan B is the monthly fee, $5, plus the per-text charge, $0.10, times the number of text messages, x.

The monthly cost for plan A	must equal	the monthly cost for plan B.

$$20 + 0.05x = 5 + 0.10x$$

Notice that each side of this equation defines a linear function:

Plan A

$$f(x) = 0.05x + 20$$

The slope, 0.05, indicates a rate of increase of $0.05 per text.

Plan B

$$g(x) = 0.10x + 5.$$

The slope, 0.10, indicates a rate of increase of $0.10 per text.

These functions, $f(x) = 0.05x + 20$ and $g(x) = 0.10x + 5$, model monthly costs, $f(x)$ and $g(x)$, for x text messages. We are interested in the input, x, that results in the same output (the same cost) for the two functions, or $f(x) = g(x)$.

Step 4 Solve the equation and answer the question.

$$20 + 0.05x = 5 + 0.10x \quad \text{This is the equation that models the problem's conditions.}$$

$$20 = 5 + 0.05x \quad \text{Subtract 0.05x from both sides.}$$

$$15 = 0.05x \quad \text{Subtract 5 from both sides.}$$

$$\frac{15}{0.05} = \frac{0.05x}{0.05} \quad \text{Divide both sides by 0.05.}$$

$$300 = x \quad \text{Simplify.}$$

Because x represents the number of text messages for which the two plans cost the same, the costs will be the same for 300 texts per month.

Step 5 Check the proposed solution in the original wording of the problem. The problem states that the costs for the two plans should be the same. Let's see if they are with 300 text messages:

Cost for plan A = $20 + $0.05(300) = $20 + $15 = $35

Monthly fee Per-text charge

Cost for plan B = $5 + $0.10(300) = $5 + $30 = $35.

With 300 text messages, both plans cost $35 for the month. Thus, the proposed solution, 300 minutes, satisfies the problem's conditions. ●●●

TECHNOLOGY

Numeric and Graphic Connections

We can use a graphing utility to numerically or graphically verify our work in Example 3. Enter the linear function defined by each side of the equation that models the problems conditions.

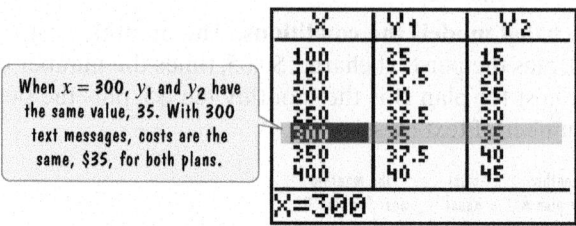

| The monthly cost for plan A | must equal | the monthly cost for plan B. |

$$20 + 0.05x \quad = \quad 5 + 0.10x$$

Enter $y_1 = 20 + .05x$. Enter $y_2 = 5 + .10x$.

Numeric Check

Display a table for y_1 and y_2.

When $x = 300$, y_1 and y_2 have the same value, 35. With 300 text messages, costs are the same, $35, for both plans.

X	Y1	Y2
100	25	15
150	27.5	20
200	30	25
250	32.5	30
300	35	35
350	37.5	40
400	40	45

X=300

Graphic Check

Display graphs for y_1 and y_2. Use the intersection feature.

$y_1 = 20 + 0.05x$

$y_2 = 5 + 0.10x$

Graphs intersect at (300, 35). With 300 text messages, costs are the same, $35, for both plans.

Intersection
X=300 _____ Y=35 _____

[0, 500, 100] by [0, 50, 5]

☑ Check Point **3** You are choosing between two texting plans. Plan A has a monthly fee of $15 with a charge of $0.08 per text. Plan B has a monthly fee of $3 with a charge of $0.12 per text. For how many text messages will the costs for the two plans be the same?

EXAMPLE 4 A Price Reduction on a Digital Camera

Your local computer store is having a terrific sale on digital cameras. After a 40% price reduction, you purchase a digital camera for $276. What was the camera's price before the reduction?

SOLUTION

Step 1 Let x represent one of the unknown quantities. We will let

$x =$ the price of the digital camera prior to the reduction.

Step 2 Represent other unknown quantities in terms of x. There are no other unknown quantities to find, so we can skip this step.

Step 3 Write an equation in x that models the conditions. The camera's original price minus the 40% reduction is the reduced price, $276.

| Original price | minus | the reduction (40% of the original price) | is | the reduced price, $276. |

$$x \quad - \quad 0.4x \quad = \quad 276$$

GREAT QUESTION!

Why is the 40% reduction written as 0.4x in Example 4?

- 40% is written 0.40 or 0.4.
- "Of" represents multiplication, so 40% of the original price is 0.4x.

Notice that the original price, x, reduced by 40% is $x - 0.4x$ and *not* $x - 0.4$.

Step 4 Solve the equation and answer the question.

$$x - 0.4x = 276 \quad \text{This is the equation that models}$$
the problem's conditions.

$$0.6x = 276 \quad \text{Combine like terms:}$$
$x - 0.4x = 1x - 0.4x = (1 - 0.4)x = 0.6x.$

$$\frac{0.6x}{0.6} = \frac{276}{0.6} \quad \text{Divide both sides by 0.6.}$$

$$x = 460 \quad \text{Simplify: } 0.6 \overline{)\, 276.0\,} \begin{smallmatrix}460.\end{smallmatrix}$$

The digital camera's price before the reduction was $460.

Step 5 Check the proposed solution in the original wording of the problem. The price before the reduction, $460, minus the 40% reduction should equal the reduced price given in the original wording, $276:

$$460 - 40\% \text{ of } 460 = 460 - 0.4(460) = 460 - 184 = 276.$$

This verifies that the digital camera's price before the reduction was $460. •••

⊘ Check Point **4** After a 30% price reduction, you purchase a new computer for $840. What was the computer's price before the reduction?

Our next example is about simple interest. Simple interest involves interest calculated only on the amount of money that we invest, called the **principal**. The formula $I = Pr$ is used to find the simple interest, I, earned for one year when the principal, P, is invested at an annual interest rate, r. Dual investment problems involve different amounts of money in two or more investments, each paying a different rate.

EXAMPLE 5 Solving a Dual Investment Problem

Your grandmother needs your help. She has $50,000 to invest. Part of this money is to be invested in noninsured bonds paying 15% annual interest. The rest of this money is to be invested in a government-insured certificate of deposit paying 7% annual interest. She told you that she requires $6000 per year in extra income from the combination of these investments. How much money should be placed in each investment?

SOLUTION

Step 1 Let x represent one of the unknown quantities. We will let

$x = $ the amount invested in the noninsured bonds at 15%.

Step 2 Represent other unknown quantities in terms of x. The other quantity that we seek is the amount invested at 7% in the certificate of deposit. Because the total amount Grandma has to invest is $50,000 and we already used up x,

$50,000 - x = $ the amount invested in the certificate of deposit at 7%.

Step 3 Write an equation in x that models the conditions. Because Grandma requires $6000 in total interest, the interest for the two investments combined must be $6000. Interest is Pr or rP for each investment.

Interest from the 15% investment	plus	interest from the 7% investment	is	$6000.
$0.15x$	$+$	$0.07(50,000 - x)$	$=$	6000
rate times principal		rate times principal		

Step 4 Solve the equation and answer the question.

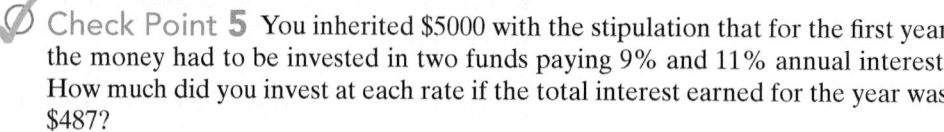

$$0.15x + 0.07(50{,}000 - x) = 6000$$ This is the equation that models the problem's conditions.

$$0.15x + 3500 - 0.07x = 6000$$ Use the distributive property.

$$0.08x + 3500 = 6000$$ Combine like terms.

$$0.08x = 2500$$ Subtract 3500 from both sides.

$$\frac{0.08x}{0.08} = \frac{2500}{0.08}$$ Divide both sides by 0.08.

$$x = 31{,}250$$ Simplify.

Thus,

the amount invested at 15% $= x = 31{,}250.$

the amount invested at 7% $= 50{,}000 - 31{,}250 = 18{,}750.$

Grandma should invest $31,250 at 15% and $18,750 at 7%.

Step 5 Check the proposed solution in the original wording of the problem. The problem states that the total interest from the dual investments should be $6000. Can Grandma count on $6000 interest? The interest earned on $31,250 at 15% is ($31,250) (0.15), or $4687.50. The interest earned on $18,750 at 7% is ($18,750)(0.07), or $1312.50. The total interest is $4687.50 + $1312.50, or $6000, exactly as it should be. You've made your grandmother happy. (Now if you would just visit her more often …) ● ● ●

✐ Check Point **5** You inherited $5000 with the stipulation that for the first year the money had to be invested in two funds paying 9% and 11% annual interest. How much did you invest at each rate if the total interest earned for the year was $487?

Solving geometry problems usually requires a knowledge of basic geometric ideas and formulas. Formulas for area, perimeter, and volume are given in **Table 2.1**.

Table 2.1 Common Formulas for Area, Perimeter, and Volume

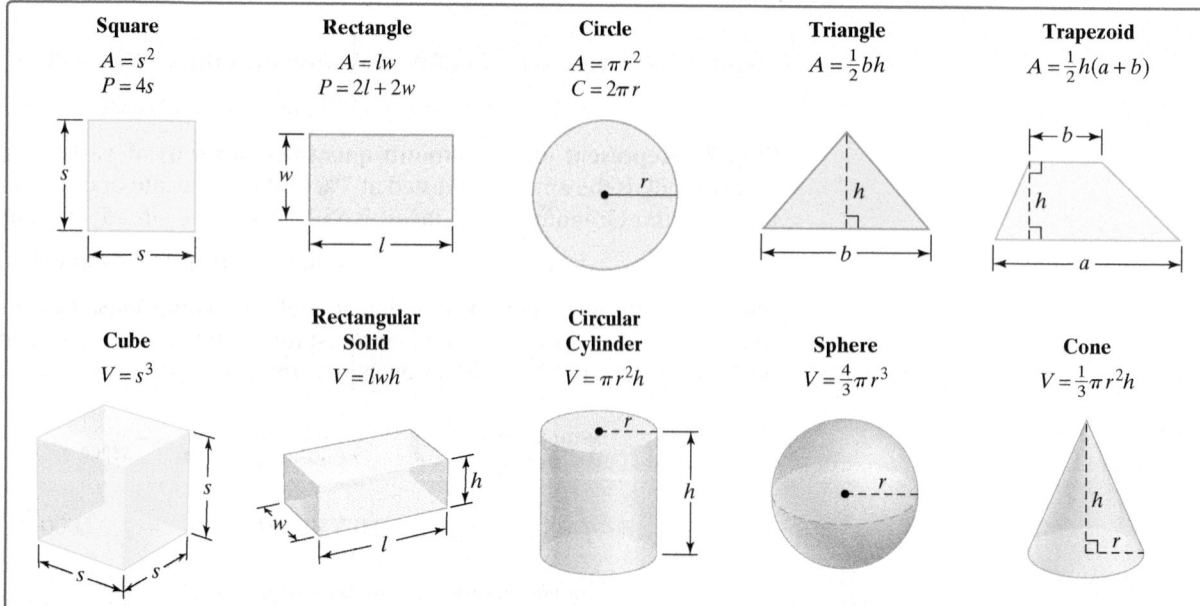

We will be using the formula for the perimeter of a rectangle, $P = 2l + 2w$, in our next example. The formula states that a rectangle's perimeter is the sum of twice its length and twice its width.

EXAMPLE 6 Finding the Dimensions of an American Football Field

The length of an American football field is 200 feet more than the width. If the perimeter of the field is 1040 feet, what are its dimensions?

SOLUTION

Step 1 Let x represent one of the unknown quantities. We know something about the length; the length is 200 feet more than the width. We will let

$$x = \text{the width.}$$

Step 2 Represent other unknown quantities in terms of x. Because the length is 200 feet more than the width, we add 200 to the width to represent the length. Thus,

$$x + 200 = \text{the length.}$$

Figure 2.5 illustrates an American football field and its dimensions.

Step 3 Write an equation in x that models the conditions. Because the perimeter of the field is 1040 feet,

Twice the length	plus	twice the width	is	the perimeter.
$2(x + 200)$	$+$	$2x$	$=$	$1040.$

Step 4 Solve the equation and answer the question.

$2(x + 200) + 2x = 1040$	This is the equation that models the problem's conditions.
$2x + 400 + 2x = 1040$	Apply the distributive property.
$4x + 400 = 1040$	Combine like terms: $2x + 2x = 4x$.
$4x = 640$	Subtract 400 from both sides.
$x = 160$	Divide both sides by 4.

Thus,

$$\text{width} = x = 160.$$
$$\text{length} = x + 200 = 160 + 200 = 360.$$

The dimensions of an American football field are 160 feet by 360 feet. (The 360-foot length is usually described as 120 yards.)

Step 5 Check the proposed solution in the original wording of the problem. The perimeter of the football field using the dimensions that we found is

$$2(160 \text{ feet}) + 2(360 \text{ feet}) = 320 \text{ feet} + 720 \text{ feet} = 1040 \text{ feet.}$$

Because the problem's wording tells us that the perimeter is 1040 feet, our dimensions are correct. • • •

FIGURE 2.5 An American football field

GREAT QUESTION!

Should I draw pictures like Figure 2.5 when solving geometry problems?

When solving word problems, particularly problems involving geometric figures, drawing a picture of the situation is often helpful. Label x on your drawing and, where appropriate, label other parts of the drawing in terms of x.

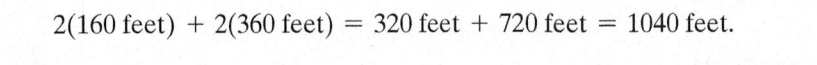 Check Point **6** The length of a rectangular basketball court is 44 feet more than the width. If the perimeter of the basketball court is 288 feet, what are its dimensions?

CONCEPT AND VOCABULARY CHECK

Fill in each blank so that the resulting statement is true.

1. According to the U.S. Office of Management and Budget, the 2011 budget for defense exceeded the budget for education by $658.6 billion. If x represents the budget for education, in billions of dollars, the budget for defense can be represented by _____.

2. In 2000, 31% of U.S. adults viewed a college education as essential for success. For the period from 2000 through 2010, this percentage increased by approximately 2.4 each year. The percentage of U.S. adults who viewed a college education as essential for success x years after 2000 can be represented by_____.

3. A text message plan costs $4 per month plus $0.15 per text. The monthly cost for x text messages can be represented by _____.

4. I purchased a computer after a 15% price reduction. If x represents the computer's original price, the reduced price can be represented by _____.

5. The combined yearly interest for x dollars invested at 12% and $30{,}000 - x$ dollars invested at 9% is _____.

EXERCISE SET 2.2

Practice Exercises

Use the five-step strategy for solving word problems to find the number or numbers described in Exercises 1–10.

1. When five times a number is decreased by 4, the result is 26. What is the number?

2. When two times a number is decreased by 3, the result is 11. What is the number?

3. When a number is decreased by 20% of itself, the result is 20. What is the number?

4. When a number is decreased by 30% of itself, the result is 28. What is the number?

5. When 60% of a number is added to the number, the result is 192. What is the number?

6. When 80% of a number is added to the number, the result is 252. What is the number?

7. 70% of what number is 224?

8. 70% of what number is 252?

9. One number exceeds another by 26. The sum of the numbers is 64. What are the numbers?

10. One number exceeds another by 24. The sum of the numbers is 58. What are the numbers?

Practice Plus

In Exercises 11–18, find all values of x satisfying the given conditions.

11. $f(x) = 13x - 4$, $g(x) = 5x + 10$, and the outputs of f exceed those of g by 2.

12. $f(x) = 10x + 6$, $g(x) = 12x - 7$, and the ouputs of f exceed those of g by 3.

13. $f(x) = 10(2x - 1)$, $g(x) = 2x + 1$, and the function values of f are 14 more than 8 times those of g.

14. $f(x) = 9(3x - 5)$, $g(x) = 3x - 1$, and the function values of f are 51 less than 12 times those of g.

15. $f(x) = 2x + 6$, $g(x) = x + 8$, $h(x) = x$, and the difference between 3 times the outputs of f and 5 times those of g is 22 less than the outputs of h.

16. $f(x) = 2.5$, $g(x) = 2x + 1$, $h(x) = x$, and the difference between 2 times the outputs of f and 3 times those of g is 8 less than 4 times the outputs of h.

17. $f(x) = \dfrac{1}{x}$, $g(x) = \dfrac{1}{2x}$, $h(x) = \dfrac{1}{x - 1}$, and the sum of 3 times the function values of f and 4 times the function values of g is 4 times those of h.

18. $f(x) = \dfrac{1}{x}$, $g(x) = \dfrac{1}{x^2 - x}$, $h(x) = \dfrac{1}{x - 1}$, and the difference between 6 times the function values of f and 3 times the function values of g is 7 times those of h.

Application Exercises

How will you spend your average life expectancy of 78 years? The bar graph shows the average number of years you will devote to each of your most time-consuming activities. Exercises 19–20 are based on the data displayed by the graph.

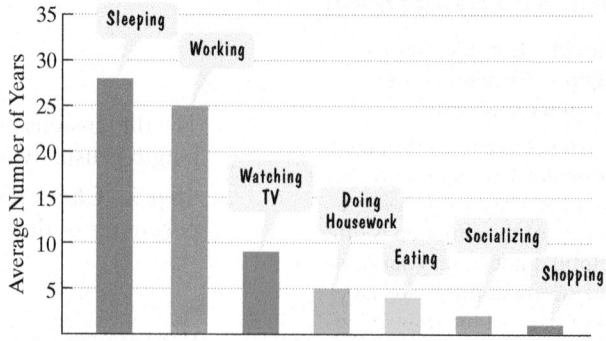

How You Will Spend Your Average Life Expectancy of 78 Years

Source: U.S. Bureau of Labor Statistics

19. According to the American Bureau of Labor Statistics, you will devote 37 years to sleeping and watching TV. The number of years sleeping will exceed the number of years watching TV by 19. Over your lifetime, how many years will you spend on each of these activities?

20. According to the American Bureau of Labor Statistics, you will devote 32 years to sleeping and eating. The number of years sleeping will exceed the number of years eating by 24. Over your lifetime, how many years will you spend on each of these activities?

The bar graph shows average yearly earnings in the United States for people with a college education, by final degree earned. Exercises 21–22 are based on the data displayed by the graph.

Average Earnings of Full-Time Workers in the U.S., by Final Degree Earned

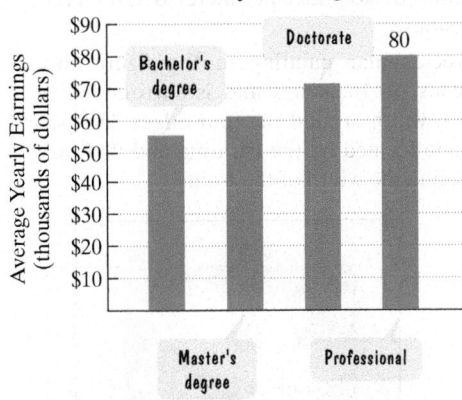

Source: U.S. Census Bureau

21. The average yearly salary of an American whose final degree is a master's is $49 thousand less than twice that of an American whose final degree is a bachelor's. Combined, two people with each of these educational attainments earn $116 thousand. Find the average yearly salary of Americans with each of these final degrees.

22. The average yearly salary of an American whose final degree is a doctorate is $39 thousand less than twice that of an American whose final degree is a bachelor's. Combined, two people with each of these educational attainments earn $126 thousand. Find the average yearly salary of Americans with each of these final degrees.

Even as Americans increasingly view a college education as essential for success, many believe that a college education is becoming less available to qualified students. Exercises 23–24 are based on the data displayed by the graph.

Changing Attitudes Toward College in the United States

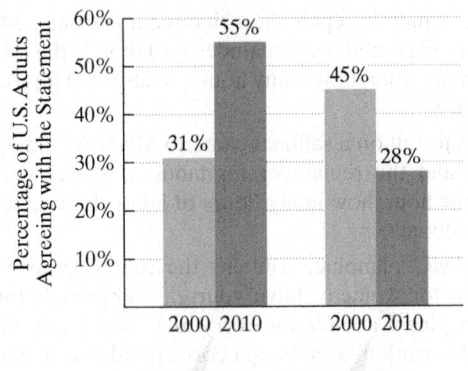

Source: *Public Agenda*

23. In 2000, 31% of U.S. adults viewed a college education as essential for success. For the period 2000 through 2010, the percentage viewing a college education as essential for success increased on average by approximately 2.4 each year. If this trend continues, by which year will 67% of all American adults view a college education as essential for success?

24. The data displayed by the graph at the bottom of the previous column indicate that in 2000, 45% of U.S. adults believed most qualified students get to attend college. For the period from 2000 through 2010, the percentage who believed that a college education is available to most qualified students decreased by approximately 1.7 each year. If this trend continues, by which year will only 11% of all American adults believe that most qualified students get to attend college?

25. A new car worth $24,000 is depreciating in value by $3000 per year.
 a. Write a linear function that models the car's value, $V(x)$, in dollars, after x years.
 b. Use the function from part (a) to determine after how many years the car's value will be $9000.
 c. Graph the function from part (a) in the first quadrant of a rectangular coordinate system. Then show your solution to part (b) on the graph.

26. A new car worth $45,000 is depreciating in value by $5000 per year.
 a. Write a linear function that models the car's value, $V(x)$, in dollars, after x years.
 b. Use the function from part (a) to determine after how many years the car's value will be $10,000.
 c. Graph the function from part (a) in the first quadrant of a rectangular coordinate system. Then show your solution to part (b) on the graph.

27. You are choosing between two health clubs. Club A offers membership for a fee of $40 plus a monthly fee of $25. Club B offers membership for a fee of $15 plus a monthly fee of $30. After how many months will the total cost at each health club be the same? What will be the total cost for each club?

28. You need to rent a rug cleaner. Company A will rent the machine you need for $22 plus $6 per hour. Company B will rent the same machine for $28 plus $4 per hour. After how many hours of use will the total amount spent at each company be the same? What will be the total amount spent at each company?

29. The bus fare in a city is $1.25. People who use the bus have the option of purchasing a monthly discount pass for $15.00. With the discount pass, the fare is reduced to $0.75. Determine the number of times in a month the bus must be used so that the total monthly cost without the discount pass is the same as the total monthly cost with the discount pass.

30. A discount pass for a bridge costs $30 per month. The toll for the bridge is normally $5.00, but it is reduced to $3.50 for people who have purchased the discount pass. Determine the number of times in a month the bridge must be crossed so that the total monthly cost without the discount pass is the same as the total monthly cost with the discount pass.

31. In 2010, there were 13,300 students at college A, with a projected enrollment increase of 1000 students per year. In the same year, there were 26,800 students at college B, with a projected enrollment decline of 500 students per year.

 a. According to these projections, when will the colleges have the same enrollment? What will be the enrollment in each college at that time?

 b. Use the following table to check your work in part (a) numerically. What equations were entered for Y_1 and Y_2 to obtain this table?

X	Y1	Y2
7	20300	23300
8	21300	22800
9	22300	22300
10	23300	21800
11	24300	21300
12	25300	20800
13	26300	20300

X=7

32. In 2000, the population of Greece was 10,600,000, with projections of a population decrease of 28,000 people per year. In the same year, the population of Belgium was 10,200,000, with projections of a population decrease of 12,000 people per year. (*Source*: United Nations) According to these projections, when will the two countries have the same population? What will be the population at that time?

33. After a 20% reduction, you purchase a television for $336. What was the television's price before the reduction?

34. After a 30% reduction, you purchase a dictionary for $30.80. What was the dictionary's price before the reduction?

35. Including 8% sales tax, an inn charges $162 per night. Find the inn's nightly cost before the tax is added.

36. Including 5% sales tax, an inn charges $252 per night. Find the inn's nightly cost before the tax is added.

Exercises 37–38 involve markup, the amount added to the dealer's cost of an item to arrive at the selling price of that item.

37. The selling price of a refrigerator is $584. If the markup is 25% of the dealer's cost, what is the dealer's cost of the refrigerator?

38. The selling price of a scientific calculator is $15. If the markup is 25% of the dealer's cost, what is the dealer's cost of the calculator?

39. You invested $7000 in two accounts paying 6% and 8% annual interest. If the total interest earned for the year was $520, how much was invested at each rate?

40. You invested $11,000 in two accounts paying 5% and 8% annual interest. If the total interest earned for the year was $730, how much was invested at each rate?

41. Things did not go quite as planned. You invested $8000, part of it in stock that paid 12% annual interest. However, the rest of the money suffered a 5% loss. If the total annual income from both investments was $620, how much was invested at each rate?

42. Things did not go quite as planned. You invested $12,000, part of it in stock that paid 14% annual interest. However, the rest of the money suffered a 6% loss. If the total annual income from both investments was $680, how much was invested at each rate?

43. A rectangular soccer field is twice as long as it is wide. If the perimeter of the soccer field is 300 yards, what are its dimensions?

44. A rectangular swimming pool is three times as long as it is wide. If the perimeter of the pool is 320 feet, what are its dimensions?

45. The length of the rectangular tennis court at Wimbledon is 6 feet longer than twice the width. If the court's perimeter is 228 feet, what are the court's dimensions?

46. The length of a rectangular pool is 6 meters less than twice the width. If the pool's perimeter is 126 meters, what are its dimensions?

47. The rectangular painting in the figure shown measures 12 inches by 16 inches and is surrounded by a frame of uniform width around the four edges. The perimeter of the rectangle formed by the painting and its frame is 72 inches. Determine the width of the frame.

48. The rectangular swimming pool in the figure shown measures 40 feet by 60 feet and is surrounded by a path of uniform width around the four edges. The perimeter of the rectangle formed by the pool and the surrounding path is 248 feet. Determine the width of the path.

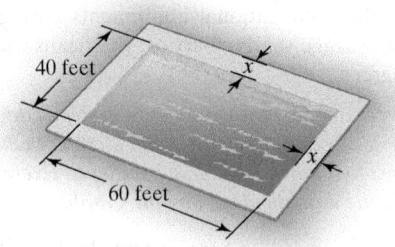

49. An automobile repair shop charged a customer $448, listing $63 for parts and the remainder for labor. If the cost of labor is $35 per hour, how many hours of labor did it take to repair the car?

50. A repair bill on a sailboat came to $1603, including $532 for parts and the remainder for labor. If the cost of labor is $63 per hour, how many hours of labor did it take to repair the sailboat?

51. An HMO pamphlet contains the following recommended weight for women: "Give yourself 100 pounds for the first 5 feet plus 5 pounds for every inch over 5 feet tall." Using this description, what height corresponds to a recommended weight of 135 pounds?

52. A job pays an annual salary of $33,150, which includes a holiday bonus of $750. If paychecks are issued twice a month, what is the gross amount for each paycheck?

53. Answer the question in the following *Peanuts* cartoon strip. (*Note:* You may not use the answer given in the cartoon!)

Peanuts © 1978 Peanuts Worldwide LLC. Used by permission of Universal Uclick. All rights reserved.

54. The rate for a particular international person-to-person telephone call is $0.43 for the first minute, $0.32 for each additional minute, and a $2.10 service charge. If the cost of a call is $5.73, how long did the person talk?

Writing in Mathematics

55. In your own words, describe a step-by-step approach for solving algebraic word problems.

56. Write an original word problem that can be solved using a linear equation. Then solve the problem.

57. Did you have difficulties solving some of the problems that were assigned in this Exercise Set? Discuss what you did if this happened to you. Did your course of action enhance your ability to solve algebraic word problems?

Technology Exercises

58. Use a graphing utility to numerically or graphically verify your work in any one exercise from Exercises 27–30. For assistance on how to do this, refer to the Technology box on page 258.

59. A tennis club offers two payment options. Members can pay a monthly fee of $30 plus $5 per hour for court rental time. The second option has no monthly fee, but court time costs $7.50 per hour.

 a. Write a linear function that models total monthly costs for each option for x hours of court rental time.

 b. Use a graphing utility to graph the two functions in a [0, 15, 1] by [0, 120, 20] viewing rectangle.

 c. Use your utility's trace or intersection feature to determine where the two graphs intersect. Describe what the coordinates of this intersection point represent in practical terms.

 d. Verify part (c) using an algebraic approach by setting the two functions equal to one another and determining how many hours one has to rent the court so that the two plans result in identical monthly costs.

Critical Thinking Exercises

Make Sense? *In Exercises 60–63, determine whether each statement makes sense or does not make sense, and explain your reasoning.*

60. By modeling attitudes of college freshmen from 1969 through 2009, I can make precise predictions about the attitudes of the freshman class of 2020.

61. I find the hardest part in solving a word problem is writing the equation that models the verbal conditions.

62. I encountered a number of word problems where I had to determine the input that resulted in the same output for two linear functions.

63. After a 35% reduction, a computer's price is $780, so I determined the original price, x, by solving $x - 0.35 = 780$.

64. At the north campus of a performing arts school, 10% of the students are music majors. At the south campus, 90% of the students are music majors. The campuses are merged into one east campus. If 42% of the 1000 students at the east campus are music majors, how many students did the north and south campuses have before the merger?

65. The price of a dress is reduced by 40%. When the dress still does not sell, it is reduced by 40% of the reduced price. If the price of the dress after both reductions is $72, what was the original price?

66. In a film, the actor Charles Coburn plays an elderly "uncle" character criticized for marrying a woman when he is 3 times her age. He wittily replies, "Ah, but in 20 years time I shall only be twice her age." How old is the "uncle" and the woman?

67. Suppose that we agree to pay you 8¢ for every problem in this chapter that you solve correctly and fine you 5¢ for every problem done incorrectly. If at the end of 26 problems we do not owe each other any money, how many problems did you solve correctly?

68. It was wartime when the Ricardos found out Mrs. Ricardo was pregnant. Ricky Ricardo was drafted and made out a will, deciding that $14,000 in a savings account was to be divided between his wife and his child-to-be. Rather strangely, and certainly with gender bias, Ricky stipulated that if the child were a boy, he would get twice the amount of the mother's portion. If it were a girl, the mother would get twice the amount the girl was to receive. We'll never know what Ricky was thinking of, for (as fate would have it) he did not return from war. Mrs. Ricardo gave birth to twins—a boy and a girl. How was the money divided?

69. A thief steals a number of rare plants from a nursery. On the way out, the thief meets three security guards, one after another. To each security guard, the thief is forced to give one-half the plants that he still has, plus 2 more. Finally, the thief leaves the nursery with 1 lone palm. How many plants were originally stolen?

Group Exercise

70. One of the best ways to learn how to *solve* a word problem in algebra is to *design* word problems of your own. Creating a word problem makes you very aware of precisely how much information is needed to solve the problem. You must also focus on the best way to present information to a reader and on how much information to give. As you write your problem, you gain skills that will help you solve problems created by others.

 The group should design five different word problems that can be solved using linear equations. All of the problems should be on different topics. For example, the group should not have more than one problem on simple interest. The group should turn in both the problems and their algebraic solutions.

Preview Exercises

Exercises 71–73 will help you prepare for the material covered in the next section.

71. Multiply: $(7 - 3x)(-2 - 5x)$.

72. Simplify: $\sqrt{18} - \sqrt{8}$.

73. Rationalize the denominator: $\dfrac{7 + 4\sqrt{2}}{2 - 5\sqrt{2}}$.

Objectives

1. Add and subtract complex numbers.
2. Multiply complex numbers.
3. Divide complex numbers.
4. Perform operations with square roots of negative numbers.

Who is this kid warning us about our eyeballs turning black if we attempt to find the square root of -9? Don't believe what you hear on the street. Although square roots of negative numbers are not real numbers, they do play a significant role in algebra. In this section, we move beyond the real numbers and discuss square roots with negative radicands.

Copyright © 2011 by Roz Chast/The New Yorker Collection/ The Cartoon Bank

The Imaginary Unit *i*

In the next section, we will study equations whose solutions may involve the square roots of negative numbers. Because the square of a real number is never negative, there is no real number x such that $x^2 = -1$. To provide a setting in which such equations have solutions, mathematicians have invented an expanded system of numbers, the complex numbers. The *imaginary number i*, defined to be a solution of the equation $x^2 = -1$, is the basis of this new number system.

> The Imaginary Unit *i*
>
> The **imaginary unit** *i* is defined as
> $$i = \sqrt{-1}, \quad \text{where } i^2 = -1.$$

Using the imaginary unit *i*, we can express the square root of any negative number as a real multiple of *i*. For example,
$$\sqrt{-25} = \sqrt{-1}\sqrt{25} = i\sqrt{25} = 5i.$$
We can check this result by squaring $5i$ and obtaining -25.
$$(5i)^2 = 5^2 i^2 = 25(-1) = -25$$

A new system of numbers, called *complex numbers*, is based on adding multiples of *i*, such as $5i$, to real numbers.

> Complex Numbers and Imaginary Numbers
>
> The set of all numbers in the form
> $$a + bi,$$
> with real numbers a and b, and i, the imaginary unit, is called the set of **complex numbers**. The real number a is called the **real part** and the real number b is called the **imaginary part** of the complex number $a + bi$. If $b \neq 0$, then the complex number is called an **imaginary number** (**Figure 2.6**). An imaginary number in the form bi is called a **pure imaginary number**.

FIGURE 2.6 The complex number system

Here are some examples of complex numbers. Each number can be written in the form $a + bi$.

$$-4 + 6i \qquad\qquad 2i = 0 + 2i \qquad\qquad 3 = 3 + 0i$$

| a, the real part, is −4. | b, the imaginary part, is 6. | a, the real part, is 0. | b, the imaginary part, is 2. | a, the real part, is 3. | b, the imaginary part, is 0. |

Can you see that b, the imaginary part, is not zero in the first two complex numbers? Because $b \neq 0$, these complex numbers are imaginary numbers. Furthermore, the imaginary number $2i$ is a pure imaginary number. By contrast, the imaginary part of the complex number on the right is zero. This complex number is not an imaginary number. The number 3, or $3 + 0i$, is a real number.

A complex number is said to be **simplified** if it is expressed in the **standard form** $a + bi$. If b contains a radical, we usually write i before the radical. For example, we write $7 + 3i\sqrt{5}$ rather than $7 + 3\sqrt{5}i$, which could easily be confused with $7 + 3\sqrt{5i}$.

Expressed in standard form, two complex numbers are equal if and only if their real parts are equal and their imaginary parts are equal.

Equality of Complex Numbers

$a + bi = c + di$ if and only if $a = c$ and $b = d$.

① **Add and subtract complex numbers.**

Operations with Complex Numbers

The form of a complex number $a + bi$ is like the binomial $a + bx$. Consequently, we can add, subtract, and multiply complex numbers using the same methods we used for binomials, remembering that $i^2 = -1$.

Adding and Subtracting Complex Numbers

1. $(a + bi) + (c + di) = (a + c) + (b + d)i$

 In words, this says that you add complex numbers by adding their real parts, adding their imaginary parts, and expressing the sum as a complex number.

2. $(a + bi) - (c + di) = (a - c) + (b - d)i$

 In words, this says that you subtract complex numbers by subtracting their real parts, subtracting their imaginary parts, and expressing the difference as a complex number.

EXAMPLE 1 Adding and Subtracting Complex Numbers

Perform the indicated operations, writing the result in standard form:

a. $(5 - 11i) + (7 + 4i)$ **b.** $(-5 + i) - (-11 - 6i)$.

SOLUTION

a. $(5 - 11i) + (7 + 4i)$

$= 5 - 11i + 7 + 4i$ Remove the parentheses.

$= 5 + 7 - 11i + 4i$ Group real and imaginary terms.

$= (5 + 7) + (-11 + 4)i$ Add real parts and add imaginary parts.

$= 12 - 7i$ Simplify.

b. $(-5 + i) - (-11 - 6i)$

$= -5 + i + 11 + 6i$ Remove the parentheses. Change signs of real and imaginary parts in the complex number being subtracted.

$= -5 + 11 + i + 6i$ Group real and imaginary terms.

$= (-5 + 11) + (1 + 6)i$ Add real parts and add imaginary parts.

$= 6 + 7i$ Simplify.

• • •

GREAT QUESTION!

Are operations with complex numbers similar to operations with polynomials?

Yes. The following examples, using the same integers as in Example 1, show how operations with complex numbers are just like operations with polynomials.

a. $(5 - 11x) + (7 + 4x)$
$= 12 - 7x$

b. $(-5 + x) - (-11 - 6x)$
$= -5 + x + 11 + 6x$
$= 6 + 7x$

⊘ Check Point **1** Perform the indicated operations, writing the result in standard form:

a. $(5 - 2i) + (3 + 3i)$ **b.** $(2 + 6i) - (12 - i)$.

Multiplication of complex numbers is performed the same way as multiplication of polynomials, using the distributive property and the FOIL method. After completing the multiplication, we replace any occurrences of i^2 with -1. This idea is illustrated in the next example.

② Multiply complex numbers.

EXAMPLE 2 Multiplying Complex Numbers

Find the products:

a. $4i(3 - 5i)$ **b.** $(7 - 3i)(-2 - 5i)$.

SOLUTION

a. $4i(3 - 5i)$

$\qquad = 4i \cdot 3 - 4i \cdot 5i$ Distribute 4i throughout the parentheses.

$\qquad = 12i - 20i^2$ Multiply.

$\qquad = 12i - 20(-1)$ Replace i^2 with -1.

$\qquad = 20 + 12i$ Simplify to 12i + 20 and write in standard form.

b. $(7 - 3i)(-2 - 5i)$

$$\overset{\text{F}}{} \quad \overset{\text{O}}{} \quad \overset{\text{I}}{} \quad \overset{\text{L}}{}$$

$\qquad = -14 - 35i + 6i + 15i^2$ Use the FOIL method.

$\qquad = -14 - 35i + 6i + 15(-1)$ $i^2 = -1$

$\qquad = -14 - 15 - 35i + 6i$ Group real and imaginary terms.

$\qquad = -29 - 29i$ Combine real and imaginary terms. • • •

⊘ Check Point **2** Find the products:

a. $7i(2 - 9i)$ **b.** $(5 + 4i)(6 - 7i)$.

③ Divide complex numbers.

Complex Conjugates and Division

It is possible to multiply imaginary numbers and obtain a real number. This occurs when we multiply $a + bi$ and $a - bi$.

$$\overset{\text{F}}{} \quad \overset{\text{O}}{} \quad \overset{\text{I}}{} \quad \overset{\text{L}}{}$$

$(a + bi)(a - bi) = a^2 - abi + abi - b^2i^2$ Use the FOIL method.

$\qquad\qquad\qquad = a^2 - b^2(-1)$ $i^2 = -1$

$\qquad\qquad\qquad = a^2 + b^2$ Notice that this product eliminates i.

For the complex number $a + bi$, we define its *complex conjugate* to be $a - bi$. The multiplication of complex conjugates results in a real number.

Conjugate of a Complex Number

The **complex conjugate** of the number $a + bi$ is $a - bi$, and the complex conjugate of $a - bi$ is $a + bi$. The multiplication of complex conjugates gives a real number.

$$(a + bi)(a - bi) = a^2 + b^2$$

$$(a - bi)(a + bi) = a^2 + b^2$$

Complex conjugates are used to divide complex numbers. The goal of the division procedure is to obtain a real number in the denominator. This real number becomes the denominator of a and b in the quotient $a + bi$. By multiplying the numerator and the denominator of the division by the complex conjugate of the denominator, you will obtain this real number in the denominator.

EXAMPLE 3 Using Complex Conjugates to Divide Complex Numbers

Divide and express the result in standard form: $\dfrac{7 + 4i}{2 - 5i}$.

SOLUTION

The complex conjugate of the denominator, $2 - 5i$, is $2 + 5i$. Multiplication of both the numerator and the denominator by $2 + 5i$ will eliminate i from the denominator while maintaining the value of the expression.

$$\frac{7 + 4i}{2 - 5i} = \frac{(7 + 4i)}{(2 - 5i)} \cdot \frac{(2 + 5i)}{(2 + 5i)}$$

Multiply the numerator and the denominator by the complex conjugate of the denominator.

| F | O | I | L |

$$= \frac{14 + 35i + 8i + 20i^2}{2^2 + 5^2}$$

Use the FOIL method in the numerator and $(a - bi)(a + bi) = a^2 + b^2$ in the denominator.

$$= \frac{14 + 43i + 20(-1)}{29}$$

In the numerator, combine imaginary terms and replace i^2 with -1. In the denominator, $2^2 + 5^2 = 4 + 25 = 29$.

$$= \frac{-6 + 43i}{29}$$

Combine real terms in the numerator: $14 + 20(-1) = 14 - 20 = -6$.

$$= -\frac{6}{29} + \frac{43}{29}i$$

Express the answer in standard form.

Observe that the quotient is expressed in the standard form $a + bi$, with $a = -\frac{6}{29}$ and $b = \frac{43}{29}$. • • •

Check Point **3** Divide and express the result in standard form: $\dfrac{5 + 4i}{4 - i}$.

4 Perform operations with square roots of negative numbers.

Roots of Negative Numbers

The square of $4i$ and the square of $-4i$ both result in -16:

$$(4i)^2 = 16i^2 = 16(-1) = -16 \qquad (-4i)^2 = 16i^2 = 16(-1) = -16.$$

Consequently, in the complex number system -16 has two square roots, namely, $4i$ and $-4i$. We call $4i$ the **principal square root** of -16.

Principal Square Root of a Negative Number

For any positive real number b, the **principal square root** of the negative number $-b$ is defined by

$$\sqrt{-b} = i\sqrt{b}.$$

Consider the multiplication problem

$$5i \cdot 2i = 10i^2 = 10(-1) = -10.$$

This problem can also be given in terms of principal square roots of negative numbers:

$$\sqrt{-25} \cdot \sqrt{-4}.$$

Because the product rule for radicals only applies to real numbers, multiplying radicands is incorrect. **When performing operations with square roots of negative numbers, begin by expressing all square roots in terms of *i*.** Then perform the indicated operation.

Correct:

$$\sqrt{-25} \cdot \sqrt{-4} = i\sqrt{25} \cdot i\sqrt{4}$$
$$= 5i \cdot 2i$$
$$= 10i^2 = 10(-1) = -10$$

Incorrect:

$$\sqrt{-25} \cdot \sqrt{-4} = \sqrt{(-25)(-4)}$$
$$= \sqrt{100}$$
$$= 10$$

EXAMPLE 4 Operations Involving Square Roots of Negative Numbers

Perform the indicated operations and write the result in standard form:

a. $\sqrt{-18} - \sqrt{-8}$ **b.** $\left(-1 + \sqrt{-5}\right)^2$ **c.** $\dfrac{-25 + \sqrt{-50}}{15}$.

SOLUTION

Begin by expressing all square roots of negative numbers in terms of *i*.

a. $\sqrt{-18} - \sqrt{-8} = i\sqrt{18} - i\sqrt{8} = i\sqrt{9 \cdot 2} - i\sqrt{4 \cdot 2}$
$$= 3i\sqrt{2} - 2i\sqrt{2} = i\sqrt{2}$$

b. $\left(-1 + \sqrt{-5}\right)^2 = \left(-1 + i\sqrt{5}\right)^2 = (-1)^2 + 2(-1)\left(i\sqrt{5}\right) + \left(i\sqrt{5}\right)^2$
$$= 1 - 2i\sqrt{5} + 5i^2$$
$$= 1 - 2i\sqrt{5} + 5(-1)$$
$$= -4 - 2i\sqrt{5}$$

c. $\dfrac{-25 + \sqrt{-50}}{15}$

$= \dfrac{-25 + i\sqrt{50}}{15}$ $\sqrt{-b} = i\sqrt{b}$

$= \dfrac{-25 + 5i\sqrt{2}}{15}$ $\sqrt{50} = \sqrt{25 \cdot 2} = 5\sqrt{2}$

$= \dfrac{-25}{15} + \dfrac{5i\sqrt{2}}{15}$ Write the complex number in standard form.

$= -\dfrac{5}{3} + i\dfrac{\sqrt{2}}{3}$ Simplify. $\bullet\bullet\bullet$

⌀ Check Point **4** Perform the indicated operations and write the result in standard form:

a. $\sqrt{-27} + \sqrt{-48}$ **b.** $\left(-2 + \sqrt{-3}\right)^2$ **c.** $\dfrac{-14 + \sqrt{-12}}{2}$.

CONCEPT AND VOCABULARY CHECK

Fill in each blank so that the resulting statement is true.

1. The imaginary unit i is defined as $i =$ _____, where $i^2 =$ ____.
2. The set of all numbers in the form $a + bi$ is called the set of _____ numbers. If $b \neq 0$, then the number is also called a/an _____ number. If $b = 0$, then the number is also called a/an _____ number.
3. $-9i + 3i =$ _____
4. $10i - (-4i) =$ _____
5. Consider the following multiplication problem:
$$(3 + 2i)(6 - 5i).$$
Using the FOIL method, the product of the first terms is ____, the product of the outside terms is _____, and

the product of the inside terms is _____. The product of the last terms in terms of i^2 is _____, which simplifies to ____.
6. The conjugate of $2 - 9i$ is _____.
7. The division
$$\frac{7 + 4i}{2 - 5i}$$
is performed by multiplying the numerator and denominator by _____.
8. $\sqrt{-20} =$ ___ $\sqrt{20} =$ ___ $\sqrt{4 \cdot 5} =$ _____

EXERCISE SET 2.3

Practice Exercises

In Exercises 1–8, add or subtract as indicated and write the result in standard form.

1. $(7 + 2i) + (1 - 4i)$
2. $(-2 + 6i) + (4 - i)$
3. $(3 + 2i) - (5 - 7i)$
4. $(-7 + 5i) - (-9 - 11i)$
5. $6 - (-5 + 4i) - (-13 - i)$
6. $7 - (-9 + 2i) - (-17 - i)$
7. $8i - (14 - 9i)$
8. $15i - (12 - 11i)$

In Exercises 9–20, find each product and write the result in standard form.

9. $-3i(7i - 5)$
10. $-8i(2i - 7)$
11. $(-5 + 4i)(3 + i)$
12. $(-4 - 8i)(3 + i)$
13. $(7 - 5i)(-2 - 3i)$
14. $(8 - 4i)(-3 + 9i)$
15. $(3 + 5i)(3 - 5i)$
16. $(2 + 7i)(2 - 7i)$
17. $(-5 + i)(-5 - i)$
18. $(-7 - i)(-7 + i)$
19. $(2 + 3i)^2$
20. $(5 - 2i)^2$

In Exercises 21–28, divide and express the result in standard form.

21. $\dfrac{2}{3 - i}$
22. $\dfrac{3}{4 + i}$
23. $\dfrac{2i}{1 + i}$
24. $\dfrac{5i}{2 - i}$
25. $\dfrac{8i}{4 - 3i}$
26. $\dfrac{-6i}{3 + 2i}$
27. $\dfrac{2 + 3i}{2 + i}$
28. $\dfrac{3 - 4i}{4 + 3i}$

In Exercises 29–44, perform the indicated operations and write the result in standard form.

29. $\sqrt{-64} - \sqrt{-25}$
30. $\sqrt{-81} - \sqrt{-144}$
31. $5\sqrt{-16} + 3\sqrt{-81}$
32. $5\sqrt{-8} + 3\sqrt{-18}$
33. $\left(-2 + \sqrt{-4}\right)^2$
34. $\left(-5 - \sqrt{-9}\right)^2$
35. $\left(-3 - \sqrt{-7}\right)^2$
36. $\left(-2 + \sqrt{-11}\right)^2$
37. $\dfrac{-8 + \sqrt{-32}}{24}$
38. $\dfrac{-12 + \sqrt{-28}}{32}$
39. $\dfrac{-6 - \sqrt{-12}}{48}$
40. $\dfrac{-15 - \sqrt{-18}}{33}$
41. $\sqrt{-8}\left(\sqrt{-3} - \sqrt{5}\right)$
42. $\sqrt{-12}\left(\sqrt{-4} - \sqrt{2}\right)$
43. $\left(3\sqrt{-5}\right)\left(-4\sqrt{-12}\right)$
44. $\left(3\sqrt{-7}\right)\left(2\sqrt{-8}\right)$

Practice Plus

In Exercises 45–50, perform the indicated operation(s) and write the result in standard form.

45. $(2 - 3i)(1 - i) - (3 - i)(3 + i)$
46. $(8 + 9i)(2 - i) - (1 - i)(1 + i)$
47. $(2 + i)^2 - (3 - i)^2$
48. $(4 - i)^2 - (1 + 2i)^2$
49. $5\sqrt{-16} + 3\sqrt{-81}$
50. $5\sqrt{-8} + 3\sqrt{-18}$
51. Let $f(x) = x^2 - 2x + 2$. Find $f(1 + i)$.
52. Let $f(x) = x^2 - 2x + 5$. Find $f(1 - 2i)$.

In Exercises 53–56, simplify each evaluation to standard form.

53. Let $f(x) = x - 3i$ and $g(x) = 4x + 2i$. Find $(fg)(-1)$.

54. Let $f(x) = 12x - i$ and $g(x) = 6x + 3i$. Find $(fg)(-\frac{1}{3})$.

55. Let $f(x) = \dfrac{x^2 + 19}{2 - x}$. Find $f(3i)$.

56. Let $f(x) = \dfrac{x^2 + 11}{3 - x}$. Find $f(4i)$.

Application Exercises

Complex numbers are used in electronics to describe the current in an electric circuit. Ohm's law relates the current in a circuit, I, in amperes, the voltage of the circuit, E, in volts, and the resistance of the circuit, R, in ohms, by the formula $E = IR$. Use this formula to solve Exercises 57–58.

57. Find E, the voltage of a circuit, if $I = (4 - 5i)$ amperes and $R = (3 + 7i)$ ohms.

58. Find E, the voltage of a circuit, if $I = (2 - 3i)$ amperes and $R = (3 + 5i)$ ohms.

59. The mathematician Girolamo Cardano is credited with the first use (in 1545) of negative square roots in solving the now-famous problem, "Find two numbers whose sum is 10 and whose product is 40." Show that the complex numbers $5 + i\sqrt{15}$ and $5 - i\sqrt{15}$ satisfy the conditions of the problem. (Cardano did not use the symbolism $i\sqrt{15}$ or even $\sqrt{-15}$. He wrote R.m 15 for $\sqrt{-15}$, meaning "radix minus 15." He regarded the numbers $5 +$ R.m 15 and $5 -$ R.m 15 as "fictitious" or "ghost numbers," and considered the problem "manifestly impossible." But in a mathematically adventurous spirit, he exclaimed, "Nevertheless, we will operate.")

Writing in Mathematics

60. What is i?

61. Explain how to add complex numbers. Provide an example with your explanation.

62. Explain how to multiply complex numbers and give an example.

63. What is the complex conjugate of $2 + 3i$? What happens when you multiply this complex number by its complex conjugate?

64. Explain how to divide complex numbers. Provide an example with your explanation.

65. Explain each of the three jokes in the cartoon on page 266.

66. A stand-up comedian uses algebra in some jokes, including one about a telephone recording that announces "You have just reached an imaginary number. Please multiply by i and dial again." Explain the joke.

Explain the error in Exercises 67–68.

67. $\sqrt{-9} + \sqrt{-16} = \sqrt{-25} = i\sqrt{25} = 5i$

68. $\left(\sqrt{-9}\right)^2 = \sqrt{-9} \cdot \sqrt{-9} = \sqrt{81} = 9$

Critical Thinking Exercises

Make Sense? *In Exercises 69–72, determine whether each statement makes sense or does not make sense, and explain your reasoning.*

69. The humor in the cartoon is based on the fact that "rational" and "real" have different meanings in mathematics and in everyday speech.

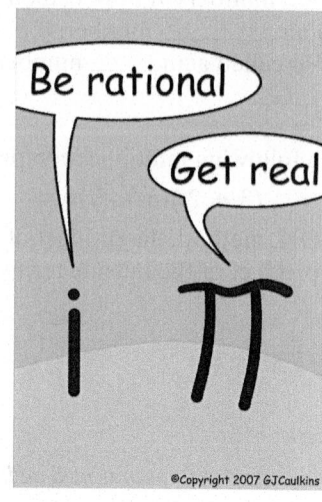

© 2007 GJ Caulkins

70. The word *imaginary* in imaginary numbers tells me that these numbers are undefined.

71. By writing the imaginary number $5i$, I can immediately see that 5 is the constant and i is the variable.

72. When I add or subtract complex numbers, I am basically combining like terms.

In Exercises 73–76, determine whether each statement is true or false. If the statement is false, make the necessary change(s) to produce a true statement.

73. Some irrational numbers are not complex numbers.

74. $(3 + 7i)(3 - 7i)$ is an imaginary number.

75. $\dfrac{7 + 3i}{5 + 3i} = \dfrac{7}{5}$

76. In the complex number system, $x^2 + y^2$ (the sum of two squares) can be factored as $(x + yi)(x - yi)$.

In Exercises 77–79, perform the indicated operations and write the result in standard form.

77. $\dfrac{4}{(2 + i)(3 - i)}$

78. $\dfrac{1 + i}{1 + 2i} + \dfrac{1 - i}{1 - 2i}$

79. $\dfrac{8}{1 + \frac{2}{i}}$

Preview Exercises

Exercises 80–82 will help you prepare for the material covered in the next section.

80. Factor: $2x^2 + 7x - 4$.

81. Factor: $x^2 - 6x + 9$.

82. Evaluate

$$\frac{-b - \sqrt{b^2 - 4ac}}{2a}$$

for $a = 2$, $b = 9$, and $c = -5$.

SECTION 2.4

Quadratic Equations

Objectives

1. Solve quadratic equations by factoring.
2. Solve quadratic equations by the square root property.
3. Solve quadratic equations by completing the square.
4. Solve quadratic equations using the quadratic formula.
5. Use the discriminant to determine the number and type of solutions.
6. Determine the most efficient method to use when solving a quadratic equation.
7. Solve problems modeled by quadratic equations.

*I'm very well acquainted, too, with matters mathematical,
I understand equations, both simple and quadratical.
About binomial theorem I'm teeming with a lot of news,
With many cheerful facts about the square of the
hypotenuse.*

—Gilbert and Sullivan,
The Pirates of Penzance

Equations quadratical? Cheerful news about the square of the hypotenuse? You've come to the right place. In this section, we study a number of methods for solving *quadratic equations*, equations in which the highest exponent on the variable is 2. (Yes, it's *quadratic* and not *quadratical*, despite the latter's rhyme with mathematical.) We also look at applications of quadratic equations, introducing (cheerfully, of course) the Pythagorean Theorem and the square of the hypotenuse.

We begin by defining a quadratic equation.

Definition of a Quadratic Equation

A **quadratic equation** in x is an equation that can be written in the **general form**

$$ax^2 + bx + c = 0,$$

where a, b, and c are real numbers, with $a \neq 0$. A quadratic equation in x is also called a **second-degree polynomial equation** in x.

① Solve quadratic equations by factoring.

Solving Quadratic Equations by Factoring

Here is an example of a quadratic equation in general form:

$$x^2 - 7x + 10 = 0.$$

$a = 1$ $b = -7$ $c = 10$

We can factor the left side of this equation. We obtain $(x - 5)(x - 2) = 0$. If a quadratic equation has zero on one side and a factored expression on the other side, it can be solved using the **zero-product principle**.

The Zero-Product Principle

If the product of two algebraic expressions is zero, then at least one of the factors is equal to zero.

If $AB = 0$, then $A = 0$ or $B = 0$.

For example, consider the equation $(x - 5)(x - 2) = 0$. According to the zero-product principle, this product can be zero only if at least one of the factors is zero. We set each individual factor equal to zero and solve each resulting equation for x.

$$(x - 5)(x - 2) = 0$$
$$x - 5 = 0 \quad \text{or} \quad x - 2 = 0$$
$$x = 5 \qquad\qquad x = 2$$

We can check each of these proposed solutions, 5 and 2, in the original quadratic equation, $x^2 - 7x + 10 = 0$. Substitute each one separately for x in the equation.

Check 5:

$$x^2 - 7x + 10 = 0$$
$$5^2 - 7 \cdot 5 + 10 \stackrel{?}{=} 0$$
$$25 - 35 + 10 \stackrel{?}{=} 0$$
$$0 = 0, \quad \text{true}$$

Check 2:

$$x^2 - 7x + 10 = 0$$
$$2^2 - 7 \cdot 2 + 10 \stackrel{?}{=} 0$$
$$4 - 14 + 10 \stackrel{?}{=} 0$$
$$0 = 0, \quad \text{true}$$

The resulting true statements indicate that the solutions are 2 and 5. The solution set is {2, 5}. Note that with a quadratic equation, we can have two solutions, compared to the conditional linear equation that had one.

Solving a Quadratic Equation by Factoring

1. If necessary, rewrite the equation in the general form $ax^2 + bx + c = 0$, moving all nonzero terms to one side, thereby obtaining zero on the other side.
2. Factor completely.
3. Apply the zero-product principle, setting each factor containing a variable equal to zero.
4. Solve the equations in step 3.
5. Check the solutions in the original equation.

EXAMPLE 1 Solving Quadratic Equations by Factoring

Solve by factoring:

 a. $4x^2 - 2x = 0$ **b.** $2x^2 + 7x = 4$.

SOLUTION

 a. We begin with $4x^2 - 2x = 0$.

Step 1 Move all nonzero terms to one side and obtain zero on the other side. All nonzero terms are already on the left and zero is on the other side, so we can skip this step.

Step 2 Factor. We factor out $2x$ from the two terms on the left side.

$$4x^2 - 2x = 0 \quad \text{This is the given equation.}$$
$$2x(2x - 1) = 0 \quad \text{Factor.}$$

Steps 3 and 4 Set each factor equal to zero and solve the resulting equations.

$$2x = 0 \quad \text{or} \quad 2x - 1 = 0$$
$$x = 0 \qquad\qquad 2x = 1$$
$$x = \tfrac{1}{2}$$

Step 5 Check the solutions in the original equation.

<div align="center">

Check 0:

$$4x^2 - 2x = 0$$
$$4 \cdot 0^2 - 2 \cdot 0 \stackrel{?}{=} 0$$
$$0 - 0 \stackrel{?}{=} 0$$
$$0 = 0, \quad \text{true}$$

</div>

<div align="center">

Check $\frac{1}{2}$:

$$4x^2 - 2x = 0$$
$$4\left(\tfrac{1}{2}\right)^2 - 2\left(\tfrac{1}{2}\right) \stackrel{?}{=} 0$$
$$4\left(\tfrac{1}{4}\right) - 2\left(\tfrac{1}{2}\right) \stackrel{?}{=} 0$$
$$1 - 1 \stackrel{?}{=} 0$$
$$0 = 0, \quad \text{true}$$

</div>

The solution set is $\left\{0, \tfrac{1}{2}\right\}$.

b. Next, we solve $2x^2 + 7x = 4$.

Step 1 Move all nonzero terms to one side and obtain zero on the other side. Subtract 4 from both sides and write the equation in general form.

$$
\begin{aligned}
2x^2 + 7x &= 4 && \text{This is the given equation.}\\
2x^2 + 7x - 4 &= 4 - 4 && \text{Subtract 4 from both sides.}\\
2x^2 + 7x - 4 &= 0 && \text{Simplify.}
\end{aligned}
$$

Step 2 Factor.

$$2x^2 + 7x - 4 = 0$$
$$(2x - 1)(x + 4) = 0$$

Steps 3 and 4 Set each factor equal to zero and solve the resulting equations.

$$
\begin{array}{lcl}
2x - 1 = 0 & \quad\text{or}\quad & x + 4 = 0\\
2x = 1 & & x = -4\\
x = \tfrac{1}{2} & &
\end{array}
$$

Step 5 Check the solutions in the original equation.

<div align="center">

Check $\frac{1}{2}$:

$$2x^2 + 7x = 4$$
$$2\left(\tfrac{1}{2}\right)^2 + 7\left(\tfrac{1}{2}\right) \stackrel{?}{=} 4$$
$$\tfrac{1}{2} + \tfrac{7}{2} \stackrel{?}{=} 4$$
$$4 = 4, \quad \text{true}$$

</div>

<div align="center">

Check -4:

$$2x^2 + 7x = 4$$
$$2(-4)^2 + 7(-4) \stackrel{?}{=} 4$$
$$32 + (-28) \stackrel{?}{=} 4$$
$$4 = 4, \quad \text{true}$$

</div>

The solution set is $\left\{-4, \tfrac{1}{2}\right\}$. • • •

⌀ **Check Point 1** Solve by factoring:

 a. $3x^2 - 9x = 0$ **b.** $2x^2 + x = 1$.

There are important connections between quadratic functions, $f(x) = ax^2 + bx + c$, and quadratic equations in general form, $ax^2 + bx + c = 0$ or $f(x) = 0$. **The real solutions, or real roots, of the quadratic equation correspond to the x-intercepts of the graph of the quadratic function.** For example, the solutions of $2x^2 + 7x - 4 = 0$

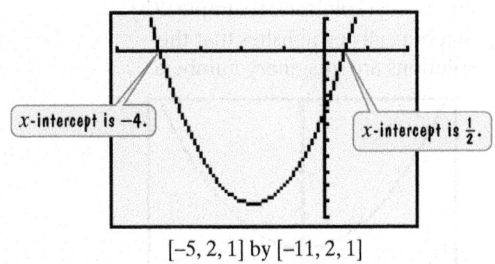

$[-5, 2, 1]$ by $[-11, 2, 1]$

FIGURE 2.7

are -4 and $\tfrac{1}{2}$. These solutions appear as the x-intercepts of the graph of $f(x) = 2x^2 + 7x - 4$, as shown in **Figure 2.7**.

When solving all types of equations, including quadratic equations, always keep in mind the relationship among zeros, roots, and x-intercepts. **The zeros of function f are the roots, or solutions, of the equation $f(x) = 0$. Furthermore, the real zeros, or real solutions, are the x-intercepts of the graph of f.**

② Solve quadratic equations by the square root property.

Solving Quadratic Equations by the Square Root Property

In Chapter P, we introduced the *square root property*. Now that we have studied square roots of negative numbers and represented them in terms of i, we can extend the property to include such roots. The **square root property** states that if u is an algebraic expression and d is a nonzero number, then $u^2 = d$ has exactly two solutions:

$$\text{If } u^2 = d, \text{ then } u = \sqrt{d} \text{ or } u = -\sqrt{d}, \text{ abbreviated } u = \pm\sqrt{d}.$$

EXAMPLE 2 Solving Quadratic Equations by the Square Root Property

Solve by the square root property:

a. $3x^2 - 15 = 0$ **b.** $9x^2 + 25 = 0$ **c.** $(x - 2)^2 = 6.$

SOLUTION

To apply the square root property, we need a squared expression by itself on one side of the equation.

$$3x^2 - 15 = 0 \qquad 9x^2 + 25 = 0 \qquad (x - 2)^2 = 6$$

We want x^2 by itself. We want x^2 by itself. The squared expression is by itself.

a.
$$3x^2 - 15 = 0 \qquad \text{This is the original equation.}$$
$$3x^2 = 15 \qquad \text{Add 15 to both sides.}$$
$$x^2 = 5 \qquad \text{Divide both sides by 3.}$$
$$x = \sqrt{5} \text{ or } x = -\sqrt{5} \qquad \text{Apply the square root property. Equivalently, } x = \pm\sqrt{5}.$$

By checking both proposed solutions in the original equation, we can confirm that the solution set is $\{-\sqrt{5}, \sqrt{5}\}$ or $\{\pm\sqrt{5}\}$.

b.
$$9x^2 + 25 = 0 \qquad \text{This is the original equation.}$$
$$9x^2 = -25 \qquad \text{Subtract 25 from both sides.}$$
$$x^2 = -\frac{25}{9} \qquad \text{Divide both sides by 9.}$$
$$x = \pm\sqrt{-\frac{25}{9}} \qquad \text{Apply the square root property.}$$
$$x = \pm i\sqrt{\frac{25}{9}} = \pm\frac{5}{3}i \qquad \text{Express solutions in terms of } i.$$

Because the equation has an x^2-term and no x-term, we can check both proposed solutions, $\pm\frac{5}{3}i$, at once.

$$\textbf{Check } \frac{5}{3}i \textbf{ and } -\frac{5}{3}i\text{:}$$

$$9x^2 + 25 = 0$$

$$9\left(\pm\frac{5}{3}i\right)^2 + 25 \stackrel{?}{=} 0$$

TECHNOLOGY

Graphic Connections

The graph of

$$y = 9x^2 + 25$$

has no x-intercepts. This shows that

$$9x^2 + 25 = 0$$

has no real solutions. Example 2(b) algebraically establishes that the solutions are imaginary numbers.

$[-3, 3, 1]$ by $[-5, 100, 5]$

$$9\left(\frac{25}{9}i^2\right) + 25 \stackrel{?}{=} 0$$

$$25i^2 + 25 \stackrel{?}{=} 0$$

$$i^2 = -1$$

$$25(-1) + 25 \stackrel{?}{=} 0$$

$$0 = 0 \quad \text{true}$$

The solutions are $-\frac{5}{3}i$ and $\frac{5}{3}i$. The solution set is $\left\{-\frac{5}{3}i, \frac{5}{3}i\right\}$ or $\left\{\pm\frac{5}{3}i\right\}$.

c. $(x - 2)^2 = 6$ This is the original equation.

 $x - 2 = \pm\sqrt{6}$ Apply the square root property.

 $x = 2 \pm \sqrt{6}$ Add 2 to both sides.

By checking both values in the original equation, we can confirm that the solution set is $\{2 + \sqrt{6}, 2 - \sqrt{6}\}$ or $\{2 \pm \sqrt{6}\}$. • • •

✓ Check Point **2** Solve by the square root property:

 a. $3x^2 - 21 = 0$ **b.** $5x^2 + 45 = 0$ **c.** $(x + 5)^2 = 11$.

③ Solve quadratic equations by completing the square.

Completing the Square

How do we solve an equation in the form $ax^2 + bx + c = 0$ if the trinomial $ax^2 + bx + c$ cannot be factored? We cannot use the zero-product principle in such a case. However, we can convert the equation into an equivalent equation that can be solved using the square root property. This is accomplished by **completing the square**.

> *This box was first presented in Chapter P. We've repeated it to remind you of the procedure.*

Completing the Square

If $x^2 + bx$ is a binomial, then by adding $\left(\frac{b}{2}\right)^2$, which is the square of half the coefficient of x, a perfect square trinomial will result. That is,

$$x^2 + bx + \left(\frac{b}{2}\right)^2 = \left(x + \frac{b}{2}\right)^2.$$

We can solve *any* quadratic equation by completing the square. If the coefficient of the x^2-term is one, we add the square of half the coefficient of x to both sides of the equation. **When you add a constant term to one side of the equation to complete the square, be certain to add the same constant to the other side of the equation.** These ideas are illustrated in Example 3.

EXAMPLE 3 Solving a Quadratic Equation by Completing the Square

Solve by completing the square: $x^2 - 6x + 4 = 0$.

SOLUTION

We begin by subtracting 4 from both sides. This is done to isolate the binomial $x^2 - 6x$, so that we can complete the square.

$$x^2 - 6x + 4 = 0 \quad \text{This is the original equation.}$$
$$x^2 - 6x = -4 \quad \text{Subtract 4 from both sides.}$$

GREAT QUESTION!

When I solve a quadratic equation by completing the square, doesn't this result in a new equation? How do I know that the solutions of this new equation are the same as those of the given equation?

When you complete the square for the binomial expression $x^2 + bx$, you obtain a different polynomial. When you solve a quadratic equation by completing the square, you obtain an equation with the same solution set because the constant needed to complete the square is added to *both sides*.

Next, we work with $x^2 - 6x = -4$ and complete the square. Find half the coefficient of the x-term and square it. The coefficient of the x-term is -6. Half of -6 is -3 and $(-3)^2 = 9$. Thus, we add 9 to both sides of the equation.

$$x^2 - 6x + 9 = -4 + 9 \qquad \text{Add 9 to both sides to complete the square.}$$

$$(x - 3)^2 = 5 \qquad \text{Factor and simplify.}$$

$$x - 3 = \sqrt{5} \quad \text{or} \quad x - 3 = -\sqrt{5} \qquad \text{Apply the square root property.}$$

$$x = 3 + \sqrt{5} \qquad x = 3 - \sqrt{5} \qquad \text{Add 3 to both sides in each equation.}$$

The solutions are $3 \pm \sqrt{5}$ and the solution set is $\{3 + \sqrt{5}, 3 - \sqrt{5}\}$, or $\{3 \pm \sqrt{5}\}$. • • •

✓ **Check Point 3** Solve by completing the square: $x^2 + 4x - 1 = 0$.

If the coefficient of the x^2-term in a quadratic equation is not 1, you must divide each side of the equation by this coefficient before completing the square. For example, to solve $9x^2 - 6x - 4 = 0$ by completing the square, first divide every term by 9:

$$\frac{9x^2}{9} - \frac{6x}{9} - \frac{4}{9} = \frac{0}{9}$$

$$x^2 - \frac{6}{9}x - \frac{4}{9} = 0$$

$$x^2 - \frac{2}{3}x - \frac{4}{9} = 0.$$

Now that the coefficient of the x^2-term is 1, we can solve by completing the square.

EXAMPLE 4 Finding a Function's Zeros by Completing the Square

Find all zeros of $f(x) = 9x^2 - 6x - 4$.

SOLUTION

By definition, the zeros are the values of x for which $f(x)$ is equal to 0. Thus, we set $f(x)$ equal to 0:

$$f(x) = 9x^2 - 6x - 4 = 0.$$

We solve the quadratic equation $9x^2 - 6x - 4 = 0$ by completing the square.

$$9x^2 - 6x - 4 = 0 \qquad \text{This is the equation needed to find the function's zeros.}$$

$$x^2 - \frac{2}{3}x - \frac{4}{9} = 0 \qquad \text{Divide both sides by 9.}$$

$$x^2 - \frac{2}{3}x = \frac{4}{9} \qquad \text{Add } \tfrac{4}{9} \text{ to both sides to isolate the binomial.}$$

$$x^2 - \frac{2}{3}x + \frac{1}{9} = \frac{4}{9} + \frac{1}{9} \qquad \text{Complete the square: Half of } -\tfrac{2}{3} \text{ is } -\tfrac{2}{6}, \text{ or } -\tfrac{1}{3}, \text{ and } \left(-\tfrac{1}{3}\right)^2 = \tfrac{1}{9}.$$

$$\left(x - \frac{1}{3}\right)^2 = \frac{5}{9} \qquad \text{Factor and simplify.}$$

$$x - \frac{1}{3} = \sqrt{\frac{5}{9}} \quad \text{or} \quad x - \frac{1}{3} = -\sqrt{\frac{5}{9}} \qquad \text{Apply the square root property.}$$

$$x - \frac{1}{3} = \frac{\sqrt{5}}{3} \qquad x - \frac{1}{3} = -\frac{\sqrt{5}}{3} \qquad \sqrt{\frac{5}{9}} = \frac{\sqrt{5}}{\sqrt{9}} = \frac{\sqrt{5}}{3}$$

$$x = \frac{1}{3} + \frac{\sqrt{5}}{3} \qquad x = \frac{1}{3} - \frac{\sqrt{5}}{3} \qquad \text{Add } \tfrac{1}{3} \text{ to both sides and solve for } x.$$

TECHNOLOGY

Graphic Connections

Obtain a decimal approximation for each zero of

$$f(x) = 9x^2 - 6x - 4,$$

the function in Example 4.

$$\frac{1 + \sqrt{5}}{3} \approx 1.1; \quad \frac{1 - \sqrt{5}}{3} \approx -0.4$$

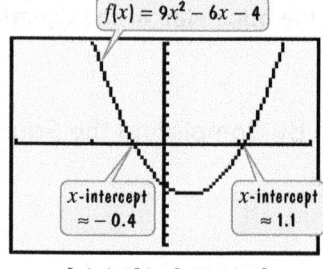

$f(x) = 9x^2 - 6x - 4$

x-intercept ≈ -0.4 x-intercept ≈ 1.1

$[-2, 2, 1]$ by $[-10, 10, 1]$

The x-intercepts of the graph of $f(x) = 9x^2 - 6x - 4$ verify the zeros that we found algebraically.

$$x = \frac{1 + \sqrt{5}}{3} \qquad\qquad x = \frac{1 - \sqrt{5}}{3}$$

Express solutions with a common denominator.

The zeros of f are $\dfrac{1 + \sqrt{5}}{3}$ and $\dfrac{1 - \sqrt{5}}{3}$. The graph of f in the technology box on the previous page shows that each zero is an x-intercept.

• • •

⊘ Check Point **4** Find all zeros of: $f(x) = 2x^2 + 3x - 4$.

Solving Quadratic Equations Using the Quadratic Formula

We can use the method of completing the square to derive a formula that can be used to solve all quadratic equations. The derivation given below also shows a particular quadratic equation, $3x^2 - 2x - 4 = 0$, to specifically illustrate each of the steps.

Deriving the Quadratic Formula

General Form of a Quadratic Equation	Comment	A Specific Example
$ax^2 + bx + c = 0, a > 0$	This is the given equation.	$3x^2 - 2x - 4 = 0$
$x^2 + \dfrac{b}{a}x + \dfrac{c}{a} = 0$	Divide both sides by the coefficient of x^2.	$x^2 - \dfrac{2}{3}x - \dfrac{4}{3} = 0$
$x^2 + \dfrac{b}{a}x = -\dfrac{c}{a}$	Isolate the binomial by adding $-\dfrac{c}{a}$ on both sides of the equation.	$x^2 - \dfrac{2}{3}x = \dfrac{4}{3}$
$x^2 + \dfrac{b}{a}x + \left(\dfrac{b}{2a}\right)^2 = -\dfrac{c}{a} + \left(\dfrac{b}{2a}\right)^2$ $\underbrace{\qquad}_{(\text{half})^2}$	Complete the square. Add the square of half the coefficient of x to both sides.	$x^2 - \dfrac{2}{3}x + \left(-\dfrac{1}{3}\right)^2 = \dfrac{4}{3} + \left(-\dfrac{1}{3}\right)^2$ $\underbrace{\qquad}_{(\text{half})^2}$
$x^2 + \dfrac{b}{a}x + \dfrac{b^2}{4a^2} = -\dfrac{c}{a} + \dfrac{b^2}{4a^2}$		$x^2 - \dfrac{2}{3}x + \dfrac{1}{9} = \dfrac{4}{3} + \dfrac{1}{9}$
$\left(x + \dfrac{b}{2a}\right)^2 = -\dfrac{c}{a}\cdot\dfrac{4a}{4a} + \dfrac{b^2}{4a^2}$	Factor on the left side and obtain a common denominator on the right side.	$\left(x - \dfrac{1}{3}\right)^2 = \dfrac{4}{3}\cdot\dfrac{3}{3} + \dfrac{1}{9}$
$\left(x + \dfrac{b}{2a}\right)^2 = \dfrac{-4ac + b^2}{4a^2}$	Add fractions on the right side.	$\left(x - \dfrac{1}{3}\right)^2 = \dfrac{12 + 1}{9}$
$\left(x + \dfrac{b}{2a}\right)^2 = \dfrac{b^2 - 4ac}{4a^2}$		$\left(x - \dfrac{1}{3}\right)^2 = \dfrac{13}{9}$
$x + \dfrac{b}{2a} = \pm\sqrt{\dfrac{b^2 - 4ac}{4a^2}}$	Apply the square root property.	$x - \dfrac{1}{3} = \pm\sqrt{\dfrac{13}{9}}$
$x + \dfrac{b}{2a} = \pm\dfrac{\sqrt{b^2 - 4ac}}{2a}$	Take the square root of the quotient, simplifying the denominator.	$x - \dfrac{1}{3} = \pm\dfrac{\sqrt{13}}{3}$
$x = \dfrac{-b}{2a} \pm \dfrac{\sqrt{b^2 - 4ac}}{2a}$	Solve for x by subtracting $\dfrac{b}{2a}$ from both sides.	$x = \dfrac{1}{3} \pm \dfrac{\sqrt{13}}{3}$
$x = \dfrac{-b \pm \sqrt{b^2 - 4ac}}{2a}$	Combine fractions on the right side.	$x = \dfrac{1 \pm \sqrt{13}}{3}$

The formula shown at the bottom of the left column is called the *quadratic formula*. A similar proof shows that the same formula can be used to solve quadratic equations if a, the coefficient of the x^2-term, is negative.

④ Solve quadratic equations using the quadratic formula.

The Quadratic Formula

The solutions of a quadratic equation in general form $ax^2 + bx + c = 0$, with $a \neq 0$, are given by the **quadratic formula**:

$$x = \frac{-b \pm \sqrt{b^2 - 4ac}}{2a}.$$

> *x equals negative b plus or minus the square root of $b^2 - 4ac$, all divided by 2a.*

To use the quadratic formula, rewrite the quadratic equation in general form if necessary. Then determine the numerical values for a (the coefficient of the x^2-term), b (the coefficient of the x-term), and c (the constant term). Substitute the values of a, b, and c into the quadratic formula and evaluate the expression. The \pm sign indicates that there are two (not necessarily distinct) solutions of the equation.

EXAMPLE 5 Solving a Quadratic Equation Using the Quadratic Formula

Solve using the quadratic formula: $2x^2 - 6x + 1 = 0$.

SOLUTION

The given equation is in general form. Begin by identifying the values for a, b, and c.

$$2x^2 - 6x + 1 = 0$$

$$a = 2 \qquad b = -6 \qquad c = 1$$

Substituting these values into the quadratic formula and simplifying gives the equation's solutions.

$$x = \frac{-b \pm \sqrt{b^2 - 4ac}}{2a}$$ Use the quadratic formula.

$$= \frac{-(-6) \pm \sqrt{(-6)^2 - 4(2)(1)}}{2 \cdot 2}$$ Substitute the values for a, b, and c: $a = 2$, $b = -6$, and $c = 1$.

$$= \frac{6 \pm \sqrt{36 - 8}}{4}$$ $-(-6) = 6$, $(-6)^2 = (-6)(-6) = 36$, and $4(2)(1) = 8$.

$$= \frac{6 \pm \sqrt{28}}{4}$$ Complete the subtraction under the radical.

$$= \frac{6 \pm 2\sqrt{7}}{4}$$ $\sqrt{28} = \sqrt{4 \cdot 7} = \sqrt{4}\sqrt{7} = 2\sqrt{7}$

$$= \frac{2(3 \pm \sqrt{7})}{4}$$ Factor out 2 from the numerator.

$$= \frac{3 \pm \sqrt{7}}{2}$$ Divide the numerator and denominator by 2.

The solution set is $\left\{ \dfrac{3 + \sqrt{7}}{2}, \dfrac{3 - \sqrt{7}}{2} \right\}$ or $\left\{ \dfrac{3 \pm \sqrt{7}}{2} \right\}$.

● ● ●

TECHNOLOGY

You can use a graphing utility to verify that the solutions of $2x^2 - 6x + 1 = 0$ are $\dfrac{3 \pm \sqrt{7}}{2}$. Begin by entering $y_1 = 2x^2 - 6x + 1$ in the $\boxed{Y=}$ screen. Then evaluate this equation at each of the proposed solutions.

```
Y₁((3+√(7))/2)
                    0
Y₁((3-√(7))/2)
                    0
■
```

> In each case, the value is 0, verifying that the solutions satisfy $2x^2 - 6x + 1 = 0$.

GREAT QUESTION!

The simplification of the irrational solutions in Example 5 is kind of tricky. Any suggestions to guide the process?

Many students use the quadratic formula correctly until the last step, where they make an error in simplifying the solutions. Be sure to factor the numerator before dividing the numerator and denominator by the greatest common factor:

$$\frac{6 \pm 2\sqrt{7}}{4} = \frac{2(3 \pm \sqrt{7})}{4} = \frac{\overset{1}{\cancel{2}}(3 \pm \sqrt{7})}{\underset{2}{\cancel{4}}} = \frac{3 \pm \sqrt{7}}{2}.$$

Factor first. Then divide by the GCF.

You cannot divide just one term in the numerator and the denominator by their greatest common factor.

Incorrect!

$$\frac{6 \pm 2\sqrt{7}}{4} = \frac{\overset{3}{\cancel{6}} \pm 2\sqrt{7}}{\underset{2}{\cancel{4}}} = \frac{3 \pm 2\sqrt{7}}{2} \qquad \frac{6 \pm 2\sqrt{7}}{4} = \frac{6 \pm \overset{1}{\cancel{2}}\sqrt{7}}{\underset{2}{\cancel{4}}} = \frac{6 \pm \sqrt{7}}{2}$$

Can all irrational solutions of quadratic equations be simplified? No. The following solutions cannot be simplified.

$$\frac{5 \pm 2\sqrt{7}}{2}$$ Other than 1, terms in each numerator have no common factor. $$\frac{-4 \pm 3\sqrt{7}}{2}$$

✓ Check Point **5** Solve using the quadratic formula:

$$2x^2 + 2x - 1 = 0.$$

EXAMPLE 6 Finding a Function's Zeros Using the Quadratic Formula

Find all zeros of $f(x) = 3x^2 - 2x + 4$.

SOLUTION

We find the zeros of f by setting $f(x)$ equal to 0 and solving the resulting quadratic equation.

$$f(x) = 3x^2 - 2x + 4 = 0$$

$$a = 3 \quad b = -2 \quad c = 4$$

Substituting these values into the quadratic formula and simplifying gives the function's zeros.

GREAT QUESTION!

Should I check irrational and imaginary solutions by substitution in the given quadratic equation?

No. Checking irrational and imaginary solutions can be time-consuming. The solutions given by the quadratic formula are always correct, unless you have made a careless error. Checking for computational errors or errors in simplification is sufficient.

$$x = \frac{-b \pm \sqrt{b^2 - 4ac}}{2a}$$ Use the quadratic formula.

$$= \frac{-(-2) \pm \sqrt{(-2)^2 - 4(3)(4)}}{2(3)}$$ Substitute the values for a, b, and c: $a = 3, b = -2$, and $c = 4$.

$$= \frac{2 \pm \sqrt{4 - 48}}{6}$$ $-(-2) = 2$ and $(-2)^2 = (-2)(-2) = 4$.

$$= \frac{2 \pm \sqrt{-44}}{6}$$ Subtract under the radical. Because the number under the radical sign is negative, the solutions will not be real numbers.

$$= \frac{2 \pm 2i\sqrt{11}}{6}$$ $\sqrt{-44} = \sqrt{4(11)(-1)} = 2i\sqrt{11}$

$$= \frac{2(1 \pm i\sqrt{11})}{6}$$ Factor 2 from the numerator.

$$= \frac{1 \pm i\sqrt{11}}{3}$$ Divide the numerator and denominator by 2.

$$= \frac{1}{3} \pm i\frac{\sqrt{11}}{3}$$ Write the complex numbers in standard form.

Because $x = \frac{1}{3} \pm i\frac{\sqrt{11}}{3}$, the zeros of f are $\frac{1}{3} + i\frac{\sqrt{11}}{3}$ and $\frac{1}{3} - i\frac{\sqrt{11}}{3}$. These imaginary zeros are complex conjugates.

•••

TECHNOLOGY

Graphic Connections

The graph of $f(x) = 3x^2 - 2x + 4$, the function in Example 6, has no x-intercepts. This shows that the function has no real zeros.

$f(x) = 3x^2 - 2x + 4$

$[-2, 2, 1]$ by $[-1, 10, 1]$

The graph has no x-intercepts.

If a function has no real zeros, its graph will not have x-intercepts. Only real zeros correspond to x-intercepts. Thus, if $ax^2 + bx + c = 0$ has complex imaginary solutions, the graph of $f(x) = ax^2 + bx + c$ will lie entirely above or entirely below the x-axis.

⊘ Check Point **6** Find all zeros of $f(x) = x^2 - 2x + 2$.

⑤ Use the discriminant to determine the number and type of solutions.

The Discriminant

The quantity $b^2 - 4ac$, which appears under the radical sign in the quadratic formula, is called the **discriminant**. **Table 2.2** shows how the discriminant of the quadratic equation $ax^2 + bx + c = 0$ determines the number and type of solutions.

Table 2.2 The Discriminant and the Kinds of Solutions to $ax^2 + bx + c = 0$

Discriminant $b^2 - 4ac$	Kinds of Solutions to $ax^2 + bx + c = 0$	Graph of $y = ax^2 + bx + c$
$b^2 - 4ac > 0$	**Two unequal real solutions:** If a, b, and c are rational numbers and the discriminant is a perfect square, the solutions are *rational*. If the discriminant is not a perfect square, the solutions are *irrational* conjugates.	Two x-intercepts
$b^2 - 4ac = 0$	**One solution (a repeated solution) that is a real number:** If a, b, and c are rational numbers, the repeated solution is also a rational number.	One x-intercept
$b^2 - 4ac < 0$	**No real solution; two imaginary solutions:** The solutions are complex conjugates.	No x-intercepts

GREAT QUESTION!

Is the square root sign part of the discriminant?

No. The discriminant is $b^2 - 4ac$. It is not $\sqrt{b^2 - 4ac}$, so do not give the discriminant as a radical.

EXAMPLE 7 Using the Discriminant

For each equation, compute the discriminant. Then determine the number and type of solutions:

a. $3x^2 + 4x - 5 = 0$ **b.** $9x^2 - 6x + 1 = 0$ **c.** $3x^2 - 8x + 7 = 0.$

SOLUTION

Begin by identifying the values for $a, b,$ and c in each equation. Then compute $b^2 - 4ac,$ the discriminant.

a. $3x^2 + 4x - 5 = 0$

 $a = 3$ $b = 4$ $c = -5$

Substitute and compute the discriminant:

$$b^2 - 4ac = 4^2 - 4 \cdot 3(-5) = 16 - (-60) = 16 + 60 = 76.$$

The discriminant, 76, is a positive number that is not a perfect square. Thus, there are two irrational solutions. (These solutions are conjugates of each other.)

b. $9x^2 - 6x + 1 = 0$

 $a = 9$ $b = -6$ $c = 1$

Substitute and compute the discriminant:

$$b^2 - 4ac = (-6)^2 - 4 \cdot 9 \cdot 1 = 36 - 36 = 0.$$

The discriminant, 0, shows that there is only one real solution. This real solution is a rational number.

c. $3x^2 - 8x + 7 = 0$

 $a = 3$ $b = -8$ $c = 7$

$$b^2 - 4ac = (-8)^2 - 4 \cdot 3 \cdot 7 = 64 - 84 = -20$$

The negative discriminant, -20, shows that there are two imaginary solutions. (These solutions are complex conjugates of each other.) • • •

⊘ **Check Point 7** For each equation, compute the discriminant. Then determine the number and type of solutions:

a. $x^2 + 6x + 9 = 0$ **b.** $2x^2 - 7x - 4 = 0$ **c.** $3x^2 - 2x + 4 = 0.$

⑥ Determine the most efficient method to use when solving a quadratic equation.

Determining Which Method to Use

All quadratic equations can be solved by the quadratic formula. However, if an equation is in the form $u^2 = d,$ such as $x^2 = 5$ or $(2x + 3)^2 = 8,$ it is faster to use the square root property, taking the square root of both sides. If the equation is not in the form $u^2 = d,$ write the quadratic equation in general form ($ax^2 + bx + c = 0$). Try to solve the equation by factoring. If $ax^2 + bx + c$ cannot be factored, then solve the quadratic equation by the quadratic formula.

Because we used the method of completing the square to derive the quadratic formula, we no longer need it for solving quadratic equations. However, we will use completing the square later in the book to help graph equations.

GREAT QUESTION!

Is factoring the most important technique to use when solving a quadratic equation?

No. Even though you first learned to solve quadratic equations by factoring, most quadratic equations in general form cannot be factored. Be sure to apply the quadratic formula when this occurs.

Table 2.3 summarizes our observations about which technique to use when solving a quadratic equation.

Table 2.3 Determining the Most Efficient Technique to Use when Solving a Quadratic Equation

Description and Form of the Quadratic Equation	Most Efficient Solution Method	Example
$ax^2 + bx + c = 0$ and $ax^2 + bx + c$ can be factored easily.	Factor and use the zero-product principle.	$3x^2 + 5x - 2 = 0$ $(3x - 1)(x + 2) = 0$ $3x - 1 = 0 \quad$ or $\quad x + 2 = 0$ $x = \dfrac{1}{3} \qquad\qquad x = -2$
$ax^2 + bx = 0$ The quadratic equation has no constant term. ($c = 0$)	Factor and use the zero-product principle.	$6x^2 + 9x = 0$ $3x(2x + 3) = 0$ $3x = 0 \quad$ or $\quad 2x + 3 = 0$ $x = 0 \qquad\qquad 2x = -3$ $\qquad\qquad\qquad x = -\frac{3}{2}$
$ax^2 + c = 0$ The quadratic equation has no x-term. ($b = 0$)	Solve for x^2 and apply the square root property.	$7x^2 - 4 = 0$ $7x^2 = 4$ $x^2 = \dfrac{4}{7}$ $x = \pm\sqrt{\dfrac{4}{7}}$ $= \pm\dfrac{2}{\sqrt{7}} = \pm\dfrac{2}{\sqrt{7}} \cdot \dfrac{\sqrt{7}}{\sqrt{7}} = \pm\dfrac{2\sqrt{7}}{7}$
$u^2 = d$; u is a first-degree polynomial.	Use the square root property.	$(x + 4)^2 = 5$ $x + 4 = \pm\sqrt{5}$ $x = -4 \pm \sqrt{5}$
$ax^2 + bx + c = 0$ and $ax^2 + bx + c$ cannot be factored or the factoring is too difficult.	Use the quadratic formula: $x = \dfrac{-b \pm \sqrt{b^2 - 4ac}}{2a}$.	$x^2 - 2x - 6 = 0$ $\boxed{a = 1} \quad \boxed{b = -2} \quad \boxed{c = -6}$ $x = \dfrac{-(-2) \pm \sqrt{(-2)^2 - 4(1)(-6)}}{2(1)}$ $= \dfrac{2 \pm \sqrt{4 + 24}}{2(1)}$ $= \dfrac{2 \pm \sqrt{28}}{2} = \dfrac{2 \pm \sqrt{4}\sqrt{7}}{2}$ $= \dfrac{2 \pm 2\sqrt{7}}{2} = \dfrac{2(1 \pm \sqrt{7})}{2}$ $= 1 \pm \sqrt{7}$

⑦ Solve problems modeled by quadratic equations.

Applications

EXAMPLE 8 Blood Pressure and Age

The graphs in **Figure 2.8** illustrate that a person's normal systolic blood pressure, measured in millimeters of mercury (mm Hg), depends on his or her age. (The symbol ⌇ on the vertical axis shows there is a break in values between 0 and 100. Thus, the first tick mark on the vertical axis represents a blood pressure of 100 mm Hg.) The function

$$P(A) = 0.006A^2 - 0.02A + 120$$

models a man's normal systolic pressure, $P(A)$, at age A.

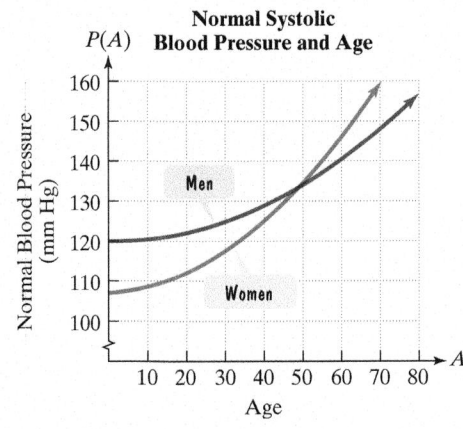

FIGURE 2.8

a. Find the age, to the nearest year, of a man whose normal systolic blood pressure is 125 mm Hg.

b. Use the graphs in **Figure 2.8** to describe the differences between the normal systolic blood pressures of men and women as they age.

SOLUTION

a. We are interested in the age of a man with a normal systolic blood pressure of 125 millimeters of mercury. Thus, we substitute 125 for $P(A)$ in the given function for men. Then we solve for A, the man's age.

$P(A) = 0.006A^2 - 0.02A + 120$	This is the given function for men.
$125 = 0.006A^2 - 0.02A + 120$	Substitute 125 for $P(A)$.
$0 = 0.006A^2 - 0.02A - 5$	Subtract 125 from both sides and write the quadratic equation in general form.

$$a = 0.006 \quad b = -0.02 \quad c = -5$$

Because the trinomial on the right side of the equation is prime, we solve using the quadratic formula.

Notice that the variable is A, rather than the usual x.

$$A = \frac{-b \pm \sqrt{b^2 - 4ac}}{2a}$$

Use the quadratic formula.

$$= \frac{-(-0.02) \pm \sqrt{(-0.02)^2 - 4(0.006)(-5)}}{2(0.006)}$$

Substitute the values for a, b, and c: $a = 0.006$, $b = -0.02$, and $c = -5$.

$$= \frac{0.02 \pm \sqrt{0.1204}}{0.012}$$

Use a calculator to simplify the radicand.

$$\approx \frac{0.02 \pm 0.347}{0.012}$$

Use a calculator: $\sqrt{0.1204} \approx 0.347$.

$$A \approx \frac{0.02 + 0.347}{0.012} \quad \text{or} \quad A \approx \frac{0.02 - 0.347}{0.012}$$

$$A \approx 31 \qquad\qquad\qquad A \approx -27$$

Use a calculator and round to the nearest integer.

Reject this solution.
Age cannot be negative.

TECHNOLOGY

On most calculators, here is how to approximate

$$\frac{0.02 + \sqrt{0.1204}}{0.012}.$$

Many Scientific Calculators

(.02 + .1204 √)

÷ .012 =

Many Graphing Calculators

(.02 + √ .1204)

÷ .012 ENTER

If your calculator displays an open parenthesis after √, you'll need to enter another closed parenthesis here.

The positive solution, $A \approx 31$, indicates that 31 is the approximate age of a man whose normal systolic blood pressure is 125 mm Hg. This is illustrated by the black lines with the arrows on the red graph representing men in **Figure 2.9**.

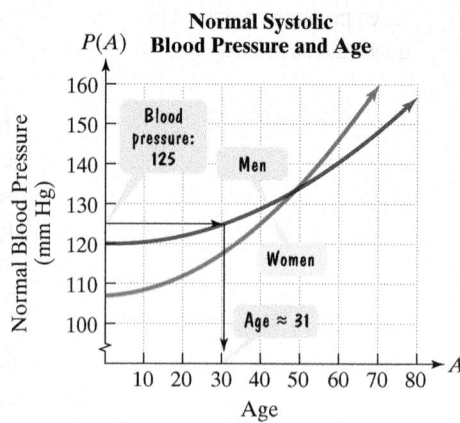

FIGURE 2.9

b. Take a second look at the graphs in **Figure 2.9**. Before approximately age 50, the blue graph representing women's normal systolic blood pressure lies below the red graph representing men's normal systolic blood pressure. Thus, up to age 50, women's normal systolic blood pressure is lower than men's, although it is increasing at a faster rate. After age 50, women's normal systolic blood pressure is higher than men's. •••

⊘ Check Point **8** The function $P(A) = 0.01A^2 + 0.05A + 107$ models a woman's normal systolic blood pressure, $P(A)$, at age A. Use this function to find the age, to the nearest year, of a woman whose normal systolic blood pressure is 115 mm Hg. Use the blue graph in **Figure 2.9** to verify your solution.

In section 1.9, we used the Pythagorean Theorem to derive the distance formula. Recall that in a right triangle, the side opposite the 90° angle is the hypotenuse. and the other sides are legs.

The Pythagorean Theorem

The sum of the squares of the lengths of the legs of a right triangle equals the square of the length of the hypotenuse.

If the legs have lengths a and b, and the hypotenuse has length c, then

$$a^2 + b^2 = c^2.$$

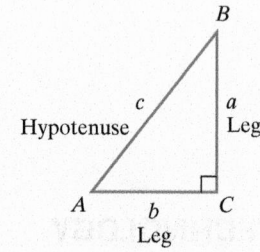

EXAMPLE 9 Using the Pythagorean Theorem

a. A wheelchair ramp with a length of 122 inches has a horizontal distance of 120 inches. What is the ramp's vertical distance?

b. Construction laws are very specific when it comes to access ramps for people with disabilities. Every vertical rise of 1 inch requires a horizontal run of 12 inches. Does this ramp satisfy the requirement?

SOLUTION

a. **Figure 2.10** shows the right triangle that is formed by the ramp, the wall, and the ground. We can find x, the ramp's vertical distance, using the Pythagorean Theorem.

FIGURE 2.10

(leg)²	plus	(leg)²	equals	(hypotenuse)²
x^2	$+$	120^2	$=$	122^2

We solve this equation using the square root property.

$$x^2 + 120^2 = 122^2$$ This is the equation resulting from the Pythagorean Theorem.

$$x^2 + 14,400 = 14,884$$ Square 120 and 122.

$$x^2 = 484$$ Isolate x^2 by subtracting 14,400 from both sides.

$$x = \sqrt{484} \quad \text{or} \quad x = -\sqrt{484}$$ Apply the square root property.
$$x = 22 \qquad\qquad x = -22$$

Because x represents the ramp's vertical distance, we reject the negative value. Thus, the ramp's vertical distance is 22 inches.

b. Every vertical rise of 1 inch requires a horizontal run of 12 inches. Because the ramp has a vertical distance of 22 inches, it requires a horizontal distance of 22(12) inches, or 264 inches. The horizontal distance is only 120 inches, so this ramp does not satisfy construction laws for access ramps for people with disabilities. • • •

Ø **Check Point 9** In a 25-inch television set, the length of the screen's diagonal is 25 inches. If the screen's height is 15 inches, what is its width?

Blitzer Bonus ‖ Art, Nature, and Quadratic Equations

In *The Bathers at Asnières*, by the French impressionist Georges Seurat (1859–1891), the artist positions parts of the painting as though they were inside golden rectangles. *Bathers at Asnières* (1884), Georges Seurat, Art Resource.

A **golden rectangle** can be a rectangle of any size, but its long side must be Φ times as long as its short side, where $\Phi \approx 1.6$. Artists often use golden rectangles in their work because they are considered to be more visually pleasing than other rectangles.

If a golden rectangle is divided into a square and a rectangle, as in **Figure 2.11(a)**, the smaller rectangle is a golden rectangle. If the smaller golden rectangle is divided again, the same is true of the yet smaller rectangle, and so on. The process of repeatedly dividing each golden rectangle in this manner is illustrated in **Figure 2.11(b)**. We've also created a spiral by connecting the opposite corners of all the squares with a smooth curve. This spiral matches the spiral shape of the chambered nautilus shell shown in **Figure 2.11(c)**. The shell spirals out at an ever-increasing rate that is governed by this geometry.

Golden Rectangle *A*

| Square | Golden Rectangle *B* |

FIGURE 2.11(a)

FIGURE 2.11(b)

FIGURE 2.11(c)

In the Exercise Set that follows, you will use the golden rectangles in **Figure 2.11(a)** to obtain an exact value for Φ, the ratio of the long side to the short side in a golden rectangle of any size. Your model will involve a quadratic equation that can be solved by the quadratic formula. (See Exercise 131.)

CONCEPT AND VOCABULARY CHECK

Fill in each blank so that the resulting statement is true.

1. An equation that can be written in the general form $ax^2 + bx + c = 0$, $a \neq 0$, is called a/an _____ equation.

2. The zero-product principle states that if $AB = 0$, then _____.

3. The solutions of $ax^2 + bx + c = 0$ correspond to the _____ for the graph of $y = ax^2 + bx + c$.

4. The square root property states that if $u^2 = d$, then $u =$ _____.

5. If $x^2 = 7$, then $x =$ _____.

6. To solve $x^2 + 6x = 7$ by completing the square, add _____ to both sides of the equation.

7. To solve $x^2 - \dfrac{2}{3}x = \dfrac{4}{9}$ by completing the square, add _____ to both sides of the equation.

8. The solutions of a quadratic equation in the general form $ax^2 + bx + c = 0$, $a \neq 0$, are given by the quadratic formula $x =$ _____.

9. In order to solve $2x^2 + 9x - 5 = 0$ by the quadratic formula, we use $a =$ _____, $b =$ _____, and $c =$ _____.

10. In order to solve $x^2 = 4x + 1$ by the quadratic formula, we use $a =$ _____, $b =$ _____, and $c =$ _____.

11. $x = \dfrac{-(-4) \pm \sqrt{(-4)^2 - 4(1)(2)}}{2(1)}$ simplifies to

 $x =$ _____.

12. $x = \dfrac{-4 \pm \sqrt{4^2 - 4 \cdot 2 \cdot 5}}{2 \cdot 2}$ simplifies to $x =$ _____.

13. The discriminant of $ax^2 + bx + c = 0$ is defined by _____.

14. If the discriminant of $ax^2 + bx + c = 0$ is negative, the quadratic equation has _____ real solutions.

15. If the discriminant of $ax^2 + bx + c = 0$ is positive, the quadratic equation has _____ real solutions.

16. The most efficient technique for solving $(2x + 7)^2 = 25$ is by using _____.

17. The most efficient technique for solving $x^2 + 5x - 10 = 0$ is by using _____.

18. The most efficient technique for solving $x^2 + 8x + 15 = 0$ is by using _____.

19. A triangle with one angle measuring $90°$ is called a/an _____ triangle. The side opposite the $90°$ angle is called the _____. The other sides are called _____.

20. The Pythagorean Theorem states that in any _____ triangle, the sum of the squares of the lengths of the _____ equals _____.

EXERCISE SET 2.4

Practice Exercises

Solve each equation in Exercises 1–14 by factoring.

1. $x^2 - 3x - 10 = 0$
2. $x^2 - 13x + 36 = 0$
3. $x^2 = 8x - 15$
4. $x^2 = -11x - 10$
5. $6x^2 + 11x - 10 = 0$
6. $9x^2 + 9x + 2 = 0$
7. $3x^2 - 2x = 8$
8. $4x^2 - 13x = -3$
9. $3x^2 + 12x = 0$
10. $5x^2 - 20x = 0$
11. $2x(x - 3) = 5x^2 - 7x$
12. $16x(x - 2) = 8x - 25$
13. $7 - 7x = (3x + 2)(x - 1)$
14. $10x - 1 = (2x + 1)^2$

Solve each equation in Exercises 15–34 by the square root property.

15. $3x^2 = 27$
16. $5x^2 = 45$
17. $5x^2 + 1 = 51$
18. $3x^2 - 1 = 47$
19. $2x^2 - 5 = -55$
20. $2x^2 - 7 = -15$
21. $(x + 2)^2 = 25$
22. $(x - 3)^2 = 36$
23. $3(x - 4)^2 = 15$
24. $3(x + 4)^2 = 21$
25. $(x + 3)^2 = -16$
26. $(x - 1)^2 = -9$
27. $(x - 3)^2 = -5$
28. $(x + 2)^2 = -7$
29. $(3x + 2)^2 = 9$
30. $(4x - 1)^2 = 16$
31. $(5x - 1)^2 = 7$
32. $(8x - 3)^2 = 5$
33. $(3x - 4)^2 = 8$
34. $(2x + 8)^2 = 27$

Solve each equation in Exercises 35–52 by completing the square.

35. $x^2 + 6x = 7$
36. $x^2 + 6x = -8$
37. $x^2 - 2x = 2$
38. $x^2 + 4x = 12$
39. $x^2 - 6x - 11 = 0$
40. $x^2 - 2x - 5 = 0$
41. $x^2 + 4x + 1 = 0$
42. $x^2 + 6x - 5 = 0$
43. $x^2 - 5x + 6 = 0$
44. $x^2 + 7x - 8 = 0$
45. $x^2 + 3x - 1 = 0$
46. $x^2 - 3x - 5 = 0$
47. $2x^2 - 7x + 3 = 0$
48. $2x^2 + 5x - 3 = 0$
49. $4x^2 - 4x - 1 = 0$
50. $2x^2 - 4x - 1 = 0$
51. $3x^2 - 2x - 2 = 0$
52. $3x^2 - 5x - 10 = 0$

Solve each equation in Exercises 53–62 using the quadratic formula.

53. $x^2 + 8x + 15 = 0$
54. $x^2 + 8x + 12 = 0$
55. $x^2 + 5x + 3 = 0$
56. $x^2 + 5x + 2 = 0$
57. $3x^2 - 3x - 4 = 0$
58. $5x^2 + x - 2 = 0$
59. $4x^2 = 2x + 7$
60. $3x^2 = 6x - 1$
61. $x^2 - 6x + 10 = 0$
62. $x^2 - 2x + 17 = 0$

In Exercises 63–70, compute the discriminant. Then determine the number and type of solutions for the given equation.

63. $x^2 - 4x - 5 = 0$

64. $4x^2 - 2x + 3 = 0$

65. $2x^2 - 11x + 3 = 0$

66. $2x^2 + 11x - 6 = 0$

67. $x^2 - 2x + 1 = 0$

68. $3x^2 = 2x - 1$

69. $x^2 - 3x - 7 = 0$

70. $3x^2 + 4x - 2 = 0$

Solve each equation in Exercises 71–96 by the method of your choice.

71. $2x^2 - x = 1$

72. $3x^2 - 4x = 4$

73. $5x^2 + 2 = 11x$

74. $5x^2 = 6 - 13x$

75. $3x^2 = 60$

76. $2x^2 = 250$

77. $x^2 - 2x = 1$

78. $2x^2 + 3x = 1$

79. $(2x + 3)(x + 4) = 1$

80. $(2x - 5)(x + 1) = 2$

81. $(3x - 4)^2 = 16$

82. $(2x + 7)^2 = 25$

83. $3x^2 - 12x + 12 = 0$

84. $9 - 6x + x^2 = 0$

85. $4x^2 - 16 = 0$

86. $3x^2 - 27 = 0$

87. $x^2 - 6x + 13 = 0$

88. $x^2 - 4x + 29 = 0$

89. $x^2 = 4x - 7$

90. $5x^2 = 2x - 3$

91. $2x^2 - 7x = 0$

92. $2x^2 + 5x = 3$

93. $\dfrac{1}{x} + \dfrac{1}{x + 2} = \dfrac{1}{3}$

94. $\dfrac{1}{x} + \dfrac{1}{x + 3} = \dfrac{1}{4}$

95. $\dfrac{2x}{x - 3} + \dfrac{6}{x + 3} = -\dfrac{28}{x^2 - 9}$

96. $\dfrac{3}{x - 3} + \dfrac{5}{x - 4} = \dfrac{x^2 - 20}{x^2 - 7x + 12}$

In Exercises 97–102, find the zeros of each function. Use the zeros to match the function with its graph. The graphs are shown in $[-10, 10, 1]$ by $[-10, 10, 1]$ viewing rectangles and labeled (a) through (f).

97. $y = x^2 - 4x - 5$

98. $y = x^2 - 6x + 7$

99. $y = -(x + 1)^2 + 4$

100. $y = -(x + 3)^2 + 1$

101. $y = x^2 - 2x + 2$

102. $y = x^2 + 6x + 9$

a.

b.

c.

d.

e.

f.

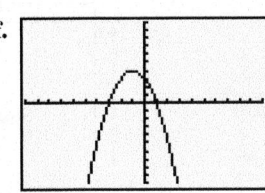

In Exercises 103–110, find all values of x satisfying the given conditions.

103. $f(x) = 2x^2 - 3x$ and $f(x) = 2$.

104. $f(x) = 5x^2 + 3x$ and $f(x) = 2$.

105. $f(x) = x - 1, g(x) = x + 4,$ and $(fg)(x) = 14$.

106. $f(x) = x - 3, g(x) = x + 8,$ and $(fg)(x) = -30$.

107. $f(x) = \dfrac{2x}{x + 2}, g(x) = \dfrac{3}{x + 4},$ and $(f + g)(x) = 1$.

108. $f(x) = \dfrac{3}{x - 1}, g(x) = \dfrac{8}{x},$ and $(f + g)(x) = 3$.

109. $f(x) = 2x^2 + 5x - 4, g(x) = -x^2 + 15x - 10,$ and $(f - g)(x) = 0$.

110. $f(x) = -x^2 + 4x - 2, g(x) = -3x^2 + x - 1,$ and $(f - g)(x) = 0$.

Practice Plus

In Exercises 111–112, list all numbers that must be excluded from the domain of each function.

111. $f(x) = \dfrac{3}{2x^2 + 4x - 9}$

112. $f(x) = \dfrac{7}{2x^2 - 8x + 5}$

In Exercises 113–116, find all values of x satisfying the given conditions.

113. $f(x) = x - 3, g(x) = 2x^2,$ and $(g \circ f)(x) = 50$.

114. $f(x) = x + 4, g(x) = 3x^2,$ and $(g \circ f)(x) = 12$.

115. $f(x) = 2x - 5, g(x) = x^2 - 3x + 8,$ and $(f \circ g)(x) = 7$.

116. $f(x) = 1 - 2x, g(x) = 3x^2 + x - 1,$ and $(f \circ g)(x) = -5$.

117. When the sum of 6 and twice a positive number is subtracted from the square of the number, 0 results. Find the number.

118. When the sum of 1 and twice a negative number is subtracted from twice the square of the number, 0 results. Find the number.

In Exercises 119–122, solve each equation by the method of your choice.

119. $\dfrac{1}{x^2 - 3x + 2} = \dfrac{1}{x + 2} + \dfrac{5}{x^2 - 4}$

120. $\dfrac{x - 1}{x - 2} + \dfrac{x}{x - 3} = \dfrac{1}{x^2 - 5x + 6}$

121. $\sqrt{2}x^2 + 3x - 2\sqrt{2} = 0$

122. $\sqrt{3}x^2 + 6x + 7\sqrt{3} = 0$

Application Exercises

In a round-robin chess tournament, each player is paired with every other player once. The function

$$N(x) = \frac{x^2 - x}{2}$$

models the number of chess games, $N(x)$, that must be played in a round-robin tournament with x chess players. Use this function to solve Exercises 123–124.

123. In a round-robin chess tournament, 21 games were played. How many players were entered in the tournament?

124. In a round-robin chess tournament, 36 games were played. How many players were entered in the tournament?

The graph of the function in Exercises 123–124 is shown. Use the graph to solve Exercises 125–126.

125. Identify your solution to Exercise 123 as a point on the graph.

126. Identify your solution to Exercise 124 as a point on the graph.

A driver's age has something to do with his or her chance of getting into a fatal car crash. The bar graph shows the number of fatal vehicle crashes per 100 million miles driven for drivers of various age groups. For example, 25-year-old drivers are involved in 4.1 fatal crashes per 100 million miles driven. Thus, when a group of 25-year-old Americans have driven a total of 100 million miles, approximately 4 have been in accidents in which someone died.

Age of United States Drivers and Fatal Crashes

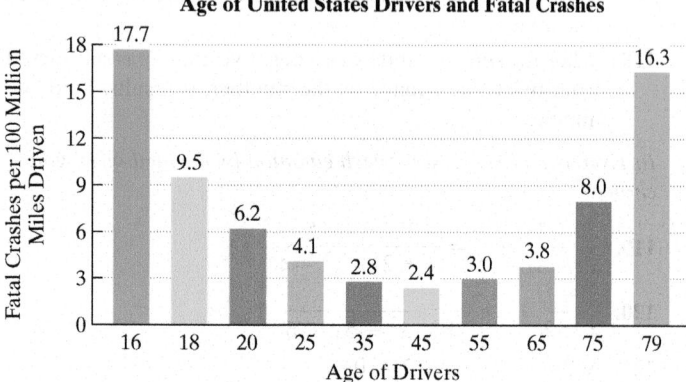

Source: Insurance Institute for Highway Safety

The number of fatal vehicle crashes per 100 million miles, $N(x)$, for drivers of age x can be modeled by the quadratic function

$$N(x) = 0.013x^2 - 1.19x + 28.24.$$

Use the function to solve Exercises 127–128.

127. What age groups are expected to be involved in 3 fatal crashes per 100 million miles driven? How well does the function model the trend in the actual data shown in the bar graph?

128. What age groups are expected to be involved in 10 fatal crashes per 100 million miles driven? How well does the function model the trend in the actual data shown in the bar graph?

Throwing events in track and field include the shot put, the discus throw, the hammer throw, and the javelin throw. The distance that an athlete can achieve depends on the initial velocity of the object thrown and the angle above the horizontal at which the object leaves the hand.

In Exercises 129–130, an athlete whose event is the shot put releases the shot with the same initial velocity, but at different angles.

129. When the shot is released at an angle of 35°, its path can be modeled by the function

$$f(x) = -0.01x^2 + 0.7x + 6.1,$$

in which x is the shot's horizontal distance, in feet, and $f(x)$ is its height, in feet. This function is shown by one of the graphs, (a) or (b), in the figure. Use the function to determine the shot's maximum distance. Use a calculator and round to the nearest tenth of a foot. Which graph, (a) or (b), shows the shot's path?

Horizontal Distance
[0, 80, 10] by [0, 40, 10]

130. (Refer to the preceding information and the graphs shown in Exercise 129.) When the shot is released at an angle of 65°, its path can be modeled by the function

$$f(x) = -0.04x^2 + 2.1x + 6.1,$$

in which x is the shot's horizontal distance, in feet, and $f(x)$ is its height, in feet. This function is shown by one of the graphs, (a) or (b), in the figure in Exercise 129. Use the function to determine the shot's maximum distance. Use a calculator and round to the nearest tenth of a foot. Which graph, (a) or (b), shows the shot's path?

131. If you have not yet done so, read the Blitzer Bonus on page 287. In this exercise, you will use the golden rectangles shown to obtain an exact value for Φ, the ratio of the long side to the short side in a golden rectangle of any size.

Golden Rectangle *A*

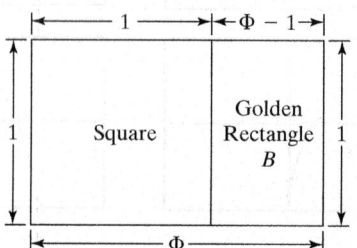

a. The golden ratio in rectangle A, or the ratio of the long side to the short side, can be modeled by $\dfrac{\Phi}{1}$. Write a fractional expression that models the golden ratio in rectangle B.

b. Set the expression for the golden ratio in rectangle A equal to the expression for the golden ratio in rectangle B. Multiply both sides of the equation by the least common denominator and solve the resulting quadratic equation using the quadratic formula. Express Φ as an exact value in simplified radical form.

c. Use your solution from part (b) to complete this statement: The ratio of the long side to the short side in a golden rectangle of any size is _____ to 1.

Use the Pythagorean Theorem and the square root property to solve Exercises 132–135. Express answers in simplified radical form. Then find a decimal approximation to the nearest tenth.

132 A rectangular park is 6 miles long and 3 miles wide. How long is a pedestrian route that runs diagonally across the park?

133. A rectangular park is 4 miles long and 2 miles wide. How long is a pedestrian route that runs diagonally across the park?

134. The base of a 30-foot ladder is 10 feet from a building. If the ladder reaches the flat roof, how tall is the building?

135. A baseball diamond is actually a square with 90-foot sides. What is the distance from home plate to second base?

136. An **isosceles right triangle** has legs that are the same length and acute angles each measuring 45°.

a. Write an expression in terms of a that represents the length of the hypotenuse.

b. Use your result from part (a) to write a sentence that describes the length of the hypotenuse of an isosceles right triangle in terms of the length of a leg.

137. The length of a rectangular sign is 3 feet longer than the width. If the sign's area is 54 square feet, find its length and width.

138. A rectangular parking lot has a length that is 3 yards greater than the width. The area of the parking lot is 180 square yards. Find the length and the width.

139. Each side of a square is lengthened by 3 inches. The area of this new, larger square is 64 square inches. Find the length of a side of the original square.

140. Each side of a square is lengthened by 2 inches. The area of this new, larger square is 36 square inches. Find the length of a side of the original square.

141. A pool measuring 10 meters by 20 meters is surrounded by a path of uniform width, as shown in the figure at the top of the next column. If the area of the pool and the path combined is 600 square meters, what is the width of the path?

142. A vacant rectangular lot is being turned into a community vegetable garden measuring 15 meters by 12 meters. A path of uniform width is to surround the garden, as shown in the figure. If the area of the garden and path combined is 378 square meters, find the width of the path.

143. A machine produces open boxes using square sheets of metal. The figure illustrates that the machine cuts equal-sized squares measuring 2 inches on a side from the corners and then shapes the metal into an open box by turning up the sides. If each box must have a volume of 200 cubic inches, find the length and width of the open box.

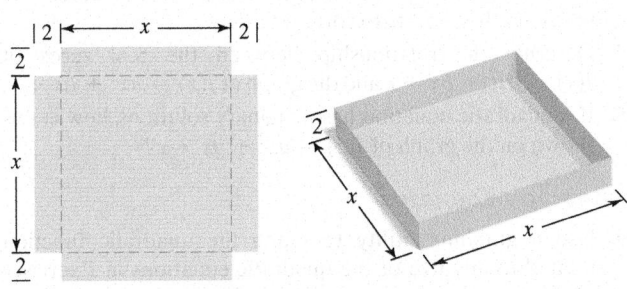

144. A machine produces open boxes using square sheets of metal. The machine cuts equal-sized squares measuring 3 inches on a side from the corners and then shapes the metal into an open box by turning up the sides. If each box must have a volume of 75 cubic inches, find the length and width of the open box.

145. A rain gutter is made from sheets of aluminum that are 20 inches wide. As shown in the figure, the edges are turned up to form right angles. Determine the depth of the gutter that will allow a cross-sectional area of 13 square inches. Show that there are two different solutions to the problem. Round to the nearest tenth of an inch.

Flat sheet 20 inches wide

146. A piece of wire is 8 inches long. The wire is cut into two pieces and then each piece is bent into a square. Find the length of each piece if the sum of the areas of these squares is to be 2 square inches.

Writing in Mathematics

147. What is a quadratic equation?

148. Explain how to solve $x^2 + 6x + 8 = 0$ using factoring and the zero-product principle.

149. Explain how to solve $x^2 + 6x + 8 = 0$ by completing the square.

150. Explain how to solve $x^2 + 6x + 8 = 0$ using the quadratic formula.

151. How is the quadratic formula derived?

152. What is the discriminant and what information does it provide about a quadratic equation?

153. If you are given a quadratic equation, how do you determine which method to use to solve it?

154. Describe the relationship between the real zeros of $f(x) = ax^2 + bx + c$ and the graph of $f(x) = ax^2 + bx + c$.

155. If a quadratic equation has imaginary solutions, how is this shown on the graph of $f(x) = ax^2 + bx + c$?

Technology Exercises

156. Use a graphing utility to graph the quadratic function related to any five of the quadratic equations in Exercises 63–70. How does each graph illustrate what you determined algebraically using the discriminant?

157. Reread Exercise 145. The cross-sectional area of the gutter is given by the quadratic function
$$f(x) = x(20 - 2x).$$
Graph the function in a $[0, 10, 1]$ by $[0, 60, 5]$ viewing rectangle. Then ⎡TRACE⎤ along the curve or use the maximum function feature to determine the depth of the gutter that will maximize its cross-sectional area and allow the greatest amount of water to flow. What is the maximum area? Does the situation described in Exercise 145 take full advantage of the sheets of aluminum?

Critical Thinking Exercises

Make Sense? *In Exercises 158–161, determine whether each statement makes sense or does not make sense, and explain your reasoning.*

158. Because I want to solve $25x^2 - 169 = 0$ fairly quickly, I'll use the quadratic formula.

159. I'm looking at a graph with one x-intercept, so it must be the graph of a linear function.

160. I obtained -17 for the discriminant, so there are two imaginary irrational solutions.

161. When I use the square root property to determine the length of a right triangle's side, I don't even bother to list the negative square root.

In Exercises 162–165, determine whether each statement is true or false. If the statement is false, make the necessary change(s) to produce a true statement.

162. The equation $(2x - 3)^2 = 25$ is equivalent to $2x - 3 = 5$.

163. Any quadratic equation that can be solved by completing the square can be solved by the quadratic formula.

164. The quadratic formula is developed by applying factoring and the zero-product principle to the quadratic equation $ax^2 + bx + c = 0$.

165. In using the quadratic formula to solve the quadratic equation $5x^2 = 2x - 7$, we have $a = 5$, $b = 2$, and $c = -7$.

166. Write a quadratic equation in general form whose solution set is $\{-3, 5\}$.

167. Solve for t: $s = -16t^2 + v_0t$.

168. A rectangular swimming pool is 12 meters long and 8 meters wide. A tile border of uniform width is to be built around the pool using 120 square meters of tile. The tile is from a discontinued stock (so no additional materials are available) and all 120 square meters are to be used. How wide should the border be? Round to the nearest tenth of a meter. If zoning laws require at least a 2-meter-wide border around the pool, can this be done with the available tile?

Preview Exercises

Exercises 169–171 will help you prepare for the material covered in the next section.

169. Factor completely: $x^3 + x^2 - 4x - 4$.

170. Use the special product $(A + B)^2 = A^2 + 2AB + B^2$ to multiply: $\left(\sqrt{x + 4} + 1\right)^2$.

171. If -8 is substituted for x in the equation $5x^{\frac{2}{3}} + 11x^{\frac{1}{3}} + 2 = 0$, is the resulting statement true or false?

CHAPTER 2 Mid-Chapter Check Point

WHAT YOU KNOW: Zeros of linear functions, $f(x) = mx + b$, $m \neq 0$, are solutions, or roots, of linear equations, $mx + b = 0$. Zeros of quadratic functions, $f(x) = ax^2 + bx + c$, $a \neq 0$, are solutions, or roots, of quadratic equations, $ax^2 + bx + c = 0$. We solved linear equations. We saw that some equations have no solution, whereas others have all real numbers as solutions. We solved quadratic equations using factoring,

the square root property, completing the square, and the quadratic formula. We saw that the discriminant of $ax^2 + bx + c = 0$, $b^2 - 4ac$, determines the number and type of solutions. We performed operations with complex numbers and used the imaginary unit i ($i = \sqrt{-1}$, where $i^2 = -1$) to represent solutions of quadratic equations with negative discriminants. Only real solutions correspond to x-intercepts We also solved rational equations by multiplying both sides by the least common denominator and clearing fractions. We developed a strategy for solving a variety of applied problems, using equations to model verbal conditions.

In Exercises 1–12, solve each equation.

1. $-5 + 3(x + 5) = 2(3x - 4)$

2. $5x^2 - 2x = 7$

3. $\dfrac{x - 3}{5} - 1 = \dfrac{x - 5}{4}$

4. $3x^2 - 6x - 2 = 0$

5. $4x - 2(1 - x) = 3(2x + 1) - 5$

6. $5x^2 + 1 = 37$

7. $x(2x - 3) = -4$

8. $\dfrac{3x}{4} - \dfrac{x}{3} + 1 = \dfrac{4x}{5} - \dfrac{3}{20}$

9. $(x + 3)^2 = 24$

10. $\dfrac{1}{x^2} - \dfrac{4}{x} + 1 = 0$

11. $3x + 1 - (x - 5) = 2x - 4$

12. $\dfrac{2x}{x^2 + 6x + 8} = \dfrac{x}{x + 4} - \dfrac{2}{x + 2}$

In Exercises 13–17, find the zeros of each function.

13. $f(x) = x^2 + 6x + 2$

14. $f(x) = 4(x + 1) - 3x - (6 - x)$

15. $f(x) = 2x^2 + 26$

16. $f(x) = \dfrac{x^2}{3} + \dfrac{x}{2} - \dfrac{2}{3}$

17. $f(x) = x^2 - 5x + 8$

In Exercises 18–19, find all values of x satisfying the given conditions.

18. $f(x) = 3(2x - 5) - 2(4x + 1)$, $g(x) = -5(x + 3) - 2$, and $f(x) = g(x)$.

19. $f(x) = 2x + 3$, $g(x) = x + 2$, and $(fg)(x) = 10$.

20. Solve by completing the square: $x^2 + 10x - 3 = 0$.

In Exercises 21–22, without solving the equation, determine the number and type of solutions.

21. $2x^2 + 5x + 4 = 0$

22. $10x(x + 4) = 15x - 15$

23. What's the last word in capital punishment? An analysis of the final statements of all men and women Texas has executed since the Supreme Court reinstated the death penalty in 1976 revealed that "love" is by far the most frequently uttered word. The bar graph shows the number of times various words were used in final statements by Texas death-row inmates.

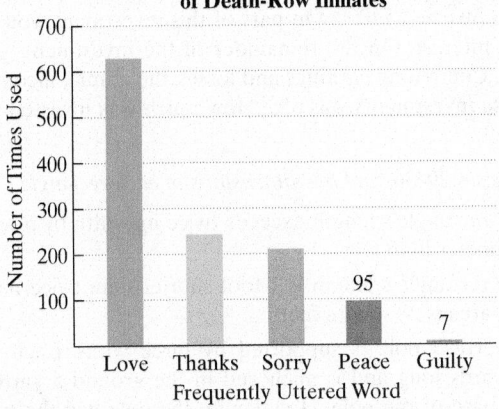

Frequently Uttered Words in Final Statements of Death-Row Inmates

Source: *Texas Department of Criminal Justice*

The number of times "love" wats used exceeded the number of times "sorry" was used by 419. The number of utterances of "thanks" exceeded the number of utterances of "sorry" by 32. Combined, these three words were used 1084 times. Determine the number of times each of these words was used in final statements by Texas inmates.

24. The line graph indicates that in 1960, 23% of U.S. taxes came from corporate income tax. For the period from 1960 through 2010, this percentage decreased by approximately 0.28 each year. If this trend continues, by which year will corporations pay zero taxes? Round to the nearest year.

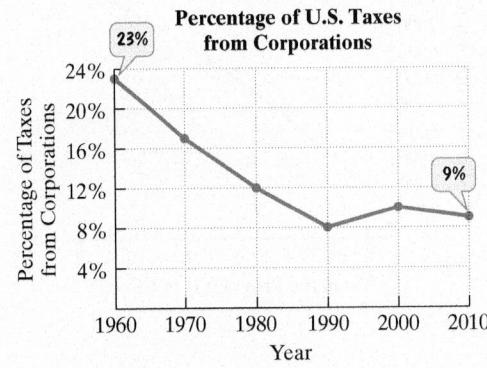

Percentage of U.S. Taxes from Corporations

Source: *Office of Management and Budget*

25. You invested $25,000 in two accounts paying 8% and 9% annual interest. At the end of the year, the total interest from these investments was $2135. How much was invested at each rate?

26. You are choosing between two texting plans. One plan has a monthly fee of $15 with a charge of $0.05 per text. The other plan has a monthly fee of $10 with a charge of $0.075 per text. For how many text messages will the costs for the two plans be the same? What will be the cost for each plan?

27. After a 40% price reduction, you purchase a camcorder for $468. What was the camcorder's price before the reduction?

28. You invested $4000. On part of this investment, you earned 4% interest. On the remainder of the investment, you lost 3%. Combining earnings and losses, the annual income from these investments was $55. How much was invested at each rate?

In Exercises 29–30, find the dimensions of each rectangle.

29. The rectangle's length exceeds twice its width by 5 feet. The perimeter is 46 feet.

30. The rectangle's length is 1 foot shorter than twice its width. The area is 28 square feet.

31. A vertical pole is supported by three wires. Each wire is 13 yards long and is anchored in the ground 5 yards from the base of the pole. How far up the pole will the wires be attached?

32. a. The alligator, at one time an endangered species, was the subject of a protection program. The function

$$P(x) = -10x^2 + 475x + 3500$$

models the alligator population, $P(x)$, after x years of the protection program, where $0 \le x \le 12$. How long did it take the population to reach 5990 alligators?

 b. The graph of the function modeling the alligator population is shown below. Identify your solution from part (a) as a point on the graph.

Years the Program Is in Effect

33. A substantial percentage of the United States population is foreign-born. The bar graph shows the percentage of foreign-born Americans for selected years from 1920 through 2008.

Percentage of the United States Population That Was Foreign-Born, 1920–2008

Source: *U.S. Census Bureau*

The percentage, $p(x)$, of the United States population that was foreign-born x years after 1920 can be modeled by the function

$$p(x) = 0.004x^2 - 0.37x + 14.1.$$

According to this function, in which year will 25% of the United States population be foreign-born? Round to the nearest year.

In Exercises 34–39, perform the indicated operations and write the result in standard form.

34. $(6 - 2i) - (7 - i)$

35. $3i(2 + i)$

36. $(1 + i)(4 - 3i)$

37. $\dfrac{1 + i}{1 - i}$

38. $\sqrt{-75} - \sqrt{-12}$

39. $\left(2 - \sqrt{-3}\right)^2$

SECTION 2.5 Other Types of Equations

Objectives

1. Solve polynomial equations by factoring.
2. Solve radical equations.
3. Solve equations with rational exponents.
4. Solve equations that are quadratic in form.
5. Solve equations involving absolute value.
6. Solve problems modeled by equations.

"Bad television is three things: a bullet train to a morally bankrupt youth, a slow spiral into an intellectual void, and of course, a complete blast to watch."

—Dennis Miller

A blast, indeed. Americans spend more of their free time watching TV than any other activity.

What does algebra have to say about this? In particular, what sort of mathematical model can be used to describe the hours per week we watch TV, by annual income? In this section, we answer the question after studying procedures for solving a variety of different types of equations.

Polynomial Equations

1 Solve polynomial equations by factoring.

A **polynomial equation** is the result of setting two polynomials equal to each other. The equation is in **general form** if one side is 0 and the polynomial on the other side is in descending powers of the variable. The **degree of a polynomial equation** is the same as the highest degree of any term in the equation. Here are examples of three polynomial equations:

$$3x + 5 = 14 \qquad 2x^2 + 7x = 4 \qquad x^3 + x^2 = 4x + 4.$$

This equation is of degree 1 because 1 is the highest degree.

This equation is of degree 2 because 2 is the highest degree.

This equation is of degree 3 because 3 is the highest degree.

Notice that a polynomial equation of degree 1 is a linear equation. A polynomial equation of degree 2 is a quadratic equation.

Some polynomial equations of degree 3 or higher can be solved by moving all terms to one side, thereby obtaining 0 on the other side. Once the equation is in general form, factor and then set each factor equal to 0.

EXAMPLE 1 Solving a Polynomial Equation by Factoring

Solve by factoring: $3x^4 = 27x^2$.

SOLUTION

Step 1 Move all nonzero terms to one side and obtain zero on the other side. Subtract $27x^2$ from both sides.

$$3x^4 = 27x^2 \qquad \text{This is the given equation.}$$
$$3x^4 - 27x^2 = 27x^2 - 27x^2 \qquad \text{Subtract } 27x^2 \text{ from both sides.}$$
$$3x^4 - 27x^2 = 0 \qquad \text{Simplify.}$$

Step 2 Factor. We can factor $3x^2$ from each term.

$$3x^4 - 27x^2 = 0$$
$$3x^2(x^2 - 9) = 0$$

GREAT QUESTION!

Can I solve $3x^4 = 27x^2$ by first dividing both sides by x^2?

No. If you divide both sides of the equation by x^2, you'll lose 0 as a solution. In general, do not divide both sides of an equation by a variable expression because that expression might take on the value 0 and you cannot divide by 0.

Steps 3 and 4 Set each factor equal to zero and solve the resulting equations.

$$3x^2 = 0 \qquad \text{or} \qquad x^2 - 9 = 0$$
$$x^2 = 0 \qquad\qquad\qquad x^2 = 9$$
$$x = \pm\sqrt{0} \qquad\qquad\quad x = \pm\sqrt{9}$$
$$x = 0 \qquad\qquad\qquad\quad x = \pm 3$$

Step 5 Check the solutions in the original equation. Check the three solutions, $0, -3$, and 3, by substituting them into the original equation, $3x^4 = 27x^2$. Can you verify that the solution set is $\{-3, 0, 3\}$? •••

⊘ Check Point **1** Solve by factoring: $4x^4 = 12x^2$.

EXAMPLE 2 Solving a Polynomial Equation by Factoring

Solve by factoring: $x^3 + x^2 = 4x + 4$.

SOLUTION

Step 1 Move all nonzero terms to one side and obtain zero on the other side. Subtract $4x$ and subtract 4 from both sides.

$$x^3 + x^2 = 4x + 4 \qquad\qquad \text{This is the given equation.}$$
$$x^3 + x^2 - 4x - 4 = 4x + 4 - 4x - 4 \qquad \text{Subtract } 4x \text{ and } 4 \text{ from both sides.}$$
$$x^3 + x^2 - 4x - 4 = 0 \qquad\qquad \text{Simplify.}$$

Step 2 Factor. Use factoring by grouping. Group terms that have a common factor.

$$\boxed{x^3 + x^2} \; + \; \boxed{-4x - 4} = 0$$

Common factor is x^2. Common factor is -4.

$$x^2(x + 1) - 4(x + 1) = 0 \qquad \text{Factor } x^2 \text{ from the first two terms and } -4 \text{ from the last two terms.}$$

$$(x + 1)(x^2 - 4) = 0 \qquad \text{Factor out the common binomial factor, } x + 1, \text{ from each term.}$$

Steps 3 and 4 Set each factor equal to zero and solve the resulting equations.

$$x + 1 = 0 \qquad \text{or} \qquad x^2 - 4 = 0$$
$$x = -1 \qquad\qquad\qquad x^2 = 4$$
$$x = \pm\sqrt{4} = \pm 2$$

Step 5 Check the solutions in the original equation. Check the three solutions, $-1, -2$, and 2, by substituting them into $x^3 + x^2 = 4x + 4$, the original equation. Can you verify that the solution set is $\{-2, -1, 2\}$? •••

TECHNOLOGY

Numeric Connections

A graphing utility's $\boxed{\text{TABLE}}$ feature can be used to verify numerically that $\{-2, -1, 2\}$ is the solution set of

$$x^3 + x^2 = 4x + 4.$$

Enter $y_1 = x^3 + x^2$. Enter $y_2 = 4x + 4$.

y_1 and y_2 are equal when $x = -2$, $x = -1$, and $x = 2$.

TECHNOLOGY

Graphic Connections

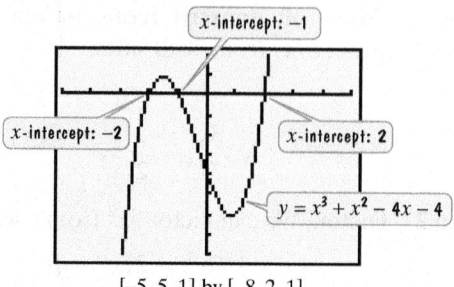

x-intercept: -1
x-intercept: -2
x-intercept: 2
$y = x^3 + x^2 - 4x - 4$
$[-5, 5, 1]$ by $[-8, 2, 1]$

You can use a graphing utility to check the solutions of $x^3 + x^2 - 4x - 4 = 0$. Graph $y = x^3 + x^2 - 4x - 4$, as shown on the left. The x-intercepts are $-2, -1$, and 2, corresponding to the equation's solutions. Equivalently, the zeros of the polynomial function $f(x) = x^3 + x^2 - 4x - 4$ are $-2, -1$, and 2.

DISCOVERY

Suggest a method involving intersecting graphs that can be used with a graphing utility to verify that $\{-2, -1, 2\}$ is the solution set of

$$x^3 + x^2 = 4x + 4.$$

Apply this method to verify the solution set.

⊘ Check Point **2** Solve by factoring: $2x^3 + 3x^2 = 8x + 12$.

2 Solve radical equations.

Radical Equations

A **radical equation** is an equation in which the variable occurs in a square root, cube root, or any higher root. An example of a radical equation is

$$\sqrt{x} = 9.$$

We solve this equation by squaring both sides:

Squaring both sides eliminates the square root.

$$\left(\sqrt{x}\right)^2 = 9^2$$
$$x = 81.$$

The proposed solution, 81, can be checked in the original equation, $\sqrt{x} = 9$. Because $\sqrt{81} = 9$, the solution is 81 and the solution set is {81}.

In general, we solve radical equations with square roots by squaring both sides of the equation. We solve radical equations with nth roots by raising both sides of the equation to the nth power. Unfortunately, if n is even, all the solutions of the equation raised to the even power may not be solutions of the original equation. Consider, for example, the equation

$$x = 4.$$

If we square both sides, we obtain

$$x^2 = 16.$$

Solving this equation using the square root property, we obtain

$$x = \pm\sqrt{16} = \pm 4.$$

The new equation $x^2 = 16$ has two solutions, -4 and 4. By contrast, only 4 is a solution of the original equation, $x = 4$. For this reason, **when raising both sides of an equation to an even power, always check proposed solutions in the original equation.**

Here is a general method for solving radical equations with nth roots:

Solving Radical Equations Containing nth Roots

1. If necessary, arrange terms so that one radical is isolated on one side of the equation.
2. Raise both sides of the equation to the nth power to eliminate the isolated nth root.
3. Solve the resulting equation. If this equation still contains radicals, repeat steps 1 and 2.
4. Check all proposed solutions in the original equation.

Extra solutions may be introduced when you raise both sides of a radical equation to an even power. Such solutions, which are not solutions of the given equation, are called **extraneous solutions** or **extraneous roots**.

EXAMPLE 3 Solving a Radical Equation

Solve: $\sqrt{2x - 1} + 2 = x$.

SOLUTION

Step 1 Isolate a radical on one side. We isolate the radical, $\sqrt{2x - 1}$, by subtracting 2 from both sides.

$$\sqrt{2x - 1} + 2 = x \qquad \text{This is the given equation.}$$
$$\sqrt{2x - 1} = x - 2 \qquad \text{Subtract 2 from both sides.}$$

GREAT QUESTION!

Can I square the right side of $\sqrt{2x - 1} = x - 2$ by first squaring x and then squaring 2?

No. Be sure to square *both sides* of an equation. Do *not* square each term.

Correct:

$$\left(\sqrt{2x - 1}\right)^2 = (x - 2)^2$$

Incorrect!

$$\left(\sqrt{2x - 1}\right)^2 = x^2 - 2^2$$

Step 2 Raise both sides to the nth power. Because n, the index of $\sqrt{2x - 1} = x - 2$, is 2, we square both sides.

$$\left(\sqrt{2x - 1}\right)^2 = (x - 2)^2$$
$$2x - 1 = x^2 - 4x + 4$$

Simplify. Use the formula $(A - B)^2 = A^2 - 2AB + B^2$ on the right side.

Step 3 Solve the resulting equation. Because of the x^2-term, the resulting equation is a quadratic equation. We can obtain 0 on the left side by subtracting $2x$ and adding 1 on both sides.

$$2x - 1 = x^2 - 4x + 4 \qquad \text{The resulting equation is quadratic.}$$
$$0 = x^2 - 6x + 5 \qquad \begin{array}{l}\text{Write in general form, subtracting } 2x \\ \text{and adding 1 on both sides.}\end{array}$$
$$0 = (x - 1)(x - 5) \qquad \text{Factor.}$$
$$x - 1 = 0 \quad \text{or} \quad x - 5 = 0 \qquad \text{Set each factor equal to 0.}$$
$$x = 1 \qquad\qquad x = 5 \qquad \text{Solve the resulting equations.}$$

Step 4 Check the proposed solutions in the original equation.

Check 1:	**Check 5:**

$$\sqrt{2x - 1} + 2 = x \qquad\qquad \sqrt{2x - 1} + 2 = x$$
$$\sqrt{2 \cdot 1 - 1} + 2 \overset{?}{=} 1 \qquad\qquad \sqrt{2 \cdot 5 - 1} + 2 \overset{?}{=} 5$$
$$\sqrt{1} + 2 \overset{?}{=} 1 \qquad\qquad \sqrt{9} + 2 \overset{?}{=} 5$$
$$1 + 2 \overset{?}{=} 1 \qquad\qquad 3 + 2 \overset{?}{=} 5$$
$$3 = 1, \quad \text{false} \qquad\qquad 5 = 5, \quad \text{true}$$

Thus, 1 is an extraneous solution. The only solution is 5, and the solution set is {5}. • • •

TECHNOLOGY

Numeric Connections

A graphing utility's $\boxed{\text{TABLE}}$ feature provides a numeric check that 1 is not a solution and 5 is a solution of $\sqrt{2x - 1} + 2 = x$.

⊘ Check Point **3** Solve: $\sqrt{x + 3} + 3 = x$.

Solving a radical equation with two or more square root expressions involves isolating a radical, squaring both sides, and then repeating this process. Let's consider an equation containing two square root expressions.

EXAMPLE 4 Solving an Equation That Has Two Radicals

Solve: $\sqrt{3x + 1} - \sqrt{x + 4} = 1$.

SOLUTION

Step 1 Isolate a radical on one side. We can isolate the radical $\sqrt{3x + 1}$ by adding $\sqrt{x + 4}$ to both sides. We obtain

$$\sqrt{3x + 1} = \sqrt{x + 4} + 1.$$

Step 2 Square both sides.

$$\left(\sqrt{3x + 1}\right)^2 = \left(\sqrt{x + 4} + 1\right)^2$$

Squaring the expression on the right side of the equation can be a bit tricky. We have to use the formula

$$(A + B)^2 = A^2 + 2AB + B^2.$$

Focusing on just the right side, here is how the squaring is done:

$$(A + B)^2 \quad = \quad A^2 \quad + \quad 2 \cdot A \cdot B + B^2$$

$$\left(\sqrt{x + 4} + 1\right)^2 = \left(\sqrt{x + 4}\right)^2 + 2 \cdot \sqrt{x + 4} \cdot 1 + 1^2 = x + 4 + 2\sqrt{x + 4} + 1.$$

Now let's return to squaring both sides.

$$\left(\sqrt{3x + 1}\right)^2 = \left(\sqrt{x + 4} + 1\right)^2 \qquad \text{Square both sides of the equation with an isolated radical.}$$

$$3x + 1 = x + 4 + 2\sqrt{x + 4} + 1 \qquad \left(\sqrt{3x + 1}\right)^2 = 3x + 1; \text{square the right side using the formula for } (A + B)^2.$$

$$3x + 1 = x + 5 + 2\sqrt{x + 4} \qquad \text{Combine numerical terms on the right side: } 4 + 1 = 5.$$

Can you see that the resulting equation still contains a radical, namely, $\sqrt{x + 4}$? Thus, we need to repeat the first two steps.

Repeat Step 1 Isolate a radical on one side. We isolate $2\sqrt{x + 4}$, the radical term, by subtracting $x + 5$ from both sides. We obtain

$$3x + 1 = x + 5 + 2\sqrt{x + 4} \qquad \text{This is the equation from our last step.}$$

$$2x - 4 = 2\sqrt{x + 4}. \qquad \text{Subtract } x \text{ and subtract 5 from both sides.}$$

Although we can simplify the equation by dividing both sides by 2, this sort of simplification is not always helpful. Thus, we will work with the equation in this form.

Repeat Step 2 Square both sides.

> Be careful in squaring both sides. Use $(A - B)^2 = A^2 - 2AB + B^2$ to square the left side. Use $(AB)^2 = A^2 B^2$ to square the right side.

$$(2x - 4)^2 = \left(2\sqrt{x + 4}\right)^2 \quad \text{Square both sides.}$$

$$4x^2 - 16x + 16 = 4(x + 4) \qquad \text{Square both 2 and } \sqrt{x + 4} \text{ on the right side.}$$

Step 3 Solve the resulting equation. We solve this quadratic equation by writing it in general form.

$$4x^2 - 16x + 16 = 4x + 16 \qquad \text{Use the distributive property.}$$

$$4x^2 - 20x = 0 \qquad \text{Subtract } 4x + 16 \text{ from both sides.}$$

$$4x(x - 5) = 0 \qquad \text{Factor.}$$

$$4x = 0 \quad \text{or} \quad x - 5 = 0 \qquad \text{Set each factor equal to zero.}$$

$$x = 0 \qquad\qquad x = 5 \qquad \text{Solve for } x.$$

TECHNOLOGY

Graphic Connections

The graph of

$$y = \sqrt{3x + 1} - \sqrt{x + 4} - 1$$

has only one x-intercept at 5. This verifies that the solution set of $\sqrt{3x + 1} - \sqrt{x + 4} = 1$ is $\{5\}$.

$[0, 7, 1]$ by $[-1, 1, 1]$

Step 4 Check the proposed solutions in the original equation.

Check 0:	**Check 5:**
$\sqrt{3x + 1} - \sqrt{x + 4} = 1$	$\sqrt{3x + 1} - \sqrt{x + 4} = 1$
$\sqrt{3 \cdot 0 + 1} - \sqrt{0 + 4} \stackrel{?}{=} 1$	$\sqrt{3 \cdot 5 + 1} - \sqrt{5 + 4} \stackrel{?}{=} 1$
$\sqrt{1} - \sqrt{4} \stackrel{?}{=} 1$	$\sqrt{16} - \sqrt{9} \stackrel{?}{=} 1$
$1 - 2 \stackrel{?}{=} 1$	$4 - 3 \stackrel{?}{=} 1$
$-1 = 1$, false	$1 = 1$, true

The check indicates that 0 is not a solution. It is an extraneous solution brought about by squaring each side of the equation. The only solution is 5 and the solution set is $\{5\}$. •••

GREAT QUESTION!

In Example 4, do I have to use the formula for $(A + B)^2$ in order to square $\sqrt{x + 4} + 1$?

No. You can also use the FOIL method.

$$(\sqrt{x + 4} + 1)^2 = (\sqrt{x + 4} + 1)(\sqrt{x + 4} + 1)$$

$$= (\sqrt{x + 4})^2 + 1 \cdot \sqrt{x + 4} + 1 \cdot \sqrt{x + 4} + 1^2$$
$$= x + 4 + 2\sqrt{x + 4} + 1$$

✓ **Check Point 4** Solve: $\sqrt{x + 5} - \sqrt{x - 3} = 2$.

③ Solve equations with rational exponents.

Equations with Rational Exponents

We know that expressions with rational exponents represent radicals:

$$a^{\frac{m}{n}} = \left(\sqrt[n]{a} \right)^m = \sqrt[n]{a^m}.$$

For example, the radical equation $3\sqrt[4]{x^3} - 6 = 0$ can be expressed as $3x^{\frac{3}{4}} - 6 = 0$.
 A radical equation with a rational exponent can be solved by

 1. isolating the expression with the rational exponent, and
 2. raising both sides of the equation to a power that is the reciprocal of the rational exponent.

Solving Radical Equations of the Form $x^{\frac{m}{n}} = k$

Assume that m and n are positive integers, $\frac{m}{n}$ is in lowest terms, and k is a real number.

 1. Isolate the expression with the rational exponent.
 2. Raise both sides of the equation to the $\frac{n}{m}$ power.

If m is even:	If m is odd:

$$x^{\frac{m}{n}} = k$$

$$\left(x^{\frac{m}{n}}\right)^{\frac{n}{m}} = \pm k^{\frac{n}{m}}$$

$$x = \pm k^{\frac{n}{m}}$$

$$x^{\frac{m}{n}} = k$$

$$\left(x^{\frac{m}{n}}\right)^{\frac{n}{m}} = k^{\frac{n}{m}}$$

$$x = k^{\frac{n}{m}}$$

It is incorrect to insert the \pm symbol when the numerator of the exponent is odd. An odd index has only one root.

3. Check all proposed solutions in the original equation to find out if they are actual solutions or extraneous solutions.

EXAMPLE 5 Solving Equations Involving Rational Exponents

Solve:

a. $3x^{\frac{3}{4}} - 6 = 0$ **b.** $x^{\frac{2}{3}} - \dfrac{3}{4} = -\dfrac{1}{2}$.

SOLUTION

a. To solve $3x^{\frac{3}{4}} - 6 = 0$, our goal is to isolate $x^{\frac{3}{4}}$. Then we raise both sides of the equation to the $\frac{4}{3}$ power because $\frac{4}{3}$ is the reciprocal of $\frac{3}{4}$.

$$3x^{\frac{3}{4}} - 6 = 0 \qquad \text{This is the given equation; we will isolate } x^{\frac{3}{4}}.$$

$$3x^{\frac{3}{4}} = 6 \qquad \text{Add 6 to both sides.}$$

$$\frac{3x^{\frac{3}{4}}}{3} = \frac{6}{3} \qquad \text{Divide both sides by 3.}$$

$$x^{\frac{3}{4}} = 2 \qquad \text{Simplify.}$$

$$\left(x^{\frac{3}{4}}\right)^{\frac{4}{3}} = 2^{\frac{4}{3}} \qquad \begin{array}{l}\text{Raise both sides to the } \frac{4}{3} \text{ power. Because } \frac{m}{n} = \frac{3}{4} \text{ and } m \text{ is}\\ \text{odd, we do not use the } \pm \text{ symbol.}\end{array}$$

$$x = 2^{\frac{4}{3}} \qquad \text{Simplify the left side: } \left(x^{\frac{3}{4}}\right)^{\frac{4}{3}} = x^{\frac{3 \cdot 4}{4 \cdot 3}} = x^{\frac{12}{12}} = x^1 = x.$$

The proposed solution is $2^{\frac{4}{3}}$. Complete the solution process by checking this value in the given equation.

$$3x^{\frac{3}{4}} - 6 = 0 \qquad \text{This is the original equation.}$$

$$3\left(2^{\frac{4}{3}}\right)^{\frac{3}{4}} - 6 \stackrel{?}{=} 0 \qquad \text{Substitute the proposed solution.}$$

$$3 \cdot 2 - 6 \stackrel{?}{=} 0 \qquad \left(2^{\frac{4}{3}}\right)^{\frac{3}{4}} = 2^{\frac{4 \cdot 3}{3 \cdot 4}} = 2^{\frac{12}{12}} = 2^1 = 2$$

$$0 = 0, \quad \text{true} \qquad 3 \cdot 2 - 6 = 6 - 6 = 0$$

The solution is $2^{\frac{4}{3}} = \sqrt[3]{2^4} \approx 2.52$. The solution set is $\left\{2^{\frac{4}{3}}\right\}$.

GREAT QUESTION!

Why didn't you eliminate fractions in Example 5(b) by first multiplying both sides by the LCD, 4?

Our goal is to write the equation in the form $x^{\frac{m}{n}} = k$, where the coefficient of the variable term is 1. Multiplying by 4 results in a coefficient of 4 rather than 1.

b. To solve $x^{\frac{2}{3}} - \frac{3}{4} = -\frac{1}{2}$, our goal is to isolate $x^{\frac{2}{3}}$. Then we raise both sides of the equation to the $\frac{3}{2}$ power because $\frac{3}{2}$ is the reciprocal of $\frac{2}{3}$.

$$x^{\frac{2}{3}} - \frac{3}{4} = -\frac{1}{2}$$ This is the given equation.

$$x^{\frac{2}{3}} = \frac{1}{4}$$ Add $\frac{3}{4}$ to both sides. $\frac{3}{4} - \frac{1}{2} = \frac{3}{4} - \frac{2}{4} = \frac{1}{4}$.

$$\left(x^{\frac{2}{3}}\right)^{\frac{3}{2}} = \pm\left(\frac{1}{4}\right)^{\frac{3}{2}}$$ Raise both sides to the $\frac{3}{2}$ power. Because $\frac{m}{n} = \frac{2}{3}$ and m is even, the \pm symbol is necessary.

$$x = \pm\frac{1}{8}$$ $\left(\frac{1}{4}\right)^{\frac{3}{2}} = \left(\sqrt{\frac{1}{4}}\right)^3 = \left(\frac{1}{2}\right)^3 = \frac{1}{8}$

Take a moment to verify that the solution set is $\left\{-\frac{1}{8}, \frac{1}{8}\right\}$. • • •

⌀ **Check Point 5** Solve:

 a. $5x^{\frac{3}{2}} - 25 = 0$ **b.** $x^{\frac{2}{3}} - 8 = -4$.

 Solve equations that are quadratic in form.

Equations That Are Quadratic in Form

Some equations that are not quadratic can be written as quadratic equations using an appropriate substitution. Here are some examples:

GREAT QUESTION!

Do I have to use the letter u in the substitution?

No. Although we selected u, any letter can be used.

Given Equation	Substitution	New Equation
$x^4 - 10x^2 + 9 = 0$ or $(x^2)^2 - 10x^2 + 9 = 0$	$u = x^2$	$u^2 - 10u + 9 = 0$
$5x^{\frac{2}{3}} + 11x^{\frac{1}{3}} + 2 = 0$ or $5\left(x^{\frac{1}{3}}\right)^2 + 11x^{\frac{1}{3}} + 2 = 0$	$u = x^{\frac{1}{3}}$	$5u^2 + 11u + 2 = 0$

An equation that is **quadratic in form** is one that can be expressed as a quadratic equation using an appropriate substitution. Both of the preceding given equations are quadratic in form.

Equations that are quadratic in form contain an expression to a power, the same expression to that power squared, and a constant term. By letting u equal the expression to the power, a quadratic equation in u will result. Now it's easy. Solve this quadratic equation for u. Finally, use your substitution to find the values for the variable in the given equation. Example 6 shows how this is done.

 EXAMPLE 6 Solving an Equation Quadratic in Form

Solve: $x^4 - 8x^2 - 9 = 0$.

SOLUTION

Notice that the equation contains an expression to a power, x^2, the same expression to that power squared, x^4 or $(x^2)^2$, and a constant term, -9. We let u equal the expression to the power. Thus,

$$\text{let } u = x^2.$$

TECHNOLOGY

Graphic Connections

The graph of

$$f(x) = x^4 - 8x^2 - 9$$

has x-intercepts at -3 and 3. This verifies that the real zeros of f are -3 and 3. The imaginary zeros, $-i$ and i, are not shown as intercepts.

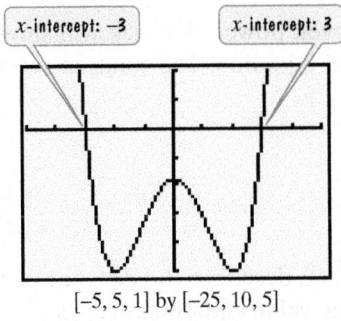

x-intercept: -3

x-intercept: 3

$[-5, 5, 1]$ by $[-25, 10, 5]$

Now we write the given equation as a quadratic equation in u and solve for u.

$x^4 - 8x^2 - 9 = 0$	This is the given equation.
$(x^2)^2 - 8x^2 - 9 = 0$	The given equation contains x^2 and x^2 squared.
$u^2 - 8u - 9 = 0$	Replace x^2 with u.
$(u - 9)(u + 1) = 0$	Factor.
$u - 9 = 0$ or $u + 1 = 0$	Apply the zero-product principle.
$u = 9$ $u = -1$	Solve for u.

We're not done! Why not? We were asked to solve for x and we have values for u. We use the original substitution, $u = x^2$, to solve for x. Replace u with x^2 in each equation shown, namely $u = 9$ and $u = -1$.

$u = 9$	$u = -1$	These are the equations in u.
$x^2 = 9$	$x^2 = -1$	Replace u with x^2.
$x = \pm\sqrt{9}$	$x = \pm\sqrt{-1}$	Apply the square root property.
$x = \pm 3$	$x = \pm i$	Simplify.

The solution set is $\{-3, 3, -i, i\}$. The graph in the technology box shows that only the real solutions, -3 and 3, appear as x-intercepts. • • •

⊘ Check Point **6** Solve: $x^4 - 5x^2 + 6 = 0$.

EXAMPLE 7 Solving an Equation Quadratic in Form

Solve: $5x^{\frac{2}{3}} + 11x^{\frac{1}{3}} + 2 = 0$.

SOLUTION

Notice that the equation contains an expression to a power, $x^{\frac{1}{3}}$, the same expression to that power squared, $x^{\frac{2}{3}}$ or $\left(x^{\frac{1}{3}}\right)^2$, and a constant term, 2. We let u equal the expression to the power. Thus,

$$\text{let } u = x^{\frac{1}{3}}.$$

Now we write the given equation as a quadratic equation in u and solve for u.

$5x^{\frac{2}{3}} + 11x^{\frac{1}{3}} + 2 = 0$	This is the given equation.
$5\left(x^{\frac{1}{3}}\right)^2 + 11x^{\frac{1}{3}} + 2 = 0$	The given equation contains $x^{\frac{1}{3}}$ and $x^{\frac{1}{3}}$ squared.
$5u^2 + 11u + 2 = 0$	Let $u = x^{\frac{1}{3}}$.
$(5u + 1)(u + 2) = 0$	Factor.
$5u + 1 = 0$ or $u + 2 = 0$	Set each factor equal to 0.
$5u = -1$ $u = -2$	Solve for u.
$u = -\dfrac{1}{5}$	

Use the original substitution, $u = x^{\frac{1}{3}}$, to solve for x. Replace u with $x^{\frac{1}{3}}$ in each of the preceding equations, namely $u = -\frac{1}{5}$ and $u = -2$.

$$u = -\frac{1}{5} \qquad\qquad u = -2 \qquad \text{These are the equations in } u.$$

$$x^{\frac{1}{3}} = -\frac{1}{5} \qquad\qquad x^{\frac{1}{3}} = -2 \qquad \text{Replace } u \text{ with } x^{\frac{1}{3}}.$$

$$\left(x^{\frac{1}{3}}\right)^3 = \left(-\frac{1}{5}\right)^3 \qquad \left(x^{\frac{1}{3}}\right)^3 = (-2)^3 \qquad \text{Solve for } x \text{ by cubing both sides of each equation.}$$

$$x = -\frac{1}{125} \qquad\qquad x = -8 \qquad \text{Simplify.}$$

We did not raise both sides of an equation to an even power. A check will show that both -8 and $-\frac{1}{125}$ are solutions of the original equation. The solution set is $\left\{-8, -\frac{1}{125}\right\}$.

• • •

⊘ Check Point **7** Solve: $3x^{\frac{2}{3}} - 11x^{\frac{1}{3}} - 4 = 0$.

⑤ Solve equations involving absolute value.

Equations Involving Absolute Value

We have seen that the absolute value of x, denoted $|x|$, describes the distance of x from zero on a number line. Now consider an **absolute value equation**, such as

$$|x| = 2.$$

This means that we must determine real numbers whose distance from the origin on a number line is 2. **Figure 2.12** shows that there are two numbers such that $|x| = 2$, namely, 2 and -2. We write $x = 2$ or $x = -2$. This observation can be generalized as follows:

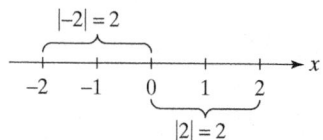

FIGURE 2.12

Rewriting an Absolute Value Equation without Absolute Value Bars

If c is a positive real number and u represents any algebraic expression, then $|u| = c$ is equivalent to $u = c$ or $u = -c$.

TECHNOLOGY

Graphic Connections

You can use a graphing utility to verify the solution of an absolute value equation. Consider, for example,

$$|2x - 3| = 11.$$

Graph $y_1 = |2x - 3|$ and $y_2 = 11$. The graphs are shown in a $[-10, 10, 1]$ by $[-1, 15, 1]$ viewing rectangle. The x-coordinates of the intersection points are -4 and 7, verifying that $\{-4, 7\}$ is the solution set.

$[-10, 10, 1]$ by $[-1, 15, 1]$

EXAMPLE 8 Solving an Equation Involving Absolute Value

Solve: $|2x - 3| = 11$.

SOLUTION

$$|2x - 3| = 11 \qquad \text{This is the given equation.}$$

$$2x - 3 = 11 \quad \text{or} \quad 2x - 3 = -11 \qquad \text{Rewrite the equation without absolute value bars: } |u| = c \text{ is equivalent to } u = c \text{ or } u = -c.$$

$$2x = 14 \qquad\qquad 2x = -8 \qquad \text{Add 3 to both sides of each equation.}$$

$$x = 7 \qquad\qquad x = -4 \qquad \text{Divide both sides of each equation by 2.}$$

Check 7:	**Check -4:**	
$\|2x - 3\| = 11$	$\|2x - 3\| = 11$	This is the original equation.
$\|2(7) - 3\| \stackrel{?}{=} 11$	$\|2(-4) - 3\| \stackrel{?}{=} 11$	Substitute the proposed solutions.
$\|14 - 3\| \stackrel{?}{=} 11$	$\|-8 - 3\| \stackrel{?}{=} 11$	Perform operations inside the absolute value bars.
$\|11\| \stackrel{?}{=} 11$	$\|-11\| \stackrel{?}{=} 11$	
$11 = 11,$ true	$11 = 11,$ true	These true statements indicate that 7 and -4 are solutions.

The solution set is $\{-4, 7\}$.

• • •

⊘ Check Point **8** Solve: $|2x - 1| = 5$.

EXAMPLE 9 Solving an Equation Involving Absolute Value

Find all zeros of $f(x) = 5|1 - 4x| - 15$.

SOLUTION

We find the zeros by solving the equation $f(x) = 0$:

$$f(x) = 5|1 - 4x| - 15 = 0.$$

We need to isolate $|1 - 4x|$, the absolute value expression.

$5	1 - 4x	= 15$		Add 15 to both sides.
$	1 - 4x	= 3$		Divide both sides by 5.
$1 - 4x = 3$ or $1 - 4x = -3$		Rewrite $	u	= c$ as $u = c$ or $u = -c$.
$-4x = 2$ $-4x = -4$		Subtract 1 from both sides of each equation.		
$x = -\dfrac{1}{2}$ $x = 1$		Divide both sides of each equation by -4.		

The zeros of $f(x) = 5|1 - 4x| - 15$ are $-\frac{1}{2}$ and 1. These zeros appear as x-intercepts for the graph of f, as shown in **Figure 2.13**. Take a moment to verify the zeros by finding $f(-\frac{1}{2})$ and $f(1)$. You should obtain 0 in each case. • • •

x-intercept: $-\frac{1}{2}$ x-intercept: 1

[−2, 2, 1] by [−16, 4, 1]

FIGURE 2.13 The graph of $f(x) = 5|1 - 4x| - 15$

✓ **Check Point 9** Find all zeros of $f(x) = 4|1 - 2x| - 20$.

The absolute value of a number is never negative. Thus, if u is an algebraic expression and c is a negative number, then $|u| = c$ has no solution. For example, the equation $|3x - 6| = -2$ has no solution because $|3x - 6|$ cannot be negative. The solution set is \varnothing, the empty set. Equivalently, $f(x) = |3x - 6| + 2$ has no zeros. The graph of f, shown in **Figure 2.14**, lies above the x-axis and has no x-intercepts.

The absolute value of 0 is 0. Thus, if u is an algebraic expression and $|u| = 0$, the solution is found by solving $u = 0$. For example, the solution of $|x - 2| = 0$ is obtained by solving $x - 2 = 0$. The solution is 2 and the solution set is $\{2\}$.

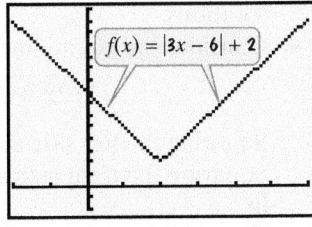

$f(x) = |3x - 6| + 2$

[−2, 6, 1] by [−2, 15, 1]

FIGURE 2.14 An absolute value function whose graph has no x-intercepts

⑥ Solve problems modeled by equations.

Applications

EXAMPLE 10 Television Viewing, by Annual Income

The graph in **Figure 2.15** shows the average number of hours per week spent watching TV, by annual income. The formula

$$H = -2.3\sqrt{I} + 67.6$$

models weekly television viewing time, H, in hours, by annual income, I, in thousands of dollars. What annual income corresponds to 44.6 hours per week watching TV?

SOLUTION

Because we are interested in finding the annual income that corresponds to 44.6 hours of weekly viewing time, we substitute 44.6 for H in the given formula. Then we solve for I, the annual income, in thousands of dollars.

$H = -2.3\sqrt{I} + 67.6$	This is the given formula.
$44.6 = -2.3\sqrt{I} + 67.6$	Substitute 44.6 for H.
$-23 = -2.3\sqrt{I}$	Subtract 67.6 from both sides.
$\dfrac{-23}{-2.3} = \dfrac{-2.3\sqrt{I}}{-2.3}$	Divide both sides by −2.3.

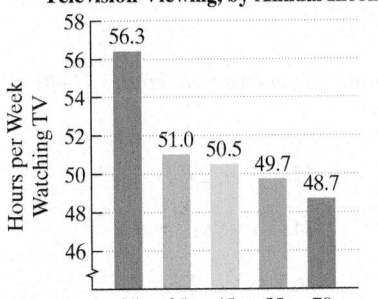

Television Viewing, by Annual Income

56.3, 51.0, 50.5, 49.7, 48.7

Hours per Week Watching TV

Annual Income (thousands of dollars)

25 35 45 55 70

FIGURE 2.15
Source: *Nielsen Media Research*

$$10 = \sqrt{I}$$

$$10^2 = \left(\sqrt{I}\right)^2 \qquad \text{Square both sides.}$$

$$100 = I \qquad \text{Simplify.}$$

Simplify $\dfrac{-23}{-2.3} = \dfrac{-2.3\sqrt{I}}{-2.3}; \dfrac{-23}{-2.3} = 10.$

The model indicates that an annual income of 100 thousand dollars, or \$100,000, corresponds to 44.6 hours of weekly viewing time.　　● ● ●

⊘ Check Point **10** Use the formula in Example 10 on the previous page to find the annual income that corresponds to 33.1 hours per week watching TV.

CONCEPT AND VOCABULARY CHECK

Fill in each blank so that the resulting statement is true.

1. The first step in solving the polynomial equation

$$2x^3 + 3x^2 = 8x + 12$$

is to _____.

2. An equation in which the variable occurs in a square root, cube root, or any higher root is called a/an _____ equation.

3. Solutions of a squared equation that are not solutions of the original equation are called _____ solutions.

4. Consider the equation

$$\sqrt{2x + 1} = x - 7.$$

Squaring the left side and simplifying results in _____. Squaring the right side and simplifying results in _____.

5. Consider the equation

$$\sqrt{x + 2} = 3 - \sqrt{x - 1}.$$

Squaring the left side and simplifying results in _____. Squaring the right side and simplifying results in _____.

6. If $x^{\frac{3}{4}} = 5$, then $x = $ ____.

7. If $x^{\frac{2}{3}} = 5$, then $x = $ _____.

8. We solve $x^4 - 13x^2 + 36 = 0$ by letting $u = $ ____. We then rewrite the equation in terms of u as _____.

9. We solve $x^{\frac{2}{3}} + 2x^{\frac{1}{3}} - 3 = 0$ by letting $u = $ ____. We then rewrite the equation in terms of u as _____.

10. If $c > 0$, $|u| = c$ is equivalent to $u = $ __ or $u = $ __.

11. $|3x - 1| = 7$ is equivalent to _____ or _____.

EXERCISE SET 2.5

Practice Exercises

Solve each polynomial equation in Exercises 1–10 by factoring and then using the zero-product principle.

1. $3x^4 - 48x^2 = 0$
2. $5x^4 - 20x^2 = 0$
3. $3x^3 + 2x^2 = 12x + 8$
4. $4x^3 - 12x^2 = 9x - 27$
5. $2x - 3 = 8x^3 - 12x^2$
6. $x + 1 = 9x^3 + 9x^2$
7. $4y^3 - 2 = y - 8y^2$
8. $9y^3 + 8 = 4y + 18y^2$
9. $2x^4 = 16x$
10. $3x^4 = 81x$

Solve each radical equation in Exercises 11–30. Check all proposed solutions.

11. $\sqrt{3x + 18} = x$
12. $\sqrt{20 - 8x} = x$
13. $\sqrt{x + 3} = x - 3$
14. $\sqrt{x + 10} = x - 2$
15. $\sqrt{2x + 13} = x + 7$
16. $\sqrt{6x + 1} = x - 1$
17. $x - \sqrt{2x + 5} = 5$
18. $x - \sqrt{x + 11} = 1$
19. $\sqrt{2x + 19} - 8 = x$
20. $\sqrt{2x + 15} - 6 = x$
21. $\sqrt{3x + 10} = x + 4$
22. $\sqrt{x - 3} = x - 9$
23. $\sqrt{x + 8} - \sqrt{x - 4} = 2$
24. $\sqrt{x + 5} - \sqrt{x - 3} = 2$
25. $\sqrt{x - 5} - \sqrt{x - 8} = 3$
26. $\sqrt{2x - 3} - \sqrt{x - 2} = 1$
27. $\sqrt{2x + 3} + \sqrt{x - 2} = 2$
28. $\sqrt{x + 2} + \sqrt{3x + 7} = 1$
29. $\sqrt{3\sqrt{x + 1}} = \sqrt{3x - 5}$
30. $\sqrt{1 + 4\sqrt{x}} = 1 + \sqrt{x}$

Solve each equation with rational exponents in Exercises 31–40. Check all proposed solutions.

31. $x^{\frac{3}{2}} = 8$
32. $x^{\frac{3}{2}} = 27$
33. $(x - 4)^{\frac{3}{2}} = 27$
34. $(x + 5)^{\frac{3}{2}} = 8$
35. $6x^{\frac{5}{2}} - 12 = 0$
36. $8x^{\frac{5}{3}} - 24 = 0$
37. $(x - 4)^{\frac{2}{3}} = 16$
38. $(x + 5)^{\frac{2}{3}} = 4$
39. $(x^2 - x - 4)^{\frac{3}{4}} - 2 = 6$
40. $(x^2 - 3x + 3)^{\frac{3}{2}} - 1 = 0$

Solve each equation in Exercises 41–60 by making an appropriate substitution.

41. $x^4 - 5x^2 + 4 = 0$

42. $x^4 - 13x^2 + 36 = 0$

43. $9x^4 = 25x^2 - 16$

44. $4x^4 = 13x^2 - 9$

45. $x - 13\sqrt{x} + 40 = 0$

46. $2x - 7\sqrt{x} - 30 = 0$

47. $x^{-2} - x^{-1} - 20 = 0$

48. $x^{-2} - x^{-1} - 6 = 0$

49. $x^{\frac{2}{3}} - x^{\frac{1}{3}} - 6 = 0$

50. $2x^{\frac{2}{3}} + 7x^{\frac{1}{3}} - 15 = 0$

51. $x^{\frac{3}{2}} - 2x^{\frac{3}{4}} + 1 = 0$

52. $x^{\frac{2}{5}} + x^{\frac{1}{5}} - 6 = 0$

53. $2x - 3x^{\frac{1}{2}} + 1 = 0$

54. $x + 3x^{\frac{1}{2}} - 4 = 0$

55. $(x - 5)^2 - 4(x - 5) - 21 = 0$

56. $(x + 3)^2 + 7(x + 3) - 18 = 0$

57. $(x^2 - x)^2 - 14(x^2 - x) + 24 = 0$

58. $(x^2 - 2x)^2 - 11(x^2 - 2x) + 24 = 0$

59. $\left(y - \dfrac{8}{y}\right)^2 + 5\left(y - \dfrac{8}{y}\right) - 14 = 0$

60. $\left(y - \dfrac{10}{y}\right)^2 + 6\left(y - \dfrac{10}{y}\right) - 27 = 0$

In Exercises 61–78, solve each absolute value equation or indicate that the equation has no solution.

61. $|x| = 8$

62. $|x| = 6$

63. $|x - 2| = 7$

64. $|x + 1| = 5$

65. $|2x - 1| = 5$

66. $|2x - 3| = 11$

67. $2|3x - 2| = 14$

68. $3|2x - 1| = 21$

69. $7|5x| + 2 = 16$

70. $7|3x| + 2 = 16$

71. $2\left|4 - \dfrac{5}{2}x\right| + 6 = 18$

72. $4\left|1 - \dfrac{3}{4}x\right| + 7 = 10$

73. $|x + 1| + 5 = 3$

74. $|x + 1| + 6 = 2$

75. $|2x - 1| + 3 = 3$

76. $|3x - 2| + 4 = 4$

Hint for Exercises 77–78: Absolute value expressions are equal when the expressions inside the absolute value bars are equal to or opposites of each other.

77. $|3x - 1| = |x + 5|$

78. $|2x - 7| = |x + 3|$

In Exercises 79–84, find the zeros of the given function, f. Then use the zeros to match the function with its graph. [The graphs are labeled (a) through (f).]

79. $f(x) = \sqrt{x + 2} + \sqrt{x - 1} - 3$

80. $f(x) = \sqrt{x - 4} + \sqrt{x + 4} - 4$

81. $f(x) = x^{\frac{1}{3}} + 2x^{\frac{1}{6}} - 3$

82. $f(x) = x^{-2} - x^{-1} - 6$

83. $f(x) = (x + 2)^2 - 9(x + 2) + 20$

84. $f(x) = 2(x + 2)^2 + 5(x + 2) - 3$

a.

$[-1, 10, 1]$ by $[-3, 3, 1]$

b.

$[-3, 3, 1]$ by $[-10, 10, 1]$

c.

$[-1, 10, 1]$ by $[-4, 4, 1]$

d.

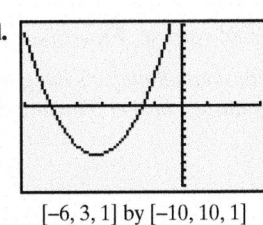

$[-6, 3, 1]$ by $[-10, 10, 1]$

e.

$[-1, 10, 1]$ by $[-3, 3, 1]$

f.

$[-1, 6, 1]$ by $[-1, 10, 1]$

In Exercises 85–94, find all values of x satisfying the given conditions.

85. $f(x) = |5 - 4x|$ and $f(x) = 11$.

86. $f(x) = |2 - 3x|$ and $f(x) = 13$.

87. $f(x) = x + \sqrt{x + 5}$ and $f(x) = 7$.

88. $f(x) = x - \sqrt{x - 2}$ and $f(x) = 4$.

89. $f(x) = 2x^3 + x^2 - 8x + 2$ and $f(x) = 6$.

90. $f(x) = x^3 + 4x^2 - x + 6$ and $f(x) = 10$.

91. $f(x) = (x + 4)^{\frac{3}{2}}$ and $f(x) = 8$.

92. $f(x) = (x - 5)^{\frac{3}{2}}$ and $f(x) = 125$.

93. $f(x) = (x^2 - 1)^2$, $g(x) = 2(x^2 - 1)$, and $f(x)$ exceeds $g(x)$ by 3.

94. $f(x) = 6\left(\dfrac{2x}{x - 3}\right)^2$, $g(x) = 5\left(\dfrac{2x}{x - 3}\right)$, and $f(x)$ exceeds $g(x)$ by 6.

Practice Plus

In Exercises 95–98, solve each equation.

95. $|x^2 + 2x - 36| = 12$

96. $|x^2 + 6x + 1| = 8$

97. $x(x + 1)^3 - 42(x + 1)^2 = 0$

98. $x(x - 2)^3 - 35(x - 2)^2 = 0$

99. If 5 times a number is decreased by 4, the principal square root of this difference is 2 less than the number. Find the number(s).

100. If a number is decreased by 3, the principal square root of this difference is 5 less than the number. Find the number(s).

101. Solve for V: $r = \sqrt{\dfrac{3V}{\pi h}}$.

102. Solve for A: $r = \sqrt{\dfrac{A}{4\pi}}$.

In Exercises 103–104, list all numbers that must be excluded from the domain of each function.

103. $f(x) = \dfrac{|x - 1| - 3}{|x + 2| - 14}$

104. $f(x) = \dfrac{x^3 - 2x^2 - 9x + 18}{x^3 + 3x^2 - x - 3}$

Application Exercises

A basketball player's hang time is the time spent in the air when shooting a basket. The function

$$t(d) = \frac{\sqrt{d}}{2}$$

models hang time, t(d), in seconds, in terms of the vertical distance of a player's jump, d, in feet. Use this function to solve Exercises 105–106.

105. When Michael Wilson of the Harlem Globetrotters slamdunked a basketball, his hang time for the shot was approximately 1.16 seconds. What was the vertical distance of his jump, rounded to the nearest tenth of a foot?

106. If hang time for a shot by a professional basketball player is 0.85 second, what is the vertical distance of the jump, rounded to the nearest tenth of a foot?

Use the graph of the function for hang time to solve Exercises 107–108.

107. How is your answer to Exercise 105 shown on the graph?

108. How is your answer to Exercise 106 shown on the graph?

By the end of 2010, women made up more than half of the labor force in the United States for the first time in the country's history. The graphs show the percentage of jobs in the U.S. labor force held by men and by women from 1972 through 2009. Exercises 109–110 are based on the data displayed by the graphs.

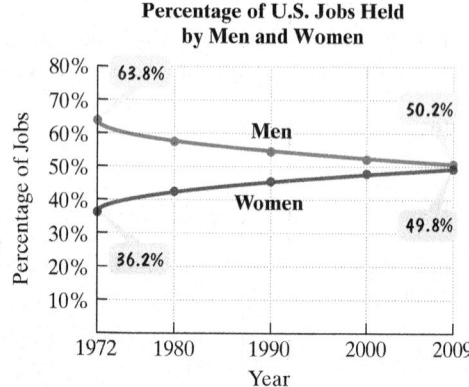

Percentage of U.S. Jobs Held by Men and Women

Source: Bureau of Labor Statistics

109. The function

$$p(t) = 2.2\sqrt{t} + 36.2$$

models the percentage of jobs in the U.S. labor force, $p(t)$, held by women t years after 1972.

a. Use the appropriate graph to estimate the percentage of jobs in the U.S. labor force held by women in 2000. Give your estimation to the nearest percent.

b. Use the function to determine the percentage of jobs in the U.S. labor force held by women in 2000. Round to the nearest tenth of a percent.

c. According to the function, when will 52% of jobs in the U.S. labor force be held by women? Round to the nearest year.

110. The function

$$p(t) = -2.2\sqrt{t} + 63.8$$

models the percentage of jobs in the U.S. labor force, $p(t)$, held by men t years after 1972.

a. Use the appropriate graph to estimate the percentage of jobs in the U.S. labor force held by men in 2000. Give your estimation to the nearest percent.

b. Use the function to determine the percentage of jobs in the U.S. labor force held by men in 2000. Round to the nearest tenth of a percent.

c. According to the function, when will 48% of jobs in the U.S. labor force be held by men? Round to the nearest year.

For each planet in our solar system, its year is the time it takes the planet to revolve once around the sun. The function

$$E(x) = 0.2x^{\frac{3}{2}}$$

models the number of Earth days in a planet's year, E(x), where x is the average distance of the planet from the sun, in millions of kilometers. Use the equation to solve Exercises 111–112.

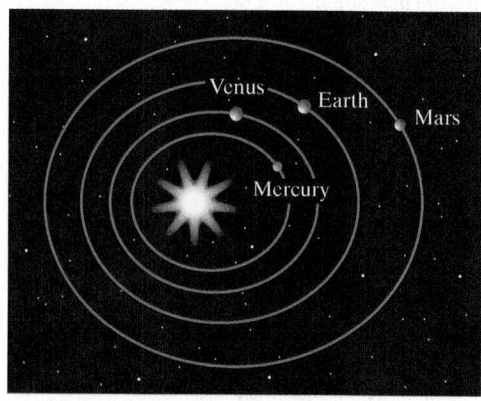

111. We, of course, have 365 Earth days in our year. What is the average distance of Earth from the sun? Use a calculator and round to the nearest million kilometers.

112. There are approximately 88 Earth days in the year of the planet Mercury. What is the average distance of Mercury from the sun? Use a calculator and round to the nearest million kilometers.

Use the Pythagorean Theorem to solve Exercises 113–114.

113. Two vertical poles of lengths 6 feet and 8 feet stand 10 feet apart. A cable reaches from the top of one pole to some point on the ground between the poles and then to the top of the other pole. Where should this point be located to use 18 feet of cable?

114. Towns A and B are located 6 miles and 3 miles, respectively, from a major expressway. The point on the expressway closest to town A is 12 miles from the point on the expressway closest to town B. Two new roads are to be built from A to the expressway and then to B.

a. Express the combined lengths of the new roads in terms of x as shown in the figure.

b. If the combined length of the new roads is 15 miles, what distance does x represent?

Writing in Mathematics

115. Without actually solving the equation, give a general description of how to solve $x^3 - 5x^2 - x + 5 = 0$.

116. In solving $\sqrt{3x + 4} - \sqrt{2x + 4} = 2$, why is it a good idea to isolate a radical term? What if we don't do this and simply square each side? Describe what happens.

117. What is an extraneous solution to a radical equation?

118. Explain how to recognize an equation that is quadratic in form. Provide two original examples with your explanation.

119. Describe two methods for solving this equation: $x - 5\sqrt{x} + 4 = 0$.

120. Explain how to solve an equation involving absolute value.

121. Explain why the procedure that you explained in Exercise 120 does not apply to the equation $|x - 2| = -3$. What is the solution set for this equation?

Technology Exercises

In Exercises 122–124, use a graphing utility and the graph's x-intercepts to solve each equation. Check by direct substitution. A viewing rectangle is given.

122. $x^3 + 3x^2 - x - 3 = 0$
$[-6, 6, 1]$ by $[-6, 6, 1]$

123. $-x^4 + 4x^3 - 4x^2 = 0$
$[-6, 6, 1]$ by $[-9, 2, 1]$

124. $\sqrt{2x + 13} - x - 5 = 0$
$[-5, 5, 1]$ by $[-5, 5, 1]$

Critical Thinking Exercises

Make Sense? *In Exercises 125–128, determine whether each statement makes sense or does not make sense, and explain your reasoning.*

125. When checking a radical equation's proposed solution, I can substitute into the original equation or any equation that is part of the solution process.

126. After squaring both sides of a radical equation, the only solution that I obtained was extraneous, so \varnothing must be the solution set of the original equation.

127. The equation $5x^{\frac{2}{3}} + 11x^{\frac{1}{3}} + 2 = 0$ is quadratic in form, but when I reverse the variable terms and obtain $11x^{\frac{1}{3}} + 5x^{\frac{2}{3}} + 2 = 0$, the resulting equation is no longer quadratic in form.

128. When I solve an equation that is quadratic in form, it's important to write down the substitution that I am making.

In Exercises 129–132, determine whether each statement is true or false. If the statement is false, make the necessary change(s) to produce a true statement.

129. The first step in solving $\sqrt{x + 6} = x + 2$ is to square both sides, obtaining $x + 6 = x^2 + 4$.

130. The equations $\sqrt{x + 4} = -5$ and $x + 4 = 25$ have the same solution set.

131. To solve $x - 9\sqrt{x} + 14 = 0$, we let $\sqrt{u} = x$.

132. The equation $|x| = -6$ is equivalent to $x = 6$ or $x = -6$.

133. Solve: $\sqrt{6x - 2} = \sqrt{2x + 3} - \sqrt{4x - 1}$.

134. Solve without squaring both sides:

$$5 - \frac{2}{x} = \sqrt{5 - \frac{2}{x}}.$$

135. Solve for x: $\sqrt[3]{x\sqrt{x}} = 9$.

136. Solve for x: $x^{\frac{5}{6}} + x^{\frac{2}{3}} - 2x^{\frac{1}{2}} = 0$.

Preview Exercises

Exercises 137–139 will help you prepare for the material covered in the next section.

137. Is -1 a solution of $3 - 2x \le 11$?

138. Solve: $-2x - 4 = x + 5$.

139. Solve: $\dfrac{x + 3}{4} = \dfrac{x - 2}{3} + \dfrac{1}{4}$.

Objectives

① Solve linear inequalities.

② Recognize inequalities with no solution or all real numbers as solutions.

③ Solve compound inequalities.

④ Use boundary points to solve absolute value inequalities.

⑤ Use equivalent compound inequalities to solve absolute value inequalities.

Rent-a-Heap, a car rental company, charges $125 per week plus $0.20 per mile to rent one of their cars. If x represents the number of miles you drive their heap in a week, we can write a linear function that models the weekly cost, $C(x)$, for driving x miles:

Weekly cost	is	the weekly charge of $125	plus	the charge of $0.20 per mile for x miles.

$$C(x) \quad = \quad 125 \quad + \quad 0.20x.$$

Suppose you are limited by how much money you can spend for the week: You can spend at most $335. We can write an inequality that models your limitation:

The weekly charge of $125	plus	the charge of $0.20 per mile for x miles	must be less than or equal to	$335.

$$125 \quad + \quad 0.20x \quad \leq \quad 335.$$

Placing an inequality symbol between a linear function and a constant results in a *linear inequality in one variable*. In this section, we will study how to solve linear inequalities such as $125 + 0.20x \leq 335$. **Solving an inequality** is the process of finding the set of numbers that make the inequality a true statement. These numbers are called the **solutions** of the inequality and we say that they **satisfy** the inequality. The set of all solutions is called the **solution set** of the inequality. Set-builder notation and interval notation are used to represent these solution sets.

① Solve linear inequalities.

Solving Linear Inequalities in One Variable

We know that a linear function in x can be expressed as $f(x) = mx + b$. A **linear inequality in x** can be written in one of the following forms:

$$mx + b < 0, \, mx + b \leq 0, \, mx + b > 0, \, mx + b \geq 0.$$

In each form, $m \neq 0$.

Back to our question that opened this section: How many miles can you drive your Rent-a-Heap car if you can spend at most $335? We answer the question by solving

$$0.20x + 125 \leq 335$$

for x. The solution procedure is nearly identical to that for solving

$$0.20x + 125 = 335.$$

Our goal is to get x by itself on the left side. We do this by subtracting 125 from both sides to isolate $0.20x$:

$$0.20x + 125 \leq 335 \qquad \text{This is the given inequality.}$$

$$0.20x + 125 - 125 \leq 335 - 125 \qquad \text{Subtract 125 from both sides.}$$

$$0.20x \leq 210. \qquad \text{Simplify.}$$

Finally, we isolate x from $0.20x$ by dividing both sides of the inequality by 0.20:

$$\frac{0.20x}{0.20} \leq \frac{210}{0.20} \qquad \text{Divide both sides by 0.20.}$$

$$x \leq 1050. \qquad \text{Simplify.}$$

With at most $335 to spend, you can travel at most 1050 miles.

We started with the inequality $0.20x + 125 \leq 335$ and obtained the inequality $x \leq 1050$ in the final step. These inequalities have the same solution set, namely $\{x \mid x \leq 1050\}$. Inequalities such as these, with the same solution set, are said to be **equivalent**.

We isolated x from $0.20x$ by dividing both sides of $0.20x \leq 210$ by 0.20, a positive number. Let's see what happens if we divide both sides of an inequality by a negative number. Consider the inequality $10 < 14$. Divide 10 and 14 by -2:

$$\frac{10}{-2} = -5 \quad \text{and} \quad \frac{14}{-2} = -7.$$

Because -5 lies to the right of -7 on the number line, -5 is greater than -7:

$$-5 > -7.$$

Notice that the direction of the inequality symbol is reversed:

$$10 < 14$$

Dividing by -2 changes the direction of the inequality symbol.

$$-5 > -7.$$

In general, **when we multiply or divide both sides of an inequality by a negative number, the direction of the inequality symbol is reversed**. When we reverse the direction of the inequality symbol, we say that we change the *sense* of the inequality.

We can isolate a variable in a linear inequality in the same way we isolate a variable in a linear equation. The properties on the next page are used to create equivalent inequalities.

GREAT QUESTION!

What are some common English phrases and sentences that I can model with inequalities?

English phrases such as "at least" and "at most" can be represented by inequalities.

English Sentence	Inequality
x is at least 5.	$x \geq 5$
x is at most 5.	$x \leq 5$
x is between 5 and 7.	$5 < x < 7$
x is no more than 5.	$x \leq 5$
x is no less than 5.	$x \geq 5$

Properties of Inequalities

Property	The Property in Words	Example
The Addition Property of Inequality If $a < b$, then $a + c < b + c$. If $a < b$, then $a - c < b - c$.	If the same quantity is added to or subtracted from both sides of an inequality, the resulting inequality is equivalent to the original one.	$2x + 3 < 7$ Subtract 3: $2x + 3 - 3 < 7 - 3.$ Simplify: $2x < 4.$
The Positive Multiplication Property of Inequality If $a < b$ and c is positive, then $ac < bc$. If $a < b$ and c is positive, then $\dfrac{a}{c} < \dfrac{b}{c}$.	If we multiply or divide both sides of an inequality by the same positive quantity, the resulting inequality is equivalent to the original one.	$2x < 4$ Divide by 2: $\dfrac{2x}{2} < \dfrac{4}{2}.$ Simplify: $x < 2.$
The Negative Multiplication Property of Inequality If $a < b$ and c is negative, then $ac > bc$. If $a < b$ and c is negative, then $\dfrac{a}{c} > \dfrac{b}{c}$.	If we multiply or divide both sides of an inequality by the same negative quantity and reverse the direction of the inequality symbol, the resulting inequality is equivalent to the original one.	$-4x < 20$ Divide by -4 and change the sense of the inequality: $\dfrac{-4x}{-4} > \dfrac{20}{-4}.$ Simplify: $x > -5.$

EXAMPLE 1 Solving a Linear Inequality

Solve and graph the solution set on a number line:

$$3 - 2x \le 11.$$

SOLUTION

$$
\begin{aligned}
3 - 2x &\le 11 & &\text{This is the given inequality.}\\
3 - 2x - 3 &\le 11 - 3 & &\text{Subtract 3 from both sides.}\\
-2x &\le 8 & &\text{Simplify.}\\
\frac{-2x}{-2} &\ge \frac{8}{-2} & &\text{Divide both sides by } -2 \text{ and}\\
& & &\text{change the sense of the inequality.}\\
x &\ge -4 & &\text{Simplify.}
\end{aligned}
$$

The solution set consists of all real numbers that are greater than or equal to -4, expressed as $\{x \mid x \ge -4\}$ in set-builder notation. The interval notation for this solution set is $[-4, \infty)$. The graph of the solution set is shown as follows:

DISCOVERY

As a partial check, select one number from the solution set of $3 - 2x \le 11$. Substitute that number into the original inequality. Perform the resulting computations. You should obtain a true statement.

Is it possible to perform a partial check using a number that is not in the solution set? What should happen in this case? Try doing this.

⊘ Check Point **1** Solve and graph the solution set on a number line:

$$2 - 3x \le 5.$$

EXAMPLE 2 Solving a Linear Inequality

Solve and graph the solution set on a number line:

$$-2x - 4 > x + 5.$$

SOLUTION

Step 1 Simplify each side. Because each side is already simplified, we can skip this step.

GREAT QUESTION!

Do I have to solve the inequality in Example 2 by isolating the variable on the left?

No. You can solve

$$-2x - 4 > x + 5$$

by isolating x on the right side. Add $2x$ to both sides.

$$-2x - 4 + 2x > x + 5 + 2x$$
$$-4 > 3x + 5$$

Now subtract 5 from both sides.

$$-4 - 5 > 3x + 5 - 5$$
$$-9 > 3x$$

Finally, divide both sides by 3.

$$\frac{-9}{3} > \frac{3x}{3}$$
$$-3 > x$$

This last inequality means the same thing as $x < -3$.

Step 2 Collect variable terms on one side and constant terms on the other side. We will collect variable terms of $-2x - 4 > x + 5$ on the left and constant terms on the right.

$-2x - 4 > x + 5$	This is the given inequality.
$-2x - 4 - x > x + 5 - x$	Subtract x from both sides.
$-3x - 4 > 5$	Simplify.
$-3x - 4 + 4 > 5 + 4$	Add 4 to both sides.
$-3x > 9$	Simplify.

Step 3 Isolate the variable and solve. We isolate the variable, x, by dividing both sides by -3. Because we are dividing by a negative number, we must reverse the inequality symbol.

$\dfrac{-3x}{-3} < \dfrac{9}{-3}$	Divide both sides by -3 and change the sense of the inequality.
$x < -3$	Simplify.

Step 4 Express the solution set in set-builder or interval notation and graph the set on a number line. The solution set consists of all real numbers that are less than -3, expressed in set-builder notation as $\{x \mid x < -3\}$. The interval notation for this solution set is $(-\infty, -3)$. The graph of the solution set is shown as follows:

$$\xleftarrow{\hspace{1cm}}\overset{\displaystyle \;\;\;\;\;\;\;\; -5 \;\;\; -4 \;\;\; -3 \;\;\; -2 \;\;\; -1 \;\;\;\; 0 \;\;\;\; 1 \;\;\;\; 2 \;\;\;\; 3 \;\;\;\; 4 \;\;\;\; 5}{}\xrightarrow{\hspace{1cm}} x$$

• • •

TECHNOLOGY

Numeric and Graphic Connections

You can use a graphing utility to check the solution set of a linear inequality. Enter the linear function defined by each side of the inequality separately under y_1 and y_2. Then use the table or the graphs. To use the table, first locate the x-value for which the y-values are the same. Then scroll up or down to locate x values for which y_1 is greater than y_2 or for which y_1 is less than y_2. To use the graphs, locate the intersection point and then find the x-values for which the graph of y_1 lies above the graph of $y_2 (y_1 > y_2)$ or for which the graph of y_1 lies below the graph of $y_2 (y_1 < y_2)$.

Let's verify our work in Example 2 and show that $(-\infty, -3)$ is the solution set of

$$-2x - 4 > x + 5.$$

Enter $y_1 = -2x - 4$ in the $\boxed{y =}$ screen. Enter $y_2 = x + 5$ in the $\boxed{y =}$ screen.

We are looking for values of x for which y_1 is greater than y_2.

Numeric Check	Graphic Check

Scrolling through the table shows that $y_1 > y_2$ for values of x that are less than -3 (when $x = -3$, $y_1 = y_2$). This verifies $(-\infty, -3)$ is the solution set of $-2x - 4 > x + 5$.

Display the graphs for y_1 and y_2. Use the intersection feature. The solution set is the set of x-values for which the graph of y_1 lies above the graph of y_2.

Graphs intersect at $(-3, 2)$. When x is less than -3, the graph of y_1 lies above the graph of y_2. This graphically verifies $(-\infty, -3)$ is the solution set of $-2x - 4 > x + 5$.

$[-10, 10, 1]$ by $[-10, 10, 1]$

⊘ Check Point **2** Solve and graph the solution set on a number line: $3x + 1 > 7x - 15$.

If an inequality contains fractions with constants in the denominators, begin by multiplying both sides by the least common denominator. This will clear the inequality of fractions.

EXAMPLE 3 Solving a Linear Inequality Containing Fractions

Solve and graph the solution set on a number line:

$$\frac{x + 3}{4} \geq \frac{x - 2}{3} + \frac{1}{4}.$$

SOLUTION

The denominators are 4, 3, and 4. The least common denominator is 12. We begin by multiplying both sides of the inequality by 12.

$$\frac{x + 3}{4} \geq \frac{x - 2}{3} + \frac{1}{4}$$ This is the given inequality.

$$12\left(\frac{x + 3}{4}\right) \geq 12\left(\frac{x - 2}{3} + \frac{1}{4}\right)$$ Multiply both sides by 12. Multiplying by a positive number preserves the sense of the inequality.

$$\frac{12}{1} \cdot \frac{x + 3}{4} \geq \frac{12}{1} \cdot \frac{x - 2}{3} + \frac{12}{1} \cdot \frac{1}{4}$$ Multiply each term by 12. Use the distributive property on the right side.

$$\frac{\overset{3}{12}}{1} \cdot \frac{x + 3}{\underset{1}{4}} \geq \frac{\overset{4}{12}}{1} \cdot \frac{x - 2}{\underset{1}{3}} + \frac{\overset{3}{12}}{1} \cdot \frac{1}{\underset{1}{4}}$$ Divide out common factors in each multiplication.

$$3(x + 3) \geq 4(x - 2) + 3$$ The fractions are now cleared.

Now that the fractions have been cleared, we follow the four steps that we used in the previous example.

Step 1 Simplify each side.

$$3(x + 3) \geq 4(x - 2) + 3$$ This is the inequality with the fractions cleared.

$$3x + 9 \geq 4x - 8 + 3$$ Use the distributive property.

$$3x + 9 \geq 4x - 5$$ Simplify.

Step 2 Collect variable terms on one side and constant terms on the other side. We will collect variable terms on the left and constant terms on the right.

$$3x + 9 - 4x \geq 4x - 5 - 4x$$ Subtract 4x from both sides.

$$-x + 9 \geq -5$$ Simplify.

$$-x + 9 - 9 \geq -5 - 9$$ Subtract 9 from both sides.

$$-x \geq -14$$ Simplify.

Step 3 Isolate the variable and solve. To isolate x, we must eliminate the negative sign in front of the x. Because $-x$ means $-1x$, we can do this by multiplying (or dividing) both sides of the inequality by -1. We are multiplying by a negative number. Thus, we must reverse the direction of the inequality symbol.

$$(-1)(-x) \leq (-1)(-14)$$ Multiply both sides by −1 and change the sense of the inequality.

$$x \leq 14$$ Simplify.

Step 4 Express the solution set in set-builder or interval notation and graph the set on a number line. The solution set consists of all real numbers that are less

than or equal to 14, expressed in set-builder notation as $\{x \mid x \leq 14\}$. The interval notation for this solution set is $(-\infty, 14]$. The graph of the solution set is shown as follows:

● Check Point **3** Solve and graph the solution set on a number line:

$$\frac{x-4}{2} \geq \frac{x-2}{3} + \frac{5}{6}.$$

② Recognize inequalities with no solution or all real numbers as solutions.

Inequalities with Unusual Solution Sets

We have seen that some equations have no solution. This is also true for some inequalities. An example of such an inequality is

$$x > x + 1.$$

There is no number that is greater than itself plus 1. This inequality has no solution and its solution set is \varnothing, the empty set.

By contrast, some inequalities are true for all real numbers. An example of such an inequality is

$$x < x + 1.$$

Every real number is less than itself plus 1. The solution set is $\{x \mid x$ is a real number$\}$ or \mathbb{R}. In interval notation, the solution set is $(-\infty, \infty)$.

If you attempt to solve an inequality that has no solution, you will eliminate the variable and obtain a false statement, such as $0 > 1$. If you attempt to solve an inequality that is true for all real numbers, you will eliminate the variable and obtain a true statement, such as $0 < 1$.

EXAMPLE 4 Solving Linear Inequalities

Solve each inequality:

a. $2(x + 4) > 2x + 3$ **b.** $x + 7 \leq x - 2$.

SOLUTION

a.
$2(x + 4) > 2x + 3$	This is the given inequality.
$2x + 8 > 2x + 3$	Apply the distributive property.
$2x + 8 - 2x > 2x + 3 - 2x$	Subtract 2x from both sides.
$8 > 3$	Simplify. The statement 8 > 3 is true.

The inequality $8 > 3$ is true for all values of x. Because this inequality is equivalent to the original inequality, the original inequality is true for all real numbers. The solution set is

$$\{x \mid x \text{ is a real number}\} \text{ or } \mathbb{R} \text{ or } (-\infty, \infty).$$

b.
$x + 7 \leq x - 2$	This is the given inequality.
$x + 7 - x \leq x - 2 - x$	Subtract x from both sides.
$7 \leq -2$	Simplify. The statement 7 ≤ −2 is false.

The inequality $7 \leq -2$ is false for all values of x. Because this inequality is equivalent to the original inequality, the original inequality has no solution. The solution set is \varnothing. ● ● ●

● Check Point **4** Solve each inequality:

a. $3(x + 1) > 3x + 2$ **b.** $x + 1 \leq x - 1$.

TECHNOLOGY

Graphic Connections

The graphs of the linear functions

$$y_1 = 2(x + 4) \text{ and } y_2 = 2x + 3$$

have the same slope, 2, but different y-intercepts, 8 and 3. Thus, the graphs are parallel lines.

$[-10, 10, 1]$ by $[-10, 10, 1]$

The graph of y_1 is always above the graph of y_2. Every value of x satisfies the inequality $y_1 > y_2$. Thus, the solution set of the inequality

$$2(x + 4) > 2x + 3$$

is $(-\infty, \infty)$.

③ Solve compound inequalities.

Solving Compound Inequalities

We now consider two inequalities such as

$$-3 < 2x + 1 \quad \text{and} \quad 2x + 1 \leq 3,$$

expressed as a **compound inequality**

$$-3 < 2x + 1 \leq 3.$$

The word *and* does not appear when the inequality is written in the shorter form, although intersection is implied. The shorter form enables us to solve both inequalities at once. By performing each operation on all three parts of the inequality, our goal is to **isolate x in the middle**.

EXAMPLE 5 Solving a Compound Inequality

Solve and graph the solution set on a number line:

$$-3 < 2x + 1 \leq 3.$$

SOLUTION

We would like to isolate x in the middle. We can do this by first subtracting 1 from all three parts of the compound inequality. Then we isolate x from $2x$ by dividing all three parts of the inequality by 2.

$-3 < 2x + 1 \leq 3$	This is the given inequality.
$-3 - 1 < 2x + 1 - 1 \leq 3 - 1$	Subtract 1 from all three parts.
$-4 < 2x \leq 2$	Simplify.
$\dfrac{-4}{2} < \dfrac{2x}{2} \leq \dfrac{2}{2}$	Divide each part by 2.
$-2 < x \leq 1$	Simplify.

The solution set consists of all real numbers greater than -2 and less than or equal to 1, represented by $\{x \mid -2 < x \leq 1\}$ in set-builder notation and $(-2, 1]$ in interval notation. The graph is shown as follows:

Check Point **5** Solve and graph the solution set on a number line:
$1 \leq 2x + 3 < 11$.

TECHNOLOGY

Numeric and Graphic Connections

Let's verify our work in Example 5 and show that $(-2, 1]$ is the solution set of $-3 < 2x + 1 \leq 3$.

Numeric Check	Graphic Check

Numeric Check

To check numerically, enter $y_1 = 2x + 1$.

The shaded part of the table shows that values of $y_1 = 2x + 1$ are greater than -3 and less than or equal to 3 when x is in the interval $(-2, 1]$.

X	Y1
-4	-7
-3	-5
-2	-3
-1	-1
0	1
1	3
2	5

$y_1 = 2x + 1$

$Y_1 = -2$

Graphic Check

To check graphically, graph the functions defined by each part of

$$-3 < 2x + 1 \leq 3.$$

Enter Enter Enter
$y_1 = -3$. $y_2 = 2x + 1$. $y_3 = 3$.

The figure shows that the graph of $y_2 = 2x + 1$ lies above the graph of $y_1 = -3$ and on or below the graph of $y_3 = 3$ when x is in the interval $(-2, 1]$.

$[-3, 3, 1]$ by $[-5, 5, 1]$

4 Use boundary points to solve absolute value inequalities.

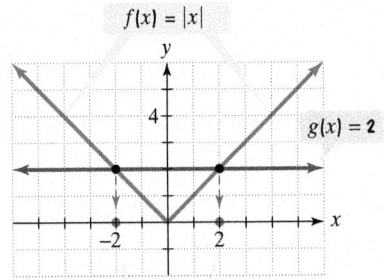

$f(x) = g(x)$ at **−2** and **2**. The solution set of $|x| = 2$ is {**−2, 2**}.

FIGURE 2.16

Solving Inequalities Involving Absolute Value Using Boundary Points

Graphs can help us visualize the solution sets of equations and inequalities involving absolute value. Let's first consider the equation

$$|x| = 2.$$

Figure 2.16 shows the graphs of $f(x) = |x|$ and $g(x) = 2$. The x-coordinates of the intersection points illustrate that $\{-2, 2\}$ is the solution set of $|x| = 2$.

Now let's see what the graphs of $f(x) = |x|$ and $g(x) = 2$ can tell us about the solution sets of the following inequalities involving absolute value:

$$|x| < 2 \qquad \text{and} \qquad |x| > 2.$$

Look for values of x where the graph of $f(x) = |x|$ lies *below* the graph of $g(x) = 2$.

Look for values of x where the graph of $f(x) = |x|$ lies *above* the graph of $g(x) = 2$.

Figure 2.17(a) illustrates the solution set of $|x| < 2$. **Figure 2.17(b)** illustrates the solution set of $|x| > 2$.

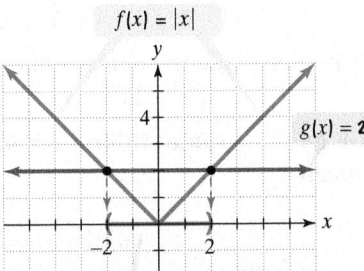

The blue graph of f lies below the red graph of g for all x in (−2, 2). The solution set of $|x| < 2$ is (−2, 2).

FIGURE 2.17(a)

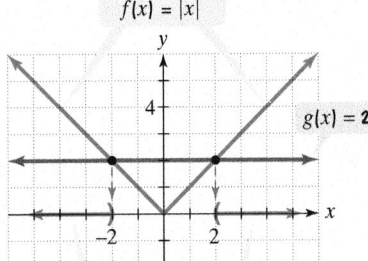

The blue graph of f lies above the red graph of g for all x in (−∞, −2) or (2, ∞). The solution set of $|x| > 2$ is (−∞, −2) ∪ (2, ∞).

FIGURE 2.17(b)

Can you see how -2 and 2, the solutions of $|x| = 2$, serve as boundary points that divide the x-axis into intervals? On each interval, the blue graph of f is either below the red graph of g or above the red graph of g.

Boundary points play a fundamental role in solving absolute value inequalities in the form $|u| < c$ or $|u| > c$, where u is an algebraic expression and c is a positive number. The boundary points are found by solving the equation $|u| = c$.

Using Boundary Points to Solve Absolute Value Inequalities

If u is an algebraic expression and c is a positive number, the inequalities $|u| < c$ and $|u| > c$ can be solved by the following procedure.

1. Solve the equation $|u| = c$. The solutions are the **boundary points**.
2. Locate these boundary points on a number line, thereby dividing the number line into intervals.
3. Choose one representative number, called a **test value**, within each interval and substitute that number into the given inequality.
 a. If a true statement results, then all numbers, x, in the interval satisfy the given inequality.
 b. If a false statement results, then no numbers, x, in the interval satisfy the given inequality.
4. Write the solution set, selecting the interval or intervals that satisfy the given inequality.

This procedure is valid if $<$ is replaced by \leq, or $>$ is replaced by \geq. However, if the inequality involves \leq or \geq, include the boundary points (the solutions of $|u| = c$) in the solution set.

GREAT QUESTION!

Can I use the endpoints of intervals as test values?

No. Each test value must be chosen from the *interior* of an interval. Test values should not be endpoints of intervals.

EXAMPLE 6 Solving an Absolute Value Inequality Using Boundary Points

Solve and graph the solution set on a number line:

$$|x - 4| < 3.$$

SOLUTION

Step 1 Solve the equation $|u| = c$. We find the boundary points for $|x - 4| < 3$ by solving $|x - 4| = 3$.

$$|x - 4| = 3 \qquad \text{\small This is the equation needed to find the boundary points.}$$

$$x - 4 = 3 \quad \text{or} \quad x - 4 = -3 \qquad \text{\small Rewrite } |u| = c \text{ as } u = c \text{ or } u = \ \ c.$$

$$x = 7 \qquad\qquad x = 1 \qquad \text{\small Add 4 to both sides of each equation and solve for x.}$$

The boundary points are 1 and 7.

Step 2 Locate the boundary points on a number line and separate the line into intervals. The number line with the boundary points is shown as follows:

```
                 1                 7
 ──┼──┼──┼──┼──●──┼──┼──┼──┼──┼──●──┼──→ x
   -2  -1   0   1   2   3   4   5   6   7   8
```

The boundary points divide the number line into three intervals:

$$(-\infty, 1) \qquad (1, 7) \qquad (7, \infty).$$

Step 3 Choose one test value within each interval and substitute that value into the given inequality.

Interval	Test Value	Substitute into $\lvert x - 4\rvert < 3$	Conclusion
$(-\infty, 1)$	0	$\lvert 0 - 4\rvert \overset{?}{<} 3$ $\lvert -4\rvert \overset{?}{<} 3$ $4 < 3, \ \text{\small false}$	$(-\infty, 1)$ does not belong to the solution set.
$(1, 7)$	2	$\lvert 2 - 4\rvert \overset{?}{<} 3$ $\lvert -2\rvert \overset{?}{<} 3$ $2 < 3, \ \text{\small true}$	$(1, 7)$ belongs to the solution set.
$(7, \infty)$	8	$\lvert 8 - 4\rvert \overset{?}{<} 3$ $\lvert 4\rvert \overset{?}{<} 3$ $4 < 3, \ \text{\small false}$	$(7, \infty)$ does not belong to the solution set.

Step 4 Write the solution set, selecting the interval or intervals that satisfy the given inequality. Based on our work in step 3, we see that the solution set of the given inequality, $|x - 4| < 3$, is $(1, 7)$. The solution set can be expressed in set-builder notation as $\{x \mid 1 < x < 7\}$. The graph of the solution set is shown as follows:

We can use the rectangular coordinate system to visualize the solution set of

$$|x - 4| < 3.$$

Figure 2.18 shows the graphs of $f(x) = |x - 4|$ and $g(x) = 3$. The solution set of $|x - 4| < 3$ consists of all values of x for which the blue graph of f lies below the red graph of g. These x-values make up the interval $(1, 7)$, which is the solution set.

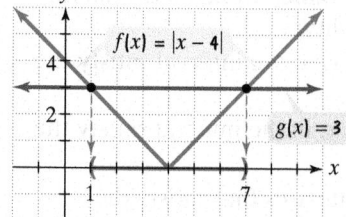

FIGURE 2.18 The solution set of $|x - 4| < 3$ is $(1, 7)$.

☑ Check Point **6** Solve and graph the solution set on a number line: $|x - 2| < 5$.

EXAMPLE 7 Solving an Absolute Value Inequality Using Boundary Points

Solve and graph the solution set on a number line:

$$|2x + 3| \geq 5.$$

SOLUTION

Step 1 Solve the equation $|u| = c$. We find the boundary points by solving $|2x + 3| = 5$.

$	2x + 3	= 5$	This is the equation needed to find the boundary points.
$2x + 3 = 5$ or $2x + 3 = -5$	Rewrite $	u	= c$ as $u = c$ or $u = -c$.
$2x = 2 \qquad\qquad 2x = -8$	Subtract 3 from both sides of each equation.		
$x = 1 \qquad\qquad x = -4$	Divide both sides of each equation by 2 and solve for x.		

The boundary points are -4 and 1.

Step 2 Locate the boundary points on a number line and separate the line into intervals. The number line with the boundary points is shown as follows:

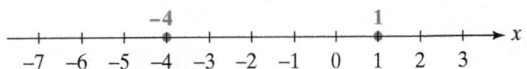

The boundary points divide the number line into three intervals:

$$(-\infty, -4) \qquad (-4, 1) \qquad (1, \infty).$$

Step 3 Choose one test value within each interval and substitute that value into the given inequality.

Interval	Test Value	Substitute into $\|2x + 3\| \geq 5$	Conclusion
$(-\infty, -4)$	-5	$\|2(-5) + 3\| \overset{?}{\geq} 5$ $\|-10 + 3\| \overset{?}{\geq} 5$ $\|-7\| \overset{?}{\geq} 5$ $7 \geq 5$, true	$(-\infty, -4)$ belongs to the solution set.
$(-4, 1)$	0	$\|2 \cdot 0 + 3\| \overset{?}{\geq} 5$ $\|0 + 3\| \overset{?}{\geq} 5$ $\|3\| \overset{?}{\geq} 5$ $3 \geq 5$, false	$(-4, 1)$ does not belong to the solution set.
$(1, \infty)$	2	$\|2 \cdot 2 + 3\| \overset{?}{\geq} 5$ $\|4 + 3\| \overset{?}{\geq} 5$ $\|7\| \overset{?}{\geq} 5$ $7 \geq 5$ true	$(1, \infty)$ belongs to the solution set.

Step 4 Write the solution set, selecting the interval or intervals that satisfy the given inequality. Based on our work in step 3, we see that all x in $(-\infty, -4)$ or $(1, \infty)$ belong to the solution set of $|2x + 3| \geq 5$. However, because the inequality

involves ≥ (greater than or *equal to*), we must also include the solutions of $|2x + 3| = 5$, namely the boundary points −4 and 1, in the solution set. Thus, the solution set of $|2x + 3| \geq 5$ is

$$(-\infty, -4] \cup [1, \infty)$$

$$\text{or} \quad \{x \mid x \leq -4 \quad \text{or} \quad x \geq 1\}.$$

The graph of the solution set on a number line is shown as follows:

● ● ●

✓ Check Point **7** Solve and graph the solution set on a number line: $|2x - 5| \geq 3$.

EXAMPLE 8 Solving an Absolute Value Inequality Using Boundary Points

Solve and graph the solution set on a number line: $-2|3x + 5| + 7 \geq -13$.

SOLUTION

$$-2|3x + 5| + 7 \geq -13 \qquad \text{This is the given inequality.}$$

We need to isolate $|3x + 5|$, the absolute value expression.

$$-2|3x + 5| + 7 - 7 \geq -13 - 7 \quad \text{Subtract 7 from both sides.}$$

$$-2|3x + 5| \geq -20 \qquad \text{Simplify.}$$

$$\frac{-2|3x + 5|}{-2} \leq \frac{-20}{-2} \qquad \begin{array}{l}\text{Divide both sides by −2 and change the sense of}\\\text{the inequality.}\end{array}$$

$$|3x + 5| \leq 10 \qquad \text{Simplify.}$$

Now that we have isolated the absolute value expression, we will use boundary points to solve $|3x + 5| \leq 10$.

Step 1 Solve the equation $|u| = c$. We find the boundary points by solving $|3x + 5| = 10$.

$$|3x + 5| = 10 \qquad \text{This is the equation needed to find the boundary points.}$$

$$3x + 5 = 10 \quad \text{or} \quad 3x + 5 = -10 \quad \text{Rewrite } |u| = c \text{ as } u = c \text{ or } u = -c.$$

$$3x = 5 \qquad\qquad 3x = -15 \qquad \text{Subtract 5 from both sides of each equation.}$$

$$x = \tfrac{5}{3} \qquad\qquad x = -5 \qquad \text{Divide both sides of each equation by 3 and solve for x.}$$

The boundary points are −5 and $\frac{5}{3}$ (or $1\frac{2}{3}$).

Step 2 Locate the boundary points on a number line and separate the line into intervals. The number line with the boundary points is shown as follows:

The boundary points divide the number line into three intervals:

$$(-\infty, -5) \qquad (-5, \tfrac{5}{3}) \qquad (\tfrac{5}{3}, \infty).$$

Step 3 Choose one test value within each interval and substitute that value into the given inequality. We will continue to use the equivalent inequality with the isolated absolute value expression.

Interval	Test Value	Substitute into $\|3x + 5\| \leq 10$	Conclusion
$(-\infty, -5)$	-6	$\|3(-6) + 5\| \overset{?}{\leq} 10$ $\|-18 + 5\| \overset{?}{\leq} 10$ $\|-13\| \overset{?}{\leq} 10$ $13 \leq 10,$ false	$(-\infty, -5)$ does not belong to the solution set.
$(-5, \frac{5}{3})$	0	$\|3 \cdot 0 + 5\| \overset{?}{\leq} 10$ $\|0 + 5\| \overset{?}{\leq} 10$ $\|5\| \overset{?}{\leq} 10$ $5 \leq 10,$ true	$(-5, \frac{5}{3})$ belongs to the solution set.
$(\frac{5}{3}, \infty)$	2	$\|3 \cdot 2 + 5\| \overset{?}{\leq} 10$ $\|6 + 5\| \overset{?}{\leq} 10$ $\|11\| \overset{?}{\leq} 10$ $11 \leq 10,$ false	$(\frac{5}{3}, \infty)$ does not belong to the solution set.

Step 4 Write the solution set, selecting the interval or intervals that satisfy the given inequality. Based on our work in step 3, we see that all x in $(-5, \frac{5}{3})$ belong to the solution set of $\|3x + 5\| \leq 10$. However, because the inequality involves \leq (less than or *equal to*), we must also include the solutions of $\|3x + 5\| = 10$, namely the boundary points -5 and $\frac{5}{3}$, in the solution set. Thus, the solution set of $\|3x + 5\| \leq 10$ is

$$[-5, \tfrac{5}{3}] \quad \text{or} \quad \{x | -5 \leq x \leq \tfrac{5}{3}\}.$$

This is also the solution set of the given (equivalent) inequality $-2|3x + 5| + 7 \geq -13$. The graph of the solution set on a number line is shown as follows:

✓ **Check Point 8** Solve and graph the solution set on a number line: $-3|5x - 2| + 20 \geq -19$.

⑤ Use equivalent compound inequalities to solve absolute value inequalities.

Solving Inequalities Involving Absolute Value Using Equivalent Compound Inequalities

In **Figure 2.17** on page 317, we used graphs of $f(x) = |x|$ and $g(x) = 2$ a to visualize the solution sets of two absolute value inequalities:

- The solution set of $|x| < 2$ is $(-2, 2)$ or $\{x | -2 < x < 2\}$.
- The solution set of $|x| > 2$ is $(-\infty, -2) \cup (2, \infty)$ or $\{x | x < -2 \text{ or } x > 2\}$.

We can verify these results by interpreting $|x|$ as the distance from 0 to x on a number line.

Generalizing from these observations gives us a second method for solving inequalities with absolute value. This method involves rewriting the given inequality without absolute value bars.

Using Equivalent Compound Inequalities to Solve Absolute Value Inequalities

If u is an algebraic expression and c is a positive number,
 1. The solutions of $|u| < c$ are the numbers that satisfy $-c < u < c$.
 2. The solutions of $|u| > c$ are the numbers that satisfy $u < -c$ or $u > c$.

These rules are valid if $<$ is replaced by \leq and $>$ is replaced by \geq.

Let's rework Examples 6 and 7 by converting to equivalent compound inequalities. Although this method is faster than using boundary points, it's easy to make an error when rewriting absolute value inequalities without absolute value bars. Ask your instructor if one method is preferred over another.

EXAMPLE 9 Solving Absolute Value Inequalities Using Equivalent Compound Inequalities

Solve each inequality:

 a. $|x - 4| < 3$　　　　**b.** $|2x + 3| \geq 5$.

SOLUTION

 a. We rewrite the inequality without absolute value bars.

$$|u| < c \text{ means } -c < u < c.$$

$$|x - 4| < 3 \quad \text{means} \quad -3 < x - 4 < 3.$$

We solve the compound inequality by adding 4 to all three parts.
$$-3 < x - 4 < 3$$
$$-3 + 4 < x - 4 + 4 < 3 + 4$$
$$1 < x < 7$$

The solution set is all real numbers greater than 1 and less than 7, denoted by $\{x | 1 < x < 7\}$ or $(1, 7)$.

 b. We rewrite the inequality without absolute value bars.

$$|u| \geq c \quad \text{means} \quad u \leq -c \quad \text{or} \quad u \geq c.$$

$$|2x + 3| \geq 5 \quad \text{means} \quad 2x + 3 \leq -5 \quad \text{or} \quad 2x + 3 \geq 5.$$

We solve this compound inequality by solving each of these inequalities separately. Then we take the union of their solution sets.

$2x + 3 \leq -5$ or $2x + 3 \geq 5$		These are the inequalities without absolute value bars.
$2x \leq -8$	$2x \geq 2$	Subtract 3 from both sides.
$x \leq -4$	$x \geq 1$	Divide both sides by 2.

The solution set consists of all numbers that are less than or equal to -4 or greater than or equal to 1. The solution set is $\{x | x \leq -4 \text{ or } x \geq 1\}$, or, in interval notation, $(-\infty, -4] \cup [1, \infty)$.

● ● ●

GREAT QUESTION!

When is the graph of the solution set of an absolute value inequality a single interval and when is it the union of two intervals?

If u is a linear expression and $c > 0$, the graph of the solution set for $|u| > c$ will be divided into two intervals whose union cannot be represented as a single interval. The graph of the solution set for $|u| < c$ will be a single interval. Avoid the common error of rewriting $|u| > c$ as $-c < u > c$.

Ⓞ Check Point **9** Solve each inequality using equivalent compound inequalities. (These are the inequalities that you solved in Check Points 6 and 7 using boundary points.)

a. $|x - 2| < 5$ **b.** $|2x - 5| \geq 3$

Now let's rework Example 8 by using an equivalent compound inequality.

EXAMPLE 10 Solving an Absolute Value Inequality Using an Equivalent Compound Inequality

Solve and graph the solution set on a number line: $-2|3x + 5| + 7 \geq -13$.

SOLUTION

As with the boundary point method (see Example 8), we first need to isolate the absolute value expression.

$$-2|3x + 5| + 7 \geq -13 \qquad \text{This is the given inequality.}$$

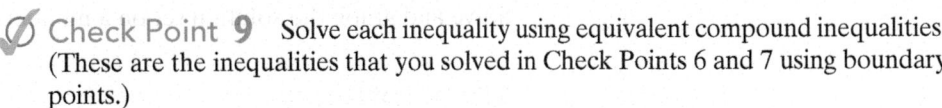

We need to isolate $|3x + 5|$, the absolute value expression.

$$-2|3x + 5| + 7 - 7 \geq -13 - 7 \qquad \text{Subtract 7 from both sides.}$$

$$-2|3x + 5| \geq -20 \qquad \text{Simplify.}$$

$$\frac{-2|3x + 5|}{-2} \leq \frac{-20}{-2} \qquad \text{Divide both sides by } -2 \text{ and change the sense of the inequality.}$$

$$|3x + 5| \leq 10 \qquad \text{Simplify.}$$

$$-10 \leq 3x + 5 \leq 10 \qquad \text{Rewrite without absolute value bars:} \\ |u| \leq c \text{ means } -c \leq u \leq c.$$

Now we need to isolate x in the middle.

$$-10 - 5 \leq 3x + 5 - 5 \leq 10 - 5 \qquad \text{Subtract 5 from all three parts.}$$

$$-15 \leq 3x \leq 5 \qquad \text{Simplify.}$$

$$\frac{-15}{3} \leq \frac{3x}{3} \leq \frac{5}{3} \qquad \text{Divide each part by 3.}$$

$$-5 \leq x \leq \frac{5}{3} \qquad \text{Simplify.}$$

The solution set is $\left\{ x \mid -5 \leq x \leq \frac{5}{3} \right\}$ in set-builder notation and $\left[-5, \frac{5}{3} \right]$ in interval notation. The graph is shown as follows:

```
        [———————————————]
  ┼———┼———┼———┼———┼———┼———┼———┼———┼———┼→ x
 −7  −6  −5  −4  −3  −2  −1   0   1   2   3
                                  1⅔
```

●●●

⬛ Check Point **10** Solve and graph the solution set on a number line: $-3|5x - 2| + 20 \geq -19$. (This is the inequality that you solved in Check Point 8 using boundary points.)

EXAMPLE 11 Solving an Absolute Value Inequality Using an Equivalent Compound Inequality

Solve and graph the solution set on a number line: $7 < |5 - 2x|$.

SOLUTION

We begin by expressing the inequality with the absolute value expression on the left side:

$$|5 - 2x| > 7.$$

> $c < |u|$ means the same thing as $|u| > c$. In both cases, the inequality symbol points to c.

We rewrite this inequality without absolute value bars.

$$|u| \;>\; c \quad \text{means} \quad u \;<\; -c \quad \text{or} \quad u \;>\; c.$$

$|5 - 2x| > 7$ means $5 - 2x < -7$ or $5 - 2x > 7$.

Because $|5 - 2x| > 7$ means $5 - 2x < -7$ or $5 - 2x > 7$, we solve $5 - 2x < -7$ and $5 - 2x > 7$ separately. Then we take the union of their solution sets.

$5 - 2x < -7$ or	$5 - 2x > 7$	These are the inequalities without absolute value bars.
$5 - 5 - 2x < -7 - 5$	$5 - 5 - 2x > 7 - 5$	Subtract 5 from both sides.
$-2x < -12$	$-2x > 2$	Simplify.
$\dfrac{-2x}{-2} > \dfrac{-12}{-2}$	$\dfrac{-2x}{-2} < \dfrac{2}{-2}$	Divide both sides by -2 and change the sense of each inequality.
$x > 6$	$x < -1$	Simplify.

The solution set consists of all numbers that are less than -1 or greater than 6. The solution set is $\{x \mid x < -1 \text{ or } x > 6\}$, or, in interval notation $(-\infty, -1) \cup (6, \infty)$. The graph of the solution set is shown as follows:

⬛ Check Point **11** Solve and graph the solution set on a number line: $18 < |6 - 3x|$.

Applications

Our next example shows how to use an inequality to select the better deal between two pricing options. We use our strategy for solving word problems, modeling the verbal conditions of the problem with a linear inequality.

EXAMPLE 12 Selecting the Better Deal

Acme Car rental agency charges $4 a day plus $0.15 per mile, whereas Interstate rental agency charges $20 a day and $0.05 per mile. How many miles must be driven to make the daily cost of an Acme rental a better deal than an Interstate rental?

SOLUTION

Step 1 Let x represent one of the unknown quantities. We are looking for the number of miles that must be driven in a day to make Acme the better deal. Thus,

$$\text{let } x = \text{the number of miles driven in a day.}$$

Step 2 Represent other unknown quantities in terms of x. We are not asked to find another quantity, so we can skip this step.

Step 3 Write an inequality in x that models the conditions. Acme is a better deal than Interstate if the daily cost of Acme is less than the daily cost of Interstate.

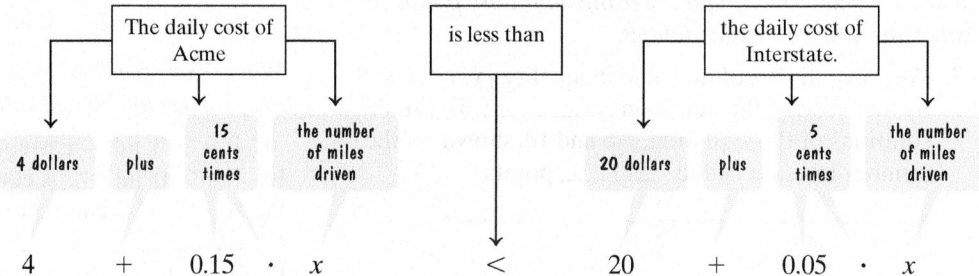

$$4 \quad + \quad 0.15 \quad \cdot \quad x \qquad < \qquad 20 \quad + \quad 0.05 \quad \cdot \quad x$$

TECHNOLOGY

Graphic Connections

The graphs of the daily cost models for the car rental agencies

$$y_1 = 4 + 0.15x$$

and $y_2 = 20 + 0.05x$

are shown in a $[0, 300, 10]$ by $[0, 40, 4]$ viewing rectangle. The graphs intersect at $(160, 28)$.

To the left of $x = 160$, the graph of Acme's daily cost lies below that of Interstate's daily cost. This shows that for fewer than 160 miles per day, Acme offers the better deal.

Step 4 Solve the inequality and answer the question.

$4 + 0.15x < 20 + 0.05x$	This is the inequality that models the verbal conditions.
$4 + 0.15x - 0.05x < 20 + 0.05x - 0.05x$	Subtract 0.05x from both sides.
$4 + 0.1x < 20$	Simplify.
$4 + 0.1x - 4 < 20 - 4$	Subtract 4 from both sides.
$0.1x < 16$	Simplify.
$\dfrac{0.1x}{0.1} < \dfrac{16}{0.1}$	Divide both sides by 0.1.
$x < 160$	Simplify.

Thus, driving fewer than 160 miles per day makes Acme the better deal.

Step 5 Check the proposed solution in the original wording of the problem. One way to do this is to take a mileage less than 160 miles per day to see if Acme is the better deal. Suppose that 150 miles are driven in a day.

$$\text{Cost for Acme} = 4 + 0.15(150) = 26.50$$
$$\text{Cost for Interstate} = 20 + 0.05(150) = 27.50$$

Acme has a lower daily cost, making Acme the better deal. • • •

✓ Check Point **12** A car can be rented from Basic Rental for $260 per week with no extra charge for mileage. Continental charges $80 per week plus 25 cents for each mile driven to rent the same car. How many miles must be driven in a week to make the rental cost for Basic Rental a better deal than Continental's?

CONCEPT AND VOCABULARY CHECK

Fill in each blank so that the resulting statement is true.

1. The linear inequality $-3x - 4 > 5$ can be solved by first _____ to both sides and then _____ both sides by _____, which changes the _____ of the inequality symbol from _____ to _____.

2. In solving an inequality, if you eliminate the variable and obtain a false statement such as $7 < -2$, the solution set is ___.

3. In solving an inequality, if you eliminate the variable and obtain a true statement such as $8 > 3$, the solution set is _____.

4. The way to solve $-7 < 3x - 4 \leq 5$ is to isolate x in the _____.

Work Exercises 5–8 if you are using boundary points to solve absolute value inequalities.

5. We solve the absolute value inequality $|x - 2| > 8$ by first solving the equation _____. The real solutions of this equation, -6 and 10, shown on the number line, are called _____ points.

6. True or false: A test value for the leftmost interval on the number line shown in the previous column could be -10. _____.

7. True or false: A test value for the rightmost interval on the number line shown in the previous column could be 10. _____.

8. Using test values to solve $|x - 2| > 8$, we obtain the following results:

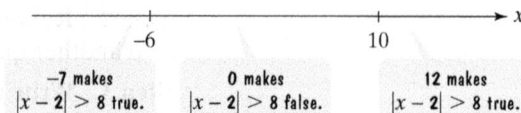

Based on the information shown above, the solution set of $|x - 2| > 8$ is _____.

Work Exercises 9–12 if you are using equivalent compound inequalities to solve absolute value inequalities.

9. If $c > 0$, $|u| < c$ is equivalent to ___ $< u <$ ___.

10. If $c > 0$, $|u| > c$ is equivalent to $u <$ ___ or $u >$ ___.

11. $|x - 7| < 2$ can be rewritten without absolute value bars as _____.

12. $|x - 7| > 2$ can be rewritten without absolute value bars as _____.

EXERCISE SET 2.6

Practice Exercises

In all exercises, other than \varnothing, use interval notation to express solution sets and graph each solution set on a number line.

In Exercises 1–24, solve each linear inequality.

1. $5x + 11 < 26$
2. $2x + 5 < 17$
3. $3x - 7 \geq 13$
4. $8x - 2 \geq 14$
5. $-9x \geq 36$
6. $-5x \leq 30$
7. $8x - 11 \leq 3x - 13$
8. $18x + 45 \leq 12x - 8$
9. $4(x + 1) + 2 \geq 3x + 6$
10. $8x + 3 > 3(2x + 1) + x + 5$
11. $2x - 11 < -3(x + 2)$
12. $-4(x + 2) > 3x + 20$
13. $1 - (x + 3) \geq 4 - 2x$
14. $5(3 - x) \leq 3x - 1$
15. $\dfrac{x}{4} - \dfrac{3}{2} \leq \dfrac{x}{2} + 1$
16. $\dfrac{3x}{10} + 1 \geq \dfrac{1}{5} - \dfrac{x}{10}$
17. $1 - \dfrac{x}{2} > 4$
18. $7 - \dfrac{4}{5}x < \dfrac{3}{5}$
19. $\dfrac{x - 4}{6} \geq \dfrac{x - 2}{9} + \dfrac{5}{18}$
20. $\dfrac{4x - 3}{6} + 2 \geq \dfrac{2x - 1}{12}$
21. $4(3x - 2) - 3x < 3(1 + 3x) - 7$
22. $3(x - 8) - 2(10 - x) > 5(x - 1)$
23. $5(x - 2) - 3(x + 4) \geq 2x - 20$
24. $6(x - 1) - (4 - x) \geq 7x - 8$

In Exercises 25–32, solve each compound inequality.

25. $6 < x + 3 < 8$
26. $7 < x + 5 < 11$
27. $-3 \leq x - 2 < 1$
28. $-6 < x - 4 \leq 1$
29. $-11 < 2x - 1 \leq -5$
30. $3 \leq 4x - 3 < 19$
31. $-3 \leq \dfrac{2}{3}x - 5 < -1$
32. $-6 \leq \dfrac{1}{2}x - 4 < -3$

In Exercises 33–68, solve each absolute value inequality. Use either boundary points or an equivalent compound inequality, or the method specified by your instructor.

33. $|x| < 3$
34. $|x| < 5$
35. $|x - 1| \leq 2$
36. $|x + 3| \leq 4$
37. $|2x - 6| < 8$
38. $|3x + 5| < 17$
39. $|2(x - 1) + 4| \leq 8$
40. $|3(x - 1) + 2| \leq 20$
41. $\left|\dfrac{2x + 6}{3}\right| < 2$
42. $\left|\dfrac{3(x - 1)}{4}\right| < 6$
43. $|x| > 3$
44. $|x| > 5$
45. $|x - 1| \geq 2$
46. $|x + 3| \geq 4$
47. $|3x - 8| > 7$
48. $|5x - 2| > 13$
49. $\left|\dfrac{2x + 2}{4}\right| \geq 2$
50. $\left|\dfrac{3x - 3}{9}\right| \geq 1$
51. $\left|3 - \dfrac{2}{3}x\right| > 5$
52. $\left|3 - \dfrac{3}{4}x\right| > 9$
53. $3|x - 1| + 2 \geq 8$
54. $5|2x + 1| - 3 \geq 9$

55. $-2|x - 4| \geq -4$

56. $-3|x + 7| \geq -27$

57. $-4|1 - x| < -16$

58. $-2|5 - x| < -6$

59. $3 \leq |2x - 1|$

60. $9 \leq |4x + 7|$

61. $5 > |4 - x|$

62. $2 > |11 - x|$

63. $1 < |2 - 3x|$

64. $4 < |2 - x|$

65. $12 < \left|-2x + \dfrac{6}{7}\right| + \dfrac{3}{7}$

66. $1 < \left|x - \dfrac{11}{3}\right| + \dfrac{7}{3}$

67. $4 + \left|3 - \dfrac{x}{3}\right| \geq 9$

68. $\left|2 - \dfrac{x}{2}\right| - 1 \leq 1$

In Exercises 69–76, use interval notation to represent all values of x satisfying the given conditions.

69. $f(x) = \dfrac{x}{2} + 3$, $g(x) = \dfrac{x}{3} + \dfrac{5}{2}$, and $f(x) \leq g(x)$.

70. $f(x) = \dfrac{2}{3}(6x - 9) + 4$, $g(x) = 5x + 1$, and $f(x) > g(x)$.

71. $f(x) = 1 - (x + 3) + 2x$ and $f(x)$ is at least 4.

72. $f(x) = 2x - 11 + 3(x + 2)$ and $f(x)$ is at most 0.

73. $f(x) = |3x - 4| + 2$ and $f(x) < 8$.

74. $f(x) = |2x - 5| + 1$ and $f(x) > 9$.

75. $f(x) = 7 - \left|\dfrac{x}{2} + 2\right|$ and $f(x)$ is at most 4.

76. $f(x) = 8 - |5x + 3|$ and $f(x)$ is at least 6.

Practice Plus

In Exercises 77–78, use the graph of $f(x) = |4 - x|$ to solve each inequality.

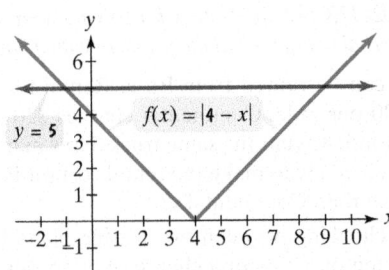

77. $|4 - x| < 5$

78. $|4 - x| \geq 5$

In Exercises 79–80, use the table to solve each inequality.

79. $-2 \leq 5x + 3 < 13$

80. $-3 < 2x - 5 \leq 3$

81. When 3 times a number is subtracted from 4, the absolute value of the difference is at least 5. Use interval notation to express the set of all numbers that satisfy this condition.

82. When 4 times a number is subtracted from 5, the absolute value of the difference is at most 13. Use interval notation to express the set of all numbers that satisfy this condition.

In Exercises 83–88, find the domain of each function.

83. $f(x) = \sqrt{20 - 4x}$

84. $f(x) = \sqrt{15 - 3x}$

85. $f(x) = \dfrac{20 - 4x}{\sqrt{3(2x - 5) - 8x + 5}}$

86. $f(x) = \dfrac{15 - 3x}{\sqrt{5(2x + 7) + 4x + 7}}$

87. $f(x) = \sqrt{2x - 8} + \sqrt{4x - 4}$

88. $f(x) = \sqrt{-4x + 2} + \sqrt{-2x - 8}$

Application Exercises

The graphs show that the three components of love, namely, passion, intimacy, and commitment, progress differently over time. Passion peaks early in a relationship and then declines. By contrast, intimacy and commitment build gradually. Use the graphs to solve Exercises 89–96.

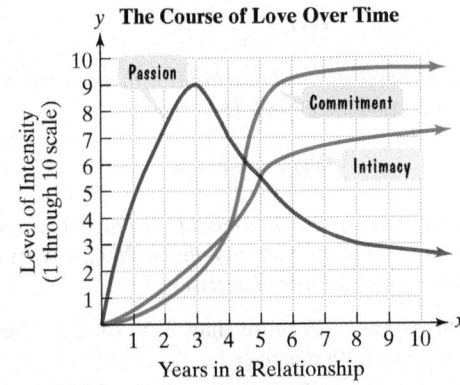

Source: *R. J. Sternberg. A Triangular Theory of Love, Psychological Review, 93, 119–135.*

89. Use interval notation to write an inequality that expresses for which years in a relationship intimacy is greater than commitment.

90. Use interval notation to write an inequality that expresses for which years in a relationship passion is greater than or equal to intimacy.

91. What is the relationship between passion and intimacy on the interval [5, 7)?

92. What is the relationship between intimacy and commitment on the interval [4, 7)?

93. What is the relationship between passion and commitment for $\{x \mid 6 < x < 8\}$?

94. What is the relationship between passion and commitment for $\{x \mid 7 < x < 9\}$?

95. What is the maximum level of intensity for passion? After how many years in a relationship does this occur?

96. After approximately how many years do levels of intensity for commitment exceed the maximum level of intensity for passion?

In more U.S. marriages, spouses have different faiths. The bar graph shows the percentage of households with an interfaith marriage in 1988 and 2008. Also shown is the percentage of households in which a person of faith is married to someone with no religion.

Percentage of U.S. Households in Which Married Couples Do Not Share the Same Faith

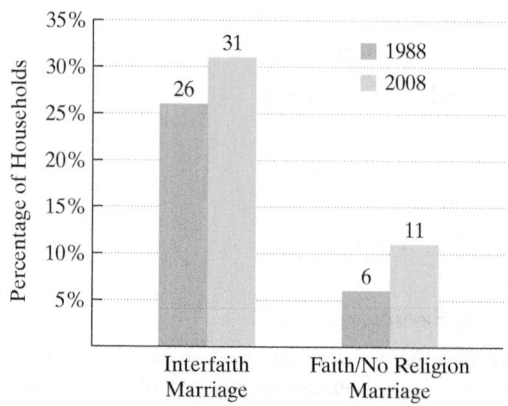

Source: *General Social Survey, University of Chicago*

The function

$$I(x) = \frac{1}{4}x + 26$$

models the percentage of U.S. households with an interfaith marriage, I(x), x years after 1988. The function

$$N(x) = \frac{1}{4}x + 6$$

models the percentage of U.S. households in which a person of faith is married to someone with no religion, N(x), x years after 1988. Use these models to solve Exercises 97–98.

97. a. In which years will more than 33% of U.S. households have an interfaith marriage?

 b. In which years will more than 14% of U.S. households have a person of faith married to someone with no religion?

 c. Based on your answers to parts (a) and (b), in which years will more than 33% of households have an interfaith marriage and more than 14% have a faith/no religion marriage?

 d. Based on your answers to parts (a) and (b), in which years will more than 33% of households have an interfaith marriage or more than 14% have a faith/no religion marriage?

98. a. In which years will more than 34% of U.S. households have an interfaith marriage?

 b. In which years will more than 15% of U.S. households have a person of faith married to someone with no religion?

 c. Based on your answers to parts (a) and (b), in which years will more than 34% of households have an interfaith marriage and more than 15% have a faith/no religion marriage?

 d. Based on your answers to parts (a) and (b), in which years will more than 34% of households have an interfaith marriage or more than 15% have a faith/no religion marriage?

99. A basic cellphone plan costs $20 per month for 60 calling minutes. Additional time costs $0.40 per minute. The formula

$$C = 20 + 0.40(x - 60)$$

gives the monthly cost for this plan, C, for x calling minutes, where $x > 60$. How many calling minutes are possible for a monthly cost of at least $28 and at most $40?

100. The formula for converting Fahrenheit temperature, F, to Celsius temperature, C, is

$$C = \frac{5}{9}(F - 32).$$

If Celsius temperature ranges from 15° to 35°, inclusive, what is the range for the Fahrenheit temperature? Use interval notation to express this range.

101. If a coin is tossed 100 times, we would expect approximately 50 of the outcomes to be heads. It can be demonstrated that a coin is unfair if h, the number of outcomes that result in heads, satisfies $\left|\dfrac{h - 50}{5}\right| \geq 1.645$. Describe the number of outcomes that determine an unfair coin that is tossed 100 times.

In Exercises 102–113, use the strategy for solving word problems, modeling the verbal conditions of the problem with a linear inequality.

102. A truck can be rented from Basic Rental for $50 per day plus $0.20 per mile. Continental charges $20 per day plus $0.50 per mile to rent the same truck. How many miles must be driven in a day to make the rental cost for Basic Rental a better deal than Continental's?

103. You are choosing between two texting plans. Plan A has a monthly fee of $15 with a charge of $0.08 per text. Plan B has a monthly fee of $3 with a charge of $0.12 per text. How many text messages in a month make plan A the better deal?

104. A city commission has proposed two tax bills. The first bill requires that a homeowner pay $1800 plus 3% of the assessed home value in taxes. The second bill requires taxes of $200 plus 8% of the assessed home value. What price range of home assessment would make the first bill a better deal?

105. A local bank charges $8 per month plus 5¢ per check. The credit union charges $2 per month plus 8¢ per check. How many checks should be written each month to make the credit union a better deal?

106. A company manufactures and sells blank audiocassette tapes. The weekly fixed cost is $10,000 and it costs $0.40 to produce each tape. The selling price is $2.00 per tape. How many tapes must be produced and sold each week for the company to generate a profit?

107. A company manufactures and sells personalized stationery. The weekly fixed cost is $3000 and it costs $3.00 to produce each package of stationery. The selling price is $5.50 per package. How many packages of stationery must be produced and sold each week for the company to generate a profit?

108. An elevator at a construction site has a maximum capacity of 2800 pounds. If the elevator operator weighs 265 pounds and each cement bag weighs 65 pounds, how many bags of cement can be safely lifted on the elevator in one trip?

109. An elevator at a construction site has a maximum capacity of 3000 pounds. If the elevator operator weighs 245 pounds and each cement bag weighs 95 pounds, how many bags of cement can be safely lifted on the elevator in one trip?

110. To earn an A in a course, you must have a final average of at least 90%. On the first four examinations, you have grades of 86%, 88%, 92%, and 84%. If the final examination counts as two grades, what must you get on the final to earn an A in the course?

111. On two examinations, you have grades of 86 and 88. There is an optional final examination, which counts as one grade. You decide to take the final in order to get a course grade of A, meaning a final average of at least 90.
 a. What must you get on the final to earn an A in the course?
 b. By taking the final, if you do poorly, you might risk the B that you have in the course based on the first two exam grades. If your final average is less than 80, you will lose your B in the course. Describe the grades on the final that will cause this to happen.

112. Parts for an automobile repair cost $175. The mechanic charges $34 per hour. If you receive an estimate for at least $226 and at most $294 for fixing the car, what is the time interval that the mechanic will be working on the job?

113. The toll to a bridge is $3.00. A three-month pass costs $7.50 and reduces the toll to $0.50. A six-month pass costs $30 and permits crossing the bridge for no additional fee. How many crossings per three-month period does it take for the three-month pass to be the best deal?

Writing in Mathematics

114. When graphing the solutions of an inequality, what does a parenthesis signify? What does a square bracket signify?

115. Describe ways in which solving a linear inequality is similar to solving a linear equation.

116. Describe ways in which solving a linear inequality is different than solving a linear equation.

117. What is a compound inequality and how is it solved?

118. Describe one method for solving an absolute value inequality.

119. Explain why $|x| < -4$ has no solution.

120. Describe the solution set of $|x| > -4$.

Technology Exercises

In Exercises 121–122, solve each inequality using a graphing utility. Graph the linear function on each side separately. Then determine the values of x for which the graph of the function on the left side lies above the graph of the function on the right side.

121. $-3(x - 6) > 2x - 2$ **122.** $-2(x + 4) > 6x + 16$

123. Use a graphing utility's $\boxed{\text{TABLE}}$ feature to verify your work in Exercises 121–122.

124. A bank offers two checking account plans. Plan A has a base service charge of $4.00 per month plus 10¢ per check. Plan B charges a base service charge of $2.00 per month plus 15¢ per check.
 a. Write linear functions that model the total monthly costs $C(x)$ for each plan if x checks are written.
 b. Use a graphing utility to graph the models in the same $[0, 50, 10]$ by $[0, 10, 1]$ viewing rectangle.
 c. Use the graphs (and the intersection feature) to determine for what number of checks per month plan A will be better than plan B.
 d. Verify the result of part (c) algebraically by solving an inequality.

Critical Thinking Exercises

Make Sense? *In Exercises 125–128, determine whether each statement makes sense or does not make sense, and explain your reasoning.*

125. I can check inequalities by substituting 0 for the variable: When 0 belongs to the solution set, I should obtain a true statement, and when 0 does not belong to the solution set, I should obtain a false statement.

126. In an inequality such as $5x + 4 < 8x - 5$, I can avoid division by a negative number depending on which side I collect the variable terms and on which side I collect the constant terms.

127. I have problems setting up the appropriate compound inequality for an absolute value inequality, so I use the boundary point method.

128. I noticed that the graph of $f(x) = |x - 3|$ lies below the graph of $g(x) = 4$ in $(-1, 7)$, so the solution set of $|x - 3| > 4$ must be $(-\infty, -1) \cup (7, \infty)$.

In Exercises 129–132, determine whether each statement is true or false. If the statement is false, make the necessary change(s) to produce a true statement.

129. The smallest real number in the solution set of $2x > 6$ is 4.

130. Values of -5 and 5 satisfy $|x| = 5, |x| \le 5$, and $|x| \ge -5$.

131. The inequality $3x > 6$ is equivalent to $2 > x$.

132. All irrational numbers satisfy $|x - 4| > 0$.

133. What's wrong with this argument? Suppose x and y represent two real numbers, where $x > y$.

$2 > 1$	This is a true statement.
$2(y - x) > 1(y - x)$	Multiply both sides by $y - x$.
$2y - 2x > y - x$	Use the distributive property.
$y - 2x > -x$	Subtract y from both sides.
$y > x$	Add 2x to both sides.

The final inequality, $y > x$, is impossible because we were initially given $x > y$.

134. Write an absolute value inequality for which the interval shown is the solution.

Solutions lie within
3 units of 4.

a.

-2 -1 0 1 2 3 4 5 6 7 8

b.

-2 -1 0 1 2 3 4 5 6 7 8

Group Exercise

135. Each group member should research one situation that provides two different pricing options. These can involve areas such as public transportation options (with or without discount passes), cellphone plans, long-distance telephone plans, or anything of interest. Be sure to bring in all the details for each option. At a second group meeting, select the two pricing situations that are most interesting and relevant. Using each situation, write a word problem about selecting the better of the two options. The word problem should be one that can be solved using a linear inequality. The group should turn in the two problems and their solutions.

Preview Exercises

Exercises 136–138 will help you prepare for the material covered in the first section of the next chapter.

In Exercises 136–137, solve each quadratic equation by the method of your choice.

136. $0 = -2(x - 3)^2 + 8$ **137.** $-x^2 - 2x + 1 = 0$

138. Use the graph of $f(x) = x^2$ to graph $g(x) = (x + 3)^2 + 1$.

CHAPTER 2 Summary, Review, and Test

DEFINITIONS AND CONCEPTS EXAMPLES

2.1 Linear Equations and Rational Equations

a. A linear equation in one variable x can be written in the form $mx + b = 0, m \neq 0$.

b. The procedure for solving a linear equation is given in the box on page 239. Ex. 1, p. 240

c. If an equation contains fractions, begin by multiplying both sides by the least common denominator, thereby clearing fractions. Ex. 2, p. 241

d. A rational equation is an equation containing one or more rational expressions. If an equation contains rational expressions with variable denominators, avoid in the solution set any values of the variable that make a denominator zero. Ex. 3, p. 242; Ex. 4, p. 243; Ex. 5, p. 244

e. An identity is an equation that is true for all real numbers for which both sides are defined. When solving an identity, the variable is eliminated and a true statement, such as 3 = 3, results. A conditional equation is not an identity and is true for at least one real number. An inconsistent equation is an equation that is not true for even one real number. A false statement, such as 3 = 7, results when solving such an equation, whose solution set is \varnothing, the empty set. Ex. 6, p. 246; Ex. 7, p. 247

2.2 Models and Applications

a. A five-step procedure for solving word problems using equations that model verbal conditions is given in the box on page 253. Ex. 1, p. 254; Ex. 2, p. 255; Ex. 3, p. 257; Ex. 4, p. 258; Ex. 5, p. 259; Ex. 6, p. 261

2.3 Complex Numbers

a. The imaginary unit i is defined as

$$i = \sqrt{-1}, \text{ where } i^2 = -1.$$

The set of numbers in the form $a + bi$ is called the set of complex numbers; a is the real part and b is the imaginary part. If $b = 0$, the complex number is a real number. If $b \neq 0$, the complex number is an imaginary number. Complex numbers in the form bi are called pure imaginary numbers.

b. Rules for adding and subtracting complex numbers are given in the box on page 267. Ex. 1, p. 267

DEFINITIONS AND CONCEPTS

c. To multiply complex numbers, multiply as if they are polynomials. After completing the multiplication, replace i^2 with -1 and simplify.

Ex. 2, p. 268

d. The complex conjugate of $a + bi$ is $a - bi$ and vice versa. The multiplication of complex conjugates gives a real number:

$$(a + bi)(a - bi) = a^2 + b^2.$$

e. To divide complex numbers, multiply the numerator and the denominator by the complex conjugate of the denominator.

Ex. 3, p. 269

f. When performing operations with square roots of negative numbers, begin by expressing all square roots in terms of i. The principal square root of $-b$ is defined by

$$\sqrt{-b} = i\sqrt{b}.$$

Ex. 4, p. 270

2.4 Quadratic Equations

a. The zeros of function f are the roots, or solutions, of the equation $f(x) = 0$. The real zeros, or real solutions, are the x-intercepts of the graph of f.

Figure 2.7, p. 275

b. A quadratic function in x can be written in the form $f(x) = ax^2 + bx + c$, $\ a \neq 0$.

A quadratic equation in x can be written in the general form $ax^2 + bx + c = 0$, $\ a \neq 0$.

c. The procedure for solving a quadratic equation by factoring and the zero-product principle is given in the box on page 274.

Ex. 1, p. 274

d. The procedure for solving a quadratic equation by the square root property is described on page 276. If $u^2 = d$, then $u = \pm\sqrt{d}$.

Ex. 2, p. 276

e. All quadratic equations can be solved by completing the square. Isolate the binomial with the two variable terms on one side of the equation. If the coefficient of the x^2-term is not one, divide each side of the equation by this coefficient. Then add the square of half the coefficient of x to both sides.

Ex. 3, p. 277;
Ex. 4, p. 278

f. All quadratic equations can be solved by the quadratic formula

Ex. 5, p. 280;
Ex. 6, p. 281

$$x = \frac{-b \pm \sqrt{b^2 - 4ac}}{2a}.$$

The formula is derived by completing the square of the equation $ax^2 + bx + c = 0$.

g. The discriminant, $b^2 - 4ac$, indicates the number and type of solutions to the quadratic equation $ax^2 + bx + c = 0$, shown in Table 2.2 on page 282.

Ex. 7, p. 283

h. Table 2.3 on page 284 shows the most efficient technique to use when solving a quadratic equation.

Ex. 8, p. 285;
Ex. 9, p. 286

2.5 Other Types of Equations

a. Some polynomial equations of degree 3 or greater can be solved by moving all terms to one side, obtaining zero on the other side, factoring, and using the zero-product principle. Factoring by grouping is often used.

Ex. 1, p. 295;
Ex. 2, p. 296

b. A radical equation is an equation in which the variable occurs in a square root, a cube root, and so on. A radical equation can be solved by isolating the radical and raising both sides of the equation to a power equal to the radical's index. When raising both sides to an even power, check all proposed solutions in the original equation. Eliminate extraneous solutions from the solution set.

Ex. 3, p. 297;
Ex. 4, p. 299

c. A radical equation with a rational exponent can be solved by isolating the expression with the rational exponent and raising both sides of the equation to a power that is the reciprocal of the rational exponent. See the details in the box on pages 300–301.

Ex. 5, p. 301

d. An equation is quadratic in form if it can be written in the form $au^2 + bu + c = 0$, where u is an algebraic expression and $a \neq 0$. Solve for u and use the substitution that resulted in this equation to find the values for the variable in the given equation.

Ex. 6, p. 302;
Ex. 7, p. 303

e. Absolute value equations in the form $|u| = c, c > 0$, can be solved by rewriting the equation without absolute value bars: $u = c$ or $u = -c$.

Ex. 8, p. 304;
Ex. 9, p. 305

DEFINITIONS AND CONCEPTS

EXAMPLES

2.6 Linear Inequalities and Absolute Value Inequalities

a. A linear inequality in one variable x can be expressed as $mx + b \leq 0, mx + b < 0, mx + b \geq 0$, or $mx + b > 0, m \neq 0$.

b. A linear inequality is solved using a procedure similar to solving a linear equation. However, when multiplying or dividing by a negative number, change the sense of the inequality.

Ex. 1, p. 312;
Ex. 2, p. 312;
Ex. 3, p. 314

c. A compound inequality with three parts can be solved by isolating the variable in the middle.

Ex. 5, p. 316

d. Inequalities involving absolute value can be solved using boundary points. The procedure is described in the box on page 317.

Ex. 6, p. 318;
Ex. 7, p. 319;
Ex. 8, p. 320

e. Inequalities involving absolute value can be solved by rewriting the inequalities without absolute value bars. If c is a positive number,

Ex. 9, p. 322;
Ex. 10, p. 323;
Ex. 11, p. 324

1. The solutions of $|u| < c$ are the numbers that satisfy $-c < u < c$.

2. The solutions of $|u| > c$ are the numbers that satisfy $u < -c$ or $u > c$.

REVIEW EXERCISES

2.1

In Exercises 1–19, solve each equation. Then state whether the equation is an identity, a conditional equation, or an inconsistent equation.

1. $2x - 5 = 7$

2. $5x + 20 = 3x$

3. $7(x - 4) = x + 2$

4. $1 - 2(6 - x) = 3x + 2$

5. $2(x - 4) + 3(x + 5) = 2x - 2$

6. $2x - 4(5x + 1) = 3x + 17$

7. $7x + 5 = 5(x + 3) + 2x$

8. $7x + 13 = 2(2x - 5) + 3x + 23$

9. $\dfrac{2x}{3} = \dfrac{x}{6} + 1$

10. $\dfrac{x}{2} - \dfrac{1}{10} = \dfrac{x}{5} + \dfrac{1}{2}$

11. $\dfrac{2x}{3} = 6 - \dfrac{x}{4}$

12. $\dfrac{x}{4} = 2 - \dfrac{x - 3}{3}$

13. $\dfrac{3x + 1}{3} - \dfrac{13}{2} = \dfrac{1 - x}{4}$

14. $\dfrac{9}{4} - \dfrac{1}{2x} = \dfrac{4}{x}$

15. $\dfrac{7}{x - 5} + 2 = \dfrac{x + 2}{x - 5}$

16. $\dfrac{1}{x - 1} - \dfrac{1}{x + 1} = \dfrac{2}{x^2 - 1}$

17. $\dfrac{5}{x + 3} + \dfrac{1}{x - 2} = \dfrac{8}{x^2 + x - 6}$

18. $\dfrac{1}{x + 5} = 0$

19. $\dfrac{4}{x + 2} + \dfrac{3}{x} = \dfrac{10}{x^2 + 2x}$

In Exercises 20–21, find the zeros of each function.

20. $f(x) = 3 - 5(2x + 1) - 2(x - 4)$

21. $f(x) = \dfrac{x + 2}{x + 3} + \dfrac{1}{x^2 + 2x - 3} - 1$

2.2

In Exercises 22–29, use the five-step strategy for solving word problems.

22. Destined for Gory. As sequels to horror films increase, so does the body count. Wes Craven's slasher *Scream* series adheres to that axiom.

Body Count in Wes Craven's
***Scream* Series**

Scream Scream Scream Scream
 2 3 4

Whether it's knife to the back, knife to the gut, or knife to the head, the body count in *Scream 2* exceeds the departed in *Scream* by 2. Appropriately, the number of characters killed off in *Scream 3* exceeds the departed in *Scream* by 3. The total body count in the four *Scream* films shown in the graphic is 33. Find the body count in *Scream*, *Scream 2*, and *Scream 3*.

23. The bar graph shows the average price of a movie ticket for selected years from 1980 through 2010. The graph indicates that in 1980, the average movie ticket price was $2.69. For the period from 1980 through 2010, the price increased by approximately $0.15 per year. If this trend continues, by which year will the average price of a movie ticket be $8.69?

1980
Ordinary People
Ticket Price $2.69

2010
Alice in Wonderland
Ticket Price $7.85

Sources: *Motion Picture Association of America, National Association of Theater Owners (NATO), and Bureau of Labor Statistics (BLS)*

24. You are choosing between two texting plans. One plan has a monthly fee of $15 with a charge of $0.05 per text. The other plan has a monthly fee of $5 with a charge of $0.07 per text. For how many text messages will the costs for the two plans be the same?

25. After a 20% price reduction, a cordless phone sold for $48. What was the phone's price before the reduction?

26. A salesperson earns $300 per week plus 5% commission of sales. How much must be sold to earn $800 in a week?

27. You invested $9000 in two funds paying 4% and 7% annual interest. At the end of the year, the total interest from these investments was $555. How much was invested at each rate?

28. You invested $8000 in two funds paying 2% and 5% annual interest. At the end of the year, the interest from the 5% investment exceeded the interest from the 2% investment by $85. How much money was invested at each rate?

29. The length of a rectangular field is 6 yards less than triple the width. If the perimeter of the field is 340 yards, what are its dimensions?

30. In 2010, there were 14,100 students at college A, with a projected enrollment increase of 1500 students per year. In the same year, there were 41,700 students at college B, with a projected enrollment decline of 800 students per year.

 a. Let x represent the number of years after 2010. Write, but do not solve, an equation that can be used to find how many years after 2010 the colleges will have the same enrollment.

 b. The table in the next column is based on your equation in part (a). Y_1 represents one side of the equation and Y_2 represents the other side of the equation. Use the table to answer the following questions: In which year will the colleges have the same enrollment? What will be the enrollment in each college at that time?

X	Y1	Y2
7	24600	36100
8	26100	35300
9	27600	34500
10	29100	33700
11	30600	32900
12	32100	32100
13	33600	31300

X=7

2.3

In Exercises 31–40, perform the indicated operations and write the result in standard form.

31. $(8 - 3i) - (17 - 7i)$

32. $4i(3i - 2)$

33. $(7 - i)(2 + 3i)$

34. $(3 - 4i)^2$

35. $(7 + 8i)(7 - 8i)$

36. $\dfrac{6}{5 + i}$

37. $\dfrac{3 + 4i}{4 - 2i}$

38. $\sqrt{-32} - \sqrt{-18}$

39. $\left(-2 + \sqrt{-100}\right)^2$

40. $\dfrac{4 + \sqrt{-8}}{2}$

2.4

Solve each equation in Exercises 41–42 by factoring.

41. $2x^2 + 15x = 8$

42. $5x^2 + 20x = 0$

Solve each equation in Exercises 43–46 by the square root property.

43. $2x^2 - 3 = 125$

44. $\dfrac{x^2}{2} + 5 = -3$

45. $(x + 3)^2 = -10$

46. $(3x - 4)^2 = 18$

Solve each equation in Exercises 47–48 by completing the square.

47. $x^2 - 12x + 27 = 0$

48. $3x^2 - 12x + 11 = 0$

Solve each equation in Exercises 49–51 using the quadratic formula.

49. $x^2 = 2x + 4$

50. $x^2 - 2x + 19 = 0$

51. $2x^2 = 3 - 4x$

In Exercises 52–53, without solving the given quadratic equation, determine the number and type of solutions.

52. $x^2 - 4x + 13 = 0$

53. $9x^2 = 2 - 3x$

Solve each equation in Exercises 54–59 by the method of your choice.

54. $2x^2 - 11x + 5 = 0$

55. $(3x + 5)(x - 3) = 5$

56. $3x^2 - 7x + 1 = 0$

57. $x^2 - 9 = 0$

58. $(x - 3)^2 - 25 = 0$

59. $3x^2 - x + 2 = 0$

In Exercises 60–62, find all zeros of each function.

60. $f(x) = 3x^2 - 10x - 8$

61. $f(x) = (x + 2)^2 + 4$

62. $f(x) = \dfrac{5}{x + 1} + \dfrac{x - 1}{4} - 2$

63. The function $W(t) = 3t^2$ models the weight of a human fetus, $W(t)$, in grams, after t weeks, where $0 \le t \le 39$. After how many weeks does the fetus weigh 588 grams?

64. As gas prices surge, more Americans are cycling as a way to save money, stay fit, or both. In 2010, Boston installed 20 miles of bike lanes and New York City added more than 50 miles. The bar graph shows the number of bicycle friendly U.S. communities, as designated by the League of American Bicyclists, for selected years from 2003 through 2011.

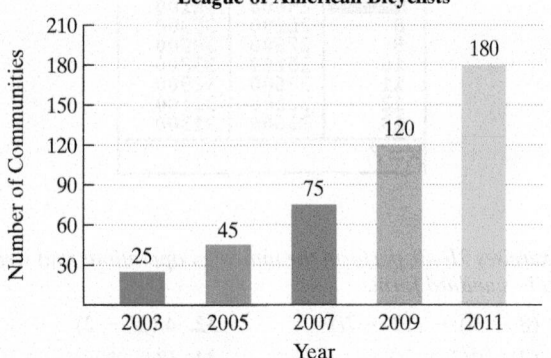

Number of U.S. Communities Designated "Bicycle Friendly" by the League of American Bicyclists

Source: *League of American Bicyclists*

The quadratic function

$$B(x) = 1.7x^2 + 6x + 26$$

models the number of bicycle-friendly communities, $B(x)$, x years after 2003.

a. Use the function to find the number of bicycle-friendly communities in 2011. Round to the nearest whole number. Does this rounded value underestimate or overestimate the number shown by the graph? By how much?

b. Use the function to determine the year in which 826 U.S. communities will be bicycle friendly.

65. An architect is allowed 15 square yards of floor space to add a small bedroom to a house. Because of the room's design in relationship to the existing structure, the width of the rectangular floor must be 7 yards less than two times the length. Find the length and width of the rectangular floor that the architect is permitted.

66. A building casts a shadow that is double the height of the building. If the distance from the end of the shadow to the top of the building is 300 meters, how high is the building? Round to the nearest meter.

2.5

Solve each polynomial equation in Exercises 67–68.

67. $2x^4 = 50x^2$

68. $2x^3 - x^2 - 18x + 9 = 0$

Solve each radical equation in Exercises 69–70.

69. $\sqrt{2x - 3} + x = 3$

70. $\sqrt{x - 4} + \sqrt{x + 1} = 5$

Solve the equations with rational exponents in Exercises 71–72.

71. $3x^{\frac{3}{4}} - 24 = 0$

72. $(x - 7)^{\frac{2}{3}} = 25$

Solve each equation in Exercises 73–74 by making an appropriate substitution.

73. $x^4 - 5x^2 + 4 = 0$

74. $x^{\frac{1}{2}} + 3x^{\frac{1}{4}} - 10 = 0$

Solve the equations containing absolute value in Exercises 75–76.

75. $|2x + 1| = 7$

76. $2|x - 3| - 6 = 10$

Solve each equation in Exercises 77–80 by the method of your choice.

77. $3x^{\frac{4}{3}} - 5x^{\frac{2}{3}} + 2 = 0$

78. $2\sqrt{x - 1} = x$

79. $|2x - 5| - 3 = 0$

80. $x^3 + 2x^2 = 9x + 18$

In Exercises 81–84, find all zeros of each function.

81. $f(x) = \sqrt{10 - 3x} - x$

82. $f(x) = 4x^{\frac{4}{3}} - 13x^{\frac{2}{3}} + 9$

83. $f(x) = x^3 + 3x^2 - 2x - 6$

84. $f(x) = -4|x + 1| + 12$

85. Long before the passage of laws prohibiting marriage between people of the same sex, a substantial percentage of Americans were in favor of laws prohibiting interracial marriage. Since 1972, the General Social Survey has asked

> Do you think that there should be laws against marriages between negroes/blacks/African Americans and whites?

In 1972, 39.3% of the adult U.S. population was in favor of such laws. The bar graph shows the percentage of Americans in favor of legislation prohibiting interracial marriage for five selected years from 1993 through 2002, the last year the data were collected.

Percentage of Americans in Favor of Laws Prohibiting Interracial Marriage

Source: *Ben Schott*, Schott's Almanac 2007, *Donnelley and Sons*

The function

$$p(t) = -2.5\sqrt{t} + 17$$

models the percentage of Americans, $p(t)$, in favor of laws prohibiting interracial marriage t years after 1993. If trends indicated by this model continue, in what year did the percentage favoring such legislation decrease to 7%?

2.6

In Exercises 86–96, solve each inequality. Other than \varnothing, use interval notation to express solution sets and graph each solution set on a number line.

86. $-6x + 3 \leq 15$

87 $6x - 9 \geq -4x - 3$

88. $\dfrac{x}{3} - \dfrac{3}{4} - 1 > \dfrac{x}{2}$

89. $6x + 5 > -2(x - 3) - 25$

90. $3(2x - 1) - 2(x - 4) \geq 7 + 2(3 + 4x)$

91. $5(x - 2) - 3(x + 4) \geq 2x - 20$

92. $7 < 2x + 3 \leq 9$

93. $|2x + 3| \leq 15$

94. $\left|\dfrac{2x + 6}{3}\right| > 2$

95. $|2x + 5| - 7 \geq -6$

96. $-4|x + 2| + 5 \leq -7$

In Exercises 97–98, use interval notation to represent all values of x satisfying the given conditions.

97 $f(x) = -10 - 3(2x + 1)$, $g(x) = 8x + 1$, and $f(x) > g(x)$.

98. $f(x) = 3 - |2x - 5|$ and $f(x)$ is at least -6.

99. A car rental agency rents a certain car for \$40 per day with unlimited mileage or \$24 per day plus \$0.20 per mile. How far can a customer drive this car per day for the \$24 option to cost no more than the unlimited mileage option?

100. To receive a B in a course, you must have an average of at least 80% but less than 90% on five exams. Your grades on the first four exams were 95%, 79%, 91%, and 86%. What range of grades on the fifth exam will result in a B for the course?

101. A retiree requires an annual income of at least \$9000 from an investment paying 7.5% annual interest. How much should the retiree invest to achieve the desired return?

CHAPTER 2 TEST

In Exercises 1–17, solve each equation or inequality. Other than \varnothing, use interval notation to express solution sets of inequalities and graph these solution sets on a number line.

1. $7(x - 2) = 4(x + 1) - 21$

2. $\dfrac{2x - 3}{4} = \dfrac{x - 4}{2} - \dfrac{x + 1}{4}$

3. $\dfrac{2}{x - 3} - \dfrac{4}{x + 3} = \dfrac{8}{x^2 - 9}$

4. $2x^2 - 3x - 2 = 0$

5. $(3x - 1)^2 = 75$

6. $x(x - 2) = 4$

7. $4x^2 = 8x - 5$

8. $x^3 - 4x^2 - x + 4 = 0$

9. $\sqrt{x - 3} + 5 = x$

10. $\sqrt{x + 4} + \sqrt{x - 1} = 5$

11. $5x^{\frac{3}{2}} - 10 = 0$

12. $x^{\frac{2}{3}} - 9x^{\frac{1}{3}} + 8 = 0$

13. $\left|\dfrac{2}{3}x - 6\right| = 2$

14. $3(x + 4) \geq 5x - 12$

15. $\dfrac{x}{6} + \dfrac{1}{8} \leq \dfrac{x}{2} - \dfrac{3}{4}$

16. $-3 \leq \dfrac{2x + 5}{3} < 6$

17. $|3x + 2| \geq 3$

In Exercises 18–23, find all zeros of each function.

18. $f(x) = \sqrt{8 - 2x} - x$

19. $f(x) = -3|4x - 7| + 15$

20. $f(x) = -10 - 3(2x + 1) - 8x - 1$

21. $f(x) = \dfrac{1}{x^2} - \dfrac{4}{x} + 1$

22. $f(x) = (x + 3)^2 - 24$

23. $f(x) = \dfrac{2x}{x^2 + 6x + 8} - \dfrac{x}{x + 4} + \dfrac{2}{x + 2}$

In Exercises 24–25, use interval notation to represent all values of x satisfying the given conditions.

24. $f(x) = 2x - 5$, and $f(x)$ is at least -3 and no more than 7.

25. $f(x) = \left|\dfrac{2 - x}{4}\right|$ and $f(x)$ is at least 1.

In Exercises 26–28, perform the indicated operations and write the result in standard form.

26. $(6 - 7i)(2 + 5i)$

27. $\dfrac{5}{2 - i}$

28. $2\sqrt{-49} + 3\sqrt{-64}$

The graphs show the amount being paid in Social Security benefits and the amount going into the system. All data are expressed in billions of dollars. Amounts from 2012 through 2024 are projections.

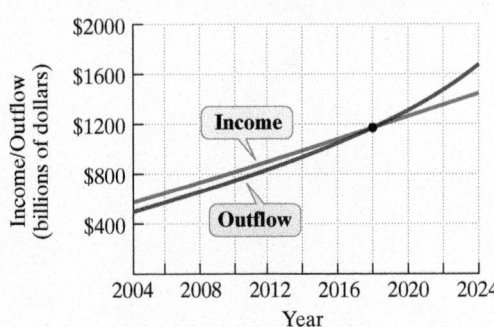

Social Insecurity: Income and Outflow of the Social Security System

Source: *2004 Social Security Trustees Report*

Exercises 29–31 are based on the data shown by the graphs.

29. In 2004, the system's income was \$575 billion, projected to increase at an average rate of \$43 billion per year. In which year will the system's income be \$1177 billion?

30. The data for the system's outflow can be modeled by the quadratic function

$$B(x) = 0.07x^2 + 47.4x + 500,$$

where $B(x)$ represents the amount paid in benefits, in billions of dollars, x years after 2004. According to this model, when will the amount paid in benefits be $1177 billion? Round to the nearest year.

31. How well do your answers to Exercises 29 and 30 model the data shown by the graphs on the previous page?

32. Here's Looking at You. According to University of Texas economist Daniel Hamermesh (*Beauty Pays: Why Attractive People Are More Successful*), strikingly attractive and good-looking men and women can expect to earn an average of $230,000 more in a lifetime than a person who is homely or plain. (Your author feels the need to start affirmative action for the beauty-bereft, consoled by the reality that looks are only one of many things that matter.) The bar graph shows the distribution of looks for American men and women, ranging from homely to strikingly attractive.

The percentage of average-looking men exceeds the percentage of strikingly attractive men by 57. The percentage of good-looking men exceeds the percentage of strikingly attractive men by 25. A total of 88% of American men range between average looking, good-looking, and strikingly attractive. Find the percentage of men who fall within each of these three categories of looks.

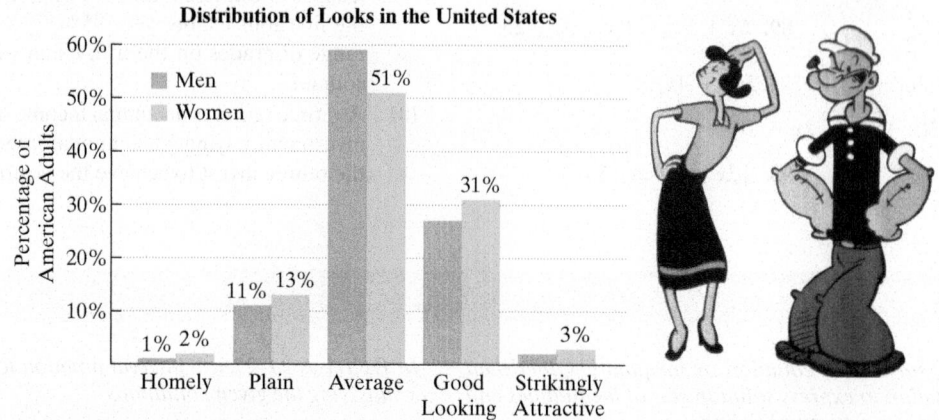

Distribution of Looks in the United States

Source: *Time*, August 22, 2011

33. The costs for two different kinds of heating systems for a small home are given in the following table. After how many years will total costs for solar heating and electric heating be the same? What will be the cost at that time?

System	Cost to Install	Operating Cost/Year
Solar	$29,700	$150
Electric	$5000	$1100

34. You invested $10,000 in two accounts paying 8% and 10% annual interest. At the end of the year, the total interest from these investments was $940. How much was invested at each rate?

35. The length of a rectangular carpet is 4 feet greater than twice its width. If the area is 48 square feet, find the carpet's length and width.

36. A vertical pole is to be supported by a wire that is 26 feet long and anchored 24 feet from the base of the pole. How far up the pole should the wire be attached?

37. After a 60% reduction, a jacket sold for $20. What was the jacket's price before the reduction?

38. You are choosing between two texting plans. Plan A charges $25 per month for unlimited texting. Plan B has a monthly fee of $13 with a charge of $0.06 per text. How many text messages in a month make plan A the better deal?

CUMULATIVE REVIEW EXERCISES (CHAPTERS 1–2)

Use the graph of $y = f(x)$ to solve Exercises 1–5.

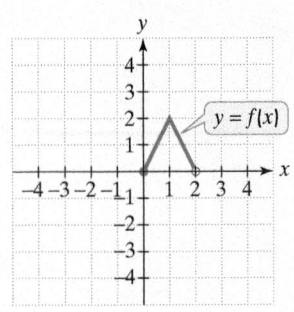

1. Find the domain and the range of f.
2. For what value(s) of x is $f(x) = 1$?
3. Find the relative maximum.
4. Graph $g(x) = f(x - 1) + 1$.
5. Graph $h(x) = -2f\left(\frac{1}{2}x\right)$.

In Exercises 6–10, solve each equation or inequality.

6. $(x + 3)(x - 4) = 8$
7. $3(4x - 1) = 4 - 6(x - 3)$
8. $\sqrt{x} + 2 = x$
9. $x^{\frac{2}{3}} - x^{\frac{1}{3}} - 6 = 0$
10. $\frac{x}{2} - 3 \le \frac{x}{4} + 2$

In Exercises 11–14, graph each equation in a rectangular coordinate system. If two functions are indicated, graph both in the same system. Then use your graphs to identify each relation's domain and range.

11. $3x - 6y - 12 = 0$ 12. $(x - 2)^2 + (y + 1)^2 = 4$
13. $f(x) = \sqrt[3]{x}$ and $g(x) = \sqrt[3]{x - 3} + 4$
14. $f(x) = \sqrt{x - 3} + 2$ and f^{-1}

In Exercises 15–16, let $f(x) = 4 - x^2$ and $g(x) = x + 5$.

15. Find $\dfrac{f(x + h) - f(x)}{h}$ and simplify.
16. Find all values of x satisfying $(f \circ g)(x) = 0$.
17. Write equations in point-slope form, slope-intercept form, and general form for the line passing through $(-2, 5)$ and perpendicular to the line whose equation is $y = -\frac{1}{4}x + \frac{1}{3}$.
18. You invested \$6000 in two accounts paying 7% and 9% annual interest. At the end of the year, the total interest from these investments was \$510. How much was invested at each rate?
19. For a summer sales job, you are choosing between two pay arrangements: a weekly salary of \$200 plus 5% commission on sales, or a straight 15% commission. For how many dollars of sales will the earnings be the same regardless of the pay arrangement?
20. The length of a rectangular garden is 2 feet more than twice its width. If 22 feet of fencing is needed to enclose the garden, what are its dimensions?

POLYNOMIAL AND RATIONAL FUNCTIONS

CHAPTER 3

We are surrounded by evidence that the world is profoundly mathematical. After turning a somersault, a diver's path can be modeled by a quadratic function, $f(x) = ax^2 + bx + c$, as can the path of a football tossed from quarterback to receiver or the path of a flipped coin. Even if you throw an object directly upward, although its path is straight and vertical, its changing height over time is described by a quadratic function. And tailgaters beware: Whether you're driving a car, a motorcycle, or a truck on dry or wet roads, an array of quadratic functions that model your required stopping distances at various speeds are available to help you become a safer driver.

HERE'S WHERE YOU'LL FIND THESE APPLICATIONS:

The quadratic functions surrounding our long history of throwing things appear throughout the chapter, including Example 5 in Section 3.1 and Example 5 in Section 3.6.

Tailgaters should pay close attention to the Section 3.6 opener, Exercises 77–78 and 92–93 in Exercise Set 3.6, and Exercise 75 in the Chapter Review Exercises.

SECTION 3.1 Quadratic Functions

Objectives

① Recognize characteristics of parabolas.

② Graph parabolas.

③ Determine a quadratic function's minimum or maximum value.

④ Solve problems involving a quadratic function's minimum or maximum value.

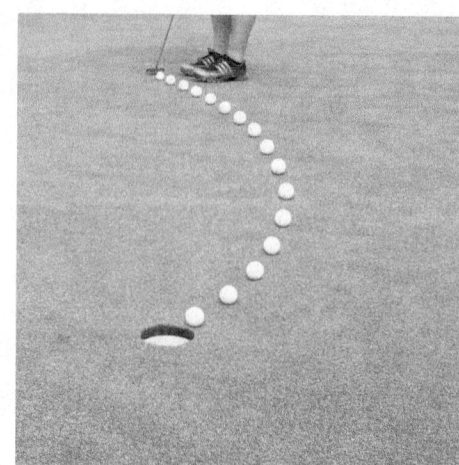

Many sports involve objects that are thrown, kicked, or hit, and then proceed with no additional force of their own. Such objects are called **projectiles**. Paths of projectiles, as well as their heights over time, can be modeled by quadratic functions. We have seen that a **quadratic function** is any function of the form

$$f(x) = ax^2 + bx + c,$$

where $a, b,$ and c are real numbers, with $a \neq 0$. A quadratic function is a polynomial function whose greatest exponent is 2. In this section, you will learn to use graphs of quadratic functions to gain a visual understanding of the algebra that describes football, baseball, basketball, the shot put, and other projectile sports.

① Recognize characteristics of parabolas.

Graphs of Quadratic Functions

The graph of any quadratic function is called a **parabola**. Parabolas are shaped like bowls or inverted bowls, as shown in **Figure 3.1**. If the coefficient of x^2 (the value of a in $ax^2 + bx + c$) is positive, the parabola opens upward. If the coefficient of x^2 is negative, the parabola opens downward. The **vertex** (or turning point) of the parabola is the lowest point on the graph when it opens upward and the highest point on the graph when it opens downward.

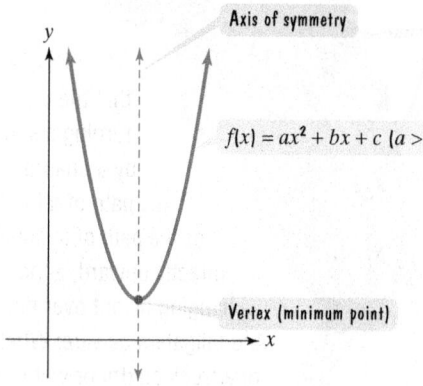

$a > 0$: Parabola opens upward.

$a < 0$: Parabola opens downward.

FIGURE 3.1 Characteristics of graphs of quadratic functions

Look at the unusual image of the word *mirror* shown here. The artist, Scott Kim, has created the image so that the two halves of the whole are mirror images of each other. A parabola shares this kind of symmetry, in which a vertical line through the vertex divides the figure in half. Parabolas are symmetric with respect to this line, called the **axis of symmetry**. If a parabola is folded along its axis of symmetry, the two halves match exactly.

② Graph parabolas.

Graphing Quadratic Functions in Standard Form

In our earlier work with transformations, we applied a series of transformations to the graph of $f(x) = x^2$. The graph of this function is a parabola. The vertex for this parabola is $(0, 0)$. In **Figure 3.2(a)**, the graph of $f(x) = ax^2$ for $a > 0$ is shown in black; it opens *upward*. In **Figure 3.2(b)**, the graph of $f(x) = ax^2$ for $a < 0$ is shown in black; it opens *downward*.

Transformations of $f(x) = ax^2$

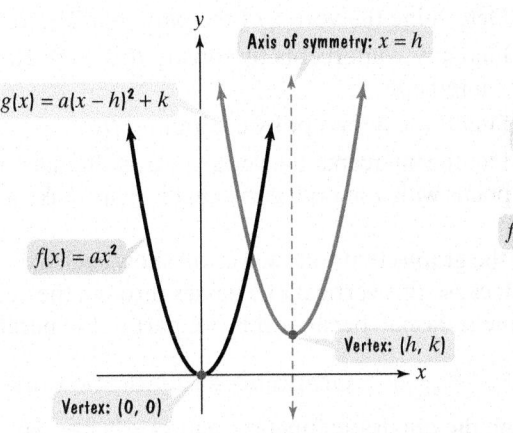

FIGURE 3.2(a) $a > 0$: Parabola opens upward.

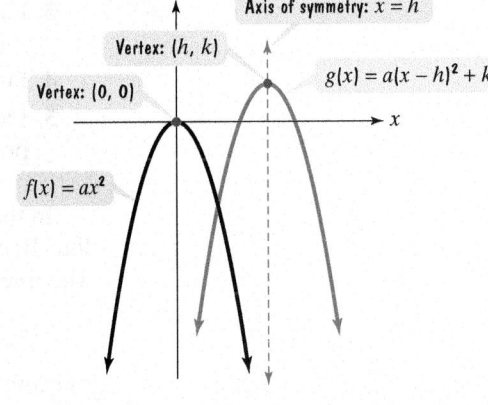

FIGURE 3.2(b) $a < 0$: Parabola opens downward.

Figure 3.2(a) and **3.2(b)** also show the graph of $g(x) = a(x - h)^2 + k$ in blue. Compare these graphs to those of $f(x) = ax^2$. Observe that h determines a horizontal shift and k determines a vertical shift of the graph of $f(x) = ax^2$:

$$g(x) = a(x - h)^2 + k.$$

| If $h > 0$, the graph of $f(x) = ax^2$ is shifted h units to the right. | If $k > 0$, the graph of $y = a(x - h)^2$ is shifted k units up. |

Consequently, the vertex $(0, 0)$ on the black graph of $f(x) = ax^2$ moves to the point (h, k) on the blue graph of $g(x) = a(x - h)^2 + k$. The axis of symmetry is the vertical line whose equation is $x = h$.

The form of the expression for g is convenient because it immediately identifies the vertex of the parabola as (h, k). This is the **standard form** of a quadratic function.

The Standard Form of a Quadratic Function

The quadratic function

$$f(x) = a(x - h)^2 + k, \qquad a \neq 0$$

is in **standard form**. The graph of f is a parabola whose vertex is the point (h, k). The parabola is symmetric with respect to the line $x = h$. If $a > 0$, the parabola opens upward; if $a < 0$, the parabola opens downward.

The sign of a in $f(x) = a(x - h)^2 + k$ determines whether the parabola opens upward or downward. Furthermore, if $|a|$ is small, the parabola opens more flatly than if $|a|$ is large. Here is a general procedure for graphing parabolas whose equations are in standard form:

Graphing Quadratic Functions with Equations in Standard Form

To graph $f(x) = a(x - h)^2 + k$,

1. Determine whether the parabola opens upward or downward. If $a > 0$, it opens upward. If $a < 0$, it opens downward.
2. Determine the vertex of the parabola. The vertex is (h, k).
3. Find any x-intercepts by solving $f(x) = 0$. The function's real zeros are the x-intercepts.
4. Find the y-intercept by computing $f(0)$.
5. Plot the intercepts, the vertex, and additional points as necessary. Connect these points with a smooth curve that is shaped like a bowl or an inverted bowl.

In the graphs that follow, we will show each axis of symmetry as a dashed vertical line. Because this vertical line passes through the vertex, (h, k), its equation is $x = h$. The line is dashed because it is not part of the parabola.

EXAMPLE 1 Graphing a Quadratic Function in Standard Form

Graph the quadratic function $f(x) = -2(x - 3)^2 + 8$.

SOLUTION

We can graph this function by following the steps in the preceding box. We begin by identifying values for a, h, and k.

Standard form
$$f(x) = a(x - h)^2 + k$$

$$a = -2 \quad h = 3 \quad k = 8$$

Given function
$$f(x) = -2(x - 3)^2 + 8$$

Step 1 Determine how the parabola opens. Note that a, the coefficient of x^2, is -2. Thus, $a < 0$; this negative value tells us that the parabola opens downward.

Step 2 Find the vertex. The vertex of the parabola is at (h, k). Because $h = 3$ and $k = 8$, the parabola has its vertex at $(3, 8)$.

Step 3 Find the x-intercepts by solving $f(x) = 0$. Replace $f(x)$ with 0 in $f(x) = -2(x - 3)^2 + 8$.

$0 = -2(x - 3)^2 + 8$	Find x-intercepts, setting f(x) equal to 0.
$2(x - 3)^2 = 8$	Solve for x. Add 2(x − 3)² to both sides of the equation.
$(x - 3)^2 = 4$	Divide both sides by 2.
$x - 3 = \sqrt{4}$ or $x - 3 = -\sqrt{4}$	Apply the square root property.
$x - 3 = 2 \qquad\qquad x - 3 = -2$	$\sqrt{4} = 2$
$x = 5 \qquad\qquad\quad x = 1$	Add 3 to both sides in each equation.

The x-intercepts are 1 and 5. The parabola passes through $(1, 0)$ and $(5, 0)$.

Step 4 Find the y-intercept by computing $f(0)$. Replace x with 0 in $f(x) = -2(x - 3)^2 + 8$.

$$f(0) = -2(0 - 3)^2 + 8 = -2(-3)^2 + 8 = -2(9) + 8 = -10$$

The y-intercept is -10. The parabola passes through $(0, -10)$.

Step 5 Graph the parabola. With a vertex at $(3, 8)$, x-intercepts at 5 and 1, and a y-intercept at -10, the graph of f is shown in **Figure 3.3**. The axis of symmetry is the vertical line whose equation is $x = 3$.

• • •

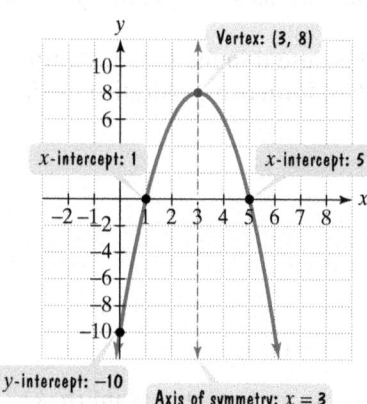

FIGURE 3.3 The graph of $f(x) = -2(x - 3)^2 + 8$

⟁ Check Point **1** Graph the quadratic function $f(x) = -(x - 1)^2 + 4$.

EXAMPLE 2 Graphing a Quadratic Function in Standard Form

Graph the quadratic function $f(x) = (x + 3)^2 + 1$.

SOLUTION

We begin by finding values for a, h, and k.

$$f(x) = a(x - h)^2 + k \qquad \text{Standard form of quadratic function}$$
$$f(x) = (x + 3)^2 + 1 \qquad \text{Given function}$$
$$f(x) = 1(x - (-3))^2 + 1$$

$$a = 1 \qquad h = -3 \qquad k = 1$$

Step 1 Determine how the parabola opens. Note that a, the coefficient of x^2, is 1. Thus, $a > 0$; this positive value tells us that the parabola opens upward.

Step 2 Find the vertex. The vertex of the parabola is at (h, k). Because $h = -3$ and $k = 1$, the parabola has its vertex at $(-3, 1)$.

Step 3 Find the x-intercepts by solving $f(x) = 0$. Replace $f(x)$ with 0 in $f(x) = (x + 3)^2 + 1$. Because the vertex is at $(-3, 1)$, which lies above the x-axis, and the parabola opens upward, it appears that this parabola has no x-intercepts. We can verify this observation algebraically.

$$0 = (x + 3)^2 + 1 \qquad \text{Find possible } x\text{-intercepts, setting } f(x) \text{ equal to 0.}$$

$$-1 = (x + 3)^2 \qquad \text{Solve for } x. \text{ Subtract 1 from both sides.}$$

$$x + 3 = \sqrt{-1} \quad \text{or} \quad x + 3 = -\sqrt{-1} \qquad \text{Apply the square root property.}$$
$$x + 3 = i \qquad\qquad x + 3 = -i \qquad \sqrt{-1} = i$$
$$x = -3 + i \qquad\qquad x = -3 - i \qquad \text{The solutions are } -3 \pm i.$$

Because this equation has no real solutions, the parabola has no x-intercepts.

Step 4 Find the y-intercept by computing $f(0)$. Replace x with 0 in $f(x) = (x + 3)^2 + 1$.

$$f(0) = (0 + 3)^2 + 1 = 3^2 + 1 = 9 + 1 = 10$$

The y-intercept is 10. The parabola passes through $(0, 10)$.

Step 5 Graph the parabola. With a vertex at $(-3, 1)$, no x-intercepts, and a y-intercept at 10, the graph of f is shown in **Figure 3.4**. The axis of symmetry is the vertical line whose equation is $x = -3$. • • •

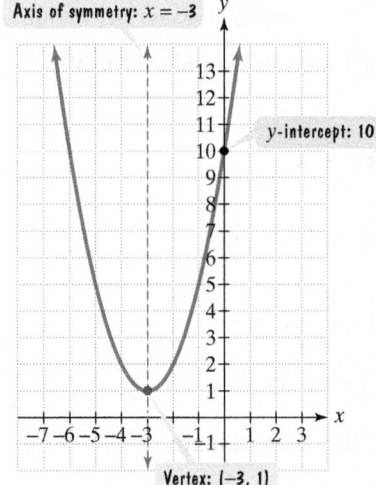

FIGURE 3.4 The graph of $f(x) = (x + 3)^2 + 1$

GREAT QUESTION!

I'm confused about finding h from the equation $f(x) = a(x - h)^2 + k$. Can you help me out?

It's easy to make a sign error when finding h, the x-coordinate of the vertex. In

$$f(x) = a(x - h)^2 + k,$$

h is the number that follows the subtraction sign.

• $f(x) = -2(x - 3)^2 + 8$

 The number *after* the subtraction is 3: $h = 3$.

• $f(x) = (x + 3)^2 + 1$
 $\quad = (x - (-3))^2 + 1$

 The number *after* the subtraction is -3: $h = -3$.

⬭ Check Point **2** Graph the quadratic function $f(x) = (x - 2)^2 + 1$.

Graphing Quadratic Functions in the Form $f(x) = ax^2 + bx + c$

Quadratic functions are frequently expressed in the form $f(x) = ax^2 + bx + c$. How can we identify the vertex of a parabola whose equation is in this form? Completing the square provides the answer to this question.

$$f(x) = ax^2 + bx + c$$

$$= a\left(x^2 + \frac{b}{a}x\right) + c \qquad \text{Factor out } a \text{ from } ax^2 + bx.$$

$$= a\left(x^2 + \frac{b}{a}x + \frac{b^2}{4a^2}\right) + c - a\left(\frac{b^2}{4a^2}\right)$$

> Complete the square by adding the square of half the coefficient of x.

> By completing the square, we added $a \cdot \dfrac{b^2}{4a^2}$. To avoid changing the function's equation, we must subtract this term.

$$= a\left(x + \frac{b}{2a}\right)^2 + c - \frac{b^2}{4a} \qquad \text{\textit{Write the trinomial as the square of a binomial and simplify the constant term.}}$$

Compare this form of the equation with a quadratic function's **standard form**.

Standard form
$$f(x) = a(x - h)^2 + k$$
$$h = -\frac{b}{2a} \qquad k = c - \frac{b^2}{4a}$$

Equation under discussion
$$f(x) = a\left(x - \left(-\frac{b}{2a}\right)\right)^2 + c - \frac{b^2}{4a}$$

The important part of this observation is that h, the x-coordinate of the vertex, is $-\dfrac{b}{2a}$. The y-coordinate can be found by evaluating the function at $-\dfrac{b}{2a}$.

The Vertex of a Parabola Whose Equation Is $f(x) = ax^2 + bx + c$

Consider the parabola defined by the quadratic function $f(x) = ax^2 + bx + c$. The parabola's vertex is $\left(-\dfrac{b}{2a}, f\left(-\dfrac{b}{2a}\right)\right)$. The x-coordinate is $-\dfrac{b}{2a}$. The y-coordinate is found by substituting the x-coordinate into the parabola's equation and evaluating the function at this value of x.

We can apply our five-step procedure to graph parabolas in the form $f(x) = ax^2 + bx + c$.

Graphing Quadratic Functions with Equations in the Form $f(x) = ax^2 + bx + c$

To graph $f(x) = ax^2 + bx + c$,

1. Determine whether the parabola opens upward or downward. If $a > 0$, it opens upward. If $a < 0$, it opens downward.

2. Determine the vertex of the parabola. The vertex is $\left(-\dfrac{b}{2a}, f\left(-\dfrac{b}{2a}\right)\right)$.

3. Find any x-intercepts by solving $f(x) = 0$. The real solutions of $ax^2 + bx + c = 0$ are the x-intercepts.

4. Find the y-intercept by computing $f(0)$. Because $f(0) = c$ (the constant term in the function's equation), the y-intercept is c and the parabola passes through $(0, c)$.

5. Plot the intercepts, the vertex, and additional points as necessary. Connect these points with a smooth curve.

EXAMPLE 3 Graphing a Quadratic Function in the Form $f(x) = ax^2 + bx + c$

Graph the quadratic function $f(x) = -x^2 - 2x + 1$. Use the graph to identify the function's domain and its range.

SOLUTION

Step 1 Determine how the parabola opens. Note that a, the coefficient of x^2, is -1. Thus, $a < 0$; this negative value tells us that the parabola opens downward.

Step 2 Find the vertex. We know that the x-coordinate of the vertex is $x = -\dfrac{b}{2a}$. We identify a, b, and c in $f(x) = ax^2 + bx + c$.

$$f(x) = -x^2 - 2x + 1$$

$$a = -1 \qquad b = -2 \qquad c = 1$$

Substitute the values of a and b into the equation for the x-coordinate:

$$x = -\frac{b}{2a} = -\frac{-2}{2(-1)} = -\left(\frac{-2}{-2}\right) = -1.$$

The x-coordinate of the vertex is -1. We substitute -1 for x in the equation of the function, $f(x) = -x^2 - 2x + 1$, to find the y-coordinate:

$$f(-1) = -(-1)^2 - 2(-1) + 1 = -1 + 2 + 1 = 2.$$

The vertex is at $(-1, 2)$.

Step 3 Find the x-intercepts by solving $f(x) = 0$. Replace $f(x)$ with 0 in $f(x) = -x^2 - 2x + 1$. We obtain $0 = -x^2 - 2x + 1$. This equation cannot be solved by factoring. We will use the quadratic formula to solve it.

$$-x^2 - 2x + 1 = 0$$

$$a = -1 \qquad b = -2 \qquad c = 1$$

$$x = \frac{-b \pm \sqrt{b^2 - 4ac}}{2a} = \frac{-(-2) \pm \sqrt{(-2)^2 - 4(-1)(1)}}{2(-1)} = \frac{2 \pm \sqrt{4 - (-4)}}{-2}$$

To locate the x-intercepts, we need decimal approximations. Thus, there is no need to simplify the radical form of the solutions.

$$x = \frac{2 + \sqrt{8}}{-2} \approx -2.4 \qquad \text{or} \qquad x = \frac{2 - \sqrt{8}}{-2} \approx 0.4$$

The x-intercepts are approximately -2.4 and 0.4. The parabola passes through $(-2.4, 0)$ and $(0.4, 0)$.

Step 4 Find the *y*-intercept by computing *f*(0). Replace *x* with 0 in $f(x) = -x^2 - 2x + 1$.

$$f(0) = -0^2 - 2 \cdot 0 + 1 = 1$$

The *y*-intercept is 1, which is the constant term in the function's equation. The parabola passes through $(0, 1)$.

Step 5 Graph the parabola. With a vertex at $(-1, 2)$, *x*-intercepts at approximately -2.4 and 0.4, and a *y*-intercept at 1, the graph of *f* is shown in **Figure 3.5(a)**. The axis of symmetry is the vertical line whose equation is $x = -1$.

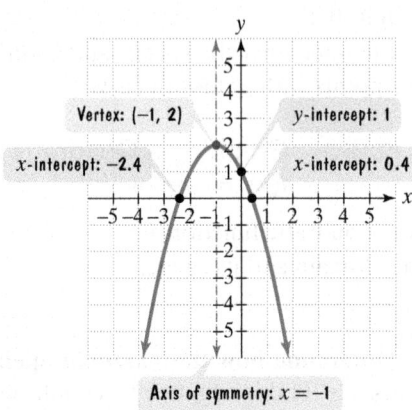

FIGURE 3.5(a) The graph of $f(x) = -x^2 - 2x + 1$

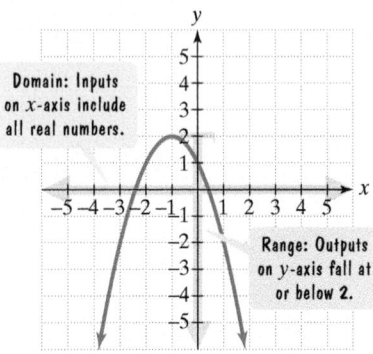

FIGURE 3.5(b) Determining the domain and range of $f(x) = -x^2 - 2x + 1$

GREAT QUESTION!

Are there rules to find domains and ranges of quadratic functions?

Yes. The domain of any quadratic function includes all real numbers. If the vertex is the graph's highest point, the range includes all real numbers at or below the *y*-coordinate of the vertex. If the vertex is the graph's lowest point, the range includes all real numbers at or above the *y*-coordinate of the vertex.

Now we are ready to determine the domain and range of $f(x) = -x^2 - 2x + 1$. We can use the parabola, shown again in **Figure 3.5(b)**, to do so. To find the domain, look for all the inputs on the *x*-axis that correspond to points on the graph. As the graph widens and continues to fall at both ends, can you see that these inputs include all real numbers?

Domain of *f* is $\{x \mid x$ is a real number$\}$ or $(-\infty, \infty)$.

To find the range, look for all the outputs on the *y*-axis that correspond to points on the graph. **Figure 3.5(b)** shows that the parabola's vertex, $(-1, 2)$, is the highest point on the graph. Because the *y*-coordinate of the vertex is 2, outputs on the *y*-axis fall at or below 2.

Range of *f* is $\{y \mid y \le 2\}$ or $(-\infty, 2]$. ●●●

✓ Check Point **3** Graph the quadratic function $f(x) = -x^2 + 4x + 1$. Use the graph to identify the function's domain and its range.

③ Determine a quadratic function's minimum or maximum value.

Minimum and Maximum Values of Quadratic Functions

Consider the quadratic function $f(x) = ax^2 + bx + c$. If $a > 0$, the parabola opens upward and the vertex is its lowest point. If $a < 0$, the parabola opens downward and the vertex is its highest point. The *x*-coordinate of the vertex is $-\dfrac{b}{2a}$. Thus, we can find the minimum or maximum value of *f* by evaluating the quadratic function at $x = -\dfrac{b}{2a}$.

Minimum and Maximum: Quadratic Functions

Consider the quadratic function $f(x) = ax^2 + bx + c$.

1. If $a > 0$, then f has a minimum that occurs at $x = -\dfrac{b}{2a}$. This minimum value is $f\left(-\dfrac{b}{2a}\right)$.

2. If $a < 0$, then f has a maximum that occurs at $x = -\dfrac{b}{2a}$. This maximum value is $f\left(-\dfrac{b}{2a}\right)$.

In each case, the value of x gives the location of the minimum or maximum value. The value of y, or $f\left(-\dfrac{b}{2a}\right)$, gives that minimum or maximum value.

EXAMPLE 4 Obtaining Information about a Quadratic Function from Its Equation

Consider the quadratic function $f(x) = -3x^2 + 6x - 13$.

 a. Determine, without graphing, whether the function has a minimum value or a maximum value.

 b. Find the minimum or maximum value and determine where it occurs.

 c. Identify the function's domain and its range.

SOLUTION

We begin by identifying a, b, and c in the function's equation:

$$f(x) = -3x^2 + 6x - 13.$$

$$a = -3 \qquad b = 6 \qquad c = -13$$

 a. Because $a < 0$, the function has a maximum value.

 b. The maximum value occurs at

$$x = -\frac{b}{2a} = -\frac{6}{2(-3)} = -\frac{6}{-6} = -(-1) = 1.$$

The maximum value occurs at $x = 1$ and the maximum value of $f(x) = -3x^2 + 6x - 13$ is

$$f(1) = -3 \cdot 1^2 + 6 \cdot 1 - 13 = -3 + 6 - 13 = -10.$$

We see that the maximum is -10 at $x = 1$.

 c. Like all quadratic functions, the domain is $(-\infty, \infty)$. Because the function's maximum value is -10, the range includes all real numbers at or below -10. The range is $(-\infty, -10]$. •••

We can use the graph of $f(x) = -3x^2 + 6x - 13$ to visualize the results of Example 4. **Figure 3.6** shows the graph in a $[-6, 6, 1]$ by $[-50, 20, 10]$ viewing rectangle. The maximum function feature verifies that the function's maximum is -10 at $x = 1$. Notice that x gives the location of the maximum and y gives the maximum value. Notice, too, that the maximum value is -10 and not the ordered pair $(1, -10)$.

Range is $(-\infty, -10]$.

Maximum
X=1 Y=-10

$[-6, 6, 1]$ by $[-50, 20, 10]$

FIGURE 3.6

⊘ Check Point **4** Repeat parts (a) through (c) of Example 4 using the quadratic function $f(x) = 4x^2 - 16x + 1000$.

④ Solve problems involving a quadratic function's minimum or maximum value.

Applications of Quadratic Functions

Many applied problems involve finding the maximum or minimum value of a quadratic function, as well as where this value occurs.

EXAMPLE 5 The Parabolic Path of a Punted Football

Figure 3.7 shows that when a football is kicked, the nearest defensive player is 6 feet from the point of impact with the kicker's foot. The height of the punted football, $f(x)$, in feet, can be modeled by

$$f(x) = -0.01x^2 + 1.18x + 2,$$

where x is the ball's horizontal distance, in feet, from the point of impact with the kicker's foot.

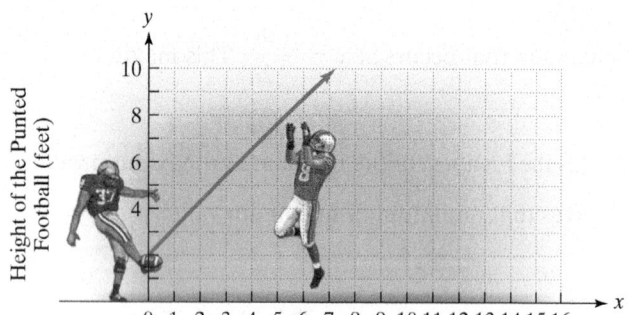

Height of the Punted Football (feet)

Distance from the Point of Impact (feet)

FIGURE 3.7

a. What is the maximum height of the punt and how far from the point of impact does this occur?

b. How far must the nearest defensive player, who is 6 feet from the kicker's point of impact, reach to block the punt?

c. If the ball is not blocked by the defensive player, how far down the field will it go before hitting the ground?

d. Graph the function that models the football's parabolic path.

SOLUTION

a. We begin by identifying the numbers a, b, and c in the function's equation.

$$f(x) = -0.01x^2 + 1.18x + 2$$

$$a = -0.01 \qquad b = 1.18 \qquad c = 2$$

Because $a < 0$, the function has a maximum that occurs at $x = -\dfrac{b}{2a}$.

$$x = -\frac{b}{2a} = -\frac{1.18}{2(-0.01)} = -(-59) = 59$$

This means that the maximum height of the punt occurs 59 feet from the kicker's point of impact. The maximum height of the punt is

$$f(59) = -0.01(59)^2 + 1.18(59) + 2 = 36.81,$$

or 36.81 feet.

b. **Figure 3.7** shows that the defensive player is 6 feet from the kicker's point of impact. To block the punt, he must touch the football along its parabolic path. This means that we must find the height of the ball 6 feet from the kicker. Replace x with 6 in the given function, $f(x) = -0.01x^2 + 1.18x + 2$.

$$f(6) = -0.01(6)^2 + 1.18(6) + 2 = -0.36 + 7.08 + 2 = 8.72$$

The defensive player must reach 8.72 feet above the ground to block the punt.

c. Assuming that the ball is not blocked by the defensive player, we are interested in how far down the field it will go before hitting the ground. We are looking for the ball's horizontal distance, x, when its height above the ground, $f(x)$, is 0 feet. To find this x-intercept, replace $f(x)$ with 0 in $f(x) = -0.01x^2 + 1.18x + 2$. We obtain $0 = -0.01x^2 + 1.18x + 2$, or $-0.01x^2 + 1.18x + 2 = 0$. The equation cannot be solved by factoring. We will use the quadratic formula to solve it.

$$-0.01x^2 + 1.18x + 2 = 0$$

$$a = -0.01 \quad b = 1.18 \quad c = 2$$

The equation for determining the ball's maximum horizontal distance

Use a calculator to evaluate the radicand.

$$x = \frac{-b \pm \sqrt{b^2 - 4ac}}{2a} = \frac{-1.18 \pm \sqrt{(1.18)^2 - 4(-0.01)(2)}}{2(-0.01)} = \frac{-1.18 \pm \sqrt{1.4724}}{-0.02}$$

$$x = \frac{-1.18 + \sqrt{1.4724}}{-0.02} \quad \text{or} \quad x = \frac{-1.18 - \sqrt{1.4724}}{-0.02}$$

$$x \approx -1.7 \qquad\qquad\qquad x \approx 119.7$$

Use a calculator and round to the nearest tenth.

Reject this value. We are interested in the football's height corresponding to horizontal distances from its point of impact onward, or $x \geq 0$.

If the football is not blocked by the defensive player, it will go approximately 119.7 feet down the field before hitting the ground.

d. In terms of graphing the model for the football's parabolic path, $f(x) = -0.01x^2 + 1.18x + 2$, we have already determined the vertex and the approximate x-intercept.

vertex: $(59, 36.81)$

The ball's maximum height, 36.81 feet, occurs at a horizontal distance of 59 feet.

x-intercept: 119.7

The ball's maximum horizontal distance is approximately 119.7 feet.

Figure 3.7 indicates that the y-intercept is 2, meaning that the ball is kicked from a height of 2 feet. Let's verify this value by replacing x with 0 in $f(x) = -0.01x^2 + 1.18x + 2$.

$$f(0) = -0.01 \cdot 0^2 + 1.18 \cdot 0 + 2 = 0 + 0 + 2 = 2$$

Using the vertex, $(59, 36.81)$, the x-intercept, 119.7, and the y-intercept, 2, the graph of the equation that models the football's parabolic path is shown in **Figure 3.8**. The graph is shown only for $x \geq 0$, indicating horizontal distances that begin at the football's impact with the kicker's foot and end with the ball hitting the ground.

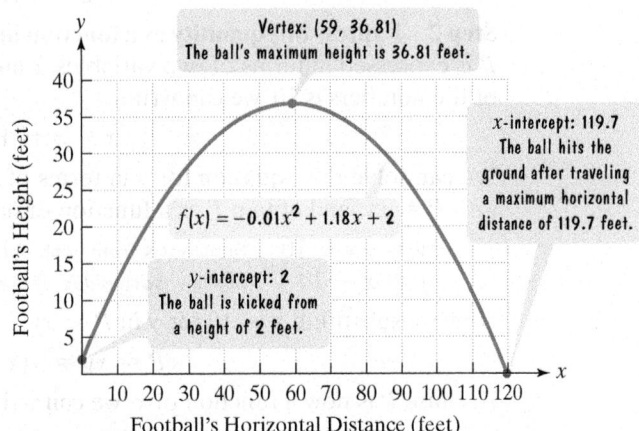

FIGURE 3.8 The parabolic path of a punted football

✓ Check Point **5** An archer's arrow follows a parabolic path. The height of the arrow, $f(x)$, in feet, can be modeled by

$$f(x) = -0.005x^2 + 2x + 5,$$

where x is the arrow's horizontal distance, in feet.

a. What is the maximum height of the arrow and how far from its release does this occur?

b. Find the horizontal distance the arrow travels before it hits the ground. Round to the nearest foot.

c. Graph the function that models the arrow's parabolic path.

Quadratic functions can also be used to model verbal conditions. Once we have obtained a quadratic function, we can then use the x-coordinate of the vertex to determine its maximum or minimum value. Here is a step-by-step strategy for solving these kinds of problems:

Strategy for Solving Problems Involving Maximizing or Minimizing Quadratic Functions

1. Read the problem carefully and decide which quantity is to be maximized or minimized.

2. Use the conditions of the problem to express the quantity as a function in one variable.

3. Rewrite the function in the form $f(x) = ax^2 + bx + c$.

4. Calculate $-\dfrac{b}{2a}$. If $a > 0$, f has a minimum at $x = -\dfrac{b}{2a}$. This minimum value is $f\left(-\dfrac{b}{2a}\right)$. If $a < 0$, f has a maximum at $x = -\dfrac{b}{2a}$. This maximum value is $f\left(-\dfrac{b}{2a}\right)$.

5. Answer the question posed in the problem.

EXAMPLE 6 Minimizing a Product

Among all pairs of numbers whose difference is 10, find a pair whose product is as small as possible. What is the minimum product?

SOLUTION

Step 1 Decide what must be maximized or minimized. We must minimize the product of two numbers. Calling the numbers x and y, and calling the product P, we must minimize

$$P = xy.$$

Step 2 Express this quantity as a function in one variable. In the formula $P = xy$, P is expressed in terms of two variables, x and y. However, because the difference of the numbers is 10, we can write

$$x - y = 10.$$

We can solve this equation for y in terms of x (or vice versa), substitute the result into $P = xy$, and obtain P as a function of one variable.

$-y = -x + 10$ Subtract x from both sides of x − y = 10.

$y = x - 10$ Multiply both sides of the equation by −1 and solve for y.

Now we substitute $x - 10$ for y in $P = xy$.

$$P = xy = x(x - 10)$$

Because P is now a function of x, we can write

$$P(x) = x(x - 10)$$

Step 3 Write the function in the form $f(x) = ax^2 + bx + c$**.** We apply the distributive property to obtain

$$P(x) = x(x - 10) = x^2 - 10x.$$

$$a = 1 \qquad b = -10$$

Step 4 Calculate $-\dfrac{b}{2a}$**. If** $a > 0$**, the function has a minimum at this value.** The voice balloons show that $a = 1$ and $b = -10$.

$$x = -\frac{b}{2a} = -\frac{-10}{2(1)} = -(-5) = 5$$

This means that the product, P, of two numbers whose difference is 10 is a minimum when one of the numbers, x, is 5.

Step 5 Answer the question posed by the problem. The problem asks for the two numbers and the minimum product. We found that one of the numbers, x, is 5. Now we must find the second number, y.

$$y = x - 10 = 5 - 10 = -5$$

The number pair whose difference is 10 and whose product is as small as possible is 5, −5. The minimum product is 5(−5), or −25. •••

⊘ **Check Point 6** Among all pairs of numbers whose difference is 8, find a pair whose product is as small as possible. What is the minimum product?

TECHNOLOGY

Numeric Connections

The ⎢TABLE⎥ feature of a graphing utility can be used to verify our work in Example 6.

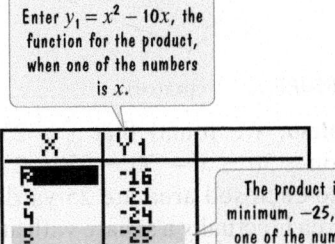

Enter $y_1 = x^2 - 10x$, the function for the product, when one of the numbers is x.

X	Y₁
2	-16
3	-21
4	-24
5	-25
6	-24
7	-21
8	-16

X=2

The product is a minimum, −25, when one of the numbers is 5.

EXAMPLE 7 Maximizing Area

You have 100 yards of fencing to enclose a rectangular region. Find the dimensions of the rectangle that maximize the enclosed area. What is the maximum area?

SOLUTION

Step 1 Decide what must be maximized or minimized. We must maximize area. What we do not know are the rectangle's dimensions, x and y.

Step 2 Express this quantity as a function in one variable. Because we must maximize the area of a rectangle, we have $A = xy$. We need to transform this into a function in which A is represented by one variable. Because you have 100 yards of fencing, the perimeter of the rectangle is 100 yards. This means that

$$2x + 2y = 100.$$

We can solve this equation for y in terms of x, substitute the result into $A = xy$, and obtain A as a function in one variable. We begin by solving for y.

$2y = 100 - 2x$ Subtract 2x from both sides of 2x + 2y = 100.

$y = \dfrac{100 - 2x}{2}$ Divide both sides by 2.

$y = 50 - x$ Divide each term in the numerator by 2.

Now we substitute $50 - x$ for y in $A = xy$.

$$A = xy = x(50 - x)$$

The rectangle and its dimensions are illustrated in **Figure 3.9**. Because A is now a function of x, we can write

$$A(x) = x(50 - x).$$

This function models the area, $A(x)$, of any rectangle whose perimeter is 100 yards in terms of one of its dimensions, x.

FIGURE 3.9 What value of x will maximize the rectangle's area?

Step 3 **Write the function in the form** $f(x) = ax^2 + bx + c$. We apply the distributive property to obtain

$$A(x) = x(50 - x) = 50x - x^2 = -x^2 + 50x.$$

$$\boxed{a = -1} \quad \boxed{b = 50}$$

TECHNOLOGY

Graphic Connections

The graph of the area function

$$A(x) = x(50 - x)$$

was obtained with a graphing utility using a $[0, 50, 2]$ by $[0, 700, 25]$ viewing rectangle. The maximum function feature verifies that a maximum area of 625 square yards occurs when one of the dimensions is 25 yards.

Step 4 **Calculate** $-\dfrac{b}{2a}$. **If** $a < 0$, **the function has a maximum at this value.** The voice balloons show that $a = -1$ and $b = 50$.

$$x = -\frac{b}{2a} = -\frac{50}{2(-1)} = 25$$

This means that the area, $A(x)$, of a rectangle with perimeter 100 yards is a maximum when one of the rectangle's dimensions, x, is 25 yards.

FIGURE 3.9 (repeated)

Step 5 **Answer the question posed by the problem.** We found that $x = 25$. **Figure 3.9** shows that the rectangle's other dimension is $50 - x = 50 - 25 = 25$. The dimensions of the rectangle that maximize the enclosed area are 25 yards by 25 yards. The rectangle that gives the maximum area is actually a square with an area of 25 yards · 25 yards, or 625 square yards. $\bullet\bullet\bullet$

⊘ Check Point **7** You have 120 feet of fencing to enclose a rectangular region. Find the dimensions of the rectangle that maximize the enclosed area. What is the maximum area?

Blitzer Bonus ‖ Addressing Stress Parabolically

FIGURE 3.10
Source: Herbert Benson, *Your Maximum Mind*, Random House, 1987.

Stress levels can help or hinder performance. The parabola in **Figure 3.10** serves as a model that shows people under both low stress and high stress perform worse than their moderate-stress counterparts.

CONCEPT AND VOCABULARY CHECK

Fill in each blank so that the resulting statement is true.

1. The quadratic function $f(x) = a(x - h)^2 + k, a \neq 0$, is in _____ form. The graph of f is called a/an _____ whose vertex is the point _____. The graph opens upward if a _____ and opens downward if a _____.

2. Consider the quadratic function $f(x) = ax^2 + bx + c, a \neq 0$. If $a > 0$, then f has a minimum that occurs at $x =$ _____. This minimum value is _____. If $a < 0$, then f has a maximum that occurs at $x =$ _____. This maximum value is _____.

3. True or false: The graph of $f(x) = (x - 2)^2 + 1$ opens upward. _____

4. True or false: The graph of $f(x) = (x + 5)^2 + 3$ has its vertex at $(5, 3)$. _____

5. True or false: The y-coordinate of the vertex of $f(x) = 4x^2 - 16x + 300$ is $f(2)$. _____

6. The difference between two numbers is 8. If one number is represented by x, the other number can be expressed as _____. The product of the numbers, $P(x)$, expressed in the form $P(x) = ax^2 + bx + c$, is $P(x) =$ _____.

7. The perimeter of a rectangle is 80 feet. If the length of the rectangle is represented by x, its width can be expressed as _____. The area of the rectangle, $A(x)$, expressed in the form $A(x) = ax^2 + bx + c$, is $A(x) =$ _____.

EXERCISE SET 3.1

Practice Exercises

In Exercises 1–4, the graph of a quadratic function is given. Write the function's equation, selecting from the following options.

$$f(x) = (x + 1)^2 - 1 \qquad g(x) = (x + 1)^2 + 1$$
$$h(x) = (x - 1)^2 + 1 \qquad j(x) = (x - 1)^2 - 1$$

1.

2.

3.

4.

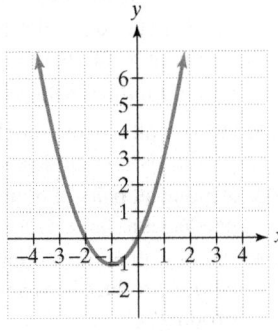

In Exercises 5–8, the graph of a quadratic function is given. Write the function's equation, selecting from the following options.

$$f(x) = x^2 + 2x + 1 \qquad g(x) = x^2 - 2x + 1$$
$$h(x) = x^2 - 1 \qquad j(x) = -x^2 - 1$$

5.

6.

7.

8.

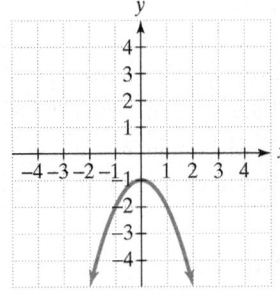

In Exercises 9–16, find the coordinates of the vertex for the parabola defined by the given quadratic function.

9. $f(x) = 2(x - 3)^2 + 1$
10. $f(x) = -3(x - 2)^2 + 12$
11. $f(x) = -2(x + 1)^2 + 5$
12. $f(x) = -2(x + 4)^2 - 8$
13. $f(x) = 2x^2 - 8x + 3$
14. $f(x) = 3x^2 - 12x + 1$
15. $f(x) = -x^2 - 2x + 8$
16. $f(x) = -2x^2 + 8x - 1$

In Exercises 17–38, use the vertex and intercepts to sketch the graph of each quadratic function. Give the equation of the parabola's axis of symmetry. Use the graph to determine the function's domain and range.

17. $f(x) = (x - 4)^2 - 1$
18. $f(x) = (x - 1)^2 - 2$
19. $f(x) = (x - 1)^2 + 2$
20. $f(x) = (x - 3)^2 + 2$
21. $y - 1 = (x - 3)^2$
22. $y - 3 = (x - 1)^2$
23. $f(x) = 2(x + 2)^2 - 1$
24. $f(x) = \frac{5}{4} - \left(x - \frac{1}{2}\right)^2$
25. $f(x) = 4 - (x - 1)^2$
26. $f(x) = 1 - (x - 3)^2$
27. $f(x) = x^2 - 2x - 3$
28. $f(x) = x^2 - 2x - 15$
29. $f(x) = x^2 + 3x - 10$
30. $f(x) = 2x^2 - 7x - 4$
31. $f(x) = 2x - x^2 + 3$
32. $f(x) = 5 - 4x - x^2$
33. $f(x) = x^2 + 6x + 3$
34. $f(x) = x^2 + 4x - 1$
35. $f(x) = 2x^2 + 4x - 3$
36. $f(x) = 3x^2 - 2x - 4$
37. $f(x) = 2x - x^2 - 2$
38. $f(x) = 6 - 4x + x^2$

In Exercises 39–44, an equation of a quadratic function is given.
 a. Determine, without graphing, whether the function has a minimum value or a maximum value.
 b. Find the minimum or maximum value and determine where it occurs.
 c. Identify the function's domain and its range.

39. $f(x) = 3x^2 - 12x - 1$
40. $f(x) = 2x^2 - 8x - 3$
41. $f(x) = -4x^2 + 8x - 3$
42. $f(x) = -2x^2 - 12x + 3$
43. $f(x) = 5x^2 - 5x$
44. $f(x) = 6x^2 - 6x$

Practice Plus

In Exercises 45–48, give the domain and the range of each quadratic function whose graph is described.

45. The vertex is $(-1, -2)$ and the parabola opens up.

46. The vertex is $(-3, -4)$ and the parabola opens down.

47. Maximum $= -6$ at $x = 10$

48. Minimum $= 18$ at $x = -6$

In Exercises 49–52, write an equation in standard form of the parabola that has the same shape as the graph of $f(x) = 2x^2$, but with the given point as the vertex.

49. $(5, 3)$ **50.** $(7, 4)$

51. $(-10, -5)$ **52.** $(-8, -6)$

In Exercises 53–56, write an equation in standard form of the parabola that has the same shape as the graph of $f(x) = 3x^2$ or $g(x) = -3x^2$, but with the given maximum or minimum.

53. Maximum $= 4$ at $x = -2$ **54.** Maximum $= -7$ at $x = 5$

55. Minimum $= 0$ at $x = 11$ **56.** Minimum $= 0$ at $x = 9$

Application Exercises

An athlete whose event is the shot put releases the shot with the same initial velocity but at different angles. The figure shows the parabolic paths for shots released at angles of 35° and 65°. Exercises 57–58 are based on the functions that model the parabolic paths.

57. When the shot whose path is shown by the blue graph is released at an angle of 35°, its height, $f(x)$, in feet, can be modeled by

$$f(x) = -0.01x^2 + 0.7x + 6.1,$$

where x is the shot's horizontal distance, in feet, from its point of release. Use this model to solve parts (a) through (c) and verify your answers using the blue graph.

 a. What is the maximum height of the shot and how far from its point of release does this occur?

 b. What is the shot's maximum horizontal distance, to the nearest tenth of a foot, or the distance of the throw?

 c. From what height was the shot released?

58. When the shot whose path is shown by the red graph in the previous column is released at an angle of 65°, its height, $g(x)$, in feet, can be modeled by

$$g(x) = -0.04x^2 + 2.1x + 6.1,$$

where x is the shot's horizontal distance, in feet, from its point of release. Use this model to solve parts (a) through (c) and verify your answers using the red graph.

 a. What is the maximum height, to the nearest tenth of a foot, of the shot and how far from its point of release does this occur?

 b. What is the shot's maximum horizontal distance, to the nearest tenth of a foot, or the distance of the throw?

 c. From what height was the shot released?

59. A ball is thrown upward and outward from a height of 6 feet. The height of the ball, $f(x)$, in feet, can be modeled by

$$f(x) = -0.8x^2 + 2.4x + 6,$$

where x is the ball's horizontal distance, in feet, from where it was thrown.

 a. What is the maximum height of the ball and how far from where it was thrown does this occur?

 b. How far does the ball travel horizontally before hitting the ground? Round to the nearest tenth of a foot.

 c. Graph the function that models the ball's parabolic path.

60. A ball is thrown upward and outward from a height of 6 feet. The height of the ball, $f(x)$, in feet, can be modeled by

$$f(x) = -0.8x^2 + 3.2x + 6,$$

where x is the ball's horizontal distance, in feet, from where it was thrown.

 a. What is the maximum height of the ball and how far from where it was thrown does this occur?

 b. How far does the ball travel horizontally before hitting the ground? Round to the nearest tenth of a foot.

 c. Graph the function that models the ball's parabolic path.

61. Among all pairs of numbers whose sum is 16, find a pair whose product is as large as possible. What is the maximum product?

62. Among all pairs of numbers whose sum is 20, find a pair whose product is as large as possible. What is the maximum product?

63. Among all pairs of numbers whose difference is 16, find a pair whose product is as small as possible. What is the minimum product?

64. Among all pairs of numbers whose difference is 24, find a pair whose product is as small as possible. What is the minimum product?

65. You have 600 feet of fencing to enclose a rectangular plot that borders on a river. If you do not fence the side along the river, find the length and width of the plot that will maximize the area. What is the largest area that can be enclosed?

66. You have 200 feet of fencing to enclose a rectangular plot that borders on a river. If you do not fence the side along the river, find the length and width of the plot that will maximize the area. What is the largest area that can be enclosed?

$200 - 2x$

67. You have 50 yards of fencing to enclose a rectangular region. Find the dimensions of the rectangle that maximize the enclosed area. What is the maximum area?

68. You have 80 yards of fencing to enclose a rectangular region. Find the dimensions of the rectangle that maximize the enclosed area. What is the maximum area?

69. A rectangular playground is to be fenced off and divided in two by another fence parallel to one side of the playground. Six hundred feet of fencing is used. Find the dimensions of the playground that maximize the total enclosed area. What is the maximum area?

70. A rectangular playground is to be fenced off and divided in two by another fence parallel to one side of the playground. Four hundred feet of fencing is used. Find the dimensions of the playground that maximize the total enclosed area. What is the maximum area?

71. A rain gutter is made from sheets of aluminum that are 20 inches wide by turning up the edges to form right angles. Determine the depth of the gutter that will maximize its cross-sectional area and allow the greatest amount of water to flow. What is the maximum cross-sectional area?

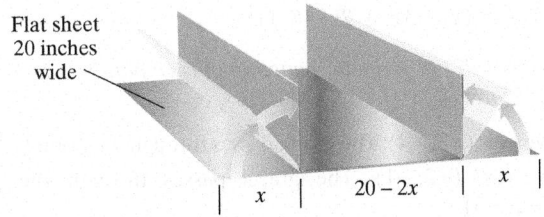

Flat sheet 20 inches wide

x $20 - 2x$ x

72. A rain gutter is made from sheets of aluminum that are 12 inches wide by turning up the edges to form right angles. Determine the depth of the gutter that will maximize its cross-sectional area and allow the greatest amount of water to flow. What is the maximum cross-sectional area?

73. Hunky Beef, a local sandwich store, has a fixed weekly cost of $525.00, and variable costs for making a roast beef sandwich are $0.55.

 a. Let x represent the number of roast beef sandwiches made and sold each week. Write the weekly cost function, C, for Hunky Beef. (*Hint:* The cost function is the sum of fixed and variable costs.)

 b. The function $R(x) = -0.001x^2 + 3x$ describes the money, in dollars, that Hunky Beef takes in each week from the sale of x roast beef sandwiches. Use this revenue function and the cost function from part (a) to write the store's weekly profit function, P. (*Hint:* The profit function is the difference between the revenue and cost functions.)

 c. Use the store's profit function to determine the number of roast beef sandwiches it should make and sell each week to maximize profit. What is the maximum weekly profit?

Writing in Mathematics

74. What is a quadratic function?

75. What is a parabola? Describe its shape.

76. Explain how to decide whether a parabola opens upward or downward.

77. Describe how to find a parabola's vertex if its equation is expressed in standard form. Give an example.

78. Describe how to find a parabola's vertex if its equation is in the form $f(x) = ax^2 + bx + c$. Use $f(x) = x^2 - 6x + 8$ as an example.

79. A parabola that opens upward has its vertex at (1, 2). Describe as much as you can about the parabola based on this information. Include in your discussion the number of x-intercepts (if any) for the parabola.

Technology Exercises

80. Use a graphing utility to verify any five of your hand-drawn graphs in Exercises 17–38.

81. a. Use a graphing utility to graph $y = 2x^2 - 82x + 720$ in a standard viewing rectangle. What do you observe?

 b. Find the coordinates of the vertex for the given quadratic function.

 c. The answer to part (b) is $(20.5, -120.5)$. Because the leading coefficient, 2, of the given function is positive, the vertex is a minimum point on the graph. Use this fact to help find a viewing rectangle that will give a relatively complete picture of the parabola. With an axis of symmetry at $x = 20.5$, the setting for x should extend past this, so try Xmin = 0 and Xmax = 30. The setting for y should include (and probably go below) the y-coordinate of the graph's minimum y-value, so try Ymin = −130. Experiment with Ymax until your utility shows the parabola's major features.

 d. In general, explain how knowing the coordinates of a parabola's vertex can help determine a reasonable viewing rectangle on a graphing utility for obtaining a complete picture of the parabola.

In Exercises 82–85, find the vertex for each parabola. Then determine a reasonable viewing rectangle on your graphing utility and use it to graph the quadratic function.

82. $y = -0.25x^2 + 40x$

83. $y = -4x^2 + 20x + 160$

84. $y = 5x^2 + 40x + 600$

85. $y = 0.01x^2 + 0.6x + 100$

86. The bar graph shows the ratings of *American Idol* from season 1 (2002) through season 9 (2010).

American Idol: Each Season's Champion and Average Number of Viewers

Source: Nielsen

a. Let x represent *American Idol's* season number and let y represent the average number of viewers, in millions. Use a graphing utility to draw a scatter plot of the data. Explain why a quadratic function is appropriate for modeling these data.

b. Use the quadratic regression feature to find the quadratic function that best fits the data. Round all numbers to two decimal places.

c. Use the model in part (b) to determine the season in which *American Idol* had the greatest number of viewers. Round to the nearest whole number. According to the model, how many millions of viewers were there in that season? Round to one decimal place.

d. How do the results obtained from the model in part (c) compare with the data displayed by the graph?

e. Use a graphing utility to draw a scatter plot of the data and graph the quadratic function of best fit on the scatter plot. Can you see why projections based on the graph have the producers of *American Idol* looking for a shake-up?

Critical Thinking Exercises

Make Sense? *In Exercises 87–90, determine whether each statement makes sense or does not make sense, and explain your reasoning.*

87. I must have made an error when graphing this parabola because its axis of symmetry is the *y*-axis.

88. I like to think of a parabola's vertex as the point where it intersects its axis of symmetry.

89. I threw a baseball vertically upward and its path was a parabola.

90. **Figure 3.7** on page 348 shows that a linear function provides a better description of the football's path than a quadratic function.

In Exercises 91–94, determine whether each statement is true or false. If the statement is false, make the necessary change(s) to produce a true statement.

91. No quadratic functions have a range of $(-\infty, \infty)$.

92. The vertex of the parabola described by $f(x) = 2(x - 5)^2 - 1$ is at $(5, 1)$.

93. The graph of $f(x) = -2(x + 4)^2 - 8$ has one *y*-intercept and two *x*-intercepts.

94. The maximum value of y for the quadratic function $f(x) = -x^2 + x + 1$ is 1.

In Exercises 95–96, find the axis of symmetry for each parabola whose equation is given. Use the axis of symmetry to find a second point on the parabola whose y-coordinate is the same as the given point.

95. $f(x) = 3(x + 2)^2 - 5$; $(-1, -2)$

96. $f(x) = (x - 3)^2 + 2$; $(6, 11)$

In Exercises 97–98, write the equation of each parabola in standard form.

97. Vertex: $(-3, -4)$; The graph passes through the point $(1, 4)$.

98. Vertex: $(-3, -1)$; The graph passes through the point $(-2, -3)$.

99. A rancher has 1000 feet of fencing to construct six corrals, as shown in the figure. Find the dimensions that maximize the enclosed area. What is the maximum area?

100. The annual yield per lemon tree is fairly constant at 320 pounds when the number of trees per acre is 50 or fewer. For each additional tree over 50, the annual yield per tree for all trees on the acre decreases by 4 pounds due to overcrowding. Find the number of trees that should be planted on an acre to produce the maximum yield. How many pounds is the maximum yield?

Group Exercise

101. Each group member should consult an almanac, newspaper, magazine, or the Internet to find data that initially increase and then decrease, or vice versa, and therefore can be modeled by a quadratic function. Group members should select the two sets of data that are most interesting and relevant. For each data set selected,

 a. Use the quadratic regression feature of a graphing utility to find the quadratic function that best fits the data.

 b. Use the equation of the quadratic function to make a prediction from the data. What circumstances might affect the accuracy of your prediction?

 c. Use the equation of the quadratic function to write and solve a problem involving maximizing or minimizing the function.

Preview Exercises

Exercises 102–104 will help you prepare for the material covered in the next section.

102. Factor: $x^3 + 3x^2 - x - 3$.

103. If $f(x) = x^3 - 2x - 5$, find $f(2)$ and $f(3)$. Then explain why the continuous graph of f must cross the x-axis between 2 and 3.

104. Determine whether $f(x) = x^4 - 2x^2 + 1$ is even, odd, or neither. Describe the symmetry, if any, for the graph of f.

SECTION 3.2 Polynomial Functions and Their Graphs

Objectives

1. Identify polynomial functions.
2. Recognize characteristics of graphs of polynomial functions.
3. Determine end behavior.
4. Use factoring to find zeros of polynomial functions.
5. Identify zeros and their multiplicities.
6. Use the Intermediate Value Theorem.
7. Understand the relationship between degree and turning points.
8. Graph polynomial functions.

In 1980, U.S. doctors diagnosed 41 cases of a rare form of cancer, Kaposi's sarcoma, that involved skin lesions, pneumonia, and severe immunological deficiencies. All cases involved gay men ranging in age from 26 to 51. By the end of 2008, approximately 1.1 million Americans, straight and gay, male and female, old and young, were infected with the HIV virus.

Modeling AIDS-related data and making predictions about the epidemic's havoc is serious business. **Figure 3.11** on the next page shows the number of AIDS cases diagnosed in the United States from 1983 through 2008.

Basketball player Magic Johnson (1959–) tested positive for HIV in 1991.

AIDS Cases Diagnosed in the United States, 1983–2008

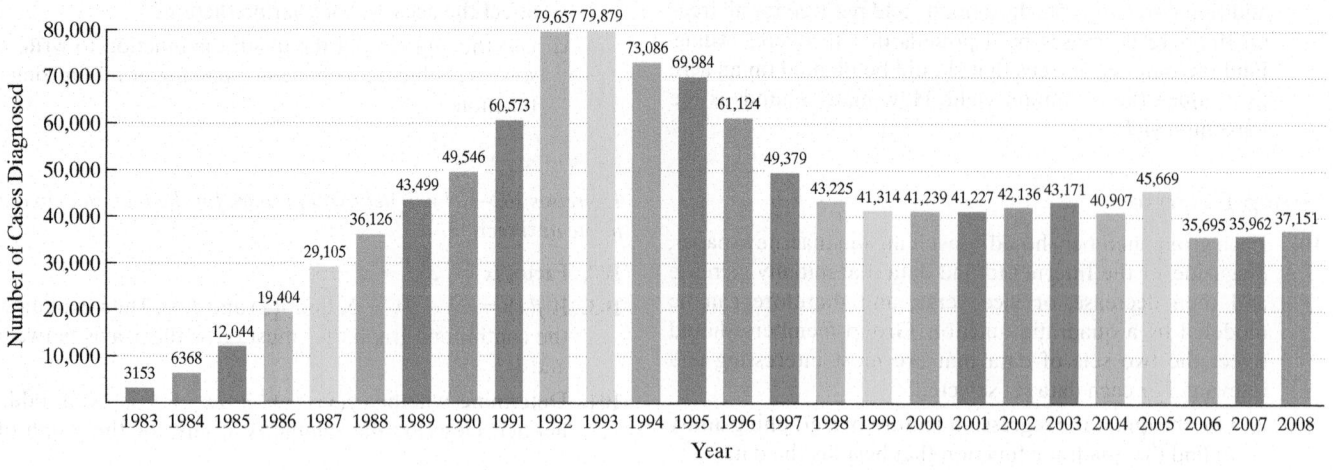

FIGURE 3.11
Source: Department of Health and Human Services

① Identify polynomial functions.

Changing circumstances and unforeseen events can result in models for AIDS-related data that are not particularly useful over long periods of time. For example, the function

$$f(x) = -49x^3 + 806x^2 + 3776x + 2503$$

models the number of AIDS cases diagnosed in the United States x years after 1983. The model was obtained using a portion of the data shown in **Figure 3.11**, namely cases diagnosed from 1983 through 1991, inclusive. **Figure 3.12** shows the graph of f from 1983 through 1991. This function is an example of a *polynomial function of degree 3.*

$f(x) = -49x^3 + 806x^2 + 3776x + 2503$

Years after 1983
$[0, 8, 1]$ by $[0, 60{,}000, 5000]$

FIGURE 3.12 The graph of a function modeling the number of AIDS diagnoses from 1983 through 1991

Definition of a Polynomial Function

Let n be a nonnegative integer and let $a_n, a_{n-1}, \ldots, a_2, a_1, a_0$ be real numbers, with $a_n \neq 0$. The function defined by

$$f(x) = a_n x^n + a_{n-1} x^{n-1} + \cdots + a_2 x^2 + a_1 x + a_0$$

is called a **polynomial function of degree n**. The number a_n, the coefficient of the variable to the highest power, is called the **leading coefficient**.

Polynomial Functions	**Not Polynomial Functions**

Polynomial Functions

$$f(x) = -3x^5 + \sqrt{2}x^2 + 5$$

Polynomial function of degree 5

$$g(x) = -3x^4(x - 2)(x + 3)$$
$$= -3x^4(x^2 + x - 6)$$
$$= -3x^6 - 3x^5 + 18x^4$$

Polynomial function of degree 6

Not Polynomial Functions

$$F(x) = -3\sqrt{x} + \sqrt{2}x^2 + 5$$
$$= -3x^{\frac{1}{2}} + \sqrt{2}x^2 + 5$$

The exponent on the variable is not an integer.

$$G(x) = -\frac{3}{x^2} + \sqrt{2}x^2 + 5$$
$$= -3x^{-2} + \sqrt{2}x^2 + 5$$

The exponent on the variable is not a nonnegative integer.

A constant function $f(x) = c$, where $c \neq 0$, is a polynomial function of degree 0. A linear function $f(x) = mx + b$, where $m \neq 0$, is a polynomial function of degree 1. A quadratic function $f(x) = ax^2 + bx + c$, where $a \neq 0$, is a polynomial function of degree 2. In this section, we focus on polynomial functions of degree 3 or higher.

② Recognize characteristics of graphs of polynomial functions.

Smooth, Continuous Graphs

Polynomial functions of degree 2 or higher have graphs that are *smooth* and *continuous*. By **smooth**, we mean that the graphs contain only rounded curves with no sharp corners. By **continuous**, we mean that the graphs have no breaks and can be drawn without lifting your pencil from the rectangular coordinate system. These ideas are illustrated in **Figure 3.13**.

Graphs of Polynomial Functions

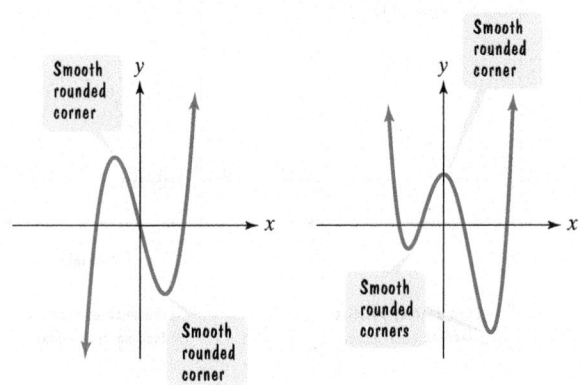

Not Graphs of Polynomial Functions

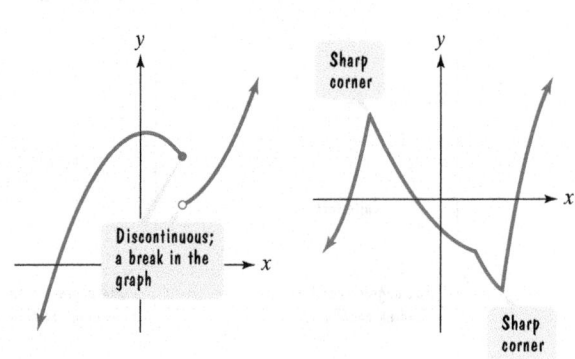

FIGURE 3.13 Recognizing graphs of polynomial functions

③ Determine end behavior.

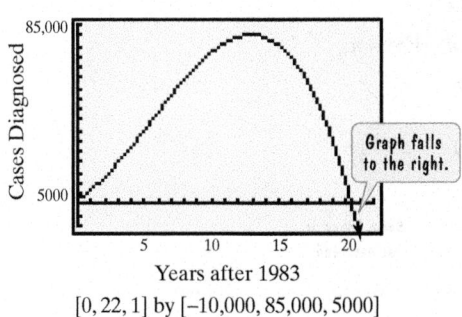

[0, 22, 1] by [−10,000, 85,000, 5000]

FIGURE 3.14 By extending the viewing rectangle, we see that *y* is eventually negative and the function no longer models the number of AIDS cases.

End Behavior of Polynomial Functions

Figure 3.14 shows the graph of the function

$$f(x) = -49x^3 + 806x^2 + 3776x + 2503,$$

which models the number of U.S. AIDS diagnoses from 1983 through 1991. Look what happens to the graph when we extend the year up through 2005. By year 21 (2004), the values of *y* are negative and the function no longer models AIDS diagnoses. We've added an arrow to the graph at the far right to emphasize that it continues to decrease without bound. It is this far-right *end behavior* of the graph that makes it inappropriate for modeling AIDS cases into the future.

The behavior of the graph of a function to the far left or the far right is called its **end behavior**. Although the graph of a polynomial function may have intervals where it increases or decreases, the graph will eventually rise or fall without bound as it moves far to the left or far to the right.

How can you determine whether the graph of a polynomial function goes up or down at each end? The end behavior of a polynomial function

$$f(x) = a_n x^n + a_{n-1} x^{n-1} + \cdots + a_1 x + a_0$$

depends upon the leading term $a_n x^n$, because when $|x|$ is large, the other terms are relatively insignificant in size. In particular, the sign of the leading coefficient, a_n, and the degree, n, of the polynomial function reveal its end behavior. In terms of end behavior, only the term of highest degree counts, as summarized by the **Leading Coefficient Test**.

The Leading Coefficient Test

As x increases or decreases without bound, the graph of the polynomial function

$$f(x) = a_n x^n + a_{n-1} x^{n-1} + a_{n-2} x^{n-2} + \cdots + a_1 x + a_0 \quad (a_n \neq 0)$$

eventually rises or falls. In particular,

1. For n odd:

If the leading coefficient is positive, the graph falls to the left and rises to the right. (\swarrow, \nearrow)

$a_n > 0$

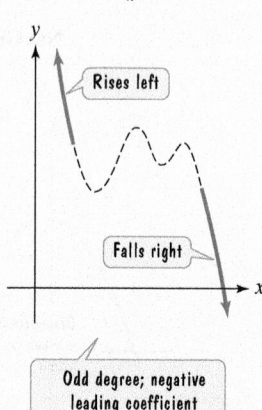

Odd degree; positive leading coefficient

If the leading coefficient is negative, the graph rises to the left and falls to the right. (\nwarrow, \searrow)

$a_n < 0$

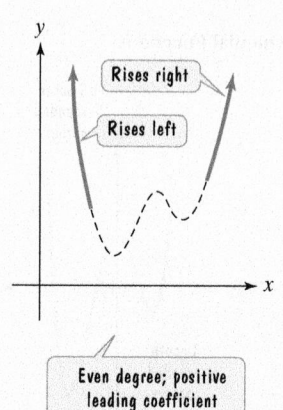

Odd degree; negative leading coefficient

2. For n even:

If the leading coefficient is positive, the graph rises to the left and rises to the right. (\nwarrow, \nearrow)

$a_n > 0$

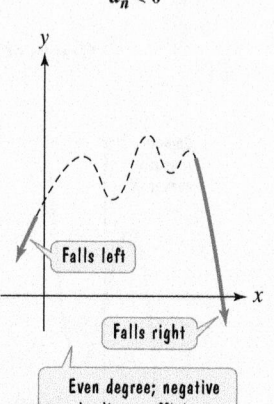

Even degree; positive leading coefficient

If the leading coefficient is negative, the graph falls to the left and falls to the right. (\swarrow, \searrow)

$a_n < 0$

Even degree; negative leading coefficient

DISCOVERY

Verify each of the four cases of the Leading Coefficient Test by using a graphing utility to graph $f(x) = x^3$, $f(x) = -x^3$, $f(x) = x^2$, and $f(x) = -x^2$.

GREAT QUESTION!

What's the bottom line on the Leading Coefficient Test?

Odd-degree polynomial functions have graphs with opposite behavior at each end. Even-degree polynomial functions have graphs with the same behavior at each end. Here's a table to help you remember the details:

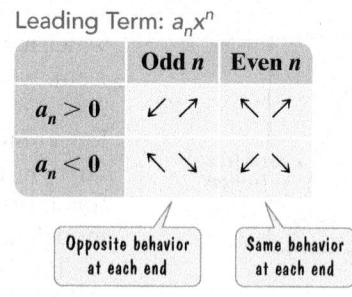

Leading Term: $a_n x^n$

	Odd n	Even n
$a_n > 0$	$\swarrow \nearrow$	$\nwarrow \nearrow$
$a_n < 0$	$\nwarrow \searrow$	$\swarrow \searrow$

Opposite behavior at each end / Same behavior at each end

EXAMPLE 1 Using the Leading Coefficient Test

Use the Leading Coefficient Test to determine the end behavior of the graph of

$$f(x) = x^3 + 3x^2 - x - 3.$$

SOLUTION

We begin by identifying the sign of the leading coefficient and the degree of the polynomial.

$$f(x) = x^3 + 3x^2 - x - 3$$

The leading coefficient, 1, is positive.

The degree of the polynomial, 3, is odd.

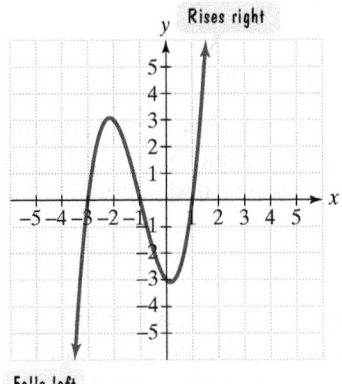

FIGURE 3.15 The graph of
$f(x) = x^3 + 3x^2 - x - 3$

The degree of the function f is 3, which is odd. Odd-degree polynomial functions have graphs with opposite behavior at each end. The leading coefficient, 1, is positive. Thus, the graph falls to the left and rises to the right (\swarrow, \nearrow). The graph of f is shown in **Figure 3.15**. •••

✓ Check Point **1** Use the Leading Coefficient Test to determine the end behavior of the graph of $f(x) = x^4 - 4x^2$.

EXAMPLE 2 Using the Leading Coefficient Test

Use the Leading Coefficient Test to determine the end behavior of the graph of

$$f(x) = -4x^3(x - 1)^2(x + 5).$$

SOLUTION

Although the equation for f is in factored form, it is not necessary to multiply to determine the degree of the function.

$$f(x) = -4x^3(x - 1)^2(x + 5)$$

Degree of this factor is 3. Degree of this factor is 2. Degree of this factor is 1.

When multiplying exponential expressions with the same base, we add the exponents. This means that the degree of f is $3 + 2 + 1$, or 6, which is even. Even-degree polynomial functions have graphs with the same behavior at each end. Without multiplying out, you can see that the leading coefficient is -4, which is negative. Thus, the graph of f falls to the left and falls to the right (\swarrow, \searrow). •••

✓ Check Point **2** Use the Leading Coefficient Test to determine the end behavior of the graph of $f(x) = 2x^3(x - 1)(x + 5)$.

EXAMPLE 3 Using the Leading Coefficient Test

Use end behavior to explain why

$$f(x) = -49x^3 + 806x^2 + 3776x + 2503$$

is only an appropriate model for AIDS diagnoses for a limited time period.

SOLUTION

We begin by identifying the sign of the leading coefficient and the degree of the polynomial.

$$f(x) = -49x^3 + 806x^2 + 3776x + 2503$$

The leading coefficient, -49, is negative. The degree of the polynomial, 3, is odd.

The degree of f is 3, which is odd. Odd-degree polynomial functions have graphs with opposite behavior at each end. The leading coefficient, -49, is negative. Thus, the graph rises to the left and falls to the right (\nwarrow, \searrow). The fact that the graph falls to the right indicates that at some point the number of AIDS diagnoses will be negative, an impossibility. If a function has a graph that decreases without bound over time, it will not be capable of modeling nonnegative phenomena over long time periods. Model breakdown will eventually occur. •••

✓ Check Point **3** The polynomial function

$$f(x) = -0.27x^3 + 9.2x^2 - 102.9x + 400$$

models the ratio of students to computers in U.S. public schools x years after 1980. Use end behavior to determine whether this function could be an appropriate model for computers in the classroom well into the twenty-first century. Explain your answer.

If you use a graphing utility to graph a polynomial function, it is important to select a viewing rectangle that accurately reveals the graph's end behavior. If the viewing rectangle, or window, is too small, it may not accurately show a complete graph with the appropriate end behavior.

EXAMPLE 4 Using the Leading Coefficient Test

The graph of $f(x) = -x^4 + 8x^3 + 4x^2 + 2$ was obtained with a graphing utility using a $[-8, 8, 1]$ by $[-10, 10, 1]$ viewing rectangle. The graph is shown in **Figure 3.16**. Is this a complete graph that shows the end behavior of the function?

SOLUTION

We begin by identifying the sign of the leading coefficient and the degree of the polynomial.

$$f(x) = -x^4 + 8x^3 + 4x^2 + 2$$

The leading coefficient, −1, is negative. The degree of the polynomial, 4, is even.

The degree of f is 4, which is even. Even-degree polynomial functions have graphs with the same behavior at each end. The leading coefficient, −1, is negative. Thus, the graph should fall to the left and fall to the right (↙, ↘). The graph in **Figure 3.16** is falling to the left, but it is not falling to the right. Therefore, the graph is not complete enough to show end behavior. A more complete graph of the function is shown in a larger viewing rectangle in **Figure 3.17**. ● ● ●

✓ Check Point **4** The graph of $f(x) = x^3 + 13x^2 + 10x - 4$ is shown in a standard viewing rectangle in **Figure 3.18**. Use the Leading Coefficient Test to determine whether this is a complete graph that shows the end behavior of the function. Explain your answer.

$[-8, 8, 1]$ by $[-10, 10, 1]$

FIGURE 3.16

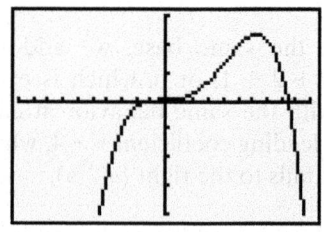

$[-10, 10, 1]$ by $[-1000, 750, 250]$

FIGURE 3.17

FIGURE 3.18

④ Use factoring to find zeros of polynomial functions.

Zeros of Polynomial Functions

If f is a polynomial function, then the values of x for which $f(x)$ is equal to 0 are called the **zeros** of f. These values of x are the **roots**, or **solutions**, of the polynomial equation $f(x) = 0$. Each real root of the polynomial equation appears as an x-intercept of the graph of the polynomial function.

EXAMPLE 5 Finding Zeros of a Polynomial Function

Find all zeros of $f(x) = x^3 + 3x^2 - x - 3$.

SOLUTION

By definition, the zeros are the values of x for which $f(x)$ is equal to 0. Thus, we set $f(x)$ equal to 0:

$$f(x) = x^3 + 3x^2 - x - 3 = 0.$$

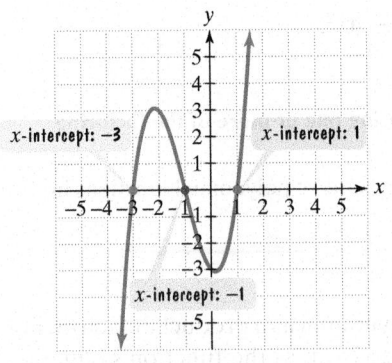

FIGURE 3.19

We solve the polynomial equation $x^3 + 3x^2 - x - 3 = 0$ for x as follows:

$$x^3 + 3x^2 - x - 3 = 0 \qquad \text{This is the equation needed to find the function's zeros.}$$

$$x^2(x + 3) - 1(x + 3) = 0 \qquad \text{Factor } x^2 \text{ from the first two terms and } -1 \text{ from the last two terms.}$$

$$(x + 3)(x^2 - 1) = 0 \qquad \text{A common factor of } x + 3 \text{ is factored from the expression.}$$

$$x + 3 = 0 \quad \text{or} \quad x^2 - 1 = 0 \qquad \text{Set each factor equal to 0.}$$

$$x = -3 \qquad\qquad x^2 = 1 \qquad \text{Solve for x.}$$

$$x = \pm 1 \qquad \text{Remember that if } x^2 = d, \text{ then } x = \pm\sqrt{d}.$$

The zeros of f are $-3, -1$, and 1. The graph of f in **Figure 3.19** shows that each zero is an x-intercept. The graph passes through the points $(-3, 0)$, $(-1, 0)$, and $(1, 0)$. ●●●

TECHNOLOGY

Graphic and Numeric Connections

A graphing utility can be used to verify that $-3, -1$, and 1 are the three real zeros of $f(x) = x^3 + 3x^2 - x - 3$.

Numeric Check

Display a table for the function.

Graphic Check

Display a graph for the function. The x-intercepts indicate that $-3, -1$, and 1 are the real zeros.

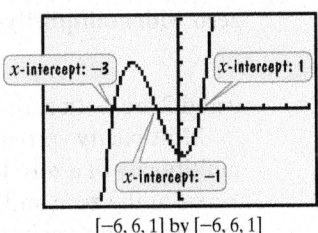

$[-6, 6, 1]$ by $[-6, 6, 1]$

The utility's $\boxed{\text{ZERO}}$ feature on the graph of f also verifies that $-3, -1$, and 1 are the function's real zeros.

✓ **Check Point 5** Find all zeros of $f(x) = x^3 + 2x^2 - 4x - 8$.

EXAMPLE 6 Finding Zeros of a Polynomial Function

Find all zeros of $f(x) = -x^4 + 4x^3 - 4x^2$.

SOLUTION

We find the zeros of f by setting $f(x)$ equal to 0 and solving the resulting equation.

$$-x^4 + 4x^3 - 4x^2 = 0 \qquad \text{We now have a polynomial equation.}$$

$$x^4 - 4x^3 + 4x^2 = 0 \qquad \text{Multiply both sides by } -1. \text{ This step is optional.}$$

$$x^2(x^2 - 4x + 4) = 0 \qquad \text{Factor out } x^2.$$

$$x^2(x - 2)^2 = 0 \qquad \text{Factor completely.}$$

$$x^2 = 0 \quad \text{or} \quad (x - 2)^2 = 0 \qquad \text{Set each factor equal to 0.}$$

$$x = 0 \qquad\qquad x = 2 \qquad \text{Solve for x.}$$

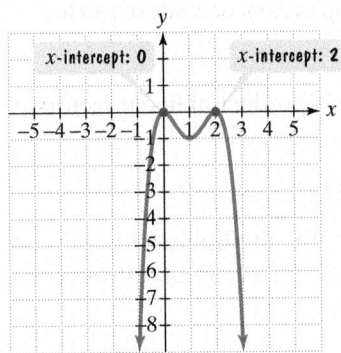

FIGURE 3.20 The zeros of $f(x) = -x^4 + 4x^3 - 4x^2$, namely 0 and 2, are the x-intercepts for the graph of f.

The zeros of $f(x) = -x^4 + 4x^3 - 4x^2$ are 0 and 2. The graph of f, shown in **Figure 3.20**, has x-intercepts at 0 and 2. The graph passes through the points $(0, 0)$ and $(2, 0)$. ●●●

⊘ Check Point **6** Find all zeros of $f(x) = x^4 - 4x^2$.

GREAT QUESTION!

Can zeros of polynomial functions always be found using one or more of the factoring techniques that were reviewed in Section P.6?

No. You'll be learning additional strategies for finding zeros of polynomial functions in the next two sections of this chapter.

⑤ Identify zeros and their multiplicities.

Multiplicities of Zeros

We can use the results of factoring to express a polynomial as a product of factors. For instance, in Example 6, we can use our factoring to express the function's equation as follows:

$$f(x) = -x^4 + 4x^3 - 4x^2 = -(x^4 - 4x^3 + 4x^2) = -x^2(x - 2)^2.$$

| The factor x occurs twice: $x^2 = x \cdot x.$ | The factor $(x - 2)$ occurs twice: $(x - 2)^2 = (x - 2)(x - 2).$ |

Notice that each factor occurs twice. In factoring the equation for the polynomial function f, if the same factor $x - r$ occurs k times, but not $k + 1$ times, we call r a **zero with multiplicity k**. For the polynomial function

$$f(x) = -x^2(x - 2)^2,$$

0 and 2 are both zeros with multiplicity 2.

Multiplicity provides another connection between zeros and graphs. The multiplicity of a zero tells us whether the graph of a polynomial function touches the x-axis at the zero and turns around or if the graph crosses the x-axis at the zero. For example, look again at the graph of $f(x) = -x^4 + 4x^3 - 4x^2$ in **Figure 3.20**. Each zero, 0 and 2, is a zero with multiplicity 2. The graph of f touches, but does not cross, the x-axis at each of these zeros of even multiplicity. By contrast, a graph crosses the x-axis at zeros of odd multiplicity.

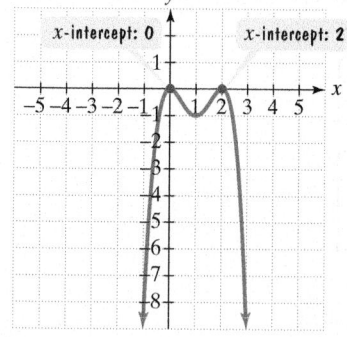

FIGURE 3.20 (repeated) The graph of $f(x) = -x^4 + 4x^3 - 4x^2$

Multiplicity and x-Intercepts

If r is a zero of **even multiplicity**, then the graph **touches** the x-axis **and turns around** at r. If r is a zero of **odd multiplicity**, then the graph **crosses** the x-axis at r. Regardless of whether the multiplicity of a zero is even or odd, graphs tend to flatten out near zeros with multiplicity greater than one.

GREAT QUESTION!

If r is a zero of even multiplicity, how come the graph of f doesn't just cross the x-axis at r?

Because r is a zero of even multiplicity, the sign of $f(x)$ does not change from one side of r to the other side of r. This means that the graph must turn around at r. On the other hand, if r is a zero of odd multiplicity, the sign of $f(x)$ changes from one side of r to the other side. That's why the graph crosses the x-axis at r.

If a polynomial function's equation is expressed as a product of linear factors, we can quickly identify zeros and their multiplicities.

EXAMPLE 7 Finding Zeros and Their Multiplicities

Find the zeros of $f(x) = \frac{1}{2}(x + 1)(2x - 3)^2$ and give the multiplicity of each zero. State whether the graph crosses the x-axis or touches the x-axis and turns around at each zero.

SOLUTION

We find the zeros of f by setting $f(x)$ equal to 0:

$$\tfrac{1}{2}(x + 1)(2x - 3)^2 = 0.$$

Set each variable factor equal to 0.

$$x + 1 = 0$$
$$x = -1$$

$$2x - 3 = 0$$
$$x = \frac{3}{2}$$

$$\frac{1}{2}(x + 1)^1(2x - 3)^2 = 0$$

This exponent is 1.
Thus, the multiplicity
of −1 is 1.

This exponent is 2.
Thus, the multiplicity
of $\frac{3}{2}$ is 2.

−1 is a zero of odd multiplicity.
Graph crosses x-axis.

$\frac{3}{2}$ is a zero of even multiplicity.
Graph touches x-axis, flattens,
and turns around.

$[-3, 3, 1]$ by $[-10, 10, 1]$

FIGURE 3.21 The graph of
$f(x) = \frac{1}{2}(x + 1)(2x - 3)^2$

The zeros of $f(x) = \frac{1}{2}(x + 1)(2x - 3)^2$ are −1, with multiplicity 1, and $\frac{3}{2}$, with multiplicity 2. Because the multiplicity of −1 is odd, the graph crosses the x-axis at this zero. Because the multiplicity of $\frac{3}{2}$ is even, the graph touches the x-axis and turns around at this zero. These relationships are illustrated by the graph of f in **Figure 3.21**. • • •

✓ Check Point **7** Find the zeros of $f(x) = -4\left(x + \frac{1}{2}\right)^2(x - 5)^3$ and give the multiplicity of each zero. State whether the graph crosses the x-axis or touches the x-axis and turns around at each zero.

⑥ Use the Intermediate Value Theorem.

The Intermediate Value Theorem

The *Intermediate Value Theorem* tells us of the existence of real zeros. The idea behind the theorem is illustrated in **Figure 3.22**. The figure shows that if $(a, f(a))$ lies below the x-axis and $(b, f(b))$ lies above the x-axis, the smooth, continuous graph of the polynomial function f must cross the x-axis at some value c between a and b. This value is a real zero for the function.

These observations are summarized in the **Intermediate Value Theorem**.

The Intermediate Value Theorem for Polynomial Functions

Let f be a polynomial function with real coefficients. If $f(a)$ and $f(b)$ have opposite signs, then there is at least one value of c between a and b for which $f(c) = 0$. Equivalently, the equation $f(x) = 0$ has at least one real root between a and b.

y
$(b, f(b))$
$f(b) > 0$
$f(c) = 0$
a
c
b
x
$(a, f(a))$
$f(a) < 0$

FIGURE 3.22 The graph must cross the x-axis at some value between a and b.

EXAMPLE 8 Using the Intermediate Value Theorem

Show that the polynomial function $f(x) = x^3 - 2x - 5$ has a real zero between 2 and 3.

SOLUTION

Let us evaluate f at 2 and at 3. If $f(2)$ and $f(3)$ have opposite signs, then there is at least one real zero between 2 and 3. Using $f(x) = x^3 - 2x - 5$, we obtain

$$f(2) = 2^3 - 2 \cdot 2 - 5 = 8 - 4 - 5 = -1$$

$f(2)$ is negative.

and

$$f(3) = 3^3 - 2 \cdot 3 - 5 = 27 - 6 - 5 = 16.$$

$f(3)$ is positive.

$y = x^3 - 2x - 5$

Zero
X=2.0945515 Y=0

$[-3, 3, 1]$ by $[-10, 10, 1]$

FIGURE 3.23

Because $f(2) = -1$ and $f(3) = 16$, the sign change shows that the polynomial function has a real zero between 2 and 3. This zero is actually irrational and is approximated using a graphing utility's ZERO feature as 2.0945515 in **Figure 3.23**. • • •

⌀ Check Point **8** Show that the polynomial function $f(x) = 3x^3 - 10x + 9$ has a real zero between -3 and -2.

⑦ Understand the relationship between degree and turning points.

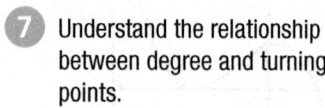

Turning points: from increasing to decreasing

FIGURE 3.24 Graph with four turning points

⑧ Graph polynomial functions.

Turning Points of Polynomial Functions

The graph of $f(x) = x^5 - 6x^3 + 8x + 1$ is shown in **Figure 3.24**. The graph has four smooth **turning points**.

At each turning point in **Figure 3.24**, the graph changes direction from increasing to decreasing or vice versa. The given equation has 5 as its greatest exponent and is therefore a polynomial function of degree 5. Notice that the graph has four turning points. In general, **if f is a polynomial function of degree n, then the graph of f has at most $n - 1$ turning points.**

Figure 3.24 illustrates that the y-coordinate of each turning point is either a relative maximum or a relative minimum of f. Without the aid of a graphing utility or a knowledge of calculus, it is difficult and often impossible to locate turning points of polynomial functions with degrees greater than 2. If necessary, test values can be taken between the x-intercepts to get a general idea of how high the graph rises or how low the graph falls. For the purpose of graphing in this section, a general estimate is sometimes appropriate and necessary.

A Strategy for Graphing Polynomial Functions

Here's a general strategy for graphing a polynomial function. A graphing utility is a valuable complement, but not a necessary component, to this strategy. If you are using a graphing utility, some of the steps listed in the following box will help you to select a viewing rectangle that shows the important parts of the graph.

GREAT QUESTION!

When graphing a polynomial function, how do I determine the location of its turning points?

Without calculus, it is often impossible to give the exact location of turning points. However, you can obtain additional points satisfying the function to estimate how high the graph rises or how low it falls. To obtain these points, use values of x between (and to the left and right of) the x-intercepts.

Graphing a Polynomial Function
$$f(x) = a_n x^n + a_{n-1} x^{n-1} + a_{n-2} x^{n-2} + \cdots + a_1 x + a_0, a_n \neq 0$$

1. Use the Leading Coefficient Test to determine the graph's end behavior.
2. Find x-intercepts by setting $f(x) = 0$ and solving the resulting polynomial equation. If there is an x-intercept at r as a result of $(x - r)^k$ in the complete factorization of $f(x)$, then
 a. If k is even, the graph touches the x-axis at r and turns around.
 b. If k is odd, the graph crosses the x-axis at r.
 c. If $k > 1$, the graph flattens out near $(r, 0)$.
3. Find the y-intercept by computing $f(0)$.
4. Use symmetry, if applicable, to help draw the graph:
 a. y-axis symmetry: $f(-x) = f(x)$
 b. Origin symmetry: $f(-x) = -f(x)$.
5. Use the fact that the maximum number of turning points of the graph is $n - 1$, where n is the degree of the polynomial function, to check whether it is drawn correctly.

EXAMPLE 9 Graphing a Polynomial Function

Graph: $f(x) = x^4 - 2x^2 + 1$.

SOLUTION

Step 1 Determine end behavior. Identify the sign of a_n, the leading coefficient, and the degree, n, of the polynomial function.

$$f(x) = x^4 - 2x^2 + 1$$

The leading coefficient, 1, is positive. The degree of the polynomial function, 4, is even.

Because the degree, 4, is even, the graph has the same behavior at each end. The leading coefficient, 1, is positive. Thus, the graph rises to the left and rises to the right.

Step 2 Find x-intercepts (zeros of the function) by setting f(x) = 0.

$$x^4 - 2x^2 + 1 = 0 \quad \text{Set } f(x) \text{ equal to } 0.$$
$$(x^2 - 1)(x^2 - 1) = 0 \quad \text{Factor.}$$
$$(x + 1)(x - 1)(x + 1)(x - 1) = 0 \quad \text{Factor completely.}$$
$$(x + 1)^2(x - 1)^2 = 0 \quad \text{Express the factorization in a more compact form.}$$
$$(x + 1)^2 = 0 \quad \text{or} \quad (x - 1)^2 = 0 \quad \text{Set each factorization equal to } 0.$$
$$x = -1 \qquad\qquad x = 1 \quad \text{Solve for } x.$$

We see that −1 and 1 are both repeated zeros with multiplicity 2. Because of the even multiplicity, the graph touches the x-axis at −1 and 1 and turns around. Furthermore, the graph tends to flatten out near these zeros with multiplicity greater than one.

Step 3 Find the y-intercept by computing f(0). We use $f(x) = x^4 - 2x^2 + 1$ and compute $f(0)$.

$$f(0) = 0^4 - 2 \cdot 0^2 + 1 = 1$$

There is a y-intercept at 1, so the graph passes through $(0, 1)$.

It appears that 1 is a relative maximum, but we need more information to be certain.

Step 4 Use possible symmetry to help draw the graph. Our partial graph suggests y-axis symmetry. Let's verify this by finding $f(-x)$.

$$f(x) = x^4 - 2x^2 + 1$$

Replace x with −x.

$$f(-x) = (-x)^4 - 2(-x)^2 + 1 = x^4 - 2x^2 + 1$$

Because $f(-x) = f(x)$, the graph of f is symmetric with respect to the y-axis. **Figure 3.25** shows the graph of $f(x) = x^4 - 2x^2 + 1$.

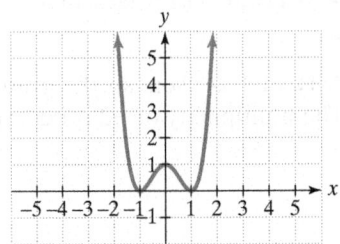

FIGURE 3.25 The graph of $f(x) = x^4 - 2x^2 + 1$

Step 5 Use the fact that the maximum number of turning points of the graph is n − 1 to check whether it is drawn correctly. Because $n = 4$, the maximum number of turning points is $4 - 1$, or 3. Because the graph in **Figure 3.25** has three turning points, we have not violated the maximum number possible. Can you see how this verifies that 1 is indeed a relative maximum and $(0, 1)$ is a turning point? If the graph rose above 1 on either side of $x = 0$, it would have to rise above 1 on the other side as well because of symmetry. This would require additional turning points to smoothly curve back to the x-intercepts. The graph already has three turning points, which is the maximum number for a fourth-degree polynomial function.

• • •

⊘ Check Point **9** Use the five-step strategy to graph $f(x) = x^3 - 3x^2$.

EXAMPLE 10 Graphing a Polynomial Function

Graph: $f(x) = -2(x - 1)^2(x + 2)$.

SOLUTION

Step 1 Determine end behavior. Identify the sign of a_n, the leading coefficient, and the degree, n, of the polynomial function.

$$f(x) = \underbrace{-2(x - 1)^2(x + 2)}$$

The leading term is $-2x^2 \cdot x$, or $-2x^3$.

> The leading coefficient, **−2**, is negative.

> The degree of the polynomial function, **3**, is odd.

Because the degree, 3, is odd, the graph has opposite behavior at each end. The leading coefficient, −2, is negative. Thus, the graph rises to the left and falls to the right.

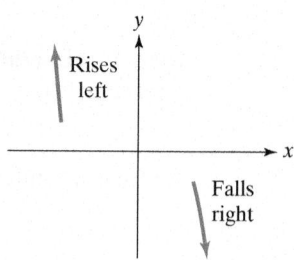

Step 2 Find x-intercepts (zeros of the function) by setting $f(x) = 0$.

$$-2(x - 1)^2(x + 2) = 0 \qquad \text{Set } f(x) \text{ equal to } 0.$$
$$(x - 1)^2 = 0 \quad \text{or} \quad x + 2 = 0 \qquad \text{Set each variable factor equal to } 0.$$
$$x = 1 \qquad\qquad x = -2 \qquad \text{Solve for } x.$$

We see that the zeros are 1 and −2. The multiplicity of 1 is even, so the graph touches the x-axis at 1, flattens, and turns around. The multiplicity of −2 is odd, so the graph crosses the x-axis at −2.

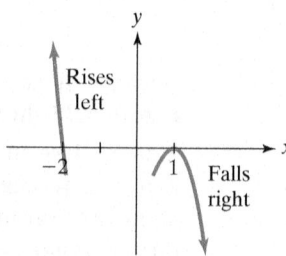

Step 3 Find the y-intercept by computing $f(0)$. We use $f(x) = -2(x - 1)^2(x + 2)$ and compute $f(0)$.

$$f(0) = -2(0 - 1)^2(0 + 2) = -2(1)(2) = -4$$

There is a y-intercept at -4, so the graph passes through $(0, -4)$.

We're not sure about the graph's behavior for x between −2 and 0 or between 0 and 1. Additional points will be helpful. Furthermore, we may need to rescale the y-axis.

Based on the note in the voice balloon, let's evaluate the function at $-1.5, -1, -0.5$, and 0.5, as well as at $-3, 2$, and 3.

x	-3	-1.5	-1	-0.5	0.5	2	3
$f(x) = -2(x-1)^2(x+2)$	32	-6.25	-8	-6.75	-1.25	-8	-40

In order to accommodate these points, we'll scale the y-axis from -50 to 50, with each tick mark representing 10 units.

Step 4 Use possible symmetry to help draw the graph. Our partial graph illustrates that we have neither y-axis symmetry nor origin symmetry. Using end behavior, intercepts, and the points from our table, **Figure 3.26** shows the graph of $f(x) = -2(x-1)^2(x+2)$.

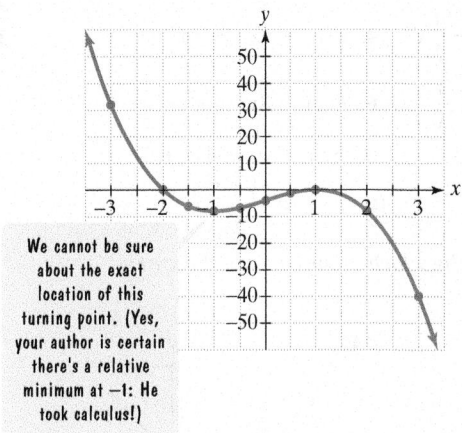

We cannot be sure about the exact location of this turning point. (Yes, your author is certain there's a relative minimum at −1: He took calculus!)

FIGURE 3.26 The graph of $f(x) = -2(x-1)^2(x+2)$

Step 5 Use the fact that the maximum number of turning points of the graph is $n - 1$ to check whether it is drawn correctly. The leading term is $-2x^3$, so $n = 3$. The maximum number of turning points is $3 - 1$, or 2. Because the graph in **Figure 3.26** has two turning points, we have not violated the maximum number possible. • • •

⊘ Check Point **10** Use the five-step strategy to graph
$f(x) = 2(x+2)^2(x-3)$.

CONCEPT AND VOCABULARY CHECK

Fill in each blank so that the resulting statement is true.

1. The degree of the polynomial function
$f(x) = -2x^3(x - 1)(x + 5)$ is _____. The leading
coefficient is _____.

2. True or false: Some polynomial functions of degree 2
or higher have breaks in their graphs. _____

3. The behavior of the graph of a polynomial function
to the far left or the far right is called its _____
behavior, which depends upon the _____ term.

4. The graph of $f(x) = x^3$ _____ to the left and
_____ to the right.

5. The graph of $f(x) = -x^3$ _____ to the left and
_____ to the right.

6. The graph of $f(x) = x^2$ _____ to the left and _____
to the right.

7. The graph of $f(x) = -x^2$ _____ to the left and
_____ to the right.

8. True or false: Odd-degree polynomial functions have
graphs with opposite behavior at each end. _____

9. True or false: Even-degree polynomial functions have
graphs with the same behavior at each end. _____

10. Every real zero of a polynomial function appears as
a/an _____ of the graph.

11. If r is a zero of even multiplicity, then the graph
touches the x-axis and _____ at r. If r is
a zero of odd multiplicity, then the graph _____
the x-axis at r.

12. If f is a polynomial function and $f(a)$ and $f(b)$ have
opposite signs, then there must be at least one value
of c between a and b for which $f(c) =$ _____. This
result is called the _____ Theorem.

13. If f is a polynomial function of degree n, then the
graph of f has at most _____ turning points.

EXERCISE SET 3.2

Practice Exercises

In Exercises 1–10, determine which functions are polynomial functions. For those that are, identify the degree.

1. $f(x) = 5x^2 + 6x^3$

2. $f(x) = 7x^2 + 9x^4$

3. $g(x) = 7x^5 - \pi x^3 + \frac{1}{5}x$

4. $g(x) = 6x^7 + \pi x^5 + \frac{2}{3}x$

5. $h(x) = 7x^3 + 2x^2 + \frac{1}{x}$

6. $h(x) = 8x^3 - x^2 + \frac{2}{x}$

7. $f(x) = x^{\frac{1}{2}} - 3x^2 + 5$

8. $f(x) = x^{\frac{1}{3}} - 4x^2 + 7$

9. $f(x) = \frac{x^2 + 7}{x^3}$

10. $f(x) = \frac{x^2 + 7}{3}$

In Exercises 11–14, identify which graphs are not those of polynomial functions.

11.

12.

13.

14.
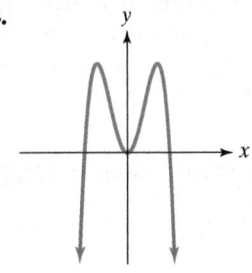

In Exercises 15–18, use the Leading Coefficient Test to determine the end behavior of the graph of the given polynomial function. Then use this end behavior to match the polynomial function with its graph. [The graphs are labeled (a) through (d).]

15. $f(x) = -x^4 + x^2$

16. $f(x) = x^3 - 4x^2$

17. $f(x) = (x - 3)^2$

18. $f(x) = -x^3 - x^2 + 5x - 3$

a.

b.

c.

d.

In Exercises 19–24, use the Leading Coefficient Test to determine the end behavior of the graph of the polynomial function.

19. $f(x) = 5x^3 + 7x^2 - x + 9$

20. $f(x) = 11x^3 - 6x^2 + x + 3$

21. $f(x) = 5x^4 + 7x^2 - x + 9$

22. $f(x) = 11x^4 - 6x^2 + x + 3$

23. $f(x) = -5x^4 + 7x^2 - x + 9$

24. $f(x) = -11x^4 - 6x^2 + x + 3$

In Exercises 25–32, find the zeros for each polynomial function and give the multiplicity for each zero. State whether the graph crosses the x-axis, or touches the x-axis and turns around, at each zero.

25. $f(x) = 2(x - 5)(x + 4)^2$

26. $f(x) = 3(x + 5)(x + 2)^2$

27. $f(x) = 4(x - 3)(x + 6)^3$

28. $f(x) = -3\left(x + \frac{1}{2}\right)(x - 4)^3$

29. $f(x) = x^3 - 2x^2 + x$

30. $f(x) = x^3 + 4x^2 + 4x$

31. $f(x) = x^3 + 7x^2 - 4x - 28$

32. $f(x) = x^3 + 5x^2 - 9x - 45$

In Exercises 33–40, use the Intermediate Value Theorem to show that each polynomial has a real zero between the given integers.

33. $f(x) = x^3 - x - 1$; between 1 and 2

34. $f(x) = x^3 - 4x^2 + 2$; between 0 and 1

35. $f(x) = 2x^4 - 4x^2 + 1$; between −1 and 0

36. $f(x) = x^4 + 6x^3 - 18x^2$; between 2 and 3

37. $f(x) = x^3 + x^2 - 2x + 1$; between −3 and −2

38. $f(x) = x^5 - x^3 - 1$; between 1 and 2

39. $f(x) = 3x^3 - 10x + 9$; between −3 and −2

40. $f(x) = 3x^3 - 8x^2 + x + 2$; between 2 and 3

In Exercises 41–64,

a. *Use the Leading Coefficient Test to determine the graph's end behavior.*

b. *Find the x-intercepts. State whether the graph crosses the x-axis, or touches the x-axis and turns around, at each intercept.*

c. *Find the y-intercept.*

d. *Determine whether the graph has y-axis symmetry, origin symmetry, or neither.*

e. *If necessary, find a few additional points and graph the function. Use the maximum number of turning points to check whether it is drawn correctly.*

41. $f(x) = x^3 + 2x^2 - x - 2$

42. $f(x) = x^3 + x^2 - 4x - 4$

43. $f(x) = x^4 - 9x^2$

44. $f(x) = x^4 - x^2$

45. $f(x) = -x^4 + 16x^2$

46. $f(x) = -x^4 + 4x^2$

47. $f(x) = x^4 - 2x^3 + x^2$

48. $f(x) = x^4 - 6x^3 + 9x^2$

49. $f(x) = -2x^4 + 4x^3$

50. $f(x) = -2x^4 + 2x^3$

51. $f(x) = 6x^3 - 9x - x^5$

52. $f(x) = 6x - x^3 - x^5$

53. $f(x) = 3x^2 - x^3$

54. $f(x) = \frac{1}{2} - \frac{1}{2}x^4$

55. $f(x) = -3(x - 1)^2(x^2 - 4)$

56. $f(x) = -2(x - 4)^2(x^2 - 25)$

57. $f(x) = x^2(x - 1)^3(x + 2)$

58. $f(x) = x^3(x + 2)^2(x + 1)$

59. $f(x) = -x^2(x - 1)(x + 3)$

60. $f(x) = -x^2(x + 2)(x - 2)$

61. $f(x) = -2x^3(x - 1)^2(x + 5)$

62. $f(x) = -3x^3(x - 1)^2(x + 3)$

63. $f(x) = (x - 2)^2(x + 4)(x - 1)$

64. $f(x) = (x + 3)(x + 1)^3(x + 4)$

Practice Plus

In Exercises 65–72, complete graphs of polynomial functions whose zeros are integers are shown.

a. *Find the zeros and state whether the multiplicity of each zero is even or odd.*

b. *Write an equation, expressed as the product of factors, of a polynomial function that might have each graph. Use a leading coefficient of 1 or −1, and make the degree of f as small as possible.*

c. *Use both the equation in part (b) and the graph to find the y-intercept.*

65.

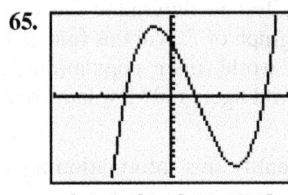

[−5, 5, 1] by [−12, 12, 1]

66.

[−6, 6, 1] by [−40, 40, 10]

67.

[−3, 6, 1] by [−10, 10, 1]

68.

[−3, 3, 1] by [−10, 10, 1]

69.

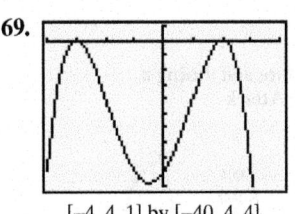

[−4, 4, 1] by [−40, 4, 4]

70.

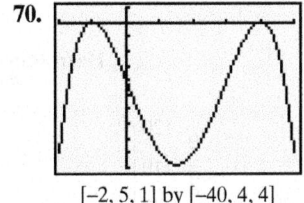

[−2, 5, 1] by [−40, 4, 4]

71.

[−3, 3, 1] by [−5, 10, 1]

72.

[−3, 3, 1] by [−5, 10, 1]

Application Exercises

Experts fear that without conservation efforts, tigers could disappear from the wild by 2022. Just one hundred years ago, there were at least 100,000 wild tigers. By 2010, the estimated world tiger population was 3200. The bar graph shows the estimated world tiger population for selected years from 1970 through 2010. Also shown is a polynomial function, with its graph, that models the data. Use this information to solve Exercises 73–74.

Vanishing Tigers

Estimated World Tiger Population

Graph of a Polynomial Model for the Data

$$f(x) = 0.76x^3 - 30x^2 - 882x + 37,807$$

Years after 1970

Source: World Wildlife Fund

73. a. Find and interpret $f(40)$. Identify this information as a point on the graph of f.

 b. Does $f(40)$ overestimate or underestimate the actual data shown by the bar graph? By how much?

 c. Use the Leading Coefficient Test to determine the end behavior to the right for the graph of f. Will this function be useful in modeling the world tiger population if conservation efforts to save wild tigers fail? Explain your answer.

74. a. Find and interpret $f(10)$. Identify this information as a point on the graph of f.

 d. Does $f(10)$ overestimate or underestimate the actual data shown by the bar graph? By how much?

 c. Use the Leading Coefficient Test to determine the end behavior to the right for the graph of f. Might this function be useful in modeling the world tiger population if conservation efforts to save wild tigers are successful? Explain your answer.

75. During a diagnostic evaluation, a 33-year-old woman experienced a panic attack a few minutes after she had been asked to relax her whole body. The graph shows the rapid increase in heart rate during the panic attack.

a. For which time periods during the diagnostic evaluation was the woman's heart rate increasing?

b. For which time periods during the diagnostic evaluation was the woman's heart rate decreasing?

c. How many turning points (from increasing to decreasing or from decreasing to increasing) occurred for the woman's heart rate during the first 12 minutes of the diagnostic evaluation?

d. Suppose that a polynomial function is used to model the data displayed by the graph using

 (time during the evaluation, heart rate).

 Use the number of turning points to determine the degree of the polynomial function of best fit.

e. For the model in part (d), should the leading coefficient of the polynomial function be positive or negative? Explain your answer.

f. Use the graph to estimate the woman's maximum heart rate during the first 12 minutes of the diagnostic evaluation. After how many minutes did this occur?

g. Use the graph to estimate the woman's minimum heart rate during the first 12 minutes of the diagnostic evaluation. After how many minutes did this occur?

76. Volatility at the Pump The graph shows the average price per gallon of gasoline in the United States in January for the period from 2005 through 2011.

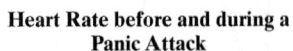

Heart Rate before and during a Panic Attack

Source: Davis and Palladino, *Psychology*, Fifth Edition, Prentice Hall, 2007.

Average January Gasoline Price in the United States

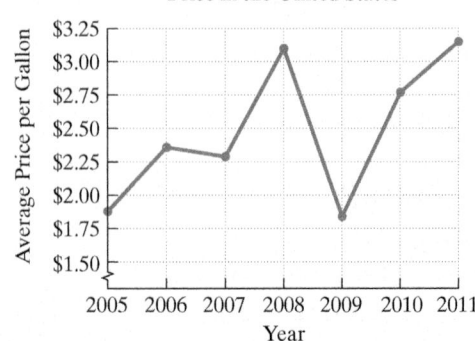

Source: U.S. Energy Information Administration

a. For which years was the average price per gallon in January increasing?

b. For which years was the average price per gallon in January decreasing?

c. How many turning points (from increasing to decreasing or from decreasing to increasing) does the graph have for the period shown?

d. Suppose that a polynomial function is used to model the data displayed by the graph using

(number of years after 2005, average January price per gallon).

Use the number of turning points to determine the degree of the polynomial function of best fit.

e. For the model in part (d), should the leading coefficient of the polynomial function be positive or negative? Explain your answer.

f. Use the graph to estimate the maximum average January price per gallon. In which year did this occur?

g. Use the graph to estimate the minimum average January price per gallon. In which year did this occur?

Writing in Mathematics

77. What is a polynomial function?

78. What do we mean when we describe the graph of a polynomial function as smooth and continuous?

79. What is meant by the end behavior of a polynomial function?

80. Explain how to use the Leading Coefficient Test to determine the end behavior of a polynomial function.

81. Why is a third-degree polynomial function with a negative leading coefficient not appropriate for modeling nonnegative real-world phenomena over a long period of time?

82. What are the zeros of a polynomial function and how are they found?

83. Explain the relationship between the multiplicity of a zero and whether or not the graph crosses or touches the x-axis at that zero.

84. If f is a polynomial function, and $f(a)$ and $f(b)$ have opposite signs, what must occur between a and b? If $f(a)$ and $f(b)$ have the same sign, does it necessarily mean that this will not occur? Explain your answer.

85. Explain the relationship between the degree of a polynomial function and the number of turning points on its graph.

86. Can the graph of a polynomial function have no x-intercepts? Explain.

87. Can the graph of a polynomial function have no y-intercept? Explain.

88. Describe a strategy for graphing a polynomial function. In your description, mention intercepts, the polynomial's degree, and turning points.

Technology Exercises

89. Use a graphing utility to verify any five of the graphs that you drew by hand in Exercises 41–64.

Write a polynomial function that imitates the end behavior of each graph in Exercises 90–93. The dashed portions of the graphs indicate that you should focus only on imitating the left and right behavior of the graph and can be flexible about what occurs between the left and right ends. Then use your graphing utility to graph the polynomial function and verify that you imitated the end behavior shown in the given graph.

90. **91.**

92. **93.**

In Exercises 94–97, use a graphing utility with a viewing rectangle large enough to show end behavior to graph each polynomial function.

94. $f(x) = x^3 + 13x^2 + 10x - 4$

95. $f(x) = -2x^3 + 6x^2 + 3x - 1$

96. $f(x) = -x^4 + 8x^3 + 4x^2 + 2$

97. $f(x) = -x^5 + 5x^4 - 6x^3 + 2x + 20$

In Exercises 98–99, use a graphing utility to graph f and g in the same viewing rectangle. Then use the $\boxed{\text{ZOOM OUT}}$ feature to show that f and g have identical end behavior.

98. $f(x) = x^3 - 6x + 1$, $g(x) = x^3$

99. $f(x) = -x^4 + 2x^3 - 6x$, $g(x) = -x^4$

Critical Thinking Exercises

Make Sense? *In Exercises 100–103, determine whether each statement makes sense or does not make sense, and explain your reasoning.*

100. When I'm trying to determine end behavior, it's the coefficient of the leading term of a polynomial function that I should inspect.

101. I graphed $f(x) = (x + 2)^3(x - 4)^2$, and the graph touched the x-axis and turned around at -2.

102. I'm graphing a fourth-degree polynomial function with four turning points.

103. Although I have not yet learned techniques for finding the x-intercepts of $f(x) = x^3 + 2x^2 - 5x - 6$, I can easily determine the y-intercept.

In Exercises 104–107, determine whether each statement is true or false. If the statement is false, make the necessary change(s) to produce a true statement.

104. If $f(x) = -x^3 + 4x$, then the graph of f falls to the left and falls to the right.

105. A mathematical model that is a polynomial of degree n whose leading term is $a_n x^n$, n odd and $a_n < 0$, is ideally suited to describe phenomena that have positive values over unlimited periods of time.

106. There is more than one third-degree polynomial function with the same three x-intercepts.

107. The graph of a function with origin symmetry can rise to the left and rise to the right.

Use the descriptions in Exercises 108–109 to write an equation of a polynomial function with the given characteristics. Use a graphing utility to graph your function to see if you are correct. If not, modify the function's equation and repeat this process.

108. Crosses the x-axis at $-4, 0,$ and 3; lies above the x-axis between -4 and 0; lies below the x-axis between 0 and 3

109. Touches the x-axis at 0 and crosses the x-axis at 2; lies below the x-axis between 0 and 2

Preview Exercises

Exercises 110–112 will help you prepare for the material covered in the next section.

110. Divide 737 by 21 without using a calculator. Write the answer as

$$\text{quotient} + \frac{\text{remainder}}{\text{divisor}}.$$

111. Rewrite $4 - 5x - x^2 + 6x^3$ in descending powers of x.

112. Use

$$\frac{2x^3 - 3x^2 - 11x + 6}{x - 3} = 2x^2 + 3x - 2$$

to factor $2x^3 - 3x^2 - 11x + 6$ completely.

SECTION 3.3 Dividing Polynomials; Remainder and Factor Theorems

Objectives

① Use long division to divide polynomials.

② Use synthetic division to divide polynomials.

③ Evaluate a polynomial using the Remainder Theorem.

④ Use the Factor Theorem to solve a polynomial equation.

A moth has moved into your closet. She appeared in your bedroom at night, but somehow her relatively stout body escaped your clutches. Within a few weeks, swarms of moths in your tattered wardrobe suggest that Mama Moth was in the family way. There must be at least 200 critters nesting in every crevice of your clothing.

Two hundred plus moth-tykes from one female moth—is this possible? Indeed it is. The number of eggs, $f(x)$, in a female moth is a function of her abdominal width, x, in millimeters, modeled by

$$f(x) = 14x^3 - 17x^2 - 16x + 34, \quad 1.5 \leq x \leq 3.5.$$

Because there are 200 moths feasting on your favorite sweaters, Mama's abdominal width can be estimated by finding the solutions of the polynomial equation

$$14x^3 - 17x^2 - 16x + 34 = 200.$$

How can we solve such an equation? You might begin by subtracting 200 from both sides to obtain zero on one side. But then what? The factoring that we used in the previous section will not work in this situation.

In Section 3.4, we will present techniques for solving certain kinds of polynomial equations. These techniques will further enhance your ability to manipulate algebraically the polynomial functions that model your world. Because these techniques are based on understanding polynomial division, in this section we look at two methods for dividing polynomials. (We'll return to Mama Moth's abdominal width in the Exercise Set.)

① Use long division to divide polynomials.

Long Division of Polynomials and the Division Algorithm

We begin by looking at division by a polynomial containing more than one term, such as

$$x + 3 \overline{)x^2 + 10x + 21}.$$

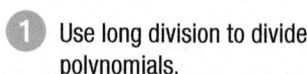

| Divisor has two terms and is a binomial. | The polynomial dividend has three terms and is a trinomial. |

When a divisor has more than one term, the four steps used to divide whole numbers—**divide, multiply, subtract, bring down the next term**—form the repetitive procedure for polynomial long division.

EXAMPLE 1 Long Division of Polynomials

Divide $x^2 + 10x + 21$ by $x + 3$.

SOLUTION

The following steps illustrate how polynomial division is very similar to numerical division.

$$x + 3 \overline{)x^2 + 10x + 21}$$

Arrange the terms of the dividend ($x^2 + 10x + 21$) and the divisor ($x + 3$) in descending powers of x.

$$\begin{array}{r} x \phantom{{}+ 10x + 21} \\ x + 3 \overline{)x^2 + 10x + 21} \end{array}$$

Divide x^2 (the first term in the dividend) by x (the first term in the divisor): $\dfrac{x^2}{x} = x$. Align like terms.

$$x(x + 3) = x^2 + 3x \qquad \begin{array}{r} x \phantom{{}+ 10x + 21} \\ x + 3 \overline{)x^2 + 10x + 21} \\ x^2 + 3x \phantom{{}+ 21} \end{array}$$

Multiply each term in the divisor (x + 3) by x, aligning terms of the product under like terms in the dividend.

$$\begin{array}{r} x \phantom{{}+ 10x + 21} \\ x + 3 \overline{)x^2 + 10x + 21} \\ \ominus x^2 \ominus 3x \phantom{{}+ 21} \\ \hline 7x \phantom{{}+ 21} \end{array}$$

Subtract $x^2 + 3x$ from $x^2 + 10x$ by changing the sign of each term in the lower expression and adding.

Change signs of the polynomial being subtracted.

$$\begin{array}{r} x \phantom{{}+ 10x + 21} \\ x + 3 \overline{)x^2 + 10x + 21} \\ x^2 + 3x \phantom{{}} \downarrow \\ \hline 7x + 21 \end{array}$$

Bring down 21 from the original dividend and add algebraically to form a new dividend.

$$\begin{array}{r} x + 7 \\ x + 3 \overline{)x^2 + 10x + 21} \\ x^2 + 3x \phantom{{}+ 21} \\ \hline 7x + 21 \end{array}$$

Find the second term of the quotient. Divide the first term of $7x + 21$ by x, the first term of the divisor: $\dfrac{7x}{x} = 7$.

$$7(x + 3) = 7x + 21 \qquad \begin{array}{r} x + 7 \\ x + 3 \overline{)x^2 + 10x + 21} \\ x^2 + 3x \phantom{{}+ 21} \\ \hline 7x + 21 \\ \ominus 7x \ominus 21 \\ \hline 0 \end{array}$$

Multiply the divisor (x + 3) by 7, aligning under like terms in the new dividend. Then subtract to obtain the remainder of 0.

Remainder

The quotient is $x + 7$. Because the remainder is 0, we can conclude that $x + 3$ is a factor of $x^2 + 10x + 21$ and

$$\frac{x^2 + 10x + 21}{x + 3} = x + 7.$$

● ● ●

⊘ **Check Point 1** Divide $x^2 + 14x + 45$ by $x + 9$.

Before considering additional examples, let's summarize the general procedure for dividing one polynomial by another.

Long Division of Polynomials

1. **Arrange** the terms of both the dividend and the divisor in descending powers of any variable.
2. **Divide** the first term in the dividend by the first term in the divisor. The result is the first term of the quotient.
3. **Multiply** every term in the divisor by the first term in the quotient. Write the resulting product beneath the dividend with like terms lined up.
4. **Subtract** the product from the dividend.
5. **Bring down** the next term in the original dividend and write it next to the remainder to form a new dividend.
6. Use this new expression as the dividend and repeat this process until the remainder can no longer be divided. This will occur when the degree of the remainder (the highest exponent on a variable in the remainder) is less than the degree of the divisor.

In our next long division, we will obtain a nonzero remainder.

EXAMPLE 2 Long Division of Polynomials

Divide $4 - 5x - x^2 + 6x^3$ by $3x - 2$.

SOLUTION

We begin by writing the dividend in descending powers of x.

$$4 - 5x - x^2 + 6x^3 = 6x^3 - x^2 - 5x + 4$$

Now we divide $3x^2$ by $3x$ to obtain x, multiply x and the divisor, and subtract.

Now we divide $-3x$ by $3x$ to obtain -1, multiply -1 and the divisor, and subtract.

$$-1(3x - 2) = -3x + 2$$

$$
\begin{array}{r}
2x^2 + x - 1 \\
3x - 2 \overline{)6x^3 - x^2 - 5x + 4} \\
6x^3 - 4x^2 \\
\hline
3x^2 - 5x \\
3x^2 - 2x \\
\hline
-3x + 4 \\
-3x + 2 \\
\hline
2
\end{array}
$$

Divide: $\dfrac{-3x}{3x} = -1.$

Multiply: $-1(3x - 2) = -3x + 2.$

Subtract $-3x + 2$ from $-3x + 4$, leaving a remainder of 2.

Change signs of the polynomial being subtracted.

Remainder

The quotient is $2x^2 + x - 1$ and the remainder is 2. When there is a nonzero remainder, as in this example, list the quotient, plus the remainder above the divisor. Thus,

$$\frac{6x^3 - x^2 - 5x + 4}{3x - 2} = \underbrace{2x^2 + x - 1}_{\text{Quotient}} + \underbrace{\frac{2}{3x - 2}}_{\substack{\text{Remainder} \\ \text{above divisor}}}.$$

An important property of division can be illustrated by clearing fractions in the equation that concluded Example 2. Multiplying both sides of this equation by $3x - 2$ results in the following equation:

$$6x^3 - x^2 - 5x + 4 = (3x - 2)(2x^2 + x - 1) + 2.$$

Dividend Divisor Quotient Remainder

Polynomial long division is checked by multiplying the divisor with the quotient and then adding the remainder. This should give the dividend. The process illustrates the **Division Algorithm.**

The Division Algorithm

If $f(x)$ and $d(x)$ are polynomials, with $d(x) \neq 0$, and the degree of $d(x)$ is less than or equal to the degree of $f(x)$, then there exist unique polynomials $q(x)$ and $r(x)$ such that

$$f(x) = d(x) \cdot q(x) + r(x).$$

Dividend Divisor Quotient Remainder

The remainder, $r(x)$, equals 0 or it is of degree less than the degree of $d(x)$. If $r(x) = 0$, we say that $d(x)$ **divides evenly** into $f(x)$ and that $d(x)$ and $q(x)$ are **factors** of $f(x)$.

\varnothing **Check Point 2** Divide $7 - 11x - 3x^2 + 2x^3$ by $x - 3$. Express the result in the form quotient, plus remainder divided by divisor.

If a power of x is missing in either a dividend or a divisor, add that power of x with a coefficient of 0 and then divide. In this way, like terms will be aligned as you carry out the long division.

EXAMPLE 3 Long Division of Polynomials

Divide $6x^4 + 5x^3 + 3x - 5$ by $3x^2 - 2x$.

SOLUTION

We write the dividend, $6x^4 + 5x^3 + 3x - 5$, as $6x^4 + 5x^3 + 0x^2 + 3x - 5$ to keep all like terms aligned.

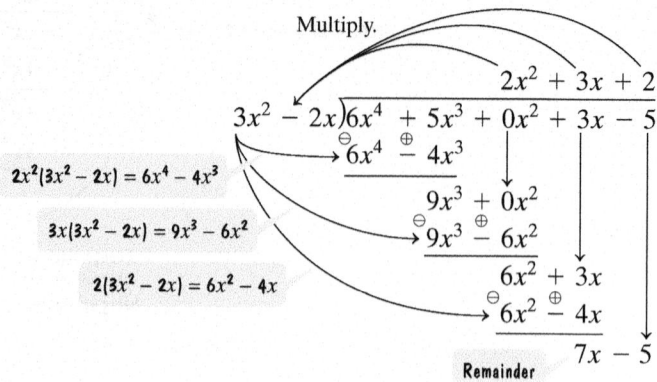

$2x^2(3x^2 - 2x) = 6x^4 - 4x^3$

$3x(3x^2 - 2x) = 9x^3 - 6x^2$

$2(3x^2 - 2x) = 6x^2 - 4x$

The division process is finished because the degree of $7x - 5$, which is 1, is less than the degree of the divisor $3x^2 - 2x$, which is 2. The answer is

$$\frac{6x^4 + 5x^3 + 3x - 5}{3x^2 - 2x} = 2x^2 + 3x + 2 + \frac{7x - 5}{3x^2 - 2x}.$$

• • •

✓ Check Point **3** Divide $2x^4 + 3x^3 - 7x - 10$ by $x^2 - 2x$.

② Use synthetic division to divide polynomials.

Dividing Polynomials Using Synthetic Division

We can use **synthetic division** to divide polynomials if the divisor is of the form $x - c$. This method provides a quotient more quickly than long division. Let's compare the two methods showing $x^3 + 4x^2 - 5x + 5$ divided by $x - 3$.

Long Division

Quotient

$$
\begin{array}{r}
x^2 + 7x + 16 \\
x - 3\overline{)x^3 + 4x^2 - 5x + 5} \\
\ominus\ \oplus \\
x^3 - 3x^2 \\
\hline
7x^2 - 5x \\
\ominus\ \oplus \\
7x^2 - 21x \\
\hline
16x + 5 \\
\ominus\ \oplus \\
16x - 48 \\
\hline
53
\end{array}
$$

Divisor
$x - c;$
$c = 3$

Dividend

Remainder

Synthetic Division

$$
\begin{array}{r|rrrr}
3 & 1 & 4 & -5 & 5 \\
& & 3 & 21 & 48 \\
\hline
& 1 & 7 & 16 & 53
\end{array}
$$

Notice the relationship between the polynomials in the long division process and the numbers that appear in synthetic division.

These are the coefficients of the dividend $x^3 + 4x^2 - 5x + 5$.

The divisor is $x - 3$.
This is 3, or c, in $x - c$.

$$
\begin{array}{r|rrrr}
3 & 1 & 4 & -5 & 5 \\
& & 3 & 21 & 48 \\
\hline
& 1 & 7 & 16 & 53
\end{array}
$$

These are the coefficients of the quotient $x^2 + 7x + 16$.

This is the remainder.

Now let's look at the steps involved in synthetic division.

Synthetic Division

To divide a polynomial by $x - c$:

	Example

1. Arrange the polynomial in descending powers, with a 0 coefficient for any missing term.

$$x - 3\overline{)x^3 + 4x^2 - 5x + 5}$$

2. Write c for the divisor, $x - c$. To the right, write the coefficients of the dividend.

$$\underline{3|}\quad 1\quad 4\quad -5\quad 5$$

3. Write the leading coefficient of the dividend on the bottom row.

$$\underline{3|}\quad 1\quad 4\quad -5\quad 5$$
$$\qquad \downarrow\quad \text{Bring down 1.}$$
$$\quad\ 1$$

4. Multiply c (in this case, 3) times the value just written on the bottom row. Write the product in the next column in the second row.

$$\underline{3|}\quad 1\quad 4\quad -5\quad 5$$
$$\qquad\qquad 3$$
$$\quad\ 1$$

Multiply by 3: $3 \cdot 1 = 3$.

5. Add the values in this new column, writing the sum in the bottom row.

$$\underline{3|}\quad 1\quad 4\quad -5\quad 5$$
$$\qquad\qquad 3\quad \text{Add.}$$
$$\quad\ 1\quad 7$$

6. Repeat this series of multiplications and additions until all columns are filled in.

$$\underline{3|}\quad 1\quad 4\quad -5\quad 5$$
$$\qquad\qquad 3\quad 21\quad \text{Add.}$$
$$\quad\ 1\quad 7\quad 16$$

Multiply by 3: $3 \cdot 7 = 21$.

$$\underline{3|}\quad 1\quad 4\quad -5\quad 5$$
$$\qquad\qquad 3\quad 21\quad 48\quad \text{Add.}$$
$$\quad\ 1\quad 7\quad 16\quad 53$$

Multiply by 3: $3 \cdot 16 = 48$.

7. Use the numbers in the last row to write the quotient, plus the remainder above the divisor. **The degree of the first term of the quotient is one less than the degree of the first term of the dividend.** The final value in this row is the remainder.

Written from
1 7 16 53
the last row of the synthetic division

$$1x^2 + 7x + 16 + \frac{53}{x - 3}$$
$$x - 3\overline{)x^3 + 4x^2 - 5x + 5}$$

EXAMPLE 4 Using Synthetic Division

Use synthetic division to divide $5x^3 + 6x + 8$ by $x + 2$.

SOLUTION

The divisor must be in the form $x - c$. Thus, we write $x + 2$ as $x - (-2)$. This means that $c = -2$. Writing a 0 coefficient for the missing x^2-term in the dividend, we can express the division as follows:

$$x - (-2)\overline{)5x^3 + 0x^2 + 6x + 8.}$$

Now we are ready to set up the problem so that we can use synthetic division.

Use the coefficients of the dividend
$5x^3 + 0x^2 + 6x + 8$ in descending powers of x.

$$x - (-2)\overline{)5x^3 + 0x^2 + 6x + 8}$$

The given division problem (repeated)

This is c in
$x - (-2)$. $-2\,|\ \ 5\ \ \ 0\ \ \ 6\ \ \ 8$

We begin the synthetic division process by bringing down 5. This is followed by a series of multiplications and additions.

1. Bring down 5.

$-2\,|\ \ 5\ \ \ 0\ \ \ 6\ \ \ 8$
$\quad\quad\downarrow$
$\quad\quad 5$

2. Multiply: $-2(5) = -10$.

$-2\,|5\ \ \ \ 0\ \ \ 6\ \ \ 8$
$\qquad\quad -10$
$\quad 5$

Multiply 5 by –2.

3. Add: $0 + (-10) = -10$.

$-2\,|\ 5\ \ \ \ 0\ |\ 6\ \ 8$
$\qquad\quad -10\ \downarrow\ \text{Add.}$
$\quad\ \ 5\ \ -10$

4. Multiply: $-2(-10) = 20$.

$-2\,|5\ \ \ \ 0\ \ \ 6\ \ \ 8$
$\qquad\ -10\ \ 20$
$\quad 5\ \ \ -10$

Multiply –10 by –2.

5. Add: $6 + 20 = 26$.

$-2\,|\ 5\ \ \ \ 0\ \ \ 6\ |\ \ 8$
$\qquad\quad -10\ \ 20\ \downarrow\ \text{Add.}$
$\quad\ \ 5\ \ -10\ \ 26$

6. Multiply: $-2(26) = -52$.

$-2\,|5\ \ \ \ 0\ \ \ 6\ \ \ \ 8$
$\qquad\ -10\ \ 20\ \ -52$
$\quad 5\ \ \ -10\ \ 26$

Multiply 26 by –2.

7. Add: $8 + (-52) = -44$.

$-2\,|\ 5\ \ \ \ 0\ \ \ 6\ \ \ \ 8\ |$
$\qquad\quad -10\ \ 20\ \ -52\ \downarrow\ \text{Add.}$
$\quad\ \ 5\ \ -10\ \ 26\ \ -44$

The numbers in the last row represent the coefficients of the quotient and the remainder. The degree of the first term of the quotient is one less than that of the dividend. Because the degree of the dividend, $5x^3 + 6x + 8$, is 3, the degree of the quotient is 2. This means that the 5 in the last row represents $5x^2$.

$$-2\,|\ 5\ \ \ \ 0\ \ \ 6\ \ \ \ 8$$
$$\qquad\quad -10\ \ 20\ \ -52$$
$$\quad\ \ 5\ \ -10\ \ 26\ \ -44$$

The quotient is The remainder
$5x^2 - 10x + 26$. is **–44.**

Thus,

$$\frac{5x^2 - 10x + 26 - \dfrac{44}{x + 2}}{x + 2\overline{)5x^3 + 6x + 8}}.$$

$\bullet\bullet\bullet$

✓ Check Point **4** Use synthetic division to divide $x^3 - 7x - 6$ by $x + 2$.

3 Evaluate a polynomial using the Remainder Theorem.

The Remainder Theorem

Let's consider the Division Algorithm when the dividend, $f(x)$, is divided by $x - c$. In this case, the remainder must be a constant because its degree is less than one, the degree of $x - c$.

$$f(x)\quad =\quad (x - c)q(x)\quad +\quad r$$

Dividend Divisor Quotient The remainder, r, is a constant when
dividing by $x - c$.

Now let's evaluate f at c.

$$f(c) = (c - c)q(c) + r \quad \text{Find } f(c) \text{ by letting } x = c \text{ in } f(x) = (x - c)q(x) + r.$$
$$\text{This will give an expression for } r.$$

$$f(c) = 0 \cdot q(c) + r \qquad c - c = 0$$
$$f(c) = r \qquad 0 \cdot q(c) = 0 \text{ and } 0 + r = r.$$

What does this last equation mean? If a polynomial is divided by $x - c$, the remainder is the value of the polynomial at c. This result is called the **Remainder Theorem**.

The Remainder Theorem

If the polynomial $f(x)$ is divided by $x - c$, then the remainder is $f(c)$.

Example 5 shows how we can use the Remainder Theorem to evaluate a polynomial function at 2. Rather than substituting 2 for x, we divide the function by $x - 2$. The remainder is $f(2)$.

EXAMPLE 5 Using the Remainder Theorem
 to Evaluate a Polynomial Function

Given $f(x) = x^3 - 4x^2 + 5x + 3$, use the Remainder Theorem to find $f(2)$.

SOLUTION

By the Remainder Theorem, if $f(x)$ is divided by $x - 2$, then the remainder is $f(2)$. We'll use synthetic division to divide.

$$\begin{array}{r|rrrr} 2 & 1 & -4 & 5 & 3 \\ & & 2 & -4 & 2 \\ \hline & 1 & -2 & 1 & 5 \end{array} \quad \text{Remainder}$$

The remainder, 5, is the value of $f(2)$. Thus, $f(2) = 5$. We can verify that this is correct by evaluating $f(2)$ directly. Using $f(x) = x^3 - 4x^2 + 5x + 3$, we obtain

$$f(2) = 2^3 - 4 \cdot 2^2 + 5 \cdot 2 + 3 = 8 - 16 + 10 + 3 = 5. \qquad \bullet\bullet\bullet$$

⊘ Check Point **5** Given $f(x) = 3x^3 + 4x^2 - 5x + 3$, use the Remainder Theorem to find $f(-4)$.

④ Use the Factor Theorem to solve a polynomial equation.

The Factor Theorem

Let's look again at the Division Algorithm when the divisor is of the form $x - c$.

$$f(x) \quad = \quad (x - c)q(x) \quad + \quad r$$

Dividend Divisor Quotient Constant remainder

By the Remainder Theorem, the remainder r is $f(c)$, so we can substitute $f(c)$ for r:

$$f(x) = (x - c)q(x) + f(c).$$

Notice that if $f(c) = 0$, then

$$f(x) = (x - c)q(x)$$

so that $x - c$ is a factor of $f(x)$. This means that for the polynomial function $f(x)$, if $f(c) = 0$, then $x - c$ is a factor of $f(x)$.

Let's reverse directions and see what happens if $x - c$ is a factor of $f(x)$. This means that

$$f(x) = (x - c)q(x).$$

If we replace x in $f(x) = (x - c)q(x)$ with c, we obtain

$$f(c) = (c - c)q(c) = 0 \cdot q(c) = 0.$$

Thus, if $x - c$ is a factor of $f(x)$, then $f(c) = 0$.

We have proved a result known as the **Factor Theorem**.

> The Factor Theorem
>
> Let $f(x)$ be a polynomial.
>
> **a.** If $f(c) = 0$, then $x - c$ is a factor of $f(x)$.
> **b.** If $x - c$ is a factor of $f(x)$, then $f(c) = 0$.

The example that follows shows how the Factor Theorem can be used to solve a polynomial equation.

EXAMPLE 6 Using the Factor Theorem

Solve the equation $2x^3 - 3x^2 - 11x + 6 = 0$ given that 3 is a zero of $f(x) = 2x^3 - 3x^2 - 11x + 6$.

SOLUTION

We are given that 3 is a zero of $f(x) = 2x^3 - 3x^2 - 11x + 6$. This means that $f(3) = 0$. Because $f(3) = 0$, the Factor Theorem tells us that $x - 3$ is a factor of $f(x)$. We'll use synthetic division to divide $f(x)$ by $x - 3$.

TECHNOLOGY

Graphic Connections

Because the solution set of

$$2x^3 - 3x^2 - 11x + 6 = 0$$

is $\left\{-2, \frac{1}{2}, 3\right\}$, this implies that the polynomial function

$$f(x) = 2x^3 - 3x^2 - 11x + 6$$

has x-intercepts (or zeros) at $-2, \frac{1}{2}$, and 3. This is verified by the graph of f.

$[-10, 10, 1]$ by $[-15, 15, 1]$

$$
\begin{array}{r|rrrr}
3 & 2 & -3 & -11 & 6 \\
 & & 6 & 9 & -6 \\
\hline
 & 2 & 3 & -2 & 0
\end{array}
$$

The remainder, 0, verifies that $x - 3$ is a factor of $2x^3 - 3x^2 - 11x + 6$.

$$
\begin{array}{r}
2x^2 + 3x - 2 \\
x - 3 \overline{)2x^3 - 3x^2 - 11x + 6}
\end{array}
$$

Equivalently,

$$2x^3 - 3x^2 - 11x + 6 = (x - 3)(2x^2 + 3x - 2).$$

Now we can solve the polynomial equation.

$2x^3 - 3x^2 - 11x + 6 = 0$	This is the given equation.
$(x - 3)(2x^2 + 3x - 2) = 0$	Factor using the result from the synthetic division.
$(x - 3)(2x - 1)(x + 2) = 0$	Factor the trinomial.
$x - 3 = 0$ or $2x - 1 = 0$ or $x + 2 = 0$	Set each factor equal to 0.
$x = 3$ $\qquad x = \frac{1}{2}$ $\qquad x = -2$	Solve for x.

The solution set is $\left\{-2, \frac{1}{2}, 3\right\}$. $\bullet\bullet\bullet$

Based on the Factor Theorem, the following statements are useful in solving polynomial equations:

1. If $f(x)$ is divided by $x - c$ and the remainder is zero, then c is a zero of f and c is a root of the polynomial equation $f(x) = 0$.

2. If $f(x)$ is divided by $x - c$ and the remainder is zero, then $x - c$ is a factor of $f(x)$.

 Check Point 6 Solve the equation $15x^3 + 14x^2 - 3x - 2 = 0$ given that -1 is a zero of $f(x) = 15x^3 + 14x^2 - 3x - 2$.

CONCEPT AND VOCABULARY CHECK

Fill in each blank so that the resulting statement is true.

1. Consider the following long division problem:
$$x + 4\overline{)6x - 4 + 2x^3}.$$
We begin the division process by rewriting the dividend as _____.

2. Consider the following long division problem:
$$3x - 1\overline{)6x^3 + 7x^2 + 12x - 5}.$$
We begin the division process by dividing _____ by _____. We obtain _____. We write this result above _____ in the dividend.

3. In the following long division problem, the first step has been completed:
$$\begin{array}{r} 2x^2 \\ 5x - 2\overline{)10x^3 + 6x^2 - 9x + 10} \end{array}.$$
The next step is to multiply _____ and _____. We obtain _____. We write this result below _____.

4. In the following long division problem, the first two steps have been completed:
$$\begin{array}{r} 2x \\ 3x - 5\overline{)6x^2 + 8x - 4} \\ 6x^2 - 10x \end{array}.$$
The next step is to subtract _____ from _____. We obtain _____. Then we bring down _____ and form the new dividend _____.

5. In the following long division problem, most of the steps have been completed:
$$\begin{array}{r} 3x - 5 \\ 2x + 1\overline{)6x^2 - 7x + 4} \\ 6x^2 + 3x \\ \overline{-10x + 4} \\ -10x - 5 \\ \overline{} \\ ? \end{array}.$$
Completing the step designated by the question mark, we obtain _____. Thus, the quotient is _____ and the remainder is _____. The answer to this long division problem is _____.

6. After performing polynomial long division, the answer may be checked by multiplying the _____ by the _____, and then adding the _____. You should obtain the _____.

7. To divide $x^3 + 5x^2 - 7x + 1$ by $x - 4$ using synthetic division, the first step is to write
$$\underline{}\rfloor \quad \underline{} \quad \underline{} \quad \underline{} \quad \underline{}.$$

8. To divide $4x^3 - 8x - 2$ by $x + 5$ using synthetic division, the first step is to write
$$\underline{}\rfloor \quad \underline{} \quad \underline{} \quad \underline{} \quad \underline{}.$$

9. True or false:
$$\begin{array}{r|rrrr} -1 & 3 & -4 & 2 & -1 \\ & & -3 & 7 & -9 \\ \hline & 3 & -7 & 9 & -10 \end{array} \text{ means}$$
$$\frac{3x^3 - 4x^2 + 2x - 1}{x + 1} = 3x^2 - 7x + 9 - \frac{10}{x + 1}. _____$$

10. The Remainder Theorem states that if the polynomial $f(x)$ is divided by $x - c$, then the remainder is _____.

11. The Factor Theorem states that if f is a polynomial function and $f(c) = 0$, then _____ is a factor of $f(x)$.

EXERCISE SET 3.3

Practice Exercises

In Exercises 1–16, divide using long division. State the quotient, $q(x)$, and the remainder, $r(x)$.

1. $(x^2 + 8x + 15) \div (x + 5)$

2. $(x^2 + 3x - 10) \div (x - 2)$

3. $(x^3 + 5x^2 + 7x + 2) \div (x + 2)$

4. $(x^3 - 2x^2 - 5x + 6) \div (x - 3)$

5. $(6x^3 + 7x^2 + 12x - 5) \div (3x - 1)$

6. $(6x^3 + 17x^2 + 27x + 20) \div (3x + 4)$

7. $(12x^2 + x - 4) \div (3x - 2)$

8. $(4x^2 - 8x + 6) \div (2x - 1)$

9. $\dfrac{2x^3 + 7x^2 + 9x - 20}{x + 3}$

10. $\dfrac{3x^2 - 2x + 5}{x - 3}$

11. $\dfrac{4x^4 - 4x^2 + 6x}{x - 4}$

12. $\dfrac{x^4 - 81}{x - 3}$

13. $\dfrac{6x^3 + 13x^2 - 11x - 15}{3x^2 - x - 3}$

14. $\dfrac{x^4 + 2x^3 - 4x^2 - 5x - 6}{x^2 + x - 2}$

15. $\dfrac{18x^4 + 9x^3 + 3x^2}{3x^2 + 1}$

16. $\dfrac{2x^5 - 8x^4 + 2x^3 + x^2}{2x^3 + 1}$

In Exercises 17–32, divide using synthetic division.

17. $(2x^2 + x - 10) \div (x - 2)$ 18. $(x^2 + x - 2) \div (x - 1)$

19. $(3x^2 + 7x - 20) \div (x + 5)$

20. $(5x^2 - 12x - 8) \div (x + 3)$

21. $(4x^3 - 3x^2 + 3x - 1) \div (x - 1)$

22. $(5x^3 - 6x^2 + 3x + 11) \div (x - 2)$

23. $(6x^5 - 2x^3 + 4x^2 - 3x + 1) \div (x - 2)$

24. $(x^5 + 4x^4 - 3x^2 + 2x + 3) \div (x - 3)$

25. $(x^2 - 5x - 5x^3 + x^4) \div (5 + x)$

26. $(x^2 - 6x - 6x^3 + x^4) \div (6 + x)$

27. $\dfrac{x^5 + x^3 - 2}{x - 1}$

28. $\dfrac{x^7 + x^5 - 10x^3 + 12}{x + 2}$

29. $\dfrac{x^4 - 256}{x - 4}$

30. $\dfrac{x^7 - 128}{x - 2}$

31. $\dfrac{2x^5 - 3x^4 + x^3 - x^2 + 2x - 1}{x + 2}$

32. $\dfrac{x^5 - 2x^4 - x^3 + 3x^2 - x + 1}{x - 2}$

In Exercises 33–40, use synthetic division and the Remainder Theorem to find the indicated function value.

33. $f(x) = 2x^3 - 11x^2 + 7x - 5;\ \ f(4)$

34. $f(x) = x^3 - 7x^2 + 5x - 6;\ \ f(3)$

35. $f(x) = 3x^3 - 7x^2 - 2x + 5;\ \ f(-3)$

36. $f(x) = 4x^3 + 5x^2 - 6x - 4;\ \ f(-2)$

37. $f(x) = x^4 + 5x^3 + 5x^2 - 5x - 6;\ \ f(3)$

38. $f(x) = x^4 - 5x^3 + 5x^2 + 5x - 6;\ \ f(2)$

39. $f(x) = 2x^4 - 5x^3 - x^2 + 3x + 2;\ \ f\left(-\dfrac{1}{2}\right)$

40. $f(x) = 6x^4 + 10x^3 + 5x^2 + x + 1;\ \ f\left(-\dfrac{2}{3}\right)$

41. Use synthetic division to divide
$$f(x) = x^3 - 4x^2 + x + 6 \text{ by } x + 1.$$
Use the result to find all zeros of f.

42. Use synthetic division to divide
$$f(x) = x^3 - 2x^2 - x + 2 \text{ by } x + 1.$$
Use the result to find all zeros of f.

43. Solve the equation $2x^3 - 5x^2 + x + 2 = 0$ given that 2 is a zero of $f(x) = 2x^3 - 5x^2 + x + 2$.

44. Solve the equation $2x^3 - 3x^2 - 11x + 6 = 0$ given that -2 is a zero of $f(x) = 2x^3 - 3x^2 - 11x + 6$.

45. Solve the equation $12x^3 + 16x^2 - 5x - 3 = 0$ given that $-\dfrac{3}{2}$ is a root.

46. Solve the equation $3x^3 + 7x^2 - 22x - 8 = 0$ given that $-\dfrac{1}{3}$ is a root.

Practice Plus

In Exercises 47–50, use the graph or the table to determine a solution of each equation. Use synthetic division to verify that this number is a solution of the equation. Then solve the polynomial equation.

47. $x^3 + 2x^2 - 5x - 6 = 0$

$[0, 4, 1]$ by $[-25, 25, 5]$

48. $2x^3 + x^2 - 13x + 6 = 0$

$[-4, 0, 1]$ by $[-25, 25, 5]$

49. $6x^3 - 11x^2 + 6x - 1 = 0$

50. $2x^3 + 11x^2 - 7x - 6 = 0$

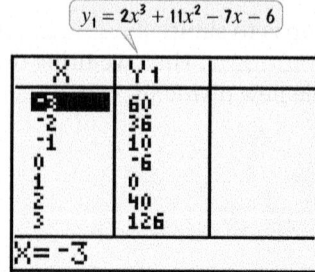

Application Exercises

51. **a.** Use synthetic division to show that 3 is a solution of the polynomial equation
$$14x^3 - 17x^2 - 16x - 177 = 0.$$

b. Use the solution from part (a) to solve this problem. The number of eggs, $f(x)$, in a female moth is a function of her abdominal width, x, in millimeters, modeled by

$$f(x) = 14x^3 - 17x^2 - 16x + 34.$$

What is the abdominal width when there are 211 eggs?

52. a. Use synthetic division to show that 2 is a solution of the polynomial equation

$$2h^3 + 14h^2 - 72 = 0.$$

b. Use the solution from part (a) to solve this problem. The width of a rectangular box is twice the height and the length is 7 inches more than the height. If the volume is 72 cubic inches, find the dimensions of the box.

In Exercises 53–54, write a polynomial that represents the length of each rectangle.

53.

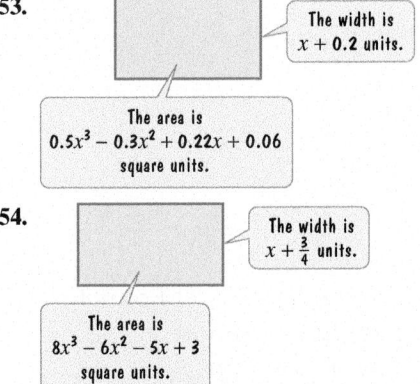

The width is $x + 0.2$ units.

The area is $0.5x^3 - 0.3x^2 + 0.22x + 0.06$ square units.

54.

The width is $x + \frac{3}{4}$ units.

The area is $8x^3 - 6x^2 - 5x + 3$ square units.

During the 1980s, the controversial economist Arthur Laffer promoted the idea that tax increases lead to a reduction in government revenue. Called supply-side economics, the theory uses functions such as

$$f(x) = \frac{80x - 8000}{x - 110}, \ 30 \le x \le 100.$$

This function models the government tax revenue, $f(x)$, in tens of billions of dollars, in terms of the tax rate, x. The graph of the function is shown. It illustrates tax revenue decreasing quite dramatically as the tax rate increases. At a tax rate of (gasp) 100%, the government takes all our money and no one has an incentive to work. With no income earned, zero dollars in tax revenue is generated.

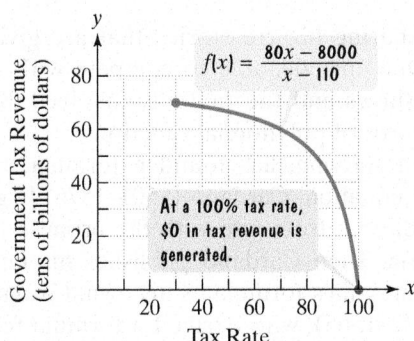

Use function f and its graph to solve Exercises 55–56.

55. a. Find and interpret $f(30)$. Identify the solution as a point on the graph of the function.

b. Rewrite the function by using long division to perform

$$(80x - 8000) \div (x - 110).$$

Then use this new form of the function to find $f(30)$. Do you obtain the same answer as you did in part (a)?

c. Is f a polynomial function? Explain your answer.

56. a. Find and interpret $f(40)$. Identify the solution as a point on the graph of the function.

b. Rewrite the function by using long division to perform

$$(80x - 8000) \div (x - 110).$$

Then use this new form of the function to find $f(40)$. Do you obtain the same answer as you did in part (a)?

c. Is f a polynomial function? Explain your answer.

Writing in Mathematics

57. Explain how to perform long division of polynomials. Use $2x^3 - 3x^2 - 11x + 7$ divided by $x - 3$ in your explanation.

58. In your own words, state the Division Algorithm.

59. How can the Division Algorithm be used to check the quotient and remainder in a long division problem?

60. Explain how to perform synthetic division. Use the division problem in Exercise 57 to support your explanation.

61. State the Remainder Theorem.

62. Explain how the Remainder Theorem can be used to find $f(-6)$ if $f(x) = x^4 + 7x^3 + 8x^2 + 11x + 5$. What advantage is there to using the Remainder Theorem in this situation rather than evaluating $f(-6)$ directly?

63. How can the Factor Theorem be used to determine if $x - 1$ is a factor of $x^3 - 2x^2 - 11x + 12$?

64. If you know that -2 is a zero of

$$f(x) = x^3 + 7x^2 + 4x - 12,$$

explain how to solve the equation

$$x^3 + 7x^2 + 4x - 12 = 0.$$

Technology Exercise

65. For each equation that you solved in Exercises 43–46, use a graphing utility to graph the polynomial function defined by the left side of the equation. Use end behavior to obtain a complete graph. Then use the graph's x-intercepts to verify your solutions.

Critical Thinking Exercises

Make Sense? *In Exercises 66–69, determine whether each statement makes sense or does not make sense, and explain your reasoning.*

66. When performing the division $(x^5 + 1) \div (x + 1)$, there's no need for me to follow all the steps involved in polynomial long division because I can work the problem in my head and see that the quotient must be $x^4 + 1$.

67. Every time I divide polynomials using synthetic division, I am using a highly condensed form of the long division procedure where omitting the variables and exponents does not involve the loss of any essential data.

68. The only nongraphic method that I have for evaluating a function at a given value is to substitute that value into the function's equation.

69. I found the zeros of function f, but I still need to find the solutions of the equation $f(x) = 0$.

In Exercises 70–73, determine whether each statement is true or false. If the statement is false, make the necessary change(s) to produce a true statement.

70. If a trinomial in x of degree 6 is divided by a trinomial in x of degree 3, the degree of the quotient is 2.

71. Synthetic division can be used to find the quotient of $10x^3 - 6x^2 + 4x - 1$ and $x - \frac{1}{2}$.

72. Any problem that can be done by synthetic division can also be done by the method for long division of polynomials.

73. If a polynomial long-division problem results in a remainder that is a whole number, then the divisor is a factor of the dividend.

74. Find k so that $4x + 3$ is a factor of
$$20x^3 + 23x^2 - 10x + k.$$

75. When $2x^2 - 7x + 9$ is divided by a polynomial, the quotient is $2x - 3$ and the remainder is 3. Find the polynomial.

76. Find the quotient of $x^{3n} + 1$ and $x^n + 1$.

77. Synthetic division is a process for dividing a polynomial by $x - c$. The coefficient of x in the divisor is 1. How might synthetic division be used if you are dividing by $2x - 4$?

78. Use synthetic division to show that 5 is a solution of
$$x^4 - 4x^3 - 9x^2 + 16x + 20 = 0.$$
Then solve the polynomial equation.

Preview Exercises

Exercises 79–81 will help you prepare for the material covered in the next section.

79. Solve: $x^2 + 4x - 1 = 0$.

80. Solve: $x^2 + 4x + 6 = 0$.

81. Let $f(x) = a_n(x^4 - 3x^2 - 4)$. If $f(3) = -150$, determine the value of a_n.

SECTION 3.4 Zeros of Polynomial Functions

Objectives

① Use the Rational Zero Theorem to find possible rational zeros.

② Find zeros of a polynomial function.

③ Solve polynomial equations.

④ Use the Linear Factorization Theorem to find polynomials with given zeros.

⑤ Use Descartes's Rule of Signs.

HIERONYMVS CARDANVS.

You stole my formula!

Tartaglia's Secret Formula for One Solution of $x^3 + mx = n$

$$x = \sqrt[3]{\sqrt{\left(\frac{n}{2}\right)^2 + \left(\frac{m}{3}\right)^3} + \frac{n}{2}}$$
$$- \sqrt[3]{\sqrt{\left(\frac{n}{2}\right)^2 + \left(\frac{m}{3}\right)^3} - \frac{n}{2}}$$

Popularizers of mathematics are sharing bizarre stories that are giving math a secure place in popular culture. One episode, able to compete with the wildest fare served up by television talk shows and the tabloids, involves three Italian mathematicians and, of all things, zeros of polynomial functions.

Tartaglia (1499–1557), poor and starving, has found a formula that gives a root for a third-degree polynomial equation. Cardano (1501–1576) begs Tartaglia to reveal the secret formula, wheedling it from him with the promise he will find the impoverished Tartaglia a patron. Then Cardano publishes his famous work *Ars Magna*, in which he presents Tartaglia's formula as his own. Cardano uses his most talented student, Ferrari (1522–1565), who derived a formula for a root of a fourth-degree polynomial equation, to falsely accuse Tartaglia of plagiarism. The dispute becomes violent and Tartaglia is fortunate to escape alive.

The noise from this "You Stole My Formula" episode is quieted by the work of French mathematician Evariste Galois (1811–1832). Galois proved that there is no general formula for finding roots of polynomial equations of degree 5 or higher. There are, however, methods for finding roots. In this section, we study methods for finding zeros of polynomial functions. We begin with a theorem that plays an important role in this process.

GREAT QUESTION!

When finding zeros of polynomial functions, what kinds of numbers can I get?

Here's a quick example that involves all possible kinds of zeros:

$$f(x) = \underbrace{(x + 3)(2x - 1)}(x + \sqrt{2})(x - \sqrt{2})(x - 4 + 5i)(x - 4 - 5i).$$

Zeros: $-3,$ $\dfrac{1}{2},$ $-\sqrt{2},$ $\sqrt{2},$ $4 - 5i,$ $4 + 5i$

Rational zeros Irrational zeros Complex imaginary zeros

Real zeros Nonreal zeros

① Use the Rational Zero Theorem to find possible rational zeros.

The Rational Zero Theorem

The *Rational Zero Theorem* provides us with a tool that we can use to make a list of all possible rational zeros of a polynomial function. Equivalently, the theorem gives all possible rational roots of a polynomial equation. Not every number in the list will be a zero of the function, but every rational zero of the polynomial function will appear somewhere in the list.

> ### The Rational Zero Theorem
>
> If $f(x) = a_n x^n + a_{n-1} x^{n-1} + \cdots + a_1 x + a_0$ has *integer* coefficients and $\dfrac{p}{q}$ (where $\dfrac{p}{q}$ is reduced to lowest terms) is a rational zero of f, then p is a factor of the constant term, a_0, and q is a factor of the leading coefficient, a_n.

You can explore the "why" behind the Rational Zero Theorem in Exercise 92 of Exercise Set 3.4. For now, let's see if we can figure out what the theorem tells us about possible rational zeros. To use the theorem, list all the integers that are factors of the constant term, a_0. Then list all the integers that are factors of the leading coefficient, a_n. Finally, list all possible rational zeros:

$$\text{Possible rational zeros} = \frac{\text{Factors of the constant term}}{\text{Factors of the leading coefficient}}.$$

EXAMPLE 1 Using the Rational Zero Theorem

List all possible rational zeros of $f(x) = -x^4 + 3x^2 + 4$.

SOLUTION

The constant term is 4. We list all of its factors: $\pm 1, \pm 2, \pm 4$. The leading coefficient is -1. Its factors are ± 1.

Factors of the constant term, 4: $\pm 1,$ $\pm 2,$ ± 4
Factors of the leading coefficient, -1: ± 1

Because

$$\text{Possible rational zeros} = \frac{\text{Factors of the constant term}}{\text{Factors of the leading coefficient}},$$

we must take each number in the first row, $\pm1,\ \pm2,\ \pm4$, and divide by each number in the second row, ±1.

$$\text{Possible rational zeros} = \frac{\text{Factors of } 4}{\text{Factors of } -1} = \frac{\pm1,\ \pm2,\ \pm4}{\pm1} = \pm1,\quad \pm2,\quad \pm4$$

| Divide ±1 by ±1. | Divide ±2 by ±1. | Divide ±4 by ±1. |

GREAT QUESTION!

What's the relationship among zeros, roots, and x-intercepts?

The zeros of a function f are the roots, or solutions, of the equation $f(x) = 0$. Furthermore, the real zeros, or real roots, are the x-intercepts of the graph of f.

There are six possible rational zeros. The graph of $f(x) = -x^4 + 3x^2 + 4$ is shown in **Figure 3.27**. The x-intercepts are -2 and 2. Thus, -2 and 2 are the actual rational zeros.

• • •

FIGURE 3.27 The graph of $f(x) = -x^4 + 3x^2 + 4$ shows that -2 and 2 are rational zeros.

⊘ Check Point **1** List all possible rational zeros of

$$f(x) = x^3 + 2x^2 - 5x - 6.$$

EXAMPLE 2 Using the Rational Zero Theorem

List all possible rational zeros of $f(x) = 15x^3 + 14x^2 - 3x - 2$.

SOLUTION

The constant term is -2 and the leading coefficient is 15.

$$\text{Possible rational zeros} = \frac{\text{Factors of the constant term, } -2}{\text{Factors of the leading coefficient, } 15} = \frac{\pm1,\ \pm2}{\pm1,\ \pm3,\ \pm5,\ \pm15}$$

$$= \pm1,\quad \pm2,\quad \pm\frac{1}{3},\quad \pm\frac{2}{3},\quad \pm\frac{1}{5},\quad \pm\frac{2}{5},\quad \pm\frac{1}{15},\quad \pm\frac{2}{15}$$

| Divide ±1 and ±2 by ±1. | Divide ±1 and ±2 by ±3. | Divide ±1 and ±2 by ±5. | Divide ±1 and ±2 by ±15. |

There are 16 possible rational zeros. The actual solution set of

$$15x^3 + 14x^2 - 3x - 2 = 0$$

is $\left\{-1, -\frac{1}{3}, \frac{2}{5}\right\}$, which contains three of the 16 possible zeros. • • •

⊘ Check Point **2** List all possible rational zeros of

$$f(x) = 4x^5 + 12x^4 - x - 3.$$

② Find zeros of a polynomial function.

How do we determine which (if any) of the possible rational zeros are rational zeros of the polynomial function? To find the first rational zero, we can use a trial-and-error process involving synthetic division: If $f(x)$ is divided by $x - c$ and the remainder is zero, then c is a zero of f. After we identify the first rational zero, we use the result of the synthetic division to factor the original polynomial. Then we set each factor equal to zero to identify any additional rational zeros.

EXAMPLE 3 Finding Zeros of a Polynomial Function

Find all zeros of $f(x) = x^3 + 2x^2 - 5x - 6$.

SOLUTION

We begin by listing all possible rational zeros.

Possible rational zeros

$$= \frac{\text{Factors of the constant term, } -6}{\text{Factors of the leading coefficient, } 1} = \frac{\pm 1, \pm 2, \pm 3, \pm 6}{\pm 1} = \pm 1, \pm 2, \pm 3, \pm 6$$

Divide the eight numbers in the numerator by ±1.

Now we will use synthetic division to see if we can find a rational zero among the possible rational zeros $\pm 1, \pm 2, \pm 3, \pm 6$. Keep in mind that if $f(x)$ is divided by $x - c$ and the remainder is zero, then c is a zero of f. Let's start by testing 1. If 1 is not a rational zero, then we will test other possible rational zeros.

Test 1.

Coefficients of
$f(x) = x^3 + 2x^2 - 5x - 6$

Possible
rational zero

$$\underline{1}\,|\ \ 1 \quad 2 \ -5 \ -6$$
$$\quad \ \ \ 1 \quad 3 \ -2$$
$$\overline{\ \ 1 \quad 3 \ -2 \ -8}$$

*The nonzero remainder
shows that 1 is not a zero.*

Test 2.

Coefficients of
$f(x) = x^3 + 2x^2 - 5x - 6$

Possible
rational zero

$$\underline{2}\,|\ \ 1 \quad 2 \ -5 \ -6$$
$$\quad \ \ \ 2 \quad 8 \ \ \ 6$$
$$\overline{\ \ 1 \quad 4 \quad 3 \ \ \ 0}$$

*The zero remainder
shows that 2 is a zero.*

The zero remainder tells us that 2 is a zero of the polynomial function $f(x) = x^3 + 2x^2 - 5x - 6$. Equivalently, 2 is a solution, or root, of the polynomial equation $x^3 + 2x^2 - 5x - 6 = 0$. Thus, $x - 2$ is a factor of the polynomial. The first three numbers in the bottom row of the synthetic division on the right, 1, 4, and 3, give the coefficients of the other factor. This factor is $x^2 + 4x + 3$.

$$x^3 + 2x^2 - 5x - 6 = 0 \quad \text{\small Finding the zeros of } f(x) = x^3 + 2x^2 - 5x - 6 \text{ is}$$
$$\text{\small the same as finding the roots of this equation.}$$

$$(x - 2)(x^2 + 4x + 3) = 0 \quad \text{\small Factor using the result from the synthetic division.}$$
$$(x - 2)(x + 3)(x + 1) = 0 \quad \text{\small Factor completely.}$$
$$x - 2 = 0 \quad \text{or} \quad x + 3 = 0 \quad \text{or} \quad x + 1 = 0 \quad \text{\small Set each factor equal to zero.}$$
$$x = 2 \qquad\qquad x = -3 \qquad\qquad x = -1 \quad \text{\small Solve for x.}$$

The solution set is $\{-3, -1, 2\}$. The zeros of f are $-3, -1$, and 2. • • •

 Check Point 3 Find all zeros of
$$f(x) = x^3 + 8x^2 + 11x - 20.$$

Our work in Example 3 involved finding zeros of a third-degree polynomial function. The Rational Zero Theorem is a tool that allows us to rewrite such functions as products of two factors, one linear and one quadratic. Zeros of the quadratic factor are found by factoring, the quadratic formula, or the square root property.

EXAMPLE 4 Finding Zeros of a Polynomial Function

Find all zeros of $f(x) = x^3 + 7x^2 + 11x - 3$.

SOLUTION

We begin by listing all possible rational zeros.

$$\text{Possible rational zeros} = \frac{\text{Factors of the constant term, } -3}{\text{Factors of the leading coefficient, } 1} = \frac{\pm 1, \pm 3}{\pm 1} = \pm 1, \pm 3$$

Now we will use synthetic division to see if we can find a rational zero among the four possible rational zeros, $\pm 1, \pm 3$, of $f(x) = x^3 + 7x^2 + 11x - 3$.

Test 1.	Test −1.	Test 3.	Test −3.
$\underline{1}\rvert$ 1 7 11 −3	$\underline{-1}\rvert$ 1 7 11 −3	$\underline{3}\rvert$ 1 7 11 −3	$\underline{-3}\rvert$ 1 7 11 −3
1 8 19	−1 −6 −5	3 30 123	−3 −12 3
1 8 19 16	1 6 5 −8	1 10 41 120	1 4 −1 0

The zero remainder when testing −3 tells us that −3 is a zero of the polynomial function $f(x) = x^3 + 7x^2 + 11x - 3$. To find all zeros of f, we proceed as follows:

$$x^3 + 7x^2 + 11x - 3 = 0$$
Finding the zeros of f is the same thing as finding the roots of f(x) = 0.

$$(x + 3)(x^2 + 4x - 1) = 0$$
This result is from the last synthetic division, shown above. The first three numbers in the bottom row, 1, 4, and −1, give the coefficients of the second factor.

$$x + 3 = 0 \quad \text{or} \quad x^2 + 4x - 1 = 0$$
$$x = -3.$$
Set each factor equal to 0.
Solve the linear equation.

We can use the quadratic formula to solve $x^2 + 4x - 1 = 0$.

$$x = \frac{-b \pm \sqrt{b^2 - 4ac}}{2a}$$
We use the quadratic formula because $x^2 + 4x - 1$ cannot be factored.

$$= \frac{-4 \pm \sqrt{4^2 - 4(1)(-1)}}{2(1)}$$
Let $a = 1, b = 4,$ and $c = -1$.

$$= \frac{-4 \pm \sqrt{20}}{2}$$
Multiply and subtract under the radical: $4^2 - 4(1)(-1) = 16 - (-4) = 16 + 4 = 20.$

$$= \frac{-4 \pm 2\sqrt{5}}{2}$$
$\sqrt{20} = \sqrt{4 \cdot 5} = 2\sqrt{5}$

$$= -2 \pm \sqrt{5}$$
Divide the numerator and the denominator by 2.

The solution set is $\{-3, -2 - \sqrt{5}, -2 + \sqrt{5}\}$. The zeros of $f(x) = x^3 + 7x^2 + 11x - 3$ are $-3, -2 - \sqrt{5},$ and $-2 + \sqrt{5}$. Among these three real zeros, one zero is rational and two are irrational. • • •

✓ **Check Point 4** Find all zeros of $f(x) = x^3 + x^2 - 5x - 2$.

If the degree of a polynomial function or equation is 4 or higher, it is often necessary to find more than one linear factor by synthetic division.

One way to speed up the process of finding the first zero is to graph the function. Any x-intercept is a zero.

③ Solve polynomial equations.

EXAMPLE 5 Solving a Polynomial Equation

Solve: $x^4 - 6x^2 - 8x + 24 = 0$.

SOLUTION

Recall that we refer to the *zeros* of a polynomial function and the *roots* of a polynomial equation. Because we are given an equation, we will use the word "roots," rather than "zeros," in the solution process. We begin by listing all possible rational roots.

$$\text{Possible rational roots} = \frac{\text{Factors of the constant term, 24}}{\text{Factors of the leading coefficient, 1}}$$

$$= \frac{\pm1, \pm2, \pm3, \pm4, \pm6, \pm8, \pm12, \pm24}{\pm1}$$

$$= \pm1, \pm2, \pm3, \pm4, \pm6, \pm8, \pm12, \pm24$$

Part of the graph of $f(x) = x^4 - 6x^2 - 8x + 24$ is shown in **Figure 3.28**. Because the x-intercept is 2, we will test 2 by synthetic division and show that it is a root of the given equation. Without the graph, the procedure would be to start the trial-and-error synthetic division with 1 and proceed until a zero remainder is found, as we did in Example 4.

FIGURE 3.28 The graph of $f(x) = x^4 - 6x^2 - 8x + 24$ in a $[-1, 5, 1]$ by $[-2, 10, 1]$ viewing rectangle

$$\begin{array}{r|rrrrr} 2 & 1 & 0 & -6 & -8 & 24 \\ & & 2 & 4 & -4 & -24 \\ \hline & 1 & 2 & -2 & -12 & 0 \end{array}$$

Careful!
$x^4 - 6x^2 - 8x + 24 = x^4 + 0x^3 - 6x^2 - 8x + 24$

The zero remainder indicates that 2 is a root of
$x^4 - 6x^2 - 8x + 24 = 0.$

Now we can rewrite the given equation in factored form.

$$x^4 - 6x^2 - 8x + 24 = 0 \qquad \text{This is the given equation.}$$

$$(x - 2)(x^3 + 2x^2 - 2x - 12) = 0 \qquad \text{This is the result obtained from the synthetic division. The first four numbers in the bottom row, 1, 2, −2, and −12, give the coefficients of the second factor.}$$

$$x - 2 = 0 \quad \text{or} \quad x^3 + 2x^2 - 2x - 12 = 0 \qquad \text{Set each factor equal to 0.}$$

We can use the same approach to look for rational roots of the polynomial equation $x^3 + 2x^2 - 2x - 12 = 0$, listing all possible rational roots. Without the graph in **Figure 3.28**, the procedure would be to start testing possible rational roots by trial-and-error synthetic division. However, take a second look at the graph in **Figure 3.28**. Because the graph turns around at 2, this means that 2 is a root of even multiplicity. Thus, 2 must also be a root of $x^3 + 2x^2 - 2x - 12 = 0$, confirmed by the following synthetic division.

$$\begin{array}{r|rrrr} 2 & 1 & 2 & -2 & -12 \\ & & 2 & 8 & 12 \\ \hline & 1 & 4 & 6 & 0 \end{array}$$

These are the coefficients of
$x^3 + 2x^2 - 2x - 12 = 0.$

The zero remainder indicates that 2 is a root of $x^3 + 2x^2 - 2x - 12 = 0.$

Now we can solve the original equation as follows:

$$x^4 - 6x^2 - 8x + 24 = 0 \qquad \text{This is the given equation.}$$

$$(x - 2)(x^3 + 2x^2 - 2x - 12) = 0 \qquad \text{This factorization was obtained from the first synthetic division.}$$

$$(x - 2)(x - 2)(x^2 + 4x + 6) = 0 \qquad \text{This factorization was obtained from the second synthetic division. The first three numbers in the bottom row, 1, 4, and 6, give the coefficients of the third factor.}$$

$$x - 2 = 0 \quad \text{or} \quad x - 2 = 0 \quad \text{or} \quad x^2 + 4x + 6 = 0 \qquad \text{Set each factor equal to 0.}$$

$$x = 2 \qquad\qquad x = 2. \qquad\qquad\qquad \text{Solve the linear equations.}$$

We can use the quadratic formula to solve $x^2 + 4x + 6 = 0$.

$$x = \frac{-b \pm \sqrt{b^2 - 4ac}}{2a}$$ We use the quadratic formula because $x^2 + 4x + 6$ cannot be factored.

$$= \frac{-4 \pm \sqrt{4^2 - 4(1)(6)}}{2(1)}$$ Let $a = 1$, $b = 4$, and $c = 6$.

$$= \frac{-4 \pm \sqrt{-8}}{2}$$ Multiply and subtract under the radical: $4^2 - 4(1)(6) = 16 - 24 = -8$.

$$= \frac{-4 \pm 2i\sqrt{2}}{2}$$ $\sqrt{-8} = \sqrt{4(2)(-1)} = 2i\sqrt{2}$

$$= -2 \pm i\sqrt{2}$$ Simplify.

FIGURE 3.28 (repeated) The graph of $f(x) = x^4 - 6x^2 - 8x + 24$ in a $[-1, 5, 1]$ by $[-2, 10, 1]$ viewing rectangle

The solution set of the original equation, $x^4 - 6x^2 - 8x + 24 = 0$, is $\{2, -2 - i\sqrt{2}, -2 + i\sqrt{2}\}$. **Figure 3.28**, repeated in the margin, shows that a graphing utility does not reveal the two imaginary roots. ●●●

In Example 5, 2 is a repeated root of the equation with multiplicity 2. Counting this multiple root separately, the fourth-degree equation $x^4 - 6x^2 - 8x + 24 = 0$ has four roots: 2, 2, $-2 + i\sqrt{2}$, and $-2 - i\sqrt{2}$. The equation and its roots illustrate two general properties:

Properties of Roots of Polynomial Equations

1. If a polynomial equation is of degree n, then counting multiple roots separately, the equation has n roots.
2. If $a + bi$ is a root of a polynomial equation with real coefficients ($b \neq 0$), then the imaginary number $a - bi$ is also a root. Imaginary roots, if they exist, occur in conjugate pairs.

 Check Point 5 Solve: $x^4 - 6x^3 + 22x^2 - 30x + 13 = 0$.

The Fundamental Theorem of Algebra

The fact that a polynomial equation of degree n has n roots is a consequence of a theorem proved in 1799 by a 22-year-old student named Carl Friedrich Gauss in his doctoral dissertation. His result is called the **Fundamental Theorem of Algebra**.

The Fundamental Theorem of Algebra

If $f(x)$ is a polynomial of degree n, where $n \geq 1$, then the equation $f(x) = 0$ has at least one complex root.

Suppose, for example, that $f(x) = 0$ represents a polynomial equation of degree n. By the Fundamental Theorem of Algebra, we know that this equation has at least one complex root; we'll call it c_1. By the Factor Theorem, we know that $x - c_1$ is a factor of $f(x)$. Therefore, we obtain

$$(x - c_1)q_1(x) = 0$$ The degree of the polynomial $q_1(x)$ is $n - 1$.

$$x - c_1 = 0 \quad \text{or} \quad q_1(x) = 0.$$ Set each factor equal to 0.

If the degree of $q_1(x)$ is at least 1, by the Fundamental Theorem of Algebra, the equation $q_1(x) = 0$ has at least one complex root. We'll call it c_2. The Factor Theorem gives us

$$q_1(x) = 0$$ The degree of $q_1(x)$ is $n - 1$.

$$(x - c_2)q_2(x) = 0$$ The degree of $q_2(x)$ is $n - 2$.

$$x - c_2 = 0 \quad \text{or} \quad q_2(x) = 0.$$ Set each factor equal to 0.

GREAT QUESTION!

Do all polynomial equations have at least one imaginary root?

No. As you read the Fundamental Theorem of Algebra, don't confuse *complex root* with *imaginary root* and conclude that every polynomial equation has at least one imaginary root. Recall that complex numbers, $a + bi$, include both real numbers ($b = 0$) and imaginary numbers ($b \neq 0$).

Let's see what we have up to this point and then continue the process.

$$f(x) = 0 \qquad \text{This is the original polynomial equation of degree } n.$$
$$(x - c_1)q_1(x) = 0 \qquad \text{This is the result from our first application of the}$$
$$\text{Fundamental Theorem.}$$
$$(x - c_1)(x - c_2)q_2(x) = 0 \qquad \text{This is the result from our second application of}$$
$$\text{the Fundamental Theorem.}$$

By continuing this process, we will obtain the product of n linear factors. Setting each of these linear factors equal to zero results in n complex roots. Thus, if $f(x)$ is a polynomial of degree n, where $n \geq 1$, then $f(x) = 0$ has exactly n roots, where roots are counted according to their multiplicity.

The Linear Factorization Theorem

4 Use the Linear Factorization Theorem to find polynomials with given zeros.

In Example 5, we found that $x^4 - 6x^2 - 8x + 24 = 0$ has $\left\{ 2, -2 \pm i\sqrt{2} \right\}$ as a solution set, where 2 is a repeated root with multiplicity 2. The polynomial can be factored over the complex nonreal numbers as follows:

$$f(x) = x^4 - 6x^2 - 8x + 24 \qquad \text{These are the four zeros.}$$
$$= \underbrace{[x - (-2 + i\sqrt{2})]}\underbrace{[x - (-2 - i\sqrt{2})]}\underbrace{(x - 2)}\underbrace{(x - 2)}.$$

These are the linear factors.

This fourth-degree polynomial has four linear factors. Just as an nth-degree polynomial equation has n roots, an nth-degree polynomial has n linear factors. This is formally stated as the **Linear Factorization Theorem**.

> ### The Linear Factorization Theorem
>
> If $f(x) = a_n x^n + a_{n-1}x^{n-1} + \cdots + a_1 x + a_0$, where $n \geq 1$ and $a_n \neq 0$, then
>
> $$f(x) = a_n(x - c_1)(x - c_2) \cdots (x - c_n),$$
>
> where c_1, c_2, \ldots, c_n are complex numbers (possibly real and not necessarily distinct). In words: An nth-degree polynomial can be expressed as the product of a nonzero constant and n linear factors, where each linear factor has a leading coefficient of 1.

Many of our problems involving polynomial functions and polynomial equations dealt with the process of finding zeros and roots. The Linear Factorization Theorem enables us to reverse this process, finding a polynomial function when the zeros are given.

EXAMPLE 6 Finding a Polynomial Function with Given Zeros

Find a fourth-degree polynomial function $f(x)$ with real coefficients that has -2, 2, and i as zeros and such that $f(3) = -150$.

SOLUTION

Because i is a zero and the polynomial has real coefficients, the conjugate, $-i$, must also be a zero. We can now use the Linear Factorization Theorem.

$$f(x) = a_n(x - c_1)(x - c_2)(x - c_3)(x - c_4) \qquad \text{This is the linear factorization for a}$$
$$\text{fourth-degree polynomial.}$$
$$= a_n(x + 2)(x - 2)(x - i)(x + i) \qquad \text{Use the given zeros:}$$
$$c_1 = -2, c_2 = 2, c_3 = i, \text{ and, from above,}$$
$$c_4 = -i.$$
$$= a_n(x^2 - 4)(x^2 + 1) \qquad \text{Multiply: } (x - i)(x + i) = x^2 - i^2 =$$
$$x^2 - (-1) = x^2 + 1.$$

TECHNOLOGY

Graphic Connections

The graph of
$f(x) = -3x^4 + 9x^2 + 12$, shown
in a $[-3, 3, 1]$ by $[-200, 20, 20]$
viewing rectangle, verifies that -2
and 2 are real zeros. By tracing
along the curve, we can check that
$f(3) = -150$.

| $f(x) = a_n(x^4 - 3x^2 - 4)$ | Complete the multiplication for $f(x) = a_n(x^2 - 4)(x^2 + 1)$. |

$$f(3) = a_n(3^4 - 3 \cdot 3^2 - 4) = -150 \qquad \text{To find } a_n, \text{ use the fact that } f(3) = -150.$$

$$a_n(81 - 27 - 4) = -150 \qquad \text{Solve for } a_n.$$

$$50a_n = -150 \qquad \text{Simplify: } 81 - 27 - 4 = 50.$$

$$a_n = -3 \qquad \text{Divide both sides by 50.}$$

Substituting -3 for a_n in the formula for $f(x)$, we obtain

$$f(x) = -3(x^4 - 3x^2 - 4).$$

Equivalently,

$$f(x) = -3x^4 + 9x^2 + 12. \qquad \bullet\bullet\bullet$$

 Check Point **6** Find a third-degree polynomial function $f(x)$ with real coefficients that has -3 and i as zeros and such that $f(1) = 8$.

⑤ Use Descartes's Rule of Signs.

"An equation can have as many true [positive] roots as it contains changes of sign, from plus to minus or from minus to plus." René Descartes (1596–1650) in *La Géométrie* (1637)

GREAT QUESTION!

Does Descartes's Rule of Signs include both rational and irrational zeros?

Yes. The number of real zeros given by Descartes's Rule of Signs includes rational zeros from a list of possible rational zeros, as well as irrational zeros not on the list. It does not include any imaginary zeros.

Descartes's Rule of Signs

Because an nth-degree polynomial equation might have roots that are imaginary numbers, we should note that such an equation can have *at most n* real roots. **Descartes's Rule of Signs** provides even more specific information about the number of real zeros that a polynomial can have. The rule is based on considering *variations in sign* between consecutive coefficients. For example, the function $f(x) = 3x^7 - 2x^5 - x^4 + 7x^2 + x - 3$ has three sign changes:

$$f(x) = 3x^7 \underset{\text{sign change}}{-} 2x^5 - x^4 \underset{\text{sign change}}{+} 7x^2 + x \underset{\text{sign change}}{-} 3.$$

Descartes's Rule of Signs

Let $f(x) = a_n x^n + a_{n-1}x^{n-1} + \cdots + a_2 x^2 + a_1 x + a_0$ be a polynomial with real coefficients.

1. The number of *positive real zeros* of f is either
 a. the same as the number of sign changes of $f(x)$

 or

 b. less than the number of sign changes of $f(x)$ by a positive even integer. If $f(x)$ has only one variation in sign, then f has exactly one positive real zero.

2. The number of *negative real zeros* of f is either
 a. the same as the number of sign changes of $f(-x)$

 or

 b. less than the number of sign changes of $f(-x)$ by a positive even integer. If $f(-x)$ has only one variation in sign, then f has exactly one negative real zero.

GREAT QUESTION!

Does a polynomial function have to be written in descending powers of x when counting sign changes?

Yes. Be sure that the function's equation is in this form before using Descartes's Rule of signs.

Table 3.1 illustrates what Descartes's Rule of Signs tells us about the positive real zeros of various polynomial functions.

Table 3.1 Descartes's Rule of Signs and Positive Real Zeros

Polynomial Function	Sign Changes	Conclusion
$f(x) = 3x^7 - 2x^5 - x^4 + 7x^2 + x - 3$ sign change sign change sign change	3	There are 3 positive real zeros. or There is $3 - 2 = 1$ positive real zero.
$f(x) = 4x^5 + 2x^4 - 3x^2 + x + 5$ sign change sign change	2	There are 2 positive real zeros. or There are $2 - 2 = 0$ positive real zeros.
$f(x) = -7x^6 - 5x^4 + x + 9$ sign change	1	There is 1 positive real zero.

EXAMPLE 7 Using Descartes's Rule of Signs

Determine the possible numbers of positive and negative real zeros of $f(x) = x^3 + 2x^2 + 5x + 4$.

SOLUTION

1. To find possibilities for positive real zeros, count the number of sign changes in the equation for $f(x)$. Because all the coefficients are positive, there are no variations in sign. Thus, there are no positive real zeros.

2. To find possibilities for negative real zeros, count the number of sign changes in the equation for $f(-x)$. We obtain this equation by replacing x with $-x$ in the given function.

$$f(x) = x^3 + 2x^2 + 5x + 4$$

Replace x with $-x$.

$$f(-x) = (-x)^3 + 2(-x)^2 + 5(-x) + 4$$
$$= -x^3 + 2x^2 - 5x + 4$$

Now count the sign changes.

$$f(-x) = -x^3 + 2x^2 - 5x + 4$$

sign change sign change sign change

There are three variations in sign. The number of negative real zeros of f is either equal to the number of sign changes, 3, or is less than this number by an even integer. This means that either there are 3 negative real zeros or there is $3 - 2 = 1$ negative real zero. • • •

What do the results of Example 7 mean in terms of solving

$$x^3 + 2x^2 + 5x + 4 = 0?$$

Without using Descartes's Rule of Signs, we list the possible rational roots as follows:

Possible rational roots

$$= \frac{\text{Factors of the constant term, 4}}{\text{Factors of the leading coefficient, 1}} = \frac{\pm 1, \pm 2, \pm 4}{\pm 1} = \pm 1, \pm 2, \pm 4.$$

We see that the possible rational roots of $x^3 + 2x^2 + 5x + 4 = 0$ are $\pm 1, \pm 2$, and ± 4. However, Descartes's Rule of Signs informed us that $f(x) = x^3 + 2x^2 + 5x + 4$ has no positive real zeros. Thus, the polynomial equation $x^3 + 2x^2 + 5x + 4 = 0$ has no positive real roots. This means that we can eliminate the positive numbers 1, 2, and 4 from our list of possible rational roots. Possible rational roots include only $-1, -2,$ and -4. We can use synthetic division and test the first of these three possible rational roots of $x^3 + 2x^2 + 5x + 4 = 0$ as follows:

Test
−1.

$$\begin{array}{r|rrrr} -1 & 1 & 2 & 5 & 4 \\ & & -1 & -1 & -4 \\ \hline & 1 & 1 & 4 & 0. \end{array}$$

The zero remainder shows
that −1 is a root.

By solving the equation $x^3 + 2x^2 + 5x + 4 = 0$, you will find that this equation of degree 3 has three roots. One root is -1 and the other two roots are imaginary numbers in a conjugate pair. Verify this by completing the solution process.

✅ Check Point **7** Determine the possible numbers of positive and negative real zeros of $f(x) = x^4 - 14x^3 + 71x^2 - 154x + 120$.

CONCEPT AND VOCABULARY CHECK

Fill in each blank so that the resulting statement is true.

1. Consider the polynomial function with integer coefficients
$$f(x) = a_n x^n + a_{n-1} x^{n-1} + \cdots + a_1 x + a_0, a_n \neq 0.$$
The Rational Zero Theorem states that if $\dfrac{p}{q}$ is a rational zero of f (where $\dfrac{p}{q}$ is reduced to lowest terms), then p is a factor of _____ and q is a factor of _____.

2. True or false: $\dfrac{3}{2}$ is a possible rational zero of $f(x) = 2x^3 + 11x^2 - 7x - 6.$ _____

3. True or false: $\dfrac{1}{2}$ is a possible rational zero of $f(x) = 3x^4 - 3x^3 + x^2 - x + 1.$ _____

4. If a polynomial equation is of degree n, then counting multiple roots separately, the equation has _____ roots.

5. If $a + bi$ is a root of a polynomial equation with real coefficients, $b \neq 0$, then _____ is also a root of the equation.

6. Consider solving $2x^3 + 11x^2 - 7x - 6 = 0$. The synthetic division shown below indicates that _____ is a root.

$$\begin{array}{r|rrrr} -6 & 2 & 11 & -7 & -6 \\ & & -12 & 6 & 6 \\ \hline & 2 & -1 & -1 & 0 \end{array}$$
Based on the synthetic division, $2x^3 + 11x^2 - 7x - 6 = 0$ can be written in factored form as _____.

7. The Linear Factorization Theorem states that an nth-degree polynomial can be expressed as the product of a nonzero constant and _____ linear factors, where each linear factor has a leading coefficient of ____.

Use Descartes's Rule of Signs to determine whether each statement is true of false.

8. A polynomial function with four sign changes must have four positive real zeros. _____

9. A polynomial function with one sign change must have one positive real zero. _____

10. A polynomial function with seven sign changes can have one, three, five, or seven positive real zeros. _____

EXERCISE SET 3.4

Practice Exercises

In Exercises 1–8, use the Rational Zero Theorem to list all possible rational zeros for each given function.

1. $f(x) = x^3 + x^2 - 4x - 4$
2. $f(x) = x^3 + 3x^2 - 6x - 8$
3. $f(x) = 3x^4 - 11x^3 - x^2 + 19x + 6$

4. $f(x) = 2x^4 + 3x^3 - 11x^2 - 9x + 15$
5. $f(x) = 4x^4 - x^3 + 5x^2 - 2x - 6$
6. $f(x) = 3x^4 - 11x^3 - 3x^2 - 6x + 8$
7. $f(x) = x^5 - x^4 - 7x^3 + 7x^2 - 12x - 12$
8. $f(x) = 4x^5 - 8x^4 - x + 2$

In Exercises 9–16,

 a. List all possible rational zeros.

 b. Use synthetic division to test the possible rational zeros and find an actual zero.

 c. Use the quotient from part (b) to find the remaining zeros of the polynomial function.

9. $f(x) = x^3 + x^2 - 4x - 4$

10. $f(x) = x^3 - 2x^2 - 11x + 12$

11. $f(x) = 2x^3 - 3x^2 - 11x + 6$

12. $f(x) = 2x^3 - 5x^2 + x + 2$

13. $f(x) = x^3 + 4x^2 - 3x - 6$

14. $f(x) = 2x^3 + x^2 - 3x + 1$

15. $f(x) = 2x^3 + 6x^2 + 5x + 2$

16. $f(x) = x^3 - 4x^2 + 8x - 5$

In Exercises 17–24,

 a. List all possible rational roots.

 b. Use synthetic division to test the possible rational roots and find an actual root.

 c. Use the quotient from part (b) to find the remaining roots and solve the equation.

17. $x^3 - 2x^2 - 11x + 12 = 0$

18. $x^3 - 2x^2 - 7x - 4 = 0$

19. $x^3 - 10x - 12 = 0$

20. $x^3 - 5x^2 + 17x - 13 = 0$

21. $6x^3 + 25x^2 - 24x + 5 = 0$

22. $2x^3 - 5x^2 - 6x + 4 = 0$

23. $x^4 - 2x^3 - 5x^2 + 8x + 4 = 0$

24. $x^4 - 2x^2 - 16x - 15 = 0$

In Exercises 25–32, find an nth-degree polynomial function with real coefficients satisfying the given conditions. If you are using a graphing utility, use it to graph the function and verify the real zeros and the given function value.

25. $n = 3$; 1 and $5i$ are zeros; $f(-1) = -104$

26. $n = 3$; 4 and $2i$ are zeros; $f(-1) = -50$

27. $n = 3$; -5 and $4 + 3i$ are zeros; $f(2) = 91$

28. $n = 3$; 6 and $-5 + 2i$ are zeros; $f(2) = -636$

29. $n = 4$; i and $3i$ are zeros; $f(-1) = 20$

30. $n = 4$; $-2, -\frac{1}{2}$, and i are zeros; $f(1) = 18$

31. $n = 4$; $-2, 5$, and $3 + 2i$ are zeros; $f(1) = -96$

32. $n = 4$; $-4, \frac{1}{3}$, and $2 + 3i$ are zeros; $f(1) = 100$

In Exercises 33–38, use Descartes's Rule of Signs to determine the possible number of positive and negative real zeros for each given function.

33. $f(x) = x^3 + 2x^2 + 5x + 4$

34. $f(x) = x^3 + 7x^2 + x + 7$

35. $f(x) = 5x^3 - 3x^2 + 3x - 1$

36. $f(x) = -2x^3 + x^2 - x + 7$

37. $f(x) = 2x^4 - 5x^3 - x^2 - 6x + 4$

38. $f(x) = 4x^4 - x^3 + 5x^2 - 2x - 6$

In Exercises 39–52, find all zeros of the polynomial function or solve the given polynomial equation. Use the Rational Zero Theorem, Descartes's Rule of Signs, and possibly the graph of the polynomial function shown by a graphing utility as an aid in obtaining the first zero or the first root.

39. $f(x) = x^3 - 4x^2 - 7x + 10$

40. $f(x) = x^3 + 12x^2 + 21x + 10$

41. $2x^3 - x^2 - 9x - 4 = 0$

42. $3x^3 - 8x^2 - 8x + 8 = 0$

43. $f(x) = x^4 - 2x^3 + x^2 + 12x + 8$

44. $f(x) = x^4 - 4x^3 - x^2 + 14x + 10$

45. $x^4 - 3x^3 - 20x^2 - 24x - 8 = 0$

46. $x^4 - x^3 + 2x^2 - 4x - 8 = 0$

47. $f(x) = 3x^4 - 11x^3 - x^2 + 19x + 6$

48. $f(x) = 2x^4 + 3x^3 - 11x^2 - 9x + 15$

49. $4x^4 - x^3 + 5x^2 - 2x - 6 = 0$

50. $3x^4 - 11x^3 - 3x^2 - 6x + 8 = 0$

51. $2x^5 + 7x^4 - 18x^2 - 8x + 8 = 0$

52. $4x^5 + 12x^4 - 41x^3 - 99x^2 + 10x + 24 = 0$

Practice Plus

Exercises 53–60 show incomplete graphs of given polynomial functions.

 a. Find all the zeros of each function.

 b. Without using a graphing utility, draw a complete graph of the function.

53. $f(x) = -x^3 + x^2 + 16x - 16$

$[-5, 0, 1]$ by $[-40, 25, 5]$

54. $f(x) = -x^3 + 3x^2 - 4$

$[-2, 0, 1]$ by $[-10, 10, 1]$

55. $f(x) = 4x^3 - 8x^2 - 3x + 9$

$[-2, 0, 1]$ by $[-10, 10, 1]$

56. $f(x) = 3x^3 + 2x^2 + 2x - 1$

$[0, 2, \frac{1}{6}]$ by $[-3, 15, 1]$

57. $f(x) = 2x^4 - 3x^3 - 7x^2 - 8x + 6$

$[0, 1, \frac{1}{4}]$ by $[-10, 10, 1]$

58. $f(x) = 2x^4 + 2x^3 - 22x^2 - 18x + 36$

$[0, 4, 1]$ by $[-50, 50, 10]$

59. $f(x) = 3x^5 + 2x^4 - 15x^3 - 10x^2 + 12x + 8$

$[0, 4, 1]$ by $[-20, 25, 5]$

60. $f(x) = -5x^4 + 4x^3 - 19x^2 + 16x + 4$

$[0, 2, 1]$ by $[-10, 10, 1]$

Application Exercises

A popular model of carry-on luggage has a length that is 10 inches greater than its depth. Airline regulations require that the sum of the length, width, and depth cannot exceed 40 inches. These conditions, with the assumption that this sum is 40 inches, can be modeled by a function that gives the volume of the luggage, V, in cubic inches, in terms of its depth, x, in inches.

Volume	=	depth	·	length	·	width: 40 − (depth + length)

$$V(x) = x \cdot (x + 10) \cdot [40 - (x + x + 10)]$$
$$V(x) = x(x + 10)(30 - 2x)$$

Use function V to solve Exercises 61–62.

61. If the volume of the carry-on luggage is 2000 cubic inches, determine two possibilities for its depth. Where necessary, round to the nearest tenth of an inch.

62. If the volume of the carry-on luggage is 1500 cubic inches, determine two possibilities for its depth. Where necessary, round to the nearest tenth of an inch.

Use the graph of the function modeling the volume of the carry-on luggage to solve Exercises 63–64.

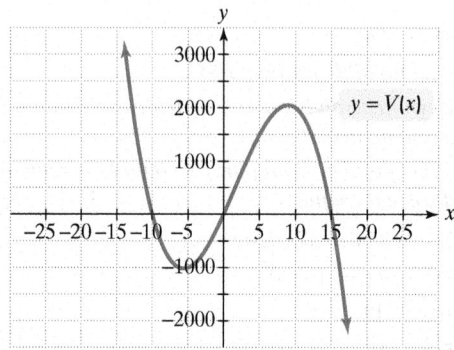

63. a. Identify your answers from Exercise 61 as points on the graph.

 b. Use the graph to describe a realistic domain, x, for the volume function, where x represents the depth of the carry-on luggage.

64. a. Identify your answers from Exercise 62 as points on the graph.

 b. Use the graph to describe a realistic domain, x, for the volume function, where x represents the depth of the carry-on luggage.

Writing in Mathematics

65. Describe how to find the possible rational zeros of a polynomial function.

66. How does the linear factorization of $f(x)$, that is,

$$f(x) = a_n(x - c_1)(x - c_2) \cdots (x - c_n),$$

show that a polynomial equation of degree n has n roots?

67. Describe how to use Descartes's Rule of Signs to determine the possible number of positive real zeros of a polynomial function.

68. Describe how to use Descartes's Rule of Signs to determine the possible number of negative roots of a polynomial equation.

69. Why must every polynomial equation with real coefficients of degree 3 have at least one real root?

70. Explain why the equation $x^4 + 6x^2 + 2 = 0$ has no rational roots.

71. Suppose $\frac{3}{4}$ is a root of a polynomial equation. What does this tell us about the leading coefficient and the constant term in the equation?

Technology Exercises

The equations in Exercises 72–75 have real roots that are rational. Use the Rational Zero Theorem to list all possible rational roots. Then graph the polynomial function in the given viewing rectangle to determine which possible rational roots are actual roots of the equation.

72. $2x^3 - 15x^2 + 22x + 15 = 0$; $[-1, 6, 1]$ by $[-50, 50, 10]$
73. $6x^3 - 19x^2 + 16x - 4 = 0$; $[0, 2, 1]$ by $[-3, 2, 1]$
74. $2x^4 + 7x^3 - 4x^2 - 27x - 18 = 0$; $[-4, 3, 1]$ by $[-45, 45, 15]$
75. $4x^4 + 4x^3 + 7x^2 - x - 2 = 0$; $[-2, 2, 1]$ by $[-5, 5, 1]$

76. Use Descartes's Rule of Signs to determine the possible number of positive and negative real zeros of $f(x) = 3x^4 + 5x^2 + 2$. What does this mean in terms of the graph of f? Verify your result by using a graphing utility to graph f.

77. Use Descartes's Rule of Signs to determine the possible number of positive and negative real zeros of $f(x) = x^5 - x^4 + x^3 - x^2 + x - 8$. Verify your result by using a graphing utility to graph f.

78. Write equations for several polynomial functions of odd degree and graph each function. Is it possible for the graph to have no real zeros? Explain. Try doing the same thing for polynomial functions of even degree. Now is it possible to have no real zeros?

Use a graphing utility to obtain a complete graph for each polynomial function in Exercises 79–82. Then determine the number of real zeros and the number of imaginary zeros for each function.

79. $f(x) = x^3 - 6x - 9$
80. $f(x) = 3x^5 - 2x^4 + 6x^3 - 4x^2 - 24x + 16$
81. $f(x) = 3x^4 + 4x^3 - 7x^2 - 2x - 3$
82. $f(x) = x^6 - 64$

Critical Thinking Exercises

Make Sense? *In Exercises 83–86, determine whether each statement makes sense or does not make sense, and explain your reasoning.*

83. I've noticed that $f(-x)$ is used to explore the number of negative real zeros of a polynomial function, as well as to determine whether a function is even, odd, or neither.

84. By using the quadratic formula, I do not need to bother with synthetic division when solving polynomial equations of degree 3 or higher.

85. I'm working with a fourth-degree polynomial function with integer coefficients and zeros at 1 and $3 + \sqrt{5}$. I'm certain that $3 + \sqrt{2}$ cannot also be a zero of this function.

86. I'm working with the polynomial function $f(x) = x^4 + 3x^2 + 2$ that has four possible rational zeros but no actual rational zeros.

In Exercises 87–90, determine whether each statement is true or false. If the statement is false, make the necessary change(s) to produce a true statement.

87. The equation $x^3 + 5x^2 + 6x + 1 = 0$ has one positive real root.

88. Descartes's Rule of Signs gives the exact number of positive and negative real roots for a polynomial equation.

89. Every polynomial equation of degree 3 with real coefficients has at least one real root.

90. Every polynomial equation of degree n has n distinct solutions.

91. If the volume of the solid shown in the figure is 208 cubic inches, find the value of x.

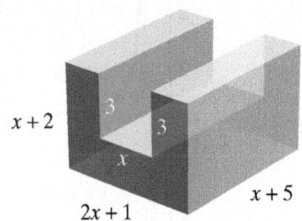

92. In this exercise, we lead you through the steps involved in the proof of the Rational Zero Theorem. Consider the polynomial equation

$$a_n x^n + a_{n-1}x^{n-1} + a_{n-2}x^{n-2} + \cdots + a_1 x + a_0 = 0,$$

and let $\dfrac{p}{q}$ be a rational root reduced to lowest terms.

a. Substitute $\dfrac{p}{q}$ for x in the equation and show that the equation can be written as

$$a_n p^n + a_{n-1}p^{n-1}q + a_{n-2}p^{n-2}q^2 + \cdots + a_1 pq^{n-1} = -a_0 q^n.$$

b. Why is p a factor of the left side of the equation?

c. Because p divides the left side, it must also divide the right side. However, because $\dfrac{p}{q}$ is reduced to lowest terms, p and q have no common factors other than -1 and 1. Because p does divide the right side and has no factors in common with q^n, what can you conclude?

d. Rewrite the equation from part (a) with all terms containing q on the left and the term that does not have a factor of q on the right. Use an argument that parallels parts (b) and (c) to conclude that q is a factor of a_n.

In Exercises 93–96, the graph of a polynomial function is given. What is the smallest degree that each polynomial could have?

97. Explain why a polynomial function of degree 20 cannot cross the x-axis exactly once.

Preview Exercises

Exercises 98–100 will help you prepare for the material covered in the next section. Use the graph of function f to solve each exercise.

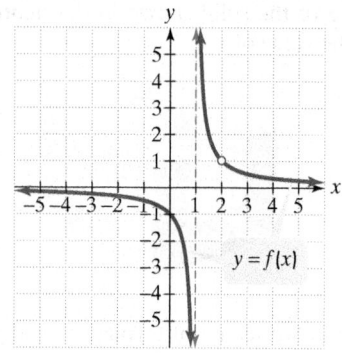

$y = f(x)$

98. For what values of x is the function undefined?

99. Write the equation of the vertical asymptote, or the vertical line that the graph of f approaches but does not touch.

100. Write the equation of the horizontal asymptote, or the horizontal line that the graph of f approaches but does not touch.

CHAPTER 3 Mid-Chapter Check Point

WHAT YOU KNOW: We graphed quadratic functions using vertices, intercepts, and additional points, as necessary. We learned that the vertex of $f(x) = a(x - h)^2 + k$ is (h, k) and the vertex of $f(x) = ax^2 + bx + c$ is $\left(-\frac{b}{2a}, f\left(-\frac{b}{2a}\right)\right)$. We used the vertex to solve problems that involved minimizing or maximizing quadratic functions. We learned a number of techniques for finding the zeros of a polynomial function f of degree 3 or higher or, equivalently, finding the roots, or solutions, of the equation $f(x) = 0$. For some functions, the zeros were found by factoring $f(x)$. For other functions, we listed possible rational zeros and used synthetic division and the Factor Theorem to determine the zeros. We saw that graphs cross the x-axis at zeros of odd multiplicity and touch the x-axis and turn around at zeros of even multiplicity. We learned to graph polynomial functions using zeros, the Leading Coefficient Test, intercepts, and symmetry. We checked graphs using the fact that a polynomial function of degree n has a graph with at most $n - 1$ turning points. After finding zeros of polynomial functions, we reversed directions by using the Linear Factorization Theorem to find functions with given zeros.

In Exercises 1–4, graph the given quadratic function. Give each function's domain and range.

1. $f(x) = (x - 3)^2 - 4$ **2.** $f(x) = 5 - (x + 2)^2$
3. $f(x) = -x^2 - 4x + 5$ **4.** $f(x) = 3x^2 - 6x + 1$

In Exercises 5–13, find all zeros of each polynomial function. Then graph the function.

5. $f(x) = (x - 2)^2(x + 1)^3$
6. $f(x) = -(x - 2)^2(x + 1)^2$
7. $f(x) = x^3 - x^2 - 4x + 4$
8. $f(x) = x^4 - 5x^2 + 4$
9. $f(x) = -(x + 1)^6$
10. $f(x) = -6x^3 + 7x^2 - 1$

11. $f(x) = 2x^3 - 2x$
12. $f(x) = x^3 - 2x^2 + 26x$
13. $f(x) = -x^3 + 5x^2 - 5x - 3$

In Exercises 14–19, solve each polynomial equation.

14. $x^3 - 3x + 2 = 0$
15. $6x^3 - 11x^2 + 6x - 1 = 0$
16. $(2x + 1)(3x - 2)^3(2x - 7) = 0$
17. $2x^3 + 5x^2 - 200x - 500 = 0$
18. $x^4 - x^3 - 11x^2 = x + 12$
19. $2x^4 + x^3 - 17x^2 - 4x + 6 = 0$

20. A company manufactures and sells bath cabinets. The function
$$P(x) = -x^2 + 150x - 4425$$
models the company's daily profit, $P(x)$, when x cabinets are manufactured and sold per day. How many cabinets should be manufactured and sold per day to maximize the company's profit? What is the maximum daily profit?

21. Among all pairs of numbers whose sum is -18, find a pair whose product is as large as possible. What is the maximum product?

22. The base of a triangle measures 40 inches minus twice the measure of its height. For what measure of the height does the triangle have a maximum area? What is the maximum area?

In Exercises 23–24, divide, using synthetic division if possible.

23. $(6x^4 - 3x^3 - 11x^2 + 2x + 4) \div (3x^2 - 1)$
24. $(2x^4 - 13x^3 + 17x^2 + 18x - 24) \div (x - 4)$

In Exercises 25–26, find an nth-degree polynomial function with real coefficients satisfying the given conditions.

25. $n = 3$; 1 and i are zeros; $f(-1) = 8$
26. $n = 4$; 2 (with multiplicity 2) and $3i$ are zeros; $f(0) = 36$
27. Does $f(x) = x^3 - x - 5$ have a real zero between 1 and 2?

SECTION 3.5 Rational Functions and Their Graphs

Objectives

1. Find the domains of rational functions.
2. Use arrow notation.
3. Identify vertical asymptotes.
4. Identify horizontal asymptotes.
5. Use transformations to graph rational functions.
6. Graph rational functions.
7. Identify slant asymptotes.
8. Solve applied problems involving rational functions.

Your grandmother appears to be slowing down. Enter ... Mechanical-Grandma! Japanese researchers have developed the robotic exoskeleton shown here to help the elderly and disabled walk and even lift heavy objects like the three 22-pound bags of rice in the photo. It's called the Hybrid Assistive Limb, or HAL. (The inventor has obviously never seen *2001: A Space Odyssey*.) HAL's brain is a computer housed in a backpack that learns to mimic the wearer's gait and posture. Bioelectric sensors pick up signals transmitted from the brain to the muscles, so it can anticipate movements the moment the wearer thinks of them. A commercial version is available at a hefty cost ranging between $14,000 and $20,000. (*Source:* sanlab. kz.tsukuba.ac.jp)

The cost of manufacturing robotic exoskeletons can be modeled by rational functions. In this section, you will see that high production levels of HAL can eventually make this amazing invention more affordable for the elderly and people with disabilities.

Rational Functions

1. Find the domains of rational functions.

Rational functions are quotients of polynomial functions. This means that rational functions can be expressed as

$$f(x) = \frac{p(x)}{q(x)},$$

where p and q are polynomial functions and $q(x) \neq 0$. The **domain** of a rational function is the set of all real numbers except the x-values that make the denominator zero. For example, the domain of the rational function

$$f(x) = \frac{x^2 + 7x + 9}{x(x-2)(x+5)} \quad \begin{array}{l} \text{This is } p(x). \\ \text{This is } q(x). \end{array}$$

is the set of all real numbers except $0, 2,$ and -5.

EXAMPLE 1 Finding the Domain of a Rational Function

Find the domain of each rational function:

a. $f(x) = \dfrac{x^2 - 9}{x - 3}$ **b.** $g(x) = \dfrac{x}{x^2 - 9}$ **c.** $h(x) = \dfrac{x + 3}{x^2 + 9}$.

SOLUTION

Rational functions contain division. Because division by 0 is undefined, we must exclude from the domain of each function values of x that cause the polynomial function in the denominator to be 0.

a. The denominator of $f(x) = \dfrac{x^2 - 9}{x - 3}$ is 0 if $x = 3$. Thus, x cannot equal 3.

The domain of f consists of all real numbers except 3. We can express the domain in set-builder or interval notation:

Domain of $f = \{x \mid x \neq 3\}$

Domain of $f = (-\infty, 3) \cup (3, \infty)$.

GREAT QUESTION!

Other than observing the denominator, is there a procedure I can use to find the domain of a rational function?

Yes. Because the domain of a rational function is the set of all real numbers except those for which the denominator is 0, you can identify such numbers by setting the denominator equal to 0 and solving for x. Exclude the resulting real values of x from the domain.

 Use arrow notation.

b. The denominator of $g(x) = \dfrac{x}{x^2 - 9}$ is 0 if $x = -3$ or $x = 3$. Thus, the domain of g consists of all real numbers except -3 and 3. We can express the domain in set-builder or interval notation:

$$\text{Domain of } g = \{x \mid x \neq -3, x \neq 3\}$$
$$\text{Domain of } g = (-\infty, -3) \cup (-3, 3) \cup (3, \infty).$$

c. No real numbers cause the denominator of $h(x) = \dfrac{x + 3}{x^2 + 9}$ to equal 0. The domain of h consists of all real numbers.

$$\text{Domain of } h = (-\infty, \infty) \qquad \bullet \bullet \bullet$$

✓ **Check Point 1** Find the domain of each rational function:

a. $f(x) = \dfrac{x^2 - 25}{x - 5}$ **b.** $g(x) = \dfrac{x}{x^2 - 25}$ **c.** $h(x) = \dfrac{x + 5}{x^2 + 25}.$

The most basic rational function is the **reciprocal function**, defined by $f(x) = \dfrac{1}{x}$. The denominator of the reciprocal function is zero when $x = 0$, so the domain of f is the set of all real numbers except 0.

Let's look at the behavior of f near the excluded value 0. We start by evaluating $f(x)$ to the left of 0.

 x approaches 0 from the left. 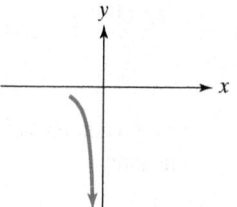

x	-1	-0.5	-0.1	-0.01	-0.001
$f(x) = \dfrac{1}{x}$	-1	-2	-10	-100	-1000

Mathematically, we say that "x approaches 0 from the left." From the table and the accompanying graph, it appears that as x approaches 0 from the left, the function values, $f(x)$, decrease without bound. We say that "$f(x)$ approaches negative infinity." We use a special arrow notation to describe this situation symbolically:

$$\text{As } x \to 0^-,\ f(x) \to -\infty.$$

As x approaches 0 from the left, $f(x)$ approaches negative infinity (that is, the graph falls).

Observe that the minus $(-)$ superscript on the 0 $(x \to 0^-)$ is read "from the left." Next, we evaluate $f(x)$ to the right of 0.

 x approaches 0 from the right. 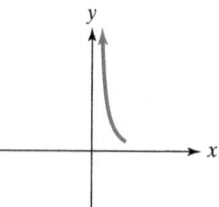

x	0.001	0.01	0.1	0.5	1
$f(x) = \dfrac{1}{x}$	1000	100	10	2	1

Mathematically, we say that "x approaches 0 from the right." From the table and the accompanying graph, it appears that as x approaches 0 from the right, the function values, $f(x)$, increase without bound. We say that "$f(x)$ approaches infinity." We again use a special arrow notation to describe this situation symbolically:

$$\text{As } x \to 0^+,\ f(x) \to \infty.$$

As x approaches 0 from the right, $f(x)$ approaches infinity (that is, the graph rises).

Observe that the plus $(+)$ superscript on the 0 $(x \to 0^+)$ is read "from the right."

Now let's see what happens to the function values of $f(x) = \dfrac{1}{x}$ as x gets farther away from the origin. The following tables suggest what happens to $f(x)$ as x increases or decreases without bound.

x increases without bound:					x decreases without bound:				
x	1	10	100	1000	x	−1	−10	−100	−1000
$f(x) = \dfrac{1}{x}$	1	0.1	0.01	0.001	$f(x) = \dfrac{1}{x}$	−1	−0.1	−0.01	−0.001

It appears that as x increases or decreases without bound, the function values, $f(x)$, are getting progressively closer to 0.

Figure 3.29 illustrates the end behavior of $f(x) = \dfrac{1}{x}$ as x increases or decreases without bound. The graph shows that the function values, $f(x)$, are approaching 0. This means that as x increases or decreases without bound, the graph of f is approaching the horizontal line $y = 0$ (that is, the x-axis). We use arrow notation to describe this situation:

$$\text{As } x \to \infty,\ f(x) \to 0 \qquad \text{and} \qquad \text{as } x \to -\infty,\ f(x) \to 0.$$

As x approaches infinity
(that is, increases without bound),
$f(x)$ approaches 0.

As x approaches negative infinity
(that is, decreases without bound),
$f(x)$ approaches 0.

Thus, as x approaches infinity $(x \to \infty)$ or as x approaches negative infinity $(x \to -\infty)$, the function values are approaching zero: $f(x) \to 0$.

The graph of the reciprocal function $f(x) = \dfrac{1}{x}$ is shown in **Figure 3.30**. Unlike the graph of a polynomial function, the graph of the reciprocal function has a break and is composed of two branches.

FIGURE 3.29 $f(x)$ approaches 0 as x increases or decreases without bound.

FIGURE 3.30 The graph of the reciprocal function $f(x) = \dfrac{1}{x}$

GREAT QUESTION!

What is the relationship between x and $\dfrac{1}{x}$ when x is far from 0? What happens if x is close to 0?

If x is far from 0, then $\dfrac{1}{x}$ is close to 0. By contrast, if x is close to 0, then $\dfrac{1}{x}$ is far from 0.

Blitzer Bonus || The Reciprocal Function as a Knuckle Tattoo

"I got the tattoo because I like the idea of math not being well behaved. That sounds lame and I really don't mean that in some kind of anarchy-type way. I just think that it's kind of nice that something as perfectly functional as math can kink up around the edges."

Kink up around the edges? On the next page, we'll describe the graphic behavior of the reciprocal function using *asymptotes* rather than kink. Asymptotes are lines that graphs approach but never touch. *Asymptote* comes from the Greek word *asymptotos*, meaning "not meeting."

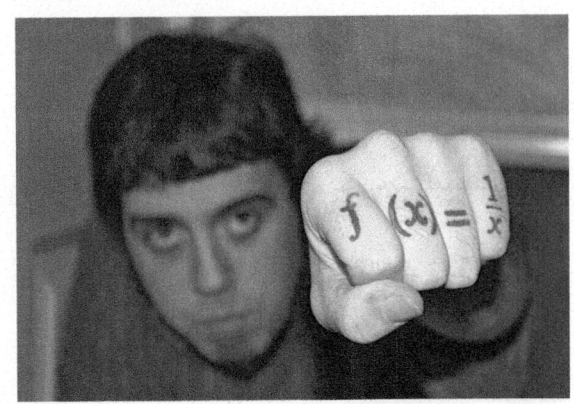

The arrow notation used throughout our discussion of the reciprocal function is summarized in the following box:

Arrow Notation

Symbol	Meaning
$x \to a^+$	x approaches a from the right.
$x \to a^-$	x approaches a from the left.
$x \to \infty$	x approaches infinity; that is, x increases without bound.
$x \to -\infty$	x approaches negative infinity; that is, x decreases without bound.

Another basic rational function is $f(x) = \dfrac{1}{x^2}$. The graph of this even function, with y-axis symmetry and positive function values, is shown in **Figure 3.31**. Like the reciprocal function, the graph has a break and is composed of two distinct branches.

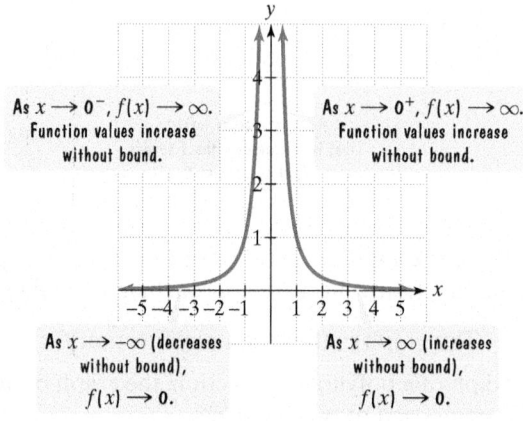

As $x \to 0^-$, $f(x) \to \infty$.
Function values increase without bound.

As $x \to 0^+$, $f(x) \to \infty$.
Function values increase without bound.

As $x \to -\infty$ (decreases without bound), $f(x) \to 0$.

As $x \to \infty$ (increases without bound), $f(x) \to 0$.

FIGURE 3.31 The graph of $f(x) = \dfrac{1}{x^2}$

3 Identify vertical asymptotes.

Vertical Asymptotes of Rational Functions

Look again at the graph of $f(x) = \dfrac{1}{x^2}$ in **Figure 3.31**. The curve approaches, but does not touch, the y-axis. The y-axis, or $x = 0$, is said to be a *vertical asymptote* of the graph. A rational function may have no vertical asymptotes, one vertical asymptote, or several vertical asymptotes. The graph of a rational function never intersects a vertical asymptote. We will use dashed lines to show asymptotes.

Definition of a Vertical Asymptote

The line $x = a$ is a **vertical asymptote** of the graph of a function f if $f(x)$ increases or decreases without bound as x approaches a.

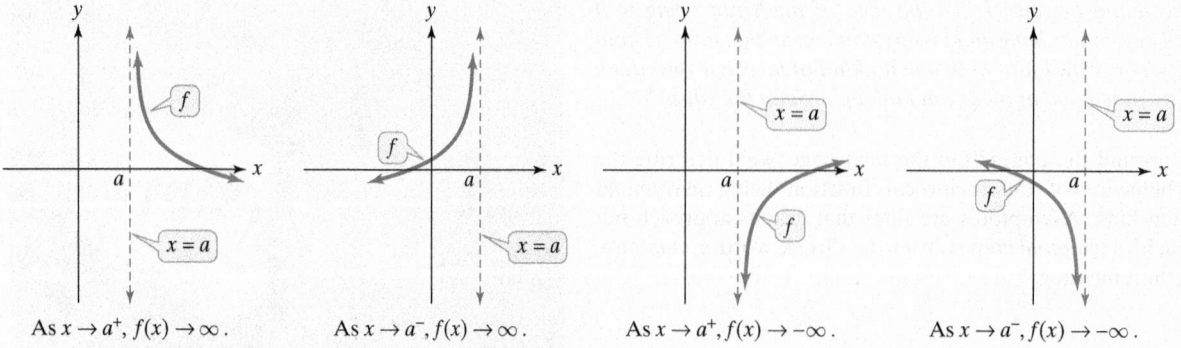

As $x \to a^+$, $f(x) \to \infty$.

As $x \to a^-$, $f(x) \to \infty$.

As $x \to a^+$, $f(x) \to -\infty$.

As $x \to a^-$, $f(x) \to -\infty$.

Thus, as x approaches a from either the left or the right, $f(x) \to \infty$ or $f(x) \to -\infty$.

If the graph of a rational function has vertical asymptotes, they can be located using the following theorem:

Locating Vertical Asymptotes

If $f(x) = \dfrac{p(x)}{q(x)}$ is a rational function in which $p(x)$ and $q(x)$ have no common factors and a is a zero of $q(x)$, the denominator, then $x = a$ is a vertical asymptote of the graph of f.

EXAMPLE 2 Finding the Vertical Asymptotes of a Rational Function

Find the vertical asymptotes, if any, of the graph of each rational function:

a. $f(x) = \dfrac{x}{x^2 - 9}$ **b.** $g(x) = \dfrac{x + 3}{x^2 - 9}$ **c.** $h(x) = \dfrac{x + 3}{x^2 + 9}$.

SOLUTION

Factoring is usually helpful in identifying zeros of denominators and any common factors in the numerators and denominators.

a. $f(x) = \dfrac{x}{x^2 - 9} = \dfrac{x}{(x + 3)(x - 3)}$

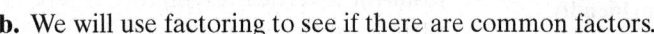

| This factor is 0 if $x = -3$. | This factor is 0 if $x = 3$. |

There are no common factors in the numerator and the denominator. The zeros of the denominator are -3 and 3. Thus, the lines $x = -3$ and $x = 3$ are the vertical asymptotes for the graph of f. [See **Figure 3.32(a)**.]

b. We will use factoring to see if there are common factors.

$$g(x) = \dfrac{x + 3}{x^2 - 9} = \dfrac{(x + 3)}{(x + 3)(x - 3)} = \dfrac{1}{x - 3}, \text{ provided } x \neq -3$$

There is a common factor, $x + 3$, so simplify. This denominator is 0 if $x = 3$.

The only zero of the denominator of $g(x)$ in simplified form is 3. Thus, the line $x = 3$ is the only vertical asymptote of the graph of g. [See **Figure 3.32(b)**.]

c. We cannot factor the denominator of $h(x)$ over the real numbers.

$$h(x) = \dfrac{x + 3}{x^2 + 9}$$

No real numbers make this denominator 0.

The denominator has no real zeros. Thus, the graph of h has no vertical asymptotes. [See **Figure 3.32(c)**.]

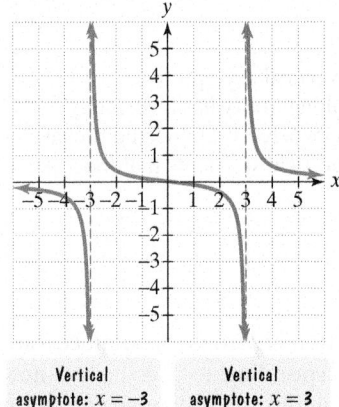

Vertical asymptote: $x = -3$ Vertical asymptote: $x = 3$

FIGURE 3.32(a) The graph of $f(x) = \dfrac{x}{x^2 - 9}$ has two vertical asymptotes.

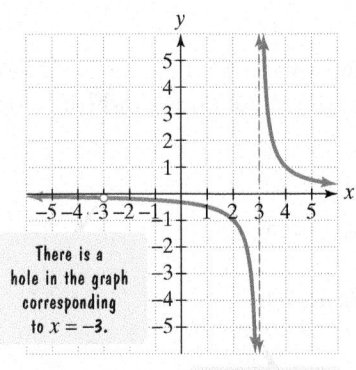

There is a hole in the graph corresponding to $x = -3$.

Vertical asymptote: $x = 3$

FIGURE 3.32(b) The graph of $g(x) = \dfrac{x + 3}{x^2 - 9}$ has one vertical asymptote.

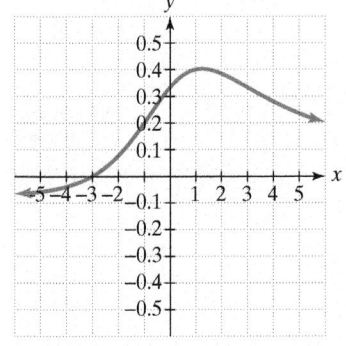

FIGURE 3.32(c) The graph of $h(x) = \dfrac{x + 3}{x^2 + 9}$ has no vertical asymptotes.

• • •

⊘ **Check Point 2** Find the vertical asymptotes, if any, of the graph of each rational function:

a. $f(x) = \dfrac{x}{x^2 - 1}$ **b.** $g(x) = \dfrac{x - 1}{x^2 - 1}$ **c.** $h(x) = \dfrac{x - 1}{x^2 + 1}$.

TECHNOLOGY

The graph of the rational function $f(x) = \dfrac{x}{x^2 - 9}$, drawn by hand in **Figure 3.32(a)** on the previous page, is graphed below in a $[-5, 5, 1]$ by $[-4, 4, 1]$ viewing rectangle. The graph is shown in connected mode and in dot mode. In connected mode, the graphing utility plots many points and connects the points with curves. In dot mode, the utility plots the same points but does not connect them.

Connected Mode

Dot Mode

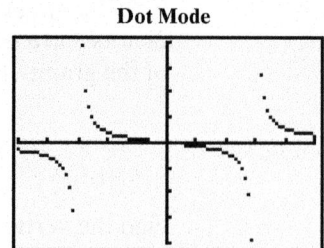

This might appear to be the vertical asymptote $x = -3$, but it is neither vertical nor an asymptote.

This might appear to be the vertical asymptote $x = 3$, but it is neither vertical nor an asymptote.

The steep lines that may appear on some graphing utilities in connected mode and seem to be the vertical asymptotes $x = -3$ and $x = 3$ are not part of the graph and do not represent the vertical asymptotes. The graphing utility has incorrectly connected the last point to the left of $x = -3$ with the first point to the right of $x = -3$. It has also incorrectly connected the last point to the left of $x = 3$ with the first point to the right of $x = 3$. The effect is to create two near-vertical segments that look like asymptotes. This erroneous effect does not appear using dot mode.

GREAT QUESTION!

Do I have to factor to identify a rational function's possible vertical asymptotes or holes?

Yes. It is essential to factor the numerator and the denominator of the rational function.

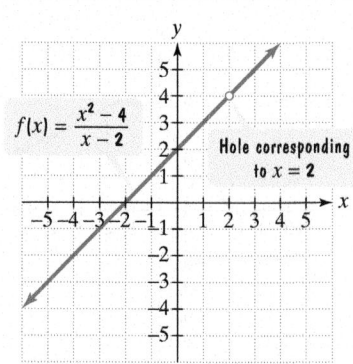

FIGURE 3.33 A graph with a hole corresponding to the denominator's zero

A value where the denominator of a rational function is zero does not necessarily result in a vertical asymptote. There is a hole corresponding to $x = a$, and not a vertical asymptote, in the graph of a rational function under the following conditions: The value a causes the denominator to be zero, but there is a reduced form of the function's equation in which a does not cause the denominator to be zero.

Consider, for example, the function

$$f(x) = \frac{x^2 - 4}{x - 2}.$$

Because the denominator is zero when $x = 2$, the function's domain is all real numbers except 2. However, there is a reduced form of the equation in which 2 does not cause the denominator to be zero:

$$f(x) = \frac{x^2 - 4}{x - 2} = \frac{(x + 2)(x - 2)}{x - 2} = x + 2, \ x \neq 2.$$

Denominator is zero at $x = 2$.

In this reduced form, 2 does not result in a zero denominator.

Figure 3.33 shows that the graph has a hole corresponding to $x = 2$. Graphing utilities do not show this feature of the graph.

④ Identify horizontal asymptotes.

Horizontal Asymptotes of Rational Functions

Figure 3.30, repeated in the margin at the top of the next page, shows the graph of the reciprocal function $f(x) = \dfrac{1}{x}$. As $x \to \infty$ and as $x \to -\infty$, the function values are approaching 0: $f(x) \to 0$. The line $y = 0$ (that is, the x-axis) is a *horizontal asymptote* of the graph. Many, but not all, rational functions have horizontal asymptotes.

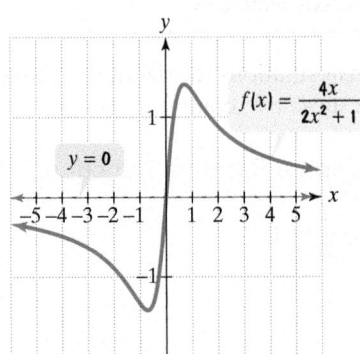

FIGURE 3.30 The graph of $f(x) = \dfrac{1}{x}$ (repeated)

The line $y = b$ is a **horizontal asymptote** of the graph of a function f if $f(x)$ approaches b as x increases or decreases without bound.

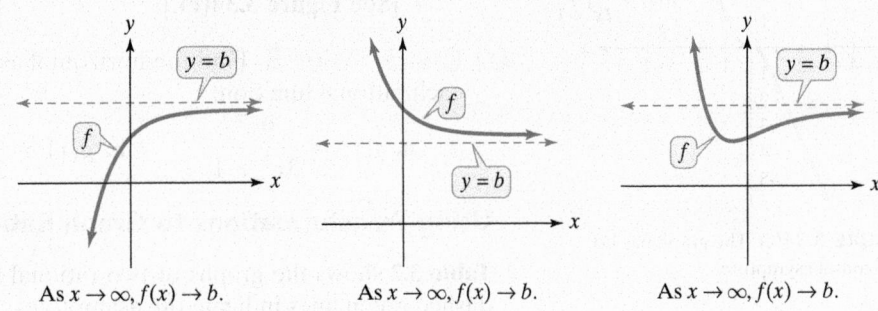

As $x \to \infty, f(x) \to b$. As $x \to \infty, f(x) \to b$. As $x \to \infty, f(x) \to b$.

Recall that a rational function may have several vertical asymptotes. By contrast, it can have at most one horizontal asymptote. Although a graph can never intersect a vertical asymptote, it may cross its horizontal asymptote.

If the graph of a rational function has a horizontal asymptote, it can be located using the following theorem:

GREAT QUESTION!

Do I have to factor to identify a rational function's possible horizontal asymptote?

No. Unlike identifying possible vertical asymptotes or holes, we do not use factoring to determine a possible horizontal asymptote.

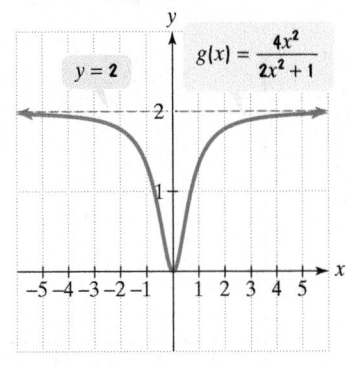

FIGURE 3.34(a) The horizontal asymptote of the graph is $y = 0$.

FIGURE 3.34(b) The horizontal asymptote of the graph is $y = 2$.

Locating Horizontal Asymptotes

Let f be the rational function given by

$$f(x) = \frac{a_n x^n + a_{n-1}x^{n-1} + \cdots + a_1 x + a_0}{b_m x^m + b_{m-1}x^{m-1} + \cdots + b_1 x + b_0}, \quad a_n \neq 0, b_m \neq 0.$$

The degree of the numerator is n. The degree of the denominator is m.

1. If $n < m$, the x-axis, or $y = 0$, is the horizontal asymptote of the graph of f.

2. If $n = m$, the line $y = \dfrac{a_n}{b_m}$ is the horizontal asymptote of the graph of f.

3. If $n > m$, the graph of f has no horizontal asymptote.

EXAMPLE 3 Finding the Horizontal Asymptote of a Rational Function

Find the horizontal asymptote, if there is one, of the graph of each rational function:

a. $f(x) = \dfrac{4x}{2x^2 + 1}$ **b.** $g(x) = \dfrac{4x^2}{2x^2 + 1}$ **c.** $h(x) = \dfrac{4x^3}{2x^2 + 1}$.

SOLUTION

a. $f(x) = \dfrac{4x}{2x^2 + 1}$

The degree of the numerator, 1, is less than the degree of the denominator, 2. Thus, the graph of f has the x-axis as a horizontal asymptote. [See **Figure 3.34(a)**.] The equation of the horizontal asymptote is $y = 0$.

b. $g(x) = \dfrac{4x^2}{2x^2 + 1}$

The degree of the numerator, 2, is equal to the degree of the denominator, 2. The leading coefficients of the numerator and denominator, 4 and 2, are used to obtain the equation of the horizontal asymptote. The equation of the horizontal asymptote is $y = \frac{4}{2}$ or $y = 2$. [See **Figure 3.34(b)**.]

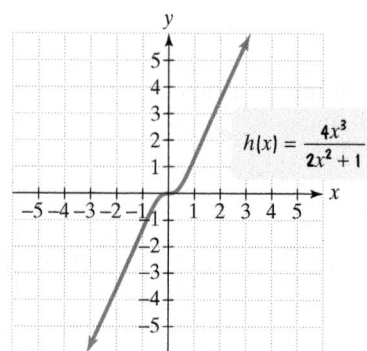

FIGURE 3.34(c) The graph has no horizontal asymptote.

c. $h(x) = \dfrac{4x^3}{2x^2 + 1}$

The degree of the numerator, 3, is greater than the degree of the denominator, 2. Thus, the graph of h has no horizontal asymptote. [See **Figure 3.34(c)**.]

●●●

✓ Check Point **3** Find the horizontal asymptote, if there is one, of the graph of each rational function:

a. $f(x) = \dfrac{9x^2}{3x^2 + 1}$ **b.** $g(x) = \dfrac{9x}{3x^2 + 1}$ **c.** $h(x) = \dfrac{9x^3}{3x^2 + 1}$.

Using Transformations to Graph Rational Functions

⑤ Use transformations to graph rational functions.

Table 3.2 shows the graphs of two rational functions, $f(x) = \dfrac{1}{x}$ and $f(x) = \dfrac{1}{x^2}$. The dashed green lines indicate the asymptotes.

Table 3.2 Graphs of Common Rational Functions

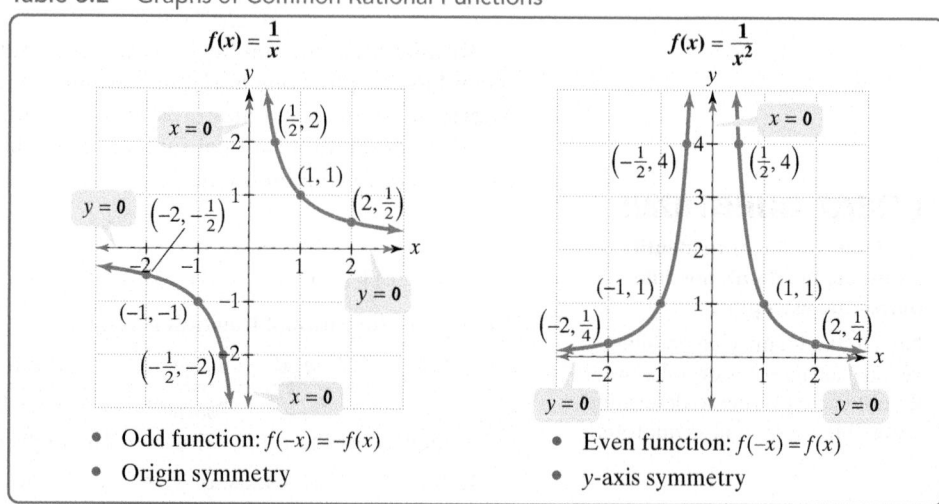

- Odd function: $f(-x) = -f(x)$
- Origin symmetry

- Even function: $f(-x) = f(x)$
- y-axis symmetry

Some rational functions can be graphed using transformations (horizontal shifting, stretching or shrinking, reflecting, vertical shifting) of these two common graphs.

EXAMPLE 4 Using Transformations to Graph a Rational Function

Use the graph of $f(x) = \dfrac{1}{x^2}$ to graph $g(x) = \dfrac{1}{(x-2)^2} + 1$.

SOLUTION

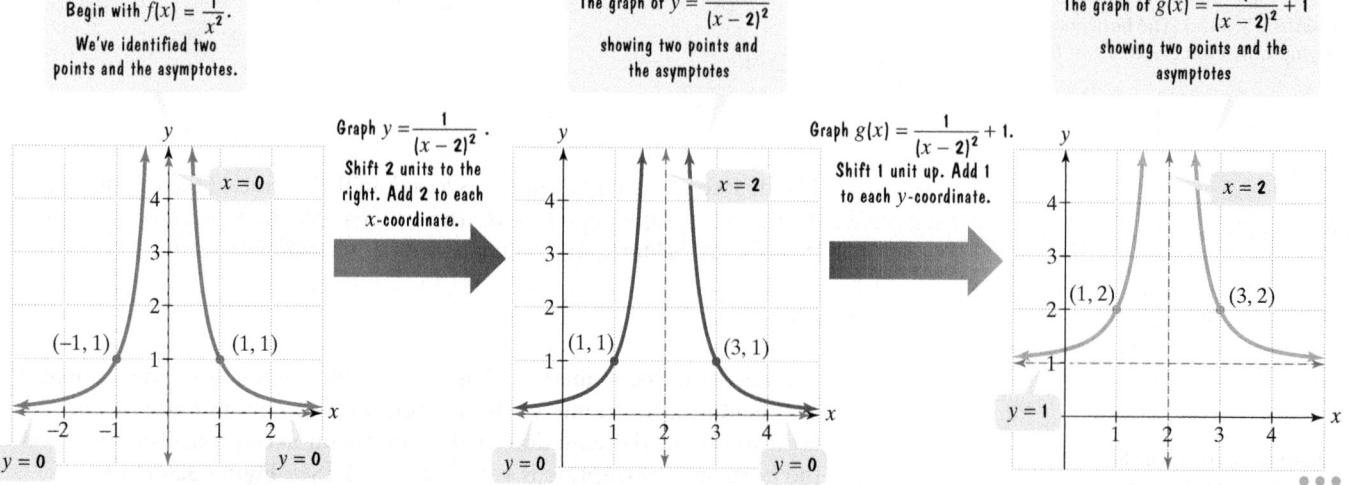

●●●

⊘ **Check Point 4** Use the graph of $f(x) = \dfrac{1}{x}$ to graph $g(x) = \dfrac{1}{x+2} - 1$.

⑥ Graph rational functions.

Graphing Rational Functions

Rational functions that are not transformations of $f(x) = \dfrac{1}{x}$ or $f(x) = \dfrac{1}{x^2}$ can be graphed using the following procedure:

> Strategy for Graphing a Rational Function
>
> The following strategy can be used to graph
>
> $$f(x) = \frac{p(x)}{q(x)},$$
>
> where p and q are polynomial functions with no common factors.
>
> 1. Determine whether the graph of f has symmetry.
>
> $f(-x) = f(x)$: y-axis symmetry
>
> $f(-x) = -f(x)$: origin symmetry
>
> 2. Find the y-intercept (if there is one) by evaluating $f(0)$.
> 3. Find the x-intercepts (if there are any) by solving the equation $p(x) = 0$.
> 4. Find any vertical asymptote(s) by solving the equation $q(x) = 0$.
> 5. Find the horizontal asymptote (if there is one) using the rule for determining the horizontal asymptote of a rational function.
> 6. Plot at least one point between and beyond each x-intercept and vertical asymptote.
> 7. Use the information obtained previously to graph the function between and beyond the vertical asymptotes.

EXAMPLE 5 Graphing a Rational Function

Graph: $f(x) = \dfrac{2x - 1}{x - 1}$.

SOLUTION

Step 1 Determine symmetry.

$$f(-x) = \frac{2(-x) - 1}{-x - 1} = \frac{-2x - 1}{-x - 1} = \frac{2x + 1}{x + 1}$$

Because $f(-x)$ does not equal either $f(x)$ or $-f(x)$, the graph has neither y-axis symmetry nor origin symmetry.

Step 2 Find the y-intercept. Evaluate $f(0)$.

$$f(0) = \frac{2 \cdot 0 - 1}{0 - 1} = \frac{-1}{-1} = 1$$

The y-intercept is 1, so the graph passes through $(0, 1)$.

Step 3 Find x-intercept(s). This is done by solving $p(x) = 0$, where $p(x)$ is the numerator of $f(x)$.

$$2x - 1 = 0 \quad \text{Set the numerator equal to 0.}$$
$$2x = 1 \quad \text{Add 1 to both sides.}$$
$$x = \frac{1}{2} \quad \text{Divide both sides by 2.}$$

The x-intercept is $\frac{1}{2}$, so the graph passes through $\left(\frac{1}{2}, 0\right)$.

Step 4 Find the vertical asymptote(s). Solve $q(x) = 0$, where $q(x)$ is the denominator of $f(x)$, thereby finding zeros of the denominator. (Note that the numerator and denominator of $f(x) = \dfrac{2x - 1}{x - 1}$ have no common factors.)

$$x - 1 = 0 \qquad \text{Set the denominator equal to 0.}$$
$$x = 1 \qquad \text{Add 1 to both sides.}$$

The equation of the vertical asymptote is $x = 1$.

Step 5 Find the horizontal asymptote. Because the numerator and denominator of $f(x) = \dfrac{2x - 1}{x - 1}$ have the same degree, 1, the leading coefficients of the numerator and denominator, 2 and 1, respectively, are used to obtain the equation of the horizontal asymptote. The equation is

$$y = \frac{2}{1} = 2.$$

The equation of the horizontal asymptote is $y = 2$.

Step 6 Plot points between and beyond each x-intercept and vertical asymptote. With an x-intercept at $\frac{1}{2}$ and a vertical asymptote at $x = 1$, we evaluate the function at $-2, -1, \dfrac{3}{4}, 2,$ and 4.

x	-2	-1	$\dfrac{3}{4}$	2	4
$f(x) = \dfrac{2x - 1}{x - 1}$	$\dfrac{5}{3}$	$\dfrac{3}{2}$	-2	3	$\dfrac{7}{3}$

Figure 3.35 shows these points, the y-intercept, the x-intercept, and the asymptotes.

Step 7 Graph the function. The graph of $f(x) = \dfrac{2x - 1}{x - 1}$ is shown in **Figure 3.36**.

TECHNOLOGY

The graph of $y = \dfrac{2x - 1}{x - 1}$, obtained using the dot mode in a $[-6, 6, 1]$ by $[-6, 6, 1]$ viewing rectangle, verifies that our hand-drawn graph in **Figure 3.36** is correct.

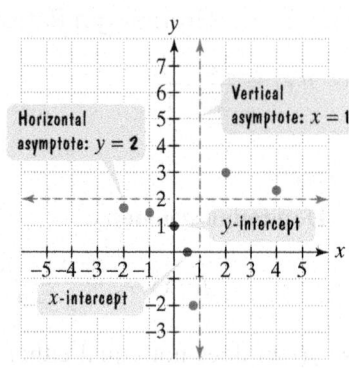

FIGURE 3.35 Preparing to graph the rational function $f(x) = \dfrac{2x - 1}{x - 1}$

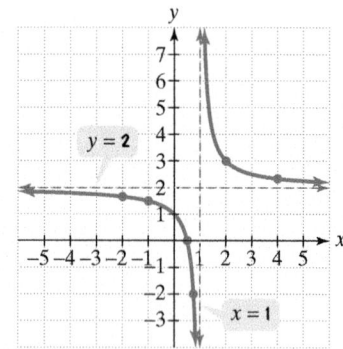

FIGURE 3.36 The graph of $f(x) = \dfrac{2x - 1}{x - 1}$

● ● ●

☑ Check Point **5** Graph: $f(x) = \dfrac{3x - 3}{x - 2}$.

EXAMPLE 6 Graphing a Rational Function

Graph: $f(x) = \dfrac{3x^2}{x^2 - 4}$.

SOLUTION

Step 1 Determine symmetry. $f(-x) = \dfrac{3(-x)^2}{(-x)^2 - 4} = \dfrac{3x^2}{x^2 - 4} = f(x)$: The graph of f is symmetric with respect to the y-axis.

Step 2 Find the *y*-intercept. $f(0) = \dfrac{3 \cdot 0^2}{0^2 - 4} = \dfrac{0}{-4} = 0$: The *y*-intercept is 0, so the graph passes through the origin.

Step 3 Find the *x*-intercept(s). $3x^2 = 0$, so $x = 0$: The *x*-intercept is 0, verifying that the graph passes through the origin.

Step 4 Find the vertical asymptote(s). Set $q(x) = 0$. (Note that the numerator and denominator of $f(x) = \dfrac{3x^2}{x^2 - 4}$ have no common factors.)

$$x^2 - 4 = 0 \qquad \text{Set the denominator equal to 0.}$$
$$x^2 = 4 \qquad \text{Add 4 to both sides.}$$
$$x = \pm 2 \qquad \text{Use the square root property.}$$

The vertical asymptotes are $x = -2$ and $x = 2$.

Step 5 Find the horizontal asymptote. Because the numerator and denominator of $f(x) = \dfrac{3x^2}{x^2 - 4}$ have the same degree, 2, their leading coefficients, 3 and 1, are used to determine the equation of the horizontal asymptote. The equation is $y = \frac{3}{1} = 3$.

Step 6 Plot points between and beyond each *x*-intercept and vertical asymptote. With an *x*-intercept at 0 and vertical asymptotes at $x = -2$ and $x = 2$, we evaluate the function at $-3, -1, 1, 3,$ and 4.

x		-3	-1	1	3	4
$f(x) = \dfrac{3x^2}{x^2 - 4}$		$\dfrac{27}{5}$	-1	-1	$\dfrac{27}{5}$	4

Figure 3.37 shows the points $\left(-3, \dfrac{27}{5}\right), (-1, -1), (1, -1), \left(3, \dfrac{27}{5}\right)$, and $(4, 4)$, the *y*-intercept, the *x*-intercept, and the asymptotes.

Step 7 Graph the function. The graph of $f(x) = \dfrac{3x^2}{x^2 - 4}$ is shown in **Figure 3.38**. The *y*-axis symmetry is now obvious.

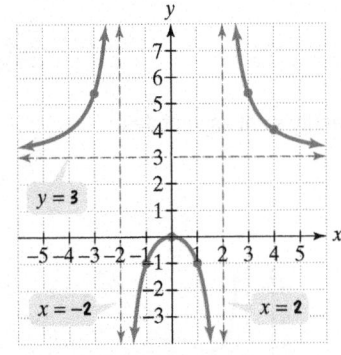

FIGURE 3.37 Preparing to graph $f(x) = \dfrac{3x^2}{x^2 - 4}$

FIGURE 3.38 The graph of $f(x) = \dfrac{3x^2}{x^2 - 4}$

● ● ●

GREAT QUESTION!

Do I have to evaluate $f(x)$ at all five of the values shown in the table for Step 6?

No. Because the graph has *y*-axis symmetry, it is not necessary to evaluate the even function at -3 and again at 3.

$$f(-3) = f(3) = \tfrac{27}{5}$$

This also applies to evaluation at -1 and 1.

TECHNOLOGY

The graph of $y = \dfrac{3x^2}{x^2 - 4}$, generated by a graphing utility, verifies that our hand-drawn graph is correct.

$[-6, 6, 1]$ by $[-6, 6, 1]$

✓ Check Point **6** Graph: $f(x) = \dfrac{2x^2}{x^2 - 9}$.

Example 7 illustrates that not every rational function has vertical and horizontal asymptotes.

EXAMPLE 7 Graphing a Rational Function

Graph: $f(x) = \dfrac{x^4}{x^2 + 1}$.

SOLUTION

Step 1 Determine symmetry. $f(-x) = \dfrac{(-x)^4}{(-x)^2 + 1} = \dfrac{x^4}{x^2 + 1} = f(x)$

The graph of f is symmetric with respect to the y-axis.

Step 2 Find the y-intercept. $f(0) = \dfrac{0^4}{0^2 + 1} = \dfrac{0}{1} = 0$: The y-intercept is 0.

Step 3 Find the x-intercept(s). $x^4 = 0$, so $x = 0$: The x-intercept is 0.

Step 4 Find the vertical asymptote(s). Set $q(x) = 0$.

$$x^2 + 1 = 0 \qquad \text{Set the denominator equal to 0.}$$
$$x^2 = -1 \qquad \text{Subtract 1 from both sides.}$$

Although this equation has imaginary roots ($x = \pm i$), there are no real roots. Thus, the graph of f has no vertical asymptotes.

Step 5 Find the horizontal asymptote. Because the degree of the numerator, 4, is greater than the degree of the denominator, 2, there is no horizontal asymptote.

Step 6 Plot points between and beyond each x-intercept and vertical asymptote. With an x-intercept at 0 and no vertical asymptotes, let's look at function values at $-2, -1, 1$, and 2. You can evaluate the function at 1 and 2. Use y-axis symmetry to obtain function values at -1 and -2:

$$f(-1) = f(1) \text{ and } f(-2) = f(2).$$

x		-2	-1	1	2
$f(x) = \dfrac{x^4}{x^2 + 1}$		$\dfrac{16}{5}$	$\dfrac{1}{2}$	$\dfrac{1}{2}$	$\dfrac{16}{5}$

Step 7 Graph the function. Figure 3.39 shows the graph of f using the points obtained from the table and y-axis symmetry. Notice that as x approaches infinity or negative infinity ($x \to \infty$ or $x \to -\infty$), the function values, $f(x)$, are getting larger without bound [$f(x) \to \infty$]. ● ● ●

☑ **Check Point 7** Graph: $f(x) = \dfrac{x^4}{x^2 + 2}$.

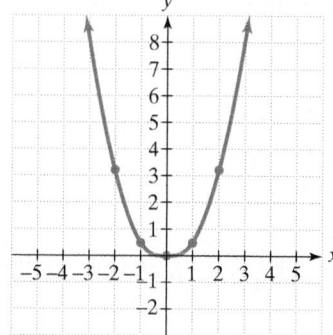

FIGURE 3.39 The graph of $f(x) = \dfrac{x^4}{x^2 + 1}$

⑦ Identify slant asymptotes.

Slant Asymptotes

Examine the graph of

$$f(x) = \frac{x^2 + 1}{x - 1},$$

shown in **Figure 3.40**. Note that the degree of the numerator, 2, is greater than the degree of the denominator, 1. Thus, the graph of this function has no horizontal asymptote. However, the graph has a **slant asymptote**, $y = x + 1$.

The graph of a rational function has a slant asymptote if the degree of the numerator is one more than the degree of the denominator. The equation of the slant asymptote can be found by division. For example, to find the slant asymptote for the graph of $f(x) = \dfrac{x^2 + 1}{x - 1}$, divide $x - 1$ into $x^2 + 1$:

$$\underline{1)\ \ 1 \quad 0 \quad 1}$$
$$\phantom{\underline{1)\ \ 1}} 1 \quad 1$$
$$\overline{\ 1 \quad 1 \quad 2}$$

$$1x + 1 + \frac{2}{x - 1}.$$
$$x - 1 \overline{)x^2 + 0x + 1}$$

Remainder

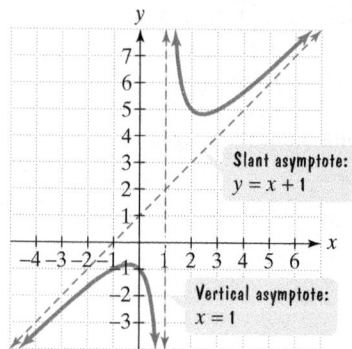

FIGURE 3.40 The graph of $f(x) = \dfrac{x^2 + 1}{x - 1}$ with a slant asymptote

Observe that

$$f(x) = \frac{x^2 + 1}{x - 1} = \underbrace{x + 1} + \frac{2}{x - 1}.$$

The equation of the slant asymptote is $y = x + 1$.

As $|x| \to \infty$, the value of $\frac{2}{x - 1}$ is approximately 0. Thus, when $|x|$ is large, the function is very close to $y = x + 1 + 0$. This means that as $x \to \infty$ or as $x \to -\infty$, the graph of f gets closer and closer to the line whose equation is $y = x + 1$. The line $y = x + 1$ is a slant asymptote of the graph.

In general, if $f(x) = \frac{p(x)}{q(x)}$, p and q have no common factors, and the degree of p is one greater than the degree of q, find the slant asymptote by dividing $q(x)$ into $p(x)$. The division will take the form

$$\frac{p(x)}{q(x)} = mx + b + \frac{\text{remainder}}{q(x)}.$$

Slant asymptote:
$y = mx + b$

The equation of the slant asymptote is obtained by dropping the term with the remainder. Thus, the equation of the slant asymptote is $y = mx + b$.

EXAMPLE 8 Finding the Slant Asymptote of a Rational Function

Find the slant asymptote of $f(x) = \dfrac{x^2 - 4x - 5}{x - 3}$.

SOLUTION

Because the degree of the numerator, 2, is exactly one more than the degree of the denominator, 1, and $x - 3$ is not a factor of $x^2 - 4x - 5$, the graph of f has a slant asymptote. To find the equation of the slant asymptote, divide $x - 3$ into $x^2 - 4x - 5$:

$$\underline{3}\ \begin{array}{ccc} 1 & -4 & -5 \\ & 3 & -3 \\ \hline 1 & -1 & -8 \end{array} \quad \text{Remainder}$$

$$\begin{array}{r} 1x - 1 - \dfrac{8}{x - 3} \\ x - 3 \overline{)x^2 - 4x - 5} \end{array}$$

Drop the remainder term and you'll have the equation of the slant asymptote.

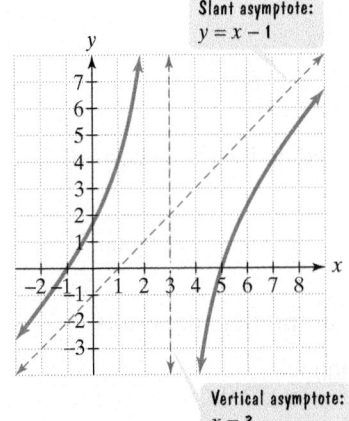

Slant asymptote:
$y = x - 1$

Vertical asymptote:
$x = 3$

FIGURE 3.41 The graph of $f(x) = \dfrac{x^2 - 4x - 5}{x - 3}$

The equation of the slant asymptote is $y = x - 1$. Using our strategy for graphing rational functions, the graph of $f(x) = \dfrac{x^2 - 4x - 5}{x - 3}$ is shown in **Figure 3.41**. •••

⊘ Check Point **8** Find the slant asymptote of $f(x) = \dfrac{2x^2 - 5x + 7}{x - 2}$.

⑧ Solve applied problems involving rational functions.

Applications

There are numerous examples of asymptotic behavior in functions that model real-world phenomena. Let's consider an example from the business world. The **cost function**, C, for a business is the sum of its fixed and variable costs:

$$C(x) = (\text{fixed cost}) + cx.$$

Cost per unit times the number of units produced, x

The **average cost** per unit for a company to produce x units is the sum of its fixed and variable costs divided by the number of units produced. The **average cost function** is a rational function that is denoted by \overline{C}. Thus,

Cost of producing x units: fixed plus variable costs

$$\overline{C}(x) = \frac{(\text{fixed cost}) + cx}{x}.$$

Number of units produced

EXAMPLE 9 Average Cost for a Business

We return to the robotic exoskeleton described in the section opener. Suppose a company that manufactures this invention has a fixed monthly cost of $1,000,000 and that it costs $5000 to produce each robotic system.

a. Write the cost function, C, of producing x robotic systems.
b. Write the average cost function, \overline{C}, of producing x robotic systems.
c. Find and interpret $\overline{C}(1000)$, $\overline{C}(10{,}000)$, and $\overline{C}(100{,}000)$.
d. What is the horizontal asymptote for the graph of the average cost function, \overline{C}? Describe what this represents for the company.

SOLUTION

a. The cost function, C, is the sum of the fixed cost and the variable costs.

$$C(x) = 1{,}000{,}000 + 5000x$$

Fixed cost is $1,000,000. Variable cost: $5000 for each robotic system produced

b. The average cost function, \overline{C}, is the sum of fixed and variable costs divided by the number of robotic systems produced.

$$\overline{C}(x) = \frac{1{,}000{,}000 + 5000x}{x} \quad \text{or} \quad \overline{C}(x) = \frac{5000x + 1{,}000{,}000}{x}$$

c. We evaluate \overline{C} at 1000, 10,000, and 100,000, interpreting the results.

$$\overline{C}(1000) = \frac{5000(1000) + 1{,}000{,}000}{1000} = 6000$$

The average cost per robotic system of producing 1000 systems per month is $6000.

$$\overline{C}(10{,}000) = \frac{5000(10{,}000) + 1{,}000{,}000}{10{,}000} = 5100$$

The average cost per robotic system of producing 10,000 systems per month is $5100.

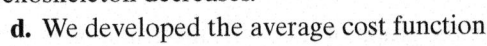

$$\overline{C}(100,000) = \frac{5000(100,000) + 1,000,000}{100,000} = 5010$$

The average cost per robotic system of producing 100,000 systems per month is $5010. Notice that with higher production levels, the cost of producing each robotic exoskeleton decreases.

d. We developed the average cost function

$$\overline{C}(x) = \frac{5000x + 1,000,000}{x}$$

in which the degree of the numerator, 1, is equal to the degree of the denominator, 1. The leading coefficients of the numerator and denominator, 5000 and 1, are used to obtain the equation of the horizontal asymptote. The equation of the horizontal asymptote is

$$y = \frac{5000}{1} \quad \text{or} \quad y = 5000.$$

The horizontal asymptote is shown in **Figure 3.42**. This means that the more robotic systems produced each month, the closer the average cost per system for the company comes to $5000. The least possible cost per robotic exoskeleton is approaching $5000. Competitively low prices take place with high production levels, posing a major problem for small businesses. •••

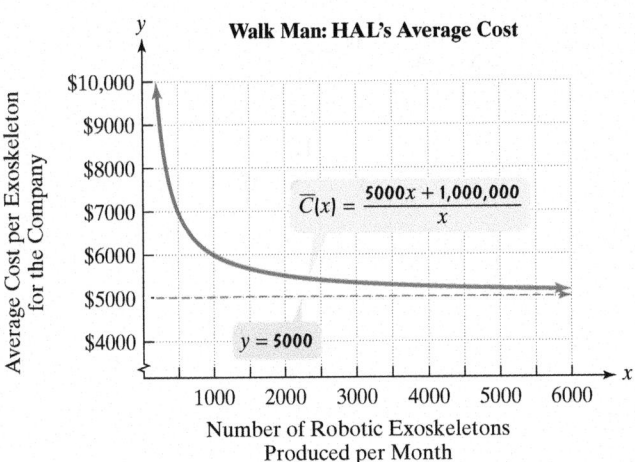

FIGURE 3.42

✓ Check Point **9** A company is planning to manufacture wheelchairs that are light, fast, and beautiful. The fixed monthly cost will be $500,000 and it will cost $400 to produce each radically innovative chair.

a. Write the cost function, C, of producing x wheelchairs.

b. Write the average cost function, \overline{C}, of producing x wheelchairs.

c. Find and interpret $\overline{C}(1000)$, $\overline{C}(10,000)$, and $\overline{C}(100,000)$.

d. What is the horizontal asymptote for the graph of the average cost function, \overline{C}? Describe what this represents for the company.

CONCEPT AND VOCABULARY CHECK

Fill in each blank so that the resulting statement is true.

1. All rational functions can be expressed as

$$f(x) = \frac{p(x)}{q(x)},$$

where p and q are _____ functions and $q(x) \neq 0$.

2. True or false: The domain of every rational function is the set of all real numbers. _____

3. True or false: The graph of the reciprocal function $f(x) = \dfrac{1}{x}$ has a break and is composed of two distinct branches. _____

4. If the graph of a function f increases or decreases without bound as x approaches a, then the line $x = a$ is a/an _____ of the graph of f. The equation of such a line for the graph of $f(x) = \dfrac{2}{x + 5}$ is _____.

5. If the graph of a function f approaches b as x increases or decreases without bound, then the line $y = b$ is a/an _____ of the graph of f. The equation of such a line for the graph of $y = \dfrac{x - 10}{3x^2 + x + 1}$ is _____. The equation of such a line for the graph of $y = \dfrac{x^2 - 10}{3x^2 + x + 1}$ is _____.

6. True or false: If the degree of the numerator of a rational function equals the degree of the denominator, then setting y equal to the ratio of the leading coefficients gives the equation of the horizontal asymptote. _____

7. Compared with the graph of $f(x) = \dfrac{1}{x}$, the graph of $g(x) = \dfrac{1}{x+2} - 1$ is shifted 2 units _____ and 1 unit _____.

8. The graph of a rational function has a slant asymptote if the degree of the numerator is _____ the degree of the denominator.

9. Based on the synthetic division shown below, the equation of the slant asymptote of

$$f(x) = \frac{3x^2 - 7x + 5}{x - 4} \text{ is } \underline{\hspace{1.5cm}}.$$

$$\underline{4}\begin{array}{|rrr} 3 & -7 & 5 \\ & 12 & 20 \\ \hline 3 & 5 & 25 \end{array}$$

EXERCISE SET 3.5

Practice Exercises

In Exercises 1–8, find the domain of each rational function.

1. $f(x) = \dfrac{5x}{x-4}$

2. $f(x) = \dfrac{7x}{x-8}$

3. $g(x) = \dfrac{3x^2}{(x-5)(x+4)}$

4. $g(x) = \dfrac{2x^2}{(x-2)(x+6)}$

5. $h(x) = \dfrac{x+7}{x^2 - 49}$

6. $h(x) = \dfrac{x+8}{x^2 - 64}$

7. $f(x) = \dfrac{x+7}{x^2 + 49}$

8. $f(x) = \dfrac{x+8}{x^2 + 64}$

Use the graph of the rational function in the figure shown to complete each statement in Exercises 9–14.

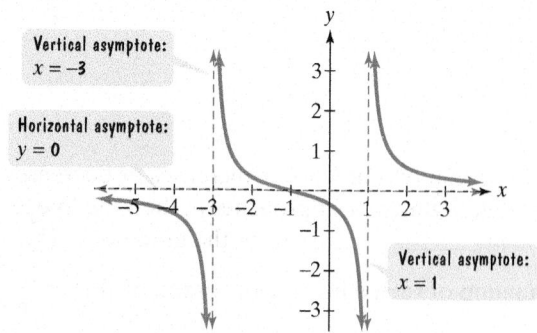

9. As $x \to -3^-$, $f(x) \to$ _____.

10. As $x \to -3^+$, $f(x) \to$ _____.

11. As $x \to 1^-$, $f(x) \to$ _____.

12. As $x \to 1^+$, $f(x) \to$ _____.

13. As $x \to -\infty$, $f(x) \to$ _____.

14. As $x \to \infty$, $f(x) \to$ _____.

Use the graph of the rational function in the figure shown to complete each statement in Exercises 15–20.

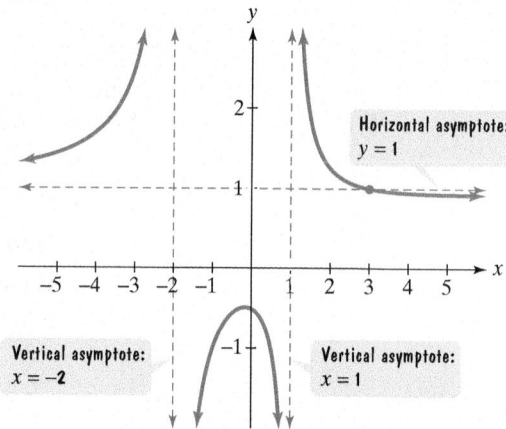

15. As $x \to 1^+$, $f(x) \to$ _____.

16. As $x \to 1^-$, $f(x) \to$ _____.

17. As $x \to -2^+$, $f(x) \to$ _____.

18. As $x \to -2^-$, $f(x) \to$ _____.

19. As $x \to \infty$, $f(x) \to$ _____.

20. As $x \to -\infty$, $f(x) \to$ _____.

In Exercises 21–36, find the vertical asymptotes, if any, and the values of x corresponding to holes, if any, of the graph of each rational function.

21. $f(x) = \dfrac{x}{x+4}$

22. $f(x) = \dfrac{x}{x-3}$

23. $g(x) = \dfrac{x+3}{x(x+4)}$

24. $g(x) = \dfrac{x+3}{x(x-3)}$

25. $h(x) = \dfrac{x}{x(x+4)}$

26. $h(x) = \dfrac{x}{x(x-3)}$

27. $r(x) = \dfrac{x}{x^2 + 4}$

28. $r(x) = \dfrac{x}{x^2 + 3}$

29. $f(x) = \dfrac{x^2 - 9}{x - 3}$

30. $f(x) = \dfrac{x^2 - 25}{x - 5}$

31. $g(x) = \dfrac{x - 3}{x^2 - 9}$

32. $g(x) = \dfrac{x - 5}{x^2 - 25}$

33. $h(x) = \dfrac{x + 7}{x^2 + 4x - 21}$

34. $h(x) = \dfrac{x + 6}{x^2 + 2x - 24}$

35. $r(x) = \dfrac{x^2 + 4x - 21}{x + 7}$

36. $r(x) = \dfrac{x^2 + 2x - 24}{x + 6}$

In Exercises 37–44, find the horizontal asymptote, if there is one, of the graph of each rational function.

37. $f(x) = \dfrac{12x}{3x^2 + 1}$

38. $f(x) = \dfrac{15x}{3x^2 + 1}$

39. $g(x) = \dfrac{12x^2}{3x^2 + 1}$

40. $g(x) = \dfrac{15x^2}{3x^2 + 1}$

41. $h(x) = \dfrac{12x^3}{3x^2 + 1}$

42. $h(x) = \dfrac{15x^3}{3x^2 + 1}$

43. $f(x) = \dfrac{-2x + 1}{3x + 5}$

44. $f(x) = \dfrac{-3x + 7}{5x - 2}$

In Exercises 45–56, use transformations of $f(x) = \dfrac{1}{x}$ or $f(x) = \dfrac{1}{x^2}$ to graph each rational function.

45. $g(x) = \dfrac{1}{x - 1}$

46. $g(x) = \dfrac{1}{x - 2}$

47. $h(x) = \dfrac{1}{x} + 2$

48. $h(x) = \dfrac{1}{x} + 1$

49. $g(x) = \dfrac{1}{x + 1} - 2$

50. $g(x) = \dfrac{1}{x + 2} - 2$

51. $g(x) = \dfrac{1}{(x + 2)^2}$

52. $g(x) = \dfrac{1}{(x + 1)^2}$

53. $h(x) = \dfrac{1}{x^2} - 4$

54. $h(x) = \dfrac{1}{x^2} - 3$

55. $h(x) = \dfrac{1}{(x - 3)^2} + 1$

56. $h(x) = \dfrac{1}{(x - 3)^2} + 2$

In Exercises 57–80, follow the seven steps on page 409 to graph each rational function.

57. $f(x) = \dfrac{4x}{x - 2}$

58. $f(x) = \dfrac{3x}{x - 1}$

59. $f(x) = \dfrac{2x}{x^2 - 4}$

60. $f(x) = \dfrac{4x}{x^2 - 1}$

61. $f(x) = \dfrac{2x^2}{x^2 - 1}$

62. $f(x) = \dfrac{4x^2}{x^2 - 9}$

63. $f(x) = \dfrac{-x}{x + 1}$

64. $f(x) = \dfrac{-3x}{x + 2}$

65. $f(x) = -\dfrac{1}{x^2 - 4}$

66. $f(x) = -\dfrac{2}{x^2 - 1}$

67. $f(x) = \dfrac{2}{x^2 + x - 2}$

68. $f(x) = \dfrac{-2}{x^2 - x - 2}$

69. $f(x) = \dfrac{2x^2}{x^2 + 4}$

70. $f(x) = \dfrac{4x^2}{x^2 + 1}$

71. $f(x) = \dfrac{x + 2}{x^2 + x - 6}$

72. $f(x) = \dfrac{x - 4}{x^2 - x - 6}$

73. $f(x) = \dfrac{x - 2}{x^2 - 4}$

74. $f(x) = \dfrac{x - 3}{x^2 - 9}$

75. $f(x) = \dfrac{x^4}{x^2 + 2}$

76. $f(x) = \dfrac{2x^4}{x^2 + 1}$

77. $f(x) = \dfrac{x^2 + x - 12}{x^2 - 4}$

78. $f(x) = \dfrac{x^2}{x^2 + x - 6}$

79. $f(x) = \dfrac{3x^2 + x - 4}{2x^2 - 5x}$

80. $f(x) = \dfrac{x^2 - 4x + 3}{(x + 1)^2}$

*In Exercises 81–88, **a.** Find the slant asymptote of the graph of each rational function and **b.** Follow the seven-step strategy and use the slant asymptote to graph each rational function.*

81. $f(x) = \dfrac{x^2 - 1}{x}$

82. $f(x) = \dfrac{x^2 - 4}{x}$

83. $f(x) = \dfrac{x^2 + 1}{x}$

84. $f(x) = \dfrac{x^2 + 4}{x}$

85. $f(x) = \dfrac{x^2 + x - 6}{x - 3}$

86. $f(x) = \dfrac{x^2 - x + 1}{x - 1}$

87. $f(x) = \dfrac{x^3 + 1}{x^2 + 2x}$

88. $f(x) = \dfrac{x^3 - 1}{x^2 - 9}$

Practice Plus

In Exercises 89–94, the equation for f is given by the simplified expression that results after performing the indicated operation. Write the equation for f and then graph the function.

89. $\dfrac{5x^2}{x^2 - 4} \cdot \dfrac{x^2 + 4x + 4}{10x^3}$

90. $\dfrac{x - 5}{10x - 2} \div \dfrac{x^2 - 10x + 25}{25x^2 - 1}$

91. $\dfrac{x}{2x + 6} - \dfrac{9}{x^2 - 9}$

92. $\dfrac{2}{x^2 + 3x + 2} - \dfrac{4}{x^2 + 4x + 3}$

93. $\dfrac{1 - \dfrac{3}{x + 2}}{1 + \dfrac{1}{x - 2}}$

94. $\dfrac{x - \dfrac{1}{x}}{x + \dfrac{1}{x}}$

In Exercises 95–98, use long division to rewrite the equation for g in the form

$$quotient + \dfrac{remainder}{divisor}.$$

Then use this form of the function's equation and transformations of $f(x) = \dfrac{1}{x}$ to graph g.

95. $g(x) = \dfrac{2x + 7}{x + 3}$

96. $g(x) = \dfrac{3x + 7}{x + 2}$

97. $g(x) = \dfrac{3x - 7}{x - 2}$

98. $g(x) = \dfrac{2x - 9}{x - 4}$

Application Exercises

99. A company is planning to manufacture mountain bikes. The fixed monthly cost will be $100,000 and it will cost $100 to produce each bicycle.

 a. Write the cost function, C, of producing x mountain bikes.

 b. Write the average cost function, \overline{C}, of producing x mountain bikes.

 c. Find and interpret $\overline{C}(500)$, $\overline{C}(1000)$, $\overline{C}(2000)$, and $\overline{C}(4000)$.

 d. What is the horizontal asymptote for the graph of the average cost function, \overline{C}? Describe what this means in practical terms.

100. A company that manufactures running shoes has a fixed monthly cost of $300,000. It costs $30 to produce each pair of shoes.

a. Write the cost function, C, of producing x pairs of shoes.

b. Write the average cost function, \overline{C}, of producing x pairs of shoes.

c. Find and interpret $\overline{C}(1000)$, $\overline{C}(10,000)$, and $\overline{C}(100,000)$.

d. What is the horizontal asymptote for the graph of the average cost function, \overline{C}? Describe what this represents for the company.

101. The function

$$f(x) = \frac{6.5x^2 - 20.4x + 234}{x^2 + 36}$$

models the pH level, $f(x)$, of the human mouth x minutes after a person eats food containing sugar. The graph of this function is shown in the figure.

Number of Minutes after
Eating Food Containing Sugar

a. Use the graph to obtain a reasonable estimate, to the nearest tenth, of the pH level of the human mouth 42 minutes after a person eats food containing sugar.

b. After eating sugar, when is the pH level the lowest? Use the function's equation to determine the pH level, to the nearest tenth, at this time.

c. According to the graph, what is the normal pH level of the human mouth?

d. What is the equation of the horizontal asymptote associated with this function? Describe what this means in terms of the mouth's pH level over time.

e. Use the graph to describe what happens to the pH level during the first hour.

102. A drug is injected into a patient and the concentration of the drug in the bloodstream is monitored. The drug's concentration, $C(t)$, in milligrams per liter, after t hours is modeled by

$$C(t) = \frac{5t}{t^2 + 1}.$$

The graph of this rational function, obtained with a graphing utility, is shown in the figure.

$[0, 10, 1]$ by $[0, 3, 1]$

a. Use the preceding graph to obtain a reasonable estimate of the drug's concentration after 3 hours.

b. Use the function's equation displayed in the voice balloon by the graph to determine the drug's concentration after 3 hours.

c. Use the function's equation to find the horizontal asymptote for the graph. Describe what this means about the drug's concentration in the patient's bloodstream as time increases.

Among all deaths from a particular disease, the percentage that is smoking related (21–39 cigarettes per day) is a function of the disease's **incidence ratio**. *The incidence ratio describes the number of times more likely smokers are than nonsmokers to die from the disease. The following table shows the incidence ratios for heart disease and lung cancer for two age groups.*

Incidence Ratios		
	Heart Disease	**Lung Cancer**
Ages 55–64	1.9	10
Ages 65–74	1.7	9

Source: Alexander M. Walker, *Observations and Inference*, Epidemiology Resources Inc., 1991.

For example, the incidence ratio of 9 in the table means that smokers between the ages of 65 and 74 are 9 times more likely than nonsmokers in the same age group to die from lung cancer. The rational function

$$P(x) = \frac{100(x - 1)}{x}$$

models the percentage of smoking-related deaths among all deaths from a disease, $P(x)$, in terms of the disease's incidence ratio, x. The graph of the rational function is shown. Use this function to solve Exercises 103–106.

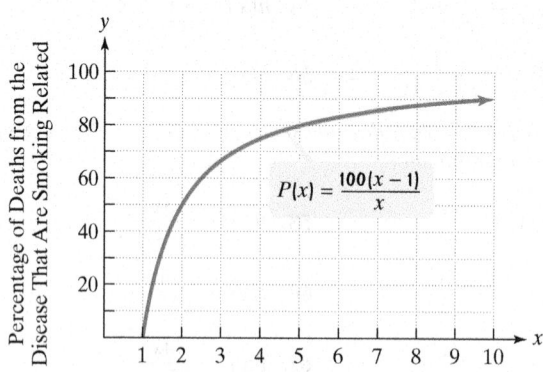

The Disease's Incidence Ratio:
The number of times more likely smokers are than nonsmokers to die from the disease

103. Find $P(10)$. Describe what this means in terms of the incidence ratio, 10, given in the table. Identify your solution as a point on the graph.

104. Find $P(9)$. Round to the nearest percent. Describe what this means in terms of the incidence ratio, 9, given in the table. Identify your solution as a point on the graph.

105. What is the horizontal asymptote of the graph? Describe what this means about the percentage of deaths caused by smoking with increasing incidence ratios.

106. According to the model and its graph, is there a disease for which all deaths are caused by smoking? Explain your answer.

107. The bar graph shows the amount, in billions of dollars, that the United States government spent on human resources and total budget outlays for six selected years. (Human resources include education, health, Medicare, Social Security, and veterans benefits and services.)

Federal Budget Expenditures on Human Resources

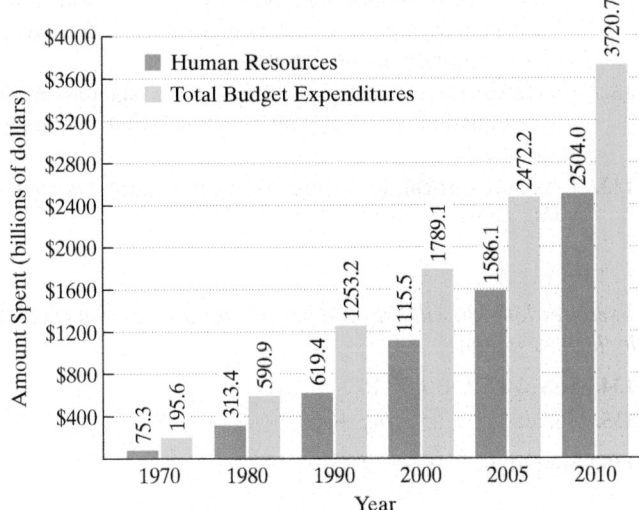

Source: Office of Management and Budget

The function $p(x) = 1.75x^2 - 15.9x + 160$ models the amount, $p(x)$, in billions of dollars, that the United States government spent on human resources x years after 1970. The function $q(x) = 2.1x^2 - 3.5x + 296$ models total budget expenditures, $q(x)$, in billions of dollars, x years after 1970.

a. Use p and q to write a rational function that models the fraction of total budget outlays spent on human resources x years after 1970.

b. Use the data displayed by the bar graph to find the percentage of federal expenditures spent on human resources in 2010. Round to the nearest percent.

c. Use the rational function from part (a) to find the percentage of federal expenditures spent on human resources in 2010. Round to the nearest percent. Does this underestimate or overestimate the actual percent that you found in part (b)? By how much?

d. What is the equation of the horizontal asymptote associated with the rational function in part (a)? If trends modeled by the function continue, what percentage of the federal budget will be spent on human resources over time? Round to the nearest percent.

Writing in Mathematics

108. What is a rational function?

109. Use everyday language to describe the graph of a rational function f such that as $x \to -\infty, f(x) \to 3$.

110. Use everyday language to describe the behavior of a graph near its vertical asymptote if $f(x) \to \infty$ as $x \to -2^-$ and $f(x) \to -\infty$ as $x \to -2^+$.

111. If you are given the equation of a rational function, explain how to find the vertical asymptotes, if there is one, of the function's graph.

112. If you are given the equation of a rational function, explain how to find the horizontal asymptote, if any, of the function's graph.

113. Describe how to graph a rational function.

114. If you are given the equation of a rational function, how can you tell if the graph has a slant asymptote? If it does, how do you find its equation?

115. Is every rational function a polynomial function? Why or why not? Does a true statement result if the two adjectives *rational* and *polynomial* are reversed? Explain.

116. Although your friend has a family history of heart disease, he smokes, on average, 25 cigarettes per day. He sees the table showing incidence ratios for heart disease (see Exercises 103–106) and feels comfortable that they are less than 2, compared to 9 and 10 for lung cancer. He claims that all family deaths have been from heart disease and decides not to give up smoking. Use the given function and its graph to describe some additional information not given in the table that might influence his decision.

Technology Exercises

117. Use a graphing utility to verify any five of your hand-drawn graphs in Exercises 45–88.

118. Use a graphing utility to graph $y = \dfrac{1}{x}, y = \dfrac{1}{x^3}$, and $\dfrac{1}{x^5}$ in the same viewing rectangle. For odd values of n, how does changing n affect the graph of $y = \dfrac{1}{x^n}$?

119. Use a graphing utility to graph $y = \dfrac{1}{x^2}, y = \dfrac{1}{x^4}$, and $y = \dfrac{1}{x^6}$ in the same viewing rectangle. For even values of n, how does changing n affect the graph of $y = \dfrac{1}{x^n}$?

120. Use a graphing utility to graph

$$f(x) = \frac{x^2 - 4x + 3}{x - 2} \quad \text{and} \quad g(x) = \frac{x^2 - 5x + 6}{x - 2}.$$

What differences do you observe between the graph of f and the graph of g? How do you account for these differences?

121. The rational function

$$f(x) = \frac{27,725(x - 14)}{x^2 + 9} - 5x$$

models the number of arrests, $f(x)$, per 100,000 drivers, for driving under the influence of alcohol, as a function of a driver's age, x.

a. Graph the function in a $[0, 70, 5]$ by $[0, 400, 20]$ viewing rectangle.

b. Describe the trend shown by the graph.

c. Use the $\boxed{\text{ZOOM}}$ and $\boxed{\text{TRACE}}$ features or the maximum function feature of your graphing utility to find the age that corresponds to the greatest number of arrests. How many arrests, per 100,000 drivers, are there for this age group?

Critical Thinking Exercises

Make Sense? *In Exercises 122–125, determine whether each statement makes sense or does not make sense, and explain your reasoning.*

122. I've graphed a rational function that has two vertical asymptotes and two horizontal asymptotes.

123. My graph of $y = \dfrac{x - 1}{(x - 1)(x - 2)}$ has vertical asymptotes at $x = 1$ and $x = 2$.

124. The function

$$f(x) = \frac{1.96x + 3.14}{3.04x + 21.79}$$

models the fraction of nonviolent prisoners in New York State prisons x years after 1980. I can conclude from this equation that over time the percentage of nonviolent prisoners will exceed 60%.

125. As production level increases, the average cost for a company to produce each unit of its product also increases.

In Exercises 126–129, determine whether each statement is true or false. If the statement is false, make the necessary change(s) to produce a true statement.

126. The graph of a rational function cannot have both a vertical asymptote and a horizontal asymptote.

127. It is possible to have a rational function whose graph has no y-intercept.

128. The graph of a rational function can have three vertical asymptotes.

129. The graph of a rational function can never cross a vertical asymptote.

In Exercises 130–133, write the equation of a rational function $f(x) = \dfrac{p(x)}{q(x)}$ having the indicated properties, in which the degrees of p and q are as small as possible. More than one correct function may be possible. Graph your function using a graphing utility to verify that it has the required properties.

130. f has a vertical asymptote given by $x = 3$, a horizontal asymptote $y = 0$, y-intercept at -1, and no x-intercept.

131. f has vertical asymptotes given by $x = -2$ and $x = 2$, a horizontal asymptote $y = 2$, y-intercept at $\frac{9}{2}$, x-intercepts at -3 and 3, and y-axis symmetry.

132. f has a vertical asymptote given by $x = 1$, a slant asymptote whose equation is $y = x$, y-intercept at 2, and x-intercepts at -1 and 2.

133. f has no vertical, horizontal, or slant asymptotes, and no x-intercepts.

Preview Exercises

Exercises 134–136 will help you prepare for the material covered in the next section.

134. Solve: $2x^2 + x = 15$.

135. Solve: $x^3 + x^2 = 4x + 4$.

136. Simplify: $\dfrac{x + 1}{x + 3} - 2$.

SECTION 3.6 Polynomial and Rational Inequalities

Objectives

① Solve polynomial inequalities.

② Solve rational inequalities.

③ Solve problems modeled by polynomial or rational inequalities.

© Warren Miller/The New Yorker Collection/Cartoonbank

Tailgaters beware: If your car is going 35 miles per hour on dry pavement, your required stopping distance is 160 feet, or the width of a football field. At 65 miles per hour, the distance required is 410 feet, or approximately the length of one and one-tenth football fields. **Figure 3.43** at the top of the next page shows stopping distances for cars at various speeds on dry roads and on wet roads.

Stopping Distances for Cars at Selected Speeds

Dry Pavement Wet Pavement

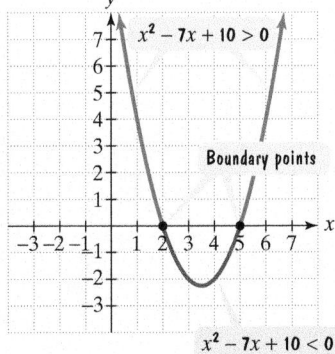

FIGURE 3.43
Source: National Highway Traffic Safety Administration

TECHNOLOGY

We used the statistical menu of a graphing utility and the quadratic regression program to obtain the quadratic function that models stopping distance on dry pavement. After entering the appropriate data from **Figure 3.43**, namely,

$(35, 160), (45, 225), (55, 310), (65, 410),$

we obtained the results shown in the screen.

```
QuadReg
y=ax²+bx+c
a=.0875
b=-.4
c=66.5625
```

A car's required stopping distance, $f(x)$, in feet, on dry pavement traveling at x miles per hour can be modeled by the quadratic function

$$f(x) = 0.0875x^2 - 0.4x + 66.6.$$

How can we use this function to determine speeds on dry pavement requiring stopping distances that exceed the length of one and one-half football fields, or 540 feet? We must solve the inequality

$$0.0875x^2 - 0.4x + 66.6 > 540.$$

Required stopping distance exceeds 540 feet.

We begin by subtracting 540 from both sides. This will give us zero on the right:

$$0.0875x^2 - 0.4x + 66.6 - 540 > 540 - 540$$
$$0.0875x^2 - 0.4x - 473.4 > 0.$$

The form of this inequality is $ax^2 + bx + c > 0$. Such a quadratic inequality is called a *polynomial inequality*.

Definition of a Polynomial Inequality

A polynomial inequality is any inequality that can be put into one of the forms

$$f(x) < 0, \quad f(x) > 0, \quad f(x) \le 0, \quad \text{or} \quad f(x) \ge 0,$$

where f is a polynomial function.

In this section, we establish the basic techniques for solving polynomial inequalities. We will also use these techniques to solve inequalities involving rational functions.

Solving Polynomial Inequalities

Graphs can help us visualize the solutions of polynomial inequalities. For example, the graph of $f(x) = x^2 - 7x + 10$ is shown in **Figure 3.44**. The x-intercepts, 2 and 5, are **boundary points** between where the graph lies above the x-axis, shown in blue, and where the graph lies below the x-axis, shown in red.

FIGURE 3.44

Locating the x-intercepts of a polynomial function, f, is an important step in finding the solution set for polynomial inequalities in the form $f(x) < 0$ or $f(x) > 0$. We use the x-intercepts of f as boundary points that divide the real number line into intervals. On each interval, the graph of f is either above the x-axis $[f(x) > 0]$ or below the x-axis $[f(x) < 0]$. For this reason, x-intercepts play a fundamental role in solving polynomial inequalities. The x-intercepts are found by solving the equation $f(x) = 0$.

1 Solve polynomial inequalities.

Procedure for Solving Polynomial Inequalities

1. Express the inequality in the form

$$f(x) < 0 \quad \text{or} \quad f(x) > 0,$$

where f is a polynomial function.
2. Solve the equation $f(x) = 0$. The real solutions are the **boundary points**.
3. Locate these boundary points on a number line, thereby dividing the number line into intervals.
4. Choose one representative number, called a **test value**, within each interval and evaluate f at that number.
 a. If the value of f is positive, then $f(x) > 0$ for all numbers, x, in the interval.
 b. If the value of f is negative, then $f(x) < 0$ for all numbers, x, in the interval.
5. Write the solution set, selecting the interval or intervals that satisfy the given inequality.

This procedure is valid if $<$ is replaced by \leq or $>$ is replaced by \geq. However, if the inequality involves \leq or \geq, include the boundary points [the solutions of $f(x) = 0$] in the solution set.

EXAMPLE 1 Solving a Polynomial Inequality

Solve and graph the solution set on a real number line: $2x^2 + x > 15$.

SOLUTION

Step 1 Express the inequality in the form $f(x) < 0$ or $f(x) > 0$. We begin by rewriting the inequality so that 0 is on the right side.

$2x^2 + x > 15$	This is the given inequality.
$2x^2 + x - 15 > 15 - 15$	Subtract 15 from both sides.
$2x^2 + x - 15 > 0$	Simplify.

This inequality is equivalent to the one we wish to solve. It is in the form $f(x) > 0$, where $f(x) = 2x^2 + x - 15$.

Step 2 Solve the equation $f(x) = 0$. We find the x-intercepts of $f(x) = 2x^2 + x - 15$ by solving the equation $2x^2 + x - 15 = 0$.

$2x^2 + x - 15 = 0$	This polynomial equation is a quadratic equation.
$(2x - 5)(x + 3) = 0$	Factor.
$2x - 5 = 0 \quad \text{or} \quad x + 3 = 0$	Set each factor equal to 0.
$x = \frac{5}{2} \qquad\qquad x = -3$	Solve for x.

The x-intercepts of f are -3 and $\frac{5}{2}$. We will use these x-intercepts as boundary points on a number line.

Step 3 Locate the boundary points on a number line and separate the line into intervals. The number line with the boundary points is shown as follows:

The boundary points divide the number line into three intervals:

$$(-\infty, -3) \quad \left(-3, \tfrac{5}{2}\right) \quad \left(\tfrac{5}{2}, \infty\right).$$

Step 4 Choose one test value within each interval and evaluate f at that number.

Interval	Test Value	Substitute into $f(x) = 2x^2 + x - 15$	Conclusion
$(-\infty, -3)$	-4	$f(-4) = 2(-4)^2 + (-4) - 15$ $= 13$, positive	$f(x) > 0$ for all x in $(-\infty, -3)$.
$\left(-3, \frac{5}{2}\right)$	0	$f(0) = 2 \cdot 0^2 + 0 - 15$ $= -15$, negative	$f(x) < 0$ for all x in $\left(-3, \frac{5}{2}\right)$.
$\left(\frac{5}{2}, \infty\right)$	3	$f(3) = 2 \cdot 3^2 + 3 - 15$ $= 6$, positive	$f(x) > 0$ for all x in $\left(\frac{5}{2}, \infty\right)$.

TECHNOLOGY

Graphic Connections

The solution set for

$$2x^2 + x > 15$$

or, equivalently,

$$2x^2 + x - 15 > 0$$

can be verified with a graphing utility. The graph of $f(x) = 2x^2 + x - 15$ was obtained using a $[-10, 10, 1]$ by $[-16, 6, 1]$ viewing rectangle. The graph lies above the x-axis, representing $>$, for all x in $(-\infty, -3)$ or $\left(\frac{5}{2}, \infty\right)$.

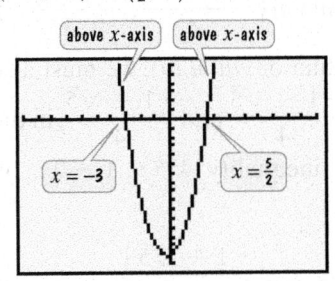

Step 5 Write the solution set, selecting the interval or intervals that satisfy the given inequality. We are interested in solving $f(x) > 0$, where $f(x) = 2x^2 + x - 15$. Based on our work in step 4, we see that $f(x) > 0$ for all x in $(-\infty, -3)$ or $\left(\frac{5}{2}, \infty\right)$. Thus, the solution set of the given inequality, $2x^2 + x > 15$, or, equivalently, $2x^2 + x - 15 > 0$, is

$$(-\infty, -3) \cup \left(\tfrac{5}{2}, \infty\right) \text{or} \left\{x \mid x < -3 \text{ or } x > \tfrac{5}{2}\right\}.$$

The graph of the solution set on a number line is shown as follows:

✅ **Check Point 1** Solve and graph the solution set: $x^2 - x > 20$.

EXAMPLE 2 Solving a Polynomial Inequality

Solve and graph the solution set on a real number line: $4x^2 \le 1 - 2x$.

SOLUTION

Step 1 Express the inequality in the form $f(x) \le 0$ or $f(x) \ge 0$. We begin by rewriting the inequality so that 0 is on the right side.

$4x^2 \le 1 - 2x$	This is the given inequality.
$4x^2 + 2x - 1 \le 1 - 2x + 2x - 1$	Add 2x and subtract 1 on both sides.
$4x^2 + 2x - 1 \le 0$	Simplify.

This inequality is equivalent to the one we wish to solve. It is in the form $f(x) \le 0$, where $f(x) = 4x^2 + 2x - 1$.

Step 2 Solve the equation $f(x) = 0$. We will find the x-intercepts of $f(x) = 4x^2 + 2x - 1$ by solving the equation $4x^2 + 2x - 1 = 0$. This equation cannot be solved by factoring. We will use the quadratic formula to solve it.

$$4x^2 + 2x - 1 = 0$$

$$\boxed{a = 4} \quad \boxed{b = 2} \quad \boxed{c = -1}$$

$$x = \frac{-b \pm \sqrt{b^2 - 4ac}}{2a} = \frac{-2 \pm \sqrt{2^2 - 4 \cdot 4(-1)}}{2 \cdot 4} = \frac{-2 \pm \sqrt{4 - (-16)}}{8}$$

$$= \frac{-2 \pm \sqrt{20}}{8} = \frac{-2 \pm \sqrt{4}\sqrt{5}}{8} = \frac{-2 \pm 2\sqrt{5}}{8} = \frac{2(-1 \pm \sqrt{5})}{8} = \frac{-1 \pm \sqrt{5}}{4}$$

$$x = \frac{-1 + \sqrt{5}}{4} \approx 0.3 \qquad x = \frac{-1 - \sqrt{5}}{4} \approx -0.8$$

The x-intercepts of f are $\dfrac{-1 + \sqrt{5}}{4}$ (approximately 0.3) and $\dfrac{-1 - \sqrt{5}}{4}$ (approximately -0.8). We will use these x-intercepts as boundary points on a number line.

Step 3 **Locate the boundary points on a number line and separate the line into intervals.** The number line with the boundary points is shown as follows:

$$\frac{-1-\sqrt{5}}{4} \approx -0.8 \qquad \frac{-1+\sqrt{5}}{4} \approx 0.3$$

The boundary points divide the number line into three intervals:

$$\left(-\infty, \frac{-1-\sqrt{5}}{4}\right) \quad \left(\frac{-1-\sqrt{5}}{4}, \frac{-1+\sqrt{5}}{4}\right) \quad \left(\frac{-1+\sqrt{5}}{4}, \infty\right).$$

Step 4 **Choose one test value within each interval and evaluate f at that number.**

Interval	Test Value	Substitute into $f(x) = 4x^2 + 2x - 1$	Conclusion
$\left(-\infty, \dfrac{-1-\sqrt{5}}{4}\right)$	-1	$\begin{aligned}f(-1) &= 4(-1)^2 + 2(-1) - 1\\ &= 1,\text{ positive}\end{aligned}$	$f(x) > 0$ for all x in $\left(-\infty, \dfrac{-1-\sqrt{5}}{4}\right)$.
$\left(\dfrac{-1-\sqrt{5}}{4}, \dfrac{-1+\sqrt{5}}{4}\right)$	0	$\begin{aligned}f(0) &= 4\cdot 0^2 + 2\cdot 0 - 1\\ &= -1,\text{ negative}\end{aligned}$	$f(x) < 0$ for all x in $\left(\dfrac{-1-\sqrt{5}}{4}, \dfrac{-1+\sqrt{5}}{4}\right)$.
$\left(\dfrac{-1+\sqrt{5}}{4}, \infty\right)$	1	$\begin{aligned}f(1) &= 4\cdot 1^2 + 2\cdot 1 - 1\\ &= 5,\text{ positive}\end{aligned}$	$f(x) > 0$ for all x in $\left(\dfrac{-1+\sqrt{5}}{4}, \infty\right)$.

TECHNOLOGY

Graphic Connections

The solution set for

$$4x^2 \le 1 - 2x$$

or, equivalently,

$$4x^2 + 2x - 1 \le 0$$

can be verified with a graphing utility. The graph of $f(x) = 4x^2 + 2x - 1$ was obtained using a $[-2, 2, 1]$ by $[-10, 10, 1]$ viewing rectangle. The graph lies on or below the x-axis, representing \le, for all x in

$$\left[\frac{-1-\sqrt{5}}{4}, \frac{-1+\sqrt{5}}{4}\right]$$

$$\approx [-0.8, 0.3].$$

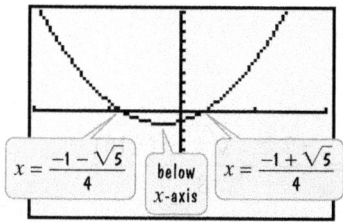

$x = \dfrac{-1-\sqrt{5}}{4}$ \quad below x-axis \quad $x = \dfrac{-1+\sqrt{5}}{4}$

$[-2, 2, 1]$ by $[-10, 10, 1]$

Step 5 **Write the solution set, selecting the interval or intervals that satisfy the given inequality.** We are interested in solving $f(x) \le 0$, where $f(x) = 4x^2 + 2x - 1$. Based on our work in step 4, we see that $f(x) < 0$ for all x in $\left(\dfrac{-1-\sqrt{5}}{4}, \dfrac{-1+\sqrt{5}}{4}\right)$.

However, because the inequality involves \le (less than or *equal to*), we must also include the solutions of $4x^2 + 2x - 1 = 0$, namely, $\dfrac{-1-\sqrt{5}}{4}$ and $\dfrac{-1+\sqrt{5}}{4}$, in the solution set. Thus, the solution set of the given inequality $4x^2 \le 1 - 2x$, or, equivalently, $4x^2 + 2x - 1 \le 0$, is

$$\left[\frac{-1-\sqrt{5}}{4}, \frac{-1+\sqrt{5}}{4}\right] \quad \text{or} \quad \left\{x \,\middle|\, \frac{-1-\sqrt{5}}{4} \le x \le \frac{-1+\sqrt{5}}{4}\right\}.$$

The graph of the solution set on a number line is shown as follows:

✓ Check Point **2** Solve and graph the solution set on a real number line: $2x^2 \le -6x - 1$.

EXAMPLE 3 Solving a Polynomial Inequality

Solve and graph the solution set on a real number line: $x^3 + x^2 \le 4x + 4$.

SOLUTION

Step 1 **Express the inequality in the form $f(x) \le 0$ or $f(x) \ge 0$.** We begin by rewriting the inequality so that 0 is on the right side.

$$x^3 + x^2 \le 4x + 4 \qquad \text{This is the given inequality.}$$
$$x^3 + x^2 - 4x - 4 \le 4x + 4 - 4x - 4 \qquad \text{Subtract } 4x + 4 \text{ from both sides.}$$
$$x^3 + x^2 - 4x - 4 \le 0 \qquad \text{Simplify.}$$

This inequality is equivalent to the one we wish to solve. It is in the form $f(x) \le 0$, where $f(x) = x^3 + x^2 - 4x - 4$.

Step 2 Solve the equation $f(x) = 0$. We find the x-intercepts of $f(x) = x^3 + x^2 - 4x - 4$ by solving the equation $x^3 + x^2 - 4x - 4 = 0$.

$$x^3 + x^2 - 4x - 4 = 0 \quad \text{This polynomial equation is of degree 3.}$$

$$x^2(x + 1) - 4(x + 1) = 0 \quad \begin{array}{l}\text{Factor } x^2 \text{ from the first two terms and} \\ -4 \text{ from the last two terms.}\end{array}$$

$$(x + 1)(x^2 - 4) = 0 \quad \begin{array}{l}\text{A common factor of } x + 1 \text{ is factored} \\ \text{from the expression.}\end{array}$$

$$(x + 1)(x + 2)(x - 2) = 0 \quad \text{Factor completely.}$$

$$x + 1 = 0 \quad \text{or} \quad x + 2 = 0 \quad \text{or} \quad x - 2 = 0 \quad \text{Set each factor equal to 0.}$$

$$x = -1 \qquad\qquad x = -2 \qquad\qquad x = 2 \quad \text{Solve for x.}$$

The x-intercepts of f are $-2, -1$, and 2. We will use these x-intercepts as boundary points on a number line.

Step 3 Locate the boundary points on a number line and separate the line into intervals. The number line with the boundary points is shown as follows:

The boundary points divide the number line into four intervals:

$$(-\infty, -2) \quad (-2, -1) \quad (-1, 2) \quad (2, \infty).$$

Step 4 Choose one test value within each interval and evaluate f at that number.

Interval	Test Value	Substitute into $f(x) = x^3 + x^2 - 4x - 4$	Conclusion
$(-\infty, -2)$	-3	$f(-3) = (-3)^3 + (-3)^2 - 4(-3) - 4$ $= -10$, negative	$f(x) < 0$ for all x in $(-\infty, -2)$.
$(-2, -1)$	-1.5	$f(-1.5) = (-1.5)^3 + (-1.5)^2 - 4(-1.5) - 4$ $= 0.875$, positive	$f(x) > 0$ for all x in $(-2, -1)$.
$(-1, 2)$	0	$f(0) = 0^3 + 0^2 - 4 \cdot 0 - 4$ $= -4$, negative	$f(x) < 0$ for all x in $(-1, 2)$.
$(2, \infty)$	3	$f(3) = 3^3 + 3^2 - 4 \cdot 3 - 4$ $= 20$, positive	$f(x) > 0$ for all x in $(2, \infty)$.

TECHNOLOGY

Graphic Connections

The solution set for

$$x^3 + x^2 \le 4x + 4$$

or, equivalently,

$$x^3 + x^2 - 4x - 4 \le 0$$

can be verified with a graphing utility. The graph of $f(x) = x^3 + x^2 - 4x - 4$ lies on or below the x-axis, representing \le, for all x in $(-\infty, -2]$ or $[-1, 2]$.

$[-4, 4, 1]$ by $[-7, 3, 1]$

Step 5 Write the solution set, selecting the interval or intervals that satisfy the given inequality. We are interested in solving $f(x) \le 0$, where $f(x) = x^3 + x^2 - 4x - 4$. Based on our work in step 4, we see that $f(x) < 0$ for all x in $(-\infty, -2)$ or $(-1, 2)$. However, because the inequality involves \le (less than or *equal to*), we must also include the solutions of $x^3 + x^2 - 4x - 4 = 0$, namely $-2, -1$, and 2, in the solution set. Thus, the solution set of the given inequality, $x^3 + x^2 \le 4x + 4$, or, equivalently, $x^3 + x^2 - 4x - 4 \le 0$, is

$$(-\infty, -2] \cup [-1, 2]$$

$$\text{or} \quad \{x \mid x \le -2 \text{ or} -1 \le x \le 2\}.$$

The graph of the solution set on a number line is shown as follows:

● ● ●

✓ **Check Point 3** Solve and graph the solution set on a real number line: $x^3 + 3x^2 \le x + 3$.

② Solve rational inequalities.

Solving Rational Inequalities

A **rational inequality** is any inequality that can be put into one of the forms

$$f(x) < 0, \quad f(x) > 0, \quad f(x) \le 0, \quad \text{or} \quad f(x) \ge 0,$$

where f is a rational function. An example of a rational inequality is

$$\frac{3x + 3}{2x + 4} > 0.$$

This inequality is in the form $f(x) > 0$, where f is the rational function given by

$$f(x) = \frac{3x + 3}{2x + 4}.$$

The graph of f is shown in **Figure 3.45**.

We can find the x-intercept of f by setting the numerator equal to 0:

$$3x + 3 = 0$$
$$3x = -3 \qquad \boxed{\text{f has an x-intercept}}$$
$$x = -1. \qquad \text{at -1 and passes through $(-1, 0)$.}$$

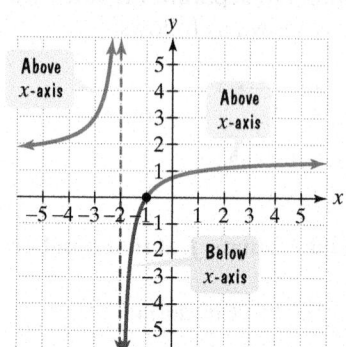

FIGURE 3.45 The graph of $f(x) = \dfrac{3x + 3}{2x + 4}$

We can determine where f is undefined by setting the denominator equal to 0:

$$2x + 4 = 0$$
$$2x = -4 \qquad \boxed{\text{f is undefined at -2.}}$$
$$x = -2. \qquad \text{\textbf{Figure 3.45} shows that the function's vertical asymptote is $x = -2$.}$$

By setting both the numerator and the denominator of f equal to 0, we obtained -2 and -1. These numbers separate the x-axis into three intervals: $(-\infty, -2), (-2, -1)$, and $(-1, \infty)$. On each interval, the graph of f is either above the x-axis $[f(x) > 0]$ or below the x-axis $[f(x) < 0]$.

Examine the graph in **Figure 3.45** carefully. Can you see that it is above the x-axis for all x in $(-\infty, -2)$ or $(-1, \infty)$, shown in blue? Thus, the solution set of $\dfrac{3x + 3}{2x + 4} > 0$ is $(-\infty, -2) \cup (-1, \infty)$. By contrast, the graph of f lies below the x-axis for all x in $(-2, -1)$, shown in red. Thus, the solution set of $\dfrac{3x + 3}{2x + 4} < 0$ is $(-2, -1)$.

The first step in solving a rational inequality is to bring all terms to one side, obtaining zero on the other side. Then express the rational function on the nonzero side as a single quotient. The second step is to set the numerator and the denominator of the rational function f equal to zero. The solutions of these equations serve as boundary points that separate the real number line into intervals. At this point, the procedure is the same as the one we used for solving polynomial inequalities.

GREAT QUESTION!

Can I begin solving

$$\frac{x + 1}{x + 3} \ge 2$$

by multiplying both sides by $x + 3$?

No. We do not know if $x + 3$ is positive or negative. Thus, we do not know whether or not to change the sense of the inequality.

EXAMPLE 4 Solving a Rational Inequality

Solve and graph the solution set: $\dfrac{x + 1}{x + 3} \ge 2$.

SOLUTION

Step 1 Express the inequality so that one side is zero and the other side is a single quotient. We subtract 2 from both sides to obtain zero on the right.

$$\frac{x+1}{x+3} \ge 2 \qquad \text{This is the given inequality.}$$

$$\frac{x+1}{x+3} - 2 \ge 0 \qquad \text{Subtract 2 from both sides, obtaining 0 on the right.}$$

$$\frac{x+1}{x+3} - \frac{2(x+3)}{x+3} \ge 0 \qquad \text{The least common denominator is } x+3. \text{ Express 2 in terms of this denominator.}$$

$$\frac{x+1-2(x+3)}{x+3} \ge 0 \qquad \text{Subtract rational expressions.}$$

$$\frac{x+1-2x-6}{x+3} \ge 0 \qquad \text{Apply the distributive property.}$$

$$\frac{-x-5}{x+3} \ge 0 \qquad \text{Simplify.}$$

This inequality is equivalent to the one we wish to solve. It is in the form $f(x) \ge 0$, where $f(x) = \dfrac{-x-5}{x+3}$.

Step 2 Set the numerator and the denominator of f equal to zero. The real solutions are the boundary points.

$$-x - 5 = 0 \qquad x + 3 = 0 \qquad \text{Set the numerator and denominator equal to 0. These are the values that make the previous quotient zero or undefined.}$$

$$x = -5 \qquad x = -3 \qquad \text{Solve for x.}$$

We will use these solutions as boundary points on a number line.

Step 3 Locate the boundary points on a number line and separate the line into intervals. The number line with the boundary points is shown as follows:

The boundary points divide the number line into three intervals:

$$(-\infty, -5) \quad (-5, -3) \quad (-3, \infty).$$

Step 4 Choose one test value within each interval and evaluate f at that number.

Interval	Test Value	Substitute into $f(x) = \dfrac{-x-5}{x+3}$	Conclusion
$(-\infty, -5)$	-6	$f(-6) = \dfrac{-(-6)-5}{-6+3}$ $= -\frac{1}{3}$, negative	$f(x) < 0$ for all x in $(-\infty, -5)$.
$(-5, -3)$	-4	$f(-4) = \dfrac{-(-4)-5}{-4+3}$ $= 1$, positive	$f(x) > 0$ for all x in $(-5, -3)$.
$(-3, \infty)$	0	$f(0) = \dfrac{-0-5}{0+3}$ $= -\frac{5}{3}$, negative	$f(x) < 0$ for all x in $(-3, \infty)$.

Step 5 Write the solution set, selecting the interval or intervals that satisfy the given inequality. We are interested in solving $f(x) \ge 0$, where $f(x) = \dfrac{-x-5}{x+3}$.

Based on our work in step 4, we see that $f(x) > 0$ for all x in $(-5, -3)$. However, because the inequality involves \ge (greater than or *equal to*), we must also include the solution of $f(x) = 0$, namely, the value that we obtained when we set the

GREAT QUESTION!

Which boundary points must I always exclude from the solution set of a rational inequality?

Never include values that cause a rational function's denominator to equal zero. Division by zero is undefined.

TECHNOLOGY

Graphic Connections

The solution set for

$$\frac{x + 1}{x + 3} \ge 2$$

or, equivalently,

$$\frac{-x - 5}{x + 3} \ge 0$$

can be verified with a graphing utility. The graph of $f(x) = \dfrac{-x - 5}{x + 3}$ lies on or above the x-axis, representing \ge, for all x in $[-5, -3)$.

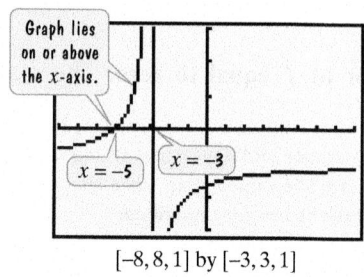

$[-8, 8, 1]$ by $[-3, 3, 1]$

3 Solve problems modeled by polynomial or rational inequalities.

$t = 0$
$s_0 = 190$
$v_0 = 96$

190 feet

FIGURE 3.46 Throwing a ball from 190 feet with a velocity of 96 feet per second

numerator of f equal to zero. Thus, we must include -5 in the solution set. The solution set of the given inequality is

$$[-5, -3) \text{ or } \{x \mid -5 \le x < -3\}.$$

−3 causes the denominator of f to equal zero. It must be excluded from the solution set.

The graph of the solution set on a number line is shown as follows:

```
    +   +   [———)   +   +   +   +   +   +   → x
   -7  -6  -5  -4  -3  -2  -1   0   1   2   3
```

⊘ Check Point **4** Solve and graph the solution set: $\dfrac{2x}{x + 1} \ge 1$.

Applications

If you throw an object directly upward, although its path is straight and vertical, its changing height over time can be described by a quadratic function.

The Position Function for a Free-Falling Object Near Earth's Surface

An object that is falling or vertically projected into the air has its height above the ground, $s(t)$, in feet, given by

$$s(t) = -16t^2 + v_0 t + s_0,$$

where v_0 is the original velocity (initial velocity) of the object, in feet per second, t is the time that the object is in motion, in seconds, and s_0 is the original height (initial height) of the object, in feet.

In Example 5, we solve a polynomial inequality in a problem about the position of a free-falling object.

EXAMPLE 5 Using the Position Function

A ball is thrown vertically upward from the top of the Leaning Tower of Pisa (190 feet high) with an initial velocity of 96 feet per second (**Figure 3.46**). During which time period will the ball's height exceed that of the tower?

SOLUTION

$s(t) = -16t^2 + v_0 t + s_0$ This is the position function for a free-falling object.

$s(t) = -16t^2 + 96t + 190$ Because v_0 (initial velocity) = 96 and s_0 (initial position) = 190, substitute these values into the formula.

When will $s(t)$, the ball's height, exceed that of the tower?

$$-16t^2 + 96t + 190 > 190$$

$-16t^2 + 96t + 190 > 190$ This is the inequality that models the problem's question. We must find t.

$-16t^2 + 96t > 0$ Subtract 190 from both sides. This inequality is in the form $f(t) > 0$, where $f(t) = -16t^2 + 96t$.

$-16t^2 + 96t = 0$ Solve the equation $f(t) = 0$.

$-16t(t - 6) = 0$ Factor.

$-16t = 0 \quad \text{or} \quad t - 6 = 0$ Set each factor equal to 0.

$t = 0 \qquad\qquad t = 6$ Solve for t. The boundary points are 0 and 6. Locate these values on a number line.

```
  +   +   ●   +   +   +   +   ●   +   +   → t
 -2  -1   0   1   2   3   4   5   6   7   8
```

The intervals are $(-\infty, 0), (0, 6)$, and $(6, \infty)$. For our purposes, the mathematical model is useful only from $t = 0$ until the ball hits the ground. (By setting $-16t^2 + 96t + 190$ equal to zero, we find $t \approx 7.57$; the ball hits the ground after approximately 7.57 seconds.) Thus, we use $(0, 6)$ and $(6, 7.57)$ for our intervals.

Interval	Test Value	Substitute into $f(t) = -16t^2 + 96t$	Conclusion
$(0, 6)$	1	$f(1) = -16 \cdot 1^2 + 96 \cdot 1$ $= 80$, positive	$f(t) > 0$ for all t in $(0, 6)$.
$(6, 7.57)$	7	$f(7) = -16 \cdot 7^2 + 96 \cdot 7$ $= -112$, negative	$f(t) < 0$ for all t in $(6, 7.57)$.

We are interested in solving $f(t) > 0$, where $f(t) = -16t^2 + 96t$. We see that $f(t) > 0$ for all t in $(0, 6)$. This means that the ball's height exceeds that of the tower between 0 and 6 seconds.　• • •

TECHNOLOGY

Graphic Connections

The graphs of

$$y_1 = -16x^2 + 96x + 190$$

and

$$y_2 = 190$$

are shown in a

$$[0, 8, 1] \text{ by } [0, 360, 36]$$

<table><tr><td>seconds
in motion</td><td>height,
in feet</td></tr></table>

viewing rectangle. The graphs show that the ball's height exceeds that of the tower between 0 and 6 seconds.

✓ Check Point **5** An object is propelled straight up from ground level with an initial velocity of 80 feet per second. Its height at time t is modeled by

$$s(t) = -16t^2 + 80t,$$

where the height, $s(t)$, is measured in feet and the time, t, is measured in seconds. In which time interval will the object be more than 64 feet above the ground?

CONCEPT AND VOCABULARY CHECK

Fill in each blank so that the resulting statement is true.

1. We solve the polynomial inequality $x^2 + 8x + 15 > 0$ by first solving the equation _____. The real solutions of this equation, -5 and -3, shown on the number line, are called _____ points.

2. The points at -5 and -3 shown in Exercise 1 divide the number line into three intervals:

_____, _____, _____.

3. True or false: A test value for the leftmost interval on the number line shown in Exercise 1 could be -10. _____

4. True or false: A test value for the rightmost interval on the number line shown in Exercise 1 could be 0. _____

5. Consider the rational inequality

$$\frac{x - 1}{x + 2} \geq 0.$$

Setting the numerator and the denominator of $\dfrac{x - 1}{x + 2}$ equal to zero, we obtain $x = 1$ and $x = -2$. These values are shown as points on the number line. Also shown is information about three test values.

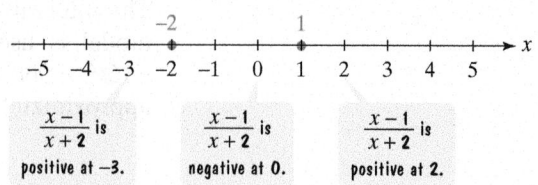

Based on the information shown above, the solution set of $\dfrac{x - 1}{x + 2} \geq 0$ is _____.

EXERCISE SET 3.6

Practice Exercises

Solve each polynomial inequality in Exercises 1–42 and graph the solution set on a real number line. Express each solution set in interval notation.

1. $(x - 4)(x + 2) > 0$
2. $(x + 3)(x - 5) > 0$
3. $(x - 7)(x + 3) \leq 0$
4. $(x + 1)(x - 7) \leq 0$
5. $x^2 - 5x + 4 > 0$
6. $x^2 - 4x + 3 < 0$
7. $x^2 + 5x + 4 > 0$
8. $x^2 + x - 6 > 0$
9. $x^2 - 6x + 9 < 0$
10. $x^2 - 2x + 1 > 0$
11. $3x^2 + 10x - 8 \leq 0$
12. $9x^2 + 3x - 2 \geq 0$
13. $2x^2 + x < 15$
14. $6x^2 + x > 1$
15. $4x^2 + 7x < -3$
16. $3x^2 + 16x < -5$
17. $5x \leq 2 - 3x^2$
18. $4x^2 + 1 \geq 4x$
19. $x^2 - 4x \geq 0$
20. $x^2 + 2x < 0$
21. $2x^2 + 3x > 0$
22. $3x^2 - 5x \leq 0$
23. $-x^2 + x \geq 0$
24. $-x^2 + 2x \geq 0$
25. $x^2 \leq 4x - 2$
26. $x^2 \leq 2x + 2$
27. $9x^2 - 6x + 1 < 0$
28. $4x^2 - 4x + 1 \geq 0$
29. $(x - 1)(x - 2)(x - 3) \geq 0$
30. $(x + 1)(x + 2)(x + 3) \geq 0$
31. $x(3 - x)(x - 5) \leq 0$
32. $x(4 - x)(x - 6) \leq 0$
33. $(2 - x)^2\left(x - \dfrac{7}{2}\right) < 0$
34. $(5 - x)^2\left(x - \dfrac{13}{2}\right) < 0$
35. $x^3 + 2x^2 - x - 2 \geq 0$
36. $x^3 + 2x^2 - 4x - 8 \geq 0$
37. $x^3 - 3x^2 - 9x + 27 < 0$
38. $x^3 + 7x^2 - x - 7 < 0$
39. $x^3 + x^2 + 4x + 4 > 0$
40. $x^3 - x^2 + 9x - 9 > 0$
41. $x^3 \geq 9x^2$
42. $x^3 \leq 4x^2$

Solve each rational inequality in Exercises 43–60 and graph the solution set on a real number line. Express each solution set in interval notation.

43. $\dfrac{x - 4}{x + 3} > 0$
44. $\dfrac{x + 5}{x - 2} > 0$
45. $\dfrac{x + 3}{x + 4} < 0$
46. $\dfrac{x + 5}{x + 2} < 0$
47. $\dfrac{-x + 2}{x - 4} \geq 0$
48. $\dfrac{-x - 3}{x + 2} \leq 0$
49. $\dfrac{4 - 2x}{3x + 4} \leq 0$
50. $\dfrac{3x + 5}{6 - 2x} \geq 0$
51. $\dfrac{x}{x - 3} > 0$
52. $\dfrac{x + 4}{x} > 0$
53. $\dfrac{(x + 4)(x - 1)}{x + 2} \leq 0$
54. $\dfrac{(x + 3)(x - 2)}{x + 1} \leq 0$
55. $\dfrac{x + 1}{x + 3} < 2$
56. $\dfrac{x}{x - 1} > 2$

57. $\dfrac{x + 4}{2x - 1} \leq 3$
58. $\dfrac{1}{x - 3} < 1$
59. $\dfrac{x - 2}{x + 2} \leq 2$
60. $\dfrac{x}{x + 2} \geq 2$

Practice Plus

In Exercises 61–64, find the domain of each function.

61. $f(x) = \sqrt{2x^2 - 5x + 2}$
62. $f(x) = \dfrac{1}{\sqrt{4x^2 - 9x + 2}}$
63. $f(x) = \sqrt{\dfrac{2x}{x + 1} - 1}$
64. $f(x) = \sqrt{\dfrac{x}{2x - 1} - 1}$

Solve each inequality in Exercises 65–70 and graph the solution set on a real number line.

65. $|x^2 + 2x - 36| > 12$
66. $|x^2 + 6x + 1| > 8$
67. $\dfrac{3}{x + 3} > \dfrac{3}{x - 2}$
68. $\dfrac{1}{x + 1} > \dfrac{2}{x - 1}$
69. $\dfrac{x^2 - x - 2}{x^2 - 4x + 3} > 0$
70. $\dfrac{x^2 - 3x + 2}{x^2 - 2x - 3} > 0$

In Exercises 71–72, use the graph of the polynomial function to solve each inequality.

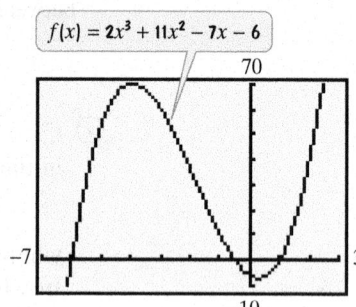

71. $2x^3 + 11x^2 \geq 7x + 6$
72. $2x^3 + 11x^2 < 7x + 6$

In Exercises 73–74, use the graph of the rational function to solve each inequality.

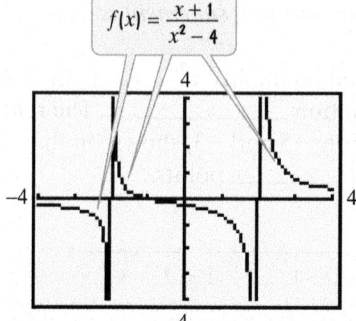

73. $\dfrac{1}{4(x + 2)} \leq -\dfrac{3}{4(x - 2)}$

74. $\dfrac{1}{4(x + 2)} > -\dfrac{3}{4(x - 2)}$

Application Exercises

Use the position function

$$s(t) = -16t^2 + v_0 t + s_0$$

$(v_0 = $ *initial velocity,* $s_0 = $ *initial position,* $t = $ *time*)

to answer Exercises 75–76.

75. Divers in Acapulco, Mexico, dive headfirst at 8 feet per second from the top of a cliff 87 feet above the Pacific Ocean. During which time period will a diver's height exceed that of the cliff?

76. You throw a ball straight up from a rooftop 160 feet high with an initial velocity of 48 feet per second. During which time period will the ball's height exceed that of the rooftop?

The functions

$$f(x) = 0.0875x^2 - 0.4x + 66.6$$

Dry pavement

and

Wet pavement

$$g(x) = 0.0875x^2 + 1.9x + 11.6$$

model a car's stopping distance, $f(x)$ or $g(x)$, in feet, traveling at x miles per hour. Function f models stopping distance on dry pavement and function g models stopping distance on wet pavement. The graphs of these functions are shown for $\{x \mid x \geq 30\}$. Notice that the figure does not specify which graph is the model for dry roads and which is the model for wet roads. Use this information to solve Exercises 77–78.

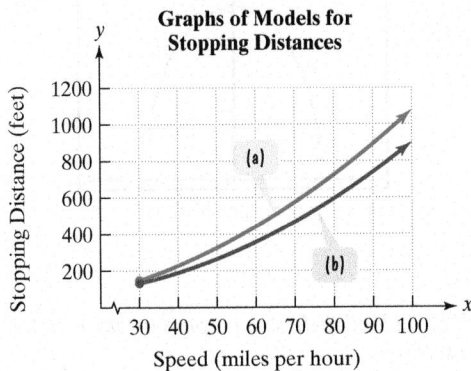

Graphs of Models for Stopping Distances

77. **a.** Use the given functions to find the stopping distance on dry pavement and the stopping distance on wet pavement for a car traveling at 35 miles per hour. Round to the nearest foot.

 b. Based on your answers to part (a), which rectangular coordinate graph shows stopping distances on dry pavement and which shows stopping distances on wet pavement?

 c. How well do your answers to part (a) model the actual stopping distances shown in **Figure 3.43** on page 421?

 d. Determine speeds on dry pavement requiring stopping distances that exceed the length of one and one-half football fields, or 540 feet. Round to the nearest mile per hour. How is this shown on the appropriate graph of the models?

78. **a.** Use the given functions to find the stopping distance on dry pavement and the stopping distance on wet pavement for a car traveling at 55 miles per hour. Round to the nearest foot.

 b. Based on your answers to part (a), which rectangular coordinate graph shows stopping distances on dry pavement and which shows stopping distances on wet pavement?

 c. How well do your answers to part (a) model the actual stopping distances shown in **Figure 3.43** on page 421?

 d. Determine speeds on wet pavement requiring stopping distances that exceed the length of one and one-half football fields, or 540 feet. Round to the nearest mile per hour. How is this shown on the appropriate graph of the models?

79. The perimeter of a rectangle is 50 feet. Describe the possible lengths of a side if the area of the rectangle is not to exceed 114 square feet.

80. The perimeter of a rectangle is 180 feet. Describe the possible lengths of a side if the area of the rectangle is not to exceed 800 square feet.

Writing in Mathematics

81. What is a polynomial inequality?

82. What is a rational inequality?

83. If f is a polynomial or rational function, explain how the graph of f can be used to visualize the solution set of the inequality $f(x) < 0$.

Technology Exercises

84. Use a graphing utility to verify your solution sets to any three of the polynomial inequalities that you solved algebraically in Exercises 1–42.

85. Use a graphing utility to verify your solution sets to any three of the rational inequalities that you solved algebraically in Exercises 43–60.

Solve each inequality in Exercises 86–91 using a graphing utility.

86. $x^2 + 3x - 10 > 0$

87. $2x^2 + 5x - 3 \leq 0$

88. $x^3 + x^2 - 4x - 4 > 0$

89. $\dfrac{x - 4}{x - 1} \leq 0$

90. $\dfrac{x + 2}{x - 3} \leq 2$

91. $\dfrac{1}{x + 1} \leq \dfrac{2}{x + 4}$

The graph shows stopping distances for trucks at various speeds on dry roads and on wet roads. Use this information to solve Exercises 92–93.

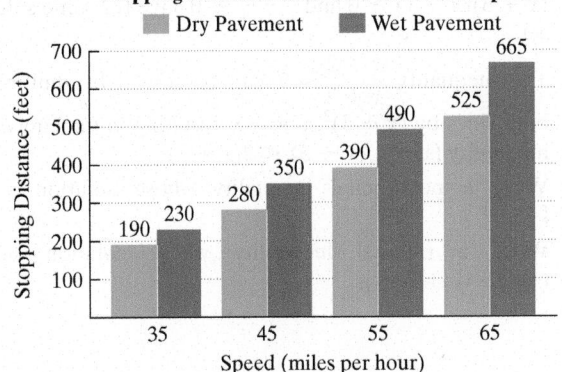

Stopping Distances for Trucks at Selected Speeds

Source: National Highway Traffic Safety Administration

(In Exercises 92–93, be sure to refer to the graph at the bottom of the previous page.)

92. a. Use the statistical menu of your graphing utility and the quadratic regression program to obtain the quadratic function that models a truck's stopping distance, $f(x)$, in feet, on dry pavement traveling at x miles per hour. Round the x-coefficient and the constant term to one decimal place.

b. Use the function from part (a) to determine speeds on dry pavement requiring stopping distances that exceed 455 feet. Round to the nearest mile per hour.

93. a. Use the statistical menu of your graphing utility and the quadratic regression program to obtain the quadratic function that models a truck's stopping distance, $f(x)$, in feet, on wet pavement traveling at x miles per hour. Round the x-coefficient and the constant term to one decimal place.

b. Use the function from part (a) to determine speeds on wet pavement requiring stopping distances that exceed 446 feet.

Critical Thinking Exercises

Make Sense? *In Exercises 94–97, determine whether each statement makes sense or does not make sense, and explain your reasoning.*

94. When solving $f(x) > 0$, where f is a polynomial function, I only pay attention to the sign of f at each test value and not the actual function value.

95. I'm solving a polynomial inequality that has a value for which the polynomial function is undefined.

96. Because it takes me longer to come to a stop on a wet road than on a dry road, graph (a) for Exercises 77–78 is the model for stopping distances on wet pavement and graph (b) is the model for stopping distances on dry pavement.

97. I began the solution of the rational inequality $\dfrac{x+1}{x+3} \geq 2$ by setting both $x + 1$ and $x + 3$ equal to zero.

In Exercises 98–101, determine whether each statement is true or false. If the statement is false, make the necessary change(s) to produce a true statement.

98. The solution set of $x^2 > 25$ is $(5, \infty)$.

99. The inequality $\dfrac{x-2}{x+3} < 2$ can be solved by multiplying both sides by $x + 3$, resulting in the equivalent inequality $x - 2 < 2(x + 3)$.

100. $(x + 3)(x - 1) \geq 0$ and $\dfrac{x+3}{x-1} \geq 0$ have the same solution set.

101. The inequality $\dfrac{x-2}{x+3} < 2$ can be solved by multiplying both sides by $(x + 3)^2$, $x \neq -3$, resulting in the equivalent inequality $(x - 2)(x + 3) < 2(x + 3)^2$.

102. Write a polynomial inequality whose solution set is $[-3, 5]$.

103. Write a rational inequality whose solution set is $(-\infty, -4) \cup [3, \infty)$.

In Exercises 104–107, use inspection to describe each inequality's solution set. Do not solve any of the inequalities.

104. $(x - 2)^2 > 0$

105. $(x - 2)^2 \leq 0$

106. $(x - 2)^2 < -1$

107. $\dfrac{1}{(x-2)^2} > 0$

108. The graphing utility screen shows the graph of $y = 4x^2 - 8x + 7$.

[−2, 6, 1] by [−2, 8, 1]

a. Use the graph to describe the solution set of $4x^2 - 8x + 7 > 0$.

b. Use the graph to describe the solution set of $4x^2 - 8x + 7 < 0$.

c. Use an algebraic approach to verify each of your descriptions in parts (a) and (b).

109. The graphing utility screen shows the graph of $y = \sqrt{27 - 3x^2}$. Write and solve a quadratic inequality that explains why the graph only appears for $-3 \leq x \leq 3$.

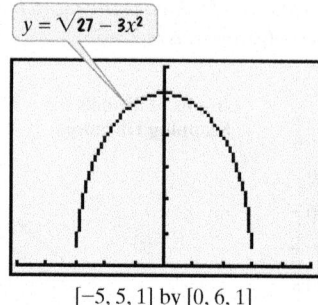

[−5, 5, 1] by [0, 6, 1]

Preview Exercises

Exercises 110–112 will help you prepare for the material covered in the next section.

110. a. If $y = kx^2$, find the value of k using $x = 2$ and $y = 64$.

b. Substitute the value for k into $y = kx^2$ and write the resulting equation.

c. Use the equation from part (b) to find y when $x = 5$.

111. a. If $y = \dfrac{k}{x}$, find the value of k using $x = 8$ and $y = 12$.

b. Substitute the value for k into $y = \dfrac{k}{x}$ and write the resulting equation.

c. Use the equation from part (b) to find y when $x = 3$.

112. If $S = \dfrac{kA}{P}$, find the value of k using $A = 60{,}000$, $P = 40$, and $S = 12{,}000$.

SECTION 3.7 Modeling Using Variation

Objectives

1 Solve direct variation problems.
2 Solve inverse variation problems.
3 Solve combined variation problems.
4 Solve problems involving joint variation.

Have you ever wondered how telecommunication companies estimate the number of phone calls expected per day between two cities? The formula

$$C = \frac{0.02 P_1 P_2}{d^2}$$

shows that the daily number of phone calls, C, increases as the populations of the cities, P_1 and P_2, in thousands, increase and decreases as the distance, d, between the cities increases.

Certain formulas occur so frequently in applied situations that they are given special names. Variation formulas show how one quantity changes in relation to other quantities. Quantities can vary *directly, inversely,* or *jointly.* In this section, we look at situations that can be modeled by each of these kinds of variation.

1 Solve direct variation problems.

Direct Variation

When you swim underwater, the pressure in your ears depends on the depth at which you are swimming. The formula

$$p = 0.43d$$

describes the water pressure, p, in pounds per square inch, at a depth of d feet. We can use this linear function to determine the pressure in your ears at various depths:

If $d = 20$, $p = 0.43(20) = 8.6$. At a depth of 20 feet, water pressure is 8.6 pounds per square inch.

Doubling the depth doubles the pressure.

If $d = 40$, $p = 0.43(40) = 17.2$. At a depth of 40 feet, water pressure is 17.2 pounds per square inch.

Doubling the depth doubles the pressure.

If $d = 80$, $p = 0.43(80) = 34.4$. At a depth of 80 feet, water pressure is 34.4 pounds per square inch.

The formula $p = 0.43d$ illustrates that water pressure is a constant multiple of your underwater depth. If your depth is doubled, the pressure is doubled; if your depth is tripled, the pressure is tripled; and so on. Because of this, the pressure in your ears is said to **vary directly** as your underwater depth. The **equation of variation** is

$$p = 0.43d.$$

Generalizing our discussion of pressure and depth, we obtain the following statement:

Direct Variation

If a situation is described by an equation in the form

$$y = kx,$$

where k is a nonzero constant, we say that **y varies directly as x** or **y is directly proportional to x**. The number k is called the **constant of variation** or the **constant of proportionality**.

Can you see that **the direct variation equation, $y = kx$, is a special case of the linear function $y = mx + b$?** When $m = k$ and $b = 0$, $y = mx + b$ becomes $y = kx$. Thus, the slope of a direct variation equation is k, the constant of variation. Because b, the y-intercept, is 0, the graph of a variation equation is a line passing through the origin. This is illustrated in **Figure 3.47**, which shows the graph of $p = 0.43d$: Water pressure varies directly as depth.

FIGURE 3.47 Water pressure at various depths

Problems involving direct variation can be solved using the following procedure. This procedure applies to direct variation problems, as well as to the other kinds of variation problems that we will discuss.

Solving Variation Problems

1. Write an equation that models the given English statement.
2. Substitute the given pair of values into the equation in step 1 and find the value of k, the constant of variation.
3. Substitute the value of k into the equation in step 1.
4. Use the equation from step 3 to answer the problem's question.

EXAMPLE 1 Solving a Direct Variation Problem

The volume of blood, B, in a person's body varies directly as body weight, W. A person who weighs 160 pounds has approximately 5 quarts of blood. Estimate the number of quarts of blood in a person who weighs 200 pounds.

SOLUTION

Step 1 Write an equation. We know that y *varies directly as x* is expressed as

$$y = kx.$$

By changing letters, we can write an equation that models the following English statement: The volume of blood, B, varies directly as body weight, W.

$$B = kW$$

Step 2 Use the given values to find k. A person who weighs 160 pounds has approximately 5 quarts of blood. Substitute 160 for W and 5 for B in the direct variation equation. Then solve for k.

$B = kW$	The volume of blood varies directly as body weight.
$5 = k \cdot 160$	Substitute 160 for W and 5 for B.
$\dfrac{5}{160} = \dfrac{k \cdot 160}{160}$	Divide both sides by 160.
$0.03125 = k$	Express $\frac{5}{160}$, or $\frac{1}{32}$, in decimal form.

Step 3 Substitute the value of k into the equation.

$B = kW$	Use the equation from step 1.
$B = 0.03125W$	Replace k, the constant of variation, with 0.03125.

Step 4 Answer the problem's question. We are interested in estimating the number of quarts of blood in a person who weighs 200 pounds. Substitute 200 for W in $B = 0.03125W$ and solve for B.

$$B = 0.03125W \qquad \text{This is the equation from step 3.}$$
$$B = 0.03125(200) \qquad \text{Substitute 200 for W.}$$
$$ = 6.25 \qquad \text{Multiply.}$$

A person who weighs 200 pounds has approximately 6.25 quarts of blood. • • •

⌀ Check Point **1** The number of gallons of water, W, used when taking a shower varies directly as the time, t, in minutes, in the shower. A shower lasting 5 minutes uses 30 gallons of water. How much water is used in a shower lasting 11 minutes?

The direct variation equation $y = kx$ is a linear function. If $k > 0$, then the slope of the line is positive. Consequently, as x increases, y also increases.
A direct variation situation can involve variables to higher powers. For example, y can vary directly as $x^2 (y = kx^2)$ or as $x^3 (y = kx^3)$.

Direct Variation with Powers

y* varies directly as the *n*th power of *x if there exists some nonzero constant k such that

$$y = kx^n.$$

We also say that ***y* is directly proportional to the *n*th power of *x*.**

Direct variation with whole number powers is modeled by polynomial functions. In our next example, the graph of the variation equation is the familiar parabola.

EXAMPLE 2 Solving a Direct Variation Problem

The distance, s, that a body falls from rest varies directly as the square of the time, t, of the fall. If skydivers fall 64 feet in 2 seconds, how far will they fall in 4.5 seconds?

SOLUTION

Step 1 Write an equation. We know that *y varies directly as the square of x* is expressed as

$$y = kx^2.$$

By changing letters, we can write an equation that models the following English statement: Distance, s, varies directly as the square of time, t, of the fall.

$$s = kt^2$$

Step 2 Use the given values to find k. Skydivers fall 64 feet in 2 seconds. Substitute 64 for s and 2 for t in the direct variation equation. Then solve for k.

$$s = kt^2 \qquad \text{Distance varies directly as the square of time.}$$
$$64 = k \cdot 2^2 \qquad \text{Skydivers fall 64 feet in 2 seconds.}$$
$$64 = 4k \qquad \text{Simplify: } 2^2 = 4.$$
$$\frac{64}{4} = \frac{4k}{4} \qquad \text{Divide both sides by 4.}$$
$$16 = k \qquad \text{Simplify.}$$

Distance Fallen by Skydivers over Time

FIGURE 3.48 The graph of $s(t) = 16t^2$

Step 3 Substitute the value of k into the equation.

$$s = kt^2 \quad \text{Use the equation from step 1.}$$
$$s = 16t^2 \quad \text{Replace } k, \text{ the constant of variation, with 16.}$$

Step 4 Answer the problem's question. How far will the skydivers fall in 4.5 seconds? Substitute 4.5 for t in $s = 16t^2$ and solve for s.

$$s = 16(4.5)^2 = 16(20.25) = 324$$

Thus, in 4.5 seconds, the skydivers will fall 324 feet. • • •

We can express the variation equation from Example 2 in function notation, writing

$$s(t) = 16t^2.$$

The distance that a body falls from rest is a function of the time, t, of the fall. The parabola that is the graph of this quadratic function is shown in **Figure 3.48**. The graph increases rapidly from left to right, showing the effects of the acceleration of gravity.

✓ Check Point **2** The weight of a great white shark varies directly as the cube of its length. A great white shark caught off Catalina Island, California, was 15 feet long and weighed 2025 pounds. What was the weight of the 25-foot-long shark in the novel *Jaws*?

② Solve inverse variation problems.

Inverse Variation

The distance from San Francisco to Los Angeles is 420 miles. The time that it takes to drive from San Francisco to Los Angeles depends on the rate at which one drives and is given by

$$\text{Time} = \frac{420}{\text{Rate}}.$$

For example, if you average 30 miles per hour, the time for the drive is

$$\text{Time} = \frac{420}{30} = 14,$$

or 14 hours. If you average 50 miles per hour, the time for the drive is

$$\text{Time} = \frac{420}{50} = 8.4,$$

or 8.4 hours. As your rate (or speed) increases, the time for the trip decreases and vice versa. This is illustrated by the graph in **Figure 3.49**.

We can express the time for the San Francisco–Los Angeles trip using t for time and r for rate:

$$t = \frac{420}{r}.$$

This equation is an example of an **inverse variation** equation. Time, t, **varies inversely** as rate, r. When two quantities vary inversely, one quantity increases as the other decreases and vice versa.

Generalizing, we obtain the following statement:

Inverse Variation

If a situation is described by an equation in the form

$$y = \frac{k}{x},$$

where k is a nonzero constant, we say that **y varies inversely as x** or **y is inversely proportional to x.** The number k is called the **constant of variation**.

$t = \frac{420}{r}$

Averaging 30 mph, the trip takes 14 hours.

Averaging 50 mph, the trip takes 8.4 hours.

FIGURE 3.49

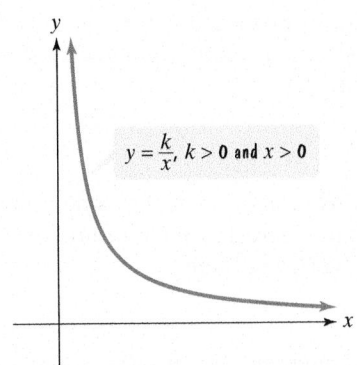

$$y = \frac{k}{x}, \; k > 0 \text{ and } x > 0$$

FIGURE 3.50 The graph of the inverse variation equation

Doubling the pressure halves the volume.

Notice that the **inverse variation equation**

$$y = \frac{k}{x}, \quad \text{or} \quad f(x) = \frac{k}{x},$$

is a rational function. For $k > 0$ and $x > 0$, the graph of the function takes on the shape shown in **Figure 3.50**.

We use the same procedure to solve inverse variation problems as we did to solve direct variation problems. Example 3 illustrates this procedure.

EXAMPLE 3 Solving an Inverse Variation Problem

When you use a spray can and press the valve at the top, you decrease the pressure of the gas in the can. This decrease of pressure causes the volume of the gas in the can to increase. Because the gas needs more room than is provided in the can, it expands in spray form through the small hole near the valve. In general, if the temperature is constant, the pressure, P, of a gas in a container varies inversely as the volume, V, of the container. The pressure of a gas sample in a container whose volume is 8 cubic inches is 12 pounds per square inch. If the sample expands to a volume of 22 cubic inches, what is the new pressure of the gas?

SOLUTION

Step 1 Write an equation. We know that y *varies inversely as* x is expressed as

$$y = \frac{k}{x}.$$

By changing letters, we can write an equation that models the following English statement: The pressure, P, of a gas in a container varies inversely as the volume, V.

$$P = \frac{k}{V}$$

Step 2 Use the given values to find k. The pressure of a gas sample in a container whose volume is 8 cubic inches is 12 pounds per square inch. Substitute 12 for P and 8 for V in the inverse variation equation. Then solve for k.

$$P = \frac{k}{V} \qquad \text{Pressure varies inversely as volume.}$$

$$12 = \frac{k}{8} \qquad \text{The pressure in an 8 cubic-inch} \\ \text{container is 12 pounds per square inch.}$$

$$12 \cdot 8 = \frac{k}{8} \cdot 8 \qquad \text{Multiply both sides by 8.}$$

$$96 = k \qquad \text{Simplify.}$$

Step 3 Substitute the value of k into the equation.

$$P = \frac{k}{V} \qquad \text{Use the equation from step 1.}$$

$$P = \frac{96}{V} \qquad \text{Replace } k, \text{ the constant of variation,} \\ \text{with 96.}$$

Step 4 Answer the problem's question. We need to find the pressure when the volume expands to 22 cubic inches. Substitute 22 for V and solve for P.

$$P = \frac{96}{V} = \frac{96}{22} = \frac{48}{11} = 4\frac{4}{11}$$

When the volume is 22 cubic inches, the pressure of the gas is $4\frac{4}{11}$ pounds per square inch. ● ● ●

✓ Check Point **3** The length of a violin string varies inversely as the frequency of its vibrations. A violin string 8 inches long vibrates at a frequency of 640 cycles per second. What is the frequency of a 10-inch string?

③ Solve combined variation problems.

Combined Variation

In **combined variation**, direct variation and inverse variation occur at the same time. For example, as the advertising budget, A, of a company increases, its monthly sales, S, also increase. Monthly sales vary directly as the advertising budget:

$$S = kA.$$

By contrast, as the price of the company's product, P, increases, its monthly sales, S, decrease. Monthly sales vary inversely as the price of the product:

$$S = \frac{k}{P}.$$

We can combine these two variation equations into one combined equation:

$$S = \frac{kA}{P}.$$

> Monthly sales, S, vary directly as the advertising budget, A, and inversely as the price of the product, P.

The following example illustrates an application of combined variation.

EXAMPLE 4 Solving a Combined Variation Problem

The owners of Rollerblades Plus determine that the monthly sales, S, of their skates vary directly as their advertising budget, A, and inversely as the price of the skates, P. When $60,000 is spent on advertising and the price of the skates is $40, the monthly sales are 12,000 pairs of rollerblades.

a. Write an equation of variation that describes this situation.

b. Determine monthly sales if the amount of the advertising budget is increased to $70,000.

SOLUTION

a. Write an equation.

$$S = \frac{kA}{P}$$

> Translate "sales vary directly as the advertising budget and inversely as the skates' price."

Use the given values to find k.

$$12,000 = \frac{k(60,000)}{40}$$ When $60,000 is spent on advertising ($A = 60,000$) and the price is $40 ($P = 40$), monthly sales are 12,000 units ($S = 12,000$).

$$12,000 = k \cdot 1500$$ Divide 60,000 by 40.

$$\frac{12,000}{1500} = \frac{k \cdot 1500}{1500}$$ Divide both sides of the equation by 1500.

$$8 = k$$ Simplify.

Therefore, the equation of variation that models monthly sales is

$$S = \frac{8A}{P}.$$ Substitute 8 for k in $S = \frac{kA}{P}$.

b. The advertising budget is increased to $70,000, so $A = 70{,}000$. The skates' price is still $40, so $P = 40$.

$$S = \frac{8A}{P}$$ *This is the equation from part (a).*

$$S = \frac{8(70{,}000)}{40}$$ *Substitute 70,000 for A and 40 for P.*

$$S = 14{,}000$$ *Simplify.*

With a $70,000 advertising budget and $40 price, the company can expect to sell 14,000 pairs of rollerblades in a month (up from 12,000). • • •

⬚ Check Point **4** The number of minutes needed to solve an Exercise Set of variation problems varies directly as the number of problems and inversely as the number of people working to solve the problems. It takes 4 people 32 minutes to solve 16 problems. How many minutes will it take 8 people to solve 24 problems?

④ Solve problems involving joint variation.

Joint Variation

Joint variation is a variation in which a variable varies directly as the product of two or more other variables. Thus, the equation $y = kxz$ is read "y varies jointly as x and z."

Joint variation plays a critical role in Isaac Newton's formula for gravitation:

$$F = G\frac{m_1 m_2}{d^2}.$$

The formula states that the force of gravitation, F, between two bodies varies jointly as the product of their masses, m_1 and m_2, and inversely as the square of the distance between them, d. (G is the gravitational constant.) The formula indicates that gravitational force exists between any two objects in the universe, increasing as the distance between the bodies decreases. One practical result is that the pull of the moon on the oceans is greater on the side of Earth closer to the moon. This gravitational imbalance is what produces tides.

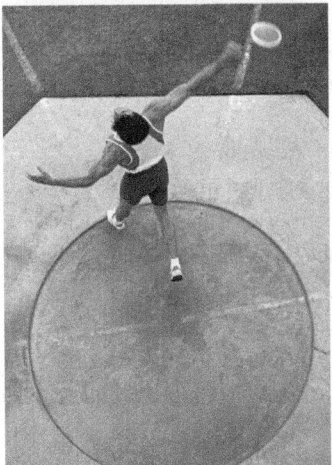

EXAMPLE 5 Modeling Centrifugal Force

The centrifugal force, C, of a body moving in a circle varies jointly with the radius of the circular path, r, and the body's mass, m, and inversely with the square of the time, t, it takes to move about one full circle. A 6-gram body moving in a circle with radius 100 centimeters at a rate of 1 revolution in 2 seconds has a centrifugal force of 6000 dynes. Find the centrifugal force of an 18-gram body moving in a circle with radius 100 centimeters at a rate of 1 revolution in 3 seconds.

SOLUTION

$$C = \frac{krm}{t^2}$$ *Translate "Centrifugal force, C, varies jointly with radius, r, and mass, m, and inversely with the square of time, t."*

$$6000 = \frac{k(100)(6)}{2^2}$$ *A 6-gram body (m = 6) moving in a circle with radius 100 centimeters (r = 100) at 1 revolution in 2 seconds (t = 2) has a centrifugal force of 6000 dynes (C = 6000).*

$$6000 = 150k$$ *Simplify.*

$$40 = k$$ *Divide both sides by 150 and solve for k.*

$$C = \frac{40rm}{t^2}$$ *Substitute 40 for k in the model for centrifugal force.*

$$C = \frac{40(100)(18)}{3^2}$$ Find centrifugal force, C, of an 18-gram body
$(m = 18)$ moving in a circle with radius
100 centimeters $(r = 100)$ at 1 revolution
in 3 seconds $(t = 3)$.

$$= 8000$$ Simplify.

The centrifugal force is 8000 dynes. •••

⊘ Check Point **5** The volume of a cone, V, varies jointly as its height, h, and the square of its radius, r. A cone with a radius measuring 6 feet and a height measuring 10 feet has a volume of 120π cubic feet. Find the volume of a cone having a radius of 12 feet and a height of 2 feet.

CONCEPT AND VOCABULARY CHECK

Fill in each blank so that the resulting statement is true.

1. y varies directly as x can be modeled by the equation _____, where k is called the

_____.

2. y varies directly as the nth power of x can be modeled by the equation _____.

3. y varies inversely as x can be modeled by the equation

_____.

4. y varies directly as x and inversely as z can be

modeled by the equation _____.

5. y varies jointly as x and z can be modeled by the equation _____.

6. In the equation $S = \dfrac{8A}{P}$, S varies _____ as A

and _____ as P.

7. In the equation $C = \dfrac{0.02P_1P_2}{d^2}$, C varies _____ as

P_1 and P_2 and _____ as the square of d.

EXERCISE SET 3.7

Practice Exercises

Use the four-step procedure for solving variation problems given on page 434 to solve Exercises 1–10.

1. y varies directly as x. $y = 65$ when $x = 5$. Find y when $x = 12$.

2. y varies directly as x. $y = 45$ when $x = 5$. Find y when $x = 13$.

3. y varies inversely as x. $y = 12$ when $x = 5$. Find y when $x = 2$.

4. y varies inversely as x. $y = 6$ when $x = 3$. Find y when $x = 9$.

5. y varies directly as x and inversely as the square of z. $y = 20$ when $x = 50$ and $z = 5$. Find y when $x = 3$ and $z = 6$.

6. a varies directly as b and inversely as the square of c. $a = 7$ when $b = 9$ and $c = 6$. Find a when $b = 4$ and $c = 8$.

7. y varies jointly as x and z. $y = 25$ when $x = 2$ and $z = 5$. Find y when $x = 8$ and $z = 12$.

8. C varies jointly as A and T. $C = 175$ when $A = 2100$ and $T = 4$. Find C when $A = 2400$ and $T = 6$.

9. y varies jointly as a and b and inversely as the square root of c. $y = 12$ when $a = 3, b = 2$, and $c = 25$. Find y when $a = 5, b = 3$, and $c = 9$.

10. y varies jointly as m and the square of n and inversely as p. $y = 15$ when $m = 2, n = 1$, and $p = 6$. Find y when $m = 3, n = 4$, and $p = 10$.

Practice Plus

In Exercises 11–20, write an equation that expresses each relationship. Then solve the equation for y.

11. x varies jointly as y and z.

12. x varies jointly as y and the square of z.

13. x varies directly as the cube of z and inversely as y.

14. x varies directly as the cube root of z and inversely as y.

15. x varies jointly as y and z and inversely as the square root of w.

16. x varies jointly as y and z and inversely as the square of w.

17. x varies jointly as z and the sum of y and w.

18. x varies jointly as z and the difference between y and w.

19. x varies directly as z and inversely as the difference between y and w.

20. x varies directly as z and inversely as the sum of y and w.

Application Exercises

Use the four-step procedure for solving variation problems given on page 434 to solve Exercises 21–36.

21. An alligator's tail length, T, varies directly as its body length, B. An alligator with a body length of 4 feet has a tail length of 3.6 feet. What is the tail length of an alligator whose body length is 6 feet?

|← ——— Body length, B ——— →|← ——— Tail length, T ——— →|

22. An object's weight on the moon, M, varies directly as its weight on Earth, E. Neil Armstrong, the first person to step on the moon on July 20, 1969, weighed 360 pounds on Earth (with all of his equipment on) and 60 pounds on the moon. What is the moon weight of a person who weighs 186 pounds on Earth?

23. The height that a ball bounces varies directly as the height from which it was dropped. A tennis ball dropped from 12 inches bounces 8.4 inches. From what height was the tennis ball dropped if it bounces 56 inches?

24. The distance that a spring will stretch varies directly as the force applied to the spring. A force of 12 pounds is needed to stretch a spring 9 inches. What force is required to stretch the spring 15 inches?

25. If all men had identical body types, their weight would vary directly as the cube of their height. Shown below is Robert Wadlow, who reached a record height of 8 feet 11 inches (107 inches) before his death at age 22. If a man who is 5 feet 10 inches tall (70 inches) with the same body type as Mr. Wadlow weighs 170 pounds, what was Robert Wadlow's weight shortly before his death?

26. The number of houses that can be served by a water pipe varies directly as the square of the diameter of the pipe. A water pipe that has a 10-centimeter diameter can supply 50 houses.

 a. How many houses can be served by a water pipe that has a 30-centimeter diameter?

 b. What size water pipe is needed for a new subdivision of 1250 houses?

27. The figure shows that a bicyclist tips the cycle when making a turn. The angle B, formed by the vertical direction and the bicycle, is called the banking angle. The banking angle varies inversely as the cycle's turning radius. When the turning radius is 4 feet, the banking angle is 28°. What is the banking angle when the turning radius is 3.5 feet?

$B°$

28. The water temperature of the Pacific Ocean varies inversely as the water's depth. At a depth of 1000 meters, the water temperature is 4.4° Celsius. What is the water temperature at a depth of 5000 meters?

29. Radiation machines, used to treat tumors, produce an intensity of radiation that varies inversely as the square of the distance from the machine. At 3 meters, the radiation intensity is 62.5 milliroentgens per hour. What is the intensity at a distance of 2.5 meters?

30. The illumination provided by a car's headlight varies inversely as the square of the distance from the headlight. A car's headlight produces an illumination of 3.75 foot-candles at a distance of 40 feet. What is the illumination when the distance is 50 feet?

31. Body-mass index, or BMI, takes both weight and height into account when assessing whether an individual is underweight or overweight. BMI varies directly as one's weight, in pounds, and inversely as the square of one's height, in inches. In adults, normal values for the BMI are between 20 and 25, inclusive. Values below 20 indicate that an individual is underweight and values above 30 indicate that an individual is obese. A person who weighs 180 pounds and is 5 feet, or 60 inches, tall has a BMI of 35.15. What is the BMI, to the nearest tenth, for a 170-pound person who is 5 feet 10 inches tall? Is this person overweight?

32. One's intelligence quotient, or IQ, varies directly as a person's mental age and inversely as that person's chronological age. A person with a mental age of 25 and a chronological age of 20 has an IQ of 125. What is the chronological age of a person with a mental age of 40 and an IQ of 80?

33. The heat loss of a glass window varies jointly as the window's area and the difference between the outside and inside temperatures. A window 3 feet wide by 6 feet long loses 1200 Btu per hour when the temperature outside is 20° colder than the temperature inside. Find the heat loss through a glass window that is 6 feet wide by 9 feet long when the temperature outside is 10° colder than the temperature inside.

34. Kinetic energy varies jointly as the mass and the square of the velocity. A mass of 8 grams and velocity of 3 centimeters per second has a kinetic energy of 36 ergs. Find the kinetic energy for a mass of 4 grams and velocity of 6 centimeters per second.

35. Sound intensity varies inversely as the square of the distance from the sound source. If you are in a movie theater and you change your seat to one that is twice as far from the speakers, how does the new sound intensity compare to that of your original seat?

36. Many people claim that as they get older, time seems to pass more quickly. Suppose that the perceived length of a period of time is inversely proportional to your age. How long will a year seem to be when you are three times as old as you are now?

37. The average number of daily phone calls, C, between two cities varies jointly as the product of their populations, P_1 and P_2, and inversely as the square of the distance, d, between them.

 a. Write an equation that expresses this relationship.

 b. The distance between San Francisco (population: 777,000) and Los Angeles (population: 3,695,000) is 420 miles. If the average number of daily phone calls between the cities is 326,000, find the value of k to two decimal places and write the equation of variation.

 c. Memphis (population: 650,000) is 400 miles from New Orleans (population: 490,000). Find the average number of daily phone calls, to the nearest whole number, between these cities.

38. The force of wind blowing on a window positioned at a right angle to the direction of the wind varies jointly as the area of the window and the square of the wind's speed. It is known that a wind of 30 miles per hour blowing on a window measuring 4 feet by 5 feet exerts a force of 150 pounds. During a storm with winds of 60 miles per hour, should hurricane shutters be placed on a window that measures 3 feet by 4 feet and is capable of withstanding 300 pounds of force?

39. The table shows the values for the current, I, in an electric circuit and the resistance, R, of the circuit.

I (amperes)	0.5	1.0	1.5	2.0	2.5	3.0	4.0	5.0
R (ohms)	12.0	6.0	4.0	3.0	2.4	2.0	1.5	1.2

 a. Graph the ordered pairs in the table of values, with values of I along the x-axis and values of R along the y-axis. Connect the eight points with a smooth curve.

 b. Does current vary directly or inversely as resistance? Use your graph and explain how you arrived at your answer.

 c. Write an equation of variation for I and R, using one of the ordered pairs in the table to find the constant of variation. Then use your variation equation to verify the other seven ordered pairs in the table.

Writing in Mathematics

40. What does it mean if two quantities vary directly?

41. In your own words, explain how to solve a variation problem.

42. What does it mean if two quantities vary inversely?

43. Explain what is meant by combined variation. Give an example with your explanation.

44. Explain what is meant by joint variation. Give an example with your explanation.

In Exercises 45–46, describe in words the variation shown by the given equation.

45. $z = \dfrac{k\sqrt{x}}{y^2}$ 46. $z = kx^2\sqrt{y}$

47. We have seen that the daily number of phone calls between two cities varies jointly as their populations and inversely as the square of the distance between them. This model, used by telecommunication companies to estimate the line capacities needed among various cities, is called the *gravity model*. Compare the model to Newton's formula for gravitation on page 439 and describe why the name *gravity model* is appropriate.

Technology Exercise

48. Use a graphing utility to graph any three of the variation equations in Exercises 21–30. Then ⊏TRACE⊐ along each curve and identify the point that corresponds to the problem's solution.

Critical Thinking Exercises

Make Sense? *In Exercises 49–52, determine whether each statement makes sense or does not make sense, and explain your reasoning.*

49. I'm using an inverse variation equation and I need to determine the value of the dependent variable when the independent variable is zero.

50. The graph of this direct variation equation that has a positive constant of variation shows one variable increasing as the other variable decreases.

51. When all is said and done, it seems to me that direct variation equations are special kinds of linear functions and inverse variation equations are special kinds of rational functions.

52. Using the language of variation, I can now state the formula for the area of a trapezoid, $A = \frac{1}{2}h(b_1 + b_2)$, as, "A trapezoid's area varies jointly with its height and the sum of its bases."

53. In a hurricane, the wind pressure varies directly as the square of the wind velocity. If wind pressure is a measure of a hurricane's destructive capacity, what happens to this destructive power when the wind speed doubles?

54. The illumination from a light source varies inversely as the square of the distance from the light source. If you raise a lamp from 15 inches to 30 inches over your desk, what happens to the illumination?

55. The heat generated by a stove element varies directly as the square of the voltage and inversely as the resistance. If the voltage remains constant, what needs to be done to triple the amount of heat generated?

56. Galileo's telescope brought about revolutionary changes in astronomy. A comparable leap in our ability to observe the universe took place as a result of the Hubble Space Telescope. The space telescope was able to see stars and galaxies whose brightness is $\frac{1}{50}$ of the faintest objects observable using ground-based telescopes. Use the fact that the brightness of a point source, such as a star, varies inversely as the square of its distance from an observer to show that the space telescope was able to see about seven times farther than a ground-based telescope.

Group Exercise

57. Begin by deciding on a product that interests the group because you are now in charge of advertising this product. Members were told that the demand for the product varies directly as the amount spent on advertising and inversely as the price of the product. However, as more money is spent on advertising, the price of your product rises. Under what conditions would members recommend an increased expense in advertising? Once you've determined what your product is, write formulas for the given conditions and experiment with hypothetical numbers. What other factors might you take into consideration in terms of your recommendation? How do these factors affect the demand for your product?

Preview Exercises

Exercises 58–60 will help you prepare for the material covered in the first section of the next chapter.

58. Use point plotting to graph $f(x) = 2^x$. Begin by setting up a partial table of coordinates, selecting integers from -3 to 3, inclusive, for x. Because $y = 0$ is a horizontal asymptote, your graph should approach, but never touch, the negative portion of the x-axis.

In Exercises 59–60, use transformations of your graph from Exercise 58 to graph each function.

59. $g(x) = f(-x) = 2^{-x}$ **60.** $h(x) = f(x) + 1 = 2^x + 1$

CHAPTER 3 Summary, Review, and Test

SUMMARY

DEFINITIONS AND CONCEPTS	EXAMPLES

3.1 Quadratic Functions

 a. A quadratic function is of the form $f(x) = ax^2 + bx + c, a \neq 0$.

 b. The standard form of a quadratic function is $f(x) = a(x - h)^2 + k, a \neq 0$.

 c. The graph of a quadratic function is a parabola. The vertex is (h, k) or $\left(-\dfrac{b}{2a}, f\left(-\dfrac{b}{2a}\right)\right)$. A procedure for graphing a quadratic function in standard form is given in the box on page 342. A procedure for graphing a quadratic function in the form $f(x) = ax^2 + bx + c$ is given in the box on page 345.

 Ex. 1, p. 342;
 Ex. 2, p. 343;
 Ex. 3, p. 345

 d. See the box on page 347 for minimum or maximum values of quadratic functions.

 Ex. 4, p. 347;
 Ex. 5, p. 348

 e. A strategy for solving problems involving maximizing or minimizing quadratic functions is given in the box on page 350.

 Ex. 6, p. 350;
 Ex. 7, p. 351

3.2 Polynomial Functions and Their Graphs

 a. Polynomial Function of Degree n: $f(x) = a_n x^n + a_{n-1}x^{n-1} + \cdots + a_2 x^2 + a_1 x + a_0, a_n \neq 0$

 b. The graphs of polynomial functions are smooth and continuous.

 Fig. 3.13, p. 359

 c. The end behavior of the graph of a polynomial function depends on the leading term, given by the Leading Coefficient Test in the box on page 360. Odd-degree polynomial functions have graphs with opposite behavior at each end. Even-degree polynomial functions have graphs with the same behavior at each end.

 Ex. 1, p. 360;
 Ex. 2, p. 361;
 Ex. 3, p. 361;
 Ex. 4, p. 362

 d. The values of x for which $f(x)$ is equal to 0 are the zeros of the polynomial function f. These values are the roots, or solutions, of the polynomial equation $f(x) = 0$.

 Ex. 5, p. 362;
 Ex. 6, p. 363

 e. If $x - r$ occurs k times in a polynomial function's factorization, r is a repeated zero with multiplicity k. If k is even, the graph touches the x-axis and turns around at r. If k is odd, the graph crosses the x-axis at r.

 Ex. 7, p. 364

 f. The Intermediate Value Theorem: If f is a polynomial function and $f(a)$ and $f(b)$ have opposite signs, there is at least one value of c between a and b for which $f(c) = 0$.

 Ex. 8, p. 365

DEFINITIONS AND CONCEPTS	EXAMPLES

g. If f is a polynomial of degree n, the graph of f has at most $n - 1$ turning points.

Fig. 3.24, p. 366

h. A strategy for graphing a polynomial function is given in the box on page 366.

Ex. 9, p. 366;
Ex. 10, p. 368

3.3 Dividing Polynomials; Remainder and Factor Theorems

a. Long division of polynomials is performed by dividing, multiplying, subtracting, bringing down the next term, and repeating this process until the degree of the remainder is less than the degree of the divisor. The details are given in the box on page 376.

Ex. 1, p. 375;
Ex. 2, p. 376;
Ex. 3, p. 377

b. The Division Algorithm: $f(x) = d(x)q(x) + r(x)$. The dividend is the product of the divisor and the quotient plus the remainder.

c. Synthetic division is used to divide a polynomial by $x - c$. The details are given in the box on page 379.

Ex. 4, p. 379

d. The Remainder Theorem: If a polynomial $f(x)$ is divided by $x - c$, then the remainder is $f(c)$.

Ex. 5, p. 381

e. The Factor Theorem: If $x - c$ is a factor of a polynomial function $f(x)$, then c is a zero of f and a root of $f(x) = 0$. If c is a zero of f or a root of $f(x) = 0$, then $x - c$ is a factor of $f(x)$.

Ex. 6, p. 382

3.4 Zeros of Polynomial Functions

a. The Rational Zero Theorem states that the possible rational zeros of a polynomial

$$\text{function} = \frac{\text{Factors of the constant term}}{\text{Factors of the leading coefficient}}.$$ The theorem is stated in the box on page 387.

Ex. 1, p. 387;
Ex. 2, p. 388;
Ex. 3, p. 388;
Ex. 4, p. 389;
Ex. 5, p. 390

b. Number of roots: If $f(x)$ is a polynomial of degree $n \geq 1$, then, counting multiple roots separately, the equation $f(x) = 0$ has n roots.

c. If $a + bi$ is a root of $f(x) = 0$ with real coefficients, then $a - bi$ is also a root.

d. The Linear Factorization Theorem: An nth-degree polynomial can be expressed as the product of n linear factors. Thus, $f(x) = a_n(x - c_1)(x - c_2) \cdots (x - c_n)$.

Ex. 6, p. 393

e. Descartes's Rule of Signs: The number of positive real zeros of f equals the number of sign changes of $f(x)$ or is less than that number by an even integer. The number of negative real zeros of f applies a similar statement to $f(-x)$.

Table 3.1, p. 395;
Ex. 7, p. 395

3.5 Rational Functions and Their Graphs

a. Rational function: $f(x) = \dfrac{p(x)}{q(x)}$; p and q are polynomial functions and $q(x) \neq 0$. The domain of f is the set of all real numbers excluding values of x that make $q(x)$ zero.

Ex. 1, p. 401

b. Arrow notation is summarized in the box on page 404.

c. The line $x = a$ is a vertical asymptote of the graph of f if $f(x)$ increases or decreases without bound as x approaches a. Vertical asymptotes are located using the theorem in the box on page 405.

Ex. 2, p. 405

d. The line $y = b$ is a horizontal asymptote of the graph of f if $f(x)$ approaches b as x increases or decreases without bound. Horizontal asymptotes are located using the theorem in the box on page 407.

Ex. 3, p. 407

e. Table 3.2 on page 408 shows the graphs of $f(x) = \dfrac{1}{x}$ and $f(x) = \dfrac{1}{x^2}$. Some rational functions can be graphed using transformations of these common graphs.

Ex. 4, p. 408

f. A strategy for graphing rational functions is given in the box on page 409.

Ex. 5, p. 409;
Ex. 6, p. 410;
Ex. 7, p. 412

g. The graph of a rational function has a slant asymptote when the degree of the numerator is one more than the degree of the denominator. The equation of the slant asymptote is found using division and dropping the remainder term.

Ex. 8, p. 413

3.6 Polynomial and Rational Inequalities

a. A polynomial inequality can be expressed as $f(x) < 0, f(x) > 0, f(x) \leq 0,$ or $f(x) \geq 0$, where f is a polynomial function. A procedure for solving polynomial inequalities is given in the box on page 422.

Ex. 1, p. 422;
Ex. 2, p. 423;
Ex. 3, p. 424

DEFINITIONS AND CONCEPTS

EXAMPLES

b. A rational inequality can be expressed as $f(x) < 0, f(x) > 0, f(x) \leq 0,$ or $f(x) \geq 0,$ where f is a rational function. The procedure for solving such inequalities begins with expressing them so that one side is zero and the other side is a single quotient. Find boundary points by setting the numerator and denominator equal to zero. Then follow a procedure similar to that for solving polynomial inequalities.

Ex. 4, p. 426

3.7 Modeling Using Variation

a. A procedure for solving variation problems is given in the box on page 434.

b.

English Statement	Equation	
y varies directly as x. y is directly proportional to x.	$y = kx$	Ex. 1, p. 434
y varies directly as x^n. y is directly proportional to x^n.	$y = kx^n$	Ex. 2, p. 435
y varies inversely as x. y is inversely proportional to x.	$y = \dfrac{k}{x}$	Ex. 3, p. 437; Ex. 4, p. 438
y varies inversely as x^n. y is inversely proportional to x^n.	$y = \dfrac{k}{x^n}$	
y varies jointly as x and z.	$y = kxz$	Ex. 5, p. 439

REVIEW EXERCISES

3.1

In Exercises 1–4, use the vertex and intercepts to sketch the graph of each quadratic function. Give the equation for the parabola's axis of symmetry. Use the graph to determine the function's domain and range.

1. $f(x) = -(x + 1)^2 + 4$ **2.** $f(x) = (x + 4)^2 - 2$

3. $f(x) = -x^2 + 2x + 3$ **4.** $f(x) = 2x^2 - 4x - 6$

In Exercises 5–6, use the function's equation, and not its graph, to find

a. *the minimum or maximum value and where it occurs.*

b. *the function's domain and its range.*

5. $f(x) = -x^2 + 14x - 106$ **6.** $f(x) = 2x^2 + 12x + 703$

7. A quarterback tosses a football to a receiver 40 yards downfield. The height of the football, $f(x)$, in feet, can be modeled by

$$f(x) = -0.025x^2 + x + 6,$$

where x is the ball's horizontal distance, in yards, from the quarterback.

a. What is the ball's maximum height and how far from the quarterback does this occur?

b. From what height did the quarterback toss the football?

c. If the football is not blocked by a defensive player nor caught by the receiver, how far down the field will it go before hitting the ground? Round to the nearest tenth of a yard.

d. Graph the function that models the football's parabolic path.

8. A field bordering a straight stream is to be enclosed. The side bordering the stream is not to be fenced. If 1000 yards of fencing material is to be used, what are the dimensions of the largest rectangular field that can be fenced? What is the maximum area?

9. Among all pairs of numbers whose difference is 14, find a pair whose product is as small as possible. What is the minimum product?

3.2

In Exercises 10–13, use the Leading Coefficient Test to determine the end behavior of the graph of the given polynomial function. Then use this end behavior to match the polynomial function with its graph. [The graphs are labeled (a) through (d).]

10. $f(x) = -x^3 + x^2 + 2x$ **11.** $f(x) = x^6 - 6x^4 + 9x^2$

12. $f(x) = x^5 - 5x^3 + 4x$ **13.** $f(x) = -x^4 + 1$

a.

b.

c. **d.**

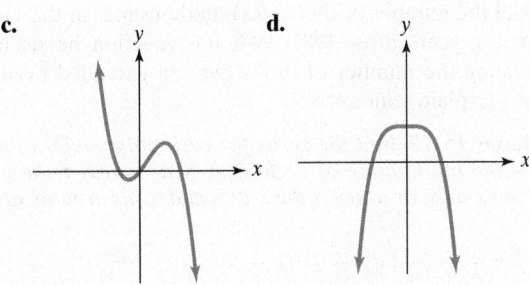

14. The Brazilian Amazon rain forest is the world's largest tropical rain forest, with some of the greatest biodiversity of any region. In 2009, the number of trees cut down in the Amazon dropped to its lowest level in 20 years. The line graph shows the number of square kilometers cleared from 2001 through 2009.

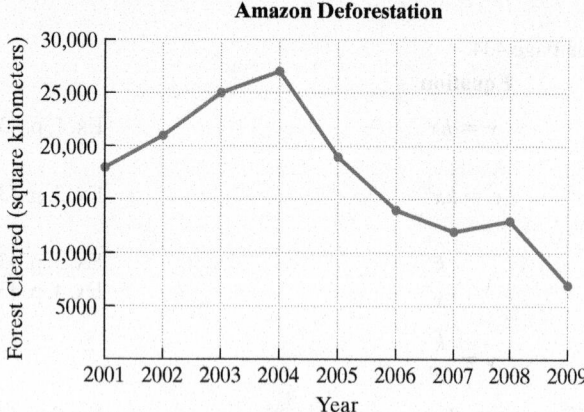

Amazon Deforestation

Source: Brazil's National Institute for Space Research

The data in the line graph can be modeled by the following third- and fourth-degree polynomial functions:

$$f(x) = 158x^3 - 2845x^2 + 12,926x + 7175$$
$$g(x) = -17x^4 + 508x^3 - 5180x^2 + 18,795x + 2889.$$

Amazon deforestation, in square kilometers, x years after 2000

a. Use the Leading Coefficient Test to determine the end behavior to the right for the graph of f.

b. Assume that the rate at which the Amazon rain forest is being cut down continues to decline. Based on your answer to part (a), will f be useful in modeling Amazon deforestation over an extended period of time? Explain your answer.

c. Use the Leading Coefficient Test to determine the end behavior to the right for the graph of g

d. Assume that the rate at which the Amazon rain forest is being cut down continues to decline. Based on your answer to part (c), will g be useful in modeling Amazon deforestation over an extended period of time? Explain your answer.

15. The polynomial function

$$f(x) = -0.87x^3 + 0.35x^2 + 81.62x + 7684.94$$

model the number of thefts, $f(x)$, in thousands in the United States x years after 1987. Will this function be useful in modeling the number of thefts over an extended period of time? Explain your answer.

In Exercises 16–17, find the zeros for each polynomial function and give the multiplicity of each zero. State whether the graph crosses the x-axis, or touches the x-axis and turns around, at each zero.

16. $f(x) = -2(x - 1)(x + 2)^2(x + 5)^3$

17. $f(x) = x^3 - 5x^2 - 25x + 125$

18. Show that $f(x) = x^3 - 2x - 1$ has a real zero between 1 and 2.

In Exercises 19–24,
 a. *Use the Leading Coefficient Test to determine the graph's end behavior.*
 b. *Determine whether the graph has y-axis symmetry, origin symmetry, or neither.*
 c. *Graph the function.*

19. $f(x) = x^3 - x^2 - 9x + 9$

20. $f(x) = 4x - x^3$

21. $f(x) = 2x^3 + 3x^2 - 8x - 12$

22. $f(x) = -x^4 + 25x^2$

23. $f(x) = -x^4 + 6x^3 - 9x^2$

24. $f(x) = 3x^4 - 15x^3$

In Exercises 25–26, graph each polynomial function.

25. $f(x) = 2x^2(x - 1)^3(x + 2)$

26. $f(x) = -x^3(x + 4)^2(x - 1)$

3.3

In Exercises 27–29, divide using long division.

27. $(4x^3 - 3x^2 - 2x + 1) \div (x + 1)$

28. $(10x^3 - 26x^2 + 17x - 13) \div (5x - 3)$

29. $(4x^4 + 6x^3 + 3x - 1) \div (2x^2 + 1)$

In Exercises 30–31, divide using synthetic division.

30. $(3x^4 + 11x^3 - 20x^2 + 7x + 35) \div (x + 5)$

31. $(3x^4 - 2x^2 - 10x) \div (x - 2)$

32. Given $f(x) = 2x^3 - 7x^2 + 9x - 3$, use the Remainder Theorem to find $f(-13)$.

33. Use synthetic division to divide $f(x) = 2x^3 + x^2 - 13x + 6$ by $x - 2$. Use the result to find all zeros of f.

34. Solve the equation $x^3 - 17x + 4 = 0$ given that 4 is a root.

3.4

In Exercises 35–36, use the Rational Zero Theorem to list all possible rational zeros for each given function.

35. $f(x) = x^4 - 6x^3 + 14x^2 - 14x + 5$

36. $f(x) = 3x^5 - 2x^4 - 15x^3 + 10x^2 + 12x - 8$

In Exercises 37–38, use Descartes's Rule of Signs to determine the possible number of positive and negative real zeros for each given function.

37. $f(x) = 3x^4 - 2x^3 - 8x + 5$

38. $f(x) = 2x^5 - 3x^3 - 5x^2 + 3x - 1$

39. Use Descartes's Rule of Signs to explain why $2x^4 + 6x^2 + 8 = 0$ has no real roots.

For Exercises 40–46,
 a. *List all possible rational roots or rational zeros.*
 b. *Use Descartes's Rule of Signs to determine the possible number of positive and negative real roots or real zeros.*
 c. *Use synthetic division to test the possible rational roots or zeros and find an actual root or zero.*
 d. *Use the quotient from part (c) to find all the remaining roots or zeros.*

40. $f(x) = x^3 + 3x^2 - 4$

41. $f(x) = 6x^3 + x^2 - 4x + 1$

42. $8x^3 - 36x^2 + 46x - 15 = 0$

43. $2x^3 + 9x^2 - 7x + 1 = 0$

44. $x^4 - x^3 - 7x^2 + x + 6 = 0$

45. $4x^4 + 7x^2 - 2 = 0$

46. $f(x) = 2x^4 + x^3 - 9x^2 - 4x + 4$

In Exercises 47–48, find an nth-degree polynomial function with real coefficients satisfying the given conditions. If you are using a graphing utility, graph the function and verify the real zeros and the given function value.

47. $n = 3$; 2 and $2 - 3i$ are zeros; $f(1) = -10$

48. $n = 4$; i is a zero; -3 is a zero of multiplicity 2; $f(-1) = 16$

In Exercises 49–50, find all the zeros of each polynomial function and write the polynomial as a product of linear factors.

49. $f(x) = 2x^4 + 3x^3 + 3x - 2$

50. $g(x) = x^4 - 6x^3 + x^2 + 24x + 16$

In Exercises 51–54, graphs of fifth-degree polynomial functions are shown. In each case, specify the number of real zeros and the number of imaginary zeros. Indicate whether there are any real zeros with multiplicity other than 1.

51.

52.

53.

54.

3.5

In Exercises 55–56, use transformations of $f(x) = \dfrac{1}{x}$ or $f(x) = \dfrac{1}{x^2}$ to graph each rational function.

55. $g(x) = \dfrac{1}{(x + 2)^2} - 1$ **56.** $h(x) = \dfrac{1}{x - 1} + 3$

In Exercises 57–64, find the vertical asymptotes, if any, the horizontal asymptote, if one exists, and the slant asymptote, if there is one, of the graph of each rational function. Then graph the rational function.

57. $f(x) = \dfrac{2x}{x^2 - 9}$ **58.** $g(x) = \dfrac{2x - 4}{x + 3}$

59. $h(x) = \dfrac{x^2 - 3x - 4}{x^2 - x - 6}$ **60.** $r(x) = \dfrac{x^2 + 4x + 3}{(x + 2)^2}$

61. $y = \dfrac{x^2}{x + 1}$ **62.** $y = \dfrac{x^2 + 2x - 3}{x - 3}$

63. $f(x) = \dfrac{-2x^3}{x^2 + 1}$ **64.** $g(x) = \dfrac{4x^2 - 16x + 16}{2x - 3}$

65. A company is planning to manufacture affordable graphing calculators. The fixed monthly cost will be $50,000 and it will cost $25 to produce each calculator.

 a. Write the cost function, C, of producing x graphing calculators.

 b. Write the average cost function, \overline{C}, of producing x graphing calculators.

 c. Find and interpret $\overline{C}(50)$, $\overline{C}(100)$, $\overline{C}(1000)$, and $\overline{C}(100,000)$.

 d. What is the horizontal asymptote for the graph of this function and what does it represent?

Exercises 66–67 involve rational functions that model the given situations. In each case, find the horizontal asymptote as $x \to \infty$ and then describe what this means in practical terms.

66. $f(x) = \dfrac{150x + 120}{0.05x + 1}$; the number of bass, $f(x)$, after x months in a lake that was stocked with 120 bass

67. $P(x) = \dfrac{72,900}{100x^2 + 729}$; the percentage, $P(x)$, of people in the United States with x years of education who are unemployed

68. The bar graph shows the population of the United States, in millions, for six selected years.

Source: U.S. Census Bureau

Here are two functions that model the data:

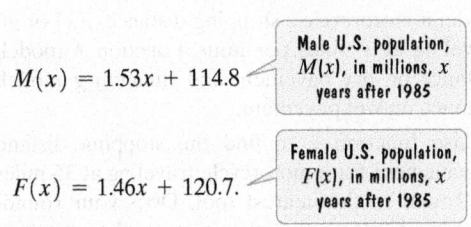

$$M(x) = 1.53x + 114.8$$ Male U.S. population, $M(x)$, in millions, x years after 1985

$$F(x) = 1.46x + 120.7.$$ Female U.S. population, $F(x)$, in millions, x years after 1985

 a. Write a function that models the total U.S. population, $P(x)$, in millions, x years after 1985.

 b. Write a rational function that models the fraction of men in the U.S. population, $R(x)$, x years after 1985.

 c. What is the equation of the horizontal asymptote associated with the function in part (b)? Round to two decimal places. What does this mean about the percentage of men in the U.S. population over time?

3.6

In Exercises 69–74, solve each inequality and graph the solution set on a real number line.

69. $2x^2 + 5x - 3 < 0$

70. $2x^2 + 9x + 4 \geq 0$

71. $x^3 + 2x^2 > 3x$

72. $\dfrac{x - 6}{x + 2} > 0$

73. $\dfrac{(x + 1)(x - 2)}{x - 1} \geq 0$

74. $\dfrac{x + 3}{x - 4} \leq 5$

75. The graph shows stopping distances for motorcycles at various speeds on dry roads and on wet roads.

Stopping Distances for Motorcycles at Selected Speeds

Source: National Highway Traffic Safety Administration

The functions

$$f(x) = 0.125x^2 - 0.8x + 99$$

Dry pavement

and

Wet pavement

$$g(x) = 0.125x^2 + 2.3x + 27$$

model a motorcycle's stopping distance, $f(x)$ or $g(x)$, in feet, traveling at x miles per hour. Function f models stopping distance on dry pavement and function g models stopping distance on wet pavement.

a. Use function g to find the stopping distance on wet pavement for a motorcycle traveling at 35 miles per hour. Round to the nearest foot. Does your rounded answer overestimate or underestimate the stopping distance shown by the graph? By how many feet?

b. Use function f to determine speeds on dry pavement requiring stopping distances that exceed 267 feet.

76. Use the position function

$$s(t) = -16t^2 + v_0 t + s_0$$

to solve this problem. A projectile is fired vertically upward from ground level with an initial velocity of 48 feet per second. During which time period will the projectile's height exceed 32 feet?

3.7

Solve the variation problems in Exercises 77–82.

77. Many areas of Northern California depend on the snowpack of the Sierra Nevada mountain range for their water supply. The volume of water produced from melting snow varies directly as the volume of snow. Meteorologists have determined that 250 cubic centimeters of snow will melt to 28 cubic centimeters of water. How much water does 1200 cubic centimeters of melting snow produce?

78. The distance that a body falls from rest is directly proportional to the square of the time of the fall. If skydivers fall 144 feet in 3 seconds, how far will they fall in 10 seconds?

79. The pitch of a musical tone varies inversely as its wavelength. A tone has a pitch of 660 vibrations per second and a wavelength of 1.6 feet. What is the pitch of a tone that has a wavelength of 2.4 feet?

80. The loudness of a stereo speaker, measured in decibels, varies inversely as the square of your distance from the speaker. When you are 8 feet from the speaker, the loudness is 28 decibels. What is the loudness when you are 4 feet from the speaker?

81. The time required to assemble computers varies directly as the number of computers assembled and inversely as the number of workers. If 30 computers can be assembled by 6 workers in 10 hours, how long would it take 5 workers to assemble 40 computers?

82. The volume of a pyramid varies jointly as its height and the area of its base. A pyramid with a height of 15 feet and a base with an area of 35 square feet has a volume of 175 cubic feet. Find the volume of a pyramid with a height of 20 feet and a base with an area of 120 square feet.

83. Heart rates and life spans of most mammals can be modeled using inverse variation. The bar graph shows the average heart rate and the average life span of five mammals.

Heart Rate and Life Span

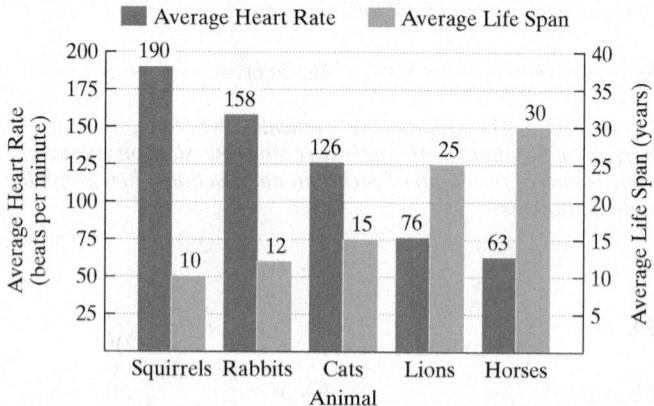

Source: *The Handy Science Answer Book,* Visible Ink Press, 2003.

a. A mammal's average life span, L, in years, varies inversely as its average heart rate, R, in beats per minute. Use the data shown for horses to write the equation that models this relationship.

b. Is the inverse variation equation in part (a) an exact model or an approximate model for the data shown for lions?

c. Elephants have an average heart rate of 27 beats per minute. Determine their average life span.

CHAPTER 3 TEST

In Exercises 1–2, use the vertex and intercepts to sketch the graph of each quadratic function. Give the equation for the parabola's axis of symmetry. Use the graph to determine the function's domain and range.

1. $f(x) = (x + 1)^2 + 4$

2. $f(x) = x^2 - 2x - 3$

3. Determine, without graphing, whether the quadratic function $f(x) = -2x^2 + 12x - 16$ has a minimum value or a maximum value. Then find

 a. the minimum or maximum value and where it occurs.

 b. the function's domain and its range.

4. The function $f(x) = -x^2 + 46x - 360$ models the daily profit, $f(x)$, in hundreds of dollars, for a company that manufactures x computers daily. How many computers should be manufactured each day to maximize profit? What is the maximum daily profit?

5. Among all pairs of numbers whose sum is 14, find a pair whose product is as large as possible. What is the maximum product?

6. Consider the function $f(x) = x^3 - 5x^2 - 4x + 20$.

 a. Use factoring to find all zeros of f.

 b. Use the Leading Coefficient Test and the zeros of f to graph the function.

7. Use end behavior to explain why the following graph cannot be the graph of $f(x) = x^5 - x$. Then use intercepts to explain why the graph cannot represent $f(x) = x^5 - x$.

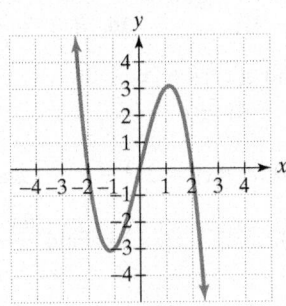

8. The graph of $f(x) = 6x^3 - 19x^2 + 16x - 4$ is shown in the figure.

 a. Based on the graph of f, find the root of the equation $6x^3 - 19x^2 + 16x - 4 = 0$ that is an integer.

 b. Use synthetic division to find the other two roots of $6x^3 - 19x^2 + 16x - 4 = 0$.

9. Use the Rational Zero Theorem to list all possible rational zeros of $f(x) = 2x^3 + 11x^2 - 7x - 6$.

10. Use Descartes's Rule of Signs to determine the possible number of positive and negative real zeros of

$$f(x) = 3x^5 - 2x^4 - 2x^2 + x - 1.$$

11. Solve: $x^3 + 9x^2 + 16x - 6 = 0$.

12. Consider the function whose equation is given by $f(x) = 2x^4 - x^3 - 13x^2 + 5x + 15$.

 a. List all possible rational zeros.

 b. Use the graph of f in the figure shown and synthetic division to find all zeros of the function.

13. Use the graph of $f(x) = x^3 + 3x^2 - 4$ in the figure shown to factor $x^3 + 3x^2 - 4$.

14. Find a fourth-degree polynomial function $f(x)$ with real coefficients that has $-1, 1,$ and i as zeros and such that $f(3) = 160$.

15. The figure shows an incomplete graph of $f(x) = -3x^3 - 4x^2 + x + 2$. Find all the zeros of the function. Then draw a complete graph.

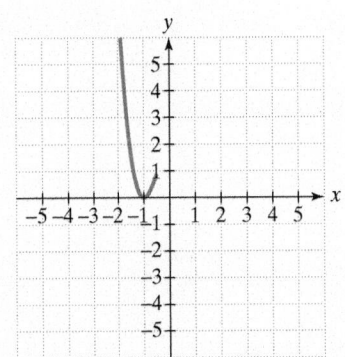

In Exercises 16–21, find the domain of each rational function and graph the function.

16. $f(x) = \dfrac{1}{(x + 3)^2}$

17. $f(x) = \dfrac{1}{x - 1} + 2$

18. $f(x) = \dfrac{x}{x^2 - 16}$

19. $f(x) = \dfrac{x^2 - 9}{x - 2}$

20. $f(x) = \dfrac{x + 1}{x^2 + 2x - 3}$

21. $f(x) = \dfrac{4x^2}{x^2 + 3}$

22. A company is planning to manufacture portable satellite radio players. The fixed monthly cost will be $300,000 and it will cost $10 to produce each player.

 a. Write the average cost function, \overline{C}, of producing x players.

 b. What is the horizontal asymptote for the graph of this function and what does it represent?

Solve each inequality in Exercises 23–24 and graph the solution set on a real number line. Express each solution set in interval notation.

23. $x^2 < x + 12$

24. $\dfrac{2x + 1}{x - 3} \le 3$

25. The intensity of light received at a source varies inversely as the square of the distance from the source. A particular light has an intensity of 20 foot-candles at 15 feet. What is the light's intensity at 10 feet?

CUMULATIVE REVIEW EXERCISES (CHAPTERS 1–3)

Use the graph of $y = f(x)$ to solve Exercises 1–6.

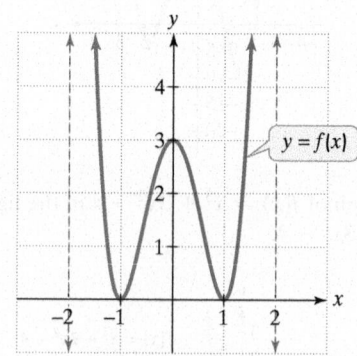

1. Find the domain and the range of f.

2. Find the zeros and the least possible multiplicity of each zero.

3. Where does the relative maximum occur?

4. Find $(f \circ f)(-1)$.

5. Use arrow notation to complete this statement: $f(x) \to \infty$ as _____ or as _____.

6. Graph $g(x) = f(x + 2) + 1$.

In Exercises 7–12, solve each equation or inequality.

7. $|2x - 1| = 3$

8. $3x^2 - 5x + 1 = 0$

9. $9 + \dfrac{3}{x} = \dfrac{2}{x^2}$

10. $x^3 + 2x^2 - 5x - 6 = 0$

11. $|2x - 5| > 3$

12. $3x^2 > 2x + 5$

In Exercises 13–18, graph each equation in a rectangular coordinate system. If two functions are given, graph both in the same system.

13. $f(x) = x^3 - 4x^2 - x + 4$

14. $f(x) = x^2 + 2x - 8$

15. $f(x) = x^2(x - 3)$

16. $f(x) = \dfrac{x - 1}{x - 2}$

17. $f(x) = |x|$ and $g(x) = -|x| - 1$

18. $x^2 + y^2 - 2x + 4y - 4 = 0$

In Exercises 19–20, let $f(x) = 2x^2 - x - 1$ and $g(x) = 4x - 1$.

19. Find $(f \circ g)(x)$.

20. Find $\dfrac{f(x + h) - f(x)}{h}$.

EXPONENTIAL AND LOGARITHMIC FUNCTIONS

Can I put aside $25,000 when I'm 20 and wind up sitting on half a million dollars by my early fifties? Will population growth lead to a future without comfort or individual choice? Why did I feel I was walking too slowly on my visit to New York City? Are Californians at greater risk from drunk drivers than from earthquakes? What is the difference between earthquakes measuring 6 and 7 on the Richter scale? Can I live forever? And what can possibly be causing merchants at our local shopping mall to grin from ear to ear as they watch the browsers?

The functions that you will be learning about in this chapter will provide you with the mathematics for answering these questions. You will see how these remarkable functions enable us to predict the future and rediscover the past.

HERE'S WHERE YOU'LL FIND THESE APPLICATIONS:

- You'll be sitting on $500,000 in Example 10 of Section 4.4.

- World population growth: Section 4.5, Example 5

- Population and walking speed: Section 4.5, Check Point 4

- Alcohol and risk of a car accident: Section 4.4, Example 9

- Earthquake intensity: Section 4.2, Example 9

- Immortality: Blitzer Bonus, page 456

We open the chapter with those grinning merchants and the sound of ka-ching!

451

SECTION 4.1

Exponential Functions

Objectives

① Evaluate exponential functions.

② Graph exponential functions.

③ Evaluate functions with base *e*.

④ Use compound interest formulas.

Just browsing? Take your time. Researchers know, to the dollar, the average amount the typical consumer spends per minute at the shopping mall. And the longer you stay, the more you spend. So if you say you're just browsing, that's just fine with the mall merchants. Browsing is time and, as shown in **Figure 4.1**, time is money.

The data in **Figure 4.1** can be modeled by the function

$$f(x) = 42.2(1.56)^x,$$

where $f(x)$ is the average amount spent, in dollars, at a shopping mall after x hours. Can you see how this function is different from a polynomial function? The variable x is in the exponent. Functions whose equations contain a variable in the exponent are called **exponential functions**. Many real-life situations, including population growth, growth of epidemics, radioactive decay, and other changes that involve rapid increase or decrease, can be described using exponential functions.

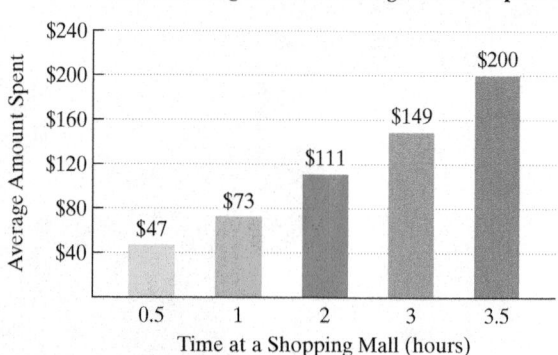

Mall Browsing Time and Average Amount Spent

FIGURE 4.1
Source: International Council of Shopping Centers Research, 2006

Definition of the Exponential Function

The **exponential function** f **with base** b is defined by

$$f(x) = b^x \quad \text{or} \quad y = b^x,$$

where b is a positive constant other than 1 ($b > 0$ and $b \neq 1$) and x is any real number.

Here are some examples of exponential functions:

$$f(x) = 2^x \qquad g(x) = 10^x \qquad h(x) = 3^{x+1} \qquad j(x) = \left(\frac{1}{2}\right)^{x-1}.$$

Base is 2. Base is 10. Base is 3. Base is $\frac{1}{2}$.

Each of these functions has a constant base and a variable exponent.

By contrast, the following functions are not exponential functions:

$$F(x) = x^2 \qquad G(x) = 1^x \qquad H(x) = (-1)^x \qquad J(x) = x^x.$$

Variable is the base and not the exponent. The base of an exponential function must be a positive constant other than 1. The base of an exponential function must be positive. Variable is both the base and the exponent.

Why is $G(x) = 1^x$ not classified as an exponential function? The number 1 raised to any power is 1. Thus, the function G can be written as $G(x) = 1$, which is a constant function.

Why is $H(x) = (-1)^x$ not an exponential function? The base of an exponential function must be positive to avoid having to exclude many values of x from the domain that result in nonreal numbers in the range:

$$H(x) = (-1)^x \qquad H\left(\frac{1}{2}\right) = (-1)^{\frac{1}{2}} = \sqrt{-1} = i.$$

Not an exponential function

All values of x resulting in even roots of negative numbers produce nonreal numbers.

① Evaluate exponential functions.

You will need a calculator to evaluate exponential expressions. Most scientific calculators have a $\boxed{y^x}$ key. Graphing calculators have a $\boxed{\wedge}$ key. To evaluate expressions of the form b^x, enter the base b, press $\boxed{y^x}$ or $\boxed{\wedge}$, enter the exponent x, and finally press $\boxed{=}$ or $\boxed{\text{ENTER}}$.

EXAMPLE 1 Evaluating an Exponential Function

The exponential function $f(x) = 42.2(1.56)^x$ models the average amount spent, $f(x)$, in dollars, at a shopping mall after x hours. What is the average amount spent, to the nearest dollar, after four hours?

SOLUTION

Because we are interested in the amount spent after four hours, substitute 4 for x and evaluate the function.

$$f(x) = 42.2(1.56)^x \quad \text{This is the given function.}$$
$$f(4) = 42.2(1.56)^4 \quad \text{Substitute 4 for x.}$$

Use a scientific or graphing calculator to evaluate $f(4)$. Press the following keys on your calculator to do this:

Scientific calculator: 42.2 $\boxed{\times}$ 1.56 $\boxed{y^x}$ 4 $\boxed{=}$

Graphing calculator: 42.2 $\boxed{\times}$ 1.56 $\boxed{\wedge}$ 4 $\boxed{\text{ENTER}}$.

The display should be approximately 249.92566.

$$f(4) = 42.2(1.56)^4 \approx 249.92566 \approx 250$$

Thus, the average amount spent after four hours at a mall is $250. •••

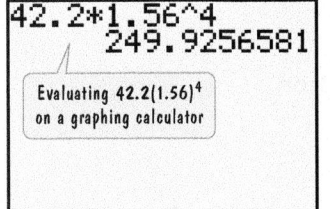

42.2*1.56^4
 249.9256581

Evaluating $42.2(1.56)^4$
on a graphing calculator

✅ **Check Point 1** Use the exponential function in Example 1 to find the average amount spent, to the nearest dollar, after three hours at a shopping mall. Does this rounded function value underestimate or overestimate the amount shown in **Figure 4.1**? By how much?

② Graph exponential functions.

Graphing Exponential Functions

We are familiar with expressions involving b^x, where x is a rational number. For example,

$$b^{1.7} = b^{\frac{17}{10}} = \sqrt[10]{b^{17}} \quad \text{and} \quad b^{1.73} = b^{\frac{173}{100}} = \sqrt[100]{b^{173}}.$$

However, note that the definition of $f(x) = b^x$ includes all real numbers for the domain x. You may wonder what b^x means when x is an irrational number, such as $b^{\sqrt{3}}$ or b^{π}. Using closer and closer approximations for $\sqrt{3}$ ($\sqrt{3} \approx 1.73205$), we can think of $b^{\sqrt{3}}$ as the value that has the successively closer approximations

$$b^{1.7}, b^{1.73}, b^{1.732}, b^{1.73205}, \dots .$$

In this way, we can graph exponential functions with no holes, or points of discontinuity, at the irrational domain values.

●

EXAMPLE 2 Graphing an Exponential Function

Graph: $f(x) = 2^x$.

SOLUTION

We begin by setting up a table of coordinates.

x	$f(x) = 2^x$
-3	$f(-3) = 2^{-3} = \dfrac{1}{8}$
-2	$f(-2) = 2^{-2} = \dfrac{1}{4}$
-1	$f(-1) = 2^{-1} = \dfrac{1}{2}$
0	$f(0) = 2^0 = 1$
1	$f(1) = 2^1 = 2$
2	$f(2) = 2^2 = 4$
3	$f(3) = 2^3 = 8$

> We selected integers from **−3** to **3**, inclusive, to include three negative numbers, **0**, and three positive numbers. We also wanted to keep the resulting computations for y relatively simple.

FIGURE 4.2 The graph of $f(x) = 2^x$

We plot these points, connecting them with a continuous curve. **Figure 4.2** shows the graph of $f(x) = 2^x$. Observe that the graph approaches, but never touches, the negative portion of the x-axis. Thus, the x-axis, or $y = 0$, is a horizontal asymptote. The range is the set of all positive real numbers. Although we used integers for x in our table of coordinates, you can use a calculator to find additional points. For example, $f(0.3) = 2^{0.3} \approx 1.231$ and $f(0.95) = 2^{0.95} \approx 1.932$. The points $(0.3, 1.231)$ and $(0.95, 1.932)$ approximately fit the graph. ●●●

✓ **Check Point 2** Graph: $f(x) = 3^x$.

EXAMPLE 3 Graphing an Exponential Function

Graph: $g(x) = \left(\dfrac{1}{2}\right)^x$.

SOLUTION

We begin by setting up a table of coordinates. We compute the function values by noting that

$$g(x) = \left(\frac{1}{2}\right)^x = (2^{-1})^x = 2^{-x}.$$

x	$g(x) = \left(\dfrac{1}{2}\right)^x$ or 2^{-x}
-3	$g(-3) = 2^{-(-3)} = 2^3 = 8$
-2	$g(-2) = 2^{-(-2)} = 2^2 = 4$
-1	$g(-1) = 2^{-(-1)} = 2^1 = 2$
0	$g(0) = 2^{-0} = 1$
1	$g(1) = 2^{-1} = \dfrac{1}{2^1} = \dfrac{1}{2}$
2	$g(2) = 2^{-2} = \dfrac{1}{2^2} = \dfrac{1}{4}$
3	$g(3) = 2^{-3} = \dfrac{1}{2^3} = \dfrac{1}{8}$

We plot these points, connecting them with a continuous curve. **Figure 4.3** shows the graph of $g(x) = \left(\frac{1}{2}\right)^x$. This time the graph approaches, but never touches, the *positive* portion of the x-axis. Once again, the x-axis, or $y = 0$, is a horizontal asymptote. The range consists of all positive real numbers.

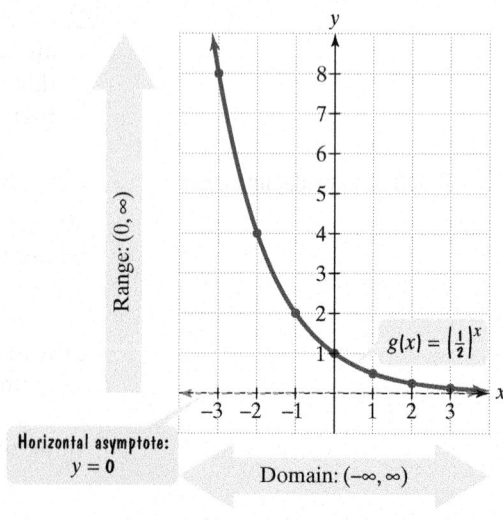

FIGURE 4.3 The graph of $g(x) = \left(\dfrac{1}{2}\right)^x$

●●●

Do you notice a relationship between the graphs of $f(x) = 2^x$ and $g(x) = \left(\frac{1}{2}\right)^x$ in **Figures 4.2** and **4.3**? The graph of $g(x) = \left(\frac{1}{2}\right)^x$ is the graph of $f(x) = 2^x$ reflected about the y-axis:

$$g(x) = \left(\frac{1}{2}\right)^x = 2^{-x} = f(-x).$$

> Recall that the graph of $y = f(-x)$ is the graph of $y = f(x)$ reflected about the y-axis.

✓ **Check Point 3** Graph: $f(x) = \left(\frac{1}{3}\right)^x$. Note that $f(x) = \left(\frac{1}{3}\right)^x = (3^{-1})^x = 3^{-x}$.

Four exponential functions have been graphed in **Figure 4.4**. Compare the black and green graphs, where $b > 1$, to those in blue and red, where $b < 1$. When $b > 1$, the value of y increases as the value of x increases. When $b < 1$, the value of y decreases as the value of x increases. Notice that all four graphs pass through $(0, 1)$.

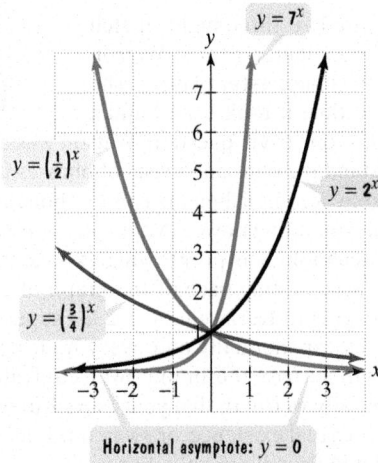

Horizontal asymptote: $y = 0$

FIGURE 4.4 Graphs of four exponential functions

These graphs illustrate the following general characteristics of exponential functions:

Characteristics of Exponential Functions of the Form $f(x) = b^x$

1. The domain of $f(x) = b^x$ consists of all real numbers: $(-\infty, \infty)$. The range of $f(x) = b^x$ consists of all positive real numbers: $(0, \infty)$.

2. The graphs of all exponential functions of the form $f(x) = b^x$ pass through the point $(0, 1)$ because $f(0) = b^0 = 1 (b \neq 0)$. The y-intercept is 1. There is no x-intercept.

3. If $b > 1$, $f(x) = b^x$ has a graph that goes up to the right and is an increasing function. The greater the value of b, the steeper the increase.

4. If $0 < b < 1$, $f(x) = b^x$ has a graph that goes down to the right and is a decreasing function. The smaller the value of b, the steeper the decrease.

5. $f(x) = b^x$ is one-to-one and has an inverse that is a function.

6. The graph of $f(x) = b^x$ approaches, but does not touch, the x-axis. The x-axis, or $y = 0$, is a horizontal asymptote.

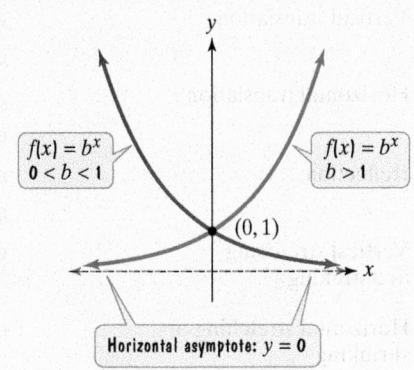

Blitzer Bonus || Exponential Growth: The Year Humans Become Immortal

In 2011, *Jeopardy!* aired a three-night match between a personable computer named Watson and the show's two most successful players. The winner: Watson. In the time it took each human contestant to respond to one trivia question, Watson was able to scan the content of one million books. It was also trained to understand the puns and twists of phrases unique to *Jeopardy!* clues.

Watson's remarkable accomplishments can be thought of as a single data point on an exponential curve that models growth in computing power. According to inventor, author, and computer scientist Ray Kurzweil (1948–), computer technology is progressing exponentially, doubling in power each year. What does this mean in terms of the accelerating pace of the graph of $y = 2^x$ that starts slowly and then rockets skyward toward infinity? According to Kurzweil, by 2023, a supercomputer will surpass the brainpower of a human. As progress accelerates exponentially and every hour brings a century's worth of scientific breakthroughs, by 2045, computers will surpass the brainpower equivalent to that of all human brains combined. Here's where it gets exponentially weird: In that year (says Kurzweil), we will be able to scan our consciousness into computers and enter a virtual existence, or swap our bodies for immortal robots. Indefinite life extension will become a reality and people will die only if they choose to.

Transformations of Exponential Functions

The graphs of exponential functions can be translated vertically or horizontally, reflected, stretched, or shrunk. These transformations are summarized in **Table 4.1**.

Table 4.1 Transformations Involving Exponential Functions

In each case, c represents a positive real number.

Transformation	Equation	Description
Vertical translation	$g(x) = b^x + c$	• Shifts the graph of $f(x) = b^x$ upward c units.
	$g(x) = b^x - c$	• Shifts the graph of $f(x) = b^x$ downward c units.
Horizontal translation	$g(x) = b^{x+c}$	• Shifts the graph of $f(x) = b^x$ to the left c units.
	$g(x) = b^{x-c}$	• Shifts the graph of $f(x) = b^x$ to the right c units.
Reflection	$g(x) = -b^x$	• Reflects the graph of $f(x) = b^x$ about the x-axis.
	$g(x) = b^{-x}$	• Reflects the graph of $f(x) = b^x$ about the y-axis.
Vertical stretching or shrinking	$g(x) = cb^x$	• Vertically stretches the graph of $f(x) = b^x$ if $c > 1$. • Vertically shrinks the graph of $f(x) = b^x$ if $0 < c < 1$.
Horizontal stretching or shrinking	$g(x) = b^{cx}$	• Horizontally shrinks the graph of $f(x) = b^x$ if $c > 1$. • Horizontally stretches the graph of $f(x) = b^x$ if $0 < c < 1$.

EXAMPLE 4 Transformations Involving Exponential Functions

Use the graph of $f(x) = 3^x$ to obtain the graph of $g(x) = 3^{x+1}$.

SOLUTION

The graph of $g(x) = 3^{x+1}$ is the graph of $f(x) = 3^x$ shifted 1 unit to the left.

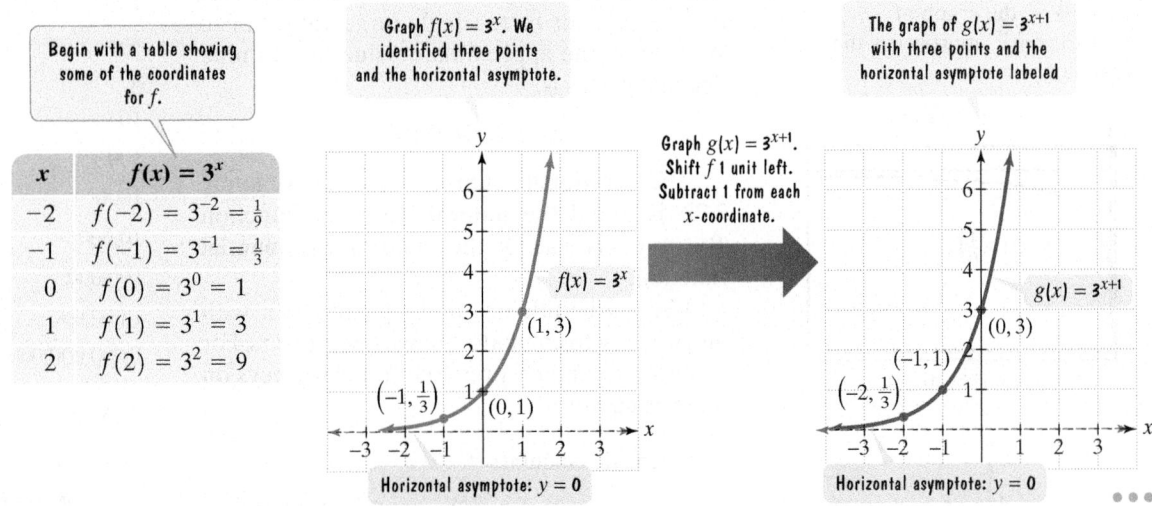

Begin with a table showing some of the coordinates for f.

x	$f(x) = 3^x$
-2	$f(-2) = 3^{-2} = \frac{1}{9}$
-1	$f(-1) = 3^{-1} = \frac{1}{3}$
0	$f(0) = 3^0 = 1$
1	$f(1) = 3^1 = 3$
2	$f(2) = 3^2 = 9$

Graph $f(x) = 3^x$. We identified three points and the horizontal asymptote.

Graph $g(x) = 3^{x+1}$. Shift f 1 unit left. Subtract 1 from each x-coordinate.

The graph of $g(x) = 3^{x+1}$ with three points and the horizontal asymptote labeled

Horizontal asymptote: $y = 0$

✓ **Check Point 4** Use the graph of $f(x) = 3^x$ to obtain the graph of $g(x) = 3^{x-1}$.

If an exponential function is translated upward or downward, the horizontal asymptote is shifted by the amount of the vertical shift.

EXAMPLE 5 Transformations Involving Exponential Functions

Use the graph of $f(x) = 2^x$ to obtain the graph of $g(x) = 2^x - 3$.

SOLUTION

The graph of $g(x) = 2^x - 3$ is the graph of $f(x) = 2^x$ shifted down 3 units.

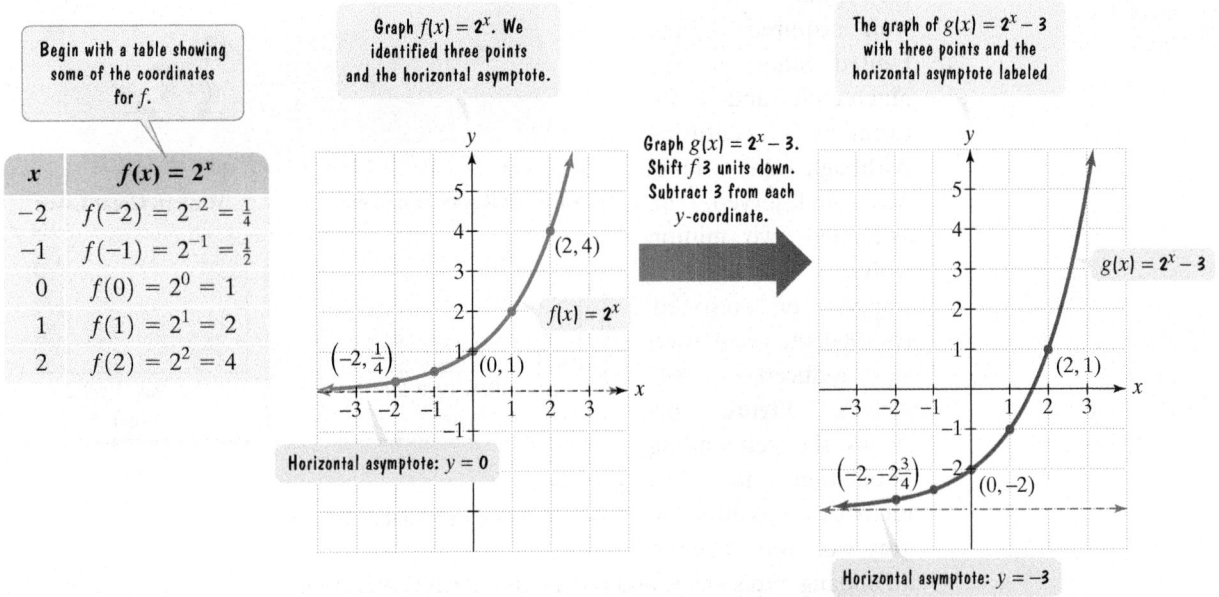

Begin with a table showing some of the coordinates for f.

x	$f(x) = 2^x$
-2	$f(-2) = 2^{-2} = \frac{1}{4}$
-1	$f(-1) = 2^{-1} = \frac{1}{2}$
0	$f(0) = 2^0 = 1$
1	$f(1) = 2^1 = 2$
2	$f(2) = 2^2 = 4$

Graph $f(x) = 2^x$. We identified three points and the horizontal asymptote.

Graph $g(x) = 2^x - 3$. Shift f 3 units down. Subtract 3 from each y-coordinate.

The graph of $g(x) = 2^x - 3$ with three points and the horizontal asymptote labeled

Horizontal asymptote: $y = 0$

Horizontal asymptote: $y = -3$

✓ **Check Point 5** Use the graph of $f(x) = 2^x$ to obtain the graph of $g(x) = 2^x + 1$.

③ Evaluate functions with base *e*.

TECHNOLOGY

Graphic Connections

As $n \to \infty$, the graph of $y = \left(1 + \frac{1}{n}\right)^n$ approaches the graph of $y = e$.

[0, 15, 1] by [0, 3, 1]

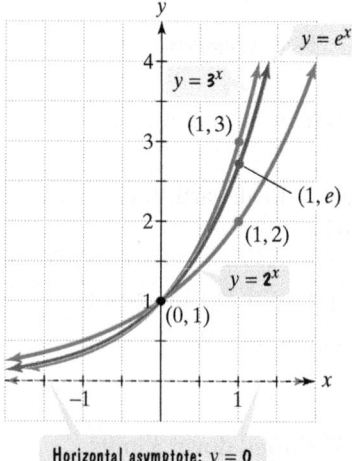

Horizontal asymptote: $y = 0$

FIGURE 4.5 Graphs of three exponential functions

The Natural Base *e*

An irrational number, symbolized by the letter *e*, appears as the base in many applied exponential functions. The number *e* is defined as the value that $\left(1 + \frac{1}{n}\right)^n$ approaches as *n* gets larger and larger. **Table 4.2** shows values of $\left(1 + \frac{1}{n}\right)^n$ for increasingly large values of *n*. As $n \to \infty$, the approximate value of *e* to nine decimal places is

$$e \approx 2.718281827.$$

The irrational number *e*, approximately 2.72, is called the **natural base**. The function $f(x) = e^x$ is called the **natural exponential function**.

Use a scientific or graphing calculator with an $\boxed{e^x}$ key to evaluate *e* to various powers. For example, to find e^2, press the following keys on most calculators:

Scientific calculator: $2 \boxed{e^x}$

Graphing calculator: $\boxed{e^x} \ 2 \ \boxed{\text{ENTER}}$.

The display should be approximately 7.389.

$$e^2 \approx 7.389$$

The number *e* lies between 2 and 3. Because $2^2 = 4$ and $3^2 = 9$, it makes sense that e^2, approximately 7.389, lies between 4 and 9.

Because $2 < e < 3$, the graph of $y = e^x$ is between the graphs of $y = 2^x$ and $y = 3^x$, shown in **Figure 4.5**.

Table 4.2

n	$\left(1 + \dfrac{1}{n}\right)^n$
1	2
2	2.25
5	2.48832
10	2.59374246
100	2.704813829
1000	2.716923932
10,000	2.718145927
100,000	2.718268237
1,000,000	2.718280469
1,000,000,000	2.718281827

As $n \to \infty$, $\left(1 + \dfrac{1}{n}\right)^n \to e$.

In calculus, this is expressed as

$$\lim_{n \to \infty} \left(1 + \frac{1}{n}\right)^n = e.$$

EXAMPLE 6 Gray Wolf Population

Insatiable killer. That's the reputation the gray wolf acquired in the United States in the nineteenth and early twentieth centuries. Although the label was undeserved, an estimated two million wolves were shot, trapped, or poisoned. By 1960, the population was reduced to 800 wolves. **Figure 4.6** shows the rebounding population in two recovery areas after the gray wolf was declared an endangered species and received federal protection.

Gray Wolf Population in Two Recovery Areas for Selected Years

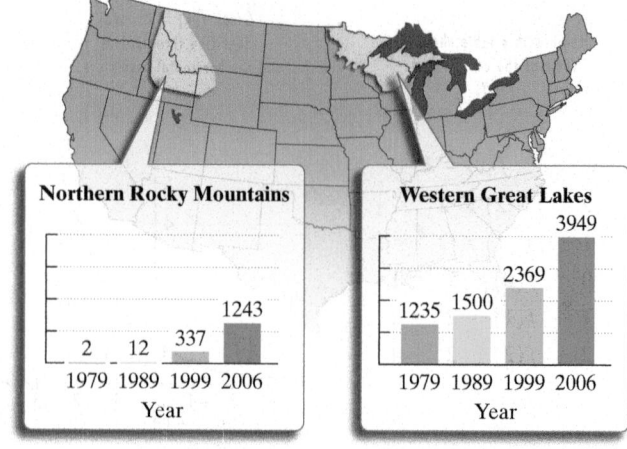

FIGURE 4.6
Source: U.S. Fish and Wildlife Service

The exponential function

$$f(x) = 1.26e^{0.247x}$$

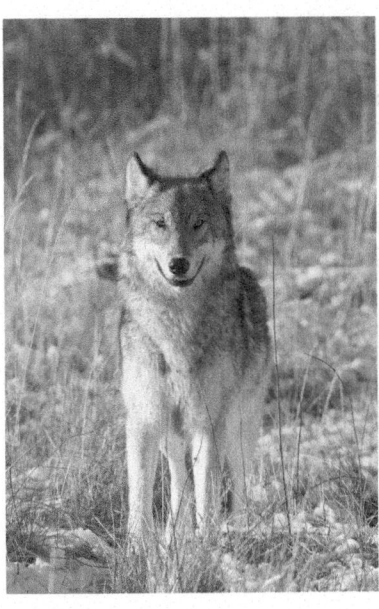

models the gray wolf population of the Northern Rocky Mountains, $f(x)$, x years after 1978. If the wolf is not removed from the endangered species list and trends shown in **Figure 4.6** continue, project its population in the recovery area in 2010.

SOLUTION

Because 2010 is 32 years after 1978, we substitute 32 for x in the given function.

$$f(x) = 1.26e^{0.247x} \quad \text{This is the given function.}$$

$$f(32) = 1.26e^{0.247(32)} \quad \text{Substitute 32 for x.}$$

Perform this computation on your calculator.

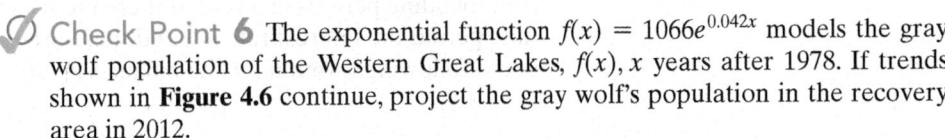

Scientific calculator:

Graphing calculator:

The display should be approximately 3412.1973.

Thus, $\qquad f(32) = 1.26e^{0.247(32)} \approx 3412.$

> This parenthesis is given on some calculators.

This indicates that the gray wolf population of the Northern Rocky Mountains in the year 2010 is projected to have been 3412. •••

✓ Check Point **6** The exponential function $f(x) = 1066e^{0.042x}$ models the gray wolf population of the Western Great Lakes, $f(x)$, x years after 1978. If trends shown in **Figure 4.6** continue, project the gray wolf's population in the recovery area in 2012.

In 2008, using exponential functions and projections like those in Example 6 and Check Point 6, the U.S. Fish and Wildlife Service removed the gray wolf from the endangered species list, a ruling environmentalists vowed to appeal.

④ Use compound interest formulas.

Compound Interest

We all want a wonderful life with fulfilling work, good health, and loving relationships. And let's be honest: Financial security wouldn't hurt! Achieving this goal depends on understanding how money in savings accounts grows in remarkable ways as a result of *compound interest*. **Compound interest** is interest computed on your original investment as well as on any accumulated interest.

Suppose a sum of money, called the **principal**, P, is invested at an annual percentage rate r, in decimal form, compounded once per year. Because the interest is added to the principal at year's end, the accumulated value, A, is

$$A = P + Pr = P(1 + r).$$

The accumulated amount of money follows this pattern of multiplying the previous principal by $(1 + r)$ for each successive year, as indicated in **Table 4.3**.

Table 4.3

Time in Years	Accumulated Value after Each Compounding
0	$A = P$
1	$A = P(1 + r)$
2	$A = P(1 + r)(1 + r) = P(1 + r)^2$
3	$A = P(1 + r)^2(1 + r) = P(1 + r)^3$
4	$A = P(1 + r)^3(1 + r) = P(1 + r)^4$
⋮	⋮
t	$A = P(1 + r)^t$

> This formula gives the balance, A, that a principal, P, is worth after t years at interest rate r, compounded once a year.

Most savings institutions have plans in which interest is paid more than once a year. If compound interest is paid twice a year, the compounding period is six months. We say that the interest is **compounded semiannually**. When compound interest is paid four times a year, the compounding period is three months and the interest is said to be **compounded quarterly**. Some plans allow for monthly compounding or daily compounding.

In general, when compound interest is paid n times a year, we say that there are **n compounding periods per year**. **Table 4.4** shows the three most frequently used plans in which interest is paid more than once a year.

Table 4.4 Interest Plans

Name	Number of Compounding Periods per Year	Length of Each Compounding Period
Semiannual Compounding	$n = 2$	6 months
Quarterly Compounding	$n = 4$	3 months
Monthly Compounding	$n = 12$	1 month

The formula $A = P(1 + r)^t$ can be adjusted to take into account the number of compounding periods in a year. If there are n compounding periods per year, in each time period the interest rate is $\frac{r}{n}$ and there are nt time periods in t years. This results in the following formula for the balance, A, after t years:

$$A = P\left(1 + \frac{r}{n}\right)^{nt}.$$

Some banks use **continuous compounding**, where the number of compounding periods increases infinitely (compounding interest every trillionth of a second, every quadrillionth of a second, etc.). Let's see what happens to the balance, A, as $n \to \infty$.

$$\frac{n}{r} \cdot rt = nt$$

$$A = P\left(1 + \frac{r}{n}\right)^{nt} = P\left[\left(1 + \frac{1}{\frac{n}{r}}\right)^{\frac{n}{r}}\right]^{rt} = P\left[\left(1 + \frac{1}{h}\right)^{h}\right]^{rt} = Pe^{rt}$$

Let $h = \frac{n}{r}$.
As $n \to \infty$, $h \to \infty$.

As $h \to \infty$, by definition
$\left(1 + \frac{1}{h}\right)^{h} \to e$.

We see that the formula for continuous compounding is $A = Pe^{rt}$. Although continuous compounding sounds terrific, it yields only a fraction of a percent more interest over a year than daily compounding.

Formulas for Compound Interest

After t years, the balance, A, in an account with principal P and annual interest rate r (in decimal form) is given by the following formulas:

1. For n compounding periods per year: $A = P\left(1 + \frac{r}{n}\right)^{nt}$

2. For continuous compounding: $A = Pe^{rt}$.

EXAMPLE 7 Choosing between Investments

You decide to invest $8000 for 6 years and you have a choice between two accounts. The first pays 7% per year, compounded monthly. The second pays 6.85% per year, compounded continuously. Which is the better investment?

SOLUTION

The better investment is the one with the greater balance in the account after 6 years. Let's begin with the account with monthly compounding. We use the compound interest model with $P = 8000, r = 7\% = 0.07, n = 12$ (monthly compounding means 12 compoundings per year), and $t = 6$.

$$A = P\left(1 + \frac{r}{n}\right)^{nt} = 8000\left(1 + \frac{0.07}{12}\right)^{12 \cdot 6} \approx 12{,}160.84$$

The balance in this account after 6 years is \$12,160.84.

For the second investment option, we use the model for continuous compounding with $P = 8000, r = 6.85\% = 0.0685$, and $t = 6$.

$$A = Pe^{rt} = 8000e^{0.0685(6)} \approx 12{,}066.60$$

The balance in this account after 6 years is \$12,066.60, slightly less than the previous amount. Thus, the better investment is the 7% monthly compounding option. ●●●

⊘ Check Point **7** A sum of \$10,000 is invested at an annual rate of 8%. Find the balance in the account after 5 years subject to **a.** quarterly compounding and **b.** continuous compounding.

CONCEPT AND VOCABULARY CHECK

Fill in each blank so that the resulting statement is true.

1. The exponential function f with base b is defined by $f(x) =$ _____, $b > 0$ and $b \neq 1$. Using interval notation, the domain of this function is _____ and the range is _____.

2. The graph of the exponential function f with base b approaches, but does not touch, the _____-axis. This axis, whose equation is _____, is a/an _____ asymptote.

3. The value that $\left(1 + \dfrac{1}{n}\right)^n$ approaches as n gets larger and larger is the irrational number _____, called the _____ base. This irrational number is approximately equal to _____.

4. Consider the compound interest formula

$$A = P\left(1 + \frac{r}{n}\right)^{nt}.$$

This formula gives the balance, _____, in an account with principal _____ and annual interest rate _____, in decimal form, subject to compound interest paid _____ times per year.

5. If compound interest is paid twice a year, we say that the interest is compounded _____. If compound interest is paid four times a year, we say that the interest is compounded _____. If the number of compounding periods increases infinitely, we call this _____ compounding.

EXERCISE SET 4.1

Practice Exercises

In Exercises 1–10, approximate each number using a calculator. Round your answer to three decimal places.

1. $2^{3.4}$ **2.** $3^{2.4}$ **3.** $3^{\sqrt{5}}$ **4.** $5^{\sqrt{3}}$ **5.** $4^{-1.5}$

6. $6^{-1.2}$ **7.** $e^{2.3}$ **8.** $e^{3.4}$ **9.** $e^{-0.95}$ **10.** $e^{-0.75}$

In Exercises 11–18, graph each function by making a table of coordinates. If applicable, use a graphing utility to confirm your hand-drawn graph.

11. $f(x) = 4^x$

13. $g(x) = \left(\frac{3}{2}\right)^x$

15. $h(x) = \left(\frac{1}{2}\right)^x$

17. $f(x) = (0.6)^x$

12. $f(x) = 5^x$

14. $g(x) = \left(\frac{4}{3}\right)^x$

16. $h(x) = \left(\frac{1}{3}\right)^x$

18. $f(x) = (0.8)^x$

In Exercises 19–24, the graph of an exponential function is given. Select the function for each graph from the following options:

$$f(x) = 3^x, g(x) = 3^{x-1}, h(x) = 3^x - 1,$$

$$F(x) = -3^x, G(x) = 3^{-x}, H(x) = -3^{-x}.$$

19.

20.

21. 　　**22.**

23. 　　**24.**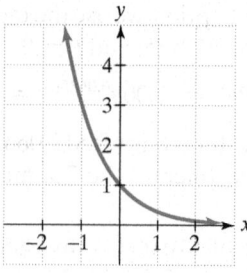

In Exercises 25–34, begin by graphing $f(x) = 2^x$. Then use transformations of this graph to graph the given function. Be sure to graph and give equations of the asymptotes. Use the graphs to determine each function's domain and range. If applicable, use a graphing utility to confirm your hand-drawn graphs.

25. $g(x) = 2^{x+1}$　　　　　　**26.** $g(x) = 2^{x+2}$

27. $g(x) = 2^x - 1$　　　　　　**28.** $g(x) = 2^x + 2$

29. $h(x) = 2^{x+1} - 1$　　　　**30.** $h(x) = 2^{x+2} - 1$

31. $g(x) = -2^x$　　　　　　　**32.** $g(x) = 2^{-x}$

33. $g(x) = 2 \cdot 2^x$　　　　　　**34.** $g(x) = \frac{1}{2} \cdot 2^x$

The figure shows the graph of $f(x) = e^x$. In Exercises 35–46, use transformations of this graph to graph each function. Be sure to give equations of the asymptotes. Use the graphs to determine each function's domain and range. If applicable, use a graphing utility to confirm your hand-drawn graphs.

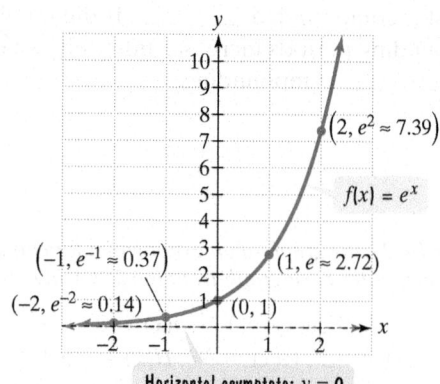

35. $g(x) = e^{x-1}$　　　　　　　**36.** $g(x) = e^{x+1}$

37. $g(x) = e^x + 2$　　　　　　**38.** $g(x) = e^x - 1$

39. $h(x) = e^{x-1} + 2$　　　　**40.** $h(x) = e^{x+1} - 1$

41. $h(x) = e^{-x}$　　　　　　　**42.** $h(x) = -e^x$

43. $g(x) = 2e^x$　　　　　　　**44.** $g(x) = \frac{1}{2}e^x$

45. $h(x) = e^{2x} + 1$　　　　　**46.** $h(x) = e^{\frac{x}{2}} + 2$

In Exercises 47–52, graph functions f and g in the same rectangular coordinate system. Graph and give equations of all asymptotes. If applicable, use a graphing utility to confirm your hand-drawn graphs.

47. $f(x) = 3^x$ and $g(x) = 3^{-x}$

48. $f(x) = 3^x$ and $g(x) = -3^x$

49. $f(x) = 3^x$ and $g(x) = \frac{1}{3} \cdot 3^x$

50. $f(x) = 3^x$ and $g(x) = 3 \cdot 3^x$

51. $f(x) = \left(\frac{1}{2}\right)^x$ and $g(x) = \left(\frac{1}{2}\right)^{x-1} + 1$

52. $f(x) = \left(\frac{1}{2}\right)^x$ and $g(x) = \left(\frac{1}{2}\right)^{x-1} + 2$

Use the compound interest formulas $A = P\left(1 + \dfrac{r}{n}\right)^{nt}$ and $A = Pe^{rt}$ to solve Exercises 53–56. Round answers to the nearest cent.

53. Find the accumulated value of an investment of $10,000 for 5 years at an interest rate of 5.5% if the money is **a.** compounded semiannually; **b.** compounded quarterly; **c.** compounded monthly; **d.** compounded continuously.

54. Find the accumulated value of an investment of $5000 for 10 years at an interest rate of 6.5% if the money is **a.** compounded semiannually; **b.** compounded quarterly; **c.** compounded monthly; **d.** compounded continuously.

55. Suppose that you have $12,000 to invest. Which investment yields the greater return over 3 years: 7% compounded monthly or 6.85% compounded continuously?

56. Suppose that you have $6000 to invest. Which investment yields the greater return over 4 years: 8.25% compounded quarterly or 8.3% compounded semiannually?

Practice Plus

In Exercises 57–58, graph f and g in the same rectangular coordinate system. Then find the point of intersection of the two graphs.

57. $f(x) = 2^x$, $g(x) = 2^{-x}$

58. $f(x) = 2^{x+1}$, $g(x) = 2^{-x+1}$

59. Graph $y = 2^x$ and $x = 2^y$ in the same rectangular coordinate system.

60. Graph $y = 3^x$ and $x = 3^y$ in the same rectangular coordinate system.

In Exercises 61–64, give the equation of each exponential function whose graph is shown.

61. 　　**62.**

63.

64.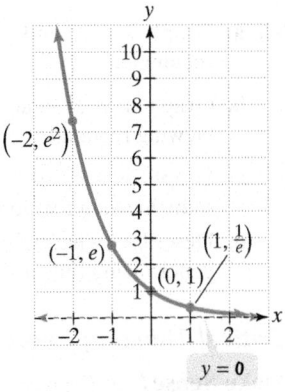

Application Exercises

Use a calculator with a $\boxed{y^x}$ key or a $\boxed{\wedge}$ key to solve Exercises 65–70.

65. India is currently one of the world's fastest-growing countries. By 2040, the population of India will be larger than the population of China; by 2050, nearly one-third of the world's population will live in these two countries alone. The exponential function $f(x) = 574(1.026)^x$ models the population of India, $f(x)$, in millions, x years after 1974.

 a. Substitute 0 for x and, without using a calculator, find India's population in 1974.

 b. Substitute 27 for x and use your calculator to find India's population, to the nearest million, in the year 2001 as modeled by this function.

 c. Find India's population, to the nearest million, in the year 2028 as predicted by this function.

 d. Find India's population, to the nearest million, in the year 2055 as predicted by this function.

 e. What appears to be happening to India's population every 27 years?

66. The 1986 explosion at the Chernobyl nuclear power plant in the former Soviet Union sent about 1000 kilograms of radioactive cesium-137 into the atmosphere. The function $f(x) = 1000(0.5)^{\frac{x}{30}}$ describes the amount, $f(x)$, in kilograms, of cesium-137 remaining in Chernobyl x years after 1986. If even 100 kilograms of cesium-137 remain in Chernobyl's atmosphere, the area is considered unsafe for human habitation. Find $f(80)$ and determine if Chernobyl will be safe for human habitation by 2066.

The formula $S = C(1 + r)^t$ models inflation, where $C =$ the value today, $r =$ the annual inflation rate, and $S =$ the inflated value t years from now. Use this formula to solve Exercises 67–68. Round answers to the nearest dollar.

67. If the inflation rate is 6%, how much will a house now worth $465,000 be worth in 10 years?

68. If the inflation rate is 3%, how much will a house now worth $510,000 be worth in 5 years?

69. A decimal approximation for $\sqrt{3}$ is 1.7320508. Use a calculator to find $2^{1.7}$, $2^{1.73}$, $2^{1.732}$, $2^{1.73205}$, and $2^{1.7320508}$. Now find $2^{\sqrt{3}}$. What do you observe?

70. A decimal approximation for π is 3.141593. Use a calculator to find 2^3, $2^{3.1}$, $2^{3.14}$, $2^{3.141}$, $2^{3.1415}$, $2^{3.14159}$, and $2^{3.141593}$. Now find 2^{π}. What do you observe?

Use a calculator with an $\boxed{e^x}$ key to solve Exercises 71–76.

Average annual premiums for employer-sponsored family health insurance policies more than doubled over 11 years. The bar graph shows the average cost of a family health insurance plan in the United States for six selected years from 2000 through 2011.

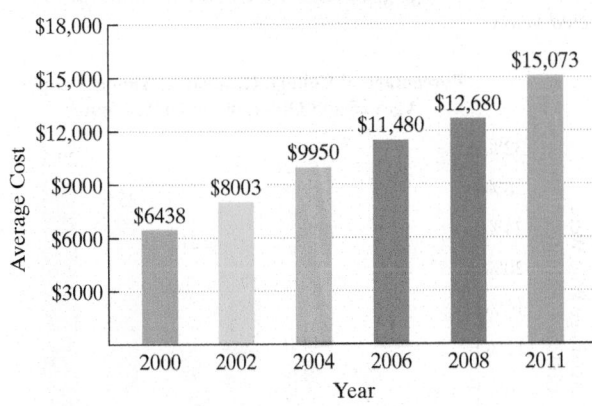

Source: Kaiser Family Foundation

The data can be modeled by

$$f(x) = 782x + 6564 \quad \text{and} \quad g(x) = 6875e^{0.077x},$$

in which $f(x)$ and $g(x)$ represent the average cost of a family health insurance plan x years after 2000. Use these functions to solve Exercises 71–72. Where necessary, round answers to the nearest whole dollar.

71. a. According to the linear model, what was the average cost of a family health insurance plan in 2011?

 b. According to the exponential model, what was the average cost of a family health insurance plan in 2011?

 c. Which function is a better model for the data in 2011?

72. a. According to the linear model, what was the average cost of a family health insurance plan in 2008?

 b. According to the exponential model, what was the average cost of a family health insurance plan in 2008?

 c. Which function is a better model for the data in 2008?

73. In college, we study large volumes of information—information that, unfortunately, we do not often retain for very long. The function

$$f(x) = 80e^{-0.5x} + 20$$

describes the percentage of information, $f(x)$, that a particular person remembers x weeks after learning the information.

 a. Substitute 0 for x and, without using a calculator, find the percentage of information remembered at the moment it is first learned.

 b. Substitute 1 for x and find the percentage of information that is remembered after 1 week.

 c. Find the percentage of information that is remembered after 4 weeks.

 d. Find the percentage of information that is remembered after one year (52 weeks).

74. In 1626, Peter Minuit convinced the Wappinger Indians to sell him Manhattan Island for $24. If the Native Americans had put the $24 into a bank account paying 5% interest, how much would the investment have been worth in the year 2010 if interest were compounded

 a. monthly? **b.** continuously?

The bar graph shows the percentage of people 25 years of age and older who were college graduates in the United States for seven selected years.

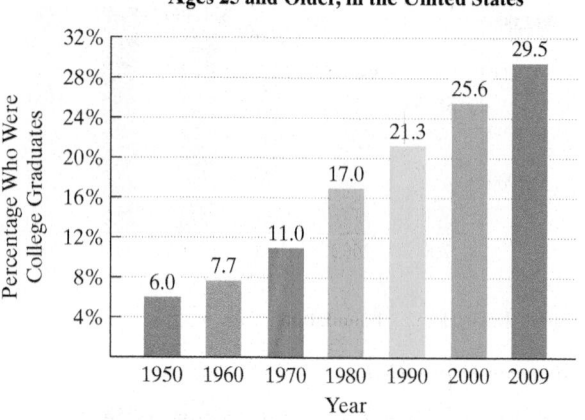

Percentage of College Graduates, Among People Ages 25 and Older, in the United States

Source: U.S. Census Bureau

The functions

$$f(x) = 6.25(1.029)^x \quad \text{and} \quad g(x) = \frac{38.8}{1 + 6.3e^{-0.051x}}$$

model the percentage of college graduates, among people ages 25 and older, $f(x)$ or $g(x)$, x years after 1950. Use these functions to solve Exercises 75–76.

75. Which function is a better model for the percentage who were college graduates in 2009?

76. Which function is a better model for the percentage who were college graduates in 1990?

Writing in Mathematics

77. What is an exponential function?

78. What is the natural exponential function?

79. Use a calculator to evaluate $\left(1 + \dfrac{1}{x}\right)^x$ for $x = 10, 100, 1000,$ 10,000, 100,000, and 1,000,000. Describe what happens to the expression as x increases.

80. Describe how you could use the graph of $f(x) = 2^x$ to obtain a decimal approximation for $\sqrt{2}$.

Technology Exercises

81. You have $10,000 to invest. One bank pays 5% interest compounded quarterly and a second bank pays 4.5% interest compounded monthly.

 a. Use the formula for compound interest to write a function for the balance in each bank at any time t.

 b. Use a graphing utility to graph both functions in an appropriate viewing rectangle. Based on the graphs, which bank offers the better return on your money?

82. a. Graph $y = e^x$ and $y = 1 + x + \dfrac{x^2}{2}$ in the same viewing rectangle.

 b. Graph $y = e^x$ and $y = 1 + x + \dfrac{x^2}{2} + \dfrac{x^3}{6}$ in the same viewing rectangle.

 c. Graph $y = e^x$ and $y = 1 + x + \dfrac{x^2}{2} + \dfrac{x^3}{6} + \dfrac{x^4}{24}$ in the same viewing rectangle.

 d. Describe what you observe in parts (a)–(c). Try generalizing this observation.

Critical Thinking Exercises

Make Sense? *In Exercises 83–86, determine whether each statement makes sense or does not make sense, and explain your reasoning.*

83. My graph of $f(x) = 3 \cdot 2^x$ shows that the horizontal asymptote for f is $x = 3$.

84. I'm using a photocopier to reduce an image over and over by 50%, so the exponential function $f(x) = \left(\frac{1}{2}\right)^x$ models the new image size, where x is the number of reductions.

85. I'm looking at data that show the number of new college programs in green studies, and a linear function appears to be a better choice than an exponential function for modeling the number of new college programs from 2005 through 2009.

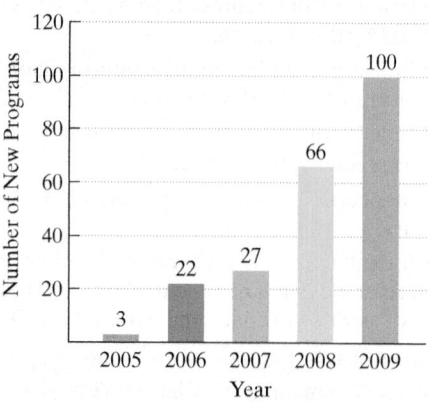

Number of New Programs in Green Studies at U.S. Colleges

Source: Association for the Advancement of Sustainability in Higher Education

86. I use the natural base e when determining how much money I'd have in a bank account that earns compound interest subject to continuous compounding.

In Exercises 87–90, determine whether each statement is true or false. If the statement is false, make the necessary change(s) to produce a true statement.

87. As the number of compounding periods increases on a fixed investment, the amount of money in the account over a fixed interval of time will increase without bound.

88. The functions $f(x) = 3^{-x}$ and $g(x) = -3^x$ have the same graph.

89. If $f(x) = 2^x$, then $f(a + b) = f(a) + f(b)$.

90. The functions $f(x) = \left(\frac{1}{3}\right)^x$ and $g(x) = 3^{-x}$ have the same graph.

91. The graphs labeled (a)–(d) in the figure represent $y = 3^x$, $y = 5^x$, $y = \left(\frac{1}{3}\right)^x$, and $y = \left(\frac{1}{5}\right)^x$, but not necessarily in that order. Which is which? Describe the process that enables you to make this decision.

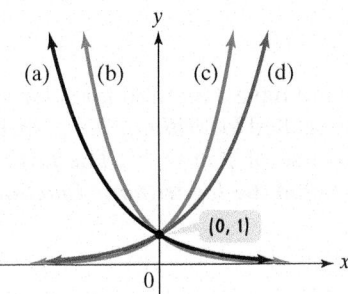

92. Graph $f(x) = 2^x$ and its inverse function in the same rectangular coordinate system.

93. The *hyperbolic cosine* and *hyperbolic sine* functions are defined by

$$\cosh x = \frac{e^x + e^{-x}}{2} \quad \text{and} \quad \sinh x = \frac{e^x - e^{-x}}{2}.$$

a. Show that $\cosh x$ is an even function.

b. Show that $\sinh x$ is an odd function.

c. Prove that $(\cosh x)^2 - (\sinh x)^2 = 1$.

Preview Exercises

Exercises 94–96 will help you prepare for the material covered in the next section.

94. What problem do you encounter when using the switch-and-solve strategy to find the inverse of $f(x) = 2^x$? (The switch-and-solve strategy is described in the box on page 211.)

95. 25 to what power gives 5? ($25^? = 5$)

96. Solve: $(x - 3)^2 > 0$.

SECTION 4.2 Logarithmic Functions

Objectives

① Change from logarithmic to exponential form.

② Change from exponential to logarithmic form.

③ Evaluate logarithms.

④ Use basic logarithmic properties.

⑤ Graph logarithmic functions.

⑥ Find the domain of a logarithmic function.

⑦ Use common logarithms.

⑧ Use natural logarithms.

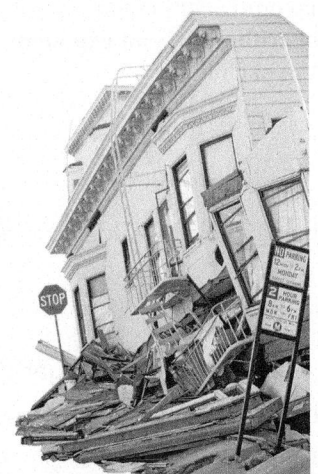

The earthquake that ripped through northern California on October 17, 1989, measured 7.1 on the Richter scale, killed more than 60 people, and injured more than 2400. Shown here is San Francisco's Marina district, where shock waves tossed houses off their foundations and into the street.

A higher measure on the Richter scale is more devastating than it seems because for each increase in one unit on the scale, there is a tenfold increase in the intensity of an earthquake. In this section, our focus is on the inverse of the exponential function, called the logarithmic function. The logarithmic function will help you to understand diverse phenomena, including earthquake intensity, human memory, and the pace of life in large cities.

GREAT QUESTION!

You mentioned that the inverse of the exponential function is called the logarithmic function. We haven't discussed inverses of functions since Section 1.8. What should I already know about functions and their inverses?

Here's a brief summary:

1. Only one-to-one functions have inverses that are functions. A function, f, has an inverse function, f^{-1}, if there is no horizontal line that intersects the graph of f at more than one point.

2. If a function is one-to-one, its inverse function can be found by interchanging x and y in the function's equation and solving for y.

3. If $f(a) = b$, then $f^{-1}(b) = a$. The domain of f is the range of f^{-1}. The range of f is the domain of f^{-1}.

4. $f(f^{-1}(x)) = x$ and $f^{-1}(f(x)) = x$.

5. The graph of f^{-1} is the reflection of the graph of f about the line $y = x$.

The Definition of Logarithmic Functions

No horizontal line can be drawn that intersects the graph of an exponential function at more than one point. This means that the exponential function is one-to-one and has an inverse. Let's use our switch-and-solve strategy from Section 1.8 to find the inverse.

All exponential functions have
inverse functions.

$$f(x) = b^x$$

Step 1 **Replace $f(x)$ with y:** $\quad y = b^x$.

Step 2 **Interchange x and y:** $\quad x = b^y$.

Step 3 **Solve for y:** $\qquad\qquad$?

The question mark indicates that we do not have a method for solving $b^y = x$ for y. To isolate the exponent y, a new notation, called *logarithmic notation*, is needed. This notation gives us a way to name the inverse of $f(x) = b^x$. **The inverse function of the exponential function with base b is called the *logarithmic function with base b*.**

> ### Definition of the Logarithmic Function
>
> For $x > 0$ and $b > 0$, $b \neq 1$,
>
> $$y = \log_b x \text{ is equivalent to } b^y = x.$$
>
> The function $f(x) = \log_b x$ is the **logarithmic function with base b**.

The equations

$$y = \log_b x \quad \text{and} \quad b^y = x$$

are different ways of expressing the same thing. The first equation is in **logarithmic form** and the second equivalent equation is in **exponential form**.

Notice that a **logarithm, y, is an exponent.** You should learn the location of the base and exponent in each form.

> ### Location of Base and Exponent in Exponential and Logarithmic Forms
>
> Exponent
>
> Logarithmic Form: $y = \log_b x$ \qquad Exponential Form: $b^y = x$
>
> Exponent
>
> Base $\qquad\qquad\qquad$ Base

① Change from logarithmic to exponential form.

EXAMPLE 1 Changing from Logarithmic to Exponential Form

Write each equation in its equivalent exponential form:

a. $2 = \log_5 x$ \qquad **b.** $3 = \log_b 64$ \qquad **c.** $\log_3 7 = y$.

SOLUTION

We use the fact that $y = \log_b x$ means $b^y = x$.

a. $2 = \log_5 x$ means $5^2 = x$. \qquad **b.** $3 = \log_b 64$ means $b^3 = 64$.

Logarithms are exponents. $\qquad\qquad\qquad$ Logarithms are exponents.

c. $\log_3 7 = y$ or $y = \log_3 7$ means $3^y = 7$. \qquad • • •

GREAT QUESTION!

Much of what you've discussed so far involves changing from logarithmic form to the more familiar exponential form. Is there a pattern I can use to help me remember how to do this?

Yes. To change from logarithmic form to exponential form, use this pattern:

$$y = \log_b x \quad \text{means} \quad b^y = x.$$

✓ Check Point **1** Write each equation in its equivalent exponential form:

a. $3 = \log_7 x$ \qquad **b.** $2 = \log_b 25$ \qquad **c.** $\log_4 26 = y$.

② Change from exponential to logarithmic form.

EXAMPLE 2 Changing from Exponential to Logarithmic Form

Write each equation in its equivalent logarithmic form:

a. $12^2 = x$ **b.** $b^3 = 8$ **c.** $e^y = 9$.

SOLUTION

We use the fact that $b^y = x$ means $y = \log_b x$.

a. $12^2 = x$ means $2 = \log_{12} x$. **b.** $b^3 = 8$ means $3 = \log_b 8$.

Exponents are logarithms.

Exponents are logarithms.

c. $e^y = 9$ means $y = \log_e 9$. •••

 Check Point **2** Write each equation in its equivalent logarithmic form:

a. $2^5 = x$ **b.** $b^3 = 27$ **c.** $e^y = 33$.

③ Evaluate logarithms.

Remembering that logarithms are exponents makes it possible to evaluate some logarithms by inspection. The logarithm of x with base b, $\log_b x$, is the exponent to which b must be raised to get x. For example, suppose we want to evaluate $\log_2 32$. We ask, 2 to what power gives 32? Because $2^5 = 32$, $\log_2 32 = 5$.

EXAMPLE 3 Evaluating Logarithms

Evaluate:

a. $\log_2 16$ **b.** $\log_7 \frac{1}{49}$ **c.** $\log_{25} 5$ **d.** $\log_2 \sqrt[5]{2}$.

SOLUTION

Logarithmic Expression	Question Needed for Evaluation	Logarithmic Expression Evaluated
a. $\log_2 16$	2 to what power gives 16? $2^? = 16$	$\log_2 16 = 4$ because $2^4 = 16$.
b. $\log_7 \frac{1}{49}$	7 to what power gives $\frac{1}{49}$? $7^? = \frac{1}{49}$	$\log_7 \frac{1}{49} = -2$ because $7^{-2} = \frac{1}{7^2} = \frac{1}{49}$.
c. $\log_{25} 5$	25 to what power gives 5? $25^? = 5$	$\log_{25} 5 = \frac{1}{2}$ because $25^{\frac{1}{2}} = \sqrt{25} = 5$.
d. $\log_2 \sqrt[5]{2}$	2 to what power gives $\sqrt[5]{2}$, or $2^{\frac{1}{5}}$? $2^? = \sqrt[5]{2}$	$\log_2 \sqrt[5]{2} = \frac{1}{5}$ because $2^{\frac{1}{5}} = \sqrt[5]{2}$.

•••

 Check Point **3** Evaluate:

a. $\log_{10} 100$ **b.** $\log_5 \frac{1}{125}$ **c.** $\log_{36} 6$ **d.** $\log_3 \sqrt[3]{3}$.

④ Use basic logarithmic properties.

Basic Logarithmic Properties

Because logarithms are exponents, they have properties that can be verified using properties of exponents.

Basic Logarithmic Properties Involving One

1. $\log_b b = 1$ because 1 is the exponent to which b must be raised to obtain b. $(b^1 = b)$

2. $\log_b 1 = 0$ because 0 is the exponent to which b must be raised to obtain 1. $(b^0 = 1)$

EXAMPLE 4 Using Properties of Logarithms

Evaluate:

a. $\log_7 7$ b. $\log_5 1$.

SOLUTION

a. Because $\log_b b = 1$, we conclude $\log_7 7 = 1$. This means that $7^1 = 7$.
b. Because $\log_b 1 = 0$, we conclude $\log_5 1 = 0$. This means that $5^0 = 1$.

● ● ●

◯ Check Point 4 Evaluate:

a. $\log_9 9$ b. $\log_8 1$.

Now that we are familiar with logarithmic notation, let's resume and finish the switch-and-solve strategy for finding the inverse of $f(x) = b^x$.

Step 1 Replace $f(x)$ with y: $y = b^x$.
Step 2 Interchange x and y: $x = b^y$.
Step 3 Solve for y: $y = \log_b x$.
Step 4 Replace y with $f^{-1}(x)$: $f^{-1}(x) = \log_b x$.

The completed switch-and-solve strategy illustrates that if $f(x) = b^x$, then $f^{-1}(x) = \log_b x$. The inverse of an exponential function is the logarithmic function with the same base.

In Section 1.8, we saw how inverse functions "undo" one another. In particular,

$$f(f^{-1}(x)) = x \quad \text{and} \quad f^{-1}(f(x)) = x.$$

Applying these relationships to exponential and logarithmic functions, we obtain the following **inverse properties of logarithms**:

Inverse Properties of Logarithms

For $b > 0$ and $b \neq 1$,

$\log_b b^x = x$ The logarithm with base b of b raised to a power equals that power.

$b^{\log_b x} = x$. b raised to the logarithm with base b of a number equals that number.

EXAMPLE 5 Using Inverse Properties of Logarithms

Evaluate:

a. $\log_4 4^5$ b. $6^{\log_6 9}$.

SOLUTION

a. Because $\log_b b^x = x$, we conclude $\log_4 4^5 = 5$.
b. Because $b^{\log_b x} = x$, we conclude $6^{\log_6 9} = 9$.

● ● ●

◯ Check Point 5 Evaluate:

a. $\log_7 7^8$ b. $3^{\log_3 17}$.

⑤ Graph logarithmic functions.

Graphs of Logarithmic Functions

How do we graph logarithmic functions? We use the fact that a logarithmic function is the inverse of an exponential function. This means that the logarithmic function reverses the coordinates of the exponential function. It also means that the graph of the logarithmic function is a reflection of the graph of the exponential function about the line $y = x$.

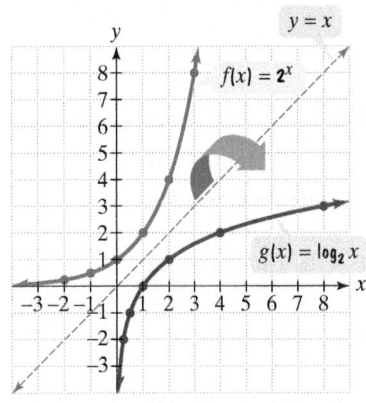

FIGURE 4.7 The graphs of $f(x) = 2^x$ and its inverse function

EXAMPLE 6 Graphs of Exponential and Logarithmic Functions

Graph $f(x) = 2^x$ and $g(x) = \log_2 x$ in the same rectangular coordinate system.

SOLUTION

We first set up a table of coordinates for $f(x) = 2^x$. Reversing these coordinates gives the coordinates for the inverse function $g(x) = \log_2 x$.

x	−2	−1	0	1	2	3
$f(x) = 2^x$	$\frac{1}{4}$	$\frac{1}{2}$	1	2	4	8

Reverse coordinates.

x	$\frac{1}{4}$	$\frac{1}{2}$	1	2	4	8
$g(x) = \log_2 x$	−2	−1	0	1	2	3

We now plot the ordered pairs from each table, connecting them with smooth curves. **Figure 4.7** shows the graphs of $f(x) = 2^x$ and its inverse function $g(x) = \log_2 x$. The graph of the inverse can also be drawn by reflecting the graph of $f(x) = 2^x$ about the line $y = x$. ● ● ●

GREAT QUESTION!

You found the coordinates of $g(x) = \log_2 x$ by reversing the coordinates of $f(x) = 2^x$. Do I have to do it that way?

Not necessarily. You can obtain a partial table of coordinates for $g(x) = \log_2 x$ without having to obtain and reverse coordinates for $f(x) = 2^x$. Because $g(x) = \log_2 x$ means $2^{g(x)} = x$, we begin with values for $g(x)$ and compute corresponding values for x:

Use $x = 2^{g(x)}$ to compute x. For example, if $g(x) = -2$, $x = 2^{-2} = \frac{1}{2^2} = \frac{1}{4}$.

Start with values for $g(x)$.

x	$\frac{1}{4}$	$\frac{1}{2}$	1	2	4	8
$g(x) = \log_2 x$	−2	−1	0	1	2	3

✓ Check Point **6** Graph $f(x) = 3^x$ and $g(x) = \log_3 x$ in the same rectangular coordinate system.

Figure 4.8 illustrates the relationship between the graph of an exponential function, shown in blue, and its inverse, a logarithmic function, shown in red, for bases greater than 1 and for bases between 0 and 1. Also shown and labeled are the exponential function's horizontal asymptote ($y = 0$) and the logarithmic function's vertical asymptote ($x = 0$).

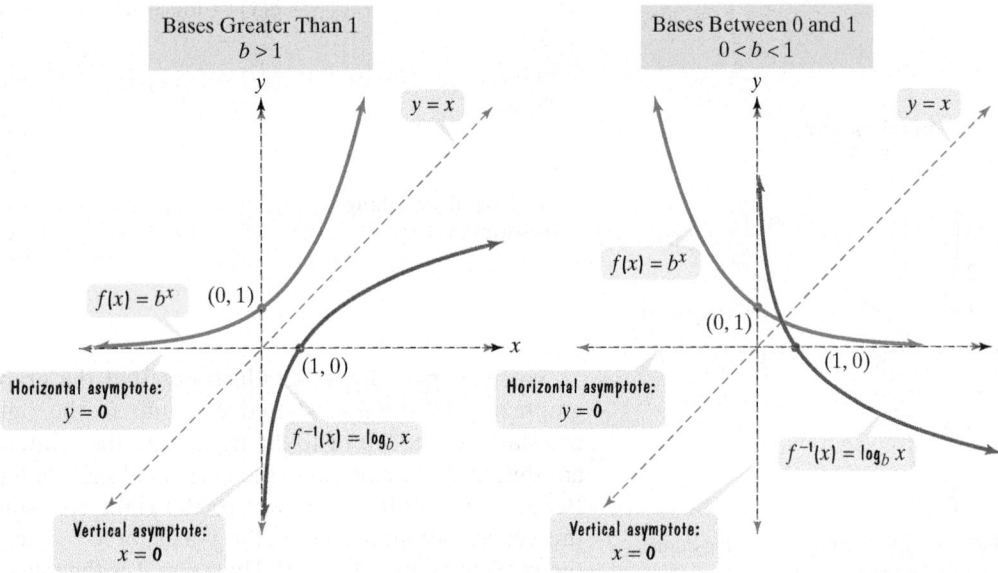

FIGURE 4.8 Graphs of exponential and logarithmic functions

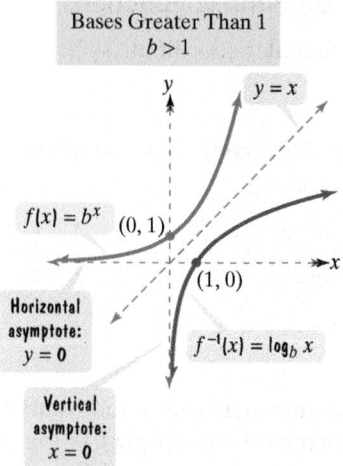

Bases Greater Than 1
$b > 1$

Horizontal asymptote: $y = 0$

Vertical asymptote: $x = 0$

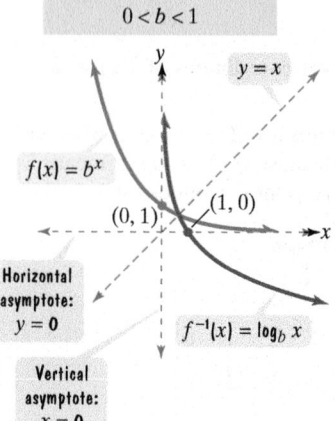

Bases Between 0 and 1
$0 < b < 1$

Horizontal asymptote: $y = 0$

Vertical asymptote: $x = 0$

FIGURE 4.8 (repeated)

The red graphs in **Figure 4.8** illustrate the following general characteristics of logarithmic functions:

> **Characteristics of Logarithmic Functions of the Form** $f(x) = \log_b x$
>
> 1. The domain of $f(x) = \log_b x$ consists of all positive real numbers: $(0, \infty)$. The range of $f(x) = \log_b x$ consists of all real numbers: $(-\infty, \infty)$.
> 2. The graphs of all logarithmic functions of the form $f(x) = \log_b x$ pass through the point $(1, 0)$ because $f(1) = \log_b 1 = 0$. The x-intercept is 1. There is no y-intercept.
> 3. If $b > 1$, $f(x) = \log_b x$ has a graph that goes up to the right and is an increasing function.
> 4. If $0 < b < 1$, $f(x) = \log_b x$ has a graph that goes down to the right and is a decreasing function.
> 5. The graph of $f(x) = \log_b x$ approaches, but does not touch, the y-axis. The y-axis, or $x = 0$, is a vertical asymptote.

The graphs of logarithmic functions can be translated vertically or horizontally, reflected, stretched, or shrunk. These transformations are summarized in **Table 4.5**.

Table 4.5 Transformations Involving Logarithmic Functions

In each case, c represents a positive real number.

Transformation	Equation	Description
Vertical translation	$g(x) = \log_b x + c$	• Shifts the graph of $f(x) = \log_b x$ upward c units.
	$g(x) = \log_b x - c$	• Shifts the graph of $f(x) = \log_b x$ downward c units.
Horizontal translation	$g(x) = \log_b(x + c)$	• Shifts the graph of $f(x) = \log_b x$ to the left c units. Vertical asymptote: $x = -c$
	$g(x) = \log_b(x - c)$	• Shifts the graph of $f(x) = \log_b x$ to the right c units. Vertical asymptote: $x = c$
Reflection	$g(x) = -\log_b x$	• Reflects the graph of $f(x) = \log_b x$ about the x-axis.
	$g(x) = \log_b(-x)$	• Reflects the graph of $f(x) = \log_b x$ about the y-axis.
Vertical stretching or shrinking	$g(x) = c \log_b x$	• Vertically stretches the graph of $f(x) = \log_b x$ if $c > 1$. • Vertically shrinks the graph of $f(x) = \log_b x$ if $0 < c < 1$.
Horizontal stretching or shrinking	$g(x) = \log_b(cx)$	• Horizontally shrinks the graph of $f(x) = \log_b x$ if $c > 1$. • Horizontally stretches the graph of $f(x) = \log_b x$ if $0 < c < 1$.

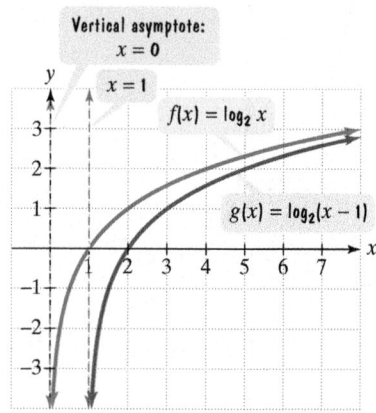

FIGURE 4.9 Shifting $f(x) = \log_2 x$ one unit to the right

For example, **Figure 4.9** illustrates that the graph of $g(x) = \log_2(x - 1)$ is the graph of $f(x) = \log_2 x$ shifted one unit to the right. If a logarithmic function is translated to the left or to the right, both the x-intercept and the vertical asymptote are shifted by the amount of the horizontal shift. In **Figure 4.9**, the x-intercept of f is 1. Because g is shifted one unit to the right, its x-intercept is 2. Also observe that the vertical asymptote for f, the y-axis, or $x = 0$, is shifted one unit to the right for the vertical asymptote for g. Thus, $x = 1$ is the vertical asymptote for g.

Here are some other examples of transformations of graphs of logarithmic functions:

- The graph of $g(x) = 3 + \log_4 x$ is the graph of $f(x) = \log_4 x$ shifted up three units, shown in **Figure 4.10**.
- The graph of $h(x) = -\log_2 x$ is the graph of $f(x) = \log_2 x$ reflected about the x-axis, shown in **Figure 4.11**.
- The graph of $r(x) = \log_2(-x)$ is the graph of $f(x) = \log_2 x$ reflected about the y-axis, shown in **Figure 4.12**.

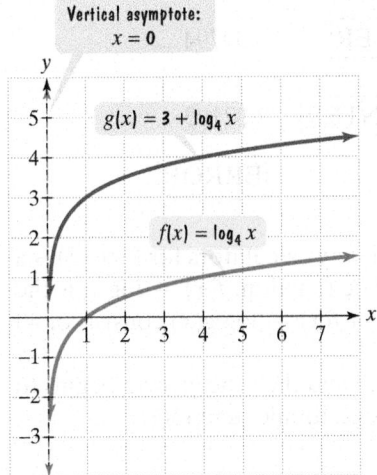

FIGURE 4.10 Shifting vertically up three units

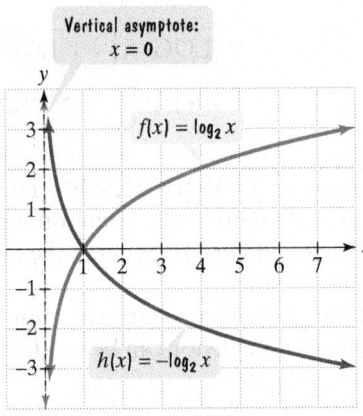

FIGURE 4.11 Reflection about the x-axis

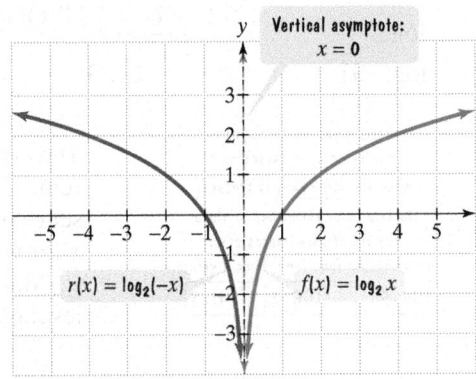

FIGURE 4.12 Reflection about the y-axis

⑥ Find the domain of a logarithmic function.

The Domain of a Logarithmic Function

In Section 4.1, we learned that the domain of an exponential function of the form $f(x) = b^x$ includes all real numbers and its range is the set of positive real numbers. Because the logarithmic function reverses the domain and the range of the exponential function, the **domain of a logarithmic function of the form $f(x) = \log_b x$ is the set of all positive real numbers**. Thus, $\log_2 8$ is defined because the value of x in the logarithmic expression, 8, is greater than zero and therefore is included in the domain of the logarithmic function $f(x) = \log_2 x$. However, $\log_2 0$ and $\log_2(-8)$ are not defined because 0 and -8 are not positive real numbers and therefore are excluded from the domain of the logarithmic function $f(x) = \log_2 x$. In general, **the domain of $f(x) = \log_b g(x)$ consists of all x for which $g(x) > 0$.**

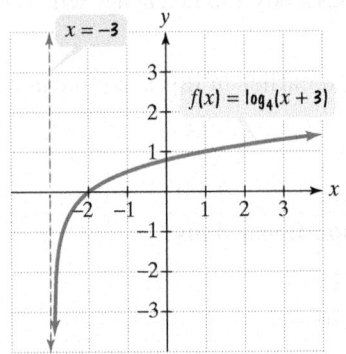

FIGURE 4.13 The domain of $f(x) = \log_4(x + 3)$ is $(-3, \infty)$.

EXAMPLE 7 Finding the Domain of a Logarithmic Function

Find the domain of $f(x) = \log_4(x + 3)$.

SOLUTION

The domain of f consists of all x for which $x + 3 > 0$. Solving this inequality for x, we obtain $x > -3$. Thus, the domain of f is $(-3, \infty)$. This is illustrated in **Figure 4.13**. The vertical asymptote is $x = -3$ and all points on the graph of f have x-coordinates that are greater than -3. ● ● ●

☑ Check Point **7** Find the domain of $f(x) = \log_4(x - 5)$.

 Use common logarithms.

Common Logarithms

The logarithmic function with base 10 is called the **common logarithmic function**. The function $f(x) = \log_{10} x$ is usually expressed as $f(x) = \log x$. A calculator with a [LOG] key can be used to evaluate common logarithms. Here are some examples:

Logarithm	Most Scientific Calculator Keystrokes	Most Graphing Calculator Keystrokes	Display (or Approximate Display)
$\log 1000$	1000 [LOG]	[LOG] 1000 [ENTER]	3
$\log \dfrac{5}{2}$	[(] 5 [÷] 2 [)] [LOG]	[LOG] [(] 5 [÷] 2 [)] [ENTER]	0.39794
$\dfrac{\log 5}{\log 2}$	5 [LOG] [÷] 2 [LOG] [=]	[LOG] 5 [÷] [LOG] 2 [ENTER]	2.32193
$\log(-3)$	3 [+/−] [LOG]	[LOG] [(−)] 3 [ENTER]	[ERROR]

Some graphing calculators display an open parenthesis when the [LOG] key is pressed. In this case, remember to close the set of parentheses after entering the function's domain value: [LOG] 5 [)] [÷] [LOG] 2 [)] [ENTER].

The error message or [NONREAL ANS] given by many calculators for $\log(-3)$ is a reminder that the domain of the common logarithmic function, $f(x) = \log x$, is the set of positive real numbers. In general, the domain of $f(x) = \log g(x)$ consists of all x for which $g(x) > 0$.

Many real-life phenomena start with rapid growth and then the growth begins to level off. This type of behavior can be modeled by logarithmic functions.

EXAMPLE 8 Modeling Height of Children

The percentage of adult height attained by a boy who is x years old can be modeled by

$$f(x) = 29 + 48.8 \log(x + 1),$$

where x represents the boy's age (from 5 to 15) and $f(x)$ represents the percentage of his adult height. Approximately what percentage of his adult height has a boy attained at age eight?

SOLUTION

We substitute the boy's age, 8, for x and evaluate the function.

$$f(x) = 29 + 48.8 \log(x + 1)$$ This is the given function.
$$f(8) = 29 + 48.8 \log(8 + 1)$$ Substitute 8 for x.
$$= 29 + 48.8 \log 9$$ Graphing calculator keystrokes:
 29 [+] 48.8 [LOG] 9 [ENTER]
$$\approx 76$$

Thus, an 8-year-old boy has attained approximately 76% of his adult height. • • •

✓ **Check Point 8** Use the function in Example 8 to answer this question: Approximately what percentage of his adult height has a boy attained at age ten?

The basic properties of logarithms that were listed earlier in this section can be applied to common logarithms.

Properties of Common Logarithms

General Properties	Common Logarithms
1. $\log_b 1 = 0$	**1.** $\log 1 = 0$
2. $\log_b b = 1$	**2.** $\log 10 = 1$
3. $\log_b b^x = x$ Inverse	**3.** $\log 10^x = x$
4. $b^{\log_b x} = x$ properties	**4.** $10^{\log x} = x$

The property $\log 10^x = x$ can be used to evaluate common logarithms involving powers of 10. For example,

$$\log 100 = \log 10^2 = 2, \quad \log 1000 = \log 10^3 = 3, \quad \text{and} \quad \log 10^{7.1} = 7.1.$$

EXAMPLE 9 Earthquake Intensity

The magnitude, R, on the Richter scale of an earthquake of intensity I is given by

$$R = \log\frac{I}{I_0},$$

where I_0 is the intensity of a barely felt zero-level earthquake. The earthquake that destroyed San Francisco in 1906 was $10^{8.3}$ times as intense as a zero-level earthquake. What was its magnitude on the Richter scale?

SOLUTION

Because the earthquake was $10^{8.3}$ times as intense as a zero-level earthquake, the intensity, I, is $10^{8.3}I_0$.

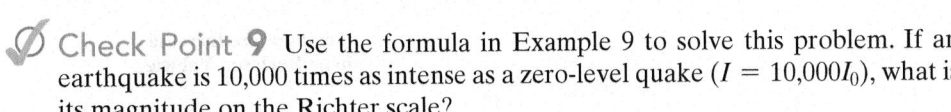

$$R = \log\frac{I}{I_0} \qquad \text{This is the formula for magnitude on the Richter scale.}$$

$$R = \log\frac{10^{8.3}I_0}{I_0} \qquad \text{Substitute } 10^{8.3}I_0 \text{ for } I.$$

$$= \log 10^{8.3} \qquad \text{Simplify.}$$

$$= 8.3 \qquad \text{Use the property } \log 10^x = x.$$

San Francisco's 1906 earthquake registered 8.3 on the Richter scale. • • •

Check Point **9** Use the formula in Example 9 to solve this problem. If an earthquake is 10,000 times as intense as a zero-level quake ($I = 10,000I_0$), what is its magnitude on the Richter scale?

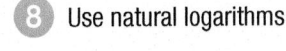

8 Use natural logarithms.

Natural Logarithms

The logarithmic function with base e is called the **natural logarithmic function**. The function $f(x) = \log_e x$ is usually expressed as $f(x) = \ln x$, read "el en of x." A calculator with an $\boxed{\text{LN}}$ key can be used to evaluate natural logarithms. Keystrokes are identical to those shown for common logarithmic evaluations on page 472.

Like the domain of all logarithmic functions, the domain of the natural logarithmic function $f(x) = \ln x$ is the set of all positive real numbers. Thus, the domain of $f(x) = \ln g(x)$ consists of all x for which $g(x) > 0$.

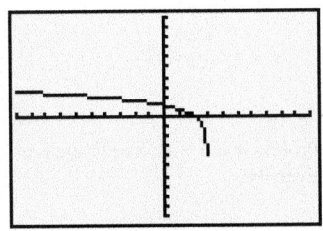

[–10, 10, 1] by [–10, 10, 1]

FIGURE 4.14 The domain of $f(x) = \ln(3 - x)$ is $(-\infty, 3)$.

EXAMPLE 10 Finding Domains of Natural Logarithmic Functions

Find the domain of each function:

a. $f(x) = \ln(3 - x)$ **b.** $h(x) = \ln(x - 3)^2.$

SOLUTION

a. The domain of f consists of all x for which $3 - x > 0$. Solving this inequality for x, we obtain $x < 3$. Thus, the domain of f is $\{x \mid x < 3\}$ or $(-\infty, 3)$. This is verified by the graph in **Figure 4.14**.

b. The domain of h consists of all x for which $(x - 3)^2 > 0$. It follows that the domain of h is all real numbers except 3. Thus, the domain of h is $\{x \mid x \neq 3\}$ or $(-\infty, 3) \cup (3, \infty)$. This is shown by the graph in **Figure 4.15**. To make it more obvious that 3 is excluded from the domain, we used a $\boxed{\text{DOT}}$ format. • • •

[–10, 10, 1] by [–10, 10, 1]

FIGURE 4.15 3 is excluded from the domain of $h(x) = \ln(x - 3)^2$.

Check Point **10** Find the domain of each function:

a. $f(x) = \ln(4 - x)$ **b.** $h(x) = \ln x^2.$

The basic properties of logarithms that were listed earlier in this section can be applied to natural logarithms.

Properties of Natural Logarithms

General Properties	**Natural Logarithms**
1. $\log_b 1 = 0$	**1.** $\ln 1 = 0$
2. $\log_b b = 1$	**2.** $\ln e = 1$
3. $\log_b b^x = x$ — Inverse	**3.** $\ln e^x = x$
4. $b^{\log_b x} = x$ — properties	**4.** $e^{\ln x} = x$

Examine the inverse properties, $\ln e^x = x$ and $e^{\ln x} = x$. Can you see how \ln and e "undo" one another? For example,

$$\ln e^2 = 2, \quad \ln e^{7x^2} = 7x^2, \quad e^{\ln 2} = 2, \quad \text{and} \quad e^{\ln 7x^2} = 7x^2.$$

EXAMPLE 11 Dangerous Heat: Temperature in an Enclosed Vehicle

When the outside air temperature is anywhere from 72° to 96° Fahrenheit, the temperature in an enclosed vehicle climbs by 43° in the first hour. The bar graph in **Figure 4.16** shows the temperature increase throughout the hour. The function

$$f(x) = 13.4 \ln x - 11.6$$

models the temperature increase, $f(x)$, in degrees Fahrenheit, after x minutes. Use the function to find the temperature increase, to the nearest degree, after 50 minutes. How well does the function model the actual increase shown in **Figure 4.16**?

Temperature Increase in an Enclosed Vehicle

FIGURE 4.16
Source: Professor Jan Null, San Francisco State University

SOLUTION

We find the temperature increase after 50 minutes by substituting 50 for x and evaluating the function at 50.

$$f(x) = 13.4 \ln x - 11.6 \qquad \text{This is the given function.}$$
$$f(50) = 13.4 \ln 50 - 11.6 \qquad \text{Substitute 50 for x.}$$
$$\approx 41$$

Graphing calculator keystrokes:
13.4 |ln| 50 |−| 11.6 |ENTER|. On some calculators, a parenthesis is needed after 50.

According to the function, the temperature will increase by approximately 41° after 50 minutes. Because the increase shown in **Figure 4.16** is 41°, the function models the actual increase extremely well. • • •

⊘ Check Point **11** Use the function in Example 11 to find the temperature increase, to the nearest degree, after 30 minutes. How well does the function model the actual increase shown in **Figure 4.16**?

Blitzer Bonus ‖ The Curious Number e

FIGURE 4.17

You will learn more about each curiosity mentioned below if you take calculus.

- The number e was named by the Swiss mathematician Leonhard Euler (1707–1783), who proved that it is the limit as $n \to \infty$ of $\left(1 + \dfrac{1}{n}\right)^n$.
- e features in Euler's remarkable relationship $e^{i\pi} = -1$, in which $i = \sqrt{-1}$.
- The first few decimal places of e are fairly easy to remember:
 $e = 2.7\ 1828\ 1828\ 45\ 90\ 45\ldots.$
- The best approximation of e using numbers less than 1000 is also easy to remember:
 $e \approx \dfrac{878}{323} \approx 2.71826\ldots.$
- Isaac Newton (1642–1727), one of the cofounders of calculus, showed that

$$e^x = 1 + x + \frac{x^2}{2!} + \frac{x^3}{3!} + \frac{x^4}{4!} + \cdots,$$ from which we obtain $e = 1 + 1 + \dfrac{1}{2!} + \dfrac{1}{3!} + \dfrac{1}{4!} + \cdots,$ an infinite

sum suitable for calculation because its terms decrease so rapidly. (*Note:* $n!$ (n factorial) is the product of all the consecutive integers from n down to 1: $n! = n(n-1)(n-2)(n-3)\cdots \cdot 3 \cdot 2 \cdot 1$.)

- The area of the region bounded by $y = \dfrac{1}{x}$, the x-axis, $x = 1$, and $x = t$ (shaded in **Figure 4.17**) is a function of

t, designated by $A(t)$. Grégoire de Saint-Vincent, a Belgian Jesuit (1584–1667), spent his entire professional life attempting to find a formula for $A(t)$. With his student, he showed that $A(t) = \ln t$, becoming one of the first mathematicians to make use of the logarithmic function for something other than a computational device.

CONCEPT AND VOCABULARY CHECK

Fill in each blank so that the resulting statement is true.

1. $y = \log_b x$ is equivalent to the exponential form _____, $x > 0, b > 0, b \neq 1$.
2. The function $f(x) = \log_b x$ is the _____ function with base _____.
3. $\log_b b = $ _____
4. $\log_b 1 = $ _____
5. $\log_b b^x = $ _____
6. $b^{\log_b x} = $ _____
7. Using interval notation, the domain of $f(x) = \log_b x$ is _____ and the range is _____.
8. The graph of $f(x) = \log_b x$ approaches, but does not touch, the _____-axis. This axis, whose equation is _____, is a/an _____ asymptote.
9. The graph of $g(x) = 5 + \log_2 x$ is the graph of $f(x) = \log_2 x$ shifted _____.

10. The graph of $g(x) = \log_3(x + 5)$ is the graph of $f(x) = \log_3 x$ shifted _____.
11. The graph of $g(x) = -\log_4 x$ is the graph of $f(x) = \log_4 x$ reflected about the _____.
12. The graph of $g(x) = \log_5(-x)$ is the graph of $f(x) = \log_5 x$ reflected about the _____.
13. The domain of $g(x) = \log_2(5 - x)$ can be found by solving the inequality _____.
14. The logarithmic function with base 10 is called the _____ logarithmic function. The function $f(x) = \log_{10} x$ is usually expressed as $f(x) = $ _____.
15. The logarithmic function with base e is called the _____ logarithmic function. The function $f(x) = \log_e x$ is usually expressed as $f(x) = $ _____.

EXERCISE SET 4.2

Practice Exercises

In Exercises 1–8, write each equation in its equivalent exponential form.

1. $4 = \log_2 16$
2. $6 = \log_2 64$
3. $2 = \log_3 x$
4. $2 = \log_9 x$
5. $5 = \log_b 32$
6. $3 = \log_b 27$
7. $\log_6 216 = y$
8. $\log_5 125 = y$

In Exercises 9–20, write each equation in its equivalent logarithmic form.

9. $2^3 = 8$
10. $5^4 = 625$
11. $2^{-4} = \frac{1}{16}$
12. $5^{-3} = \frac{1}{125}$
13. $\sqrt[3]{8} = 2$
14. $\sqrt[3]{64} = 4$
15. $13^2 = x$
16. $15^2 = x$
17. $b^3 = 1000$
18. $b^3 = 343$
19. $7^y = 200$
20. $8^y = 300$

In Exercises 21–42, evaluate each expression without using a calculator.

21. $\log_4 16$ **22.** $\log_7 49$ **23.** $\log_2 64$ **24.** $\log_3 27$

25. $\log_5 \frac{1}{5}$ **26.** $\log_6 \frac{1}{6}$ **27.** $\log_2 \frac{1}{8}$ **28.** $\log_3 \frac{1}{9}$

29. $\log_7 \sqrt{7}$ **30.** $\log_6 \sqrt{6}$ **31.** $\log_2 \frac{1}{\sqrt{2}}$ **32.** $\log_3 \frac{1}{\sqrt{3}}$

33. $\log_{64} 8$ **34.** $\log_{81} 9$ **35.** $\log_5 5$ **36.** $\log_{11} 11$

37. $\log_4 1$ **38.** $\log_6 1$ **39.** $\log_5 5^7$ **40.** $\log_4 4^6$

41. $8^{\log_8 19}$ **42.** $7^{\log_7 23}$

43. Graph $f(x) = 4^x$ and $g(x) = \log_4 x$ in the same rectangular coordinate system.

44. Graph $f(x) = 5^x$ and $g(x) = \log_5 x$ in the same rectangular coordinate system.

45. Graph $f(x) = \left(\frac{1}{2}\right)^x$ and $g(x) = \log_{\frac{1}{2}} x$ in the same rectangular coordinate system.

46. Graph $f(x) = \left(\frac{1}{4}\right)^x$ and $g(x) = \log_{\frac{1}{4}} x$ in the same rectangular coordinate system.

In Exercises 47–52, the graph of a logarithmic function is given. Select the function for each graph from the following options:

$$f(x) = \log_3 x, \, g(x) = \log_3(x - 1), \, h(x) = \log_3 x - 1,$$

$$F(x) = -\log_3 x, \, G(x) = \log_3(-x), \, H(x) = 1 - \log_3 x.$$

47.

48.

49.

50.

51.

52.
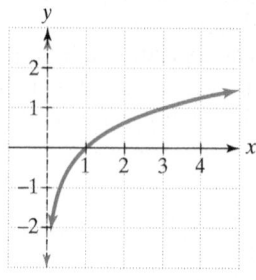

In Exercises 53–58, begin by graphing $f(x) = \log_2 x$. Then use transformations of this graph to graph the given function. What is the vertical asymptote? Use the graphs to determine each function's domain and range.

53. $g(x) = \log_2(x + 1)$ **54.** $g(x) = \log_2(x + 2)$

55. $h(x) = 1 + \log_2 x$ **56.** $h(x) = 2 + \log_2 x$

57. $g(x) = \frac{1}{2}\log_2 x$ **58.** $g(x) = -2\log_2 x$

The figure shows the graph of $f(x) = \log x$. In Exercises 59–64, use transformations of this graph to graph each function. Graph and give equations of the asymptotes. Use the graphs to determine each function's domain and range.

59. $g(x) = \log(x - 1)$ **60.** $g(x) = \log(x - 2)$

61. $h(x) = \log x - 1$ **62.** $h(x) = \log x - 2$

63. $g(x) = 1 - \log x$ **64.** $g(x) = 2 - \log x$

The figure shows the graph of $f(x) = \ln x$. In Exercises 65–74, use transformations of this graph to graph each function. Graph and give equations of the asymptotes. Use the graphs to determine each function's domain and range.

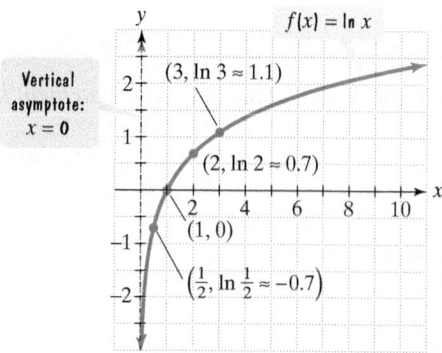

65. $g(x) = \ln(x + 2)$ **66.** $g(x) = \ln(x + 1)$

67. $h(x) = \ln(2x)$ **68.** $h(x) = \ln\left(\frac{1}{2}x\right)$

69. $g(x) = 2\ln x$ **70.** $g(x) = \frac{1}{2}\ln x$

71. $h(x) = -\ln x$ **72.** $h(x) = \ln(-x)$

73. $g(x) = 2 - \ln x$ **74.** $g(x) = 1 - \ln x$

In Exercises 75–80, find the domain of each logarithmic function.

75. $f(x) = \log_5(x + 4)$ **76.** $f(x) = \log_5(x + 6)$

77. $f(x) = \log(2 - x)$ **78.** $f(x) = \log(7 - x)$

79. $f(x) = \ln(x - 2)^2$ **80.** $f(x) = \ln(x - 7)^2$

In Exercises 81–100, evaluate or simplify each expression without using a calculator.

81. $\log 100$ **82.** $\log 1000$ **83.** $\log 10^7$ **84.** $\log 10^8$

85. $10^{\log 33}$ **86.** $10^{\log 53}$ **87.** $\ln 1$ **88.** $\ln e$

89. $\ln e^6$ **90.** $\ln e^7$ **91.** $\ln \frac{1}{e^6}$ **92.** $\ln \frac{1}{e^7}$

93. $e^{\ln 125}$ **94.** $e^{\ln 300}$ **95.** $\ln e^{9x}$ **96.** $\ln e^{13x}$

97. $e^{\ln 5x^2}$ **98.** $e^{\ln 7x^2}$ **99.** $10^{\log \sqrt{x}}$ **100.** $10^{\log \sqrt[3]{x}}$

Practice Plus

In Exercises 101–104, write each equation in its equivalent exponential form. Then solve for x.

101. $\log_3(x - 1) = 2$

102. $\log_5(x + 4) = 2$

103. $\log_4 x = -3$

104. $\log_{64} x = \dfrac{2}{3}$

In Exercises 105–108, evaluate each expression without using a calculator.

105. $\log_3(\log_7 7)$

106. $\log_5(\log_2 32)$

107. $\log_2(\log_3 81)$

108. $\log(\ln e)$

In Exercises 109–112, find the domain of each logarithmic function.

109. $f(x) = \ln(x^2 - x - 2)$

110. $f(x) = \ln(x^2 - 4x - 12)$

111. $f(x) = \log\left(\dfrac{x + 1}{x - 5}\right)$

112. $f(x) = \log\left(\dfrac{x - 2}{x + 5}\right)$

Application Exercises

The percentage of adult height attained by a girl who is x years old can be modeled by

$$f(x) = 62 + 35 \log(x - 4),$$

where x represents the girl's age (from 5 to 15) and f(x) represents the percentage of her adult height. Use the function to solve Exercises 113–114. Round answers to the nearest tenth of a percent.

113. Approximately what percentage of her adult height has a girl attained at age 13?

114. Approximately what percentage of her adult height has a girl attained at age ten?

The bar graph indicates that the percentage of first-year college students expressing antifeminist views declined after 1970. Use this information to solve Exercises 115–116.

Opposition to Feminism among First-Year United States College Students, 1970–2008

Statement: "The activities of married women are best confined to the home and family."

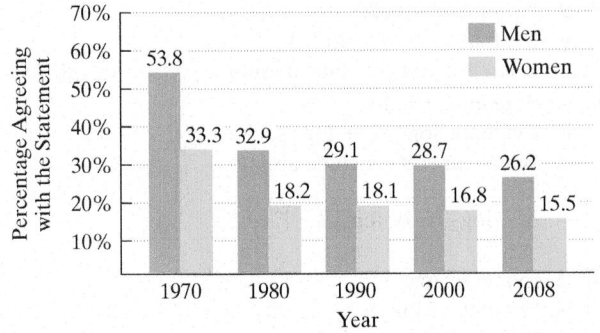

Source: John Macionis, *Sociology, Thirteenth Edition*, Prentice Hall, 2010.

115. The function

$$f(x) = -7.52 \ln x + 53$$

models the percentage of first-year college men, $f(x)$, expressing antifeminist views (by agreeing with the statement) x years after 1969.

a. Use the function to find the percentage of first-year college men expressing antifeminist views in 2008. Round to one decimal place. Does this function value overestimate or underestimate the percentage displayed by the graph? By how much?

b. Use the function to project the percentage of first-year college men who will express antifeminist views in 2015. Round to one decimal place.

116. The function

$$f(x) = -4.82 \ln x + 32.5$$

models the percentage of first-year college women, $f(x)$, expressing antifeminist views (by agreeing with the statement) x years after 1969.

a. Use the function to find the percentage of first-year college women expressing antifeminist views in 2008. Round to one decimal place. Does this function value overestimate or underestimate the percentage displayed by the graph? By how much?

b. Use the function to project the percentage of first-year college women who will express antifeminist views in 2015. Round to one decimal place.

The loudness level of a sound, D, in decibels, is given by the formula

$$D = 10 \log(10^{12} I),$$

where I is the intensity of the sound, in watts per meter². Decibel levels range from 0, a barely audible sound, to 160, a sound resulting in a ruptured eardrum. (Any exposure to sounds of 130 decibels or higher puts a person at immediate risk for hearing damage.) Use the formula to solve Exercises 117–118.

117. The sound of a blue whale can be heard 500 miles away, reaching an intensity of 6.3×10^6 watts per meter². Determine the decibel level of this sound. At close range, can the sound of a blue whale rupture the human eardrum?

118. What is the decibel level of a normal conversation, 3.2×10^{-6} watt per meter²?

119. Students in a psychology class took a final examination. As part of an experiment to see how much of the course content they remembered over time, they took equivalent forms of the exam in monthly intervals thereafter. The average score for the group, $f(t)$, after t months was modeled by the function

$$f(t) = 88 - 15 \ln(t + 1), \qquad 0 \le t \le 12.$$

a. What was the average score on the original exam?

b. What was the average score after 2 months? 4 months? 6 months? 8 months? 10 months? one year?

c. Sketch the graph of f (either by hand or with a graphing utility). Describe what the graph indicates in terms of the material retained by the students.

Writing in Mathematics

120. Describe the relationship between an equation in logarithmic form and an equivalent equation in exponential form.

121. What question can be asked to help evaluate $\log_3 81$?

122. Explain why the logarithm of 1 with base b is 0.

123. Describe the following property using words: $\log_b b^x = x$.

124. Explain how to use the graph of $f(x) = 2^x$ to obtain the graph of $g(x) = \log_2 x$.

125. Explain how to find the domain of a logarithmic function.

126. Logarithmic models are well suited to phenomena in which growth is initially rapid but then begins to level off. Describe something that is changing over time that can be modeled using a logarithmic function.

127. Suppose that a girl is 4 feet 6 inches at age 10. Explain how to use the function in Exercises 113–114 to determine how tall she can expect to be as an adult.

Technology Exercises

In Exercises 128–131, graph f and g in the same viewing rectangle. Then describe the relationship of the graph of g to the graph of f.

128. $f(x) = \ln x, g(x) = \ln(x + 3)$

129. $f(x) = \ln x, g(x) = \ln x + 3$

130. $f(x) = \log x, g(x) = -\log x$

131. $f(x) = \log x, g(x) = \log(x - 2) + 1$

132. Students in a mathematics class took a final examination. They took equivalent forms of the exam in monthly intervals thereafter. The average score, $f(t)$, for the group after t months was modeled by the human memory function $f(t) = 75 - 10\log(t + 1)$, where $0 \le t \le 12$. Use a graphing utility to graph the function. Then determine how many months elapsed before the average score fell below 65.

133. In parts (a)–(c), graph f and g in the same viewing rectangle.
 a. $f(x) = \ln(3x), g(x) = \ln 3 + \ln x$
 b. $f(x) = \log(5x^2), g(x) = \log 5 + \log x^2$
 c. $f(x) = \ln(2x^3), g(x) = \ln 2 + \ln x^3$
 d. Describe what you observe in parts (a)–(c). Generalize this observation by writing an equivalent expression for $\log_b(MN)$, where $M > 0$ and $N > 0$.
 e. Complete this statement: The logarithm of a product is equal to _____.

134. Graph each of the following functions in the same viewing rectangle and then place the functions in order from the one that increases most slowly to the one that increases most rapidly.
$$y = x, y = \sqrt{x}, y = e^x, y = \ln x, y = x^x, y = x^2$$

Critical Thinking Exercises

Make Sense? *In Exercises 135–138, determine whether each statement makes sense or does not make sense, and explain your reasoning.*

135. I've noticed that exponential functions and logarithmic functions exhibit inverse, or opposite, behavior in many ways. For example, a vertical translation shifts an exponential function's horizontal asymptote and a horizontal translation shifts a logarithmic function's vertical asymptote.

136. I estimate that $\log_8 16$ lies between 1 and 2 because $8^1 = 8$ and $8^2 = 64$.

137. I can evaluate some common logarithms without having to use a calculator.

138. An earthquake of magnitude 8 on the Richter scale is twice as intense as an earthquake of magnitude 4.

In Exercises 139–142, determine whether each statement is true or false. If the statement is false, make the necessary change(s) to produce a true statement.

139. $\dfrac{\log_2 8}{\log_2 4} = \dfrac{8}{4}$

140. $\log(-100) = -2$

141. The domain of $f(x) = \log_2 x$ is $(-\infty, \infty)$.

142. $\log_b x$ is the exponent to which b must be raised to obtain x.

143. Without using a calculator, find the exact value of
$$\frac{\log_3 81 - \log_\pi 1}{\log_{2\sqrt{2}} 8 - \log 0.001}.$$

144. Without using a calculator, find the exact value of $\log_4[\log_3(\log_2 8)]$.

145. Without using a calculator, determine which is the greater number: $\log_4 60$ or $\log_3 40$.

Group Exercise

146. This group exercise involves exploring the way we grow. Group members should create a graph for the function that models the percentage of adult height attained by a boy who is x years old, $f(x) = 29 + 48.8\log(x + 1)$. Let $x = 5, 6, 7, \ldots, 15$, find function values, and connect the resulting points with a smooth curve. Then create a graph for the function that models the percentage of adult height attained by a girl who is x years old, $g(x) = 62 + 35\log(x - 4)$. Let $x = 5, 6, 7, \ldots, 15$, find function values, and connect the resulting points with a smooth curve. Group members should then discuss similarities and differences in the growth patterns for boys and girls based on the graphs.

Preview Exercises

Exercises 147–149 will help you prepare for the material covered in the next section. In each exercise, evaluate the indicated logarithmic expressions without using a calculator.

147. a. Evaluate: $\log_2 32$.
 b. Evaluate: $\log_2 8 + \log_2 4$.
 c. What can you conclude about $\log_2 32$, or $\log_2(8 \cdot 4)$?

148. a. Evaluate: $\log_2 16$.
 b. Evaluate: $\log_2 32 - \log_2 2$.
 c. What can you conclude about
$$\log_2 16, \text{ or } \log_2\left(\frac{32}{2}\right)?$$

149. a. Evaluate: $\log_3 81$.
 b. Evaluate: $2\log_3 9$.
 c. What can you conclude about
$$\log_3 81, \text{ or } \log_3 9^2?$$

SECTION 4.3

Properties of Logarithms

Objectives

1. Use the product rule.
2. Use the quotient rule.
3. Use the power rule.
4. Expand logarithmic expressions.
5. Condense logarithmic expressions.
6. Use the change-of-base property.

We all learn new things in different ways. In this section, we consider important properties of logarithms. What would be the most effective way for you to learn these properties? Would it be helpful to use your graphing utility and discover one of these properties for yourself? To do so, work Exercise 133 in Exercise Set 4.2 before continuing. Would it be helpful to evaluate certain logarithmic expressions that suggest three of the properties? If this is the case, work Preview Exercises 147–149 in Exercise Set 4.2 before continuing. Would the properties become more meaningful if you could see exactly where they come from? If so, you will find details of the proofs of many of these properties in the appendix. The remainder of our work in this chapter will be based on the properties of logarithms that you learn in this section.

① Use the product rule.

DISCOVERY

We know that log 100,000 = 5. Show that you get the same result by writing 100,000 as 1000 · 100 and then using the product rule. Then verify the product rule by using other numbers whose logarithms are easy to find.

The Product Rule

Properties of exponents correspond to properties of logarithms. For example, when we multiply with the same base, we add exponents:

$$b^m \cdot b^n = b^{m+n}.$$

This property of exponents, coupled with an awareness that a logarithm is an exponent, suggests the following property, called the **product rule:**

> **The Product Rule**
>
> Let b, M, and N be positive real numbers with $b \neq 1$.
> $$\log_b(MN) = \log_b M + \log_b N$$
> The logarithm of a product is the sum of the logarithms.

You can find the proof of the product rule in the appendix.

When we use the product rule to write a single logarithm as the sum of two logarithms, we say that we are **expanding a logarithmic expression**. For example, we can use the product rule to expand $\ln(7x)$:

$$\ln(7x) \ = \ \ln 7 \ + \ \ln x.$$

The logarithm of a product — is — the sum of the logarithms.

EXAMPLE 1 Using the Product Rule

Use the product rule to expand each logarithmic expression:

a. $\log_4(7 \cdot 5)$ **b.** $\log(10x)$.

SOLUTION

a. $\log_4(7 \cdot 5) = \log_4 7 + \log_4 5$ The logarithm of a product is the sum of the logarithms.

b. $\log(10x) = \log 10 + \log x$ The logarithm of a product is the sum of the logarithms. These are common logarithms with base 10 understood.

$= 1 + \log x$ Because $\log_b b = 1$, then $\log 10 = 1$. • • •

✓ Check Point **1** Use the product rule to expand each logarithmic expression:

a. $\log_6(7 \cdot 11)$ **b.** $\log(100x)$.

② Use the quotient rule.

The Quotient Rule

When we divide with the same base, we subtract exponents:

$$\frac{b^m}{b^n} = b^{m-n}.$$

This property suggests the following property of logarithms, called the **quotient rule**:

DISCOVERY

We know that $\log_2 16 = 4$. Show that you get the same result by writing 16 as $\dfrac{32}{2}$ and then using the quotient rule. Then verify the quotient rule using other numbers whose logarithms are easy to find.

> **The Quotient Rule**
>
> Let b, M, and N be positive real numbers with $b \neq 1$.
>
> $$\log_b\left(\frac{M}{N}\right) = \log_b M - \log_b N$$
>
> The logarithm of a quotient is the difference of the logarithms.

When we use the quotient rule to write a single logarithm as the difference of two logarithms, we say that we are **expanding a logarithmic expression**. For example, we can use the quotient rule to expand $\log \dfrac{x}{2}$:

$$\log\left(\frac{x}{2}\right) = \log x - \log 2.$$

The logarithm of a quotient is the difference of the logarithms.

EXAMPLE 2 Using the Quotient Rule

Use the quotient rule to expand each logarithmic expression:

a. $\log_7\left(\dfrac{19}{x}\right)$ **b.** $\ln\left(\dfrac{e^3}{7}\right)$.

SOLUTION

a. $\log_7\left(\dfrac{19}{x}\right) = \log_7 19 - \log_7 x$ The logarithm of a quotient is the difference of the logarithms.

b. $\ln\left(\dfrac{e^3}{7}\right) = \ln e^3 - \ln 7$ The logarithm of a quotient is the difference of the logarithms. These are natural logarithms with base e understood.

$= 3 - \ln 7$ Because $\ln e^x = x$, then $\ln e^3 = 3$. • • •

⊘ Check Point **2** Use the quotient rule to expand each logarithmic expression:

a. $\log_8\left(\dfrac{23}{x}\right)$ **b.** $\ln\left(\dfrac{e^5}{11}\right)$.

③ Use the power rule.

The Power Rule

When an exponential expression is raised to a power, we multiply exponents:

$$(b^m)^n = b^{mn}.$$

This property suggests the following property of logarithms, called the **power rule**:

> ### The Power Rule
>
> Let b and M be positive real numbers with $b \neq 1$, and let p be any real number.
> $$\log_b M^p = p \log_b M$$
> The logarithm of a number with an exponent is the product of the exponent and the logarithm of that number.

When we use the power rule to "pull the exponent to the front," we say that we are **expanding a logarithmic expression**. For example, we can use the power rule to expand $\ln x^2$:

$$\ln x^2 = 2 \ln x.$$

| The logarithm of a number with an exponent | is | the product of the exponent and the logarithm of that number. |

Figure 4.18 shows the graphs of $y = \ln x^2$ and $y = 2 \ln x$ in $[-5, 5, 1]$ by $[-5, 5, 1]$ viewing rectangles. Are $\ln x^2$ and $2 \ln x$ the same? The graphs illustrate that $y = \ln x^2$ and $y = 2 \ln x$ have different domains. The graphs are only the same if $x > 0$. Thus, we should write

$$\ln x^2 = 2 \ln x \text{ for } x > 0.$$

 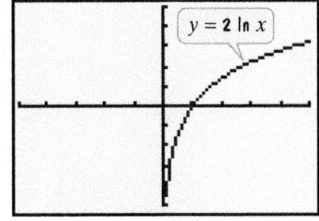

Domain: $(-\infty, 0) \cup (0, \infty)$ Domain: $(0, \infty)$

FIGURE 4.18 $\ln x^2$ and $2 \ln x$ have different domains.

When expanding a logarithmic expression, you might want to determine whether the rewriting has changed the domain of the expression. For the rest of this section, assume that all variables and variable expressions represent positive numbers.

EXAMPLE 3 Using the Power Rule

Use the power rule to expand each logarithmic expression:

a. $\log_5 7^4$ **b.** $\ln \sqrt{x}$ **c.** $\log(4x)^5$.

SOLUTION

a. $\log_5 7^4 = 4 \log_5 7$ The logarithm of a number with an exponent is the exponent times the logarithm of the number.

b. $\ln \sqrt{x} = \ln x^{\frac{1}{2}}$ Rewrite the radical using a rational exponent.

 $= \frac{1}{2} \ln x$ Use the power rule to bring the exponent to the front.

c. $\log(4x)^5 = 5 \log(4x)$ We immediately apply the power rule because the entire variable expression, 4x, is raised to the 5th power. • • •

✓ Check Point **3** Use the power rule to expand each logarithmic expression:

a. $\log_6 3^9$ **b.** $\ln \sqrt[3]{x}$ **c.** $\log(x + 4)^2$.

④ Expand logarithmic expressions.

Expanding Logarithmic Expressions

It is sometimes necessary to use more than one property of logarithms when you expand a logarithmic expression. Properties for expanding logarithmic expressions are as follows:

Properties for Expanding Logarithmic Expressions

For $M > 0$ and $N > 0$:

1. $\log_b(MN) = \log_b M + \log_b N$ Product rule

2. $\log_b\left(\dfrac{M}{N}\right) = \log_b M - \log_b N$ Quotient rule

3. $\log_b M^p = p \log_b M$ Power rule

GREAT QUESTION!

Are there some common screw-ups that I can avoid when using properties of logarithms?

The graphs show that

$\ln(x + 3) \neq \ln x + \ln 3.$

$y = \ln x$ shifted 3 units left $y = \ln x$ shifted ln 3 units up

In general,

$\log_b(M + N) \neq \log_b M + \log_b N.$

$y_1 = \ln(x + 3)$

$y_2 = \ln x + \ln 3$

$[-4, 5, 1]$ by $[-3, 3, 1]$

Try to avoid the following errors:

Incorrect!

$\log_b(M + N) = \log_b M + \log_b N$

$\log_b(M - N) = \log_b M - \log_b N$

$\log_b(M \cdot N) = \log_b M \cdot \log_b N$

$\log_b\left(\dfrac{M}{N}\right) = \dfrac{\log_b M}{\log_b N}$

$\dfrac{\log_b M}{\log_b N} = \log_b M - \log_b N$

$\log_b(MN^p) = p \log_b(MN)$

EXAMPLE 4 Expanding Logarithmic Expressions

Use logarithmic properties to expand each expression as much as possible:

a. $\log_b(x^2\sqrt{y})$ **b.** $\log_6\left(\dfrac{\sqrt[3]{x}}{36y^4}\right)$.

SOLUTION

We will have to use two or more of the properties for expanding logarithms in each part of this example.

a. $\log_b\left(x^2\sqrt{y}\right) = \log_b\left(x^2 y^{\frac{1}{2}}\right)$ Use exponential notation.

$\qquad\qquad\quad = \log_b x^2 + \log_b y^{\frac{1}{2}}$ Use the product rule.

$\qquad\qquad\quad = 2\log_b x + \dfrac{1}{2}\log_b y$ Use the power rule.

b. $\log_6\left(\dfrac{\sqrt[3]{x}}{36y^4}\right) = \log_6\left(\dfrac{x^{\frac{1}{3}}}{36y^4}\right)$ Use exponential notation.

$\qquad\qquad\quad = \log_6 x^{\frac{1}{3}} - \log_6(36y^4)$ Use the quotient rule.

$\qquad\qquad\quad = \log_6 x^{\frac{1}{3}} - (\log_6 36 + \log_6 y^4)$ Use the product rule on $\log_6(36y^4)$.

$\qquad\qquad\quad = \dfrac{1}{3}\log_6 x - (\log_6 36 + 4\log_6 y)$ Use the power rule.

$\qquad\qquad\quad = \dfrac{1}{3}\log_6 x - \log_6 36 - 4\log_6 y$ Apply the distributive property.

$\qquad\qquad\quad = \dfrac{1}{3}\log_6 x - 2 - 4\log_6 y$ $\log_6 36 = 2$ because 2 is the power to which we must raise 6 to get 36. ($6^2 = 36$) • • •

☑ **Check Point 4** Use logarithmic properties to expand each expression as much as possible:

a. $\log_b\left(x^4\sqrt[3]{y}\right)$ **b.** $\log_5\left(\dfrac{\sqrt{x}}{25y^3}\right)$.

⑤ Condense logarithmic expressions.

Condensing Logarithmic Expressions

To **condense a logarithmic expression**, we write the sum or difference of two or more logarithmic expressions as a single logarithmic expression. We use the properties of logarithms to do so.

GREAT QUESTION!

Are the properties listed on the right the same as those in the box on page 482?

Yes. The only difference is that we've reversed the sides in each property from the previous box.

Properties for Condensing Logarithmic Expressions

For $M > 0$ and $N > 0$:

1. $\log_b M + \log_b N = \log_b(MN)$ Product rule

2. $\log_b M - \log_b N = \log_b\left(\dfrac{M}{N}\right)$ Quotient rule

3. $p\log_b M = \log_b M^p$ Power rule

EXAMPLE 5 Condensing Logarithmic Expressions

Write as a single logarithm:

a. $\log_4 2 + \log_4 32$ **b.** $\log(4x - 3) - \log x$.

SOLUTION

a. $\log_4 2 + \log_4 32 = \log_4(2 \cdot 32)$ Use the product rule.

$\qquad\qquad\qquad\quad = \log_4 64$ We now have a single logarithm. However, we can simplify.

$\qquad\qquad\qquad\quad = 3$ $\log_4 64 = 3$ because $4^3 = 64$.

b. $\log(4x - 3) - \log x = \log\left(\dfrac{4x - 3}{x}\right)$ Use the quotient rule. • • •

☑ Check Point **5** Write as a single logarithm:

a. $\log 25 + \log 4$ **b.** $\log (7x + 6) - \log x$.

Coefficients of logarithms must be 1 before you can condense them using the product and quotient rules. For example, to condense

$$2 \ln x + \ln(x + 1),$$

the coefficient of the first term must be 1. We use the power rule to rewrite the coefficient as an exponent:

> 1. Use the power rule to make the number in front an exponent.

$$2 \ln x + \ln(x + 1) = \ln x^2 + \ln(x + 1) = \ln[x^2(x + 1)].$$

> 2. Use the product rule. The sum of logarithms with coefficients of 1 is the logarithm of the product.

EXAMPLE 6 Condensing Logarithmic Expressions

Write as a single logarithm:

a. $\frac{1}{2} \log x + 4 \log (x - 1)$ **b.** $3 \ln (x + 7) - \ln x$

c. $4 \log_b x - 2 \log_b 6 - \frac{1}{2} \log_b y$.

SOLUTION

a. $\frac{1}{2} \log x + 4 \log (x - 1)$

$= \log x^{\frac{1}{2}} + \log (x - 1)^4$ Use the power rule so that all coefficients are 1.

$= \log\left[x^{\frac{1}{2}}(x - 1)^4 \right]$ Use the product rule. The condensed form can be expressed as $\log[\sqrt{x}\,(x - 1)^4]$.

b. $3 \ln(x + 7) - \ln x$

$= \ln(x + 7)^3 - \ln x$ Use the power rule so that all coefficients are 1.

$= \ln\left[\dfrac{(x + 7)^3}{x} \right]$ Use the quotient rule.

c. $4 \log_b x - 2 \log_b 6 - \frac{1}{2} \log_b y$

$= \log_b x^4 - \log_b 6^2 - \log_b y^{\frac{1}{2}}$ Use the power rule so that all coefficients are 1.

$= \log_b x^4 - \left(\log_b 36 + \log_b y^{\frac{1}{2}} \right)$ Rewrite as a single subtraction.

$= \log_b x^4 - \log_b \left(36y^{\frac{1}{2}} \right)$ Use the product rule.

$= \log_b \left(\dfrac{x^4}{36y^{\frac{1}{2}}} \right)$ or $\log_b \left(\dfrac{x^4}{36\sqrt{y}} \right)$ Use the quotient rule. • • •

☑ Check Point **6** Write as a single logarithm:

a. $2 \ln x + \frac{1}{3} \ln (x + 5)$ **b.** $2 \log (x - 3) - \log x$

c. $\frac{1}{4}\log_b x - 2 \log_b 5 - 10 \log_b y$.

⑥ Use the change-of-base property.

The Change-of-Base Property

We have seen that calculators give the values of both common logarithms (base 10) and natural logarithms (base e). To find a logarithm with any other base, we can use the following change-of-base property:

The Change-of-Base Property

For any logarithmic bases a and b, and any positive number M,

$$\log_b M = \frac{\log_a M}{\log_a b}.$$

The logarithm of M with base b is equal to the logarithm of M with any new base divided by the logarithm of b with that new base.

In the change-of-base property, base b is the base of the original logarithm. Base a is a new base that we introduce. Thus, the change-of-base property allows us to change from base b to *any* new base a, as long as the newly introduced base is a positive number not equal to 1.

The change-of-base property is used to write a logarithm in terms of quantities that can be evaluated with a calculator. Because calculators contain keys for common (base 10) and natural (base e) logarithms, we will frequently introduce base 10 or base e.

Change-of-Base Property	**Introducing Common Logarithms**	**Introducing Natural Logarithms**
$\log_b M = \dfrac{\log_a M}{\log_a b}$	$\log_b M = \dfrac{\log_{10} M}{\log_{10} b}$	$\log_b M = \dfrac{\log_e M}{\log_e b}$
a is the new introduced base.	10 is the new introduced base.	e is the new introduced base.

Using the notations for common logarithms and natural logarithms, we have the following results:

The Change-of-Base Property: Introducing Common and Natural Logarithms

Introducing Common Logarithms

$$\log_b M = \frac{\log M}{\log b}$$

Introducing Natural Logarithms

$$\log_b M = \frac{\ln M}{\ln b}$$

EXAMPLE 7 Changing Base to Common Logarithms

Use common logarithms to evaluate $\log_5 140$.

SOLUTION

Because $\log_b M = \dfrac{\log M}{\log b}$,

$$\log_5 140 = \frac{\log 140}{\log 5}$$

$$\approx 3.07.$$ Use a calculator: 140 ⎡LOG⎤ ⎡÷⎤ 5 ⎡LOG⎤ ⎡=⎤
or ⎡LOG⎤ 140 ⎡÷⎤ ⎡LOG⎤ 5 ⎡ENTER⎤. On some calculators, parentheses are needed after 140 and 5.

This means that $\log_5 140 \approx 3.07$. • • •

DISCOVERY

Find a reasonable estimate of $\log_5 140$ to the nearest whole number. To what power can you raise 5 in order to get 140? Compare your estimate to the value obtained in Example 7.

✓ Check Point **7** Use common logarithms to evaluate $\log_7 2506$.

EXAMPLE 8 Changing Base to Natural Logarithms

Use natural logarithms to evaluate $\log_5 140$.

SOLUTION

Because $\log_b M = \dfrac{\ln M}{\ln b}$,

$$\log_5 140 = \dfrac{\ln 140}{\ln 5}$$

$$\approx 3.07. \quad \text{Use a calculator: } 140 \;\boxed{\text{LN}}\; \boxed{\div}\; 5\; \boxed{\text{LN}}\; \boxed{=}$$

$$\text{or } \boxed{\text{LN}}\; 140\; \boxed{\div}\; \boxed{\text{LN}}\; 5\; \boxed{\text{ENTER}} . \text{ On some calculators,}$$
$$\text{parentheses are needed after 140 and 5.}$$

We have again shown that $\log_5 140 \approx 3.07$. • • •

✓ **Check Point 8** Use natural logarithms to evaluate $\log_7 2506$.

TECHNOLOGY

We can use the change-of-base property to graph logarithmic functions with bases other than 10 or e on a graphing utility. For example, **Figure 4.19** shows the graphs of

$$y = \log_2 x \quad \text{and} \quad y = \log_{20} x$$

in a $[0, 10, 1]$ by $[-3, 3, 1]$ viewing rectangle. Because $\log_2 x = \dfrac{\ln x}{\ln 2}$ and $\log_{20} x = \dfrac{\ln x}{\ln 20}$, the functions are entered as

$$y_1 = \boxed{\text{LN}} x\; \boxed{\div}\, \boxed{\text{LN}}\, 2$$
$$\text{and } y_2 = \boxed{\text{LN}} x\; \boxed{\div}\, \boxed{\text{LN}}\, 20.$$

> On some calculators, parentheses are needed after x, **2**, and **20**.

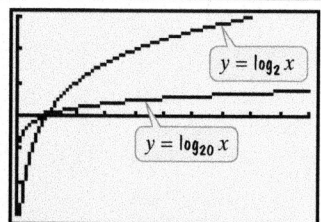

FIGURE 4.19 Using the change-of-base property to graph logarithmic functions

CONCEPT AND VOCABULARY CHECK

Fill in each blank so that the resulting statement is true.

1. The product rule for logarithms states that $\log_b(MN) =$ _____. The logarithm of a product is the _____ of the logarithms.
2. The quotient rule for logarithms states that $\log_b\left(\dfrac{M}{N}\right) =$ _____. The logarithm of a quotient is the _____ of the logarithms.
3. The power rule for logarithms states that $\log_b M^p =$ _____. The logarithm of a number with an exponent is the _____ of the exponent and the logarithm of that number.
4. The change-of-base property for logarithms allows us to write logarithms with base b in terms of a new base a. Introducing base a, the property states that

$$\log_b M = \dfrac{\rule{1cm}{0.4pt}}{\rule{1cm}{0.4pt}}.$$

EXERCISE SET 4.3

Practice Exercises

In Exercises 1–40, use properties of logarithms to expand each logarithmic expression as much as possible. Where possible, evaluate logarithmic expressions without using a calculator.

1. $\log_5(7 \cdot 3)$ **2.** $\log_8(13 \cdot 7)$ **3.** $\log_7(7x)$

4. $\log_9(9x)$ **5.** $\log(1000x)$ **6.** $\log(10{,}000x)$

7. $\log_7\left(\dfrac{7}{x}\right)$ **8.** $\log_9\left(\dfrac{9}{x}\right)$ **9.** $\log\left(\dfrac{x}{100}\right)$

10. $\log\left(\dfrac{x}{1000}\right)$ **11.** $\log_4\left(\dfrac{64}{y}\right)$ **12.** $\log_5\left(\dfrac{125}{y}\right)$

13. $\ln\left(\dfrac{e^2}{5}\right)$ **14.** $\ln\left(\dfrac{e^4}{8}\right)$ **15.** $\log_b x^3$

16. $\log_b x^7$ **17.** $\log N^{-6}$ **18.** $\log M^{-8}$

19. $\ln\sqrt[5]{x}$ **20.** $\ln\sqrt[7]{x}$ **21.** $\log_b(x^2 y)$

22. $\log_b(xy^3)$ **23.** $\log_4\left(\dfrac{\sqrt{x}}{64}\right)$ **24.** $\log_5\left(\dfrac{\sqrt{x}}{25}\right)$

25. $\log_6\left(\dfrac{36}{\sqrt{x+1}}\right)$ **26.** $\log_8\left(\dfrac{64}{\sqrt{x+1}}\right)$ **27.** $\log_b\left(\dfrac{x^2 y}{z^2}\right)$

28. $\log_b\left(\dfrac{x^3 y}{z^2}\right)$ **29.** $\log\sqrt{100x}$ **30.** $\ln\sqrt{ex}$

31. $\log\sqrt[3]{\dfrac{x}{y}}$ **32.** $\log\sqrt[5]{\dfrac{x}{y}}$ **33.** $\log_b\left(\dfrac{\sqrt{x}y^3}{z^3}\right)$

34. $\log_b\left(\dfrac{\sqrt[3]{x}y^4}{z^5}\right)$ **35.** $\log_5\sqrt[3]{\dfrac{x^2 y}{25}}$ **36.** $\log_2\sqrt[5]{\dfrac{xy^4}{16}}$

37. $\ln\left[\dfrac{x^3\sqrt{x^2+1}}{(x+1)^4}\right]$ **38.** $\ln\left[\dfrac{x^4\sqrt{x^2+3}}{(x+3)^5}\right]$

39. $\log\left[\dfrac{10x^2\sqrt[3]{1-x}}{7(x+1)^2}\right]$ **40.** $\log\left[\dfrac{100x^3\sqrt[3]{5-x}}{3(x+7)^2}\right]$

In Exercises 41–70, use properties of logarithms to condense each logarithmic expression. Write the expression as a single logarithm whose coefficient is 1. Where possible, evaluate logarithmic expressions without using a calculator.

41. $\log 5 + \log 2$ **42.** $\log 250 + \log 4$

43. $\ln x + \ln 7$ **44.** $\ln x + \ln 3$

45. $\log_2 96 - \log_2 3$ **46.** $\log_3 405 - \log_3 5$

47. $\log(2x+5) - \log x$ **48.** $\log(3x+7) - \log x$

49. $\log x + 3\log y$ **50.** $\log x + 7\log y$

51. $\frac{1}{2}\ln x + \ln y$ **52.** $\frac{1}{3}\ln x + \ln y$

53. $2\log_b x + 3\log_b y$ **54.** $5\log_b x + 6\log_b y$

55. $5\ln x - 2\ln y$ **56.** $7\ln x - 3\ln y$

57. $3\ln x - \frac{1}{3}\ln y$ **58.** $2\ln x - \frac{1}{2}\ln y$

59. $4\ln(x+6) - 3\ln x$ **60.** $8\ln(x+9) - 4\ln x$

61. $3\ln x + 5\ln y - 6\ln z$ **62.** $4\ln x + 7\ln y - 3\ln z$

63. $\frac{1}{2}(\log x + \log y)$ **64.** $\frac{1}{3}(\log_4 x - \log_4 y)$

65. $\frac{1}{2}(\log_5 x + \log_5 y) - 2\log_5(x+1)$

66. $\frac{1}{3}(\log_4 x - \log_4 y) + 2\log_4(x+1)$

67. $\frac{1}{3}[2\ln(x+5) - \ln x - \ln(x^2-4)]$

68. $\frac{1}{3}[5\ln(x+6) - \ln x - \ln(x^2-25)]$

69. $\log x + \log(x^2-1) - \log 7 - \log(x+1)$

70. $\log x + \log(x^2-4) - \log 15 - \log(x+2)$

In Exercises 71–78, use common logarithms or natural logarithms and a calculator to evaluate to four decimal places.

71. $\log_5 13$ **72.** $\log_6 17$ **73.** $\log_{14} 87.5$

74. $\log_{16} 57.2$ **75.** $\log_{0.1} 17$ **76.** $\log_{0.3} 19$

77. $\log_\pi 63$ **78.** $\log_\pi 400$

In Exercises 79–82, use a graphing utility and the change-of-base property to graph each function.

79. $y = \log_3 x$ **80.** $y = \log_{15} x$

81. $y = \log_2(x+2)$ **82.** $y = \log_3(x-2)$

Practice Plus

In Exercises 83–88, let $\log_b 2 = A$ and $\log_b 3 = C$. Write each expression in terms of A and C.

83. $\log_b \frac{3}{2}$ **84.** $\log_b 6$ **85.** $\log_b 8$

86. $\log_b 81$ **87.** $\log_b\sqrt{\dfrac{2}{27}}$ **88.** $\log_b\sqrt{\dfrac{3}{16}}$

In Exercises 89–102, determine whether each equation is true or false. Where possible, show work to support your conclusion. If the statement is false, make the necessary change(s) to produce a true statement.

89. $\ln e = 0$ **90.** $\ln 0 = e$

91. $\log_4(2x^3) = 3\log_4(2x)$ **92.** $\ln(8x^3) = 3\ln(2x)$

93. $x\log 10^x = x^2$ **94.** $\ln(x+1) = \ln x + \ln 1$

95. $\ln(5x) + \ln 1 = \ln(5x)$ **96.** $\ln x + \ln(2x) = \ln(3x)$

97. $\log(x+3) - \log(2x) = \dfrac{\log(x+3)}{\log(2x)}$

98. $\dfrac{\log(x+2)}{\log(x-1)} = \log(x+2) - \log(x-1)$

99. $\log_6\left(\dfrac{x-1}{x^2+4}\right) = \log_6(x-1) - \log_6(x^2+4)$

100. $\log_6[4(x+1)] = \log_6 4 + \log_6(x+1)$

101. $\log_3 7 = \dfrac{1}{\log_7 3}$ **102.** $e^x = \dfrac{1}{\ln x}$

Application Exercises

103. The loudness level of a sound can be expressed by comparing the sound's intensity to the intensity of a sound barely audible to the human ear. The formula

$$D = 10(\log I - \log I_0)$$

describes the loudness level of a sound, D, in decibels, where I is the intensity of the sound, in watts per meter2, and I_0 is the intensity of a sound barely audible to the human ear.

a. Express the formula so that the expression in parentheses is written as a single logarithm.

b. Use the form of the formula from part (a) to answer this question: If a sound has an intensity 100 times the intensity of a softer sound, how much larger on the decibel scale is the loudness level of the more intense sound?

104. The formula

$$t = \frac{1}{c}[\ln A - \ln(A - N)]$$

describes the time, t, in weeks, that it takes to achieve mastery of a portion of a task, where A is the maximum learning possible, N is the portion of the learning that is to be achieved, and c is a constant used to measure an individual's learning style.

a. Express the formula so that the expression in brackets is written as a single logarithm.

b. The formula is also used to determine how long it will take chimpanzees and apes to master a task. For example, a typical chimpanzee learning sign language can master a maximum of 65 signs. Use the form of the formula from part (a) to answer this question: How many weeks will it take a chimpanzee to master 30 signs if c for that chimp is 0.03?

Writing in Mathematics

105. Describe the product rule for logarithms and give an example.

106. Describe the quotient rule for logarithms and give an example.

107. Describe the power rule for logarithms and give an example.

108. Without showing the details, explain how to condense $\ln x - 2\ln(x + 1)$.

109. Describe the change-of-base property and give an example.

110. Explain how to use your calculator to find $\log_{14} 283$.

111. You overhear a student talking about a property of logarithms in which division becomes subtraction. Explain what the student means by this.

112. Find ln 2 using a calculator. Then calculate each of the following: $1 - \frac{1}{2}$; $\quad 1 - \frac{1}{2} + \frac{1}{3}$; $\quad 1 - \frac{1}{2} + \frac{1}{3} - \frac{1}{4}$; $1 - \frac{1}{2} + \frac{1}{3} - \frac{1}{4} + \frac{1}{5}$; Describe what you observe.

Technology Exercises

113. a. Use a graphing utility (and the change-of-base property) to graph $y = \log_3 x$.

 b. Graph $y = 2 + \log_3 x$, $y = \log_3(x + 2)$, and $y = -\log_3 x$ in the same viewing rectangle as $y = \log_3 x$. Then describe the change or changes that need to be made to the graph of $y = \log_3 x$ to obtain each of these three graphs.

114. Graph $y = \log x$, $y = \log(10x)$, and $y = \log(0.1x)$ in the same viewing rectangle. Describe the relationship among the three graphs. What logarithmic property accounts for this relationship?

115. Use a graphing utility and the change-of-base property to graph $y = \log_3 x$, $y = \log_{25} x$, and $y = \log_{100} x$ in the same viewing rectangle.

 a. Which graph is on the top in the interval $(0, 1)$? Which is on the bottom?

 b. Which graph is on the top in the interval $(1, \infty)$? Which is on the bottom?

 c. Generalize by writing a statement about which graph is on top, which is on the bottom, and in which intervals, using $y = \log_b x$ where $b > 1$.

Disprove each statement in Exercises 116–120 by

 a. *letting y equal a positive constant of your choice, and*

 b. *using a graphing utility to graph the function on each side of the equal sign. The two functions should have different graphs, showing that the equation is not true in general.*

116. $\log(x + y) = \log x + \log y$

117. $\log\left(\dfrac{x}{y}\right) = \dfrac{\log x}{\log y}$

118. $\ln(x - y) = \ln x - \ln y$

119. $\ln(xy) = (\ln x)(\ln y)$

120. $\dfrac{\ln x}{\ln y} = \ln x - \ln y$

Critical Thinking Exercises

Make Sense? *In Exercises 121–124, determine whether each statement makes sense or does not make sense, and explain your reasoning.*

121. Because I cannot simplify the expression $b^m + b^n$ by adding exponents, there is no property for the logarithm of a sum.

122. Because logarithms are exponents, the product, quotient, and power rules remind me of properties for operations with exponents.

123. I can use any positive number other than 1 in the change-of-base property, but the only practical bases are 10 and e because my calculator gives logarithms for these two bases.

124. I expanded $\log_4 \sqrt{\dfrac{x}{y}}$ by writing the radical using a rational exponent and then applying the quotient rule, obtaining $\dfrac{1}{2}\log_4 x - \log_4 y$.

In Exercises 125–128, determine whether each statement is true or false. If the statement is false, make the necessary change(s) to produce a true statement.

125. $\ln\sqrt{2} = \dfrac{\ln 2}{2}$

126. $\dfrac{\log_7 49}{\log_7 7} = \log_7 49 - \log_7 7$

127. $\log_b(x^3 + y^3) = 3\log_b x + 3\log_b y$

128. $\log_b(xy)^5 = (\log_b x + \log_b y)^5$

129. Use the change-of-base property to prove that

$$\log e = \frac{1}{\ln 10}.$$

130. If $\log 3 = A$ and $\log 7 = B$, find $\log_7 9$ in terms of A and B.

131. Write as a single term that does not contain a logarithm:

$$e^{\ln 8x^5 - \ln 2x^2}.$$

132. If $f(x) = \log_b x$, show that

$$\frac{f(x + h) - f(x)}{h} = \log_b\left(1 + \frac{h}{x}\right)^{\frac{1}{h}}, h \neq 0.$$

133. Use the proof of the product rule on page 805 of the appendix to prove the quotient rule.

Preview Exercises

Exercises 134–136 will help you prepare for the material covered in the next section.

134. Solve for x: $a(x - 2) = b(2x + 3)$.

135. Solve: $x(x - 7) = 3$.

136. Solve: $\dfrac{x + 2}{4x + 3} = \dfrac{1}{x}$.

CHAPTER 4 Mid-Chapter Check Point

WHAT YOU KNOW: We evaluated and graphed exponential functions $[f(x) = b^x, b > 0$ and $b \neq 1]$, including the natural exponential function $[f(x) = e^x, e \approx 2.718]$. A function has an inverse that is a function if there is no horizontal line that intersects the function's graph more than once. The exponential function passes this horizontal line test and we called the inverse of the exponential function with base b the logarithmic function with base b. We learned that $y = \log_b x$ is equivalent to $b^y = x$. We evaluated and graphed logarithmic functions, including the common logarithmic function $[f(x) = \log_{10} x$ or $f(x) = \log x]$ and the natural logarithmic function $[f(x) = \log_e x$ or $f(x) = \ln x]$. We learned to use transformations to graph exponential and logarithmic functions. Finally, we used properties of logarithms to expand and condense logarithmic expressions.

In Exercises 1–5, graph f and g in the same rectangular coordinate system. Graph and give equations of all asymptotes. Give each function's domain and range.

1. $f(x) = 2^x$ and $g(x) = 2^x - 3$
2. $f(x) = \left(\frac{1}{2}\right)^x$ and $g(x) = \left(\frac{1}{2}\right)^{x-1}$
3. $f(x) = e^x$ and $g(x) = \ln x$
4. $f(x) = \log_2 x$ and $g(x) = \log_2(x - 1) + 1$
5. $f(x) = \log_{\frac{1}{2}} x$ and $g(x) = -2 \log_{\frac{1}{2}} x$

In Exercises 6–9, find the domain of each function.

6. $f(x) = \log_3(x + 6)$
7. $g(x) = \log_3 x + 6$
8. $h(x) = \log_3(x + 6)^2$
9. $f(x) = 3^{x+6}$

In Exercises 10–20, evaluate each expression without using a calculator. If evaluation is not possible, state the reason.

10. $\log_2 8 + \log_5 25$
11. $\log_3 \frac{1}{9}$
12. $\log_{100} 10$
13. $\log \sqrt[3]{10}$
14. $\log_2(\log_3 81)$
15. $\log_3\left(\log_2 \frac{1}{8}\right)$
16. $6^{\log_6 5}$
17. $\ln e^{\sqrt{7}}$
18. $10^{\log 13}$
19. $\log_{100} 0.1$
20. $\log_\pi \pi^{\sqrt{\pi}}$

In Exercises 21–22, expand and evaluate numerical terms.

21. $\log\left(\dfrac{\sqrt{xy}}{1000}\right)$
22. $\ln(e^{19} x^{20})$

In Exercises 23–25, write each expression as a single logarithm.

23. $8 \log_7 x - \frac{1}{3}\log_7 y$
24. $7 \log_5 x + 2 \log_5 x$
25. $\frac{1}{2}\ln x - 3 \ln y - \ln(z - 2)$

26. Use the formulas
$$A = P\left(1 + \frac{r}{n}\right)^{nt} \quad \text{and} \quad A = Pe^{rt}$$
to solve this exercise. You decide to invest $8000 for 3 years at an annual rate of 8%. How much more is the return if the interest is compounded continuously than if it is compounded monthly? Round to the nearest dollar.

SECTION 4.4 Exponential and Logarithmic Equations

Objectives

1. Use like bases to solve exponential equations.
2. Use logarithms to solve exponential equations.
3. Use the definition of a logarithm to solve logarithmic equations.
4. Use the one-to-one property of logarithms to solve logarithmic equations.
5. Solve applied problems involving exponential and logarithmic equations.

At age 20, you inherit $30,000. You'd like to put aside $25,000 and eventually have over half a million dollars for early retirement. Is this possible? In Example 10 in this section, you will see how techniques for solving equations with variable exponents provide an answer to this question.

Exponential Equations

An **exponential equation** is an equation containing a variable in an exponent. Examples of exponential equations include

$$2^{3x-8} = 16, \quad 4^x = 15, \quad \text{and} \quad 40e^{0.6x} = 240.$$

Some exponential equations can be solved by expressing each side of the equation as a power of the same base. All exponential functions are one-to-one—that is, no two different ordered pairs have the same second component. Thus, if b is a positive number other than 1 and $b^M = b^N$, then $M = N$.

1 Use like bases to solve exponential equations.

Solving Exponential Equations by Expressing Each Side as a Power of the Same Base

$$\text{If } b^M = b^N, \text{ then } M = N.$$

| Express each side as a power of the same base. | Set the exponents equal to each other. |

1. Rewrite the equation in the form $b^M = b^N$.
2. Set $M = N$.
3. Solve for the variable.

Graphic Connections

The graphs of

$$y_1 = 2^{3x-8}$$

and $y_2 = 16$

have an intersection point whose x-coordinate is 4. This verifies that $\{4\}$ is the solution set of $2^{3x-8} = 16$.

$[-1, 5, 1]$ by $[0, 20, 1]$

EXAMPLE 1 Solving Exponential Equations

Solve:

a. $2^{3x-8} = 16$ **b.** $27^{x+3} = 9^{x-1}$.

SOLUTION

In each equation, express both sides as a power of the same base. Then set the exponents equal to each other and solve for the variable.

a. Because 16 is 2^4, we express each side of $2^{3x-8} = 16$ in terms of base 2.

$2^{3x-8} = 16$	This is the given equation.
$2^{3x-8} = 2^4$	Write each side as a power of the same base.
$3x - 8 = 4$	If $b^M = b^N$, $b > 0$ and $b \neq 1$, then $M = N$.
$3x = 12$	Add 8 to both sides.
$x = 4$	Divide both sides by 3.

Substituting 4 for x into the original equation produces the true statement $16 = 16$. The solution set is $\{4\}$.

b. Because $27 = 3^3$ and $9 = 3^2$, we can express both sides of $27^{x+3} = 9^{x-1}$ in terms of base 3.

$27^{x+3} = 9^{x-1}$	This is the given equation.
$\left(3^3\right)^{x+3} = \left(3^2\right)^{x-1}$	Write each side as a power of the same base.
$3^{3(x+3)} = 3^{2(x-1)}$	When an exponential expression is raised to a power, multiply exponents.
$3(x + 3) = 2(x - 1)$	If two powers of the same base are equal, then the exponents are equal.
$3x + 9 = 2x - 2$	Apply the distributive property.
$x + 9 = -2$	Subtract 2x from both sides.
$x = -11$	Subtract 9 from both sides.

Substituting -11 for x into the original equation produces $27^{-8} = 9^{-12}$, which simplifies to the true statement $3^{-24} = 3^{-24}$. The solution set is $\{-11\}$. • • •

⟡ Check Point **1** Solve:

a. $5^{3x-6} = 125$ **b.** $8^{x+2} = 4^{x-3}$.

2 Use logarithms to solve exponential equations.

Most exponential equations cannot be rewritten so that each side has the same base. Here are two examples:

$$4^x = 15 \qquad\qquad 10^x = 120{,}000.$$

| We cannot rewrite both sides in terms of base 2 or base 4. | We cannot rewrite both sides in terms of base 10. |

Logarithms are extremely useful in solving these equations. The solution begins with isolating the exponential expression. Notice that the exponential expression is already isolated in both $4^x = 15$ and $10^x = 120,000$. Then we take the logarithm on both sides. Why can we do this? All logarithmic relations are functions. Thus, if M and N are positive real numbers and $M = N$, then $\log_b M = \log_b N$.

The base that is used when taking the logarithm on both sides of an equation can be any base at all. If the exponential equation involves base 10, as in $10^x = 120,000$, we'll take the common logarithm on both sides. If the exponential equation involves any other base, as in $4^x = 15$, we'll take the natural logarithm on both sides.

Using Logarithms to Solve Exponential Equations

1. Isolate the exponential expression.
2. Take the common logarithm on both sides of the equation for base 10. Take the natural logarithm on both sides of the equation for bases other than 10.
3. Simplify using one of the following properties:

$$\ln b^x = x \ln b \quad \text{or} \quad \ln e^x = x \quad \text{or} \quad \log 10^x = x.$$

4. Solve for the variable.

EXAMPLE 2 Solving Exponential Equations

Solve:

 a. $4^x = 15$ **b.** $10^x = 120,000$.

SOLUTION

We will use the natural logarithmic function to solve $4^x = 15$ and the common logarithmic function to solve $10^x = 120,000$.

 a. Because the exponential expression, 4^x, is already isolated on the left side of $4^x = 15$, we begin by taking the natural logarithm on both sides of the equation.

$4^x = 15$	This is the given equation.
$\ln 4^x = \ln 15$	Take the natural logarithm on both sides.
$x \ln 4 = \ln 15$	Use the power rule and bring the variable exponent to the front: $\ln b^x = x \ln b$.
$x = \dfrac{\ln 15}{\ln 4}$	Solve for x by dividing both sides by $\ln 4$. Equivalently, $x = \log_4 15$.

We now have an exact value for x. We use the exact value for x in the equation's solution set. Thus, the equation's solution is $\dfrac{\ln 15}{\ln 4}$ and the solution set is $\left\{\dfrac{\ln 15}{\ln 4}\right\}$. We can obtain a decimal approximation by using a calculator: $x \approx 1.95$. Because $4^2 = 16$, it seems reasonable that the solution to $4^x = 15$ is approximately 1.95.

 b. Because the exponential expression, 10^x, is already isolated on the left side of $10^x = 120,000$, we begin by taking the common logarithm on both sides of the equation.

$10^x = 120,000$	This is the given equation.
$\log 10^x = \log 120,000$	Take the common logarithm on both sides.
$x = \log 120,000$	Use the inverse property $\log 10^x = x$ on the left.

The equation's solution is $\log 120,000$ and the solution set is $\{\log 120,000\}$. We can obtain a decimal approximation by using a calculator: $x \approx 5.08$. Because $10^5 = 100,000$, it seems reasonable that the solution to $10^x = 120,000$ is approximately 5.08.

 • • •

DISCOVERY

Keep in mind that the base used when taking the logarithm on both sides of an equation can be any base at all. Solve $4^x = 15$ by taking the common logarithm on both sides. Solve again, this time taking the logarithm with base 4 on both sides. Use the change-of-base property to show that the solutions are the same as the one obtained in Example 2(a).

☑ Check Point **2** Solve:

a. $5^x = 134$ b. $10^x = 8000$.

Find each solution set and then use a calculator to obtain a decimal approximation to two decimal places for the solution.

EXAMPLE 3 Solving an Exponential Equation

Solve: $40e^{0.6x} - 3 = 237$.

SOLUTION

We begin by adding 3 to both sides and dividing both sides by 40 to isolate the exponential expression, $e^{0.6x}$. Then we take the natural logarithm on both sides of the equation.

$40e^{0.6x} - 3 = 237$	This is the given equation.
$40e^{0.6x} = 240$	Add 3 to both sides.
$e^{0.6x} = 6$	Isolate the exponential factor by dividing both sides by 40.
$\ln e^{0.6x} = \ln 6$	Take the natural logarithm on both sides.
$0.6x = \ln 6$	Use the inverse property $\ln e^x = x$ on the left.
$x = \dfrac{\ln 6}{0.6} \approx 2.99$	Divide both sides by 0.6 and solve for x.

Thus, the solution of the equation is $\dfrac{\ln 6}{0.6} \approx 2.99$. Try checking this approximate solution in the original equation to verify that $\left\{ \dfrac{\ln 6}{0.6} \right\}$ is the solution set. • • •

☑ Check Point **3** Solve: $7e^{2x} - 5 = 58$. Find the solution set and then use a calculator to obtain a decimal approximation to two decimal places for the solution.

EXAMPLE 4 Solving an Exponential Equation

Solve: $5^{x-2} = 4^{2x+3}$.

SOLUTION

Because each exponential expression is isolated on one side of the equation, we begin by taking the natural logarithm on both sides.

$5^{x-2} = 4^{2x+3}$	This is the given equation.
$\ln 5^{x-2} = \ln 4^{2x+3}$	Take the natural logarithm on both sides.

Be sure to insert parentheses around the binomials.

$(x - 2)\ln 5 = (2x + 3)\ln 4$	Use the power rule and bring the variable exponents to the front: $\ln b^x = x \ln b$.

Remember that ln 5 and ln 4 are constants, not variables.

$x \ln 5 - 2 \ln 5 = 2x \ln 4 + 3 \ln 4$	Use the distributive property to distribute ln 5 and ln 4 to both terms in parentheses.
$x \ln 5 - 2x \ln 4 = 2 \ln 5 + 3 \ln 4$	Collect variable terms involving x on the left by subtracting 2x ln 4 and adding 2 ln 5 on both sides.

DISCOVERY

Use properties of logarithms to show that the solution in Example 4 can be expressed as

$$\frac{\ln 1600}{\ln\left(\frac{5}{16}\right)}.$$

$$x(\ln 5 - 2\ln 4) = 2\ln 5 + 3\ln 4 \qquad \text{Factor out } x \text{ from the two terms on the left.}$$

$$x = \frac{2\ln 5 + 3\ln 4}{\ln 5 - 2\ln 4} \qquad \text{Isolate } x \text{ by dividing both sides by } \ln 5 - 2\ln 4.$$

The solution set is $\left\{\dfrac{2\ln 5 + 3\ln 4}{\ln 5 - 2\ln 4}\right\}$. The solution is approximately -6.34. • • •

✅ **Check Point 4** Solve: $3^{2x-1} = 7^{x+1}$. Find the solution set and then use a calculator to obtain a decimal approximation to two decimal places for the solution.

EXAMPLE 5 Solving an Exponential Equation

Solve: $e^{2x} - 4e^x + 3 = 0$.

SOLUTION

The given equation is quadratic in form. If $u = e^x$, the equation can be expressed as $u^2 - 4u + 3 = 0$. Because this equation can be solved by factoring, we factor to isolate the exponential term.

$$e^{2x} - 4e^x + 3 = 0 \qquad \text{This is the given equation.}$$

$$(e^x - 3)(e^x - 1) = 0 \qquad \text{Factor on the left. Notice that if } u = e^x, u^2 - 4u + 3 = (u - 3)(u - 1).$$

$$e^x - 3 = 0 \quad \text{or} \quad e^x - 1 = 0 \qquad \text{Set each factor equal to 0.}$$

$$e^x = 3 \qquad\qquad e^x = 1 \qquad \text{Solve for } e^x.$$

$$\ln e^x = \ln 3 \qquad\qquad x = 0 \qquad \text{Take the natural logarithm on both sides of the first equation. The equation on the right can be solved by inspection.}$$

$$x = \ln 3 \qquad\qquad\qquad\qquad \ln e^x = x$$

The solution set is $\{0, \ln 3\}$. The solutions are 0 and $\ln 3$, which is approximately 1.10. • • •

TECHNOLOGY

Graphic Connections

Shown below is the graph of $y = e^{2x} - 4e^x + 3$. There are two x-intercepts, one at 0 and one at approximately 1.10. These intercepts verify our algebraic solution.

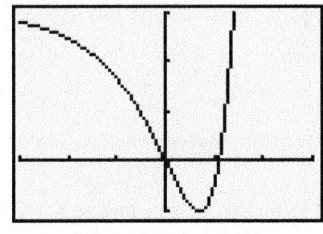

$[-3, 3, 1]$ by $[-1, 3, 1]$

✅ **Check Point 5** Solve: $e^{2x} - 8e^x + 7 = 0$. Find the solution set and then use a calculator to obtain a decimal approximation to two decimal places, if necessary, for the solutions.

③ Use the definition of a logarithm to solve logarithmic equations.

Logarithmic Equations

A **logarithmic equation** is an equation containing a variable in a logarithmic expression. Examples of logarithmic equations include

$$\log_4(x + 3) = 2 \quad \text{and} \quad \ln(x + 2) - \ln(4x + 3) = \ln\left(\frac{1}{x}\right).$$

Some logarithmic equations can be expressed in the form $\log_b M = c$. We can solve such equations by rewriting them in exponential form.

1. Express the equation in the form $\log_b M = c$.

2. Use the definition of a logarithm to rewrite the equation in exponential form:

$$\log_b M = c \quad \text{means} \quad b^c = M.$$

> Logarithms are exponents.

3. Solve for the variable.

4. Check proposed solutions in the original equation. Include in the solution set only values for which $M > 0$.

EXAMPLE 6 Solving Logarithmic Equations

Solve:

 a. $\log_4(x + 3) = 2$ **b.** $3\ln(2x) = 12$.

SOLUTION

The form $\log_b M = c$ involves a single logarithm whose coefficient is 1 on one side and a constant on the other side. Equation (a) is already in this form. We will need to divide both sides of equation (b) by 3 to obtain this form.

TECHNOLOGY

Graphic Connections

The graphs of

 $y_1 = \log_4(x + 3)$ and $y_2 = 2$

have an intersection point whose x-coordinate is 13. This verifies that $\{13\}$ is the solution set for $\log_4(x + 3) = 2$.

$[-3, 17, 1]$ by $[-2, 3, 1]$

a. $\log_4(x + 3) = 2$ *This is the given equation.*

 $4^2 = x + 3$ *Rewrite in exponential form: $\log_b M = c$ means $b^c = M$.*

 $16 = x + 3$ *Square 4.*

 $13 = x$ *Subtract 3 from both sides.*

Check 13:

 $\log_4(x + 3) = 2$ *This is the given logarithmic equation.*

 $\log_4(13 + 3) \overset{?}{=} 2$ *Substitute 13 for x.*

 $\log_4 16 \overset{?}{=} 2$

 $2 = 2$, true *$\log_4 16 = 2$ because $4^2 = 16$.*

This true statement indicates that the solution set is $\{13\}$.

b. $3\ln(2x) = 12$ *This is the given equation.*

 $\ln(2x) = 4$ *Divide both sides by 3.*

 $\log_e(2x) = 4$ *Rewrite the natural logarithm showing base e. This step is optional.*

 $e^4 = 2x$ *Rewrite in exponential form: $\log_b M = c$ means $b^c = M$.*

 $\dfrac{e^4}{2} = x$ *Divide both sides by 2.*

Check $\dfrac{e^4}{2}$:

 $3\ln(2x) = 12$ *This is the given logarithmic equation.*

 $3\ln\left[2\left(\dfrac{e^4}{2}\right)\right] \overset{?}{=} 12$ *Substitute $\dfrac{e^4}{2}$ for x.*

 $3\ln e^4 \overset{?}{=} 12$ *Simplify: $\dfrac{\cancel{2}}{1} \cdot \dfrac{e^4}{\cancel{2}} = e^4$.*

 $3 \cdot 4 \overset{?}{=} 12$ *Because $\ln e^x = x$, we conclude $\ln e^4 = 4$.*

 $12 = 12$, true

This true statement indicates that the solution set is $\left\{\dfrac{e^4}{2}\right\}$. •••

⍉ Check Point **6** Solve:

 a. $\log_2(x - 4) = 3$ **b.** $4\ln(3x) = 8$.

 Logarithmic expressions are defined only for logarithms of positive real numbers. **Always check proposed solutions of a logarithmic equation in the original equation. Exclude from the solution set any proposed solution that produces the logarithm of a negative number or the logarithm of 0.**

GREAT QUESTION!

Can a negative number belong to the solution set of a logarithmic equation?

Yes. Here's an example.

$$\log_2(x + 20) = 3 \qquad \text{Solve this equation.}$$
$$2^3 = x + 20 \qquad \text{Rewrite in exponential form.}$$
$$8 = x + 20 \qquad \text{Cube 2.}$$
$$-12 = x \qquad \text{Subtract 20 from both sides.}$$

Check −12:
$$\log_2(-12 + 20) \stackrel{?}{=} 3 \qquad \text{Substitute −12 for x.}$$
$$\log_2 8 \stackrel{?}{=} 3$$
$$3 = 3, \text{ true} \quad \log_2 8 = 3 \text{ because } 2^3 = 8.$$

The solution set is {−12}. Although −12 is negative, it does not produce the logarithm of a negative number in $\log_2(x + 20) = 3$, the given equation. Note that the domain of the expression $\log_2(x + 20)$ is $(-20, \infty)$, which includes negative numbers such as −12.

To rewrite the logarithmic equation $\log_b M = c$ in the equivalent exponential form $b^c = M$, we need a single logarithm whose coefficient is one. It is sometimes necessary to use properties of logarithms to condense logarithms into a single logarithm. In the next example, we use the product rule for logarithms to obtain a single logarithmic expression on the left side.

EXAMPLE 7 Solving a Logarithmic Equation

Solve: $\log_2 x + \log_2(x - 7) = 3$.

SOLUTION

$$\log_2 x + \log_2(x - 7) = 3 \qquad \text{This is the given equation.}$$
$$\log_2[x(x - 7)] = 3 \qquad \text{Use the product rule to obtain a single logarithm: } \log_b M + \log_b N = \log_b(MN).$$
$$2^3 = x(x - 7) \qquad \text{Rewrite in exponential form: } \log_b M = c \text{ means } b^c = M.$$
$$8 = x^2 - 7x \qquad \text{Evaluate } 2^3 \text{ on the left and apply the distributive property on the right.}$$
$$0 = x^2 - 7x - 8 \qquad \text{Set the equation equal to 0.}$$
$$0 = (x - 8)(x + 1) \qquad \text{Factor.}$$
$$x - 8 = 0 \quad \text{or} \quad x + 1 = 0 \qquad \text{Set each factor equal to 0.}$$
$$x = 8 \qquad\qquad x = -1 \qquad \text{Solve for x.}$$

Check 8:	Check −1:
$\log_2 x + \log_2(x - 7) = 3$	$\log_2 x + \log_2(x - 7) = 3$
$\log_2 8 + \log_2(8 - 7) \stackrel{?}{=} 3$	$\log_2(-1) + \log_2(-1 - 7) \stackrel{?}{=} 3$
$\log_2 8 + \log_2 1 \stackrel{?}{=} 3$	The number −1 does not check.
$3 + 0 \stackrel{?}{=} 3$	The domain of a logarithmic function consists of positive numbers.
$3 = 3, \text{ true}$	

The solution set is {8}. ● ● ●

✓ **Check Point 7** Solve: $\log x + \log(x - 3) = 1$.

④ Use the one-to-one property of logarithms to solve logarithmic equations.

Some logarithmic equations can be expressed in the form $\log_b M = \log_b N$, where the bases on both sides of the equation are the same. Because all logarithmic functions are one-to-one, we can conclude that $M = N$.

> **Using the One-to-One Property of Logarithms to Solve Logarithmic Equations**
>
> **1.** Express the equation in the form $\log_b M = \log_b N$. This form involves a single logarithm whose coefficient is 1 on each side of the equation.
> **2.** Use the one-to-one property to rewrite the equation without logarithms: If $\log_b M = \log_b N$, then $M = N$.
> **3.** Solve for the variable.
> **4.** Check proposed solutions in the original equation. Include in the solution set only values for which $M > 0$ and $N > 0$.

EXAMPLE 8 Solving a Logarithmic Equation

Solve: $\ln(x + 2) - \ln(4x + 3) = \ln\left(\dfrac{1}{x}\right)$.

SOLUTION

In order to apply the one-to-one property of logarithms, we need a single logarithm whose coefficient is 1 on each side of the equation. The right side is already in this form. We can obtain a single logarithm on the left side by applying the quotient rule.

$$\ln(x + 2) - \ln(4x + 3) = \ln\left(\frac{1}{x}\right)$$

This is the given equation.

$$\ln\left(\frac{x + 2}{4x + 3}\right) = \ln\left(\frac{1}{x}\right)$$

Use the quotient rule to obtain a single logarithm on the left side: $\log_b M - \log_b N = \log_b\left(\dfrac{M}{N}\right)$.

$$\frac{x + 2}{4x + 3} = \frac{1}{x}$$

Use the one-to-one property: If $\log_b M = \log_b N$, then $M = N$.

$$x(4x + 3)\left(\frac{x + 2}{4x + 3}\right) = x(4x + 3)\left(\frac{1}{x}\right)$$

Multiply both sides by $x(4x + 3)$, the LCD.

$$x(x + 2) = 4x + 3$$

Simplify.

$$x^2 + 2x = 4x + 3$$

Apply the distributive property.

$$x^2 - 2x - 3 = 0$$

Subtract $4x + 3$ from both sides and set the equation equal to 0.

$$(x - 3)(x + 1) = 0$$

Factor.

$$x - 3 = 0 \quad \text{or} \quad x + 1 = 0$$

Set each factor equal to 0.

$$x = 3 \qquad\qquad x = -1$$

Solve for x.

Substituting 3 for x into $\ln(x + 2) - \ln(4x + 3) = \ln\left(\frac{1}{x}\right)$ produces the true statement $\ln\left(\frac{1}{3}\right) = \ln\left(\frac{1}{3}\right)$. However, substituting -1 produces logarithms of negative numbers. Thus, -1 is not a solution. The solution set is $\{3\}$. • • •

☑ Check Point 8 Solve: $\ln(x - 3) = \ln(7x - 23) - \ln(x + 1)$.

⑤ Solve applied problems involving exponential and logarithmic equations.

Applications

Our first applied example provides a mathematical perspective on the old slogan "Alcohol and driving don't mix." In California, where 38% of fatal traffic crashes involve drinking drivers, it is illegal to drive with a blood alcohol concentration of 0.08 or higher. At these levels, drivers may be arrested and charged with driving under the influence.

TECHNOLOGY

Numeric Connections

A graphing utility's ⟨TABLE⟩ feature can be used to verify that {3} is the solution set of

$$\ln(x + 2) - \ln(4x + 3) = \ln\left(\frac{1}{x}\right).$$

y_1 and y_2 are equal when $x = 3$.

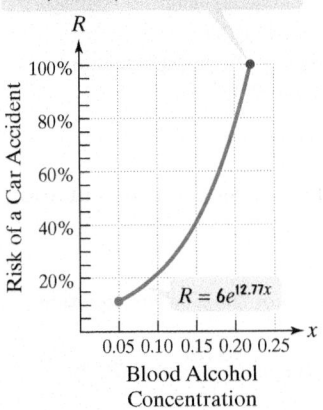

A blood alcohol concentration of 0.22 corresponds to near certainty, or a 100% probability, of a car accident.

FIGURE 4.20

EXAMPLE 9 Alcohol and Risk of a Car Accident

Medical research indicates that the risk of having a car accident increases exponentially as the concentration of alcohol in the blood increases. The risk is modeled by

$$R = 6e^{12.77x},$$

where x is the blood alcohol concentration and R, given as a percent, is the risk of having a car accident. What blood alcohol concentration corresponds to a 17% risk of a car accident? How is this shown on the graph of R in **Figure 4.20**?

SOLUTION

For a risk of 17%, we let $R = 17$ in the equation and solve for x, the blood alcohol concentration.

$R = 6e^{12.77x}$	This is the given equation.
$6e^{12.77x} = 17$	Substitute 17 for R and (optional) reverse the two sides of the equation.
$e^{12.77x} = \dfrac{17}{6}$	Isolate the exponential factor by dividing both sides by 6.
$\ln e^{12.77x} = \ln\left(\dfrac{17}{6}\right)$	Take the natural logarithm on both sides.
$12.77x = \ln\left(\dfrac{17}{6}\right)$	Use the inverse property $\ln e^x = x$ on the left side.
$x = \dfrac{\ln\left(\dfrac{17}{6}\right)}{12.77} \approx 0.08$	Divide both sides by 12.77.

For a blood alcohol concentration of 0.08, the risk of a car accident is 17%. This is shown on the graph of R in **Figure 4.20** by the point $(0.08, 17)$ that lies on the blue curve. Take a moment to locate this point on the curve. In many states, it is illegal to drive with a blood alcohol concentration of 0.08. • • •

⊘ Check Point **9** Use the formula in Example 9 to answer this question: What blood alcohol concentration corresponds to a 7% risk of a car accident? (In many states, drivers under the age of 21 can lose their licenses for driving at this level.)

Suppose that you inherit $30,000 at age 20. Is it possible to invest $25,000 and have over half a million dollars for early retirement? Our next example illustrates the power of compound interest.

EXAMPLE 10 Revisiting the Formula for Compound Interest

The formula

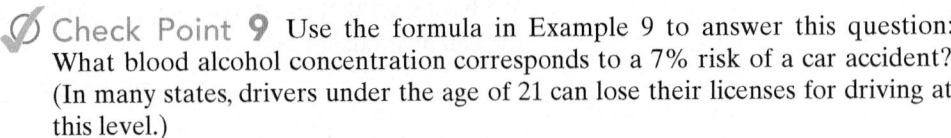

$$A = P\left(1 + \frac{r}{n}\right)^{nt}$$

describes the accumulated value, A, of a sum of money, P, the principal, after t years at annual percentage rate r (in decimal form) compounded n times a year. How long will it take $25,000 to grow to $500,000 at 9% annual interest compounded monthly?

SOLUTION

$$A = P\left(1 + \frac{r}{n}\right)^{nt}$$

This is the given formula.

$$500{,}000 = 25{,}000\left(1 + \frac{0.09}{12}\right)^{12t}$$

A(the desired accumulated value) = $500,000,
P(the principal) = $25,000,
r(the interest rate) = 9% = 0.09, and n = 12
(monthly compounding).

Because of the exponent, 12*t*, do not distribute
25,000 over the terms in parentheses.

Blitzer Bonus
**Playing Doubles:
Interest Rates
and Doubling Time**

One way to calculate what your
savings will be worth at some point
in the future is to consider doubling
time. The following table shows
how long it takes for your money
to double at different annual
interest rates subject to continuous
compounding.

Annual Interest Rate	Years to Double
5%	13.9 years
7%	9.9 years
9%	7.7 years
11%	6.3 years

Of course, the first problem is
collecting some money to invest.
The second problem is finding a
reasonably safe investment with a
return of 9% or more.

Our goal is to solve the equation for t. Let's reverse the two sides of the equation
and then simplify within parentheses.

$$25{,}000\left(1 + \frac{0.09}{12}\right)^{12t} = 500{,}000$$

Reverse the two sides of the previous equation.

$$25{,}000(1 + 0.0075)^{12t} = 500{,}000$$

Divide within parentheses: $\frac{0.09}{12} = 0.0075$.

$$25{,}000(1.0075)^{12t} = 500{,}000$$

Add within parentheses.

$$(1.0075)^{12t} = 20$$

Divide both sides by 25,000.

$$\ln(1.0075)^{12t} = \ln 20$$

Take the natural logarithm on both sides.

$$12t \ln(1.0075) = \ln 20$$

Use the power rule to bring the exponent to the
front: $\ln b^x = x \ln b$.

$$t = \frac{\ln 20}{12 \ln 1.0075}$$

Solve for t, dividing both sides by 12 ln 1.0075.

$$\approx 33.4$$

Use a calculator.

After approximately 33.4 years, the $25,000 will grow to an accumulated value of
$500,000. If you set aside the money at age 20, you can begin enjoying a life as a
half-millionaire at about age 53. •••

✓ Check Point **10** How long, to the nearest tenth of a year, will it take $1000 to
grow to $3600 at 8% annual interest compounded quarterly?

EXAMPLE 11 Revisiting the Model for Height of Children

We have seen that the percentage of adult height attained by a boy who is x years
old can be modeled by

$$f(x) = 29 + 48.8 \log(x + 1),$$

where x represents the boy's age (from 5 to 15) and $f(x)$ represents the percentage
of his adult height. At what age, rounded to the nearest year, has a boy attained
85% of his adult height?

SOLUTION

To find at what age a boy has attained 85% of his adult height, we substitute 85 for
$f(x)$ and solve for x, the boy's age.

$$f(x) = 29 + 48.8 \log(x + 1)$$

This is the given function.

$$85 = 29 + 48.8 \log(x + 1)$$

Substitute 85 for f(x).

Our goal is to isolate log(*x* + 1) and then
rewrite the equation in exponential form.

$$56 = 48.8 \log(x + 1)$$

Subtract 29 from both sides.

$$\frac{56}{48.8} = \log(x + 1)$$

Divide both sides by 48.8.

$$\frac{56}{48.8} = \log_{10}(x + 1)$$

Rewrite the common logarithm showing base 10. This step is optional.

$$10^{\frac{56}{48.8}} = x + 1$$

Rewrite in exponential form.

$$10^{\frac{56}{48.8}} - 1 = x$$

Subtract 1 from both sides.

$$13 \approx x$$

Use a calculator:
10 ∧ ((56 ÷ 48.8)) − 1 ENTER.

At approximately age 13, a boy has attained 85% of his adult height. This is shown on the graph of the model in **Figure 4.21** by the point $(13, 85)$.

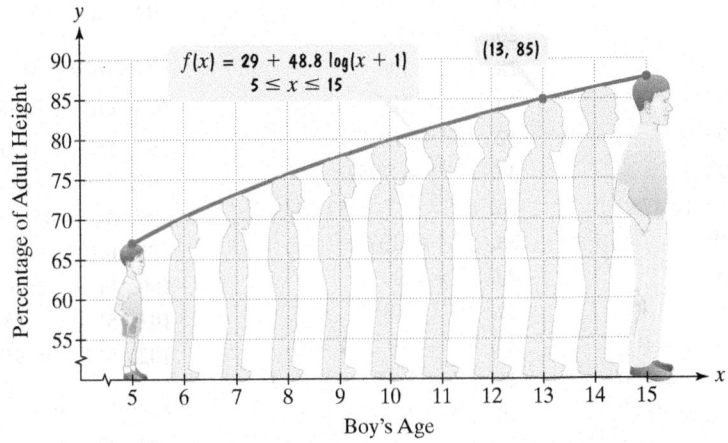

FIGURE 4.21 Graph of a model for the percentage of adult height attained by a boy • • •

⌀ Check Point **11** The percentage of adult height attained by a girl who is x years old can be modeled by

$$f(x) = 62 + 35 \log(x - 4),$$

where x represents the girl's age (from 5 to 15) and $f(x)$ represents the percentage of her adult height. At what age has a girl attained 97% of her adult height?

CONCEPT AND VOCABULARY CHECK

Fill in each blank so that the resulting statement is true.

1. If $b^M = b^N$, then _____.

2. If $2^{4x-1} = 2^7$, then _____ = 7.

3. If $x \ln 9 = \ln 20$, then $x =$ _____.

4. If $e^{0.6x} = 6$, then $0.6x =$ _____.

5. If $\log_5(x + 1) = 3$, then _____ = $x + 1$.

6. If $\log_3 x + \log_3(x + 1) = 2$, then \log_3 _____ = 2.

7. If $\ln\left(\dfrac{7x - 23}{x + 1}\right) = \ln(x - 3)$, then _____ = $x - 3$.

8. True or false: $x^4 = 15$ is an exponential equation. _____

9. True or false: $4^x = 15$ is an exponential equation. _____

10. True or false: -3 is a solution of $\log_5 9 = 2 \log_5 x$. _____

11. True or false: -10 is a solution of $\log_5(x + 35) = 2$. _____

EXERCISE SET 4.4

Practice Exercises

Solve each exponential equation in Exercises 1–22 by expressing each side as a power of the same base and then equating exponents.

1. $2^x = 64$

2. $3^x = 81$

3. $5^x = 125$

4. $5^x = 625$

5. $2^{2x-1} = 32$

6. $3^{2x+1} = 27$

7. $4^{2x-1} = 64$

8. $5^{3x-1} = 125$

9. $32^x = 8$

10. $4^x = 32$

11. $9^x = 27$

12. $125^x = 625$

13. $3^{1-x} = \frac{1}{27}$

14. $5^{2-x} = \frac{1}{125}$

15. $6^{\frac{x-3}{4}} = \sqrt{6}$

16. $7^{\frac{x-2}{6}} = \sqrt{7}$

17. $4^x = \dfrac{1}{\sqrt{2}}$

18. $9^x = \dfrac{1}{\sqrt[3]{3}}$

19. $8^{x+3} = 16^{x-1}$

20. $8^{1-x} = 4^{x+2}$

21. $e^{x+1} = \dfrac{1}{e}$

22. $e^{x+4} = \dfrac{1}{e^{2x}}$

Solve each exponential equation in Exercises 23–48. Express the solution set in terms of natural logarithms or common logarithms. Then use a calculator to obtain a decimal approximation, correct to two decimal places, for the solution.

23. $10^x = 3.91$

24. $10^x = 8.07$

25. $e^x = 5.7$

26. $e^x = 0.83$

27. $5^x = 17$

28. $19^x = 143$

29. $5e^x = 23$

30. $9e^x = 107$

31. $3e^{5x} = 1977$

32. $4e^{7x} = 10{,}273$

33. $e^{1-5x} = 793$

34. $e^{1-8x} = 7957$

35. $e^{5x-3} - 2 = 10{,}476$

36. $e^{4x-5} - 7 = 11{,}243$

37. $7^{x+2} = 410$

38. $5^{x-3} = 137$

39. $7^{0.3x} = 813$

40. $3^{\frac{x}{7}} = 0.2$

41. $5^{2x+3} = 3^{x-1}$

42. $7^{2x+1} = 3^{x+2}$

43. $e^{2x} - 3e^x + 2 = 0$

44. $e^{2x} - 2e^x - 3 = 0$

45. $e^{4x} + 5e^{2x} - 24 = 0$

46. $e^{4x} - 3e^{2x} - 18 = 0$

47. $3^{2x} + 3^x - 2 = 0$

48. $2^{2x} + 2^x - 12 = 0$

Solve each logarithmic equation in Exercises 49–92. Be sure to reject any value of x that is not in the domain of the original logarithmic expressions. Give the exact answer. Then, where necessary, use a calculator to obtain a decimal approximation, correct to two decimal places, for the solution.

49. $\log_3 x = 4$

50. $\log_5 x = 3$

51. $\ln x = 2$

52. $\ln x = 3$

53. $\log_4(x + 5) = 3$

54. $\log_5(x - 7) = 2$

55. $\log_2(x + 25) = 4$

56. $\log_2(x + 50) = 5$

57. $\log_3(x + 4) = -3$

58. $\log_7(x + 2) = -2$

59. $\log_4(3x + 2) = 3$

60. $\log_2(4x + 1) = 5$

61. $5 \ln(2x) = 20$

62. $6 \ln(2x) = 30$

63. $6 + 2 \ln x = 5$

64. $7 + 3 \ln x = 6$

65. $\ln\sqrt{x + 3} = 1$

66. $\ln\sqrt{x + 4} = 1$

67. $\log_5 x + \log_5(4x - 1) = 1$

68. $\log_6(x + 5) + \log_6 x = 2$

69. $\log_3(x + 6) + \log_3(x + 4) = 1$

70. $\log_6(x + 3) + \log_6(x + 4) = 1$

71. $\log_2(x + 2) - \log_2(x - 5) = 3$

72. $\log_4(x + 2) - \log_4(x - 1) = 1$

73. $2 \log_3(x + 4) = \log_3 9 + 2$

74. $3 \log_2(x - 1) = 5 - \log_2 4$

75. $\log_2(x - 6) + \log_2(x - 4) - \log_2 x = 2$

76. $\log_2(x - 3) + \log_2 x - \log_2(x + 2) = 2$

77. $\log(x + 4) = \log x + \log 4$

78. $\log(5x + 1) = \log(2x + 3) + \log 2$

79. $\log(3x - 3) = \log(x + 1) + \log 4$

80. $\log(2x - 1) = \log(x + 3) + \log 3$

81. $2 \log x = \log 25$

82. $3 \log x = \log 125$

83. $\log(x + 4) - \log 2 = \log(5x + 1)$

84. $\log(x + 7) - \log 3 = \log(7x + 1)$

85. $2 \log x - \log 7 = \log 112$

86. $\log(x - 2) + \log 5 = \log 100$

87. $\log x + \log(x + 3) = \log 10$

88. $\log(x + 3) + \log(x - 2) = \log 14$

89. $\ln(x - 4) + \ln(x + 1) = \ln(x - 8)$

90. $\log_2(x - 1) - \log_2(x + 3) = \log_2\left(\dfrac{1}{x}\right)$

91. $\ln(x - 2) - \ln(x + 3) = \ln(x - 1) - \ln(x + 7)$

92. $\ln(x - 5) - \ln(x + 4) = \ln(x - 1) - \ln(x + 2)$

Practice Plus

In Exercises 93–102, solve each equation.

93. $5^{2x} \cdot 5^{4x} = 125$

94. $3^{x+2} \cdot 3^x = 81$

95. $2|\ln x| - 6 = 0$

96. $3|\log x| - 6 = 0$

97. $3^{x^2} = 45$

98. $5^{x^2} = 50$

99. $\ln(2x + 1) + \ln(x - 3) - 2 \ln x = 0$

100. $\ln 3 - \ln(x + 5) - \ln x = 0$

101. $5^{x^2-12} = 25^{2x}$

102. $3^{x^2-12} = 9^{2x}$

Application Exercises

103. The formula $A = 37.3e^{0.0095t}$ models the population of California, A, in millions, t years after 2010.

 a. What was the population of California in 2010?

 b. When will the population of California reach 40 million?

104. The formula $A = 25.1e^{0.0187t}$ models the population of Texas, A, in millions, t years after 2010.

 a. What was the population of Texas in 2010?

 b. When will the population of Texas reach 28 million?

The function $f(x) = 20(0.975)^x$ models the percentage of surface sunlight, $f(x)$, that reaches a depth of x feet beneath the surface of the ocean. The figure shows the graph of this function. Use this information to solve Exercises 105–106.

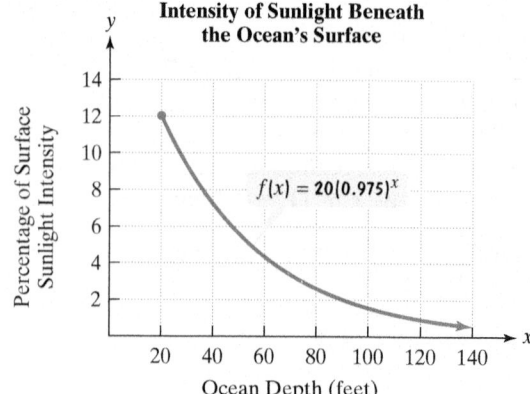

105. Use the function to determine at what depth, to the nearest foot, there is 1% of surface sunlight. How is this shown on the graph of f?

106. Use the function to determine at what depth, to the nearest foot, there is 3% of surface sunlight. How is this shown on the graph of f?

In Exercises 107–110, complete the table for a savings account subject to n compoundings yearly $\left[A = P\left(1 + \dfrac{r}{n}\right)^{nt} \right]$. Round answers to one decimal place.

	Amount Invested	Number of Compounding Periods	Annual Interest Rate	Accumulated Amount	Time t in Years
107.	$12,500	4	5.75%	$20,000	
108.	$7250	12	6.5%	$15,000	
109.	$1000	360	16.8%	$1400	
110.	$5000	360	14.7%	$9000	

In Exercises 111–114, complete the table for a savings account subject to continuous compounding ($A = Pe^{rt}$). Round answers to one decimal place.

	Amount Invested	Annual Interest Rate	Accumulated Amount	Time t in Years
111.	$8000	8%	Double the amount invested	
112.	$8000	20.3%	$12,000	
113.	$2350	15.7%	Triple the amount invested	
114.	$17,425	4.25%	$25,000	

By 2019, nearly $1 out of every $5 spent in the U.S. economy is projected to go for health care. The bar graph shows the percentage of the U.S. gross domestic product (GDP) going toward health care from 2007 through 2010, with projections for 2014 and 2019.

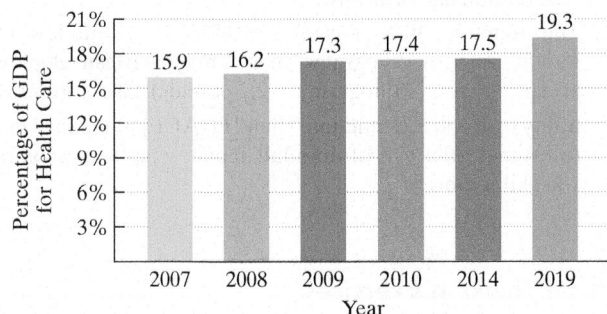

Percentage of the Gross Domestic Product in the United States Going Toward Health Care

Source: Health Affairs (healthaffairs.org)

The data can be modeled by the function $f(x) = 1.2 \ln x + 15.7$, where $f(x)$ is the percentage of the U.S. gross domestic product going toward health care x years after 2006. Use this information to solve Exercises 115–116.

115. a. Use the function to determine the percentage of the U.S. gross domestic product that went toward health care in 2009. Round to the nearest tenth of a percent. Does this underestimate or overestimate the percent displayed by the graph? By how much?

 b. According to the model, when will 18.5% of the U.S. gross domestic product go toward health care? Round to the nearest year.

116. a. Use the function to determine the percentage of the U.S. gross domestic product that went toward health care in 2008. Round to the nearest tenth of a percent. Does this underestimate or overestimate the percent displayed by the graph? By how much?

 b. According to the model, when will 18.6% of the U.S. gross domestic product go toward health care? Round to the nearest year.

The function $P(x) = 95 - 30 \log_2 x$ models the percentage, $P(x)$, of students who could recall the important features of a classroom lecture as a function of time, where x represents the number of days that have elapsed since the lecture was given. The figure shows the graph of the function. Use this information to solve Exercises 117–118. Round answers to one decimal place.

117. After how many days do only half the students recall the important features of the classroom lecture? (Let $P(x) = 50$ and solve for x.) Locate the point on the graph that conveys this information.

118. After how many days have all students forgotten the important features of the classroom lecture? (Let $P(x) = 0$ and solve for x.) Locate the point on the graph that conveys this information.

The pH scale is used to measure the acidity or alkalinity of a solution. The scale ranges from 0 to 14. A neutral solution, such as pure water, has a pH of 7. An acid solution has a pH less than 7 and an alkaline solution has a pH greater than 7. The lower the pH below 7, the more acidic is the solution. Each whole-number decrease in pH represents a tenfold increase in acidity.

The pH Scale

The pH of a solution is given by

$$pH = -\log x,$$

where x represents the concentration of the hydrogen ions in the solution, in moles per liter. Use the formula to solve Exercises 119–120. Express answers as powers of 10.

119. a. Normal, unpolluted rain has a pH of about 5.6. What is the hydrogen ion concentration?

b. An environmental concern involves the destructive effects of acid rain. The most acidic rainfall ever had a pH of 2.4. What was the hydrogen ion concentration?

c. How many times greater is the hydrogen ion concentration of the acidic rainfall in part (b) than the normal rainfall in part (a)?

120. a. The figure indicates that lemon juice has a pH of 2.3. What is the hydrogen ion concentration?

b. Stomach acid has a pH that ranges from 1 to 3. What is the hydrogen ion concentration of the most acidic stomach?

c. How many times greater is the hydrogen ion concentration of the acidic stomach in part (b) than the lemon juice in part (a)?

Writing in Mathematics

121. Explain how to solve an exponential equation when both sides can be written as a power of the same base.

122. Explain how to solve an exponential equation when both sides cannot be written as a power of the same base. Use $3^x = 140$ in your explanation.

123. Explain the differences between solving $\log_3(x - 1) = 4$ and $\log_3(x - 1) = \log_3 4$.

124. In many states, a 17% risk of a car accident with a blood alcohol concentration of 0.08 is the lowest level for charging a motorist with driving under the influence. Do you agree with the 17% risk as a cutoff percentage, or do you feel that the percentage should be lower or higher? Explain your answer. What blood alcohol concentration corresponds to what you believe is an appropriate percentage?

Technology Exercises

In Exercises 125–132, use your graphing utility to graph each side of the equation in the same viewing rectangle. Then use the x-coordinate of the intersection point to find the equation's solution set. Verify this value by direct substitution into the equation.

125. $2^{x+1} = 8$

126. $3^{x+1} = 9$

127. $\log_3(4x - 7) = 2$

128. $\log_3(3x - 2) = 2$

129. $\log(x + 3) + \log x = 1$

130. $\log(x - 15) + \log x = 2$

131. $3^x = 2x + 3$

132. $5^x = 3x + 4$

Hurricanes are one of nature's most destructive forces. These low-pressure areas often have diameters of over 500 miles. The function $f(x) = 0.48 \ln(x + 1) + 27$ models the barometric air pressure, $f(x)$, in inches of mercury, at a distance of x miles from the eye of a hurricane. Use this function to solve Exercises 133–134.

133. Graph the function in a $[0, 500, 50]$ by $[27, 30, 1]$ viewing rectangle. What does the shape of the graph indicate about barometric air pressure as the distance from the eye increases?

134. Use an equation to answer this question: How far from the eye of a hurricane is the barometric air pressure 29 inches of mercury? Use the TRACE and ZOOM features or the intersect command of your graphing utility to verify your answer.

135. The function $P(t) = 145e^{-0.092t}$ models a runner's pulse, $P(t)$, in beats per minute, t minutes after a race, where $0 \le t \le 15$. Graph the function using a graphing utility. TRACE along the graph and determine after how many minutes the runner's pulse will be 70 beats per minute. Round to the nearest tenth of a minute. Verify your observation algebraically.

136. The function $W(t) = 2600(1 - 0.51e^{-0.075t})^3$ models the weight, $W(t)$, in kilograms, of a female African elephant at age t years. (1 kilogram \approx 2.2 pounds) Use a graphing utility to graph the function. Then TRACE along the curve to estimate the age of an adult female elephant weighing 1800 kilograms.

Critical Thinking Exercises

Make Sense? *In Exercises 137–140, determine whether each statement makes sense or does not make sense, and explain your reasoning.*

137. Because the equations $2^x = 15$ and $2^x = 16$ are similar, I solved them using the same method.

138. Because the equations

$$\log(3x + 1) = 5 \text{ and } \log(3x + 1) = \log 5$$

are similar, I solved them using the same method.

139. I can solve $4^x = 15$ by writing the equation in logarithmic form.

140. It's important for me to check that the proposed solution of an equation with logarithms gives only logarithms of positive numbers in the original equation.

In Exercises 141–144, determine whether each statement is true or false. If the statement is false, make the necessary change(s) to produce a true statement.

141. If $\log(x + 3) = 2$, then $e^2 = x + 3$.

142. If $\log(7x + 3) - \log(2x + 5) = 4$, then the equation in exponential form is $10^4 = (7x + 3) - (2x + 5)$.

143. If $x = \dfrac{1}{k}\ln y$, then $y = e^{kx}$.

144. Examples of exponential equations include $10^x = 5.71$, $e^x = 0.72$, and $x^{10} = 5.71$.

145. If $4000 is deposited into an account paying 3% interest compounded annually and at the same time $2000 is deposited into an account paying 5% interest compounded annually, after how long will the two accounts have the same balance? Round to the nearest year.

Solve each equation in Exercises 146–148. Check each proposed solution by direct substitution or with a graphing utility.

146. $(\ln x)^2 = \ln x^2$

147. $(\log x)(2 \log x + 1) = 6$

148. $\ln(\ln x) = 0$

Group Exercise

149. Research applications of logarithmic functions as mathematical models and plan a seminar based on your group's research. Each group member should research one of the following areas or any other area of interest: pH (acidity of solutions), intensity of sound (decibels), brightness of stars, human memory, progress over time in a sport, profit over time. For the area that you select, explain how logarithmic functions are used and provide examples.

Preview Exercises

Exercises 150–152 will help you prepare for the material covered in the next section.

150. The formula $A = 10e^{-0.003t}$ models the population of Hungary, A, in millions, t years after 2006.

 a. Find Hungary's population, in millions, for 2006, 2007, 2008, and 2009. Round to two decimal places.

 b. Is Hungary's population increasing or decreasing?

151. a. Simplify: $e^{\ln 3}$.

 b. Use your simplification from part (a) to rewrite 3^x in terms of base e.

152. U.S. soldiers fight Russian troops who have invaded New York City. Incoming missiles from Russian submarines and warships ravage the Manhattan skyline. It's just another scenario for the multi-billion-dollar video games *Call of Duty*, which have sold more than 100 million games since the franchise's birth in 2003.

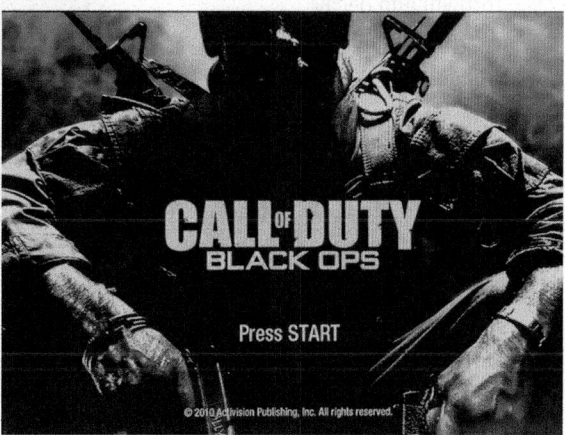

The table shows the annual retail sales for *Call of Duty* video games from 2004 through 2010. Create a scatter plot for the data. Based on the shape of the scatter plot, would a logarithmic function, an exponential function, or a linear function be the best choice for modeling the data?

Annual Retail Sales for *Call of Duty* Games

Year	Retail Sales (millions of dollars)
2004	56
2005	101
2006	196
2007	352
2008	436
2009	778
2010	980

Source: The NPD Group

SECTION 4.5

SECTION 4.5 Exponential Growth and Decay; Modeling Data

Objectives

1. Model exponential growth and decay.
2. Use logistic growth models.
3. Choose an appropriate model for data.
4. Express an exponential model in base *e*.

On October 31, 2011, the world marked a major milestone: According to the United Nations, the number of people on Earth reached 7 billion—and counting. Since the dawn of humankind some 50,000 years ago, an estimated total of 108 billion people have lived on our planet, which means that about 6.5% of all humans ever born are alive today. That's a lot of bodies to feed, clothe, and shelter. Scientists, politicians, economists, and demographers have long disagreed when it comes to making predictions about the effects of the world's growing population. Debates about entities that are growing exponentially can be approached mathematically: We can create functions that model data and use these functions to make predictions. In this section, we will show you how this is done.

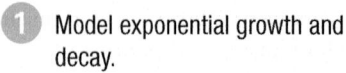

① Model exponential growth and decay.

Exponential Growth and Decay

One of algebra's many applications is to predict the behavior of variables. This can be done with *exponential growth* and *decay models*. With exponential growth or decay, quantities grow or decay at a rate directly proportional to their size. Populations that are growing exponentially grow extremely rapidly as they get larger because there are more adults to have offspring. For example, world population is increasing at approximately 1.2% per year. This means that each year world population is 1.2% more than what it was in the previous year. In 2010, world population was 6.9 billion. Thus, we compute the world population in 2011 as follows:

$$6.9 \text{ billion} + 1.2\% \text{ of } 6.9 \text{ billion} = 6.9 + (0.012)(6.9) = 6.9828.$$

This computation indicates that 6.9828 billion people populated the world in 2011. The 0.0828 billion represents an increase of 82.8 million people from 2010 to 2011, the equivalent of the population of Germany. Using 1.2% as the annual rate of increase, world population for 2012 is found in a similar manner:

$$6.9828 + 1.2\% \text{ of } 6.9828 = 6.9828 + (0.012)(6.9828) \approx 7.067.$$

This computation indicates that approximately 7.1 billion people populated the world in 2012.

The explosive growth of world population may remind you of the growth of money in an account subject to compound interest. Just as the growth rate for world population is multiplied by the population plus any increase in the population, a compound interest rate is multiplied by the original investment plus any accumulated interest. The balance in an account subject to continuous compounding and world population are special cases of *exponential growth models*.

GREAT QUESTION!

Why does the formula for exponential growth look familiar?

You have seen the formula for exponential growth before, but with different letters. It is the formula for compound interest with continuous compounding.

$$A = Pe^{rt}$$

| Amount at time t | Principal is the original amount. | Interest rate is the growth rate. |

$$A = A_0 e^{kt}$$

Exponential Growth and Decay Models

The mathematical model for **exponential growth** or **decay** is given by

$$f(t) = A_0 e^{kt} \quad \text{or} \quad A = A_0 e^{kt}.$$

- **If $k > 0$, the function models the amount, or size, of a *growing* entity.** A_0 is the original amount, or size, of the growing entity at time $t = 0$, A is the amount at time t, and k is a constant representing the growth rate.
- **If $k < 0$, the function models the amount, or size, of a *decaying* entity.** A_0 is the original amount, or size, of the decaying entity at time $t = 0$, A is the amount at time t, and k is a constant representing the decay rate.

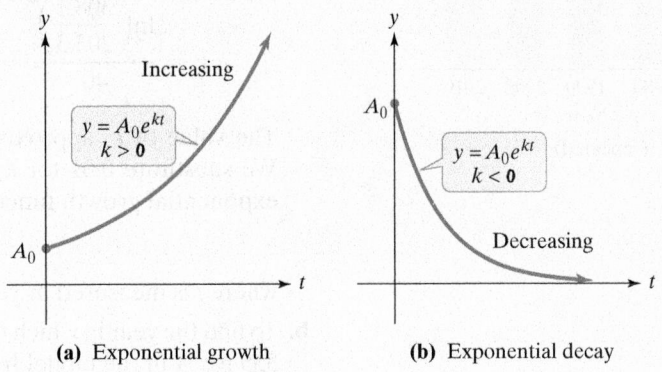

(a) Exponential growth **(b)** Exponential decay

Sometimes we need to use given data to determine k, the rate of growth or decay. After we compute the value of k, we can use the formula $A = A_0 e^{kt}$ to make predictions. This idea is illustrated in our first two examples.

EXAMPLE 1 Modeling the Growth of the U.S. Population

The graph in **Figure 4.22** shows the U.S. population, in millions, for five selected years from 1970 through 2010. In 1970, the U.S. population was 203.3 million. By 2010, it had grown to 308.7 million.

a. Find an exponential growth function that models the data for 1970 through 2010.

b. By which year will the U.S. population reach 335 million?

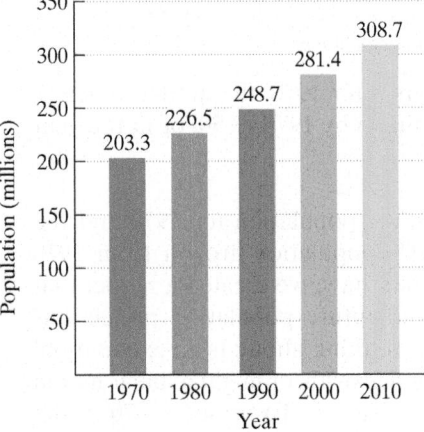

FIGURE 4.22
Source: U.S. Census Bureau

SOLUTION

a. We use the exponential growth model

$$A = A_0 e^{kt}$$

in which t is the number of years after 1970. This means that 1970 corresponds to $t = 0$. At that time the U.S. population was 203.3 million, so we substitute 203.3 for A_0 in the growth model:

$$A = 203.3 e^{kt}.$$

We are given that 308.7 million was the population in 2010. Because 2010 is 40 years after 1970, when $t = 40$ the value of A is 308.7. Substituting these numbers into the growth model will enable us to find k, the growth rate. We know that $k > 0$ because the problem involves growth.

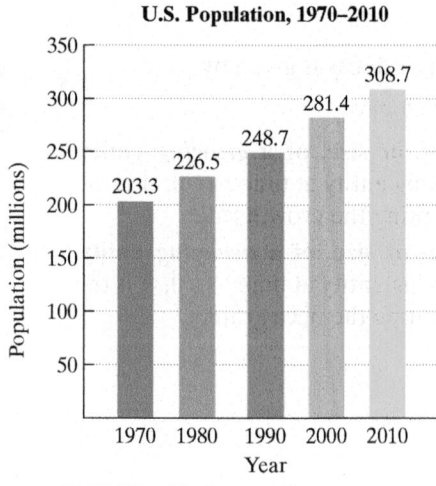

U.S. Population, 1970–2010

FIGURE 4.22 (repeated)

$$A = 203.3e^{kt}$$
$$308.7 = 203.3e^{k \cdot 40}$$

Use the growth model with $A_0 = 203.3$. When $t = 40$, $A = 308.7$. Substitute these numbers into the model.

$$e^{40k} = \frac{308.7}{203.3}$$

Isolate the exponential factor by dividing both sides by 203.3. We also reversed the sides.

$$\ln e^{40k} = \ln\left(\frac{308.7}{203.3}\right)$$

Take the natural logarithm on both sides.

$$40k = \ln\left(\frac{308.7}{203.3}\right)$$

Simplify the left side using $\ln e^x = x$.

$$k = \frac{\ln\left(\frac{308.7}{203.3}\right)}{40} \approx 0.01$$

Divide both sides by 40 and solve for k. Then use a calculator.

The value of k, approximately 0.01, indicates a growth rate of about 1%. We substitute 0.01 for k in the growth model, $A = 203.3e^{kt}$, to obtain an exponential growth function for the U.S. population. It is

$$A = 203.3e^{0.01t},$$

where t is measured in years after 1970.

b. To find the year in which the U.S. population will reach 335 million, substitute 335 for A in the model from part (a) and solve for t.

$$A = 203.3e^{0.01t}$$

This is the model from part (a).

$$335 = 203.3e^{0.01t}$$

Substitute 335 for A.

$$e^{0.01t} = \frac{335}{203.3}$$

Divide both sides by 203.3. We also reversed the sides.

$$\ln e^{0.01t} = \ln\left(\frac{335}{203.3}\right)$$

Take the natural logarithm on both sides.

$$0.01t = \ln\left(\frac{335}{203.3}\right)$$

Simplify on the left using $\ln e^x = x$.

$$t = \frac{\ln\left(\frac{335}{203.3}\right)}{0.01} \approx 50$$

Divide both sides by 0.01 and solve for t. Then use a calculator.

Because t represents the number of years after 1970, the model indicates that the U.S. population will reach 335 million by 1970 + 50, or in the year 2020. • • •

In Example 1, we used only two data values, the population for 1970 and the population for 2010, to develop a model for U.S. population growth from 1970 through 2010. By not using data for any other years, have we created a model that inaccurately describes both the existing data and future population projections given by the U.S. Census Bureau? Something else to think about: Is an exponential model the best choice for describing U.S. population growth, or might a linear model provide a better description? We return to these issues in Exercises 66–70 in the Exercise Set.

☑ Check Point **1** In 2000, the population of Africa was 807 million and by 2011 it had grown to 1052 million.

a. Use the exponential growth model $A = A_0e^{kt}$, in which t is the number of years after 2000, to find the exponential growth function that models the data.

b. By which year will Africa's population reach 2000 million, or two billion?

Our next example involves exponential decay and its use in determining the age of fossils and artifacts. The method is based on considering the percentage of carbon-14 remaining in the fossil or artifact. Carbon-14 decays exponentially with a *half-life* of approximately 5715 years. The **half-life** of a substance is the time required for half of a given sample to disintegrate. Thus, after 5715 years a given amount of carbon-14 will have decayed to half the original amount. Carbon dating is useful for artifacts or fossils up to 80,000 years old. Older objects do not have enough carbon-14 left to determine age accurately.

EXAMPLE 2 Carbon-14 Dating: The Dead Sea Scrolls

a. Use the fact that after 5715 years a given amount of carbon-14 will have decayed to half the original amount to find the exponential decay model for carbon-14.

b. In 1947, earthenware jars containing what are known as the Dead Sea Scrolls were found by an Arab Bedouin herdsman. Analysis indicated that the scroll wrappings contained 76% of their original carbon-14. Estimate the age of the Dead Sea Scrolls.

SOLUTION

a. We begin with the exponential decay model $A = A_0 e^{kt}$. We know that $k < 0$ because the problem involves the decay of carbon-14. After 5715 years ($t = 5715$), the amount of carbon-14 present, A, is half the original amount, A_0.

Thus, we can substitute $\dfrac{A_0}{2}$ for A in the exponential decay model. This will enable us to find k, the decay rate.

$A = A_0 e^{kt}$	Begin with the exponential decay model.
$\dfrac{A_0}{2} = A_0 e^{k5715}$	After 5715 years ($t = 5715$), $A = \dfrac{A_0}{2}$ (because the amount present, A, is half the original amount, A_0).
$\dfrac{1}{2} = e^{5715k}$	Divide both sides of the equation by A_0.
$\ln\left(\dfrac{1}{2}\right) = \ln e^{5715k}$	Take the natural logarithm on both sides.
$\ln\left(\dfrac{1}{2}\right) = 5715k$	Simplify the right side using $\ln e^x = x$.
$k = \dfrac{\ln\left(\dfrac{1}{2}\right)}{5715} \approx -0.000121$	Divide both sides by 5715 and solve for k.

Substituting -0.000121 for k in the decay model, $A = A_0 e^{kt}$, the model for carbon-14 is

$$A = A_0 e^{-0.000121t}.$$

b. In 1947, the Dead Sea Scrolls contained 76% of their original carbon-14. To find their age in 1947, substitute $0.76A_0$ for A in the model from part (a) and solve for t.

$A = A_0 e^{-0.000121t}$	This is the decay model for carbon-14.
$0.76A_0 = A_0 e^{-0.000121t}$	A, the amount present, is 76% of the original amount, so $A = 0.76A_0$.
$0.76 = e^{-0.000121t}$	Divide both sides of the equation by A_0.
$\ln 0.76 = \ln e^{-0.000121t}$	Take the natural logarithm on both sides.
$\ln 0.76 = -0.000121t$	Simplify the right side using $\ln e^x = x$.
$t = \dfrac{\ln 0.76}{-0.000121} \approx 2268$	Divide both sides by -0.000121 and solve for t.

The Dead Sea Scrolls are approximately 2268 years old plus the number of years between 1947 and the current year.

● ● ●

Blitzer Bonus

Carbon Dating and Artistic Development

The artistic community was electrified by the discovery in 1995 of spectacular cave paintings in a limestone cavern in France. Carbon dating of the charcoal from the site showed that the images, created by artists of remarkable talent, were 30,000 years old, making them the oldest cave paintings ever found. The artists seemed to have used the cavern's natural contours to heighten a sense of perspective. The quality of the painting suggests that the art of early humans did not mature steadily from primitive to sophisticated in any simple linear fashion.

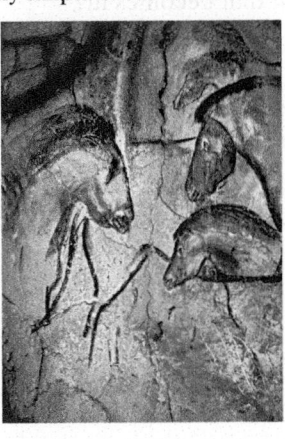

⊘ Check Point **2** Strontium-90 is a waste product from nuclear reactors. As a consequence of fallout from atmospheric nuclear tests, we all have a measurable amount of strontium-90 in our bones.

a. The half-life of strontium-90 is 28 years, meaning that after 28 years a given amount of the substance will have decayed to half the original amount. Find the exponential decay model for strontium-90.

b. Suppose that a nuclear accident occurs and releases 60 grams of strontium-90 into the atmosphere. How long will it take for strontium-90 to decay to a level of 10 grams?

② Use logistic growth models.

Logistic Growth Models

From population growth to the spread of an epidemic, nothing on Earth can grow exponentially indefinitely. Growth is always limited. This is shown in **Figure 4.23** by the horizontal asymptote. The *logistic growth model* is a function used to model situations of this type.

Logistic Growth Model

The mathematical model for limited logistic growth is given by

$$f(t) = \frac{c}{1 + ae^{-bt}} \quad \text{or} \quad A = \frac{c}{1 + ae^{-bt}},$$

where a, b, and c are constants, with $c > 0$ and $b > 0$.

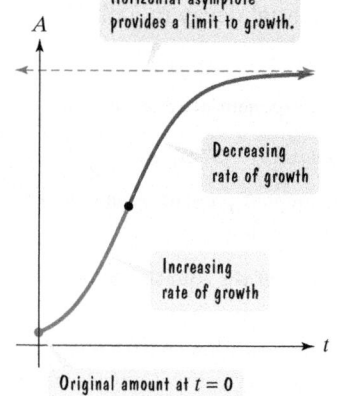

Horizontal asymptote provides a limit to growth.

Decreasing rate of growth

Increasing rate of growth

Original amount at $t = 0$

FIGURE 4.23 The logistic growth curve has a horizontal asymptote that identifies the limit of the growth of A over time.

As time increases ($t \to \infty$), the expression ae^{-bt} in the model approaches 0, and A gets closer and closer to c. This means that $y = c$ is a horizontal asymptote for the graph of the function. Thus, the value of A can never exceed c and c represents the limiting size that A can attain.

EXAMPLE 3 Modeling the Spread of the Flu

The function

$$f(t) = \frac{30,000}{1 + 20e^{-1.5t}}$$

describes the number of people, $f(t)$, who have become ill with influenza t weeks after its initial outbreak in a town with 30,000 inhabitants.

a. How many people became ill with the flu when the epidemic began?

b. How many people were ill by the end of the fourth week?

c. What is the limiting size of $f(t)$, the population that becomes ill?

SOLUTION

a. The time at the beginning of the flu epidemic is $t = 0$. Thus, we can find the number of people who were ill at the beginning of the epidemic by substituting 0 for t.

$$f(t) = \frac{30,000}{1 + 20e^{-1.5t}} \qquad \text{This is the given logistic growth function.}$$

$$f(0) = \frac{30,000}{1 + 20e^{-1.5(0)}} \qquad \text{When the epidemic began, } t = 0.$$

$$= \frac{30,000}{1 + 20} \qquad e^{-1.5(0)} = e^0 = 1$$

$$\approx 1429$$

Approximately 1429 people were ill when the epidemic began.

TECHNOLOGY

The graph of the logistic growth function for the flu epidemic

$$y = \frac{30,000}{1 + 20e^{-1.5x}}$$

can be obtained using a graphing utility. We started x at 0 and ended at 10. This takes us to week 10. (In Example 3, we found that by week 4 approximately 28,583 people were ill.) We also know that 30,000 is the limiting size, so we took values of y up to 30,000. Using a [0, 10, 1] by [0, 30,000, 3000] viewing rectangle, the graph of the logistic growth function is shown below.

b. We find the number of people who were ill at the end of the fourth week by substituting 4 for t in the logistic growth function.

$$f(t) = \frac{30,000}{1 + 20e^{-1.5t}} \quad \text{Use the given logistic growth function.}$$

$$f(4) = \frac{30,000}{1 + 20e^{-1.5(4)}} \quad \text{To find the number of people ill by the end of week four, let } t = 4.$$

$$\approx 28,583 \quad \text{Use a calculator.}$$

Approximately 28,583 people were ill by the end of the fourth week. Compared with the number of people who were ill initially, 1429, this illustrates the virulence of the epidemic.

c. Recall that in the logistic growth model, $f(t) = \dfrac{c}{1 + ae^{-bt}}$, the constant c represents the limiting size that $f(t)$ can attain. Thus, the number in the numerator, 30,000, is the limiting size of the population that becomes ill. •••

✓ Check Point **3** In a learning theory project, psychologists discovered that

$$f(t) = \frac{0.8}{1 + e^{-0.2t}}$$

is a model for describing the proportion of correct responses, $f(t)$, after t learning trials.

a. Find the proportion of correct responses prior to learning trials taking place.

b. Find the proportion of correct responses after 10 learning trials.

c. What is the limiting size of $f(t)$, the proportion of correct responses, as continued learning trials take place?

③ Choose an appropriate model for data.

Modeling Data

Throughout this chapter, we have been working with models that were given. However, we can create functions that model data by observing patterns in scatter plots. **Figure 4.24** shows scatter plots for data that are exponential, logarithmic, and linear.

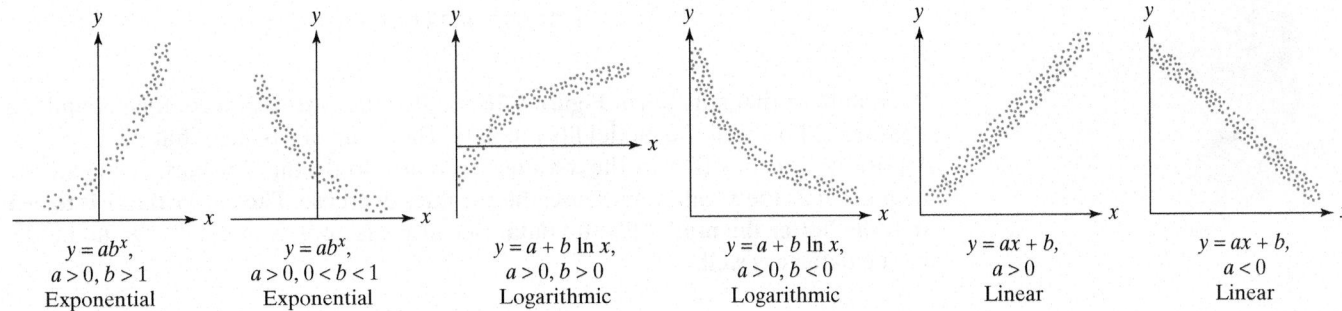

FIGURE 4.24 Scatter plots for exponential, logarithmic, and linear models

Table 4.6 Growth of the Human Brain

Age	Percentage of Adult Size Brain
1	30%
2	50%
4	78%
6	88%
8	92%
10	95%
11	99%

Source: Gerrig and Zimbardo, *Psychology and Life*, 18th Edition, Allyn and Bacon, 2008.

EXAMPLE 4 Choosing a Model for Data

The data in **Table 4.6** indicate that between the ages of 1 and 11, the human brain does not grow linearly, or steadily. A scatter plot for the data is shown in **Figure 4.25**. What type of function would be a good choice for modeling the data?

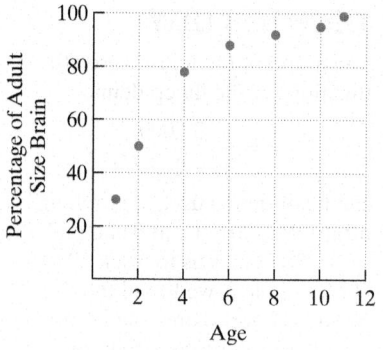

FIGURE 4.25

SOLUTION

Because the data in the scatter plot increase rapidly at first and then begin to level off a bit, the shape suggests that a logarithmic function is a good choice for modeling the data.

● ● ●

☑ Check Point **4** **Table 4.7** shows the populations of various cities, in thousands, and the average walking speed, in feet per second, of a person living in the city. Create a scatter plot for the data. Based on the scatter plot, what type of function would be a good choice for modeling the data?

Table 4.7 Population and Walking Speed

Population (thousands)	Walking Speed (feet per second)
5.5	0.6
14	1.0
71	1.6
138	1.9
342	2.2

Source: Mark H. Bornstein and Helen G. Bornstein, "The Pace of Life." *Nature*, 259, Feb. 19, 1976, pp. 557–559

Table 4.8

x, Age	y, Percentage of Adult Size Brain
1	30
2	50
4	78
6	88
8	92
10	95
11	99

```
LnReg
y=a+blnx
a=31.95404756
b=28.94733911
r²=.9806647799
r=.9902852013
```

FIGURE 4.26 A logarithmic model for the data in **Table 4.8**

How can we obtain a logarithmic function that models the data for the growth of the human brain? A graphing utility can be used to obtain a logarithmic model of the form $y = a + b \ln x$. **Because the domain of the logarithmic function is the set of positive numbers, zero must not be a value for x.** This is not a problem for the data giving the percentage of an adult size brain because the data begin at age 1. We will assign x to represent age and y to represent the percentage of an adult size brain. This gives us the data shown in **Table 4.8**. Using the logarithmic regression option, we obtain the equation in **Figure 4.26**.

From **Figure 4.26**, we see that the logarithmic model of the data, with numbers rounded to three decimal places, is

$$y = 31.954 + 28.947 \ln x.$$

The number r that appears in **Figure 4.26** is called the **correlation coefficient** and is a measure of how well the model fits the data. The value of r is such that $-1 \le r \le 1$. A positive r means that as the x-values increase, so do the y-values. A negative r means that as the x-values increase, the y-values decrease. **The closer that r is to −1 or 1, the better the model fits the data.** Because r is approximately 0.99, the model fits the data very well.

EXAMPLE 5 Choosing a Model for Data

Figure 4.27(a) shows world population, in billions, for seven selected years from 1950 through 2010. A scatter plot is shown in **Figure 4.27(b)**. Suggest two types of functions that would be good choices for modeling the data.

World Population, 1950–2010

FIGURE 4.27(a)

FIGURE 4.27(b)

Source: U.S. Census Bureau, International Database

Table 4.9 Percentage of U.S. Men Who Are Married or Who Have Been Married, by Age

Age	Percent
18	2
20	7
25	36
30	61
35	75

Source: National Center for Health Statistics

SOLUTION

Because the data in the scatter plot appear to increase more and more rapidly, the shape suggests that an exponential model might be a good choice. Furthermore, we can probably draw a line that passes through or near the seven points. Thus, a linear function would also be a good choice for modeling the data. • • •

✓ Check Point **5** **Table 4.9** shows the percentage of U.S. men who are married or who have been married, by age. Create a scatter plot for the data. Based on the scatter plot, what type of function would be a good choice for modeling the data?

EXAMPLE 6 Comparing Linear and Exponential Models

The data for world population are shown in **Table 4.10**. Using a graphing utility's linear regression feature and exponential regression feature, we enter the data and obtain the models shown in **Figure 4.28**.

Table 4.10

> Although the domain of $y = ab^x$ is the set of all real numbers, some graphing utilities only accept positive values for x. That's why we assigned x to represent the number of years after 1949.

x, Number of Years after 1949	y, World Population (billions)
1 (1950)	2.6
11 (1960)	3.0
21 (1970)	3.7
31 (1980)	4.5
41 (1990)	5.3
51 (2000)	6.1
61 (2010)	6.9

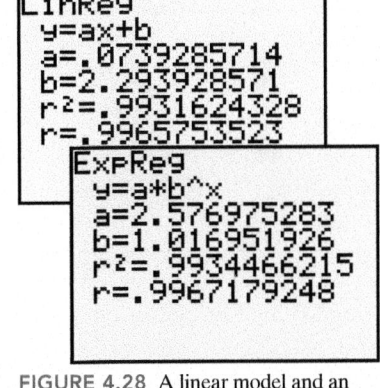

FIGURE 4.28 A linear model and an exponential model for the data in **Table 4.10**

Because r, the correlation coefficient, is close to 1 in each screen in **Figure 4.28**, the models fit the data very well.

a. Use **Figure 4.28** to express each model in function notation, with numbers rounded to three decimal places.

b. How well do the functions model world population in 2000?

c. By one projection, world population is expected to reach 8 billion in the year 2026. Which function serves as a better model for this prediction?

FIGURE 4.28 (repeated) A linear model and an exponential model for world population

GREAT QUESTION!

How can I use a graphing utility to see how well my models describe the data?

Once you have obtained one or more models for the data, you can use a graphing utility's $\boxed{\text{TABLE}}$ feature to numerically see how well each model describes the data. Enter the models as y_1, y_2, and so on. Create a table, scroll through the table, and compare the table values given by the models to the actual data.

SOLUTION

a. Using **Figure 4.28** and rounding to three decimal places, the functions

$$f(x) = 0.074x + 2.294 \quad \text{and} \quad g(x) = 2.577(1.017)^x$$

model world population, in billions, x years after 1949. We named the linear function f and the exponential function g, although any letters can be used.

b. **Table 4.10** on the previous page shows that world population in 2000 was 6.1 billion. The year 2000 is 51 years after 1949. Thus, we substitute 51 for x in each function's equation and then evaluate the resulting expressions with a calculator to see how well the functions describe world population in 2000.

$f(x) = 0.074x + 2.294$	This is the linear model.
$f(51) = 0.074(51) + 2.294$	Substitute 51 for x.
≈ 6.1	Use a calculator.
$g(x) = 2.577(1.017)^x$	This is the exponential model.
$g(51) = 2.577(1.017)^{51}$	Substitute 51 for x.
≈ 6.1	Use a calculator: 2.577 $\boxed{\times}$ 1.017 $\boxed{y^x}$ (or $\boxed{\wedge}$) 51 $\boxed{=}$.

Because 6.1 billion was the actual world population in 2000, both functions model world population in 2000 extremely well.

c. Let's see which model comes closer to projecting a world population of 8 billion in the year 2026. Because 2026 is 77 years after 1949 $(2026 - 1949 = 77)$, we substitute 77 for x in each function's equation.

$f(x) = 0.074x + 2.294$	This is the linear model.
$f(77) = 0.074(77) + 2.294$	Substitute 77 for x.
≈ 8.0	Use a calculator.
$g(x) = 2.577(1.017)^x$	This is the exponential model.
$g(77) = 2.577(1.017)^{77}$	Substitute 77 for x.
≈ 9.4	Use a calculator: 2.577 $\boxed{\times}$ 1.017 $\boxed{y^x}$ (or $\boxed{\wedge}$) 77 $\boxed{=}$.

The linear function $f(x) = 0.074x + 2.294$ serves as a better model for a projected world population of 8 billion by 2026. ● ● ●

\oslash Check Point **6** Use the models $f(x) = 0.074x + 2.294$ and $g(x) = 2.577(1.017)^x$ to solve this problem.

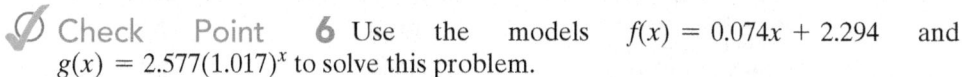

a. World population in 1970 was 3.7 billion. Which function serves as a better model for this year?

b. By one projection, world population is expected to reach 9.3 billion by 2050. Which function serves as a better model for this projection?

When using a graphing utility to model data, begin with a scatter plot, drawn either by hand or with the graphing utility, to obtain a general picture for the shape of the data. It might be difficult to determine which model best fits the data— linear, logarithmic, exponential, quadratic, or something else. If necessary, use your graphing utility to fit several models to the data. The best model is the one that yields the value r, the correlation coefficient, closest to 1 or -1. Finding a proper fit for data can be almost as much art as it is mathematics. In this era of technology, the process of creating models that best fit data is one that involves more decision making than computation.

④ Express an exponential model in base e.

Expressing $y = ab^x$ in Base e

Graphing utilities display exponential models in the form $y = ab^x$. However, our discussion of exponential growth involved base e. Because of the inverse property $b = e^{\ln b}$, we can rewrite any model in the form $y = ab^x$ in terms of base e.

> ### Expressing an Exponential Model in Base e
>
> $$y = ab^x \quad \text{is equivalent to} \quad y = ae^{(\ln b) \cdot x}.$$

EXAMPLE 7 Rewriting the Model for World Population in Base e

We have seen that the function

$$g(x) = 2.577(1.017)^x$$

models world population, $g(x)$, in billions, x years after 1949. Rewrite the model in terms of base e.

SOLUTION

We use the two equivalent equations shown in the voice balloons to rewrite the model in terms of base e.

$$y = ab^x \qquad\qquad y = ae^{(\ln b) \cdot x}$$

$$g(x) = 2.577(1.017)^x \quad \text{is equivalent to} \quad g(x) = 2.577e^{(\ln 1.017)x}.$$

Using $\ln 1.017 \approx 0.017$, the exponential growth model for world population, $g(x)$, in billions, x years after 1949 is

$$g(x) = 2.577e^{0.017x}. \qquad \bullet\bullet\bullet$$

In Example 7, we can replace $g(x)$ with A and x with t so that the model has the same letters as those in the exponential growth model $A = A_0e^{kt}$.

$$A = \quad A_0 \quad e^{kt} \qquad \text{This is the exponential growth model.}$$

$$A = 2.577e^{0.017t} \qquad \text{This is the model for world population.}$$

The value of k, 0.017, indicates a growth rate of 1.7%. Although this is an excellent model for the data, we must be careful about making projections about world population using this growth function. Why? World population growth rate is now 1.2%, not 1.7%, so our model will overestimate future populations.

 Check Point **7** Rewrite $y = 4(7.8)^x$ in terms of base e. Express the answer in terms of a natural logarithm and then round to three decimal places.

Blitzer Bonus

Global Population Increase

By the time you finish reading Example 7 and working Check Point 7, more than 1000 people will have been added to our planet. By this time tomorrow, world population will have increased by more than 220,000.

CONCEPT AND VOCABULARY CHECK

Fill in each blank so that the resulting statement is true.

1. Consider the model for exponential growth or decay given by
$$A = A_0e^{kt}.$$
 If k _____, the function models the amount, or size, of a growing entity. If k _____, the function models the amount, or size, of a decaying entity.

2. In the model for exponential growth or decay, the amount, or size, at $t = 0$ is represented by _____. The amount, or size, at time t is represented by _____.

3. Consider the model for limited logistic growth given by
$$A = \frac{c}{1 + ae^{-bt}}.$$
 The amount, or size, at time t is represented by _____. This value can never exceed _____.

For each of the scatter plots in Exercises 4–6, determine whether an exponential function, a logarithmic function, or a linear function is the best choice for modeling the data.

4.

5.

6.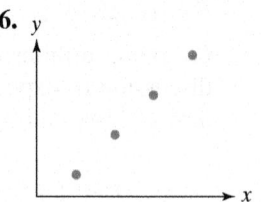

7. $y = 3(5)^x$ can be written in terms of base e as $y = 3e^{(-\quad)\cdot x}$.

EXERCISE SET 4.5

Practice Exercises and Application Exercises

The exponential models describe the population of the indicated country, A, in millions, t years after 2010. Use these models to solve Exercises 1–6.

India	$A = 1173.1e^{0.008t}$
Iraq	$A = 31.5e^{0.019t}$
Japan	$A = 127.3e^{-0.006t}$
Russia	$A = 141.9e^{-0.005t}$

1. What was the population of Japan in 2010?

2. What was the population of Iraq in 2010?

3. Which country has the greatest growth rate? By what percentage is the population of that country increasing each year?

4. Which countries have a decreasing population? By what percentage is the population of these countries decreasing each year?

5. When will India's population be 1377 million?

6. When will India's population be 1491 million?

About the size of New Jersey, Israel has seen its population soar to more than 6 million since it was established. The graphs show that by 2050, Palestinians in the West Bank, Gaza Strip, and East Jerusalem will outnumber Israelis. Exercises 7–8 involve the projected growth of these two populations.

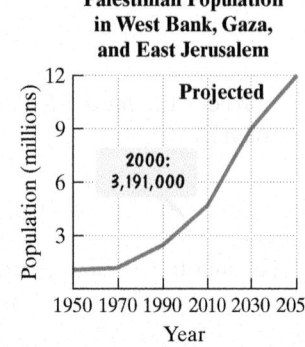

Source: *Newsweek*

7. a. In 2000, the population of Israel was approximately 6.04 million and by 2050 it is projected to grow to 10 million. Use the exponential growth model $A = A_0e^{kt}$, in which t is the number of years after 2000, to find an exponential growth function that models the data.

b. In which year will Israel's population be 9 million?

8. a. In 2000, the population of the Palestinians in the West Bank, Gaza Strip, and East Jerusalem was approximately 3.2 million and by 2050 it is projected to grow to 12 million. Use the exponential growth model $A = A_0e^{kt}$, in which t is the number of years after 2000, to find the exponential growth function that models the data.

b. In which year will the Palestinian population be 9 million?

In Exercises 9–14, complete the table. Round projected populations to one decimal place and values of k to four decimal places.

	Country	2010 Population (millions)	Projected 2050 Population (millions)	Projected Growth Rate, k
9.	Philippines	99.9		0.0095
10.	Pakistan	184.4		0.0149
11.	Colombia	44.2	62.9	
12.	Madagascar	21.3	42.7	
13.	Germany	82.3	70.5	
14.	Bulgaria	7.1	5.4	

Source: International Programs Center, U.S. Census Bureau

An artifact originally had 16 grams of carbon-14 present. The decay model $A = 16e^{-0.000121t}$ describes the amount of carbon-14 present after t years. Use this model to solve Exercises 15–16.

15. How many grams of carbon-14 will be present in 5715 years?

16. How many grams of carbon-14 will be present in 11,430 years?

17. The half-life of the radioactive element krypton-91 is 10 seconds. If 16 grams of krypton-91 are initially present, how many grams are present after 10 seconds? 20 seconds? 30 seconds? 40 seconds? 50 seconds?

18. The half-life of the radioactive element plutonium-239 is 25,000 years. If 16 grams of plutonium-239 are initially present, how many grams are present after 25,000 years? 50,000 years? 75,000 years? 100,000 years? 125,000 years?

Use the exponential decay model for carbon-14, $A = A_0e^{-0.000121t}$, to solve Exercises 19–20.

19. Prehistoric cave paintings were discovered in a cave in France. The paint contained 15% of the original carbon-14. Estimate the age of the paintings.

20. Skeletons were found at a construction site in San Francisco in 1989. The skeletons contained 88% of the expected amount of carbon-14 found in a living person. In 1989, how old were the skeletons?

In Exercises 21–26, complete the table. Round half-lives to one decimal place and values of k to six decimal places.

	Radioactive Substance	Half-Life	Decay Rate, k
21.	Tritium		5.5% per year = −0.055
22.	Krypton-85		6.3% per year = −0.063
23.	Radium-226	1620 years	
24.	Uranium-238	4560 years	
25.	Arsenic-74	17.5 days	
26.	Calcium-47	113 hours	

27. The August 1978 issue of *National Geographic* described the 1964 find of bones of a newly discovered dinosaur weighing 170 pounds, measuring 9 feet, with a 6-inch claw on one toe of each hind foot. The age of the dinosaur was estimated using potassium-40 dating of rocks surrounding the bones.

 a. Potassium-40 decays exponentially with a half-life of approximately 1.31 billion years. Use the fact that after 1.31 billion years a given amount of potassium-40 will have decayed to half the original amount to show that the decay model for potassium-40 is given by $A = A_0e^{-0.52912t}$, where *t* is in billions of years.

 b. Analysis of the rocks surrounding the dinosaur bones indicated that 94.5% of the original amount of potassium-40 was still present. Let $A = 0.945A_0$ in the model in part (a) and estimate the age of the bones of the dinosaur.

Use the exponential decay model, $A = A_0e^{kt}$, to solve Exercises 28–31. Round answers to one decimal place.

28. The half-life of thorium-229 is 7340 years. How long will it take for a sample of this substance to decay to 20% of its original amount?

29. The half-life of lead is 22 years. How long will it take for a sample of this substance to decay to 80% of its original amount?

30. The half-life of aspirin in your bloodstream is 12 hours. How long will it take for the aspirin to decay to 70% of the original dosage?

31. Xanax is a tranquilizer used in the short-term relief of symptoms of anxiety. Its half-life in the bloodstream is 36 hours. How long will it take for Xanax to decay to 90% of the original dosage?

32. A bird species in danger of extinction has a population that is decreasing exponentially ($A = A_0e^{kt}$). Five years ago the population was at 1400 and today only 1000 of the birds are alive. Once the population drops below 100, the situation will be irreversible. When will this happen?

33. Use the exponential growth model, $A = A_0e^{kt}$, to show that the time it takes a population to double (to grow from A_0 to $2A_0$) is given by $t = \dfrac{\ln 2}{k}$.

34. Use the exponential growth model, $A = A_0e^{kt}$, to show that the time it takes a population to triple (to grow from A_0 to $3A_0$) is given by $t = \dfrac{\ln 3}{k}$.

Use the formula $t = \dfrac{\ln 2}{k}$ that gives the time for a population with a growth rate k to double to solve Exercises 35–36. Express each answer to the nearest whole year.

35. The growth model $A = 4.3e^{0.01t}$ describes New Zealand's population, A, in millions, t years after 2010.

 a. What is New Zealand's growth rate?

 b. How long will it take New Zealand to double its population?

36. The growth model $A = 112.5e^{0.012t}$ describes Mexico's population, A, in millions, t years after 2010.

 a. What is Mexico's growth rate?

 b. How long will it take Mexico to double its population?

37. The logistic growth function

$$f(t) = \frac{100{,}000}{1 + 5000e^{-t}}$$

describes the number of people, $f(t)$, who have become ill with influenza *t* weeks after its initial outbreak in a particular community.

 a. How many people became ill with the flu when the epidemic began?

 b. How many people were ill by the end of the fourth week?

 c. What is the limiting size of the population that becomes ill?

Shown, again, in the following table is world population, in billions, for seven selected years from 1950 through 2010. Using a graphing utility's logistic regression option, we obtain the equation shown on the screen.

x, Number of Years after 1949	y, World Population (billions)
1 (1950)	2.6
11 (1960)	3.0
21 (1970)	3.7
31 (1980)	4.5
41 (1990)	5.3
51 (2000)	6.1
61 (2010)	6.9

```
Logistic
y=c/(1+ae^(-bx))
a=4.110936638
b=.0264832525
c=12.57374318
```

We see from the calculator screen at the bottom of the previous page that a logistic growth model for world population, f(x), in billions, x years after 1949 is

$$f(x) = \frac{12.57}{1 + 4.11e^{-0.026x}}.$$

Use this function to solve Exercises 38–42.

38. How well does the function model the data showing a world population of 6.1 billion for 2000?

39. How well does the function model the data showing a world population of 6.9 billion for 2010?

40. When will world population reach 7 billion?

41. When will world population reach 8 billion?

42. According to the model, what is the limiting size of the population that Earth will eventually sustain?

The logistic growth function

$$P(x) = \frac{90}{1 + 271e^{-0.122x}}$$

models the percentage, P(x), of Americans who are x years old with some coronary heart disease. Use the function to solve Exercises 43–46.

43. What percentage of 20-year-olds have some coronary heart disease?

44. What percentage of 80-year-olds have some coronary heart disease?

45. At what age is the percentage of some coronary heart disease 50%?

46. At what age is the percentage of some coronary heart disease 70%?

Exercises 47–52 present data in the form of tables. For each data set shown by the table,

 a. Create a scatter plot for the data.

 b. Use the scatter plot to determine whether an exponential function, a logarithmic function, or a linear function is the best choice for modeling the data. (If applicable, in Exercise 72, you will use your graphing utility to obtain these functions.)

47. Percent of Miscarriages, by Age

Woman's Age	Percent of Miscarriages
22	9%
27	10%
32	13%
37	20%
42	38%
47	52%

Source: *Time*

48. Savings Needed for Health-Care Expenses during Retirement

Age at Death	Savings Needed
80	$219,000
85	$307,000
90	$409,000
95	$524,000
100	$656,000

Source: Employee Benefit Research Institute

49. Intensity and Loudness Level of Various Sounds

Intensity (watts per meter2)	Loudness Level (decibels)
0.1 (loud thunder)	110
1 (rock concert, 2 yd from speakers)	120
10 (jackhammer)	130
100 (jet takeoff, 40 yd away)	140

50. Temperature Increase in an Enclosed Vehicle

Minutes	Temperature Increase (°F)
10	19°
20	29°
30	34°
40	38°
50	41°
60	43°

51. Dads Raising Kids Alone

Year	Number of Single U.S. Fathers Heading Households with Children Younger Than 18 (millions)
1980	0.6
1990	1.2
2000	1.8
2010	2.5

Source: U.S. Census Bureau

52. Percentage of Births to Unmarried Women in the United States

Year	Percentage of Births
1960	5%
1970	11%
1980	18%
1990	28%
2000	33%
2009	41%

Source: National Center for Health Statistics

In Exercises 53–56, rewrite the equation in terms of base e. Express the answer in terms of a natural logarithm and then round to three decimal places.

53. $y = 100(4.6)^x$ **54.** $y = 1000(7.3)^x$

55. $y = 2.5(0.7)^x$ **56.** $y = 4.5(0.6)^x$

Writing in Mathematics

57. Nigeria has a growth rate of 0.025 or 2.5%. Describe what this means.

58. How can you tell whether an exponential model describes exponential growth or exponential decay?

59. Suppose that a population that is growing exponentially increases from 800,000 people in 2007 to 1,000,000 people in 2010. Without showing the details, describe how to obtain the exponential growth function that models the data.

60. What is the half-life of a substance?

61. Describe a difference between exponential growth and logistic growth.

62. Describe the shape of a scatter plot that suggests modeling the data with an exponential function.

63. You take up weightlifting and record the maximum number of pounds you can lift at the end of each week. You start off with rapid growth in terms of the weight you can lift from week to week, but then the growth begins to level off. Describe how to obtain a function that models the number of pounds you can lift at the end of each week. How can you use this function to predict what might happen if you continue the sport?

64. Would you prefer that your salary be modeled exponentially or logarithmically? Explain your answer.

65. One problem with all exponential growth models is that nothing can grow exponentially forever. Describe factors that might limit the size of a population.

Technology Exercises

In Example 1 on page 505, we used two data points and an exponential function to model the population of the United States from 1970 through 2010. The data are shown again in the table. Use all five data points to solve Exercises 66–70.

x, Number of Years after 1969	y, U.S. Population (millions)
1 (1970)	203.3
11 (1980)	226.5
21 (1990)	248.7
31 (2000)	281.4
41 (2010)	308.7

66. a. Use your graphing utility's exponential regression option to obtain a model of the form $y = ab^x$ that fits the data. How well does the correlation coefficient, r, indicate that the model fits the data?

 b. Rewrite the model in terms of base e. By what percentage is the population of the United States increasing each year?

67. Use your graphing utility's logarithmic regression option to obtain a model of the form $y = a + b \ln x$ that fits the data. How well does the correlation coefficient, r, indicate that the model fits the data?

68. Use your graphing utility's linear regression option to obtain a model of the form $y = ax + b$ that fits the data. How well does the correlation coefficient, r, indicate that the model fits the data?

69. Use your graphing utility's power regression option to obtain a model of the form $y = ax^b$ that fits the data. How well does the correlation coefficient, r, indicate that the model fits the data?

70. Use the values of r in Exercises 66–69 to select the two models of best fit. Use each of these models to predict by which year the U.S. population will reach 335 million. How do these answers compare to the year we found in Example 1, namely 2020? If you obtained different years, how do you account for this difference?

71. The figure shows the number of people in the United States age 65 and over, with projected figures for the year 2020 and beyond.

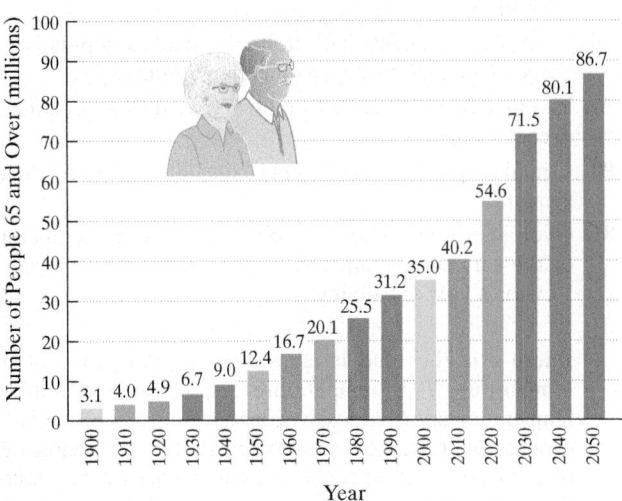

United States Population Age 65 and Over

Source: U.S. Census Bureau

 a. Let x represent the number of years after 1899 and let y represent the U.S. population age 65 and over, in millions. Use your graphing utility to find the model that best fits the data in the bar graph.

 b. Rewrite the model in terms of base e. By what percentage is the 65 and over population increasing each year?

72. In Exercises 47–52, you determined the best choice for the kind of function that modeled the data in the table. For each of the exercises that you worked, use a graphing utility to find the actual function that best fits the data. Then use the model to make a reasonable prediction for a value that exceeds those shown in the table's first column.

Critical Thinking Exercises

Make Sense? *In Exercises 73–76, determine whether each statement makes sense or does not make sense, and explain your reasoning.*

73. I used an exponential model with a positive growth rate to describe the depreciation in my car's value over four years.

74. After 100 years, a population whose growth rate is 3% will have three times as many people as a population whose growth rate is 1%.

75. When I used an exponential function to model Russia's declining population, the growth rate k was negative.

76. Because carbon-14 decays exponentially, carbon dating can determine the ages of ancient fossils.

The exponential growth models describe the population of the indicated country, A, in millions, t years after 2006.

Canada $A = 33.1e^{0.009t}$

Uganda $A = 28.2e^{0.034t}$

In Exercises 77–80, use this information to determine whether each statement is true or false. If the statement is false, make the necessary change(s) to produce a true statement.

(In Exercises 77–80, be sure to refer to the information at the bottom of the previous page.)

77. In 2006, Canada's population exceeded Uganda's by 4.9 million.

78. By 2009, the models indicate that Canada's population will exceed Uganda's by approximately 2.8 million.

79. The models indicate that in 2013, Uganda's population will exceed Canada's.

80. Uganda's growth rate is approximately 3.8 times that of Canada's.

81. Over a period of time, a hot object cools to the temperature of the surrounding air. This is described mathematically by Newton's Law of Cooling:

$$T = C + (T_0 - C)e^{-kt},$$

where t is the time it takes for an object to cool from temperature T_0 to temperature T, C is the surrounding air temperature, and k is a positive constant that is associated with the cooling object. A cake removed from the oven has a temperature of 210°F and is left to cool in a room that has a temperature of 70°F. After 30 minutes, the temperature of the cake is 140°F. What is the temperature of the cake after 40 minutes?

Group Exercise

82. Each group member should consult an almanac, newspaper, magazine, or the Internet to find data that can be modeled by exponential or logarithmic functions. Group members should select the two sets of data that are most interesting and relevant. For each set selected, find a model that best fits the data. Each group member should make one prediction based on the model and then discuss a consequence of this prediction. What factors might change the accuracy of the prediction?

Preview Exercises

Exercises 83–85 will help you prepare for the material covered in the first section of the next chapter.

83. a. Does $(4, -1)$ satisfy $x + 2y = 2$?
 b. Does $(4, -1)$ satisfy $x - 2y = 6$?

84. Graph $x + 2y = 2$ and $x - 2y = 6$ in the same rectangular coordinate system. At what point do the graphs intersect?

85. Solve: $5(2x - 3) - 4x = 9$.

CHAPTER 4 Summary, Review, and Test

SUMMARY

DEFINITIONS AND CONCEPTS	EXAMPLES

4.1 Exponential Functions

a. The exponential function with base b is defined by $f(x) = b^x$, where $b > 0$ and $b \neq 1$. — Ex. 1, p. 453

b. Characteristics of exponential functions and graphs for $0 < b < 1$ and $b > 1$ are shown in the box on page 455. — Ex. 2, p. 454; Ex. 3, p. 454

c. Transformations involving exponential functions are summarized in Table 4.1 on page 456. — Ex. 4, p. 457; Ex. 5, p. 457

d. The natural exponential function is $f(x) = e^x$. The irrational number e is called the natural base, where $e \approx 2.7183$. e is the value that $\left(1 + \dfrac{1}{n}\right)^n$ approaches as $n \to \infty$. — Ex. 6, p. 458

e. Formulas for compound interest: After t years, the balance, A, in an account with principal P and annual interest rate r (in decimal form) is given by one of the following formulas: — Ex. 7, p. 460

 1. For n compoundings per year: $A = P\left(1 + \dfrac{r}{n}\right)^{nt}$

 2. For continuous compounding: $A = Pe^{rt}$.

4.2 Logarithmic Functions

a. Definition of the logarithmic function: For $x > 0$ and $b > 0$, $b \neq 1$, $y = \log_b x$ is equivalent to $b^y = x$. The function $f(x) = \log_b x$ is the logarithmic function with base b. This function is the inverse function of the exponential function with base b. — Ex. 1, p. 466; Ex. 2, p. 467; Ex. 3, p. 467

b. Graphs of logarithmic functions for $b > 1$ and $0 < b < 1$ are shown in Figure 4.8 on page 469. Characteristics of the graphs are summarized in the box on page 470. — Ex. 6, p. 469

c. Transformations involving logarithmic functions are summarized in Table 4.5 on page 470. — Figures 4.9–4.12, pp. 470–471

d. The domain of a logarithmic function of the form $f(x) = \log_b x$ is the set of all positive real numbers. The domain of $f(x) = \log_b g(x)$ consists of all x for which $g(x) > 0$. — Ex. 7, p. 471; Ex. 10, p. 473

DEFINITIONS AND CONCEPTS

EXAMPLES

e. Common and natural logarithms: $f(x) = \log x$ means $f(x) = \log_{10} x$ and is the common logarithmic function. $f(x) = \ln x$ means $f(x) = \log_e x$ and is the natural logarithmic function.

Ex. 8, p. 472;
Ex. 9, p. 473;
Ex. 11, p. 474

f. Basic Logarithmic Properties

Base b ($b > 0, b \neq 1$)	Base 10 (Common Logarithms)	Base e (Natural Logarithms)
$\log_b 1 = 0$	$\log 1 = 0$	$\ln 1 = 0$
$\log_b b = 1$	$\log 10 = 1$	$\ln e = 1$
$\log_b b^x = x$	$\log 10^x = x$	$\ln e^x = x$
$b^{\log_b x} = x$	$10^{\log x} = x$	$e^{\ln x} = x$

Ex. 4, p. 468

Ex. 5, p. 468

4.3 Properties of Logarithms

a. *The Product Rule:* $\log_b(MN) = \log_b M + \log_b N$

Ex. 1, p. 479

b. *The Quotient Rule:* $\log_b\left(\dfrac{M}{N}\right) = \log_b M - \log_b N$

Ex. 2, p. 480

c. *The Power Rule:* $\log_b M^p = p \log_b M$

Ex. 3, p. 481

d. *The Change-of-Base Property:*

The General Property	Introducing Common Logarithms	Introducing Natural Logarithms
$\log_b M = \dfrac{\log_a M}{\log_a b}$	$\log_b M = \dfrac{\log M}{\log b}$	$\log_b M = \dfrac{\ln M}{\ln b}$

Ex. 7, p. 485;
Ex. 8, p. 486

e. Properties for expanding logarithmic expressions are given in the box on page 482.

Ex. 4, p. 482

f. Properties for condensing logarithmic expressions are given in the box on page 483.

Ex. 5, p. 483;
Ex. 6, p. 484

4.4 Exponential and Logarithmic Equations

a. An exponential equation is an equation containing a variable in an exponent. Some exponential equations can be solved by expressing each side as a power of the same base: If $b^M = b^N$, then $M = N$. Details are in the box on page 490.

Ex. 1, p. 490

b. The procedure for using logarithms to solve exponential equations is given in the box on page 491. The solution procedure involves isolating the exponential expression. Take the common logarithm on both sides for base 10. Take the natural logarithm on both sides for bases other than 10. Simplify using

$$\ln b^x = x \ln b \quad \text{or} \quad \ln e^x = x \quad \text{or} \quad \log 10^x = x.$$

Ex. 2, p. 491;
Ex. 3, p. 492;
Ex. 4, p. 492;
Ex. 5, p. 493

c. A logarithmic equation is an equation containing a variable in a logarithmic expression. Some logarithmic equations can be expressed in the form $\log_b M = c$. The definition of a logarithm is used to rewrite the equation in exponential form: $b^c = M$. See the box on page 493. When checking logarithmic equations, reject proposed solutions that produce the logarithm of a negative number or the logarithm of 0 in the original equation.

Ex. 6, p. 494;
Ex. 7, p. 495

d. Some logarithmic equations can be expressed in the form $\log_b M = \log_b N$. Use the one-to-one property to rewrite the equation without logarithms: $M = N$. See the box on page 496.

Ex. 8, p. 496

4.5 Exponential Growth and Decay; Modeling Data

a. Exponential growth and decay models are given by $A = A_0 e^{kt}$ in which t represents time, A_0 is the amount present at $t = 0$, and A is the amount present at time t. If $k > 0$, the model describes growth and k is the growth rate. If $k < 0$, the model describes decay and k is the decay rate.

Ex. 1, p. 505;
Ex. 2, p. 507

b. The logistic growth model, given by $A = \dfrac{c}{1 + ae^{-bt}}$, describes situations in which growth is limited. $y = c$ is a horizontal asymptote for the graph, and growth, A, can never exceed c.

Ex. 3, p. 508

c. Scatter plots for exponential, logarithmic, and linear models are shown in Figure 4.24 on page 509. When using a graphing utility to model data, the closer that the correlation coefficient, r, is to -1 or 1, the better the model fits the data.

Ex. 4, p. 510;
Ex. 5, p. 511;
Ex. 6, p. 511

d. Expressing an Exponential Model in Base e: $y = ab^x$ is equivalent to $y = ae^{(\ln b) \cdot x}$.

Ex. 7, p. 513

REVIEW EXERCISES

4.1

In Exercises 1–4, the graph of an exponential function is given. Select the function for each graph from the following options:

$$f(x) = 4^x, g(x) = 4^{-x},$$

$$h(x) = -4^{-x}, r(x) = -4^{-x} + 3.$$

1.

2.

3.

4.
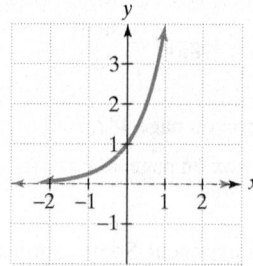

In Exercises 5–9, graph f and g in the same rectangular coordinate system. Use transformations of the graph of f to obtain the graph of g. Graph and give equations of all asymptotes. Use the graphs to determine each function's domain and range.

5. $f(x) = 2^x$ and $g(x) = 2^{x-1}$

6. $f(x) = 3^x$ and $g(x) = 3^x - 1$

7. $f(x) = 3^x$ and $g(x) = -3^x$

8. $f(x) = \left(\frac{1}{2}\right)^x$ and $g(x) = \left(\frac{1}{2}\right)^{-x}$

9. $f(x) = e^x$ and $g(x) = 2e^{\frac{x}{2}}$

Use the compound interest formulas to solve Exercises 10–11.

10. Suppose that you have $5000 to invest. Which investment yields the greater return over 5 years: 5.5% compounded semiannually or 5.25% compounded monthly?

11. Suppose that you have $14,000 to invest. Which investment yields the greater return over 10 years: 7% compounded monthly or 6.85% compounded continuously?

12. A cup of coffee is taken out of a microwave oven and placed in a room. The temperature, T, in degrees Fahrenheit, of the coffee after t minutes is modeled by the function $T = 70 + 130e^{-0.04855t}$. The graph of the function is shown in the figure.

Use the graph to answer each of the following questions.

a. What was the temperature of the coffee when it was first taken out of the microwave?

b. What is a reasonable estimate of the temperature of the coffee after 20 minutes? Use your calculator to verify this estimate.

c. What is the limit of the temperature to which the coffee will cool? What does this tell you about the temperature of the room?

4.2

In Exercises 13–15, write each equation in its equivalent exponential form.

13. $\frac{1}{2} = \log_{49} 7$ **14.** $3 = \log_4 x$ **15.** $\log_3 81 = y$

In Exercises 16–18, write each equation in its equivalent logarithmic form.

16. $6^3 = 216$ **17.** $b^4 = 625$ **18.** $13^y = 874$

In Exercises 19–29, evaluate each expression without using a calculator. If evaluation is not possible, state the reason.

19. $\log_4 64$ **20.** $\log_5 \frac{1}{25}$ **21.** $\log_3(-9)$

22. $\log_{16} 4$ **23.** $\log_{17} 17$ **24.** $\log_3 3^8$

25. $\ln e^5$ **26.** $\log_3 \frac{1}{\sqrt{3}}$ **27.** $\ln \frac{1}{e^2}$

28. $\log \frac{1}{1000}$ **29.** $\log_3(\log_8 8)$

30. Graph $f(x) = 2^x$ and $g(x) = \log_2 x$ in the same rectangular coordinate system. Use the graphs to determine each function's domain and range.

31. Graph $f(x) = \left(\frac{1}{3}\right)^x$ and $g(x) = \log_{\frac{1}{3}} x$ in the same rectangular coordinate system. Use the graphs to determine each function's domain and range.

In Exercises 32–35, the graph of a logarithmic function is given. Select the function for each graph from the following options:

$$f(x) = \log x, g(x) = \log(-x),$$

$$h(x) = \log(2 - x), r(x) = 1 + \log(2 - x).$$

32. **33.**

34. **35.**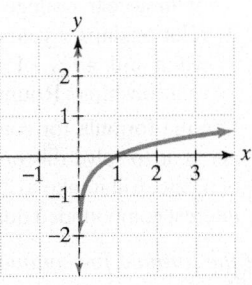

In Exercises 36–38, begin by graphing $f(x) = \log_2 x$. Then use transformations of this graph to graph the given function. What is the graph's x-intercept? What is the vertical asymptote? Use the graphs to determine each function's domain and range.

36. $g(x) = \log_2(x - 2)$ **37.** $h(x) = -1 + \log_2 x$
38. $r(x) = \log_2(-x)$

In Exercises 39–40, graph f and g in the same rectangular coordinate system. Use transformations of the graph of f to obtain the graph of g. Graph and give equations of all asymptotes. Use the graphs to determine each function's domain and range.

39. $f(x) = \log x$ and $g(x) = -\log(x + 3)$
40. $f(x) = \ln x$ and $g(x) = -\ln(2x)$

In Exercises 41–43, find the domain of each logarithmic function.

41. $f(x) = \log_8(x + 5)$ **42.** $f(x) = \log(3 - x)$
43. $f(x) = \ln(x - 1)^2$

In Exercises 44–46, use inverse properties of logarithms to simplify each expression.

44. $\ln e^{6x}$ **45.** $e^{\ln \sqrt{x}}$ **46.** $10^{\log 4x^2}$

47. On the Richter scale, the magnitude, R, of an earthquake of intensity I is given by $R = \log \dfrac{I}{I_0}$, where I_0 is the intensity of a barely felt zero-level earthquake. If the intensity of an earthquake is $1000I_0$, what is its magnitude on the Richter scale?

48. Students in a psychology class took a final examination. As part of an experiment to see how much of the course content they remembered over time, they took equivalent forms of the exam in monthly intervals thereafter. The average score, $f(t)$, for the group after t months is modeled by the function $f(t) = 76 - 18 \log(t + 1)$, where $0 \le t \le 12$.

 a. What was the average score when the exam was first given?

 b. What was the average score after 2 months? 4 months? 6 months? 8 months? one year?

 c. Use the results from parts (a) and (b) to graph f. Describe what the shape of the graph indicates in terms of the material retained by the students.

49. The formula

$$t = \frac{1}{c}\ln\left(\frac{A}{A - N}\right)$$

describes the time, t, in weeks, that it takes to achieve mastery of a portion of a task. In the formula, A represents maximum learning possible, N is the portion of the learning that is to be achieved, and c is a constant used to measure an individual's learning style. A 50-year-old man decides to start running as a way to maintain good health. He feels that the maximum rate he could ever hope to achieve is 12 miles per hour. How many weeks will it take before the man can run 5 miles per hour if $c = 0.06$ for this person?

4.3

In Exercises 50–53, use properties of logarithms to expand each logarithmic expression as much as possible. Where possible, evaluate logarithmic expressions without using a calculator.

50. $\log_6(36x^3)$ **51.** $\log_4\left(\dfrac{\sqrt{x}}{64}\right)$

52. $\log_2\left(\dfrac{xy^2}{64}\right)$ **53.** $\ln \sqrt[3]{\dfrac{x}{e}}$

In Exercises 54–57, use properties of logarithms to condense each logarithmic expression. Write the expression as a single logarithm whose coefficient is 1.

54. $\log_b 7 + \log_b 3$ **55.** $\log 3 - 3 \log x$

56. $3 \ln x + 4 \ln y$ **57.** $\frac{1}{2} \ln x - \ln y$

In Exercises 58–59, use common logarithms or natural logarithms and a calculator to evaluate to four decimal places.

58. $\log_6 72{,}348$ **59.** $\log_4 0.863$

In Exercises 60–63, determine whether each equation is true or false. Where possible, show work to support your conclusion. If the statement is false, make the necessary change(s) to produce a true statement.

60. $(\ln x)(\ln 1) = 0$

61. $\log(x + 9) - \log(x + 1) = \dfrac{\log(x + 9)}{\log(x + 1)}$

62. $(\log_2 x)^4 = 4 \log_2 x$ **63.** $\ln e^x = x \ln e$

4.4

In Exercises 64–73, solve each exponential equation. Where necessary, express the solution set in terms of natural or common logarithms and use a calculator to obtain a decimal approximation, correct to two decimal places, for the solution.

64. $2^{4x-2} = 64$ **65.** $125^x = 25$

66. $10^x = 7000$ **67.** $9^{x+2} = 27^{-x}$

68. $8^x = 12{,}143$ **69.** $9e^{5x} = 1269$

70. $e^{12-5x} - 7 = 123$ **71.** $5^{4x+2} = 37{,}500$

72. $3^{x+4} = 7^{2x-1}$ **73.** $e^{2x} - e^x - 6 = 0$

In Exercises 74–79, solve each logarithmic equation.

74. $\log_4(3x - 5) = 3$

75. $3 + 4\ln(2x) = 15$

76. $\log_2(x + 3) + \log_2(x - 3) = 4$

77. $\log_3(x - 1) - \log_3(x + 2) = 2$

78. $\ln(x + 4) - \ln(x + 1) = \ln x$

79. $\log_4(2x + 1) = \log_4(x - 3) + \log_4(x + 5)$

80. The function $P(x) = 14.7e^{-0.21x}$ models the average atmospheric pressure, $P(x)$, in pounds per square inch, at an altitude of x miles above sea level. The atmospheric pressure at the peak of Mt. Everest, the world's highest mountain, is 4.6 pounds per square inch. How many miles above sea level, to the nearest tenth of a mile, is the peak of Mt. Everest?

81. The bar graph shows the number of smartphones sold in the United States from 2004 through 2010.

Number of Smartphones Sold in the United States

Source: *Newsweek*, April 11, 2011

The function $S(t) = 21.4(1.56)^t$ models the number of smartphones sold in the United States, $S(t)$, in millions, t years after 2004. When does this model project that 1170.9 million smartphones will be sold? Round to the nearest year.

82. Researchers have surveyed attitudes of college freshmen every year since 1969. The bar graph shows that since 1980, there has been a decline in first-year college students' opposition to homosexual relationships.

Opposition to Homosexual Relationships among First-Year United States College Students, 1980–2008

Statement: "It is important to have laws prohibiting homosexual relationships."

Source: John Macionis, *Sociology, Thirteenth Edition*, Prentice Hall, 2010.

a. The function

$$f(x) = -6.2 \ln x + 40.5$$

models the percentage of first-year college women, $f(x)$, opposed to homosexual relationships x years after 1979. If trends modeled by the function continue, when will opposition to homosexual relationships among first-year college women diminish to 16%? Round to the nearest year.

b. The function $g(x) = -7\ln x + 59$ models the percentage of first-year college men, $g(x)$, opposed to homosexual relationships x years after 1979. According to this model, when did 40% of male freshmen oppose homosexual relationships? Round to the nearest year.

83. Use the formula for compound interest with n compoundings per year to solve this problem. How long, to the nearest tenth of a year, will it take $12,500 to grow to $20,000 at 6.5% annual interest compounded quarterly?

Use the formula for continuous compounding to solve Exercises 84–85.

84. How long, to the nearest tenth of a year, will it take $50,000 to triple in value at 7.5% annual interest compounded continuously?

85. What interest rate, to the nearest percent, is required for an investment subject to continuous compounding to triple in 5 years?

4.5

86. According to the U.S. Bureau of the Census, in 2000 there were 35.3 million residents of Hispanic origin living in the United States. By 2010, the number had increased to 50.5 million. The exponential growth function $A = 35.3e^{kt}$ describes the U.S. Hispanic population, A, in millions, t years after 2000.

a. Find k, correct to three decimal places.

b. Use the resulting model to project the Hispanic resident population in 2015.

c. In which year will the Hispanic resident population reach 70 million?

87. Use the exponential decay model, $A = A_0e^{kt}$, to solve this exercise. The half-life of polonium-210 is 140 days. How long will it take for a sample of this substance to decay to 20% of its original amount?

88. The function

$$f(t) = \frac{500,000}{1 + 2499e^{-0.92t}}$$

models the number of people, $f(t)$, in a city who have become ill with influenza t weeks after its initial outbreak.

a. How many people became ill with the flu when the epidemic began?

b. How many people were ill by the end of the sixth week?

c. What is the limiting size of $f(t)$, the population that becomes ill?

Exercises 89–91 present data in the form of tables. For each data set shown by the table,

a. *Create a scatter plot for the data.*

b. *Use the scatter plot to determine whether an exponential function, a logarithmic function, or a linear function is the best choice for modeling the data.*

89. U.S. Ice Cream Consumption

Year	Average Annual Consumption (pounds)
1968	18.5
1978	17.6
1988	17.3
1998	16.3
2008	13.9

Source: USDA, Feb. 2010

90. Percentage of Bachelor's Degrees Awarded to Women in the United States

Year	Percentage of Degrees
1975	45%
1985	51%
1995	55%
2005	57%

Source: U.S. Department of Education

91. E-Filing of Federal Taxes

Year	Percentage of E-Filed Returns
1990	4%
1995	10%
2000	28%
2005	51%
2010	69%

Source: IRS

In Exercises 92–93, rewrite the equation in terms of base e. Express the answer in terms of a natural logarithm and then round to three decimal places.

92. $y = 73(2.6)^x$

93. $y = 6.5(0.43)^x$

94. The figure shows world population projections through the year 2150. The data are from the United Nations Family Planning Program and are based on optimistic or pessimistic expectations for successful control of human population growth. Suppose that you are interested in modeling these data using exponential, logarithmic, linear, and quadratic functions. Which function would you use to model each of the projections? Explain your choices. For the choice corresponding to a quadratic model, would your formula involve one with a positive or negative leading coefficient? Explain.

Projections in
World Population Growth

Source: United Nations

CHAPTER 4 TEST

1. Graph $f(x) = 2^x$ and $g(x) = 2^{x+1}$ in the same rectangular coordinate system.

2. Graph $f(x) = \log_2 x$ and $g(x) = \log_2(x - 1)$ in the same rectangular coordinate system.

3. Write in exponential form: $\log_5 125 = 3$.

4. Write in logarithmic form: $\sqrt{36} = 6$.

5. Find the domain: $f(x) = \ln(3 - x)$.

In Exercises 6–7, use properties of logarithms to expand each logarithmic expression as much as possible. Where possible, evaluate logarithmic expressions without using a calculator.

6. $\log_4(64x^5)$

7. $\log_3\left(\dfrac{\sqrt[3]{x}}{81}\right)$

In Exercises 8–9, write each expression as a single logarithm.

8. $6 \log x + 2 \log y$

9. $\ln 7 - 3 \ln x$

10. Use a calculator to evaluate $\log_{15} 71$ to four decimal places.

In Exercises 11–18, solve each equation.

11. $3^{x-2} = 9^{x+4}$

12. $5^x = 1.4$

13. $400e^{0.005x} = 1600$

14. $e^{2x} - 6e^x + 5 = 0$

15. $\log_6(4x - 1) = 3$

16. $2 \ln(3x) = 8$

17. $\log x + \log(x + 15) = 2$

18. $\ln(x - 4) - \ln(x + 1) = \ln 6$

19. On the decibel scale, the loudness of a sound, D, in decibels, is given by $D = 10 \log \dfrac{I}{I_0}$, where I is the intensity of the sound, in watts per meter2, and I_0 is the intensity of a sound barely audible to the human ear. If the intensity of a sound is $10^{12}I_0$, what is its loudness in decibels? (Such a sound is potentially damaging to the ear.)

In Exercises 20–22, simplify each expression.

20. $\ln e^{5x}$

21. $\log_b b$

22. $\log_6 1$

Use the compound interest formulas to solve Exercises 23–25.

23. Suppose you have $3000 to invest. Which investment yields the greater return over 10 years: 6.5% compounded semiannually or 6% compounded continuously? How much more (to the nearest dollar) is yielded by the better investment?

24. How long, to the nearest tenth of a year, will it take $4000 to grow to $8000 at 5% annual interest compounded quarterly?

25. What interest rate, to the nearest tenth of a percent, is required for an investment subject to continuous compounding to double in 10 years?

26. The function

$$A = 82.3e^{-0.004t}$$

models the population of Germany, A, in millions, t years after 2010.

a. What was the population of Germany in 2010?

b. Is the population of Germany increasing or decreasing? Explain.

c. In which year will the population of Germany be 79.1 million?

27. The 2010 population of Asia was 4121 million; in 2050, it is projected to be 5231 million. Write the exponential growth function that describes the population of Asia, in millions, t years after 2010.

28. Use the exponential decay model, $A = A_0e^{kt}$, to solve this exercise. The half-life of iodine-131 is 7.2 days. How long will it take for a sample of this substance to decay to 30% of its original amount? Round to one decimal place.

29. The logistic growth function

$$f(t) = \frac{140}{1 + 9e^{-0.165t}}$$

describes the population, $f(t)$, of an endangered species of elk t years after they were introduced to a nonthreatening habitat.

a. How many elk were initially introduced to the habitat?

b. How many elk are expected in the habitat after 10 years?

c. What is the limiting size of the elk population that the habitat will sustain?

In Exercises 30–33, determine whether the values in each table belong to an exponential function, a logarithmic function, a linear function, or a quadratic function.

30.

x	y
0	3
1	1
2	−1
3	−3
4	−5

31.

x	y
$\frac{1}{3}$	−1
1	0
3	1
9	2
27	3

32.

x	y
0	1
1	5
2	25
3	125
4	625

33.

x	y
0	12
1	3
2	0
3	3
4	12

34. Rewrite $y = 96(0.38)^x$ in terms of base e. Express the answer in terms of a natural logarithm and then round to three decimal places.

CUMULATIVE REVIEW EXERCISES (CHAPTERS 1–4)

In Exercises 1–8, solve each equation or inequality.

1. $|3x - 4| = 2$

2. $\sqrt{2x - 5} - \sqrt{x - 3} = 1$

3. $x^4 + x^3 - 3x^2 - x + 2 = 0$

4. $e^{5x} - 32 = 96$

5. $\log_2(x + 5) + \log_2(x - 1) = 4$

6. $\ln(x + 4) + \ln(x + 1) = 2\ln(x + 3)$

7. $14 - 5x \geq -6$

8. $|2x - 4| \leq 2$

In Exercises 9–14, graph each equation in a rectangular coordinate system. If two functions are indicated, graph both in the same system.

9. $(x - 3)^2 + (y + 2)^2 = 4$

10. $f(x) = (x - 2)^2 - 1$

11. $f(x) = \dfrac{x^2 - 1}{x^2 - 4}$

12. $f(x) = (x - 2)^2(x + 1)$

13. $f(x) = 2x - 4$ and $f^{-1}(x)$

14. $f(x) = \ln x$ and $g(x) = \ln(x - 2) + 1$

15. Write the point-slope form and the slope-intercept form of the line passing through $(1, 3)$ and $(3, -3)$.

16. If $f(x) = x^2$ and $g(x) = x + 2$, find $(f \circ g)(x)$ and $(g \circ f)(x)$.

17. You discover that the number of hours you sleep each night varies inversely as the square of the number of cups of coffee consumed during the early evening. If 2 cups of coffee are consumed, you get 8 hours of sleep. If the number of cups of coffee is doubled, how many hours should you expect to sleep?

A baseball player hits a pop fly into the air. The function

$$s(t) = -16t^2 + 64t + 5$$

models the ball's height above the ground, $s(t)$, in feet, t seconds after it is hit. Use the function to solve Exercises 18–19.

18. When does the baseball reach its maximum height? What is that height?

19. After how many seconds does the baseball hit the ground? Round to the nearest tenth of a second.

20. You are paid time-and-a-half for each hour worked over 40 hours a week. Last week you worked 50 hours and earned $660. What is your normal hourly salary?

SYSTEMS OF EQUATIONS AND INEQUALITIES

CHAPTER 5

Television, movies, and magazines place great emphasis on physical beauty. Our culture emphasizes physical appearance to such an extent that it is a central factor in the perception and judgment of others. The modern emphasis on thinness as the ideal body shape has been suggested as a major cause of eating disorders among adolescent women.

Cultural values of physical attractiveness change over time. During the 1950s, actress Jayne Mansfield embodied the postwar ideal: curvy, buxom, and big-hipped. Men, too, have been caught up in changes of how they "ought" to look. The 1960s ideal was the soft and scrawny hippie. Today's ideal man is tough and muscular.

Given the importance of culture in setting standards of attractiveness, how can you establish a healthy weight range for your age and height? In this chapter, we will use systems of inequalities to explore these skin-deep issues.

HERE'S WHERE YOU'LL FIND THESE APPLICATIONS:

You'll find a weight that fits you using the models (mathematical, not fashion) in Example 5 of Section 5.5 and Exercises 77–80 in Exercise Set 5.5. Exercises 85–86 use graphs and a formula for body-mass index to indicate whether you are obese, overweight, borderline overweight, normal weight, or underweight.

525

SECTION 5.1

Systems of Linear Equations in Two Variables

Objectives

1. Decide whether an ordered pair is a solution of a linear system.
2. Solve linear systems by substitution.
3. Solve linear systems by addition.
4. Identify systems that do not have exactly one ordered-pair solution.
5. Solve problems using systems of linear equations.

FIGURE 5.1
Source: Gerrig and Zimbardo, *Psychology and Life*, 18th Edition, Allyn and Bacon, 2008.

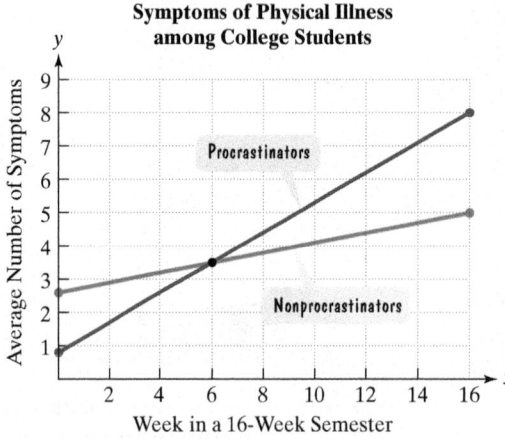

Symptoms of Physical Illness among College Students

Researchers identified college students who generally were procrastinators or nonprocrastinators. The students were asked to report throughout the semester how many symptoms of physical illness they had experienced. **Figure 5.1** shows that by late in the semester, all students experienced increases in symptoms. Early in the semester, procrastinators reported fewer symptoms, but late in the semester, as work came due, they reported more symptoms than their nonprocrastinating peers.

The data in **Figure 5.1** can be analyzed using a pair of linear models in two variables. The figure shows that by week 6, both groups reported the same number of symptoms of illness, an average of approximately 3.5 symptoms per group. In this section, you will learn two algebraic methods, called *substitution* and *addition*, that will reinforce this graphic observation, verifying (6, 3.5) as the point of intersection.

1. Decide whether an ordered pair is a solution of a linear system.

Systems of Linear Equations and Their Solutions

All equations in the form $Ax + By = C$ are straight lines when graphed. Two such equations are called a **system of linear equations** or a **linear system. A solution to a system of linear equations in two variables** is an ordered pair that satisfies both equations in the system. For example, $(3, 4)$ satisfies the system

$$\begin{cases} x + y = 7 & (3 + 4 \text{ is, indeed, } 7.) \\ x - y = -1. & (3 - 4 \text{ is, indeed, } -1.) \end{cases}$$

Thus, $(3, 4)$ satisfies both equations and is a solution of the system. The solution can be described by saying that $x = 3$ and $y = 4$. The solution can also be described using set notation. The solution set to the system is $\{(3, 4)\}$ — that is, the set consisting of the ordered pair $(3, 4)$.

A system of linear equations can have exactly one solution, no solution, or infinitely many solutions. We begin with systems that have exactly one solution.

EXAMPLE 1 Determining Whether Ordered Pairs Are Solutions of a System

Consider the system:

$$\begin{cases} x + 2y = 2 \\ x - 2y = 6. \end{cases}$$

Determine if each ordered pair is a solution of the system:

 a. $(4, -1)$ **b.** $(-4, 3)$.

SOLUTION

a. We begin by determining whether $(4, -1)$ is a solution. Because 4 is the x-coordinate and -1 is the y-coordinate of $(4, -1)$, we replace x with 4 and y with -1.

$$
\begin{array}{ll}
x + 2y = 2 & x - 2y = 6 \\
4 + 2(-1) \stackrel{?}{=} 2 & 4 - 2(-1) \stackrel{?}{=} 6 \\
4 + (-2) \stackrel{?}{=} 2 & 4 - (-2) \stackrel{?}{=} 6 \\
2 = 2, \quad \text{true} & 4 + 2 \stackrel{?}{=} 6 \\
& 6 = 6, \quad \text{true}
\end{array}
$$

The pair $(4, -1)$ satisfies both equations: It makes each equation true. Thus, the ordered pair is a solution of the system.

b. To determine whether $(-4, 3)$ is a solution, we replace x with -4 and y with 3.

$$
\begin{array}{ll}
x + 2y = 2 & x - 2y = 6 \\
-4 + 2 \cdot 3 \stackrel{?}{=} 2 & -4 - 2 \cdot 3 \stackrel{?}{=} 6 \\
-4 + 6 \stackrel{?}{=} 2 & -4 - 6 \stackrel{?}{=} 6 \\
2 = 2, \quad \text{true} & -10 = 6, \quad \text{false}
\end{array}
$$

The pair $(-4, 3)$ fails to satisfy *both* equations: It does not make both equations true. Thus, the ordered pair is not a solution of the system. • • •

GREAT QUESTION!

Can I use a rough sketch on scratch paper to solve a linear system by graphing?

No. When solving linear systems by graphing, neatly drawn graphs are essential for determining points of intersection.

- Use rectangular coordinate graph paper.
- Use a ruler or straightedge.
- Use a pencil with a sharp point.

The solution of a system of linear equations can sometimes be found by graphing both of the equations in the same rectangular coordinate system. For a system with one solution, the **coordinates of the point of intersection give the system's solution**. For example, the system in Example 1,

$$
\begin{cases} x + 2y = 2 \\ x - 2y = 6, \end{cases}
$$

is graphed in **Figure 5.2**. The solution of the system, $(4, -1)$, corresponds to the point of intersection of the lines.

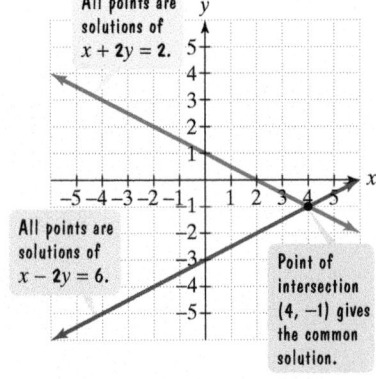

FIGURE 5.2 Visualizing a system's solution

✓ Check Point **1** Consider the system:

$$
\begin{cases} 2x - 3y = -4 \\ 2x + y = 4. \end{cases}
$$

Determine if each ordered pair is a solution of the system:
 a. $(1, 2)$ **b.** $(7, 6)$.

❷ Solve linear systems by substitution.

Eliminating a Variable Using the Substitution Method

Finding the solution to a linear system by graphing equations may not be easy to do. For example, a solution of $\left(-\frac{2}{3}, \frac{157}{29}\right)$ would be difficult to "see" as an intersection point on a graph.

Let's consider a method that does not depend on finding a system's solution visually: the substitution method. This method involves converting the system to one equation in one variable by an appropriate substitution.

GREAT QUESTION!

In the first step of the substitution method, how do I know which variable to isolate and in which equation?

You can choose both the variable and the equation. If possible, solve for a variable whose coefficient is 1 or −1 to avoid working with fractions.

Solving Linear Systems by Substitution

1. Solve either of the equations for one variable in terms of the other. (If one of the equations is already in this form, you can skip this step.)
2. Substitute the expression found in step 1 into the *other* equation. This will result in an equation in one variable.
3. Solve the equation containing one variable.
4. Back-substitute the value found in step 3 into one of the original equations. Simplify and find the value of the remaining variable.
5. Check the proposed solution in both of the system's given equations.

EXAMPLE 2 Solving a System by Substitution

Solve by the substitution method:

$$\begin{cases} 5x - 4y = 9 \\ x - 2y = -3. \end{cases}$$

SOLUTION

Step 1 Solve either of the equations for one variable in terms of the other. We begin by isolating one of the variables in either of the equations. By solving for x in the second equation, which has a coefficient of 1, we can avoid fractions.

$$x - 2y = -3 \qquad \text{This is the second equation in the given system.}$$
$$x = 2y - 3 \qquad \text{Solve for } x \text{ by adding } 2y \text{ to both sides.}$$

Step 2 Substitute the expression from step 1 into the other equation. We substitute $2y - 3$ for x in the first equation.

$$x = \boxed{2y - 3} \qquad\qquad 5\boxed{x} - 4y = 9$$

This gives us an equation in one variable, namely

$$5(2y - 3) - 4y = 9.$$

The variable x has been eliminated.

Step 3 Solve the resulting equation containing one variable.

$$5(2y - 3) - 4y = 9 \qquad \text{This is the equation containing one variable.}$$
$$10y - 15 - 4y = 9 \qquad \text{Apply the distributive property.}$$
$$6y - 15 = 9 \qquad \text{Combine like terms.}$$
$$6y = 24 \qquad \text{Add 15 to both sides.}$$
$$y = 4 \qquad \text{Divide both sides by 6.}$$

GREAT QUESTION!

When I back-substitute the value for one of the variables, which equation should I use?

The equation from step 1, in which one variable is expressed in terms of the other, is equivalent to one of the original equations. It is often easiest to back-substitute the obtained value into this equation to find the value of the other variable. After obtaining both values, get into the habit of checking the ordered-pair solution in *both* equations of the system.

Step 4 Back-substitute the obtained value into one of the original equations. Now that we have the y-coordinate of the solution, we back-substitute 4 for y into one of the original equations to find x. Let's use both equations to show that we obtain the same value for x in either case.

Using the first equation:	Using the second equation:
$5x - 4y = 9$	$x - 2y = -3$
$5x - 4(4) = 9$	$x - 2(4) = -3$
$5x - 16 = 9$	$x - 8 = -3$
$5x = 25$	$x = 5$
$x = 5$	

With $x = 5$ and $y = 4$, the proposed solution is $(5, 4)$.

Step 5 Check. Take a moment to show that $(5, 4)$ satisfies both given equations. The solution set is $\{(5, 4)\}$.

• • •

TECHNOLOGY

Graphic Connections

A graphing utility can be used to solve the system in Example 2. Solve each equation for y, graph the equations, and use the intersection feature. The utility displays the solution $(5, 4)$ as $x = 5, y = 4$.

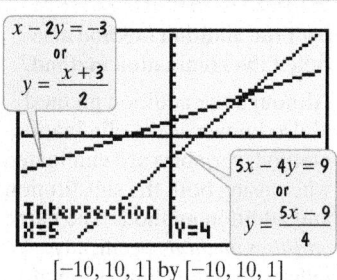

$[-10, 10, 1]$ by $[-10, 10, 1]$

✓ Check Point **2** Solve by the substitution method:

$$\begin{cases} 3x + 2y = 4 \\ 2x + \ y = 1. \end{cases}$$

③ Solve linear systems by addition.

Eliminating a Variable Using the Addition Method

The substitution method is most useful if one of the given equations has an isolated variable. A second, and frequently the easiest, method for solving a linear system is the addition method. Like the substitution method, the addition method involves eliminating a variable and ultimately solving an equation containing only one variable. However, this time we eliminate a variable by adding the equations.

For example, consider the following system of linear equations:

$$\begin{cases} 3x - 4y = 11 \\ -3x + 2y = -7. \end{cases}$$

When we add these two equations, the x-terms are eliminated. This occurs because the coefficients of the x-terms, 3 and -3, are opposites (additive inverses) of each other:

$$\begin{cases} \ \ \ 3x - 4y = \ \ 11 \\ \underline{-3x + 2y = \ -7} \end{cases}$$

The sum is an equation in one variable.

$$-2y = \ \ \ 4$$
$$y = -2.$$

Divide both sides by -2 and solve for y.

Now we can back-substitute -2 for y into one of the original equations to find x. It does not matter which equation you use; you will obtain the same value for x in either case. If we use either equation, we can show that $x = 1$ and the solution $(1, -2)$ satisfies both equations in the system.

When we use the addition method, we want to obtain two equations whose sum is an equation containing only one variable. The key step is to **obtain, for one of the variables, coefficients that differ only in sign**. To do this, we may need to multiply one or both equations by some nonzero number so that the coefficients of one of the variables, x or y, become opposites. Then when the two equations are added, this variable is eliminated.

GREAT QUESTION!

Isn't the addition method also called the elimination method?

Although the addition method is also known as the elimination method, variables are eliminated when using both the substitution and addition methods. The name *addition method* specifically tells us that the elimination of a variable is accomplished by adding two equations.

Solving Linear Systems by Addition

1. If necessary, rewrite both equations in the form $Ax + By = C$.
2. If necessary, multiply either equation or both equations by appropriate nonzero numbers so that the sum of the x-coefficients or the sum of the y-coefficients is 0.
3. Add the equations in step 2. The sum is an equation in one variable.
4. Solve the equation in one variable.
5. Back-substitute the value obtained in step 4 into either of the given equations and solve for the other variable.
6. Check the solution in both of the original equations.

EXAMPLE 3 Solving a System by the Addition Method

Solve by the addition method:

$$\begin{cases} 3x + 2y = 48 \\ 9x - 8y = -24. \end{cases}$$

SOLUTION

Step 1 Rewrite both equations in the form $Ax + By = C$. Both equations are already in this form. Variable terms appear on the left and constants appear on the right.

Step 2 If necessary, multiply either equation or both equations by appropriate numbers so that the sum of the x-coefficients or the sum of the y-coefficients is 0. We can eliminate x or y. Let's eliminate x. Consider the terms in x in each equation, that is, $3x$ and $9x$. To eliminate x, we can multiply each term of the first equation by -3 and then add the equations.

$$\begin{cases} 3x + 2y = 48 \quad \xrightarrow{\text{Multiply by } -3.} \\ 9x - 8y = -24 \quad \xrightarrow{\text{No change}} \end{cases} \begin{cases} -9x - 6y = -144 \\ \underline{9x - 8y = -24} \end{cases}$$

Step 3 Add the equations. Add: $-14y = -168$

Step 4 Solve the equation in one variable. We solve $-14y = -168$ by dividing both sides by -14.

$$\frac{-14y}{-14} = \frac{-168}{-14} \quad \text{Divide both sides by } -14.$$

$$y = 12 \quad \text{Simplify.}$$

Step 5 Back-substitute and find the value for the other variable. We can back-substitute 12 for y into either one of the given equations. We'll use the first one.

$$3x + 2y = 48 \quad \text{This is the first equation in the given system.}$$
$$3x + 2(12) = 48 \quad \text{Substitute 12 for } y.$$
$$3x + 24 = 48 \quad \text{Multiply.}$$
$$3x = 24 \quad \text{Subtract 24 from both sides.}$$
$$x = 8 \quad \text{Divide both sides by 3.}$$

We found that $y = 12$ and $x = 8$. The proposed solution is $(8, 12)$.

Step 6 Check. Take a few minutes to show that $(8, 12)$ satisfies both of the original equations in the system. The solution set is $\{(8, 12)\}$. ● ● ●

✅ Check Point **3** Solve by the addition method:

$$\begin{cases} 4x + 5y = 3 \\ 2x - 3y = 7. \end{cases}$$

Some linear systems have solutions that are not integers. If the value of one variable turns out to be a "messy" fraction, back-substitution might lead to cumbersome arithmetic. If this happens, you can return to the original system and use the addition method to find the value of the other variable.

EXAMPLE 4 Solving a System by the Addition Method

Solve by the addition method:

$$\begin{cases} 2x = 7y - 17 \\ 5y = 17 - 3x. \end{cases}$$

SOLUTION

Step 1 Rewrite both equations in the form $Ax + By = C$. We first arrange the system so that variable terms appear on the left and constants appear on the right. We obtain

$$\begin{cases} 2x - 7y = -17 \quad \text{Subtract } 7y \text{ from both sides of the first equation.} \\ 3x + 5y = 17. \quad \text{Add } 3x \text{ to both sides of the second equation.} \end{cases}$$

Step 2 If necessary, multiply either equation or both equations by appropriate numbers so that the sum of the x-coefficients or the sum of the y-coefficients is 0. We can eliminate x or y. Let's eliminate x by multiplying the first equation by 3 and the second equation by -2.

$$\begin{cases} 2x - 7y = -17 \xrightarrow{\text{Multiply by 3.}} \\ 3x + 5y = 17 \xrightarrow{\text{Multiply by } -2.} \end{cases} \begin{cases} 6x - 21y = -51 \\ -6x - 10y = -34 \end{cases}$$

Step 3 Add the equations. Add: $-31y = -85$

Step 4 Solve the equation in one variable. We solve $-31y = -85$ by dividing both sides by -31.

$$\frac{-31y}{-31} = \frac{-85}{-31} \quad \text{Divide both sides by } -31.$$

$$y = \frac{85}{31} \quad \text{Simplify.}$$

Step 5 Back-substitute and find the value for the other variable. Back-substitution of $\frac{85}{31}$ for y into either of the given equations results in cumbersome arithmetic. Instead, let's use the addition method on the given system in the form $Ax + By = C$ to find the value for x. Thus, we eliminate y by multiplying the first equation by 5 and the second equation by 7.

$$\begin{cases} 2x - 7y = -17 \xrightarrow{\text{Multiply by 5.}} \\ 3x + 5y = 17 \xrightarrow{\text{Multiply by 7.}} \end{cases} \begin{cases} 10x - 35y = -85 \\ 21x + 35y = 119 \end{cases}$$

$$\text{Add:} \quad 31x = 34$$

$$x = \frac{34}{31} \quad \text{Divide both sides by 31.}$$

We found that $y = \dfrac{85}{31}$ and $x = \dfrac{34}{31}$. The proposed solution is $\left(\dfrac{34}{31}, \dfrac{85}{31}\right)$.

Step 6 Check. For this system, a calculator is helpful in showing that $\left(\frac{34}{31}, \frac{85}{31}\right)$ satisfies both of the original equations in the system. The solution set is $\left\{\left(\frac{34}{31}, \frac{85}{31}\right)\right\}$.

• • •

☑ Check Point **4** Solve by the addition method:

$$\begin{cases} 2x = 9 + 3y \\ 4y = 8 - 3x. \end{cases}$$

④ Identify systems that do not have exactly one ordered-pair solution.

Linear Systems Having No Solution or Infinitely Many Solutions

We have seen that a system of linear equations in two variables represents a pair of lines. The lines either intersect at one point, are parallel, or are identical. Thus, there are three possibilities for the number of solutions to a system of two linear equations in two variables.

The Number of Solutions to a System of Two Linear Equations

The number of solutions to a system of two linear equations in two variables is given by one of the following. (See **Figure 5.3**.)

Number of Solutions	What This Means Graphically
Exactly one ordered-pair solution	The two lines intersect at one point.
No solution	The two lines are parallel.
Infinitely many solutions	The two lines are identical.

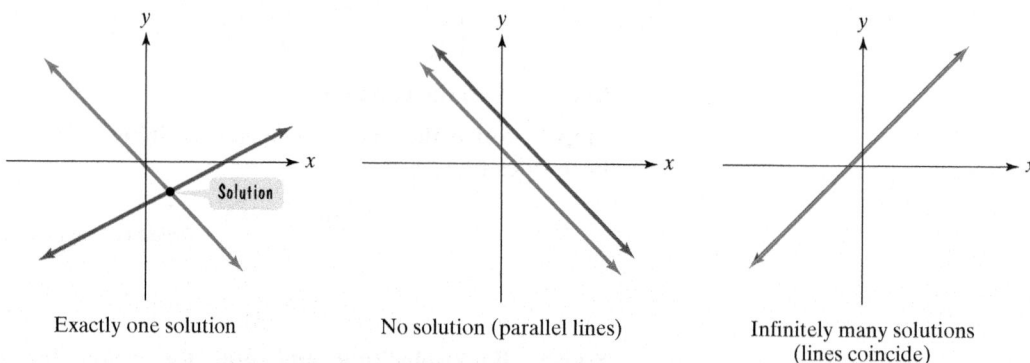

| Exactly one solution | No solution (parallel lines) | Infinitely many solutions (lines coincide) |

FIGURE 5.3 Possible graphs for a system of two linear equations in two variables

A linear system with no solution is called an **inconsistent system**. If you attempt to solve such a system by substitution or addition, you will eliminate both variables. A false statement, such as $0 = 12$, will be the result.

EXAMPLE 5 A System with No Solution

Solve the system:

$$\begin{cases} 4x + 6y = 12 \\ 6x + 9y = 12. \end{cases}$$

SOLUTION

Because no variable is isolated, we will use the addition method. To obtain coefficients of x that differ only in sign, we multiply the first equation by 3 and multiply the second equation by -2.

$$\begin{cases} 4x + 6y = 12 \\ 6x + 9y = 12 \end{cases} \quad \xrightarrow[\text{Multiply by } -2.}]{\text{Multiply by 3.}} \quad \begin{cases} 12x + 18y = 36 \\ -12x - 18y = -24 \end{cases}$$

Add: $\qquad\qquad\qquad\qquad 0 = 12$

There are no values of x and y for which $0 = 12$. No values of x and y satisfy $0x + 0y = 12$.

The false statement $0 = 12$ indicates that the system is inconsistent and has no solution. The solution set is the empty set, \varnothing. • • •

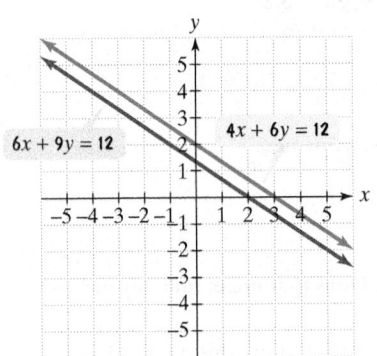

FIGURE 5.4 The graph of an inconsistent system

The lines corresponding to the two equations in Example 5 are shown in **Figure 5.4.** The lines are parallel and have no point of intersection.

DISCOVERY

Show that the graphs of $4x + 6y = 12$ and $6x + 9y = 12$ must be parallel lines by solving each equation for y. What are the slope and y-intercept for each line? What does this mean? If a linear system is inconsistent, what must be true about the slopes and y-intercepts for the system's graphs?

✓ **Check Point 5** Solve the system:

$$\begin{cases} 5x - 2y = 4 \\ -10x + 4y = 7. \end{cases}$$

A linear system that has at least one solution is called a **consistent system**. Lines that intersect and lines that coincide both represent consistent systems. If the lines coincide, then the consistent system has infinitely many solutions, represented by every point on either line.

The equations in a linear system with infinitely many solutions are called **dependent**. If you attempt to solve such a system by substitution or addition, you will eliminate both variables. However, a true statement, such as $10 = 10$, will be the result.

EXAMPLE 6 A System with Infinitely Many Solutions

Solve the system:

$$\begin{cases} y = 3x - 2 \\ 15x - 5y = 10. \end{cases}$$

SOLUTION

Because the variable y is isolated in $y = 3x - 2$, the first equation, we can use the substitution method. We substitute the expression for y into the second equation.

$$y = \boxed{3x - 2} \qquad 15x - 5\boxed{y} = 10 \qquad \text{Substitute } 3x - 2 \text{ for } y.$$

$$15x - 5(3x - 2) = 10 \qquad \text{The substitution results in an equation in one variable.}$$

$$15x - 15x + 10 = 10 \qquad \text{Apply the distributive property.}$$

$$10 = 10 \qquad \text{Simplify.}$$

This statement is true for all values of x and y.

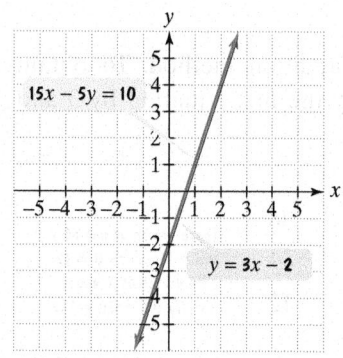

$15x - 5y = 10$

$y = 3x - 2$

FIGURE 5.5 The graph of a system with infinitely many solutions

In our final step, both variables have been eliminated and the resulting statement, $10 = 10$, is true. This true statement indicates that the system has infinitely many solutions. The solution set consists of all points (x, y) lying on either of the coinciding lines, $y = 3x - 2$ or $15x - 5y = 10$, as shown in **Figure 5.5**.

We express the solution set for the system in one of two equivalent ways:

$$\{(x, y) \mid y = 3x - 2\} \quad \text{or} \quad \{(x, y) \mid 15x - 5y = 10\}.$$

The set of all ordered pairs (x, y) such that $y = 3x - 2$

The set of all ordered pairs (x, y) such that $15x - 5y = 10$

• • •

GREAT QUESTION!

The system in Example 6 has infinitely many solutions. Does that mean that any ordered pair of numbers is a solution?

No. Although the system in Example 6 has infinitely many solutions, this does not mean that any ordered pair of numbers you can form will be a solution. The ordered pair (x, y) must satisfy one of the system's equations, $y = 3x - 2$ or $15x - 5y = 10$, and there are infinitely many such ordered pairs. Because the graphs are coinciding lines, the ordered pairs that are solutions of one of the equations are also solutions of the other equation.

⊘ Check Point **6** Solve the system:

$$\begin{cases} x = 4y - 8 \\ 5x - 20y = -40. \end{cases}$$

⑤ Solve problems using systems of linear equations.

Functions of Business: Break-Even Analysis

Suppose that a company produces and sells x units of a product. Its *revenue* is the money generated by selling x units of the product. Its *cost* is the cost of producing x units of the product.

Revenue and Cost Functions

A company produces and sells x units of a product.

Revenue Function

$$R(x) = (\text{price per unit sold})x$$

Cost Function

$$C(x) = \text{fixed cost} + (\text{cost per unit produced})x$$

The point of intersection of the graphs of the revenue and cost functions is called the **break-even point**. The x-coordinate of the point reveals the number of units that a company must produce and sell so that money coming in, the revenue, is equal to money going out, the cost. The y-coordinate of the break-even point gives the amount of money coming in and going out. Example 7 illustrates the use of the substitution method in determining a company's break-even point.

EXAMPLE 7 Finding a Break-Even Point

Technology is now promising to bring light, fast, and beautiful wheelchairs to millions of people with disabilities. A company is planning to manufacture these radically different wheelchairs. Fixed cost will be $500,000 and it will cost $400 to produce each wheelchair. Each wheelchair will be sold for $600.

a. Write the cost function, C, of producing x wheelchairs.

b. Write the revenue function, R, from the sale of x wheelchairs.

c. Determine the break-even point. Describe what this means.

SOLUTION

a. The cost function is the sum of the fixed cost and variable cost.

> Fixed cost of $500,000 plus Variable cost: $400 for each chair produced

$$C(x) = 500,000 + 400x$$

b. The revenue function is the money generated from the sale of x wheelchairs.

> Revenue per chair, $600, times the number of chairs sold

$$R(x) = 600x$$

c. The break-even point occurs where the graphs of C and R intersect. Thus, we find this point by solving the system

$$\begin{cases} C(x) = 500,000 + 400x \\ R(x) = 600x \end{cases} \quad \text{or} \quad \begin{cases} y = 500,000 + 400x \\ y = 600x. \end{cases}$$

Using substitution, we can substitute $600x$ for y in the first equation:

$$600x = 500,000 + 400x \qquad \text{Substitute } 600x \text{ for } y \text{ in } y = 500,000 + 400x.$$

$$200x = 500,000 \qquad \text{Subtract } 400x \text{ from both sides.}$$

$$x = 2500. \qquad \text{Divide both sides by } 200.$$

Back-substituting 2500 for x in either of the system's equations (or functions), we obtain

$$R(2500) = 600(2500) = 1,500,000.$$

> We used $R(x) = 600x$.

The break-even point is $(2500, 1,500,000)$. This means that the company will break even if it produces and sells 2500 wheelchairs. At this level, the money coming in is equal to the money going out: $1,500,000. • • •

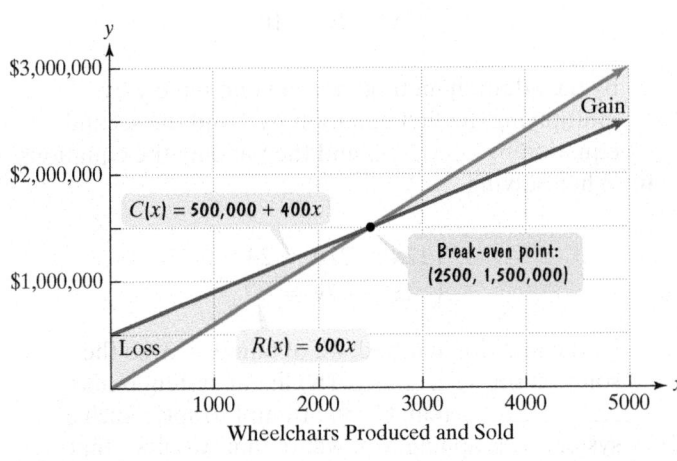

FIGURE 5.6

Figure 5.6 shows the graphs of the revenue and cost functions for the wheelchair business. Similar graphs and models apply no matter how small or large a business venture may be.

The intersection point confirms that the company breaks even by producing and selling 2500 wheelchairs. Can you see what happens for $x < 2500$? The red cost graph lies above the blue revenue graph. The cost is greater than the revenue and the business is losing money. Thus, if the company sells fewer than 2500 wheelchairs, the result is a *loss*. By contrast, look at what happens for $x > 2500$. The blue revenue graph lies above the red cost graph. The revenue is greater than the cost and the business is making money. Thus, if the company sells more than 2500 wheelchairs, the result is a *gain*.

✓ Check Point **7** A company that manufactures running shoes has a fixed cost of $300,000. Additionally, it costs $30 to produce each pair of shoes. They are sold at $80 per pair.

 a. Write the cost function, C, of producing x pairs of running shoes.

 b. Write the revenue function, R, from the sale of x pairs of running shoes.

 c. Determine the break-even point. Describe what this means.

What does every entrepreneur, from a kid selling lemonade to Mark Zuckerberg, want to do? Generate profit, of course. The *profit* made is the money taken in, or the revenue, minus the money spent, or the cost. This relationship between revenue and cost allows us to define the *profit function, $P(x)$*.

The Profit Function

The profit, $P(x)$, generated after producing and selling x units of a product is given by the **profit function**

$$P(x) = R(x) - C(x),$$

where R and C are the revenue and cost functions, respectively.

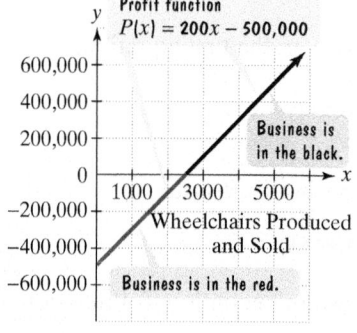

The profit function for the wheelchair business in Example 7 is

$$P(x) = R(x) - C(x)$$
$$= 600x - (500{,}000 + 400x)$$
$$= 200x - 500{,}000.$$

The graph of this profit function is shown in **Figure 5.7**. The red portion lies below the x-axis and shows a loss when fewer than 2500 wheelchairs are sold. The business is "in the red." The black portion lies above the x-axis and shows a gain when more than 2500 wheelchairs are sold. The wheelchair business is "in the black."

FIGURE 5.7

CONCEPT AND VOCABULARY CHECK

Fill in each blank so that the resulting statement is true.

1. A solution to a system of linear equations in two variables is an ordered pair that _____

_____.

2. When solving a system of linear equations by graphing, the system's solution is determined by locating _____.

3. When solving

$$\begin{cases} 3x - 2y = 5 \\ y = 3x - 3 \end{cases}$$

by the substitution method, we obtain $x = \frac{1}{3}$, so the solution set is _____.

4. When solving

$$\begin{cases} 2x + 10y = 9 \\ 8x + 5y = 7 \end{cases}$$

by the addition method, we can eliminate y by multiplying the second equation by _____ and then adding the equations.

5. When solving

$$\begin{cases} 4x - 3y = 15 \\ 3x - 2y = 10 \end{cases}$$

by the addition method, we can eliminate y by multiplying the first equation by 2 and the second equation by _____, and then adding the equations.

6. When solving

$$\begin{cases} 12x - 21y = 24 \\ 4x - 7y = 7 \end{cases}$$

by the addition method, we obtain $0 = 3$, so the solution set is _____. The linear system is a/an _____ system. If you attempt to solve such a system by graphing, you will obtain two lines that are _____.

7. When solving

$$\begin{cases} x = 3y + 2 \\ 5x - 15y = 10 \end{cases}$$

by the substitution method, we obtain $10 = 10$, so the solution set is _____.
The equations in this system are called _____. If you attempt to solve such a system by graphing, you will obtain two lines that _____.

8. A company's _____ function is the money generated by selling x units of its product. The difference between this function and the company's cost function is called its _____ function.

9. A company has a graph that shows the money it generates by selling x units of its product. It also has a graph that shows its cost of producing x units of its product. The point of intersection of these graphs is called the company's _____.

EXERCISE SET 5.1

Practice Exercises

In Exercises 1–4, determine whether the given ordered pair is a solution of the system.

1. $(2, 3)$
$$\begin{cases} x + 3y = 11 \\ x - 5y = -13 \end{cases}$$

2. $(-3, 5)$
$$\begin{cases} 9x + 7y = 8 \\ 8x - 9y = -69 \end{cases}$$

3. $(2, 5)$
$$\begin{cases} 2x + 3y = 17 \\ x + 4y = 16 \end{cases}$$

4. $(8, 5)$
$$\begin{cases} 5x - 4y = 20 \\ 3y = 2x + 1 \end{cases}$$

In Exercises 5–18, solve each system by the substitution method.

5. $\begin{cases} x + y = 4 \\ y = 3x \end{cases}$

6. $\begin{cases} x + y = 6 \\ y = 2x \end{cases}$

7. $\begin{cases} x + 3y = 8 \\ y = 2x - 9 \end{cases}$

8. $\begin{cases} 2x - 3y = -13 \\ y = 2x + 7 \end{cases}$

9. $\begin{cases} x = 4y - 2 \\ x = 6y + 8 \end{cases}$

10. $\begin{cases} x = 3y + 7 \\ x = 2y - 1 \end{cases}$

11. $\begin{cases} 5x + 2y = 0 \\ x - 3y = 0 \end{cases}$

12. $\begin{cases} 4x + 3y = 0 \\ 2x - y = 0 \end{cases}$

13. $\begin{cases} 2x + 5y = -4 \\ 3x - y = 11 \end{cases}$

14. $\begin{cases} 2x + 5y = 1 \\ -x + 6y = 8 \end{cases}$

15. $\begin{cases} 2x - 3y = 8 - 2x \\ 3x + 4y = x + 3y + 14 \end{cases}$

16. $\begin{cases} 3x - 4y = x - y + 4 \\ 2x + 6y = 5y - 4 \end{cases}$

17. $\begin{cases} y = \dfrac{1}{3}x + \dfrac{2}{3} \\ y = \dfrac{5}{7}x - 2 \end{cases}$

18. $\begin{cases} y = -\dfrac{1}{2}x + 2 \\ y = \dfrac{3}{4}x + 7 \end{cases}$

In Exercises 19–30, solve each system by the addition method.

19. $\begin{cases} x + y = 1 \\ x - y = 3 \end{cases}$

20. $\begin{cases} x + y = 6 \\ x - y = -2 \end{cases}$

21. $\begin{cases} 2x + 3y = 6 \\ 2x - 3y = 6 \end{cases}$

22. $\begin{cases} 3x + 2y = 14 \\ 3x - 2y = 10 \end{cases}$

23. $\begin{cases} x + 2y = 2 \\ -4x + 3y = 25 \end{cases}$

24. $\begin{cases} 2x - 7y = 2 \\ 3x + y = -20 \end{cases}$

25. $\begin{cases} 4x + 3y = 15 \\ 2x - 5y = 1 \end{cases}$

26. $\begin{cases} 3x - 7y = 13 \\ 6x + 5y = 7 \end{cases}$

27. $\begin{cases} 3x - 4y = 11 \\ 2x + 3y = -4 \end{cases}$

28. $\begin{cases} 2x + 3y = -16 \\ 5x - 10y = 30 \end{cases}$

29. $\begin{cases} 3x = 4y + 1 \\ 3y = 1 - 4x \end{cases}$

30. $\begin{cases} 5x = 6y + 40 \\ 2y = 8 - 3x \end{cases}$

In Exercises 31–42, solve by the method of your choice. Identify systems with no solution and systems with infinitely many solutions, using set notation to express their solution sets.

31. $\begin{cases} x = 9 - 2y \\ x + 2y = 13 \end{cases}$

32. $\begin{cases} 6x + 2y = 7 \\ y = 2 - 3x \end{cases}$

33. $\begin{cases} y = 3x - 5 \\ 21x - 35 = 7y \end{cases}$

34. $\begin{cases} 9x - 3y = 12 \\ y = 3x - 4 \end{cases}$

35. $\begin{cases} 3x - 2y = -5 \\ 4x + y = 8 \end{cases}$

36. $\begin{cases} 2x + 5y = -4 \\ 3x - y = 11 \end{cases}$

37. $\begin{cases} x + 3y = 2 \\ 3x + 9y = 6 \end{cases}$

38. $\begin{cases} 4x - 2y = 2 \\ 2x - y = 1 \end{cases}$

39. $\begin{cases} \dfrac{x}{4} - \dfrac{y}{4} = -1 \\ x + 4y = -9 \end{cases}$

40. $\begin{cases} \dfrac{x}{6} - \dfrac{y}{2} = \dfrac{1}{3} \\ x + 2y = -3 \end{cases}$

41. $\begin{cases} 2x = 3y + 4 \\ 4x = 3 - 5y \end{cases}$

42. $\begin{cases} 4x = 3y + 8 \\ 2x = -14 + 5y \end{cases}$

In Exercises 43–46, let x represent one number and let y represent the other number. Use the given conditions to write a system of equations. Solve the system and find the numbers.

43. The sum of two numbers is 7. If one number is subtracted from the other, their difference is -1. Find the numbers.

44. The sum of two numbers is 2. If one number is subtracted from the other, their difference is 8. Find the numbers.

45. Three times a first number decreased by a second number is 1. The first number increased by twice the second number is 12. Find the numbers.

46. The sum of three times a first number and twice a second number is 8. If the second number is subtracted from twice the first number, the result is 3. Find the numbers.

Practice Plus

In Exercises 47–48, solve each system by the method of your choice.

47. $\begin{cases} \dfrac{x + 2}{2} - \dfrac{y + 4}{3} = 3 \\ \dfrac{x + y}{5} = \dfrac{x - y}{2} - \dfrac{5}{2} \end{cases}$

48. $\begin{cases} \dfrac{x - y}{3} = \dfrac{x + y}{2} - \dfrac{1}{2} \\ \dfrac{x + 2}{2} - 4 = \dfrac{y + 4}{3} \end{cases}$

In Exercises 49–50, solve each system for x and y, expressing either value in terms of a or b, if necessary. Assume that $a \neq 0$ and $b \neq 0$.

49. $\begin{cases} 5ax + 4y = 17 \\ ax + 7y = 22 \end{cases}$ **50.** $\begin{cases} 4ax + by = 3 \\ 6ax + 5by = 8 \end{cases}$

51. For the linear function $f(x) = mx + b, f(-2) = 11$ and $f(3) = -9$. Find m and b.

52. For the linear function $f(x) = mx + b, f(-3) = 23$ and $f(2) = -7$. Find m and b.

Use the graphs of the linear functions to solve Exercises 53–54.

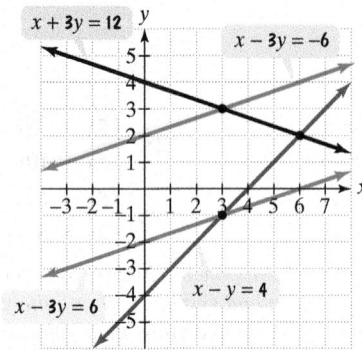

53. Write the linear system whose solution set is $\{(6, 2)\}$. Express each equation in the system in slope-intercept form.

54. Write the linear system whose solution set is \varnothing. Express each equation in the system in slope-intercept form.

Application Exercises

The figure shows the graphs of the cost and revenue functions for a company that manufactures and sells small radios. Use the information in the figure to solve Exercises 55–60.

Radios Produced and Sold

55. How many radios must be produced and sold for the company to break even?

56. More than how many radios must be produced and sold for the company to have a profit?

57. Use the formulas shown in the voice balloons to find $R(200) - C(200)$. Describe what this means for the company.

58. Use the formulas shown in the voice balloons to find $R(300) - C(300)$. Describe what this means for the company.

59. a. Use the formulas shown in the voice balloons to write the company's profit function, P, from producing and selling x radios.

 b. Find the company's profit if 10,000 radios are produced and sold.

60. a. Use the formulas shown in the voice balloons to write the company's profit function, P, from producing and selling x radios.

 b. Find the company's profit if 20,000 radios are produced and sold.

Exercises 61–64 describe a number of business ventures. For each exercise,

 a. *Write the cost function, C.*

 b. *Write the revenue function, R.*

 c. *Determine the break-even point. Describe what this means.*

61. A company that manufactures small canoes has a fixed cost of $18,000. It costs $20 to produce each canoe. The selling price is $80 per canoe. (In solving this exercise, let x represent the number of canoes produced and sold.)

62. A company that manufactures bicycles has a fixed cost of $100,000. It costs $100 to produce each bicycle. The selling price is $300 per bike. (In solving this exercise, let x represent the number of bicycles produced and sold.)

63. You invest in a new play. The cost includes an overhead of $30,000, plus production costs of $2500 per performance. A sold-out performance brings in $3125. (In solving this exercise, let x represent the number of sold-out performances.)

64. You invested $30,000 and started a business writing greeting cards. Supplies cost 2¢ per card and you are selling each card for 50¢. (In solving this exercise, let x represent the number of cards produced and sold.)

*An important application of systems of equations arises in connection with supply and demand. As the price of a product increases, the demand for that product decreases. However, at higher prices, suppliers are willing to produce greater quantities of the product. The price at which supply and demand are equal is called the **equilibrium price**. The quantity supplied and demanded at that price is called the **equilibrium quantity**. Exercises 65–66 involve supply and demand.*

65. The following models describe wages for low-skilled labor.

Demand Model	**Supply Model**
$p = -0.325x + 5.8$	$p = 0.375x + 3$

Price of labor (per hour)	Millions of workers employers will hire	Price of labor (per hour)	Millions of available workers

Source: O'Sullivan and Sheffrin, *Economics,* Prentice Hall, 2007.

 a. Solve the system and find the equilibrium number of workers, in millions, and the equilibrium hourly wage.

 b. Use your answer from part (a) to complete this statement:

 If workers are paid _____ per hour, there will be _____ million available workers and _____ million workers will be hired.

 c. In 2007, the federal minimum wage was set at $5.15 per hour. Substitute 5.15 for p in the demand model, $p = -0.325x + 5.8$, and determine the millions of workers employers will hire at this price.

 d. At a minimum wage of $5.15 per hour, use the supply model, $p = 0.375x + 3$, to determine the millions of available workers. Round to one decimal place.

 e. At a minimum wage of $5.15 per hour, use your answers from parts (c) and (d) to determine how many more people are looking for work than employers are willing to hire.

66. The following models describe demand and supply for three-bedroom rental apartments.

Demand Model

$p = -50x + 2000$

Monthly rental price

Number of apartments demanded, in thousands

Supply Model

$p = 50x$

Monthly rental price

Number of apartments supplied, in thousands

a. Solve the system and find the equilibrium quantity and the equilibrium price.

b. Use your answer from part (a) to complete this statement:

When rents are _____ per month, consumers will demand _____ apartments and suppliers will offer _____ apartments for rent.

67. The bar graph indicates that fewer U.S. adults are getting married.

Marital Status of U.S. Adults

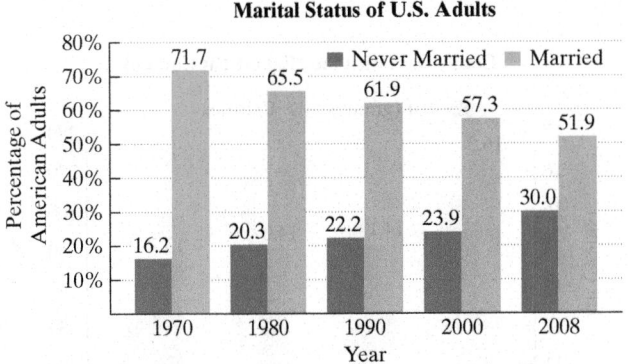

Source: U.S. Census Bureau

The data can be modeled by the following system of linear equations:

$$\begin{cases} -3x + 10y = 160 \\ x + 2y = 142. \end{cases}$$

Percentage of never-married American adults, y, x years after 1970

Percentage of married American adults, y, x years after 1970

a. Use these models to determine the year, rounded to the nearest year, when the percentage of never-married adults will be the same as the percentage of married adults. For that year, approximately what percentage of Americans, rounded to the nearest percent, will belong to each group?

b. How is your approximate solution from part (a) shown by the following graphs?

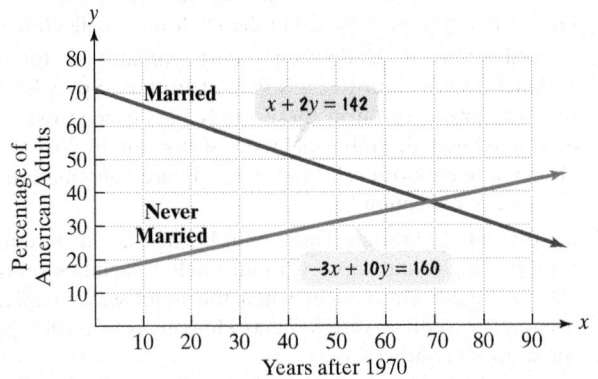

68. The bar graph shows the percentage of Americans for and against the death penalty for a person convicted of murder.

Are You in Favor of the Death Penalty for a Person Convicted of Murder?

Source: *Newsweek* poll

The data can be modeled by the following system of equations:

$$\begin{cases} 13x + 12y = 992 \\ -x + y = 16. \end{cases}$$

The percent, y, in favor of the death penalty x years after 1988

The percent, y, against the death penalty x years after 1988

In which year will the percentage of Americans in favor of the death penalty be the same as the percentage of Americans who oppose it? For that year, what percent will be for the death penalty and what percent will be against it?

69. We opened this section with a study showing that late in the semester, procrastinating students reported more symptoms of physical illness than their nonprocrastinating peers.

a. At the beginning of the semester, procrastinators reported an average of 0.8 symptoms, increasing at a rate of 0.45 symptoms per week. Write a function that models the average number of symptoms after x weeks.

b. At the beginning of the semester, nonprocrastinators reported an average of 2.6 symptoms, increasing at a rate of 0.15 symptoms per week. Write a function that models the average number of symptoms after x weeks.

c. By which week in the semester did both groups report the same number of symptoms of physical illness? For that week, how many symptoms were reported by each group? How is this shown in **Figure 5.1** on page 526?

70. Although Social Security is a problem, some projections indicate that there's a much bigger time bomb ticking in the federal budget, and that's Medicare. In 2000, the cost of Social Security was 5.48% of the gross domestic product, increasing by 0.04% of the GDP per year. In 2000, the cost of Medicare was 1.84% of the gross domestic product, increasing by 0.17% of the GDP per year.

(Source: Congressional Budget Office)

(Exercise 70 continues on the next page.)

a. Write a function that models the cost of Social Security as a percentage of the GDP x years after 2000.

b. Write a function that models the cost of Medicare as a percentage of the GDP x years after 2000.

c. In which year will the cost of Medicare and Social Security be the same? For that year, what will be the cost of each program as a percentage of the GDP? Which program will have the greater cost after that year?

71. The graphs show the percentage of American adults in two living arrangements from 1960 through 2008.

Living Arrangements of U.S. Adults

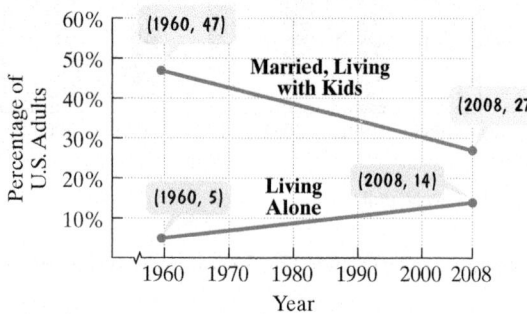

Source: U.S. Census Bureau

a. Write the slope-intercept equation of the line that models the percentage of U.S. adults living alone, y, x years after 1960. Round the value of the slope, m, to one decimal place.

b. Write the slope-intercept equation of the line that models the percentage of married U.S. adults living with kids, y, x years after 1960. Round the value of the slope, m, to one decimal place.

c. Use the models from parts (a) and (b) to project the year in which the percentage of adults living alone will be the same as the percentage of married adults living with kids. What percentage of U.S. adults will belong to each group during that year?

72. The graphs show that the U.S. public has taken a more conservative stance on gun control for the period from 1999 through 2009.

U.S. Public Opinion on Gun Control

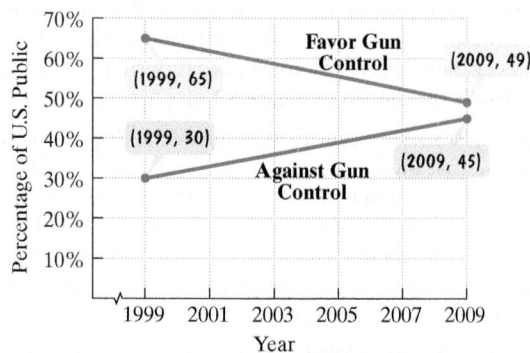

Source: pewresearch.org

a. Write the slope-intercept equation of the line that models the percentage of the U.S. public that favors gun control, y, x years after 1999.

b. Write the slope-intercept equation of the line that models the percentage of the U.S. public against gun control, y, x years after 1999.

c. Use the models from parts (a) and (b) to project the year, to the nearest whole year, during which the percentage in favor of gun control was the same as the percentage against gun control. What percentage of the U.S. public, to the nearest percent, shared each of these opinions during that year?

Use a system of linear equations to solve Exercises 73–84.

Looking for Mr. Goodbar? It's probably not a good idea if you want to look like Mr. Universe or Julia Roberts. The graph shows the four candy bars with the highest fat content, representing grams of fat and calories in each bar. Exercises 73–76 are based on the graph.

Candy Bars with the Highest Fat Content

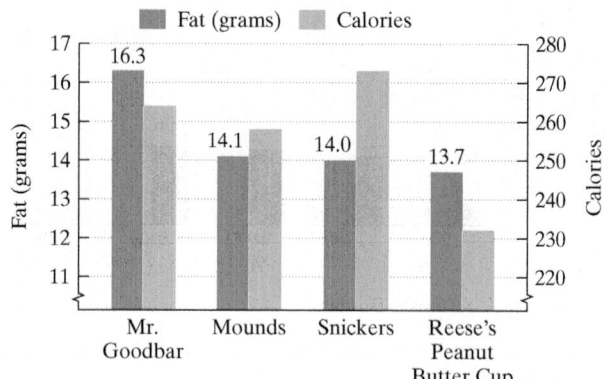

Source: Krantz and Sveum, *The World's Worsts*, HarperCollins, 2005

73. One Mr. Goodbar and two Mounds bars contain 780 calories. Two Mr. Goodbars and one Mounds bar contain 786 calories. Find the caloric content of each candy bar.

74. One Snickers bar and two Reese's Peanut Butter Cups contain 737 calories. Two Snickers bars and one Reese's Peanut Butter Cup contain 778 calories. Find the caloric content of each candy bar.

75. A collection of Halloween candy contains a total of five Mr. Goodbars and Mounds bars. Chew on this: The grams of fat in these candy bars exceed the daily maximum desirable fat intake of 70 grams by 7.1 grams. How many bars of each kind of candy are contained in the Halloween collection?

76. A collection of Halloween candy contains a total of 12 Snickers bars and Reese's Peanut Butter Cups. Chew on this: The grams of fat in these candy bars exceed twice the daily maximum desirable fat intake of 70 grams by 26.5 grams. How many bars of each kind of candy are contained in the Halloween collection?

77. A hotel has 200 rooms. Those with kitchen facilities rent for $100 per night and those without kitchen facilities rent for $80 per night. On a night when the hotel was completely occupied, revenues were $17,000. How many of each type of room does the hotel have?

78. A new restaurant is to contain two-seat tables and four-seat tables. Fire codes limit the restaurant's maximum occupancy to 56 customers. If the owners have hired enough servers to handle 17 tables of customers, how many of each kind of table should they purchase?

79. A rectangular lot whose perimeter is 360 feet is fenced along three sides. An expensive fencing along the lot's length costs $20 per foot and an inexpensive fencing along the two side widths costs only $8 per foot. The total cost of the fencing along the three sides comes to $3280. What are the lot's dimensions?

80. A rectangular lot whose perimeter is 320 feet is fenced along three sides. An expensive fencing along the lot's length costs $16 per foot and an inexpensive fencing along the two side widths costs only $5 per foot. The total cost of the fencing along the three sides comes to $2140. What are the lot's dimensions?

81. When a crew rows with the current, it travels 16 miles in 2 hours. Against the current, the crew rows 8 miles in 2 hours. Let x = the crew's rowing rate in still water and let y = the rate of the current. The following chart summarizes this information:

	Rate	×	Time	=	Distance
Rowing with current	$x + y$		2		16
Rowing against current	$x - y$		2		8

Find the rate of rowing in still water and the rate of the current.

82. When an airplane flies with the wind, it travels 800 miles in 4 hours. Against the wind, it takes 5 hours to cover the same distance. Find the plane's rate in still air and the rate of the wind.

In Exercises 83–84, an isosceles triangle containing two angles with equal measure is shown. The degree measure of each triangle's three interior angles and an exterior angle is represented with variables. Find the measure of the three interior angles.

83.

84.
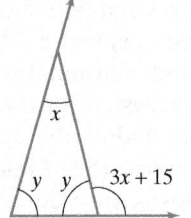

Writing in Mathematics

85. What is a system of linear equations? Provide an example with your description.

86. What is the solution of a system of linear equations?

87. Explain how to solve a system of equations using the substitution method. Use $y = 3 - 3x$ and $3x + 4y = 6$ to illustrate your explanation.

88. Explain how to solve a system of equations using the addition method. Use $3x + 5y = -2$ and $2x + 3y = 0$ to illustrate your explanation.

89. When is it easier to use the addition method rather than the substitution method to solve a system of equations?

90. When using the addition or substitution method, how can you tell if a system of linear equations has infinitely many solutions? What is the relationship between the graphs of the two equations?

91. When using the addition or substitution method, how can you tell if a system of linear equations has no solution? What is the relationship between the graphs of the two equations?

92. Describe the break-even point for a business.

Technology Exercise

93. Verify your solutions to any five exercises in Exercises 5–42 by using a graphing utility to graph the two equations in the system in the same viewing rectangle. Then use the intersection feature to display the solution.

Critical Thinking Exercises

Make Sense? *In Exercises 94–97, determine whether each statement makes sense or does not make sense, and explain your reasoning.*

94. Even if a linear system has a solution set involving fractions, such as $\{(\frac{8}{11}, \frac{43}{11})\}$, I can use graphs to determine if the solution set is reasonable.

95. Each equation in a system of linear equations has infinitely many ordered-pair solutions.

96. Every linear system has infinitely many ordered-pair solutions.

97. If I know the perimeter of this rectangle and triangle, each in the same unit of measure, I can use a system of linear equations to determine values for x and y.

 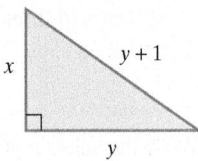

98. Write a system of equations having $\{(-2, 7)\}$ as a solution set. (More than one system is possible.)

99. Solve the system for x and y in terms of $a_1, b_1, c_1, a_2, b_2,$ and c_2:

$$\begin{cases} a_1 x + b_1 y = c_1 \\ a_2 x + b_2 y = c_2. \end{cases}$$

100. Two identical twins can only be distinguished by the characteristic that one always tells the truth and the other always lies. One twin tells you of a lucky number pair: "When I multiply my first lucky number by 3 and my second lucky number by 6, the addition of the resulting numbers produces a sum of 12. When I add my first lucky number and twice my second lucky number, the sum is 5." Which twin is talking?

101. A marching band has 52 members, and there are 24 in the pom-pom squad. They wish to form several hexagons and squares like those diagrammed below. Can it be done with no people left over?

B = Band Member
P = Pom-pom Person

Group Exercise

102. The group should write four different word problems that can be solved using a system of linear equations in two variables. All of the problems should be on different topics. The group should turn in the four problems and their algebraic solutions.

Preview Exercises

Exercises 103–105 will help you prepare for the material covered in the next section.

103. If $x = 3, y = 2,$ and $z = -3,$ does the ordered triple (x, y, z) satisfy the equation $2x - y + 4z = -8$?

104. Consider the following equations:

$$\begin{cases} 5x - 2y - 4z = 3 & \text{Equation 1} \\ 3x + 3y + 2z = -3. & \text{Equation 2} \end{cases}$$

Use this process to eliminate z: Copy Equation 1 and multiply Equation 2 by 2. Then add the equations.

105. Write an equation involving $a, b,$ and c based on the following description:

When the value of x in $y = ax^2 + bx + c$ is 4, the value of y is 1682.

SECTION 5.2 Systems of Linear Equations in Three Variables

Objectives

1. Verify the solution of a system of linear equations in three variables.
2. Solve systems of linear equations in three variables.
3. Solve problems using systems in three variables.

All animals sleep, but the length of time they sleep varies widely: Cattle sleep for only a few minutes at a time. We humans seem to need more sleep than other animals, up to eight hours a day. Without enough sleep, we have difficulty concentrating, make mistakes in routine tasks, lose energy, and feel bad-tempered. There is a relationship between hours of sleep and death rate per year per 100,000 people. How many hours of sleep will put you in the group with the minimum death rate? In this section, we will answer this question by solving a system of linear equations with more than two variables.

① Verify the solution of a system of linear equations in three variables.

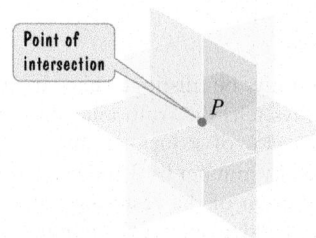

FIGURE 5.8

Systems of Linear Equations in Three Variables and Their Solutions

An equation such as $x + 2y - 3z = 9$ is called a *linear equation in three variables*. In general, any equation of the form

$$Ax + By + Cz = D,$$

where $A, B, C,$ and D are real numbers such that $A, B,$ and C are not all 0, is a **linear equation in three variables: $x, y,$ and z**. The graph of this linear equation in three variables is a plane in three-dimensional space.

The process of solving a system of three linear equations in three variables is geometrically equivalent to finding the point of intersection (assuming that there is one) of three planes in space. (See **Figure 5.8.**) A **solution** of a system of linear equations in three variables is an ordered triple of real numbers that satisfies all equations of the system. The **solution set** of the system is the set of all its solutions.

EXAMPLE 1 Determining Whether an Ordered Triple Satisfies a System

Show that the ordered triple $(-1, 2, -2)$ is a solution of the system:

$$\begin{cases} x + 2y - 3z = 9 \\ 2x - y + 2z = -8 \\ -x + 3y - 4z = 15. \end{cases}$$

SOLUTION

Because -1 is the x-coordinate, 2 is the y-coordinate, and -2 is the z-coordinate of $(-1, 2, -2)$, we replace x with -1, y with 2, and z with -2 in each of the three equations.

$$x + 2y - 3z = 9$$
$$-1 + 2(2) - 3(-2) \overset{?}{=} 9$$
$$-1 + 4 + 6 \overset{?}{=} 9$$
$$9 = 9, \quad \text{true}$$

$$2x - y + 2z = -8$$
$$2(-1) - 2 + 2(-2) \overset{?}{=} -8$$
$$-2 - 2 - 4 \overset{?}{=} -8$$
$$-8 = -8, \quad \text{true}$$

$$-x + 3y - 4z = 15$$
$$-(-1) + 3(2) - 4(-2) \overset{?}{=} 15$$
$$1 + 6 + 8 \overset{?}{=} 15$$
$$15 = 15, \quad \text{true}$$

The ordered triple $(-1, 2, -2)$ satisfies the three equations: It makes each equation true. Thus, the ordered triple is a solution of the system. ● ● ●

⌀ **Check Point 1** Show that the ordered triple $(-1, -4, 5)$ is a solution of the system:

$$\begin{cases} x - 2y + 3z = 22 \\ 2x - 3y - z = 5 \\ 3x + y - 5z = -32. \end{cases}$$

② Solve systems of linear equations in three variables.

Solving Systems of Linear Equations in Three Variables by Eliminating Variables

The method for solving a system of linear equations in three variables is similar to that used on systems of linear equations in two variables. We use addition to eliminate any variable, reducing the system to two equations in two variables. Once we obtain a system of two equations in two variables, we use addition or substitution to eliminate a variable. The result is a single equation in one variable. We solve this equation to get the value of the remaining variable. Other variable values are found by back-substitution.

GREAT QUESTION!

When solving a linear system in three variables, which variable should I eliminate first?

It does not matter which variable you eliminate first, as long as you eliminate the same variable in two different pairs of equations.

Solving Linear Systems in Three Variables by Eliminating Variables

1. Reduce the system to two equations in two variables. This is usually accomplished by taking two different pairs of equations and using the addition method to eliminate the same variable from both pairs.
2. Solve the resulting system of two equations in two variables using addition or substitution. The result is an equation in one variable that gives the value of that variable.
3. Back-substitute the value of the variable found in step 2 into either of the equations in two variables to find the value of the second variable.
4. Use the values of the two variables from steps 2 and 3 to find the value of the third variable by back-substituting into one of the original equations.
5. Check the proposed solution in each of the original equations.

EXAMPLE 2 Solving a System in Three Variables

Solve the system:

$$\begin{cases} 5x - 2y - 4z = 3 & \text{Equation 1} \\ 3x + 3y + 2z = -3 & \text{Equation 2} \\ -2x + 5y + 3z = 3. & \text{Equation 3} \end{cases}$$

SOLUTION

There are many ways to proceed. Because our initial goal is to reduce the system to two equations in two variables, **the central idea is to take two different pairs of equations and eliminate the same variable from both pairs.**

Step 1 Reduce the system to two equations in two variables. We choose any two equations and use the addition method to eliminate a variable. Let's eliminate z using Equations 1 and 2. We do so by multiplying Equation 2 by 2. Then we add equations.

(Equation 1) $\begin{cases} 5x - 2y - 4z = 3 \end{cases}$ $\xrightarrow{\text{No change}}$ $\begin{cases} 5x - 2y - 4z = 3 \end{cases}$
(Equation 2) $\begin{cases} 3x + 3y + 2z = -3 \end{cases}$ $\xrightarrow{\text{Multiply by 2.}}$ $\begin{cases} 6x + 6y + 4z = -6 \end{cases}$

$$\text{Add:} \quad 11x + 4y = -3 \quad \text{Equation 4}$$

$$\begin{cases} 5x - 2y - 4z = 3 & \text{Equation 1} \\ 3x + 3y + 2z = -3 & \text{Equation 2} \\ -2x + 5y + 3z = 3 & \text{Equation 3} \end{cases}$$

The given system (repeated)

Now we must eliminate the *same* variable using another pair of equations. We can eliminate z from Equations 2 and 3. First, we multiply Equation 2 by -3. Next, we multiply Equation 3 by 2. Finally, we add equations.

(Equation 2)
(Equation 3)
$$\begin{cases} 3x + 3y + 2z = -3 \\ -2x + 5y + 3z = 3 \end{cases} \xrightarrow[\text{Multiply by 2.}]{\text{Multiply by }-3.} \begin{cases} -9x - 9y - 6z = 9 \\ -4x + 10y + 6z = 6 \end{cases}$$

Add: $\quad -13x + \quad y \quad = 15 \quad$ Equation 5

Equation 4, $11x + 4y = -3$, obtained on the previous page, and Equation 5, $-13x + y = 15$, obtained above, give us a system of two equations in two variables:

$$\begin{cases} 11x + 4y = -3 & \text{Equation 4} \\ -13x + y = 15. & \text{Equation 5} \end{cases}$$

Step 2 Solve the resulting system of two equations in two variables. We will use the addition method to solve Equations 4 and 5 for x and y. To do so, we multiply Equation 5 on both sides by -4 and add this to Equation 4.

(Equation 4)
(Equation 5)
$$\begin{cases} 11x + 4y = -3 \\ -13x + y = 15 \end{cases} \xrightarrow[\text{Multiply by }-4.]{\text{No change}} \begin{cases} 11x + 4y = -3 \\ 52x - 4y = -60 \end{cases}$$

Add: $\quad 63x \qquad = -63$

$\qquad\qquad x \qquad = -1 \quad$ Divide both sides by 63.

Step 3 Use back-substitution in one of the equations in two variables to find the value of the second variable. We back-substitute -1 for x in either Equation 4 or 5 to find the value of y.

$$\begin{aligned} -13x + y &= 15 & \text{Equation 5} \\ -13(-1) + y &= 15 & \text{Substitute }-1\text{ for }x. \\ 13 + y &= 15 & \text{Multiply.} \\ y &= 2 & \text{Subtract 13 from both sides.} \end{aligned}$$

Step 4 Back-substitute the values found for two variables into one of the original equations to find the value of the third variable. We can now use any one of the original equations and back-substitute the values of x and y to find the value for z. We will use Equation 2.

$$\begin{aligned} 3x + 3y + 2z &= -3 & \text{Equation 2} \\ 3(-1) + 3(2) + 2z &= -3 & \text{Substitute }-1\text{ for }x\text{ and 2 for }y. \\ 3 + 2z &= -3 & \text{Multiply and then add:} \\ & & 3(-1) + 3(2) = -3 + 6 = 3. \\ 2z &= -6 & \text{Subtract 3 from both sides.} \\ z &= -3 & \text{Divide both sides by 2.} \end{aligned}$$

With $x = -1$, $y = 2$, and $z = -3$, the proposed solution is the ordered triple $(-1, 2, -3)$.

Step 5 Check. Check the proposed solution, $(-1, 2, -3)$, by substituting the values for x, y, and z into each of the three original equations. These substitutions yield three true statements. Thus, the solution set is $\{(-1, 2, -3)\}$. • • •

✓ Check Point **2** Solve the system:

$$\begin{cases} x + 4y - z = 20 \\ 3x + 2y + z = 8 \\ 2x - 3y + 2z = -16. \end{cases}$$

In some examples, one of the variables is already eliminated from a given equation. In this case, the missing variable should be eliminated from the other two equations, thereby making it possible to omit one of the elimination steps. We illustrate this idea in Example 3.

EXAMPLE 3 Solving a System of Equations with a Missing Term

Solve the system:

$$\begin{cases} x + z = 8 & \text{Equation 1} \\ x + y + 2z = 17 & \text{Equation 2} \\ x + 2y + z = 16. & \text{Equation 3} \end{cases}$$

DISCOVERY

Solve Equation 1 for z and then replace z with $8 - x$ in Equations 2 and 3. Solve the resulting system of two equations in two variables. Compare your solution method with the one shown on the right. Which method do you prefer?

SOLUTION

Step 1 Reduce the system to two equations in two variables. Because Equation 1 contains only x and z, we could omit one of the elimination steps by eliminating y using Equations 2 and 3. This will give us two equations in x and z. To eliminate y using Equations 2 and 3, we multiply Equation 2 by -2 and add Equation 3.

$$\begin{array}{ll} \text{(Equation 2)} & \begin{cases} x + y + 2z = 17 \\ x + 2y + z = 16 \end{cases} \end{array} \xrightarrow[\text{No change}]{\text{Multiply by } -2.} \begin{cases} -2x - 2y - 4z = -34 \\ \underline{x + 2y + z = 16} \end{cases}$$
$$\text{Add:} \quad -x - 3z = -18 \quad \text{Equation 4}$$

Equation 4 and the given Equation 1 provide us with a system of two equations in two variables:

$$\begin{cases} x + z = 8 & \text{Equation 1} \\ -x - 3z = -18. & \text{Equation 4} \end{cases}$$

Step 2 Solve the resulting system of two equations in two variables. We will solve Equations 1 and 4 for x and z.

$$\begin{cases} x + z = 8 & \text{Equation 1} \\ -x - 3z = -18 & \text{Equation 4} \end{cases}$$
$$\text{Add:} \quad -2z = -10$$
$$z = 5 \quad \text{Divide both sides by } -2.$$

Step 3 Use back-substitution in one of the equations in two variables to find the value of the second variable. To find x, we back-substitute 5 for z in either Equation 1 or 4. We will use Equation 1.

$$\begin{array}{ll} x + z = 8 & \text{Equation 1} \\ x + 5 = 8 & \text{Substitute 5 for z.} \\ x = 3 & \text{Subtract 5 from both sides.} \end{array}$$

Step 4 Back-substitute the values found for two variables into one of the original equations to find the value of the third variable. To find y, we back-substitute 3 for x and 5 for z into Equation 2 or 3. We cannot use Equation 1 because y is missing in this equation. We will use Equation 2.

$$\begin{array}{ll} x + y + 2z = 17 & \text{Equation 2} \\ 3 + y + 2(5) = 17 & \text{Substitute 3 for x and 5 for z.} \\ y + 13 = 17 & \text{Multiply and add.} \\ y = 4 & \text{Subtract 13 from both sides.} \end{array}$$

We found that $z = 5$, $x = 3$, and $y = 4$. Thus, the proposed solution is the ordered triple $(3, 4, 5)$.

Step 5 Check. Substituting 3 for x, 4 for y, and 5 for z into each of the three original equations yields three true statements. Consequently, the solution set is $\{(3, 4, 5)\}$. • • •

✓ Check Point **3** Solve the system:

$$\begin{cases} 2y - z = 7 \\ x + 2y + z = 17 \\ 2x - 3y + 2z = -1. \end{cases}$$

A system of linear equations in three variables represents three planes. The three planes may not always intersect at one point. The planes may have no common point of intersection and represent an inconsistent system with no solution. By contrast, the planes may coincide or intersect along a line. In these cases, the planes have infinitely many points in common and represent systems with infinitely many solutions. Systems of linear equations in three variables that are inconsistent or that contain dependent equations will be discussed in Chapter 6.

③ Solve problems using systems in three variables.

Applications

Systems of equations may allow us to find models for data without using a graphing utility. Three data points that do not lie on or near a line determine the graph of a quadratic function of the form $y = ax^2 + bx + c, a \neq 0$. Quadratic functions often model situations in which values of y are decreasing and then increasing, suggesting the bowl-like shape of a parabola.

EXAMPLE 4 Modeling Data Relating Sleep and Death Rate

In a study relating sleep and death rate, the following data were obtained. Use the function $y = ax^2 + bx + c$ to model the data.

x (Average Number of Hours of Sleep)	y (Death Rate per Year per 100,000 Males)
4	1682
7	626
9	967

SOLUTION

We need to find values for a, b, and c in $y = ax^2 + bx + c$. We can do so by solving a system of three linear equations in a, b, and c. We obtain the three equations by using the values of x and y from the data as follows:

$y = ax^2 + bx + c$ *Use the quadratic function to model the data.*

When $x = 4, y = 1682$: $\begin{cases} 1682 = a \cdot 4^2 + b \cdot 4 + c \\ 626 = a \cdot 7^2 + b \cdot 7 + c \\ 967 = a \cdot 9^2 + b \cdot 9 + c \end{cases}$ or $\begin{cases} 16a + 4b + c = 1682 \\ 49a + 7b + c = 626 \\ 81a + 9b + c = 967. \end{cases}$

When $x = 7, y = 626$:

When $x = 9, y = 967$:

The easiest way to solve this system is to eliminate c from two pairs of equations, obtaining two equations in a and b. Solving this system gives $a = 104.5$, $b = -1501.5$, and $c = 6016$. We now substitute the values for a, b, and c into $y = ax^2 + bx + c$. The function that models the given data is

$$y = 104.5x^2 - 1501.5x + 6016.$$ • • •

We can use the model that we obtained in Example 4 to find the death rate of males who average, say, 6 hours of sleep. First, write the model in function notation:

$$f(x) = 104.5x^2 - 1501.5x + 6016.$$

Substitute 6 for x:

$$f(6) = 104.5(6)^2 - 1501.5(6) + 6016 = 769.$$

According to the model, the death rate for males who average 6 hours of sleep is 769 deaths per 100,000 males.

TECHNOLOGY

The graph of
$$y = 104.5x^2 - 1501.5x + 6016$$
is displayed in a [3, 12, 1] by [500, 2000, 100] viewing rectangle. The minimum function feature shows that the lowest point on the graph, the vertex, is approximately (7.2, 622.5). Men who average 7.2 hours of sleep are in the group with the lowest death rate, approximately 622.5 deaths per 100,000 males.

☑ Check Point **4** Find the quadratic function $y = ax^2 + bx + c$ whose graph passes through the points $(1, 4)$, $(2, 1)$, and $(3, 4)$.

CONCEPT AND VOCABULARY CHECK

Fill in each blank so that the resulting statement is true.

1. A solution of a system of linear equations in three variables is an ordered _____ of real numbers that satisfies <u>all/some</u> of the equations in the system.

Circle the correct choice.

2. Consider the following system:

$$\begin{cases} x + y - z = -1 & \text{Equation 1} \\ 2x - 2y - 5z = 7 & \text{Equation 2} \\ 4x + y - 2z = 7. & \text{Equation 3} \end{cases}$$

We can eliminate x from Equations 1 and 2 by multiplying Equation 1 by _____ and adding equations. We can eliminate x from Equations 1 and 3 by multiplying Equation 1 by _____ and adding equations.

3. Consider the following system:

$$\begin{cases} x + y + z = 2 & \text{Equation 1} \\ 2x - 3y = 3 & \text{Equation 2} \\ 10y - z = 12. & \text{Equation 3} \end{cases}$$

Equation 2 does not contain the variable ____. To obtain a second equation that does not contain this variable, we can _____.

EXERCISE SET 5.2

Practice Exercises

In Exercises 1–4, determine if the given ordered triple is a solution of the system.

1. $(2, -1, 3)$
$$\begin{cases} x + y + z = 4 \\ x - 2y - z = 1 \\ 2x - y - 2z = -1 \end{cases}$$

2. $(5, -3, -2)$
$$\begin{cases} x + y + z = 0 \\ x + 2y - 3z = 5 \\ 3x + 4y + 2z = -1 \end{cases}$$

3. $(4, 1, 2)$
$$\begin{cases} x - 2y = 2 \\ 2x + 3y = 11 \\ y - 4z = -7 \end{cases}$$

4. $(-1, 3, 2)$
$$\begin{cases} x - 2z = -5 \\ y - 3z = -3 \\ 2x - z = -4 \end{cases}$$

Solve each system in Exercises 5–18.

5. $\begin{cases} x + y + 2z = 11 \\ x + y + 3z = 14 \\ x + 2y - z = 5 \end{cases}$

6. $\begin{cases} 2x + y - 2z = -1 \\ 3x - 3y - z = 5 \\ x - 2y + 3z = 6 \end{cases}$

7. $\begin{cases} 4x - y + 2z = 11 \\ x + 2y - z = -1 \\ 2x + 2y - 3z = -1 \end{cases}$

8. $\begin{cases} x - y + 3z = 8 \\ 3x + y - 2z = -2 \\ 2x + 4y + z = 0 \end{cases}$

9. $\begin{cases} 3x + 2y - 3z = -2 \\ 2x - 5y + 2z = -2 \\ 4x - 3y + 4z = 10 \end{cases}$

10. $\begin{cases} 2x + 3y + 7z = 13 \\ 3x + 2y - 5z = -22 \\ 5x + 7y - 3z = -28 \end{cases}$

11. $\begin{cases} 2x - 4y + 3z = 17 \\ x + 2y - z = 0 \\ 4x - y - z = 6 \end{cases}$

12. $\begin{cases} x + z = 3 \\ x + 2y - z = 1 \\ 2x - y + z = 3 \end{cases}$

13. $\begin{cases} 2x + y = 2 \\ x + y - z = 4 \\ 3x + 2y + z = 0 \end{cases}$

14. $\begin{cases} x + 3y + 5z = 20 \\ y - 4z = -16 \\ 3x - 2y + 9z = 36 \end{cases}$

15. $\begin{cases} x + y = -4 \\ y - z = 1 \\ 2x + y + 3z = -21 \end{cases}$

16. $\begin{cases} x + y = 4 \\ x + z = 4 \\ y + z = 4 \end{cases}$

17. $\begin{cases} 3(2x + y) + 5z = -1 \\ 2(x - 3y + 4z) = -9 \\ 4(1 + x) = -3(z - 3y) \end{cases}$

18. $\begin{cases} 7z - 3 = 2(x - 3y) \\ 5y + 3z - 7 = 4x \\ 4 + 5z = 3(2x - y) \end{cases}$

In Exercises 19–22, find the quadratic function $y = ax^2 + bx + c$ whose graph passes through the given points.

19. $(-1, 6), (1, 4), (2, 9)$ **20.** $(-2, 7), (1, -2), (2, 3)$
21. $(-1, -4), (1, -2), (2, 5)$ **22.** $(1, 3), (3, -1), (4, 0)$

In Exercises 23–24, let x represent the first number, y the second number, and z the third number. Use the given conditions to write a system of equations. Solve the system and find the numbers.

23. The sum of three numbers is 16. The sum of twice the first number, 3 times the second number, and 4 times the third number is 46. The difference between 5 times the first number and the second number is 31. Find the three numbers.

24. The following is known about three numbers: Three times the first number plus the second number plus twice the third number is 5. If 3 times the second number is subtracted from the sum of the first number and 3 times the third number, the result is 2. If the third number is subtracted from 2 times the first number and 3 times the second number, the result is 1. Find the numbers.

Practice Plus

Solve each system in Exercises 25–26.

25. $\begin{cases} \dfrac{x+2}{6} - \dfrac{y+4}{3} + \dfrac{z}{2} = 0 \\[2mm] \dfrac{x+1}{2} + \dfrac{y-1}{2} - \dfrac{z}{4} = \dfrac{9}{2} \\[2mm] \dfrac{x-5}{4} + \dfrac{y+1}{3} + \dfrac{z-2}{2} = \dfrac{19}{4} \end{cases}$

26. $\begin{cases} \dfrac{x+3}{2} - \dfrac{y-1}{2} + \dfrac{z+2}{4} = \dfrac{3}{2} \\[2mm] \dfrac{x-5}{2} + \dfrac{y+1}{3} - \dfrac{z}{4} = -\dfrac{25}{6} \\[2mm] \dfrac{x-3}{4} - \dfrac{y+1}{2} + \dfrac{z-3}{2} = -\dfrac{5}{2} \end{cases}$

In Exercises 27–28, find the equation of the quadratic function $y = ax^2 + bx + c$ whose graph is shown. Select three points whose coordinates appear to be integers.

27.

28.

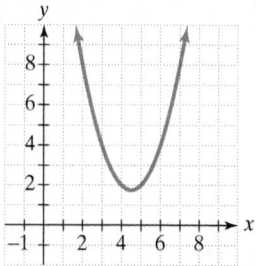

In Exercises 29–30, solve each system for (x, y, z) in terms of the nonzero constants a, b, and c.

29. $\begin{cases} ax - by - 2cz = 21 \\ ax + by + cz = 0 \\ 2ax - by + cz = 14 \end{cases}$ **30.** $\begin{cases} ax - by + 2cz = -4 \\ ax + 3by - cz = 1 \\ 2ax + by + 3cz = 2 \end{cases}$

Application Exercises

31. You throw a ball straight up from a rooftop. The ball misses the rooftop on its way down and eventually strikes the ground. A mathematical model can be used to describe the relationship for the ball's height above the ground, y, after x seconds. Consider the following data:

x, seconds after the ball is thrown	y, ball's height, in feet, above the ground
1	224
3	176
4	104

a. Find the quadratic function $y = ax^2 + bx + c$ whose graph passes through the given points.

b. Use the function in part (a) to find the value for y when $x = 5$. Describe what this means.

32. A mathematical model can be used to describe the relationship between the number of feet a car travels once the brakes are applied, y, and the number of seconds the car is in motion after the brakes are applied, x. A research firm collects the following data:

x, seconds in motion after brakes are applied	y, feet car travels once the brakes are applied
1	46
2	84
3	114

a. Find the quadratic function $y = ax^2 + bx + c$ whose graph passes through the given points.

b. Use the function in part (a) to find the value for y when $x = 6$. Describe what this means.

Use a system of linear equations in three variables to solve Exercises 33–41.

The bar graph shows the average annual spending per person on selected items in 1980 and 2010. All dollar amounts are adjusted for inflation. Use this display to solve Exercises 33–34.

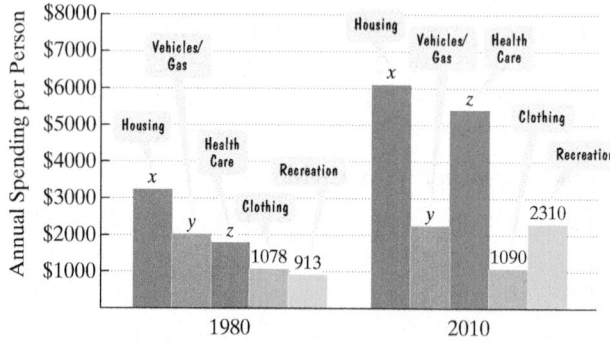

Annual Spending per Person in the United States, Adjusted for Inflation

Source: U.S. Bureau of Economic Analysis

33. In this exercise, we refer to annual spending per person in 2010. The combined spending on housing, vehicles/gas, and health care was \$13,840. The difference between spending on housing and spending on vehicles/gas was \$3864. The difference between spending on housing and spending on health care was \$695. Find the average per-person spending on housing, vehicles/gas, and health care in 2010.

34. In this exercise, we refer to annual spending per person in 1980. The combined spending on housing, vehicles/gas, and health care was \$7073. The difference between spending on housing and spending on vehicles/gas was \$1247. The difference between spending on housing and spending on health care was \$1466. Find the average per-person spending on housing, vehicles/gas, and health care in 1980.

35. On a recent trip to the convenience store, you picked up 2 gallons of milk, 5 bottles of water, and 6 snack-size bags of chips. Your total bill (before tax) was $19.00. If a bottle of water costs twice as much as a bag of chips, and a gallon of milk costs $2.00 more than a bottle of water, how much does each item cost?

36. On a recent trip to the convenience store, you picked up 1 gallon of milk, 7 bottles of water, and 4 snack-size bags of chips. Your total bill (before tax) was $17.00. If a bottle of water costs twice as much as a bag of chips, and a gallon of milk costs $2.00 more than a bottle of water, how much does each item cost?

37. At a college production of *A Streetcar Named Desire*, 400 tickets were sold. The ticket prices were $8, $10, and $12, and the total income from ticket sales was $3700. How many tickets of each type were sold if the combined number of $8 and $10 tickets sold was 7 times the number of $12 tickets sold?

38. A certain brand of razor blades comes in packages of 6, 12, and 24 blades, costing $2, $3, and $4 per package, respectively. A store sold 12 packages containing a total of 162 razor blades and took in $35. How many packages of each type were sold?

39. A person invested $6700 for one year, part at 8%, part at 10%, and the remainder at 12%. The total annual income from these investments was $716. The amount of money invested at 12% was $300 more than the amount invested at 8% and 10% combined. Find the amount invested at each rate.

40. A person invested $17,000 for one year, part at 10%, part at 12%, and the remainder at 15%. The total annual income from these investments was $2110. The amount of money invested at 12% was $1000 less than the amount invested at 10% and 15% combined. Find the amount invested at each rate.

41. In the following triangle, the degree measures of the three interior angles and two of the exterior angles are represented with variables. Find the measure of each interior angle.

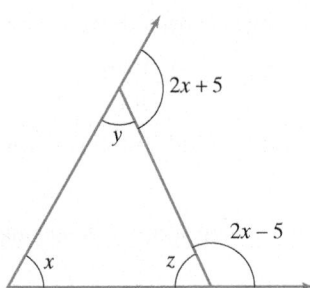

Writing in Mathematics

42. What is a system of linear equations in three variables?

43. How do you determine whether a given ordered triple is a solution of a system in three variables?

44. Describe in general terms how to solve a system in three variables.

45. AIDS is taking a deadly toll on southern Africa. Describe how to use the techniques that you learned in this section to obtain a model for African life span using projections with AIDS, shown by the red graph in the figure. Let x represent the number of years after 1985 and let y represent African life span in that year.

African Life Span

Source: United Nations

Technology Exercises

46. Does your graphing utility have a feature that allows you to solve linear systems by entering coefficients and constant terms? If so, use this feature to verify the solutions to any five exercises that you worked by hand from Exercises 5–16.

47. Verify your results in Exercises 19–22 by using a graphing utility to graph the resulting parabola. Trace along the curve and convince yourself that the three points given in the exercise lie on the parabola.

Critical Thinking Exercises

Make Sense? *In Exercises 48–51, determine whether each statement makes sense or does not make sense, and explain your reasoning.*

48. Solving a system in three variables, I found that $x = 3$ and $y = -1$. Because z represents a third variable, z cannot equal 3 or -1.

49. A system of linear equations in three variables, x, y, and z, cannot contain an equation in the form $y = mx + b$.

50. I'm solving a three-variable system in which one of the given equations has a missing term, so it will not be necessary to use any of the original equations twice when I reduce the system to two equations in two variables.

51. Because the percentage of the U.S. population that was foreign-born decreased from 1910 through 1970 and then increased after that, a quadratic function of the form $f(x) = ax^2 + bx + c$, rather than a linear function of the form $f(x) = mx + b$, should be used to model the data.

52. Describe how the system

$$\begin{cases} x + y - z - 2w = -8 \\ x - 2y + 3z + w = 18 \\ 2x + 2y + 2z - 2w = 10 \\ 2x + y - z + w = 3 \end{cases}$$

could be solved. Is it likely that in the near future a graphing utility will be available to provide a geometric solution (using intersecting graphs) to this system? Explain.

53. A modernistic painting consists of triangles, rectangles, and pentagons, all drawn so as to not overlap or share sides. Within each rectangle are drawn 2 red roses and each pentagon contains 5 carnations. How many triangles, rectangles, and pentagons appear in the painting if the painting contains a total of 40 geometric figures, 153 sides of geometric figures, and 72 flowers?

Group Exercise

54. Group members should develop appropriate functions that model each of the projections shown in Exercise 45.

Preview Exercises

Exercises 55–57 will help you prepare for the material covered in the next section.

55. Subtract: $\dfrac{3}{x-4} - \dfrac{2}{x+2}$.

56. Add: $\dfrac{5x-3}{x^2+1} + \dfrac{2x}{(x^2+1)^2}$.

57. Solve:

$$\begin{cases} A + B = 3 \\ 2A - 2B + C = 17 \\ 4A - 2C = 14. \end{cases}$$

SECTION 5.3 Partial Fractions

Objectives

1. Decompose $\dfrac{P}{Q}$, where Q has only distinct linear factors.

2. Decompose $\dfrac{P}{Q}$, where Q has repeated linear factors.

3. Decompose $\dfrac{P}{Q}$, where Q has a nonrepeated prime quadratic factor.

4. Decompose $\dfrac{P}{Q}$, where Q has a prime, repeated quadratic factor.

The rising and setting of the sun suggest the obvious: Things change over time. Calculus is the study of rates of change, allowing the motion of the rising sun to be measured by "freezing the frame" at one instant in time. If you are given a function, calculus reveals its rate of change at any "frozen" instant. In this section, you will learn an algebraic technique used in calculus to find a function if its rate of change is known. The technique involves expressing a given function in terms of simpler functions.

The Idea behind Partial Fraction Decomposition

We know how to use common denominators to write a sum or difference of rational expressions as a single rational expression. For example,

$$\frac{3}{x-4} - \frac{2}{x+2} = \frac{3(x+2) - 2(x-4)}{(x-4)(x+2)}$$

$$= \frac{3x+6 - 2x+8}{(x-4)(x+2)} = \frac{x+14}{(x-4)(x+2)}.$$

For solving the kind of calculus problem described in the section opener, we must reverse this process:

$\dfrac{x+14}{(x-4)(x+2)}$ is expressed as the sum of two simpler fractions.

Partial fraction ‹ Partial fraction

$$\frac{x+14}{(x-4)(x+2)} = \frac{3}{x-4} + \frac{-2}{x+2}.$$

This is the partial fraction decomposition of $\dfrac{x+14}{(x-4)(x+2)}$.

Each of the two fractions on the right is called a **partial fraction**. The sum of these fractions is called the **partial fraction decomposition** of the rational expression on the left-hand side.

Partial fraction decompositions can be written for rational expressions of the form $\dfrac{P(x)}{Q(x)}$, where P and Q have no common factors and the highest power in the numerator is less than the highest power in the denominator. In this section, we will show you how to write the partial fraction decompositions for each of the following rational expressions:

$$\dfrac{9x^2 - 9x + 6}{(2x - 1)(x + 2)(x - 2)}$$

$P(x) = 9x^2 - 9x + 6$; highest power $= 2$

$Q(x) = (2x - 1)(x + 2)(x - 2)$; multiplying factors, highest power $= 3$.

$$\dfrac{5x^3 - 3x^2 + 7x - 3}{(x^2 + 1)^2}.$$

$P(x) = 5x^3 - 3x^2 + 7x - 3$; highest power $= 3$

$Q(x) = (x^2 + 1)^2$; squaring the expression, highest power $= 4$.

The partial fraction decomposition of a rational expression depends on the factors of the denominator. We consider four cases involving different kinds of factors in the denominator:

1. The denominator is a product of distinct linear factors.

2. The denominator is a product of linear factors, some of which are repeated.

3. The denominator has prime quadratic factors, none of which is repeated.

4. The denominator has a repeated prime quadratic factor.

① Decompose $\dfrac{P}{Q}$, where Q has only distinct linear factors.

The Partial Fraction Decomposition of a Rational Expression with Distinct Linear Factors in the Denominator

If the denominator of a rational expression has a linear factor of the form $ax + b$, then the partial fraction decomposition will contain a term of the form

$$\dfrac{A}{ax + b}.$$

Constant

Linear factor

Each distinct linear factor in the denominator produces a partial fraction of the form *constant over linear factor*. For example,

$$\dfrac{9x^2 - 9x + 6}{(2x - 1)(x + 2)(x - 2)} = \dfrac{A}{2x - 1} + \dfrac{B}{x + 2} + \dfrac{C}{x - 2}.$$

We write a constant over each linear factor in the denominator.

The Partial Fraction Decomposition of $\dfrac{P(x)}{Q(x)}$: $Q(x)$ Has Distinct Linear Factors

The form of the partial fraction decomposition for a rational expression with distinct linear factors in the denominator is

$$\dfrac{P(x)}{(a_1x + b_1)(a_2x + b_2)(a_3x + b_3) \cdots (a_nx + b_n)}$$

$$= \dfrac{A_1}{a_1x + b_1} + \dfrac{A_2}{a_2x + b_2} + \dfrac{A_3}{a_3x + b_3} + \cdots + \dfrac{A_n}{a_nx + b_n}.$$

EXAMPLE 1 Partial Fraction Decomposition with Distinct Linear Factors

Find the partial fraction decomposition of

$$\frac{x + 14}{(x - 4)(x + 2)}.$$

SOLUTION

We begin by setting up the partial fraction decomposition with the unknown constants. Write a constant over each of the two distinct linear factors in the denominator.

$$\frac{x + 14}{(x - 4)(x + 2)} = \frac{A}{x - 4} + \frac{B}{x + 2}$$

Our goal is to find A and B. We do this by multiplying both sides of the equation by the least common denominator, $(x - 4)(x + 2)$.

$$(x - 4)(x + 2)\frac{x + 14}{(x - 4)(x + 2)} = (x - 4)(x + 2)\left(\frac{A}{x - 4} + \frac{B}{x + 2}\right)$$

We use the distributive property on the right side.

$$(x - 4)(x + 2)\frac{x + 14}{(x - 4)(x + 2)}$$

$$= (x - 4)(x + 2)\frac{A}{(x - 4)} + (x - 4)(x + 2)\frac{B}{(x + 2)}$$

Dividing out common factors in numerators and denominators, we obtain

$$x + 14 = A(x + 2) + B(x - 4).$$

To find values for A and B that make both sides equal, we will express the sides in exactly the same form by writing the variable x-terms and then writing the constant terms. Apply the distributive property on the right side.

$x + 14 = Ax + 2A + Bx - 4B$	Distribute A and B over the parentheses.
$x + 14 = Ax + Bx + 2A - 4B$	Rearrange terms.
$1x + 14 = (A + B)x + (2A - 4B)$	Rewrite to identify the coefficient of x and the constant term.

As shown by the arrows, if two polynomials are equal, coefficients of like powers of x must be equal ($A + B = 1$) and their constant terms must be equal ($2A - 4B = 14$). Consequently, A and B satisfy the following two equations:

$$\begin{cases} A + B = 1 \\ 2A - 4B = 14. \end{cases}$$

We can use the addition method to solve this linear system in two variables. By multiplying the first equation by -2 and adding equations, we obtain $A = 3$ and $B = -2$. Thus,

$$\frac{x + 14}{(x - 4)(x + 2)} = \frac{A}{x - 4} + \frac{B}{x + 2} = \frac{3}{x - 4} + \frac{-2}{x + 2} \left(\text{or } \frac{3}{x - 4} - \frac{2}{x + 2} \right).$$

●●●

Steps in Partial Fraction Decomposition

1. Set up the partial fraction decomposition with the unknown constants A, B, C, etc., in the numerators of the decomposition.
2. Multiply both sides of the resulting equation by the least common denominator.
3. Simplify the right side of the equation.
4. Write both sides in descending powers, equate coefficients of like powers of x, and equate constant terms.
5. Solve the resulting linear system for A, B, C, etc.
6. Substitute the values for A, B, C, etc., into the equation in step 1 and write the partial fraction decomposition.

GREAT QUESTION!

Is there any way to speed up the process of finding partial fraction decompositions?

Sometimes. In certain situations, you can use the equation in step 2, the identity obtained after clearing fractions, to determine the constants A, B, C, etc., needed to write the partial fraction decomposition. In Example 1, the identity obtained after fractions were cleared was

$$x + 14 = A(x + 2) + B(x - 4).$$

Using suitable choices for x, we can quickly find A and B.

$$x + 14 = A(x + 2) + B(x - 4) \qquad x + 14 = A(x + 2) + B(x - 4)$$

If we let $x = -2$, the term containing A drops out.
$12 = -6B$
$-2 = B$

If we let $x = 4$, the term containing B drops out.
$18 = 6A$
$3 = A$

As in the solution to Example 1, $A = 3$ and $B = -2$. This shortcut sometimes makes life easier by letting you skip steps 3, 4, and 5 in the box. The downside: Using suitable choices for x to determine constants does not always work.

Ø Check Point **1** Find the partial fraction decomposition of $\dfrac{5x - 1}{(x - 3)(x + 4)}$.

2 Decompose $\dfrac{P}{Q}$, where Q has repeated linear factors.

The Partial Fraction Decomposition of a Rational Expression with Linear Factors in the Denominator, Some of Which Are Repeated

Suppose that $(ax + b)^n$ is a factor of the denominator. This means that the linear factor $ax + b$ is repeated n times. When this occurs, the partial fraction decomposition will contain a sum of n fractions for this factor of the denominator.

The Partial Fraction Decomposition of $\dfrac{P(x)}{Q(x)}$: $Q(x)$ Has Repeated Linear Factors

The form of the partial fraction decomposition for a rational expression containing the linear factor $ax + b$ occurring n times as its denominator is

$$\frac{P(x)}{(ax + b)^n} = \frac{A_1}{ax + b} + \frac{A_2}{(ax + b)^2} + \frac{A_3}{(ax + b)^3} + \cdots + \frac{A_n}{(ax + b)^n}.$$

Include one fraction with a constant numerator for each power of $ax + b$.

EXAMPLE 2 Partial Fraction Decomposition with a Repeated Linear Factor

Find the partial fraction decomposition of $\dfrac{x - 18}{x(x - 3)^2}$.

SOLUTION

Step 1 Set up the partial fraction decomposition with the unknown constants. Because the linear factor $x - 3$ occurs twice, we must include one fraction with a constant numerator for each power of $x - 3$.

$$\frac{x - 18}{x(x - 3)^2} = \frac{A}{x} + \frac{B}{x - 3} + \frac{C}{(x - 3)^2}$$

Step 2 Multiply both sides of the resulting equation by the least common denominator. We clear fractions, multiplying both sides by $x(x - 3)^2$, the least common denominator.

$$x(x - 3)^2 \left[\frac{x - 18}{x(x - 3)^2} \right] = x(x - 3)^2 \left[\frac{A}{x} + \frac{B}{x - 3} + \frac{C}{(x - 3)^2} \right]$$

We use the distributive property on the right side.

$$x(x - 3)^2 \cdot \frac{x - 18}{x(x - 3)^2} = x(x - 3)^2 \cdot \frac{A}{x} + x(x - 3)^2 \cdot \frac{B}{(x - 3)} + x(x - 3)^2 \cdot \frac{C}{(x - 3)^2}$$

Dividing out common factors in numerators and denominators, we obtain

$$x - 18 = A(x - 3)^2 + Bx(x - 3) + Cx.$$

You can use this equation to find C (let $x = 3$) and to find A (let $x = 0$).

Step 3 Simplify the right side of the equation. Square $x - 3$. Then apply the distributive property.

$$x - 18 = A(x^2 - 6x + 9) + Bx(x - 3) + Cx \qquad \text{Square } x - 3 \text{ using } (A - B)^2 = A^2 - 2AB + B^2.$$

$$x - 18 = Ax^2 - 6Ax + 9A + Bx^2 - 3Bx + Cx \qquad \text{Apply the distributive property.}$$

GREAT QUESTION!

When setting up the partial fraction decomposition for

$$\frac{x - 18}{x(x - 3)^2},$$

can I just use a constant over $x - 3$ and another constant over $x - 3$? After all,

$$(x - 3)^2 = (x - 3)(x - 3).$$

No. Avoid this common error:

$$\frac{x - 18}{x(x - 3)^2} = \frac{A}{x} + \frac{B}{x - 3} + \frac{C}{x - 3}$$

Listing $x - 3$ twice does not take into account $(x - 3)^2$.

Step 4 Write both sides in descending powers, equate coefficients of like powers of x, and equate constant terms. The left side, $x - 18$, is in descending powers of x: $x - 18x^0$. We will write the right side in descending powers of x.

$x - 18 = Ax^2 + Bx^2 - 6Ax - 3Bx + Cx + 9A$ Rearrange terms on the right side.

Express both sides in the same form.

$0x^2 + 1x - 18 = (A + B)x^2 + (-6A - 3B + C)x + 9A$ Rewrite to identify coefficients and the constant term.

Equating coefficients of like powers of x and equating constant terms results in the following system of linear equations:

$$\begin{cases} A + B = 0 \\ -6A - 3B + C = 1 \\ 9A = -18. \end{cases}$$

Step 5 Solve the resulting system for A, B, and C. Dividing both sides of the last equation by 9, we obtain $A = -2$. Substituting -2 for A in the first equation, $A + B = 0$, gives $-2 + B = 0$, so $B = 2$. We find C by substituting -2 for A and 2 for B in the middle equation, $-6A - 3B + C = 1$. We obtain $C = -5$.

Step 6 Substitute the values of A, B, and C, and write the partial fraction decomposition. With $A = -2$, $B = 2$, and $C = -5$, the required partial fraction decomposition is

$$\frac{x - 18}{x(x - 3)^2} = \frac{A}{x} + \frac{B}{x - 3} + \frac{C}{(x - 3)^2} = -\frac{2}{x} + \frac{2}{x - 3} - \frac{5}{(x - 3)^2}. \quad \bullet\bullet\bullet$$

✓ **Check Point 2** Find the partial fraction decomposition of $\dfrac{x + 2}{x(x - 1)^2}$.

GREAT QUESTION!

In Examples 1 and 2, the denominators of the given rational expressions were factored. What do I do when I have to determine a partial fraction decomposition in which the denominator of the given rational expression is not already factored?

Begin by factoring the denominator. Then apply the steps needed to obtain the partial fraction decomposition.

③ Decompose $\dfrac{P}{Q}$, where Q has a nonrepeated prime quadratic factor.

The Partial Fraction Decomposition of a Rational Expression with Prime, Nonrepeated Quadratic Factors in the Denominator

Our final two cases of partial fraction decomposition involve prime quadratic factors of the form $ax^2 + bx + c$. Based on our work with the discriminant, we know that $ax^2 + bx + c$ is prime and cannot be factored over the integers if $b^2 - 4ac < 0$ or if $b^2 - 4ac$ is not a perfect square.

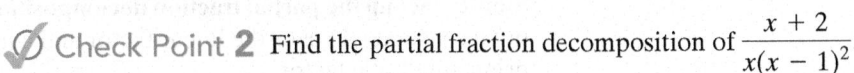

The Partial Fraction Decomposition of $\dfrac{P(x)}{Q(x)}$: $Q(x)$ Has a Nonrepeated, Prime Quadratic Factor

If $ax^2 + bx + c$ is a prime quadratic factor of $Q(x)$, the partial fraction decomposition will contain a term of the form

$$\frac{Ax + B}{ax^2 + bx + c}.$$

Linear numerator

Quadratic factor

The voice balloons in the box show that each distinct prime quadratic factor in the denominator produces a partial fraction of the form *linear numerator over quadratic factor*.

Here's an example of a partial fraction decomposition containing a linear numerator over a quadratic factor:

$$\frac{3x^2 + 17x + 14}{(x - 2)(x^2 + 2x + 4)} = \frac{A}{x - 2} + \frac{Bx + C}{x^2 + 2x + 4}.$$

| We write a constant over the linear factor in the denominator. | We write a linear numerator over the prime quadratic factor in the denominator. |

Our next example illustrates how a linear system in three variables is used to determine values for A, B, and C.

EXAMPLE 3 Partial Fraction Decomposition

Find the partial fraction decomposition of

$$\frac{3x^2 + 17x + 14}{(x - 2)(x^2 + 2x + 4)}.$$

SOLUTION

Step 1 Set up the partial fraction decomposition with the unknown constants. We put a constant (A) over the linear factor and a linear expression ($Bx + C$) over the prime quadratic factor.

$$\frac{3x^2 + 17x + 14}{(x - 2)(x^2 + 2x + 4)} = \frac{A}{x - 2} + \frac{Bx + C}{x^2 + 2x + 4}$$

Step 2 Multiply both sides of the resulting equation by the least common denominator. We clear fractions, multiplying both sides by $(x - 2)(x^2 + 2x + 4)$, the least common denominator.

$$(x - 2)(x^2 + 2x + 4)\left[\frac{3x^2 + 17x + 14}{(x - 2)(x^2 + 2x + 4)}\right] = (x - 2)(x^2 + 2x + 4)\left[\frac{A}{x - 2} + \frac{Bx + C}{x^2 + 2x + 4}\right]$$

We use the distributive property on the right side.

$$(x - 2)(x^2 + 2x + 4) \cdot \frac{3x^2 + 17x + 14}{(x - 2)(x^2 + 2x + 4)}$$

$$= (x - 2)(x^2 + 2x + 4) \cdot \frac{A}{x - 2} + (x - 2)(x^2 + 2x + 4) \cdot \frac{Bx + C}{x^2 + 2x + 4}$$

Dividing out common factors in numerators and denominators, we obtain

$$3x^2 + 17x + 14 = A(x^2 + 2x + 4) + (Bx + C)(x - 2).$$

Step 3 Simplify the right side of the equation. We simplify on the right side by distributing A over each term in parentheses and multiplying $(Bx + C)(x - 2)$ using the FOIL method.

$$3x^2 + 17x + 14 = Ax^2 + 2Ax + 4A + Bx^2 - 2Bx + Cx - 2C$$

Step 4 Write both sides in descending powers, equate coefficients of like powers of x, and equate constant terms. The left side, $3x^2 + 17x + 14$, is in descending powers of x. We write the right side in descending powers of x

$$3x^2 + 17x + 14 = Ax^2 + Bx^2 + 2Ax - 2Bx + Cx + 4A - 2C$$

and express both sides in the same form.

$$3x^2 + 17x + 14 = (A + B)x^2 + (2A - 2B + C)x + (4A - 2C)$$

Equating coefficients of like powers of x and equating constant terms results in the following system of linear equations:

$$\begin{cases} A + B = 3 \\ 2A - 2B + C = 17 \\ 4A - 2C = 14. \end{cases}$$

Step 5 Solve the resulting system for $A, B,$ and C. Because the first equation involves A and B, we can obtain another equation in A and B by eliminating C from the second and third equations. Multiply the second equation by 2 and add equations. Solving in this manner, we obtain $A = 5, B = -2,$ and $C = 3$.

Step 6 Substitute the values of $A, B,$ and C, and write the partial fraction decomposition. With $A = 5, B = -2,$ and $C = 3$, the required partial fraction decomposition is

$$\frac{3x^2 + 17x + 14}{(x - 2)(x^2 + 2x + 4)} = \frac{A}{x - 2} + \frac{Bx + C}{x^2 + 2x + 4} = \frac{5}{x - 2} + \frac{-2x + 3}{x^2 + 2x + 4}. \quad \bullet\bullet\bullet$$

TECHNOLOGY

Numeric Connections

You can use the TABLE feature of a graphing utility to check a partial fraction decomposition. To check the result of Example 3, enter the given rational function and its partial fraction decomposition:

$$y_1 = \frac{3x^2 + 17x + 14}{(x - 2)(x^2 + 2x + 4)}$$

$$y_2 = \frac{5}{x - 2} + \frac{-2x + 3}{x^2 + 2x + 4}.$$

X	Y1	Y2
-3	.28571	.28571
-2	.5	.5
-1	0	0
0	-1.75	-1.75
1	-4.857	-4.857
2	ERROR	ERROR
3	4.8421	4.8421

X=-3

No matter how far up or down we scroll, $y_1 = y_2$, so the decomposition appears to be correct.

✓ **Check Point 3** Find the partial fraction decomposition of

$$\frac{8x^2 + 12x - 20}{(x + 3)(x^2 + x + 2)}.$$

④ Decompose $\dfrac{P}{Q}$, where Q has a prime, repeated quadratic factor.

The Partial Fraction Decomposition of a Rational Expression with a Prime, Repeated Quadratic Factor in the Denominator

Suppose that $(ax^2 + bx + c)^n$ is a factor of the denominator and that $ax^2 + bx + c$ cannot be factored further. This means that the quadratic factor $ax^2 + bx + c$ occurs n times. When this occurs, the partial fraction decomposition will contain a linear numerator for each power of $ax^2 + bx + c$.

The Partial Fraction Decomposition of $\dfrac{P(x)}{Q(x)}$: $Q(x)$ Has a Prime, Repeated Quadratic Factor

The form of the partial fraction decomposition for a rational expression containing the prime factor $ax^2 + bx + c$ occurring n times as its denominator is

$$\frac{P(x)}{(ax^2 + bx + c)^n} = \frac{A_1x + B_1}{ax^2 + bx + c} + \frac{A_2x + B_2}{(ax^2 + bx + c)^2} + \frac{A_3x + B_3}{(ax^2 + bx + c)^3} + \cdots + \frac{A_nx + B_n}{(ax^2 + bx + c)^n}.$$

Include one fraction with a linear numerator for each power of $ax^2 + bx + c$.

EXAMPLE 4 Partial Fraction Decomposition with a Repeated Quadratic Factor

Find the partial fraction decomposition of

$$\frac{5x^3 - 3x^2 + 7x - 3}{(x^2 + 1)^2}.$$

SOLUTION

Step 1 Set up the partial fraction decomposition with the unknown constants. Because the quadratic factor $x^2 + 1$ occurs twice, we must include one fraction with a linear numerator for each power of $x^2 + 1$.

$$\frac{5x^3 - 3x^2 + 7x - 3}{(x^2 + 1)^2} = \frac{Ax + B}{x^2 + 1} + \frac{Cx + D}{(x^2 + 1)^2}$$

Step 2 Multiply both sides of the resulting equation by the least common denominator. We clear fractions, multiplying both sides by $(x^2 + 1)^2$, the least common denominator.

$$(x^2 + 1)^2 \left[\frac{5x^3 - 3x^2 + 7x - 3}{(x^2 + 1)^2} \right] = (x^2 + 1)^2 \left[\frac{Ax + B}{x^2 + 1} + \frac{Cx + D}{(x^2 + 1)^2} \right]$$

Now we multiply and simplify.

$$5x^3 - 3x^2 + 7x - 3 = (x^2 + 1)(Ax + B) + Cx + D$$

Step 3 Simplify the right side of the equation. We multiply $(x^2 + 1)(Ax + B)$ using the FOIL method.

$$5x^3 - 3x^2 + 7x - 3 = Ax^3 + Bx^2 + Ax + B + Cx + D$$

Step 4 Write both sides in descending powers, equate coefficients of like powers of x, and equate constant terms.

$$5x^3 - 3x^2 + 7x - 3 = Ax^3 + Bx^2 + Ax + Cx + B + D$$

$$5x^3 - 3x^2 + 7x - 3 = Ax^3 + Bx^2 + (A + C)x + (B + D)$$

Equating coefficients of like powers of x and equating constant terms results in the following system of linear equations:

$$\begin{cases} A = 5 \\ B = -3 \\ A + C = 7 \qquad \text{With } A = 5, \text{ we immediately obtain } C = 2. \\ B + D = -3. \qquad \text{With } B = -3, \text{ we immediately obtain } D = 0. \end{cases}$$

GREAT QUESTION!

When setting up partial fraction decompositions, when should I use constant numerators and when should I use linear numerators?

When the denominator of a rational expression contains a **power of a linear factor**, set up the partial fraction decomposition with **constant numerators** (A, B, C, etc.). When the denominator of a rational expression contains a **power of a prime quadratic factor**, set up the partial fraction decomposition with **linear numerators** ($Ax + B, Cx + D$, etc.).

Step 5 **Solve the resulting system for A, B, C, and D.** Based on our observations in step 4, $A = 5$, $B = -3$, $C = 2$, and $D = 0$.

Step 6 **Substitute the values of A, B, C, and D, and write the partial fraction decomposition.**

$$\frac{5x^3 - 3x^2 + 7x - 3}{(x^2 + 1)^2} = \frac{Ax + B}{x^2 + 1} + \frac{Cx + D}{(x^2 + 1)^2} = \frac{5x - 3}{x^2 + 1} + \frac{2x}{(x^2 + 1)^2} \quad \bullet \bullet \bullet$$

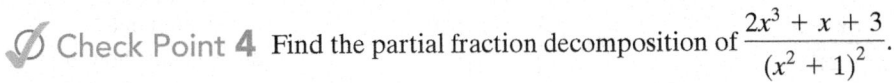 Check Point **4** Find the partial fraction decomposition of $\dfrac{2x^3 + x + 3}{(x^2 + 1)^2}$.

GREAT QUESTION!

When a rational expression contains a power of a factor in the denominator, how many fractions do I have to include in the partial fraction decomposition?

When a rational expression contains a power of a factor in the denominator, be sure to set up the partial fraction decomposition to allow for every natural-number power of that factor less than or equal to the power. Example:

$$\frac{2x + 1}{(x - 5)^2 x^3}$$

$$= \frac{A}{x - 5} + \frac{B}{(x - 5)^2} + \frac{C}{x} + \frac{D}{x^2} + \frac{E}{x^3}$$

> Although $(x - 5)^2$ and x^2 are quadratic, they are still expressed as powers of linear factors, $x - 5$ and x. Thus, the numerators are constant.

CONCEPT AND VOCABULARY CHECK

Determine whether each partial fraction decomposition is set up correctly. If the setup is incorrect, make the necessary changes to produce the correct decomposition.

1. Correct or incorrect:

$$\frac{7x}{(x + 2)(x - 3)} = \frac{A}{x + 2} + \frac{B}{x - 3} \underline{\hspace{2cm}}$$

2. Correct or incorrect:

$$\frac{3x}{(x + 5)(x - 4)^2} = \frac{A}{x + 5} + \frac{B}{x - 4} + \frac{C}{x - 4} \underline{\hspace{2cm}}$$

3. Correct or incorrect:

$$\frac{1}{(x + 1)(x^2 + 4)} = \frac{A}{x + 1} + \frac{B}{x^2 + 4} \underline{\hspace{2cm}}$$

4. Correct or incorrect:

$$\frac{7x - 5}{(x^2 + x + 1)^2} = \frac{Ax + B}{x^2 + x + 1} + \frac{Cx + D}{(x^2 + x + 1)^2} \underline{\hspace{2cm}}$$

EXERCISE SET 5.3

Practice Exercises

In Exercises 1–8, write the form of the partial fraction decomposition of the rational expression. It is not necessary to solve for the constants.

1. $\dfrac{11x - 10}{(x - 2)(x + 1)}$

2. $\dfrac{5x + 7}{(x - 1)(x + 3)}$

3. $\dfrac{6x^2 - 14x - 27}{(x + 2)(x - 3)^2}$

4. $\dfrac{3x + 16}{(x + 1)(x - 2)^2}$

5. $\dfrac{5x^2 - 6x + 7}{(x - 1)(x^2 + 1)}$

6. $\dfrac{5x^2 - 9x + 19}{(x - 4)(x^2 + 5)}$

7. $\dfrac{x^3 + x^2}{(x^2 + 4)^2}$

8. $\dfrac{7x^2 - 9x + 3}{(x^2 + 7)^2}$

In Exercises 9–42, write the partial fraction decomposition of each rational expression.

9. $\dfrac{x}{(x - 3)(x - 2)}$

10. $\dfrac{1}{x(x - 1)}$

11. $\dfrac{3x + 50}{(x - 9)(x + 2)}$

12. $\dfrac{5x - 1}{(x - 2)(x + 1)}$

13. $\dfrac{7x - 4}{x^2 - x - 12}$

14. $\dfrac{9x + 21}{x^2 + 2x - 15}$

15. $\dfrac{4}{2x^2 - 5x - 3}$

16. $\dfrac{x}{x^2 + 2x - 3}$

17. $\dfrac{4x^2 + 13x - 9}{x(x - 1)(x + 3)}$

18. $\dfrac{4x^2 - 5x - 15}{x(x + 1)(x - 5)}$

19. $\dfrac{4x^2 - 7x - 3}{x^3 - x}$

20. $\dfrac{2x^2 - 18x - 12}{x^3 - 4x}$

21. $\dfrac{6x - 11}{(x - 1)^2}$

22. $\dfrac{x}{(x + 1)^2}$

23. $\dfrac{x^2 - 6x + 3}{(x - 2)^3}$

24. $\dfrac{2x^2 + 8x + 3}{(x + 1)^3}$

25. $\dfrac{x^2 + 2x + 7}{x(x - 1)^2}$

26. $\dfrac{3x^2 + 49}{x(x + 7)^2}$

27. $\dfrac{x^2}{(x - 1)^2(x + 1)}$

28. $\dfrac{x^2}{(x - 1)^2(x + 1)^2}$

29. $\dfrac{5x^2 - 6x + 7}{(x - 1)(x^2 + 1)}$

30. $\dfrac{5x^2 - 9x + 19}{(x - 4)(x^2 + 5)}$

31. $\dfrac{5x^2 + 6x + 3}{(x + 1)(x^2 + 2x + 2)}$

32. $\dfrac{9x + 2}{(x - 2)(x^2 + 2x + 2)}$

33. $\dfrac{x + 4}{x^2(x^2 + 4)}$

34. $\dfrac{10x^2 + 2x}{(x - 1)^2(x^2 + 2)}$

35. $\dfrac{6x^2 - x + 1}{x^3 + x^2 + x + 1}$

36. $\dfrac{3x^2 - 2x + 8}{x^3 + 2x^2 + 4x + 8}$

37. $\dfrac{x^3 + x^2 + 2}{(x^2 + 2)^2}$

38. $\dfrac{x^2 + 2x + 3}{(x^2 + 4)^2}$

39. $\dfrac{x^3 - 4x^2 + 9x - 5}{(x^2 - 2x + 3)^2}$

40. $\dfrac{3x^3 - 6x^2 + 7x - 2}{(x^2 - 2x + 2)^2}$

41. $\dfrac{4x^2 + 3x + 14}{x^3 - 8}$

42. $\dfrac{3x - 5}{x^3 - 1}$

Practice Plus

In Exercises 43–46, perform each long division and write the partial fraction decomposition of the remainder term.

43. $\dfrac{x^5 + 2}{x^2 - 1}$

44. $\dfrac{x^5}{x^2 - 4x + 4}$

45. $\dfrac{x^4 - x^2 + 2}{x^3 - x^2}$

46. $\dfrac{x^4 + 2x^3 - 4x^2 + x - 3}{x^2 - x - 2}$

In Exercises 47–50, write the partial fraction decomposition of each rational expression.

47. $\dfrac{1}{x^2 - c^2}$ $(c \neq 0)$

48. $\dfrac{ax + b}{x^2 - c^2}$ $(c \neq 0)$

49. $\dfrac{ax + b}{(x - c)^2}$ $(c \neq 0)$

50. $\dfrac{1}{x^2 - ax - bx + ab}$ $(a \neq b)$

Application Exercises

51. Find the partial fraction decomposition for $\dfrac{1}{x(x + 1)}$ and use the result to find the following sum:

$$\dfrac{1}{1 \cdot 2} + \dfrac{1}{2 \cdot 3} + \dfrac{1}{3 \cdot 4} + \cdots + \dfrac{1}{99 \cdot 100}.$$

52. Find the partial fraction decomposition for $\dfrac{2}{x(x + 2)}$ and use the result to find the following sum:

$$\dfrac{2}{1 \cdot 3} + \dfrac{2}{3 \cdot 5} + \dfrac{2}{5 \cdot 7} + \cdots + \dfrac{2}{99 \cdot 101}.$$

Writing in Mathematics

53. Explain what is meant by the partial fraction decomposition of a rational expression.

54. Explain how to find the partial fraction decomposition of a rational expression with distinct linear factors in the denominator.

55. Explain how to find the partial fraction decomposition of a rational expression with a repeated linear factor in the denominator.

56. Explain how to find the partial fraction decomposition of a rational expression with a prime quadratic factor in the denominator.

57. Explain how to find the partial fraction decomposition of a rational expression with a repeated, prime quadratic factor in the denominator.

58. How can you verify your result for the partial fraction decomposition for a given rational expression without using a graphing utility?

Technology Exercise

59. Use the $\boxed{\text{TABLE}}$ feature of a graphing utility to verify any three of the decompositions that you obtained in Exercises 9–42.

Critical Thinking Exercises

Make Sense? *In Exercises 60–63, determine whether each statement makes sense or does not make sense, and explain your reasoning.*

60. Partial fraction decomposition involves finding a single rational expression for a given sum or difference of rational expressions.

61. I apply partial fraction decompositions for rational expressions of the form $\dfrac{P(x)}{Q(x)}$, where P and Q have no common factors and the degree of P is greater than the degree of Q.

62. Because $x + 5$ is linear and $x^2 - 3x + 2$ is quadratic, I set up the following partial fraction decomposition:

$$\dfrac{7x^2 + 9x + 3}{(x + 5)(x^2 - 3x + 2)} = \dfrac{A}{x + 5} + \dfrac{Bx + C}{x^2 - 3x + 2}.$$

63. Because $(x + 3)^2$ consists of two factors of $x + 3$, I set up the following partial fraction decomposition:

$$\dfrac{5x + 2}{(x + 3)^2} = \dfrac{A}{x + 3} + \dfrac{B}{x + 3}.$$

64. Use an extension of the Great Question! on page 558 to describe how to set up the partial fraction decomposition of a rational expression that contains powers of a prime cubic factor in the denominator. Give an example of such a decomposition.

65. Find the partial fraction decomposition of

$$\dfrac{4x^2 + 5x - 9}{x^3 - 6x - 9}.$$

Preview Exercises

Exercises 66–68 will help you prepare for the material covered in the next section.

66. Solve by the substitution method:

$$\begin{cases} 4x + 3y = 4 \\ y = 2x - 7. \end{cases}$$

67. Solve by the addition method:

$$\begin{cases} 2x + 4y = -4 \\ 3x + 5y = -3. \end{cases}$$

68. Graph $x - y = 3$ and $(x - 2)^2 + (y + 3)^2 = 4$ in the same rectangular coordinate system. What are the two intersection points? Show that each of these ordered pairs satisfies both equations.

SECTION 5.4 Systems of Nonlinear Equations in Two Variables

Objectives

1. Recognize systems of nonlinear equations in two variables.
2. Solve nonlinear systems by substitution.
3. Solve nonlinear systems by addition.
4. Solve problems using systems of nonlinear equations.

Scientists debate the probability that a "doomsday rock" will collide with Earth. It has been estimated that an asteroid, a tiny planet that revolves around the sun, crashes into Earth about once every 250,000 years, and that such a collision would have disastrous results. In 1908, a small fragment struck Siberia, leveling thousands of acres of trees. One theory about the extinction of dinosaurs 65 million years ago involves Earth's collision with a large asteroid and the resulting drastic changes in Earth's climate.

Understanding the path of Earth and the path of a comet is essential to detecting threatening space debris. Orbits about the sun are not described by linear equations in the form $Ax + By = C$. The ability to solve systems that contain nonlinear equations provides NASA scientists watching for troublesome asteroids with a way to locate possible collision points with Earth's orbit.

① Recognize systems of nonlinear equations in two variables.

Systems of Nonlinear Equations and Their Solutions

A **system of** two **nonlinear equations** in two variables, also called a **nonlinear system**, contains at least one equation that cannot be expressed in the form $Ax + By = C$. Here are two examples:

$$\begin{cases} x^2 = 2y + 10 \\ 3x - y = 9 \end{cases}$$
Not in the form $Ax + By = C$. The term x^2 is not linear.

$$\begin{cases} y = x^2 + 3 \\ x^2 + y^2 = 9. \end{cases}$$
Neither equation is in the form $Ax + By = C$. The terms x^2 and y^2 are not linear.

A **solution** of a nonlinear system in two variables is an ordered pair of real numbers that satisfies both equations in the system. The **solution set** of the system is the set of all such ordered pairs. As with linear systems in two variables, the solution of a nonlinear system (if there is one) corresponds to the intersection point(s) of the graphs of the equations in the system. Unlike linear systems, the graphs can be circles, parabolas, or anything other than two lines. We will solve nonlinear systems using the substitution method and the addition method.

② Solve nonlinear systems by substitution.

Eliminating a Variable Using the Substitution Method

The substitution method involves converting a nonlinear system into one equation in one variable by an appropriate substitution. The steps in the solution process are exactly the same as those used to solve a linear system by substitution. However, when you obtain an equation in one variable, this equation may not be linear. In our first example, this equation is quadratic.

EXAMPLE 1 Solving a Nonlinear System by the Substitution Method

Solve by the substitution method:

$$\begin{cases} x^2 = 2y + 10 & \text{(The graph is a parabola.)} \\ 3x - y = 9. & \text{(The graph is a line.)} \end{cases}$$

SOLUTION

Step 1 Solve one of the equations for one variable in terms of the other. We begin by isolating one of the variables raised to the first power in either of the equations. By solving for y in the second equation, which has a coefficient of -1, we can avoid fractions.

$$3x - y = 9 \qquad \text{This is the second equation in the given system.}$$
$$3x = y + 9 \qquad \text{Add } y \text{ to both sides.}$$
$$3x - 9 = y \qquad \text{Subtract 9 from both sides.}$$

Step 2 Substitute the expression from step 1 into the other equation. We substitute $3x - 9$ for y in the first equation.

$$y = \boxed{3x - 9} \qquad x^2 = 2\boxed{y} + 10$$

This gives us an equation in one variable, namely,

$$x^2 = 2(3x - 9) + 10.$$

The variable y has been eliminated.

Step 3 Solve the resulting equation containing one variable.

$$x^2 = 2(3x - 9) + 10 \qquad \text{This is the equation containing one variable.}$$
$$x^2 = 6x - 18 + 10 \qquad \text{Use the distributive property.}$$
$$x^2 = 6x - 8 \qquad \text{Combine numerical terms on the right.}$$
$$x^2 - 6x + 8 = 0 \qquad \text{Move all terms to one side and set the quadratic equation equal to 0.}$$
$$(x - 4)(x - 2) = 0 \qquad \text{Factor.}$$
$$x - 4 = 0 \quad \text{or} \quad x - 2 = 0 \qquad \text{Set each factor equal to 0.}$$
$$x = 4 \qquad\qquad x = 2 \qquad \text{Solve for x.}$$

Step 4 Back-substitute the obtained values into the equation from step 1. Now that we have the x-coordinates of the solutions, we back-substitute 4 for x and 2 for x into the equation $y = 3x - 9$.

If x is 4, $y = 3(4) - 9 = 3,$ so $(4, 3)$ is a solution.
If x is 2, $y = 3(2) - 9 = -3,$ so $(2, -3)$ is a solution.

Step 5 Check the proposed solutions in both of the system's given equations. We begin by checking $(4, 3)$. Replace x with 4 and y with 3.

$$x^2 = 2y + 10 \qquad\qquad 3x - y = 9 \qquad \text{These are the given equations.}$$
$$4^2 \overset{?}{=} 2(3) + 10 \qquad 3(4) - 3 \overset{?}{=} 9 \qquad \text{Let } x = 4 \text{ and } y = 3.$$
$$16 \overset{?}{=} 6 + 10 \qquad\quad 12 - 3 \overset{?}{=} 9 \qquad \text{Simplify.}$$
$$16 = 16, \quad \text{true} \qquad\qquad 9 = 9, \quad \text{true} \quad \text{True statements result.}$$

The ordered pair $(4, 3)$ satisfies both equations. Thus, $(4, 3)$ is a solution of the system. Now let's check $(2, -3)$. Replace x with 2 and y with -3 in both given equations.

$$x^2 = 2y + 10 \qquad\qquad 3x - y = 9 \qquad \text{These are the given equations.}$$
$$2^2 \overset{?}{=} 2(-3) + 10 \qquad 3(2) - (-3) \overset{?}{=} 9 \qquad \text{Let } x = 2 \text{ and } y = -3.$$
$$4 \overset{?}{=} -6 + 10 \qquad\qquad 6 + 3 \overset{?}{=} 9 \qquad \text{Simplify.}$$
$$4 = 4, \quad \text{true} \qquad\qquad 9 = 9, \quad \text{true} \quad \text{True statements result.}$$

The ordered pair $(2, -3)$ also satisfies both equations and is a solution of the system. The solutions are $(4, 3)$ and $(2, -3)$, and the solution set is $\{(4, 3), (2, -3)\}$.

Figure 5.9 shows the graphs of the equations in the system and the solutions as intersection points. •••

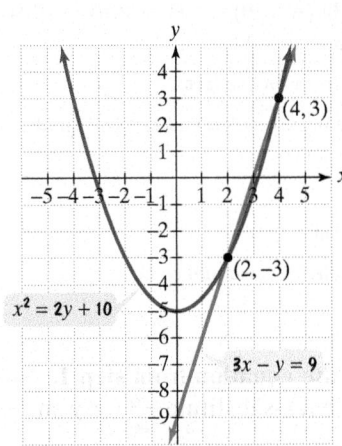

FIGURE 5.9 Points of intersection illustrate the nonlinear system's solutions.

 Check Point **1** Solve by the substitution method:

$$\begin{cases} x^2 = y - 1 \\ 4x - y = -1. \end{cases}$$

EXAMPLE 2 Solving a Nonlinear System by the Substitution Method

Solve by the substitution method:

$$\begin{cases} x - y = 3 & \text{(The graph is a line.)} \\ (x - 2)^2 + (y + 3)^2 = 4. & \text{(The graph is a circle.)} \end{cases}$$

GREAT QUESTION!

Why is the graph of
$(x - 2)^2 + (y + 3)^2 = 4$ **a circle?**

Recall that

$$(x - h)^2 + (y - k)^2 = r^2$$

describes a circle with center (h, k) and radius r.

SOLUTION

Graphically, we are finding the intersection of a line and a circle with center $(2, -3)$ and radius 2.

Step 1 Solve one of the equations for one variable in terms of the other. We will solve for x in the linear equation—that is, the first equation. (We could also solve for y.)

$$x - y = 3 \qquad \text{This is the first equation in the given system.}$$
$$x = y + 3 \qquad \text{Add } y \text{ to both sides.}$$

Step 2 Substitute the expression from step 1 into the other equation. We substitute $y + 3$ for x in the second equation.

$$x = \boxed{y + 3} \qquad (\boxed{x} - 2)^2 + (y + 3)^2 = 4$$

This gives an equation in one variable, namely,

$$(y + 3 - 2)^2 + (y + 3)^2 = 4.$$

The variable x has been eliminated.

Step 3 Solve the resulting equation containing one variable.

$$(y + 3 - 2)^2 + (y + 3)^2 = 4 \qquad \text{This is the equation containing one variable.}$$
$$(y + 1)^2 + (y + 3)^2 = 4 \qquad \text{Combine numerical terms in the first parentheses.}$$
$$y^2 + 2y + 1 + y^2 + 6y + 9 = 4 \qquad \text{Use the formula } (A + B)^2 = A^2 + 2AB + B^2 \text{ to square } y + 1 \text{ and } y + 3.$$
$$2y^2 + 8y + 10 = 4 \qquad \text{Combine like terms on the left.}$$
$$2y^2 + 8y + 6 = 0 \qquad \text{Subtract 4 from both sides and set the quadratic equation equal to 0.}$$
$$2(y^2 + 4y + 3) = 0 \qquad \text{Factor out 2.}$$
$$2(y + 3)(y + 1) = 0 \qquad \text{Factor completely.}$$
$$y + 3 = 0 \quad \text{or} \quad y + 1 = 0 \qquad \text{Set each variable factor equal to 0.}$$
$$y = -3 \qquad\qquad y = -1 \qquad \text{Solve for } y.$$

Step 4 Back-substitute the obtained values into the equation from step 1. Now that we have the y-coordinates of the solutions, we back-substitute -3 for y and -1 for y in the equation $x = y + 3$.

If $y = -3$: $x = -3 + 3 = 0$, so $(0, -3)$ is a solution.
If $y = -1$: $x = -1 + 3 = 2$, so $(2, -1)$ is a solution.

Step 5 Check the proposed solutions in both of the system's given equations. Take a moment to show that each ordered pair satisfies both given equations, $x - y = 3$ and $(x - 2)^2 + (y + 3)^2 = 4$. The solutions are $(0, -3)$ and $(2, -1)$, and the solution set of the given system is $\{(0, -3), (2, -1)\}$.

Figure 5.10 shows the graphs of the equations in the system and the solutions as intersection points. •••

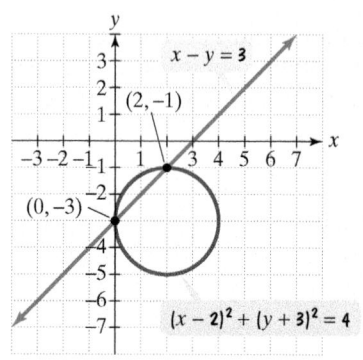

FIGURE 5.10 Points of intersection illustrate the nonlinear system's solutions.

✓ Check Point **2** Solve by the substitution method:

$$\begin{cases} x + 2y = 0 \\ (x - 1)^2 + (y - 1)^2 = 5. \end{cases}$$

③ Solve nonlinear systems by addition.

Eliminating a Variable Using the Addition Method

In solving linear systems with two variables, we learned that the addition method works well when each equation is in the form $Ax + By = C$. For nonlinear systems, the addition method can be used when each equation is in the form $Ax^2 + By^2 = C$. If necessary, we will multiply either equation or both equations by appropriate numbers so that the coefficients of x^2 or y^2 will have a sum of 0. We then add equations. The sum will be an equation in one variable.

EXAMPLE 3 Solving a Nonlinear System by the Addition Method

Solve the system:

$$\begin{cases} 4x^2 + y^2 = 13 & \text{Equation 1} \\ x^2 + y^2 = 10. & \text{Equation 2} \end{cases}$$

SOLUTION

We can use the same steps that we did when we solved linear systems by the addition method.

Step 1 Write both equations in the form $Ax^2 + By^2 = C$. Both equations are already in this form, so we can skip this step.

Step 2 If necessary, multiply either equation or both equations by appropriate numbers so that the sum of the x^2-coefficients or the sum of the y^2-coefficients is 0. We can eliminate y^2 by multiplying Equation 2 by -1.

$$\begin{cases} 4x^2 + y^2 = 13 \\ x^2 + y^2 = 10 \end{cases} \xrightarrow[\text{Multiply by } -1.]{\text{No change}} \begin{cases} 4x^2 + y^2 = 13 \\ -x^2 - y^2 = -10 \end{cases}$$

Steps 3 and 4 Add equations and solve for the remaining variable.

$$\begin{cases} 4x^2 + y^2 = 13 \\ \underline{-x^2 - y^2 = -10} \end{cases}$$

$$3x^2 \quad\quad = 3 \qquad \text{Add equations.}$$

$$x^2 = 1 \qquad \text{Divide both sides by 3.}$$

$$x = \pm 1 \qquad \text{Use the square root property:}$$
$$\text{If } x^2 = c, \text{ then } x = \pm \sqrt{c}.$$

Step 5 Back-substitute and find the values for the other variable. We must back-substitute each value of x into either one of the original equations. Let's use $x^2 + y^2 = 10$, Equation 2. If $x = 1$,

$$1^2 + y^2 = 10 \qquad \text{Replace } x \text{ with 1 in Equation 2.}$$

$$y^2 = 9 \qquad \text{Subtract 1 from both sides.}$$

$$y = \pm 3. \qquad \text{Apply the square root property.}$$

$(1, 3)$ and $(1, -3)$ are solutions. If $x = -1$,

$$(-1)^2 + y^2 = 10 \qquad \text{Replace } x \text{ with } -1 \text{ in Equation 2.}$$

$$y^2 = 9 \qquad \text{The steps are the same as before.}$$

$$y = \pm 3.$$

$(-1, 3)$ and $(-1, -3)$ are solutions.

Step 6 Check. Take a moment to show that each of the four ordered pairs satisfies the given equations, $4x^2 + y^2 = 13$ and $x^2 + y^2 = 10$. The solution set of the given system is $\{(1, 3), (1, -3), (-1, 3), (-1, -3)\}$.

Figure 5.11 shows the graphs of the equations in the system and the solutions as intersection points. • • •

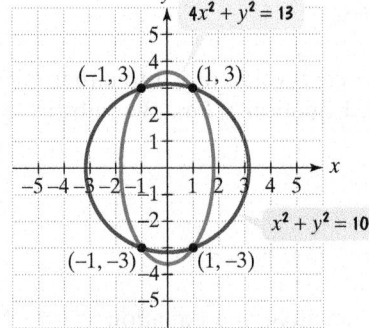

FIGURE 5.11 A system with four solutions

✐ Check Point **3** Solve the system:

$$\begin{cases} 3x^2 + 2y^2 = 35 \\ 4x^2 + 3y^2 = 48. \end{cases}$$

In solving nonlinear systems, we include only ordered pairs with real numbers in the solution set. We have seen that each of these ordered pairs corresponds to a point of intersection of the system's graphs.

GREAT QUESTION!

When solving nonlinear systems, do I really have to go to the hassle of checking my solutions?

Yes. Extra solutions may be introduced that do not satisfy both equations in the system. Therefore, you should get into the habit of checking all proposed pairs in each of the system's two equations.

EXAMPLE 4 Solving a Nonlinear System by the Addition Method

Solve the system:

$$\begin{cases} y = x^2 + 3 & \text{Equation 1 (The graph is a parabola.)} \\ x^2 + y^2 = 9. & \text{Equation 2 (The graph is a circle.)} \end{cases}$$

SOLUTION

We could use substitution because Equation 1, $y = x^2 + 3$, has y expressed in terms of x, but substituting $x^2 + 3$ for y in $x^2 + y^2 = 9$ would result in a fourth-degree equation. However, we can rewrite Equation 1 by subtracting x^2 from both sides and adding the equations to eliminate the x^2-terms.

Notice how like terms are arranged in columns.

$$\begin{cases} -x^2 + y = 3 & \text{Subtract } x^2 \text{ from both sides of Equation 1.} \\ x^2 + y^2 = 9 & \text{This is Equation 2.} \end{cases}$$
$$\overline{\; y + y^2 = 12} \quad \text{Add the equations.}$$

We now solve this quadratic equation.

$$y + y^2 = 12 \quad \text{This is the equation containing one variable.}$$

$$y^2 + y - 12 = 0 \quad \text{Subtract 12 from both sides and set the quadratic equation equal to 0.}$$

$$(y + 4)(y - 3) = 0 \quad \text{Factor.}$$

$$y + 4 = 0 \quad \text{or} \quad y - 3 = 0 \quad \text{Set each factor equal to 0.}$$

$$y = -4 \qquad\qquad y = 3 \quad \text{Solve for } y.$$

To complete the solution, we must back-substitute each value of y into either one of the original equations. We will use $y = x^2 + 3$, Equation 1. First, we substitute -4 for y.

$$-4 = x^2 + 3$$

$$-7 = x^2 \quad \text{Subtract 3 from both sides.}$$

Because the square of a real number cannot be negative, the equation $x^2 = -7$ does not have real-number solutions. We will not include the imaginary solutions, $x = \pm\sqrt{-7}$, or $i\sqrt{7}$ and $-i\sqrt{7}$, in the ordered pairs that make up the solution set. Thus, we move on to our other value for y, 3, and substitute this value into Equation 1.

$$y = x^2 + 3 \quad \text{This is Equation 1.}$$

$$3 = x^2 + 3 \quad \text{Back-substitute 3 for } y.$$

$$0 = x^2 \quad \text{Subtract 3 from both sides.}$$

$$0 = x \quad \text{Solve for } x.$$

We showed that if $y = 3$, then $x = 0$. Thus, $(0, 3)$ is the solution with a real ordered pair. Take a moment to show that $(0, 3)$ satisfies the given equations, $y = x^2 + 3$ and $x^2 + y^2 = 9$. The solution set of the system is $\{(0, 3)\}$. **Figure 5.12** shows the system's graphs and the solution as an intersection point. ● ● ●

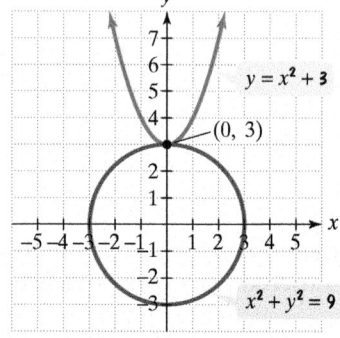

FIGURE 5.12 A system with one real solution

⊘ Check Point **4** Solve the system:

$$\begin{cases} y = x^2 + 5 \\ x^2 + y^2 = 25. \end{cases}$$

④ Solve problems using systems of nonlinear equations.

Applications

Many geometric problems can be modeled and solved by the use of systems of nonlinear equations. We will use our step-by-step strategy for solving problems using mathematical models that are created from verbal conditions.

EXAMPLE 5 An Application of a Nonlinear System

You have 36 yards of fencing to build the enclosure in **Figure 5.13**. Some of this fencing is to be used to build an internal divider. If you'd like to enclose 54 square yards, what are the dimensions of the enclosure?

FIGURE 5.13 Building an enclosure

SOLUTION

Step 1 Use variables to represent unknown quantities. Let x = the enclosure's length and y = the enclosure's width. These variables are shown in **Figure 5.13**.

Step 2 Write a system of equations modeling the problem's conditions. The first condition is that you have 36 yards of fencing.

Fencing along both lengths	plus	fencing along both widths	plus	fencing for the internal divider	equals	36 yards.
$2x$	$+$	$2y$	$+$	y	$=$	36

Adding like terms, we can express the equation that models the verbal conditions for the fencing as $2x + 3y = 36$.

The second condition is that you'd like to enclose 54 square yards. The rectangle's area, the product of its length and its width, must be 54 square yards.

Length	times	width	is	54 square yards.
x	\cdot	y	$=$	54

Step 3 Solve the system and answer the problem's question. We must solve the system

$$\begin{cases} 2x + 3y = 36 & \text{Equation 1} \\ xy = 54. & \text{Equation 2} \end{cases}$$

We will use substitution. Because Equation 1 has no coefficients of 1 or −1, we will work with Equation 2 and solve for y. Dividing both sides of $xy = 54$ by x, we obtain

$$y = \frac{54}{x}.$$

Now we substitute $\frac{54}{x}$ for y in Equation 1 and solve for x.

$$2x + 3y = 36 \qquad \text{This is Equation 1.}$$
$$2x + 3 \cdot \frac{54}{x} = 36 \qquad \text{Substitute } \frac{54}{x} \text{ for } y.$$
$$2x + \frac{162}{x} = 36 \qquad \text{Multiply.}$$

$$x\left(2x + \frac{162}{x}\right) = 36 \cdot x \qquad \text{Clear fractions by multiplying both sides by x.}$$

$$2x^2 + 162 = 36x \qquad \text{Use the distributive property on the left side.}$$

$$2x^2 - 36x + 162 = 0 \qquad \text{Subtract 36x from both sides and set the quadratic equation equal to 0.}$$

$$2(x^2 - 18x + 81) = 0 \qquad \text{Factor out 2.}$$

$$2(x - 9)^2 = 0 \qquad \text{Factor completely using } A^2 - 2AB + B^2 = (A - B)^2.$$

$$x - 9 = 0 \qquad \text{Set the repeated factor equal to zero.}$$

$$x = 9 \qquad \text{Solve for } x.$$

We back-substitute this value of x into $y = \dfrac{54}{x}$.

$$\text{If } x = 9, \quad y = \frac{54}{9} = 6.$$

This means that the dimensions of the enclosure in **Figure 5.13** are 9 yards by 6 yards.

FIGURE 5.13 (repeated)

Step 4 Check the proposed solution in the original wording of the problem. Take a moment to check that a length of 9 yards and a width of 6 yards results in 36 yards of fencing and an area of 54 square yards. • • •

 Check Point **5** Find the length and width of a rectangle whose perimeter is 20 feet and whose area is 21 square feet.

CONCEPT AND VOCABULARY CHECK

Fill in each blank so that the resulting statement is true.

1. A system of two equations in two variables that contains at least one equation that cannot be expressed in the form $Ax + By = C$ is called a system of _____ equations.

2. When solving

$$\begin{cases} x^2 - 4y = 4 \\ x + y = -1 \end{cases}$$

by the substitution method, we obtain $x = -4$ or $x = 0$, so the solution set is _____.

3. When solving

$$\begin{cases} 3x^2 + 2y^2 = 35 \\ 4x^2 + 3y^2 = 48 \end{cases}$$

by the addition method, we can eliminate x^2 by multiplying the first equation by -4 and the second equation by _____ and then adding the equations.

4. When solving

$$\begin{cases} x^2 + 4y^2 = 16 \\ x^2 - y^2 = 1 \end{cases}$$

by the addition method, we obtain $y^2 = 3$, so the solution set is _____.

5. When solving

$$\begin{cases} x^2 + y^2 = 13 \\ x^2 - y = 7 \end{cases}$$

by the addition method, we can eliminate x^2 by multiplying the second equation by _____ and then adding the equations. We obtain _____, a quadratic equation in y.

6. When solving

$$\begin{cases} x^2 + 4y^2 = 20 \\ xy = 4 \end{cases}$$

by the substitution method, we can eliminate y by solving the second equation for y. We obtain

$y =$ _____. Then we substitute _____ for _____ in the first equation.

EXERCISE SET 5.4

Practice Exercises

In Exercises 1–18, solve each system by the substitution method.

1. $\begin{cases} x + y = 2 \\ y = x^2 - 4 \end{cases}$

2. $\begin{cases} x - y = -1 \\ y = x^2 + 1 \end{cases}$

3. $\begin{cases} x + y = 2 \\ y = x^2 - 4x + 4 \end{cases}$

4. $\begin{cases} 2x + y = -5 \\ y = x^2 + 6x + 7 \end{cases}$

5. $\begin{cases} y = x^2 - 4x - 10 \\ y = -x^2 - 2x + 14 \end{cases}$

6. $\begin{cases} y = x^2 + 4x + 5 \\ y = x^2 + 2x - 1 \end{cases}$

7. $\begin{cases} x^2 + y^2 = 25 \\ x - y = 1 \end{cases}$

8. $\begin{cases} x^2 + y^2 = 5 \\ 3x - y = 5 \end{cases}$

9. $\begin{cases} xy = 6 \\ 2x - y = 1 \end{cases}$

10. $\begin{cases} xy = -12 \\ x - 2y + 14 = 0 \end{cases}$

11. $\begin{cases} y^2 = x^2 - 9 \\ 2y = x - 3 \end{cases}$

12. $\begin{cases} x^2 + y = 4 \\ 2x + y = 1 \end{cases}$

13. $\begin{cases} xy = 3 \\ x^2 + y^2 = 10 \end{cases}$

14. $\begin{cases} xy = 4 \\ x^2 + y^2 = 8 \end{cases}$

15. $\begin{cases} x + y = 1 \\ x^2 + xy - y^2 = -5 \end{cases}$

16. $\begin{cases} x + y = -3 \\ x^2 + 2y^2 = 12y + 18 \end{cases}$

17. $\begin{cases} x + y = 1 \\ (x - 1)^2 + (y + 2)^2 = 10 \end{cases}$

18. $\begin{cases} 2x + y = 4 \\ (x + 1)^2 + (y - 2)^2 = 4 \end{cases}$

In Exercises 19–28, solve each system by the addition method.

19. $\begin{cases} x^2 + y^2 = 13 \\ x^2 - y^2 = 5 \end{cases}$

20. $\begin{cases} 4x^2 - y^2 = 4 \\ 4x^2 + y^2 = 4 \end{cases}$

21. $\begin{cases} x^2 - 4y^2 = -7 \\ 3x^2 + y^2 = 31 \end{cases}$

22. $\begin{cases} 3x^2 - 2y^2 = -5 \\ 2x^2 - y^2 = -2 \end{cases}$

23. $\begin{cases} 3x^2 + 4y^2 - 16 = 0 \\ 2x^2 - 3y^2 - 5 = 0 \end{cases}$

24. $\begin{cases} 16x^2 - 4y^2 - 72 = 0 \\ x^2 - y^2 - 3 = 0 \end{cases}$

25. $\begin{cases} x^2 + y^2 = 25 \\ (x - 8)^2 + y^2 = 41 \end{cases}$

26. $\begin{cases} x^2 + y^2 = 5 \\ x^2 + (y - 8)^2 = 41 \end{cases}$

27. $\begin{cases} y^2 - x = 4 \\ x^2 + y^2 = 4 \end{cases}$

28. $\begin{cases} x^2 - 2y = 8 \\ x^2 + y^2 = 16 \end{cases}$

In Exercises 29–42, solve each system by the method of your choice.

29. $\begin{cases} 3x^2 + 4y^2 = 16 \\ 2x^2 - 3y^2 = 5 \end{cases}$

30. $\begin{cases} x + y^2 = 4 \\ x^2 + y^2 = 16 \end{cases}$

31. $\begin{cases} 2x^2 + y^2 = 18 \\ xy = 4 \end{cases}$

32. $\begin{cases} x^2 + 4y^2 = 20 \\ xy = 4 \end{cases}$

33. $\begin{cases} x^2 + 4y^2 = 20 \\ x + 2y = 6 \end{cases}$

34. $\begin{cases} 3x^2 - 2y^2 = 1 \\ 4x - y = 3 \end{cases}$

35. $\begin{cases} x^3 + y = 0 \\ x^2 - y = 0 \end{cases}$

36. $\begin{cases} x^3 + y = 0 \\ 2x^2 - y = 0 \end{cases}$

37. $\begin{cases} x^2 + (y - 2)^2 = 4 \\ x^2 - 2y = 0 \end{cases}$

38. $\begin{cases} x^2 - y^2 - 4x + 6y - 4 = 0 \\ x^2 + y^2 - 4x - 6y + 12 = 0 \end{cases}$

39. $\begin{cases} y = (x + 3)^2 \\ x + 2y = -2 \end{cases}$

40. $\begin{cases} (x - 1)^2 + (y + 1)^2 = 5 \\ 2x - y = 3 \end{cases}$

41. $\begin{cases} x^2 + y^2 + 3y = 22 \\ 2x + y = -1 \end{cases}$

42. $\begin{cases} x - 3y = -5 \\ x^2 + y^2 - 25 = 0 \end{cases}$

In Exercises 43–46, let x represent one number and let y represent the other number. Use the given conditions to write a system of nonlinear equations. Solve the system and find the numbers.

43. The sum of two numbers is 10 and their product is 24. Find the numbers.

44. The sum of two numbers is 20 and their product is 96. Find the numbers.

45. The difference between the squares of two numbers is 3. Twice the square of the first number increased by the square of the second number is 9. Find the numbers.

46. The difference between the squares of two numbers is 5. Twice the square of the second number subtracted from three times the square of the first number is 19. Find the numbers.

Practice Plus

In Exercises 47–52, solve each system by the method of your choice.

47. $\begin{cases} 2x^2 + xy = 6 \\ x^2 + 2xy = 0 \end{cases}$

48. $\begin{cases} 4x^2 + xy = 30 \\ x^2 + 3xy = -9 \end{cases}$

49. $\begin{cases} -4x + y = 12 \\ y = x^3 + 3x^2 \end{cases}$

50. $\begin{cases} -9x + y = 45 \\ y = x^3 + 5x^2 \end{cases}$

51. $\begin{cases} \dfrac{3}{x^2} + \dfrac{1}{y^2} = 7 \\ \dfrac{5}{x^2} - \dfrac{2}{y^2} = -3 \end{cases}$

52. $\begin{cases} \dfrac{2}{x^2} + \dfrac{1}{y^2} = 11 \\ \dfrac{4}{x^2} - \dfrac{2}{y^2} = -14 \end{cases}$

In Exercises 53–54, make a rough sketch in a rectangular coordinate system of the graphs representing the equations in each system.

53. The system, whose graphs are a line with positive slope and a parabola whose equation has a positive leading coefficient, has two solutions.

54. The system, whose graphs are a line with negative slope and a parabola whose equation has a negative leading coefficient, has one solution.

Application Exercises

55. A planet's orbit follows a path described by $16x^2 + 4y^2 = 64$. A comet follows the parabolic path $y = x^2 - 4$. Where might the comet intersect the orbiting planet?

56. A system for tracking ships indicates that a ship lies on a path described by $2y^2 - x^2 = 1$. The process is repeated and the ship is found to lie on a path described by $2x^2 - y^2 = 1$. If it is known that the ship is located in the first quadrant of the coordinate system, determine its exact location.

57. Find the length and width of a rectangle whose perimeter is 36 feet and whose area is 77 square feet.

58. Find the length and width of a rectangle whose perimeter is 40 feet and whose area is 96 square feet.

Use the formula for the area of a rectangle and the Pythagorean Theorem to solve Exercises 59–60.

59. A small television has a picture with a diagonal measure of 10 inches and a viewing area of 48 square inches. Find the length and width of the screen.

60. The area of a rug is 108 square feet and the length of its diagonal is 15 feet. Find the length and width of the rug.

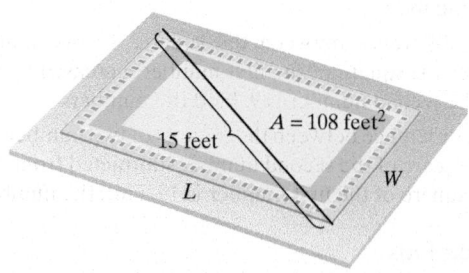

61. The figure shows a square floor plan with a smaller square area that will accommodate a combination fountain and pool. The floor with the fountain-pool area removed has an area of 21 square meters and a perimeter of 24 meters. Find the dimensions of the floor and the dimensions of the square that will accommodate the pool.

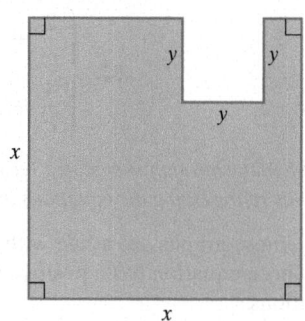

62. The area of the rectangular piece of cardboard shown below is 216 square inches. The cardboard is used to make an open box by cutting a 2-inch square from each corner and turning up the sides. If the box is to have a volume of 224 cubic inches, find the length and width of the cardboard that must be used.

63. The bar graph shows that compared to a century ago, work in the United States now involves mostly white-collar service jobs.

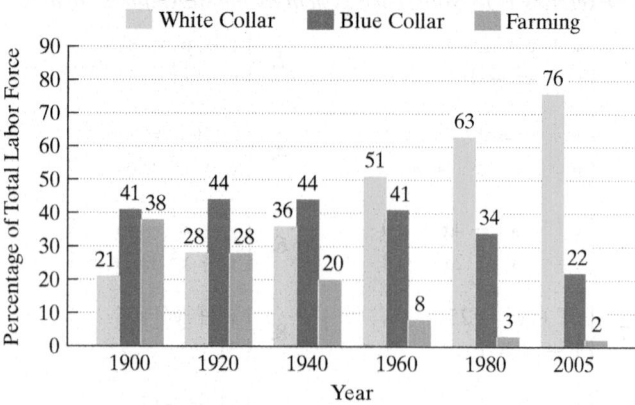

Source: U.S. Department of Labor

The data can be modeled by linear and quadratic functions.

White collar $\quad 0.5x - y = -18$

Blue collar $\quad y = -0.004x^2 + 0.23x + 41$

Farming $\quad 0.4x + y = 35$

In each function, x represents the number of years after 1900 and y represents the percentage of the total U.S. labor force.

a. Based on the information in the graph, it appears that there was a year when the percentage of white-collar workers in the labor force was the same as the percentage of blue-collar workers in the labor force. According to the graph, between which two decades did this occur?

b. Solve a nonlinear system to determine the year described in part (a). Round to the nearest year. What percentage of the labor force, to the nearest percent, consisted of white-collar workers and what percentage consisted of blue-collar workers?

c. According to the graph, for which year was the percentage of white-collar workers the same as the percentage of farmers? What percentage of U.S. workers were in each of these groups?

d. Solve a linear system to determine the year described in part (c). Round to the nearest year. Use the models to find the percentage of the labor force consisting of white-collar workers and the percentage consisting of farmers. How well do your answers model the actual data specified in part (c)?

Writing in Mathematics

64. What is a system of nonlinear equations? Provide an example with your description.

65. Explain how to solve a nonlinear system using the substitution method. Use $x^2 + y^2 = 9$ and $2x - y = 3$ to illustrate your explanation.

66. Explain how to solve a nonlinear system using the addition method. Use $x^2 - y^2 = 5$ and $3x^2 - 2y^2 = 19$ to illustrate your explanation.

Technology Exercises

67. Verify your solutions to any five exercises from Exercises 1–42 by using a graphing utility to graph the two equations in the system in the same viewing rectangle. Then use the intersection feature to verify the solutions.

68. Write a system of equations, one equation whose graph is a line and the other whose graph is a parabola, that has no ordered pairs that are real numbers in its solution set. Graph the equations using a graphing utility and verify that you are correct.

Critical Thinking Exercises

Make Sense? *In Exercises 69–72, determine whether each statement makes sense or does not make sense, and explain your reasoning.*

69. I use the same steps to solve nonlinear systems as I did to solve linear systems, although I don't obtain linear equations when a variable is eliminated.

70. I graphed a nonlinear system that modeled the orbits of Earth and Mars, and the graphs indicated the system had a solution with a real ordered pair.

71. Without using any algebra, it's obvious that the nonlinear system consisting of $x^2 + y^2 = 4$ and $x^2 + y^2 = 25$ does not have real-number solutions.

72. I think that the nonlinear system consisting of $x^2 + y^2 = 36$ and $y = (x - 2)^2 - 3$ is easier to solve graphically than by using the substitution method or the addition method.

In Exercises 73–76, determine whether each statement is true or false. If the statement is false, make the necessary change(s) to produce a true statement.

73. A system of two equations in two variables whose graphs are a circle and a line can have four real ordered-pair solutions.

74. A system of two equations in two variables whose graphs are a parabola and a circle can have four real ordered-pair solutions.

75. A system of two equations in two variables whose graphs are two circles must have at least two real ordered-pair solutions.

76. A system of two equations in two variables whose graphs are a parabola and a circle cannot have only one real ordered-pair solution.

77. The points of intersection of the graphs of $xy = 20$ and $x^2 + y^2 = 41$ are joined to form a rectangle. Find the area of the rectangle.

78. Find a and b in this figure.

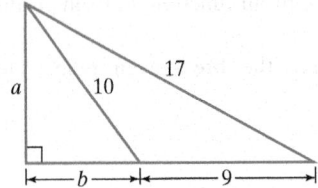

Solve the systems in Exercises 79–80.

79. $\begin{cases} \log_y x = 3 \\ \log_y(4x) = 5 \end{cases}$

80. $\begin{cases} \log x^2 = y + 3 \\ \log x = y - 1 \end{cases}$

Preview Exercises

Exercises 81–83 will help you prepare for the material covered in the next section. In each exercise, graph the linear function.

81. $2x - 3y = 6$

82. $f(x) = -\frac{2}{3}x$

83. $f(x) = -2$

CHAPTER 5 Mid-Chapter Check Point

WHAT YOU KNOW: We learned to solve systems of equations. We solved linear and nonlinear systems in two variables by the substitution method and by the addition method. We solved linear systems in three variables by eliminating a variable, reducing the system to two equations in two variables. We saw that some linear systems, called inconsistent systems, have no solution, whereas other linear systems, called dependent systems, have infinitely many solutions. We applied systems to a variety of situations, including finding the break-even point for a business, finding a quadratic function from three points on its graph, and finding a rational function's partial fraction decomposition.

In Exercises 1–12, solve each system by the method of your choice.

1. $\begin{cases} x = 3y - 7 \\ 4x + 3y = 2 \end{cases}$

2. $\begin{cases} 3x + 4y = -5 \\ 2x - 3y = 8 \end{cases}$

3. $\begin{cases} \dfrac{2x}{3} + \dfrac{y}{5} = 6 \\ \dfrac{x}{6} - \dfrac{y}{2} = -4 \end{cases}$

4. $\begin{cases} y = 4x - 5 \\ 8x - 2y = 10 \end{cases}$

5. $\begin{cases} 2x + 5y = 3 \\ 3x - 2y = 1 \end{cases}$

6. $\begin{cases} \dfrac{x}{12} - y = \dfrac{1}{4} \\ 4x - 48y = 16 \end{cases}$

7. $\begin{cases} 2x - y + 2z = -8 \\ x + 2y - 3z = 9 \\ 3x - y - 4z = 3 \end{cases}$

8. $\begin{cases} x - 3z = -5 \\ 2x - y + 2z = 16 \\ 7x - 3y - 5z = 19 \end{cases}$

9. $\begin{cases} x^2 + y^2 = 9 \\ x + 2y - 3 = 0 \end{cases}$

10. $\begin{cases} 3x^2 + 2y^2 = 14 \\ 2x^2 - y^2 = 7 \end{cases}$

11. $\begin{cases} y = x^2 - 6 \\ x^2 + y^2 = 8 \end{cases}$

12. $\begin{cases} x - 2y = 4 \\ 2y^2 + xy = 8 \end{cases}$

In Exercises 13–16, write the partial fraction decomposition of each rational expression.

13. $\dfrac{x^2 - 6x + 3}{(x - 2)^3}$

14. $\dfrac{10x^2 + 9x - 7}{(x + 2)(x^2 - 1)}$

15. $\dfrac{x^2 + 4x - 23}{(x + 3)(x^2 + 4)}$

16. $\dfrac{x^3}{(x^2 + 4)^2}$

17. A company is planning to manufacture PDAs (personal digital assistants). The fixed cost will be $400,000 and it will cost $20 to produce each PDA. Each PDA will be sold for $100.

 a. Write the cost function, C, of producing x PDAs.

 b. Write the revenue function, R, from the sale of x PDAs.

 c. Write the profit function, P, from producing and selling x PDAs.

 d. Determine the break-even point. Describe what this means.

18. Roses sell for $3 each and carnations for $1.50 each. If a mixed bouquet of 20 flowers consisting of roses and carnations costs $39, how many of each type of flower is in the bouquet?

19. Find the measure of each angle whose degree measure is represented with a variable.

20. Find the quadratic function $y = ax^2 + bx + c$ whose graph passes through the points $(-1, 0)$, $(1, 4)$, and $(2, 3)$.

21. Find the length and width of a rectangle whose perimeter is 21 meters and whose area is 20 square meters.

SECTION 5.5 Systems of Inequalities

Objectives

1. Graph a linear inequality in two variables.
2. Graph a nonlinear inequality in two variables.
3. Use mathematical models involving linear inequalities.
4. Graph a system of inequalities.

We opened the chapter noting that the modern emphasis on thinness as the ideal body shape has been suggested as a major cause of eating disorders. In this section (Example 5), as well as in the Exercise Set (Exercises 77–80), we use systems of linear inequalities in two variables that will enable you to establish a healthy weight range for your height and age.

Linear Inequalities in Two Variables and Their Solutions

We have seen that equations in the form $Ax + By = C$ are straight lines when graphed. If we change the symbol $=$ to $>$, $<$, \geq, or \leq, we obtain a **linear inequality in two variables**. Some examples of linear inequalities in two variables are $x + y > 2$, $3x - 5y \leq 15$, and $2x - y < 4$.

 A **solution of an inequality in two variables**, x and y, is an ordered pair of real numbers with the following property: When the x-coordinate is substituted for x and the y-coordinate is substituted for y in the inequality, we obtain a true statement. For example, $(3, 2)$ is a solution of the inequality $x + y > 1$. When 3 is substituted for x and 2 is substituted for y, we obtain the true statement $3 + 2 > 1$, or $5 > 1$. Because there are infinitely many pairs of numbers that have a sum greater than 1, the inequality $x + y > 1$ has infinitely many solutions. Each ordered-pair solution is said to **satisfy** the inequality. Thus, $(3, 2)$ satisfies the inequality $x + y > 1$.

① Graph a linear inequality in two variables.

The Graph of a Linear Inequality in Two Variables

We know that the graph of an equation in two variables is the set of all points whose coordinates satisfy the equation. Similarly, the **graph of an inequality in two variables** is the set of all points whose coordinates satisfy the inequality.

Let's use **Figure 5.14** to get an idea of what the graph of a linear inequality in two variables looks like. Part of the figure shows the graph of the linear equation $x + y = 2$. The line divides the points in the rectangular coordinate system into three sets. First, there is the set of points along the line, satisfying $x + y = 2$. Next, there is the set of points in the green region above the line. Points in the green region satisfy the linear inequality $x + y > 2$. Finally, there is the set of points in the purple region below the line. Points in the purple region satisfy the linear inequality $x + y < 2$.

A **half-plane** is the set of all the points on one side of a line. In **Figure 5.14**, the green region is a half-plane. The purple region is also a half-plane.

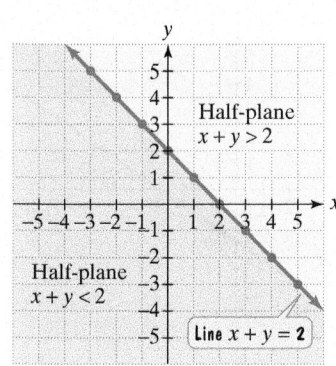

FIGURE 5.14

A half-plane is the graph of a linear inequality that involves $>$ or $<$. The graph of a linear inequality that involves \geq or \leq is a half-plane and a line. A solid line is used to show that a line is part of a solution set. A dashed line is used to show that a line is not part of a solution set.

Graphing a Linear Inequality in Two Variables

1. Replace the inequality symbol with an equal sign and graph the corresponding linear equation. Draw a solid line if the original inequality contains a \leq or \geq symbol. Draw a dashed line if the original inequality contains a $<$ or $>$ symbol.

2. Choose a test point from one of the half-planes. (Do not choose a point on the line.) Substitute the coordinates of the test point into the inequality.

3. If a true statement results, shade the half-plane containing this test point. If a false statement results, shade the half-plane not containing this test point.

EXAMPLE 1 Graphing a Linear Inequality in Two Variables

Graph: $2x - 3y \geq 6$.

SOLUTION

Step 1 Replace the inequality symbol by = and graph the linear equation. We need to graph $2x - 3y = 6$. We can use intercepts to graph this line.

We set $y = 0$ to find the x-intercept.	We set $x = 0$ to find the y-intercept.
$2x - 3y = 6$	$2x - 3y = 6$
$2x - 3 \cdot 0 = 6$	$2 \cdot 0 - 3y = 6$
$2x = 6$	$-3y = 6$
$x = 3$	$y = -2$

The x-intercept is 3, so the line passes through $(3, 0)$. The y-intercept is -2, so the line passes through $(0, -2)$. Using the intercepts, the line is shown in **Figure 5.15** as a solid line. This is because the inequality $2x - 3y \geq 6$ contains a \geq symbol, in which equality is included.

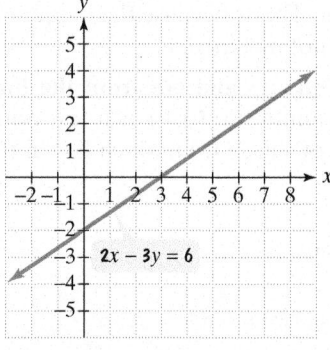

FIGURE 5.15 Preparing to graph $2x - 3y \geq 6$

Step 2 Choose a test point from one of the half-planes and not from the line. Substitute its coordinates into the inequality. The line $2x - 3y = 6$ divides the plane into three parts—the line itself and two half-planes. The points in one half-plane satisfy $2x - 3y > 6$. The points in the other half-plane satisfy $2x - 3y < 6$. We need to find which half-plane belongs to the solution of $2x - 3y \geq 6$. To do so, we test a point from either half-plane. The origin, $(0, 0)$, is the easiest point to test.

$$2x - 3y \geq 6 \quad \text{This is the given inequality.}$$
$$2 \cdot 0 - 3 \cdot 0 \overset{?}{\geq} 6 \quad \text{Test } (0, 0) \text{ by substituting 0 for } x \text{ and 0 for } y.$$
$$0 - 0 \overset{?}{\geq} 6 \quad \text{Multiply.}$$
$$0 \geq 6 \quad \text{This statement is false.}$$

Step 3 If a false statement results, shade the half-plane not containing the test point. Because 0 is not greater than or equal to 6, the test point, $(0, 0)$, is not part of the solution set. Thus, the half-plane below the solid line $2x - 3y = 6$ is part of the solution set. The solution set is the line and the half-plane that does not contain the point $(0, 0)$, indicated by shading this half-plane. The graph is shown using green shading and a blue line in **Figure 5.16**. ●●●

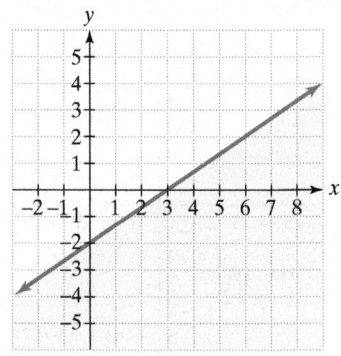

FIGURE 5.16 The graph of $2x - 3y \geq 6$

✓ Check Point **1** Graph: $4x - 2y \geq 8$.

When graphing a linear inequality, choose a test point that lies in one of the half-planes and *not on the line dividing the half-planes*. The test point $(0, 0)$ is convenient because it is easy to calculate when 0 is substituted for each variable. However, if $(0, 0)$ lies on the dividing line and not in a half-plane, a different test point must be selected.

EXAMPLE 2 Graphing a Linear Inequality in Two Variables

Graph: $y > -\frac{2}{3}x.$

SOLUTION

Step 1 Replace the inequality symbol by = and graph the linear equation. Because we are interested in graphing $y > -\frac{2}{3}x$, we begin by graphing $y = -\frac{2}{3}x$. We can use the slope and the y-intercept to graph this linear function.

$$y = -\frac{2}{3}x + 0$$

$$\text{Slope} = \frac{-2}{3} = \frac{\text{rise}}{\text{run}} \qquad y\text{-intercept} = 0$$

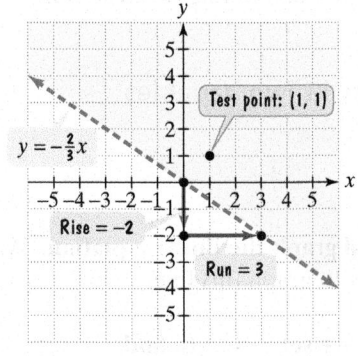

FIGURE 5.17 The graph of $y > -\frac{2}{3}x$

The y-intercept is 0, so the line passes through $(0, 0)$. Using the y-intercept and the slope, the line is shown in **Figure 5.17** as a dashed line. This is because the inequality $y > -\frac{2}{3}x$ contains a $>$ symbol, in which equality is not included.

Step 2 Choose a test point from one of the half-planes and not from the line. Substitute its coordinates into the inequality. We cannot use $(0, 0)$ as a test point because it lies on the line and not in a half-plane. Let's use $(1, 1)$, which lies in the half-plane above the line.

$$y > -\frac{2}{3}x \quad \text{This is the given inequality.}$$
$$1 \overset{?}{>} -\frac{2}{3} \cdot 1 \quad \text{Test } (1, 1) \text{ by substituting 1 for } x \text{ and 1 for } y.$$
$$1 > -\frac{2}{3} \quad \text{This statement is true.}$$

Step 3 If a true statement results, shade the half-plane containing the test point. Because 1 is greater than $-\frac{2}{3}$, the test point $(1,1)$ is part of the solution set. All the points on the same side of the line $y = -\frac{2}{3}x$ as the point $(1,1)$ are members of the solution set. The solution set is the half-plane that contains the point $(1,1)$, indicated by shading this half-plane. The graph is shown using green shading and a dashed blue line in **Figure 5.17**. ● ● ●

TECHNOLOGY

Most graphing utilities can graph inequalities in two variables with the ⬚SHADE⬚ feature. The procedure varies by model, so consult your manual. For most graphing utilities, you must first solve for y if it is not already isolated. The figure shows the graph of $y > -\frac{2}{3}x$. Most displays do not distinguish between dashed and solid boundary lines.

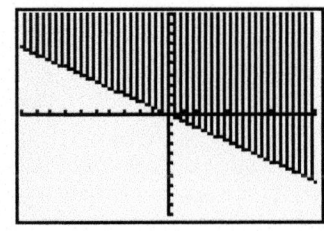

 ✓ **Check Point 2** Graph: $y > -\dfrac{3}{4}x$.

Graphing Linear Inequalities without Using Test Points

You can graph inequalities in the form $y > mx + b$ or $y < mx + b$ without using test points. The inequality symbol indicates which half-plane to shade.

- If $y > mx + b$, shade the half-plane above the line $y = mx + b$.
- If $y < mx + b$, shade the half-plane below the line $y = mx + b$.

Observe how this is illustrated in **Figure 5.17** in the margin on the previous page. The graph of $y > -\frac{2}{3}x$ is the half-plane above the line $y = -\frac{2}{3}x$.

It is also not necessary to use test points when graphing inequalities involving half-planes on one side of a vertical or a horizontal line.

For the Vertical Line $x = a$:	For the Horizontal Line $y = b$:
• If $x > a$, shade the half-plane to the right of $x = a$.	• If $y > b$, shade the half-plane above $y = b$.
• If $x < a$, shade the half-plane to the left of $x = a$.	• If $y < b$, shade the half-plane below $y = b$.

GREAT QUESTION!

When is it important to use test points to graph linear inequalities?

Continue using test points to graph inequalities in the form $Ax + By > C$ or $Ax + By < C$. The graph of $Ax + By > C$ can lie above or below the line given by $Ax + By = C$, depending on the values of A and B. The same comment applies to the graph of $Ax + By < C$.

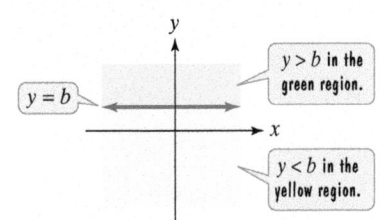

EXAMPLE 3 Graphing Inequalities without Using Test Points

Graph each inequality in a rectangular coordinate system:

a. $y \le -3$ **b.** $x > 2$.

SOLUTION

a. $y \le -3$

Graph $y = -3$, a horizontal line with y-intercept -3. The line is solid because equality is included in $y \le -3$. Because of the less than part of \le, shade the half-plane below the horizontal line.

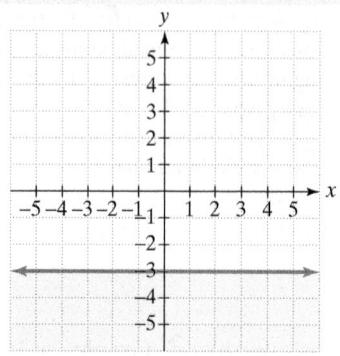

b. $x > 2$

Graph $x = 2$, a vertical line with x-intercept **2**. The line is dashed because equality is not included in $x > 2$. Because of $>$, the greater than symbol, shade the half-plane to the right of the vertical line.

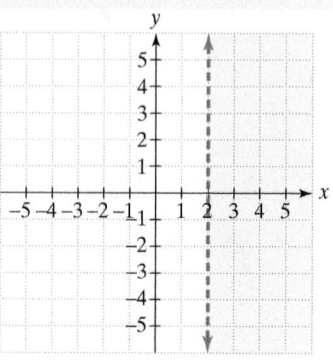

• • •

✓ **Check Point 3** Graph each inequality in a rectangular coordinate system:

a. $y > 1$ **b.** $x \le -2$.

② Graph a nonlinear inequality in two variables.

Graphing a Nonlinear Inequality in Two Variables

Example 4 illustrates that a nonlinear inequality in two variables is graphed in the same way that we graph a linear inequality.

EXAMPLE 4 Graphing a Nonlinear Inequality in Two Variables

Graph: $x^2 + y^2 \le 9$.

SOLUTION

Step 1 Replace the inequality symbol with = and graph the nonlinear equation. We need to graph $x^2 + y^2 = 9$. The graph is a circle of radius 3 with its center at the origin. The graph is shown in **Figure 5.18** as a solid circle because equality is included in the \le symbol.

Step 2 Choose a test point from one of the regions and not from the circle. Substitute its coordinates into the inequality. The circle divides the plane into three parts—the circle itself, the region inside the circle, and the region outside the circle. We need to determine whether the region inside or outside the circle is included in the solution. To do so, we will use the test point $(0, 0)$ from inside the circle.

$$x^2 + y^2 \le 9 \quad \text{This is the given inequality.}$$

$$0^2 + 0^2 \overset{?}{\le} 9 \quad \text{Test } (0, 0) \text{ by substituting 0 for } x \text{ and 0 for } y.$$

$$0 + 0 \overset{?}{\le} 9 \quad \text{Square 0: } 0^2 = 0.$$

$$0 \le 9 \quad \text{Add. This statement is true.}$$

Step 3 If a true statement results, shade the region containing the test point. The true statement tells us that all the points inside the circle satisfy $x^2 + y^2 \le 9$. The graph is shown using green shading and a solid blue circle in **Figure 5.19**.

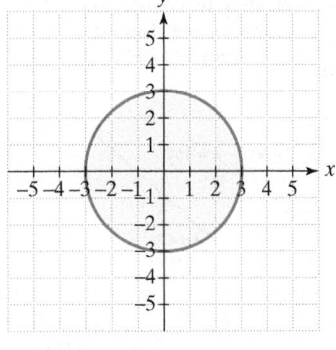

FIGURE 5.18 Preparing to graph
$x^2 + y^2 \le 9$

FIGURE 5.19 The graph of
$x^2 + y^2 \le 9$

$\bullet\bullet\bullet$

\varnothing Check Point **4** Graph: $x^2 + y^2 \ge 16$.

③ Use mathematical models involving linear inequalities.

Modeling with Systems of Linear Inequalities

Just as two or more linear equations make up a system of linear equations, two or more linear inequalities make up a **system of linear inequalities**. A **solution of a system of linear inequalities** in two variables is an ordered pair that satisfies each inequality in the system.

EXAMPLE 5 Does Your Weight Fit You?

The latest guidelines, which apply to both men and women, give healthy weight ranges, rather than specific weights, for your height. **Figure 5.20** shows the healthy weight region for various heights for people between the ages of 19 and 34, inclusive.

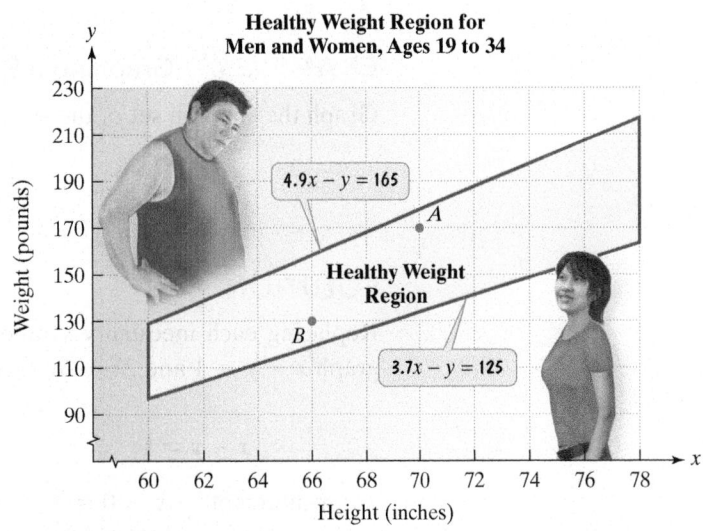

FIGURE 5.20
Source: U.S. Department of Health and Human Services

If x represents height, in inches, and y represents weight, in pounds, the healthy weight region in **Figure 5.20** can be modeled by the following system of linear inequalities:

$$\begin{cases} 4.9x - y \ge 165 \\ 3.7x - y \le 125. \end{cases}$$

Show that point A in **Figure 5.20** is a solution of the system of inequalities that describes healthy weight.

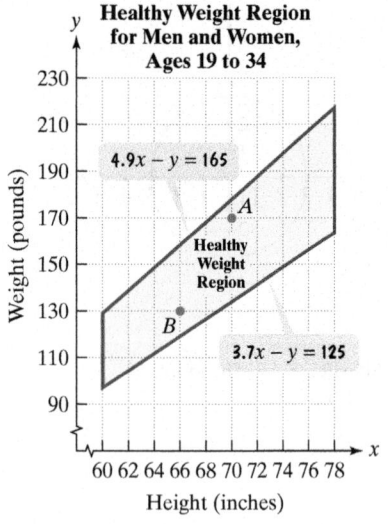

Healthy Weight Region for Men and Women, Ages 19 to 34

FIGURE 5.20 (repeated)

SOLUTION

Point *A* has coordinates (70, 170). This means that if a person is 70 inches tall, or 5 feet 10 inches, and weighs 170 pounds, then that person's weight is within the healthy weight region. We can show that (70, 170) satisfies the system of inequalities by substituting 70 for *x* and 170 for *y* in each inequality in the system.

$$4.9x - y \geq 165$$
$$4.9(70) - 170 \overset{?}{\geq} 165$$
$$343 - 170 \overset{?}{\geq} 165$$
$$173 \geq 165, \quad \text{true}$$

$$3.7x - y \leq 125$$
$$3.7(70) - 170 \overset{?}{\leq} 125$$
$$259 - 170 \overset{?}{\leq} 125$$
$$89 \leq 125, \quad \text{true}$$

The coordinates (70, 170) make each inequality true. Thus, (70, 170) satisfies the system for the healthy weight region and is a solution of the system. ● ● ●

✓ Check Point **5** Show that point *B* in **Figure 5.20** is a solution of the system of inequalities that describes healthy weight.

④ Graph a system of inequalities.

Graphing Systems of Linear Inequalities

The **solution set of a system of linear inequalities in two variables** is the set of all ordered pairs that satisfy each inequality in the system. Thus, to graph a system of inequalities in two variables, begin by graphing each individual inequality in the same rectangular coordinate system. Then find the region, if there is one, that is common to every graph in the system. This region of intersection gives a picture of the system's solution set.

EXAMPLE 6 Graphing a System of Linear Inequalities

Graph the solution set of the system:

$$\begin{cases} x - y < 1 \\ 2x + 3y \geq 12. \end{cases}$$

SOLUTION

Replacing each inequality symbol with an equal sign indicates that we need to graph $x - y = 1$ and $2x + 3y = 12$. We can use intercepts to graph these lines.

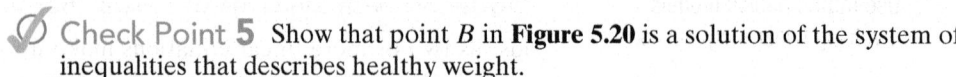

x − y = 1

x-intercept: $x - 0 = 1$ (Set y = 0 in each equation.)
 $x = 1$
The line passes through (1, 0).

y-intercept: $0 - y = 1$ (Set x = 0 in each equation.)
 $-y = 1$
 $y = -1$
The line passes through (0, −1).

2x + 3y = 12

x-intercept: $2x + 3 \cdot 0 = 12$
 $2x = 12$
 $x = 6$
The line passes through (6, 0).

y-intercept: $2 \cdot 0 + 3y = 12$
 $3y = 12$
 $y = 4$
The line passes through (0, 4).

Now we are ready to graph the solution set of the system of linear inequalities.

Graph $x - y < 1$. The blue line, $x - y = 1$, is dashed: Equality is not included in $x - y < 1$. Because $(0, 0)$ makes the inequality true $(0 - 0 < 1,$ or $0 < 1,$ is true$)$, shade the half-plane containing $(0, 0)$ in yellow.

Add the graph of $2x + 3y \geq 12$. The red line, $2x + 3y = 12$, is solid: Equality is included in $2x + 3y \geq 12$. Because $(0, 0)$ makes the inequality false $(2 \cdot 0 + 3 \cdot 0 \geq 12,$ or $0 \geq 12,$ is false$)$, shade the half-plane not containing $(0, 0)$ using green vertical shading.

The solution set of the system is graphed as the intersection (the overlap) of the two half-planes. This is the region in which the yellow shading and the green vertical shading overlap.

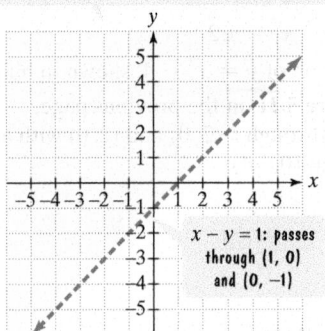

The graph of $x - y < 1$

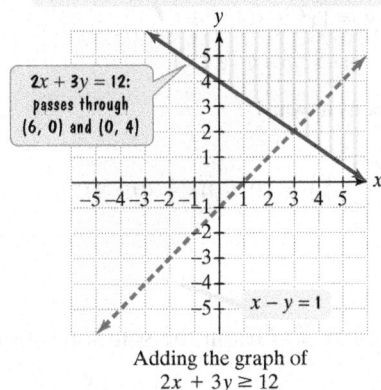

Adding the graph of $2x + 3y \geq 12$

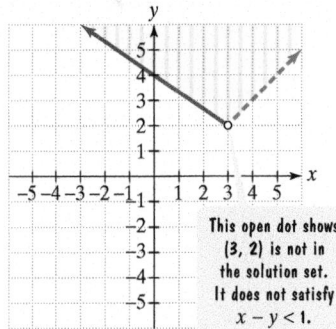

The graph of $x - y < 1$ and $2x + 3y \geq 12$

✓ Check Point **6** Graph the solution set of the system:

$$\begin{cases} x - 3y < 6 \\ 2x + 3y \geq -6. \end{cases}$$

EXAMPLE 7 Graphing a System of Inequalities

Graph the solution set of the system:

$$\begin{cases} y \geq x^2 - 4 \\ x - y \geq 2. \end{cases}$$

SOLUTION

We begin by graphing $y \geq x^2 - 4$. Because equality is included in \geq, we graph $y = x^2 - 4$ as a solid parabola. Because $(0, 0)$ makes the inequality $y \geq x^2 - 4$ true (we obtain $0 \geq -4$), we shade the interior portion of the parabola containing $(0, 0)$, shown in yellow in **Figure 5.21**.

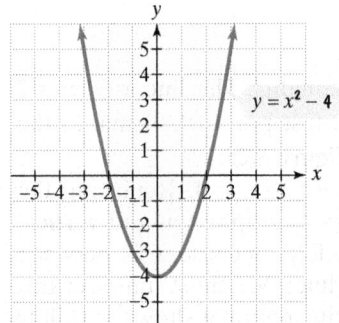

FIGURE 5.21 The graph of $y \geq x^2 - 4$

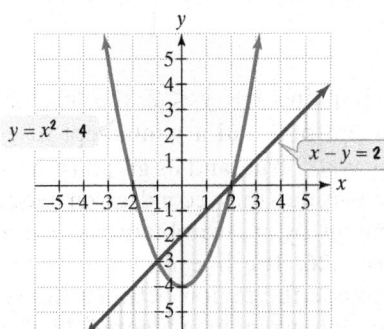

FIGURE 5.22 Adding the graph of $x - y \geq 2$

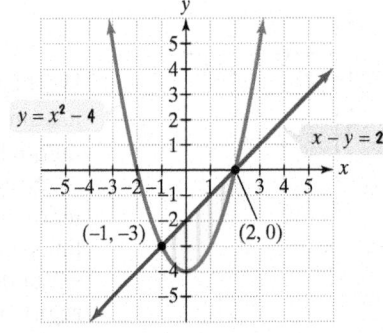

FIGURE 5.23 The graph of $y \geq x^2 - 4$ and $x - y \geq 2$

Now we graph $x - y \geq 2$ in the same rectangular coordinate system. First we graph the line $x - y = 2$ using its x-intercept, 2, and its y-intercept, -2. Because $(0, 0)$ makes the inequality $x - y \geq 2$ false (we obtain $0 \geq 2$), we shade the half-plane below the line. This is shown in **Figure 5.22** using green vertical shading.

The solution of the system is shown in **Figure 5.23** by the intersection (the overlap) of the solid yellow and green vertical shadings. The graph of the system's solution set consists of the region enclosed by the parabola and the line. To find the points of intersection of the parabola and the line, use the substitution method to solve the nonlinear system containing $y = x^2 - 4$ and $x - y = 2$. Take a moment to show that the solutions are $(-1, -3)$ and $(2, 0)$, as shown in **Figure 5.23**. • • •

GREAT QUESTION!

In Example 7, how did you graph $y = x^2 - 4$?

We used the parabola's vertex and its x-intercepts.

$$y = 1x^2 + 0x - 4$$

$$a = 1 \quad b = 0 \quad c = -4$$

- $a > 0$: opens upward
- Vertex:

$$\left(-\frac{b}{2a}, f\left(-\frac{b}{2a}\right)\right) = (0, f(0)) = (0, -4)$$

- x-intercepts:

$$x^2 - 4 = 0$$
$$x^2 = 4$$
$$x = \pm 2$$

The graph of $y = x^2 - 4$, shown in blue in **Figure 5.21** on the previous page, passes through $(-2, 0)$ and $(2, 0)$ with a vertex at $(0, -4)$.

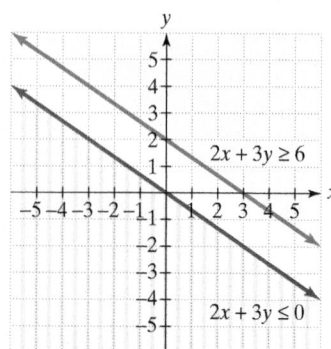

FIGURE 5.24 A system of inequalities with no solution

⊘ Check Point **7** Graph the solution set of the system:

$$\begin{cases} y \geq x^2 - 4 \\ x + y \leq 2. \end{cases}$$

A system of inequalities has no solution if there are no points in the rectangular coordinate system that simultaneously satisfy each inequality in the system. For example, the system

$$\begin{cases} 2x + 3y \geq 6 \\ 2x + 3y \leq 0, \end{cases}$$

whose separate graphs are shown in **Figure 5.24**, has no overlapping region. Thus, the system has no solution. The solution set is ∅, the empty set.

EXAMPLE 8 Graphing a System of Inequalities

Graph the solution set of the system:

$$\begin{cases} x - y < 2 \\ -2 \leq x < 4 \\ y < 3. \end{cases}$$

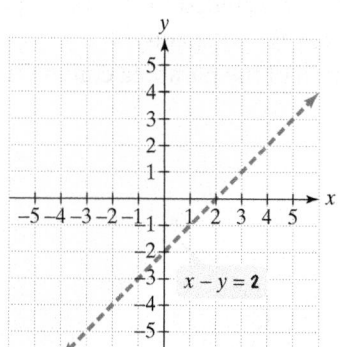

FIGURE 5.25 The graph of $x - y < 2$

SOLUTION

We begin by graphing $x - y < 2$, the first given inequality. The line $x - y = 2$ has an x-intercept of 2 and a y-intercept of -2. The test point $(0, 0)$ makes the inequality $x - y < 2$ true, and its graph is shown in **Figure 5.25**.

Now, let's consider the second given inequality, $-2 \leq x < 4$. Replacing the inequality symbols by $=$, we obtain $x = -2$ and $x = 4$, graphed as red vertical lines in **Figure 5.26**. The line of $x = 4$ is not included. Because x is between -2 and 4, we shade the region between the vertical lines. We must intersect this region with the yellow region in **Figure 5.25**. The resulting region is shown in yellow and green vertical shading in **Figure 5.26**.

Finally, let's consider the third given inequality, $y < 3$. Replacing the inequality symbol by $=$, we obtain $y = 3$, which graphs as a horizontal line. Because of the less than symbol in $y < 3$, the graph consists of the half-plane below the line $y = 3$. We must intersect this half-plane with the region in **Figure 5.26**. The resulting region is shown in yellow and green vertical shading in **Figure 5.27**. This region represents the graph of the solution set of the given system.

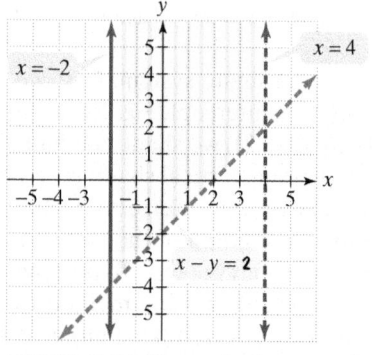

FIGURE 5.26 The graph of $x - y < 2$ and $-2 \leq x < 4$

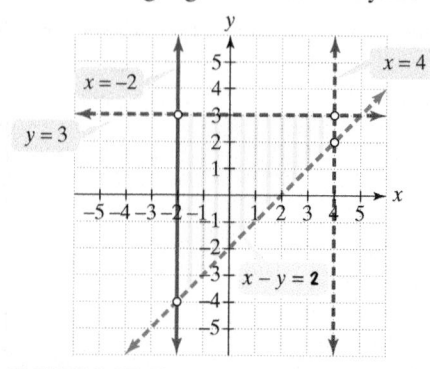

FIGURE 5.27 The graph of $x - y < 2$ and $-2 \leq x < 4$ and $y < 3$

In **Figure 5.27** it may be difficult to tell where the graph of $x - y = 2$ intersects the vertical line $x = 4$. Using the substitution method, it can be determined that this intersection point is $(4, 2)$. Take a moment to verify that the four intersection points in **Figure 5.27** are, clockwise from upper left, $(-2, 3)$, $(4, 3)$, $(4, 2)$, and $(-2, -4)$. These points are shown as open dots because none satisfies all three of the system's inequalities. • • •

☑ Check Point **8** Graph the solution set of the system:

$$\begin{cases} x + y < 2 \\ -2 \le x < 1 \\ y > -3. \end{cases}$$

CONCEPT AND VOCABULARY CHECK

Fill in each blank so that the resulting statement is true.

1. The ordered pair $(3, 2)$ is a/an _____ of the inequality $x + y > 1$ because when 3 is substituted for _____ and 2 is substituted for _____, the true statement _____ is obtained.
2. The set of all points that satisfy a linear inequality in two variables is called the _____ of the inequality.
3. The set of all points on one side of a line is called a/an _____.
4. True or false: The graph of $2x - 3y > 6$ includes the line $2x - 3y = 6$. _____
5. True or false: The graph of the linear equation $2x - 3y = 6$ is used to graph the linear inequality $2x - 3y > 6$. _____
6. True or false: When graphing $4x - 2y \ge 8$, to determine which side of the line to shade, choose a test point on $4x - 2y = 8$. _____

7. When graphing $x^2 + y^2 > 25$, to determine whether to shade the region inside the circle or the region outside the circle, we can use _____ as a test point.
8. The solution set of the system

$$\begin{cases} x - y < 1 \\ 2x + 3y \ge 12. \end{cases}$$

is the set of ordered pairs that satisfy _____ and _____.
9. True or false: The graph of the solution set of the system

$$\begin{cases} x - 3y < 6 \\ 2x + 3y \ge -6. \end{cases}$$

includes the intersection point of $x - 3y = 6$ and $2x + 3y = -6$. _____

EXERCISE SET 5.5

Practice Exercises

In Exercises 1–26, graph each inequality.

1. $x + 2y \le 8$
2. $3x - 6y \le 12$
3. $x - 2y > 10$
4. $2x - y > 4$
5. $y \le \dfrac{1}{3}x$
6. $y \le \dfrac{1}{4}x$
7. $y > 2x - 1$
8. $y > 3x + 2$
9. $x \le 1$
10. $x \le -3$
11. $y > 1$
12. $y > -3$
13. $x^2 + y^2 \le 1$
14. $x^2 + y^2 \le 4$
15. $x^2 + y^2 > 25$
16. $x^2 + y^2 > 36$
17. $(x - 2)^2 + (y + 1)^2 < 9$
18. $(x + 2)^2 + (y - 1)^2 < 16$
19. $y < x^2 - 1$
20. $y < x^2 - 9$
21. $y \ge x^2 - 9$
22. $y \ge x^2 - 1$
23. $y > 2^x$
24. $y \le 3^x$
25. $y \ge \log_2(x + 1)$
26. $y \ge \log_3(x - 1)$

In Exercises 27–62, graph the solution set of each system of inequalities or indicate that the system has no solution.

27. $\begin{cases} 3x + 6y \le 6 \\ 2x + y \le 8 \end{cases}$
28. $\begin{cases} x - y \ge 4 \\ x + y \le 6 \end{cases}$
29. $\begin{cases} 2x - 5y \le 10 \\ 3x - 2y > 6 \end{cases}$
30. $\begin{cases} 2x - y \le 4 \\ 3x + 2y > -6 \end{cases}$
31. $\begin{cases} y > 2x - 3 \\ y < -x + 6 \end{cases}$
32. $\begin{cases} y < -2x + 4 \\ y < x - 4 \end{cases}$
33. $\begin{cases} x + 2y \le 4 \\ y \ge x - 3 \end{cases}$
34. $\begin{cases} x + y \le 4 \\ y \ge 2x - 4 \end{cases}$
35. $\begin{cases} x \le 2 \\ y \ge -1 \end{cases}$
36. $\begin{cases} x \le 3 \\ y \le -1 \end{cases}$
37. $-2 \le x < 5$
38. $-2 < y \le 5$
39. $\begin{cases} x - y \le 1 \\ x \ge 2 \end{cases}$
40. $\begin{cases} 4x - 5y \ge -20 \\ x \ge -3 \end{cases}$
41. $\begin{cases} x + y > 4 \\ x + y < -1 \end{cases}$
42. $\begin{cases} x + y > 3 \\ x + y < -2 \end{cases}$

43. $\begin{cases} x + y > 4 \\ x + y > -1 \end{cases}$

44. $\begin{cases} x + y > 3 \\ x + y > -2 \end{cases}$

45. $\begin{cases} y \geq x^2 - 1 \\ x - y \geq -1 \end{cases}$

46. $\begin{cases} y \geq x^2 - 4 \\ x - y \geq 2 \end{cases}$

47. $\begin{cases} x^2 + y^2 \leq 16 \\ x + y > 2 \end{cases}$

48. $\begin{cases} x^2 + y^2 \leq 4 \\ x + y > 1 \end{cases}$

49. $\begin{cases} x^2 + y^2 > 1 \\ x^2 + y^2 < 16 \end{cases}$

50. $\begin{cases} x^2 + y^2 > 1 \\ x^2 + y^2 < 9 \end{cases}$

51. $\begin{cases} (x - 1)^2 + (y + 1)^2 < 25 \\ (x - 1)^2 + (y + 1)^2 \geq 16 \end{cases}$

52. $\begin{cases} (x + 1)^2 + (y - 1)^2 < 16 \\ (x + 1)^2 + (y - 1)^2 \geq 4 \end{cases}$

53. $\begin{cases} x^2 + y^2 \leq 1 \\ y - x^2 > 0 \end{cases}$

54. $\begin{cases} x^2 + y^2 < 4 \\ y - x^2 \geq 0 \end{cases}$

55. $\begin{cases} x^2 + y^2 < 16 \\ y \geq 2^x \end{cases}$

56. $\begin{cases} x^2 + y^2 \leq 16 \\ y < 2^x \end{cases}$

57. $\begin{cases} x - y \leq 2 \\ x > -2 \\ y \leq 3 \end{cases}$

58. $\begin{cases} 3x + y \leq 6 \\ x > -2 \\ y \leq 4 \end{cases}$

59. $\begin{cases} x \geq 0 \\ y \geq 0 \\ 2x + 5y < 10 \\ 3x + 4y \leq 12 \end{cases}$

60. $\begin{cases} x \geq 0 \\ y \geq 0 \\ 2x + y < 4 \\ 2x - 3y \leq 6 \end{cases}$

61. $\begin{cases} 3x + y \leq 6 \\ 2x - y \leq -1 \\ x > -2 \\ y < 4 \end{cases}$

62. $\begin{cases} 2x + y \leq 6 \\ x + y > 2 \\ 1 \leq x \leq 2 \\ y < 3 \end{cases}$

Practice Plus

In Exercises 63–64, write each sentence as an inequality in two variables. Then graph the inequality.

63. The *y*-variable is at least 4 more than the product of −2 and the *x*-variable.

64. The *y*-variable is at least 2 more than the product of −3 and the *x*-variable.

In Exercises 65–68, write the given sentences as a system of inequalities in two variables. Then graph the system.

65. The sum of the *x*-variable and the *y*-variable is at most 4. The *y*-variable added to the product of 3 and the *x*-variable does not exceed 6.

66. The sum of the *x*-variable and the *y*-variable is at most 3. The *y*-variable added to the product of 4 and the *x*-variable does not exceed 6.

67. The sum of the *x*-variable and the *y*-variable is no more than 2. The *y*-variable is no less than the difference between the square of the *x*-variable and 4.

68. The sum of the squares of the *x*-variable and the *y*-variable is no more than 25. The sum of twice the *y*-variable and the *x*-variable is no less than 5.

In Exercises 69–70, rewrite each inequality in the system without absolute value bars. Then graph the rewritten system in rectangular coordinates.

69. $\begin{cases} |x| \leq 2 \\ |y| \leq 3 \end{cases}$

70. $\begin{cases} |x| \leq 1 \\ |y| \leq 2 \end{cases}$

*The graphs of solution sets of systems of inequalities involve finding the intersection of the solution sets of two or more inequalities. By contrast, in Exercises 71–72, you will be graphing the **union** of the solution sets of two inequalities.*

71. Graph the union of $y > \frac{3}{2}x - 2$ and $y < 4$.

72. Graph the union of $x - y \geq -1$ and $5x - 2y \leq 10$.

Without graphing, in Exercises 73–76, determine if each system has no solution or infinitely many solutions.

73. $\begin{cases} 3x + y < 9 \\ 3x + y > 9 \end{cases}$

74. $\begin{cases} 6x - y \leq 24 \\ 6x - y > 24 \end{cases}$

75. $\begin{cases} (x + 4)^2 + (y - 3)^2 \leq 9 \\ (x + 4)^2 + (y - 3)^2 \geq 9 \end{cases}$

76. $\begin{cases} (x - 4)^2 + (y + 3)^2 \leq 24 \\ (x - 4)^2 + (y + 3)^2 \geq 24 \end{cases}$

Application Exercises

The figure shows the healthy weight region for various heights for people ages 35 and older.

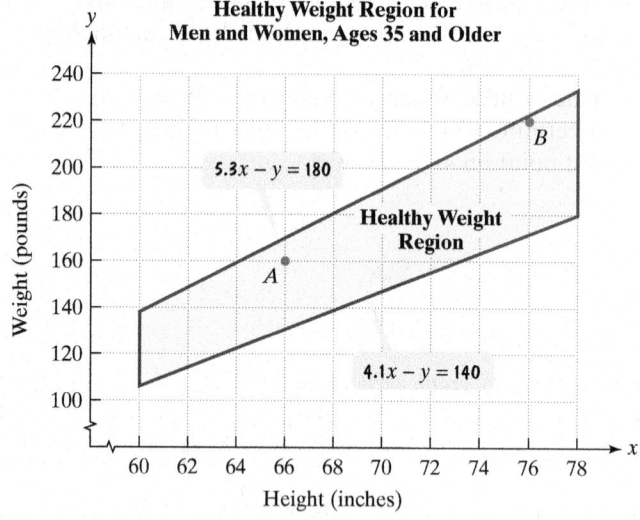

Source: U.S. Department of Health and Human Services

If x represents height, in inches, and y represents weight, in pounds, the healthy weight region can be modeled by the following system of linear inequalities:

$$\begin{cases} 5.3x - y \geq 180 \\ 4.1x - y \leq 140. \end{cases}$$

Use this information to solve Exercises 77–80.

77. Show that point *A* is a solution of the system of inequalities that describes healthy weight for this age group.

78. Show that point *B* is a solution of the system of inequalities that describes healthy weight for this age group.

79. Is a person in this age group who is 6 feet tall weighing 205 pounds within the healthy weight region?

80. Is a person in this age group who is 5 feet 8 inches tall weighing 135 pounds within the healthy weight region?

81. Many elevators have a capacity of 2000 pounds.

 a. If a child averages 50 pounds and an adult 150 pounds, write an inequality that describes when x children and y adults will cause the elevator to be overloaded.

 b. Graph the inequality. Because x and y must be positive, limit the graph to quadrant I only.

 c. Select an ordered pair satisfying the inequality. What are its coordinates and what do they represent in this situation?

82. A patient is not allowed to have more than 330 milligrams of cholesterol per day from a diet of eggs and meat. Each egg provides 165 milligrams of cholesterol. Each ounce of meat provides 110 milligrams.

 a. Write an inequality that describes the patient's dietary restrictions for x eggs and y ounces of meat.

 b. Graph the inequality. Because x and y must be positive, limit the graph to quadrant I only.

 c. Select an ordered pair satisfying the inequality. What are its coordinates and what do they represent in this situation?

83. On your next vacation, you will divide lodging between large resorts and small inns. Let x represent the number of nights spent in large resorts. Let y represent the number of nights spent in small inns.

 a. Write a system of inequalities that models the following conditions:

 > You want to stay at least 5 nights. At least one night should be spent at a large resort. Large resorts average $200 per night and small inns average $100 per night. Your budget permits no more than $700 for lodging.

 b. Graph the solution set of the system of inequalities in part (a).

 c. Based on your graph in part (b), what is the greatest number of nights you could spend at a large resort and still stay within your budget?

84. A person with no more than $15,000 to invest plans to place the money in two investments. One investment is high risk, high yield; the other is low risk, low yield. At least $2000 is to be placed in the high-risk investment. Furthermore, the amount invested at low risk should be at least three times the amount invested at high risk. Find and graph a system of inequalities that describes all possibilities for placing the money in the high- and low-risk investments.

The graph of an inequality in two variables is a region in the rectangular coordinate system. Regions in coordinate systems have numerous applications. For example, the regions in the following two graphs indicate whether a person is obese, overweight, borderline overweight, normal weight, or underweight.

Source: Centers for Disease Control and Prevention

The horizontal axis shows a person's age. The vertical axis shows that person's body-mass index (BMI), computed using the following formula:

$$\text{BMI} = \frac{703W}{H^2}.$$

The variable W represents weight, in pounds. The variable H represents height, in inches. Use this information to solve Exercises 85–86.

85. A man is 20 years old, 72 inches (6 feet) tall, and weighs 200 pounds.

 a. Compute the man's BMI. Round to the nearest tenth.

 b. Use the man's age and his BMI to locate this information as a point in the coordinate system for males. Is this person obese, overweight, borderline overweight, normal weight, or underweight?

86. A woman is 25 years old, 66 inches (5 feet, 6 inches) tall, and weighs 105 pounds.

 a. Compute the woman's BMI. Round to the nearest tenth.

 b. Use the woman's age and her BMI to locate this information as a point in the coordinate system for females. Is this person obese, overweight, borderline overweight, normal weight, or underweight?

Writing in Mathematics

87. What is a linear inequality in two variables? Provide an example with your description.

88. How do you determine if an ordered pair is a solution of an inequality in two variables, x and y?

89. What is a half-plane?

90. What does a solid line mean in the graph of an inequality?

91. What does a dashed line mean in the graph of an inequality?

92. Compare the graphs of $3x - 2y > 6$ and $3x - 2y \leq 6$. Discuss similarities and differences between the graphs.

93. What is a system of linear inequalities?

94. What is a solution of a system of linear inequalities?

95. Explain how to graph the solution set of a system of inequalities.

96. What does it mean if a system of linear inequalities has no solution?

Technology Exercises

Graphing utilities can be used to shade regions in the rectangular coordinate system, thereby graphing an inequality in two variables. Read the section of the user's manual for your graphing utility that describes how to shade a region. Then use your graphing utility to graph the inequalities in Exercises 97–102.

97. $y \leq 4x + 4$

98. $y \geq \dfrac{2}{3}x - 2$

99. $y \geq x^2 - 4$

100. $y \geq \dfrac{1}{2}x^2 - 2$

101. $2x + y \leq 6$

102. $3x - 2y \geq 6$

103. Does your graphing utility have any limitations in terms of graphing inequalities? If so, what are they?

104. Use a graphing utility with a ⬛SHADE⬛ feature to verify any five of the graphs that you drew by hand in Exercises 1–26.

105. Use a graphing utility with a ⬛SHADE⬛ feature to verify any five of the graphs that you drew by hand for the systems in Exercises 27–62.

Critical Thinking Exercises

Make Sense? *In Exercises 106–109, determine whether each statement makes sense or does not make sense, and explain your reasoning.*

106. When graphing a linear inequality, I should always use $(0, 0)$ as a test point because it's easy to perform the calculations when 0 is substituted for each variable.

107. When graphing $3x - 4y < 12$, it's not necessary for me to graph the linear equation $3x - 4y = 12$ because the inequality contains a $<$ symbol, in which equality is not included.

108. The reason that systems of linear inequalities are appropriate for modeling healthy weight is because guidelines give healthy weight ranges, rather than specific weights, for various heights.

109. I graphed the solution set of $y \geq x + 2$ and $x \geq 1$ without using test points.

In Exercises 110–113, write a system of inequalities for each graph.

110.

111.

112. **113.**

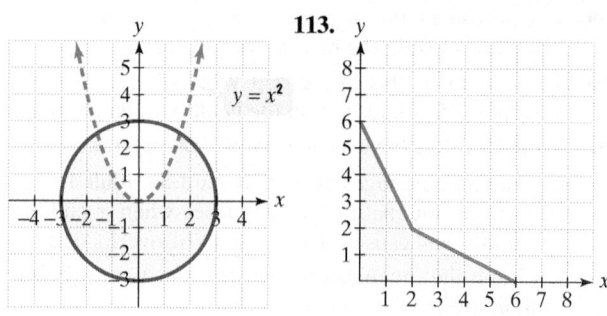

114. Write a system of inequalities whose solution set includes every point in the rectangular coordinate system.

115. Sketch the graph of the solution set for the following system of inequalities:

$$\begin{cases} y \geq nx + b \ (n < 0, b > 0) \\ y \leq mx + b \ (m > 0, b > 0). \end{cases}$$

Preview Exercises

Exercises 116–118 will help you prepare for the material covered in the next section.

116. a. Graph the solution set of the system:

$$\begin{cases} x + y \geq 6 \\ x \leq 8 \\ y \geq 5. \end{cases}$$

b. List the points that form the corners of the graphed region in part (a).

c. Evaluate $3x + 2y$ at each of the points obtained in part (b).

117. a. Graph the solution set of the system:

$$\begin{cases} x \geq 0 \\ y \geq 0 \\ 3x - 2x \leq 6 \\ y \leq -x + 7. \end{cases}$$

b. List the points that form the corners of the graphed region in part (a).

c. Evaluate $2x + 5y$ at each of the points obtained in part (b).

118. Bottled water and medical supplies are to be shipped to survivors of an earthquake by plane. The bottled water weighs 20 pounds per container and medical kits weigh 10 pounds per kit. Each plane can carry no more than 80,000 pounds. If x represents the number of bottles of water to be shipped per plane and y represents the number of medical kits per plane, write an inequality that models each plane's 80,000-pound weight restriction.

SECTION 5.6 Linear Programming

Objectives

1. Write an objective function describing a quantity that must be maximized or minimized.
2. Use inequalities to describe limitations in a situation.
3. Use linear programming to solve problems.

West Berlin children at Tempelhof airport watch fleets of U.S. airplanes bringing in supplies to circumvent the Soviet blockade. The airlift began June 28, 1948, and continued for 15 months.

The Berlin Airlift (1948–1949) was an operation by the United States and Great Britain in response to military action by the former Soviet Union: Soviet troops closed all roads and rail lines between West Germany and Berlin, cutting off supply routes to the city. The Allies used a mathematical technique developed during World War II to maximize the amount of supplies transported. During the 15-month airlift, 278,228 flights provided basic necessities to blockaded Berlin, saving one of the world's great cities.

In this section, we will look at an important application of systems of linear inequalities. Such systems arise in **linear programming**, a method for solving problems in which a particular quantity that must be maximized or minimized is limited by other factors. Linear programming is one of the most widely used tools in management science. It helps businesses allocate resources to manufacture products in a way that will maximize profit. Linear programming accounts for more than 50% and perhaps as much as 90% of all computing time used for management decisions in business. The Allies used linear programming to save Berlin.

1 Write an objective function describing a quantity that must be maximized or minimized.

Objective Functions in Linear Programming

Many problems involve quantities that must be maximized or minimized. Businesses are interested in maximizing profit. An operation in which bottled water and medical kits are shipped to earthquake survivors needs to maximize the number of survivors helped by this shipment. An **objective function** is an algebraic expression in two or more variables describing a quantity that must be maximized or minimized.

EXAMPLE 1 Writing an Objective Function

Bottled water and medical supplies are to be shipped to survivors of an earthquake by plane. Each container of bottled water will serve 10 people and each medical kit will aid 6 people. If x represents the number of bottles of water to be shipped and y represents the number of medical kits, write the objective function that models the number of people that can be helped.

SOLUTION

Because each bottle of water serves 10 people and each medical kit aids 6 people, we have

The number of people helped	is	10 times the number of bottles of water	plus	6 times the number of medical kits.
$=$		$10x$	$+$	$6y.$

Using z to represent the number of people helped, the objective function is

$$z = 10x + 6y.$$

Unlike the functions that we have seen so far, the objective function is an equation in three variables. For a value of x and a value of y, there is one and only one value of z. Thus, z is a function of x and y. •••

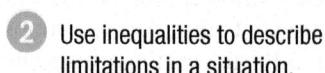 Check Point **1** A company manufactures bookshelves and desks for computers. Let x represent the number of bookshelves manufactured daily and y the number of desks manufactured daily. The company's profits are $25 per bookshelf and $55 per desk. Write the objective function that models the company's total daily profit, z, from x bookshelves and y desks. (Check Points 2 through 4 are related to this situation, so keep track of your answers.)

② Use inequalities to describe limitations in a situation.

Constraints in Linear Programming

Ideally, the number of earthquake survivors helped in Example 1 should increase without restriction so that every survivor receives water and medical supplies. However, the planes that ship these supplies are subject to weight and volume restrictions. In linear programming problems, such restrictions are called **constraints**. Each constraint is expressed as a linear inequality. The list of constraints forms a system of linear inequalities.

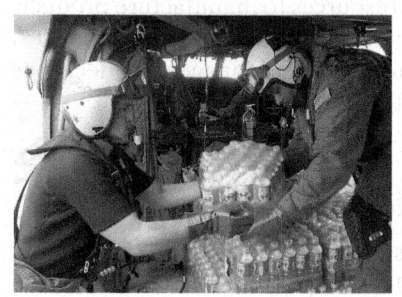

EXAMPLE 2 Writing a Constraint

Each plane can carry no more than 80,000 pounds. The bottled water weighs 20 pounds per container and each medical kit weighs 10 pounds. Let x represent the number of bottles of water to be shipped and y the number of medical kits. Write an inequality that models this constraint.

SOLUTION

Because each plane can carry no more than 80,000 pounds, we have

The total weight of the water bottles	plus	the total weight of the medical kits	must be less than or equal to	80,000 pounds.
$20x$	$+$	$10y$	\leq	$80,000.$

Each bottle weighs 20 pounds. Each kit weighs 10 pounds.

The plane's weight constraint is modeled by the inequality

$$20x + 10y \leq 80,000.$$ •••

 Check Point **2** To maintain high quality, the company in Check Point 1 should not manufacture more than a total of 80 bookshelves and desks per day. Write an inequality that models this constraint.

In addition to a weight constraint on its cargo, each plane has a limited amount of space in which to carry supplies. Example 3 demonstrates how to express this constraint.

EXAMPLE 3 Writing a Constraint

Each plane can carry a total volume of supplies that does not exceed 6000 cubic feet. Each water bottle is 1 cubic foot and each medical kit also has a volume of 1 cubic foot. With x still representing the number of water bottles and y the number of medical kits, write an inequality that models this second constraint.

SOLUTION

Because each plane can carry a volume of supplies that does not exceed 6000 cubic feet, we have

$$1x \quad + \quad 1y \quad \le \quad 6000.$$

Each bottle is 1 cubic foot. Each kit is 1 cubic foot.

The plane's volume constraint is modeled by the inequality $x + y \le 6000$. • • •

In summary, here's what we have described so far in this aid-to-earthquake-survivors situation:

$$z = 10x + 6y$$ This is the objective function modeling the number of people helped with x bottles of water and y medical kits.

$$\begin{cases} 20x + 10y \le 80{,}000 \\ x + y \le 6000. \end{cases}$$ These are the constraints based on each plane's weight and volume limitations.

✓ Check Point **3** To meet customer demand, the company in Check Point 1 must manufacture between 30 and 80 bookshelves per day, inclusive. Furthermore, the company must manufacture at least 10 and no more than 30 desks per day. Write an inequality that models each of these sentences. Then summarize what you have described about this company by writing the objective function for its profits and the three constraints.

③ Use linear programming to solve problems.

Solving Problems with Linear Programming

The problem in the earthquake situation described previously is to maximize the number of survivors who can be helped, subject to each plane's weight and volume constraints. The process of solving this problem is called *linear programming*, based on a theorem that was proven during World War II.

Solving a Linear Programming Problem

Let $z = ax + by$ be an objective function that depends on x and y. Furthermore, z is subject to a number of constraints on x and y. If a maximum or minimum value of z exists, it can be determined as follows:

1. Graph the system of inequalities representing the constraints.
2. Find the value of the objective function at each corner, or **vertex**, of the graphed region. The maximum and minimum of the objective function occur at one or more of the corner points.

EXAMPLE 4 Solving a Linear Programming Problem

Determine how many bottles of water and how many medical kits should be sent on each plane to maximize the number of earthquake survivors who can be helped.

SOLUTION

We must maximize $z = 10x + 6y$ subject to the following constraints:

$$\begin{cases} 20x + 10y \leq 80,000 \\ x + y \leq 6000. \end{cases}$$

Step 1 Graph the system of inequalities representing the constraints. Because x (the number of bottles of water per plane) and y (the number of medical kits per plane) must be nonnegative, we need to graph the system of inequalities in quadrant I and its boundary only.

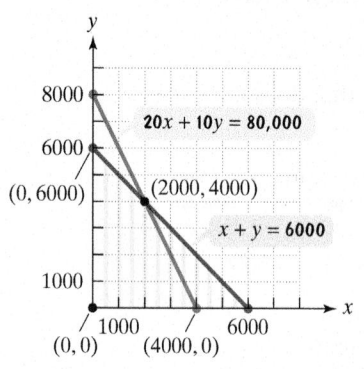

To graph the inequality $20x + 10y \leq 80,000$, we graph the equation $20x + 10y = 80,000$ as a solid blue line (**Figure 5.28**). Setting $y = 0$, the x-intercept is 4000 and setting $x = 0$, the y-intercept is 8000. Using $(0, 0)$ as a test point, the inequality is satisfied, so we shade below the blue line, as shown in yellow in **Figure 5.28**.

Now we graph $x + y \leq 6000$ by first graphing $x + y = 6000$ as a solid red line. Setting $y = 0$, the x-intercept is 6000. Setting $x = 0$, the y-intercept is 6000. Using $(0, 0)$ as a test point, the inequality is satisfied, so we shade below the red line, as shown using green vertical shading in **Figure 5.28**.

FIGURE 5.28 The region in quadrant I representing the constraints $20x + 10y \leq 80,000$ and $x + y \leq 6000$

We use the addition method to find where the lines $20x + 10y = 80,000$ and $x + y = 6000$ intersect.

$$\begin{cases} 20x + 10y = 80,000 \\ x + y = 6000 \end{cases} \xrightarrow[\text{Multiply by } -10.]{\text{No change}} \begin{cases} 20x + 10y = 80,000 \\ -10x - 10y = -60,000 \end{cases}$$
$$\text{Add:} \qquad 10x = 20,000$$
$$x = 2000$$

Back-substituting 2000 for x in $x + y = 6000$, we find $y = 4000$, so the intersection point is $(2000, 4000)$.

The system of inequalities representing the constraints is shown by the region in which the yellow shading and the green vertical shading overlap in **Figure 5.28**. The graph of the system of inequalities is shown again in **Figure 5.29**. The red and blue line segments are included in the graph.

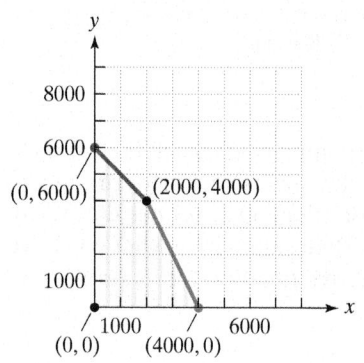

FIGURE 5.29

Step 2 Find the value of the objective function at each corner of the graphed region. The maximum and minimum of the objective function occur at one or more of the corner points. We must evaluate the objective function, $z = 10x + 6y$, at the four corners, or vertices, of the region in **Figure 5.29**.

Corner (x, y)	Objective Function $z = 10x + 6y$
$(0, 0)$	$z = 10(0) + 6(0) = 0$
$(4000, 0)$	$z = 10(4000) + 6(0) = 40,000$
$(2000, 4000)$	$z = 10(2000) + 6(4000) = 44,000 \leftarrow$ maximum
$(0, 6000)$	$z = 10(0) + 6(6000) = 36,000$

Thus, the maximum value of z is 44,000 and this occurs when $x = 2000$ and $y = 4000$. In practical terms, this means that the maximum number of earthquake survivors who can be helped with each plane shipment is 44,000. This can be accomplished by sending 2000 water bottles and 4000 medical kits per plane. • • •

✓ Check Point **4** For the company in Check Points 1–3, how many bookshelves and how many desks should be manufactured per day to obtain maximum profit? What is the maximum daily profit?

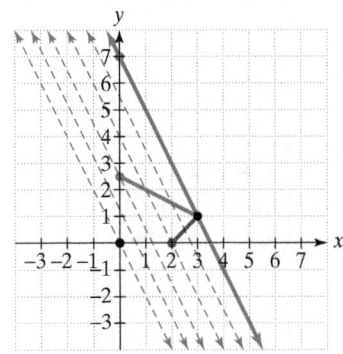

FIGURE 5.30 The graph of $x + 2y \leq 5$ and $x - y \leq 2$ in quadrant I

EXAMPLE 5 Solving a Linear Programming Problem

Find the maximum value of the objective function

$$z = 2x + y$$

subject to the following constraints:

$$\begin{cases} x \geq 0, y \geq 0 \\ x + 2y \leq 5 \\ x - y \leq 2. \end{cases}$$

SOLUTION

We begin by graphing the region in quadrant I ($x \geq 0, y \geq 0$) formed by the constraints. The graph is shown in **Figure 5.30**.

Now we evaluate the objective function at the four vertices of this region.

Objective function: $z = 2x + y$

At $(0, 0)$: $z = 2 \cdot 0 + 0 = 0$

At $(2, 0)$: $z = 2 \cdot 2 + 0 = 4$ Maximum

At $(3, 1)$: $z = 2 \cdot 3 + 1 = 7$ value of z

At $(0, 2.5)$: $z = 2 \cdot 0 + 2.5 = 2.5$

Thus, the maximum value of z is 7, and this occurs when $x = 3$ and $y = 1$. • • •

We can see why the objective function in Example 5 has a maximum value that occurs at a vertex by solving the equation for y.

$$z = 2x + y$$ This is the objective function of Example 5.

$$y = -2x + z$$ Solve for y. Recall that the slope-intercept form of a line is $y = mx + b$.

In this form, z represents the y-intercept of the objective function. The equation describes infinitely many parallel lines, each with slope -2. The process in linear programming involves finding the maximum z-value for all lines that intersect the region determined by the constraints. Of all the lines whose slope is -2, we're looking for the one with the greatest y-intercept that intersects the given region. As we see in **Figure 5.31**, such a line will pass through one (or possibly more) of the vertices of the region.

FIGURE 5.31 The line with slope -2 with the greatest y-intercept that intersects the shaded region passes through one of its vertices.

✓ Check Point **5** Find the maximum value of the objective function $z = 3x + 5y$ subject to the constraints $x \geq 0, y \geq 0, x + y \geq 1, x + y \leq 6$.

CONCEPT AND VOCABULARY CHECK

Fill in each blank so that the resulting statement is true.

1. A method for finding the maximum or minimum value of a quantity that is subject to various limitations is called _____.

2. An algebraic expression in two or more variables describing a quantity that must be maximized or minimized is called a/an _____ function.

3. A system of linear inequalities is used to represent restrictions, or _____, on a function that must be maximized or minimized. Using the graph of such a system of inequalities, the maximum and minimum values of the function occur at one or more of the _____ points.

EXERCISE SET 5.6

Practice Exercises

In Exercises 1–4, find the value of the objective function at each corner of the graphed region. What is the maximum value of the objective function? What is the minimum value of the objective function?

1. Objective Function
$z = 5x + 6y$

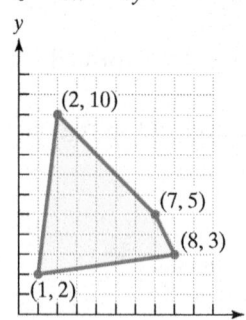

2. Objective Function
$z = 3x + 2y$

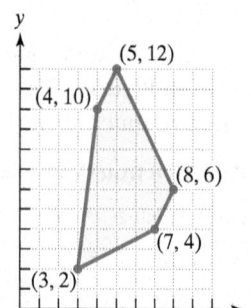

3. Objective Function
$z = 40x + 50y$

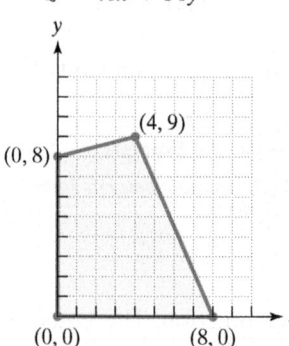

4. Objective Function
$z = 30x + 45y$

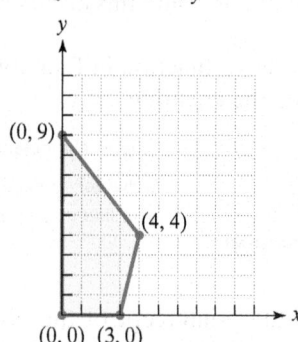

In Exercises 5–14, an objective function and a system of linear inequalities representing constraints are given.

 a. *Graph the system of inequalities representing the constraints.*

 b. *Find the value of the objective function at each corner of the graphed region.*

 c. *Use the values in part (b) to determine the maximum value of the objective function and the values of x and y for which the maximum occurs.*

5. Objective Function
Constraints

$z = 3x + 2y$
$\begin{cases} x \geq 0, y \geq 0 \\ 2x + y \leq 8 \\ x + y \geq 4 \end{cases}$

6. Objective Function
Constraints

$z = 2x + 3y$
$\begin{cases} x \geq 0, y \geq 0 \\ 2x + y \leq 8 \\ 2x + 3y \leq 12 \end{cases}$

7. Objective Function
Constraints

$z = 4x + y$
$\begin{cases} x \geq 0, y \geq 0 \\ 2x + 3y \leq 12 \\ x + y \geq 3 \end{cases}$

8. Objective Function
Constraints

$z = x + 6y$
$\begin{cases} x \geq 0, y \geq 0 \\ 2x + y \leq 10 \\ x - 2y \geq -10 \end{cases}$

9. Objective Function
Constraints

$z = 3x - 2y$
$\begin{cases} 1 \leq x \leq 5 \\ y \geq 2 \\ x - y \geq -3 \end{cases}$

10. Objective Function
Constraints

$z = 5x - 2y$
$\begin{cases} 0 \leq x \leq 5 \\ 0 \leq y \leq 3 \\ x + y \geq 2 \end{cases}$

11. Objective Function
Constraints

$z = 4x + 2y$
$\begin{cases} x \geq 0, y \geq 0 \\ 2x + 3y \leq 12 \\ 3x + 2y \leq 12 \\ x + y \geq 2 \end{cases}$

12. Objective Function
Constraints

$z = 2x + 4y$
$\begin{cases} x \geq 0, y \geq 0 \\ x + 3y \geq 6 \\ x + y \geq 3 \\ x + y \leq 9 \end{cases}$

13. Objective Function
Constraints

$z = 10x + 12y$
$\begin{cases} x \geq 0, y \geq 0 \\ x + y \leq 7 \\ 2x + y \leq 10 \\ 2x + 3y \leq 18 \end{cases}$

14. Objective Function
Constraints

$z = 5x + 6y$
$\begin{cases} x \geq 0, y \geq 0 \\ 2x + y \geq 10 \\ x + 2y \geq 10 \\ x + y \leq 10 \end{cases}$

Application Exercises

15. A television manufacturer makes rear-projection and plasma televisions. The profit per unit is $125 for the rear-projection televisions and $200 for the plasma televisions.

 a. Let $x =$ the number of rear-projection televisions manufactured in a month and let $y =$ the number of plasma televisions manufactured in a month. Write the objective function that models the total monthly profit.

 b. The manufacturer is bound by the following constraints:

 • Equipment in the factory allows for making at most 450 rear-projection televisions in one month.

 • Equipment in the factory allows for making at most 200 plasma televisions in one month.

 • The cost to the manufacturer per unit is $600 for the rear-projection televisions and $900 for the plasma televisions. Total monthly costs cannot exceed $360,000.

 Write a system of three inequalities that models these constraints.

c. Graph the system of inequalities in part (b). Use only the first quadrant and its boundary, because *x* and *y* must both be nonnegative.

d. Evaluate the objective function for total monthly profit at each of the five vertices of the graphed region. [The vertices should occur at $(0, 0)$, $(0, 200)$, $(300, 200)$, $(450, 100)$, and $(450, 0)$.]

e. Complete the missing portions of this statement: The television manufacturer will make the greatest profit by manufacturing _____ rear-projection televisions each month and _____ plasma televisions each month. The maximum monthly profit is $ _____.

16. a. A student earns $10 per hour for tutoring and $7 per hour as a teacher's aide. Let *x* = the number of hours each week spent tutoring and let *y* = the number of hours each week spent as a teacher's aide. Write the objective function that models total weekly earnings.

b. The student is bound by the following constraints:

• To have enough time for studies, the student can work no more than 20 hours per week.

• The tutoring center requires that each tutor spend at least three hours per week tutoring.

• The tutoring center requires that each tutor spend no more than eight hours per week tutoring.

Write a system of three inequalities that models these constraints.

c. Graph the system of inequalities in part (b). Use only the first quadrant and its boundary, because *x* and *y* are nonnegative.

d. Evaluate the objective function for total weekly earnings at each of the four vertices of the graphed region. [The vertices should occur at $(3, 0)$, $(8, 0)$, $(3, 17)$, and $(8, 12)$.]

e. Complete the missing portions of this statement: The student can earn the maximum amount per week by tutoring for ____ hours per week and working as a teacher's aide for ____ hours per week. The maximum amount that the student can earn each week is $____.

Use the two steps for solving a linear programming problem, given in the box on page 587, to solve the problems in Exercises 17–23.

17. A manufacturer produces two models of mountain bicycles. The times (in hours) required for assembling and painting each model are given in the following table:

	Model *A*	Model *B*
Assembling	5	4
Painting	2	3

The maximum total weekly hours available in the assembly department and the paint department are 200 hours and 108 hours, respectively. The profits per unit are $25 for model *A* and $15 for model *B*. How many of each type should be produced to maximize profit?

18. A large institution is preparing lunch menus containing foods A and B. The specifications for the two foods are given in the following table:

Food	Units of Fat per Ounce	Units of Carbohydrates per Ounce	Units of Protein per Ounce
A	1	2	1
B	1	1	1

Each lunch must provide at least 6 units of fat per serving, no more than 7 units of protein, and at least 10 units of carbohydrates. The institution can purchase food A for $0.12 per ounce and food B for $0.08 per ounce. How many ounces of each food should a serving contain to meet the dietary requirements at the least cost?

19. Food and clothing are shipped to survivors of a natural disaster. Each carton of food will feed 12 people, while each carton of clothing will help 5 people. Each 20-cubic-foot box of food weighs 50 pounds and each 10-cubic-foot box of clothing weighs 20 pounds. The commercial carriers transporting food and clothing are bound by the following constraints:

• The total weight per carrier cannot exceed 19,000 pounds.

• The total volume must be less than 8000 cubic feet.

How many cartons of food and clothing should be sent with each plane shipment to maximize the number of people who can be helped?

20. On June 24, 1948, the former Soviet Union blocked all land and water routes through East Germany to Berlin. A gigantic airlift was organized using American and British planes to bring food, clothing, and other supplies to the more than 2 million people in West Berlin. The cargo capacity was 30,000 cubic feet for an American plane and 20,000 cubic feet for a British plane. To break the Soviet blockade, the Western Allies had to maximize cargo capacity but were subject to the following restrictions:

• No more than 44 planes could be used.

• The larger American planes required 16 personnel per flight, double that of the requirement for the British planes. The total number of personnel available could not exceed 512.

• The cost of an American flight was $9000 and the cost of a British flight was $5000. Total weekly costs could not exceed $300,000.

Find the number of American and British planes that were used to maximize cargo capacity.

21. A theater is presenting a program for students and their parents on drinking and driving. The proceeds will be donated to a local alcohol information center. Admission is $2.00 for parents and $1.00 for students. However, the situation has two constraints: The theater can hold no more than 150 people and every two parents must bring at least one student. How many parents and students should attend to raise the maximum amount of money?

22. You are about to take a test that contains computation problems worth 6 points each and word problems worth 10 points each. You can do a computation problem in 2 minutes and a word problem in 4 minutes. You have 40 minutes to take the test and may answer no more than 12 problems. Assuming you answer all the problems attempted correctly, how many of each type of problem must you answer to maximize your score? What is the maximum score?

23. In 1978, a ruling by the Civil Aeronautics Board allowed Federal Express to purchase larger aircraft. Federal Express's options included 20 Boeing 727s that United Airlines was retiring and/or the French-built Dassault Fanjet Falcon 20. To aid in their decision, executives at Federal Express analyzed the following data:

	Boeing 727	Falcon 20
Direct Operating Cost	$1400 per hour	$500 per hour
Payload	42,000 pounds	6000 pounds

Federal Express was faced with the following constraints:
* Hourly operating cost was limited to $35,000.
* Total payload had to be at least 672,000 pounds.
* Only twenty 727s were available.

Given the constraints, how many of each kind of aircraft should Federal Express have purchased to maximize the number of aircraft?

Writing in Mathematics

24. What kinds of problems are solved using the linear programming method?

25. What is an objective function in a linear programming problem?

26. What is a constraint in a linear programming problem? How is a constraint represented?

27. In your own words, describe how to solve a linear programming problem.

28. Describe a situation in your life in which you would really like to maximize something, but you are limited by at least two constraints. Can linear programming be used in this situation? Explain your answer.

Critical Thinking Exercises

Make Sense? *In Exercises 29–32, determine whether each statement makes sense or does not make sense, and explain your reasoning.*

29. In order to solve a linear programming problem, I use the graph representing the constraints and the graph of the objective function.

30. I use the coordinates of each vertex from my graph representing the constraints to find the values that maximize or minimize an objective function.

31. I need to be able to graph systems of linear inequalities in order to solve linear programming problems.

32. An important application of linear programming for businesses involves maximizing profit.

33. Suppose that you inherit $10,000. The will states how you must invest the money. Some (or all) of the money must be invested in stocks and bonds. The requirements are that at least $3000 be invested in bonds, with expected returns of $0.08 per dollar, and at least $2000 be invested in stocks, with expected returns of $0.12 per dollar. Because the stocks are medium risk, the final stipulation requires that the investment in bonds should never be less than the investment in stocks. How should the money be invested so as to maximize your expected returns?

34. Consider the objective function $z = Ax + By$ ($A > 0$ and $B > 0$) subject to the following constraints: $2x + 3y \leq 9, x - y \leq 2, x \geq 0$, and $y \geq 0$. Prove that the objective function will have the same maximum value at the vertices $(3, 1)$ and $(0, 3)$ if $A = \frac{2}{3}B$.

Group Exercises

35. Group members should choose a particular field of interest. Research how linear programming is used to solve problems in that field. If possible, investigate the solution of a specific practical problem. Present a report on your findings, including the contributions of George Dantzig, Narendra Karmarkar, and L. G. Khachion to linear programming.

36. Members of the group should interview a business executive who is in charge of deciding the product mix for a business. How are production policy decisions made? Are other methods used in conjunction with linear programming? What are these methods? What sort of academic background, particularly in mathematics, does this executive have? Present a group report addressing these questions, emphasizing the role of linear programming for the business.

Preview Exercises

Exercises 37–39 will help you prepare for the material covered in the first section of the next chapter.

37. Solve the system:

$$\begin{cases} x + y + 2z = 19 \\ y + 2z = 13 \\ z = 5. \end{cases}$$

What makes it fairly easy to find the solution?

38. Solve the system:

$$\begin{cases} w - x + 2y - 2z = -1 \\ x - \frac{1}{3}y + z = \frac{8}{3} \\ y - z = 1 \\ z = 3. \end{cases}$$

Express the solution set in the form $\{(w, x, y, z)\}$. What makes it fairly easy to find the solution?

39. Consider the following array of numbers:

$$\begin{bmatrix} 1 & 2 & -1 \\ 4 & -3 & -15 \end{bmatrix}.$$

Rewrite the array as follows: Multiply each number in the top row by -4 and add this product to the corresponding number in the bottom row. Do not change the numbers in the top row.

CHAPTER 5 Summary, Review, and Test

SUMMARY

DEFINITIONS AND CONCEPTS	EXAMPLES

5.1 Systems of Linear Equations in Two Variables

a. Two equations in the form $Ax + By = C$ are called a system of linear equations. A solution of the system is an ordered pair that satisfies both equations in the system.

Ex. 1, p. 526

b. Systems of linear equations in two variables can be solved by eliminating a variable, using the substitution method (see the box on page 528) or the addition method (see the box on page 530).

Ex. 2, p. 528;
Ex. 3, p. 530;
Ex. 4, p. 531

c. Some linear systems have no solution and are called inconsistent systems; others have infinitely many solutions. The equations in a linear system with infinitely many solutions are called dependent. For details, see the box on page 532.

Ex. 5, p. 532;
Ex. 6, p. 533

d. Functions of Business

Ex. 7, p. 534;
Figure 5.7, p. 536

Revenue Function

$$R(x) = (\text{price per unit sold})x$$

Cost Function

$$C(x) = \text{fixed cost} + (\text{cost per unit produced})x$$

Profit Function

$$P(x) = R(x) - C(x)$$

The point of intersection of the graphs of R and C is the break-even point. The x-coordinate of the point reveals the number of units that a company must produce and sell so that the money coming in, the revenue, is equal to the money going out, the cost. The y-coordinate gives the amount of money coming in and going out.

5.2 Systems of Linear Equations in Three Variables

a. Three equations in the form $Ax + By + Cz = D$ are called a system of linear equations in three variables. A solution of the system is an ordered triple that satisfies all three equations in the system.

Ex. 1, p. 542

b. A system of linear equations in three variables can be solved by eliminating variables. Use the addition method to eliminate any variable, reducing the system to two equations in two variables. Use substitution or the addition method to solve the resulting system in two variables. Details are found in the box on page 543.

Ex. 2, p. 543;
Ex. 3, p. 545

c. Three points that do not lie on a line determine the graph of a quadratic function $y = ax^2 + bx + c$. Use the three given points to create a system of three equations. Solve the system to find a, b, and c.

Ex. 4, p. 546

5.3 Partial Fractions

a. Partial fraction decomposition is used on rational expressions in which the numerator and denominator have no common factors and the highest power in the numerator is less than the highest power in the denominator. The steps in partial fraction decomposition are given in the box on page 553.

b. Include one partial fraction with a constant numerator for each distinct linear factor in the denominator. Include one partial fraction with a constant numerator for each power of a repeated linear factor in the denominator.

Ex. 1, p. 552;
Ex. 2, p. 554

c. Include one partial fraction with a linear numerator for each distinct prime quadratic factor in the denominator. Include one partial fraction with a linear numerator for each power of a prime, repeated quadratic factor in the denominator.

Ex. 3, p. 556;
Ex. 4, p. 558

5.4 Systems of Nonlinear Equations in Two Variables

a. A system of two nonlinear equations in two variables contains at least one equation that cannot be expressed as $Ax + By = C$.

b. Systems of nonlinear equations in two variables can be solved algebraically by eliminating all occurrences of one of the variables by the substitution or addition methods.

Ex. 1, p. 562;
Ex. 2, p. 563;
Ex. 3, p. 564;
Ex. 4, p. 566

DEFINITIONS AND CONCEPTS

EXAMPLES

5.5 Systems of Inequalities

a. A linear inequality in two variables can be written in the form
$Ax + By > C$, $Ax + By \geq C$, $Ax + By < C$, or $Ax + By \leq C$.

b. The procedure for graphing a linear inequality in two variables is given in the box on page 573. A nonlinear inequality in two variables is graphed using the same procedure.

Ex. 1, p. 573;
Ex. 2, p. 574;
Ex. 3, p. 576;
Ex. 4, p. 576

c. To graph the solution set of a system of inequalities, graph each inequality in the system in the same rectangular coordinate system. Then find the region, if there is one, that is common to every graph in the system.

Ex. 6, p. 578;
Ex. 7, p. 579;
Ex. 8, p. 580

5.6 Linear Programming

a. An objective function is an algebraic expression in three variables describing a quantity that must be maximized or minimized.

Ex. 1, p. 585

b. Constraints are restrictions, expressed as linear inequalities.

Ex. 2, p. 586;
Ex. 3, p. 587

c. Linear programming is a method for solving problems in which an objective function that must be maximized or minimized is limited by constraints. Steps for solving a linear programming problem are given in the box on page 587.

Ex. 4, p. 588;
Ex. 5, p. 589

REVIEW EXERCISES

5.1

In Exercises 1–5, solve by the method of your choice. Identify systems with no solution and systems with infinitely many solutions, using set notation to express their solution sets.

1. $\begin{cases} y = 4x + 1 \\ 3x + 2y = 13 \end{cases}$

2. $\begin{cases} x + 4y = 14 \\ 2x - y = 1 \end{cases}$

3. $\begin{cases} 5x + 3y = 1 \\ 3x + 4y = -6 \end{cases}$

4. $\begin{cases} 2y - 6x = 7 \\ 3x - y = 9 \end{cases}$

5. $\begin{cases} 4x - 8y = 16 \\ 3x - 6y = 12 \end{cases}$

6. A company is planning to manufacture computer desks. The fixed cost will be $60,000 and it will cost $200 to produce each desk. Each desk will be sold for $450.

a. Write the cost function, C, of producing x desks.

b. Write the revenue function, R, from the sale of x desks.

c. Determine the break-even point. Describe what this means.

7.

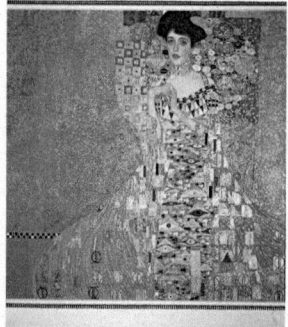

Adele Bloch-Bauer, I (1907), Gustav Klimt/Neue Galerie/Art Resource, NY

Boy with a Pipe (1905), Pablo Picasso. © 2011 Picasso Estate/ARS

Talk about paintings by numbers: In 2007, this glittering Klimt set the record for the most ever paid for a painting, a title that had been held by Picasso's *Boy with a Pipe*. Combined, the two paintings sold for $239 million. The difference between the selling price for Klimt's work and the selling price for Picasso's work was $31 million. Find the amount paid for each painting.

8. The music business is evolving into a digital marketplace. The bar graph shows that from 2004 through 2009, CD album sales declined, while digital track sales grew.

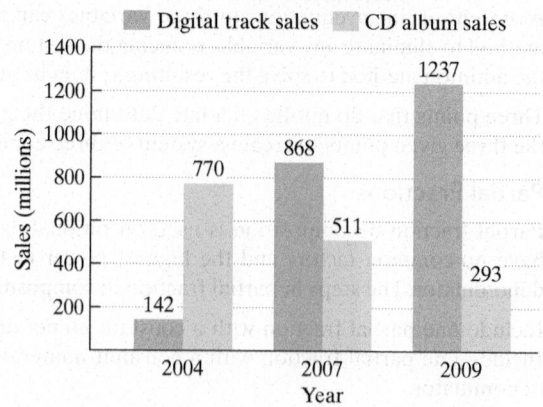

Sales of Albums and Digital Tracks in the United States

Source: RIAA

a. In 2004, 142 million digital tracks were sold. For the period shown by the graph, this has increased at an average rate of 219 million digital tracks per year. Write a function that models the sale of digital tracks, y, in millions, x years after 2004.

b. The function $95x + y = 770$ models the number of albums sold, y, in millions, x years after 2004. Use this model and the model you obtained in part (a) to determine when digital track sales caught up with album sales. How many digital tracks and how many albums were sold in that year?

9. The perimeter of a rectangular table top is 34 feet. The difference between 4 times the length and 3 times the width is 33 feet. Find the dimensions.

10. A travel agent offers two package vacation plans. The first plan costs $360 and includes 3 days at a hotel and a rental car for 2 days. The second plan costs $500 and includes 4 days at a hotel and a rental car for 3 days. The daily charge for the hotel is the same under each plan, as is the daily charge for the car. Find the cost per day for the hotel and for the car.

11. The calorie-nutrient information for an apple and an avocado is given in the table. How many of each should be eaten to get exactly 1000 calories and 100 grams of carbohydrates?

	One Apple	One Avocado
Calories	100	350
Carbohydrates (grams)	24	14

5.2

Solve each system in Exercises 12–13.

12. $\begin{cases} 2x - y + z = 1 \\ 3x - 3y + 4z = 5 \\ 4x - 2y + 3z = 4 \end{cases}$ **13.** $\begin{cases} x + 2y - z = 5 \\ 2x - y + 3z = 0 \\ 2y + z = 1 \end{cases}$

14. Find the quadratic function $y = ax^2 + bx + c$ whose graph passes through the points $(1, 4), (3, 20),$ and $(-2, 25)$.

15. 20th Century Death The greatest cause of death in the 20th century was disease, killing 1390 million people. The bar graph shows the five leading causes of death in that century, excluding disease.

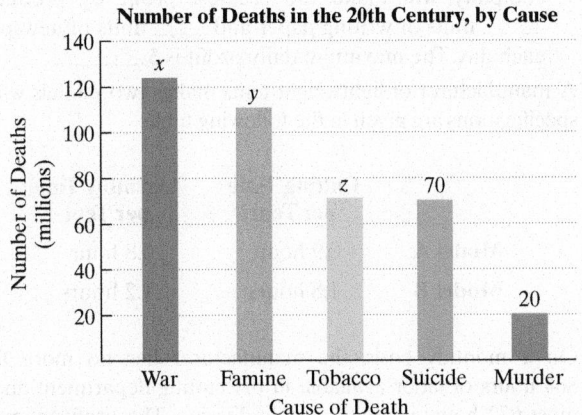

Number of Deaths in the 20th Century, by Cause

Source: Wikipedia

War, famine, and tobacco combined resulted in 306 million deaths. The difference between the number of deaths from war and famine was 13 million. The difference between the number of deaths from war and tobacco was 53 million. Find the number of 20th century deaths from war, famine, and tobacco.

5.3

In Exercises 16–24, write the partial fraction decomposition of each rational expression.

16. $\dfrac{x}{(x - 3)(x + 2)}$

17. $\dfrac{11x - 2}{x^2 - x - 12}$

18. $\dfrac{4x^2 - 3x - 4}{x(x + 2)(x - 1)}$

19. $\dfrac{2x + 1}{(x - 2)^2}$

20. $\dfrac{2x - 6}{(x - 1)(x - 2)^2}$

21. $\dfrac{3x}{(x - 2)(x^2 + 1)}$

22. $\dfrac{7x^2 - 7x + 23}{(x - 3)(x^2 + 4)}$

23. $\dfrac{x^3}{(x^2 + 4)^2}$

24. $\dfrac{4x^3 + 5x^2 + 7x - 1}{(x^2 + x + 1)^2}$

5.4

In Exercises 25–35, solve each system by the method of your choice.

25. $\begin{cases} 5y = x^2 - 1 \\ x - y = 1 \end{cases}$ **26.** $\begin{cases} y = x^2 + 2x + 1 \\ x + y = 1 \end{cases}$

27. $\begin{cases} x^2 + y^2 = 2 \\ x + y = 0 \end{cases}$ **28.** $\begin{cases} 2x^2 + y^2 = 24 \\ x^2 + y^2 = 15 \end{cases}$

29. $\begin{cases} xy - 4 = 0 \\ y - x = 0 \end{cases}$ **30.** $\begin{cases} y^2 = 4x \\ x - 2y + 3 = 0 \end{cases}$

31. $\begin{cases} x^2 + y^2 = 10 \\ y = x + 2 \end{cases}$ **32.** $\begin{cases} xy = 1 \\ y = 2x + 1 \end{cases}$

33. $\begin{cases} x + y + 1 = 0 \\ x^2 + y^2 + 6y - x = -5 \end{cases}$ **34.** $\begin{cases} x^2 + y^2 = 13 \\ x^2 - y = 7 \end{cases}$

35. $\begin{cases} 2x^2 + 3y^2 = 21 \\ 3x^2 - 4y^2 = 23 \end{cases}$

36. The perimeter of a rectangle is 26 meters and its area is 40 square meters. Find its dimensions.

37. Find the coordinates of all points (x, y) that lie on the line whose equation is $2x + y = 8$, so that the area of the rectangle shown in the figure is 6 square units.

38. Two adjoining square fields with an area of 2900 square feet are to be enclosed with 240 feet of fencing. The situation is represented in the figure. Find the length of each side where a variable appears.

5.5

In Exercises 39–45, graph each inequality.

39. $3x - 4y > 12$

40. $y \le -\dfrac{1}{2}x + 2$

41. $x < -2$

42. $y \ge 3$

43. $x^2 + y^2 > 4$

44. $y \le x^2 - 1$

45. $y \le 2^x$

In Exercises 46–55, graph the solution set of each system of inequalities or indicate that the system has no solution.

46. $\begin{cases} 3x + 2y \ge 6 \\ 2x + y \ge 6 \end{cases}$

47. $\begin{cases} 2x - y \ge 4 \\ x + 2y < 2 \end{cases}$

48. $\begin{cases} y < x \\ y \le 2 \end{cases}$

49. $\begin{cases} x + y \le 6 \\ y \ge 2x - 3 \end{cases}$

50. $\begin{cases} 0 \le x \le 3 \\ y > 2 \end{cases}$

51. $\begin{cases} 2x + y < 4 \\ 2x + y > 6 \end{cases}$

52. $\begin{cases} x^2 + y^2 \le 16 \\ x + y < 2 \end{cases}$

53. $\begin{cases} x^2 + y^2 \le 9 \\ y < -3x + 1 \end{cases}$

54. $\begin{cases} y > x^2 \\ x + y < 6 \\ y < x + 6 \end{cases}$

55. $\begin{cases} y \ge 0 \\ 3x + 2y \ge 4 \\ x - y \le 3 \end{cases}$

5.6

56. Find the value of the objective function $z = 2x + 3y$ at each corner of the graphed region shown. What is the maximum value of the objective function? What is the minimum value of the objective function?

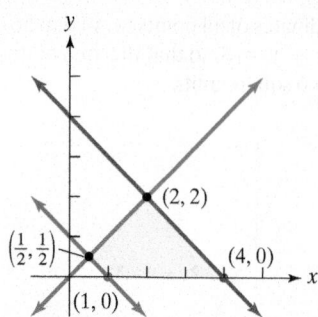

In Exercises 57–59, graph the region determined by the constraints. Then find the maximum value of the given objective function, subject to the constraints.

57. Objective Function
Constraints

$z = 2x + 3y$
$\begin{cases} x \ge 0, y \ge 0 \\ x + y \le 8 \\ 3x + 2y \ge 6 \end{cases}$

58. Objective Function
Constraints

$z = x + 4y$
$\begin{cases} 0 \le x \le 5, 0 \le y \le 7 \\ x + y \ge 3 \end{cases}$

59. Objective Function
Constraints

$z = 5x + 6y$
$\begin{cases} x \ge 0, y \ge 0 \\ y \le x \\ 2x + y \le 12 \\ 2x + 3y \ge 6 \end{cases}$

60. A paper manufacturing company converts wood pulp to writing paper and newsprint. The profit on a unit of writing paper is $500 and the profit on a unit of newsprint is $350.

a. Let x represent the number of units of writing paper produced daily. Let y represent the number of units of newsprint produced daily. Write the objective function that models total daily profit.

b. The manufacturer is bound by the following constraints:
 - Equipment in the factory allows for making at most 200 units of paper (writing paper and newsprint) in a day.
 - Regular customers require at least 10 units of writing paper and at least 80 units of newsprint daily.

 Write a system of inequalities that models these constraints.

c. Graph the inequalities in part (b). Use only the first quadrant, because x and y must both be positive. (*Suggestion:* Let each unit along the x- and y-axes represent 20.)

d. Evaluate the objective function at each of the three vertices of the graphed region.

e. Complete the missing portions of this statement: The company will make the greatest profit by producing _____ units of writing paper and _____ units of newsprint each day. The maximum daily profit is $_____.

61. A manufacturer of lightweight tents makes two models whose specifications are given in the following table:

	Cutting Time per Tent	Assembly Time per Tent
Model A	0.9 hour	0.8 hour
Model B	1.8 hours	1.2 hours

On a monthly basis, the manufacturer has no more than 864 hours of labor available in the cutting department and at most 672 hours in the assembly division. The profits come to $25 per tent for model A and $40 per tent for model B. How many of each should be manufactured monthly to maximize the profit?

CHAPTER 5 TEST

In Exercises 1–5, solve the system.

1. $\begin{cases} x = y + 4 \\ 3x + 7y = -18 \end{cases}$

2. $\begin{cases} 2x + 5y = -2 \\ 3x - 4y = 20 \end{cases}$

3. $\begin{cases} x + y + z = 6 \\ 3x + 4y - 7z = 1 \\ 2x - y + 3z = 5 \end{cases}$

4. $\begin{cases} x^2 + y^2 = 25 \\ x + y = 1 \end{cases}$

5. $\begin{cases} 2x^2 - 5y^2 = -2 \\ 3x^2 + 2y^2 = 35 \end{cases}$

6. Find the partial fraction decomposition for $\dfrac{x}{(x + 1)(x^2 + 9)}$.

In Exercises 7–10, graph the solution set of each inequality or system of inequalities.

7. $x - 2y < 8$

8. $\begin{cases} x \geq 0, y \geq 0 \\ 3x + y \leq 9 \\ 2x + 3y \geq 6 \end{cases}$

9. $\begin{cases} x^2 + y^2 > 1 \\ x^2 + y^2 < 4 \end{cases}$

10. $\begin{cases} y \leq 1 - x^2 \\ x^2 + y^2 \leq 9 \end{cases}$

11. Find the maximum value of the objective function $z = 3x + 5y$ subject to the following constraints: $x \geq 0, y \geq 0, x + y \leq 6, x \geq 2$.

12. Health experts agree that cholesterol intake should be limited to 300 mg or less each day. Three ounces of shrimp and 2 ounces of scallops contain 156 mg of cholesterol. Five ounces of shrimp and 3 ounces of scallops contain 45 mg of cholesterol less than the suggested maximum daily intake. Determine the cholesterol content in an ounce of each item.

13. A company is planning to produce and sell a new line of computers. The fixed cost will be \$360,000 and it will cost \$850 to produce each computer. Each computer will be sold for \$1150.

 a. Write the cost function, C, of producing x computers.

 b. Write the revenue function, R, from the sale of x computers.

 c. Determine the break-even point. Describe what this means.

14. Find the quadratic function whose graph passes through the points $(-1, -2), (2, 1),$ and $(-2, 1)$.

15. The rectangular plot of land shown in the figure is to be fenced along three sides using 39 feet of fencing. No fencing is to be placed along the river's edge. The area of the plot is 180 square feet. What are its dimensions?

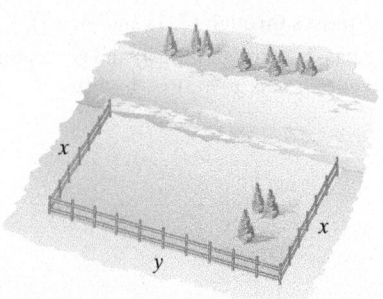

16. A manufacturer makes two types of jet skis, regular and deluxe. The profit on a regular jet ski is \$200 and the profit on the deluxe model is \$250. To meet customer demand, the company must manufacture at least 50 regular jet skis per week and at least 75 deluxe models. To maintain high quality, the total number of both models of jet skis manufactured by the company should not exceed 150 per week. How many jet skis of each type should be manufactured per week to obtain maximum profit? What is the maximum weekly profit?

CUMULATIVE REVIEW EXERCISES (CHAPTERS 1–5)

The figure shows the graph of $y = f(x)$ and its two vertical asymptotes. Use the graph to solve Exercises 1–10.

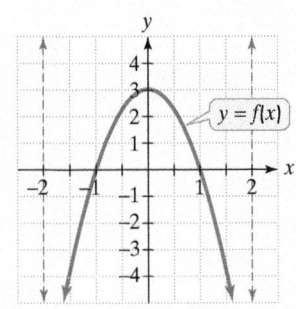

1. Find the domain and the range of f.

2. Find the zeros.

3. What is the relative maximum and where does it occur?

4. Find the interval(s) on which f is decreasing.

5. Is $f(-0.7)$ positive or negative?

6. Find $(f \circ f)(-1)$.

7. Use arrow notation to complete this statement:

$$f(x) \to -\infty \text{ as } _____ \text{ or as } _____.$$

8. Does f appear to be even, odd, or neither?

9. Graph $g(x) = f(x + 2) - 1$.

10. Graph $h(x) = \frac{1}{2}f\left(\frac{1}{2}x\right)$.

In Exercises 11–21, solve each equation, inequality, or system of equations.

11. $\sqrt{x^2 - 3x} = 2x - 6$

12. $4x^2 = 8x - 7$

13. $\left|\frac{x}{3} + 2\right| < 4$

14. $\frac{x + 5}{x - 1} > 2$

15. $2x^3 + x^2 - 13x + 6 = 0$

16. $6x - 3(5x + 2) = 4(1 - x)$

17. $\log(x + 3) + \log x = 1$

18. $3^{x+2} = 11$

19. $x^{\frac{1}{2}} - 2x^{\frac{1}{4}} - 15 = 0$

20. $\begin{cases} 3x - y = -2 \\ 2x^2 - y = 0 \end{cases}$

21. $\begin{cases} x + 2y + 3z = -2 \\ 3x + 3y + 10z = -2 \\ 2y - 5z = 6 \end{cases}$

In Exercises 22–28, graph each equation, function, or inequality in a rectangular coordinate system. If two functions are indicated, graph both in the same system.

22. $f(x) = (x + 2)^2 - 4$

23. $2x - 3y \leq 6$

24. $y = 3^{x-2}$

25. $f(x) = \frac{x^2 - x - 6}{x + 1}$

26. $f(x) = 2x - 4$ and f^{-1}

27. $(x - 2)^2 + (y - 4)^2 > 9$

28. $f(x) = |x|$ and $g(x) = -|x - 2|$

In Exercises 29–30, let $f(x) = 2x^2 - x - 1$ *and* $g(x) = 1 - x$.

29. Find $(f \circ g)(x)$ and $(g \circ f)(x)$.

30. Find $\dfrac{f(x + h) - f(x)}{h}$ and simplify.

In Exercises 31–32, write the linear function in slope-intercept form satisfying the given conditions.

31. Graph of f passes through $(2, 4)$ and $(4, -2)$.

32. Graph of g passes through $(-1, 0)$ and is perpendicular to the line whose equation is $x + 3y - 6 = 0$.

33. You invested $4000 in two stocks paying 12% and 14% annual interest. At the end of the year, the total interest from these investments was $508. How much was invested at each rate?

34. The length of a rectangle is 1 meter more than twice the width. If the rectangle's area is 36 square meters, find its dimensions.

35. What interest rate is required for an investment of $6000 subject to continuous compounding to grow to $18,000 in 10 years?

MATRICES AND DETERMINANTS

You are being drawn deeper into cyberspace, spending more time online each week. With constantly improving high-resolution images, cyberspace is reshaping your life by nourishing shared enthusiasms. The people who built your computer talk of bandwidth that will give you the visual experience, in high-definition 3-D format, of being in the same room with a person who is actually in another city. Rectangular arrays of numbers, called matrices, play a central role in representing computer images and in the forthcoming technology of tele-immersion.

HERE'S WHERE YOU'LL FIND THESE APPLICATIONS:

The use of rectangular arrays of numbers in the digital representation of images and the manipulation of images on a computer screen is discussed in Examples 8 and 9 in Section 6.3.

SECTION 6.1

Matrix Solutions to Linear Systems

Objectives

1. Write the augmented matrix for a linear system.
2. Perform matrix row operations.
3. Use matrices and Gaussian elimination to solve systems.
4. Use matrices and Gauss-Jordan elimination to solve systems.

In Chapter 5, we used systems of inequalities to establish healthy weight ranges for men and women of various ages and heights. Now it's time to step on the scale. The data below show the average weight of American adults, by age.

Average Weight by Age

	Ages 20–29	Ages 30–39	Ages 40–49	Ages 50–59	Ages 60–69	Ages 70–79	Ages 80+
Men	188	194	202	199	198	187	168
Women	156	165	171	172	171	156	142

Source: National Center for Health Statistics

The 14 numbers inside the brackets are arranged in two rows and seven columns. This rectangular array of 14 numbers, arranged in rows and columns and placed in brackets, is an example of a **matrix** (plural: **matrices**). The numbers inside the brackets are called **elements** of the matrix. Matrices are used to display information and to solve systems of linear equations. Because systems involving two equations in two variables can easily be solved by substitution or addition, we will focus on matrix solutions to systems of linear equations in three or more variables.

1. Write the augmented matrix for a linear system.

Augmented Matrices

A matrix gives us a shortened way of writing a system of equations. The first step in solving a system of linear equations using matrices is to write the *augmented matrix*. An **augmented matrix** has a vertical bar separating the columns of the matrix into two groups. The coefficients of each variable are placed to the left of the vertical line and the constants are placed to the right. If any variable is missing, its coefficient is 0. Here are two examples:

GREAT QUESTION!

Do linear systems have to be expressed in a special form when writing augmented matrices?

Yes. Variable terms must be on one side of each equation and constants must be on the other side. Furthermore, the variables must be in the same order in each equation.

System of Linear Equations	Augmented Matrix

$$\begin{cases} 3x + y + 2z = 31 \\ x + y + 2z = 19 \\ x + 3y + 2z = 25 \end{cases} \qquad \begin{bmatrix} 3 & 1 & 2 & | & 31 \\ 1 & 1 & 2 & | & 19 \\ 1 & 3 & 2 & | & 25 \end{bmatrix}$$

$$\begin{cases} x + 2y - 5z = -19 \\ y + 3z = 9 \\ z = 4 \end{cases} \qquad \begin{bmatrix} 1 & 2 & -5 & | & -19 \\ 0 & 1 & 3 & | & 9 \\ 0 & 0 & 1 & | & 4 \end{bmatrix}.$$

Our goal in solving a system of linear equations in three variables using matrices is to produce a matrix with 1s down the diagonal from upper left to lower right on the left side of the vertical bar, called the **main diagonal**, and 0s below the 1s. In general, the matrix will be of the form

$$\begin{bmatrix} 1 & a & b & | & c \\ 0 & 1 & d & | & e \\ 0 & 0 & 1 & | & f \end{bmatrix},$$

where *a* through *f* represent real numbers. The third row of this matrix gives us the value of one variable. The other variables can then be found by back-substitution.

② Perform matrix row operations.

Matrix Row Operations

A matrix with 1s down the main diagonal and 0s below the 1s is said to be in **row-echelon form**. How do we produce a matrix in this form? We use **row operations** on the augmented matrix. These row operations are just like what you did when solving a linear system by the addition method. The difference is that we no longer write the variables, usually represented by x, y, and z.

Matrix Row Operations

The following row operations produce matrices that represent systems with the same solution set:

1. Two rows of a matrix may be interchanged. This is the same as interchanging two equations in a linear system.

2. The elements in any row may be multiplied by a nonzero number. This is the same as multiplying both sides of an equation by a nonzero number.

3. The elements in any row may be multiplied by a nonzero number, and these products may be added to the corresponding elements in any other row. This is the same as multiplying both sides of an equation by a nonzero number and then adding equations to eliminate a variable.

Two matrices are **row equivalent** if one can be obtained from the other by a sequence of row operations.

Each matrix row operation in the preceding box can be expressed symbolically as follows:

1. Interchange the elements in the ith and jth rows: $R_i \leftrightarrow R_j$.

2. Multiply each element in the ith row by k: kR_i.

3. Add k times the elements in row i to the corresponding elements in row j: $kR_i + R_j$.

GREAT QUESTION!

Can you clarify what I'm supposed to do to find $kR_i + R_j$? Which row do I work with and which row do I replace?

When performing the row operation

$$kR_i + R_j,$$

you use row i to find the products. However, **elements in row i do not change. It is the elements in row j that change:** Add k times the elements in row i to the corresponding elements in row j. Replace elements in row j by these sums.

EXAMPLE 1 Performing Matrix Row Operations

Use the matrix

$$\begin{bmatrix} 3 & 18 & -12 & 21 \\ 1 & 2 & -3 & 5 \\ -2 & -3 & 4 & -6 \end{bmatrix}$$

and perform each indicated row operation:

a. $R_1 \leftrightarrow R_2$ **b.** $\frac{1}{3}R_1$ **c.** $2R_2 + R_3$.

SOLUTION

a. The notation $R_1 \leftrightarrow R_2$ means to interchange the elements in row 1 and row 2. This results in the row-equivalent matrix

$$\begin{bmatrix} 1 & 2 & -3 & 5 \\ 3 & 18 & -12 & 21 \\ -2 & -3 & 4 & -6 \end{bmatrix}.$$

This was row 2; now it's row 1.

This was row 1; now it's row 2.

b. The notation $\frac{1}{3}R_1$ means to multiply each element in row 1 by $\frac{1}{3}$. This results in the row-equivalent matrix

$$\begin{bmatrix} \frac{1}{3}(3) & \frac{1}{3}(18) & \frac{1}{3}(-12) & \frac{1}{3}(21) \\ 1 & 2 & -3 & 5 \\ -2 & -3 & 4 & -6 \end{bmatrix} = \begin{bmatrix} 1 & 6 & -4 & 7 \\ 1 & 2 & -3 & 5 \\ -2 & -3 & 4 & -6 \end{bmatrix}.$$

$$\begin{bmatrix} 3 & 18 & -12 & | & 21 \\ 1 & 2 & -3 & | & 5 \\ -2 & -3 & 4 & | & -6 \end{bmatrix}$$

The given matrix (repeated)

c. The notation $2R_2 + R_3$ means to add 2 times the elements in row 2 to the corresponding elements in row 3. Replace the elements in row 3 by these sums. First, we find 2 times the elements in row 2, namely, 1, 2, −3, and 5:

$$2(1) \text{ or } 2, \quad 2(2) \text{ or } 4, \quad 2(-3) \text{ or } -6, \quad 2(5) \text{ or } 10.$$

Now we add these products to the corresponding elements in row 3. Although we use row 2 to find the products, row 2 does not change. It is the elements in row 3 that change, resulting in the row-equivalent matrix

Replace row 3 by the sum of itself and 2 times row 2.

$$\begin{bmatrix} 3 & 18 & -12 & | & 21 \\ 1 & 2 & -3 & | & 5 \\ -2+2=0 & -3+4=1 & 4+(-6)=-2 & | & -6+10=4 \end{bmatrix} = \begin{bmatrix} 3 & 18 & -12 & | & 21 \\ 1 & 2 & -3 & | & 5 \\ 0 & 1 & -2 & | & 4 \end{bmatrix}.$$

• • •

⊘ Check Point 1 Use the matrix

$$\begin{bmatrix} 4 & 12 & -20 & | & 8 \\ 1 & 6 & -3 & | & 7 \\ -3 & -2 & 1 & | & -9 \end{bmatrix}$$

and perform each indicated row operation:

a. $R_1 \leftrightarrow R_2$ **b.** $\frac{1}{4}R_1$ **c.** $3R_2 + R_3$.

③ Use matrices and Gaussian elimination to solve systems.

Solving Linear Systems Using Gaussian Elimination

The process that we use to solve linear systems using matrix row operations is called **Gaussian elimination**, after the German mathematician Carl Friedrich Gauss (1777–1855). Here are the steps used in Gaussian elimination:

Solving Linear Systems of Three Equations with Three Variables Using Gaussian Elimination

1. Write the augmented matrix for the system.
2. Use matrix row operations to simplify the matrix to a row-equivalent matrix in row-echelon form, with 1s down the main diagonal from upper left to lower right, and 0s below the 1s in the first and second columns.

$$\begin{bmatrix} 1 & * & * & | & * \\ * & * & * & | & * \\ * & * & * & | & * \end{bmatrix} \rightarrow \begin{bmatrix} 1 & * & * & | & * \\ 0 & * & * & | & * \\ 0 & * & * & | & * \end{bmatrix} \rightarrow \begin{bmatrix} 1 & * & * & | & * \\ 0 & 1 & * & | & * \\ 0 & * & * & | & * \end{bmatrix} \rightarrow \begin{bmatrix} 1 & * & * & | & * \\ 0 & 1 & * & | & * \\ 0 & 0 & * & | & * \end{bmatrix} \rightarrow \begin{bmatrix} 1 & * & * & | & * \\ 0 & 1 & * & | & * \\ 0 & 0 & 1 & | & * \end{bmatrix}$$

Get 1 in the upper left-hand corner. Use the 1 in the first column to get 0s below it. Get 1 in the second row, second column position. Use the 1 in the second column to get 0 below it. Get 1 in the third row, third column position.

3. Write the system of linear equations corresponding to the matrix in step 2 and use back-substitution to find the system's solution.

EXAMPLE 2 Gaussian Elimination with Back-Substitution

Use matrices to solve the system:

$$\begin{cases} 3x + y + 2z = 31 \\ x + y + 2z = 19 \\ x + 3y + 2z = 25. \end{cases}$$

SOLUTION

Step 1 Write the augmented matrix for the system.

Linear System	Augmented Matrix

$$\begin{cases} 3x + y + 2z = 31 \\ x + y + 2z = 19 \\ x + 3y + 2z = 25 \end{cases} \qquad \begin{bmatrix} 3 & 1 & 2 & | & 31 \\ 1 & 1 & 2 & | & 19 \\ 1 & 3 & 2 & | & 25 \end{bmatrix}$$

Step 2 Use matrix row operations to simplify the matrix to row-echelon form, with 1s down the main diagonal from upper left to lower right, and 0s below the 1s in the first and second columns. Our first step in achieving this goal is to get 1 in the top position of the first column.

$$\begin{array}{c} \text{We want 1 in} \\ \text{this position.} \end{array} \quad \begin{bmatrix} 3 & 1 & 2 & | & 31 \\ 1 & 1 & 2 & | & 19 \\ 1 & 3 & 2 & | & 25 \end{bmatrix}$$

To get 1 in this position, we interchange row 1 and row 2: $R_1 \leftrightarrow R_2$. (We could also interchange row 1 and row 3 to attain our goal.)

$$\begin{bmatrix} 1 & 1 & 2 & | & 19 \\ 3 & 1 & 2 & | & 31 \\ 1 & 3 & 2 & | & 25 \end{bmatrix} \quad \begin{array}{l} \text{This was row 2; now it's row 1.} \\ \\ \text{This was row 1; now it's row 2.} \end{array}$$

Now we want to get 0s below the 1 in the first column.

$$\begin{array}{c} \text{We want 0 in} \\ \text{these positions.} \end{array} \quad \begin{bmatrix} 1 & 1 & 2 & | & 19 \\ 3 & 1 & 2 & | & 31 \\ 1 & 3 & 2 & | & 25 \end{bmatrix}$$

To get a 0 where there is now a 3, multiply the top row of numbers by -3 and add these products to the second row of numbers: $-3R_1 + R_2$. To get a 0 where there is now a 1, multiply the top row of numbers by -1 and add these products to the third row of numbers: $-1R_1 + R_3$. Although we are using row 1 to find the products, the numbers in row 1 do not change.

Replace row 2 by
$-3R_1 + R_2$.

Replace row 3 by
$-1R_1 + R_3$.

$$\begin{bmatrix} 1 & 1 & 2 & | & 19 \\ -3(1) + 3 & -3(1) + 1 & -3(2) + 2 & | & -3(19) + 31 \\ -1(1) + 1 & -1(1) + 3 & -1(2) + 2 & | & -1(19) + 25 \end{bmatrix} = \begin{bmatrix} 1 & 1 & 2 & | & 19 \\ 0 & -2 & -4 & | & -26 \\ 0 & 2 & 0 & | & 6 \end{bmatrix}$$

We want 1 in this position.

We move on to the second column. To get 1 in the desired position, we multiply -2 by its reciprocal, $-\frac{1}{2}$. Therefore, we multiply all the numbers in the second row by $-\frac{1}{2}$: $-\frac{1}{2}R_2$.

$$-\frac{1}{2}R_2 \quad \begin{bmatrix} 1 & 1 & 2 & | & 19 \\ -\frac{1}{2}(0) & -\frac{1}{2}(-2) & -\frac{1}{2}(-4) & | & -\frac{1}{2}(-26) \\ 0 & 2 & 0 & | & 6 \end{bmatrix} = \begin{bmatrix} 1 & 1 & 2 & | & 19 \\ 0 & 1 & 2 & | & 13 \\ 0 & 2 & 0 & | & 6 \end{bmatrix}.$$

We want 0 in this position.

$$\begin{bmatrix} 1 & 1 & 2 & | & 19 \\ 0 & 1 & 2 & | & 13 \\ 0 & 2 & 0 & | & 6 \end{bmatrix}$$

We want 0 in this position.

The matrix from the bottom
of the previous page (repeated)

So far, our matrix row operations have resulted in the matrix that we repeated in the margin. We are not yet done with the second column. The voice balloon shows that we want to get a 0 where there is now a 2. If we multiply the second row of numbers by -2 and add these products to the third row of numbers, we will get 0 in this position: $-2R_2 + R_3$. Although we are using the numbers in row 2 to find the products, the numbers in row 2 do not change.

Replace row 3 by
$-2R_2 + R_3$.

$$\begin{bmatrix} 1 & 1 & 2 & | & 19 \\ 0 & 1 & 2 & | & 13 \\ -2(0) + 0 & -2(1) + 2 & -2(2) + 0 & | & -2(13) + 6 \end{bmatrix} = \begin{bmatrix} 1 & 1 & 2 & | & 19 \\ 0 & 1 & 2 & | & 13 \\ 0 & 0 & -4 & | & -20 \end{bmatrix}$$

We want 1 in this position.

We move on to the third column. To get 1 in the desired position, we multiply -4 by its reciprocal, $-\frac{1}{4}$. Therefore, we multiply all the numbers in the third row by $-\frac{1}{4}$: $-\frac{1}{4}R_3$.

$-\frac{1}{4}R_3$

$$\begin{bmatrix} 1 & 1 & 2 & | & 19 \\ 0 & 1 & 2 & | & 13 \\ -\frac{1}{4}(0) & -\frac{1}{4}(0) & -\frac{1}{4}(-4) & | & -\frac{1}{4}(-20) \end{bmatrix} = \begin{bmatrix} 1 & 1 & 2 & | & 19 \\ 0 & 1 & 2 & | & 13 \\ 0 & 0 & 1 & | & 5 \end{bmatrix}$$

We now have the desired matrix in row-echelon form, with 1s down the main diagonal and 0s below the 1s in the first and second columns.

Step 3 Write the system of linear equations corresponding to the matrix in step 2 and use back-substitution to find the system's solution. The system represented by the matrix in step 2 is

$$\begin{bmatrix} 1 & 1 & 2 & | & 19 \\ 0 & 1 & 2 & | & 13 \\ 0 & 0 & 1 & | & 5 \end{bmatrix} \rightarrow \begin{cases} 1x + 1y + 2z = 19 \\ 0x + 1y + 2z = 13 \\ 0x + 0y + 1z = 5 \end{cases} \text{ or } \begin{cases} x + y + 2z = 19 & (1) \\ y + 2z = 13. & (2) \\ z = 5 & (3) \end{cases}$$

We immediately see from equation (3) that the value for z is 5. To find y, we back-substitute 5 for z in the second equation.

$$\begin{aligned} y + 2z &= 13 && \text{Equation (2)} \\ y + 2(5) &= 13 && \text{Substitute 5 for z.} \\ y + 10 &= 13 && \text{Multiply.} \\ y &= 3 && \text{Subtract 10 from both sides and solve for y.} \end{aligned}$$

TECHNOLOGY

Most graphing utilities can convert an augmented matrix to row-echelon form, with 1s down the main diagonal and 0s below the 1s. However, row-echelon form is not unique. Your graphing utility might give a row-echelon form different from the one you obtained by hand. However, all row-echelon forms for a given system's augmented matrix produce the same solution to the system. Enter the augmented matrix and name it A. Then use the REF (row-echelon form) command on matrix A.

Finally, back-substitute 3 for y and 5 for z in the first equation.

$$\begin{aligned} x + y + 2z &= 19 && \text{Equation (1)} \\ x + 3 + 2(5) &= 19 && \text{Substitute 3 for y and 5 for z.} \\ x + 13 &= 19 && \text{Multiply and add.} \\ x &= 6 && \text{Subtract 13 from both sides and solve for x.} \end{aligned}$$

With $z = 5$, $y = 3$, and $x = 6$, the solution set of the original system is $\{(6, 3, 5)\}$. Check to see that the solution satisfies all three equations in the given system. • • •

⊘ Check Point **2** Use matrices to solve the system:

$$\begin{cases} 2x + y + 2z = 18 \\ x - y + 2z = 9 \\ x + 2y - z = 6. \end{cases}$$

Modern supercomputers are capable of solving systems with more than 600,000 variables. The augmented matrices for such systems are huge, but the solution using matrices is exactly like what we did in Example 2. Work with the augmented matrix, one column at a time. Get 1s down the main diagonal from upper left to lower right and 0s below the 1s. Let's see how this works for a linear system involving four equations in four variables.

EXAMPLE 3 Gaussian Elimination with Back-Substitution

Use matrices to solve the system:

$$\begin{cases} 2w + x + 3y - z = 6 \\ w - x + 2y - 2z = -1 \\ w - x - y + z = -4 \\ -w + 2x - 2y - z = -7. \end{cases}$$

SOLUTION

Step 1 Write the augmented matrix for the system.

Linear System

$$\begin{cases} 2w + x + 3y - z = 6 \\ w - x + 2y - 2z = -1 \\ w - x - y + z = -4 \\ -w + 2x - 2y - z = -7 \end{cases}$$

Augmented Matrix

$$\left[\begin{array}{cccc|c} 2 & 1 & 3 & -1 & 6 \\ 1 & -1 & 2 & -2 & -1 \\ 1 & -1 & -1 & 1 & -4 \\ -1 & 2 & -2 & -1 & -7 \end{array}\right]$$

Step 2 Use matrix row operations to simplify the matrix to row-echelon form, with 1s down the main diagonal from upper left to lower right, and 0s below the 1s in the first, second, and third columns. Our first step in achieving this goal is to get 1 in the top position of the first column. To do this, we interchange row 1 and row 2: $R_1 \leftrightarrow R_2$.

We want Os in these positions.

$$\left[\begin{array}{cccc|c} 1 & -1 & 2 & -2 & -1 \\ 2 & 1 & 3 & -1 & 6 \\ 1 & -1 & -1 & 1 & -4 \\ -1 & 2 & -2 & -1 & -7 \end{array}\right]$$

This was row 2; now it's row 1.

This was row 1; now it's row 2.

Now we use the 1 at the top of the first column to get 0s below it.

Use the previous matrix and:
Replace row 2 by $-2R_1 + R_2$.
Replace row 3 by $-1R_1 + R_3$.
Replace row 4 by $1R_1 + R_4$.

$$\left[\begin{array}{cccc|c} 1 & -1 & 2 & -2 & -1 \\ 0 & 3 & -1 & 3 & 8 \\ 0 & 0 & -3 & 3 & -3 \\ 0 & 1 & 0 & -3 & -8 \end{array}\right]$$

We want 1 in this position.

We move on to the second column. We can obtain 1 in the desired position by multiplying the numbers in the second row by $\frac{1}{3}$, the reciprocal of 3.

$$\left[\begin{array}{cccc|c} 1 & -1 & 2 & -2 & -1 \\ \frac{1}{3}(0) & \frac{1}{3}(3) & \frac{1}{3}(-1) & \frac{1}{3}(3) & \frac{1}{3}(8) \\ 0 & 0 & -3 & 3 & -3 \\ 0 & 1 & 0 & -3 & -8 \end{array}\right] = \left[\begin{array}{cccc|c} 1 & -1 & 2 & -2 & -1 \\ 0 & 1 & -\frac{1}{3} & 1 & \frac{8}{3} \\ 0 & 0 & -3 & 3 & -3 \\ 0 & 1 & 0 & -3 & -8 \end{array}\right] \quad \frac{1}{3}R_2$$

We want Os in these positions.
The top position already has a 0.

$$\begin{bmatrix} 1 & -1 & 2 & -2 & | & -1 \\ 0 & 1 & -\frac{1}{3} & 1 & | & \frac{8}{3} \\ 0 & 0 & -3 & 3 & | & -3 \\ 0 & 1 & 0 & -3 & | & -8 \end{bmatrix}$$

We want 0s in these positions.
The top position already has a 0.

The matrix from the bottom of the previous page (repeated)

So far, our matrix row operations have resulted in the matrix that we repeated in the margin. Now we use the 1 in the second row, second column position to get 0s below it.

$$\begin{bmatrix} 1 & -1 & 2 & -2 & | & -1 \\ 0 & 1 & -\frac{1}{3} & 1 & | & \frac{8}{3} \\ 0 & 0 & -3 & 3 & | & -3 \\ 0 & 0 & \frac{1}{3} & -4 & | & -\frac{32}{3} \end{bmatrix}$$

We want 1 in this position.

Replace row 4 in the previous matrix by $-1R_2 + R_4$.

We move on to the third column. We can obtain 1 in the desired position by multiplying the numbers in the third row by $-\frac{1}{3}$, the reciprocal of -3.

$$\begin{bmatrix} 1 & -1 & 2 & -2 & | & -1 \\ 0 & 1 & -\frac{1}{3} & 1 & | & \frac{8}{3} \\ -\frac{1}{3}(0) & -\frac{1}{3}(0) & -\frac{1}{3}(-3) & -\frac{1}{3}(3) & | & -\frac{1}{3}(-3) \\ 0 & 0 & \frac{1}{3} & -4 & | & -\frac{32}{3} \end{bmatrix} = \begin{bmatrix} 1 & -1 & 2 & -2 & | & -1 \\ 0 & 1 & -\frac{1}{3} & 1 & | & \frac{8}{3} \\ 0 & 0 & 1 & -1 & | & 1 \\ 0 & 0 & \frac{1}{3} & -4 & | & -\frac{32}{3} \end{bmatrix} \quad -\frac{1}{3}R_3$$

We want 0 in this position.

Now we use the 1 in the third column to get 0 below it.

$$\begin{bmatrix} 1 & -1 & 2 & -2 & | & -1 \\ 0 & 1 & -\frac{1}{3} & 1 & | & \frac{8}{3} \\ 0 & 0 & 1 & -1 & | & 1 \\ 0 & 0 & 0 & -\frac{11}{3} & | & -11 \end{bmatrix}$$

We want 1 in this position.

Replace row 4 in the previous matrix by $-\frac{1}{3}R_3 + R_4$.

We move on to the fourth column. Because we want 1s down the diagonal from upper left to lower right, we want 1 where there is now $-\frac{11}{3}$. We can obtain 1 in this position by multiplying the numbers in the fourth row by $-\frac{3}{11}$.

$$\begin{bmatrix} 1 & -1 & 2 & -2 & | & -1 \\ 0 & 1 & -\frac{1}{3} & 1 & | & \frac{8}{3} \\ 0 & 0 & 1 & -1 & | & 1 \\ -\frac{3}{11}(0) & -\frac{3}{11}(0) & -\frac{3}{11}(0) & -\frac{3}{11}(-\frac{11}{3}) & | & -\frac{3}{11}(-11) \end{bmatrix}$$

$$= \begin{bmatrix} 1 & -1 & 2 & -2 & | & -1 \\ 0 & 1 & -\frac{1}{3} & 1 & | & \frac{8}{3} \\ 0 & 0 & 1 & -1 & | & 1 \\ 0 & 0 & 0 & 1 & | & 3 \end{bmatrix} \quad -\frac{3}{11}R_4$$

We now have the desired matrix in row-echelon form, with 1s down the main diagonal and 0s below the 1s. An equivalent row-echelon matrix can be obtained using a graphing utility and the ⎢REF⎢ command on the augmented matrix.

Step 3 Write the system of linear equations corresponding to the matrix in step 2 and use back-substitution to find the system's solution. The system represented by the matrix in step 2 is

$$\begin{bmatrix} 1 & -1 & 2 & -2 & | & -1 \\ 0 & 1 & -\frac{1}{3} & 1 & | & \frac{8}{3} \\ 0 & 0 & 1 & -1 & | & 1 \\ 0 & 0 & 0 & 1 & | & 3 \end{bmatrix} \rightarrow \begin{cases} 1w - 1x + 2y - 2z = -1 \\ 0w + 1x - \frac{1}{3}y + 1z = \frac{8}{3} \\ 0w + 0x + 1y - 1z = 1 \\ 0w + 0x + 0y + 1z = 3 \end{cases} \text{ or } \begin{cases} w - x + 2y - 2z = -1 \\ x - \frac{1}{3}y + z = \frac{8}{3} \\ y - z = 1 \\ z = 3. \end{cases}$$

We immediately see that the value for z is 3. We can now use back-substitution to find the values for y, x, and w.

These are the
four equations
from the last column.

$$z = 3 \quad \Big| \quad y - z = 1 \quad \Big| \quad x - \frac{1}{3}y + z = \frac{8}{3} \quad \Big| \quad w - x + 2y - 2z = -1$$

$$y - 3 = 1 \quad \Big| \quad x - \frac{1}{3}(4) + 3 = \frac{8}{3} \quad \Big| \quad w - 1 + 2(4) - 2(3) = -1$$

$$y = 4 \quad \Big| \quad x + \frac{5}{3} = \frac{8}{3} \quad \Big| \quad w - 1 + 8 - 6 = -1$$

$$x = 1 \quad \Big| \quad w + 1 = -1$$

$$w = -2$$

Let's agree to write the solution for the system in the alphabetical order of the variables from left to right, namely (w, x, y, z). Thus, the solution set is $\{(-2, 1, 4, 3)\}$. We can verify the solution by substituting the value for each variable into the original system of equations and obtaining four true statements. •••

⬦ Check Point **3** Use matrices to solve the system:

$$\begin{cases} w - 3x - 2y + z = -3 \\ 2w - 7x - y + 2z = 1 \\ 3w - 7x - 3y + 3z = -5 \\ 5w + x + 4y - 2z = 18. \end{cases}$$

④ Use matrices and Gauss-Jordan elimination to solve systems.

Solving Linear Systems Using Gauss-Jordan Elimination

Using Gaussian elimination, we obtain a matrix in row-echelon form, with 1s down the main diagonal and 0s below the 1s. A second method, called **Gauss-Jordan elimination**, after Carl Friedrich Gauss and Wilhelm Jordan (1842–1899), continues the process until a matrix with 1s down the main diagonal and 0s in every position *above and below* each 1 is found. Such a matrix is said to be in **reduced row-echelon form**. For a system of three linear equations in three variables, x, y, and z, we must get the augmented matrix into the form

$$\begin{bmatrix} 1 & 0 & 0 & | & a \\ 0 & 1 & 0 & | & b \\ 0 & 0 & 1 & | & c \end{bmatrix}.$$

Based on this matrix, we conclude that $x = a$, $y = b$, and $z = c$.

Solving Linear Systems Using Gauss-Jordan Elimination

1. Write the augmented matrix for the system.

2. Use matrix row operations to simplify the matrix to a row-equivalent matrix in reduced row-echelon form, with 1s down the main diagonal from upper left to lower right, and 0s above and below the 1s.

 a. Get 1 in the upper left-hand corner.

 b. Use the 1 in the first column to get 0s below it.

 c. Get 1 in the second row, second column.

 d. Use the 1 in the second column to make the remaining entries in the second column 0.

 e. Get 1 in the third row, third column.

 f. Use the 1 in the third column to make the remaining entries in the third column 0.

 g. Continue this procedure as far as possible.

3. Use the reduced row-echelon form of the matrix in step 2 to write the system's solution set. (Back-substitution is not necessary.)

TECHNOLOGY

Most graphing utilities can convert a matrix to reduced row-echelon form. Enter the system's augmented matrix and name it A. Then use the ⎡RREF⎤ (reduced row-echelon form) command on matrix A.

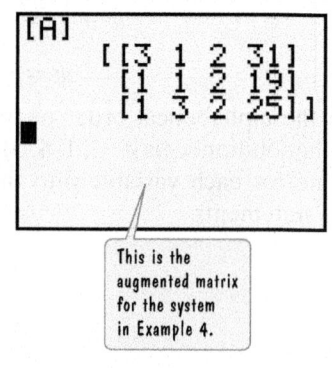

This is the augmented matrix for the system in Example 4.

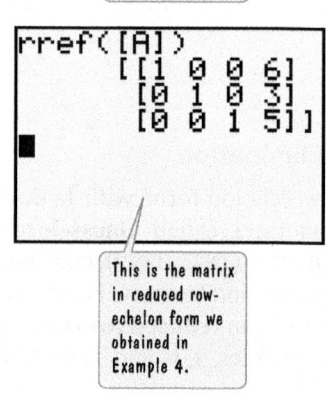

This is the matrix in reduced row-echelon form we obtained in Example 4.

EXAMPLE 4 Using Gauss-Jordan Elimination

Use Gauss-Jordan elimination to solve the system:

$$\begin{cases} 3x + y + 2z = 31 \\ x + y + 2z = 19 \\ x + 3y + 2z = 25. \end{cases}$$

SOLUTION

In Example 2, we used Gaussian elimination to obtain the following matrix:

We want
0s in these
positions.
$$\begin{bmatrix} 1 & 1 & 2 & | & 19 \\ 0 & 1 & 2 & | & 13 \\ 0 & 0 & 1 & | & 5 \end{bmatrix}.$$

To use Gauss-Jordan elimination, we need 0s both above and below the 1s in the main diagonal. We use the 1 in the second row, second column to get a 0 above it.

Replace row 1
in the previous
matrix by
$-1R_2 + R_1$
$$\begin{bmatrix} 1 & 0 & 0 & | & 6 \\ 0 & 1 & 2 & | & 13 \\ 0 & 0 & 1 & | & 5 \end{bmatrix}$$
We want
0s in these
positions.

We use the 1 in the third column to get 0s above it.

$$\begin{bmatrix} 1 & 0 & 0 & | & 6 \\ 0 & 1 & 0 & | & 3 \\ 0 & 0 & 1 & | & 5 \end{bmatrix}$$
Replace row 2 in the previous
matrix by $-2R_3 + R_2$.

This last matrix corresponds to

$$x = 6, \quad y = 3, \quad z = 5.$$

As we found in Example 2, the solution set is $\{(6, 3, 5)\}$. • • •

✓ Check Point **4** Solve the system in Check Point 2 using Gauss-Jordan elimination. Begin by working with the matrix that you obtained in Check Point 2.

CONCEPT AND VOCABULARY CHECK

Fill in each blank so that the resulting statement is true.

1. A rectangular array of numbers, arranged in rows and columns and placed in brackets, is called a/an _____. The numbers inside the brackets are called _____.

2. Consider the matrix

$$\begin{bmatrix} 1 & 1 & -1 & | & -2 \\ -3 & -4 & 2 & | & 4 \\ 2 & 1 & 1 & | & 6 \end{bmatrix}.$$

We can obtain 0 in the position shaded by a rectangle if we multiply the top row of numbers by _____ and add these products to the _____ row of numbers. We can obtain 0 in the position shaded by an oval if we multiply the top row of numbers by _____ and add these products to the _____ row of numbers.

3. The augmented matrix for the system

$$\begin{cases} 2x + y + 4z = -4 \\ 3x \quad\quad + z = 1 \\ 4x + 3y + z = 8 \end{cases}$$

is

4. Using Gauss-Jordan elimination to solve the system

$$\begin{cases} x - y + z = -4 \\ 5x + y - 2z = 12, \\ 2x - 3y + 4z = -15 \end{cases}$$

we obtain the matrix

$$\begin{bmatrix} 1 & 0 & 0 & | & 1 \\ 0 & 1 & 0 & | & 3 \\ 0 & 0 & 1 & | & -2 \end{bmatrix}.$$

The system's solution set is _____.

5. True or false: Back-substitution is required to solve linear systems using Gaussian elimination. _____

6. True or false: Back-substitution is required to solve linear systems using Gauss-Jordan elimination. _____

EXERCISE SET 6.1

Practice Exercises

In Exercises 1–8, write the augmented matrix for each system of linear equations.

1. $\begin{cases} 2x + y + 2z = 2 \\ 3x - 5y - z = 4 \\ x - 2y - 3z = -6 \end{cases}$

2. $\begin{cases} 3x - 2y + 5z = 31 \\ x + 3y - 3z = -12 \\ -2x - 5y + 3z = 11 \end{cases}$

3. $\begin{cases} x - y + z = 8 \\ y - 12z = -15 \\ z = 1 \end{cases}$

4. $\begin{cases} x - 2y + 3z = 9 \\ y + 3z = 5 \\ z = 2 \end{cases}$

5. $\begin{cases} 5x - 2y - 3z = 0 \\ x + y = 5 \\ 2x - 3z = 4 \end{cases}$

6. $\begin{cases} x - 2y + z = 10 \\ 3x + y = 5 \\ 7x + 2z = 2 \end{cases}$

7. $\begin{cases} 2w + 5x - 3y + z = 2 \\ 3x + y = 4 \\ w - x + 5y = 9 \\ 5w - 5x - 2y = 1 \end{cases}$

8. $\begin{cases} 4w + 7x - 8y + z = 3 \\ 5x + y = 5 \\ w - x - y = 17 \\ 2w - 2x + 11y = 4 \end{cases}$

In Exercises 9–12, write the system of linear equations represented by the augmented matrix. Use x, y, and z, or, if necessary, w, x, y, and z, for the variables.

9. $\begin{bmatrix} 5 & 0 & 3 & | & -11 \\ 0 & 1 & -4 & | & 12 \\ 7 & 2 & 0 & | & 3 \end{bmatrix}$

10. $\begin{bmatrix} 7 & 0 & 4 & | & -13 \\ 0 & 1 & -5 & | & 11 \\ 2 & 7 & 0 & | & 6 \end{bmatrix}$

11. $\begin{bmatrix} 1 & 1 & 4 & 1 & | & 3 \\ -1 & 1 & -1 & 0 & | & 7 \\ 2 & 0 & 0 & 5 & | & 11 \\ 0 & 0 & 12 & 4 & | & 5 \end{bmatrix}$

12. $\begin{bmatrix} 4 & 1 & 5 & 1 & | & 6 \\ 1 & -1 & 0 & -1 & | & 8 \\ 3 & 0 & 0 & 7 & | & 4 \\ 0 & 0 & 11 & 5 & | & 3 \end{bmatrix}$

In Exercises 13–18, perform each matrix row operation and write the new matrix.

13. $\begin{bmatrix} 2 & -6 & 4 & | & 10 \\ 1 & 5 & -5 & | & 0 \\ 3 & 0 & 4 & | & 7 \end{bmatrix}$ $\frac{1}{2}R_1$

14. $\begin{bmatrix} 3 & -12 & 6 & | & 9 \\ 1 & -4 & 4 & | & 0 \\ 2 & 0 & 7 & | & 4 \end{bmatrix}$ $\frac{1}{3}R_1$

15. $\begin{bmatrix} 1 & -3 & 2 & | & 0 \\ 3 & 1 & -1 & | & 7 \\ 2 & -2 & 1 & | & 3 \end{bmatrix}$ $-3R_1 + R_2$

16. $\begin{bmatrix} 1 & -1 & 5 & | & -6 \\ 3 & 3 & -1 & | & 10 \\ 1 & 3 & 2 & | & 5 \end{bmatrix}$ $-3R_1 + R_2$

17. $\begin{bmatrix} 1 & -1 & 1 & 1 & | & 3 \\ 0 & 1 & -2 & -1 & | & 0 \\ 2 & 0 & 3 & 4 & | & 11 \\ 5 & 1 & 2 & 4 & | & 6 \end{bmatrix}$ $\begin{matrix} -2R_1 + R_3 \\ -5R_1 + R_4 \end{matrix}$

18. $\begin{bmatrix} 1 & -5 & 2 & -2 & | & 4 \\ 0 & 1 & -3 & -1 & | & 0 \\ 3 & 0 & 2 & -1 & | & 6 \\ -4 & 1 & 4 & 2 & | & -3 \end{bmatrix}$ $\begin{matrix} -3R_1 + R_3 \\ 4R_1 + R_4 \end{matrix}$

In Exercises 19–20, a few steps in the process of simplifying the given matrix to row-echelon form, with 1s down the diagonal from upper left to lower right, and 0s below the 1s, are shown. Fill in the missing numbers in the steps that are shown.

19. $\begin{bmatrix} 1 & -1 & 1 & | & 8 \\ 2 & 3 & -1 & | & -2 \\ 3 & -2 & -9 & | & 9 \end{bmatrix} \rightarrow \begin{bmatrix} 1 & -1 & 1 & | & 8 \\ 0 & 5 & \square & | & \square \\ 0 & 1 & \square & | & \square \end{bmatrix}$

$\rightarrow \begin{bmatrix} 1 & -1 & 1 & | & 8 \\ 0 & 1 & \square & | & \square \\ 0 & 1 & \square & | & \square \end{bmatrix}$

20. $\begin{bmatrix} 1 & -2 & 3 & | & 4 \\ 2 & 1 & -4 & | & 3 \\ -3 & 4 & -1 & | & -2 \end{bmatrix} \rightarrow \begin{bmatrix} 1 & -2 & 3 & | & 4 \\ 0 & 5 & \square & | & \square \\ 0 & -2 & \square & | & \square \end{bmatrix}$

$\rightarrow \begin{bmatrix} 1 & -2 & 3 & | & 4 \\ 0 & 1 & \square & | & \square \\ 0 & -2 & \square & | & \square \end{bmatrix}$

In Exercises 21–38, solve each system of equations using matrices. Use Gaussian elimination with back-substitution or Gauss-Jordan elimination.

21. $\begin{cases} x + y - z = -2 \\ 2x - y + z = 5 \\ -x + 2y + 2z = 1 \end{cases}$

22. $\begin{cases} x - 2y - z = 2 \\ 2x - y + z = 4 \\ -x + y - 2z = -4 \end{cases}$

23. $\begin{cases} x + 3y = 0 \\ x + y + z = 1 \\ 3x - y - z = 11 \end{cases}$

24. $\begin{cases} 3y - z = -1 \\ x + 5y - z = -4 \\ -3x + 6y + 2z = 11 \end{cases}$

25. $\begin{cases} 2x - y - z = 4 \\ x + y - 5z = -4 \\ x - 2y = 4 \end{cases}$ **26.** $\begin{cases} x - 3z = -2 \\ 2x + 2y + z = 4 \\ 3x + y - 2z = 5 \end{cases}$

27. $\begin{cases} x + y + z = 4 \\ x - y - z = 0 \\ x - y + z = 2 \end{cases}$ **28.** $\begin{cases} 3x + y - z = 0 \\ x + y + 2z = 6 \\ 2x + 2y + 3z = 10 \end{cases}$

29. $\begin{cases} x + 2y = z - 1 \\ x = 4 + y - z \\ x + y - 3z = -2 \end{cases}$ **30.** $\begin{cases} 2x + y = z + 1 \\ 2x = 1 + 3y - z \\ x + y + z = 4 \end{cases}$

31. $\begin{cases} 3a - b - 4c = 3 \\ 2a - b + 2c = -8 \\ a + 2b - 3c = 9 \end{cases}$ **32.** $\begin{cases} 3a + b - c = 0 \\ 2a + 3b - 5c = 1 \\ a - 2b + 3c = -4 \end{cases}$

33. $\begin{cases} 2x + 2y + 7z = -1 \\ 2x + y + 2z = 2 \\ 4x + 6y + z = 15 \end{cases}$ **34.** $\begin{cases} 3x + 2y + 3z = 3 \\ 4x - 5y + 7z = 1 \\ 2x + 3y - 2z = 6 \end{cases}$

35. $\begin{cases} w + x + y + z = 4 \\ 2w + x - 2y - z = 0 \\ w - 2x - y - 2z = -2 \\ 3w + 2x + y + 3z = 4 \end{cases}$

36. $\begin{cases} w + x + y + z = 5 \\ w + 2x - y - 2z = -1 \\ w - 3x - 3y - z = -1 \\ 2w - x + 2y - z = -2 \end{cases}$

37. $\begin{cases} 3w - 4x + y + z = 9 \\ w + x - y - z = 0 \\ 2w + x + 4y - 2z = 3 \\ -w + 2x + y - 3z = 3 \end{cases}$

38. $\begin{cases} 2w + y - 3z = 8 \\ w - x + 4z = -10 \\ 3w + 5x - y - z = 20 \\ w + x - y - z = 6 \end{cases}$

Practice Plus

39. Find the quadratic function $f(x) = ax^2 + bx + c$ for which $f(-2) = -4, f(1) = 2$, and $f(2) = 0$.

40. Find the quadratic function $f(x) = ax^2 + bx + c$ for which $f(-1) = 5, f(1) = 3$, and $f(2) = 5$.

41. Find the cubic function $f(x) = ax^3 + bx^2 + cx + d$ for which $f(-1) = 0, f(1) = 2, f(2) = 3$, and $f(3) = 12$.

42. Find the cubic function $f(x) = ax^3 + bx^2 + cx + d$ for which $f(-1) = 3, f(1) = 1, f(2) = 6$, and $f(3) = 7$.

43. Solve the system:
$$\begin{cases} 2 \ln w + \ln x + 3 \ln y - 2 \ln z = -6 \\ 4 \ln w + 3 \ln x + \ln y - \ln z = -2 \\ \ln w + \ln x + \ln y + \ln z = -5 \\ \ln w + \ln x - \ln y - \ln z = 5. \end{cases}$$

(*Hint:* Let $A = \ln w, B = \ln x, C = \ln y$, and $D = \ln z$. Solve the system for A, B, C, and D. Then use the logarithmic equations to find w, x, y, and z.)

44. Solve the system:
$$\begin{cases} \ln w + \ln x + \ln y + \ln z = -1 \\ -\ln w + 4 \ln x + \ln y - \ln z = 0 \\ \ln w - 2 \ln x + \ln y - 2 \ln z = 11 \\ -\ln w - 2 \ln x + \ln y + 2 \ln z = -3. \end{cases}$$

(*Hint:* Let $A = \ln w, B = \ln x, C = \ln y$, and $D = \ln z$. Solve the system for A, B, C, and D. Then use the logarithmic equations to find w, x, y, and z.)

Application Exercises

45. A ball is thrown straight upward. A position function
$$s(t) = \tfrac{1}{2}at^2 + v_0 t + s_0$$
can be used to describe the ball's height, $s(t)$, in feet, after t seconds.

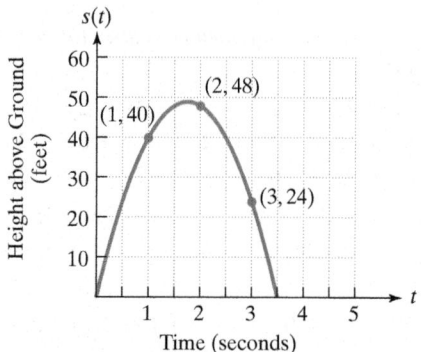

a. Use the points labeled in the graph to find the values of a, v_0, and s_0. Solve the system of linear equations involving a, v_0, and s_0 using matrices.

b. Find and interpret $s(3.5)$. Identify your solution as a point on the graph shown.

c. After how many seconds does the ball reach its maximum height? What is its maximum height?

46. A football is kicked straight upward. A position function
$$s(t) = \tfrac{1}{2}at^2 + v_0 t + s_0$$
can be used to describe the ball's height, $s(t)$, in feet, after t seconds.

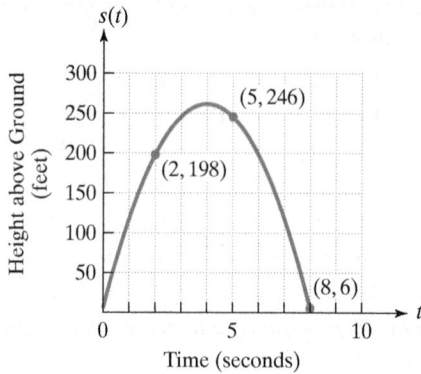

a. Use the points labeled in the graph to find the values of a, v_0, and s_0. Solve the system of linear equations involving a, v_0, and s_0 using matrices.

b. Find and interpret $s(7)$. Identify your solution as a point on the graph shown.

c. After how many seconds does the ball reach its maximum height? What is its maximum height?

Write a system of linear equations in three or four variables to solve Exercises 47–50. Then use matrices to solve the system.

47. Three foods have the following nutritional content per ounce.

	Calories	Protein (in grams)	Vitamin C (in milligrams)
Food A	40	5	30
Food B	200	2	10
Food C	400	4	300

If a meal consisting of the three foods allows exactly 660 calories, 25 grams of protein, and 425 milligrams of vitamin C, how many ounces of each kind of food should be used?

48. A furniture company produces three types of desks: a children's model, an office model, and a deluxe model. Each desk is manufactured in three stages: cutting, construction, and finishing. The time requirements for each model and manufacturing stage are given in the following table.

	Children's Model	Office Model	Deluxe Model
Cutting	2 hr	3 hr	2 hr
Construction	2 hr	1 hr	3 hr
Finishing	1 hr	1 hr	2 hr

Each week the company has available a maximum of 100 hours for cutting, 100 hours for construction, and 65 hours for finishing. If all available time must be used, how many of each type of desk should be produced each week?

49. Imagine the entire global population as a village of precisely 200 people. The bar graph shows some numeric observations based on this scenario.

Earth's Population as a Village of 200 People

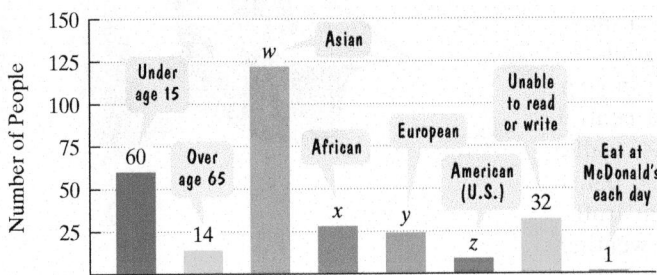

Source: Gary Rimmer, *Number Freaking,* The Disinformation Company Ltd., 2006

Combined, there are 183 Asians, Africans, Europeans, and Americans in the village. The number of Asians exceeds the number of Africans and Europeans by 70. The difference between the number of Europeans and Americans is 15. If the number of Africans is doubled, their population exceeds the number of Europeans and Americans by 23. Determine the number of Asians, Africans, Europeans, and Americans in the global village.

50. The bar graph shows the number of rooms, bathrooms, fireplaces, and elevators in the U.S. White House.

The U.S. White House by the Numbers

Source: The White House

Combined, there are 198 rooms, bathrooms, fireplaces, and elevators. The number of rooms exceeds the number of bathrooms and fireplaces by 69. The difference between the number of fireplaces and elevators is 25. If the number of bathrooms is doubled, it exceeds the number of fireplaces and elevators by 39. Determine the number of rooms, bathrooms, fireplaces, and elevators in the U.S. White House.

Writing in Mathematics

51. What is a matrix?

52. Describe what is meant by the augmented matrix of a system of linear equations.

53. In your own words, describe each of the three matrix row operations. Give an example with each of the operations.

54. Describe how to use row operations and matrices to solve a system of linear equations.

55. What is the difference between Gaussian elimination and Gauss-Jordan elimination?

Technology Exercises

56. Most graphing utilities can perform row operations on matrices. Consult the owner's manual for your graphing utility to learn proper keystrokes for performing these operations. Then duplicate the row operations of any three exercises that you solved from Exercises 13–18.

57. If your graphing utility has a ⎡REF⎤ (row-echelon form) command or a ⎡RREF⎤ (reduced row-echelon form) command, use this feature to verify your work with any five systems that you solved from Exercises 21–38.

58. Solve using a graphing utility's ⎡REF⎤ or ⎡RREF⎤ command:

$$\begin{cases} 2x_1 - 2x_2 + 3x_3 - x_4 = 12 \\ x_1 + 2x_2 - x_3 + 2x_4 - x_5 = -7 \\ x_1 + x_3 + x_4 - 5x_5 = 1 \\ -x_1 + x_2 - x_3 - 2x_4 - 3x_5 = 0 \\ x_1 - x_2 - x_4 + x_5 = 4. \end{cases}$$

Critical Thinking Exercises

Make Sense? *In Exercises 59–62, determine whether each statement makes sense or does not make sense, and explain your reasoning.*

59. Matrix row operations remind me of what I did when solving a linear system by the addition method, although I no longer write the variables.

60. When I use matrices to solve linear systems, the only arithmetic involves multiplication or a combination of multiplication and addition.

61. When I use matrices to solve linear systems, I spend most of my time using row operations to express the system's augmented matrix in row-echelon form.

62. Using row operations on an augmented matrix, I obtain a row in which 0s appear to the left of the vertical bar, but 6 appears on the right, so the system I'm working with has no solution.

In Exercises 63–66, determine whether each statement is true or false. If the statement is false, make the necessary change(s) to produce a true statement.

63. A matrix row operation such as $-\frac{4}{5}R_1 + R_2$ is not permitted because of the negative fraction.

64. The augmented matrix for the system

$$\begin{cases} x - 3y = 5 \\ y - 2z = 7 \\ 2x + z = 4 \end{cases} \text{ is } \begin{bmatrix} 1 & -3 & | & 5 \\ 1 & -2 & | & 7 \\ 2 & 1 & | & 4 \end{bmatrix}.$$

65. In solving a linear system of three equations in three variables, we begin with the augmented matrix and use row operations to obtain a row-equivalent matrix with 0s down the diagonal from left to right and 1s below each 0.

66. The row operation $kR_i + R_j$ indicates that it is the elements in row i that change.

67. The table shows the daily production level and profit for a business.

x (Number of Units Produced Daily)	30	50	100
y (Daily Profit)	$5900	$7500	$4500

Use the quadratic function $y = ax^2 + bx + c$ to determine the number of units that should be produced each day for maximum profit. What is the maximum daily profit?

Preview Exercises

Exercises 68–70 will help you prepare for the material covered in the next section. In each exercise, refer to the following system:

$$\begin{cases} 3x - 4y + 4z = 7 \\ x - y - 2z = 2 \\ 2x - 3y + 6z = 5. \end{cases}$$

68. Show that $(12z + 1, 10z - 1, z)$ satisfies the system for $z = 0$.

69. Show that $(12z + 1, 10z - 1, z)$ satisfies the system for $z = 1$.

70. a. Select a value for z other than 0 or 1 and show that $(12z + 1, 10z - 1, z)$ satisfies the system.
b. Based on your work in Exercises 68–70(a), how does this system differ from those in Exercises 21–34?

SECTION 6.2

Inconsistent and Dependent Systems and Their Applications

Objectives

1 Apply Gaussian elimination to systems without unique solutions.

2 Apply Gaussian elimination to systems with more variables than equations.

3 Solve problems involving systems without unique solutions.

Traffic jams getting you down? Powerful computers, able to solve systems with hundreds of thousands of variables in a single bound, may promise a gridlock-free future. The computer in your car could be linked to a central computer that manages traffic flow by controlling traffic lights, rerouting you away from traffic congestion, issuing weather reports, and selecting the best route to your destination. New technologies could eventually drive your car at a steady 75 miles per hour along automated highways as you comfortably nap. In this section, we look at the role of linear systems without unique solutions in a future free of traffic jams.

1 Apply Gaussian elimination to systems without unique solutions.

Linear systems can have one solution, no solution, or infinitely many solutions. We can use Gaussian elimination on systems with three or more variables to determine how many solutions such systems may have. In the case of systems with no solution or infinitely many solutions, it is impossible to rewrite the augmented matrix in the desired form with 1s down the main diagonal from upper left to lower right, and 0s below the 1s. Let's see what this means by looking at a system that has no solution.

EXAMPLE 1 A System with No Solution

EXAMPLE 1 A System with No Solution

Use Gaussian elimination to solve the system:

$$\begin{cases} x - y - 2z = 2 \\ 2x - 3y + 6z = 5 \\ 3x - 4y + 4z = 12. \end{cases}$$

SOLUTION

Step 1 Write the augmented matrix for the system.

<table>
<tr><td align="center">**Linear System**</td><td align="center">**Augmented Matrix**</td></tr>
<tr><td>$\begin{cases} x - y - 2z = 2 \\ 2x - 3y + 6z = 5 \\ 3x - 4y + 4z = 12 \end{cases}$</td><td>$\begin{bmatrix} 1 & -1 & -2 & 2 \\ 2 & -3 & 6 & 5 \\ 3 & -4 & 4 & 12 \end{bmatrix}$</td></tr>
</table>

Step 2 Attempt to simplify the matrix to row-echelon form, with 1s down the main diagonal and 0s below the 1s. Notice that the augmented matrix already has a 1 in the top position of the first column. Now we want 0s below the 1. To get the first 0, multiply row 1 by −2 and add these products to row 2. To get the second 0, multiply row 1 by −3 and add these products to row 3. Performing these operations, we obtain the following matrix:

We want 1 in this position. $\begin{bmatrix} 1 & -1 & -2 & 2 \\ 0 & -1 & 10 & 1 \\ 0 & -1 & 10 & 6 \end{bmatrix}$.

Use the augmented matrix and:
Replace row 2 by $-2R_1 + R_2$.
Replace row 3 by $-3R_1 + R_3$.

Moving on to the second column, we obtain 1 in the desired position by multiplying row 2 by −1.

$$\begin{bmatrix} 1 & -1 & -2 & 2 \\ -1(0) & -1(-1) & -1(10) & -1(1) \\ 0 & -1 & 10 & 6 \end{bmatrix} = \begin{bmatrix} 1 & -1 & -2 & 2 \\ 0 & 1 & -10 & -1 \\ 0 & -1 & 10 & 6 \end{bmatrix} \quad -1R_2$$

We want 0 in this position.

Now we want a 0 below the 1 in column 2. To get the 0, multiply row 2 by 1 and add these products to row 3. (Equivalently, add row 2 to row 3.) We obtain the following matrix:

$$\begin{bmatrix} 1 & -1 & -2 & 2 \\ 0 & 1 & -10 & -1 \\ 0 & 0 & 0 & 5 \end{bmatrix}.$$

Replace row 3 in the previous matrix by $1R_2 + R_3$.

It is impossible to convert this last matrix to the desired form of 1s down the main diagonal. If we translate the last row back into equation form, we get

$$0x + 0y + 0z = 5,$$

There are no values of x, y, and z for which $0 = 5$.

which is false. Regardless of which values we select for x, y, and z, the last equation can never be a true statement. Consequently, the system has no solution. The solution set is \varnothing, the empty set. • • •

✓ Check Point **1** Use Gaussian elimination to solve the system:

$$\begin{cases} x - 2y - z = -5 \\ 2x - 3y - z = 0 \\ 3x - 4y - z = 1. \end{cases}$$

DISCOVERY

Use the addition method to solve Example 1. Describe what happens. Why does this mean that there is no solution?

Recall that the graph of a system of three linear equations in three variables consists of three planes. When these planes intersect in a single point, the system has precisely one ordered-triple solution. When the planes have no point in common, the system has no solution, like the one in Example 1. **Figure 6.1** illustrates some of the geometric possibilities for these inconsistent systems.

Three planes are parallel with no common intersection point.

Two planes are parallel with no common intersection point.

Planes intersect two at a time. There is no intersection point common to all three planes.

FIGURE 6.1 Three planes may have no common point of intersection.

Now let's see what happens when we apply Gaussian elimination to a system with infinitely many solutions. Representing the solution set for these systems can be a bit tricky.

EXAMPLE 2 A System with an Infinite Number of Solutions

Use Gaussian elimination to solve the following system:

$$\begin{cases} 3x - 4y + 4z = 7 \\ x - y - 2z = 2 \\ 2x - 3y + 6z = 5. \end{cases}$$

SOLUTION

As always, we start with the augmented matrix.

$$\begin{bmatrix} 3 & -4 & 4 & | & 7 \\ 1 & -1 & -2 & | & 2 \\ 2 & -3 & 6 & | & 5 \end{bmatrix} \xrightarrow[\substack{\text{Interchange rows} \\ \text{1 and 2.}}]{R_1 \leftrightarrow R_2} \begin{bmatrix} 1 & -1 & -2 & | & 2 \\ 3 & -4 & 4 & | & 7 \\ 2 & -3 & 6 & | & 5 \end{bmatrix} \xrightarrow[\substack{\text{Replace row 3} \\ \text{by} -2R_1 + R_3.}]{\substack{\text{Replace row 2} \\ \text{by} -3R_1 + R_2.}}$$

$$\begin{bmatrix} 1 & -1 & -2 & | & 2 \\ 0 & -1 & 10 & | & 1 \\ 0 & -1 & 10 & | & 1 \end{bmatrix} \xrightarrow[\substack{\text{Multiply row 2} \\ \text{by} -1.}]{-1R_2} \begin{bmatrix} 1 & -1 & -2 & | & 2 \\ 0 & 1 & -10 & | & -1 \\ 0 & -1 & 10 & | & 1 \end{bmatrix} \xrightarrow[\substack{\text{Replace row 3} \\ \text{by} 1R_2 + R_3.}]{}$$

$$\begin{bmatrix} 1 & -1 & -2 & | & 2 \\ 0 & 1 & -10 & | & -1 \\ 0 & 0 & 0 & | & 0 \end{bmatrix}$$

If we translate row 3 of the matrix into equation form, we obtain

$$0x + 0y + 0z = 0$$

or

$$0 = 0.$$

This equation results in a true statement regardless of which values we select for x, y, and z. Consequently, the equation $0x + 0y + 0z = 0$ is *dependent* on the other two equations in the system in the sense that it adds no new information about the variables. Thus, we can drop it from the system, which can now be expressed in the form

$$\begin{bmatrix} 1 & -1 & -2 & | & 2 \\ 0 & 1 & -10 & | & -1 \end{bmatrix}.$$

This is the last matrix from above with row 3 omitted.

The original system is equivalent to the system

$$\begin{cases} x - y - 2z = 2 \\ \quad\quad y - 10z = -1. \end{cases}$$ This is the system represented by the matrix at the bottom of the previous page.

Although neither of these equations gives a value for z, we can use them to express x and y in terms of z. From the last equation, we obtain

$$y = 10z - 1.$$ Add 10z to both sides and isolate y.

Back-substituting for y into the first equation obtained from the final matrix, we can find x in terms of z.

$$x - y - 2z = 2$$ This is the first equation obtained from the final matrix.

$$x - (10z - 1) - 2z = 2$$ Because y = 10z − 1, substitute 10z − 1 for y.

$$x - 10z + 1 - 2z = 2$$ Apply the distributive property.

$$x - 12z + 1 = 2$$ Combine like terms.

$$x = 12z + 1$$ Solve for x in terms of z.

We have now found two equations expressing x and y in terms of z:

$$x = 12z + 1$$
$$y = 10z - 1.$$

Because no value is determined for z, we can find a solution of the system by letting z equal any real number and then using these equations to obtain x and y. For example, if $z = 1$, then

$$x = 12z + 1 = 12(1) + 1 = 13 \text{ and}$$
$$y = 10z - 1 = 10(1) - 1 = 9.$$

Consequently, $(13, 9, 1)$ is a solution of the system. On the other hand, if we let $z = -1$, then

$$x = 12z + 1 = 12(-1) + 1 = -11 \text{ and}$$
$$y = 10z - 1 = 10(-1) - 1 = -11.$$

Thus, $(-11, -11, -1)$ is another solution of the system.

We see that for any arbitrary choice of z, every ordered triple of the form $(12z + 1, 10z - 1, z)$ is a solution of the system. The solution set of this system with dependent equations is

$$\{(12z + 1, 10z - 1, z)\}.$$ • • •

We have seen that when three planes have no point in common, the corresponding system has no solution. When the system has infinitely many solutions, like the one in Example 2, the three planes intersect in more than one point. **Figure 6.2** illustrates geometric possibilities for systems with dependent equations.

The planes intersect along a common line. The planes coincide.

FIGURE 6.2 Three planes may intersect at infinitely many points.

☑ Check Point **2** Use Gaussian elimination to solve the following system:

$$\begin{cases} x - 2y - z = 5 \\ 2x - 5y + 3z = 6 \\ x - 3y + 4z = 1. \end{cases}$$

② Apply Gaussian elimination to systems with more variables than equations.

Nonsquare Systems

Up to this point, we have encountered only *square* systems in which the number of equations is equal to the number of variables. In a **nonsquare system**, the number of variables differs from the number of equations. In Example 3, we have two equations and three variables.

EXAMPLE 3 A System with Fewer Equations Than Variables

Use Gaussian elimination to solve the system:

$$\begin{cases} 3x + 7y + 6z = 26 \\ x + 2y + z = 8. \end{cases}$$

SOLUTION

We begin with the augmented matrix.

$$\begin{bmatrix} 3 & 7 & 6 & | & 26 \\ 1 & 2 & 1 & | & 8 \end{bmatrix} \xrightarrow{R_1 \leftrightarrow R_2} \begin{bmatrix} 1 & 2 & 1 & | & 8 \\ 3 & 7 & 6 & | & 26 \end{bmatrix} \xrightarrow[\text{by } -3R_1 + R_2.]{\text{Replace row 2}} \begin{bmatrix} 1 & 2 & 1 & | & 8 \\ 0 & 1 & 3 & | & 2 \end{bmatrix}$$

Because the matrix $\begin{bmatrix} 1 & 2 & 1 & | & 8 \\ 0 & 1 & 3 & | & 2 \end{bmatrix}$ has 1s down the diagonal that begins with the upper-left entry and a 0 below the leading 1, we translate this matrix back into equation form.

$$\begin{cases} x + 2y + z = 8 & \text{Equation 1} \\ y + 3z = 2 & \text{Equation 2} \end{cases}$$

We can let z equal any real number and use back-substitution to express x and y in terms of z.

DISCOVERY

Let $z = 1$ for the solution

$$(5z + 4, -3z + 2, z).$$

What solution do you obtain? Substitute these three values in the two original equations:

$$\begin{cases} 3x + 7y + 6z = 26 \\ x + 2y + z = 8. \end{cases}$$

Show that each equation is satisfied. Repeat this process for two other values of z.

Equation 2

$$y + 3z = 2$$
$$y = -3z + 2$$

Equation 1

$$x + 2y + z = 8$$
$$x + 2(-3z + 2) + z = 8$$
$$x - 6z + 4 + z = 8$$
$$x - 5z + 4 = 8$$
$$x = 5z + 4$$

For any arbitrary choice of z, every ordered triple of the form $(5z + 4, -3z + 2, z)$ is a solution of the system. We can express the system's solution set as

$$\{(5z + 4, -3z + 2, z)\}. \qquad \bullet\bullet\bullet$$

☑ Check Point **3** Use Gaussian elimination to solve the system:

$$\begin{cases} x + 2y + 3z = 70 \\ x + y + z = 60. \end{cases}$$

③ Solve problems involving systems without unique solutions.

Applications

How will computers be programmed to control traffic flow and avoid congestion? They will be required to solve systems continually based on the following premise: If traffic is to keep moving, during any period of time the number of cars entering an intersection must equal the number of cars leaving that intersection. Let's see what this means by looking at the intersections of four one-way city streets.

EXAMPLE 4 Traffic Control

Figure 6.3 shows the intersections of four one-way streets. As you study the figure, notice that 300 cars per hour want to enter intersection I_1 from the north on 27th Avenue. Also, 200 cars per hour want to head east from intersection I_2 on Palm Drive. The letters $w, x, y,$ and z stand for the number of cars passing between the intersections.

a. If the traffic is to keep moving, at each intersection the number of cars entering per hour must equal the number of cars leaving per hour. Use this idea to set up a linear system of equations involving $w, x, y,$ and z.

b. Use Gaussian elimination to solve the system.

c. If construction on 27th Avenue limits z to 50 cars per hour, how many cars per hour must pass between the other intersections to keep traffic flowing?

FIGURE 6.3 The intersections of four one-way streets

SOLUTION

a. Set up the system by considering one intersection at a time, referring to **Figure 6.3**.

For Intersection I_1: Because $300 + 700 = 1000$ cars enter I_1 and $w + z$ cars leave the intersection, then $w + z = 1000$.

For Intersection I_2: Because $w + x$ cars enter the intersection and $200 + 900 = 1100$ cars leave I_2, then $w + x = 1100$.

For Intersection I_3: **Figure 6.3** indicates that $300 + 400 = 700$ cars enter and $x + y$ leave, so $x + y = 700$.

For Intersection I_4: With $y + z$ cars entering and $200 + 400 = 600$ cars exiting, traffic will keep flowing if $y + z = 600$.

The system of equations that models this situation is given by

$$\begin{cases} w + z = 1000 \\ w + x = 1100 \\ x + y = 700 \\ y + z = 600. \end{cases}$$

b. To solve this system using Gaussian elimination, we begin with the augmented matrix.

System of Linear Equations (showing missing variables with 0 coefficients)

Augmented Matrix

$$\begin{cases} 1w + 0x + 0y + 1z = 1000 \\ 1w + 1x + 0y + 0z = 1100 \\ 0w + 1x + 1y + 0z = 700 \\ 0w + 0x + 1y + 1z = 600 \end{cases} \qquad \begin{bmatrix} 1 & 0 & 0 & 1 & | & 1000 \\ 1 & 1 & 0 & 0 & | & 1100 \\ 0 & 1 & 1 & 0 & | & 700 \\ 0 & 0 & 1 & 1 & | & 600 \end{bmatrix}$$

We can now use row operations to obtain the following matrix:

$$\begin{bmatrix} 1 & 0 & 0 & 1 & | & 1000 \\ 0 & 1 & 0 & -1 & | & 100 \\ 0 & 0 & 1 & 1 & | & 600 \\ 0 & 0 & 0 & 0 & | & 0 \end{bmatrix}$$

$w + z = 1000$
$x - z = 100$
$y + z = 600$

The last row of the matrix shows that the system in the voice balloons has dependent equations and infinitely many solutions. To write the solution set containing these infinitely many solutions, let z equal any real number. Use the three equations in the voice balloons to express w, x, and y in terms of z:

$$w = 1000 - z, \quad x = 100 + z, \quad \text{and} \quad y = 600 - z.$$

With z arbitrary, the alphabetically ordered solution (w, x, y, z) enables us to express the system's solution set as

$$\{(1000 - z, 100 + z, 600 - z, z)\}.$$

c. We are given that construction limits z to 50 cars per hour. Because $z = 50$, we substitute 50 for z in the system's ordered solution:

$$(1000 - z, 100 + z, 600 - z, z) \qquad \text{Use the system's solution.}$$
$$= (1000 - 50, 100 + 50, 600 - 50, 50) \qquad z = 50$$
$$= (950, 150, 550, 50).$$

Thus, $w = 950$, $x = 150$, and $y = 550$. (See **Figure 6.4**.) With construction on 27th Avenue, this means that to keep traffic flowing, 950 cars per hour must be routed between I_1 and I_2, 150 per hour between I_3 and I_2, and 550 per hour between I_3 and I_4. •••

FIGURE 6.4 With z limited to 50 cars per hour, values for w, x, and y are determined.

✓ Check Point **4** **Figure 6.5** shows a system of four one-way streets. The numbers in the figure denote the number of cars per minute that travel in the direction shown.

a. Use the requirement that the number of cars entering each of the intersections per minute must equal the number of cars leaving per minute to set up a system of equations in w, x, y, and z.

b. Use Gaussian elimination to solve the system.

c. If construction limits z to 10 cars per minute, how many cars per minute must pass between the other intersections to keep traffic flowing?

FIGURE 6.5

CONCEPT AND VOCABULARY CHECK

Fill in each blank so that the resulting statement is true.

Using Gaussian elimination on linear systems in three variables, we obtained each of the matrices shown in Exercises 1 through 3. State whether the linear system has one solution, no solution, or infinitely many solutions.

1. $\begin{bmatrix} 1 & 1 & 1 & | & 6 \\ 0 & 1 & 1 & | & -6 \\ 0 & 0 & 0 & | & -27 \end{bmatrix}$ _____

2. $\begin{bmatrix} 1 & 1 & 2 & | & 19 \\ 0 & 1 & 2 & | & 13 \\ 0 & 0 & 1 & | & 5 \end{bmatrix}$ _____

3. $\begin{bmatrix} 1 & -1 & -2 & | & 2 \\ 0 & 1 & -10 & | & -1 \\ 0 & 0 & 0 & | & 0 \end{bmatrix}$ _____

4. True or false: If $\{(2z + 3, 5z - 1, z)\}$ is the solution set of a system with dependent equations, then $(5, 4, 1)$ is a solution of this system. _____

5. Using Gaussian elimination to solve

$$\begin{cases} 3x + 2y - 2z = 0 \\ x - y + z = 5, \end{cases}$$

we obtain the matrix

$$\begin{bmatrix} 1 & -1 & 1 & | & 5 \\ 0 & 1 & -1 & | & -3 \end{bmatrix}.$$

Translating this matrix back into equation form gives

$$\begin{cases} \underline{\hspace{3cm}} & \text{Equation 1} \\ \underline{\hspace{3cm}}. & \text{Equation 2} \end{cases}$$

Solving Equation 2 for y in terms of z results in $y = \underline{\hspace{1cm}}$. Substituting this expression for y in Equation 1 gives $x = \underline{\hspace{1cm}}$. The system's solution set is \underline{\hspace{1.5cm}}.

EXERCISE SET 6.2

Practice Exercises

In Exercises 1–24, use Gaussian elimination to find the complete solution to each system of equations, or show that none exists.

1. $\begin{cases} 5x + 12y + z = 10 \\ 2x + 5y + 2z = -1 \\ x + 2y - 3z = 5 \end{cases}$

2. $\begin{cases} 2x - 4y + z = 3 \\ x - 3y + z = 5 \\ 3x - 7y + 2z = 12 \end{cases}$

3. $\begin{cases} 5x + 8y - 6z = 14 \\ 3x + 4y - 2z = 8 \\ x + 2y - 2z = 3 \end{cases}$

4. $\begin{cases} 5x - 11y + 6z = 12 \\ -x + 3y - 2z = -4 \\ 3x - 5y + 2z = 4 \end{cases}$

5. $\begin{cases} 3x + 4y + 2z = 3 \\ 4x - 2y - 8z = -4 \\ x + y - z = 3 \end{cases}$

6. $\begin{cases} 2x - y - z = 0 \\ x + 2y + z = 3 \\ 3x + 4y + 2z = 8 \end{cases}$

7. $\begin{cases} 8x + 5y + 11z = 30 \\ -x - 4y + 2z = 3 \\ 2x - y + 5z = 12 \end{cases}$

8. $\begin{cases} x + y - 10z = -4 \\ x \quad - 7z = -5 \\ 3x + 5y - 36z = -10 \end{cases}$

9. $\begin{cases} w - 2x - y - 3z = -9 \\ w + x - y = 0 \\ 3w + 4x + z = 6 \\ 2x - 2y + z = 3 \end{cases}$

10. $\begin{cases} 2w + x - 2y - z = 3 \\ w - 2x + y + z = 4 \\ -w - 8x + 7y + 5z = 13 \\ 3w + x - 2y + 2z = 6 \end{cases}$

11. $\begin{cases} 2w + x - y = 3 \\ w - 3x + 2y = -4 \\ 3w + x - 3y + z = 1 \\ w + 2x - 4y - z = -2 \end{cases}$

12. $\begin{cases} 2w - x + 3y + z = 0 \\ 3w + 2x + 4y - z = 0 \\ 5w - 2x - 2y - z = 0 \\ 2w + 3x - 7y - 5z = 0 \end{cases}$

13. $\begin{cases} w - 3x + y - 4z = 4 \\ -2w + x + 2y = -2 \\ 3w - 2x + y - 6z = 2 \\ -w + 3x + 2y - z = -6 \end{cases}$

14. $\begin{cases} 3w + 2x - y + 2z = -12 \\ 4w - x + y + 2z = 1 \\ w + x + y + z = -2 \\ -2w + 3x + 2y - 3z = 10 \end{cases}$

15. $\begin{cases} 2x + y - z = 2 \\ 3x + 3y - 2z = 3 \end{cases}$

16. $\begin{cases} 3x + 2y - z = 5 \\ x + 2y - z = 1 \end{cases}$

17. $\begin{cases} x + 2y + 3z = 5 \\ y - 5z = 0 \end{cases}$

18. $\begin{cases} 3x - y + 4z = 8 \\ y + 2z = 1 \end{cases}$

19. $\begin{cases} x + y - 2z = 2 \\ 3x - y - 6z = -7 \end{cases}$

20. $\begin{cases} -2x - 5y + 10z = 19 \\ x + 2y - 4z = 12 \end{cases}$

21. $\begin{cases} w + x - y + z = -2 \\ 2w - x + 2y - z = 7 \\ -w + 2x + y + 2z = -1 \end{cases}$

22. $\begin{cases} 2w - 3x + 4y + z = 7 \\ w - x + 3y - 5z = 10 \\ 3w + x - 2y - 2z = 6 \end{cases}$

23. $\begin{cases} w + 2x + 3y - z = 7 \\ 2x - 3y + z = 4 \\ w - 4x + y = 3 \end{cases}$

24. $\begin{cases} w - x + z = 0 \\ w - 4x + y + 2z = 0 \\ 3w - y + 2z = 0 \end{cases}$

Practice Plus

In Exercises 25–28, the first screen shows the augmented matrix, A, for a nonsquare linear system of three equations in four variables, w, x, y, and z. The second screen shows the reduced row-echelon form of matrix A. For each exercise,

 a. Write the system represented by A.

 b. Use the reduced row-echelon form of A to find the system's complete solution.

25.

26.

27.

28.

Application Exercises

The figure for Exercises 29–32 shows the intersections of three one-way streets. To keep traffic moving, the number of cars per minute entering an intersection must equal the number exiting that intersection. For intersection I_1, $x + 10$ cars enter and $y + 14$ cars exit per minute. Thus, $x + 10 = y + 14$.

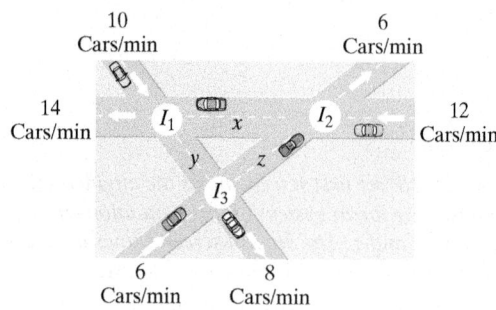

29. Write an equation for intersection I_2 that keeps traffic moving.

30. Write an equation for intersection I_3 that keeps traffic moving.

31. Use Gaussian elimination to solve the system formed by the equation given prior to Exercise 29 and the two equations that you obtained in Exercises 29–30.

32. Use your ordered solution obtained in Exercise 31 to solve this exercise. If construction limits z to 4 cars per minute, how many cars per minute must pass between the other intersections to keep traffic flowing?

33. The figure shows the intersections of four one-way streets.

 a. Set up a system of equations that keeps traffic moving.

 b. Use Gaussian elimination to solve the system.

 c. If construction limits z to 50 cars per hour, how many cars per hour must pass between the other intersections to keep traffic moving?

34. The vitamin content per ounce for three foods is given in the following table.

	Milligrams per Ounce		
	Thiamin	Riboflavin	Niacin
Food A	3	7	1
Food B	1	5	3
Food C	3	8	2

 a. Use matrices to show that no combination of these foods can provide exactly 14 mg of thiamin, 32 mg of riboflavin, and 9 mg of niacin.

 b. Use matrices to describe in practical terms what happens if the riboflavin requirement is increased by 5 mg and the other requirements stay the same.

35. Three foods have the following nutritional content per ounce.

	Units per Ounce		
	Vitamin A	Iron	Calcium
Food 1	20	20	10
Food 2	30	10	10
Food 3	10	10	30

 a. A diet must consist precisely of 220 units of vitamin A, 180 units of iron, and 340 units of calcium. However, the dietician runs out of Food 1. Use a matrix approach to show that under these conditions the dietary requirements cannot be met.

 b. Now suppose that all three foods are available. Use matrices to give two possible ways to meet the iron and calcium requirements with the three foods.

36. A company that manufactures products A, B, and C does both manufacturing and testing. The hours needed to manufacture and test each product are shown in the table.

	Hours Needed Weekly to Manufacture	Hours Needed Weekly to Test
Product A	7	2
Product B	6	2
Product C	3	1

The company has exactly 67 hours per week available for manufacturing and 20 hours per week available for testing. Give two different combinations for the number of products that can be manufactured and tested weekly.

Writing in Mathematics

37. Describe what happens when Gaussian elimination is used to solve an inconsistent system.

38. Describe what happens when Gaussian elimination is used to solve a system with dependent equations.

39. In solving a system of dependent equations in three variables, one student simply said that there are infinitely many solutions. A second student expressed the solution set as $\{(4z + 3, 5z - 1, z)\}$. Which is the better form of expressing the solution set and why?

Technology Exercise

40. a. The figure shows the intersections of a number of one-way streets. The numbers given represent traffic flow at a peak period (from 4 P.M. to 5:30 P.M.). Use the figure to write a linear system of six equations in seven variables based on the idea that at each intersection the number of cars entering must equal the number of cars leaving.

 b. Use a graphing utility with a $\boxed{\text{REF}}$ or $\boxed{\text{RREF}}$ command to find the complete solution to the system.

Critical Thinking Exercises

Make Sense? *In Exercises 41–44, determine whether each statement makes sense or does not make sense, and explain your reasoning.*

41. I omitted row 3 from $\begin{bmatrix} 1 & -1 & -2 & 2 \\ 0 & 1 & -10 & -1 \\ 0 & 0 & 0 & 5 \end{bmatrix}$ and expressed the system in the form $\begin{bmatrix} 1 & -1 & -2 & 2 \\ 0 & 1 & -10 & -1 \end{bmatrix}$.

42. I omitted row 3 from $\begin{bmatrix} 1 & -1 & -2 & 2 \\ 0 & 1 & -10 & -1 \\ 0 & 0 & 0 & 0 \end{bmatrix}$ and expressed the system in the form $\begin{bmatrix} 1 & -1 & -2 & 2 \\ 0 & 1 & -10 & -1 \end{bmatrix}$.

43. I solved a nonsquare system in which the number of equations was the same as the number of variables.

44. Models for controlling traffic flow are based on an equal number of cars entering an intersection and leaving that intersection.

45. Consider the linear system
$$\begin{cases} x + 3y + z = a^2 \\ 2x + 5y + 2az = 0 \\ x + y + a^2z = -9. \end{cases}$$

For which values of a will the system be inconsistent?

Group Exercise

46. Before beginning this exercise, the group needs to read and solve Exercise 40.

 a. A political group is planning a demonstration on 95th Street between 113th Place and 117th Court for 5 P.M. Wednesday. The problem becomes one of minimizing traffic flow on 95th Street (between 113th and 117th) without causing traffic tie-ups on other streets. One possible solution is to close off traffic on 95th Street between 113th and 117th (let $x_6 = 0$). What can group members conclude about x_7 under these conditions?

 b. Working with a matrix allows us to simplify the problem caused by the political demonstration, but it did not actually solve the problem. There are an infinite number of solutions; each value of x_7 we choose gives us a new picture. We also assumed x_6 was equal to 0; changing that assumption would also lead to different solutions. With your group, design another solution to the traffic flow problem caused by the political demonstration.

Preview Exercises

Exercises 47–49 will help you prepare for the material covered in the next section. In each exercise, perform the indicated operation or operations.

47. $-6 - (-5)$

48. $1(-4) + 2(5) + 3(-6)$

49. $\frac{1}{2}[8 - (-8)]$

Objectives

1 Use matrix notation.
2 Understand what is meant by equal matrices.
3 Add and subtract matrices.
4 Perform scalar multiplication.
5 Solve matrix equations.
6 Multiply matrices.
7 Model applied situations with matrix operations.

Use your smartphone to read your e-mail. Turn on your computer to write a paper. When you need to do research, use the Internet to browse through art museums and photography exhibits. When you need a break, load a flight simulator program and fly through a photorealistic computer world. As different as these experiences may be, they all share one thing — you're looking at images based on

matrices. Matrices have applications in numerous fields, including the technology of digital photography in which pictures are represented by numbers rather than film. In this section, we turn our attention to matrix algebra and some of its applications.

1 Use matrix notation.

Notations for Matrices

We have seen that an array of numbers, arranged in rows and columns and placed in brackets, is called a matrix. We can represent a matrix in two different ways.

- A capital letter, such as A, B, or C, can denote a matrix.
- A lowercase letter enclosed in brackets, such as that shown below, can denote a matrix.

$$A = [a_{ij}] \quad \text{Matrix } A \text{ with elements } a_{ij}$$

A general element in matrix A is denoted by a_{ij}. This refers to the element in the ith row and jth column. For example, a_{32} is the element of A located in the third row, second column.

A matrix of **order $m \times n$** has m rows and n columns. If $m = n$, a matrix has the same number of rows as columns and is called a **square matrix**.

EXAMPLE 1 Matrix Notation

Let

$$A = \begin{bmatrix} 3 & 2 & 0 \\ -4 & -5 & -\frac{1}{5} \end{bmatrix}.$$

a. What is the order of A?
b. If $A = [a_{ij}]$, identify a_{23} and a_{12}.

SOLUTION

a. The matrix has 2 rows and 3 columns, so it is of order 2×3.

b. The element a_{23} is in the second row and third column. Thus, $a_{23} = -\frac{1}{5}$. The element a_{12} is in the first row and second column. Consequently, $a_{12} = 2$.

• • •

✓ Check Point **1** Let

$$A = \begin{bmatrix} 5 & -2 \\ -3 & \pi \\ 1 & 6 \end{bmatrix}.$$

a. What is the order of A? **b.** Identify a_{12} and a_{31}.

2 Understand what is meant by equal matrices.

Equality of Matrices

Two matrices are **equal** if and only if they have the same order and corresponding elements are equal.

Definition of Equality of Matrices

Two matrices A and B are **equal** if and only if they have the same order $m \times n$ and $a_{ij} = b_{ij}$ for $i = 1, 2, \ldots, m$ and $j = 1, 2, \ldots, n$.

For example, if $A = \begin{bmatrix} x & y + 1 \\ z & 6 \end{bmatrix}$ and $B = \begin{bmatrix} 1 & 5 \\ 3 & 6 \end{bmatrix}$, then $A = B$ if and only if $x = 1, y + 1 = 5$ (so $y = 4$), and $z = 3$.

3 Add and subtract matrices.

Matrix Addition and Subtraction

Table 6.1 shows that matrices of the same order can be added or subtracted by simply adding or subtracting corresponding elements.

Table 6.1 Adding and Subtracting Matrices

Let $A = [a_{ij}]$ and $B = [b_{ij}]$ be matrices of order $m \times n$.

Definition	The Definition in Words	Example
Matrix Addition $A + B = [a_{ij} + b_{ij}]$	Matrices of the same order are added by adding the elements in corresponding positions.	$\begin{bmatrix} 1 & -2 \\ 3 & 5 \end{bmatrix} + \begin{bmatrix} -1 & 6 \\ 0 & 4 \end{bmatrix}$ $= \begin{bmatrix} 1 + (-1) & -2 + 6 \\ 3 + 0 & 5 + 4 \end{bmatrix} = \begin{bmatrix} 0 & 4 \\ 3 & 9 \end{bmatrix}$
Matrix Subtraction $A - B = [a_{ij} - b_{ij}]$	Matrices of the same order are subtracted by subtracting the elements in corresponding positions.	$\begin{bmatrix} 1 & -2 \\ 3 & 5 \end{bmatrix} - \begin{bmatrix} -1 & 6 \\ 0 & 4 \end{bmatrix}$ $= \begin{bmatrix} 1 - (-1) & -2 - 6 \\ 3 - 0 & 5 - 4 \end{bmatrix} = \begin{bmatrix} 2 & -8 \\ 3 & 1 \end{bmatrix}$

The sum or difference of two matrices of different orders is undefined. For example, consider the matrices

$$A = \begin{bmatrix} 0 & 3 \\ 4 & 3 \end{bmatrix} \quad \text{and} \quad B = \begin{bmatrix} 1 & 9 \\ 4 & 5 \\ 2 & 3 \end{bmatrix}.$$

The order of A is 2×2; the order of B is 3×2. These matrices are of different orders and cannot be added or subtracted.

TECHNOLOGY

Graphing utilities can add and subtract matrices. Enter the matrices and name them $[A]$ and $[B]$. Then use a keystroke sequence similar to

$[A]$ $+$ $[B]$ ENTER

$[A]$ $-$ $[B]$ ENTER

Consult your manual and verify the results in Example 2.

EXAMPLE 2 Adding and Subtracting Matrices

Perform the indicated matrix operations:

a. $\begin{bmatrix} 0 & 5 & 3 \\ -2 & 6 & -8 \end{bmatrix} + \begin{bmatrix} -2 & 3 & 5 \\ 7 & -9 & 6 \end{bmatrix}$

b. $\begin{bmatrix} -6 & 7 \\ 2 & -3 \end{bmatrix} - \begin{bmatrix} -5 & 6 \\ 0 & -4 \end{bmatrix}$.

SOLUTION

a.

$\begin{bmatrix} 0 & 5 & 3 \\ -2 & 6 & -8 \end{bmatrix} + \begin{bmatrix} -2 & 3 & 5 \\ 7 & -9 & 6 \end{bmatrix}$

$= \begin{bmatrix} 0 + (-2) & 5 + 3 & 3 + 5 \\ -2 + 7 & 6 + (-9) & -8 + 6 \end{bmatrix}$ Add the corresponding elements in the 2 × 3 matrices.

$= \begin{bmatrix} -2 & 8 & 8 \\ 5 & -3 & -2 \end{bmatrix}$ Simplify.

b.

$\begin{bmatrix} -6 & 7 \\ 2 & -3 \end{bmatrix} - \begin{bmatrix} -5 & 6 \\ 0 & -4 \end{bmatrix}$

$= \begin{bmatrix} -6 - (-5) & 7 - 6 \\ 2 - 0 & -3 - (-4) \end{bmatrix}$ Subtract the corresponding elements in the 2 × 2 matrices.

$= \begin{bmatrix} -1 & 1 \\ 2 & 1 \end{bmatrix}$ Simplify. • • •

✓ **Check Point 2** Perform the indicated matrix operations:

a. $\begin{bmatrix} -4 & 3 \\ 7 & -6 \end{bmatrix} + \begin{bmatrix} 6 & -3 \\ 2 & -4 \end{bmatrix}$ b. $\begin{bmatrix} 5 & 4 \\ -3 & 7 \\ 0 & 1 \end{bmatrix} - \begin{bmatrix} -4 & 8 \\ 6 & 0 \\ -5 & 3 \end{bmatrix}$.

A matrix whose elements are all equal to 0 is called a **zero matrix**. If A is an $m \times n$ matrix and 0 is the $m \times n$ zero matrix, then $A + 0 = A$. For example,

$$\begin{bmatrix} -5 & 2 \\ 3 & 6 \end{bmatrix} + \begin{bmatrix} 0 & 0 \\ 0 & 0 \end{bmatrix} = \begin{bmatrix} -5 & 2 \\ 3 & 6 \end{bmatrix}.$$

The $m \times n$ zero matrix is called the **additive identity** for $m \times n$ matrices.

For any matrix A, the **additive inverse** of A, written $-A$, is the matrix with the same order as A such that every element of $-A$ is the opposite of the corresponding element of A. Because corresponding elements are added in matrix addition, $A + (-A)$ is a zero matrix. For example,

$$\begin{bmatrix} -5 & 2 \\ 3 & 6 \end{bmatrix} + \begin{bmatrix} 5 & -2 \\ -3 & -6 \end{bmatrix} = \begin{bmatrix} 0 & 0 \\ 0 & 0 \end{bmatrix}.$$

Properties of matrix addition are similar to properties for adding real numbers.

Properties of Matrix Addition

If A, B, and C are $m \times n$ matrices and 0 is the $m \times n$ zero matrix, then the following properties are true.

1. $A + B = B + A$ Commutative property of addition

2. $(A + B) + C = A + (B + C)$ Associative property of addition

3. $A + 0 = 0 + A = A$ Additive identity property

4. $A + (-A) = (-A) + A = 0$ Additive inverse property

④ Perform scalar multiplication.

Scalar Multiplication

A matrix of order 1×1, such as [6], contains only one entry. To distinguish this matrix from the number 6, we refer to 6 as a **scalar**. In general, in our work with matrices, we will refer to real numbers as scalars.

To multiply a matrix A by a scalar c, we multiply each entry in A by c. For example,

$$4 \begin{bmatrix} 2 & 5 \\ -3 & 0 \end{bmatrix} = \begin{bmatrix} 4(2) & 4(5) \\ 4(-3) & 4(0) \end{bmatrix} = \begin{bmatrix} 8 & 20 \\ -12 & 0 \end{bmatrix}.$$

Scalar Matrix

Definition of Scalar Multiplication

If $A = [a_{ij}]$ is a matrix of order $m \times n$ and c is a scalar, then the matrix cA is the $m \times n$ matrix given by

$$cA = [ca_{ij}].$$

This matrix is obtained by multiplying each element of A by the real number c. We call cA a **scalar multiple** of A.

EXAMPLE 3 Scalar Multiplication

If $A = \begin{bmatrix} -1 & 4 \\ 3 & 0 \end{bmatrix}$ and $B = \begin{bmatrix} 2 & -3 \\ 5 & -6 \end{bmatrix}$, find the following matrices:

 a. $-5B$ **b.** $2A + 3B$.

SOLUTION

a. $-5B = -5 \begin{bmatrix} 2 & -3 \\ 5 & -6 \end{bmatrix} = \begin{bmatrix} -5(2) & -5(-3) \\ -5(5) & -5(-6) \end{bmatrix} = \begin{bmatrix} -10 & 15 \\ -25 & 30 \end{bmatrix}$

Multiply each element by −5.

b. $2A + 3B = 2 \begin{bmatrix} -1 & 4 \\ 3 & 0 \end{bmatrix} + 3 \begin{bmatrix} 2 & -3 \\ 5 & -6 \end{bmatrix}$

$\qquad\qquad = \begin{bmatrix} 2(-1) & 2(4) \\ 2(3) & 2(0) \end{bmatrix} + \begin{bmatrix} 3(2) & 3(-3) \\ 3(5) & 3(-6) \end{bmatrix}$

Multiply each element in A by 2. Multiply each element in B by 3.

$\qquad\qquad = \begin{bmatrix} -2 & 8 \\ 6 & 0 \end{bmatrix} + \begin{bmatrix} 6 & -9 \\ 15 & -18 \end{bmatrix} = \begin{bmatrix} -2 + 6 & 8 + (-9) \\ 6 + 15 & 0 + (-18) \end{bmatrix}$

Perform the addition of these 2 x 2 matrices by adding corresponding elements.

$\qquad\qquad = \begin{bmatrix} 4 & -1 \\ 21 & -18 \end{bmatrix}$ • • •

TECHNOLOGY

You can verify the algebraic solution in Example 3(b) by first entering the matrices $[A]$ and $[B]$ into your graphing utility. The screen below shows the required computation.

```
2[A]+3[B]
      [[4   -1 ]
       [21  -18]]
■
```

✓ **Check Point 3** If $A = \begin{bmatrix} -4 & 1 \\ 3 & 0 \end{bmatrix}$ and $B = \begin{bmatrix} -1 & -2 \\ 8 & 5 \end{bmatrix}$, find the following matrices:

 a. $-6B$ **b.** $3A + 2B$.

DISCOVERY

Verify each of the four properties listed in the box using

$$A = \begin{bmatrix} 2 & -4 \\ -5 & 3 \end{bmatrix},$$

$$B = \begin{bmatrix} 4 & 0 \\ 1 & -6 \end{bmatrix},$$

$c = 4$, and $d = 2$.

Properties of scalar multiplication are similar to properties for multiplying real numbers.

Properties of Scalar Multiplication

If A and B are $m \times n$ matrices, and c and d are scalars, then the following properties are true.

1. $(cd)A = c(dA)$ Associative property of scalar multiplication
2. $1A = A$ Scalar identity property
3. $c(A + B) = cA + cB$ Distributive property
4. $(c + d)A = cA + dA$ Distributive property

⑤ Solve matrix equations.

Have you noticed the many similarities between addition of real numbers and matrix addition, subtraction of real numbers and matrix subtraction, and multiplication of real numbers and scalar multiplication? Example 4 shows how these similarities can be used to solve matrix equations involving matrix addition, matrix subtraction, and scalar multiplication.

EXAMPLE 4 Solving a Matrix Equation

Solve for X in the matrix equation

$$2X + A = B,$$

where $A = \begin{bmatrix} 1 & -5 \\ 0 & 2 \end{bmatrix}$ and $B = \begin{bmatrix} -6 & 5 \\ 9 & 1 \end{bmatrix}$.

SOLUTION

We begin by solving the matrix equation for X.

$$2X + A = B$$ This is the given matrix equation.

$$2X = B - A$$ Subtract matrix A from both sides.

We multiply both sides by $\frac{1}{2}$ rather than divide both sides by 2. This is in anticipation of performing scalar multiplication.

$$X = \frac{1}{2}(B - A)$$ Multiply both sides by $\frac{1}{2}$ and solve for matrix X.

Now we use the matrices A and B to find the matrix X.

$$X = \frac{1}{2}\left(\begin{bmatrix} -6 & 5 \\ 9 & 1 \end{bmatrix} - \begin{bmatrix} 1 & -5 \\ 0 & 2 \end{bmatrix} \right)$$ Substitute the matrices into $X = \frac{1}{2}(B - A)$.

$$= \frac{1}{2}\begin{bmatrix} -7 & 10 \\ 9 & -1 \end{bmatrix}$$ Subtract matrices by subtracting corresponding elements.

$$= \begin{bmatrix} -\dfrac{7}{2} & 5 \\ \dfrac{9}{2} & -\dfrac{1}{2} \end{bmatrix}$$ Perform the scalar multiplication by multiplying each element by $\frac{1}{2}$.

Take a few minutes to show that this matrix satisfies the given equation $2X + A = B$. Substitute the matrix for X and the given matrices for A and B into the equation. The matrices on each side of the equal sign, $2X + A$ and B, should be equal. • • •

\emptyset Check Point **4** Solve for X in the matrix equation $3X + A = B$, where

$$A = \begin{bmatrix} 2 & -8 \\ 0 & 4 \end{bmatrix} \quad \text{and} \quad B = \begin{bmatrix} -10 & 1 \\ -9 & 17 \end{bmatrix}.$$

⑥ Multiply matrices.

Matrix Multiplication

We do not multiply two matrices by multiplying the corresponding entries of the matrices. Instead, we must think of matrix multiplication as *row-by-column multiplication*. To better understand how this works, let's begin with the definition of matrix multiplication for matrices of order 2×2.

Definition of Matrix Multiplication: 2 × 2 Matrices

Row 1 of A × Column 1 of B Row 1 of A × Column 2 of B

$$AB = \begin{bmatrix} a & b \\ c & d \end{bmatrix} \begin{bmatrix} e & f \\ g & h \end{bmatrix} = \begin{bmatrix} ae + bg & af + bh \\ ce + dg & cf + dh \end{bmatrix}$$

Row 2 of A × Column 1 of B Row 2 of A × Column 2 of B

Notice that we obtain the element in the ith row and jth column in AB by performing computations with elements in the ith row of A and the jth column of B. For example, we obtain the element in the first row and first column of AB by performing computations with elements in the first row of A and the first column of B.

First row of A First column of B

$$\begin{bmatrix} a & b \end{bmatrix} \begin{bmatrix} e \\ g \end{bmatrix} = \begin{bmatrix} ae + bg \end{bmatrix}$$

1. Multiply each element in row 1 of A by the corresponding element in column 1 of B.

2. Add these products.

3. Record the sum as the element in row 1, column 1 of the product matrix.

Corresponding elements

$$\begin{bmatrix} a & b \end{bmatrix} \begin{bmatrix} e \\ g \end{bmatrix}$$

Corresponding elements

FIGURE 6.6 Finding corresponding elements when multiplying matrices

You may wonder how to find the corresponding elements in step 1 in the voice balloon. The element at the far left of row 1 corresponds to the element at the top of column 1. The second element from the left of row 1 corresponds to the second element from the top of column 1. This is illustrated in **Figure 6.6**.

EXAMPLE 5 Multiplying Matrices

Find AB, given

$$A = \begin{bmatrix} 2 & 3 \\ 4 & 7 \end{bmatrix} \quad \text{and} \quad B = \begin{bmatrix} 0 & 1 \\ 5 & 6 \end{bmatrix}.$$

SOLUTION

We will perform a row-by-column computation.

$$AB = \begin{bmatrix} 2 & 3 \\ 4 & 7 \end{bmatrix} \begin{bmatrix} 0 & 1 \\ 5 & 6 \end{bmatrix}$$

Row 1 of A × Column 1 of B　　Row 1 of A × Column 2 of B

$$= \begin{bmatrix} 2(0) + 3(5) & 2(1) + 3(6) \\ 4(0) + 7(5) & 4(1) + 7(6) \end{bmatrix} = \begin{bmatrix} 15 & 20 \\ 35 & 46 \end{bmatrix}$$

Row 2 of A × Column 1 of B　　Row 2 of A × Column 2 of B

• • •

✓ Check Point **5** Find AB, given $A = \begin{bmatrix} 1 & 3 \\ 2 & 5 \end{bmatrix}$ and $B = \begin{bmatrix} 4 & 6 \\ 1 & 0 \end{bmatrix}$.

We can generalize the process of Example 5 to multiply an $m \times n$ matrix and an $n \times p$ matrix. **For the product of two matrices to be defined, the number of columns of the first matrix must equal the number of rows of the second matrix.**

First Matrix　　　Second Matrix

$m \times n$　　　　$n \times p$

The number of columns in the first matrix must be the same as the number of rows in the second matrix.

GREAT QUESTION!

Is there a diagram I can use to determine the order of the matrix product AB?

Yes. The following diagram illustrates the first sentence in the box defining matrix multiplication.

Definition of Matrix Multiplication

The **product** of an $m \times n$ matrix, A, and an $n \times p$ matrix, B, is an $m \times p$ matrix, AB, whose elements are found as follows: The element in the ith row and jth column of AB is found by multiplying each element in the ith row of A by the corresponding element in the jth column of B and adding the products.

To find a product AB, each row of A must have the same number of elements as each column of B. We obtain p_{ij}, the element in the ith row and jth column in AB, by performing computations with elements in the ith row of A and the jth column of B:

When multiplying corresponding elements, keep in mind that the element at the far left of row i corresponds to the element at the top of column j. The element second from the left in row i corresponds to the element second from the top in column j. Likewise, the element third from the left in row i corresponds to the element third from the top in column j, and so on.

EXAMPLE 6　Multiplying Matrices

Matrices A and B are defined as follows:

$$A = \begin{bmatrix} 1 & 2 & 3 \end{bmatrix} \qquad B = \begin{bmatrix} 4 \\ 5 \\ 6 \end{bmatrix}.$$

Find each product:　　**a.** AB　　**b.** BA.

SOLUTION

a. Matrix A is a 1×3 matrix and matrix B is a 3×1 matrix. Thus, the product AB is a 1×1 matrix.

$$A = [1 \quad 2 \quad 3] \quad B = \begin{bmatrix} 4 \\ 5 \\ 6 \end{bmatrix}$$

$$AB = [1 \quad 2 \quad 3]\begin{bmatrix} 4 \\ 5 \\ 6 \end{bmatrix} \qquad \text{We will perform a row-by-column computation.}$$

$$= [(1)(4) + (2)(5) + (3)(6)] \qquad \text{Multiply elements in row 1 of } A \text{ by corresponding elements in column 1 of } B \text{ and add the products.}$$

$$= [4 + 10 + 18] \qquad \text{Perform the multiplications.}$$

$$= [32] \qquad \text{Add.}$$

b. Matrix B is a 3×1 matrix and matrix A is a 1×3 matrix. Thus, the product BA is a 3×3 matrix.

$$B = \begin{bmatrix} 4 \\ 5 \\ 6 \end{bmatrix} \quad A = [1 \quad 2 \quad 3]$$

$$BA = \begin{bmatrix} 4 \\ 5 \\ 6 \end{bmatrix}[1 \quad 2 \quad 3] \qquad \text{We perform a row-by-column computation.}$$

Row 1 of $B \times$ Column 1 of A Row 1 of $B \times$ Column 2 of A Row 1 of $B \times$ Column 3 of A

$$= \begin{bmatrix} (4)(1) & (4)(2) & (4)(3) \\ (5)(1) & (5)(2) & (5)(3) \\ (6)(1) & (6)(2) & (6)(3) \end{bmatrix}$$

Row 2 of $B \times$ Column 1 of A Row 2 of $B \times$ Column 2 of A Row 2 of $B \times$ Column 3 of A

Row 3 of $B \times$ Column 1 of A Row 3 of $B \times$ Column 2 of A Row 3 of $B \times$ Column 3 of A

$$= \begin{bmatrix} 4 & 8 & 12 \\ 5 & 10 & 15 \\ 6 & 12 & 18 \end{bmatrix} \qquad \text{Simplify.}$$

• • •

In Example 6, did you notice that AB and BA are different matrices? For most matrices A and B, $AB \neq BA$. Because **matrix multiplication is not commutative**, be careful about the order in which matrices appear when performing this operation.

✓ Check Point **6** If $A = [2 \quad 0 \quad 4]$ and $B = \begin{bmatrix} 1 \\ 3 \\ 7 \end{bmatrix}$, find AB and BA.

TECHNOLOGY

The screens illustrate the solution of Example 6 using a graphing utility.

```
[A]
        [[1 2 3]]
[B]
         [[4]
          [5]
          [6]]
■
```

```
[A][B]
         [[32]]
[B][A]
         [[4 8  12]
          [5 10 15]
          [6 12 18]]
■
```

Blitzer Bonus

Arthur Cayley

Matrices were first studied intensively by the English mathematician Arthur Cayley (1821–1895). Before reaching the age of 25, he published 25 papers, setting a pattern of prolific creativity that lasted throughout his life. Cayley was a lawyer, painter, mountaineer, and Cambridge professor whose greatest invention was that of matrices and matrix theory. Cayley's matrix algebra, especially the noncommutativity of multiplication ($AB \neq BA$), opened up a new area of mathematics called abstract algebra.

EXAMPLE 7 Multiplying Matrices

Where possible, find each product:

a. $\begin{bmatrix} 4 & 2 \\ 1 & 3 \end{bmatrix}\begin{bmatrix} 1 & 2 & 3 & 4 \\ 0 & 2 & -1 & 6 \end{bmatrix}$

b. $\begin{bmatrix} 1 & 2 & 3 & 4 \\ 0 & 2 & -1 & 6 \end{bmatrix}\begin{bmatrix} 4 & 2 \\ 1 & 3 \end{bmatrix}.$

SOLUTION

a. The first matrix is a 2×2 matrix and the second is a 2×4 matrix. The product will be a 2×4 matrix.

$$\begin{bmatrix} 4 & 2 \\ 1 & 3 \end{bmatrix}\begin{bmatrix} 1 & 2 & 3 & 4 \\ 0 & 2 & -1 & 6 \end{bmatrix}$$ We perform a row-by-column computation.

$$= \begin{bmatrix} 4(1) + 2(0) & 4(2) + 2(2) & 4(3) + 2(-1) & 4(4) + 2(6) \\ 1(1) + 3(0) & 1(2) + 3(2) & 1(3) + 3(-1) & 1(4) + 3(6) \end{bmatrix}$$

$$= \begin{bmatrix} 4 + 0 & 8 + 4 & 12 - 2 & 16 + 12 \\ 1 + 0 & 2 + 6 & 3 - 3 & 4 + 18 \end{bmatrix}$$

$$= \begin{bmatrix} 4 & 12 & 10 & 28 \\ 1 & 8 & 0 & 22 \end{bmatrix}$$

b. $\begin{bmatrix} 1 & 2 & 3 & 4 \\ 0 & 2 & -1 & 6 \end{bmatrix}\begin{bmatrix} 4 & 2 \\ 1 & 3 \end{bmatrix}$

First matrix Second matrix
2×4 2×2

These numbers must be the same to multiply the matrices.

The number of columns in the first matrix does not equal the number of rows in the second matrix. Thus, the product of these two matrices is undefined. • • •

✓ Check Point **7** Where possible, find each product:

a. $\begin{bmatrix} 1 & 3 \\ 0 & 2 \end{bmatrix}\begin{bmatrix} 2 & 3 & -1 & 6 \\ 0 & 5 & 4 & 1 \end{bmatrix}$

b. $\begin{bmatrix} 2 & 3 & -1 & 6 \\ 0 & 5 & 4 & 1 \end{bmatrix}\begin{bmatrix} 1 & 3 \\ 0 & 2 \end{bmatrix}.$

Although matrix multiplication is not commutative, it does obey many of the properties of real numbers.

DISCOVERY

Verify the properties listed in the box using

$$A = \begin{bmatrix} 3 & 2 \\ -1 & 4 \end{bmatrix},$$

$$B = \begin{bmatrix} 1 & 0 \\ 3 & 2 \end{bmatrix},$$

$$C = \begin{bmatrix} 1 & 2 \\ -1 & 1 \end{bmatrix},$$

and $c = 3$.

7 Model applied situations with matrix operations.

Properties of Matrix Multiplication

If A, B, and C are matrices and c is a scalar, then the following properties are true. (Assume the order of each matrix is such that all operations in these properties are defined.)

1. $(AB)C = A(BC)$ Associative property of matrix multiplication

2. $A(B + C) = AB + AC$ Distributive properties of matrix multiplication
$(A + B)C = AC + BC$

3. $c(AB) = (cA)B$ Associative property of scalar multiplication

Applications

All of the still images that you see on the Web have been created or manipulated on a computer in a digital format—made up of hundreds of thousands, or even millions, of tiny squares called **pixels**. Pixels are created by dividing an image into a grid. The computer can change the brightness of every square or pixel in this grid. A digital camera captures photos in this digital format. Also, you can scan pictures to convert them into digital format. Example 8 illustrates the role that matrices play in this technology.

EXAMPLE 8 Matrices and Digital Photography

The letter L in **Figure 6.7** is shown using 9 pixels in a 3×3 grid. The colors possible in the grid are shown in **Figure 6.8**. Each color is represented by a specific number: 0, 1, 2, or 3.

FIGURE 6.7 The letter L FIGURE 6.8 Color levels

a. Find a matrix that represents a digital photograph of this letter L.

b. Increase the contrast of the letter L by changing the dark gray to black and the light gray to white. Use matrix addition to accomplish this.

SOLUTION

a. Look at the L and the background in **Figure 6.7**. Because the L is dark gray, color level 2, and the background is light gray, color level 1, a digital photograph of **Figure 6.7** can be represented by the matrix

$$\begin{bmatrix} 2 & 1 & 1 \\ 2 & 1 & 1 \\ 2 & 2 & 1 \end{bmatrix}.$$

b. We can make the L black, color level 3, by increasing each 2 in the above matrix to 3. We can make the background white, color level 0, by decreasing each 1 in the above matrix to 0. This is accomplished using the following matrix addition:

$$\begin{bmatrix} 2 & 1 & 1 \\ 2 & 1 & 1 \\ 2 & 2 & 1 \end{bmatrix} + \begin{bmatrix} 1 & -1 & -1 \\ 1 & -1 & -1 \\ 1 & 1 & -1 \end{bmatrix} = \begin{bmatrix} 3 & 0 & 0 \\ 3 & 0 & 0 \\ 3 & 3 & 0 \end{bmatrix}.$$

The picture corresponding to the matrix sum to the right of the equal sign is shown in **Figure 6.9**. •••

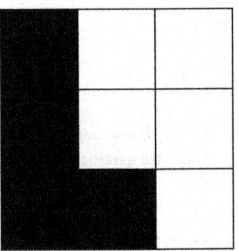

FIGURE 6.9 Changing contrast: the letter L

Check Point **8** Change the contrast of the letter L in **Figure 6.7** on the previous page by making the L light gray and the background black. Use matrix addition to accomplish this.

Blitzer Bonus ‖ Images of Space

Photographs sent back from space use matrices with thousands of pixels. Each pixel is assigned a number from 0 to 63 representing its color—0 for pure white and 63 for pure black. In the image of Saturn shown here, matrix operations provide false colors that emphasize the banding of the planet's upper atmosphere.

We have seen how functions can be transformed using translations, reflections, stretching, and shrinking. In a similar way, matrix operations are used to transform and manipulate computer graphics.

EXAMPLE 9 Transformations of an Image

The quadrilateral in **Figure 6.10** can be represented by the matrix

Coordinates of vertices

$$A = \begin{bmatrix} -2 & -1 & 3 & 1 \\ -3 & 2 & 4 & -2 \end{bmatrix}.$$ *x*-coordinates *y*-coordinates

Each column in the matrix gives the coordinates of a vertex, or corner, of the quadrilateral. Use matrix operations to perform the following transformations:

a. Move the quadrilateral 4 units to the right and 1 unit down.

b. Shrink the quadrilateral to half its perimeter.

c. Let $B = \begin{bmatrix} -1 & 0 \\ 0 & 1 \end{bmatrix}$. Find BA. What effect does this have on the quadrilateral in **Figure 6.10**?

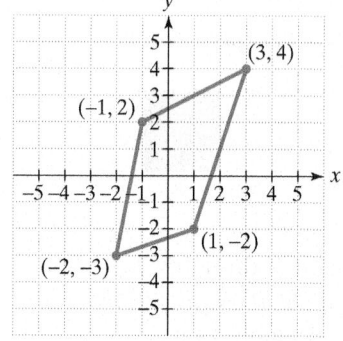

FIGURE 6.10

SOLUTION

a. We translate the quadrilateral 4 units right and 1 unit down by adding 4 to each *x*-coordinate and subtracting 1 from each *y*-coordinate. This is accomplished using the following matrix addition:

$$\begin{bmatrix} -2 & -1 & 3 & 1 \\ -3 & 2 & 4 & -2 \end{bmatrix} + \begin{bmatrix} 4 & 4 & 4 & 4 \\ -1 & -1 & -1 & -1 \end{bmatrix} = \begin{bmatrix} 2 & 3 & 7 & 5 \\ -4 & 1 & 3 & -3 \end{bmatrix}.$$

This matrix represents the original quadrilateral. Shift 4 units to the right and 1 unit down. This matrix represents the translated quadrilateral.

Each column in the matrix on the right gives the coordinates of a vertex of the translated quadrilateral. The original quadrilateral and the translated image are shown in **Figure 6.11**.

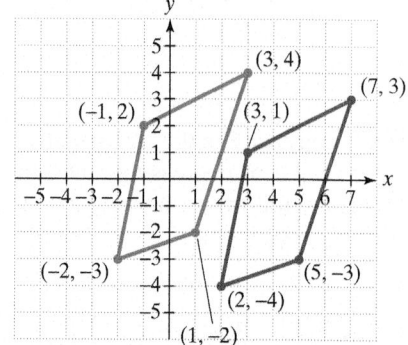

FIGURE 6.11 Shifting the quadrilateral 4 units right and 1 unit down

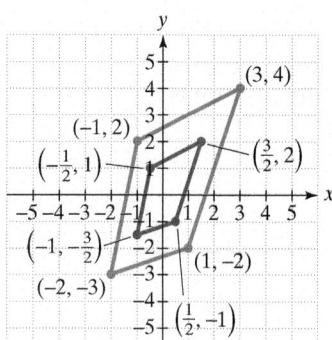

FIGURE 6.12 Shrinking the quadrilateral to half the original perimeter

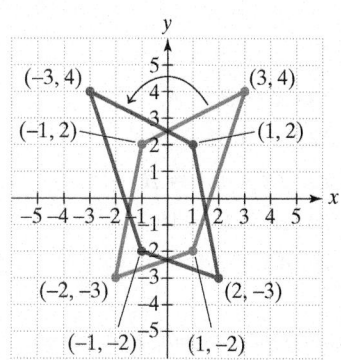

FIGURE 6.13

b. We shrink the quadrilateral in **Figure 6.10**, shown in blue in **Figure 6.12**, to half its perimeter by multiplying each x-coordinate and each y-coordinate by $\frac{1}{2}$. This is accomplished using the following scalar multiplication:

$$\frac{1}{2}\begin{bmatrix} -2 & -1 & 3 & 1 \\ -3 & 2 & 4 & -2 \end{bmatrix} = \begin{bmatrix} -1 & -\frac{1}{2} & \frac{3}{2} & \frac{1}{2} \\ -\frac{3}{2} & 1 & 2 & -1 \end{bmatrix}.$$

This matrix represents the original quadrilateral. This matrix represents the quadrilateral with half the original perimeter.

Each column in the matrix on the right gives the coordinates of a vertex of the reduced quadrilateral. The original quadrilateral and the reduced image are shown in **Figure 6.12**.

c. We begin by finding BA. Keep in mind that A represents the original quadrilateral, shown in blue in **Figure 6.13**.

$$BA = \begin{bmatrix} -1 & 0 \\ 0 & 1 \end{bmatrix}\begin{bmatrix} -2 & -1 & 3 & 1 \\ -3 & 2 & 4 & -2 \end{bmatrix}$$

$$= \begin{bmatrix} (-1)(-2)+0(-3) & (-1)(-1)+0(2) & (-1)(3)+0(4) & (-1)(1)+0(-2) \\ 0(-2)+1(-3) & 0(-1)+1(2) & 0(3)+1(4) & 0(1)+1(-2) \end{bmatrix}$$

$$= \begin{bmatrix} 2 & 1 & -3 & -1 \\ -3 & 2 & 4 & -2 \end{bmatrix}$$

Each column in the matrix multiplication gives the coordinates of a vertex of the transformed image. The original quadrilateral and this transformed image are shown in **Figure 6.13**. Notice that each x-coordinate on the original blue image is replaced with its opposite on the transformed red image.

We can conclude that multiplication by $\begin{bmatrix} -1 & 0 \\ 0 & 1 \end{bmatrix}$ reflected the blue quadrilateral about the y-axis. • • •

✓ Check Point **9** Consider the triangle represented by the matrix

$$A = \begin{bmatrix} 0 & 3 & 4 \\ 0 & 5 & 2 \end{bmatrix}.$$

Use matrix operations to perform the following transformations:

 a. Move the triangle 3 units to the left and 1 unit down.

 b. Enlarge the triangle to twice its original perimeter.

Illustrate your results in parts (a) and (b) by showing the original triangle and the transformed image in a rectangular coordinate system.

 c. Let $B = \begin{bmatrix} 1 & 0 \\ 0 & -1 \end{bmatrix}$. Find BA. What effect does this have on the original triangle?

CONCEPT AND VOCABULARY CHECK

Fill in each blank so that the resulting statement is true.

1. The notation a_{34} refers to the element in the _____ row and _____ column of a matrix.

2. The order of $A = [2 \quad 3 \quad 7]$ is _____.

3. A matrix that has the same number of rows as columns is called a/an _____ matrix.

4. If $\begin{bmatrix} 3 & x \\ y & -7 \end{bmatrix} = \begin{bmatrix} 3 & -10 \\ 6 & -7 \end{bmatrix}$, then $x = $ _____ and $y = $ _____.

5. True or false: Matrix addition is commutative. _____

6. True or false: Matrices of different orders can be added. _____

7. True or false: The scalar multiple $-4A$ is obtained by multiplying each element of A by -4. _____

8. If A is an $m \times n$ matrix and B is an $n \times p$ matrix, then AB is defined as an ____ × ____ matrix. To find the product AB, the number of _____ in matrix A must equal the number of _____ in matrix B.

9. True or false: Matrices of different orders can sometimes be multiplied. _____

10. True or false: Matrix multiplication is commutative. _____

EXERCISE SET 6.3

Practice Exercises

In Exercises 1–4,

 a. *Give the order of each matrix.*

 b. *If $A = [a_{ij}]$, identify a_{32} and a_{23}, or explain why identification is not possible.*

1. $\begin{bmatrix} 4 & -7 & 5 \\ -6 & 8 & -1 \end{bmatrix}$ **2.** $\begin{bmatrix} -6 & 4 & -1 \\ -9 & 0 & \frac{1}{2} \end{bmatrix}$

3. $\begin{bmatrix} 1 & -5 & \pi & e \\ 0 & 7 & -6 & -\pi \\ -2 & \frac{1}{2} & 11 & -\frac{1}{5} \end{bmatrix}$ **4.** $\begin{bmatrix} -4 & 1 & 3 & -5 \\ 2 & -1 & \pi & 0 \\ 1 & 0 & -e & \frac{1}{5} \end{bmatrix}$

In Exercises 5–8, find values for the variables so that the matrices in each exercise are equal.

5. $\begin{bmatrix} x \\ 4 \end{bmatrix} = \begin{bmatrix} 6 \\ y \end{bmatrix}$ **6.** $\begin{bmatrix} x \\ 7 \end{bmatrix} = \begin{bmatrix} 11 \\ y \end{bmatrix}$

7. $\begin{bmatrix} x & 2y \\ z & 9 \end{bmatrix} = \begin{bmatrix} 4 & 12 \\ 3 & 9 \end{bmatrix}$ **8.** $\begin{bmatrix} x & y+3 \\ 2z & 8 \end{bmatrix} = \begin{bmatrix} 12 & 5 \\ 6 & 8 \end{bmatrix}$

In Exercises 9–16, find the following matrices:

 a. $A + B$ **b.** $A - B$

 c. $-4A$ **d.** $3A + 2B$.

9. $A = \begin{bmatrix} 4 & 1 \\ 3 & 2 \end{bmatrix}$, $B = \begin{bmatrix} 5 & 9 \\ 0 & 7 \end{bmatrix}$

10. $A = \begin{bmatrix} -2 & 3 \\ 0 & 1 \end{bmatrix}$, $B = \begin{bmatrix} 8 & 1 \\ 5 & 4 \end{bmatrix}$

11. $A = \begin{bmatrix} 1 & 3 \\ 3 & 4 \\ 5 & 6 \end{bmatrix}$, $B = \begin{bmatrix} 2 & -1 \\ 3 & -2 \\ 0 & 1 \end{bmatrix}$

12. $A = \begin{bmatrix} 3 & 1 & 1 \\ -1 & 2 & 5 \end{bmatrix}$, $B = \begin{bmatrix} 2 & -3 & 6 \\ -3 & 1 & -4 \end{bmatrix}$

13. $A = \begin{bmatrix} 2 \\ -4 \\ 1 \end{bmatrix}$, $B = \begin{bmatrix} -5 \\ 3 \\ -1 \end{bmatrix}$

14. $A = [6 \quad 2 \quad -3]$, $B = [4 \quad -2 \quad 3]$

15. $A = \begin{bmatrix} 2 & -10 & -2 \\ 14 & 12 & 10 \\ 4 & -2 & 2 \end{bmatrix}$, $B = \begin{bmatrix} 6 & 10 & -2 \\ 0 & -12 & -4 \\ -5 & 2 & -2 \end{bmatrix}$

16. $A = \begin{bmatrix} 6 & -3 & 5 \\ 6 & 0 & -2 \\ -4 & 2 & -1 \end{bmatrix}$, $B = \begin{bmatrix} -3 & 5 & 1 \\ -1 & 2 & -6 \\ 2 & 0 & 4 \end{bmatrix}$

In Exercises 17–26, let

$$A = \begin{bmatrix} -3 & -7 \\ 2 & -9 \\ 5 & 0 \end{bmatrix} \text{ and } B = \begin{bmatrix} -5 & -1 \\ 0 & 0 \\ 3 & -4 \end{bmatrix}.$$

Solve each matrix equation for X.

17. $X - A = B$ **18.** $X - B = A$

19. $2X + A = B$ **20.** $3X + A = B$

21. $3X + 2A = B$ **22.** $2X + 5A = B$

23. $B - X = 4A$ **24.** $A - X = 4B$

25. $4A + 3B = -2X$ **26.** $4B + 3A = -2X$

In Exercises 27–36, find (if possible) the following matrices:

 a. AB **b.** BA.

27. $A = \begin{bmatrix} 1 & 3 \\ 5 & 3 \end{bmatrix}$, $B = \begin{bmatrix} 3 & -2 \\ -1 & 6 \end{bmatrix}$

28. $A = \begin{bmatrix} 3 & -2 \\ 1 & 5 \end{bmatrix}$, $B = \begin{bmatrix} 0 & 0 \\ 5 & -6 \end{bmatrix}$

29. $A = [1 \quad 2 \quad 3 \quad 4]$, $B = \begin{bmatrix} 1 \\ 2 \\ 3 \\ 4 \end{bmatrix}$

30. $A = \begin{bmatrix} -1 \\ -2 \\ -3 \end{bmatrix}$, $B = [1 \quad 2 \quad 3]$

31. $A = \begin{bmatrix} 1 & -1 & 4 \\ 4 & -1 & 3 \\ 2 & 0 & -2 \end{bmatrix}$, $B = \begin{bmatrix} 1 & 1 & 0 \\ 1 & 2 & 4 \\ 1 & -1 & 3 \end{bmatrix}$

32. $A = \begin{bmatrix} 1 & -1 & 1 \\ 5 & 0 & -2 \\ 3 & -2 & 2 \end{bmatrix}$, $B = \begin{bmatrix} 1 & 1 & 0 \\ 1 & -4 & 5 \\ 3 & -1 & 2 \end{bmatrix}$

33. $A = \begin{bmatrix} 4 & 2 \\ 6 & 1 \\ 3 & 5 \end{bmatrix}$, $B = \begin{bmatrix} 2 & 3 & 4 \\ -1 & -2 & 0 \end{bmatrix}$

34. $A = \begin{bmatrix} 2 & 4 \\ 3 & 1 \\ 4 & 2 \end{bmatrix}$, $B = \begin{bmatrix} 3 & 2 & 0 \\ -1 & -3 & 5 \end{bmatrix}$

35. $A = \begin{bmatrix} 2 & -3 & 1 & -1 \\ 1 & 1 & -2 & 1 \end{bmatrix}$, $B = \begin{bmatrix} 1 & 2 \\ -1 & 1 \\ 5 & 4 \\ 10 & 5 \end{bmatrix}$

36. $A = \begin{bmatrix} 2 & -1 & 3 & 2 \\ 1 & 0 & -2 & 1 \end{bmatrix}$, $B = \begin{bmatrix} -1 & 2 \\ 1 & 1 \\ 3 & -4 \\ 6 & 5 \end{bmatrix}$

In Exercises 37–44, perform the indicated matrix operations given that A, B, and C are defined as follows. If an operation is not defined, state the reason.

$$A = \begin{bmatrix} 4 & 0 \\ -3 & 5 \\ 0 & 1 \end{bmatrix} \quad B = \begin{bmatrix} 5 & 1 \\ -2 & -2 \end{bmatrix} \quad C = \begin{bmatrix} 1 & -1 \\ -1 & 1 \end{bmatrix}$$

37. $4B - 3C$

38. $5C - 2B$

39. $BC + CB$

40. $A(B + C)$

41. $A - C$

42. $B - A$

43. $A(BC)$

44. $A(CB)$

Practice Plus

In Exercises 45–50, let

$$A = \begin{bmatrix} 1 & 0 \\ 0 & 1 \end{bmatrix}, \quad B = \begin{bmatrix} 1 & 0 \\ 0 & -1 \end{bmatrix}, \quad C = \begin{bmatrix} -1 & 0 \\ 0 & 1 \end{bmatrix},$$

$$D = \begin{bmatrix} -1 & 0 \\ 0 & -1 \end{bmatrix}.$$

45. Find the product of the sum of A and B and the difference between C and D.

46. Find the product of the difference between A and B and the sum of C and D.

47. Use any three of the matrices to verify a distributive property.

48. Use any three of the matrices to verify an associative property.

In Exercises 49–50, suppose that the vertices of a computer graphic are points, (x, y), represented by the matrix

$$Z = \begin{bmatrix} x \\ y \end{bmatrix}.$$

49. Find BZ and explain why this reflects the graphic about the x-axis.

50. Find CZ and explain why this reflects the graphic about the y-axis.

Application Exercises

The $+$ sign in the figure is shown using 9 pixels in a 3×3 grid. The color levels are given to the right of the figure. Each color is represented by a specific number: 0, 1, 2, or 3. Use this information to solve Exercises 51–52.

White	Light gray	Dark gray	Black
0	1	2	3

51. a. Find a matrix that represents a digital photograph of the $+$ sign.

b. Adjust the contrast by changing the black to dark gray and the light gray to white. Use matrix addition to accomplish this.

c. Adjust the contrast by changing the black to light gray and the light gray to dark gray. Use matrix addition to accomplish this.

52. a. Find a matrix that represents a digital photograph of the $+$ sign.

b. Adjust the contrast by changing the black to dark gray and the light gray to black. Use matrix addition to accomplish this.

c. Adjust the contrast by leaving the black alone and changing the light gray to white. Use matrix addition to accomplish this.

The figure shows the letter L in a rectangular coordinate system.

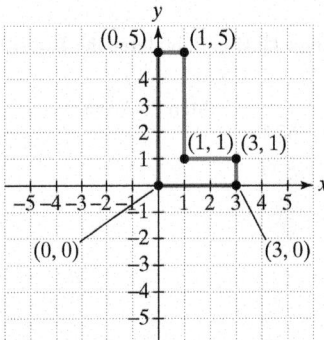

The figure can be represented by the matrix

$$B = \begin{bmatrix} 0 & 3 & 3 & 1 & 1 & 0 \\ 0 & 0 & 1 & 1 & 5 & 5 \end{bmatrix}.$$

Each column in the matrix describes a point on the letter. The order of the columns shows the direction in which a pencil must move to draw the letter. The L is completed by connecting the last point in the matrix, $(0, 5)$, to the starting point, $(0, 0)$. Use these ideas to solve Exercises 53–60.

53. Use matrix operations to move the L 2 units to the left and 3 units down. Then graph the letter and its transformation in a rectangular coordinate system.

54. Use matrix operations to move the L 2 units to the right and 3 units down. Then graph the letter and its transformation in a rectangular coordinate system.

55. Reduce the L to half its perimeter and move the reduced image 1 unit up. Then graph the letter and its transformation.

56. Reduce the L to half its perimeter and move the reduced image 2 units up. Then graph the letter and its transformation.

57. a. If $A = \begin{bmatrix} 1 & 0 \\ 0 & -1 \end{bmatrix}$, find AB.

b. Graph the object represented by matrix AB. What effect does the matrix multiplication have on the letter L represented by matrix B?

58. a. If $A = \begin{bmatrix} -1 & 0 \\ 0 & 1 \end{bmatrix}$, find AB.

b. Graph the object represented by matrix AB. What effect does the matrix multiplication have on the letter L represented by matrix B?

(In Exercises 59–60, be sure to refer to matrix B described in the second column on the previous page.)

59. a. If $A = \begin{bmatrix} 0 & -1 \\ 1 & 0 \end{bmatrix}$, find AB.

 b. Graph the object represented by matrix AB. What effect does the matrix multiplication have on the letter L represented by matrix B?

60. a. If $A = \begin{bmatrix} 2 & 0 \\ 0 & 1 \end{bmatrix}$, find AB.

 b. Graph the object represented by matrix AB. What effect does the matrix multiplication have on the letter L represented by matrix B?

61. Completing the transition to adulthood is measured by one or more of the following: leaving home, finishing school, getting married, having a child, or being financially independent. The bar graph shows the percentage of Americans, ages 20 and 30, who had completed the transition to adulthood in 1960 and in 2000.

Percentage Having Completed the Transition to Adulthood

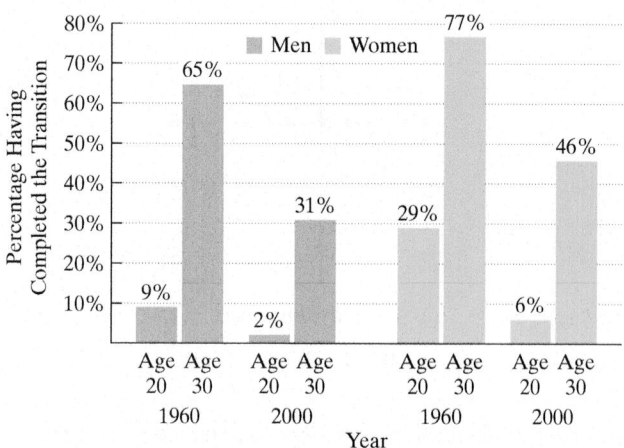

Source: James M. Henslin, *Sociology,* Eleventh Edition, Pearson, 2012.

 a. Use a 2 × 2 matrix to represent the data for 2000. Entries in the matrix should be percents that are organized as follows:

$$\begin{array}{c} \\ \text{Age 20} \\ \text{Age 30} \end{array} \overset{\displaystyle \text{Men} \quad \text{Women}}{\begin{bmatrix} & \\ & \end{bmatrix}}$$

 Call this matrix A.

 b. Use a 2 × 2 matrix to represent the data for 1960. Call this matrix B.

 c. Find $B - A$. What does this matrix represent?

62. The table gives an estimate of basic caloric needs for different age groups and activity levels.

Age Range	Sedentary		Moderately Active		Active	
19 − 30	2400	2000	2700	2100	3000	2400
31 − 50	2200	1800	2500	2000	2900	2200
51 +	2000	1600	2300	1800	2600	2100
	Men	Women	Men	Women	Men	Women

Source: *USA Today*

a. Use a 3 × 3 matrix to represent the daily caloric needs, by age and activity level, for men. Call this matrix M.

b. Use a 3 × 3 matrix to represent the daily caloric needs, by age and activity level, for women. Call this matrix W.

c. Find $M - W$. What does this matrix represent?

63. The final grade in a particular course is determined by grades on the midterm and final. The grades for five students and the two grading systems are modeled by the following matrices. Call the first matrix A and the second B.

	Midterm	Final
Student 1	76	92
Student 2	74	84
Student 3	94	86
Student 4	84	62
Student 5	58	80

	System 1	System 2
Midterm	0.5	0.3
Final	0.5	0.7

a. Describe the grading system that is represented by matrix B.

b. Compute the matrix AB and assign each of the five students a final course grade first using system 1 and then using system 2. ($89.5 - 100 = A, 79.5 - 89.4 = B, 69.5 - 79.4 = C, 59.5 - 69.4 = D$, below $59.5 = F$)

64. In a certain county, the proportion of voters in each age group registered as Republicans, Democrats, or Independents is given by the following matrix, which we'll call A.

	Age		
	18–30	**31–50**	**Over 50**
Republicans	0.40	0.30	0.70
Democrats	0.30	0.60	0.25
Independents	0.30	0.10	0.05

The distribution, by age and gender, of this county's voting population is given by the following matrix, which we'll call B.

		Male	**Female**
	18–30	6000	8000
Age	**31–50**	12,000	14,000
	Over 50	14,000	16,000

a. Calculate the product AB.

b. How many female Democrats are there?

c. How many male Republicans are there?

Writing in Mathematics

65. What is meant by the order of a matrix? Give an example with your explanation.

66. What does a_{ij} mean?

67. What are equal matrices?

68. How are matrices added?

69. Describe how to subtract matrices.

70. Describe matrices that cannot be added or subtracted.

71. Describe how to perform scalar multiplication. Provide an example with your description.

72. Describe how to multiply matrices.

73. Describe when the multiplication of two matrices is not defined.

74. If two matrices can be multiplied, describe how to determine the order of the product.

75. Low-resolution digital photographs use 262,144 pixels in a 512 × 512 grid. If you enlarge a low-resolution digital photograph enough, describe what will happen.

Technology Exercise

76. Use the matrix feature of a graphing utility to verify each of your answers to Exercises 37–44.

Critical Thinking Exercises

Make Sense? *In Exercises 77–80, determine whether each statement makes sense or does not make sense, and explain your reasoning.*

77. I added matrices of the same order by adding corresponding elements.

78. I multiplied an $m \times n$ matrix and an $n \times p$ matrix by multiplying corresponding elements.

79. I'm working with two matrices that can be added but not multiplied.

80. I'm working with two matrices that can be multiplied but not added.

81. Find two matrices A and B such that $AB = BA$.

82. Consider a square matrix such that each element that is not on the diagonal from upper left to lower right is zero. Experiment with such matrices (call each matrix A) by finding AA. Then write a sentence or two describing a method for multiplying this kind of matrix by itself.

83. If $AB = -BA$, then A and B are said to be anticommutative. Are $A = \begin{bmatrix} 0 & -1 \\ 1 & 0 \end{bmatrix}$ and $B = \begin{bmatrix} 1 & 0 \\ 0 & -1 \end{bmatrix}$ anticommutative?

Group Exercise

84. The interesting and useful applications of matrix theory are nearly unlimited. Applications of matrices range from representing digital photographs to predicting long-range trends in the stock market. Members of the group should research an application of matrices that they find intriguing. The group should then present a seminar to the class about this application.

Preview Exercises

Exercises 85–87 will help you prepare for the material covered in the next section.

85. Multiply:

$$\begin{bmatrix} a_{11} & a_{12} \\ a_{21} & a_{22} \end{bmatrix}\begin{bmatrix} 1 & 0 \\ 0 & 1 \end{bmatrix}.$$

After performing the multiplication, describe what happens to the elements in the first matrix.

86. Use Gauss-Jordan elimination to solve the system:

$$\begin{cases} -x - y - z = 1 \\ 4x + 5y = 0 \\ y - 3z = 0. \end{cases}$$

87. Multiply and write the linear system represented by the following matrix multiplication:

$$\begin{bmatrix} a_1 & b_1 & c_1 \\ a_2 & b_2 & c_2 \\ a_3 & b_3 & c_3 \end{bmatrix}\begin{bmatrix} x \\ y \\ z \end{bmatrix} = \begin{bmatrix} d_1 \\ d_2 \\ d_3 \end{bmatrix}.$$

CHAPTER 6 Mid-Chapter Check Point

WHAT YOU KNOW: We learned to use matrices to solve systems of linear equations. Gaussian elimination required simplifying the augmented matrix to one with 1s down the main diagonal and 0s below the 1s. Gauss-Jordan elimination simplified the augmented matrix to one with 1s down the main diagonal and 0s above and below each 1. Such a matrix, in reduced row-echelon form, did not require back-substitution to solve the system. We applied Gaussian elimination to systems with no solution, as well as to represent the solution set for systems with infinitely many solutions, including nonsquare systems. We learned how to perform operations with matrices, including matrix addition, matrix subtraction, scalar multiplication, and matrix multiplication.

In Exercises 1–5, use matrices to find the complete solution to each system of equations, or show that none exists.

1. $\begin{cases} x + 2y - 3z = -7 \\ 3x - y + 2z = 8 \\ 2x - y + z = 5 \end{cases}$

2. $\begin{cases} 2x + 4y + 5z = 2 \\ x + y + 2z = 1 \\ 3x + 5y + 7z = 4 \end{cases}$

3. $\begin{cases} x - 2y + 2z = -2 \\ 2x + 3y - z = 1 \end{cases}$

4. $\begin{cases} w + x + y + z = 6 \\ w - x + 3y + z = -14 \\ w + 2x - 3z = 12 \\ 2w + 3x + 6y + z = 1 \end{cases}$

5. $\begin{cases} 2x - 2y + 2z = 5 \\ x - y + z = 2 \\ 2x + y - z = 1 \end{cases}$

In Exercises 6–10, perform the indicated matrix operations or solve the matrix equation for X given that A, B, and C are defined as follows. If an operation is not defined, state the reason.

$$A = \begin{bmatrix} 0 & 2 \\ -1 & 3 \\ 1 & 0 \end{bmatrix} \quad B = \begin{bmatrix} 4 & 1 \\ -6 & -2 \end{bmatrix} \quad C = \begin{bmatrix} -1 & 0 \\ 0 & 1 \end{bmatrix}$$

6. $2C - \frac{1}{2}B$ **7.** $A(B + C)$ **8.** $A(BC)$

9. $A + C$ **10.** $2X - 3C = B$

SECTION 6.4

Multiplicative Inverses of Matrices and Matrix Equations

Objectives

1 Find the multiplicative inverse of a square matrix.

2 Use inverses to solve matrix equations.

3 Encode and decode messages.

In 1939, Britain's secret service hired top chess players, mathematicians, and other masters of logic to break the code used by the Nazis in communications between headquarters and troops. The project, which employed over 10,000 people, broke the code less than a year later, providing the Allies with information about Nazi troop movements throughout World War II.

Messages must often be sent in such a way that the real meaning is hidden from everyone but the sender and the recipient. In this section, we will look at the role that matrices and their inverses play in this process.

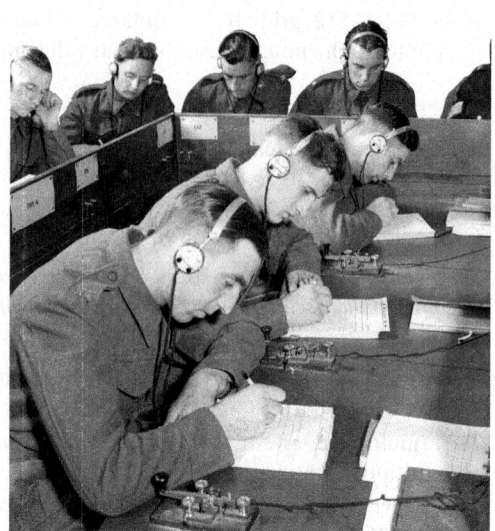

The Multiplicative Identity Matrix

For the real numbers, we know that 1 is the multiplicative identity because $a \cdot 1 = 1 \cdot a = a$. Is there a similar property for matrix multiplication? That is, is there a matrix I such that $AI = A$ and $IA = A$? The answer is yes. A square matrix with 1s down the main diagonal from upper left to lower right and 0s elsewhere does not change the elements in a matrix in products with that matrix. In the case of 2×2 matrices,

$$\begin{bmatrix} a_{11} & a_{12} \\ a_{21} & a_{22} \end{bmatrix} \begin{bmatrix} 1 & 0 \\ 0 & 1 \end{bmatrix} = \begin{bmatrix} a_{11} & a_{12} \\ a_{21} & a_{22} \end{bmatrix}$$

The elements in the matrix do not change.

and $\begin{bmatrix} 1 & 0 \\ 0 & 1 \end{bmatrix} \begin{bmatrix} a_{11} & a_{12} \\ a_{21} & a_{22} \end{bmatrix} = \begin{bmatrix} a_{11} & a_{12} \\ a_{21} & a_{22} \end{bmatrix}.$

The elements in the matrix do not change.

The $n \times n$ square matrix with 1s down the main diagonal from upper left to lower right and 0s elsewhere is called the **multiplicative identity matrix of order n**. This matrix is designated by I_n. For example,

$$I_2 = \begin{bmatrix} 1 & 0 \\ 0 & 1 \end{bmatrix}, \quad I_3 = \begin{bmatrix} 1 & 0 & 0 \\ 0 & 1 & 0 \\ 0 & 0 & 1 \end{bmatrix},$$

and so on.

1 Find the multiplicative inverse of a square matrix.

The Multiplicative Inverse of a Matrix

The multiplicative identity matrix, I_n, will help us to define a new concept: the multiplicative inverse of a matrix. To do so, let's consider a similar concept, the multiplicative inverse of a nonzero number, a. Recall that the multiplicative inverse of a is $\frac{1}{a}$. The multiplicative inverse has the following property:

$$a \cdot \tfrac{1}{a} = 1 \quad \text{and} \quad \tfrac{1}{a} \cdot a = 1.$$

We can define the multiplicative inverse of a square matrix in a similar manner.

> ### Definition of the Multiplicative Inverse of a Square Matrix
>
> Let A be an $n \times n$ matrix. If there exists an $n \times n$ matrix A^{-1} (read: "A inverse") such that
>
> $$AA^{-1} = I_n \quad \text{and} \quad A^{-1}A = I_n,$$
>
> then A^{-1} is the **multiplicative inverse** of A.

We have seen that matrix multiplication is not commutative. Thus, to show that a matrix B is the multiplicative inverse of the matrix A, find both AB and BA. If B is the multiplicative inverse of A, both products (AB and BA) will be the multiplicative identity matrix, I_n.

EXAMPLE 1 The Multiplicative Inverse of a Matrix

Show that B is the multiplicative inverse of A, where

$$A = \begin{bmatrix} -1 & 3 \\ 2 & -5 \end{bmatrix} \quad \text{and} \quad B = \begin{bmatrix} 5 & 3 \\ 2 & 1 \end{bmatrix}.$$

SOLUTION

To show that B is the multiplicative inverse of A, we must find the products AB and BA. If B is the multiplicative inverse of A, then AB will be the multiplicative identity matrix and BA will be the multiplicative identity matrix. Because A and B are 2×2 matrices, $n = 2$. Thus, we denote the multiplicative identity matrix as I_2; it is also a 2×2 matrix. We must show that

- $AB = I_2 = \begin{bmatrix} 1 & 0 \\ 0 & 1 \end{bmatrix}$ and

- $BA = I_2 = \begin{bmatrix} 1 & 0 \\ 0 & 1 \end{bmatrix}.$

Let's first show $AB = I_2$.

$$AB = \begin{bmatrix} -1 & 3 \\ 2 & -5 \end{bmatrix}\begin{bmatrix} 5 & 3 \\ 2 & 1 \end{bmatrix}$$

$$= \begin{bmatrix} -1(5) + 3(2) & -1(3) + 3(1) \\ 2(5) + (-5)(2) & 2(3) + (-5)(1) \end{bmatrix} = \begin{bmatrix} 1 & 0 \\ 0 & 1 \end{bmatrix}$$

Let's now show $BA = I_2$.

$$BA = \begin{bmatrix} 5 & 3 \\ 2 & 1 \end{bmatrix}\begin{bmatrix} -1 & 3 \\ 2 & -5 \end{bmatrix}$$

$$= \begin{bmatrix} 5(-1) + 3(2) & 5(3) + 3(-5) \\ 2(-1) + 1(2) & 2(3) + 1(-5) \end{bmatrix} = \begin{bmatrix} 1 & 0 \\ 0 & 1 \end{bmatrix}$$

Both products give the multiplicative identity matrix. Thus, B is the multiplicative inverse of A and we can designate B as $A^{-1} = \begin{bmatrix} 5 & 3 \\ 2 & 1 \end{bmatrix}.$ • • •

⬤ Check Point **1** Show that B is the multiplicative inverse of A, where

$$A = \begin{bmatrix} 2 & 1 \\ 1 & 1 \end{bmatrix} \quad \text{and} \quad B = \begin{bmatrix} 1 & -1 \\ -1 & 2 \end{bmatrix}.$$

One method for finding the multiplicative inverse of a matrix A is to begin by denoting the elements in A^{-1} with variables. Using the equation $AA^{-1} = I_n$, we can find a value for each element in the multiplicative inverse that is represented by a variable. Example 2 shows how this is done.

EXAMPLE 2 Finding the Multiplicative Inverse of a Matrix

Find the multiplicative inverse of

$$A = \begin{bmatrix} 2 & 1 \\ 5 & 3 \end{bmatrix}.$$

SOLUTION

Let us denote the multiplicative inverse by

$$A^{-1} = \begin{bmatrix} w & x \\ y & z \end{bmatrix}.$$

Because A is a 2×2 matrix, we use the equation $AA^{-1} = I_2$ to find values for $w, x, y,$ and z.

$$\begin{matrix} A & A^{-1} & I_2 \end{matrix}$$

$$\begin{bmatrix} 2 & 1 \\ 5 & 3 \end{bmatrix} \begin{bmatrix} w & x \\ y & z \end{bmatrix} = \begin{bmatrix} 1 & 0 \\ 0 & 1 \end{bmatrix}$$

$$\begin{bmatrix} 2w + y & 2x + z \\ 5w + 3y & 5x + 3z \end{bmatrix} = \begin{bmatrix} 1 & 0 \\ 0 & 1 \end{bmatrix}$$

Use row-by-column matrix multiplication on the left side of $\begin{bmatrix} 2 & 1 \\ 5 & 3 \end{bmatrix} \begin{bmatrix} w & x \\ y & z \end{bmatrix} = \begin{bmatrix} 1 & 0 \\ 0 & 1 \end{bmatrix}$.

We now equate corresponding elements to obtain the following two systems of linear equations:

$$\begin{cases} 2w + y = 1 \\ 5w + 3y = 0 \end{cases} \text{ and } \begin{cases} 2x + z = 0 \\ 5x + 3z = 1. \end{cases}$$

Each of these systems can be solved using the addition method.

$$\begin{cases} 2w + y = 1 \\ 5w + 3y = 0 \end{cases}$$

Multiply by -3.

No change

$$\begin{aligned} -6w - 3y &= -3 \\ \underline{5w + 3y} &= \underline{\ 0} \end{aligned}$$

Add: $-w\ \ \ \ \ \ = -3$

$w = 3$

Use back-substitution.

$y = -5$

$$\begin{cases} 2x + z = 0 \\ 5x + 3z = 1 \end{cases}$$

Multiply by -3.

No change

$$\begin{aligned} -6x - 3z &= 0 \\ \underline{5x + 3z} &= \underline{\ 1} \end{aligned}$$

Add: $-x\ \ \ \ \ \ = 1$

$x = -1$

Use back-substitution.

$z = 2$

Using these values, we have

$$A^{-1} = \begin{bmatrix} w & x \\ y & z \end{bmatrix} = \begin{bmatrix} 3 & -1 \\ -5 & 2 \end{bmatrix}.$$

$\bullet\bullet\bullet$

TECHNOLOGY

You can use a graphing utility to find the inverse of the matrix in Example 2. Enter the matrix and name it A. The screens show A and A^{-1}. Verify that this is correct by showing that

$AA^{-1} = I_2$ and $A^{-1}A = I_2$.

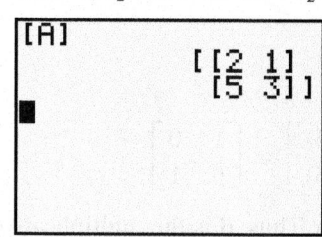

[A]
$$\begin{bmatrix} [2 & 1] \\ [5 & 3] \end{bmatrix}$$

[A]⁻¹
$$\begin{bmatrix} [3 & -1] \\ [-5 & 2] \end{bmatrix}$$

✐ **Check Point 2** Find the multiplicative inverse of $A = \begin{bmatrix} 5 & 7 \\ 2 & 3 \end{bmatrix}$.

Only square matrices of order $n \times n$ have multiplicative inverses, but not every square matrix possesses a multiplicative inverse. For example, suppose that you apply the procedure of Example 2 to $A = \begin{bmatrix} -6 & 4 \\ -3 & 2 \end{bmatrix}$:

This is
A.

This represents
A^{-1}.

This is the
multiplicative
identity matrix.

$$\begin{bmatrix} -6 & 4 \\ -3 & 2 \end{bmatrix} \begin{bmatrix} w & x \\ y & z \end{bmatrix} = \begin{bmatrix} 1 & 0 \\ 0 & 1 \end{bmatrix}.$$

Multiplying matrices on the left and equating corresponding elements results in inconsistent systems with no solutions. There are no values for $w, x, y,$ and z. This shows that matrix A does not have a multiplicative inverse.

A nonsquare matrix, one with a different number of rows than columns, cannot have a multiplicative inverse. If A is an $m \times n$ matrix and B is an $n \times m$ matrix $(n \neq m)$, then the products AB and BA are of different orders. This means that they could not be equal to each other, so that AB and BA could not both equal the multiplicative identity matrix.

If a square matrix has a multiplicative inverse, that inverse is unique. This means that the square matrix has no more than one inverse. If a square matrix has a multiplicative inverse, it is said to be **invertible** or **nonsingular**. If a square matrix has no multiplicative inverse, it is called **singular**.

A Quick Method for Finding the Multiplicative Inverse of a 2 × 2 Matrix

The same method used in Example 2 can be used to develop the general form of the multiplicative inverse of a 2×2 matrix. The following rule enables us to calculate the multiplicative inverse, if there is one:

Multiplicative Inverse of a 2 × 2 Matrix

If $A = \begin{bmatrix} a & b \\ c & d \end{bmatrix}$, then $A^{-1} = \dfrac{1}{ad - bc} \begin{bmatrix} d & -b \\ -c & a \end{bmatrix}$.

The matrix A is invertible if and only if $ad - bc \neq 0$. If $ad - bc = 0$, then A does not have a multiplicative inverse.

EXAMPLE 3 Using the Quick Method to Find a Multiplicative Inverse

Find the multiplicative inverse of

$$A = \begin{bmatrix} -1 & -2 \\ 3 & 4 \end{bmatrix}.$$

SOLUTION

a b

$$A = \begin{bmatrix} -1 & -2 \\ 3 & 4 \end{bmatrix}$$

c d

This is the given matrix.
We've designated the
elements a, b, c, and d.

GREAT QUESTION!

Can you provide any suggestions that will help me remember the formula for the multiplicative inverse of a 2 × 2 matrix?

Yes. To find the matrix that appears as the second factor for the inverse of

$$A = \begin{bmatrix} a & b \\ c & d \end{bmatrix}:$$

• Reverse a and d, the numbers in the diagonal from upper left to lower right.

• Negate b and c, the numbers in the other diagonal.

GREAT QUESTION!

When using the formula to find the multiplicative inverse of

$$\begin{bmatrix} a & b \\ c & d \end{bmatrix},$$

what should I do first?

Start by computing $ad - bc$. If the computed value is 0, there is no need to continue. The given matrix is singular—that is, it does not have a multiplicative inverse.

$$A = \begin{bmatrix} -1 & -2 \\ 3 & 4 \end{bmatrix}$$

a *b*

c *d*

The given matrix (repeated)

$$A^{-1} = \frac{1}{ad - bc} \begin{bmatrix} d & -b \\ -c & a \end{bmatrix}$$

This is the formula for the inverse of $\begin{bmatrix} a & b \\ c & d \end{bmatrix}$.

$$= \frac{1}{(-1)(4) - (-2)(3)} \begin{bmatrix} 4 & -(-2) \\ -3 & -1 \end{bmatrix}$$

Apply the formula with $a = -1$, $b = -2$, $c = 3$, and $d = 4$.

$$= \frac{1}{2} \begin{bmatrix} 4 & 2 \\ -3 & -1 \end{bmatrix}$$

Simplify.

$$= \begin{bmatrix} 2 & 1 \\ -\frac{3}{2} & -\frac{1}{2} \end{bmatrix}$$

Perform the scalar multiplication by multiplying each element in the matrix by $\frac{1}{2}$.

The inverse of $A = \begin{bmatrix} -1 & -2 \\ 3 & 4 \end{bmatrix}$ is $A^{-1} = \begin{bmatrix} 2 & 1 \\ -\frac{3}{2} & -\frac{1}{2} \end{bmatrix}$.

We can verify this result by showing that $AA^{-1} = I_2$ and $A^{-1}A = I_2$. ● ● ●

⊘ Check Point **3** Find the multiplicative inverse of

$$A = \begin{bmatrix} 3 & -2 \\ -1 & 1 \end{bmatrix}.$$

Finding Multiplicative Inverses of $n \times n$ Matrices with n Greater Than 2

To find the multiplicative inverse of a 3×3 invertible matrix, we begin by denoting the elements in the multiplicative inverse with variables. Here is an example:

$$\begin{bmatrix} -1 & -1 & -1 \\ 4 & 5 & 0 \\ 0 & 1 & -3 \end{bmatrix} \begin{bmatrix} x_1 & x_2 & x_3 \\ y_1 & y_2 & y_3 \\ z_1 & z_2 & z_3 \end{bmatrix} = \begin{bmatrix} 1 & 0 & 0 \\ 0 & 1 & 0 \\ 0 & 0 & 1 \end{bmatrix}.$$

This is matrix A whose inverse we wish to find. This represents A^{-1}. This is the multiplicative identity matrix, I_3.

We multiply the matrices on the left, using the row-by-column definition of matrix multiplication.

$$\begin{bmatrix} -x_1 - y_1 - z_1 & -x_2 - y_2 - z_2 & -x_3 - y_3 - z_3 \\ 4x_1 + 5y_1 + 0z_1 & 4x_2 + 5y_2 + 0z_2 & 4x_3 + 5y_3 + 0z_3 \\ 0x_1 + 1y_1 - 3z_1 & 0x_2 + 1y_2 - 3z_2 & 0x_3 + 1y_3 - 3z_3 \end{bmatrix} = \begin{bmatrix} 1 & 0 & 0 \\ 0 & 1 & 0 \\ 0 & 0 & 1 \end{bmatrix}$$

We now equate corresponding entries to obtain the following three systems of linear equations:

$$\begin{cases} -x_1 - y_1 - z_1 = 1 \\ 4x_1 + 5y_1 + 0z_1 = 0 \\ 0x_1 + y_1 - 3z_1 = 0 \end{cases} \quad \begin{cases} -x_2 - y_2 - z_2 = 0 \\ 4x_2 + 5y_2 + 0z_2 = 1 \\ 0x_2 + y_2 - 3z_2 = 0 \end{cases} \quad \begin{cases} -x_3 - y_3 - z_3 = 0 \\ 4x_3 + 5y_3 + 0z_3 = 0 \\ 0x_3 + y_3 - 3z_3 = 1. \end{cases}$$

Notice that the variables on the left of the equal signs have the same coefficients in each system. We can use Gauss-Jordan elimination to solve all three systems at once. Form an augmented matrix that contains the coefficients of the three systems to the left of the vertical line and the constants for the systems to the right.

$$\left[\begin{array}{ccc|ccc} -1 & -1 & -1 & 1 & 0 & 0 \\ 4 & 5 & 0 & 0 & 1 & 0 \\ 0 & 1 & -3 & 0 & 0 & 1 \end{array}\right]$$

| Coefficients of the three systems | Constants on the right in each of the three systems |

To solve all three systems using Gauss-Jordan elimination, we must obtain $\left[\begin{array}{ccc} 1 & 0 & 0 \\ 0 & 1 & 0 \\ 0 & 0 & 1 \end{array}\right]$

to the left of the vertical line. Use matrix row operations, working one column at a time. Obtain 1 in the required position. Then obtain 0s in the other two positions. Using these operations, we obtain the matrix

$$\left[\begin{array}{ccc|ccc} 1 & 0 & 0 & 15 & 4 & -5 \\ 0 & 1 & 0 & -12 & -3 & 4 \\ 0 & 0 & 1 & -4 & -1 & 1 \end{array}\right].$$

This augmented matrix provides the solutions to the three systems of equations. They are given by

$$\left[\begin{array}{ccc|c} 1 & 0 & 0 & 15 \\ 0 & 1 & 0 & -12 \\ 0 & 0 & 1 & -4 \end{array}\right] \quad \begin{array}{l} x_1 = 15 \\ y_1 = -12 \\ z_1 = -4 \end{array}$$

and $\left[\begin{array}{ccc|c} 1 & 0 & 0 & 4 \\ 0 & 1 & 0 & -3 \\ 0 & 0 & 1 & -1 \end{array}\right] \quad \begin{array}{l} x_2 = 4 \\ y_2 = -3 \\ z_2 = -1 \end{array}$ and $\left[\begin{array}{ccc|c} 1 & 0 & 0 & -5 \\ 0 & 1 & 0 & 4 \\ 0 & 0 & 1 & 1 \end{array}\right] \quad \begin{array}{l} x_3 = -5 \\ y_3 = 4 \\ z_3 = 1. \end{array}$

Using the preceding nine values, the inverse matrix is

$$\left[\begin{array}{ccc} x_1 & x_2 & x_3 \\ y_1 & y_2 & y_3 \\ z_1 & z_2 & z_3 \end{array}\right] = \left[\begin{array}{ccc} 15 & 4 & -5 \\ -12 & -3 & 4 \\ -4 & -1 & 1 \end{array}\right].$$

Take a second look at the matrix obtained at the point where Gauss-Jordan elimination was completed. This matrix is shown, again, below. Notice that the 3×3 matrix to the right of the vertical bar is the multiplicative inverse of A. Also notice that the multiplicative identity matrix, I_3 is the matrix that appears to the left of the vertical bar.

$$\left[\begin{array}{ccc|ccc} 1 & 0 & 0 & 15 & 4 & -5 \\ 0 & 1 & 0 & -12 & -3 & 4 \\ 0 & 0 & 1 & -4 & -1 & 1 \end{array}\right]$$

| This is the multiplicative identity, I_3. | This is the multiplicative inverse of A. |

The observations in the voice balloons and the procedures followed above give us a general method for finding the multiplicative inverse of an invertible matrix. This method is given at the top of the next page.

GREAT QUESTION!

Should I use the procedure in the box for finding the multiplicative inverse of a 2 × 2 matrix?

No. Because we have a quick method for finding the multiplicative inverse of a 2 × 2 matrix, the procedure on the right is recommended for matrices of order 3 × 3 or greater when a graphing utility is not being used.

Procedure for Finding the Multiplicative Inverse of an Invertible Matrix

To find A^{-1} for any $n \times n$ matrix A for which A^{-1} exists,

1. Form the augmented matrix $[A \mid I_n]$, where I_n is the multiplicative identity matrix of the same order as the given matrix A.
2. Perform row operations on $[A \mid I_n]$ to obtain a matrix of the form $[I_n \mid B]$. This is equivalent to using Gauss-Jordan elimination to change A into the identity matrix.
3. Matrix B is A^{-1}.
4. Verify the result by showing that $AA^{-1} = I_n$ and $A^{-1}A = I_n$.

EXAMPLE 4 Finding the Multiplicative Inverse of a 3 × 3 Matrix

Find the multiplicative inverse of

$$A = \begin{bmatrix} 1 & -1 & 1 \\ 0 & -2 & 1 \\ -2 & -3 & 0 \end{bmatrix}.$$

SOLUTION

Step 1 Form the augmented matrix $[A \mid I_3]$.

$$\begin{bmatrix} 1 & -1 & 1 & 1 & 0 & 0 \\ 0 & -2 & 1 & 0 & 1 & 0 \\ -2 & -3 & 0 & 0 & 0 & 1 \end{bmatrix}$$

This is matrix A.

This is I_3, the multiplicative identity matrix, with 1s down the main diagonal and 0s elsewhere.

Step 2 Perform row operations on $[A \mid I_3]$ to obtain a matrix of the form $[I_3 \mid B]$. To the left of the vertical dividing line, we want 1s down the diagonal from upper left to lower right and 0s elsewhere.

$$\begin{bmatrix} 1 & -1 & 1 & 1 & 0 & 0 \\ 0 & -2 & 1 & 0 & 1 & 0 \\ -2 & -3 & 0 & 0 & 0 & 1 \end{bmatrix}$$

Replace row 3 by $2R_1 + R_3$.

$$\xrightarrow{} \begin{bmatrix} 1 & -1 & 1 & 1 & 0 & 0 \\ 0 & -2 & 1 & 0 & 1 & 0 \\ 0 & -5 & 2 & 2 & 0 & 1 \end{bmatrix} \xrightarrow{-\frac{1}{2}R_2}$$

$$\begin{bmatrix} 1 & -1 & 1 & 1 & 0 & 0 \\ 0 & 1 & -\frac{1}{2} & 0 & -\frac{1}{2} & 0 \\ 0 & -5 & 2 & 2 & 0 & 1 \end{bmatrix}$$

Replace row 1 by $1R_2 + R_1$.
Replace row 3 by $5R_2 + R_3$.

$$\xrightarrow{} \begin{bmatrix} 1 & 0 & \frac{1}{2} & 1 & -\frac{1}{2} & 0 \\ 0 & 1 & -\frac{1}{2} & 0 & -\frac{1}{2} & 0 \\ 0 & 0 & -\frac{1}{2} & 2 & -\frac{5}{2} & 1 \end{bmatrix} \xrightarrow{-2R_3}$$

$$\begin{bmatrix} 1 & 0 & \frac{1}{2} & 1 & -\frac{1}{2} & 0 \\ 0 & 1 & -\frac{1}{2} & 0 & -\frac{1}{2} & 0 \\ 0 & 0 & 1 & -4 & 5 & -2 \end{bmatrix}$$

Replace row 1 by $-\frac{1}{2}R_3 + R_1$.
Replace row 2 by $\frac{1}{2}R_3 + R_2$.

$$\xrightarrow{} \begin{bmatrix} 1 & 0 & 0 & 3 & -3 & 1 \\ 0 & 1 & 0 & -2 & 2 & -1 \\ 0 & 0 & 1 & -4 & 5 & -2 \end{bmatrix}$$

This is the multiplicative identity, I_3.

This is the multiplicative inverse of A.

TECHNOLOGY

We can use a graphing utility to verify the inverse matrix we found in Example 4. Enter the elements in matrix A and press $\boxed{x^{-1}}$ to display A^{-1}.

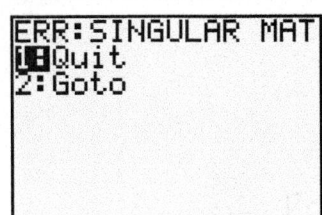

TECHNOLOGY

The matrix

$$A = \begin{bmatrix} 4 & 6 \\ 2 & 3 \end{bmatrix} = \begin{bmatrix} a & b \\ c & d \end{bmatrix}$$

has no multiplicative inverse because

$$ad - bc = 4 \cdot 3 - 6 \cdot 2$$
$$= 12 - 12 = 0.$$

When we try to find the inverse with a graphing utility, an ERROR message occurs, indicating the matrix is singular.

```
[A]
        [[4 6]
         [2 3]]
[A]-1
■
```

```
ERR:SINGULAR MAT
1■Quit
2:Goto
```

Step 3 **Matrix B is A^{-1}.** The last matrix shown at the bottom of the previous page is in the form $[I_3 \mid B]$. The multiplicative identity matrix is on the left of the vertical bar. Matrix B, the multiplicative inverse of A, is on the right. Thus, the multiplicative inverse of A is

$$A^{-1} = \begin{bmatrix} 3 & -3 & 1 \\ -2 & 2 & -1 \\ -4 & 5 & -2 \end{bmatrix}.$$

Step 4 **Verify the result by showing that $AA^{-1} = I_3$ and $A^{-1}A = I_3$.** Try confirming the result by multiplying A and A^{-1} to obtain I_3. Do you obtain I_3 if you reverse the order of the multiplication? • • •

We have seen that not all square matrices have multiplicative inverses. If the row operations in step 2 result in all zeros in a row or column to the left of the vertical line, the given matrix does not have a multiplicative inverse.

⌀ Check Point **4** Find the multiplicative inverse of

$$A = \begin{bmatrix} 1 & 0 & 2 \\ -1 & 2 & 3 \\ 1 & -1 & 0 \end{bmatrix}.$$

Summary: Finding Multiplicative Inverses for Invertible Matrices

Use a graphing utility with matrix capabilities, or

a. If the matrix is 2×2: The inverse of $A = \begin{bmatrix} a & b \\ c & d \end{bmatrix}$ is

$$A^{-1} = \frac{1}{ad - bc} \begin{bmatrix} d & -b \\ -c & a \end{bmatrix}.$$

b. If the matrix A is $n \times n$ where $n > 2$: Use the procedure on page 644. Form $[A \mid I_n]$ and use row transformations to obtain $[I_n \mid B]$. Then $A^{-1} = B$.

Solving Systems of Equations Using Multiplicative Inverses of Matrices

Matrix multiplication can be used to represent a system of linear equations.

Linear System

$$\begin{cases} a_1x + b_1y + c_1z = d_1 \\ a_2x + b_2y + c_2z = d_2 \\ a_3x + b_3y + c_3z = d_3 \end{cases}$$

Matrix Form of the System

$$\begin{bmatrix} a_1 & b_1 & c_1 \\ a_2 & b_2 & c_2 \\ a_3 & b_3 & c_3 \end{bmatrix} \begin{bmatrix} x \\ y \\ z \end{bmatrix} = \begin{bmatrix} d_1 \\ d_2 \\ d_3 \end{bmatrix}$$

| The matrix contains the system's coefficients. | The matrix contains the system's variables. | The matrix contains the system's constants. |

You can work with the matrix form of the system and obtain the form of the linear system on the left. To do so, perform the matrix multiplication on the left side of the matrix equation. Then equate the corresponding elements.

The matrix equation

$$\begin{bmatrix} a_1 & b_1 & c_1 \\ a_2 & b_2 & c_2 \\ a_3 & b_3 & c_3 \end{bmatrix} \begin{bmatrix} x \\ y \\ z \end{bmatrix} = \begin{bmatrix} d_1 \\ d_2 \\ d_3 \end{bmatrix}$$

$$\downarrow \qquad\quad \downarrow \quad\ \downarrow$$
$$A \qquad\quad X \ = \ B$$

is abbreviated as $AX = B$, where A is the **coefficient matrix** of the system, and X and B are matrices containing one column, called **column matrices**. The matrix B is called the **constant matrix**.

Here is a specific example of a linear system and its matrix form:

Linear System

$$\begin{cases} x - y + z = 2 \\ \quad - 2y + z = 2 \\ -2x - 3y \quad\quad = \frac{1}{2} \end{cases}$$

Matrix Form

Coefficients

$$\begin{bmatrix} 1 & -1 & 1 \\ 0 & -2 & 1 \\ -2 & -3 & 0 \end{bmatrix} \begin{bmatrix} x \\ y \\ z \end{bmatrix} = \begin{bmatrix} 2 \\ 2 \\ \frac{1}{2} \end{bmatrix}$$

Constants

A, the coefficient matrix X B, the constant matrix

The matrix equation $AX = B$ can be solved using A^{-1} if it exists.

$AX = B$ This is the matrix equation.

$A^{-1}AX = A^{-1}B$ Multiply both sides by A^{-1}. Because matrix multiplication is not commutative, put A^{-1} in the same left position on both sides.

$I_nX = A^{-1}B$ The multiplicative inverse property tells us that $A^{-1}A = I_n$.

$X = A^{-1}B$ Because I_n is the multiplicative identity, $I_nX = X$.

We see that if $AX = B$, then $X = A^{-1}B$.

② Use inverses to solve matrix equations.

Solving a System Using A^{-1}

If $AX = B$ has a unique solution, then $X = A^{-1}B$. To solve a linear system of equations, multiply A^{-1} and B to find X.

EXAMPLE 5 Using the Inverse of a Matrix to Solve a System

Solve the system by using A^{-1}, the inverse of the coefficient matrix:

$$\begin{cases} x - y + z = 2 \\ \quad - 2y + z = 2 \\ -2x - 3y \quad\quad = \frac{1}{2}. \end{cases}$$

TECHNOLOGY

We can use a graphing utility to solve a linear system with a unique solution by entering the elements in A, the coefficient matrix, and B, the column matrix. Then find the product of A^{-1} and B. The screen below verifies our solution in Example 5.

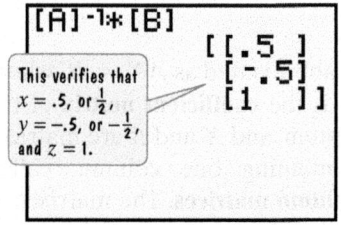

```
[A]⁻¹*[B]
            [[.5 ]
             [-.5]
             [1  ]]
```
This verifies that
$x = .5$, or $\frac{1}{2}$,
$y = -.5$, or $-\frac{1}{2}$,
and $z = 1$.

SOLUTION

The linear system can be written as

$$\begin{bmatrix} 1 & -1 & 1 \\ 0 & -2 & 1 \\ -2 & -3 & 0 \end{bmatrix} \begin{bmatrix} x \\ y \\ z \end{bmatrix} = \begin{bmatrix} 2 \\ 2 \\ \frac{1}{2} \end{bmatrix}.$$

$\quad\quad A \quad\quad\quad X \quad\quad B$

The solution is given by $X = A^{-1}B$. Consequently, we must find A^{-1}. We found the inverse of matrix A in Example 4. Using this result,

$$X = A^{-1}B = \begin{bmatrix} 3 & -3 & 1 \\ -2 & 2 & -1 \\ -4 & 5 & -2 \end{bmatrix} \begin{bmatrix} 2 \\ 2 \\ \frac{1}{2} \end{bmatrix} = \begin{bmatrix} 3\cdot2 + (-3)\cdot2 + 1\cdot\frac{1}{2} \\ -2\cdot2 + 2\cdot2 + (-1)\cdot\frac{1}{2} \\ -4\cdot2 + 5\cdot2 + (-2)\cdot\frac{1}{2} \end{bmatrix} = \begin{bmatrix} \frac{1}{2} \\ -\frac{1}{2} \\ 1 \end{bmatrix}.$$

Thus, $x = \frac{1}{2}$, $y = -\frac{1}{2}$, and $z = 1$. The solution set is $\left\{\left(\frac{1}{2}, -\frac{1}{2}, 1\right)\right\}$. • • •

 Check Point **5** Solve the system by using A^{-1}, the inverse of the coefficient matrix that you found in Check Point 4:

$$\begin{cases} x & + 2z = 6 \\ -x + 2y + 3z = -5 \\ x - y & = 6. \end{cases}$$

 Encode and decode messages.

Applications of Matrix Inverses to Coding

A **cryptogram** is a message written so that no one other than the intended recipient can understand it. To encode a message, we begin by assigning a number to each letter in the alphabet: $A = 1, B = 2, C = 3,\ldots, Z = 26$, and a space $= 0$. For example, the numerical equivalent of the word MATH is $13, 1, 20, 8$. The numerical equivalent of the message is then converted into a matrix. Finally, an invertible matrix can be used to convert the message into code. The multiplicative inverse of this matrix can be used to decode the message.

Encoding a Word or Message

1. Express the word or message numerically.
2. List the numbers in step 1 by columns and form a square matrix. If you do not have enough numbers to form a square matrix, put zeros in any remaining spaces in the last column.
3. Select any square invertible matrix, called the **coding matrix**, the same size as the matrix in step 2. Multiply the coding matrix by the square matrix that expresses the message numerically. The resulting matrix is the **coded matrix**.
4. Use the numbers, by columns, from the coded matrix in step 3 to write the encoded message.

EXAMPLE 6 Encoding a Word

Use matrices to encode the word MATH.

SOLUTION

Step 1 Express the word numerically. As shown previously, the numerical equivalent of MATH is $13, 1, 20, 8$.

Step 2 List the numbers in step 1 by columns and form a square matrix. The 2×2 matrix for the numerical equivalent of MATH, $13, 1, 20, 8$, is

$$\begin{bmatrix} 13 & 20 \\ 1 & 8 \end{bmatrix}.$$

Step 3 Multiply the matrix in step 2 by a square invertible matrix. We will use $\begin{bmatrix} -2 & -3 \\ 3 & 4 \end{bmatrix}$ as the coding matrix.

$$\begin{bmatrix} -2 & -3 \\ 3 & 4 \end{bmatrix}\begin{bmatrix} 13 & 20 \\ 1 & 8 \end{bmatrix} = \begin{bmatrix} -2(13) - 3(1) & -2(20) - 3(8) \\ 3(13) + 4(1) & 3(20) + 4(8) \end{bmatrix}$$

Coding matrix · Numerical representation of MATH

$$= \begin{bmatrix} -29 & -64 \\ 43 & 92 \end{bmatrix}$$

Coded matrix

Step 4 Use the numbers, by columns, from the coded matrix in step 3 to write the encoded message. The encoded message is $-29, 43, -64, 92$. ● ● ●

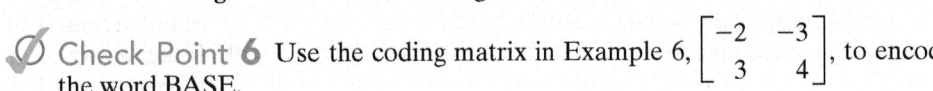 Check Point **6** Use the coding matrix in Example 6, $\begin{bmatrix} -2 & -3 \\ 3 & 4 \end{bmatrix}$, to encode the word BASE.

The inverse of a coding matrix can be used to decode a word or message that was encoded.

Decoding a Word or Message That Was Encoded

1. Find the multiplicative inverse of the coding matrix.
2. Multiply the multiplicative inverse of the coding matrix and the coded matrix.
3. Express the numbers, by columns, from the matrix in step 2 as letters.

EXAMPLE 7 Decoding a Word

Decode $-29, 43, -64, 92$ from Example 6.

SOLUTION

Step 1 Find the inverse of the coding matrix. The coding matrix in Example 6 was $\begin{bmatrix} -2 & -3 \\ 3 & 4 \end{bmatrix}$. We use the formula for the multiplicative inverse of a 2×2 matrix to find the multiplicative inverse of this matrix. It is $\begin{bmatrix} 4 & 3 \\ -3 & -2 \end{bmatrix}$.

Step 2 Multiply the multiplicative inverse of the coding matrix and the coded matrix.

$$\underset{\substack{\text{Multiplicative inverse} \\ \text{of the coding matrix}}}{\begin{bmatrix} 4 & 3 \\ -3 & -2 \end{bmatrix}} \underset{\substack{\text{Coded} \\ \text{matrix}}}{\begin{bmatrix} -29 & -64 \\ 43 & 92 \end{bmatrix}} = \begin{bmatrix} 4(-29) + 3(43) & 4(-64) + 3(92) \\ -3(-29) - 2(43) & -3(-64) - 2(92) \end{bmatrix}$$

$$= \begin{bmatrix} 13 & 20 \\ 1 & 8 \end{bmatrix}$$

Step 3 Express the numbers, by columns, from the matrix in step 2 as letters. The numbers are $13, 1, 20$, and 8. Using letters, the decoded message is MATH. • • •

⊘ **Check Point 7** Decode the word that you encoded in Check Point 6.

Decoding is simple for an authorized receiver who knows the coding matrix. Because any invertible matrix can be used for the coding matrix, decoding a cryptogram for an unauthorized receiver who does not know this matrix is extremely difficult.

CONCEPT AND VOCABULARY CHECK

Fill in each blank so that the resulting statement is true.

1. The multiplicative identity matrix of order 2 is
$$I_2 = \underline{\hspace{2cm}}.$$

2. The multiplicative identity matrix of order 3 is
$$I_3 = \underline{\hspace{2cm}}.$$

3. For $n \times n$ matrices A and B, if $AB = I_n$ and $BA = I_n$, then B is called the _____ of A.

4. True or false: Only square matrices have multiplicative inverses. _____

5. If $A = \begin{bmatrix} a & b \\ c & d \end{bmatrix}$, the matrix A is invertible if and only if _____.

6. If a square matrix does not have a multiplicative inverse, it is called _____.

7. True or false: $A = \begin{bmatrix} 3 & 2 \\ 9 & 6 \end{bmatrix}$ is invertible. _____

8. To find the multiplicative inverse of an invertible matrix A, we perform row operations on $[A \,|\, I_n]$ to obtain a matrix of the form $[I_n \,|\, B]$, where $B = \underline{\hspace{1cm}}$.

9. If the matrix equation $AX = B$ has a unique solution, then we can solve the equation using $X = \underline{\hspace{1cm}}$.

EXERCISE SET 6.4

Practice Exercises

In Exercises 1–12, find the products AB and BA to determine whether B is the multiplicative inverse of A.

1. $A = \begin{bmatrix} 4 & -3 \\ -5 & 4 \end{bmatrix}$, $B = \begin{bmatrix} 4 & 3 \\ 5 & 4 \end{bmatrix}$

2. $A = \begin{bmatrix} -2 & -1 \\ -1 & 1 \end{bmatrix}$, $B = \begin{bmatrix} 1 & 1 \\ 1 & 2 \end{bmatrix}$

3. $A = \begin{bmatrix} -4 & 0 \\ 1 & 3 \end{bmatrix}$, $B = \begin{bmatrix} -2 & 4 \\ 0 & 1 \end{bmatrix}$

4. $A = \begin{bmatrix} -2 & 4 \\ 1 & -2 \end{bmatrix}$, $B = \begin{bmatrix} 1 & 2 \\ -1 & -2 \end{bmatrix}$

5. $A = \begin{bmatrix} -2 & 1 \\ \frac{3}{2} & -\frac{1}{2} \end{bmatrix}$, $B = \begin{bmatrix} 1 & 2 \\ 3 & 4 \end{bmatrix}$

6. $A = \begin{bmatrix} 4 & 5 \\ 2 & 3 \end{bmatrix}$, $B = \begin{bmatrix} \frac{3}{2} & -\frac{5}{2} \\ -1 & 2 \end{bmatrix}$

7. $A = \begin{bmatrix} 0 & 1 & 0 \\ 0 & 0 & 1 \\ 1 & 0 & 0 \end{bmatrix}$, $B = \begin{bmatrix} 0 & 0 & 1 \\ 1 & 0 & 0 \\ 0 & 1 & 0 \end{bmatrix}$

8. $A = \begin{bmatrix} -2 & 1 & -1 \\ -5 & 2 & -1 \\ 3 & -1 & 1 \end{bmatrix}$, $B = \begin{bmatrix} 1 & 0 & 1 \\ 2 & 1 & 3 \\ -1 & 1 & 1 \end{bmatrix}$

9. $A = \begin{bmatrix} 1 & 2 & 3 \\ 1 & 3 & 4 \\ 1 & 4 & 3 \end{bmatrix}$, $B = \begin{bmatrix} \frac{7}{2} & -3 & \frac{1}{2} \\ -\frac{1}{2} & 0 & \frac{1}{2} \\ -\frac{1}{2} & 1 & -\frac{1}{2} \end{bmatrix}$

10. $A = \begin{bmatrix} 0 & 2 & 0 \\ 3 & 3 & 2 \\ 2 & 5 & 1 \end{bmatrix}$, $B = \begin{bmatrix} -3.5 & -1 & 2 \\ 0.5 & 0 & 0 \\ 4.5 & 2 & -3 \end{bmatrix}$

11. $A = \begin{bmatrix} 0 & 0 & -2 & 1 \\ -1 & 0 & 1 & 1 \\ 0 & 1 & -1 & 0 \\ 1 & 0 & 0 & -1 \end{bmatrix}$, $B = \begin{bmatrix} 1 & 2 & 0 & 3 \\ 0 & 1 & 1 & 1 \\ 0 & 1 & 0 & 1 \\ 1 & 2 & 0 & 2 \end{bmatrix}$

12. $A = \begin{bmatrix} 1 & -2 & 1 & 0 \\ 0 & 1 & -2 & 1 \\ 0 & 0 & 1 & -2 \\ 0 & 0 & 0 & 1 \end{bmatrix}$, $B = \begin{bmatrix} 1 & 2 & 3 & 4 \\ 0 & 1 & 2 & 3 \\ 0 & 0 & 1 & 2 \\ 0 & 0 & 0 & 1 \end{bmatrix}$

In Exercises 13–18, use the fact that if $A = \begin{bmatrix} a & b \\ c & d \end{bmatrix}$, then

$A^{-1} = \dfrac{1}{ad - bc}\begin{bmatrix} d & -b \\ -c & a \end{bmatrix}$ *to find the inverse of each matrix, if possible. Check that $AA^{-1} = I_2$ and $A^{-1}A = I_2$.*

13. $A = \begin{bmatrix} 2 & 3 \\ -1 & 2 \end{bmatrix}$

14. $A = \begin{bmatrix} 0 & 3 \\ 4 & -2 \end{bmatrix}$

15. $A = \begin{bmatrix} 3 & -1 \\ -4 & 2 \end{bmatrix}$

16. $A = \begin{bmatrix} 2 & -6 \\ 1 & -2 \end{bmatrix}$

17. $A = \begin{bmatrix} 10 & -2 \\ -5 & 1 \end{bmatrix}$

18. $A = \begin{bmatrix} 6 & -3 \\ -2 & 1 \end{bmatrix}$

In Exercises 19–28, find A^{-1} by forming $[A\,|\,I]$ and then using row operations to obtain $[I\,|\,B]$, where $A^{-1} = [B]$. Check that $AA^{-1} = I$ and $A^{-1}A = I$.

19. $A = \begin{bmatrix} 2 & 0 & 0 \\ 0 & 4 & 0 \\ 0 & 0 & 6 \end{bmatrix}$

20. $A = \begin{bmatrix} 3 & 0 & 0 \\ 0 & 6 & 0 \\ 0 & 0 & 9 \end{bmatrix}$

21. $A = \begin{bmatrix} 1 & 2 & -1 \\ -2 & 0 & 1 \\ 1 & -1 & 0 \end{bmatrix}$

22. $A = \begin{bmatrix} 1 & -1 & 1 \\ 0 & 2 & -1 \\ 2 & 3 & 0 \end{bmatrix}$

23. $A = \begin{bmatrix} 2 & 2 & -1 \\ 0 & 3 & -1 \\ -1 & -2 & 1 \end{bmatrix}$

24. $A = \begin{bmatrix} 2 & 4 & -4 \\ 1 & 3 & -4 \\ 2 & 4 & -3 \end{bmatrix}$

25. $A = \begin{bmatrix} 5 & 0 & 2 \\ 2 & 2 & 1 \\ -3 & 1 & -1 \end{bmatrix}$

26. $A = \begin{bmatrix} 3 & 2 & 6 \\ 1 & 1 & 2 \\ 2 & 2 & 5 \end{bmatrix}$

27. $A = \begin{bmatrix} 1 & 0 & 0 & 0 \\ 0 & -1 & 0 & 0 \\ 0 & 0 & 3 & 0 \\ 1 & 0 & 0 & 1 \end{bmatrix}$

28. $A = \begin{bmatrix} 2 & 0 & 0 & 1 \\ 0 & 1 & 0 & 0 \\ 0 & 0 & -1 & 0 \\ 0 & 0 & 0 & 2 \end{bmatrix}$

In Exercises 29–32, write each linear system as a matrix equation in the form $AX = B$, where A is the coefficient matrix and B is the constant matrix.

29. $\begin{cases} 6x + 5y = 13 \\ 5x + 4y = 10 \end{cases}$

30. $\begin{cases} 7x + 5y = 23 \\ 3x + 2y = 10 \end{cases}$

31. $\begin{cases} x + 3y + 4z = -3 \\ x + 2y + 3z = -2 \\ x + 4y + 3z = -6 \end{cases}$

32. $\begin{cases} x + 4y - z = 3 \\ x + 3y - 2z = 5 \\ 2x + 7y - 5z = 12 \end{cases}$

In Exercises 33–36, write each matrix equation as a system of linear equations without matrices.

33. $\begin{bmatrix} 4 & -7 \\ 2 & -3 \end{bmatrix}\begin{bmatrix} x \\ y \end{bmatrix} = \begin{bmatrix} -3 \\ 1 \end{bmatrix}$

34. $\begin{bmatrix} 3 & 0 \\ -3 & 1 \end{bmatrix}\begin{bmatrix} x \\ y \end{bmatrix} = \begin{bmatrix} 6 \\ -7 \end{bmatrix}$

35. $\begin{bmatrix} 2 & 0 & -1 \\ 0 & 3 & 0 \\ 1 & 1 & 0 \end{bmatrix}\begin{bmatrix} x \\ y \\ z \end{bmatrix} = \begin{bmatrix} 6 \\ 9 \\ 5 \end{bmatrix}$

36. $\begin{bmatrix} -1 & 0 & 1 \\ 0 & -1 & 0 \\ 0 & 1 & 1 \end{bmatrix}\begin{bmatrix} x \\ y \\ z \end{bmatrix} = \begin{bmatrix} -4 \\ 2 \\ 4 \end{bmatrix}$

In Exercises 37–42,

a. *Write each linear system as a matrix equation in the form $AX = B$.*

b. *Solve the system using the inverse that is given for the coefficient matrix.*

37. $\begin{cases} 2x + 6y + 6z = 8 \\ 2x + 7y + 6z = 10 \\ 2x + 7y + 7z = 9 \end{cases}$ The inverse of $\begin{bmatrix} 2 & 6 & 6 \\ 2 & 7 & 6 \\ 2 & 7 & 7 \end{bmatrix}$ is $\begin{bmatrix} \frac{7}{2} & 0 & -3 \\ -1 & 1 & 0 \\ 0 & -1 & 1 \end{bmatrix}$.

The inverse of

38. $\begin{cases} x + 2y + 5z = 2 \\ 2x + 3y + 8z = 3 \\ -x + y + 2z = 3 \end{cases}$ $\begin{bmatrix} 1 & 2 & 5 \\ 2 & 3 & 8 \\ -1 & 1 & 2 \end{bmatrix}$ is $\begin{bmatrix} 2 & -1 & -1 \\ 12 & -7 & -2 \\ -5 & 3 & 1 \end{bmatrix}$.

The inverse of

39. $\begin{cases} x - y + z = 8 \\ 2y - z = -7 \\ 2x + 3y = 1 \end{cases}$ $\begin{bmatrix} 1 & -1 & 1 \\ 0 & 2 & -1 \\ 2 & 3 & 0 \end{bmatrix}$ is $\begin{bmatrix} 3 & 3 & -1 \\ -2 & -2 & 1 \\ -4 & -5 & 2 \end{bmatrix}$.

The inverse of

40. $\begin{cases} x - 6y + 3z = 11 \\ 2x - 7y + 3z = 14 \\ 4x - 12y + 5z = 25 \end{cases}$ $\begin{bmatrix} 1 & -6 & 3 \\ 2 & -7 & 3 \\ 4 & -12 & 5 \end{bmatrix}$ is $\begin{bmatrix} 1 & -6 & 3 \\ 2 & -7 & 3 \\ 4 & -12 & 5 \end{bmatrix}$.

41. $\begin{cases} w - x + 2y \quad\;\; = -3 \\ x - y + z = 4 \\ -w + x - y + 2z = 2 \\ -x + y - 2z = -4 \end{cases}$

The inverse of

$\begin{bmatrix} 1 & -1 & 2 & 0 \\ 0 & 1 & -1 & 1 \\ -1 & 1 & -1 & 2 \\ 0 & -1 & 1 & -2 \end{bmatrix}$ is $\begin{bmatrix} 0 & 0 & -1 & -1 \\ 1 & 4 & 1 & 3 \\ 1 & 2 & 1 & 2 \\ 0 & -1 & 0 & -1 \end{bmatrix}$.

42. $\begin{cases} 2w + y + z = 6 \\ 3w + z = 9 \\ -w + x - 2y + z = 4 \\ 4w - x + y = 6 \end{cases}$

The inverse of

$\begin{bmatrix} 2 & 0 & 1 & 1 \\ 3 & 0 & 0 & 1 \\ -1 & 1 & -2 & 1 \\ 4 & -1 & 1 & 0 \end{bmatrix}$ is $\begin{bmatrix} -1 & 2 & -1 & -1 \\ -4 & 9 & -5 & -6 \\ 0 & 1 & -1 & -1 \\ 3 & -5 & 3 & 3 \end{bmatrix}$.

Practice Plus

In Exercises 43–44, find A^{-1} and check.

43. $A = \begin{bmatrix} e^x & e^{3x} \\ -e^{3x} & e^{5x} \end{bmatrix}$

44. $A = \begin{bmatrix} e^{2x} & -e^x \\ e^{3x} & e^{2x} \end{bmatrix}$

In Exercises 45–46, if I is the multiplicative identity matrix of order 2, find $(I - A)^{-1}$ for the given matrix A.

45. $\begin{bmatrix} 8 & -5 \\ -3 & 2 \end{bmatrix}$

46. $\begin{bmatrix} 7 & -5 \\ -4 & 3 \end{bmatrix}$

In Exercises 47–48, find $(AB)^{-1}$, $A^{-1}B^{-1}$, and $B^{-1}A^{-1}$. What do you observe?

47. $A = \begin{bmatrix} 2 & 1 \\ 3 & 1 \end{bmatrix}$ $B = \begin{bmatrix} 4 & 7 \\ 1 & 2 \end{bmatrix}$

48. $A = \begin{bmatrix} 2 & -9 \\ 1 & -4 \end{bmatrix}$ $B = \begin{bmatrix} 9 & 5 \\ 7 & 4 \end{bmatrix}$

49. Prove the following statement:

If $A = \begin{bmatrix} a & 0 & 0 \\ 0 & b & 0 \\ 0 & 0 & c \end{bmatrix}$, $a \neq 0, b \neq 0, c \neq 0,$

then $A^{-1} = \begin{bmatrix} \frac{1}{a} & 0 & 0 \\ 0 & \frac{1}{b} & 0 \\ 0 & 0 & \frac{1}{c} \end{bmatrix}$.

50. Prove the following statement:

If $A = \begin{bmatrix} a & b \\ c & d \end{bmatrix}$ and $ad - bc \neq 0$,

then $A^{-1} = \dfrac{1}{ad - bc}\begin{bmatrix} d & -b \\ -c & a \end{bmatrix}$.

(*Hint:* Use the method of Example 2 on page 640.)

Application Exercises

In Exercises 51–52, use the coding matrix

$$A = \begin{bmatrix} 4 & -1 \\ -3 & 1 \end{bmatrix} \text{ and its inverse } A^{-1} = \begin{bmatrix} 1 & 1 \\ 3 & 4 \end{bmatrix}$$

to encode and then decode the given message.

51. HELP

52. LOVE

In Exercises 53–54, use the coding matrix

$$A = \begin{bmatrix} 1 & -1 & 0 \\ 3 & 0 & 2 \\ -1 & 0 & -1 \end{bmatrix} \text{ and its inverse}$$

$$A^{-1} = \begin{bmatrix} 0 & 1 & 2 \\ -1 & 1 & 2 \\ 0 & -1 & -3 \end{bmatrix} \text{ to write a cryptogram for each}$$

message. Check your result by decoding the cryptogram.

53. S E N D _ C A S H
 19 5 14 4 0 3 1 19 8

Use $\begin{bmatrix} 19 & 4 & 1 \\ 5 & 0 & 19 \\ 14 & 3 & 8 \end{bmatrix}$.

54. S T A Y _ W E L L
 19 20 1 25 0 23 5 12 12

Use $\begin{bmatrix} 19 & 25 & 5 \\ 20 & 0 & 12 \\ 1 & 23 & 12 \end{bmatrix}$.

Writing in Mathematics

55. What is the multiplicative identity matrix?

56. If you are given two matrices, A and B, explain how to determine if B is the multiplicative inverse of A.

57. Explain why a matrix that does not have the same number of rows and columns cannot have a multiplicative inverse.

58. Explain how to find the multiplicative inverse for a 2×2 invertible matrix.

59. Explain how to find the multiplicative inverse for a 3×3 invertible matrix.

60. Explain how to write a linear system of three equations in three variables as a matrix equation.

61. Explain how to solve the matrix equation $AX = B$.

62. What is a cryptogram?

63. It's January 1, and you've written down your major goal for the year. You do not want those closest to you to see what you've written in case you do not accomplish your objective. Consequently, you decide to use a coding matrix to encode your goal. Explain how this can be accomplished.

64. A year has passed since Exercise 63. (Time flies when you're solving exercises in algebra books.) It's been a terrific year and so many wonderful things have happened that you can't remember your goal from a year ago. You consult your personal journal and you find the encoded message and the coding matrix. How can you use these to find your original goal?

Technology Exercises

In Exercises 65–70, use a graphing utility to find the multiplicative inverse of each matrix. Check that the displayed inverse is correct.

65. $\begin{bmatrix} 3 & -1 \\ -2 & 1 \end{bmatrix}$

66. $\begin{bmatrix} -4 & 1 \\ 6 & -2 \end{bmatrix}$

67. $\begin{bmatrix} -2 & 1 & -1 \\ -5 & 2 & -1 \\ 3 & -1 & 1 \end{bmatrix}$

68. $\begin{bmatrix} 1 & 1 & -1 \\ -3 & 2 & -1 \\ 3 & -3 & 2 \end{bmatrix}$

69. $\begin{bmatrix} 7 & -3 & 0 & 2 \\ -2 & 1 & 0 & -1 \\ 4 & 0 & 1 & -2 \\ -1 & 1 & 0 & -1 \end{bmatrix}$

70. $\begin{bmatrix} 1 & 2 & 0 & 0 \\ 0 & 0 & 1 & 0 \\ 1 & 3 & 0 & 1 \\ 4 & 0 & 0 & 2 \end{bmatrix}$

In Exercises 71–76, write each system in the form $AX = B$. Then solve the system by entering A and B into your graphing utility and computing $A^{-1}B$.

71. $\begin{cases} x - y + z = -6 \\ 4x + 2y + z = 9 \\ 4x - 2y + z = -3 \end{cases}$

72. $\begin{cases} y + 2z = 0 \\ -x + y = 1 \\ 2x - y + z = -1 \end{cases}$

73. $\begin{cases} 3x - 2y + z = -2 \\ 4x - 5y + 3z = -9 \\ 2x - y + 5z = -5 \end{cases}$

74. $\begin{cases} x - y = 1 \\ 6x + y + 20z = 14 \\ y + 3z = 1 \end{cases}$

75. $\begin{cases} v - 3x + z = -3 \\ w + y = -1 \\ x + z = 7 \\ v + w - x + 4y = -8 \\ v + w + x + y + z = 8 \end{cases}$

76. $\begin{cases} w + x + y + z = 4 \\ w + 3x - 2y + 2z = 7 \\ 2w + 2x + y + z = 3 \\ w - x + 2y + 3z = 5 \end{cases}$

In Exercises 77–78, use a coding matrix A of your choice. Use a graphing utility to find the multiplicative inverse of your coding matrix. Write a cryptogram for each message. Check your result by decoding the cryptogram. Use your graphing utility to perform all necessary matrix multiplications.

77.
A	R	R	I	V	E	D	_	S	A	F	E	L	Y
1	18	18	9	22	5	4	0	19	1	6	5	12	25

78.
A	R	T	_	E	N	R	I	C	H	E	S
1	18	20	0	5	14	18	9	3	8	5	19

Critical Thinking Exercises

Make Sense? *In Exercises 79–82, determine whether each statement makes sense or does not make sense, and explain your reasoning.*

79. I found the multiplicative inverse of a 2×3 matrix.

80. I used Gauss-Jordan elimination to find the multiplicative inverse of a 3×3 matrix.

81. I used matrix multiplication to represent a system of linear equations.

82. I made an encoding error by selecting the wrong square invertible matrix.

In Exercises 83–88, determine whether each statement is true or false. If the statement is false, make the necessary change(s) to produce a true statement.

83. All square 2×2 matrices have inverses because there is a formula for finding these inverses.

84. Two 2×2 invertible matrices can have a matrix sum that is not invertible.

85. To solve the matrix equation $AX = B$ for X, multiply A and the inverse of B.

86. $(AB)^{-1} = A^{-1}B^{-1}$, assuming A, B, and AB are invertible.

87. $(A + B)^{-1} = A^{-1} + B^{-1}$, assuming A, B, and $A + B$ are invertible.

88. $\begin{bmatrix} 1 & -3 \\ -1 & 3 \end{bmatrix}$ is an invertible matrix.

89. Give an example of a 2×2 matrix that is its own inverse.

90. If $A = \begin{bmatrix} 3 & 5 \\ 2 & 4 \end{bmatrix}$, find $(A^{-1})^{-1}$.

91. Find values of a for which the following matrix is not invertible:

$$\begin{bmatrix} 1 & a + 1 \\ a - 2 & 4 \end{bmatrix}.$$

Group Exercise

92. Each person in the group should work with one partner. Send a coded word or message to each other by giving your partner the coded matrix and the coding matrix that you selected. Once messages are sent, each person should decode the message received.

Preview Exercises

Exercises 93–95 will help you prepare for the material covered in the next section. Simplify the expression in each exercise.

93. $2(-5) - (-3)(4)$

94. $\dfrac{2(-5) - 1(-4)}{5(-5) - 6(-4)}$

95. $2(-30 - (-3)) - 3(6 - 9) + (-1)(1 - 15)$

Objectives

1 Evaluate a second-order determinant.

2 Solve a system of linear equations in two variables using Cramer's Rule.

3 Evaluate a third-order determinant.

4 Solve a system of linear equations in three variables using Cramer's Rule.

5 Evaluate higher-order determinants.

A portion of Charles Babbage's unrealized difference engine

As cyberspace absorbs more and more of our work, play, shopping, and socializing, where will it all end? Which activities will still be offline in 2025?

Our technologically transformed lives can be traced back to the English inventor Charles Babbage (1792–1871). Babbage knew of a method for solving linear systems called *Cramer's Rule*, in honor of the Swiss geometer Gabriel Cramer (1704–1752). Cramer's Rule was simple but involved numerous multiplications for large systems. Babbage designed a machine, called the "difference engine" that consisted of toothed wheels on shafts for performing these multiplications. Despite the fact that only one-seventh of the functions ever worked, Babbage's invention demonstrated how complex calculations could be handled mechanically. In 1944, scientists at IBM used the lessons of the difference engine to create the world's first computer.

Those who invented computers hoped to relegate the drudgery of repeated computation to a machine. In this section, we look at a method for solving linear systems that played a critical role in this process. The method uses real numbers, called *determinants*, that are associated with arrays of numbers. As with matrix methods, solutions are obtained by writing down the coefficients and constants of a linear system and performing operations with them.

1 Evaluate a second-order determinant.

The Determinant of a 2 × 2 Matrix

Associated with every square matrix is a real number, called its **determinant**. The determinant for a 2 × 2 square matrix is defined as follows:

GREAT QUESTION!

What does the definition of a determinant mean? What am I supposed to do?

To evaluate a second-order determinant, find the difference of the product of the two diagonals.

$$\begin{vmatrix} a_1 & b_1 \\ a_2 & b_2 \end{vmatrix} = a_1b_2 - a_2b_1$$

Definition of the Determinant of a 2 × 2 Matrix

The determinant of the matrix $\begin{bmatrix} a_1 & b_1 \\ a_2 & b_2 \end{bmatrix}$ is denoted by $\begin{vmatrix} a_1 & b_1 \\ a_2 & b_2 \end{vmatrix}$ and is defined by

$$\begin{vmatrix} a_1 & b_1 \\ a_2 & b_2 \end{vmatrix} = a_1b_2 - a_2b_1.$$

We also say that the **value** of the **second-order determinant** $\begin{vmatrix} a_1 & b_1 \\ a_2 & b_2 \end{vmatrix}$ is $a_1b_2 - a_2b_1$.

Example 1 illustrates that the determinant of a matrix may be positive or negative. A determinant can also have 0 as its value.

EXAMPLE 1 Evaluating the Determinant of a 2 × 2 Matrix

Evaluate the determinant of each of the following matrices:

a. $\begin{bmatrix} 5 & 6 \\ 7 & 3 \end{bmatrix}$ **b.** $\begin{bmatrix} 2 & 4 \\ -3 & -5 \end{bmatrix}$.

DISCOVERY

Write and then evaluate three determinants, one whose value is positive, one whose value is negative, and one whose value is 0.

SOLUTION

We multiply and subtract as indicated.

a. $\begin{vmatrix} 5 & 6 \\ 7 & 3 \end{vmatrix} = 5 \cdot 3 - 7 \cdot 6 = 15 - 42 = -27$ *The value of the second-order determinant is −27.*

b. $\begin{vmatrix} 2 & 4 \\ -3 & -5 \end{vmatrix} = 2(-5) - (-3)(4) = -10 + 12 = 2$ *The value of the second-order determinant is 2.* • • •

☑ Check Point **1** Evaluate the determinant of each of the following matrices:

a. $\begin{bmatrix} 10 & 9 \\ 6 & 5 \end{bmatrix}$ **b.** $\begin{bmatrix} 4 & 3 \\ -5 & -8 \end{bmatrix}$.

② Solve a system of linear equations in two variables using Cramer's Rule.

Solving Systems of Linear Equations in Two Variables Using Determinants

Determinants can be used to solve a linear system in two variables. In general, such a system appears as

$$\begin{cases} a_1 x + b_1 y = c_1 \\ a_2 x + b_2 y = c_2. \end{cases}$$

Let's first solve this system for x using the addition method. We can solve for x by eliminating y from the equations. Multiply the first equation by b_2 and the second equation by $-b_1$. Then add the two equations:

$$\begin{cases} a_1 x + b_1 y = c_1 & \xrightarrow{\text{Multiply by } b_2.} \\ a_2 x + b_2 y = c_2 & \xrightarrow{\text{Multiply by } -b_1.} \end{cases} \quad \begin{cases} a_1 b_2 x + b_1 b_2 y = c_1 b_2 \\ -a_2 b_1 x - b_1 b_2 y = -c_2 b_1 \end{cases}$$
$$\text{Add:} \quad (a_1 b_2 - a_2 b_1)x = c_1 b_2 - c_2 b_1$$
$$x = \frac{c_1 b_2 - c_2 b_1}{a_1 b_2 - a_2 b_1}.$$

Because

$$\begin{vmatrix} c_1 & b_1 \\ c_2 & b_2 \end{vmatrix} = c_1 b_2 - c_2 b_1 \quad \text{and} \quad \begin{vmatrix} a_1 & b_1 \\ a_2 & b_2 \end{vmatrix} = a_1 b_2 - a_2 b_1,$$

we can express our answer for x as the quotient of two determinants:

$$x = \frac{c_1 b_2 - c_2 b_1}{a_1 b_2 - a_2 b_1} = \frac{\begin{vmatrix} c_1 & b_1 \\ c_2 & b_2 \end{vmatrix}}{\begin{vmatrix} a_1 & b_1 \\ a_2 & b_2 \end{vmatrix}}.$$

Similarly, we could use the addition method to solve our system for y, again expressing y as the quotient of two determinants. This method of using determinants to solve the linear system, called **Cramer's Rule**, is summarized in the box at the top of the next page.

Solving a Linear System in Two Variables Using Determinants
Cramer's Rule

If

$$\begin{cases} a_1 x + b_1 y = c_1 \\ a_2 x + b_2 y = c_2, \end{cases}$$

then

$$x = \frac{\begin{vmatrix} c_1 & b_1 \\ c_2 & b_2 \end{vmatrix}}{\begin{vmatrix} a_1 & b_1 \\ a_2 & b_2 \end{vmatrix}} \quad \text{and} \quad y = \frac{\begin{vmatrix} a_1 & c_1 \\ a_2 & c_2 \end{vmatrix}}{\begin{vmatrix} a_1 & b_1 \\ a_2 & b_2 \end{vmatrix}},$$

where

$$\begin{vmatrix} a_1 & b_1 \\ a_2 & b_2 \end{vmatrix} \neq 0.$$

Here are some helpful tips when solving

$$\begin{cases} a_1 x + b_1 y = c_1 \\ a_2 x + b_2 y = c_2 \end{cases}$$

using determinants:

1. Three different determinants are used to find x and y. The determinants in the denominators for x and y are identical. The determinants in the numerators for x and y differ. In abbreviated notation, we write

$$x = \frac{D_x}{D} \quad \text{and} \quad y = \frac{D_y}{D}, \text{where } D \neq 0.$$

2. The elements of D, the determinant in the denominator, are the coefficients of the variables in the system.

$$D = \begin{vmatrix} a_1 & b_1 \\ a_2 & b_2 \end{vmatrix}$$

3. D_x, the determinant in the numerator of x, is obtained by replacing the x-coefficients, in D, a_1 and a_2, with the constants on the right sides of the equations, c_1 and c_2.

$$D = \begin{vmatrix} a_1 & b_1 \\ a_2 & b_2 \end{vmatrix} \quad \text{and} \quad D_x = \begin{vmatrix} c_1 & b_1 \\ c_2 & b_2 \end{vmatrix} \qquad \text{Replace the column with } a_1 \text{ and } a_2 \\ \text{with the constants } c_1 \text{ and } c_2 \text{ to get } D_x.$$

4. D_y, the determinant in the numerator for y, is obtained by replacing the y-coefficients, in D, b_1 and b_2, with the constants on the right sides of the equations, c_1 and c_2.

$$D = \begin{vmatrix} a_1 & b_1 \\ a_2 & b_2 \end{vmatrix} \quad \text{and} \quad D_y = \begin{vmatrix} a_1 & c_1 \\ a_2 & c_2 \end{vmatrix} \qquad \text{Replace the column with } b_1 \text{ and } b_2 \\ \text{with the constants } c_1 \text{ and } c_2 \text{ to get } D_y.$$

EXAMPLE 2 Using Cramer's Rule to Solve a Linear System

Use Cramer's Rule to solve the system:

$$\begin{cases} 5x - 4y = 2 \\ 6x - 5y = 1. \end{cases}$$

SOLUTION

Because

$$x = \frac{D_x}{D} \quad \text{and} \quad y = \frac{D_y}{D},$$

we will set up and evaluate the three determinants D, D_x, and D_y.

1. D, the determinant in both denominators, consists of the x- and y-coefficients.

$$D = \begin{vmatrix} 5 & -4 \\ 6 & -5 \end{vmatrix} = (5)(-5) - (6)(-4) = -25 + 24 = -1$$

Because this determinant is not zero, we continue to use Cramer's Rule to solve the system.

2. D_x, the determinant in the numerator for x, is obtained by replacing the x-coefficients in D, 5 and 6, by the constants on the right sides of the equations, 2 and 1.

$$D_x = \begin{vmatrix} 2 & -4 \\ 1 & -5 \end{vmatrix} = (2)(-5) - (1)(-4) = -10 + 4 = -6$$

3. D_y, the determinant in the numerator for y, is obtained by replacing the y-coefficients in D, -4 and -5, by the constants on the right sides of the equations, 2 and 1.

$$D_y = \begin{vmatrix} 5 & 2 \\ 6 & 1 \end{vmatrix} = (5)(1) - (6)(2) = 5 - 12 = -7$$

4. Thus,

$$x = \frac{D_x}{D} = \frac{-6}{-1} = 6 \quad \text{and} \quad y = \frac{D_y}{D} = \frac{-7}{-1} = 7.$$

As always, the solution $(6, 7)$ can be checked by substituting these values into the original equations. The solution set is $\{(6, 7)\}$. • • •

⬚ Check Point **2** Use Cramer's Rule to solve the system:

$$\begin{cases} 5x + 4y = 12 \\ 3x - 6y = 24. \end{cases}$$

③ Evaluate a third-order determinant.

The Determinant of a 3 × 3 Matrix

Associated with every square matrix is a real number called its determinant. The determinant for a 3 × 3 matrix is defined as follows:

Definition of a Third-Order Determinant

$$\begin{vmatrix} a_1 & b_1 & c_1 \\ a_2 & b_2 & c_2 \\ a_3 & b_3 & c_3 \end{vmatrix} = a_1b_2c_3 + b_1c_2a_3 + c_1a_2b_3 - a_3b_2c_1 - b_3c_2a_1 - c_3a_2b_1$$

The six terms and the three factors in each term in this complicated evaluation formula, $a_1b_2c_3 + b_1c_2a_3 + c_1a_2b_3 - a_3b_2c_1 - b_3c_2a_1 - c_3a_2b_1$, can be rearranged, and then we can apply the distributive property. We obtain

$$a_1b_2c_3 - a_1b_3c_2 - a_2b_1c_3 + a_2b_3c_1 + a_3b_1c_2 - a_3b_2c_1$$
$$= a_1(b_2c_3 - b_3c_2) - a_2(b_1c_3 - b_3c_1) + a_3(b_1c_2 - b_2c_1)$$
$$= a_1 \begin{vmatrix} b_2 & c_2 \\ b_3 & c_3 \end{vmatrix} - a_2 \begin{vmatrix} b_1 & c_1 \\ b_3 & c_3 \end{vmatrix} + a_3 \begin{vmatrix} b_1 & c_1 \\ b_2 & c_2 \end{vmatrix}.$$

You can evaluate each of the second-order determinants and obtain the three expressions in parentheses in the second step.

In summary, we now have arranged the definition of a third-order determinant as follows:

Definition of the Determinant of a 3 × 3 Matrix

A third-order determinant is defined by

$$\begin{vmatrix} a_1 & b_1 & c_1 \\ a_2 & b_2 & c_2 \\ a_3 & b_3 & c_3 \end{vmatrix} = a_1 \begin{vmatrix} b_2 & c_2 \\ b_3 & c_3 \end{vmatrix} \overset{\text{Subtract.}}{-} a_2 \begin{vmatrix} b_1 & c_1 \\ b_3 & c_3 \end{vmatrix} \overset{\text{Add.}}{+} a_3 \begin{vmatrix} b_1 & c_1 \\ b_2 & c_2 \end{vmatrix}.$$

Each a on the right comes from the first column.

Here are some tips that may be helpful when evaluating the determinant of a 3 × 3 matrix:

Evaluating the Determinant of a 3 × 3 Matrix

1. Each of the three terms in the definition contains two factors—a numerical factor and a second-order determinant.
2. The numerical factor in each term is an element from the first column of the third-order determinant.
3. The minus sign precedes the second term.
4. The second-order determinant that appears in each term is obtained by crossing out the row and the column containing the numerical factor.

$$a_1 \begin{vmatrix} b_2 & c_2 \\ b_3 & c_3 \end{vmatrix} - a_2 \begin{vmatrix} b_1 & c_1 \\ b_3 & c_3 \end{vmatrix} + a_3 \begin{vmatrix} b_1 & c_1 \\ b_2 & c_2 \end{vmatrix}$$

$$\begin{vmatrix} \cancel{a_1} & \cancel{b_1} & \cancel{c_1} \\ a_2 & b_2 & c_2 \\ a_3 & b_3 & c_3 \end{vmatrix} \quad \begin{vmatrix} a_1 & b_1 & c_1 \\ \cancel{a_2} & \cancel{b_2} & \cancel{c_2} \\ a_3 & b_3 & c_3 \end{vmatrix} \quad \begin{vmatrix} a_1 & b_1 & c_1 \\ a_2 & b_2 & c_2 \\ \cancel{a_3} & \cancel{b_3} & \cancel{c_3} \end{vmatrix}$$

The **minor** of an element is the determinant that remains after deleting the row and column of that element. For this reason, we call this method **expansion by minors**.

EXAMPLE 3 Evaluating the Determinant of a 3 × 3 Matrix

Evaluate the determinant of the following matrix:

$$\begin{bmatrix} 4 & 1 & 0 \\ -9 & 3 & 4 \\ -3 & 8 & 1 \end{bmatrix}.$$

SOLUTION

We know that each of the three terms in the determinant contains a numerical factor and a second-order determinant. The numerical factors are from the first column of the given matrix. They are highlighted in red in the following matrix:

$$\begin{bmatrix} 4 & 1 & 0 \\ -9 & 3 & 4 \\ -3 & 8 & 1 \end{bmatrix}.$$

We find the minor for each numerical factor by deleting the row and column of that element:

The minor for 4 is $\begin{vmatrix} 3 & 4 \\ 8 & 1 \end{vmatrix}$. The minor for -9 is $\begin{vmatrix} 1 & 0 \\ 8 & 1 \end{vmatrix}$. The minor for -3 is $\begin{vmatrix} 1 & 0 \\ 3 & 4 \end{vmatrix}$.

Now we have three numerical factors, 4, −9, and −3, and three second-order determinants. We multiply each numerical factor by its second-order determinant to find the three terms of the third-order determinant:

$$4\begin{vmatrix} 3 & 4 \\ 8 & 1 \end{vmatrix}, \quad -9\begin{vmatrix} 1 & 0 \\ 8 & 1 \end{vmatrix}, \quad -3\begin{vmatrix} 1 & 0 \\ 3 & 4 \end{vmatrix}.$$

Based on the preceding definition, we subtract the second term from the first term and add the third term:

> **Don't forget to supply the minus sign.**

$$\begin{vmatrix} 4 & 1 & 0 \\ -9 & 3 & 4 \\ -3 & 8 & 1 \end{vmatrix} = 4\begin{vmatrix} 3 & 4 \\ 8 & 1 \end{vmatrix} - (-9)\begin{vmatrix} 1 & 0 \\ 8 & 1 \end{vmatrix} - 3\begin{vmatrix} 1 & 0 \\ 3 & 4 \end{vmatrix}$$ *Begin by evaluating the three second-order determinants.*

$$= 4(3 \cdot 1 - 8 \cdot 4) + 9(1 \cdot 1 - 8 \cdot 0) - 3(1 \cdot 4 - 3 \cdot 0)$$

$$= 4(3 - 32) + 9(1 - 0) - 3(4 - 0)$$ *Multiply within parentheses.*

$$= 4(-29) + 9(1) - 3(4)$$ *Subtract within parentheses.*

$$= -116 + 9 - 12$$ *Multiply.*

$$= -119$$ *Add and subtract as indicated.* •••

✓ **Check Point 3** Evaluate the determinant of the following matrix:

$$\begin{bmatrix} 2 & 1 & 7 \\ -5 & 6 & 0 \\ -4 & 3 & 1 \end{bmatrix}.$$

The six terms in the definition of a third-order determinant can be rearranged and factored in a variety of ways. Thus, it is possible to expand a determinant by minors about any row or any column. *Minus signs must be supplied preceding any element appearing in a position where the sum of its row and its column is an odd number.* For example, expanding about the elements in column 2 gives us

$$\begin{vmatrix} a_1 & b_1 & c_1 \\ a_2 & b_2 & c_2 \\ a_3 & b_3 & c_3 \end{vmatrix} = -b_1\begin{vmatrix} a_2 & c_2 \\ a_3 & c_3 \end{vmatrix} + b_2\begin{vmatrix} a_1 & c_1 \\ a_3 & c_3 \end{vmatrix} - b_3\begin{vmatrix} a_1 & c_1 \\ a_2 & c_2 \end{vmatrix}.$$

Minus sign is supplied because b_1 appears in row 1 and column 2; $1 + 2 = 3$, an odd number.

Minus sign is supplied because b_3 appears in row 3 and column 2; $3 + 2 = 5$, an odd number.

TECHNOLOGY

A graphing utility can be used to evaluate the determinant of a matrix. Enter the matrix and call it A. Then use the determinant command. The screen below verifies our result in Example 3.

```
[A]
        [[4   1   0]
         [-9  3   4]
         [-3  8   1]]
det([A])
                -119
■
```

GREAT QUESTION!

Is there a way I can remember the signs for the numerical factors of the minors?

Yes. Keep in mind that you can expand a determinant by minors about any row or column. Use alternating plus and minus signs to precede the numerical factors of the minors according to the following sign array:

$$\begin{vmatrix} + & - & + \\ - & + & - \\ + & - & + \end{vmatrix}.$$

Expanding by minors about column 3, we obtain

$$\begin{vmatrix} a_1 & b_1 & c_1 \\ a_2 & b_2 & c_2 \\ a_3 & b_3 & c_3 \end{vmatrix} = c_1 \begin{vmatrix} a_2 & b_2 \\ a_3 & b_3 \end{vmatrix} - c_2 \begin{vmatrix} a_1 & b_1 \\ a_3 & b_3 \end{vmatrix} + c_3 \begin{vmatrix} a_1 & b_1 \\ a_2 & b_2 \end{vmatrix}.$$

Minus sign must be supplied because c_2 appears in row 2 and column 3; $2 + 3 = 5$, an odd number.

When evaluating a 3×3 determinant using expansion by minors, you can expand about any row or column. To simplify the arithmetic, if a row or column contains one or more 0s, expand about that row or column.

EXAMPLE 4 Evaluating a Third-Order Determinant

Evaluate:

$$\begin{vmatrix} 9 & 5 & 0 \\ -2 & -3 & 0 \\ 1 & 4 & 2 \end{vmatrix}.$$

SOLUTION

Note that the last column has two 0s. We will expand the determinant about the elements in that column.

$$\begin{vmatrix} 9 & 5 & 0 \\ -2 & -3 & 0 \\ 1 & 4 & 2 \end{vmatrix} = 0 \begin{vmatrix} -2 & -3 \\ 1 & 4 \end{vmatrix} - 0 \begin{vmatrix} 9 & 5 \\ 1 & 4 \end{vmatrix} + 2 \begin{vmatrix} 9 & 5 \\ -2 & -3 \end{vmatrix}$$

$$= 0 - 0 + 2[9(-3) - (-2)5] \quad \text{Evaluate the second-order}$$
$$= 2(-27 + 10) \quad \text{determinant whose numerical}$$
$$= 2(-17) \quad \text{factor is not 0.}$$
$$= -34$$

● ● ●

⬙ Check Point **4** Evaluate:

$$\begin{vmatrix} 6 & 4 & 0 \\ -3 & -5 & 3 \\ 1 & 2 & 0 \end{vmatrix}.$$

④ Solve a system of linear equations in three variables using Cramer's Rule.

Solving Systems of Linear Equations in Three Variables Using Determinants

Cramer's Rule can be applied to solving systems of linear equations in three variables. The determinants in the numerator and denominator of all variables are third-order determinants.

Solving Three Equations in Three Variables Using Determinants
Cramer's Rule

If

$$\begin{cases} a_1x + b_1y + c_1z = d_1 \\ a_2x + b_2y + c_2z = d_2 \\ a_3x + b_3y + c_3z = d_3 \end{cases}$$

then

$$x = \frac{D_x}{D}, y = \frac{D_y}{D}, \text{ and } z = \frac{D_z}{D}, \text{ where } D \neq 0.$$

These four third-order determinants are given by

$$D = \begin{vmatrix} a_1 & b_1 & c_1 \\ a_2 & b_2 & c_2 \\ a_3 & b_3 & c_3 \end{vmatrix}$$ These are the coefficients of the variables x, y, and z.

$$D_x = \begin{vmatrix} d_1 & b_1 & c_1 \\ d_2 & b_2 & c_2 \\ d_3 & b_3 & c_3 \end{vmatrix}$$ Replace x-coefficients in D with the constants on the right of the three equations.

$$D_y = \begin{vmatrix} a_1 & d_1 & c_1 \\ a_2 & d_2 & c_2 \\ a_3 & d_3 & c_3 \end{vmatrix}$$ Replace y-coefficients in D with the constants on the right of the three equations.

$$D_z = \begin{vmatrix} a_1 & b_1 & d_1 \\ a_2 & b_2 & d_2 \\ a_3 & b_3 & d_3 \end{vmatrix}$$ Replace z-coefficients in D with the constants on the right of the three equations.

EXAMPLE 5 Using Cramer's Rule to Solve a Linear System in Three Variables

Use Cramer's Rule to solve:

$$\begin{cases} x + 2y - z = -4 \\ x + 4y - 2z = -6 \\ 2x + 3y + z = 3. \end{cases}$$

SOLUTION
Because

$$x = \frac{D_x}{D}, \quad y = \frac{D_y}{D}, \quad \text{and} \quad z = \frac{D_z}{D},$$

we need to set up and evaluate four determinants.

Step 1 Set up the determinants.

1. D, the determinant in all three denominators, consists of the x-, y-, and z-coefficients.

$$D = \begin{vmatrix} 1 & 2 & -1 \\ 1 & 4 & -2 \\ 2 & 3 & 1 \end{vmatrix}$$

2. D_x, the determinant in the numerator for x, is obtained by replacing the x-coefficients in D, 1, 1, and 2, with the constants on the right sides of the equations, $-4, -6$, and 3.

$$D_x = \begin{vmatrix} -4 & 2 & -1 \\ -6 & 4 & -2 \\ 3 & 3 & 1 \end{vmatrix}$$

3. D_y, the determinant in the numerator for y, is obtained by replacing the y-coefficients in D, 2, 4, and 3, with the constants on the right sides of the equations, $-4, -6$, and 3.

$$D_y = \begin{vmatrix} 1 & -4 & -1 \\ 1 & -6 & -2 \\ 2 & 3 & 1 \end{vmatrix}$$

4. D_z, the determinant in the numerator for z, is obtained by replacing the z-coefficients in D, $-1, -2$, and 1, with the constants on the right sides of the equations, $-4, -6$, and 3.

$$D_z = \begin{vmatrix} 1 & 2 & -4 \\ 1 & 4 & -6 \\ 2 & 3 & 3 \end{vmatrix}$$

Step 2 Evaluate the four determinants.

$$D = \begin{vmatrix} 1 & 2 & -1 \\ 1 & 4 & -2 \\ 2 & 3 & 1 \end{vmatrix} = 1\begin{vmatrix} 4 & -2 \\ 3 & 1 \end{vmatrix} - 1\begin{vmatrix} 2 & -1 \\ 3 & 1 \end{vmatrix} + 2\begin{vmatrix} 2 & -1 \\ 4 & -2 \end{vmatrix}$$

$$= 1(4+6) - 1(2+3) + 2(-4+4)$$

$$= 1(10) - 1(5) + 2(0) = 5$$

Using the same technique to evaluate each determinant, we obtain

$$D_x = -10, \quad D_y = 5, \quad \text{and} \quad D_z = 20.$$

Step 3 Substitute these four values and solve the system.

$$x = \frac{D_x}{D} = \frac{-10}{5} = -2$$

$$y = \frac{D_y}{D} = \frac{5}{5} = 1$$

$$z = \frac{D_z}{D} = \frac{20}{5} = 4$$

The solution $(-2, 1, 4)$ can be checked by substitution into the original three equations. The solution set is $\{(-2, 1, 4)\}$. ● ● ●

GREAT QUESTION!

Can I use D, the determinant in each denominator, to find the determinants in the numerators?

No. To find D_x, D_y, and D_z, you'll need to apply the evaluation process for a 3×3 determinant three more times. The values of D_x, D_y, and D_z cannot be obtained from the number that occurs in the computation of D.

☑ Check Point **5** Use Cramer's Rule to solve the system:

$$\begin{cases} 3x - 2y + z = 16 \\ 2x + 3y - z = -9 \\ x + 4y + 3z = 2. \end{cases}$$

GREAT QUESTION!

What should I do if D, the determinant in the denominator of Cramer's Rule, is zero?

If $D = 0$, the system is inconsistent or contains dependent equations. Use a method other than Cramer's Rule to determine the solution set. For systems in two variables, use the substitution method or the addition method. For systems in three variables, use Gaussian elimination.

Speaking of Cramer meltdown, there's another situation in which Cramer's Rule is useless. This involves nonsquare systems such as

$$\begin{cases} 3x + 7y + 6z = 26 \\ x + 2y + z = 8. \end{cases}$$

Determinants are only associated with square matrices. Gaussian elimination must be used to solve systems with fewer equations than variables. (This system is solved in Example 3 on page 616.)

In essence, the use of Cramer's Rule to determine solution sets is only an alternative to Gaussian elimination in some situations.

⑤ Evaluate higher-order determinants.

The Determinant of Any $n \times n$ Matrix

The determinant of a matrix with n rows and n columns is said to be an ***n*th-order determinant**. The value of an nth-order determinant ($n > 2$) can be found in terms of determinants of order $n - 1$. For example, we found the value of a third-order determinant in terms of determinants of order 2.

We can generalize this idea for fourth-order determinants and higher. We have seen that the **minor** of the element a_{ij} is the determinant obtained by deleting the ith row and the jth column in the given array of numbers. The **cofactor** of the element a_{ij} is $(-1)^{i+j}$ times the minor of a_{ij}. If the sum of the row and column $(i + j)$ is even, the cofactor is the same as the minor. If the sum of the row and column $(i + j)$ is odd, the cofactor is the opposite of the minor.

Let's see what this means in the case of a fourth-order determinant.

EXAMPLE 6 Evaluating the Determinant of a 4 × 4 Matrix

Evaluate the determinant of the following matrix:

$$A = \begin{bmatrix} 1 & -2 & 3 & 0 \\ -1 & 1 & 0 & 2 \\ 0 & 2 & 0 & -3 \\ 2 & 3 & -4 & 1 \end{bmatrix}.$$

SOLUTION

$$|A| = \begin{vmatrix} 1 & -2 & 3 & 0 \\ -1 & 1 & 0 & 2 \\ 0 & 2 & 0 & -3 \\ 2 & 3 & -4 & 1 \end{vmatrix}$$

With two 0s in the third column, we will expand along the third column.

$$= (-1)^{1+3}(3) \begin{vmatrix} -1 & 1 & 2 \\ 0 & 2 & -3 \\ 2 & 3 & 1 \end{vmatrix} + (-1)^{4+3}(-4) \begin{vmatrix} 1 & -2 & 0 \\ -1 & 1 & 2 \\ 0 & 2 & -3 \end{vmatrix}$$

3 is in row 1, column 3. −4 is in row 4, column 3.

$$= 3 \begin{vmatrix} -1 & 1 & 2 \\ 0 & 2 & -3 \\ 2 & 3 & 1 \end{vmatrix} + 4 \begin{vmatrix} 1 & -2 & 0 \\ -1 & 1 & 2 \\ 0 & 2 & -3 \end{vmatrix}$$

The determinant that follows 3 is obtained by crossing out the row and the column (row 1, column 3) in the original determinant. The minor for −4 is obtained in a similar manner.

Evaluate the two third-order determinants to get

$$|A| = 3(-25) + 4(-1) = -79.$$

•••

⊘ **Check Point 6** Evaluate the determinant of the following matrix:

$$A = \begin{bmatrix} 0 & 4 & 0 & -3 \\ -1 & 1 & 5 & 2 \\ 1 & -2 & 0 & 6 \\ 3 & 0 & 0 & 1 \end{bmatrix}.$$

If a linear system has n equations, Cramer's Rule requires you to compute $n + 1$ determinants of nth order. The excessive number of calculations required to perform Cramer's Rule for systems with four or more equations makes it an inefficient method for solving large systems.

CONCEPT AND VOCABULARY CHECK

Fill in each blank so that the resulting statement is true.

1. $\begin{vmatrix} 5 & 4 \\ 2 & 3 \end{vmatrix} = __ \cdot __ - __ \cdot __ = __ - __ = __$

 The value of this second-order _____
 is _____.

2. Using Cramer's Rule to solve

 $$\begin{cases} x + y = 8 \\ x - y = -2 \end{cases}$$

 we obtain

 $$x = \dfrac{\begin{vmatrix} _ & _ \\ _ & _ \end{vmatrix}}{\begin{vmatrix} _ & _ \\ _ & _ \end{vmatrix}} \quad \text{and} \quad y = \dfrac{\begin{vmatrix} _ & _ \\ _ & _ \end{vmatrix}}{\begin{vmatrix} _ & _ \\ _ & _ \end{vmatrix}}.$$

3. $\begin{vmatrix} 3 & 2 & 1 \\ 4 & 3 & 1 \\ 5 & 1 & 1 \end{vmatrix} = 3\begin{vmatrix} _ & _ \\ _ & _ \end{vmatrix} - 4\begin{vmatrix} _ & _ \\ _ & _ \end{vmatrix} + 5\begin{vmatrix} _ & _ \\ _ & _ \end{vmatrix}$

4. Using Cramer's Rule to solve

 $$\begin{cases} 3x + y + 4z = -8 \\ 2x + 3y - 2z = 11 \\ x - 3y - 2z = 4 \end{cases}$$

 for y, we obtain

 $$y = \dfrac{\begin{vmatrix} _ & _ & _ \\ _ & _ & _ \\ _ & _ & _ \end{vmatrix}}{\begin{vmatrix} _ & _ & _ \\ _ & _ & _ \\ _ & _ & _ \end{vmatrix}}.$$

 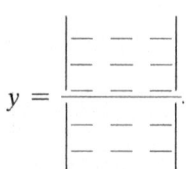

5. The easiest way to evaluate

 $$\begin{vmatrix} 3 & 2 & 8 \\ 5 & -4 & 0 \\ -6 & 7 & 0 \end{vmatrix}$$

 is to expand about the elements in
 _____.

EXERCISE SET 6.5

Practice Exercises

Evaluate each determinant in Exercises 1–10.

1. $\begin{vmatrix} 5 & 7 \\ 2 & 3 \end{vmatrix}$

2. $\begin{vmatrix} 4 & 8 \\ 5 & 6 \end{vmatrix}$

3. $\begin{vmatrix} -4 & 1 \\ 5 & 6 \end{vmatrix}$

4. $\begin{vmatrix} 7 & 9 \\ -2 & -5 \end{vmatrix}$

5. $\begin{vmatrix} -7 & 14 \\ 2 & -4 \end{vmatrix}$

6. $\begin{vmatrix} 1 & -3 \\ -8 & 2 \end{vmatrix}$

7. $\begin{vmatrix} -5 & -1 \\ -2 & -7 \end{vmatrix}$

8. $\begin{vmatrix} \frac{1}{5} & \frac{1}{6} \\ -6 & 5 \end{vmatrix}$

9. $\begin{vmatrix} \frac{1}{2} & \frac{1}{2} \\ \frac{1}{8} & -\frac{3}{4} \end{vmatrix}$

10. $\begin{vmatrix} \frac{2}{3} & \frac{1}{3} \\ -\frac{1}{2} & \frac{3}{4} \end{vmatrix}$

For Exercises 11–22, use Cramer's Rule to solve each system.

11. $\begin{cases} x + y = 7 \\ x - y = 3 \end{cases}$

12. $\begin{cases} 2x + y = 3 \\ x - y = 3 \end{cases}$

13. $\begin{cases} 12x + 3y = 15 \\ 2x - 3y = 13 \end{cases}$

14. $\begin{cases} x - 2y = 5 \\ 5x - y = -2 \end{cases}$

15. $\begin{cases} 4x - 5y = 17 \\ 2x + 3y = 3 \end{cases}$

16. $\begin{cases} 3x + 2y = 2 \\ 2x + 2y = 3 \end{cases}$

17. $\begin{cases} x + 2y = 3 \\ 3x - 4y = 4 \end{cases}$

18. $\begin{cases} 2x - 9y = 5 \\ 3x - 3y = 11 \end{cases}$

19. $\begin{cases} 3x - 4y = 4 \\ 2x + 2y = 12 \end{cases}$

20. $\begin{cases} 3x = 7y + 1 \\ 2x = 3y - 1 \end{cases}$

21. $\begin{cases} 2x = 3y + 2 \\ 5x = 51 - 4y \end{cases}$

22. $\begin{cases} y = -4x + 2 \\ 2x = 3y + 8 \end{cases}$

Evaluate each determinant in Exercises 23–28.

23. $\begin{vmatrix} 3 & 0 & 0 \\ 2 & 1 & -5 \\ 2 & 5 & -1 \end{vmatrix}$

24. $\begin{vmatrix} 4 & 0 & 0 \\ 3 & -1 & 4 \\ 2 & -3 & 5 \end{vmatrix}$

25. $\begin{vmatrix} 3 & 1 & 0 \\ -3 & 4 & 0 \\ -1 & 3 & -5 \end{vmatrix}$

26. $\begin{vmatrix} 2 & -4 & 2 \\ -1 & 0 & 5 \\ 3 & 0 & 4 \end{vmatrix}$

27. $\begin{vmatrix} 1 & 1 & 1 \\ 2 & 2 & 2 \\ -3 & 4 & -5 \end{vmatrix}$

28. $\begin{vmatrix} 1 & 2 & 3 \\ 2 & 2 & -3 \\ 3 & 2 & 1 \end{vmatrix}$

In Exercises 29–36, use Cramer's Rule to solve each system.

29. $\begin{cases} x + y + z = 0 \\ 2x - y + z = -1 \\ -x + 3y - z = -8 \end{cases}$

30. $\begin{cases} x - y + 2z = 3 \\ 2x + 3y + z = 9 \\ -x - y + 3z = 11 \end{cases}$

31. $\begin{cases} 4x - 5y - 6z = -1 \\ x - 2y - 5z = -12 \\ 2x - y = 7 \end{cases}$

32. $\begin{cases} x - 3y + z = -2 \\ x + 2y = 8 \\ 2x - y = 1 \end{cases}$

33. $\begin{cases} x + y + z = 4 \\ x - 2y + z = 7 \\ x + 3y + 2z = 4 \end{cases}$

34. $\begin{cases} 2x + 2y + 3z = 10 \\ 4x - y + z = -5 \\ 5x - 2y + 6z = 1 \end{cases}$

35. $\begin{cases} x + 2z = 4 \\ 2y - z = 5 \\ 2x + 3y = 13 \end{cases}$

36. $\begin{cases} 3x + 2z = 4 \\ 5x - y = -4 \\ 4y + 3z = 22 \end{cases}$

Evaluate each determinant in Exercises 37–40.

37. $\begin{vmatrix} 4 & 2 & 8 & -7 \\ -2 & 0 & 4 & 1 \\ 5 & 0 & 0 & 5 \\ 4 & 0 & 0 & -1 \end{vmatrix}$

38. $\begin{vmatrix} 3 & -1 & 1 & 2 \\ -2 & 0 & 0 & 0 \\ 2 & -1 & -2 & 3 \\ 1 & 4 & 2 & 3 \end{vmatrix}$

39. $\begin{vmatrix} -2 & -3 & 3 & 5 \\ 1 & -4 & 0 & 0 \\ 1 & 2 & 2 & -3 \\ 2 & 0 & 1 & 1 \end{vmatrix}$ **40.** $\begin{vmatrix} 1 & -3 & 2 & 0 \\ -3 & -1 & 0 & -2 \\ 2 & 1 & 3 & 1 \\ 2 & 0 & -2 & 0 \end{vmatrix}$

Practice Plus

In Exercises 41–42, evaluate each determinant.

41. $\begin{vmatrix} \begin{vmatrix} 3 & 1 \\ -2 & 3 \end{vmatrix} & \begin{vmatrix} 7 & 0 \\ 1 & 5 \end{vmatrix} \\ \begin{vmatrix} 3 & 0 \\ 0 & 7 \end{vmatrix} & \begin{vmatrix} 9 & -6 \\ 3 & 5 \end{vmatrix} \end{vmatrix}$ **42.** $\begin{vmatrix} \begin{vmatrix} 5 & 0 \\ 4 & -3 \end{vmatrix} & \begin{vmatrix} -1 & 0 \\ 0 & -1 \end{vmatrix} \\ \begin{vmatrix} 7 & -5 \\ 4 & 6 \end{vmatrix} & \begin{vmatrix} 4 & 1 \\ -3 & 5 \end{vmatrix} \end{vmatrix}$

In Exercises 43–44, write the system of linear equations for which Cramer's Rule yields the given determinants.

43. $D = \begin{vmatrix} 2 & -4 \\ 3 & 5 \end{vmatrix}, \quad D_x = \begin{vmatrix} 8 & -4 \\ -10 & 5 \end{vmatrix}$

44. $D = \begin{vmatrix} 2 & -3 \\ 5 & 6 \end{vmatrix}, \quad D_x = \begin{vmatrix} 8 & -3 \\ 11 & 6 \end{vmatrix}$

In Exercises 45–48, solve each equation for x.

45. $\begin{vmatrix} -2 & x \\ 4 & 6 \end{vmatrix} = 32$ **46.** $\begin{vmatrix} x+3 & -6 \\ x-2 & -4 \end{vmatrix} = 28$

47. $\begin{vmatrix} 1 & x & -2 \\ 3 & 1 & 1 \\ 0 & -2 & 2 \end{vmatrix} = -8$ **48.** $\begin{vmatrix} 2 & x & 1 \\ -3 & 1 & 0 \\ 2 & 1 & 4 \end{vmatrix} = 39$

Application Exercises

Determinants are used to find the area of a triangle whose vertices are given by three points in a rectangular coordinate system. The area of a triangle with vertices (x_1, y_1), (x_2, y_2), and (x_3, y_3) is

$$\text{Area} = \pm \frac{1}{2} \begin{vmatrix} x_1 & y_1 & 1 \\ x_2 & y_2 & 1 \\ x_3 & y_3 & 1 \end{vmatrix},$$

where the \pm symbol indicates that the appropriate sign should be chosen to yield a positive area. Use this information to work Exercises 49–50.

49. Use determinants to find the area of the triangle whose vertices are $(3, -5)$, $(2, 6)$, and $(-3, 5)$.

50. Use determinants to find the area of the triangle whose vertices are $(1, 1)$, $(-2, -3)$, and $(11, -3)$.

Determinants are used to show that three points lie on the same line (are collinear). If

$$\begin{vmatrix} x_1 & y_1 & 1 \\ x_2 & y_2 & 1 \\ x_3 & y_3 & 1 \end{vmatrix} = 0,$$

then the points (x_1, y_1), (x_2, y_2), and (x_3, y_3) are collinear. If the determinant does not equal 0, then the points are not collinear. Use this information to work Exercises 51–52.

51. Are the points $(3, -1)$, $(0, -3)$, and $(12, 5)$ collinear?

52. Are the points $(-4, -6)$, $(1, 0)$, and $(11, 12)$ collinear?

Determinants are used to write an equation of a line passing through two points. An equation of the line passing through the distinct points (x_1, y_1) and (x_2, y_2) is given by

$$\begin{vmatrix} x & y & 1 \\ x_1 & y_1 & 1 \\ x_2 & y_2 & 1 \end{vmatrix} = 0.$$

Use this information to work Exercises 53–54.

53. Use the determinant to write an equation of the line passing through $(3, -5)$ and $(-2, 6)$. Then expand the determinant, expressing the line's equation in slope-intercept form.

54. Use the determinant to write an equation of the line passing through $(-1, 3)$ and $(2, 4)$. Then expand the determinant, expressing the line's equation in slope-intercept form.

Writing in Mathematics

55. Explain how to evaluate a second-order determinant.

56. Describe the determinants D_x and D_y in terms of the coefficients and constants in a system of two equations in two variables.

57. Explain how to evaluate a third-order determinant.

58. When expanding a determinant by minors, when is it necessary to supply minus signs?

59. Without going into too much detail, describe how to solve a linear system in three variables using Cramer's Rule.

60. In applying Cramer's Rule, what should you do if $D = 0$?

61. The process of solving a linear system in three variables using Cramer's Rule can involve tedious computation. Is there a way of speeding up this process, perhaps using Cramer's Rule to find the value for only one of the variables? Describe how this process might work, presenting a specific example with your description. Remember that your goal is still to find the value for each variable in the system.

62. If you could use only one method to solve linear systems in three variables, which method would you select? Explain why this is so.

Technology Exercises

63. Use the feature of your graphing utility that evaluates the determinant of a square matrix to verify any five of the determinants that you evaluated by hand in Exercises 1–10, 23–28, or 37–40.

In Exercises 64–65, use a graphing utility to evaluate the determinant for the given matrix.

64. $\begin{bmatrix} 3 & -2 & -1 & 4 \\ -5 & 1 & 2 & 7 \\ 2 & 4 & 5 & 0 \\ -1 & 3 & -6 & 5 \end{bmatrix}$

65. $\begin{bmatrix} 8 & 2 & 6 & -1 & 0 \\ 2 & 0 & -3 & 4 & 7 \\ 2 & 1 & -3 & 6 & -5 \\ -1 & 2 & 1 & 5 & -1 \\ 4 & 5 & -2 & 3 & -8 \end{bmatrix}$

66. What is the fastest method for solving a linear system with your graphing utility?

Critical Thinking Exercises

Make Sense? *In Exercises 67–70, determine whether each statement makes sense or does not make sense, and explain your reasoning.*

67. I'm solving a linear system using a determinant that contains two rows and three columns.

68. I can speed up the tedious computations required by Cramer's Rule by using the value of D to determine the value of D_x.

69. When using Cramer's Rule to solve a linear system, the number of determinants that I set up and evaluate is the same as the number of variables in the system.

70. Using Cramer's Rule to solve a linear system, I found the value of D to be zero, so the value of each variable is zero.

71. a. Evaluate: $\begin{vmatrix} a & a \\ 0 & a \end{vmatrix}$.

 b. Evaluate: $\begin{vmatrix} a & a & a \\ 0 & a & a \\ 0 & 0 & a \end{vmatrix}$.

 c. Evaluate: $\begin{vmatrix} a & a & a & a \\ 0 & a & a & a \\ 0 & 0 & a & a \\ 0 & 0 & 0 & a \end{vmatrix}$.

 d. Describe the pattern in the given determinants.

 e. Describe the pattern in the evaluations.

72. Evaluate: $\begin{vmatrix} 2 & 0 & 0 & 0 & 0 \\ 0 & 3 & 0 & 0 & 0 \\ 0 & 0 & 2 & 0 & 0 \\ 0 & 0 & 0 & 1 & 0 \\ 0 & 0 & 0 & 0 & 4 \end{vmatrix}$.

73. What happens to the value of a second-order determinant if the two columns are interchanged?

74. Consider the system

$$\begin{cases} a_1 x + b_1 y = c_1 \\ a_2 x + b_2 y = c_2. \end{cases}$$

Use Cramer's Rule to prove that if the first equation of the system is replaced by the sum of the two equations, the resulting system has the same solution as the original system.

75. Show that the equation of a line through (x_1, y_1) and (x_2, y_2) is given by the determinant equation in Exercises 53–54.

Group Exercise

76. We have seen that determinants can be used to solve linear equations, give areas of triangles in rectangular coordinates, and determine equations of lines. Not impressed with these applications? Members of the group should research an application of determinants that they find intriguing. The group should then present a seminar to the class about this application.

Preview Exercises

Exercises 77–79 will help you prepare for the material covered in the first section of the next chapter.

77. Consider the equation $\dfrac{x^2}{9} + \dfrac{y^2}{4} = 1$.

 a. Set $y = 0$ and find the x-intercepts.

 b. Set $x = 0$ and find the y-intercepts.

78. Divide both sides of $25x^2 + 16y^2 = 400$ by 400 and simplify.

79. Complete the square and write the circle's equation in standard form:

$$x^2 + y^2 - 2x + 4y = 4.$$

Then give the center and radius of the circle and graph the equation.

CHAPTER 6 Summary, Review, and Test

SUMMARY

DEFINITIONS AND CONCEPTS	EXAMPLES
6.1 Matrix Solutions to Linear Systems	
a. Matrix row operations are described in the box on page 601.	Ex. 1, p. 601
b. To solve a linear system using Gaussian elimination, begin with the system's augmented matrix. Use matrix row operations to get 1s down the main diagonal from upper left to lower right, and 0s below the 1s. Such a matrix is in row-echelon form. Details are in the box on page 602.	Ex. 2, p. 602; Ex. 3, p. 605
c. To solve a linear system using Gauss-Jordan elimination, use the procedure of Gaussian elimination, but obtain 0s above and below the 1s in the main diagonal from upper left to lower right. Such a matrix is in reduced row-echelon form. Details are in the box on page 607.	Ex. 4, p. 608

6.2 Inconsistent and Dependent Systems and Their Applications

a. If Gaussian elimination results in a matrix with a row containing all 0s to the left of the vertical line and a nonzero number to the right, the system has no solution (is inconsistent).

Ex. 1, p. 613

b. In a square system, if Gaussian elimination results in a matrix with a row with all 0s, but not a row like the one in part (a), the system has an infinite number of solutions (contains dependent equations).

Ex. 2, p. 614

c. In nonsquare systems, the number of variables differs from the number of equations.

Ex. 3, p. 616

6.3 Matrix Operations and Their Applications

a. A matrix of order $m \times n$ has m rows and n columns. Two matrices are equal if and only if they have the same order and corresponding elements are equal.

Ex. 1, p. 622

b. Matrix Addition and Subtraction: Matrices of the same order are added or subtracted by adding or subtracting corresponding elements. Properties of matrix addition are given in the box on page 624.

Ex. 2, p. 624

c. Scalar Multiplication: If A is a matrix and c is a scalar, then cA is the matrix formed by multiplying each element in A by c. Properties of scalar multiplication are given in the box on page 626.

Ex. 3, p. 625;
Ex. 4, p. 626

d. Matrix Multiplication: The product of an $m \times n$ matrix A and an $n \times p$ matrix B is an $m \times p$ matrix AB. The element in the ith row and jth column of AB is found by multiplying each element in the ith row of A by the corresponding element in the jth column of B and adding the products. Matrix multiplication is not commutative: $AB \neq BA$. Properties of matrix multiplication are given in the box on page 631.

Ex. 5, p. 627;
Ex. 6, p. 628;
Ex. 7, p. 630

6.4 Multiplicative Inverses of Matrices and Matrix Equations

a. The multiplicative identity matrix I_n is an $n \times n$ matrix with 1s down the main diagonal from upper left to lower right and 0s elsewhere.

b. Let A be an $n \times n$ square matrix. If there is a square matrix A^{-1} such that $AA^{-1} = I_n$ and $A^{-1}A = I_n$, then A^{-1} is the multiplicative inverse of A.

Ex. 1, p. 639

c. If a square matrix has a multiplicative inverse, it is invertible or nonsingular. Methods for finding multiplicative inverses for invertible matrices, including a formula for 2×2 matrices, are given in the box on page 645.

Ex. 2, p. 640;
Ex. 3, p. 641;
Ex. 4, p. 644

d. Linear systems can be represented by matrix equations of the form $AX = B$ in which A is the coefficient matrix and B is the constant matrix. If $AX = B$ has a unique solution, then $X = A^{-1}B$.

Ex. 5, p. 646

6.5 Determinants and Cramer's Rule

a. Value of a second-order determinant:

Ex. 1, p. 653

$$\begin{vmatrix} a_1 & b_1 \\ a_2 & b_2 \end{vmatrix} = a_1b_2 - a_2b_1$$

b. Cramer's Rule for solving systems of linear equations in two variables uses three second-order determinants and is stated in the box on page 654.

Ex. 2, p. 655

c. To evaluate an nth-order determinant, where $n > 2$,

Ex. 3, p. 656;
Ex. 4, p. 658;
Ex. 6, p. 661

1. Select a row or column about which to expand.

2. Multiply each element a_{ij} in the row or column by $(-1)^{i+j}$ times the determinant obtained by deleting the ith row and the jth column in the given array of numbers.

3. The value of the determinant is the sum of the products found in step 2.

d. Cramer's Rule for solving systems of linear equations in three variables uses four third-order determinants and is stated in the box beginning on page 658.

Ex. 5, p. 659

e. Cramer's Rule cannot be used to determine solution sets with inconsistent or dependent systems.

REVIEW EXERCISES

6.1

In Exercises 1–2, perform each matrix row operation and write the new matrix.

1. $\begin{bmatrix} 1 & 2 & 2 & | & 2 \\ 0 & 1 & -1 & | & 2 \\ 0 & 5 & 4 & | & 1 \end{bmatrix}$ $-5R_2 + R_3$

2. $\begin{bmatrix} 2 & -2 & 1 & | & -1 \\ 1 & 2 & -1 & | & 2 \\ 6 & 4 & 3 & | & 5 \end{bmatrix}$ $\frac{1}{2}R_1$

In Exercises 3–5, solve each system of equations using matrices. Use Gaussian elimination with back-substitution or Gauss-Jordan elimination.

3. $\begin{cases} x + 2y + 3z = -5 \\ 2x + y + z = 1 \\ x + y - z = 8 \end{cases}$

4. $\begin{cases} x - 2y + z = 0 \\ y - 3z = -1 \\ 2y + 5z = -2 \end{cases}$

5. $\begin{cases} 3x_1 + 5x_2 - 8x_3 + 5x_4 = -8 \\ x_1 + 2x_2 - 3x_3 + x_4 = -7 \\ 2x_1 + 3x_2 - 7x_3 + 3x_4 = -11 \\ 4x_1 + 8x_2 - 10x_3 + 7x_4 = -10 \end{cases}$

6. The table shows the pollutants in the air in a city on a typical summer day.

x (Hours after 6 A.M.)	y (Amount of Pollutants in the Air, in parts per million)
2	98
4	138
10	162

a. Use the function $y = ax^2 + bx + c$ to model the data. Use either Gaussian elimination with back-substitution or Gauss-Jordan elimination to find the values for a, b, and c.

b. Use the function to find the time of day at which the city's air pollution level is at a maximum. What is the maximum level?

7. Sociologists Joseph Kahl and Dennis Gilbert developed a six-tier model to portray the class structure of the United States. The bar graph represents the percentage of Americans who are members of each of the six social classes.

The United States Social Class Ladder

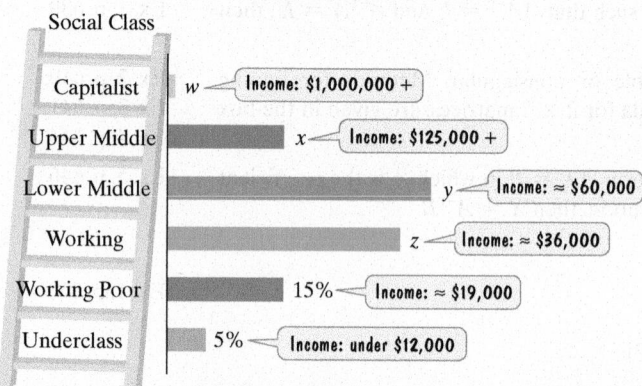

Percentage of the Population

Source: James Henslin, *Sociology,* Ninth Edition, Allyn and Bacon, 2011.

Combined, members of the capitalist class, the upper-middle class, the lower-middle class, and the working class make up 80% of the U.S. population. The percentage of the population belonging to the lower-middle class exceeds the percentage belonging to capitalist and upper-middle classes by 18%. The difference between the percentage belonging to the lower-middle class and the working class is 4%. If the percentage belonging to the upper-middle class is tripled, it exceeds the percentage belonging to the capitalist and lower-middle classes by 10%. Determine the percentage of the U.S. population who are members of the capitalist class, the upper-middle class, the lower-middle class, and the working class.

6.2

In Exercises 8–11, use Gaussian elimination to find the complete solution to each system, or show that none exists.

8. $\begin{cases} 2x - 3y + z = 1 \\ x - 2y + 3z = 2 \\ 3x - 4y - z = 1 \end{cases}$

9. $\begin{cases} x - 3y + z = 1 \\ -2x + y + 3z = -7 \\ x - 4y + 2z = 0 \end{cases}$

10. $\begin{cases} x_1 + 4x_2 + 3x_3 - 6x_4 = 5 \\ x_1 + 3x_2 + x_3 - 4x_4 = 3 \\ 2x_1 + 8x_2 + 7x_3 - 5x_4 = 11 \\ 2x_1 + 5x_2 \quad\quad - 6x_4 = 4 \end{cases}$

11. $\begin{cases} 2x + 3y - 5z = 15 \\ x + 2y - z = 4 \end{cases}$

12. The figure shows the intersections of three one-way streets. The numbers given represent traffic flow, in cars per hour, at a peak period (from 4 P.M. to 6 P.M.).

a. Use the idea that the number of cars entering each intersection per hour must equal the number of cars leaving per hour to set up a system of linear equations involving x, y, and z.

b. Use Gaussian elimination to solve the system.

c. If construction limits the value of z to 400, how many cars per hour must pass between the other intersections to keep traffic flowing?

6.3

13. Find values for x, y, and z so that the following matrices are equal:

$$\begin{bmatrix} 2x & y + 7 \\ z & 4 \end{bmatrix} = \begin{bmatrix} -10 & 13 \\ 6 & 4 \end{bmatrix}.$$

In Exercises 14–27, perform the indicated matrix operations given that A, B, C, and D are defined as follows. If an operation is not defined, state the reason.

$$A = \begin{bmatrix} 2 & -1 & 2 \\ 5 & 3 & -1 \end{bmatrix} \quad B = \begin{bmatrix} 0 & -2 \\ 3 & 2 \\ 1 & -5 \end{bmatrix}$$

$$C = \begin{bmatrix} 1 & 2 & 3 \\ -1 & 1 & 2 \\ -1 & 2 & 1 \end{bmatrix} \quad D = \begin{bmatrix} -2 & 3 & 1 \\ 3 & -2 & 4 \end{bmatrix}$$

14. $A + D$

15. $2B$

16. $D - A$

17. $B + C$

18. $3A + 2D$

19. $-2A + 4D$

20. $-5(A + D)$

21. AB

22. BA

23. BD

24. DB

25. $AB - BA$

26. $(A - D)C$

27. $B(AC)$

28. Solve for X in the matrix equation

$$3X + A = B,$$

where $A = \begin{bmatrix} 4 & 6 \\ -5 & 0 \end{bmatrix}$ and $B = \begin{bmatrix} -2 & -12 \\ 4 & 1 \end{bmatrix}.$

In Exercises 29–30, use nine pixels in a 3 × 3 grid and the color levels shown.

29. Write a 3×3 matrix that represents a digital photograph of the letter T in dark gray on a light gray background.

30. Find a matrix B so that $A + B$ increases the contrast of the letter T by changing the dark gray to black and the light gray to white.

The figure shows a right triangle in a rectangular coordinate system.

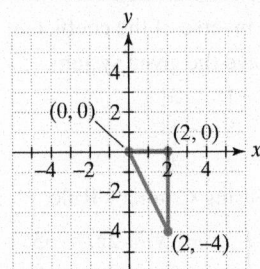

The figure can be represented by the matrix

$$B = \begin{bmatrix} 0 & 2 & 2 \\ 0 & 0 & -4 \end{bmatrix}.$$

Use the triangle and the matrix that represents it to solve Exercises 31–36.

31. Use matrix operations to move the triangle 2 units to the left and 1 unit up. Then graph the triangle and its transformation in a rectangular coordinate system.

32. Use matrix operations to reduce the triangle to half its perimeter and move the reduced image 2 units down. Then graph the triangle and its transformation in a rectangular coordinate system.

In Exercises 33–36, find AB and graph the resulting image. What effect does the multiplication have on the triangle represented by matrix B?

33. $A = \begin{bmatrix} 1 & 0 \\ 0 & -1 \end{bmatrix}$

34. $A = \begin{bmatrix} -1 & 0 \\ 0 & 1 \end{bmatrix}$

35. $A = \begin{bmatrix} 0 & -1 \\ 1 & 0 \end{bmatrix}$

36. $A = \begin{bmatrix} 2 & 0 \\ 0 & 1 \end{bmatrix}$

6.4

In Exercises 37–38, find the products AB and BA to determine whether B is the multiplicative inverse of A.

37. $A = \begin{bmatrix} 2 & 7 \\ 1 & 4 \end{bmatrix}$, $B = \begin{bmatrix} 4 & -7 \\ -1 & 3 \end{bmatrix}$

38. $A = \begin{bmatrix} 1 & 0 & 0 \\ 0 & 2 & -7 \\ 0 & -1 & 4 \end{bmatrix}$, $B = \begin{bmatrix} 1 & 0 & 0 \\ 0 & 4 & 7 \\ 0 & 1 & 2 \end{bmatrix}$

In Exercises 39–42, find A^{-1}. Check that $AA^{-1} = I$ and $A^{-1}A = I$.

39. $A = \begin{bmatrix} 1 & -1 \\ -2 & 3 \end{bmatrix}$

40. $A = \begin{bmatrix} 0 & 1 \\ 5 & 3 \end{bmatrix}$

41. $A = \begin{bmatrix} 1 & 0 & -2 \\ 2 & 1 & 0 \\ 1 & 0 & -3 \end{bmatrix}$

42. $A = \begin{bmatrix} 1 & 3 & -2 \\ 4 & 13 & -7 \\ 5 & 16 & -8 \end{bmatrix}$

In Exercises 43–44,

a. *Write each linear system as a matrix equation in the form $AX = B$.*

b. *Solve the system using the inverse that is given for the coefficient matrix.*

43. $\begin{cases} x + y + 2z = 7 \\ \quad\;\; y + 3z = -2 \\ 3x \quad\;\; - 2z = 0 \end{cases}$ The inverse of $\begin{bmatrix} 1 & 1 & 2 \\ 0 & 1 & 3 \\ 3 & 0 & -2 \end{bmatrix}$ is $\begin{bmatrix} -2 & 2 & 1 \\ 9 & -8 & -3 \\ -3 & 3 & 1 \end{bmatrix}$.

44. $\begin{cases} x - y + 2z = 12 \\ \quad\;\; y - z = -5 \\ x \quad\;\; + 2z = 10 \end{cases}$ The inverse of $\begin{bmatrix} 1 & -1 & 2 \\ 0 & 1 & -1 \\ 1 & 0 & 2 \end{bmatrix}$ is $\begin{bmatrix} 2 & 2 & -1 \\ -1 & 0 & 1 \\ -1 & -1 & 1 \end{bmatrix}$.

45. Use the coding matrix $A = \begin{bmatrix} 3 & 2 \\ 4 & 3 \end{bmatrix}$ and its inverse $A^{-1} = \begin{bmatrix} 3 & -2 \\ -4 & 3 \end{bmatrix}$ to encode and then decode the word RULE.

6.5

In Exercises 46–51, evaluate each determinant.

46. $\begin{vmatrix} 3 & 2 \\ -1 & 5 \end{vmatrix}$

47. $\begin{vmatrix} -2 & -3 \\ -4 & -8 \end{vmatrix}$

48. $\begin{vmatrix} 2 & 4 & -3 \\ 1 & -1 & 5 \\ -2 & 4 & 0 \end{vmatrix}$

49. $\begin{vmatrix} 4 & 7 & 0 \\ -5 & 6 & 0 \\ 3 & 2 & -4 \end{vmatrix}$

50. $\begin{vmatrix} 1 & 1 & 0 & 2 \\ 0 & 3 & 2 & 1 \\ 0 & -2 & 4 & 0 \\ 0 & 3 & 0 & 1 \end{vmatrix}$

51. $\begin{vmatrix} 2 & 2 & 2 & 2 \\ 0 & 2 & 2 & 2 \\ 0 & 0 & 2 & 2 \\ 0 & 0 & 0 & 2 \end{vmatrix}$

In Exercises 52–55, use Cramer's Rule to solve each system.

52. $\begin{cases} x - 2y = 8 \\ 3x + 2y = -1 \end{cases}$

53. $\begin{cases} 7x + 2y = 0 \\ 2x + y = -3 \end{cases}$

54. $\begin{cases} x + 2y + 2z = 5 \\ 2x + 4y + 7z = 19 \\ -2x - 5y - 2z = 8 \end{cases}$

55. $\begin{cases} 2x + y = -4 \\ \quad\; y - 2z = 0 \\ 3x \quad\; - 2z = -11 \end{cases}$

56. Use the quadratic function $y = ax^2 + bx + c$ to model the following data:

x (Age of a Driver)	y (Average Number of Automobile Accidents per Day in the United States)
20	400
40	150
60	400

Use Cramer's Rule to determine values for a, b, and c. Then use the model to write a statement about the average number of automobile accidents in which 30-year-olds and 50-year-olds are involved daily.

CHAPTER 6 TEST

In Exercises 1–2, solve each system of equations using matrices.

1. $\begin{cases} x + 2y - z = -3 \\ 2x - 4y + z = -7 \\ -2x + 2y - 3z = 4 \end{cases}$

2. $\begin{cases} x - 2y + z = 2 \\ 2x - y - z = 1 \end{cases}$

In Exercises 3–6, let

$$A = \begin{bmatrix} 3 & 1 \\ 1 & 0 \\ 2 & 1 \end{bmatrix}, \quad B = \begin{bmatrix} 1 & -1 \\ 2 & 1 \end{bmatrix}, \quad \text{and} \quad C = \begin{bmatrix} 1 & 2 \\ -1 & 3 \end{bmatrix}.$$

Carry out the indicated operations.

3. $2B + 3C$ **4.** AB

5. C^{-1} **6.** $BC - 3B$

7. If $A = \begin{bmatrix} 1 & 2 & 2 \\ 2 & 3 & 3 \\ 1 & -1 & -2 \end{bmatrix}$ and $B = \begin{bmatrix} -3 & 2 & 0 \\ 7 & -4 & 1 \\ -5 & 3 & -1 \end{bmatrix}$, show that B is the inverse of A.

8. Consider the system

$$\begin{cases} 3x + 5y = 9 \\ 2x - 3y = -13. \end{cases}$$

a. Express the system in the form $AX = B$, where A, X, and B are appropriate matrices.

b. Find A^{-1}, the inverse of the coefficient matrix.

c. Use A^{-1} to solve the given system.

9. Evaluate: $\begin{vmatrix} 4 & -1 & 3 \\ 0 & 5 & -1 \\ 5 & 2 & 4 \end{vmatrix}$.

10. Solve for x only using Cramer's Rule:

$$\begin{cases} 3x + y - 2z = -3 \\ 2x + 7y + 3z = 9 \\ 4x - 3y - z = 7. \end{cases}$$

CUMULATIVE REVIEW EXERCISES (CHAPTERS 1–6)

Solve each equation or inequality in Exercises 1–6.

1. $2x^2 = 4 - x$ **2.** $5x + 8 \leq 7(1 + x)$

3. $\sqrt{2x + 4} - \sqrt{x + 3} - 1 = 0$

4. $3x^3 + 8x^2 - 15x + 4 = 0$

5. $e^{2x} - 14e^x + 45 = 0$

6. $\log_3 x + \log_3(x + 2) = 1$

7. Use matrices to solve this system:

$$\begin{cases} x - y + z = 17 \\ 2x + 3y + z = 8 \\ -4x + y + 5z = -2. \end{cases}$$

8. Solve for y using Cramer's Rule:

$$\begin{cases} x - 2y + z = 7 \\ 2x + y - z = 0 \\ 3x + 2y - 2z = -2. \end{cases}$$

9. If $f(x) = \sqrt{4x - 7}$, find $f^{-1}(x)$.

10. Graph: $f(x) = \dfrac{x}{x^2 - 16}$.

11. Use the graph of $f(x) = 4x^4 - 4x^3 - 25x^2 + x + 6$ shown in the figure to factor the polynomial completely.

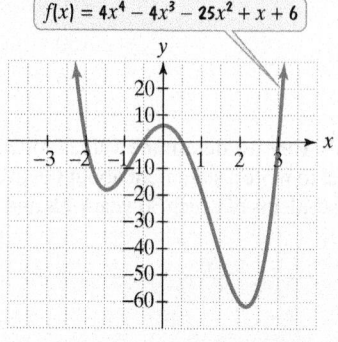

$f(x) = 4x^4 - 4x^3 - 25x^2 + x + 6$

12. Graph $y = \log_2 x$ and $y = \log_2(x + 1)$ in the same rectangular coordinate system.

13. Use the exponential decay model $A = A_0 e^{kt}$ to solve this problem. A radioactive substance has a half-life of 40 days. There are initially 900 grams of the substance.

a. Find the decay model for this substance. Round k to the nearest thousandth.

b. How much of the substance will remain after 10 days? Round to the nearest hundredth of a gram.

14. Multiply the matrices: $\begin{bmatrix} 1 & -1 & 0 \\ 2 & 1 & 3 \end{bmatrix} \begin{bmatrix} 4 & -1 \\ 2 & 0 \\ 1 & 1 \end{bmatrix}$.

15. Find the partial fraction decomposition of

$$\frac{3x^2 + 17x - 38}{(x - 3)(x - 2)(x + 2)}.$$

In Exercises 16–19, graph each equation, function, or inequality in a rectangular coordinate system.

16. $y = -\frac{2}{3}x - 1$

17. $3x - 5y < 15$

18. $f(x) = x^2 - 2x - 3$

19. $(x - 1)^2 + (y + 1)^2 = 9$

20. Use synthetic division to divide $x^3 - 6x + 4$ by $x - 2$.

CONIC SECTIONS

From ripples in water to the path on which humanity journeys through space, certain curves occur naturally throughout the universe. Over two thousand years ago, the ancient Greeks studied these curves, called *conic sections*, without regard to their immediate usefulness simply because studying them elicited ideas that were exciting, challenging, and interesting. The ancient Greeks could not have imagined the applications of these curves in the twenty-first century. They enable the Hubble Space Telescope, a large satellite about the size of a school bus orbiting 375 miles above Earth, to gather distant rays of light and focus them into spectacular images of our evolving universe. They provide doctors with a procedure for dissolving kidney stones painlessly without invasive surgery. In this chapter, we use the rectangular coordinate system to study the conic sections and the mathematics behind their surprising applications.

HERE'S WHERE YOU'LL FIND
THESE APPLICATIONS THAT
MOVE BEYOND PLANET EARTH:

- Planetary orbits: Section 7.1, page 678; Exercise Set 7.1, Exercise 78
- Halley's Comet: Blitzer Bonus on page 679
- Hubble Space Telescope: Section 7.3, pages 699 and 707.

For a kidney stone here on Earth, see Section 7.1, page 678.

SECTION 7.1

The Ellipse

Objectives

1. Graph ellipses centered at the origin.
2. Write equations of ellipses in standard form.
3. Graph ellipses not centered at the origin.
4. Solve applied problems involving ellipses.

You took on a summer job driving a truck, delivering books that were ordered online. You're an avid reader, so just being around books sounded appealing. However, now you're feeling a bit shaky driving the truck for the first time. It's 10 feet wide and 9 feet high; compared to your compact car, it feels like you're behind the wheel of a tank. Up ahead you see a sign at the semielliptical entrance to a tunnel: Caution! Tunnel is 10 Feet High at Center Peak. Then you see another sign: Caution! Tunnel Is 40 Feet Wide. Will your truck clear the opening of the tunnel's archway?

Mathematics is present in the movements of planets, bridge and tunnel construction, navigational systems used to keep track of a ship's location, manufacture of lenses for telescopes, and even in a procedure for disintegrating kidney stones. The mathematics behind these applications involves conic sections. Conic sections are curves that result from the intersection of a right circular cone and a plane. **Figure 7.1** illustrates the four conic sections: the circle, the ellipse, the parabola, and the hyperbola.

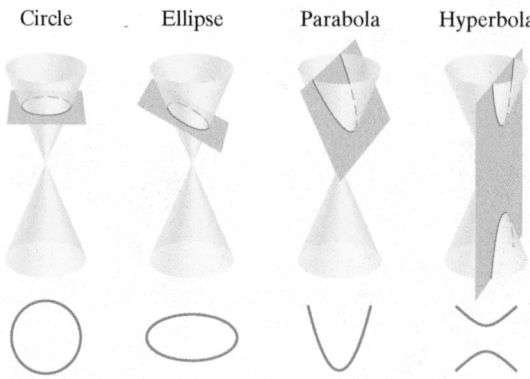

FIGURE 7.1 Obtaining the conic sections by intersecting a plane and a cone

In this section, we study the symmetric oval-shaped curve known as the ellipse. We will use a geometric definition for an ellipse to derive its equation. With this equation, we will determine if your delivery truck will clear the tunnel's entrance.

Definition of an Ellipse

Figure 7.2 illustrates how to draw an ellipse. Place pins at two fixed points, each of which is called a *focus* (plural: *foci*). If the ends of a fixed length of string are fastened to the pins and we draw the string taut with a pencil, the path traced by the pencil will be an ellipse. Notice that the sum of the distances of the pencil point from the foci remains constant because the length of the string is fixed. This procedure for drawing an ellipse illustrates its geometric definition.

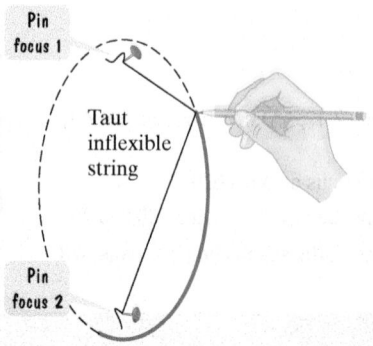

FIGURE 7.2 Drawing an ellipse

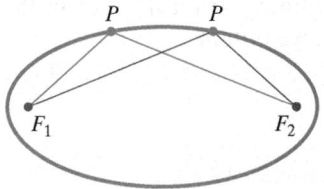

FIGURE 7.3

Definition of an Ellipse

An **ellipse** is the set of all points, P, in a plane the sum of whose distances from two fixed points, F_1 and F_2, is constant (see **Figure 7.3**). These two fixed points are called the **foci** (plural of **focus**). The midpoint of the segment connecting the foci is the **center** of the ellipse.

Figure 7.4 illustrates that an ellipse can be elongated in any direction. In this section, we will limit our discussion to ellipses that are elongated horizontally or vertically. The line through the foci intersects the ellipse at two points, called the **vertices** (singular: **vertex**). The line segment that joins the vertices is the **major axis**. Notice that the midpoint of the major axis is the center of the ellipse. The line segment whose endpoints are on the ellipse and that is perpendicular to the major axis at the center is called the **minor axis** of the ellipse.

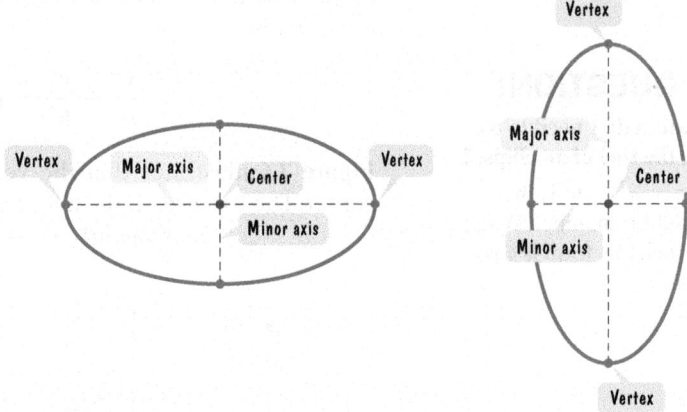

FIGURE 7.4 Horizontal and vertical elongations of an ellipse

Standard Form of the Equation of an Ellipse

The rectangular coordinate system gives us a unique way of describing an ellipse. It enables us to translate an ellipse's geometric definition into an algebraic equation.

We start with **Figure 7.5** to obtain an ellipse's equation. We've placed an ellipse that is elongated horizontally into a rectangular coordinate system. The foci are on the x-axis at $(-c, 0)$ and $(c, 0)$, as in **Figure 7.5**. In this way, the center of the ellipse is at the origin. We let (x, y) represent the coordinates of any point on the ellipse.

What does the definition of an ellipse tell us about the point (x, y) in **Figure 7.5**? For any point (x, y) on the ellipse, the sum of the distances to the two foci, $d_1 + d_2$, must be constant. As we shall see, it is convenient to denote this constant by $2a$. Thus, the point (x, y) is on the ellipse if and only if

$$d_1 + d_2 = 2a.$$

$$\sqrt{(x + c)^2 + y^2} + \sqrt{(x - c)^2 + y^2} = 2a \qquad \text{Use the distance formula.}$$

After eliminating radicals and simplifying, we obtain

$$(a^2 - c^2)x^2 + a^2y^2 = a^2(a^2 - c^2).$$

Look at the triangle in **Figure 7.5**. Notice that the distance from F_1 to F_2 is $2c$. Because the length of any side of a triangle is less than the sum of the lengths of the other two sides, $2c < d_1 + d_2$. Equivalently, $2c < 2a$ and $c < a$. Consequently,

FIGURE 7.5

GREAT QUESTION!

How did you go from
$$\sqrt{(x + c)^2 + y^2} +$$
$$\sqrt{(x - c)^2 + y^2} = 2a$$
to $(a^2 - c^2)x^2 + a^2y^2 =$
$$a^2(a^2 - c^2)?$$

The algebraic details behind eliminating the radicals can be found in the appendix. There you will find a step-by-step derivation of the ellipse's equation.

$a^2 - c^2 > 0$. For convenience, let $b^2 = a^2 - c^2$. Substituting b^2 for $a^2 - c^2$ in the preceding equation, $(a^2 - c^2)x^2 + a^2y^2 = a^2(a^2 - c^2)$, we obtain

$$b^2x^2 + a^2y^2 = a^2b^2$$

$$\frac{b^2x^2}{a^2b^2} + \frac{a^2y^2}{a^2b^2} = \frac{a^2b^2}{a^2b^2} \quad \text{Divide both sides by } a^2b^2.$$

$$\frac{x^2}{a^2} + \frac{y^2}{b^2} = 1. \quad \text{Simplify.}$$

This last equation is the **standard form of the equation of an ellipse centered at the origin**. There are two such equations, one for a horizontal major axis and one for a vertical major axis.

GREAT QUESTION!

Which equation do I need to use for locating the foci of an ellipse?

The form $c^2 = a^2 - b^2$ is the one you should remember. When finding the foci, this form is easy to manipulate.

Standard Forms of the Equations of an Ellipse

The **standard form of the equation of an ellipse** with center at the origin, and major and minor axes of lengths $2a$ and $2b$ (where a and b are positive, and $a^2 > b^2$) is

$$\frac{x^2}{a^2} + \frac{y^2}{b^2} = 1 \quad \text{or} \quad \frac{x^2}{b^2} + \frac{y^2}{a^2} = 1.$$

Figure 7.6 illustrates that the vertices are on the major axis, a units from the center. The foci are on the major axis, c units from the center. For both equations, $b^2 = a^2 - c^2$. Equivalently, $c^2 = a^2 - b^2$.

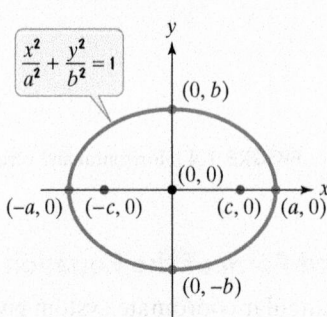

FIGURE 7.6(a) Major axis is horizontal with length $2a$.

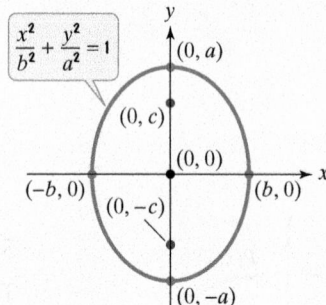

FIGURE 7.6(b) Major axis is vertical with length $2a$.

The intercepts shown in **Figure 7.6(a)** can be obtained algebraically. Let's do this for

$$\frac{x^2}{a^2} + \frac{y^2}{b^2} = 1.$$

x-intercepts: Set $y = 0$. **y-intercepts: Set $x = 0$.**

$$\frac{x^2}{a^2} = 1 \qquad\qquad \frac{y^2}{b^2} = 1$$

$$x^2 = a^2 \qquad\qquad y^2 = b^2$$

$$x = \pm a \qquad\qquad y = \pm b$$

x-intercepts are $-a$ and a. The graph passes through $(-a, 0)$ and $(a, 0)$, which are the vertices.

y-intercepts are $-b$ and b. The graph passes through $(0, -b)$ and $(0, b)$.

Using the Standard Form of the Equation of an Ellipse

We can use the standard form of an ellipse's equation to graph the ellipse. Although the definition of the ellipse is given in terms of its foci, the foci are not part of the graph. A complete graph of an ellipse can be obtained without graphing the foci.

① Graph ellipses centered at the origin.

EXAMPLE 1 Graphing an Ellipse Centered at the Origin

Graph and locate the foci: $\dfrac{x^2}{9} + \dfrac{y^2}{4} = 1.$

SOLUTION

The given equation is the standard form of an ellipse's equation with $a^2 = 9$ and $b^2 = 4$.

$$\frac{x^2}{9} + \frac{y^2}{4} = 1$$

$a^2 = 9$. This is the larger of the two denominators. $b^2 = 4$. This is the smaller of the two denominators.

TECHNOLOGY

We graph $\dfrac{x^2}{9} + \dfrac{y^2}{4} = 1$ with a graphing utility by solving for y.

$$\frac{y^2}{4} = 1 - \frac{x^2}{9}$$

$$y^2 = 4\left(1 - \frac{x^2}{9}\right)$$

$$y = \pm 2\sqrt{1 - \frac{x^2}{9}}$$

Notice that the square root property requires us to define two functions. Enter

$y_1 = 2\ \boxed{\sqrt{\ }}\ (1\ \boxed{-}\ x\ \boxed{\wedge}\ 2\ \boxed{\div}\ 9)$

and

$$y_2 = -y_1.$$

To see the true shape of the ellipse, use the

$\boxed{\text{ZOOM SQUARE}}$

feature so that one unit on the y-axis is the same length as one unit on the x-axis.

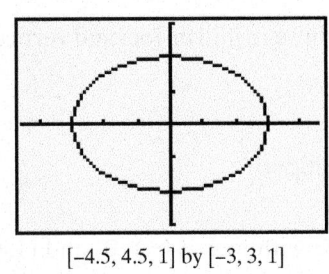

[-4.5, 4.5, 1] by [-3, 3, 1]

Because the denominator of the x^2-term is greater than the denominator of the y^2-term, the major axis is horizontal. Based on the standard form of the equation, we know the vertices are $(-a, 0)$ and $(a, 0)$. Because $a^2 = 9$, $a = 3$. Thus, the vertices are $(-3, 0)$ and $(3, 0)$, shown in **Figure 7.7**.

Now let us find the endpoints of the vertical minor axis. According to the standard form of the equation, these endpoints are $(0, -b)$ and $(0, b)$. Because $b^2 = 4$, $b = 2$. Thus, the endpoints of the minor axis are $(0, -2)$ and $(0, 2)$. They are shown in **Figure 7.7**.

Finally, we find the foci, which are located at $(-c, 0)$ and $(c, 0)$. We can use the formula $c^2 = a^2 - b^2$ to do so. We know that $a^2 = 9$ and $b^2 = 4$. Thus,

$$c^2 = a^2 - b^2 = 9 - 4 = 5.$$

Because $c^2 = 5$, $c = \sqrt{5}$. The foci, $(-c, 0)$ and $(c, 0)$, are located at $\left(-\sqrt{5}, 0\right)$ and $\left(\sqrt{5}, 0\right)$. They are shown in **Figure 7.7**.

You can sketch the ellipse in **Figure 7.7** by locating endpoints on the major and minor axes.

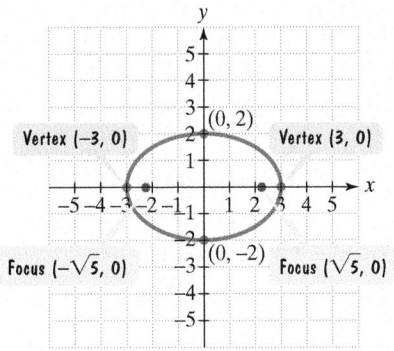

FIGURE 7.7 The graph of $\dfrac{x^2}{9} + \dfrac{y^2}{4} = 1$

$$\frac{x^2}{3^2} + \frac{y^2}{2^2} = 1$$

Endpoints of the major axis are 3 units to the right and left of the center. Endpoints of the minor axis are 2 units up and down from the center.

• • •

✓ Check Point 1 Graph and locate the foci: $\dfrac{x^2}{36} + \dfrac{y^2}{9} = 1.$

EXAMPLE 2 Graphing an Ellipse Centered at the Origin

Graph and locate the foci: $25x^2 + 16y^2 = 400$.

SOLUTION

We begin by expressing the equation in standard form. Because we want 1 on the right side, we divide both sides by 400.

$$\frac{25x^2}{400} + \frac{16y^2}{400} = \frac{400}{400}$$

$$\frac{x^2}{16} + \frac{y^2}{25} = 1$$

$b^2 = 16$. This is the smaller of the two denominators.　　$a^2 = 25$. This is the larger of the two denominators.

The equation is the standard form of an ellipse's equation with $a^2 = 25$ and $b^2 = 16$. Because the denominator of the y^2-term is greater than the denominator of the x^2-term, the major axis is vertical. Based on the standard form of the equation, we know the vertices are $(0, -a)$ and $(0, a)$. Because $a^2 = 25, a = 5$. Thus, the vertices are $(0, -5)$ and $(0, 5)$, shown in **Figure 7.8**.

Now let us find the endpoints of the horizontal minor axis. According to the standard form of the equation, these endpoints are $(-b, 0)$ and $(b, 0)$. Because $b^2 = 16, b = 4$. Thus, the endpoints of the minor axis are $(-4, 0)$ and $(4, 0)$. They are shown in **Figure 7.8**.

Finally, we find the foci, which are located at $(0, -c)$ and $(0, c)$. We can use the formula $c^2 = a^2 - b^2$ to do so. We know that $a^2 = 25$ and $b^2 = 16$. Thus,

$$c^2 = a^2 - b^2 = 25 - 16 = 9.$$

Because $c^2 = 9, c = 3$. The foci, $(0, -c)$ and $(0, c)$, are located at $(0, -3)$ and $(0, 3)$. They are shown in **Figure 7.8**.

You can sketch the ellipse in **Figure 7.8** by locating endpoints on the major and minor axes.

$$\frac{x^2}{4^2} + \frac{y^2}{5^2} = 1$$

Endpoints of the minor axis are 4 units to the right and left of the center.　　Endpoints of the major axis are 5 units up and down from the center.

• • •

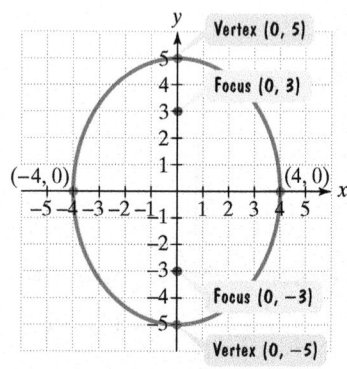

FIGURE 7.8 The graph of $25x^2 + 16y^2 = 400$, or $\frac{x^2}{16} + \frac{y^2}{25} = 1$

✓ Check Point **2** Graph and locate the foci: $16x^2 + 9y^2 = 144$.

❷ Write equations of ellipses in standard form.

In Examples 1 and 2, we used the equation of an ellipse to find its foci and vertices. In the next example, we reverse this procedure.

EXAMPLE 3 Finding the Equation of an Ellipse from Its Foci and Vertices

Find the standard form of the equation of an ellipse with foci at $(-1, 0)$ and $(1, 0)$ and vertices $(-2, 0)$ and $(2, 0)$.

SOLUTION

Because the foci are located at $(-1, 0)$ and $(1, 0)$, on the x-axis, the major axis is horizontal. The center of the ellipse is midway between the foci, located at $(0, 0)$. Thus, the form of the equation is

$$\frac{x^2}{a^2} + \frac{y^2}{b^2} = 1.$$

We need to determine the values for a^2 and b^2. The distance from the center, $(0, 0)$, to either vertex, $(-2, 0)$ or $(2, 0)$, is 2. Thus, $a = 2$.

$$\frac{x^2}{2^2} + \frac{y^2}{b^2} = 1 \quad \text{or} \quad \frac{x^2}{4} + \frac{y^2}{b^2} = 1$$

We must still find b^2. The distance from the center, $(0, 0)$, to either focus, $(-1, 0)$ or $(1, 0)$, is 1, so $c = 1$. Using $c^2 = a^2 - b^2$, we have

$$1^2 = 2^2 - b^2$$

and

$$b^2 = 2^2 - 1^2 = 4 - 1 = 3.$$

Substituting 3 for b^2 in $\dfrac{x^2}{4} + \dfrac{y^2}{b^2} = 1$ gives us the standard form of the ellipse's equation. The equation is

$$\frac{x^2}{4} + \frac{y^2}{3} = 1. \qquad \bullet\bullet\bullet$$

∅ **Check Point 3** Find the standard form of the equation of an ellipse with foci at $(-2, 0)$ and $(2, 0)$ and vertices $(-3, 0)$ and $(3, 0)$.

③ Graph ellipses not centered at the origin.

Translations of Ellipses

Horizontal and vertical translations can be used to graph ellipses that are not centered at the origin. **Figure 7.9** illustrates that the graphs of

$$\frac{(x - h)^2}{a^2} + \frac{(y - k)^2}{b^2} = 1 \quad \text{and} \quad \frac{x^2}{a^2} + \frac{y^2}{b^2} = 1$$

have the same size and shape. However, the graph of the first equation is centered at (h, k) rather than at the origin.

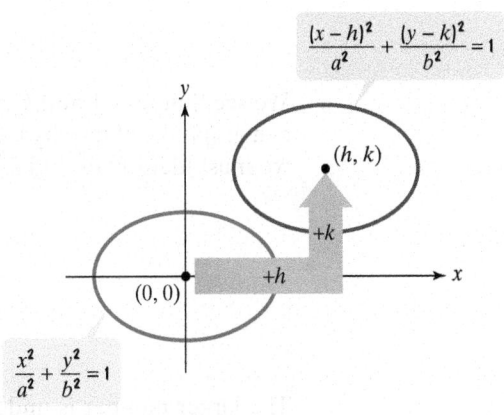

FIGURE 7.9 Translating an ellipse's graph

Table 7.1 gives the standard forms of equations of ellipses centered at (h, k) and shows their graphs.

Table 7.1 Standard Forms of Equations of Ellipses Centered at (h, k)

Equation	Center	Major Axis	Vertices	Graph
$$\frac{(x-h)^2}{a^2} + \frac{(y-k)^2}{b^2} = 1$$ Endpoints of major axis are a units right and a units left of center. $\quad a^2 > b^2$ Foci are c units right and c units left of center, where $c^2 = a^2 - b^2$.	(h, k)	Parallel to the x-axis, horizontal	$(h - a, k)$ $(h + a, k)$	
$$\frac{(x-h)^2}{b^2} + \frac{(y-k)^2}{a^2} = 1$$ $a^2 > b^2$ \quad Endpoints of the major axis are a units above and a units below the center. Foci are c units above and c units below the center, where $c^2 = a^2 - b^2$.	(h, k)	Parallel to the y-axis, vertical	$(h, k - a)$ $(h, k + a)$	

EXAMPLE 4 Graphing an Ellipse Centered at (h, k)

Graph: $\dfrac{(x-1)^2}{4} + \dfrac{(y+2)^2}{9} = 1$. Where are the foci located?

SOLUTION

To graph the ellipse, we need to know its center, (h, k). In the standard forms of equations centered at (h, k), h is the number subtracted from x and k is the number subtracted from y.

This is $(x - h)^2$, with $h = 1$. \qquad This is $(y - k)^2$, with $k = -2$.

$$\frac{(x-1)^2}{4} + \frac{(y-(-2))^2}{9} = 1$$

We see that $h = 1$ and $k = -2$. Thus, the center of the ellipse, (h, k), is $(1, -2)$. We can graph the ellipse by locating endpoints on the major and minor axes. To do this, we must identify a^2 and b^2.

$$\frac{(x-1)^2}{4} + \frac{(y+2)^2}{9} = 1$$

$b^2 = 4$. This is the smaller of the two denominators. \qquad $a^2 = 9$. This is the larger of the two denominators.

The larger number is under the expression involving y. This means that the major axis is vertical and parallel to the y-axis.

We can sketch the ellipse by locating endpoints on the major and minor axes.

$$\frac{(x-1)^2}{2^2} + \frac{(y+2)^2}{3^2} = 1$$

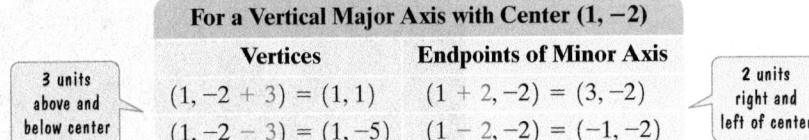

We categorize the observations in the voice balloons as follows:

For a Vertical Major Axis with Center (1, −2)		
Vertices	**Endpoints of Minor Axis**	
$(1, -2 + 3) = (1, 1)$	$(1 + 2, -2) = (3, -2)$	
$(1, -2 - 3) = (1, -5)$	$(1 - 2, -2) = (-1, -2)$	

3 units above and below center

2 units right and left of center

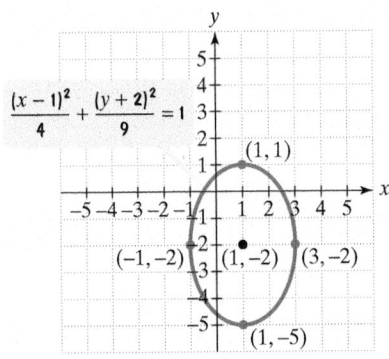

FIGURE 7.10 The graph of an ellipse centered at (1, −2)

Using the center and these four points, we can sketch the ellipse shown in **Figure 7.10**.

With $c^2 = a^2 - b^2$, we have $c^2 = 9 - 4 = 5$. So the foci are located $\sqrt{5}$ units above and below the center, at $\left(1, -2 + \sqrt{5}\right)$ and $\left(1, -2 - \sqrt{5}\right)$. $\bullet\bullet\bullet$

⊘ **Check Point 4** Graph: $\dfrac{(x+1)^2}{9} + \dfrac{(y-2)^2}{4} = 1$. Where are the foci located?

In some cases, it is necessary to convert the equation of an ellipse to standard form by completing the square on x and y. For example, suppose that we wish to graph the ellipse whose equation is

$$9x^2 + 4y^2 - 18x + 16y - 11 = 0.$$

Because we plan to complete the square on both x and y, we need to rearrange terms so that

- x-terms are arranged in descending order.
- y-terms are arranged in descending order.
- the constant term appears on the right.

$9x^2 + 4y^2 - 18x + 16y - 11 = 0$ This is the given equation.

$(9x^2 - 18x) + (4y^2 + 16y) = 11$ Group terms and add 11 to both sides.

$9(x^2 - 2x + \square) + 4(y^2 + 4y + \square) = 11$ To complete the square, coefficients of x^2 and y^2 must be 1. Factor out 9 and 4, respectively.

We added 9 · 1, or 9, to the left side.

We also added 4 · 4, or 16, to the left side.

$9(x^2 - 2x + 1) + 4(y^2 + 4y + 4) = 11 + 9 + 16$ Complete each square by adding the square of half the coefficient of x and y, respectively.

9 and 16, added on the left side, must also be added on the right side.

$9(x-1)^2 + 4(y+2)^2 = 36$ Factor.

$\dfrac{9(x-1)^2}{36} + \dfrac{4(y+2)^2}{36} = \dfrac{36}{36}$ Divide both sides by 36.

$\dfrac{(x-1)^2}{4} + \dfrac{(y+2)^2}{9} = 1$ Simplify.

The equation is now in standard form. This is precisely the form of the equation that we graphed in Example 4.

GREAT QUESTION!

Can you remind me how to complete the square?

To complete the square on $x^2 + bx$, take half the coefficient of x. Then square this number. By adding the square of half the coefficient of x, a perfect square trinomial will result. Once you've completed the square, remember that changes made on the left side of the equation must also be made on the right side of the equation.

4 Solve applied problems involving ellipses.

Applications

Ellipses have many applications. German scientist Johannes Kepler (1571–1630) showed that the planets in our solar system move in elliptical orbits, with the sun at a focus. Earth satellites also travel in elliptical orbits, with Earth at a focus.

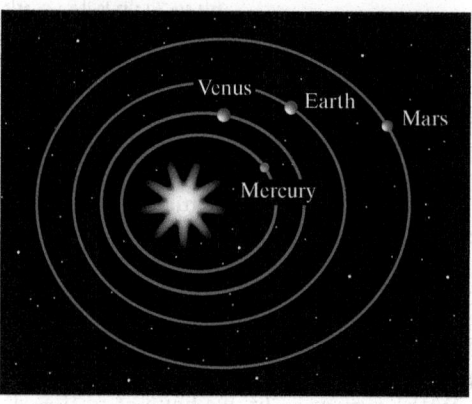

Planets move in elliptical orbits.

Whispering in an elliptical dome

One intriguing aspect of the ellipse is that a ray of light or a sound wave emanating from one focus will be reflected from the ellipse to exactly the other focus. A whispering gallery is an elliptical room with an elliptical, dome-shaped ceiling. People standing at the foci can whisper and hear each other quite clearly, while persons in other locations in the room cannot hear them. Statuary Hall in the U.S. Capitol Building is elliptical. President John Quincy Adams, while a member of the House of Representatives, was aware of this acoustical phenomenon. He situated his desk at a focal point of the elliptical ceiling, easily eavesdropping on the private conversations of other House members located near the other focus.

The elliptical reflection principle is used in a procedure for disintegrating kidney stones. The patient is placed within a device that is elliptical in shape. The patient is placed so the kidney is centered at one focus, while ultrasound waves from the other focus hit the walls and are reflected to the kidney stone. The convergence of the ultrasound waves at the kidney stone causes vibrations that shatter it into fragments. The small pieces can then be passed painlessly through the patient's system. The patient recovers in days, as opposed to up to six weeks if surgery is used instead.

Disintegrating kidney stones

Ellipses are often used for supporting arches of bridges and in tunnel construction. This application forms the basis of our next example.

Blitzer Bonus

Halley's Comet

Halley's Comet has an elliptical orbit with the sun at one focus. The comet returns every 76.3 years. The first recorded sighting was in 239 B.C. It was last seen in 1986. At that time, spacecraft went close to the comet, measuring its nucleus to be 7 miles long and 4 miles wide. By 2024, Halley's Comet will have reached the farthest point in its elliptical orbit before returning to be next visible from Earth in 2062.

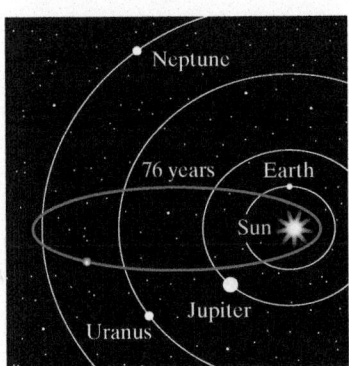

The elliptical orbit of Halley's Comet

EXAMPLE 5 An Application Involving an Ellipse

A semielliptical archway over a one-way road has a height of 10 feet and a width of 40 feet (see **Figure 7.11**). Your truck has a width of 10 feet and a height of 9 feet. Will your truck clear the opening of the archway?

FIGURE 7.11
A semielliptical archway

SOLUTION

Because your truck's width is 10 feet, to determine the clearance, we must find the height of the archway 5 feet from the center. If that height is 9 feet or less, the truck will not clear the opening.

In **Figure 7.12**, we've constructed a coordinate system with the x-axis on the ground and the origin at the center of the archway. Also shown is the truck, whose height is 9 feet.

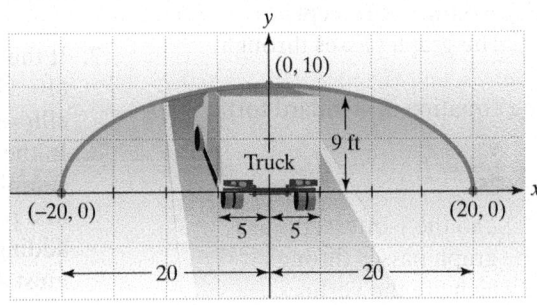

FIGURE 7.12

Using the equation $\dfrac{x^2}{a^2} + \dfrac{y^2}{b^2} = 1$, we can express the equation of the blue archway in **Figure 7.12** as $\dfrac{x^2}{20^2} + \dfrac{y^2}{10^2} = 1$, or $\dfrac{x^2}{400} + \dfrac{y^2}{100} = 1$.

As shown in **Figure 7.12**, the edge of the 10-foot-wide truck corresponds to $x = 5$. We find the height of the archway 5 feet from the center by substituting 5 for x and solving for y.

$$\frac{5^2}{400} + \frac{y^2}{100} = 1 \qquad \text{Substitute 5 for x in } \frac{x^2}{400} + \frac{y^2}{100} = 1.$$

$$\frac{25}{400} + \frac{y^2}{100} = 1 \qquad \text{Square 5.}$$

$$400\left(\frac{25}{400} + \frac{y^2}{100}\right) = 400(1) \qquad \text{Clear fractions by multiplying both sides by 400.}$$

$$25 + 4y^2 = 400 \qquad \text{Use the distributive property and simplify.}$$

$$4y^2 = 375 \qquad \text{Subtract 25 from both sides.}$$

$$y^2 = \frac{375}{4} \qquad \text{Divide both sides by 4.}$$

$$y = \sqrt{\frac{375}{4}} \qquad \text{Take only the positive square root. The archway is above the x-axis, so y is nonnegative.}$$

$$\approx 9.68 \qquad \text{Use a calculator.}$$

Thus, the height of the archway 5 feet from the center is approximately 9.68 feet. Because your truck's height is 9 feet, there is enough room for the truck to clear the archway. • • •

⊘ Check Point **5** Will a truck that is 12 feet wide and has a height of 9 feet clear the opening of the archway described in Example 5?

CONCEPT AND VOCABULARY CHECK

Fill in each blank so that the resulting statement is true.

1. The set of all points in a plane the sum of whose distances from two fixed points is constant is a/an _____. The two fixed points are called the _____. The midpoint of the line segment connecting the two fixed points is the _____.

2. Consider the following equation in standard form:
$$\frac{x^2}{25} + \frac{y^2}{9} = 1.$$
The value of a^2 is _____, so the x-intercepts are _____ and _____. The graph passes through _____ and _____, which are the vertices.
The value of b^2 is _____, so the y-intercepts are _____ and _____. The graph passes through _____ and _____.

3. Consider the following equation in standard form:
$$\frac{x^2}{9} + \frac{y^2}{25} = 1.$$
The value of a^2 is _____, so the y-intercepts are _____ and _____. The graph passes through _____ and _____, which are the vertices.
The value of b^2 is _____, so the x-intercepts are _____ and _____. The graph passes through _____ and _____.

4. Consider an ellipse centered at the origin whose major axis is vertical. The equation of this ellipse in standard form indicates that $a^2 = 9$ and $b^2 = 4$. Thus, $c^2 =$ _____. The foci are located at _____ and _____.

5. The graph of $\dfrac{(x+1)^2}{25} + \dfrac{(y-4)^2}{9} = 1$ has its center at _____.

6. If the center of an ellipse is $(3, -2)$, the major axis is horizontal and parallel to the x-axis, and the distance from the center of the ellipse to its vertices is $a = 5$ units, then the coordinates of the vertices are _____ and _____.

7. If the foci of an ellipse are located at $(-8, 6)$ and $(10, 12)$, then the coordinates of the center of the ellipse are _____.

8. In the equation $3(x^2 + 4x) + 4(y^2 - 2y) = 32$, we complete the square on x by adding _____ within the first parentheses. We complete the square on y by adding _____ within the second parentheses. Thus, we must add _____ to the right side of the equation.

EXERCISE SET 7.1

Practice Exercises

In Exercises 1–18, graph each ellipse and locate the foci.

1. $\dfrac{x^2}{16} + \dfrac{y^2}{4} = 1$

2. $\dfrac{x^2}{25} + \dfrac{y^2}{16} = 1$

3. $\dfrac{x^2}{9} + \dfrac{y^2}{36} = 1$

4. $\dfrac{x^2}{16} + \dfrac{y^2}{49} = 1$

5. $\dfrac{x^2}{25} + \dfrac{y^2}{64} = 1$

6. $\dfrac{x^2}{49} + \dfrac{y^2}{36} = 1$

7. $\dfrac{x^2}{49} + \dfrac{y^2}{81} = 1$

8. $\dfrac{x^2}{64} + \dfrac{y^2}{100} = 1$

9. $\dfrac{x^2}{\frac{9}{4}} + \dfrac{y^2}{\frac{25}{4}} = 1$

10. $\dfrac{x^2}{\frac{81}{4}} + \dfrac{y^2}{\frac{25}{16}} = 1$

11. $x^2 = 1 - 4y^2$

12. $y^2 = 1 - 4x^2$

13. $25x^2 + 4y^2 = 100$

14. $9x^2 + 4y^2 = 36$

15. $4x^2 + 16y^2 = 64$

16. $4x^2 + 25y^2 = 100$

17. $7x^2 = 35 - 5y^2$

18. $6x^2 = 30 - 5y^2$

In Exercises 19–24, find the standard form of the equation of each ellipse and give the location of its foci.

19.

20.

21.

22.

23.

24.

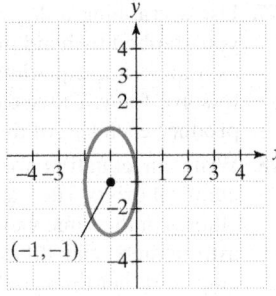

48. $\dfrac{(x+1)^2}{2} + \dfrac{(y-3)^2}{5} = 1$

49. $9(x-1)^2 + 4(y+3)^2 = 36$

50. $36(x+4)^2 + (y+3)^2 = 36$

In Exercises 51–56, convert each equation to standard form by completing the square on x and y. Then graph the ellipse and give the location of its foci.

51. $9x^2 + 25y^2 - 36x + 50y - 164 = 0$

52. $4x^2 + 9y^2 - 32x + 36y + 64 = 0$

53. $9x^2 + 16y^2 - 18x + 64y - 71 = 0$

54. $x^2 + 4y^2 + 10x - 8y + 13 = 0$

55. $4x^2 + y^2 + 16x - 6y - 39 = 0$

56. $4x^2 + 25y^2 - 24x + 100y + 36 = 0$

Practice Plus

In Exercises 57–62, find the solution set for each system by graphing both of the system's equations in the same rectangular coordinate system and finding points of intersection. Check all solutions in both equations.

57. $\begin{cases} x^2 + y^2 = 1 \\ x^2 + 9y^2 = 9 \end{cases}$

58. $\begin{cases} x^2 + y^2 = 25 \\ 25x^2 + y^2 = 25 \end{cases}$

59. $\begin{cases} \dfrac{x^2}{25} + \dfrac{y^2}{9} = 1 \\ \\ y = 3 \end{cases}$

60. $\begin{cases} \dfrac{x^2}{4} + \dfrac{y^2}{36} = 1 \\ \\ x = -2 \end{cases}$

61. $\begin{cases} 4x^2 + y^2 = 4 \\ 2x - y = 2 \end{cases}$

62. $\begin{cases} 4x^2 + y^2 = 4 \\ x + y = 3 \end{cases}$

In Exercises 63–64, graph each semiellipse.

63. $y = -\sqrt{16 - 4x^2}$

64. $y = -\sqrt{4 - 4x^2}$

Application Exercises

65. Will a truck that is 8 feet wide carrying a load that reaches 7 feet above the ground clear the semielliptical arch on the one-way road that passes under the bridge shown in the figure?

66. A semielliptic archway has a height of 20 feet and a width of 50 feet, as shown in the figure. Can a truck 14 feet high and 10 feet wide drive under the archway without going into the other lane?

In Exercises 25–36, find the standard form of the equation of each ellipse satisfying the given conditions.

25. Foci: $(-5, 0), (5, 0)$; vertices: $(-8, 0), (8, 0)$

26. Foci: $(-2, 0), (2, 0)$; vertices: $(-6, 0), (6, 0)$

27. Foci: $(0, -4), (0, 4)$; vertices: $(0, -7), (0, 7)$

28. Foci: $(0, -3), (0, 3)$; vertices: $(0, -4), (0, 4)$

29. Foci: $(-2, 0), (2, 0)$; y-intercepts: -3 and 3

30. Foci: $(0, -2), (0, 2)$; x-intercepts: -2 and 2

31. Major axis horizontal with length 8; length of minor axis $= 4$; center: $(0, 0)$

32. Major axis horizontal with length 12; length of minor axis $= 6$; center: $(0, 0)$

33. Major axis vertical with length 10; length of minor axis $= 4$; center: $(-2, 3)$

34. Major axis vertical with length 20; length of minor axis $= 10$; center: $(2, -3)$

35. Endpoints of major axis: $(7, 9)$ and $(7, 3)$
Endpoints of minor axis: $(5, 6)$ and $(9, 6)$

36. Endpoints of major axis: $(2, 2)$ and $(8, 2)$
Endpoints of minor axis: $(5, 3)$ and $(5, 1)$

In Exercises 37–50, graph each ellipse and give the location of its foci.

37. $\dfrac{(x-2)^2}{9} + \dfrac{(y-1)^2}{4} = 1$

38. $\dfrac{(x-1)^2}{16} + \dfrac{(y+2)^2}{9} = 1$

39. $(x+3)^2 + 4(y-2)^2 = 16$

40. $(x-3)^2 + 9(y+2)^2 = 18$

41. $\dfrac{(x-4)^2}{9} + \dfrac{(y+2)^2}{25} = 1$

42. $\dfrac{(x-3)^2}{9} + \dfrac{(y+1)^2}{16} = 1$

43. $\dfrac{x^2}{25} + \dfrac{(y-2)^2}{36} = 1$

44. $\dfrac{(x-4)^2}{4} + \dfrac{y^2}{25} = 1$

45. $\dfrac{(x+3)^2}{9} + (y-2)^2 = 1$

46. $\dfrac{(x+2)^2}{16} + (y-3)^2 = 1$

47. $\dfrac{(x-1)^2}{2} + \dfrac{(y+3)^2}{5} = 1$

67. The elliptical ceiling in Statuary Hall in the U.S. Capitol Building is 96 feet long and 23 feet tall.

a. Using the rectangular coordinate system in the figure shown, write the standard form of the equation of the elliptical ceiling.

b. John Quincy Adams discovered that he could overhear the conversations of opposing party leaders near the left side of the chamber if he situated his desk at the focus at the right side of the chamber. How far from the center of the ellipse along the major axis did Adams situate his desk? (Round to the nearest foot.)

68. If an elliptical whispering room has a height of 30 feet and a width of 100 feet, where should two people stand if they would like to whisper back and forth and be heard?

Writing in Mathematics

69. What is an ellipse?

70. Describe how to graph $\dfrac{x^2}{25} + \dfrac{y^2}{16} = 1$.

71. Describe how to locate the foci for $\dfrac{x^2}{25} + \dfrac{y^2}{16} = 1$.

72. Describe one similarity and one difference between the graphs of $\dfrac{x^2}{25} + \dfrac{y^2}{16} = 1$ and $\dfrac{x^2}{16} + \dfrac{y^2}{25} = 1$.

73. Describe one similarity and one difference between the graphs of $\dfrac{x^2}{25} + \dfrac{y^2}{16} = 1$ and $\dfrac{(x-1)^2}{25} + \dfrac{(y-1)^2}{16} = 1$.

74. An elliptipool is an elliptical pool table with only one pocket. A pool shark places a ball on the table, hits it in what appears to be a random direction, and yet it bounces off the edge, falling directly into the pocket. Explain why this happens.

Technology Exercises

75. Use a graphing utility to graph any five of the ellipses that you graphed by hand in Exercises 1–18.

76. Use a graphing utility to graph any three of the ellipses that you graphed by hand in Exercises 37–50. First solve the given equation for y by using the square root property. Enter each of the two resulting equations to produce each half of the ellipse.

77. Use a graphing utility to graph any one of the ellipses that you graphed by hand in Exercises 51–56. Write the equation as a quadratic equation in y and use the quadratic formula to solve for y. Enter each of the two resulting equations to produce each half of the ellipse.

78. Write an equation for the path of each of the following elliptical orbits. Then use a graphing utility to graph the two ellipses in the same viewing rectangle. Can you see why early astronomers had difficulty detecting that these orbits are ellipses rather than circles?

- Earth's orbit: Length of major axis: 186 million miles

 Length of minor axis: 185.8 million miles

- Mars's orbit: Length of major axis: 283.5 million miles

 Length of minor axis: 278.5 million miles

Critical Thinking Exercises

Make Sense? *In Exercises 79–82, determine whether each statement makes sense or does not make sense, and explain your reasoning.*

79. I graphed an ellipse with a horizontal major axis and foci on the y-axis.

80. I graphed an ellipse that was symmetric about its major axis but not symmetric about its minor axis.

81. You told me that an ellipse centered at the origin has vertices at $(-5, 0)$ and $(5, 0)$, so I was able to graph the ellipse.

82. In a whispering gallery at our science museum, I stood at one focus, my friend stood at the other focus, and we had a clear conversation, very little of which was heard by the 25 museum visitors standing between us.

83. Find the standard form of the equation of an ellipse with vertices at $(0, -6)$ and $(0, 6)$, passing through $(2, -4)$.

84. An Earth satellite has an elliptical orbit described by

$$\frac{x^2}{(5000)^2} + \frac{y^2}{(4750)^2} = 1.$$

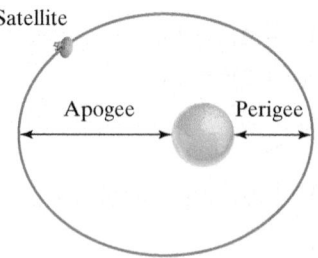

(All units are in miles.) The coordinates of the center of Earth are $(16, 0)$.

a. The perigee of the satellite's orbit is the point that is nearest Earth's center. If the radius of Earth is approximately 4000 miles, find the distance of the perigee above Earth's surface.

b. The apogee of the satellite's orbit is the point that is the greatest distance from Earth's center. Find the distance of the apogee above Earth's surface.

85. The equation of the red ellipse in the figure shown is

$$\frac{x^2}{25} + \frac{y^2}{9} = 1.$$

Write the equation for each circle shown in the figure.

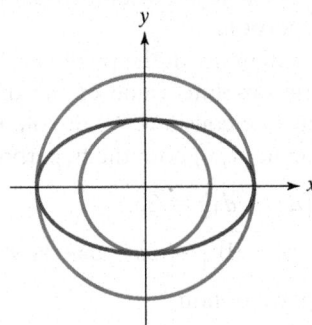

86. What happens to the shape of the graph of $\frac{x^2}{a^2} + \frac{y^2}{b^2} = 1$ as $\frac{c}{a} \rightarrow 0$, where $c^2 = a^2 - b^2$?

Preview Exercises

Exercises 87–89 will help you prepare for the material covered in the next section.

87. Divide both sides of $4x^2 - 9y^2 = 36$ by 36 and simplify. How does the simplified equation differ from that of an ellipse?

88. Consider the equation $\frac{x^2}{16} - \frac{y^2}{9} = 1$.

 a. Find the *x*-intercepts.
 b. Explain why there are no *y*-intercepts.

89. Consider the equation $\frac{y^2}{9} - \frac{x^2}{16} = 1$.

 a. Find the *y*-intercepts.
 b. Explain why there are no *x*-intercepts.

SECTION 7.2 The Hyperbola

Objectives

1. Locate a hyperbola's vertices and foci.
2. Write equations of hyperbolas in standard form.
3. Graph hyperbolas centered at the origin.
4. Graph hyperbolas not centered at the origin.
5. Solve applied problems involving hyperbolas.

Conic sections are often used to create unusual architectural designs. The top of St. Mary's Cathedral in San Francisco is a 2135-cubic-foot dome with walls rising 200 feet above the floor and supported by four massive concrete pylons that extend 94 feet into the ground. Cross sections of the roof are parabolas and hyperbolas. In this section, we study the curve with two parts known as the hyperbola.

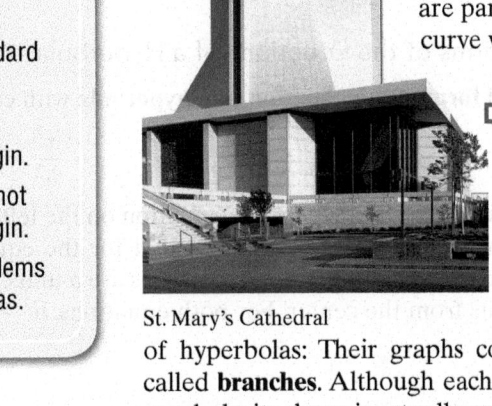

St. Mary's Cathedral

Definition of a Hyperbola

Figure 7.13 shows a cylindrical lampshade casting two shadows on a wall. These shadows indicate the distinguishing feature of hyperbolas: Their graphs contain two disjoint parts, called **branches**. Although each branch might look like a parabola, its shape is actually quite different.

The definition of a hyperbola is similar to that of an ellipse. For an ellipse, the *sum* of the distances to the foci is a constant. By contrast, for a hyperbola the *difference* of the distances to the foci is a constant.

FIGURE 7.13 Casting hyperbolic shadows

Definition of a Hyperbola

A **hyperbola** is the set of points in a plane the difference of whose distances from two fixed points, called foci, is constant.

Figure 7.14 illustrates the two branches of a hyperbola. The line through the foci intersects the hyperbola at two points, called the **vertices**. The line segment that joins the vertices is the **transverse axis**. The midpoint of the transverse axis is the **center** of the hyperbola. Notice that the center lies midway between the vertices, as well as midway between the foci.

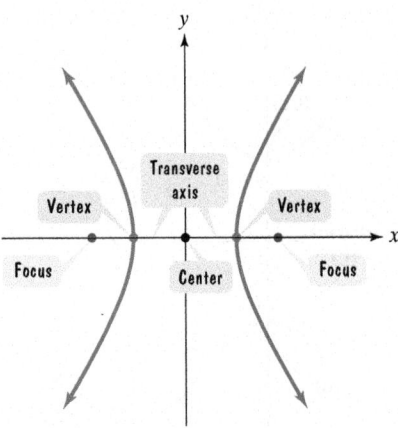

FIGURE 7.14 The two branches of a hyperbola

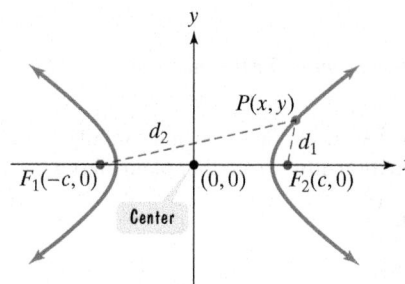

FIGURE 7.15

Standard Form of the Equation of a Hyperbola

The rectangular coordinate system enables us to translate a hyperbola's geometric definition into an algebraic equation. **Figure 7.15** is our starting point for obtaining an equation. We place the foci, F_1 and F_2, on the x-axis at the points $(-c, 0)$ and $(c, 0)$. Note that the center of this hyperbola is at the origin. We let (x, y) represent the coordinates of any point, P, on the hyperbola.

What does the definition of a hyperbola tell us about the point (x, y) in **Figure 7.15**? For any point (x, y) on the hyperbola, the absolute value of the difference of the distances from the two foci, $|d_2 - d_1|$, must be constant. We denote this constant by $2a$, just as we did for the ellipse. Thus, the point (x, y) is on the hyperbola if and only if

$$|d_2 - d_1| = 2a.$$

$$\left| \sqrt{(x + c)^2 + (y - 0)^2} - \sqrt{(x - c)^2 + (y - 0)^2} \right| = 2a \quad \text{Use the distance formula.}$$

After eliminating radicals and simplifying, we obtain

$$(c^2 - a^2)x^2 - a^2y^2 = a^2(c^2 - a^2).$$

For convenience, let $b^2 = c^2 - a^2$. Substituting b^2 for $c^2 - a^2$ in the preceding equation, we obtain

$$b^2x^2 - a^2y^2 = a^2b^2.$$

$$\frac{b^2x^2}{a^2b^2} - \frac{a^2y^2}{a^2b^2} = \frac{a^2b^2}{a^2b^2} \quad \text{Divide both sides by } a^2b^2.$$

$$\frac{x^2}{a^2} - \frac{y^2}{b^2} = 1 \quad \text{Simplify.}$$

This last equation is called the **standard form of the equation of a hyperbola centered at the origin**. There are two such equations. The first is for a hyperbola in which the transverse axis lies on the x-axis. The second is for a hyperbola in which the transverse axis lies on the y-axis.

Standard Forms of the Equations of a Hyperbola

The **standard form of the equation of a hyperbola** with center at the origin is

$$\frac{x^2}{a^2} - \frac{y^2}{b^2} = 1 \quad \text{or} \quad \frac{y^2}{a^2} - \frac{x^2}{b^2} = 1.$$

Figure 7.16(a) illustrates that for the equation on the left, the transverse axis lies on the x-axis. **Figure 7.16(b)** illustrates that for the equation on the right, the transverse axis lies on the y-axis. The vertices are a units from the center and the foci are c units from the center. For both equations, $b^2 = c^2 - a^2$. Equivalently, $c^2 = a^2 + b^2$.

GREAT QUESTION!

Which equation do I need to use for locating the foci of a hyperbola?

The form $c^2 = a^2 + b^2$ is the one you should remember. When finding the foci, this form is easy to manipulate.

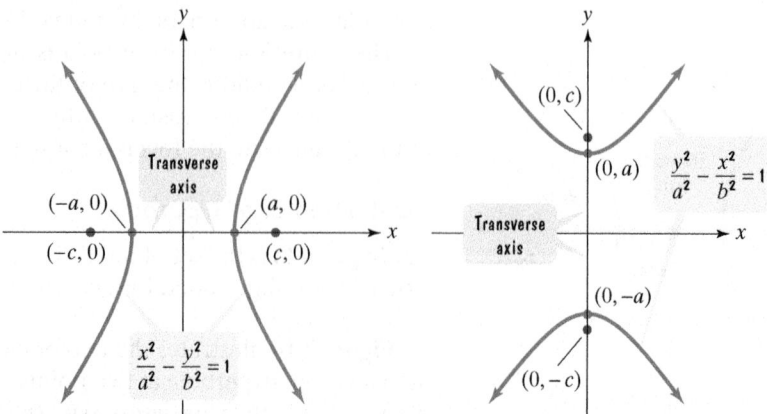

FIGURE 7.16(a) Transverse axis lies on the x-axis.

FIGURE 7.16(b) Transverse axis lies on the y-axis.

GREAT QUESTION!

How can I tell from a hyperbola's equation whether the transverse axis is horizontal or vertical?

When the x^2-term is preceded by a plus sign, the transverse axis is horizontal. When the y^2-term is preceded by a plus sign, the transverse axis is vertical.

① Locate a hyperbola's vertices and foci.

GREAT QUESTION!

What's the difference between the equations for locating an ellipse's foci and locating a hyperbola's foci?

Notice the sign difference between the following equations:

Finding an ellipse's foci:

$$c^2 = a^2 - b^2$$

Finding a hyperbola's foci:

$$c^2 = a^2 + b^2.$$

Using the Standard Form of the Equation of a Hyperbola

We can use the standard form of the equation of a hyperbola to find its vertices and locate its foci. Because the vertices are a units from the center, begin by identifying a^2 in the equation. In the standard form of a hyperbola's equation, **a^2 is the number under the variable whose term is preceded by a plus sign (+).** If the x^2-term is preceded by a plus sign, the transverse axis lies along the x-axis. Thus, the vertices are a units to the left and right of the origin. If the y^2-term is preceded by a plus sign, the transverse axis lies along the y-axis. Thus, the vertices are a units above and below the origin.

We know that the foci are c units from the center. The substitution that is used to derive the hyperbola's equation, $c^2 = a^2 + b^2$, is needed to locate the foci when a^2 and b^2 are known.

EXAMPLE 1 Finding Vertices and Foci from a Hyperbola's Equation

Find the vertices and locate the foci for each of the following hyperbolas with the given equation:

a. $\dfrac{x^2}{16} - \dfrac{y^2}{9} = 1$ **b.** $\dfrac{y^2}{9} - \dfrac{x^2}{16} = 1.$

SOLUTION

Both equations are in standard form. We begin by identifying a^2 and b^2 in each equation.

a. The first equation is in the form $\dfrac{x^2}{a^2} - \dfrac{y^2}{b^2} = 1.$

$$\frac{x^2}{16} - \frac{y^2}{9} = 1$$

> $a^2 = 16.$ This is the denominator of the term preceded by a plus sign. $b^2 = 9.$ This is the denominator of the term preceded by a minus sign.

Because the x^2-term is preceded by a plus sign, the transverse axis lies along the x-axis. Thus, the vertices are a units to the *left* and *right* of the origin. Based on the standard form of the equation, we know the vertices are $(-a, 0)$ and $(a, 0)$. Because $a^2 = 16$, $a = 4$. Thus, the vertices are $(-4, 0)$ and $(4, 0)$, shown in **Figure 7.17**.

We use $c^2 = a^2 + b^2$ to find the foci, which are located at $(-c, 0)$ and $(c, 0)$. We know that $a^2 = 16$ and $b^2 = 9$; we need to find c^2 in order to find c.

$$c^2 = a^2 + b^2 = 16 + 9 = 25$$

Because $c^2 = 25$, $c = 5$. The foci are located at $(-5, 0)$ and $(5, 0)$. They are shown in **Figure 7.17**.

b. The second given equation is in the form $\dfrac{y^2}{a^2} - \dfrac{x^2}{b^2} = 1.$

$$\frac{y^2}{9} - \frac{x^2}{16} = 1$$

> $a^2 = 9.$ This is the denominator of the term preceded by a plus sign. $b^2 = 16.$ This is the denominator of the term preceded by a minus sign.

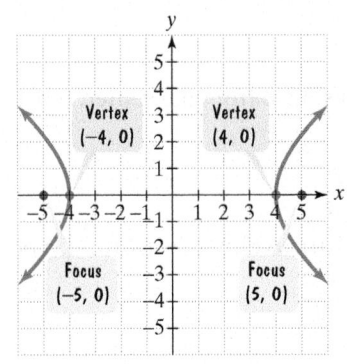

FIGURE 7.17 The graph of $\dfrac{x^2}{16} - \dfrac{y^2}{9} = 1$

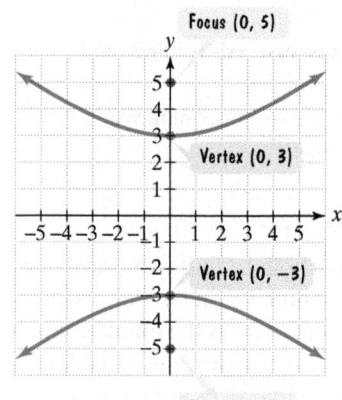

FIGURE 7.18 The graph of
$\dfrac{y^2}{9} - \dfrac{x^2}{16} = 1$

Because the y^2-term in $\dfrac{y^2}{9} - \dfrac{x^2}{16} = 1$ is preceded by a plus sign, the transverse axis lies along the y-axis. Thus, the vertices are a units *above* and *below* the origin. Based on the standard form of the equation, we know the vertices are $(0, -a)$ and $(0, a)$. Because $a^2 = 9$, $a = 3$. Thus, the vertices are $(0, -3)$ and $(0, 3)$, shown in **Figure 7.18**.

We use $c^2 = a^2 + b^2$ to find the foci, which are located at $(0, -c)$ and $(0, c)$.

$$c^2 = a^2 + b^2 = 9 + 16 = 25$$

Because $c^2 = 25$, $c = 5$. The foci are located at $(0, -5)$ and $(0, 5)$. They are shown in **Figure 7.18**. • • •

✓ Check Point **1** Find the vertices and locate the foci for each of the following hyperbolas with the given equation:

a. $\dfrac{x^2}{25} - \dfrac{y^2}{16} = 1$ **b.** $\dfrac{y^2}{25} - \dfrac{x^2}{16} = 1$.

In Example 1, we used equations of hyperbolas to find their foci and vertices. In the next example, we reverse this procedure.

2 Write equations of hyperbolas in standard form.

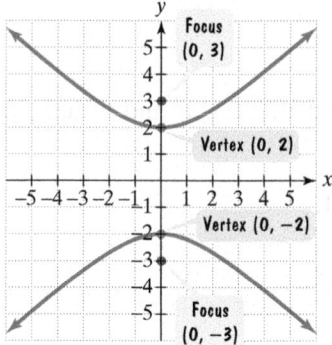

FIGURE 7.19

EXAMPLE 2 Finding the Equation of a Hyperbola from Its Foci and Vertices

Find the standard form of the equation of a hyperbola with foci at $(0, -3)$ and $(0, 3)$ and vertices $(0, -2)$ and $(0, 2)$, shown in **Figure 7.19**.

SOLUTION

Because the foci are located at $(0, -3)$ and $(0, 3)$, on the y-axis, the transverse axis lies on the y-axis. The center of the hyperbola is midway between the foci, located at $(0, 0)$. Thus, the form of the equation is

$$\dfrac{y^2}{a^2} - \dfrac{x^2}{b^2} = 1.$$

We need to determine the values for a^2 and b^2. The distance from the center, $(0, 0)$, to either vertex, $(0, -2)$ or $(0, 2)$, is 2, so $a = 2$.

$$\dfrac{y^2}{2^2} - \dfrac{x^2}{b^2} = 1 \quad \text{or} \quad \dfrac{y^2}{4} - \dfrac{x^2}{b^2} = 1$$

We must still find b^2. The distance from the center, $(0, 0)$, to either focus, $(0, -3)$ or $(0, 3)$, is 3. Thus, $c = 3$. Using $c^2 = a^2 + b^2$, we have

$$3^2 = 2^2 + b^2$$

and

$$b^2 = 3^2 - 2^2 = 9 - 4 = 5.$$

Substituting 5 for b^2 in $\dfrac{y^2}{4} - \dfrac{x^2}{b^2} = 1$ gives us the standard form of the hyperbola's equation. The equation is

$$\dfrac{y^2}{4} - \dfrac{x^2}{5} = 1.$$ • • •

✓ Check Point **2** Find the standard form of the equation of a hyperbola with foci at $(0, -5)$ and $(0, 5)$ and vertices $(0, -3)$ and $(0, 3)$.

The Asymptotes of a Hyperbola

As x and y get larger, the two branches of the graph of a hyperbola approach a pair of intersecting straight lines, called **asymptotes**. The asymptotes pass through the center of the hyperbola and are helpful in graphing hyperbolas.

Figure 7.20 shows the asymptotes for the graphs of hyperbolas centered at the origin. The asymptotes pass through the corners of a rectangle. Note that the dimensions of this rectangle are $2a$ by $2b$. The line segment of length $2b$ is the **conjugate axis** of the hyperbola and is perpendicular to the transverse axis through the center of the hyperbola.

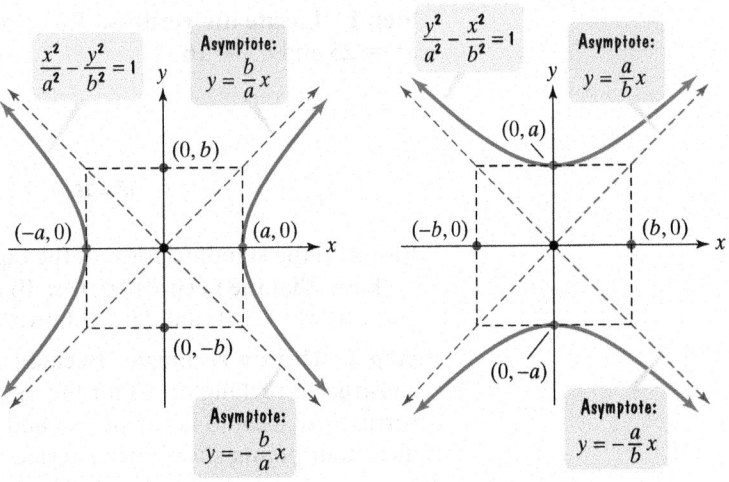

FIGURE 7.20 Asymptotes of a hyperbola

The Asymptotes of a Hyperbola Centered at the Origin

The hyperbola $\dfrac{x^2}{a^2} - \dfrac{y^2}{b^2} = 1$ has a horizontal transverse axis and two asymptotes

$$y = \frac{b}{a}x \qquad \text{and} \qquad y = -\frac{b}{a}x.$$

The hyperbola $\dfrac{y^2}{a^2} - \dfrac{x^2}{b^2} = 1$ has a vertical transverse axis and two asymptotes

$$y = \frac{a}{b}x \qquad \text{and} \qquad y = -\frac{a}{b}x.$$

Why are $y = \pm\dfrac{b}{a}x$ the asymptotes for a hyperbola whose transverse axis is horizontal? The proof can be found in the appendix.

③ Graph hyperbolas centered at the origin.

Graphing Hyperbolas Centered at the Origin

Hyperbolas are graphed using vertices and asymptotes.

Graphing Hyperbolas

1. Locate the vertices.
2. Use dashed lines to draw the rectangle centered at the origin with sides parallel to the axes, crossing one axis at $\pm a$ and the other at $\pm b$.
3. Use dashed lines to draw the diagonals of this rectangle and extend them to obtain the asymptotes.
4. Draw the two branches of the hyperbola by starting at each vertex and approaching the asymptotes.

EXAMPLE 3 Graphing a Hyperbola

Graph and locate the foci: $\dfrac{x^2}{25} - \dfrac{y^2}{16} = 1$. What are the equations of the asymptotes?

SOLUTION

Step 1 Locate the vertices. The given equation is in the form $\dfrac{x^2}{a^2} - \dfrac{y^2}{b^2} = 1$, with $a^2 = 25$ and $b^2 = 16$.

$$\underset{a^2 = 25}{} \frac{x^2}{25} - \frac{y^2}{16} = 1 \underset{b^2 = 16}{}$$

Based on the standard form of the equation with the transverse axis on the x-axis, we know that the vertices are $(-a, 0)$ and $(a, 0)$. Because $a^2 = 25$, $a = 5$. Thus, the vertices are $(-5, 0)$ and $(5, 0)$, shown in **Figure 7.21**.

Step 2 Draw a rectangle. Because $a^2 = 25$ and $b^2 = 16$, $a = 5$ and $b = 4$. We construct a rectangle to find the asymptotes, using -5 and 5 on the x-axis (the vertices are located here) and -4 and 4 on the y-axis. The rectangle passes through these four points, shown using dashed lines in **Figure 7.21**.

Step 3 Draw extended diagonals for the rectangle to obtain the asymptotes. We draw dashed lines through the opposite corners of the rectangle, shown in **Figure 7.21**, to obtain the graph of the asymptotes. Based on the standard form of the hyperbola's equation, the equations for these asymptotes are

$$y = \pm \frac{b}{a}x \quad \text{or} \quad y = \pm \frac{4}{5}x.$$

TECHNOLOGY

Graph $\dfrac{x^2}{25} - \dfrac{y^2}{16} = 1$ by solving for y:

$$y_1 = \frac{\sqrt{16x^2 - 400}}{5}$$

$$y_2 = -\frac{\sqrt{16x^2 - 400}}{5} = -y_1.$$

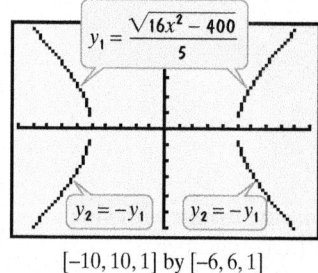

$[-10, 10, 1]$ by $[-6, 6, 1]$

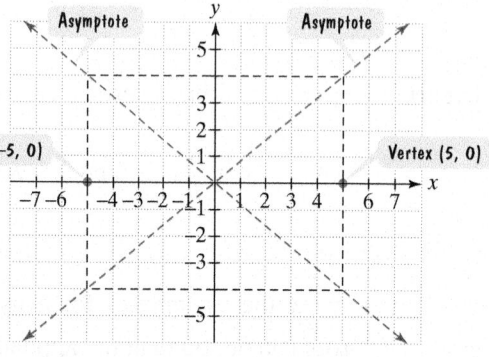

FIGURE 7.21 Preparing to graph $\dfrac{x^2}{25} - \dfrac{y^2}{16} = 1$

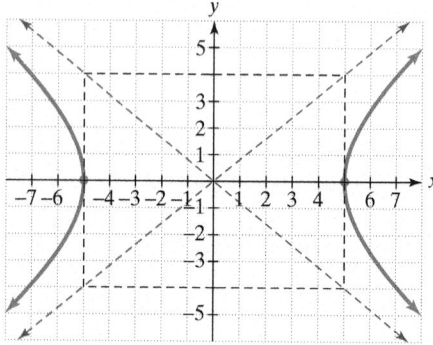

FIGURE 7.22 The graph of $\dfrac{x^2}{25} - \dfrac{y^2}{16} = 1$

Step 4 Draw the two branches of the hyperbola by starting at each vertex and approaching the asymptotes. The hyperbola is shown in **Figure 7.22**.

We now consider the foci, located at $(-c, 0)$ and $(c, 0)$. We find c using $c^2 = a^2 + b^2$.

$$c^2 = 25 + 16 = 41$$

Because $c^2 = 41$, $c = \sqrt{41}$. The foci are located at $\left(-\sqrt{41}, 0\right)$ and $\left(\sqrt{41}, 0\right)$, approximately $(-6.4, 0)$ and $(6.4, 0)$. ● ● ●

Ø **Check Point 3** Graph and locate the foci: $\dfrac{x^2}{36} - \dfrac{y^2}{9} = 1$. What are the equations of the asymptotes?

EXAMPLE 4 Graphing a Hyperbola

Graph and locate the foci: $9y^2 - 4x^2 = 36$. What are the equations of the asymptotes?

SOLUTION

We begin by writing the equation in standard form. The right side should be 1, so we divide both sides by 36.

$$\frac{9y^2}{36} - \frac{4x^2}{36} = \frac{36}{36}$$

$$\frac{y^2}{4} - \frac{x^2}{9} = 1 \quad \text{Simplify. The right side is now 1.}$$

Now we are ready to use our four-step procedure for graphing hyperbolas.

Step 1 Locate the vertices. The equation that we obtained is in the form $\frac{y^2}{a^2} - \frac{x^2}{b^2} = 1$, with $a^2 = 4$ and $b^2 = 9$.

$$\underbrace{\frac{y^2}{4}}_{a^2 = 4} - \underbrace{\frac{x^2}{9}}_{b^2 = 9} = 1$$

Based on the standard form of the equation with the transverse axis on the y-axis, we know that the vertices are $(0, -a)$ and $(0, a)$. Because $a^2 = 4$, $a = 2$. Thus, the vertices are $(0, -2)$ and $(0, 2)$, shown in **Figure 7.23**.

Step 2 Draw a rectangle. Because $a^2 = 4$ and $b^2 = 9$, $a = 2$ and $b = 3$. We construct a rectangle to find the asymptotes, using -2 and 2 on the y-axis (the vertices are located here) and -3 and 3 on the x-axis. The rectangle passes through these four points, shown using dashed lines in **Figure 7.23**.

Step 3 Draw extended diagonals of the rectangle to obtain the asymptotes. We draw dashed lines through the opposite corners of the rectangle, shown in **Figure 7.23**, to obtain the graph of the asymptotes. Based on the standard form of the hyperbola's equation, the equations of these asymptotes are

$$y = \pm\frac{a}{b}x \quad \text{or} \quad y = \pm\frac{2}{3}x.$$

Step 4 Draw the two branches of the hyperbola by starting at each vertex and approaching the asymptotes. The hyperbola is shown in **Figure 7.24**.

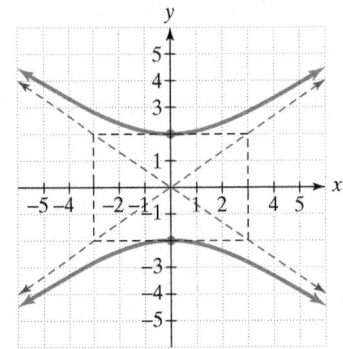

FIGURE 7.23 Preparing to graph $\frac{y^2}{4} - \frac{x^2}{9} = 1$

FIGURE 7.24 The graph of $\frac{y^2}{4} - \frac{x^2}{9} = 1$

We now consider the foci, located at $(0, -c)$ and $(0, c)$. We find c using $c^2 = a^2 + b^2$.

$$c^2 = 4 + 9 = 13$$

Because $c^2 = 13$, $c = \sqrt{13}$. The foci are located at $\left(0, -\sqrt{13}\right)$ and $\left(0, \sqrt{13}\right)$, approximately $(0, -3.6)$ and $(0, 3.6)$. • • •

✓ Check Point **4** Graph and locate the foci: $y^2 - 4x^2 = 4$. What are the equations of the asymptotes?

④ Graph hyperbolas not centered at the origin.

Translations of Hyperbolas

The graph of a hyperbola can be centered at (h, k), rather than at the origin. Horizontal and vertical translations are accomplished by replacing x with $x - h$ and y with $y - k$ in the standard form of the hyperbola's equation.

Table 7.2 gives the standard forms of equations of hyperbolas centered at (h, k) and shows their graphs.

Table 7.2 Standard Forms of Equations of Hyperbolas Centered at (h, k)

Equation	Center	Transverse Axis	Vertices	Graph
$\dfrac{(x - h)^2}{a^2} - \dfrac{(y - k)^2}{b^2} = 1$ Vertices are a units right and a units left of center. Foci are c units right and c units left of center, where $c^2 = a^2 + b^2$.	(h, k)	Parallel to the x-axis; horizontal	$(h - a, k)$ $(h + a, k)$	
$\dfrac{(y - k)^2}{a^2} - \dfrac{(x - h)^2}{b^2} = 1$ Vertices are a units above and a units below the center. Foci are c units above and c units below the center, where $c^2 = a^2 + b^2$.	(h, k)	Parallel to the y-axis; vertical	$(h, k - a)$ $(h, k + a)$	

EXAMPLE 5 Graphing a Hyperbola Centered at (h, k)

Graph: $\dfrac{(x - 2)^2}{16} - \dfrac{(y - 3)^2}{9} = 1$. Where are the foci located? What are the equations of the asymptotes?

SOLUTION

In order to graph the hyperbola, we need to know its center, (h, k). In the standard forms of equations centered at (h, k), h is the number subtracted from x and k is the number subtracted from y.

This is $(x - h)^2$, with $h = 2$. This is $(y - k)^2$, with $k = 3$.

$$\frac{(x - 2)^2}{16} - \frac{(y - 3)^2}{9} = 1$$

We see that $h = 2$ and $k = 3$. Thus, the center of the hyperbola, (h, k), is $(2, 3)$. We can graph the hyperbola by using vertices, asymptotes, and our four-step graphing procedure.

Step 1 Locate the vertices. To do this, we must identify a^2.

$$\frac{(x - 2)^2}{16} - \frac{(y - 3)^2}{9} = 1 \quad \text{The form of this equation is } \frac{(x - h)^2}{a^2} - \frac{(y - k)^2}{b^2} = 1.$$

$a^2 = 16$ $b^2 = 9$

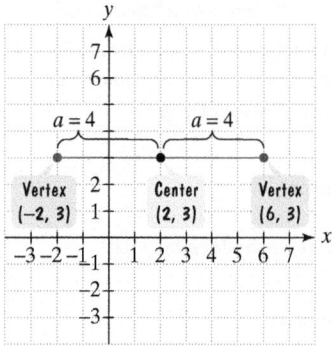

FIGURE 7.25 Locating a hyperbola's center and vertices

Based on the standard form of the equation with a horizontal transverse axis, the vertices are a units to the left and right of the center. Because $a^2 = 16$, $a = 4$. This means that the vertices are 4 units to the left and right of the center, $(2, 3)$. Four units to the left of $(2, 3)$ puts one vertex at $(2 - 4, 3)$, or $(-2, 3)$. Four units to the right of $(2, 3)$ puts the other vertex at $(2 + 4, 3)$, or $(6, 3)$. The vertices are shown in **Figure 7.25**.

Step 2 Draw a rectangle. Because $a^2 = 16$ and $b^2 = 9$, $a = 4$ and $b = 3$. The rectangle passes through points that are 4 units to the right and left of the center (the vertices are located here) and 3 units above and below the center. The rectangle is shown using dashed lines in **Figure 7.26**.

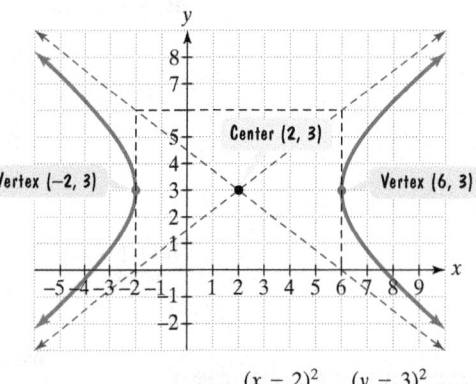

FIGURE 7.26 The graph of $\dfrac{(x - 2)^2}{16} - \dfrac{(y - 3)^2}{9} = 1$

Step 3 Draw extended diagonals of the rectangle to obtain the asymptotes. We draw dashed lines through the opposite corners of the rectangle, shown in **Figure 7.26**, to obtain the graph of the asymptotes. The equations of the asymptotes of the unshifted hyperbola $\dfrac{x^2}{16} - \dfrac{y^2}{9} = 1$ are $y = \pm\dfrac{b}{a}x$, or $y = \pm\dfrac{3}{4}x$. Thus, the asymptotes for the hyperbola that is shifted two units to the right and three units up, namely,

$$\frac{(x - 2)^2}{16} - \frac{(y - 3)^2}{9} = 1$$

have equations that can be expressed as

$$y - 3 = \pm\frac{3}{4}(x - 2).$$

Step 4 Draw the two branches of the hyperbola by starting at each vertex and approaching the asymptotes. The hyperbola is shown in **Figure 7.26**.

We now consider the foci, located c units to the right and left of the center. We find c using $c^2 = a^2 + b^2$.

$$c^2 = 16 + 9 = 25$$

Because $c^2 = 25$, $c = 5$. This means that the foci are 5 units to the left and right of the center, $(2, 3)$. Five units to the left of $(2, 3)$ puts one focus at $(2 - 5, 3)$, or $(-3, 3)$. Five units to the right of $(2, 3)$ puts the other focus at $(2 + 5, 3)$, or $(7, 3)$. $\bullet\bullet\bullet$

GREAT QUESTION!

Do I have to use the equations for the asymptotes of a hyperbola centered at the origin,

$$y = \pm\frac{b}{a}x \text{ or } y = \pm\frac{a}{b}x, \text{ to}$$

determine the equations of the asymptotes when the hyperbola's center is (h, k)?

No. You can also use the point-slope form of a line's equation

$$y - y_1 = m(x - x_1)$$

to find the equations of the asymptotes. The center of the hyperbola, (h, k), is a point on each asymptote, so $x_1 = h$ and $y_1 = k$. The slopes, m, are $\pm\dfrac{b}{a}$ for a horizontal transverse axis and $\pm\dfrac{a}{b}$ for a vertical transverse axis.

✓ Check Point **5** Graph: $\dfrac{(x-3)^2}{4} - \dfrac{(y-1)^2}{1} = 1$. Where are the foci located? What are the equations of the asymptotes?

In our next example, it is necessary to convert the equation of a hyperbola to standard form by completing the square on x and y.

EXAMPLE 6 Graphing a Hyperbola Centered at (h, k)

Graph: $4x^2 - 24x - 25y^2 + 250y - 489 = 0$. Where are the foci located? What are the equations of the asymptotes?

SOLUTION

We begin by completing the square on x and y.

$$4x^2 - 24x - 25y^2 + 250y - 489 = 0$$ This is the given equation.

$$(4x^2 - 24x) + (-25y^2 + 250y) = 489$$ Group terms and add 489 to both sides.

$$4(x^2 - 6x + \square) - 25(y^2 - 10y + \square) = 489$$ Factor out 4 and -25, respectively, so coefficients of x^2 and y^2 are 1.

$$4(x^2 - 6x + 9) - 25(y^2 - 10y + 25) = 489 + 36 + (-625)$$ Complete each square by adding the square of half the coefficient of x and y, respectively.

We added 4 · 9, or 36, to the left side. We added $-25 \cdot 25$, or -625, to the left side. Add $36 + (-625)$ to the right side.

$$4(x-3)^2 - 25(y-5)^2 = -100$$ Factor.

$$\dfrac{4(x-3)^2}{-100} - \dfrac{25(y-5)^2}{-100} = \dfrac{-100}{-100}$$ Divide both sides by -100.

$$\dfrac{(x-3)^2}{-25} + \dfrac{(y-5)^2}{4} = 1$$ Simplify.

This is $(y-k)^2$, with $k = 5$. This is $(x-h)^2$, with $h = 3$.

$$\dfrac{(y-5)^2}{4} - \dfrac{(x-3)^2}{25} = 1$$ Write the equation in standard form, $\dfrac{(y-k)^2}{a^2} - \dfrac{(x-h)^2}{b^2} = 1$.

GREAT QUESTION!

Shouldn't the center of $\dfrac{(y-5)^2}{4} - \dfrac{(x-3)^2}{25} = 1$ be $(5, 3)$?

No. The hyperbola's center is $(3, 5)$ because the equation shows that 3 is subtracted from x and 5 is subtracted from y. Many students tend to read the equation from left to right and get the center backward. The hyperbola's center is *not* $(5, 3)$.

We see that $h = 3$ and $k = 5$. Thus, the center of the hyperbola, (h, k), is $(3, 5)$. Because the x^2-term is being subtracted, the transverse axis is vertical and the hyperbola opens upward and downward.

We use our four-step procedure to obtain the graph of

$$\dfrac{(y-5)^2}{4} - \dfrac{(x-3)^2}{25} = 1.$$

$a^2 = 4$ $b^2 = 25$

Step 1 Locate the vertices. Based on the standard form of the equation with a vertical transverse axis, the vertices are a units above and below the center. Because $a^2 = 4$, $a = 2$. This means that the vertices are 2 units above and below the center, $(3, 5)$. This puts the vertices at $(3, 7)$ and $(3, 3)$, shown in **Figure 7.27**.

Step 2 Draw a rectangle. Because $a^2 = 4$ and $b^2 = 25$, $a = 2$ and $b = 5$. The rectangle passes through points that are 2 units above and below the center (the vertices are located here) and 5 units to the right and left of the center. The rectangle is shown using dashed lines in **Figure 7.27**.

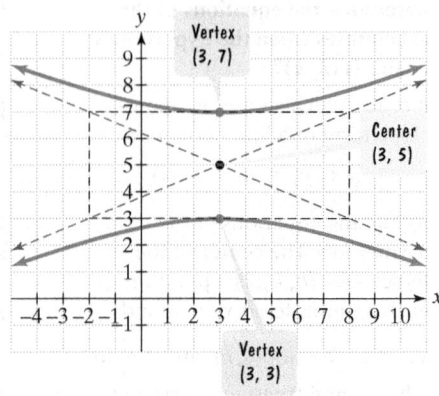

FIGURE 7.27 The graph of $\dfrac{(y-5)^2}{4} - \dfrac{(x-3)^2}{25} = 1$

Step 3 Draw extended diagonals of the rectangle to obtain the asymptotes.
We draw dashed lines through the opposite corners of the rectangle, shown in
Figure 7.27, to obtain the graph of the asymptotes. The equations of the asymptotes
of the unshifted hyperbola $\dfrac{y^2}{4} - \dfrac{x^2}{25} = 1$ are $y = \pm\dfrac{a}{b}x$, or $y = \pm\dfrac{2}{5}x$. Thus, the
asymptotes for the hyperbola that is shifted three units to the right and five units
up, namely,

$$\frac{(y-5)^2}{4} - \frac{(x-3)^2}{25} = 1$$

have equations that can be expressed as

$$y - 5 = \pm\frac{2}{5}(x-3).$$

**Step 4 Draw the two branches of the hyperbola by starting at each vertex and
approaching the asymptotes.** The hyperbola is shown in **Figure 7.27**.
 We now consider the foci, located c units above and below the center, $(3,5)$. We
find c using $c^2 = a^2 + b^2$.

$$c^2 = 4 + 25 = 29$$

Because $c^2 = 29$, $c = \sqrt{29}$. The foci are located at $\left(3, 5 + \sqrt{29}\right)$ and
$\left(3, 5 - \sqrt{29}\right)$.

● ● ●

⊘ **Check Point 6** Graph: $4x^2 - 24x - 9y^2 - 90y - 153 = 0$. Where are the
foci located? What are the equations of the asymptotes?

⑤ Solve applied problems involving
hyperbolas.

Applications

Hyperbolas have many applications. When a jet flies at a speed greater than the speed of
sound, the shock wave that is created is heard as a sonic boom. The wave has the shape
of a cone. The shape formed as the cone hits the ground is one branch of a hyperbola.
 Halley's Comet, a permanent part of our solar system, travels around the sun in
an elliptical orbit. Other comets pass through the solar system only once, following
a hyperbolic path with the sun as a focus.
 Hyperbolas are of practical importance in fields ranging from architecture to
navigation. Cooling towers used in the design for nuclear power plants have cross
sections that are both ellipses and hyperbolas. Three-dimensional solids whose cross
sections are hyperbolas are used in some rather unique architectural creations,
including the TWA building at Kennedy Airport in New York City and the St. Louis
Science Center Planetarium.

The hyperbolic shape of a sonic boom

EXAMPLE 7 An Application Involving Hyperbolas

An explosion is recorded by two microphones that are 2 miles apart. Microphone
M_1 received the sound 4 seconds before microphone M_2. Assuming sound travels
at 1100 feet per second, determine the possible locations of the explosion relative to
the location of the microphones.

SOLUTION

We begin by putting the
microphones in a coordinate
system. Because 1 mile = 5280
feet, we place M_1 5280 feet on
a horizontal axis to the right of
the origin and M_2 5280 feet on a
horizontal axis to the left of the
origin. **Figure 7.28** illustrates that
the two microphones are 2 miles
apart.

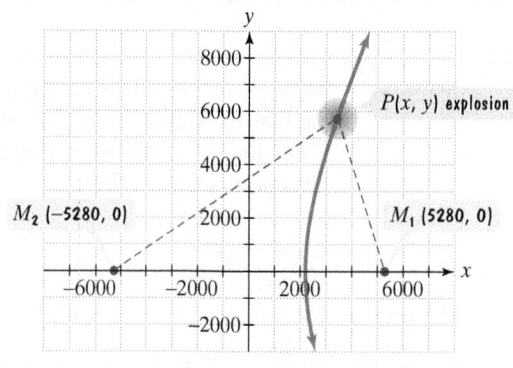

FIGURE 7.28 Locating an explosion on the branch of a hyperbola

Orbits of comets

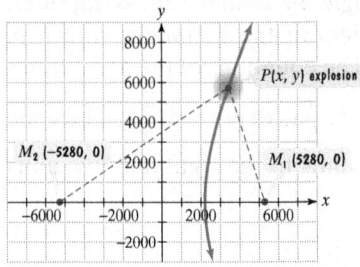

FIGURE 7.28 (repeated) Locating an explosion on the branch of a hyperbola

We know that M_2 received the sound 4 seconds after M_1. Because sound travels at 1100 feet per second, the difference between the distance from P to M_1 and the distance from P to M_2 is 4400 feet. The set of all points P (or locations of the explosion) satisfying these conditions fits the definition of a hyperbola, with microphones M_1 and M_2 at the foci.

$$\frac{x^2}{a^2} - \frac{y^2}{b^2} = 1 \quad \text{Use the standard form of the hyperbola's equation. } P(x, y), \text{ the explosion point, lies on this hyperbola. We must find } a^2 \text{ and } b^2.$$

The difference between the distances, represented by $2a$ in the derivation of the hyperbola's equation, is 4400 feet. Thus, $2a = 4400$ and $a = 2200$.

$$\frac{x^2}{(2200)^2} - \frac{y^2}{b^2} = 1 \quad \text{Substitute 2200 for } a.$$

$$\frac{x^2}{4,840,000} - \frac{y^2}{b^2} = 1 \quad \text{Square 2200.}$$

We must still find b^2. We know that $a = 2200$. The distance from the center, $(0, 0)$, to either focus, $(-5280, 0)$ or $(5280, 0)$, is 5280. Thus, $c = 5280$. Using $c^2 = a^2 + b^2$, we have

$$5280^2 = 2200^2 + b^2$$

and

$$b^2 = 5280^2 - 2200^2 = 23,038,400.$$

The equation of the hyperbola with a microphone at each focus is

$$\frac{x^2}{4,840,000} - \frac{y^2}{23,038,400} = 1. \quad \text{Substitute 23,038,400 for } b^2.$$

We can conclude that the explosion occurred somewhere on the right branch (the branch closer to M_1) of the hyperbola given by this equation. • • •

In Example 7, we determined that the explosion occurred somewhere along one branch of a hyperbola, but not exactly where on the hyperbola. If, however, we had received the sound from another pair of microphones, we could locate the sound along a branch of another hyperbola. The exact location of the explosion would be the point where the two hyperbolas intersect.

⊘ Check Point **7** Rework Example 7 assuming microphone M_1 receives the sound 3 seconds before microphone M_2.

CONCEPT AND VOCABULARY CHECK

Fill in each blank so that the resulting statement is true.

1. The set of all points in a plane the difference of whose distances from two fixed points is constant is a/an _____. The two fixed points are called the _____. The line through these points intersects the graph at two points, called the _____, which are joined by the _____ axis.

2. The vertices of $\dfrac{x^2}{25} - \dfrac{y^2}{9} = 1$ are _____ and _____. The foci are located at _____ and _____.

3. The vertices of $\dfrac{y^2}{25} - \dfrac{x^2}{9} = 1$ are _____ and _____. The foci are located at _____ and _____.

4. The two branches of the graph of a hyperbola approach a pair of intersecting lines, called _____. These intersecting lines pass through the _____ of the hyperbola.

5. The equation $9x^2 - 4y^2 = 36$ can be written in standard form by _____ both sides by _____.

6. The equations for the asymptotes of $\dfrac{x^2}{4} - \dfrac{y^2}{9} = 1$ are _____ and _____.

7. The equations for the asymptotes of $\dfrac{y^2}{4} - x^2 = 1$ are _____ and _____.

8. If the center of a hyperbola with a horizontal transverse axis is $(2, 3)$ and $a^2 = 25$, then the coordinates of the vertices are _____ and _____.

9. The center of $\dfrac{(y + 2)^2}{4} - \dfrac{(x - 7)^2}{36} = 1$ is _____.

10. In the equation $9(x^2 - 8x) - 16(y^2 + 2y) = 16$, we complete the square on x by adding _____ within the first parentheses. We complete the square on y by adding _____ within the second parentheses. Thus, we must add _____ to the right side of the equation.

EXERCISE SET 7.2

Practice Exercises

In Exercises 1–4, find the vertices and locate the foci of each hyperbola with the given equation. Then match each equation to one of the graphs that are shown and labeled (a)–(d).

1. $\dfrac{x^2}{4} - \dfrac{y^2}{1} = 1$

2. $\dfrac{x^2}{1} - \dfrac{y^2}{4} = 1$

3. $\dfrac{y^2}{4} - \dfrac{x^2}{1} = 1$

4. $\dfrac{y^2}{1} - \dfrac{x^2}{4} = 1$

a.

b.

c.

d.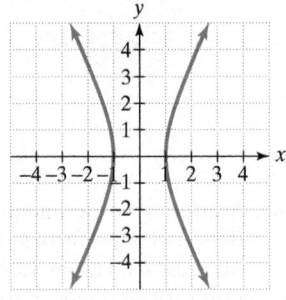

In Exercises 5–12, find the standard form of the equation of each hyperbola satisfying the given conditions.

5. Foci: $(0, -3), (0, 3)$; vertices: $(0, -1), (0, 1)$

6. Foci: $(0, -6), (0, 6)$; vertices: $(0, -2), (0, 2)$

7. Foci: $(-4, 0), (4, 0)$; vertices: $(-3, 0), (3, 0)$

8. Foci: $(-7, 0), (7, 0)$; vertices: $(-5, 0), (5, 0)$

9. Endpoints of transverse axis: $(0, -6), (0, 6)$; asymptote: $y = 2x$

10. Endpoints of transverse axis: $(-4, 0), (4, 0)$; asymptote: $y = 2x$

11. Center: $(4, -2)$; Focus: $(7, -2)$; vertex: $(6, -2)$

12. Center: $(-2, 1)$; Focus: $(-2, 6)$; vertex: $(-2, 4)$

In Exercises 13–26, use vertices and asymptotes to graph each hyperbola. Locate the foci and find the equations of the asymptotes.

13. $\dfrac{x^2}{9} - \dfrac{y^2}{25} = 1$

14. $\dfrac{x^2}{16} - \dfrac{y^2}{25} = 1$

15. $\dfrac{x^2}{100} - \dfrac{y^2}{64} = 1$

16. $\dfrac{x^2}{144} - \dfrac{y^2}{81} = 1$

17. $\dfrac{y^2}{16} - \dfrac{x^2}{36} = 1$

18. $\dfrac{y^2}{25} - \dfrac{x^2}{64} = 1$

19. $4y^2 - x^2 = 1$

20. $9y^2 - x^2 = 1$

21. $9x^2 - 4y^2 = 36$

22. $4x^2 - 25y^2 = 100$

23. $9y^2 - 25x^2 = 225$

24. $16y^2 - 9x^2 = 144$

25. $y = \pm\sqrt{x^2 - 2}$

26. $y = \pm\sqrt{x^2 - 3}$

In Exercises 27–32, find the standard form of the equation of each hyperbola.

27.

28.

29.

30.

31.

32.

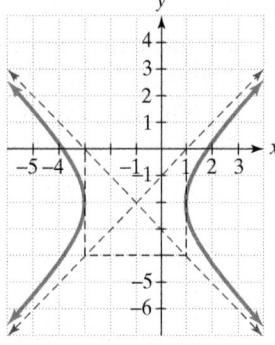

In Exercises 33–42, use the center, vertices, and asymptotes to graph each hyperbola. Locate the foci and find the equations of the asymptotes.

33. $\dfrac{(x+4)^2}{9} - \dfrac{(y+3)^2}{16} = 1$ **34.** $\dfrac{(x+2)^2}{9} - \dfrac{(y-1)^2}{25} = 1$

35. $\dfrac{(x+3)^2}{25} - \dfrac{y^2}{16} = 1$ **36.** $\dfrac{(x+2)^2}{9} - \dfrac{y^2}{25} = 1$

37. $\dfrac{(y+2)^2}{4} - \dfrac{(x-1)^2}{16} = 1$

38. $\dfrac{(y-2)^2}{36} - \dfrac{(x+1)^2}{49} = 1$

39. $(x-3)^2 - 4(y+3)^2 = 4$

40. $(x+3)^2 - 9(y-4)^2 = 9$

41. $(x-1)^2 - (y-2)^2 = 3$

42. $(y-2)^2 - (x+3)^2 = 5$

In Exercises 43–50, convert each equation to standard form by completing the square on x and y. Then graph the hyperbola. Locate the foci and find the equations of the asymptotes.

43. $x^2 - y^2 - 2x - 4y - 4 = 0$

44. $4x^2 - y^2 + 32x + 6y + 39 = 0$

45. $16x^2 - y^2 + 64x - 2y + 67 = 0$

46. $9y^2 - 4x^2 - 18y + 24x - 63 = 0$

47. $4x^2 - 9y^2 - 16x + 54y - 101 = 0$

48. $4x^2 - 9y^2 + 8x - 18y - 6 = 0$

49. $4x^2 - 25y^2 - 32x + 164 = 0$

50. $9x^2 - 16y^2 - 36x - 64y + 116 = 0$

Practice Plus

In Exercises 51–56, graph each relation. Use the relation's graph to determine its domain and range.

51. $\dfrac{x^2}{9} - \dfrac{y^2}{16} = 1$ **52.** $\dfrac{x^2}{25} - \dfrac{y^2}{4} = 1$

53. $\dfrac{x^2}{9} + \dfrac{y^2}{16} = 1$ **54.** $\dfrac{x^2}{25} + \dfrac{y^2}{4} = 1$

55. $\dfrac{y^2}{16} - \dfrac{x^2}{9} = 1$ **56.** $\dfrac{y^2}{4} - \dfrac{x^2}{25} = 1$

In Exercises 57–60, find the solution set for each system by graphing both of the system's equations in the same rectangular coordinate system and finding points of intersection. Check all solutions in both equations.

57. $\begin{cases} x^2 - y^2 = 4 \\ x^2 + y^2 = 4 \end{cases}$ **58.** $\begin{cases} x^2 - y^2 = 9 \\ x^2 + y^2 = 9 \end{cases}$

59. $\begin{cases} 9x^2 + y^2 = 9 \\ y^2 - 9x^2 = 9 \end{cases}$ **60.** $\begin{cases} 4x^2 + y^2 = 4 \\ y^2 - 4x^2 = 4 \end{cases}$

Application Exercises

61. An explosion is recorded by two microphones that are 1 mile apart. Microphone M_1 received the sound 2 seconds before microphone M_2. Assuming sound travels at 1100 feet per second, determine the possible locations of the explosion relative to the location of the microphones.

62. Radio towers A and B, 200 kilometers apart, are situated along the coast, with A located due west of B. Simultaneous radio signals are sent from each tower to a ship, with the signal from B received 500 microseconds before the signal from A.

 a. Assuming that the radio signals travel 300 meters per microsecond, determine the equation of the hyperbola on which the ship is located.

 b. If the ship lies due north of tower B, how far out at sea is it?

63. An architect designs two houses that are shaped and positioned like a part of the branches of the hyperbola whose equation is $625y^2 - 400x^2 = 250{,}000$, where x and y are in yards. How far apart are the houses at their closest point?

64. Scattering experiments, in which moving particles are deflected by various forces, led to the concept of the nucleus of an atom. In 1911, the physicist Ernest Rutherford (1871–1937) discovered that when alpha particles are directed toward the nuclei of gold atoms, they are eventually deflected along hyperbolic paths, illustrated in the figure. If a particle gets as close as 3 units to the nucleus along a hyperbolic path with an asymptote given by $y = \frac{1}{2}x$, what is the equation of its path?

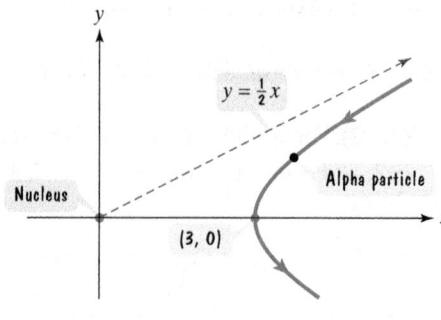

Moiré patterns, such as those shown in Exercises 65–66, can appear when two repetitive patterns overlap to produce a third, sometimes unintended, pattern.

 a. *In each exercise, use the name of a conic section to describe the moiré pattern.*

 b. *Select one of the following equations that can possibly describe a conic section within the moiré pattern:*

$$x^2 + y^2 = 1; \quad x^2 - y^2 = 1; \quad x^2 + 4y^2 = 4.$$

65. **66.**

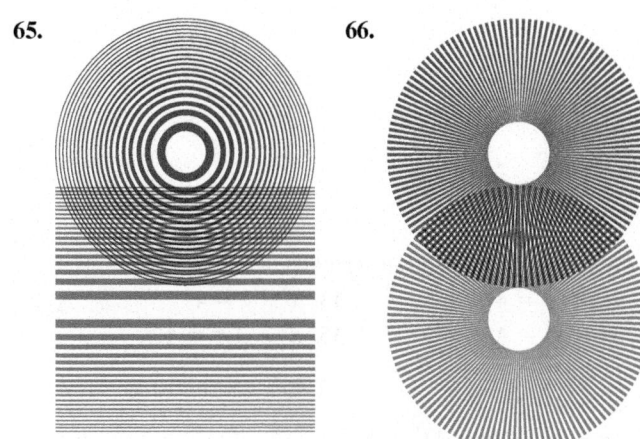

Writing in Mathematics

67. What is a hyperbola?

68. Describe how to graph $\dfrac{x^2}{9} - \dfrac{y^2}{1} = 1$.

69. Describe how to locate the foci of the graph of $\dfrac{x^2}{9} - \dfrac{y^2}{1} = 1$.

70. Describe one similarity and one difference between the graphs of $\dfrac{x^2}{9} - \dfrac{y^2}{1} = 1$ and $\dfrac{y^2}{9} - \dfrac{x^2}{1} = 1$.

71. Describe one similarity and one difference between the graphs of $\dfrac{x^2}{9} - \dfrac{y^2}{1} = 1$ and $\dfrac{(x-3)^2}{9} - \dfrac{(y+3)^2}{1} = 1$.

72. How can you distinguish an ellipse from a hyperbola by looking at their equations?

73. In 1992, a NASA team began a project called Spaceguard Survey, calling for an international watch for comets that might collide with Earth. Why is it more difficult to detect a possible "doomsday comet" with a hyperbolic orbit than one with an elliptical orbit?

Technology Exercises

74. Use a graphing utility to graph any five of the hyperbolas that you graphed by hand in Exercises 13–26.

75. Use a graphing utility to graph any three of the hyperbolas that you graphed by hand in Exercises 33–42. First solve the given equation for y by using the square root property. Enter each of the two resulting equations to produce each branch of the hyperbola.

76. Use a graphing utility to graph any one of the hyperbolas that you graphed by hand in Exercises 43–50. Write the equation as a quadratic equation in y and use the quadratic formula to solve for y. Enter each of the two resulting equations to produce each branch of the hyperbola.

77. Use a graphing utility to graph $\dfrac{x^2}{4} - \dfrac{y^2}{9} = 0$. Is the graph a hyperbola? In general, what is the graph of $\dfrac{x^2}{a^2} - \dfrac{y^2}{b^2} = 0$?

78. Graph $\dfrac{x^2}{a^2} - \dfrac{y^2}{b^2} = 1$ and $\dfrac{x^2}{a^2} - \dfrac{y^2}{b^2} = -1$ in the same viewing rectangle for values of a^2 and b^2 of your choice. Describe the relationship between the two graphs.

79. Write $4x^2 - 6xy + 2y^2 - 3x + 10y - 6 = 0$ as a quadratic equation in y and then use the quadratic formula to express y in terms of x. Graph the resulting two equations using a graphing utility in a $[-50, 70, 10]$ by $[-30, 50, 10]$ viewing rectangle. What effect does the xy-term have on the graph of the resulting hyperbola? What problems would you encounter if you attempted to write the given equation in standard form by completing the square?

80. Graph $\dfrac{x^2}{16} - \dfrac{y^2}{9} = 1$ and $\dfrac{x|x|}{16} - \dfrac{y|y|}{9} = 1$ in the same viewing rectangle. Explain why the graphs are not the same.

Critical Thinking Exercises

Make Sense? *In Exercises 81–84, determine whether each statement makes sense or does not make sense, and explain your reasoning.*

81. I changed the addition in an ellipse's equation to subtraction and this changed its elongation from horizontal to vertical.

82. I noticed that the definition of a hyperbola closely resembles that of an ellipse in that it depends on the distances between a set of points in a plane to two fixed points, the foci.

83. I graphed a hyperbola centered at the origin that had y-intercepts but no x-intercepts.

84. I graphed a hyperbola centered at the origin that was symmetric with respect to the x-axis and also symmetric with respect to the y-axis.

In Exercises 85–88, determine whether each statement is true or false. If the statement is false, make the necessary change(s) to produce a true statement.

85. If one branch of a hyperbola is removed from a graph, then the branch that remains must define y as a function of x.

86. All points on the asymptotes of a hyperbola also satisfy the hyperbola's equation.

87. The graph of $\dfrac{x^2}{9} - \dfrac{y^2}{4} = 1$ does not intersect the line $y = -\dfrac{2}{3}x$.

88. Two different hyperbolas can never share the same asymptotes.

89. What happens to the shape of the graph of $\dfrac{x^2}{a^2} - \dfrac{y^2}{b^2} = 1$ as $\dfrac{c}{a} \to \infty$, where $c^2 = a^2 + b^2$?

90. Find the standard form of the equation of the hyperbola with vertices $(5, -6)$ and $(5, 6)$, passing through $(0, 9)$.

91. Find the equation of a hyperbola whose asymptotes are perpendicular.

Exercises 92–94 will help you prepare for the material covered in the next section.

In Exercises 92–93, graph each parabola with the given equation.

92. $y = x^2 + 4x - 5$

93. $y = -3(x - 1)^2 + 2$

94. Isolate the terms involving y on the left side of the equation:

$$y^2 + 2y + 12x - 23 = 0.$$

Then write the equation in an equivalent form by completing the square on the left side.

CHAPTER 7 Mid-Chapter Check Point

WHAT YOU KNOW: We learned that the four conic sections are the circle, the ellipse, the hyperbola, and the parabola. Prior to this chapter, we graphed circles with center (h, k) and radius r:

$$(x - h)^2 + (y - k)^2 = r^2.$$

In this chapter, you learned to graph ellipses centered at the origin and ellipses centered at (h, k):

$$\frac{(x-h)^2}{a^2} + \frac{(y-k)^2}{b^2} = 1 \text{ or } \frac{(x-h)^2}{b^2} + \frac{(y-k)^2}{a^2} = 1, a^2 > b^2.$$

We saw that the larger denominator (a^2) determines whether the major axis is horizontal or vertical. We used vertices and asymptotes to graph hyperbolas centered at the origin and hyperbolas centered at (h, k):

$$\frac{(x - h)^2}{a^2} - \frac{(y - k)^2}{b^2} = 1 \text{ or } \frac{(y - k)^2}{a^2} - \frac{(x - h)^2}{b^2} = 1.$$

We used $c^2 = a^2 - b^2$ to locate the foci of an ellipse. We used $c^2 = a^2 + b^2$ to locate the foci of a hyperbola.

In Exercises 1–5, graph each ellipse. Give the location of the foci.

1. $\dfrac{x^2}{25} + \dfrac{y^2}{4} = 1$

2. $9x^2 + 4y^2 = 36$

3. $\dfrac{(x - 2)^2}{16} + \dfrac{(y + 1)^2}{25} = 1$

4. $\dfrac{(x + 2)^2}{25} + \dfrac{(y - 1)^2}{16} = 1$

5. $x^2 + 9y^2 - 4x + 54y + 49 = 0$

In Exercises 6–11, graph each hyperbola. Give the location of the foci and the equations of the asymptotes.

6. $\dfrac{x^2}{9} - y^2 = 1$

7. $\dfrac{y^2}{9} - x^2 = 1$

8. $y^2 - 4x^2 = 16$

9. $4x^2 - 49y^2 = 196$

10. $\dfrac{(x - 2)^2}{9} - \dfrac{(y + 2)^2}{16} = 1$

11. $4x^2 - y^2 + 8x + 6y + 11 = 0$

In Exercises 12–18, graph each equation.

12. $x^2 + y^2 = 4$

13. $x + y = 4$

14. $x^2 - y^2 = 4$

15. $x^2 + 4y^2 = 4$

16. $(x + 1)^2 + (y - 1)^2 = 4$

17. $x^2 + 4(y - 1)^2 = 4$

18. $(x - 1)^2 - (y - 1)^2 = 4$

In Exercises 19–22, find the standard form of the equation of the conic section satisfying the given conditions.

19. Ellipse; Foci: $(-4, 0), (4, 0)$; Vertices: $(-5, 0), (5, 0)$

20. Ellipse; Endpoints of major axis: $(-8, 2), (10, 2)$; Foci: $(-4, 2), (6, 2)$

21. Hyperbola; Foci: $(0, -3), (0, 3)$; Vertices: $(0, -2), (0, 2)$

22. Hyperbola; Foci: $(-4, 5), (2, 5)$; Vertices: $(-3, 5), (1, 5)$

23. A semielliptical archway over a one-way road has a height of 10 feet and a width of 30 feet. A truck has a width of 10 feet and a height of 9.5 feet. Will this truck clear the opening of the archway?

24. A lithotriper is used to disintegrate kidney stones. The patient is placed within an elliptical device with the kidney centered at one focus, while ultrasound waves from the other focus hit the walls and are reflected to the kidney stone, shattering the stone. Suppose that the length of the major axis of the ellipse is 40 centimeters and the length of the minor axis is 20 centimeters. How far from the kidney stone should the electrode that sends the ultrasound waves be placed in order to shatter the stone?

25. An explosion is recorded by two forest rangers, one at a primary station and the other at an outpost 6 kilometers away. The ranger at the primary station hears the explosion 6 seconds before the ranger at the outpost.

 a. Assuming sound travels at 0.35 kilometer per second, write an equation in standard form that gives all the possible locations of the explosion. Use a coordinate system with the two ranger stations on the x-axis and the midpoint between the stations at the origin.

 b. Graph the equation that gives the possible locations of the explosion. Show the locations of the ranger stations in your drawing.

SECTION 7.3 The Parabola

Objectives

1. Graph parabolas with vertices at the origin.
2. Write equations of parabolas in standard form.
3. Graph parabolas with vertices not at the origin.
4. Solve applied problems involving parabolas.

At first glance, this image looks like columns of smoke rising from a fire into a starry sky. Those are, indeed, stars in the background, but you are not looking at ordinary smoke columns. These stand almost 6 trillion miles high and are 7000 light-years from Earth—more than 400 million times as far away as the sun.

This NASA photograph is one of a series of stunning images captured from the ends of the universe by the Hubble Space Telescope. The image shows infant star systems the size of our solar system emerging from the gas and dust that shrouded their creation. Using a parabolic mirror that is 94.5 inches in diameter, the Hubble has provided answers to many of the profound mysteries of the cosmos: How big and how old is the universe? How did the galaxies come to exist? Do other Earth-like planets orbit other sun-like stars? In this section, we study parabolas and their applications, including parabolic shapes that gather distant rays of light and focus them into spectacular images.

Definition of a Parabola

In Chapter 3, we studied parabolas, viewing them as graphs of quadratic functions in the form

$$y = a(x - h)^2 + k \quad \text{or} \quad y = ax^2 + bx + c.$$

GREAT QUESTION!

What should I already know about graphing parabolas?

Here's a brief summary:

Graphing $y = a(x - h)^2 + k$ and $y = ax^2 + bx + c$

1. If $a > 0$, the graph opens upward. If $a < 0$, the graph opens downward.
2. The vertex of $y = a(x - h)^2 + k$ is (h, k).

3. The x-coordinate of the vertex of $y = ax^2 + bx + c$ is $x = -\dfrac{b}{2a}$.

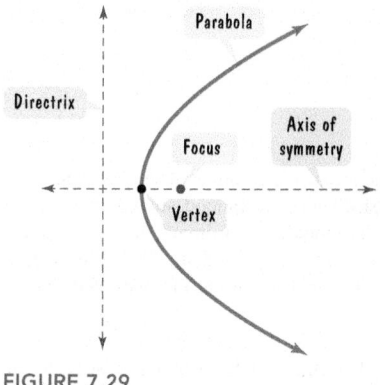

FIGURE 7.29

Parabolas can be given a geometric definition that enables us to include graphs that open to the left or to the right, as well as those that open obliquely. The definitions of ellipses and hyperbolas involved two fixed points, the foci. By contrast, the definition of a parabola is based on one point and a line.

Definition of a Parabola

A **parabola** is the set of all points in a plane that are equidistant from a fixed line, the **directrix**, and a fixed point, the **focus**, that is not on the line (see **Figure 7.29**).

In **Figure 7.29**, find the line passing through the focus and perpendicular to the directrix. This is the **axis of symmetry** of the parabola. The point of intersection of the parabola with its axis of symmetry is called the **vertex**. Notice that the vertex is midway between the focus and the directrix.

Standard Form of the Equation of a Parabola

The rectangular coordinate system enables us to translate a parabola's geometric definition into an algebraic equation. **Figure 7.30** is our starting point for obtaining an equation. We place the focus on the x-axis at the point $(p, 0)$. The directrix has an equation given by $x = -p$. The vertex, located midway between the focus and the directrix, is at the origin.

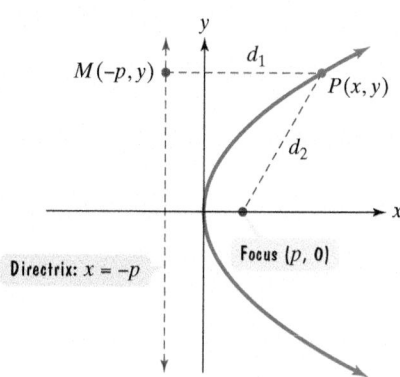

FIGURE 7.30

What does the definition of a parabola tell us about the point (x, y) in **Figure 7.30**? For any point (x, y) on the parabola, the distance d_1 to the directrix is equal to the distance d_2 to the focus. Thus, the point (x, y) is on the parabola if and only if

$$d_1 = d_2.$$

$$\sqrt{(x + p)^2 + (y - y)^2} = \sqrt{(x - p)^2 + (y - 0)^2} \quad \text{Use the distance formula.}$$

$$(x + p)^2 = (x - p)^2 + y^2 \quad \text{Square both sides of the equation.}$$

$$x^2 + 2px + p^2 = x^2 - 2px + p^2 + y^2 \quad \text{Square } x + p \text{ and } x - p.$$

$$2px = -2px + y^2 \quad \text{Subtract } x^2 + p^2 \text{ from both sides of the equation.}$$

$$y^2 = 4px \quad \text{Solve for } y^2.$$

This last equation is called the **standard form of the equation of a parabola with its vertex at the origin**. There are two such equations, one for a focus on the x-axis and one for a focus on the y-axis.

Standard Forms of the Equations of a Parabola

The **standard form of the equation of a parabola** with vertex at the origin is

$$y^2 = 4px \quad \text{or} \quad x^2 = 4py.$$

Figure 7.31(a) at the top of the next page illustrates that for the equation on the left, the focus is on the x-axis, which is the axis of symmetry. **Figure 7.31(b)** illustrates that for the equation on the right, the focus is on the y-axis, which is the axis of symmetry.

GREAT QUESTION!

What is the relationship between the sign of p and the position of the focus from the vertex?

It is helpful to think of p as the *directed distance* from the vertex to the focus. If $p > 0$, the focus lies p units to the right of the vertex or p units above the vertex. If $p < 0$, the focus lies $|p|$ units to the left of the vertex or $|p|$ units below the vertex.

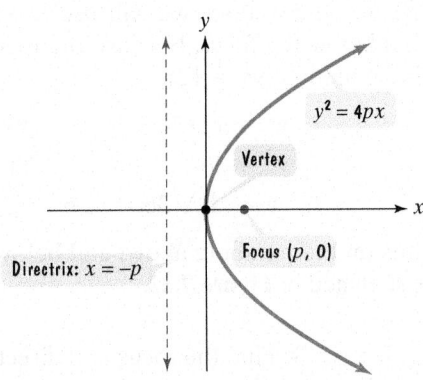

FIGURE 7.31(a) Parabola with the x-axis as the axis of symmetry. If $p > 0$, the graph opens to the right. If $p < 0$, the graph opens to the left.

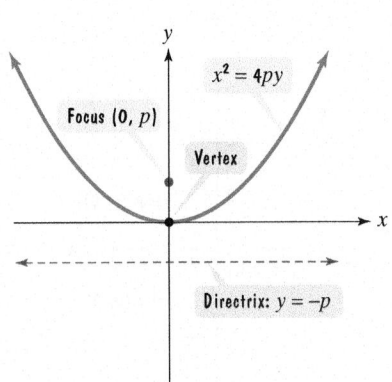

FIGURE 7.31(b) Parabola with the y-axis as the axis of symmetry. If $p > 0$, the graph opens upward. If $p < 0$, the graph opens downward.

① Graph parabolas with vertices at the origin.

Using the Standard Form of the Equation of a Parabola

We can use the standard form of the equation of a parabola to find its focus and directrix. Observing the graph's symmetry from its equation is helpful in locating the focus.

$$y^2 = 4px \qquad\qquad x^2 = 4py$$

The equation does not change if y is replaced with $-y$. There is x-axis symmetry and the focus is on the x-axis at $(p, 0)$.

The equation does not change if x is replaced with $-x$. There is y-axis symmetry and the focus is on the y-axis at $(0, p)$.

Although the definition of a parabola is given in terms of its focus and its directrix, the focus and directrix are not part of the graph. The vertex, located at the origin, is a point on the graph of $y^2 = 4px$ and $x^2 = 4py$. Example 1 illustrates how you can find two additional points on the parabola.

EXAMPLE 1 Finding the Focus and Directrix of a Parabola

Find the focus and directrix of the parabola given by $y^2 = 12x$. Then graph the parabola.

SOLUTION

The given equation, $y^2 = 12x$, is in the standard form $y^2 = 4px$, so $4p = 12$.

No change if y is replaced with $-y$. The parabola has x-axis symmetry. $y^2 = 12x$

This is $4p$.

We can find both the focus and the directrix by finding p.

$$4p = 12$$
$$p = 3 \quad \text{Divide both sides by 4.}$$

Because p is positive, the parabola, with its x-axis symmetry, opens to the right. The focus is 3 units to the right of the vertex, $(0, 0)$.

Focus: $(p, 0) = (3, 0)$
Directrix: $x = -p; x = -3$

FIGURE 7.32 The graph of $y^2 = 12x$

The focus, $(3, 0)$, and directrix, $x = -3$, are shown in **Figure 7.32**.

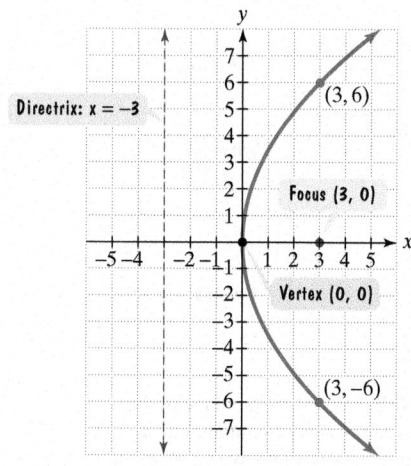

FIGURE 7.32 (repeated) The graph of $y^2 = 12x$

TECHNOLOGY

We graph $y^2 = 12x$ with a graphing utility by first solving for y. The screen shows the graphs of $y = \sqrt{12x}$ and $y = -\sqrt{12x}$. The graph fails the vertical line test. Because $y^2 = 12x$ is not a function, you were not familiar with this form of the parabola's equation in Chapter 3.

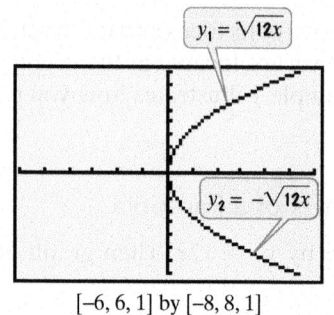

$[-6, 6, 1]$ by $[-8, 8, 1]$

To graph the parabola, we will use two points on the graph that lie directly above and below the focus. Because the focus is at $(3, 0)$, substitute 3 for x in the parabola's equation, $y^2 = 12x$.

$$y^2 = 12 \cdot 3 \qquad \text{Replace } x \text{ with 3 in } y^2 = 12x.$$
$$y^2 = 36 \qquad \text{Simplify.}$$
$$y = \pm\sqrt{36} = \pm 6 \qquad \text{Apply the square root property.}$$

The points on the parabola above and below the focus are $(3, 6)$ and $(3, -6)$. The graph is sketched in **Figure 7.32**. • • •

⊘ Check Point **1** Find the focus and directrix of the parabola given by $y^2 = 8x$. Then graph the parabola.

In general, the points on a parabola $y^2 = 4px$ that lie above and below the focus, $(p, 0)$, are each at a distance $|2p|$ from the focus. This is because if $x = p$, then $y^2 = 4px = 4p^2$, so $y = \pm 2p$. The line segment joining these two points is called the *latus rectum*; its length is $|4p|$.

The Latus Rectum and Graphing Parabolas

The **latus rectum** of a parabola is a line segment that passes through its focus, is parallel to its directrix, and has its endpoints on the parabola. **Figure 7.33** shows that the length of the latus rectum for the graphs of $y^2 = 4px$ and $x^2 = 4py$ is $|4p|$.

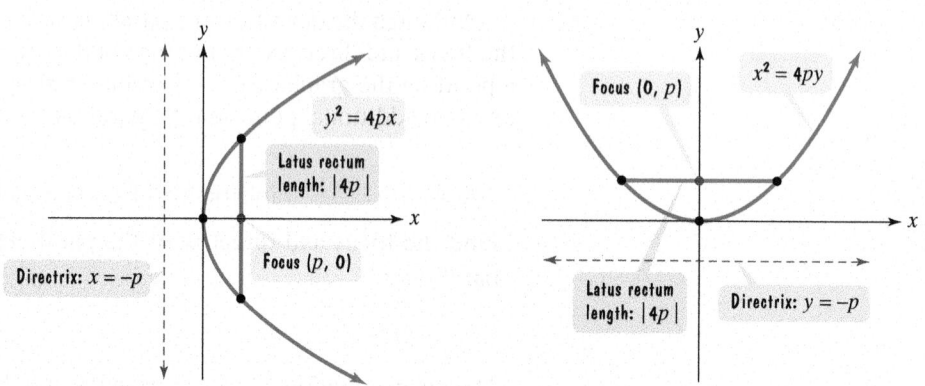

FIGURE 7.33 Endpoints of the latus rectum are helpful in determining a parabola's "width," or how it opens.

EXAMPLE 2 Finding the Focus and Directrix of a Parabola

Find the focus and directrix of the parabola given by $x^2 = -8y$. Then graph the parabola.

SOLUTION

The given equation, $x^2 = -8y$, is in the standard form $x^2 = 4py$, so $4p = -8$.

No change if x is replaced with $-x$. The parabola has y-axis symmetry.

$$x^2 = -8y$$

This is $4p$.

FIGURE 7.34 The graph of $x^2 = -8y$

TECHNOLOGY

Graph $x^2 = -8y$ by first solving

for y: $y = -\dfrac{x^2}{8}$. The graph passes

the vertical line test. Because
$x^2 = -8y$ is a function, you were
familiar with the parabola's
alternate algebraic form,

$y = -\dfrac{1}{8}x^2$, in Chapter 3. The

form is $y = ax^2 + bx + c$, with

$a = -\dfrac{1}{8}, b = 0$, and $c = 0$.

$[-6, 6, 1]$ by $[-6, 6, 1]$

② Write equations of parabolas in standard form.

We can find both the focus and the directrix by finding p.

$$4p = -8$$
$$p = -2 \quad \text{Divide both sides by 4.}$$

Because p is negative, the parabola, with its y-axis symmetry, opens downward. The focus is 2 units below the vertex, $(0,0)$.

$$\text{Focus:} \qquad (0, p) = (0, -2)$$
$$\text{Directrix:} \quad y = -p; y = 2$$

The focus and directrix are shown in **Figure 7.34**.

To graph the parabola, we will use the vertex, $(0, 0)$, and the two endpoints of the latus rectum. The length of the latus rectum is

$$|4p| = |4(-2)| = |-8| = 8.$$

Because the graph has y-axis symmetry, the latus rectum extends 4 units to the left and 4 units to the right of the focus, $(0, -2)$. The endpoints of the latus rectum are $(-4, -2)$ and $(4, -2)$. Passing a smooth curve through the vertex and these two points, we sketch the parabola, shown in **Figure 7.34**. • • •

⊘ Check Point **2** Find the focus and directrix of the parabola given by $x^2 = -12y$. Then graph the parabola.

In Examples 1 and 2, we used the equation of a parabola to find its focus and directrix. In the next example, we reverse this procedure.

EXAMPLE 3 Finding the Equation of a Parabola from Its Focus and Directrix

Find the standard form of the equation of a parabola with focus $(5, 0)$ and directrix $x = -5$, shown in **Figure 7.35**.

SOLUTION

The focus is $(5, 0)$. Thus, the focus is on the x-axis. We use the standard form of the equation in which there is x-axis symmetry, namely, $y^2 = 4px$.

We need to determine the value of p. **Figure 7.35** shows that the focus is 5 units to the right of the vertex, $(0, 0)$. Thus, p is positive and $p = 5$. We substitute 5 for p in $y^2 = 4px$ to obtain the standard form of the equation of the parabola. The equation is

$$y^2 = 4 \cdot 5x \qquad \text{or} \qquad y^2 = 20x.$$

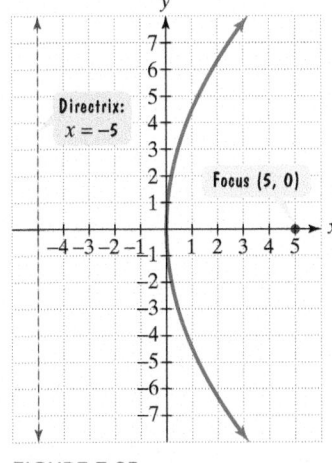

FIGURE 7.35 • • •

⊘ Check Point **3** Find the standard form of the equation of a parabola with focus $(8, 0)$ and directrix $x = -8$.

③ Graph parabolas with vertices not at the origin.

Translations of Parabolas

The graph of a parabola can have its vertex at (h, k), rather than at the origin. Horizontal and vertical translations are accomplished by replacing x with $x - h$ and y with $y - k$ in the standard form of the parabola's equation.

Table 7.3 gives the standard forms of equations of parabolas with vertex at (h, k). **Figure 7.36** shows their graphs.

Table 7.3 Standard Forms of Equations of Parabolas with Vertex at (h, k)

Equation	Vertex	Axis of Symmetry	Focus	Directrix	Description
$(y - k)^2 = 4p(x - h)$	(h, k)	Horizontal	$(h + p, k)$	$x = h - p$	If $p > 0$, opens to the right. If $p < 0$, opens to the left.
$(x - h)^2 = 4p(y - k)$	(h, k)	Vertical	$(h, k + p)$	$y = k - p$	If $p > 0$, opens upward. If $p < 0$, opens downward.

GREAT QUESTION!

What are the main differences between parabolas whose equations contain y^2 and parabolas whose equations contain x^2?

If y is the squared term, there is horizontal symmetry and the parabola's equation is not a function. If x is the squared term, there is vertical symmetry and the parabola's equation is a function. Continue to think of p as the directed distance from the vertex, (h, k), to the focus.

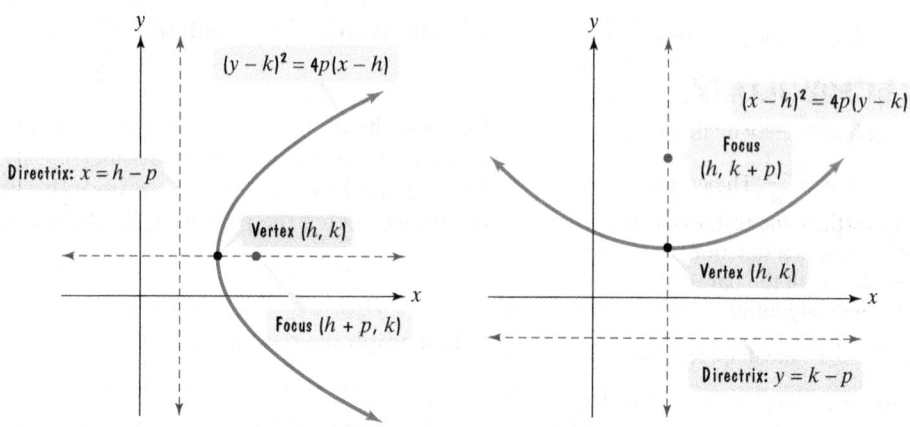

FIGURE 7.36 Graphs of parabolas with vertex at (h, k) and $p > 0$

The two parabolas shown in **Figure 7.36** illustrate standard forms of equations for $p > 0$. If $p < 0$, a parabola with a horizontal axis of symmetry will open to the left and the focus will lie to the left of the directrix. If $p < 0$, a parabola with a vertical axis of symmetry will open downward and the focus will lie below the directrix.

EXAMPLE 4 Graphing a Parabola with Vertex at (h, k)

Find the vertex, focus, and directrix of the parabola given by

$$(x - 3)^2 = 8(y + 1).$$

Then graph the parabola.

SOLUTION

In order to find the focus and directrix, we need to know the vertex. In the standard forms of equations with vertex at (h, k), h is the number subtracted from x and k is the number subtracted from y.

$$(x - 3)^2 = 8(y - (-1))$$

This is $(x - h)^2$, with $h = 3$. This is $y - k$, with $k = -1$.

We see that $h = 3$ and $k = -1$. Thus, the vertex of the parabola is $(h, k) = (3, -1)$. Now that we have the vertex, we can find both the focus and directrix by finding p.

$$(x - 3)^2 = 8(y + 1)$$

This is $4p$.

The equation is in the standard form $(x - h)^2 = 4p(y - k)$. Because x is the squared term, there is vertical symmetry and the parabola's equation is a function.

Because $4p = 8$, $p = 2$. Based on the standard form of the equation, the axis of symmetry is vertical. With a positive value for p and a vertical axis of symmetry,

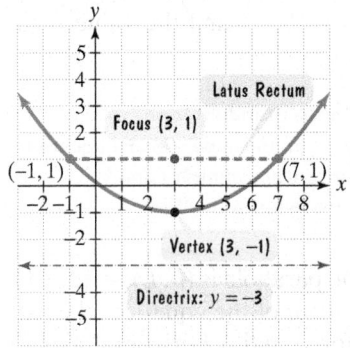

FIGURE 7.37 The graph of $(x - 3)^2 = 8(y + 1)$

TECHNOLOGY

Graph $(x - 3)^2 = 8(y + 1)$ by first solving for y:

$$\frac{1}{8}(x - 3)^2 = y + 1$$

$$y = \frac{1}{8}(x - 3)^2 - 1.$$

The graph passes the vertical line test. Because $(x - 3)^2 = 8(y + 1)$ is a function, you were familiar with the parabola's alternate algebraic form, $y = \frac{1}{8}(x - 3)^2 - 1$, in Chapter 3. The form is $y = a(x - h)^2 + k$ with $a = \frac{1}{8}, h = 3$, and $k = -1$.

$[-3, 9, 1]$ by $[-6, 6, 1]$

the parabola opens upward. Because $p = 2$, the focus is located 2 units above the vertex, $(3, -1)$. Likewise, the directrix is located 2 units below the vertex.

Focus: $\qquad (h, k + p) = (3, -1 + 2) = (3, 1)$

The vertex, (h, k), is $(3, -1)$. 　 The focus is 2 units above the vertex, $(3, -1)$.

Directrix: $\qquad y = k - p$

$$y = -1 - 2 = -3$$

The directrix is 2 units below the vertex, $(3, -1)$.

Thus, the focus is $(3, 1)$ and the directrix is $y = -3$. They are shown in **Figure 7.37**. To graph the parabola, we will use the vertex, $(3, -1)$, and the two endpoints of the latus rectum. The length of the latus rectum is

$$|4p| = |4 \cdot 2| = |8| = 8.$$

Because the graph has vertical symmetry, the latus rectum extends 4 units to the left and 4 units to the right of the focus, $(3, 1)$. The endpoints of the latus rectum are $(3 - 4, 1)$, or $(-1, 1)$, and $(3 + 4, 1)$, or $(7, 1)$. Passing a smooth curve through the vertex and these two points, we sketch the parabola, shown in **Figure 7.37**. ● ● ●

⊘ Check Point **4** Find the vertex, focus, and directrix of the parabola given by $(x - 2)^2 = 4(y + 1)$. Then graph the parabola.

In some cases, we need to convert the equation of a parabola to standard form by completing the square on x or y, whichever variable is squared. Let's see how this is done.

EXAMPLE 5 Graphing a Parabola with Vertex at (h, k)

Find the vertex, focus, and directrix of the parabola given by

$$y^2 + 2y + 12x - 23 = 0.$$

Then graph the parabola.

SOLUTION

We convert the given equation to standard form by completing the square on the variable y. We isolate the terms involving y on the left side.

$y^2 + 2y + 12x - 23 = 0$	This is the given equation.
$y^2 + 2y = -12x + 23$	Isolate the terms involving y.
$y^2 + 2y + 1 = -12x + 23 + 1$	Complete the square by adding the square of half the coefficient of y.
$(y + 1)^2 = -12x + 24$	Factor.

To express the equation $(y + 1)^2 = -12x + 24$ in the standard form $(y - k)^2 = 4p(x - h)$, we factor out -12 on the right. The standard form of the parabola's equation is

$$(y + 1)^2 = -12(x - 2).$$

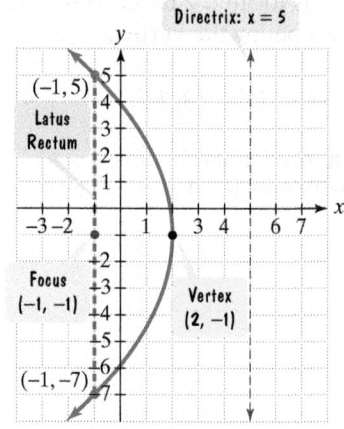

FIGURE 7.38 The graph of $y^2 + 2y + 12x - 23 = 0$, or $(y + 1)^2 = -12(x - 2)$

TECHNOLOGY

Graph $y^2 + 2y + 12x - 23 = 0$ by solving the equation for y.

$$y^2 + 2y + (12x - 23) = 0$$

$$a = 1 \quad b = 2 \quad c = 12x - 23$$

Use the quadratic formula to solve for y and enter the resulting equations.

$$y_1 = \frac{-2 + \sqrt{4 - 4(12x - 23)}}{2}$$

$$y_2 = \frac{-2 - \sqrt{4 - 4(12x - 23)}}{2}$$

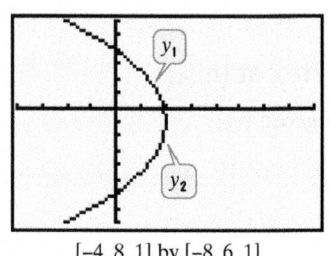

$[-4, 8, 1]$ by $[-8, 6, 1]$

④ Solve applied problems involving parabolas.

We use $(y + 1)^2 = -12(x - 2)$ to identify the vertex, (h, k), and the value for p needed to locate the focus and the directrix.

$$[y - (-1)]^2 = -12(x - 2)$$

> This is $(y - k)^2$, with $k = -1$. This is $4p$. This is $x - h$, with $h = 2$.

> The equation is in the standard form $(y - k)^2 = 4p(x - h)$. Because y is the squared term, there is horizontal symmetry and the parabola's equation is not a function.

We see that $h = 2$ and $k = -1$. Thus, the vertex of the parabola is $(h, k) = (2, -1)$. Because $4p = -12, p = -3$. Based on the standard form of the equation, the axis of symmetry is horizontal. With a negative value for p and a horizontal axis of symmetry, the parabola opens to the left. Because $p = -3$, the focus is located 3 units to the left of the vertex, $(2, -1)$. Likewise, the directrix is located 3 units to the right of the vertex.

Focus: $\qquad (h + p, k) = (2 + (-3), -1) = (-1, -1)$

> The vertex, (h, k), is $(2, -1)$. The focus is 3 units to the left of the vertex, $(2, -1)$.

Directrix: $\qquad x = h - p$
$$\qquad\qquad x = 2 - (-3) = 5$$

> The directrix is 3 units to the right of the vertex, $(2, -1)$.

Thus, the focus is $(-1, -1)$ and the directrix is $x = 5$. They are shown in **Figure 7.38**.

To graph the parabola, we will use the vertex, $(2, -1)$, and the two endpoints of the latus rectum. The length of the latus rectum is

$$|4p| = |4(-3)| = |-12| = 12.$$

Because the graph has horizontal symmetry, the latus rectum extends 6 units above and 6 units below the focus, $(-1, -1)$. The endpoints of the latus rectum are $(-1, -1 + 6)$, or $(-1, 5)$, and $(-1, -1 - 6)$, or $(-1, -7)$. Passing a smooth curve through the vertex and these two points, we sketch the parabola shown in **Figure 7.38**.　•••

✓ Check Point **5** Find the vertex, focus, and directrix of the parabola given by $y^2 + 2y + 4x - 7 = 0$. Then graph the parabola.

Applications

Parabolas have many applications. Cables hung between structures to form suspension bridges form parabolas. Arches constructed of steel and concrete, whose main purpose is strength, are usually parabolic in shape.

Suspension bridge　　　　　　Arch bridge

We have seen that comets in our solar system travel in orbits that are ellipses and hyperbolas. Some comets follow parabolic paths. Only comets with elliptical orbits, such as Halley's Comet, return to our part of the galaxy.

Blitzer Bonus

The Hubble Space Telescope

The Hubble Space Telescope

For decades, astronomers hoped to create an observatory above the atmosphere that would provide an unobscured view of the universe. This vision was realized with the 1990 launching of the Hubble Space Telescope. The telescope initially had blurred vision due to problems with its parabolic mirror. The mirror had been ground two millionths of a meter smaller than design specifications. In 1993, astronauts from the Space Shuttle *Endeavor* equipped the telescope with optics to correct the blurred vision. "A small change for a mirror, a giant leap for astronomy," Christopher J. Burrows of the Space Telescope Science Institute said when clear images from the ends of the universe were presented to the public after the repair mission.

If a parabola is rotated about its axis of symmetry, a parabolic surface is formed. **Figure 7.39(a)** shows how a parabolic surface can be used to reflect light. Light originates at the focus. Note how the light is reflected by the parabolic surface, so that the outgoing light is parallel to the axis of symmetry. The reflective properties of parabolic surfaces are used in the design of searchlights [see **Figure 7.39(b)**] and automobile headlights.

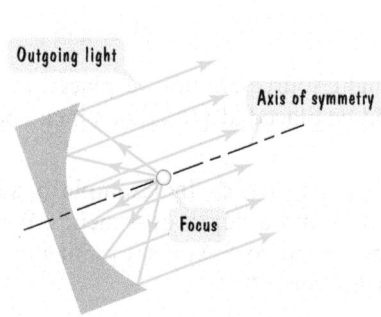

FIGURE 7.39(a) Parabolic surface reflecting light

FIGURE 7.39(b) Light from the focus is reflected parallel to the axis of symmetry.

Figure 7.40(a) shows how a parabolic surface can be used to reflect *incoming* light. Note that light rays strike the surface and are reflected *to the focus*. This principle is used in the design of reflecting telescopes, radar, and television satellite dishes. Reflecting telescopes magnify the light from distant stars by reflecting the light from these bodies to the focus of a parabolic mirror [see **Figure 7.40(b)**].

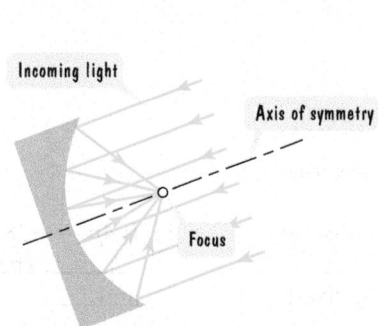

FIGURE 7.40(a) Parabolic surface reflecting incoming light

FIGURE 7.40(b) Incoming light rays are reflected to the focus.

EXAMPLE 6 Using the Reflection Property of Parabolas

An engineer is designing a flashlight using a parabolic reflecting mirror and a light source, shown in **Figure 7.41**. The casting has a diameter of 4 inches and a depth of 2 inches. What is the equation of the parabola used to shape the mirror? At what point should the light source be placed relative to the mirror's vertex?

FIGURE 7.41 Designing a flashlight

SOLUTION

We position the parabola with its vertex at the origin and opening upward (**Figure 7.42**). Thus, the focus is on the *y*-axis, located at $(0, p)$. We use the standard

FIGURE 7.42

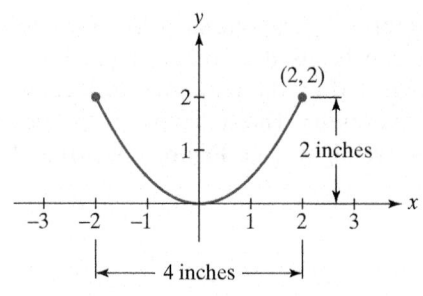

FIGURE 7.42 (repeated)

Blitzer Bonus

Five Things Scientists Learned from the Hubble Space Telescope

• The universe is approximately 13.7 billion years old.

• There is a high probability that the universe is expanding at ever-accelerating rates.

• New stars are created from clouds of gas and dust when giant galaxies collide.

• Dust rings around stars transform into planets.

• Planets are far more common than previously thought.

Source: *Listomania*, Harper Design, 2011

form of the equation in which there is *y*-axis symmetry, namely, $x^2 = 4py$. We need to find *p*. Because $(2, 2)$ lies on the parabola, we let $x = 2$ and $y = 2$ in $x^2 = 4py$.

$$2^2 = 4p \cdot 2 \quad \text{Substitute 2 for } x \text{ and 2 for } y \text{ in } x^2 = 4py.$$

$$4 = 8p \quad \text{Simplify.}$$

$$p = \tfrac{1}{2} \quad \text{Divide both sides of the equation by 8 and reduce the resulting fraction.}$$

We substitute $\tfrac{1}{2}$ for *p* in $x^2 = 4py$ to obtain the standard form of the equation of the parabola. The equation of the parabola used to shape the mirror is

$$x^2 = 4 \cdot \tfrac{1}{2} y \quad \text{or} \quad x^2 = 2y.$$

The light source should be placed at the focus, $(0, p)$. Because $p = \tfrac{1}{2}$, the light should be placed at $\left(0, \tfrac{1}{2}\right)$, or $\tfrac{1}{2}$ inch above the vertex. •••

 Check Point **6** In Example 6, suppose that the casting has a diameter of 6 inches and a depth of 4 inches. What is the equation of the parabola used to shape the mirror? At what point should the light source be placed relative to the mirror's vertex?

Degenerate Conic Sections

We opened the chapter by noting that conic sections are curves that result from the intersection of a cone and a plane. However, these intersections might not result in a conic section. Three degenerate cases occur when the cutting plane passes through the vertex. These **degenerate conic sections** are a point, a line, and a pair of intersecting lines, illustrated in **Figure 7.43**.

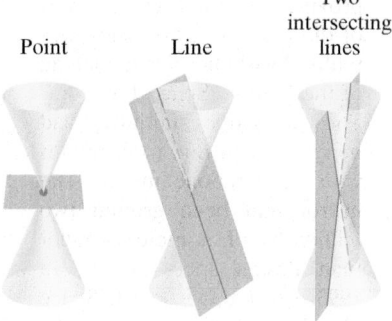

FIGURE 7.43 Degenerate conics

CONCEPT AND VOCABULARY CHECK

Fill in each blank so that the resulting statement is true.

1. The set of all points in a plane that are equidistant from a fixed line and a fixed point is a/an _____. The fixed line is called the _____ and the fixed point is called the _____.

Use the graph shown to answer Exercises 2–5.

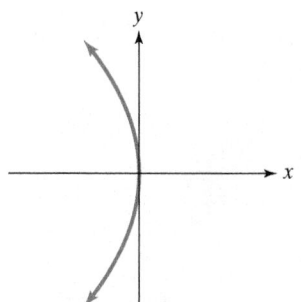

2. The equation of the parabola is of the form
 a. $y^2 = 4px$
 b. $x^2 = 4py$. ____

3. If $4p = -28$, then the coordinates of the focus are _____.

4. If $4p = -28$, then the equation of the directrix is _____.

5. If $4p = -28$, then the length of the latus rectum is _____. The endpoints of the latus rectum are _____ and _____.

Use the graph shown to answer Exercises 6–9.

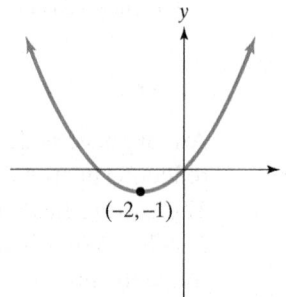

6. The equation of the parabola is of the form
 a. $(y - 1)^2 = 4p(x + 2)$
 b. $(y + 1)^2 = 4p(x + 2)$
 c. $(x - 2)^2 = 4p(y - 1)$
 d. $(x + 2)^2 = 4p(y + 1)$.

7. If $4p = 4$, then the coordinates of the focus are _____.

8. If $4p = 4$, then the equation of the directrix is _____.

9. If $4p = 4$, then the length of the latus rectum is _____. The endpoints of the latus rectum are _____ and _____.

EXERCISE SET 7.3

Practice Exercises

In Exercises 1–4, find the focus and directrix of each parabola with the given equation. Then match each equation to one of the graphs that are shown and labeled (a)–(d).

1. $y^2 = 4x$

3. $x^2 = -4y$

2. $x^2 = 4y$

4. $y^2 = -4x$

a.

b.

c.

d.

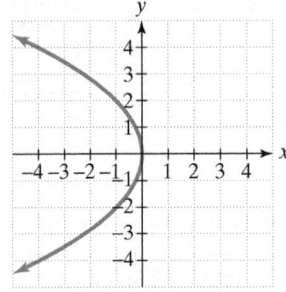

In Exercises 5–16, find the focus and directrix of the parabola with the given equation. Then graph the parabola.

5. $y^2 = 16x$

7. $y^2 = -8x$

9. $x^2 = 12y$

11. $x^2 = -16y$

13. $y^2 - 6x = 0$

15. $8x^2 + 4y = 0$

6. $y^2 = 4x$

8. $y^2 = -12x$

10. $x^2 = 8y$

12. $x^2 = -20y$

14. $x^2 - 6y = 0$

16. $8y^2 + 4x = 0$

In Exercises 17–30, find the standard form of the equation of each parabola satisfying the given conditions.

17. Focus: $(7, 0)$; Directrix: $x = -7$

18. Focus: $(9, 0)$; Directrix: $x = -9$

19. Focus: $(-5, 0)$; Directrix: $x = 5$

20. Focus: $(-10, 0)$; Directrix: $x = 10$

21. Focus: $(0, 15)$; Directrix: $y = -15$

22. Focus: $(0, 20)$; Directrix: $y = -20$

23. Focus: $(0, -25)$; Directrix: $y = 25$

24. Focus: $(0, -15)$; Directrix: $y = 15$

25. Vertex: $(2, -3)$; Focus: $(2, -5)$

26. Vertex: $(5, -2)$; Focus: $(7, -2)$

27. Focus: $(3, 2)$; Directrix: $x = -1$

28. Focus: $(2, 4)$; Directrix: $x = -4$

29. Focus: $(-3, 4)$; Directrix: $y = 2$

30. Focus: $(7, -1)$; Directrix: $y = -9$

In Exercises 31–34, find the vertex, focus, and directrix of each parabola with the given equation. Then match each equation to one of the graphs that are shown and labeled (a)–(d).

31. $(y - 1)^2 = 4(x - 1)$

32. $(x + 1)^2 = 4(y + 1)$

33. $(x + 1)^2 = -4(y + 1)$

34. $(y - 1)^2 = -4(x - 1)$

a.

b.

c.

d.

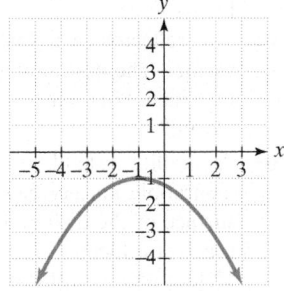

In Exercises 35–42, find the vertex, focus, and directrix of each parabola with the given equation. Then graph the parabola.

35. $(x - 2)^2 = 8(y - 1)$

37. $(x + 1)^2 = -8(y + 1)$

39. $(y + 3)^2 = 12(x + 1)$

41. $(y + 1)^2 = -8x$

36. $(x + 2)^2 = 4(y + 1)$

38. $(x + 2)^2 = -8(y + 2)$

40. $(y + 4)^2 = 12(x + 2)$

42. $(y - 1)^2 = -8x$

In Exercises 43–48, convert each equation to standard form by completing the square on x or y. Then find the vertex, focus, and directrix of the parabola. Finally, graph the parabola.

43. $x^2 - 2x - 4y + 9 = 0$

45. $y^2 - 2y + 12x - 35 = 0$

47. $x^2 + 6x - 4y + 1 = 0$

44. $x^2 + 6x + 8y + 1 = 0$

46. $y^2 - 2y - 8x + 1 = 0$

48. $x^2 + 8x - 4y + 8 = 0$

Practice Plus

In Exercises 49–54, use the vertex and the direction in which the parabola opens to determine the relation's domain and range. Is the relation a function?

49. $y^2 + 6y - x + 5 = 0$

50. $y^2 - 2y - x - 5 = 0$

51. $y = -x^2 + 4x - 3$

52. $y = -x^2 - 4x + 4$

53. $x = -4(y - 1)^2 + 3$

54. $x = -3(y - 1)^2 - 2$

In Exercises 55–60, find the solution set for each system by graphing both of the system's equations in the same rectangular coordinate system and finding points of intersection. Check all solutions in both equations.

55. $\begin{cases} (y - 2)^2 = x + 4 \\ \qquad y = -\dfrac{1}{2}x \end{cases}$

56. $\begin{cases} (y - 3)^2 = x - 2 \\ \quad x + y = 5 \end{cases}$

57. $\begin{cases} x = y^2 - 3 \\ x = y^2 - 3y \end{cases}$

58. $\begin{cases} x = y^2 - 5 \\ x^2 + y^2 = 25 \end{cases}$

59. $\begin{cases} x = (y + 2)^2 - 1 \\ (x - 2)^2 + (y + 2)^2 = 1 \end{cases}$

60. $\begin{cases} x = 2y^2 + 4y + 5 \\ (x + 1)^2 + (y - 2)^2 = 1 \end{cases}$

Application Exercises

61. The reflector of a flashlight is in the shape of a parabolic surface. The casting has a diameter of 4 inches and a depth of 1 inch. How far from the vertex should the light bulb be placed?

62. The reflector of a flashlight is in the shape of a parabolic surface. The casting has a diameter of 8 inches and a depth of 1 inch. How far from the vertex should the light bulb be placed?

63. A satellite dish, like the one shown below, is in the shape of a parabolic surface. Signals coming from a satellite strike the surface of the dish and are reflected to the focus, where the receiver is located. The satellite dish shown has a diameter of 12 feet and a depth of 2 feet. How far from the base of the dish should the receiver be placed?

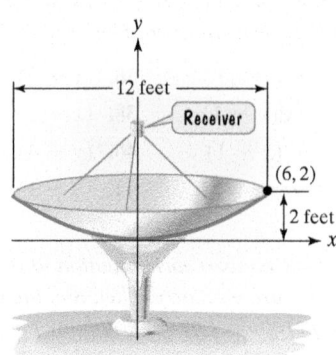

64. In Exercise 63, if the diameter of the dish is halved and the depth stays the same, how far from the base of the smaller dish should the receiver be placed?

65. The towers of the Golden Gate Bridge connecting San Francisco to Marin County are 1280 meters apart and rise 160 meters above the road. The cable between the towers has the shape of a parabola and the cable just touches the sides of the road midway between the towers. What is the height of the cable 200 meters from a tower? Round to the nearest meter.

66. The towers of a suspension bridge are 800 feet apart and rise 160 feet above the road. The cable between the towers has the shape of a parabola and the cable just touches the sides of the road midway between the towers. What is the height of the cable 100 feet from a tower?

67. The parabolic arch shown in the figure is 50 feet above the water at the center and 200 feet wide at the base. Will a boat that is 30 feet tall clear the arch 30 feet from the center?

68. A satellite dish in the shape of a parabolic surface has a diameter of 20 feet. If the receiver is to be placed 6 feet from the base, how deep should the dish be?

Writing in Mathematics

69. What is a parabola?

70. Explain how to use $y^2 = 8x$ to find the parabola's focus and directrix.

71. If you are given the standard form of the equation of a parabola with vertex at the origin, explain how to determine if the parabola opens to the right, left, upward, or downward.

72. Describe one similarity and one difference between the graphs of $y^2 = 4x$ and $(y - 1)^2 = 4(x - 1)$.

73. How can you distinguish parabolas from other conic sections by looking at their equations?

74. Look at the satellite dish shown in Exercise 63. Why must the receiver for a shallow dish be farther from the base of the dish than for a deeper dish of the same diameter?

Technology Exercises

75. Use a graphing utility to graph any five of the parabolas that you graphed by hand in Exercises 5–16.

76. Use a graphing utility to graph any three of the parabolas that you graphed by hand in Exercises 35–42. First solve the given equation for y, possibly using the square root property. Enter each of the two resulting equations to produce the complete graph.

Use a graphing utility to graph the parabolas in Exercises 77–78. Write the given equation as a quadratic equation in y and use the quadratic formula to solve for y. Enter each of the equations to produce the complete graph.

77. $y^2 + 2y - 6x + 13 = 0$

78. $y^2 + 10y - x + 25 = 0$

In Exercises 79–80, write each equation as a quadratic equation in y and then use the quadratic formula to express y in terms of x. Graph the resulting two equations using a graphing utility. What effect does the xy-term have on the graph of the resulting parabola?

79. $16x^2 - 24xy + 9y^2 - 60x - 80y + 100 = 0$

80. $x^2 + 2\sqrt{3}xy + 3y^2 + 8\sqrt{3}x - 8y + 32 = 0$

Critical Thinking Exercises

Make Sense? *In Exercises 81–84, determine whether each statement makes sense or does not make sense, and explain your reasoning.*

81. I graphed a parabola that opened to the right and contained a maximum point.

82. Knowing that a parabola opening to the right has a vertex at $(-1, 1)$ gives me enough information to determine its graph.

83. I noticed that depending on the values for A and B, assuming that they are both not zero, the graph of $Ax^2 + By^2 = C$ can represent any of the conic sections other than a parabola.

84. I'm using a telescope in which light from distant stars is reflected to the focus of a parabolic mirror.

In Exercises 85–88, determine whether each statement is true or false. If the statement is false, make the necessary change(s) to produce a true statement.

85. The parabola whose equation is $x = 2y - y^2 + 5$ opens to the right.

86. If the parabola whose equation is $x = ay^2 + by + c$ has its vertex at $(3, 2)$ and $a > 0$, then it has no y-intercepts.

87. Some parabolas that open to the right have equations that define y as a function of x.

88. The graph of $x = a(y - k) + h$ is a parabola with vertex at (h, k).

89. Find the focus and directrix of a parabola whose equation is of the form $Ax^2 + Ey = 0, A \neq 0, E \neq 0$.

90. Write the standard form of the equation of a parabola whose points are equidistant from $y = 4$ and $(-1, 0)$.

Group Exercise

91. Consult the research department of your library or the Internet to find an example of architecture that incorporates one or more conic sections in its design. Share this example with other group members. Explain precisely how conic sections are used. Do conic sections enhance the appeal of the architecture? In what ways?

Preview Exercises

Exercises 92–94 will help you prepare for the material covered in the first section of the next chapter.

92. Evaluate $\dfrac{(-1)^n}{3^n - 1}$ for $n = 1, 2, 3$, and 4.

93. Find the product of all positive integers from n down through 1 for $n = 5$.

94. Evaluate $i^2 + 1$ for all consecutive integers from 1 to 6, inclusive. Then find the sum of the six evaluations.

CHAPTER 7 Summary, Review, and Test

SUMMARY

DEFINITIONS AND CONCEPTS	EXAMPLES

7.1 The Ellipse

a. An ellipse is the set of all points in a plane the sum of whose distances from two fixed points, the foci, is constant.

b. Standard forms of the equations of an ellipse with center at the origin are $\dfrac{x^2}{a^2} + \dfrac{y^2}{b^2} = 1$ [foci: $(-c, 0), (c, 0)$] and $\dfrac{x^2}{b^2} + \dfrac{y^2}{a^2} = 1$ [foci: $(0, -c), (0, c)$], where $c^2 = a^2 - b^2$ and $a^2 > b^2$. See the box on page 672 containing Figure 7.6. *Ex. 1, p. 673; Ex. 2, p. 674; Ex. 3, p. 674*

c. Standard forms of the equations of an ellipse centered at (h, k) are $\dfrac{(x - h)^2}{a^2} + \dfrac{(y - k)^2}{b^2} = 1$ and $\dfrac{(x - h)^2}{b^2} + \dfrac{(y - k)^2}{a^2} = 1, a^2 > b^2$. See Table 7.1 on page 676. *Ex. 4, p. 676*

DEFINITIONS AND CONCEPTS EXAMPLES

7.2 The Hyperbola

a. A hyperbola is the set of all points in a plane the difference of whose distances from two fixed points, the foci, is constant.

b. Standard forms of the equations of a hyperbola with center at the origin are $\dfrac{x^2}{a^2} - \dfrac{y^2}{b^2} = 1$ [foci: $(-c, 0), (c, 0)$] Ex. 1, p. 685;
Ex. 2, p. 686

and $\dfrac{y^2}{a^2} - \dfrac{x^2}{b^2} = 1$ [foci: $(0, c), (0, -c)$], where $c^2 = a^2 + b^2$. See the box on page 684 and Figure 7.16.

c. Asymptotes for $\dfrac{x^2}{a^2} - \dfrac{y^2}{b^2} = 1$ are $y = \pm\dfrac{b}{a}x$. Asymptotes for $\dfrac{y^2}{a^2} - \dfrac{x^2}{b^2} = 1$ are $y = \pm\dfrac{a}{b}x$.

d. A procedure for graphing hyperbolas is given in the box on page 687. Ex. 3, p. 688;
Ex. 4, p. 689

e. Standard forms of the equations of a hyperbola centered at (h, k) are $\dfrac{(x - h)^2}{a^2} - \dfrac{(y - k)^2}{b^2} = 1$ and Ex. 5, p. 690;
Ex. 6, p. 692
$\dfrac{(y - k)^2}{a^2} - \dfrac{(x - h)^2}{b^2} = 1$. See Table 7.2 on page 690.

7.3 The Parabola

a. A parabola is the set of all points in a plane that are equidistant from a fixed line, the directrix, and a fixed point, the focus.

b. Standard forms of the equations of parabolas with vertex at the origin are $y^2 = 4px$ [focus: $(p, 0)$] and Ex. 1, p. 701;
$x^2 = 4py$ [focus: $(0, p)$]. See the box on page 700 and Figure 7.31 on page 701. Ex. 3, p. 703

c. A parabola's latus rectum is a line segment that passes through its focus, is parallel to its directrix, and has Ex. 2, p. 702
its endpoints on the parabola. The length of the latus rectum for $y^2 = 4px$ and $x^2 = 4py$ is $|4p|$. A parabola
can be graphed using the vertex and endpoints of the latus rectum.

d. Standard forms of the equations of a parabola with vertex at (h, k) are $(y - k)^2 = 4p(x - h)$ and Ex. 4, p. 704;
$(x - h)^2 = 4p(y - k)$. See Table 7.3 on page 704 and Figure 7.36. Ex. 5, p. 705

REVIEW EXERCISES

7.1

In Exercises 1–8, graph each ellipse and locate the foci.

1. $\dfrac{x^2}{36} + \dfrac{y^2}{25} = 1$

2. $\dfrac{y^2}{25} + \dfrac{x^2}{16} = 1$

3. $4x^2 + y^2 = 16$

4. $4x^2 + 9y^2 = 36$

5. $\dfrac{(x - 1)^2}{16} + \dfrac{(y + 2)^2}{9} = 1$

6. $\dfrac{(x + 1)^2}{9} + \dfrac{(y - 2)^2}{16} = 1$

7. $4x^2 + 9y^2 + 24x - 36y + 36 = 0$

8. $9x^2 + 4y^2 - 18x + 8y - 23 = 0$

In Exercises 9–11, find the standard form of the equation of each ellipse satisfying the given conditions.

9. Foci: $(-4, 0), (4, 0)$; Vertices: $(-5, 0), (5, 0)$

10. Foci: $(0, -3), (0, 3)$; Vertices: $(0, -6), (0, 6)$

11. Major axis horizontal with length 12; length of minor axis = 4; center: $(-3, 5)$

12. A semielliptical arch supports a bridge that spans a river 20 yards wide. The center of the arch is 6 yards above the river's center. Write an equation for the ellipse so that the x-axis coincides with the water level and the y-axis passes through the center of the arch.

13. A semielliptic archway has a height of 15 feet at the center and a width of 50 feet, as shown in the figure. The 50-foot width consists of a two-lane road. Can a truck that is 12 feet high and 14 feet wide drive under the archway without going into the other lane?

14. An elliptical pool table has a ball placed at each focus. If one ball is hit toward the edge of the table, explain what will occur.

7.2

In Exercises 15–22, graph each hyperbola. Locate the foci and find the equations of the asymptotes.

15. $\dfrac{x^2}{16} - y^2 = 1$ **16.** $\dfrac{y^2}{16} - x^2 = 1$

17. $9x^2 - 16y^2 = 144$ **18.** $4y^2 - x^2 = 16$

19. $\dfrac{(x-2)^2}{25} - \dfrac{(y+3)^2}{16} = 1$

20. $\dfrac{(y+2)^2}{25} - \dfrac{(x-3)^2}{16} = 1$

21. $y^2 - 4y - 4x^2 + 8x - 4 = 0$

22. $x^2 - y^2 - 2x - 2y - 1 = 0$

In Exercises 23–24, find the standard form of the equation of each hyperbola satisfying the given conditions.

23. Foci: $(0, -4), (0, 4)$; Vertices: $(0, -2), (0, 2)$
24. Foci: $(-8, 0), (8, 0)$; Vertices: $(-3, 0), (3, 0)$
25. Explain why it is not possible for a hyperbola to have foci at $(0, -2)$ and $(0, 2)$ and vertices at $(0, -3)$ and $(0, 3)$.
26. Radio tower M_2 is located 200 miles due west of radio tower M_1. The situation is illustrated in the figure shown, where a coordinate system has been superimposed. Simultaneous radio signals are sent from each tower to a ship, with the signal from M_2 received 500 microseconds before the signal from M_1. Assuming that radio signals travel at 0.186 mile per microsecond, determine the equation of the hyperbola on which the ship is located.

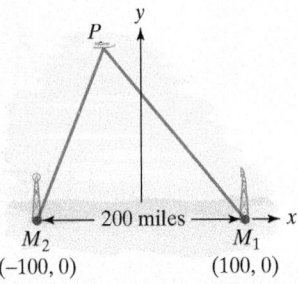

7.3

In Exercises 27–33, find the vertex, focus, and directrix of each parabola with the given equation. Then graph the parabola.

27. $y^2 = 8x$ **28.** $x^2 + 16y = 0$
29. $(y-2)^2 = -16x$ **30.** $(x-4)^2 = 4(y+1)$
31. $x^2 + 4y = 4$ **32.** $y^2 - 4x - 10y + 21 = 0$
33. $x^2 - 4x - 2y = 0$

In Exercises 34–35, find the standard form of the equation of each parabola satisfying the given conditions.

34. Focus: $(12, 0)$; Directrix: $x = -12$
35. Focus: $(0, -11)$; Directrix: $y = 11$

36. An engineer is designing headlight units for automobiles. The unit has a parabolic surface with a diameter of 12 inches and a depth of 3 inches. The situation is illustrated in the figure, where a coordinate system has been superimposed. What is the equation of the parabola in this system? Where should the light source be placed? Describe this placement relative to the vertex.

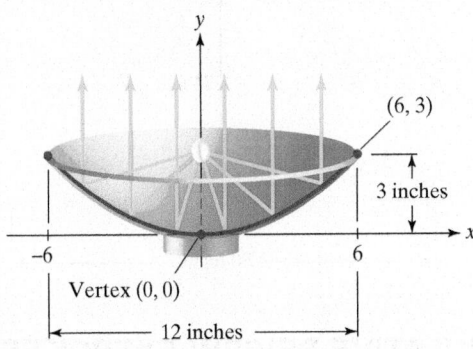

37. The George Washington Bridge spans the Hudson River from New York to New Jersey. Its two towers are 3500 feet apart and rise 316 feet above the road. As shown in the figure, the cable between the towers has the shape of a parabola and the cable just touches the sides of the road midway between the towers. What is the height of the cable 1000 feet from a tower?

38. The giant satellite dish in the figure shown is in the shape of a parabolic surface. Signals strike the surface and are reflected to the focus, where the receiver is located. The diameter of the dish is 300 feet and its depth is 44 feet. How far, to the nearest foot, from the base of the dish should the receiver be placed?

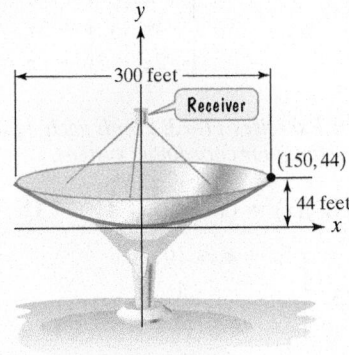

CHAPTER 7 TEST

In Exercises 1–5, graph the conic section with the given equation. For ellipses, find the foci. For hyperbolas, find the foci and give the equations of the asymptotes. For parabolas, find the vertex, focus, and directrix.

1. $9x^2 - 4y^2 = 36$ **2.** $x^2 = -8y$

3. $\dfrac{(x+2)^2}{25} + \dfrac{(y-5)^2}{9} = 1$

4. $4x^2 - y^2 + 8x + 2y + 7 = 0$

5. $(x+5)^2 = 8(y-1)$

In Exercises 6–8, find the standard form of the equation of the conic section satisfying the given conditions.

6. Ellipse; Foci: $(-7, 0), (7, 0)$; Vertices: $(-10, 0), (10, 0)$
7. Hyperbola; Foci: $(0, -10), (0, 10)$; Vertices: $(0, -7), (0, 7)$
8. Parabola; Focus: $(50, 0)$; Directrix: $x = -50$

9. A sound whispered at one focus of a whispering gallery can be heard at the other focus. The figure shows a whispering gallery whose cross section is a semielliptical arch with a height of 24 feet and a width of 80 feet. How far from the room's center should two people stand so that they can whisper back and forth and be heard?

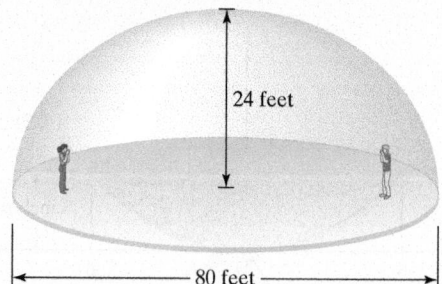

24 feet

80 feet

10. An engineer is designing headlight units for cars. The unit shown in the figure has a parabolic surface with a diameter of 6 inches and a depth of 3 inches.

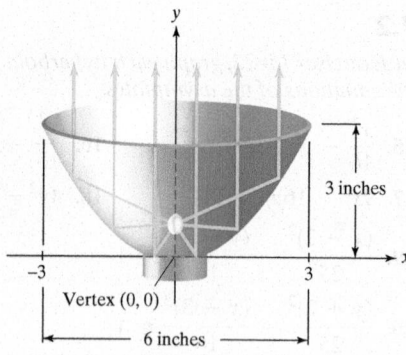

Vertex (0, 0)

3 inches

6 inches

a. Using the coordinate system that has been positioned on the unit, find the parabola's equation.

b. If the light source is located at the focus, describe its placement relative to the vertex.

CUMULATIVE REVIEW EXERCISES (CHAPTERS 1–7)

Solve each equation or inequality in Exercises 1–7.

1. $2(x - 3) + 5x = 8(x - 1)$

2. $-3(2x - 4) > 2(6x - 12)$

3. $x - 5 = \sqrt{x + 7}$ **4.** $(x - 2)^2 = 20$

5. $|2x - 1| \geq 7$ **6.** $3x^3 + 4x^2 - 7x + 2 = 0$

7. $\log_2(x + 1) + \log_2(x - 1) = 3$

Solve each system in Exercises 8–10.

8. $\begin{cases} 3x + 4y = 2 \\ 2x + 5y = -1 \end{cases}$ **9.** $\begin{cases} 2x^2 - y^2 = -8 \\ x - y = 6 \end{cases}$

10. (Use matrices.)

$$\begin{cases} x - y + z = 17 \\ -4x + y + 5z = -2 \\ 2x + 3y + z = 8 \end{cases}$$

In Exercises 11–13, graph each equation, function, or system in a rectangular coordinate system.

11. $f(x) = (x - 1)^2 - 4$ **12.** $\dfrac{x^2}{9} + \dfrac{y^2}{4} = 1$

13. $\begin{cases} 5x + y \leq 10 \\ y \geq \dfrac{1}{4}x + 2 \end{cases}$

14. a. List all possible rational roots of

$$32x^3 - 52x^2 + 17x + 3 = 0.$$

b. The graph of $f(x) = 32x^3 - 52x^2 + 17x + 3$ is shown in a $[-1, 3, 1]$ by $[-2, 6, 1]$ viewing rectangle. Use the graph of f and synthetic division to solve the equation in part (a).

15. The figure shows the graph of $y = f(x)$ and its two vertical asymptotes.

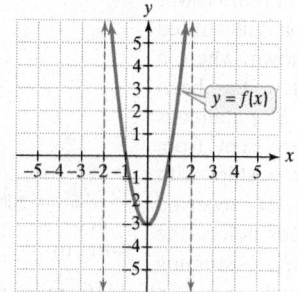

$y = f(x)$

a. Find the domain and the range of f.

b. What is the relative minimum and where does it occur?

c. Find the interval on which f is increasing.

d. Find $f(-1) - f(0)$.

e. Find $(f \circ f)(1)$.

f. Use arrow notation to complete this statement:
$$f(x) \to \infty \text{ as } \underline{\hspace{1cm}} \text{ or as } \underline{\hspace{1cm}}.$$

g. Graph $g(x) = f(x - 2) + 1$.

h. Graph $h(x) = -f(2x)$.

16. If $f(x) = x^2 - 4$ and $g(x) = x + 2$, find $(g \circ f)(x)$.

17. Expand using logarithmic properties. Where possible, evaluate logarithmic expressions.

$$\log_5\left(\frac{x^3\sqrt{y}}{125}\right)$$

18. Write the slope-intercept form of the equation of the line passing through $(1, -4)$ and $(-5, 8)$.

19. Rent-a-Truck charges a daily rental rate for a truck of $39 plus $0.16 a mile. A competing agency, Ace Truck Rentals, charges $25 a day plus $0.24 a mile for the same truck. How many miles must be driven in a day to make the daily cost of both agencies the same? What will be the cost?

20. The local cable television company offers two deals. Basic cable service with one movie channel costs $35 per month. Basic service with two movie channels costs $45 per month. Find the charge for the basic cable service and the charge for each movie channel.

SEQUENCES, INDUCTION, AND PROBABILITY

Something incredible has happened. Your college roommate, a gifted athlete, has been given a six-year contract with a professional baseball team. He will be playing against the likes of Albert Pujols and Justin Verlander. Management offers him three options. One is a beginning salary of $1,700,000 with annual increases of $70,000 per year starting in the second year. A second option is $1,700,000 the first year with an annual increase of 2% per year beginning in the second year. The third option involves less money the first year—$1,500,000—but there is an annual increase of 9% yearly after that. Which option offers the most money over the six-year contract?

HERE'S WHERE YOU'LL FIND THESE APPLICATIONS:

A similar problem appears as Exercise 67 in Exercise Set 8.3, and this problem appears as the Group Exercise on page 753.

SECTION 8.1

Sequences and Summation Notation

Objectives

1. Find particular terms of a sequence from the general term.
2. Use recursion formulas.
3. Use factorial notation.
4. Use summation notation.

Blitzer Bonus

Fibonacci Numbers on the Piano Keyboard

One Octave

Numbers in the Fibonacci sequence can be found in an octave on the piano keyboard. The octave contains 2 black keys in one cluster and 3 black keys in another cluster, for a total of 5 black keys. It also has 8 white keys, for a total of 13 keys. The numbers 2, 3, 5, 8, and 13 are the third through seventh terms of the Fibonacci sequence.

Sequences

Many creations in nature involve intricate mathematical designs, including a variety of spirals. For example, the arrangement of the individual florets in the head of a sunflower forms spirals. In some species, there are 21 spirals in the clockwise direction and 34 in the counterclockwise direction. The precise numbers depend on the species of sunflower: 21 and 34, or 34 and 55, or 55 and 89, or even 89 and 144.

This observation becomes even more interesting when we consider a sequence of numbers investigated by Leonardo of Pisa, also known as Fibonacci, an Italian mathematician of the thirteenth century. The **Fibonacci sequence** of numbers is an infinite sequence that begins as follows:

$$1, 1, 2, 3, 5, 8, 13, 21, 34, 55, 89, 144, 233, \ldots.$$

The first two terms are 1. Every term thereafter is the sum of the two preceding terms. For example, the third term, 2, is the sum of the first and second terms: $1 + 1 = 2$. The fourth term, 3, is the sum of the second and third terms: $1 + 2 = 3$, and so on. Did you know that the number of spirals in a daisy or a sunflower, 21 and 34, are two Fibonacci numbers? The number of spirals in a pine cone, 8 and 13, and a pineapple, 8 and 13, are also Fibonacci numbers.

We can think of the Fibonacci sequence as a function. The terms of the sequence

$$1, 1, 2, 3, 5, 8, 13, 21, 34, 55, 89, 144, 233, \ldots$$

are the range values for a function f whose domain is the set of positive integers.

Domain:	1,	2,	3,	4,	5,	6,	7,	...
	↓	↓	↓	↓	↓	↓	↓	
Range:	1,	1,	2,	3,	5,	8,	13,	...

Thus, $f(1) = 1, f(2) = 1, f(3) = 2, f(4) = 3, f(5) = 5, f(6) = 8, f(7) = 13$, and so on.

The letter a with a subscript is used to represent function values of a sequence, rather than the usual function notation. The subscripts make up the domain of the sequence and they identify the location of a term. Thus, a_1 represents the first term of the sequence, a_2 represents the second term, a_3 the third term, and so on. This notation is shown for the first six terms of the Fibonacci sequence:

$$1, \quad 1, \quad 2, \quad 3, \quad 5, \quad 8.$$

$a_1 = 1 \quad a_2 = 1 \quad a_3 = 2 \quad a_4 = 3 \quad a_5 = 5 \quad a_6 = 8$

The notation a_n represents the nth term, or **general term**, of a sequence. The entire sequence is represented by $\{a_n\}$.

Definition of a Sequence

An **infinite sequence** $\{a_n\}$ is a function whose domain is the set of positive integers. The function values, or **terms**, of the sequence are represented by

$$a_1, a_2, a_3, a_4, \ldots, a_n, \ldots.$$

Sequences whose domains consist only of the first n positive integers are called **finite sequences**.

① Find particular terms of a sequence from the general term.

EXAMPLE 1 Writing Terms of a Sequence from the General Term

Write the first four terms of the sequence whose nth term, or general term, is given:

a. $a_n = 3n + 4$ **b.** $a_n = \dfrac{(-1)^n}{3^n - 1}$.

SOLUTION

a. We need to find the first four terms of the sequence whose general term is $a_n = 3n + 4$. To do so, we replace n in the formula with $1, 2, 3$, and 4.

a_1, 1st term $3 \cdot 1 + 4 = 3 + 4 = 7$ a_2, 2nd term $3 \cdot 2 + 4 = 6 + 4 = 10$

a_3, 3rd term $3 \cdot 3 + 4 = 9 + 4 = 13$ a_4, 4th term $3 \cdot 4 + 4 = 12 + 4 = 16$

The first four terms are $7, 10, 13,$ and 16. The sequence defined by $a_n = 3n + 4$ can be written as

$$7, 10, 13, 16, \ldots, 3n + 4, \ldots.$$

b. We need to find the first four terms of the sequence whose general term is $a_n = \dfrac{(-1)^n}{3^n - 1}$. To do so, we replace each occurrence of n in the formula with $1, 2, 3$, and 4.

a_1, 1st term $\dfrac{(-1)^1}{3^1 - 1} = \dfrac{-1}{3 - 1} = -\dfrac{1}{2}$ a_2, 2nd term $\dfrac{(-1)^2}{3^2 - 1} = \dfrac{1}{9 - 1} = \dfrac{1}{8}$

a_3, 3rd term $\dfrac{(-1)^3}{3^3 - 1} = \dfrac{-1}{27 - 1} = -\dfrac{1}{26}$ a_4, 4th term $\dfrac{(-1)^4}{3^4 - 1} = \dfrac{1}{81 - 1} = \dfrac{1}{80}$

The first four terms are $-\frac{1}{2}, \frac{1}{8}, -\frac{1}{26},$ and $\frac{1}{80}$. The sequence defined by $\dfrac{(-1)^n}{3^n - 1}$ can be written as

$$-\frac{1}{2}, \frac{1}{8}, -\frac{1}{26}, \frac{1}{80}, \ldots, \frac{(-1)^n}{3^n - 1}, \ldots$$ ● ● ●

GREAT QUESTION!

What effect does $(-1)^n$ have on the terms of a sequence?

The factor $(-1)^n$ in the general term of a sequence causes the signs of the terms to alternate between positive and negative, depending on whether n is even or odd.

✓ Check Point **1** Write the first four terms of the sequence whose nth term, or general term, is given:

a. $a_n = 2n + 5$ **b.** $a_n = \dfrac{(-1)^n}{2^n + 1}$.

Although sequences are usually named with the letter a, any lowercase letter can be used. For example, the first four terms of the sequence $\{b_n\} = \left\{\left(\frac{1}{2}\right)^n\right\}$ are $b_1 = \frac{1}{2}, b_2 = \frac{1}{4}, b_3 = \frac{1}{8},$ and $b_4 = \frac{1}{16}$.

Because a sequence is a function whose domain is the set of positive integers, the **graph of a sequence** is a set of discrete points. For example, consider the sequence whose general term is $a_n = \frac{1}{n}$. How does the graph of this sequence differ from the

TECHNOLOGY

Graphing utilities can write the terms of a sequence and graph them. For example, to find the first six terms of $\{a_n\} = \left\{\dfrac{1}{n}\right\}$, enter

General term — Stop at a_6.

$\boxed{\text{SEQ}}\,(1\,\boxed{\div}\,x, x, 1, 6, 1).$

Variable used in general term — Start at a_1. — The "step" from a_1 to a_2, a_2 to a_3, etc., is 1.

The first few terms of the sequence are shown in the viewing rectangle. By pressing the right arrow key to scroll right, you can see the remaining terms.

```
seq(1/X,X,1,6,1)
{1 .5 .33333333…
Ans▶Frac
{1 1/2 1/3 1/4 …
```

graph of the function $f(x) = \frac{1}{x}$? The graph of $f(x) = \frac{1}{x}$ is shown in **Figure 8.1(a)** for positive values of x. To obtain the graph of the sequence $\{a_n\} = \left\{\frac{1}{n}\right\}$, remove all the points from the graph of f except those whose x-coordinates are positive integers. Thus, we remove all points except $(1, 1)$, $\left(2, \frac{1}{2}\right)$, $\left(3, \frac{1}{3}\right)$, $\left(4, \frac{1}{4}\right)$, and so on. The remaining points are the graph of the sequence $\{a_n\} = \left\{\frac{1}{n}\right\}$, shown in **Figure 8.1(b)**. Notice that the horizontal axis is labeled n and the vertical axis is labeled a_n.

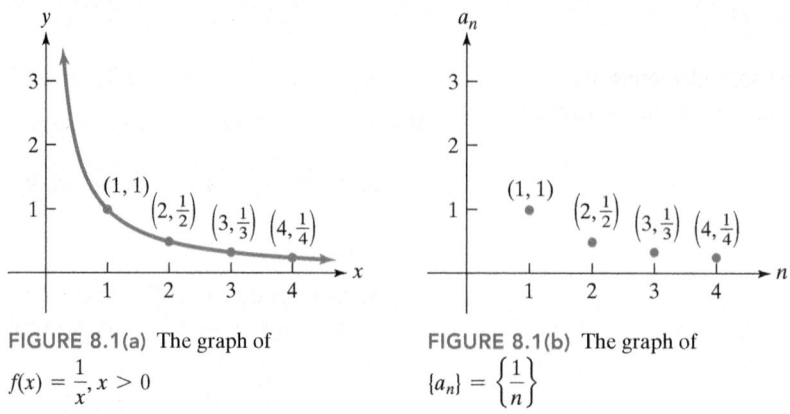

FIGURE 8.1(a) The graph of
$$f(x) = \frac{1}{x}, x > 0$$

FIGURE 8.1(b) The graph of
$$\{a_n\} = \left\{\frac{1}{n}\right\}$$

Comparing a continuous graph to the graph of a sequence

Recursion Formulas

In Example 1, the formulas used for the nth term of a sequence expressed the term as a function of n, the number of the term. Sequences can also be defined using **recursion formulas**. A recursion formula defines the nth term of a sequence as a function of the previous term. Our next example illustrates that if the first term of a sequence is known, then the recursion formula can be used to determine the remaining terms.

② Use recursion formulas.

EXAMPLE 2 Using a Recursion Formula

Find the first four terms of the sequence in which $a_1 = 5$ and $a_n = 3a_{n-1} + 2$ for $n \geq 2$.

SOLUTION

Let's be sure we understand what is given.

$$a_1 = 5 \quad \text{and} \quad a_n = 3a_{n-1} + 2$$

The first term is 5. — Each term after the first — is — 3 times the previous term — plus 2.

Now let's write the first four terms of this sequence.

$a_1 = 5$	This is the given first term.
$a_2 = 3a_1 + 2$	Use $a_n = 3a_{n-1} + 2$, with $n = 2$. Thus, $a_2 = 3a_{2-1} + 2 = 3a_1 + 2$.
$= 3(5) + 2 = 17$	Substitute 5 for a_1.
$a_3 = 3a_2 + 2$	Again use $a_n = 3a_{n-1} + 2$, with $n = 3$.
$= 3(17) + 2 = 53$	Substitute 17 for a_2.
$a_4 = 3a_3 + 2$	Notice that a_4 is defined in terms of a_3. We used $a_n = 3a_{n-1} + 2$, with $n = 4$.
$= 3(53) + 2 = 161$	Use the value of a_3, the third term, obtained above.

The first four terms are $5, 17, 53$, and 161. • • •

✅ **Check Point 2** Find the first four terms of the sequence in which $a_1 = 3$ and $a_n = 2a_{n-1} + 5$ for $n \geq 2$.

③ Use factorial notation.

Factorial Notation

Products of consecutive positive integers occur quite often in sequences. These products can be expressed in a special notation, called **factorial notation**.

Blitzer Bonus

Factorials from 0 through 20

0!	1
1!	1
2!	2
3!	6
4!	24
5!	120
6!	720
7!	5040
8!	40,320
9!	362,880
10!	3,628,800
11!	39,916,800
12!	479,001,600
13!	6,227,020,800
14!	87,178,291,200
15!	1,307,674,368,000
16!	20,922,789,888,000
17!	355,687,428,096,000
18!	6,402,373,705,728,000
19!	121,645,100,408,832,000
20!	2,432,902,008,176,640,000

As n increases, $n!$ grows very rapidly. Factorial growth is more explosive than exponential growth discussed in Chapter 4.

Factorial Notation

If n is a positive integer, the notation $n!$ (read "n factorial") is the product of all positive integers from n down through 1.

$$n! = n(n-1)(n-2)\cdots(3)(2)(1)$$

0! (zero factorial), by definition, is 1.

$$0! = 1$$

The values of $n!$ for the first six positive integers are

$$1! = 1$$
$$2! = 2 \cdot 1 = 2$$
$$3! = 3 \cdot 2 \cdot 1 = 6$$
$$4! = 4 \cdot 3 \cdot 2 \cdot 1 = 24$$
$$5! = 5 \cdot 4 \cdot 3 \cdot 2 \cdot 1 = 120$$
$$6! = 6 \cdot 5 \cdot 4 \cdot 3 \cdot 2 \cdot 1 = 720.$$

Factorials affect only the number or variable that they follow unless grouping symbols appear. For example,

$$2 \cdot 3! = 2(3 \cdot 2 \cdot 1) = 2 \cdot 6 = 12$$

whereas

$$(2 \cdot 3)! = 6! = 6 \cdot 5 \cdot 4 \cdot 3 \cdot 2 \cdot 1 = 720.$$

In this sense, factorials are similar to exponents.

EXAMPLE 3 Finding Terms of a Sequence Involving Factorials

Write the first four terms of the sequence whose nth term is

$$a_n = \frac{2^n}{(n-1)!}.$$

SOLUTION

We need to find the first four terms of the sequence. To do so, we replace each n in $\frac{2^n}{(n-1)!}$ with $1, 2, 3$, and 4.

a_1, 1st term $\quad \dfrac{2^1}{(1-1)!} = \dfrac{2}{0!} = \dfrac{2}{1} = 2$

a_2, 2nd term $\quad \dfrac{2^2}{(2-1)!} = \dfrac{4}{1!} = \dfrac{4}{1} = 4$

a_3, 3rd term $\quad \dfrac{2^3}{(3-1)!} = \dfrac{8}{2!} = \dfrac{8}{2 \cdot 1} = 4$

a_4, 4th term $\quad \dfrac{2^4}{(4-1)!} = \dfrac{16}{3!} = \dfrac{16}{3 \cdot 2 \cdot 1} = \dfrac{16}{6} = \dfrac{8}{3}$

The first four terms are $2, 4, 4$, and $\frac{8}{3}$. • • •

TECHNOLOGY

Most calculators have factorial keys. To find 5!, most calculators use one of the following:

Many Scientific Calculators

$5 \boxed{x!}$

Many Graphing Calculators

$5 \boxed{!} \boxed{\text{ENTER}}$.

Because $n!$ becomes quite large as n increases, your calculator will display these larger values in scientific notation.

⊘ Check Point **3** Write the first four terms of the sequence whose *n*th term is

$$a_n = \frac{20}{(n+1)!}.$$

When evaluating fractions with factorials in the numerator and the denominator, try to reduce the fraction before performing the multiplications. For example, consider $\frac{26!}{21!}$. Rather than write out 26! as the product of all integers from 26 down to 1, we can express 26! as

$$26! = 26 \cdot 25 \cdot 24 \cdot 23 \cdot 22 \cdot 21!.$$

In this way, we can divide both the numerator and the denominator by the common factor, 21!.

$$\frac{26!}{21!} = \frac{26 \cdot 25 \cdot 24 \cdot 23 \cdot 22 \cdot \cancel{21!}}{\cancel{21!}} = 26 \cdot 25 \cdot 24 \cdot 23 \cdot 22 = 7,893,600$$

EXAMPLE 4 Evaluating Fractions with Factorials

Evaluate each factorial expression:

a. $\dfrac{10!}{2!8!}$ **b.** $\dfrac{(n+1)!}{n!}$.

SOLUTION

a. $\dfrac{10!}{2!8!} = \dfrac{10 \cdot 9 \cdot \cancel{8!}}{2 \cdot 1 \cdot \cancel{8!}} = \dfrac{90}{2} = 45$

b. $\dfrac{(n+1)!}{n!} = \dfrac{(n+1) \cdot \cancel{n!}}{\cancel{n!}} = n+1$

• • •

⊘ Check Point **4** Evaluate each factorial expression:

a. $\dfrac{14!}{2!12!}$ **b.** $\dfrac{n!}{(n-1)!}$.

④ Use summation notation.

Summation Notation

It is sometimes useful to find the sum of the first *n* terms of a sequence. For example, consider the cost of raising a child born in the United States in 2006 to a middle-income ($43,200–$72,600 per year) family, shown in **Table 8.1**.

Table 8.1 The Cost of Raising a Child Born in the U.S. in 2006 to a Middle-Income Family

Year	2006	2007	2008	2009	2010	2011	2012	2013	2014
Average Cost	$10,600	$10,930	$11,270	$11,960	$12,330	$12,710	$12,950	$13,350	$13,760
	Child is under 1.	Child is 1.	Child is 2.	Child is 3.	Child is 4.	Child is 5.	Child is 6.	Child is 7.	Child is 8.

Year	2015	2016	2017	2018	2019	2020	2021	2022	2023
Average Cost	$13,970	$14,400	$14,840	$16,360	$16,860	$17,390	$18,430	$19,000	$19,590
	Child is 9.	Child is 10.	Child is 11.	Child is 12.	Child is 13.	Child is 14.	Child is 15.	Child is 16.	Child is 17.

Source: U.S. Department of Agriculture

We can let a_n represent the cost of raising a child in year n, where $n = 1$ corresponds to 2006, $n = 2$ to 2007, $n = 3$ to 2008, and so on. The terms of the finite sequence in **Table 8.1** are given as follows:

10,600, 10,930, 11,270, 11,960, 12,330, 12,710, 12,950, 13,350, 13,760,

a_1 a_2 a_3 a_4 a_5 a_6 a_7 a_8 a_9

13,970, 14,400, 14,840, 16,360, 16,860, 17,390, 18,430, 19,000, 19,590.

a_{10} a_{11} a_{12} a_{13} a_{14} a_{15} a_{16} a_{17} a_{18}

Why might we want to add the terms of this sequence? We do this to find the total cost of raising a child born in 2006 from birth through age 17. Thus,

$$a_1 + a_2 + a_3 + a_4 + a_5 + a_6 + a_7 + a_8 + a_9 + a_{10} + a_{11} + a_{12} + a_{13} + a_{14} + a_{15} + a_{16} + a_{17} + a_{18}$$
$$= 10{,}600 + 10{,}930 + 11{,}270 + 11{,}960 + 12{,}330 + 12{,}710 + 12{,}950 + 13{,}350 + 13{,}760$$
$$+ 13{,}970 + 14{,}400 + 14{,}840 + 16{,}360 + 16{,}860 + 17{,}390 + 18{,}430 + 19{,}000 + 19{,}590$$
$$= 260{,}700.$$

We see that the total cost of raising a child born in 2006 from birth through age 17 is \$260,700.

There is a compact notation for expressing the sum of the first n terms of a sequence. For example, rather than write

$$a_1 + a_2 + a_3 + a_4 + a_5 + a_6 + a_7 + a_8 + a_9 + a_{10} + a_{11} + a_{12} + a_{13} + a_{14} + a_{15} + a_{16} + a_{17} + a_{18},$$

we can use *summation notation* to express the sum as

$$\sum_{i=1}^{18} a_i.$$

We read this expression as "the sum as i goes from 1 to 18 of a_i." The letter i is called the *index of summation* and is not related to the use of i to represent $\sqrt{-1}$.

You can think of the symbol Σ (the uppercase Greek letter sigma) as an instruction to add up the terms of a sequence.

Summation Notation

The sum of the first n terms of a sequence is represented by the **summation notation**

$$\sum_{i=1}^{n} a_i = a_1 + a_2 + a_3 + a_4 + \cdots + a_n,$$

where i is the **index of summation**, n is the **upper limit of summation**, and 1 is the **lower limit of summation**.

Any letter can be used for the index of summation. The letters i, j, and k are used commonly. Furthermore, the lower limit of summation can be an integer other than 1.

When we write out a sum that is given in summation notation, we are **expanding the summation notation**. Example 5 shows how to do this.

EXAMPLE 5 Using Summation Notation

Expand and evaluate the sum:

a. $\displaystyle\sum_{i=1}^{6} (i^2 + 1)$ **b.** $\displaystyle\sum_{k=4}^{7} [(-2)^k - 5]$ **c.** $\displaystyle\sum_{i=1}^{5} 3.$

TECHNOLOGY

Graphing utilities can calculate the sum of a sequence. For example, to find the sum of the sequence in Example 5(a), enter

SUM SEQ $(x^2 + 1, x, 1, 6, 1)$.

Then press ENTER ; 97 should be displayed. Use this capability to verify Example 5(b).

SOLUTION

a. To find $\sum_{i=1}^{6}(i^2 + 1)$, we must replace i in the expression $i^2 + 1$ with all consecutive integers from 1 to 6, inclusive. Then we add.

$$\sum_{i=1}^{6}(i^2 + 1) = (1^2 + 1) + (2^2 + 1) + (3^2 + 1) + (4^2 + 1)$$
$$+ (5^2 + 1) + (6^2 + 1)$$
$$= 2 + 5 + 10 + 17 + 26 + 37$$
$$= 97$$

b. The index of summation in $\sum_{k=4}^{7}[(-2)^k - 5]$ is k. First we evaluate $(-2)^k - 5$ for all consecutive integers from 4 through 7, inclusive. Then we add.

$$\sum_{k=4}^{7}[(-2)^k - 5] = [(-2)^4 - 5] + [(-2)^5 - 5]$$
$$+ [(-2)^6 - 5] + [(-2)^7 - 5]$$
$$= (16 - 5) + (-32 - 5) + (64 - 5) + (-128 - 5)$$
$$= 11 + (-37) + 59 + (-133)$$
$$= -100$$

c. To find $\sum_{i=1}^{5} 3$, we observe that every term of the sum is 3. The notation $i = 1$ through 5 indicates that we must add the first five terms of a sequence in which every term is 3.

$$\sum_{i=1}^{5} 3 = 3 + 3 + 3 + 3 + 3 = 15$$ • • •

✓ Check Point **5** Expand and evaluate the sum:

a. $\sum_{i=1}^{6} 2i^2$ **b.** $\sum_{k=3}^{5} (2^k - 3)$ **c.** $\sum_{i=1}^{5} 4.$

Although the domain of a sequence is the set of positive integers, any integers can be used for the limits of summation. For a given sum, we can vary the upper and lower limits of summation, as well as the letter used for the index of summation. By doing so, we can produce different-looking summation notations for the same sum. For example, the sum of the squares of the first four positive integers, $1^2 + 2^2 + 3^2 + 4^2$, can be expressed in a number of equivalent ways:

$$\sum_{i=1}^{4} i^2 = 1^2 + 2^2 + 3^2 + 4^2 = 30$$

$$\sum_{i=0}^{3} (i + 1)^2 = (0 + 1)^2 + (1 + 1)^2 + (2 + 1)^2 + (3 + 1)^2$$
$$= 1^2 + 2^2 + 3^2 + 4^2 = 30$$

$$\sum_{k=2}^{5} (k - 1)^2 = (2 - 1)^2 + (3 - 1)^2 + (4 - 1)^2 + (5 - 1)^2$$
$$= 1^2 + 2^2 + 3^2 + 4^2 = 30.$$

EXAMPLE 6 Writing Sums in Summation Notation

Express each sum using summation notation:

a. $1^3 + 2^3 + 3^3 + \cdots + 7^3$ **b.** $1 + \dfrac{1}{3} + \dfrac{1}{9} + \dfrac{1}{27} + \cdots + \dfrac{1}{3^{n-1}}.$

SOLUTION

In each case, we will use 1 as the lower limit of summation and i for the index of summation.

a. The sum $1^3 + 2^3 + 3^3 + \cdots + 7^3$ has seven terms, each of the form i^3, starting at $i = 1$ and ending at $i = 7$. Thus,

$$1^3 + 2^3 + 3^3 + \cdots + 7^3 = \sum_{i=1}^{7} i^3.$$

b. The sum

$$1 + \frac{1}{3} + \frac{1}{9} + \frac{1}{27} + \cdots + \frac{1}{3^{n-1}}$$

has n terms, each of the form $\dfrac{1}{3^{i-1}}$, starting at $i = 1$ and ending at $i = n$. Thus,

$$1 + \frac{1}{3} + \frac{1}{9} + \frac{1}{27} + \cdots + \frac{1}{3^{n-1}} = \sum_{i=1}^{n} \frac{1}{3^{i-1}}.$$ $\bullet\bullet\bullet$

✓ **Check Point 6** Express each sum using summation notation:

a. $1^2 + 2^2 + 3^2 + \cdots + 9^2$ **b.** $1 + \dfrac{1}{2} + \dfrac{1}{4} + \dfrac{1}{8} + \cdots + \dfrac{1}{2^{n-1}}.$

Table 8.2 contains some important properties of sums expressed in summation notation.

Table 8.2 Properties of Sums

Property	Example
1. $\displaystyle\sum_{i=1}^{n} ca_i = c\sum_{i=1}^{n} a_i,\, c$ any real number	$\displaystyle\sum_{i=1}^{4} 3i^2 = 3\cdot 1^2 + 3\cdot 2^2 + 3\cdot 3^2 + 3\cdot 4^2$ $3\displaystyle\sum_{i=1}^{4} i^2 = 3(1^2 + 2^2 + 3^2 + 4^2) = 3\cdot 1^2 + 3\cdot 2^2 + 3\cdot 3^2 + 3\cdot 4^2$ Conclusion: $\displaystyle\sum_{i=1}^{4} 3i^2 = 3\sum_{i=1}^{4} i^2$
2. $\displaystyle\sum_{i=1}^{n} (a_i + b_i) = \sum_{i=1}^{n} a_i + \sum_{i=1}^{n} b_i$	$\displaystyle\sum_{i=1}^{4} (i + i^2) = (1 + 1^2) + (2 + 2^2) + (3 + 3^2) + (4 + 4^2)$ $\displaystyle\sum_{i=1}^{4} i + \sum_{i=1}^{4} i^2 = (1 + 2 + 3 + 4) + (1^2 + 2^2 + 3^2 + 4^2)$ $= (1 + 1^2) + (2 + 2^2) + (3 + 3^2) + (4 + 4^2)$ Conclusion: $\displaystyle\sum_{i=1}^{4} (i + i^2) = \sum_{i=1}^{4} i + \sum_{i=1}^{4} i^2$
3. $\displaystyle\sum_{i=1}^{n} (a_i - b_i) = \sum_{i=1}^{n} a_i - \sum_{i=1}^{n} b_i$	$\displaystyle\sum_{i=3}^{5} (i^2 - i^3) = (3^2 - 3^3) + (4^2 - 4^3) + (5^2 - 5^3)$ $\displaystyle\sum_{i=3}^{5} i^2 - \sum_{i=3}^{5} i^3 = (3^2 + 4^2 + 5^2) - (3^3 + 4^3 + 5^3)$ $= (3^2 - 3^3) + (4^2 - 4^3) + (5^2 - 5^3)$ Conclusion: $\displaystyle\sum_{i=3}^{5} (i^2 - i^3) = \sum_{i=3}^{5} i^2 - \sum_{i=3}^{5} i^3$

CONCEPT AND VOCABULARY CHECK

Fill in each blank so that the resulting statement is true.

1. $\{a_n\} = a_1, a_2, a_3, a_4, \ldots, a_n, \ldots$ represents an infinite _____, a function whose domain is the set of positive _____. The function values a_1, a_2, a_3, \ldots are called the _____.

2. The nth term of a sequence, represented by a_n, is called the _____ term.

Write the second term of each sequence.

3. $a_n = 5n - 6$ _____

4. $a_n = \dfrac{(-1)^n}{4^n - 1}$ _____

5. $a_n = 2a_{n-1} - 4, a_1 = 3$ _____

6. $5!$, called 5 _____, is the product of all positive integers from _____ down through _____. By definition, $0! = $ _____.

7. $\dfrac{(n + 3)!}{(n + 2)!} = $ _____

8. $\displaystyle\sum_{i=1}^{n} a_i = $ ____ + ____ + ____ + \cdots + ____. In this summation notation, i is called the _____ of summation, n is the _____ of summation, and 1 is the _____ of summation.

EXERCISE SET 8.1

Practice Exercises

In Exercises 1–12, write the first four terms of each sequence whose general term is given.

1. $a_n = 3n + 2$

2. $a_n = 4n - 1$

3. $a_n = 3^n$

4. $a_n = \left(\dfrac{1}{3}\right)^n$

5. $a_n = (-3)^n$

6. $a_n = \left(-\dfrac{1}{3}\right)^n$

7. $a_n = (-1)^n(n + 3)$

8. $a_n = (-1)^{n+1}(n + 4)$

9. $a_n = \dfrac{2n}{n + 4}$

10. $a_n = \dfrac{3n}{n + 5}$

11. $a_n = \dfrac{(-1)^{n+1}}{2^n - 1}$

12. $a_n = \dfrac{(-1)^{n+1}}{2^n + 1}$

The sequences in Exercises 13–18 are defined using recursion formulas. Write the first four terms of each sequence.

13. $a_1 = 7$ and $a_n = a_{n-1} + 5$ for $n \geq 2$

14. $a_1 = 12$ and $a_n = a_{n-1} + 4$ for $n \geq 2$

15. $a_1 = 3$ and $a_n = 4a_{n-1}$ for $n \geq 2$

16. $a_1 = 2$ and $a_n = 5a_{n-1}$ for $n \geq 2$

17. $a_1 = 4$ and $a_n = 2a_{n-1} + 3$ for $n \geq 2$

18. $a_1 = 5$ and $a_n = 3a_{n-1} - 1$ for $n \geq 2$

In Exercises 19–22, the general term of a sequence is given and involves a factorial. Write the first four terms of each sequence.

19. $a_n = \dfrac{n^2}{n!}$

20. $a_n = \dfrac{(n + 1)!}{n^2}$

21. $a_n = 2(n + 1)!$

22. $a_n = -2(n - 1)!$

In Exercises 23–28, evaluate each factorial expression.

23. $\dfrac{17!}{15!}$

24. $\dfrac{18!}{16!}$

25. $\dfrac{16!}{2!14!}$

26. $\dfrac{20!}{2!18!}$

27. $\dfrac{(n + 2)!}{n!}$

28. $\dfrac{(2n + 1)!}{(2n)!}$

In Exercises 29–42, find each indicated sum.

29. $\displaystyle\sum_{i=1}^{6} 5i$

30. $\displaystyle\sum_{i=1}^{6} 7i$

31. $\displaystyle\sum_{i=1}^{4} 2i^2$

32. $\displaystyle\sum_{i=1}^{5} i^3$

33. $\displaystyle\sum_{k=1}^{5} k(k + 4)$

34. $\displaystyle\sum_{k=1}^{4} (k - 3)(k + 2)$

35. $\displaystyle\sum_{i=1}^{4} \left(-\dfrac{1}{2}\right)^i$

36. $\displaystyle\sum_{i=2}^{4} \left(-\dfrac{1}{3}\right)^i$

37. $\displaystyle\sum_{i=5}^{9} 11$

38. $\displaystyle\sum_{i=3}^{7} 12$

39. $\displaystyle\sum_{i=0}^{4} \dfrac{(-1)^i}{i!}$

40. $\displaystyle\sum_{i=0}^{4} \dfrac{(-1)^{i+1}}{(i + 1)!}$

41. $\displaystyle\sum_{i=1}^{5} \dfrac{i!}{(i - 1)!}$

42. $\displaystyle\sum_{i=1}^{5} \dfrac{(i + 2)!}{i!}$

In Exercises 43–54, express each sum using summation notation. Use 1 as the lower limit of summation and i for the index of summation.

43. $1^2 + 2^2 + 3^2 + \cdots + 15^2$

44. $1^4 + 2^4 + 3^4 + \cdots + 12^4$

45. $2 + 2^2 + 2^3 + \cdots + 2^{11}$

46. $5 + 5^2 + 5^3 + \cdots + 5^{12}$

47. $1 + 2 + 3 + \cdots + 30$

48. $1 + 2 + 3 + \cdots + 40$

49. $\dfrac{1}{2} + \dfrac{2}{3} + \dfrac{3}{4} + \cdots + \dfrac{14}{14 + 1}$

50. $\dfrac{1}{3} + \dfrac{2}{4} + \dfrac{3}{5} + \cdots + \dfrac{16}{16 + 2}$

51. $4 + \dfrac{4^2}{2} + \dfrac{4^3}{3} + \cdots + \dfrac{4^n}{n}$

52. $\dfrac{1}{9} + \dfrac{2}{9^2} + \dfrac{3}{9^3} + \cdots + \dfrac{n}{9^n}$

53. $1 + 3 + 5 + \cdots + (2n - 1)$

54. $a + ar + ar^2 + \cdots + ar^{n-1}$

In Exercises 55–60, express each sum using summation notation. Use a lower limit of summation of your choice and k for the index of summation.

55. $5 + 7 + 9 + 11 + \cdots + 31$

56. $6 + 8 + 10 + 12 + \cdots + 32$

57. $a + ar + ar^2 + \cdots + ar^{12}$

58. $a + ar + ar^2 + \cdots + ar^{14}$

59. $a + (a + d) + (a + 2d) + \cdots + (a + nd)$

60. $(a + d) + (a + d^2) + \cdots + (a + d^n)$

Practice Plus

In Exercises 61–68, use the graphs of $\{a_n\}$ and $\{b_n\}$ to find each indicated sum.

The Graph of $\{a_n\}$ The Graph of $\{b_n\}$

 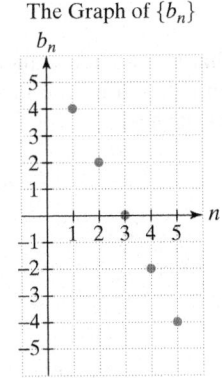

61. $\displaystyle\sum_{i=1}^{5} (a_i^2 + 1)$ **62.** $\displaystyle\sum_{i=1}^{5} (b_i^2 - 1)$

63. $\displaystyle\sum_{i=1}^{5} (2a_i + b_i)$ **64.** $\displaystyle\sum_{i=1}^{5} (a_i + 3b_i)$

65. $\displaystyle\sum_{i=4}^{5} \left(\dfrac{a_i}{b_i}\right)^2$ **66.** $\displaystyle\sum_{i=4}^{5} \left(\dfrac{a_i}{b_i}\right)^3$

67. $\displaystyle\sum_{i=1}^{5} a_i^2 + \sum_{i=1}^{5} b_i^2$ **68.** $\displaystyle\sum_{i=1}^{5} a_i^2 - \sum_{i=3}^{5} b_i^2$

Application Exercises

69. One in one hundred is the ratio of individuals in the U.S. population with autistic disorder, better known as autism. The disorder involves severe deficits in language, social bonding, and imagination, and is often accompanied by mental retardation. The bar graph at the top of the next column shows the number of autism cases diagnosed in the United States from 2001 through 2008.

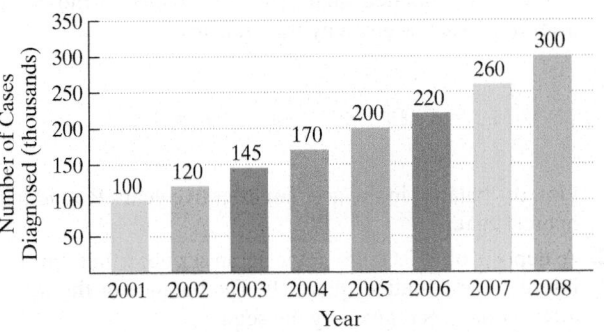

Autism Cases Diagnosed in the United States

Source: Lilienfeld et al., *Psychology,* Second Edition, Pearson, 2011.

Let a_n represent the number of autism cases diagnosed in the United States, in thousands, n years after 2000.

a. Use the numbers given in the graph to find and interpret

$$\sum_{i=1}^{8} a_i.$$

b. The finite sequence whose general term is

$$a_n = 28n + 63,$$

where $n = 1, 2, 3, \ldots, 8$, models the number of autism cases diagnosed in the United States, in thousands, n years after 2000. Use this model to find $\displaystyle\sum_{i=1}^{8} a_i$. Does this underestimate or overestimate the actual sum in part (a)? By how much?

70. The bar graph shows the average state cigarette tax per pack from 2005 through 2010.

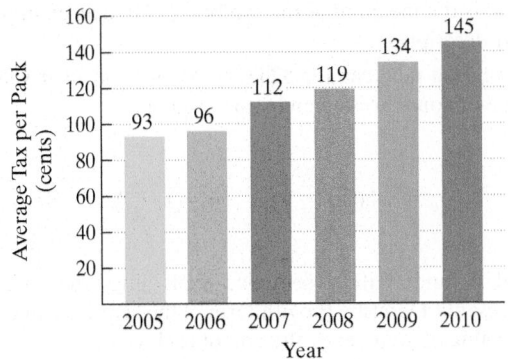

Average State Cigarette Tax per Pack

Source: "Trends in Average State Cigarette Tax Rates." Campaign for Tobacco-Free Kids

Let a_n represent the average state cigarette tax per pack, in cents, n years after 2004.

a. Use the numbers given in the graph to find and interpret

$$\dfrac{1}{6}\sum_{i=1}^{6} a_i.$$

b. The finite sequence whose general term is

$$a_n = 11n + 78,$$

where $n = 1, 2, 3, 4, 5, 6$, models the average state cigarette tax per pack, in cents, n years after 2004. Use this model to find $\dfrac{1}{6}\displaystyle\sum_{i=1}^{6} a_i$. How does this compare with the actual value you obtained in part (a)?

71. A deposit of $6000 is made in an account that earns 6% interest compounded quarterly. The balance in the account after n quarters is given by the sequence

$$a_n = 6000\left(1 + \frac{0.06}{4}\right)^n, \qquad n = 1, 2, 3, \ldots.$$

Find the balance in the account after five years. Round to the nearest cent.

72. A deposit of $10,000 is made in an account that earns 8% interest compounded quarterly. The balance in the account after n quarters is given by the sequence

$$a_n = 10,000\left(1 + \frac{0.08}{4}\right)^n, \qquad n = 1, 2, 3, \ldots.$$

Find the balance in the account after six years. Round to the nearest cent.

Writing in Mathematics

73. What is a sequence? Give an example with your description.

74. Explain how to write terms of a sequence if the formula for the general term is given.

75. What does the graph of a sequence look like? How is it obtained?

76. What is a recursion formula?

77. Explain how to find $n!$ if n is a positive integer.

78. Explain the best way to evaluate $\dfrac{900!}{899!}$ without a calculator.

79. What is the meaning of the symbol Σ? Give an example with your description.

80. You buy a new car for $24,000. At the end of n years, the value of your car is given by the sequence

$$a_n = 24,000\left(\frac{3}{4}\right)^n, \qquad n = 1, 2, 3, \ldots.$$

Find a_5 and write a sentence explaining what this value represents. Describe the nth term of the sequence in terms of the value of your car at the end of each year.

Technology Exercises

In Exercises 81–85, use a calculator's factorial key to evaluate each expression.

81. $\dfrac{200!}{198!}$ **82.** $\left(\dfrac{300}{20}\right)!$ **83.** $\dfrac{20!}{300}$

84. $\dfrac{20!}{(20-3)!}$ **85.** $\dfrac{54!}{(54-3)!3!}$

86. Use the $\boxed{\text{SEQ}}$ (sequence) capability of a graphing utility to verify the terms of the sequences you obtained for any five sequences from Exercises 1–12 or 19–22.

87. Use the $\boxed{\text{SUM}}$ $\boxed{\text{SEQ}}$ (sum of the sequence) capability of a graphing utility to verify any five of the sums you obtained in Exercises 29–42.

88. As n increases, the terms of the sequence

$$a_n = \left(1 + \frac{1}{n}\right)^n$$

get closer and closer to the number e (where $e \approx 2.7183$). Use a calculator to find $a_{10}, a_{100}, a_{1000}, a_{10,000},$ and $a_{100,000},$ comparing these terms to your calculator's decimal approximation for e.

Many graphing utilities have a sequence-graphing mode that plots the terms of a sequence as points on a rectangular coordinate system. Consult your manual; if your graphing utility has this capability, use it to graph each of the sequences in Exercises 89–92. What appears to be happening to the terms of each sequence as n gets larger?

89. $a_n = \dfrac{n}{n+1}$ n: [0, 10, 1] by a_n: [0, 1, 0.1]

90. $a_n = \dfrac{100}{n}$ n: [0, 1000, 100] by a_n: [0, 1, 0.1]

91. $a_n = \dfrac{2n^2 + 5n - 7}{n^3}$ n: [0, 10, 1] by a_n: [0, 2, 0.2]

92. $a_n = \dfrac{3n^4 + n - 1}{5n^4 + 2n^2 + 1}$ n: [0, 10, 1] by a_n: [0, 1, 0.1]

Critical Thinking Exercises

Make Sense? *In Exercises 93–96, determine whether each statement makes sense or does not make sense, and explain your reasoning.*

93. Now that I've studied sequences, I realize that the joke in this cartoon is based on the fact that you can't have a negative number of sheep.

WHEN MATHEMATICIANS CAN'T SLEEP Harry Blair

94. By writing $a_1, a_2, a_3, a_4, \ldots, a_n, \ldots,$ I can see that the range of a sequence is the set of positive integers.

95. It makes a difference whether or not I use parentheses around the expression following the summation symbol, because the value of $\displaystyle\sum_{i=1}^{8} (i + 7)$ is 92, but the value of $\displaystyle\sum_{i=1}^{8} i + 7$ is 43.

96. Without writing out the terms, I can see that $(-1)^{2n}$ in $a_n = \dfrac{(-1)^{2n}}{3n}$ causes the terms to alternate in sign.

In Exercises 97–100, determine whether each statement is true or false. If the statement is false, make the necessary change(s) to produce a true statement.

97. $\dfrac{n!}{(n-1)!} = \dfrac{1}{n-1}$

98. The Fibonacci sequence 1, 1, 2, 3, 5, 8, 13, 21, 34, 55, 89, 144, ... can be defined recursively using $a_0 = 1, a_1 = 1$; $a_n = a_{n-2} + a_{n-1}$, where $n \geq 2$.

99. $\displaystyle\sum_{i=1}^{2} (-1)^i 2^i = 0$

100. $\displaystyle\sum_{i=1}^{2} a_i b_i = \sum_{i=1}^{2} a_i \sum_{i=1}^{2} b_i$

101. Write the first five terms of the sequence whose first term is 9 and whose general term is

$$a_n = \begin{cases} \dfrac{a_{n-1}}{2} & \text{if } a_{n-1} \text{ is even} \\ 3a_{n-1} + 5 & \text{if } a_{n-1} \text{ is odd} \end{cases}$$

for $n \geq 2$.

Group Exercise

102. Enough curiosities involving the Fibonacci sequence exist to warrant a flourishing Fibonacci Association, which publishes a quarterly journal. Do some research on the Fibonacci sequence by consulting the Internet or the research department of your library, and find one property that interests you. After doing this research, get together with your group to share these intriguing properties.

Preview Exercises

Exercises 103–105 will help you prepare for the material covered in the next section.

103. Consider the sequence 8, 3, −2, −7, −12, Find $a_2 - a_1$, $a_3 - a_2, a_4 - a_3$, and $a_5 - a_4$. What do you observe?

104. Consider the sequence whose nth term is $a_n = 4n - 3$. Find $a_2 - a_1, a_3 - a_2, a_4 - a_3$, and $a_5 - a_4$. What do you observe?

105. Use the formula $a_n = 4 + (n-1)(-7)$ to find the eighth term of the sequence 4, −3, −10,

SECTION 8.2 Arithmetic Sequences

Objectives

1 Find the common difference for an arithmetic sequence.

2 Write terms of an arithmetic sequence.

3 Use the formula for the general term of an arithmetic sequence.

4 Use the formula for the sum of the first n terms of an arithmetic sequence.

Your grandmother and her financial counselor are looking at options in case an adult residential facility is needed in the future. The good news is that your grandmother's total assets are $500,000. The bad news is that adult residential community costs average $64,130 annually, increasing by $1800 each year. In this section, we will see how sequences can be used to model your grandmother's situation and help her to identify realistic options.

Arithmetic Sequences

1 Find the common difference for an arithmetic sequence.

The bar graph in **Figure 8.2** shows how much Americans spent on their pets, rounded to the nearest billion dollars, each year from 2001 through 2007.

The graph illustrates that each year spending increased by $2 billion. The sequence of annual spending

$$29, 31, 33, 35, 37, 39, 41, \ldots$$

shows that each term after the first, 29, differs from the

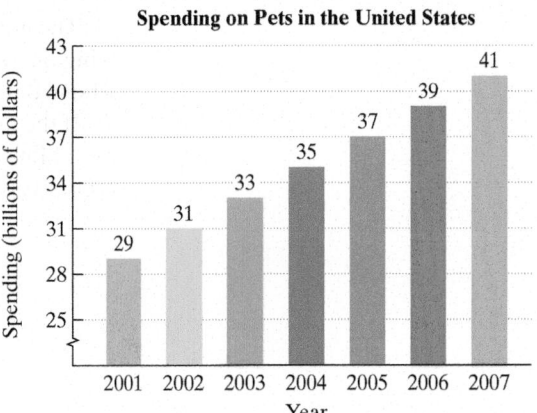

Spending on Pets in the United States

FIGURE 8.2
Source: American Pet Products Manufacturers Association

preceding term by a constant amount, namely, 2. This sequence is an example of an *arithmetic sequence*.

Definition of an Arithmetic Sequence

An **arithmetic sequence** is a sequence in which each term after the first differs from the preceding term by a constant amount. The difference between consecutive terms is called the **common difference** of the sequence.

The common difference, d, is found by subtracting any term from the term that directly follows it. In the following examples, the common difference is found by subtracting the first term from the second term, $a_2 - a_1$.

Arithmetic Sequence	Common Difference
$142, 146, 150, 154, 158, \ldots$	$d = 146 - 142 = 4$
$-5, -2, 1, 4, 7, \ldots$	$d = -2 - (-5) = -2 + 5 = 3$
$8, 3, -2, -7, -12, \ldots$	$d = 3 - 8 = -5$

Figure 8.3 shows the graphs of the last two arithmetic sequences in our list. The common difference for the increasing sequence in **Figure 8.3(a)** is 3. The common difference for the decreasing sequence in **Figure 8.3(b)** is -5.

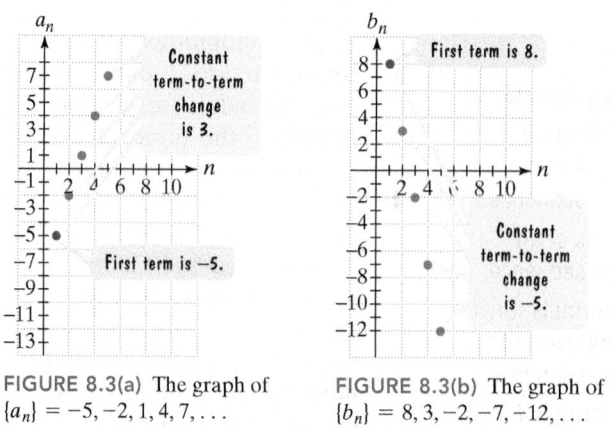

FIGURE 8.3(a) The graph of $\{a_n\} = -5, -2, 1, 4, 7, \ldots$

FIGURE 8.3(b) The graph of $\{b_n\} = 8, 3, -2, -7, -12, \ldots$

The graph of each arithmetic sequence in **Figure 8.3** forms a set of discrete points lying on a straight line. This illustrates that **an arithmetic sequence is a linear function whose domain is the set of positive integers**.

If the first term of an arithmetic sequence is a_1, each term after the first is obtained by adding d, the common difference, to the previous term. This can be expressed recursively as follows:

$$a_n = a_{n-1} + d.$$

Add d to the term in any position to get the next term.

To use this recursion formula, we must be given the first term.

② Write terms of an arithmetic sequence.

EXAMPLE 1 Writing the Terms of an Arithmetic Sequence

Write the first six terms of the arithmetic sequence in which $a_1 = 6$ and $a_n = a_{n-1} - 2$.

SOLUTION

The recursion formula $a_1 = 6$ and $a_n = a_{n-1} - 2$ indicates that each term after the first, 6, is obtained by adding -2 to the previous term.

$$a_1 = 6 \qquad\qquad\qquad\qquad \text{This is given.}$$
$$a_2 = a_1 - 2 = 6 - 2 \quad = 4 \qquad \text{Use } a_n = a_{n-1} - 2 \text{ with } n = 2.$$
$$a_3 = a_2 - 2 = 4 - 2 \quad = 2 \qquad \text{Use } a_n = a_{n-1} - 2 \text{ with } n = 3.$$
$$a_4 = a_3 - 2 = 2 - 2 \quad = 0 \qquad \text{Use } a_n = a_{n-1} - 2 \text{ with } n = 4.$$
$$a_5 = a_4 - 2 = 0 - 2 \quad = -2 \quad \text{Use } a_n = a_{n-1} - 2 \text{ with } n = 5.$$
$$a_6 = a_5 - 2 = -2 - 2 = -4 \quad \text{Use } a_n = a_{n-1} - 2 \text{ with } n = 6.$$

The first six terms are

$$6, 4, 2, 0, -2, \text{ and } -4.$$

● ● ●

⊘ Check Point 1 Write the first six terms of the arithmetic sequence in which $a_1 = 100$ and $a_n = a_{n-1} - 30$.

③ Use the formula for the general term of an arithmetic sequence.

The General Term of an Arithmetic Sequence

Consider an arithmetic sequence whose first term is a_1 and whose common difference is d. We are looking for a formula for the general term, a_n. Let's begin by writing the first six terms. The first term is a_1. The second term is $a_1 + d$. The third term is $a_1 + d + d$, or $a_1 + 2d$. Thus, we start with a_1 and add d to each successive term. The first six terms are

$$a_1, \qquad a_1 + d, \qquad a_1 + 2d, \qquad a_1 + 3d, \qquad a_1 + 4d, \qquad a_1 + 5d.$$

a_1, first term	a_2, second term	a_3, third term	a_4, fourth term	a_5, fifth term	a_6, sixth term

Compare the coefficient of d and the subscript of a denoting the term number. Can you see that the coefficient of d is 1 less than the subscript of a denoting the term number?

$$a_3: \text{third term} = a_1 + 2d \qquad\qquad a_4: \text{fourth term} = a_1 + 3d$$

One less than 3, or 2, is the coefficient of d.

One less than 4, or 3, is the coefficient of d.

Thus, the formula for the nth term is

$$a_n: n\text{th term} = a_1 + (n - 1)d.$$

One less than n, or $n - 1$, is the coefficient of d.

General Term of an Arithmetic Sequence

The nth term (the general term) of an arithmetic sequence with first term a_1 and common difference d is

$$a_n = a_1 + (n - 1)d.$$

EXAMPLE 2 Using the Formula for the General Term of an Arithmetic Sequence

Find the eighth term of the arithmetic sequence whose first term is 4 and whose common difference is −7.

SOLUTION

To find the eighth term, a_8, we replace n in the formula with 8, a_1 with 4, and d with −7.

$$a_n = a_1 + (n - 1)d$$

$$a_8 = 4 + (8 - 1)(-7) = 4 + 7(-7) = 4 + (-49) = -45$$

The eighth term is −45. We can check this result by writing the first eight terms of the sequence:

$$4, -3, -10, -17, -24, -31, -38, -45.$$

• • •

✓ Check Point **2** Find the ninth term of the arithmetic sequence whose first term is 6 and whose common difference is −5.

EXAMPLE 3 Using an Arithmetic Sequence to Model Changes in the U.S. Population

The graph in **Figure 8.4** shows the percentage of the U.S. population by race/ethnicity for 2010, with projections by the U.S. Census Bureau for 2050.

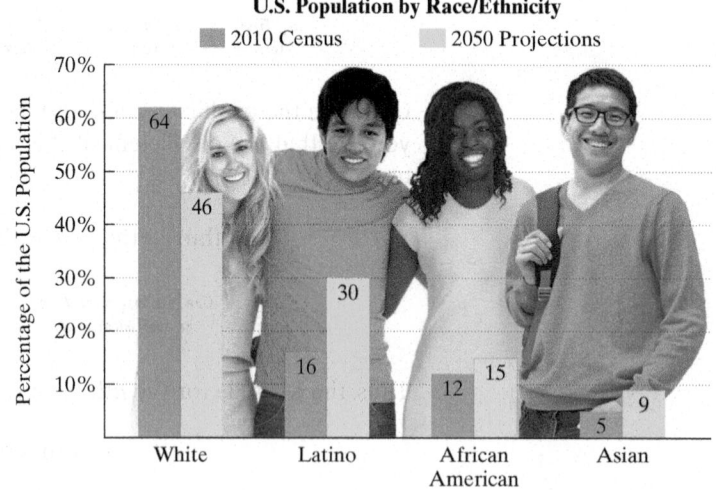

U.S. Population by Race/Ethnicity

FIGURE 8.4
Source: U.S. Census Bureau

The data show that in 2010, 64% of the U.S. population was white. On average, this is projected to decrease by approximately 0.45% per year.

a. Write a formula for the nth term of the arithmetic sequence that describes the percentage of the U.S. population that will be white n years after 2009.

b. What percentage of the U.S. population is projected to be white in 2030?

SOLUTION

a. With a yearly decrease of 0.45%, we can express the percentage of the white population by the following arithmetic sequence:

$$64, \qquad 64 - 0.45 = 63.55, \qquad 63.55 - 0.45 = 63.10, \ldots.$$

a_1: percentage of whites in the population in 2010, 1 year after 2009	a_2: percentage of whites in the population in 2011, 2 years after 2009	a_3: percentage of whites in the population in 2012, 3 years after 2009

In the sequence $64, 63.55, 63.10, \ldots$, the first term, a_1, represents the percentage of the population that was white in 2010. Each subsequent year this amount decreases by 0.45%, so $d = -0.45$. We use the formula for the general term of an arithmetic sequence to write the nth term of the sequence that describes the percentage of whites in the population n years after 2009.

$$a_n = a_1 + (n - 1)d \qquad \text{This is the formula for the general term of an arithmetic sequence.}$$

$$a_n = 64 + (n - 1)(-0.45) \qquad a_1 = 64 \text{ and } d = -0.45.$$
$$a_n = 64 - 0.45n + 0.45 \qquad \text{Distribute } -0.45 \text{ to each term in parentheses.}$$

$$a_n = -0.45n + 64.45 \qquad \text{Simplify.}$$

Thus, the percentage of the U.S. population that will be white n years after 2009 can be described by

$$a_n = -0.45n + 64.45.$$

b. Now we need to project the percentage of the population that will be white in 2030. The year 2030 is 21 years after 2009. Thus, $n = 21$. We substitute 21 for n in $a_n = -0.45n + 64.45$.

$$a_{21} = -0.45(21) + 64.45 = 55$$

The 21st term of the sequence is 55. Thus, 55% of the U.S. population is projected to be white in 2030. ● ● ●

⊘ Check Point **3** The data in **Figure 8.4** show that in 2010, 16% of the U.S. population was Latino. On average, this is projected to increase by approximately 0.35% per year.

a. Write a formula for the nth term of the arithmetic sequence that describes the percentage of the U.S. population that will be Latino n years after 2009.

b. What percentage of the U.S. population is projected to be Latino in 2030?

④ Use the formula for the sum of the first n terms of an arithmetic sequence.

The Sum of the First n Terms of an Arithmetic Sequence

The sum of the first n terms of an arithmetic sequence, denoted by S_n, and called the **nth partial sum**, can be found without having to add up all the terms. Let

$$S_n = a_1 + a_2 + a_3 + \cdots + a_n$$

be the sum of the first n terms of an arithmetic sequence. Because d is the common difference between terms, S_n can be written forward and backward as follows:

Forward: Start with the first term, a_1. Keep adding d.

Backward: Start with the last term, a_n. Keep subtracting d.

$$
\begin{aligned}
S_n &= a_1 && + (a_1 + d) && + (a_1 + 2d) && + \cdots + a_n \\
S_n &= a_n && + (a_n - d) && + (a_n - 2d) && + \cdots + a_1 \\
\hline
2S_n &= (a_1 + a_n) && + (a_1 + a_n) && + (a_1 + a_n) && + \cdots + (a_1 + a_n). \quad \text{Add the two equations.}
\end{aligned}
$$

Consider the last equation

$$2S_n = (a_1 + a_n) + (a_1 + a_n) + (a_1 + a_n) + \cdots + (a_1 + a_n).$$

Because there are n sums of $(a_1 + a_n)$ on the right side, we can express this side as $n(a_1 + a_n)$. Thus, the last equation can be written as follows:

$$2S_n = n(a_1 + a_n)$$

$$S_n = \frac{n}{2}(a_1 + a_n). \quad \text{Solve for } S_n, \text{ dividing both sides by 2.}$$

We have proved the following result:

The Sum of the First n Terms of an Arithmetic Sequence

The sum, S_n, of the first n terms of an arithmetic sequence is given by

$$S_n = \frac{n}{2}(a_1 + a_n),$$

in which a_1 is the first term and a_n is the nth term.

To find the sum of the terms of an arithmetic sequence using $S_n = \frac{n}{2}(a_1 + a_n)$, we need to know the first term, a_1, the last term, a_n, and the number of terms, n. The following examples illustrate how to use this formula.

EXAMPLE 4 Finding the Sum of n Terms of an Arithmetic Sequence

Find the sum of the first 100 terms of the arithmetic sequence: $1, 3, 5, 7, \ldots$.

SOLUTION

By finding the sum of the first 100 terms of $1, 3, 5, 7, \ldots$, we are finding the sum of the first 100 odd numbers. To find the sum of the first 100 terms, S_{100}, we replace n in the formula with 100.

$$S_n = \frac{n}{2}(a_1 + a_n)$$

$$S_{100} = \frac{100}{2}(a_1 + a_{100})$$

The first term, a_1, is 1. We must find a_{100}, the 100th term.

We use the formula for the general term of a sequence to find a_{100}. The common difference, d, of $1, 3, 5, 7, \ldots$, is 2.

$$a_n = a_1 + (n - 1)d \qquad \text{This is the formula for the nth term of an arithmetic sequence. Use it to find the 100th term.}$$

$$a_{100} = 1 + (100 - 1) \cdot 2 \qquad \text{Substitute 100 for n, 2 for d, and 1 (the first term) for a_1.}$$

$$= 1 + 99 \cdot 2 \qquad \text{Perform the subtraction in parentheses.}$$

$$= 1 + 198 = 199 \qquad \text{Multiply (99 · 2 = 198) and then add.}$$

Now we are ready to find the sum of the 100 terms $1, 3, 5, 7, \ldots, 199$.

$$S_n = \frac{n}{2}(a_1 + a_n) \qquad \text{Use the formula for the sum of the first n terms of an arithmetic sequence. Let $n = 100$, $a_1 = 1$, and $a_{100} = 199$.}$$

$$S_{100} = \frac{100}{2}(1 + 199) = 50(200) = 10{,}000$$

The sum of the first 100 odd numbers is 10,000. Equivalently, the 100th partial sum of the sequence $1, 3, 5, 7, \ldots$ is 10,000. ● ● ●

Check Point 4 Find the sum of the first 15 terms of the arithmetic sequence: $3, 6, 9, 12, \ldots$.

EXAMPLE 5 Using S_n to Evaluate a Summation

Find the following sum: $\displaystyle\sum_{i=1}^{25}(5i - 9)$.

SOLUTION

$$\sum_{i=1}^{25}(5i - 9) = (5 \cdot 1 - 9) + (5 \cdot 2 - 9) + (5 \cdot 3 - 9) + \cdots + (5 \cdot 25 - 9)$$

$$= -4 \qquad + 1 \qquad + 6 \qquad + \cdots + 116$$

By evaluating the first three terms and the last term, we see that $a_1 = -4$; d, the common difference, is $1 - (-4)$, or 5; and a_{25}, the last term, is 116.

$$S_n = \frac{n}{2}(a_1 + a_n)$$
Use the formula for the sum of the first n terms of an arithmetic sequence. Let $n = 25$, $a_1 = -4$, and $a_{25} = 116$.

$$S_{25} = \frac{25}{2}(-4 + 116) = \frac{25}{2}(112) = 1400$$

Thus,

$$\sum_{i=1}^{25}(5i - 9) = 1400. \qquad \bullet \bullet \bullet$$

✓ **Check Point 5** Find the following sum: $\displaystyle\sum_{i=1}^{30}(6i - 11)$.

EXAMPLE 6 Modeling Total Residential Community Costs over a Six-Year Period

Your grandmother has assets of $500,000. One option that she is considering involves an adult residential community for a six-year period beginning in 2014. The model

$$a_n = 1800n + 64{,}130$$

describes yearly adult residential community costs n years after 2013. Does your grandmother have enough to pay for the facility?

SOLUTION

We must find the sum of an arithmetic sequence whose general term is $a_n = 1800n + 64{,}130$. The first term of the sequence corresponds to the facility's costs in the year 2014. The last term corresponds to costs in the year 2019. Because the model describes costs n years after 2013, $n = 1$ describes the year 2014 and $n = 6$ describes the year 2019.

$$a_n = 1800n + 64{,}130$$
This is the given formula for the general term of the sequence.

$$a_1 = 1800 \cdot 1 + 64{,}130 = 65{,}930$$
Find a_1 by replacing n with 1.

$$a_6 = 1800 \cdot 6 + 64{,}130 = 74{,}930$$
Find a_6 by replacing n with 6.

The first year the facility will cost $65,930. By year 6, the facility will cost $74,930. Now we must find the sum of the costs for all six years. We focus on the sum of the first six terms of the arithmetic sequence

$$65{,}930, \ 67{,}730, \ \ldots \ , \ 74{,}930.$$

$$\underset{a_1}{\phantom{65{,}930}} \qquad \underset{a_2}{\phantom{67{,}730}} \qquad \qquad \underset{a_6}{\phantom{74{,}930}}$$

TECHNOLOGY

To find

$$\sum_{i=1}^{25}(5i - 9)$$

on a graphing utility, enter

[SUM] [SEQ] $(5x - 9, x, 1, 25, 1)$.

Then press [ENTER].

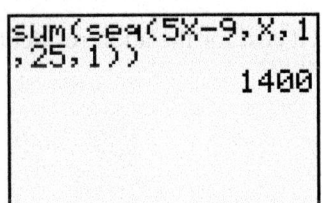

```
sum(seq(5X-9,X,1
,25,1))
              1400
```

We find the sum of the first six terms of

$$65{,}930, 67{,}730, \ldots, 74{,}930$$

using the formula for the sum of the first n terms of an arithmetic sequence. We are adding six terms: $n = 6$. The first term is 65,930: $a_1 = 65{,}930$. The last term—that is, the sixth term—is 74,930: $a_6 = 74{,}930$.

$$S_n = \frac{n}{2}(a_1 + a_n)$$

$$S_6 = \frac{6}{2}(65{,}930 + 74{,}930) = 3(140{,}860) = 422{,}580$$

Total adult residential community costs for your grandmother are predicted to be $422,580. Because your grandmother's assets are $500,000, she has enough to pay for the facility for the six-year period. ●●●

 Check Point **6** In Example 6, how much would it cost for the adult residential community for a ten-year period beginning in 2014?

CONCEPT AND VOCABULARY CHECK

Fill in each blank so that the resulting statement is true.

1. A sequence in which each term after the first differs from the preceding term by a constant amount is called a/an _____ sequence. The difference between consecutive terms is called the _____ of the sequence.

2. The nth term of the sequence described in Exercise 1 is given by the formula $a_n = $ _____, where a_1 is the _____ and d is the _____ of the sequence.

3. The sum, S_n, of the first n terms of the sequence described in Exercise 1 is given by the formula $S_n = $ _____, where a_1 is the _____ and a_n is the _____.

4. The first term of $\sum\limits_{i=1}^{20}(6i - 4)$ is _____ and the last term is _____.

5. The first three terms of $\sum\limits_{i=1}^{17}(5i + 3)$ are _____, _____, and _____. The common difference is _____.

EXERCISE SET 8.2

Practice Exercises

In Exercises 1–14, write the first six terms of each arithmetic sequence.

1. $a_1 = 200, d = 20$
2. $a_1 = 300, d = 50$
3. $a_1 = -7, d = 4$
4. $a_1 = -8, d = 5$
5. $a_1 = 300, d = -90$
6. $a_1 = 200, d = -60$
7. $a_1 = \frac{5}{2}, d = -\frac{1}{2}$
8. $a_1 = \frac{3}{4}, d = -\frac{1}{4}$
9. $a_n = a_{n-1} + 6, a_1 = -9$
10. $a_n = a_{n-1} + 4, a_1 = -7$
11. $a_n = a_{n-1} - 10, a_1 = 30$
12. $a_n = a_{n-1} - 20, a_1 = 50$
13. $a_n = a_{n-1} - 0.4, a_1 = 1.6$
14. $a_n = a_{n-1} - 0.3, a_1 = -1.7$

In Exercises 15–22, find the indicated term of the arithmetic sequence with first term, a_1, and common difference, d.

15. Find a_6 when $a_1 = 13, d = 4$.
16. Find a_{16} when $a_1 = 9, d = 2$.
17. Find a_{50} when $a_1 = 7, d = 5$.
18. Find a_{60} when $a_1 = 8, d = 6$.
19. Find a_{200} when $a_1 = -40, d = 5$.
20. Find a_{150} when $a_1 = -60, d = 5$.
21. Find a_{60} when $a_1 = 35, d = -3$.
22. Find a_{70} when $a_1 = -32, d = 4$.

In Exercises 23–34, write a formula for the general term (the nth term) of each arithmetic sequence. Do not use a recursion formula. Then use the formula for a_n to find a_{20}, the 20th term of the sequence.

23. $1, 5, 9, 13, \ldots$
24. $2, 7, 12, 17, \ldots$
25. $7, 3, -1, -5, \ldots$
26. $6, 1, -4, -9, \ldots$
27. $a_1 = 9, d = 2$
28. $a_1 = 6, d = 3$
29. $a_1 = -20, d = -4$
30. $a_1 = -70, d = -5$
31. $a_n = a_{n-1} + 3, a_1 = 4$
32. $a_n = a_{n-1} + 5, a_1 = 6$
33. $a_n = a_{n-1} - 10, a_1 = 30$
34. $a_n = a_{n-1} - 12, a_1 = 24$
35. Find the sum of the first 20 terms of the arithmetic sequence: $4, 10, 16, 22, \ldots$.
36. Find the sum of the first 25 terms of the arithmetic sequence: $7, 19, 31, 43, \ldots$.
37. Find the sum of the first 50 terms of the arithmetic sequence: $-10, -6, -2, 2, \ldots$.
38. Find the sum of the first 50 terms of the arithmetic sequence: $-15, -9, -3, 3, \ldots$.
39. Find $1 + 2 + 3 + 4 + \cdots + 100$, the sum of the first 100 natural numbers.
40. Find $2 + 4 + 6 + 8 + \cdots + 200$, the sum of the first 100 positive even integers.

41. Find the sum of the first 60 positive even integers.

42. Find the sum of the first 80 positive even integers.

43. Find the sum of the even integers between 21 and 45.

44. Find the sum of the odd integers between 30 and 54.

For Exercises 45–50, write out the first three terms and the last term. Then use the formula for the sum of the first n terms of an arithmetic sequence to find the indicated sum.

45. $\displaystyle\sum_{i=1}^{17} (5i + 3)$

46. $\displaystyle\sum_{i=1}^{20} (6i - 4)$

47. $\displaystyle\sum_{i=1}^{30} (-3i + 5)$

48. $\displaystyle\sum_{i=1}^{40} (-2i + 6)$

49. $\displaystyle\sum_{i=1}^{100} 4i$

50. $\displaystyle\sum_{i=1}^{50} (-4i)$

Practice Plus

Use the graphs of the arithmetic sequences $\{a_n\}$ and $\{b_n\}$ to solve Exercises 51–58.

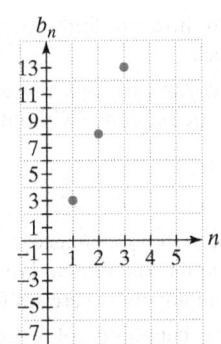

51. Find $a_{14} + b_{12}$.

52. Find $a_{16} + b_{18}$.

53. If $\{a_n\}$ is a finite sequence whose last term is -83, how many terms does $\{a_n\}$ contain?

54. If $\{b_n\}$ is a finite sequence whose last term is 93, how many terms does $\{b_n\}$ contain?

55. Find the difference between the sum of the first 14 terms of $\{b_n\}$ and the sum of the first 14 terms of $\{a_n\}$.

56. Find the difference between the sum of the first 15 terms of $\{b_n\}$ and the sum of the first 15 terms of $\{a_n\}$.

57. Write a linear function $f(x) = mx + b$, whose domain is the set of positive integers, that represents $\{a_n\}$.

58. Write a linear function $g(x) = mx + b$, whose domain is the set of positive integers, that represents $\{b_n\}$.

Use a system of two equations in two variables, a_1 and d, to solve Exercises 59–60.

59. Write a formula for the general term (the nth term) of the arithmetic sequence whose second term, a_2, is 4 and whose sixth term, a_6, is 16.

60. Write a formula for the general term (the nth term) of the arithmetic sequence whose third term, a_3, is 7 and whose eighth term, a_8, is 17.

Application Exercises

The bar graphs show changes in educational attainment for Americans ages 25 and older from 1970 to 2007. Exercises 61–62 involve developing arithmetic sequences that model the data.

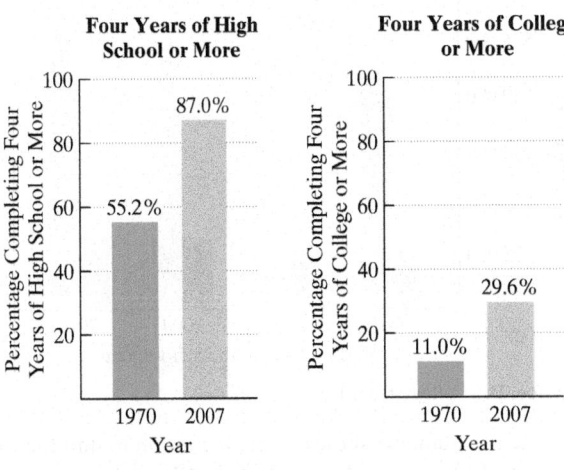

Educational Attainment for Americans Ages 25 and Older

Source: U.S. Census Bureau

61. In 1970, 11.0% of Americans ages 25 and older had completed four years of college or more. On average, this percentage has increased by approximately 0.5 each year.

a. Write a formula for the nth term of the arithmetic sequence that models the percentage of Americans ages 25 and older who had or will have completed four years of college or more n years after 1969.

b. Use the model from part (a) to project the percentage of Americans ages 25 and older who will have completed four years of college or more by 2019.

62. In 1970, 55.2% of Americans ages 25 and older had completed four years of high school or more. On average, this percentage has increased by approximately 0.86 each year.

a. Write a formula for the nth term of the arithmetic sequence that models the percentage of Americans ages 25 and older who had or will have completed four years of high school or more n years after 1969.

b. Use the model from part (a) to project the percentage of Americans ages 25 and older who will have completed four years of high school or more by 2019.

63. Company A pays $24,000 yearly with raises of $1600 per year. Company B pays $28,000 yearly with raises of $1000 per year. Which company will pay more in year 10? How much more?

64. Company A pays $23,000 yearly with raises of $1200 per year. Company B pays $26,000 yearly with raises of $800 per year. Which company will pay more in year 10? How much more?

In Exercises 65–66, we revisit the data from Chapter P showing the average cost of tuition and fees at public and private four-year U.S. colleges.

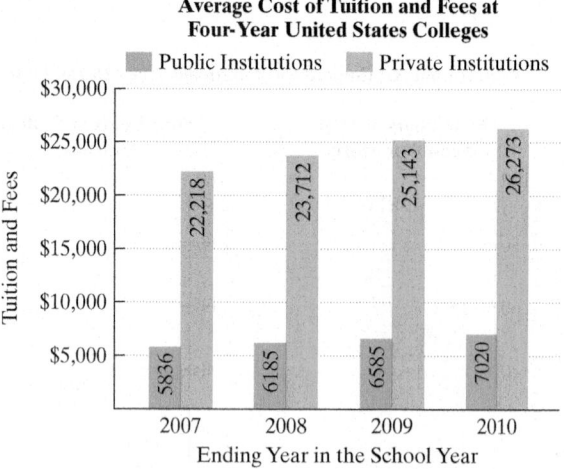

Average Cost of Tuition and Fees at Four-Year United States Colleges

Source: The College Board

65. a. Use the numbers shown in the bar graph to find the total cost of tuition and fees at public colleges for a four-year period from the school year ending in 2007 through the school year ending in 2010.

b. The model

$$a_n = 395n + 5419$$

describes the cost of tuition and fees at public colleges in academic year n, where $n = 1$ corresponds to the school year ending in 2007, $n = 2$ to the school year ending in 2008, and so on. Use this model and the formula for S_n to find the total cost of tuition and fees at public colleges for a four-year period from the school year ending in 2007 through the school year ending in 2010. How does this compare with the actual sum you obtained in part (a)?

66. a. Use the numbers shown in the bar graph to find the total cost of tuition and fees at private colleges for a four-year period from the school year ending in 2007 through the school year ending in 2010.

b. The model

$$a_n = 1360n + 20{,}938$$

describes the cost of tuition and fees at private colleges in academic year n, where $n = 1$ corresponds to the school year ending in 2007, $n = 2$ to the school year ending in 2008, and so on. Use this model and the formula for S_n to find the total cost of tuition and fees at private colleges for a four-year period from the school year ending in 2007 through the school year ending in 2010. Does the model underestimate or overestimate the actual sum that you obtained in part (a)? By how much?

67. Use one of the models in Exercises 65–66 and the formula for S_n to find the total cost of tuition and fees for your undergraduate education. How well does the model describe your anticipated costs?

68. A company offers a starting yearly salary of $33,000 with raises of $2500 per year. Find the total salary over a ten-year period.

69. You are considering two job offers. Company A will start you at $19,000 a year and guarantee a raise of $2600 per year. Company B will start you at a higher salary, $27,000 a year but will only guarantee a raise of $1200 per year. Find the total salary that each company will pay over a ten-year period. Which company pays the greater total amount?

70. A theater has 30 seats in the first row, 32 seats in the second row, increasing by 2 seats per row for a total of 26 rows. How many seats are there in the theater?

71. A section in a stadium has 20 seats in the first row, 23 seats in the second row, increasing by 3 seats each row for a total of 38 rows. How many seats are in this section of the stadium?

Writing in Mathematics

72. What is an arithmetic sequence? Give an example with your explanation.

73. What is the common difference in an arithmetic sequence?

74. Explain how to find the general term of an arithmetic sequence.

75. Explain how to find the sum of the first n terms of an arithmetic sequence without having to add up all the terms.

Technology Exercises

76. Use the $\boxed{\text{SEQ}}$ (sequence) capability of a graphing utility and the formula you obtained for a_n to verify the value you found for a_{20} in any five exercises from Exercises 23–34.

77. Use the capability of a graphing utility to calculate the sum of a sequence to verify any five of your answers to Exercises 45–50.

Critical Thinking Exercises

Make Sense? *In Exercises 78–81, determine whether each statement makes sense or does not make sense, and explain your reasoning.*

78. Rather than performing the addition, I used the formula $S_n = \frac{n}{2}(a_1 + a_n)$ to find the sum of the first 30 terms of the sequence 2, 4, 8, 16, 32,

79. I was able to find the sum of the first 50 terms of an arithmetic sequence even though I did not identify every term.

80. The sequence for the number of seats per row in our movie theater as the rows move toward the back is arithmetic with $d = 1$ so people don't block the view of those in the row behind them.

81. Beginning at 6:45 A.M., a bus stops on my block every 23 minutes, so I used the formula for the nth term of an arithmetic sequence to describe the stopping time for the nth bus of the day.

82. In the sequence 21,700, 23,172, 24,644, 26,116, . . . , which term is 314,628?

83. A *degree-day* is a unit used to measure the fuel requirements of buildings. By definition, each degree that the average daily temperature is below 65°F is 1 degree-day. For example, an average daily temperature of 42°F constitutes 23 degree-days. If the average temperature on January 1 was 42°F and fell 2°F for each subsequent day up to and including January 10, how many degree-days are included from January 1 to January 10?

84. Show that the sum of the first n positive odd integers,

$$1 + 3 + 5 + \cdots + (2n - 1),$$

is n^2.

Preview Exercises

Exercises 85–87 will help you prepare for the material covered in the next section.

85. Consider the sequence $1, -2, 4, -8, 16, \ldots$. Find $\dfrac{a_2}{a_1}, \dfrac{a_3}{a_2}, \dfrac{a_4}{a_3}$, and $\dfrac{a_5}{a_4}$. What do you observe?

86. Consider the sequence whose nth term is $a_n = 3 \cdot 5^n$. Find $\dfrac{a_2}{a_1}, \dfrac{a_3}{a_2}, \dfrac{a_4}{a_3}$, and $\dfrac{a_5}{a_4}$. What do you observe?

87. Use the formula $a_n = a_1 3^{n-1}$ to find the seventh term of the sequence $11, 33, 99, 297, \ldots$.

SECTION 8.3 Geometric Sequences and Series

Objectives

1. Find the common ratio of a geometric sequence.
2. Write terms of a geometric sequence.
3. Use the formula for the general term of a geometric sequence.
4. Use the formula for the sum of the first n terms of a geometric sequence.
5. Find the value of an annuity.
6. Use the formula for the sum of an infinite geometric series.

Here we are at the closing moments of a job interview. You're shaking hands with the manager. You managed to answer all the tough questions without losing your poise, and now you've been offered a job. As a matter of fact, your qualifications are so terrific that you've been offered two jobs—one just the day before, with a rival company in the same field! One company offers $30,000 the first year, with increases of 6% per year for four years after that. The other offers $32,000 the first year, with annual increases of 3% per year after that. Over a five-year period, which is the better offer?

If salary raises amount to a certain percent each year, the yearly salaries over time form a geometric sequence. In this section, we investigate geometric sequences and their properties. After studying the section, you will be in a position to decide which job offer to accept: You will know which company will pay you more over five years.

1 Find the common ratio of a geometric sequence.

Geometric Sequences

Figure 8.5 shows a sequence in which the number of squares is increasing. From left to right, the number of squares is 1, 5, 25, 125, and 625. In this sequence, each term after the first, 1, is obtained by multiplying the preceding term by a constant amount, namely, 5. This sequence of increasing numbers of squares is an example of a *geometric sequence*.

FIGURE 8.5 A geometric sequence of squares

Definition of a Geometric Sequence

A **geometric sequence** is a sequence in which each term after the first is obtained by multiplying the preceding term by a fixed nonzero constant. The amount by which we multiply each time is called the **common ratio** of the sequence.

The common ratio, r, is found by dividing any term after the first term by the term that directly precedes it. In the following examples, the common ratio is found by dividing the second term by the first term, $\frac{a_2}{a_1}$.

Geometric sequence	Common ratio
$1, 5, 25, 125, 625, \ldots$	$r = \frac{5}{1} = 5$
$4, 8, 16, 32, 64, \ldots$	$r = \frac{8}{4} = 2$
$6, -12, 24, -48, 96, \ldots$	$r = \frac{-12}{6} = -2$
$9, -3, 1, -\frac{1}{3}, \frac{1}{9}, \ldots$	$r = \frac{-3}{9} = -\frac{1}{3}$

Figure 8.6 shows a partial graph of the first geometric sequence in our list. The graph forms a set of discrete points lying on the exponential function $f(x) = 5^{x-1}$. This illustrates that **a geometric sequence with a positive common ratio other than 1 is an exponential function whose domain is the set of positive integers**.

How do we write out the terms of a geometric sequence when the first term and the common ratio are known? We multiply the first term by the common ratio to get the second term, multiply the second term by the common ratio to get the third term, and so on.

GREAT QUESTION!

What happens to the terms of a geometric sequence when the common ratio is negative?

When the common ratio is negative, the signs of the terms alternate.

FIGURE 8.6 The graph of $\{a_n\} = 1, 5, 25, 125, \ldots$

② Write terms of a geometric sequence.

EXAMPLE 1 Writing the Terms of a Geometric Sequence

Write the first six terms of the geometric sequence with first term 6 and common ratio $\frac{1}{3}$.

SOLUTION

The first term is 6. The second term is $6 \cdot \frac{1}{3}$, or 2. The third term is $2 \cdot \frac{1}{3}$, or $\frac{2}{3}$. The fourth term is $\frac{2}{3} \cdot \frac{1}{3}$, or $\frac{2}{9}$, and so on. The first six terms are

$$6, 2, \frac{2}{3}, \frac{2}{9}, \frac{2}{27}, \text{ and } \frac{2}{81}.$$ • • •

✅ **Check Point 1** Write the first six terms of the geometric sequence with first term 12 and common ratio $\frac{1}{2}$.

③ Use the formula for the general term of a geometric sequence.

The General Term of a Geometric Sequence

Consider a geometric sequence whose first term is a_1 and whose common ratio is r. We are looking for a formula for the general term, a_n. Let's begin by writing the first six terms. The first term is a_1. The second term is $a_1 r$. The third term is $a_1 r \cdot r$, or $a_1 r^2$. The fourth term is $a_1 r^2 \cdot r$, or $a_1 r^3$, and so on. Starting with a_1 and multiplying each successive term by r, the first six terms are

a_1,	$a_1 r$,	$a_1 r^2$,	$a_1 r^3$,	$a_1 r^4$,	$a_1 r^5$.
a_1, first term	a_2, second term	a_3, third term	a_4, fourth term	a_5, fifth term	a_6, sixth term

Can you see that the exponent on r is 1 less than the subscript of a denoting the term number?

$$a_3: \text{third term} = a_1r^2 \qquad a_4: \text{fourth term} = a_1r^3$$

One less than 3, or 2, is the exponent on r.

One less than 4, or 3, is the exponent on r.

Thus, the formula for the nth term is

$$a_n = a_1r^{n-1}.$$

One less than n, or $n - 1$, is the exponent on r.

GREAT QUESTION!

When using a_1r^{n-1} to find the nth term of a geometric sequence, what should I do first?

Be careful with the order of operations when evaluating

$$a_1r^{n-1}.$$

First subtract 1 in the exponent and then raise r to that power. Finally, multiply the result by a_1.

General Term of a Geometric Sequence

The nth term (the general term) of a geometric sequence with first term a_1 and common ratio r is

$$a_n = a_1r^{n-1}.$$

EXAMPLE 2 Using the Formula for the General Term of a Geometric Sequence

Find the eighth term of the geometric sequence whose first term is -4 and whose common ratio is -2.

SOLUTION

To find the eighth term, a_8, we replace n in the formula with 8, a_1 with -4, and r with -2.

$$a_n = a_1r^{n-1}$$
$$a_8 = -4(-2)^{8-1} = -4(-2)^7 = -4(-128) = 512$$

The eighth term is 512. We can check this result by writing the first eight terms of the sequence:

$$-4, 8, -16, 32, -64, 128, -256, 512. \qquad \bullet\bullet\bullet$$

Check Point **2** Find the seventh term of the geometric sequence whose first term is 5 and whose common ratio is -3.

In Chapter 4, we studied exponential functions of the form $f(x) = b^x$ and used an exponential function to model the growth of the U.S. population from 1970 through 2010 (Example 1 on page 505). From 2000 through 2010, the nation's population grew at the slowest rate since the 1930s. Consequently, in our next example, we consider the country's population growth from 2000 through 2010. Because a geometric sequence is an exponential function whose domain is the set of positive integers, geometric and exponential growth mean the same thing.

Blitzer Bonus

Geometric Population Growth

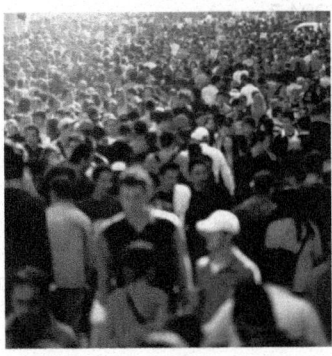

Economist Thomas Malthus (1766–1834) predicted that population would increase as a geometric sequence and food production would increase as an arithmetic sequence. He concluded that eventually population would exceed food production. If two sequences, one geometric and one arithmetic, are increasing, the geometric sequence will eventually overtake the arithmetic sequence, regardless of any head start that the arithmetic sequence might initially have.

EXAMPLE 3 Geometric Population Growth

The table shows the population of the United States in 2000 and 2010, with estimates given by the Census Bureau for 2001 through 2009.

Year	2000	2001	2002	2003	2004	2005	2006	2007	2008	2009	2010
Population (millions)	281.4	284.0	286.6	289.3	292.0	294.7	297.4	300.2	303.0	305.8	308.7

a. Show that the population is increasing geometrically.

b. Write the general term for the geometric sequence modeling the population of the United States, in millions, n years after 1999.

c. Project the U.S. population, in millions, for the year 2020.

SOLUTION

a. First, we use the sequence of population growth, 281.4, 284.0, 286.6, 289.3, and so on, to divide the population for each year by the population in the preceding year.

$$\frac{284.0}{281.4} \approx 1.009, \quad \frac{286.6}{284.0} \approx 1.009, \quad \frac{289.3}{286.6} \approx 1.009$$

Continuing in this manner, we will keep getting approximately 1.009. This means that the population is increasing geometrically with $r \approx 1.009$. The population of the United States in any year shown in the sequence is approximately 1.009 times the population the year before.

b. The sequence of the U.S. population growth is

$$281.4, 284.0, 286.6, 289.3, 292.0, 294.7, \ldots .$$

Because the population is increasing geometrically, we can find the general term of this sequence using

$$a_n = a_1 r^{n-1}.$$

In this sequence, $a_1 = 281.4$ and [from part (a)] $r \approx 1.009$. We substitute these values into the formula for the general term. This gives the general term for the geometric sequence modeling the U.S. population, in millions, n years after 1999.

$$a_n = 281.4(1.009)^{n-1}$$

c. We can use the formula for the general term, a_n, in part (b) to project the U.S. population for the year 2020. The year 2020 is 21 years after 1999—that is, $2020 - 1999 = 21$. Thus, $n = 21$. We substitute 21 for n in $a_n = 281.4(1.009)^{n-1}$.

$$a_{21} = 281.4(1.009)^{21-1} = 281.4(1.009)^{20} \approx 336.6$$

The model projects that the United States will have a population of approximately 336.6 million in the year 2020. • • •

☑ Check Point **3** Write the general term for the geometric sequence

$$3, 6, 12, 24, 48, \ldots .$$

Then use the formula for the general term to find the eighth term.

Blitzer Bonus || Ponzi Schemes and Geometric Sequences

A ponzi scheme is an investment fraud that pays returns to existing investors from funds contributed by new investors rather than from legitimate investment activity. Here's a simplified example:

$$a_1 \quad , \quad a_2 \quad , \quad a_3 \quad , \quad a_4 \quad , \ldots$$

| Round 1 | Round 2 | Round 3 | Round 4 |

The schemer takes $100 each from the first two investors.

The schemer takes $100 each from 4 new investors, pocketing $200 and paying the first two investors $100 each.

The schemer takes $100 each from 8 new investors, pocketing $200 and paying the previous 6 investors $100 each.

The schemer takes $100 each from 16 new investors, pocketing $200 and paying the previous 14 investors $100 each.

The number of investors needed to continue this Ponzi scheme,

$$2, 4, 8, 16, \ldots,$$

and the money collected in each round,

$$\$200, \$400, \$800, \$1600, \ldots,$$

form rapidly growing geometric sequences. With no legitimate earnings, the scheme requires a consistent geometric flow of money from new investors to continue. Ponzi schemes tend to collapse when it becomes difficult to recruit new investors or when a large number of investors ask to cash out.

④ Use the formula for the sum of the first n terms of a geometric sequence.

The Sum of the First n Terms of a Geometric Sequence

The sum of the first n terms of a geometric sequence, denoted by S_n and called the **nth partial sum**, can be found without having to add up all the terms. Recall that the first n terms of a geometric sequence are

$$a_1, a_1r, a_1r^2, \ldots, a_1r^{n-2}, a_1r^{n-1}.$$

We proceed as follows:

$$S_n = a_1 + a_1r + a_1r^2 + \cdots + a_1r^{n-2} + a_1r^{n-1} \qquad \text{\small S_n is the sum of the first n terms of the sequence.}$$

$$rS_n = a_1r + a_1r^2 + a_1r^3 + \cdots + a_1r^{n-1} + a_1r^n \qquad \text{\small Multiply both sides of the equation by r.}$$

$$S_n - rS_n = a_1 - a_1r^n \qquad \text{\small Subtract the second equation from the first equation.}$$

$$S_n(1 - r) = a_1(1 - r^n) \qquad \text{\small Factor out S_n on the left and a_1 on the right.}$$

$$S_n = \frac{a_1(1 - r^n)}{1 - r}. \qquad \text{\small Solve for S_n by dividing both sides by $1 - r$ (assuming that $r \neq 1$).}$$

We have proved the following result:

GREAT QUESTION!

What is the sum of the first n terms of a geometric sequence if the common ratio is 1?

If the common ratio is 1, the geometric sequence is

$$a_1, a_1, a_1, a_1, \ldots.$$

The sum of the first n terms of this sequence is na_1:

$$S_n = \underbrace{a_1 + a_1 + a_1 + \cdots + a_1.}_{\text{There are } n \text{ terms}}$$

$$= na_1.$$

The Sum of the First n Terms of a Geometric Sequence

The sum, S_n, of the first n terms of a geometric sequence is given by

$$S_n = \frac{a_1(1 - r^n)}{1 - r},$$

in which a_1 is the first term and r is the common ratio ($r \neq 1$).

To find the sum of the terms of a geometric sequence using $S_n = \dfrac{a_1(1 - r^n)}{1 - r}$, we need to know the first term, a_1, the common ratio, r, and the number of terms, n. The following examples illustrate how to use this formula.

EXAMPLE 4 Finding the Sum of the First n Terms of a Geometric Sequence

Find the sum of the first 18 terms of the geometric sequence: $2, -8, 32, -128, \ldots$.

SOLUTION

To find the sum of the first 18 terms, S_{18}, we replace n in the formula with 18.

$$S_n = \frac{a_1(1 - r^n)}{1 - r}$$

$$S_{18} = \frac{a_1(1 - r^{18})}{1 - r}$$

The first term, a_1, is 2. We must find r, the common ratio.

We can find the common ratio by dividing the second term of $2, -8, 32, -128, \ldots$ by the first term.

$$r = \frac{a_2}{a_1} = \frac{-8}{2} = -4$$

Now we are ready to find the sum of the first 18 terms of $2, -8, 32, -128, \ldots$.

$$S_n = \frac{a_1(1 - r^n)}{1 - r}$$

Use the formula for the sum of the first n terms of a geometric sequence.

$$S_{18} = \frac{2[1 - (-4)^{18}]}{1 - (-4)}$$

a_1 (the first term) $= 2$, $r = -4$, and $n = 18$ because we want the sum of the first 18 terms.

$$= -27{,}487{,}790{,}694$$ Use a calculator.

The sum of the first 18 terms is $-27{,}487{,}790{,}694$. Equivalently, this number is the 18th partial sum of the sequence $2, -8, 32, -128, \ldots$. $\bullet\bullet\bullet$

✓ Check Point **4** Find the sum of the first nine terms of the geometric sequence: $2, -6, 18, -54, \ldots$.

EXAMPLE 5 Using S_n to Evaluate a Summation

Find the following sum: $\displaystyle\sum_{i=1}^{10} 6 \cdot 2^i$.

SOLUTION

Let's write out a few terms in the sum.

$$\sum_{i=1}^{10} 6 \cdot 2^i = 6 \cdot 2 + 6 \cdot 2^2 + 6 \cdot 2^3 + \cdots + 6 \cdot 2^{10}$$

To find

$$\sum_{i=1}^{10} 6 \cdot 2^i$$

on a graphing utility, enter

$\boxed{\text{SUM}}\ \boxed{\text{SEQ}}\ (6 \times 2^x, x, 1, 10, 1)$.

Then press $\boxed{\text{ENTER}}$.

Do you see that each term after the first is obtained by multiplying the preceding term by 2? To find the sum of the 10 terms ($n = 10$), we need to know the first term, a_1, and the common ratio, r. The first term is $6 \cdot 2$ or 12: $a_1 = 12$. The common ratio is 2.

$$S_n = \frac{a_1(1 - r^n)}{1 - r} \qquad \text{Use the formula for the sum of the first } n \text{ terms of a geometric sequence.}$$

$$S_{10} = \frac{12(1 - 2^{10})}{1 - 2} \qquad a_1 \text{ (the first term)} = 12, r = 2, \text{ and } n = 10 \text{ because we are adding 10 terms.}$$

$$= 12{,}276 \qquad \text{Use a calculator.}$$

Thus,

$$\sum_{i=1}^{10} 6 \cdot 2^i = 12{,}276. \qquad \bullet\bullet\bullet$$

✓ **Check Point 5** Find the following sum: $\displaystyle\sum_{i=1}^{8} 2 \cdot 3^i$.

Some of the exercises in the previous Exercise Set involved situations in which salaries increased by a fixed amount each year. A more realistic situation is one in which salary raises increase by a certain percent each year. Example 6 shows how such a situation can be modeled using a geometric sequence.

EXAMPLE 6 Computing a Lifetime Salary

A union contract specifies that each worker will receive a 5% pay increase each year for the next 30 years. One worker is paid $20,000 the first year. What is this person's total lifetime salary over a 30-year period?

SOLUTION

The salary for the first year is $20,000. With a 5% raise, the second-year salary is computed as follows:

Salary for year 2 = $20{,}000 + 20{,}000(0.05) = 20{,}000(1 + 0.05) = 20{,}000(1.05)$.

Each year, the salary is 1.05 times what it was in the previous year. Thus, the salary for year 3 is 1.05 times $20{,}000(1.05)$, or $20{,}000(1.05)^2$. The salaries for the first five years are given in the table.

Yearly Salaries

Year 1	Year 2	Year 3	Year 4	Year 5	...
20,000	20,000(1.05)	$20{,}000(1.05)^2$	$20{,}000(1.05)^3$	$20{,}000(1.05)^4$...

The numbers in the bottom row form a geometric sequence with $a_1 = 20{,}000$ and $r = 1.05$. To find the total salary over 30 years, we use the formula for the sum of the first n terms of a geometric sequence, with $n = 30$.

$$S_n = \frac{a_1(1 - r^n)}{1 - r}$$

$$S_{30} = \frac{20{,}000[1 - (1.05)^{30}]}{1 - 1.05}$$

Total salary over 30 years

$$= \frac{20{,}000[1 - (1.05)^{30}]}{-0.05}$$

$$\approx 1{,}328{,}777 \qquad \text{Use a calculator.}$$

The total salary over the 30-year period is approximately $1,328,777. $\bullet\bullet\bullet$

⊘ Check Point **6** A job pays a salary of $30,000 the first year. During the next 29 years, the salary increases by 6% each year. What is the total lifetime salary over the 30-year period?

⑤ Find the value of an annuity.

Annuities

The compound interest formula

$$A = P(1 + r)^t$$

gives the future value, A, after t years, when a fixed amount of money, P, the principal, is deposited in an account that pays an annual interest rate r (in decimal form) compounded once a year. However, money is often invested in small amounts at periodic intervals. For example, to save for retirement, you might decide to place $1000 into an Individual Retirement Account (IRA) at the end of each year until you retire. An **annuity** is a sequence of equal payments made at equal time periods. An IRA is an example of an annuity.

Suppose P dollars is deposited into an account at the end of each year. The account pays an annual interest rate, r, compounded annually. At the end of the first year, the account contains P dollars. At the end of the second year, P dollars is deposited again. At the time of this deposit, the first deposit has received interest earned during the second year. The **value of the annuity** is the sum of all deposits made plus all interest paid. Thus, the value of the annuity after two years is

$$P + P(1 + r).$$

| Deposit of P dollars at end of second year | First-year deposit of P dollars with interest earned for a year |

The value of the annuity after three years is

$$P \quad + \quad P(1 + r) \quad + \quad P(1 + r)^2.$$

| Deposit of P dollars at end of third year | Second-year deposit of P dollars with interest earned for a year | First-year deposit of P dollars with interest earned over two years |

The value of the annuity after t years is

$$P + P(1 + r) + P(1 + r)^2 + P(1 + r)^3 + \cdots + P(1 + r)^{t-1}.$$

| Deposit of P dollars at end of year t | | First-year deposit of P dollars with interest earned over $t - 1$ years |

This is the sum of the terms of a geometric sequence with first term P and common ratio $1 + r$. We use the formula

$$S_n = \frac{a_1(1 - r^n)}{1 - r}$$

to find the sum of the terms:

$$S_t = \frac{P[1 - (1 + r)^t]}{1 - (1 + r)} = \frac{P[1 - (1 + r)^t]}{-r} = \frac{P[(1 + r)^t - 1]}{r}.$$

This formula gives the value of an annuity after t years if interest is compounded once a year. We can adjust the formula to find the value of an annuity if equal payments are made at the end of each of n yearly compounding periods.

Value of an Annuity: Interest Compounded *n* Times per Year

If *P* is the deposit made at the end of each compounding period for an annuity at *r* percent annual interest compounded *n* times per year, the value, *A*, of the annuity after *t* years is

$$A = \frac{P\left[\left(1 + \frac{r}{n}\right)^{nt} - 1\right]}{\frac{r}{n}}.$$

Blitzer Bonus

Stashing Cash and Making Taxes Less Taxing

As you prepare for your future career, retirement probably seems very far away. Making regular deposits into an IRA may not be fun, but there is a special incentive from Uncle Sam that makes it far more appealing. Traditional IRAs are **tax-deferred savings plans**. This means that you do not pay taxes on deposits and interest until you begin withdrawals, typically at retirement. Before then, yearly deposits count as adjustments to gross income and are not part of your taxable income. Not only do you get a tax break now, but also you ultimately earn more. This is because you do not pay taxes on interest from year to year, allowing earnings to accumulate until you start withdrawals. With a tax code that encourages long-term savings, opening an IRA early in your career is a smart way to gain more control over how you will spend a large part of your life.

EXAMPLE 7 Determining the Value of an Annuity

At age 25, to save for retirement, you decide to deposit $200 at the end of each month into an IRA that pays 7.5% compounded monthly.

a. How much will you have from the IRA when you retire at age 65?
b. Find the interest.

SOLUTION

a. Because you are 25, the amount that you will have from the IRA when you retire at 65 is its value after 40 years.

$$A = \frac{P\left[\left(1 + \frac{r}{n}\right)^{nt} - 1\right]}{\frac{r}{n}}$$

Use the formula for the value of an annuity.

$$A = \frac{200\left[\left(1 + \frac{0.075}{12}\right)^{12\cdot40} - 1\right]}{\frac{0.075}{12}}$$

The annuity involves month-end deposits of $200: *P* = 200. The interest rate is 7.5%: *r* = 0.075. The interest is compounded monthly: *n* = 12. The number of years is 40: *t* = 40.

$$= \frac{200[(1 + 0.00625)^{480} - 1]}{0.00625}$$

Using parentheses keys, this can be performed in a single step on a graphing calculator.

$$= \frac{200[(1.00625)^{480} - 1]}{0.00625}$$

$$\approx \frac{200(19.8989 - 1)}{0.00625}$$

Use a calculator to find $(1.00625)^{480}$:
1.00625 $\boxed{y^x}$ 480 $\boxed{=}$.

$$\approx 604{,}765$$

After 40 years, you will have approximately $604,765 when retiring at age 65.

b. Interest = Value of the IRA − Total deposits

$$\approx \$604{,}765 - \$200 \cdot 12 \cdot 40$$

$200 per month × 12 months per year × 40 years

$$= \$604{,}765 - \$96{,}000 = \$508{,}765$$

The interest is approximately $508,765, more than five times the amount of your contributions to the IRA.

•••

\oslash Check Point **7** At age 30, to save for retirement, you decide to deposit $100 at the end of each month into an IRA that pays 9.5% compounded monthly.

a. How much will you have from the IRA when you retire at age 65?

b. Find the interest.

⑥ Use the formula for the sum of an infinite geometric series.

Geometric Series

An infinite sum of the form

$$a_1 + a_1 r + a_1 r^2 + a_1 r^3 + \cdots + a_1 r^{n-1} + \cdots$$

with first term a_1 and common ratio r is called an **infinite geometric series**. How can we determine which infinite geometric series have sums and which do not? We look at what happens to r^n as n gets larger in the formula for the sum of the first n terms of this series, namely,

$$S_n = \frac{a_1(1 - r^n)}{1 - r}.$$

If r is any number between -1 and 1, that is, $-1 < r < 1$, the term r^n approaches 0 as n gets larger. For example, consider what happens to r^n for $r = \frac{1}{2}$:

$$\left(\frac{1}{2}\right)^1 = \frac{1}{2} \quad \left(\frac{1}{2}\right)^2 = \frac{1}{4} \quad \left(\frac{1}{2}\right)^3 = \frac{1}{8} \quad \left(\frac{1}{2}\right)^4 = \frac{1}{16} \quad \left(\frac{1}{2}\right)^5 = \frac{1}{32} \quad \left(\frac{1}{2}\right)^6 = \frac{1}{64}.$$

These numbers are approaching 0 as n gets larger.

Take another look at the formula for the sum of the first n terms of a geometric sequence.

$$S_n = \frac{a_1(1 - r^n)}{1 - r}$$

If $-1 < r < 1$, r^n approaches 0 as n gets larger.

Let us replace r^n with 0 in the formula for S_n. This change gives us a formula for the sum of an infinite geometric series with a common ratio between -1 and 1.

The Sum of an Infinite Geometric Series

If $-1 < r < 1$ (equivalently, $|r| < 1$), then the sum of the infinite geometric series

$$a_1 + a_1 r + a_1 r^2 + a_1 r^3 + \cdots ,$$

in which a_1 is the first term and r is the common ratio, is given by

$$S = \frac{a_1}{1 - r}.$$

If $|r| \geq 1$, the infinite series does not have a sum.

To use the formula for the sum of an infinite geometric series, we need to know the first term and the common ratio. For example, consider

First term, a_1, is $\frac{1}{2}$.
$$\frac{1}{2} + \frac{1}{4} + \frac{1}{8} + \frac{1}{16} + \frac{1}{32} + \cdots.$$

Common ratio, r, is $\frac{a_2}{a_1}$.
$$r = \frac{1}{4} \div \frac{1}{2} = \frac{1}{4} \cdot 2 = \frac{1}{2}$$

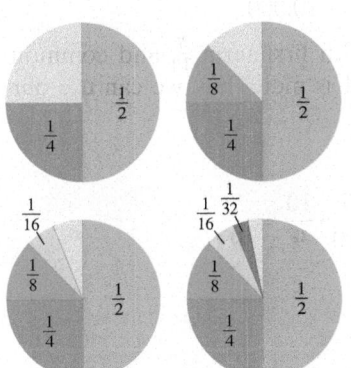

FIGURE 8.7 The sum $\frac{1}{2} + \frac{1}{4} + \frac{1}{8} + \frac{1}{16} + \frac{1}{32} + \cdots$ is approaching 1.

With $r = \dfrac{1}{2}$, the condition that $|r| < 1$ is met, so the infinite geometric series has a sum given by $S = \dfrac{a_1}{1-r}$. The sum of the series is found as follows:

$$\frac{1}{2} + \frac{1}{4} + \frac{1}{8} + \frac{1}{16} + \frac{1}{32} + \cdots = \frac{a_1}{1-r} = \frac{\frac{1}{2}}{1 - \frac{1}{2}} = \frac{\frac{1}{2}}{\frac{1}{2}} = 1.$$

Thus, the sum of the infinite geometric series is 1. Notice how this is illustrated in **Figure 8.7**. As more terms are included, the sum is approaching the area of one complete circle.

EXAMPLE 8 Finding the Sum of an Infinite Geometric Series

Find the sum of the infinite geometric series: $\frac{3}{8} - \frac{3}{16} + \frac{3}{32} - \frac{3}{64} + \cdots$.

SOLUTION

Before finding the sum, we must find the common ratio.

$$r = \frac{a_2}{a_1} = \frac{-\dfrac{3}{16}}{\dfrac{3}{8}} = -\frac{3}{16} \cdot \frac{8}{3} = -\frac{1}{2}$$

Because $r = -\frac{1}{2}$, the condition that $|r| < 1$ is met. Thus, the infinite geometric series has a sum.

$$S = \frac{a_1}{1-r} \qquad \text{This is the formula for the sum of an infinite geometric series. Let } a_1 = \frac{3}{8} \text{ and } r = -\frac{1}{2}.$$

$$= \frac{\dfrac{3}{8}}{1 - \left(-\dfrac{1}{2}\right)} = \frac{\dfrac{3}{8}}{\dfrac{3}{2}} = \frac{3}{8} \cdot \frac{2}{3} = \frac{1}{4}$$

Thus, the sum of $\frac{3}{8} - \frac{3}{16} + \frac{3}{32} - \frac{3}{64} + \cdots$ is $\frac{1}{4}$. Put in an informal way, as we continue to add more and more terms, the sum is approximately $\frac{1}{4}$. • • •

⊘ Check Point **8** Find the sum of the infinite geometric series:
$3 + 2 + \frac{4}{3} + \frac{8}{9} + \cdots$.

We can use the formula for the sum of an infinite geometric series to express a repeating decimal as a fraction in lowest terms.

EXAMPLE 9 Writing a Repeating Decimal as a Fraction

Express $0.\overline{7}$ as a fraction in lowest terms.

SOLUTION

$$0.\overline{7} = 0.7777\ldots = \frac{7}{10} + \frac{7}{100} + \frac{7}{1000} + \frac{7}{10,000} + \cdots$$

Observe that $0.\overline{7}$ is an infinite geometric series with first term $\frac{7}{10}$ and common ratio $\frac{1}{10}$. Because $r = \frac{1}{10}$, the condition that $|r| < 1$ is met. Thus, we can use our formula to find the sum. Therefore,

$$0.\overline{7} = \frac{a_1}{1-r} = \frac{\dfrac{7}{10}}{1 - \dfrac{1}{10}} = \frac{\dfrac{7}{10}}{\dfrac{9}{10}} = \frac{7}{10} \cdot \frac{10}{9} = \frac{7}{9}.$$

An equivalent fraction for $0.\overline{7}$ is $\frac{7}{9}$. •••

☑ **Check Point 9** Express $0.\overline{9}$ as a fraction in lowest terms.

Infinite geometric series have many applications, as illustrated in Example 10.

EXAMPLE 10 Tax Rebates and the Multiplier Effect

$1400

70% is spent.

$980

70% is spent.

$686

A tax rebate that returns a certain amount of money to taxpayers can have a total effect on the economy that is many times this amount. In economics, this phenomenon is called the **multiplier effect**. Suppose, for example, that the government reduces taxes so that each consumer has $2000 more income. The government assumes that each person will spend 70% of this (= $1400). The individuals and businesses receiving this $1400 in turn spend 70% of it (= $980), creating extra income for other people to spend, and so on. Determine the total amount spent on consumer goods from the initial $2000 tax rebate.

SOLUTION

The total amount spent is given by the infinite geometric series

$$1400 + 980 + 686 + \cdots.$$

70% of 1400 70% of 980

The first term is 1400: $a_1 = 1400$. The common ratio is 70%, or 0.7: $r = 0.7$. Because $r = 0.7$, the condition that $|r| < 1$ is met. Thus, we can use our formula to find the sum. Therefore,

$$1400 + 980 + 686 + \cdots = \frac{a_1}{1-r} = \frac{1400}{1-0.7} \approx 4667.$$

This means that the total amount spent on consumer goods from the initial $2000 rebate is approximately $4667. •••

☑ **Check Point 10** Rework Example 10 and determine the total amount spent on consumer goods with a $1000 tax rebate and 80% spending down the line.

CONCEPT AND VOCABULARY CHECK

Fill in each blank so that the resulting statement is true.

1. A sequence in which each term after the first is obtained by multiplying the preceding term by a fixed nonzero constant is called a/an _____ sequence. The amount by which we multiply each time is called the _____ of the sequence.

2. The nth term of the sequence described in Exercise 1 is given by the formula $a_n = $ _____, where a_1 is the _____ and r is the _____ of the sequence.

3. The sum, S_n, of the first n terms of the sequence described in Exercise 1 is given by the formula $S_n = $ _____, where a_1 is the _____ and r is the _____, $r \neq 1$.

4. A sequence of equal payments made at equal time periods is called a/an _____. Its value, A, after t years is given by the formula

$$A = \frac{P[(1 + \frac{r}{n})^{nt} - 1]}{\frac{r}{n}},$$

where _____ is the deposit made at the end of each compounding period at _____ percent annual interest compounded _____ times per year.

5. An infinite sum of the form
$$a_1 + a_1 r + a_1 r^2 + a_1 r^3 + \cdots$$
is called a/an _____.
If $-1 < r < $ _____, its sum, S, is given by the formula $S = $ _____. The series does not have a sum if _____.

6. The first four terms of $\sum\limits_{i=1}^{6} 2^i$ are _____, _____, _____, and _____. The common ratio is _____.

Determine whether each sequence is arithmetic or geometric.

7. $4, 8, 12, 16, 20, \ldots$ _____
8. $4, 8, 16, 32, 64, \ldots$ _____
9. $1, -3, 9, -27, 81, \ldots$ _____
10. $-1, 1, 3, 5, 7, \ldots$ _____

EXERCISE SET 8.3

Practice Exercises

In Exercises 1–8, write the first five terms of each geometric sequence.

1. $a_1 = 5, \quad r = 3$
2. $a_1 = 4, \quad r = 3$
3. $a_1 = 20, \quad r = \frac{1}{2}$
4. $a_1 = 24, \quad r = \frac{1}{3}$
5. $a_n = -4a_{n-1}, \quad a_1 = 10$
6. $a_n = -3a_{n-1}, \quad a_1 = 10$
7. $a_n = -5a_{n-1}, \quad a_1 = -6$
8. $a_n = -6a_{n-1}, \quad a_1 = -2$

In Exercises 9–16, use the formula for the general term (the nth term) of a geometric sequence to find the indicated term of each sequence with the given first term, a_1, and common ratio, r.

9. Find a_8 when $a_1 = 6, r = 2$.
10. Find a_8 when $a_1 = 5, r = 3$.
11. Find a_{12} when $a_1 = 5, r = -2$.
12. Find a_{12} when $a_1 = 4, r = -2$.
13. Find a_{40} when $a_1 = 1000, r = -\frac{1}{2}$.
14. Find a_{30} when $a_1 = 8000, r = -\frac{1}{2}$.
15. Find a_8 when $a_1 = 1{,}000{,}000, r = 0.1$.
16. Find a_8 when $a_1 = 40{,}000, r = 0.1$.

In Exercises 17–24, write a formula for the general term (the nth term) of each geometric sequence. Then use the formula for a_n to find a_7, the seventh term of the sequence.

17. $3, 12, 48, 192, \ldots$
18. $3, 15, 75, 375, \ldots$
19. $18, 6, 2, \frac{2}{3}, \ldots$
20. $12, 6, 3, \frac{3}{2}, \ldots$

21. $1.5, -3, 6, -12, \ldots$
22. $5, -1, \frac{1}{5}, -\frac{1}{25}, \ldots$
23. $0.0004, -0.004, 0.04, -0.4, \ldots$
24. $0.0007, -0.007, 0.07, -0.7, \ldots$

Use the formula for the sum of the first n terms of a geometric sequence to solve Exercises 25–30.

25. Find the sum of the first 12 terms of the geometric sequence: $2, 6, 18, 54, \ldots$.
26. Find the sum of the first 12 terms of the geometric sequence: $3, 6, 12, 24, \ldots$.
27. Find the sum of the first 11 terms of the geometric sequence: $3, -6, 12, -24, \ldots$.
28. Find the sum of the first 11 terms of the geometric sequence: $4, -12, 36, -108, \ldots$.
29. Find the sum of the first 14 terms of the geometric sequence: $-\frac{3}{2}, 3, -6, 12, \ldots$.
30. Find the sum of the first 14 terms of the geometric sequence: $-\frac{1}{24}, \frac{1}{12}, -\frac{1}{6}, \frac{1}{3}, \ldots$.

In Exercises 31–36, find the indicated sum. Use the formula for the sum of the first n terms of a geometric sequence.

31. $\sum\limits_{i=1}^{8} 3^i$
32. $\sum\limits_{i=1}^{6} 4^i$
33. $\sum\limits_{i=1}^{10} 5 \cdot 2^i$
34. $\sum\limits_{i=1}^{7} 4(-3)^i$
35. $\sum\limits_{i=1}^{6} \left(\frac{1}{2}\right)^{i+1}$
36. $\sum\limits_{i=1}^{6} \left(\frac{1}{3}\right)^{i+1}$

In Exercises 37–44, find the sum of each infinite geometric series.

37. $1 + \dfrac{1}{3} + \dfrac{1}{9} + \dfrac{1}{27} + \cdots$ **38.** $1 + \dfrac{1}{4} + \dfrac{1}{16} + \dfrac{1}{64} + \cdots$

39. $3 + \dfrac{3}{4} + \dfrac{3}{4^2} + \dfrac{3}{4^3} + \cdots$ **40.** $5 + \dfrac{5}{6} + \dfrac{5}{6^2} + \dfrac{5}{6^3} + \cdots$

41. $1 - \dfrac{1}{2} + \dfrac{1}{4} - \dfrac{1}{8} + \cdots$ **42.** $3 - 1 + \dfrac{1}{3} - \dfrac{1}{9} + \cdots$

43. $\displaystyle\sum_{i=1}^{\infty} 8(-0.3)^{i-1}$ **44.** $\displaystyle\sum_{i=1}^{\infty} 12(-0.7)^{i-1}$

In Exercises 45–50, express each repeating decimal as a fraction in lowest terms.

45. $0.\overline{5} = \dfrac{5}{10} + \dfrac{5}{100} + \dfrac{5}{1000} + \dfrac{5}{10,000} + \cdots$

46. $0.\overline{1} = \dfrac{1}{10} + \dfrac{1}{100} + \dfrac{1}{1000} + \dfrac{1}{10,000} + \cdots$

47. $0.\overline{47} = \dfrac{47}{100} + \dfrac{47}{10,000} + \dfrac{47}{1,000,000} + \cdots$

48. $0.\overline{83} = \dfrac{83}{100} + \dfrac{83}{10,000} + \dfrac{83}{1,000,000} + \cdots$

49. $0.\overline{257}$

50. $0.\overline{529}$

In Exercises 51–56, the general term of a sequence is given. Determine whether the sequence is arithmetic, geometric, or neither. If the sequence is arithmetic, find the common difference; if it is geometric, find the common ratio.

51. $a_n = n + 5$ **52.** $a_n = n - 3$

53. $a_n = 2^n$ **54.** $a_n = \left(\frac{1}{2}\right)^n$

55. $a_n = n^2 + 5$ **56.** $a_n = n^2 - 3$

Practice Plus

In Exercises 57–62, let

$$\{a_n\} = -5, 10, -20, 40, \ldots,$$
$$\{b_n\} = 10, -5, -20, -35, \ldots,$$

and

$$\{c_n\} = -2, 1, -\tfrac{1}{2}, \tfrac{1}{4}, \ldots.$$

57. Find $a_{10} + b_{10}$.

58. Find $a_{11} + b_{11}$.

59. Find the difference between the sum of the first 10 terms of $\{a_n\}$ and the sum of the first 10 terms of $\{b_n\}$.

60. Find the difference between the sum of the first 11 terms of $\{a_n\}$ and the sum of the first 11 terms of $\{b_n\}$.

61. Find the product of the sum of the first 6 terms of $\{a_n\}$ and the sum of the infinite series containing all the terms of $\{c_n\}$.

62. Find the product of the sum of the first 9 terms of $\{a_n\}$ and the sum of the infinite series containing all the terms of $\{c_n\}$.

In Exercises 63–64, find a_2 and a_3 for each geometric sequence.

63. $8, a_2, a_3, 27$

64. $2, a_2, a_3, -54$

Application Exercises

Use the formula for the general term (the nth term) of a geometric sequence to solve Exercises 65–68.

In Exercises 65–66, suppose you save $1 the first day of a month, $2 the second day, $4 the third day, and so on. That is, each day you save twice as much as you did the day before.

65. What will you put aside for savings on the fifteenth day of the month?

66. What will you put aside for savings on the thirtieth day of the month?

67. A professional baseball player signs a contract with a beginning salary of $3,000,000 for the first year and an annual increase of 4% per year beginning in the second year. That is, beginning in year 2, the athlete's salary will be 1.04 times what it was in the previous year. What is the athlete's salary for year 7 of the contract? Round to the nearest dollar.

68. You are offered a job that pays $30,000 for the first year with an annual increase of 5% per year beginning in the second year. That is, beginning in year 2, your salary will be 1.05 times what it was in the previous year. What can you expect to earn in your sixth year on the job?

In Exercises 69–70, you will develop geometric sequences that model the population growth for California and Texas, the two most-populated U.S. states.

69. The table shows the population of California for 2000 and 2010, with estimates given by the U.S. Census Bureau for 2001 through 2009.

Year	2000	2001	2002	2003	2004	2005
Population in millions	33.87	34.21	34.55	34.90	35.25	35.60

Year	2006	2007	2008	2009	2010
Population in millions	36.00	36.36	36.72	37.09	37.25

a. Divide the population for each year by the population in the preceding year. Round to two decimal places and show that California has a population increase that is approximately geometric.

b. Write the general term of the geometric sequence modeling California's population, in millions, n years after 1999.

c. Use your model from part (b) to project California's population, in millions, for the year 2020. Round to two decimal places.

70. The table shows the population of Texas for 2000 and 2010, with estimates given by the U.S. Census Bureau for 2001 through 2009.

Year	2000	2001	2002	2003	2004	2005
Population in millions	20.85	21.27	21.70	22.13	22.57	23.02

Year	2006	2007	2008	2009	2010
Population in millions	23.48	23.95	24.43	24.92	25.15

a. Divide the population for each year by the population in the preceding year. Round to two decimal places and show that Texas has a population increase that is approximately geometric.

b. Write the general term of the geometric sequence modeling Texas's population, in millions, n years after 1999.

c. Use your model from part (b) to project Texas's population, in millions, for the year 2020. Round to two decimal places.

Use the formula for the sum of the first n terms of a geometric sequence to solve Exercises 71–76.

In Exercises 71–72, you save $1 the first day of a month, $2 the second day, $4 the third day, continuing to double your savings each day.

71. What will your total savings be for the first 15 days?

72. What will your total savings be for the first 30 days?

73. A job pays a salary of $24,000 the first year. During the next 19 years, the salary increases by 5% each year. What is the total lifetime salary over the 20-year period? Round to the nearest dollar.

74. You are investigating two employment opportunities. Company A offers $30,000 the first year. During the next four years, the salary is guaranteed to increase by 6% per year. Company B offers $32,000 the first year, with guaranteed annual increases of 3% per year after that. Which company offers the better total salary for a five-year contract? By how much? Round to the nearest dollar.

75. A pendulum swings through an arc of 20 inches. On each successive swing, the length of the arc is 90% of the previous length.

$$20, \quad 0.9(20), \quad 0.9^2(20), \quad 0.9^3(20), \ldots$$

| 1st swing | 2nd swing | 3rd swing | 4th swing |

After 10 swings, what is the total length of the distance the pendulum has swung?

76. A pendulum swings through an arc of 16 inches. On each successive swing, the length of the arc is 96% of the previous length.

$$16, \quad 0.96(16), \quad (0.96)^2(16), \quad (0.96)^3(16), \ldots$$

| 1st swing | 2nd swing | 3rd swing | 4th swing |

After 10 swings, what is the total length of the distance the pendulum has swung?

Use the formula for the value of an annuity to solve Exercises 77–84. Round answers to the nearest dollar.

77. To save money for a sabbatical to earn a master's degree, you deposit $2000 at the end of each year in an annuity that pays 7.5% compounded annually.

a. How much will you have saved at the end of five years?

b. Find the interest.

78. To save money for a sabbatical to earn a master's degree, you deposit $2500 at the end of each year in an annuity that pays 6.25% compounded annually.

a. How much will you have saved at the end of five years?

b. Find the interest.

79. At age 25, to save for retirement, you decide to deposit $50 at the end of each month in an IRA that pays 5.5% compounded monthly.

a. How much will you have from the IRA when you retire at age 65?

b. Find the interest.

80. At age 25, to save for retirement, you decide to deposit $75 at the end of each month in an IRA that pays 6.5% compounded monthly.

a. How much will you have from the IRA when you retire at age 65?

b. Find the interest.

81. To offer scholarship funds to children of employees, a company invests $10,000 at the end of every three months in an annuity that pays 10.5% compounded quarterly.

a. How much will the company have in scholarship funds at the end of ten years?

b. Find the interest.

82. To offer scholarship funds to children of employees, a company invests $15,000 at the end of every three months in an annuity that pays 9% compounded quarterly.

a. How much will the company have in scholarship funds at the end of ten years?

b. Find the interest.

83. Here are two ways of investing $30,000 for 20 years.

Lump-Sum Deposit	Rate	Time
$30,000	5% compounded annually	20 years

Periodic Deposits	Rate	Time
$1500 at the end of each year	5% compounded annually	20 years

After 20 years, how much more will you have from the lump-sum investment than from the annuity?

84. Here are two ways of investing $40,000 for 25 years.

Lump-Sum Deposit	Rate	Time
$40,000	6.5% compounded annually	25 years

Periodic Deposits	Rate	Time
$1600 at the end of each year	6.5% compounded annually	25 years

After 25 years, how much more will you have from the lump-sum investment than from the annuity?

Use the formula for the sum of an infinite geometric series to solve Exercises 85–87.

85. A new factory in a small town has an annual payroll of $6 million. It is expected that 60% of this money will be spent in the town by factory personnel. The people in the town who receive this money are expected to spend 60% of what they receive in the town, and so on. What is the total of all this spending, called the *total economic impact* of the factory, on the town each year?

86. How much additional spending will be generated by a $10 billion tax rebate if 60% of all income is spent?

87. If the shading process shown in the figure is continued indefinitely, what fractional part of the largest square will eventually be shaded?

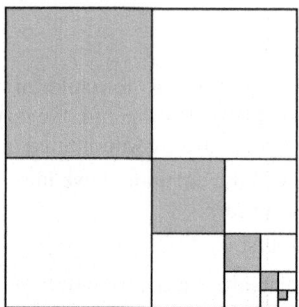

Writing in Mathematics

88. What is a geometric sequence? Give an example with your explanation.

89. What is the common ratio in a geometric sequence?

90. Explain how to find the general term of a geometric sequence.

91. Explain how to find the sum of the first n terms of a geometric sequence without having to add up all the terms.

92. What is an annuity?

93. What is the difference between a geometric sequence and an infinite geometric series?

94. How do you determine if an infinite geometric series has a sum? Explain how to find the sum of such an infinite geometric series.

95. Would you rather have $10,000,000 and a brand new BMW, or 1¢ today, 2¢ tomorrow, 4¢ on day 3, 8¢ on day 4, 16¢ on day 5, and so on, for 30 days? Explain.

96. For the first 30 days of a flu outbreak, the number of students on your campus who become ill is increasing. Which is worse: The number of students with the flu is increasing arithmetically or is increasing geometrically? Explain your answer.

Technology Exercises

97. Use the $\boxed{\text{SEQ}}$ (sequence) capability of a graphing utility and the formula you obtained for a_n to verify the value you found for a_7 in any three exercises from Exercises 17–24.

98. Use the capability of a graphing utility to calculate the sum of a sequence to verify any three of your answers to Exercises 31–36.

In Exercises 99–100, use a graphing utility to graph the function. Determine the horizontal asymptote for the graph of f and discuss its relationship to the sum of the given series.

99. Function Series

$$f(x) = \frac{2\left[1 - \left(\frac{1}{3}\right)^x\right]}{1 - \frac{1}{3}}$$ $2 + 2\left(\frac{1}{3}\right) + 2\left(\frac{1}{3}\right)^2 + 2\left(\frac{1}{3}\right)^3 + \cdots$

100. Function Series

$$f(x) = \frac{4[1 - (0.6)^x]}{1 - 0.6}$$ $4 + 4(0.6) + 4(0.6)^2 + 4(0.6)^3 + \cdots$

Critical Thinking Exercises

Make Sense? *In Exercises 101–104, determine whether each statement makes sense or does not make sense, and explain your reasoning.*

101. There's no end to the number of geometric sequences that I can generate whose first term is 5 if I pick nonzero numbers r and multiply 5 by each value of r repeatedly.

102. I've noticed that the big difference between arithmetic and geometric sequences is that arithmetic sequences are based on addition and geometric sequences are based on multiplication.

103. I modeled California's population growth with a geometric sequence, so my model is an exponential function whose domain is the set of natural numbers.

104. I used a formula to find the sum of the infinite geometric series $3 + 1 + \frac{1}{3} + \frac{1}{9} + \cdots$ and then checked my answer by actually adding all the terms.

In Exercises 105–108, determine whether each statement is true or false. If the statement is false, make the necessary change(s) to produce a true statement.

105. The sequence $2, 6, 24, 120, \ldots$ is an example of a geometric sequence.

106. The sum of the geometric series $\frac{1}{2} + \frac{1}{4} + \frac{1}{8} + \cdots + \frac{1}{512}$ can only be estimated without knowing precisely what terms occur between $\frac{1}{8}$ and $\frac{1}{512}$.

107. $10 - 5 + \frac{5}{2} - \frac{5}{4} + \cdots = \dfrac{10}{1 - \frac{1}{2}}$

108. If the nth term of a geometric sequence is $a_n = 3(0.5)^{n-1}$, the common ratio is $\frac{1}{2}$.

109. In a pest-eradication program, sterilized male flies are released into the general population each day. Ninety percent of those flies will survive a given day. How many flies should be released each day if the long-range goal of the program is to keep 20,000 sterilized flies in the population?

110. You are now 25 years old and would like to retire at age 55 with a retirement fund of $1,000,000. How much should you deposit at the end of each month for the next 30 years in an IRA paying 10% annual interest compounded monthly to achieve your goal? Round to the nearest dollar.

Group Exercise

111. Group members serve as a financial team analyzing the three options given to the professional baseball player described in the chapter opener on page 715. As a group, determine which option provides the most amount of money over the six-year contract and which provides the least. Describe one advantage and one disadvantage to each option.

Preview Exercises

Exercises 112–114 will help you prepare for the material covered in the next section.

In Exercises 112–113, show that

$$1 + 2 + 3 + \cdots + n = \frac{n(n + 1)}{2}$$

is true for the given value of n.

112. $n = 3$: Show that $1 + 2 + 3 = \dfrac{3(3 + 1)}{2}$.

113. $n = 5$: Show that $1 + 2 + 3 + 4 + 5 = \dfrac{5(5 + 1)}{2}$.

114. Simplify: $\dfrac{k(k + 1)(2k + 1)}{6} + (k + 1)^2$.

CHAPTER 8 Mid-Chapter Check Point

WHAT YOU KNOW: We learned that a sequence is a function whose domain is the set of positive integers. In an arithmetic sequence, each term after the first differs from the preceding term by a constant, the common difference, d. In a geometric sequence, each term after the first is obtained by multiplying the preceding term by a nonzero constant, the common ratio, r. We found the general term of arithmetic sequences $[a_n = a_1 + (n - 1)d]$ and geometric sequences $[a_n = a_1 r^{n-1}]$ and used these formulas to find particular terms. We determined the sum of the first n terms of arithmetic sequences $\left[S_n = \dfrac{n}{2}(a_1 + a_n) \right]$ and geometric sequences $\left[S_n = \dfrac{a_1(1 - r^n)}{1 - r} \right]$. Finally, we determined the sum of an infinite geometric series,

$$a_1 + a_1 r + a_1 r^2 + a_1 r^3 + \cdots, \text{ if } -1 < r < 1 \left(S = \frac{a_1}{1 - r} \right).$$

In Exercises 1–4, write the first five terms of each sequence. Assume that d represents the common difference of an arithmetic sequence and r represents the common ratio of a geometric sequence.

1. $a_n = (-1)^{n+1} \dfrac{n}{(n - 1)!}$

2. $a_1 = 5, d = -3$

3. $a_1 = 5, r = -3$

4. $a_1 = 3, a_n = -a_{n-1} + 4$

In Exercises 5–7, write a formula for the general term (the nth term) of each sequence. Then use the formula to find the indicated term.

5. $2, 6, 10, 14, \ldots; a_{20}$

6. $3, 6, 12, 24, \ldots; a_{10}$

7. $\dfrac{3}{2}, 1, \dfrac{1}{2}, 0, \ldots; a_{30}$

8. Find the sum of the first ten terms of the sequence:

$$5, 10, 20, 40, \ldots.$$

9. Find the sum of the first 50 terms of the sequence:

$$-2, 0, 2, 4, \ldots.$$

10. Find the sum of the first ten terms of the sequence:

$$-20, 40, -80, 160, \ldots.$$

11. Find the sum of the first 100 terms of the sequence:

$$4, -2, -8, -14, \ldots.$$

In Exercises 12–15, find each indicated sum.

12. $\displaystyle\sum_{i=1}^{4} (i + 4)(i - 1)$

13. $\displaystyle\sum_{i=1}^{50} (3i - 2)$

14. $\displaystyle\sum_{i=1}^{6} \left(\frac{3}{2} \right)^i$

15. $\displaystyle\sum_{i=1}^{\infty} \left(-\frac{2}{5} \right)^{i-1}$

16. Express $0.\overline{45}$ as a fraction in lowest terms.

17. Express the sum using summation notation. Use i for the index of summation.

$$\frac{1}{3} + \frac{2}{4} + \frac{3}{5} + \cdots + \frac{18}{20}$$

18. A skydiver falls 16 feet during the first second of a dive, 48 feet during the second second, 80 feet during the third second, 112 feet during the fourth second, and so on. Find the distance that the skydiver falls during the 15th second and the total distance the skydiver falls in 15 seconds.

19. If the average value of a house increases 10% per year, how much will a house costing \$120,000 be worth in 10 years? Round to the nearest dollar.

Mathematical Induction

Objectives

1. Understand the principle of mathematical induction.

2. Prove statements using mathematical induction.

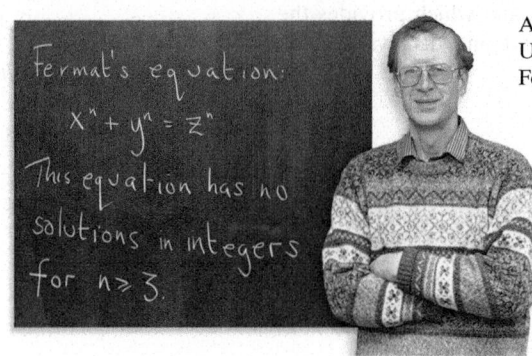

After ten years of work, Princeton University's Andrew Wiles proved Fermat's Last Theorem.

Pierre de Fermat (1601–1665) was a lawyer who enjoyed studying mathematics. In a margin of one of his books, he claimed that no positive integers satisfy

$$x^n + y^n = z^n$$

if n is an integer greater than or equal to 3.

If $n = 2$, we can find positive integers satisfying $x^n + y^n = z^n$, or $x^2 + y^2 = z^2$:

$$3^2 + 4^2 = 5^2.$$

However, Fermat claimed that no positive integers satisfy

$$x^3 + y^3 = z^3, \quad x^4 + y^4 = z^4, \quad x^5 + y^5 = z^5,$$

and so on. Fermat claimed to have a proof of his conjecture but added, "The margin of my book is too narrow to write it down." Some believe that he never had a proof and intended to frustrate his colleagues.

In 1994, 40-year-old Princeton math professor Andrew Wiles proved Fermat's Last Theorem using a principle called *mathematical induction*. In this section, you will learn how to use this powerful method to prove statements about the positive integers.

1 Understand the principle of mathematical induction.

The Principle of Mathematical Induction

How do we prove statements using mathematical induction? Let's consider an example. We will prove a statement that appears to give a correct formula for the sum of the first n positive integers:

$$S_n: 1 + 2 + 3 + \cdots + n = \frac{n(n + 1)}{2}.$$

We can verify S_n for, say, the first four positive integers. If $n = 1$, the statement S_1 is

Take the first term on the left. $1 \stackrel{?}{=} \dfrac{1(1 + 1)}{2}$ Substitute 1 for n on the right.

$$1 \stackrel{?}{=} \frac{1 \cdot 2}{2}$$

$$1 = 1 \checkmark.$$ This true statement shows that S_1 is true.

If $n = 2$, the statement S_2 is

Add the first two terms on the left. $1 + 2 \stackrel{?}{=} \dfrac{2(2 + 1)}{2}$ Substitute 2 for n on the right.

$$3 \stackrel{?}{=} \frac{2 \cdot 3}{2}$$

$$3 = 3 \checkmark.$$ This true statement shows that S_2 is true.

If $n = 3$, the statement S_3 is

Add the first three terms on the left. $1 + 2 + 3 \overset{?}{=} \dfrac{3(3 + 1)}{2}$ Substitute 3 for n on the right.

$$6 \overset{?}{=} \dfrac{3 \cdot 4}{2}$$

$6 = 6 \checkmark.$ This true statement shows that S_3 is true.

Finally, if $n = 4$, the statement S_4 is

Add the first four terms on the left. $1 + 2 + 3 + 4 \overset{?}{=} \dfrac{4(4 + 1)}{2}$ Substitute 4 for n on the right.

$$10 \overset{?}{=} \dfrac{4 \cdot 5}{2}$$

$10 = 10 \checkmark.$ This true statement shows that S_4 is true.

This approach does *not* prove that the given statement S_n is true for every positive integer n. The fact that the formula produces true statements for $n = 1, 2, 3$, and 4 does not guarantee that it is valid for all positive integers n. Thus, we need to be able to verify the truth of S_n without verifying the statement for each and every one of the positive integers.

A legitimate proof of the given statement S_n involves a technique called **mathematical induction**.

The Principle of Mathematical Induction

Let S_n be a statement involving the positive integer n. If

1. S_1 is true, and
2. the truth of the statement S_k implies the truth of the statement S_{k+1}, for every positive integer k,

then the statement S_n is true for all positive integers n.

The principle of mathematical induction can be illustrated using an unending line of dominoes, as shown in **Figure 8.8**. If the first domino is pushed over, it knocks down the next, which knocks down the next, and so on, in a chain reaction. To topple all the dominoes in the infinite sequence, two conditions must be satisfied:

1. The first domino must be knocked down.
2. If the domino in position k is knocked down, then the domino in position $k + 1$ must be knocked down.

FIGURE 8.8 Falling dominoes illustrate the principle of mathematical induction.

If the second condition is not satisfied, it does not follow that all the dominoes will topple. For example, suppose the dominoes are spaced far enough apart so that a falling domino does not push over the next domino in the line.

The domino analogy provides the two steps that are required in a proof by mathematical induction.

The Steps in a Proof by Mathematical Induction

Let S_n be a statement involving the positive integer n. To prove that S_n is true for all positive integers n requires two steps.

Step 1 Show that S_1 is true.
Step 2 Show that if S_k is assumed to be true, then S_{k+1} is also true, for every positive integer k.

Notice that to prove S_n, we work only with the statements S_1, S_k, and S_{k+1}. Our first example provides practice in writing these statements.

EXAMPLE 1 Writing S_1, S_k, and S_{k+1}

For the given statement S_n, write the three statements S_1, S_k, and S_{k+1}.

a. S_n: $1 + 2 + 3 + \cdots + n = \dfrac{n(n + 1)}{2}$

b. S_n: $1^2 + 2^2 + 3^2 + \cdots + n^2 = \dfrac{n(n + 1)(2n + 1)}{6}$

SOLUTION

a. We begin with

$$S_n: 1 + 2 + 3 + \cdots + n = \frac{n(n + 1)}{2}.$$

Write S_1 by taking the first term on the left and replacing n with 1 on the right.

$$S_1: 1 = \frac{1(1 + 1)}{2}$$

Write S_k by taking the sum of the first k terms on the left and replacing n with k on the right.

$$S_k: 1 + 2 + 3 + \cdots + k = \frac{k(k + 1)}{2}$$

Write S_{k+1} by taking the sum of the first $k + 1$ terms on the left and replacing n with $k + 1$ on the right.

$$S_{k+1}: 1 + 2 + 3 + \cdots + (k + 1) = \frac{(k + 1)[(k + 1) + 1]}{2}$$

$$S_{k+1}: 1 + 2 + 3 + \cdots + (k + 1) = \frac{(k + 1)(k + 2)}{2} \qquad \text{Simplify on the right.}$$

b. We begin with

$$S_n: 1^2 + 2^2 + 3^2 + \cdots + n^2 = \frac{n(n + 1)(2n + 1)}{6}.$$

Write S_1 by taking the first term on the left and replacing n with 1 on the right.

$$S_1: 1^2 = \frac{1(1 + 1)(2 \cdot 1 + 1)}{6}$$

Using S_n: $1^2 + 2^2 + 3^2 + \cdots + n^2 = \dfrac{n(n + 1)(2n + 1)}{6}$, we write S_k by taking the sum of the first k terms on the left and replacing n with k on the right.

$$S_k: 1^2 + 2^2 + 3^2 + \cdots + k^2 = \frac{k(k + 1)(2k + 1)}{6}$$

Write S_{k+1} by taking the sum of the first $k + 1$ terms on the left and replacing n with $k + 1$ on the right.

$$S_{k+1}: 1^2 + 2^2 + 3^2 + \cdots + (k + 1)^2 = \frac{(k + 1)[(k + 1) + 1][2(k + 1) + 1]}{6}$$

$$S_{k+1}: 1^2 + 2^2 + 3^2 + \cdots + (k + 1)^2 = \frac{(k + 1)(k + 2)(2k + 3)}{6} \qquad \text{Simplify on the right.} \qquad \bullet\bullet\bullet$$

✓ Check Point **1** For the given statement S_n, write the three statements S_1, S_k, and S_{k+1}.

a. $2 + 4 + 6 + \cdots + 2n = n(n + 1)$

b. $1^3 + 2^3 + 3^3 + \cdots + n^3 = \dfrac{n^2(n + 1)^2}{4}$

Always simplify S_{k+1} before trying to use mathematical induction to prove that S_n is true. For example, consider

$$S_n: 1^2 + 3^2 + 5^2 + \cdots + (2n - 1)^2 = \frac{n(2n - 1)(2n + 1)}{3}.$$

Begin by writing S_{k+1} as follows:

$$S_{k+1}: 1^2 + 3^2 + 5^2 + \cdots + [2(k + 1) - 1]^2$$

$$= \frac{(k + 1)[2(k + 1) - 1][2(k + 1) + 1]}{3}.$$

The sum of the first $k + 1$ terms

Replace n with $k + 1$ on the right side of S_n.

Now simplify both sides of the equation.

$$S_{k+1}: 1^2 + 3^2 + 5^2 + \cdots + (2k + 2 - 1)^2 = \frac{(k + 1)(2k + 2 - 1)(2k + 2 + 1)}{3}$$

$$S_{k+1}: 1^2 + 3^2 + 5^2 + \cdots + (2k + 1)^2 = \frac{(k + 1)(2k + 1)(2k + 3)}{3}$$

② Prove statements using mathematical induction.

Proving Statements about Positive Integers Using Mathematical Induction

Now that we know how to find S_1, S_k, and S_{k+1}, let's see how we can use these statements to carry out the two steps in a proof by mathematical induction. In Examples 2 and 3, we will use the statements S_1, S_k, and S_{k+1} to prove each of the statements S_n that we worked with in Example 1.

EXAMPLE 2 Proving a Formula by Mathematical Induction

Use mathematical induction to prove that

$$1 + 2 + 3 + \cdots + n = \frac{n(n + 1)}{2}$$

for all positive integers n.

SOLUTION

Step 1 Show that S_1 is true. Statement S_1 is

$$1 = \frac{1(1 + 1)}{2}.$$

Simplifying on the right, we obtain $1 = 1$. This true statement shows that S_1 is true.

Step 2 Show that if S_k is true, then S_{k+1} is true. Using S_k and S_{k+1} from Example 1(a), show that the truth of S_k,

$$1 + 2 + 3 + \cdots + k = \frac{k(k + 1)}{2},$$

implies the truth of S_{k+1},

$$1 + 2 + 3 + \cdots + (k + 1) = \frac{(k + 1)(k + 2)}{2}.$$

S_k:

$$1 + 2 + 3 + \cdots + k = \frac{k(k + 1)}{2}$$

S_{k+1}:

$$1 + 2 + 3 + \cdots + (k + 1) = \frac{(k + 1)(k + 2)}{2}$$

The statements S_k and S_{k+1} (repeated)

We will work with S_k. Because we assume that S_k is true, we add the next consecutive integer after k—namely, $k + 1$—to both sides.

$$1 + 2 + 3 + \cdots + k = \frac{k(k + 1)}{2}$$ This is S_k, which we assume is true.

$$1 + 2 + 3 + \cdots + k + (k + 1) = \frac{k(k + 1)}{2} + (k + 1)$$ Add $k + 1$ to both sides of the equation.

 We do not have to write this k because k is understood to be the integer that precedes $k + 1$.

$$1 + 2 + 3 + \cdots + (k + 1) = \frac{k(k + 1)}{2} + \frac{2(k + 1)}{2}$$ Write the right side with a common denominator of 2.

$$1 + 2 + 3 + \cdots + (k + 1) = \frac{(k + 1)}{2}(k + 2)$$ Factor out the common factor $\frac{k + 1}{2}$ on the right.

$$1 + 2 + 3 + \cdots + (k + 1) = \frac{(k + 1)(k + 2)}{2}$$ This final result is the statement S_{k+1}.

We have shown that if we assume that S_k is true and we add $k + 1$ to both sides of S_k, then S_{k+1} is also true. By the principle of mathematical induction, the statement S_n, namely,

$$1 + 2 + 3 + \cdots + n = \frac{n(n + 1)}{2},$$

is true for every positive integer n. •••

✓ Check Point **2** Use mathematical induction to prove that

$$2 + 4 + 6 + \cdots + 2n = n(n + 1)$$

for all positive integers n.

EXAMPLE 3 Proving a Formula by Mathematical Induction

Use mathematical induction to prove that

$$1^2 + 2^2 + 3^2 + \cdots + n^2 = \frac{n(n + 1)(2n + 1)}{6}$$

for all positive integers n.

SOLUTION

Step 1 **Show that S_1 is true.** Statement S_1 is

$$1^2 = \frac{1(1 + 1)(2 \cdot 1 + 1)}{6}.$$

Simplifying, we obtain $1 = \dfrac{1 \cdot 2 \cdot 3}{6}$. Further simplification on the right gives the statement $1 = 1$. This true statement shows that S_1 is true.

Step 2 **Show that if S_k is true, then S_{k+1} is true.** Using S_k and S_{k+1} from Example 1(b), show that the truth of

$$S_k: 1^2 + 2^2 + 3^2 + \cdots + k^2 = \frac{k(k + 1)(2k + 1)}{6}$$

implies the truth of

$$S_{k+1}: 1^2 + 2^2 + 3^2 + \cdots + (k + 1)^2 = \frac{(k + 1)(k + 2)(2k + 3)}{6}.$$

We will work with S_k. Because we assume that S_k is true, we add the square of the next consecutive integer after k, namely, $(k + 1)^2$, to both sides of the equation.

$$1^2 + 2^2 + 3^2 + \cdots + k^2 = \frac{k(k + 1)(2k + 1)}{6}$$

This is S_k, assumed to be true. We must work with this and show S_{k+1} is true.

$$1^2 + 2^2 + 3^2 + \cdots + k^2 + (k + 1)^2 = \frac{k(k + 1)(2k + 1)}{6} + (k + 1)^2$$

Add $(k + 1)^2$ to both sides.

$$1^2 + 2^2 + 3^2 + \cdots + (k + 1)^2 = \frac{k(k + 1)(2k + 1)}{6} + \frac{6(k + 1)^2}{6}$$

It is not necessary to write k^2 on the left. Express the right side with the least common denominator, 6.

$$= \frac{(k + 1)}{6}[k(2k + 1) + 6(k + 1)]$$

Factor out the common factor $\frac{k + 1}{6}$.

$$= \frac{(k + 1)}{6}(2k^2 + 7k + 6)$$

Multiply and combine like terms.

$$= \frac{(k + 1)}{6}(k + 2)(2k + 3)$$

Factor $2k^2 + 7k + 6$.

$$= \frac{(k + 1)(k + 2)(2k + 3)}{6}$$

This final statement is S_{k+1}.

We have shown that if we assume that S_k is true, and we add $(k + 1)^2$ to both sides of S_k, then S_{k+1} is also true. By the principle of mathematical induction, the statement S_n, namely,

$$1^2 + 2^2 + 3^2 + \cdots + n^2 = \frac{n(n + 1)(2n + 1)}{6},$$

is true for every positive integer n. • • •

◉ Check Point **3** Use mathematical induction to prove that

$$1^3 + 2^3 + 3^3 + \cdots + n^3 = \frac{n^2(n + 1)^2}{4}$$

for all positive integers n.

Example 4 illustrates how mathematical induction can be used to prove statements about positive integers that do not involve sums.

EXAMPLE 4 Using the Principle of Mathematical Induction

Prove that 2 is a factor of $n^2 + 5n$ for all positive integers n.

SOLUTION

Step 1 Show that S_1 is true. Statement S_1 reads

2 is a factor of $1^2 + 5 \cdot 1$.

Simplifying the arithmetic, the statement reads

2 is a factor of 6.

This statement is true: That is, $6 = 2 \cdot 3$. This shows that S_1 is true.

Step 2 Show that if S_k is true, then S_{k+1} is true. Let's use S_n: 2 is a factor of $n^2 + S_n$ to write S_k and S_{k+1}:

$$S_k: \quad \text{2 is a factor of } k^2 + 5k.$$
$$S_{k+1}: \quad \text{2 is a factor of } (k+1)^2 + 5(k+1).$$

We can rewrite statement S_{k+1} by simplifying the algebraic expression in the statement as follows:

$$(k+1)^2 + 5(k+1) = k^2 + 2k + 1 + 5k + 5 = k^2 + 7k + 6.$$

> Use the formula $(A + B)^2 = A^2 + 2AB + B^2$.

Statement S_{k+1} now reads

$$\text{2 is a factor of } k^2 + 7k + 6.$$

We need to use statement S_k—that is, 2 is a factor of $k^2 + 5k$—to prove statement S_{k+1}. We do this as follows:

$$k^2 + 7k + 6 = (k^2 + 5k) + (2k + 6) = (k^2 + 5k) + 2(k + 3).$$

> We assume that 2 is a factor of $k^2 + 5k$ because we assume S_k is true.

> Factoring the last two terms shows that 2 is a factor of $2k + 6$.

The voice balloons show that 2 is a factor of $k^2 + 5k$ and of $2(k + 3)$. Thus, if S_k is true, 2 is a factor of the sum $(k^2 + 5k) + 2(k + 3)$, or of $k^2 + 7k + 6$. This is precisely the statement S_{k+1}. We have shown that if we assume that S_k is true, then S_{k+1} is also true. By the principle of mathematical induction, the statement S_n, namely 2 is a factor of $n^2 + 5n$, is true for every positive integer n. • • •

 Check Point 4 Prove that 2 is a factor of $n^2 + n$ for all positive integers n.

CONCEPT AND VOCABULARY CHECK

Fill in each blank so that the resulting statement is true.

1. The principle of mathematical _____ states that a statement involving positive integers is true for all positive integers when two conditions have been satisfied:

 The first condition states that the statement is true for the positive integer _____.
 The second condition states that if the statement is true for some positive integer k, it is also true for the next positive integer _____.

2. Consider the statement
$$2 + 4 + 6 + \cdots + 2n = n(n + 1).$$
If $n = 1$, the statement is $2 = 1(1 + 1)$.
If $n = 2$, the statement is $2 + 4 = 2(2 + 1)$.
If $n = 3$, the statement is _____.
If $n = k + 1$, the statement is _____.

3. Consider the statement
$$3 + 7 + 11 + \cdots + (4n - 1) = n(2n + 1).$$
If $n = 1$, the statement is $3 = 1(2 + 1)$.
If $n = 2$, the statement is $3 + 7 = 2(4 + 1)$.
If $n = 3$, the statement is _____.
If $n = k + 1$, the statement before the algebra is simplified is

_____.

If $n = k + 1$, the statement after the algebra is simplified is

_____.

4. Consider the statement
$$\text{2 is a factor of } n^2 + 3n.$$
If $n = 1$, the statement is 2 is a factor of _____.
If $n = 2$, the statement is 2 is a factor of _____.
If $n = 3$, the statement is 2 is a factor of _____.
If $n = k + 1$, the statement before the algebra is simplified is 2 is a factor of _____.
If $n = k + 1$, the statement after the algebra is simplified is 2 is a factor of _____.

5. $k^2 + 3k + 2 = (k^2 + k) + 2\,(\underline{\quad\quad})$

EXERCISE SET 8.4

Practice Exercises

In Exercises 1–4, a statement S_n about the positive integers is given. Write statements S_1, S_2, and S_3, and show that each of these statements is true.

1. S_n: $1 + 3 + 5 + \cdots + (2n - 1) = n^2$

2. S_n: $3 + 4 + 5 + \cdots + (n + 2) = \dfrac{n(n + 5)}{2}$

3. S_n: 2 is a factor of $n^2 - n$.

4. S_n: 3 is a factor of $n^3 - n$.

In Exercises 5–10, a statement S_n about the positive integers is given. Write statements S_k and S_{k+1}, simplifying statement S_{k+1} completely.

5. S_n: $4 + 8 + 12 + \cdots + 4n = 2n(n + 1)$

6. S_n: $3 + 4 + 5 + \cdots + (n + 2) = \dfrac{n(n + 5)}{2}$

7. S_n: $3 + 7 + 11 + \cdots + (4n - 1) = n(2n + 1)$

8. S_n: $2 + 7 + 12 + \cdots + (5n - 3) = \dfrac{n(5n - 1)}{2}$

9. S_n: 2 is a factor of $n^2 - n + 2$.

10. S_n: 2 is a factor of $n^2 - n$.

In Exercises 11–24, use mathematical induction to prove that each statement is true for every positive integer n.

11. $4 + 8 + 12 + \cdots + 4n = 2n(n + 1)$

12. $3 + 4 + 5 + \cdots + (n + 2) = \dfrac{n(n + 5)}{2}$

13. $1 + 3 + 5 + \cdots + (2n - 1) = n^2$

14. $3 + 6 + 9 + \cdots + 3n = \dfrac{3n(n + 1)}{2}$

15. $3 + 7 + 11 + \cdots + (4n - 1) = n(2n + 1)$

16. $2 + 7 + 12 + \cdots + (5n - 3) = \dfrac{n(5n - 1)}{2}$

17. $1 + 2 + 2^2 + \cdots + 2^{n-1} = 2^n - 1$

18. $1 + 3 + 3^2 + \cdots + 3^{n-1} = \dfrac{3^n - 1}{2}$

19. $2 + 4 + 8 + \cdots + 2^n = 2^{n+1} - 2$

20. $\dfrac{1}{2} + \dfrac{1}{4} + \dfrac{1}{8} + \cdots + \dfrac{1}{2^n} = 1 - \dfrac{1}{2^n}$

21. $1 \cdot 2 + 2 \cdot 3 + 3 \cdot 4 + \cdots + n(n + 1) = \dfrac{n(n + 1)(n + 2)}{3}$

22. $1 \cdot 3 + 2 \cdot 4 + 3 \cdot 5 + \cdots + n(n + 2) = \dfrac{n(n + 1)(2n + 7)}{6}$

23. $\dfrac{1}{1 \cdot 2} + \dfrac{1}{2 \cdot 3} + \dfrac{1}{3 \cdot 4} + \cdots + \dfrac{1}{n(n + 1)} = \dfrac{n}{n + 1}$

24. $\dfrac{1}{2 \cdot 3} + \dfrac{1}{3 \cdot 4} + \dfrac{1}{4 \cdot 5} + \cdots + \dfrac{1}{(n + 1)(n + 2)} = \dfrac{n}{2n + 4}$

Practice Plus

In Exercises 25–34, use mathematical induction to prove that each statement is true for every positive integer n.

25. 2 is a factor of $n^2 - n$.

26. 2 is a factor of $n^2 + 3n$.

27. 6 is a factor of $n(n + 1)(n + 2)$.

28. 3 is a factor of $n(n + 1)(n - 1)$.

29. $\displaystyle\sum_{i=1}^{n} 5 \cdot 6^i = 6(6^n - 1)$ 30. $\displaystyle\sum_{i=1}^{n} 7 \cdot 8^i = 8(8^n - 1)$

31. $n + 2 > n$

32. If $0 < x < 1$, then $0 < x^n < 1$.

33. $(ab)^n = a^n b^n$ 34. $\left(\dfrac{a}{b}\right)^n = \dfrac{a^n}{b^n}$

Writing in Mathematics

35. Explain how to use mathematical induction to prove that a statement is true for every positive integer n.

36. Consider the statement S_n given by

$$n^2 - n + 41 \text{ is prime.}$$

Although S_1, S_2, ..., S_{40} are true, S_{41} is false. Verify that S_{41} is false. Then describe how this is illustrated by the dominoes in the figure. What does this tell you about a pattern, or formula, that seems to work for several values of n?

Critical Thinking Exercises

Make Sense? In Exercises 37–40, determine whether each statement makes sense or does not make sense, and explain your reasoning.

37. I use mathematical induction to prove that statements are true for all real numbers n.

38. I begin proofs by mathematical induction by writing S_k and S_{k+1}, both of which I assume to be true.

39. When a line of falling dominoes is used to illustrate the principle of mathematical induction, it is not necessary for all the dominoes to topple.

40. This triangular arrangement of 36 circles illustrates that

$$1 + 2 + 3 + \cdots + n = \dfrac{n(n + 1)}{2}$$

is true for $n = 8$.

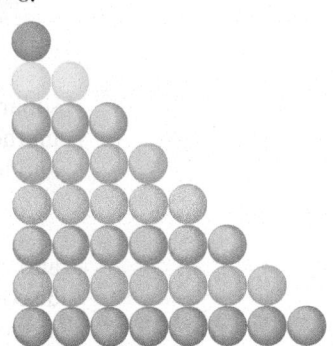

Some statements are false for the first few positive integers, but true for some positive integer m on. In these instances, you can prove S_n for $n \geq m$ by showing that S_m is true and that S_k implies S_{k+1} when $k \geq m$. Use this extended principle of mathematical induction to prove that each statement in Exercises 41–42 is true.

41. Prove that $n^2 > 2n + 1$ for $n \geq 3$. Show that the formula is true for $n = 3$ and then use step 2 of mathematical induction.

42. Prove that $2^n > n^2$ for $n \geq 5$. Show that the formula is true for $n = 5$ and then use step 2 of mathematical induction.

In Exercises 43–44, find S_1 through S_5 and then use the pattern to make a conjecture about S_n. Prove the conjectured formula for S_n by mathematical induction.

43. $S_n: \dfrac{1}{4} + \dfrac{1}{12} + \dfrac{1}{24} + \cdots + \dfrac{1}{2n(n + 1)} = ?$

44. $S_n: \left(1 - \dfrac{1}{2}\right)\left(1 - \dfrac{1}{3}\right)\left(1 - \dfrac{1}{4}\right) \cdots \left(1 - \dfrac{1}{n + 1}\right) = ?$

Group Exercise

45. Fermat's most notorious theorem, described in the section opener on page 754, baffled the greatest minds for more than three centuries. In 1994, after ten years of work, Princeton University's Andrew Wiles proved Fermat's Last Theorem. *People* magazine put him on its list of "the 25 most intriguing people of the year," the Gap asked him to model jeans, and Barbara Walters chased him for an interview. "Who's Barbara Walters?" asked the bookish Wiles, who had somehow gone through life without a television.

Using the 1993 PBS documentary "Solving Fermat: Andrew Wiles" or information about Andrew Wiles on the Internet, research and present a group seminar on what Wiles did to prove Fermat's Last Theorem, problems along the way, and the role of mathematical induction in the proof.

Preview Exercises

Exercises 46–48 will help you prepare for the material covered in the next section. Each exercise involves observing a pattern in the expanded form of the binomial expression $(a + b)^n$.

$$(a + b)^1 = a + b$$
$$(a + b)^2 = a^2 + 2ab + b^2$$
$$(a + b)^3 = a^3 + 3a^2b + 3ab^2 + b^3$$
$$(a + b)^4 = a^4 + 4a^3b + 6a^2b^2 + 4ab^3 + b^4$$
$$(a + b)^5 = a^5 + 5a^4b + 10a^3b^2 + 10a^2b^3 + 5ab^4 + b^5$$

46. Describe the pattern for the exponents on a.

47. Describe the pattern for the exponents on b.

48. Describe the pattern for the sum of the exponents on the variables in each term.

SECTION 8.5 The Binomial Theorem

Objectives

1. Evaluate a binomial coefficient.
2. Expand a binomial raised to a power.
3. Find a particular term in a binomial expansion.

Galaxies are groupings of billions of stars bound together by gravity. Some galaxies, such as the Centaurus galaxy shown here, are elliptical in shape.

Is mathematics discovered or invented? For example, planets revolve in elliptical orbits. Does that mean that the ellipse is out there, waiting for the mind to discover it? Or do people create the definition of an ellipse just as they compose a song? And is it possible for the same mathematics to be discovered/invented by independent researchers separated by time, place, and culture? This is precisely what occurred when mathematicians attempted to find efficient methods for raising binomials to higher and higher powers, such as

$$(x + 2)^3, (x + 2)^4, (x + 2)^5, (x + 2)^6,$$

and so on. In this section, we study higher powers of binomials and a method first discovered/invented by great minds in Eastern and Western cultures working independently.

① Evaluate a binomial coefficient.

Binomial Coefficients

Before turning to powers of binomials, we introduce a special notation that uses factorials.

Definition of a Binomial Coefficient $\binom{n}{r}$

For nonnegative integers n and r, with $n \geq r$, the expression $\binom{n}{r}$ (read "n above r") is called a **binomial coefficient** and is defined by

$$\binom{n}{r} = \frac{n!}{r!(n-r)!}.$$

The symbol $_nC_r$ is often used in place of $\binom{n}{r}$ to denote binomial coefficients.

EXAMPLE 1 Evaluating Binomial Coefficients

Evaluate:

a. $\binom{6}{2}$ **b.** $\binom{3}{0}$ **c.** $\binom{9}{3}$ **d.** $\binom{4}{4}$.

TECHNOLOGY

Graphing utilities can compute binomial coefficients. For example, to find $\binom{6}{2}$, many utilities require the sequence

6 $_nC_r$ 2 ENTER .

The graphing utility will display 15. Consult your manual and verify the other evaluations in Example 1.

SOLUTION

In each case, we apply the definition of the binomial coefficient.

a. $\binom{6}{2} = \dfrac{6!}{2!(6-2)!} = \dfrac{6!}{2!4!} = \dfrac{6 \cdot 5 \cdot 4!}{2 \cdot 1 \cdot 4!} = 15$

b. $\binom{3}{0} = \dfrac{3!}{0!(3-0)!} = \dfrac{3!}{0!3!} = \dfrac{1}{1} = 1$

> Remember that 0! = 1.

c. $\binom{9}{3} = \dfrac{9!}{3!(9-3)!} = \dfrac{9!}{3!6!} = \dfrac{9 \cdot 8 \cdot 7 \cdot 6!}{3 \cdot 2 \cdot 1 \cdot 6!} = 84$

d. $\binom{4}{4} = \dfrac{4!}{4!(4-4)!} = \dfrac{4!}{4!0!} = \dfrac{1}{1} = 1$ • • •

☑ **Check Point 1** Evaluate:

a. $\binom{6}{3}$ **b.** $\binom{6}{0}$ **c.** $\binom{8}{2}$ **d.** $\binom{3}{3}$.

② Expand a binomial raised to a power.

The Binomial Theorem

When we write out the *binomial expression* $(a + b)^n$, where n is a positive integer, a number of patterns begin to appear.

$$(a + b)^1 = a + b$$
$$(a + b)^2 = a^2 + 2ab + b^2$$
$$(a + b)^3 = a^3 + 3a^2b + 3ab^2 + b^3$$
$$(a + b)^4 = a^4 + 4a^3b + 6a^2b^2 + 4ab^3 + b^4$$
$$(a + b)^5 = a^5 + 5a^4b + 10a^3b^2 + 10a^2b^3 + 5ab^4 + b^5$$

$(a + b)^1 = a + b$

$(a + b)^2 = a^2 + 2ab + b^2$

$(a + b)^3 = a^3 + 3a^2b + 3ab^2 + b^3$

$(a + b)^4 = a^4 + 4a^3b + 6a^2b^2 + 4ab^3 + b^4$

$(a + b)^5 = a^5 + 5a^4b + 10a^3b^2 + 10a^2b^3 + 5ab^4 + b^5$

Observing patterns in the expansion of $(a + b)^n$ (repeated)

Each expanded form of the binomial expression is a polynomial. Observe the following patterns:

1. The first term in the expansion of $(a + b)^n$ is a^n. The exponents on a decrease by 1 in each successive term.

2. The exponents on b in the expansion of $(a + b)^n$ increase by 1 in each successive term. In the first term, the exponent on b is 0. (Because $b^0 = 1, b$ is not shown in the first term.) The last term is b^n.

3. The sum of the exponents on the variables in any term in the expansion of $(a + b)^n$ is equal to n.

4. The number of terms in the polynomial expansion is one greater than the power of the binomial, n. There are $n + 1$ terms in the expanded form of $(a + b)^n$.

Using these observations, the variable parts of the expansion of $(a + b)^6$ are

$$a^6, \quad a^5b, \quad a^4b^2, \quad a^3b^3, \quad a^2b^4, \quad ab^5, \quad b^6.$$

The first term is a^6, with the exponents on a decreasing by 1 in each successive term. The exponents on b increase from 0 to 6, with the last term being b^6. The sum of the exponents in each term is equal to 6.

We can generalize from these observations to obtain the variable parts of the expansion of $(a + b)^n$. They are

$$a^n, \quad a^{n-1}b, \quad a^{n-2}b^2, \quad a^{n-3}b^3, \dots, \quad ab^{n-1}, \quad b^n.$$

Exponents on a are decreasing by 1.
Exponents on b are increasing by 1.

Sum of exponents: $n - 1 + 1 = n$

Sum of exponents: $n - 3 + 3 = n$

Sum of exponents: $1 + n - 1 = n$

If we use binomial coefficients and the pattern for the variable part of each term, a formula called the **Binomial Theorem** can be used to expand any positive integral power of a binomial.

A Formula for Expanding Binomials: The Binomial Theorem

For any positive integer n,

$$(a + b)^n = \binom{n}{0}a^n + \binom{n}{1}a^{n-1}b + \binom{n}{2}a^{n-2}b^2 + \binom{n}{3}a^{n-3}b^3 + \cdots + \binom{n}{n}b^n$$

$$= \sum_{r=0}^{n} \binom{n}{r}a^{n-r}b^r.$$

EXAMPLE 2 Using the Binomial Theorem

Expand: $(x + 2)^4$.

SOLUTION

We use the Binomial Theorem

$$(a + b)^n = \binom{n}{0}a^n + \binom{n}{1}a^{n-1}b + \binom{n}{2}a^{n-2}b^2 + \binom{n}{3}a^{n-3}b^3 + \cdots + \binom{n}{n}b^n$$

to expand $(x + 2)^4$. In $(x + 2)^4$, $a = x$, $b = 2$, and $n = 4$. In the expansion, powers of x are in descending order, starting with x^4. Powers of 2 are in ascending order, starting with 2^0. (Because $2^0 = 1$, a 2 is not shown in the first term.) The sum of the exponents on x and 2 in each term is equal to 4, the exponent in the expression $(x + 2)^4$.

TECHNOLOGY

You can use a graphing utility's table feature to find the five binomial coefficients in Example 2.

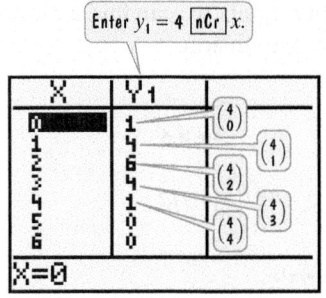

$$(x + 2)^4 = \binom{4}{0}x^4 + \binom{4}{1}x^3 \cdot 2 + \binom{4}{2}x^2 \cdot 2^2 + \binom{4}{3}x \cdot 2^3 + \binom{4}{4}2^4$$

These binomial coefficients are evaluated using $\binom{n}{r} = \frac{n!}{r!(n-r)!}$.

$$= \frac{4!}{0!4!}x^4 + \frac{4!}{1!3!}x^3 \cdot 2 + \frac{4!}{2!2!}x^2 \cdot 4 + \frac{4!}{3!1!}x \cdot 8 + \frac{4!}{4!0!} \cdot 16$$

$\frac{4!}{2!2!} = \frac{4 \cdot 3 \cdot 2!}{2! \cdot 2 \cdot 1} = \frac{12}{2} = 6$

Take a few minutes to verify the other factorial evaluations.

$$= 1 \cdot x^4 + 4x^3 \cdot 2 + 6x^2 \cdot 4 + 4x \cdot 8 + 1 \cdot 16$$
$$= x^4 + 8x^3 + 24x^2 + 32x + 16 \qquad \bullet\bullet\bullet$$

⌀ Check Point **2** Expand: $(x + 1)^4$.

EXAMPLE 3 Using the Binomial Theorem

Expand: $(2x - y)^5$.

SOLUTION

Because the Binomial Theorem involves the addition of two terms raised to a power, we rewrite $(2x - y)^5$ as $[2x + (-y)]^5$. We use the Binomial Theorem

$$(a + b)^n = \binom{n}{0}a^n + \binom{n}{1}a^{n-1}b + \binom{n}{2}a^{n-2}b^2 + \binom{n}{3}a^{n-3}b^3 + \cdots + \binom{n}{n}b^n$$

to expand $[2x + (-y)]^5$. In $[2x + (-y)]^5$, $a = 2x$, $b = -y$, and $n = 5$. In the expansion, powers of $2x$ are in descending order, starting with $(2x)^5$. Powers of $-y$ are in ascending order, starting with $(-y)^0$. [Because $(-y)^0 = 1$, a $-y$ is not shown in the first term.] The sum of the exponents on $2x$ and $-y$ in each term is equal to 5, the exponent in the expression $(2x - y)^5$.

$$(2x - y)^5 = [2x + (-y)]^5$$

$$= \binom{5}{0}(2x)^5 + \binom{5}{1}(2x)^4(-y) + \binom{5}{2}(2x)^3(-y)^2 + \binom{5}{3}(2x)^2(-y)^3 + \binom{5}{4}(2x)(-y)^4 + \binom{5}{5}(-y)^5$$

Evaluate binomial coefficients using $\binom{n}{r} = \frac{n!}{r!(n-r)!}$.

$$= \frac{5!}{0!5!}(2x)^5 + \frac{5!}{1!4!}(2x)^4(-y) + \frac{5!}{2!3!}(2x)^3(-y)^2 + \frac{5!}{3!2!}(2x)^2(-y)^3 + \frac{5!}{4!1!}(2x)(-y)^4 + \frac{5!}{5!0!}(-y)^5$$

$\frac{5!}{2!3!} = \frac{5 \cdot 4 \cdot 3!}{2 \cdot 1 \cdot 3!} = 10$

Take a few minutes to verify the other factorial evaluations.

$$= 1(2x)^5 + 5(2x)^4(-y) + 10(2x)^3(-y)^2 + 10(2x)^2(-y)^3 + 5(2x)(-y)^4 + 1(-y)^5$$

Raise both factors in these parentheses to the indicated powers.

$$= 1(32x^5) + 5(16x^4)(-y) + 10(8x^3)(-y)^2 + 10(4x^2)(-y)^3 + 5(2x)(-y)^4 + 1(-y)^5$$

Now raise $-y$ to the indicated powers.

$$= 1(32x^5) + 5(16x^4)(-y) + 10(8x^3)y^2 + 10(4x^2)(-y^3) + 5(2x)y^4 + 1(-y^5)$$

Multiplying factors in each of the six terms gives us the desired expansion:
$$(2x - y)^5 = 32x^5 - 80x^4y + 80x^3y^2 - 40x^2y^3 + 10xy^4 - y^5. \qquad \bullet\bullet\bullet$$

✅ **Check Point 3** Expand: $(x - 2y)^5$.

③ Find a particular term in a binomial expansion.

Finding a Particular Term in a Binomial Expansion

By observing the terms in the formula for expanding binomials, we can find a formula for finding a particular term without writing the entire expansion.

The exponent on b is 1 less than the term number.

Based on the observation in the bottom voice balloon, the $(r + 1)$st term of the expansion of $(a + b)^n$ is the term that contains b^r.

Finding a Particular Term in a Binomial Expansion

The $(r + 1)$st term of the expansion of $(a + b)^n$ is

$$\binom{n}{r} a^{n-r} b^r.$$

EXAMPLE 4 Finding a Single Term of a Binomial Expansion

Find the fourth term in the expansion of $(3x + 2y)^7$.

SOLUTION

The fourth term in the expansion of $(3x + 2y)^7$ contains $(2y)^3$. To find the fourth term, first note that $4 = 3 + 1$. Equivalently, the fourth term of $(3x + 2y)^7$ is the $(3 + 1)$st term. Thus, $r = 3$, $a = 3x$, $b = 2y$, and $n = 7$. The fourth term is

$$\binom{7}{3}(3x)^{7-3}(2y)^3 = \binom{7}{3}(3x)^4(2y)^3 = \frac{7!}{3!(7-3)!}(3x)^4(2y)^3.$$

Use the formula for the $(r + 1)$st term of $(a + b)^n$: $\binom{n}{r}a^{n-r}b^r$.

We use $\binom{n}{r} = \frac{n!}{r!(n-r)!}$ to evaluate $\binom{7}{3}$.

Now we need to evaluate the factorial expression and raise $3x$ and $2y$ to the indicated powers. We obtain

$$\frac{7!}{3!4!}(81x^4)(8y^3) = \frac{7 \cdot 6 \cdot 5 \cdot 4!}{3 \cdot 2 \cdot 1 \cdot 4!}(81x^4)(8y^3) = 35(81x^4)(8y^3) = 22{,}680x^4y^3.$$

The fourth term of $(3x + 2y)^7$ is $22{,}680x^4y^3$. • • •

✅ **Check Point 4** Find the fifth term in the expansion of $(2x + y)^9$.

Blitzer Bonus || The Universality of Mathematics

Pascal's triangle is an array of numbers showing coefficients of the terms in the expansions of $(a + b)^n$. Although credited to French mathematician Blaise Pascal (1623–1662), the triangular array of numbers appeared in a Chinese document printed in 1303. The Binomial Theorem was known in Eastern cultures prior to its discovery in Europe. The same mathematics is often discovered/invented by independent researchers separated by time, place, and culture.

Binomial Expansions	**Pascal's Triangle**	**Chinese Document: 1303**

Coefficients in the Expansions

$(a + b)^0 = 1$

$(a + b)^1 = a + b$

$(a + b)^2 = a^2 + 2ab + b^2$

$(a + b)^3 = a^3 + 3a^2b + 3ab^2 + b^3$

$(a + b)^4 = a^4 + 4a^3b + 6a^2b^2 + 4ab^3 + b^4$

$(a + b)^5 = a^5 + 5a^4b + 10a^3b^2 + 10a^2b^3 + 5ab^4 + b^5$

```
                1
              1   1
            1   2   1
          1   3   3   1
        1   4   6   4   1
      1   5  10  10   5   1
    1   6  15  20  15   6   1
  1   7  21  35  35  21   7   1
1   8  28  56  70  56  28   8   1
```

CONCEPT AND VOCABULARY CHECK

Fill in each blank so that the resulting statement is true.

1. $\binom{n}{r}$ is called a/an _____ coefficient.

2. $\binom{8}{2} = \dfrac{_!}{_!_!}$

3. $\binom{n}{r} =$ _____

4. $(x + 2)^5 = \binom{5}{0}x^5 +$ _____ $x^4 \cdot 2 +$ _____ $x^3 \cdot 2^2 +$ _____ $x^2 \cdot 2^3 +$ _____ $x \cdot 2^4 +$ _____ $\cdot 2^5$

5. $(a + b)^n = \binom{n}{0}a^n +$ _____ $a^{n-1}b +$ _____ $a^{n-2}b^2 +$ _____ $a^{n-3}b^3 + \cdots +$ _____ b^n

The sum of the exponents on a and b in each term is _____.

6. The formula in Exercise 5 is called the _____ Theorem.

7. The $(r + 1)$st term of the expansion of $(a + b)^n$ is $\binom{n}{r}$ _____.

EXERCISE SET 8.5

Practice Exercises

In Exercises 1–8, evaluate the given binomial coefficient.

1. $\binom{8}{3}$ **2.** $\binom{7}{2}$ **3.** $\binom{12}{1}$

4. $\binom{11}{1}$ **5.** $\binom{6}{6}$ **6.** $\binom{15}{2}$

7. $\binom{100}{2}$ **8.** $\binom{100}{98}$

In Exercises 9–30, use the Binomial Theorem to expand each binomial and express the result in simplified form.

9. $(x + 2)^3$ **10.** $(x + 4)^3$

11. $(3x + y)^3$ **12.** $(x + 3y)^3$

13. $(5x - 1)^3$ **14.** $(4x - 1)^3$

15. $(2x + 1)^4$ **16.** $(3x + 1)^4$

17. $(x^2 + 2y)^4$ **18.** $(x^2 + y)^4$

19. $(y - 3)^4$ **20.** $(y - 4)^4$

21. $(2x^3 - 1)^4$ **22.** $(2x^5 - 1)^4$

23. $(c + 2)^5$

24. $(c + 3)^5$

25. $(x - 1)^5$

26. $(x - 2)^5$

27. $(3x - y)^5$

28. $(x - 3y)^5$

29. $(2a + b)^6$

30. $(a + 2b)^6$

In Exercises 31–38, write the first three terms in each binomial expansion, expressing the result in simplified form.

31. $(x + 2)^8$

32. $(x + 3)^8$

33. $(x - 2y)^{10}$

34. $(x - 2y)^9$

35. $(x^2 + 1)^{16}$

36. $(x^2 + 1)^{17}$

37. $(y^3 - 1)^{20}$

38. $(y^3 - 1)^{21}$

In Exercises 39–48, find the term indicated in each expansion.

39. $(2x + y)^6$; third term

40. $(x + 2y)^6$; third term

41. $(x - 1)^9$; fifth term

42. $(x - 1)^{10}$; fifth term

43. $(x^2 + y^3)^8$; sixth term

44. $(x^3 + y^2)^8$; sixth term

45. $\left(x - \frac{1}{2}\right)^9$; fourth term

46. $\left(x + \frac{1}{2}\right)^8$; fourth term

47. $(x^2 + y)^{22}$; the term containing y^{14}

48. $(x + 2y)^{10}$; the term containing y^6

Practice Plus

In Exercises 49–52, use the Binomial Theorem to expand each expression and write the result in simplified form.

49. $(x^3 + x^{-2})^4$

50. $(x^2 + x^{-3})^4$

51. $\left(x^{\frac{1}{3}} - x^{-\frac{1}{3}}\right)^3$

52. $\left(x^{\frac{2}{3}} - \frac{1}{\sqrt[3]{x}}\right)^3$

In Exercises 53–54, find $\dfrac{f(x + h) - f(x)}{h}$ and simplify.

53. $f(x) = x^4 + 7$

54. $f(x) = x^5 + 8$

55. Find the middle term in the expansion of $\left(\dfrac{3}{x} + \dfrac{x}{3}\right)^{10}$.

56. Find the middle term in the expansion of $\left(\dfrac{1}{x} - x^2\right)^{12}$.

Application Exercises

The graph shows that U.S. smokers have a greater probability of suffering from some ailments than the general adult population. Exercises 57–58 are based on some of the probabilities, expressed as decimals, shown to the right of the bars. In each exercise, use a calculator to determine the probability, correct to four decimal places.

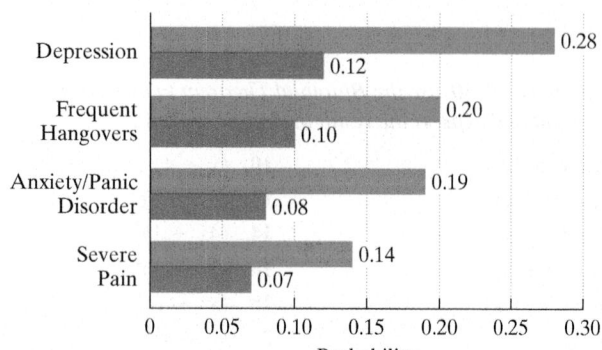

Probability That United States Adults Suffer from Various Ailments

Source: MARS 2005 OTC/DTC Pharmaceutical Study

If the probability an event will occur is p and the probability it will not occur is q, then each term in the expansion of $(p + q)^n$ represents a probability.

57. The probability that a smoker suffers from depression is 0.28. If five smokers are randomly selected, the probability that three of them will suffer from depression is the third term of the binomial expansion of

$$(0.28 + 0.72)^5.$$

| 5 smokers are selected. |

Probability a smoker suffers from depression

Probability a smoker does not suffer from depression

What is this probability?

58. The probability that a person in the general population suffers from depression is 0.12. If five people from the general population are randomly selected, the probability that three of them will suffer from depression is the third term of the binomial expansion of

$$(0.12 + 0.88)^5.$$

| 5 people from the general population are selected. |

Probability a person in the general population suffers from depression

Probability a person in the general population does not suffer from depression

What is this probability?

Writing in Mathematics

59. Explain how to evaluate $\binom{n}{r}$. Provide an example with your explanation.

60. Describe the pattern in the exponents on a in the expansion of $(a + b)^n$.

61. Describe the pattern in the exponents on b in the expansion of $(a + b)^n$.

62. What is true about the sum of the exponents on a and b in any term in the expansion of $(a + b)^n$?

63. How do you determine how many terms there are in a binomial expansion?

64. Explain how to use the Binomial Theorem to expand a binomial. Provide an example with your explanation.

65. Explain how to find a particular term in a binomial expansion without having to write out the entire expansion.

66. Describe how you would use mathematical induction to prove

$$(a + b)^n = \binom{n}{0}a^n + \binom{n}{1}a^{n-1}b + \binom{n}{2}a^{n-2}b^2$$

$$+ \cdots + \binom{n}{n-1}ab^{n-1} + \binom{n}{n}b^n.$$

What happens when $n = 1$? Write the statement that we assume to be true. Write the statement that we must prove. What must be done to the left side of the assumed statement to make it look like the left side of the statement that must be proved? (More detail on the actual proof is found in Exercise 85.)

Technology Exercises

67. Use the $\boxed{_nC_r}$ key on a graphing utility to verify your answers in Exercises 1–8.

In Exercises 68–69, graph each of the functions in the same viewing rectangle. Describe how the graphs illustrate the Binomial Theorem.

68. $f_1(x) = (x + 2)^3$ $\qquad f_2(x) = x^3$
$f_3(x) = x^3 + 6x^2 \qquad f_4(x) = x^3 + 6x^2 + 12x$
$f_5(x) = x^3 + 6x^2 + 12x + 8$
Use a $[-10, 10, 1]$ by $[-30, 30, 10]$ viewing rectangle.

69. $f_1(x) = (x + 1)^4 \qquad f_2(x) = x^4$
$f_3(x) = x^4 + 4x^3 \qquad f_4(x) = x^4 + 4x^3 + 6x^2$
$f_5(x) = x^4 + 4x^3 + 6x^2 + 4x$
$f_6(x) = x^4 + 4x^3 + 6x^2 + 4x + 1$
Use a $[-5, 5, 1]$ by $[-30, 30, 10]$ viewing rectangle.

In Exercises 70–72, use the Binomial Theorem to find a polynomial expansion for each function. Then use a graphing utility and an approach similar to the one in Exercises 68 and 69 to verify the expansion.

70. $f_1(x) = (x - 1)^3$ \qquad **71.** $f_1(x) = (x - 2)^4$
72. $f_1(x) = (x + 2)^6$

Critical Thinking Exercises

Make Sense? *In Exercises 73–76, determine whether each statement makes sense or does not make sense, and explain your reasoning.*

73. In order to expand $(x^3 - y^4)^5$, I find it helpful to rewrite the expression inside the parentheses as $x^3 + (-y^4)$.

74. Without writing the expansion of $(x - 1)^6$, I can see that the terms have alternating positive and negative signs.

75. I use binomial coefficients to expand $(a + b)^n$, where $\binom{n}{1}$ is the coefficient of the first term, $\binom{n}{2}$ is the coefficient of the second term, and so on.

76. One of the terms in my binomial expansion is $\binom{7}{5}x^2y^4$.

In Exercises 77–80, determine whether each statement is true or false. If the statement is false, make the necessary change(s) to produce a true statement.

77. The binomial expansion for $(a + b)^n$ contains n terms.

78. The Binomial Theorem can be written in condensed form as
$$(a + b)^n = \sum_{r=0}^{n} \binom{n}{r}a^{n-r}b^r.$$

79. The sum of the binomial coefficients in $(a + b)^n$ cannot be 2^n.

80. There are no values of a and b such that
$$(a + b)^4 = a^4 + b^4.$$

81. Use the Binomial Theorem to expand and then simplify the result: $(x^2 + x + 1)^3$.
Hint: Write $x^2 + x + 1$ as $x^2 + (x + 1)$.

82. Find the term in the expansion of $(x^2 + y^2)^5$ containing x^4 as a factor.

83. Prove that
$$\binom{n}{r} = \binom{n}{n - r}.$$

84. Show that
$$\binom{n}{r} + \binom{n}{r + 1} = \binom{n + 1}{r + 1}.$$

Hints:
$$(n - r)! = (n - r)(n - r - 1)!$$
$$(r + 1)! = (r + 1)r!$$

85. Follow the outline below and use mathematical induction to prove the Binomial Theorem:
$$(a + b)^n = \binom{n}{0}a^n + \binom{n}{1}a^{n-1}b + \binom{n}{2}a^{n-2}b^2$$
$$+ \cdots + \binom{n}{n - 1}ab^{n-1} + \binom{n}{n}b^n.$$

a. Verify the formula for $n = 1$.

b. Replace n with k and write the statement that is assumed true. Replace n with $k + 1$ and write the statement that must be proved.

c. Multiply both sides of the statement assumed to be true by $a + b$. Add exponents on the left. On the right, distribute a and b, respectively.

d. Collect like terms on the right. At this point, you should have
$$(a + b)^{k+1} = \binom{k}{0}a^{k+1} + \left[\binom{k}{0} + \binom{k}{1}\right]a^k b$$
$$+ \left[\binom{k}{1} + \binom{k}{2}\right]a^{k-1}b^2 + \left[\binom{k}{2} + \binom{k}{3}\right]a^{k-2}b^3$$
$$+ \cdots + \left[\binom{k}{k - 1} + \binom{k}{k}\right]ab^k + \binom{k}{k}b^{k+1}.$$

e. Use the result of Exercise 84 to add the binomial sums in brackets. For example, because $\binom{n}{r} + \binom{n}{r + 1}$
$$= \binom{n + 1}{r + 1}, \text{ then } \binom{k}{0} + \binom{k}{1} = \binom{k + 1}{1} \text{ and}$$
$$\binom{k}{1} + \binom{k}{2} = \binom{k + 1}{2}.$$

f. Because $\binom{k}{0} = \binom{k + 1}{0}$ (why?) and $\binom{k}{k} = \binom{k + 1}{k + 1}$ (why?), substitute these results and the results from part (e) into the equation in part (d). This should give the statement that we were required to prove in the second step of the mathematical induction process.

Exercises 86–88 will help you prepare for the material covered in the next section.

86. Evaluate $\dfrac{n!}{(n-r)!}$ for $n = 20$ and $r = 3$.

87. Evaluate $\dfrac{n!}{(n-r)!r!}$ for $n = 8$ and $r = 3$.

88. You can choose from two pairs of jeans (one blue, one black) and three T-shirts (one beige, one yellow, and one blue), as shown in the diagram.

True or false: The diagram shows that you can form 2×3, or 6, different outfits.

SECTION 8.6 Counting Principles, Permutations, and Combinations

Objectives

1. Use the Fundamental Counting Principle.
2. Use the permutations formula.
3. Distinguish between permutation problems and combination problems.
4. Use the combinations formula.

① Use the Fundamental Counting Principle.

Have you ever imagined what your life would be like if you won the lottery? What changes would you make? Before you fantasize about becoming a person of leisure with a staff of obedient elves, think about this: The probability of winning top prize in the lottery is about the same as the probability of being struck by lightning. There are millions of possible number combinations in lottery games and only one way of winning the grand prize. Determining the probability of winning involves calculating the chance of getting the winning combination from all possible outcomes. In this section, we begin preparing for the surprising world of probability by looking at methods for counting possible outcomes.

The Fundamental Counting Principle

It's early morning, you're groggy, and you have to select something to wear for your 8 A.M. class. (What *were* you thinking of when you signed up for a class at that hour?!) Fortunately, your "lecture wardrobe" is rather limited—just two pairs of jeans to choose from (one blue, one black), three T-shirts to choose from (one beige, one yellow, and one blue), and two pairs of sneakers to select from (one black pair, one red pair). Your possible outfits are shown in **Figure 8.9**.

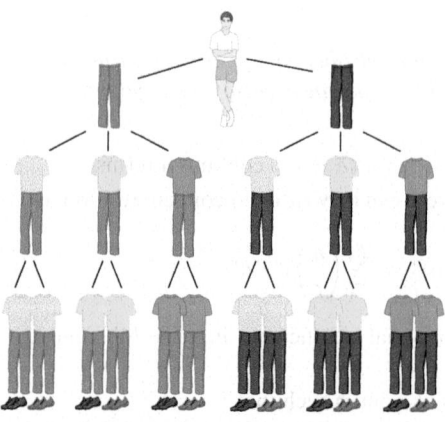

FIGURE 8.9 Selecting a wardrobe

The **tree diagram**, so named because of its branches, shows that you can form 12 outfits from your two pairs of jeans, three T-shirts, and two pairs of sneakers. Notice that the number of outfits can be obtained by multiplying the number of choices for jeans, 2, the number of choices for the T-shirts, 3, and the number of choices for the sneakers, 2:

$$2 \cdot 3 \cdot 2 = 12.$$

We can generalize this idea to any two or more groups of items—not just jeans, T-shirts, and sneakers—with the **Fundamental Counting Principle**:

The Fundamental Counting Principle

The number of ways in which a series of successive things can occur is found by multiplying the number of ways in which each thing can occur.

For example, if you own 30 pairs of jeans, 20 T-shirts, and 12 pairs of sneakers, you have

$$30 \cdot 20 \cdot 12 = 7200$$

choices for your wardrobe!

EXAMPLE 1 Options in Planning a Course Schedule

Next semester you are planning to take three courses—math, English, and humanities. Based on time blocks and highly recommended professors, there are 8 sections of math, 5 of English, and 4 of humanities that you find suitable. Assuming no scheduling conflicts, how many different three-course schedules are possible?

SOLUTION

This situation involves making choices with three groups of items.

We use the Fundamental Counting Principle to find the number of three-course schedules. Multiply the number of choices for each of the three groups:

$$8 \cdot 5 \cdot 4 = 160.$$

Thus, there are 160 different three-course schedules. • • •

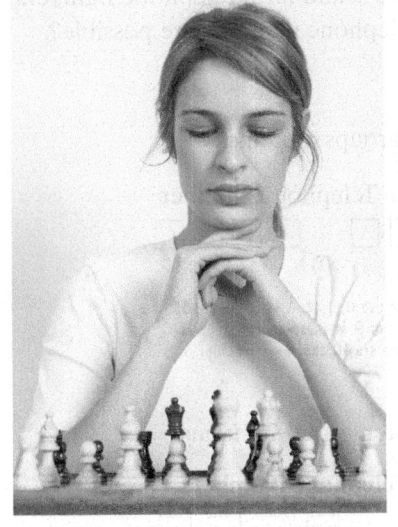

The number of possible ways of playing the first four moves on each side in a game of chess is 318,979,564,000.

✓ Check Point **1** A pizza can be ordered with three choices of size (small, medium, or large), four choices of crust (thin, thick, crispy, or regular), and six choices of toppings (ground beef, sausage, pepperoni, bacon, mushrooms, or onions). How many different one-topping pizzas can be ordered?

EXAMPLE 2 A Multiple-Choice Test

You are taking a multiple-choice test that has ten questions. Each of the questions has four answer choices, with one correct answer per question. If you select one of these four choices for each question and leave nothing blank, in how many ways can you answer the questions?

SOLUTION

This situation involves making choices with ten questions.

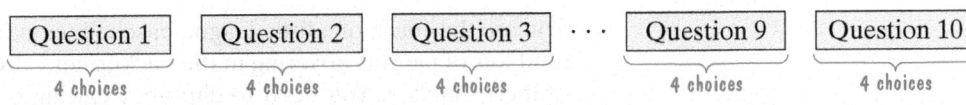

We use the Fundamental Counting Principle to determine the number of ways that you can answer the questions on the test. Multiply the number of choices, 4, for each of the ten questions.

$$4 \cdot 4 \cdot 4 \cdot 4 \cdot 4 \cdot 4 \cdot 4 \cdot 4 \cdot 4 \cdot 4 = 4^{10} = 1,048,576$$

Thus, you can answer the questions in 1,048,576 different ways.

Are you surprised that there are over one million ways of answering a ten-question multiple-choice test? Of course, there is only one way to answer the test and receive a perfect score. The probability of guessing your way into a perfect score involves calculating the chance of getting a perfect score, just one way from all 1,048,576 possible outcomes. In short, prepare for the test and do not rely on guessing!

Check Point 2 You are taking a multiple-choice test that has six questions. Each of the questions has three answer choices, with one correct answer per question. If you select one of these three choices for each question and leave nothing blank, in how many ways can you answer the questions?

EXAMPLE 3 Telephone Numbers in the United States

Telephone numbers in the United States begin with three-digit area codes followed by seven-digit local telephone numbers. Area codes and local telephone numbers cannot begin with 0 or 1. How many different telephone numbers are possible?

SOLUTION

This situation involves making choices with ten groups of items.

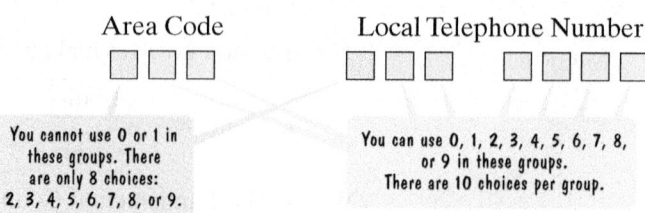

Here are the numbers of choices for each of the ten groups of items:

Area Code	Local Telephone Number
8 10 10	8 10 10 10 10 10 10 .

We use the Fundamental Counting Principle to determine the number of different telephone numbers that are possible. The total number of telephone numbers possible is

$$8 \cdot 10 \cdot 10 \cdot 8 \cdot 10 \cdot 10 \cdot 10 \cdot 10 \cdot 10 \cdot 10 = 6,400,000,000.$$

There are six billion four hundred million different telephone numbers that are possible.

Check Point 3 License plates in a particular state display two letters followed by three numbers, such as AT-887 or BB-013. How many different license plates can be manufactured?

② Use the permutations formula.

Permutations

You are the coach of a little league baseball team. There are 13 players on the team (and lots of parents hovering in the background, dreaming of stardom for their little "Albert Pujols"). You need to choose a batting order having 9 players. The order

Blitzer Bonus

Running Out of Telephone Numbers

By the year 2020, portable telephones used for business and pleasure will all be videophones. At that time, U.S. population is expected to be 323 million. Faxes, beepers, cellphones, computer phone lines, and business lines may result in certain areas running out of phone numbers. Solution: Add more digits!

With or without extra digits, we expect that the 2020 videophone greeting will still be "hello," a word created by Thomas Edison in 1877. Phone inventor Alexander Graham Bell preferred "ahoy," but "hello" won out, appearing in the *Oxford English Dictionary* in 1883.

(Source: *New York Times*)

makes a difference, because, for instance, if bases are loaded and "Little Albert" is fourth or fifth at bat, his possible home run will drive in three additional runs. How many batting orders can you form?

You can choose any of 13 players for the first person at bat. Then you will have 12 players from which to choose the second batter, then 11 from which to choose the third batter, and so on. The situation can be shown as follows:

We use the Fundamental Counting Principle to find the number of batting orders. The total number of batting orders is

$$13 \cdot 12 \cdot 11 \cdot 10 \cdot 9 \cdot 8 \cdot 7 \cdot 6 \cdot 5 = 259{,}459{,}200.$$

Nearly 260 million batting orders are possible for your 13-player little league team. Each batting order is called a *permutation* of 13 players taken 9 at a time. The number of permutations of 13 players taken 9 at a time is 259,459,200.

A **permutation** is an ordered arrangement of items that occurs when

- No item is used more than once. (Each of the 9 players in the batting order bats exactly once.)
- The order of arrangement makes a difference.

We can obtain a formula for finding the number of permutations of 13 players taken 9 at a time by rewriting our computation:

$$13 \cdot 12 \cdot 11 \cdot 10 \cdot 9 \cdot 8 \cdot 7 \cdot 6 \cdot 5$$

$$= \frac{13 \cdot 12 \cdot 11 \cdot 10 \cdot 9 \cdot 8 \cdot 7 \cdot 6 \cdot 5 \cdot \boxed{4 \cdot 3 \cdot 2 \cdot 1}}{\boxed{4 \cdot 3 \cdot 2 \cdot 1}} = \frac{13!}{4!} = \frac{13!}{(13-9)!}.$$

Thus, the number of permutations of 13 things taken 9 at a time is $\frac{13!}{(13-9)!}$. The special notation $_{13}P_9$ is used to replace the phrase "the number of permutations of 13 things taken 9 at a time." Using this new notation, we can write

$$_{13}P_9 = \frac{13!}{(13-9)!}.$$

The numerator of this expression is the number of items, 13 team members, expressed as a factorial: 13!. The denominator is also a factorial. It is the factorial of the difference between the number of items, 13, and the number of items in each permutation, 9 batters: $(13-9)!$.

The notation $_nP_r$ means the **number of permutations of n things taken r at a time**. We can generalize from the situation in which 9 batters were taken from 13 players. By generalizing, we obtain the following formula for the number of permutations if r items are taken from n items.

Permutations of n Things Taken r at a Time

The number of possible permutations if r items are taken from n items is

$$_nP_r = \frac{n!}{(n-r)!}.$$

GREAT QUESTION!

Do I have to use the formula for $_nP_r$ to solve permutation problems?

No. Because all permutation problems are also fundamental counting problems, they can be solved using the formula for $_nP_r$ or using the Fundamental Counting Principle.

TECHNOLOGY

Graphing utilities have a menu item for calculating permutations, usually labeled $\boxed{_nP_r}$. For example, to find $_{20}P_3$, the keystrokes are

20 $\boxed{_nP_r}$ 3 $\boxed{\text{ENTER}}$.

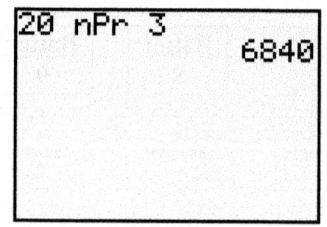

If you are using a scientific calculator, check your manual for the location of the menu item for calculating permutations and the required keystrokes.

Blitzer Bonus

How to Pass the Time for $2\frac{1}{2}$ Million Years

If you were to arrange 15 different books on a shelf and it took you one minute for each permutation, the entire task would take 2,487,965 years.

Source: Isaac Asimov's *Book of Facts*.

③ Distinguish between permutation problems and combination problems.

EXAMPLE 4 Using the Formula for Permutations

You and 19 of your friends have decided to form an Internet marketing consulting firm. The group needs to choose three officers—a CEO, an operating manager, and a treasurer. In how many ways can those offices be filled?

SOLUTION

Your group is choosing $r = 3$ officers from a group of $n = 20$ people (you and 19 friends). The order in which the officers are chosen matters because the CEO, the operating manager, and the treasurer each have different responsibilities. Thus, we are looking for the number of permutations of 20 things taken 3 at a time. We use the formula

$$_nP_r = \frac{n!}{(n-r)!}$$

with $n = 20$ and $r = 3$.

$$_{20}P_3 = \frac{20!}{(20-3)!} = \frac{20!}{17!} = \frac{20 \cdot 19 \cdot 18 \cdot 17!}{17!} = \frac{20 \cdot 19 \cdot 18 \cdot \cancel{17!}}{\cancel{17!}} = 20 \cdot 19 \cdot 18 = 6840$$

Thus, there are 6840 different ways of filling the three offices. •••

✓ **Check Point 4** A corporation has seven members on its board of directors. In how many different ways can it elect a president, vice-president, secretary, and treasurer?

EXAMPLE 5 Using the Formula for Permutations

You need to arrange seven of your favorite books along a small shelf. How many different ways can you arrange the books, assuming that the order of the books makes a difference to you?

SOLUTION

Because you are using all seven of your books in every possible arrangement, you are arranging $r = 7$ books from a group of $n = 7$ books. Thus, we are looking for the number of permutations of 7 things taken 7 at a time. We use the formula

$$_nP_r = \frac{n!}{(n-r)!}$$

with $n = 7$ and $r = 7$.

$$_7P_7 = \frac{7!}{(7-7)!} = \frac{7!}{0!} = \frac{7!}{1} = 5040$$

Thus, you can arrange the books in 5040 ways. There are 5040 different possible permutations. •••

✓ **Check Point 5** In how many ways can six books be lined up along a shelf?

Combinations

Throughout the history of entertainment, performers have featured choreography in their acts. Singers who are known for their serious moves include Beyoncé, Lady Gaga, Shakira, Justin Timberlake, and Usher.

Imagine that you ask your friends the following question: "Of these five entertainers, which three would you select to be included in a documentary on singers and choreography?" You are not asking your friends to rank their three favorite

artists in any kind of order—they should merely select the three to be included in the documentary.

One friend answers, "Beyoncé, Lady Gaga, and Usher." Another responds, "Usher, Lady Gaga, and Beyoncé." These two people have the same artists in their group of selections, even if they are named in a different order. We are interested *in which artists are named, not the order in which they are named,* for the documentary. Because the items are taken without regard to order, this is not a permutation problem. No ranking of any sort is involved.

Later on, you ask your roommate which three artists she would select for the documentary. She names Justin Timberlake, Beyoncé, and Usher. Her selection is different from those of your two other friends because different entertainers are cited.

Mathematicians describe the group of artists given by your roommate as a *combination.* A **combination** of items occurs when

- The items are selected from the same group (the five entertainers who are known for their choreography).
- No item is used more than once. (You may view Beyoncé as a phenomenal performer, but your three selections cannot be Beyoncé, Beyoncé, and Beyoncé.)
- The order of the items makes no difference. (Beyoncé, Lady Gaga, Usher is the same group in the documentary as Usher, Lady Gaga, Beyoncé.)

Do you see the difference between a permutation and a combination? A permutation is an ordered arrangement of a given group of items. A combination is a group of items taken without regard to their order. **Permutation** problems involve situations in which **order matters. Combination** problems involve situations in which the **order** of the items **makes no difference.**

EXAMPLE 6 Distinguishing between Permutations and Combinations

For each of the following problems, determine whether the problem is one involving permutations or combinations. (It is not necessary to solve the problem.)

- **a.** Six students are running for student government president, vice president, and treasurer. The student with the greatest number of votes becomes the president, the second highest vote-getter becomes vice president, and the student who gets the third largest number of votes will be treasurer. How many different outcomes are possible for these three positions?
- **b.** Six people are on the board of supervisors for your neighborhood park. A three-person committee is needed to study the possibility of expanding the park. How many different committees could be formed from the six people?
- **c.** Baskin-Robbins offers 31 different flavors of ice cream. One of its items is a bowl consisting of three scoops of ice cream, each a different flavor. How many such bowls are possible?

SOLUTION

- **a.** Students are choosing three student government officers from six candidates. The order in which the officers are chosen makes a difference because each of the offices (president, vice president, treasurer) is different. Order matters. This is a problem involving permutations.

b. A three-person committee is to be formed from the six-person board of supervisors. The order in which the three people are selected does not matter because they are not filling different roles on the committee. Because order makes no difference, this is a problem involving combinations.

c. A three-scoop bowl of three different flavors is to be formed from Baskin-Robbins's 31 flavors. The order in which the three scoops of ice cream are put into the bowl is irrelevant. A bowl with chocolate, vanilla, and strawberry is exactly the same as a bowl with vanilla, strawberry, and chocolate. Different orderings do not change things, and so this is a problem involving combinations.

● ● ●

Check Point **6** For each of the following problems, explain if the problem is one involving permutations or combinations. (It is not necessary to solve the problem.)

a. How many ways can you select 6 free DVDs from a list of 200 DVDs?

b. In a race in which there are 50 runners and no ties, in how many ways can the first three finishers come in?

4 Use the combinations formula.

A Formula for Combinations

We have seen that the notation $_nP_r$ means the number of permutations of n things taken r at a time. Similarly, the notation $_nC_r$ **means the number of combinations of n things taken r at a time.**

We can develop a formula for $_nC_r$ by comparing permutations and combinations. Consider the letters A, B, C, and D. The number of permutations of these four letters taken three at a time is

$$_4P_3 = \frac{4!}{(4-3)!} = \frac{4!}{1!} = \frac{4 \cdot 3 \cdot 2 \cdot 1}{1} = 24.$$

Here are the 24 permutations:

ABC,	ABD,	ACD,	BCD,
ACB,	ADB,	ADC,	BDC,
BAC,	BAD,	CAD,	CBD,
BCA,	BDA,	CDA,	CDB,
CAB,	DAB,	DAC,	DBC,
CBA,	DBA,	DCA,	DCB.

This column contains only one combination, ABC.	This column contains only one combination, ABD.	This column contains only one combination, ACD.	This column contains only one combination, BCD.

Because the order of items makes no difference in determining combinations, each column of six permutations represents one combination. There is a total of four combinations:

ABC, ABD, ACD, BCD.

Thus, $_4C_3 = 4$: The number of combinations of 4 things taken 3 at a time is 4. With 24 permutations and only four combinations, there are 6, or 3!, times as many permutations as there are combinations.

In general, there are $r!$ times as many permutations of n things taken r at a time as there are combinations of n things taken r at a time. Thus, we find the number of combinations of n things taken r at a time by dividing the number of permutations of n things taken r at a time by $r!$.

$$_nC_r = \frac{_nP_r}{r!} = \frac{\dfrac{n!}{(n-r)!}}{r!} = \frac{n!}{(n-r)!r!}$$

GREAT QUESTION!

Do I have to use the formula for $_nC_r$ to solve combination problems?

Yes. The number of combinations if r items are taken from n items cannot be found using the Fundamental Counting Principle and requires the use of the formula shown on the right.

Combinations of n Things Taken r at a Time

The number of possible combinations if r items are taken from n items is

$$_nC_r = \frac{n!}{(n-r)!r!}.$$

Notice that the formula for $_nC_r$ is the same as the formula for the binomial coefficient $\begin{pmatrix} n \\ r \end{pmatrix}$.

EXAMPLE 7 Using the Formula for Combinations

A three-person committee is needed to study ways of improving public transportation. How many committees could be formed from the eight people on the board of supervisors?

SOLUTION

The order in which the three people are selected does not matter. This is a problem of selecting $r = 3$ people from a group of $n = 8$ people. We are looking for the number of combinations of eight things taken three at a time. We use the formula

$$_nC_r = \frac{n!}{(n-r)!r!}$$

with $n = 8$ and $r = 3$.

$$_8C_3 = \frac{8!}{(8-3)!3!} = \frac{8!}{5!3!} = \frac{8\cdot7\cdot6\cdot5!}{5!\cdot3\cdot2\cdot1} = \frac{8\cdot7\cdot6\cdot5!}{5!\cdot3\cdot2\cdot1} = 56$$

Thus, 56 committees of three people each can be formed from the eight people on the board of supervisors. • • •

☑ **Check Point 7** From a group of 10 physicians, in how many ways can four people be selected to attend a conference on acupuncture?

EXAMPLE 8 Using the Formula for Combinations

In poker, a person is dealt 5 cards from a standard 52-card deck. The order in which you are dealt the 5 cards does not matter. How many different 5-card poker hands are possible?

SOLUTION

Because the order in which the 5 cards are dealt does not matter, this is a problem involving combinations. We are looking for the number of combinations of $n = 52$ cards dealt $r = 5$ at a time. We use the formula

$$_nC_r = \frac{n!}{(n-r)!r!}$$

with $n = 52$ and $r = 5$.

$$_{52}C_5 = \frac{52!}{(52-5)!5!} = \frac{52!}{47!5!} = \frac{52\cdot51\cdot50\cdot49\cdot48\cdot47!}{47!\cdot5\cdot4\cdot3\cdot2\cdot1} = 2{,}598{,}960$$

Thus, there are 2,598,960 different 5-card poker hands possible. It surprises many people that more than 2.5 million 5-card hands can be dealt from a mere 52 cards. • • •

TECHNOLOGY

Graphing utilities have a menu item for calculating combinations, usually labeled $_nC_r$. For example, to find $_8C_3$, the keystrokes on most graphing utilities are

8 $\boxed{_nC_r}$ 3 $\boxed{\text{ENTER}}$.

If you are using a scientific calculator, check your manual to see whether there is a menu item for calculating combinations.

If you use your calculator's factorial key to find $\frac{8!}{5!3!}$, be sure to enclose the factorials in the denominator with parentheses

8 $\boxed{!}$ $\boxed{\div}$ $\boxed{(}$ $\boxed{5}$ $\boxed{!}$ $\boxed{\times}$ $\boxed{3}$ $\boxed{!}$ $\boxed{)}$

pressing $\boxed{=}$ or $\boxed{\text{ENTER}}$ to obtain the answer.

FIGURE 8.10 A royal flush

If you are a card player, it does not get any better than to be dealt the 5-card poker hand shown in **Figure 8.10**. This hand is called a *royal flush*. It consists of an ace, king, queen, jack, and 10, all of the same suit: all hearts, all diamonds, all clubs, or all spades. The probability of being dealt a royal flush involves calculating the number of ways of being dealt such a hand: just 4 of all 2,598,960 possible hands. In the next section, we move from counting possibilities to computing probabilities.

 Check Point **8** How many different 4-card hands can be dealt from a deck that has 16 different cards?

CONCEPT AND VOCABULARY CHECK

Fill in each blank so that the resulting statement is true.

1. If you can choose one item from a group of M items and a second item from a group of N items, then the total number of two-item choices is _____.

2. The number of ways in which a series of successive things can occur is found by _____ the number of ways in which each thing can occur. This is called the _____ Principle.

3. The number of possible permutations if r objects are taken from n items is $_nP_r =$ _____.

4. The number of possible combinations if r objects are taken from n items is $_nC_r =$ _____.

5. The formula for $_nC_r$ has the same numerator as the formula for $_nP_r$ but contains an extra factor of _____ in the denominator.

EXERCISE SET 8.6

Practice Exercises

In Exercises 1–8, use the formula for $_nP_r$ to evaluate each expression.

1. $_9P_4$ **2.** $_7P_3$

3. $_8P_5$ **4.** $_{10}P_4$

5. $_6P_6$ **6.** $_9P_9$

7. $_8P_0$ **8.** $_6P_0$

In Exercises 9–16, use the formula for $_nC_r$ to evaluate each expression.

9. $_9C_5$ **10.** $_{10}C_6$

11. $_{11}C_4$ **12.** $_{12}C_5$

13. $_7C_7$ **14.** $_4C_4$

15. $_5C_0$ **16.** $_6C_0$

In Exercises 17–20, does the problem involve permutations or combinations? Explain your answer. (It is not necessary to solve the problem.)

17. A medical researcher needs 6 people to test the effectiveness of an experimental drug. If 13 people have volunteered for the test, in how many ways can 6 people be selected?

18. Fifty people purchase raffle tickets. Three winning tickets are selected at random. If first prize is $1000, second prize is $500, and third prize is $100, in how many different ways can the prizes be awarded?

19. How many different four-letter passwords can be formed from the letters A, B, C, D, E, F, and G if no repetition of letters is allowed?

20. Fifty people purchase raffle tickets. Three winning tickets are selected at random. If each prize is $500, in how many different ways can the prizes be awarded?

Practice Plus

In Exercises 21–28, evaluate each expression.

21. $\dfrac{_7P_3}{3!} - _7C_3$ **22.** $\dfrac{_{20}P_2}{2!} - _{20}C_2$

23. $1 - \dfrac{_3P_2}{_4P_3}$ **24.** $1 - \dfrac{_5P_3}{_{10}P_4}$

25. $\dfrac{_7C_3}{_5C_4} - \dfrac{98!}{96!}$ **26.** $\dfrac{_{10}C_3}{_6C_4} - \dfrac{46!}{44!}$

27. $\dfrac{_4C_2 \cdot _6C_1}{_{18}C_3}$ **28.** $\dfrac{_5C_1 \cdot _7C_2}{_{12}C_3}$

Application Exercises

Use the Fundamental Counting Principle to solve Exercises 29–40.

29. The model of the car you are thinking of buying is available in nine different colors and three different styles (hatchback, sedan, or station wagon). In how many ways can you order the car?

30. A popular brand of pen is available in three colors (red, green, or blue) and four writing tips (bold, medium, fine, or micro). How many different choices of pens do you have with this brand?

31. An ice cream store sells two drinks (sodas or milk shakes), in four sizes (small, medium, large, or jumbo), and five flavors (vanilla, strawberry, chocolate, coffee, or pistachio). In how many ways can a customer order a drink?

32. A restaurant offers the following lunch menu.

Main Course	Vegetables	Beverages	Desserts
Ham	Potatoes	Coffee	Cake
Chicken	Peas	Tea	Pie
Fish	Green beans	Milk	Ice cream
Beef		Soda	

If one item is selected from each of the four groups, in how many ways can a meal be ordered? Describe two such orders.

33. You are taking a multiple-choice test that has five questions. Each of the questions has three answer choices, with one correct answer per question. If you select one of these three choices for each question and leave nothing blank, in how many ways can you answer the questions?

34. You are taking a multiple-choice test that has eight questions. Each of the questions has three answer choices, with one correct answer per question. If you select one of these three choices for each question and leave nothing blank, in how many ways can you answer the questions?

35. In the original plan for area codes in 1945, the first digit could be any number from 2 through 9, the second digit was either 0 or 1, and the third digit could be any number except 0. With this plan, how many different area codes were possible?

36. How many different four-letter radio station call letters can be formed if the first letter must be W or K?

37. Six performers are to present their comedy acts on a weekend evening at a comedy club. One of the performers insists on being the last stand-up comic of the evening. If this performer's request is granted, how many different ways are there to schedule the appearances?

38. Five singers are to perform at a night club. One of the singers insists on being the last performer of the evening. If this singer's request is granted, how many different ways are there to schedule the appearances?

39. In the *Cambridge Encyclopedia of Language* (Cambridge University Press, 1987), author David Crystal presents five sentences that make a reasonable paragraph regardless of their order. The sentences are as follows:

- Mark had told him about the foxes.
- John looked out the window.
- Could it be a fox?
- However, nobody had seen one for months.
- He thought he saw a shape in the bushes.

How many different five-sentence paragraphs can be formed if the paragraph begins with "He thought he saw a shape in the bushes" and ends with "John looked out of the window"?

40. A television programmer is arranging the order that five movies will be seen between the hours of 6 P.M. and 4 A.M. Two of the movies have a G rating and they are to be shown in the first two time blocks. One of the movies is rated NC-17 and it is to be shown in the last of the time blocks, from 2 A.M. until 4 A.M. Given these restrictions, in how many ways can the five movies be arranged during the indicated time blocks?

Use the formula for $_nP_r$ to solve Exercises 41–48.

41. A club with ten members is to choose three officers—president, vice president, and secretary-treasurer. If each office is to be held by one person and no person can hold more than one office, in how many ways can those offices be filled?

42. A corporation has ten members on its board of directors. In how many different ways can it elect a president, vice president, secretary, and treasurer?

43. For a segment of a radio show, a disc jockey can play 7 songs. If there are 13 songs to select from, in how many ways can the program for this segment be arranged?

44. Suppose you are asked to list, in order of preference, the three best movies you have seen this year. If you saw 20 movies during the year, in how many ways can the three best be chosen and ranked?

45. In a race in which six automobiles are entered and there are no ties, in how many ways can the first three finishers come in?

46. In a production of *West Side Story*, eight actors are considered for the male roles of Tony, Riff, and Bernardo. In how many ways can the director cast the male roles?

47. Nine bands have volunteered to perform at a benefit concert, but there is only enough time for five of the bands to play. How many lineups are possible?

48. How many arrangements can be made using four of the letters of the word COMBINE if no letter is to be used more than once?

Use the formula for $_nC_r$ to solve Exercises 49–56.

49. An election ballot asks voters to select three city commissioners from a group of six candidates. In how many ways can this be done?

50. A four-person committee is to be elected from an organization's membership of 11 people. How many different committees are possible?

51. Of 12 possible books, you plan to take 4 with you on vacation. How many different collections of 4 books can you take?

52. There are 14 standbys who hope to get seats on a flight, but only 6 seats are available on the plane. How many different ways can the 6 people be selected?

53. You volunteer to help drive children at a charity event to the zoo, but you can fit only 8 of the 17 children present in your van. How many different groups of 8 children can you drive?

54. Of the 100 people in the U.S. Senate, 18 serve on the Foreign Relations Committee. How many ways are there to select Senate members for this committee (assuming party affiliation is not a factor in selection)?

55. To win at LOTTO in the state of Florida, one must correctly select 6 numbers from a collection of 53 numbers (1 through 53). The order in which the selection is made does not matter. How many different selections are possible?

56. To win in the New York State lottery, one must correctly select 6 numbers from 59 numbers. The order in which the selection is made does not matter. How many different selections are possible?

In Exercises 57–66, solve by the method of your choice.

57. In a race in which six automobiles are entered and there are no ties, in how many ways can the first four finishers come in?

58. A book club offers a choice of 8 books from a list of 40. In how many ways can a member make a selection?

59. A medical researcher needs 6 people to test the effectiveness of an experimental drug. If 13 people have volunteered for the test, in how many ways can 6 people be selected?

60. Fifty people purchase raffle tickets. Three winning tickets are selected at random. If first prize is $1000, second prize is $500, and third prize is $100, in how many different ways can the prizes be awarded?

61. From a club of 20 people, in how many ways can a group of three members be selected to attend a conference?

62. Fifty people purchase raffle tickets. Three winning tickets are selected at random. If each prize is $500, in how many different ways can the prizes be awarded?

63. How many different four-letter passwords can be formed from the letters A, B, C, D, E, F, and G if no repetition of letters is allowed?

64. Nine comedy acts will perform over two evenings. Five of the acts will perform on the first evening and the order in which the acts perform is important. How many ways can the schedule for the first evening be made?

65. Using 15 flavors of ice cream, how many cones with three different flavors can you create if it is important to you which flavor goes on the top, middle, and bottom?

66. Baskin-Robbins offers 31 different flavors of ice cream. One of its items is a bowl consisting of three scoops of ice cream, each a different flavor. How many such bowls are possible?

Exercises 67–72 are based on the following jokes about books:

- *"Outside of a dog, a book is man's best friend. Inside of a dog, it's too dark to read."—Groucho Marx*
- *"I recently bought a book of free verse. For $12."—George Carlin*
- *"If a word in the dictionary was misspelled, how would we know?"—Steven Wright*
- *"Encyclopedia is a Latin term. It means 'to paraphrase a term paper.' "—Greg Ray*
- *"A bookstore is one of the only pieces of evidence we have that people are still thinking."—Jerry Seinfeld*
- *"I honestly believe there is absolutely nothing like going to bed with a good book. Or a friend who's read one."—Phyllis Diller*

67. In how many ways can these six jokes be ranked from best to worst?

68. If Phyllis Diller's joke about books is excluded, in how many ways can the remaining five jokes be ranked from best to worst?

69. In how many ways can people select their three favorite jokes from these comments about books?

70. In how many ways can people select their two favorite jokes from these comments about books?

71. If the order in which these jokes are told makes a difference in terms of how they are received, how many ways can they be delivered if George Carlin's joke is delivered first and Jerry Seinfeld's joke is told last?

72. If the order in which these jokes are told makes a difference in terms of how they are received, how many ways can they be delivered if a joke by a man is told first?

Writing in Mathematics

73. Explain the Fundamental Counting Principle.

74. Write an original problem that can be solved using the Fundamental Counting Principle. Then solve the problem.

75. What is a permutation?

76. Describe what $_nP_r$ represents.

77. Write a word problem that can be solved by evaluating $_7P_3$.

78. What is a combination?

79. Explain how to distinguish between permutation and combination problems.

80. Write a word problem that can be solved by evaluating $_7C_3$.

Technology Exercises

81. Use a graphing utility with an $\boxed{_nP_r}$ menu item to verify your answers in Exercises 1–8.

82. Use a graphing utility with an $\boxed{_nC_r}$ menu item to verify your answers in Exercises 9–16.

Critical Thinking Exercises

Make Sense? *In Exercises 83–86, determine whether each statement makes sense or does not make sense, and explain your reasoning.*

83. I used the Fundamental Counting Principle to determine the number of five-digit ZIP codes that are available to the U.S. Postal Service.

84. I used the permutations formula to determine the number of ways the manager of a baseball team can form a 9-player batting order from a team of 25 players.

85. I used the combinations formula to determine how many different four-note sound sequences can be created from the notes C, D, E, F, G, A, and B.

86. I used the permutations formula to determine the number of ways people can select their 9 favorite baseball players from a team of 25 players.

In Exercises 87–90, determine whether each statement is true or false. If the statement is false, make the necessary change(s) to produce a true statement.

87. The number of ways to choose four questions out of ten questions on an essay test is $_{10}P_4$.

88. If $r > 1$, $_nP_r$ is less than $_nC_r$.

89. $_7P_3 = 3!_7C_3$

90. The number of ways to pick a winner and first runner-up in a talent contest with 20 contestants is $_{20}C_2$.

91. Five men and five women line up at a checkout counter in a store. In how many ways can they line up if the first person in line is a woman and the people in line alternate woman, man, woman, man, and so on?

92. How many four-digit odd numbers less than 6000 can be formed using the digits 2, 4, 6, 7, 8, and 9?

93. A mathematics exam consists of 10 multiple-choice questions and 5 open-ended problems in which all work must be shown. If an examinee must answer 8 of the multiple-choice questions and 3 of the open-ended problems, in how many ways can the questions and problems be chosen?

Group Exercise

94. The group should select real-world situations where the Fundamental Counting Principle can be applied. These could involve the number of possible student ID numbers on your campus, the number of possible phone numbers in your community, the number of meal options at a local restaurant, the number of ways a person in the group can select outfits for class, the number of ways a condominium can be purchased in a nearby community, and so on. Once situations have been selected, group members should determine in how many ways each part of the task can be done. Group members will need to obtain menus, find out about telephone-digit requirements in the community, count shirts, pants, shoes in closets, visit condominium sales offices, and so on. Once the group reassembles, apply the Fundamental Counting Principle to determine the number of available options in each situation. Because these numbers may be quite large, use a calculator.

Preview Exercises

Exercises 95–97 will help you prepare for the material covered in the next section.

The figure shows that when a die is rolled, there are six equally likely outcomes: 1, 2, 3, 4, 5, or 6. Use this information to solve each exercise.

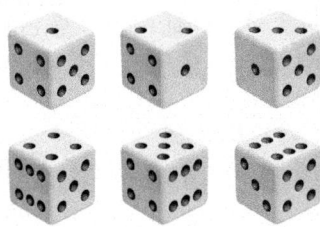

95. What fraction of the outcomes is less than 5?

96. What fraction of the outcomes is not less than 5?

97. What fraction of the outcomes is even or greater than 3?

SECTION 8.7 Probability

Objectives

1. Compute empirical probability.
2. Compute theoretical probability.
3. Find the probability that an event will not occur.
4. Find the probability of one event or a second event occurring.
5. Find the probability of one event and a second event occurring.

Table 8.3 The Hours of Sleep Americans Get on a Typical Night

Hours of Sleep	Number of Americans, in millions
4 or less	12
5	27
6	75
7	90
8	81
9	9
10 or more	6
	Total: 300

Source: Discovery Health Media

How many hours of sleep do you typically get each night? **Table 8.3** indicates that 75 million out of 300 million Americans are getting six hours of sleep on a typical night. The *probability* of an American getting six hours of sleep on a typical night is $\frac{75}{300}$. This fraction can be reduced to $\frac{1}{4}$, or expressed as 0.25, or 25%. Thus, 25% of Americans get six hours of sleep each night.

We find a probability by dividing one number by another. Probabilities are assigned to an *event*, such as getting six hours of sleep on a typical night. Events that are certain to occur are assigned probabilities of 1, or 100%. For example, the probability that a given individual will eventually die is 1. Although Woody Allen whined, "I don't want to achieve immortality through my work. I want to achieve it through not dying," death (and taxes) are always certain. By contrast, if an event cannot occur, its probability is 0. Regrettably, the probability that Elvis will return and serenade us with one final reprise of "Don't Be Cruel" (and we hope we're not) is 0.

Possible Values for Probabilities

① Compute empirical probability.

Probabilities of events are expressed as numbers ranging from 0 to 1, or 0% to 100%. The closer the probability of a given event is to 1, the more likely it is that the event will occur. The closer the probability of a given event is to 0, the less likely it is that the event will occur.

Empirical Probability

Empirical probability applies to situations in which we observe how frequently an event occurs. We use the following formula to compute the empirical probability of an event:

Computing Empirical Probability

The **empirical probability** of event E, denoted by $P(E)$, is

$$P(E) = \frac{\text{observed number of times } E \text{ occurs}}{\text{total number of observed occurrences}}.$$

EXAMPLE 1 Empirical Probabilities with Real-World Data

When women turn 40, their gynecologists typically remind them that it is time to undergo mammography screening for breast cancer. The data in **Table 8.4** are based on 100,000 U.S. women, ages 40 to 50, who participated in mammography screening.

Table 8.4 Mammography Screening on 100,000 U.S. Women, Ages 40 to 50

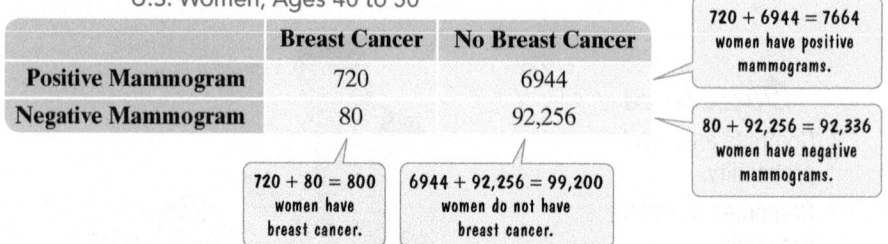

	Breast Cancer	**No Breast Cancer**
Positive Mammogram	720	6944
Negative Mammogram	80	92,256

720 + 6944 = 7664 women have positive mammograms.

80 + 92,256 = 92,336 women have negative mammograms.

720 + 80 = 800 women have breast cancer.

6944 + 92,256 = 99,200 women do not have breast cancer.

Source: Gerd Gigerenzer, *Calculated Risks*, Simon and Schuster, 2002

a. Use **Table 8.4** to find the probability that a woman aged 40 to 50 has breast cancer.

b. Among women without breast cancer, find the probability of a positive mammogram.

c. Among women with positive mammograms, find the probability of not having breast cancer.

SOLUTION

a. We begin with the probability that a woman aged 40 to 50 has breast cancer. The probability of having breast cancer is the number of women with breast cancer divided by the total number of women.

$$P \text{ (breast cancer)} = \frac{\text{number of women with breast cancer}}{\text{total number of women}}$$

$$= \frac{800}{100,000} = \frac{1}{125} = 0.008$$

The empirical probability that a woman aged 40 to 50 has breast cancer is $\frac{1}{125}$, or 0.008.

b. Now, we find the probability of a positive mammogram among women without breast cancer. Thus, we restrict the data to women without breast cancer:

	No Breast Cancer
Positive Mammogram	6944
Negative Mammogram	92,256

Within the restricted data, the probability of a positive mammogram is the number of women with positive mammograms divided by the total number of women.

$$P \text{ (positive mammogram)} = \frac{\text{number of women with positive mammograms}}{\text{total number of women in the restricted data}}$$

$$= \frac{6944}{6944 + 92{,}256} = \frac{6944}{99{,}200} = 0.07$$

This is the total number of women without breast cancer.

Among women without breast cancer, the empirical probability of a positive mammogram is $\frac{6944}{99,200}$, or 0.07.

c. Now, we find the probability of not having breast cancer among women with positive mammograms. Thus, we restrict the data to women with positive mammograms:

	Breast Cancer	No Breast Cancer
Positive Mammogram	720	6944

Within the restricted data, the probability of not having breast cancer is the number of women with no breast cancer divided by the total number of women.

$$P \text{ (no breast cancer)} = \frac{\text{number of women with no breast cancer}}{\text{total number of women in the restricted data}}$$

$$= \frac{6944}{720 + 6944} = \frac{6944}{7664} \approx 0.906$$

This is the total number of women with positive mammograms.

Among women with positive mammograms, the probability of not having breast cancer is $\frac{6944}{7664}$, or approximately 0.906. • • •

⊘ Check Point **1** Use the data in **Table 8.4** to solve this exercise. Express probabilities as fractions and as decimals rounded to three decimal places.

a. Find the probability that a woman aged 40 to 50 has a positive mammogram.

b. Among women with breast cancer, find the probability of a positive mammogram.

c. Among women with positive mammograms, find the probability of having breast cancer.

② Compute theoretical probability.

Theoretical Probability

You toss a coin. Although it is equally likely to land either heads up, denoted by H, or tails up, denoted by T, the actual outcome is uncertain. Any occurrence for which the outcome is uncertain is called an **experiment**. Thus, tossing a coin is an example of an experiment. The set of all possible outcomes of an experiment is the **sample space** of the experiment, denoted by S. The sample space for the coin-tossing experiment is

$$S = \{H, T\}.$$

Lands heads up Lands tails up

We can define an event more formally using these concepts. An **event**, denoted by E, is any subcollection, or subset, of a sample space. For example, the subset $E = \{T\}$ is the event of landing tails up when a coin is tossed.

Theoretical probability applies to situations like this, in which the sample space only contains equally likely outcomes, all of which are known. To calculate the theoretical probability of an event, we divide the number of outcomes resulting in the event by the number of outcomes in the sample space.

Computing Theoretical Probability

If an event E has $n(E)$ equally likely outcomes and its sample space S has $n(S)$ equally likely outcomes, the **theoretical probability** of event E, denoted by $P(E)$, is

$$P(E) = \frac{\text{number of outcomes in event } E}{\text{number of outcomes in sample space } S} = \frac{n(E)}{n(S)}.$$

The sum of the theoretical probabilities of all possible outcomes in the sample space is 1.

How can we use this formula to compute the probability of a coin landing tails up? We use the following sets:

$$E = \{T\} \qquad S = \{H, T\}.$$

This is the event This is the sample space with
of landing tails up. all equally likely outcomes.

The probability of a coin landing tails up is

$$P(E) = \frac{\text{number of outcomes that result in tails up}}{\text{total number of possible outcomes}} = \frac{n(E)}{n(S)} = \frac{1}{2}.$$

Theoretical probability applies to many games of chance, including rolling dice, lotteries, card games, and roulette. The next example deals with the experiment of rolling a die. **Figure 8.11** illustrates that when a die is rolled, there are six equally likely outcomes. The sample space can be shown as

$$S = \{1, 2, 3, 4, 5, 6\}.$$

FIGURE 8.11 Outcomes when a die is rolled

EXAMPLE 2 Computing Theoretical Probability

A die is rolled. Find the probability of getting a number less than 5.

SOLUTION

The sample space of equally likely outcomes is $S = \{1, 2, 3, 4, 5, 6\}$. There are six outcomes in the sample space, so $n(S) = 6$.

We are interested in the probability of getting a number less than 5. The event of getting a number less than 5 can be represented by

$$E = \{1, 2, 3, 4\}.$$

There are four outcomes in this event, so $n(E) = 4$.
The probability of rolling a number less than 5 is

$$P(E) = \frac{n(E)}{n(S)} = \frac{4}{6} = \frac{2}{3}.$$

•••

⊘ Check Point **2** A die is rolled. Find the probability of getting a number greater than 4.

EXAMPLE 3 Computing Theoretical Probability

Two ordinary six-sided dice are rolled. What is the probability of getting a sum of 8?

SOLUTION

Each die has six equally likely outcomes. By the Fundamental Counting Principle, there are $6 \cdot 6$, or 36, equally likely outcomes in the sample space. That is, $n(S) = 36$. The 36 outcomes are shown below as ordered pairs. The five ways of rolling a sum of 8 appear in the green highlighted diagonal.

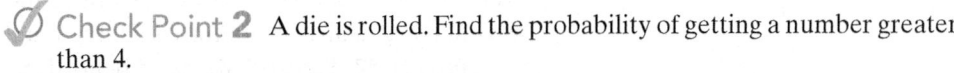

		Second Die				
	•	••	•••	••••	•••••	••••••
•	(1,1)	(1,2)	(1,3)	(1,4)	(1,5)	(1,6)
••	(2,1)	(2,2)	(2,3)	(2,4)	(2,5)	(2,6)
•••	(3,1)	(3,2)	(3,3)	(3,4)	(3,5)	(3,6)
••••	(4,1)	(4,2)	(4,3)	(4,4)	(4,5)	(4,6)
•••••	(5,1)	(5,2)	(5,3)	(5,4)	(5,5)	(5,6)
••••••	(6,1)	(6,2)	(6,3)	(6,4)	(6,5)	(6,6)

$$
\begin{aligned}
S = \{ &(1,1), (1,2), (1,3), (1,4), \\
&(1,5), (1,6), (2,1), (2,2), \\
&(2,3), (2,4), (2,5), (2,6), \\
&(3,1), (3,2), (3,3), (3,4), \\
&(3,5), (3,6), (4,1), (4,2), \\
&(4,3), (4,4), (4,5), (4,6), \\
&(5,1), (5,2), (5,3), (5,4), \\
&(5,5), (5,6), (6,1), (6,2), \\
&(6,3), (6,4), (6,5), (6,6) \}
\end{aligned}
$$

The phrase "getting a sum of 8" describes the event

$$E = \{(6, 2), (5, 3), (4, 4), (3, 5), (2, 6)\}.$$

This event has 5 outcomes, so $n(E) = 5$. Thus, the probability of getting a sum of 8 is

$$P(E) = \frac{n(E)}{n(S)} = \frac{5}{36}.$$

•••

⊘ Check Point **3** What is the probability of getting a sum of 5 when two six-sided dice are rolled?

FIGURE 8.12 A standard 52-card bridge deck

Computing Theoretical Probability without Listing an Event and the Sample Space

In some situations, we can compute theoretical probability without having to write out each event and each sample space. For example, suppose you are dealt one card from a standard 52-card deck, illustrated in **Figure 8.12**. The deck has four suits: Hearts and diamonds are red, and clubs and spades are black. Each suit has 13 different face values—A(ace), 2, 3, 4, 5, 6, 7, 8, 9, 10, J(jack), Q(queen), and K(king). Jacks, queens, and kings are called **picture cards** or **face cards**.

FIGURE 8.12 A standard 52-card bridge deck (repeated)

Hearts
Clubs
Diamonds
Spades

EXAMPLE 4 Probability and a Deck of 52 Cards

You are dealt one card from a standard 52-card deck. Find the probability of being dealt a heart.

SOLUTION

Let E be the event of being dealt a heart. Because there are 13 hearts in the deck, the event of being dealt a heart can occur in 13 ways. The number of outcomes in event E is 13: $n(E) = 13$. With 52 cards in the deck, the total number of possible ways of being dealt a single card is 52. The number of outcomes in the sample space is 52: $n(S) = 52$. The probability of being dealt a heart is

$$P(E) = \frac{n(E)}{n(S)} = \frac{13}{52} = \frac{1}{4}.$$

• • •

✓ Check Point **4** If you are dealt one card from a standard 52-card deck, find the probability of being dealt a king.

If your state has a lottery drawing each week, the probability that someone will win the top prize is relatively high. If there is no winner this week, it is virtually certain that eventually someone will be graced with millions of dollars. So, why are you so unlucky compared to this undisclosed someone? In Example 5, we provide an answer to this question, using the counting principles discussed in Section 8.6.

EXAMPLE 5 Probability and Combinations: Winning the Lottery

Florida's lottery game, LOTTO, is set up so that each player chooses six different numbers from 1 to 53. If the six numbers chosen match the six numbers drawn randomly, the player wins (or shares) the top cash prize. (As of this writing, the top cash prize has ranged from \$7 million to \$106.5 million.) With one LOTTO ticket, what is the probability of winning this prize?

SOLUTION

Because the order of the six numbers does not matter, this is a situation involving combinations. Let E be the event of winning the lottery with one ticket. With one LOTTO ticket, there is only one way of winning. Thus, $n(E) = 1$. The sample space is the set of all possible six-number combinations. We can use the combinations formula

$$_nC_r = \frac{n!}{(n-r)!r!}$$

to find the total number of possible combinations. We are selecting $r = 6$ numbers from a collection of $n = 53$ numbers.

$$_{53}C_6 = \frac{53!}{(53-6)!6!} = \frac{53!}{47!6!} = \frac{53 \cdot 52 \cdot 51 \cdot 50 \cdot 49 \cdot 48 \cdot 47!}{47! \cdot 6 \cdot 5 \cdot 4 \cdot 3 \cdot 2 \cdot 1} = 22,957,480$$

Blitzer Bonus

Comparing the Probability of Dying to the Probability of Winning Florida's LOTTO

As a healthy nonsmoking 30-year-old, your probability of dying this year is approximately 0.001. Divide this probability by the probability of winning LOTTO with one ticket:

$$\frac{0.001}{0.0000000436} \approx 22{,}936.$$

A healthy 30-year-old is nearly 23,000 times more likely to die this year than to win Florida's lottery.

There are nearly 23 million number combinations possible in LOTTO. If a person buys one LOTTO ticket, the probability of winning is

$$P(E) = \frac{n(E)}{n(S)} = \frac{1}{22{,}957{,}480} \approx 0.0000000436.$$

The probability of winning the top prize with one LOTTO ticket is $\frac{1}{22{,}957{,}480}$, or about 1 in 23 million.

Suppose that a person buys 5000 different tickets in Florida's LOTTO. Because that person has selected 5000 different combinations of the six numbers, the probability of winning is

$$\frac{5000}{22{,}957{,}480} \approx 0.000218.$$

The chances of winning top prize are about 218 in a million. At $1 per LOTTO ticket, it is highly probable that our LOTTO player will be $5000 poorer. Knowing a little probability helps a lotto.

✓ Check Point **5** People lose interest when they do not win at games of chance, including Florida's LOTTO. With drawings twice weekly instead of once, the game described in Example 5 was brought in to bring back lost players and increase ticket sales. The original LOTTO was set up so that each player chose six different numbers from 1 to 49, rather than from 1 to 53, with a lottery drawing only once a week. With one LOTTO ticket, what was the probability of winning the top cash prize in Florida's original LOTTO? Express the answer as a fraction and as a decimal correct to ten places.

③ Find the probability that an event will not occur.

Probability of an Event Not Occurring

If we know $P(E)$, the probability of an event E, we can determine the probability that the event will not occur, denoted by $P(\text{not } E)$. Because the sum of the probabilities of all possible outcomes in any situation is 1,

$$P(E) + P(\text{not } E) = 1.$$

We now solve this equation for $P(\text{not } E)$, the probability that event E will not occur, by subtracting $P(E)$ from both sides. The resulting formula is given in the following box.

The Probability of an Event Not Occurring

The probability that an event E will not occur is equal to 1 minus the probability that it will occur.

$$P(\text{not } E) = 1 - P(E)$$

World Population, by Continent/Region

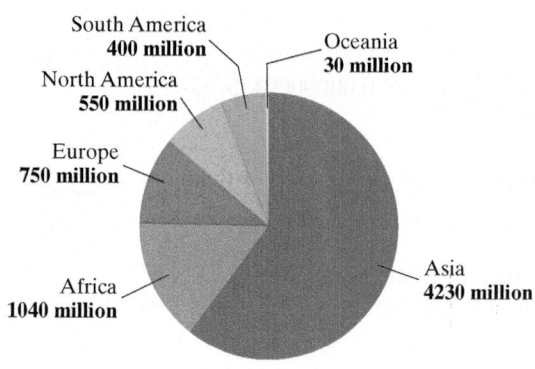

FIGURE 8.13
Source: U.S. Census Bureau

EXAMPLE 6 The Probability of an Event Not Occurring

The circle graph in **Figure 8.13** shows the distribution, by continent, of the world's 7 billion, or 7000 million, people. If one person is randomly selected, find the probability that the person does not live in Asia. Express the probability as a simplified fraction and as a decimal rounded to the nearest thousandth.

SOLUTION

We use the probability that the selected person *does* live in Asia to find the probability that the selected person does not live in this region.

P (does not live in Asia)

$$= 1 - P(\text{lives in Asia})$$

> The graph shows **4230 million** people living in Asia.

$$= 1 - \frac{4230}{7000}$$

> World population, **7000 million**, was given, but can be obtained by adding the numbers in the six sectors.

$$= \frac{7000}{7000} - \frac{4230}{7000} = \frac{2770}{7000} = \frac{277}{700} \approx 0.396$$

The probability that a randomly selected person does not live in Asia is $\frac{277}{700}$, or approximately 0.396.

⊘ Check Point **6** If one person is randomly selected from the world population represented in **Figure 8.13**, find the probability that the person does not live in North America. Express the probability as a simplified fraction and as a decimal rounded to the nearest thousandth.

④ Find the probability of one event or a second event occurring.

Or Probabilities with Mutually Exclusive Events

Suppose that you randomly select one card from a deck of 52 cards. Let A be the event of selecting a king and let B be the event of selecting a queen. Only one card is selected, so it is impossible to get both a king and a queen. The events of selecting a king and a queen cannot occur simultaneously. They are called *mutually exclusive events*. If it is impossible for any two events, A and B, to occur simultaneously, they are said to be **mutually exclusive**. If A and B are mutually exclusive events, the probability that either A or B will occur is determined by adding their individual probabilities.

Or Probabilities with Mutually Exclusive Events

If A and B are mutually exclusive events, then

$$P(A \text{ or } B) = P(A) + P(B).$$

Using set notation, $P(A \cup B) = P(A) + P(B).$

EXAMPLE 7 The Probability of Either of Two Mutually Exclusive Events Occurring

If one card is randomly selected from a deck of cards, what is the probability of selecting a king or a queen?

SOLUTION

We find the probability that either of these mutually exclusive events will occur by adding their individual probabilities.

$$P(\text{king or queen}) = P(\text{king}) + P(\text{queen}) = \frac{4}{52} + \frac{4}{52} = \frac{8}{52} = \frac{2}{13}$$

The probability of selecting a king or a queen is $\frac{2}{13}$. •••

⊘ Check Point 7 If you roll a single, six-sided die, what is the probability of getting either a 4 or a 5?

Or Probabilities with Events That Are Not Mutually Exclusive

Consider the deck of 52 cards shown in **Figure 8.14**. Suppose that these cards are shuffled and you randomly select one card from the deck. What is the probability of selecting a diamond or a picture card (jack, queen, king)? Begin by adding their individual probabilities.

There are 13 diamonds in the deck of 52 cards.

$$P(\text{diamond}) + P(\text{picture card}) = \frac{13}{52} + \frac{12}{52}$$

There are 12 picture cards in the deck of 52 cards.

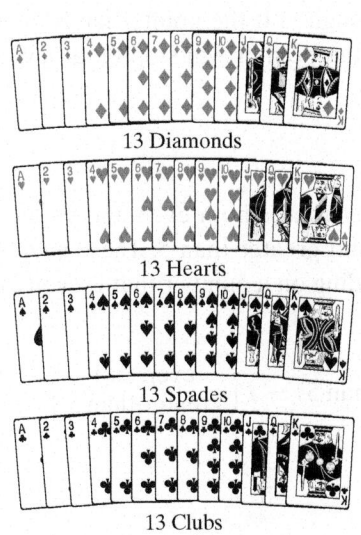

13 Diamonds

13 Hearts

13 Spades

13 Clubs

FIGURE 8.14 A deck of 52 cards

However, this sum is not the probability of selecting a diamond or a picture card. The problem is that there are three cards that are *simultaneously* diamonds and picture cards, shown in **Figure 8.15**. The events of selecting a diamond and selecting a picture card are not mutually exclusive. It is possible to select a card that is both a diamond and a picture card.

The situation is illustrated in the diagram in **Figure 8.16**. Why can't we find the probability of selecting a diamond or a picture card by adding their individual probabilities? The diagram shows that three of the cards, the three diamonds that are picture cards, get counted twice when we add the individual probabilities. First the three cards get counted as diamonds and then they get counted as picture cards. In order to avoid the error of counting the three cards twice, we need to subtract the probability of getting a diamond and a picture card, $\frac{3}{52}$, as follows:

FIGURE 8.15 Three diamonds are picture cards.

FIGURE 8.16

$P(\text{diamond or picture card})$

$$= P(\text{diamond}) + P(\text{picture card}) - P(\text{diamond and picture card})$$

$$= \frac{13}{52} + \frac{12}{52} - \frac{3}{52} = \frac{13 + 12 - 3}{52} = \frac{22}{52} = \frac{11}{26}.$$

Thus, the probability of selecting a diamond or a picture card is $\frac{11}{26}$.

In general, if A and B are events that are not mutually exclusive, the probability that A or B will occur is determined by adding their individual probabilities and then subtracting the probability that A and B occur simultaneously.

Or Probabilities with Events That Are Not Mutually Exclusive

If A and B are not mutually exclusive events, then

$$P(A \text{ or } B) = P(A) + P(B) - P(A \text{ and } B).$$

Using set notation,

$$P(A \cup B) = P(A) + P(B) - P(A \cap B).$$

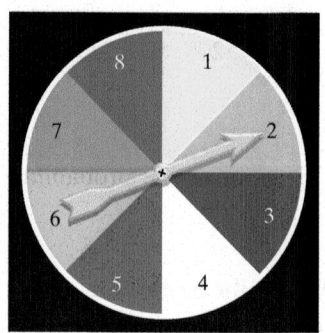

FIGURE 8.17 It is equally probable that the pointer will land on any one of the eight regions.

EXAMPLE 8 An *Or* Probability with Events That Are Not Mutually Exclusive

Figure 8.17 illustrates a spinner. It is equally probable that the pointer will land on any one of the eight regions, numbered 1 through 8. If the pointer lands on a borderline, spin again. Find the probability that the pointer will stop on an even number or a number greater than 5.

SOLUTION

It is possible for the pointer to land on a number that is both even and greater than 5. Two of the numbers, 6 and 8, are even and greater than 5. These events are not mutually exclusive. The probability of landing on a number that is even or greater than 5 is calculated as follows:

$$P\binom{\text{even or}}{\text{greater than 5}} = P(\text{even}) + P(\text{greater than 5}) - P\binom{\text{even and}}{\text{greater than 5}}$$

$$= \frac{4}{8} + \frac{3}{8} - \frac{2}{8}$$

> Four of the eight numbers, 2, 4, 6, and 8, are even.

> Three of the eight numbers, 6, 7, and 8, are greater than 5.

> Two of the eight numbers, 6 and 8, are even and greater than 5.

$$= \frac{4 + 3 - 2}{8} = \frac{5}{8}.$$

The probability that the pointer will stop on an even number or a number greater than 5 is $\frac{5}{8}$. •••

○ **Check Point 8** Use **Figure 8.17** to find the probability that the pointer will stop on an odd number or a number less than 5.

EXAMPLE 9 An *Or* Probability with Real-World Data

Table 8.5 shows the marital status of the U.S. population in 2010. Numbers in the table are expressed in millions.

Table 8.5 Marital Status of the U.S. Population, Ages 15 or Older, 2010, in Millions

	Married	Never Married	Divorced	Widowed	Total
Male	65	40	10	3	118
Female	65	34	14	11	124
Total	130	74	24	14	242

Source: U.S. Census Bureau

If one person is randomly selected from the population represented in **Table 8.5**, find the probability that

 a. the person is divorced or male.

 b. the person is married or divorced.

Express probabilities as simplified fractions and as decimals rounded to the nearest hundredth.

SOLUTION

a. It is possible to select a person who is both divorced and male. Thus, these events are not mutually exclusive.

P(divorced or male)

$$= P(\text{divorced}) + P(\text{male}) - P(\text{divorced and male})$$

$$= \frac{24}{242} + \frac{118}{242} - \frac{10}{242}$$

Of the 242 million Americans, 24 million are divorced.	Of the 242 million Americans, 118 million are male.	Of the 242 million Americans, 10 million are divorced and male.

$$= \frac{24 + 118 - 10}{242} = \frac{132}{242} = \frac{22 \cdot 6}{22 \cdot 11} = \frac{6}{11} \approx 0.55$$

The probability of selecting a person who is divorced or male is $\frac{6}{11}$, or approximately 0.55.

b. It is impossible to select a person who is both married and divorced. These events are mutually exclusive.

P(married or divorced)

$$= P(\text{married}) + P(\text{divorced})$$

$$= \frac{130}{242} + \frac{24}{242}$$

Of the 242 million Americans, 130 million are married.	Of the 242 million Americans, 24 million are divorced.

$$= \frac{130 + 24}{242} = \frac{154}{242} = \frac{22 \cdot 7}{22 \cdot 11} = \frac{7}{11} \approx 0.64$$

The probability of selecting a person who is married or divorced is $\frac{7}{11}$, or approximately 0.64.　　　　●●●

⊘ Check Point **9** If one person is randomly selected from the population represented in **Table 8.5**, find the probability that

 a. the person is married or female.

 b. the person is divorced or widowed.

Express probabilities as simplified fractions and as decimals rounded to the nearest hundredth.

⑤ Find the probability of one event and a second event occurring.

And Probabilities with Independent Events

Suppose that you toss a fair coin two times in succession. The outcome of the first toss, heads or tails, does not affect what happens when you toss the coin a second time. For example, the occurrence of tails on the first toss does not make tails more likely or less likely to occur on the second toss. The repeated toss of a coin produces

independent events because the outcome of one toss does not influence the outcome of others. Two events are **independent events** if the occurrence of either of them has no effect on the probability of the other.

If two events are independent, we can calculate the probability of the first occurring and the second occurring by multiplying their probabilities.

And Probabilities with Independent Events

If A and B are independent events, then

$$P(A \text{ and } B) = P(A) \cdot P(B).$$

EXAMPLE 10 Independent Events on a Roulette Wheel

Figure 8.18 shows a U.S. roulette wheel that has 38 numbered slots (1 through 36, 0, and 00). Of the 38 compartments, 18 are black, 18 are red, and 2 are green. A play has the dealer spin the wheel and a small ball in opposite directions. As the ball slows to a stop, it can land with equal probability on any one of the 38 numbered slots. Find the probability of red occurring on two consecutive plays.

SOLUTION

FIGURE 8.18 A U.S. roulette wheel

The wheel has 38 equally likely outcomes and 18 are red. Thus, the probability of red occurring on a play is $\frac{18}{38}$, or $\frac{9}{19}$. The result that occurs on each play is independent of all previous results. Thus,

$$P(\text{red and red}) = P(\text{red}) \cdot P(\text{red}) = \frac{9}{19} \cdot \frac{9}{19} = \frac{81}{361} \approx 0.224.$$

The probability of red occurring on two consecutive plays is $\frac{81}{361}$. • • •

Some roulette players incorrectly believe that if red occurs on two consecutive plays, then another color is "due." Because the events are independent, the outcomes of previous spins have no effect on any other spins.

⊘ Check Point **10** Find the probability of green occurring on two consecutive plays on a roulette wheel.

The *and* rule for independent events can be extended to cover three or more events. Thus, if A, B, and C are independent events, then

$$P(A \text{ and } B \text{ and } C) = P(A) \cdot P(B) \cdot P(C).$$

EXAMPLE 11 Independent Events in a Family

The picture in the margin shows a family that has had nine girls in a row. Find the probability of this occurrence.

SOLUTION

If two or more events are independent, we can find the probability of them all occurring by multiplying their probabilities. The probability of a baby girl is $\frac{1}{2}$, so the probability of nine girls in a row is $\frac{1}{2}$ used as a factor nine times.

$$P(\text{nine girls in a row}) = \frac{1}{2} \cdot \frac{1}{2} \cdot \frac{1}{2} \cdot \frac{1}{2} \cdot \frac{1}{2} \cdot \frac{1}{2} \cdot \frac{1}{2} \cdot \frac{1}{2} \cdot \frac{1}{2}$$

$$= \left(\frac{1}{2}\right)^9 = \frac{1}{512}$$

The probability of a run of nine girls in a row is $\frac{1}{512}$. (If another child is born into the family, this event is independent of the other nine, and the probability of a girl is still $\frac{1}{2}$.) • • •

☑ Check Point **11** Find the probability of a family having four boys in a row.

CONCEPT AND VOCABULARY CHECK

Fill in each blank so that the resulting statement is true.

1. Probability that is based on situations in which we observe how frequently an event occurs is called _____ probability.
2. The set of all possible outcomes of an experiment is called the _____ of the experiment.
3. The theoretical probability of event E, denoted by _____, is the _____ divided by the _____.
4. A standard bridge deck has _____ cards with four suits: _____ and _____ are red, and _____ and _____ are black.
5. The probability of winning a lottery with one lottery ticket is the number of ways of winning, which is precisely _____, divided by the total number of possible _____.

6. Because $P(E) + P(\text{not } E) = 1$, then $P(\text{not } E) =$ _____ and $P(E) =$ _____.
7. If it is impossible for events A and B to occur simultaneously, the events are said to be _____. For such events, $P(A \text{ or } B) =$ _____.
8. If it is possible for events A and B to occur simultaneously, then $P(A \text{ or } B) =$ _____.
9. If the occurrence of one event has no effect on the probability of another event, the events are said to be _____. For such events $P(A \text{ and } B) =$ _____.

EXERCISE SET 8.7

Practice and Application Exercises

Shown again is the table indicating the marital status of the U.S. population in 2010. Numbers in the table are expressed in millions. Use the data in the table to solve Exercises 1–10. Express probabilities as simplified fractions and as decimals rounded to the nearest hundredth.

Marital Status of the U.S. Population, Ages 15 or Older, 2010, in Millions

	Married	Never Married	Divorced	Widowed	Total
Male	65	40	10	3	118
Female	65	34	14	11	124
Total	130	74	24	14	242

If one person is randomly selected from the population described in the table, find the probability that the person

1. is divorced.
2. has never been married.
3. is female.
4. is male.
5. is a widowed male.
6. is a widowed female.
7. Among those who are divorced, find the probability of selecting a woman.

8. Among those who are divorced, find the probability of selecting a man.
9. Among men, find the probability of selecting a married person.
10. Among women, find the probability of selecting a married person.

In Exercises 11–16, a die is rolled. Find the probability of getting

11. a 4.
12. a 5.
13. an odd number.
14. a number greater than 3.
15. a number greater than 4.
16. a number greater than 7.

In Exercises 17–20, you are dealt one card from a standard 52-card deck. Find the probability of being dealt

17. a queen. **18.** a diamond.

19. a picture card.

20. a card greater than 3 and less than 7.

In Exercises 21–22, a fair coin is tossed two times in succession. The sample space of equally likely outcomes is {HH, HT, TH, TT}. Find the probability of getting

21. two heads.

22. the same outcome on each toss.

In Exercises 23–24, you select a family with three children. If M represents a male child and F a female child, the sample space of equally likely outcomes is {MMM, MMF, MFM, MFF, FMM, FMF, FFM, FFF}. Find the probability of selecting a family with

23. at least one male child.

24. at least two female children.

In Exercises 25–26, a single die is rolled twice. The 36 equally likely outcomes are shown as follows:

	Second Roll					
	⚀	⚁	⚂	⚃	⚄	⚅
⚀	(1,1)	(1,2)	(1,3)	(1,4)	(1,5)	(1,6)
⚁	(2,1)	(2,2)	(2,3)	(2,4)	(2,5)	(2,6)
⚂	(3,1)	(3,2)	(3,3)	(3,4)	(3,5)	(3,6)
⚃	(4,1)	(4,2)	(4,3)	(4,4)	(4,5)	(4,6)
⚄	(5,1)	(5,2)	(5,3)	(5,4)	(5,5)	(5,6)
⚅	(6,1)	(6,2)	(6,3)	(6,4)	(6,5)	(6,6)

First Roll

Find the probability of getting

25. two numbers whose sum is 4.

26. two numbers whose sum is 6.

27. To play the California lottery, a person has to select 6 out of 51 numbers, paying $1 for each six-number selection. If you pick six numbers that are the same as the ones drawn by the lottery, you win mountains of money. What is the probability that a person with one combination of six numbers will win? What is the probability of winning if 100 different lottery tickets are purchased?

28. A state lottery is designed so that a player chooses six numbers from 1 to 30 on one lottery ticket. What is the probability that a player with one lottery ticket will win? What is the probability of winning if 100 different lottery tickets are purchased?

*Exercises 29–30 involve a deck of 52 cards. If necessary, refer to the picture of a deck of cards, **Figure 8.12** on page 785.*

29. A poker hand consists of five cards.

 a. Find the total number of possible five-card poker hands.

 b. A diamond flush is a five-card hand consisting of all diamonds. Find the number of possible diamond flushes.

 c. Find the probability of being dealt a diamond flush.

30. If you are dealt 3 cards from a shuffled deck of 52 cards, find the probability that all 3 cards are picture cards.

The table shows the educational attainment of the U.S. population, ages 25 and over. Use the data in the table, expressed in millions, to solve Exercises 31–36.

Educational Attainment, in Millions, of the United States Population, Ages 25 and Over

	Less Than 4 Years High School	4 Years High School Only	Some College [Less Than 4 Years]	4 Years College [or More]	Total
Male	14	25	20	23	82
Female	15	31	24	22	92
Total	29	56	44	45	174

Source: U.S. Census Bureau

Find the probability, expressed as a simplified fraction, that a randomly selected American, age 25 or over,

31. has not completed four years (or more) of college.

32. has not completed four years of high school.

33. has completed four years of high school only or less than four years of college.

34. has completed less than four years of high school or four years of high school only.

35. has completed four years of high school only or is a man.

36. has completed four years of high school only or is a woman.

In Exercises 37–42, you are dealt one card from a 52-card deck. Find the probability that

37. you are not dealt a king.

38. you are not dealt a picture card.

39. you are dealt a 2 or a 3.

40. you are dealt a red 7 or a black 8.

41. you are dealt a 7 or a red card.

42. you are dealt a 5 or a black card.

In Exercises 43–44, it is equally probable that the pointer on the spinner shown will land on any one of the eight regions, numbered 1 through 8. If the pointer lands on a borderline, spin again.

Find the probability that the pointer will stop on

43. an odd number or a number less than 6.

44. an odd number or a number greater than 3.

Use this information to solve Exercises 45–46. The mathematics department of a college has 8 male professors, 11 female professors, 14 male teaching assistants, and 7 female teaching assistants. If a person is selected at random from the group, find the probability that the selected person is

45. a professor or a male.

46. a professor or a female.

In Exercises 47–50, a single die is rolled twice. Find the probability of rolling

47. a 2 the first time and a 3 the second time.

48. a 5 the first time and a 1 the second time.

49. an even number the first time and a number greater than 2 the second time.

50. an odd number the first time and a number less than 3 the second time.

51. If you toss a fair coin six times, what is the probability of getting all heads?

52. If you toss a fair coin seven times, what is the probability of getting all tails?

53. The probability that South Florida will be hit by a major hurricane (category 4 or 5) in any single year is $\frac{1}{16}$. (*Source:* National Hurricane Center)

 a. What is the probability that South Florida will be hit by a major hurricane two years in a row?

 b. What is the probability that South Florida will be hit by a major hurricane in three consecutive years?

 c. What is the probability that South Florida will not be hit by a major hurricane in the next ten years?

 d. What is the probability that South Florida will be hit by a major hurricane at least once in the next ten years?

Writing in Mathematics

54. Describe the difference between theoretical probability and empirical probability.

55. Give an example of an event whose probability must be determined empirically rather than theoretically.

56. Write a probability word problem whose answer is one of the following fractions: $\frac{1}{6}$ or $\frac{1}{4}$ or $\frac{1}{3}$.

57. Explain how to find the probability of an event not occurring. Give an example.

58. What are mutually exclusive events? Give an example of two events that are mutually exclusive.

59. Explain how to find *or* probabilities with mutually exclusive events. Give an example.

60. Give an example of two events that are not mutually exclusive.

61. Explain how to find *or* probabilities with events that are not mutually exclusive. Give an example.

62. Explain how to find *and* probabilities with independent events. Give an example.

63. The president of a large company with 10,000 employees is considering mandatory cocaine testing for every employee. The test that would be used is 90% accurate, meaning that it will detect 90% of the cocaine users who are tested, and that 90% of the nonusers will test negative. This also means that the test gives 10% false positive. Suppose that 1% of the employees actually use cocaine. Find the probability that someone who tests positive for cocaine use is, indeed, a user.

Hint: Find the following probability fraction:

$$\frac{\text{the number of employees who test positive and are cocaine users}}{\text{the number of employees who test positive}}.$$

This fraction is given by

$$\frac{90\% \text{ of } 1\% \text{ of } 10{,}000}{\text{the number who test positive who actually use cocaine plus the number who test positive who do not use cocaine}}.$$

What does this probability indicate in terms of the percentage of employees who test positive who are not actually users? Discuss these numbers in terms of the issue of mandatory drug testing. Write a paper either in favor of or against mandatory drug testing, incorporating the actual percentage accuracy for such tests.

Critical Thinking Exercises

Make Sense? *In Exercises 64–67, determine whether each statement makes sense or does not make sense, and explain your reasoning.*

64. The probability that Jill will win the election is 0.7 and the probability that she will not win is 0.4.

65. Assuming the next U.S. president will be a Democrat or a Republican, the probability of a Republican president is 0.5.

66. The probability that I will go to graduate school is 1.5.

67. When I toss a coin, the probability of getting heads *or* tails is 1, but the probability of getting heads *and* tails is 0.

68. The target in the figure shown contains four squares. If a dart thrown at random hits the target, find the probability that it will land in a yellow region.

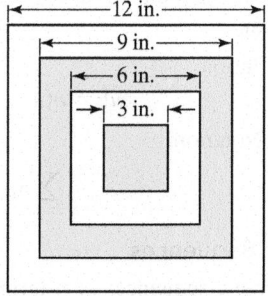

69. Suppose that it is a week in which the cash prize in Florida's LOTTO is promised to exceed $50 million. If a person purchases 22,957,480 tickets in LOTTO at $1 per ticket (all possible combinations), isn't this a guarantee of winning the lottery? Because the probability in this situation is 1, what's wrong with doing this?

70. Some three-digit numbers, such as 101 and 313, read the same forward and backward. If you select a number from all three-digit numbers, find the probability that it will read the same forward and backward.

71. In a class of 50 students, 29 are Democrats, 11 are business majors, and 5 of the business majors are Democrats. If one student is randomly selected from the class, find the probability of choosing

 a. a Democrat who is not a business major.

 b. a student who is neither a Democrat nor a business major.

72. On New Year's Eve, the probability of a person driving while intoxicated or having a driving accident is 0.35. If the probability of driving while intoxicated is 0.32 and the probability of having a driving accident is 0.09, find the probability of a person having a driving accident while intoxicated.

73. a. If two people are selected at random, the probability that they do not have the same birthday (day and month) is $\frac{365}{365} \cdot \frac{364}{365}$. Explain why this is so. (Ignore leap years and assume 365 days in a year.)

 b. If three people are selected at random, find the probability that they all have different birthdays.

 c. If three people are selected at random, find the probability that at least two of them have the same birthday.

 d. If 20 people are selected at random, find the probability that at least 2 of them have the same birthday.

 e. How large a group is needed to give a 0.5 chance of at least two people having the same birthday?

Group Exercise

74. Research and present a group report on state lotteries. Include answers to some or all of the following questions: Which states do not have lotteries? Why not? How much is spent per capita on lotteries? What are some of the lottery games? What is the probability of winning top prize in these games? What income groups spend the greatest amount of money on lotteries? If your state has a lottery, what does it do with the money it makes? Is the way the money is spent what was promised when the lottery first began?

CHAPTER 8 Summary, Review, and Test

SUMMARY

DEFINITIONS AND CONCEPTS	**EXAMPLES**
8.1 Sequences and Summation Notation	
a. An infinite sequence $\{a_n\}$ is a function whose domain is the set of positive integers. The function values, or terms, are represented by $$a_1, a_2, a_3, a_4, \ldots, a_n, \ldots$$	Ex. 1, p. 717
b. Sequences can be defined using recursion formulas that define the nth term as a function of the previous term.	Ex. 2, p. 718
c. Factorial Notation: $$n! = n(n-1)(n-2) \cdots (3)(2)(1) \quad \text{and} \quad 0! = 1$$	Ex. 3, p. 719; Ex. 4, p. 720
d. Summation Notation: $$\sum_{i=1}^{n} a_i = a_1 + a_2 + a_3 + a_4 + \cdots + a_n$$	Ex. 5, p. 721; Ex. 6, p. 723
8.2 Arithmetic Sequences	
a. In an arithmetic sequence, each term after the first differs from the preceding term by a constant, the common difference. Subtract any term from the term that directly follows to find the common difference.	List of arithmetic sequences and common differences, p. 728; Ex. 1, p. 729
b. General term or nth term: $a_n = a_1 + (n-1)d$. The first term is a_1 and the common difference is d.	Ex. 2, p. 730; Ex. 3, p. 730
c. Sum of the first n terms: $S_n = \dfrac{n}{2}(a_1 + a_n)$	Ex. 4, p. 732; Ex. 5, p. 733; Ex. 6, p. 733

DEFINITIONS AND CONCEPTS

8.3 Geometric Sequences and Series

a. In a geometric sequence, each term after the first is obtained by multiplying the preceding term by a nonzero constant, the common ratio. Divide any term after the first by the term that directly precedes it to find the common ratio.

List of geometric sequences and common ratios, p. 738; Ex. 1, p. 738

b. General term or nth term: $a_n = a_1 r^{n-1}$. The first term is a_1 and the common ratio is r.

Ex. 2, p. 739; Ex. 3, p. 740

c. Sum of the first n terms: $S_n = \dfrac{a_1(1 - r^n)}{1 - r}, r \neq 1$

Ex. 4, p. 742; Ex. 5, p. 742; Ex. 6, p. 743

d. An annuity is a sequence of equal payments made at equal time periods. The value of an annuity, A, is the sum of all deposits made plus all interest paid, given by

Ex. 7, p. 745

$$A = \dfrac{P\left[\left(1 + \dfrac{r}{n}\right)^{nt} - 1\right]}{\dfrac{r}{n}}.$$

The deposit made at the end of each period is P, the annual interest rate is r, compounded n times per year, and t is the number of years deposits have been made.

e. The sum of the infinite geometric series $a_1 + a_1 r + a_1 r^2 + a_1 r^3 + \cdots$ is $S = \dfrac{a_1}{1 - r}; |r| < 1$. If $|r| \geq 1$, the infinite series does not have a sum.

Ex. 8, p. 747; Ex. 9, p. 748; Ex. 10, p. 748

8.4 Mathematical Induction

To prove that S_n is true for all positive integers n,

1. Show that S_1 is true.

2. Show that if S_k is assumed true, then S_{k+1} is also true, for every positive integer k.

Ex. 2, p. 757; Ex. 3, p. 758; Ex. 4, p. 759

8.5 The Binomial Theorem

a. Binomial coefficient: $\dbinom{n}{r} = \dfrac{n!}{r!(n - r)!}$

Ex. 1, p. 763

b. Binomial Theorem:

Ex. 2, p. 764; Ex. 3, p. 765

$$(a + b)^n = \binom{n}{0}a^n + \binom{n}{1}a^{n-1}b + \binom{n}{2}a^{n-2}b^2 + \cdots + \binom{n}{n}b^n$$

c. The $(r + 1)$st term in the expansion of $(a + b)^n$ is

Ex. 4, p. 766

$$\binom{n}{r}a^{n-r}b^r.$$

8.6 Counting Principles, Permutations, and Combinations

a. The Fundamental Counting Principle: The number of ways in which a series of successive things can occur is found by multiplying the number of ways in which each thing can occur.

Ex. 1, p. 771; Ex. 2, p. 771; Ex. 3, p. 772

b. A permutation from a group of items occurs when no item is used more than once and the order of arrangement makes a difference.

c. Permutations Formula: The number of possible permutations if r items are taken from n items is

Ex. 4, p. 774; Ex. 5, p. 774

$$_nP_r = \dfrac{n!}{(n - r)!}.$$

d. A combination from a group of items occurs when no item is used more than once and the order of items makes no difference.

Ex. 6, p. 775

e. Combinations Formula: The number of possible combinations if r items are taken from n items is

Ex. 7, p. 777; Ex. 8, p. 777

$$_nC_r = \dfrac{n!}{(n - r)!r!}.$$

DEFINITIONS AND CONCEPTS

EXAMPLES

8.7 Probability

a. Empirical probability applies to situations in which we observe the frequency of the occurrence of an event. The empirical probability of event E is

$$P(E) = \frac{\text{observed number of times } E \text{ occurs}}{\text{total number of observed occurrences}}.$$

Ex. 1, p. 782

b. Theoretical probability applies to situations in which the sample space of all equally likely outcomes is known. The theoretical probability of event E is

$$P(E) = \frac{\text{number of outcomes in event } E}{\text{number of outcomes in sample space } S} = \frac{n(E)}{n(S)}.$$

Ex. 2, p. 784;
Ex. 3, p. 785;
Ex. 4, p. 786;
Ex. 5, p. 786

c. Probability of an event not occurring: $P(\text{not } E) = 1 - P(E)$.

Ex. 6, p. 788

d. If it is impossible for events A and B to occur simultaneously, the events are mutually exclusive.

e. If A and B are mutually exclusive events, then $P(A \text{ or } B) = P(A) + P(B)$.

Ex. 7, p. 789

f. If A and B are not mutually exclusive events, then

$$P(A \text{ or } B) = P(A) + P(B) - P(A \text{ and } B).$$

Ex. 8, p. 790;
Ex. 9, p. 790

g. Two events are independent if the occurrence of either of them has no effect on the probability of the other.

h. If A and B are independent events, then

$$P(A \text{ and } B) = P(A) \cdot P(B).$$

Ex. 10, p. 792

i. The probability of a succession of independent events is the product of each of their probabilities.

Ex. 11, p. 792

REVIEW EXERCISES

8.1

In Exercises 1–6, write the first four terms of each sequence whose general term is given.

1. $a_n = 7n - 4$

2. $a_n = (-1)^n \dfrac{n+2}{n+1}$

3. $a_n = \dfrac{1}{(n-1)!}$

4. $a_n = \dfrac{(-1)^{n+1}}{2^n}$

5. $a_1 = 9$ and $a_n = \dfrac{2}{3a_{n-1}}$ for $n \geq 2$

6. $a_1 = 4$ and $a_n = 2a_{n-1} + 3$ for $n \geq 2$

7. Evaluate: $\dfrac{40!}{4!38!}$.

In Exercises 8–9, find each indicated sum.

8. $\displaystyle\sum_{i=1}^{5} (2i^2 - 3)$

9. $\displaystyle\sum_{i=0}^{4} (-1)^{i+1} i!$

In Exercises 10–11, express each sum using summation notation. Use i for the index of summation.

10. $\dfrac{1}{3} + \dfrac{2}{4} + \dfrac{3}{5} + \cdots + \dfrac{15}{17}$

11. $4^3 + 5^3 + 6^3 + \cdots + 13^3$

8.2

In Exercises 12–15, write the first six terms of each arithmetic sequence.

12. $a_1 = 7, d = 4$

13. $a_1 = -4, d = -5$

14. $a_1 = \frac{3}{2}, d = -\frac{1}{2}$

15. $a_{n+1} = a_n + 5, a_1 = -2$

In Exercises 16–18, find the indicated term of the arithmetic sequence with first term, a_1, and common difference, d.

16. Find a_6 when $a_1 = 5, d = 3$.

17. Find a_{12} when $a_1 = -8, d = -2$.

18. Find a_{14} when $a_1 = 14, d = -4$.

In Exercises 19–21, write a formula for the general term (the nth term) of each arithmetic sequence. Do not use a recursion formula. Then use the formula for a_n to find a_{20}, the 20th term of the sequence.

19. $-7, -3, 1, 5, \ldots$

20. $a_1 = 200, d = -20$

21. $a_n = a_{n-1} - 5, a_1 = 3$

22. Find the sum of the first 22 terms of the arithmetic sequence: $5, 12, 19, 26, \ldots$.

23. Find the sum of the first 15 terms of the arithmetic sequence: $-6, -3, 0, 3, \ldots$.

24. Find $3 + 6 + 9 + \cdots + 300$, the sum of the first 100 positive multiples of 3.

In Exercises 25–27, use the formula for the sum of the first n terms of an arithmetic sequence to find the indicated sum.

25. $\displaystyle\sum_{i=1}^{16} (3i + 2)$

26. $\displaystyle\sum_{i=1}^{25} (-2i + 6)$

27. $\displaystyle\sum_{i=1}^{30} (-5i)$

28. The bar graph shows the number of hours per week devoted to housework by wives and husbands in 1965 and 2010.

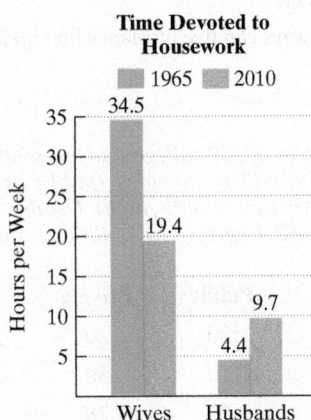

Time Devoted to Housework

■ 1965　■ 2010

Source: James Henslin, *Sociology*, Eleventh Edition, Pearson, 2012.

In 1965, wives averaged 34.5 hours per week doing housework. On average, this has decreased by approximately 0.3 hour per year since then.

a. Write a formula for the nth term of the arithmetic sequence that describes the number of hours per week devoted to housework by wives n years after 1964.

b. Use the model to project the number of hours per week wives will devote to housework in 2020.

29. A company offers a starting salary of $31,500 with raises of $2300 per year. Find the total salary over a ten-year period.

30. A theater has 25 seats in the first row and 35 rows in all. Each successive row contains one additional seat. How many seats are in the theater?

8.3

In Exercises 31–34, write the first five terms of each geometric sequence.

31. $a_1 = 3, r = 2$

32. $a_1 = \frac{1}{2}, r = \frac{1}{2}$

33. $a_1 = 16, r = -\frac{1}{2}$

34. $a_n = -5a_{n-1}, a_1 = -1$

In Exercises 35–37, use the formula for the general term (the nth term) of a geometric sequence to find the indicated term of each sequence.

35. Find a_7 when $a_1 = 2, r = 3$.

36. Find a_6 when $a_1 = 16, r = \frac{1}{2}$.

37. Find a_5 when $a_1 = -3, r = 2$.

In Exercises 38–40, write a formula for the general term (the nth term) of each geometric sequence. Then use the formula for a_n to find a_8, the eighth term of the sequence.

38. $1, 2, 4, 8, \ldots$

39. $100, 10, 1, \frac{1}{10}, \ldots$

40. $12, -4, \frac{4}{3}, -\frac{4}{9}, \ldots$

41. Find the sum of the first 15 terms of the geometric sequence: $5, -15, 45, -135, \ldots$.

42. Find the sum of the first 7 terms of the geometric sequence: $8, 4, 2, 1, \ldots$.

In Exercises 43–45, use the formula for the sum of the first n terms of a geometric sequence to find the indicated sum.

43. $\sum_{i=1}^{6} 5^i$

44. $\sum_{i=1}^{7} 3(-2)^i$

45. $\sum_{i=1}^{5} 2\left(\frac{1}{4}\right)^{i-1}$

In Exercises 46–49, find the sum of each infinite geometric series.

46. $9 + 3 + 1 + \frac{1}{3} + \cdots$

47. $2 - 1 + \frac{1}{2} - \frac{1}{4} + \cdots$

48. $-6 + 4 - \frac{8}{3} + \frac{16}{9} - \cdots$

49. $\sum_{i=1}^{\infty} 5(0.8)^i$

In Exercises 50–51, express each repeating decimal as a fraction in lowest terms.

50. $0.\overline{6}$

51. $0.\overline{47}$

52. The table shows the population of Florida for 2000 and 2010, with estimates given by the U.S. Census Bureau for 2001 through 2009.

Year	2000	2001	2002	2003	2004	2005
Population in millions	15.98	16.24	16.50	16.76	17.03	17.30

Year	2006	2007	2008	2009	2010
Population in millions	17.58	17.86	18.15	18.44	18.80

a. Divide the population, for each year by the population in the preceding year. Round to two decimal places and show that Florida has a population increase that is approximately geometric.

b. Write the general term of the geometric sequence modeling Florida's population, in millions, n years after 1999.

c. Use your model from part (b) to project Florida's population, in millions, for the year 2030. Round to two decimal places.

53. A job pays $32,000 for the first year with an annual increase of 6% per year beginning in the second year. What is the salary in the sixth year? What is the total salary paid over this six-year period? Round answers to the nearest dollar.

In Exercises 54–55, use the formula for the value of an annuity and round to the nearest dollar.

54. You spend $10 per week on lottery tickets, averaging $520 per year. Instead of buying tickets, if you deposited the $520 at the end of each year in an annuity paying 6% compounded annually,

a. How much would you have after 20 years?

b. Find the interest.

55. To save for retirement, you decide to deposit $100 at the end of each month in an IRA that pays 5.5% compounded monthly.

a. How much will you have from the IRA after 30 years?

b. Find the interest.

56. A factory in an isolated town has an annual payroll of $4 million. It is estimated that 70% of this money is spent within the town, that people in the town receiving this money will again spend 70% of what they receive in the town, and so on. What is the total of all this spending in the town each year?

8.4

In Exercises 57–61, use mathematical induction to prove that each statement is true for every positive integer n.

57. $5 + 10 + 15 + \cdots + 5n = \dfrac{5n(n + 1)}{2}$

58. $1 + 4 + 4^2 + \cdots + 4^{n-1} = \dfrac{4^n - 1}{3}$

59. $2 + 6 + 10 + \cdots + (4n - 2) = 2n^2$

60. $1 \cdot 3 + 2 \cdot 4 + 3 \cdot 5 + \cdots + n(n + 2) = \dfrac{n(n + 1)(2n + 7)}{6}$

61. 2 is a factor of $n^2 + 5n$.

8.5

In Exercises 62–63, evaluate the given binomial coefficient.

62. $\dbinom{11}{8}$

63. $\dbinom{90}{2}$

In Exercises 64–67, use the Binomial Theorem to expand each binomial and express the result in simplified form.

64. $(2x + 1)^3$

65. $(x^2 - 1)^4$

66. $(x + 2y)^5$

67. $(x - 2)^6$

In Exercises 68–69, write the first three terms in each binomial expansion, expressing the result in simplified form.

68. $(x^2 + 3)^8$

69. $(x - 3)^9$

In Exercises 70–71, find the term indicated in each expansion.

70. $(x + 2)^5$; fourth term

71. $(2x - 3)^6$; fifth term

8.6

In Exercises 72–75, evaluate each expression.

72. $_8P_3$

73. $_9P_5$

74. $_8C_3$

75. $_{13}C_{11}$

In Exercises 76–82, solve by the method of your choice.

76. A popular brand of pen comes in red, green, blue, or black ink. The writing tip can be chosen from extra bold, bold, regular, fine, or micro. How many different choices of pens do you have with this brand?

77. A stock can go up, go down, or stay unchanged. How many possibilities are there if you own five stocks?

78. A club with 15 members is to choose four officers—president, vice president, secretary, and treasurer. In how many ways can these offices be filled?

79. How many different ways can a director select 4 actors from a group of 20 actors to attend a workshop on performing in rock musicals?

80. From the 20 CDs that you've bought during the past year, you plan to take 3 with you on vacation. How many different sets of three CDs can you take?

81. How many different ways can a director select from 20 male actors and cast the roles of Mark, Roger, Angel, and Collins in the musical *Rent*?

82. In how many ways can five airplanes line up for departure on a runway?

8.7

Suppose that a survey of 350 college students is taken. Each student is asked the type of college attended (public or private) and the family's income level (low, middle, high). Use the data in the table to solve Exercises 83–88. Express probabilities as simplified fractions.

	Public	Private	Total
Low	120	20	140
Middle	110	50	160
High	22	28	50
Total	252	98	350

Find the probability that a randomly selected student in the survey

83. attends a public college.

84. is not from a high-income family.

85. is from a middle-income or a high-income family.

86. attends a private college or is from a high-income family.

87. Among people who attend a public college, find the probability that a randomly selected student is from a low-income family.

88. Among people from a middle-income family, find the probability that a randomly selected student attends a private college.

In Exercises 89–90, a die is rolled. Find the probability of

89. getting a number less than 5.

90. getting a number less than 3 or greater than 4.

In Exercises 91–92, you are dealt one card from a 52-card deck. Find the probability of

91. getting an ace or a king.

92. getting a queen or a red card.

In Exercises 93–95, it is equally probable that the pointer on the spinner shown will land on any one of the six regions, numbered 1 through 6, and colored as shown. If the pointer lands on a borderline, spin again. Find the probability of

93. not stopping on yellow.

94. stopping on red or a number greater than 3.

95. stopping on green on the first spin and stopping on a number less than 4 on the second spin.

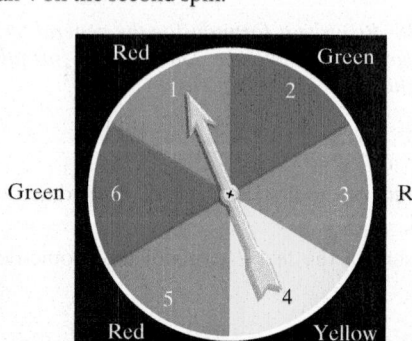

96. A lottery game is set up so that each player chooses five different numbers from 1 to 20. If the five numbers match the five numbers drawn in the lottery, the player wins (or shares) the top cash prize. What is the probability of winning the prize

 a. with one lottery ticket?

 b. with 100 different lottery tickets?

97. What is the probability of a family having five boys born in a row?

98. The probability of a flood in any given year in a region prone to floods is 0.2.

 a. What is the probability of a flood two years in a row?

 b. What is the probability of a flood for three consecutive years?

 c. What is the probability of no flooding for four consecutive years?

CHAPTER 8 TEST

1. Write the first five terms of the sequence whose general term is $a_n = \dfrac{(-1)^{n+1}}{n^2}$.

In Exercises 2–4, find each indicated sum.

2. $\displaystyle\sum_{i=1}^{5} (i^2 + 10)$

3. $\displaystyle\sum_{i=1}^{20} (3i - 4)$

4. $\displaystyle\sum_{i=1}^{15} (-2)^i$

In Exercises 5–7, evaluate each expression.

5. $\dbinom{9}{2}$

6. $_{10}P_3$

7. $_{10}C_3$

8. Express the sum using summation notation. Use i for the index of summation.

$$\frac{2}{3} + \frac{3}{4} + \frac{4}{5} + \cdots + \frac{21}{22}$$

In Exercises 9–10, write a formula for the general term (the nth term) of each sequence. Do not use a recursion formula. Then use the formula to find the twelfth term of the sequence.

9. $4, 9, 14, 19, \ldots$

10. $16, 4, 1, \frac{1}{4}, \ldots$

In Exercises 11–12, use a formula to find the sum of the first ten terms of each sequence.

11. $7, -14, 28, -56, \ldots$

12. $-7, -14, -21, -28, \ldots$

13. Find the sum of the infinite geometric series:

$$4 + \frac{4}{2} + \frac{4}{2^2} + \frac{4}{2^3} + \cdots.$$

14. Express $0.\overline{73}$ in fractional notation.

15. A job pays $30,000 for the first year with an annual increase of 4% per year beginning in the second year. What is the total salary paid over an eight-year period? Round to the nearest dollar.

16. Use mathematical induction to prove that for every positive integer n,

$$1 + 4 + 7 + \cdots + (3n - 2) = \frac{n(3n - 1)}{2}.$$

17. Use the Binomial Theorem to expand and simplify: $(x^2 - 1)^5$.

18. Use the Binomial Theorem to write the first three terms in the expansion and simplify: $(x + y^2)^8$.

19. A human resource manager has 11 applicants to fill three different positions. Assuming that all applicants are equally qualified for any of the three positions, in how many ways can this be done?

20. From the ten books that you've recently bought but not read, you plan to take four with you on vacation. How many different sets of four books can you take?

21. How many seven-digit local telephone numbers can be formed if the first three digits are 279?

A class is collecting data on eye color and gender. They organize the data they collected into the table shown. Numbers in the table represent the number of students from the class that belong to each of the categories. Use the data to solve Exercises 22–25. Express probabilities as simplified fractions.

	Brown	Blue	Green
Male	22	18	10
Female	18	20	12

Find the probability that a randomly selected student from this class

22. does not have brown eyes.

23. has brown eyes or blue eyes.

24. is female or has green eyes.

25. Among the students with blue eyes, find the probability of selecting a male.

26. A lottery game is set up so that each player chooses six different numbers from 1 to 15. If the six numbers match the six numbers drawn in the lottery, the player wins (or shares) the top cash prize. What is the probability of winning the prize with 50 different lottery tickets?

27. One card is randomly selected from a deck of 52 cards. Find the probability of selecting a black card or a picture card.

28. A group of students consists of 10 male freshmen, 15 female freshmen, 20 male sophomores, and 5 female sophomores. If one person is randomly selected from the group, find the probability of selecting a freshman or a female.

29. A quiz consisting of four multiple-choice questions has four available options (a, b, c, or d) for each question. If a person guesses at every question, what is the probability of answering all questions correctly?

30. If the spinner shown is spun twice, find the probability that the pointer lands on red on the first spin and blue on the second spin.

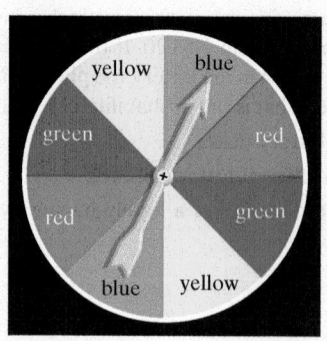

CUMULATIVE REVIEW EXERCISES (CHAPTERS 1–8)

The figure shows the graph of $y = f(x)$ and its vertical asymptote. Use the graph to solve Exercises 1–9.

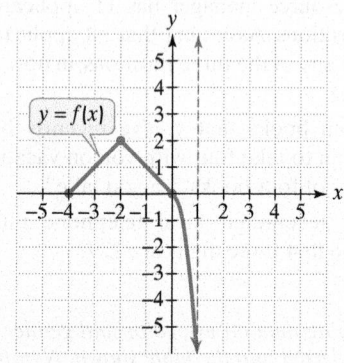

1. Find the domain and the range of f.

2. Does f have a relative maximum or a relative minimum? What is this relative maximum or minimum and where does it occur?

3. Find the interval on which f is decreasing.

4. Is f even, odd, or neither?

5. For what value(s) of x is $f(x) = 1$?

6. Find $(f \circ f)(-4)$.

7. Use arrow notation to complete this statement:

$$f(x) \to -\infty \quad \text{as} \quad \underline{\quad\quad}$$

8. Graph $g(x) = f(x - 2) + 1$.

9. Graph $h(x) = -f(2x)$.

In Exercises 10–22, solve each equation, inequality, or system of equations.

10. $-2(x - 5) + 10 = 3(x + 2)$

11. $3x^2 - 6x + 2 = 0$

12. $\log_2 x + \log_2(2x - 3) = 1$

13. $x^{\frac{1}{2}} - 6x^{\frac{1}{4}} + 8 = 0$

14. $\sqrt{2x + 4} - \sqrt{x + 3} - 1 = 0$

15. $|2x + 1| \le 1$

16. $6x^2 - 6 < 5x$

17. $\dfrac{x - 1}{x + 3} \le 0$

18. $30e^{0.7x} = 240$

19. $2x^3 + 3x^2 - 8x + 3 = 0$

20. $\begin{cases} 4x^2 + 3y^2 = 48 \\ 3x^2 + 2y^2 = 35 \end{cases}$

21. (Use matrices.)

$$\begin{cases} x - 2y + z = 16 \\ 2x - y - z = 14 \\ 3x + 5y - 4z = -10 \end{cases}$$

22. $\begin{cases} x - y = 1 \\ x^2 - x - y = 1 \end{cases}$

In Exercises 23–29, graph each equation, function, or system in a rectangular coordinate system. If two functions are indicated, graph both in the same system.

23. $100x^2 + y^2 = 25$

24. $4x^2 - 9y^2 - 16x + 54y - 29 = 0$

25. $f(x) = \dfrac{x^2 - 1}{x - 2}$

26. $\begin{cases} 2x - y \ge 4 \\ x \le 2 \end{cases}$

27. $f(x) = x^2 - 4x - 5$

28. $f(x) = \sqrt[3]{x + 4}$ and f^{-1}

29. $f(x) = \log_2 x$ and $g(x) = -\log_2(x + 1)$

In Exercises 30–31, let $f(x) = -x^2 - 2x + 1$ and $g(x) = x - 1$.

30. Find $(f \circ g)(x)$ and $(g \circ f)(x)$.

31. Find $\dfrac{f(x + h) - f(x)}{h}$ and simplify.

32. If $A = \begin{bmatrix} 4 & 2 \\ 1 & -1 \\ 0 & 5 \end{bmatrix}$ and $B = \begin{bmatrix} 2 & 4 \\ 3 & 1 \end{bmatrix}$, find $AB - 4A$.

33. Find the partial fraction decomposition for

$$\frac{2x^2 - 10x + 2}{(x - 2)(x^2 + 2x + 2)}.$$

34. Expand and simplify: $(x^3 + 2y)^5$.

35. Use the formula for the sum of the first n terms of an arithmetic sequence to find $\displaystyle\sum_{i=1}^{50} (4i - 25)$.

In Exercises 36–37, write the linear function in slope-intercept form satisfying the given conditions.

36. Graph of f passes through $(6, 3)$ and $(-2, 1)$.

37. Graph of g passes through $(0, -2)$ and is perpendicular to the line whose equation is $x - 5y - 20 = 0$.

38. For a summer sales job, you are choosing between two pay arrangements: a weekly salary of $200 plus 5% commission on sales, or a straight 15% commission. For how many dollars of sales will the earnings be the same regardless of the pay arrangement?

39. The perimeter of a soccer field is 300 yards. If the length is 50 yards longer than the width, what are the field's dimensions?

40. If 10 pens and 12 pads cost $42, and 5 of the same pens and 10 of the same pads cost $29, find the cost of a pen and a pad.

41. A ball is thrown vertically upward from the top of a 96-foot-tall building with an initial velocity of 80 feet per second. The height of the ball above ground, $s(t)$, in feet, after t seconds is modeled by the position function

$$s(t) = -16t^2 + 80t + 96.$$

a. After how many seconds will the ball strike the ground?

b. When does the ball reach its maximum height? What is the maximum height?

42. The current, I, in amperes, flowing in an electrical circuit varies inversely as the resistance, R, in ohms, in the circuit. When the resistance of an electric percolator is 22 ohms, it draws 5 amperes of current. How much current is needed when the resistance is 10 ohms?

43. The bar graph shows online ad spending worldwide, in billions of dollars, for 2010 and 2011, with projections from 2012 through 2015. Develop a linear function that models the data. Then use the function to make a projection about what might occur after 2015.

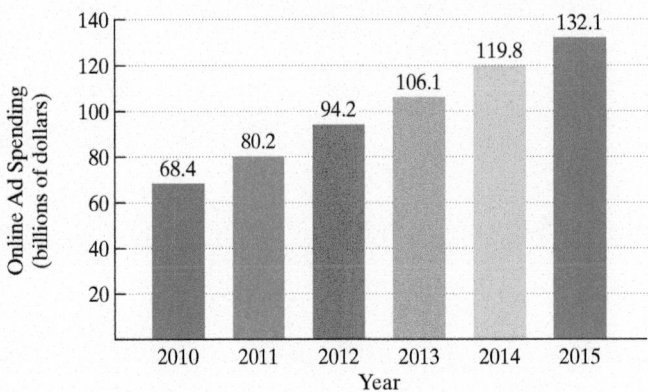

Worldwide Online Ad Spending

Source: eMarketer

APPENDIX

Where Did That Come From?
Selected Proofs

SECTION 4.3 Properties of Logarithms

The Product Rule

Let b, M, and N be positive real numbers with $b \neq 1$.

$$\log_b(MN) = \log_b M + \log_b N$$

Proof

We begin by letting $\log_b M = R$ and $\log_b N = S$.
Now we write each logarithm in exponential form.

$$\log_b M = R \quad \text{means} \quad b^R = M.$$
$$\log_b N = S \quad \text{means} \quad b^S = N.$$

By substituting and using a property of exponents, we see that

$$MN = b^R b^S = b^{R+S}.$$

Now we change $MN = b^{R+S}$ to logarithmic form.

$$MN = b^{R+S} \quad \text{means} \quad \log_b(MN) = R + S.$$

Finally, substituting $\log_b M$ for R and $\log_b N$ for S gives us

$$\log_b(MN) = \log_b M + \log_b N,$$

the property that we wanted to prove.

The quotient and power rules for logarithms are proved using similar procedures.

The Change-of-Base Property

For any logarithmic bases a and b, and any positive number M,

$$\log_b M = \frac{\log_a M}{\log_a b}.$$

Proof

To prove the change-of-base property, we let x equal the logarithm on the left side:

$$\log_b M = x.$$

Now we rewrite this logarithm in exponential form.

$$\log_b M = x \quad \text{means} \quad b^x = M.$$

Because b^x and M are equal, the logarithms with base a for each of these expressions must be equal. This means that

$$\log_a b^x = \log_a M$$

$$x \log_a b = \log_a M \qquad \text{Apply the power rule for logarithms on the left side.}$$

$$x = \frac{\log_a M}{\log_a b} \qquad \text{Solve for x by dividing both sides by } \log_a b.$$

In our first step we let x equal $\log_b M$. Replacing x on the left side by $\log_b M$ gives us

$$\log_b M = \frac{\log_a M}{\log_a b},$$

which is the change-of-base property.

SECTION 7.1 The Ellipse

The Standard Form of the Equation of an Ellipse with a Horizontal Major Axis Centered at the Origin

Proof

Refer to **Figure A.1**.

FIGURE A.1

$$d_1 + d_2 = 2a$$
The sum of the distances from P to the foci equals a constant, $2a$.

$$\sqrt{(x+c)^2 + y^2} + \sqrt{(x-c)^2 + y^2} = 2a$$
Use the distance formula.

$$\sqrt{(x+c)^2 + y^2} = 2a - \sqrt{(x-c)^2 + y^2}$$
Isolate a radical.

$$(x+c)^2 + y^2 = 4a^2 - 4a\sqrt{(x-c)^2 + y^2} + (x-c)^2 + y^2$$
Square both sides.

$$x^2 + 2cx + c^2 + y^2 = 4a^2 - 4a\sqrt{(x-c)^2 + y^2} + x^2 - 2cx + c^2 + y^2$$
Square $x + c$ and $x - c$.

$$4cx - 4a^2 = -4a\sqrt{(x-c)^2 + y^2}$$
Simplify and isolate the radical.

$$cx - a^2 = -a\sqrt{(x-c)^2 + y^2}$$
Divide both sides by 4.

$$(cx - a^2)^2 = a^2[(x-c)^2 + y^2]$$
Square both sides.

$$c^2x^2 - 2a^2cx + a^4 = a^2(x^2 - 2cx + c^2 + y^2)$$
Square $cx - a^2$ and $x - c$.

$$c^2x^2 - 2a^2cx + a^4 = a^2x^2 - 2a^2cx + a^2c^2 + a^2y^2$$
Use the distributive property.

$$c^2x^2 + a^4 = a^2x^2 + a^2c^2 + a^2y^2$$
Add $2a^2cx$ to both sides.

$$c^2x^2 - a^2x^2 - a^2y^2 = a^2c^2 - a^4$$
Rearrange the terms.

$$(c^2 - a^2)x^2 - a^2y^2 = a^2(c^2 - a^2)$$
Factor out x^2 and a^2, respectively.

$$(a^2 - c^2)x^2 + a^2y^2 = a^2(a^2 - c^2)$$
Multiply both sides by -1.

Refer to the discussion on page 672. Let $b^2 = a^2 - c^2$ in the preceding equation.

$$b^2x^2 + a^2y^2 = a^2b^2$$

$$\frac{x^2}{a^2} + \frac{y^2}{b^2} = 1$$
Divide both sides by a^2b^2.

The Asymptotes of a Hyperbola Centered at the Origin
The hyperbola

$$\frac{x^2}{a^2} - \frac{y^2}{b^2} = 1$$

with a horizontal transverse axis has the two asymptotes

$$y = \frac{b}{a}x \quad \text{and} \quad y = -\frac{b}{a}x.$$

Proof
Begin by solving the hyperbola's equation for y.

$\dfrac{x^2}{a^2} - \dfrac{y^2}{b^2} = 1$	*This is the standard form of the equation of a hyperbola.*
$\dfrac{y^2}{b^2} = \dfrac{x^2}{a^2} - 1$	*We isolate the term involving y^2 to solve for y.*
$y^2 = \dfrac{b^2 x^2}{a^2} - b^2$	*Multiply both sides by b^2.*
$y^2 = \dfrac{b^2 x^2}{a^2}\left(1 - \dfrac{a^2}{x^2}\right)$	*Factor out $\dfrac{b^2 x^2}{a^2}$ on the right. Verify that this result is correct by multiplying using the distributive property and obtaining the previous step.*
$y = \pm\sqrt{\dfrac{b^2 x^2}{a^2}\left(1 - \dfrac{a^2}{x^2}\right)}$	*Solve for y using the square root property: If $u^2 = d$, then $u = \pm\sqrt{d}$.*
$y = \pm\dfrac{b}{a}x\sqrt{1 - \dfrac{a^2}{x^2}}$	*Simplify.*

As $|x| \to \infty$, the value of $\dfrac{a^2}{x^2}$ approaches 0. Consequently, the value of y can be approximated by

$$y = \pm\frac{b}{a}x.$$

This means that the lines whose equations are $y = \dfrac{b}{a}x$ and $y = -\dfrac{b}{a}x$ are asymptotes for the graph of the hyperbola.

ANSWERS TO SELECTED EXERCISES

CHAPTER P

Section P.1

Check Point Exercises

1. 608 **2.** $9209 **3.** $\{3, 7\}$ **4.** $\{3, 4, 5, 6, 7, 8, 9\}$ **5. a.** $\sqrt{9}$ **b.** $0, \sqrt{9}$ **c.** $-9, 0, \sqrt{9}$ **d.** $-9, -1.3, 0, 0.\overline{3}, \sqrt{9}$ **e.** $\frac{\pi}{2}, \sqrt{10}$
f. $-9, -1.3, 0, 0.\overline{3}, \frac{\pi}{2}, \sqrt{9}, \sqrt{10}$ **6. a.** $\{x | -2 \le x < 5\}$ **b.** $\{x | 1 \le x \le 3.5\}$ **c.** $\{x | x < -1\}$

7. a. $(2, 3]$ **b.** $[1, 6)$ **8. a.** $\sqrt{2} - 1$ **b.** $\pi - 3$ **c.** 1 **9.** 9

Concept and Vocabulary Check

1. expression **2.** b to the nth power; base; exponent **3.** formula; modeling; models **4.** intersection; $A \cap B$ **5.** union; $A \cup B$
6. natural **7.** whole **8.** integers **9.** rational **10.** irrational **11.** rational; irrational **12.** 2; 5; 2; 5 **13.** greater than **14.** less than
or equal to **15.** $(-\infty, 9)$; intersection **16.** $(-\infty, 12)$; union **17.** absolute value; $x; -x$

Exercise Set P.1

1. 57 **3.** 10 **5.** 88 **7.** 10 **9.** 44 **11.** 46 **13.** 10 **15.** -8 **17.** $10°C$ **19.** 60 ft **21.** $\{2, 4\}$ **23.** $\{s, e, t\}$ **25.** \varnothing
27. \varnothing **29.** $\{1, 2, 3, 4, 5\}$ **31.** $\{1, 2, 3, 4, 5, 6, 7, 8, 10\}$ **33.** $\{a, e, i, o, u\}$ **35. a.** $\sqrt{100}$ **b.** $0, \sqrt{100}$ **c.** $-9, 0, \sqrt{100}$
d. $-9, -\frac{4}{5}, 0, 0.25, 9.2, \sqrt{100}$ **e.** $\sqrt{3}$ **f.** $-9, -\frac{4}{5}, 0, 0.25, \sqrt{3}, 9.2, \sqrt{100}$ **37. a.** $\sqrt{64}$ **b.** $0, \sqrt{64}$ **c.** $-11, 0, \sqrt{64}$
d. $-11, -\frac{5}{6}, 0, 0.75, \sqrt{64}$ **e.** $\sqrt{5}, \pi$ **f.** $-11, -\frac{5}{6}, 0, 0.75, \sqrt{5}, \pi, \sqrt{64}$ **39.** 0 **41.** Answers may vary; an example is 2. **43.** true
45. true **47.** true **49.** true **51.** $\{1, 2, 3, 4\}$ **53.** $\{-7, -6, -5, -4\}$ **55.** $\{7, 8, 9, 10, \ldots\}$ **57.** $\{1, 3, 5, 7, 9\}$
59. $\{x | 1 < x \le 6\}$ **61.** $\{x | -5 \le x < 2\}$ **63.** $\{x | -3 \le x \le 1\}$ **65.** $\{x | x > 2\}$

67. $\{x | x \ge -3\}$ **69.** $\{x | x < 3\}$ **71.** $\{x | x < 5.5\}$ **73.** $[-1, 0)$ **75.** $(-3, 2]$

77. $[1, 5)$ **79.** $(-\infty, 8)$ **81.** $(6, \infty)$ **83.** $[3, \infty)$ **85.** 300 **87.** $12 - \pi$ **89.** $5 - \sqrt{2}$ **91.** -1 **93.** 4 **95.** 3 **97.** 7
99. -1 **101.** $|17 - 2|$; 15 **103.** $|5 - (-2)|$; 7 **105.** $|-4 - (-19)|$; 15 **107.** $|-1.4 - (-3.6)|$; 2.2 **109.** $>$ **111.** $=$ **113.** $<$
115. $=$ **117.** 45 **119.** $\frac{1}{121}$ **121.** 14 **123.** $-\frac{8}{3}$ **125.** $-\frac{1}{2}$ **127. a.** 140 beats per minute **b.** 160 beats per minute
129. a. $26,317 **b.** overestimates by $44 **c.** $30,568 **145.** does not make sense **147.** does not make sense **149.** false **151.** false
153. $>$ **155. a.** 36 **b.** 36 **c.** $3(5 + 7) = 3 \cdot 5 + 3 \cdot 7$ **156.** true **157.** 23%

Section P.2

Check Point Exercises

1. $38x^2 + 23x$ **2.** $42 - 4x$ **3.** $\{6\}$ **4.** $w = \dfrac{P - 2l}{2}$ **5.** $C = \dfrac{P}{1 + M}$ **6.** 55 years after 1970, or in 2025

Concept and Vocabulary Check

1. $b + a; ba$ **2.** $a + (b + c); (ab)c$ **3.** $ab + ac$ **4.** 0; inverse; 0; identity **5.** inverse; 1; identity **6.** simplified **7.** a
8. equivalent **9.** add 5 to both sides **10.** divide both sides by -3 or multiply both sides by $-\dfrac{1}{3}$ **11.** isolated on one side **12.** distributive

Exercise Set P.2

1. commutative property of addition **3.** associative property of addition **5.** commutative property of addition **7.** distributive property of
multiplication over addition **9.** inverse property of multiplication **11.** $15x + 16$ **13.** $27x - 10$ **15.** $29y - 29$ **17.** $8y - 12$
19. $16y - 25$ **21.** $12x^2 + 11$ **23.** $14x$ **25.** $-2x + 3y + 6$ **27.** x **29.** $\{3\}$ **31.** $\{11\}$ **33.** $\{11\}$ **35.** $\{7\}$ **37.** $\{6\}$
39. $x - 16$ **41.** $\{9\}$ **43.** $\{19\}$ **45.** $8y - 4$ **47.** $\left\{\dfrac{5}{2}\right\}$ **49.** $r = \dfrac{d}{t}$ **51.** $P = \dfrac{I}{rt}$ **53.** $r = \dfrac{C}{2\pi}$ **55.** $m = \dfrac{E}{c^2}$

57. $m = \dfrac{y-b}{x}$ **59.** $p = \dfrac{T-D}{m}$ **61.** $x = \dfrac{-C-By}{A}$ **63.** $m = \dfrac{y-y_1}{x-x_1}$ **65.** $n = \dfrac{L-a}{d} + 1$ **67.** $l = \dfrac{A-2wh}{2w+2h}$

69. $x - (x+4)$; -4 **71.** $6(-5x)$; $-30x$ **73.** $5x - 2x$; $3x$ **75.** $8x - (3x+6)$; $5x - 6$ **77. a.** $1200 - 0.07x$ **b.** \$780
79. 48 years after 1970, or in 2018 **87.** makes sense **89.** does not make sense **91.** false **93.** true **95. a.** b^7 **b.** b^{10} **c.** Add the
exponents. **96. a.** b^4 **b.** b^6 **c.** Subtract the exponents. **97.** It moves the decimal point 3 places to the right.

Section P.3

Check Point Exercises

1. a. 3^5 or 243 **b.** $40x^5y^{10}$ **2. a.** $(-3)^3$ or -27 **b.** $9x^{11}y^3$ **3. a.** $\dfrac{1}{25}$ **b.** $-\dfrac{1}{27}$ **c.** 16 **d.** $\dfrac{3y^4}{x^6}$ **4. a.** 3^6 or 729 **b.** $\dfrac{1}{y^{14}}$

c. b^{12} **5.** $-64x^3$ **6. a.** $-\dfrac{32}{y^5}$ **b.** $\dfrac{x^{15}}{27}$ **7. a.** $16x^{12}y^{24}$ **b.** $-18x^3y^8$ **c.** $\dfrac{5y^6}{x^4}$ **d.** $\dfrac{y^8}{25x^2}$ **8. a.** $-2{,}600{,}000{,}000$ **b.** 0.000003017

9. a. 5.21×10^9 **b.** -6.893×10^{-8} **10.** 4.1×10^9 **11. a.** 3.55×10^{-1} **b.** 4×10^8 **12.** \$8300

Concept and Vocabulary Check

1. b^{m+n}; add **2.** b^{m-n}; subtract **3.** 1 **4.** $\dfrac{1}{b^n}$ **5.** false **6.** b^n **7.** true **8.** a number greater than or equal to 1 and less than 10; integer
9. true **10.** false

Exercise Set P.3

1. 50 **3.** 64 **5.** -64 **7.** 1 **9.** -1 **11.** $\dfrac{1}{64}$ **13.** 32 **15.** 64 **17.** 16 **19.** $\dfrac{1}{9}$ **21.** $\dfrac{1}{16}$ **23.** $\dfrac{y}{x^2}$ **25.** y^5 **27.** x^{10}

29. x^5 **31.** x^{21} **33.** $\dfrac{1}{x^{15}}$ **35.** x^7 **37.** x^{21} **39.** $64x^6$ **41.** $-\dfrac{64}{x^3}$ **43.** $9x^4y^{10}$ **45.** $6x^{11}$ **47.** $18x^9y^5$ **49.** $4x^{16}$ **51.** $-5a^{11}b$

53. $\dfrac{2}{b^7}$ **55.** $\dfrac{1}{16x^6}$ **57.** $\dfrac{3y^{14}}{4x^4}$ **59.** $\dfrac{y^2}{25x^6}$ **61.** $-\dfrac{27b^{15}}{a^{18}}$ **63.** 1 **65.** 380 **67.** 0.0006 **69.** $-7{,}160{,}000$ **71.** 0.79 **73.** -0.00415
75. $-60{,}000{,}100{,}000$ **77.** 3.2×10^4 **79.** 6.38×10^{17} **81.** -5.716×10^3 **83.** 2.7×10^{-3} **85.** -5.04×10^{-9} **87.** 6.3×10^7
89. 6.4×10^4 **91.** 1.22×10^{-11} **93.** 2.67×10^{13} **95.** 2.1×10^3 **97.** 4×10^5 **99.** 2×10^{-8} **101.** 5×10^3 **103.** 4×10^{15}

105. 9×10^{-3} **107.** 1 **109.** $\dfrac{y}{16x^8z^6}$ **111.** $\dfrac{1}{x^{12}y^{16}z^{20}}$ **113.** $\dfrac{x^{18}y^6}{4}$ **115. a.** 2.17×10^{12} **b.** 3.09×10^8 **c.** \$7023

117. $\$1.0586 \times 10^{10}$ **119.** 1.06×10^{-18} g **121.** 4.064×10^9 **133.** makes sense **135.** makes sense **137.** true **139.** false **141.** false
143. true **145.** $A = C + D$ **148. a.** 8 **b.** 8 **c.** $\sqrt{16} \cdot \sqrt{4} = \sqrt{16 \cdot 4}$ **149. a.** 17.32 **b.** 17.32 **c.** $\sqrt{300} = 10\sqrt{3}$
150. a. $31x$ **b.** $31\sqrt{2}$

Section P.4

Check Point Exercises

1. a. 9 **b.** -3 **c.** $\dfrac{1}{5}$ **d.** 10 **e.** 14 **2. a.** $\{-6, 6\}$ **b.** $\{\pm\sqrt{3}\}$ **3. a.** $5\sqrt{3}$ **b.** $5x\sqrt{2}$ **4. a.** $\dfrac{5}{4}$ **b.** $5x\sqrt{3}$ **5. a.** $17\sqrt{13}$
b. $-19\sqrt{17x}$ **6. a.** $17\sqrt{3}$ **b.** $10\sqrt{2x}$ **7. a.** $\dfrac{5\sqrt{3}}{3}$ **b.** $\sqrt{3}$ **8.** $\dfrac{8(4-\sqrt{5})}{11}$ or $\dfrac{32-8\sqrt{5}}{11}$ **9. a.** $2\sqrt[3]{5}$ **b.** $2\sqrt[5]{2}$ **c.** $\dfrac{5}{3}$

10. $5\sqrt[3]{3}$ **11. a.** 5 **b.** 2 **c.** -3 **d.** -2 **e.** $\dfrac{1}{3}$ **12. a.** 81 **b.** 8 **c.** $\dfrac{1}{4}$ **13. a.** $10x^4$ **b.** $4x^{5/2}$ **14.** \sqrt{x}

Concept and Vocabulary Check

1. principal **2.** 8^2 **3.** $\pm\sqrt{d}$ **4.** $\pm\sqrt{17}$ **5.** $|a|$ **6.** $\sqrt{a} \cdot \sqrt{b}$ **7.** $\dfrac{\sqrt{a}}{\sqrt{b}}$ **8.** $18\sqrt{3}$ **9.** $5; 6\sqrt{3}$ **10.** $7 - \sqrt{3}$

11. $\sqrt{10} + \sqrt{2}$ **12.** index; radicand **13.** $(-2)^5$ **14.** $a; |a|$ **15.** $\sqrt[n]{a}$ **16.** $2; 8$

Exercise Set P.4

1. 6 **3.** -6 **5.** not a real number **7.** 3 **9.** 1 **11.** 13 **13.** $\{-1, 1\}$ **15.** $\{-\sqrt{7}, \sqrt{7}\}$ or $\{\pm\sqrt{7}\}$ **17.** $\{-2, 2\}$ **19.** $\{-5, 5\}$
21. $\left\{-\dfrac{7}{2}, \dfrac{7}{2}\right\}$ **23.** $\{-\sqrt{10}, \sqrt{10}\}$ or $\{\pm\sqrt{10}\}$ **25.** $5\sqrt{2}$ **27.** $3|x|\sqrt{5}$ **29.** $2x\sqrt{3}$ **31.** $x\sqrt{x}$ **33.** $2x\sqrt{3x}$ **35.** $\dfrac{1}{9}$ **37.** $\dfrac{7}{4}$

39. $4x$ **41.** $5x\sqrt{2x}$ **43.** $2x^2\sqrt{5}$ **45.** $13\sqrt{3}$ **47.** $-2\sqrt{17x}$ **49.** $5\sqrt{2}$ **51.** $3\sqrt{2x}$ **53.** $34\sqrt{2}$ **55.** $20\sqrt{2} - 5\sqrt{3}$ **57.** $\dfrac{\sqrt{7}}{7}$

59. $\dfrac{\sqrt{10}}{5}$ **61.** $\dfrac{13(3-\sqrt{11})}{-2}$ **63.** $7(\sqrt{5} + 2)$ **65.** $3(\sqrt{5} - \sqrt{3})$ **67.** 5 **69.** -2 **71.** not a real number **73.** 3 **75.** -3

77. $-\dfrac{1}{2}$ **79.** $2\sqrt[3]{4}$ **81.** $x\sqrt[3]{x}$ **83.** $3\sqrt[3]{2}$ **85.** $2x$ **87.** $7\sqrt[5]{2}$ **89.** $13\sqrt[3]{2}$ **91.** $-y\sqrt[3]{2x}$ **93.** $\sqrt{2} + 2$ **95.** 6 **97.** 2

99. 25 **101.** $\dfrac{1}{16}$ **103.** $14x^{7/12}$ **105.** $4x^{1/4}$ **107.** x^2 **109.** $5x^2|y|^3$ **111.** $27y^{2/3}$ **113.** $\sqrt{5}$ **115.** x^2 **117.** $\sqrt[3]{x^2}$ **119.** $\sqrt[3]{x^2y}$

121. 3 **123.** $\dfrac{x^2}{7y^{3/2}}$ **125.** $\dfrac{x^3}{y^2}$ **127.** Paige Fox is bad at math. **129.** $\dfrac{\sqrt{5}+1}{2}$; 1.62 to 1 **131.** $P = 18\sqrt{5}$ ft; $A = 100$ ft^2

143. makes sense **145.** does not make sense **147.** false **149.** false **151.** first box: 25; second box : 14 **153. a.** $>$ **b.** $>$
155. $10x^7y^9$ **156.** $16x^8 + 6x^5$ **157.** $2x^3 + 11x^2 + 22x + 15$

Section P.5
Check Point Exercises

1. a. $-x^3 + x^2 - 8x - 20$ **b.** $20x^3 - 11x^2 - 2x - 8$ **2.** $15x^3 - 31x^2 + 30x - 8$ **3.** $28x^2 - 41x + 15$ **4. a.** $49x^2 - 64$ **b.** $4y^6 - 25$
5. a. $x^2 + 20x + 100$ **b.** $25x^2 + 40x + 16$ **6. a.** $x^2 - 18x + 81$ **b.** $49x^2 - 42x + 9$ **7.** $2x^2y + 5xy^2 - 2y^3$ **8. a.** $21x^2 - 25xy + 6y^2$
b. $4x^2 + 16xy + 16y^2$

Concept and Vocabulary Check

1. whole **2.** standard **3.** monomial **4.** binomial **5.** trinomial **6.** n **7.** like **8.** distributive; $4x^5 - 8x^2 + 6; 7x^3$ **9.** $5x; 3;$ like
10. $3x^2; 5x; 21x; 35$ **11.** $A^2 - B^2;$ minus **12.** $A^2 + 2AB + B^2;$ squared; product of the terms; squared **13.** $A^2 - 2AB + B^2;$ minus; product
of the terms; plus **14.** $n + m$

Exercise Set P.5

1. yes; $3x^2 + 2x - 5$ **3.** no **5.** 2 **7.** 4 **9.** $11x^3 + 7x^2 - 12x - 4; 3$ **11.** $12x^3 + 4x^2 + 12x - 14; 3$ **13.** $6x^2 - 6x + 2; 2$
15. $x^3 + 1$ **17.** $2x^3 - 9x^2 + 19x - 15$ **19.** $x^2 + 10x + 21$ **21.** $x^2 - 2x - 15$ **23.** $6x^2 + 13x + 5$ **25.** $10x^2 - 9x - 9$
27. $15x^4 - 47x^2 + 28$ **29.** $8x^5 - 40x^3 + 3x^2 - 15$ **31.** $x^2 - 9$ **33.** $9x^2 - 4$ **35.** $25 - 49x^2$ **37.** $16x^4 - 25x^2$ **39.** $1 - y^{10}$
41. $x^2 + 4x + 4$ **43.** $4x^2 + 12x + 9$ **45.** $x^2 - 6x + 9$ **47.** $16x^4 - 8x^2 + 1$ **49.** $4x^2 - 28x + 49$ **51.** $x^3 + 3x^2 + 3x + 1$
53. $8x^3 + 36x^2 + 54x + 27$ **55.** $x^3 - 9x^2 + 27x - 27$ **57.** $27x^3 - 108x^2 + 144x - 64$ **59.** $7x^2y - 4xy$ is of degree 3
61. $2x^2y + 13xy + 13$ is of degree 3 **63.** $-5x^3 + 8xy - 9y^2$ is of degree 3 **65.** $x^4y^2 + 8x^3y + y - 6x$ is of degree 6 **67.** $7x^2 + 38xy + 15y^2$
69. $2x^2 + xy - 21y^2$ **71.** $15x^2y^2 + xy - 2$ **73.** $49x^2 + 70xy + 25y^2$ **75.** $x^4y^4 - 6x^2y^2 + 9$ **77.** $x^3 - y^3$ **79.** $9x^2 - 25y^2$
81. $49x^2y^4 - 100y^2$ **83.** $48xy$ **85.** $-9x^2 + 3x + 9$ **87.** $16x^4 - 625$ **89.** $4x^2 - 28x + 49$ **91. a.** $\$54,647;$ overestimates by $\$556$
b. $M - W = 1.8x^3 - 82x^2 + 2644x - 11,449$ **c.** $\$14,434$ **d.** $\$15,136;$ underestimates by $\$702$ **93.** $4x^3 - 36x^2 + 80x$ **95.** $6x + 22$
103. makes sense **105.** makes sense **107.** false **109.** false **111.** $49x^2 + 70x + 25 - 16y^2$ **113.** $6x^n - 13$ **115.** 4 **116.** 2 **117.** 3

Mid-Chapter P Check Point

1. $12x^2 - x - 35$ **2.** $-x + 12$ **3.** $\{9\}$ **4.** $\{-3, 3\}$ **5.** $10\sqrt{6}$ **6.** $3\sqrt{3}$ **7.** $x + 45$ **8.** $64x^2 - 48x + 9$ **9.** $\dfrac{x^2}{y^3}$ **10.** $\dfrac{3}{4}$
11. $-x^2 + 5x - 6$ **12.** $2x^3 - 11x^2 + 17x - 5$ **13.** $-x^6 + 2x^3$ **14.** $18a^2 - 11ab - 10b^2$ **15.** $\{a, c, d, e, f, h\}$ **16.** $\{c, d\}$ **17.** $[3, 5)$
18. $(2, 9]$ **19.** $5x^2y^3 + 2xy - y^2$ **20.** $-\dfrac{12y^{15}}{x^3}$ **21.** $\dfrac{6y^3}{x^7}$ **22.** $\sqrt[3]{x}$ **23.** $\{6\}$ **24.** 1.2×10^{-2} **25.** $2\sqrt[3]{2}$ **26.** $x^6 - 4$
27. $x^4 + 4x^2 + 4$ **28.** $10\sqrt{3}$ **29.** $\dfrac{77 + 11\sqrt{3}}{46}$ **30.** $\dfrac{11\sqrt{3}}{3}$ **31.** $-11, -\dfrac{3}{7}, 0, 0.45, \sqrt{25}$ **32.** $\{x | -2 \le x < 6\}$

33. $\{x | x \le 6\}$ **34.** $\sqrt{13} - 2$ **35.** $-x^3$ **36.** $x = \dfrac{By - C}{A}$ **37.** 1.38×10^9 pounds **38.** 4 times

38. a. model 2 **b.** overestimates by $\$5$ **c.** 2015

Section P.6
Check Point Exercises

1. a. $2x^2(5x - 2)$ **b.** $(x - 7)(2x + 3)$ **2.** $(x + 5)(x^2 - 2)$ **3.** $(x + 8)(x + 5)$ or $(x + 5)(x + 8)$ **4.** $(x - 7)(x + 2)$ or $(x + 2)(x - 7)$
5. $(3x - 1)(2x + 7)$ or $(2x + 7)(3x - 1)$ **6.** $(3x - y)(x - 4y)$ or $(x - 4y)(3x - y)$ **7. a.** $(x + 9)(x - 9)$ **b.** $(6x + 5)(6x - 5)$
8. $(9x^2 + 4)(3x + 2)(3x - 2)$ **9. a.** $(x + 7)^2$ **b.** $(4x - 7)^2$ **10. a.** $(x + 1)(x^2 - x + 1)$ **b.** $(5x - 2)(25x^2 + 10x + 4)$ **11.** $3x(x - 5)^2$
12. $(x + 10 + 6a)(x + 10 - 6a)$ **13.** $\dfrac{2x - 1}{(x - 1)^{1/2}}$ **14. a.** $9; x^2 + 6x + 9 = (x + 3)^2$ **b.** $\dfrac{25}{4}; x^2 - 5x + \dfrac{25}{4} = \left(x - \dfrac{5}{2}\right)^2$
c. $\dfrac{1}{9}; x^2 + \dfrac{2}{3}x + \dfrac{1}{9} = \left(x + \dfrac{1}{3}\right)^2$

Concept and Vocabulary Check

1. d **2.** g **3.** b **4.** c **5.** e **6.** a **7.** f **8.** $(x + 1)^{1/2}$ **9.** $\dfrac{9}{4}$ **10.** $\dfrac{4}{25}$

Exercise Set P.6

1. $9(2x + 3)$ **3.** $3x(x + 2)$ **5.** $9x^2(x^2 - 2x + 3)$ **7.** $(x + 5)(x + 3)$ **9.** $(x - 3)(x^2 + 12)$ **11.** $(x - 2)(x^2 + 5)$ **13.** $(x - 1)(x^2 + 2)$
15. $(3x - 2)(x^2 - 2)$ **17.** $(x + 2)(x + 3)$ **19.** $(x - 5)(x + 3)$ **21.** $(x - 5)(x - 3)$ **23.** $(3x + 2)(x - 1)$ **25.** $(3x - 28)(x + 1)$
27. $(2x - 1)(3x - 4)$ **29.** $(2x + 3)(2x + 5)$ **31.** $(3x - 2)(3x - 1)$ **33.** $(5x + 8)(4x - 1)$ **35.** $(2x + y)(x + y)$ **37.** $(3x + 2y)(2x - 3y)$
39. $(x + 10)(x - 10)$ **41.** $(6x + 7)(6x - 7)$ **43.** $(3x + 5y)(3x - 5y)$ **45.** $(x^2 + 4)(x + 2)(x - 2)$ **47.** $(4x^2 + 9)(2x + 3)(2x - 3)$

49. $(x + 1)^2$ **51.** $(x - 7)^2$ **53.** $(2x + 1)^2$ **55.** $(3x - 1)^2$ **57.** $(x + 3)(x^2 - 3x + 9)$ **59.** $(x - 4)(x^2 + 4x + 16)$

61. $(2x - 1)(4x^2 + 2x + 1)$ **63.** $(4x + 3)(16x^2 - 12x + 9)$ **65.** $3x(x + 1)(x - 1)$ **67.** $4(x + 2)(x - 3)$ **69.** $2(x^2 + 9)(x + 3)(x - 3)$

71. $(x - 3)(x + 3)(x + 2)$ **73.** $2(x - 8)(x + 7)$ **75.** $x(x - 2)(x + 2)$ **77.** prime **79.** $(x - 2)(x + 2)^2$ **81.** $y(y^2 + 9)(y + 3)(y - 3)$

83. $5y^2(2y + 3)(2y - 3)$ **85.** $(x - 6 + 7y)(x - 6 - 7y)$ **87.** $(x + y)(3b + 4)(3b - 4)$ **89.** $(y - 2)(x + 4)(x - 4)$

91. $2x(x + 6 + 2a)(x + 6 - 2a)$ **93.** $x^{1/2}(x - 1)$ **95.** $\dfrac{4(1 + 2x)}{x^{2/3}}$ **97.** $-(x + 3)^{1/2}(x + 2)$ **99.** $\dfrac{x + 4}{(x + 5)^{3/2}}$ **101.** $-\dfrac{4(4x - 1)^{1/2}(x - 1)}{3}$

103. $1; x^2 + 2x + 1 = (x + 1)^2$ **105.** $49; x^2 - 14x + 49 = (x - 7)^2$ **107.** $\dfrac{49}{4}; x^2 + 7x + \dfrac{49}{4} = \left(x + \dfrac{7}{2}\right)^2$

109. $\dfrac{1}{16}; x^2 - \dfrac{1}{2}x + \dfrac{1}{16} = \left(x - \dfrac{1}{4}\right)^2$ **111.** $\dfrac{4}{9}; x^2 + \dfrac{4}{3}x + \dfrac{4}{9} = \left(x + \dfrac{2}{3}\right)^2$ **113.** $(x + 1)(5x - 6)(2x + 1)$ **115.** $(x^2 + 6)(6x^2 - 1)$

117. $y(y^2 + 1)(y^4 - y^2 + 1)$ **119.** $(x + 2y)(x - 2y)(x + y)(x - y)$ **121.** $(x - y)^2(x - y + 2)(x - y - 2)$ **123.** $(2x - y^2)(x - 3y^2)$

125. a. $(x - 0.4x)(1 - 0.4) = (0.6x)(0.6) = 0.36x$ **b.** no; 36% **127. a.** $9x^2 - 16$ **b.** $(3x + 4)(3x - 4)$ **129. a.** $x(x + y) - y(x + y)$

b. $(x + y)(x - y)$ **131.** $4a^3 - 4ab^2 = 4a(a + b)(a - b)$ **141.** makes sense **143.** does not make sense **145.** false **147.** false

149. $(x^n + 4)(x^n + 2)$ **151.** $(x - y)^3(x + y)$ **153.** $b = 8, -8, 16, -16$ **155.** $\dfrac{(x + 5)(x + 1)}{(x + 5)(x - 5)} = \dfrac{x + 1}{x - 5}$ **156.** $\dfrac{2}{3}$ **157.** $\dfrac{7}{6}$

Section P.7

Check Point Exercises

1. a. -5 **b.** $6, -6$ **c.** $-2, 7$ **2. a.** $x^2, x \neq -3$ **b.** $\dfrac{x - 1}{x + 1}, x \neq -1$ **3.** $\dfrac{x - 3}{(x - 2)(x + 3)}, x \neq 2, x \neq -2, x \neq -3$

4. $\dfrac{3(x - 1)}{x(x + 2)}, x \neq 1, x \neq 0, x \neq -2$ **5.** $-2, x \neq -1$ **6.** $\dfrac{2(4x + 1)}{(x + 1)(x - 1)}, x \neq 1, x \neq -1$ **7.** $(x - 3)(x - 3)(x + 3)$ or $(x - 3)^2(x + 3)$

8. $\dfrac{-x^2 + 11x - 20}{2(x - 5)^2}, x \neq 5$ **9.** $\dfrac{2(2 - 3x)}{4 + 3x}, x \neq 0, x \neq -\dfrac{4}{3}$ **10.** $-\dfrac{1}{x(x + 7)}, x \neq 0, x \neq -7$

Concept and Vocabulary Check

1. polynomials **2.** domain; 0 **3.** factoring; common factors **4.** $\dfrac{x^2}{15}$ **5.** $\dfrac{3}{5}$ **6.** $\dfrac{x^2 - x + 4}{3}$ **7.** $x + 3$ and $x - 2$; $x + 3$ and $x + 1$;

$(x + 3)(x - 2)(x + 1)$ **8.** $3x + 4$ **9.** complex; complex **10.** $\dfrac{x - (x + 3)}{3x(x + 3)} = \dfrac{-3}{3x(x + 3)} = -\dfrac{1}{x(x + 3)}$

Exercise Set P.7

1. 3 **3.** $5, -5$ **5.** $-1, -10$ **7.** $\dfrac{3}{x - 3}, x \neq 3$ **9.** $\dfrac{x - 6}{4}, x \neq 6$ **11.** $\dfrac{y + 9}{y - 1}, y \neq 1, 2$ **13.** $\dfrac{x + 6}{x - 6}, x \neq 6, -6$ **15.** $\dfrac{1}{3}, x \neq 2, -3$

17. $\dfrac{(x - 3)(x + 3)}{x(x + 4)}, x \neq 0, -4, 3$ **19.** $\dfrac{x - 1}{x + 2}, x \neq -2, -1, 2, 3$ **21.** $\dfrac{x^2 + 2x + 4}{3x}, x \neq -2, 0, 2$ **23.** $\dfrac{7}{9}, x \neq -1$ **25.** $\dfrac{(x - 2)^2}{x}, x \neq 0, -2, 2$

27. $\dfrac{2(x + 3)}{3}, x \neq 3, -3$ **29.** $\dfrac{x - 5}{2}, x \neq 1, -5$ **31.** $\dfrac{(x + 2)(x + 4)}{x - 5}, x \neq -6, -3, -1, 3, 5$ **33.** $2, x \neq -\dfrac{5}{6}$ **35.** $\dfrac{2x - 1}{x + 3}, x \neq 0, -3$

37. $3, x \neq 2$ **39.** $\dfrac{3}{x - 3}, x \neq 3, -4$ **41.** $\dfrac{9x + 39}{(x + 4)(x + 5)}, x \neq -4, -5$ **43.** $-\dfrac{3}{x(x + 1)}, x \neq -1, 0$ **45.** $\dfrac{3x^2 + 4}{(x + 2)(x - 2)}, x \neq -2, 2$

47. $\dfrac{2x^2 + 50}{(x - 5)(x + 5)}, x \neq -5, 5$ **49.** $\dfrac{13}{6(x + 2)}, x \neq -2$ **51.** $\dfrac{4x + 16}{(x + 3)^2}, x \neq -3$ **53.** $\dfrac{x^2 - x}{(x + 5)(x - 2)(x + 3)}, x \neq -5, 2, -3$

55. $\dfrac{-x^2 - 2x + 1}{(x + 1)(x - 1)}, x \neq -1, 1$ **57.** $\dfrac{x - 1}{x + 2}, x \neq -2, -1$ **59.** $\dfrac{1}{3}, x \neq 3$ **61.** $\dfrac{x + 1}{3x - 1}, x \neq 0, \dfrac{1}{3}$ **63.** $\dfrac{1}{xy}, x \neq 0, y \neq 0, x \neq -y$

65. $\dfrac{x}{x + 3}, x \neq -2, -3$ **67.** $-\dfrac{x - 14}{7}, x \neq -2, 2$ **69.** $\dfrac{x - 3}{x + 2}, x \neq -2, -1, 3$ **71.** $-\dfrac{2x + h}{x^2(x + h)^2}, x \neq 0, h \neq 0, x \neq -h$ **73.** $\dfrac{x^2 + 5x + 8}{(x + 2)(x + 1)}$

75. 2 **77.** $\dfrac{1}{y(y + 5)}$ **79.** $\dfrac{2d}{a^2 + ab + b^2}$ **81. a.** 86.67, 520, 1170; It costs \$86,670,000 to inoculate 40% of the population against this strain of

flu, \$520,000,000 to inoculate 80% of the population, and \$1,170,000,000 to inoculate 90% of the population. **b.** $x = 100$ **c.** The cost increases
rapidly; it is impossible to inoculate 100% of the population. **83. a.** 2078; underestimates by 22 calories **b.** 2662; underestimates by 38 calories

c. $\dfrac{-33x^2 + 263x + 515}{-60x^2 + 499x + 295}$ **85.** $\dfrac{4x^2 + 22x}{(x + 5)(x + 6)}$ **99.** does not make sense **101.** makes sense **103.** true **105.** false **107.** $\dfrac{x - 1}{x + 3}$

109. It cubes x. **110.** $-5; 0; 3; 4; 3; 0; -5$ **111.** $-8; -3; 0; 1; 0; -3; -8$ **112.** $3; 2; 1; 0; 1; 2; 3$

Chapter P Review Exercises

1. 51 **2.** 16 **3.** 124 ft **4.** $\{a, c\}$ **5.** $\{a, b, c, d, e\}$ **6.** $\{a, b, c, d, f, g\}$ **7.** $\{a\}$ **8. a.** $\sqrt{81}$ **b.** $0, \sqrt{81}$

c. $-17, 0, \sqrt{81}$ **d.** $-17, -\dfrac{9}{13}, 0, 0.75, \sqrt{81}$ **e.** $\sqrt{2}, \pi$ **f.** $-17, -\dfrac{9}{13}, 0, 0.75, \sqrt{2}, \pi, \sqrt{81}$

9. $\{x \mid -3 \le x < 5\}$

10. $\{x \mid x > -2\}$

11. $\{x \mid x \le 0\}$

12. $[-1, 1]$ **13.** $(-2, 3)$
14. $[1, 3)$ **15.** $(0, 4)$

16. 103 **17.** $\sqrt{2} - 1$ **18.** $\sqrt{17} - 3$ **19.** $|4 - (-17)|; 21$ **20.** commutative property of addition **21.** associative property of multiplication **22.** distributive property of multiplication over addition **23.** commutative property of multiplication **24.** commutative property of multiplication **25.** commutative property of addition **26.** $17x - 15$ **27.** $2x$ **28.** $\{6\}$ **29.** $\{5\}$ **30.** $5y - 17$ **31.** $10x$ **32.** $\{-3\}$

33. $g = \dfrac{s - vt}{t^2}$ **34.** $g = \dfrac{T}{r + vt}$ **35.** 55; It's the same. **36.** -108 **37.** $\dfrac{5}{16}$ **38.** $\dfrac{1}{25}$ **39.** $\dfrac{1}{27}$ **40.** $-8x^{12}y^9$ **41.** $\dfrac{10}{x^8}$ **42.** $\dfrac{1}{16x^{12}}$

43. $\dfrac{y^8}{4x^{10}}$ **44.** 37,400 **45.** 0.0000745 **46.** 3.59×10^6 **47.** 7.25×10^{-3} **48.** 390,000 **49.** 0.023 **50.** 1.35×10^{12} **51.** 3.2×10^7

52. 42,188 years **53.** $\{-\sqrt{19}, \sqrt{19}\}$ or $\{\pm \sqrt{19}\}$ **54.** $\{-2, 2\}$ **55.** $10\sqrt{3}$ **56.** $2|x|\sqrt{3}$ **57.** $2x\sqrt{5}$ **58.** $r\sqrt{r}$ **59.** $\dfrac{11}{2}$

60. $4x\sqrt{3}$ **61.** $20\sqrt{5}$ **62.** $16\sqrt{2}$ **63.** $24\sqrt{2} - 8\sqrt{3}$ **64.** $6\sqrt{5}$ **65.** $\dfrac{\sqrt{6}}{3}$ **66.** $\dfrac{5(6 - \sqrt{3})}{33}$ **67.** $7(\sqrt{7} + \sqrt{5})$ **68.** 5 **69.** -2

70. not a real number **71.** 5 **72.** $3\sqrt[3]{3}$ **73.** $y\sqrt[3]{y^2}$ **74.** $2\sqrt[4]{5}$ **75.** $13\sqrt[3]{2}$ **76.** $x\sqrt[4]{2}$ **77.** 4 **78.** $\dfrac{1}{5}$ **79.** 5 **80.** $\dfrac{1}{3}$ **81.** 16

82. $\dfrac{1}{81}$ **83.** $20x^{11/12}$ **84.** $3x^{1/4}$ **85.** $25x^4$ **86.** \sqrt{y} **87.** $8x^3 + 10x^2 - 20x - 4$; degree 3 **88.** $8x^4 - 5x^3 + 6$; degree 4

89. $12x^3 + x^2 - 21x + 10$ **90.** $6x^2 - 7x - 5$ **91.** $16x^2 - 25$ **92.** $4x^2 + 20x + 25$ **93.** $9x^2 - 24x + 16$ **94.** $8x^3 + 12x^2 + 6x + 1$

95. $125x^3 - 150x^2 + 60x - 8$ **96.** $-x^2 - 17xy - 3y^2$; degree 2 **97.** $24x^3y^2 + x^2y - 12x^2 + 4$; degree 5 **98.** $3x^2 + 16xy - 35y^2$

99. $9x^2 - 30xy + 25y^2$ **100.** $9x^4 + 12x^2y + 4y^2$ **101.** $49x^2 - 16y^2$ **102.** $a^3 - b^3$ **103.** $3x^2(5x + 1)$ **104.** $(x - 4)(x - 7)$

105. $(3x + 1)(5x - 2)$ **106.** $(8 - x)(8 + x)$ **107.** prime **108.** $3x^2(x - 5)(x + 2)$ **109.** $4x^3(5x^4 - 9)$ **110.** $(x + 3)(x - 3)^2$

111. $(4x - 5)^2$ **112.** $(x^2 + 4)(x + 2)(x - 2)$ **113.** $(y - 2)(y^2 + 2y + 4)$ **114.** $(x + 4)(x^2 - 4x + 16)$ **115.** $3x^2(x - 2)(x + 2)$

116. $(3x - 5)(9x^2 + 15x + 25)$ **117.** $x(x - 1)(x + 1)(x^2 + 1)$ **118.** $(x^2 - 2)(x + 5)$ **119.** $(x + 9 + y)(x + 9 - y)$ **120.** $\dfrac{16(1 + 2x)}{x^{3/4}}$

121. $(x + 2)(x - 2)(x^2 + 3)^{1/2}(-x^4 + x^2 + 13)$ **122.** $\dfrac{6(2x + 1)}{x^{3/2}}$ **123.** $16; x^2 - 8x + 16 = (x - 4)^2$ **124.** $\dfrac{1}{16}; x^2 + \dfrac{1}{2}x + \dfrac{1}{16} = \left(x + \dfrac{1}{4}\right)^2$

125. $x^2, x \ne -2$ **126.** $\dfrac{x - 3}{x - 6}, x \ne -6, 6$ **127.** $\dfrac{x}{x + 2}, x \ne -2$ **128.** $\dfrac{(x + 3)^3}{(x - 2)^2(x + 2)}, x \ne 2, -2$ **129.** $\dfrac{2}{x(x + 1)}, x \ne 0, 1, -1, -\dfrac{1}{3}$

130. $\dfrac{x + 3}{x - 4}, x \ne -3, 4, 2, 8$ **131.** $\dfrac{1}{x - 3}, x \ne 3, -3$ **132.** $\dfrac{4x(x - 1)}{(x + 2)(x - 2)}, x \ne 2, -2$ **133.** $\dfrac{2x^2 - 3}{(x - 3)(x + 3)(x - 2)}, x \ne 3, -3, 2$

134. $\dfrac{11x^2 - x - 11}{(2x - 1)(x + 3)(3x + 2)}, x \ne \dfrac{1}{2}, -3, -\dfrac{2}{3}$ **135.** $\dfrac{3}{x}, x \ne 0, 2$ **136.** $\dfrac{3x}{x - 4}, x \ne 0, 4, -4$ **137.** $\dfrac{3x + 8}{3x + 10}, x \ne -3, -\dfrac{10}{3}$

Chapter P Test

1. $6x^2 - 27x$ **2.** $\{2\}$ **3.** $\{-5, 5\}$ **4.** $[-1, 5]$ **5.** $(0, 2)$ **6.** $\dfrac{5y^8}{x^6}$ **7.** $3r\sqrt{2}$ **8.** $11\sqrt{2}$ **9.** $\dfrac{3(5 - \sqrt{2})}{23}$

10. $2x\sqrt[3]{2x}$ **11.** $\dfrac{x + 3}{x - 2}, x \ne 2, 1$ **12.** 2.5×10^1 **13.** $2x^3 - 13x^2 + 26x - 15$ **14.** $25x^2 + 30xy + 9y^2$ **15.** $\dfrac{2(x + 3)}{x + 1}, x \ne 3, -1, -4, -3$

16. $\dfrac{x^2 + 2x + 15}{(x + 3)(x - 3)}, x \ne 3, -3$ **17.** $\dfrac{11}{(x - 3)(x - 4)}, x \ne 3, 4$ **18.** $\dfrac{3 - x}{3}, x \ne 0$ **19.** $(x - 3)(x - 6)$ **20.** $(x^2 + 3)(x + 2)$

21. $(5x - 3)(5x + 3)$ **22.** $(6x - 7)^2$ **23.** $(y - 5)(y^2 + 5y + 25)$ **24.** $(x + 5 + 3y)(x + 5 - 3y)$ **25.** $\dfrac{2x + 3}{(x + 3)^{3/5}}$

26. $-7, -\dfrac{4}{5}, 0, 0.25, \sqrt{4}, \dfrac{22}{7}$ **27.** commutative property of addition **28.** distributive property of multiplication over addition **29.** 7.6×10^{-4}

30. $\dfrac{1}{243}$ **31.** $y = \dfrac{C - Ax}{B}$ **32.** 1.32×10^{10} **33. a.** 43.08%; overestimates by 0.08% **b.** $R = \dfrac{-0.28n + 47}{0.28n + 53}$ **c.** $\dfrac{2}{3}$; Three women will receive bachelor's degrees for every two men.; It describes the projections exactly.

CHAPTER 1

Section 1.1
Check Point Exercises

1. **2.** **3.** **4.** minimum x-value: -100; maximum x-value: 100; distance between tick marks on x-axis: 50; minimum y-value: -100; maximum y-value: 100; distance between tick marks on y-axis: 10

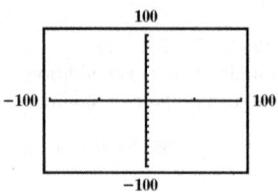

5. a. x-intercept: -3; y-intercept: 5
 b. no x-intercept; y-intercept: 4
 c. x-intercept: 0; y-intercept: 0
6. a. 65% **b.** 60% **c.** overestimates by 5

Concept and Vocabulary Check

1. x-axis **2.** y-axis **3.** origin **4.** quadrants; four **5.** x-coordinate; y-coordinate **6.** solution; satisfies
7. x-intercept; zero **8.** y-intercept; zero

Exercise Set 1.1

1. **3.** **5.** **7.** **9.** **11.**

13. **15.** **17.** **19.** **21.** **23.**

25. **27.** 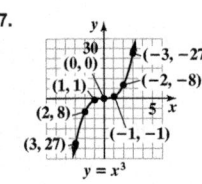 **29.** c **31.** b **33.** c **35.** no **37.** $(2, 0)$ **39.** $(-2, 4)$ and $(1, 1)$ **41. a.** 2 **b.** -4

43. a. $1, -2$ **b.** 2 **45. a.** -1 **b.** none **47.** $y = 2x + 4$

49. $y = 3 - x^2$ **51.** **53.** 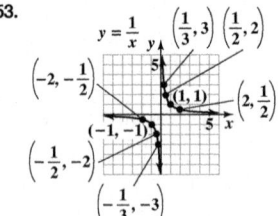 **55. a.** 20% **b.** 19.5%; underestimates by 0.5%

c. Answers will vary.; approximately 47% **d.** 46.5%; It's less than the estimate, although answers may vary. **e.** 1990; 14%

57. 8; 1 **59.** about 1.9 **67.** makes sense **69.** does not make sense **71.** false **73.** true **75.** I, III **77.** IV **79.** a **81.** b
83. b **85.** c **87.** set 1
88. **89. a.** 3 **b.** -3 and 3 **c.** all real numbers **d.** all real numbers greater than or equal to 1

Section 1.2

Check Point Exercises

1. domain: {0, 10, 20, 30, 40}; range: {9.1, 6.7, 10.7, 13.2, 21.2} **2. a.** not a function **b.** function **3. a.** $y = 6 - 2x$; function
b. $y = \pm\sqrt{1 - x^2}$; not a function **4. a.** 42 **b.** $x^2 + 6x + 15$ **c.** $x^2 + 2x + 7$
5. $f(x) = 2x$; The graph of g is the graph of f shifted down by 3 units.

$g(x) = 2x - 3$

(2, 4) (2, 1)
(1, 2) (2, 1)
(0, 0)
(−1, −2) (1, −1)
(−2, −4) (0, −3)
(−2, −7) (−1, −5)

6. a. function **b.** not a function **c.** function **d.** not a function **7. a.** 400 **b.** 9 **c.** approximately 425
8. a. domain: $\{x \mid -2 \le x \le 1\}$ or $[-2, 1]$; range: $\{y \mid 0 \le y \le 3\}$ or $[0, 3]$
b. domain: $\{x \mid -2 < x \le 1\}$ or $(-2, 1]$; range: $\{y \mid -1 \le y < 2\}$ or $[-1, 2)$
c. domain: $\{x \mid -3 \le x < 0\}$ or $[-3, 0)$; range: $\{-3, -2, -1\}$

Concept and Vocabulary Check

1. relation; domain; range **2.** function **3.** f; x **4.** true **5.** false **6.** x; $x + 6$ **7.** ordered pairs **8.** more than once; function
9. $[0, 3)$; domain **10.** $[1, \infty)$; range **11.** 0; 0; zeros **12.** false

Exercise Set 1.2

1. function; {1, 3, 5}; {2, 4, 5} **3.** not a function; {3, 4}; {4, 5} **5.** function; {3, 4, 5, 7}; {−2, 1, 9} **7.** function; {−3, −2, −1, 0}; {−3, −2, −1, 0}
9. not a function; {1}; {4, 5, 6} **11.** y is a function of x. **13.** y is a function of x. **15.** y is not a function of x. **17.** y is not a function of x.
19. y is a function of x. **21.** y is a function of x. **23.** y is a function of x. **25.** y is a function of x. **27. a.** 29 **b.** $4x + 9$ **c.** $-4x + 5$
29. **a.** 2 **b.** $x^2 + 12x + 38$ **c.** $x^2 - 2x + 3$ **31. a.** 13 **b.** 1 **c.** $x^4 - x^2 + 1$ **d.** $81a^4 - 9a^2 + 1$ **33. a.** 3 **b.** 7 **c.** $\sqrt{x} + 3$
35. **a.** $\dfrac{15}{4}$ **b.** $\dfrac{15}{4}$ **c.** $\dfrac{4x^2 - 1}{x^2}$ **37. a.** 1 **b.** −1 **c.** 1

39.

$g(x) = x + 3$
(2, 5) $f(x) = x$
(1, 4) (2, 2)
(0, 3) (1, 1)
(−1, 2) (0, 0)
(−2, 1) (−1, −1)
(−2, −2)

The graph of g is the graph of f
shifted up by 3 units.

41.

$f(x) = -2x$
$g(x) = -2x - 1$
(−2, 3) (−2, 4)
(−1, 1) (−1, 2)
(0, 0)
(0, −1) (1, −2)
(1, −3) (2, −4)
(2, −5)

The graph of g is the graph of f
shifted down by 1 unit.

43.

$g(x) = x^2 + 1$
$f(x) = x^2$
(−2, 5) (2, 5)
(−2, 4) (2, 4)
(−1, 2) (1, 2)
(−1, 1) (1, 1)
(0, 0)

The graph of g is the graph of f
shifted up by 1 unit.

45.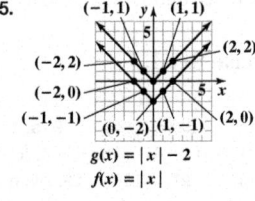

(−1, 1) (1, 1)
(−2, 2) (2, 2)
(−2, 0) (2, 0)
(−1, −1) (1, −1)
(0, −2)
$g(x) = |x| - 2$
$f(x) = |x|$

The graph of g is the graph of f
shifted down by 2 units.

47.

$f(x) = x^3$
$g(x) = x^3 + 2$
(2, 10)
(0, 2) (2, 8)
(−1, 1) (1, 3)
(1, 1)
(0, 0)
(−2, −6) (−1, −1)
(−2, −8)

The graph of g is the graph of f
shifted up by 2 units.

49.

(0, 5)
(−1, 5) (1, 5)
(−2, 5) (2, 5)
$g(x) = 5$ (2, 3)
$f(x) = 3$
(−2, 3)
(−1, 3)
(0, 3) (1, 3)

The graph of g is the graph of f
shifted up by 2 units.

51.

(4, 2) (9, 3)
(1, 1) $f(x) = \sqrt{x}$
(0, 0) $g(x) = \sqrt{x} - 1$
(0, −1)
(1, 0) (4, 1) (9, 2)

The graph of g is the graph of f
shifted down by 1 unit.

53.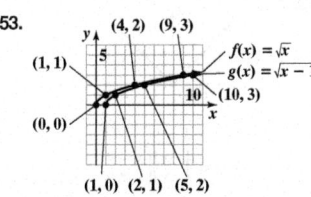

(4, 2) (9, 3)
(1, 1) $f(x) = \sqrt{x}$
$g(x) = \sqrt{x - 1}$
(0, 0) (10, 3)
(1, 0) (2, 1) (5, 2)

The graph of g is the graph of f
shifted to the right by 1 unit.

55. function **57.** function **59.** not a function **61.** function **63.** function **65.** −4 **67.** 4 **69.** 0 **71.** 2 **73.** 2
75. −2 **77. a.** $(-\infty, \infty)$ **b.** $[-4, \infty)$ **c.** −3 and 1 **d.** −3 **e.** $f(-2) = -3$ and $f(2) = 5$ **79. a.** $(-\infty, \infty)$ **b.** $[1, \infty)$ **c.** none
d. 1 **e.** $f(-1) = 2$ and $f(3) = 4$ **81. a.** $[0, 5)$ **b.** $[-1, 5)$ **c.** 2 **d.** −1 **e.** $f(3) = 1$ **83. a.** $[0, \infty)$ **b.** $[1, \infty)$ **c.** none
d. 1 **e.** $f(4) = 3$ **85. a.** $[-2, 6]$ **b.** $[-2, 6]$ **c.** 4 **d.** 4 **e.** $f(-1) = 5$ **87. a.** $(-\infty, \infty)$ **b.** $(-\infty, -2]$ **c.** none
d. −2 **e.** $f(-4) = -5$ and $f(4) = -2$ **89. a.** $(-\infty, \infty)$ **b.** $(0, \infty)$ **c.** none **d.** 1.5 **e.** $f(4) = 6$ **91. a.** {−5, −2, 0, 1, 3}
b. {2} **c.** none **d.** 2 **e.** $f(-5) + f(3) = 4$ **93.** −2; 10 **95.** −38 **97.** $-2x^3 - 2x$
99. **a.** {(Philippines, 12), (Spain, 13), (Italy, 14), (Germany, 14), (Russia, 16)} **b.** yes; Each country corresponds to exactly one age.
c. {(12, Philippines), (13, Spain), (14, Italy), (14, Germany), (16, Russia)} **d.** no; 14 in the domain corresponds to two members of the range, Italy and
Germany. **101. a.** 83; The chance that a 60-year-old will survive to age 70 is 83%. **b.** 76; The chance that a 60-year-old will survive to age 70 is 76%. **c.** f

103. a. 81; In 2010, the wage gap was 81%.; $(30, 81)$ **b.** underestimates by 2% **105.** $C = 100,000 + 100x$, where x is the number of bicycles produced; $C(90) = 109,000$; It costs \$109,000 to produce 90 bicycles. **107.** $T = \dfrac{40}{x} + \dfrac{40}{x + 30}$, where x is the rate on the outgoing trip; $T(30) = 2$; It takes 2 hours, traveling 30 mph outgoing and 60 mph returning. **119.** does not make sense **121.** does not make sense **123.** false **125.** false **127.** Answers will vary; an example is $\{(1, 1), (2, 1)\}$. **129.** 36; For 100 calling minutes, the monthly cost is \$36.

130. $f(x) = x + 2, \; x \le 1$ **131.** $4xh + 2h^2 + 3h$

Section 1.3

Check Point Exercises

1. increasing on $(-\infty, -1)$, decreasing on $(-1, 1)$, increasing on $(1, \infty)$ **2. a.** even **b.** odd **c.** neither
3. a. 20; With 40 calling minutes, the cost is \$20.; $(40, 20)$ **b.** 28; With 80 calling minutes, the cost is \$28.; $(80, 28)$
4. 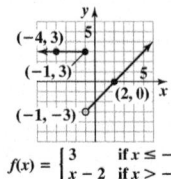 **5. a.** $-2x^2 - 4xh - 2h^2 + x + h + 5$ **b.** $-4x - 2h + 1, \; h \ne 0$

$f(x) = \begin{cases} 3 & \text{if } x \le -1 \\ x - 2 & \text{if } x > -1 \end{cases}$

Concept and Vocabulary Check

1. $< f(x_2); > f(x_2); = f(x_2)$ **2.** maximum; minimum **3.** $f(x)$; y-axis **4.** $-f(x)$; origin **5.** piecewise **6.** less than or equal to x; 2; -3; 0
7. difference quotient; $x + h$; $f(x)$; h; h **8.** false **9.** false

Exercise Set 1.3

1. a. $(-1, \infty)$ **b.** $(-\infty, -1)$ **c.** none **3. a.** $(0, \infty)$ **b.** none **c.** none **5. a.** none **b.** $(-2, 6)$ **c.** none **7. a.** $(-\infty, -1)$
b. none **c.** $(-1, \infty)$ **9. a.** $(-\infty, 0)$ or $(1.5, 3)$ **b.** $(0, 1.5)$ or $(3, \infty)$ **c.** none **11. a.** $(-2, 4)$ **b.** none **c.** $(-\infty, -2)$ or $(4, \infty)$
13. a. 0; $f(0) = 4$ **b.** $-3, 3; f(-3) = f(3) = 0$ **15. a.** $-2; f(-2) = 21$ **b.** 1; $f(1) = -6$ **17.** odd **19.** neither **21.** even **23.** even
25. even **27.** odd **29.** even **31.** odd **33. a.** $(-\infty, \infty)$ **b.** $[-4, \infty)$ **c.** 1 and 7 **d.** 4 **e.** $(4, \infty)$ **f.** $(0, 4)$ **g.** $(-\infty, 0)$
h. 4 **i.** -4 **j.** 4 **k.** 2 and 6 **l.** neither **35. a.** $(-\infty, 3]$ **b.** $(-\infty, 4]$ **c.** -3 and 3 **d.** 3 **e.** $(-\infty, 1)$ **f.** $(1, 3)$ **g.** $(-\infty, -3]$
h. A relative maximum of 4 occurs at 1. **i.** 1 **j.** positive **37. a.** -1 **b.** 7 **c.** 19 **39. a.** 3 **b.** 3 **c.** 0 **41. a.** 8 **b.** 3 **c.** 6

43. a.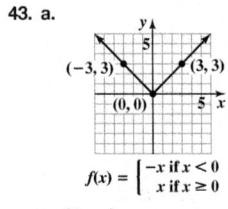

$f(x) = \begin{cases} -x & \text{if } x < 0 \\ x & \text{if } x \ge 0 \end{cases}$

b. $[0, \infty)$

45. a.

$f(x) = \begin{cases} 2x & \text{if } x \le 0 \\ 2 & \text{if } x > 0 \end{cases}$

b. $(-\infty, 0] \cup \{2\}$

47. a.

$f(x) = \begin{cases} x + 3 & \text{if } x < -2 \\ x - 3 & \text{if } x \ge -2 \end{cases}$

b. $(-\infty, \infty)$

49. a.

$f(x) = \begin{cases} 3 & \text{if } x \le -1 \\ -3 & \text{if } x > -1 \end{cases}$

b. $\{-3, 3\}$

51. a.

$f(x) = \begin{cases} \frac{1}{2}x^2 & \text{if } x < 1 \\ 2x - 1 & \text{if } x \ge 1 \end{cases}$

b. $[0, \infty)$

53. a.

$f(x) = \begin{cases} 0 & \text{if } x < -4 \\ -x & \text{if } -4 \le x < 0 \\ x^2 & \text{if } x \ge 0 \end{cases}$

b. $[0, \infty)$

55. $4, h \neq 0$ **57.** $3, h \neq 0$ **59.** $2x + h, h \neq 0$ **61.** $2x + h - 4, h \neq 0$ **63.** $4x + 2h + 1, h \neq 0$ **65.** $-2x - h + 2, h \neq 0$
67. $-4x - 2h + 5, h \neq 0$ **69.** $-4x - 2h - 1, h \neq 0$ **71.** $0, h \neq 0$ **73.** $-\dfrac{1}{x(x + h)}, h \neq 0$ **75.** $\dfrac{1}{\sqrt{x + h} + \sqrt{x}}, h \neq 0$ **77.** -18

79. $0.30t - 6$ **81.** $C(t) = \begin{cases} 50 & \text{if } 0 \leq t \leq 400 \\ 50 + 0.30(t - 400) & \text{if } t > 400 \end{cases}$

83. increasing: $(25, 55)$; decreasing: $(55, 75)$ **85.** 55 years old; 38% **87.** domain: $[25, 75]$; range: $[34, 38]$ **89.** men
91. 2575; A single taxpayer with taxable income of $20,000 owes $2575. **93.** $42{,}449 + 0.33(x - 174{,}400)$
95. 0.85; It costs $0.85 to mail a 3-ounce first-class letter. **97.** $0.65

99.

107.

increasing: $(-\infty, 1)$ or $(3, \infty)$
decreasing: $(1, 3)$

109.

increasing: $(2, \infty)$
decreasing: $(-\infty, -2)$
constant: $(-2, 2)$

111.

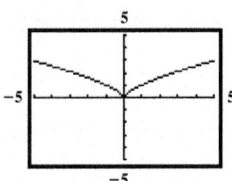

increasing: $(0, \infty)$
decreasing: $(-\infty, 0)$

c. increasing: $(0, \infty)$; decreasing: $(-\infty, 0)$

113. a.

b.

d. $f(x) = x^n$ is increasing for $(-\infty, \infty)$ when n is positive and odd. **e.**

As n increases the steepness increases.

115. makes sense **117.** makes sense **122.** 3 **123.** $\left(\dfrac{3}{2}, 0\right)$ and $(0, -2)$
124. $y = \dfrac{-3x + 4}{2}$ or $y = -\dfrac{3}{2}x + 2$

Section 1.4

Check Point Exercises

1. a. 6 **b.** $-\dfrac{7}{5}$ **2.** $y + 5 = 6(x - 2); y = 6x - 17$ **3.** $y + 1 = -5(x + 2)$ or $y + 6 = -5(x + 1); y = -5x - 11$

4.

$f(x) = \dfrac{3}{5}x + 1$

5.

6.

$3x + 6y - 12 = 0$

7. slope: $-\dfrac{1}{2}$; y-intercept: 2

8.

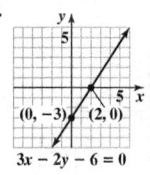

$3x - 2y - 6 = 0$

9. $f(x) = 0.016x + 52.0; 61.6°F$

Concept and Vocabulary Check

1. scatter plot; regression **2.** $\dfrac{y_2 - y_1}{x_2 - x_1}$ **3.** positive **4.** negative **5.** zero **6.** undefined **7.** $y - y_1 = m(x - x_1)$

8. $y = mx + b$; slope; y-intercept **9.** $(0, 3)$; 2; 5 **10.** horizontal **11.** vertical **12.** general

Exercise Set 1.4

1. $\dfrac{3}{4}$; rises **3.** $\dfrac{1}{4}$; rises **5.** 0; horizontal **7.** -5; falls **9.** undefined; vertical **11.** $y - 5 = 2(x - 3)$; $y = 2x - 1$

13. $y - 5 = 6(x + 2)$; $y = 6x + 17$ **15.** $y + 3 = -3(x + 2)$; $y = -3x - 9$ **17.** $y - 0 = -4(x + 4)$; $y = -4x - 16$

19. $y + 2 = -1\left(x + \dfrac{1}{2}\right)$; $y = -x - \dfrac{5}{2}$ **21.** $y - 0 = \dfrac{1}{2}(x - 0)$; $y = \dfrac{1}{2}x$ **23.** $y + 2 = -\dfrac{2}{3}(x - 6)$; $y = -\dfrac{2}{3}x + 2$

25. using $(1, 2)$, $y - 2 = 2(x - 1)$; $y = 2x$ **27.** using $(-3, 0)$, $y - 0 = 1(x + 3)$; $y = x + 3$ **29.** using $(-3, -1)$, $y + 1 = 1(x + 3)$; $y = x + 2$

31. using $(-3, -2)$, $y + 2 = \dfrac{4}{3}(x + 3)$; $y = \dfrac{4}{3}x + 2$ **33.** using $(-3, -1)$, $y + 1 = 0(x + 3)$; $y = -1$ **35.** using $(2, 4)$, $y - 4 = 1(x - 2)$; $y = x + 2$

37. using $(0, 4)$, $y - 4 = 8(x - 0)$; $y = 8x + 4$

39. $m = 2$; $b = 1$

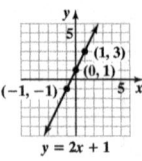

41. $m = -2$; $b = 1$

43. $m = \dfrac{3}{4}$; $b = -2$

45. $m = -\dfrac{3}{5}$; $b = 7$

47. $m = -\dfrac{1}{2}$; $b = 0$

49.

51.

53.

55.

57.

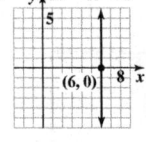

59. a. $y = -3x + 5$
b. $m = -3$; $b = 5$
c.

61. a. $y = -\dfrac{2}{3}x + 6$
b. $m = -\dfrac{2}{3}$; $b = 6$
c.

63. a. $y = 2x - 3$
b. $m = 2$; $b = -3$
c.

65. a. $y = 3$
b. $m = 0$; $b = 3$
c.

67.

69.

71.

73. $m = -\dfrac{a}{b}$; falls

75. undefined slope; vertical

77. $m = -\dfrac{A}{B}$; $b = \dfrac{C}{B}$

79. -2

81.

83. 5

85. m_1, m_3, m_2, m_4

87. a. $y - 31.1 = 0.78(x - 10)$ or $y - 38.9 = 0.78(x - 20)$ **b.** $f(x) = 0.78x + 23.3$ **c.** 54.5%

89. a & b. Life Expectancy for United States; $E(x) = 0.215x + 65.7$

Males, by Year of Birth

91. Answers will vary; an example is $y = -2.3x + 255$, where x is the percentage of adult females who are literate and y is under-five mortality per thousand.; Predictions will vary.

c. 78.6 yr

101. $m = -3$

103. $m = \dfrac{3}{4}$

105. does not make sense **107.** does not make sense
109. false **111.** true
113. coefficient of x: -6; coefficient of y: 3
118. $y = 2x + 7$ or $f(x) = 2x + 7$
119. $4x - y - 17 = 0$ **120.** 5

Section 1.5

Check Point Exercises

1. $y - 5 = 3(x + 2); y = 3x + 11$ or $f(x) = 3x + 11$ **2. a.** 3 **b.** $3x - y = 0$ **3.** $m \approx 0.32$; The number of men living alone increased at a rate of 0.32 million per year. The rate of change is 0.32 million men per year. **4. a.** 1 **b.** 7 **c.** 4 **5.** 0.01 mg per 100 ml per hr

Concept and Vocabulary Check

1. the same **2.** -1 **3.** $-\dfrac{1}{3}; 3$ **4.** $-2; \dfrac{1}{2}$ **5.** $y; x$ **6.** $\dfrac{f(x_2) - f(x_1)}{x_2 - x_1}$

Exercise Set 1.5

1. $y - 2 = 2(x - 4); y = 2x - 6$ or $f(x) = 2x - 6$ **3.** $y - 4 = -\dfrac{1}{2}(x - 2); y = -\dfrac{1}{2}x + 5$ or $f(x) = -\dfrac{1}{2}x + 5$

5. $y + 10 = -4(x + 8); y = -4x - 42$ **7.** $y + 3 = -5(x - 2); y = -5x + 7$ **9.** $y - 2 = \dfrac{2}{3}(x + 2); 2x - 3y + 10 = 0$

11. $y + 7 = -2(x - 4); 2x + y - 1 = 0$ **13.** 3 **15.** 10 **17.** $\dfrac{1}{5}$ **19.** $f(x) = 5$ **21.** $f(x) = -\dfrac{1}{2}x + 1$ **23.** $f(x) = -\dfrac{2}{3}x - 2$

25. $p(x) = -0.25x + 22$ **27.** 137; There was an average increase of approximately 137 discharges per year.
29. a. 142 **b.** overestimates by 5 discharges per year
37. a. The product of their slopes is -1.

b. ; no

c. ; The lines now appear to be perpendicular.

39. makes sense **41.** makes sense **43.** $-\dfrac{3}{7}$

44. a.

b.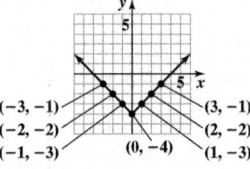

c. The graph in part (b) is the graph in part (a) shifted down 4 units.

45. a.

b.

c. The graph in part (b) is the graph in part (a) shifted to the right 2 units.

46. a. **b.** 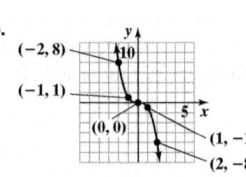 **c.** The graph in part (b) is the graph in part (a) reflected across the y-axis.

Mid-Chapter 1 Check Point

1. not a function; domain: $\{1, 2\}$; range: $\{-6, 4, 6\}$ **2.** function; domain: $\{0, 2, 3\}$; range: $\{1, 4\}$ **3.** function; domain: $\{x | -2 \le x < 2\}$ or $[-2, 2)$;
range: $\{y | 0 \le y \le 3\}$ or $[0, 3]$ **4.** not a function; domain: $\{x | -3 < x \le 4\}$ or $(-3, 4]$; range: $\{y | -1 \le y \le 2\}$ or $[-1, 2]$
5. not a function; domain: $\{-2, -1, 0, 1, 2\}$; range: $\{-2, -1, 1, 3\}$ **6.** function; domain: $\{x | x \le 1\}$ or $(-\infty, 1]$; range: $\{y | y \ge -1\}$ or $[-1, \infty)$
7. y is a function of x. **8.** y is not a function of x. **9.** No vertical line intersects the graph in more than one point. **10.** $(-\infty, \infty)$
11. $(-\infty, 4]$ **12.** -6 and 2 **13.** 3 **14.** $(-\infty, -2)$ **15.** $(-2, \infty)$ **16.** -2 **17.** 4 **18.** 3 **19.** -7 and 3 **20.** -6 and 2
21. $(-6, 2)$ **22.** negative **23.** neither **24.** -1

25. $y = -2x$ **26.** $y = -2$ **27.** $x + y = -2$ 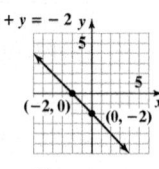 **28.** $y = \frac{1}{3}x - 2$ **29.** $x = 3.5$

30.
$4x - 2y = 8$ **31.**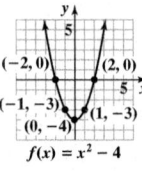
$f(x) = x^2 - 4$ **32.**
$f(x) = x - 4$ **33.**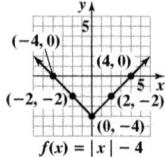
$f(x) = |x| - 4$ **34.**
$5y = -3x$

35.
$5y = 20$ **36.**
$f(x) = \begin{cases} -1 \text{ if } x \le 0 \\ 2x + 1 \text{ if } x > 0 \end{cases}$ **37. a.** $f(-x) = -2x^2 - x - 5$; neither **b.** $-4x - 2h + 1, h \ne 0$ **38. a.** 30 **b.** 50
39. $f(x) = -2x - 5$ **40.** $f(x) = 2x - 3$ **41.** $f(x) = 3x - 13$
42. $f(x) = -\frac{5}{2}x - 13$ **43.** The lines are parallel.
44. a. 0.16 **b.** 0.16; 0.16; minute of brisk walking **45.** 2

Section 1.6

Check Point Exercises

1. **2.** **3.** **4.** **5.**

6. $(-2, 4)$ **7. a.** **b.** **8.** **9.**
$g(x) = f(2x)$ $h(x) = f\left(\frac{1}{2}x\right)$ $y = -\frac{1}{3}f(x + 1) - 2$ $g(x) = 2(x - 1)^2 + 3$

Concept and Vocabulary Check

1. vertical; down **2.** horizontal; to the right **3.** x-axis **4.** y-axis **5.** vertical; y **6.** horizontal; x **7.** false

Exercise Set 1.6

1.
$g(x) = f(x) + 1$ **3.**
$g(x) = f(x + 1)$ **5.**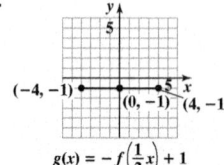
$g(x) = f(x - 1) - 2$ **7.**
$g(x) = f(-x)$ **9.**
$g(x) = -f(x) + 3$

11.
$g(x) = \frac{1}{2}f(x)$ **13.** $g(x) = f\left(\frac{1}{2}x\right)$ **15.**
$g(x) = -f\left(\frac{1}{2}x\right) + 1$ **17.**
$g(x) = f(x) - 1$ **19.**
$g(x) = f(x - 1)$

21.

$g(x) = f(x - 1) + 2$

23.

$g(x) = -f(x)$

25.

$g(x) = f(-x) + 1$

27.

$g(x) = 2f(x)$

29.

$g(x) = f(2x)$

31.

$g(x) = 2f(x + 2) + 1$

33.

$g(x) = f(x) + 2$

35.

$g(x) = f(x + 2)$

37.

$g(x) = -f(x + 2)$

39.

$g(x) = -\frac{1}{2}f(x + 2)$

41.

$g(x) = -\frac{1}{2}f(x + 2) - 2$

43.

$g(x) = \frac{1}{2}f(2x)$

45.

$g(x) = f(x - 1) - 1$

47.

$g(x) = -f(x - 1) + 1$

49.

$g(x) = 2f\left(\frac{1}{2}x\right)$

51.

$g(x) = \frac{1}{2}f(x + 1)$

53.

55.

57.

59.

61.

63.

65.

67.

69.

71.

73.

75.

77.

79.

81.

83.

85.

87.

89.

91.

93.

95.

97.

99.

101.

103.

105.

107.

109.

111.

113.

115.

117.

119.

$g(x) = 2\,\text{int}\,(x + 1)$

121. (−1, 2)

$h(x) = \text{int}(−x) + 1$

123. $y = \sqrt{x − 2}$ **125.** $y = (x + 1)^2 − 4$
127. a. First, vertically stretch the graph of $f(x) = \sqrt{x}$ by the factor 2.9; then, shift the result up 20.1 units. **b.** 40.2 in.; very well
c. 0.9 in. per month **d.** 0.2 in. per month; This is a much smaller rate of change; The graph is not as steep between 50 and 60 as it is between 0 and 10.

135. a.

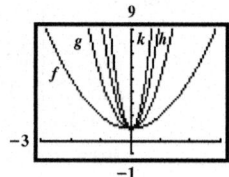

b.

137. makes sense **139.** does not make sense **141.** false **143.** false **145.** $g(x) = −(x + 4)^2$ **147.** $g(x) = −\sqrt{x − 2} + 2$
149. (−a, b) **151.** (a + 3, b) **153.** $2x^3 + x^2 − 5x + 2$ **154.** $9x^2 − 30x + 30$ **155.** $\dfrac{2x}{3 − x}$

Section 1.7

Check Point Exercises

1. a. (−∞, ∞) **b.** (−∞, −5) ∪ (−5, ∞) **c.** [2, ∞) **2. a.** $(f + g)(x) = x^2 + x − 6; (−∞, ∞)$
b. $(f − g)(x) = −x^2 + x − 4; (−∞, ∞)$ **c.** $(fg)(x) = x^3 − 5x^2 − x + 5; (−∞, ∞)$ **d.** $\left(\dfrac{f}{g}\right)(x) = \dfrac{x − 5}{x^2 − 1}; (−∞, −1) ∪ (−1, 1) ∪ (1, ∞)$
3. a. $(f + g)(x) = \sqrt{x − 3} + \sqrt{x + 1}$ **b.** [3, ∞) **4. a.** $(B + D)(x) = −3.2x^2 + 56x + 6406$ **b.** 6545.2 thousand **c.** overestimates by
7.2 thousand **5. a.** $(f \circ g)(x) = 10x^2 − 5x + 1$ **b.** $(g \circ f)(x) = 50x^2 + 115x + 65$ **c.** 16 **6. a.** $(f \circ g)(x) = \dfrac{4x}{1 + 2x}$
b. $\left(−∞, −\dfrac{1}{2}\right) ∪ \left(−\dfrac{1}{2}, 0\right) ∪ (0, ∞)$ **7.** If $f(x) = \sqrt{x}$ and $g(x) = x^2 + 5$, then $h(x) = (f \circ g)(x)$.

Concept and Vocabulary Check

1. zero **2.** negative **3.** $f(x) + g(x)$ **4.** $f(x) − g(x)$ **5.** $f(x) \cdot g(x)$ **6.** $\dfrac{f(x)}{g(x)}; g(x)$ **7.** (−∞, ∞) **8.** (2, ∞) **9.** (0, 3); (3, ∞)
10. composition; $f(g(x))$ **11.** $f; g(x)$ **12.** composition; $g(f(x))$ **13.** $g; f(x)$ **14.** false **15.** false **16.** 2

Exercise Set 1.7

1. (−∞, ∞) **3.** (−∞, −4) ∪ (−4, ∞) **5.** (−∞, 3) ∪ (3, ∞) **7.** (−∞, −7) ∪ (−7, 9) ∪ (9, ∞) **9.** (−∞, 0) ∪ (0, 3) ∪ (3, ∞)
11. (−∞, 1) ∪ (1, 3) ∪ (3, ∞) **13.** [−4, ∞) **15.** (4, ∞) **17.** (−∞, 4] **19.** [2, ∞)
21. $(f + g)(x) = 3x + 2$; domain: (−∞, ∞); $(f − g)(x) = x + 4$; domain: (−∞, ∞); $(fg)(x) = 2x^2 + x − 3$; domain: (−∞, ∞); $\left(\dfrac{f}{g}\right)(x) = \dfrac{2x + 3}{x − 1}$;
domain: (−∞, 1) ∪ (1, ∞) **23.** $(f + g)(x) = 3x^2 + x − 5$; domain: (−∞, ∞); $(f − g)(x) = −3x^2 + x − 5$; domain: (−∞, ∞); $(fg)(x) = 3x^3 − 15x^2$;
domain: (−∞, ∞); $\left(\dfrac{f}{g}\right)(x) = \dfrac{x − 5}{3x^2}$; domain: (−∞, 0) ∪ (0, ∞) **25.** $(f + g)(x) = 2x^2 − 2$; domain: (−∞, ∞); $(f − g)(x) = 2x^2 − 2x − 4$;
domain: (−∞, ∞); $(fg)(x) = 2x^3 + x^2 − 4x − 3$; domain: (−∞, ∞); $\left(\dfrac{f}{g}\right)(x) = 2x − 3$; domain: (−∞, −1) ∪ (−1, ∞)
27. $(f + g)(x) = 2x − 12$; domain: (−∞, ∞); $(f − g)(x) = −2x^2 − 2x + 18$; domain: (−∞, ∞); $(fg)(x) = −x^4 − 2x^3 + 18x^2 + 6x − 45$;
domain: (−∞, ∞); $\left(\dfrac{f}{g}\right)(x) = \dfrac{3 − x^2}{x^2 + 2x − 15}$; domain: (−∞, −5) ∪ (−5, 3) ∪ (3, ∞)
29. $(f + g)(x) = \sqrt{x} + x − 4$; domain: [0, ∞); $(f − g)(x) = \sqrt{x} − x + 4$; domain: [0, ∞); $(fg)(x) = \sqrt{x}(x − 4)$; domain: [0, ∞); $\left(\dfrac{f}{g}\right)(x) = \dfrac{\sqrt{x}}{x − 4}$;
domain: [0, 4) ∪ (4, ∞) **31.** $(f + g)(x) = \dfrac{2x + 2}{x}$; domain: (−∞, 0) ∪ (0, ∞); $(f − g)(x) = 2$; domain: (−∞, 0) ∪ (0, ∞); $(fg)(x) = \dfrac{2x + 1}{x^2}$;
domain: (−∞, 0) ∪ (0, ∞); $\left(\dfrac{f}{g}\right)(x) = 2x + 1$; domain: (−∞, 0) ∪ (0, ∞) **33.** $(f + g)(x) = \dfrac{9x − 1}{x^2 − 9}$; domain:
(−∞, −3) ∪ (−3, 3) ∪ (3, ∞); $(f − g)(x) = \dfrac{x + 3}{x^2 − 9} = \dfrac{1}{x − 3}$; domain: (−∞, −3) ∪ (−3, 3) ∪ (3, ∞); $(fg)(x) = \dfrac{20x^2 − 6x − 2}{(x^2 − 9)^2}$;
domain: (−∞, −3) ∪ (−3, 3) ∪ (3, ∞); $\left(\dfrac{f}{g}\right)(x) = \dfrac{5x + 1}{4x − 2}$; domain: (−∞, −3) ∪ $\left(−3, \dfrac{1}{2}\right)$ ∪ $\left(\dfrac{1}{2}, 3\right)$ ∪ (3, ∞) **35.** $(f + g)(x) = \sqrt{x + 4} + \sqrt{x − 1}$;
domain: [1, ∞); $(f − g)(x) = \sqrt{x + 4} − \sqrt{x − 1}$; domain: [1, ∞); $(fg)(x) = \sqrt{x^2 + 3x − 4}$; domain: [1, ∞); $\left(\dfrac{f}{g}\right)(x) = \dfrac{\sqrt{x + 4}}{\sqrt{x − 1}}$; domain: (1, ∞)
37. $(f + g)(x) = \sqrt{x − 2} + \sqrt{2 − x}$; domain: {2}; $(f − g)(x) = \sqrt{x − 2} − \sqrt{2 − x}$; domain: {2}; $(fg)(x) = \sqrt{x − 2} \cdot \sqrt{2 − x}$; domain:
{2}; $\left(\dfrac{f}{g}\right)(x) = \dfrac{\sqrt{x − 2}}{\sqrt{2 − x}}$; domain: ∅ **39. a.** $(f \circ g)(x) = 2x + 14$ **b.** $(g \circ f)(x) = 2x + 7$ **c.** $(f \circ g)(2) = 18$ **d.** $(g \circ f)(2) = 11$

41. a. $(f \circ g)(x) = 2x + 5$ **b.** $(g \circ f)(x) = 2x + 9$ **c.** $(f \circ g)(2) = 9$ **d.** $(g \circ f)(2) = 13$ **43. a.** $(f \circ g)(x) = 20x^2 - 11$
b. $(g \circ f)(x) = 80x^2 - 120x + 43$ **c.** $(f \circ g)(2) = 69$ **d.** $(g \circ f)(2) = 123$ **45. a.** $(f \circ g)(x) = x^4 - 4x^2 + 6$ **b.** $(g \circ f)(x) = x^4 + 4x^2 + 2$
c. $(f \circ g)(2) = 6$ **d.** $(g \circ f)(2) = 34$ **47. a.** $(f \circ g)(x) = -2x^2 - x - 1$ **b.** $(g \circ f)(x) = 2x^2 - 17x + 41$ **c.** $(f \circ g)(2) = -11$ **d.** $(g \circ f)(2) = 15$
49. a. $(f \circ g)(x) = \sqrt{x} - 1$ **b.** $(g \circ f)(x) = \sqrt{x - 1}$ **c.** $(f \circ g)(2) = 1$ **d.** $(g \circ f)(2) = \sqrt{2} - 1$ **51. a.** $(f \circ g)(x) = x$ **b.** $(g \circ f)(x) = x$
c. $(f \circ g)(2) = 2$ **d.** $(g \circ f)(2) = 2$ **53. a.** $(f \circ g)(x) = x$ **b.** $(g \circ f)(x) = x$ **c.** 2 **d.** $(g \circ f)(2) = 2$ **55. a.** $(f \circ g)(x) = \dfrac{2x}{1 + 3x}$
b. $\left(-\infty, -\dfrac{1}{3}\right) \cup \left(-\dfrac{1}{3}, 0\right) \cup (0, \infty)$ **57. a.** $(f \circ g)(x) = \dfrac{4}{4 + x}$ **b.** $(-\infty, -4) \cup (-4, 0) \cup (0, \infty)$ **59. a.** $(f \circ g)(x) = \sqrt{x - 2}$ **b.** $[2, \infty)$
61. a. $(f \circ g)(x) = 5 - x$ **b.** $(-\infty, 1]$ **63.** $f(x) = x^4, g(x) = 3x - 1$ **65.** $f(x) = \sqrt[3]{x}, g(x) = x^2 - 9$ **67.** $f(x) = |x|, g(x) = 2x - 5$
69. $f(x) = \dfrac{1}{x}, g(x) = 2x - 3$ **71.** 5 **73.** -1 **75.** $[-4, 3]$

77. **79.** 1 **81.** -6 **83. a.** $(M + F)(x) = 2.99x + 235.5$ **b.** 295.3 million **c.** underestimates by 2.7 million

85. $(R - C)(20,000) = -200,000$; The company loses $200,000 when 20,000 radios are sold.; $(R - C)(30,000) = 0$; The company breaks even when 30,000 radios are sold.; $(R - C)(40,000) = 200,000$; The company makes a profit of $200,000 when 40,000 radios are sold. **87. a.** f gives the price of the computer after a $400 discount. g gives the price of the computer after a 25% discount. **b.** $(f \circ g)(x) = 0.75x - 400$; This models the price of a computer after first a 25% discount and then a $400 discount. **c.** $(g \circ f)(x) = 0.75(x - 400)$; This models the price of a computer after first a $400 discount and then a 25% discount. **d.** $f \circ g$ because $0.75x - 400 < 0.75(x - 400)$

95.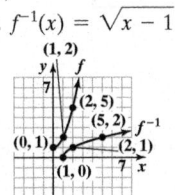
Domain: $[0, 4]$

97. makes sense **99.** does not make sense **101.** false **103.** true
105. Answers will vary; One possible answer is $f(x) = x + 1$ and $g(x) = x - 1$.
106. $\{(4, -2), (1, -1), (1, 1), (4, 2)\}$; no **107.** $y = \dfrac{5}{x - 4}$ **108.** $\{\pm\sqrt{5}\}$

Section 1.8

Check Point Exercises

1. $f(g(x)) = 4\left(\dfrac{x + 7}{4}\right) - 7 = x + 7 - 7 = x; g(f(x)) = \dfrac{(4x - 7) + 7}{4} = \dfrac{4x}{4} = x$ **2.** $f^{-1}(x) = \dfrac{x - 7}{2}$ **3.** $f^{-1}(x) = \sqrt[3]{\dfrac{x + 1}{4}}$
4. $f^{-1}(x) = \dfrac{3}{x + 1}$ **5.** (b) and (c)
6. **7.** $f^{-1}(x) = \sqrt{x} - 1$

Concept and Vocabulary Check

1. inverse **2.** $x; x$ **3.** horizontal; one-to-one **4.** $y = x$

Exercise Set 1.8

1. $f(g(x)) = x; g(f(x)) = x; f$ and g are inverses. **3.** $f(g(x)) = x; g(f(x)) = x; f$ and g are inverses.
5. $f(g(x)) = \dfrac{5x - 56}{9}; g(f(x)) = \dfrac{5x - 4}{9}; f$ and g are not inverses. **7.** $f(g(x)) = x; g(f(x)) = x; f$ and g are inverses.
9. $f(g(x)) = x; g(f(x)) = x; f$ and g are inverses. **11.** $f^{-1}(x) = x - 3$ **13.** $f^{-1}(x) = \dfrac{x}{2}$ **15.** $f^{-1}(x) = \dfrac{x - 3}{2}$ **17.** $f^{-1}(x) = \sqrt[3]{x} - 2$
19. $f^{-1}(x) = \sqrt[3]{x - 2}$ **21.** $f^{-1}(x) = \dfrac{1}{x}$ **23.** $f^{-1}(x) = x^2, x \geq 0$ **25.** $f^{-1}(x) = \dfrac{7}{x + 3}$ **27.** $f^{-1}(x) = \dfrac{3x + 1}{x - 2}; x \neq 2$
29. The function is not one-to-one, so it does not have an inverse function. **31.** The function is not one-to-one, so it does not have an inverse function.
33. The function is one-to-one, so it does have an inverse function.
35. **37.**

39. a. $f^{-1}(x) = \dfrac{x+1}{2}$

b.

c. domain of f = range of $f^{-1} = (-\infty, \infty)$; range of f = domain of $f^{-1} = (-\infty, \infty)$

41. a. $f^{-1}(x) = \sqrt{x+4}$

b.

c. domain of f = range of $f^{-1} = [0, \infty)$; range of f = domain of $f^{-1} = [-4, \infty)$

43. a. $f^{-1}(x) = -\sqrt{x} + 1$

b.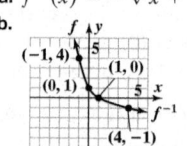

d. domain of f = range of $f^{-1} = (-\infty, 1]$; range of f = domain of $f^{-1} = [0, \infty)$

45. a. $f^{-1}(x) = \sqrt[3]{x} + 1$

b.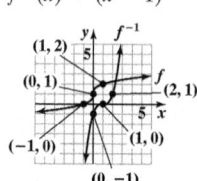

c. domain of f = range of $f^{-1} = (-\infty, \infty)$; range of f = domain of $f^{-1} = (-\infty, \infty)$

47. a. $f^{-1}(x) = \sqrt[3]{x} - 2$

b.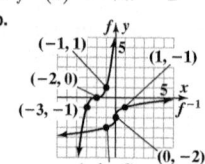

c. domain of f = range of $f^{-1} = (-\infty, \infty)$; range of f = domain of $f^{-1} = (-\infty, \infty)$

49. a. $f^{-1}(x) = x^2 + 1, x \geq 0$

b.

c. domain of f = range of $f^{-1} = [1, \infty)$; range of f = domain of $f^{-1} = [0, \infty)$

51. a. $f^{-1}(x) = (x-1)^3$

b.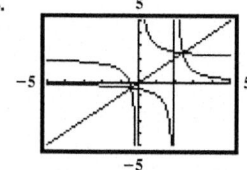

c. domain of f = range of $f^{-1} = (-\infty, \infty)$; range of f = domain of $f^{-1} = (-\infty, \infty)$

53. 5 **55.** 1 **57.** 2 **59.** −7 **61.** 3 **63.** 11
65. a. {(Zambia, 4.2), (Colombia, 4.5), (Poland, 3.3), (Italy, 3.3), (United States, 2.5)}
b. {(4.2, Zambia), (4.5, Colombia), (3.3, Poland), (3.3, Italy), (2.5, United States)}; no; The inverse of f is not a function. **67. a.** f is a one-to-one function. **b.** $f^{-1}(0.25)$ is the number of people in a room for a 25% probability of two people sharing a birthday. $f^{-1}(0.5)$ is the number of people in a room for a 50% probability of two people sharing a birthday. $f^{-1}(0.7)$ is the number of people in a room for a 70% probability of two people sharing a birthday.

69. $f(g(x)) = \dfrac{9}{5}\left[\dfrac{5}{9}(x-32)\right] + 32 = x$ and $g(f(x)) = \dfrac{5}{9}\left[\left(\dfrac{9}{5}x + 32\right) - 32\right] = x$

77.

one-to-one

79.

not one-to-one

81.

not one-to-one

83.

not one-to-one

85.

f and g are inverses.

87. makes sense **89.** makes sense **91.** false **93.** false

95. $(f \circ g)^{-1}(x) = \dfrac{x-15}{3}$; $(g^{-1} \circ f^{-1})(x) = \dfrac{x}{3} - 5 = \dfrac{x-15}{3}$

97. No; The spacecraft was at the same height, $s(t)$, for two different values of t-once during the ascent and once again during the descent.

100. $3\sqrt{5}$ **101.** **102. a.** $4; x^2 + 4x + 4 = (x+2)^2$ **b.** $9; y^2 - 6y + 9 = (y-3)^2$

Section 1.9

Check Point Exercises

1. $3\sqrt{5} \approx 6.71$ **2.** $\left(4, -\dfrac{1}{2}\right)$ **3.** $x^2 + y^2 = 16$ **4.** $x^2 + (y+6)^2 = 100$

5. a. center: $(-3, 1)$; radius: 2

b.

$(x+3)^2 + (y-1)^2 = 4$

c. domain: $[-5, -1]$
range: $[-1, 3]$

6. $(x+2)^2 + (y-2)^2 = 9$

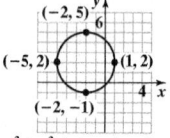

$x^2 + y^2 + 4x - 4y - 1 = 0$

Concept and Vocabulary Check

1. $\sqrt{(x_2 - x_1)^2 + (y_2 - y_1)^2}$ **2.** $\dfrac{x_1 + x_2}{2}, \dfrac{y_1 + y_2}{2}$ **3.** circle; center; radius **4.** $(x - h)^2 + (y - k)^2 = r^2$ **5.** general **6.** 4; 16

Exercise Set 1.9

1. 13 **3.** $2\sqrt{29} \approx 10.77$ **5.** 5 **7.** $\sqrt{29} \approx 5.39$ **9.** $4\sqrt{2} \approx 5.66$ **11.** $2\sqrt{5} \approx 4.47$ **13.** $2\sqrt{2} \approx 2.83$ **15.** $\sqrt{93} \approx 9.64$

17. $\sqrt{5} \approx 2.24$ **19.** $(4, 6)$ **21.** $(-4, -5)$ **23.** $\left(\dfrac{3}{2}, -6\right)$ **25.** $(-3, -2)$ **27.** $(1, 5\sqrt{5})$ **29.** $(2\sqrt{2}, 0)$ **31.** $x^2 + y^2 = 49$

33. $(x - 3)^2 + (y - 2)^2 = 25$ **35.** $(x + 1)^2 + (y - 4)^2 = 4$ **37.** $(x + 3)^2 + (y + 1)^2 = 3$ **39.** $(x + 4)^2 + y^2 = 100$

41. center: $(0, 0)$ **43.** center: $(3, 1)$ **45.** center: $(-3, 2)$ **47.** center: $(-2, -2)$ **49.** center: $(0, 1)$
 radius: 4 radius: 6 radius: 2 radius: 2 radius: 1
 domain: $[-4, 4]$ domain: $[-3, 9]$ domain: $[-5, -1]$ domain: $[-4, 0]$ domain: $[-1, 1]$
 range: $[-4, 4]$ range: $[-5, 7]$ range: $[0, 4]$ range: $[-4, 0]$ range: $[0, 2]$

51. center: $(-1, 0)$ **53.** $(x + 3)^2 + (y + 1)^2 = 4$ **55.** $(x - 5)^2 + (y - 3)^2 = 64$ **57.** $(x + 4)^2 + (y - 1)^2 = 25$
 radius: 5 center: $(-3, -1)$ center: $(5, 3)$ center: $(-4, 1)$
 domain: $[-6, 4]$ radius: 2 radius: 8 radius: 5
 range: $[-5, 5]$

 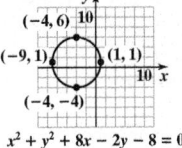

59. $(x - 1)^2 + y^2 = 16$ **61.** $\left(x - \dfrac{1}{2}\right)^2 + (y + 1)^2 = \dfrac{1}{4}$ **63.** $\left(x + \dfrac{3}{2}\right)^2 + (y - 1)^2 = \dfrac{17}{4}$ **65. a.** $(5, 10)$ **b.** $\sqrt{5}$
 c. $(x - 5)^2 + (y - 10)^2 = 5$
 center: $(1, 0)$ center: $\left(\dfrac{1}{2}, -1\right)$ center: $\left(-\dfrac{3}{2}, 1\right)$ **67.** $\{(0, -4), (4, 0)\}$

 radius: 4 radius: $\dfrac{1}{2}$ radius: $\dfrac{\sqrt{17}}{2}$

69. $\{(0, -3), (2, -1)\}$ **71.** 2693 mi **73.** $(x + 2.4)^2 + (y + 2.7)^2 = 900$ **83.** **85.** makes sense
87. does not makes sense
89. false **91.** false
93. $2\sqrt{2} + 3\sqrt{2} = 5\sqrt{2}$
95. 11π **97.** $\{9\}$
98. $-x + 10$
99. $9x - 24$

Chapter 1 Review Exercises

1. **2.** **3.** **4.** **5.**

6. x-intercept: -2; y-intercept: 2 **7.** x-intercepts: $-2, 2$; y-intercept: -4 **8.** x-intercept: 5; no y-intercept
9. $(20, 8)$; 8% of college students anticipated a starting salary of $20 thousand. **10.** $30 thousand; 22% **11.** $70 thousand; 2%
12. $25 thousand and $30 thousand **13.** 14% **14.** 19.7%; It's greater than the estimate. **15.** function; domain: $\{2, 3, 5\}$; range: $\{7\}$

16. function; domain: $\{1, 2, 13\}$; range: $\{10, 500, \pi\}$ **17.** not a function; domain: $\{12, 14\}$; range: $\{13, 15, 19\}$ **18.** y is a function of x.
19. y is a function of x. **20.** y is not a function of x. **21. a.** $f(4) = -23$ **b.** $f(x + 3) = -7x - 16$ **c.** $f(-x) = 5 + 7x$
22. a. $g(0) = 2$ **b.** $g(-2) = 24$ **c.** $g(x - 1) = 3x^2 - 11x + 10$ **d.** $g(-x) = 3x^2 + 5x + 2$ **23. a.** $g(13) = 3$ **b.** $g(0) = 4$
c. $g(-3) = 7$ **24. a.** -1 **b.** 12 **c.** 3 **25.** not a function **26.** function **27.** function **28.** not a function
29. not a function **30.** function **31. a.** $[-3, 5)$ **b.** $[-5, 0]$ **c.** -3 **d.** -2 **e.** increasing: $(-2, 0)$ or $(3, 5)$; decreasing: $(-3, -2)$
or $(0, 3)$ **f.** $f(-2) = -3$ and $f(3) = -5$ **32. a.** $(-\infty, \infty)$ **b.** $(-\infty, 3]$ **c.** -2 and 3 **d.** 3 **e.** increasing: $(-\infty, 0)$;
decreasing: $(0, \infty)$ **f.** $f(-2) = 0$ and $f(6) = -3$ **33. a.** $(-\infty, \infty)$ **b.** $[-2, 2]$ **c.** 0 **d.** 0 **e.** increasing: $(-2, 2)$;
constant: $(-\infty, -2)$ or $(2, \infty)$ **f.** $f(-9) = -2$ and $f(14) = 2$ **34. a.** $0; f(0) = -2$ **b.** $-2, 3; f(-2) = -3, f(3) = -5$
35. a. $0; f(0) = 3$ **b.** none **36.** odd; symmetric with respect to the origin **37.** even; symmetric with respect to the y-axis
38. odd; symmetric with respect to the origin

39. $; \{-3, 5\}$ **40.** $; (-\infty, 0]$

$$f(x) = \begin{cases} 5 \text{ if } x \le -1 \\ -3 \text{ if } x > -1 \end{cases}$$

$$f(x) = \begin{cases} 2x \text{ if } x < 0 \\ -x \text{ if } x \ge 0 \end{cases}$$

41. 8 **42.** $-4x - 2h + 1$ **43. a.** yes; The graph passes the vertical line test. **b.** $(3, 12)$; The eagle descended.
c. $(0, 3)$ and $(12, 17)$; The eagle's height held steady during the first 3 seconds and the eagle was on the ground for 5 seconds.
d. $(17, 30)$; The eagle was ascending.

44.

45. $-\dfrac{1}{2}$; falls **46.** 1; rises **47.** 0; horizontal **48.** undefined; vertical
49. $y - 2 = -6(x + 3); y = -6x - 16$ **50.** using $(1, 6), y - 6 = 2(x - 1); y = 2x + 4$
51. $y + 7 = -3(x - 4); y = -3x + 5$ **52.** $y - 6 = -3(x + 3); y = -3x - 3$ **53.** $x + 6y + 18 = 0$

54. Slope: $\dfrac{2}{5}$; y-intercept: -1 **55.** Slope: -4; y-intercept: 5 **56.** Slope: $-\dfrac{2}{3}$; y-intercept: -2 **57.** Slope: 0; y-intercept: 4

$y = \dfrac{2}{5}x - 1$

$f(x) = -4x + 5$

$2x + 3y + 6 = 0$

$2y - 8 = 0$

58.

$2x - 5y - 10 = 0$

59. $2x - 10 = 0$

60. a. $y - 2.3 = 0.116(x - 15)$ or $y - 11 = 0.116(x - 90)$ **b.** $f(x) = 0.116x + 0.56$ **c.** approximately 5 deaths per 100,000 persons
d. 4.3 deaths per 100,000 persons; underestimates by 0.7 death per 100,000 persons; The line passes below the point for France.
61. a. -0.48 **b.** $0.48; -0.48\%$; year **62.** 10

63. **64.** **65.** **66.** **67.**

68. **69.** **70.** **71.** **72.**

73.

74.

75.

76.

77.

78.

79.

80.

81.

82. $\left(-\dfrac{1}{2}, 1\right)$

83.

84. $(-\infty, \infty)$ **85.** $(-\infty, 7) \cup (7, \infty)$ **86.** $(-\infty, 4]$ **87.** $(-\infty, -7) \cup (-7, 3) \cup (3, \infty)$

88. $[2, 5) \cup (5, \infty)$ **89.** $[1, \infty)$ **90.** $(f + g)(x) = 4x - 6$; domain: $(-\infty, \infty)$; $(f - g)(x) = 2x + 4$;

domain: $(-\infty, \infty)$; $(fg)(x) = 3x^2 - 16x + 5$; domain: $(-\infty, \infty)$; $\left(\dfrac{f}{g}\right)(x) = \dfrac{3x - 1}{x - 5}$; domain: $(-\infty, 5) \cup (5, \infty)$

91. $(f + g)(x) = 2x^2 + x$; domain: $(-\infty, \infty)$; $(f - g)(x) = x + 2$; domain: $(-\infty, \infty)$; $(fg)(x) = x^4 + x^3 - x - 1$;

domain: $(-\infty, \infty)$; $\left(\dfrac{f}{g}\right)(x) = \dfrac{x^2 + x + 1}{x^2 - 1}$; domain: $(-\infty, -1) \cup (-1, 1) \cup (1, \infty)$

92. $(f + g)(x) = \sqrt{x + 7} + \sqrt{x - 2}$; domain: $[2, \infty)$; $(f - g)(x) = \sqrt{x + 7} - \sqrt{x - 2}$; domain: $[2, \infty)$; $(fg)(x) = \sqrt{x^2 + 5x - 14}$; domain:

$[2, \infty)$; $\left(\dfrac{f}{g}\right)(x) = \dfrac{\sqrt{x + 7}}{\sqrt{x - 2}}$; domain: $(2, \infty)$ **93. a.** $(f \circ g)(x) = 16x^2 - 8x + 4$ **b.** $(g \circ f)(x) = 4x^2 + 11$ **c.** $(f \circ g)(3) = 124$

94. a. $(f \circ g)(x) = \sqrt{x + 1}$ **b.** $(g \circ f)(x) = \sqrt{x} + 1$ **c.** $(f \circ g)(3) = 2$ **95. a.** $(f \circ g)(x) = \dfrac{1 + x}{1 - 2x}$ **b.** $(-\infty, 0) \cup \left(0, \dfrac{1}{2}\right) \cup \left(\dfrac{1}{2}, \infty\right)$

96. a. $(f \circ g)(x) = \sqrt{x + 2}$ **b.** $[-2, \infty)$ **97.** $f(x) = x^4, g(x) = x^2 + 2x - 1$ **98.** $f(x) = \sqrt[3]{x}, g(x) = 7x + 4$ **99.** $f(g(x)) = x - \dfrac{7}{10}$;

$g(f(x)) = x - \dfrac{7}{6}$; f and g are not inverses of each other. **100.** $f(g(x)) = x$; $g(f(x)) = x$; f and g are inverses of each other. **101.** $f^{-1}(x) = \dfrac{x + 3}{4}$

102. $f^{-1}(x) = \sqrt[3]{\dfrac{x - 1}{8}}$ or $\dfrac{\sqrt[3]{x - 1}}{2}$ **103.** $f^{-1}(x) = \dfrac{2}{x - 5}$ **104.** Inverse function exists. **105.** Inverse function does not exist.

106. Inverse function exists. **107.** Inverse function does not exist.

108.

109. $f^{-1}(x) = \sqrt{1 - x}$

110. $f^{-1}(x) = (x - 1)^2, x \geq 1$

111. 13 **112.** $2\sqrt{2} \approx 2.83$

113. $(-5, 5)$ **114.** $\left(-\dfrac{11}{2}, -2\right)$

115. $x^2 + y^2 = 9$

116. $(x + 2)^2 + (y - 4)^2 = 36$

117. center: $(0, 0)$; radius: 1
domain: $[-1, 1]$; range: $[-1, 1]$

$x^2 + y^2 = 1$

118. center: $(-2, 3)$; radius: 3
domain: $[-5, 1]$; range: $[0, 6]$

$(x + 2)^2 + (y - 3)^2 = 9$

119. center: $(2, -1)$; radius: 3
domain: $[-1, 5]$; range: $[-4, 2]$

$x^2 + y^2 - 4x + 2y - 4 = 0$

Chapter 1 Test

1. b, c, d **2. a.** $f(4) - f(-3) = 5$ **b.** $(-5, 6]$ **c.** $[-4, 5]$ **d.** $(-1, 2)$ **e.** $(-5, -1)$ or $(2, 6)$ **f.** 2; $f(2) = 5$ **g.** -1; $f(-1) = -4$
h. $-4, 1$, and 5 **i.** -3 **3. a.** -2 and 2 **b.** -1 and 1 **c.** 0 **d.** even **e.** no **f.** relative minimum

g.

h.

i.

j. $-\dfrac{1}{3}$

4. $x + y = 4$

domain: $(-\infty, \infty)$;
range: $(-\infty, \infty)$

5. $x^2 + y^2 = 4$

domain: $[-2, 2]$;
range: $[-2, 2]$

6. $f(x) = 4$

domain: $(-\infty, \infty)$;
range: $\{4\}$

7.

$f(x) = -\dfrac{1}{3}x + 2$

domain: $(-\infty, \infty)$;
range: $(-\infty, \infty)$

8.

$(x + 2)^2 + (y - 1)^2 = 9$

domain: $[-5, 1]$;
range: $[-2, 4]$

9.

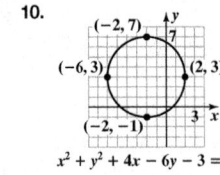

$f(x) = \begin{cases} 2 \text{ if } x \le 0 \\ -1 \text{ if } x > 0 \end{cases}$

domain: $(-\infty, \infty)$;
range: $\{-1, 2\}$

10.

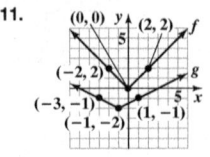

$x^2 + y^2 + 4x - 6y - 3 = 0$

domain: $[-6, 2]$;
range: $[-1, 7]$

11.

domain of f = domain of $g = (-\infty, \infty)$;
range of $f = [0, \infty)$; range of $g = [-2, \infty)$

12.

domain of f = domain of $g = (-\infty, \infty)$;
range of $f = [0, \infty)$; range of $g = (-\infty, 4]$

13.

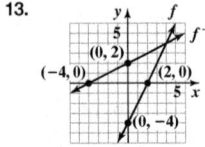

domain of f = range of $f^{-1} = (-\infty, \infty)$;
range of f = domain of $f^{-1} = (-\infty, \infty)$

14.

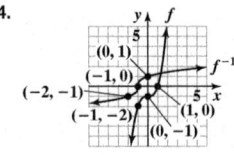

domain of f = range of $f^{-1} = (-\infty, \infty)$;
range of f = domain of $f^{-1} = (-\infty, \infty)$

15.

domain of f = range of $f^{-1} = [0, \infty)$;
range of f = domain of $f^{-1} = [-1, \infty)$

16. $f(x - 1) = x^2 - 3x - 2$ **17.** $2x + h - 1$ **18.** $(g - f)(x) = -x^2 + 3x - 2$

19. $\left(\dfrac{f}{g}\right)(x) = \dfrac{x^2 - x - 4}{2x - 6}; (-\infty, 3) \cup (3, \infty)$ **20.** $(f \circ g)(x) = 4x^2 - 26x + 38$ **21.** $(g \circ f)(x) = 2x^2 - 2x - 14$

22. -10 **23.** $f(-x) = x^2 + x - 4$; neither **24.** using $(2, 1)$, $y - 1 = 3(x - 2)$; $y = 3x - 5$

25. $y - 6 = 4(x + 4)$; $y = 4x + 22$ **26.** $2x + y + 24 = 0$ **27. a.** using $(1, 4571)$, $y - 4571 = 433(x - 1)$

b. $f(x) = 433x + 4138$ **c.** 8468 deaths **28.** 48 **29.** $g(-1) = 4$; $g(7) = 2$ **30.** $(-\infty, -5) \cup (-5, 1) \cup (1, \infty)$

31. $[1, \infty)$ **32.** $\dfrac{7x}{2 - 4x}$; domain: $(-\infty, 0) \cup \left(0, \dfrac{1}{2}\right) \cup \left(\dfrac{1}{2}, \infty\right)$ **33.** $f(x) = x^7, g(x) = 2x + 3$ **34.** $5; \left(\dfrac{7}{2}, 0\right)$ or $(3.5, 0)$

CHAPTER 2

Section 2.1

Check Point Exercises

1. $\{5\}$ **2.** $\{1\}$ **3.** $\{3\}$ **4.** \varnothing **5.** 11 **6.** \varnothing; inconsistent equation **7.** $\{x | x \text{ is a real number}\}$; identity **8.** 3.7; by the point $(3.7, 10)$

Concept and Vocabulary Check

1. linear **2.** equivalent **3.** apply the distributive property **4.** least common denominator; 12 **5.** 0 **6.** $2x$ **7.** $(x + 5)(x + 1)$
8. $x \ne 2; x \ne -4$ **9.** $5(x + 3) + 3(x + 4) = 12x + 9$ **10.** identity **11.** inconsistent

Exercise Set 2.1

1. $\{11\}$ **3.** $\{7\}$ **5.** $\{13\}$ **7.** $\{2\}$ **9.** $\{9\}$ **11.** $\{-5\}$ **13.** $\{6\}$ **15.** $\{-2\}$ **17.** $\{12\}$ **19.** $\{24\}$ **21.** $\{-15\}$ **23.** $\{5\}$ **25.** $\left\{\dfrac{33}{2}\right\}$

27. $\{-12\}$ **29.** $\left\{\dfrac{46}{5}\right\}$ **31. a.** 0 **b.** $\left\{\dfrac{1}{2}\right\}$ **33. a.** 0 **b.** $\{-2\}$ **35. a.** 0 **b.** $\{2\}$ **37. a.** 0 **b.** $\{4\}$ **39. a.** 1 **b.** $\{3\}$

41. a. -1 **b.** \varnothing **43. a.** 1 **b.** $\{2\}$ **45. a.** $-2, 2$ **b.** \varnothing **47. a.** $-1, 1$ **b.** $\{-3\}$ **49. a.** $-2, 4$ **b.** \varnothing **51.** 6
53. -7 **55.** 2 **57.** 19 **59.** -1 **61.** $\{x | x \text{ is a real number}\}$; identity **63.** \varnothing; inconsistent equation **65.** $\{0\}$; conditional equation
67. \varnothing; inconsistent equation **69.** $\{-7\}$; conditional equation **71.** \varnothing; inconsistent equation **73.** $\{-4\}$; conditional equation
75. $\{8\}$; conditional equation **77.** $\{-1\}$; conditional equation **79.** \varnothing; inconsistent equation **81.** $3(x - 4) = 3(2 - 2x)$; $\{2\}$
83. $-3(x - 3) = 5(2 - x)$; $\{0.5\}$ **85.** 2 **87.** -4 **89.** 7 **91. a.** underestimates by 2% **b.** 2030 **93. a.** \$52,000
b. \$52,723; reasonably well **c.** \$52,628; reasonably well **95.** 2018 **97.** 11 learning trials; $(11, 0.95)$ **99.** 125 liters

111. $\{5\}$

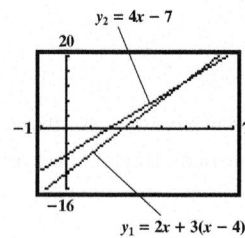

$y_2 = 4x - 7$

$y_1 = 2x + 3(x - 4)$

113. $\{-5\}$

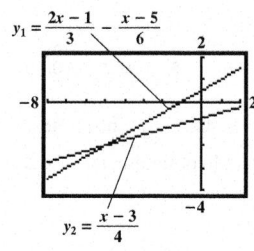

$y_1 = \dfrac{2x - 1}{3} - \dfrac{x - 5}{6}$

$y_2 = \dfrac{x - 3}{4}$

115. makes sense **117.** makes sense **119.** false
121. false **123.** 2 **125.** $x + 150$ **126.** $20 + 0.05x$
127. $4x + 400$

Section 2.2

Check Point Exercises

1. computer science: $56 thousand; economics: $49 thousand; education: $35 thousand **2.** by 67 years after 1969, or in 2036 **3.** 300 text messages
4. $1200 **5.** $3150 at 9%; $1850 at 11% **6.** 50 ft by 94 ft

Concept and Vocabulary Check

1. $x + 658.6$ **2.** $31 + 2.4x$ **3.** $4 + 0.15x$ **4.** $x - 0.15x$ or $0.85x$ **5.** $0.12x + 0.09(30,000 - x)$

Exercise Set 2.2

1. 6 **3.** 25 **5.** 120 **7.** 320 **9.** 19 and 45 **11.** 2 **13.** 8 **15.** all real numbers **17.** 5 **19.** TV: 9 years; sleeping: 28 years
21. bachelor's: $55 thousand; master's: $61 thousand **23.** by 2015 **25. a.** $V(x) = 24,000 - 3000x$ **b.** after 5 years

c.

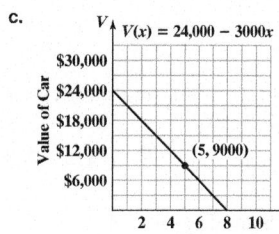

$V(x) = 24,000 - 3000x$
$(5, 9000)$
Value of Car
Years

27. after 5 months; $165 **29.** 30 times **31. a.** 2019; 22,300 students
b. $y_1 = 13,300 + 1000x$; $y_2 = 26,800 - 500x$ **33.** $420 **35.** $150 **37.** $467.20 **39.** $2000 at 6%;
$5000 at 8% **41.** $6000 at 12%; $2000 at a 5% loss **43.** 50 yd by 100 yd **45.** 36 ft by 78 ft
47. 2 in. **49.** 11 hr **51.** 5 ft 7 in. **53.** 7 oz
59. a. $f(x) = 30 + 5x$; $g(x) = 7.5x$ **b.**

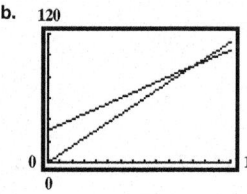

120

0 15
0

c. $(12, 90)$; For 12 hours, both options cost the same, $90.

61. makes sense **63.** does not make sense **65.** $200 **67.** 10 correct answers **69.** 36 plants **71.** $-14 - 29x + 15x^2$ or $15x^2 - 29x - 14$
72. $\sqrt{2}$ **73.** $-\dfrac{54 + 43\sqrt{2}}{46}$

Section 2.3

Check Point Exercises

1. a. $8 + i$ **b.** $-10 + 7i$ **2. a.** $63 + 14i$ **b.** $58 - 11i$ **3.** $\dfrac{16}{17} + \dfrac{21}{17}i$ **4. a.** $7i\sqrt{3}$ **b.** $1 - 4i\sqrt{3}$ **c.** $-7 + i\sqrt{3}$

Concept and Vocabulary Check

1. $\sqrt{-1}$; -1 **2.** complex; imaginary; real **3.** $-6i$ **4.** $14i$ **5.** 18; $-15i$; $12i$; $-10i^2$; 10 **6.** $2 + 9i$ **7.** $2 + 5i$ **8.** i; i; $2i\sqrt{5}$

Exercise Set 2.3

1. $8 - 2i$ **3.** $-2 + 9i$ **5.** $24 - 3i$ **7.** $-14 + 17i$ **9.** $21 + 15i$ **11.** $-19 + 7i$ **13.** $-29 - 11i$ **15.** 34 **17.** 26 **19.** $-5 + 12i$
21. $\dfrac{3}{5} + \dfrac{1}{5}i$ **23.** $1 + i$ **25.** $-\dfrac{24}{25} + \dfrac{32}{25}i$ **27.** $\dfrac{7}{5} + \dfrac{4}{5}i$ **29.** $3i$ **31.** $47i$ **33.** $-8i$ **35.** $2 + 6i\sqrt{7}$ **37.** $-\dfrac{1}{3} + i\dfrac{\sqrt{2}}{6}$ **39.** $-\dfrac{1}{8} - i\dfrac{\sqrt{3}}{24}$
41. $-2\sqrt{6} - 2i\sqrt{10}$ **43.** $24\sqrt{15}$ **45.** $-11 - 5i$ **47.** $-5 + 10i$ **49.** $0 + 47i$ or $47i$ **51.** 0 **53.** $10 + 10i$ **55.** $\dfrac{20}{13} + \dfrac{30}{13}i$
57. $(47 + 13i)$ volts **59.** $(5 + i\sqrt{15}) + (5 - i\sqrt{15}) = 10$; $(5 + i\sqrt{15})(5 - i\sqrt{15}) = 25 - 15i^2 = 25 + 15 = 40$ **69.** makes sense
71. does not make sense **73.** false **75.** false **77.** $\dfrac{14}{25} - \dfrac{2}{25}i$ **79.** $\dfrac{8}{5} + \dfrac{16}{5}i$ **80.** $(2x - 1)(x + 4)$ **81.** $(x - 3)(x - 3)$ or $(x - 3)^2$
82. -5

Section 2.4

Check Point Exercises

1. a. $\{0, 3\}$ **b.** $\left\{-1, \dfrac{1}{2}\right\}$ **2. a.** $\{-\sqrt{7}, \sqrt{7}\}$ **b.** $\{-3i, 3i\}$ **c.** $\{-5 + \sqrt{11}, -5 - \sqrt{11}\}$ **3.** $\{-2 \pm \sqrt{5}\}$ **4.** $\dfrac{-3 + \sqrt{41}}{4}, \dfrac{-3 - \sqrt{41}}{4}$
5. $\left\{\dfrac{-1 + \sqrt{3}}{2}, \dfrac{-1 - \sqrt{3}}{2}\right\}$ **6.** $1 + i, 1 - i$ **7. a.** 0; one real solution that is rational **b.** 81; two rational solutions **c.** -44; two
imaginary solutions that are complex conjugates **8.** approximately 26 years old **9.** 20 in.

Concept and Vocabulary Check

1. quadratic **2.** $A = 0$ or $B = 0$ **3.** x-intercepts **4.** $\pm\sqrt{d}$ **5.** $\pm\sqrt{7}$ **6.** 9 **7.** $\dfrac{1}{9}$ **8.** $\dfrac{-b \pm \sqrt{b^2 - 4ac}}{2a}$

9. $2; 9; -5$ **10.** $1; -4; -1$ **11.** $2 \pm \sqrt{2}$ **12.** $-1 \pm i\dfrac{\sqrt{6}}{2}$ **13.** $b^2 - 4ac$ **14.** no **15.** two **16.** the square root property **17.** the quadratic formula **18.** factoring and the zero-product principle **19.** right; hypotenuse; legs **20.** right; legs; the square of the length of the hypotenuse

Exercise Set 2.4

1. $\{-2, 5\}$ **3.** $\{3, 5\}$ **5.** $\left\{-\dfrac{5}{2}, \dfrac{2}{3}\right\}$ **7.** $\left\{-\dfrac{4}{3}, 2\right\}$ **9.** $\{-4, 0\}$ **11.** $\left\{0, \dfrac{1}{3}\right\}$ **13.** $\{-3, 1\}$ **15.** $\{-3, 3\}$ **17.** $\{-\sqrt{10}, \sqrt{10}\}$ **19.** $\{\pm 5i\}$

21. $\{-7, 3\}$ **23.** $\{4 \pm \sqrt{5}\}$ **25.** $\{-3 \pm 4i\}$ **27.** $\{3 \pm i\sqrt{5}\}$ **29.** $\left\{-\dfrac{5}{3}, \dfrac{1}{3}\right\}$ **31.** $\left\{\dfrac{1 - \sqrt{7}}{5}, \dfrac{1 + \sqrt{7}}{5}\right\}$ **33.** $\left\{\dfrac{4 - 2\sqrt{2}}{3}, \dfrac{4 + 2\sqrt{2}}{3}\right\}$

35. $\{-7, 1\}$ **37.** $\{1 + \sqrt{3}, 1 - \sqrt{3}\}$ **39.** $\{3 + 2\sqrt{5}, 3 - 2\sqrt{5}\}$ **41.** $\{-2 + \sqrt{3}, -2 - \sqrt{3}\}$ **43.** $\{2, 3\}$ **45.** $\left\{\dfrac{-3 + \sqrt{13}}{2}, \dfrac{-3 - \sqrt{13}}{2}\right\}$

47. $\left\{\dfrac{1}{2}, 3\right\}$ **49.** $\left\{\dfrac{1 + \sqrt{2}}{2}, \dfrac{1 - \sqrt{2}}{2}\right\}$ **51.** $\left\{\dfrac{1 + \sqrt{7}}{3}, \dfrac{1 - \sqrt{7}}{3}\right\}$ **53.** $\{-5, -3\}$ **55.** $\left\{\dfrac{-5 + \sqrt{13}}{2}, \dfrac{-5 - \sqrt{13}}{2}\right\}$ **57.** $\left\{\dfrac{3 + \sqrt{57}}{6}, \dfrac{3 - \sqrt{57}}{6}\right\}$

59. $\left\{\dfrac{1 + \sqrt{29}}{4}, \dfrac{1 - \sqrt{29}}{4}\right\}$ **61.** $\{3 + i, 3 - i\}$ **63.** 36; 2 unequal real solutions that are rational **65.** 97; 2 unequal real solutions that are

irrational **67.** 0; 1 real solution that is rational **69.** 37; 2 unequal real solutions that are irrational **71.** $\left\{-\dfrac{1}{2}, 1\right\}$ **73.** $\left\{\dfrac{1}{5}, 2\right\}$

75. $\{-2\sqrt{5}, 2\sqrt{5}\}$ **77.** $\{1 + \sqrt{2}, 1 - \sqrt{2}\}$ **79.** $\left\{\dfrac{-11 + \sqrt{33}}{4}, \dfrac{-11 - \sqrt{33}}{4}\right\}$ **81.** $\left\{0, \dfrac{8}{3}\right\}$ **83.** $\{2\}$ **85.** $\{-2, 2\}$ **87.** $\{3 + 2i, 3 - 2i\}$

89. $\{2 + i\sqrt{3}, 2 - i\sqrt{3}\}$ **91.** $\left\{0, \dfrac{7}{2}\right\}$ **93.** $\{2 + \sqrt{10}, 2 - \sqrt{10}\}$ **95.** $\{-5, -1\}$ **97.** -1 and 5; d **99.** -3 and 1; f **101.** no x-intercepts; b

103. $-\dfrac{1}{2}$ and 2 **105.** -6 and 3 **107.** $\dfrac{-5 - \sqrt{33}}{2}$ and $\dfrac{-5 + \sqrt{33}}{2}$ **109.** $\dfrac{5 - \sqrt{7}}{3}$ and $\dfrac{5 + \sqrt{7}}{3}$ **111.** $\dfrac{-2 - \sqrt{22}}{2}$ and $\dfrac{-2 + \sqrt{22}}{2}$ **113.** $-2, 8$

115. 1, 2 **117.** $1 + \sqrt{7}$ **119.** $\left\{\dfrac{-1 \pm \sqrt{21}}{2}\right\}$ **121.** $\left\{-2\sqrt{2}, \dfrac{\sqrt{2}}{2}\right\}$ **123.** 7 **125.** $(7, 21)$ **127.** 33-year-olds and 58-year-olds;

The function models the actual data well. **129.** 77.8 ft; (b) **131. a.** $\dfrac{1}{\Phi - 1}$ **b.** $\Phi = \dfrac{1 + \sqrt{5}}{2}$ **c.** $\dfrac{1 + \sqrt{5}}{2}$ **133.** $2\sqrt{5}$ mi, or about 4.5 mi

135. $90\sqrt{2}$ ft, or about 127.3 ft **137.** length: 9 ft; width: 6 ft **139.** 5 in. **141.** 5 m **143.** 10 in. **145.** 9.3 in. and 0.7 in.

157. depth: 5 in.; area: 50 in.2; no **159.** does not make sense **161.** makes sense **163.** true **165.** false **167.** $t = \dfrac{v_0 \pm \sqrt{v_0^2 - 64s}}{32}$

169. $(x + 1)(x + 2)(x - 2)$ **170.** $x + 5 + 2\sqrt{x + 4}$ **171.** true

Mid-Chapter 1 Check Point

1. $\{6\}$ **2.** $\left\{-1, \dfrac{7}{5}\right\}$ **3.** $\{-7\}$ **4.** $\left\{\dfrac{3 \pm \sqrt{15}}{3}\right\}$ **5.** all real numbers **6.** $\left\{\pm\dfrac{6\sqrt{5}}{5}\right\}$ **7.** $\left\{\dfrac{3}{4} \pm i\dfrac{\sqrt{23}}{4}\right\}$ **8.** $\{3\}$ **9.** $\{-3 \pm 2\sqrt{6}\}$

10. $\{2 \pm \sqrt{3}\}$ **11.** \varnothing **12.** $\{4\}$ **13.** $-3 - \sqrt{7}$ and $-3 + \sqrt{7}$ **14.** 1 **15.** $i\sqrt{13}$ and $-i\sqrt{13}$ **16.** $\dfrac{-3 - \sqrt{41}}{4}$ and $\dfrac{-3 + \sqrt{41}}{4}$

17. $\dfrac{5}{2} + i\dfrac{\sqrt{7}}{2}$ and $\dfrac{5}{2} - i\dfrac{\sqrt{7}}{2}$ **18.** 0 **19.** -4 and $\dfrac{1}{2}$ **20.** $\{-5 \pm 2\sqrt{7}\}$ **21.** two complex imaginary solutions that are complex conjugates

22. two unequal real solutions that are rational **23.** love: 630; thanks: 243; sorry: 211 **24.** by 2042 **25.** $11,500 at 8%; $13,500 at 9%

26. 200 text messages; $25 **27.** $780 **28.** $2500 at 4%; $1500 at 3% loss **29.** length: 17 ft; width: 6 ft **30.** length: 7 ft; width: 4 ft

31. 12 yd **32. a.** 6 years **b.** $(6, 5990)$ **33.** about 116 years after 1920, or in 2036 **34.** $-1 - i$ **35.** $-3 + 6i$ **36.** $7 + i$ **37.** i

38. $3i\sqrt{3}$ **39.** $1 - 4i\sqrt{3}$

Section 2.5

Check Point Exercises

1. $\{-\sqrt{3}, 0, \sqrt{3}\}$ **2.** $\left\{-2, -\dfrac{3}{2}, 2\right\}$ **3.** $\{6\}$ **4.** $\{4\}$ **5. a.** $\{\sqrt[3]{25}\}$ or $\{5^{2/3}\}$ **b.** $\{-8, 8\}$ **6.** $\{-\sqrt{3}, -\sqrt{2}, \sqrt{2}, \sqrt{3}\}$ **7.** $\left\{-\dfrac{1}{27}, 64\right\}$

8. $\{-2, 3\}$ **9.** -2 and 3 **10.** $225,000

Concept and Vocabulary Check

1. subtract $8x$ and subtract 12 from both sides **2.** radical **3.** extraneous **4.** $2x + 1; x^2 - 14x + 49$ **5.** $x + 2; x + 8 - 6\sqrt{x - 1}$ **6.** $5^{4/3}$
7. $\pm 5^{3/2}$ **8.** $x^2; u^2 - 13u + 36 = 0$ **9.** $x^{1/3}; u^2 + 2u - 3 = 0$ **10.** $c; -c$ **11.** $3x - 1 = 7; 3x - 1 = -7$

Exercise Set 2.5

1. $\{-4, 0, 4\}$ **3.** $\left\{-2, -\dfrac{2}{3}, 2\right\}$ **5.** $\left\{-\dfrac{1}{2}, \dfrac{1}{2}, \dfrac{3}{2}\right\}$ **7.** $\left\{-2, -\dfrac{1}{2}, \dfrac{1}{2}\right\}$ **9.** $\{0, 2, -1 + i\sqrt{3}, -1 - i\sqrt{3}\}$ **11.** $\{6\}$ **13.** $\{6\}$ **15.** $\{-6\}$ **17.** $\{10\}$

19. $\{-5\}$ **21.** $\{12\}$ **23.** $\{8\}$ **25.** \varnothing **27.** \varnothing **29.** $\left\{\dfrac{13 + \sqrt{105}}{6}\right\}$ **31.** $\{4\}$ **33.** $\{13\}$ **35.** $\{\sqrt[5]{4}\}$ **37.** $\{-60, 68\}$ **39.** $\{-4, 5\}$

41. $\{-2, -1, 1, 2\}$ **43.** $\left\{-\dfrac{4}{3}, -1, 1, \dfrac{4}{3}\right\}$ **45.** $\{25, 64\}$ **47.** $\left\{-\dfrac{1}{4}, \dfrac{1}{5}\right\}$ **49.** $\{-8, 27\}$ **51.** $\{1\}$ **53.** $\left\{\dfrac{1}{4}, 1\right\}$ **55.** $\{2, 12\}$ **57.** $\{-3, -1, 2, 4\}$

59. $\{-8, -2, 1, 4\}$ **61.** $\{-8, 8\}$ **63.** $\{-5, 9\}$ **65.** $\{-2, 3\}$ **67.** $\left\{-\dfrac{5}{3}, 3\right\}$ **69.** $\left\{-\dfrac{2}{5}, \dfrac{2}{5}\right\}$ **71.** $\left\{-\dfrac{4}{5}, 4\right\}$ **73.** \varnothing **75.** $\left\{\dfrac{1}{2}\right\}$ **77.** $\{-1, 3\}$

79. 2; c **81.** 1; e **83.** 2 and 3; f **85.** $-\dfrac{3}{2}$ and 4 **87.** 4 **89.** $-2, -\dfrac{1}{2}$, and 2 **91.** 0 **93.** $-2, 0$, and 2 **95.** $\{-8, -6, 4, 6\}$

97. $\{-7, -1, 6\}$ **99.** 8 **101.** $V = \dfrac{\pi r^2 h}{3}$ or $\dfrac{1}{3}\pi r^2 h$ **103.** -16 and 12 **105.** 5.4 ft **107.** $(5.4, 1.16)$ **109. a.** $48\% \pm 1\%$ **b.** 47.8% **c.** 2024

111. 149 million km **113.** either 1.2 feet or 7.5 feet from the base of the 6-foot pole **123.** $\{0, 2\}$ **125.** does not make sense

127. does not make sense **129.** false **131.** false **133.** $\left\{\dfrac{1}{2}\right\}$ **135.** $\{81\}$ **137.** yes **138.** $\{-3\}$ **139.** $\{14\}$

Section 2.6
Check Point Exercises

1. $[-1, \infty)$ or $\{x | x \geq -1\}$

2. $\{x | x < 4\}$ or $(-\infty, 4)$

3. $\{x | x \geq 13\}$ or $[13, \infty)$

4. a. $\{x | x \text{ is a real number}\}$ or or $(-\infty, \infty)$
b. \varnothing

5. $[-1, 4)$ or $\{x | -1 \leq x < 4\}$

6. $(-3, 7)$ or $\{x | -3 < x < 7\}$

7. $\{x | x \leq 1 \text{ or } x \geq 4\}$ or $(-\infty, 1] \cup [4, \infty)$

8. $\left\{x \left| -\dfrac{11}{5} \leq x \leq 3 \right.\right\}$ or $\left[-\dfrac{11}{5}, 3\right]$

9. a. $\{x | -3 < x < 7\}$ or $(-3, 7)$
b. $\{x | x \leq 1 \text{ or } x \geq 4\}$ or $(-\infty, 1] \cup [4, \infty)$

10. $\left\{x \left| -\dfrac{11}{5} \leq x \leq 3 \right.\right\}$ or $\left[-\dfrac{11}{5}, 3\right]$

11. $\{x | x < -4 \text{ or } x > 8\}$ or $(-\infty, -4) \cup (8, \infty)$ **12.** more than 720 mi per week

Concept and Vocabulary Check

1. adding 4; dividing; -3; direction; $>$; $<$ **2.** \varnothing **3.** $(-\infty, \infty)$ **4.** middle **5.** $|x - 2| = 8$; boundary **6.** true **7.** false
8. $(-\infty, -6) \cup (10, \infty)$ **9.** $-c; c$ **10.** $-c; c$ **11.** $-2 < x - 7 < 2$ **12.** $x - 7 < -2$ or $x - 7 > 2$

Exercise Set 2.6

1. $(-\infty, 3)$

3. $\left[\dfrac{20}{3}, \infty\right)$

5. $(-\infty, -4]$

7. $\left(-\infty, -\dfrac{2}{5}\right]$

9. $[0, \infty)$

11. $(-\infty, 1)$

13. $[6, \infty)$

15. $[-10, \infty)$

17. $(-\infty, -6)$

19. $[13, \infty)$

21. $(-\infty, \infty)$

23. \varnothing **25.** $(3, 5)$ **27.** $[-1, 3)$

29. $(-5, -2]$ **31.** $[3, 6)$ **33.** $(-3, 3)$ **35.** $[-1, 3]$ **37.** $(-1, 7)$ **39.** $[-5, 3]$ **41.** $(-6, 0)$ **43.** $(-\infty, -3)$ or $(3, \infty)$
45. $(-\infty, -1] \cup [3, \infty)$ **47.** $\left(-\infty, \dfrac{1}{3}\right) \cup (5, \infty)$ **49.** $(-\infty, -5] \cup [3, \infty)$ **51.** $(-\infty, -3) \cup (12, \infty)$ **53.** $(-\infty, -1] \cup [3, \infty)$ **55.** $[2, 6]$
57. $(-\infty, -3) \cup (5, \infty)$ **59.** $(-\infty, -1] \cup [2, \infty)$ **61.** $(-1, 9)$ **63.** $\left(-\infty, \dfrac{1}{3}\right) \cup (1, \infty)$ **65.** $\left(-\infty, -\dfrac{75}{14}\right) \cup \left(\dfrac{87}{14}, \infty\right)$
67. $(-\infty, -6] \cup [24, \infty)$ **69.** $(-\infty, -3]$ **71.** $[6, \infty)$ **73.** $\left(-\dfrac{2}{3}, \dfrac{10}{3}\right)$ **75.** $(-\infty, -10] \cup [2, \infty)$ **77.** $(-1, 9)$ **79.** $[-1, 2)$
81. $\left(-\infty, -\dfrac{1}{3}\right] \cup [3, \infty)$ **83.** $(-\infty, 5]$ **85.** $(-\infty, -5)$ **87.** $[4, \infty)$ **89.** $(0, 4)$ **91.** intimacy \geq passion or passion \leq intimacy

93. commitment $>$ passion or passion $<$ commitment **95.** 9; after 3 years **97. a.** years after 2016 **b.** years after 2020
c. years after 2020 **d.** years after 2016 **99.** between 80 and 110 minutes, inclusive **101.** $h \leq 41$ or $h \geq 59$ **103.** $15 + 0.08x < 3 + 0.12x$;
more than 300 texts **105.** $2 + 0.08x < 8 + 0.05x$; 199 checks or fewer **107.** $5.50x > 3000 + 3x$; more than 1200 packages **109.** $245 + 95x \leq 3000$;
at most 29 bags **111. a.** $\dfrac{86 + 88 + x}{3} \geq 90$; at least a 96 **b.** $\dfrac{86 + 88 + x}{3} < 80$; a grade less than 66 **113.** more than 3 and fewer than
15 crossings per three-month period

121.

$\{x \mid x < 4\}$ or $(-\infty, 4)$

123. a.

b.

125. makes sense **127.** makes sense **129.** false **131.** false **133.** Because $x > y$, $y - x$ represents a negative number, so when both sides
are multiplied by $(y - x)$, the inequality must be reversed. **136.** $\{1, 5\}$ **137.** $\{-1 \pm \sqrt{2}\}$ **138.**

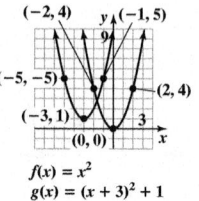

$f(x) = x^2$
$g(x) = (x + 3)^2 + 1$

Chapter 2 Review Exercises

1. $\{6\}$; conditional equation **2.** $\{-10\}$; conditional equation **3.** $\{5\}$; conditional equation **4.** $\{-13\}$; conditional equation
5. $\{-3\}$; conditional equation **6.** $\{-1\}$; conditional equation **7.** \varnothing; inconsistent equation **8.** $\{x \mid x$ is a real number$\}$; identity
9. $\{2\}$; conditional equation **10.** $\{2\}$; conditional equation **11.** $\left\{\dfrac{72}{11}\right\}$; conditional equation **12.** $\left\{\dfrac{36}{7}\right\}$; conditional equation
13. $\left\{\dfrac{77}{15}\right\}$; conditional equation **14.** $\{2\}$; conditional equation **15.** \varnothing; inconsistent equation **16.** all real numbers except -1 and 1; identity
17. $\left\{\dfrac{5}{2}\right\}$; conditional equation **18.** \varnothing; inconsistent equation **19.** $\left\{\dfrac{4}{7}\right\}$; conditional equation **20.** $\dfrac{1}{2}$ **21.** 2
22. *Scream*: 6; *Scream 2*: 8; *Scream 3*: 9 **23.** by 40 years after 1980, or in 2020 **24.** 500 text messages **25.** \$60 **26.** \$10,000 in sales
27. \$2500 at 4%; \$6500 at 7% **28.** \$4500 at 2%; \$3500 at 5% **29.** 44 yd by 126 yd **30. a.** $14,100 + 1500x = 41,700 - 800x$ **b.** 2022; 32,100
31. $-9 + 4i$ **32.** $-12 - 8i$ **33.** $17 + 19i$ **34.** $-7 - 24i$ **35.** 113 **36.** $\dfrac{15}{13} - \dfrac{3}{13}i$ **37.** $\dfrac{1}{5} + \dfrac{11}{10}i$ **38.** $i\sqrt{2}$ **39.** $-96 - 40i$
40. $2 + i\sqrt{2}$ **41.** $\left\{-8, \dfrac{1}{2}\right\}$ **42.** $\{-4, 0\}$ **43.** $\{-8, 8\}$ **44.** $\{-4i, 4i\}$ **45.** $\{-3 - i\sqrt{10}, -3 + i\sqrt{10}\}$ **46.** $\left\{\dfrac{4 - 3\sqrt{2}}{3}, \dfrac{4 + 3\sqrt{2}}{3}\right\}$
47. $\{3, 9\}$ **48.** $\left\{2 + \dfrac{\sqrt{3}}{3}, 2 - \dfrac{\sqrt{3}}{3}\right\}$ **49.** $\{1 + \sqrt{5}, 1 - \sqrt{5}\}$ **50.** $\{1 + 3i\sqrt{2}, 1 - 3i\sqrt{2}\}$ **51.** $\left\{\dfrac{-2 + \sqrt{10}}{2}, \dfrac{-2 - \sqrt{10}}{2}\right\}$
52. 2 complex imaginary solutions **53.** 2 unequal real solutions that are rational **54.** $\left\{\dfrac{1}{2}, 5\right\}$ **55.** $\left\{-2, \dfrac{10}{3}\right\}$ **56.** $\left\{\dfrac{7 + \sqrt{37}}{6}, \dfrac{7 - \sqrt{37}}{6}\right\}$
57. $\{-3, 3\}$ **58.** $\{-2, 8\}$ **59.** $\left\{\dfrac{1}{6} + i\dfrac{\sqrt{23}}{6}, \dfrac{1}{6} - i\dfrac{\sqrt{23}}{6}\right\}$ **60.** $-\dfrac{2}{3}, 4$ **61.** $-2 - 2i, -2 + 2i$ **62.** $4 + \sqrt{5}, 4 - \sqrt{5}$ **63.** 14 weeks
64. a. $182.8 \approx 183$; overestimates by 3 **b.** 2023 **65.** length $= 5$ yd; width $= 3$ yd **66.** approximately 134 m **67.** $\{-5, 0, 5\}$
68. $\left\{-3, \dfrac{1}{2}, 3\right\}$ **69.** $\{2\}$ **70.** $\{8\}$ **71.** $\{16\}$ **72.** $\{-118, 132\}$ **73.** $\{-2, -1, 1, 2\}$ **74.** $\{16\}$ **75.** $\{-4, 3\}$ **76.** $\{-5, 11\}$
77. $\left\{-1, -\dfrac{2\sqrt{6}}{9}, \dfrac{2\sqrt{6}}{9}, 1\right\}$ **78.** $\{2\}$ **79.** $\{1, 4\}$ **80.** $\{-3, -2, 3\}$ **81.** 2 **82.** $-\dfrac{27}{8}, 1, \dfrac{27}{8}$ **83.** $-3, -\sqrt{2}, \sqrt{2}$ **84.** $-4, 2$
85. by 16 years after 1993, or in 2009

86. $[-2, \infty)$

87. $\left[\dfrac{3}{5}, \infty\right)$

88. $\left(-\infty, -\dfrac{21}{2}\right)$

89. $(-3, \infty)$

90. $(-\infty, -2]$

91. \varnothing

92. $(2, 3]$

93. $[-9, 6]$

94. $(-\infty, -6) \cup (0, \infty)$

95. $(-\infty, -3] \cup [-2, \infty)$

96. $(-\infty, -5] \cup [1, \infty)$;

97. $(-\infty, -1)$ **98.** $[-2, 7]$ **99.** no more than 80 miles per day **100.** $[49\%, 99\%)$ **101.** at least $120,000

Chapter 2 Test

1. $\{-1\}$ **2.** $\{-6\}$ **3.** $\{5\}$ **4.** $\left\{-\dfrac{1}{2}, 2\right\}$ **5.** $\left\{\dfrac{1 - 5\sqrt{3}}{3}, \dfrac{1 + 5\sqrt{3}}{3}\right\}$ **6.** $\{1 - \sqrt{5}, 1 + \sqrt{5}\}$ **7.** $\left\{1 + \dfrac{1}{2}i, 1 - \dfrac{1}{2}i\right\}$ **8.** $\{-1, 1, 4\}$

9. $\{7\}$ **10.** $\{5\}$ **11.** $\{\sqrt[3]{4}\}$ **12.** $\{1, 512\}$ **13.** $\{6, 12\}$

14. $(-\infty, 12]$

15. $\left[\dfrac{21}{8}, \infty\right)$

16. $\left[-7, \dfrac{13}{2}\right)$

17. $\left(-\infty, -\dfrac{5}{3}\right] \cup \left[\dfrac{1}{3}, \infty\right)$

18. 2 **19.** $\dfrac{1}{2}, 3$ **20.** -1 **21.** $2 + \sqrt{3}, 2 - \sqrt{3}$ **22.** $-3 + 2\sqrt{6}, -3 - 2\sqrt{6}$ **23.** 4 **24.** $[1, 6]$ **25.** $(-\infty, -2] \cup [6, \infty)$

26. $47 + 16i$ **27.** $2 + i$ **28.** $38i$ **29.** 2018 **30.** 2018 **31.** quite well **32.** average: 59%; good-looking: 27%; strikingly attractive: 2%
33. 26 yr; $33,600 **34.** $3000 at 8%; $7000 at 10% **35.** length = 12 ft; width = 4 ft **36.** 10 ft **37.** $50 **38.** more than 200 text messages

Cumulative Review Exercises (Chapters 1–2)

1. domain: $[0, 2)$; range: $[0, 2]$ **2.** $\dfrac{1}{2}$ and $\dfrac{3}{2}$ **3.** 2

4.

$g(x) = f(x - 1) + 1$

5.

$h(x) = -2f\left(\dfrac{1}{2}x\right)$

6. $\{-4, 5\}$

7. $\left\{\dfrac{25}{18}\right\}$

8. $\{4\}$

9. $\{-8, 27\}$

10. $(-\infty, 20]$

11.

$3x - 6y - 12 = 0$

domain: $(-\infty, \infty)$;
range: $(-\infty, \infty)$

12.

$(x - 2)^2 + (y + 1)^2 = 4$

domain: $[0, 4]$;
range: $[-3, 1]$

13.

domain of f = domain of $g = (-\infty, \infty)$; range of f = range of $g = (-\infty, \infty)$

14.

domain of f = range of $f^{-1} = [3, \infty)$; range of f = domain of $f^{-1} = [2, \infty)$

15. $-2x - h$ **16.** -7 and -3 **17.** $y - 5 = 4(x + 2); y = 4x + 13; 4x - y + 13 = 0$ **18.** $1500 at 7%; $4500 at 9% **19.** $2000
20. 3 ft by 8 ft

CHAPTER 3

Section 3.1

Check Point Exercises

1.

$f(x) = -(x - 1)^2 + 4$

2.

$f(x) = (x - 2)^2 + 1$

3.

$f(x) = -x^2 + 4x + 1$
domain: $(-\infty, \infty)$; range: $(-\infty, 5]$

4. a. minimum **b.** Minimum is 984 at $x = 2$.
c. domain: $(-\infty, \infty)$; range: $[984, \infty)$
5. a. 205 ft; 200 ft **b.** 402 ft
c.

6. $4, -4; -16$ **7.** 30 ft by 30 ft; 900 sq ft

Concept and Vocabulary Check

1. standard; parabola; (h, k); > 0; < 0 **2.** $-\dfrac{b}{2a}; f\left(-\dfrac{b}{2a}\right); -\dfrac{b}{2a}; f\left(-\dfrac{b}{2a}\right)$ **3.** true **4.** false **5.** true **6.** $x - 8; x^2 - 8x$
7. $40 - x; -x^2 + 40x$

Exercise Set 3.1

1. $h(x) = (x - 1)^2 + 1$ **3.** $j(x) = (x - 1)^2 - 1$ **5.** $h(x) = x^2 - 1$ **7.** $g(x) = x^2 - 2x + 1$ **9.** $(3, 1)$ **11.** $(-1, 5)$
13. $(2, -5)$ **15.** $(-1, 9)$

17. domain: $(-\infty, \infty)$
range: $[-1, \infty)$
axis of symmetry: $x = 4$

$f(x) = (x - 4)^2 - 1$

19. domain: $(-\infty, \infty)$
range: $[2, \infty)$
axis of symmetry: $x = 1$

$f(x) = (x - 1)^2 + 2$

21. domain: $(-\infty, \infty)$
range: $[1, \infty)$
axis of symmetry: $x = 3$

$y - 1 = (x - 3)^2$

23. domain: $(-\infty, \infty)$
range: $[-1, \infty)$
axis of symmetry: $x = -2$

$f(x) = 2(x + 2)^2 - 1$

25. domain: $(-\infty, \infty)$
range: $(-\infty, 4]$
axis of symmetry: $x = 1$
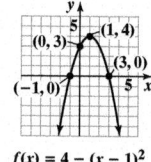
$f(x) = 4 - (x - 1)^2$

27. domain: $(-\infty, \infty)$
range: $[-4, \infty)$
axis of symmetry: $x = 1$

$f(x) = x^2 - 2x - 3$

29. domain: $(-\infty, \infty)$
range: $\left[-\dfrac{49}{4}, \infty\right)$
axis of symmetry: $x = -\dfrac{3}{2}$

$f(x) = x^2 + 3x - 10$

31. domain: $(-\infty, \infty)$
range: $(-\infty, 4]$
axis of symmetry: $x = 1$
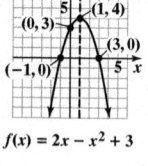
$f(x) = 2x - x^2 + 3$

33. domain: $(-\infty, \infty)$
range: $[-6, \infty)$
axis of symmetry: $x = -3$

$f(x) = x^2 + 6x + 3$

35. domain: $(-\infty, \infty)$
range: $[-5, \infty)$
axis of symmetry: $x = -1$

$f(x) = 2x^2 + 4x - 3$

37. domain: $(-\infty, \infty)$
range: $(-\infty, -1]$
axis of symmetry: $x = 1$

$f(x) = 2x - x^2 - 2$

39. a. minimum **b.** Minimum is -13 at $x = 2$. **c.** domain: $(-\infty, \infty)$; range: $[-13, \infty)$ **41. a.** maximum **b.** Maximum is 1 at $x = 1$.
c. domain: $(-\infty, \infty)$; range: $(-\infty, 1]$ **43. a.** minimum **b.** Minimum is $-\dfrac{5}{4}$ at $x = \dfrac{1}{2}$. **c.** domain: $(-\infty, \infty)$; range: $\left[-\dfrac{5}{4}, \infty\right)$
45. domain: $(-\infty, \infty)$; range: $[-2, \infty)$ **47.** domain: $(-\infty, \infty)$; range: $(-\infty, -6]$ **49.** $f(x) = 2(x - 5)^2 + 3$ **51.** $f(x) = 2(x + 10)^2 - 5$
53. $f(x) = -3(x + 2)^2 + 4$ **55.** $f(x) = 3(x - 11)^2$ **57. a.** 18.35 ft; 35 ft **b.** 77.8 ft **c.** 6.1 ft

59. a. 7.8 ft; 1.5 ft **b.** 4.6 ft **c.** **61.** 8 and 8; 64 **63.** 8, −8; −64

65. length: 300 ft; width: 150 ft; maximum area: 45,000 sq ft **67.** 12.5 yd by 12.5 yd; 156.25 sq yd **69.** 150 ft by 100 ft; 15,000 sq ft
71. 5 in.; 50 sq in. **73. a.** $C(x) = 0.55x + 525$ **b.** $P(x) = -0.001x^2 + 2.45x - 525$ **c.** 1225 roast beef sandwiches; ≈$975.63

81. a. **b.** $(20.5, -120.5)$ **83.** $(2.5, 185)$ **85.** $(-30, 91)$
c. Ymax = 750
d. You can choose Xmin and
Xmax so the x-value of the
vertex is in the center of the
graph. Choose Ymin to include
the y-value of the vertex.

You can only see a little of the parabola.

87. does not make sense **89.** does not make sense **91.** true **93.** false **95.** $x = -2; (-3, -2)$ **97.** $f(x) = \frac{1}{2}(x + 3)^2 - 4$

99. $x = 166\frac{2}{3}$ ft, $y = 125$ ft; approximately 20,833 ft^2 **102.** $(x + 3)(x + 1)(x - 1)$ **103.** $f(2) = -1; f(3) = 16$; The graph passes through $(2, -1)$,
which is below the x-axis, and $(3, 16)$, which is above the x-axis. Since the graph of f is continuous, it must cross the x-axis somewhere between 2 and 3
to get from one of these points to the other. **104.** even; symmetric with respect to the y-axis

Section 3.2

Check Point Exercises

1. The graph rises to the left and to the right. **2.** The graph falls to the left and rises to the right. **3.** Since n is odd and the leading coefficient is
negative, the graph falls to the right. Since the ratio cannot be negative, the model won't be appropriate. **4.** No; the graph should fall to the left but
doesn't appear to do so. **5.** −2 and 2 **6.** −2, 0, and 2 **7.** $-\frac{1}{2}$ with multiplicity 2 and 5 with multiplicity 3; touches and turns at $-\frac{1}{2}$ and crosses at 5
8. $f(-3) = -42; f(-2) = 5$ **9.** **10.**

Concept and Vocabulary Check

1. 5; −2 **2.** false **3.** end; leading **4.** falls; rises **5.** rises; falls **6.** rises; rises **7.** falls; falls **8.** true **9.** true
10. x-intercept **11.** turns around; crosses **12.** 0; Intermediate Value **13.** $n - 1$

Exercise Set 3.2

1. polynomial function; degree: 3 **3.** polynomial function; degree: 5 **5.** not a polynomial function **7.** not a polynomial function
9. not a polynomial function **11.** could be polynomial function **13.** not a polynomial function **15.** b **17.** a
19. falls to the left and rises to the right **21.** rises to the left and to the right **23.** falls to the left and to the right
25. $x = 5$ has multiplicity 1; The graph crosses the x-axis; $x = -4$ has multiplicity 2; The graph touches the x-axis and turns around.
27. $x = 3$ has multiplicity 1; The graph crosses the x-axis; $x = -6$ has multiplicity 3; The graph crosses the x-axis.
29. $x = 0$ has multiplicity 1; The graph crosses the x-axis; $x = 1$ has multiplicity 2; The graph touches the x-axis and turns around.
31. $x = 2, x = -2$ and $x = -7$ have multiplicity 1; The graph crosses the x-axis. **33.** $f(1) = -1; f(2) = 5$ **35.** $f(-1) = -1; f(0) = 1$
37. $f(-3) = -11; f(-2) = 1$ **39.** $f(-3) = -42; f(-2) = 5$
41. a. $f(x)$ rises to the right and falls to the left. **43. a.** $f(x)$ rises to the left and the right.
b. $x = -2, x = 1, x = -1$; **b.** $x = 0, x = 3, x = -3$;
$f(x)$ crosses the x-axis at each. $f(x)$ crosses the x-axis at −3 and 3;
c. The y-intercept is −2. $f(x)$ touches the x-axis at 0.
d. neither **c.** The y-intercept is 0.
e. **d.** y-axis symmetry
e.

45. a. $f(x)$ falls to the left and the right.
 b. $x = 0, x = 4, x = -4$;
 $f(x)$ crosses the x-axis at -4 and 4;
 $f(x)$ touches the x-axis at 0.
 c. The y-intercept is 0.
 d. y-axis symmetry
 e.

$f(x) = -x^4 + 16x^2$

49. a. $f(x)$ falls to the left and the right.
 b. $x = 0, x = 2$;
 $f(x)$ crosses the x-axis at 0 and 2.
 c. The y-intercept is 0.
 d. neither
 e.

$f(x) = -2x^4 + 4x^3$

53. a. $f(x)$ rises to the left and falls to the right.
 b. $x = 0, x = 3$;
 $f(x)$ crosses the x-axis at 3;
 $f(x)$ touches the x-axis at 0.
 c. The y-intercept is 0.
 d. neither
 e.

$f(x) = 3x^2 - x^3$

57. a. $f(x)$ rises to the left and the right.
 b. $x = -2, x = 0, x = 1$;
 $f(x)$ crosses the x-axis at -2 and 1;
 $f(x)$ touches the x-axis at 0.
 c. The y-intercept is 0.
 d. neither
 e.

$f(x) = x^2 (x - 1)^3 (x + 2)$

61. a. $f(x)$ falls to the left and the right.
 b. $x = -5, x = 0, x = 1$;
 $f(x)$ crosses the x-axis at -5 and 0;
 $f(x)$ touches the x-axis at 1.
 c. The y-intercept is 0.
 d. neither
 e.

$f(x) = -2x^3 (x - 1)^2 (x + 5)$

47. a. $f(x)$ rises to the left and the right.
 b. $x = 0, x = 1$;
 $f(x)$ touches the x-axis at 0 and 1.
 c. The y-intercept is 0.
 d. neither
 e.

$f(x) = x^4 - 2x^3 + x^2$

51. a. $f(x)$ rises to the left and falls to the right.
 b. $x = 0, x = \pm\sqrt{3}$;
 $f(x)$ crosses the x-axis at 0;
 $f(x)$ touches the x-axis at $\sqrt{3}$ and $-\sqrt{3}$.
 c. The y-intercept is 0.
 d. origin symmetry
 e.

$f(x) = 6x^3 - 9x - x^5$

55. a. $f(x)$ falls to the left and the right.
 b. $x = 1, x = -2, x = 2$;
 $f(x)$ crosses the x-axis at -2 and 2;
 $f(x)$ touches the x-axis at 1.
 c. The y-intercept is 12.
 d. neither
 e.

$f(x) = -3(x - 1)^2 (x^2 - 4)$

59. a. $f(x)$ falls to the left and the right.
 b. $x = -3, x = 0, x = 1$;
 $f(x)$ crosses the x-axis at -3 and 1;
 $f(x)$ touches the x-axis at 0.
 c. The y-intercept is 0.
 d. neither
 e.

$f(x) = -x^2 (x - 1)(x + 3)$

63. a. $f(x)$ rises to the left and the right.
 b. $x = -4, x = 1, x = 2$;
 $f(x)$ crosses the x-axis at -4 and 1;
 $f(x)$ touches the x-axis at 2.
 c. The y-intercept is -16.
 d. neither
 e.

$f(x) = (x - 2)^2 (x + 4)(x - 1)$

65. a. -2, odd; 1, odd; 4, odd **b.** $f(x) = (x + 2)(x - 1)(x - 4)$ **c.** 8 **67. a.** -1, odd; 3, even **b.** $f(x) = (x + 1)(x - 3)^2$ **c.** 9
69. a. -3, even; 2, even **b.** $f(x) = -(x + 3)^2(x - 2)^2$ **c.** -36 **71. a.** -2, even; -1, odd; 1, odd **b.** $f(x) = (x + 2)^2(x + 1)(x - 1)^3$
c. -4 **73. a.** 3167; The world tiger population in 2010 was approximately 3167.; (40, 3167) **b.** underestimates by 33 **c.** rises to the right; no;
The model indicates an increasing world tiger population that will actually decrease without conservation efforts. **75. a.** from 1 through 4 min and
from 8 through 10 min **b.** from 4 through 8 min and from 10 through 12 min **c.** 3 **d.** 4 **e.** negative; The graph falls to the left and falls to
the right. **f.** 116 ± 1 beats per min; 10 min **g.** 64 ± 1 beats per min; 8 min

95.

97.

99.

101. does not make sense **103.** makes sense **105.** false **107.** false **109.** $f(x) = x^3 - 2x^2$

110. $35\dfrac{2}{21}$ **111.** $6x^3 - x^2 - 5x + 4$ **112.** $(x - 3)(2x - 1)(x + 2)$

Section 3.3

Check Point Exercises

1. $x + 5$ **2.** $2x^2 + 3x - 2 + \dfrac{1}{x - 3}$ **3.** $2x^2 + 7x + 14 + \dfrac{21x - 10}{x^2 - 2x}$ **4.** $x^2 - 2x - 3$ **5.** -105 **6.** $\left\{-1, -\dfrac{1}{3}, \dfrac{2}{5}\right\}$

Concept and Vocabulary Check

1. $2x^3 + 0x^2 + 6x - 4$ **2.** $6x^3; 3x; 2x^2; 7x^2$ **3.** $2x^2; 5x - 2; 10x^3 - 4x^2; 10x^3 + 6x^2$ **4.** $6x^2 - 10x; 6x^2 + 8x; 18x; -4; 18x - 4$

5. $9; 3x - 5; 9; 3x - 5 + \dfrac{9}{2x + 1}$ **6.** divisor; quotient; remainder; dividend **7.** $4; 1; 5; -7; 1$ **8.** $-5; 4; 0; -8; -2$ **9.** true

10. $f(c)$ **11.** $x - c$

Exercise Set 3.3

1. $x + 3$ **3.** $x^2 + 3x + 1$ **5.** $2x^2 + 3x + 5$ **7.** $4x + 3 + \dfrac{2}{3x - 2}$ **9.** $2x^2 + x + 6 - \dfrac{38}{x + 3}$ **11.** $4x^3 + 16x^2 + 60x + 246 + \dfrac{984}{x - 4}$

13. $2x + 5$ **15.** $6x^2 + 3x - 1 - \dfrac{3x - 1}{3x^2 + 1}$ **17.** $2x + 5$ **19.** $3x - 8 + \dfrac{20}{x + 5}$ **21.** $4x^2 + x + 4 + \dfrac{3}{x - 1}$

23. $6x^4 + 12x^3 + 22x^2 + 48x + 93 + \dfrac{187}{x - 2}$ **25.** $x^3 - 10x^2 + 51x - 260 + \dfrac{1300}{x + 5}$ **27.** $x^4 + x^3 + 2x^2 + 2x + 2$ **29.** $x^3 + 4x^2 + 16x + 64$

31. $2x^4 - 7x^3 + 15x^2 - 31x + 64 - \dfrac{129}{x + 2}$ **33.** -25 **35.** -133 **37.** 240 **39.** 1 **41.** $x^2 - 5x + 6; x = -1, x = 2, x = 3$

43. $\left\{-\dfrac{1}{2}, 1, 2\right\}$ **45.** $\left\{-\dfrac{3}{2}, -\dfrac{1}{3}, \dfrac{1}{2}\right\}$ **47.** 2; The remainder is zero; $\{-3, -1, 2\}$ **49.** 1; The remainder is zero; $\left\{\dfrac{1}{3}, \dfrac{1}{2}, 1\right\}$

51. a. The remainder is 0. **b.** 3 mm **53.** $0.5x^2 - 0.4x + 0.3$ **55. a.** 70; When the tax rate is 30%, tax revenue is \$700 billion.; (30, 70)

b. $80 + \dfrac{800}{x - 110}; f(30) = 70;$ yes **c.** No, f is a rational function because it is a quotient of two polynomials. **67.** makes sense

69. does not make sense **71.** true **73.** false **75.** $x - 2$ **79.** $\{-2 \pm \sqrt{5}\}$ **80.** $\{-2 \pm i\sqrt{2}\}$ **81.** -3

Section 3.4

Check Point Exercises

1. $\pm 1, \pm 2, \pm 3, \pm 6$ **2.** $\pm 1, \pm 3, \pm\dfrac{1}{2}, \pm\dfrac{1}{4}, \pm\dfrac{3}{2}, \pm\dfrac{3}{4}$ **3.** $-5, -4,$ and 1 **4.** $2, \dfrac{-3 - \sqrt{5}}{2},$ and $\dfrac{-3 + \sqrt{5}}{2}$ **5.** $\{1, 2 - 3i, 2 + 3i\}$
6. $f(x) = x^3 + 3x^2 + x + 3$ **7.** 4, 2, or 0 positive zeros, no possible negative zeros

Concept and Vocabulary Check

1. $a_0; a_n$ **2.** true **3.** false **4.** n **5.** $a - bi$ **6.** $-6; (x + 6)(2x^2 - x - 1) = 0$ **7.** $n; 1$ **8.** false **9.** true **10.** true

Exercise Set 3.4

1. $\pm 1, \pm 2, \pm 4$ **3.** $\pm 1, \pm 2, \pm 3, \pm 6, \pm\dfrac{1}{3}, \pm\dfrac{2}{3}$ **5.** $\pm 1, \pm 2, \pm 3, \pm 6, \pm\dfrac{1}{2}, \pm\dfrac{1}{4}, \pm\dfrac{3}{2}, \pm\dfrac{3}{4}$ **7.** $\pm 1, \pm 2, \pm 3, \pm 4, \pm 6, \pm 12$ **9. a.** $\pm 1, \pm 2, \pm 4$

b. $-2, -1,$ or 2 **c.** $-2, -1,$ and 2 **11. a.** $\pm 1, \pm 2, \pm 3, \pm 6, \pm\dfrac{1}{2}, \pm\dfrac{3}{2}$ **b.** $-2, \dfrac{1}{2},$ or 3 **c.** $-2, \dfrac{1}{2},$ and 3 **13. a.** $\pm 1, \pm 2, \pm 3, \pm 6$

b. -1 **c.** $-1, \dfrac{-3 - \sqrt{33}}{2},$ and $\dfrac{-3 + \sqrt{33}}{2}$ **15. a.** $\pm 1, \pm\dfrac{1}{2}, \pm 2$ **b.** -2 **c.** $-2, \dfrac{-1 + i}{2},$ and $\dfrac{-1 - i}{2}$ **17. a.** $\pm 1, \pm 2, \pm 3, \pm 4, \pm 6, \pm 12$

b. $-3, 1,$ or 4 **c.** $\{-3, 1, 4\}$ **19. a.** $\pm 1, \pm 2, \pm 3, \pm 4, \pm 6, \pm 12$ **b.** -2 **c.** $\{-2, 1 + \sqrt{7}, 1 - \sqrt{7}\}$ **21. a.** $\pm 1, \pm 5, \pm\dfrac{1}{2}, \pm\dfrac{5}{2}, \pm\dfrac{1}{3}, \pm\dfrac{5}{3}, \pm\dfrac{1}{6}, \pm\dfrac{5}{6}$

b. $-5, \dfrac{1}{3},$ or $\dfrac{1}{2}$ **c.** $\left\{-5, \dfrac{1}{3}, \dfrac{1}{2}\right\}$ **23. a.** $\pm 1, \pm 2, \pm 4$ **b.** -2 or 2 **c.** $\{-2, 2, 1 + \sqrt{2}, 1 - \sqrt{2}\}$ **25.** $f(x) = 2x^3 - 2x^2 + 50x - 50$

27. $f(x) = x^3 - 3x^2 - 15x + 125$ **29.** $f(x) = x^4 + 10x^2 + 9$ **31.** $f(x) = x^4 - 9x^3 + 21x^2 + 21x - 130$ **33.** no positive real roots;
3 or 1 negative real roots **35.** 3 or 1 positive real roots; no negative real roots **37.** 2 or 0 positive real roots; 2 or 0 negative real roots

39. $-2, 5$, and 1 **41.** $\left\{-\dfrac{1}{2}, \dfrac{1 + \sqrt{17}}{2}, \dfrac{1 - \sqrt{17}}{2}\right\}$ **43.** $-1, 2 + 2i$, and $2 - 2i$ **45.** $\{-1, -2, 3 + \sqrt{13}, 3 - \sqrt{13}\}$

47. $-1, 2, -\dfrac{1}{3}$, and 3 **49.** $\left\{1, -\dfrac{3}{4}, i\sqrt{2}, -i\sqrt{2}\right\}$ **51.** $\left\{-2, \dfrac{1}{2}, \sqrt{2}, -\sqrt{2}\right\}$

53. a. $-4, 1$, and 4 **55. a.** -1 and $\dfrac{3}{2}$ **57. a.** $\dfrac{1}{2}, 3, -1 \pm i$ **59. a.** $-2, -1, -\dfrac{2}{3}, 1$, and 2

b.

$f(x) = -x^3 + x^2 + 16x - 16$

b.

$f(x) = 4x^3 - 8x^2 - 3x + 9$

b.

$f(x) = 2x^4 - 3x^3 - 7x^2 - 8x + 6$

b.

$f(x) = 3x^5 + 2x^4 - 15x^3 - 10x^2 + 12x + 8$

61. 7.8 in., 10 in. **63. a.** $(7.8, 2000), (10, 2000)$ **b.** $(0, 15)$ **73.** $\left\{\dfrac{1}{2}, \dfrac{2}{3}, 2\right\}$ **75.** $\left\{\pm\dfrac{1}{2}\right\}$ **77.** $5, 3$, or 1 positive real zeros; no negative real zeros

79.

1 real zero, 2 nonreal complex zeros

81.

2 real zeros, 2 nonreal complex zeros

83. makes sense **85.** makes sense **87.** false
89. true **91.** 3 in. **93.** 3 **95.** 5
98. $x = 1$ and $x = 2$ **99.** $x = 1$ **100.** $y = 0$

Mid-Chapter 3 Check Point

1.

$f(x) = (x - 3)^2 - 4$
domain: $(-\infty, \infty)$
range: $[-4, \infty)$

2.

$f(x) = 5 - (x + 2)^2$
domain: $(-\infty, \infty)$
range: $(-\infty, 5]$

3.

$f(x) = -x^2 - 4x + 5$
domain: $(-\infty, \infty)$
range: $(-\infty, 9]$

4.
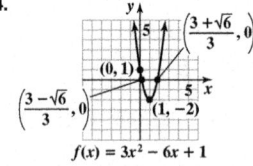
$f(x) = 3x^2 - 6x + 1$
domain: $(-\infty, \infty)$
range: $[-2, \infty)$

5. -1 and 2 **6.** -1 and 2 **7.** $-2, 1$, and 2 **8.** $-2, -1, 1$, and 2

$f(x) = (x - 2)^2 (x + 1)^3$

$f(x) = -(x - 2)^2 (x + 1)^2$

$f(x) = x^3 - x^2 - 4x + 4$

$f(x) = x^4 - 5x^2 + 4$

9. -1 **10.** $-\dfrac{1}{3}, \dfrac{1}{2}$, and 1 **11.** $-1, 0$, and 1 **12.** $0, 1 \pm 5i$

$f(x) = -(x + 1)^6$

$f(x) = -6x^3 + 7x^2 - 1$

$f(x) = 2x^3 - 2x$

$f(x) = x^3 - 2x^2 + 26x$

13. $3, 1 \pm \sqrt{2}$

$f(x) = -x^3 + 5x^2 - 5x - 3$

14. $\{-2, 1\}$ **15.** $\left\{\dfrac{1}{3}, \dfrac{1}{2}, 1\right\}$ **16.** $\left\{-\dfrac{1}{2}, \dfrac{2}{3}, \dfrac{7}{2}\right\}$ **17.** $\left\{-10, -\dfrac{5}{2}, 10\right\}$ **18.** $\{-3, 4, \pm i\}$

19. $\left\{-3, \dfrac{1}{2}, 1 \pm \sqrt{3}\right\}$ **20.** 75 cabinets per day; $1200 **21.** $-9, -9; 81$ **22.** 10 in.; 100 sq in.

23. $2x^2 - x - 3 + \dfrac{x + 1}{3x^2 - 1}$ **24.** $2x^3 - 5x^2 - 3x + 6$ **25.** $f(x) = -2x^3 + 2x^2 - 2x + 2$

26. $f(x) = x^4 - 4x^3 + 13x^2 - 36x + 36$ **27.** yes

Section 3.5

Check Point Exercises

1. a. $\{x \mid x \neq 5\}$ or $(-\infty, 5) \cup (5, \infty)$ **b.** $\{x \mid x \neq -5, x \neq 5\}$ or $(-\infty, -5) \cup (-5, 5) \cup (5, \infty)$ **c.** all real numbers or $(-\infty, \infty)$
2. a. $x = 1, x = -1$ **b.** $x = -1$ **c.** none **3. a.** $y = 3$ **b.** $y = 0$ **c.** none

4.
$g(x) = \dfrac{1}{x+2} - 1$

5.
$f(x) = \dfrac{3x-3}{x-2}$

6.
$f(x) = \dfrac{2x^2}{x^2-9}$

7.
$f(x) = \dfrac{x^4}{x^2+2}$

8. $y = 2x - 1$ **9. a.** $C(x) = 500,000 + 400x$ **b.** $\overline{C}(x) = \dfrac{500,000 + 400x}{x}$ **c.** $\overline{C}(1000) = 900$: The average cost per wheelchair of producing 1000 wheelchairs per month is \$900.; $\overline{C}(10,000) = 450$: The average cost per wheelchair of producing 10,000 wheelchairs per month is \$450.; $\overline{C}(100,000) = 405$: The average cost per wheelchair of producing 100,000 wheelchairs per month is \$405. **d.** $y = 400$; The cost per wheelchair approaches \$400 as more wheelchairs are produced.

Concept and Vocabulary Check

1. polynomial **2.** false **3.** true **4.** vertical asymptote; $x = -5$ **5.** horizontal asymptote; $y = 0$; $y = \dfrac{1}{3}$ **6.** true
7. left; down **8.** one more than **9.** $y = 3x + 5$

Exercise Set 3.5

1. $\{x \mid x \neq 4\}$ **3.** $\{x \mid x \neq 5, x \neq -4\}$ **5.** $\{x \mid x \neq 7, x \neq -7\}$ **7.** all real numbers **9.** $-\infty$ **11.** $-\infty$ **13.** 0 **15.** $+\infty$
17. $-\infty$ **19.** 1 **21.** vertical asymptote: $x = -4$; no holes **23.** vertical asymptotes: $x = -4, x = 0$; no holes **25.** vertical asymptote: $x = -4$; hole at $x = 0$ **27.** no vertical asymptotes; no holes **29.** no vertical asymptotes; hole at $x = 3$ **31.** vertical asymptote: $x = -3$; hole at $x = 3$
33. vertical asymptote: $x = 3$; hole at $x = -7$ **35.** no vertical asymptotes; hole at $x = -7$ **37.** $y = 0$ **39.** $y = 4$

41. no horizontal asymptote **43.** $y = -\dfrac{2}{3}$

45.
$g(x) = \dfrac{1}{x-1}$

47.
$h(x) = \dfrac{1}{x} + 2$

49.
$g(x) = \dfrac{1}{x+1} - 2$

51.
$g(x) = \dfrac{1}{(x+2)^2}$

53.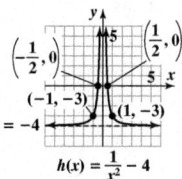
$h(x) = \dfrac{1}{x^2} - 4$

55.
$h(x) = \dfrac{1}{(x-3)^2} + 1$

57.
$f(x) = \dfrac{4x}{x-2}$

59.
$f(x) = \dfrac{2x}{x^2-4}$

61.
$f(x) = \dfrac{2x^2}{x^2-1}$

63.
$f(x) = \dfrac{-x}{x+1}$

65.
$f(x) = -\dfrac{1}{x^2-4}$

67.
$f(x) = \dfrac{2}{x^2+x-2}$

69.
$f(x) = \dfrac{2x^2}{x^2+4}$

71.
$f(x) = \dfrac{x+2}{x^2+x-6}$

73.
$f(x) = \dfrac{x-2}{x^2-4}$

75.
$f(x) = \dfrac{x^4}{x^2+2}$

77.
$f(x) = \dfrac{x^2+x-12}{x^2-4}$

79.
$f(x) = \dfrac{3x^2+x-4}{2x^2-5x}$

81. a. Slant asymptote: $y = x$
b.
$f(x) = \dfrac{x^2-1}{x}$

83. a. Slant asymptote: $y = x$
b.
$f(x) = \dfrac{x^2+1}{x}$

85. a. Slant asymptote: $y = x + 4$ **87. a.** Slant asymptote: $y = x - 2$ **89.** **91.**

b.

$$f(x) = \frac{x^2 + x - 6}{x - 3}$$

b.

$$f(x) = \frac{x^3 + 1}{x^2 + 2x}$$

$$f(x) = \frac{x + 2}{2x(x - 2)}$$

$$f(x) = \frac{x - 6}{2(x - 3)}$$

93.

$$f(x) = \frac{x - 2}{x + 2}$$

95. $g(x) = \dfrac{1}{x + 3} + 2$

$$g(x) = \frac{1}{x + 3} + 2$$

97. $g(x) = \dfrac{-1}{x - 2} + 3$

$$g(x) = \frac{-1}{x - 2} + 3$$

99. a. $C(x) = 100x + 100{,}000$ **b.** $\overline{C}(x) = \dfrac{100x + 100{,}000}{x}$ **c.** $\overline{C}(500) = 300$, when 500 bicycles are produced, it costs $300 to produce each bicycle; $\overline{C}(1000) = 200$, when 1000 bicycles are produced, it costs $200 to produce each bicycle; $\overline{C}(2000) = 150$, when 2000 bicycles are produced, it costs $150 to produce each bicycle; $\overline{C}(4000) = 125$, when 4000 bicycles are produced, it costs $125 to produce each bicycle. **d.** $y = 100$; The cost per bicycle approaches $100 as more bicycles are produced. **101. a.** 6.0 **b.** after 6 minutes; about 4.8 **c.** 6.5 **d.** $y = 6.5$; Over time, the pH level rises back to normal. **e.** It quickly drops below normal and then slowly begins to approach the normal level.
103. 90; An incidence ratio of 10 means 90% of the deaths are smoking related. **105.** $y = 100$; The percentage of deaths cannot exceed 100% as the incidence ratios increase. **107. a.** $f(x) = \dfrac{1.75x^2 - 15.9x + 160}{2.1x^2 - 3.5x + 296}$ **b.** 67% **c.** 66%; underestimates by 1% **d.** $y = \dfrac{1.75}{2.1}$; 83%

119.

The graph approaches the horizontal asymptote faster and the vertical asymptote slower as n increases.

121. a.

b. The graph increases and reaches a maximum of about 356 arrests per 100,000 drivers at age 25. **c.** at age 25, about 356 arrests

123. does not make sense **125.** does not make sense **127.** true **129.** true **134.** $\left\{-3, \dfrac{5}{2}\right\}$ **135.** $\{-2, -1, 2\}$ **136.** $\dfrac{-x - 5}{x + 3}$ or $-\dfrac{x + 5}{x + 3}$

Section 3.6

Check Point Exercises

1. $\{x \mid x < -4 \text{ or } x > 5\}$ or $(-\infty, -4) \cup (5, \infty)$
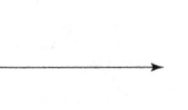

2. $\left\{x \left| \dfrac{-3 - \sqrt{7}}{2} \le x \le \dfrac{-3 + \sqrt{7}}{2}\right.\right\}$ or $\left[\dfrac{-3 - \sqrt{7}}{2}, \dfrac{-3 + \sqrt{7}}{2}\right]$

3. $\{x \mid x \le -3 \text{ or } -1 \le x \le 1\}$ or $(-\infty, -3] \cup [-1, 1]$
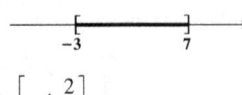
4. $\{x \mid x < -1 \text{ or } x \ge 1\}$ or $(-\infty, -1) \cup [1, \infty)$

5. between 1 and 4 seconds, excluding $t = 1$ and $t = 4$

Concept and Vocabulary Check

1. $x^2 + 8x + 15 = 0$; boundary **2.** $(-\infty, -5); (-5, -3); (-3, \infty)$ **3.** true **4.** true **5.** $(-\infty, -2) \cup [1, \infty)$

Exercise Set 3.6

1. $(-\infty, -2) \cup (4, \infty)$

3. $[-3, 7]$

5. $(-\infty, 1) \cup (4, \infty)$

7. $(-\infty, -4) \cup (-1, \infty)$

9. \varnothing

11. $\left[-4, \dfrac{2}{3}\right]$

13. $\left(-3, \dfrac{5}{2}\right)$

15. $\left(-1, -\dfrac{3}{4}\right)$

17. $\left[-2, \dfrac{1}{3}\right]$

19. $(-\infty, 0] \cup [4, \infty)$

21. $\left(-\infty, -\dfrac{3}{2}\right) \cup (0, \infty)$

23. $[0, 1]$

25. $[2 - \sqrt{2}, 2 + \sqrt{2}]$

27. \varnothing

29. $[1, 2] \cup [3, \infty)$

31. $[0, 3] \cup [5, \infty)$

33. $(-\infty, 2) \cup \left(2, \dfrac{7}{2}\right)$

35. $[-2, -1] \cup [1, \infty)$

37. $(-\infty, -3)$

39. $(-1, \infty)$

41. $\{0\} \cup [9, \infty)$

43. $(-\infty, -3) \cup (4, \infty)$

45. $(-4, -3)$

47. $[2, 4)$

49. $\left(-\infty, -\dfrac{4}{3}\right) \cup [2, \infty)$

51. $(-\infty, 0) \cup (3, \infty)$

53. $(-\infty, -4] \cup (-2, 1]$

55. $(-\infty, -5) \cup (-3, \infty)$

57. $\left(-\infty, \dfrac{1}{2}\right) \cup \left[\dfrac{7}{5}, \infty\right)$

59. $(-\infty, -6] \cup (-2, \infty)$

61. $\left(-\infty, \dfrac{1}{2}\right] \cup [2, \infty)$

63. $(-\infty, -1) \cup [1, \infty)$

65. $(-\infty, -8) \cup (-6, 4) \cup (6, \infty)$

67. $(-3, 2)$

69. $(-\infty, -1) \cup (1, 2) \cup (3, \infty)$

71. $\left[-6, -\dfrac{1}{2}\right] \cup [1, \infty)$ **73.** $(-\infty, -2) \cup [-1, 2)$ **75.** between 0 and $\dfrac{1}{2}$ second **77. a.** dry: 160 ft; wet: 185 ft **b.** dry pavement: graph (b); wet pavement: graph (a) **c.** extremely well; Function values and data are identical. **d.** speeds exceeding 76 miles per hour; points on (b) to the right of $(76, 540)$ **79.** The sides (in feet) are in $(0, 6]$ or $[19, 25)$. **87.** $\left[-3, \dfrac{1}{2}\right]$ **89.** $(1, 4]$ **91.** $(-4, -1) \cup [2, \infty)$
93. a. $f(x) = 0.1375x^2 + 0.7x + 37.8$ **b.** speeds exceeding 52 miles per hour **95.** does not make sense **97.** does not make sense
99. false **101.** true **103.** Answers may vary. One possible solution is $\dfrac{x - 3}{x + 4} \geq 0$. **105.** $\{2\}$ **107.** $(-\infty, 2) \cup (2, \infty)$
109. $27 - 3x^2 \geq 0$
$\quad 3x^2 \leq 27$
$\quad\ \ x^2 \leq 9$
$\ -3 \leq x \leq 3$

110. a. 16 **b.** $y = 16x^2$ **c.** 400 **111. a.** 96 **b.** $y = \dfrac{96}{x}$ **c.** 32 **112.** 8

Section 3.7

Check Point Exercises

1. 66 gal **2.** 9375 lb **3.** 512 cycles per second **4.** 24 min **5.** 96π cubic feet

Concept and Vocabulary Check

1. $y = kx$; constant of variation **2.** $y = kx^n$ **3.** $y = \dfrac{k}{x}$ **4.** $y = \dfrac{kx}{z}$ **5.** $y = kxz$ **6.** directly; inversely **7.** jointly; inversely

Exercise Set 3.7

1. 156 **3.** 30 **5.** $\dfrac{5}{6}$ **7.** 240 **9.** 50 **11.** $x = kyz; y = \dfrac{x}{kz}$ **13.** $x = \dfrac{kz^3}{y}; y = \dfrac{kz^3}{x}$ **15.** $x = \dfrac{kyz}{\sqrt{w}}; y = \dfrac{x\sqrt{w}}{kz}$

17. $x = kz(y + w); y = \dfrac{x - kzw}{kz}$ **19.** $x = \dfrac{kz}{y - w}; y = \dfrac{xw + kz}{x}$ **21.** 5.4 ft **23.** 80 in. **25.** about 607 lb **27.** 32°

29. 90 milliroentgens per hour **31.** This person has a BMI of 24.4 and is not overweight. **33.** 1800 Btu per hour

35. $\dfrac{1}{4}$ of what it was originally **37. a.** $C = \dfrac{kP_1P_2}{d^2}$ **b.** $k \approx 0.02$; $C = \dfrac{0.02\,P_1P_2}{d^2}$ **c.** 39,813 daily phone calls

39. a.

b. Current varies inversely as resistance. **c.** $R = \dfrac{6}{I}$

49. does not make sense **51.** makes sense **53.** The destructive power is four times as much. **55.** Reduce the resistance by a factor of $\frac{1}{3}$.

58.

$f(x) = 2^x$

59.

$g(x) = f(-x) = 2^{-x}$

60.
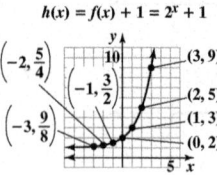
$h(x) = f(x) + 1 = 2^x + 1$

Chapter 3 Review Exercises

1.

$f(x) = -(x+1)^2 + 4$
axis of symmetry; $x = -1$
domain: $(-\infty, \infty)$; range: $(-\infty, 4]$

2.

$f(x) = (x+4)^2 - 2$
axis of symmetry: $x = -4$
domain: $(-\infty, \infty)$; range: $[-2, \infty)$

3.

$f(x) = -x^2 + 2x + 3$
axis of symmetry: $x = 1$
domain: $(-\infty, \infty)$; range: $(-\infty, 4]$

4.

$f(x) = 2x^2 - 4x - 6$
axis of symmetry: $x = 1$
domain: $(-\infty, \infty)$; range: $[-8, \infty)$

5. a. maximum is -57 at $x = 7$ **b.** domain: $(-\infty, \infty)$; range: $(-\infty, -57]$ **6. a.** minimum is 685 at $x = -3$ **b.** domain: $(-\infty, \infty)$; range: $[685, \infty)$
7. a. 16 ft; 20 yd **b.** 6 ft **c.** 45.3 yd **d.** **8.** 250 yd by 500 yd; 125,000 sq yard **9.** -7 and 7; -49

$f(x) = -0.025x^2 + x + 6$

10. c **11.** b **12.** a **13.** d **14. a.** rises to the right **b.** no; The model indicates increasing deforestation despite a declining rate in which the forest is being cut down. **c.** falls to the right **d.** no; The model indicates the amount of forest cleared, in square kilometers, will eventually be negative, which is not possible. **15.** No; the graph falls to the right, so eventually there would be a negative number of thefts, which is not possible.
16. $x = 1$, multiplicity 1, crosses; $x = -2$, multiplicity 2, touches; $x = -5$, multiplicity 3, crosses
17. $x = -5$, multiplicity 1, crosses; $x = 5$, multiplicity 2, touches
18. $f(1)$ is negative and $f(2)$ is positive, so by the Intermediate Value Theorem, f has a real zero between 1 and 2.
19. a. The graph falls to the left and rises to the right. **b.** no symmetry
c.

$f(x) = x^3 - x^2 - 9x + 9$

20. a. The graph rises to the left and falls to the right. **b.** origin symmetry
c.

$f(x) = 4x - x^3$

21. a. The graph falls to the left and rises to the right. **b.** no symmetry
c.

$f(x) = 2x^3 + 3x^2 - 8x - 12$

22. a. The graph falls to the left and to the right. **b.** y-axis symmetry
c.

$f(x) = -x^4 + 25x^2$

23. a. The graph falls to the left and to the right. **b.** no symmetry
c.

$f(x) = -x^4 + 6x^3 - 9x^2$

24. a. The graph rises to the left and to the right. **b.** no symmetry
c.

$f(x) = 3x^4 - 15x^3$

25.

$f(x) = 2x^2 (x-1)^3 (x+2)$

26.
$f(x) = -x^3 (x+4)^2 (x-1)$

27. $4x^2 - 7x + 5 - \dfrac{4}{x+1}$

28. $2x^2 - 4x + 1 - \dfrac{10}{5x-3}$

29. $2x^2 + 3x - 1$ **30.** $3x^3 - 4x^2 + 7$

31. $3x^3 + 6x^2 + 10x + 10 + \dfrac{20}{x-2}$ **32.** -5697

33. $2, \frac{1}{2}, -3$ **34.** $\{4, -2 \pm \sqrt{5}\}$ **35.** $\pm 1, \pm 5$ **36.** $\pm 1, \pm 2, \pm 4, \pm 8, \pm \frac{8}{3}, \pm \frac{4}{3}, \pm \frac{2}{3}, \pm \frac{1}{3}$ **37.** 2 or 0 positive real zeros; no negative real zeros

38. 3 or 1 positive real zeros; 2 or 0 negative real zeros **39.** No sign variations exist for either $f(x)$ or $f(-x)$, so no real roots exist.

40. a. $\pm 1, \pm 2, \pm 4$ **b.** 1 positive real zero; 2 or no negative real zeros **c.** -2 or 1 **d.** $-2, 1$

41. a. $\pm 1, \pm \frac{1}{2}, \pm \frac{1}{3}, \pm \frac{1}{6}$ **b.** 2 or 0 positive real zeros; 1 negative real zero **c.** $-1, \frac{1}{3}$, or $\frac{1}{2}$ **d.** $-1, \frac{1}{3}, \frac{1}{2}$

42. a. $\pm 1, \pm 3, \pm 5, \pm 15, \pm \frac{1}{2}, \pm \frac{1}{4}, \pm \frac{1}{8}, \pm \frac{3}{2}, \pm \frac{3}{4}, \pm \frac{3}{8}, \pm \frac{5}{2}, \pm \frac{5}{4}, \pm \frac{5}{8}, \pm \frac{15}{2}, \pm \frac{15}{4}, \pm \frac{15}{8}$ **b.** 3 or 1 positive real solutions; no negative real solutions

c. $\frac{1}{2}, \frac{3}{2}$, or $\frac{5}{2}$ **d.** $\left\{ \frac{1}{2}, \frac{3}{2}, \frac{5}{2} \right\}$ **43. a.** $\pm 1, \pm \frac{1}{2}$ **b.** 2 or 0 positive real solutions; 1 negative solution **c.** $\frac{1}{2}$ **d.** $\left\{ \frac{1}{2}, \frac{-5 - \sqrt{29}}{2}, \frac{-5 + \sqrt{29}}{2} \right\}$

44. a. $\pm 1, \pm 2, \pm 3, \pm 6$ **b.** 2 or zero positive real solutions; 2 or zero negative real solutions **c.** $-2, -1, 1$, or 3 **d.** $\{-2, -1, 1, 3\}$

45. a. $\pm 1, \pm 2, \pm \frac{1}{2}, \pm \frac{1}{4}$ **b.** 1 positive real root; 1 negative real root **c.** $-\frac{1}{2}$ or $\frac{1}{2}$ **d.** $\left\{ -\frac{1}{2}, \frac{1}{2}, i\sqrt{2}, -i\sqrt{2} \right\}$

46. a. $\pm 1, \pm 2, \pm 4, \pm \frac{1}{2}$ **b.** 2 or no positive zeros; 2 or no negative zeros **c.** $-2, -1, \frac{1}{2}$, or 2 **d.** $-2, -1, \frac{1}{2}, 2$

47. $f(x) = x^3 - 6x^2 + 21x - 26$ **48.** $f(x) = 2x^4 + 12x^3 + 20x^2 + 12x + 18$ **49.** $-2, \frac{1}{2}, \pm i; f(x) = (x - i)(x + i)(x + 2)(2x - 1)$

50. $-1, 4; g(x) = (x + 1)^2 (x - 4)^2$ **51.** 4 real zeros, one with multiplicity two **52.** 3 real zeros; 2 nonreal complex zeros

53. 2 real zeros, one with multiplicity two; 2 nonreal complex zeros **54.** 1 real zero; 4 nonreal complex zeros

55.

$g(x) = \frac{1}{(x + 2)^2} - 1$

56.

$h(x) = \frac{1}{x - 1} + 3$

57. Vertical asymptotes: $x = 3$ and $x = -3$
horizontal asymptote: $y = 0$

$f(x) = \frac{2x}{x^2 - 9}$

58. Vertical asymptote: $x = -3$
horizontal asymptote: $y = 2$

$f(x) = \frac{2x - 4}{x + 3}$

59. Vertical asymptotes: $x = 3$ and $x = -2$
horizontal asymptote: $y = 1$

$h(x) = \frac{x^2 - 3x - 4}{x^2 - x - 6}$

60. Vertical asymptote: $x = -2$
horizontal asymptote: $y = 1$

$r(x) = \frac{x^2 + 4x + 3}{(x + 2)^2}$

61. Vertical asymptote: $x = -1$
no horizontal asymptote
slant asymptote: $y = x - 1$

$y = \frac{x^2}{x + 1}$

62. Vertical asymptote: $x = 3$
no horizontal asymptote
slant asymptote: $y = x + 5$

$y = \frac{x^2 + 2x - 3}{x - 3}$

63. No vertical asymptote
no horizontal asymptote
slant asymptote: $y = -2x$

$f(x) = \frac{-2x^3}{x^2 + 1}$

64. Vertical asymptote: $x = \frac{3}{2}$
no horizontal asymptote
slant asymptote: $y = 2x - 5$

$g(x) = \frac{4x^2 - 16x + 16}{2x - 3}$

65. a. $C(x) = 25x + 50{,}000$
b. $\overline{C}(x) = \frac{25x + 50{,}000}{x}$
c. $\overline{C}(50) = 1025$, when 50 calculators are manufactured, it costs \$1025 to manufacture each; $\overline{C}(100) = 525$, when 100 calculators are manufactured, it costs \$525 to manufacture each; $\overline{C}(1000) = 75$, when 1000 calculators are manufactured, it costs \$75 to manufacture each; $\overline{C}(100{,}000) = 25.5$, when 100,000 calculators are manufactured, it costs \$25.50 to manufacture each.
d. $y = 25$; costs will approach \$25.

66. $y = 3000$; The number of fish in the pond approaches 3000.

67. $y = 0$; As the number of years of education increases the percentage rate of unemployment approaches zero.

68. a. $P(x) = 2.99x + 235.5$ **b.** $R(x) = \frac{1.53x + 114.8}{2.99x + 235.5}$ **c.** $y = 0.51$; Over time the percentage of men in the U.S. population will approach 51%.

69. $\left(-3, \dfrac{1}{2}\right)$

70. $(-\infty, -4] \cup \left[-\dfrac{1}{2}, \infty\right)$

71. $(-3, 0) \cup (1, \infty)$

72. $(-\infty, -2) \cup (6, \infty)$

73. $[-1, 1) \cup [2, \infty)$

74. $(-\infty, 4) \cup \left[\dfrac{23}{4}, \infty\right)$

75. a. 261 ft; overestimates by 1 ft **b.** speeds exceeding 40 miles per hour **76.** from 1 to 2 sec **77.** 134.4 cm^3 **78.** 1600 ft

79. 440 vibrations per second **80.** 112 decibels **81.** 16 hr **82.** 800 ft^3 **83. a.** $L = \dfrac{1890}{R}$ **b.** an approximate model **c.** 70 yr

Chapter 3 Test

1.

$f(x) = (x + 1)^2 + 4$

axis of symmetry: $x = -1$
domain: $(-\infty, \infty)$; range: $[4, \infty)$

2.

$f(x) = x^2 - 2x - 3$

axis of symmetry: $x = 1$
domain: $(-\infty, \infty)$; range: $[-4, \infty)$

3. a. maximum of 2 at $x = 3$;
 b. domain: $(-\infty, \infty)$; range: $(-\infty, 2]$
4. 23 computers;
 maximum daily profit $= \$16,900$
5. 7 and 7; 49

6. a. $5, 2, -2$
b.

$f(x) = x^3 - 5x^2 - 4x + 20$

7. Since the degree of the polynomial is odd and the leading coefficient is positive, the graph of f should fall to the left and rise to the right. The x-intercepts should be $-1, 0,$ and 1.

8. a. 2 **b.** $\dfrac{1}{2}, \dfrac{2}{3}$ **9.** $\pm 1, \pm 2, \pm 3, \pm 6, \pm \dfrac{1}{2}, \pm \dfrac{3}{2}$

10. 3 or 1 positive real zeros; no negative real zeros

11. $\{-3, -3 - \sqrt{11}, -3 + \sqrt{11}\}$ **12. a.** $\pm 1, \pm 3, \pm 5, \pm 15, \pm \dfrac{1}{2}, \pm \dfrac{3}{2}, \pm \dfrac{5}{2}, \pm \dfrac{15}{2}$ **b.** $-\sqrt{5}, -1, \dfrac{3}{2},$ and $\sqrt{5}$ **13.** $(x - 1)(x + 2)^2$

14. $f(x) = 2x^4 - 2$ **15.** -1 and $\dfrac{2}{3}$ **16.** domain: $\{x \mid x \neq -3\}$ **17.** domain: $\{x \mid x \neq 1\}$

$f(x) = -3x^3 - 4x^2 + x + 2$

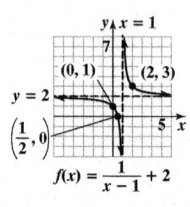

$f(x) = \dfrac{1}{(x + 3)^2}$

$f(x) = \dfrac{1}{x - 1} + 2$

18. domain: $\{x \mid x \neq 4, x \neq -4\}$ **19.** domain: $\{x \mid x \neq 2\}$ **20.** domain: $\{x \mid x \neq -3, x \neq 1\}$ **21.** domain: all real numbers

$f(x) = \dfrac{x}{x^2 - 16}$

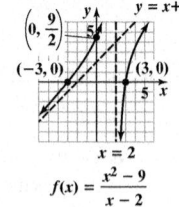

$f(x) = \dfrac{x^2 - 9}{x - 2}$

$f(x) = \dfrac{x + 1}{x^2 + 2x - 3}$

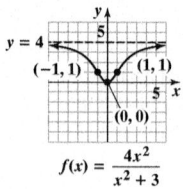

$f(x) = \dfrac{4x^2}{x^2 + 3}$

22. a. $\overline{C}(x) = \dfrac{300,000 + 10x}{x}$ **b.** $y = 10$; As the number of satellite radio players increases, the average cost approaches $10.

23. $(-3, 4)$

24. $(-\infty, 3) \cup [10, \infty)$

25. 45 foot-candles

Cumulative Review Exercises (Chapters 1–3)

1. domain: $(-2, 2)$; range: $[0, \infty)$ **2.** -1 and 1, both of multiplicity 2 **3.** 0 **4.** 3 **5.** $x \to -2^+$; $x \to 2^-$

6.
$g(x) = f(x + 2) + 1$

7. $\{2, -1\}$ **8.** $\left\{ \dfrac{5 + \sqrt{13}}{6}, \dfrac{5 - \sqrt{13}}{6} \right\}$ **9.** $\left\{ \dfrac{1}{3}, -\dfrac{2}{3} \right\}$ **10.** $\{-3, -1, 2\}$ **11.** $(-\infty, 1) \cup (4, \infty)$

12. $(-\infty, -1) \cup \left(\dfrac{5}{3}, \infty \right)$

13.
$f(x) = x^3 - 4x^2 - x + 4$

14.
$f(x) = x^2 + 2x - 8$

15.
$f(x) = x^2(x - 3)$

16.
$f(x) = \dfrac{x - 1}{x - 2}$

17.

18.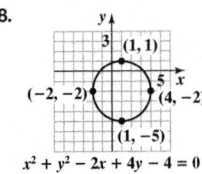
$x^2 + y^2 - 2x + 4y - 4 = 0$

19. $(f \circ g)(x) = 32x^2 - 20x + 2$ **20.** $4x + 2h - 1$

CHAPTER 4

Section 4.1

Check Point Exercises

1. approximately \$160; overestimates by \$11

2.
$f(x) = 3^x$

3.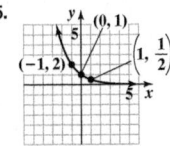
$f(x) = \left(\dfrac{1}{3} \right)^x$

4.
$f(x) = 3^x$
$g(x) = 3^{x-1}$

5.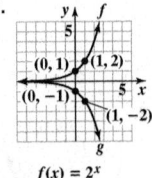
$f(x) = 2^x$
$g(x) = 2^x + 1$

6. approximately 4446 **7. a.** \$14,859.47 **b.** \$14,918.25

Concept and Vocabulary Check

1. b^x; $(-\infty, \infty)$; $(0, \infty)$ **2.** x; $y = 0$; horizontal **3.** e; natural; 2.72 **4.** A; P; r; n **5.** semiannually; quarterly; continuous

Exercise Set 4.1

1. 10.556 **3.** 11.665 **5.** 0.125 **7.** 9.974 **9.** 0.387

11.
$f(x) = 4^x$

13.
$g(x) = \left(\dfrac{3}{2} \right)^x$

15.
$h(x) = \left(\dfrac{1}{2} \right)^x$

17.
$f(x) = (0.6)^x$

19. $H(x) = -3^{-x}$
21. $F(x) = -3^x$
23. $h(x) = 3^x - 1$

25.
$f(x) = 2^x$
$g(x) = 2^x + 1$
asymptote: $y = 0$
domain: $(-\infty, \infty)$
range: $(0, \infty)$

27.
$f(x) = 2^x$
$g(x) = 2^x - 1$
asymptote: $y = -1$
domain: $(-\infty, \infty)$
range: $(-1, \infty)$

29.
$f(x) = 2^x$
$h(x) = 2^{x+1} - 1$
asymptote: $y = -1$
domain: $(-\infty, \infty)$
range: $(-1, \infty)$

31.
$f(x) = 2^x$
$g(x) = -2^x$
asymptote: $y = 0$
domain: $(-\infty, \infty)$
range: $(-\infty, 0)$

33.
$f(x) = 2^x$
$g(x) = 2 \cdot 2^x$
asymptote: $y = 0$
domain: $(-\infty, \infty)$
range: $(0, \infty)$

35.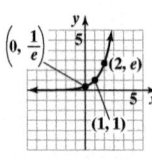
$g(x) = e^{x-1}$

asymptote: $y = 0$
domain: $(-\infty, \infty)$
range: $(0, \infty)$

37.
$g(x) = e^x + 2$

asymptote: $y = 2$
domain: $(-\infty, \infty)$
range: $(2, \infty)$

39.
$h(x) = e^{x-1} + 2$

asymptote: $y = 2$
domain: $(-\infty, \infty)$
range: $(2, \infty)$

41.
$h(x) = e^{-x}$

asymptote: $y = 0$
domain: $(-\infty, \infty)$
range: $(0, \infty)$

43.
$g(x) = 2e^x$

asymptote: $y = 0$
domain: $(-\infty, \infty)$
range: $(0, \infty)$

45.
$h(x) = e^{2x} + 1$

asymptote: $y = 1$
domain: $(-\infty, \infty)$
range: $(1, \infty)$

47.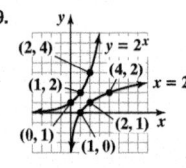
$f(x) = 3^x$
$g(x) = 3^{-x}$

asymptote of f: $y = 0$
asymptote of g: $y = 0$

49.
$f(x) = 3^x$
$g(x) = \frac{1}{3} \cdot 3^x$

asymptote of f: $y = 0$
asymptote of g: $y = 0$

51.
$f(x) = \left(\frac{1}{2}\right)^x$
$g(x) = \left(\frac{1}{2}\right)^{x-1} + 1$

asymptote of f: $y = 0$
asymptote of g: $y = 1$

53. a. $13,116.51
b. $13,140.67
c. $13,157.04
d. $13,165.31
55. 7% compounded monthly

57. $(0, 1)$
59.
$g(x) = 2^{-x}$ $f(x) = 2^x$

61. $y = 4^x$ **63.** $y = -e^x$ **65. a.** 574 million **b.** 1148 million
c. 2295 million **d.** 4590 million **e.** It appears to double. **67.** $832,744
69. 3.249009585; 3.317278183; 3.321880096; 3.321995226; 3.321997068; $2^{\sqrt{3}} \approx 3.321997085$;
The closer the exponent is to $\sqrt{3}$, the closer the value is to $2^{\sqrt{3}}$.
71. a. $15,166 **b.** $16,037 **c.** linear model **73. a.** 100% **b.** $\approx 68.5\%$
c. $\approx 30.8\%$ **d.** $\approx 20\%$ **75.** the function g

81. a. $A = 10,000\left(1 + \dfrac{0.05}{4}\right)^{4t}$; $A = 10,000\left(1 + \dfrac{0.045}{12}\right)^{12t}$
b.

5% interest compounded quarterly

83. does not make sense
85. does not make sense
87. false
89. false

91. $y = 3^x$ is (d); $y = 5^x$ is (c); $y = \left(\dfrac{1}{3}\right)^x$ is (a); $y = \left(\dfrac{1}{5}\right)^x$ is (b).

93. a. $\cosh(-x) = \dfrac{e^{-x} + e^{-(-x)}}{2} = \dfrac{e^{-x} + e^x}{2} = \dfrac{e^x + e^{-x}}{2} = \cosh x$

b. $\sinh(-x) = \dfrac{e^{-x} - e^{-(-x)}}{2} = \dfrac{e^{-x} - e^x}{2} = -\dfrac{e^x - e^{-x}}{2} = -\sinh x$

c. $\left(\dfrac{e^x + e^{-x}}{2}\right)^2 - \left(\dfrac{e^x - e^{-x}}{2}\right)^2 \overset{?}{=} 1$

$\dfrac{e^{2x} + 2 + e^{-2x}}{4} - \dfrac{e^{2x} - 2 + e^{-2x}}{4} \overset{?}{=} 1$

$\dfrac{e^{2x} + 2 + e^{-2x} - e^{2x} + 2 - e^{-2x}}{4} \overset{?}{=} 1$

$\dfrac{4}{4} \overset{?}{=} 1$

$1 = 1$

94. We don't know how to solve $x = 2^y$ for y. **95.** $\dfrac{1}{2}$ **96.** $(-\infty, 3) \cup (3, \infty)$

Section 4.2

Check Point Exercises

1. a. $7^3 = x$ **b.** $b^2 = 25$ **c.** $4^y = 26$ **2. a.** $5 = \log_2 x$ **b.** $3 = \log_b 27$ **c.** $y = \log_e 33$ **3. a.** 2 **b.** -3 **c.** $\dfrac{1}{2}$ **d.** $\dfrac{1}{7}$
4. a. 1 **b.** 0 **5. a.** 8 **b.** 17 **6.** **7.** $(5, \infty)$ **8.** 80% **9.** 4.0 **10. a.** $(-\infty, 4)$ **b.** $(-\infty, 0) \cup (0, \infty)$
$f(x) = 3^x$
$g(x) = \log_3 x$
11. 34°; extremely well

Concept and Vocabulary Check

1. $b^y = x$　**2.** logarithmic; b　**3.** 1　**4.** 0　**5.** x　**6.** x　**7.** $(0, \infty)$; $(-\infty, \infty)$　**8.** y; $x = 0$; vertical　**9.** up 5 units
10. to the left 5 units　**11.** x-axis　**12.** y-axis　**13.** $5 - x > 0$　**14.** common; $\log x$　**15.** natural; $\ln x$

Exercise Set 4.2

1. $2^4 = 16$　**3.** $3^2 = x$　**5.** $b^5 = 32$　**7.** $6^y = 216$　**9.** $\log_2 8 = 3$　**11.** $\log_2 \dfrac{1}{16} = -4$　**13.** $\log_8 2 = \dfrac{1}{3}$　**15.** $\log_{13} x = 2$

17. $\log_b 1000 = 3$　**19.** $\log_7 200 = y$　**21.** 2　**23.** 6　**25.** -1　**27.** -3　**29.** $\dfrac{1}{2}$　**31.** $-\dfrac{1}{2}$　**33.** $\dfrac{1}{2}$　**35.** 1　**37.** 0　**39.** 7　**41.** 19

43.

$f(x) = 4^x$
$g(x) = \log_4 x$

45.

$f(x) = \left(\dfrac{1}{2}\right)^x$
$g(x) = \log_{1/2} x$

47. $H(x) = 1 - \log_3 x$　**49.** $h(x) = \log_3 x - 1$　**51.** $g(x) = \log_3(x - 1)$

53.

$f(x) = \log_2 x$
$g(x) = \log_2(x+1)$

vertical asymptote: $x = -1$
domain: $(-1, \infty)$
range: $(-\infty, \infty)$

55.

$f(x) = \log_2 x$
$h(x) = 1 + \log_2 x$

vertical asymptote: $x = 0$
domain: $(0, \infty)$
range: $(-\infty, \infty)$

57.

$f(x) = \log_2 x$
$g(x) = \dfrac{1}{2} \log_2 x$

vertical asymptote: $x = 0$
domain: $(0, \infty)$
range: $(-\infty, \infty)$

59.

$g(x) = \log(x - 1)$

asymptote: $x = 1$
domain: $(1, \infty)$
range: $(-\infty, \infty)$

61.

$h(x) = \log x - 1$

asymptote: $x = 0$
domain: $(0, \infty)$
range: $(-\infty, \infty)$

63.

$g(x) = 1 - \log x$

asymptote: $x = 0$
domain: $(0, \infty)$
range: $(-\infty, \infty)$

65.

$g(x) = \ln(x + 2)$

asymptote: $x = -2$
domain: $(-2, \infty)$
range: $(-\infty, \infty)$

67.

$h(x) = \ln(2x)$

asymptote: $x = 0$
domain: $(0, \infty)$
range: $(-\infty, \infty)$

69.

$g(x) = 2 \ln x$

asymptote: $x = 0$
domain: $(0, \infty)$
range: $(-\infty, \infty)$

71.

$h(x) = -\ln x$

asymptote: $x = 0$
domain: $(0, \infty)$
range: $(-\infty, \infty)$

73.

$g(x) = 2 - \ln x$

asymptote: $x = 0$
domain: $(0, \infty)$
range: $(-\infty, \infty)$

75. $(-4, \infty)$　**77.** $(-\infty, 2)$　**79.** $(-\infty, 2) \cup (2, \infty)$　**81.** 2　**83.** 7　**85.** 33　**87.** 0　**89.** 6　**91.** -6　**93.** 125　**95.** $9x$

97. $5x^2$　**99.** \sqrt{x}　**101.** $3^2 = x - 1$; $\{10\}$　**103.** $4^{-3} = x$; $\left\{\dfrac{1}{64}\right\}$　**105.** 0　**107.** 2　**109.** $(-\infty, -1) \cup (2, \infty)$　**111.** $(-\infty, -1) \cup (5, \infty)$

113. 95.4%　**115. a.** 25.5%; underestimates by 0.7%　**b.** 24.2%　**117.** ≈ 188 db; yes

119. a. 88
b. 71.5; 63.9; 58.8; 55; 52; 49.5
c.

Material retention decreases as
time passes.

129.

$g(x)$ is $f(x)$ shifted upward
3 units.

131.

$g(x)$ is $f(x)$ shifted right 2 units
and upward 1 unit.

133. a.

b.

c.

d. They are the same.
$\log_b M + \log_b N$
e. the sum of the logarithms
of its factors

135. makes sense **137.** makes sense **139.** false **141.** false **143.** $\dfrac{4}{5}$ **145.** $\log_3 40 > \log_4 60$

147. a. 5 **b.** 5 **c.** $\log_2(8 \cdot 4) = \log_2 8 + \log_2 4$ **148. a.** 4 **b.** 4 **c.** $\log_2\left(\dfrac{32}{2}\right) = \log_2 32 - \log_2 2$

149. a. 4 **b.** 4 **c.** $\log_3 9^2 = 2 \log_3 9$

Section 4.3
Check Point Exercises

1. a. $\log_6 7 + \log_6 11$ **b.** $2 + \log x$ **2. a.** $\log_8 23 - \log_8 x$ **b.** $5 - \ln 11$ **3. a.** $9 \log_6 3$ **b.** $\dfrac{1}{3} \ln x$ **c.** $2 \log (x + 4)$

4. a. $4 \log_b x + \dfrac{1}{3} \log_b y$ **b.** $\dfrac{1}{2} \log_5 x - 2 - 3 \log_5 y$ **5. a.** $\log 100 = 2$ **b.** $\log \dfrac{7x + 6}{x}$ **6. a.** $\ln x^2 \sqrt[3]{x + 5}$ **b.** $\log \dfrac{(x - 3)^2}{x}$

c. $\log_b \dfrac{\sqrt[4]{x}}{25 y^{10}}$ **7.** 4.02 **8.** 4.02

Concept and Vocabulary Check

1. $\log_b M + \log_b N$; sum **2.** $\log_b M - \log_b N$; difference **3.** $p \log_b M$; product **4.** $\dfrac{\log_a M}{\log_a b}$

Exercise Set 4.3

1. $\log_5 7 + \log_5 3$ **3.** $1 + \log_7 x$ **5.** $3 + \log x$ **7.** $1 - \log_7 x$ **9.** $\log x - 2$ **11.** $3 - \log_4 y$ **13.** $2 - \ln 5$ **15.** $3 \log_b x$

17. $-6 \log N$ **19.** $\dfrac{1}{5} \ln x$ **21.** $2 \log_b x + \log_b y$ **23.** $\dfrac{1}{2} \log_4 x - 3$ **25.** $2 - \dfrac{1}{2} \log_6 (x + 1)$ **27.** $2 \log_b x + \log_b y - 2 \log_b z$

29. $1 + \dfrac{1}{2} \log x$ **31.** $\dfrac{1}{3} \log x - \dfrac{1}{3} \log y$ **33.** $\dfrac{1}{2} \log_b x + 3 \log_b y - 3 \log_b z$ **35.** $\dfrac{2}{3} \log_5 x + \dfrac{1}{3} \log_5 y - \dfrac{2}{3}$ **37.** $3 \ln x + \dfrac{1}{2} \ln(x^2 + 1) - 4 \ln(x + 1)$

39. $1 + 2 \log x + \dfrac{1}{3} \log (1 - x) - \log 7 - 2 \log (x + 1)$ **41.** 1 **43.** $\ln(7x)$ **45.** 5 **47.** $\log\left(\dfrac{2x + 5}{x}\right)$ **49.** $\log (xy^3)$

51. $\ln(x^{1/2} y)$ or $\ln(y \sqrt{x})$ **53.** $\log_b (x^2 y^3)$ **55.** $\ln\left(\dfrac{x^5}{y^2}\right)$ **57.** $\ln\left(\dfrac{x^3}{y^{1/3}}\right)$ or $\ln\left(\dfrac{x^3}{\sqrt[3]{y}}\right)$ **59.** $\ln \dfrac{(x + 6)^4}{x^3}$ **61.** $\ln\left(\dfrac{x^3 y^5}{z^6}\right)$ **63.** $\log \sqrt{xy}$

65. $\log_5\left(\dfrac{\sqrt{xy}}{(x + 1)^2}\right)$ **67.** $\ln \sqrt[3]{\dfrac{(x + 5)^2}{x(x^2 - 4)}}$ **69.** $\log \dfrac{x(x^2 - 1)}{7(x + 1)} = \log \dfrac{x(x - 1)}{7}$ **71.** 1.5937 **73.** 1.6944 **75.** -1.2304 **77.** 3.6193

79.

81.

83. $C - A$ **85.** $3A$ **87.** $\dfrac{1}{2} A - \dfrac{3}{2} C$ **89.** false; $\ln e = 1$ **91.** false; $\log_4(2x)^3 = 3 \log_4(2x)$ **93.** true **95.** true

97. false; $\log(x + 3) - \log(2x) = \log\left(\dfrac{x + 3}{2x}\right)$ **99.** true **101.** true **103. a.** $D = 10 \log \dfrac{I}{I_0}$ **b.** 20 decibels louder

113. a.

b.

$y = 2 + \log_3 x$ shifts the graph of $y = \log_3 x$ two units upward;
$y = \log_3(x + 2)$ shifts the graph of $y = \log_3 x$ two units left; $y = -\log_3 x$
reflects the graph of $y = \log_3 x$ about the x-axis.

115.

a. top graph: $y = \log_{100} x$; bottom graph: $y = \log_3 x$
b. top graph: $y = \log_3 x$; bottom graph: $y = \log_{100} x$
c. The graph of the equation with the largest b will be on the top in the interval $(0, 1)$ and on the bottom in the interval $(1, \infty)$.

121. makes sense **123.** makes sense **125.** true **127.** false **129.** $\log e = \log_{10} e = \dfrac{\ln e}{\ln 10} = \dfrac{1}{\ln 10}$ **131.** $4x^3$

133. Let $\log_b M = R$ and $\log_b N = S$.
Then $\log_b M = R$ means $b^R = M$ and $\log_b N = S$ means $b^S = N$.

$$\frac{M}{N} = \frac{b^R}{b^S} = b^{R-S}$$

$$\log_b \frac{M}{N} = \log_b b^{R-S} = R - S = \log_b M - \log_b N$$

134. $x = \dfrac{2a + 3b}{a - 2b}$ **135.** $\left\{\dfrac{7 \pm \sqrt{61}}{2}\right\}$ **136.** $\{-1, 3\}$

Mid-Chapter 4 Check Point

1.

$f(x) = 2^x$
$g(x) = 2^x - 3$

asymptote of f: $y = 0$
asymptote of g: $y = -3$
domain of $f =$ domain of $g = (-\infty, \infty)$
range of $f = (0, \infty)$; range of $g = (-3, \infty)$

2.

$f(x) = \left(\dfrac{1}{2}\right)^x$

$g(x) = \left(\dfrac{1}{2}\right)^{x-1}$

asymptote of f: $y = 0$
asymptote of g: $y = 0$
domain of $f =$ domain of $g = (-\infty, \infty)$
range of $f =$ range of $g = (0, \infty)$

3.

$f(x) = e^x$
$g(x) = \ln x$

asymptote of f: $y = 0$
asymptote of g: $x = 0$
domain of $f =$ range of $g = (-\infty, \infty)$
range of $f =$ domain of $g = (0, \infty)$

4.

$f(x) = \log_2 x$
$g(x) = \log_2 (x - 1) + 1$

asymptote of f: $x = 0$
asymptote of g: $x = 1$
domain of $f = (0, \infty)$; domain of $g = (1, \infty)$
range of $f =$ range of $g = (-\infty, \infty)$

5.

$f(x) = \log_{1/2} x$
$g(x) = -2\log_{1/2} x$

asymptote of f: $x = 0$
asymptote of g: $x = 0$
domain of $f =$ domain of $g = (0, \infty)$
range of $f =$ range of $g = (-\infty, \infty)$

6. $(-6, \infty)$ **7.** $(0, \infty)$ **8.** $(-\infty, -6) \cup (-6, \infty)$ **9.** $(-\infty, \infty)$ **10.** 5 **11.** -2 **12.** $\dfrac{1}{2}$ **13.** $\dfrac{1}{3}$ **14.** 2

15. Evaluation not possible; $\log_2 \dfrac{1}{8} = -3$ and $\log_3(-3)$ is undefined. **16.** 5 **17.** $\sqrt{7}$ **18.** 13 **19.** $-\dfrac{1}{2}$ **20.** $\sqrt{\pi}$

21. $\dfrac{1}{2}\log x + \dfrac{1}{2}\log y - 3$ **22.** $19 + 20 \ln x$ **23.** $\log_7\left(\dfrac{x^8}{\sqrt[3]{y}}\right)$ **24.** $\log_5 x^9$ **25.** $\ln\left[\dfrac{\sqrt{x}}{y^3(z - 2)}\right]$ **26.** \$8

Section 4.4

Check Point Exercises

1. a. $\{3\}$ **b.** $\{-12\}$ **2. a.** $\left\{\dfrac{\ln 134}{\ln 5}\right\}$; ≈ 3.04 **b.** $\{\log 8000\}$; ≈ 3.90 **3.** $\left\{\dfrac{\ln 9}{2}\right\}$; ≈ 1.10 **4.** $\left\{\dfrac{\ln 3 + \ln 7}{2\ln 3 - \ln 7}\right\}$; ≈ 12.11

5. $\{0, \ln 7\}$; $\ln 7 \approx 1.95$ **6. a.** $\{12\}$ **b.** $\left\{\dfrac{e^2}{3}\right\}$ **7.** $\{5\}$ **8.** $\{4, 5\}$ **9.** 0.01 **10.** 16.2 yr **11.** 14

Concept and Vocabulary Check

1. $M = N$ **2.** $4x - 1$ **3.** $\dfrac{\ln 20}{\ln 9}$ **4.** $\ln 6$ **5.** 5^3 **6.** $(x^2 + x)$ **7.** $\dfrac{7x - 23}{x + 1}$

8. false **9.** true **10.** false **11.** true

Exercise Set 4.4

1. $\{6\}$ **3.** $\{3\}$ **5.** $\{3\}$ **7.** $\{2\}$ **9.** $\left\{\dfrac{3}{5}\right\}$ **11.** $\left\{\dfrac{3}{2}\right\}$ **13.** $\{4\}$ **15.** $\{5\}$ **17.** $\left\{-\dfrac{1}{4}\right\}$ **19.** $\{13\}$ **21.** $\{-2\}$

23. $\left\{\dfrac{\ln 3.91}{\ln 10}\right\}$; ≈ 0.59 **25.** $\{\ln 5.7\}$; ≈ 1.74 **27.** $\left\{\dfrac{\ln 17}{\ln 5}\right\}$; ≈ 1.76 **29.** $\left\{\ln \dfrac{23}{5}\right\}$; ≈ 1.53 **31.** $\left\{\dfrac{\ln 659}{5}\right\}$; ≈ 1.30

33. $\left\{\dfrac{\ln 793 - 1}{-5}\right\}$; ≈ -1.14 **35.** $\left\{\dfrac{\ln 10,478 + 3}{5}\right\}$; ≈ 2.45 **37.** $\left\{\dfrac{\ln 410}{\ln 7} - 2\right\}$; ≈ 1.09 **39.** $\left\{\dfrac{\ln 813}{0.3 \ln 7}\right\}$; ≈ 11.48 **41.** $\left\{\dfrac{3 \ln 5 + \ln 3}{\ln 3 - 2 \ln 5}\right\}$; ≈ -2.80

43. $\{0, \ln 2\}$; $\ln 2 \approx 0.69$ **45.** $\left\{\dfrac{\ln 3}{2}\right\}$; ≈ 0.55 **47.** $\{0\}$ **49.** $\{81\}$ **51.** $\{e^2\}$; ≈ 7.39 **53.** $\{59\}$ **55.** $\{-9\}$ **57.** $\left\{-\dfrac{107}{27}\right\}$ **59.** $\left\{\dfrac{62}{3}\right\}$

61. $\left\{\dfrac{e^4}{2}\right\}$; ≈ 27.30 **63.** $\{e^{-1/2}\}$; ≈ 0.61 **65.** $\{e^2 - 3\}$; ≈ 4.39 **67.** $\left\{\dfrac{5}{4}\right\}$ **69.** $\{-3\}$ **71.** $\{6\}$ **73.** $\{5\}$ **75.** $\{12\}$ **77.** $\left\{\dfrac{4}{3}\right\}$ **79.** \varnothing

81. $\{5\}$ **83.** $\left\{\dfrac{2}{9}\right\}$ **85.** $\{28\}$ **87.** $\{2\}$ **89.** \varnothing **91.** $\left\{\dfrac{11}{3}\right\}$ **93.** $\left\{\dfrac{1}{2}\right\}$ **95.** $\{e^3, e^{-3}\}$ **97.** $\left\{\pm\sqrt{\dfrac{\ln 45}{\ln 3}}\right\}$ **99.** $\left\{\dfrac{5 + \sqrt{37}}{2}\right\}$

101. $\{-2, 6\}$ **103. a.** 37.3 million **b.** 2017 **105.** 118 ft; by the point $(118, 1)$ **107.** 8.2 yr **109.** 2.0 yr **111.** 8.7 yr **113.** 7.0 yr
115. a. 17.0%; underestimates by 0.3% **b.** 2016 **117.** 2.8 days; $(2.8, 50)$ **119. a.** $10^{-5.6}$ mole per liter **b.** $10^{-2.4}$ mole per liter
c. $10^{3.2}$ times greater **125.** $\{2\}$ **127.** $\{4\}$ **129.** $\{2\}$ **131.** $\{-1.391606, 1.6855579\}$

133. 30

0 500
27

As distance from eye increases, barometric air pressure increases.

135. 150

Y1=1454^(-.092X)

0 X=7.9787 Y=69.5949 15
0

about 7.9 min

137. does not make sense **139.** makes sense **141.** false **143.** true **145.** after 36 yr **147.** $\{10^{-2}, 10^{3/2}\}$, $10^{3/2} = 10\sqrt{10} \approx 31.62$
150. a. 10 million; 9.97 million; 9.94 million; 9.91 million **b.** decreasing **151. a.** 3 **b.** $e^{(\ln 3)x}$
152.

Annual Retail Sales of Call of Duty Games

Retail Sales (millions of dollars)

- (2010, 980)
- (2009, 778)
- (2008, 436)
- (2007, 352)
- (2006, 196)
- (2005, 101)
- (2004, 56)

2004 2008 2012 x
Year

Section 4.5

Check Point Exercises

1. a. $A = 807e^{0.024t}$ **b.** 2038 **2. a.** $A = A_0 e^{-0.0248t}$ **b.** about 72 yr **3. a.** 0.4 correct responses **b.** 0.7 correct responses
c. 0.8 correct responses
4. ; logarithmic function **5.** ; exponential function, although answers may vary

Walking Speed (feet per second)
- (138, 1.9)
- (342, 2.2)
- (71, 1.6)
- (14, 1.0)
- (5.5, 0.6)

0 100 200 300 400 x
Population (thousands)

Percent of U.S. Men Who Are Married or Have Been Married
- (35, 75)
- (18, 2)
- (30, 61)
- (25, 36)
- (20, 7)

18 22 26 30 34 38 x
Age

6. a. the exponential function g **b.** the linear function f **7.** $y = 4e^{(\ln 7.8)x}$; $y = 4e^{2.054x}$

Concept and Vocabulary Check

1. > 0; < 0 **2.** A_0; A **3.** A; c **4.** logarithmic **5.** exponential **6.** linear **7.** $\ln 5$

Exercise Set 4.5

1. 127.3 million **3.** Iraq; 1.9% **5.** 2030 **7. a.** $A = 6.04e^{0.01t}$ **b.** 2040 **9.** 146.1 million **11.** 0.0088 **13.** -0.0039
15. approximately 8 grams **17.** 8 grams after 10 seconds; 4 grams after 20 seconds; 2 grams after 30 seconds; 1 gram after 40 seconds; 0.5 gram
after 50 seconds **19.** approximately 15,679 years old **21.** 12.6 yr **23.** -0.000428; 0.0428% per year **25.** -0.039608; 3.9608% per day

27. a. $\dfrac{1}{2} = e^{1.31k}$ yields $k = \dfrac{\ln\left(\dfrac{1}{2}\right)}{1.31} \approx -0.52912$. **b.** about 0.1069 billion or 106,900,000 years old **29.** 7.1 yr **31.** 5.5 hr

33. $2A_0 = A_0 e^{kt}$; $2 = e^{kt}$; $\ln 2 = \ln e^{kt}$; $\ln 2 = kt$; $\dfrac{\ln 2}{k} = t$ **35. a.** 1% **b.** about 69 yr **37. a.** about 20 people **b.** about 1080 people

c. 100,000 people **39.** quite well **41.** 2025 **43.** about 3.7% **45.** about 48 years old

47. a.

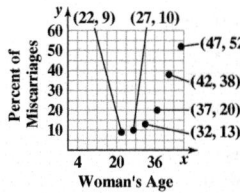

b. exponential function

49. a.

b. logarithmic function

51. a.

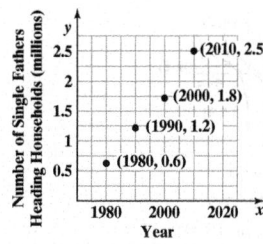

b. linear function

53. $y = 100e^{(\ln 4.6)x}$; $y = 100e^{1.526x}$ **55.** $y = 2.5e^{(\ln 0.7)x}$; $y = 2.5e^{-0.357x}$ **67.** $y = 191.9 + 24.569 \ln x$; $r \approx 0.870$; Fit is ok, but not great.
69. $y = 195.056x^{0.100}$; $r \approx 0.896$; Fits data fairly well. **71. a.** $y = 3.46(1.02)^x$ **b.** $y = 3.46e^{(\ln 1.02)x}$; $y = 3.46e^{0.02x}$; by approximately 2%
73. does not make sense **75.** makes sense **77.** true **79.** true **81.** about 126° F **83. a.** yes **b.** yes
84.

 ; $(4, -1)$ **85.** $\{4\}$

$x + 2y = 2$
$x - 2y = 6$

Chapter 4 Review Exercises

1. $g(x) = 4^{-x}$ **2.** $h(x) = -4^{-x}$ **3.** $r(x) = -4^{-x} + 3$ **4.** $f(x) = 4^x$

5.

$f(x) = 2^x$
$g(x) = 2^{x-1}$

asymptote of f: $y = 0$
asymptote of g: $y = 0$
domain of f = domain of $g = (-\infty, \infty)$
range of f = range of $g = (0, \infty)$

6.

$f(x) = 3^x$
$g(x) = 3^x - 1$

asymptote of f: $y = 0$
asymptote of g: $y = -1$
domain of f = domain of $g = (-\infty, \infty)$
range of $f = (0, \infty)$; range of $g = (-1, \infty)$

7.

$f(x) = 3^x$
$g(x) = -3^x$

asymptote of f: $y = 0$
asymptote of g: $y = 0$
domain of f = domain of $g = (-\infty, \infty)$
range of $f = (0, \infty)$; range of $g = (-\infty, 0)$

8.

$f(x) = \left(\frac{1}{2}\right)^x$

$g(x) = \left(\frac{1}{2}\right)^{-x}$

asymptote of f: $y = 0$
asymptote of g: $y = 0$
domain of f = domain of $g = (-\infty, \infty)$
range of f = range of $g = (0, \infty)$

9.

$f(x) = e^x$
$g(x) = 2e^{x/2}$

asymptote of f: $y = 0$
asymptote of g: $y = 0$
domain of f = domain of $g = (-\infty, \infty)$
range of f = range of $g = (0, \infty)$

10. 5.5% compounded semiannually **11.** 7% compounded monthly **12. a.** 200° **b.** 120°; 119° **c.** 70°; The temperature in the room is 70°.
13. $49^{1/2} = 7$ **14.** $4^3 = x$ **15.** $3^y = 81$ **16.** $\log_6 216 = 3$ **17.** $\log_b 625 = 4$ **18.** $\log_{13} 874 = y$ **19.** 3 **20.** -2 **21.** undefined;

$\log_b x$ is defined only for $x > 0$. **22.** $\frac{1}{2}$ **23.** 1 **24.** 8 **25.** 5 **26.** $-\frac{1}{2}$ **27.** -2 **28.** -3 **29.** 0

30.

$f(x) = 2^x$
$g(x) = \log_2 x$

domain of f = range of $g = (-\infty, \infty)$
range of f = domain of $g = (0, \infty)$

31.

$f(x) = \left(\frac{1}{3}\right)^x$

$g(x) = \log_{1/3} x$

domain of f = range of $g = (-\infty, \infty)$
range of f = domain of $g = (0, \infty)$

32. $g(x) = \log(-x)$
33. $r(x) = 1 + \log(2 - x)$
34. $h(x) = \log(2 - x)$
35. $f(x) = \log x$

36.

$f(x) = \log_2 x$
$g(x) = \log_2(x - 2)$

x-intercept: $(3, 0)$
vertical asymptote: $x = 2$
domain: $(2, \infty)$; range: $(-\infty, \infty)$

37.

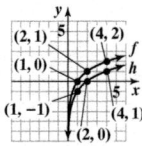

$f(x) = \log_2 x$
$h(x) = -1 + \log_2 x$

x-intercept: $(2, 0)$
vertical asymptote: $x = 0$
domain: $(0, \infty)$; range: $(-\infty, \infty)$

38.

$f(x) = \log_2 x$
$r(x) = \log_2(-x)$

x-intercept: $(-1, 0)$
vertical asymptote: $x = 0$
domain: $(-\infty, 0)$; range: $(-\infty, \infty)$

39.

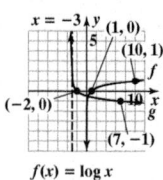

$f(x) = \log x$
$g(x) = -\log(x + 3)$

asymptote of f: $x = 0$
asymptote of g: $x = -3$
domain of $f = (0, \infty)$; domain of $g = (-3, \infty)$
range of f = range of $g = (-\infty, \infty)$

40.

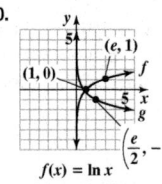

$f(x) = \ln x$
$g(x) = -\ln(2x)$

asymptote of f: $x = 0$
asymptote of g: $x = 0$
domain of f = domain of $g = (0, \infty)$
range of f = range of $g = (-\infty, \infty)$

41. $(-5, \infty)$
42. $(-\infty, 3)$
43. $(-\infty, 1) \cup (1, \infty)$
44. $6x$
45. \sqrt{x}
46. $4x^2$
47. 3.0

48. a. 76 **b.** $\approx 67, \approx 63, \approx 61, \approx 59, \approx 56$
c.

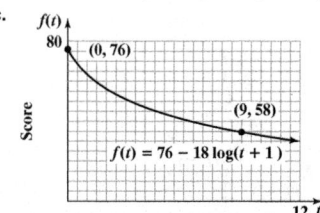

$f(t) = 76 - 18\log(t + 1)$

49. about 9 weeks **50.** $2 + 3\log_6 x$ **51.** $\frac{1}{2}\log_4 x - 3$

52. $\log_2 x + 2\log_2 y - 6$ **53.** $\frac{1}{3}\ln x - \frac{1}{3}$ **54.** $\log_b 21$ **55.** $\log\frac{3}{x^3}$

56. $\ln(x^3 y^4)$ **57.** $\ln\frac{\sqrt{x}}{y}$ **58.** 6.2448 **59.** -0.1063

60. true **61.** false; $\log(x + 9) - \log(x + 1) = \log\left(\frac{x + 9}{x + 1}\right)$

62. false; $4\log_2 x = \log_2 x^4$ **63.** true **64.** $\{2\}$ **65.** $\left\{\frac{2}{3}\right\}$ **66.** $\{\log 7000\}$; ≈ 3.85

67. $\left\{-\frac{4}{5}\right\}$ **68.** $\left\{\frac{\ln 12{,}143}{\ln 8}\right\}$; ≈ 4.52 **69.** $\left\{\frac{1}{5}\ln 141\right\}$; ≈ 0.99 **70.** $\left\{\frac{12 - \ln 130}{5}\right\}$; ≈ 1.43 **71.** $\left\{\frac{\ln 37{,}500 - 2\ln 5}{4\ln 5}\right\}$; ≈ 1.14

72. $\left\{\frac{\ln 7 + 4\ln 3}{2\ln 7 - \ln 3}\right\}$; ≈ 2.27 **73.** $\{\ln 3\}$; ≈ 1.10 **74.** $\{23\}$ **75.** $\left\{\frac{e^3}{2}\right\}$; ≈ 10.04 **76.** $\{5\}$ **77.** \varnothing **78.** $\{2\}$ **79.** $\{4\}$

80. 5.5 mi **81.** 2013 **82. a.** 2031 **b.** 1994 **83.** 7.3 yr **84.** 14.6 yr **85.** about 22% **86. a.** 0.036 **b.** 60.6 million **c.** 2019
87. 325 days **88. a.** 200 people **b.** about 45,411 people **c.** 500,000 people

89. a.

b. linear function

90. a.

b. logarithmic function

91. a.

b. exponential function

92. $y = 73e^{(\ln 2.6)x}$; $y = 73e^{0.956x}$ **93.** $y = 6.5e^{(\ln 0.43)x}$; $y = 6.5e^{-0.844x}$

Chapter 4 Test

1.

$f(x) = 2^x$
$g(x) = 2^x + 1$

2.

$f(x) = \log_2 x$
$g(x) = \log_2(x - 1)$

3. $5^3 = 125$ **4.** $\log_{36} 6 = \frac{1}{2}$ **5.** $(-\infty, 3)$ **6.** $3 + 5\log_4 x$

7. $\frac{1}{3}\log_3 x - 4$ **8.** $\log(x^6 y^2)$ **9.** $\ln\frac{7}{x^3}$ **10.** 1.5741 **11.** $\{-10\}$

12. $\left\{\frac{\ln 1.4}{\ln 5}\right\}$ **13.** $\left\{\frac{\ln 4}{0.005}\right\}$ **14.** $\{0, \ln 5\}$ **15.** $\{54.25\}$

16. $\left\{\frac{e^4}{3}\right\}$ **17.** $\{5\}$ **18.** \varnothing **19.** 120 db **20.** $5x$ **21.** 1 **22.** 0

23. 6.5% compounded semiannually; $221 more **24.** 13.9 years **25.** about 6.9%

26. a. 82.3 million **b.** decreasing; The growth rate, -0.004, is negative. **c.** 2020 **27.** $A = 4121e^{0.006t}$ **28.** 12.5 days **29. a.** 14 elk
b. about 51 elk **c.** 140 elk **30.** linear **31.** logarithmic **32.** exponential **33.** quadratic **34.** $y = 96e^{(\ln 0.38)x}$; $y = 96e^{-0.968x}$

Cumulative Review Exercises (Chapters 1–4)

1. $\left\{\frac{2}{3}, 2\right\}$ 2. $\{3, 7\}$ 3. $\{-2, -1, 1\}$ 4. $\left\{\frac{\ln 128}{5}\right\}$ 5. $\{3\}$ 6. \varnothing 7. $(-\infty, 4]$ 8. $[1, 3]$

9.

$(x - 3)^2 + (y + 2)^2 = 4$

10.

$f(x) = (x - 2)^2 - 1$

11.

$f(x) = \frac{x^2 - 1}{x^2 - 4}$

12.

$f(x) = (x - 2)^2(x + 1)$

13.

$f(x) = 2x - 4$
$f^{-1}(x) = \frac{x + 4}{2}$

14.

$f(x) = \ln x$
$g(x) = \ln(x - 2) + 1$

15. using $(1, 3)$, $y - 3 = -3(x - 1)$; $y = -3x + 6$
16. $(f \circ g)(x) = (x + 2)^2$; $(g \circ f)(x) = x^2 + 2$
17. You can expect to sleep 2 hours.
18. after 2 sec; 69 ft
19. 4.1 sec
20. $12 per hr

CHAPTER 5

Section 5.1

Check Point Exercises

1. **a.** solution **b.** not a solution 2. $\{(-2, 5)\}$ 3. $\{(2, -1)\}$ 4. $\left\{\left(\frac{60}{17}, -\frac{11}{17}\right)\right\}$ 5. no solution or \varnothing
6. $\{(x, y) \mid x = 4y - 8\}$ or $\{(x, y) \mid 5x - 20y = -40\}$ 7. **a.** $C(x) = 300,000 + 30x$ **b.** $R(x) = 80x$ **c.** $(6000, 480,000)$; The company will break even if it produces and sells 6000 pairs of shoes.

Concept and Vocabulary Check

1. satisfies both equations in the system 2. the intersection point 3. $\left\{\left(\frac{1}{3}, -2\right)\right\}$ 4. -2 5. -3 6. \varnothing; inconsistent; parallel
7. $\{(x, y) \mid x = 3y + 2\}$ or $\{(x, y) \mid 5x - 15y = 10\}$; dependent; are identical or coincide 8. revenue; profit 9. break-even point

Exercise Set 5.1

1. solution 3. not a solution 5. $\{(1, 3)\}$ 7. $\{(5, 1)\}$ 9. $\{(-22, -5)\}$ 11. $\{(0, 0)\}$ 13. $\{(3, -2)\}$ 15. $\{(5, 4)\}$ 17. $\{(7, 3)\}$

19. $\{(2, -1)\}$ 21. $\{(3, 0)\}$ 23. $\{(-4, 3)\}$ 25. $\{(3, 1)\}$ 27. $\{(1, -2)\}$ 29. $\left\{\left(\frac{7}{25}, -\frac{1}{25}\right)\right\}$ 31. \varnothing 33. $\{(x, y) \mid y = 3x - 5\}$

35. $\{(1, 4)\}$ 37. $\{(x, y) \mid x + 3y = 2\}$ 39. $\{(-5, -1)\}$ 41. $\left\{\left(\frac{29}{22}, -\frac{5}{11}\right)\right\}$ 43. $x + y = 7$; $x - y = -1$; 3 and 4

45. $3x - y = 1$; $x + 2y = 12$; 2 and 5 47. $\{(6, -1)\}$ 49. $\left\{\left(\frac{1}{a}, 3\right)\right\}$ 51. $m = -4$, $b = 3$ 53. $y = x - 4$; $y = -\frac{1}{3}x + 4$

55. 500 radios 57. -6000; When the company produces and sells 200 radios, the loss is $6000. 59. **a.** $P(x) = 20x - 10,000$ **b.** $190,000
61. **a.** $C(x) = 18,000 + 20x$ **b.** $R(x) = 80x$ **c.** $(300, 24,000)$; When 300 canoes are produced and sold, both revenue and cost are $24,000.
63. **a.** $C(x) = 30,000 + 2500x$ **b.** $R(x) = 3125x$ **c.** $(48, 150,000)$; For 48 sold-out performances, both cost and revenue are $150,000.
65. **a.** 4 million workers; $4.50 per hour **b.** $4.50; 4; 4 **c.** 2 million **d.** 5.7 million **e.** 3.7 million 67. **a.** 2039; 37%
b. by the intersection point $(69, 37)$ 69. **a.** $y = 0.45x + 0.8$ **b.** $y = 0.15x + 2.6$ **c.** week 6; 3.5 symptoms; by the intersection point $(6, 3.5)$
71. **a.** $y = 0.2x + 5$ **b.** $y = -0.4x + 47$ **c.** 2030; 19% 73. Mr. Goodbar: 264 cal; Mounds: 258 cal 75. 3 Mr. Goodbars and 2 Mounds bars
77. 50 rooms with kitchen facilities, 150 rooms without kitchen facilities 79. 100 ft long by 80 ft wide 81. rate rowing in still water: 6 mph;

rate of the current: 2 mph 83. $80°, 50°, 50°$ 95. makes sense 97. makes sense 99. $y = \frac{a_1 c_2 - a_2 c_1}{a_1 b_2 - a_2 b_1}$; $x = \frac{b_2 c_1 - b_1 c_2}{a_1 b_2 - a_2 b_1}$

101. Yes; 8 hexagons and 4 squares 103. yes 104. $11x + 4y = -3$ 105. $1682 = 16a + 4b + c$

Section 5.2

Check Point Exercises

1. $(-1) - 2(-4) + 3(5) = 22$; $2(-1) - 3(-4) - 5 = 5$; $3(-1) + (-4) - 5(5) = -32$ 2. $\{(1, 4, -3)\}$ 3. $\{(4, 5, 3)\}$ 4. $y = 3x^2 - 12x + 13$

Concept and Vocabulary Check

1. triple; all 2. -2; -4 3. z; add Equations 1 and 3

Exercise Set 5.2

1. solution **3.** solution **5.** $\{(2, 3, 3)\}$ **7.** $\{(2, -1, 1)\}$ **9.** $\{(1, 2, 3)\}$ **11.** $\{(3, 1, 5)\}$ **13.** $\{(1, 0, -3)\}$ **15.** $\{(1, -5, -6)\}$

17. $\left\{\left(\dfrac{1}{2}, \dfrac{1}{3}, -1\right)\right\}$ **19.** $y = 2x^2 - x + 3$ **21.** $y = 2x^2 + x - 5$ **23.** 7, 4, and 5 **25.** $\{(4, 8, 6)\}$ **27.** $y = -\dfrac{3}{4}x^2 + 6x - 11$

29. $\left\{\left(\dfrac{8}{a}, -\dfrac{3}{b}, -\dfrac{5}{c}\right)\right\}$ **31. a.** $y = -16x^2 + 40x + 200$ **b.** $y = 0$ when $x = 5$; The ball hits the ground after 5 seconds.

33. housing: \$6133; vehicles/gas: \$2269; health care: \$5438 **35.** milk: \$3.50; water: \$1.50; chips: \$0.75 **37.** 200 \$8 tickets; 150 \$10 tickets; 50 \$12 tickets

39. \$1200 at 8%, \$2000 at 10%, and \$3500 at 12% **41.** $x = 60, y = 55, z = 65$ **49.** does not make sense **51.** makes sense

53. 13 triangles, 21 rectangles, and 6 pentagons **55.** $\dfrac{x + 14}{(x - 4)(x + 2)}$ **56.** $\dfrac{5x^3 - 3x^2 + 7x - 3}{(x^2 + 1)^2}$ **57.** $\{(5, -2, 3)\}$

Section 5.3

Check Point Exercises

1. $\dfrac{2}{x - 3} + \dfrac{3}{x + 4}$ **2.** $\dfrac{2}{x} - \dfrac{2}{x - 1} + \dfrac{3}{(x - 1)^2}$ **3.** $\dfrac{2}{x + 3} + \dfrac{6x - 8}{x^2 + x + 2}$ **4.** $\dfrac{2x}{x^2 + 1} + \dfrac{-x + 3}{(x^2 + 1)^2}$

Concept and Vocabulary Check

1. correct **2.** incorrect **3.** incorrect **4.** correct

Exercise Set 5.3

1. $\dfrac{A}{x - 2} + \dfrac{B}{x + 1}$ **3.** $\dfrac{A}{x + 2} + \dfrac{B}{x - 3} + \dfrac{C}{(x - 3)^2}$ **5.** $\dfrac{A}{x - 1} + \dfrac{Bx + C}{x^2 + 1}$ **7.** $\dfrac{Ax + B}{x^2 + 4} + \dfrac{Cx + D}{(x^2 + 4)^2}$ **9.** $\dfrac{3}{x - 3} - \dfrac{2}{x - 2}$

11. $\dfrac{7}{x - 9} - \dfrac{4}{x + 2}$ **13.** $\dfrac{24}{7(x - 4)} + \dfrac{25}{7(x + 3)}$ **15.** $\dfrac{4}{7(x - 3)} - \dfrac{8}{7(2x + 1)}$ **17.** $\dfrac{3}{x} + \dfrac{2}{x - 1} - \dfrac{1}{x + 3}$ **19.** $\dfrac{3}{x} + \dfrac{4}{x + 1} - \dfrac{3}{x - 1}$

21. $\dfrac{6}{x - 1} - \dfrac{5}{(x - 1)^2}$ **23.** $\dfrac{1}{x - 2} - \dfrac{2}{(x - 2)^2} - \dfrac{5}{(x - 2)^3}$ **25.** $\dfrac{7}{x} - \dfrac{6}{x - 1} + \dfrac{10}{(x - 1)^2}$ **27.** $\dfrac{1}{4(x + 1)} + \dfrac{3}{4(x - 1)} + \dfrac{1}{2(x - 1)^2}$

29. $\dfrac{3}{x - 1} + \dfrac{2x - 4}{x^2 + 1}$ **31.** $\dfrac{2}{x + 1} + \dfrac{3x - 1}{x^2 + 2x + 2}$ **33.** $\dfrac{1}{4x} + \dfrac{1}{x^2} - \dfrac{x + 4}{4(x^2 + 4)}$ **35.** $\dfrac{4}{x + 1} + \dfrac{2x - 3}{x^2 + 1}$ **37.** $\dfrac{x + 1}{x^2 + 2} - \dfrac{2x}{(x^2 + 2)^2}$

39. $\dfrac{x - 2}{x^2 - 2x + 3} + \dfrac{2x + 1}{(x^2 - 2x + 3)^2}$ **41.** $\dfrac{3}{x - 2} + \dfrac{x - 1}{x^2 + 2x + 4}$ **43.** $x^3 + x - \dfrac{1}{2(x + 1)} + \dfrac{3}{2(x - 1)}$ **45.** $x + 1 - \dfrac{2}{x} - \dfrac{2}{x^2} + \dfrac{2}{x - 1}$

47. $\dfrac{\frac{1}{2c}}{x - c} - \dfrac{\frac{1}{2c}}{x + c}$ **49.** $\dfrac{a}{x - c} + \dfrac{ac + b}{(x - c)^2}$ **51.** $\dfrac{1}{x} - \dfrac{1}{x + 1}; \dfrac{99}{100}$ **61.** does not make sense **63.** does not make sense

65. $\dfrac{2}{x - 3} + \dfrac{2x + 5}{x^2 + 3x + 3}$ **66.** $\{(2.5, -2)\}$ **67.** $\{(4, -3)\}$

68.

; $(0, -3)$ and $(2, -1)$; $0 - (-3) = 3$ and $(0 - 2)^2 + (-3 + 3)^2 = 4$ are true;
$2 - (-1) = 3$ and $(2 - 2)^2 + (-1 + 3)^2 = 4$ are true.

Section 5.4

Check Point Exercises

1. $\{(0, 1), (4, 17)\}$ **2.** $\left\{\left(-\dfrac{6}{5}, \dfrac{3}{5}\right), (2, -1)\right\}$ **3.** $\{(3, 2), (3, -2), (-3, 2), (-3, -2)\}$ **4.** $\{(0, 5)\}$ **5.** length: 7 ft; width: 3 ft or length: 3 ft; width: 7 ft

Concept and Vocabulary Check

1. nonlinear **2.** $\{(-4, 3), (0, 1)\}$ **3.** 3 **4.** $\{(2, \sqrt{3}), (2, -\sqrt{3}), (-2, \sqrt{3}), (-2, -\sqrt{3})\}$ **5.** $-1; y^2 + y = 6$ **6.** $\dfrac{4}{x}; \dfrac{4}{x}; y$

Exercise Set 5.4

1. $\{(-3, 5), (2, 0)\}$ **3.** $\{(1, 1), (2, 0)\}$ **5.** $\{(4, -10), (-3, 11)\}$ **7.** $\{(4, 3), (-3, -4)\}$ **9.** $\left\{\left(-\dfrac{3}{2}, -4\right), (2, 3)\right\}$ **11.** $\{(-5, -4), (3, 0)\}$

13. $\{(3, 1), (-3, -1), (1, 3), (-1, -3)\}$ **15.** $\{(4, -3), (-1, 2)\}$ **17.** $\{(0, 1), (4, -3)\}$ **19.** $\{(3, 2), (3, -2), (-3, 2), (-3, -2)\}$

21. $\{(3, 2), (3, -2), (-3, 2), (-3, -2)\}$ **23.** $\{(2, 1), (2, -1), (-2, 1), (-2, -1)\}$ **25.** $\{(3, 4), (3, -4)\}$

27. $\{(0, 2), (0, -2), (-1, \sqrt{3}), (-1, -\sqrt{3})\}$ **29.** $\{(2, 1), (2, -1), (-2, 1), (-2, -1)\}$ **31.** $\{(-2\sqrt{2}, -\sqrt{2}), (-1, -4), (1, 4), (2\sqrt{2}, \sqrt{2})\}$

33. $\{(2, 2), (4, 1)\}$ **35.** $\{(0, 0), (-1, 1)\}$ **37.** $\{(0, 0), (-2, 2), (2, 2)\}$ **39.** $\left\{(-4, 1), \left(-\dfrac{5}{2}, \dfrac{1}{4}\right)\right\}$ **41.** $\left\{\left(\dfrac{12}{5}, -\dfrac{29}{5}\right), (-2, 3)\right\}$

43. 4 and 6 **45.** 2 and 1, 2 and -1, -2 and 1, or -2 and -1

47. $\{(2, -1), (-2, 1)\}$ **49.** $\{(2, 20), (-2, 4), (-3, 0)\}$ **51.** $\left\{\left(-1, -\dfrac{1}{2}\right), \left(-1, \dfrac{1}{2}\right), \left(1, -\dfrac{1}{2}\right), \left(1, \dfrac{1}{2}\right)\right\}$

53.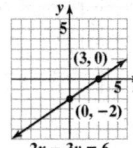
$y = x^2 - 4$
$y = 2x - 4$

55. $(0, -4), (-2, 0), (2, 0)$ **57.** 11 ft and 7 ft **59.** width: 6 in.; length: 8 in.
61. $x = 5$ m, $y = 2$ m

63. a. between the 1940s and the 1960s **b.** 1949; 43%; 43% **c.** 1920; 28% **d.** 1919; white collar: 27.5%; farmers: 27.4%; fairly well, although
answers may vary. **69.** makes sense **71.** makes sense **73.** false **75.** false **77.** 18 sq units **79.** $\{(8, 2)\}$

81.
$2x - 3y = 6$

82.

83.

Mid-Chapter 5 Check Point

1. $\{(-1, 2)\}$ **2.** $\{(1, -2)\}$ **3.** $\{(6, 10)\}$ **4.** $\{(x, y) \mid y = 4x - 5\}$ or $\{(x, y) \mid 8x - 2y = 10\}$ **5.** $\left\{\left(\dfrac{11}{19}, \dfrac{7}{19}\right)\right\}$ **6.** \varnothing **7.** $\{(-1, 2, -2)\}$

8. $\{(4, -2, 3)\}$ **9.** $\left\{\left(-\dfrac{9}{5}, \dfrac{12}{5}\right), (3, 0)\right\}$ **10.** $\{(-2, -1), (-2, 1), (2, -1), (2, 1)\}$ **11.** $\{(-\sqrt{7}, 1), (-2, -2), (2, -2), (\sqrt{7}, 1)\}$

12. $\{(0, -2), (6, 1)\}$ **13.** $\dfrac{1}{x-2} - \dfrac{2}{(x-2)^2} - \dfrac{5}{(x-2)^3}$ **14.** $\dfrac{5}{x+2} + \dfrac{3}{x+1} + \dfrac{2}{x-1}$ **15.** $-\dfrac{2}{x+3} + \dfrac{3x-5}{x^2+4}$ **16.** $\dfrac{x}{x^2+4} - \dfrac{4x}{(x^2+4)^2}$

17. a. $C(x) = 400{,}000 + 20x$ **b.** $R(x) = 100x$ **c.** $P(x) = 80x - 400{,}000$ **d.** $(5000, 500{,}000)$; The company will break even when it
produces and sells 5000 PDAs. At this level, both revenue and cost are $500,000. **18.** 6 roses and 14 carnations **19.** $x = 55°, y = 35°$
20. $y = -x^2 + 2x + 3$ **21.** length: 8 m; width: 2.5 m

Section 5.5

Check Point Exercises

1. $4x - 2y \geq 8$

2. $y > -\dfrac{3}{4}x$

3. a. $y > 1$ **b.** $x \leq -2$

4.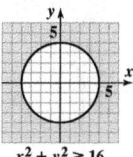
$x^2 + y^2 \geq 16$

5. Point $B = (66, 130)$; $4.9(66) - 130 \geq 165$, or $193.4 \geq 165$, is true; $3.7(66) - 130 \leq 125$, or $114.2 \leq 125$, is true.

6. $\begin{cases} x - 3y < 6 \\ 2x + 3y \geq -6 \end{cases}$

7.
$\begin{cases} y \geq x^2 - 4 \\ x + y \leq 2 \end{cases}$

8. $\begin{cases} x + y < 2 \\ -2 \leq x < 1 \\ y > -3 \end{cases}$

Concept and Vocabulary Check

1. solution; x; y; $5 > 1$ **2.** graph **3.** half-plane **4.** false **5.** true **6.** false **7.** $(0, 0)$, although answers will vary
8. $x - y < 1$; $2x + 3y \geq 12$ **9.** false

Exercise Set 5.5

1.
$x + 2y \leq 8$

3.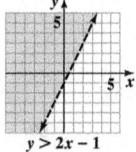
$x - 2y > 10$

5.
$y \leq \dfrac{1}{3}x$

7.
$y > 2x - 1$

9.
$x \leq 1$

11.
$y > 1$

13.
$x^2 + y^2 \leq 1$

15.
$x^2 + y^2 > 25$

17.
$(x - 2)^2 + (y + 1)^2 < 9$

19.
$y < x^2 - 1$

21.
$y \geq x^2 - 9$

23.
$y > 2^x$

25. $x = -1$
$y \geq \log_2(x + 1)$

27. $\begin{cases} 3x + 6y \leq 6 \\ 2x + y \leq 8 \end{cases}$

29. $\begin{cases} 2x - 5y \leq 10 \\ 3x - 2y > 6 \end{cases}$

31. $\begin{cases} y > 2x - 3 \\ y < -x + 6 \end{cases}$

33. $\begin{cases} x + 2y \leq 4 \\ y \geq x - 3 \end{cases}$

35. $\begin{cases} x \leq 2 \\ y \geq -1 \end{cases}$
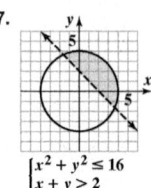

37. $-2 \leq x < 5$

39. $\begin{cases} x - y \leq 1 \\ x \geq 2 \end{cases}$

41. \varnothing

43. $\begin{cases} x + y > 4 \\ x + y > -1 \end{cases}$

45.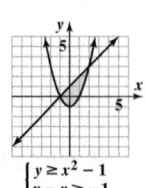
$\begin{cases} y \geq x^2 - 1 \\ x - y \geq -1 \end{cases}$

47.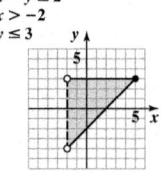
$\begin{cases} x^2 + y^2 \leq 16 \\ x + y > 2 \end{cases}$

49.
$\begin{cases} x^2 + y^2 > 1 \\ x^2 + y^2 < 16 \end{cases}$

51.
$\begin{cases} (x-1)^2 + (y+1)^2 < 25 \\ (x-1)^2 + (y+1)^2 \geq 16 \end{cases}$

53.
$\begin{cases} x^2 + y^2 \leq 1 \\ y - x^2 > 0 \end{cases}$

55.
$\begin{cases} x^2 + y^2 < 16 \\ y \geq 2^x \end{cases}$

57. $\begin{cases} x - y \leq 2 \\ x > -2 \\ y \leq 3 \end{cases}$
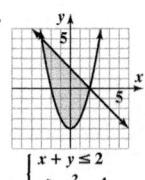

59. $\begin{cases} x \geq 0 \\ y \geq 0 \\ 2x + 5y < 10 \\ 3x + 4y \leq 12 \end{cases}$

61. $\begin{cases} 3x + y \leq 6 \\ 2x - y \leq -1 \\ x > -2 \\ y < 4 \end{cases}$

63. $y \geq -2x + 4$
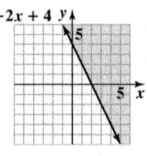

65. $\begin{cases} x + y \leq 4 \\ 3x + y \leq 6 \end{cases}$

67.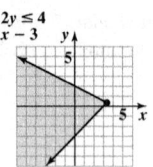
$\begin{cases} x + y \leq 2 \\ y \geq x^2 - 4 \end{cases}$

69. $\begin{cases} -2 \leq x \leq 2 \\ -3 \leq y \leq 3 \end{cases}$

71.
$y > \dfrac{3}{2}x - 2$ or $y < 4$

73. no solution

75. infinitely many solutions

77. Point $A = (66, 160)$; $5.3(66) - 160 \geq 180$, or $189.8 \geq 180$, is true; $4.1(66) - 160 \leq 140$, or $110.6 \leq 140$, is true.

79. no

81. a. $50x + 150y > 2000$
b.
$50x + 150y > 2000$
c. Answers may vary. Example:
$(20, 20)$: 20 children and 20 adults
will cause the elevator to be overloaded.

83. a. $\begin{cases} y \geq 0 \\ x + y \geq 5 \\ x \geq 1 \\ 200x + 100y \leq 700 \end{cases}$
b. $\begin{cases} y \geq 0 \\ x + y \geq 5 \\ x \geq 1 \\ 200x + 100y \leq 700 \end{cases}$

c. 2 nights

85. a. 27.1 **b.** overweight

97.

99.

101.

107. does not make sense **109.** makes sense **111.** $\begin{cases} y > x - 3 \\ y \leq x \end{cases}$ **113.** $x + 2y \leq 6$ or $2x + y \leq 6$

115. $y \geq nx + b$ $y = mx + b$
$y \leq mx + b$

$(0, b)$

$y = nx + b$

116. a.

$7 (1, 5)$ $(8, 5)$

$(8, -2)$

$x + y \geq 6$
$x \leq 8$
$y \leq 5$

b. $(1, 5), (8, 5), (8, -2)$

c. at $(1, 5)$: 13; at $(8, 5)$: 34; at $(8, -2)$: 20

117. a.

$(0, 7)$

$(4, 3)$

$(0, 0)$ $(2, 0)$ 9

$x \geq 0$
$y \geq 0$
$3x - 2y \leq 6$
$y \leq -x + 7$

b. $(0, 0), (2, 0), (4, 3), (0, 7)$

c. at $(0, 0)$: 0; at $(2, 0)$: 4; at $(4, 3)$: 23; $(0, 7)$: 35

118. $20x + 10y \leq 80{,}000$

Section 5.6

Check Point Exercises

1. $z = 25x + 55y$ **2.** $x + y \leq 80$ **3.** $30 \leq x \leq 80$; $10 \leq y \leq 30$; objective function: $z = 25x + 55y$; constraints: $\begin{cases} x + y \leq 80 \\ 30 \leq x \leq 80 \\ 10 \leq y \leq 30 \end{cases}$

4. 50 bookshelves and 30 desks; $2900 **5.** 30

Concept and Vocabulary Check

1. linear programming **2.** objective **3.** constraints; corner

Exercise Set 5.6

1. $(1, 2)$: 17; $(2, 10)$: 70; $(7, 5)$: 65; $(8, 3)$: 58; maximum: $z = 70$; minimum: $z = 17$
3. $(0, 0)$: 0; $(0, 8)$: 400; $(4, 9)$: 610; $(8, 0)$: 320; maximum: $z = 610$; minimum: $z = 0$

5. a.

$(0, 8)$

$(0, 4)$ 4

$(4, 0)$

b. $(0, 8)$: 16; $(0, 4)$: 8; $(4, 0)$: 12
c. maximum value: 16 at $x = 0$
and $y = 8$

7. a.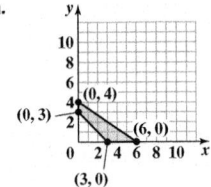

$(0, 4)$

$(0, 3)$

$(6, 0)$

$(3, 0)$

b. $(0, 4)$: 4; $(0, 3)$: 3; $(3, 0)$: 12; $(6, 0)$: 24
c. maximum value: 24 at $x = 6$ and
$y = 0$

9. a.

$(5, 8)$

$(1, 4)$

$(5, 2)$

$(1, 2)$

b. $(1, 2)$: -1; $(1, 4)$: -5; $(5, 8)$: -1; $(5, 2)$: 11
c. maximum value: 11 at $x = 5$ and $y = 2$

11. a.

$\left(\dfrac{12}{5}, \dfrac{12}{5}\right)$

$(0, 4)$ $(4, 0)$

$(0, 2)$ $(2, 0)$

b. $(0, 4)$: 8; $(0, 2)$: 4; $(2, 0)$: 8; $(4, 0)$: 16;
$\left(\dfrac{12}{5}, \dfrac{12}{5}\right)$: $\dfrac{72}{5}$
c. maximum value: 16 at $x = 4$ and $y = 0$

13. a.

$(0, 6)$

$(3, 4)$

$(0, 0)$ $(5, 0)$

b. $(0, 6)$: 72; $(0, 0)$: 0; $(5, 0)$: 50; $(3, 4)$: 78
c. maximum value: 78 at $x = 3$ and $y = 4$

15. a. $z = 125x + 200y$ **b.** $\begin{cases} x \leq 450 \\ y \leq 200 \\ 600x + 900y \leq 360{,}000 \end{cases}$

c.

$(0, 200)$

$(300, 200)$

$(450, 100)$

$(450, 0)$

$(0, 0)$

d. $(0, 0)$: 0; $(0, 200)$: 40,000;
$(300, 200)$: 77,500; $(450, 100)$: 76,250;
$(450, 0)$: 56,250
e. 300; 200; $77,500

17. 40 model A bicycles and no model B bicycles
19. 300 cartons of food and 200 cartons of clothing
21. 50 students and 100 parents
23. 10 Boeing 727s and 42 Falcon 20s

29. does not make sense **31.** makes sense **33.** $5000 in stocks and $5000 in bonds **37.** $\{(6, 3, 5)\}$; Answers may vary.

38. $\{(-2, 1, 4, 3)\}$; Answers may vary. **39.** $\begin{bmatrix} 1 & 2 & -1 \\ 0 & -11 & -11 \end{bmatrix}$

Chapter 5 Review Exercises

1. $\{(1, 5)\}$ **2.** $\{(2, 3)\}$ **3.** $\{(2, -3)\}$ **4.** \varnothing **5.** $\{(x, y) \mid 3x - 6y = 12\}$ **6. a.** $C(x) = 60{,}000 + 200x$ **b.** $R(x) = 450x$
c. $(240, 108{,}000)$; This means the company will break even if it produces and sells 240 desks. **7.** Klimt: $135 million; Picasso: $104 million
8. a. $y = 219x + 142$ **b.** 2006; 580 million of each **9.** 12 ft by 5 ft **10.** $80 per day for the room, $60 per day for the car
11. 3 apples and 2 avocados **12.** $\{(0, 1, 2)\}$ **13.** $\{(2, 1, -1)\}$ **14.** $y = 3x^2 - 4x + 5$ **15.** war: 124 million; famine: 111 million; tobacco: 71 million

16. $\dfrac{3}{5(x-3)} + \dfrac{2}{5(x+2)}$ **17.** $\dfrac{6}{x-4} + \dfrac{5}{x+3}$ **18.** $\dfrac{2}{x} + \dfrac{3}{x+2} - \dfrac{1}{x-1}$ **19.** $\dfrac{2}{x-2} + \dfrac{5}{(x-2)^2}$ **20.** $-\dfrac{4}{x-1} + \dfrac{4}{x-2} - \dfrac{2}{(x-2)^2}$

21. $\dfrac{6}{5(x-2)} + \dfrac{-6x+3}{5(x^2+1)}$ **22.** $\dfrac{5}{x-3} + \dfrac{2x-1}{x^2+4}$ **23.** $\dfrac{x}{x^2+4} - \dfrac{4x}{(x^2+4)^2}$ **24.** $\dfrac{4x+1}{x^2+x+1} + \dfrac{2x-2}{(x^2+x+1)^2}$ **25.** $\{(4,3),(1,0)\}$

26. $\{(0,1),(-3,4)\}$ **27.** $\{(1,-1),(-1,1)\}$ **28.** $\{(3,\sqrt{6}),(3,-\sqrt{6}),(-3,\sqrt{6}),(-3,-\sqrt{6})\}$ **29.** $\{(2,2),(-2,-2)\}$ **30.** $\{(9,6),(1,2)\}$

31. $\{(-3,-1),(1,3)\}$ **32.** $\left\{\left(\dfrac{1}{2},2\right),(-1,-1)\right\}$ **33.** $\left\{\left(\dfrac{5}{2},-\dfrac{7}{2}\right),(0,-1)\right\}$ **34.** $\{(2,-3),(-2,-3),(3,2),(-3,2)\}$

35. $\{(3,1),(3,-1),(-3,1),(-3,-1)\}$ **36.** 8 m and 5 m **37.** $(1,6),(3,2)$ **38.** $x=46$ and $y=28$ or $x=50$ and $y=20$

39.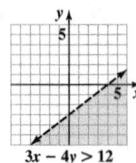
$3x - 4y > 12$

40.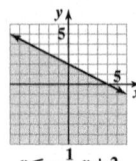
$y \le -\dfrac{1}{2}x + 2$

41.
$x < -2$

42.
$y \ge 3$

43.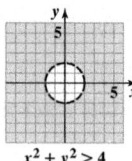
$x^2 + y^2 > 4$

44.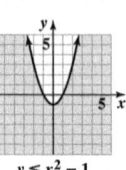
$y \le x^2 - 1$

45.
$y \le 2^x$

46.
$\begin{cases} 3x + 2y \ge 6 \\ 2x + y \ge 6 \end{cases}$

47.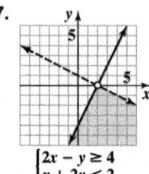
$\begin{cases} 2x - y \ge 4 \\ x + 2y < 2 \end{cases}$

48.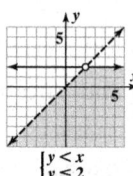
$\begin{cases} y < x \\ y \le 2 \end{cases}$

49.
$\begin{cases} x + y \le 6 \\ y \ge 2x - 3 \end{cases}$

50.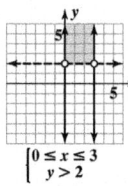
$\begin{cases} 0 \le x \le 3 \\ y > 2 \end{cases}$

51. no solution

52.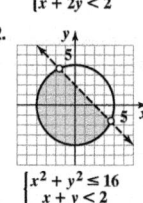
$\begin{cases} x^2 + y^2 \le 16 \\ x + y < 2 \end{cases}$

53.
$\begin{cases} x^2 + y^2 \le 9 \\ y < -3x + 1 \end{cases}$

54.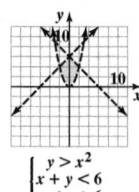
$\begin{cases} y > x^2 \\ x + y < 6 \\ y < x + 6 \end{cases}$

55.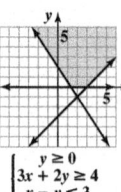
$\begin{cases} y \ge 0 \\ 3x + 2y \ge 4 \\ x - y \le 3 \end{cases}$

56. $(2,2)$: 10; $(4,0)$: 8; $\left(\dfrac{1}{2},\dfrac{1}{2}\right)$: $\dfrac{5}{2}$; $(1,0)$: 2; maximum value: 10; minimum value: 2

57.
$x \ge 0,\ y \ge 0$
$x + y \le 8$
$3x + 2y \ge 6$
Maximum is 24 at $x = 0, y = 8$.

58.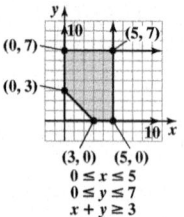
$0 \le x \le 5$
$0 \le y \le 7$
$x + y \ge 3$
Maximum is 33 at $x = 5, y = 7$.

59.
$x \ge 0, y \ge 0$
$y \le x$
$2x + y \le 12$
$2x + 3y \ge 6$
Maximum is 44 at $x = y = 4$.

60. a. $z = 500x + 350y$
b. $\begin{cases} x + y \le 200 \\ x \ge 10 \\ y \ge 80 \end{cases}$
c.
$x + y \le 200$
$x \ge 10, y \ge 80$
d. $(10, 80)$: 33,000; $(10, 190)$: 71,500; $(120, 80)$: 88,000
e. 120; 80; 88,000

61. 480 of model A and 240 of model B

Chapter 5 Test

1. $\{(1,-3)\}$ **2.** $\{(4,-2)\}$ **3.** $\{(1,3,2)\}$ **4.** $\{(4,-3),(-3,4)\}$ **5.** $\{(3,2),(3,-2),(-3,2),(-3,-2)\}$ **6.** $\dfrac{-1}{10(x+1)} + \dfrac{x+9}{10(x^2+9)}$

7.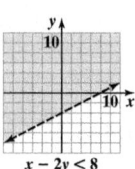
$x - 2y < 8$

8.
$\begin{cases} x \ge 0,\ y \ge 0 \\ 3x + y \le 9 \\ 2x + 3y \ge 6 \end{cases}$

9.
$\begin{cases} x^2 + y^2 > 1 \\ x^2 + y^2 < 4 \end{cases}$

10.
$\begin{cases} y \le 1 - x^2 \\ x^2 + y^2 \le 9 \end{cases}$

11. 26 **12.** shrimp: 42 mg; scallops: 15 mg **13. a.** $C(x) = 360{,}000 + 850x$ **b.** $R(x) = 1150x$ **c.** $(1200, 1{,}380{,}000)$; The company will break even if it produces and sells 1200 computers. **14.** $y = x^2 - 3$ **15.** $x = 7.5$ ft and $y = 24$ ft or $x = 12$ ft and $y = 15$ ft **16.** 50 regular and 100 deluxe jet skis; $35,000

Cumulative Review Exercises (Chapters 1–5)

1. domain: $(-2, 2)$; range: $(-\infty, 3]$ **2.** -1 and 1 **3.** maximum of 3 at $x = 0$ **4.** $(0, 2)$
5. positive **6.** 3 **7.** $x \to -2^+$; $x \to 2^-$ **8.** even

9.
$g(x) = f(x + 2) - 1$

10.
$h(x) = \frac{1}{2} f\left(\frac{1}{2} x\right)$

11. $\{3, 4\}$ **12.** $\left\{\dfrac{2 + i\sqrt{3}}{2}, \dfrac{2 - i\sqrt{3}}{2}\right\}$

13. $(-18, 6)$ **14.** $(1, 7)$

15. $\left\{-3, \dfrac{1}{2}, 2\right\}$ **16.** $\{-2\}$

17. $\{2\}$ **18.** $\{-2 + \log_3 11\}$

19. $\{625\}$ **20.** $\left\{\left(-\dfrac{1}{2}, \dfrac{1}{2}\right), (2, 8)\right\}$ **21.** $\{(8, -2, -2)\}$

22.
$f(x) = (x + 2)^2 - 4$

23.
$2x - 3y \leq 6$

24.
$y = 3^{x-2}$

25.
$f(x) = \dfrac{x^2 - x - 6}{x + 1}$

26.
$f(x) = 2x - 4$
$f^{-1}(x) = \dfrac{x + 4}{2}$

27.
$(x - 2)^2 + (y - 4)^2 > 9$

28.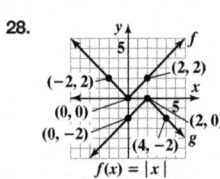
$f(x) = |x|$
$g(x) = -|x - 2|$

29. $(f \circ g)(x) = 2x^2 - 3x$; $(g \circ f)(x) = -2x^2 + x + 2$
30. $4x + 2h - 1$
31. $y = -3x + 10$
32. $y = 3x + 3$
33. $2600 at 12%; $1400 at 14%
34. 4 m by 9 m
35. 10.99%

CHAPTER 6

Section 6.1

Check Point Exercises

1. a. $\begin{bmatrix} 1 & 6 & -3 & | & 7 \\ 4 & 12 & -20 & | & 8 \\ -3 & -2 & 1 & | & -9 \end{bmatrix}$ **b.** $\begin{bmatrix} 1 & 3 & -5 & | & 2 \\ 1 & 6 & -3 & | & 7 \\ -3 & -2 & 1 & | & -9 \end{bmatrix}$ **c.** $\begin{bmatrix} 4 & 12 & -20 & | & 8 \\ 1 & 6 & -3 & | & 7 \\ 0 & 16 & -8 & | & 12 \end{bmatrix}$ **2.** $\{(5, 2, 3)\}$ **3.** $\{(1, -1, 2, -3)\}$ **4.** $\{(5, 2, 3)\}$

Concept and Vocabulary Check

1. matrix; elements **2.** 3; second; -2; third **3.** $\begin{bmatrix} 2 & 1 & 4 & | & -4 \\ 3 & 0 & 1 & | & 1 \\ 4 & 3 & 1 & | & 8 \end{bmatrix}$ **4.** $\{(1, 3, -2)\}$ **5.** true **6.** false

Exercise Set 6.1

1. $\begin{bmatrix} 2 & 1 & 2 & | & 2 \\ 3 & -5 & -1 & | & 4 \\ 1 & -2 & -3 & | & -6 \end{bmatrix}$ **3.** $\begin{bmatrix} 1 & -1 & 1 & | & 8 \\ 0 & 1 & -12 & | & -15 \\ 0 & 0 & 1 & | & 1 \end{bmatrix}$ **5.** $\begin{bmatrix} 5 & -2 & -3 & | & 0 \\ 1 & 1 & 0 & | & 5 \\ 2 & 0 & -3 & | & 4 \end{bmatrix}$ **7.** $\begin{bmatrix} 2 & 5 & -3 & 1 & | & 2 \\ 0 & 3 & 1 & 0 & | & 4 \\ 1 & -1 & 5 & 0 & | & 9 \\ 5 & -5 & -2 & 0 & | & 1 \end{bmatrix}$ **9.** $\begin{cases} 5x + 3z = -11 \\ y - 4z = 12 \\ 7x + 2y = 3 \end{cases}$

11. $\begin{cases} w + x + 4y + z = 3 \\ -w + x - y = 7 \\ 2w + 5z = 11 \\ 12y + 4z = 5 \end{cases}$ **13.** $\begin{bmatrix} 1 & -3 & 2 & | & 5 \\ 1 & 5 & -5 & | & 0 \\ 3 & 0 & 4 & | & 7 \end{bmatrix}$ **15.** $\begin{bmatrix} 1 & -3 & 2 & | & 0 \\ 0 & 10 & -7 & | & 7 \\ 2 & -2 & 1 & | & 3 \end{bmatrix}$ **17.** $\begin{bmatrix} 1 & -1 & 1 & 1 & | & 3 \\ 0 & 1 & -2 & -1 & | & 0 \\ 0 & 2 & 1 & 2 & | & 5 \\ 0 & 6 & -3 & -1 & | & -9 \end{bmatrix}$

19. $R_2: -3, -18$; $R_3: -12, -15$; $R_2: -\dfrac{3}{5}, -\dfrac{18}{5}$; $R_3: -12, -15$ **21.** $\{(1, -1, 2)\}$ **23.** $\{(3, -1, -1)\}$ **25.** $\{(2, -1, 1)\}$ **27.** $\{(2, 1, 1)\}$
29. $\{(2, -1, 1)\}$ **31.** $\{(-1, 2, -2)\}$ **33.** $\{(1, 2, -1)\}$ **35.** $\{(1, 2, 3, -2)\}$ **37.** $\{(0, -3, 0, -3)\}$ **39.** $f(x) = -x^2 + x + 2$
41. $f(x) = x^3 - 2x^2 + 3$ **43.** $\{(e^{-1}, e, e^{-3}, e^{-2})\}$ **45. a.** $a = -32, v_0 = 56, s_0 = 0$ **b.** 0; The ball hits the ground 3.5 seconds after it is thrown; $(3.5, 0)$.
c. 1.75 sec; 49 ft **47.** 4 oz of Food A; $\dfrac{1}{2}$ oz of Food B; 1 oz of Food C **49.** Asians: 122; Africans: 28; Europeans: 24; Americans: 9 **59.** makes sense
61. makes sense **63.** false **65.** false **67.** 60 units: $7700
68. For $z = 0$, $(12z + 1, 10z - 1, z)$ is $(1, -1, 0)$; $3(1) - 4(-1) + 4(0) = 7$ is true; $1 - (-1) - 2(0) = 2$ is true; $2(1) - 3(-1) + 6(0) = 5$ is true.
69. For $z = 1$, $(12z + 1, 10z - 1, z)$ is $(13, 9, 1)$; $3(13) - 4(9) + 4(1) = 7$ is true; $13 - 9 - 2(1) = 2$ is true; $2(13) - 3(9) + 6(1) = 5$ is true.
70. a. Answers may vary. **b.** This system has more than one solution.

Section 6.2

Check Point Exercises

1. \varnothing **2.** $\{(11t + 13, 5t + 4, t)\}$ **3.** $\{(t + 50, -2t + 10, t)\}$

4. a. $\begin{cases} w + z = 15 \\ w + x = 30 \\ x + y = 45 \\ y + z = 30 \end{cases}$ **b.** $\{(-t + 15, t + 15, -t + 30, t)\}$ **c.** $w = 5; x = 25; y = 20$

Concept and Vocabulary Check

1. no solution **2.** one solution **3.** infinitely many solutions **4.** true

5. $\begin{cases} x - y + z = 5 \\ y - z = -3 \end{cases}; z - 3; 2; \{(2, z - 3, z)\}$

Exercise Set 6.2

1. \varnothing **3.** $\left\{ \left(-2t + 2, 2t + \frac{1}{2}, t \right) \right\}$ **5.** $\{(-3, 4, -2)\}$ **7.** $\{(5 - 2t, -2 + t, t)\}$ **9.** $\{(-1, 2, 1, 1)\}$ **11.** $\{(1, 3, 2, 1)\}$

13. $\{(1, -2, 1, 1)\}$ **15.** $\left\{ \left(1 + \frac{1}{3}t, \frac{1}{3}t, t \right) \right\}$ **17.** $\{(-13t + 5, 5t, t)\}$ **19.** $\left\{ \left(2t - \frac{5}{4}, \frac{13}{4}, t \right) \right\}$ **21.** $\{(1, -t - 1, 2, t)\}$

23. $\left\{ \left(-\frac{2}{11}t + \frac{81}{11}, \frac{1}{22}t + \frac{10}{11}, \frac{4}{11}t - \frac{8}{11}, t \right) \right\}$

25. a. $4w - 2x + 2y - 3z = 0; 7w - x - y - 3z = 0; w + x + y - z = 0$ **b.** $\{(0.5t, 0, 0.5t, t)\}$

27. a. $w + 2x + 5y + 5z = -3; w + x + 3y + 4z = -1; w - x - y + 2z = 3$ **b.** $\{(1 - 3s - t, -2 - s - 2t, t, s)\}$

29. $z + 12 = x + 6$ **31.** $\{(t + 6, t + 2, t)\}$

33. a. $\begin{cases} w + z = 380 \\ w + x = 600 \\ x - y = 170 \\ y - z = 50 \end{cases}$ **b.** $\{(380 - t, 220 + t, 50 + t, t)\}$ **c.** $w = 330, x = 270, y = 100$

35. a. The system has no solution, so there is no way to satisfy these dietary requirements with no Food 1 available.
 b. 4 oz of Food 1, 0 oz of Food 2, 10 oz of Food 3; 2 oz of Food 1, 5 oz of Food 2, 9 oz of Food 3 (other answers are possible).

41. does not make sense **43.** does not make sense **45.** $a = 1$ or $a = 3$ **47.** -1 **48.** -12 **49.** 8

Section 6.3

Check Point Exercises

1. a. 3×2 **b.** $a_{12} = -2; a_{31} = 1$ **2. a.** $\begin{bmatrix} 2 & 0 \\ 9 & -10 \end{bmatrix}$ **b.** $\begin{bmatrix} 9 & -4 \\ -9 & 7 \\ 5 & -2 \end{bmatrix}$ **3. a.** $\begin{bmatrix} 6 & 12 \\ -48 & -30 \end{bmatrix}$ **b.** $\begin{bmatrix} -14 & -1 \\ 25 & 10 \end{bmatrix}$ **4.** $\begin{bmatrix} -4 & 3 \\ -3 & \frac{13}{3} \end{bmatrix}$

5. $\begin{bmatrix} 7 & 6 \\ 13 & 12 \end{bmatrix}$ **6.** $[30]; \begin{bmatrix} 2 & 0 & 4 \\ 6 & 0 & 12 \\ 14 & 0 & 28 \end{bmatrix}$ **7. a.** $\begin{bmatrix} 2 & 18 & 11 & 9 \\ 0 & 10 & 8 & 2 \end{bmatrix}$ **b.** The product is undefined.

8. $\begin{bmatrix} 2 & 1 & 1 \\ 2 & 1 & 1 \\ 2 & 2 & 1 \end{bmatrix} + \begin{bmatrix} -1 & 2 & 2 \\ -1 & 2 & 2 \\ -1 & -1 & 2 \end{bmatrix} = \begin{bmatrix} 1 & 3 & 3 \\ 1 & 3 & 3 \\ 1 & 1 & 3 \end{bmatrix}$

9. a. $\begin{bmatrix} 0 & 3 & 4 \\ 0 & 5 & 2 \end{bmatrix} + \begin{bmatrix} -3 & -3 & -3 \\ -1 & -1 & -1 \end{bmatrix} = \begin{bmatrix} -3 & 0 & 1 \\ -1 & 4 & 1 \end{bmatrix}$ **b.** $2\begin{bmatrix} 0 & 3 & 4 \\ 0 & 5 & 2 \end{bmatrix} = \begin{bmatrix} 0 & 6 & 8 \\ 0 & 10 & 4 \end{bmatrix}$

c. $\begin{bmatrix} 0 & 3 & 4 \\ 0 & -5 & -2 \end{bmatrix}$; Multiplying by B reflects the triangle over the x-axis.

Concept and Vocabulary Check

1. third; fourth **2.** 1×3 **3.** square **4.** $-10; 6$ **5.** true **6.** false **7.** true **8.** $m; p$; columns; rows **9.** true **10.** false

Exercise Set 6.3

1. a. 2×3 **b.** a_{32} does not exist; $a_{23} = -1$ **3. a.** 3×4 **b.** $a_{32} = \frac{1}{2}; a_{23} = -6$ **5.** $x = 6; y = 4$ **7.** $x = 4; y = 6; z = 3$

9. a. $\begin{bmatrix} 9 & 10 \\ 3 & 9 \end{bmatrix}$ **b.** $\begin{bmatrix} -1 & -8 \\ 3 & -5 \end{bmatrix}$ **c.** $\begin{bmatrix} -16 & -4 \\ -12 & -8 \end{bmatrix}$ **d.** $\begin{bmatrix} 22 & 21 \\ 9 & 20 \end{bmatrix}$ **11. a.** $\begin{bmatrix} 3 & 2 \\ 6 & 2 \\ 5 & 7 \end{bmatrix}$ **b.** $\begin{bmatrix} -1 & 4 \\ 0 & 6 \\ 5 & 5 \end{bmatrix}$ **c.** $\begin{bmatrix} -4 & -12 \\ -12 & -16 \\ -20 & -24 \end{bmatrix}$ **d.** $\begin{bmatrix} 7 & 7 \\ 15 & 8 \\ 15 & 20 \end{bmatrix}$

13. a. $\begin{bmatrix} -3 \\ -1 \\ 0 \end{bmatrix}$ **b.** $\begin{bmatrix} 7 \\ -7 \\ 2 \end{bmatrix}$ **c.** $\begin{bmatrix} -8 \\ 16 \\ -4 \end{bmatrix}$ **d.** $\begin{bmatrix} -4 \\ -6 \\ 1 \end{bmatrix}$ **15. a.** $\begin{bmatrix} 8 & 0 & -4 \\ 14 & 0 & 6 \\ -1 & 0 & 0 \end{bmatrix}$ **b.** $\begin{bmatrix} -4 & -20 & 0 \\ 14 & 24 & 14 \\ 9 & -4 & 4 \end{bmatrix}$ **c.** $\begin{bmatrix} -8 & 40 & 8 \\ -56 & -48 & -40 \\ -16 & 8 & -8 \end{bmatrix}$ **d.** $\begin{bmatrix} 18 & -10 & -10 \\ 42 & 12 & 22 \\ 2 & -2 & 2 \end{bmatrix}$

17. $\begin{bmatrix} -8 & -8 \\ 2 & -9 \\ 8 & -4 \end{bmatrix}$ **19.** $\begin{bmatrix} -1 & 3 \\ -1 & \frac{9}{2} \\ -1 & -2 \end{bmatrix}$ **21.** $\begin{bmatrix} \frac{1}{3} & \frac{13}{3} \\ -\frac{4}{3} & 6 \\ -\frac{7}{3} & -\frac{4}{3} \end{bmatrix}$ **23.** $\begin{bmatrix} 7 & 27 \\ -8 & 36 \\ -17 & -4 \end{bmatrix}$ **25.** $\begin{bmatrix} \frac{27}{2} & \frac{31}{2} \\ -4 & 18 \\ -\frac{29}{2} & 6 \end{bmatrix}$ **27. a.** $\begin{bmatrix} 0 & 16 \\ 12 & 8 \end{bmatrix}$ **b.** $\begin{bmatrix} -7 & 3 \\ 29 & 15 \end{bmatrix}$

29. a. $[30]$ **b.** $\begin{bmatrix} 1 & 2 & 3 & 4 \\ 2 & 4 & 6 & 8 \\ 3 & 6 & 9 & 12 \\ 4 & 8 & 12 & 16 \end{bmatrix}$ **31. a.** $\begin{bmatrix} 4 & -5 & 8 \\ 6 & -1 & 5 \\ 0 & 4 & -6 \end{bmatrix}$ **b.** $\begin{bmatrix} 5 & -2 & 7 \\ 17 & -3 & 2 \\ 3 & 0 & -5 \end{bmatrix}$ **33. a.** $\begin{bmatrix} 6 & 8 & 16 \\ 11 & 16 & 24 \\ 1 & -1 & 12 \end{bmatrix}$ **b.** $\begin{bmatrix} 38 & 27 \\ -16 & -4 \end{bmatrix}$

35. a. $\begin{bmatrix} 0 & 0 \\ 0 & 0 \end{bmatrix}$ **b.** $\begin{bmatrix} 4 & -1 & -3 & 1 \\ -1 & 4 & -3 & 2 \\ 14 & -11 & -3 & -1 \\ 25 & -25 & 0 & -5 \end{bmatrix}$ **37.** $\begin{bmatrix} 17 & 7 \\ -5 & -11 \end{bmatrix}$ **39.** $\begin{bmatrix} 11 & -1 \\ -7 & -3 \end{bmatrix}$

41. $A - C$ is not defined because A is 3×2 and C is 2×2. **43.** $\begin{bmatrix} 16 & -16 \\ -12 & 12 \\ 0 & 0 \end{bmatrix}$ **45.** $\begin{bmatrix} 0 & 0 \\ 0 & 0 \end{bmatrix}$

47. Answers will vary.; Example:

$$A(B + C) = \begin{bmatrix} 1 & 0 \\ 0 & 1 \end{bmatrix}\left(\begin{bmatrix} 1 & 0 \\ 0 & -1 \end{bmatrix} + \begin{bmatrix} -1 & 0 \\ 0 & 1 \end{bmatrix}\right) = \begin{bmatrix} 1 & 0 \\ 0 & 1 \end{bmatrix}\begin{bmatrix} 0 & 0 \\ 0 & 0 \end{bmatrix} = \begin{bmatrix} 0 & 0 \\ 0 & 0 \end{bmatrix}$$

$$AB + AC = \begin{bmatrix} 1 & 0 \\ 0 & 1 \end{bmatrix}\begin{bmatrix} 1 & 0 \\ 0 & -1 \end{bmatrix} + \begin{bmatrix} 1 & 0 \\ 0 & 1 \end{bmatrix}\begin{bmatrix} -1 & 0 \\ 0 & 1 \end{bmatrix} = \begin{bmatrix} 1 & 0 \\ 0 & -1 \end{bmatrix} + \begin{bmatrix} -1 & 0 \\ 0 & 1 \end{bmatrix} = \begin{bmatrix} 0 & 0 \\ 0 & 0 \end{bmatrix}$$

So, $A(B + C) = AB + AC$.

49. $\begin{bmatrix} x \\ -y \end{bmatrix}$; It changes the sign of the y-coordinate.

51. a. $\begin{bmatrix} 1 & 3 & 1 \\ 3 & 3 & 3 \\ 1 & 3 & 1 \end{bmatrix}$ **b.** $\begin{bmatrix} 1 & 3 & 1 \\ 3 & 3 & 3 \\ 1 & 3 & 1 \end{bmatrix} + \begin{bmatrix} -1 & -1 & -1 \\ -1 & -1 & -1 \\ -1 & -1 & -1 \end{bmatrix} = \begin{bmatrix} 0 & 2 & 0 \\ 2 & 2 & 2 \\ 0 & 2 & 0 \end{bmatrix}$ **c.** $\begin{bmatrix} 1 & 3 & 1 \\ 3 & 3 & 3 \\ 1 & 3 & 1 \end{bmatrix} + \begin{bmatrix} 1 & -2 & 1 \\ -2 & -2 & -2 \\ 1 & -2 & 1 \end{bmatrix} = \begin{bmatrix} 2 & 1 & 2 \\ 1 & 1 & 1 \\ 2 & 1 & 2 \end{bmatrix}$

53. $\begin{bmatrix} -2 & 1 & 1 & -1 & -1 & -2 \\ -3 & -3 & -2 & -2 & 2 & 2 \end{bmatrix}$ **55.** $\begin{bmatrix} 0 & \frac{3}{2} & \frac{3}{2} & \frac{1}{2} & \frac{1}{2} & 0 \\ 1 & 1 & \frac{3}{2} & \frac{3}{2} & \frac{7}{2} & \frac{7}{2} \end{bmatrix}$ **57. a.** $\begin{bmatrix} 0 & 3 & 3 & 1 & 1 & 0 \\ 0 & 0 & -1 & -1 & -5 & -5 \end{bmatrix}$ **b.** The effect is a reflection across the x-axis.

59. a. $\begin{bmatrix} 0 & 0 & -1 & -1 & -5 & -5 \\ 0 & 3 & 3 & 1 & 1 & 0 \end{bmatrix}$ **61. a.** $A = \begin{bmatrix} 2 & 6 \\ 31 & 46 \end{bmatrix}$ **b.** $B = \begin{bmatrix} 9 & 29 \\ 65 & 77 \end{bmatrix}$

b.

The effect is a 90° counterclockwise rotation about the origin.

c. $B - A = \begin{bmatrix} 7 & 23 \\ 34 & 31 \end{bmatrix}$
The difference between the percentage of people completing the transition to adulthood in 1960 and 2000 by age and gender

63. a. System 1: The midterm and final both count for 50% of the course grade. System 2: The midterm counts for 30% of the course grade and the final counts for 70%.

b. $\begin{bmatrix} 84 & 87.2 \\ 79 & 81 \\ 90 & 88.4 \\ 73 & 68.6 \\ 69 & 73.4 \end{bmatrix}$ System 1 grades are listed first (if different). Student 1: B; Student 2: C or B; Student 3: A or B; Student 4: C or D; Student 5: D or C

77. makes sense **79.** makes sense **83.** $AB = -BA$ so they are anticommutative.

85. $\begin{bmatrix} a_{11} & a_{12} \\ a_{21} & a_{22} \end{bmatrix}$; Nothing happens to the elements in the first matrix. **86.** $\{(15, -12, -4)\}$ **87.** $\begin{cases} a_1x + b_1y + c_1z = d_1 \\ a_2x + b_2y + c_2z = d_2 \\ a_3x + b_3y + c_3z = d_3 \end{cases}$

Mid-Chapter 6 Check Point

1. $\{(1, -1, 2)\}$ **2.** \varnothing **3.** $\left\{ \left(-\dfrac{4}{7} - \dfrac{4}{7}t, \dfrac{5}{7} + \dfrac{5}{7}t, t \right) \right\}$ **4.** $\{(3, 6, -4, 1)\}$ **5.** \varnothing **6.** $\begin{bmatrix} -4 & -\dfrac{1}{2} \\ 3 & 3 \end{bmatrix}$ **7.** $\begin{bmatrix} -12 & -2 \\ -21 & -4 \\ 3 & 1 \end{bmatrix}$ **8.** $\begin{bmatrix} 12 & -4 \\ 22 & -7 \\ -4 & 1 \end{bmatrix}$

9. $A + C$ does not exist because A is 3×2 and C is 2×2. **10.** $\begin{bmatrix} \dfrac{1}{2} & \dfrac{1}{2} \\ -3 & \dfrac{1}{2} \end{bmatrix}$

Section 6.4

Check Point Exercises

1. $AB = I_2$; $BA = I_2$ **2.** $\begin{bmatrix} 3 & -7 \\ -2 & 5 \end{bmatrix}$ **3.** $\begin{bmatrix} 1 & 2 \\ 1 & 3 \end{bmatrix}$ **4.** $\begin{bmatrix} 3 & -2 & -4 \\ 3 & -2 & -5 \\ -1 & 1 & 2 \end{bmatrix}$ **5.** $\{(4, -2, 1)\}$ **6.** The encoded message is $-7, 10, -53, 77$.

7. The decoded message is $2, 1, 19, 5$ or BASE.

Concept and Vocabulary Check

1. $\begin{bmatrix} 1 & 0 \\ 0 & 1 \end{bmatrix}$ **2.** $\begin{bmatrix} 1 & 0 & 0 \\ 0 & 1 & 0 \\ 0 & 0 & 1 \end{bmatrix}$ **3.** multiplicative inverse **4.** true **5.** $ad - bc \neq 0$

6. singular **7.** false **8.** A^{-1} **9.** $A^{-1}B$

Exercise Set 6.4

1. $AB = I_2$; $BA = I_2$; $B = A^{-1}$ **3.** $AB = \begin{bmatrix} 8 & -16 \\ -2 & 7 \end{bmatrix}$; $BA = \begin{bmatrix} 12 & 12 \\ 1 & 3 \end{bmatrix}$; $B \neq A^{-1}$ **5.** $AB = I_2$; $BA = I_2$; $B = A^{-1}$

7. $AB = I_3$; $BA = I_3$; $B = A^{-1}$ **9.** $AB = I_3$; $BA = I_3$; $B = A^{-1}$ **11.** $AB = I_4$; $BA = I_4$; $B = A^{-1}$

13. $\begin{bmatrix} \dfrac{2}{7} & -\dfrac{3}{7} \\ \dfrac{1}{7} & \dfrac{2}{7} \end{bmatrix}$ **15.** $\begin{bmatrix} 1 & \dfrac{1}{2} \\ 2 & \dfrac{3}{2} \end{bmatrix}$ **17.** A does not have an inverse. **19.** $\begin{bmatrix} \dfrac{1}{2} & 0 & 0 \\ 0 & \dfrac{1}{4} & 0 \\ 0 & 0 & \dfrac{1}{6} \end{bmatrix}$ **21.** $\begin{bmatrix} 1 & 1 & 2 \\ 1 & 1 & 1 \\ 2 & 3 & 4 \end{bmatrix}$ **23.** $\begin{bmatrix} 1 & 0 & 1 \\ 1 & 1 & 2 \\ 3 & 2 & 6 \end{bmatrix}$

25. $\begin{bmatrix} -3 & 2 & -4 \\ -1 & 1 & -1 \\ 8 & -5 & 10 \end{bmatrix}$ **27.** $\begin{bmatrix} 1 & 0 & 0 & 0 \\ 0 & -1 & 0 & 0 \\ 0 & 0 & \dfrac{1}{3} & 0 \\ -1 & 0 & 0 & 1 \end{bmatrix}$ **29.** $\begin{bmatrix} 6 & 5 \\ 5 & 4 \end{bmatrix} \begin{bmatrix} x \\ y \end{bmatrix} = \begin{bmatrix} 13 \\ 10 \end{bmatrix}$ **31.** $\begin{bmatrix} 1 & 3 & 4 \\ 1 & 2 & 3 \\ 1 & 4 & 3 \end{bmatrix} \begin{bmatrix} x \\ y \\ z \end{bmatrix} = \begin{bmatrix} -3 \\ -2 \\ -6 \end{bmatrix}$ **33.** $\begin{cases} 4x - 7y = -3 \\ 2x - 3y = 1 \end{cases}$

35. $\begin{cases} 2x - z = 6 \\ 3y = 9 \\ x + y = 5 \end{cases}$ **37. a.** $\begin{bmatrix} 2 & 6 & 6 \\ 2 & 7 & 6 \\ 2 & 7 & 7 \end{bmatrix} \begin{bmatrix} x \\ y \\ z \end{bmatrix} = \begin{bmatrix} 8 \\ 10 \\ 9 \end{bmatrix}$ **b.** $\{(1, 2, -1)\}$ **39. a.** $\begin{bmatrix} 1 & -1 & 1 \\ 0 & 2 & -1 \\ 2 & 3 & 0 \end{bmatrix} \begin{bmatrix} x \\ y \\ z \end{bmatrix} = \begin{bmatrix} 8 \\ -7 \\ 1 \end{bmatrix}$ **b.** $\{(2, -1, 5)\}$

41. a. $\begin{bmatrix} 1 & -1 & 2 & 0 \\ 0 & 1 & -1 & 1 \\ -1 & 1 & -1 & 2 \\ 0 & -1 & 1 & -2 \end{bmatrix} \begin{bmatrix} w \\ x \\ y \\ z \end{bmatrix} = \begin{bmatrix} -3 \\ 4 \\ 2 \\ -4 \end{bmatrix}$ **b.** $\{(2, 3, -1, 0)\}$ **43.** $\begin{bmatrix} \dfrac{1}{2}e^{-x} & -\dfrac{1}{2}e^{-3x} \\ \dfrac{1}{2}e^{-3x} & \dfrac{1}{2}e^{-5x} \end{bmatrix}$ **45.** $\begin{bmatrix} \dfrac{1}{8} & \dfrac{5}{8} \\ \dfrac{3}{8} & \dfrac{7}{8} \end{bmatrix}$

47. $(AB)^{-1} = \begin{bmatrix} -23 & 16 \\ 13 & -9 \end{bmatrix}$; $A^{-1}B^{-1} = \begin{bmatrix} -3 & 11 \\ 8 & -29 \end{bmatrix}$; $B^{-1}A^{-1} = \begin{bmatrix} -23 & 16 \\ 13 & -9 \end{bmatrix}$; $(AB)^{-1} = B^{-1}A^{-1}$ **49.** $AA^{-1} = I_3$ and $A^{-1}A = I_3$

51. The encoded message is $27, -19, 32, -20$.; The decoded message is $8, 5, 12, 16$ or HELP.
53. The encoded message is $14, 85, -33, 4, 18, -7, -18, 19, -9$.

65. $\begin{bmatrix} 1 & 1 \\ 2 & 3 \end{bmatrix}$ **67.** $\begin{bmatrix} 1 & 0 & 1 \\ 2 & 1 & 3 \\ -1 & 1 & 1 \end{bmatrix}$ **69.** $\begin{bmatrix} 0 & -1 & 0 & 1 \\ -1 & -5 & 0 & 3 \\ -2 & -4 & 1 & -2 \\ -1 & -4 & 0 & 1 \end{bmatrix}$ **71.** $\{(2, 3, -5)\}$ **73.** $\{(1, 2, -1)\}$ **75.** $\{(2, 1, 3, -2, 4)\}$

79. does not make sense **81.** makes sense **83.** false **85.** false **87.** false **91.** $a = 3$ or $a = -2$ **93.** 2 **94.** 6 **95.** -31

Section 6.5

Check Point Exercises

1. a. -4 **b.** -17 **2.** $\{(4, -2)\}$ **3.** 80 **4.** -24 **5.** $\{(2, -3, 4)\}$ **6.** -250

Concept and Vocabulary Check

1. $5 \cdot 3 - 2 \cdot 4 = 15 - 8 = 7$; determinant; 7

2. $x = \dfrac{\begin{vmatrix} 8 & 1 \\ -2 & -1 \end{vmatrix}}{\begin{vmatrix} 1 & 1 \\ 1 & -1 \end{vmatrix}}; y = \dfrac{\begin{vmatrix} 1 & 8 \\ 1 & -2 \end{vmatrix}}{\begin{vmatrix} 1 & 1 \\ 1 & -1 \end{vmatrix}}$

3. $3 \begin{vmatrix} 3 & 1 \\ 1 & 1 \end{vmatrix} - 4 \begin{vmatrix} 2 & 1 \\ 1 & 1 \end{vmatrix} + 5 \begin{vmatrix} 2 & 1 \\ 3 & 1 \end{vmatrix}$

4. $\dfrac{\begin{vmatrix} 3 & -8 & 4 \\ 2 & 11 & -2 \\ 1 & 4 & -2 \end{vmatrix}}{\begin{vmatrix} 3 & 1 & 4 \\ 2 & 3 & -2 \\ 1 & -3 & -2 \end{vmatrix}}$

5. column 3/the last column

Exercise Set 6.5

1. 1 **3.** -29 **5.** 0 **7.** 33 **9.** $-\dfrac{7}{16}$ **11.** $\{(5, 2)\}$ **13.** $\{(2, -3)\}$ **15.** $\{(3, -1)\}$ **17.** $\left\{\left(2, \dfrac{1}{2}\right)\right\}$ **19.** $\{(4, 2)\}$

21. $\{(7, 4)\}$ **23.** 72 **25.** -75 **27.** 0 **29.** $\{(-5, -2, 7)\}$ **31.** $\{(2, -3, 4)\}$ **33.** $\{(3, -1, 2)\}$ **35.** $\{(2, 3, 1)\}$ **37.** -200 **39.** 195

41. -42 **43.** $2x - 4y = 8; 3x + 5y = -10$ **45.** -11 **47.** 4 **49.** 28 sq units **51.** yes **53.** The equation of the line is $y = -\dfrac{11}{5}x + \dfrac{8}{5}$.

65. 13,200 **67.** does not make sense **69.** does not make sense **71. a.** a^2 **b.** a^3 **c.** a^4 **d.** Each determinant has zeros below the main diagonal and a's everywhere else. **e.** Each determinant equals a raised to the power equal to the order of the determinant.

73. The sign of the value is changed when 2 columns are interchanged in a 2nd order determinant.

75. $\begin{vmatrix} x & y & 1 \\ x_1 & y_1 & 1 \\ x_2 & y_2 & 1 \end{vmatrix} = x(y_1 - y_2) - y(x_1 - x_2) + (x_1 y_2 - x_2 y_1) = 0$; solving for y, $y = \dfrac{y_1 - y_2}{x_1 - x_2}x + \dfrac{x_1 y_2 - x_2 y_1}{x_1 - x_2}$, and $m = \dfrac{y_1 - y_2}{x_1 - x_2}$ and $b = \dfrac{x_1 y_2 - x_2 y_1}{x_1 - x_2}$.

77. a. -3 and 3 **b.** -2 and 2 **78.** $\dfrac{x^2}{16} + \dfrac{y^2}{25} = 1$ **79.** $(x - 1)^2 + (y + 2)^2 = 9$; center: $(1, -2)$; radius: 3;

$x^2 + y^2 - 2x + 4y = 4$

Chapter 6 Review Exercises

1. $\left[\begin{array}{ccc|c} 1 & 2 & 2 & 2 \\ 0 & 1 & -1 & 2 \\ 0 & 0 & 9 & -9 \end{array}\right]$ **2.** $\left[\begin{array}{ccc|c} 1 & -1 & \frac{1}{2} & -\frac{1}{2} \\ 1 & 2 & -1 & 2 \\ 6 & 4 & 3 & 5 \end{array}\right]$ **3.** $\{(1, 3, -4)\}$ **4.** $\{(-2, -1, 0)\}$ **5.** $\{(2, -2, 3, 4)\}$

6. a. $a = -2; b = 32; c = 42$ **b.** 2:00 p.m.; 170 parts per million **7.** capitalist: 1%; upper middle: 15%; lower middle: 34%; working: 30%

8. \varnothing **9.** $\{(2t + 4, t + 1, t)\}$ **10.** $\{(-37t + 2, 16t, -7t + 1, t)\}$ **11.** $\{(7t + 18, -3t - 7, t)\}$

12. a. $\begin{cases} x + z = 750 \\ y - z = -250 \\ x + y = 500 \end{cases}$ **b.** $\{(-t + 750, t - 250, t)\}$ **c.** $x = 350; y = 150$ **13.** $x = -5; y = 6; z = 6$ **14.** $\begin{bmatrix} 0 & 2 & 3 \\ 8 & 1 & 3 \end{bmatrix}$ **15.** $\begin{bmatrix} 0 & -4 \\ 6 & 4 \\ 2 & -10 \end{bmatrix}$

16. $\begin{bmatrix} -4 & 4 & -1 \\ -2 & -5 & 5 \end{bmatrix}$ **17.** Not possible since B is 3×2 and C is 3×3. **18.** $\begin{bmatrix} 2 & 3 & 8 \\ 21 & 5 & 5 \end{bmatrix}$ **19.** $\begin{bmatrix} -12 & 14 & 0 \\ 2 & -14 & 18 \end{bmatrix}$ **20.** $\begin{bmatrix} 0 & -10 & -15 \\ -40 & -5 & -15 \end{bmatrix}$

21. $\begin{bmatrix} -1 & -16 \\ 8 & 1 \end{bmatrix}$ **22.** $\begin{bmatrix} -10 & -6 & 2 \\ 16 & 3 & 4 \\ -23 & -16 & 7 \end{bmatrix}$ **23.** $\begin{bmatrix} -6 & 4 & -8 \\ 0 & 5 & 11 \\ -17 & 13 & -19 \end{bmatrix}$ **24.** $\begin{bmatrix} 10 & 5 \\ -2 & -30 \end{bmatrix}$ **25.** Not possible since AB is 2×2 and BA is 3×3.

26. $\begin{bmatrix} 7 & 6 & 5 \\ 2 & -1 & 11 \end{bmatrix}$ **27.** $\begin{bmatrix} -6 & -22 & -40 \\ 9 & 43 & 58 \\ -14 & -48 & -94 \end{bmatrix}$ **28.** $\begin{bmatrix} -2 & -6 \\ 3 & \frac{1}{3} \end{bmatrix}$ **29.** $\begin{bmatrix} 2 & 2 & 2 \\ 1 & 2 & 1 \\ 1 & 2 & 1 \end{bmatrix}$ **30.** $\begin{bmatrix} 1 & 1 & 1 \\ -1 & 1 & -1 \\ -1 & 1 & -1 \end{bmatrix}$

31. $\begin{bmatrix} -2 & 0 & 0 \\ 1 & 1 & -3 \end{bmatrix}$ **32.** $\begin{bmatrix} 0 & 1 & 1 \\ -2 & -2 & -4 \end{bmatrix}$ **33.** $\begin{bmatrix} 0 & 2 & 2 \\ 0 & 0 & 4 \end{bmatrix}$

The effect is a reflection over the x-axis

34. $\begin{bmatrix} 0 & -2 & -2 \\ 0 & 0 & -4 \end{bmatrix}$

The effect is a reflection
over the y-axis.

35. $\begin{bmatrix} 0 & 0 & 4 \\ 0 & 2 & 2 \end{bmatrix}$

The effect is a 90° counterclockwise
rotation about the origin.

36. $\begin{bmatrix} 0 & 4 & 4 \\ 0 & 0 & -4 \end{bmatrix}$

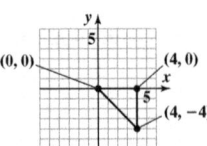

The effect is a horizontal stretch
by a factor of 2.

37. $AB = \begin{bmatrix} 1 & 7 \\ 0 & 5 \end{bmatrix}$; $BA = \begin{bmatrix} 1 & 0 \\ 1 & 5 \end{bmatrix}$; $B \neq A^{-1}$ **38.** $AB = I_3$; $BA = I_3$; $B = A^{-1}$ **39.** $\begin{bmatrix} 3 & 1 \\ 2 & 1 \end{bmatrix}$ **40.** $\begin{bmatrix} -\dfrac{3}{5} & \dfrac{1}{5} \\ 1 & 0 \end{bmatrix}$ **41.** $\begin{bmatrix} 3 & 0 & -2 \\ -6 & 1 & 4 \\ 1 & 0 & -1 \end{bmatrix}$

42. $\begin{bmatrix} 8 & -8 & 5 \\ -3 & 2 & -1 \\ -1 & -1 & 1 \end{bmatrix}$ **43. a.** $\begin{bmatrix} 1 & 1 & 2 \\ 0 & 1 & 3 \\ 3 & 0 & -2 \end{bmatrix} \begin{bmatrix} x \\ y \\ z \end{bmatrix} = \begin{bmatrix} 7 \\ -2 \\ 0 \end{bmatrix}$ **b.** $\{(-18, 79, -27)\}$ **44. a.** $\begin{bmatrix} 1 & -1 & 2 \\ 0 & 1 & -1 \\ 1 & 0 & 2 \end{bmatrix} \begin{bmatrix} x \\ y \\ z \end{bmatrix} = \begin{bmatrix} 12 \\ -5 \\ 10 \end{bmatrix}$

b. $\{(4, -2, 3)\}$ **45.** The encoded message is 96, 135, 46, 63; The decoded message is 18, 21, 12, 5 or RULE. **46.** 17 **47.** 4

48. -86 **49.** -236 **50.** 4 **51.** 16 **52.** $\left\{\left(\dfrac{7}{4}, -\dfrac{25}{8}\right)\right\}$ **53.** $\{(2, -7)\}$ **54.** $\{(23, -12, 3)\}$ **55.** $\{(-3, 2, 1)\}$

56. $a = \dfrac{5}{8}$; $b = -50$; $c = 1150$; 30- and 50-year-olds are involved in an average of 212.5 automobile accidents per day.

Chapter 6 Test

1. $\left\{\left(-3, \dfrac{1}{2}, 1\right)\right\}$ **2.** $\{(t, t - 1, t)\}$ **3.** $\begin{bmatrix} 5 & 4 \\ 1 & 11 \end{bmatrix}$ **4.** $\begin{bmatrix} 5 & -2 \\ 1 & -1 \\ 4 & -1 \end{bmatrix}$ **5.** $\begin{bmatrix} \dfrac{3}{5} & -\dfrac{2}{5} \\ \dfrac{1}{5} & \dfrac{1}{5} \end{bmatrix}$ **6.** $\begin{bmatrix} -1 & 2 \\ -5 & 4 \end{bmatrix}$ **7.** $AB = I_3$; $BA = I_3$

8. a. $\begin{bmatrix} 3 & 5 \\ 2 & -3 \end{bmatrix} \begin{bmatrix} x \\ y \end{bmatrix} = \begin{bmatrix} 9 \\ -13 \end{bmatrix}$ **b.** $\begin{bmatrix} \dfrac{3}{19} & \dfrac{5}{19} \\ \dfrac{2}{19} & -\dfrac{3}{19} \end{bmatrix}$ **c.** $\{(-2, 3)\}$ **9.** 18 **10.** $x = 2$

Cumulative Review Exercises (Chapters 1–6)

1. $\left\{\dfrac{-1 + \sqrt{33}}{4}, \dfrac{-1 - \sqrt{33}}{4}\right\}$ **2.** $\left[\dfrac{1}{2}, \infty\right)$ **3.** $\{6\}$ **4.** $\left\{-4, \dfrac{1}{3}, 1\right\}$ **5.** $\{\ln 5, \ln 9\}$ **6.** $\{1\}$ **7.** $\{(7, -4, 6)\}$

8. $y = -1$ **9.** $f^{-1}(x) = \dfrac{x^2 + 7}{4} (x \geq 0)$

10.

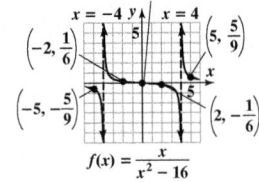

$f(x) = \dfrac{x}{x^2 - 16}$

11. $f(x) = (x + 2)(x - 3)(2x + 1)(2x - 1)$ **12.**

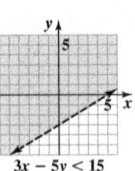

13. a. $A = 900e^{-0.017t}$ **b.** 759.30 g

14. $\begin{bmatrix} 2 & -1 \\ 13 & 1 \end{bmatrix}$ **15.** $\dfrac{8}{x - 3} + \dfrac{-2}{x - 2} + \dfrac{-3}{x + 2}$ **16.**

$y = -\dfrac{2}{3}x - 1$

17.

$3x - 5y < 15$

18.

$f(x) = x^2 - 2x - 3$

19.

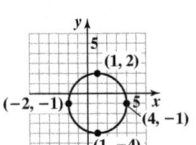

$(x - 1)^2 + (y + 1)^2 = 9$

20. $x^2 + 2x - 2$

CHAPTER 7

Section 7.1

Check Point Exercises

1. foci at $(-3\sqrt{3}, 0)$ and $(3\sqrt{3}, 0)$ **2.** foci at $(0, -\sqrt{7})$ and $(0, \sqrt{7})$ **3.** $\dfrac{x^2}{9} + \dfrac{y^2}{5} = 1$ **4.** foci at $(-1 - \sqrt{5}, 2)$ and $(-1 + \sqrt{5}, 2)$

$\dfrac{x^2}{36} + \dfrac{y^2}{9} = 1$

$16x^2 + 9y^2 = 144$

$\dfrac{(x+1)^2}{9} + \dfrac{(y-2)^2}{4} = 1$

5. Yes

Concept and Vocabulary Check

1. ellipse; foci; center **2.** 25; -5; 5; $(-5, 0)$; $(5, 0)$; 9; -3; 3; $(0, -3)$; $(0, 3)$ **3.** 25; -5; 5; $(0, -5)$; $(0, 5)$; 9; -3; 3; $(-3, 0)$; $(3, 0)$
4. 5; $(0, -\sqrt{5})$; $(0, \sqrt{5})$ **5.** $(-1, 4)$ **6.** $(-2, -2)$; $(8, -2)$ **7.** $(1, 9)$ **8.** 4; 1; 16

Exercise Set 7.1

1. foci at $(-2\sqrt{3}, 0)$ and $(2\sqrt{3}, 0)$ **3.** foci at $(0, -3\sqrt{3})$ and $(0, 3\sqrt{3})$ **5.** foci at $(0, -\sqrt{39})$ and $(0, \sqrt{39})$

$\dfrac{x^2}{16} + \dfrac{y^2}{4} = 1$

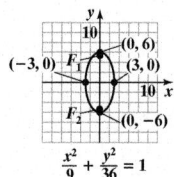

$\dfrac{x^2}{9} + \dfrac{y^2}{36} = 1$

$\dfrac{x^2}{25} + \dfrac{y^2}{64} = 1$

7. foci at $(0, -4\sqrt{2})$ and $(0, 4\sqrt{2})$ **9.** foci at $(0, -2)$ and $(0, 2)$ **11.** foci at $\left(-\dfrac{\sqrt{3}}{2}, 0\right)$ and $\left(\dfrac{\sqrt{3}}{2}, 0\right)$

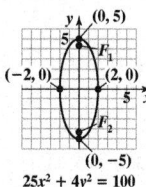

$\dfrac{x^2}{49} + \dfrac{y^2}{81} = 1$

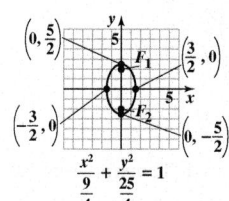

$\dfrac{x^2}{\frac{9}{4}} + \dfrac{y^2}{\frac{25}{4}} = 1$

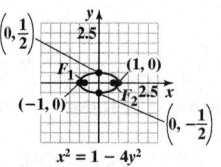

$x^2 = 1 - 4y^2$

13. foci at $(0, -\sqrt{21})$ and $(0, \sqrt{21})$ **15.** foci at $(-2\sqrt{3}, 0)$ and $(2\sqrt{3}, 0)$ **17.** foci at $(0, -\sqrt{2})$ and $(0, \sqrt{2})$

$25x^2 + 4y^2 = 100$

$4x^2 + 16y^2 = 64$

$7x^2 = 35 - 5y^2$

19. $\dfrac{x^2}{4} + \dfrac{y^2}{1} = 1$; foci at $(-\sqrt{3}, 0)$ and $(\sqrt{3}, 0)$ **21.** $\dfrac{x^2}{1} + \dfrac{y^2}{4} = 1$; foci at $(0, \sqrt{3})$ and $(0, -\sqrt{3})$

23. $\dfrac{(x+1)^2}{4} + \dfrac{(y-1)^2}{1} = 1$; foci at $(-1 - \sqrt{3}, 1)$ and $(-1 + \sqrt{3}, 1)$ **25.** $\dfrac{x^2}{64} + \dfrac{y^2}{39} = 1$ **27.** $\dfrac{x^2}{33} + \dfrac{y^2}{49} = 1$

29. $\dfrac{x^2}{13} + \dfrac{y^2}{9} = 1$ **31.** $\dfrac{x^2}{16} + \dfrac{y^2}{4} = 1$ **33.** $\dfrac{(x+2)^2}{4} + \dfrac{(y-3)^2}{25} = 1$ **35.** $\dfrac{(x-7)^2}{4} + \dfrac{(y-6)^2}{9} = 1$

37. foci at $(2 - \sqrt{5}, 1)$ and $(2 + \sqrt{5}, 1)$ **39.** foci at $(-3 - 2\sqrt{3}, 2)$ and $(-3 + 2\sqrt{3}, 2)$ **41.** foci at $(4, 2)$ and $(4, -6)$

$\dfrac{(x-2)^2}{9} + \dfrac{(y-1)^2}{4} = 1$

$(x+3)^2 + 4(y-2)^2 = 16$

$\dfrac{(x-4)^2}{9} + \dfrac{(y+2)^2}{25} = 1$

43. foci at $(0, 2 + \sqrt{11}), (0, 2 - \sqrt{11})$

$$\frac{x^2}{25} + \frac{(y - 2)^2}{36} = 1$$

45. foci at $(-3 - 2\sqrt{2}, 2)$ and $(-3 + 2\sqrt{2}, 2)$

$$\frac{(x + 3)^2}{9} + (y - 2)^2 = 1$$

47. foci at $(1, -3 + \sqrt{3})$ and $(1, -3 - \sqrt{3})$

$$\frac{(x - 1)^2}{2} + \frac{(y + 3)^2}{5} = 1$$

49. foci at $(1, -3 + \sqrt{5})$ and $(1, -3 - \sqrt{5})$

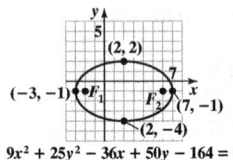

$9(x - 1)^2 + 4(y + 3)^2 = 36$

51. $\dfrac{(x - 2)^2}{25} + \dfrac{(y + 1)^2}{9} = 1$

foci at $(-2, -1)$ and $(6, -1)$

$9x^2 + 25y^2 - 36x + 50y - 164 = 0$

53. $\dfrac{(x - 1)^2}{16} + \dfrac{(y + 2)^2}{9} = 1$

foci at $(1 - \sqrt{7}, -2)$ and $(1 + \sqrt{7}, -2)$

$9x^2 + 16y^2 - 18x + 64y - 71 = 0$

55. $\dfrac{(x + 2)^2}{16} + \dfrac{(y - 3)^2}{64} = 1$

foci at $(-2, 3 + 4\sqrt{3})$ and $(-2, 3 - 4\sqrt{3})$

$4x^2 + y^2 + 16x - 6y - 39 = 0$

57. $\{(0, -1), (0, 1)\}$ **59.** $\{(0, 3)\}$ **61.** $\{(0, -2), (1, 0)\}$ **63.**

$y = -\sqrt{16 - 4x^2}$

65. Yes **67. a.** $\dfrac{x^2}{2304} + \dfrac{y^2}{529} = 1$ **b.** about 42 feet

79. does not make sense **81.** does not make sense **83.** $\dfrac{x^2}{\frac{36}{5}} + \dfrac{y^2}{36} = 1$

85. large circle: $x^2 + y^2 = 25$; small circle: $x^2 + y^2 = 9$ **87.** $\dfrac{x^2}{9} - \dfrac{y^2}{4} = 1$; Terms are separated by subtraction rather than addition.

88. a. -4 and 4 **b.** The equation $y^2 = -9$ has no real solutions. **89. a.** -3 and 3 **b.** The equation $x^2 = -16$ has no real solutions.

Section 7.2

Check Point Exercises

1. a. vertices at $(5, 0)$ and $(-5, 0)$; foci at $(\sqrt{41}, 0)$ and $(-\sqrt{41}, 0)$ **b.** vertices at $(0, 5)$ and $(0, -5)$; foci at $(0, \sqrt{41})$ and $(0, -\sqrt{41})$

2. $\dfrac{y^2}{9} - \dfrac{x^2}{16} = 1$

3. foci at $(-3\sqrt{5}, 0)$ and $(3\sqrt{5}, 0)$

asymptotes: $y = \pm\dfrac{1}{2}x$

$\dfrac{x^2}{36} - \dfrac{y^2}{9} = 1$

4. foci at $(0, \sqrt{5})$ and $(0, -\sqrt{5})$

asymptotes: $y = \pm 2x$

$y^2 - 4x^2 = 4$

5. foci at $(3 - \sqrt{5}, 1)$ and $(3 + \sqrt{5}, 1)$

asymptotes: $(y - 1) = \pm\dfrac{1}{2}(x - 3)$

$\dfrac{(x - 3)^2}{4} - \dfrac{(y - 1)^2}{1} = 1$

6. foci at $(3, -5 + \sqrt{13})$ and $(3, -5 - \sqrt{13})$

asymptotes: $(y + 5) = \pm\dfrac{2}{3}(x - 3)$

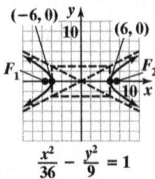

$4x^2 - 24x - 9y^2 - 90y - 153 = 0$

7. $\dfrac{x^2}{2,722,500} - \dfrac{y^2}{25,155,900} = 1$

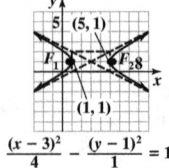

Concept and Vocabulary Check

1. hyperbola; foci; vertices; transverse
2. $(-5, 0)$; $(5, 0)$; $(-\sqrt{34}, 0)$; $(\sqrt{34}, 0)$
3. $(0, -5)$; $(0, 5)$; $(0, -\sqrt{34})$; $(0, \sqrt{34})$
4. asymptotes; center
5. dividing; 36
6. $y = -\dfrac{3}{2}x$; $y = \dfrac{3}{2}x$
7. $y = -2x$; $y = 2x$
8. $(-3, 3)$; $(7, 3)$
9. $(7, -2)$
10. 16; 1; 128

Exercise Set 7.2

1. vertices at $(2, 0)$ and $(-2, 0)$; foci at $(\sqrt{5}, 0)$ and $(-\sqrt{5}, 0)$; graph (b)
3. vertices at $(0, 2)$ and $(0, -2)$; foci at $(0, \sqrt{5})$ and $(0, -\sqrt{5})$; graph (a)

5. $y^2 - \dfrac{x^2}{8} = 1$
7. $\dfrac{x^2}{9} - \dfrac{y^2}{7} = 1$
9. $\dfrac{y^2}{36} - \dfrac{x^2}{9} = 1$
11. $\dfrac{(x - 4)^2}{4} - \dfrac{(y + 2)^2}{5} = 1$

13. foci: $(\pm\sqrt{34}, 0)$

asymptotes: $y = \pm\dfrac{5}{3}x$

$\dfrac{x^2}{9} - \dfrac{y^2}{25} = 1$

15. foci: $(\pm 2\sqrt{41}, 0)$

asymptotes: $y = \pm\dfrac{4}{5}x$

$\dfrac{x^2}{100} - \dfrac{y^2}{64} = 1$

17. foci: $(0, \pm 2\sqrt{13})$

asymptotes: $y = \pm\dfrac{2}{3}x$

$\dfrac{y^2}{16} - \dfrac{x^2}{36} = 1$

19. foci: $\left(0, \pm\dfrac{\sqrt{5}}{2}\right)$

asymptotes: $y = \pm\dfrac{1}{2}x$

$4y^2 - x^2 = 1$

21. foci: $(\pm\sqrt{13}, 0)$

asymptotes: $y = \pm\dfrac{3}{2}x$

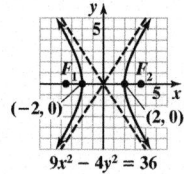

$9x^2 - 4y^2 = 36$

23. foci: $(0, \pm\sqrt{34})$

asymptotes: $y = \pm\dfrac{5}{3}x$

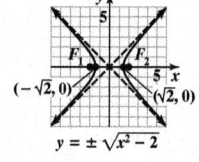

$9y^2 - 25x^2 = 225$

25. foci: $(\pm 2, 0)$

asymptotes: $y = \pm x$

$y = \pm\sqrt{x^2 - 2}$

27. $\dfrac{x^2}{9} - \dfrac{y^2}{25} = 1$
29. $\dfrac{y^2}{4} - \dfrac{x^2}{9} = 1$
31. $\dfrac{(x - 2)^2}{4} - \dfrac{(y + 3)^2}{9} = 1$

33. foci: $(-9, -3)$, $(1, -3)$

asymptotes: $(y + 3) = \pm\dfrac{4}{3}(x + 4)$

$\dfrac{(x + 4)^2}{9} - \dfrac{(y + 3)^2}{16} = 1$

35. foci: $(-3 \pm\sqrt{41}, 0)$

asymptotes: $y = \pm\dfrac{4}{5}(x + 3)$

$\dfrac{(x + 3)^2}{25} - \dfrac{y^2}{16} = 1$

37. foci: $(1, -2 \pm 2\sqrt{5})$

asymptotes: $(y + 2) = \pm\dfrac{1}{2}(x - 1)$

$\dfrac{(y + 2)^2}{4} - \dfrac{(x - 1)^2}{16} = 1$

39. foci: $(3 \pm\sqrt{5}, -3)$

asymptotes: $(y + 3) = \pm\dfrac{1}{2}(x - 3)$

$(x - 3)^2 - 4(y + 3)^2 = 4$

41. foci: $(1 \pm\sqrt{6}, 2)$

asymptotes: $(y - 2) = \pm(x - 1)$

$(x - 1)^2 - (y - 2)^2 = 3$

43. $(x - 1)^2 - (y + 2)^2 = 1$

foci: $(1 \pm\sqrt{2}, -2)$

asymptotes: $(y + 2) = \pm(x - 1)$

$x^2 - y^2 - 2x - 4y - 4 = 0$

45. $\dfrac{(y + 1)^2}{4} - \dfrac{(x + 2)^2}{0.25} = 1$

foci: $(-2, -1 \pm\sqrt{4.25})$

asymptotes: $(y + 1) = \pm 4(x + 2)$

$16x^2 - y^2 + 64x - 2y + 67 = 0$

47. $\dfrac{(x - 2)^2}{9} - \dfrac{(y - 3)^2}{4} = 1$

foci: $(2 \pm\sqrt{13}, 3)$

asymptotes: $(y - 3) = \pm\dfrac{2}{3}(x - 2)$

$4x^2 - 9y^2 - 16x + 5y - 101 = 0$

49. $\dfrac{y^2}{4} - \dfrac{(x - 4)^2}{25} = 1$

foci: $(4, \pm\sqrt{29})$

asymptotes: $y = \pm\dfrac{2}{5}(x - 4)$

$4x^2 - 25y^2 - 32x + 164 = 0$

51. domain: $(-\infty, -3] \cup [3, \infty)$
range: $(-\infty, \infty)$

$$\frac{x^2}{9} - \frac{y^2}{16} = 1$$

53. domain: $[-3, 3]$
range: $[-4, 4]$

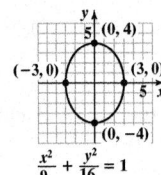

$$\frac{x^2}{9} + \frac{y^2}{16} = 1$$

55. domain: $(-\infty, \infty)$
range: $(-\infty, -4] \cup [4, \infty)$

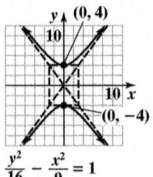

$$\frac{y^2}{16} - \frac{x^2}{9} = 1$$

57. $\{(-2, 0), (2, 0)\}$
59. $\{(0, -3), (0, 3)\}$
61. If M_1 is located 2640 feet to the right of the origin on the x-axis, the explosion is located on the right branch of the hyperbola given

by the equation $\dfrac{x^2}{1{,}210{,}000} - \dfrac{y^2}{5{,}759{,}600} = 1$. **63.** 40 yd **65. a.** ellipse **b.** $x^2 + 4y^2 = 4$

77. ; no; two intersecting lines

79. $2y^2 + (10 - 6x)y + (4x^2 - 3x - 6) = 0$

$$y = \frac{3x - 5 \pm \sqrt{x^2 - 24x + 37}}{2}$$

The xy-term rotates the hyperbola.

81. does not make sense **83.** makes sense **85.** false **87.** true **89.** $\dfrac{c}{a}$ will be large when a is small. When this happens, the asymptotes will

be nearly vertical. **91.** Any hyperbola where $a = b$, such as $\dfrac{x^2}{4} - \dfrac{y^2}{4} = 1$, has perpendicular asymptotes.

92.

93.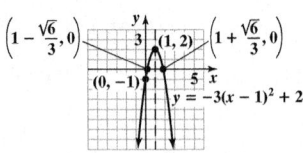

94. $(y + 1)^2 = -12x + 24$

Mid-Chapter 7 Check Point

1. foci: $(\pm\sqrt{21}, 0)$

$$\frac{x^2}{25} + \frac{y^2}{4} = 1$$

2. foci: $(0, \pm\sqrt{5})$

$$9x^2 + 4y^2 = 36$$

3. foci: $(2, 2), (2, -4)$

$$\frac{(x-2)^2}{16} + \frac{(y+1)^2}{25} = 1$$

4. foci: $(-5, 1), (1, 1)$

$$\frac{(x+2)^2}{25} + \frac{(y-1)^2}{16} = 1$$

5. foci: $(2 \pm 4\sqrt{2}, -3)$

$$x^2 + 9y^2 - 4x + 54y + 49 = 0$$

6. foci: $(\pm\sqrt{10}, 0)$

asymptotes: $y = \pm\dfrac{1}{3}x$

$$\frac{x^2}{9} - y^2 = 1$$

7. foci: $(0, \pm\sqrt{10})$

asymptotes: $y = \pm 3x$

$$\frac{y^2}{9} - x^2 = 1$$

8. foci: $(0, \pm 2\sqrt{5})$

asymptotes: $y = \pm 2x$

$$y^2 - 4x^2 = 16$$

9. foci: $(\pm\sqrt{53}, 0)$

asymptotes: $y = \pm\dfrac{2}{7}x$

$$4x^2 - 49y^2 = 196$$

10. foci: $(-3, -2), (7, -2)$

asymptotes: $(y + 2) = \pm\dfrac{4}{3}(x - 2)$

$$\frac{(x-2)^2}{9} - \frac{(y+2)^2}{16} = 1$$

11. foci: $(-1, 3 \pm 2\sqrt{5})$
asymptotes:
$(y - 3) = \pm 2(x + 1)$

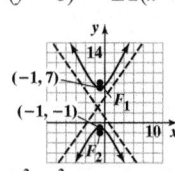

$4x^2 - y^2 + 8x + 6y + 11 = 0$

12.
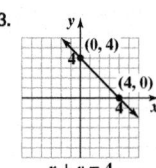

$x^2 + y^2 = 4$

13.

$x + y = 4$

14.
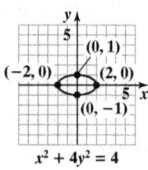

$x^2 - y^2 = 4$

15.
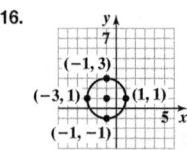

$x^2 + 4y^2 = 4$

16.

$(x + 1)^2 + (y - 1)^2 = 4$

17.

$x^2 + 4(y - 1)^2 = 4$

18.

$(x - 1)^2 - (y - 1)^2 = 4$

19. $\dfrac{x^2}{25} + \dfrac{y^2}{9} = 1$

20. $\dfrac{(x - 1)^2}{81} + \dfrac{(y - 2)^2}{56} = 1$

21. $\dfrac{y^2}{4} - \dfrac{x^2}{5} = 1$

22. $\dfrac{(x + 1)^2}{4} - \dfrac{(y - 5)^2}{5} = 1$

23. no **24.** $20\sqrt{3}$ cm

25. a. $\dfrac{x^2}{1.1025} - \dfrac{y^2}{7.8975} = 1$ **b.**

Section 7.3

Check Point Exercises

1. focus: $(2, 0)$
directrix: $x = -2$

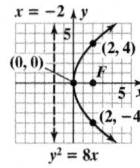

$y^2 = 8x$

2. focus: $(0, -3)$
directrix: $y = 3$

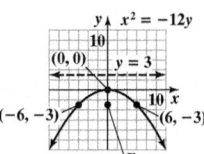

3. $y^2 = 32x$

4. vertex: $(2, -1)$; focus: $(2, 0)$
directrix: $y = -2$

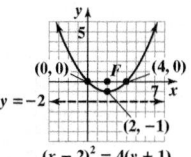

$(x - 2)^2 = 4(y + 1)$

5. vertex: $(2, -1)$; focus: $(1, -1)$
directrix: $x = 3$

$y^2 + 2y + 4x - 7 = 0$

6. $x^2 = \dfrac{9}{4}y$; The light should be placed at $\left(0, \dfrac{9}{16}\right)$, or $\dfrac{9}{16}$ inch above the vertex.

Concept and Vocabulary Check

1. parabola; directrix; focus **2.** a **3.** $(-7, 0)$ **4.** $x = 7$ **5.** 28; $(-7, -14)$; $(-7, 14)$ **6.** d **7.** $(-2, 0)$ **8.** $y = -2$ **9.** 4; $(-4, 0)$; $(0, 0)$

Exercise Set 7.3

1. focus: $(1, 0)$; directrix: $x = -1$; graph (c) **3.** focus: $(0, -1)$, directrix: $y = 1$; graph (b)

5. focus: $(4, 0)$; directrix: $x = -4$

$y^2 = 16x$

7. focus: $(-2, 0)$; directrix: $x = 2$

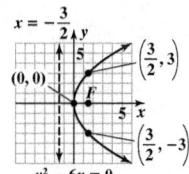

$y^2 = -8x$

9. focus: $(0, 3)$; directrix: $y = -3$

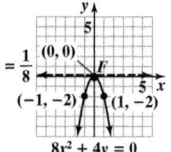

$x^2 = 12y$

11. focus: $(0, -4)$; directrix: $y = 4$

$x^2 = -16y$

13. focus: $\left(\dfrac{3}{2}, 0\right)$; directrix: $x = -\dfrac{3}{2}$

$y^2 - 6x = 0$

15. focus: $\left(0, -\dfrac{1}{8}\right)$; directrix: $y = \dfrac{1}{8}$

$8x^2 + 4y = 0$

17. $y^2 = 28x$ **19.** $y^2 = -20x$ **21.** $x^2 = 60y$ **23.** $x^2 = -100y$ **25.** $(x - 2)^2 = -8(y + 3)$ **27.** $(y - 2)^2 = 8(x - 1)$
29. $(x + 3)^2 = 4(y - 3)$ **31.** vertex: $(1, 1)$; focus: $(2, 1)$; directrix: $x = 0$; graph (c) **33.** vertex: $(-1, -1)$; focus: $(-1, -2)$; directrix: $y = 0$; graph (d)

35. vertex: (2, 1); focus: (2, 3)
directrix: $y = -1$

$(x - 2)^2 = 8(y - 1)$

37. vertex: (−1, −1); focus: (−1, −3)
directrix: $y = 1$

$(x + 1)^2 = -8(y + 1)$

39. vertex: (−1, −3); focus: (2, −3)
directrix: $x = -4$

$(y + 3)^2 = 12(x + 1)$

41. vertex: (0, −1); focus: (−2, −1)
directrix: $x = 2$

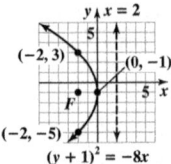

$(y + 1)^2 = -8x$

43. $(x - 1)^2 = 4(y - 2)$; vertex: (1, 2)
focus: (1, 3); directrix: $y = 1$

$x^2 - 2x - 4y + 9 = 0$

45. $(y - 1)^2 = -12(x - 3)$
vertex: (3, 1); focus: (0, 1)
directrix: $x = 6$

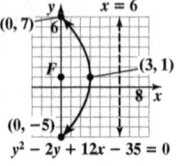

$y^2 - 2y + 12x - 35 = 0$

47. $(x + 3)^2 = 4(y + 2)$
vertex: (−3, −2); focus: (−3, −1)
directrix: $y = -3$

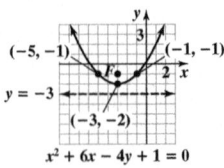

$x^2 + 6x - 4y + 1 = 0$

49. domain: $[-4, \infty)$
range: $(-\infty, \infty)$; not a function
51. domain: $(-\infty, \infty)$
range: $(-\infty, 1]$; function
53. domain: $(-\infty, 3]$
range: $(-\infty, \infty)$; not a function

55. $\{(-4, 2), (0, 0)\}$

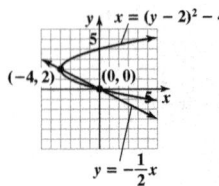

$y = -\frac{1}{2}x$

57. $\{(-2, 1)\}$

59. \varnothing

$x = (y + 2)^2 - 1$

61. 1 inch above the vertex **63.** 4.5 feet from the base of the dish **65.** 76 m **67.** yes

77. $y = -1 \pm \sqrt{6x - 12}$

79. $9y^2 + (-24x - 80)y + 16x^2 - 60x + 100 = 0$

$$y = \frac{12x + 40 \pm 10\sqrt{15x + 7}}{9}$$

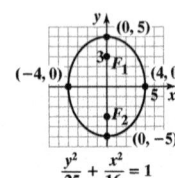

81. does not make sense
83. makes sense
85. false
87. false
89. focus: $\left(0, -\dfrac{E}{4A}\right)$; directrix: $y = \dfrac{E}{4A}$

92. $-\dfrac{1}{2}, \dfrac{1}{8}, -\dfrac{1}{26}, \dfrac{1}{80}$

93. 120 **94.** 2; 5; 10; 17; 26; 37; Sum is 97.

Chapter 7 Review Exercises

1. foci: $(\pm\sqrt{11}, 0)$

$\dfrac{x^2}{36} + \dfrac{y^2}{25} = 1$

2. foci: $(0, \pm 3)$

$\dfrac{y^2}{25} + \dfrac{x^2}{16} = 1$

3. foci: $(0, \pm 2\sqrt{3})$

$4x^2 + y^2 = 16$

4. foci: $(\pm\sqrt{5}, 0)$

$4x^2 + 9y^2 = 36$

5. foci: $(1 \pm \sqrt{7}, -2)$

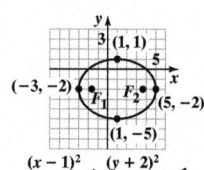

$$\frac{(x-1)^2}{16} + \frac{(y+2)^2}{9} = 1$$

6. foci: $(-1, 2 \pm \sqrt{7})$

$$\frac{(x+1)^2}{9} + \frac{(y-2)^2}{16} = 1$$

7. foci: $(-3 \pm \sqrt{5}, 2)$

$4x^2 + 9y^2 + 24x - 36y + 36 = 0$

8. foci: $(1, -1 \pm \sqrt{5})$

$9x^2 + 4y^2 - 18x + 8y - 23 = 0$

9. $\dfrac{x^2}{25} + \dfrac{y^2}{9} = 1$ **10.** $\dfrac{x^2}{27} + \dfrac{y^2}{36} = 1$ **11.** $\dfrac{(x+3)^2}{36} + \dfrac{(y-5)^2}{4} = 1$ **12.** $\dfrac{x^2}{100} + \dfrac{y^2}{36} = 1$ **13.** yes

14. The hit ball will collide with the other ball.

15. foci: $(\pm\sqrt{17}, 0)$; $y = \pm\dfrac{1}{4}x$

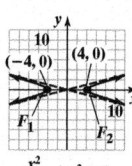

$$\frac{x^2}{16} - y^2 = 1$$

16. foci: $(0, \pm\sqrt{17})$; $y = \pm 4x$

$$\frac{y^2}{16} - x^2 = 1$$

17. foci: $(\pm 5, 0)$; $y = \pm\dfrac{3}{4}x$

$9x^2 - 16y^2 = 144$

18. foci: $(0, \pm 2\sqrt{5})$; $y = \pm\dfrac{1}{2}x$

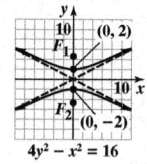

$4y^2 - x^2 = 16$

19. foci: $(2 \pm \sqrt{41}, -3)$

$$y + 3 = \pm\frac{4}{5}(x - 2)$$

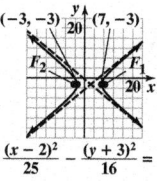

$$\frac{(x-2)^2}{25} - \frac{(y+3)^2}{16} = 1$$

20. foci: $(3, -2 \pm \sqrt{41})$

$$y + 2 = \pm\frac{5}{4}(x - 3)$$

$$\frac{(y+2)^2}{25} - \frac{(x-3)^2}{16} = 1$$

21. foci: $(1, 2 \pm \sqrt{5})$

$$y - 2 = \pm 2(x - 1)$$

$y^2 - 4y - 4x^2 + 8x - 4 = 0$

22. foci: $(1 \pm \sqrt{2}, -1)$

$$y + 1 = \pm(x - 1)$$

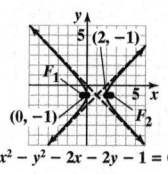

$x^2 - y^2 - 2x - 2y - 1 = 0$

23. $\dfrac{y^2}{4} - \dfrac{x^2}{12} = 1$ **24.** $\dfrac{x^2}{9} - \dfrac{y^2}{55} = 1$ **25.** c must be greater than a. **26.** $\dfrac{x^2}{2162.25} - \dfrac{y^2}{7837.75} = 1$

27. vertex: $(0, 0)$; focus: $(2, 0)$
directrix: $x = -2$

$y^2 = 8x$

28. vertex: $(0, 0)$; focus: $(0, -4)$
directrix: $y = 4$

$x^2 + 16y = 0$

29. vertex: $(0, 2)$; focus: $(-4, 2)$
directrix: $x = 4$

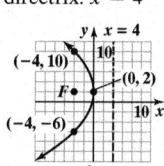

$(y - 2)^2 = -16x$

30. vertex: $(4, -1)$; focus: $(4, 0)$
directrix: $y = -2$

$(x - 4)^2 = 4(y + 1)$

31. vertex: $(0, 1)$; focus: $(0, 0)$
directrix: $y = 2$

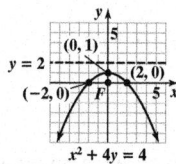

$x^2 + 4y = 4$

32. vertex: $(-1, 5)$; focus: $(0, 5)$
directrix: $x = -2$

$y^2 - 4x - 10y + 21 = 0$

33. vertex: $(2, -2)$; focus: $\left(2, -\dfrac{3}{2}\right)$; directrix: $y = -\dfrac{5}{2}$

$x^2 - 4x - 2y = 0$

34. $y^2 = 48x$ **35.** $x^2 = -44y$ **36.** $x^2 = 12y$; Place the light 3 inches from the vertex at $(0, 3)$. **37.** approximately 58 ft
38. approximately 128 ft

Chapter 7 Test

1. foci: $(\pm\sqrt{13}, 0)$

asymptotes: $y = \pm\dfrac{3}{2}x$

$9x^2 - 4y^2 = 36$

2. vertex: $(0, 0)$; focus: $(0, -2)$

directrix: $y = 2$

$x^2 = -8y$

3. foci: $(-6, 5)$, $(2, 5)$

$$\frac{(x+2)^2}{25} + \frac{(y-5)^2}{9} = 1$$

4. foci: $(-1, 1 \pm \sqrt{5})$

asymptotes: $y - 1 = \pm 2(x + 1)$

$4x^2 - y^2 + 8x + 2y + 7 = 0$

5. vertex: $(-5, 1)$; focus: $(-5, 3)$
directrix: $y = -1$

$(x + 5)^2 = 8(y - 1)$

6. $\dfrac{x^2}{100} + \dfrac{y^2}{51} = 1$ **7.** $\dfrac{y^2}{49} - \dfrac{x^2}{51} = 1$ **8.** $y^2 = 200x$ **9.** 32 ft

10. a. $x^2 = 3y$ **b.** Light is placed $\dfrac{3}{4}$ inch above the vertex.

Cumulative Review Exercises (Chapters 1–7)

1. $\{2\}$ **2.** $\{x \mid x < 2\}$ or $(-\infty, 2)$ **3.** $\{9\}$ **4.** $\{2 + 2\sqrt{5}, 2 - 2\sqrt{5}\}$ **5.** $\{x \mid x \geq 4 \text{ or } x \leq -3\}$ or $(-\infty, -3] \cup [4, \infty)$ **6.** $\left\{\dfrac{2}{3}, -1 \pm \sqrt{2}\right\}$

7. $\{3\}$ **8.** $\{(2, -1)\}$ **9.** $\{(2, -4), (-14, -20)\}$ **10.** $\{(7, -4, 6)\}$

11.

$f(x) = (x - 1)^2 - 4$

12.
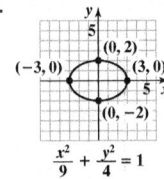
$\dfrac{x^2}{9} + \dfrac{y^2}{4} = 1$

13.

14. a. $\pm 1, \pm 3, \pm\dfrac{1}{2}, \pm\dfrac{3}{2}, \pm\dfrac{1}{4}, \pm\dfrac{3}{4}, \pm\dfrac{1}{8}, \pm\dfrac{3}{8}, \pm\dfrac{1}{16}, \pm\dfrac{3}{16}, \pm\dfrac{1}{32}, \pm\dfrac{3}{32}$ **b.** $\left\{-\dfrac{1}{8}, \dfrac{3}{4}, 1\right\}$

15. a. domain: $(-2, 2)$; range: $[-3, \infty)$ **b.** minimum of -3 at $x = 0$ **c.** $(0, 2)$ **d.** 3 **e.** -3 **f.** $x \to -2^+$; $x \to 2^-$

g.

$g(x) = f(x - 2) + 1$

h.

$h(x) = -f(2x)$

16. $(g \circ f)(x) = x^2 - 2$ **17.** $3\log_5 x + \dfrac{1}{2}\log_5 y - 3$ **18.** $y = -2x - 2$

19. The costs will be the same when the number of miles driven is 175 miles. The cost will be $67.

20. $25 for basic cable service and $10 for each movie channel

CHAPTER 8

Section 8.1

Check Point Exercises

1. a. $7, 9, 11, 13$ **b.** $-\dfrac{1}{3}, \dfrac{1}{5}, -\dfrac{1}{9}, \dfrac{1}{17}$ **2.** $3, 11, 27, 59$ **3.** $10, \dfrac{10}{3}, \dfrac{5}{6}, \dfrac{1}{6}$ **4. a.** 91 **b.** n **5. a.** 182 **b.** 47 **c.** 20

6. a. $\displaystyle\sum_{i=1}^{9} i^2$ **b.** $\displaystyle\sum_{i=1}^{n} \dfrac{1}{2^{i-1}}$

Concept and Vocabulary Check

1. sequence; integers; terms **2.** general **3.** 4 **4.** $\dfrac{1}{15}$ **5.** 2 **6.** factorial; 5; 1; 1 **7.** $n + 3$ **8.** $a_1 + a_2 + a_3 + \cdots + a_n$; index; upper limit; lower limit

Exercise Set 8.1

1. $5, 8, 11, 14$ **3.** $3, 9, 27, 81$ **5.** $-3, 9, -27, 81$ **7.** $-4, 5, -6, 7$ **9.** $\dfrac{2}{5}, \dfrac{2}{3}, \dfrac{6}{7}, 1$ **11.** $1, -\dfrac{1}{3}, \dfrac{1}{7}, -\dfrac{1}{15}$ **13.** $7, 12, 17, 22$ **15.** $3, 12, 48, 192$

17. $4, 11, 25, 53$ **19.** $1, 2, \dfrac{3}{2}, \dfrac{2}{3}$ **21.** $4, 12, 48, 240$ **23.** 272 **25.** 120 **27.** $(n + 2)(n + 1)$ **29.** 105 **31.** 60 **33.** 115 **35.** $-\dfrac{5}{16}$

37. 55 **39.** $\dfrac{3}{8}$ **41.** 15 **43.** $\displaystyle\sum_{i=1}^{15} i^2$ **45.** $\displaystyle\sum_{i=1}^{11} 2^i$ **47.** $\displaystyle\sum_{i=1}^{30} i$ **49.** $\displaystyle\sum_{i=1}^{14} \dfrac{i}{i + 1}$ **51.** $\displaystyle\sum_{i=1}^{n} \dfrac{4^i}{i}$ **53.** $\displaystyle\sum_{i=1}^{n} (2i - 1)$ **55.** $\displaystyle\sum_{k=1}^{14} (2k + 3)$ **57.** $\displaystyle\sum_{k=0}^{12} ar^k$

59. $\displaystyle\sum_{k=0}^{n} (a + kd)$ **61.** 45 **63.** 0 **65.** 2 **67.** 80 **69. a.** 1515; A total of 1515 thousand, or 1,515,000, autism cases were diagnosed in the United States from 2001 through 2008. **b.** 1512; underestimates by 3 thousand **71.** $8081.13 **81.** 39,800 **83.** 8.109673361 E15

85. 24,804 **89.**

91.

93. does not make sense **95.** makes sense **97.** false **99.** false **101.** $9, 32, 16, 8, 4$ **103.** $-5; -5; -5; -5$; The difference between consecutive terms is always -5. **104.** $4; 4; 4; 4$; The difference between consecutive terms is always 4. **105.** -45

As n gets larger, a_n approaches 1. As n gets larger, a_n approaches 0.

Section 8.2

Check Point Exercises

1. $100, 70, 40, 10, -20, -50$ **2.** -34 **3. a.** $a_n = 0.35n + 15.65$ **b.** 23% **4.** 360 **5.** 2460 **6.** $\$740,300$

Concept and Vocabulary Check

1. arithmetic; common difference **2.** $a_1 + (n - 1)d$; first term; common difference **3.** $\dfrac{n}{2}(a_1 + a_n)$; first term; nth term **4.** $2; 116$
5. $8; 13; 18; 5$

Exercise Set 8.2

1. $200, 220, 240, 260, 280, 300$ **3.** $-7, -3, 1, 5, 9, 13$ **5.** $300, 210, 120, 30, -60, -150$ **7.** $\dfrac{5}{2}, 2, \dfrac{3}{2}, 1, \dfrac{1}{2}, 0$ **9.** $-9, -3, 3, 9, 15, 21$

11. $30, 20, 10, 0, -10, -20$ **13.** $1.6, 1.2, 0.8, 0.4, 0, -0.4$ **15.** 33 **17.** 252 **19.** 955 **21.** -142 **23.** $a_n = 4n - 3; a_{20} = 77$
25. $a_n = 11 - 4n; a_{20} = -69$ **27.** $a_n = 7 + 2n; a_{20} = 47$ **29.** $a_n = -16 - 4n; a_{20} = -96$ **31.** $a_n = 1 + 3n; a_{20} = 61$
33. $a_n = 40 - 10n; a_{20} = -160$ **35.** 1220 **37.** 4400 **39.** 5050 **41.** 3660 **43.** 396 **45.** $8 + 13 + 18 + \cdots + 88; 816$
47. $2 - 1 - 4 - \cdots - 85; -1245$ **49.** $4 + 8 + 12 + \cdots + 400; 20,200$ **51.** 7 **53.** 22 **55.** 847 **57.** $f(x) = -4x + 5$
59. $a_n = 3n - 2$ **61. a.** $a_n = 0.5n + 10.5$ **b.** 35.5% **63.** Company A will pay $\$1400$ more in year 10. **65. a.** $\$25,626$
b. $\$25,626$; It's the same. **69.** Company A: $\$307,000$; Company B: $\$324,000$; Company B pays the greater total amount. **71.** 2869 seats
79. makes sense **81.** makes sense **83.** 320 degree-days **85.** $-2; -2; -2; -2$; The ratio of a term to the term that directly precedes it is
always -2. **86.** $5; 5; 5; 5$; The ratio of a term to the term that directly precedes it is always 5. **87.** 8019

Section 8.3

Check Point Exercises

1. $12, 6, 3, \dfrac{3}{2}, \dfrac{3}{4}, \dfrac{3}{8}$ **2.** 3645 **3.** $a_n = 3(2)^{n-1}; a_8 = 384$ **4.** 9842 **5.** $19,680$ **6.** $\$2,371,746$ **7. a.** $\$333,946$ **b.** $\$291,946$
8. 9 **9.** 1 **10.** $\$4000$

Concept and Vocabulary Check

1. geometric; common ratio **2.** $a_1 r^{n-1}$; first term; common ratio **3.** $\dfrac{a_1(1 - r^n)}{1 - r}$; first term; common ratio **4.** annuity; $P; r; n$
5. infinite geometric series; $1; \dfrac{a_1}{1 - r}$; $|r| \ge 1$ **6.** $2; 4; 8; 16; 2$ **7.** arithmetic **8.** geometric **9.** geometric **10.** arithmetic

Exercise Set 8.3

1. $5, 15, 45, 135, 405$ **3.** $20, 10, 5, \dfrac{5}{2}, \dfrac{5}{4}$ **5.** $10, -40, 160, -640, 2560$ **7.** $-6, 30, -150, 750, -3750$ **9.** $a_8 = 768$ **11.** $a_{12} = -10,240$

13. $a_{40} \approx -0.000000002$ **15.** $a_8 = 0.1$ **17.** $a_n = 3(4)^{n-1}; a_7 = 12,288$ **19.** $a_n = 18\left(\dfrac{1}{3}\right)^{n-1}; a_7 = \dfrac{2}{81}$ **21.** $a_n = 1.5(-2)^{n-1}; a_7 = 96$

23. $a_n = 0.0004(-10)^{n-1}; a_7 = 400$ **25.** $531,440$ **27.** 2049 **29.** $\dfrac{16,383}{2}$ **31.** 9840 **33.** $10,230$ **35.** $\dfrac{63}{128}$ **37.** $\dfrac{3}{2}$ **39.** 4 **41.** $\dfrac{2}{3}$
43. $\dfrac{80}{13} \approx 6.15385$ **45.** $\dfrac{5}{9}$ **47.** $\dfrac{47}{99}$ **49.** $\dfrac{257}{999}$ **51.** arithmetic, $d = 1$ **53.** geometric, $r = 2$ **55.** neither **57.** 2435 **59.** 2280
61. -140 **63.** $a_2 = 12, a_3 = 18$ **65.** $\$16,384$ **67.** $\$3,795,957$ **69. a.** approximately 1.01 for each division **b.** $a_n = 33.87(1.01)^{n-1}$
c. approximately 41.33 million **71.** $\$32,767$ **73.** $\$793,583$ **75.** 130.26 in. **77. a.** $\$11,617$ **b.** $\$1617$ **79. a.** $\$87,052$ **b.** $\$63,052$

81. a. $\$693,031$ **b.** $\$293,031$ **83.** $\$30,000$ **85.** $\$9$ million **87.** $\dfrac{1}{3}$ **99.**

horizontal asymptote: $y = 3$; sum of series: 3

101. makes sense **103.** makes sense **105.** false **107.** false
109. Release 2000 flies each day. **112.** $6 = 6$ **113.** $15 = 15$
114. $\dfrac{(k + 1)(k + 2)(2k + 3)}{6}$

Mid-Chapter 8 Check Point

1. $1, -2, \dfrac{3}{2}, -\dfrac{2}{3}, \dfrac{5}{24}$ **2.** $5, 2, -1, -4, -7$ **3.** $5, -15, 45, -135, 405$ **4.** $3, 1, 3, 1, 3$ **5.** $a_n = 4n - 2; a_{20} = 78$ **6.** $a_n = 3(2)^{n-1}; a_{10} = 1536$

7. $a_n = -\dfrac{1}{2}n + 2; a_{30} = -13$ **8.** 5115 **9.** 2350 **10.** 6820 **11.** $-29,300$ **12.** 44 **13.** 3725 **14.** $\dfrac{1995}{64}$ **15.** $\dfrac{5}{7}$ **16.** $\dfrac{5}{11}$

17. Answers will vary. An example is $\displaystyle\sum_{i=1}^{18} \dfrac{i}{i + 2}$. **18.** 464 ft; 3600 ft **19.** $\$311,249$

Section 8.4

Check Point Exercises

1. a. $S_1: 2 = 1(1 + 1); S_k: 2 + 4 + 6 + \cdots + 2k = k(k + 1); S_{k+1}: 2 + 4 + 6 + \cdots + 2(k + 1) = (k + 1)(k + 2)$

b. $S_1: 1^3 = \dfrac{1^2(1 + 1)^2}{4}; S_k: 1^3 + 2^3 + 3^3 + \cdots + k^3 = \dfrac{k^2(k + 1)^2}{4}; S_{k+1}: 1^3 + 2^3 + 3^3 + \cdots + (k + 1)^3 = \dfrac{(k + 1)^2(k + 2)^2}{4}$

2. $S_1: 2 = 1(1 + 1); S_k: 2 + 4 + 6 + \cdots + 2k = k(k + 1); S_{k+1}: 2 + 4 + 6 + \cdots + 2k + 2(k + 1) = (k + 1)(k + 2); S_{k+1}$ can be obtained by adding $2k + 2$ to both sides of S_k.

3. $S_1: 1^3 = \dfrac{1^2(1+1)^2}{4}$; $S_k: 1^3 + 2^3 + 3^3 + \cdots + k^3 = \dfrac{k^2(k+1)^2}{4}$; $S_{k+1}: 1^3 + 2^3 + 3^3 + \cdots + k^3 + (k+1)^3 = \dfrac{(k+1)^2(k+2)^2}{4}$; S_{k+1} can be obtained by adding $k^3 + 3k^2 + 3k + 1$ to both sides of S_k.

4. $S_1: 2$ is a factor of $1^2 + 1$; $S_k: 2$ is a factor of $k^2 + k$; $S_{k+1}: 2$ is a factor of $(k+1)^2 + (k+1) = k^2 + 3k + 2$; S_{k+1} can be obtained from S_k by writing $k^2 + 3k + 2$ as $(k^2 + k) + 2(k+1)$.

Concept and Vocabulary Check

1. induction; 1; $k + 1$ **2.** $2 + 4 + 6 = 3(3+1)$; $2 + 4 + 6 + \cdots + 2(k+1) = (k+1)(k+2)$

3. $3 + 7 + 11 = 3(6+1)$; $3 + 7 + 11 + \cdots + [4(k+1) - 1] = (k+1)[2(k+1) + 1]$; $3 + 7 + 11 + \cdots + (4k+3) = (k+1)(2k+3)$

4. 4; 10; 18; $(k+1)^2 + 3(k+1)$; $k^2 + 5k + 4$ **5.** $(k+1)$

Exercise Set 8.4

1. $S_1: 1 = 1^2$; $S_2: 1 + 3 = 2^2$; $S_3: 1 + 3 + 5 = 3^2$ **3.** $S_1: 2$ is a factor of $1 - 1 = 0$; $S_2: 2$ is a factor of $2^2 - 2 = 2$; $S_3: 2$ is a factor of $3^2 - 3 = 6$.

5. $S_k: 4 + 8 + 12 + \cdots + 4k = 2k(k+1)$; $S_{k+1}: 4 + 8 + 12 + \cdots + (4k+4) = 2(k+1)(k+2)$

7. $S_k: 3 + 7 + 11 + \cdots + (4k-1) = k(2k+1)$; $S_{k+1}: 3 + 7 + 11 + \cdots + (4k+3) = (k+1)(2k+3)$

9. $S_k: 2$ is a factor of $k^2 - k + 2$; $S_{k+1}: 2$ is a factor of $k^2 + k + 2$.

11. $S_1: 4 = 2(1)(1+1)$; $S_k: 4 + 8 + 12 + \cdots + 4k = 2k(k+1)$; $S_{k+1}: 4 + 8 + 12 + \cdots + 4(k+1) = 2(k+1)(k+2)$; S_{k+1} can be obtained by adding $4k + 4$ to both sides of S_k.

13. $S_1: 1 = 1^2$; $S_k: 1 + 3 + 5 + \cdots + (2k-1) = k^2$; $S_{k+1}: 1 + 3 + 5 + \cdots + (2k+1) = (k+1)^2$; S_{k+1} can be obtained by adding $2k + 1$ to both sides of S_k.

15. $S_1: 3 = 1[2(1) + 1]$; $S_k: 3 + 7 + 11 + \cdots + (4k-1) = k(2k+1)$; $S_{k+1}: 3 + 7 + 11 + \cdots + (4k+3) = (k+1)(2k+3)$; S_{k+1} can be obtained by adding $4k + 3$ to both sides of S_k.

17. $S_1: 1 = 2^1 - 1$; $S_k: 1 + 2 + 2^2 + \cdots + 2^{k-1} = 2^k - 1$; $S_{k+1}: 1 + 2 + 2^2 + \cdots + 2^k = 2^{k+1} - 1$; S_{k+1} can be obtained by adding 2^k to both sides of S_k.

19. $S_1: 2 = 2^{1+1} - 2$; $S_k: 2 + 4 + 8 + \cdots + 2^k = 2^{k+1} - 2$; $S_{k+1}: 2 + 4 + 8 + \cdots + 2^{k+1} = 2^{k+2} - 2$; S_{k+1} can be obtained by adding 2^{k+1} to both sides of S_k.

21. $S_1: 1 \cdot 2 = \dfrac{1(1+1)(1+2)}{3}$; $S_k: 1 \cdot 2 + 2 \cdot 3 + 3 \cdot 4 + \cdots + k(k+1) = \dfrac{k(k+1)(k+2)}{3}$;

$S_{k+1}: 1 \cdot 2 + 2 \cdot 3 + 3 \cdot 4 + \cdots + (k+1)(k+2) = \dfrac{(k+1)(k+2)(k+3)}{3}$; S_{k+1} can be obtained by adding $(k+1)(k+2)$ to both sides of S_k.

23. $S_1: \dfrac{1}{1 \cdot 2} = \dfrac{1}{1+1}$; $S_k: \dfrac{1}{1 \cdot 2} + \dfrac{1}{2 \cdot 3} + \dfrac{1}{3 \cdot 4} + \cdots + \dfrac{1}{k(k+1)} = \dfrac{k}{k+1}$; $S_{k+1}: \dfrac{1}{1 \cdot 2} + \dfrac{1}{2 \cdot 3} + \dfrac{1}{3 \cdot 4} + \cdots + \dfrac{1}{(k+1)(k+2)} = \dfrac{k+1}{k+2}$; S_{k+1} can be

obtained by adding $\dfrac{1}{(k+1)(k+2)}$ to both sides of S_k.

25. $S_1: 2$ is a factor of 0; $S_k: 2$ is a factor of $k^2 - k$; $S_{k+1}: 2$ is a factor of $k^2 + k$; S_{k+1} can be obtained from S_k by rewriting $k^2 + k$ as $(k^2 - k) + 2k$.

27. $S_1: 6$ is a factor of 6; $S_k: 6$ is a factor of $k(k+1)(k+2)$; $S_{k+1}: 6$ is a factor of $(k+1)(k+2)(k+3)$; S_{k+1} can be obtained from S_k by rewriting $(k+1)(k+2)(k+3)$ as $k(k+1)(k+2) + 3(k+1)(k+2)$ and noting that either $k+1$ or $k+2$ is even, so 6 is a factor of $3(k+1)(k+2)$.

29. $S_1: 5 \cdot 6^1 = 6(6^1 - 1)$; $S_k: \displaystyle\sum_{i=1}^{k} 5 \cdot 6^i = 6(6^k - 1)$; $S_{k+1}: \displaystyle\sum_{i=1}^{k+1} 5 \cdot 6^i = 6(6^{k+1} - 1)$; S_{k+1} can be obtained by adding $5 \cdot 6^{k+1}$ to both sides of S_k.

31. $S_1: 1 + 2 > 1$; $S_k: k + 2 > k$; $S_{k+1}: k + 3 > k + 1$; S_{k+1} can be obtained by adding 1 to both sides of S_k.

33. $S_1: (ab)^1 = a^1 b^1$; $S_k: (ab)^k = a^k b^k$; $S_{k+1}: (ab)^{k+1} = a^{k+1} b^{k+1}$; S_{k+1} can be obtained by multiplying both sides of S_k by (ab).

37. does not make sense **39.** does not make sense

41. $S_3: 3^2 > 2(3) + 1$; $S_k: k^2 > 2k + 1$ for $k \geq 3$; $S_{k+1}: (k+1)^2 > 2(k+1) + 1$ or $k^2 + 2k + 1 > 2k + 3$; S_{k+1} can be obtained from S_k by noting that S_{k+1} is the same as $k^2 > 2$ which is true for $k \geq 3$.

43. $S_1: \dfrac{1}{4}$; $S_2: \dfrac{1}{3}$; $S_3: \dfrac{3}{8}$; $S_4: \dfrac{2}{5}$; $S_5: \dfrac{5}{12}$; $S_n: \dfrac{n}{2n+2}$; Use S_k to obtain the conjectured formula.

46. The exponents begin with the exponent on $a + b$ and decrease by 1 in each successive term. **47.** The exponents begin with 0, increase by 1 in each successive term, and end with the exponent on $a + b$. **48.** The sum of the exponents is the exponent on $a + b$.

Section 8.5

Check Point Exercises

1. a. 20 **b.** 1 **c.** 28 **d.** 1 **2.** $x^4 + 4x^3 + 6x^2 + 4x + 1$ **3.** $x^5 - 10x^4 y + 40x^3 y^2 - 80x^2 y^3 + 80xy^4 - 32y^5$ **4.** $4032x^5 y^4$

Concept and Vocabulary Check

1. binomial **2.** $\dfrac{8!}{2!6!}$ **3.** $\dfrac{n!}{r!(n-r)!}$ **4.** $\dbinom{5}{0}x^5 + \dbinom{5}{1}x^4 \cdot 2 + \dbinom{5}{2}x^3 \cdot 2^2 + \dbinom{5}{3}x^2 \cdot 2^3 + \dbinom{5}{4}x \cdot 2^4 + \dbinom{5}{5}2^5$

5. $\dbinom{n}{0}a^n + \dbinom{n}{1}a^{n-1}b + \dbinom{n}{2}a^{n-2}b^2 + \dbinom{n}{3}a^{n-3}b^3 + \cdots + \dbinom{n}{n}b^n$; n **6.** Binomial **7.** $a^{n-r}b^r$

Exercise Set 8.5

1. 56 **3.** 12 **5.** 1 **7.** 4950 **9.** $x^3 + 6x^2 + 12x + 8$ **11.** $27x^3 + 27x^2 y + 9xy^2 + y^3$ **13.** $125x^3 - 75x^2 + 15x - 1$

15. $16x^4 + 32x^3 + 24x^2 + 8x + 1$ **17.** $x^8 + 8x^6 y + 24x^4 y^2 + 32x^2 y^3 + 16y^4$ **19.** $y^4 - 12y^3 + 54y^2 - 108y + 81$

21. $16x^{12} - 32x^9 + 24x^6 - 8x^3 + 1$ **23.** $c^5 + 10c^4 + 40c^3 + 80c^2 + 80c + 32$ **25.** $x^5 - 5x^4 + 10x^3 - 10x^2 + 5x - 1$
27. $243x^5 - 405x^4y + 270x^3y^2 - 90x^2y^3 + 15xy^4 - y^5$ **29.** $64a^6 + 192a^5b + 240a^4b^2 + 160a^3b^3 + 60a^2b^4 + 12ab^5 + b^6$
31. $x^8 + 16x^7 + 112x^6 + \cdots$ **33.** $x^{10} - 20x^9y + 180x^8y^2 - \cdots$ **35.** $x^{32} + 16x^{30} + 120x^{28} + \cdots$ **37.** $y^{60} - 20y^{57} + 190y^{54} - \cdots$
39. $240x^4y^2$ **41.** $126x^5$ **43.** $56x^6y^{15}$ **45.** $-\dfrac{21}{2}x^6$ **47.** $319{,}770x^{16}y^{14}$ **49.** $x^{12} + 4x^7 + 6x^2 + \dfrac{4}{x^3} + \dfrac{1}{x^8}$ **51.** $x - 3x^{1/3} + \dfrac{3}{x^{1/3}} - \dfrac{1}{x}$
53. $4x^3 + 6x^2h + 4xh^2 + h^3$ **55.** 252 **57.** 0.1138 **69.** **71.** $f_1(x) = x^4 - 8x^3 + 24x^2 - 32x + 16$
73. makes sense **75.** does not make sense **77.** false
79. false **81.** $x^6 + 3x^5 + 6x^4 + 7x^3 + 6x^2 + 3x + 1$

$f_2, f_3, f_4,$ and f_5 are approaching $f_1 = f_6.$

83. $\dbinom{n}{r} = \dfrac{n!}{r!(n-r)!} = \dfrac{n!}{(n-r)!r!} = \dfrac{n!}{(n-r)![n-(n-r)]!} = \dbinom{n}{n-r}$

85. a. $(a+b)^1 = a + b = \dbinom{1}{0}a^1 + \dbinom{1}{1}b^1$ **b.** Assume: $(a+b)^k = \dbinom{k}{0}a^k + \dbinom{k}{1}a^{k-1}b + \dbinom{k}{2}a^{k-2}b^2 + \cdots + \dbinom{k}{k-1}ab^{k-1}$

$+ \dbinom{k}{k}b^k;$ Prove: $(a+b)^{k+1} = \dbinom{k+1}{0}a^{k+1} + \dbinom{k+1}{1}a^{k+1-1}b + \dbinom{k+1}{2}a^{k+1-2}b^2 + \cdots + \dbinom{k+1}{k+1-1}ab^{k+1-1} + \dbinom{k+1}{k+1}b^{k+1}$

c. $(a+b)(a+b)^k = (a+b)\left[\dbinom{k}{0}a^k + \dbinom{k}{1}a^{k-1}b + \dbinom{k}{2}a^{k-2}b^2 + \cdots + \dbinom{k}{k-1}ab^{k-1} + \dbinom{k}{k}b^k\right]$ or $(a+b)^{k+1} = \dbinom{k}{0}a^{k+1}$

$+ \dbinom{k}{0}a^kb + \dbinom{k}{1}a^kb + \dbinom{k}{1}a^{k-1}b^2 + \dbinom{k}{2}a^{k-1}b^2 + \dbinom{k}{2}a^{k-2}b^3 + \cdots + \dbinom{k}{k-1}a^2b^{k-1} + \dbinom{k}{k-1}ab^k + \dbinom{k}{k}ab^k + \dbinom{k}{k}b^{k+1}$

d. $(a+b)^{k+1} = \dbinom{k}{0}a^{k+1} + \left[\dbinom{k}{0} + \dbinom{k}{1}\right]a^kb + \left[\dbinom{k}{1} + \dbinom{k}{2}\right]a^{k-1}b^2 + \left[\dbinom{k}{2} + \dbinom{k}{3}\right]a^{k-2}b^3 + \cdots + \left[\dbinom{k}{k-1} + \dbinom{k}{k}\right]ab^k + \dbinom{k}{k}b^{k+1}$

e. $(a+b)^{k+1} = \dbinom{k}{0}a^{k+1} + \dbinom{k+1}{1}a^kb + \dbinom{k+1}{2}a^{k-1}b^2 + \dbinom{k+1}{3}a^{k-2}b^3 + \cdots + \dbinom{k+1}{k}ab^k + \dbinom{k}{k}b^{k+1}$

f. $(a+b)^{k+1} = \dbinom{k+1}{0}a^{k+1} + \dbinom{k+1}{1}a^kb + \dbinom{k+1}{2}a^{k-1}b^2 + \dbinom{k+1}{3}a^{k-2}b^3 + \cdots + \dbinom{k+1}{k}ab^k + \dbinom{k+1}{k+1}b^{k+1}$

86. 6840 **87.** 56 **88.** true

Section 8.6

Check Point Exercises

1. 72 **2.** 729 **3.** 676,000 **4.** 840 **5.** 720 **6. a.** combinations **b.** permutations **7.** 210 **8.** 1820

Concept and Vocabulary Check

1. $M \cdot N$ **2.** multiplying; Fundamental Counting **3.** $\dfrac{n!}{(n-r)!}$ **4.** $\dfrac{n!}{(n-r)!r!}$ **5.** $r!$

Exercise Set 8.6

1. 3024 **3.** 6720 **5.** 720 **7.** 1 **9.** 126 **11.** 330 **13.** 1 **15.** 1 **17.** combinations **19.** permutations **21.** 0 **23.** $\dfrac{3}{4}$
25. -9499 **27.** $\dfrac{3}{68}$ **29.** 27 ways **31.** 40 ways **33.** 243 ways **35.** 144 area codes **37.** 120 ways **39.** 6 paragraphs **41.** 720 ways
43. 8,648,640 ways **45.** 120 ways **47.** 15,120 lineups **49.** 20 ways **51.** 495 collections **53.** 24,310 groups **55.** 22,957,480 selections
57. 360 ways **59.** 1716 ways **61.** 1140 ways **63.** 840 passwords **65.** 2730 cones **67.** 720 **69.** 20 **71.** 24 **83.** makes sense
85. does not make sense **87.** false **89.** true **91.** 14,400 ways **93.** 450 ways **95.** $\dfrac{2}{3}$ **96.** $\dfrac{1}{3}$ **97.** $\dfrac{2}{3}$

Section 8.7

Check Point Exercises

1. a. $\dfrac{7664}{100{,}000}$ or $\dfrac{479}{6250} \approx 0.077$ **b.** $\dfrac{720}{800}$ or $\dfrac{9}{10} = 0.9$ **c.** $\dfrac{720}{7664}$ or $\dfrac{45}{479} \approx 0.094$ **2.** $\dfrac{1}{3}$ **3.** $\dfrac{1}{9}$ **4.** $\dfrac{1}{13}$ **5.** $\dfrac{1}{13{,}983{,}816} \approx 0.0000000715$
6. $\dfrac{129}{140} \approx 0.921$ **7.** $\dfrac{1}{3}$ **8.** $\dfrac{3}{4}$ **9. a.** $\dfrac{189}{242} \approx 0.78$ **b.** $\dfrac{19}{121} \approx 0.16$ **10.** $\dfrac{1}{361} \approx 0.00277$ **11.** $\dfrac{1}{16}$

Concept and Vocabulary Check

1. empirical **2.** sample space **3.** $P(E)$; number of outcomes in E; total number of possible outcomes **4.** 52; hearts; diamonds; clubs; spades
5. 1; combinations **6.** $1 - P(E); 1 - P(\text{not } E)$ **7.** mutually exclusive; $P(A) + P(B)$ **8.** $P(A) + P(B) - P(A \text{ and } B)$
9. independent; $P(A) \cdot P(B)$

Exercise Set 8.7

1. $\frac{12}{121} \approx 0.10$ **3.** $\frac{62}{121} \approx 0.51$ **5.** $\frac{3}{242} \approx 0.01$ **7.** $\frac{7}{12} \approx 0.58$ **9.** $\frac{65}{118} \approx 0.55$ **11.** $\frac{1}{6}$ **13.** $\frac{1}{2}$ **15.** $\frac{1}{3}$ **17.** $\frac{1}{13}$ **19.** $\frac{3}{13}$

21. $\frac{1}{4}$ **23.** $\frac{7}{8}$ **25.** $\frac{1}{12}$ **27.** $\frac{1}{18,009,460}; \frac{5}{900,473}$ **29. a.** 2,598,960 **b.** 1287 **c.** $\frac{1287}{2,598,960} \approx 0.0005$ **31.** $\frac{43}{58}$ **33.** $\frac{50}{87}$ **35.** $\frac{113}{174}$

37. $\frac{12}{13}$ **39.** $\frac{2}{13}$ **41.** $\frac{7}{13}$ **43.** $\frac{3}{4}$ **45.** $\frac{33}{40}$ **47.** $\frac{1}{36}$ **49.** $\frac{1}{3}$ **51.** $\frac{1}{64}$ **53. a.** $\frac{1}{256}$ **b.** $\frac{1}{4096}$ **c.** $\left(\frac{15}{16}\right)^{10} \approx 0.524$

d. $1 - \left(\frac{15}{16}\right)^{10} \approx 0.476$ **65.** does not make sense **67.** makes sense **71. a.** $\frac{12}{25}$ **b.** $\frac{3}{10}$ **73. a.** The first person can have any birthday in

the year. The second person can have all but one birthday. **b.** $\frac{365}{365} \cdot \frac{364}{365} \cdot \frac{363}{365} \approx 0.99$ **c.** ≈ 0.01 **d.** ≈ 0.41 **e.** 23 people

Chapter 8 Review Exercises

1. $a_1 = 3; a_2 = 10; a_3 = 17; a_4 = 24$ **2.** $a_1 = -\frac{3}{2}; a_2 = \frac{4}{3}; a_3 = -\frac{5}{4}; a_4 = \frac{6}{5}$ **3.** $a_1 = 1; a_2 = 1; a_3 = \frac{1}{2}; a_4: \frac{1}{6}$

4. $a_1 = \frac{1}{2}; a_2 = -\frac{1}{4}; a_3 = \frac{1}{8}; a_4 = -\frac{1}{16}$ **5.** $a_1 = 9; a_2 = \frac{2}{27}; a_3 = 9; a_4 = \frac{2}{27}$ **6.** $a_1 = 4; a_2 = 11; a_3 = 25; a_4 = 53$ **7.** 65 **8.** 95

9. -20 **10.** $\sum_{i=1}^{15} \frac{i}{i+2}$ **11.** $\sum_{i=4}^{13} i^3$ or $\sum_{i=1}^{10}(i+3)^3$ **12.** 7, 11, 15, 19, 23, 27 **13.** $-4, -9, -14, -19, -24, -29$ **14.** $\frac{3}{2}, 1, \frac{1}{2}, 0, -\frac{1}{2}, -1$

15. $-2, 3, 8, 13, 18, 23$ **16.** $a_6 = 20$ **17.** $a_{12} = -30$ **18.** $a_{14} = -38$ **19.** $a_n = 4n - 11; a_{20} = 69$ **20.** $a_n = 220 - 20n; a_{20} = -180$
21. $a_n = 8 - 5n; a_{20} = -92$ **22.** 1727 **23.** 225 **24.** 15,150 **25.** 440 **26.** -500 **27.** -2325 **28. a.** $a_n = 34.8 - 0.3n$

b. 18 hours per week **29.** \$418,500 **30.** 1470 seats **31.** 3, 6, 12, 24, 48 **32.** $\frac{1}{2}, \frac{1}{4}, \frac{1}{8}, \frac{1}{16}, \frac{1}{32}$ **33.** $16, -8, 4, -2, 1$ **34.** $-1, 5, -25, 125, -625$

35. $a_7 = 1458$ **36.** $a_6 = \frac{1}{2}$ **37.** $a_5 = -48$ **38.** $a_n = 2^{n-1}; a_8 = 128$ **39.** $a_n = 100\left(\frac{1}{10}\right)^{n-1}; a_8 = \frac{1}{100,000}$

40. $a_n = 12\left(-\frac{1}{3}\right)^{n-1}; a_8 = -\frac{4}{729}$ **41.** 17,936,135 **42.** $\frac{127}{8}$ **43.** 19,530 **44.** -258 **45.** $\frac{341}{128}$ **46.** $\frac{27}{2}$ **47.** $\frac{4}{3}$ **48.** $-\frac{18}{5}$ **49.** 20

50. $\frac{2}{3}$ **51.** $\frac{47}{99}$ **52. a.** approximately 1.02 for each division **b.** $a_n = 15.98(1.02)^{n-1}$ **c.** 28.95 million **53.** \$42,823; \$223,210

54. a. \$19,129 **b.** \$8729 **55. a.** \$91,361 **b.** \$55,361 **56.** $9\frac{1}{3}$ million
57. $S_1: 5 = \frac{5(1)(1+1)}{2}; S_k: 5 + 10 + 15 + \cdots + 5k = \frac{5k(k+1)}{2}; S_{k+1}: 5 + 10 + 15 + \cdots + 5(k+1) = \frac{5(k+1)(k+2)}{2}; S_{k+1}$
can be obtained by adding $5(k+1)$ to both sides of S_k.
58. $S_1: 1 = \frac{4^1 - 1}{3}; S_k: 1 + 4 + 4^2 + \cdots + 4^{k-1} = \frac{4^k - 1}{3}; S_{k+1}: 1 + 4 + 4^2 + \cdots + 4^k = \frac{4^{k+1} - 1}{3}; S_{k+1}$ can be obtained by adding 4^k
to both sides of S_k.
59. $S_1: 2 = 2(1)^2; S_k: 2 + 6 + 10 + \cdots + (4k - 2) = 2k^2; S_{k+1}: 2 + 6 + 10 + \cdots + (4k + 2) = 2k^2 + 4k + 2; S_{k+1}$ can be obtained
by adding $4k + 2$ to both sides of S_k.
60. $S_1: 1 \cdot 3 = \frac{1(1+1)[2(1)+7]}{6}; S_k: 1 \cdot 3 + 2 \cdot 4 + 3 \cdot 5 + \cdots + k(k+2) = \frac{k(k+1)(2k+7)}{6};$
$S_{k+1}: 1 \cdot 3 + 2 \cdot 4 + 3 \cdot 5 + \cdots + (k+1)(k+3) = \frac{(k+1)(k+2)(2k+9)}{6}; S_{k+1}$ can be obtained by adding $(k+1)(k+3)$ to both sides of S_k.
61. $S_1: 2$ is a factor of 6; $S_k: 2$ is a factor of $k^2 + 5k; S_{k+1}: 2$ is a factor of $k^2 + 7k + 6; S_{k+1}$ can be obtained from S_k by rewriting $k^2 + 7k + 6$
as $(k^2 + 5k) + 2(k + 3)$.
62. 165 **63.** 4005 **64.** $8x^3 + 12x^2 + 6x + 1$ **65.** $x^8 - 4x^6 + 6x^4 - 4x^2 + 1$ **66.** $x^5 + 10x^4y + 40x^3y^2 + 80x^2y^3 + 80xy^4 + 32y^5$
67. $x^6 - 12x^5 + 60x^4 - 160x^3 + 240x^2 - 192x + 64$ **68.** $x^{16} + 24x^{14} + 252x^{12} + \cdots$ **69.** $x^9 - 27x^8 + 324x^7 - \cdots$ **70.** $80x^2$ **71.** $4860x^2$
72. 336 **73.** 15,120 **74.** 56 **75.** 78 **76.** 20 choices **77.** 243 possibilities **78.** 32,760 ways **79.** 4845 ways **80.** 1140 sets

81. 116,280 ways **82.** 120 ways **83.** $\frac{18}{25}$ **84.** $\frac{6}{7}$ **85.** $\frac{3}{5}$ **86.** $\frac{12}{35}$ **87.** $\frac{10}{21}$ **88.** $\frac{5}{16}$ **89.** $\frac{2}{3}$ **90.** $\frac{2}{3}$ **91.** $\frac{2}{13}$ **92.** $\frac{7}{13}$ **93.** $\frac{5}{6}$

94. $\frac{5}{6}$ **95.** $\frac{1}{6}$ **96. a.** $\frac{1}{15,504}$ **b.** $\frac{1}{3876}$ **97.** $\frac{25}{32}$ **98. a.** 0.04 **b.** 0.008 **c.** 0.4096

Chapter 8 Test

1. $a_1 = 1; a_2 = -\frac{1}{4}; a_3 = \frac{1}{9}; a_4 = -\frac{1}{16}; a_5 = \frac{1}{25}$ **2.** 105 **3.** 550 **4.** $-21,846$ **5.** 36 **6.** 720 **7.** 120 **8.** $\sum_{i=1}^{20} \frac{i+1}{i+2}$

9. $a_n = 5n - 1; a_{12} = 59$ **10.** $a_n = 16\left(\frac{1}{4}\right)^{n-1}; a_{12} = \frac{1}{262,144}$ **11.** -2387 **12.** -385 **13.** 8 **14.** $\frac{73}{99}$ **15.** \$276,427

16. $S_1: 1 = \frac{1[3(1) - 1]}{2}; S_k: 1 + 4 + 7 + \cdots + (3k - 2) = \frac{k(3k - 1)}{2}; S_{k+1}: 1 + 4 + 7 + \cdots + (3k + 1) = \frac{(k+1)(3k+2)}{2}; S_{k+1}$
can be obtained by adding $3k + 1$ to both sides of S_k.

17. $x^{10} - 5x^8 + 10x^6 - 10x^4 + 5x^2 - 1$ **18.** $x^8 + 8x^7y^2 + 28x^6y^4$ **19.** 990 ways **20.** 210 sets **21.** $10^4 = 10,000$ **22.** $\frac{3}{5}$ **23.** $\frac{39}{50}$
24. $\frac{3}{5}$ **25.** $\frac{9}{19}$ **26.** $\frac{10}{1001}$ **27.** $\frac{8}{13}$ **28.** $\frac{3}{5}$ **29.** $\frac{1}{256}$ **30.** $\frac{1}{16}$

Cumulative Review Exercises (Chapters 1–8)

1. domain: $[-4, 1)$; range: $(-\infty, 2]$ **2.** maximum of 2 at $x = -2$ **3.** $(-2, 1)$ **4.** neither **5.** -3 and -1 **6.** 0 **7.** $x \to 1^-$

8.

$g(x) = f(x - 2) + 1$

9.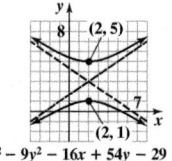

$h(x) = -f(2x)$

10. $\left\{ \dfrac{14}{5} \right\}$ **11.** $\left\{ \dfrac{3 + \sqrt{3}}{3}, \dfrac{3 - \sqrt{3}}{3} \right\}$ **12.** $\{2\}$

13. $\{16, 256\}$ **14.** $\{6\}$ **15.** $-1 \le x \le 0$ or $[-1, 0]$

16. $\left(-\dfrac{2}{3}, \dfrac{3}{2} \right)$ **17.** $(-3, 1]$ **18.** $\{2.9706\}$ **19.** $\left\{ 1, \dfrac{1}{2}, -3 \right\}$

20. $\{(3, 2), (3, -2), (-3, 2), (-3, -2)\}$ **21.** $\{(6, -4, 2)\}$ **22.** $\{(0, -1), (2, 1)\}$

23.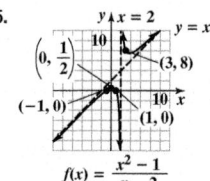

$100x^2 + y^2 = 25$

24.

$4x^2 - 9y^2 - 16x + 54y - 29 = 0$

25.

$f(x) = \dfrac{x^2 - 1}{x - 2}$

26.

$2x - y \ge 4$
$x \le 2$

27.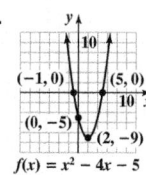

$f(x) = x^2 - 4x - 5$

28.

$f(x) = \sqrt[3]{x + 4}$
$f^{-1}(x) = x^3 - 4$

29.

$f(x) = \log_2 x$
$g(x) = -\log_2 (x + 1)$

30. $(f \circ g)(x) = -x^2 + 2$; $(g \circ f)(x) = -x^2 - 2x$ **31.** $-2x - h - 2$

32. $\begin{bmatrix} -2 & 10 \\ -5 & 7 \\ 15 & -15 \end{bmatrix}$ **33.** $\dfrac{-1}{x - 2} + \dfrac{3x - 2}{x^2 + 2x + 2}$

34. $x^{15} + 10x^{12}y + 40x^9y^2 + 80x^6y^3 + 80x^3y^4 + 32y^5$ **35.** 3850

36. $f(x) = \dfrac{1}{4}x + \dfrac{3}{2}$ **37.** $g(x) = -5x - 2$ **38.** \$2000 **39.** length: 100 yd; width: 50 yd **40.** pen: \$1.80; pad: \$2 **41. a.** 6 sec

b. 2.5 sec; 196 ft **42.** 11 amps **43.** Answers will vary. An example is: $f(x) = 11.8x + 68.4$ where x is the number of years after 2010.

SUBJECT INDEX

PHOTO CREDITS

FM COVER Mustache: Pelonmaker/Shutterstock p. iii (left) Courtesy of author p. iii (right) Courtesy of author p. iv (baby) Jose Luis Pelaez Inc/Blend Images/Alamy p. iv (cap) Discpicture/Shutterstock p. ix Zoonar/E Lam/AGE Fotostock p. v (bottom) SuperStock p. v (top) K.Magnusson/Alamy p. vi Big Cheese Photo LLC/Alamy p. vii Charles Tsevis p. vii (bottom) Irochka/Dreamstime p. x Alamy p. xii Pearson Education, Inc. p. xiii Pearson Education, Inc. p. xv Courtesy of author

CHAPTER P p. 1 (baby) Jose Luis Pelaez Inc/Blend Images/Alamy p. 1 (cap) Discpicture/Shutterstock p. 2 William Scott/Alamy p. 3 (left) Image Source/AGE Fotostock p. 3 (right) Stockbroker/SuperStock p. 4 (cap) Stephen Coburn/Shutterstock p. 4 (child) Iofoto/Shutterstock p. 8 Martin Allinger/Shutterstock p. 15 Ikon Images/SuperStock p. 19 *Foxtrot* copyright © 2003, 2009 by Bill Amend/Distributed by Universal Uclick p. 20 Photodisc/Getty Images p. 42 Mike Kiev/Fotolia p. 45 Daily Mirror/Mirrorpix/Newscom p. 58 *The Persistence of Memory* (1931) Salvador Dali. Oil on canvas, 9 1/2 × 13 in. (24.1 × 33 cm). (162.1934). Given anonoymously to the Museum of Modern Art/Digital image copyright © the Museum of Modern Art/Licensed by Scala/Art Resource, New York/Artists Rights Society (ARS), New York p. 60 *Foxtrot* © 2003, 2009 by Bill Amend/Used by permission of Universal Uclick. All rights reserved. p. 61 (left) John G. Ross/Photo Researchers, Inc. p. 61 (right) *Peanuts* © 1978 Peanuts Worldwide LLC. Used by permission of Universal Uclick. All rights reserved. p. 62 (left) Jan Tyler/iStockphoto p. 62 (right) Courtesy of Robert F. Blitzer p. 71 Fancy Collection/SuperStock p. 75 Steve Shot/Dorling Kindersley, Ltd. p. 88 Comstock/Thinkstock

CHAPTER 1 p. 105 Michael Nolan/Specialist Stock/AGE Fotostock p. 106 The Everett Collection p. 112 Mike Kemp/RubberBall/AGE Fotostock p. 118 (HIV virus) NIBSC/Photo Researchers, Inc. p. 118 (Cowell) Joy Scheller/iPhoto/Newscom p. 118 (DeGene res) Nancy Kaszerman/ZUMApress/Newscom p. 118 (McGraw) K03/ZUMA Press/Newscom p. 118 (Seacrest) Debbie VanStory/RockinExposures/Newscom p. 118 (Winfrey) Dan Herrick-KPA/ZUMA KPA/Newscom p. 137 (top) Matthew Ward/Dorling Kindersley, Ltd. p. 151 Copyright © 2011 Scott Kim, scottkim.com p. 152 Image Source/AGE Fotostock p. 167 (left) SuperStock p. 167 (right) Nomad Soul/Shutterstock p. 176 John Serafin/French Government Tourist Office p. 178 Everett Collection p. 183 Gene Ahrens/SuperStock p. 194 Günay Mutlu/iStockphoto p. 208 (left) Archives du 7eme Art/Photos 12/Alamy p. 208 (right) The Everett Collection p. 219 By permission of John L. Hart FLP and Creators Syndicate, Inc. p. 220 Millard H. Sharp/Photo Researchers, Inc.

CHAPTER 2 p. 237 20th Century Fox TV/Album/Newscom p. 238 Zuma Press/Newscom p. 253 (Bryant) Icon Sports Media/Newscom p. 253 (CEO) ER Productions/Corbis Premium RF/Alamy p. 253 (doctor) Michael N. Paras/AGE Fotostock p. 253 (janitor) Kurhan/Shutterstock p. 253 (McGraw) S. Bukley/Newscom p. 253 (Pitt) WENN/Newscom p. 253 (Stern) Stuart Ramson/AP Images p. 253 (teacher) Image Source/AGE Fotostock p. 255 Courtesy of Robert F. Blitzer p. 265 *Peanuts* © 1978 Peanuts Worldwide LLC. Used by permission of Universal Uclick. All rights reserved. p. 266 Copyright © 2011 by Roz Chast/The New Yorker Collection/The Cartoon Bank p. 272 Gordon Caulkins p. 273 Brian Goodman/Shutterstock p. 287 *Bathers at Asnières* (1884), Georges Seurat. Oil on canvas, 201 cm × 300 cm (79 in × 118 in). National Gallery, London/Art Resource, New York p. 295 Ouzounova/Splash News/Newscom p. 308 Steven Chen/Surf/Corbis p. 310 (left) Craft Alan King/Alamy p. 310 (right) Hire Image Picture Library/Alamy

p. 333 (left) Paramount Pictures/Everett Collection p. 333 (right) Archives du 7e Art/Photos 12/Alamy p. 336 Alamy

CHAPTER 3 p. 339 K.Magnusson/Alamy p. 340 (top) Marcin Janiec/Fotolia p. 340 (bottom left) Pascal Rondeau/Getty Images p. 340 (bottom right) Copyright © 2011 Scott Kim, scottkim.com p. 356 (Allen) James Diddick/Globe Photos/ZUMA Press/Newscom p. 356 (Sparks) Pictorial Press Ltd/Alamy p. 356 (Barrino) Amanda M. Parks/iPhoto Inc/Newscom p. 356 (Clarkson) Dee Cercone/Everett Collection/Alamy p. 356 (Cook) Roth Stock/Everett Collection/Alamy p. 356 (DeWyze) Jen Lowery/Splash News/Newscom p. 356 (Hicks) Allstar Picture Library/Alamy p. 356 (Studdard) Allstar Picture Library/Alamy p. 356 (Underwood) Debbie VanStory/iPhoto Inc/Newscom p. 357 Andrew D. Bernstein/Getty Images p. 374 Dorling Kindersley, Ltd. p. 386 Photo Researchers, Inc. p. 394 Library of Congress Prints and Photographs Division [LC-USZ62-61365] p. 401 Kyodo News/Newscom p. 403 Courtesy of KnuckleTattoos.com p. 414 Kyodo News/Newscom p. 420 Warren Miller/Cartoonbank p. 433 (left) Alamy p. 433 (right) Alamy p. 434 NBC/Getty Images p. 435 Exactostock/SuperStock p. 439 (top) Ty Allison/Getty Images p. 439 (bottom) David Madison/Photographer's Choice/Getty Images p. 441 Ullstein Bild/The Image Works

CHAPTER 4 p. 451 SuperStock p. 452 SuperStock p. 456 (left) Seth Wenig/AP Images p. 456 (right) Rolffimages/Dreamstime p. 459 Tim Davis/Corbis p. 465 John Trotter/Sacramento Bee/ZUMA Press/Newscom p. 479 Colin Anderson/Alamy p. 489 Mitchel Gray/SuperStock p. 503 Jamaway/Alamy p. 504 Chlorophylle/Fotolia p. 507 Courtesy of the French Ministry of Culture and Communication p. 510 Science Photo Library/SuperStock

CHAPTER 5 p. 525 Big Cheese Photo LLC/Alamy p. 526 Vienetta/Shutterstock p. 534 SUN/Newscom p. 542 Alamy p. 550 Alamy p. 561 Dave King/Graham High/Centaur Studios/Dorling Kindersley, Ltd. p. 572 Terry J. Alcorn/iStockphoto p. 585 AP Images p. 586 David Crawford/US Navy/AP Images p. 594 (left) *Adele Bloch-Bauer I* (1907), Gustav Klimt/Neue Galerie/Art Resource, New York p. 594 (right) *Boy with a Pipe (Garçon à la Pipe)* (1905), Pablo Picasso. Oil on canvas, 100 cm × 81.3 cm (39.4 in × 32.0 in). Private collection. © 2011 Picasso Estate/ARS/Photo by BridgemanArt Library/Superstock

CHAPTER 6 p. 599 Charles Tsevis p. 600 Zoonar/E Lam/AGE Fotostock p. 612 Alamy p. 622 Ian Shaw/Alamy p. 630 Photo Researchers, Inc. p. 632 Tim Furniss/Genesis Space Photo Library/NASA p. 638 Popperfoto/Getty Images p. 652 SSPL/The Image Works

CHAPTER 7 p. 669 SuperStock p. 670 Kevin Fleming/Corbis p. 679 NASA p. 683 Gunter Marx/Alamy p. 699 NASA/Jeff Hester and Paul Scowen Arizona State University p. 707 NASA Johnson Space Center Collection

CHAPTER 8 p. 715 Louis Lopez/Cal Sport Media/Newscom p. 716 Irochka/Dreamstime p. 726 © 2004 Harry Blair. All rights reserved. p. 727 SuperStock p. 737 SuperStock p. 739 Shutterstock p. 743 U.S. Bureau of Engraving and Printing p. 754 Charles Rex Arbogast/AP Images p. 762 Dr. Rudolph Schild/Photo Researchers, Inc. p. 767 Reprinted with the permission of Cambridge University Press p. 770 SuperStock p. 771 Halfdark/Alamy p. 772 Oleksiy Maksymenko/Alamy p. 775 (Usher) Alamy p. 775 (Beyonce) Sven Hoogerhuis/Alamy p. 775 (Lady GaGa) JM5 WENN Photos/Newscom p. 775 (Shakira) Edd Westmacott/Alamy p. 775 (Timberlake) Sven Hoogerhuis/Alamy p. 781 Thinkstock p. 792 Corbis

Definitions, Rules, and Formulas

THE REAL NUMBERS

Natural Numbers: $\{1, 2, 3, \ldots\}$

Whole Numbers: $\{0, 1, 2, 3, \ldots\}$

Integers: $\{\ldots, -3, -2, -1, 0, 1, 2, 3, \ldots\}$

Rational Numbers: $\left\{\frac{a}{b} \mid a \text{ and } b \text{ are integers}, b \neq 0\right\}$

Irrational Numbers: $\{x \mid x \text{ is real and not rational}\}$

PROPERTIES OF ADDITION AND MULTIPLICATION

Commutative: $a + b = b + a$; $ab = ba$

Associative: $(a + b) + c = a + (b + c)$;
$$(ab)c = a(bc)$$

Distributive: $a(b + c) = ab + ac$; $a(b - c) = ab - ac$

Identity: $a + 0 = a$; $a \cdot 1 = a$

Inverse: $a + (-a) = 0$; $a \cdot \frac{1}{a} = 1 \ (a \neq 0)$

Multiplication Properties: $(-1)a = -a$;
$(-1)(-a) = a$; $a \cdot 0 = 0$; $(-a)(b) = (a)(-b) = -ab$;
$(-a)(-b) = ab$

EXPONENTS

Definitions of Rational Exponents

1. $a^{\frac{1}{n}} = \sqrt[n]{a}$ **2.** $a^{\frac{m}{n}} = \left(\sqrt[n]{a}\right)^m \text{ or } \sqrt[n]{a^m}$

3. $a^{-\frac{m}{n}} = \dfrac{1}{a^{\frac{m}{n}}}$

Properties of Rational Exponents

If m and n are rational exponents, and a and b are real numbers for which the following expressions are defined, then

1. $b^m \cdot b^n = b^{m+n}$ **2.** $\dfrac{b^m}{b^n} = b^{m-n}$

3. $(b^m)^n = b^{mn}$ **4.** $(ab)^n = a^n b^n$

5. $\left(\dfrac{a}{b}\right)^n = \dfrac{a^n}{b^n}$

RADICALS

If $\sqrt[n]{a}$ and $\sqrt[n]{b}$ are real numbers, then

1. If n is even, then $\sqrt[n]{a^n} = |a|$.

2. If n is odd, then $\sqrt[n]{a^n} = a$.

3. The product rule: $\sqrt[n]{a} \cdot \sqrt[n]{b} = \sqrt[n]{ab}$

4. The quotient rule: $\dfrac{\sqrt[n]{a}}{\sqrt[n]{b}} = \sqrt[n]{\dfrac{a}{b}}$

INTERVAL NOTATION, SET-BUILDER NOTATION, AND GRAPHS

$(a, b) = \{x \mid a < x < b\}$

$[a, b) = \{x \mid a \leq x < b\}$

$(a, b] = \{x \mid a < x \leq b\}$

$[a, b] = \{x \mid a \leq x \leq b\}$

$(-\infty, b) = \{x \mid x < b\}$

$(-\infty, b] = \{x \mid x \leq b\}$

$(a, \infty) = \{x \mid x > a\}$

$[a, \infty) = \{x \mid x \geq a\}$

$(-\infty, \infty) = \{x \mid x \text{ is a real number}\} = \{x \mid x \in R\}$

SLOPE FORMULA

$$\text{slope } (m) = \frac{\text{change in } y}{\text{change in } x} = \frac{y_2 - y_1}{x_2 - x_1}, \quad (x_1 \neq x_2)$$

EQUATIONS OF LINES

1. *Slope-intercept form:* $y = mx + b$
m is the line's slope and b is its y-intercept.

2. *General form:* $Ax + By + C = 0$

3. *Point-slope form:* $y - y_1 = m(x - x_1)$
m is the line's slope and (x_1, y_1) is a fixed point on the line.

4. *Horizontal line parallel to the x-axis:* $y = b$

5. *Vertical line parallel to the y-axis:* $x = a$

ABSOLUTE VALUE

1. $|x| = \begin{cases} x & \text{if } x \geq 0 \\ -x & \text{if } x < 0 \end{cases}$

2. If $|x| = c$, then $x = c$ or $x = -c$. $(c > 0)$

3. If $|x| < c$, then $-c < x < c$. $(c > 0)$

4. If $|x| > c$, then $x < -c$ or $x > c$. $(c > 0)$

SPECIAL FACTORIZATIONS

1. *Difference of two squares:*
$$A^2 - B^2 = (A + B)(A - B)$$

2. *Perfect square trinomials:*
$$A^2 + 2AB + B^2 = (A + B)^2$$
$$A^2 - 2AB + B^2 = (A - B)^2$$

3. *Sum of two cubes:*
$$A^3 + B^3 = (A + B)(A^2 - AB + B^2)$$

4. *Difference of two cubes:*
$$A^3 - B^3 = (A - B)(A^2 + AB + B^2)$$

ALGEBRA'S COMMON GRAPHS

Identity Function

Absolute Value Function

Standard Quadratic Function

Square Root Function

Standard Cubic Function

Cube Root Function

Greatest Integer Function

Reciprocal Function

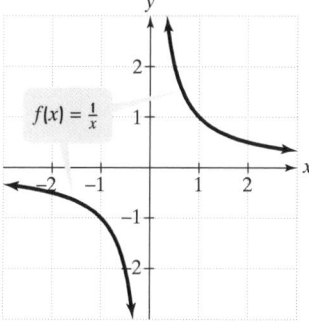

TRANSFORMATIONS

In each case, c represents a positive real number.

	Function	**Draw the graph of f and:**
Vertical translations	$\begin{cases} y = f(x) + c \\ y = f(x) - c \end{cases}$	Shift f upward c units. Shift f downward c units.
Horizontal translations	$\begin{cases} y = f(x - c) \\ y = f(x + c) \end{cases}$	Shift f to the right c units. Shift f to the left c units.
Reflections	$\begin{cases} y = -f(x) \\ y = f(-x) \end{cases}$	Reflect f about the x-axis. Reflect f about the y-axis.
Vertical Stretching or Shrinking	$\begin{cases} y = cf(x); c > 1 \\ y = cf(x); 0 < c < 1 \end{cases}$	Vertically stretch f, multiplying each of its y-coordinates by c. Vertically shrink f, multiplying each of its y-coordinates by c.
Horizontal Stretching or Shrinking	$\begin{cases} y = f(cx); c > 1 \\ y = f(cx); 0 < c < 1 \end{cases}$	Horizontally shrink f, dividing each of its x-coordinates by c. Horizontally stretch f, dividing each of its x-coordinates by c.

DISTANCE AND MIDPOINT FORMULAS

1. The distance from (x_1, y_1) to (x_2, y_2) is
$$\sqrt{(x_2 - x_1)^2 + (y_2 - y_1)^2}.$$

2. The midpoint of the line segment with endpoints (x_1, y_1) and (x_2, y_2) is
$$\left(\frac{x_1 + x_2}{2}, \frac{y_1 + y_2}{2}\right).$$

QUADRATIC FORMULA

The solutions of $ax^2 + bx + c = 0$ with $a \neq 0$ are
$$x = \frac{-b \pm \sqrt{b^2 - 4ac}}{2a}.$$

FUNCTIONS

1. Linear Function: $f(x) = mx + b$
 Graph is a line with slope m and y-intercept b.

2. Quadratic Function: $f(x) = ax^2 + bx + c, a \neq 0$

 Graph is a parabola with vertex at $x = -\dfrac{b}{2a}$.

 Quadratic Function: $f(x) = a(x - h)^2 + k$
 In this form, the parabola's vertex is (h, k).

3. nth-Degree Polynomial Function:
 $$f(x) = a_n x^n + a_{n-1}x^{n-1} + a_{n-2}x^{n-2} + \cdots + a_1 x + a_0, a_n \neq 0$$

 For n odd and $a_n > 0$, graph falls to the left and rises to the right.

 For n odd and $a_n < 0$, graph rises to the left and falls to the right.

 For n even and $a_n > 0$, graph rises to the left and rises to the right.

 For n even and $a_n < 0$, graph falls to the left and falls to the right.

4. Rational Function: $f(x) = \dfrac{p(x)}{q(x)}, p(x)$ and $q(x)$ are
 polynomials, $q(x) \neq 0$

5. Exponential Function: $f(x) = b^x, b > 0, b \neq 1$

 Graphs:

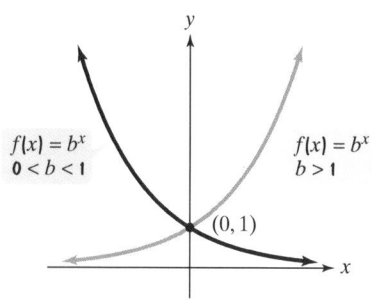

6. Logarithmic Function: $f(x) = \log_b x, b > 0, b \neq 1$
 $y = \log_b x$ is equivalent to $x = b^y$.

 Graph:

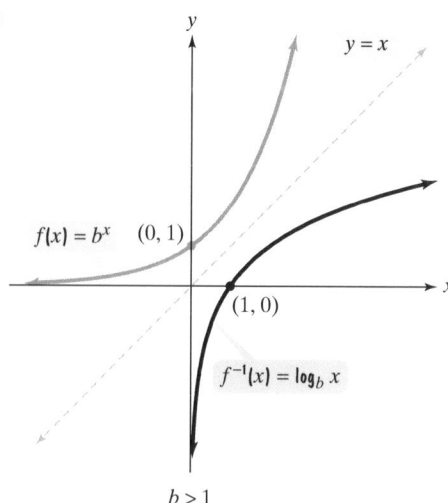

PROPERTIES OF LOGARITHMS

1. $\log_b(MN) = \log_b M + \log_b N$

2. $\log_b\left(\dfrac{M}{N}\right) = \log_b M - \log_b N$

3. $\log_b M^p = p \log_b M$

4. $\log_b M = \dfrac{\log_a M}{\log_a b} = \dfrac{\ln M}{\ln b} = \dfrac{\log M}{\log b}$

5. $\log_b b^x = x;\quad \ln e^x = x$

6. $b^{\log_b x} = x;\quad e^{\ln x} = x$

INVERSE OF A 2 × 2 MATRIX

If $A = \begin{bmatrix} a & b \\ c & d \end{bmatrix}$, then $A^{-1} = \dfrac{1}{ad - bc}\begin{bmatrix} d & -b \\ -c & a \end{bmatrix}$, where $ad - bc \neq 0$.

CRAMER'S RULE

If
$$a_{11}x_1 + a_{12}x_2 + a_{13}x_3 + \cdots + a_{1n}x_n = b_1$$
$$a_{21}x_1 + a_{22}x_2 + a_{23}x_3 + \cdots + a_{2n}x_n = b_2$$
$$a_{31}x_1 + a_{32}x_2 + a_{33}x_3 + \cdots + a_{3n}x_n = b_3$$
$$\vdots$$
$$a_{n1}x_1 + a_{n2}x_2 + a_{n3}x_3 + \cdots + a_{nn}x_n = b_n$$

then $x_i = \dfrac{D_i}{D}, D \neq 0$.

D: determinant of the system's coefficients

D_i: determinant in which coefficients of x_1 are replaced by $b_1, b_2, b_3, \ldots, b_n$.

CONIC SECTIONS

Circle

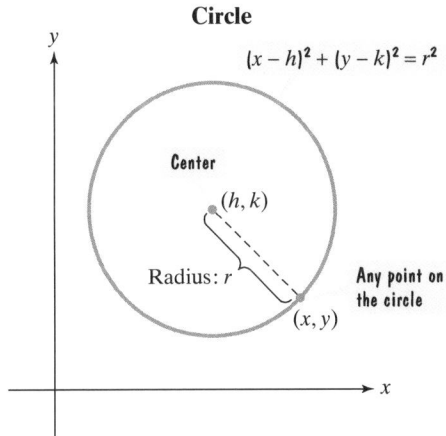

$$(x - h)^2 + (y - k)^2 = r^2$$

Ellipse

$$\frac{(x-h)^2}{a^2} + \frac{(y-k)^2}{b^2} = 1$$

$$\frac{(x-h)^2}{b^2} + \frac{(y-k)^2}{a^2} = 1$$

$$a^2 > b^2 \text{ and } c^2 = a^2 - b^2$$

Hyperbola

$$\frac{(x-h)^2}{a^2} - \frac{(y-k)^2}{b^2} = 1$$

$$\frac{(y-k)^2}{a^2} - \frac{(x-h)^2}{b^2} = 1$$

$$c^2 = a^2 + b^2$$

Parabolas

$$(y - k)^2 = 4p(x - h)$$

$$(x - h)^2 = 4p(y - k)$$

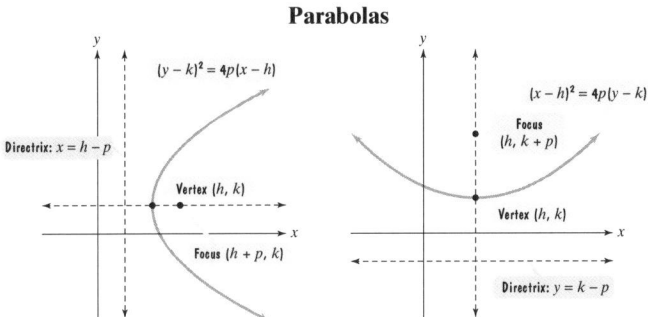

SEQUENCES

1. Infinite Sequence:

$$\{a_n\} = a_1, a_2, a_3, \ldots, a_n, \ldots$$

2. Summation Notation:

$$\sum_{i=1}^{n} a_i = a_1 + a_2 + a_3 + \cdots + a_n$$

3. nth Term of an Arithmetic Sequence:

$$a_n = a_1 + (n - 1)d$$

4. Sum of First n Terms of an Arithmetic Sequence:

$$S_n = \frac{n}{2}(a_1 + a_n)$$

5. nth Term of a Geometric Sequence: $a_n = a_1 r^{n-1}$

6. Sum of First n Terms of a Geometric Sequence:

$$S_n = \frac{a_1(1 - r^n)}{1 - r} (r \neq 1)$$

7. Sum of an Infinite Geometric Series with $|r| < 1$:

$$S = \frac{a_1}{1 - r}$$

THE BINOMIAL THEOREM

1. $n! = n(n - 1)(n - 2) \cdots 3 \cdot 2 \cdot 1; 0! = 1$

2. $\binom{n}{r} = \dfrac{n!}{r!(n - r)!}$

3. Binomial Theorem:

$$(a + b)^n = \binom{n}{0}a^n + \binom{n}{1}a^{n-1}b +$$
$$\binom{n}{2}a^{n-2}b^2 + \cdots + \binom{n}{n}b^n.$$

PERMUTATIONS, COMBINATIONS, AND PROBABILITY

1. $_nP_r$, the number of permutations of n elements taken r at a time, is given by

$$_nP_r = \frac{n!}{(n - r)!}.$$

2. $_nC_r$, the number of combinations of n elements taken r at a time, is given by

$$_nC_r = \frac{n!}{(n - r)!r!}.$$

3. *Probability of an Event:* $P(E) = \dfrac{n(E)}{n(S)}$, where

$n(E) = $ the number of outcomes in event E and $n(S) = $ the number of outcomes in the sample space.

STATISTICS

A Tool for Social Research and Data Analysis

Eleventh Edition

Joseph F. Healey
Christopher Newport University

Christopher Donoghue
Montclair State University

CENGAGE

Australia • Brazil • Japan • Mexico • Singapore • United Kingdom • United States

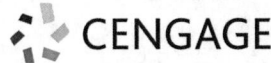

Statistics: A Tool for Social Research and Data Analysis, Eleventh Edition

Joseph F. Healey, Christopher Donoghue

Product Director: Laura Ross

Product Manager: Kori Alexander

Learning Designer: Emma Guiton

Content Manager: Meaghan Ford

Marketing Manager: Tricia Salata

Product Assistant: Shelby Blakey

Digital Delivery Lead: Matt Altieri

Intellectual Property Analyst: Deanna Ettinger

Senior Intellectual Property Manager: Betsy Hathaway

Designer, Creative Studio: Nadine Ballard

Production Service/Composition: MPS Limited

© 2021, 2015 Cengage Learning, Inc.

ALL RIGHTS RESERVED. No part of this work covered by the copyright herein may be reproduced or distributed in any form or by any means, except as permitted by U.S. copyright law, without the prior written permission of the copyright owner.

For product information and technology assistance, contact us at **Cengage Customer & Sales Support, 1-800-354-9706.**

For permission to use material from this text or product, submit all requests online at **www.cengage.com/permissions.** Further permissions questions can be e-mailed to **permissionrequest@cengage.com.**

Library of Congress Control Number: 2020902062

Student Edition:
ISBN: 978-0-357-37107-7

Loose-leaf Edition:
ISBN: 978-0-357-37109-1

Cengage
200 Pier 4 Boulevard
Boston, MA 02210
USA

Cengage is a leading provider of customized learning solutions with employees residing in nearly 40 different countries and sales in more than 125 countries around the world. Find your local representative at **www.cengage.com**.

Cengage products are represented in Canada by Nelson Education, Ltd.

To learn more about Cengage platforms and services, register or access your online learning solution, or purchase materials for your course, visit **www.cengage.com**.

Printed in the United States of America
Print Number: 01 Print Year: 2020

Brief Contents

Detailed Contents

Preface

Society has become increasingly data driven. At colleges and universities, statistics are an essential part of sociology and other social sciences (including political science, social work, psychology, public administration, criminal justice, urban studies, and others). In social policy work, researchers are challenged to use data (on poverty, health care, policing, and other topics) to demonstrate why policy reform will be beneficial to society. In the private sector, and in sports, decisions are increasingly based on data analytics.

This means that understanding data has become an essential skill that college students can bring to the job market when they graduate. Nevertheless, learning statistics is challenging, especially for students that do not have an extensive background in mathematics. Social science students, in particular, possess a wide range of skills and interests but still find themselves puzzled about the relevance of statistics for them, and frequently, this leads to math anxiety. This text introduces statistical analysis for the social sciences in a way that is intended to reduce those anxieties.

The Authors and the New Title of the Book

We begin by telling you a little bit about the authors and the new approach they have taken to this edition. **Joseph Healey** taught sociology and statistics for more than 40 years. His research and teaching interests centered on ethnic and race relations, immigration, sports, violence, and globalization. He began this project because of his dissatisfaction with texts that were either too mathematical and technical or too simplistic. He sought to find a middle approach that was accessible to beginning students and that prepared them to appreciate the professional literature and to do actual social research themselves.

New to the 11th edition is the addition of **Christopher Donoghue** as coauthor. Chris is a sociologist with research interests in peer aggression and bullying, school climate, ethnic and racial social distance, and sexual behavior among adolescents. He has also conducted extensive research on the quality of care in nursing homes. As a student, he used a much earlier edition of this book and since he began teaching the course in 2001, it is the only book he has used. Although he has taught the course for many years now, every semester is different (students come with new needs and interests, trends in using data change quite often). Changing the course to add new techniques, technologies, and strategies is something that he greatly enjoys. Joe and Chris also share an interest in effectively teaching statistics to students that lack confidence in mathematics, so their collaboration on this text has been a natural fit.

Together, Joe and Chris set out to make the 11th edition of the book more current and reflective of the increased emphasis in society on data and evidence-based decision making. The new title, **Statistics: A Tool for Social Research and Data Analysis**, is indicative of the book's new approach to use real data and demonstrate real applications in all chapters. Although the previous editions of the book used real data, they also contained many boxes on Applied Statistics that used fictional data. In the latest edition, these boxes have been replaced by boxes on **Social Research and Data Analysis**, nearly all of which contain real problems with actual data. Many other updates were made to the text as well, with the consistent goal of infusing both real and current data to the learning experience.

Who Is This Book For?

Most undergraduate students of sociology and the other social sciences take a course in social statistics. This book is rich with examples from those fields and it is designed for that audience, but its many additional examples from social policy research, market research and sports, and its applied orientation make it flexible enough to be used in statistics courses for all kinds of undergraduate majors as well as graduate courses.

Many students dread that first course in statistics, especially if they have not had success in mathematics in the past. This book was designed with that sort of student in mind. Since 1984, students have been using this text to learn about statistics in an approachable and straightforward fashion. The book contains simple instructions, language that is easy to understand, and data from research and everyday life that students will enjoy reading about. The book makes few assumptions about mathematical background. In fact, if you can *add, subtract, multiply,* and *divide,* you will find that almost all of the operations in the text can be accomplished without much additional mathematical learning. They just take some concentration and the willingness to learn new concepts and methods.

The book is written at an intermediate level and its goal is to show the relevance and value of statistics for the social sciences. We emphasize interpretation and understanding statistics in the context of social science research, but we have not sacrificed comprehensive coverage or statistical correctness. Mathematical explanations are kept at an elementary level, as is appropriate in a first exposure to social statistics. The text does not claim that statistics are always "fun" or that the material can be mastered without considerable effort. At the same time, students are not overwhelmed with abstract proofs, formula derivations, or mathematical theory, which can frustrate the learning experience at this level.

Changes to the 11th Edition

Previous users will find that the structure of the book and the instructional content have not changed. What is new is the extensive addition of new data on current problems throughout the text from sociology and the other social sciences (e.g., political science, social work, and gerontology), market research, social

policy, sports, and more. The influx of data analytics to professional industries is also discussed and students are encouraged to learn more by exploring statistical software packages such as SPSS, SAS, and R.

The following are the most important changes to this edition:

- Current examples and information can be found in all chapters on topics such as the 2016 presidential election in the United States, the mass shooting epidemic, contemporary uses of the term "racism," debates about the appropriateness of abstinence teaching in sexual education, the "fake news" phenomenon, and more.

- Social Research and Data Analysis boxes now contain examples in which students are encouraged to use the formulas in the chapter to explore real and current data.

- New boxes on Reading Statistics have been added, containing interesting findings from real academic and professional studies.

- The Pew Research Center's 2018 Core Trends data set has been added to provide students with more variables to explore than in previous editions.

- All of the data on current events, the SPSS exercises, and the Statistics in Everyday Life boxes have been updated.

- New end-of-chapter problems have been added on the topics of Market Research and Social Policy.

- In Chapter 15, new terms for multiple regression statistics have been added (e.g., unstandardized regression coefficients and constants) to make the chapter more consistent with academic research articles and more usable to students.

- New commentary now appears in Chapter 3 on bimodal distributions, Chapter 4 on skewness and kurtosis, and Chapter 6 on sampling frames.

- Several new Step-by-Step boxes have been added to Chapter 6 to teach students how to employ different random sampling techniques.

- The Using Statistics boxes now direct students to specific pages in the chapters where the subject material is used to solve an interesting problem or to address an important topic.

- SPSS problems have been added to the end-of-chapter problems throughout the text. In some chapters (e.g., Chapters 13 and 15), the SPSS problems replace problems using hand calculators.

- The data sets used in the text have been expanded and updated. The data sets are used throughout the text in the new "Using SPSS" sections, in the end-of-chapter problems, and in the "You Are the Researcher" projects at the end of most chapters. The data sets are available for downloading at the Web site for this text: **www.cengagebrain.com**; they include:

 - A General Social Survey (GSS) data set (*GSS2018.sav*), which has been updated to 2018.

 - A smaller version of the 2018 GSS (*GSS2018 Sample*) for students using the student version of SPSS. This data set has only 50 variables and fewer than 1500 cases.

- The Pew Research Center's 2018 Core Trends Data Set.
- A data set that includes census and crime data on the 50 states (*States.sav*).
- A data set that includes demographic data for 99 nations (*International.sav*).
- A fourth data set (*CrimeTrends84-17.sav*) is used only for the graphing exercises in Chapter 2.

Numerous other changes have been made throughout the text. All are intended to clarify explanations and make the material more accessible to students. As with previous editions, our goal is to offer a comprehensive, flexible, and student-oriented book that will provide a challenging first exposure to social statistics.

Goal of the Text

Although the book is evolving, its goal is still to introduce students to basic statistics and demonstrate their usefulness in social research of all kinds. The statistically literate person understands the role of statistics in the research process, is competent to perform basic calculations, and can appreciate any analyses of data that she or he may encounter in everyday life. Let's consider these three competencies in a little more detail.

1. **An Appreciation of Statistics.** Appreciating statistics means understanding their relevance and being capable of employing statistical tools when they are needed, and in the right situations. This textbook develops these qualities in several ways.

 - *The relevance of statistics.* Throughout the text, each example is framed in the context of a research situation. A question is posed and then, with the aid of a statistic, answered. This central theme of usefulness is further reinforced by a series of "Social Research and Data Analysis" boxes, each of which illustrates some specific way statistics can be used to answer questions, such as those posed in the "Using Statistics" feature that opens every chapter.

 - *Employing techniques and strategies.* At the end of each chapter, students are challenged to use what they have learned to solve problems relevant to specific fields: **SOC for sociology, SW for social work, PS for political science, SP for social policy (new to this edition), MR for market research (new to this edition), CJ for criminal justice, PA for public administration, and GER for gerontology.** These problems show the relevance of statistics to students with a wide array of academic and career interests.

 - *Interpreting statistics.* For most students, interpretation—saying what statistics mean—is a big challenge. The ability to interpret statistics can be developed only by exposure and experience. To provide exposure, we have been careful to express the meaning of the statistics in terms of the original research question. To provide experience, the Using SPSS boxes and the end-of-chapter problems call for an interpretation of the statistics that are calculated. Students can also learn these lessons from the "Interpreting Statistics" sections that appear in many chapters.

- *Using statistics: Ideas for research projects.* Appendix E and the You Are the Researcher boxes offer ideas for independent data-analysis projects for students. The projects require students to use SPSS to analyze a data set. They can be assigned at intervals throughout the semester or at the end of the course. Each project provides an opportunity for students to practice and apply their statistical skills and to exercise their ability to interpret the statistics they produce.

2. **Computational Competence.** Students should emerge from their first course in statistics with the ability to perform elementary forms of data analysis. While computers have made computation less of an issue today, computation is inseparable from statistics, so we have included a number of features to help students cope with these mathematical challenges.

 - *"One Step at a Time:"* boxes for each statistic break down computation into individual steps for maximum clarity and ease.

 - *Extensive problem sets* are provided at the end of each chapter. For the most part, these problems use fictitious data and are designed for ease of computation.

 - *Cumulative exercises* are included at the end of each Part of the book to provide practice in choosing, computing, and analyzing statistics. These exercises present only data sets and research questions. Students must choose appropriate statistics as part of the exercise.

 - *Solutions* to odd-numbered computational problems are provided so that students may check their answers.

 - *SPSS* gives students access to the computational power of the computer. This is explained in more detail later.

3. **The Ability to Read the Professional Social Science Literature.** Comprehending and critically appraising statistical reports is another essential skill that can come out of a basic statistics course. The development of this skill is a particular problem for many students since (1) the vocabulary of professional researchers is so much more concise than the language of the textbook, and (2) the statistics featured in the literature are more advanced than those covered at the introductory level. To help bridge this gap, we have included a series of boxes labeled "**Reading Statistics.**" In each such box, we briefly describe the reporting style typically used for the statistic in question and try to alert students about what to expect when they approach the professional literature. These inserts have been updated in this edition and most include excerpts from the research literature that illustrate how statistics are actually applied and interpreted by social scientists.

Additional Features

A number of other features make the text more meaningful for students and more useful for instructors.

- *Readability and clarity.* The writing style is informal and accessible to students without ignoring the traditional vocabulary of statistics. For the more difficult material (such as hypothesis testing), students are first walked through

an example problem before being confronted by formal terminology and concepts. Each chapter ends with a summary of major points and formulas and a glossary of important concepts. A list of frequently used formulas inside the front cover and a glossary of symbols inside the back cover can be used for quick reference.

- *Organization and coverage.* The text is divided into four parts, with most of the coverage devoted to univariate descriptive statistics, inferential statistics, and bivariate measures of association. The distinction between description and inference is introduced in the first chapter and maintained throughout the text. In selecting statistics for inclusion, we have tried to strike a balance between the essential concepts with which students must be familiar and the amount of material students can reasonably be expected to learn in their first (and perhaps only) statistics course, all the while bearing in mind that different instructors will naturally wish to stress different aspects of the subject. Thus, the text covers a full gamut of the usual statistics, with each chapter broken into subsections so that instructors may choose the particular statistics they wish to include.

- *Learning objectives.* Learning objectives are stated at the beginning of each chapter. These are intended to serve as "study guides" and to help students identify and focus on the most important material.

- *Review of mathematical skills.* A comprehensive review of all of the mathematical skills that will be used in this text is included as a Prologue. Students who are inexperienced or out of practice with mathematics can study this review early in the course and/or refer to it as needed. A self-test is included so that students can check their level of preparation for the course.

- *Statistical techniques and end-of-chapter problems are explicitly linked.* After a technique is introduced, students are directed to specific problems for practice and review. The "how-to-do-it" aspects of calculation are immediately and clearly reinforced.

- *End-of-chapter problems are organized progressively.* Simpler problems with small data sets are presented first. Often, explicit instructions or hints accompany the first several problems in a set. The problems gradually become more challenging and require more decision making by the student (e.g., choosing the most appropriate statistic for a certain situation). Thus, each problem set develops problem-solving abilities gradually and progressively.

- *Computer applications.* This text integrates SPSS, the leading social science statistics package, to help students take advantage of the power of the computer. Appendix F provides an introduction to SPSS, and demonstrations are integrated into the chapters. SPSS-based problems are included at the end of chapters, and research projects using SPSS are presented in the "You Are the Researcher" feature.

- *Realistic, up-to-date data.* The databases for computer applications in the text include the 2018 General Social Survey, the Pew Research Center's 2018 Core Trends data set, a data set that includes census and crime data for the 50 states, and a data set that includes demographic data for 99 nations. These data sets will give students the opportunity to practice their statistical skills on "real-life" data. All databases are described in Appendix G.

Supplements

Several ancillaries accompany the text:

MindTap™: The Personal Learning Experience The redesigned Mind-Tap for *Statistics*: *A Tool for Social Research and Data Analysis*, 11th Edition, from Cengage represents a new approach to a highly personalized, online learning platform. MindTap combines all of a student's learning tools, readings, and multimedia activities into a "Learning Path" that guides students through their statistics course. Four new, highly interactive activities challenge students to think critically by exploring, analyzing, and creating content, developing their sociological imagination through personal, local, and global lenses.

MindTap for Statistics: *A Tool for Social Research and Data Analysis* is easy to use and saves instructors time by allowing you to

- Break course content down into manageable modules to promote personalization, encourage interactivity, and ensure student engagement.

- Customize the course—from tools to text—and make adjustments "on the fly," making it possible to intertwine breaking news into your lessons and incorporate today's teachable moments.

- Bring interactivity into learning through the integration of multimedia assets (apps from Cengage and other providers) and numerous in-context exercises and supplements; student engagement will increase, leading to better student outcomes.

- Track students' use, activities, and comprehension in real time, which provides opportunities for early intervention to influence progress and outcomes. Grades are visible and archived so students and instructors always have access to current standings in the class.

- Assess knowledge throughout each section: after readings, and in automatically graded activities and assignments.

- A new digital implementation guide will help you integrate the new MindTap Learning Path into your course.

- Learn more at **www.cengage.com/mindtap**.

- *Online PowerPoint® Slides.* A revised series of PowerPoint slides allows instructors to deliver class lectures and presentations discussing chapter-by-chapter content.

- *Online Instructor's Manual/Testbank.* The Instructor's Manual includes chapter summaries, a test item file of multiple-choice questions, answers to even-numbered computational problems, and step-by-step solutions to selected problems. In addition, the Instructor's Manual includes cumulative exercises (with answers) that can be used for testing purposes. To access these instructor resources, please log in to your account at **http://login.cengage.com**.

- *Cengage Learning Testing Powered by Cognero®.* Cengage Learning Testing powered by Cognero is a flexible online system that allows instructors to author, edit, and manage Test Bank content and quickly create multiple test versions. You can deliver tests from your LMS, your classroom—or wherever you want.

Acknowledgments

Joseph F. Healey

I would like to thank the instructors who made statistics understandable to me (Professors Satoshi Ito, Noelie Herzog, and Ed Erikson) and all of my colleagues at Christopher Newport University for their support and encouragement (especially Professors F. Samuel Bauer, Stephanie Byrd, Cheryl Chambers, Robert Durel, Marcus Griffin, Mai Lan Gustafsson, Kai Heiddemann, Ruth Kernodle, Michael Lewis, Marion Manton, Eileen O'Brien, Lea Pellet, Eduardo Perez, Virginia Purtle, Andrea Timmer, and Linda Waldron). Also, I thank all of my students for their patience and thoughtful feedback.

I want to thank Professor Christopher Donoghue for his countless contributions as coauthor of the 11th edition. His understanding of the material and his years of experience in teaching and research have greatly strengthened this text, and I look forward to many years of collaboration on this project.

I also want to acknowledge the support of my family and rededicate this work to them. I have the extreme good fortune to be a member of an extended family that is remarkable in many ways and that continues to increase in size. Although I cannot list everyone, I would especially like to thank the older generation (my mother, Alice T. Healey), my wife Patricia A. Healey, the next generation (my sons, Kevin and Christopher, my daughters-in-law, Jennifer and Jessica, my stepson Christopher Schroen, and stepdaughters Kate Cowell and Jennifer Schroen), and the youngest generation (Benjamin, Caroline, and Isabelle Healey and Abigail Cowell).

Christopher Donoghue

I owe a great deal of debt to three mentors that I have been fortunate to have in Social Statistics. The first is Dr. Rosemary Cooney of Fordham University. I like to say that Rosemary taught me the "language of statistics." To this day, I can hear her voice whenever I am teaching the course. Dr. Joseph Healey was also a mentor to me in statistics even though I did not meet him in person until 2013. He wrote this book to help students who struggled with mathematics, and that is exactly the kind of student I was as an undergraduate in the 1990s. This book made statistics easy for me and it helped me immensely in communicating the subject matter to my students. Finally, Dr. Yong Wang of Montclair State University has been a constant and reliable source of information about statistics for me over the last several years. Yong has pushed me to incorporate many new techniques that have improved my courses. He is also the first person I talk to when I need the answer to a question about statistics.

Joe and I also would like to express a special thanks to our Writer's Assistant, Taulant Asani. Taulant was a great reviewer of our work and he was excellent at checking our math in Excel. We are also in debt to Stephen Shahin, and undergraduate student Montclair State University. Stephen was a careful reader of the prior edition. He caught a number of our errors and he made great suggestions for the current edition.

I would also like to thank my wife, Lisa Donoghue, my daughter, Amanda Donoghue, and my mother Teresa Donoghue. All three of them have supported me throughout my academic career, and in this endeavor, they have been a constant source of motivation that I could not do without.

Finally, we are grateful to the Literary Executor of the late Sir Ronald A. Fisher, F.R.S., to Dr. Frank Yates, F.R.S., and to Longman Group Ltd., London, for permission to reprint Appendixes B, C, and D from their book *Statistical Tables for Biological, Agricultural and Medical Research* (6th edition, 1974).

Prologue:
Basic Mathematics Review

You will probably be relieved to hear that this text does not place a great emphasis on mathematical computations. While you will encounter many numbers to work with and formulas to solve, the main focus will be on understanding the role of statistics in research and the logic we use to answer research questions empirically. You will also find that, in this text, the sample problems have been simplified so that the computations will not impede your understanding of the statistics themselves.

Statistics do involve numbers, however, so you cannot avoid the mathematics altogether. If you have anxieties about math, we recommend you focus on the four simplest operations with which you are already familiar: ***adding, subtracting, multiplying,*** and ***dividing***. When it comes down to it, every formula in this book can be completed by carrying out those four simple tasks. The greater challenges in the work of statistics are instead about figuring out which of these tasks to employ, and when to employ them. So to make your work easier, we have included boxes called "Step by Step" that tell you exactly which tasks to perform and when to perform them. We recommend you try these boxes out whenever a formula gives you trouble.

Calculators and Computers

A calculator is a virtual necessity for this text. Even the simplest, least expensive model will save you time and effort and is definitely worth the investment. However, if you have experience using Microsoft Excel or Google Sheets, you will find that these software packages can help you even more because they will enable you to perform simple tasks (like adding and multiplying) over and over again in a clear and efficient manner. In the text, you will also learn how to use SPSS for this purpose. One benefit of SPSS is that it enables you to perform more sophisticated techniques with a simple point and click method, but unfortunately, it does not show you how the statistics are computed. For this reason, there is also a benefit to using Microsoft Excel or Google Sheets because these packages require you to set the math up yourself.

Variables and Symbols

Statistics are a set of techniques by which we can describe, analyze, and manipulate variables. A **variable** is a trait that can change value from case to case or from time to time. Examples of variables include height, weight, level of

prejudice, and political party preference. The possible values or scores associated with a given variable might be numerous (e.g., income) or relatively few (e.g., gender). We will often use symbols, usually the letter X, to refer to variables in general or to a specific variable.

Sometimes we will need to refer to a specific value or a set of values of a variable. This is usually done with the aid of subscripts. So the symbol X_1 (read "X-sub-one") would refer to the first score in a set of scores, X_2 ("X-sub-two") to the second score, and so forth. Also, we will use the subscript i to refer to all the scores in a set. Thus, the symbol X_i ("X-sub-i") refers to all of the scores associated with a given variable (for example, the test grades of a particular class).

Operations

You are all familiar with the four basic mathematical operations of addition, subtraction, multiplication, and division and the standard symbols $(+, -, \times, \div)$ used to denote them. We should remind you that multiplication and division can be symbolized in a variety of ways. For example, the operation of multiplying some number a by some number b may be symbolized in (at least) six different ways:

$$a \times b$$
$$a \cdot b$$
$$a * b$$
$$ab$$
$$a(b)$$
$$(a)(b)$$

In this text, we will commonly use the "adjacent symbols" format (that is, ab), the conventional times sign (\times), or adjacent parentheses to indicate multiplication. On most calculators and computers, the asterisk (*) is the symbol for multiplication.

The operation of division can also be expressed in several different ways. In this text, in addition to the standard symbol for division, we will use either of these two methods:

$$a/b \text{ or } \frac{a}{b}$$

Several formulas require us to find the square of a number. To do this, multiply the number by itself. This operation is symbolized as X^2 (read "X squared"), which is the same thing as $(X)(X)$. If X has a value of 4, then

$$X^2 = (X)(X) = (4)(4) = 16$$

or we could say that "4 squared is 16."

The square root of a number is the value that, when multiplied by itself, results in the original number. So the square root of 16 is 4 because (4)(4) is 16. The operation of finding the square root of a number is symbolized as

$$\sqrt{X}$$

A final operation with which you should be familiar is summation, or the addition of the scores associated with a particular variable. When a formula requires the addition of a series of scores, this operation is usually symbolized as ΣX_i.

"Σ" is uppercase Greek letter sigma and stands for "the summation of." So the combination of symbols ΣX_i means "the summation of all the scores" and directs us to add all the scores for that variable. If four people had family sizes of 2, 4, 5, and 7, then the summation of these four scores for this variable could be symbolized as

$$\Sigma X_i = 2 + 4 + 5 + 7 = 18$$

The symbol Σ is an operator, just like the $+$ and \times signs. It directs us to add all of the scores on the variable indicated by the X symbol.

There are two other common uses of the summation sign, and, unfortunately, the symbols denoting these uses are not, at first glance, sharply different from each other or from the symbol used earlier. Some careful attention to these various meanings should minimize the confusion.

The first set of symbols is ΣX^2, which means "the sum of the squared scores." This quantity is found by *first* squaring each of the scores and *then* adding the squared scores together. A second common set of symbols will be $(\Sigma X_i)^2$, which means "the sum of the scores, squared." This quantity is found by *first* summing the scores and *then* squaring the total.

These distinctions might be confusing at first, so let's use an example to help clarify the situation. Suppose we had a set of three scores: 10, 12, and 13. So

$$X_i = 10, 12, 13$$

The sum of these scores is

$$\Sigma X_i = 10 + 12 + 13 = 35$$

The sum of the squared scores would be

$$\Sigma X^2 = (10)^2 + (12)^2 + (13)^2 = 100 + 144 + 169 = 413$$

Take careful note of the order of operations here. First the scores are squared one at a time, and then the squared scores are added. This is a completely different operation from squaring the sum of the scores:

$$(\Sigma X_i)^2 = (10 + 12 + 13)^2 = (35)^2 = 1225$$

To find this quantity, first the scores are summed and then the total of all the scores is squared. The squared sum of the scores (1225) is *not* the same as the sum of the squared scores (413).

In summary, the operations associated with each set of symbols can be summarized as follows:

Symbol	Operations
ΣX_i	Add the scores
ΣX_i^2	First square the scores and then add the squared scores
$(\Sigma X_i)^2$	First add the scores and then square the total

Operations with Negative Numbers

A number can be either positive (if it is preceded by a $+$ sign or by no sign at all) or negative (if it is preceded by a $-$ sign). Positive numbers are greater than 0, and negative numbers are less than 0. It is important to keep track of signs

because they will affect the outcome of virtually every mathematical operation. This section briefly summarizes the relevant rules for dealing with negative numbers.

First, adding a negative number is the same as subtraction. For example,

$$3 + (-1) = 3 - 1 = 2$$

Second, subtraction changes the sign of a negative number:

$$3 - (-1) = 3 + 1 = 4$$

Note the importance of keeping track of signs here. If you neglected to change the sign of the negative number in the second expression, you would get the wrong answer.

For multiplication and division, you need to be aware of various combinations of negative and positive numbers. Ignoring the case of all positive numbers, this leaves several possible combinations. A negative number times a positive number results in a negative value:

$$(-3)(4) = -12$$
$$(3)(-4) = -12$$

A negative number multiplied by a negative number is always positive:

$$(-3)(-4) = 12$$

Division follows the same patterns. If there is a negative number in the calculations, the answer will be negative. If both numbers are negative, the answer will be positive. So

$$\frac{-4}{2} = -2$$

and

$$\frac{4}{-2} = -2$$

but

$$\frac{-4}{-2} = 2$$

Negative numbers do not have square roots, because multiplying a number by itself cannot result in a negative value. Squaring a negative number always results in a positive value (see the multiplication rules earlier).

Accuracy and Rounding Off

A possible source of confusion in computation involves accuracy and rounding off. People work at different levels of precision and, for this reason alone, may arrive at different answers to problems. This is important because our answers can be at least slightly different if you work at one level of precision and we (or your instructor or your study partner) work at another. You may sometimes think you've gotten the wrong answer when all you've really done is round off at a different place in the calculations or in a different way.

There are two issues here: when to round off and how to round off. Our practice is to work in as much accuracy as our calculator or statistics package will allow and then round off to two places of accuracy (two places beyond, or to the right of, the decimal point) only at the end. If a set of calculations is lengthy and requires the reporting of intermediate sums or subtotals, we will round the subtotals off to two places as we go.

In terms of how to round off, begin by looking at the digit immediately to the right of the last digit you want to retain. If you want to round off to 100ths (two places beyond the decimal point), look at the digit in the 1000ths place (three places beyond the decimal point). If that digit is 5 or more, round up. For example, 23.346 would round off to 23.35. If the digit to the right is less than 5, round down. So, 23.343 would become 23.34.

Let's look at an additional example of how to follow these rules of rounding. If you are calculating the mean value of a set of test scores and your calculator shows a final value of 83.459067 and you want to round off to two places, look at the digit three places beyond the decimal point. In this case the value is 9 (greater than 5), so we would round the second digit beyond the decimal point up and report the mean as 83.46. If the value had been 83.453067, we would have reported our final answer as 83.45.

Formulas, Complex Operations, and the Order of Operations

A mathematical formula is a set of directions, stated in general symbols, for calculating a particular statistic. To "solve a formula," you replace the symbols with the proper values and then perform a series of calculations. Even the most complex formula can be simplified by breaking the operations down into smaller steps.

Working through the steps requires some knowledge of general procedure and the rules of precedence of mathematical operations. This is because the order in which you perform calculations may affect your final answer. Consider the following expression:

$$2 + 3(4)$$

If you add first, you will evaluate the expression as

$$5(4) = 20$$

but if you multiply first, the expression becomes

$$2 + 12 = 14$$

Obviously, it is crucial to complete the steps of a calculation in the correct order.

The basic rules of precedence are to find all squares and square roots first, then do all multiplication and division, and finally complete all addition and subtraction. So the expression

$$8 + 2 \times 2^2/2$$

would be evaluated as

$$8 + 2 \times \frac{4}{2} = 8 + \frac{8}{2} = 8 + 4 = 12$$

The rules of precedence may be overridden by parentheses. Solve all expressions within parentheses before applying the rules stated earlier. For most formulas in this text, the order of calculations will be controlled by the parentheses. Consider the following expression:

$$(8 + 2) - \frac{4}{(5 - 1)}$$

Resolving the parenthetical expressions first, we would have

$$(8 + 2) - \frac{4}{(5 - 1)} = (10) - \frac{4}{4} = 10 - 1 = 9$$

A final operation you will encounter in some formulas in this text involves denominators of fractions that themselves contain fractions. In this situation, solve the fraction in the denominator first and then complete the division. For example,

$$\frac{15 - 9}{6/2}$$

would become

$$\frac{15 - 9}{6/2} = \frac{6}{3} = 2$$

When you are confronted with complex expressions such as these, don't be intimidated. If you're patient with yourself and work through them step by step, beginning with the parenthetical expressions, even the most imposing formulas can be managed.

Exercises

You can use the following problems as a self-test on the material presented in this review. If you can handle these problems, you're ready to do all of the arithmetic in this text. If you have difficulty with any of these problems, please review the appropriate section of this prologue. You might also want to use this section as an opportunity to become more familiar with your calculator. The answers are given immediately following these exercises, along with commentary and some reminders.

1. Complete each of the following:
 a. $17 \times 3 =$
 b. $17(3) =$
 c. $(17)(3) =$
 d. $17/3 =$
 e. $(42)^2 =$
 f. $\sqrt{113} =$

2. For the set of scores (X_i) of 50, 55, 60, 65, and 70, evaluate each of the following expressions:

$\Sigma X_i =$

$\Sigma X_i^2 =$

$(\Sigma X_i)^2 =$

3. Complete each of the following:

a. $17 + (-3) + (4) + (-2) =$

b. $(-27)(54) =$

c. $(-14)(-100) =$

d. $-34/(-2) =$

e. $322/(-11) =$

f. $\sqrt{-2} =$

g. $(-17)^2 =$

4. Round off each of the following to two places beyond the decimal point:

a. 17.17532

b. 43.119

c. 1076.77337

d. 32.4641152301

e. 32.4751152301

5. Evaluate each of the following:

a. $(3 + 7)/10 =$

b. $3 + 7/10 =$

c. $\dfrac{(4 - 3) + (7 + 2)(3)}{(4 + 5)(10)}$

d. $\dfrac{22 + 44}{15/3}$

Answers to Exercises

1. **a.** 51 **b.** 51 **c.** 51

 (The obvious purpose of these first three problems is to remind you that there are several different ways of expressing multiplication.)

 d. 5.67 (Note the rounding off.)

 e. 1764

 f. 10.63

2. The first expression is "the sum of the scores," so this operation would be

 $$\Sigma X_i = 50 + 55 + 60 + 65 + 70 = 300$$

The second expression is the "sum of the squared scores." So

$$\Sigma X_i^2 = (50)^2 + (55)^2 + (60)^2 + (65)^2 + (70)^2$$

$$\Sigma X_i^2 = 2500 + 3025 + 3600 + 4225 + 4900$$

$$\Sigma X_i^2 = 18{,}250$$

The third expression is "the sum of the scores, squared":

$$(\Sigma X_i)^2 = (50 + 55 + 60 + 65 + 70)^2$$

$$(\Sigma X_i)^2 = (300)^2$$

$$(\Sigma X_i)^2 = 90{,}000$$

Remember that ΣX_i^2 and $(\Sigma X_i)^2$ are two completely different expressions with very different values.

3. **a.** 16
 b. −1458
 c. 1400
 d. 17
 e. −29.27

f. Your calculator probably gave you some sort of error message for this problem, because negative numbers do not have square roots.

g. 289

4. **a.** 17.18 **b.** 43.12 **c.** 1076.77
 d. 32.46 **e.** 32.48

5. **a.** 1 **b.** 3.7 (Note the importance of parentheses.) **c.** 0.31 **d.** 13.2

1 Introduction

LEARNING OBJECTIVES By the end of this chapter, you will be able to:

1. Identify the key stages and terms in social scientific research.
2. Distinguish between descriptive and inferential statistics.
3. Provide examples of discrete and continuous variables.
4. Describe three levels of measurement and cite examples of each.

USING STATISTICS

One of the most important themes of this text is that statistics are useful tools for analyzing and understanding information and for communicating our conclusions to others.

To stress this theme, each chapter will begin with a list of situations in which statistics can be applied to good advantage. In some chapters, you can find the page numbers on which we use statistics for these purposes.

Statistics can be used to

- Describe the relationship between Internet use and social class. **See page 8**
- Make estimates about the percentage of Americans that are in favor of same-sex marriage and those that are against it. **See page 9**
- Measure the percentage of Americans that identify as members of the lower class over time. **See page 13**
- Compare the cost of living (housing prices and rents, the cost of groceries and gas, health care, and so forth) between different localities (cities, states, and even nations).

Why Study Statistics?

Students sometimes wonder about the value of studying statistics. What, after all, do numbers and formulas have to do with understanding people and society? The value of statistics will become clear as we move from chapter to chapter, but, for now, we can demonstrate the importance of statistics by considering **research** or disciplined and careful inquiries. Scientists conduct

research to answer questions, examine ideas, and test theories. Research can take numerous forms, and statistics are relevant for **quantitative research** projects, or projects that collect information in the form of numbers or **data**. **Statistics** are mathematical techniques used by social scientists to analyze data in order to answer questions and test theories.

What is so important about learning how to manipulate data? On one hand, some of the most important and enlightening works in the social sciences do not utilize statistics at all. There is nothing magical about data and statistics. The mere presence of numbers guarantees nothing about the quality of a scientific inquiry.

On the other hand, data can be the most trustworthy information available to the researcher and, consequently, deserve special attention. Data that have been carefully collected and thoughtfully analyzed can be the strongest, most objective foundations for building theory and enhancing our understanding of the social world.

Let us be clear about one point: It is never enough merely to gather data (or, for that matter, any kind of information). Even the most carefully collected numerical information does not and cannot speak for itself. The researcher must be able to use statistics effectively to organize, evaluate, and analyze the data. Without a good understanding of statistical analysis, the researcher will be unable to make sense of the data. Without an appropriate application of statistics, data are useless.

Statistics are an indispensable tool for the social sciences. They provide the scientist with some of the most useful techniques for evaluating ideas, testing theory, and making evidence-based decisions. The next section describes the relationships among theory, research, and statistics in more detail.

The Role of Statistics in Scientific Inquiry

Figure 1.1 graphically represents the role of statistics in the research process. The diagram is based on the thinking of Walter Wallace and illustrates a useful conception of how any scientific enterprise grows and develops. One point the diagram makes is that scientific theory and research continually shape each other. Statistics are one of the most important means by which research and theory interact. Let's take a closer look at the process.

FIGURE 1.1 The Wheel of Science

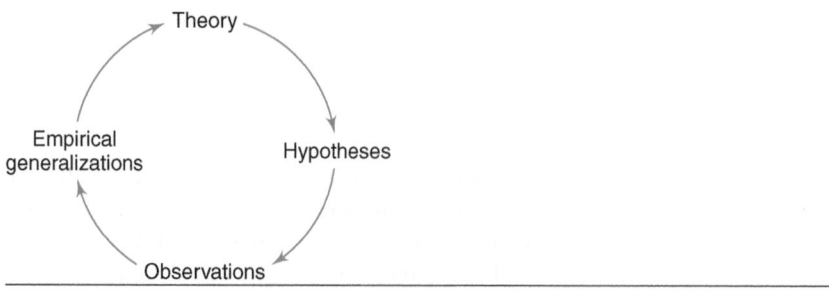

Source: Adapted from Walter Wallace, *The Logic of Science in Sociology* (Chicago: Aldine-Atherton, 1971).

A JOURNEY THROUGH THE SCIENTIFIC PROCESS

Because Figure 1.1 is circular, with no beginning or end, we could start our journey at any point. For the sake of convenience, let's begin at the top, with theory, and follow the arrows around the circle. A **theory** is an explanation of the relationships among phenomena. People naturally (and endlessly) wonder about problems in society (such as prejudice, poverty, child abuse, and serial murders), and they develop explanations ("low levels of education cause prejudice") in their attempts to understand. Unlike our everyday, informal explanations, scientific theory is subject to a rigorous testing process. Let's take prejudice as an example to illustrate how the research process works.

Theory. Why are some people prejudiced against other groups? One possible answer to this question is provided by a theory of prejudice stated more than 70 years ago by social psychologist Gordon Allport and tested on a number of occasions since that time.[1]

The theory states that prejudice will decrease in situations in which members of different groups have equal status and cooperate with each other to work toward mutual goals. The more equal and cooperative the contact, the more likely people will see each other as individuals and not as representatives of a particular group. For example, we might predict that members of a racially mixed athletic team who work together to achieve victory will experience a decline in prejudice. However, when different groups compete for jobs, housing, or other resources, prejudice will tend to increase. Allport's theory is not a complete explanation of prejudice, of course, but it will serve to illustrate a sociological theory.

Variables and Causation. Note that Allport's theory is stated in terms of causal relationships between two variables. A **variable** is any trait that can change values from case to case; examples include gender, age, income, and political party affiliation. A theory may identify some variables as causes and others as effects or results. In the language of science, the causes are **independent variables** and the effects or results are **dependent variables**. In our theory, equal-status contact would be the independent variable (or the cause) and level of prejudice would be the dependent variable (the result or effect). In other words, the theory argues that the level of prejudice *depends on* (or is caused by) the extent to which a person participates in equal-status, cooperative contacts with members of other groups.

[1]Allport, Gordon, 1954. *The Nature of Prejudice*. Reading, MA: Addison-Wesley. This theory is often called "the contact hypothesis." For recent attempts to test this theory, see: McLaren, Lauren. 2003. "Anti-Immigrant Prejudice in Europe: Contact, Threat Perception, and Preferences for the Exclusion of Migrants." *Social Forces*, 81:909–937; Pettigrew, Thomas. 1997. "Generalized Intergroup Contact Effects on Prejudice." *Personality and Social Psychology Bulletin*, 23:173–185; and Pettigrew, T. F., Tropp, L. R., Wagner, U., & Christ, O. 2011. "Recent Advances in Intergroup Contact Theory." *International Journal of Intercultural Relations*, 35: 271–280.

Diagrams can be a useful way of representing the relationships between variables:

$$\text{Equal-status contact} \rightarrow \text{Prejudice}$$
$$\text{Independent variable} \rightarrow \text{Dependent variable}$$
$$X \rightarrow Y$$

The arrow represents the direction of the causal relationship, and X and Y are general symbols for the independent and dependent variables, respectively.

Hypotheses. So far, we have a theory of prejudice and independent and dependent variables. Is the theory true or false? To find out, we need to compare our theory with the facts; we need to do some research. Our next steps would be to define our terms and ideas. One problem we often face in research is that theories are complex and abstract, and we need to be very specific to conduct a valid test. Often, we do this by deriving a hypothesis from the theory: A **hypothesis** is a statement about the relationship between variables that is specific and exact.

For example, if we wished to test Allport's theory, we would have to say exactly what we mean by prejudice and we would need to describe "equal-status, cooperative contact" in detail. We would also review the research literature to help develop and clarify our definitions and our understanding of these concepts.

As our hypothesis takes shape, we begin the next step of the research process, during which we decide exactly how we will gather data. We must decide how cases will be selected and tested, how variables will be measured, and a host of related matters. Ultimately, these plans will lead to the observation phase (the bottom of the wheel of science), where we actually measure social reality. Before we can do this, we must have a clear idea of what we are looking for and a well-defined strategy for conducting the search.

Making Observations and Using Statistics. To test Allport's theory of prejudice, we might begin with people from different racial or ethnic groups. We might place some subjects in cooperative situations and others in competitive situations. We would need to measure levels of prejudice before and after each type of contact. We might do this by administering a survey that asks subjects to agree or disagree with statements such as "Greater efforts must be made to racially integrate the public school system." Our goal would be to see whether the people exposed to the cooperative-contact situations actually become less prejudiced.

Now, finally, we come to the use of statistics in our research. During the observation phase, we will collect a great deal of numerical information or data. If we have a sample of 100 people, we will have 200 completed surveys measuring prejudice: 100 completed before the contact situation and 100 filled out afterward. Try to imagine dealing with 200 completed surveys. If we ask each respondent just five questions to measure their prejudice, we will have a total of 1000 separate pieces of information to deal with. What will we do? We'll have to organize and analyze this information, and statistics will become very helpful at this point. They will supply us with many ideas about "what to do" with the data. We will begin to look at some of the options in the next chapter; for now, let us stress two points about statistics.

First, statistics are crucial. Simply put, without statistics, quantitative research is impossible. We need statistics to analyze data and to shape and refine our theories of the social world.

Second, and somewhat paradoxically, the role of statistics is limited. As Figure 1.1 makes clear, scientific research proceeds through several mutually interdependent stages. Statistics become directly relevant only at the end of the observation stage. Before any statistical analysis can be legitimately applied, however, earlier phases must have been successfully completed. If the researcher has asked poorly conceived questions or has made serious errors of design or method, then even the most sophisticated statistical analysis is valueless. As useful as they can be, statistics cannot substitute for rigorous conceptualization, detailed and careful planning, or creative use of theory. Statistics cannot salvage a poorly conceived or designed research project. They cannot make sense out of garbage.

On the other hand, inappropriate statistical applications can limit the usefulness of an otherwise carefully done project. Only by successfully completing *all* phases of the process can a quantitative research project hope to contribute to understanding. A reasonable knowledge of the uses and limitations of statistics is as essential to the education of the social scientist as is training in theory and methodology.

Empirical Generalizations. Our statistical analysis would focus on assessing our theory, but we would also look for other trends in the data. For example, if we found that equal-status, cooperative contact reduces prejudice in general, we might go on to ask whether the pattern applies to males as well as to females and to the well educated as well as to the poorly educated. As we probed the data, we might begin to develop some generalizations based on the empirical patterns we observe. For example, what if we found that contact reduced prejudice for younger respondents but not for older respondents? Could it be that younger people are less "set in their ways" and have attitudes and feelings that are more open to change? As we developed tentative explanations, we would begin to revise or elaborate the theory that guides the research project.

STATISTICS In Everyday Life

Push Polls

Political campaigns sometimes use "push polls" to sway public opinion. These polls are designed to influence opinions, sometimes by starting or circulating rumors and innuendo. For example, they may try to brand an opponent as untrustworthy by asking questions like "Would you still support the candidate if you found out that he is an alcoholic?" Even when completely fabricated, such a question can create negative associations in the minds of voters. Statistics may be used to "analyze" data gathered by push polls (or by marketing campaigns that use similar techniques), but the results have little or no scientific validity.

A New Theory? If we changed our theory because of our empirical generalizations, a new research project would probably be needed to test the revised theory, and the wheel of science would begin to turn again. We (or perhaps other researchers) would go through the entire process once again with this new—and, hopefully, improved—theory. This second project might result in further revisions that would require still more research, and the wheel of science would continue to turn as long as scientists were able to suggest revisions or develop new insights. Every time the wheel turned, our understanding of the phenomenon under consideration would (hopefully) improve.

Summary: The Dialog Between Theory and Research. This description of the research process does not include white-coated, clipboard-carrying scientists who, in a blinding flash of inspiration, discover some fundamental truth about reality and shout "Eureka!" The truth is that, in the normal course of science, we rarely are in a position to declare a theory true or false. Rather, evidence will gradually accumulate over time, and ultimate judgments of truth will be the result of many years of hard work, research, and debate. It is for this reason that social scientists usually report that their results point to "evidence" for something, rather than saying that it "proves" something.

Let's briefly review our imaginary research project. We began with a theory and examined the various stages of the research project that would be needed to test that theory. We wound up back at the top of the wheel, ready to begin a new project guided by a revised theory. We saw how theory can motivate research and how our observations might cause us to revise theory and, thus, motivate a new research project: Theory stimulates research and research shapes theory. This constant interaction is the lifeblood of science and the key to enhancing our understanding of the social world.

The dialog between theory and research occurs at many levels and in multiple forms. Statistics are one of the most important links between these two realms. Statistics permit us to analyze data, to identify and probe trends and relationships, to develop generalizations, and to revise and improve our theories. As you will see throughout this text, statistics are limited in many ways. They are also an indispensable part of the research enterprise, an essential tool for conducting research and shaping theory. *(For practice in describing the relationship between theory and research and the role of statistics in research, see problems 1.1 and 1.2.)*

The Goals of This Text

Society is becoming increasingly data driven. Social scientific researchers, as well as other professionals that work in social policy, market research, sports, and many other fields, use data to aide their decision making all the time. The goals of this book are to assist students in becoming skilled at using basic statistics as well as to introduce them to the ways that social scientists and other professionals read, analyze, and interpret data. In this section, we address some of the key outcomes that students can expect after taking this course.

First, this textbook takes the view that statistics are tools: useful, but not ends in themselves. Thus, we will not take a "mathematical" approach to the subject, although we will cover enough material so that you can develop a basic understanding of why statistics do what they do. The operations in this text can be carried out manually, but since calculators are rarely used for this purpose in real-world settings, we also demonstrate how to carry out most operations using SPSS, a valuable statistical software package. Many professionals use SPSS for the same purposes, but others use Microsoft Excel, SAS, STATA, R (a free and open sourced program), and other software packages.

Second, you will soon become involved in advanced coursework in your major field of study, and you will find that much of the professional literature assumes at least basic statistical literacy. Furthermore, after graduation, many of you will find yourselves in positions—either in a career or in graduate school—where some understanding of statistics will be helpful or perhaps even required. Very few of you will become statisticians, but you must have a grasp of statistics in order to read and critically appreciate the research literature of your discipline. As a student in the social sciences, you simply cannot realize your full potential without a background in statistics.

This textbook is an introduction to statistics as they are utilized in the social sciences and applied in many professional settings. The general objective of the text is to develop an appreciation—a healthy respect—for statistics and their place in the research process. You should emerge from this experience with the ability to use statistics intelligently and to know when other people have done so—and when they have not! Over time, you will become familiar with the advantages and limitations of the more commonly used statistical techniques, and you will learn which operations are appropriate for a given purpose. The overall experience should give you insight into how statistics are used to address social problems, as well as to make informed decisions in most areas of work and life.

Descriptive and Inferential Statistics

As noted earlier, statistics are tools used to analyze data and answer research questions. Two general classes of statistical techniques, introduced in this section, are available to accomplish this task.

DESCRIPTIVE STATISTICS

The first class of techniques, known as **descriptive statistics**, is relevant in several different situations:

1. When a researcher needs to summarize or describe the distribution of a single variable. These statistics are called *univariate* ("one variable") descriptive statistics.

2. When the researcher wishes to describe the relationship between two or more variables. These statistics are called *bivariate* ("two variable") or *multivariate* (more than two variable) descriptive statistics.

STATISTICS　　In Everyday Life

Using Descriptive Statistics

American society is increasingly connected to the Internet. In 2016, about 82% of American households had access to the Internet, a dramatic increase from 53% in 2000. However, connectedness is dependent on social class: 97% of the most affluent Americans used the Internet vs. only 59% of the least affluent. What are the implications of this pattern? Are the less affluent being left behind? Will their lower use of this increasingly essential resource affect the education and job prospects of their children and thus perpetuate their lower income over the generations? What additional information would you need to answer these questions?

Source: U.S. Bureau of the Census. 2018. Computer and Internet Use in the United States, 2016. Available at https://www.census.gov/content/dam/Census/library/publications/2018/acs/ACS-39.pdf

Univariate Descriptive Statistics. Many of these statistics are probably familiar to you. For example, percentages, averages, and graphs can all be used to describe single variables.

To illustrate the usefulness of univariate descriptive statistics, consider the following problem: Suppose you wanted to summarize the income distribution of a community of 10,000 families. How would you do it? Obviously, you couldn't simply list all the incomes and let it go at that. You would want to use some summary measures of the overall distribution—perhaps a graph, an average, or the percentage of incomes that are low, moderate, or high. Whatever method you choose, its function is to reduce these thousands of items of information into a few clear, concise, and easily understood numbers. The process of using a few numbers to summarize many, called **data reduction**, is the basic goal of univariate descriptive statistics. Part I of this text is devoted to these statistics.

Bivariate and Multivariate Descriptive Statistics. The second type of descriptive statistics is designed to help us understand the relationship between two or more variables. These statistics, called **measures of association**, allow us to quantify the strength and direction of a relationship. We can use these statistics to investigate two matters of central theoretical and practical importance to any science: causation and prediction. Measures of association help us disentangle the connections between variables and trace the ways in which some variables might affect others. We can also use them to predict scores on one variable from scores on another.

For example, suppose you were interested in the relationship between two variables—amount of study time and grades—and had gathered data from a group of college students. By calculating the appropriate measure of association, you could determine the strength of the relationship and its direction. Suppose you found a strong, positive relationship. This would indicate that "study time" and "grade" were closely related (strength of the relationship) and that as one increased in value, the other also increased (direction of the relationship). You could make predictions from one variable to the other ("the longer the study time, the higher the grade").

Measures of association can give us valuable information about relationships between variables and help us understand how one variable causes another.

One important point to keep in mind about these statistics, however, is that they cannot, by themselves, *prove* that two variables are causally related. Even if a measure of association shows a very strong relationship between study time and grades, we cannot conclude that one variable causes the other. Correlation is not the same thing as causation, and the mere existence of a correlation cannot prove that a causal relationship exists between variables. We will consider bivariate associations or correlations in Part III of this text, and we will cover multivariate analysis in Part IV.

INFERENTIAL STATISTICS

This second class of statistical techniques becomes relevant when we wish to generalize to a **population**, the total collection of all cases in which the researcher is interested and wishes to understand better. Examples of possible populations would be voters in the United States, all parliamentary democracies, or all unemployed people in Atlanta.

Populations can theoretically range from enormous ("all humanity") to quite small (all sophomores on your campus) but are usually fairly large. Social scientists almost never have the resources or time to test every case in a population, hence the need for **inferential statistics**. This class of techniques involves using information from **samples** (carefully chosen subsets of the population) to make inferences about populations. Because they have fewer cases, samples are cheaper to assemble, and—if the proper techniques are followed—generalizations based on samples can be very accurate representations of populations.

Many of the concepts and procedures involved in inferential statistics may be unfamiliar, but most of us are experienced consumers of inferential statistics—most familiarly, perhaps, in the form of public-opinion polls and election projections. When a public-opinion poll reports that 42% of the American electorate plans to vote for a certain presidential candidate, it is essentially reporting a generalization to a population ("the American electorate"—which numbers over 130 million people) from a carefully drawn sample (usually between 1000 and 3000 respondents). Inferential statistics will occupy our attention in Part II of this book. *(For practice in describing different statistical applications, see problems 1.3 and 1.7.)*

STATISTICS In Everyday Life

Using Inferential Statistics

In 2019, a sample of adult Americans was asked about same-sex marriage. About 61% were in favor of allowing gays and lesbians to legally marry, a clear majority of the population. Further, surveys show that the level of support has steadily increased since the turn of the century: in 2001, only 35% of the public supported same-sex marriage. Is American society reaching a consensus on this issue? Before coming to a conclusion, remember that the level of support varies dramatically from group to group. For example, in 2019, 79% of Americans with no religious affiliation supported same-sex marriages vs. only 29% of white Evangelical Protestants.

Source: Pew Research Center. 2019. "Religion and Public Life." Available at https://www.pewforum.org/fact-sheet/changing-attitudes-on-gay-marriage/

Discrete and Continuous Variables

In the next chapter, you will begin to encounter the many statistics available to social scientists. One aspect of using statistics that can be puzzling is deciding when to use which statistic. You will learn specific guidelines as you go along, but we will introduce some basic and general guidelines at this point. The first of these concerns discrete and continuous variables; the second, covered in the next section, concerns level of measurement.

DISCRETE VARIABLES

A variable is said to be **discrete** if it has a basic unit of measurement that cannot be subdivided. For example, number of people per household is a discrete variable. The basic unit is people, a variable that will always be measured in whole numbers; you'll never find 2.7 people living in a specific household. The scores of discrete variables will be 0, 1, 2, 3, or some other whole integer. Other examples of discrete variables include number of siblings, children, or cars. To measure these variables, we count the number of units (people, cars, siblings) for each case (household, person) and record results in whole numbers.

CONTINUOUS VARIABLES

A variable is **continuous** if it has scores that can be subdivided infinitely (at least theoretically). One example of a continuous variable is time, which can be measured in minutes, seconds, milliseconds (thousandths of a second), nanoseconds (billionths of a second), or even smaller units. In a sense, when we measure a continuous variable, we are always approximating and rounding off the scores. We could report somebody's time in the 100-yard dash as 10.7 seconds or 10.732451 seconds, but, because time can be infinitely subdivided, we will never be able to report the exact time elapsed. Because we cannot work with infinitely long numbers, we must report the scores on continuous variables as if they were discrete.

The scores of discrete and continuous variables may look the same even though we measure and process them differently. This distinction is one of the most basic in statistics and will be one of the criteria we use to choose among various statistics and graphs. *(For practice in distinguishing between discrete and continuous variables, see problems 1.4 through 1.9.)*

Level of Measurement

A second basic and very important guideline for the selection of statistics is the **level of measurement**, or the mathematical nature of the variables under consideration. Some variables, such as age and income, have numerical scores (years, dollars) and can be analyzed in a variety of ways, using many different statistics. For example, we could summarize these variables with a mean or average and make statements like "The average income of residents of this city is $43,000" or "The average age of students on this campus is 19.7."

Other variables, such as gender and zip code, have "scores" that are really just labels, not numbers at all, and we have fewer options for analyzing them. The average would be meaningless as a way of describing these variables. Your personal zip code might *look* like a number, but it is merely an arbitrary label that happens to be expressed in digits. These "numbers" cannot be added or divided, and statistics like the average zip code of a group of people are pointless.

Level of measurement is crucial because our statistical analysis must match the mathematical characteristics of our variables. One of the first steps in any quantitative research project is to determine the level of measurement of our variables, and we will consider the level of measurement of variables throughout this text and each time a statistical technique is introduced.

There are three levels of measurement. In order of increasing sophistication, they are nominal, ordinal, and interval-ratio. Each will be discussed separately.

THE NOMINAL LEVEL OF MEASUREMENT

Variables measured at the nominal level (such as gender) have non-numerical "scores" or categories. Other examples of variables at this level include zip code, political party affiliation, and religious preference. Statistical analysis with nominal-level variables is limited to comparing the relative sizes of the categories and making statements such as "There are more females than males in this dorm" or "The most common zip code on this campus is 20158."

Let's look at some examples of nominal-level variables and take a moment to consider some terminology. Earlier in this chapter, we defined a variable as any trait that can vary from case to case. A variable consists of categories or scores—the traits that can vary. We "measure" a variable (e.g., gender) by classifying a case into one of the categories (e.g., male, female, transgender, other) or assigning a score to it. Several examples of nominal-level variables and their scores or categories are presented in Table 1.1.

Note that there is a number assigned to each of the categories in Table 1.1 (e.g., 1 for Protestant, 2 for Catholic, etc.), as is commonly done in quantitative research, especially when the data are being prepared for computer analysis. Recall that, for nominal-level variables, these "numbers" are just labels and cannot be treated mathematically: They cannot be added, divided, or otherwise manipulated. The categories or scores do not make up a mathematical scale; they are different from each other but not "more or less" or "higher or lower" than each other. Males and females differ in terms of gender, but neither category has more

TABLE 1.1 Some Examples of Nominal-Level Variables and Their Categories

Variables →	Gender	Political Party Preference	Religious Preference
Categories →	1 Male	1 Democrat	1 Protestant
	2 Female	2 Republican	2 Catholic
	3 Transgender	3 Other	3 Jewish
	4 Other	4 Independent	4 None
			5 Other

TABLE 1.2 **Criteria for Stating the Categories of Variables**

The categories of variables must
1. Be mutually exclusive (Each case must fit into one and only one category.)
2. Be exhaustive (There must be a category for every case.)
3. Include elements that are similar (The cases in each category must be similar to each other.)

TABLE 1.3 **Four Scales for Measuring Religious Affiliation**

Scale A	Scale B	Scale C	Scale D
Protestant	Protestant	Protestant	Protestant
Episcopalian	Catholic	Non-Protestant	Catholic
Catholic	Jewish		Jewish
Jewish			None
None			Other
Other			

or less gender than the other. In the same way, a zip code of 54398 is different from but not "higher" than a zip code of 13427.

Nominal variables are rudimentary, but there are criteria that we need to observe in order to measure them adequately. These criteria, in fact, apply to variables measured at all levels. They are stated in Table 1.2 and illustrated in Table 1.3. We consider and illustrate the criteria one at a time.

The first criterion ("mutually exclusive") means that there can be no overlap between categories and no confusion or ambiguity about where a case belongs. To illustrate, consider Scale A in Table 1.3, which violates this principle because of the overlap between the categories "Protestant" and "Episcopalian."

The second criterion ("exhaustive") means that there must be a place for every case or score. Scale B in Table 1.3 violates this criterion because there is no category for people who have no religious affiliation and for those who belong to religions other than the three stated. Often, we meet this criterion by including an "Other" category (as in Scale D).

The third criterion ("similar") means that the categories should include cases that are truly comparable. To put it another way, we should avoid lumping apples with oranges. There are no hard-and-fast guidelines for judging whether a category includes similar items. The researcher must make that decision in terms of the specific purpose of the research, and categories that are too broad for some purposes may be perfectly adequate for others. Scale C in Table 1.3 violates this criterion because it uses a category (Non-Protestant, which would include Catholics, Jews, Buddhists, Atheists, and so forth) that seems too broad for meaningful research.

Scale D in Table 1.3 is probably the most common way religious preference has been measured in North America, and this scale would be adequate in many situations. However, it may be too general for some research purposes. For example, an investigation focused on moral questions such as support for

legal abortion would probably have to distinguish among the various Protestant denominations, and an effort to document religious diversity certainly would need to add categories for Buddhists, Muslims, and numerous other religious faiths.

THE ORDINAL LEVEL OF MEASUREMENT

Variables measured at the ordinal level are more sophisticated than nominal-level variables. They have scores or categories that can be ranked from high to low, so, in addition to classifying cases into categories, we can describe the categories in terms of "more or less" with respect to each other. Thus, with ordinal-level variables, not only can we say that one case is different from another; we can also say that one case is higher or lower, more or less than another.

For example, the variable socioeconomic status is usually measured at the ordinal level, often using categories such as those in Table 1.4. Individual cases can be compared in terms of the categories into which they are classified. Thus, an individual classified as a 4 (upper class) would be ranked higher than an

STATISTICS **In Everyday Life**

Changes in Socioeconomic Status in the United States

The poverty rate in the United States has fluctuated between 10% and 15% since the mid-1960s. It normally rises during periods of recession and declines as the economy recovers. It is interesting to note, however, that the percentage of Americans who define themselves as "lower class" is much less than the facts about poverty suggest, although this has been changing over time. According to the General Social Survey, only 4.2% of Americans described themselves as lower class in 1980. This rose to 5% in 1990, 8.6% in 2010, and 9% in 2018. Why does this disparity between self-identification and the realities of poverty persist? What sociological factors might help to explain this phenomenon?

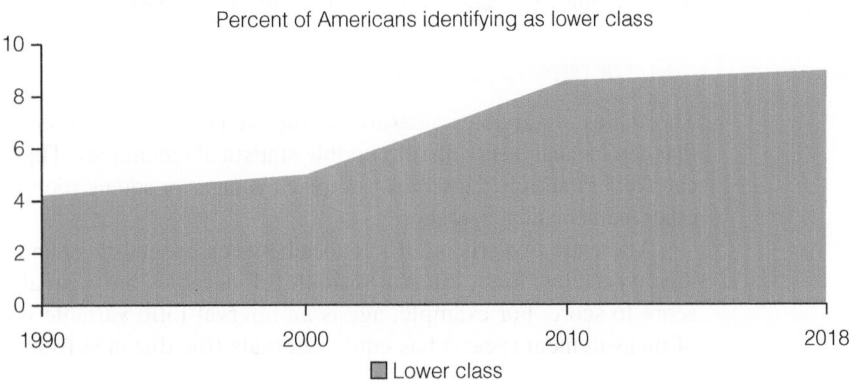

Sources: U.S. Bureau of the Census. Available at https://www.census.gov/library/stories/2018/09/poverty-rate-drops-third-consecutive-year-2017.html; National Opinion Research Center. The General Social Survey. Available at https://gssdataexplorer.norc.org/

TABLE 1.4 Measuring Socioeconomic Status

If you were asked to describe yourself, into which of the following classes would you say you belong?	
Score	Class
1	Lower class
2	Working class
3	Middle class
4	Upper class

individual classified as a 2 (working class), and a lower-class person (1) would rank lower than a middle-class person (3). Other variables that are usually measured at the ordinal level include attitude and opinion scales, such as those that measure prejudice, alienation, or political conservatism.

The major limitation of the ordinal level of measurement is that the scores have no absolute or objective meaning: They only represent position with respect to other scores. We can distinguish between high and low scores, but the distance between the scores cannot be described in precise terms. Thus, in terms of the scale shown in Table 1.4, although we know that a social class score of 4 is more than a score of 2, we do not know whether it is twice as much as 2.

Our options for statistical analysis with ordinal-level variables are limited by the fact that we don't know the exact distances from score to score. For example, addition (and other mathematical operations) assumes that the intervals between scores are exactly equal. If the distances from score to score are not equal, 2 + 2 might equal 3 or 5 or even 15.

Strictly speaking, statistics such as the average, or mean (which requires that the scores be added together and then divided by the number of scores), are not permitted with ordinal-level variables. The most sophisticated mathematical operation fully justified with an ordinal variable is ranking categories and cases (although, as we will see, it is not unusual for social scientists to violate this rule by calculating an average for an ordinal-level variable).

THE INTERVAL-RATIO LEVEL OF MEASUREMENT[2]

The scores of variables measured at the interval-ratio level are actual numbers that can be analyzed with all possible statistical techniques. This means that we can add or multiply the scores, compute averages or square roots, or perform any other mathematical operation.

There are two crucial differences between ordinal-level and interval-ratio–level variables. First, interval-ratio–level variables have equal intervals from score to score. For example, age is an interval-ratio variable because the unit of measurement (years) has equal intervals (the distance from year to year is

[2]Many statisticians distinguish between the interval level (equal intervals) and the ratio level (true zero point). This text will treat these two levels as one.

365 days). Similarly, if we ask people how many siblings they have, we would produce a variable with equal intervals: Two siblings are 1 more than 1, and 13 is 1 more than 12. Likewise, if I have four children and you have two children, I have twice as many children as you do.

Second, interval-ratio variables have true zero points. That is, the score of 0 for these variables is not arbitrary; it indicates the absence or complete lack of whatever is being measured. For example, the variable "number of siblings" has a true zero point because it is possible to have no siblings at all. Similarly, it is possible to have 0 years of education, no income at all, a score of 0 on a

ONE STEP AT A TIME | **Determining the Level of Measurement of a Variable**

Step	*Operation*
1.	Inspect the scores or categories of the variable.
2.	Can the scores be ranked from low to high?

Illustration: Marital status is a variable that may be categorized as follows:

- Married
- Never married
- Divorced
- Widowed

Do people who are widowed have *more* marital status than those who are divorced?
Your answer should be *no*, because categories of marital status cannot be ranked from low to high.

STOP HERE. The level of measurement is nominal.

For variables for which the categories *can* be ranked from low to high, proceed to step 3.

3.	Is the distance between the scores equal or defined?

Illustration: Consider the following scale that measures support for the death penalty.

- Somewhat oppose
- Strongly oppose
- Somewhat support
- Strongly support

People who "strongly" support the death penalty are more in favor than people who "somewhat" support it, but it is unclear from the categories how much additional support to expect. Your answer should be *no*, the distance is not equal or defined.

STOP HERE. The level of measurement is ordinal.

For variables for which the scores *are* equally spaced and defined, proceed to step 4.

4.	You have determined that the scores or categories of your variable can be ranked from low to high and are equally spaced and defined.

STOP HERE. The level of measurement is interval-ratio.

Illustration: Consider the following question: What is your current age?
Ages can be ranked from low to high and are defined and equally spaced. Age 4 comes before age 5. The distance between the ages of 4 and 5 is 12 months, just like the distance between the ages of 68 and 69. A person who is 20 has lived exactly twice as long as a person who is 10.

READING STATISTICS | **An Introduction**

Many students are drawn to the social sciences because they want to make a difference in society. It sometimes comes as a shock to them that they need to learn about research methods and statistics. Some students ask, What's the point? One answer to this question is that the analysis of data has become an essential part of everyday decision making in all sectors of society. To learn more about how statistics are used to make evidence-based decisions, see the boxes in this text on Social Research and Data Analysis.

A second answer to this question is that data are analyzed in all forms of published research in the social sciences. If you are interested in becoming a social science researcher, these boxes on Reading Statistics are for you. As you will see, professional researchers use a reporting style that is quite different from the statistical language you will find in this text. Space in reports and research journals can be expensive, and the typical research project requires the analysis of many variables. Thus, a large volume of information must be summarized in few words. Researchers may express in a word or two a result or an interpretation that will take us a paragraph or more to state.

Because this is an introductory textbook, we have been careful to break down the computation and logic of each calculation and to explain what we are doing when we use statistics. In this text, we will never be concerned with more than a few variables at a time. We will be able to take pages or even entire chapters to develop a statistical idea or concept. Thus, a major purpose of these boxed inserts is to summarize how our comparatively long-winded (but more careful) vocabulary is translated into the concise language of the professional researcher.

When you have difficulty reading words, you consult reference books (especially dictionaries or online search engines) to help you analyze the elements (words) of the passage. When you have difficulty reading statistics, you should do exactly the same thing. We hope you will find this text a valuable reference book, but if you learn enough to be able to use any source to help you read statistics, this text will have fulfilled one of its major goals.

multiple-choice test, and to be 0 years old (although not for very long). Other examples of interval-ratio variables normally include number of children, life expectancy, and years married. All mathematical operations are permitted for variables measured at the interval-ratio level.

SOME FINAL POINTS

Table 1.5 summarizes the discussion of the three levels of measurement. Note that the number of permitted mathematical operations increases as we move from nominal to ordinal to interval-ratio levels of measurement. Ordinal-level variables are more sophisticated and flexible than nominal-level variables, and interval-ratio–level variables permit the broadest range of mathematical operations.

Let us end this section by making four points. The first stresses the importance of level of measurement, and the next three discuss some common points of confusion in applying this concept.

1. Knowing the level of measurement of a variable is crucial because it tells us which statistics are appropriate and useful. Not all statistics can be used with all variables. As shown in Table 1.5, different statistics require different

TABLE 1.5 **Basic Characteristics of the Three Levels of Measurement**

Levels	Examples	Measurement Procedures	Mathematical Operations Permitted
Nominal	Gender, race, religion, marital status	Classification into categories	Counting number in each category, comparing sizes of categories
Ordinal	Social class, attitude and opinion scales	Classification into categories *plus* ranking of categories with respect to each other	All of the above *plus* judgments of "greater than" and "less than"
Interval-Ratio	Age, number of children, income	All of the above *plus* description of scores in terms of equal units	All of the above *plus* all other mathematical operations (addition, subtraction, multiplication, division, square roots, etc.)

mathematical operations. For example, computing an average requires addition and division, and finding a median (or middle score) requires that the scores be ranked from high to low. Addition and division are appropriate only for interval-ratio–level variables, and ranking is possible only for variables that are at least ordinal in level of measurement. Your first step in dealing with a variable and selecting appropriate statistics is *always* to determine its level of measurement.

2. The distinction made earlier between discrete and continuous variables is a concern only for interval-ratio–level variables. Nominal- and ordinal-level variables are almost always discrete. That is, researchers typically measure these variables by asking respondents to select the single category that best describes them, as illustrated in Table 1.6. The scores of these survey items, as they are stated here, are discrete because respondents must select only one category and because these scores cannot be subdivided.

 Interval-ratio variables can be either discrete (number of times you've been divorced, which must be a whole integer) or continuous (age, which could be measured to the nanosecond, or even more precisely). Remember that, since we cannot work with infinitely long numbers, continuous interval-ratio variables have to be rounded off and reported as if they were discrete. The distinction between discrete and continuous variables relates more to our options for appropriate statistics or graphs than to the appearance of the variables.

3. In determining level of measurement, always examine the way in which the scores of the variable are *actually stated*. This is particularly a problem with variables that are interval-ratio but that have been measured at the ordinal level.

 To illustrate, consider income as a variable. If we asked respondents to list their exact income in dollars, we would generate scores that are interval-ratio in level of measurement. Measured in this way, the variable would have a true zero point (no income at all) and equal intervals from score to score (1 dollar). It is more convenient for respondents, however, simply to check the appropriate category from a list of income ranges, as in Table 1.7. The four scores or

TABLE 1.6 **Measuring a Nominal Variable (Marital Status) and an Ordinal Variable (Support for Capital Punishment) as Discrete Variables**

What is your marital status? Are you presently:		Do you support the death penalty for persons convicted of homicide?	
Score	Category	Score	Category
1	Married	1	Strongly support
2	Divorced	2	Somewhat support
3	Separated	3	Neither support nor oppose
4	Widowed	4	Somewhat oppose
5	Single	5	Strongly oppose

TABLE 1.7 **Measuring Income at the Ordinal Level**

Score	Income Range
1	Less than $25,000
2	$25,000 to $49,999
3	$50,000 to $99,999
4	$100,000 or more

categories in Table 1.7 are ordinal in level of measurement because they are unequal in size. It is common for researchers to sacrifice precision (income in actual dollars) for the convenience of the respondents. You should be careful to look at the way in which the variable is measured before making a decision about its level of measurement.

4. There is a mismatch between the variables that are usually of most interest to social scientists (race, gender, marital status, attitudes, and opinions) and the most powerful and interesting statistics (such as the mean). The former are typically nominal or ordinal in level of measurement, but the more sophisticated statistics require measurement at the interval-ratio level.

This mismatch creates some real difficulties for social science researchers. On one hand, researchers will want to measure variables at the highest, most precise level of measurement possible. If income is measured in exact dollars, for example, researchers can make precise descriptive statements about the differences between people: "Ms. Parashar earns $12,547 more than Mr. Wong." If the same variable is measured in broad, unequal categories, such as those in Table 1.7, comparisons between individuals would be less precise and provide less information: "Ms. Parashar earns more than Mr. Wong."

On the other hand, given the nature of the disparity, researchers are more likely to treat variables as if they were higher in level of measurement than they actually are. In particular, variables measured at the ordinal level, especially when they have many possible categories or scores, are often treated as if they were interval-ratio and are analyzed with the more powerful, flexible, and interesting statistics available at the higher level. This practice is common, but researchers should be cautious in assessing statistical results and developing interpretations when the level of measurement criterion has been violated.

In conclusion, level of measurement is a very basic characteristic of a variable, and we will always consider it when presenting statistical procedures. Level of measurement is also a major organizing principle for the material that follows, and you should make sure that you are familiar with these guidelines. *(For practice in determining the level of measurement of a variable, see problems 1.4 through 1.9.)*

SUMMARY

1. The purpose of statistics is to organize, manipulate, and analyze data so that researchers can test theories and answer questions. Along with theory and methodology, statistics are a basic tool used by social scientists to enhance their understanding of the social world.

2. There are two general classes of statistics. Descriptive statistics are used to summarize the distribution of a single variable and the relationships between two or more variables. We use inferential statistics to generalize to populations from random samples.

3. Variables may be either discrete or continuous. They may be measured at any of three different levels. At the nominal level, we can compare category sizes. At the ordinal level, scores can be ranked from high to low. At the interval-ratio level, all mathematical operations are permitted. Interval-ratio–level variables can be either discrete or continuous. Variables at the nominal or ordinal level are almost always discrete.

GLOSSARY

Continuous variable. A variable with a unit of measurement that can be subdivided infinitely.

Data. Information expressed as numbers.

Data reduction. Summarizing many scores with a few statistics.

Dependent variable. A variable that is identified as an effect or outcome. The dependent variable is thought to be caused by the independent variable.

Descriptive statistics. The branch of statistics concerned with (1) summarizing the distribution of a single variable or (2) measuring the relationship between two or more variables.

Discrete variable. A variable with a basic unit of measurement that cannot be subdivided.

Hypothesis. A specific statement, derived from a theory, about the relationship between variables.

Independent variable. A variable that is identified as a cause. The independent variable is thought to cause the dependent variable.

Inferential statistics. The branch of statistics concerned with making generalizations from samples to populations.

Level of measurement. The mathematical characteristic of a variable and the major criterion for selecting statistical techniques. Variables can be measured at any of three levels, each permitting certain mathematical operations and statistical techniques. The characteristics of the three levels are summarized in Table 1.5.

Measures of association. Statistics that summarize the strength and direction of the relationship between variables.

Population. The total collection of all cases in which the researcher is interested.

Quantitative research. Research projects that collect data or information in the form of numbers.

Research. Any process of gathering information systematically and carefully to answer questions or test theories. Statistics are useful for research projects that collect numerical information or data.

Sample. A carefully chosen subset of a population. In inferential statistics, information is gathered from a sample and then generalized to a population.

Statistics. A set of mathematical techniques for organizing and analyzing data.

Theory. A generalized explanation of the relationship between two or more variables.

Variable. Any trait that can change values from case to case.

PROBLEMS

1.1 In your own words, describe how statistics might be used to measure changes in divorce rates over time. Using the "wheel of science" as a framework, explain how statistics can be used to link theories about family change with research.

1.2 Find a research article in any social science journal. Choose an article on a subject of interest to you, and don't worry about being able to understand all of the statistics that are reported.

 a. How much of the article is devoted to statistics?

 b. Is the research based on a sample from some population? How large is the sample? How were subjects or cases selected? Can the findings be generalized to some population?

 c. What variables are used? Which are independent and which are dependent? For each variable, determine the level of measurement and whether the variable is discrete or continuous.

 d. What statistical techniques are used? Try to follow the statistical analysis and see how much you can understand. Save the article and read it again after you finish this course and see if you do any better.

1.3 Distinguish between descriptive and inferential statistics. Describe a research situation that would use each type.

1.4 For each of the following items from a public-opinion survey, indicate the level of measurement and whether the variable will be discrete or continuous. *(HINT: Remember that only interval-ratio–level variables can be continuous.)*

 a. In what country were you born? _____

 b. What is your age? _____

 c. How many years of school have you completed? _____

 d. What is your occupation? _____

 e. If you were asked to use one of these four names for your social class, in which would you say you belonged?

 _____ Upper _____ Middle

 _____ Working _____ Lower

 f. What is your grade-point average? _____

 g. What is your major? _____

 h. The only way to deal with the drug problem is to legalize all drugs.

 _____ Strongly disagree

 _____ Disagree

 _____ Undecided

 _____ Agree

 _____ Strongly agree

 i. What is your astrological sign? _____

 j. How many brothers and sisters do you have? _____

1.5 Following are brief descriptions of how researchers measured a variable. For each situation, determine the level of measurement of the variable and whether it is continuous or discrete.

 a. Race or ethnicity. Check all that apply:

 _____ Black **or African American**

 _____ White

 _____ Hispanic

 _____ Asian

 _____ Native Hawaiian or Other Pacific Islander

 _____ American Indian or Alaskan Native

 _____ Other (Please specify: _____)

 b. Honesty. Subjects were observed as they passed by a spot where an apparently lost wallet was lying. The wallet contained money and complete identification. Subjects were classified into one of the following categories:

 _____ Returned the wallet with money

 _____ Returned the wallet but kept the money

 _____ Did not return the wallet

 c. Social class. What was your family situation when you were 16 years old?

 _____ Very well off compared to other families

 _____ About average

 _____ Not so well off

 d. Education. How many years of schooling have you completed? _____

 e. Racial integration on campus. Students were observed during lunchtime at the cafeteria for a month. The number of students sitting with students of other races was counted for each meal period.

 f. Number of children. How many children have you ever had? _____

 g. Student seating patterns in classrooms. On the first day of class, instructors noted where each student sat. Seating patterns were remeasured every two weeks until the end of the semester. Each student was classified as:

 _____ Same seat as at last measurement

 _____ Adjacent seat

 _____ Different seat, not adjacent

 _____ Absent

h. Physicians per capita. The number of physicians was counted in each of 50 cities. The researchers used population data to compute the number of physicians per capita.

i. Physical attractiveness. A panel of 10 judges rated each of 50 photos of a mixed-race sample of males and females for physical attractiveness on a scale from 0 to 20, with 20 being the highest score.

j. Number of accidents. The number of traffic accidents per year for each of 20 intersections was recorded. Also, each accident was rated as:

_____ Minor damage, no injuries

_____ Moderate damage, personal injury requiring hospitalization

_____ Severe damage and injury

1.6 For each research project listed here, identify the variables and classify them in terms of level of measurement, and whether they are discrete or continuous. Is each variable independent or dependent?

a. For a research project in a political science course, a student collected information for 50 nations. She used infant mortality rates as a measure of quality of life and the percentage of all adults who are permitted to vote as a measure of democratization. Her hypothesis was that quality of life is higher in more democratic nations.

b. A highway engineer wonders whether increasing the speed limit on a heavily traveled highway will result in more accidents. He plans to collect information on traffic volume, number of accidents, and number of fatalities for the six-month periods before and after the speed limit is changed.

c. Students are planning a program to promote "safe sex" and awareness of other health concerns on campus. To measure the effectiveness of the program, they plan to survey students about their knowledge of safe sex practices before and after the program.

d. A graduate student asks 500 female students whether they have experienced any sexual harassment on campus. Each student is asked to estimate the frequency of these incidents as either "often, sometimes, rarely, or never." The researcher also gathers data on age and major to see whether there is any connection between these variables and frequency of sexual harassment.

e. A supervisor in the solid waste management division of city government is assessing two different methods of trash collection. One area of the city is served by trucks with two-man crews who do "backyard" pickups, and the rest of the city is served by "high-tech" single-person trucks with curbside pickup. The assessment measures include the number of complaints received from the two different areas over a six-month period, the amount of time per day required to service each area, and the cost per ton of trash collected.

f. Does tolerance for diversity vary by race or ethnicity? Samples of white, black, Asian, Hispanic, and Native Americans have been given a survey that measures their interest in and appreciation of cultures other than their own.

g. States have drastically cut their budgets for mental health care. Will this increase the number of homeless people? A researcher contacts a number of agencies serving the homeless in each state and develops an estimate of the size of the homeless population before and after the cuts.

h. The adult bookstore near campus has been raided and closed by the police. Your social research class has decided to poll a sample of students to find out: whether he or she supports the store's closing, how many times each has visited the store, and whether he or she agrees that "pornography causes sexual assaults on women." The class also collects information on the gender, political philosophy, and major of the students.

1.7 For each of the following research situations, identify the level of measurement of all variables and indicate whether they are discrete or continuous. Also, decide which statistical applications are used: descriptive statistics (single variable), descriptive statistics (two or more variables), or inferential statistics. Remember that it is quite common for a given situation to require more than one type of application.

a. The administration of your university is proposing a change in parking policy. You select a random sample of students and ask each one how strongly they favor or oppose the change. You then use the results to estimate the support for the change in the entire student body.

b. You ask everyone in your social research class for their highest letter grade in any math course and their grade (percentage of items correct) on a recent statistics test. You compare the two sets of scores to see whether there is a relationship.

c. Your aunt is running for mayor and hires you to question a random sample of voters about their concerns. Specifically, she wants to use this information to characterize the entire city in terms of political party affiliation, gender, and what percentage favor widening of the main street in town.

d. Several years ago, a state reinstituted the death penalty for first-degree homicide. Did this reduce the homicide rate? A researcher has gathered information on the number of homicides in the state for the two-year periods before and after the change.

e. A local automobile dealer is concerned about customer satisfaction. He mails a survey form to all customers for the past year and asks whether they are satisfied, very satisfied, or not satisfied with their purchases.

1.8 For each of the first 20 items in the General Social Survey (see Appendix G), indicate the level of measurement and whether the variable is continuous or discrete.

1.9 Identify all variables in the research situations listed below and classify them by their level of measurement. Which variables are independent and which are dependent?

a. A researcher is wondering about racial preferences in dating among college students and asks a large sample of undergraduates about their own racial self-identification. She then asks the respondents to rank some racial-ethnic categories (stated as: white, black, Latino, Asian) in terms of desirability as potential dates.

b. For high school students, does GPA affect sexual activity? A sample has been interviewed about the number of different romantic relationships they have had, the number of times they have had sexual intercourse, and their high school grade-point average.

c. Several hundred voting precincts across the nation have been classified in terms of percentage of minority voters, voting turnout, and percentage of local elected officials who are members of minority groups. Do precincts with higher percentages of minority voters have lower turnout? Do precincts with higher percentages of minority elected officials have higher turnout?

d. As nations become more affluent (as measured by per capita income), does the percentage of children enrolled in school increase? Is this relationship different for boys and girls?

e. Does the level of support for gun control vary by number of years of schooling? Does this relationship vary by gender, region of the country, or political party preference? Support for gun control was measured by a five-point scale that ranged from "strongly in favor" to "strongly opposed."

YOU ARE THE RESEARCHER

Introduction

The best way—maybe the only way—to learn statistics and to appreciate their importance is to apply and use them. This includes selecting the correct statistic for a given purpose, doing the calculations, and interpreting the result. We have included extensive end-of-chapter problems to give you multiple opportunities to select and calculate statistics and say what they mean. Many of these problems have been written so that they can be solved with just a simple hand calculator. We've purposely kept the number of cases unrealistically low so that the tedium of lengthy calculations would not interfere with the learning process. These problems present an important and useful opportunity to develop your statistical skills.

As important as they are, these end-of-chapter problems are simplified and several steps removed from the complex realities of social science research and data analysis. To provide a more realistic statistical experience, we have included a feature called "You Are the Researcher," in which you will walk through many of the steps of a research project, making decisions about how to apply your growing knowledge of research and statistics and interpreting the statistical output you generate.

To conduct research and data analysis, you need two things. First, you need something to analyze, a database or a collection of information. These days, databases are often very large and include hundreds or thousands or more of bits of information. They are far too large to analyze with even the most sophisticated hand

calculator, and this brings us to the second thing you need: a computerized statistical analysis program. We address these two needs separately.

In the "You Are the Researcher" exercises, you will usually analyze the 2018 General Social Survey (GSS), although several other databases are supplied with this text. Please visit www.cengagebrain.com to view and download the databases. The GSS is a public-opinion poll that has been conducted on nationally representative samples of citizens of the United States since 1972. The 2018 GSS includes more than 1000 items of information (e.g., gender, race, level of education) and questions covering a broad range of social and political issues (e.g., support for legal abortion, political beliefs and values, and levels of trust in other people). The GSS and the other databases are actual, "real-life" data, so you have the opportunity to practice your statistical skills in a realistic context.

To analyze the GSS—or any large database—you will use a computerized statistical software package. There are several such programs commonly available, such as SAS, STATA, and R. The one featured in this text is called the Statistical Package for the Social Sciences (SPSS). A statistical package is a set of computer programs designed to analyze data. The advantage of these packages is that, because the programs are already written, you can capitalize on the power of the computer with minimal computer literacy and virtually no programming experience. Be sure to read Appendix F before attempting any data analysis.

Some of you may not have access to SPSS on your campus but you can still conduct these research projects by using the student version of SPSS.[3] This student version is limited to smaller available online with fewer than 1500 respondents and 50 variables.

In most of the research exercises, which begin in Chapter 2, you will make the same kinds of decisions as do professional researchers and move through some of the steps of a research project. You will select variables and statistics, generate and analyze output, and express your conclusions. When you finish these exercises, you will be well prepared to conduct your own research project (within limits, of course) and perhaps make a contribution to the ever-growing social science research literature.

[3]The student version of SPSS is available online.

Part I Descriptive Statistics

Part I consists of four chapters, each devoted to a different application of univariate descriptive statistics. Chapter 2 covers "basic" descriptive statistics, including frequency distributions, percentages, ratios, rates, and graphs. This material is relatively elementary and at least vaguely familiar to most people. Although these statistics are "basic," they are not necessarily simple or obvious, and the explanations and examples should be considered carefully before attempting the end-of-chapter problems or using them in actual research.

Chapters 3 and 4 cover measures of central tendency and variation, respectively. Measures of central tendency describe the typical case or average score (e.g., the mean), while measures of variation describe the amount of variety or diversity among the scores (e.g., the range, or the distance from the high score to the low score). These two types of statistics are presented in separate chapters to stress the point that centrality and variation are independent, separate characteristics of a variable. You should realize, however, that both measures are necessary and commonly reported together, along with some of the statistics presented in Chapter 2. To reinforce this idea, many of the problems at the end of Chapter 4 require the computation of a measure of central tendency from Chapter 3.

Chapter 5 is a pivotal chapter in the flow of the text. It takes some of the statistics from Chapters 2 through 4 and applies them to the normal curve, a concept of great importance in statistics. The normal curve is a line chart (see Chapter 2) that can be used to describe the position of scores using means (Chapter 3) and standard deviations (Chapter 4). Chapter 5 also uses proportions and percentages (Chapter 2).

In addition to its role in descriptive statistics, the normal curve is a central concept in inferential statistics, the topic of Part II of this text. Thus, Chapter 5 ends the presentation of univariate descriptive statistics and lays essential groundwork for the material to come.

2 Basic Descriptive Statistics: Tables, Percentages, Ratios and Rates, and Graphs

LEARNING OBJECTIVES By the end of this chapter, you will be able to:

1. Explain how descriptive statistics can be used to make your data understandable to others.
2. Construct frequency distributions for variables at each of the three levels of measurement.
3. Compute percentages, proportions, ratios, rates, and percentage change for numerical data.
4. Analyze pie and bar charts, histograms, and line graphs.
5. Create frequency distributions in SPSS and analyze the output.

USING STATISTICS

The statistical techniques presented in this chapter are used to summarize the scores on a single variable. They can be used to

- Organize information into easy-to-read tables, charts, and graphs.
- Express the percentage of people in the United States that belong to various religions, including people with no religious preference and atheists. **See page 29**
- Display the percentages of Americans that are in poverty by state using a map chart. **See page 30**
- Describe the changes in alcohol and marijuana use among college and noncollege students over time. **See page 58**
- Make estimates about the nature of population change for nations around the world. **See page 46**

Research results do not speak for themselves. Researchers use statistics to organize and manipulate data so that their meaning can be understood by their readers. The purpose of descriptive statistics is to clarify and express research findings effectively.

In this chapter, we consider some commonly used techniques for presenting research results, including tables, percentages, rates, and graphs. These univariate descriptive statistics are not mathematically complex (although they are not as

simple as they might seem at first glance), but they can be extremely useful tools for organizing and analyzing results and communicating conclusions.

Often, the first step in a quantitative research project is to examine the variables and see how scores are distributed. One of the most useful ways to do this is to construct tables, or **frequency distributions**, that report the number of cases in each category for all variables. These tables can be used with variables at any level of measurement. We will begin with nominal-level variables and consider not only frequency distributions but also some statistics that can increase the clarity of results.

Frequency Distributions for Nominal-Level Variables

Constructing frequency distributions for nominal-level variables is typically straightforward. Count the number of times each category or score of the variable occurs and display the frequencies in table format. Table 2.1, for example, displays the distribution of gender for 113 respondents.

Note that the table has a title and clearly labeled categories, and it reports the total number of cases (*N*) at the bottom of the frequency column. These items must be included in *all* frequency distributions.

For some tables, the researcher might have to make some choices about the number of categories to be included. For example, recall Table 1.3, which displayed several different ways of measuring religious affiliation in North America. As we noted in Chapter 1, it is probably most common to use five categories (see Scale D of Table 1.3) to display this variable, but what if a researcher wanted to emphasize the diversity of religions? We could do this by expanding the "Other" category and including more religions. But where do we stop? There are hundreds—maybe thousands—of different religions in the United States. Increasing the number of categories will present a more accurate picture of religious diversity, but clarity and ease of communication may suffer.

Tables 2.2 and 2.3 illustrate the choices that sometimes have to be made. Table 2.2 uses the "standard" five categories to display religious membership in the United States as of 2018. Table 2.3 presents more detail by adding the two largest "Other" religions. Note that the number of categories could be further increased by including more "Other" religions or by subdividing the Protestant category into the various denominations (Methodist, Lutheran, Episcopalian, and so forth).

At what point do we have "enough" detail? When does the picture presented by the table become too cluttered, too complex, and unclear? These are questions that must be answered in the context of the purpose of the research project. If you wanted

TABLE 2.1 Gender (fictitious data)

Gender	Frequency
Males	53
Females	60
	N = 113

TABLE 2.2 Self-Described Religious Affiliation of Adult Americans, 2018

Religious Group	Frequency
Protestant	124,148,263
Catholic	53,822,969
Jewish	4,315,993
None	59,154,489
Other	12,440,215
	N = 253,881,929

Source: General Social Survey, 2018. Frequencies are approximations computed by multiplying the percentage in each category by the total adult population in 2018.

TABLE 2.3 Self-Described Religious Affiliation of Adult Americans, 2018

Religious Group	Frequency
Protestant	124,148,263
Catholic	53,822,969
Jewish	4,315,993
Muslim	1,777,173
Buddhist	2,031,055
None	59,154,489
Other	8,631,985
	N = 253,881,927

Source: General Social Survey, 2018. Frequencies are approximations computed by multiplying the percentage in each category by the total adult population in 2018.

to stress the numerical predominance of Protestants in the United States, Table 2.2 might be preferred. On the other hand, if you wanted to stress the diversity of religious affiliations, Table 2.3 would be preferable. There are no hard-and-fast rules, and the choice between greater detail (more categories) and more clarity (fewer categories) may be confronted with variables at all three levels of measurement.

Enhancing Clarity: Percentages and Proportions

It is likely that you have noted a difficulty with Tables 2.2 and 2.3: The frequencies involved are large and, therefore, may be difficult to comprehend. For example, imagine yourself trying to explain these tables to someone else in terms of the frequencies: "124,148,263 of 253,881,927 adult Americans described themselves as Protestant." There is nothing wrong with this statement—it is literally true—but the same point could have been more clearly conveyed using a **percentage**: "About 49% of adult Americans describe themselves as Protestant."

Percentages are extremely useful statistics because they supply a frame of reference by standardizing the raw frequencies to the base 100. The mathematical definition of a percentage is

FORMULA 2.1
$$\text{Percentage: } \% = \left(\frac{f}{N}\right) \times 100$$

where f = frequency, or the number of cases in any category
N = the number of cases in all categories

To illustrate the computation of percentages, consider the data presented in Table 2.1. How can we find the percentage of males in the sample? Note that there are 53 males ($f = 53$) and a total of 113 cases ($N = 113$). So

$$\% = \left(\frac{f}{N}\right) \times 100 = \left(\frac{53}{113}\right) \times 100 = (0.4690) \times 100 = 46.90\%$$

Using the same procedure, we can find the percentage of females:

$$\% = \left(\frac{f}{N}\right) \times 100 = \left(\frac{60}{113}\right) \times 100 = (0.5310) \times 100 = 53.10\%$$

TABLE 2.4 **Self-Described Religious Affiliation of Adult Americans, 2018**

Religious Group	Frequency	Percentages
Protestant	124,148,263	48.9
Catholic	53,822,969	21.2
Jewish	4,315,993	1.7
Muslim	1,777,173	0.7
Buddhist	2,031,055	0.8
None	59,154,489	23.3
Other	8,631,985	3.4
	N = 253,881,927	100.0%

Percentages are easier to read and comprehend than raw frequencies, and a column for percentages is commonly added to frequency distributions for variables at all levels of measurement. For example, Table 2.4 reports the same information as Table 2.3, but the percentages make it much easier to read.

The advantage of using percentages is particularly obvious when we want to compare groups of different sizes. For example, consider the information presented in Table 2.5. Which college has the higher *relative* number of social science majors? Because the total enrollments are so different, comparisons are difficult to make from the raw frequencies. Percentages eliminate the difference in size of the two campuses by standardizing both distributions to the base of 100. The same data are presented in Table 2.6 with percentages. See the box on "Statistics in Everyday Life" for real data on college majors in the United States.

Table 2.6 makes it easier to identify differences as well as similarities. College A has a higher percentage of social science majors than College B (even

TABLE 2.5 **Declared Major Fields on Two College Campuses (fictitious data)**

Major	College A	College B
Business	103	3120
Natural sciences	82	2799
Social sciences	137	1884
Humanities	93	2176
	N = 415	N = 9979

TABLE 2.6 **Declared Major Fields on Two College Campuses (fictitious data)**

Major	College A	College B
Business	24.82%	31.27%
Natural sciences	19.76%	28.05%
Social sciences	33.01%	18.88%
Humanities	22.41%	21.81%
	100.00%	100.00%
	(415)	(9979)

FIGURE 2.1 Percentages of People Living in Poverty by State in 2017

Percentage of people in poverty

- 18.0 or more
- 16.0 to 17.9
- 13.0 to 15.9
- 11.0 to 12.9
- Less than 11.0

U.S. percentage is 13.4

US Census Bureau

Accessed at https://www.census.gov/library/visualizations/2018/comm/acs-poverty-map.html

though the absolute number of social science majors is less than at College B) and about the same percentage of humanities majors. How would you describe the differences in the remaining two major fields?

Percentages can also be arrayed over a map to allow for differences to be observed by geographic region. In Figure 2.1, it is easy to see that the states with the highest percentages of people in poverty are New Mexico, Louisiana, Mississippi, and West Virginia.

Social scientists use **proportions** as well as percentages. Proportions vary from 0.00 to 1.00: They standardize results to a base of 1.00 instead of to the base of 100 used for percentages. A proportion is the same as a percentage except that we do *not* multiply by 100.

FORMULA 2.2

$$\text{Proportion} = \left(\frac{f}{N}\right)$$

Percentages can be converted to proportions by dividing by 100 and, conversely, proportions can be converted to percentages by multiplying by 100. The two statistics are equivalent expressions of the same message and are interchangeable. For example, we could state the relative number of males in Table 2.1 as a proportion:

$$\text{Proportion} = \left(\frac{f}{N}\right) = \left(\frac{53}{113}\right) = 0.47$$

ONE STEP AT A TIME **Finding Percentages and Proportions**

1. Determine the values for f (number of cases in a category) and N (number of cases in all categories). Remember that f will be the number of cases in a *specific category* (e.g., males on your campus) and N will be the number of cases in *all* categories (e.g., students of all genders on your campus). Note that f will be smaller than N, except when the category and the entire group are the same (e.g., when all students are male). Proportions cannot exceed 1.00, and percentages cannot exceed 100%.
2. For a proportion, divide f by N.
3. For a percentage, multiply the value you calculated in step 2 by 100.

How can we choose between these statistics? Percentages are easier for most people (including statisticians) to comprehend and generally would be preferred. Proportions are used less frequently, generally when we are working with probabilities (see Chapter 5). The preference for percentages is based solely on ease of communication: Proportions and percentages are equally valid ways of expressing results. Be sure to familiarize yourself with the guidelines stated in Social Research and Data Analysis 2.1. *(For practice in computing and interpreting percentages and proportions, see problems 2.1 and 2.2.)*

Social Research and Data Analysis 2.1 **Using Proportions and Percentages in Practice**

Proportions and percentages are simple statistics, but they can be deceiving in certain situations. Follow these guidelines to make sure you are being upfront about your data and its limitations.

1. When the number of cases (N) is small (say, fewer than 20), it is usually preferable to report the actual frequencies rather than percentages or proportions. With a small number of cases, the percentages can change drastically with relatively minor changes in the number of cases. For example, if you begin with 10 employees that have a bachelor's degree and 10 that have a master's degree, the distribution is 50% of each. If you then add a new employee with a master's degree, the percentage distributions will change to 52.4% with a master's degree and 47.6% with a bachelor's degree. Of course, as the number of observations increases, each additional case will have a smaller impact. If we started with 500 employees with a bachelor's degree and 500 with a master's degree, and then added one more with a master's degree, the percentage with a master's degree would change by less than a tenth of a percent (from 50% to 50.05%).

2. *Always* report the number of observations along with proportions and percentages. This permits the reader to judge the adequacy of the sample size and, conversely, helps to prevent the researcher from lying with statistics. Statements like "Two out of three people experienced a decrease in their cold symptoms after taking just one pill" might sound impressive, but the claim would lose its gloss if you learned that only three people were tested. *You should be extremely suspicious of reports that fail to mention the number of cases that were tested.*

3. Percentages and proportions can be calculated for variables at the ordinal and nominal levels of measurement, in spite of the fact that they require division. This is not a violation of the level-of-measurement guideline (see Table 1.5). Percentages and proportions do not require the division of the *scores* of the variable (as would be the case in computing the average score on a test, for example). Rather, the *number of cases* in a particular category (f) is divided by the total number of cases in the sample (N). As such, it makes sense for Apple to report that the iPhone made up about 48% of its sales in 2019. When we make a statement like "The iPhone makes up 48% of Apple's sales," we are expressing the relative size of a category (iPhone) of the variable (total sales) in a convenient way.

Social Research and Data Analysis 2.2 | **Communicating with Statistics**

Social service agencies that administer Medicaid and Supplemental Nutrition Assistance Program (or SNAP) benefits (previously known as Food Stamps) are often overburdened with applications. Social policy research organizations like MDRC conduct research on the size of their case load and make it known to the public. Imagine if MDRC published information on their website stating that of the 177 full-time employees in a county-level social service agency, only 50 are taking SNAP applications. Yet 6231 of the 16,722 applications being handled by their agency were for SNAP benefits.

What would their readers think? Could this information be conveyed more effectively?

Since the goal is to compare two sets of numbers (the number of staff taking SNAP applications vs. the total number of applications being taken by the agency), proportions or percentages would be a more forceful way of presenting results. What if they had said, "Only 28.25% of the employees are accepting SNAP applications, but SNAP applications comprise 37.26% of the total workload of the agency"? Is this a clearer message?

The first percentage is found by

$$\% = \left(\frac{f}{n}\right) \times 100 = \left(\frac{50}{177}\right) \times 100$$

$$= (0.2825) \times 100 = 28.25\%$$

and the second percentage is found by

$$\% = \left(\frac{f}{N}\right) \times 100 = \left(\frac{6231}{16,722}\right) \times 100$$

$$= (0.3726) \times 100 = 37.26\%$$

STATISTICS | **In Everyday Life**

Tables 2.5 and 2.6 present fictional data on major fields of study of college students. What does this variable look like nationally? The table below presents the top five most common majors. Together, these five majors account for over half of all degrees.

Bachelor's Degrees Awarded in the United States by Field of Study, 2015–16

Major	Percentage of All Degrees
Business	19%
Health Professions	12%
Social Science and History	8%
Psychology	6%
Biology and Biomedical Sciences	6%

The choice of major has changed quite a bit over time! In 1970–71, the top five fields of study were Education, Social Science and History, Business, English, and Engineering.

Source: https://nces.ed.gov/programs/digest/d17/tables/dt17_322.10.asp?current=yes

Frequency Distributions for Ordinal-Level Variables

Frequency distributions for ordinal-level variables are constructed in the same way as for nominal-level variables. Table 2.7 reports the frequency distribution of responses of students to a survey item measuring support for birth control on the campus. Note that a percentage column has been added to the table to increase clarity.

TABLE 2.7 Support for Birth Control on a University Campus (fictitious data)

Do you strongly agree, agree, disagree, or strongly disagree that the University Health Center should provide condoms and other "safe sex" items on demand and at no additional cost to students?		
Response	Frequency	Percentage
Strongly Agree	350	25.55%
Agree	462	33.72%
Disagree	348	25.40%
Strongly Disagree	210	15.33%
	1370	100.00%

TABLE 2.8 Support for Birth Control on a University Campus (fictitious data)

Do you strongly agree, agree, disagree, or strongly disagree that the University Health Center should be able to provide condoms and other "safe sex" items on demand and at no additional cost to students?		
Response	Frequency	Percentage
Strongly Agree or Agree	812	59.27%
Disagree or Strongly Disagree	558	40.73%
	1370	100.00%

The table reports that opinions were fairly evenly distributed. The single most popular response was "Agree" (33.72%), and a majority of students (59.27%) agreed or strongly agreed that condoms and other "safe sex" devices should be available. If the researcher wanted to emphasize this pattern or make the table more compact, categories could be collapsed as in Table 2.8. However, there is a price to pay for the simpler, more compact expression of results: The exact breakdown of degrees of agreement and disagreement is lost. As we saw when discussing the display of religious affiliations in Tables 2.2 and 2.3, the researcher must strike a balance between more detail (more categories) and greater clarity (fewer categories). *(For practice in constructing and interpreting frequency distributions for nominal- and ordinal-level variables, see problems 2.5 and 2.12.)*

Frequency Distributions for Interval-Ratio–Level Variables

In general, the construction of frequency distributions for variables measured at the interval-ratio level is more complex than for nominal and ordinal variables. Interval-ratio variables usually have a wide range of scores, and this means that the researcher must collapse or group categories to produce reasonably compact tables. Once again, we see that the researcher needs to decide between more detail and more clarity.

For example, suppose you wished to report the distribution of the variable "age" for a community. In most communities, there would be a broad range of ages, from newborns to people in their 90s or older. If you simply reported the number of times that each year of age (or score) occurred, you could wind up with a frequency distribution that had 80, 90, or even more categories; such a table would be very difficult to read. The scores (years) must be grouped into larger

TABLE 2.9 **Age of Students in a College Class (fictitious data)**

Interval width = 1 year of age	
Ages	*Frequency*
18	5
19	6
20	3
21	2
22	1
23	1
24	1
25	0
26	1
	$N = 20$

categories to increase ease of comprehension. How large should these categories be? How many categories should be included in the table? Should we provide more information (a greater number of narrow categories) or clarity (a smaller number of wide categories)?

CONSTRUCTING FREQUENCY DISTRIBUTIONS FOR INTERVAL-RATIO–LEVEL VARIABLES

For ease of illustration, let's consider a frequency distribution for a small college class of 20 students. Because of the narrower age range of college students, we can use categories of only one year (these categories are often called **class intervals** when working with interval-ratio data). The frequency distribution is constructed by listing the ages in order, counting the number of times each score (year of age) occurs, and then totaling the number of scores for each category. Table 2.9 presents the information and reveals a concentration of scores in the 18 and 19 class intervals.

Even though this table is fairly clear, assume that you desire a more compact (less detailed) summary. To obtain this, you will have to group scores into wider class intervals. Increasing the interval width (say, to two years) will reduce the number of intervals and produce a more compact expression. The grouping of scores in Table 2.10 clearly emphasizes the predominance of younger respondents. This trend can be stressed even more by adding a column for percentages.

TABLE 2.10 **Age of Students in a College Class (fictitious data)**

Interval width = 2 years of age		
Age	*Frequency*	*Percentage*
18–19	11	55.0%
20–21	5	25.0%
22–23	2	10.0%
24–25	1	5.0%
26–27	1	5.0%
	$N = 20$	100.0%

STATED LIMITS

Note that the class intervals in Table 2.10 have been stated with an apparent gap between them. That is, the **stated class limits** are separated by a distance of one unit. At first glance, these gaps may appear to violate the principle of exhaustiveness introduced in Chapter 1 (see Table 1.2) but, because age has been measured in whole numbers, the "gaps" are actually not a problem. Given the level of precision of the measurement (in whole years as opposed to, say, tenths of a year), no case could have a score that falls between the intervals. The class intervals in Table 2.10 are exhaustive and mutually exclusive, and each of the 20 respondents can be sorted into one and only one age category.

However, consider the difficulties if age had been measured with greater precision. If age had been measured in tenths of a year, into which class interval in Table 2.10 would a 19.5-year-old subject be placed? We avoid this ambiguity by always stating the limits of the class intervals at the same level of precision as the data. Thus, if age were being measured in tenths of a year, the limits of the class intervals in Table 2.10 would be stated in tenths of a year. For example:

17.0–18.9

19.0–20.9

21.0–22.9

23.0–24.9

25.0–26.9

To maintain mutual exclusivity between categories, do not overlap the class intervals. If you state the limits of the class intervals at the same level of precision as the data (which might be in whole numbers, tenths, hundredths, etc.) and maintain a "gap" between intervals, you will always produce a frequency distribution where each case can be assigned to one and only one category.

MIDPOINTS

On occasion—when constructing certain graphs, for example—you will need to work with the **midpoints** of the class intervals. Midpoints are exactly half-way between the upper and lower limits of a class interval and can be found by dividing the sum of the upper and lower limits by 2. To illustrate, Table 2.11

TABLE 2.11 Finding Midpoints

Class interval width = 3	
Class Interval	*Midpoint*
0–2	1.0
3–5	4.0
6–8	7.0
9–11	10.0

Class interval width = 6	
100–105	102.5
106–111	108.5
112–117	114.5
118–123	120.5

ONE STEP AT A TIME | Finding Midpoints

1. Find the upper and lower limits of the lowest interval in the frequency distribution. For the top set of intervals in Table 2.11, the lowest interval (0–2) includes scores of 0, 1, and 2. The upper limit of this interval is 2 and the lower limit is 0.
2. Add the upper and lower limits and divide by 2. For the interval 0–2: (0 + 2)/2 = 1. The midpoint for this interval is 1.
3. Midpoints for other intervals can be found by repeating steps 1 and 2 for each interval. As an alternative, you can find the midpoint for any interval by adding the value of the interval width to the midpoint of the next lower interval. For example, the lowest interval in the top panel of Table 2.11 is 0–2 and the midpoint is 1. Intervals are 3 units wide (that is, they each include three scores), so the midpoint for the next higher interval (3–5) is 1 + 3, or 4. The midpoint for the interval 6–8 is 4 + 3, or 7, and so forth.

displays midpoints for two different sets of class intervals. See the "One Step at a Time" box for detailed instructions. *(For practice in finding midpoints, see problems 2.8b and 2.9b.)*

REAL LIMITS

To construct certain graphs, we must eliminate the "gap" between class intervals and treat a distribution as a continuous series of categories that border each other. We can use Table 2.10 to illustrate how to do this. The limits shown in the table, stated with the apparent "gap," organize the scores into a series of discrete, non-overlapping intervals. To treat the variable as continuous and find the **real class limits**, follow the instructions in the "One Step at a Time" box.

The stated and real limits for Table 2.10 are:

Stated Limits	Real Limits
18–19	17.5–19.5
20–21	19.5–21.5
22–23	21.5–23.5
24–25	23.5–25.5
26–27	25.5–27.5

Table 2.12 presents additional illustrations of real limits for two different sets of class intervals. In both cases, the "gap" between the stated limits is 1. *(For practice in finding real limits, see problems 2.7b and 2.8d.)*

ONE STEP AT A TIME | Finding Real Limits

1. Find the distance (the "gap") between the stated class intervals. In Table 2.10, for example, this value is 1.
2. Divide the value found in step 1 in half.
3. Add the value found in step 2 to all upper stated limits and subtract it from all lower stated limits.

TABLE 2.12 Real Class Limits

Class Intervals (stated limits)	Real Class Limits
3–5	2.5–5.5
6–8	5.5–8.5
9–11	8.5–11.5

Class Intervals (stated limits)	Real Class Limits
100–105	99.5–105.5
106–111	105.5–111.5
112–117	111.5–117.5
118–123	117.5–123.5

CUMULATIVE FREQUENCY AND CUMULATIVE PERCENTAGE

Two commonly used adjuncts to the basic frequency distribution for interval-ratio data are the **cumulative frequency** and **cumulative percentage** columns. These columns allow us to tell at a glance how many cases fall below or above a given score or class interval in the distribution. To add these columns, follow the instructions in the "One Step at a Time" box. Table 2.13 shows a cumulative frequency column and a cumulative percentage column added to Table 2.10.

These cumulative columns are quite useful in situations where the researcher wants to make a point about how cases are spread across the range of scores. For example, Table 2.13 shows clearly that the great majority of students in the class

ONE STEP AT A TIME **Adding Cumulative Frequency and Percentage Columns to Frequency Distributions**

Use the data in Table 2.13 for the illustrations below.

Cumulative Frequency Column

1. Begin with the lowest class interval (the interval with the lowest scores). The entry in the cumulative frequency column will be the same as the number of cases in this interval. In this case, the value for both columns is 11.
2. Go to the next class interval. The cumulative frequency for this interval is the number of cases in the interval plus the number of cases in the lower interval, or 11 + 5 = 16.
3. Continue adding (or accumulating) cases from interval to interval until you reach the interval with the highest scores, which will have a cumulative frequency equal to N.

Cumulative Percentage Column

1. Compute the percentage of cases in each category and then follow the pattern for the cumulative frequencies. The entry for the lowest class interval will be the same as the percentage of cases in the interval. For this table, both the percentage and cumulative percentage for the first interval are equal to 55%.
2. For the next higher interval, the cumulative percentage is the percentage of cases in the interval plus the percentage of cases in the lower interval. So, 55.0% + 25% = 80.0%.
3. Continue adding (or accumulating) percentages from interval to interval until you reach the interval with the highest scores, which will have a cumulative percentage of 100%.

TABLE 2.13 Age of Students in a College Class

Age	Frequency	Cumulative Frequency	Percentage	Cumulative Percentage
18–19	11	11	55.0%	55.0%
20–21	5	16	25.0%	80.0%
22–23	2	18	10.0%	90.0%
24–25	1	19	5.0%	95.0%
26–27	1	20	5.0%	100.0%
	$N = 20$		100.0%	

are younger than 21. If the researcher wishes to stress this fact, then these cumulative columns are quite handy.

Most realistic research situations will be concerned with many more than 20 cases and/or many more categories than our tables have, and the cumulative percentage column would usually be preferred to the cumulative frequencies column.

UNEQUAL CLASS INTERVALS

As a general rule, the class intervals of frequency distributions should be equal in size in order to maximize clarity and ease of comprehension. For example, all of the class intervals in Table 2.13 are 2 years wide. There are other possibilities for stating class intervals, and we will examine them here separately.

Open-Ended Intervals. What would happen to the frequency distribution in Table 2.13 if we added one more student who was 47 years of age? We would now have 21 cases and there would be a large gap between the oldest respondent (now 47) and the second oldest (age 26). If we simply added the older student to Table 2.13, we would have to include nine new class intervals (28–29, 30–31, 32–33, and so forth) with zero cases in them before we got to the 46–47 interval. This would waste space and probably be unclear and confusing. An alternative way to handle a few very high (or low) scores would be to add an "open-ended" interval to the frequency distribution, as in Table 2.14.

The open-ended interval in Table 2.14 allows us to present the information compactly. We could handle an extremely low score by adding an open-ended

TABLE 2.14 Age of Students in a College Class ($N = 21$)

Age	Frequency	Cumulative Frequency
18–19	11	11
20–21	5	16
22–23	2	18
24–25	1	19
26–27	1	20
28 and older	1	21
	$N = 21$	

interval as the lowest class interval (e.g., "17 and younger"). While open-ended intervals are sometimes convenient, you should also note their potential disadvantages: they omit information about the exact scores in the interval and they transform interval-ratio–level variables to ordinal level (since the categories are no longer equal in size).

Intervals of Unequal Size. On some variables, most scores are tightly clustered together but others are strewn across a broad range of scores. Consider the distribution of income in the United States, as presented in Table 2.15. In 2017, most households (about 52%) reported annual incomes between $25,000 and $100,000 and a sizeable grouping (21%) earned less than that. The problem (from a statistical point of view) comes with more affluent households, those with incomes above $100,000. There are fewer households in these categories (26.2%) but we must still account for them.

If we used equal intervals of, say, $10,000, in Table 2.15, we would need 30 or 40 or more intervals to include all of the more affluent households, and many of our intervals in the higher income ranges—especially those over $150,000—would have few or zero cases. The resulting table would be large and hard to comprehend.

In situations such as this, we can use intervals of unequal size to summarize the variable more efficiently, as in Table 2.15. Some of the intervals in Table 2.15 are $10,000 wide, others are $25,000 or $50,000 wide, and two (the lowest and highest intervals) are open-ended. Tables that use intervals of mixed widths might be a little confusing, but the trade-off in compactness and efficiency can be considerable. Note also that Table 2.15 changes income to an ordinal-level variable, since the class intervals are now unequal in size. *(For practice in constructing and interpreting frequency distributions for interval-ratio–level variables, see problems 2.5 to 2.9 and 2.12.)*

TABLE 2.15 Distribution of Income by Household, United States, 2017

Income	Percentage of Households	Cumulative Percentages
Less than $10,000	6.70%	6.7%
$10,000 to $14,999	4.90%	11.60%
$15,000 to $24,999	9.80%	21.40%
$25,000 to $34,999	9.50%	30.90%
$35,000 to $49,999	13.00%	43.90%
$50,000 to $74,999	17.70%	61.50%
$75,000 to $99,999	12.30%	73.80%
$100,000 to $149,999	14.10%	87.90%
$150,000 to $199,000	5.80%	93.70%
$200,000 and above	6.30%	100.00%
	100.0%	
	($N = 118,825,921$)	

Source: U.S. Census Bureau. 2018. American Community Survey, 2017. Five-Year Summary. Retrieved from https://factfinder.census.gov/faces/tableservices/jsf/pages/productview .xhtml?pid=ACS_17_5YR_DP03&prodType=table

SUMMARY: FREQUENCY DISTRIBUTIONS FOR INTERVAL-RATIO–LEVEL VARIABLES

Remember that there are no absolute rules for constructing frequency distributions for interval-ratio variables. As long as you state the class intervals with a "gap" and have an interval for every score, you are free to make your own decisions about the best balance between detail (more categories) and clarity (fewer categories). The "One Step at a Time" box provides some general guidelines that may be helpful in constructing a frequency distribution by hand.

Also, it is likely that you have access to computers and software such as SPSS to help you construct these tables. The "Using SPSS" section presents instructions for using the software to generate a frequency distribution.

ONE STEP AT A TIME	Constructing Frequency Distributions for Interval-Ratio Variables

Use the data in Tables 2.9 and 2.13 for the illustrations below.

1. Decide how many class intervals (k) you wish to use. One reasonable suggestion is to use about 10 categories but, in this case, with only 20 cases and a small range of scores, we will use five.
2. Find the range (R) of the scores by subtracting the low score from the high score. For the data here, the highest value is 26 and the lowest is 18. The range is $26 - 18 = 8$.
3. Find the size of the class intervals (i) by dividing R (from step 2) by k (from step 1):

$$i = R/k \text{ or in this case, } 8/5 = 1.6$$

Round the value of i to a convenient whole number. This will be the interval size or width. Let's round it here to 2.
4. State the lowest interval so that its lower limit is equal to or below the lowest score. See Table 2.13, where the lower limit for the data in Table 2.9 is set to 18–19. By the same token, your highest interval will be the one that contains the highest score, or in this case 26–27. Generally, intervals should be equal in size, but unequal and open-ended intervals may be used when convenient.
5. State the limits of the class intervals at the same level of precision as you have used to measure the data. Do not overlap intervals. You will thereby define the class intervals so that each case can be sorted into one and only one category.
6. Count the number of cases in each class interval and report these subtotals in a column labeled "Frequency." Report the total number of cases (N) at the bottom of this column. The table may also include a column for percentages, cumulative frequencies, and cumulative percentages. See Table 2.13 for the completed table.
7. Inspect the frequency distribution carefully. Has too much detail been obscured? If so, reconstruct the table with a greater number of class intervals (or use a smaller interval size). Is the table too detailed? If so, reconstruct the table with fewer class intervals (or use wider intervals). Are there too many intervals with no cases in them? If so, consider using open-ended intervals or intervals of unequal size. Remember that the frequency distribution results from a number of decisions you make in a rather arbitrary manner. If the appearance of the table seems less than optimal given the purpose of the research, redo the table until you are satisfied that you have struck the best balance between detail and conciseness.
8. Give your table a clear, concise title, and number the table if your report contains more than one. All categories and columns must also be clearly labeled.

For another example, see Social Research and Data Analysis 2.3.

Social Research and Data Analysis 2.3 | Frequency Distributions

The Bureau of Justice Statistics (BJS) is a major source of data on crime and punishment. Imagine if you were to collect data for the BJS on the ages of 50 prisoners enrolled in a work-release program that allows them to leave their facility for part of the day to work at a community job. You might choose to use a frequency distribution to provide an accurate picture of the overall age structure.

18	60	57	27	19
20	32	62	26	20
25	35	75	25	21
30	45	67	41	30
37	47	65	42	25
18	51	22	52	30
22	18	27	53	38
27	23	32	35	42
32	37	32	40	45
55	42	45	50	47

Let's use 10 intervals to display these data. By inspection we see that the youngest prisoner is 18 and the oldest is 75. The range is 57 ($75 - 18 = 57$). A good interval size would be 5.7 ($57/10 = 5.7$), which we will round to 6. Use a six-year interval beginning at 18. The limits of the lowest interval will be 18–23. Next, state the limits of all the other intervals, count the number of cases in each interval, and display these counts in a frequency distribution. Columns may optionally be added for cumulative percentages and cumulative frequencies. The complete distribution, with just a column added for percentages, is shown below.

Ages	Frequency	Percentages
18–23	10	20%
24–29	7	14%
30–35	9	18%
36–41	5	10%
42–47	8	16%
48–53	4	8%
54–59	2	4%
60–65	3	6%
66–71	1	2%
72–77	1	2%
	$N = 50$	100%

The prisoners seem to be fairly evenly spread across the age groups up to the 48–53 interval. There is a noticeable lack of prisoners in the oldest age groups and a concentration of prisoners in their 20s and 30s.

USING SPSS | Frequency Distributions

In this first instalment of Using SPSS, we will use the program to create a frequency distribution for a variable called *attend*, a measure of how often Americans attend religious services.

Follow these steps to get a frequency distribution:

1. Find and click the SPSS icon on your desktop.
2. Load the database.
 a. Find the **File** command on the far left of the menu bar and then click **File →
 Open → Data**.
 b. Find the 2018 General Social Survey (GSS) database file supplied with this text. You can download this file from the website for this text if you haven't already.
3. From the menu bar across the top of the SPSS window, click **Analyze, Descriptive Statistics**, and **Frequencies**. The screen will look like this:

(continued)

USING SPSS *(Continued)*

4. Find the variable (*attend*) in the box on the left of the **Frequencies** window and click the arrow to move the variable name to the box on the right. What you see on the screen will be similar to this:

The window on the left may display variables by name (e.g., *abany*, *abpoor*) or by label (e.g., ABORTION IF WOMAN WANTS FOR ANY REASON). If labels are displayed, it may be helpful to switch to variable names by right clicking a variable, and then selecting **Variable Names**. See Appendix F and Table F.2 for further information.

5. Click **OK** and the frequency distribution will be sent to the SPSS Output window.

The output is:

How often R attends religious services

		Frequency	Percent	Valid Percent	Cumulative Percent
Valid	NEVER	716	30.5	30.7	30.7
	LT ONCE A YEAR	137	5.8	5.9	36.6
	ONCE A YEAR	300	12.8	12.9	49.4
	SEVRL TIMES A YR	239	10.2	10.2	59.7
	ONCE A MONTH	146	6.2	6.3	66.0
	2-3X A MONTH	186	7.9	8.0	73.9
	NRLY EVERY WEEK	88	3.7	3.8	77.7
	EVERY WEEK	391	16.7	16.8	94.5
	MORE THN ONCE WK	129	5.5	5.5	100.0
	Total	2332	99.3	100.0	
Missing	DK,NA	16	.7		
Total		2348	100.0		

Note, first of all, that this variable is ordinal in level of measurement. There is a true zero point ("Never"), but attendance rates are grouped into broad, unequal categories.

Second, note that this table looks like the frequency distributions presented in this chapter. The categories (or scores or class intervals) are listed on the left and frequencies appear in the next column. We see immediately that the most common score for church attendance is "never" (716 respondents) followed by "every week" (391 respondents).

The percentages in the next column (Percent) are calculated from the entire sample, including the 16 respondents whose response was DK (Don't Know) or NA (No Answer) and who should be regarded as "missing" and excluded from all calculations. We will ignore this column and move to the Valid Percent column, the percentages that are made after eliminating the missing cases. A glance down this column reinforces the conclusion that the two most common scores are "never" (30.7%) and "every week" (16.8%). Finally, the Cumulative Percent column shows that most respondents (59.7%) attended church "Several Times a Year" or less.

Overall, how would you describe the pattern in this table? Is attendance at religious services an important value in the United States? Are Americans divided or polarized in their religious observances? How do you think that the pattern of attendance would vary by denomination or by age or gender?

Ratios, Rates, and Percentage Change

Ratios, rates, and percentage change are statistics that are used to summarize results simply and clearly. They may be used independently or with frequency distributions, and they may be computed for variables at any level of measurement. Although they are similar to each other, each statistic has a specific application and purpose, and we will consider them one at a time.

RATIOS

Ratios are computed by dividing the frequency in one category by the frequency in another. The formula for a ratio is

FORMULA 2.3

$$\text{Ratio} = \frac{f_1}{f_2}$$

where f_1 = the number of cases in the first category

f_2 = the number of cases in the second category

Ratios are especially useful for comparing the relative sizes of different categories of a variable. To illustrate, suppose that you were interested in the relative sizes of the various religious denominations and found that a particular community included 1370 Protestant families and 930 Catholic families. To find the ratio of Protestants (f_1) to Catholics (f_2), divide 1370 by 930:

$$\text{Ratio} = \frac{f_1}{f_2} = \frac{1370}{930} = 1.47$$

The ratio of 1.47 means that there are 1.47 Protestant families for every Catholic family.

Ratios can be very economical ways of expressing relative size. In our example, it is obvious from the raw data that Protestants outnumber Catholics, but ratios tell us exactly how much one category outnumbers the other.

Ratios are often multiplied by some power of 10 to eliminate decimal points. For example, the ratio of Protestants to Catholics might be multiplied by 100 and reported as 147 instead of 1.47. This would mean that there are 147 Protestant families for every 100 Catholic families in the community. To ensure clarity, the comparison units for the ratio are often expressed as well. Based on a unit of ones, the ratio of Protestants to Catholics would be expressed as 1.47:1. Based on hundreds, the same statistic might be expressed as 147:100. *(For practice in computing and interpreting ratios, see problems 2.1 and 2.2.)*

RATES

Rates provide still another way of summarizing the distribution of a single variable. A rate is defined as the number of actual occurrences of a phenomenon

Social Research and Data Analysis 2.4 **Ratios**

PRS IN VIVO is a market research company that studies consumer shopping experiences. Imagine you have collected data for them on the number of people that preferred a new flavor of Reese's Peanut Butter Cups (2799) compared to the traditional flavor (1884). This question could be answered with frequencies, but a more easily understood way of expressing the answer would be with a ratio. The ratio of people preferring the new flavor to those preferring the old flavor is described below.

$$\text{Ratio} = \frac{f_1}{f_2} = \frac{2799}{1884} = 1.49$$

For every person preferring the traditional flavor, there are 1.49 people preferring the new flavor.

divided by the number of possible occurrences per some unit of time. Rates are usually multiplied by some power of 10 to eliminate decimal points.

For example, the crude death rate for a population is defined as the total number of deaths in that population (actual occurrences) divided by the number of people in the population (possible occurrences) per year. This quantity is then multiplied by 1000. This formula can be expressed as

$$\text{Crude death rate} = \frac{\text{Number of deaths}}{\text{Total population}} \times 1000$$

If there were 100 deaths during a given year in a town of 7000, the crude death rate would be

$$\text{Crude death rate} = \frac{100}{7000} \times 1000 = (0.01429) \times 1000 = 14.29$$

Or, for every 1000 people, there were 14.29 deaths during this particular year.

Rates are commonly used to measure crime. For a particular crime, the rate is the number of occurrences divided by the size of the population, multiplied by 100,000. For example, if a city of 237,000 people experienced 120 auto thefts during a particular year, the auto theft rate would be

$$\text{Auto theft rate} = \frac{120}{237,000} \times 100,000 = (0.0005063) \times 100,000 = 50.63$$

Or, for every 100,000 people, there were 50.63 auto thefts during the year in question. *(For practice in computing and interpreting rates, see problems 2.3 and 2.4a.)*

PERCENTAGE CHANGE

Measuring social change, in all its variety, is an important task for all social sciences. One useful statistic for this purpose is the **percentage change**, which tells us how much a variable has increased or decreased over a certain span of time.

Social Research and Data Analysis 2.5 | **Rates**

Demographers calculate the crude birth rate of a region by dividing the number of births by the number of people in a population and multiplying by 1000 (to eliminate decimal points). In 2019, there were 2500 births in a city of 167,000. In 2009, when the population of the city was only 133,000, there were 2700 births. Is the birth rate rising or falling? A demographer would compare the crude birth rates from both years to find out.

Although this question can be answered from the preceding information, the trend in birth rates will be much more obvious if we compute birth rates for both years. For 2009:

$$\text{Crude Birth Rate} = \frac{2700}{133,000} \times 1000 = 20.30$$

In 2009, there were 20.30 births for every 1000 people in the city. For 2019:

$$\text{Crude Birth Rate} = \frac{2500}{167,000} \times 1000 = 14.97$$

In 2019, there were 14.97 births for every 1000 people in the city. With the help of these statistics, the decline in the birth rate is clearly expressed.

To compute this statistic, we need the scores of a variable at two different points in time. The scores could be in the form of frequencies, rates, or percentages. The percentage change will tell us how much the score has changed at the later time relative to the earlier time. Using death rates as an example once again, imagine a society suffering from a devastating outbreak of disease in which the death rate rose from 16 deaths per 1000 in the year 2000 to 24 deaths per 1000 in 2019. Clearly, the death rate is higher in 2019, but by how much relative to 2000?

The formula for the percent change is

FORMULA 2.4
$$\text{Percent change} = \left(\frac{f_2 - f_1}{f_1}\right) \times 100$$

$$\text{where} \quad f_1 = \text{first score, frequency, or value}$$
$$f_2 = \text{second score, frequency, or value}$$

In our example, f_1 is the death rate in 2000 ($f_1 = 16$) and f_2 is the death rate in 2019 ($f_2 = 24$). The formula tells us to subtract the earlier score from the later score and then divide by the earlier score. The result expresses the size of the change in scores ($f_2 - f_1$) relative to the score at the earlier time (f_1). The value is then multiplied by 100 to express the change in the form of a percentage:

$$\text{Percent change} = \left(\frac{24 - 16}{16}\right) \times 100 = \left(\frac{8}{16}\right) \times 100 = (0.50) \times 100 = 50\%$$

The death rate in 2019 is 50% higher than in 2000. This means that the 2019 rate was equal to the 2000 rate *plus* half of the earlier score. If the rate had risen to 32 deaths per 1000, the percent change would have been 100% (the rate would have doubled), and if the death rate had fallen to 8 per 1000, the percentage change would have been −50%. Note the negative sign: It means that the death rate has decreased by 50% and the 2010 rate is half the size of the 2000 rate.

An additional example should make the computation and interpretation of the percentage change clearer. Suppose we wanted to compare the projected population growth rates for various nations through 2050. The necessary information is presented in Table 2.16, which shows the actual population for each nation in 2018 and the projected population for 2050. The "Increase/Decrease" column shows how many people will be added or lost over the time span.

TABLE 2.16 Projected Population Growth for Six Nations, 2018–2050

Nation	Population, 2018 (f_1)	Population, 2050 (f_2)	Increase/Decrease ($f_1 - f_2$)	Percentage Change
China	1,393,800,000	1,343,900,000	−49,900,000	−3.58
U.S.	328,000,000	389,600,000	61,600,000	18.78
Nigeria	195,900,000	410,600,000	214,700,000	109.60
Mexico	130,800,000	164,300,000	33,500,000	25.61
U.K.	66,400,000	74,700,000	8,300,000	12.50
Canada	37,200,000	49,900,000	12,700,000	34.14

Source: Population reference Bureau, 2019. *2018 World Population Data Sheet*. Retrieved from https://www.prb.org/2018-world-population-data-sheet-with-focus-on-changing-age-structures/

Social Research and Data Analysis 2.6 | Percentage Change

The American family has been changing rapidly over the past several decades. One major change has been an increase in the number of married women and mothers with young children who hold jobs outside the home. For example, in 1970, 30.3% of married women with children under the age of 6 worked outside the home. In 2017, this percentage had risen to 65.1%. How large is this change?

Obviously, the 2017 percentage is much higher. Calculating the percentage change will tell us the exact magnitude of the change. The 1970 percentage is f_1 and the 2017 figure is f_2, so

$$\text{Percent Change} = \left(\frac{65.1 - 30.3}{30.3}\right) \times 100$$

$$= \left(\frac{34.8}{30.3}\right) \times 100 = 114.85$$

In the 47 years between 1970 and 2017, the percentage of married women with children younger than 6 who worked outside the home increased by 114.85%. This is an extremely large change—more than doubling the earlier percentage—and signals major changes in this social institution.

Sources: 1970: U.S. Bureau of the Census. 2012. *Statistical Abstract of the United States, 2012*. Washington, DC: Government Printing Office. p. 385. Retrieved from http://www.census.gov/prod/2011pubs/12statab/labor.pdf; 2017: U.S. Department of Labor. 2018. *Mothers and Families*. https://www.dol.gov/wb/stats/mother_families.htm

Which nation will grow the fastest? Casual inspection will give us some information, but we can make precise comparisons using percentage change. The values in the right-hand column of Table 2.16 were computed by subtracting the 2018 population (f_1) from the 2050 population (f_2), dividing by the 2018 population, and multiplying by 100.

China has the largest population of these six nations, but it will actually lose 3.6% of its population by 2050. The U.S. population will increase by almost 19%, Mexico will grow by over 25%, and the U.K. will increase by 12.5%. Canada will grow more rapidly—by over a third of its 2018 population—but Nigeria has by far the highest growth rate: It will add the most people and its population will increase by well over 100%. This means that the 2050 population of Nigeria will be more than twice its 2018 size. *(For practice in computing and interpreting percent change, see problem 2.4b.)*

ONE STEP AT A TIME | Finding Ratios, Rates, and Percent Change

Ratios

1. Determine the values for f_1 and f_2. The value for f_1 will be the number of cases in the first category (e.g., the number of males on your campus), and the value for f_2 will be the number of cases in the second category (e.g., the number of females on your campus).
2. Divide the value of f_1 by the value of f_2.
3. You may multiply the value you calculated in step 2 by some power of 10 when reporting results.

(continued)

ONE STEP AT A TIME *(Continued)*

Rates

1. Determine the number of actual occurrences (e.g., births, deaths, homicides, assaults). This value will be the numerator of the formula.
2. Determine the number of possible occurrences. This value will usually be the total population for the area in question and will be the denominator of the formula.
3. Divide the number of actual occurrences (step 1) by the number of possible occurrences (step 2).
4. Multiply the value you calculated in step 3 by some power of 10. Conventionally, birth rates and death rates are multiplied by 1000 and crime rates are multiplied by 100,000.

Percentage Change

1. Determine the values for f_1 and f_2. The former will be the score at time 1 (the earlier time) and the latter will be the score at time 2 (the later time).
2. Subtract f_1 from f_2.
3. Divide the quantity you found in step 2 by f_1.
4. Multiply the quantity you found in step 3 by 100.

STATISTICS **In Everyday Life**

Traffic Fatalities

Are American highways becoming more dangerous? Here are some statistical facts: In 1990 there were 46,800 motor vehicle deaths, and that number fell to 37,133 in 2017. This is a decrease of 20.7%, and that is certainly good news, but do the raw numbers tell the whole story? Wouldn't this comparison be more meaningful and informative if it took account of changes in population, number of cars and drivers on the road, or changing driving habits?

One way to make the statistic more useful would be to compute a rate of traffic deaths based on the number of miles Americans drive each year: For every 100 million miles driven, the rate of traffic deaths fell from 2.1 in 1990 to 1.2 in 2017, an even more dramatic decrease of about 42.9%. What factors might account for this increase in highway safety?

Sources: 1990: U.S. Bureau of the Census, *Statistical Abstract of the United States, 2012*, p. 693. Retrieved from http://www.census.gov/prod/2011pubs/12statab/trans.pdf; 2017: National Highway and Traffic Safety Authority. 2019. Quick Facts, 2017. Retrieved from https://crashstats.nhtsa.dot.gov/#/DocumentTypeList/11

Using Graphs to Present Data

Researchers frequently use graphs to present their data in a visually dramatic way. These devices are particularly useful for communicating the overall shape of a distribution and for highlighting any clustering of cases in a particular range of scores.

Many types of graphs are available, but we will examine just four. The first two, pie charts and bar charts, are appropriate for discrete variables at any level of

measurement. The last two, histograms and line charts (or frequency polygons), are used with both discrete and continuous interval-ratio variables but are particularly appropriate for the latter.

All of the graphs described here are easily produced with SPSS, Microsoft Excel, and other software programs commonly available on college campuses. These programs are relatively flexible and easy to use. We will demonstrate how to use SPSS to produce graphs in the next section and in the "You Are the Researcher" section at the end of this chapter.

PIE CHARTS

A **pie chart** is an excellent way of displaying the relative sizes of the categories of a variable. Figure 2.2 shows the distribution of religious identification in the United States first shown in Table 2.2. The table is reproduced here as Table 2.17 with percentages added.

The pie chart divides a circle into "slices" proportional to the relative frequencies of the categories. The largest slice represents the Protestant group, with almost 50% of the cases, and the smallest slice represents the Jewish group, the smallest category. You might think of a pie chart as a visual frequency distribution, and Figure 2.2 dramatically and clearly displays the relative sizes of the various religions.

FIGURE 2.2 Self-Described Religious Identification of Adult Americans, 2018

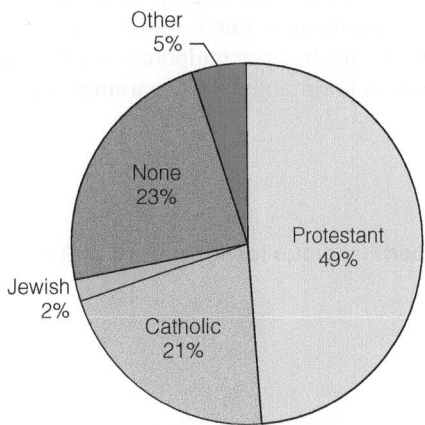

TABLE 2.17 Self-Described Religious Affiliation of Adult Americans, 2018

Religious Group	Frequency	Percentages
Protestant	124,148,263	48.90
Catholic	53,822,969	21.20
Jewish	4,315,993	1.70
None	59,154,489	23.30
Other	12,440,215	4.90
	$N = 253,881,927$	100.0%

BAR CHARTS

Like pie charts, **bar charts** are straightforward. The categories of the variable are arrayed along the horizontal axis (or *abscissa*) and frequencies, or percentages if you prefer, along the vertical axis (or *ordinate*). The heights of the bars are proportional to the relative frequencies of the categories and, like pie charts, bar charts are visual equivalents of frequency distributions. Figure 2.3 reproduces the data on religious identification displayed in Figure 2.2 and Table 2.17.

Figure 2.3 would be interpreted in exactly the same way as the pie chart in Figure 2.2, and researchers are free to choose between these two methods of displaying data. However, if a variable has more than four or five categories, the bar chart would be preferred, since the pie chart gets crowded with many categories.

Bar charts are particularly effective ways to display the relative frequencies for two or more categories of a variable when you want to emphasize some comparisons. Suppose, for example, that you wished to make a point about suicide rates for males and females of various age groups. Figure 2.4 displays the data in a dramatic and easily comprehended way. First, the bar chart shows that rates for males are considerably higher than rates for females for all age groups. Also, the suicide rate is lowest for the youngest age groups of both genders and tends to increase with age for males. The rate for females, on the other hand, peaks in the 45 to 64 age group. What sociological factors might account for these patterns? *(For practice in constructing and interpreting pie and bar charts, see problems 2.5b and 2.10.)*

Some bar charts display data on more than one variable at a time. In Figure 2.5, the unemployment rate for all levels of education is shown on the left. On the right, the median (or midpoint) usual earnings are shown. Notice that unemployment is lower and weekly earnings are higher for those with a higher education.

FIGURE 2.3 Self-Described Religious Identifications of Adult Americans, 2018

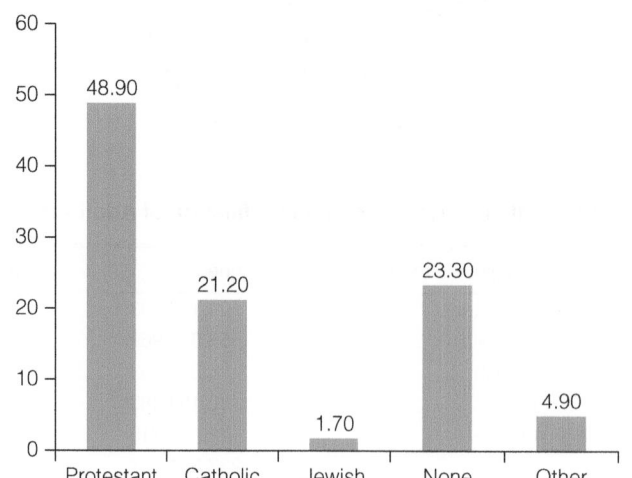

FIGURE 2.4 Suicide Rates for Males and Females by Age Group, 2017

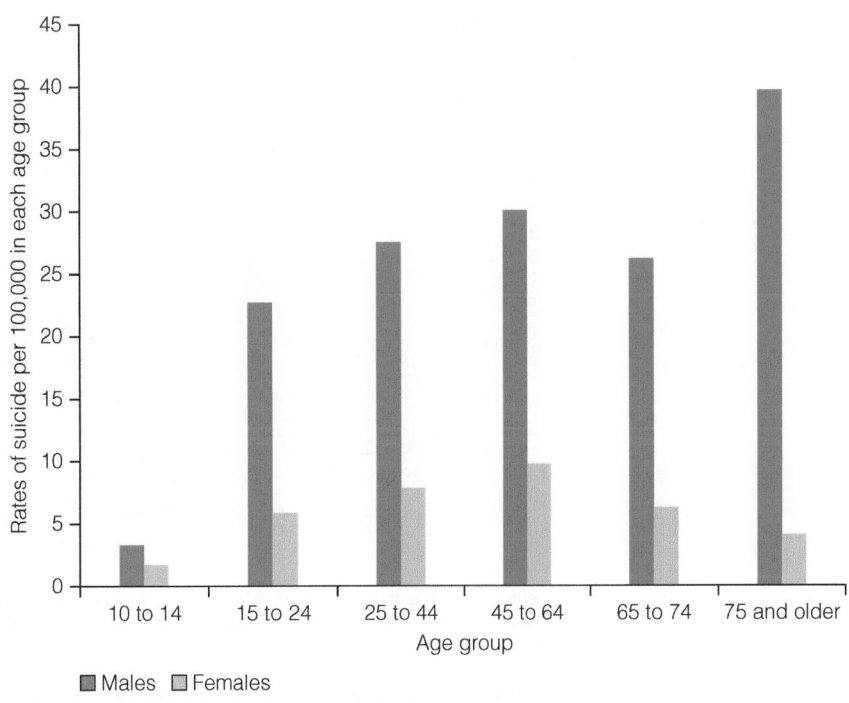

Source: National Center for Health Statistics, 2018. Suicide Mortality in the United States, 1999–2017.
https://www.cdc.gov/nchs/data/databriefs/db330-h.pdf

HISTOGRAMS

Histograms look a lot like bar charts and, in fact, are constructed in much the same way. However, histograms use real limits rather than stated limits, and the categories or scores of the variable border each other, as if they merged into each other in a continuous series. Therefore, these graphs are most appropriate for

FIGURE 2.5 Unemployment Rate and Earnings by Educational Attainment, 2018

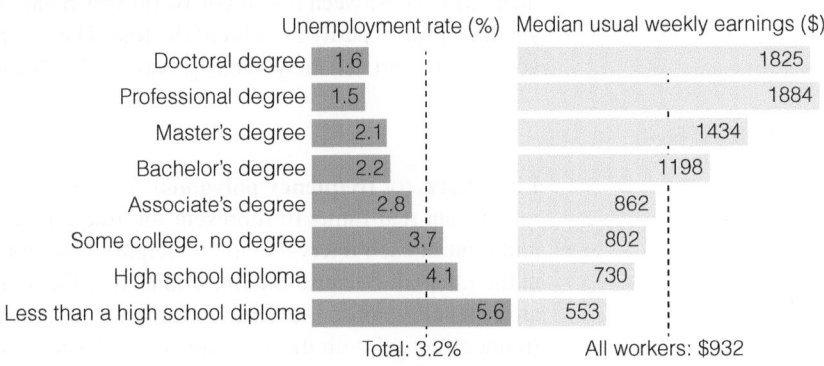

Note: Data are for persons age 25 and over. Earnings are for full-time wage and salary workers.

Source: U.S. Bureau of Labor Statistics, Current Population Survey.

FIGURE 2.6 **Age Distribution of the United States, 2017**

Source: U.S. Bureau of the Census. 2018. American Community Survey. 2013–2017, Five Year Estimates. https://factfinder.census.gov/faces/tableservices/jsf/pages/productview.xhtml?pid=ACS_17_5YR_S0101& prodType=table

continuous interval-ratio–level variables, but they are commonly used for discrete interval-ratio–level variables as well.

As an example, Figure 2.6 uses a histogram to display the distribution of age for the population of the United States in 2017. Note that age is an interval-ratio variable that has been collapsed into categories of 5 years, except for the oldest category (85 and older), which is an open-ended interval.

The bars in the graph are 5 years wide, and their uneven heights reflect the varying number of respondents for each 5-year group. As you would expect, there is a sharp drop in the graph starting at about age 60 and continuing to the oldest age group, a reflection of the higher death rate in older populations.

Note that the graph peaks for people in their early 50s and in their 20s. The older age group is the tail end of the "baby boom," produced by the unusually high birth rate between the end of World War II and the mid-1960s. The second, or younger, peak is partly an echo of the first. The two peaks are 30 years, or about a generation, apart and the younger group are the children of the older group.

LINE CHARTS

Line charts (or **frequency polygons**) are similar to histograms but use midpoints rather than real limits to represent the frequencies. Dots are placed above the midpoint of each interval, and the height of the dot reflects the number of cases in the interval. Because the line connecting the dots is continuous, these graphs are especially appropriate for continuous interval-ratio–level variables but are frequently used with discrete interval-ratio–level variables as well. Figure 2.7 displays a line chart showing the age distribution of males and females in the United States in 2017. Note that males outnumber females from the youngest age

FIGURE 2.7 Age Distribution of the Population of the United States by Gender, 2017

Source: U.S. Bureau of the Census. 2018. American Community Survey. 2013–2017, Five Year Estimates. https://factfinder.census.gov/faces/tableservices/jsf/pages/productview.xhtml?pid=ACS_17_5YR_S0101& prodType=table

groups through about age 55. At that point, females begin to outnumber males and this pattern continues through the oldest age group. This pattern reflects the higher death rate for males in the older age groups.

Histograms and line charts are alternative ways of displaying essentially the same message. Thus, the choice between the two techniques is left to the aesthetic pleasures of the researcher. *(For practice in constructing and interpreting histograms and line charts, see problems 2.10 and 2.13.)*

USING SPSS **Graphs**

SPSS can produce a variety of graphs and, in this installment of "Using SPSS," we will use the General Social Survey (GSS) to produce a pie chart and a histogram to illustrate the procedure. We will use *attend* again for the pie chart and age for the line chart. We can compare the distribution of age for the GSS sample with the population distribution presented in Figure 2.6.

Follow these steps to get a pie chart:

1. Find and click the SPSS icon on your desktop.
2. Load the database.
 a. Find the **File** command on the far left of the menu bar and then click **File, Open, Data**.
 b. Find the General Social Survey (GSS2018.sav) database file supplied with this text. You can download this file from the Web site for this text if you haven't already.
3. From the menu bar across the top of the SPSS window, click **Graphs**, **Legacy Dialog**, and **Pie**.
4. The "Pie Charts" window will appear with "Summaries for groups of cases" preselected. This is what we want, so click **Define**.

(*continued*)

USING SPSS

(Continued)

5. A new window will appear. Find attend in the list of variables on the left and click the arrow to move the variable name to the "Define Slices by:" box. When you are done, the window will look like this:

Click OK and the pie chart will be sent to the SPSS Output window:

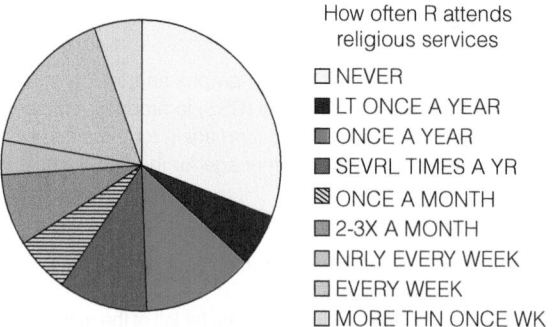

How often R attends religious services

☐ NEVER
■ LT ONCE A YEAR
■ ONCE A YEAR
■ SEVRL TIMES A YR
▧ ONCE A MONTH
■ 2-3X A MONTH
☐ NRLY EVERY WEEK
☐ EVERY WEEK
☐ MORE THN ONCE WK

Remember that a pie chart is essentially a visual frequency distribution. As we saw before, the most common score (largest "slice") is "never." The "every week" and "more than once a week" slices are about the same size.

To get a histogram:

1. Click **Graphs**, **Legacy Dialog**, and **Histogram**.
2. In the **Histogram** window, find *age* and move it to the "Variable:" box.

USING SPSS | **(Continued)**

3. Click **OK** and the graph will be sent to the SPSS Output window:

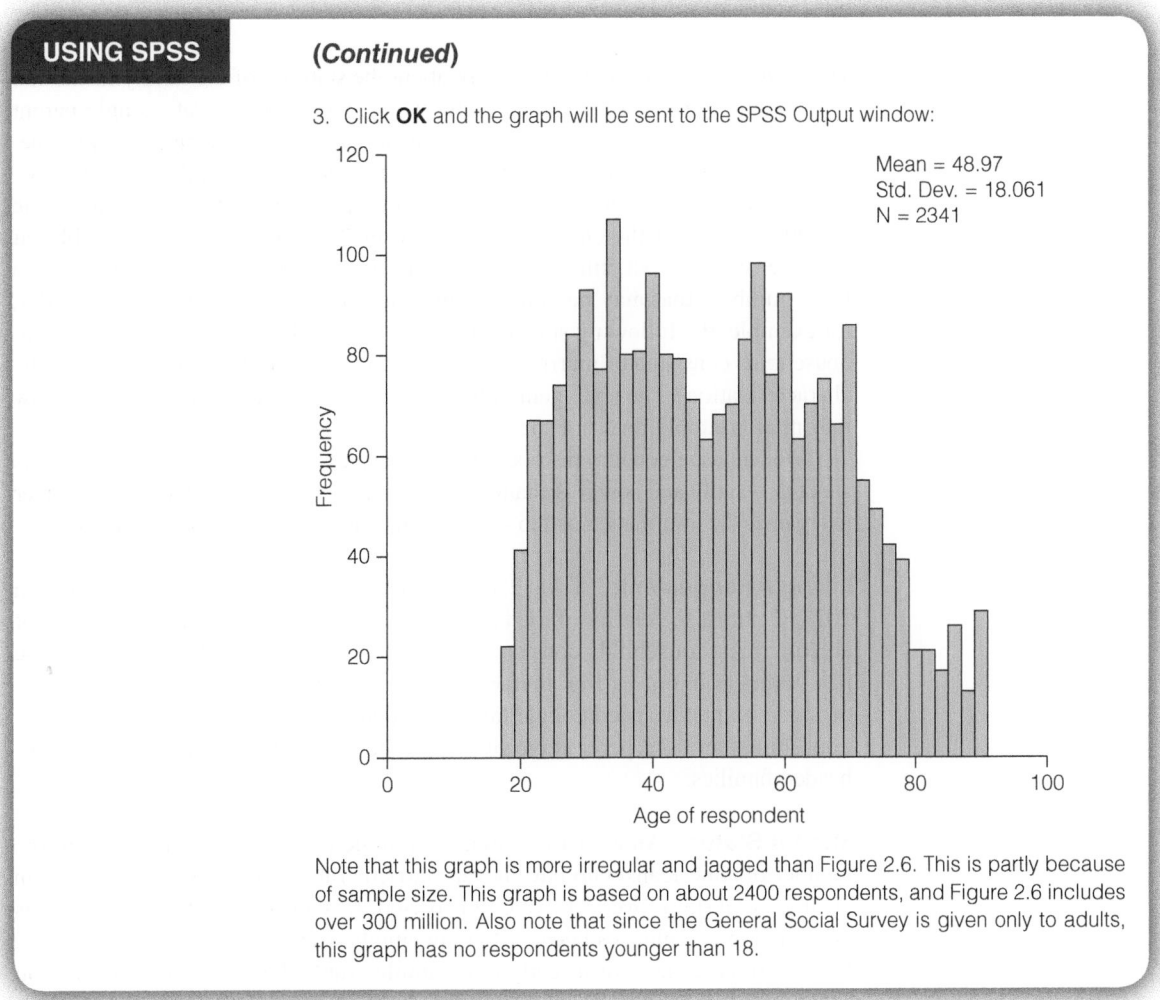

Mean = 48.97
Std. Dev. = 18.061
N = 2341

Note that this graph is more irregular and jagged than Figure 2.6. This is partly because of sample size. This graph is based on about 2400 respondents, and Figure 2.6 includes over 300 million. Also note that since the General Social Survey is given only to adults, this graph has no respondents younger than 18.

Interpreting Statistics: Changing Patterns of American Family Life

We have covered a sizeable volume of material, and it will be useful to conclude by focusing on meaning and interpretation. What can you say after you have constructed a frequency distribution or a graph and computed percentages or rates? Remember that statistics are tools to help us analyze information and answer questions. They never speak for themselves and they always have to be understood in the context of some research question or test of a hypothesis. This section provides an example of interpretation by posing and answering questions. The interpretation (words) will be explicitly linked to the statistics (numbers) so that you will be able to see how and why conclusions are developed.

THE U.S. FAMILY

There has been a great deal of concern about the stability of the family institution in the United States. People commonly worry about divorce rate, single-parent families, and the rise of new family relationships including cohabitation and same-sex marriage. Are the concerns justified? What is happening to the American family?

We can get an overview of the changing family from the U.S. Census, the decennial count of the entire population. Census data are readily available (at www.census.gov) but, since it counts everyone, census results are expressed in large numbers that may be cumbersome and difficult to understand. Consider, for example, the following statement: "In 2018, 61,241,000 of 127,586,000 U.S. households consisted of married couples." Can you distill any meaning from this characterization? The statement is literally true but illustrates the point that raw information simply does not speak for itself.

We will use frequency distributions, percentages, and rates to present some facts about the family and how it is changing, using the statistics presented in this chapter to organize the information and to state conclusions. We start with basic structures.

Living Arrangements. Table 2.18 presents information on U.S. households in 2018. Contrast the clarity of the table and the percentages ("In 2018, 48.00% of all American households consisted of married couples") with the chaos of the raw numbers in our earlier characterization. The table makes some easily understood points: Married couple families are now a slight minority of American households and the percentage of female-headed families far exceeds male-headed families.

Marital Status. Another trend that is of some concern is the increase in the number of Americans who have never married. Table 2.19 presents information about marital status over a 48-year time period. As with Table 2.18, information is presented in percentages because the raw numbers are quite large. (Incidentally, these two tables are not directly comparable. Table 2.18 presents information about households while Table 2.19 is based on individuals.)

How would you describe these changes? Three trends stand out:

1. In 1970, the great majority of adult Americans were married. By 2018, this percentage had declined to about 52%.

2. In 1970, the percentage of adult Americans who were divorced was tiny, at a little more than 3%. By 2018, this percentage had tripled and about one out of 10 adult Americans was divorced.

TABLE 2.18 U.S. Households by Type, 2018

Family Households:	65.12%
Married Couples	48.00%
Male Headed	5.03%
Female Headed	12.09%
Nonfamily Households:	34.88%

Source: U.S. Bureau of the Census. 2019. Historical Household Tables, Table HH1. https://www.census.gov/data/tables/time-series/demo/families/households.html

TABLE 2.19 Marital Status for Selected Years

Status	1970	1990	2018
Married	71.70%	62.31%	52.0%
Never Married	16.15%	22.36%	32.3%
Divorced	3.25%	7.80%	9.9%
Widowed	8.91%	7.53%	5.8%
	100.01%	100.00%	100.00%
	($N = 132,500,000$)	($N = 180,700,000$)	($N = 228,119,000$)

Sources: 1970 and 1990: U.S. Bureau of the Census. 2012. *Statistical Abstract of the United States, 2012.* https://www.census.gov/library/publications/2011/compendia/statab/131ed.html. 2018: U.S. Bureau of the Census. 2019. *America's Families and Living Arrangements: 2018.* Table A1. 2018 https://www.census.gov/data/tables/2018/demo/families/cps-2018.html

3. Between 1970 and 2018, the percentage of Americans who have never married increased dramatically from about 16% to over 32%.

Ratios provide another way to express these changes:

1. In 1970, married Americans outnumbered Americans who had never married by 4.4 to 1. By 2018, the ratio had dropped to 1.6 to 1.

2. In 1970, married Americans outnumbered divorced Americans by 22 to 1. By 2018, the ratio had dropped to 5.3 to 1.

We could also use percentage changes to state these trends:

1. Between 1970 and 2018, the percentage of Americans who were married declined by 27.5%.

2. Between 1970 and 2018, the percentage of Americans who had never married increased by 100%.

3. Between 1970 and 2018, the percentage of Americans who were divorced increased by 204.6%.

Divorce Rate. It is well known that the U.S. rate of divorce is higher than that of other Western industrialized nations, but what is generally not known is that the rate has been falling for decades, as shown in Figure 2.8.

The figure shows an increase in both marriage and divorce rates in the 1960s, a peak in the early 1980s (and another peak in the early 1970s for marriage rate), followed by slow declines over the past several decades. The divorce rate has declined about 45% since 1980, a fact that most Americans would probably be surprised to hear.

What factors would explain the decline? We cannot go into detail here, but one factor is obvious: Fewer Americans are getting married, as shown in Table 2.19 and by the falling marriage rate in Figure 2.8. You can't get divorced if you don't get married! What other factors might account for the trends displayed in Figure 2.8?

Note that the information in Figure 2.8 could have been presented in a table. What are the advantages of the "picture" presented in the graph? Are there any disadvantages of using graphs rather than tables? What kinds of variables are most appropriately displayed in line charts like Figure 2.8?

FIGURE 2.8 **U.S. Rates of Marriage and Divorce, 1950–2017 (rates per 100,000 population)**

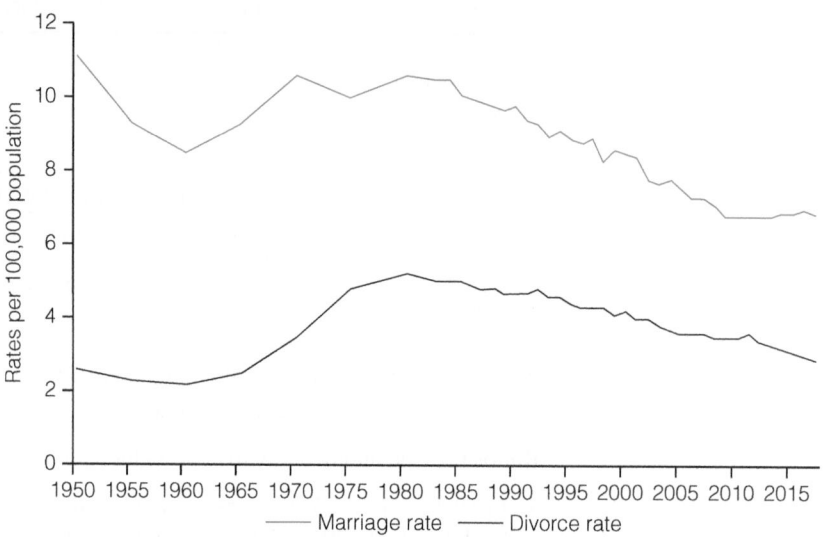

READING STATISTICS Substance Abuse Among College Students

The statistics covered in this chapter are frequently used in the research literature of the social sciences—as well as in the popular press and the media—and one of the goals of this text is to help you develop your ability to understand and critically analyze these applications. Fortunately, this task is usually quite straightforward, but these statistical tools are sometimes not as simple as they appear and they can be misused. Here are some ideas to keep in mind when reading research reports that use the statistics covered in this chapter.

First, there are many different formats for presenting results and the tables and graphs you find in the research literature will not necessarily follow the conventions used in this text. Be flexible and study the presentation carefully before drawing conclusions.

Second, because of space limitations, tables and graphs may be presented with minimum detail. For example, the researcher may present a frequency distribution with only a percentage column.

Begin your analysis by examining the statistics carefully. If you are reading a table or graph, first read the title, all labels (that is, row and/or column headings), and any footnotes. These will tell you exactly what information is being presented. Inspect the body of the table or graph with the author's analysis in mind. See whether you agree with the author's analysis. (You almost always will, but it never hurts to double-check and exercise your critical abilities.)

Finally, remember that most research projects analyze interrelationships among many variables. Because the tables and graphs covered in this chapter display variables one at a time, they are unlikely to be included in such research reports (or, perhaps, included only as background information). Even when they are not reported, you can be sure that the research began with an inspection of percentages, frequency distributions, or graphs for each variable. Univariate tables and graphs display a great deal of information about the variables in a compact, easily understood format and are almost universally used as descriptive devices.

STATISTICS IN THE PROFESSIONAL LITERATURE

A recent report published by the National Institute on Drug Abuse compared drug use among college students and noncollege students of the same age

READING STATISTICS *(Continued)*

TABLE 1 Drug Use by College Students and Nonstudents

Drug	College	Noncollege
Daily Cigarette Use	2%	14%
Used Alcohol in the Past Month	62%	56%

group. Among other differences, the researchers found striking differences in the use of cigarettes and alcohol. As displayed in Table 1, cigarette use among college students was virtually nonexistent but seven times greater among nonstudents (although still uncommon). In contrast, students were more likely to report alcohol use within the past month than nonstudents, although the behavior was common in both groups.

Perhaps unexpectedly, as shown in Figure 1, daily use of marijuana among students was much lower than among nonstudents and has declined since 2014, while use of the drug has increased among nonstudents. By 2017, daily use of marijuana among college students was almost as rare as daily cigarette use. What sociological factors might help explain these differences?

FIGURE 1 Reported Daily Use of Marijuana by College Students and Their Noncollege Age Peers

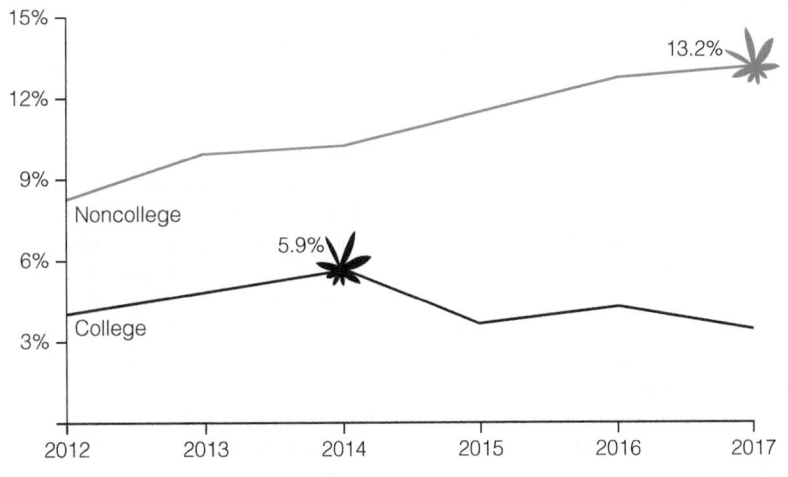

SUMMARY

1. We considered several different ways of summarizing the distribution of a single variable and, more generally, reporting the results of our research. Our emphasis throughout was on the need to communicate the results clearly and concisely.

2. Frequency distributions are tables that summarize the entire distribution of some variable. Statistical analysis almost always starts with the construction and review of these tables for each variable. Columns for percentages, cumulative frequencies, and/or cumulative percentages often enhance the readability of frequency distributions.

3. Percentages and proportions, ratios, rates, and percentage changes give us different ways of expressing our results in terms of relative frequency. Percentages

and proportions report the relative occurrence of some category of a variable compared with the distribution as a whole. Ratios compare two categories with each other, and rates report the actual occurrences of some phenomenon compared with the number of possible occurrences per some unit of time. Percentage change shows the relative increase or decrease in a variable over time.

4. Pie charts and bar charts, histograms, and line charts (or frequency polygons) are graphs that express the information contained in the frequency distribution in a compact and visually dramatic way.

SUMMARY OF FORMULAS

Percentage	**2.1**	$\% = \left(\dfrac{f}{N}\right) \times 100$	Ratio	**2.3**	$\text{Ratio} = \dfrac{f_1}{f_2}$
Proportion	**2.2**	$\text{Proportion} = \left(\dfrac{f}{N}\right)$	Percentage change	**2.4**	$\text{Percent change} = \left(\dfrac{f_2 - f_1}{f_1}\right) \times 100$

GLOSSARY

Bar chart. A graph used for discrete variables. Categories are represented by bars of equal width, the height of each corresponding to the number (or percentage) of cases in the category.

Class intervals. The categories used in the frequency distributions for interval-ratio variables.

Cumulative frequency. An optional column in a frequency distribution that displays the number of cases within an interval and all preceding intervals.

Cumulative percentage. An optional column in a frequency distribution that displays the percentage of cases within an interval and all preceding intervals.

Frequency distribution. A table that displays the number of cases in each category of a variable.

Frequency polygon. See **Line chart**.

Histogram. A graph used for interval-ratio variables. Class intervals are represented by contiguous bars of equal width (equal to the class limits), the height of each corresponding to the number (or percentage) of cases in the interval.

Line chart. A graph used for interval-ratio variables. Class intervals are represented by dots placed over the midpoints, the height of each corresponding to the number (or percentage) of cases in the interval. All dots are connected by straight lines.

Midpoint. The point exactly halfway between the upper and lower limits of a class interval.

Percentage. The number of cases in a category of a variable divided by the number of cases in all categories of the variable, the entire quantity multiplied by 100.

Percent change. A statistic that expresses the magnitude of change in a variable from time 1 to time 2.

Pie chart. A graph used for discrete variables with only a few categories. A circle (the pie) is divided into segments proportional in size to the percentage of cases in each category of the variable.

Proportion. The number of cases in one category of a variable divided by the number of cases in all categories of the variable.

Rate. The number of actual occurrences of some phenomenon or trait divided by the number of possible occurrences per some unit of time. Rates are usually multiplied by some power of 10.

Ratio. The number of cases in one category divided by the number of cases in some other category.

Real class limits. The class intervals of a frequency distribution when stated as continuous categories.

Stated class limits. The class intervals of a frequency distribution when stated as discrete categories.

PROBLEMS

2.1 SOC The table that follow report the marital status of 20 respondents in two different apartment complexes. (*HINT: As you solve these problems, make sure that you have the correct numbers in the numerator and denominator. For example, problem 2.1a asks for "the percentage of respondents in each complex who are married," and the denominators will be 20 for these two fractions. Problem 2.1d, on*

the other hand, asks for the "percentage of the single respondents who live in Complex B," and the denominator for this fraction will be 4 + 6, or 10.)

Status	Complex A	Complex B
Married	5	10
Unmarried ("living together")	8	2
Single	4	6
Separated	2	1
Widowed	0	1
Divorced	1	0
	20	20

a. What percentage of the respondents in each complex are married?

b. What is the ratio of single to married respondents at each complex?

c. What proportion of the residents of each complex are widowed?

d. What percentage of the single respondents live in Complex B?

e. What is the ratio of the "unmarried/living together" to the "married" at each complex?

2.2 At St. Winnefred Hospital, the numbers of males and females in various job categories are as follows:

Job	Males	Females	TOTALS
Doctor	83	15	98
Nurse	62	116	178
Aide	151	12	163
Lab technician	32	30	62
Administrator	12	1	13
Clerk	72	141	213
TOTALS	412	315	727

Read each question carefully before constructing the fraction and solving for the answer. *(HINT: Be sure you place the proper number in the denominator of each fraction. For example, some problems use the total number of males or females as the denominator, whereas others use the total number of subjects.)*

a. What percentage of nurses are male?

b. What proportion of aides are female?

c. For doctors, what is the ratio of males to females?

d. What percentage of the total staff are males?

e. What is the ratio of males to females for the entire sample?

f. What proportion of the clerks are male?

g. What percentage of the sample are administrators?

h. What is the ratio of nurses to doctors?

i. What is the ratio of female lab technicians to male lab technicians?

j. What proportion of the males are clerks?

2.3 [SP] According to the Centers for Disease Control, in 2017 in the state of Florida, 5088 people died from a drug overdose, 2724 from a firearm, and 1269 due to homicide. In that same year, Census records show that 20,278,447 people were living in Florida. Calculate the rate of death per 100,000 for each cause of death. *(HINT: Make sure that you set up the fraction with size of population in the denominator.)*

2.4 [CJ] The numbers of homicides in five U.S. states and five Canadian provinces for the years 1997 and 2017 (U.S.) or 2018 (Canada) are presented in two tables.

a. Calculate the homicide rate per 100,000 population for each state and each province for each year. Relatively speaking, which state and which province had the highest homicide rates in each year? Which society seems to have the higher homicide rate? Write a paragraph describing these results.

b. Using the rates you calculated in part a, calculate the percent change between 1997 and the more recent date for each state and each province. Which states and provinces had the largest increase and decrease? Which society seems to have the largest change in homicide rates? Summarize your results in a paragraph.

	1997		2017	
State	Homicides	Population	Homicides	Population
New Jersey	338	8,053,000	324	9,005,644
Iowa	52	2,852,000	79	1,875,128
Alabama	426	4,139,000	337	3,732,789
Texas	1327	19,439,000	1309	25,244,087
California	2579	32,268,000	1778	38,706,966

Source: FBI, 2019. *Uniform Crime Reports.* https://ucr.fbi.gov/crime-in-the-u.s/2017/crime-in-the-u.s.-2017/tables/table-5

	1997		2018	
Province	Homicides	Population	Homicides	Population
Nova Scotia	24	936,100	11	959,942
Quebec	132	7,323,600	83	8,390,499
Ontario	178	11,387,400	266	14,322,757
Manitoba	31	1,137,900	55	1,352,154
British Columbia	116	3,997,100	89	4,991,687

Source: Statistics Canada. https://www150.statcan.gc.ca/n1/en/subjects?MM=1

2.5 SOC The scores of 15 respondents on four variables are as reported next. These scores were taken from a public-opinion survey called the General Social Survey (GSS), the data set used for the computer exercises in this text. For the actual questions, see Appendix G. The numerical codes for the variables are as follows:

Sex	Support for Gun Control	Level of Education	Age
1 = Male	1 = In favor	0 = Less than HS	Actual years
2 = Female	2 = Opposed	1 = HS	
	2 = Jr. college		
	3 = Bachelor's		
	4 = Graduate		

Case Number	Sex	Support for Gun Control	Level of Education	Age
1	2	1	1	45
2	1	2	1	48
3	2	1	3	55
4	1	1	2	32
5	2	1	3	33
6	1	1	1	28
7	2	2	0	77
8	1	1	1	50
9	1	2	0	43
10	2	1	1	48
11	1	1	4	33
12	1	1	4	35
13	1	1	0	39
14	2	1	1	25
15	1	1	1	23

Construct a frequency distribution for each variable. Include a column for percentages.

2.6 SW A local youth service agency has begun a sex education program for teenage girls who have been referred by the juvenile courts. The girls were given a 20-item test for general knowledge about sex, contraception, and anatomy and physiology upon admission to the program and again after completion of the program. The scores of the first 15 girls to complete the program are as follows:

Case	Pretest	Posttest	Case	Pretest	Posttest
A	8	12	I	5	7
B	7	13	J	15	12
C	10	12	K	13	20
D	15	19	L	4	5
E	10	8	M	10	15
F	10	17	N	8	11
G	3	12	O	12	20
H	10	11			

Construct frequency distributions for the pretest and posttest scores. Include a column for percentages. (*HINT: There were 20 items on the test, so the maximum range for these scores is 20. If you use 10 class intervals to display these scores, the interval size will be 2. Since there are no scores of 0 or 1 for either test, you may state the first interval as 2–3. To make comparisons easier, both frequency distributions should have the same intervals.*)

2.7 MR In a survey conducted by AT&T, 15 males and 16 females were asked how many years they have been in their existing mobile phone plan. Their answers are reported below.

Males			
6	5	6	6
2	5	2	3
5	3	1	4
5	6	1	

Females			
4	1	3	7
7	2	3	8
7	2	4	2
5	2	8	8

Display each of these variables in a frequency distribution with columns for percentages and cumulative percentages.

2.8 GER Twenty-five residents of a community for senior citizens were asked to keep track of the number of hours they watched TV during the past week.

20	2	12	7	32
7	10	7	13	17
14	15	15	9	27
5	21	4	7	6
2	18	10	5	11

a. Construct a frequency distribution to display these data using the guidelines stated in the "One Step at a Time: Constructing Frequency Distributions for Interval-Ratio Variables" earlier in the chapter. (*HINT: For this problem you will need to set class intervals with a "gap," as in 1–5, 6–10, etc.*)

b. What are the midpoints of the class intervals?

c. Add columns to the table to display the percentage distribution, cumulative frequencies, and cumulative percentages.

d. What are the real limits for the intervals you selected?

e. Write a paragraph summarizing this distribution of scores.

2.9 CJ Twenty-five students completed a questionnaire that measured their attitudes toward the use of force by police. A high score indicates that the respondent believed that physical force could legitimately be used in many situations. Low scores mean that the respondent thought that violence could be justified in few or no situations.

52	47	17	8	92
53	23	28	9	90
17	63	17	17	23
19	66	10	20	47
20	66	5	25	17

a. Construct a frequency distribution to display these data using the guidelines stated in the "One Step at a Time: Constructing Frequency Distributions for Interval-Ratio Variables" earlier in the chapter.

(*HINT: For this problem you should set class intervals to 0–9, 10–19, etc.*)

b. What are the midpoints of the class intervals?

c. Add columns to the table to display the percentage distribution, cumulative frequencies, and cumulative percentages.

d. Write a paragraph summarizing this distribution of scores.

2.10 PA/CJ As part of an evaluation of the efficiency of your local police force, you have gathered the following data on police response time to calls for assistance during two different years. (Response times were rounded off to whole minutes.) Convert both frequency distributions into percentages, and write a paragraph comparing the changes in response time between the two years.

Response Time, 2000	Frequency (*f*)
21 minutes or more	20
16–20 minutes	90
11–15 minutes	185
6–10 minutes	370
Less than 6 minutes	210
	875

Response Time, 2019	Frequency (*f*)
21 minutes or more	45
16–20 minutes	95
11–15 minutes	155
6–10 minutes	350
Less than 6 minutes	250
	895

Statistical Analysis Using SPSS

2.11 SOC In this exercise, you will use SPSS to graph trends in crime in the United States. You will use the *CrimeTrends84-17* data set, which is available from the Web site for this text.

- Find and click the SPSS icon on your desktop.
- Load the *CrimeTrends84-17* data set.
- From the menu bar across the top of the SPSS window, click **Graphs**, then **Legacy Dialogs** and **Line**.
- The **Line Chart** dialog box will open. Click **Simple** in the top part of the window and **Values of Individual Cases** from the "Data in Chart Are" panel.
- Click **Define**.

- The "Define Simple Line" dialog box will open. In the "Category Labels" panel, click **Variable**. Next click *year* in the list of variables and click the arrow pointing to the **Variable** box.
- Click one of the measures of crime in the list of variables and click the arrow pointing to the "Line Represents" box. Click **OK** and the line chart will be created in the SPSS Output window.
- Repeat these steps for the other crime variables.
 - Each time you return to the "Define Simple Line" dialog box, click the variable name in the "Line Represents" box and then click the left-pointing arrow to return the variable to the list of variable names.
 - Click on a new measure of crime and click the arrow pointing to the "Line Represents" box. Click **OK** and the next line chart will be created. DO NOT change the *year* variable. Repeat until you have charts for all five measures of crime.

Write a paragraph describing each of these graphs. What similarities and differences can you observe among the three graphs? (For example, do crime rates always change in the same direction?) Note the differences in the vertical axis from chart to chart—for homicide, the axis has a much smaller range than for the other crimes, which are far more common, and a scale with smaller intervals is needed to display the rates.

2.12 SOC In this exercise, you will use the *GSS2018.sav* database supplied with this text to produce frequency distributions and bar charts or pie charts for three variables. The steps are the same as in the "USING SPSS: Frequency Distributions" section of this chapter, except that you will choose your own variables.

- Find and click the SPSS icon on your desktop.
- Load the 2018 GSS (*GSS2018.sav*) database supplied with this text.
- Use Appendix G to select one nominal-level, one ordinal-level (other than *attend*), and one interval-ratio–level variable. To determine level of measurement, use "ONE STEP AT A TIME: Determining the Level of Measurement of a Variable" from Chapter 1 and Table 1.5. Make sure that you examine the way in which the scores of the variable are *actually stated*.

- From the menu bar across the top of the SPSS window, click **Analyze**, **Descriptive Statistics**, and **Frequencies**.
- Choose your three variables from the list and then click the arrow to move each variable name—one at a time—to the box on the right.
- Click **OK** and the frequency distributions will be sent to the **Output** window.

Study the tables carefully, including the **Valid Percent** column and, if appropriate, the **Cumulative Percent** column. Write a sentence or two of interpretation for each table.

2.13 SOC In this exercise, you will use the *GSS2018. sav* database to produce pie charts and bar charts for marital status (*marital*) and a line chart and histogram for hours worked (*hrs1*). With the *GSS2018.sav* data loaded, click **Graphs** and then **Legacy Dialogs**. The submenu displays the types of graphs available. Each time you finish a graph, return to the Data Editor window and click **Graphs** and **Legacy Dialogs** again to get the next graph.

- For a pie chart:
 - Click **Pie** and then **Define** on the "Pie Chart" window.
 - On the "Define Pie:" window, move *marital* into the "Define Slices by" box.
 - Click **OK**.
- For a bar chart:
 - Click **Bar** and then **Define** on the "Bar Charts" window.
 - On the "Define Simple Bar:" window, move *marital* into the "Category Axis" box.
 - Click **OK**.
- For a line chart:
 - Click **Line** and then **Define** on the "Line Charts" window.
 - On the "Define Simple Line:" window, move *hrs1* into the "Category Axis" box.
 - Click **OK**.
- For a histogram:
 - Click **Histogram**.
 - Move *hrs1* into the "Variables" box.
 - Click **OK**.

Use the graphs to write a few sentences of description for each variable.

Is There a "Culture War" in the United States?

One of the early steps in a research project is to produce frequency distributions for all variables. If nothing else, these tables provide excellent background information and, sometimes, you can use them to begin to answer research questions. In this installment of "You Are the Researcher," you will use SPSS to produce summary tables for several variables that measure attitudes about controversial issues in U.S. society and that may map the battlefronts in what many call the American culture wars.

The United States seems to be divided on a number of religious, political, and cultural issues and values. We might characterize the opposing sides using terms like "liberal vs. conservative," "modern vs. traditional," or "progressive vs. old-school"; some of the most bitter debates across these lines include abortion, same-sex marriage, and gun control. As you know, debates over issues like these can be intense, bitter, and even violent: Adherents of one position may view their opponents with utter contempt, demonize them, and dismiss their arguments. How deep is this fault line in U.S. society?

We can begin to investigate these questions by choosing three variables from the 2018 General Social Survey (GSS) that seem to differentiate the sides in the American culture war. Before doing this, let's take a moment to consider the process of "picking variables." Technically, selecting a variable to represent a concept such as "culture war" is called operationalization, and this can be one of the most difficult steps in a research project. We discussed operationalization in Chapter 1 when we reviewed the research process (see Figure 1.1). When we move from the "Theory" stage to the "Hypothesis" stage, we identify specific variables (such as responses to survey items) that match our concepts (like prejudice). This can be challenging, as our concepts are usually quite abstract and subject to a variety of perspectives. What exactly is a culture war, and why are some positions liberal or traditional, conservative or progressive? In order to do research, we must use concrete variables to represent our abstract concepts, but which variables relate to which concepts?

Any pairing we make between variables and concepts is bound to be at least a little arbitrary. In many cases, the best strategy is to use several variables to represent the concept: If our operationalizations are reasonable, our selected variables will behave similarly and each will behave as the abstract concept would if we could measure it directly. This is why we ask you to select three different variables to represent the culture wars. Each researcher may select different variables but, if everyone makes reasonable decisions, the chosen variables should be close representations of the concept.

Begin by starting SPSS and opening the 2018 GSS. You can download this file from the Web site for this text if you haven't already.

Next, select three variables that seem relevant to the culture war by browsing through the list of variables in Appendix G. You can also see a list of variables in the database by clicking **Utilities** and then **Variables** on the menu bar of the Data Editor window. After you have made your selections, complete the following steps.

Step 1: Identify Your Three Variables

Write the names of your variables as they appear in the database (for example: *abany*, *marital*, or *sex*). To "explain what the variable measures," look at the wording of the

survey items in Appendix G or use the abbreviated statements in SPSS by clicking **Utilities** and then **Variables** on the menu bar.

SPSS variable name	Explain exactly what this variable measures

Step 2: Operationalization

Explain how each of your variables relates to the culture wars. Which position is liberal and which is conservative? For example, you might argue that liberal would be pro-choice on abortion and conservatives would be more supportive of gun control.

SPSS variable name	Explain the relationship between each variable and the culture wars. Which position is liberal and which is conservative?

Step 3: Producing Frequency Distributions

Now we are ready to generate output and get some background on the nature of disagreements over values and issues among Americans.

Generating Frequency Distributions

We produced and examined a frequency distribution for the variable *sex* in Appendix F and for *attend* in the "Using SPSS" demonstration in this chapter. Use the same procedures to produce frequency distributions for the three variables you used to represent the American culture wars. From the menu bar, click **Analyze**, then click **Descriptive Statistics**, and then **Frequencies**.

Find your three variables in the **Frequencies** window one at a time. Click each variable name and then click the arrow button in the middle of the screen to move it to the right-hand window. SPSS will process all variables listed in the right-hand box together. Click **OK** in the **Frequencies** window, and SPSS will rush off to create the frequency distributions you requested. The tables will be in the **SPSS Viewer** or Output window that will now be "closest" to you on the screen.

Reading SPSS Frequency Distributions

We discussed how to read frequency distributions produced by SPSS in the "Using SPSS" demonstration earlier in this chapter. Remember to disregard the "Percent" column.

Step 4: Interpreting Results

Summarize your results by reporting the percentage (not the frequencies) of respondents who endorsed each response. How large are the divisions in American values? Is there consensus on the issue measured by your variable (do the great

majority endorse the same response) or is there considerable disagreement? The lower the consensus, the greater the opportunity for the issue to be included in the culture war.

SPSS name of variable 1: _____

Summarize the frequency distribution in terms of the percentage of respondents who endorsed each position:

Are these results consistent with the idea that there is a "war" over American values? How?

SPSS name of variable 2: _____

Summarize the frequency distribution in terms of the percentage of respondents who endorsed each position:

Are these results consistent with the idea that there is a "war" over American values? How?

SPSS name of variable 3: _____

Summarize the frequency distribution in terms of the percentage of respondents who endorsed each position:

Are these results consistent with the idea that there is a "war" over American values? How?

3 Measures of Central Tendency

LEARNING OBJECTIVES By the end of this chapter, you will be able to:

1. Explain to others how measures of central tendency can make data understandable.
2. Calculate the appropriate measure of central tendency for variables at each of the three levels of measurement.
3. Explain the difference in the types of information provided by the mode, median, and mean.
4. Explain the mathematical characteristics of the mean, especially how it is affected by skew.
5. Produce measures of central tendency using SPSS and explain their meaning.

USING STATISTICS

The statistical techniques presented in this chapter are used to find the typical case or average score on a single variable. They can be used to

- Identify the most common opinion that teenagers have about how often they should communicate with their dating partner. **See page 70**

- Compare the teen birth rate in the United States by state to determine which states have the highest and lowest levels. **See page 74**

- Measure the median income and the mean income among workers in the United States. **See page 86**

- Measure the average duration of traffic congestion in hours per day by city. **See page 90**

One benefit of frequency distributions and graphs is that they summarize the overall shape of a distribution of scores in a way that can be quickly comprehended. However, you will almost always need to report more detailed information about the distribution. Specifically, two additional kinds of statistics are extremely useful: some idea of the typical or average case in the distribution (for example, "The average starting salary for social workers in this state is $39,000") and some idea of how much variety there is in the distribution ("In this state, starting salaries for social workers range from $33,000 to $45,000 per year"). The first kind of statistic, called a **measure of central tendency**, is the subject of this chapter. The second kind of statistic, a measure of variation, is presented in Chapter 4.

The three commonly used measures of central tendency—the mode, median, and mean—are all probably familiar to you. They summarize a distribution of scores by describing the most common score (the mode), the score of the middle case (the median), or the average score (the mean) of that distribution. These statistics are powerful because they can reduce huge arrays of data to a single, easily understood number. Remember that the central purpose of descriptive statistics is to summarize or "reduce" data.

Even though they share a common purpose, the three measures of central tendency are different statistics and they will have the same value only under certain conditions. They vary in terms of level of measurement and, perhaps more importantly, in terms of how they define central tendency. They will not necessarily identify the same score or case as "typical." Thus, your choice of an appropriate measure of central tendency will depend, in part, on what exact information you want to convey.

The Mode

The **mode** of any distribution of scores is the value that occurs most frequently. For example, in the set of scores 58, 82, 82, 90, 98, the mode is 82 because it occurs twice and the other scores occur only once.

The mode is a simple statistic, most useful when you are interested in the most common score and when you are working with nominal-level variables. In fact, the mode is the only measure of central tendency for nominal-level variables. For example, Table 3.1 displays the distribution of household living arrangements for Americans in 2018. This information is similar to that reported in Table 2.18, but it uses different categories and is for a representative sample, not for the entire population. The mode of this distribution, the single most common household type, is "married couples with children," but note that the "single adult" category is nearly as large. This table reinforces the idea, first expressed in Chapter 2, that the "nuclear family" (parents and children) has lost popularity in the United States. Finally, note that cohabitation has become a prominent feature of American family life.

You should note that the mode has several limitations. First, distributions can have no mode at all (when all scores have the same frequency) or so many modes that the statistic becomes meaningless. For example, consider the unusual but not impossible distributions of test scores in Table 3.2. Test A has no "single most common" score and, therefore, no mode. Test B has four different

TABLE 3.1 Household Living Arrangements in the United States, 2018

Household Type	Frequency	Percentage
Married Couple with Children	705	33.73
Married Couple, no Children	343	16.41
Single Adult	669	32.01
Cohabiting Couple, with and without Children	162	7.75
Other	211	10.10
	$N = 2090$	100.00%

TABLE 3.2 Distributions of Scores on Two Tests

Score (% correct)	Test A Frequency of scores	Test B Frequency of scores
97	14	22
91	14	3
90	14	4
86	14	22
77	14	3
60	14	22
55	14	22
	$N = 98$	98

TABLE 3.3 A Distribution of Test Scores

Score (% correct)	Frequency
93	8
68	3
67	4
66	2
62	7
	$N = 24$

modes—55, 60, 86, and 97—and it is unlikely that reporting all four would convey any useful information about central tendency in the distribution.

Second, the modal score of ordinal or interval-ratio variables may not be central to the distribution as a whole. That is, *most common* does not necessarily identify the center of the distribution. For example, consider the unusual but, again, not impossible distribution of test scores in Table 3.3. The mode of the distribution is 93. Is this score close to the majority of the scores? If the instructor summarized this distribution by reporting only the modal score, would he or she be conveying an accurate picture of the distribution as a whole? *(For practice in finding and interpreting the mode, see problems 3.1 to 3.7.)*

STATISTICS **In Everyday Life**

Communication Between Dating Teenagers

How often do you think teenagers in dating relationships communicate with their partners? Data from the Pew Research Center tell us that the answer is quite a lot! When asked how often they expect to hear from their dating partner, the modal (or most common) response among teenagers was "once a day" (38%) followed closely by "every few hours" (35%) and hourly (11%). It's also interesting to note that these dating teenagers think that their partners have similar expectations of them. In the same survey, when asked how much their partner expects to hear from them, the modal response was "every few hours" (38%), followed by "once a day" (35%) and "hourly" (15%).

Source: Pew Research Center at: https://www.pewinternet.org/2015/10/01/teens-technology-and-romantic-relationships/

The Median

Unlike the mode, the **median (Md)** is always at the exact center of a distribution of scores. The median is the score of the case that is at the middle of a distribution: Half the cases have scores higher than the median and half the cases have scores lower than the median. Thus, if the median family income for a community is $52,000, half the families earn more than $52,000 and half earn less.

Before finding the median, you must place the cases in order from the highest to the lowest score—or from the lowest to the highest. Then you can determine the median by locating the case that divides the distribution into two equal halves. The median is the score of the middle case. If five students received grades of 93, 87, 80, 75, and 61 on a test, the median would be 80, the score that splits the distribution into two equal halves.

When the number of cases (N) is odd, the value of the median is unambiguous because there will always be a middle case. With an even number of cases, however, there will be two middle cases and, in this situation, the median is defined as the score exactly halfway between the scores of the two middle cases.

To illustrate, assume that seven students were asked to indicate their level of support for intercollegiate athletics at their universities on a scale ranging from 10 (great support) to 0 (no support). After arranging their responses from high to low, you can find the median by locating the case that divides the distribution into two equal halves. With a total of seven cases, the middle case would be the fourth case, since there will be three cases above and three cases below this case. Table 3.4 lists the cases in order and identifies the median. With seven cases, the median is the score of the fourth case.

Now suppose we add another student, whose support for athletics was measured as a 1. This would make N an even number (8) and we would no longer have a single middle case. Table 3.5 presents the new distribution of scores, and, as you can see, any value between 7 and 5 would technically satisfy the definition of a median (that is, it would split the distribution into two equal halves of four cases each). We resolve this ambiguity by defining the median as the average of the scores of the two middle cases. In this example, the median would be defined as $(7 + 5)/2$, or 6.

TABLE 3.4 Finding the Median with Seven Cases (N is odd)

Case	Score	
A	10	
B	10	
C	8	
D	7	← Md
E	5	
F	4	
G	2	

TABLE 3.5 Finding the Median with Eight Cases (N is even)

Case	Score	
A	10	
B	10	
C	8	
D	7	
		← Md = $(7 + 5)/2 = 6$
E	5	
F	4	
G	2	
H	1	

ONE STEP AT A TIME Finding the Median

Step	Operation
1.	Array the score in order from high score to low score.
2.	Count the number of scores to determine whether N is odd or even.

IF N is ODD:
⇓

IF N is EVEN:
⇓

3.	The median will be the score of the middle case.	3.	The median is halfway between the scores of the two middle cases.
4.	To find the middle case, add 1 to N and divide the result by 2.	4.	To find the first middle case, divide N by 2.
5.	The value you calculated in Step 4 is the number of the middle case. The median is the score of this case. For example, if $N = 13$, the median will be the score of the $(13 + 1)/2$ or seventh case.	5.	To find the second middle case, increase the value you computed in Step 4 by 1.
		6.	Add the scores of the two middle cases and divide by 2. The result is the median. For example, if $N = 14$, the median is the score halfway between the seventh and eighth cases. If the middle cases have the same score, that score is defined as the median.

Remember that we follow different procedures to find the median, depending on whether N is odd or even. The procedures are stated in general terms in the "One Step at a Time" box.

The median cannot be calculated for variables measured at the nominal level because it requires that scores be ranked from high to low, and nominal-level variables cannot be ordered or ranked. The scores of nominal-level variables are different from each other but do not form a mathematical scale of any sort. The median can be found for either ordinal or interval-ratio data but is generally more appropriate for the former. *(The median may be found for any problem at the end of this chapter.)*

Other Measures of Position: Percentiles, Deciles, and Quartiles

In addition to serving as a measure of central tendency, the median belongs to a class of statistics that measure position or location. The median identifies the exact middle of a distribution, but it is sometimes useful to locate other points as well. We may want to know, for example, the scores that split the distribution into thirds or fourths, or the point below which a certain percentage of the cases fall. A familiar application of these measures would be scores on standardized tests, which are often reported in terms of location (for example, "A score of 476 is higher than 46% of the scores").

PERCENTILES

One commonly used statistic for reporting position is the **percentile**, which identifies the point below which a specific percentage of cases fall. If a score of 476 is reported as the 46th percentile (P_{46}), this means that 46% of the cases scored less than 476 and 54% scored higher.

To find a percentile, first arrange the scores in order. Next, multiply the number of cases (N) by the proportional value of the percentile. The resultant value identifies the number of the case that marks the percentile. For example, the proportional value for the 46th percentile would be 0.46. To find the 46th percentile for a sample of 78 cases, multiply 78 by 0.46. The result is 35.88, and the 46th percentile would be 88/100 of the distance between the scores of the 35th and 36th cases. In most situations, we would probably round off 35.88 to 36 and call the score of the 36th case the 46th percentile. The slight inaccuracy would be worth the savings in time and computational effort.

In terms of percentiles, the median is the 50th percentile, and in our example we would find the median by multiplying 78 by 0.50, finding the 39th case, and declaring the score of that case to be the 50th percentile. Notice again that we are cutting some corners here. Technically, the median would be the score halfway between the two middle cases (the 39th and 40th cases), but it is unlikely that this inaccuracy would be very significant.

DECILES AND QUARTILES

Some other commonly used measures of position are **deciles** and **quartiles**. Deciles divide the distribution of scores into tenths. So the first decile (D_1) is the point below which 10% of the cases fall and is equivalent to the 10th percentile (P_{10}). The fifth decile (D_5) is also the same as the 50th percentile (P_{50}), which is the same as (you guessed it) the median.

Quartiles divide the distribution into quarters, and the first quartile (Q_1) is the same as the 25th percentile (P_{25}), the second quartile (Q_2) is the 50th percentile (P_{50}), and the third quartile (Q_3) is the 75th percentile (P_{75}). Any of these measures can be found by the method described earlier for percentiles. Remember that multiplying N by the proportional value of the percentile, decile, or quartile identifies the *case*, and it's the *score* of the case that actually marks the location. Also remember that this technique cuts some (probably minor) computational corners, so use it with caution. *(Quartiles, deciles, and percentiles may be found for any ordinal or interval-ratio variable in the problems at the end of this chapter.)*

USING PERCENTILES TO FIND THE MEDIAN AND THE MODE

Percentiles in a frequency distribution can also be used to find the median and the mode. Take a look at the frequency distribution in Table 3.6. Orient your eyes to it. As you learned in Chapter 2, in the leftmost column you can see that there are 9 values, ranging from "0" for no children to "eight or more" children. The frequency column tells you how many respondents gave each answer. So, 663 people reported that they have no children. The Percent column tells you the percentage of respondents that gave each answer using the overall total (2348) as the N, including the respondents who refused to answer (NA) or claimed to not know (DK). As we

TABLE 3.6 **Frequency Distribution for the Number of Children in the 2018 General Social Survey**

Number of children

		Frequency	Percent	Valid Percent	Cumulative Percent
Valid	0	663	28.2	28.3	28.3
	1	349	14.9	14.9	43.2
	2	623	26.5	26.6	69.8
	3	383	16.3	16.3	86.1
	4	166	7.1	7.1	93.2
	5	74	3.2	3.2	96.3
	6	47	2.0	2.0	98.3
	7	23	1.0	1.0	99.3
	EIGHT OR MORE	16	.7	.7	100.0
	Total	2344	99.8	100.0	
Missing	DK NA	4	.2		
Total		2348	100.0		

Source: Smith, Tom W., Davern, Michael, Freese, Jeremy, and Morgan, Stephen L., General Social Surveys, 1972–2018 [machine-readable data file] /Principal Investigator, Smith, Tom W.; Co-Principal Investigators, Michael Davern, Jeremy Freese, and Stephen L. Morgan; Sponsored by National Science Foundation.—NORC ed.—Chicago: NORC, 2019.

discussed in the "Using SPSS" feature in Chapter 2, we will generally ignore this column. The Valid Percent column represents the percentage that gave each answer using the total number of valid cases as the N (2344). Finally, the Cumulative Percent represents the percentage of valid cases that gave each response or less (for example, 43.2% of the respondents reported having either no children or one child). Now, how can you use this information to quickly find the mode and median?

The mode is easy. Just select the value with the highest Valid Percent. The mode here is zero children because its Valid Percent (28.3) is the highest in the table. For the median, use the Cumulative Percent column. Select the value that is associated with the first Cumulative Percentage to reach or exceed 50%. In this case, the median is 2 because its Cumulative Percentage (69.8) is the first to reach or exceed 50%. This means that the person (or persons in an even numbered sample) in the middle of the distribution has exactly 2 children.

Social Research and Data Analysis 3.1 **Finding Measures of Position**

Table 3.7 presents the teen birth rate (number of births for women ages 15–19 per 1000 population) for 50 states in 2017. Scores have been ranked from low (Massachusetts with a rate of 8.1) to high (Arkansas with a rate of 32.8). The various percentiles, deciles, and quartiles have been noted in the table.

Since there are 50 cases, the 10th percentile (P_{10}) or 1st decile (D_1) is found by first multiplying 50 by 0.10. The

score of the 5th case (New Jersey) is 10.3 and this is the 10th percentile (or 1st decile).

The case that marks the first quartile (Q_1) is found by multiplying 50 by 0.25. This gives a value of 12.5, so Q_1 is the score halfway between the 12th and 13th cases or a value of 14.8 (since both Pennsylvania and Washington have the same rate). Follow the same routines to find other percentiles, deciles, and quartiles.

(continued)

Social Research and Data Analysis 3.1 *(Continued)*

Where does your state fall in the ranking? What sociological factors might help to explain these scores? For example, why do you think Massachusetts, New Hampshire, and Connecticut are ranked so low while Mississippi and Arkansas rank highest? Incidentally, the teen birth rate for the nation has fallen by a remarkable 70% since 1991.[1] What factors might account for this decline?

TABLE 3.7 Deciles, Quartiles, and Percentiles for Teen Birth Rates of 50 States, 2017

No.	State	Birth Rate		No.	State	Birth Rate	
1	MA	8.1		26	ID	18.6	
2	NH	8.4		27	HI	19.1	
3	CT	8.8		28	NC	20.6	
4	VT	10.1		29	OH	20.8	
5	NJ	10.3	←D_1, P_{10}	30	MT	21.2	← D_6, P_{60}
6	RI	11.4		31	KS	21.3	
7	MN	12.1		32	SC	21.7	
8	NY	12.5		33	GA	21.9	
9	ME	13.1		34	NV	21.9	
10	WI	13.8	← D_2, P_{20}	35	AK	22.0	← D_7, P_{70}
11	MD	14.2		36	AZ	22.0	
12	PA	14.8	← Q_1 (12.5), P_{25}	37	MO	22.5	← Q_3 (37.5), P_{75}
13	WA	14.8		38	SD	22.6	
14	OR	15.0		39	IN	22.8	
15	VA	15.0	← D_3, P_{30}	40	WY	24.6	← D_8, P_{80}
16	CA	15.1		41	TN	26.6	
17	UT	15.2		42	AL	27.0	
18	IA	16.0		43	WV	27.1	
19	CO	16.1		44	TX	27.6	
20	ND	16.2	← D_4, P_{40}	45	NM	27.9	← D_9, P_{90}
21	MI	16.4		46	KY	29.0	
22	IL	17.4		47	LA	29.1	
23	NE	18.1		48	OK	29.7	
24	FL	18.2		49	MS	31.0	
25	DE	18.5	← D_5, Q_2, P_{50}, Md	50	AR	32.8	← D_{10}, P_{100}, Q_4

Source: Centers for Disease Control, 2018. "Teen Birth Rates by State." https://www.cdc.gov/nchs/pressroom/sosmap/teen-births/teenbirths.htm

The Mean

The **mean** (\overline{X}—read this as "ex-bar"),[1] or arithmetic average, is by far the most commonly used measure of central tendency. It reports the average score of a distribution, and it is calculated by dividing the sum of the scores by the number of scores (N).

This is the symbol for the mean of a sample. The symbol for the mean of a population is Greek letter mu (μ), pronounced 'mew'.

To illustrate: A birth control clinic administered a 20-item test of general knowledge about contraception to 10 clients. The numbers of correct responses were 2, 10, 15, 11, 9, 16, 18, 10, 11, and 7. To find the mean of this distribution, add the scores (total = 109) and divide by the number of scores (10). The result (10.9) is the average score on the test.

The mathematical formula for the mean is

FORMULA 3.1

$$\overline{X} = \frac{\Sigma(X_i)}{N}$$

where \overline{X} = the mean

$\Sigma(X_i)$ = the summation of the scores

N = the number of cases

Let's take a moment to consider the new symbols introduced in this formula. First, the symbol Σ (uppercase Greek letter sigma) is a mathematical operator just like the plus sign (+) or divide sign (÷). It stands for "the summation of" and directs us to add whatever quantities are stated immediately following it.[2]

The second new symbol is X_i ("X-sub-i"), which refers to any single score—the "ith" score. If we wish to refer to the score of a particular case, we replace the subscript with the number of the case. Thus, X_1 would refer to the score of the first case, X_2 to the score of the second case, X_{26} to the 26th case, and so forth.

The operation of adding all the scores is symbolized as $\Sigma(X_i)$. This combination of symbols directs us to sum the scores, beginning with the first score and ending with the last score in the distribution. Thus, Formula 3.1 states in symbols what has already been stated in words (to calculate the mean, add the scores and divide by the number of scores), but in a succinct and precise way. *(For practice in computing the mean, see any of the problems at the end of this chapter.)*

Because computation of the mean requires addition and division, it should be used with variables measured at the interval-ratio level. However, researchers do calculate the mean for variables measured at the ordinal level, because the mean is much more flexible than the median and is a central feature of many interesting and powerful advanced statistical techniques. Thus, if the researcher plans to do any more than merely describe his or her data, the mean will probably be the preferred measure of central tendency even for ordinal-level variables.

ONE STEP AT A TIME **Finding the Mean**

1. Add up the scores (X_i)
2. Divide the quantity you found in step 1 (ΣX_i) by N.

[2]See the Prologue (Basic Review of Math) for further information on the summation sign and on summation operations.

Three Characteristics of the Mean

The mean is the most commonly used measure of central tendency, and we will consider its mathematical and statistical characteristics in some detail.

1. **The Mean Balances All the Scores.** First, the mean is an excellent measure of central tendency because it acts like a fulcrum that "balances" all the scores: The mean is the point around which all of the scores cancel out. We can express this property symbolically:

$$\Sigma(X_i - \overline{X}) = 0$$

This expression says that if we subtract the mean (\overline{X}) from each score (X_i) in a distribution and then sum the differences, the result will always be 0.

To illustrate, consider the test scores presented in Table 3.8. The mean of these five scores is 390/5, or 78. The difference between each score and the mean is listed in the right-hand column, and the sum of these differences is 0. The total of the negative differences (-19) is exactly equal to the total of the positive differences $(+19)$, as will always be the case. Thus, the mean "balances" the scores and is at the center of the distribution.

2. **The Mean Minimizes the Variation of the Scores.** A second characteristic of the mean is called the "least-squares" principle, a characteristic that is expressed in the statement

$$\Sigma(X_i - \overline{X})^2 = \text{minimum}$$

or: The mean is the point in a distribution around which the variation of the scores (as indicated by the squared differences) is minimized. If the differences between the scores and the mean are squared and then added, the resultant sum will be less than the sum of the squared differences between the scores and any other point in the distribution.

To illustrate this principle, consider Table 3.9. Column 1 of the table shows the same five scores displayed in Table 3.8, and column 2 displays the differences between the scores and the mean. In column 3, the differences between the scores and the mean are squared, and the sum of these differences is 388.

If we performed those same mathematical operations with any number other than the mean—say, 77—the result would be a sum greater than 388.

TABLE 3.8 A Demonstration Showing That All Scores Cancel Out Around the Mean

X_i	$X_i - \overline{X}$
65	$65 - 78 = -13$
73	$73 - 78 = -5$
77	$77 - 78 = -1$
85	$85 - 78 = 7$
90	$90 - 78 = 12$
$\Sigma(X_i) = 390$	$\Sigma(X_i - \overline{X}) = 0$
$\overline{X} = 390/5 = 78$	

TABLE 3.9 A Demonstration Showing That the Mean Is the Point of Minimized Variation

1	2	3	4
X_i	$X_i - \overline{X}$	$(X_i - \overline{X})^2$	$(X_i - 77)^2$
65	$65 - 78 = -13$	$(-13)^2 = 169$	$65 - 77 = (-12)^2 = 144$
73	$73 - 78 = -5$	$(-5)^2 = 25$	$73 - 77 = (-4)^2 = 16$
77	$77 - 78 = -1$	$(-1)^2 = 1$	$77 - 77 = (0)^2 = 0$
85	$85 - 78 = 7$	$(7)^2 = 49$	$85 - 77 = (8)^2 = 64$
90	$90 - 78 = 12$	$(12)^2 = 144$	$90 - 77 = (13)^2 = 169$
$\Sigma(X_i) = 390$	$\Sigma(X_i - \overline{X}) = 0$	$\Sigma(X_i - X)^2 = 388$	$\Sigma(X_i - 77)^2 = 393$

This point is illustrated in column 4 of Table 3.9, which shows that the sum of the squared differences around 77 is 393, a value greater than 388.

This least-squares principle underlines the fact that the mean is closer to all of the scores than the other measures of central tendency. Also, this characteristic of the mean is important for the statistical techniques of correlation and regression, topics we take up toward the end of this text.

3. **The Mean Is Affected by All Scores and Can Be Misleading If the Distribution Has "Outliers."** The final important characteristic of the mean is that all the scores in a distribution are included in its calculation ("to find the mean, add up *all* the scores and divide by N"). In contrast, the mode reflects only the most common score, and the median deals only with the score of the middle case.

On one hand, this characteristic is an advantage because the mean utilizes all the available information. On the other hand, when a distribution has **outliers** or some extremely high or low scores, the mean may become misleading: It may not represent the central or typical score. Distributions with outliers have a **skew**. If there are some extremely high scores, this is called a *positive skew*, and distributions with a few very low scores have a *negative skew*.

The point to remember is that the mean will be pulled in the direction of the outlying scores relative to the median. With a positive skew, the mean will be greater in value than the median, and just the opposite will occur with a negative skew.

Why is this problematic? Because the median uses *only* the middle cases, it will *always* reflect the center of the distribution. The mean, because it uses *all* cases (including outliers), may be much higher or lower than the bulk of the scores and give a false impression of centrality.

To illustrate, consider Table 3.10. The five scores in column 1 are symmetrical or unskewed: They are equidistant from each other, and both the high and low scores are an equal distance from the central score of 25. Because these scores have no skew, the median and mean are the same value (see column 2).

In column 3, an extremely high score (3500) has been added, and the distribution now has a positive skew. In column 4, we see that this very high score has no effect on the median; it remains at 25. This is because the median is based *only* on the score of the middle case and is not affected by changes in the scores of other cases. The mean, because it takes *all* scores into account, is very much affected. The mean changes from 25 to 718 solely because of the one extremely high score of 3500.

Finally, the scores in column 5 are the same as those in column 1, except the low score is changed to 0, thus creating a negative skew. Column 6

TABLE 3.10 A Demonstration Showing That the Mean Is Affected by Every Score

1	2	3	4	5	6
Scores	Measures of Central Tendency	Scores	Measures of Central Tendency	Scores	Measures of Central Tendency
15	Mean = 25	15	Mean = 718	0	Mean = 22
20		20		20	
25	Median = 25	25	Median = 25	25	Median = 25
30		30		30	
35		3500		35	

shows that the median stays at 25 but, in response to the lower score, the mean drops to 22.

We should make a few more points about these relationships:

1. The mean and median will have the same value only in distributions that are symmetrical, like the distribution in column 1.

2. Note that the mean in column 4 is very different from the five scores listed in column 3. In this case, is the mean or the median a better representation of the centrality of the scores? When a distribution has a pronounced skew, the median may be the preferred measure of central tendency even for interval-ratio variables.

3. Compare the means and medians in column 4 and column 6. In column 4, the mean is very different from the median, reflecting the fact that the "outlier" or extreme score of 3500 is very different from the other scores. The difference between the mean and median in column 6 is much less because the skew is not as extreme.

The general principle, again, is that the mean is always pulled in the direction of extreme scores relative to the median. The greater the skew (whether negative or positive), the greater will be the difference between the two measures of central tendency. The mean and median will have the same value when and only when a distribution is symmetrical. Figures 3.1 to 3.3 depict three different line charts that demonstrate these relationships.

These relationships between medians and means have a practical value. For one thing, a quick comparison of the median and mean will always tell you whether a distribution is skewed and the direction of the skew. If the mean is less than the median, the distribution has a negative skew. If the mean is greater than the median, the distribution has a positive skew.

FIGURE 3.1 A Positively Skewed Distribution (The mean is greater in value than the median.)

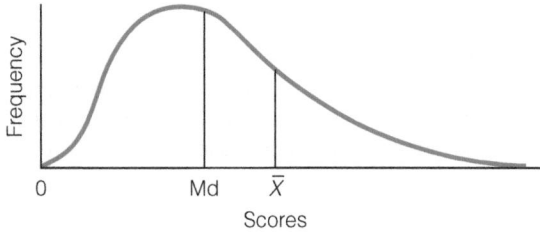

FIGURE 3.2 A Negatively Skewed Distribution (The mean is less than the median.)

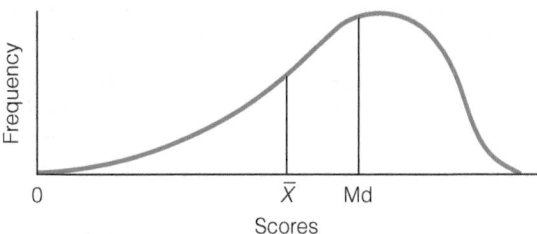

FIGURE 3.3 An Unskewed, Symmetrical Distribution (The mean and median are equal.)

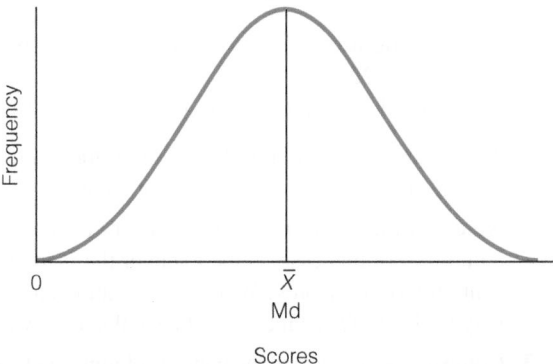

Statisticians use the term "skewness" to refer to the level of skew and its direction. Skewness is said to be "zero" when the mode, median, and mean are equal to one another, negative when the data are negatively skewed (when there are some extremely low scores), and positive when the data are positively skewed (when there are some extremely high scores). We'll learn more about cases when the median and mean are equal to one another in Chapter 5, which is devoted to the normal curve. A related term, kurtosis, can also tell us how many outliers there are relative to the center of the distribution, or how many cases are concentrated in the tails of the distribution versus the middle.

Which measure of central tendency is more appropriate for skewed distributions? This will depend on what point the researcher wishes to make but, as a rule, either both statistics or the median alone should be reported when the distribution has a few extreme scores. (*For practice in dealing with the effects of extreme scores on means and medians, see problems 3.11, 3.13, 3.14, and 3.15.*)

STATISTICS In Everyday Life

Means, Medians, and Baseball Salaries

Why are some professional baseball teams more successful than others? You don't have to be a sports fan to hypothesize that the best teams have the best athletes. At the professional level, this suggests that the best teams also have the highest payrolls because excellence is expensive!

(*continued*)

STATISTICS *(Continued)*

Major League Baseball publishes the salaries of every team and this information can create some interesting comparisons. The table below displays some payroll statistics for three of the most and least successful teams in the 2019 season. Do you see some patterns?

The most obvious message is that success costs money: the mean salaries of the more successful teams are roughly twice that of the less successful teams. Also note that the mean salaries are much higher than the median salaries for all teams. Does this suggest that these distributions are skewed? In what direction? What would account for the skew?

Here's a hint and some further information. The lowest salary on all teams was about $550,000, the major league minimum in 2019. Note that the median salaries for the less successful teams were not much higher than this minimum. Median salaries are much higher on the more successful teams, especially for the Astros. Does this suggest that the more successful teams have rosters crowded with high-priced, gifted athletes while less successful teams are more likely to be staffed by less proven (cheaper) talent?

	Mean	Median
Less Successful in 2019		
Baltimore Orioles	$2,857,573	$568,250
Detroit Tigers	$3,169,148	$570,800
Florida Marlins	$2,615,186	$570,000
More Successful in 2019		
L. A. Dodgers	$6,103,690	$2,887,500
Houston Astros	$6,135,296	$4,500,000
New York Yankees	$6,420,065	$1,593,750

Sources: https://www.usatoday.com/sports/mlb/salaries/2019/team/all/; https://www.spotrac.com/mlb/payroll/

USING SPSS **Measures of Central Tendency**

In this installment of "Using SPSS," we will get measures of central tendency for the U.S. homicide rate in two different years, using the 50 states as our cases. Our database will be the *States* file, which can be downloaded from the Web site for this text. We will continue to use **Frequencies**, the only SPSS procedure that will calculate all three measures of central tendency.

Follow these steps:

1. Click the SPSS icon on your desktop.
2. Load the database.
 a. Find the **File** command on the far left of the menu bar and then click **File, Open, and Data**.
 b. Find the *States* database. You can download this file from the Web site for this text if you haven't already.
3. From the menu bar across the top of the SPSS window, click **Analyze, Descriptive Statistics**, and **Frequencies**.

(continued)

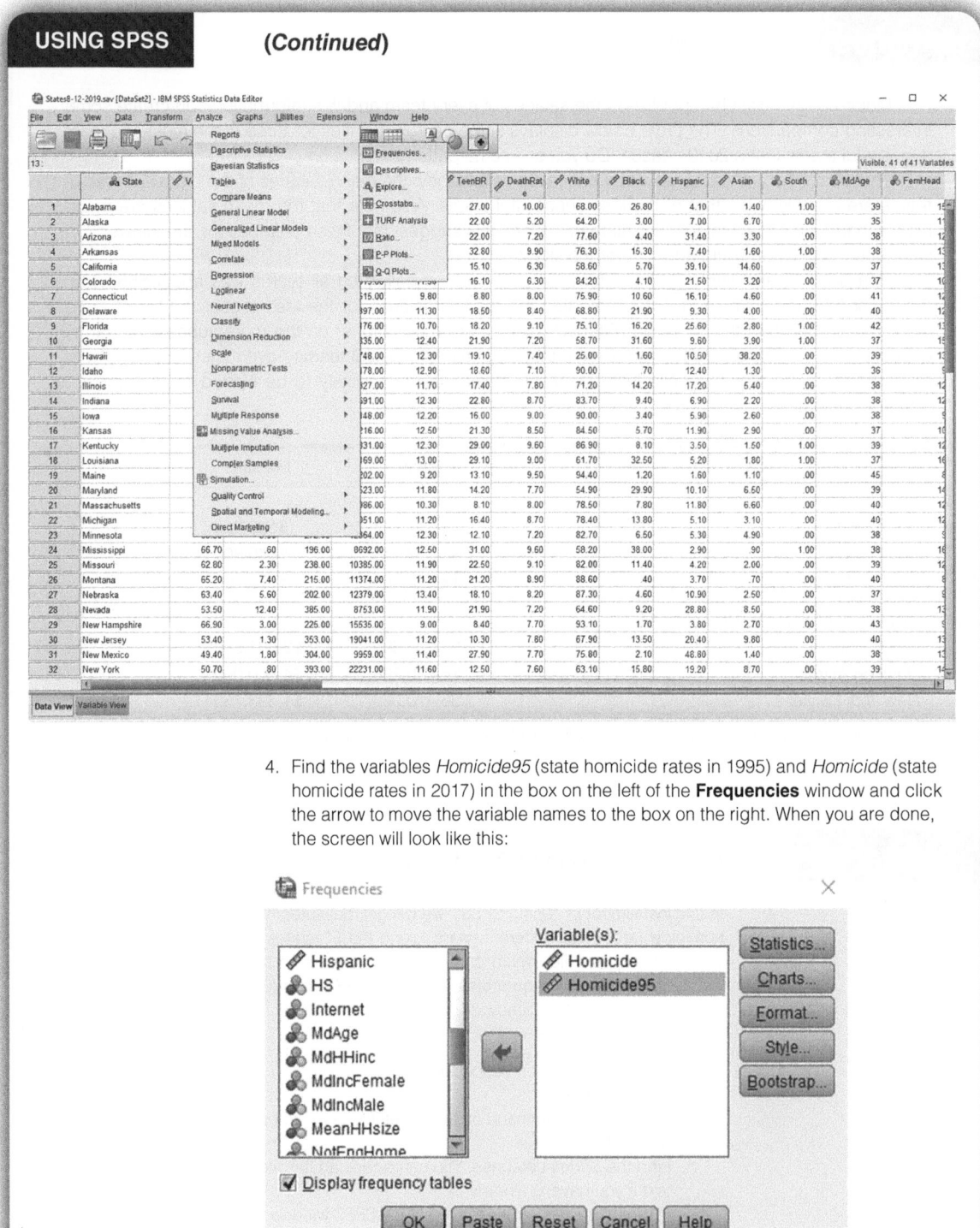

4. Find the variables *Homicide95* (state homicide rates in 1995) and *Homicide* (state homicide rates in 2017) in the box on the left of the **Frequencies** window and click the arrow to move the variable names to the box on the right. When you are done, the screen will look like this:

USING SPSS

(*Continued*)

5. Click the **Statistics** button on the upper right and the **Frequencies: Statistics** box will appear. Select the mean, median, and mode by clicking on the boxes to the left. The box will look like this when you are done:

Click Continue and you will be returned to the Frequency window.

6. Make sure you click the box next to the "Display frequencies tables" option under the variable list window to NOT select this option. We are interested only in the summary statistics, not in the tables.

7. Click **OK**.

The output will look like this:

Statistics

		Homicide	Homicide rate 1995
N	Valid	50	50
	Missing	0	0
Mean		5.010	6.7120
Median		5.000	6.2500
Mode		7.8	1.80

(*continued*)

USING SPSS

(Continued)

The mode for 2017 is 7.80 and for 1995 is 1.80, but this is not very useful information. The mean for 1995 (6.71) is higher than the median (6.25), indicating a positive skew in the data. With just 50 cases, we can find the outlier by inspection. Look at the SPSS Data Editor or the screen that displays the variables in columns (vertically) and the cases in rows (horizontally). Find the column titled Homicide95 and run your eyes down the column, Louisiana's score of 17.00 stands out because it is so much higher than the next highest score (12.90 for Mississippi).

The mean homicide rate for 2017 is lower than 1995 and comparing the mean (5.01) and median (5.00) tells us that there is no skew, or very high scores, in the distribution. By inspecting the scores in the SPSS Data Editor (find the column titled Homicide as you did before), we can see that Louisiana again has the highest homicide rate (12.4), but it is much lower than in 1995. Overall, we can conclude that the homicide rate fell between 1995 and 2017. We will defer further discussion of this trend until we have introduced the concept of variation in Chapter 4.

Choosing a Measure of Central Tendency

You should consider two main criteria when choosing a measure of central tendency. First, make sure that you know the level of measurement of the variable in question. This will generally tell you whether you should report the mode, median, or mean.

Table 3.11 shows the relationship between the level of measurement and measures of central tendency. The capitalized, boldface "**YES**" identifies the most appropriate measure of central tendency for each level of measurement and the non-boldface "Yes" indicates the levels of measurement for which the measure is also permitted. An entry of "No" in the table means that the statistic cannot be computed for that level of measurement. Finally, the "Yes (?)" entry in the bottom row indicates that the mean is often used with ordinal-level variables even though, strictly speaking, this practice violates level-of-measurement guidelines.

Second, consider the definitions of the three measures of central tendency and remember that they provide different types of information. They will be the same value only under certain specific conditions (namely, for symmetrical

TABLE 3.11 **The Relationship Between Level of Measure and Measures of Central Tendency**

Measure of Central Tendency	Level of Measurement		
	Nominal	*Ordinal*	*Interval-Ratio*
Mode	**YES**	Yes	Yes
Median	No	**YES**	Yes
Mean	No	Yes (?)	**YES**

TABLE 3.12 Choosing a Measure of Central Tendency

Use the mode when:	1	The variable is measured at the nominal level.
	2	You want a quick and easy measure for ordinal and interval-ratio variables.
	3	You want to report the most common score.
	4	You have an interval-ratio variable with two modes that are separated from one another. This is called a bimodal distribution. In this case, report both modes.
Use the median when:	1	The variable is measured at the ordinal level.
	2	An interval-ratio variable is highly skewed.
	3	You want to report the central score. The median always lies at the exact center of the distribution.
Use the mean when:	1	The variable is measured at the interval-ratio level (except when the variable is highly skewed).
	2	You want to report the typical score. The mean is the "fulcrum that exactly balances all of the scores."
	3	You anticipate additional statistical analysis.

distributions with one mode), and each has its own message to report. In many circumstances, you might want to report all three.

The guidelines in Table 3.12 stress both selection criteria and may be helpful when choosing a specific measure of central tendency. Item 4 under the top portion of the table introduces a new distribution referred to as a bimodal distribution. This is a case in which there are two modes in a distribution separated by at least a few values. For example, if you were to ask a sample of people how many years of education they have, you might find that most people would report 12 years (for a high school education) or 16 years (for a college education). Perhaps the frequencies for these two values would not be exactly the same, but they might be the two highest values, and very close to one another. In this sort of distribution, it often makes sense to report the two modes (12 and 16) because both the median and the mean would fall in between these two values and not do the best job of describing central tendency. This is illustrated in Figure 3.4.

FIGURE 3.4 A Bimodal Distribution Where the Median and the Mean Can Be Found in Between the Two Modes

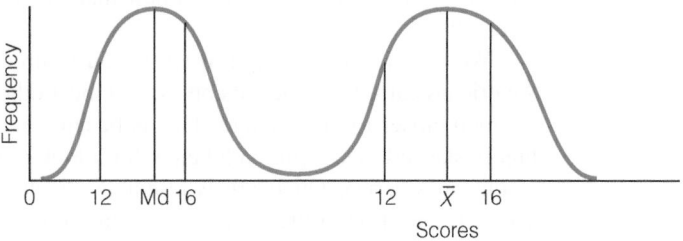

Social Research and Data Analysis 3.2 | **The Mean, Mode, and Median**

Ten students have been asked how many hours they spent in the college library during the past week. What is the average "library time" for these students? The hours are reported in the following list, and we will find the mode, the median, and the mean for these data.

Student	Hours Spent in the Library Last Week (X_i)
A	0
B	2
C	5
D	5
E	7
F	10
G	14
H	14
I	20
J	30
	$\Sigma(X_i) = 107$

By scanning the scores, we can see that two scores, 5 and 14, occurred twice, and no other score occurred more than once. This distribution has two modes, 5 and 14.

Because the number of cases is even, the median will be the average of the two middle cases after all cases have been ranked in order. With 10 cases, the first middle case will be the ($N/2$), or (10/2), or the fifth case. The second middle case is the ($N/2$) + 1, or (10/2) + 1, or the sixth case. The median will be the score halfway between the scores of the fifth and sixth cases. Counting down from the top, we find that the score of the fifth case (student E) is 7 and the score of the sixth case (student F) is 10. The median for these data is (7 + 10)/2, or (17/2), or 8.5. The median, the score that divides this distribution in half, is 8.5.

The mean is found by first adding all the scores and then dividing by the number of scores. The sum of the scores is 107, so the mean is

$$\bar{X} = \frac{\Sigma(X_i)}{N} = \frac{107}{10} = 10.7$$

These 10 students spent an average of 10.7 hours in the library during the week in question.

Note that the mean is higher than the median. This indicates a positive skew in the distribution (a few extremely high scores). By inspection we can see that the positive skew is caused by the two students who spent many more hours (20 and 30 hours) in the library than the other students.

Interpreting Statistics: Choosing a Measure of Central Tendency to Describe Income

How much income does the average American earn? There are a number of ways to answer this question, so it's wise to consider all of your options and be careful not to report deceptive or misleading data. First, let's consider what the shape of the distribution of income looks like. In other words, how many people do you expect have a low income and how many people do you expect have a high income?

Well, you're on the right track if you figured out that a small number of Americans earn the highest incomes, and most other people earn an income that is much closer to the average. This is because income in the United States is highly skewed. See Figure 3.5 below for a look at the distribution of *household* income in America, which largely resembles the distribution of *individual* income. The concentration of high bars on the left indicates that most American households have an income that is low compared to the small number of households on

FIGURE 3.5

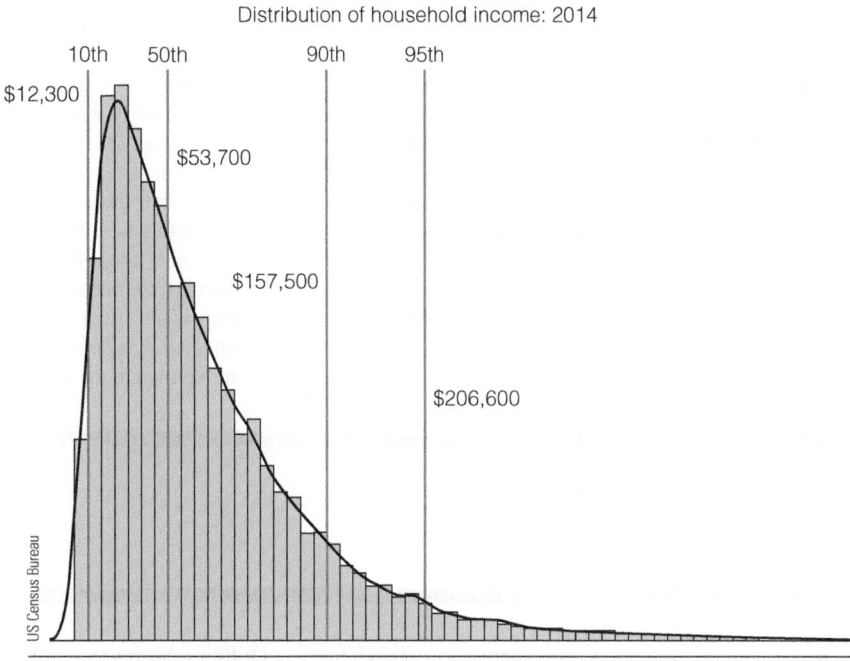

Distribution of household income: 2014

Source: https://www.ssa.gov/oact/cola/central.html

the right that have much higher incomes. Examine the position of the vertical line marked "50th." This tells you that 50% of the households had an income below $53,700 in 2014. Note that this line falls pretty far to the left of the distribution, instead of in the middle. Thus, this distribution is highly skewed to the right, with a few very high scores or incomes.

Let's use Table 3.12 to choose a measure of central tendency that will do the best job of telling us about how much Americans earn per year. The table guides us to select the median since the data are at the interval-level and highly skewed. But why is the median better than the mean? The answer can be found in the goals of reporting central tendency.

A good measure of central tendency is one that tells us the most typical or representative value among the cases. Which statistic, the median or the mean, would you expect to be closer to the most typical case in the figure above?

Let's look at some individual income data to provide the solution. In 2017, the median *individual* income was $31,561.49, and the average individual income was $48,251.57. It's no accident that the mean is higher than the median. You can always count on this being the case when we have positively skewed data. The small number of outlying incomes drags the average up and above the median (or below the median for negatively skewed data), placing it farther away from the concentration of the data.

So to answer our original question, it is best to report the median income because income is highly skewed, and as a result the median will be closer to the most typical case than the mean.

SUMMARY

1. The three measures of central tendency presented in this chapter share a common purpose. Each reports some information about the most typical or representative value in a distribution. Appropriate use of these statistics permits the researcher to report important information about an entire distribution of scores in a single, easily understood number.
2. The mode reports the most common score and is used most appropriately with nominal-level variables.
3. The median (Md) reports the score that is the exact center of the distribution. It is most appropriately used with variables measured at the ordinal level and with interval-ratio variables when the distribution is skewed.

4. Other measures of position besides the median include percentiles, deciles, and quartiles.
5. The mean (\overline{X}), the most frequently used of the three measures, reports the most typical score. It is used most appropriately with variables measured at the interval-ratio level (except when the distribution is highly skewed).
6. The mean has a number of important mathematical characteristics. First, it is the point in a distribution of scores around which all other scores cancel out. Second, the mean is the point of minimized variation (this is the "least-squares" principle). Last, the mean is affected by every score in the distribution and is therefore pulled in the direction of extreme scores.

SUMMARY OF FORMULAS

3.1 Mean $\overline{X} = \dfrac{\Sigma(X_i)}{N}$

GLOSSARY

Deciles. The points that divide a distribution of scores into tenths.
Mean. The arithmetic average of the scores. \overline{X} represents the mean of a sample, and μ, the mean of a population.
Measures of central tendency. Statistics that summarize a distribution of scores by reporting the most typical or representative value of the distribution.
Median (Md). The point in a distribution of scores above and below which exactly half of the cases fall.
Mode. The most common value in a distribution or the largest category of a variable.

Outliers. Scores that are very high or very low compared to most scores.
Percentile. A point below which a specific percentage of the cases fall.
Quartiles. The points that divide a distribution into quarters.
Σ (uppercase Greek letter sigma). "The summation of."
Skew. The extent to which a distribution of scores has a few scores that are extremely high (positive skew) or extremely low (negative skew).
X_i ("X-sub-i"). Any score in a distribution.

PROBLEMS

3.1 [SOC] A variety of information has been gathered from a sample of college freshmen and seniors, including
- Their region of birth;
- The extent to which they support legalization of marijuana (measured on a scale on which 7 = strong support, 4 = neutral, and 1 = strong opposition);
- The amount of money they spend each week out-of-pocket for food, drinks, and entertainment;
- How many movies they watched in their dorm rooms last week;
- Their opinion of cafeteria food (10 = excellent, 0 = very bad); and
- Their religious affiliation.

Some results are presented here. Find the *most appropriate* measure of central tendency for each variable for freshmen and then for seniors. Report both the measure you selected and its value for each variable (e.g., "Mode = 3" or "Median = 3.5"). *(HINT: Determine the level of measurement for each variable first. In general, this will tell you which measure of central tendency is appropriate. See the section "Choosing a Measure of Central Tendency" to review the relationship between measure of central tendency and level of measurement. Also, remember that the mode is the* most common *score, and especially remember to array scores from high to low before finding the median.)*

FRESHMEN

Student	Region of Birth	Legalization	Out-of-Pocket Expenses	Movies	Cafeteria Food	Religion
A	North	7	43	0	10	Protestant
B	North	4	49	14	7	Protestant
C	South	3	55	10	2	Catholic
D	Midwest	2	57	7	1	None
E	North	3	72	5	8	Protestant
F	North	5	58	1	6	Jew
G	South	1	62	0	10	Protestant
H	South	4	75	14	0	Other
I	Midwest	1	61	3	5	Other
J	West	2	53	4	6	Catholic

SENIORS

Student	Region of Birth	Legalization	Out-of-Pocket Expenses	Movies	Cafeteria Food	Religion
K	North	7	75	0	1	None
L	Midwest	6	72	5	2	Protestant
M	North	7	70	11	8	Protestant
N	North	5	95	3	4	Catholic
O	South	1	72	4	3	Protestant
P	South	5	67	14	6	Protestant
Q	West	6	50	0	2	Catholic
R	West	7	59	7	9	None
S	North	3	55	5	4	None
T	West	5	95	3	7	Other
U	North	4	88	5	4	None

3.2 A variety of information has been collected for all district high schools. Find the most appropriate measure of central tendency for each variable, and summarize this information in a paragraph.

(HINT: The level of measurement of the variable will generally tell you which measure of central tendency is appropriate. Remember to organize the scores from high to low before finding the median.)

High School	Enrollment	Largest Racial/ Ethnic Group	% College Bound	Most Popular Sport	Condition of Physical Plant (10 = Excellent)
A	1400	White	25	Football	10
B	1223	White	77	Baseball	7
C	876	Black	52	Football	5
D	1567	Hispanic	29	Football	8
E	778	White	43	Basketball	4
F	1690	Black	35	Basketball	5
G	1250	White	66	Soccer	6
H	970	White	54	Football	9

3.3 PS You have been observing the local Democratic Party in a large city and have compiled some information about a small sample of party regulars. Find the appropriate measure of central tendency for each variable.

Respondent	Gender	Social Class	No. of Years in Party	Education	Marital Status	No. of Children
A	M	High	32	High school	Married	5
B	M	Middle	17	High school	Married	0
C	M	Working	32	High school	Single	0
D	M	Working	50	8th grade	Widowed	7
E	M	Working	25	4th grade	Married	4
F	M	Middle	25	High school	Divorced	3
G	F	High	12	College	Divorced	3
H	F	High	10	College	Separated	2
I	F	Middle	21	College	Married	1
J	F	Middle	33	College	Married	5
K	M	Working	37	High school	Single	0
L	F	Working	15	High school	Divorced	0
M	F	Working	31	8th grade	Widowed	1

3.4 PA The following table presents the average duration of traffic congestion in hours per day for a group of cities in 2016.

City	Average Hours of Congestion
Birmingham	0.33
Boston	6.25
Charlotte	4.08
Cincinnati	3.08
Cleveland	2.54
Detroit	4.25
Hartford	2.54
Indianapolis	3.01
Kansas City	3.07
Las Vegas	4.20
Los Angeles	9.35
Louisville	2.30
Nashville	2.55
New Orleans	4.27
Oklahoma City	2.22
Phoenix	3.39
Richmond	2.08
San Antonio	3.45
San Diego	4.29

(*Continued next column*)

(*Continued*)

City	Average Hours of Congestion
St Louis	3.26
Tampa	2.51
Washington, D.C.	7.45

Source: U.S. Department of Transportation. 2017. "Urban Congestion Report." https://ops.fhwa.dot.gov/perf_measurement/ucr/reports/52urbanareas/fy2017_q1.pdf

a. Calculate the mean and median of this distribution.
b. Compare the mean and median. Which is the higher value? Why?
c. If you removed Los Angeles from this distribution and recalculated, what would happen to the mean? To the median? Why?
d. Report the mean and median as you would in a formal research report.

3.5 SOC Data have been gathered on four variables for 15 respondents (see the following table). Find and report the appropriate measure of central tendency for each variable. (On the "Attitude on Abortion" scale, a high score indicates strong opposition.)

Respondent	Marital Status	Racial/Ethnic Group	Age	Attitude on Abortion
A	Single	White	18	10
B	Single	Hispanic	20	9
C	Widowed	White	21	8
D	Married	White	30	10
E	Married	Hispanic	25	7
F	Married	White	26	7
G	Divorced	Black	19	9
H	Widowed	White	29	6
I	Divorced	White	31	10
J	Married	Black	55	5
K	Widowed	Asian American	32	4
L	Married	Native American	28	3
M	Divorced	White	23	2
N	Married	White	24	1
O	Divorced	Black	32	9

3.6 SOC The following tables list the median family and household incomes for 13 Canadian provinces and territories in 2000 and 2015 and for 13 states of the United States in 1999 and 2017. Compute the mean and median income for each year and compare the two measures of central tendency. Which measure of central tendency is greater for each year? Are the distributions skewed? In which direction?

Median Family Income for Canadian Provinces and Territories, 2000 and 2015 (Canadian dollars)

Province or Territory	2000	2015
Newfoundland and Labrador	38,800	67,272
Prince Edward Island	44,200	61,163
Nova Scotia	44,500	60,764
New Brunswick	43,200	59,347
Quebec	47,700	59,822
Ontario	55,700	74,287
Manitoba	47,300	68,147
Saskatchewan	45,800	75,412
Alberta	55,200	93,835
British Columbia	49,100	69,995
Yukon	56,000	84,521
Northwest Territories	61,000	117,688
Nunavut	37,600	97,441
Mean =		
Median =		

Source: Statistics Canada. 2019. Number of Households, Median Income, and Median Income Rank, https://www150.statcan.gc.ca/n1/daily-quotidien/170913/t001a-eng.htm

Median Household Income for Thirteen States, 1999 and 2017 (U.S. dollars)

State	1999	2017
Alabama	36,251	46,472
Alaska	51,396	76,114
Arkansas	29,682	43.813
California	43,629	67,169
Connecticut	50,593	73,781
Illinois	46,330	61,229
Kansas	37,348	55,477
Maryland	52,205	78,916
Michigan	46,089	52,668
New York	39,989	62,765
Ohio	39,489	52,407
South Dakota	35,825	54,126
Texas	38,688	57,051
Mean =		
Median =		

Sources: U.S. Census Bureau. 2019. Table H-8-a "Median Household Income by State." https://www.census.gov/data/tables/time-series/demo/income-poverty/historical-income-households.html

3.7 Find the appropriate measure of central tendency for each variable displayed in problem 2.5. Report each statistic as you would in a formal research report.

3.8 The table on the next page gives scores on four variables for 30 cases from the General Social Survey:
- "Age" is reported in years.
- "Happiness" consists of answers to the question "Taken all together, would you say that you are

1. very happy
2. pretty happy
3. not too happy

- "Number of Partners" refers to sex partners over the past five years:
 - A score of 0 to 4 is actual number of partners.
 - A score of 5 means 5–10 partners.
 - A score of 6 means 11–20 partners.
 - A score of 7 means 21–100 partners.
 - A score of 8 means more than 100 partners.
- "Religion" is self-identified religious affiliation (Protestant, Catholic, Jew, Other, or None).

For each variable, find the appropriate measure of central tendency, and write a sentence reporting this statistical information as you would in a research report.

Respondent	Age	Happiness	Number of Partners	Religion
1	20	1	2	Protestant
2	32	1	1	Protestant
3	31	1	1	Catholic
4	34	2	5	Protestant
5	34	2	3	Protestant
6	31	3	0	Jew
7	35	1	4	None
8	42	1	3	Protestant
9	48	1	1	Catholic
10	27	2	1	None
11	41	1	1	Protestant
12	42	2	0	Other
13	29	1	8	None
14	28	1	1	Jew
15	47	2	1	Protestant
16	69	2	2	Catholic
17	44	1	4	Other
18	21	3	1	Protestant
19	33	2	1	None
20	56	1	2	Protestant
21	73	2	0	Catholic
22	31	1	1	Catholic
23	53	2	3	None
24	78	1	0	Protestant
25	47	2	3	Protestant
26	88	3	0	Catholic
27	43	1	2	Protestant
28	24	1	1	None
29	24	2	3	None
30	60	1	1	Protestant

3.9 [SOC] The college administration is considering a total ban on student automobiles. You have conducted a poll on this issue of 20 fellow students and 20 of the neighbors who live around the campus and have calculated scores for your respondents. On the scale you used, a high score indicates strong opposition to the proposed ban. The scores are presented here for both groups. Calculate an appropriate measure of central tendency and compare the two groups in a sentence or two.

Students		Neighbors	
10	11	0	7
10	9	1	6
10	8	0	0
10	11	1	3
9	8	7	4
10	11	11	0
9	7	0	0
5	1	1	10
5	2	10	9
0	10	10	0

3.10 [SW] As the head of a social services agency, you believe that your staff of 20 social workers is very much overworked compared with 10 years ago. The caseloads for each worker are reported below for each of the two years in question. Has the average caseload increased? Which measure of central tendency is most appropriate to answer this question? Why?

2009		2019	
52	55	42	82
50	49	75	50
57	50	69	52
49	52	65	50
45	59	58	55
65	60	64	65
60	65	69	60
55	68	60	60
42	60	50	60
50	42	60	60

3.11 [SOC] The following table lists the number of roller coaster per capita for 15 nations. Compute the mean and median for these data. Which is greater in value? Is there a positive skew in the data? How do you know?

Nation	Number of Roller Coasters per Capita
Argentina	0.33
Belgium	2.18
Canada	1.57
China	0.05
Denmark	6.07
Ecuador	0.14
France	1.02
India	0.02
Japan	1.88
Mexico	0.20
New Zealand	0.72
Russia	0.13
Saudi Arabia	0.59
United Kingdom	2.64
United States	2.09
Mean =	
Median =	

Source: https://www.nationmaster.com/country-info/stats/Lifestyle/Roller-coasters-per-million

3.12 SP The social policy research agency you work for specializes in gathering data on teenage sexual activity and sexual education programming in Chicago. In their first survey conducted prior to the implementation of a sexual education program (the pretest), they asked the frequency in which teenagers engaged in sexual activity per week. One year later they went back to the same school after they had begun implementing the program and asked the same students the same question (the posttest). A sampling of the fictitious data are shown below. Compute the median and mean for both the pretest and posttest. Interpret these statistics. Did the program make a difference?

Case	Pretest	Posttest
A	4	2
B	1	1
C	1	2

(Continued next column)

(Continued)

Case	Pretest	Posttest
D	2	2
E	3	3
F	0	1
G	1	1
H	1	1
I	2	2
J	3	1
K	1	1
L	1	2

3.13 SOC A sample of 25 freshmen at a major university completed a survey that measured their degree of racial prejudice (the higher the score, the greater the prejudice).

a. Compute the median and mean scores for these data.

10	43	30	30	45
40	12	40	42	35
45	25	10	33	50
42	32	38	11	47
22	26	37	38	10

b. These same 25 students completed the same survey during their senior year. Compute the median and mean for this second set of scores, and compare them to the earlier set. What happened?

10	45	35	27	50
35	10	50	40	30
40	10	10	37	10
40	15	30	20	43
23	25	30	40	10

3.14 MR A market research company has gathered survey data by telephone on consumer's smartphone plans. Their goal is to measure the most commonly utilized smartphone features to use in an upcoming ad campaign. The data are shown below. For each variable, determine the best measure of central tendency and then calculate it.

Case	Present Income ($)	Type of Voice Mail	Unlimited Texting	Talk Minutes
A	104,000	Visual	Yes	800
B	68,000	Audio	No	400
C	54,000	Email	Yes	400
D	45,000	Email	No	400
E	40,000	Visual	No	400
F	85,000	Visual	Yes	800
J	48,000	Visual	Yes	400

3.15 SOC Professional athletes are threatening to strike because they claim that they are underpaid. The team owners have released a statement that says, in part, "The average salary for players was $1.2 million last year." The players counter by issuing their own statement that says, in part, "The typical player earned only $753,000 last year." Is either side necessarily lying? If you were a sports reporter and had just read Chapter 3 of this text, what questions would you ask about these statistics?

Statistical Analysis Using SPSS

3.16 SOC In this exercise, you will use SPSS to get measures of central tendency for several different variables in the *States* data set, which is available from the Web site for this text.
- Find and click the SPSS icon on your desktop.
- Load the *States* data set.
- From the menu bar across the top of the SPSS window, click **Analyze**, **Descriptive Statistics**, and **Frequencies**.
- Find these variables in the box on the left of the **Frequencies** window: *FamPoor* (percentage of all families in the state below the poverty line), *DeathRate* (number of deaths per 1000 population), and *Asian* (percentage of the state's population that is of Asian descent). Click the arrow to move the variable names into the box on the right.
- Click the **Statistics** button and, under "Central Tendency," select Mean, Median, and Mode. Click **Continue** to return to the **Frequencies** window.
- Make sure the box next to the "Display frequency tables" option is *not* checked. We are interested only in the summary statistics, not in the tables.
- Click **OK**.
 a. Write a paragraph analyzing and summarizing the three variables. Are any of the distributions skewed? In what direction?

b. Select one of the skewed variables and skim the scores of the variable in the SPSS Data Editor to find the outlier(s). That is, find the column in which the scores of the variable are stored and skim the scores looking for especially high scores (if the skew is positive) or low scores (if the skew is negative). What sociological factors might account for these very high or low scores?

3.17 SOC In this exercise, you will use SPSS to get measures of central tendency for several different variables in the *GSS2018* data set, which is available from the Web site for this text. In this data set, the cases are a representative sample (see Chapter 1) of the U.S. population.
- Find and click the SPSS icon on your desktop.
- Load the *GSS2018* data set.
- From the menu bar across the top of the SPSS window, click **Analyze**, **Descriptive Statistics**, and **Frequencies**.
- Find these variables in the box on the left of the **Frequencies** window: *helpblk*, *hrs1*, *partnrs5*, and *region*. Consult Appendix G for information on what each variable measures and the coding scheme for the scores. Click the arrow to move the variable names into the box on the right.
- Click the **Statistics** button and, under "Central Tendency," select Mean, Median, and Mode. Click **Continue** to return to the **Frequencies** window.
- Make sure the box next to the "Display frequency tables" option is *not* checked. We are interested only in the summary statistics, not in the tables.
- Click **OK**.

For each variable, select the *most appropriate* measure of central tendency (consult Table 3.11) and write a summary sentence about the variable.

YOU ARE THE RESEARCHER

The Typical American

Is there such a thing as a "typical" American? In this exercise, you will develop a profile of the average American based on measures of central tendency for 10 variables, chosen by you from the 2018 General Social Survey (*GSS2018*). Choose variables that you think are the most important in defining what it means to be a member of this society and then choose an *appropriate* measure of central tendency for each variable.

Use this information to write a description of the typical American. We will also use this opportunity to introduce a new SPSS program.

Step 1: Choosing Your Variables

Scroll through the list of available variables either in Appendix G or by using the **Utilities → Variables** command in SPSS to get the online codebook. Select 10 variables that, in your view, are central to defining or describing the "typical American" and list them in the table below. *Select at least one variable from each of the three levels of measurement.*

Variable	SPSS Name	Explain exactly what this variable measures	Level of Measurement
1			
2			
3			
4			
5			
6			
7			
8			
9			
10			

Step 2: Getting the Statistics

Find and click the SPSS icon on your desktop. Load the 2018 GSS (*GSS2018*) data set.

Using the "Frequencies" Procedure for the Mode and Median

The only SPSS procedure that will produce all three measures of central tendency is **Frequencies**. Here you will use **Frequencies** to get modes and medians for any nominal- and ordinal-level variables you selected in step 1.

Click **Analyze → Descriptive Statistics → Frequencies**. In the **Frequencies** dialog box, find the names of your nominal- and ordinal-level variables in the list on the left and click the arrow button in the middle of the screen to move the names to the **Variables** box on the right.

Click the **Statistics** button, find the **Central Tendency** box on the right, and click **Median** and **Mode**. Click **Continue**, and you will be returned to the **Frequencies** dialog box. Click (uncheck) the "Display frequency tables" box so that SPSS will *not* produce frequency distribution tables. Click **OK** and record your results in the table below. Report the mode for all nominal-level variables and the median for ordinal-level variables. Use as many lines as needed.

Variable	SPSS Name	Level of Measurement	Mode	Median
1				
2				
3				
4				
5				
6				
7				
8				
9				

Using the "Descriptives" Procedure for the Mean

The **Descriptives** command in SPSS is designed to provide summary statistics for continuous interval-ratio–level variables. By default (i.e., unless you tell it otherwise), **Descriptives** produces the mean, the minimum and maximum scores (i.e., the lowest and highest scores, which can be used to compute the range), and the standard deviation. We will introduce the range and standard deviation in the next chapter.

To use **Descriptives**, click **Analyze, Descriptive Statistics**, and **Descriptives**. The **Descriptives** dialog box will open. This dialog box looks just like the **Frequencies** dialog box and works in the same way. Find the names of your interval-ratio–level variables in the list on the left and, once they are highlighted, click the arrow button in the middle of the screen to transfer them to the **Variables** box on the right. Click **OK** and record your results in the table below, using as many lines as necessary.

Variable	SPSS Name	Level of Measurement	Mean
1			
2			
3			
4			
5			
6			
7			
8			
9			

Step 3: Interpreting Results

Examine the two tables displaying your results, and write a summary paragraph or two describing the typical American. Be sure to report all 10 variables and, as appropriate, describe the most common case (the mode), the typical case (the median), or the typical score (mean). Write as if you are reporting in a newspaper: Your goal should be clarity and accuracy. As an example, you might report that "The typical American is Protestant and married."

4 Measures of Variation

LEARNING OBJECTIVES

By the end of this chapter, you will be able to:

1. Explain the purpose of measures of variation and the information they convey.
2. Compute the range (R), the interquartile range (Q), the standard deviation (s), and the variance (s^2).
3. Select an appropriate measure of variation for a given level of measurement.
4. Interpret the components of a boxplot.
5. Explain the meaning of the standard deviation and the variance.
6. Use SPSS to calculate the range and the standard deviation.

USING STATISTICS

The statistical techniques presented in this chapter are used to describe the variability or diversity in a set of scores. They can be used to describe the

- Variation in birth rates across nations in the world in a box plot. **See page 107**

- Differences in life expectancy by gender, income level, and nation. **See page 110**

- Variation in homicide rates by state in different regions of America. **See page 112**

- Differences in pornography use and levels of religiosity among Americans. **See page 117**

In Chapters 2 and 3, you learned a variety of ways to describe a variable, including frequency distributions, graphs, and measures of central tendency. For a complete description of a distribution of scores, we must combine these with **measures of variation**, the subject of this chapter. While measures of central tendency describe the typical, average, or central score, measures of variation describe the variety, diversity, or heterogeneity of a distribution of scores.

The importance of the concept of **variation** might be easier to grasp with the aid of an example. Suppose that the director of public safety wants to evaluate two ambulance services that have contracted with the city to provide emergency medical aid. As a part of the investigation, she has collected data showing that the mean response time to calls for assistance is 7.4 minutes for Service A and 7.6 minutes for Service B.

These averages are virtually identical and provide no clear basis for judging one service as more efficient than the other. Measures of variation, however, can reveal substantial differences between distributions even when the measures of central tendency are the same. For example, consider the graphs in Figure 4.1, which display the distribution of response times for the two services.

Note that the line chart for Service B is much flatter than that for Service A. This is because the scores for Service B are more spread out, or more diverse, than the scores for Service A. In other words, Service B was much more variable in response time and had more scores in the high *and* low ranges and fewer in the middle. Service A was more consistent in its response time, and its scores are more clustered, or grouped, around the mean. Both distributions have essentially the same *average* response time, but there is considerably more *variation* in the response times for Service B.

If you were the director of public safety, would you be more likely to select an ambulance service that was always on the scene of an emergency in about the same amount of time (Service A) or one that was sometimes slow and sometimes very quick to respond (Service B)? Note that if you had not considered variation, a possibly important difference in the performance of the two ambulance services might have gone unnoticed.

Keep the two shapes in Figure 4.1 in mind as visual representations of the concept of variation. The greater clustering of scores around the mean in

FIGURE 4.1 Response Time for Two Ambulance Services

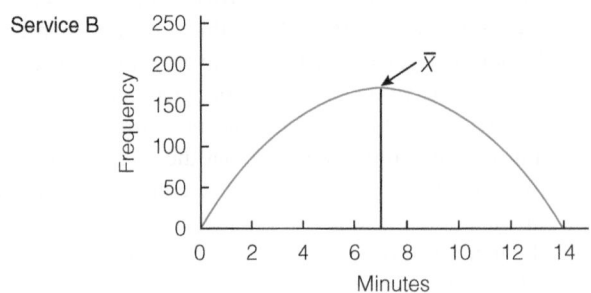

the distribution for Service A indicates *less* variation, and the flatter curve of the distribution for Service B indicates *more* variation. Any measure of variation will decrease as the scores become less variable and the distribution becomes more peaked (that is, as the distribution looks more and more like Service A's) and increase as the scores become more variable and the distribution becomes flatter (that is, as the distribution looks more and more like Service B's).

These ideas and Figure 4.1 may give you a general notion of what is meant by variation, but the concept is not easily described in words alone: We need to consider some statistics that are designed to quantify the variation in a set of scores. In this chapter, we introduce some of the more common measures of variation. We begin with the range and the interquartile range but devote most of our attention to the standard deviation. We also consider a visual representation of variation called a boxplot.

The Range (*R*) and Interquartile Range (*Q*)

The **range (*R*)** is defined as the distance between the highest and lowest scores in a distribution:

FORMULA 4.1
$$R = \text{High score} - \text{Low score}$$

The range is easy to calculate and is perhaps most useful as a quick and general indicator of variability. The statistic is also easy to interpret: The greater the value of the range, the greater the distance from high to low score, and the greater the variation in the distribution.

Unfortunately, because it is based on only the highest and lowest scores, the range has some important limitations. First, almost any sizeable distribution will contain some scores that are atypically high and/or low (outliers) compared to most of the scores (for example, see Table 3.3). Thus, *R* might exaggerate the amount of variation for most of the scores in the distribution. Also, *R* yields no information about the variation of the scores between the highest and lowest scores.

The **interquartile range (*Q*)** is another kind of range. It avoids some of the problems associated with *R* by considering only the middle 50% of the cases in a distribution. To find *Q*, first arrange the scores from highest to lowest and then divide the distribution into quarters (as distinct from halves when locating the median). We discussed quartiles in Chapter 3. The first quartile (Q_1) is the point below which 25% of the cases fall and above which 75% of the cases fall. The second quartile (Q_2) divides the distribution into halves (thus, Q_2 is equal to the median). The third quartile (Q_3) is the point below which 75% of the cases fall and above which 25% of the cases fall.

The interquartile range is defined as the distance from the third to the first quartile:

FORMULA 4.2
$$Q = Q_3 - Q_1$$

The interquartile range essentially extracts the middle 50% of the distribution and, like R, is based on only two scores. Q is interpreted in the same way as R: The greater its value, the greater the variation. Unlike the range, Q avoids the problem of being based on the most extreme scores, but it also fails to yield any information about the variation of the scores other than the two on which it is based.

Computing the Range and Interquartile Range

Table 4.1 presents birth rates (number of births per 1000 population) for a group of 40 nations. What are the range and interquartile range of these data? Note that the scores have already been ordered from high to low. This makes the range easy to calculate and is necessary for finding the interquartile range. Of these 40 nations, Niger ranked the highest with a birth rate of 48 and several nations are tied for the lowest rank with a birth rate of 8. The range is: $48 - 8$ or 40 ($R = 40$).

To find Q, we must locate the first and third quartiles (Q_1 and Q_3). We define these points in terms of the scores associated with certain cases, as we did when finding the median. Q_1 is determined by multiplying N by 0.25. Because (40) × (0.25) = 10, Q_1 is the score associated with the tenth case, counting up from the lowest score. The tenth case is China, with a score of 12. So $Q_1 = 12$. (Note that several other nations also have birth rates of 12 and could have been taken as the tenth case. This would change the *identity* of the tenth case but not the *score*.)

TABLE 4.1 **Birth Rates (Number of Births per 1000 Population) for 40 Nations, 2018**

Rank	Nation	Birth Rate	Rank	Nation	Birth Rate
40 (Highest)	Niger	48	**20**	India	20
39	Angola	45	**19**	Libya	20
38	Uganda	41	**18**	Mexico	19
37	Guinea	39	**17**	Venezuela	19
36	Mozambique	39	**16**	Vietnam	16
35	Nigeria	39	**15**	Colombia	15
34	Tanzania	39	**14**	Chile	14
33	Malawi	34	**13**	Kuwait	14
32	Togo	34	**12**	Australia	13
31	Ethiopia	33	**11**	Ireland	13
30	Senegal	33	**10**	China	12
29	Rwanda	32	**9**	Russia	12
28	Kenya	31	**8**	United Kingdom	12
27	Ghana	30	**7**	United States	12
26	Egypt	27	**6**	Canada	11
25	Pakistan	26	**5**	France	11
24	Haiti	25	**4**	Germany	9
23	Guatemala	24	**3**	Italy	8
22	Cambodia	23	**2**	Japan	8
21	Syria	22	**1** (Lowest)	Spain	8

The case that lies at the third quartile (Q_3) is given by multiplying N by 0.75, and $(40) \times (0.75) = 30$th case. The 30th case, again counting up from the lowest score, is Senegal, with a score of 33 ($Q_3 = 33$). Therefore:

$$Q = Q_3 - Q_1$$
$$Q = 33 - 12$$
$$Q = 21$$

(For practice in finding and interpreting R and Q, see problems 4.12 and 4.13. The range may be found for any of the problems at the end of this chapter.)

USING A FREQUENCY DISTRIBUTION TO FIND THE RANGE AND THE INTERQUARTILE RANGE

In Chapter 3, we used percentiles in a frequency distribution to find the mode and the median for the number of children respondents had in the 2018 General Social Survey. Let's review this by taking a look at the number of children that those same respondents thought was ideal for a family. The data can be found in Table 4.2. Compare this to Table 3.6. Are you surprised? According to Table 4.2, what are the mode and the median for the number of children the respondents think an ideal family should have?[1]

As you might recall, the most common number of children the respondents had was zero, but now we can see that the mode for the ideal number of children the respondents thought a family should have is 2, with a Valid Percent of 51.2%. The median is also 2 because it is the first value with a Valid Percent that is 50% or higher.

A frequency distribution can also be used to determine the range and the interquartile range. Using the values column, we can easily find the range. It is equal to the highest value minus the lowest value, $6 - 0 = 6$. But how about the interquartile range? To solve for it, we need to subtract the value associated with the 25th percentile from the value associated with the 75th percentile. Can you figure out what those values are? *(HINT: Use the Cumulative Percentage column.)*

TABLE 4.2 What Do You Think Is the Ideal Number of Children for a Family to Have?

		Frequency	Percent	Valid Percent	Cumulative Percent
Valid	0	18	.8	1.3	1.3
	1	34	1.4	2.5	3.9
	2	691	29.4	51.2	55.1
	3	414	17.6	30.7	85.8
	4	152	6.5	11.3	97.0
	5	27	1.1	2.0	99.0
	6	13	.6	1.0	100.0
	Total	1349	57.5	100.0	
Missing	System	999	42.5		
	Total	2348	100.0		

[1]Note that in this table we have excluded a small number of people that said "7 or more" as well as 13% of the sample that said "As many as they want."

ONE STEP AT A TIME Finding the Interquartile Range (Q)

1. Array the cases in order.
2. Find the case that lies at the first quartile (Q_1) by multiplying N by 0.25. If the result is not a whole number, round it off to the nearest whole number. This is the number of the case that marks the first quartile. Note the *score* of this case.
3. Find the case that lies at the third quartile (Q_3) by multiplying N by 0.75. If the result is not a whole number, round off to the nearest whole number. This is the number of the case that marks the third quartile. Note the *score* of this case.
4. Subtract the score of the case at the first quartile (Q_1)—see step 2—from the score of the case at the third quartile (Q_3)—see step 3. The result is the interquartile range or Q.

STATISTICS **In Everyday Life**

Increasing Racial Diversity

U.S. society is growing increasingly diverse in terms of race, ethnicity, culture, and language. Variables like race and ethnicity are nominal in level of measurement, so we can't express their variation with statistics like the range. However, the increasing diversity can be quickly appreciated from the table below. By the middle of this century, the United States is projected to become "majority-minority," a much more diverse population in which Hispanic and Asian Americans are far more prominent.

Racial and Ethnic Makeup of U.S. Society, 1980, 2010, and 2060. Percentage of Total Population

Racial and Ethnic Groups	1980	2010	2060 (Projected)
Non-Hispanic Whites	81%	65%	45%
Blacks or African Americans	11%	13%	16%
Hispanics*	6%	16%	28%
Asian and Pacific Islanders	2%	5%	10%
American Indians and Alaska Natives	1%	1%	1%
	101%	100%	100%

*Note: Hispanics may be of any race.

Sources: 1980: Heaton, T., Chadwick, B., & Jacobsen, C. 2000. *Statistical Handbook on Racial Groups in the United States.* Oryx Press; 2010: Humes, K., Jones, N., & Rameriz, R. 2011. Overview of Race and Hispanic Origin: 2010. https://www.census.gov/prod/cen2010/briefs/c2010br-02.pdf; 2060: U.S. Census Bureau. 2019. National Population Projection Tables. https://www.census.gov/data/tables/2017/demo/popproj/2017-summary-tables.html

The value associated with the 25th percentile is that first value with a Cumulative Percentage of 25% or higher. In this case, it's 2 since the Cumulative Percentage of 55.1% is the first in the table to reach 25%. As you might have guessed, the value associated with the 75th percentile is the first value with a Cumulative Percentage of 75% or higher, or 3 in this case. As such, the interquartile range is $3 - 2 = 1$.

The Standard Deviation and Variance

A basic limitation of both Q and R is that they use only two scores and, thus, do not use all of the available information. Also, neither statistic provides any information on how far the scores are from each other or from some central point, such as the mean. How can we design a measure of variation that would correct these faults? We can begin with some specifications. A good measure of variation should:

- Use all the scores in the distribution. The statistic should use all the information available.

- Describe the average or typical deviation of the scores. The statistic should give us an idea about how far the scores are from each other or from the center of the distribution.

- Increase in value as the scores became more diverse. This would be a handy feature when comparing distributions because it would permit us to tell at a glance which was more variable: The higher the numerical value of the statistic, the greater the variation.

One way to develop a statistic to meet these criteria would be to start with the distances between each score and the mean, or the **deviations** $(X_i - \overline{X})$. The value of the deviations will increase as the differences between the scores and the mean increase. If the scores are more clustered around the mean (remember the graph for Service A in Figure 4.1), the deviations will be small. If the scores are more spread out, or more varied (like the scores for Service B in Figure 4.1), the deviations will be larger. How can we use the deviations to develop a useful statistic?

One course of action would be to use the sum of the deviations, $\Sigma(X_i - \overline{X})$, as the basis for a statistic, but, as we saw in Chapter 3, the sum of deviations will always be 0. To illustrate, consider the distribution of five scores presented in Table 4.3. If we sum the deviations of any set of scores from their mean, we would always wind up with a total of 0, regardless of the amount of variety in the scores.

Still, the sum of the deviations is a logical basis for a measure of variation, and statisticians have developed a way around the fact that the positive deviations always equal the negative deviations. If we square each of the deviations, all values will be positive because a negative number multiplied by itself becomes positive. For example, $(-20) \times (-20) = +400$. For the scores in Table 4.3, the

TABLE 4.3 A Demonstration That the Sum of the Deviations of the Scores Around the Mean Will Always Total Zero

Scores (X_i)	Deviations $(X_i - \overline{X})$
10	$(10 - 30) = -20$
20	$(20 - 30) = -10$
30	$(30 - 30) = 0$
40	$(40 - 30) = 10$
50	$(50 - 30) = 20$
$\Sigma(X_i) = 150$	$\Sigma(X_i - \overline{X}) = 0$
$\overline{X} = 150/5 = 30$	

TABLE 4.4 **Computing the Standard Deviation**

Scores (X_i)	Deviations ($X_i - \overline{X}$)	Deviations Squared ($X_i - \overline{X}$)²
10	$(10 - 30) = -20$	$(-20)^2 = 400$
20	$(20 - 30) = -10$	$(-10)^2 = 100$
30	$(30 - 30) = 0$	$(- 0)^2 = 0$
40	$(40 - 30) = 10$	$(10)^2 = 100$
50	$(50 - 30) = 20$	$(20)^2 = 400$
$\Sigma(X_i) = 150$	$\Sigma(X_i - \overline{X}) = 0$	$\Sigma(X_i - \overline{X})^2 = 1000$

sum of the squared deviations would be (400 + 100 + 0 + 100 + 400), or 1000 (see Table 4.4). Thus, a statistic based on the sum of the squared deviations will have the properties we want in a good measure of variation.

Before we finish designing our measure of variation, we must deal with another problem. The sum of the squared deviations will increase with sample size: The larger the *number* of scores, the greater the value of the measure. This would make it difficult to compare the relative variability of distributions based on samples of different sizes. We can solve this problem by dividing the sum of the squared deviations by N (sample size) and thus standardizing for samples of different sizes.

These procedures yield a statistic known as the **variance**, which is symbolized as s^2. The variance is used primarily in inferential statistics, although it is a central concept in the design of some measures of association. For purposes of describing the variation of a distribution, a closely related statistic, called the **standard deviation** (symbolized as s), is typically used, and this statistic is our focus for the remainder of the chapter.

The formulas for the variance (s^2) and standard deviation (s) are

FORMULA 4.3

$$s^2 = \frac{\Sigma(X_i - \overline{X})^2}{N}$$

FORMULA 4.4

$$s = \sqrt{\frac{\Sigma(X_i - \overline{X})^2}{N}}$$

Strictly speaking, Formulas 4.3 and 4.4 are for the variance and standard deviation of a population. Slightly different formulas, with $N - 1$ instead of N in the denominator, should be used when we are working with random samples rather than entire populations. This is an important point because many of the electronic calculators and statistical software packages you might be using (including SPSS) use $N - 1$ in the denominator and, thus, produce results that are at least slightly different from those given by Formulas 4.3 and 4.4. The size of the difference will decrease as sample size increases, but many of the problems and examples in this chapter use small samples, and the differences between using N and $N - 1$ in the denominator can be considerable in such cases. Some calculators allow the user to choose either $N - 1$ or N when calculating the standard deviation. If you have such a calculator, choose "N" and your answers should match ours.

ONE STEP AT A TIME Finding the Standard Deviation (s) and the Variance (s²)

1. Construct a computing table like Table 4.4, with columns for the scores (X_i), the deviations $(X_i - \overline{X})$, and the deviations squared $(X_i - \overline{X})^2$.
2. List the scores (X_i) in the left-hand column. Add up the scores and divide by N to find the mean.
3. Find the deviations $(X_i - \overline{X})$ by subtracting the mean from each score, one at a time. List the deviations in the second column.
4. Add up the deviations. The sum must equal zero (within rounding error). If the sum of the deviations does not equal zero, you have made an error and need to repeat step 3.
5. Square each deviation and list the result in the third column.
6. Add up column 3 (the squared deviations) and substitute that value into numerator of Formula 4.4.
7. Divide the sum of the squared deviations from step 6 by N.
8. Find the square root of the quantity you found in step 7. This is the standard deviation.
9. To find the variance (s^2), square the value of the standard deviation (s).

Social Research and Data Analysis 4.1 The Standard Deviation

Recent years have been marked by an alarming number of mass shooting in the United States. Below we will use data from *Time* magazine on several mass shootings in 2019 to solve for the standard deviation. In this case, mass shootings are defined as those that took place in public and claimed the lives of at least 3 people. The death counts below include nine attacks that occurred in Sebring, Florida; State College, Pennsylvania; Aurora, Illinois; Virginia Beach, Virginia; Abbeville, South Carolina; Gilroy, California; El Paso, Texas; Dayton, Ohio; and Odessa, Texas. We will use Formula 4.4 to compute the standard deviation.

Number of Aggressive Acts

(X_i)	$(X_i - \overline{X})$	$(X_i - \overline{X})^2$
5	$5 - 7.7 = -2.7$	7.3
3	$3 - 7.7 = -4.7$	22.1
5	$5 - 7.7 = -2.7$	7.3
12	$12 - 7.7 = \;\;\;4.3$	18.5
3	$3 - 7.7 = -4.7$	22.1
3	$3 - 7.7 = -4.7$	22.1
22	$22 - 7.7 = \;\;14.3$	204.5
9	$9 - 7.7 = \;\;\;1.3$	1.7
7	$7 - 7.7 = -0.7$	0.5
$\Sigma(X_i) = 69$	$\Sigma(X_i - \overline{X}) = -0.3^*$	$\Sigma(X_i - \overline{X})^2 = 306.1$

$$\overline{X} = \frac{\Sigma X_i}{N} = \frac{69}{9} = 7.7$$

*The sum of this column does not equal zero because of rounding error.

Substituting into Formula 4.4, we have

$$s = \sqrt{\frac{\Sigma(X_i - \overline{X})^2}{N}} = \sqrt{\frac{306.1}{9}} = \sqrt{34.01} = 5.8$$

The standard deviation for these data is 5.8.

Source: https://time.com/5643553/2019-mass-shootings-list/

It is advisable to use a table such as Table 4.4 when computing the standard deviation. The five scores in our example are listed in the left-hand column of the table, the deviations are in the middle column, and the squared deviations are in the right-hand column.

The total of the last column in Table 4.4 is the sum of the squared deviations, which is substituted directly into Formula 4.4:

$$s = \sqrt{\frac{\Sigma(X_i - \overline{X})^2}{N}}$$

$$s = \sqrt{\frac{1000}{5}}$$

$$s = \sqrt{200}$$

$$s = 14.14$$

See the "One Step at a Time" box for detailed instructions on solving Formulas 4.3 and 4.4.

Computing the Standard Deviation: An Additional Example

An additional example will help to clarify the procedures for computing and interpreting the standard deviation. A researcher is comparing the student bodies of two campuses. One college is located in a small town, and almost all of the students reside on campus. The other is located in a large city, and the students are almost all part-time commuters. The researcher wishes to compare the ages of the students at the two campuses and has compiled the information presented in Table 4.5. Which student body is older and which is more diverse in age? (Needless to say, these groups are much too small for serious research and are used here only to simplify computations.)

We see from the means that the students from the residential campus are quite a bit younger, on the average, than the students from the urban campus (19 vs. 23 years of age). Which group is more diverse in age? Computing the standard deviation (Formula 4.4) will answer this question:

Residential campus:

$$s = \sqrt{\frac{\Sigma(X_i - \overline{X})^2}{N}} = \sqrt{\frac{4}{5}} = \sqrt{0.8} = 0.89$$

Urban campus:

$$s = \sqrt{\frac{\Sigma(X_i - \overline{X})^2}{N}} = \sqrt{\frac{88}{5}} = \sqrt{17.6} = 4.20$$

TABLE 4.5 **Computing the Standard Deviation for Two Campuses**

	Residential Campus	
Scores (X_i)	Deviations ($X_i - \overline{X}$)	Deviations Squared ($X_i - \overline{X}$)2
18	$(18 - 19) = -1$	$(-1)^2 = 1$
19	$(19 - 19) = 0$	$(0)^2 = 0$
20	$(20 - 19) = 1$	$(1)^2 = 1$
18	$(18 - 19) = -1$	$(-1)^2 = 1$
20	$(20 - 19) = 1$	$(1)^2 = 1$
$\Sigma(X_i) = 95$	$\Sigma(X_i - \overline{X}) = 0$	$\Sigma(X_i - \overline{X})^2 = 4$

$$\overline{X} = \frac{\Sigma X_i}{N} = \frac{95}{5} = 19$$

	Urban Campus	
Scores (X_i)	Deviations ($X_i - \overline{X}$)	Deviations Squared ($X_i - \overline{X}$)2
20	$(20 - 23) = -3$	$(-3)^2 = 9$
22	$(22 - 23) = -1$	$(-1)^2 = 1$
18	$(18 - 23) = -5$	$(-5)^2 = 25$
25	$(25 - 23) = 2$	$(2)^2 = 4$
30	$(30 - 23) = 7$	$(7)^2 = 49$
$\Sigma(X_i) = 115$	$\Sigma(X_i - \overline{X}) = 0$	$\Sigma(X_i - \overline{X})^2 = 88$

$$\overline{X} = \frac{\Sigma X_i}{N} = \frac{115}{5} = 23$$

The higher value of the standard deviation for the urban campus means that it is more diverse with respect to age. As you can see by scanning the scores, the ages of the students at the residential college fall within a narrow age range ($R = 20 - 18 = 2$), whereas the students at the urban campus are more mixed and include students of age 25 and 30 ($R = 30 - 18 = 12$). *(For practice in computing and interpreting the standard deviation, see any of the problems at the end of this chapter. Problems with smaller data sets, such as 4.1 and 4.2, are recommended for practicing computations until you are comfortable with these procedures.)*

Visualizing Variation: Boxplots

A graph called a **boxplot** or a **box and whiskers plot** provides a helpful way to visualize and analyze variation and gives us an opportunity to apply some of our growing array of statistical tools. Boxplots use the median (see Chapter 3), range (R), and interquartile range (Q) to depict both central tendency and variability. They also display any outliers or extreme scores that might be included in the distribution.

Let's begin by examining the parts of a boxplot using Figure 4.2 as an example. This graph presents the birth rates (number of births per 1000 population)

FIGURE 4.2 Boxplot for Birth Rates for 99 Nations, 2018

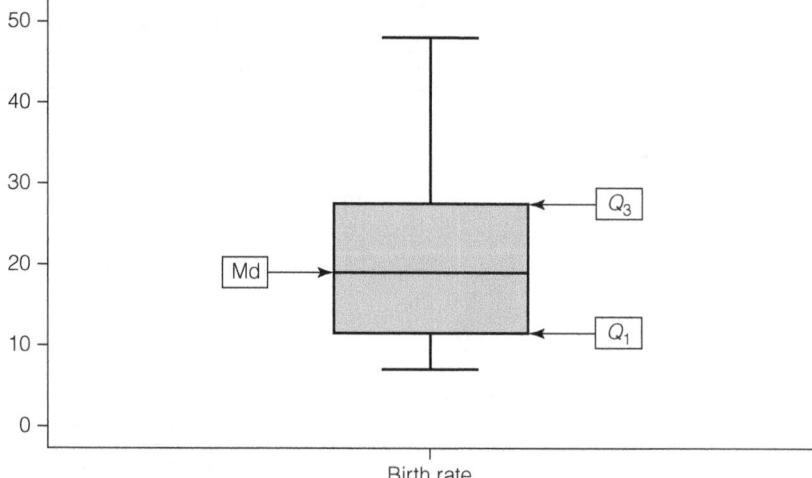

of 99 nations from a variety of levels of development and affluence. The box-plot consists of a box and two T-shaped lines, one above and one below the box. The box stretches from the 3rd quartile at the top (Q_3) to the 1st quartile at the bottom (Q_1). Thus, the height of the box reflects the value of Q or the interquartile range. The horizontal line through the box is drawn at the median (Md).

The T-shaped lines (or "whiskers") reflect the range of the scores. They extend 1.5 times the height of the box *or* to the high and low scores, which-ever is closer to the box. In this case, the highest actual birth rate is 48 births per 1000 population, and this is where the upper whisker is drawn. The lower whisker is drawn at a birth rate of 7, the lowest birth rate in this group of nations.

In a boxplot, an "outlier" is defined as any score outside the "whiskers," or any score beyond 1.5 times the height of the box in either direction. Scores that are beyond three times the height of the box in either direction are called "ex-treme outliers." The nations displayed in Figure 4.2 have no extreme scores but we shall see some examples shortly.

Boxplots are especially useful when we want to compare the distribution of a variable across different conditions or times. Figure 4.3 presents birth rates for the same 99 nations but, this time, the nations have been classified into income groups. Summary statistics for birth rate by income group are reported in Table 4.6.

The income groups are based on economic criteria developed by the World Bank (http://data.worldbank.org/about/country-classifications/country -and-lending-groups), and the nations in each income group share many social characteristics:

- Low-income nations have agricultural economies, which typically value large families. This category includes many African nations, Haiti, and some Asian nations.

FIGURE 4.3 **Boxplots for Birth Rates by Income Level for 99 Nations, 2018**

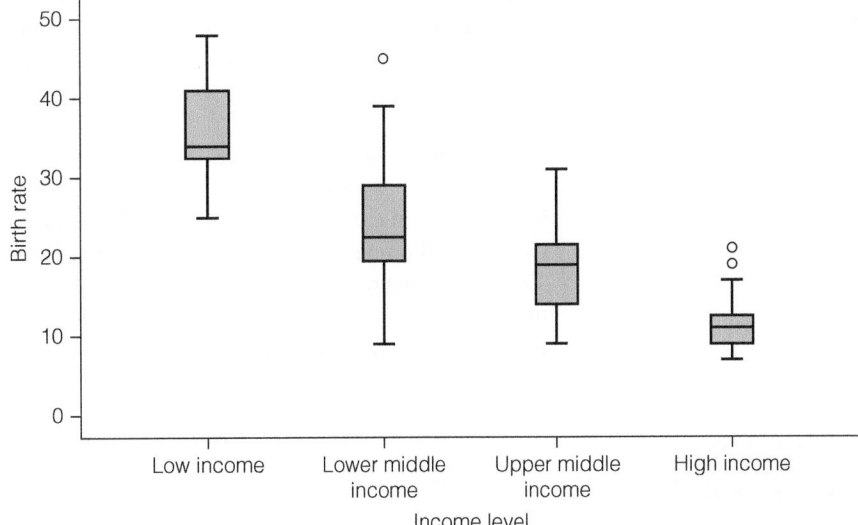

TABLE 4.6 **Summary Statistics for Birth Rate by Income Levels for 99 Nations, 2018**

	Income Level			
Statistics	Low Income	Lower Middle Income	Upper Middle Income	High Income
Low score	25	9	9	7
Q_1	32	19.25	14	9
Median	34	22.5	19	11
Q_3	41	30	21.75	12.75
High score	48	45	31	21
Q	9	10.75	7.75	3.75
R	23	36	22	14
s	6.0	8.8	5.7	3.3
\overline{X}	36.6	24.5	18.1	11.5
	$N = 19$	20	28	32

Source: Population Reference Bureau, 2019. *World Population Data Sheet.* https://www.prb.org
/2018-world-population-data-sheet-with-focus-on-changing-age-structures/

- Lower-middle–income nations are generally in the early stages of industriali-
 zation and include Egypt, Bolivia, and India.
- Upper-middle–income nations have more advanced economies and include
 Brazil, Turkey, and China.
- Almost all high-income nations are highly industrialized and urbanized, and
 they have the highest standards of living. Children are more expensive to raise
 in these economies, so families are smaller and birth rates lower. This cate-
 gory includes the United States, Canada, most Western European nations, and
 Japan, Australia, and New Zealand.

Looking first at central tendency, we see that the median birth rate (marked by the horizontal lines in the boxes) decreases sharply as income level rises, as does the mean (see Table 4.6). Low-income nations have much higher average birth rates, and high-income nations have the lowest average birth rates. Lower-middle and upper-middle nations are intermediate in average birth rates.

Turning to variation, we see several patterns. First, lower-middle–income nations are the most variable according to all three measures of variation (R, Q, and s). As you can see in the boxplot, these nations have an outlier, as noted by the small circle above the box. The outlier (which is Angola, a nation on the southwest coast of Africa) inflates the value of the range but the other two measures of variation (Q and s) reinforce the conclusion that the lower-middle–income nations are the most variable on birth rate.

STATISTICS | **In Everyday Life**

Life Expectancy Around the Globe

The citizens of more affluent societies can routinely expect to live into their 70s, 80s, or even older. This is not the case for nations with lower standards of living, nutrition, and health care. The table below presents information on life expectancy for men and women for 99 nations at four different income levels. Note that average life expectancy is higher for females than males at all income levels and increases for both sexes as income level increases. On the other hand, the measures of variation (standard deviation and range) tend to do just the opposite and are highest for lower-middle–income nations. Why are lower-income nations more diverse on life expectancy? *(Hint: Are males and females in less affluent societies more vulnerable at all ages?)*

Summary Statistics on Life Expectancy by Income Level for 99 Nations

Income Level	Mean	Standard Deviation	High	Low	Range	N
High Income						
Males	78.1	3.1	82	69	13	
Females	83.0	2.2	87	76	11	32
Upper Middle Income						
Males	71.5	4.4	77	61	16	
Females	76.6	4.5	83	65	18	28
Lower Middle Income						
Males	65.8	6.1	73	52	21	
Females	70.1	7.2	78	54	24	20
Low Income						
Males	61.4	4.3	72	52	20	
Females	64.7	4.5	75	54	21	19

Source: Population Reference Bureau. 2019. *World Population Data Sheet*. Retrieved from https://www.prb.org/2018-world-population-data-sheet-with-focus-on-changing-age-structures/

Second, high-income nations are the least variable by far. This is apparent from their short boxplot and by their much lower interquartile range (3.75), range (14), and standard deviation (3.3). These cases are most similar to each other, in spite of the two outliers, which are Israel and Panama. Why would these two nations be so different from other high-income nations in their birth rates?

In conclusion, boxplots provide useful visual and analytical information about variation and central tendency. Like R, Q, and s, boxplots are most useful when comparing variables across different conditions. *(For practice in interpreting boxplots, see problems 4.18 and 4.19.)*

Interpreting the Standard Deviation

It is possible that the meaning of the standard deviation (i.e., why we calculate it) is not completely obvious to you at this point. You might be asking, "Once I've gone to the trouble of calculating the standard deviation, what do I have?" The meaning of this measure of variation can be expressed in three ways. The first and most important involves the normal curve, and we will defer this interpretation until the next chapter.

A second way of thinking about the standard deviation is as an index of variation that increases in value as the distribution becomes more variable. In other words, the standard deviation is higher for more diverse distributions and lower for less diverse distributions. The lowest value the standard deviation can have is 0, and this would occur for distributions with no variation (i.e., if every single case in the sample had exactly the same score). Thus, 0 is the lowest value possible for the standard deviation (although there is no upper limit).

A third way to get a feel for the meaning of the standard deviation is by comparing one distribution with another. We already did this when comparing the two ambulance services in Figure 4.1 and the residential and urban campuses in Table 4.5. You might also do this when comparing one group with another (e.g., men vs. women, blacks vs. whites) or the same variable at two different times. For example, suppose we found that the ages of the students on a particular campus had changed over time as indicated by the summary statistics shown in Table 4.7.

Clearly, the student body has grown older on average, and, according to the standard deviation, it has also grown more diverse in age. The lower standard deviation for 1990 indicates that ages would be more clustered around the mean in that year (remember the distribution for Service A in Figure 4.1). In 2015, in contrast, the larger standard deviation means that the distribution is flatter and more spread out, like the distribution for Service B in Figure 4.1. In other words, compared to 2015, the students in 1990 were more similar to each other with respect to age and more clustered in a narrower age range. The standard deviation is extremely useful for making comparisons of this sort between distributions of scores.

Table 4.7 Age on a College Campus in Two Different Years (Fictitious Data)

1990	2015
$\overline{X} = 21$	$\overline{X} = 25$
$s = 1$	$s = 4$

Social Research and Data Analysis 4.2	Describing Variation

The homicide rates of five New England and five Western states are listed in the tables below. Which group of states has the higher rate? Which is more variable?

Homicide Rate for Five New England States, 2017 (Homicides per 100,000 population)

State	Homicide Rate	Deviations	Deviations Squared
Connecticut	2.8	0.58	0.34
Massachusetts	2.5	0.28	0.08
Rhode Island	2.2	−0.02	0.00
Vermont	1.9	−0.32	0.10
Maine	1.7	−0.52	0.27
	$\Sigma(X_i) = 11.1$	$\Sigma(X_i - \bar{X}) = 0.00$	$\Sigma(X_i - \bar{X})^2 = 0.79$

$$\bar{X} = \frac{\Sigma X_i}{N} = \frac{11.1}{5} = 2.22$$

$$s = \sqrt{\frac{\Sigma(X_i - \bar{X})^2}{N}} = \sqrt{\frac{0.79}{5}} = \sqrt{0.16} = 0.40$$

Homicide Rate for Five Western States, 2017 (Homicides per 100,000 population)

State	Homicide Rate	Deviations	Deviations Squared
Nevada	9.1	3.62	13.10
California	4.6	−0.88	0.77
Arizona	5.9	0.42	0.18
Montana	3.9	−1.58	2.50
Colorado	3.9	−1.58	2.50
	$\Sigma(X_i) = 27.4$	$\Sigma(X_i - \bar{X}) = 0.00$	$\Sigma(X_i - \bar{X})^2 = 19.05$

$$\bar{X} = \frac{\Sigma X_i}{N} = \frac{27.4}{5} = 5.48$$

$$s = \sqrt{\frac{\Sigma(X_i - \bar{X})^2}{N}} = \sqrt{\frac{19.05}{5}} = \sqrt{3.81} = 1.95$$

With such small groups, you can tell by inspection that the Western states have higher homicide rates. This impression is confirmed by both the median—which is 2.2 for the New England states (Vermont is the middle case) and 4.6 for the Western states (California is the middle case)—and the mean (2.22 for the New England states and 5.48 for the Western states).

The five Western states are also much more variable. The range for the Western states is 5.2 ($R = 9.1 - 3.9 = 5.2$), much higher than the range for the New England states of 1.1 ($R = 2.8 - 1.7 = 1.1$). Similarly, the standard deviation for the Western states (1.95) is more than four times greater than the standard deviation for the New England states (.40).

In summary, the five Western states average higher homicide rates and are also more variable than the five New England states.

USING SPSS

Measures of Variation

In Chapter 3, we looked at the mean and median U.S. homicide rates for 1995 and 2017. In this installment of "Using SPSS," we will expand the analysis to include measures of variation. We will use the **Descriptives** command that was introduced at the end of Chapter 3.
Follow these steps:

1. Click the SPSS icon on your desktop.
2. Load the *States* database.
 a. Find the **File** command on the far left of the menu bar and then click **File →
 Open → Data**.
 b. Find the *States* database. You can download this file from the website for this text
 if you haven't already.
3. From the menu bar across the top of the SPSS window, click **Analyze**, **Descriptive
 Statistics**, and **Descriptives**.

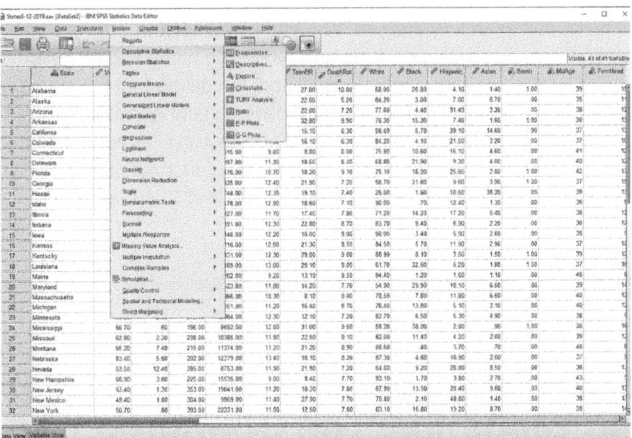

4. Find the variables *Homicide95* (state homicide rates in 1995) and *Homicide* (state
 homicide rates in 2017) in the box on the left of the **Descriptives** window and click
 the arrow to move the variable names to the box on the right. When you are done,
 the screen will look like this:

(*continued*)

USING SPSS **(Continued)**

5. Click the Options button on the **Descriptives** window and check the box next to **Range** and **Variance** to select those statistics. Other statistics (standard deviation, mean, and the low and high scores) are selected by default.
6. Click **OK**.

The output will look like this:

Descriptive Statistics							
	N	Range	Minimum	Maximum	Mean	Std. Deviation	Variance
Homicide	50	11.4	1.0	12.4	5.010	2.6266	6.899
Homicide rate 1995	50	16.10	.90	17.00	6.7120	3.62027	13.106
Valid N (listwise)	50						

We saw in Chapter 3 that the median and mean homicide rates declined between 1995 and 2017. The measures of variation (range, standard deviation, and variance) also declined, indicating that the states were more similar to each other in 2017. The standard deviation, for example, fell from 3.62 in 1995 to 2.63 in 2017. As you recall from our discussion in Chapter 3, this decrease reflects the lower score of Louisiana (an extreme outlier in 1995), the fact that the homicide rate fell in virtually every state between 1995 and 2010, and the clustering of scores in the low (less than 3.00) range in 2017.

Interpreting Statistics: The Changing Distribution of Income in the United States

In Chapter 3, we analyzed the distribution of income in the United States at a single point in time and concluded that the median was the best measure of central tendency to describe this variable. In this installment of "Interpreting Statistics," we examine changes in the distribution of income in the United States over time and add measures of variation to the analysis. We will use information from the Census Bureau to answer several questions: Is average or mean income rising or falling? What about median income: Does the average American receive more or less income than in the past? Is the distribution of income becoming more unequal (i.e., are the rich getting richer)? We will use the mean to look at changes in average income and the median to observe changes in income for the average American. The standard deviation would be the preferred measure of variation for an interval-ratio–level variable such as income, but the Census Bureau does not supply this statistic. Instead, we will use a statistic similar to the interquartile range to measure variation.

As we saw in Chapter 3, we should keep in mind that virtually any distribution of income will be positively skewed (see Figure 3.1). That is, in any group, community, state, or nation, the incomes of most people will be grouped around

the mean or median but—in the language of boxplots—some people will be extreme outliers and have incomes far higher than the average. Some people will also have incomes far below the mean or median, and their incomes may even be zero. However, low incomes will have less effect on the mean than the highest incomes. The lowest income cannot be less than 0, but the highest incomes can be in the billions of dollars. Since the mean uses *all* the scores, including the highest, and is pulled in the direction of extreme scores, mean income will always be greater than median income.

Figure 4.4 shows the changes in mean and median income for the United States across five decades. Income is expressed in 2017 dollars to eliminate the effect of inflation over the years. Without this adjustment, recent incomes would appear to be much higher than older incomes, not because of increased buying power and well-being, but rather because of the changing value of the dollar. Also, Figure 4.4 is based on total income for entire households, not individual income.

The line labeled "Median Household Income" in Figure 4.4 shows that the income of the average American household was about $46,000 in 1967. This value gradually trends upward, with some declines in the mid-1970s, early 1980s, and early 1990s, all periods of recession. There is an increase in median income during the boom economy of the 1990s, but then we see a leveling off and a decline following the most recent recession that started in 2007. Over the past several years, median income has again risen and, in 2017, it stood at $61,372, about $15,000 higher than 1967 and a little higher than the previous peak year of 1999. Overall, the median indicates that the standard of living generally increased for the average American household over this time period.

The pattern of the mean income is almost identical to that of the median. It rises and falls in almost exactly the same ways and at the same times. Notice, however, that the mean is always higher than the median, a reflection of the

FIGURE 4.4 Mean and Median Household Incomes for the United States, 1967–2017 (in 2017 dollars)

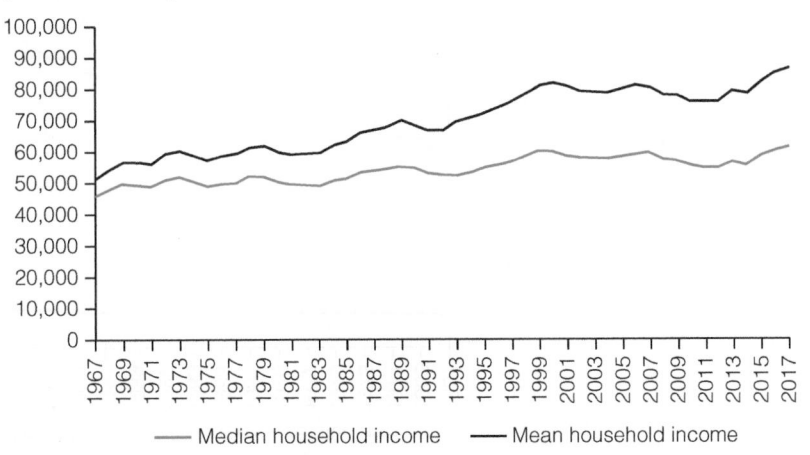

Source: U.S. Bureau of the Census, 2019. "Historical Income Tables: Table H-5." https://www.census.gov/data/tables/time-series/demo/income-poverty/historical-income-households.html

characteristic positive skew of income data. Also notice that the size of the gap between the mean and the median increases over time, indicating that the degree of skew is increasing. This is because the income of the rich is increasing relative to the income of the typical American household.

An increasing skew means that the scores are becoming more diverse, and this means, in turn, that any measure of variation computed for the distribution of income between 1967 and 2017 will increase in value. The Census Bureau provides statistics similar to the interquartile range, and these are presented in Figure 4.5. Note that, once again, incomes are presented in 2017 dollars to eliminate any changes due to inflation over the time period.

Each line in Figure 4.5 marks a certain percentile, as opposed to quartiles. For example, the 20th percentile is the income that splits the distribution into two groups: 20% of households earned less than this value and 80% earned more. The 80th percentile also splits the distribution into two groups: 20% earned more than this income and 80% earned less. Finally, the 95th percentile separates the top 5% of household incomes from the huge majority (95%) of American households.

Note that the line marking the 20th percentile (P_{20}) stayed level throughout the time period while the 80th percentile (P_{80}) line generally rises. This means that the financial situation of lower-income Americans has remained virtually unchanged while the income of more affluent Americans has increased. This increasing income for the more affluent is consistent with the increasing positive skew in Figure 4.4.

If we subtracted the P_{20} income from the P_{80} income, we would have a measure of variation similar to the interquartile range (Q) and the value of this measure would increase throughout the time period, indicating that incomes were

FIGURE 4.5 Percentiles of Household Income for the United States, 1967–2017 (in 2017 dollars)

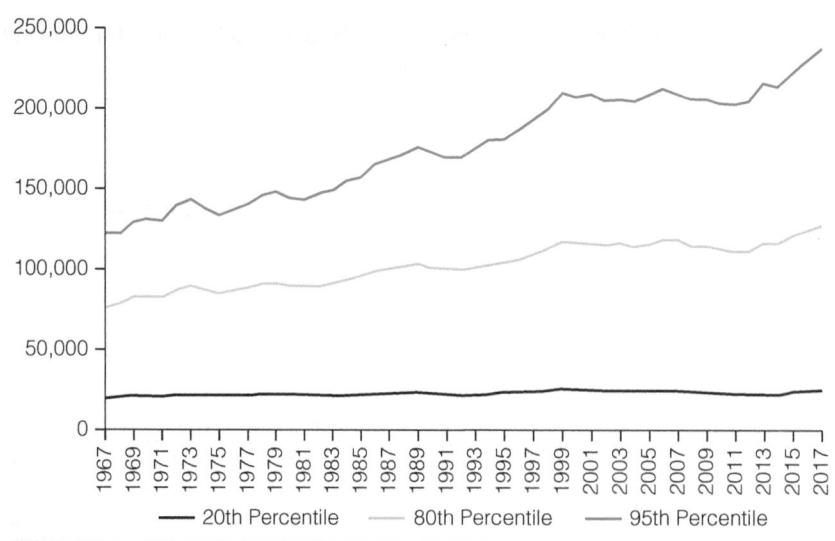

Source: U.S. Bureau of the Census, 2019. "Historical Income Tables: Table H-1." https://www.census.gov/data/tables/time-series/demo/income-poverty/historical-income-households.html

becoming more unequal. We can see from Figure 4.5 that this increase in variation is solely a result of the rising incomes associated with P_{80}, not with any changes in the incomes associated with P_{20}. The increasing variation in American household incomes would be even more dramatic if we used the 95th percentile instead of the 80th.

Taken together, these graphs show increases in the income of the average American household (median) and the average income for all American households (mean) over this time period. However, the increase was not shared equally. It was virtually nonexistent for the less affluent (the 20th percentile) and quite robust for those with higher incomes. Thus, variation increased: Households with modest incomes continued to have modest incomes, and, consistent with the ancient folk wisdom, the rich got richer.

READING STATISTICS Pornography and Religiosity

The professional research literature focuses on the relationships between variables, and univariate descriptive statistics—like means, standard deviations, and frequency distributions—are usually just the first step in the data analysis, not the ultimate concern, and probably will not be included in the final publication. On the other hand, some statistics (e.g., the mean and standard deviation) serve a dual function. They not only are valuable descriptive statistics but also form the basis for many analytical techniques. Thus, they may be reported in the latter role if not in the former.

When included in research reports, measures of central tendency and variation will most often be presented in some summary form such as a table. A fragment of such a summary table might look like this:

Characteristics of the Sample

Variable	\overline{X}	S	N
Age	33.2	1.3	1078
Number of children	2.3	0.7	1078
Years married	7.8	1.5	1052
Income	$55,786	1500	987

Tables such as this describe the overall characteristics of the sample succinctly and, if you inspect the table carefully, you will have a good sense of the nature of the sample on the traits relevant to the project. Note that the number of cases varies from variable to variable. This is normal in social science research

and is caused by missing data or incomplete information on some of the cases.

STATISTICS IN THE PROFESSIONAL LITERATURE

Two sociologists were interested in the effects of pornography on religiosity for teenage and young adult Americans. Because of the pervasiveness of the Internet, pornography is more available than in previous generations. Does viewing pornography decrease interest in religion and weaken attachments to religious institutions? The researchers used a large, representative sample of American young people to examine these complex relationships. In this installment of "Reading Statistics," we review some of the descriptive statistics reported by the researchers. As is usually the case, they reported these statistics as background information and to help identify differences and important patterns. The actual hypotheses are tested with more advanced statistics.

They present means and standard deviations for many different variables, several of which are presented below. Pornography viewing was measured by asking how many pornographic programs (movies, etc.) each respondent had watched over the past year. Responses ranged from a low of zero to a high of 300, but the variable was collapsed to "0 to 5 or more" for this analysis. The researchers measured religiosity (e.g., religious attendance) with scales ranging from zero (never) to a high of 5 or 6 (religious attendance more than once a week).

(continued)

READING STATISTICS **(Continued)**

By inspecting the table, you can see that pornography consumption increased with age while all measures of religiosity decreased. These patterns are consistent with the researcher's argument that, for youth, pornography increases secularization and decreases attachment to the religious institution.

Want to find out more? The source is noted below the table. If you are interested in the topic and wish to read the entire article, you will find that the author's arguments and conclusions are quite accessible, even though their statistical presentation might be beyond your understanding at this point in your education.

Means and Standard Deviations for Pornography Viewing and Religiosity by Age

Variables	Range	Ages 13–17		Ages 18–24	
		Mean	Std. Dev.	Mean	Std. Dev.
Pornography Viewing	0–5	0.92	1.63	1.60	2.00
Religious Attendance	0–6	3.09	2.22	2.17	2.13
Prayer Frequency	0–6	3.26	2.00	2.98	2.07
Closeness to God	0–5	2.96	1.29	2.69	1.35

Source: Perry, S., and Hayward, G. 2017. "Seeing Is (Not) Believing: How Viewing Pornography Shapes the Religious Lives of Young Americans," *Social Forces*: 95(4), pp. 1757–1788.

SUMMARY

1. Measures of variation summarize information about the heterogeneity, or variety, in a distribution of scores. When combined with a measure of central tendency, these statistics convey a large volume of information in just a few numbers. While measures of central tendency locate the central points of the distribution, measures of variation indicate the amount of diversity in the distribution.

2. The range (R) is the distance from the highest to the lowest score in the distribution. The interquartile range (Q) is the distance from the third to the first quartile (the "range" of the middle 50% of the scores). These two ranges can be used with variables measured at either the ordinal or interval-ratio level.

3. The standard deviation (s) is the most important measure of variation because of its central role in many more advanced statistical applications. The standard deviation has a minimum value of 0 (indicating no variation in the distribution) and increases in value as the variability of the distribution increases. It is used most appropriately with variables measured at the interval-ratio level but is frequently computed for ordinal-level variables.

4. The variance (s^2) is used primarily in inferential statistics and in the design of some measures of association.

5. Boxplots provide a useful way of visualizing and analyzing variation.

SUMMARY OF FORMULAS

Range	4.1	$R = \text{High score} - \text{Low score}$

Interquartile range	4.2	$Q = Q_3 - Q_1$

Variance	4.3	$s^2 = \dfrac{\Sigma(X_i - \bar{X})^2}{N}$

Standard deviation	4.4	$s = \sqrt{\dfrac{\Sigma(X_i - \bar{X})^2}{N}}$

GLOSSARY

Boxplot. A graph that presents information about the central tendency and variation of a variable.

Deviation. The distance between the score and the mean.

Interquartile range (Q). The distance from the third quartile to the first quartile.

Measures of variation. Statistics that indicate the amount of variety, or heterogeneity, in a distribution of scores.

Range (R). The highest score minus the lowest score.

Standard deviation. A statistic that quantifies the amount of variation around the mean. The statistic is computed by summing the squared deviations of the scores around the mean, dividing by N, and, finally, taking the

square root of the result. The most important and useful descriptive measure of variation; s represents the deviation of a sample; σ represents the standard deviation of a population.

Variance. The square of the standard deviation. The sum of the squared deviations of the scores around the mean, divided by N. A measure of variation used primarily in inferential statistics and also in correlation and regression techniques; s^2 represents the variance of a sample; σ^2 represents the variance of a population.

Variation. The amount of variety, or heterogeneity, in a distribution of scores.

PROBLEMS

4.1 Compute the range and standard deviation of the following 10 scores. (*HINT: It will be helpful to organize your computations as in Table 4.4.*)

10, 12, 15, 20, 25, 30, 32, 35, 40, 50

4.2 Compute the range and standard deviation of the following 10 test scores.

66, 75, 69, 72, 84, 90, 96, 70, 55, 45

4.3 SOC In your sociology class, you were assigned to carry out a survey of the relationship between age, awareness of political news, and support for the

president. You have measured age and the number of political news stories the respondents read last week as interval-level variables. Support for the president was rated on a scale of 1–10 with 10 being the most support. Technically, this last variable is at the ordinal level, but we will treat it as an interval-level variable. You have also separated the results by gender. Calculate the mode, the median, the mean, the range, and the standard deviation for each variable by gender. What information is added by the measures of variation? Write a paragraph summarizing the differences between males and females.

Age		Number of Political News Stories Read Last Week		Support for the President	
Males	*Females*	*Males*	*Females*	*Males*	*Females*
33	65	0	0	10	1
39	62	14	5	7	2
45	60	10	11	2	8
47	90	7	3	1	4
62	62	5	4	8	3
48	57	1	14	6	6
52	40	0	0	10	2
65	49	14	7	0	9
51	45	3	5	5	4
43	85	4	3	6	7
	78		5		4

4.4 SOC In problem 3.5 at the end of Chapter 3, you calculated measures of central tendency for four variables for 15 respondents. Two of those variables are reproduced here. Calculate the mean, the range, and the standard deviation for each variable. Again, we will treat the attitudinal variable as if it is at the

interval level. What information is added by the measures of variation? Write a paragraph summarizing these statistics.

Respondent	Age	Attitude on Abortion (High score = Strong opposition)
A	18	10
B	20	9
C	21	8
D	30	10
E	25	7
F	26	7
G	19	9
H	29	6
I	31	10
J	55	5
K	32	4
L	28	3
M	23	2
N	24	1
O	32	9

4.5 SOC In problem 3.6, you computed mean and median income for 13 Canadian provinces and territories and for 13 U.S. states in two separate years. Now compute the standard deviation and range for each year, and, taking account of the two measures of central tendency and the two measures of variation, write a paragraph summarizing the distributions. What do the measures of variation add to what you already knew about central tendency? Did the median income of the provinces and states become more or less variable over the period? The scores are reproduced here.

Province or Territory

	2000	2015
Newfoundland and Labrador	38,800	67,272
Prince Edward Island	44,200	61,163
Nova Scotia	44,500	60,764
New Brunswick	43,200	59,347
Quebec	47,700	59,822
Ontario	55,700	74,287
Manitoba	47,300	68,147
Saskatchewan	45,800	75,412
Alberta	55,200	93,835
British Columbia	49,100	69,995
Yukon	56,000	84,521
Northwest Territories	61,000	117,688
Nunavit	37,600	97,441

Median Household Incomes for U.S. States, 1999 and 2017 (U.S. dollars)

State	1999	2017
Alabama	36,251	46,472
Alaska	51,396	76,114
Arkansas	29,682	43,813
California	43,629	67,169
Connecticut	50,593	73,781
Illinois	46,330	61,229
Kansas	37,348	55,477
Maryland	52,205	78,916
Michigan	46,089	52,668
New York	39,989	62,765
Ohio	39,489	52,407
South Dakota	35,825	54,126
Texas	38,688	57,051

4.6 SW A licensed clinical social worker has collected data on 10 families that have started family therapy treatment to reduce conflict between parents and their teenage children. Prior to the treatment, she measured the number of times the teenagers reported having arguments with their parents in the last four weeks using data submitted by the teenagers from a diary. After the therapy, the teenagers were asked to keep a diary for four weeks again, and to report the number of arguments during that time. The data for the pretest below show the number of conflicts before the therapy. The data for the posttest below show the number of conflicts after the therapy. Compute the mean and the standard deviation for the pretest and posttest scores. Did the therapy make a difference? Use the mean to answer this question. Was there more variation in the pretest scores or the posttest scores? Use the standard deviation for this one. What might she conclude? Write a paragraph addressing these questions and citing the results.

Case	Pretest	Posttest
A	8	12
B	7	13
C	10	12
D	15	19
E	10	8
F	10	17
G	3	12
H	10	11
I	5	7
J	15	12
K	13	20
L	4	5
M	10	15
N	8	11
O	12	20

4.7 SOC Labor force participation rates (percent employed), percentage high school graduates, and mean income for males and females in 10 states are reported here. Calculate the mean and the standard deviation for both groups for each variable and describe the differences. Are males and females unequal on any of these variables? How great is the gender inequality?

| State | Labor Force Participation Rate | | Percentage HS Graduates | | Mean Income | |
	Male	Female	Male	Female	Male	Female
A	65.8	54.3	81.0	81.9	55,623	50,012
B	76.7	63.0	88.4	89.7	52,345	51,556
C	71.8	57.2	82.4	84.6	55,789	48,231
D	76.1	66.6	89.5	90.9	48,907	46,289
E	75.1	63.1	86.9	88.7	62,023	58,034
F	69.9	61.1	86.3	86.4	55,000	53,897
G	73.6	59.6	87.1	87.6	49,145	47,148
H	70.5	60.3	87.0	87.6	51,897	50,659
I	66.3	55.2	81.7	84.1	51,238	45,289
J	74.5	67.1	89.1	91.6	60,746	56,489

4.8 SOC Data on several variables measuring overall health and well-being for 11 nations are reported here for 2010, with projections to 2020. Are nations becoming more or less diverse on these variables? Calculate the mean, range, and standard deviation for each year for each variable. Summarize the results in a paragraph.

| Nation | Life Expectancy (years) | | Infant Mortality Rate* | | Fertility Rate[#] | |
	2010	2020	2010	2020	2010	2020
Canada	81	82	5.0	4.4	1.6	1.6
China	75	76	16.5	12.6	1.5	1.5
Egypt	72	75	26.2	17.9	3.0	2.7
Germany	79	81	4.0	3.6	1.4	1.5
Japan	82	83	2.8	2.7	1.2	1.3
Mali	52	57	113.7	91.9	6.5	5.5
Mexico	76	78	17.8	13.2	2.3	2.1
Peru	71	74	27.7	20.2	2.3	2.0
Ukraine	69	70	8.7	7.3	1.3	1.4
United States	78	80	6.1	5.4	2.1	2.1
Zambia	52	54	68.4	50.6	6.0	5.3

*Number of deaths of children under one year of age per 1000 live births.

[#]Average number of children per female.

Source: U.S. Bureau of the Census. 2012. *Statistical Abstract of the United States: 2012.* p. 842.

4.9 SOC The price of a liter of regular gas for 20 nations is reported below. Compute the mean, median, range, interquartile range, and standard deviation for this variable, and write a paragraph summarizing these statistics.

	Nation	Cost of a Liter of Gasoline (U.S. dollars) in 2019
1	Nigeria	0.40
2	Mexico	1.05
3	United States	0.78
4	India	1.04
5	Colombia	0.71
6	Canada	1.01
7	Ukraine	1.14
8	China	1.01
9	Kenya	1.09
10	South Africa	1.01
11	Australia	0.97
12	Brazil	1.06
13	Guatemala	0.89
14	Spain	1.45
15	Germany	1.59
16	Japan	1.32
17	Iceland	1.88
18	Sweden	1.62
19	United Kingdom	1.56
20	Italy	1.74

Source: Global Petrol Prices. 2019. https://www.globalpetrolprices.com/gasoline_prices/

4.10 SOC In problem 3.11, you computed measures of central tendency for the number of roller coasters per capita for 15 nations. The scores are reproduced here. Compute the standard deviation for this variable, and write a paragraph summarizing the mean, median, and standard deviation.

Nation	Number of Roller Coasters per Capita
Argentina	0.33
Belgium	2.18
Canada	1.57
China	0.05
Denmark	6.07
Ecuador	0.14
France	1.02
India	0.02
Japan	1.88
Mexico	0.20
New Zealand	0.72

(Continued next column)

(Continued)

Nation	Number of Roller Coasters per Capita
Russia	0.13
Saudi Arabia	0.59
United Kingdom	2.64
United States	2.09

4.11 MR The year 2018 was bad for an LGBTQ organization in a large city. Its events were poorly attended. In preparation for their schedule of events in 2019, they hired a public relations firm to improve their advertising. In 2019, they held the same 15 events and their attendance improved, but by how much? Also, how much did the variation in the number of people attending change? Compute the mean attendance at events and the standard deviation for each year, and describe the differences between 2018 and 2019.

City	2018	2019
A	52	197
B	87	124
C	101	131
D	52	197
E	115	119
F	88	87
G	100	150
H	101	209
I	95	110
J	78	140
K	103	178
L	107	94
M	85	125
N	117	200
O	167	225
P	180	210
Q	120	201
R	78	141
S	55	248
T	92	131

4.12 a. PA Compute the range, interquartile range, and standard deviation for the data presented in problem 3.4. The data are reproduced here. (*NOTE: To simplify computations for Q, round off the case numbers that mark the first and third quartiles to the nearest whole number.*)

b. What would happen to the value of the standard deviation if you removed Los Angeles (the city with the highest score) from this distribution and recalculated? Why?

City	Average Hours of Congestion
Birmingham	0.33
Boston	6.25
Charlotte	4.08
Cincinnati	3.08
Cleveland	2.54
Detroit	4.25
Hartford	2.54
Indianapolis	3.01
Kansas City	3.07
Las Vegas	4.20
Los Angeles	9.35
Louisville	2.30
Nashville	2.55
New Orleans	4.27
Oklahoma City	2.22
Phoenix	3.39
Richmond	2.08
San Antonio	3.45
San Diego	4.29
St Louis	3.26
Tampa	2.51
Washington, D.C.	7.45

4.13 SOC Listed here are the rates of abortion per 100,000 women for 20 states in 1973 and 1975. Describe what happened to these distributions over the two-year period. Did the average rate increase or decrease? What happened to the variation of this distribution? Did the range, interquartile range, and standard deviation increase or decrease? What happened between 1973 and 1975 that might explain these changes in central tendency and variation? (*HINT: It was a Supreme Court decision.*)

	State	1973	1975
1	Mississippi	0.2	0.6
2	Arkansas	2.9	6.3
3	Montana	3.1	9.9
4	Maine	3.5	9.5
5	South Carolina	3.8	10.3
6	Tennessee	4.2	19.2
7	Texas	6.8	19.1
8	Arizona	6.9	15.8

(*Continued next column*)

(*Continued*)

	State	1973	1975
9	Ohio	7.3	17.9
10	Nebraska	7.3	14.3
11	Virginia	7.8	18.0
12	Iowa	8.8	14.7
13	Massachusetts	10.0	25.7
14	Pennsylvania	12.1	18.5
15	Colorado	14.4	24.6
16	Florida	15.8	30.5
17	Michigan	18.7	20.3
18	Hawaii	26.3	31.6
19	California	30.8	33.6
20	New York	53.5	40.7

Source: U.S. Bureau of the Census. 1977. *Statistical Abstract of the United States: 1977.* Washington, D.C.: Government Printing Office, 1977.

4.14 SW One of your goals as the new chief administrator of a large social service bureau is to equalize workloads within the various divisions of the agency. You have gathered data on caseloads per worker within each division. Which division comes closest to the ideal of an equalized workload? Which is furthest away?

A	B	C	D
50	60	60	75
51	59	61	80
55	58	58	74
60	55	59	70
68	56	59	69
59	61	60	82
60	62	61	85
57	63	60	83
50	60	59	65
55	59	58	60

4.15 CJ You are the governor of the state and must decide which of four metropolitan police departments will win the annual award for efficiency. The performance of each department is summarized in monthly arrest statistics as reported here. Which department will win the award? Why?

	Department			
	A	B	C	D
Mean =	601.30	633.17	592.70	599.99
$s =$	2.30	27.32	40.17	60.23

4.16 SP A large university in the Midwest has developed a new training program for its faculty members to educate them on how to identify cases of sexual harassment. They have administered the training to half of their faculty members so far, and to determine the degree of its effectiveness they have administered a test on the subject to those that have not taken the test yet and those that have already taken the training. The scores are on a scale of 1 to 100, where 100 is a perfect score. The data are shown below. What can you conclude about the training program?

Untrained Faculty	Trained Faculty
$\bar{X} = 77.8$	$\bar{X} = 76.8$
$s = 12.3$	$s = 6.2$
$N = 478$	$N = 465$

4.17 GR As more U.S. states have legalized marijuana use, senior citizens have been among the fastest to increase their usage of cannabis. Below are fictional data from a survey showing the level of support (1–50) for further legalization of marijuana in the United States among two groups of older Americans, those between 55 and 65 and those between 65 and 75. Compare the mean and the standard deviation for each group. Describe what you have learned about the differences between the groups as well as the variation.

Ages 55–65				
10	43	30	30	45
40	12	40	42	35
45	25	10	33	50
42	32	38	11	47
22	26	37	38	10

Ages 65–75				
10	45	35	27	50
35	10	50	40	30
40	10	10	37	10
40	15	30	20	43
23	25	30	40	10

Statistical Analysis Using SPSS

4.18 SOC In this exercise, you will use SPSS to get measures of central tendency and variation for the rate of traffic fatalities in two different years (1990 and 2017) from the *States* data set. As an option, you can also get boxplots for these variables.
- Find and click the SPSS icon on your desktop.
- Load the *States* data set.

- From the menu bar across the top of the SPSS window, click **Analyze, Descriptive Statistics,** and **Frequencies**.
- In the list of variables in the **Frequencies** window, find *TrafDths17* (the rate of traffic fatalities in 2017) and *TrafDths90* (the rate of traffic fatalities in 1990). Click the arrow to move the variable names into the box on the right.
- Click the **Statistics** tab and check the appropriate boxes to select the following statistics: Quartiles, Mean, Median, Std. deviation, and Range. Click **Continue** and then click **OK**.
- *OPTIONAL*: To get boxplots for these variables:
 - Click **Graphs, Legacy Dialogs,** and **Boxplot**.
 - On the **Boxplot** window, check "Summaries of separate variables" and then click **Define**.
 - Find *TrafDths17* and *TrafDths90* in the list of variables and click the arrow to move the variable names into the "Boxes Represent" box.
 - Click **OK**.

a. Compute Q for each variable by subtracting the value for the 25th percentile from the value for the 75th percentile.

b. Write a paragraph describing the changes in the rate of traffic fatalities between 1990 and 2017. Be sure to discuss central tendency and variation separately, and clearly identify each statistic. *(OPTIONAL: Include an analysis of the boxplots.)*

It may be helpful to model your paragraph on one of the discussions presented earlier in this chapter, including the discussion of changing homicide rates in the "Using SPSS" section. What factors might explain the changing rate of traffic fatalities? What additional information would you need to test your explanations?

4.19 SOC In this exercise, you will once again use the *States* data set, this time to analyze changes in the level of poverty between 2000 and 2017. We will use the **Descriptives** command to generate output.
- Find and click the SPSS icon on your desktop.
- Load the *States* data set.
- From the menu bar across the top of the SPSS window, click **Analyze, Descriptive Statistics,** and **Descriptives**.
- In the list of variables, find *FamPoor00* (the percentage of families below the poverty line in each state in 2000) and *FamPoor17* (the percentage of families below the poverty line in 2017). Click the arrow to move the variable names into the box on the right.
- Click the **Options** tab and check the box to select the Range. Click **Continue** and then click **OK**.

- *OPTIONAL*: To get boxplots for these variables:
 - Click **Graphs**, **Legacy Dialogs**, and **Boxplot**.
 - On the **Boxplot** window, check "Summaries of separate variables" and then click **Define**.
 - Find *Fampoor17* and *Fampoor09* in the list of variables, and click the arrow to move the variable names into the "Boxes Represent" box.
 - Click **OK**.

Write a paragraph describing the changes in family poverty between 2000 and 2017. Be sure to discuss central tendency and variation separately, and clearly identify each statistic. *(OPTIONAL: Include an analysis of the boxplots.)* It may be helpful to model your paragraph on one of the discussions presented earlier in this chapter.

YOU ARE THE RESEARCHER

Two projects are presented below. The first follows up on the project presented at the end of Chapter 3, and the second continues the project presented at the end of Chapter 2. In the second project, you will be introduced to a new SPSS command. You are urged to complete both projects.

Project 1: The Typical American (Revisited)

In Chapter 3, you described the typical American using 10 variables selected from the 2018 GSS. Now you will examine variation on some of the variables you selected.

Step 1: Choosing the Variables

Select at least five of the ordinal and interval-ratio variables you used in Chapter 3. Add more variables if you had fewer than five variables at this level of measurement.

Step 2: Getting the Statistics

Use **Descriptives** to find the range and standard deviation for each of your selected variables. With the 2018 GSS loaded, click **Analyze**, **Descriptive Statistics**, and **Descriptives** from the Main Menu of SPSS. The **Descriptives** dialog box will open. Find the names of your variables in the list on the left and click the right arrow button to transfer them to the **Variables** box. Click **OK**, and SPSS will produce the same output you analyzed in Chapter 3. Now, however, we will consider variation rather than central tendency.

The standard deviation for each variable is reported in the column labeled "Std. Deviation," and the range can be computed from the values given in the Minimum and Maximum columns. At this point, the range is probably easier to understand and interpret than the standard deviation. As we have seen, the latter is more meaningful when we have a point of comparison. For example, suppose we were interested in the variable *tvhours* and how television viewing habits have changed over the years. The **Descriptives** output for 2018 shows that people watched an average of 2.94 hours a day with a standard deviation of 2.84. Suppose that data from an earlier year showed an average of 3.70 hours of television viewing a day with a standard deviation of 1.1. You could conclude that television watching had, on average, decreased over the 35-year period, but that Americans had also become much more diverse in their viewing habits.

Record your results in this table:

Variable	SPSS Name	Mean	Range	Standard Deviation
1.				
2.				
3.				
4.				
5.				

Step 3: Interpreting Results

Start with the descriptions of the variables you wrote in Chapter 3 and add information about variation, referring both to the range and the standard deviation. Your job now is both to describe the "typical" American *and* to suggest the amount of variation around this central point.

Project 2: The Culture Wars (Revisited)

In Chapter 2, you examined some of the dimensions of the "culture wars" in U.S. society. In this project, you will reexamine the topic and learn how to use a new SPSS command to combine existing variables into a new summary variable. This computed variable can be used to summarize feelings and attitudes "in general" and to explore new dimensions of the issue.

Step 1: Creating a Scale to Summarize Attitudes Toward Abortion

One of the most controversial issues in the American culture wars is abortion: Under what conditions, if any, should abortion be legal? The 2018 GSS data set supplied with this text includes two variables that measure support for legal abortion. The variables differ in context: one asks specifically if an abortion should be available if the mother is poor and wants no more children (*abpoor*). The other asks if an abortion should be available "for any reason" (*abany*). Since these two situations are distinct, each item should be analyzed in its own right. Suppose, however, that you wanted to create a summary scale that indicated a person's *overall* feelings about abortion.

One way to do this would be to add the scores on the two variables together. This would create a new variable, which we will call *abscale*, with three possible scores:

- A score of 2 means that the respondent was consistently "pro-abortion" and answered "yes" (coded as "1") to both items.
- A score of 3 means that the respondent answered "yes" (1) to one item and "no" (2) to the other. This might be labeled an "intermediate" position.
- A score of 4 means that the respondent answered "no" (coded as "2") to both items. This would be a consistent "anti-abortion" position.

The table below summarizes the scoring possibilities.

If response on *abany* is:	and	response on *abpoor* is:	score on *abscale* will be:
1 (Yes)		1 (Yes)	2 (pro-abortion)
1 (Yes)		2 (No)	3 (intermediate)
2 (No)		1 (Yes)	3 (intermediate)
2 (No)		2 (No)	4 (anti-abortion)

The new variable, *abscale*, summarizes each respondent's overall position on the issue. Once created, *abscale* could be analyzed, transformed, and manipulated exactly like a variable actually recorded in the data file.

Using Compute We will create *abscale* with the **Compute** command. To begin, click **Transform** and then **Compute Variable** from the main menu. The **Compute Variable** window will appear. Find the **Target Variable** box in the upper-left-hand corner of this window. Follow these steps to compute *abscale*:

- First, assign a name to the new variable we are about to compute. To do this, type *abscale* in the **Target Variable** box.
- Next, we need to tell SPSS how to compute the new variable. In this case, *abscale* will be computed by adding the scores of *abany* and *abpoor*.
 - Find *abany* in the variable list on the left and click the arrow button in the middle of the screen to transfer the variable name to the **Numeric Expression** box.
 - Click the plus sign (+) on the calculator pad under the **Numeric Expression** box, and the sign will appear next to *abany*.
 - Finally, highlight *abpoor* in the variable list and click the arrow button to transfer the variable name to the **Numeric Expression** box.
 - The expression in the **Numeric Expression** box should now read

 abany + abpoor

- Click **OK**, and *abscale* will be created and added to the data set.

If you want to keep this new variable permanently, click **Save** from the **File** menu, and the updated data set with *abscale* added will be saved. (If you are using the student version of SPSS, remember that your data set is limited to 50 variables and you may not be allowed to save the data set with *abscale* added.)

Examining the Variables We now have three variables that measure attitudes toward abortion—two items referring to specific situations, and a more general, summary item. It is always a good idea to check the frequency distribution for computed variables to make sure that the computations were carried out as we intended. Use the **Frequencies** procedure (click **Analyze**, **Descriptive Statistics**, and **Frequencies**) to get tables for *abany*, *abpoor*, and *abscale*. The results for the full 2018 GSS are reproduced below. *(NOTE: If you are using the smaller sample GSS, your results will differ.)*

		Frequency	Percent	Valid Percent	Cumulative Percent
		Abortion if a woman wants for any reason			
Valid	YES	764	32.5	50.1	50.1
	NO	760	32.4	49.9	100.0
	Total	1524	64.9	100.0	
Missing	IAP	774	33.0		
	DK	30	1.3		
	NA	20	.9		
	Total	824	35.1		
Total		2348	100.0		

Low income—can't afford more children					
		Frequency	Percent	Valid Percent	Cumulative Percent
Valid	YES	750	31.9	49.3	49.3
	NO	770	32.8	50.7	100.0
	Total	1520	64.7	100.0	
Missing	IAP	774	33.0		
	DK	36	1.5		
	NA	18	.8		
	Total	828	35.3		
Total		2348	100.0		

abscale					
		Frequency	Percent	Valid Percent	Cumulative Percent
Valid	2.00	649	27.6	43.4	43.4
	3.00	200	8.5	13.4	56.7
	4.00	648	27.6	43.3	100.0
	Total	1497	63.8	100.0	
Missing	System	851	36.2		
Total		2348	100.0		

Missing Cases Note that about 1520 respondents answered each specific abortion item (*abany* and *abpoor*). Remember that no respondent is given the entire GSS, and the vast majority of the "missing cases" received a form of the GSS that did not include these two items. These cases are coded as "IAP" in the tables above. Other "missing" scores are DK or "Don't Know" (the respondent said they didn't know) and NA or "No Answer" (the respondent declined to answer).

Now look at *abscale* and note that even fewer cases (1497) are included in the summary scale than in the two original items. When SPSS executes a **Compute** statement, it automatically eliminates any cases that are missing scores on any of the constituent items. If these cases were not eliminated, a variety of errors and misclassifications could result. For example, if cases with missing scores were included, a person who scored a 2 ("anti-abortion") on *abany* and then didn't respond to *abpoor* would have a total score of 2 on *abscale*. Thus, this case would be treated as "pro-abortion" when the only information we have indicates that this respondent is "anti-abortion." Cases with missing scores are deleted to avoid this type of error.

Step 2: Interpreting the Results

Compare the distributions of these three variables with each other and write a report in which you answer the following questions:

1. How does the level of approval for abortion differ from one specific situation to another? Remember to use the "Valid Percent" column.
2. What percent of the respondents approved and disapproved of abortion in both circumstances? What percentage approved in one situation but not the other? *(Use the distribution of abscale to answer this question.)*
3. What do these patterns reveal about value consensus in U.S. society? Are Americans generally in agreement about the issue of abortion?

5 The Normal Curve

6/16 7/16 9/16 11/16 12/

LEARNING OBJECTIVES By the end of this chapter, you will be able to:

1. Explain the normal curve and its characteristics.
2. Convert numbers around the mean into Z scores.
3. Use Z scores and the normal curve table (Appendix A) to find areas above, below, and between points on the curve.
4. Express areas under the curve in terms of probabilities.

USING STATISTICS

The statistical techniques presented in this chapter can be used to

- Estimate the percentage of Americans that had their first child by a particular age. **See page 140**.
- Estimate the number of books that Americans have read in the last year. **See page 148**.

The **normal curve** is a concept of great importance in statistics. In combination with the mean and standard deviation, it is used to make precise descriptive statements about empirical distributions. Also, the normal curve is central to the theory that underlies inferential statistics. Thus, this chapter concludes our treatment of Part I (Descriptive Statistics) and lays important groundwork for Part II (Inferential Statistics).

The normal curve is sometimes used to assign grades on tests, and you may be familiar with this application. In this guise, the normal curve is often called the "bell curve," and grading "on the curve" means that the instructor wants the scores to follow a specific pattern: The modal grade is a C and there will be equal numbers of As and Fs, Bs and Ds. In other words, the distribution of grades should look like a bell or a mound.

Properties of the Normal Curve

The normal curve is a theoretical model, a line chart (see Chapter 2) that is unimodal (i.e., has a single mode, or peak), perfectly smooth, and symmetrical (unskewed), so its mean, median, and mode are all exactly the same

value. As we learned in Chapter 3, the value of its skewness is zero. It is bell-shaped, and its tails extend infinitely in both directions. No empirical distribution matches this ideal model perfectly, but some variables (e.g., test results from large classes, standardized test scores such as the GRE) are close enough to permit the assumption of normality. In turn, this assumption makes possible one of the most important uses of the normal curve—the description of empirical distributions based on our knowledge of the theoretical normal curve.

The crucial point about the normal curve is that distances along the horizontal axis, when measured in standard deviations from the mean, always encompass the same proportion of the total area under the curve. In other words, the distance from any point to the mean—when measured in standard deviations—will cut off exactly the same part of the area under the curve.

To illustrate, Figures 5.1 and 5.2 present two hypothetical distributions of IQ scores, both normally distributed, for fictional groups of males and females, such that:

Males	Females
$\overline{X} = 100$	$\overline{X} = 100$
$s = 20$	$s = 10$
$N = 1000$	$N = 1000$

Figures 5.1 and 5.2 are drawn with two scales on the horizontal axis of the graph. The upper scale is stated in "IQ units" and the lower scale in standard deviations from the mean. These scales are interchangeable and we can easily shift from one to the other. For example, for the males, an IQ score of 120 is one standard deviation (remember that, for the male group, $s = 20$) above the mean and an IQ of 140 is two standard deviations above (to the right of) the mean. IQ scores to the left of the mean are marked as negative values on

FIGURE 5.1 Scores for a Group of Males

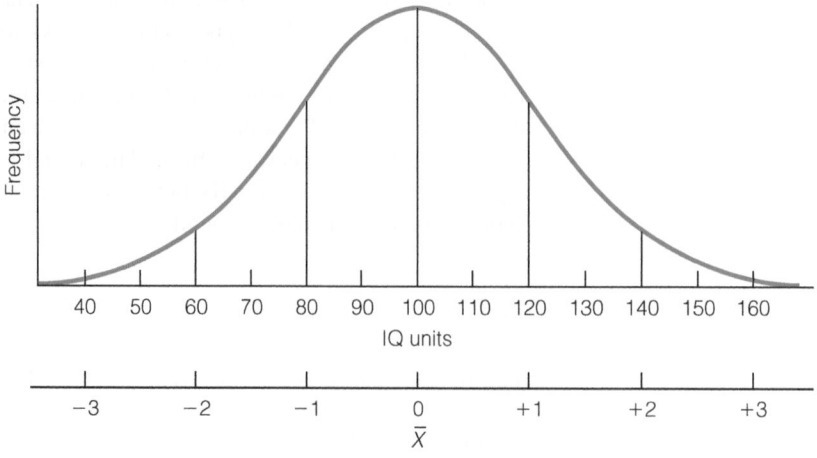

FIGURE 5.2 IQ Scores for a Group of Females

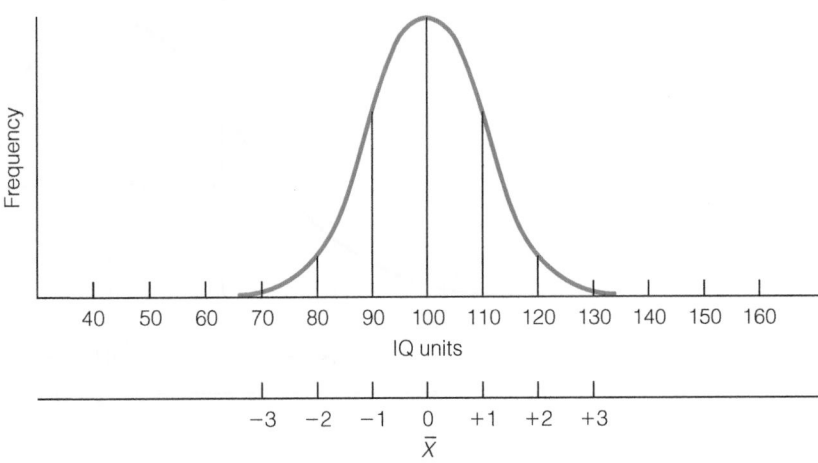

the standard deviation scale because they are less than the mean. An IQ of 80 is one standard deviation below the mean, an IQ score of 60 is two standard deviations below the mean, and so forth. Figure 5.2 is marked in a similar way, except that, because its standard deviation is a different value ($s = 10$), the markings occur at different points. For the female sample, one standard deviation above the mean is an IQ of 110, one standard deviation below the mean is an IQ of 90, and so forth.

Recall that on any normal curve, distances along the horizontal axis, when measured in standard deviations, always encompass exactly the same proportion of the total area under the curve. Specifically, the distance between one standard deviation above the mean and one standard deviation below the mean (or ± 1 standard deviation) encompasses exactly 68.26% of the total area under the curve. This means that in Figure 5.1, 68.26% of the total area lies between the score of 80 (-1 standard deviation) and 120 ($+1$ standard deviation). The standard deviation for females is 10, so the same percentage of the area (68.26%) lies between the scores of 90 and 110. On any normal distribution, 68.26% of the total area will always fall between ± 1 standard deviation, regardless of the trait being measured and the numerical values of the mean and standard deviation.

It will be useful to familiarize yourself with the following relationships between distances from the mean and areas under the normal curve:

Between	Lies
± 1 standard deviation	68.26% of the area
± 2 standard deviations	95.44% of the area
± 3 standard deviations	99.72% of the area

These relationships are displayed graphically in Figure 5.3.

FIGURE 5.3 Areas Under the Theoretical Normal Curve

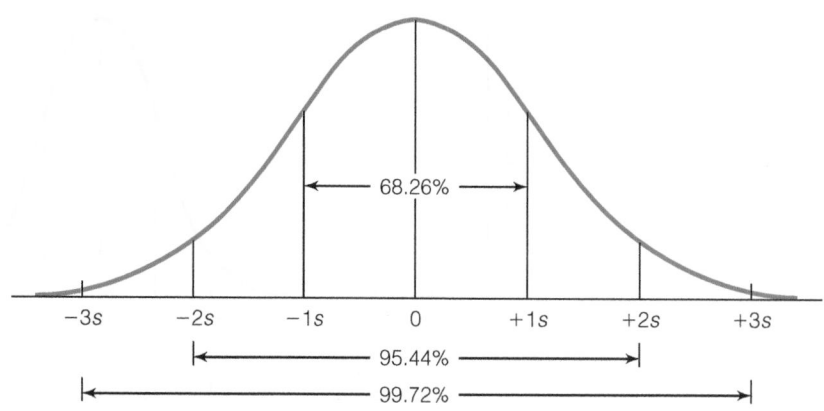

We can describe empirical distributions that are at least approximately normal using these relationships between distance from the mean and area. The position of individual scores can be described with respect to the mean, the distribution as a whole, or any other score in the distribution.

The areas between scores can also be expressed in numbers of cases rather than percentage of total area. For example, a normal distribution of 1000 cases will contain about 683 cases (68.26% of 1000 cases) between ± 1 standard deviation of the mean, about 954 between ± 2 standard deviations, and about 997 between ± 3 standard deviations. Thus, for any normal distribution, only a few cases will be farther away from the mean than ± 3 standard deviations.

STATISTICS **In Everyday Life**

Tests of Intelligence

Tests of intelligence are designed to generate distributions of scores that are approximately normal. That is, the tests mix easier and harder questions so that the mean will be about 100 and the final distribution of scores will be bell-shaped, have a mode of about 100, and have roughly equal numbers of high and low scores.

Intelligence testing can be controversial and, although we cannot go into detail here, you should be aware that there is considerable debate about the meaning of IQ scores. Tests of intelligence may have cultural and other biases and there is considerable disagreement about whether or not they actually measure native mental ability. In particular, you should be aware that the more or less normal distribution displayed by IQ scores does not necessarily mean that the underlying and elusive quality called intelligence—however that might be defined—is also normally distributed.

Using the Normal Curve

We have seen that we can find areas under the normal curve for scores that are 1, 2, or 3 standard deviations above or below the mean. To work with values that are not exact multiples of the standard deviation, we must express the original scores in units of the standard deviation or convert them into **Z scores**. The original scores could be in any unit of measurement (feet, IQ, dollars), but Z scores always have the same values for their mean (0) and standard deviation (1).

COMPUTING Z SCORES

Think of converting the original scores into Z scores as a process of changing value scales—similar to changing from meters to yards, kilometers to miles, or gallons to liters. These units are different but equally valid ways of expressing distance, length, or volume. For example, a mile is equal to 1.61 kilometers, so two towns that are 10 miles apart are also 16.1 kilometers apart and a "5k" race covers about 3.10 miles. Although you may be more familiar with miles than kilometers, either unit works perfectly well as a way of expressing distance.

In the same way, the original (or "raw") scores and Z scores are two equally valid but different ways of measuring distances under the normal curve. In Figure 5.1, for example, we could describe a particular score in terms of IQ units ("John's score was 120") or standard deviations ("John scored one standard deviation above the mean").

When we compute Z scores, we convert the original units of measurement (IQ scores, inches, dollars, etc.) to Z scores and, thus, "standardize" the normal curve to a distribution that has a mean of 0 and a standard deviation of 1. The mean of the empirical normal distribution will be converted to 0, its standard deviation to 1, and all values will be expressed in Z-score form. The formula for computing Z scores is

FORMULA 5.1

$$Z = \frac{X_i - \overline{X}}{s}$$

This formula will convert any score (X_i) from an empirical normal distribution into the equivalent Z score. To illustrate with the men's IQ data (Figure 5.1), the Z-score equivalent of a raw score of 120 would be

$$Z = \frac{120 - 100}{20} = \frac{20}{20} = +1.00$$

ONE STEP AT A TIME **Finding Z Scores**

1. Subtract the value of the mean (\overline{X}) from the value of the score (X_i).
2. Divide the quantity found in step 1 by the value of the standard deviation (s). The result is the Z-score equivalent for this raw score.

The Z score of positive 1.00 indicates that the original score lies one standard deviation unit above (to the right of) the mean. A negative score would fall below (to the left of) the mean. *(For practice in computing Z scores, see any of the problems at the end of this chapter.)*

The Normal Curve Table (Appendix A)

The theoretical normal curve has been thoroughly described by statisticians. The areas related to any Z score have been precisely determined and organized into a table format. This **normal curve table**, or Z-score table, is presented as Appendix A in this text; for purposes of illustration, a small portion of it is reproduced here as Table 5.1.

The normal curve table consists of three columns, with Z scores in the left-hand column (column a), areas between the Z score and the mean in the middle (column b), and areas beyond the Z score in the right-hand column (column c). Take note of the pictures above column b and column c in Appendix A. The shaded regions show you that column b is measuring the distance to the mean, and column c is measuring the distance from any point away from (above or below) the mean. To find the area between any Z score and the mean, go down column a until you find the Z score. Then move to your right and find the distance (or proportion) in column b that is associated with that Z score. For example, go down column a (either in Appendix A or in Table 5.1) until you find a Z score of $+1.00$. The entry in column b for that row shows that the "Area Between Mean and Z" is 0.3413.

TABLE 5.1 An Illustration of How to Find Areas Under the Normal Curve Using Appendix A

(a) Z	(b) Area Between Mean and Z	(c) Area Beyond Z
0.00	0.0000	0.5000
0.01	0.0040	0.4960
0.02	0.0080	0.4920
0.03	0.0120	0.4880
.	.	.
.	.	.
.	.	.
.	.	.
1.00	0.3413	0.1587
1.01	0.3438	0.1562
1.02	0.3461	0.1539
1.03	0.3485	0.1515
.	.	.
.	.	.
.	.	.
1.50	0.4332	0.0668

The table presents areas in the form of proportions, but we can easily translate these into percentages by multiplying them by 100 (see Chapter 2). We could say either "a proportion of 0.3413 of the total area under the curve lies between a Z score of 1.00 and the mean" or "34.13% of the total area lies between a score of 1.00 and the mean."

To illustrate further, find the Z score of 1.50 either in the Z-score column of Appendix A or the abbreviated table presented in Table 5.1. This score is 1½ standard deviations to the right of the mean and corresponds to an IQ of 130 for the men's IQ distribution. The area in column b for this score is 0.4332. This means that a proportion of 0.4332—or a percentage of 43.32%—of all the area under the curve lies between this score and the mean.

The third column in the table presents "Areas Beyond Z." These are areas above positive scores or below negative scores. The use of this column will be explained later in this chapter.

To conserve space, the normal curve table in Appendix A includes only positive Z scores. Because the normal curve is perfectly symmetrical, however, the area between the score and the mean—column b—for a negative score will be exactly the same as for a positive score of the same numerical value. For example, the area between a Z score of −1.00 and the mean will also be 34.13%, exactly the same as the area we found previously for a score of +1.00. As will be repeatedly demonstrated later, however, the sign of the Z score is extremely important and should be carefully noted.

For practice in using Appendix A to describe areas under an empirical normal curve, verify that the Z scores and areas in Table 5.2 are correct for the men's IQ distribution. For each IQ score, we compute the Z score with Formula 5.1, and then consult Appendix A to find areas between the score and the mean.

As shown in Table 5.3, the same procedures apply when the Z-score equivalent of an actual score happens to be a negative value (that is, when the raw score lies below the mean).

Remember that the areas in Appendix A will be the same for Z scores of the same numerical value regardless of sign. For the men's IQ distribution, the area between the score of 138 ($Z = +1.90$) and the mean is the same as the area between 62 ($Z = -1.90$) and the mean. *(For practice in using the normal curve table, see any of the problems at the end of this chapter.)*

TABLE 5.2 Finding Positive Z Scores with $\overline{X} = 100$ and $s = 20$

IQ Score	Z Score	Area Between Z and Mean
110	+0.50	19.15%
125	+1.25	39.44%
133	+1.65	45.05%
138	+1.90	47.13%

TABLE 5.3 Finding Negative Z Scores with $\overline{X} = 100$ and $s = 20$

IQ Score	Z Score	Area Between Z and Mean
93	−0.35	13.68%
85	−0.75	27.34%
67	−1.65	45.05%
62	−1.90	47.13%

Finding the Total Area Above and Below a Score

So far, we have seen how the normal curve table can be used to find areas between a Z score and the mean. Appendix A can also be used to find other kinds of areas in empirical distributions that are at least approximately normal in shape. For example, suppose you need to determine the total area below the scores of two male subjects in the distribution described in Figure 5.1. The first subject has a score of 117 ($X_1 = 117$), which is equivalent to a Z score of +0.85:

$$Z = \frac{X_i - \overline{X}}{s} = \frac{117 - 100}{20} = \frac{17}{20} = +0.85$$

The plus sign of the Z score indicates that the score should be placed above (to the right of) the mean. To find the area below a positive Z score, the area between the score and the mean (see column b) must be added to the area below the mean. As we noted earlier, the normal curve is symmetrical (unskewed) and its mean will be equal to its median. Therefore, the area below the mean (just like the median) will be 50%. Study Figure 5.4 carefully. We are interested in the shaded area.

By consulting the normal curve table, we find that the area between the score and the mean (see column b) is 30.23% of the total area. The area below a Z score of +0.85 is therefore 80.23% (50.00% + 30.23%). This subject scored higher than 80.23% of the persons tested.

The second subject has an IQ score of 73 ($X_2 = 73$), which is equivalent to a Z score of −1.35:

$$Z = \frac{X_i - \overline{X}}{s} = \frac{73 - 100}{20} = -\frac{27}{20} = -1.35$$

To find the area below a negative score, we use the right-hand column or "Area Beyond Z." The area in which we are interested is depicted in Figure 5.5, and we must determine the size of the shaded area. The "area beyond" (see column c) a score of −1.35 is 0.0885, which we can express as 8.85%. The second subject ($X_2 = 73$) scored higher than 8.85% of the tested group.

FIGURE 5.4 Finding the Area Below a Positive Z Score

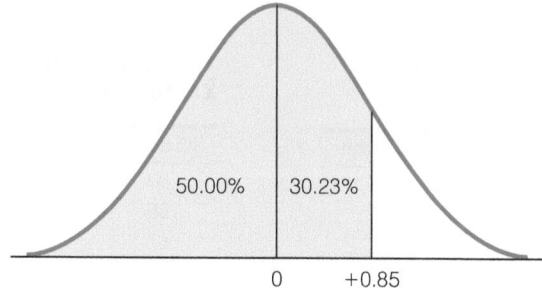

50.00% 30.23%

0 +0.85

FIGURE 5.5 Finding the Area Below a Negative Z Score

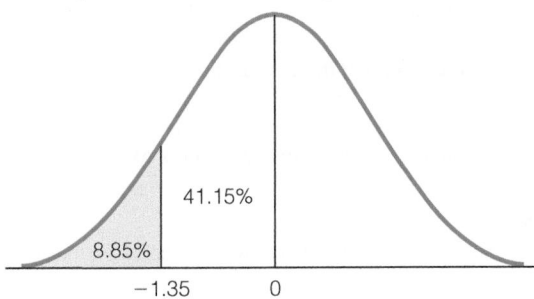

FIGURE 5.6 Finding the Area Above a Positive Z Score

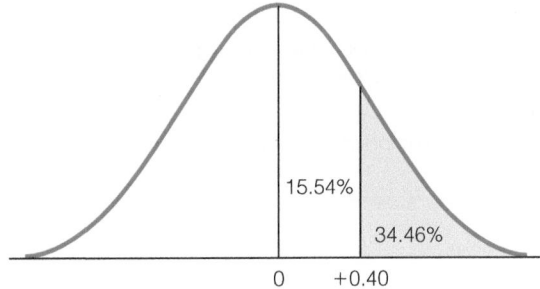

In the foregoing examples, we found the area below a score. Essentially the same techniques are used to find the area above a score. If we need to determine the area above an IQ score of 108, for example, we would first convert to a Z score,

$$Z = \frac{X_i - \overline{X}}{s} = \frac{108 - 100}{20} = \frac{8}{20} = +0.40$$

and then proceed to Appendix A. The shaded area in Figure 5.6 represents the area in which we are interested. The area above a positive score is found in the "Area Beyond Z" column, and, in this case, the area is 0.3446, or 34.46%.

These procedures are summarized in Table 5.4 and in the "One Step at a Time" box. They might be confusing at first, so you should *always* draw the curve and shade in the areas in which you are interested when starting a problem. *(For practice in finding areas above or below Z scores, see problems 5.1 to 5.7.)*

TABLE 5.4 Finding Areas Above and Below Positive and Negative Scores

	When the Z Score Is	
To Find Area	*Positive*	*Negative*
Above Z	Look in column c	Add column b area to .5000 or to 50.00%
Below Z	Add column b area to .5000 or to 50.00%	Look in column c

ONE STEP AT A TIME	Finding Areas Above and Below Positive and Negative Z Scores

1. ***Important: Start by shading in the region of the distribution that you want to measure, as in Figures 5.4, 5.5, or 5.6.***
2. Compute the Z score. Note whether the score is positive or negative.
3. Find the Z score in column a of the normal curve table (Appendix A).
4. To find the total area:

Below a Positive Z Score:
- Add the column b area to .5000. To express the area as a percentage, multiply the column b area by 100 and add it to 50.00%.

Above a Positive Z Score:
- Look in column c. This value is the area above the score expressed as a proportion. To express the area as a percentage, multiply the column c area by 100.

Below a Negative Z Score:
- Look in column c. This value is the area below the score expressed as a proportion. To express the area as a percentage, multiply the column c area by 100.

Above a Negative Z Score:
- Add the column b area to .5000. To express the areas as a percentage, multiply the column b area by 100 and add it to 50.00%.

Finding Areas Between Two Scores

On occasion, you will need to determine the area between two scores. When the scores are on opposite sides of the mean, the area between them can be found by adding the areas between each score and the mean. Using the men's IQ distribution as an example, if we wished to know the area between the IQ scores of 93 and 112, we would convert both scores to Z scores, find the area between each score and the mean in Appendix A, and add these two areas together. The first IQ score of 93 converts to a Z score of -0.35:

$$Z = \frac{X_i - \overline{X}}{s} = \frac{93 - 100}{20} = -\frac{7}{20} = -0.35$$

The second IQ score (112) converts to $+0.60$:

$$Z = \frac{X_i - \overline{X}}{s} = \frac{112 - 100}{20} = \frac{12}{20} = +0.60$$

Both scores are placed on Figure 5.7. We are interested in the total shaded area. The total area between these two scores is 13.68% + 22.57%, or 36.25%. Therefore, 36.25% of the total area (or about 363 of the 1000 cases) lies between the IQ scores of 93 and 112.

FIGURE 5.7 **Finding the Area Between Two Scores**

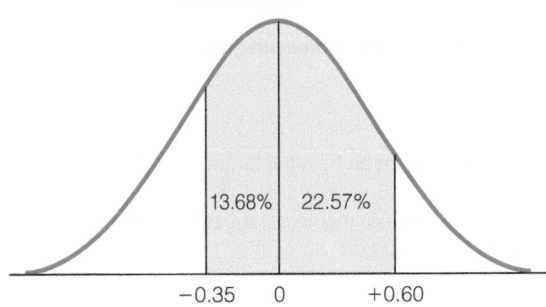

FIGURE 5.8 **Finding the Area Between Two Scores**

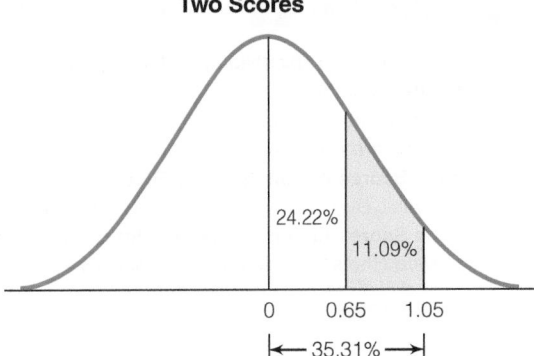

When the scores of interest are on the same side of the mean, a different procedure must be followed to determine the area between them. For example, if we were interested in the area between the scores of 113 and 121, we would begin by converting these scores into Z scores:

$$Z = \frac{X_i - \overline{X}}{s} = \frac{113 - 100}{20} = \frac{13}{20} = +0.65$$

$$Z = \frac{X_i - \overline{X}}{s} = \frac{121 - 100}{20} = \frac{21}{20} = +1.05$$

The scores are noted in Figure 5.8; we are interested in the shaded area. To find the area between two scores on the same side of the mean, find the area between each score and the mean, and then subtract the smaller area from the larger. Using column b of Appendix A, we see that 24.22% of the total area lies between the Z score of +0.65 and the mean, and that 35.31% of the area lies between the Z score of +1.05 and the mean. Therefore, the area between these two scores is 35.31% − 24.22%, or 11.09% of the total area. The same technique would be followed if both scores had been below the mean. The procedures for finding areas between two scores are summarized in Table 5.5 and in the "One Step at a Time" box. Note that you can also subtract the proportions in the c columns for each Z score and get the same result. *(For practice in finding areas between two scores, see problems 5.3, 5.4, and 5.6 to 5.9.)*

TABLE 5.5 **Finding Areas Between Scores**

Situation	Procedure
Scores are on the SAME side of the mean	Find the areas between each score and the mean (see column b). Subtract the smaller area from the larger area.
Scores are on OPPOSITE sides of the mean	Find the areas between each score and the mean (see column b). Add the areas together.

ONE STEP AT A TIME | Finding Areas Between Z Scores

1. ***Important: Start by shading in the region of the distribution that you want to measure, as in Figure 5.7 or 5.8.***
2. Compute the Z scores for both raw scores.
3. Find the areas between each score and the mean (see column b).

 If the Scores Are on the *Same* Side of the Mean → Subtract the smaller area from the larger area. Multiply this value by 100 to express it as a percentage.

 If the Scores Are on *Opposite* Sides of the Mean → Add the two areas together to get the total area between the scores. Multiply this value by 100 to express it as a percentage.

Social Research and Data Analysis 5.1 | Finding *Z* Scores and Area Below

How old do you think you will be if and when you have your first child? Or have you already had your first child? Let's see what the percentages look like nationally by using the 2018 General Social Survey data. It turns out that the mean age when a first child was born is 24.30 and the standard deviation is 5.74. So let's figure out the percentage of respondents that had their first child before the age of 26.

If you can assume that the ages are normally distributed, you can compute a Z score and find the area below the age of 26. The Z-score equivalent would be

$$Z = \frac{X_i - \overline{X}}{s} = \frac{26 - 24.30}{5.74} = \frac{1.70}{5.74} = +0.30$$

Turning to Appendix A, we find that the "Area Between Mean and Z" for this score is 0.1179, which could also be expressed as 11.79%. Since this is a positive Z score, we need to add this area to 50.00% to find the total area below. This tells us that 50% + 11.79% or 61.79% of the respondents had their first child before the age of 26.

Social Research and Data Analysis 5.2 | Finding *Z* Scores and Area Between

Using the same data as in Social Research and Data Analysis 5.1, let's estimate the percentage of people that had their first child between the ages of 19 and 21, and between 22 and 26.

The first two scores are both below the mean. Using Table 5.5 as a guide, we must first compute Z scores, find areas between each score and the mean, and then subtract the smaller area from the larger.

$$Z = \frac{X_i - \overline{X}}{s} = \frac{19 - 24.3}{5.74} = -\frac{5.3}{5.74} = -0.92$$

$$Z = \frac{X_i - \overline{X}}{s} = \frac{21 - 24.3}{5.74} = -\frac{3.3}{5.74} = -0.57$$

The area between $Z = -.92$ and the mean is 0.3212 and the area between $Z = -0.57$ and the mean is 0.2157. Subtracting the smaller from the larger (0.3212 − 0.2157) gives us 0.1055. Changing to percentage format, we can

say that 10.55% of the respondents had their first child between the ages of 19 and 21.

To find the percentage of respondents that had their first child between the ages of 22 and 26, we must add the column b areas together, since the ages (22 and 26) are on opposite sides of the mean (see Table 5.5):

$$Z = \frac{X_i - \overline{X}}{s} = \frac{22 - 24.30}{5.74} = -\frac{2.3}{5.74} = -0.40$$

$$Z = \frac{X_i - \overline{X}}{s} = \frac{26 - 24.30}{5.74} = \frac{1.7}{5.74} = +0.30$$

The area between $Z = -0.40$ and the mean is 0.1554 and the area between $Z = +0.30$ and the mean is 0.1179. Therefore, the total area between these two scores is 0.1554 + 0.1179, or 0.2733. Translating to percentages again, we can say that 27.33% of the respondents had their first child between the ages of 22 and 26.

Using the Normal Curve to Estimate Probabilities

To this point, we have thought of the theoretical normal curve as a way of describing areas above, below, and between scores. We have also seen that these areas can be converted into the number of cases above, below, and between scores.

In this section, we will explore an additional use of the normal curve: It can be used to estimate probabilities or the chances that certain events will occur. We will find these probabilities using the same techniques used to find areas earlier in this chapter. The only new idea introduced in this section is that the areas under the theoretical normal curve (or in Appendix A) can be considered probabilities. Before we consider these mechanics, however, let us examine what is meant by "probability."

ESTIMATING PROBABILITIES

Although we are rarely systematic or rigorous about it, we all attempt to deal with probabilities every day, and, indeed, we base our behavior on our estimates of the likelihood that certain events will occur. We constantly ask (and answer) questions such as: What is the chance (or probability) of rain? Of drawing an inside straight in poker? Of the worn-out tires on my car going flat? Of passing a test if I don't study?

To estimate the probability of an event, we must first be able to define what would constitute a "success." The preceding examples contain several different definitions of a success (rain, drawing a certain card, flat tires, and passing grades). To determine a probability, a fraction must be established, with the numerator equaling the number of events that would constitute a success and the denominator equaling the total number of possible events where a success could theoretically occur:

$$\text{Probability} = \frac{\#\ successes}{\#\ events}$$

To illustrate, assume that we wish to know the probability of selecting a specific card—say, the king of hearts—in one draw from a well-shuffled deck of cards. Our definition of a success is specific (drawing the king of hearts), and with the information given, we can establish a fraction. Only one card satisfies our definition of success, so the number of events that would constitute a success is 1; this value will be the numerator of the fraction. There are 52 possible events (that is, 52 cards in the deck), so the denominator will be 52. The fraction is thus 1/52, which represents the probability of selecting the king of hearts on one draw from a well-shuffled deck of cards. Our probability of success is 1 out of 52.

We can leave this fraction as it is, or we can express it in several other ways. For example, we can express it as an odds ratio by inverting the fraction, showing that the odds of selecting the king of hearts on a single draw are 52:1 (or a probability of fifty-two to one). We can express the fraction as a proportion by dividing the numerator by the denominator. For our example, the corresponding proportion is 0.0192, which is the proportion of all possible

events that would satisfy our definition of a success. In the social sciences, probabilities are usually expressed as proportions, and we will follow this convention throughout the remainder of this section. Using p to represent "probability," the probability of drawing the king of hearts (or any specific card) can be expressed as

$$p(\text{king of hearts}) = \frac{\#\ successes}{\#\ events} = \frac{1}{52} = .0192$$

As conceptualized here, probabilities have an exact meaning: Over the long run, the events that we define as successes will bear a certain proportional relationship to the total number of events. The probability of 0.0192 for selecting the king of hearts in a single draw really means that, over thousands of selections of one card at a time from a well-shuffled deck of 52 cards, the proportion of successful draws would be 0.0192. Or, for every 10,000 draws, 192 would be the king of hearts, and the remaining 9808 selections would be other cards. Thus, when we say that the probability of drawing the king of hearts in one draw is 0.0192, we are essentially applying to a single draw our knowledge of what would happen over thousands of draws.

Like proportions, probabilities range from 0.00 (meaning that the event has absolutely no chance of occurrence) to 1.00 (a certainty). As the value of the probability increases, the likelihood that the defined event will occur also increases. A probability of 0.0192 is close to 0, and this means that the event (drawing the king of hearts) is unlikely, or improbable.

These techniques can be used to establish simple probabilities in any situation in which we can specify the number of successes and the total number of events. For example, a single die has six sides, or faces, each with a different value, ranging from 1 to 6. The probability of getting any specific number (say, a 4) in a single roll of a die is therefore

$$p(\text{rolling a 4}) = \frac{1}{6} = .1667$$

PROBABILITY AND THE NORMAL CURVE

Combining this way of thinking about probability with our knowledge of the theoretical normal curve allows us to estimate the likelihood of selecting a case that has a score within a certain range. For example, suppose we wished to estimate the probability that a randomly chosen subject from the distribution of men's IQ scores would have an IQ score between 95 and the mean score of 100. Our definition of a success here would be the selection of any subject with a score in the specified range. Normally, we would next establish a fraction with the numerator equal to the number of subjects with scores in the defined range and the denominator equal to the total number of subjects. However, if the empirical distribution is normal in form, we can skip this step, because the probabilities, in proportion form, are already stated in Appendix A. That is, the areas in Appendix A can be interpreted as probabilities.

To determine the probability that a randomly selected male will have a score between 95 and the mean of 100 (from the example at the beginning of the chapter), we would convert the original score to a Z score:

$$Z = \frac{X_i - \overline{X}}{s} = \frac{95 - 100}{20} = -\frac{5}{20} = -0.25$$

Using Appendix A, we see that the area between this score and the mean is 0.0987. This is the probability we are seeking. The probability that a randomly selected case will have a score between 95 and 100 is 0.0987 (or, rounded off, 0.10, or 1 out of 10). In the same fashion, the probability of selecting a subject from any range of scores can be estimated. Note that the techniques for estimating probabilities are exactly the same as those for finding areas.

To consider an additional example, what is the probability that a randomly selected male will have an IQ less than 123? We will find probabilities in exactly the same way we found areas. The score (X_i) is above the mean and, following the directions in Table 5.4, we will find the probability we are seeking by adding the area in column b to 0.5000. First, we find the Z score:

$$Z = \frac{X_i - \overline{X}}{s} = \frac{123 - 100}{20} = \frac{23}{20} = +1.15$$

Next, look in column b of Appendix A to find the area between this score and the mean. Then add the area (0.3749) to 0.5000. The probability of selecting a male with an IQ of less than 123 is 0.3749 + 0.5000, or 0.8749. Rounding this value to 0.88, we can say that the odds are 0.88 (very high) that we will select a male with an IQ score in this range. Technically, remember that this probability expresses what would happen over the long run: For every 100 males selected from this group over an infinite number of trials, 88 would have IQ scores less than 123 and 12 would not.

Let us close by stressing an important point about probabilities and the normal curve. The probability is high that any case randomly selected from a normal distribution will have a score close in value to the mean. The shape of the normal curve is such that most cases are clustered around the mean and decline in frequency as we move farther away—either to the right or to the left—from the mean value. In fact, given what we know about the normal curve, the probability that a randomly selected case will have a score within ±1 standard deviation of the mean is 0.6826. Rounding off, we can say that 68 out of 100 cases—or a little more than two-thirds of all cases—selected over the long run will have a score between ±1 standard deviation, or Z score, of the mean. The probabilities are high that any randomly selected case will have a score close in value to the mean.

In contrast, the probability of the case having a score beyond 3 standard deviations from the mean is very low. Look in column c ("Area Beyond Z") for a Z score of 3.00 and you will find the value 0.0014. Adding the area in the upper tail (beyond +3.00) to the area in the lower tail (beyond −3.00) gives us 0.0014 + 0.0014, for a total of 0.0028. The probability of selecting a case with a very high score or a very low score is 0.0028. If we randomly select cases from a normal

distribution, we would select cases with Z scores beyond ± 3.00 only 28 times out of every 10,000 trials ($0.0028 \times 10,000$).

The general point to remember is that cases with scores close to the mean are common, and cases with scores that are far above or below the mean are rare. This relationship is central for an understanding of inferential statistics in Part II. *(For practice in using the normal curve table to find probabilities, see problems 5.8 to 5.10 and 5.13.)*

ONE STEP AT A TIME | **Finding Probabilities**

1. *Important: Start by shading in the region of the distribution that you want to measure, as in Figures 5.4–5.6.*
2. Compute the Z score (or scores). Note whether the score is positive or negative.
3. Find the Z score (or scores) in the normal curve table (Appendix A).
4. Find the area above or below the score (or between the scores) as you would normally (see the previous One Step at a Time boxes in this chapter) and express the result as a proportion. Typically, probabilities are expressed as a value between 0.00 and 1.00 rounded to two digits beyond the decimal point.

Social Research and Data Analysis 5.3 | **Finding Probabilities**

What is the probability of having your first child before the age of 20? Let's find out using the same data we cited earlier from the 2018 General Social Survey. Recall that the mean age was 24.30 and the standard deviation was 5.74. To answer this question, start by drawing the normal curve and shading in the region below the age of 20. Then calculate the Z scores and consult Appendix A. We are looking for probabilities, so we will leave the areas in proportion form. The Z score for an age of 20 is

$$Z = \frac{X_i - \overline{X}}{s} = \frac{20 - 24.30}{5.74} = -\frac{4.30}{5.74} = -.75$$

This score is a negative value (below, or to the left of, the mean) and we are looking for the area below. Using Table 5.4 as a guide, we will find the area below a negative score in column c. This area is 0.2266. Rounding off, we can say that the odds of having a child before the age of 20 is 0.2266 or 22.66 out of 100.

How about the age of 35? What is the probability of having your first child after that age?

$$Z = \frac{X_i - \overline{X}}{s} = \frac{35 - 24.30}{5.74} = \frac{10.7}{5.74} = +1.86$$

The Z score is positive, and to find the area above (greater than) 35, we look in column c of Appendix A. This value is 0.0322. The odds of having your first child at that age or older is 0.0322 or 3.22 out of 100.

Let's try one more. What is the probability of having your first child before you turn 30?

$$Z = \frac{X_i - \overline{X}}{s} = \frac{30 - 24.30}{5.74} = \frac{5.70}{5.74} = +.99$$

To find the area below a positive Z score, we add the area between the score and the mean (column b) to 0.5000 (see Table 5.4). This value is 0.3389 + 0.5000, or 0.8389. So it is very likely that a randomly selected person will have their first child before the age of 30.

Keep in mind that there are some limitations to the data we are using. The respondents in the survey that were asked this question already had their first child. Remember that some people will never have a child. Also, these probabilities refer to life chances. They don't tell us the probability that a woman will conceive of a child if she is trying. Finally, remember that this analysis assumes that the variable is normally distributed.

STATISTICS **In Everyday Life**

Probability Theory in Action

Statistics and data analytics have become extremely popular in professional sports. Oftentimes, fans read data about the probability of their sports teams making the playoffs or winning a championship without even knowing how those statistics are calculated. The Web site FiveThirtyEight uses a formula they call the Elo rating to predict the fate of teams in the NFL. Although the variables they use are much greater in number than those that we have considered in this chapter, the probabilities are still based on estimates like the ones we covered here with the normal curve. For the 2019 NFL season, the Elo model predicts that the reigning Super Bowl Champion New England Patriots will win 11 games. They also have an 80% chance of making the playoffs and a 14% chance of repeating as Super Bowl Champions. Check your local sports page to see how accurate these predictions were.

Source: https://fivethirtyeight.com/features/who-are-the-favorites-in-the-2019-nfl-season/

SUMMARY

1. The normal curve, in combination with the mean and standard deviation, can be used to construct precise descriptive statements about empirical distributions that are normally distributed. This chapter also lays some important groundwork for Part II.

2. To work with the theoretical normal curve, we must transform raw scores into their equivalent Z scores. Z scores allow us to find areas under the theoretical normal curve (Appendix A).

3. We considered three uses of the theoretical normal curve: finding total areas above and below a score, finding areas between two scores, and expressing these areas as probabilities. This last use of the normal curve is especially germane because inferential statistics are centrally concerned with estimating probabilities in a fashion similar to the process introduced in this chapter.

SUMMARY OF FORMULAS

Z Scores 5.1 $Z = \dfrac{X_i - \overline{X}}{s}$

GLOSSARY

Normal curve. A theoretical distribution of scores that is symmetrical, unimodal, and bell-shaped. The standard normal curve always has a mean of 0 and a standard deviation of 1.

Normal curve table. A detailed description of the area between a Z score and the mean of any standardized normal distribution. See Appendix A.

Z scores. Standard scores; the way scores are expressed after they have been standardized to the theoretical normal curve.

PROBLEMS

5.1 Scores on a quiz were normally distributed and had a mean of 10 and a standard deviation of 3. For each of the following scores, find the Z score and the percentage of area above and below the score.

X_i	Z Score	% Area Above	% Area Below
5			
6			
7			
8			
9			
11			
12			
14			
15			
16			
18			

5.2 Assume that the distribution of scores on a college entrance exam is normal, with a mean of 500 and a standard deviation of 100. For each of the following scores, find the equivalent Z score, the percentage of the area above the score, and the percentage of the area below the score.

X_i	Z Score	% Area Above	% Area Below
650			
400			
375			
586			
437			
526			
621			
498			
517			
398			

5.3 The senior class has been given a comprehensive examination to assess educational experience. The mean score on the test was 74 and the standard deviation was 10. What percentage of the students had scores

a. above 80?
b. above 83?
c. below 75?
d. below 77?
e. below 80?
f. below 85?
g. between 75 and 85?
h. between 80 and 85?
i. between 80 and 70?
j. between 75 and 70?

5.4 For a normal distribution where the mean is 50 and the standard deviation is 10, what percentage of the area is
a. above a score of 47?
b. below a score of 53?
c. between the scores of 40 and 47?
d. between the scores of 35 and 65?
e. above a score of 72?
f. below a score of 31 and above a score of 69?
g. between the scores of 55 and 62?
h. between the scores of 32 and 47?

5.5 MR A pension provider has hired a data analytics company to examine patterns in the demographic characteristics of its retirees. In their research on a sample (*N*) of 200 retirees, they have noted that the mean age of retirement was 72 and the standard deviation was 6. The following table presents the ages of 10 of the retirees when they retired. Convert each age into a Z score and determine the *number of people* that retired at an older or a younger age than each of the 10 retirees listed here. (*HINT: Determine the proportion of people in the usual way for each column below, multiply it by N, and round the result.*)

X_i	Z Score	Number of Retirees Above	Number of Retirees Below
60			
57			
55			
67			
70			
72			
78			
82			
90			
95			

5.6 SP In 2019, The College Board cancelled its plans to develop an adversity score that would have measured factors such as the availability of advanced coursework and the crime rate in a student's area in order to provide context for their SAT score. Assuming that the adversity scores were arrayed on a scale from 1 to 100, that the mean was 78 and the standard deviation was 11, what percentage of the area would have lied
a. below 60?
b. below 70?
c. below 80?
d. below 90?
e. between 60 and 65?
f. between 65 and 79?
g. between 70 and 95?
h. between 80 and 90?

i. above 99?
j. above 89?
k. above 75?
l. above 65?

5.7 A scale measuring prejudice has been administered to a large sample of respondents. The distribution of scores is approximately normal, with a mean of 31 and a standard deviation of 5. What percentage of the sample had scores
 a. below 20?
 b. below 40?
 c. between 30 and 40?
 d. between 35 and 45?
 e. above 25?
 f. above 35?

5.8 SOC On the scale mentioned in problem 5.7, if a score of 40 or more is considered "highly prejudiced," what is the probability that a person selected at random will have a score in that range?

5.9 The average speed on a local highway on a Monday afternoon was 59 and the standard deviation was 4. What is the probability that a randomly selected driver will be going
 a. between 55 and 65?
 b. between 60 and 65?
 c. above 65?
 d. between 60 and 50?
 e. between 55 and 50?
 f. below 55?

5.10 The average number of dropouts for a school district has been 305 per year with a standard deviation of 50. What is the probability that the number of dropouts next year will be
 a. less than 250?
 b. less than 300?
 c. more than 350?
 d. more than 400?
 e. between 250 and 350?
 f. between 300 and 350?
 g. between 350 and 375?

5.11 CJ The local police force gives all applicants an entrance exam and accepts only those applicants who score in the top 15% on this test. If the mean score this year is 87 and the standard deviation is 8, would an individual with a score of 110 be accepted?

5.12 To be accepted into an honor society, a student must have a grade-point average (GPA) in the top 10% of the school. If the mean GPA at the school is 2.78 and the standard deviation is 0.33, which of the following GPAs would qualify?

 3.20, 3.21, 3.25, 3.30, 3.35

5.13 In a distribution of scores with a mean of 35 and a standard deviation of 4, which event is more likely: that a randomly selected score will be between 29 and 31 or that a randomly selected score will be between 40 and 42?

5.14 After taking the state merit examinations for the positions of school counselor and social worker, you receive the following information on the tests and on your performance. On which of the tests did you do better?

School Counselor	Social Worker
$\overline{X} = 118$	$\overline{X} = 27$
$s = 17$	$s = 3$
Your score = 127	Your score = 29

5.15 A large university administers entrance exams to all entering students. The test is administered again as an exit exam for all graduates. The test results for one group of students are given in the table.

Freshman Year	Senior Year
$\overline{X} = 53$	$\overline{X} = 92$
$s = 7$	$s = 4$

 a. The scores for five students on both tests are listed below. For each student, calculate Z scores and determine whether the student performed better as a freshman or as a senior.

Student	Score, Freshman Year	Score, Senior Year
A	57	97
B	51	94
C	45	82
D	73	101
E	62	98

 b. Determine the probabilities that randomly selected students will have scores in each of these ranges:

Freshman Year	
Score	Probability
Less than 52	
Less than 57	
Between 40 and 50	
More than 51	
More than 62	

Senior Year	
Score	Probability
Less than 88	
Less than 98	
Between 80 and 100	
More than 97	
More than 85	

In this chapter, you learned how to estimate the percentage of cases that are above, below, or between scores in a normal distribution. We should also consider what happens when we do not have a perfectly normal distribution. Let's try working with some "non-normal" distributions and see what happens.

We will use the 2018 Core Trends Survey for these exercises: it is included on the Web site for this text. Below are two distributions from that data set, one for *age* and one for *books1* (the number of books the respondents read in the last 12 months). Each distribution (see Figures 1 and 2 below) contains contains bars showing the number of people at each age, or the number of books that people read. There is also a curved line over the distribution representing a normal curve for each figure. The mean and the standard deviation are below the figures. Neither distribution is perfectly normal, but which one looks more normal, or less skewed?

Your answer should be age, since its distribution appears more balanced and less skewed. In Chapter 3, we learned that a distribution that has a skewness value farther from zero is more skewed and less normal. In this case, the level of skewness (not shown) for the number of books is 3.04, and for age it is −0.14. The level of skewness for age is pretty close to zero, indicating that it is more normal than perhaps it appeared at first glance.

Since neither of these variables are perfectly normal, we can't make exact estimates about the proportion of cases above or below certain values, but that's almost always the case in research. Let's try making a prediction for each variable and see how close it comes to the actual distribution of scores.

A. Estimating Age (*age*)

What percentage of the sample is 40 or younger? Using the statistics below the figure for *age* (\overline{X} = 50.55 and s = 18.67), begin by calculating the Z score.

$$Z = \frac{X_i - \overline{X}}{s} = \underline{\hspace{1cm}}$$

Since the Z score is a negative value, use column c of Appendix A to find the total area below. As always, skim the left-hand column until you find the Z score and then find the value in column c. This area is _____, which can also be expressed as a percentage: _____ (multiply the column c area by 100 to find the percentage). This value is your **estimate** of the percentage of cases 40 or younger, if the distribution is normal.

Now, run the **Frequencies** program to find the actual percentage of cases 40 or younger. With the 2018 Core Trend Survey loaded, click **Analyze, Descriptive Statistics**, and then **Frequencies**. Find *age* in the list of variables on the left and click the arrow to move it to the box on the right. Click **OK**, and the frequency distribution for *age* will be produced in the SPSS Statistics Viewer window.

Find 40 in the list of scores on the left of the frequency distribution and then look in the Cumulative Percent column to find the percentage of cases 40 and younger. Write that value here: _____. This value is the **actual** percentage of cases 40 and younger for this distribution. Compare this with the **estimated** percentage found earlier: Subtract the smaller from the larger of the two values and record the result here: _____ How close was the estimate? *It should be pretty close.*

B. Estimating Number of Books Read (*books1*)

How well can we estimate the number of books read (*books1*)? We can see from the distributions and the skewness scores that *books1* is further from normal than *age* (see the figures below and recall that the skewness for *books1* was 3.04 and for *age* was −0.14), so estimates for *books1* should be less accurate than those for *age*.

What percentage of the sample read 16 or fewer books over the past year? Using the statistics below the figure for *age* (\overline{X} = 10.70 and *s* = 17.38), begin by calculating the *Z* score.

$$Z = \frac{X_i - \overline{X}}{s} = \underline{\hspace{2cm}}$$

Since the *Z* score is a positive value, use column b of Appendix A to find the total area between the score and the mean. This area is _____, which can also be expressed as a percentage: _____ (multiply the column c area by 100 to find the percentage). Add this value to 50.00% and record the result here _____. This is your **estimate** of the percentage of cases that read 16 or fewer books, if the distribution is normal.

Now, run the **Frequencies** program to find the actual percentage of cases that read 16 or fewer books for this sample. With the 2018 Core Trend Survey loaded, click **Analyze, Descriptive Statistics**, and then **Frequencies**. Find *books1* in the list of variables on the left and click the arrow to move it to the box on the right. Click **OK**, and the frequency distribution for *books1* will be produced in the SPSS Statistics Viewer window.

Find 16 in the list of scores on the left and then look in the Cumulative Percent column to find the percentage of people that read 16 or fewer books and write that value here: _____. This is the **actual** percentage of cases that read 16 or fewer books for this distribution. Compare this with the **estimated** percentage found earlier: Subtract the smaller from the larger of the two values and record the result here: _____ How close was the estimate? Compare this with the accuracy of our earlier estimate of *age*. You should have found that the estimate for age was closer.

What can we conclude? Not all variables have a normal distribution. In fact, it would be rare to find one that is perfectly normal. However, the closer a distribution is to normal, the better able we will be to make predictions about the percentage of cases above, below, and between scores.

FIGURE 1 Distribution for Age

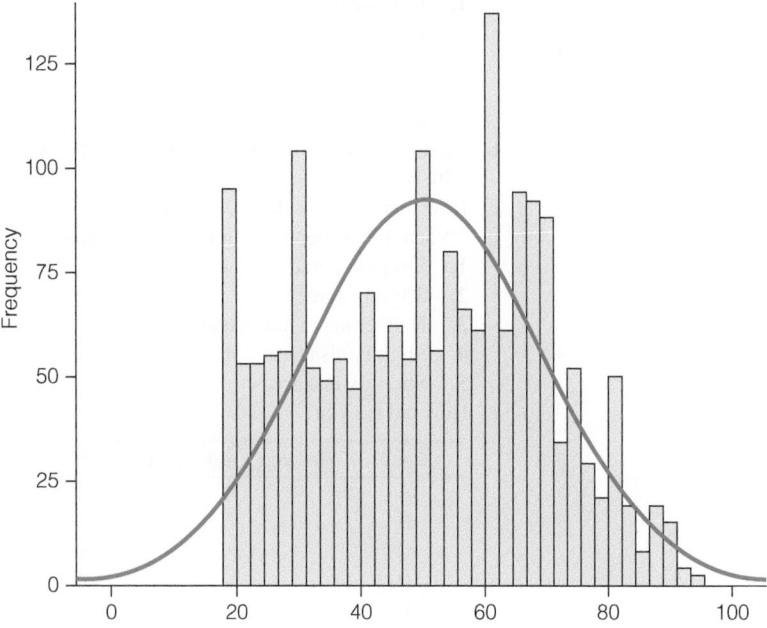

(Mean = 50.55, Standard Deviation = 18.67)

FIGURE 2 Distribution for the number of books read in the last 12 months

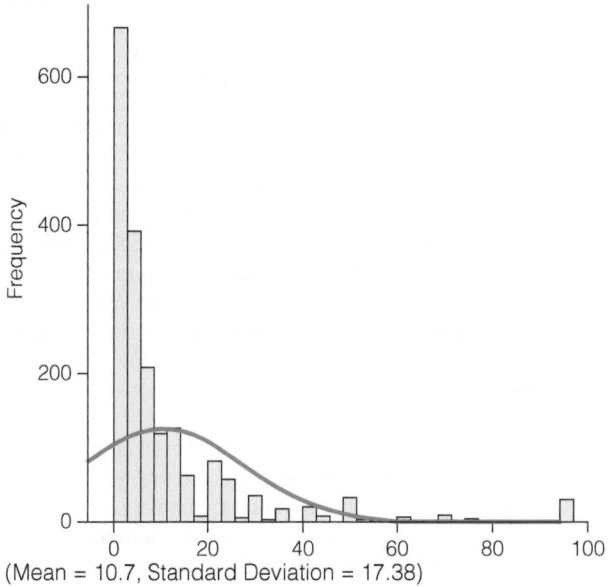

(Mean = 10.7, Standard Deviation = 17.38)

PART I CUMULATIVE EXERCISES

1. To what extent do people apply religion to the problems of everyday living? Fifteen people have responded to a series of questions:
 a. What is your religion?
 1. Protestant
 2. Catholic
 3. Jewish
 4. None
 5. Other
 b. On a scale of 1 to 10 (with 10 being the highest), how strong would you say your faith is?
 c. How many times a day do you pray?
 d. "When things aren't going well, my religion is my major source of comfort."
 1. Strongly agree
 2. Slightly agree
 3. Neither agree nor disagree
 4. Slightly disagree
 5. Strongly disagree
 e. How old are you?

Case	Religion	Strength	Pray	Comfort	Age
A	1	2	0	2	30
B	1	2	0	2	30
C	1	2	0	2	30
D	1	6	0	1	43
E	3	9	3	1	32
F	3	3	0	3	18

G	4	0	0	5	52
H	5	9	6	1	37
I	1	5	0	2	54
J	2	8	2	1	55
K	2	3	0	5	33
L	1	6	1	3	45
M	1	8	2	2	37
N	1	7	2	1	50
O	2	9	1	1	25

For each variable, construct a frequency distribution and calculate an *appropriate* measure of central tendency and variation. Write a sentence summarizing each variable.

2. A survey measuring attitudes toward interracial dating was administered to 1000 respondents. The questions were:
 a. What is your age?
 b. What is your gender?
 1. Male
 2. Female
 c. "Marriages between people of different racial groups just don't work out and should be banned by law."
 1. Strongly agree
 2. Slightly agree
 3. Neither agree nor disagree
 4. Slightly disagree
 5. Strongly disagree
 d. How many years of schooling have you completed?
 e. Which of the following categories best describes the place where you grew up?
 1. Large city
 2. Medium-size city
 3. Suburbs of a city
 4. Small town
 5. Rural area
 f. What is your marital status?
 1. Married
 2. Separated or divorced
 3. Widowed
 4. Never married

The scores of 20 respondents were:

Case	Age	Sex	Interracial Dating	Years of School	Area	Marital Status
A	17	1	5	12	1	4
B	25	2	3	12	2	1
C	55	2	3	14	2	1
D	45	1	1	12	3	1
E	38	2	1	10	3	1
F	21	1	1	16	5	1
G	29	2	2	16	2	2
H	30	2	1	12	4	1
I	37	1	1	12	2	1

J	42	2	3	18	5	4
K	57	2	4	12	2	3
L	24	2	2	12	4	1
M	27	1	2	18	3	2
N	44	1	1	15	1	1
O	37	1	1	10	5	4
P	35	1	1	12	4	1
Q	41	2	2	15	3	1
R	42	2	1	10	2	4
S	20	2	1	16	1	4
T	21	2	1	16	1	4

a. For each variable, construct a frequency distribution and select and calculate an appropriate measure of central tendency and a measure of variation. Summarize each variable in a sentence.

b. For *all* 1000 respondents, the mean age was 34.70 with a standard deviation of 3.4 years. Assuming the distribution of age is approximately normal, compute Z scores for each of the first 10 respondents and determine the percentage of the area below (younger than) each respondent.

Statistical Analysis Using SPSS

3. Start SPSS and load the General Social Survey 2018 (*GSS2018*) data set. For each of the variables listed below, get a frequency distribution and *appropriate* measures of central tendency and variation. *(HINT: Use the* **Frequencies** *command and click the* **Statistics** *button to select appropriate statistics.)* Summarize each variable in a sentence. Consult Appendix G for the actual wording of items and meaning of codes. *Make sure you understand what each variable actually measures!*

Variables: *childs, degree, racecen1, spanking, tvhours, and relig*

4. Start SPSS and load the *States* data set. For each of the variables listed below, get measures of central tendency and variation. *(HINT: Use the* **Descriptives** *command.)* Summarize each variable in a sentence. Consult Appendix G for the definitions of variables. *Make sure you understand what each variable actually measures!*

Variables: *aggass, BA, NotEnglHome, Internet, TeenBR, and WrkTrav*

Part II Inferential Statistics

The chapters in this part cover the techniques and concepts of inferential statistics. Generally speaking, these applications allow us to learn about large groups (populations) from smaller, carefully selected subgroups (samples). These statistical techniques are powerful and extremely useful. They are used to poll public opinion, to research the potential market for new consumer products, to project the winners of elections, to test scientific hypotheses and theories, and in hundreds of other ways both inside and outside the social sciences.

Chapter 6 includes a brief description of sampling, the techniques used by social scientists to select subgroups so that inferences can be made to populations. This section is intended to give you a general overview of the process, not a comprehensive or detailed treatment of the subject. The most important part of this chapter, however, concerns the sampling distribution, the single most important concept in inferential statistics. The sampling distribution is normal in shape, and it is the key link between populations and samples.

There are two main applications in inferential statistics, and Chapter 7 covers the first: using statistical information from a sample (e.g., a mean or a proportion) to estimate the characteristics of a population. The technique, called *estimation*, is most commonly used in public-opinion polling and election projection.

Chapters 8 through 11 cover the second application of inferential statistics: hypothesis testing. Most of the relevant concepts are introduced in Chapter 8, and each chapter covers a different application of hypothesis testing. For example, Chapter 9 presents the techniques used when we are comparing information from two different samples or groups (e.g., men vs. women), while Chapter 10 covers applications involving more than two groups or samples (e.g., Republicans vs. Democrats vs. Independents).

Hypothesis testing is one of the more challenging aspects of statistics for beginning students, and we have included an abundance of learning aids to ease the chore of assimilating this material. Hypothesis testing is also one of the most common and important statistical applications to be found in social science research. Mastery of this material is essential for developing your ability to read the professional literature.

Introduction to Inferential Statistics: Sampling and the Sampling Distribution

LEARNING OBJECTIVES By the end of this chapter, you will be able to:

1. Define these key terms: *population, sample, parameter, statistic, representative, EPSEM.*
2. Describe four different kinds of EPSEM (Equal Probability of Selection Method) sampling strategies.
3. Differentiate between the sampling distribution, the sample, and the population.
4. Explain how inferential statistics can be used to generalize from a sample to a population.
5. Explain the two theorems presented in the chapter.

USING STATISTICS

The statistical concepts presented in this chapter are fundamental to all applications of inferential statistics, including the estimation of the characteristics of large populations from small samples. Estimation can be used to assess

- Changes in attitudes toward homosexuality in the United States over the years. **See page 166**

- The number of men and women working in business and finance in the United States. **See page 160**

- The factors that are associated with a college professor offering extra credit. **See page 172**

- The appeal of candidates for political office in various segments of the voting population (e.g., among women or Catholics or Southerners).

- Immediate public reactions to controversial issues, such as new laws on immigration or health care.

One of the goals of social science research is to test our theories using many different people, groups, societies, and historical eras. Obviously, we can have the greatest confidence in theories that have stood up to testing against the greatest variety of cases and social settings. However, a major problem in social science research is that the most appropriate populations for our tests are too large. For example, a theory concerning political party preference among U.S. citizens

would be most suitably tested using the entire electorate, but it is impossible to interview all 253 million adult Americans. Indeed, even for theories that reasonably could be tested with smaller populations—such as a local community or the student body at a university—the logistics of gathering data from every single case in the population can be staggering to contemplate.

How can we test our theories if we can't access entire populations? To deal with this problem, social scientists select samples, or subsets of cases, from the populations of interest. Our goal in inferential statistics is to learn about the characteristics of a population (often called **parameters**) based on what we can learn from the sample. Two applications of inferential statistics are covered in this text. In estimation procedures, covered in Chapter 7, an "educated guess" of the population parameter is made, based on what is known about the sample. In hypothesis testing, covered in Chapters 8 through 11, the validity of a hypothesis about the population is tested against sample outcomes. In this chapter, we look briefly at sampling (the techniques for selecting cases for a sample) and then introduce a key concept in inferential statistics: the sampling distribution.

Probability Sampling: Basic Concepts and Terms

Social scientists have developed a variety of sampling techniques. In this chapter, we review the basic procedures for selecting probability samples, the only type of sample that fully supports the use of inferential statistics to generalize to populations. These samples are often described as *random*, and you may be more familiar with this terminology. Because of its greater familiarity, we will often use the term *random sample* in the following chapters. The term *probability sample* is preferred, however, because, in everyday language, *random* is often used to mean "by coincidence" or to give a connotation of unpredictability. As you will see, probability samples are selected carefully and methodically, in ways that leave no room for haphazardness. Interviewing the people you happen to meet in a mall one afternoon may be "random" in some sense of the word, but this technique will not result in a sample that could support inferential statistics.

Before considering probability sampling, let us point out that social scientists sometimes use **nonprobability samples**. For example, social scientists studying small-group dynamics or the structure of attitudes or personal values might use the students enrolled in their classes as subjects. These "convenience" samples are useful for a number of purposes (e.g., exploring ideas or pretesting survey forms before embarking on a more ambitious project) and are typically cheaper and easier to assemble. Their major limitation is that results cannot be generalized beyond the group being tested. If a theory of prejudice, for example, has been tested only on the students in a particular section of introductory sociology at a particular university, we cannot conclude that results would apply to other types of people. Therefore, we cannot place a lot of confidence in theories tested on nonprobability samples, even when they produce such strong evidence.

When constructing a probability sample, our goal is to select cases so that the final sample is **representative** of the population from which it was drawn. A sample is representative if it reproduces the important characteristics of the

STATISTICS In Everyday Life

Advertising and Sampling

Advertisers sometimes phrase their appeals in language that might seem scientific. Claims such as "80% of the 2000 people surveyed prefer Dasani water" might sound convincing, and a sample size of 2000 certainly seems impressive. But what if the sample consisted solely of people leaving a promotional event for Dasani water during which free samples of the product were distributed? Does the claim still have legitimacy? Only properly selected probability samples can be used to make conclusions about larger populations.

population. For example, if the population consists of 60% females and 40% males, the sample should have about the same proportional makeup. In other words, a representative sample is much like the population—only smaller.

It is crucial for inferential statistics that samples be representative. How can we assure ourselves that our samples are representative? Unfortunately, it is not possible to guarantee representativeness. However, we can maximize the chance of obtaining a representative sample by following the rule of EPSEM (the "**E**qual **P**robability of **SE**lection **M**ethod"), the fundamental principle of probability sampling. To follow EPSEM and maximize the probability that our sample will be representative, we select the sample so that every element or case in the population has an equal probability of being selected. Just as a six-sided die has an equal probability of showing any value from 1 to 6 when thrown, every case in the population must have the same chance of being included in the sample.

Remember that EPSEM and representativeness are two different things. In other words, the fact that a sample is selected according to EPSEM does not guarantee that it will be an exact representation of the population. There is a high probability that an EPSEM sample will be representative; but, just as a perfectly honest coin will sometimes show 10 heads in a row when flipped, an EPSEM sample will occasionally present an inaccurate picture of the population. One of the great strengths of inferential statistics is that they allow the researcher to estimate the probability of this type of error and interpret results accordingly.

EPSEM Sampling Techniques

This section presents four techniques for selecting random samples. The most basic technique is presented first; the remaining techniques are refinements or variations on the basic technique.

SIMPLE RANDOM SAMPLING

The most basic EPSEM sampling technique produces a **simple random sample**. To use this technique, we need a list of all cases in the population (also known as a **sampling frame**)—each with its own unique identifier, usually a number—and

ONE STEP AT A TIME	Drawing a Simple Random Sample

Step	Operation
1.	Gather a sampling frame, or a list of all the cases in a population. For example, obtain a list of all the students in your class.
2.	Number all of the cases.
3.	Draw a random number using a list of random numbers obtained from the Internet. Select the case that corresponds with that number.
4.	Repeat Step #3 until you have obtained your desired sample size.

a system of selection that guarantees that every case has an equal chance of being chosen for the sample.

The selection process could involve flipping coins, throwing dice, or drawing numbers from a hat but, in social science research, we often use computer programs to produce a list of random numbers. Try searching the Internet for a list of random numbers. You're sure to find one. Then select a random number and find the case that corresponds with that number. Repeat these steps until you have reached your desired sample size; if an identification number is selected more than once, skip it and move on to the next selection.[1] The result of these steps is what we call a simple random sample.

SYSTEMATIC RANDOM SAMPLING

Following the foregoing procedures will produce random samples as long as you select from a complete list of the population. However, this technique can be time consuming when the population is large. Consider the situation when your population numbers 10,000 cases. It is perfectly possible that the first case you select will come from the front of the list, the second from the back, the third from the front again, and so on.

To save the time required to flip back and forth through the list in such a situation, researchers often use a technique called **systematic random sampling**, where only the first case is randomly selected. Thereafter, every kth case is selected. For example, if you are drawing from a list of 10,000 cases and desire a sample of 200, select the first case randomly and, thereafter, every 50th (10,000/200) case. If you randomly start with case #13, then your second case will be case #63, your third case #113, and so on, until you reach the end of the list.

Note that systematic sampling does not strictly conform to the criterion of EPSEM. That is, once the first case has been selected, the other cases no longer have an equal probability of being chosen. In our example, cases other than the 13th, the 63rd, the 113th, and so on will not be selected for the sample. In general,

[1]Ignoring identification numbers when they repeat is called "sampling without replacement" and, technically, this practice compromises the randomness of the selection process. However, if the sample is a small fraction of the population, we will be unlikely to select the same case twice and ignoring repeats will not bias our conclusions.

ONE STEP AT A TIME	Drawing a Systematic Random Sample

Step	Operation
1.	Gather a sampling frame and make sure that the list is in no particular order.
2.	Number all of the cases.
3.	Draw the first case randomly by using a list of random numbers and select the case that corresponds with that number.
4.	Then draw every kth case by skipping down the list past the case you drew in Step #3. To determine k, divide the total number of cases (10,000, for example) by the desired sample size (200, for example).
5.	Continue to select cases until you have obtained your desired sample size.

this increased probability of error is small as long as the list from which cases are chosen is random, or at least noncyclical with respect to the traits you wish to measure. For example, there might be a problem if you are concerned with ethnicity and are drawing your sample from an alphabetical list, because there is a tendency for certain ethnic names to begin with the same letter (e.g., the Irish prefix *O*). Therefore, when using systematic sampling, you must ensure that the list of cases is not in any particular order.

STRATIFIED RANDOM SAMPLING

The above techniques are desirable because they do the best job of giving all cases an equal chance of selection. However, sometimes those methods can be problematic since a completely random sample might not include certain kinds of cases that are small in number. For example, imagine if you desired a random sample of 500 college students in the United States, but you wanted to ensure that it includes at least some people from all 50 states. If you used either of the techniques we've discussed already, you might not get anyone from Vermont or Wyoming since their populations are so small.

In a situation like this where you want to ensure the selection of at least some cases that are small in number, a third type of EPSEM sampling, **stratified random sampling**, might be in order. To apply this technique, you first *stratify* (or divide) the sampling frame into sublists according to some relevant trait and then sample from the sublists. Using the example above, you might stack or order your list by the state in which all the college students in your sampling frame go to school. Then select every kth case (as you did when systematic random sampling), and you will find that your sample includes someone from each state.

Consider another example. Suppose you are drawing a sample of 300 of your classmates and you wish to have proportional representation from every major field on campus. If only 10% of the student body is majoring in, say, zoology, random and systematic sampling could result in a sample with few (or even no) zoologists. If, however, you first divide the population into sublists by major, you can use the systematic sampling technique to select the right proportions of students from all majors, including 30 zoologists (or 10% of the

ONE STEP AT A TIME	Drawing a Stratified Random Sample

Step	Operation
1.	Gather a sampling frame and order it by the characteristic you want to proportionally represent (e.g., college major).
2.	Follow steps 3–5 for systematic random sampling as you select cases from each sublist. As you draw every kth case, you will be sure to proportionally represent the characteristic of interest.

sample) from the appropriate sublist. This works because once the cases are stacked, you can't help but select the right proportion in each major (except when the number of cases is too tiny) as you go down the list and select every kth case.

Stratified samples are guaranteed to meet the all-important criterion of representativeness (at least for the traits that are used to stratify the samples). The major limitation of stratified random sampling is that we usually lack detailed information about the exact composition of the population. Without this information, we will be unable to establish a scheme for stratification and determine how many cases should be taken from each sublist.

CLUSTER SAMPLING

The above random sampling techniques are ideal, but in reality we often find that they are impossible to carry out. This is because it is rare to find a sampling frame, or a list of all the cases in a population. Do you have a list of all Americans or a list of all the students in your college? Probably not, so let's consider a method we can use when we do not have a sampling frame.

Social scientists have devised several ways of dealing with the limitations imposed by the scarcity of lists. Probably the most significant of these is **cluster sampling**, which involves selecting groups of cases (clusters) rather than single cases. The clusters are often based on geography, and the selection of clusters often proceeds in stages. For example, you might draw a cluster sample of your city or town by first numbering all voting precincts. Next, you would use an EPSEM technique to select a sample of precincts (or clusters of cases). The second stage of selection would involve numbering the blocks within each of the selected precincts and using EPSEM to select a sample of blocks (a second cluster of cases). A third stage might involve the selection of households within each selected block. When these stages are completed, you would have a sample that has a high probability of being representative of the entire city without ever using a list of city residents.

Cluster samples are somewhat less likely to be representative of the population than simple random samples of comparable size. In part, this lower accuracy is a result of the multiple selection stages. With a simple random sample, the sample is drawn in one selection from the list of the population. In a multistage cluster sample, each stage in the selection process (e.g., first the precincts, then the blocks, and then the households) has a probability of error. That is, each time

ONE STEP AT A TIME Drawing a Cluster Sample

Step	Operation
1.	When you do not have a list of all the cases in a population, draw a list of clusters (or groupings) of cases (e.g., voting precincts, majors in your university, counties in your state).
2.	Then use any EPSEM technique to randomly select a sample of clusters from your list. If you chose "majors" as your cluster in Step #1, you would now have a list of randomly selected majors.
3.	Next, choose another cluster inside the first cluster. For example, you might choose 100-level classes offered by all of the majors you selected in Step #2.
4.	Continue selecting clusters until they get small enough for you to find. If you followed the steps above, your smallest cluster might be Tuesday morning classes at the 100-level in a selection of majors.
5.	Finally, select cases using an EPSEM technique from the smallest clusters. Your final step in the example would be to go to the classes you picked in Step #4 and randomly sample the students you find in attendance.

we sample, we run the risk of selecting an unrepresentative sample. In simple random sampling, we run this risk once; with cluster sampling we will run the risk anew at each stage.

Although we have to treat inferences about populations based on cluster samples with some additional caution, we usually have no alternative method of sampling. While it may be extremely difficult (or even impossible) to construct an accurate list of an entire city population, all you need to compile a cluster sample is a map (or a list of voting precincts, census tracts, and so forth).

Social Research and Data Analysis 6.1 **The American Community Survey**

When most people think about the Census, they have the decennial census in mind. The United States Census Bureau has conducted a count of the American population every 10 years since 1790. Back then, the Census Bureau determined that there were less than 4 million people in America. Things have changed quite a bit since then. The current population is now over 328 million people. But the decennial census, from which this count is derived, contains only 10 questions. A great deal more information is collected by the Census Bureau in its American Community Survey (ACS).

The ACS is a monthly survey that is carried out on millions of Americans that are selected randomly to participate. It asks questions about marital status, place of birth, health insurance coverage, poverty status, computer and Internet use, telephone service availability, and many other topics. By utilizing a cluster sampling technique, the Census Bureau can gather vast amounts of information and do

a good job of describing the American population without having to find everyone in the population to ask them all of these questions.

Check out the American Fact Finder on the Census Bureau website (see the URLs below). It uses ACS data to make estimates about the U.S. population, as well as the population of cities and states, on all sorts of variables. For example, the data show that in 2017, approximately 12.6 million men were employed in management, business, and financial occupations, compared to 10.2 million women. The estimates are not perfect. For the men, there is a $+/- 47,054$ margin of error at the 90% confidence level. This means they are 90% confident that their estimate is within 47,054 people. Confidence intervals like these are not hard to calculate. We will learn how to compute them with a simple formula in Chapter 7.

Sources: https://www.census.gov/; https://factfinder.census.gov/faces/nav/jsf/pages/index.xhtml

STATISTICS	In Everyday Life

Sampling and Telephone Soliciting

The sampling techniques presented in this chapter are also used for commercial purposes and for mass marketing. The next time you are annoyed by a telemarketer calling during dinnertime, remember that your phone number may have been selected by a sophisticated, complex application of random and cluster sampling. This probably will not lessen your annoyance, but it will remind you that sampling technology is a part of daily life and in many areas outside of the social sciences.

Before we leave this discussion of sampling techniques, we should stress a major point. The purpose of inferential statistics is to acquire knowledge about populations based on the information gathered from samples of that population. Each of the statistics to be presented in Part II of this text requires that samples be selected according to EPSEM. While even the most painstaking and sophisticated sampling techniques are no guarantee, the probability is high that EPSEM samples will be representative of the populations from which they are selected.

The Sampling Distribution

Once we have selected a probability sample according to some EPSEM procedure, what do we know? On one hand, we can gather a great deal of information from the cases in the sample. On the other hand, we know nothing about the population. Indeed, if we had information about the population, we probably wouldn't need the sample. Remember that we use inferential statistics to learn more about populations, and information from the sample is important primarily insofar as it allows us to generalize to the population.

When we use inferential statistics, we generally measure some variable (e.g., age, political party preference, or opinions about abortion) in the sample and then use the information from the sample to learn more about that variable in the population. In Part I of this text, you learned that three types of information are necessary to describe a variable adequately:

1. The shape of its distribution (Chapter 2),
2. A measure of central tendency (Chapter 3), and
3. A measure of variation (Chapter 4).

Clearly, all three kinds of information can be gathered (or computed) on a variable from the cases in the sample. Just as clearly, none of the information is available for the population. Except in rare situations (e.g., IQ tests are designed so that the scores of a population will be approximately normal in distribution, and the distribution of income is almost always positively skewed), nothing can be known about the exact shape of the distribution of a variable in the population.

FIGURE 6.1 Relationships Among the Sample, the Sampling Distribution, and the Population

The means and standard deviations of variables in the population are also unknown. If we had this information for the population, inferential statistics would be unnecessary.

In statistics, we link information from the sample to the population with the **sampling distribution**: *the theoretical, probabilistic distribution of a statistic for all possible samples of a certain size (N).* That is, the sampling distribution is the distribution of a statistic (e.g., a mean or a proportion) based on every conceivable sample or combination of cases from the population. A crucial point about the sampling distribution is that its characteristics are based on the laws of probability, not on empirical information, and are very well known. In other words, it is based on theory; we can't observe it in reality. In fact, the sampling distribution is the central concept in inferential statistics, and we will discuss it in depth in the remainder of this chapter.

THE THREE DISTRIBUTIONS USED IN INFERENTIAL STATISTICS

As illustrated by Figure 6.1, all applications of inferential statistics move between the sample and the population via the sampling distribution. Thus, three separate and distinct distributions of a variable are involved in every application of inferential statistics:

1. The *population distribution* of the variable, which, while empirical (i.e., it exists in reality), is unknown. For example, there is a real total population of women in the United States. We just aren't capable of determining it by counting them one by one. Amassing information about or making inferences to the population is the sole purpose of inferential statistics.

2. The *sampling distribution* of the variable, which is nonempirical, or theoretical. Because of the laws of probability, a great deal is known about this distribution. Specifically, the shape, central tendency, and variation of the distribution can be deduced and, therefore, the distribution can be adequately characterized.

3. The *sample distribution* of the variable, which is empirical and known. The shape, central tendency, and variation of the variable can be ascertained for the sample. Remember, however, that information from the sample is important primarily insofar as it allows the researcher to learn about the population.

The utility of the sampling distribution is implied by its definition. Because it includes the statistics from all possible sample outcomes, the sampling distribution enables us to estimate the probability of any particular sample outcome, a process that will occupy our attention for the next five chapters.

CONSTRUCTING A SAMPLING DISTRIBUTION

The sampling distribution is theoretical, which means that it is never actually constructed. However, to understand better the structure and function of the distribution, let's consider an example of how one might be constructed. Suppose we wanted to gather some information about the ages of a particular community of 10,000 individuals. We draw an EPSEM sample of 100 residents, ask all 100 respondents their age, and use those individual scores to compute a mean age of 27.

FIGURE 6.2 Constructing a Sampling Distribution

This score is noted on the graph in Figure 6.2. Note that this sample is one of countless possible combinations of 100 people taken from this population of 10,000, and the statistic (the mean of 27) is one of millions of possible sample outcomes.

Now draw another sample of the same size ($N = 100$), and again compute the average age. Assume that the mean for the second sample is 30 and note this sample outcome on Figure 6.2. This second sample is another of the countless combinations of 100 people taken from this population of 10,000, and the sample mean of 30 is another of the millions of possible sample statistics. Replace these respondents and draw still another sample, calculate and note the mean, replace this third sample, and draw a fourth sample, continuing these operations an infinite number of times, calculating and noting the mean of each sample. Now try to imagine what Figure 6.2 would look like after tens of thousands of individual samples had been collected and the mean had been computed for each sample. What shape, mean, and standard deviation would this distribution of sample means have after we had collected all possible combinations of 100 respondents from the population of 10,000?

For one thing, we know that each sample will be at least slightly different from every other sample, since it is extremely unlikely that we will sample the exact same 100 people twice. Because each sample will be unique, each sample mean will be at least slightly different in value.

We also know that not all the samples will be representative, even if we follow EPSEM closely. For example, if we continue taking samples of 100 people long enough, we will eventually choose a sample that includes only the youngest residents. Such a sample would have a mean much lower than the true population mean. Likewise, some of our samples will include only senior citizens and will have means that are much higher than the population mean. Common sense suggests, however, that such nonrepresentative samples will be rare and that most sample means will cluster around the true population value.

To illustrate further, assume that we somehow come to know that the true mean age of the population of 10,000 individuals is 30. Because, as we have just seen, most of the sample means will also be approximately 30, the sampling distribution of these sample means should peak at 30. Some of the samples will be nonrepresentative and their means will "miss the mark," but the frequency of such misses should decline as we get farther away from the population mean of 30. That is, the distribution should slope to the base as we move away from the population value—sample means of 29 or 31 should be common; means of 20 or 40 should be rare. The samples are random, so their means should miss an equal number of times on either side of the population value, and the distribution itself should therefore be roughly symmetrical.

Does this form sound familiar? It's the form we studied in Chapter 5, the normal distribution. The sampling distribution of all possible sample means

FIGURE 6.3 **A Sampling Distribution of Sample Means**

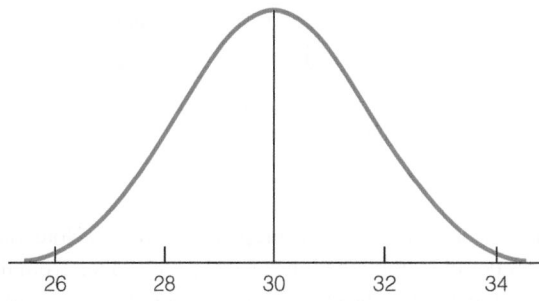

should be approximately normal and will resemble the distribution presented in Figure 6.3. Recall from Chapter 5 that, on any normal curve, cases close to the mean (say, within ±1 standard deviation) are common and cases far away from the mean (say, beyond 3 standard deviations) are rare. *(See problem 6.2 for a demonstrations of the construction of sampling distributions.)*

TWO THEOREMS

These commonsense notions about the shape of the sampling distribution and other important information about central tendency and variation are stated in two theorems. Before examining the theorems, we need to review some of the symbols we will use. Recall that the symbol for the mean of a sample is \overline{X} and that μ is the symbol for the mean of a population. For measures of variation, we use s to symbolize the standard deviation of a sample and σ (lowercase Greek letter sigma) to refer to the standard deviation of a population. We will present the theorems informally here, but they are presented in their formal, mathematical language following this discussion. The first theorem applies to variables that are normally distributed across a population: IQ scores, for example. As we discussed in the previous section, we construct a sampling distribution by taking multiple random samples of the same size (N) from the population and then plotting the sample statistic (a mean, for example) on a graph. According to the first theorem, when the population distribution is normal, the sampling distribution will have these characteristics:

1. It will be normal in shape,

2. Its mean will be the same value as the mean of the population (μ), and

3. Its standard deviation will be equal to the standard deviation of the population (σ) divided by the square root of N.

If we know that the variable is distributed normally in the population, we can assume that the sampling distribution is normal. Furthermore, the theorem states that the mean of the sampling distribution will be exactly the same value as the mean of the population. For example, if we know that the mean IQ of the entire population is 100, then we know that the mean of any sampling distribution of sample mean IQ scores will also be 100.

Exactly why these two means will be the same value cannot be fully explained at this level. Recall, however, that most sample means will cluster around

the population value over the long run. Thus, the fact that these two values are equal should have intuitive appeal.

Finally, the theorem says that the standard deviation of the sampling distribution, also called the **standard error of the mean**, will be equal to the standard deviation of the population divided by the square root of N (symbolically: σ/\sqrt{N}). If the mean and standard deviation of a normally distributed variable in a population are known, the theorem allows us to compute the mean and standard deviation of the sampling distribution.[2] Thus, we will know exactly as much about the sampling distribution (shape, central tendency, and variation) as we ever knew about any empirical distribution.

You might be wondering what happens when the distribution of the variable in question is not normal. What if the shape of the population distribution is unknown or is known *not* to be normal in shape (e.g., income, which almost always has a positive skew)? This situation is covered by a second theorem, called the **Central Limit Theorem**, which says that the sampling distribution will become normal in shape as sample size (N) increases for *any* variable, even a variable that is not normally distributed across the population. When N is large, the mean of the sampling distribution will equal the population mean, and its standard deviation (or the standard error of the mean) will be equal to σ/\sqrt{N}.

The Central Limit Theorem is important because it removes the condition that the variable be normally distributed in the population. Thus, even if we are working with a variable that is known to have a skewed or bimodal distribution, we can still assume a normal sampling distribution.

The issue remaining is to define what is meant by a large sample. A good rule of thumb is that if sample size (N) is 100 or more, the Central Limit Theorem applies, and you can assume that the sampling distribution of sample statistics is normal in shape. When N is less than 100, you must have good evidence of a normal population distribution before you can assume that the sampling distribution is normal. Thus, as long as samples are larger than 100, we can assume that the sampling distribution will be normal in shape.

SUMMARY: TWO THEOREMS

The formal or mathematical statements of the theorems are:

First Theorem: If repeated random samples of size N are drawn from a normal population with mean μ and standard deviation σ, then the sampling distribution of sample means will be normal, with a mean μ and a standard deviation of σ/\sqrt{N}.

Second Theorem (Central Limit Theorem): If repeated random samples of size N are drawn from any population with mean μ and standard deviation σ, then, as N becomes large, the sampling distribution of sample means will approach normality, with mean μ and standard deviation σ/\sqrt{N}.

[2]In the typical research situation, the values of the population mean and standard deviation are unknown. However, these values can be estimated from sample statistics, as we shall see in the chapters that follow.

| STATISTICS | In Everyday Life |

Using the GSS to Measure Changing American Attitudes

The General Social Survey has been administered since 1972, and this permits us to track change in our society over time. As one illustration, data from the GSS demonstrate a growing acceptance of homosexuality. In the 1970s, about 72% of respondents felt that homosexuality was "always wrong." In the most recent administration of the GSS in 2018, that percentage had fallen to about 33%.

The Sampling Distribution: An Additional Example Using the General Social Survey (GSS)

Developing an understanding of the sampling distribution—what it is and why it's important—can be a challenging task. It may be helpful to list briefly the most important points about the sampling distribution:

1. Its definition: *The sampling distribution is the distribution of a statistic (such as a mean or a proportion) for all possible sample outcomes of a certain size.*

2. Its shape: *normal* (see Chapter 5 and Appendix A).

3. Its central tendency and variation: *The mean of the sampling distribution is the same value as the mean of the population. The standard deviation of the sampling distribution—or the standard error—is equal to the population standard deviation divided by the square root of N (see the theorems).*

4. The role of the sampling distribution in inferential statistics: *It links the sample with the population* (see Figure 6.1).

To reinforce these points, let's consider an additional example using the General Social Survey (GSS), one of the databases used for SPSS exercises in this text. The GSS has been administered to randomly selected samples of adult Americans since 1972 and explores a broad range of characteristics and issues, including confidence in the Supreme Court, attitudes about assisted suicide, number of siblings, and level of education. The GSS has its limits, but it is a valuable resource for testing theory and for learning more about American society. Focusing on this survey, let's review the roles played by the population, the sample, and the sampling distribution when we use this database.

THE GSS AND THE POPULATION

We'll start with the population, or the group we want to learn more about. In the case of the GSS, the population consists of all adult (older than 18) Americans, which includes 253 million people. Clearly, we can never interview all of these people and learn what they are like or what they are thinking about abortion, gun control, sex education in the public schools, or any other issue. We should also note that this information is worth having. It could help inform public debates,

provide some basis in fact for the discussion of many controversial issues (e.g., the polls show consistently that the majority of Americans favor some form of gun control), and assist people in clarifying their personal beliefs. If the information is valuable, how can we learn more about this huge population?

THE GSS AND THE SAMPLE

This brings us to the sample, a carefully chosen subset of the population. The General Social Survey is administered to several thousand people, each of whom is chosen by a sophisticated technology based on the principle of EPSEM. A key point to remember is that samples chosen by this method are very likely to be representative of the populations from which they were selected. Whatever is true of the sample will also be true of the population (with some limits and qualifications, of course).

The respondents are contacted at home and asked for background information (religion, gender, years of education, and so on) as well as their opinions and attitudes. When all of this information is collated, the GSS database includes information (shape, central tendency, variation) on hundreds of variables (age, level of prejudice, marital status) for the people in the sample. Because of this, we have a lot of information about the variables for the *sample* (the people who actually responded to the survey), but no information about these variables for the *population* (the 253 million adult Americans). How do we get from the known sample to the unknown population? This is the central question of inferential statistics; the answer, as you hopefully realize by now, is "by using the sampling distribution."

THE GSS AND THE SAMPLING DISTRIBUTION

Remember that, unlike the sample and the population, the sampling distribution is theoretical, and, because of the theorems presented earlier in this chapter, we know its shape, central tendency, and variation. For any variable from the GSS, the theorems tell us that:

- The sampling distribution will be normal in shape because the sample is "large" (N is much greater than 100). This will be true regardless of the shape of the variable in the population.

- The mean of the sampling distribution will be the same value as the mean of the population. If *all* adult Americans have completed an average of 13.5 years of schooling ($\mu = 13.5$), then the mean of the sampling distribution will also be 13.5.

- The standard deviation (or standard error) of the sampling distribution is equal to the population standard deviation (σ) divided by the square root of N.

Thus, the theorems tell us the statistical characteristics of the sampling distribution (shape, central tendency, and variation), and this information allows us to link the sample to the population.

How does the sampling distribution link the sample to the population? The fact that the sampling distribution will be normal when N is large is crucial. This means that more than two-thirds (68%) of all samples will be within ± 1 Z score of the mean of the sampling distribution (which is the same value as the mean of the population), about 95% are within ± 2 Z scores, and so forth. We do not (and cannot) know the actual value of the mean of the sampling distribution, but we do know

STATISTICS In Everyday Life

Using Surveys to Compare American Attitudes with Other Nations

The World Values Survey has been administered to random samples from many nations since 1981. Although U.S. attitudes toward homosexuality are growing more tolerant, Americans are less accepting than many other advanced industrialized nations, as shown in the table below.

"Please tell me if you think homosexuality can always be justified, never be justified, or something in between."

Nation	Percent "Never Justified"
United Sates	24.0%
Japan	17.6%
Germany	17.8%
Netherlands	11.8%
Sweden	7.5%

Source: World Values Survey. Wave 6 (2010–2014) http://www.worldvaluessurvey.org/WVSOnline.jsp

that the probabilities are high that our sample statistic is approximately equal to this parameter. Similarly, the theorems give us crucial information about the mean and standard error of the sampling distribution that we can use, as you will see in the chapters that follow, to link information from the sample to the population.

To summarize, our goal is to infer information about the population (in the case of the GSS, all adult Americans). When populations are too large to test, we use information from randomly selected samples, carefully drawn from the population of interest, to estimate the characteristics of the population. In the case of the GSS, the full sample consists of several thousand adult Americans who have responded to the questions on the survey. The sampling distribution, the theoretical distribution whose characteristics are defined by the theorems, links the known sample to the unknown population.

READING STATISTICS Who Offers Extra Credit?

Have you ever asked your professor for extra credit? Who hasn't? But have you ever wondered whether there are sociological factors at work when a professor decides to say yes or no? If you wanted to probe this issue, it would be a good idea to study a random sample of college professors. So, that's exactly what two sociologists did recently in a study that was published in *Teaching Sociology*.

Before we look at the results, let's take a look at their sampling strategy. According to the methods section of the study, the authors gathered a stratified random sample. They started with a list of all U.S. colleges, from which they drew a sample of 100 four-year universities and 50 two-year colleges. Once they had selected the colleges, they used the Internet to generate a list of sociology instructors. This yielded a sample of 978 faculty members and 247 graduate student instructors. But, as you might expect, not all of these people responded to their survey, and the

(continued)

READING STATISTICS *(Continued)*

final sample consisted of 306 sociology instructors, or 25% of the original sample.

The study found that two-thirds of the sample offered extra credit in the last two years and that women were more likely to offer it than men. The authors speculated that this might be so because female professors tend to score lower on student evaluations than men, and offering extra credit might be a way to boost their popularity. Professors with less teaching experience were also more likely to offer extra credit. The authors suggested that this might be because more experienced professors have had more time to test out assignments for extra credit, and have incorporated these materials into their classes. Less experienced professors may also

be more likely to use extra credit as a way of dealing with the stresses of their new teaching assignments.

Like any study, this one has its limitations. Most obvious, perhaps, is the low response rate of 25%. This is a common problem with survey research and could limit our willingness to generalize these results to the population of all sociology instructors. When we see such results, we look for supporting evidence. If we can combine evidence from multiple studies—even though each is limited in some way—we may still be able to come to some conclusions about the subject under study.

Source: Izienicki, H., & Setchfield, S. (2019). Extra Credit in the Sociology Classroom. *Teaching Sociology, 47*, 1, 32–42.

Symbols and Terminology

In the following chapters, we will be working with three entirely different distributions. Furthermore, we will be concerned with several different kinds of sampling distributions—including the sampling distribution of sample means and the sampling distribution of sample proportions.

To distinguish among these various distributions, we will often use symbols. The symbols for the means and standard deviations of samples and populations have already been introduced in Chapters 3 and 4. For quick reference, Table 6.1 presents the symbols that will be used for the sampling distribution. Basically, the sampling distribution is denoted with Greek letters that are subscripted according to the sample statistic of interest.

Note that the mean and standard deviation of a sample are denoted with English letters (\overline{X} and s), while the mean and standard deviation of a population are denoted with Greek-letter equivalents (μ and σ). Proportions calculated on samples are symbolized as P_s (P-sub-s; s for sample), while population proportions are denoted as P_u (P-sub-u; u for "universe" or population). The symbols

TABLE 6.1 Symbols for Means and Standard Deviations of Three Distributions

	Mean	Standard Deviation	Proportion
1. Samples	\overline{X}	s	P_s
2. Populations	μ	σ	P_u
3. Sampling distributions			
of means	$\mu_{\overline{x}}$	$\sigma_{\overline{x}}$	
of proportions	μ_p	σ_p	

for the sampling distribution are Greek letters with English-letter subscripts. The mean and standard deviation of a sampling distribution of sample means are $\mu_{\overline{X}}$ ("mu-sub-x-bar") and $\sigma_{\overline{X}}$ ("sigma-sub-x-bar"). The mean and standard deviation of a sampling distribution of sample proportions are μ_p ("mu-sub-p") and σ_p ("sigma-sub-p").

SUMMARY

1. Because populations are almost always too large to test, a fundamental strategy of social science research is to select a sample from the defined population and then to use information from the sample to generalize to the population. This is done either by estimation or by hypothesis testing.

2. Several techniques are commonly used for selecting random samples. Each of these techniques involves selecting cases for the sample according to EPSEM. Even the most rigorous technique, however, cannot guarantee that the sample will be representative. One of the great strengths of inferential statistics is that the probability that the sample is nonrepresentative can be estimated.

3. The sampling distribution, the central concept in inferential statistics, is a theoretical distribution of all possible sample outcomes. Its overall shape, mean, and standard deviation are known (under the conditions specified in the two theorems), so the sampling distribution can be adequately characterized and utilized by researchers.

4. The two theorems that were introduced in this chapter state that when the variable of interest is normally distributed in the population or when sample size is large, the sampling distribution will be normal in shape, its mean will be equal to the population mean, and its standard deviation (or standard error) will be equal to the population standard deviation divided by the square root of N.

5. All applications of inferential statistics involve generalizing from the sample to the population by means of the sampling distribution. Both estimation procedures and hypothesis testing incorporate the three distributions, and it is crucial that you develop a clear understanding of each distribution and its role in inferential statistics.

GLOSSARY

Central Limit Theorem. A theorem that specifies the mean, standard deviation, and shape of the sampling distribution, given that the sample is large.

Cluster sampling. A method of sampling by which geographical units are randomly selected and all cases within each selected unit are tested.

EPSEM. The Equal Probability of SElection Method for selecting samples. Every element or case in the population must have an equal probability of selection for the sample.

μ. The mean of a population.

μ_p. The mean of a sampling distribution of sample proportions.

$\mu_{\overline{X}}$. The mean of a sampling distribution of sample means.

Nonprobability sample. Any sample that does not meet the EPSEM criterion. Such samples have a variety of uses in social science research but cannot be used to generalize to a population.

P_s. Any sample proportion.

P_u. Any population proportion.

Parameter. A characteristic of a population.

Representative sample. A sample that reproduces the major characteristics of the population from which it was drawn.

Sampling distribution. The distribution of a statistic for all possible sample outcomes of a certain size. Under conditions specified in two theorems, the sampling distribution will be normal in shape, with a mean equal to the population value and a standard deviation equal to the population standard deviation divided by the square root of N.

Sampling frame. A list of all the cases in a population. The list can be numbered and used to draw a simple random sample, a systematic random sample, or a stratified random sample.

Simple random sample. A method for choosing cases from a population by which every case and every combination of cases has an equal chance of being included.

Standard error of the mean. The standard deviation of a sampling distribution of sample means.

Stratified sample. A method of sampling by which cases are selected from sublists of the population.

Systematic sampling. A method of sampling by which the first case from a list of the population is randomly selected. Thereafter, every kth case is selected.

PROBLEMS

6.1 Imagine that you had to gather a random sample ($N = 300$) of the student body at your school. How would you acquire a list of the population? Would the list be complete and accurate? What procedure would you follow in selecting cases? Would cluster sampling be an appropriate technique (assuming that no list was available)? Describe in detail how you would construct a cluster sample. (*HINT: See the One Step at a Time box in this chapter.*)

Statistical Analysis Using SPSS

6.2 This exercise will give you a tangible understanding of sampling distributions and the two theorems presented in this chapter. *For this demonstration only*, the 2018 GSS database will be treated as a *population* and its characteristics will be treated as *parameters*. We will use SPSS to draw ten random samples and calculate statistics for *age* on each sample. Recall that a sampling distribution includes *all possible* samples, so these 10 samples are the merest beginning of an actual sampling distribution.

After drawing each sample, you will record the mean and the standard error of the mean for *age*. Recall that the standard error is the standard deviation of the sampling distribution ($\sigma_{\bar{x}}$). The means of our 10 samples should be within a standard error of the popu-

lation mean, which in this artificial situation we know is 48.97 ($\mu = 48.97$).

- Find and click the SPSS icon on your desktop.
- Load the *GSS2018* data set.
- Click **Data** from the menu bar of the Data Editor and then click **Select Cases**. Next, click the button next to **Random sample of cases** and then click the **Sample** button. On the **Select Cases: Random Sample** window, we can specify the size of the sample in two different ways. We'll use the second option and select samples of size 200 ($N = 200$). Click the box next to **Exactly** and fill in the boxes so that this line reads "Exactly <u>200</u> cases from the first <u>2348</u> cases."
- Click **Continue**, and then click **OK** on the **Select Cases** window. The sample will be selected.
- To find the mean *age* for the sample, click **Analyze**, **Descriptive Statistics**, and **Descriptives**. Find *age* in the variable list and transfer it to the **Variable(s):** window.
- On the **Descriptives** menu, click the **Options** button and select **S.E. MEAN** in addition to the usual statistics. Click **Continue** and then **OK**, and the requested statistics will appear in the output window.
- Here are the results for our first sample.
- Write your results in the table top of next page. We have entered results from our first sample.

	Descriptive Statistics					
	N	Minimum	Maximum	Mean		Std. Deviation
	Statistic	Statistic	Statistic	Statistic	Std. Error	Statistic
Age of respondent	200	18	89	48.53	1.330	18.805
Valid N (listwise)	200					

- Before getting the next sample, we need to restore the full data set. To do this, click **Data**, **Select Cases**, and then click the **Reset** button at the bottom of the window. Click **OK** and the GSS data set will be restored to its original size.
- To produce the second sample, click **Data**, **Select Cases**, and then click the button next to **Random Sample of Cases**. Click **Sample** and, as before, specify "Exactly 200 cases from the first 2348 cases."
 - Click **Continue** and **OK**, and the second sample will be selected. Use **Descriptives** to calculate statistics and record your results in the table below.
- Repeat these steps for the third sample and record your results. Make sure that you restore the full data

set (click **Data**, **Select Cases**, and **Reset**) each time before selecting the next sample. Continue until you have results for ten samples, including our sample.
- To complete the table:
 - Enter the sample means and standard errors (S.E.) of the mean in the appropriate columns for each sample.
 - The right-hand column shows an interval that is the sample mean plus and minus the standard error. Find the lower limit of the interval by subtracting the standard error from the sample mean and the upper limit by adding the standard error to the sample mean. The population mean ($\mu = 48.97$) will almost always be in this interval.

	\overline{X}	S.E.	$\overline{X} \pm$ S.E.
1.	48.53	1.33	47.30 .04 to 49.86 8.54
2.			_____ to _____
3.			_____ to _____
4.			_____ to _____
5.			_____ to _____
6.			_____ to _____
7.			_____ to _____
8.			_____ to _____
9.			_____ to _____
10.			_____ to _____

What can we conclude? First, if you had the patience to draw all 10 samples, congratulations! Second, we must point out that this exercise would work better if you drew even more samples. Remember that the theorems about the characteristics of the sampling distribution are based on an *infinite* number of sample outcomes.

Third, the 10 outcomes in the table *are* a sampling distribution, although not a fully developed one. The 10 outcomes in the table are just the beginning of a long process if we were to construct an actual sampling distribution.

Remember that the sampling distribution is normal and has a mean equal to the population mean and a standard deviation equal to the standard error (S.E.). About 68% of all sample outcomes will be within ± 1 standard error of the mean, 95% will be within ± 2 standard errors, and so forth. What percentage of your 10 sample means are in these ranges? Random chance is at work here and some of your sample means might be different from the population mean, but most should be reasonably close (i.e., within ± 1 S.E.).

We would not ordinarily know the value of the population mean, but we do know that it is the same value as the mean of the sampling distribution. Thus, the odds are in our favor that any particular sample mean will be close to the parameter. This exercise should convince you that properly drawn EPSEM samples are likely to be representative of populations, and statistics computed on these samples should be close in value to their respective parameters.

7

Estimation Procedures

LEARNING OBJECTIVES By the end of this chapter, you will be able to:

1. Explain the logic of estimation and the role of the sample, sampling distribution, and the population.
2. Define the concepts of bias and efficiency.
3. Calculate confidence intervals using sample means and sample proportions.
4. Explain the relationships between the confidence level, sample size, and the width of the confidence interval.
5. Use SPSS to construct confidence intervals with sample means and proportions.

USING STATISTICS

The statistical techniques presented in this chapter can be used to estimate the characteristics of large populations based on information from small samples. They can be used to estimate

- Changes in values and attitudes (e.g., support for capital punishment or same-sex marriage) in the United States over the years.
- Levels of happiness and well-being in nations throughout the world. **See page 196**
- The average number of science courses Americans have taken in college. **See page 181**
- The winner of the general election for President of the United States. **See page 189**
- The average approval level of the President of the United States. **See page 190**

The object of this branch of inferential statistics is to estimate population values or parameters from statistics computed from samples. You are already familiar with public-opinion polls and election projections, the most common applications of these techniques. Polls and surveys on every conceivable issue—from the sublime to the trivial—have become a staple of mass media and popular culture. The statistical techniques you will learn in this chapter are the same as those used by the most reputable, sophisticated, and scientific pollsters.

The standard procedure for estimating population values is to construct a **confidence interval**, a mathematical statement that says that the parameter lies

within a certain interval or range of values. For example, based on the 2018 GSS, a confidence interval might say "63% ± 3%—or, between 60% and 66%—of Americans approve of capital punishment." In the media, the central value of the interval (63% in this case) is usually stressed, but it is important to realize that the population parameter (the percentage of *all* Americans who approve of capital punishment) could be anywhere in the interval between 60% and 66%.

Estimation Selection Criteria: Bias and Efficiency

Estimation procedures are based on sample statistics. Which of the many available sample statistics should be used? Estimators can be selected according to two criteria: **bias** and **efficiency**. Estimates should be based on sample statistics that are unbiased and relatively efficient. We cover each of these criteria separately.

BIAS

An estimator is unbiased if the mean of its sampling distribution is equal to the population value of interest. We know from the theorems presented in Chapter 6 that sample means (e.g., average age of a sample) meet this criterion. The mean of the sampling distribution of sample means (which we will note symbolically as $\mu_{\overline{X}}$) is the same as the population mean (μ).

Sample proportions (P_s) (e.g., the proportion of a sample that is male) are also unbiased. If we calculate sample proportions from repeated random samples of size N, the sampling distribution of sample proportions will have a mean (μ_p) equal to the population proportion (P_u). Thus, if we are concerned with coin flips and sample honest coins 10 at a time ($N = 10$), the sampling distribution will have a mean equal to 0.5, which is the probability that an honest coin will show heads (or tails) when flipped. Of the statistics commonly used in social research, only sample means and sample proportions are unbiased (that is, have sampling distributions with means equal to the population value).

Because sample means and sample proportions are unbiased, we can determine the probability that they are within a certain distance of the population values we are trying to estimate. To illustrate, consider a specific problem. Assume that we wish to estimate the average income of a community. A random sample of 500 households is taken ($N = 500$), and a sample mean income of $45,000 is computed. In this example, the population mean (μ) is the average income of *all* households in the community, and the sample mean (\overline{X}) is the average income for the 500 households that happened to be selected for our sample. Note that we do not know the value of the population mean (μ)—if we did, we wouldn't need the sample—but it is μ that we are interested in. The sample mean of $45,000 is important primarily insofar as it can give us information about the population mean.

The two theorems presented in Chapter 6 give us a great deal of information about the sampling distribution of all possible sample means in this situation. Because N is large ($N > 100$), we know that the sampling distribution is normal and that it has a mean equal to the population mean ($\mu_{\overline{X}} = \mu$). We also know that all normal curves contain about 68% of the cases (the cases here are sample means) within ±1 Z, 95% of the cases within ±2 Z's, and more than 99% of the cases

FIGURE 7.1 Areas Under the Sampling Distribution of Sample Means

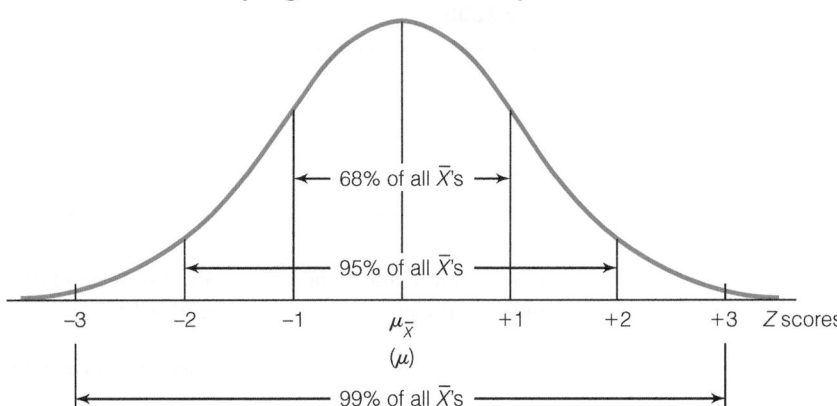

within ± 3 Z's of the mean. Remember that we are discussing the sampling distribution here—the distribution of all possible sample outcomes, or, in this instance, sample means. Thus, the probabilities are very good (approximately 68 out of 100 chances) that our sample mean of $45,000 is within ± 1 Z, excellent (95 out of 100) that it is within ± 2 Z's, and overwhelming (99 out of 100) that it is within ± 3 Z's of the mean of the sampling distribution (which is the same value as the population mean). These relationships are graphically depicted in Figure 7.1.

If an estimator is unbiased, it is probably an accurate estimate of the population parameter (μ in this case). However, in less than 1% of the cases, a sample mean will be more than ± 3 Z's away from the mean of the sampling distribution by random chance alone. We literally have no idea whether our particular sample mean of $45,000 is in this small minority. We do know, however, that the odds are high that our sample mean is considerably closer than 3 Z's to the mean of the sampling distribution and, thus, to the population mean.

EFFICIENCY

The second desirable characteristic of an estimator is efficiency, which is the extent to which the sampling distribution is clustered about its mean. Efficiency, or clustering, is essentially a matter of variation (see Figure 4.1). The smaller the standard deviation of a sampling distribution, the greater the clustering around the mean and the higher the efficiency.

Recall from the theorems stated in Chapter 6 that the standard deviation of the sampling distribution of sample means, or the standard error of the mean, is equal to the population standard deviation divided by the square root of N (i.e., $\sigma_{\bar{x}} = \sigma/\sqrt{N}$). Therefore, the standard deviation of the sampling distribution is an inverse function of N: As sample size increases, $\sigma_{\bar{x}}$ will decrease. We can improve the efficiency (or decrease the standard deviation of the sampling distribution) for any estimator by increasing sample size.

An example should make this clearer. Consider the two samples listed in Table 7.1. Both sample means are unbiased, but which is the more efficient estimator?

TABLE 7.1 Standard Deviation of the Sampling Distribution of Two Samples, Assuming $\sigma = \$500$

	Sample 1	Sample 2
Sample Mean	$\bar{X}_1 = \$45,000$	$\bar{X}_2 = \$45,000$
Sample Size	$N_1 = 100$	$N_2 = 1,000$
Standard Deviation of the Sampling Distribution (or Standard Error)	$\sigma_{\bar{x}} = \$50.00$	$\sigma_{\bar{x}} = \$15.81$

For sample 1, the standard deviation of the sampling distribution of all possible means with an N of 100 would be $50.00:

$$\sigma_{\bar{X}} = \sigma/\sqrt{N} = 500/\sqrt{100} = 500/10 = 50.00$$

For sample 2, the standard deviation of all possible sample means with an N of 1,000 would be a much smaller value of $15.81:

$$\sigma_{\bar{X}} = \sigma/\sqrt{N} = 500/\sqrt{1000} = 500/31.62 = 15.81$$

The sampling distribution based on sample 2 is much more clustered than the sampling distribution based on sample 1. In fact, it contains 68% of all possible sample means within ±15.81 of μ, while the sampling distribution based on sample 1 requires a much broader interval of ±50.00 to do the same. The estimate based on a sample with 1000 cases is much more likely to approximate the population parameter than the estimate based on a sample of 100 cases. Figures 7.2 and 7.3 illustrate these relationships graphically.

The key point to remember is that the standard deviation of all sampling distributions is an inverse function of N: The larger the sample, the greater the clustering and the higher the efficiency. In part, these relationships between sample size and the standard deviation of the sampling distribution do nothing more than underscore our commonsense notion that more confidence can be placed in large samples than in small (as long as both have been selected according to EPSEM).

FIGURE 7.2 A Sampling Distribution with $N = 100$ and $\sigma_{\bar{x}} = \$50.00$

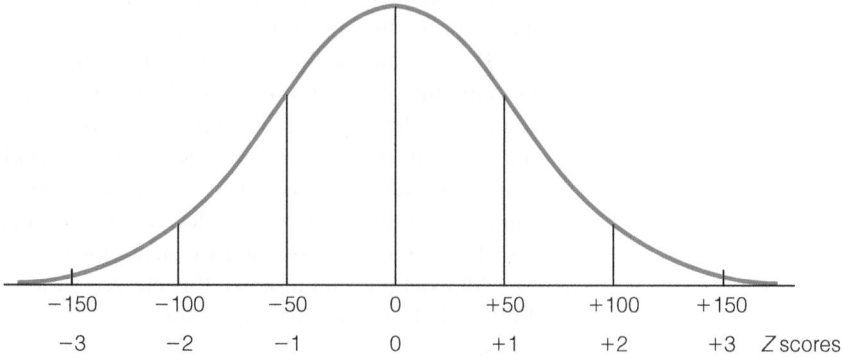

FIGURE 7.3 A Sampling Distribution with _N_ = 1000 and $\sigma_{\bar{x}}$ = \$15.81

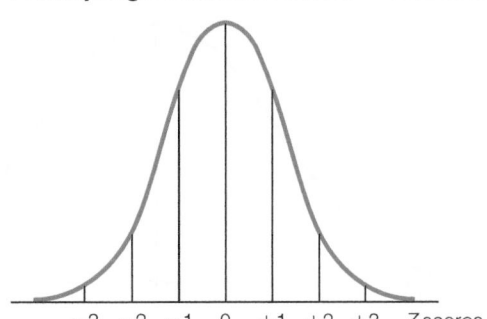

-3 -2 -1 0 +1 +2 +3 _Z_ scores

Interval Estimation Procedures

We are now ready to estimate population values based on sample statistics. We will do this by constructing **confidence intervals** or mathematical statements that say that the parameter is within a range of values. Confidence intervals are constructed in three steps.

The first step is to decide on the risk you are willing to take of being wrong. An interval estimate is wrong if it does *not* include the population parameter. This probability of error is called **alpha** (symbolized by Greek letter α). The exact value of alpha will depend on the nature of the research situation, but a 0.05 probability is commonly used. Setting alpha equal to 0.05—also called using the 95% **confidence level**—means that over the long run, the researcher is willing to be wrong only 5% of the time. Or, to put it another way, if an infinite number of intervals were constructed at this alpha level (and with all other things being equal), 95% of them would contain the population value and 5% would not. In reality, only one interval is constructed, and, by setting the probability of error so low, we are setting the odds in our favor that the interval will include the population value.

The second step is to picture the sampling distribution, divide the probability of error equally into the upper and lower tails of the distribution, and then find the corresponding Z score. For example, if we decided to set alpha equal to 0.05, we would place half (0.025) of this probability in the lower tail and half in the upper tail of the distribution. The sampling distribution would thus be divided as illustrated in Figure 7.4.

We need to find the Z score that marks the beginnings of the shaded areas. In Chapter 5, we learned how to calculate Z scores and find areas under the normal curve. Here, we will reverse that process. We need to find the Z score beyond which lies a proportion of 0.0250 of the total area. To do this, go down column c of Appendix A until you find the proportion 0.0250. The associated Z score is 1.96. Because the curve is symmetrical and we are interested in both the upper and lower tails, we designate the Z score that corresponds to an alpha of 0.05 as ±1.96 (see Figure 7.5).

We now know that 95% of all possible sample outcomes fall within ±1.96 Z scores of the population value. In reality, there is only one sample outcome,

FIGURE 7.4 The Sampling Distribution with Alpha (α) Equal to 0.05

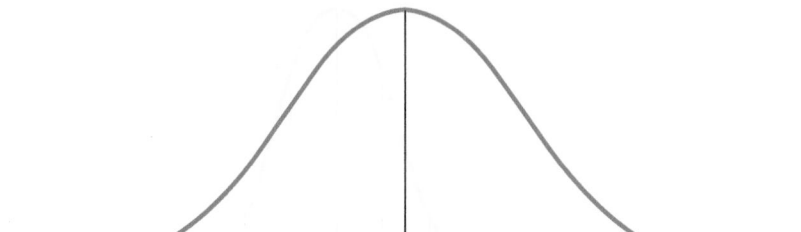

FIGURE 7.5 Finding the Z Score That Corresponds to an Alpha (α) of 0.05

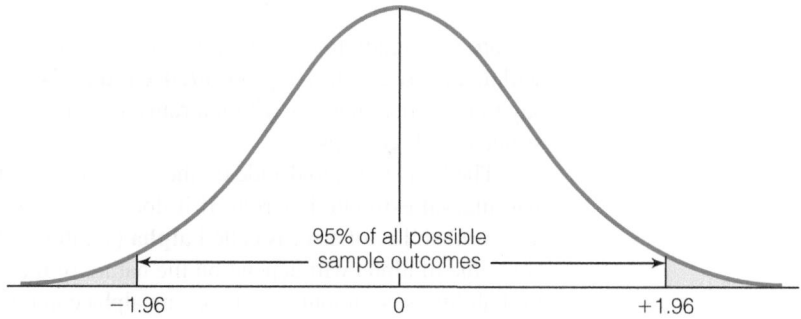

but, if we construct an interval estimate based on ± 1.96 Z's, the probabilities are that 95% of all such intervals will trap the population value. Thus, we can be 95% confident that our interval contains the population value.

Besides the 95% level, there are four other commonly used confidence levels: the 90% level ($\alpha = 0.10$), the 99% level ($\alpha = 0.01$), the 99.9% level ($\alpha = 0.001$), and the 99.99% level ($\alpha = 0.0001$). To find the corresponding Z scores for these levels, follow the procedures outlined earlier for an alpha of 0.05. Table 7.2 summarizes all the information you will need.

You should turn to Appendix A and confirm for yourself that the Z scores in Table 7.2 correspond to these alpha levels. As you do, note that, in the cases where alpha is set at 0.10 and 0.01, the precise areas we seek are not in the table. For example, with an alpha of 0.10, we would look in column c ("Area Beyond Z")

TABLE 7.2 Z Scores for Various Levels of Alpha (α)

Confidence Level	Alpha (α)	$\alpha/2$	Z Score
90%	0.10	0.05	± 1.65
95%	0.05	0.025	± 1.96
99%	0.01	0.005	± 2.58
99.9%	0.001	0.0005	± 3.32
99.99%	0.0001	0.00005	± 3.90

for the area 0.0500. Instead we find an area of 0.0505 ($Z = \pm 1.64$) and an area of 0.0495 ($Z = \pm 1.65$). The Z score we are seeking is somewhere between these two other scores. When this occurs, take the larger of the two scores as Z. This will make the interval as wide as possible under the circumstances and is thus the most conservative course of action.

For the alpha of 0.01, we encounter the same problem (the exact area 0.0050 is not in the table), resolve it the same way, and take the larger score as Z. For the alpha of 0.001, we take the largest of the several scores listed for the area as our Z score. Finally, for the lowest alpha of 0.0001, the table is not detailed enough to show exact areas, and we will use ± 3.90 as our Z score. *(For practice in finding Z scores for various levels of confidence, see problem 7.3.)*

The third step is to construct the confidence interval. In the sections that follow, we construct interval estimates of population parameters, first using sample means and then using sample proportions.

Interval Estimation Procedures Using Sample Means (Large Samples)

The formula for constructing a confidence interval based on sample means is given in Formula 7.1.

FORMULA 7.1
$$\text{c.i.} = \overline{X} \pm Z(\sigma/\sqrt{N})$$

where c.i. = confidence interval

\overline{X} = the sample mean

Z = the Z score as determined by the alpha level

(σ/\sqrt{N}) = the standard deviation of the sampling distribution, or the standard error of the mean

As an example, suppose you wanted to estimate the average IQ of a community and had randomly selected a sample of 200 residents, with a sample mean IQ of 105. Assume that the population standard deviation for IQ scores is about 15, so we can set σ equal to 15. If we are willing to run a 5% chance of being wrong and set alpha at 0.05, the corresponding Z score will be ± 1.96. These values can be substituted directly into Formula 7.1, and an interval can be constructed:

$$\text{c.i.} = \overline{X} \pm Z(\sigma/\sqrt{N})$$
$$\text{c.i.} = 105 \pm 1.96(15/\sqrt{200})$$
$$\text{c.i.} = 105 \pm 1.96(15/14.14)$$
$$\text{c.i.} = 105 \pm 1.96(1.06)$$
$$\text{c.i.} = 105 \pm 2.08$$

That is, our estimate is that the average IQ for the population in question is somewhere between 102.92 ($105 - 2.08$) and 107.08 ($105 + 2.08$). Because 95% of all possible sample means are within ± 1.96 Z's (or 2.08 IQ units in

this case) of the mean of the sampling distribution, the odds are high that our interval will contain the population mean. In fact, even if the sample mean is as far off as ± 1.96 Z's (which is unlikely), our interval will still contain the mean of the sampling distribution and, thus, the population mean (μ). Only if our sample mean is one of the few that is more than ± 1.96 Z's from the mean of the sampling distribution will we have failed to include the population mean.

Note that the value of the population standard deviation was supplied in our example. Needless to say, it is unusual to have such information about a population. In the great majority of cases, we will not know the value of σ, but we can estimate σ with s, the sample standard deviation. Unfortunately, s is a biased estimator of σ, and the formula must be changed slightly to correct for the bias. For larger samples, the bias of s will not affect the interval very much. The revised formula for cases in which σ is unknown is

FORMULA 7.2
$$\text{c.i.} = \overline{X} \pm Z(s/\sqrt{N-1})$$

In comparing this formula with Formula 7.1, note that there are two changes. First, σ is replaced by s, and, second, the denominator of the last term is the square root of $N - 1$ rather than the square root of N. The latter change is the correction for the fact that s is biased.

Let us stress here that the substitution of s for σ is permitted only for large samples (that is, samples with 100 or more cases). For smaller samples, when the value of the population standard deviation is unknown, the standardized normal distribution summarized in Appendix A cannot be used in the estimation process. To construct confidence intervals from sample means with sample sizes smaller than 100, we must use a different theoretical distribution, called the Student's t distribution, to find areas under the sampling distribution. We defer the presentation of the t distribution until Chapter 8 and confine our attention here to estimation procedures for large samples only.

We will close this section by working through a sample problem with Formula 7.2. Average income for a random sample of 500 residents of a particular community is \$45,000, with a standard deviation of \$200. What is the 95% interval estimate of the population mean income, μ?

Given that

$$\overline{X} = \$45,000$$
$$s = \$200$$
$$N = 500$$

and using an alpha of 0.05, we can construct the interval:

$$\text{c.i.} = \overline{X} \pm Z(s/\sqrt{N-1})$$
$$\text{c.i.} = 45,000 \pm (1.96)(200/\sqrt{499})$$
$$\text{c.i.} = 45,000 \pm (1.96)(200/22.34)$$
$$\text{c.i.} = 45,000 \pm (1.96)(8.95)$$
$$\text{c.i.} = 45,000 \pm 17.54$$

The average income for the community as a whole is between \$44,982.46 (45,000 − 17.54) and \$45,017.54 (45,000 + 17.54). Remember that this interval has only a 5% chance of being wrong (that is, of not containing the population mean). *(For practice with confidence intervals using sample means, see problems 7.1, 7.4–7.7, 7.18, and 7.19a.)*

ONE STEP AT A TIME Constructing Confidence Intervals for Sample Means

Using Formula 7.2

1. Select an alpha level and find the associated Z score in Table 7.2. If you use the conventional alpha level of 0.05, the Z score is ± 1.96.
2. Substitute the sample values into Formula 7.2.

To Solve Formula 7.2

1. Find the square root of $N - 1$.
2. Divide the value you found in step 1 into s, the sample standard deviation.
3. Multiply the value you found in step 2 by the value of Z.
4. The value you found in step 3 is the width of the confidence interval. To find the lower and upper limits of the interval, subtract and add this value to the sample mean.

ONE STEP AT A TIME Interpreting the Confidence Interval

1. Express the confidence interval in a sentence or two that identifies each of these elements:
 a. The sample statistic (a mean in this case)
 b. The confidence interval
 c. Sample size (N)
 d. The population to which you are estimating
 e. The confidence level (e.g., 95%)

The confidence interval we constructed in this section could be expressed as: "The average income for this community is estimated at \$45,000 ± \$17.55. The estimate is based on a sample of 500 respondents, and we can be 95% confident that it is correct."

Social Research and Data Analysis 7.1 Estimating a Population Mean

Most college students take science courses even if they are outside of their major. Have you ever wondered if you've taken more or less science courses than most people? You can try asking your friends, but that wouldn't provide you with information about most Americans. A better way to find out would be to consult the 2018 General Social Survey, which asked a random sample of Americans how many college-level science courses they have taken. After shrinking the sample down to those that took at least one science course (regardless of whether or not they graduated), we found that the average was 6.4 science courses. Let's use this statistic as well as the information below to estimate the population mean for the U.S. population in 2018.

(continued)

Social Research and Data Analysis 7.1 *(Continued)*

$$\overline{X} = 6.4$$
$$s = 9.1$$
$$N = 490$$

If we set alpha at 0.05, the corresponding Z score will be ± 1.96, and the 95% confidence interval will be

$$\text{c.i.} = \overline{X} \pm Z(s/\sqrt{N-1})$$
$$\text{c.i.} = 6.4 \pm 1.96(9.1/\sqrt{489})$$
$$\text{c.i.} = 6.4 \pm 1.96(9.1/22.11)$$
$$\text{c.i.} = 6.4 \pm 1.96(0.41)$$
$$\text{c.i.} = 6.4 \pm 0.80$$

Based on this result, we would estimate that the population mean is 6.4 ± 0.80. The lower limit of our interval estimate ($6.4 - 0.80$) is 5.6, and the upper limit ($6.4 + 0.80$) is 7.2. Thus, another way to state the interval would be

$$5.6 \le \mu \ge 7.2$$

The population mean is greater than or equal to 5.6 and less than or equal to 7.2. Because alpha was set at the 0.05 level, this estimate has a 5% chance of being wrong (that is, of not containing the population mean).

Interval Estimation Procedures Using Sample Proportions (Large Samples)

Estimation procedures using sample proportions are essentially the same as those using sample means, as long as samples are large ($N > 100$). We do not cover estimation procedures for small samples in this text.

Because proportions and means are different statistics, we must use a different sampling distribution. In fact, again based on the Central Limit Theorem (see Chapter 6), we know that sample proportions have sampling distributions that are normal in shape with means (μ_p) equal to the population value (P_u) and standard deviations (σ_p) equal to $\sqrt{\dfrac{P_u(1 - P_u)}{N}}$. The formula for constructing confidence intervals based on sample proportions is

FORMULA 7.3

$$\text{c.i.} = p_s \pm z\sqrt{\frac{P_u(1 - P_u)}{N}}$$

The values for P_s and N come directly from the sample, and the value of Z is determined by the confidence level, as was the case with sample means. Only one term in the formula (P_u) (the population value) is undefined—but this is the quantity we are trying to estimate! How can we set a value for P_u?

We can resolve this dilemma by setting the value of P_u at 0.5. Because the second factor in the numerator under the radical ($1 - P_u$) is the complement of P_u, the entire numerator $P_u(1 - P_u)$ will always have a value of 0.5×0.5, or 0.25, which is the maximum value this expression can attain. That is, if we set P_u at any value other than 0.5, the expression $P_u(1 - P_u)$ will decrease in value. If we set P_u at 0.4, for example, the second factor ($1 - P_u$) would be 0.6, and the value of $P_u(1 - P_u)$ would decrease to 0.24. Setting P_u at 0.5 ensures that the

expression $P_u(1 - P_u)$ will be at its maximum possible value and, consequently, the interval will be at maximum width. This is the most conservative solution possible to the dilemma posed by having to assign a value to P_u in the estimation equation.

To illustrate these procedures, assume that you wish to estimate the proportion of students at your university who missed at least one day of classes because of illness last semester. Out of a random sample of 200 students, 60 reported that they had been sick enough to miss classes. The sample proportion on which we will base our estimate is thus 60/200, or 0.30 (or 30%). At the 95% level, the interval estimate will be:

$$\text{c.i.} = P_s \pm Z\sqrt{\frac{P_u(1 - P_u)}{N}}$$

$$\text{c.i.} = 0.30 \pm (1.96)\sqrt{\frac{(0.5)(0.5)}{200}}$$

$$\text{c.i.} = 0.30 \pm (1.96)\sqrt{\frac{0.25}{200}}$$

$$\text{c.i.} = 0.30 \pm (1.96)\sqrt{0.00125}$$

$$\text{c.i.} = 0.30 \pm (1.96)(0.04)$$

$$\text{c.i.} = 0.30 \pm 0.08$$

You would estimate that the proportion of students who missed at least one day of classes because of illness was between 0.22 and 0.38. The estimate could also be phrased in percentages by reporting that between 22% and 38% of the student body was affected by illness at least once during the past semester. (*For practice with confidence intervals using sample proportions, see problems 7.2, 7.8–7.12, 7.16, 7.17, and 7.19b.*)

ONE STEP AT A TIME **Constructing Confidence Intervals for Sample Proportions**

Begin by selecting an alpha level and finding the associated Z score in Table 7.1. If you use the conventional alpha level of 0.05, the Z score is ± 1.96.

To Solve Formula 7.3

1. Substitute the value 0.25 for the expression $P_u(1 - P_u)$ in the numerator of the fraction under the square root sign.
2. Divide N into 0.25.
3. Find the square root of the value you found in step 2.
4. Multiply the value you found in step 3 by the value of Z.
5. The value you found in step 4 is the width of the confidence interval. To find the lower and upper limits of the interval, subtract and add this value to the sample proportion.

> **ONE STEP AT A TIME** **Interpreting the Confidence Interval**
>
> 1. Express the confidence interval in a sentence or two that identifies each of these elements:
> a. The sample statistic (a proportion in this case)
> b. The confidence interval
> c. Sample size (N)
> d. The population to which you are estimating
> e. The confidence level (e.g., 95%)
>
> The confidence interval we constructed in this section could be expressed as: "On this campus, 30% ± 8% of the students were sick enough to miss class at least once during the semester. The estimate is based on a sample of 200 respondents, and we can be 95% confident that it is correct."

Social Research and Data Analysis 7.2 **Estimating Population Proportions**

The World Values Survey interviews random samples from nations around the world about a variety of issues. The table below presents the percentage of respondents from India, the United States, Sweden, and Mexico that agree that "Most People Can Be Trusted."

Percentage of Samples Agreeing that "Most People Can Be Trusted"

Nation	Sample Size (N)	% Agreeing	95% Confidence Interval ($\alpha = 0.05$)
Mexico	2,000	12.4%	0.124 ± 0.022 or 12.4% ± 2.2%
Sweden	1,206	60.1%	0.601 ± 0.028 or 60.1% ± 2.8%
U.S.	2,232	34.8%	0.348 ± 0.021 or 34.8% ± 2.1%
India	4,078	16.7%	0.167 ± 0.015 or 16.7% ± 1.5%

Source: World Values Survey: http://www.worldvaluessurvey.org/WVSOnline.jsp

To construct the confidence intervals, we must convert the percentages to proportions. With an alpha of 0.05, the Z score is ± 1.96 and the confidence interval for Mexico is:

$$\text{c.i.} = P_s \pm Z\sqrt{\frac{P_u(1 - P_u)}{N}}$$

$$\text{c.i.} = 0.124 \pm (1.96)\sqrt{\frac{(0.5)(0.5)}{2,000}}$$

$$\text{c.i.} = 0.124 \pm (1.96)\sqrt{\frac{0.25}{2,000}}$$

$$\text{c.i.} = 0.124 \pm (1.96)\sqrt{0.00013}$$

$$\text{c.i.} = 0.124 \pm (1.96)(0.011)$$

$$\text{c.i.} = 0.124 \pm 0.022$$

The interval can also be expressed as a percentage: 12.4% ± 2.2%. We can estimate that between 10.2% and 14.6% of the population of Mexico agree that most people can be trusted. This interval has a 5% chance of not containing the population value. The same formula and procedures are used to find the confidence intervals for the other three nations.

Why would these nations be so different in their levels of trust? Why are Swedes so trusting and Mexicans and Indians so wary? What national traditions and values could be at work?

STATISTICS In Everyday Life

The National Mood

According to a 2018 Gallup Poll, Americans were angrier and more stressed and worried than at most points over the past decade. More than half (55%) of the respondents reported feeling a lot of stress, almost half (45%) reported a lot of worry, and 22% reported a lot of anger. All three of these feelings seem to be increasing, and Americans have one of the highest stress rates of the 143 nations studied in 2018. What sociological factors might account for these disturbing trends?

These results are based on a sample of 1004 Americans, ages 15 and older. They are accurate to within ± 3% and have a 95% confidence level

Source: Gallup: https://news.gallup.com/poll/249098/americans-stress-worry-anger-intensified-2018.aspx

Computing Confidence Intervals: A Summary

Table 7.3 presents the three formulas for confidence intervals we have covered, organized by the situations in which they are used. To compute a confidence interval for a mean, choose between the two formulas based on whether the population standard deviation (σ) is known or unknown. To compute a confidence interval for a proportion, always use Formula 7.3.

Controlling the Width of Interval Estimates

The width of a confidence interval for either a mean or a proportion can be partly controlled by manipulating two terms in the formula. First, the confidence level can be raised or lowered and, second, the interval can be widened or narrowed by gathering samples of different size.

The researcher alone determines the risk he or she is willing to take of being wrong (that is, of not including the population value in the interval estimate). The exact confidence level (or alpha level) will depend, in part, on the purpose of the research. For example, if potentially harmful drugs were being tested, the researcher would naturally demand near-perfect high levels of confidence

TABLE 7.3 Choosing Formulas for Confidence Intervals

If the sample statistic is a	And the population standard deviation (σ) is	Use Formula
Mean (\bar{X})	Known	7.1 c.i. $= \bar{X} \pm Z(\sigma/\sqrt{N})$
Mean (\bar{X})	Unknown	7.2 c.i. $= \bar{X} \pm Z(s/\sqrt{N-1})$
Proportion (P_s)		7.3 c.i. $= P_s \pm Z\sqrt{\dfrac{P_u(1-P_u)}{N}}$

(99.99% or even 99.999%). On the other hand, if intervals are being constructed only for loose "guesstimates," then much lower confidence levels can be tolerated (such as 90%).

INTERVAL WIDTH AND CONFIDENCE LEVEL

The relationship between interval size and confidence level is this: *Intervals widen as confidence levels increase.* This should make intuitive sense. Wider intervals are more likely to include the population value; hence, more confidence can be placed in them.

To illustrate this relationship, let us return to the example where we estimated the average income for a community. In this problem, we were working with a sample of 500 residents, and the average income for this sample was \$45,000, with a standard deviation of \$200. We constructed the 95% confidence interval and found that it extended 17.54 around the sample mean (that is, the interval was \$45,000 ± \$17.54).

If we had constructed the 90% confidence interval (a lower confidence level) based on these sample data, the Z score in the formula would have decreased to ±1.65, and the interval would have been narrower:

$$\text{c.i.} = \overline{X} \pm Z(s/\sqrt{N-1})$$
$$\text{c.i.} = 45{,}000 \pm (1.65)(200/\sqrt{499})$$
$$\text{c.i.} = 45{,}000 \pm (1.65)(8.95)$$
$$\text{c.i.} = 45{,}000 \pm 14.77$$

On the other hand, if we had constructed the 99% confidence interval, the Z score would have increased to ±2.58, and the interval would have been wider:

$$\text{c.i.} = \overline{X} \pm Z(s/\sqrt{N-1})$$
$$\text{c.i.} = 45{,}000 \pm (2.58)(200/\sqrt{499})$$
$$\text{c.i.} = 45{,}000 \pm (2.58)(8.95)$$
$$\text{c.i.} = 45{,}000 \pm 23.09$$

At the 99.9% confidence level, the Z score would be ±3.32, and the interval would be wider still:

$$\text{c.i.} = \overline{X} \pm Z(s/\sqrt{N-1})$$
$$\text{c.i.} = 45{,}000 \pm (3.32)(200/\sqrt{499})$$
$$\text{c.i.} = 45{,}000 \pm (3.32)(8.95)$$
$$\text{c.i.} = 45{,}000 \pm 29.71$$

The four intervals are grouped together in Table 7.4 and we can readily see the increase in interval size as the confidence level increases. Although we have used confidence intervals for means to illustrate, exactly the same relationships apply to confidence intervals for proportions. *(To explore further the relationship between alpha and interval width, see problem 7.13.)*

TABLE 7.4 **Confidence Intervals Grow Wider as Confidence Levels Increase**
(\bar{X} = $45,000, s = $200, N = 500 throughout)

Alpha (α)	Confidence Level	Interval	Interval Width
0.10	90%	$45,000 ± 14.77	$29.54
0.05	95%	$45,000 ± 17.54	$35.08
0.01	99%	$45,000 ± 23.09	$46.18
0.001	99.9%	$45,000 ± 29.71	$59.42

TABLE 7.5 **Confidence Intervals Grow Narrower as Sample Size Increases**
(\bar{X} = $45,000, s = $200, α = 0.05 throughout)

N	Confidence Interval	Interval Width
100	c.i. = $45,000 ± 1.96 (200/$\sqrt{99}$) = $45,000 ± $39.40	$78.80
500	c.i. = $45,000 ± 1.96 (200/$\sqrt{499}$) = $45,000 ± $17.54	$35.08
1,000	c.i. = $45,000 ± 1.96 (200/$\sqrt{999}$) = $45,000 ± $12.40	$24.80
10,000	c.i. = $45,000 ± 1.96 (200/$\sqrt{9999}$) = $45,000 ± $3.92	$ 7.84

INTERVAL WIDTH AND SAMPLE SIZE

Sample size bears the opposite relationship to interval width: *As sample size increases, interval width decreases.* Larger samples give more precise (narrower) estimates.

An example should make this clear. Table 7.5 presents confidence intervals for four samples of different size (*N*). The sample data are the same as in Table 7.4, and the confidence level is 95% throughout. The calculation of the confidence intervals is illustrated in the middle column, and the width of each interval is listed in the right-hand column. Note the decrease in interval width as sample size increases. The relationships illustrated in Table 7.5 also hold true for sample proportions.

Notice that the decrease in interval width (or increase in precision) does not bear a constant, or linear, relationship with sample size. For example, the sample with an *N* of 500 is five times larger than the sample with *N* = 100, but the interval is not five times as narrow. This is an important relationship because it means that *N* might have to be increased many times over to appreciably improve the precision of an estimate. Because the cost of a research project is a direct function of sample size, this relationship means that there is a point of diminishing returns in estimation procedures. A sample with *N* = 10,000 will cost about twice as much as a sample with *N* = 5000, but estimates based on the larger sample will not be twice as precise. *(To explore further the relationship between sample size and interval width, see problem 7.14.)*

USING SPSS	**Constructing Confidence Intervals**

SPSS does not provide a program specifically for constructing confidence intervals, although some of the procedures we'll cover in future chapters do include confidence intervals as part of the output. Rather than make use of these programs, we will get sample statistics from the 2018 GSS and then construct interval estimates to the population (U.S. society as of 2018). We will construct two confidence intervals, first with sample means

(continued)

USING SPSS

(Continued)

and then with proportions. (NOTE: If you are using the Student Version of SPSS and use the smaller *GSS2018Sample* data set, your results will be slightly different.)

1. Click the SPSS icon on your desktop.
2. Load the *GSS2018* database.
 a. Find the **File** command on the far left of the menu bar and then click **File → Open → Data**.
 b. Find the *GSS2018* database and click the file name.

To construct a confidence interval with a sample mean:

3. Click **Analyze, Descriptive Statistics**, and **Descriptives**.
4. Find *childs* in the list of variables on the left of the **Descriptives** window and click the arrow to move the variable names to the box on the right.
5. Click **OK** and the output will look like this:

Descriptive Statistics

	N	Minimum	Maximum	Mean	Std. Deviation
Number of children	2344	0	8	1.86	1.674
Valid N (listwise)	2344				

6. We can substitute these values into Formula 7.2 to construct the 95% confidence interval:

$$\text{c.i.} = \overline{X} \pm Z(s/\sqrt{N-1})$$
$$\text{c.i.} = 1.86 \pm 1.96(1.67/\sqrt{2{,}344})$$
$$\text{c.i.} = 1.86 \pm 1.96(0.03)$$
$$\text{c.i.} = 1.86 \pm 0.06$$

Based on a sample of 2344 and at the 95% confidence level, we estimate that Americans have 1.86 ± 0.06 children or between 1.80 and 1.92 children.

To construct a confidence interval with a sample proportion:

7. Click **Analyze, Descriptive Statistics**, and **Frequencies**.
8. Find *marital* in the list of variables on the left and click the arrow to move the variable names to the box on the right.
9. Click **OK** and the output will look like this:

Marital status

		Frequency	Percent	Valid Percent	Cumulative Percent
Valid	MARRIED	998	42.5	42.5	42.5
	WIDOWED	200	8.5	8.5	51.1
	DIVORCED	403	17.2	17.2	68.2
	SEPARATED	75	3.2	3.2	71.4
	NEVER MARRIED	670	28.5	28.6	100.0
	Total	2346	99.9	100.0	
Missing	NA	2	.1		
Total		2348	100.0		

USING SPSS *(Continued)*

10. We will estimate the percentage of Americans who have never married. Convert the percentage in the Valid Percent column (28.6%) to a proportion (0.29) and substitute into Formula 7.3:

$$\text{c.i.} = P_s \pm Z\sqrt{\frac{P_u(1 - P_u)}{N}}$$

$$\text{c.i.} = 0.29 \pm 1.96\sqrt{\frac{0.25}{2,346}}$$

$$\text{c.i.} = 0.29 \pm (1.96)(.01)$$

$$\text{c.i.} = 0.29 \pm 0.02$$

Converting back to percentages, we can estimate, at the 95% confidence level, that 29% ± 2% of Americans have never married. This estimate is based on a sample of 2348.

Interpreting Statistics: Predicting Election Outcomes and Judging the Performance of the President

The statistical techniques covered in this chapter have become a part of everyday life in the United States and in many other societies. In politics, for example, estimation techniques are used to track public sentiment, measure how citizens perceive the performance of our leaders, and project the likely winners of upcoming elections.

We should acknowledge that these applications of estimation techniques can be controversial. Does the easy availability of polls makes politicians overly sensitive to the whims of public sentiment? Do election projections decrease people's readiness to vote on Election Day? These are serious concerns, but we can do little more than acknowledge them in this text and hope that you have the opportunity to pursue them in other, more appropriate settings.

In this installment of "Interpreting Statistics," we examine the accuracy of election projections for the 2016 presidential election, and we'll also examine the polls that measure the approval ratings of two U.S. presidents. Both kinds of polls use the same formulas introduced in this chapter to construct confidence intervals (although the random samples were assembled by a complex and sophisticated technology that is beyond the scope of this text).

PREDICTING THE 2016 PRESIDENTIAL ELECTION

The outcome of the 2016 presidential election, in which Donald Trump defeated Hillary Clinton, was a surprise to most Americans because pollsters

TABLE 7.6 Polls and Actual Results, Presidential Race, 2016

Polls*	N	Margin of Error	Predicted for Clinton	Predicted for Trump
Bloomberg	799	±3.5	46%	43%
IBD/TIPP Tracking	1107	±3.1	43%	42%
LA Times/USC Tracking	3935	±4.5	44%	47%
ABC/Wash Post Tracking	2220	±2.5	49%	46%
FOX News	1295	±2.5	48%	44%
	ACTUAL VOTE		48%	46%

had estimated the probability of a Clinton win to be very high. For example, just a day before the election, Real Clear Politics, a popular polling organization, predicted that Clinton's chances of winning were at 88%. Needless to say, this prediction turned out to be wrong. Presidential elections in the United States are determined by the Electoral College and not by the popular vote. While he lost the popular vote, Trump won the Electoral College by a margin of 306 to 232. Actually, most polling organizations were accurate in their predictions of the popular vote. As you can see in Table 7.6, five major polling organizations predicted a popular vote outcome that was close to the actual vote.

BREAKING DOWN THE POLL NUMBERS

As shown in Table 7.6, each poll has a "margin of error," which is determined mostly by sample size. In the five polls shown, the margin of error is as low as 2.5 and as high as 4.5. Pollsters normally base these margins of error on the 95% confidence level, indicating that they estimate only a 5% chance that their prediction will miss by more than the margin of error. For example, the Bloomberg Poll predicted a 3% Clinton win, but the poll had a margin of error of ±3.5%. This means that Clinton's support could have been as low as 42.5% (46% − 3.5%) or as high as 49.5% (46% + 3.5%). Among the polls shown, only the IBD/TIPP Tracking Poll missed the mark because it estimated that Clinton's support could only be as high as 46.1% (43% + 3.1%). For more on the 2016 election, check out the "Statistics in Everyday Life" later in this chapter.

THE UPS AND DOWNS OF PRESIDENTIAL POPULARITY

Once you get to be president, you are not free of public-opinion surveys—quite the contrary, in fact. Since the middle of the 20th century, pollsters have tracked the president's popularity by asking samples of adult Americans whether they approve or disapprove of the way the president is handling the job. Approval ratings for the last two years of President Obama's term and the first two years of President Trump's term are presented in Figure 7.6. For purposes of clarity, only the sample percentages are used for this graph, but remember that the confidence

FIGURE 7.6 Presidential Popularity, Jan. 2015 to Aug. 2019

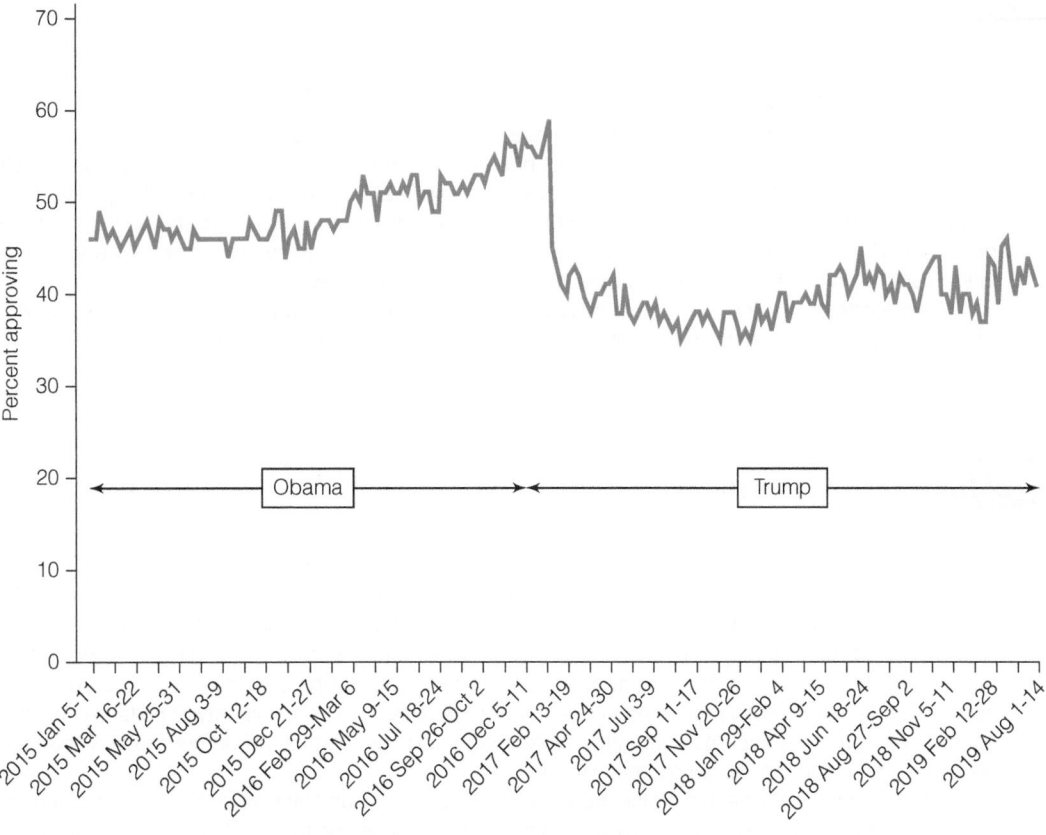

interval estimate would range about ±3% (at the 95% confidence level) around these points.

Americans responded to President Obama's inauguration in 2008 with high levels of approval (not shown in Figure 7.6). His initial ratings were in the high 70s but then gradually declined throughout his first years in office as the nation's economic woes continued and the administration was caught up in a lengthy, acrimonious debate over health care. By his last two years in office (as shown in Figure 7.6), his approval had fallen to below 50%, although he did enjoy a surge of popularity in the final months of 2016. President Trump, in contrast, started his term with an approval rating well below 50% and has fluctuated around the 40% to 45% approval level since.

Returning to the controversies surrounding public-opinion polls, many would argue that it is harmful to judge the president so continuously. These polls expose the president (and all politicians whose performance is similarly measured) to the whims of public sentiment and, at some level, pressure him or her to cater to popular opinion and shape ("spin") his or her image to maintain support.

STATISTICS **In Everyday Life**

The Classic Election Prediction Error

"Dewey Defeats Truman," read the banner headline of the morning edition of the *Chicago Tribune* the day after the 1948 presidential election. The headline became famous because the final results were just the reverse: Not only had Truman won his bid for reelection, but he won by a handy margin. What happened? Because the political "experts" were certain that Truman would lose, the pollsters had stopped taking polls early in the fall of 1948 and missed a late surge of support for the incumbent president. This major embarrassment for the *Chicago Tribune* (and many other newspapers) resulted in the current practice of polling the public right up until Election Day.

What kinds of opinions or attitudes would be more likely to change rapidly, as did support for President Truman? Can you think of attitudes or opinions that would be less likely to fluctuate?

STATISTICS **In Everyday Life**

The Big Miss by Pollsters in the 2016 Presidential Election

Most forecasters predicted a victory for Hillary Clinton over Donald Trump in the presidential race of 2016. Without question, the most celebrated forecaster was Nate Silver, the founder of FiveThirtyEight.com, who predicted that Clinton's chances of winning the presidency were 71.4%.* Silver's techniques go far beyond the statistics and formulas presented in this text, but he still relies on 95% confidence intervals and the arithmetic presented in this chapter. He uses data from a variety of polls, weighs them according to previous accuracy and other factors, and adds economic and demographic variables to his estimation algorithm. So, what went wrong?

According to an article in the Harvard Business Review by Dan Cassino, the biggest problem was one that has plagued pollsters for many years now—low response rate to polls. For pollsters to truly claim that their estimates are accurate at the 95% confidence level, their results need to be based on a random sample. Polling organizations aim for random samples by using techniques such as random digit dialing, but due to the advent of Caller ID, average response rates have fallen to less than 15%.

Statisticians can use weights to adjust for low response rates when they know which groups are less likely to respond, but it becomes more challenging when they don't know the reasons. One theory about what went wrong in the predictions for the 2016 election is that Trump voters were less likely to respond to political polls because they thought they were rigged. This might explain why more people voted for Trump than were expected, especially in key swing states that altered the expected outcome of the Electoral College vote.

*Source: https://projects.fivethirtyeight.com/2016-election-forecast/

READING STATISTICS **Public-Opinion Polls**

You are most likely to encounter the estimation techniques covered in this chapter in the form of public-opinion polls, election projections, and the like. Responsible reporting by the media will usually emphasize the estimate itself (e.g., "57% of the respondents said that they approve of the president's performance") but also will report the width of the interval ("This estimate is accurate to within ±3%" or

(*continued*)

READING STATISTICS *(Continued)*

"Figures from this poll are subject to a sampling error of ±3%"), the alpha level (usually as the confidence level of 95%), and the size of the sample ("1458 households were surveyed").

Election projections have been common since the middle of the 20th century. More recently, public-opinion polling has been used to gauge reactions to everything from the newest movies to the hottest gossip to the president's conduct during the latest crisis, and polls are routinely included in news stories in the media. We would include an example of these applications, but polls have become so pervasive that you can easily find your own. Just pick up a magazine or newspaper and leaf through it casually, and we bet you'll find at least one poll. Read the story and see whether you can identify the population, the confidence interval width, the sample size, and the confidence level. Bring the news item to class and dazzle your instructor.

As a citizen, as well as a social scientist, you should be extremely suspicious of polls that do not include such vital information as sample size and interval width (often reported as "margin of error" or "sampling error"). You should also check to see how the sample was assembled. Nonrandom samples are not representative of the American public or, for that matter, of any population other than the sample itself. You might encounter samples such as these when news or sports programs ask viewers to call in and register their opinions about some current controversy. The "samples" assembled in this way are not representative, and the results should be regarded as entertainment (at best), not as reflections of public opinion in general. You should read all polls and surveys critically and analytically and place confidence only in polls that are based on samples selected according to the rule of EPSEM (see Chapter 6) from some defined population.

Advertisements, commercials, and reports published by partisan groups sometimes report statistics that *seem* to be estimates of the population values. Often, such estimates are based on woefully inadequate sample sizes and biased

sampling procedures, and the data are collected under circumstances that evoke a desired response. "Person on the street" (or shopper in the grocery store) interviews have a certain folksy appeal, but they must not be accorded the same credibility as surveys conducted by reputable polling firms.

PUBLIC-OPINION SURVEYS IN THE PROFESSIONAL LITERATURE

For the social sciences, probably the single most important consequence of the growth in opinion polling is that many nationally representative databases are now available for research purposes. These high-quality databases are often available at no cost or for a nominal fee, and they make it possible to conduct "state-of-the-art" research without the expense and difficulty of collecting data yourself. This is an important development because we can now test our theories against very high-quality data and our conclusions can have a stronger empirical basis. Our research efforts can have greater credibility with our colleagues, with policy makers, and with the public at large.

One of the more important and widely used databases of this sort is the General Social Survey, or the GSS. Since 1972, the National Opinion Research Council has questioned a nationally representative sample of Americans about a wide variety of issues and concerns. Since many of the questions are asked every year, the GSS offers a longitudinal record of American sentiment and opinion about a large variety of topics for almost four decades. Each year, new topics of current concern are added and explored, and the variety of information available continues to expand. Like other nationally representative samples, the GSS sample is chosen by a complex probability design that resembles cluster sampling (see Chapter 6). Sample size varies from 1400 to more than 4000, and estimates based on samples this large will be accurate to within about ±3% or less. Many of the computer exercises in this text are based on the 2018 GSS; this database is described more fully in Appendix F.

SUMMARY

1. Population values can be estimated with sample values. Confidence intervals estimate that the population value falls within a certain range of values and have a known probability of being incorrect.
2. Estimates based on sample statistics must be unbiased and relatively efficient. Means and proportions are unbiased because the means of their sampling distributions are equal to the respective population values. Efficiency is largely a matter of sample size. The greater the sample size, the lower the value of the standard deviation of the sampling distribution, the more tightly clustered the sample outcomes will be around the mean of the sampling distribution, and the more efficient the estimate.
3. The width of a confidence interval is a function of the risk we are willing to take of being wrong (the alpha level) and the sample size. The interval widens as our probability of being wrong decreases and as sample size decreases.

SUMMARY OF FORMULAS

Confidence interval using sample means, large samples ($N > 100$), population standard deviation known:

$$\textbf{7.1} \quad \text{c.i.} = \bar{X} \pm Z(\sigma/\sqrt{N})$$

Confidence interval using sample means, large samples ($N > 100$), population standard deviation unknown:

$$\textbf{7.2} \quad \text{c.i.} = \bar{X} \pm Z(s/\sqrt{N-1})$$

Confidence interval using sample proportions, large samples ($N > 100$):

$$\textbf{7.3} \quad \text{c.i.} = P_s \pm Z\sqrt{\frac{P_u(1 - P_u)}{N}}$$

GLOSSARY

Alpha (α). The probability of error, or the probability that a confidence interval does not contain the population value. Alpha levels are usually set at 0.10, 0.05, 0.01, or 0.001.

Bias. A criterion used to select sample statistics as estimators. A statistic is unbiased if the mean of its sampling distribution is equal to the population value of interest.

Confidence interval. An estimate of a population value in which a range of values is specified.

Confidence level. A frequently used alternate way of expressing alpha, the probability that an interval estimate will not contain the population value. Confidence levels of 90%, 95%, 99%, and 99.9% correspond to alphas of 0.10, 0.05, 0.01, and 0.001, respectively.

Efficiency. The extent to which the sample outcomes are clustered around the mean of the sampling distribution.

PROBLEMS

7.1 For each set of sample outcomes below, use Formula 7.2 to construct the 95% confidence interval for estimating μ, the population mean.

 a. $\bar{X} = 5.2$ **b.** $\bar{X} = 100$ **c.** $\bar{X} = 20$
 $s = 0.7$ $s = 9$ $s = 3$
 $N = 157$ $N = 620$ $N = 220$

 d. $\bar{X} = 1020$ **e.** $\bar{X} = 7.3$ **f.** $\bar{X} = 33$
 $s = 50$ $s = 1.2$ $s = 6$
 $N = 329$ $N = 105$ $N = 220$

7.2 For each of the following sets of sample outcomes, use Formula 7.3 to construct the 99% confidence interval for estimating P_u.

 a. $P_s = 0.40$ **b.** $P_s = 0.37$ **c.** $P_s = 0.79$
 $N = 548$ $N = 522$ $N = 121$

 d. $P_s = 0.14$ **e.** $P_s = 0.43$ **f.** $P_s = 0.63$
 $N = 100$ $N = 1049$ $N = 300$

7.3 For each of the following confidence levels, determine the corresponding Z score.

Confidence Level	Alpha	Area Beyond Z	Z score
95%	0.05	0.0250	± 1.96
94%			
92%			
97%			
98%			

7.4 SP Non-standard work schedules involving shifts on weekends, evenings, and overnight are common among low-wage workers. To study the impact of these shifts on the lives of the working class, you have developed a series of questions about life satisfaction among those that work these shifts and administered it to a random sample of 100 of them. On a scale of 1–20, their average score came to 10.6, with a standard deviation of 2.8. What is your estimate of the average life satisfaction score for the population of workers on these schedules? Use the 95% confidence level.

7.5 SOC A researcher has gathered information from a random sample of 178 households. For each of the following variables, construct confidence intervals to estimate the population mean. Use the 90% level.

a. An average of 2.3 people resides in each household. Standard deviation is 0.35.

b. There was an average of 2.1 television sets ($s = 0.10$) and 0.78 telephones ($s = 0.55$) per household.

c. The households averaged 6.0 hours of television viewing per day ($s = 3.0$).

7.6 SP As a researcher in a social policy organization, you have collected data on literacy in the Sudan. Your random sample of 100 Sudanese people indicate that the average age of those that are illiterate is 19. At the 99% level, what is your estimate of the population value?

$$\overline{X} = 19$$
$$s = 0.85$$
$$N = 100$$

7.7 SOC A random sample of 429 college students was interviewed about a number of matters. Use the results to construct confidence interval estimates of the population mean at the 99% level.

a. They reported that they had spent an average of $478.23 on textbooks during the previous semester, with a sample standard deviation of $15.78.

b. They also reported that they had visited the health clinic an average of 1.5 times a semester, with a sample standard deviation of 0.3.

c. On the average, the sample had missed 2.8 days of classes per semester because of illness, with a sample standard deviation of 1.0.

d. On the average, the sample had missed 3.5 days of classes per semester for reasons other than illness, with a sample standard deviation of 1.5.

7.8 CJ A random sample of 100 inmates at a maximum security prison shows that exactly 10 of the respondents had been the victims of violent crime during their incarceration. Estimate the proportion of victims for the population as a whole, using the 90% confidence level. (*HINT: Calculate the sample proportion P_s before using Formula 7.3. Remember that a proportion is equal to frequency divided by* N.)

7.9 SOC The survey mentioned in problem 7.5 found that 25 of the 178 households consisted of unmarried couples who were living together. What is your estimate of the population proportion? Use the 95% level.

7.10 PA A random sample of 260 workers in a high-rise office building revealed that 30% were very satisfied with the quality of elevator service. At the 99% level, what is your estimate of the population value?

7.11 SOC A random sample of 1496 respondents of a major metropolitan area was questioned about a number of issues. Construct estimates to the population at the 95% level for each of the results reported next. Express the final confidence interval in percentages (e.g., "between 40% and 45% agreed that premarital sex was always wrong").

a. When asked to agree or disagree with the statement "Internet pornography leads to rape and other sex crimes," 823 agreed.

b. When asked to agree or disagree with the statement "Hand guns should be outlawed," 650 agreed.

c. When asked, 375 of the sample agreed that marijuana should be legalized.

d. When asked, 1023 of the sample said that they had attended a church, synagogue, mosque, or other place of worship at least once within the past month.

e. When asked, 800 agreed that public elementary schools should have sex education programs starting in the fifth grade.

7.12 SW A random sample of 100 patients treated in a program for alcoholism and drug dependency over the past 10 years was selected. It was determined that 53 of the patients had been readmitted to the

program at least once. At the 95% level, construct an estimate of the population proportion.

7.13 For the following sample data, construct five different interval estimates of the population mean, one each for the 90%, 95%, 99%, 99.9%, and 99.99% levels. What happens to the interval width as confidence level increases? Why?

$$\bar{X} = 100$$
$$s = 10$$
$$N = 500$$

7.14 For each of the following three sample sizes, construct the 95% confidence interval for the population proportion. Use a sample proportion of 0.40 throughout. What happens to interval width as sample size increases? Why?

$$P_s = 0.40$$
Sample A: $N = 100$
Sample B: $N = 1000$
Sample C: $N = 10,000$

7.15 [PS] Two individuals are running for mayor of a large city. You conduct an election survey a week before the election and find that 51% of the respondents prefer candidate A. Can you predict a winner? Use the 99% level. (*HINT: In a two-candidate race, what percentage of the vote would the winner need? Does the confidence interval indicate that candidate A has a sure margin of victory? Remember that while the population parameter is probably [a = 0.01] in the confidence interval, it may be anywhere in the interval.*)

$$P_s = 0.51$$
$$N = 578$$

7.16 [SOC] A survey of random samples from societies around the globe was administered recently. Listed here are the numbers of respondents in each nation who said that they are "very happy." Compute sample proportions, and construct confidence interval estimates for each nation at the 95% level.

Nation	Number "Very Happy"	Sample Size	Confidence Interval
Great Britain	528	1041	
Japan	311	1096	
Brazil	510	1500	
China	424	2015	
Malaysia	428	1201	
Mali	598	1534	
Russia	214	2033	

7.17 [SOC] The fraternities and sororities at St. Algebra College have been plagued by declining membership over the past several years and want to know whether the incoming freshman class will be a fertile recruiting ground. Not having enough money to survey all 1600 freshmen, they commission you to survey the interests of a random sample. You find that 35 of your 150 respondents are "extremely" interested in social clubs. At the 95% level, what is your estimate of the number of freshmen who would be extremely interested? (*HINT: The high and low values of your final confidence interval are proportions. How can proportions be expressed as numbers?*)

7.18 [SOC] You are the automotive specialist for the monthly magazine *Consumer Beware*. Part of your job is to investigate the claims made by automakers in their TV commercials. You are particularly suspicious of a new economy car that the automaker claims will get 78 miles per gallon. After checking the mileage figures for a random sample of 125 owners of this car, you find an average miles per gallon of 75.5, with a standard deviation of 3.7. At the 99% level, do your results tend to confirm or refute the manufacturer's claims?

Statistical Analysis Using SPSS

7.19 [SOC] In this exercise, you will use SPSS to generate sample statistics from the 2018 General Social Survey (*GSS2018*). You will then construct confidence intervals to estimate population values or parameters.
- Click the SPSS icon on your desktop.
- Load the *GSS2018* database.
- Use the **Descriptives** program to find means and standard deviations for hours worked (*hrs1*) and age (*age*).
 a. Use Formula 7.2 to estimate population values for each variable. Use the 95% confidence level and express the confidence interval in a few sentences.
- Use the **Frequencies** program to get frequency distributions for religious affiliation (*relig*) and support for the death penalty (*cappun*).
 b. Use Formula 7.3 to estimate population values for the percentage Catholic and the percentage that support the death penalty. Use the 95% confidence level and express the confidence intervals in a few sentences.

Estimating the Characteristics of the Typical American

As we did in the "Using SPSS" demonstration, we will use SPSS to produce sample statistics from the 2018 GSS, which you can then use as the basis for interval estimates to the population.

Step 1: Choosing the Variables

As a framework for this exercise, we will return to the task of describing the "typical American" begun in Chapter 3 and continued in Chapter 4. Select four of the variables you used in the earlier exercises and add four more that you did NOT use in the earlier chapters, in the "Using SPSS" demonstration in this chapter, or in problem 7.19. Make sure that you have at least one variable from the nominal level and at least one measured at the ordinal or interval-ratio level. List the variables, along with level of measurement, here.

	SPSS Variable Name	Explain Exactly What This Variable Measures	Levels of Measurement
1			
2			
3			
4			
5			
6			
7			
8			

Step 2: Getting Sample Statistics

For interval-ratio variables and ordinal-level variables with 4 or more scores, use **Descriptives** to get sample means, standard deviations, and sample sizes. For nominal-level variables and for ordinal-level variables with 3 or fewer scores, use **Frequencies** to produce frequency distributions. For either procedure, click **Analyze → Descriptive Statistics** and then select either **Frequencies** or **Descriptives**. Select your variables from the box on the left and click the arrow to move the variable name to the window on the right. SPSS will process all variables listed at the same time.

Step 3: Constructing Confidence Intervals

Once you have the SPSS output, use the results to construct 95% confidence intervals around the sample statistics. For nominal-level variables and ordinal-level variables with 3 or fewer scores, select one category (e.g., female for *sex* or Catholic for *relig*) and look in the **Valid Percent** column of the frequency distribution to get the percentage of cases in the sample in that category. Change this value to a proportion (divide by 100). This value is P_s and can be substituted directly into Formula 7.3. Remember to estimate P_u at 0.5. After you have found the confidence interval, change your proportion back to a percentage and record your results in the table below.

For ordinal-level variables with 4 or more scores and for interval-ratio variables, find the sample mean, standard deviation, and sample size in the output of the **Descriptives** procedure. Substitute these values into Formula 7.2 and record your results in the table below.

Step 4: Recording Results

Use the table to summarize your confidence intervals.

	SPSS Variable Name	Sample Statistic (\overline{X} or P_s)	N	95% Confidence Interval
1				
2				
3				
4				
5				
6				
7				
8				

NOTE: Some ordinal-level variables, like *attend*, use scores to stand for broad categories (see Appendix G for examples), and sample means should be interpreted in terms of these categories, not as absolute scores. For example, if the mean of *attend* is 3.7, this would indicate that the average score for this variable is somewhere between "several times a year" (a score of 3) and "once a month" (a score of 4).

Step 5: Reporting Results

For each statistic, express the confidence interval in words, as if you were reporting results in a newspaper story. Be sure to identify the population and include each of the following: the sample statistic, sample size, the upper and lower limits of the confidence interval, and the confidence level (95%). A sample sentence might read, "I estimate that 45% ± 3% of all Americans support candidate X for president. This estimate is based on a sample of 1362 adult Americans, and I am 95% confident that it is correct."

Rawpixel.com/Shutterstock.com

8 Hypothesis Testing I: The One-Sample Case

LEARNING OBJECTIVES By the end of this chapter, you will be able to:

1. Explain the logic of hypothesis testing, including the concepts of the null hypothesis, the sampling distribution, the alpha level, and the test statistic.
2. Explain what it means to "reject the null hypothesis" or "fail to reject the null hypothesis."
3. Identify and cite examples of situations in which one-sample tests of hypotheses are appropriate.
4. Use the five-step model to test the significance of single-sample means and proportions and correctly interpret the results.
5. Explain the difference between one- and two-tailed tests, and specify when each is appropriate.
6. Use the Student's *t* distribution to conduct a hypothesis test on a small sample.

USING STATISTICS

The statistical techniques presented in this chapter can be used to answer research questions when we want to compare a sample to a population. Examples include:

1. Are people who complete a rehabilitation program for alcoholics more reliable workers than people in the community as a whole? **See page 200**
2. Are sociology majors more sophisticated than the rest of the student body? **See page 212**
3. Do commuter students have higher or lower GPAs than students as whole? **See page 217**
4. Do low-income neighborhoods differ from the city as a whole in terms of the proportion of female-headed households? **See page 222**

In Chapter 7, you were introduced to inferential statistical techniques for estimating population parameters from sample statistics. In Chapters 8 through 11, we investigate a second application of inferential statistics, called **hypothesis testing** or **significance testing**. In this chapter, the techniques for hypothesis testing in the one-sample case are introduced. We use these techniques to compare a random sample to a population. Note that we are not interested in the sample per se but in the larger group from which it was selected: We want to know whether the group represented by the sample differs from the population on a specific variable.

It would be better if we could test all members of the group in which we are interested rather than a smaller sample. However, as we have seen, researchers usually do not have the resources to test large groups and must use random samples instead.

For example, consider a situation in which a random sample of senior citizens showed a higher rate of victimization than the state as a whole. We might conclude that "senior citizens are significantly more likely to be victims of crime." The word *significantly* is a key word: It means that the difference between the sample's victimization rate and the population's rate is unlikely to be caused by random chance alone. In other words, it is likely that *all* senior citizens (not just those in the sample) have a higher victimization rate than the state as a whole.

Thus, we can use samples to represent larger groups (all senior citizens, for example) and compare and contrast the characteristics of the sample with those of the population and be extremely confident in our conclusions. Remember, however, that there is no guarantee that random samples will be representative, and there will always be a small amount of uncertainty in our conclusions. One of the great advantages of inferential statistics is that we will be able to estimate the probability of error and evaluate our decisions accordingly.

An Overview of Hypothesis Testing

We'll begin with an informal overview of hypothesis testing, using the first situation mentioned in "Using Statistics" as an example. We will introduce the more technical considerations and terminology throughout the remainder of the chapter.

Let's examine this situation in some detail. First, the main question here is, "Are people treated in this program more reliable workers than people in the community in general?" In other words, what the researcher would really like to do is compare *all* clients (the *population* of alcoholics treated in the program) with the entire metropolitan area. If she had information for these groups, she could answer the question easily, completely, and finally.

The problem is that the researcher does not have the resources to gather information on the thousands of people who have been treated by the program. Instead, she has drawn a random sample of 127 clients. The absentee rates for the sample and the community are:

Community	Sample of Treated Alcoholics
$\mu = 7.2$ days per year	$\bar{X} = 6.8$ days per year
$\sigma = 1.43$	$N = 127$

We can see that the average rate of absenteeism for the sample is lower than the rate for the community. Although it's tempting, we can't make any conclusions yet because we are working with a random sample ($N = 127$), not the population itself (*all* people treated in the program).

Figure 8.1 should clarify these relationships. The community is symbolized by the largest circle because it is the largest group. The population of all treated alcoholics is also symbolized by a large circle because it is a sizeable group,

FIGURE 8.1 A Test of Hypothesis for Single-Sample Means

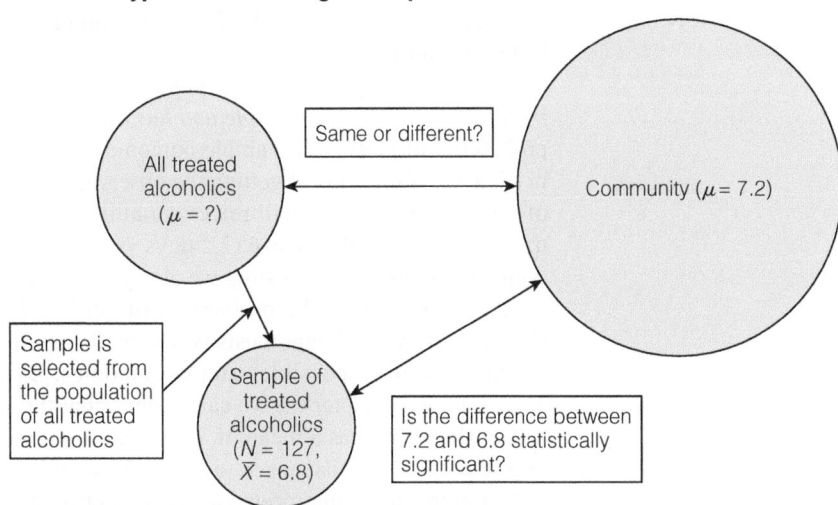

although only a fraction of the community as a whole. The random sample of 127, the smallest of the three groups, is symbolized by the smallest circle.

The labels on the arrows connecting the circles summarize the major question in this research situation: "Does the population of *all* treated alcoholics have different absentee rates than the community as a whole?" The population of treated alcoholics, too large to test, is represented by the random sample of 127.

We are interested in what caused the difference between the sample mean of 6.8 and the community mean of 7.2. There are two possible explanations for this difference:

1. The difference between the mean for the entire community (7.2) and the sample mean (6.8) is "statistically significant," which means that it is unlikely to have occurred by random chance alone. If this explanation is true, the population of *all* treated alcoholics is different from the community and the sample did *not* come from a population with a mean absentee rate of 7.2 days.

2. The second explanation is called the **null hypothesis** (symbolized as H_0, or *H*-sub-zero). It states that the difference between the sample and community is not important and was caused by mere random chance. If the null hypothesis is true, treated alcoholics are just like everyone else and have a mean absentee rate of 7.2 days.

Which explanation is correct? We cannot answer this question with absolute certainty as long as we are working with a random sample. We can, however, set up a decision-making procedure that will allow us to choose one of the explanations with a great deal of confidence. This decision-making process begins with the assumption that the second explanation—the null hypothesis—is correct and that the mean absentee rate for all treated alcoholics is the same as the rate for the community as a whole:

$$H_0: \mu = 7.2 \text{ days per year}$$

Remember that this μ is the one in Figure 8.1 that refers to the mean for *all* treated alcoholics, not just the 127 in the sample. This assumption, $\mu = 7.2$, can be tested statistically.

If the null hypothesis ("*The population of treated alcoholics is not different from the community as a whole and has a μ of 7.2*") is true, then we can find the probability of getting the sample outcome ($\overline{X} = 6.8$). Let us add a decision rule in advance. If the odds of getting the observed difference are less than 0.05 (5 out of 100, or 1 in 20), we will reject the null hypothesis. If this explanation were true, a difference of this size (7.2 days vs. 6.8 days) would be an extremely rare event, and in hypothesis testing we always bet *against* rare events.

We can estimate the probability of getting the observed sample outcome ($\overline{X} = 6.8$) if the null hypothesis is true by using our knowledge of the sampling distribution of all possible sample outcomes. By applying the Central Limit Theorem (see Chapter 6), we can assume that the sampling distribution is normal in shape and has a mean of 7.2 (because $\mu_{\overline{X}} = \mu$) and a standard deviation of $1.43/\sqrt{127}$ because $\sigma_{\overline{X}} = \sigma/\sqrt{N}$. We also know that the standard normal distribution can be interpreted as a distribution of probabilities (see Chapter 5) and that this sample outcome ($\overline{X} = 6.8$) is one of thousands of possible sample outcomes. The sampling distribution, with the sample outcome noted, is depicted in Figure 8.2.

We can add further useful information to this sampling distribution of sample means. We can depict our decision rule ("Any sample outcome with probability less than 0.05 will cause us to reject the null hypothesis") with Z scores. The probability of 0.05 can be translated into an area and divided equally into the upper and lower tails of the sampling distribution. Using Appendix A, we find that the Z-score equivalent of this area is ± 1.96. (To review finding Z scores from areas or probabilities, see Chapter 7.) The areas and Z scores are depicted in Figure 8.3.

The decision rule can now be rephrased. Any sample outcome falling in the shaded areas depicted in Figure 8.3 has a probability of less than 0.05. Such an outcome would be rare and would cause us to reject the null hypothesis.

All that remains is to translate our sample outcome into a Z score to see where it falls on the curve. To do this, we use the standard formula for locating

FIGURE 8.2 The Sampling Distribution of All Possible Sample Means

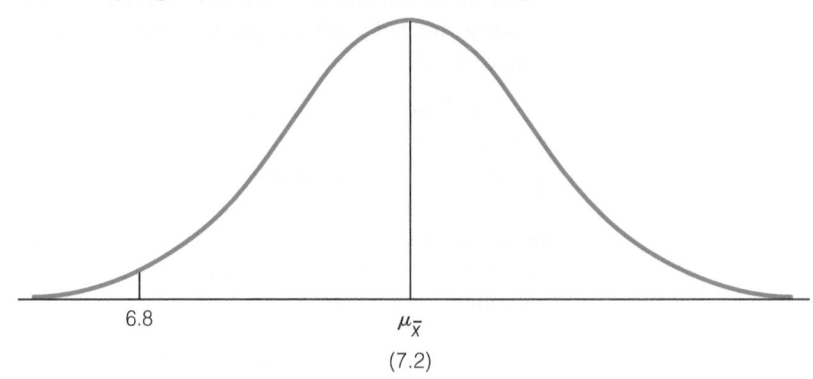

FIGURE 8.3 The Sampling Distribution of All Possible Sample Means

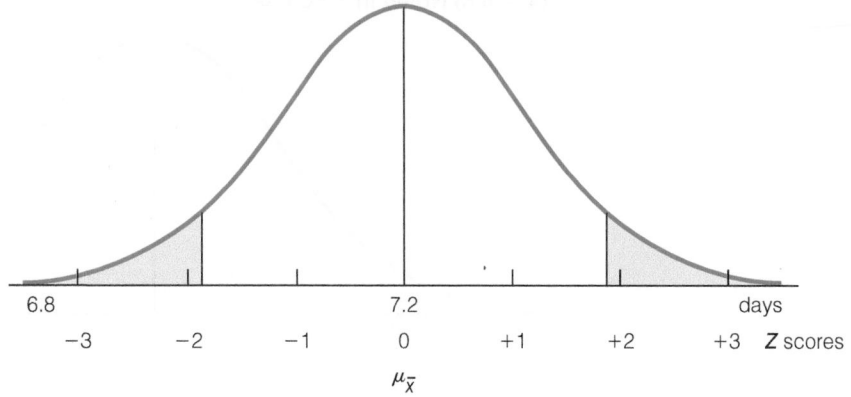

any score under a normal distribution. When we use known or empirical distributions, this formula is expressed as

$$Z = \frac{X_i - \overline{X}}{s}$$

Or, to find the equivalent Z score for any raw score, subtract the mean of the distribution from the raw score and divide by the standard deviation of the distribution.

Because we are now concerned with the sampling distribution of all sample means rather than an empirical distribution, the symbols in the formula will change, but the form remains exactly the same:

FORMULA 8.1

$$Z = \frac{\overline{X} - \mu}{\sigma/\sqrt{N}}$$

Or, to find the equivalent Z score for any sample mean, subtract the mean of the sampling distribution (which is equal to the population mean, or μ) from the sample mean and divide by the standard deviation of the sampling distribution.

Recalling the data given on this problem, we can now find the Z-score equivalent of the sample mean:

$$Z = \frac{\overline{X} - \mu}{\sigma/\sqrt{N}}$$

$$Z = \frac{6.8 - 7.2}{1.43/\sqrt{127}}$$

$$Z = \frac{-0.40}{1.43/11.27}$$

$$Z = \frac{-0.40}{0.127}$$

$$Z = -3.15$$

FIGURE 8.4 The Sampling Distribution of Sample Means with the Sample Outcome ($\overline{X} = 6.8$) Noted in Z Scores

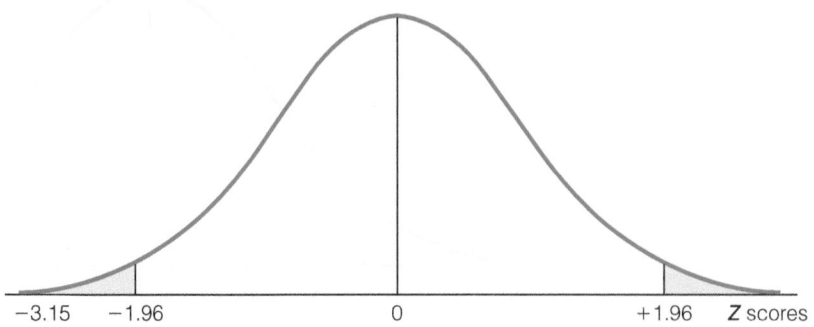

In Figure 8.4, this Z score of -3.15 is noted on the distribution of all possible sample means, and we see that the sample outcome does fall in the shaded area. If the null hypothesis is true, this sample outcome has a probability of less than 0.05. The sample outcome ($\overline{X} = 6.8$, or $Z = -3.15$) would be rare if the null hypothesis were true, and the researcher may therefore reject the null hypothesis. The sample of 127 treated alcoholics comes from a population that is significantly different from the community on absenteeism. Or, to put it another way, the sample does not come from a population that has a mean of 7.2 days of absences.

Keep in mind that significance testing is based on random samples. On rare occasions, an EPSEM sample may not be representative of the population from which it was selected. The decision-making process just outlined has a high probability of resulting in correct decisions, but we always face an element of risk. The decision to reject the null hypothesis might be incorrect if our sample happens to be one of the few that is unrepresentative of the population of alcoholics treated in this program.

One important strength of hypothesis testing is that we can estimate the probability of making an incorrect decision. In the example at hand, the null hypothesis was rejected and the probability that this decision was incorrect is 0.05—the decision rule established at the beginning of the process. To say that the probability of rejecting the null hypothesis incorrectly is 0.05 means that, if we repeated this same test an infinite number of times, we would incorrectly reject the null hypothesis only 5 times out of every 100.

STATISTICS In Everyday Life

Testing Drugs

The federal Food and Drug Administration (FDA) estimates that it takes around 10 years for a new drug to be approved for use in the United States. Before the manufacturer of a drug can apply for FDA approval, the drug must go through a rigorous testing process, which may include testing on animals, healthy volunteer humans, and samples of humans who have been diagnosed with the specific disease or condition that the manufacturer claims the drug was designed to treat. The results of these studies must show that the drug is safe and effective and that its benefits outweigh its risks. All the tests are based on null hypotheses: that the drug is ineffective in the treatment of the target disease or condition.

The Five-Step Model for Hypothesis Testing

The previous section introduced the concepts and procedures used in hypothesis testing. Next, we will examine a **five-step model** for organizing all tests of hypothesis.

Step 1. Making assumptions and meeting test requirements

Step 2. Stating the null hypothesis

Step 3. Selecting the sampling distribution and establishing the critical region

Step 4. Computing the test statistic

Step 5. Making a decision and interpreting the results of the test

We now look at each step individually, using the problem from the previous section as an example.

Step 1. Making Assumptions and Meeting Test Requirements. Any application of statistics is based on a set of assumptions. For hypothesis tests with a sample mean, three assumptions have to be satisfied:

- The sample has been selected according to the rules of EPSEM.
- The variable being tested is interval-ratio in level of measurement.
- The sampling distribution of all possible sample means is normal in shape. This will allow us to use the standardized normal distribution to find areas under the sampling distribution. We can be sure that this assumption is satisfied by using large samples (see the Central Limit Theorem in Chapter 6).

 Usually, we will state these assumptions in abbreviated form as a mathematical model for the test. For example:

> Model: Random sampling
> Level of measurement is interval-ratio
> Sampling distribution is normal

Step 2. Stating the Null Hypothesis. The null hypothesis is always a statement of "no difference," but its exact form will vary depending on the test being conducted. In the single-sample case, the null hypothesis states that the sample comes from a population with a certain characteristic. In our example, the null hypothesis is that the population of treated alcoholics is "no different" from the community as a whole, that their average days of absenteeism is also 7.2, and that the difference between 7.2 and the sample mean of 6.8 is caused by random chance. As we saw previously, the null hypothesis is stated as

$$H_0: \mu = 7.2$$

where μ refers to the mean of the population of treated alcoholics. The null hypothesis is the central element in any test of hypothesis because the entire process is aimed at rejecting or failing to reject H_0.

Usually, the researcher believes there is a difference between the sample and the population and wants to reject the null hypothesis. The researcher's belief is stated in a **research hypothesis (H_1)**, a statement that directly contradicts the null hypothesis. Often, the researcher's goal is to support the research hypothesis by rejecting the null hypothesis.

The research hypothesis can be stated in several ways. One form would state that the population from which the sample was selected did *not* have a certain characteristic or, in terms of our example, had a mean that was *not* equal to a specific value:

$$(H_1: \mu \neq 7.2)$$

where \neq means "not equal to"

The research hypothesis asserts that the population of all treated alcoholics is *different from* the community as a whole. The null hypothesis, in contrast, asserts that the population of all treated alcoholics is *the same as* the community as a whole. One (and only one) of these two statements must be true.

The research hypothesis is enclosed in parentheses to emphasize that it has no formal standing in the hypothesis-testing process (except, as we shall see in the next section, in choosing between one-tailed and two-tailed tests). It serves as a reminder of what the researcher believes to be the truth.

Step 3. Selecting the Sampling Distribution and Establishing the Critical Region.

The sampling distribution is the probabilistic yardstick against which a particular sample outcome is measured. By assuming that the null hypothesis is true, we can attach values to the mean and standard deviation of the sampling distribution and measure the probability of any specific sample outcome. We will cover several different sampling distributions in this text, but for now we will confine our attention to the sampling distribution described by the standard normal curve, as summarized in Appendix A.

The **critical region**, or **region of rejection**, consists of the areas under the sampling distribution that include unlikely sample outcomes. In our earlier example, this area began at a Z score of ± 1.96, called **Z(critical)**, that was graphically displayed in Figure 8.3. The shaded area is the critical region. Any sample outcome for which the Z-score equivalent fell in this area (that is, below -1.96 or above $+1.96$) would have caused us to reject the null hypothesis.

By convention, the size of the critical region is reported as alpha (α), the proportion of all of the area included in the critical region. In our example, our **alpha level** was 0.05. Other commonly used alphas are 0.10, 0.01, 0.001, and 0.0001.

These decisions can be stated in abbreviated form. The critical region is noted by the Z scores that mark its beginnings.

$$\text{Sampling distribution} = Z \text{ distribution}$$
$$\alpha = 0.05$$
$$Z(\text{critical}) = \pm 1.96$$

(For practice in finding Z(critical) scores, see problem 8.1a.)

Step 4. Computing the Test Statistic.

To evaluate the probability of the sample outcome, we must convert the sample value into a Z score. Finding this Z score is

> **ONE STEP AT A TIME** Completing Step 4 of the Five-Step Model: Computing Z(obtained)
>
> Use these procedures if the population standard deviation (σ) is known or sample size (N) is greater than 100. See the section titled "The Student's t Distribution" for procedures when σ is unknown and N is less than 100.
>
> ***To Compute the Test Statistic Using Formula 8.1***
>
> 1. Find the square root of N.
> 2. Divide the square root of N into the population standard deviation (σ).
> 3. Subtract the population mean (μ) from the sample mean (\overline{X}).
> 4. Divide the quantity you found in step 3 by the quantity you found in step 2. This value is Z(obtained).

called computing the **test statistic**, and the resultant value will be referred to as **Z(obtained)** to differentiate it from the score that marks the beginning of the critical region. In our example, we found a Z(obtained) of -3.15. *(For practice in computing obtained Z scores for means, see problems 8.1c, 8.2 to 8.7, and 8.15e and f.)*

Step 5. Making a Decision and Interpreting the Results of the Test. If the test statistic falls in the critical region, our decision will be to reject the null hypothesis. If the test statistic does not fall in the critical region, we fail to reject the null hypothesis. In our example, the two values were

$$Z(\text{critical}) = \pm 1.96$$

$$Z(\text{obtained}) = -3.15$$

and we saw that the Z(obtained) fell in the critical region (see Figure 8.4). Our decision was to reject the null hypothesis, which stated: "Treated alcoholics have a mean absentee rate of 7.2 days." When we reject this null hypothesis, we are saying that treated alcoholics do *not* have a mean absentee rate of 7.2 days and that there *is* a difference between them and the community. We can also say that the difference between the sample mean of 6.8 and the community mean of 7.2 is statistically significant, or unlikely to be caused by random chance alone.

Note that, to complete step 5, you have to do two things. First, you reject or fail to reject the null hypothesis (see Table 8.1). Second, you need to say what this decision means. In this case, we rejected the null hypothesis and concluded that there is a significant difference between the mean of the sample and the mean for the entire community. Therefore, we concluded that treated alcoholics are different from the community as a whole.

TABLE 8.1 Making a Decision in Step 5 and Interpreting the Results of the Test

Situation	Decision	Interpretation
The test statistic is in the critical region	Reject the null hypothesis (H_0)	The difference is statistically significant
The test statistic is not in the critical region	Fail to reject the null hypothesis (H_0)	The difference is not statistically significant

ONE STEP AT A TIME Completing Step 5 of the Five-Step Model: Making a Decision and Interpreting Results

1. Compare the Z(obtained) to the Z(critical). If Z(obtained) is *in* the critical region, *reject* the null hypothesis. If Z(obtained) is *not in* the critical region, *fail to reject* the null hypothesis.
2. Interpret your decision in terms of the original question. For example, our conclusion for the example problem was "Treated alcoholics miss significantly fewer days of work than the community as a whole."

STATISTICS In Everyday Life

Hypothesis Testing and Gambling

Think of hypothesis testing as analogous to gambling. Imagine that you have been invited to participate in a game of chance involving coin flips: heads you win, tails you lose. You would agree to participate only if you could **assume** that the coin was honest and that the probabilities of heads and tails were both 0.5. What if your opponent flips 10 tails in a row? Think of these flips as a **test** of your original assumption. At some point, as the coin shows tail after tail, you must compare the outcomes with your original assumption of honesty and make a **decision**: Either the coin is weighted toward tails or you have just witnessed a very rare series of events.

If you conclude that the game is rigged, you have rejected the null hypothesis that there is no difference in the probabilities of heads and tails. This seems like a reasonable decision but note that there is a chance that your decision was wrong: It is possible (though unlikely) that the game is not rigged and the coin is not weighted toward tails.

In hypothesis testing, we also **make assumptions** (steps 1 through 3), **test** these assumptions in step 4, and make a **decision**, based on probabilities, in step 5.

This five-step model will serve as a framework for decision making throughout the hypothesis-testing chapters. The exact nature and method of expression for our decisions will be different for different situations. However, the five-step model will provide a common frame of reference for all significance testing.

Choosing a One-Tailed or Two-Tailed Test

The five-step model for hypothesis testing allows for few choices. Nonetheless, the researcher must still deal with two options. First, he or she must decide between a one-tailed and a two-tailed test. Second, an alpha level must be selected. In this section, we discuss the former decision; we discuss the latter in the next section.

TYPES OF SIGNIFICANCE TESTS

The choice between a one- and two-tailed test is based on the researcher's expectations about the population from which the sample was selected. These expectations

are reflected in the research hypothesis (H_1), which is contradictory to the null hypothesis and usually states what the researcher believes to be "the truth." In most situations, the researcher will wish to support the research hypothesis by rejecting the null hypothesis.

The format for the research hypothesis may take either of two forms, depending on the relationship between what the null hypothesis states and what the researcher believes to be the truth. In the one-sample tests considered in this chapter, the null hypothesis states that the population has a specific characteristic. In the example that we have used in this chapter, the null hypothesis stated, in symbols, "All treated alcoholics have the *same* absentee rate (7.2 days) as the community." The researcher might believe that the population of treated alcoholics actually has *less* absenteeism (their population mean is *lower than* the value stated in the null hypothesis) or *more* absenteeism (their population mean is *greater than* the value stated in the null hypothesis), or he or she might be unsure about the direction of the difference.

If the researcher is unsure about the direction, the research hypothesis would state only that the population mean is "not equal" to the value stated in the null hypothesis. The research hypothesis stated in our example ($\mu \neq 7.2$) was in this format. This is called a **two-tailed test** of significance because the researcher will be equally concerned with the possibility that the true population value is greater than *or* less than the value specified in the null hypothesis.

In other situations, the researcher might be concerned only with differences in a specific direction. If the direction of the difference can be predicted, or if the researcher is concerned only with differences in one direction, a **one-tailed test** can be used.

A one-tailed test may take one of two forms. If the researcher believes that the true population value is greater than the value specified in the null hypothesis, the research hypothesis would use the ">" ("greater than") symbol. In our example, if we had predicted that treated alcoholics had *higher* absentee rates than the community (or averaged *more* than 7.2 days of absenteeism), our research hypothesis would have been

$$(H_1: \mu > 7.2)$$

where $>$ signifies "greater than"

If we predicted that treated alcoholics had *lower* absentee rates than the community (or averaged *fewer* than 7.2 days of absenteeism), our research hypothesis would have been

$$(H_1: \mu < 7.2)$$

where $<$ signifies "less than"

One-tailed tests are often appropriate when programs designed to solve a problem or improve a situation are being evaluated. If the program for treating alcoholics made them *less* reliable workers, for example, the program would be a failure, at least on that criterion. In this situation, the researcher may focus only on outcomes that would indicate that the program is a success (i.e., when treated

alcoholics have lower rates) and conduct a one-tailed test with a research hypothesis in the form $H_1: \mu < 7.2$.

Or consider the evaluation of a program designed to reduce unemployment. The evaluators would be concerned only with outcomes that show a decrease in the unemployment rate. If the rate shows no change or if unemployment increases, the program is a failure, and both of these outcomes might be considered equally negative by the researchers. Thus, the researchers could legitimately use a one-tailed test that stated that unemployment rates for graduates of the program would be less than ($<$) rates in the community.

ONE- VERSUS TWO-TAILED TESTS AND THE FIVE-STEP MODEL

In terms of the five-step model, the choice of a one-tailed or two-tailed test determines what we do with the critical region in step 3. In a two-tailed test, we split the critical region equally into the upper and lower tails of the sampling distribution. In a one-tailed test, we place the entire critical area in either the upper or lower tail. If we believe that the population characteristic is greater than the value stated in the null hypothesis (if H_1 includes the $>$ symbol), we place the entire critical region in the upper tail. If we believe that the characteristic is less than the value stated in the null hypothesis (if H_1 includes the $<$ symbol), the entire critical region goes in the lower tail.

For example, in a two-tailed test with alpha equal to 0.05, the critical region begins at Z(critical) = ± 1.96. In a one-tailed test at the same alpha level, the Z(critical) is $+1.65$ if the upper tail is specified and -1.65 if the lower tail is specified. Table 8.2 summarizes the procedures to follow in terms of the nature of the research hypothesis. The difference in placing the critical region is graphically summarized in Figure 8.5, and the critical Z scores for the most common alpha levels are given in Table 8.3 for both one- and two-tailed tests.

TABLE 8.2 One- vs. Two-Tailed Tests, $\alpha = 0.05$

If the Research Hypothesis (H_1) Uses	The test is	Concern is on	Z(critical) is
=	two-tailed	Both tails	± 1.96
>	one-tailed	Upper tail	$+ 1.65$
<	one-tailed	Lower tail	$- 1.65$

TABLE 8.3 Finding Critical Z Scores for One-Tailed Tests

Alpha	Two-Tailed Value	One-Tailed Value Upper Tail	One-Tailed Value Lower Tail
0.10	± 1.65	$+1.29$	-1.29
0.05	± 1.96	$+1.65$	-1.65
0.01	± 2.58	$+2.33$	-2.33
0.001	± 3.32	$+3.10$	-3.10
0.0001	± 3.90	$+3.70$	-3.70

FIGURE 8.5 Establishing the Critical Region, One-Tailed Tests Versus Two-Tailed Tests (alpha = 0.05)

A. The two-tailed test, Z(critical) $= \pm 1.96$

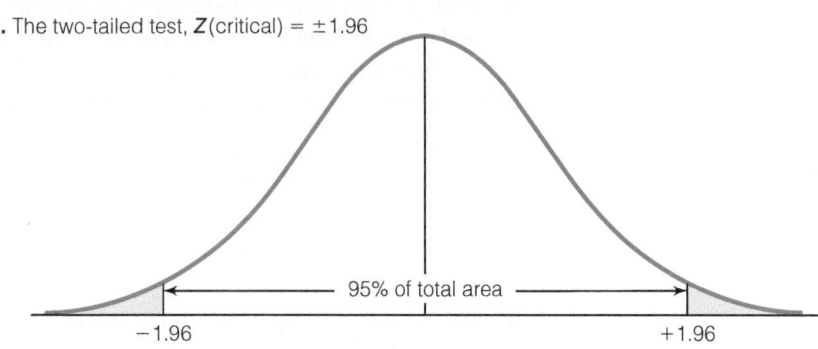

95% of total area

−1.96 +1.96

B. The one-tailed test for upper tail, Z(critical) $= +1.65$

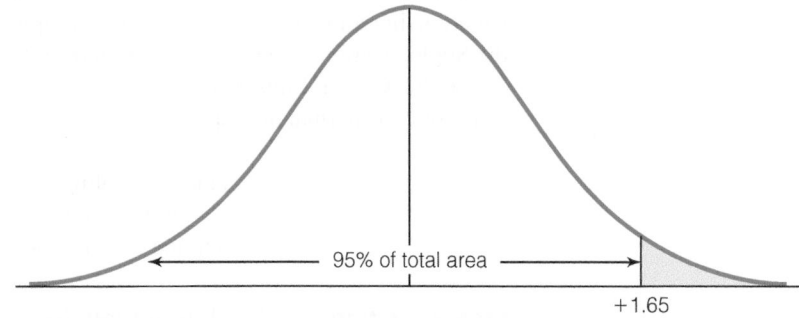

95% of total area

+1.65

C. The one-tailed test for lower tail, Z(critical) $= -1.65$

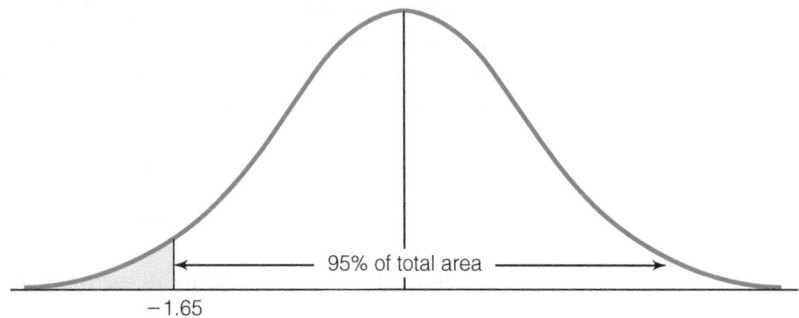

95% of total area

−1.65

Note that, for a given alpha level, the critical Z values for one-tailed tests are closer to the mean of the sampling distribution. Thus, a one-tailed test is more likely to reject H_0 without changing the alpha level (assuming that we have specified the correct tail). One-tailed tests are a way of statistically both having and eating your cake and should be used whenever (1) the direction of the difference can be confidently predicted or (2) the researcher is concerned only with differences in one tail of the sampling distribution. An example should clarify these procedures.

USING A ONE-TAILED TEST

A sociologist has noted that sociology majors seem more sophisticated, charming, and cosmopolitan than the rest of the student body. A "Sophistication Scale" test has been administered to the entire student body and to a random sample of 100 sociology majors, and these results have been obtained:

Student Body	Sociology Majors
$\mu = 17.3$	$\bar{X} = 19.2$
$\sigma = 7.4$	$N = 100$

We will use the five-step model to test the H_0 of no difference between sociology majors and the general student body.

Step 1. Making Assumptions and Meeting Test Requirements. Because we are using a mean to summarize the sample outcome, we must assume that the Sophistication Scale generates interval-ratio–level data. With a sample size of 100, the Central Limit Theorem applies, and we can assume that the sampling distribution is normal in shape.

> Model: Random sampling
> Level of measurement is interval-ratio
> Sampling distribution is normal

Step 2. Stating the Null Hypothesis (H_0). The null hypothesis states that there is no difference between sociology majors and the general student body. The research hypothesis (H_1) will also be stated at this point. The researcher has predicted a direction for the difference ("Sociology majors are *more* sophisticated"), so a one-tailed test is justified. The one-tailed research hypothesis asserts that sociology majors have a higher ($>$) score on the Sophistication Scale. The two hypotheses may be stated as

$$H_0: \mu = 17.3$$
$$(H_0: \mu > 17.3)$$

As before, the μ here represents all sociology majors.

Step 3. Selecting the Sampling Distribution and Establishing the Critical Region. We will use Appendix A to find areas under the sampling distribution. If alpha is set at 0.05, the critical region will begin at the Z score $+1.65$. That is, the researcher has predicted that sociology majors are *more* sophisticated and that this sample comes from a population that has a mean *greater than* 17.3, so he or she will be concerned only with sample outcomes in the upper tail of the sampling distribution. If sociology majors are *the same as* other students in terms of sophistication (if H_0 is true) or if they are *less* sophisticated (and come from a population with a mean less than 17.3), the theory is not supported. These decisions may be summarized as

FIGURE 8.6 Z(Obtained) Versus Z(Critical) (alpha = 0.05, one-tailed test)

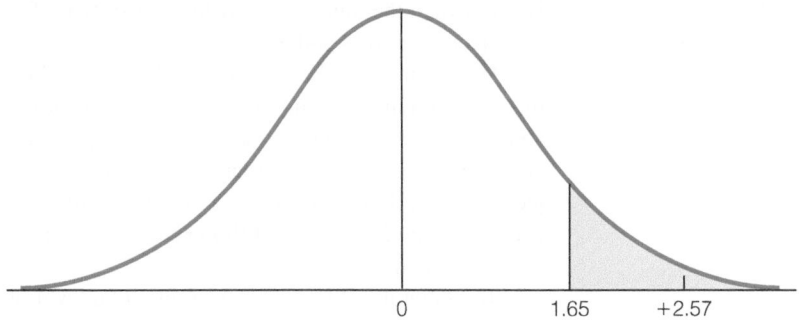

$$\text{Sampling distribution} = Z \text{ distribution}$$
$$\alpha = 0.05$$
$$Z(\text{critical}) = +1.65$$

Step 4. Computing the Test Statistic.

$$Z = \frac{\overline{X} - \mu}{\sigma/\sqrt{N}}$$
$$Z = \frac{19.2 - 17.3}{7.4/\sqrt{100}}$$
$$Z = \frac{1.90}{0.74}$$
$$Z = +2.57$$

Step 5. Making a Decision and Interpreting Test Results. In this step, we compare the Z(obtained) with the Z(critical):

$$Z(\text{critical}) = +1.65$$
$$Z(\text{obtained}) = +2.57$$

The test statistic falls in the critical region and this outcome is depicted in Figure 8.6. We will reject the null hypothesis because, if H_0 were true, a difference of this size would be highly unlikely. There is a significant difference between sociology majors and the general student body in terms of sophistication. Since the null hypothesis has been rejected, the research hypothesis (sociology majors are more sophisticated) is supported. *(For practice in dealing with tests of significance for means that may call for one-tailed tests, see problems 8.2b, 8.3, 8.6, 8.8, and 8.17.)*

Selecting an Alpha Level

In addition to deciding between one-tailed and two-tailed tests, the researcher must select an alpha level. We have seen that the alpha level plays a crucial role in hypothesis testing. When we assign a value to alpha, we define what we mean by

an "unlikely" sample outcome. If the probability of the observed sample outcome is lower than the alpha level (if the test statistic falls in the critical region), then we reject the null hypothesis.

How can we make reasonable decisions about the value of alpha? Recall that, in addition to defining what we mean by *unlikely*, the alpha level is the probability that the decision to reject the null hypothesis if the test statistic falls in the critical region will be incorrect. In hypothesis testing, the error of incorrectly rejecting the null hypothesis—or rejecting a null hypothesis that is actually true—is called **Type I error**, or **alpha error**. To minimize this type of error, use small values for alpha.

To elaborate: When we select an alpha level, we divide the sampling distribution into two sets of possible sample outcomes. One, the critical region, includes all unlikely or rare sample outcomes. The other—the remainder of the area—consists of all sample outcomes that are not rare. A lower alpha level reduces the size of the critical region and moves the critical region further from the mean of the sampling distribution.

We can illustrate this point with the alpha levels and Z(critical) scores in Table 8.4 (which were also presented in Table 7.2). Note that the critical region becomes smaller and moves farther away from the mean of the sampling distribution as alpha levels go down. Thus, lower alpha levels make it less likely that we will reject the null hypothesis and—because a Type I error can occur only if we reject the null hypothesis—less likely that we will commit a Type I error. To minimize the probability of rejecting a null hypothesis that is in fact true, use low alpha levels.

However, as the critical region decreases in size (as alpha levels decrease), the "noncritical" region becomes larger. This increases the chance that the sample outcome will not fall in the critical region and that our decision in step 5 will be to "fail to reject."

This raises the possibility of a second type of mistake, called **Type II error**, or **beta (β) error**: failing to reject a null hypothesis that is, in fact, false. As the alpha level decreases, the probability of a Type I error decreases, but the probability of a Type II error increases. The two types of error are inversely related, and we cannot minimize both in the same test. As the probability of one type of error decreases, the probability of the other increases, and vice versa.

Table 8.5 may clarify these relationships. The table lists the two decisions we can make in step 5 of the five-step model (reject or fail to reject the null hypothesis) along with the two possible conditions of the null hypothesis (it is either actually true or actually false). The table combines these possibilities into a total of four possible outcomes. The first two outcomes are desirable ("OK") but the other two indicate that an error has been made.

TABLE 8.4 The Relationship Between Alpha and Z(critical) for a Two-Tailed Test

If Alpha =	The Two-Tailed Critical Region Will Begin at Z(critical) =
0.10	±1.65
0.05	±1.96
0.01	±2.58
0.001	±3.32

TABLE 8.5 Decision Making and the Five-Step Model

	If our decision is to	And the H_0 is actually	The result is
a	Reject the H_0	False	OK
b	Fail to reject the H_0	True	OK
c	Reject the H_0	True	Type I or alpha (α) error
d	Fail to reject the H_0	False	Type II or beta (β) error

STATISTICS In Everyday Life

Alpha Levels

In social science research, the 0.05 alpha level has become the standard indicator of a significant difference. This alpha level means that we will incorrectly reject the null hypothesis only five times out of every 100 tests. These might seem like excellent odds but research that involves potentially harmful drugs would call for even lower alpha levels (0.001, 0.0001, or even lower) to minimize the possibility of making incorrect decisions and endangering health.

Let's consider the two desirable ("OK") outcomes first. Obviously, we always want to reject false null hypotheses and "fail to reject" those that are true. In other words, we always want to verify the truth and refute incorrect statements. The remaining two combinations are errors. If we reject a null hypothesis that is actually true, we are saying that a true statement is false. Likewise, if we fail to reject a null hypothesis that is actually false, we are saying that a false statement is true. We always want to reject false statements and accept the truth when we find it. Hypothesis testing always carries an element of risk and it is not possible to minimize the chances of both Type I and Type II errors simultaneously.

What all of this means, finally, is that you must think of selecting an alpha level as an attempt to balance the two types of error. Higher alpha levels will minimize the probability of Type II error (saying that false statements are true), and lower alpha levels will minimize the probability of Type I error (saying that true statements are false). Normally, in social science research, we will want to minimize Type I error, and lower alpha levels (0.05, 0.01, 0.001, or lower) will be used. The 0.05 level in particular has emerged as a generally recognized indicator of a significant result. However, the widespread use of the 0.05 level is simply a convention, and there is no reason that alpha cannot be set at virtually any sensible level (such as 0.04, 0.027, 0.083). The researcher has the responsibility of selecting the alpha level that seems most reasonable in terms of the goals of the research project.

The Student's *t* Distribution

To this point, we have considered only situations involving single-sample means where the value of the population standard deviation (σ) was known. Needless to say, in most research situations, the value of σ will be unknown. However, a value for σ is

FIGURE 8.7 The *t* Distribution and the *Z* Distribution

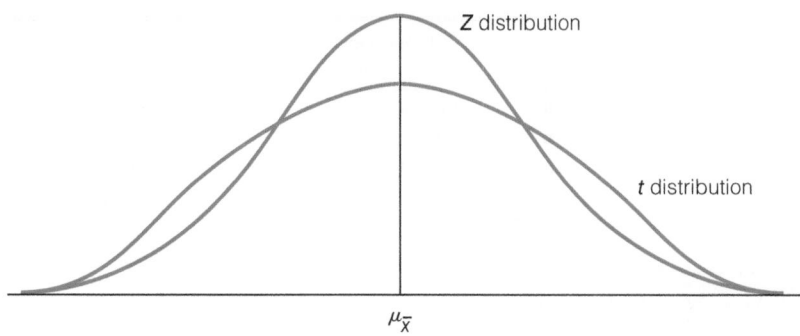

required to compute the standard error of the mean (σ/\sqrt{N}), convert our sample outcome into a *Z* score, and place the *Z*(obtained) on the sampling distribution (step 4). How can we reasonably obtain a value for the population standard deviation?

It might seem sensible to estimate σ with *s*, the sample standard deviation. As we noted in Chapter 7, *s* is a biased estimator of σ, but the degree of bias decreases as sample size increases. For large samples (that is, samples with 100 or more cases), the sample standard deviation yields an adequate estimate of σ. Thus, for large samples, we simply substitute *s* for σ in the formula for *Z*(obtained) in step 4 and continue to use the standard normal curve to find areas under the sampling distribution.[1]

For smaller samples, however, when σ is unknown, an alternative distribution called the **Student's *t* distribution** must be used to find areas under the sampling distribution and establish the critical region. The shape of the *t* distribution varies as with sample size. The relative shapes of the *t* and *Z* distributions are depicted in Figure 8.7. For small samples, the *t* distribution is much flatter than the *Z* distribution, but the *t* distribution comes to resemble the *Z* distribution more and more as sample size increases until the two are essentially identical for sample sizes greater than 120. As *N* increases, the sample standard deviation (*s*) becomes a better estimate of the population standard deviation (*s*), and the *t* distribution becomes more and more like the *Z* distribution.

THE DISTRIBUTION OF *t*: USING APPENDIX B

The *t* distribution is summarized in Appendix B. The *t* table differs from the *Z* table in three ways.

1. There is a column at the left of the table labeled *df* for "degrees of freedom."[2] Because the exact shape of the *t* distribution varies by sample size, the exact location of the critical region varies as well. Degrees of freedom, which are

[1]Even though the effect will be minor and will decrease with sample size, we will always correct for the bias in *s* by using $N - 1$ rather than *N* in the computation of the standard deviation of the sampling distribution when σ is unknown.

[2]"Degrees of freedom" refers to the number of values in a distribution that are free to vary. For a sample mean, a distribution has $N - 1$ degrees of freedom. This means that, for a specific value of the mean and of N, $N - 1$ scores are free to vary. For example, if the mean is 3 and $N = 5$, the distribution would have 4 degrees of freedom. When the values of four of the scores are known, the value of the fifth is fixed. If the four scores are 1, 2, 3, and 4, the fifth score must be 5 and no other value.

equal to $N - 1$ in the case of a single-sample mean, must be computed before the critical region for any alpha can be located.

2. Alpha levels are arrayed across the top of Appendix B in two rows, one row for one-tailed tests and one for two-tailed tests. To use the table, begin by locating the selected alpha level in the appropriate row.

3. The entries in the table are the actual scores, called **t(critical)**, that mark the beginnings of the critical regions and not areas under the sampling distribution.

To illustrate the use of this table with single-sample means, find the critical region for alpha equal to 0.05, two-tailed test, for $N = 30$. The degrees of freedom will be $N - 1$, or 29; reading down the proper column, you should find a value of 2.045. Thus, the critical region for this test will begin at t(critical) $= \pm 2.045$.

Notice that this t(critical) is larger in value than the comparable Z(critical), which for a two-tailed test at an alpha of 0.05 would be ± 1.96. This is because the t distribution is flatter than the Z distribution (see Figure 8.7). On the t distribution, the critical regions will begin farther away from the mean of the sampling distribution; therefore, the null hypothesis will be harder to reject. Furthermore, the smaller the sample size (the lower the degrees of freedom), the larger the value of **t(obtained)** necessary to reject the H_0.

Also note that the values of t(critical) decrease as degrees of freedom increase. For 1 degree of freedom, the t(critical) for an alpha of 0.05, two-tailed test is 12.706, but this score becomes smaller for larger samples. For degrees of freedom greater than 120, the value of t(critical) is the same as the comparable value of Z(critical), or ± 1.96. As sample size increases, the t distribution resembles the Z distribution more and more until, with sample sizes greater than 120, the two distributions are essentially identical.[3]

USING THE t DISTRIBUTION IN A TEST OF HYPOTHESIS

To demonstrate the uses of the t distribution in more detail, we will work through an example problem. Note that, in terms of the five-step model, the changes occur mostly in steps 3 and 4. In step 3, the sampling distribution will be the t distribution, and degrees of freedom (df) must be computed before locating the critical region or the t (critical) score. In step 4, a slightly different formula for computing the test statistic, t(obtained), will be used. As compared with the formula for Z(obtained), s will replace σ and $N - 1$ will replace N. Specifically,

FORMULA 8.2
$$t\text{(obtained)} = \frac{\overline{X} - \mu}{s/\sqrt{N - 1}}$$

A researcher wonders whether commuter students are different from the general student body in terms of academic achievement. She has gathered a random

[3]Appendix B abbreviates the t distribution by presenting a limited number of critical t scores for degrees of freedom between 31 and 120. When the exact df is not listed in the table, take the larger available score as t(critical). This is the most conservative course of action because it makes rejection of H_0 less likely.

sample of 30 commuter students and has learned from the registrar that the mean grade-point average for all students is 2.50 (μ = 2.50), but the standard deviation of the population (σ) has never been computed. Sample data are reported here. Is the sample from a population that has a mean of 2.50?

Student Body	Commuter Students
$\mu = 2.50\ (= \mu_{\bar{x}})$	$\bar{X} = 2.78$
$\sigma = ?$	$s = 1.23$
	$N = 30$

Step 1. Making Assumptions and Meeting Test Requirements.

> Model: Random sampling
> Level of measurement is interval-ratio
> Sampling distribution is normal

Step 2. Stating the Null Hypothesis.

$$H_0: \mu = 2.50$$

$$(H_0: \mu \neq 2.50)$$

You can see from the research hypothesis that the researcher has not predicted a direction for the difference. This will be a two-tailed test.

Step 3. Selecting the Sampling Distribution and Establishing the Critical Region. Since σ is unknown and the sample size is small, the t distribution will be used to find the critical region. Alpha will be set at 0.01.

> Sampling distribution = t distribution
> $\alpha = 0.01$, two-tailed test
> $df = (N - 1) = 29$
> $t(\text{critical}) = \pm 2.756$

Step 4. Computing the Test Statistic.

$$t(\text{obtained}) = \frac{\bar{X} - \mu}{s/\sqrt{N - 1}}$$

$$t(\text{obtained}) = \frac{2.78 - 2.50}{1.23/\sqrt{29}}$$

$$t(\text{obtained}) = \frac{0.28}{1.23/5.39}$$

$$t(\text{obtained}) = \frac{0.28}{0.23}$$

$$t(\text{obtained}) = 1.22$$

FIGURE 8.8 Sampling Distribution Showing *t*(Obtained) Versus *t*(Critical) ($\alpha = 0.05$, Two-Tailed Test, *df* = 29)

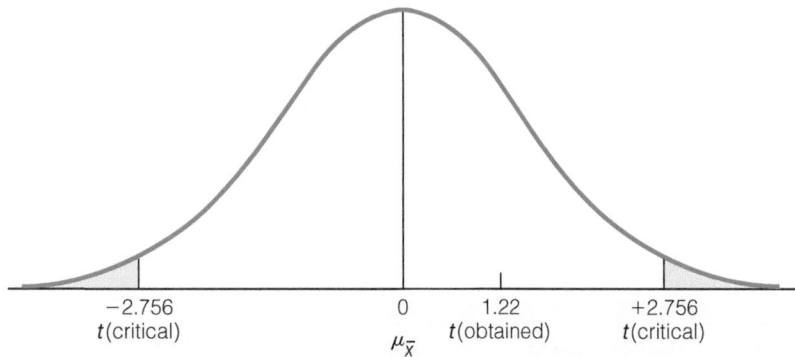

| −2.756 | 0 | 1.22 | +2.756 |
| *t*(critical) | $\mu_{\bar{X}}$ | *t*(obtained) | *t*(critical) |

Step 5. Making a Decision and Interpreting Test Results. The test statistic does not fall in the critical region. Therefore, the researcher fails to reject H_0. The difference between the sample mean (2.78) and the population mean (2.50) is not statistically significant. The difference is no greater than what would be expected if only random chance were operating. The test statistic and critical regions are displayed in Figure 8.8.

To summarize, when testing single-sample means, we must make a choice regarding the theoretical distribution we will use to establish the critical region. The choice is straightforward. If the population standard deviation (σ) is known or sample size is large, the Z distribution (summarized in Appendix A) will be used. If σ is unknown and the sample is small, the *t* distribution (summarized in Appendix B) will be used. These decisions are summarized in Table 8.6. *(For practice in using the t distribution in a test of hypothesis, see problems 8.8 to 8.10 and 8.17.)*

TABLE 8.6 Choosing a Sampling Distribution When Testing Single-Sample Means for Significance

If Population Standard Deviation (σ) is	Sampling Distribution is the
Known	Z distribution
Unknown and sample size (*N*) is large	Z distribution
Unknown and sample size (*N*) is small	*t* distribution

ONE STEP AT A TIME **Completing Step 4 of the Five-Step Model: Computing *t*(obtained)**

Follow these procedures when using the Student's *t* distribution.

To Compute the Test Statistic Using Formula 8.2

1. Find the square root of $N - 1$.
2. Divide the quantity you found in step 1 into the sample standard deviation (*s*).
3. Subtract the population mean (μ) from the sample mean (\bar{X}).
4. Divide the quantity you found in step 3 by the quantity you found in step 2. This value is *t*(obtained).

ONE STEP AT A TIME | **Completing Step 5 of the Five-Step Model: Making a Decision and Interpreting Results**

1. Compare the t(obtained) to the t(critical). If t(obtained) is *in* the critical region, *reject* the null hypothesis. If t(obtained) is *not in* the critical region, *fail to reject* the null hypothesis.
2. Interpret your decision in terms of the original question. For example, our conclusion for the example problem used in this section was "There is no significant difference between the average GPAs of commuter students and the general student body."

Social Research and Data Analysis 8.1 | **Testing a Sample Mean for Significance**

As a large business owner in Texas that aims for high productivity, one issue to consider is the average amount of time your employees spend getting to work. If their commute is longer than average for the state, you might be sacrificing valuable time by not letting them work from home more often. But what is the average commute for your workers? Is it different from the average in Texas? To find out, you gather a random sample of 152 of your employees and find that it takes them 28.6 minutes to get to work. Using American Fact Finder data from the U.S. Census Bureau, you can see that the average number of minutes it takes Texans to get to work is 26.1 minutes. Obviously, your workers have a different commute time, but we will need to use the five-step model hypothesis test to see if we can be 95% sure that it is really different. In this case, the sample size is large and the population standard deviation is unknown. We will use the Z distribution.

Step 1. Making Assumptions and Meeting Test Requirements.

> Model: Random sampling
> Level of measurement is interval-ratio
> Sampling distribution is normal

This is a large sample ($N > 100$) and length of commute in minutes is an interval-ratio variable, so we can conclude that the model assumptions are satisfied.

Step 2. Stating the Null Hypothesis (H_0). The null hypothesis states that the average commute time for your employees is the same as all of Texas. In symbols:

$$H_0: \mu = 26.1$$

Remember, the μ above refers to the population of your employees. Our research hypothesis is that it is significantly different from the overall population in Texas, but note that

it does not specify a direction; it only asks if your employees' commute time is "different from" (not higher or lower than) the statewide average. This suggests a two-tailed test:

$$(H_1: \mu \neq 26.1)$$

Step 3. Selecting the Sampling Distribution and Establishing the Critical Region. Since this is a large sample, we can use Appendix A to establish the critical region and state the critical scores as Z scores (as opposed to t scores).

> Sampling distribution = Z distribution
> $\alpha = 0.05$
> Z(critical) = ± 1.96

Step 4. Computing the Test Statistic. The necessary information for conducting a test of the null hypothesis is below.

$$\bar{X} = 28.6 \quad \mu = 26.1$$
$$s = 3.7$$
$$N = 152$$

The test statistic, Z(obtained), is

$$Z(\text{obtained}) = \frac{\bar{X} - \mu}{\sigma/\sqrt{N-1}}$$

$$Z(\text{obtained}) = \frac{28.6 - 26.1}{3.7/\sqrt{152-1}}$$

$$Z(\text{obtained}) = \frac{2.50}{3.7/\sqrt{151}}$$

$$Z(\text{obtained}) = \frac{2.50}{3.7/12.29}$$

$$Z(\text{obtained}) = \frac{2.50}{0.30}$$

$$Z(\text{obtained}) = 8.33$$

Note that, even though the sample size is large, we have used "$N - 1$" rather than "N" under the square root

(continued)

Social Research and Data Analysis 8.1	*(Continued)*

sign. We will do this routinely whenever σ is unknown, even when the effect on the value of Z(obtained) will be minimal.

Step 5. Making a Decision and Interpreting the Test Results. With alpha set at 0.05, the critical region begins at Z(critical) = ± 1.96. With an obtained Z score of 8.33, the null hypothesis is rejected. The difference between the commuting time for your employees and all of those in the state of Texas is statistically significant. The difference is so large that we may conclude that it did not occur by random chance. The decision to reject the null hypothesis has a 0.05 probability of being wrong.

Source: https://factfinder.census.gov/faces/nav/jsf/pages/community_facts.xhtml?src=bkmk

Tests of Hypothesis for Single-Sample Proportions (Large Samples)

In many cases, we work with variables that are not interval-ratio in level of measurement. One alternative in this situation would be to use a sample proportion (P_s), rather than a sample mean, as the test statistic. As we shall see, the overall procedures for testing single-sample proportions are the same as those for testing means. The central question is still "Does the population from which the sample was drawn have a certain characteristic?" We still conduct the test based on the assumption that the null hypothesis is true, and we still evaluate the probability of the obtained sample outcome against a sampling distribution of all possible sample outcomes. Our decision at the end of the test is also the same. If the obtained test statistic falls in the critical region (and thus is unlikely, given the assumption that H_0 is true), we reject H_0.

There are also some important differences in significance tests for sample proportions. These differences are best related in terms of the five-step model:

1. In step 1, we assume that the variable is measured at the nominal level.

2. In step 2, the symbols used to state the null hypothesis are different even though it is still a statement of "no difference."

3. In step 3, we will use only the Z distribution to find areas under the sampling distribution and locate the critical region. This will be appropriate as long as sample size is large. We will not consider small-sample tests of hypothesis for proportions in this text.

4. In step 4, computing the test statistic, the form of the formula remains the same. That is, the test statistic, Z(obtained), equals the sample statistic minus the mean of the sampling distribution, divided by the standard deviation of the sampling distribution. However, the symbols change because we are basing the tests on sample proportions. The formula can be stated as

FORMULA 8.3

$$Z(\text{obtained}) = \frac{P_s - P_u}{\sqrt{P_u(1 - P_u)/N}}$$

5. Step 5, making a decision, is exactly the same as before. If the test statistic, Z(obtained), falls in the critical region, reject H_0.

A TEST OF HYPOTHESIS USING SAMPLE PROPORTIONS

An example should clarify these procedures. A random sample of 122 households in a low-income neighborhood revealed that 53 (or a proportion of 0.43) of the households were headed by females. In the city as a whole, the proportion of female-headed households is 0.39. Are households in the lower-income neighborhood significantly different from the city as a whole on this characteristic?

Step 1. Making Assumptions and Meeting Test Requirements.

> Model: Random sampling
> Level of measurement is nominal
> Sampling distribution is normal in shape

Step 2. Stating the Null Hypothesis. The research question asks whether the sample proportion is *different from* the population proportion. A two-tailed test will be used because a direction for the difference has not been predicted.

$$H_0: P_u = 0.39$$
$$(H_1: P_u \neq 0.39)$$

Step 3. Selecting the Sampling Distribution and Establishing the Critical Region.

> Sampling distribution = Z distribution
> $\alpha = 0.10$, two-tailed
> $Z(\text{critical}) = \pm 1.65$

Step 4. Computing the Test Statistic.

$$Z(\text{obtained}) = \frac{P_s - P_u}{\sqrt{P_u(1 - P_u)/N}}$$

$$Z(\text{obtained}) = \frac{0.43 - 0.39}{\sqrt{(0.39)(0.61)/122}}$$

$$Z(\text{obtained}) = \frac{0.04}{\sqrt{0.24/122}}$$

$$Z(\text{obtained}) = \frac{0.04}{\sqrt{0.002}}$$

$$Z(\text{obtained}) = \frac{0.04}{0.05}$$

$$Z(\text{obtained}) = 0.80$$

Step 5. Making a Decision and Interpreting Test Results. The test statistic, $Z(\text{obtained})$, does not fall in the critical region. Therefore, we fail to reject H_0. There is no statistically significant difference between the low-income community

and the city as a whole in terms of the proportion of households headed by females. Figure 8.9 displays the sampling distribution, the critical region, and the Z(obtained). *(For practice in tests of significance using sample proportions, see problems 8.1c, 8.11 to 8.14, 8.15a to d, 8.16, and 8.19a and b.)*

FIGURE 8.9 Sampling Distribution Showing Z(Obtained) Versus Z(Critical) ($\alpha = 0.10$, Two-Tailed Test)

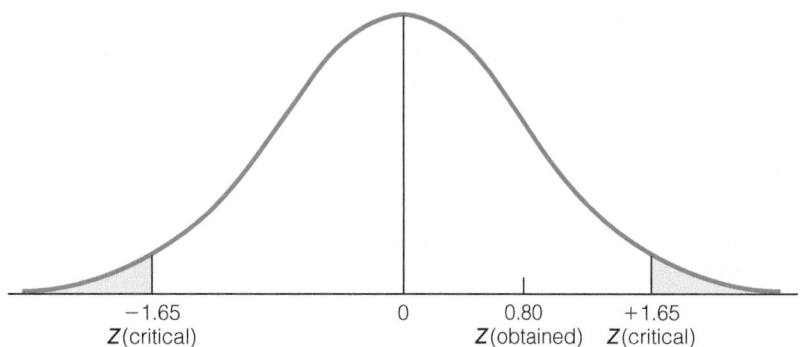

-1.65		0	0.80	$+1.65$
Z(critical)			Z(obtained)	Z(critical)

ONE STEP AT A TIME **Completing Step 4 of the Five-Step Model: Computing Z(obtained)**

To Compute the Test Statistic Using Formula 8.3

1. Start with the denominator of Formula 8.3 and substitute in the value for P_u. This value will be given in the statement of the problem.
2. Subtract the value of P_u from 1.
3. Multiply the value you found in step 2 by the value of P_u.
4. Divide the quantity you found in step 3 by N.
5. Take the square root of the value you found in step 4.
6. Subtract P_u from P_s.
7. Divide the quantity you found in step 6 by the quantity you found in step 5. This value is Z(obtained).

ONE STEP AT A TIME **Completing Step 5 of the Five-Step Model: Making a Decision and Interpreting Results**

1. Compare your Z(obtained) to your Z(critical). If Z(obtained) is *in* the critical region, *reject* the null hypothesis. If Z(obtained) is *not in* the critical region, *fail to reject* the null hypothesis.
2. Interpret the decision in terms of the original question. For example, our conclusion for the example problem used in this section was "There is no significant difference between the low-income community and the city as a whole in the proportion of households that are headed by females."

Social Research and Data Analysis 8.2 **Testing a Sample Proportion for Significance**

In the 2016 presidential election, 48.2% of the vote in Pennsylvania went to Donald Trump, giving him a narrow victory in the state and 20 crucial electoral votes. But of course, that doesn't mean that all communities in that state favored Trump. As a political sociologist specializing in race, you wonder if communities with a higher population of whites were more likely to vote for Trump. So you gather a random sample ($N = 103$) of neighborhoods in Pennsylvania with a higher proportion of whites than average and you find that 49% of the respondents in these neighborhoods voted for Trump. This is pretty close. Do you have enough evidence to conclude that residents of all neighborhoods with a higher than average proportion of whites were more likely to vote for Trump? We will need a sample proportion test of significance to figure this out.

Step 1. Making Assumptions and Meeting Test Requirements.

> Model: Random sampling
> > Level of measurement is nominal
> > Sampling distribution is normal

This is a large sample, so we may assume a normal sampling distribution.

Step 2. Stating the Null Hypothesis (H_0). The null hypothesis says that the communities with a higher proportion of whites are not different from the state as a whole.

$$H_0: P_u = 0.482$$

The original question ("Were the white neighborhoods *more* likely to vote for Trump?") suggests a one-tailed research hypothesis:

$$(H_1: P_u > 0.482)$$

Step 3. Selecting the Sampling Distribution and Establishing the Critical Region.

> Sampling distribution = Z distribution
> $$\alpha = 0.05$$
> $$Z(\text{critical}) = +1.65$$

The research hypothesis says that we will be concerned only with outcomes in which the white neighborhoods are *more* likely to vote for Trump or with sample outcomes in the upper tail of the sampling distribution.

Step 4. Computing the Test Statistic. The information necessary for a test of the null hypothesis, expressed in the form of proportions, is

Sample	State
$P_s = 0.49$	$P_u = 0.482$
$N = 103$	

The test statistic, Z(obtained), would be

$$Z(\text{obtained}) = \frac{P_s - P_u}{\sqrt{P_u(1 - P_u)/N}}$$
$$Z(\text{obtained}) = \frac{0.49 - 0.482}{\sqrt{(0.48)(0.52)/103}}$$
$$Z(\text{obtained}) = \frac{0.008}{\sqrt{0.25/103}}$$
$$Z(\text{obtained}) = \frac{0.008}{\sqrt{0.0024}}$$
$$Z(\text{obtained}) = \frac{0.008}{0.05}$$
$$Z(\text{obtained}) = 0.16$$

Step 5. Making a Decision and Interpreting Test Results. With alpha set at 0.05, one-tailed, the critical region begins at $Z(\text{critical}) = +1.65$. With an obtained Z score of 0.16, the null hypothesis is accepted. The difference between the white neighborhoods and the state as a whole is not statistically significant. Residents of the white neighborhoods were not significantly more likely to have voted for Trump in 2016.

Source: https://www.nytimes.com/elections/2016/results/pennsylvania

SUMMARY

1. All the basic concepts and techniques for testing hypotheses were presented in this chapter. We saw how to test the null hypothesis of "no difference" for single-sample means and proportions. In both cases, the central question is whether the population represented by the sample has a certain characteristic.

2. All significance tests involve finding the probability of the observed sample outcome, given that the null hypothesis is true. If the outcome has a low probability, we reject the null hypothesis. In the usual research situation, we will wish to reject the null hypothesis and thereby support the research hypothesis.

3. The five-step model will be our framework for decision making throughout the hypothesis-testing chapters. However, what we do during each step will vary, depending on the specific test being conducted.

4. If we can predict a direction for the difference in stating the research hypothesis, a one-tailed test is called for. If no direction can be predicted, a two-tailed test is appropriate.

5. There are two kinds of errors in hypothesis testing. Type I, or alpha, error is rejecting a true null hypothesis; Type II, or beta, error is failing to reject a false null hypothesis. The probabilities of committing these two types of error are inversely related and cannot be simultaneously minimized in the same test. In selecting an alpha level, we try to balance the probability of these two errors.

6. When testing sample means, we use the t distribution to find the critical region when the population standard deviation is unknown and sample size is small.

7. Tests of significance based on sample proportions assume a nominal level of measurement, use different symbols to state the null hypothesis, and use a different formula (8.3) to compute Z(obtained).

8. If you are still confused about the uses of inferential statistics described in this chapter, don't be alarmed or discouraged. A sizeable volume of complex material has been presented, and only rarely will someone fully comprehend the unique logic of hypothesis testing on first exposure. After all, it is not every day that you learn how to test a statement you don't believe (the null hypothesis) against a distribution that doesn't exist (the sampling distribution)!

SUMMARY OF FORMULAS

Single-sample means, large samples:

8.1 $$Z = \frac{\overline{X} - \mu}{\sigma/\sqrt{N}}$$

Single-sample means when samples are small and population standard deviation is unknown:

8.2 $$t\text{(obtained)} = \frac{\overline{X} - \mu}{s/\sqrt{N - 1}}$$

Single-sample proportions, large samples:

8.3 $$Z\text{(obtained)} = \frac{P_s - P_u}{\sqrt{P_u(1 - P_u)/N}}$$

GLOSSARY

Alpha level (α). The proportion of area under the sampling distribution that contains unlikely sample outcomes, given that the null hypothesis is true. Also, the probability of Type I error.

Critical region (region of rejection). The area under the sampling distribution that, in advance of the test itself, is defined as including unlikely sample outcomes, given that the null hypothesis is true.

Five-step model. A step-by-step guideline for conducting tests of hypotheses. A framework that organizes decisions and computations for all tests of significance.

Hypothesis testing. Statistical tests that estimate the probability of sample outcomes if assumptions about the population (the null hypothesis) are true.

Null hypothesis (H_0). A statement of "no difference." In the context of single-sample tests of significance, the null hypothesis states that the population from which the sample was drawn has a certain characteristic or value.

One-tailed test. A type of hypothesis test used when (1) the direction of the difference can be predicted or (2) concern focuses on outcomes in only one tail of the sampling distribution.

Research hypothesis (H_1). A statement that contradicts the null hypothesis. In the context of single-sample tests of significance, the research hypothesis says that the population from which the sample was drawn does not have a certain characteristic or value.

Significance testing. See **Hypothesis testing**.

Student's t distribution. A distribution used to find the critical region for tests of sample means when s is unknown and sample size is small.

t(critical). The t score that marks the beginning of the critical region of a t distribution.

t(obtained). The test statistic computed in step 4 of the five-step model. The sample outcome expressed as a t score.

Test statistic. The value computed in step 4 of the five-step model that converts the sample outcome into either a t score or a Z score.

Two-tailed test. A type of hypothesis test used when (1) the direction of the difference cannot be predicted or (2) concern focuses on outcomes in both tails of the sampling distribution.

Type I error (alpha error). The probability of rejecting a null hypothesis that is, in fact, true.

Type II error (beta error). The probability of failing to reject a null hypothesis that is, in fact, false.

Z(critical). The Z score that marks the beginnings of the critical region on a Z distribution.

Z(obtained). The test statistic computed in step 4 of the five-step model. The sample outcomes expressed as a Z score.

PROBLEMS

8.1 a. For each situation, find Z(critical).

Alpha (α)	Form of the test	Z(critical)
0.05	One-tailed	
0.10	Two-tailed	
0.06	Two-tailed	
0.01	One-tailed	
0.02	Two-tailed	

b. For each situation, find the critical t score.

Alpha (α)	Form of the test	N	t(critical)
0.10	Two-tailed	31	
0.02	Two-tailed	24	
0.01	Two-tailed	121	
0.01	One-tailed	31	
0.05	One-tailed	61	

c. Compute the appropriate test statistic (Z or t) for each situation.

	Population	Sample	Z(obtained) or t(obtained)
1	$\mu = 2.40$ $\sigma = 0.75$	$\bar{X} = 2.20$ $N = 200$	
2	$\mu = 17.1$	$\bar{X} = 16.8$ $s = 0.9$ $N = 45$	
3	$\mu = 10.2$	$\bar{X} = 9.4$ $s = 1.7$ $N = 150$	
4	$P_u = 0.57$	$P_s = 0.60$ $N = 117$	
5	$P_u = 0.32$	$P_s = 0.30$ $N = 322$	

8.2 a. SP In 2016, the average SNAP benefits (formerly known as Food Stamps) per individual was $121.71. As a social policy researcher interested in welfare benefits, you are wondering if Asians on SNAP benefits in Arizona are different from the national average. So you have gathered a random sample of 150 Asians in Arizona on SNAP benefits and you find that they average $116.75 per month, with a standard deviation of $18.00. Are Asians in Arizona different from all SNAP beneficiaries? Determine this using the 95% confidence level. (*HINT: The wording of the research question suggests a two-tailed test. This means that the alternative, or research, hypothesis in step 2 will be stated as H_1: $\mu \neq 3.3$ and that the critical region will be split between the upper and lower tails of the sampling distribution. See Table 8.3 for values of Z(critical) for various alpha levels.*)

b. What if the research question were changed to "Do Asians collect *less than* all beneficiaries in Arizona?" How would the test conducted in problem 8.2a change? (*HINT: This wording implies a one-tailed test of significance. How would the research hypothesis change? For the alpha you used in problem 8.2a, what would the value of Z(critical) be?*)

8.3 a. SW Statewide, social workers average 10.2 years of experience. In a random sample, 203 social workers in greater metropolitan Shinbone average only 8.7 years, with a standard deviation of 0.52. Are social workers in Shinbone significantly less experienced? (*NOTE: The wording of the research hypothesis may justify a one-tailed test of significance. For a one-tailed test, what form would the research hypothesis take, and where would the critical region begin?*)

b. The same sample of social workers reports an average annual salary of $25,782, with a standard deviation of $622. Is this figure significantly higher than the statewide average of $24,509? (*NOTE: The wording of the research hypothesis suggests a one-tailed test. What form would the research hypothesis take, and where would the critical region begin?*)

8.4 MR According to a recent article in *USA Today*, Americans spend $484 per year on alcoholic beverages. As a market research firm for a local wine distributor, you have accessed data on a random sample of Boston residents to see if their spending is higher than average, and perhaps a good area for your client to make a local investment. The average spending in the sample is $404, with a standard deviation of $102. Does this represent a significant difference?

8.5 SOC A random sample of 423 Chinese Americans drawn from the population of a particular state has

finished an average of 12.7 years of formal education, with a standard deviation of 1.7. Is this significantly different from the statewide average of 12.2 years?

8.6 SP According to the General Social Survey, the average age when a person's first child was born is 24.3. To determine how different urban working-class people are from this norm, you have gathered a random sample of 105 urban working-class people and found that their average age when their first child was born is 22, with a standard deviation of 4.6. Are working-class people in urban areas different than Americans in general? Conduct both one- and two-tailed tests.

8.7 **a.** SOC Statewide, the population as a whole watches 6.2 hours of TV per day. A random sample of 1017 senior citizens in the state report watching an average of 5.9 hours per day, with a standard deviation of 0.7. Is the difference significant?

 b. The same sample of senior citizens reports that they belong to an average of 2.1 voluntary organizations and clubs, with a standard deviation of 0.5. Statewide, the average is 1.7. Is the difference significant?

8.8 SOC A school system has assigned several hundred "chronic and severe underachievers" to an alternative educational experience. To assess the program, a random sample of 35 has been selected for comparison with all students in the system.

 a. In terms of GPA, did the program work?

Systemwide GPA	Program GPA
$\mu = 2.47$	$\bar{X} = 2.55$ $s = 0.70$ $N = 35$

 b. In terms of absenteeism (number of days missed per year), what can be said about the success of the program?

Systemwide	Program
$\mu = 6.137$	$\bar{X} = 4.78$ $s = 1.11$ $N = 35$

 c. In terms of standardized test scores in math and reading, was the program successful?

Math Test—Systemwide	Math Test—Program
$\mu = 103$	$\bar{X} = 106$ $s = 2.0$ $N = 35$

Reading Test—Systemwide	Reading Test—Program
$\mu = 110$	$\bar{X} = 113$ $s = 2.0$ $N = 35$

(*HINT: Note the wording of the research questions. Is a one-tailed test justified? Is the program a success if the students in the program are no different from students systemwide? What if the program students were performing at lower levels? If a one-tailed test is used, what form should the research hypothesis take? Where will the critical region begin?*)

8.9 SOC A random sample of 26 local sociology graduates scored an average of 458 on the GRE advanced sociology test, with a standard deviation of 20. Is this significantly different from the national average ($\mu = 440$)?

8.10 PA Nationally, the per capita monthly fuel oil bill is $110. A random sample of 36 cities in the Southeast average $78, with a standard deviation of $4. Is the difference significant? Summarize your conclusions in a sentence or two.

8.11 GER/CJ A survey shows that 10% of the population is victimized by property crime each year. A random sample of 527 older citizens (65 years or more of age) shows a victimization rate of 14%. Are older people more likely to be victimized? Conduct both one- and two-tailed tests of significance.

8.12 CJ A random sample of 113 felons convicted of nonviolent crimes in a state prison system completed a program designed to improve their employment possibilities before being released on parole. Fifty-eight eventually became repeat offenders. Is this recidivism rate significantly different from the rate for all offenders in that state (57%)? Summarize your conclusions in a sentence or two. (*HINT: You must use the information given in the problem to compute a sample proportion. Remember to convert the population percentage to a proportion.*)

8.13 PS In a recent statewide election, 55% of the voters rejected a proposal to institute a state lottery. In a random sample of 150 urban precincts, 49% of the voters rejected the proposal. Is the difference significant? Summarize your conclusions in a sentence or two.

8.14 CJ Statewide, the police clear by arrest 35% of the robberies and 42% of the aggravated assaults reported to them. A researcher takes a random sample of all the robberies ($N = 207$) and aggravated assaults ($N = 178$)

reported to a metropolitan police department in one year and finds that 83 of the robberies and 80 of the assaults were cleared by arrest. Are the local arrest rates significantly different from the statewide rates? Write a sentence or two interpreting your decision.

8.15 SOC/SW A researcher has compiled a file of information on a random sample of 317 families that have chronic, long-term patterns of child abuse. Reported here are some of the characteristics of the sample, along with values for the city as a whole. For each trait, test the null hypothesis of "no difference" and summarize your findings.

	Variable	City	Sample
a.	Mother's educational level (proportion completing high school)	$P_u = 0.65$	$P_s = 0.61$
b.	Family size (proportion of families with four or more children)	$P_u = 0.21$	$P_s = 0.26$
c.	Mother's work status (proportion of mothers with jobs outside the home)	$P_u = 0.51$	$P_s = 0.27$
d.	Relations with kin (proportion of families that have contact with kin at least once a week)	$P_u = 0.82$	$P_s = 0.43$
e.	Father's educational achievement (average years of formal schooling)	$\mu = 12.3$	$\overline{X} = 12.5$ $s = 1.7$
f.	Father's occupational stability (average years in present job)	$\mu = 5.2$	$\overline{X} = 3.7$ $s = 0.5$

8.16 SW You are the head of an agency seeking funding for a program to reduce unemployment among teenage males. Nationally, the unemployment rate for this group is 18%. A random sample of 323 teenage males in your area reveals an unemployment rate of 21.7%. Is the difference significant? Can you demonstrate a need for the program? Should you use a one-tailed test in this situation? Why? Explain the result of your test of significance as you would to a funding agency.

8.17 PA The city manager has received a complaint from the local union of firefighters to the effect that they are underpaid. Not having much time, the city manager gathers the records of a random sample of 27 firefighters and finds that their average salary is $38,073, with a standard deviation of $575. If she knows that the average salary nationally is $38,202,

how can she respond to the complaint? Should she use a one-tailed test in this situation? Why? What would she say in a memo to the union that would respond to the complaint?

8.18 The following essay questions review the basic principles and concepts of inferential statistics. The order of the questions roughly follows the five-step model.
 a. Hypothesis testing or significance testing can be conducted only with a random sample. Why?
 b. Under what specific conditions can it be assumed that the sampling distribution is normal in shape?
 c. Explain the role of the sampling distribution in a test of hypothesis.
 d. The null hypothesis is an assumption about reality that makes it possible to test sample outcomes for their significance. Explain.
 e. What is the critical region? How is the size of the critical region determined?
 f. Describe a research situation in which a one-tailed test of hypothesis would be appropriate.
 g. With regard to the shape of the sampling distribution, why does use of the t distribution (as opposed to the Z distribution) make it more difficult to reject the null hypothesis?
 h. What exactly can be concluded in the one-sample case when the test statistic falls in the critical region?

8.19 SOC A researcher is studying changes in the student body at her university and has selected a random sample of 163 from the freshman class. The table below compares their characteristics to the student body as a whole. Which differences are significant?

	Variable	Freshmen	All Students
a.	Proportion Republican	$P_s = 0.44$	$P_u = 0.49$
b.	Proportion majoring in math and science	$P_s = 0.17$	$P_u = 0.09$
c.	Average family income	$\overline{X} = 63{,}542$ $s = 1568$	$\mu = 59{,}235$
d.	Average number of siblings	$\overline{X} = 0.9$ $s = 0.2$	$\mu = 1.2$
e.	Average number of telephone calls to parents per week	$\overline{X} = 1.5$ $s = 0.4$	$\mu = 0.4$
f.	Average number of close friends living on campus	$\overline{X} = 4.5$ $s = 0.3$	$\mu = 7.8$

9

0.01 0.002 **Hypothesis Testing II:** 0.003 0.004
The Two-Sample Case

Rawpixel.com/Shutterstock.com

LEARNING OBJECTIVES By the end of this chapter, you will be able to:

1. Identify and cite examples of situations in which the two-sample test of hypothesis is appropriate.
2. Explain the logic of hypothesis testing as applied to the two-sample case.
3. Explain the concept of independent random sampling.
4. Perform a hypothesis test for two sample means and two sample proportions following the five-step model and correctly interpret the results.
5. Explain how each of the factors (especially sample size) affect the probability of rejecting the null hypothesis.
6. Use the Student's *t* distribution to conduct a hypothesis test on small samples.
7. Explain the differences between statistical significance and importance.
8. Use SPSS to conduct a test of significance for the difference between sample means.

USING STATISTICS

This chapter presents statistical techniques used to compare two random samples. If the samples are significantly different, we conclude that the populations from which they were selected are different. This conclusion has a known probability of error, often set at 0.05. Examples of these situations include:

1. A researcher is examining the differences in support for gun control between men and women in a sample that is representative of the U.S. population. If the samples are significantly different, the researcher will conclude that the populations of all U.S. men and all U.S. women differ on this issue.

2. Sociologists have been interested in the reasons why people have closer or more distant feelings toward people of different ethnicities and races for many years. By conducting independent sample means tests we can measure the level of social distance males have toward people of other ethnicities and races and compare to that of women. **See page 250**

3. Who performs more labor per week, foreign-born or native-born Americans? By gathering a random sample from each group and measuring their average hours worked last week we can see if there is a significant difference between the two. **See page 235**

FIGURE 9.1 A Test of Hypothesis for Two Sample Means

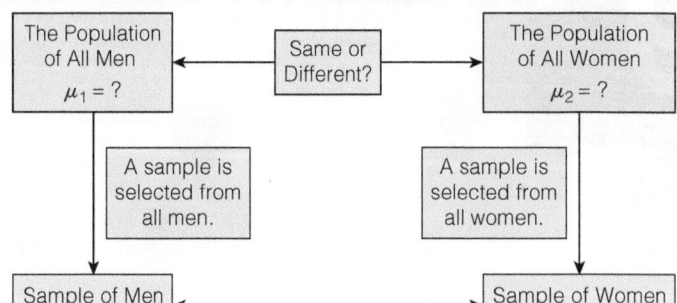

In Chapter 8, we dealt with hypothesis testing in the one-sample case and were concerned with the significance of the difference between a sample statistic and a population parameter. In this chapter, we consider a new research situation. We will be concerned with the difference between two samples and our question will be: Is the difference between the samples large enough for us to conclude that the populations represented by the samples are different?

For example, do men and women in the United States vary in their support for gun control? Obviously, we cannot ask every male and female for their opinions on this issue. Instead, we must draw random samples and use the information gathered from these samples to infer population patterns. If the difference between the samples is large enough, we will conclude (with a known probability of error) that the populations represented by the samples are different and that there is a true variance between men and women in the population.

Figure 9.1 illustrates the logic of the two-sample case, using gender differences in support for gun control as an example. It may be helpful to compare this diagram with Figure 8.1, which depicted the one-sample case.

In this chapter, we consider tests for the significance of the difference between sample means and sample proportions. In both tests, the five-step model will be the framework for our decision making. The hypothesis-testing process is similar to that of the one-sample case, but there are some important differences as well.

The One-Sample Case Versus the Two-Sample Case

There are three important differences between the one-sample case considered in Chapter 8 and the two-sample case covered in this chapter:

1. The model in step 1 will include a new assumption: **independent random sampling**. To satisfy this assumption, we select cases for each sample randomly *and* separately from each other. In our example, this means that selecting a specific male would not affect the probability of selecting any particular female: Each case must be selected independently of all other cases.

This requirement can be satisfied by drawing EPSEM samples from separate lists (for example, one for females and one for males) but it is usually more convenient to draw a single EPSEM sample from a single list of the population and then subdivide the cases into separate groups (males and females, for example). As long as the original sample is random, any subsamples created by the researcher will meet the assumption of independence.

2. The null hypothesis will be stated differently. In the one-sample case, the null hypothesis stated that the population from which the sample was drawn had a certain characteristic. Now, the null hypothesis will say that the two populations are the same (e.g., "There is no significant difference between men and women in their support of gun control"). If the test statistic falls in the critical region, the null hypothesis of no difference between the populations can be rejected, and the argument that the populations are different will be supported.

3. The sampling distribution, which is the distribution of all possible sample outcomes, will be new. In Chapter 8, the sample outcome was a single statistic, either a mean or a proportion. Now we are dealing with two samples (e.g., samples of men and women), and the sample outcome is the *difference between* the sample statistics. In our example, the sampling distribution would include all possible differences in sample means for support of gun control between men and women. If the null hypothesis is true and men and women do *not* have different views about gun control, the difference between the population means would be 0, the mean of the sampling distribution would be 0, and the huge majority of differences between sample means would be 0 or very close to 0. The greater the differences between the sample means, the further the sample outcome will be from the mean of the sampling distribution (0), and the more likely that the difference reflects a true variance between the populations represented by the samples.

Hypothesis Testing with Sample Means (Large Samples)

To illustrate the procedure for testing sample means, assume that a researcher has access to a nationally representative random sample and that the individuals in the sample have answered a survey that measures attitudes toward gun control. The sample is divided by gender, and sample statistics are computed for males and females.

As long as sample size is large (that is, as long as the combined number of cases in the two samples exceeds 100), the sampling distribution of the differences in sample means will be normal and the normal curve (Appendix A) can be used to establish the critical regions. The test statistic, Z(obtained), will be computed by the usual formula: sample outcome (the difference between the sample means) minus the mean of the sampling distribution, divided by the standard deviation of the sampling distribution.

The formula for Z(obtained) is presented as Formula 9.1. Note that numerical subscripts are used to identify the samples and populations. The "$\overline{X} - \overline{X}$" subscript attached to σ indicates that we are dealing with the sampling distribution of the *differences* in sample means.

FORMULA 9.1

$$Z(\text{obtained}) = \frac{(\overline{X}_1 - \overline{X}_2) - (\mu_1 - \mu_2)}{\sigma_{\overline{X} - \overline{X}}}$$

where $(\overline{X}_1 - \overline{X}_2)$ = the difference in the sample means
$(\mu_1 - \mu_2)$ = the difference in the population means
$\sigma_{\overline{X} - \overline{X}}$ = the standard deviation of the sampling
distribution of the differences in sample means

Recall that tests of significance are always based on the assumption that the null hypothesis is true. If the means of the two populations are equal, then the term $(\mu_1 - \mu_2)$ will be 0 and can be dropped from the equation. In effect, the formula we will ultimately use to compute the test statistic in step 4 will be

FORMULA 9.2

$$Z(\text{obtained}) = \frac{\overline{X}_1 - \overline{X}_2}{\sigma_{\overline{X} - \overline{X}}}$$

For large samples, the standard deviation of the sampling distribution of the difference in sample means is defined as

FORMULA 9.3

$$\sigma_{\overline{X} - \overline{X}} = \sqrt{\frac{\sigma_1^2}{N_1} + \frac{\sigma_2^2}{N_2}}$$

Because we will rarely, if ever, know the values of the population standard deviations (σ_1 and σ_2), we must use the sample standard deviations, corrected for bias, to estimate them. Formula 9.4 displays the equation used to estimate the standard deviation of the sampling distribution in this situation. This is called a **pooled estimate** because it combines information from both samples.

FORMULA 9.4

$$\sigma_{\overline{X} - \overline{X}} = \sqrt{\frac{s_1^2}{N_1 - 1} + \frac{s_2^2}{N_2 - 1}}$$

The sample outcomes for support of gun control are:

Sample 1 (Men)	Sample 2 (Women)
$\overline{X}_1 = 6.2$	$\overline{X}_2 = 6.5$
$s_1 = 1.3$	$s_2 = 1.4$
$N_1 = 324$	$N_2 = 317$

We see from the sample statistics that men have a lower average score and are less supportive of gun control. The test of hypothesis will tell us whether this difference is large enough to conclude that it did not occur by random chance and reflects an actual difference between all men and all women on this issue.

Step 1. Making Assumptions and Meeting Test Requirements. We now assume that the random samples are independent, but the rest of the model is the same as in the one-sample case.

> Model: Independent random samples
> Level of measurement is interval-ratio
> Sampling distribution is normal

Step 2. Stating the Null Hypothesis. The null hypothesis states that the *populations* represented by the samples are not different on this variable. No direction for the difference has been predicted, so a two-tailed test is called for:

$$H_0: \mu_1 = \mu_2$$

$$(H_1: \mu_1 \neq \mu_2)$$

Step 3. Selecting the Sampling Distribution and Establishing the Critical Region. For large samples, the Z distribution can be used to find areas under the sampling distribution and establish the critical region. Alpha will be set at 0.05.

> Sampling distribution = Z distribution
> Alpha = 0.05
> Z(critical) = ± 1.96

Step 4. Computing the Test Statistic. The population standard deviations are unknown, so Formula 9.4 will be used to estimate the standard deviation of the sampling distribution. This value will then be substituted into Formula 9.2 and Z(obtained) will be computed.

$$\sigma_{\bar{X}-\bar{X}} = \sqrt{\frac{s_1^2}{N_1 - 1} + \frac{s_2^2}{N_2 - 1}}$$

$$\sigma_{\bar{X}-\bar{X}} = \sqrt{\frac{(1.3)^2}{324 - 1} + \frac{(1.4)^2}{317 - 1}}$$

$$\sigma_{\bar{X}-\bar{X}} = \sqrt{\frac{1.69}{323} + \frac{1.96}{316}}$$

$$\sigma_{\bar{X}-\bar{X}} = \sqrt{0.0052 + 0.0062}$$

$$\sigma_{\bar{X}-\bar{X}} = \sqrt{0.0114}$$

$$\sigma_{\bar{X}-\bar{X}} = 0.107$$

$$Z(\text{obtained}) = \frac{\bar{X}_1 - \bar{X}_2}{\sigma_{\bar{X}-\bar{X}}}$$

$$Z(\text{obtained}) = \frac{6.2 - 6.5}{0.107}$$

$$Z(\text{obtained}) = \frac{-0.3}{0.107}$$

$$Z(\text{obtained}) = -2.80$$

Step 5. Making a Decision and Interpreting the Results of the Test. Compare the test statistic with the critical region:

$$Z(\text{obtained}) = -2.80$$
$$Z(\text{obtained}) = \pm 1.96$$

We see that the Z score falls in the critical region, which means that a difference as large as -0.3 ($6.2 - 6.5$) between the sample means is unlikely if the null hypothesis is true. The null hypothesis of no difference is rejected and the alternative hypothesis ("men and women differ in terms of their support of gun control") is supported. The decision to reject the null hypothesis has only a 0.05 probability (the alpha level) of being incorrect.

Note that the value for $Z(\text{obtained})$ is negative, indicating that men have significantly lower support for gun control. The sign of the test statistic reflects our arbitrary decision to label men sample 1 and women sample 2. If we had reversed the labels and called women sample 1 and men sample 2, the sign of the $Z(\text{obtained})$ would have been positive, but its value (2.80) would have been exactly the same, as would our decision in step 5. *(For practice in testing the significance of the difference between sample means for large samples, see problems 9.1 to 9.6, 9.9, 9.15d to f, 9.16d and e, and 9.17.)*

ONE STEP AT A TIME Completing Step 4 of the Five-Step Model: Computing Z(obtained)

Use these procedures when samples are large. Solve Formula 9.4 first and then solve Formula 9.2.

To Solve Formula 9.4

1. Subtract 1 from N_1.
2. Square the value of the standard deviation for the first sample (s_1^2).
3. Divide the quantity you found in step 2 by the quantity you found in step 1.
4. Subtract 1 from N_2.
5. Square the value of the standard deviation for the second sample (s_2^2).
6. Divide the quantity you found in step 5 by the quantity you found in step 4.
7. Add the quantity you found in step 6 to the quantity you found in step 3.
8. Take the square root of the quantity you found in step 7.

To Solve Formula 9.2

1. Subtract \bar{X}_2 from \bar{X}_1.
2. Divide the value you found in step 1 by the quantity you found in step 8 of "To Solve Formula 9.4." This is $Z(\text{obtained})$.

ONE STEP AT A TIME Completing Step 5 of the Five-Step Model: Making a Decision and Interpreting Results

1. Compare the $Z(\text{obtained})$ to your $Z(\text{critical})$. If $Z(\text{obtained})$ is *in* the critical region, *reject* the null hypothesis. If $Z(\text{obtained})$ is *not in* the critical region, *fail to reject* the null hypothesis.
2. Interpret the decision to reject or fail to reject the null hypothesis in terms of the original question. For example, our conclusion for the problem was "There is a significant difference between men and women in their support for gun control."

Social Research and Data Analysis 9.1 | A Test of Significance for Sample Means

A sociologist studying labor patterns in immigration is exploring differences in the number of hours that foreign-born Americans work compared to those that are native born. According to 2018 General Social Survey data, there really is not much of a difference. Those born in the United States worked 41.19 hours in the week before their interview and those that were foreign born worked 41.86 hours. Still, it is not possible for her to know how important this difference is without conducting a test for significance between sample means. So let's carry one out for her to see if there is a significant difference beween the two. (Note that you can follow the steps in *Using SPSS* for a similar problem in this chapter to let your computer do the work.)

The sample information is:

Sample 1 (Born in United States)	Sample 2 (Foreign Born)
$\bar{X}_1 = 41.19$	$\bar{X}_2 = 41.86$
$s_1 = 14.58$	$s_2 = 13.89$
$N_1 = 1190$	$N_2 = 191$

We can see from the sample results that those born in the United States worked slightly fewer hours, but let's proceed to see if the difference is significant.

Step 1. Making Assumptions and Meeting Test Requirements.

> Model: Independent random samples
> Level of measurement is interval-ratio
> Sampling distribution is normal

Step 2. Stating the Null Hypothesis.

$$H_0: \mu_1 = \mu_2$$
$$(H_1: \mu_1 \neq \mu_2)$$

Step 3. Selecting the Sampling Distribution and Establishing the Critical Region.

> Sampling distribution = Z distribution
> Alpha = 0.05, two-tailed
> $Z(\text{critical}) = \pm 1.96$

Step 4. Computing the Test Statistic.

$$\sigma_{\bar{X} - \bar{X}} = \sqrt{\frac{s_1^2}{N_1 - 1} + \frac{s_2^2}{N_2 - 1}}$$

$$\sigma_{\bar{X} - \bar{X}} = \sqrt{\frac{(14.58)^2}{1190 - 1} + \frac{(13.89)^2}{191 - 1}}$$

$$\sigma_{\bar{X} - \bar{X}} = \sqrt{\frac{212.58}{1189} + \frac{192.93}{190}}$$

$$\sigma_{\bar{X} - \bar{X}} = \sqrt{0.18 + 1.02}$$

$$\sigma_{\bar{X} - \bar{X}} = \sqrt{1.2}$$

$$\sigma_{\bar{X} - \bar{X}} = 1.10$$

$$Z(\text{obtained}) = \frac{\bar{X}_1 - \bar{X}_2}{\sigma_{\bar{X} - \bar{X}}}$$

$$Z(\text{obtained}) = \frac{41.19 - 41.86}{1.10}$$

$$Z(\text{obtained}) = \frac{-0.67}{1.10}$$

$$Z(\text{obtained}) = -0.61$$

Step 5. Making a Decision and Interpreting the Results of the Test. Compare the test statistic with the critical region,

> $Z(\text{obtained}) = -0.61$
> $Z(\text{critical}) = \pm 1.96$

We accept the null hypothesis. There is no significant difference in the number of hours worked last week between workers born in the United States and those that are foreign born.

STATISTICS | In Everyday Life

Reading and Math

The reading scores of American fourth graders have remained relatively unchanged over recent years. The average reading score declined by only 1 point (on a scale of 0 to 500) between 2015 and 2017, but this was not a significant change. Tests are administered every other year to large, randomly selected representative samples of about 150,000 fourth graders.

*Reports are available at https://www.nationsreportcard.gov/reading_math_2017_highlights/

Hypothesis Testing with Sample Means (Small Samples)

When the population standard deviation is unknown and sample size is small (combined Ns of less than 100), the Z distribution cannot be used to find areas under the sampling distribution. Instead, we will use the t distribution to find the critical region and identify unlikely sample outcomes. To do this, we need to perform one additional calculation and make one additional assumption. The calculation is for degrees of freedom, which we need in order to use the t table (Appendix B). In the two-sample case, degrees of freedom are equal to $N_1 + N_2 - 2$.

The additional assumption is a more complex matter. When samples are small, we must assume that the population variances are equal in order to justify the assumption of a normal sampling distribution and to form a pooled estimate of the standard deviation of the sampling distribution. There are ways to test the assumption of equal variances in the population, but, for our purposes, we will simply assume equal population variances without formal testing. This assumption is safe as long as sample sizes are approximately equal.

THE FIVE-STEP MODEL AND THE t DISTRIBUTION

To illustrate this procedure, assume that a researcher believes that center-city families have significantly more children than suburban families. Random samples from both areas are gathered and the following sample statistics computed.

Sample 1 (Suburban)	Sample 2 (Center-City)
$\overline{X}_1 = 2.37$	$\overline{X}_2 = 2.78$
$s_1 = 0.63$	$s_2 = 0.95$
$N_1 = 42$	$N_2 = 37$

The sample data show a difference in the predicted direction. The significance of this observed difference can be tested with the five-step model.

Step 1. Making Assumptions and Meeting Test Requirements. Sample size is small, and the population standard deviation is unknown. Hence, we must assume equal population variances in the model.

> Model: Independent random samples
> Level of measurement is interval-ratio
> Population variances are equal
> Sampling distribution is normal

Step 2. Stating the Null Hypothesis. A direction has been predicted, so a one-tailed test will be used. The research hypothesis is stated accordingly.

$$H_0: \mu_1 = \mu_2$$
$$H_1: \mu_1 < \mu_2$$

Step 3. Selecting the Sampling Distribution and Establishing the Critical Region. With small samples, the t distribution is used to establish the critical region. Alpha will be set at 0.05, and a one-tailed test will be used.

$$\text{Sampling distribution} = t \text{ distribution}$$
$$\text{Alpha} = 0.05, \text{ one-tailed}$$
$$\text{Degrees of freedom} = N_1 + N_2 - 2 = 42 + 37 - 2 = 77$$
$$t(\text{critical}) = -1.671$$

Note that the critical region is placed in the lower tail of the sampling distribution in accordance with the direction specified in H_1.

Step 4. Computing the Test Statistic. With small samples, a different formula (Formula 9.5) is used for the pooled estimate of the standard deviation of the sampling distribution. This value is then substituted directly into the denominator of the formula for t(obtained) in Formula 9.6.

FORMULA 9.5

$$\sigma_{\overline{X}-\overline{X}} = \sqrt{\frac{N_1 s_1^2 + N_2 s_2^2}{N_1 + N_2 - 2}} \sqrt{\frac{N_1 + N_2}{N_1 N_2}}$$

$$\sigma_{\overline{X}-\overline{X}} = \sqrt{\frac{(42)(0.63)^2 + (37)(0.95)^2}{42 + 37 - 2}} \sqrt{\frac{42 + 37}{(42)(37)}}$$

$$\sigma_{\overline{X}-\overline{X}} = \sqrt{\frac{(42)(0.40) + (37)(0.90)}{77}} \sqrt{\frac{79}{1554}}$$

$$\sigma_{\overline{X}-\overline{X}} = \sqrt{\frac{50.10}{77}} \sqrt{\frac{79}{1554}}$$

$$\sigma_{\overline{X}-\overline{X}} = \sqrt{0.65} \sqrt{0.05}$$

$$\sigma_{\overline{X}-\overline{X}} = (.81)(.22)$$

$$\sigma_{\overline{X}-\overline{X}} = 0.18$$

FORMULA 9.6

$$t(\text{obtained}) = \frac{\overline{X}_1 - \overline{X}_2}{\sigma_{\overline{X}-\overline{X}}}$$

$$t(\text{obtained}) = \frac{2.37 - 2.78}{0.18}$$

$$t(\text{obtained}) = \frac{-0.41}{0.18}$$

$$t(\text{obtained}) = -2.28$$

Step 5. Making a Decision and Interpreting the Results of the Test. The test statistic falls in the critical region:

$$t(\text{obtained}) = -2.28$$

$$t(\text{critical}) = -1.671$$

If the null hypothesis (H_0: $\mu_1 = \mu_2$) were true, this would be a very unlikely outcome, so the null hypothesis can be rejected. There is a statistically significant difference (a difference so large that it is unlikely to be due to random chance) in the sizes of center-city and suburban families. Furthermore, center-city families are significantly larger in size. The test statistic and sampling distribution are depicted in Figure 9.2. *(For practice in testing the significance of the difference between sample means for small samples, see problems 9.7 and 9.8.)*

ONE STEP AT A TIME Completing Step 4 of the Five-Step Model: Computing t(obtained)

Solve Formula 9.5 first.

To Solve Formula 9.5

1. Add N_1 and N_2 and then subtract 2 from this total.
2. Square the standard deviation for the first sample (s_1^2) and multiply the result by N_1.
3. Square the standard deviation for the second sample (s_2^2) and multiply the result by N_2.
4. Add the quantities you found in steps 2 and 3.
5. Divide the quantity you found in step 4 by the quantity you found in step 1 and take the square root of the result.
6. Multiply N_1 by N_2.
7. Add N_1 and N_2.
8. Divide the quantity you found in step 7 by the quantity you found in step 6 and take the square root of the result.
9. Multiply the quantity you found in step 8 by the quantity you found in step 5.

To Solve Formula 9.6

1. Subtract \bar{X}_2 from \bar{X}_1.
2. Divide the difference between the sample means by the quantity you found in step 9 of "To Solve Formula 9.5." This is t(obtained).

ONE STEP AT A TIME Completing Step 5 of the Five-Step Model: Making a Decision and Interpreting Results

1. Compare the t(obtained) to your t(critical). If t(obtained) is *in* the critical region, *reject* the null hypothesis. If t(obtained) is *not in* the critical region, *fail to reject* the null hypothesis.
2. Interpret the decision to reject or fail to reject the null hypothesis in terms of the original question. For example, our conclusion for the example problem was "There is a significant difference between the size of center city and suburban families."

FIGURE 9.2 The Sampling Distribution with Critical Region and Test Statistic Displayed

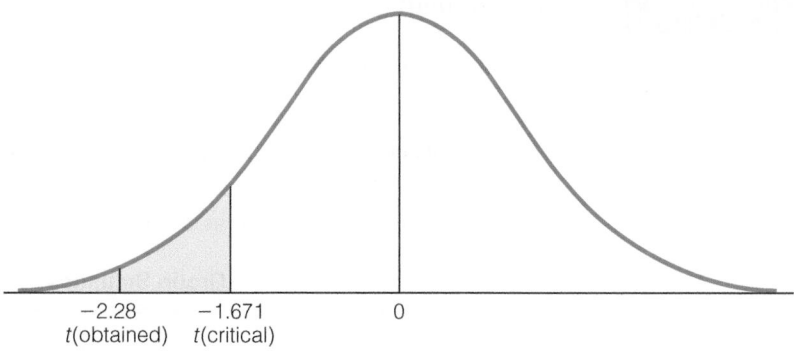

-2.28 -1.671 0

t(obtained) *t*(critical)

USING SPSS

Testing Sample Means for Significant Differences

SPSS for Windows provides several tests for the significance of the difference between means. We'll use the **Independent-Samples T Test**, which is similar to the test presented in this chapter for large samples.

To provide an illustration, we will compare high school and college graduates in terms of how often they use the Internet. Level of education is the independent variable and the dependent variable will be *wwwhr* or the number of hours per week the respondent uses the Internet. As we noted in Chapter 1 ("Statistics in Everyday Life"), there are important social class differences in access to modern technology and we expect that college graduates will have significantly higher scores on *wwwhr* than high school graduates. If the difference between the sample means is large enough, we can conclude that similar differences exist in the population. You will have the opportunity to use this procedure and conduct tests of significance for yourself later in the chapter.

To begin:

1. Click the SPSS icon on your desktop.
2. Load the *GSS2018* database.
 a. Find the **File** command on the far left of the menu bar and then click **File →
 Open → Data**.
 b. Find the *GSS2018* database and click the file name.
3. From the main menu bar, click **Analyze**, then **Compare Means**, and then
 Independent-Samples T Test. The **Independent-Samples T Test** dialog box will
 open with the usual list of variables on the left.
4. Find *wwwhr*, our dependent variable, and click the top arrow in the middle of the
 window to move the variable name to the **Test Variable(s)** box.
5. To select the independent variable, find *degree* in the list of variables and click the
 bottom arrow in the middle of the window to move *degree* to the **Grouping Variable**
 box.
6. Two question marks will appear in the **Grouping Variable** box, and the **Define
 Groups** button will become active. SPSS needs to know which cases go in which
 groups. Let's assign high school graduates (indicated by a score of 1 on *degree*) to
 group 1 and college graduates (a score of 3 on *degree*) to group 2.
7. Click the **Define Groups** button, and the **Define Groups** window will appear. The
 cursor will be blinking in the box beside Group 1. SPSS is asking for the score

(continued)

USING SPSS

(Continued)

that will determine which cases go in this group. Type a 1 in this box (for high school graduates) and then click the box next to Group 2 and type a 3 (for college graduates).

8. Click **Continue** to return to the **Independent-Samples T Test** window and click **OK** and your output will be produced.

The output is presented in two boxes, the first of which looks like this:

Group Statistics

	R's highest degree	N	Mean	Std. Deviation	Std. Error Mean
www hours per week	HIGH SCHOOL	688	13.30	17.370	.662
	BACHELOR	294	14.15	16.246	.948

NOTE: Your results will be different if you use the smaller GSS2018 Sample database for this test.

This box ("Group Statistics") presents descriptive statistics. There were 688 high school graduates, and their average score on *wwwhr* was 13.30. The 294 college graduates averaged 14.15 on *wwwhr*. We can see from this output that the sample means are different and that, on the average, college graduates have the higher average score (as we predicated). Is the difference between the sample means significant?

The results of the test for significance are reported in the next block of output. *SPSS for Windows* does a separate test for each of two assumptions about the population variance, but we will look only at the "Equal Variances Assumed" reported in the top row. This is the model used in this chapter.

		Levene's Test for Equality of Variances		t-Test of Equality of Means		
		F	Sig.	t	df	Sig. (2-tailed)
www hours per week	Equal Variances Assumed	0.369	.543	−.718	980	.473
	Equal Variances Not Assumed			−.737	589.191	.461

NOTE: This box has been edited for clarity and some output has been omitted.

Skip over the first columns of the output block (which report the results of a test for equality of the population variances). In the top row, *SPSS for Windows* reports a *t* value (−.718), the degrees of freedom (*df* = 980), and a "Sig. (2-tailed)" of .473. This last piece of information is an alpha level except that it is the *exact* probability of getting the observed difference in sample means if only chance is operating. Thus, there is no need to look up the test statistic in a *t* or *Z* table. The value for alpha is greater than our usual indicator of significance of 0.05. We conclude that the difference in *wwwhr* is not statistically significant. Contrary to our expectation, there is no significant difference in average use of the Internet per week between college and high school graduates. What do you think we would have found if we had compared college graduates with people with less than a high school degree (a score of 0 on *degree*)?

Hypothesis Testing with Sample Proportions (Large Samples)

Testing for the significance of the difference between two sample proportions follows the pattern for testing sample means. The null hypothesis states that there is no difference between the populations from which the samples are drawn. We compute a test statistic in step 4, which is then compared with the critical region. When sample sizes are large (combined Ns of more than 100), the Z distribution is used to find the critical region. In this text, we will not consider tests of significance for proportions based on small samples.

To find the value of the test statistic, several preliminary equations must be solved. Formula 9.7 uses the values of the two sample proportions (P_s) to give us an estimate of the population proportion (P_u), which is the proportion of cases in the population that have the trait under consideration assuming the null hypothesis is true.

FORMULA 9.7

$$P_u = \frac{N_1 P_{s1} + N_2 P_{s2}}{N_1 + N_2}$$

The value of P_u is then used to compute the standard deviation of the sampling distribution of the difference in sample proportions in Formula 9.8.

FORMULA 9.8

$$\sigma_{p-p} = \sqrt{P_u(1 - P_u)}\sqrt{\frac{N_1 + N_2}{N_1 N_2}}$$

This value is then substituted into the formula for computing the test statistic, presented as Formula 9.9.

FORMULA 9.9

$$Z(\text{obtained}) = \frac{(P_{s1} - P_{s2}) - (P_{u1} - P_{u2})}{\sigma_{p-p}}$$

where $(P_{s1} - P_{s2})$ = the difference between the sample proportions

$(P_{u1} - P_{u2})$ = the difference between the population proportions

σ_{p-p} = the standard deviation of the sampling distribution of the difference between sample proportions

As was the case with sample means, the second term in the numerator is assumed to be zero by the null hypothesis. Therefore, the formula reduces to

FORMULA 9.10

$$Z(\text{obtained}) = \frac{(P_{s1} - P_{s2})}{\sigma_{p-p}}$$

Remember to solve these equations in order, starting with Formula 9.7 (and skipping Formula 9.9).

CONDUCTING A TEST OF HYPOTHESIS WITH SAMPLE PROPORTIONS (LARGE SAMPLES)

An example will make these procedures clearer. Suppose we are researching social networks among senior citizens and wonder whether black people and white people differ in the number of memberships in clubs and other organizations they hold. Random samples of black and white senior citizens have been selected and classified as high or low in terms of their number of memberships. Is there a statistically significant difference in the participation patterns of black and white elderly? The proportions of each group classified as "high" in participation and sample size for both groups are:

Black Senior Citizens (Sample 1)	White Senior Citizens (Sample 2)
$P_{s1} = 0.34$	$P_{s1} = 0.25$
$N_1 = 83$	$N_1 = 103$

Step 1. Making Assumptions and Meeting Test Requirements.

Model: Independent random samples
Level of measurement is nominal
Sampling distribution is normal

Step 2. Stating the Null Hypothesis. Because no direction has been predicted, this will be a two-tailed test.

$$(H_0: P_{u1} = P_{u2})$$

$$(H_1: P_{u1} \neq P_{u2})$$

Step 3. Selecting the Sampling Distribution and Establishing the Critical Region. Because sample size is large, the Z distribution will be used to establish the critical region. Setting alpha at 0.05, we have

Sampling distribution = Z distribution

Alpha = 0.05, two-tailed

Z(critical) = ± 1.96

Step 4. Computing the Test Statistic. Solve Formula 9.7 first, substitute the resultant value into Formula 9.8, and then solve for Z(obtained) with Formula 9.10.

$$P_u = \frac{N_1 P_{s1} + N_2 P_{s2}}{N_1 + N_2}$$

$$P_u = \frac{(83)(0.34) + (103)(0.25)}{83 + 103}$$

$$P_u = \frac{28.22 + 25.75}{186}$$

$$P_u = \frac{53.97}{186}$$

$$P_u = 0.29$$

$$\sigma_{p-p} = \sqrt{P_u(1-P_u)}\sqrt{\frac{N_1 + N_2}{N_1 N_2}}$$

$$\sigma_{p-p} = \sqrt{(0.29)(0.71)}\sqrt{\frac{83 + 103}{(83)(103)}}$$

$$\sigma_{p-p} = \sqrt{0.2059}\sqrt{\frac{186}{8549}}$$

$$\sigma_{p-p} = \sqrt{0.2059}\sqrt{0.0218}$$

$$\sigma_{p-p} = (0.45)(0.15)$$

$$\sigma_{p-p} = 0.07$$

$$Z(\text{obtained}) = \frac{P_{s1} - P_{s2}}{\sigma_{p-p}}$$

$$Z(\text{obtained}) = \frac{0.34 - 0.25}{0.07}$$

$$Z(\text{obtained}) = \frac{0.09}{0.07}$$

$$Z(\text{obtained}) = 1.29$$

Step 5. Making a Decision and Interpreting the Results of the Test. The test statistic does not fall in the critical region:

$$Z(\text{critical}) = \pm 1.96$$

$$Z(\text{obtained}) = 1.29$$

We fail to reject the null hypothesis. The difference between the sample proportions is no greater than what would be expected if the null hypothesis were true and only random chance were operating. Black and white senior citizens are not significantly different in their participation in voluntary associations. *(For practice in testing the significance of the difference between sample proportions, see problems 9.10 to 9.14, 9.15a to c, and 9.16a to c.)*

ONE STEP AT A TIME **Completing Step 4 of the Five-Step Model: Computing Z(obtained)**

Solve Formulas 9.7, 9.8, and 9.10 to find the test statistic.

To Solve Formula 9.7

1. Add N_1 and N_2.
2. Multiply P_{s1} by N_1.
3. Multiply P_{s2} by N_2.

(continued)

ONE STEP AT A TIME *(Continued)*

4. Add the quantity you found in step 3 to the quantity you found in step 2.
5. Divide the quantity you found in step 4 by the quantity you found in step 1. This is P_u.

To Solve Formula 9.8

1. Multiply P_u by $(1 - P_u)$.
2. Take the square root of the quantity you found in step 1.
3. Multiply N_1 by N_2.
4. Add N_1 and N_2. *(NOTE: You already found this value when solving Formula 9.7. See step 1.)*
5. Divide the quantity you found in step 4 by the quantity you found in step 3.
6. Take the square root of the quantity you found in step 5.
7. Multiply the quantity you found in step 6 by the quantity you found in step 2.

To Solve Formula 9.10

1. Subtract the Sample 2 proportion from the Sample 1 proportion.
2. Divide the quantity you found in step 1 by the quantity you found in step 7 of "To Solve Formula 9.8." This value is Z(obtained).

ONE STEP AT A TIME **Completing Step 5 of the Five-Step Model: Making a Decision and Interpreting Results**

1. Compare Z(obtained) to Z(critical). If Z(obtained) is *in* the critical region, *reject* the null hypothesis. If Z(obtained) is *not in* the critical region, *fail to reject* the null hypothesis.
2. Interpret the decision to reject or fail to reject the null hypothesis in terms of the original question. For example, our conclusion for the example problem was "There is no significant difference between black and white senior citizens in their participation pattern."

Social Research and Data Analysis 9.2 **Testing the Significance of the Difference Between Sample Proportions**

Do attitudes toward premarital sex vary by religion? Let's compare Protestants to Catholics. According to the 2018 GSS, 29% of Protestants think it is "always wrong," compared to only 9% of Catholics. Using the data below we can determine if this difference is statistically significant.

Protestants (Sample 1)	Catholics (Sample 2)
$P_{s1} = 0.29$	$P_{s2} = 0.09$
$N_1 = 726$	$N_2 = 355$

The table presents all the information we will need to conduct a test of the null hypothesis following the familiar five-step model with alpha set at 0.05, two-tailed test.

Step 1. Making Assumptions and Meeting Test Requirements.

Model: Independent random samples
Level of measurement is nominal
Sampling distribution is normal

(continued)

Social Research and Data Analysis 9.2 *(Continued)*

Step 2. Stating the Null Hypothesis.

$$H_0: P_{u1} = P_{u2}$$

$$H_1: P_{u1} \neq P_{u2}$$

Step 3. Selecting the Sampling Distribution and Establishing the Critical Region.

$$\text{Sampling distribution} = Z \text{ distribution}$$
$$\text{Alpha} = 0.05, \text{ two-tailed}$$
$$Z(\text{critical}) = \pm 1.96$$

Step 4. Computing the Test Statistic. Remember to start with Formula 9.7, substitute the value for P_u into Formula 9.8, and then substitute that value into Formula 9.10 to solve for $Z(\text{obtained})$.

$$P_u = \frac{N_1 P_{s1} + N_2 P_{s2}}{N_1 + N_2}$$

$$P_u = \frac{(726)(0.29) + (355)(0.09)}{726 + 355}$$

$$P_u = \frac{210.54 + 31.95}{1081}$$

$$P_u = \frac{242.49}{1081}$$

$$P_u = 0.22$$

$$\sigma_{p-p} = \sqrt{P_u(1 - P_u)} \sqrt{\frac{N_1 + N_2}{N_1 N_2}}$$

$$\sigma_{p-p} = \sqrt{(0.22)(0.78)} \sqrt{\frac{726 + 355}{(726)(355)}}$$

$$\sigma_{p-p} = \sqrt{0.17} \sqrt{\frac{1081}{257,730}}$$

$$\sigma_{p-p} = (0.41)\sqrt{0.0042}$$

$$\sigma_{p-p} = (0.41)(0.06)$$

$$\sigma_{p-p} = 0.03$$

$$Z(\text{obtained}) = \frac{(P_{s1} - P_{s2})}{\sigma_{p-p}}$$

$$Z(\text{obtained}) = \frac{(0.29 - 0.09)}{0.03}$$

$$Z(\text{obtained}) = \frac{0.20}{0.03}$$

$$Z(\text{obtained}) = 6.67$$

Step 5. Making a Decision and Interpreting the Results of the Test. With an obtained Z score of 6.67, we reject the null hypothesis. There is a statistically significant difference between the proportion of Protestants and Catholics who feel that premarital sex is always wrong.

STATISTICS In Everyday Life

Abortion and Polarization in American Politics

According to Gallup polls, Republicans and Democrats have become increasingly polarized over the abortion issue. In 1975, one year after the landmark *Roe v. Wade* Supreme Court decision, 18% of Republicans and 19% of Democrats agreed that abortion should be legal "under any circumstances." Opinions fluctuated over the years since then but, in 2019, the percentage of Republicans approving abortion under any circumstances was 12% (after reaching a low of 9% in 2012) while the percentage of Democrats approving had risen to 39% (down from a high of 46% in 2017). The 1 percentage point difference in 1975 might have been statistically significant, but it was clearly not important. The 27% point difference in 2019 is statistically significant, important, and extremely consequential for the possibility of civil discourse in American politics. These results are based on random samples of about 1000 adult Americans.

How might we explain these trends? One possibility is that the Republican Party has lost many moderate, pro-choice members since the 1970s, leaving the party smaller and more ideologically homogenous. What information would you need to investigate this possibility?

Source: Gallup Polls. 2019. "Abortion Trends by Political Party." https://news.gallup.com/poll/246278/abortion-trends-party.aspx

The Limitations of Hypothesis Testing: Significance Versus Importance

We are usually interested in rejecting the null hypothesis and should take a moment to consider systematically the factors that affect our decision in step 5. Generally speaking, the probability of rejecting the null hypothesis is a function of four independent factors, only the first of which is not under the direct control of the researcher:

1. *The size of the difference between the sample statistics.* The larger the difference between the means or proportions, the greater the likelihood that the null hypothesis will be rejected.

2. *The alpha level.* As the alpha level increases, the critical region grows larger, and the probability of rejecting the null hypothesis increases. Thus, it is easier to reject H_0 at the 0.05 level than at the 0.01 level, and easier still at the 0.10 level. The danger here is that higher alpha levels increase the chance of Type I errors, and we might declare small differences to be statistically significant.

3. *One-tailed vs. two-tailed test.* A one-tailed test will increase the probability of rejecting the null hypothesis (assuming that the proper direction has been predicted).

4. *The size of the sample.* With all other factors constant, the probability of rejecting H_0 increases with sample size: The larger the sample, the more likely we are to reject the null hypothesis. With very large samples, we may declare small, unimportant differences to be statistically significant.

The relationship between sample size and the probability of rejecting the null hypothesis may be surprising and we will consider it in more detail. The probability of rejecting H_0 increases with sample size because sample size (N) is in the "denominator of the denominator" in all of the formulas for computing the test statistic (step 4). Algebraically, this is equivalent to being in the numerator of the formula and means that the value of the test statistic is directly proportional to N and that the two will increase together.

To illustrate, consider Table 9.1, which shows the value of the test statistic for single-sample means from samples of various sizes. The value of the test statistic, Z(obtained), increases as N increases, even though none of the other terms in the formula change.

With a small sample ($N = 50$), the test statistic of 1.41 is not significant at the 0.05 level. However, doubling the sample size to 100 produces a significant result, even though none of the other values in the equation change. The test statistic—Z(obtained)—continues to increase as sample size increases. This pattern holds for all tests of significance: The larger the sample, the greater the value of the test statistic and the higher the probability of rejecting the null hypothesis.

On one hand, the relationship between sample size and the probability of rejecting the null hypothesis should not alarm us unduly. Larger samples are better approximations of the populations they represent. Thus, decisions based on larger samples can be trusted more than decisions based on smaller samples.

On the other hand, this relationship clearly underlines what is perhaps the most important limitation of hypothesis testing: *Simply because a difference is*

TABLE 9.1 Test Statistics for Single-Sample Means Computed from Samples of Various Sizes ($\overline{X} = 80$, $\mu = 79$, $s = 5$ throughout)

Sample Size (N)	Computing the Test Statistic	Test Statistic, Z(obtained)
50	$Z\text{(obtained)} = \dfrac{\overline{X} - \mu}{\sigma/\sqrt{N-1}} = \dfrac{80 - 79}{5/\sqrt{49}} = \dfrac{1}{0.71} =$	1.41
100	$Z\text{(obtained)} = \dfrac{\overline{X} - \mu}{\sigma/\sqrt{N-1}} = \dfrac{80 - 79}{5/\sqrt{99}} = \dfrac{1}{0.50} =$	2.00
500	$Z\text{(obtained)} = \dfrac{\overline{X} - \mu}{\sigma/\sqrt{N-1}} = \dfrac{80 - 79}{5/\sqrt{499}} = \dfrac{1}{0.22} =$	4.55
1000	$Z\text{(obtained)} = \dfrac{\overline{X} - \mu}{\sigma/\sqrt{N-1}} = \dfrac{80 - 79}{5/\sqrt{999}} = \dfrac{1}{0.16} =$	6.25
10,000	$Z\text{(obtained)} = \dfrac{\overline{X} - \mu}{\sigma/\sqrt{N-1}} = \dfrac{80 - 79}{5/\sqrt{9999}} = \dfrac{1}{0.05} =$	20.00

statistically significant does not guarantee that it is important in any other sense. Particularly with very large samples, relatively small differences may be statistically significant. Even with small samples, differences that are otherwise trivial or uninteresting may be statistically significant.

The crucial point is that statistical significance and theoretical or practical importance can be two different things. Statistical significance is a necessary but not sufficient condition for theoretical or practical importance. A difference that is not statistically significant is almost certainly unimportant. However, significance by itself does not guarantee importance. Even when it is clear that the research results were not produced by random chance, the researcher must still assess their importance. Do they firmly support a theory or hypothesis? Are they consistent with the results of similar studies? Are they clearly consistent with a prediction or analysis? Do they strongly indicate a line of action in solving some problem? These are the kinds of questions a researcher must ask when assessing the importance of the results of a statistical test.

Also, we should note that researchers have access to some powerful ways of analyzing the importance (vs. the statistical significance) of research results. These statistics, which include bivariate measures of association and multivariate statistical techniques, are introduced in Parts III and IV of this text.

Interpreting Statistics: Is There a Statistically Significant Gender Gap in Income?

In the United States, as in many other nations around the globe, concerted efforts have been made to equalize working conditions for men and women. How successful have these efforts been? Is there a "gender gap" in income?

Note that we could answer these questions with absolute certainty only if we knew the average income of every single man and woman in the United States. Because this information is not available, we will investigate the relationship by using a random sample. If the difference between the sample means for men and women is large enough, we can infer that there is a difference in average salaries between men and women in the population. The General Social Survey, or GSS, is given to randomly selected samples of Americans and will be used as the basis for this test.

Before conducting the test, we need to deal with two issues. First, the GSS measures personal income with a series of categories rather than in actual dollars. In other words, respondents were asked to choose from a list of salary ranges (e.g., $10,000 to $12,499, $12,500 to $14,999, and so forth). Thus, income is measured at the ordinal level, and, ideally, we would like an interval-ratio dependent variable to compute sample means. To deal with this problem, we can convert the variable to a form that more closely approximates interval-ratio data by substituting the midpoints for each of the salary intervals. For example, we will use the midpoint ($13,750) for the interval "12,500 to 14,999."

This technique makes income more like an interval-ratio variable but it creates other issues and inaccuracies, and the sample means used in this test are only approximations of the population means. However, because we are concerned with the *difference* in income rather than actual dollar values, we should still be able to come to some conclusions.

Second, we shouldn't compare the incomes of *all* men and *all* women in the sample because some difference could be the result of the fact that women are less likely to be in the paid labor market (e.g., they are more likely to be occupied as wives and mothers) and, thus, less likely to have a personal income. We will deal with this by restricting the comparison to respondents who work full-time.

Once these adjustments have been made, we can test for the significance of the difference in income. The sample information calculated from the full 2018 GSS database is:

Males	Females
$\overline{X}_1 = 66,119.45$	$\overline{X}_2 = 49,500.53$
$s_1 = 45,200.59$	$s_2 = 39,283.61$
$N_1 = 509$	$N_2 = 475$

It appears that there is a large gender gap in income and that males average $16,000 a year more than females. Is this difference in sample means significant? Could it have occurred by random chance?

The null hypothesis is that males and females have the same average salary in the population ($\mu_1 = \mu_2$). We will skip the customary trip through the five-step model and simply report that the Z(obtained) calculated with Formula 9.2 (step 4) is 6.14, much greater than the customary Z(critical) score of ± 1.96 associated with an alpha level of 0.05. The difference in sample means is so large that we must reject the null hypothesis and conclude (with a probability of error of 0.05) that the population means are different: Males earn significantly more than females.

Is the gender gap in income due to differences in levels of education? If females were significantly less educated than males, this would account for at least some of the difference in income. Using the full 2018 General Social Survey

again to get information on years of education for all males and females in the sample, we have the following summary.

Males	Females
$\overline{X}_1 = 13.69$	$\overline{X}_2 = 13.76$
$s_1 = 3.01$	$s_2 = 2.94$
$N_1 = 1051$	$N_2 = 1294$

We can see that both males and females average about 13.7 years of schooling and that the sample means are almost identical. The test statistic computed in step 4 is a Z(obtained) of -0.57, which is not in the critical region, as marked by a Z(critical) score of ± 1.96 at the 0.05 level. We fail to reject the null hypothesis of no difference. Males and females have essentially equal levels of schooling in the population, and education cannot account for the differences in income.

These results suggest that females are getting a lower return in income for their education. Although females and males are essentially equal in preparation for work and careers (at least as measured by years of schooling), a large and statistically significant gender gap in income exists. Why? To answer this question in detail would carry us far beyond the bounds of a statistics text. However, we can note one part of the answer. Men and women tend to pursue different kinds of careers. In particular, men tend to dominate the more lucrative, higher-prestige occupations (e.g., lawyer, doctor), while women are concentrated in jobs that have lower levels of remuneration (e.g., elementary school teacher, nurse). For example, in 2017, women were 90% of all registered nurses and 86% of all legal assistants versus only 40% of all doctors and 37% of all lawyers (U.S. Bureau of Labor Statistics. 2017. "Employed Persons by Detailed Occupation, Sex, Race, and Hispanic or Latino Ethnicity." https://www .bls.gov/cps/cpsaat11.htm). Thus, the gender gap in income is sustained, in part, by the fact that women are more likely to be employed in occupations that command lower salaries. What sociological factors might account for these choices?

STATISTICS **In Everyday Life**

Is the Gender Gap Decreasing?

Is the gender gap closing? Comparing only full-time workers, the median income for women in 1955 was 64% of the median income for men. By 2017, the gap had closed to 80%.[*]

Note that it may not be possible to close this gap much further as long as there is a "glass ceiling" on women's chances for promotion to the highest (most lucrative) positions and relatively fewer women in the highest income brackets. For example, in 2018, 14% of men earned more than $100,000 versus only 6% of women.[#]

What does the glass ceiling imply about the overall shape of the distribution of men's and women's income? Which distribution will be more positively skewed? What effect will that have on the mean? (See Chapter 4 for a review of the effect of skew on measures of central tendency.)

[*]U.S. Bureau of the Census. Table P-36, available at https://www.census.gov/data/tables/time-series/demo/income-poverty /historical-income-people.html

[#]U.S. Bureau of the Census. Table PINC-01, available at https://www.census.gov/data/tables/2019/demo/cps/pinc-01.html

| READING STATISTICS | Comparing Ethnic and Racial Social Distance by Gender |

In research by Christopher Donoghue (coauthor) and Vincent Parrillo on ethnic and racial social distance, female college students have consistently reported closer feelings than males toward a group of 30 ethnic and racial groups in America. In their articles on this topic, they report only the sample means, the value of the test statistic (a Z or t score), the alpha level, and sample size.

In their 2011 *Journal of Social Science* paper they reported that the average level of social distance (on a scale of 1–7, where 1 is the closest) was 1.43 for females and 1.48 for men, a difference which was significant at the ($p < .05$) level or the 95% confidence level. This was based on a sample size of 2916 college students. When they replicated the study on a sample of 3166 students in a 2013 *Sociological Forum* study, ethnic and racial social distance had increased for all college students, but females continued to report closer feelings with an average of 1.60, compared to 1.72 for males. Again, this difference was significant at the ($p < .05$) level. However, in their most recent study (not yet published), they found no significant difference between males and females.

Note that the alpha level is reported as "$p < 0.05$." This is shorthand for "the probability of a difference of this magnitude occurring by chance alone. It is a good illustration of how researchers can convey a great deal of information in just a few symbols. In a similar fashion, our statement "The test statistic falls in the critical region and, therefore, the null hypothesis is rejected" is rendered tersely and simply: "The difference was found to be significant."

When researchers need to report the results of multiple tests, they will often use a summary table to report the sample information and whether the difference is significant at a certain alpha level. If you read the researcher's description and analysis of such tables, you should have little difficulty interpreting and understanding them.

Sources: Parrillo, Vincent, and Christopher Donoghue. 2013. "The National Social Distance Survey: Ten Years Later." *Sociological Forum,* 28(13):597–614;

Parrillo, Vincent, and Christopher Donoghue. 2005. "Updating the Bogardus Social Distance Studies: A New National Survey." *The Social Science Journal,* 42(2):257–271.

SUMMARY

1. A common research situation is to test for the significance of the difference between two populations. Sample statistics are calculated for random samples of each population, and then we test for the significance of the difference between the samples as a way of inferring differences between the populations.
2. When sample information is summarized in the form of sample means and N is large, the Z distribution is used to find the critical region. When N is small, the t distribution is used to establish the critical region.
3. Differences in sample proportions may also be tested for significance. For large samples, the Z distribution is used to find the critical region.

4. In all tests of hypothesis, a number of factors, including sample size, affect the probability of rejecting the null hypothesis. Statistical significance is not the same thing as theoretical or practical importance. Even after a difference is found to be statistically significant, the researcher must still demonstrate the relevance or importance of his or her findings. The statistics presented in Parts III and IV of this text will give us the tools we need to deal directly with issues beyond statistical significance.

SUMMARY OF FORMULAS

Test statistic for two sample means, large samples:

9.1 $$Z(\text{obtained}) = \frac{(\bar{X}_1 - \bar{X}_2) - (\mu_1 - \mu_2)}{\sigma_{\bar{X} - \bar{X}}}$$

Test statistic for two sample means, large samples (simplified):

9.2 $$Z(\text{obtained}) = \frac{\bar{X}_1 - \bar{X}_2}{\sigma_{\bar{X} - \bar{X}}}$$

Standard deviation of the sampling distribution of the difference in sample means, large samples:

9.3
$$\sigma_{\overline{X} - \overline{X}} = \sqrt{\frac{\sigma_1^2}{N_1} + \frac{\sigma_2^2}{N_2}}$$

Pooled estimate of the standard deviation of the sampling distribution of the difference in sample means, large samples:

9.4
$$\sigma_{\overline{X} - \overline{X}} = \sqrt{\frac{s_1^2}{N_1 - 1} + \frac{s_2^2}{N_2 - 1}}$$

Pooled estimate of the standard deviation of the sampling distribution of the difference in sample means, small samples:

9.5
$$\sigma_{\overline{X} - \overline{X}} = \sqrt{\frac{N_1 s_1^2 + N_2 s_2^2}{N_1 + N_2 - 2}} \sqrt{\frac{N_1 + N_2}{N_1 N_2}}$$

Test statistic for two sample means, small samples:

9.6
$$t(\text{obtained}) = \frac{\overline{X}_1 - \overline{X}_2}{\sigma_{\overline{X} - \overline{X}}}$$

Pooled estimate of population proportion, large samples:

9.7
$$P_u = \frac{N_1 P_{s1} + N_2 P_{s2}}{N_1 + N_2}$$

Standard deviation of the sampling distribution of the difference in sample proportions, large samples:

9.8
$$\sigma_{p - p} = \sqrt{P_u(1 - P_u)} \sqrt{\frac{N_1 + N_2}{N_1 N_2}}$$

Test statistic for two sample proportions, large samples:

9.9
$$Z(\text{obtained}) = \frac{(P_{s1} - P_{s2}) - (P_{u1} - P_{u2})}{\sigma_{p - p}}$$

Test statistic for two sample proportions, large samples (simplified):

9.10
$$Z(\text{obtained}) = \frac{P_{s1} - P_{s2}}{\sigma_{p - p}}$$

GLOSSARY

Independent random samples. Random samples gathered in such a way that the selection of a particular case for one sample has no effect on the probability that any other particular case will be selected for the other samples.

Pooled estimate. An estimate of the standard deviation of the sampling distribution of the difference in

sample means based on the standard deviations of both samples.

$\sigma_{p - p}$. Symbol for the standard deviation of the sampling distribution of the differences in sample proportions.

$\sigma_{\overline{X} - \overline{X}}$. Symbol for the standard deviation of the sampling distribution of the differences in sample means.

PROBLEMS

9.1 For each of the following, test for the significance of the difference in sample statistics using the five-step model. (*HINT: Remember to solve Formula 9.4 before attempting to solve Formula 9.2. Also, in Formula 9.4, perform the mathematical operations in the proper sequence. First square each sample standard deviation, then divide by the proper N, add the resultant values, and then find the square root of the sum.*)

a.

Sample 1	Sample 2
$\overline{X}_1 = 72.5$	$\overline{X}_2 = 76.0$
$s_1 = 14.3$	$s_2 = 10.2$
$N_1 = 136$	$N_2 = 257$

b.

Sample 1	Sample 2
$\overline{X}_1 = 107$	$\overline{X}_2 = 103$
$s_1 = 14$	$s_2 = 17$
$N_1 = 175$	$N_2 = 200$

9.2 SOC Two sociologists administered questionnaires to samples of undergraduates that measured attitudes toward interpersonal violence (higher scores indicate greater approval of interpersonal violence). Test the results as reported here for gender, racial, and social-class differences.

a.

Sample 1 (Males)	Sample 2 (Females)
$\overline{X}_1 = 2.99$	$\overline{X}_2 = 2.29$
$s_1 = 0.88$	$s_2 = 0.91$
$N_1 = 122$	$N_2 = 251$

b.

Sample 1 (Blacks)	Sample 2 (Whites)
$\overline{X}_1 = 2.76$	$\overline{X}_2 = 2.49$
$s_1 = 0.68$	$s_2 = 0.91$
$N_1 = 43$	$N_2 = 304$

c.

Sample 1 (White Collar)	Sample 2 (Blue Collar)
$\overline{X}_1 = 2.46$	$\overline{X}_2 = 2.67$
$s_1 = 0.91$	$s_2 = 0.87$
$N_1 = 249$	$N_2 = 97$

Summarize your results in terms of the significance and the direction of the differences. Which of these three factors seems to make the biggest difference in attitudes toward interpersonal violence?

9.3 SOC Do athletes in different sports vary in terms of their readiness for college? Reported here are college entrance exam scores of fictitious random samples of college basketball and football players and male and female athletes. Is there a significant difference between groups? Write a sentence or two explaining the difference.

a.

Sample 1 (Basketball Players)	Sample 2 (Football Players)
$\overline{X}_1 = 460$	$\overline{X}_2 = 442$
$s_1 = 92$	$s_2 = 57$
$N_1 = 102$	$N_2 = 117$

b.

Sample 1 (Males)	Sample 2 (Females)
$\overline{X}_1 = 452$	$\overline{X}_2 = 480$
$s_1 = 88$	$s_2 = 75$
$N_1 = 107$	$N_2 = 105$

9.4 PA A number of years ago, the road and highway maintenance department of Sin City, Nevada, began recruiting minority group members through an affirmative action program. In terms of efficiency ratings, as compiled by their superiors, how do the affirmative action employees rate? The ratings of random samples of both groups were collected, and the results are reported here (higher ratings indicate greater efficiency). Write a few sentences reporting and interpreting the results of the test.

Sample 1 (Affirmative Action)	Sample 2 (Regular)
$\overline{X}_1 = 15.2$	$\overline{X}_2 = 15.5$
$s_1 = 3.9$	$s_2 = 2.0$
$N_1 = 97$	$N_2 = 100$

9.5 SOC Are middle-class families more likely than working-class families to maintain contact with kin? Write a paragraph summarizing the results of these tests.

a. A sample of middle-class families reported an average of 7.3 visits per year with close kin, while a sample of working-class families averaged 8.2 visits. Is the difference significant?

Sample 1 (Middle Class)	Sample 2 (Working Class)
$\overline{X}_2 = 7.3$	$\overline{X}_2 = 8.2$
$s_1 = 0.3$	$s_2 = 0.5$
$N_1 = 89$	$N_2 = 55$

b. The middle-class families averaged 2.3 phone calls and 8.7 e-mail messages per month with close kin. The working-class families averaged 2.7 calls and 5.7 e-mail messages per month. Are these differences significant?

Phone Calls per Month

Sample 1 (Middle Class)	Sample 2 (Working Class)
$\overline{X}_1 = 2.3$	$\overline{X}_2 = 2.7$
$s_1 = 0.5$	$s_2 = 0.8$
$N_1 = 89$	$N_2 = 55$

E-Mail Messages per Month

Sample 1 (Middle Class)	Sample 2 (Working Class)
$\overline{X}_1 = 8.7$	$\overline{X}_2 = 5.7$
$s_1 = 0.3$	$s_2 = 1.1$
$N_1 = 89$	$N_2 = 55$

9.6 MR According to a recent article in *Business Insider*, adults between the ages of 35 and 49 watch 221 minutes of TV per day. Older adults between the ages of 50 and 64 watch 338 minutes per day. If the sample sizes (N) and sample standard deviations were as given on the next page, would the difference between the two age groups be significantly different?

Sample 1 (Ages 35–49)	Sample 2 (Ages 50–64)
$\overline{X}_1 = 221$	$\overline{X}_2 = 338$
$s_1 = 100$	$s_2 = 116$
$N_1 = 158$	$N_2 = 173$

9.7 Are senior citizens who live in retirement communities more socially active than those who live in age-integrated communities? Write a sentence or two explaining the results of these tests. *(HINT: Remember to use the proper formulas for small sample sizes.)*

a. A random sample of senior citizens living in a retirement village reported that they had an average of 1.42 face-to-face interactions per day with their neighbors. A random sample of those living in age-integrated communities reported 1.58 interactions. Is the difference significant?

Sample 1 (Retirement Community)	Sample 2 (Age-Integrated Neighborhood)
$\overline{X}_1 = 1.42$	$\overline{X}_2 = 1.58$
$s_1 = 0.10$	$s_2 = 0.78$
$N_1 = 43$	$N_2 = 37$

b. Senior citizens living in the retirement village reported that they had an average of 7.43 telephone calls with friends and relatives each week, while those in the age-integrated communities reported an average of 5.50 calls. Is the difference significant?

Sample 1 (Retirement Community)	Sample 2 (Age-Integrated Neighborhood)
$\overline{X}_1 = 7.43$	$\overline{X}_2 = 5.50$
$s_1 = 0.75$	$s_2 = 0.25$
$N_1 = 43$	$N_2 = 37$

9.8 [SW] As a school social worker, one of your tasks is to monitor student absences and be ready to seek interventions with families if they are in need. Recently, you have noticed that the students with higher absences tend to be Hispanic. The school where you work is very diverse ethnically and racially. Could it really be the case that Hispanics have higher absences than non-Hipanics? Some of your colleagues might conclude that this is the case without considering the evidence, but you decide to investigate to see if the data confirm the pattern. Below are the statistics you have gathered on student absences for two samples in the last six months. Run a Five-Step

Model Hypothesis Test to see if Hispanics have a higher mean for absences than non-Hispanics in the whole student population.

Sample 1 (Non-Hispanics)	Sample 2 (Hispanics)
$\overline{X}_1 = 10.3$	$\overline{X}_2 = 12.3$
$s_1 = 2.7$	$s_2 = 4.2$
$N_1 = 40$	$N_2 = 55$

9.9 [SOC] A survey has been administered to random samples of respondents in each of five nations. Respondents were asked: "How satisfied are you with your life as a whole?" Responses varied from 1 (very dissatisfied) to 10 (very satisfied). Conduct a test for the significance of the difference in mean scores for each nation between men and women.

Brazil

Males	Females
$\overline{X}_1 = 7.7$	$\overline{X}_2 = 7.6$
$s_1 = 2.0$	$s_2 = 2.2$
$N_1 = 712$	$N_2 = 783$

Ukraine

Males	Females
$\overline{X}_1 = 5.8$	$\overline{X}_2 = 5.5$
$s_1 = 2.3$	$s_2 = 2.4$
$N_1 = 446$	$N_2 = 549$

Vietnam

Males	Females
$\overline{X}_1 = 7.2$	$\overline{X}_2 = 7.0$
$s_1 = 1.9$	$s_2 = 1.8$
$N_1 = 762$	$N_2 = 720$

South Africa

Males	Females
$\overline{X}_1 = 6.8$	$\overline{X}_2 = 7.2$
$s_1 = 2.5$	$s_2 = 2.3$
$N_1 = 1492$	$N_2 = 1482$

Egypt

Males	Females
$\overline{X}_1 = 5.6$	$\overline{X}_2 = 5.9$
$s_1 = 2.7$	$s_2 = 2.7$
$N_1 = 1557$	$N_2 = 1493$

9.10 For each problem, test the sample proportions for the significance of the difference.

a.

Sample 1	Sample 2
$p_{s1} = 0.20$	$p_{s2} = 0.17$
$N_1 = 114$	$N_2 = 101$

b.

Sample 1	Sample 2
$p_{s1} = 0.60$	$p_{s2} = 0.62$
$N_1 = 478$	$N_2 = 532$

9.11 SW About half of the police officers in Shinbone, Kansas, have completed a special course in investigative procedures. Has the course increased their efficiency in clearing crimes by arrest? The proportions of cases cleared by arrest for samples of trained and untrained officers are reported here.

Sample 1 (Trained)	Sample 2 (Untrained)
$p_{s1} = 0.47$	$p_{s2} = 0.43$
$N_1 = 157$	$N_2 = 113$

9.12 SW A large counseling center needs to evaluate several experimental programs. Write a paragraph summarizing the results of these tests. Did the new programs work?

a. One program is designed for divorce counseling; the key feature of the program is its counselors, who are married couples working in teams. About half of all clients have been randomly assigned to this special program and half to the regular program, and the proportion of cases that eventually ended in divorce was recorded for both. The results for random samples of couples from both programs are reported here. In terms of preventing divorce, did the new program work? .

Sample 1 (Special Program)	Sample 2 (Regular Program)
$p_{s1} = 0.53$	$p_{s2} = 0.59$
$N_1 = 78$	$N_2 = 82$

b. The agency is also experimenting with a program that includes peer counseling for depressed teenagers. About half of all clients were randomly assigned to the new program. After a year, a random sample of teens from the new program was compared with a random sample that received standard counseling. In terms of the percentage of children who were judged to be "much improved," did the new program work?

Sample 1 (Peer Counseling)	Sample 2 (Standard Program)
$p_{s1} = 0.10$	$p_{s2} = 0.15$
$N_1 = 52$	$N_2 = 56$

9.13 SOC At St. Algebra College, the sociology and psychology departments have been feuding for years about the respective quality of their programs. In an attempt to resolve the dispute, you have gathered data about the graduate school experience of random samples of both groups of majors. As measured by these data, is there a significant difference in program quality?

a. Proportion of majors who applied to graduate school:

Sample 1 (Sociology)	Sample 2 (Psychology)
$p_{s1} = 0.53$	$p_{s2} = 0.40$
$N_1 = 150$	$N_2 = 175$

b. Proportion accepted by program of first choice:

Sample 1 (Sociology)	Sample 2 (Psychology)
$p_{s1} = 0.75$	$p_{s2} = 0.86$
$N_1 = 80$	$N_2 = 70$

c. Proportion completing the program:

Sample 1 (Sociology)	Sample 2 (Psychology)
$p_{s1} = 0.75$	$p_{s2} = 0.70$
$N_1 = 60$	$N_2 = 60$

9.14 CJ The mayor of a large city started a "crimeline" program some years ago and wonders whether it is working. The program publicizes unsolved violent crimes in the local media and offers cash rewards for information leading to arrests. Are "featured" crimes more likely to be cleared by arrest than other violent crimes? Results from random samples of both types of crimes are reported as follows:

Sample 1 (Crimeline Crimes Cleared by Arrest)	Sample 2 (Non-Crimeline Crimes Cleared by Arrest)
$p_{s1} = 0.35$	$p_{s2} = 0.25$
$N_1 = 178$	$N_2 = 212$

9.15 SOC Some results from a survey administered to a nationally representative sample are reported here in terms of differences by gender. Which of these

differences, if any, are significant? Write a sentence or two of interpretation for each test.

a. Proportion favoring the legalization of marijuana:

Males	Females
$p_{s1} = 0.37$	$p_{s2} = 0.31$
$N_1 = 202$	$N_2 = 246$

b. Proportion strongly agreeing that "kids are life's greatest joy":

Males	Females
$p_{s11} = 0.45$	$p_{s2} = 0.58$
$N_1 = 251$	$N_2 = 351$

c. Proportion voting for President Trump in 2016:

Males	Females
$p_{s1} = 0.51$	$p_{s2} = 0.37$
$N_1 = 399$	$N_2 = 509$

d. Average hours spent with e-mail each week:

Males	Females
$\bar{X}_1 = 4.18$	$\bar{X}_2 = 3.38$
$s_1 = 7.21$	$s_2 = 5.92$
$N_1 = 431$	$N_2 = 535$

e. Average rate of church attendance (number of times per year):

Males	Females
$\bar{X}_1 = 3.19$	$\bar{X}_2 = 3.99$
$s_1 = 2.60$	$s_2 = 2.72$
$N_1 = 641$	$N_2 = 808$

f. Number of children:

Males	Females
$\bar{X}_1 = 1.49$	$\bar{X}_2 = 1.93$
$s_1 = 1.50$	$s_2 = 1.50$
$N_1 = 635$	$N_2 = 803$

9.16 SOC A researcher is studying the effects of the college experience on attitudes, values, and behaviors and is comparing random samples of freshmen and seniors at the same university. Which of the following differences are significant?

a. Proportion calling parents at least once a day:

Freshmen	Seniors
$p_{s1} = 0.24$	$p_{s2} = 0.08$
$N_1 = 153$	$N_2 = 117$

b. Proportion with a "close friend" from another racial or ethnic group:

Freshmen	Seniors
$p_{s1} = 0.15$	$p_{s2} = 0.26$
$N_1 = 155$	$N_2 = 114$

c. Proportion saying that they are "very interested" in national politics:

Freshmen	Seniors
$p_{s1} = 0.46$	$p_{s2} = 0.48$
$N_1 = 147$	$N_2 = 111$

d. Average score on a scale that measures political ideology. On this scale, 10 means "very conservative" and 1 means "very liberal."

Freshmen	Seniors
$\bar{X}_1 = 5.23$	$\bar{X}_2 = 5.12$
$s_1 = 1.78$	$s_2 = 1.07$
$N_1 = 145$	$N_2 = 105$

e. Average score on a scale that measures support for traditional gender roles. This scale ranges from 7 ("very traditional") to 1 ("very nontraditional").

Freshmen	Seniors
$\bar{X}_1 = 2.07$	$\bar{X}_2 = 2.17$
$s_1 = 1.23$	$s_2 = 0.78$
$N_1 = 143$	$N_2 = 111$

Statistical Analysis Using SPSS

9.17 SOC In this exercise, you will use SPSS and the 2018 General Social Survey (GSS) to compare Protestants and Catholics for significant differences in their average social class standing, using level of education (*educ*) and occupational prestige (*prestg10*) as measures of social class. You will use the **Independent-Samples T Test** to conduct the tests.

- Click the SPSS icon on your desktop.
- Load the *GSS2018* database.
- Click **Analyze → Compare Means → Independent-Samples T Test**.
- Find *educ* and *prestg10* in the list of variables and move them into the **Test Variable(s)** box.
- Find *relig* in the list of variables and move it into the **Grouping Variable** box.
- Click **Define Groups** and type 1 (for Protestants) in the box for Group 1 and 2 (for Catholics) in the box for Group 2. Click **Continue**.

- Click **OK** and the results of the test will be printed to the SPSS output window. Find the group means, the t score, degrees of freedom, and the "Sig. (2-tailed)" (which is the exact probability of getting this difference if H_0 is true), and report the results of the test in a few sentences.

YOU ARE THE RESEARCHER

Gender Gaps and Support for Traditional Gender Roles

There are two projects below. The first tests for significant differences in sample means between men and women on four variables of your own choosing. The second uses the **Compute** command to explore attitudes toward abortion or traditional gender roles. You are urged to complete both projects. Follow the instructions in the "Using SPSS" demonstration presented earlier in this chapter to produce the output.

PROJECT 1: Exploring the Gender Gap

In this enlightened age, with its heavy stress on gender equality, how many important differences persist between the sexes? In this project, you will select four dependent variables and test whether the genders are significantly different on the variables you select.

Step 1: Choosing Dependent Variables

Select four dependent variables from the *GSS2018* data set. Choose *only* interval-ratio variables or ordinal variables with three or more scores or categories. As you select variables, you might keep in mind the issues at the forefront of the debate over gender equality: income, education, and other measures of equality. Or you might choose variables that relate to lifestyle choices and patterns of everyday life: religiosity, TV viewing habits, or political ideas. See Appendix G for a list of some of the variables available for this test or browse the variables in *GSS2018*. You can do this by clicking the "Variable View" tab on the Data Editor window and then reading what each variable measures in the "Label" column.

List your four dependent variables in the table below.

Variable	SPSS Name	What exactly does this variable measure?
1		
2		
3		
4		

Step 2: Stating Hypotheses

For each dependent variable, state a hypothesis about the difference you expect to find. For example, you might hypothesize that men will be more liberal or women will be more educated. Or you can hypothesize that there will be no significant difference between the genders. You can base your hypotheses on your own experiences or on the information about gender differences that you have acquired in your courses or from other sources.

Hypotheses:
1.
2.
3.
4.

Step 3: Getting the Output

Load the *GSS2018* data set and use the **Independent-Samples T Test** to produce output. See the "Using SPSS" demonstration in this chapter for detailed instructions. Find the four dependent variables you selected in the list and click the top arrow in the middle of the window to move the variable names to the **Test Variable(s)** box.

Next, highlight *sex* and click the bottom arrow in the middle of the window to move *sex* to the **Grouping Variable** box. Click the **Define Groups** button and type 1 in the first box (for males) and 2 in the second box (for females). Click **Continue** to return to the **Independent-Samples T Test** window and click **OK**.

Step 4: Reading the Output

Remember that the first block of output ("Group Statistics") presents descriptive statistics, and the second block reports the test for significance. In the top row of the second block of output, find the *t* value, the degrees of freedom, and the "Sig. (2-tailed)," which, as you recall, is the *exact* probability of getting the observed difference in sample means if only chance is operating.

Step 5: Recording Your Results

Record your results in the table below. Write the SPSS variable name in the first column and then record the descriptive statistics. Next, record the results of the test of significance, using the top row ("Equal Variance Assumed") of the second output block. Record the *t* score, the degrees of freedom (*df*), and whether the difference is significant at the 0.05 level. If the value of "Sig. (2-tailed)" is less than 0.05, write YES in this column. If the value of "Sig. (2-tailed)" is more than 0.05, write NO in the column.

Dependent Variables		\overline{X}	s	N	t score	df	Significant?
	Men						
	Women						
	Men						
	Women						
	Men						
	Women						
	Men						
	Women						

Step 6: Interpreting Your Results

Summarize your findings. For each dependent variable, write

1. At least one sentence summarizing the test, in which you identify the variables being tested, the sample means for each group, N, the t score, and the significance level. In the professional research literature, you might find the results reported as: "For a sample of 1417 respondents, there was no significant difference between the average age of men (48.21) and the average age of women (48.12) ($t = 0.95$, $df = 1416$, $p > 0.05$)."
2. A sentence relating to your hypotheses. Were they supported? How?

PROJECT 2: Using the Compute Command to Explore Gender Differences

In this project, you will use the **Compute** command, which was introduced in Chapter 4, to construct a summary scale for either support for legal abortion or support for traditional gender roles. Do these attitudes vary significantly by gender? You will also choose a second independent variable, other than gender, to test for significant differences.

Step 1: Creating Summary Scales

To refresh your memory, we used the **Compute** command in Chapter 4 to create a summary scale (*abscale*) for attitudes toward abortion by adding the scores on two items (*abany* and *abpoor*). Remember that, once created, a computed variable is added to the active file and can be used like any of the variables actually recorded in the file. If you did not save the data file with *abscale* included, you can quickly re-create the variable by following the instructions in Chapter 4.

As an alternative, you can also create a scale to measure support for traditional gender roles. The full GSS 2018 data set supplied with this text includes two variables that measure gender attitudes. One of these (*fefam*) states, "It is much better for everyone involved if the man is the achiever outside the home and the woman takes care of the home and family." There are four possible responses to this item, ranging from "Strongly agree" (1) to "Strongly disagree" (4). Note that the lowest score (1) is most supportive of traditional gender roles.

The other item (*fechld1*) states, "A working mother can establish just as warm and secure a relationship with her children as a mother who does not work." This item also has four possible responses and the response "strongly disagree" (1) is the most consistent with support of traditional gender roles.[1]

The scores of the two items vary in the same direction (for both, the lowest score indicates the strongest support for traditional gender roles), so we can create a summary scale by simply adding the two variables together. Follow the commands in Chapter 4 to add the scores of *fefam* and *fechld* and create the scale, which we will call *fescale*. The computed variable will have a total of seven possible scores, with lower scores indicating more support for traditional gender roles and higher scores indicating less support.

Step 2: Stating Hypotheses

Choose either *abscale* or *fescale* and state hypotheses about what differences you expect to find between men and women. Which gender will be more supportive (have

[1] The variable *fechld1* is *fechld* recoded to make the scoring consistent with *fefam*.

a lower average score on *abscale*) of legal abortion? Why? Will men or women be more supportive (have a lower average score on *fescale*) of traditional gender roles? Why?

Step 3: Getting and Interpreting the Output

Run the **Independent-Samples T Test** as before with your chosen scale as the **Test Variable** and *sex* as the **Grouping Variable**. See the instructions for Project 1 above.

Step 4: Interpreting Your Results

Summarize your results as in Project 1, step 6. Was your hypothesis confirmed? How?

Step 5: Extending the Test by Selecting an Additional Independent Variable

What other independent variable besides gender might be related to attitudes toward abortion or traditional gender roles? Select another independent variable besides *sex*, and conduct an additional *t* test with either *abscale* or *fescale* as the dependent variable. Remember that the *t* test requires that the independent variable have *only* two categories. For variables with more than two categories (*relig* or *racecen1*, for example), you can meet this requirement by using the **Define Groups** button in the **Grouping Variables** box to select specific categories of a variable. You could, for example, compare Protestants and Catholics on *relig* by choosing scores of 1 (Protestants) and 2 (Catholics).

Step 6: Stating Hypotheses

State a hypothesis about what differences you expect to find between the categories of your independent variable. Which category will be more supportive of legal abortion or more supportive of traditional gender roles? Why?

Step 7: Getting and Interpreting the Output

Run the **Independent-Samples T Test** as before with the scale you selected as the **Test Variable** and your independent variable as the **Grouping Variable**.

Step 8: Interpreting Your Results

Summarize your results as in Project 1, step 6. Was your hypothesis confirmed? How?

10 Hypothesis Testing III:
The Analysis of Variance

LEARNING OBJECTIVES By the end of this chapter, you will be able to:

1. Cite examples of situations in which analysis of variance (ANOVA) is appropriate.
2. Explain the logic of hypothesis testing as applied to ANOVA.
3. Perform the ANOVA test, using the five-step model as a guide and interpret the results.
4. Explain the concepts of population variance, total sum of squares, sum of squares between, sum of squares within, mean square estimates, and post hoc tests.
5. Use SPSS to conduct the analysis of variance test.

USING STATISTICS

This chapter presents statistical techniques used to compare differences between more than two random samples. If the differences are significant, we conclude that the populations from which the samples have been selected are different. This conclusion will have a known probability of error, which is often set at the 0.05 level. Here are some examples where we can use analysis of variance tests:

1. A researcher is interested in examining differences in life expectancy by the wealth level of nations. **See page 269**
2. A sociologist wants to detect differences in hours worked last week by social class. **See page 271**
3. To examine the influence of religion on public policy, you might use an ANOVA test to see if their sexual education practices vary by the degree and nature of religiosity in the state. **See page 279**

In this chapter, we examine a flexible and widely used test of significance called the **analysis of variance** (often abbreviated as **ANOVA**). This test is designed to be used with interval-ratio–level dependent variables and is a powerful tool for analyzing the most sophisticated and precise measurements you are likely to encounter.

It is perhaps easiest to think of ANOVA as an extension of the t test for the significance of the difference between two sample means, which was presented in Chapter 9. The t test is used in situations in which our independent variable has exactly two categories (e.g., Protestants and Catholics). The analysis of variance,

on the other hand, is appropriate for independent variables with more than two categories (e.g., Protestants, Catholics, Jews, people with no religious affiliation, Muslims, Hindus, and so forth).

To illustrate, suppose we were interested in analyzing the relationship between support for capital punishment (the dependent variable) and age group (the independent variable). Do attitudes change as people age? Are older people more supportive of the death penalty than younger people?

Suppose that we administered a scale that measures support for capital punishment at the interval-ratio level to a randomly selected sample and have divided the respondents into four age groups: 18–29, 30–45, 46–64, and 65 and older. We have four categories of subjects, and we want to see whether opinion varies significantly by the category (age group) into which a person is classified.

The Logic of the Analysis of Variance

For ANOVA, the null hypothesis is that the populations from which the samples are drawn have the same mean score on the dependent variable. As applied to our problem, the null hypothesis could be phrased as "People from different age groups do not vary in their support for the death penalty" or, symbolically, as $\mu_1 = \mu_2 = \mu_3 = \mu_4$. (Note that this is an extended version of the null hypothesis for the two-sample t test.) The researcher will normally want to reject the null hypothesis and, in this case, show that support is related to age group.

If the null hypothesis of "no difference" in the populations is true, then any means calculated from randomly selected samples should be roughly equal in value. The average score for the 18–29-year-olds should be about the same as the average score for each of the other age groups. Note that the averages are unlikely to be exactly the same value even if the null hypothesis really is true, because we will always encounter some error or chance fluctuations in the measurement process. We are *not* asking: "Are there differences between the samples (or, in our example, the age groups)?" Rather, we are asking: "Are the differences between the samples large enough to reject the null hypothesis and justify the conclusion that the populations from which the samples were drawn are different?"

Consider what kinds of outcomes we might encounter if we administered a "Support of Capital Punishment Scale" and organized the scores by age group. Let's focus on the two extreme outcomes presented in Tables 10.1 and 10.2.

TABLE 10.1 Support for Capital Punishment by Age Group (fictitious data)

	18–29	30–45	46–64	65+
Mean	10.3	11.0	10.1	9.9
Standard deviation	2.4	1.9	2.2	1.7

TABLE 10.2 Support for Capital Punishment by Age Group (fictitious data)

	18–29	30–45	46–64	65+
Mean	10.0	13.0	16.0	22.0
Standard deviation	2.4	1.9	2.2	1.7

In the first set of hypothetical results (Table 10.1), we see that the means and standard deviations of the groups are quite similar. The average scores are about the same, and all four groups exhibit about the same variation. These results would be consistent with the null hypothesis of no difference between the populations. Neither the average score nor the variation of the scores varies in any important way by age group.

Now consider another set of fictitious results, as displayed in Table 10.2. Here we see substantial differences in average score, with the youngest age group showing the least support and the oldest group showing the most. Also, the standard deviations are low and similar from category to category, indicating that there is not much variation within the age groups. Table 10.2 shows marked differences between age groups combined with homogeneity (or low standard deviations) within age groups. In other words, the age groups are different from each other and the variation within each age group is low. These results would contradict the null hypothesis and support the notion that support for the death penalty does vary by age group.

ANOVA proceeds by making the kinds of comparisons outlined above. The test compares the amount of variation between categories (that is, between age groups) with the amount of variation within categories (within each age group). The greater the difference between categories (as measured by the means) relative to the differences within categories (as measured by the standard deviations), the more likely that the null hypothesis of "no difference" can be rejected. If support for capital punishment does vary by age group, then the sample mean for each age group should be quite different from the others and variation within the categories should be relatively low.

The Computation of ANOVA

Even though we have been thinking of ANOVA as a test for the significance of the difference between sample means, the computational routine involves developing two separate estimates of the population variance, σ^2 (hence the name *analysis of variance*). Recall from Chapter 4 that the variance is the standard deviation squared. One estimate of the population variance is based on the amount of variation within each of the categories of the independent variable, and the other is based on the amount of variation between the categories.

Before constructing these estimates, we need to introduce some new concepts and statistics. The first new concept is the total variation of the scores, which is measured by a quantity called the **total sum of squares**, or **SST**:

FORMULA 10.1
$$\text{SST} = \Sigma X_i^2 - N\overline{X}^2$$

To solve this formula, first find the sum of the squared scores (in other words, square each score and then add up the squared scores). Next, square the mean of all scores, multiply that value by the total number of cases in the sample (*N*), and subtract that quantity from the sum of the squared scores.

Formula 10.1 may seem vaguely familiar. A similar expression, $\Sigma(X_i - \overline{X})^2$, appears in the formula for the standard deviation and variance (see Chapter 4).

All three statistics incorporate information about the variation of the scores (or, in the case of SST, the squared scores) around the mean (or, in the case of SST, the square of the mean multiplied by N). All three of these statistics are measures of the variation of the scores.

To construct the two separate estimates of the population variance, we divide the total variation (SST) into two components. One component reflects the pattern of variation within each of the categories and is called the **sum of squares within (SSW)**. In our example problem, SSW would measure the amount of variation in support for the death penalty within each of the age groups.

The other component is based on the variation between categories and is called the **sum of squares between (SSB)**. Using our example to illustrate, SSB measures how different people in each age group are from each other in their support for capital punishment. SSW and SSB are components of SST, as reflected in Formula 10.2:

FORMULA 10.2
$$SST = SSB + SSW$$

Let's start with the computation of SSB, our measure of the variation between categories. We use the category means as summary statistics to determine the size of the difference from category to category. In other words, we compare the average support for the death penalty for each age group with the average support for all age groups to determine SSB. The formula for the sum of squares between (SSB) is

FORMULA 10.3
$$SSB = \Sigma N_k(\overline{X}_k - \overline{X})^2$$

where SSB = the sum of squares between the categories
N_k = the number of cases in a category
\overline{X}_k = the mean of a category

To find SSB, subtract the overall mean of all scores (\overline{X}) from each category mean (\overline{X}_k), square the difference, multiply by the number of cases in the category (N_k), and add the results across all the categories.

The second estimate of the population variance (SSW) is based on the amount of variation within the categories. Formula 10.2 shows that the total sum of squares (SST) is equal to the sum of SSW and SSB. This relationship means that we can find SSW by simple subtraction. Formula 10.4 rearranges the symbols in Formula 10.2:

FORMULA 10.4
$$SSW = SST - SSB$$

Let's pause for a second to remember what we are after here. If the null hypothesis is true, then there should not be much variation from category to category (see Table 10.1), and SSW and SSB should be roughly equal. If the null hypothesis is not true, there will be large differences between categories (see Table 10.2) relative to the differences within categories, and SSB should be much larger than SSW. SSB will increase as the difference between category means increases, especially when there is not much variation within the categories (SSW). The larger SSB is compared to SSW, the more likely we are to reject the null hypothesis.

The next step in the computational routine is to construct the estimates of the population variance. To do this, we divide each sum of squares by its respective degrees of freedom. To find the degrees of freedom associated with SSW, subtract the number of categories (k) from the number of cases (N). The degrees of freedom associated with SSB are the number of categories minus 1. In summary,

FORMULA 10.5
$$\text{dfw} = N - k$$

where dfw = degrees of freedom associated with SSW
N = total number of cases
k = number of categories

FORMULA 10.6
$$\text{dfb} = k - 1$$

where dfb = degrees of freedom associated with SSB
k = number of categories

The actual estimates of the population variance, called the **mean square estimates**, are calculated by dividing each sum of squares by its respective degrees of freedom:

FORMULA 10.7
$$\text{Mean square within} = \frac{\text{SSW}}{\text{dfw}}$$

FORMULA 10.8
$$\text{Mean square between} = \frac{\text{SSB}}{\text{dfb}}$$

The test statistic calculated in step 4 of the five-step model is called the **F ratio**, and its value is determined by the following formula:

FORMULA 10.9
$$F(\text{obtained}) = \frac{\text{Mean square between}}{\text{Mean square within}}$$

As you can see, the value of the F ratio is a function of the amount of variation between categories (based on SSB) compared to the amount of variation within the categories (based on SSW). The greater the variation between the categories relative to the variation within, the higher the value of the F ratio and the more likely we will reject the null hypothesis.

An Example of Computing the Analysis of Variance

Assume that we have administered our Support for Capital Punishment Scale to a sample of 16 individuals who are equally divided among the four age categories. (Obviously, this sample is much too small for any serious research and is intended solely for purposes of illustration.) All scores are reported in Table 10.3 along with the other quantities needed to complete the computations. The scores (X_i) are listed for each of the four categories, and a column has been added for the squared scores (X_i^2). The sums of both X_i and X_i^2 are reported at the bottom of each column. The category means (\overline{X}_k) show that the four 18–29-year-olds

TABLE 10.3 Support for Capital Punishment by Age Group (fictitious data)

18–29		30–45		46–64		65+	
X_i	X_i^2	X_i	X_i^2	X_i	X_i^2	X_i	X_i^2
7	49	10	100	12	144	17	289
8	64	12	144	15	225	20	400
10	100	13	169	17	289	24	576
15	225	17	289	20	400	27	729
40	438	52	702	64	1058	88	1994
$\overline{X}_k = 10.0$		$\overline{X}_k = 13.0$		$\overline{X}_k = 16.0$		$\overline{X}_k = 22.0$	
			$\overline{X} = 15.25$				

averaged 10.0 on the Support for Capital Punishment Scale, the four 30–45-year-olds averaged 13.0, and so forth. Finally, the overall mean (sometimes called the *grand mean*) is reported in the bottom row of the table. This shows that the average score for all 16 respondents was 15.25.

To organize our computations, we'll follow the routine summarized in the "One Step at a Time" box. We begin by finding SST (Formula 10.1):

$$SST = \Sigma X_i^2 - N\overline{X}^2$$
$$SST = (438 + 702 + 1058 + 1994) - (16)(15.25)^2$$
$$SST = 4192 - (16)(232.56)$$
$$SST = 4192 - 3720.96$$
$$SST = 471.04$$

The sum of squares between (SSB) is found by means of Formula 10.3:

$$SSB = \Sigma N_k(\overline{X}_k - \overline{X})^2$$
$$SSB = 4(10 - 15.25)^2 + 4(13 - 15.25)^2$$
$$\qquad + 4(16 - 15.25)^2 + 4(22 - 15.25)^2$$
$$SSB = 4(-5.25)^2 + 4(-2.25)^2 + 4(0.75)^2 + 4(6.75)^2$$
$$SSB = 4(27.56) + 4(5.06) + 4(0.56) + 4(45.56)$$
$$SSB = 110.24 + 20.24 + 2.24 + 182.24$$
$$SSB = 314.96$$

Now SSW can be found by subtraction (Formula 10.4):

$$SSW = SST - SSB$$
$$SSW = 471.04 - 314.96$$
$$SSW = 156.08$$

To find the degrees of freedom for the two sums of squares, we use Formulas 10.5 and 10.6:

$$dfw = N - k = 16 - 4 = 12$$
$$dfb = k - 1 = 4 - 1 = 3$$

Finally, we are ready to construct the mean square estimates of the population variance. For the estimate based on SSW, we use Formula 10.7:

$$\text{Mean square within} = \frac{\text{SSW}}{\text{dfw}} = \frac{156.08}{12} = 13.00$$

For the between estimate, we use Formula 10.8:

$$\text{Mean square between} = \frac{\text{SSB}}{\text{dfb}} = \frac{314.96}{3} = 104.99$$

The test statistic, or *F* ratio, is found by using Formula 10.9:

$$F = \frac{\text{Mean square between}}{\text{Mean square within}} = \frac{104.99}{13.00} = 8.08$$

This statistic must still be evaluated for its significance. *(Solve any of the end-of-chapter problems to practice computing these quantities and solving these formulas.)*

A Test of Significance for ANOVA

Now we will see how to test an *F* ratio for significance. We will also look at some of the assumptions underlying the ANOVA test using the five-step model.

Step 1. Making Assumptions and Meeting Test Requirements.

> Model: Independent random samples
> Level of measurement is interval-ratio
> Populations are normally distributed
> Population variances are equal

The model assumptions are quite strict and underscore the fact that ANOVA should be used only with dependent variables that have been carefully and precisely measured. However, as long as the categories are roughly equal in size, ANOVA can tolerate some violation of the model assumptions. In situations where you are uncertain or have samples of very different sizes, it is probably advisable to use an alternative test. (Chi square, described in Chapter 11, is one option.)

Step 2. Stating the Null Hypothesis. For ANOVA, the null hypothesis always states that the means of the populations from which the samples were drawn are

equal. For our example problem, we are concerned with four different populations, or categories, so our null hypothesis would be

$$H_0: \mu_1 = \mu_2 = \mu_3 = \mu_4$$

where μ_1 represents the mean for the 18–29 age group,
μ_2 the mean for people 30–45, and so forth.

The alternative hypothesis states simply that at least one of the population means is different. The wording here is important. If we reject the null hypothesis, ANOVA does not identify which mean or means are significantly different. In the final section of the chapter, we briefly consider an advanced test that can help us identify which pairs of means are significantly different.

$(H_1:$ At least one of the population means is different.)

Step 3. Selecting the Sampling Distribution and Establishing the Critical Region. The sampling distribution for ANOVA is the F distribution, which is summarized in Appendix D. Note that there are separate tables for alphas of 0.05 and 0.01. As with the t table, the value of the critical F score will vary by degrees of freedom. For ANOVA, there are two separate degrees of freedom, one for each estimate of the population variance. The numbers across the top of the table (the n_1 values) are the degrees of freedom associated with the between estimate (dfb), and the numbers down the side of the table (the n_2 values) are those associated with the within estimate (dfw). In our example, dfb is $(k - 1)$, or 3, and dfw is $(N - k)$, or 12 (see Formulas 10.5 and 10.6). So, if we set alpha at 0.05, our critical F score will be 3.49.

Summarizing these considerations:

$$\text{Sampling distribution} = F \text{ distribution}$$
$$\text{Alpha } (\alpha) = 0.05$$
$$\text{Degrees of freedom between (dfb)} = (k - 1) = 3$$
$$\text{Degrees of freedom within (dfw)} = (N - k) = 12$$
$$F(\text{critical}) = 3.49$$

Take a moment to inspect the two F tables and you will notice that all the values are greater than 1.00. This is because ANOVA is a one-tailed test and we are concerned only with outcomes in which there is more variance between categories than within categories. F values of less than 1.00 would indicate that the between estimate was lower in value than the within estimate, and, because we would always fail to reject the null hypothesis in such cases, we simply ignore this class of outcomes.

Step 4. Computing the Test Statistic. This was done in the previous section, where we found an obtained F ratio of 8.08.

Step 5. Making a Decision and Interpreting the Results of the Test. Compare the test statistic with the critical value:

$$F\text{(critical)} = 3.49$$

$$F\text{(obtained)} = 8.08$$

The test statistic falls in the critical region, so our decision would be to reject the null hypothesis. Support for capital punishment differs significantly by age group.

ONE STEP AT A TIME **Completing Step 4 of the Five-Step Model: Computing F(obtained)**

To compute the obtained F ratio, use Formulas 10.1, 10.3, and 10.4 to find SST, SSB, and SSW. Then, calculate the degrees of freedom, mean square estimates of the population variance, and the obtained F ratio. It is strongly recommended that you use a computing table like Table 10.3 to organize the computations.

To Solve Formula 10.1 (SST)

1. To find ΣX_i^2:
 a. Square each score.
 b. Find the sum of the squared scores.
2. To find $N\overline{X}^2$:
 a. Square the value of the mean.
 b. Multiply \overline{X}^2 by N.
3. Subtract the quantity you found in step 2 from the quantity you found in step 1. This is SST.

To Solve Formula 10.3 (SSB)

1. Subtract the mean of all scores (\overline{X}) from the mean of each category (\overline{X}_k) and then square each difference.
2. Multiply each of the squared differences you found in step 1 by the number of cases in the category (N_k).
3. Add all of the quantities you found in step 2. This is SSB.

To Solve Formula 10.4 (SSW)

1. Subtract the value of SSB from the value of SST.

To Solve Formulas 10.5 and 10.6 (dfw and dfb)

1. For dfw, subtract the number of categories (k) from the number of cases (N).
2. For dfb, subtract 1 from the number of categories (k).

To Solve Formulas 10.7 and 10.8 (Mean Square Estimates)

1. To find the mean square within estimate, divide SSW by dfw.
2. To find the mean square between estimate, divide SSB by dfb.

To Solve Formula 10.9 (Obtained F ratio)

1. Divide the mean square between estimate by the mean square within estimate. The result is the obtained F ratio.

ONE STEP AT A TIME | **Completing Step 5 of the Five-Step Model: Making a Decision and Interpreting Results**

1. Compare F(obtained) to F(critical). If F(obtained) is greater than F(critical), it is *in* the critical region. *Reject* the null hypothesis. If F(obtained) is less than F(critical), it is *not in* the critical region. *Fail to reject* the null hypothesis.
2. Interpret the decision to reject or fail to reject the null hypothesis in terms of the original question. For example, our conclusion for the example problem was "Support for capital punishment varies significantly by age group."

STATISTICS | **In Everyday Life**

What Causes Obesity?

Obesity has become a significant problem in the United States. Is this a result of genetic factors? How do lifestyle choices such as diet and exercise affect weight gain? A team of researchers explored these questions in a sophisticated analysis of more than 8000 people whose health has been monitored since 1948. Among other conclusions, they found that the effects of genetic factors lessen as people age and are reduced by physical activity. Although the study used sophisticated statistics not covered in this text, the researchers relied on ANOVA to test many of the relationships they explored.

Source: Guo, G., H. Liu, L. Wang, H. Shen, and W. Hu. 2015. "The Genome-Wide Influence on Human BMI Depends on Physical Activity, Life Course, and Historical Period." *Demography*, 52, 1651–1670.

Social Research and Data Analysis 10.1 | **The Analysis of Variance**

A random sample of 20 nations from four income levels has been selected. "Low income" nations are largely agricultural and have the lowest quality of life. "High income" nations are industrial and the most affluent and modern. "Upper-middle income" and "lower-middle income" nations are between these extremes. Are these general characteristics reflected in differences in life expectancy (the number of years the average male citizen can expect to live at birth) between the categories?

Life Expectancy for Males (L.E.) by Income Level

INCOME LEVELS							
Low		Lower-Middle		Upper-Middle		High	
Nation	*L.E.*	*Nation*	*L.E.*	*Nation*	*L.E.*	*Nation*	*L.E.*
Cambodia	67	Bolivia	67	China	75	Australia	80
Malawi	61	Guatemala	69	Colombia	73	Canada	80
Nepal	70	India	67	Jordan	73	Japan	81
Niger	59	Philippines	66	Libya	69	Spain	80
Zimbabwe	58	Vietnam	71	Russia	68	U.S.	76

Source: Population Reference Bureau. 2018 World Population Data Sheet. Available at http://www.prb.org/.

(continued)

Social Research and Data Analysis 10.1 *(Continued)*

To find F(obtained) and conduct the ANOVA test, computations will be organized into table format:

	Low		Lower-Middle		Upper-Middle		High	
	X_i	X_i^2	X_i	X_i^2	X_i	X_i^2	X_i	X_i^2
	67	4489	67	4489	75	5625	80	6400
	61	3721	69	4761	73	5329	80	6400
	70	4900	67	4489	73	5329	81	6561
	59	3481	66	4356	69	4761	80	6400
	58	3364	71	5041	68	4624	76	5776
$\Sigma X_i =$	315		340		358		397	
$\Sigma X_i^2 =$		19,955		23,136		25,668		31,537
$\bar{X}_k =$	63.0		68.0		71.6		79.4	

$$\bar{X} = 70.5$$

The ANOVA test will tell us if these differences are large enough to justify the conclusion that they did not occur by chance alone. Following the usual computational routine:

$$SST = \Sigma X_i^2 - N\bar{X}^2$$

$$SST = (19{,}955 + 23{,}136 + 25{,}668 + 31{,}537) - (20)(70.5)^2$$

$$SST = (100{,}296) - (20)(4970.25)$$

$$SST = 100{,}296 - 99{,}405.00$$

$$SST = 891.00$$

$$SSB = \Sigma N_k(\bar{X}_k - \bar{X})^2$$

$$SSB = 5(63.0 - 70.5)^2 + 5(68.0 - 70.5)^2$$
$$+ 5(71.6 - 70.5)^2 + 5(79.4 - 70.5)^2$$

$$SSB = 5(-7.50)^2 + 5(-2.50)^2 + 5(1.10)^2 + 5(8.90)^2$$

$$SSB = 5(56.25) + 5(6.25) + 5(1.21) + 5(79.21)$$

$$SSB = 281.25 + 31.25 + 6.05 + 396.05$$

$$SSB = 714.60$$

$$SSW = SST - SSB$$

$$SSW = 891.00 - 714.60$$

$$SSW = 176.40$$

$$dfw = N - k = 20 - 4 = 16$$
$$dfb = k - 1 = 4 - 1 = 3$$

$$\text{Mean Square Within} = \frac{SSW}{dfw} = \frac{176.40}{16} = 11.025$$

$$\text{Mean Square Between} = \frac{SSB}{dfb} = \frac{714.60}{3} = 238.20$$

$$F(\text{obtained}) = \frac{\text{Mean Square Between}}{\text{Mean Square Within}} = \frac{238.20}{11.025} = 21.61$$

We can now conduct the test of significance.

Step 1. Making Assumptions and Meeting Test Requirements.

> Model: Independent random samples
> Level of measurement is interval-ratio
> Populations are normally distributed
> Population variances are equal

Step 2. Stating the Null Hypothesis.

$$H_0: \mu_1 = \mu_2 = \mu_3 = \mu_4$$

(H_1: At least one of the population means is different.)

Step 3. Selecting the Sampling Distribution and Establishing the Critical Region.

> Sampling distribution = F distribution
> Alpha = 0.05
> dfb = 3
> dfw = 16
> F(critical) = 3.24

Step 4. Computing the Test Statistic. We found an obtained F ratio of 21.61.

Step 5. Making a Decision and Interpreting the Results of the Test. Compare the test statistic with the critical value

> F(critical) = 3.24
> F(obtained) = 21.61

The null hypothesis ("The population means are equal") can be rejected. The differences in life expectancy between nations at different income levels are statistically significant and reflect differences in the populations from which these samples were selected.

USING SPSS

The ANOVA Test

SPSS provides several different ways of conducting the analysis of variance test. The procedure summarized below—**One-Way ANOVA**—is the most accessible of these, but it still incorporates options and capabilities that we have not covered in this chapter. If you wish to explore these possibilities, please use the online **Help** facility for SPSS.

To provide an illustration, we will test for significant differences in the number of hours worked last week by social class. ANOVA will allow us to expand the number of categories in the independent variable (*class*) beyond two. Our dependent variable will be (*hrs1*) for the number of hours worked last week.

To begin:

1. Click the SPSS icon on your desktop.
2. Load the *GSS2018* database.
 a. Find the **File** command on the far left of the menu bar and then click **File** → **Open** → **Data**.
 b. Find the *GSS2018* database and click the file name.
3. From the main menu bar, click **Analyze**, then **Compare Means**, and then **One-Way ANOVA**. The **One-Way ANOVA** window appears. Find *hrs1* in the variable list on the left and click the arrow to move the variable name into the **Dependent List** box. Note that you can request more than one dependent variable at a time, but that won't be necessary here. Next, find *class* and click the arrow to move the variable name into the **Factor** box.
4. Click **Options** and then click the box next to **Descriptive** in the **Statistics** box to request means and standard deviations along with the analysis of variance. Click **Continue**.
5. Finally, click **Post Hoc** and select LSD under Equal Variances Assumed. Post Hoc tests permit us to see which differences between means are most significant. See the "Post Hoc Analysis" section toward the end of this chapter for more information. Click **Continue** and then click **OK**, and the following output will be produced:

	N	Mean	Std. Deviation	Std. Error
LOWER CLASS	74	35.69	16.666	1.937
WORKING CLASS	710	41.63	14.162	.532
MIDDLE CLASS	547	41.34	13.874	.593
UPPER CLASS	48	44.81	19.605	2.830
Total	1379	41.31	14.473	.390

As you can see, the group that worked the most hours last week is the Upper Class group (\overline{X} = 44.81), followed by the working and middle class groups, which are close together. The Lower Class group reported the fewest hours, at 35.69. Are these differences significant? Let's check the ANOVA test results.

ANOVA

Number of hours worked last week

	Sum of Squares	df	Mean Square	F	Sig.
Between Groups	3000.532	3	1000.177	4.815	0.002
Within Groups	285641.868	1375	207.740		
Total	288642.400	1378			

(continued)

USING SPSS *(Continued)*

The output box labeled **ANOVA** reports the degrees of freedom, all of the sums of squares, the mean square estimates, the F ratio (4.815), and, at the far right, the probability ("Sig.") of getting these results if the null hypothesis is true. This is reported as 0.002, which is less than 0.05, so we can be 95% confident that the differences in the hours worked last week for these groups did not occur by chance. But let's take it a step further. We can use the post hoc tests to examine the differences by groups.

Post Hoc Tests

(I) Subjective Class Identification	(J) Subjective Class Identification	Mean Difference (I − J)	Std. Error	Sig.
LOWER CLASS	WORKING CLASS	−5.942*	1.761	0.001
	MIDDLE CLASS	−5.655*	1.785	0.002
	UPPER CLASS	−9.123*	2.671	0.001
WORKING CLASS	LOWER CLASS	5.942*	1.761	0.001
	MIDDLE CLASS	.287	.820	0.726
	UPPER CLASS	−3.182	2.150	0.139
MIDDLE CLASS	LOWER CLASS	5.655*	1.785	0.002
	WORKING CLASS	−.287	.820	0.726
	UPPER CLASS	−3.469	2.170	0.110
UPPER CLASS	LOWER CLASS	9.123*	2.671	0.001
	WORKING CLASS	3.182	2.150	0.139
	MIDDLE CLASS	3.469	2.170	0.110

The output box labeled **Post Hoc Tests/Multiple Comparisons** shows us the mean difference between groups. On the first line, we can see the difference between the Lower Class and the Working Class group is −5.942. Since it is negative, this means that the first group (Lower Class) worked fewer hours than the second group (Working Class). We can also see the standard error and the significance level. For this first row, the difference between Lower Class and Working Class is significant at the 0.001 level (note that it is less than 0.05), leading us to 95% certainty that the difference is not due to chance.

Moving down the significance column, we can see that most of the differences between groups are also significant ($p < 0.05$), especially those involving the Lower Class. The single biggest difference, more than nine hours, is between the Lower and Upper Classes.

An Additional Example of Computing and Testing the Analysis of Variance

In this section, we work through an additional example of the computation and interpretation of the ANOVA test. We first review matters of computation, find the obtained F ratio, and then test the statistic for its significance.

TABLE 10.4 Number of Days Required to Process Cases for Three Agencies (fictitious data)

Client	Agency A X_i	Agency A X_i^2	Agency B X_i	Agency B X_i^2	Agency C X_i	Agency C X_i^2
1	5	25	12	144	9	81
2	7	49	10	100	8	64
3	8	64	19	361	12	144
4	10	100	20	400	15	225
5	4	16	12	144	20	400
6	9	81	11	121	21	441
7	6	36	13	169	20	400
8	9	81	14	196	19	361
9	6	36	10	100	15	225
10	6	36	9	81	11	121
$\Sigma X_i =$	70		130		150	
$\Sigma X_i^2 =$		524		1816		2462
$\overline{X}_k =$	7.0		13.0		15.0	

$$\overline{X} = 350/30 = 11.67$$

A researcher has been asked to evaluate the efficiency of three social service agencies. One area of concern is how fast the agencies are determining the eligibility of clients for a particular program. The researcher has information on the number of days it took to process random samples of 10 cases in each agency. Is there a significant difference between the agencies? The data are reported in Table 10.4, which also includes the information we will need to complete our calculations.

To find SST by means of Formula 10.1:

$$SST = \Sigma X_i^2 - N\overline{X}^2$$
$$SST = (524 + 1816 + 2462) - (30)(11.67)^2$$
$$SST = 4802 - (30)(136.19)$$
$$SST = 4802 - 4085.70$$
$$SST = 716.30$$

To find SSB by means of Formula 10.3:

$$SSB = \Sigma N_k(\overline{X}_k - \overline{X})^2$$

$$SSB = (10)(7.0 - 11.67)^2 + (10)(13.0 - 11.67)^2 + (10)(15.0 - 11.67)^2$$

$$SSB = (10)(-4.67)^2 + (10)(1.33)^2 + (10)(3.33)^2$$

$$SSB = (10)(21.81) + (10)(1.77) + (10)(11.09)$$

$$SSB = 218.10 + 17.70 + 110.90$$

$$SSB = 346.70$$

Now we can find SSW using Formula 10.4:

$$\text{SSW} = \text{SST} - \text{SSB}$$
$$\text{SSW} = 716.30 - 346.70$$
$$\text{SSW} = 369.60$$

The degrees of freedom are found through Formulas 10.5 and 10.6:

$$\text{dfb} = k - 1 = 3 - 1 = 2$$
$$\text{dfw} = N - k = 30 - 3 = 27$$

The estimates of the population variances are found by means of Formulas 10.7 and 10.8:

$$\text{Mean square within} = \frac{\text{SSW}}{\text{dfw}} = \frac{369.6}{27} = 13.69$$

$$\text{Mean square between} = \frac{\text{SSB}}{\text{dfb}} = \frac{346.70}{2} = 173.35$$

The F ratio (Formula 10.9) is:

$$F(\text{obtained}) = \frac{\text{Mean square between}}{\text{Mean square within}} = \frac{173.35}{13.69} = 12.66$$

And we can now test this value for its significance.

Step 1. Making Assumptions and Meeting Test Requirements.

> Model: Independent random samples
> Level of measurement is interval-ratio
> Populations are normally distributed
> Population variances are equal

The researcher will always be in a position to judge the adequacy of the first two assumptions in the model. The second two assumptions are more problematic, but remember that ANOVA will tolerate some deviation from its assumptions as long as sample sizes are roughly equal.

Step 2. Stating the Null Hypothesis.

$$H_0: \mu_1 = \mu_2 = \mu_3$$

(H_1: At least one of the population means is different.)

Step 3. Selecting the Sampling Distribution and Establishing the Critical Region.

> Sampling distribution = F distribution
> Alpha = 0.05

$$\text{Degrees of freedom (between)} = (k - 1) = (3 - 1) = 2$$
$$\text{Degrees of freedom (within)} = (N - k) = (30 - 3) = 27$$
$$F\text{(critical)} = 3.35$$

Step 4. Computing the Test Statistic. We found an obtained F ratio of 12.66.

Step 5. Making a Decision and Interpreting the Results of the Test.
Compare the test statistic with the critical value:

$$F\text{(critical)} = 3.35$$
$$F\text{(obtained)} = 12.66$$

The test statistic is in the critical region, and we would reject the null hypothesis of no difference. The differences between the three agencies are very unlikely to have occurred by chance alone. The agencies are significantly different in the speed with which they process paperwork and determine eligibility. *(For practice in conducting the ANOVA test, see problems 10.2 to 10.8. Begin with the lower-numbered problems since they have smaller data sets, fewer categories, and, therefore, the simplest calculations.)*

The Limitations of the Test

ANOVA is appropriate when you want to test differences between the means of an interval-ratio–level dependent variable between three or more categories of an independent variable. This application is called a **one-way analysis of variance**, because it involves the effect of a single variable (e.g., religion) on another (e.g., support for capital punishment). This is the simplest application of ANOVA, and you should be aware that the technique has numerous more advanced and complex forms. For example, you may encounter research projects in which the effects of two separate variables (such as religion and gender) on some third variable were observed.

One important limitation of ANOVA is that it requires an interval-ratio dependent variable and roughly equal numbers of cases in each category of the independent variable. The former condition may be difficult to meet with complete confidence for many variables of interest in the social sciences. The latter condition may create problems when the research hypothesis calls for comparisons between groups that are, by their nature, unequal in numbers (for example, white versus black Americans). Neither of these limitations should be particularly crippling, because ANOVA tolerates some deviation from its model assumptions, but you should be aware of these limitations in planning your own research as well as in judging the adequacy of research conducted by others.

Another limitation of ANOVA, introduced in Chapter 9, applies to all forms of significance testing. Tests of significance are designed to detect nonrandom differences or differences so large that they are unlikely to be produced by random chance alone. The problem is that differences that are statistically significant

are not necessarily important in any other sense. Parts III and IV of this text provide some statistical techniques that can directly assess the importance of results directly.

A final limitation of ANOVA relates to the research hypothesis, which simply asserts that at least one of the population means is different from the others. When we reject the null hypothesis, we would like to know *which* differences between the sample means are significant. We can sometimes make this determination by simple inspection. In our problem involving social service agencies, it is clear from Table 10.4 that Agency A is the source of most of the differences. This informal "eyeball" method can be misleading, however, and you should exercise caution in drawing conclusions. In the next section, we provide an extended example of the interpretation of ANOVA and introduce a technique (called **post hoc**, or "after the fact," analysis) that permits us to reliably identify significant differences between the sample means. The computational routines for post hoc analysis are beyond the scope of this text, but the tests are commonly available in computerized statistical packages such as SPSS.

Interpreting Statistics: Does Sexual Activity Vary by Age?

Does age make a difference in people's level of sexual activity? Do younger individuals have sex more often? Do they report having more sexual partners? We will use the data from the full 2018 General Social Survey to answer these questions scientifically. Recall that the respondents to this survey are a representative sample (see Chapter 6) of the population of the United States.

This analysis involves three variables. The independent variable is age, and there will be two separate dependent variables: the number of different sexual partners in the past five years (*partnrs5* in the GSS 2018) and the number of times the respondent had sex during the last 12 months (*sexfreq*). We have recoded age into four groups of roughly equal size for this analysis. The two dependent variables are ordinal in level of measurement rather than the interval-ratio level required by ANOVA. Respondents were asked to describe their level of sexual activity by selecting from broad categories such as "2 or 3 times a month" (see Appendix G for a complete listing of the scores of the two variables that measure sexual activity).

We encountered a similar problem in Chapter 9 when we assessed the gender gap in income. Then, we solved the problem by substituting midpoints for each of the intervals of the variables, and we will follow a similar strategy. On the number of sex partners, we will substitute 7.5 for the category "5 to 10," and a value of 100 will be substituted for the "More than 100 partners" category. The variable measuring frequency of sexual activity has been transformed in a similar way: A value of 12 has been substituted for "about once a month," 52 for "about once a week," and so forth.

The summary statistics for the two sexual activity variables are presented in Table 10.5. The sample averaged 2.52 sex partners over the last five years and had sex at a rate of about once a week over the past year (53.81). Note that the standard deviation is high relative to the mean for both variables. This is caused by extreme positive skews: Most cases are grouped in the low range of the variable (e.g., they had only one or two sexual partners), but a few respondents have very high scores.

TABLE 10.5 Descriptive Statistics for Sexual Activity Variables

	Number of Times Respondent Had Sex During the Past Year	Number of Different Sex Partners over the Past Five Years
Mean	53.81	2.52
Standard Deviation	66.31	6.66
$N =$	1752	1767

TABLE 10.6 Number of Sexual Partners in the Last Five Years by Age Group

	18–33	34–48	49–63	64+	F Ratio	Alpha
Mean	4.29	2.10	1.46	0.70	29.24	<0.001
Standard Deviation	9.30	3.63	2.18	0.66		
$N =$	355	366	366	296		

The means and standard deviations for number of partners over the last five years for each age group are reported in Table 10.6. The F ratio for these differences is 29.24, which is significant at less than the 0.001 level. Age does make a significant difference in the number of sex partners. Comparing the category means, the results are what one would expect. The youngest respondents had the highest average number of partners (4.29), and the average number of partners decreased as age increased. The oldest age group had the lowest average score on this measure of sexual activity.

Table 10.7 reports the means and standard deviations for the number of times the respondent reported having sex in the last 12 months by age group. Again the younger age groups reported higher average levels of sexual activity, with an average rate of over 60 times for the year (more than once a week). The rate of sexual activity declined as age group increased and the oldest respondents reported an average rate of about 14 times, or a little more than once a month. The F ratio of 62.99 is significant at less than 0.001, so we can conclude that the rate of sexual activity is significantly affected by age.

POST HOC ANALYSIS

Tables 10.6 and 10.7 indicate that the relationships are significant, but which differences between sample means are most important and contribute the most to

TABLE 10.7 Frequency of Sex over the Past Year by Age Group

	18–33	34–48	49–63	64+	F Ratio	Alpha
Mean	61.28	61.67	34.21	14.42	62.99	<0.001
Standard Deviation	62.34	57.12	48.60	31.22		
$N =$	359	361	341	296		

the significant F ratios? We can see by inspection that the oldest respondents are very different from the others, but what other differences might be important? A post hoc analysis of the differences in sample means can provide objective answers to these questions.

We will not present the computational routines for post hoc tests, but the interpretation of the results of these tests is straightforward. The tests compare the means of all possible pairs of categories (i.e., the 18–33-year-olds with the 34–48-year-olds, the 49–63-year-olds with the 64 and older group, and so forth) and tell us exactly which combinations contribute the most to a significant F ratio. Comparing sample means in this way increases the probability of making an alpha error (falsely rejecting a true null hypothesis—see Chapter 8); to correct for this, post hoc tests use more stringent criteria to identify significant differences.

Table 10.8 presents the results of post hoc tests for both measures of sexual activity. The categories are listed at the left-hand side of the table and again across the top, along with the category means. The entries in the table are the differences between the means for the category listed at the left of the table and the category listed across the top. A positive difference indicates that the category listed to the left had a higher score, and a negative difference means that the category listed across the top had the higher score. For example, the average number of partners over the last five years was 4.29 for the 18–33 age group (category listed on the left) and 2.10 for the 34–48-year-olds (category listed across the top). The difference between the two categories means is $4.29 - 2.10$, or 2.19, the value reported at the intersection of both categories. The positive value means that younger respondents (category to the left) had a higher score than older respondents (category across the top). An asterisk (*) next to the value means that the difference is significant at the 0.05 level.

Looking first at the top portion of the table (number of sexual partners over the last five years), we see that all differences are significant except for two: the difference between the 34–48-year-olds and the 49–63-year-olds and the difference between the 49–63-year-olds and the oldest group. The largest difference is

TABLE 10.8 A Post Hoc Test for Differences in Sexual Activity by Age

Age Group	Mean Differences in Number of Partners		
	34–48 (2.10)	*49–63 (1.46)*	*64+ (0.70)*
18–33 (4.29)	2.19*	2.82*	3.60*
34–48 (2.10)		0.64	1.41*
49–63 (1.46)			0.77
	Mean Differences in Frequency of Sex		
	34–48 (61.67)	*49–63 (34.21)*	*64+ (14.42)*
18–33 (61.28)	−0.39	26.96*	46.86*
34–48 (61.67)		27.35*	47.25*
49–63 (34.21)			19.90*

*Mean difference is significant at the 0.05 level, Modified Least Significant Differences Test.

between the youngest and oldest respondents. In the bottom portion of the table, we see that all differences are significant except for the difference between the 18–33-year-olds and the 34–48-year-olds. Once again, the largest differences are between the two younger groups and the oldest respondents.

The post hoc analysis confirms what you probably suspected from the start: there is a significant relationship between these variables and sexual activity decreases with age.

READING STATISTICS Religion and Sex Education

We can use a recent research article to illustrate how to read ANOVA results in the professional literature. A team of researchers was concerned about how state-mandated sex education curricula are affected by the degree and nature of religiosity in the state. Are states with higher rates of church attendance and a greater presence of fundamentalism more likely to stress abstinence? Are states with lower levels of involvement in organized religion more likely to stress contraception?

The researchers used a variety of variables to measure religiosity, including rates of church attendance (measured on a scale ranging from 1 to 6), percent "literalists" (those that believe the Bible is the word of God), and percent "theists" (those who profess a belief in God). These independent variables were measured by surveys administered to random samples of each state's population. Some of their results for the prominence of abstinence in sex education are presented in the table below.

Means for Religious Predictors of State Policy on Abstinence Coverage in Sexual Education

Variable	Abstinence Is:			
	Not Required	*Covered*	*Stressed*	*F Ratio*
Average Attendance	3.40	3.61	3.82	4.83*
Percent Literalist	25.27	32.21	38.63	4.62*
Percent Theist	89.96	93.40	95.17	14.05***

*$p < 0.05$ ***$p < 0.001$

Based on Table 2, p. 242

Source: Baker, J., K. Smith, and Y. Stoss. 2015. "Theism, Secularism, and Sexual Education in the United States." *Sex Research and Social Policy*, 12: 236–247.

The results indicate that "percent theist" was by far the most significant, strongest predictor of a strong abstinence policy, while church attendance was the second strongest. Want to learn more? The citation is given below.

SUMMARY

1. One-way analysis of variance is a powerful test of significance that is commonly used when comparisons across more than two categories or samples are of interest. It is perhaps easiest to conceptualize ANOVA as an extension of the test for the difference in sample means.
2. ANOVA compares the amount of variation within the categories to the amount of variation between categories. If the null hypothesis of no difference is false, there should be relatively great variation between categories and relatively little variation within categories. The greater the differences from category to category relative to the differences within the categories, the more likely we will be able to reject the null hypothesis.

3. The computational routine for even simple applications of ANOVA can quickly become quite complex (a fact that suggests the value of computerized statistical packages such as SPSS). The basic process is to construct separate estimates of the population variance based on the variation within the categories and the variation between the categories. The test statistic is the F ratio, which is based on a comparison of these two estimates.

4. The ANOVA test can be organized into the familiar five-step model for testing the significance of sample outcomes. Although the model assumptions (step 1) require high-quality data, the test can tolerate some deviation as long as sample sizes are roughly equal. The null hypothesis takes the familiar form of stating that there is no difference of any importance among the population values, while the alternative hypothesis asserts that at least one population mean is different. The sampling distribution is the F distribution, and the test is always one-tailed. The decision to reject or to fail to reject the null hypothesis is based on a comparison of the obtained F ratio with the critical F ratio as determined for a given alpha level and degrees of freedom. The decision to reject the null hypothesis indicates only that one or more of the population means is different from the others. We can often determine which of the sample means accounts for the difference by inspecting the sample data, but this informal method should be used with caution, and post hoc tests are more reliable indicators of significant differences.

SUMMARY OF FORMULAS

Total sum of squares:

10.1 $\quad \text{SST} = \Sigma X_i^2 - N\overline{X}^2$

The two components of the total sum of squares:

10.2 $\quad \text{SST} = \text{SSB} + \text{SSW}$

Sum of squares between:

10.3 $\quad \text{SSB} = \Sigma N_k(\overline{X}_k - \overline{X})^2$

Sum of squares within:

10.4 $\quad \text{SSW} = \text{SST} - \text{SSB}$

Degrees of freedom for SSW:

10.5 $\quad \text{dfw} = N - k$

Degrees of freedom for SSB:

10.6 $\quad \text{dfb} = k - 1$

Mean square within:

10.7 $\quad \text{Mean square within} = \dfrac{\text{SSW}}{\text{dfw}}$

Mean square between:

10.8 $\quad \text{Mean square between} = \dfrac{\text{SSB}}{\text{dfb}}$

F ratio:

10.9 $\quad F(\text{obtained}) = \dfrac{\text{Mean square between}}{\text{Mean square within}}$

GLOSSARY

Analysis of variance. A test of significance appropriate for situations in which we are concerned with the differences between more than two sample means.

ANOVA. See **Analysis of variance**.

F ratio. The test statistic computed in step 4 of the ANOVA test.

Mean square estimate. An estimate of the variance calculated by dividing the sum of squares within (SSW) or the sum of squares between (SSB) by the proper degrees of freedom.

One-way analysis of variance. Applications of ANOVA in which the effect of a single independent variable on a dependent variable is observed.

Post hoc test. A technique for determining which pairs of means are significantly different.

Sum of squares between (SSB). The sum of the squared deviations of the sample means from the overall mean, weighted by sample size.

Sum of squares within (SSW). The sum of the squared deviations of scores from the category means.

Total sum of squares (SST). The sum of the squared deviations of the scores from the overall mean.

PROBLEMS

NOTE: The number of cases in these problems is very low—a fraction of the sample size necessary for any serious research—in order to simplify computations.

10.1 Conduct an ANOVA test for each of the following sets of scores. (*HINT: Follow the computational routine outlined in the "One Step at a Time" box and keep track of all sums and means by constructing computational tables like Table 10.3 or 10.4.*)

a.

Category		
A	B	C
5	10	12
7	12	16
8	14	18
9	15	20

b.

Category		
A	B	C
1	2	3
10	12	10
9	2	7
20	3	14
8	1	1

c.

Category			
A	B	C	D
13	45	23	10
15	40	78	20
10	47	80	25
11	50	34	27
10	45	30	20

10.2 [SOC] A local church has been experiencing steady growth in membership and wants to launch a membership drive. Before they begin, they would like to find out what types of people are most likely to attend services each week. A random sample of 15 existing members has been asked how many times they attended services over the past five weeks and for

certain other information. Which independent variables have significant relationships with attendance?

a. Attendance by education:

Less Than High School	High School	College
0	1	0
1	3	3
2	3	4
3	4	4
4	5	4

b. Attendance by length of residence in present community:

Less Than 2 Years	2–5 Years	More Than 5 Years
0	0	1
1	2	3
3	3	3
4	4	4
4	5	4

c. Attendance by extent of television watching:

Little or None	Moderate	High
0	3	4
0	3	4
1	3	4
1	3	4
2	4	5

d. Attendance by number of children:

None	One Child	More Than One Child
0	2	0
1	3	3
1	4	4
3	4	4
3	4	5

10.3 [SOC] In a local community, a random sample of 18 couples has been assessed on a scale that measures the extent to which power and decision making are shared (lower scores) or monopolized by one party (higher scores) and on marital happiness (higher scores indicate higher levels of happiness). The couples were also classified by type of relationship: traditional (only the husband works outside the home), dual-career (both parties work), and cohabitational (parties living together but not legally married, regardless of work patterns). Does decision making or happiness vary significantly by type of relationship?
a. Decision making

Traditional	Dual-Career	Cohabitational
7	8	2
8	5	1
2	4	3
5	4	4
7	5	1
6	5	2

b. Happiness

Traditional	Dual-Career	Cohabitational
10	12	12
14	12	14
20	12	15
22	14	17
23	15	18
24	20	22

10.4 [CJ] Two separate after-school programs for at-risk teenagers have been implemented in the city of Redland. One is a big-brother/big-sister mentoring program, in which adult volunteers are paired with troubled teenagers. The second involves extracurricular team sports competitions. In terms of the percentage change in crimes reported to the police over a one-year period, were the programs successful? The results are for random samples of 18 neighborhoods drawn from the entire city.

Mentoring Program	Competition Program	No Program
−10	−21	+30
−20	−15	−10
+10	−80	+14
+20	−10	+80
+70	−50	+50
+10	−10	−20

10.5 Are sexually active teenagers any better informed about AIDS and other potential health problems related to sex than teenagers who are sexually inactive? A 15-item test of general knowledge about sex and health was administered to random samples of teens who are sexually inactive, teens who are sexually active but with only a single partner, and teens who are sexually active with more than one partner. Is there any significant difference in the test scores?

Inactive	Active—One Partner	Active—More Than One Partner
10	11	12
12	11	12
8	6	10
10	5	4
8	15	3
5	10	15

10.6 [SOC] Does the rate of voter turnout vary significantly by the type of election? A random sample of voting precincts displays the following pattern of voter turnout by election type. Assess the results for significance.

Local Only	State	National
33	35	42
78	56	40
32	35	52
28	40	66
10	45	78
12	42	62
61	65	57
28	62	75
29	25	72
45	47	51
44	52	69
41	55	59

10.7 [GER] Do older citizens lose interest in politics and current affairs? A brief quiz on recent headline stories was administered to random samples of respondents from each of four age groups. Is there a significant difference? The following data represent numbers of correct responses.

High School (15–18)	Young Adult (21–30)	Middle-Aged (30–55)	Retired (65+)
0	0	2	5
1	0	3	6
1	2	3	6
2	2	4	6
2	4	4	7
2	4	5	7
3	4	6	8
5	6	7	10
5	7	7	10
7	7	8	10
7	7	8	10
9	10	10	10

10.8 MR A small random sample of respondents has been selected for a study of the ways that iPhone owners use their apps. Specifically, a research firm is comparing the apps for Starbucks, Amazon.com, and Yelp. Are there statistically significant differences by app for any of the variables listed here?

a. Minutes of usage per week

Starbucks App	Amazon App	Yelp App
5	5	5
5	6	6
7	10	7
8	11	10
8	12	10
10	14	11
10	15	12
11	16	12
12	18	12
13	20	13

b. iPhone screen number that the app can be found on. A "1" means it can be found on the user's first iPhone screen, followed by a "2" and so on.

Starbucks App	Amazon App	Yelp App
1	2	1
1	1	4
1	2	2
2	1	3
1	2	3
1	2	2
2	3	5
2	2	1
1	2	4
1	1	4

c. Percent of battery used by the app in the last 24 hours

Starbucks App	Amazon App	Yelp App
5	6	5
7	8	5
8	11	11
11	12	10
8	12	9
9	11	6
8	11	10
3	9	7
9	10	9
10	12	8

d. Length of time (in months) that the user has owned the app

Starbucks App	Amazon App	Yelp App
12	14	1
7	13	5
0	2	4
4	5	4
5	8	18
8	5	4
7	8	8
5	7	8
7	2	8
4	6	5

e. Satisfaction level on a scale of 1–10 with the app

Starbucks App	Amazon App	Yelp App
5	5	3
3	7	7
10	10	5
2	2	10
10	3	1
2	10	8
3	1	5
4	3	10
5	4	3
9	1	1

10.9 SOC Does support for assisted suicide ("death with dignity") vary by social class? Is this relationship different in different nations? Small samples in three nations were asked whether it is ever justified for a person with an incurable disease to take his or her own life. Respondents answered in terms of a 10-point scale, on which 10 was "always justified" (the strongest support for "death with dignity") and 1 was "never justified" (the lowest level of support). Results are reported here.

Mexico

Lower Class	Working Class	Middle Class	Upper Class
5	2	1	2
2	2	1	4
4	1	3	5
5	1	4	7
4	6	1	8
2	5	2	10
3	7	1	10
1	2	5	9
1	3	1	8
3	1	1	8

Canada

Lower Class	Working Class	Middle Class	Upper Class
7	5	1	5
7	6	3	7
6	7	4	8
4	8	5	9
7	8	7	10
8	9	8	10
9	5	8	8
9	6	9	5
6	7	9	8
5	8	5	9

United States

Lower Class	Working Class	Middle Class	Upper Class
4	4	4	1
5	5	6	5
6	1	7	8
1	4	5	9
3	3	8	9
3	3	9	9
3	4	9	8
5	2	8	6
3	1	7	9
6	1	2	9

10.10 [SP] An organization that advocates for affordable housing has gathered a random sample of working-class home owners, condo owners, apartment renters, and co-op owners to measure the differences in their need for assistance to pay their rent or mortgage. Upon surveying them, they have found that the levels of need in different areas tend to vary. Use the data below to determine which of these areas of need are significantly different by group.

a. Number of months in the last year when the respondent struggled to pay rent or mortgage:

Home Owner	Condo Owner	Apartment Renter	Co-op Owner
7	0	10	0
2	3	9	2
5	4	8	3
12	2	5	1
9	7	1	1
1	2	8	4
0	0	7	5
3	3	12	7
4	4	10	10
6	2	11	9

b. Number of times they have made late payments on their rent or mortgage in the last 12 months:

Home Owner	Condo Owner	Apartment Renter	Co-op Owner
7	7	9	1
7	3	8	1
5	7	5	5
10	8	0	6
9	5	1	5
7	4	1	3
6	1	2	5
5	2	3	6
4	4	5	1
3	2	6	5

c. Number of dependents claimed on last year's taxes:

Home Owner	Condo Owner	Apartment Renter	Co-op Owner
0	3	2	6
3	0	2	0
4	0	3	0
6	0	1	1
0	5	1	5
4	6	1	5
0	3	3	3
1	4	2	6
0	2	1	6
0	5	1	1

Statistical Analysis Using SPSS

10.11 In this exercise, you will follow up on problem 9.17, in which you compared Protestants and Catholics for significant differences in their level of education (*educ*) and occupational prestige

(*prestg10*). We will now use **One-Way ANOVA** to compare all five categories of the *religR* variable.

- Click the SPSS icon on your desktop.
- Load the *GSS2018* database.
- Click **Analyze** → **Compare Means** → **One-Way ANOVA.**
- Find *educ* and *prestg10* in the list of variables and move them into the **Dependent List** box.
- Find *religR* in the list of variables and move it into the **Factor** box.

- Click **Options** and click the box next to **Descriptive** in the **Statistics** box to get statistical information on each category of *religR*. Click **Continue**.
- Click **OK** and the results of the test will be printed to the SPSS output window.

Find the group means, the F ratio, degrees of freedom, and the "Sig." (which is the exact probability of getting this difference if H_0 is true) and report the results of the test in a few sentences.

YOU ARE THE RESEARCHER

Why Are Some People Liberal (or Conservative)?

This project investigates political ideology using *polviews*—a seven-point scale that measures how liberal or conservative a person is—as the dependent variable. This variable is ordinal in level of measurement but will be treated as interval-ratio for this test. What social forces affect political views?

One possible independent variable is *age*: People may grow more conservative as they grow older. It might be appropriate to investigate the possible causal impact of aging, but note that the variable is in the wrong form. Independent variables in ANOVA must have just a few categories, and respondents to the 2018 GSS ranged in age from 18 to over 89. How can we change age to a form that fits the test? In the next section, we introduce the **Recode** command, which enables us to collapse scores on a variable.

Recoding Variables

We will recode *age* into a variable with just a few scores, which we can use as an independent variable in ANOVA. We will create a new version of *age* that has three categories. When we are finished, we will have two versions of the same variable in the data set: the original interval-ratio version with age measured in years, and a new ordinal-level version with collapsed categories. If we wish, we can add the new version of age to the permanent data file and use it in the future.

The decision to use three categories for *age* is arbitrary; we could easily have decided on four, five, or even six categories for the new, recoded independent variable. If we find that we are unhappy with the three-category version of the variable, we can always recode the original version of *age* into a different set of categories.

We will collapse the values of *age* into three broad categories with roughly equal numbers of cases in each category. How can we define these new categories? Begin by running the **Frequencies** command for *age* to inspect the distribution of the scores. We used the cumulative percent column of the frequency distribution to find the ages that divide the variable into thirds. We found that 32.3% of the 2018 GSS sample were younger than 37 and that 67.4% were younger than 58, and we decided to use these ages as the cut points for the new categories, as summarized below:

Ages	Percent of Sample
18–37	32.3%
38–58	35.1%
59–89	32.6%
	100.0%

To recode *age* into these categories, follow these steps:

1. In the **SPSS Data Editor** window, click **Transform** from the menu bar and then click **Recode**. A window will open that gives us two choices: "**into same variable**" or "**into different variable**." If we choose **into same variable**, the new version of the variable will *replace* the old and the variable *age* (measured in years) would disappear. We do *not* want this to happen, so we will choose (click on) **into different variable**. This option will allow us to keep both the old and new versions of the variable.

2. The **Recode into Different Variable** window will open. The usual box listing the variables will be on the left. Use the cursor to highlight *age*, and then click on the arrow button to move the variable to the **Input Variable → Output Variable** box. The input variable is the old version of age, and the output variable is the new, recoded version we will soon create.

3. In the **Output Variable** box on the right, click in the **Name** box and type a name for the new (output) variable. We suggest *ager* ("**age r**ecoded") for the new variable, but you can assign any name as long as it does not duplicate the name of some other variable in the data set. Click the **Change** button and the expression *age → ager* will appear in the **Input Variable → Output Variable** box.

4. Click on the **Old and New Values** button in the middle of the screen, and a new dialog box will open. Read down the left-hand column until you find the **Range** button. Click on the button, and the cursor will move to the small box immediately below. Type **18** (the youngest age in the sample) into the left-hand **Range** dialog box and then click on the right-hand box and type **37**.

5. In the **New Value** box in the upper-right corner of the screen, click the **Value** button. Type **1** in the **Value** dialog box and then click the **Add** button directly below. The expression "**18–37 → 1**" will appear in the **Old → New** dialog box.

6. Return to the **Range** dialog boxes. Type **38** in the left-hand box and **58** in the right-hand box, and then click the **Value** button in the **New Values** box. Type **2** in the **Value** dialog box and then click the **Add** button. The expression "**38–58 → 2**" appears in the **Old → New** dialog box.

7. Finish by returning to the **Range** dialog box and entering the values **59** and **89** in the correct boxes. Click the **Value** button in the **New Values** box. Type **3** in the **Value** dialog box and then click the **Add** button. The expression "**59–89 → 3**" appears in the **Old → New** dialog box.

8. Click the **Continue** button at the bottom of the screen, and you will return to the **Recode into Different Variable** dialog box. Click **OK**, and SPSS will execute the transformation.

You now have a data set with an additional variable named *ager* (or whatever name you gave the recoded variable). You can find the new variable in the last column to the right in the data window. You can make the new variable a permanent part of the data set by saving the data file at the end of the session. If you do not wish to save the new, expanded data file, click **No** when you are asked whether you want to save the data file. If you are using the student version of SPSS, remember that you are limited to a maximum of 50 variables and you may not be able to save the new variable.

Investigating Political Ideology (*polviews*)

Why are some people more liberal and others more conservative? In this project, you will analyze the relationships between *polviews* (the dependent variable), recoded *age*, and

two more independent variables of your own choosing. As you analyze and interpret your results, remember that *polviews* is coded so that higher scores indicate greater conservatism (see Appendix G or click **Utilities** → **Variables** on the SPSS menu bar).

Step 1: Choosing Independent Variables

Select recoded *age* and two more possible independent variables from the 2018 GSS. What factors might help to explain a person's political views? Choose *only* independent variables with three to six scores or categories. Among other possibilities, you might consider education (use *degree*), religious denomination (*religR*), or social class (*class*). Use the **Recode** command to collapse independent variables with more than six categories into three or four categories. In general, variables that measure personal characteristics or traits (like gender or race) will work better than those that measure attitude or opinion (like *cappun* or *gunlaw*), which are more likely to be manifestations of political ideology, not causes.

List your independent variables in the table below:

Variable	SPSS Name	What Exactly Does This Variable Measure?
1.	*ager* (or whatever name you assigned to recoded *age*)	
2.		
3.		

Step 2: Stating Hypotheses

For each independent variable, state a hypothesis about its relationship with *polviews*. For example, you might hypothesize that people with greater education will be more liberal or that the more religious will be more conservative. You can base your hypotheses on your own experiences or on information you have acquired in your courses.

1.
2.
3.

Step 3: Getting and Reading the Output

To use the **One-Way ANOVA** procedure, follow the instructions in the "Using SPSS: The ANOVA Test" in this chapter.

1. Place *polviews* in the **Dependent List** box and your first independent variable in the **Factor** box.

2. Don't forget to click **Options** and click the box next to **Descriptive** to get summary statistics. Click **OK** and the first test will be conducted.

3. Return to the **One-Way ANOVA** window and click the arrow next to the **Factor** box to move your first independent variable back to the variable list.

4. Move your second independent variable into the **Factor** box and click **OK**.

5. Repeat steps 3 and 4 to test your third independent variable.

6. Print or save your results.

Step 4: Recording Results

Record your results in the table below, using as many rows for each independent variable as necessary.

1. Write the SPSS variable name in the first column.

2. Write the names of the categories of that independent variable in the next column.

3. Record the descriptive statistics in the designated columns.

4. Write the value of the F ratio and, in the far right-hand column, indicate whether the results are significant at the 0.05 level. If the value in the "Sig." column of the ANOVA output is less than 0.05, write YES in this column. If the value in the "Sig." column of the ANOVA output is more than 0.05, write NO in this column.

5. For each independent variable, record the grand or overall means, standard deviations, and sample sizes in the row labeled "Totals = ."

Independent Variables	Categories	Mean	Standard Deviation	N	F Ratio	Sig. at 0.05 Level?
1. *ager*	1.				____	____
	2.					
	3.					
	Totals =					
2. _____	1.				____	____
	2.					
	3.					
	4.					
	5.					
	6.					
	Totals =					
3. _____	1.				____	____
	2.					
	3.					
	4.					
	5.					
	6.					
	Totals =					

Step 6: Interpreting Your Results

Summarize your findings. For each test, write:

1. At least one sentence summarizing the test in which you identify the variables being tested, the sample means for each group, N, the F ratio, and the significance level. In the professional research literature, you might find the results reported as: "For a sample of 1417 respondents, there was no significant difference between the average age of Southerners (46.50), Northerners (44.20), Midwesterners (47.80), and Westerners (47.20) ($F = 1.77$, $p > 0.05$)."

2. A sentence relating to your hypotheses. Were they supported? How?

Now that you have tested for all significant differences by groups, you might also conduct post hoc tests to see where the intergroup differences lie. Follow the directions in the Using SPSS exercise earlier in the chapter to test these differences.

11

Hypothesis Testing IV: Chi Square

LEARNING OBJECTIVES By the end of this chapter, you will be able to:

1. Cite examples of situations in which the chi square test is appropriate.
2. Explain the structure of a bivariate table.
3. Explain the concept of independence as it relates to bivariate tables.
4. Explain the logic of hypothesis testing in terms of chi square.
5. Perform the chi square test using the five-step model and correctly interpret the results.
6. Conduct a goodness-of-fit test and correctly interpret the results.
7. Explain the limitations of the chi square test.
8. Use SPSS to conduct the chi square test of significance.

USING STATISTICS

This chapter presents statistical techniques used to test the statistical significance of the relationships between variables that have been arrayed in a table. If there is a significant difference between the frequencies in the table and what would be expected by random chance, we conclude that the variables are related in the population. This conclusion has a known probability of error, which is often set at the 0.05 level. Examples of situations in which the chi square test is appropriate include:

1. A researcher is studying the belief some LGBT youth have that there is no support for them to see if knowledge about gay and lesbian communities is related to deviant and criminal acts. **See page 310**

2. A student interested in "Fake News" wants to know if Democrats use Facebook more than Republicans leading them to have a greater exposure to political commercials on that social media site. **See page 302**

3. A sociologist wants to know if blacks and whites differ in their opinions about whether a racist should be allowed to teach in a college. **See page 309**

4. A technology researcher wants to compare views on whether the Internet is a good thing for society by age. **See page 298**

Historically, the **chi square (χ^2) test** has probably been the most frequently used test of hypothesis in the social sciences. This popularity is due largely to the fact that the assumptions and requirements in step 1 of the five-step model are easy to satisfy. The test can be conducted with variables measured at the nominal level and is **nonparametric**, which means that it requires no assumption at all about the shape of the population or sampling distribution.

Why is it an advantage to have easy-to-satisfy assumptions and requirements? The decision to reject the null hypothesis (step 5) is not specific: It means only that one statement in the model (step 1) or the null hypothesis (step 2) is wrong. Usually we single out the null hypothesis for rejection. The more certain we are of the model stated in step 1, the greater our confidence that the null hypothesis is the faulty assumption. A weak or easily satisfied model means that our decision to reject the null hypothesis can be made with greater certainty.

Chi square has also been popular for its flexibility. It can be used with variables that have many categories or scores. For example, in Chapter 9, we tested the significance of the difference in the proportions of black and white citizens who were "highly participatory" in voluntary associations. What if the researcher wished to expand the test to include Hispanic and Asian American citizens? The two-sample test would no longer be applicable, but chi square handles the more complex variable easily. Also, unlike the ANOVA test presented in Chapter 10, the chi square test can be conducted with variables at any level of measurement.

Bivariate Tables

Chi square is computed from **bivariate tables**, so called because they display the scores of cases on two different variables at the same time. Bivariate tables are used to test for significant relationships and for other purposes that we will investigate in later chapters. These tables are commonly used in research and a detailed examination of them is in order.

First, bivariate tables have two dimensions. We refer to the horizontal (across) dimension as **rows** and the vertical dimension (up and down) as **columns**. Each column or row represents a score on a variable, and the intersections of the rows and columns (**cells**) represent the combined scores on both variables.

Let's use an example to clarify. Suppose that a researcher is interested in the lives of senior citizens and wonders whether their participation in voluntary groups is affected by their marital status. To simplify the analysis, the researcher has confined the sample to people who are presently married or not married (including people who are single and divorced) and has measured involvement in voluntary associations as a simple dichotomy. People have been classified as either high or low in participation.

By convention, we array the independent variable (the variable that is taken to be the cause) in the columns and the dependent variable in the rows. In the example at hand, marital status is the causal variable (the question was "Is participation *affected by* marital status?"), and each column will represent a score on this variable. Each row, on the other hand, will represent a score on level of involvement (high or low). Table 11.1 displays the outline of the bivariate table for a sample of 100 senior citizens.

TABLE 11.1 **Rates of Participation in Voluntary Associations by Marital Status for 100 Senior Citizens**

	Marital Status		
Participation Rates	*Married*	*Unmarried*	TOTALS
High			50
Low	—	—	50
TOTALS	50	50	100

Note some further details of the table. First, subtotals are included for each column and row. These are called the row or column **marginals**, and in this case they tell us that 50 members of the sample are married and 50 are not married (the column marginals), and that 50 are high in participation and 50 are low (the row marginals).

Second, the total number of cases in the sample ($N = 100$) is reported at the intersection of the row and column marginals. Finally, take careful note of the labeling of the table. Each row and column is identified and the table has a descriptive title that includes the names of the variables, with the dependent variable listed first. Clear, complete labels and concise titles must be included in all tables, graphs, and charts.

Table 11.1 lacks some crucial information, specifically the numbers in the body of the table. To complete the table, we need to classify the marital status and level of participation of each member of the sample, keep count of how often each combination of scores occurs, and record these numbers in the appropriate cells of the table. Because each variable has two possible scores, four combinations of scores are possible, each corresponding to a cell in the table. For example, married people with high levels of participation would be counted in the upper-left cell, unmarried people with low levels of participation would be counted in the lower-right cell, and so forth. When we are finished counting, each cell will display the number of times each combination of scores occurred.

Finally, note that we could expand the table to accommodate variables with more than two possible scores. If we wished to include people with other marital statuses (widowed, separated, and so forth), we would simply add columns. More elaborate dependent variables could also be easily accommodated. If we had measured participation rates with three categories (e.g., high, moderate, and low), we would simply add a row to the table.

The Logic of Chi Square

The chi square test has several different uses and most of this chapter deals with an application called the *chi square test for independence*. We have encountered the term *independence* in connection with the requirements for the two-sample hypothesis-testing case (Chapter 9) and for the ANOVA test (Chapter 10). In the context of chi square, **independence** refers to the relationship between the variables, not between the samples. Two variables are independent if the classification of a case into a category of one variable has no effect on the

TABLE 11.2 **Rates of Participation in Voluntary Associations by Marital Status for 100 Senior Citizens**

Participation Rates	Marital Status		TOTALS
	Married	*Unmarried*	
High	25	25	50
Low	25	25	50
TOTALS	50	50	100

probability that the case will fall into any category of the second variable. The variables in Table 11.1 would be independent of each other if the classification of a case as married or not married has no effect on the classification of the case as high or low on participation. We can conclude that the variables are independent if level of participation and marital status are completely unrelated to each other.

Consider Table 11.1 again. If the variables were independent, the cell frequencies would be determined solely by random chance and we would find that about half of the married respondents would rank high on participation and half would rank low, just as an honest coin would show heads about 50% of the time. The same pattern would hold for the 50 unmarried respondents and, therefore, each of the four cells would have about 25 cases in it, as illustrated in Table 11.2. This pattern of cell frequencies indicates that marital status has no effect on a person's level of participation. The probability of being classified as high or low would be 0.50 for both marital statuses, and the variables would therefore be independent.

The null hypothesis for the chi square test for independence is that the variables are independent. Under the assumption that the null hypothesis is true, we compute the cell frequencies that we would expect to find if only random chance were operating. These frequencies are called **expected frequencies** (symbolized f_e) and we compare them, cell by cell, with the frequencies observed in the table (**observed frequencies**, symbolized f_o). If the null hypothesis is true and the variables are independent, then there should be little difference between the expected and observed frequencies. If the null hypothesis is false, however, there should be large differences between the two. The greater the differences between expected (f_e) and observed (f_o) frequencies, the less likely that the variables are independent and the more likely that we will be able to reject the null hypothesis.

The Computation of Chi Square

To conduct the chi square test we compute a test statistic, χ^2(**obtained**), from the sample data and then place that value on the sampling distribution of all possible sample outcomes. Specifically, the χ^2(obtained) will be compared with the value of χ^2(**critical**), which will be determined by consulting a chi square table (Appendix C) for a particular alpha level and degrees of freedom.

The procedure for calculating chi square is given in Formula 11.1:

FORMULA 11.1
$$\chi^2(\text{obtained}) = \sum \frac{(f_o - f_e)^2}{f_e}$$

where f_o = the cell frequencies observed in the bivariate table
f_e = the cell frequencies that would be expected if the variables were independent

For each cell subtract the expected frequency from the observed frequency, square the result, and then divide by the expected frequency for that cell. Sum the resultant values for all cells.

This formula requires an expected frequency for each cell in the table. In Table 11.2, the marginals are the same value for all rows and columns, and the expected frequencies are obvious by intuition: f_e = 25 for all four cells. In the more usual case, the expected frequencies will not be obvious, the marginals will be unequal, and we must use Formula 11.2 to find the expected frequency for each cell:

FORMULA 11.2
$$f_e = \frac{\text{Row marginal} \times \text{Column marginal}}{N}$$

The expected frequency for any cell is equal to the total number of cases in the row (the row marginal) times the total number of cases in the column (the column marginal) divided by the total number of cases in the table (N).

An example using Table 11.3 should clarify these procedures. A random sample of 100 social work majors have been classified in terms of whether the Council on Social Work Education has accredited their undergraduate programs (the column, or independent, variable) and whether they were hired in social work positions within three months of graduation (the row, or dependent, variable).

Beginning with the upper-left cell (graduates of accredited programs who are working as social workers), the expected frequency for this cell, using Formula 11.2, is (40 × 55)/100, or 22. For the other cell in this row (graduates of nonaccredited programs who are working as social workers), the expected frequency is (40 × 45)/100, or 18. For the two cells in the bottom row, the expected frequencies are (60 × 55)/100, or 33, and (60 × 45)/100, or 27, respectively. The expected frequencies for all four cells are displayed in Table 11.4.

TABLE 11.3 Employment of 100 Social Work Majors by Accreditation Status of Undergraduate Program

	Accreditation Status		
Employment Status	Accredited	Not Accredited	TOTALS
Working as a social worker	30	10	40
Not working as a social worker	25	35	60
TOTALS	55	45	100

TABLE 11.4 Expected Frequencies for Table 11.3

Employment Status	Accreditation Status		TOTALS
	Accredited	Not Accredited	
Working as a social worker	22	18	40
Not working as a social worker	33	27	60
TOTALS	55	45	100

Note that the row and column marginals as well as the total number of cases in Table 11.4 are the same as those in Table 11.3. The row and column marginals for the expected frequencies must *always* equal those of the observed frequencies, a relationship that provides a convenient way of checking your arithmetic to this point.

The value for chi square for these data can now be found by solving Formula 11.1. It will be helpful to use a computing table, such as Table 11.5, to organize the several steps required to compute chi square. The table lists the observed frequencies (f_o) in column 1 in order from the upper-left cell to the lower-right cell, moving from left to right across the table and from top to bottom. Column 2 lists the expected frequencies (f_e) in exactly the same order. Double-check to make sure you have listed the cell frequencies in the same order for both columns.

The next step is to subtract the expected frequency from the observed frequency for each cell and list these values in column 3. To complete column 4, square the value in column 3 and then, in column 5, divide the column 4 value by the expected frequency for that cell. Finally, add up column 5. The sum of this column is obtained chi square. For Table 11.3, χ^2(obtained) = 10.78.

Note that the totals for columns 1 (f_o) and 2 (f_e) are exactly the same. This will always be the case, and if the totals do not match, you have made a computational error, probably in the calculation of the expected frequencies. Also note that the sum of column 3 will always be 0, another convenient way to check your math to this point.

This sample value for chi square must still be tested for its significance. *(For practice in computing chi square, see any problem at the end of this chapter, starting with the smallest tables.)*

TABLE 11.5 Computational Table for Table 11.3

(1)	(2)	(3)	(4)	(5)
f_o	f_e	$f_o - f_e$	$(f_o - f_e)^2$	$(f_o - f_e)^2/f_e$
30	22	8	64	2.91
10	18	−8	64	3.56
25	33	−8	64	1.94
35	27	8	64	2.37
100	100	0		10.78

> **ONE STEP AT A TIME** Computing Expected Frequencies (f_e)
>
> Use Formula 11.2.
>
> 1. Start with the upper-left-hand cell of the bivariate table. Multiply the row marginal for this cell by the column marginal.
> 2. Divide the quantity you found in step 1 by N. The result is the expected frequency for that cell.
> 3. Record this f_e in a separate table (see Table 11.4) or in the second column of a computational table (see Table 11.5).
> 4. Repeat steps 1–3 for each cell.
> 5. *Double-check to make sure that you are using the correct row and column marginal when computing the expected frequencies.*

> **ONE STEP AT A TIME** Computing Chi Square
>
> Use Formula 11.1 and a computing table like Table 11.5.
>
> 1. List the observed frequencies (f_o) in column 1. The total for this column is the number of cases (N).
> 2. List the expected frequencies (f_e) in column 2. The total for this column must equal the total for column 1 (N).
> 3. For each cell, subtract the expected frequency (f_e) from the observed frequency (f_o), and list these values in the third column of the computational table, labeled ($f_o - f_e$). Find the total for this column. If this total is not zero, you have made a mistake and need to check your computations.
> 4. Square each of the values in the third column of the table, and record the result in the fourth column, labeled ($f_o - f_e$)2.
> 5. Divide each value in column 4 by the expected frequency for that cell, and record the result in the fifth column, labeled ($f_o - f_e$)2/f_e.
> 6. Find the total for the fifth column. This value is χ^2(obtained).

The Chi Square Test for Independence

We are now ready to conduct the chi square test for independence. Recall that if the variables are independent of each other, the score of a case on one variable will have no relationship with its score on the other variable. The five-step model for significance testing will provide the framework for organizing our decision making. The data presented in Table 11.3 will serve as our example.

Step 1. Making Assumptions and Meeting Test Requirements. Note that, because the test is nonparametric, we make no assumptions at all about the shape of the sampling distribution.

> Model: Independent random samples
> Level of measurement is nominal

Step 2. Stating the Null Hypothesis. The null hypothesis states that the two variables are independent. If the null hypothesis is true, the differences between the observed and expected frequencies will be small. As usual, the research hypothesis directly contradicts the null hypothesis. Thus, if we reject H_0, the research hypothesis will be supported.

$$H_0: \text{The two variables are independent.}$$
$$(H_1: \text{The two variables are dependent.})$$

Step 3. Selecting the Sampling Distribution and Establishing the Critical Region. The sampling distribution of sample chi squares, unlike the Z and t distributions, is positively skewed, with higher values of sample chi squares in the upper tail of the distribution (to the right). Thus, with the chi square test, the critical region is established in the upper tail of the sampling distribution.

Values for χ^2(critical) are given in Appendix C. This table is similar to the t table, with alpha levels arrayed across the top and degrees of freedom down the side. However, with chi square, the degrees of freedom (df) are found by Formula 11.3:

FORMULA 11.3
$$df = (r - 1)(c - 1)$$

where r = the number of rows in the table
c = the number of columns in the table

A table with two rows and two columns (a 2×2 table) has 1 degree of freedom regardless of the number of cases in the sample.[1] A table with two rows and three columns would have $(2 - 1)(3 - 1)$, or 2 degrees of freedom. Our sample problem involves a 2×2 table with 1 degree of freedom, so if we set alpha at 0.05, the critical chi square score would be 3.841. Summarizing these decisions, we have

$$\text{Sampling distribution} = \chi^2 \text{ distribution}$$
$$\text{Alpha} = 0.05$$
$$\text{Degrees of freedom} = 1$$
$$\chi^2(\text{critical}) = 3.841$$

Step 4. Computing the Test Statistic. The computation of chi square was introduced in the previous section. As you recall, we had

$$\chi^2(\text{obtained}) = 10.78$$

[1]Degrees of freedom are the number of values in a distribution that are free to vary for any statistic. A 2×2 table has 1 degree of freedom because once one cell frequency is determined, all other cell frequencies are fixed and no longer free to vary. In Table 11.3, for example, if any cell frequency is known, all others are determined. If the upper-left cell is known to be 30, the remaining cell in that row must be 10, because there are 40 cases total in the row and $40 - 30 = 10$. Once the frequencies of the cells in the top row are established, cell frequencies for the bottom row are determined by subtraction from the column marginals. *NOTE: This relationship can be used to quickly compute expected frequencies. For example, in a 2×2 table, only one expected frequency needs to be computed. The f_es for all other cells can then be found by subtraction.*

Step 5. Making a Decision and Interpreting the Results of the Test.
Comparing the test statistic with the critical region,

$$\chi^2(\text{obtained}) = 10.78$$
$$\chi^2(\text{critical}) = 3.841$$

we see that the test statistic falls in the critical region and, therefore, we reject the null hypothesis of independence. The pattern of cell frequencies observed in Table 11.3 is unlikely to have occurred by chance alone. The variables are dependent. Based on these sample data, the probability of securing employment in the field of social work is dependent on the accreditation status of the program. *(For practice in conducting and interpreting the chi square test for independence, see problems 11.2 to 11.15.)*

We should be clear about exactly what the chi square test does and does not tell us. A significant chi square means that the variables are (probably) dependent on each other in the population. In terms of our example, this means that there is a relationship between accreditation and whether a person is working as a social worker. But what exactly is the relationship between the variables? Which type of graduate is more likely to find work in the profession? To make this determination, we must perform an additional calculation.

Column Percentages

We can figure out how the independent variable (accreditation status in our example) is affecting the dependent variable (employment as a social worker) by computing **column percentages** or by calculating percentages within each column of the bivariate table. This procedure is analogous to calculating percentages for frequency distributions (see Chapter 2).

To calculate column percentages, divide each cell frequency by the total number of cases in the column (the column marginal) and multiply the result by 100. For Table 11.3, starting in the upper-left cell, we see that there are 30 cases in this cell and 55 cases in the column. So 30 of the 55 graduates of accredited programs are working as social workers. The column percentage for this cell is therefore $(30/55) \times 100 = 54.55\%$. For the lower-left cell, the column percent is $(25/55) \times 100 = 45.45\%$. For the two cells in the right-hand column (graduates of nonaccredited programs), the column percents are $(10/45) \times 100 = 22.22$ and $(35/45) \times 100 = 77.78$. Table 11.6 displays all column percents for Table 11.3.

Column percentages make the relationship between the variables more obvious, and we can see easily from Table 11.6 that it's the students from accredited programs who are more likely to be working as social workers. Nearly 55% of

TABLE 11.6 **Column Percentages for Table 11.3**

Employment Status	Accreditation Status		TOTALS
	Accredited	*Not Accredited*	
Working as a social worker	54.55%	22.22%	40.00%
Not working as a social worker	45.45%	77.78%	60.00%
TOTALS	100.00%	100.00%	100.00%
	(55)	(45)	

Social Research and Data Analysis 11.1 The Chi Square Test

Who do you think has a more favorable view of the Internet, senior citizens or younger people? Your educated guess might be right, but how big is the difference and is it significant? Let's use some real data from the 2018 Pew Research Center's Core Trends Survey to find out. In the survey, people were asked to add up all the advantages and disadvantages of the Internet and decide whether they think it has been a good thing or a bad thing for society. The results of the survey are below. For the purpose of this demonstration, we removed all of the missing cases, and the cases where people said "some of both" on the question about the Internet. We then divided the sample into those under 65 and those 65 or over. *To learn how to use SPSS for a problem like this on your own, see the Using SPSS box in this chapter.*

Opinion on the Internet	Age		TOTALS
	Under 65	*65 or over*	
Good thing	1000	338	1338
Bad thing	194	100	294
TOTALS	1194	438	1632

The expected frequencies are calculated cell by cell using Formula 11.2. Double-check to make sure that you have the correct row and column marginal when calculating expected frequencies.

Opinion on the Internet	Age		TOTALS
	Under 65	*65 or over*	
Good thing	978.90	359.10	1338
Bad thing	215.10	78.90	294
TOTALS	1194	438	1632

Organize the calculations of chi square with a computational table.

1	2	3	4	5
f_o	f_e	$f_o - f_e$	$(f_o - f_e)^2$	$(f_o - f_e)^2/f_e$
1000	978.9	21.1	445.21	0.45
338	359.1	−21.1	445.21	1.24
194	215.1	−21.1	445.21	2.07
100	78.9	21.1	445.21	5.64
1632	1632		χ^2(obtained) =	9.40

Step 1. Making Assumptions and Meeting Test Requirements.

> Model: Independent random samples
> Level of measurement is nominal

Step 2. Stating the Null Hypothesis.

> H_0: The two variables are independent.
> (H_1: The two variables are dependent.)

Step 3. Selecting the Sampling Distribution and Establishing the Critical Region.

> Sampling distribution = χ^2 distribution
> Alpha = 0.05
> Degrees of freedom = 1
> χ^2(critical) = 3.841

Step 4. Computing the Test Statistic.

$$\chi^2(\text{obtained}) = \sum \frac{(f_o - f_e)^2}{f_e}$$

$$\chi^2(\text{obtained}) = 9.40$$

Step 5. Making a Decision and Interpreting the Results of the Test. With an obtained chi square of 9.40, we would reject the null hypothesis of independence. For this sample there is a statistically significant relationship between age (under 65 or 65 and over) and one's views on the Internet.

How are these variables related? Which age group has a more favorable view of the Internet? We can determine this by computing and analyzing column percentages:

Opinion on the Internet	Age		TOTALS
	Under 65	*65 or over*	
Good thing	83.75%	77.17%	81.99%
Bad thing	16.25%	22.83%	18.01%
TOTALS	100.00%	100.00%	100.00% (1632)

The column percentages show that 83.75% of younger people in this sample see the Internet as a good thing versus 77.17% of those over 65. We have already concluded that the relationship is significant, and now we know the pattern. Those under 65 are more likely than those 65 and older to see the Internet as a good thing for society.

these students are working as social workers versus about 22% of the students from nonaccredited programs. We already know that this relationship is significant (unlikely to be caused by random chance), and with the aid of column percents, we know how the two variables are related. According to these results, graduates from accredited programs have a decided advantage in securing social work jobs.

Let's highlight two points in summary:

1. Chi square is a test of statistical significance. It tests the null hypothesis that the variables are independent in the population. If we reject the null hypothesis, we are concluding, with a known probability of error (determined by the alpha level), that the variables are dependent on each other in the population. By itself, chi square does not tell us the nature of the relationship.

2. Column percentages show us how the independent variable affects the dependent variable. In our example, they show that it's the graduates of accredited programs who are more likely to find work as social workers. We explore column percentages more extensively when we discuss bivariate association in Chapter 12.

ONE STEP AT A TIME **Computing Column Percentages**

1. Start with the upper-left-hand cell. Divide the cell frequency (the number of cases in the cell) by the total number of cases in that column (or the column marginal). Multiply the result by 100 to convert to a percentage.
2. Move down one cell and repeat step 1. Continue moving down the column, cell by cell, until you have converted all cell frequencies to percentages.
3. Move to the next column. Start with the cell in the top row and repeat step 1 (making sure that you use the correct column total in the denominator of the fraction).
4. Continue moving down the second column until you have converted all cell frequencies to percentages.
5. Continue these operations, moving from column to column one at a time, until you have converted all cell frequencies to percentages.

STATISTICS **In Everyday Life**

The Gender Gap

There seems to be a gender gap in American politics. Women tend to be more sympathetic to Democratic presidential candidates while men tend to prefer the Republican candidates. The Gender Gap is depicted in the table below for the 2016 election. These results are based on a sample of 1341 voters and are statistically significant at well below the 0.05 level.

Why do you suppose The Gender Gap exists? What additional information would you like to have to explore the gap further?

Presidential Preference	Gender	
	Male	*Female*
H. Clinton (Democrat)	49.8%	63.1%
D. Trump (Republican)	50.2%	36.9%
TOTALS	100.00% (615)	100.00% (726)

$$\chi^2 = 24.13,\ df = 1,\ p < 0.05$$

Source: 2018 General Social Survey

The Chi Square Test: An Additional Example

To this point, we have confined our attention to 2×2 tables, with two rows and two columns. For purposes of illustration, we will work through the computational routines and decision-making process for a larger table. As you will see, larger tables require more computations (because they have more cells), but in all other essentials they are dealt with in the same way as the 2×2 table.

A researcher is concerned with the possible effects of marital status on the academic progress of college students. Do married students, with their extra burden of family responsibilities, suffer academically as compared to unmarried students? Is academic performance dependent on marital status? A random sample of 453 students is gathered and each student is classified by marital status and as a good, average, or poor student, based on grade-point average (GPA). Results are presented in Table 11.7.

For the top-left cell (married students with good GPAs), the expected frequency would be $(160 \times 175)/453$, or 61.8. For the other cell in this row, expected frequency is $(160 \times 278)/453$, or 98.2. In similar fashion, all expected frequencies are computed and displayed in Table 11.8.

The next step is to solve the formula for χ^2(obtained), being sure to use the proper f_os and f_es for each cell. Once again, we will use a computational table (Table 11.9) to organize the calculations and then test the obtained chi square for its statistical significance. Remember that obtained chi square is equal to the total of column 5.

TABLE 11.7 Grade-Point Average (GPA) by Marital Status

GPA	Marital Status		TOTALS
	Married	Unmarried	
Good	70	90	160
Average	60	110	170
Poor	45	78	123
TOTALS	175	278	453

TABLE 11.8 Expected Frequencies for Table 11.7

GPA	Marital Status		TOTALS
	Married	Unmarried	
Good	61.8	98.2	160
Average	65.7	104.3	170
Poor	47.5	75.5	123
TOTALS	175	278	453

TABLE 11.9 Computational Table for Table 11.7

(1)	(2)	(3)	(4)	(5)
f_o	f_e	$f_o - f_e$	$(f_o - f_e)^2$	$(f_o - f_e)^2/f_e$
70	61.8	8.2	67.24	1.09
90	98.2	−8.2	67.24	0.69
60	65.7	−5.7	32.49	0.50
110	104.3	5.7	32.49	0.31
45	47.5	−2.5	6.25	0.13
78	75.5	2.5	6.25	0.08
453	453	0		χ^2(obtained) = 2.80

The value of the obtained chi square (2.80) can now be tested for its significance.

Step 1. Making Assumptions and Meeting Test Requirements.

Model: Independent random samples
Level of measurement is nominal

Step 2. Stating the Null Hypothesis.

H_0: The two variables are independent.
(H_1: The two variables are dependent.)

Step 3. Selecting the Sampling Distribution and Establishing the Critical Region.

Sampling distribution = χ^2 distribution
Alpha = 0.05
Degrees of freedom = $(r - 1)(c - 1) = (3 - 1)(2 - 1) = 2$
χ^2(critical) = 5.991

Step 4. Computing the Test Statistic.

$$\chi^2\text{(obtained)} = 2.80$$

Step 5. Making a Decision and Interpreting the Results of the Test. The test statistic, χ^2(obtained) = 2.80, does not fall in the critical region, which, for alpha = 0.05, $df = 2$, begins at χ^2(critical) of 5.991. Therefore, we fail to reject the null hypothesis. The observed frequencies are not significantly different from the frequencies we would expect to find if the variables were independent and only random chance were operating. Based on these sample results, we can conclude that the academic performance of college students is not dependent on their marital status. Because we failed to reject the null hypothesis, we will not examine column percentages as we did for Table 11.3.

USING SPSS

Crosstabs and Chi Square

In this installment of "Using SPSS," we will use a procedure called **Crosstabs** to see if there is an association between political party and the use of Facebook. Since the 2016 general election, much attention has been paid to the spread of fake news on social media sites. By applying this test, we learn a little bit about which party is more exposed to political advertising on Facebook. In the procedure, we will produce a bivariate table with column percentages and a chi square value.

We will use the Pew Research Center's Core Trends Survey for this test and examine the relationship between political party (*party*), the independent variable, and Facebook use (*web1c*), the dependent variable. Is there a relationship between these variables? Do Democrats use Facebook more than Republicans? Or vice versa?

To begin:

1. Click the SPSS icon on your desktop.
2. Load the *Pew Core Trends Survey* database.
 a. Find the **File** command on the far left of the menu bar and then click **File → Open → Data**.
 b. Find the *Pew Core Trends Survey* database and click the file name.
3. Next, let's clean the *party* variable by designating values 3–9 as missing. These are the Independents and the people that did not tell us their political party affiliation. In **Variable View**, find the row for *party*. Inside the **Missing** column, select **Range plus one optional discrete missing value** and enter 4 for **Low** and 9 for **High**.
4. From the main menu bar, click **Analyze**, **Descriptive Statistics**, and **Crosstabs**.
5. The **Crosstabs** dialog box appears with the variables listed in a box on the left.
 a. Highlight the dependent variable (*web1c*) and click the arrow to move the variable name into the **Rows** box.
 b. Next, find the independent variable (*party*) and move it into the **Columns** box. SPSS will process all combinations of variables in the row and column boxes at one time.
6. Click the **Statistics** button at the bottom of the window and click the box next to Chi-square.
7. Return to the **Crosstabs** window and click the **Cells** button and select "column" in the **Percentages** box. This will generate column percentages for the table.
8. Return to the **Crosstabs** window and click **OK** and the bivariate table will be produced (see next page).

Each cell displays the cell frequency and column percentages. Starting with the upper-left-hand cell, there were 295 respondents who were Republicans reporting that Yes they use Facebook online or on their cell phones. These represent 65.7% of all Republicans in the sample. In contrast, 420 of the Democrats (73.0%) also reported using Facebook. The column percentages are different, and this suggests a relationship between these variables. Is the relationship significant?

The next output block shows that there is a significant relationship. There is a lot of information in this block which is not reproduced here to conserve space. We only need to read the top line, or "Pearson Chi-Square," which reports a chi square (obtained) of 6.450 with 1 degree of freedom. The significance of this chi square, reported in the column labeled "Asymp. Sig (2-sided)," is .011. Since this is below our standard indicator of a significant result (alpha = 0.05), we may conclude, as we saw with the column percentages, that there is a statistically significant relationship between these variables. Use of Facebook is dependent on political party, and according to the column percentages, Democrats are more likely to use Facebook than Republicans.

(continued)

USING SPSS **(Continued)**

WEB1c. Please tell me if you ever use any of the following social media sites online or on your cell phone. Do you ever use... Facebook?* PARTY. In politics TODAY, do you consider yourself a Republican, Democrat, or independent? Crosstabulation

| | | | PARTY. In politics TODAY, do you consider yourself a Republican, Democrat, or independent? | | |
			Republican	Democrat	Total
WEB1c. Please tell me if you ever use any of the following social media sites online or on your cell phone. Do you ever use... Facebook?	Yes, do this	Count	295	420	715
		% within PARTY. In politics TODAY, do you consider yourself a Republican, Democrat, or independent?	65.7%	73.0%	69.8%
	No, do not do this	Count	154	155	309
		% within PARTY. In politics TODAY, do you consider yourself a Republican, Democrat, or independent?	34.3%	27.0%	30.2%
Total		Count	449	575	1024
		% within PARTY. In politics TODAY, do you consider yourself a Republican, Democrat, or independent?	100.0%	100.0%	100.0%

Chi-Square Tests

	Value	df	Asymptotic Significance (2-sided)	Exact Sig. (2-sided)	Exact Sig. (1-sided)
Pearson Chi-Square	6.450[a]	1	.011		
Continuity Correction[b]	6.106	1	.013		
Likelihood Ratio	6.426	1	.011		
Fisher's Exact Test				.013	.007
Linear-by-Linear Association	6.444	1	.011		
N of Valid Cases	1024				

a. 0 cells (0.0%) have expected count less than 5. The minimum expected count is 135.49.
b. Computed only for a 2 × 2 table

Chi Square and the Goodness-of-Fit Test[2]

In addition to the test for independence between two variables, chi square can be used to test the significance of the distribution of scores for a single variable. This application is called the **goodness-of-fit test**; it is used to compare an actual

[2]This section is optional.

or observed distribution of scores against a set of expected frequencies. The greater the difference between the observed and expected scores, the more likely that the observed pattern did not occur by random chance alone. If the observed and expected frequencies are similar, it is said that there is a "good fit" (hence the name of this application), and we would conclude that the two distributions were not significantly different.

A major difference in this new application is that the null hypothesis, not Formula 11.2, is used to determine the expected frequencies. An example should clarify this process. Suppose you were gambling on coin tosses and suspected that a particular coin was biased toward heads. Over a series of tosses, what percentage of heads would you expect to observe from an unbiased coin? Your null hypothesis would be that the coin was unbiased and that half of all tosses should be heads and half tails. Notice what we have just done: we figured out the expected frequencies (half of the flips should be heads) from the null hypothesis (the coin is unbiased) rather than by computing them.

Let's consider another example of a situation in which the goodness-of-fit test is appropriate. A school district has five high schools with roughly equal enrollments. Are the schools equally represented on the all-district sports teams? A total of 380 students have been named in all sports. How should they be distributed across the high schools if only random chance is operating?

The problem involves only one variable (being named to the all-district sports teams), and we are concerned with how that variable is distributed by high school. If the five high schools do *not* vary in their athletic proficiency, we would expect that about 1/5 of the all-stars should come from each school. Our null hypothesis would be that athletic prowess does not vary by school and our expected frequencies would be calculated by dividing the observed number of all-stars by the number of schools.

The actual distribution of all-stars by school is displayed in Table 11.10. The expected distribution is found by dividing the total number of all-stars by 5: 380/5 = 76. Note that the expected frequency will be the same for every school. With these values determined, we can calculate chi square by using Formula 11.1.

$$\chi^2(\text{obtained}) = \sum \frac{(f_o - f_e)^2}{f_e}$$

$$\chi^2(\text{obtained}) = \frac{(110 - 76)^2}{76} + \frac{(75 - 76)^2}{76} + \frac{(82 - 76)^2}{76}$$
$$+ \frac{(45 - 76)^2}{76} + \frac{(68 - 76)^2}{76}$$

$$\chi^2(\text{obtained}) = \frac{(34)^2}{76} + \frac{(-1)^2}{76} + \frac{(6)^2}{76} + \frac{(-31)^2}{76} + \frac{(-8)^2}{76}$$

$$\chi^2(\text{obtained}) = \frac{1156}{76} + \frac{1}{76} + \frac{36}{76} + \frac{961}{76} + \frac{64}{76}$$

$$\chi^2(\text{obtained}) = 15.21 + 0.01 + 0.47 + 12.64 + 0.84$$

$$\chi^2(\text{obtained}) = 29.17$$

The χ^2(obtained) of 29.17 can now be tested for significance in the usual manner.

TABLE 11.10 Number of All-Stars per High School

High School	Number of All-Stars
Millard Fillmore	110
West Lake	75
Central	82
Dunbar	45
Heritage	68
	380

Step 1. Making Assumptions and Meeting Test Requirements.

Model: Random sampling
Level of measurement is nominal

Step 2. Stating the Null Hypothesis.

H_0: There is no difference in the number of all-stars by high school.
(H_1: There is a difference in the number of all-stars by high school.)

Step 3. Selecting the Sampling Distribution and Establishing the Critical Region.
In the goodness-of-fit test, degrees of freedom are equal to the number of categories minus 1, or $df = k - 1$. In the problem under consideration, there are 5 categories and 4 degrees of freedom.

$$\text{Sampling distribution} = \chi^2 \text{ distribution}$$
$$\text{Alpha} = 0.05$$
$$\text{Degrees of freedom} = k - 1 = 5 - 1 = 4$$
$$\chi^2(\text{critical}) = 9.488$$

Step 4. Computing the Test Statistic.

$$\chi^2(\text{obtained}) = 29.17$$

Step 5. Making a Decision and Interpreting the Results of the Test.
The test statistic (29.17) clearly falls in the critical region (which begins at 9.488), so we reject the null hypothesis. These data suggest that the number of all-district athletes varies by high school nonrandomly. *(For practice in conducting and interpreting the chi square goodness-of-fit test, see problems 11.15 and 11.16.)*

The Limitations of the Chi Square Test

Like any other test, chi square has limitations, and you should be aware of several potential difficulties. First, chi square becomes difficult to interpret when the variables have many categories. For example, two variables with five categories each would generate a 5×5 table with 25 cells, which would probably be too many

combinations of scores to be easily absorbed or understood. As a rule of thumb, the chi square test is easiest to interpret and understand when both variables have four or fewer scores.

Two further limitations of the test are related to sample size. When sample size is small, we can no longer assume that the sampling distribution of all possible sample outcomes is accurately described by the chi square distribution. For chi square, a small sample is defined as one where a high percentage of the cells have expected frequencies (f_e) of 5 or less. Various rules of thumb have been developed to help the researcher decide what constitutes a "high percentage of cells." The safest course is to take corrective action whenever any of the cells have expected frequencies of 5 or less.

For 2 × 2 tables, the value of χ^2(obtained) can be adjusted when we have an expected frequency of 5 or less by applying Yates' correction for continuity, the formula for which is:

FORMULA 11.4

$$\chi_c^2 = \sum \frac{\left(\left|f_o - f_e\right| - 0.5\right)^2}{f_e}$$

where χ_c^2 = corrected chi square
$\left|f_o - f_e\right|$ = the absolute value of the difference between observed and expected frequency for each cell.
(NOTE: Absolute values ignore plus and minus signs.)

The correction factor is applied by reducing the absolute value of the term $(f_o - f_e)$ by 0.5 before squaring the difference and dividing by the expected frequency for the cell.

For tables larger than 2 × 2, there is no correction formula for computing χ^2(obtained) for small samples. It may be possible to combine some of the categories of the variables and thereby increase cell sizes. For example, in Table 11.7, the "average" and "poor" categories might be combined. This would reduce the table size to 2 × 2 and maximize the number of cases in the bottom row.

Categories should be combined only when it is sensible to do so. In other words, meaningful distinctions, like those between average and poor students, should not be erased to conform to the requirements of a statistical test.

When you feel that categories cannot be combined to build up cell frequencies and the percentage of cells with expected frequencies of 5 or less is small, it is justifiable to continue with the uncorrected chi square test as long as the results are regarded with a suitable amount of caution.

A second potential problem related to sample size occurs with large samples. We pointed out in Chapter 9 that all tests of hypothesis are sensitive to sample size. That is, the probability of rejecting the null hypothesis increases as the number of cases increases, regardless of any other factor. Chi square is especially sensitive to sample size: The value of χ^2(obtained) will increase at the same rate as sample size. If N is doubled, for example, the value of χ^2(obtained) will double. Thus, larger samples may lead to a decision to reject the null hypothesis when the actual relationship is trivial. *(For an illustration of this principle, see problem 11.14.)*

STATISTICS In Everyday Life

Who Uses Social Networking Sites?

Would men or women be more likely to use social networking sites like Twitter and Facebook? Why? The table below presents results of a 2018 survey given to a nationally representative sample. Do the results confirm your expectations? What are some possible sociological explanations for these results?

Do You Ever Use Social Networking Sites?	Gender		TOTALS
	Male	Female	
Yes	65.0%	69.7%	67.2% (1343)
No	35.0%	30.3%	32.8% (656)
TOTALS	100.00% (1080)	100.00% (919)	1999

$$\chi^2 = 5.08, \; df = 1, \; p < 0.05$$

Source: Pew Core Trend Survey, 2018

Note that the difference between males and females is not particularly large. About 2/3rds of both genders use these sites and the difference between male and females is less than 5%. Could this be an example of a result that is statistically significant but not very important? In Part III, we will introduce some statistical tools that will allow us to assess importance vs. significance.

You should be aware of the relationship between sample size and the value of chi square because it once again raises the distinction between statistical significance and theoretical importance. On one hand, tests of significance play a crucial role in research. When we are working with random samples, we must know whether our research results could have been produced by mere random chance.

On the other hand, like any other statistical technique, tests of hypothesis are limited in the range of questions they can answer. These tests will tell us whether our results are statistically significant or not. However, they will not necessarily tell us whether the results are important in any other sense. To deal more directly with questions of importance we must use an additional set of statistical techniques called *measures of association*. We previewed these techniques in this chapter when we computed column percentages, and measures of association are the subject of Part III of this text.

Interpreting Statistics: Family Values and Social Class

When sociologists examine patterns of family life in the United States, they commonly find that parents of different social classes raise their children in different ways. These differences include disciplinary techniques, expectations regarding success in school, and a host of other variables. In this installment of "Interpreting Statistics," we examine differences in the goals that parents of different social classes, as

measured by level of education, have for their children. Specifically, we look at an item from the 2018 General Social Survey that asks parents: "If you had to choose, which thing on this list would you pick as the most important for a child to learn to prepare him or her for life?" Respondents were offered a list of traits from which to choose; we focus on two of these: "To obey" and "To think for himself or herself."

Respondents could rank each trait from 1 ("most important") to 5 ("least important"). To simplify the analysis, we have collapsed the scale to three scores:

> 1 = Most important (High)
> 2 = Second or third most important (Moderate)
> 3 = Fourth or fifth most important (Low)

Why would parents of different social classes have different socialization goals? One possible causal factor is the nature of the work associated with social class. For people in the higher social classes, work tends to stress individual initiative and creativity and be less supervised. The working-class occupations of those with less education, in contrast, are more likely to be highly routinized, regimented, and closely supervised.

Thus, in a reflection of their everyday work situations, we can hypothesize that parents in the higher social classes will stress independent thinking, while parents in the lower social classes will tend to endorse obedience as the most important socialization goal. In the language of chi square, we can hypothesize that socialization goals will be dependent on social class.

Tables 11.11 and 11.12 show the relationships between socialization goals and level of education for the entire 2018 GSS. Recall that the General Social Survey is administered to random samples of adult Americans, selected in accordance with the rules of EPSEM (see Chapter 6). Thus, we can assume that the conclusions of this analysis will apply to U.S. society in general.

First, note that results are presented using column percentages instead of cell frequencies. This is consistent with the professional research literature and makes it easier to identify patterns and differences. For example, the row marginals of the tables show there is much less support for obedience as a socialization goal. Only 10.8% of all respondents ranked it highest, while 41.7% ranked independent thinking highest.

The column percentages show that our hypotheses are supported. Looking at the top row (those that ranked obedience "high") of Table 11.11, we see that support for obedience is highest for the least educated respondents and declines as level of education increases. Inspecting the same row of Table 11.12, we see

TABLE 11.11 Socialization Goal (Obedience) by Level of Education

Importance of "Obey"	Less Than High School	High School Graduate	At Least Some College	TOTALS
High	19.9%	13.6%	4.8%	10.8%
Moderate	36.4%	29.7%	19.6%	26.5%
Low	43.8%	56.7%	75.6%	62.6%
TOTALS	100.0% (176)	100.0% (760)	100.0% (606)	100.0% (1542)
Chi square = 89.68, $df = 4$, $p < 0.01$				

TABLE 11.11 Socialization Goal (Independent Thinking) by Level of Education

Importance of "Obey"	Less Than High School	High School Graduate	At Least Some College	TOTALS
High	27.8%	37.5%	51.0%	41.7%
Moderate	38.6%	38.4%	35.8%	37.4%
Low	33.5%	24.1%	13.2%	20.9%
TOTALS	99.9%	100.0%	100.0%	100.0%
	(176)	(760)	(606)	(1542)

Chi square = 58.87, $df = 4$, $p < 0.01$

that support for independent thinking is highest for the most educated and increases as level of education increases.

The results of the chi square test are reported below the table and both chi squares are statistically significant. The notation "$p < 0.01$" is the format used in the professional research literature to report significance and, for the chi square test, means "the probability of these cell frequencies, if the null hypothesis of independence is true, is less than 0.01."

What can we conclude? These results show support for our hypothesis. The relationships are statistically significant and our expectations about the pattern of the column percentages are confirmed. Support for obedience is higher among the least educated and support for independent thinking is highest in the most educated.

The findings that overall support for obedience is much lower than support for independent thinking could reflect the fact that the United States is now a post-industrial society, in which good jobs demand high levels of education and critical thinking. If people of all class backgrounds have come to recognize this economic reality, we would expect to find less difference in socialization goals by social class than in the past.

Social Research and Data Analysis 11.2 Racism and Free Speech

In recent years, debates over who is a racist and whether racists should be allowed to engage in free speech have intensified in the United States. In the example below, we will use 2018 General Social Survey data to measure the different opinions that whites, blacks, and those of other races have about whether a "racist" should be allowed to teach in a college.

Let's begin by examining the observed statistics below. You can explore the same data in the GSS2018 file by running a cross tab between *race* and *colrace*.

Should Racist Be Allowed to Teach in College	White	Black	Other	TOTALS
Allowed	498	100	53	651
Not Allowed	603	153	116	872
TOTALS	1101	253	169	1523

A quick look at the observed frequencies shows us that more whites say that a racist should be allowed to teach in college, but this is deceiving since there are many more whites in the sample than blacks and others. We will need to keep working to see if the differences by race are significant. Continuing on, below is a table of the expected frequencies.

Should Racist Be Allowed to Teach in College	White	Black	Other	TOTALS
Allowed	470.62	108.14	72.24	651.00
Not Allowed	630.38	144.86	96.76	872.00
TOTALS	1101.00	253.00	169.00	1523.00

(*continued*)

> ### Social Research and Data Analysis 11.2 (Continued)
>
> Next, we will set the data up in columns to solve for chi square.
>
1	2	3	4	5
> | f_o | f_e | $f_o - f_e$ | $(f_o - f_e)^2$ | $(f_o - f_e)^2/f_e$ |
> | 498.00 | 470.62 | 27.38 | 749.66 | 1.59 |
> | 603.00 | 630.38 | −27.38 | 749.66 | 1.19 |
> | 100.00 | 108.14 | −8.14 | 66.26 | 0.61 |
> | 153.00 | 144.86 | 8.14 | 66.26 | 0.46 |
> | 53.00 | 72.24 | −19.24 | 370.18 | 5.12 |
> | 116.00 | 96.76 | 19.24 | 370.18 | 3.83 |
> | 1523.00 | 1523.00 | | | 12.80 |
>
> Note that the cells are stated in a different order than in previous tables. This is fine as long as the observed and expected frequencies are paired correctly.
>
> What should we conclude? With $df = 2$ and an alpha of 0.05, χ^2(critical) is 5.991. Thus, the relationship is significant. Use the column percentages below to see where the differences can be found. Note that most respondents (57.26%) oppose allowing a racist to teach in college, but which race is the least supportive? Why?
>
Should Racist Be Allowed to Teach in College	Race			
> | | White | Black | Other | TOTALS |
> | Allowed | 45.23% | 39.53% | 31.36% | 42.74% |
> | Not Allowed | 54.77% | 60.47% | 68.64% | 57.26% |
> | TOTALS | 100.00% | 100.00% | 100.00% | 100.00% |
>
> Source: 2018 General Social Survey

> ### READING STATISTICS Awareness of Support for LGBT Communities Among Youth
>
> The chi square statistics is often used in categorical analyses, in which a researcher is trying to identify an association between variables that have nominal or ordinal categorical responses. In a recent study in *The Journal of LGBT Youth*, Jeffrey Dennis was interested in learning whether gay and lesbian youth were less likely to engage in deviant behaviors or commit small crimes if they had a greater awareness of the presence of LGBT communities in their society. His logic was based on the idea that most LGBT youth are raised to believe that there is "nothing out there" for them in a society that is almost entirely driven by heteronormative preferences, language, and behaviors.
>
> In his study of 224 gay and lesbian youth, he found that 41.7% of those with a low awareness of LGBT communities reported getting drunk on a regular basis, compared to 18.2% of those with a high awareness. Since the sample is not random, the study did not meet all of the assumptions that we require in this chapter so we cannot generalize the results to a population. However, this is a common limitation in academic research and we can still test to see if the association was by chance or not. According to the article, the chi square value was significant ($\chi^2 = 5.98$, $p < .05$) for the association between awareness and getting drunk regularly, which leads us to the conclusion that we can be 95% confident that it did not occur due to chance. Later in the paper, the author also reported that 75% of those with a low awareness used an above-average amount of illicit substances ($\chi^2 = 8.41$, $p < .05$). Again, we can conclude that the association did not occur by chance alone.
>
> Source: Dennis, J. (2019). Nothing out there: Community awareness and delinquency among gay and lesbian youth. *Journal of LGBT Youth, 16*:1, 1–17.

SUMMARY

1. The chi square test for independence is appropriate for situations in which the variables of interest have been organized into table format. The null hypothesis is that the variables are independent, or that the classification of a case into a particular category on one variable has no effect on the probability that the case will be classified into any particular category of the second variable.

2. Because chi square is nonparametric and requires only nominally measured variables, its model assumptions are easily satisfied. Furthermore, it is computed from bivariate tables, in which the number of rows and columns can be easily expanded, so the chi square test can be used in many situations in which other tests are inapplicable.

3. In the chi square test, we first find the frequencies that would appear in the cells if the variables were independent (f_e) and then compare those frequencies, cell by cell, with the frequencies actually observed in the cells (f_o). If the null hypothesis is true, expected and observed frequencies should be quite close in value. The greater the difference between the observed and expected frequencies, the greater the possibility of rejecting the null hypothesis.

4. The chi square test has several important limitations. It is often difficult to interpret when tables have more than four or five dimensions. Furthermore, as sample size (N) decreases, the chi square test becomes less trustworthy and corrective action may be required. Finally, with very large samples, we may declare relatively trivial relationships to be statistically significant. As is the case with all tests of hypothesis, statistical significance is not the same thing as "importance" in any other sense. It is important to remember that statistical significance is a necessary, but not sufficient, condition for theoretical or practical importance.

SUMMARY OF FORMULAS

Chi square (obtained):

11.1 $$\chi^2(\text{obtained}) = \sum \frac{(f_o - f_e)^2}{f_e}$$

Expected frequencies:

11.2 $$f_e = \frac{\text{Row marginal} \times \text{Column marginal}}{N}$$

Degrees of freedom, bivariate tables:

11.3 $$df = (r - 1)(c - 1)$$

Yates' correction for continuity:

11.4 $$\chi_c^2 = \sum \frac{(|f_o - f_e| - 0.5)^2}{f_e}$$

GLOSSARY

Bivariate table. A table that displays the joint frequency distributions of two variables.

Cells. The cross-classification categories of the variables in a bivariate table.

χ^2(critical). The score on the sampling distribution of all possible sample chi squares that marks the beginning of the critical region.

χ^2(obtained). The test statistic computed from sample results.

Chi square (χ^2) test. A nonparametric test of hypothesis for variables that have been organized into a bivariate table.

Column. The vertical dimension of a bivariate table. By convention, each column represents a score on the independent variable.

Expected frequency (f_e). The cell frequencies that would be expected in a bivariate table if the variables were independent.

Goodness-of-fit test. An additional use for chi square that tests the significance of the distribution of a single variable.

Independence. The null hypothesis in the chi square test. Two variables are independent if, for all cases, the classification of a case on one variable has no effect on the probability that the case will be classified in any particular category of the second variable.

Marginals. The row and column subtotals in a bivariate table.

Nonparametric test. A "distribution-free" test. These tests do not assume a normal sampling distribution.

Observed frequency (f_o). The cell frequencies observed in a bivariate table.

Row. The horizontal dimension of a bivariate table, conventionally representing a score on the dependent variable.

PROBLEMS ▰▰▰▰▰▰▰▰▰▰▰▰▰▰▰▰▰▰▰▰▰▰▰▰▰▰▰▰▰▰▰▰

11.1 Below are four 2 × 2 tables, presented without labels or titles, to provide simplified opportunities to practice your computational skills. Calculate chi square for each.

HINTS:

- *Calculate the expected frequencies for each cell with Formula 11.2. Double-check to make sure you are using the correct row and column marginals for each cell.*
- *It may be helpful to record the expected frequencies in table format—see Tables 11.4 and 11.8.*
- *Use a computational table to organize the calculation for Formula 11.1—see Tables 11.5 and 11.9.*
 - *Follow the step-by-step instructions.*
 - *Double-check to make sure you are using the correct values for each cell.*

a.

20	25	45
25	20	45
45	45	90

c.

25	15	40
30	30	60
55	45	100

b.

10	15	25
20	30	50
30	45	75

d.

20	45	65
15	20	35
35	65	100

11.2 |PS| As mentioned in this chapter, many analysts have noted a "gender gap" in U.S. elections, with women more likely to vote for the Democratic candidate. A sample of university faculty has been asked about their political party preference. Do their responses indicate a significant relationship between gender and party preference for this group?

Party Preference	Gender		TOTALS
	Male	*Female*	
Democrat	10	15	25
Republican	15	10	25
TOTALS	25	25	50

a. Is there a statistically significant relationship between gender and party preference?

b. Compute column percentages for the table to clarify the pattern of the relationship. Which gender is more likely to prefer the Democrats?

11.3 |SW| A local politician is concerned that a program for the homeless in her city is discriminating against blacks and other minorities. The following data were taken from a random sample of black and white homeless people.

Received Services?	Race		TOTALS
	Black	*White*	
Yes	6	7	13
No	4	9	13
TOTALS	10	16	26

a. Is there a statistically significant relationship between race and whether the person has received services from the program?

b. Compute column percents for the table to clarify the pattern of the relationship. Which group was more likely to get services?

11.4 |SOC| A sample of 25 cities have been classified as high or low on their homicide rates and on the number of handguns sold within the city limits.

a. Is there a relationship between these variables? Do cities with higher homicide rates have significantly higher handgun sales? Explain your results in a sentence or two, citing both the results of the chi square test and the patterns of the column percentages.

b. The table is constructed assuming that homicide rate is the independent variable ("higher homicide rates cause people to buy more guns"). Can you think of an argument that would make handgun sales the independent variable? Which is the more logical argument? Why?

Volume of Handgun Sales	Homicide Rate		TOTALS
	Low	*High*	
High	8	5	13
Low	4	8	12
TOTALS	12	13	25

11.5 |PA| Is there a relationship between salary levels and unionization for public employees? The following data represent this relationship for fire departments in a random sample of 100 cities of roughly the same size. Salary data have been dichotomized at the median. Explain your results in a sentence or two, citing both the results of the chi square test and the patterns of the column percentages.

Salary	Status		TOTALS
	Union	*Nonunion*	
High	21	29	50
Low	14	36	50
TOTALS	35	65	100

11.6 CJ A local judge has been allowing some individuals convicted of "driving under the influence" to work in a hospital emergency room as an alternative to fines, suspensions, and other penalties. A random sample of offenders has been drawn. Do participants in this program have lower rates of recidivism for this offense? Explain your results in a sentence or two, citing both the results of the chi square test and the patterns of the column percentages.

| | Status | | |
Recidivist?	Non-participant	Participant	TOTALS
Yes	60	123	183
No	55	108	163
TOTALS	115	231	346

11.7 SOC The state Department of Education has rated a sample of local school systems for compliance with state-mandated guidelines for quality.
 a. Is there a statistically significant relationship between the quality of a school system and the affluence of the community as measured by per capita income?
 b. Compute column percentages. Are high- or low-income communities more likely to have high-quality schools?

| | Per Capita Income | | |
Quality	Low	High	TOTALS
Low	16	8	24
High	9	17	26
TOTALS	25	25	50

11.8 SOC A program of pet therapy has been running at a local nursing home. Are the participants in the program significantly more alert and responsive than nonparticipants? *(Remember to compute column percentages to identify the pattern of the relationship.)*

| | Participation? | | |
Alertness	Yes	No	TOTALS
Low	11	18	29
High	23	15	38
TOTALS	34	33	67

11.9 SOC Is there a statistically significant relationship between length of marriage and satisfaction with marriage? The necessary information has been collected from a random sample of 100 respondents drawn from a local community. Write a sentence or two

explaining your findings, including the results of the chi square test and pattern of column percentages.

| | Length of Marriage (in years) | | | |
Satisfaction	Less Than 5	5–10	More Than 10	TOTALS
Low	10	20	20	50
High	20	20	10	50
TOTALS	30	40	30	100

11.10 PS Is there a relationship between college class standing and political orientation? Are juniors and seniors significantly different from freshmen and sophomores on this variable? The following table reports the relationship between these two variables for a random sample of 267 college students. Summarize your findings in a sentence or two, including the results of the chi square test and the pattern of column percentages. Which group is more likely to be liberal? Which group is more likely to be conservative?

| | Class Standing | | |
Political Views	Freshman and Sophomore	Junior and Senior	TOTALS
Liberal	40	43	83
Moderate	50	50	100
Conservative	44	40	84
TOTALS	134	133	267

11.11 SOC At a large urban college, about half of the students live off campus in various arrangements, and the other half live on campus. Is academic performance dependent on living arrangements? Write a sentence or two explaining your findings, including the results of the chi square test and pattern of column percentages.

| | Residential Status | | | |
GPA	Off Campus with Roommates	Off Campus with Parents	On Campus	TOTALS
Low	22	20	48	90
Moderate	36	40	54	130
High	32	10	38	80
TOTALS	90	70	140	300

11.12 |SOC| An urban planning commissioner has created a database describing a sample of the neighborhoods in their city and has developed a scale by which each area can be rated for the "quality of life" (this includes measures of pollution, noise, open space, services available, etc.). They have also asked samples of residents of these areas about their level of satisfaction with their neighborhoods. Is there statistically significant agreement between the commissioner's objective ratings of quality of life and the respondents' self-reports of satisfaction? Write a sentence or two explaining your findings, including the results of the chi square test and pattern of column percentages.

	Quality of Life			
Satisfaction	*Low*	*Moderate*	*High*	TOTALS
Low	21	15	6	42
Moderate	12	25	21	58
High	8	17	32	57
TOTALS	41	57	59	157

11.13 |SOC| Does support for the legalization of marijuana vary by region of the country? The table displays the relationship between the two variables for a random sample of 1020 adult citizens. Is the relationship significant? Describe the pattern of the relationship, using column percentages. Which region is most in support of legalization?

	Region				
Legalize?	*North*	*Midwest*	*South*	*West*	TOTALS
Yes	60	65	42	78	245
No	245	200	180	150	775
TOTALS	305	265	222	228	1020

11.14 |SOC| A researcher is concerned with the relationship between attitudes toward violence and violent behavior. Do attitudes "cause" behavior? Do people who have positive attitudes toward violence have higher rates of violent behavior? A pretest was conducted on 70 respondents; among other things, the respondents were asked, "Have you been involved in a violent incident of any kind over the past six months?" The researcher established the following relationship:

	Attitude Toward Violence		
Involvement	*Favorable*	*Unfavorable*	TOTALS
Yes	16	19	35
No	14	21	35
TOTALS	30	40	70

The chi square for this table is 0.23, which is not significant at the 0.05 level (*confirm this with your own calculations*). Undeterred by this result, the researcher proceeded with the project and gathered a random sample of 7000. In terms of percentage distributions, the results for the full sample were the same as for the pretest:

	Attitude Toward Violence		
Involvement	*Favorable*	*Unfavorable*	TOTALS
Yes	1600	1900	3500
No	1400	2100	3500
TOTALS	3000	4000	7000

However, the chi square obtained is a very healthy 23.4 (*confirm with your own calculations*). Why is the full-sample chi square significant when the chi square for the pretest was not? What happened? Do you think that the second result is important?

11.15 |SOC| (*This problem is optional.*) The director of athletics at the local high school wonders whether the sports program is getting a proportional amount of support from each of the four classes. If there are roughly equal numbers of students in each of the classes, what does the following breakdown of class membership from a random sample of students in attendance at a recent basketball game suggest?

Class	Frequency
Freshman	200
Sophomore	150
Junior	120
Senior	110
TOTALS	580

11.16 |SOC| (*This problem is optional.*) A small western town has roughly equal numbers of Hispanic, Asian, and Anglo-American residents. Are the three groups equally represented at town meetings? The breakdown of group membership for a random sample drawn from those attending a meeting is given in the table.

Group	Frequency
Hispanic	74
Asian	55
Anglo-American	53
TOTALS	182

Is there a statistically significant pattern here?

11.17 PS A random sample of 748 voters in a large city were asked how they voted in the presidential election of 2016. Calculate chi square and the column percentages for each table below, and write a brief report describing the significance of the relationships as well as the patterns you observe. As you recall, Trump was the Republican candidate and Clinton was the nominee of the Democratic Party.

a. Presidential preference and gender

	Gender		
Preference	*Male*	*Female*	TOTALS
Trump	175	163	338
Clinton	200	210	410
TOTALS	375	373	748

b. Presidential preference and race/ethnicity

	Race/Ethnicity			
Preference	*White*	*Black*	*Latino*	TOTALS
Trump	289	5	44	338
Clinton	249	95	66	410
TOTALS	538	100	110	748

c. Presidential preference by education

	Education				
Preference	*Less Than HS*	*HS Graduate*	*College Graduate*	*Postgrad. Degree*	TOTALS
Trump	30	180	118	10	338
Clinton	35	120	218	37	410
TOTALS	65	300	336	47	748

d. Presidential preference by religion

	Religion					
Preference	*Protestant*	*Catholic*	*Jewish*	*None*	*Other*	TOTALS
Trump	165	110	10	28	25	338
Clinton	245	55	20	60	30	410
TOTALS	410	165	30	88	55	748

Statistical Analysis Using SPSS

11.18 Is there a relationship between gender and fear of walking alone at night (*fear*), support for capital punishment (*cappun*), and job satisfaction (*satjob*)? Are men or women more likely to say that there is a place nearby where they are afraid to walk at night? Which gender is more supportive of the death penalty? Are men or women more satisfied with their jobs (*satjob*)? We will use the **Crosstabs** procedure, with chi square and column percentages to explore this relationship.
- Click the SPSS icon on your desktop.
- Load the *GSS2018* database.
- Click **Analyze → Descriptive → Crosstabs**.
- Find *fear, cappun,* and *satjob* in the list of variables and move them into the **Rows** box.
- Find *sex* in the list of variables and move it into the **Columns** box.
- Click the **Statistics** button and select **Chi-square**.
- Return to the **Crosstabs** window, click the **Cells** button, and select **Column** in the **Percentages** box.
- Return to the **Crosstabs** window and click **OK**. The results of all three tests will be printed to the SPSS output window.

Inspect the column percentages, chi square, and the "Asymp. Sig (2-sided)." Report these results in a few sentences.

YOU ARE THE RESEARCHER

Understanding Political Beliefs

Two projects are presented below and we urge you to apply your understanding of the chi square test to both. In the first, you will examine the sources of people's beliefs about some of the most hotly debated topics in U.S. society: assisted suicide, gay

marriage, and immigration. In the second, you will compare various independent variables to see which has the most significant relationship with your chosen dependent variable.

Using Crosstabs

In these projects you will use **Crosstabs**, the procedure introduced in the "Using SPSS" installment and in Problem 11.18. To run **Crosstabs** for both projects:

* Click **Analyze**, **Descriptive Statistics**, and **Crosstabs**.
* Place your dependent variable(s) in the **Rows** box.
* Place your independent variable(s) in the **Columns** box.
* Click the **Statistics** button and select chi-square.
* Click the **Cells** button and select "column" in the **Percentages** box.

PROJECT 1: Explaining Beliefs

In this project, you will analyze beliefs about assisted suicide (*letdie1*), gay marriage (*marhomo*), and immigration (*letin1*). You will select an independent variable, use SPSS to generate chi squares and column percentages, and analyze and interpret your results.

Step 1: Choose an Independent Variable

Select an independent variable that seems likely to be an important cause of people's attitudes about these issues. *(Be sure to select an independent variable that has only 2–5 categories.)* If you would like to use an independent variable with more than 5 scores, use the **recode** command (see "You Are the Researcher" in Chapter 10 for instructions in using the recode command) to reduce the number of categories. You might consider gender, level of education (use *degree*), religion, or age (the recoded version—see Chapter 10) as possible independent variables, but there are many others. Record the variable name and state exactly what the variable measures in the box below.

SPSS Name (e.g., *degree*)	What Exactly Does This Variable Measure?

Step 2: Stating Hypotheses

State hypotheses about the relationships you expect to find between your independent variable and each of the four dependent variables. State these hypotheses in terms of which category of the independent variable you expect to be associated with which category of the dependent variable (e.g., "I expect that men will be more supportive of the legal right to an abortion for any reason").

Dependent Variable	Hypothesis
letdie1	
marhomo	
letin1	

Step 3: Running Crosstabs

All three dependent variables can be placed in the **Rows**: box together. Move the independent variable you selected in the **Columns**: box. Don't forget to click the **Statistics** button to get chi square and the **Cells** button for column percentages.

Step 4: Recording Results

There will be a lot of output and it will be helpful to summarize your results in the following table. You will find the value of chi square, the degrees of freedom, and the statistical significance listed in the first line ("Pearson Chi Square") of the "Chi-Square Tests" block of output. If the value in the column labeled "Asymp. Sig (2-sided)" is less than 0.05, write "Yes" in the right-hand column of the table below.

Dep. Var.	Chi Square	Degrees of Freedom	Significant at 0.05 Level?
letdie1			
marhomo			
letin1			

Step 5: Analyzing and Interpreting Results

Write a short summary of results of each test in which you:

1. Identify the variables, the value and significance of chi square, *N*, and the pattern (if any) of the column percentages. In the professional research literature, you might find the results reported as: "For a sample of 949 respondents, there was no significant relationship between gender and support for abortion (chi square = 2.863, *df* = 1, *p* > 0.05). About 46% of the men supported the legal right to an abortion vs. about 44% of the women."

2. Explain if your hypotheses were supported.

PROJECT 2: Exploring the Impact of Various Independent Variables

In this project, you will examine the relative ability of different independent variables to account for a single dependent variable. You will again use the **Crosstabs** procedure in SPSS to generate chi squares and column percentages, as well as the value of alpha to judge which independent variable has the most significant relationship with your dependent variable.

Step 1: Choosing Variables

Select a dependent variable. You may use any of the variables from Project 1 or select a new dependent variable. *(Be sure that your dependent variable has no more than five values or scores.)* Use the **recode** command (see "You Are the Researcher" in Chapter 10 for instructions in using the recode command) as necessary to reduce the number of categories. Good choices for dependent variables include any measure of attitudes or opinions. Do not select characteristics like race, sex, or religion as dependent variables.

Select three independent variables that seem likely to be important causes of the dependent variable you selected. *(Be sure your independent variable has no more than*

five values or scores.) Use the **recode** command as necessary to reduce the number of categories. You might consider gender, level of education (use *degree*), religion, or age (the recoded version—see Chapter 10) as possibilities, but there are many others.

Record the variable names and state exactly what each variable measures in the table below.

SPSS Name	What Exactly Does This Variable Measure?
Dependent Variable	
Independent Variables	

Step 2: Stating Hypotheses

State hypotheses about the relationships you expect to find between your independent variables and the dependent variable. State these hypotheses in terms of which category of the independent variable you expect to be associated with which category of the dependent variable (for example, "I expect that men will be more supportive of the legal right to an abortion for any reason").

Independent Variables (SPSS Names)	Hypothesis
1. _____	
2. _____	
3. _____	

Step 3: Running Crosstabs

Place your dependent variable in the **Rows**: box and all three of your independent variables in the **Columns**: box. Click the **Statistics** button to get chi square and the **Cells** button for column percentages.

Step 4: Recording Results

There will be a lot of output and it will be helpful to summarize your results in the following table. You will find the value of chi square, the degrees of freedom, and the statistical significance listed in the first line ("Pearson Chi Square") of the "Chi-Square Tests" block of output. If the value in the column labeled "Asymp. Sig (2-sided)" is less than 0.05, write "Yes" in the right-hand column of the table below.

Indep. Var.	Chi Square	Degrees of Freedom	Significant at 0.05 Level?
1. _____			
2. _____			
3. _____			

Step 5: Analyzing and Interpreting Results

Write a short summary of results of each test in which you:

1. Identify the variables being tested, the value and significance of chi square, N, and the pattern (if any) of the column percentages.

2. Explain whether your hypotheses were supported.

3. Explain which independent variable had the most significant relationship (lowest value in the "Asymp. Sig 2-tailed" column) with your dependent variable.

PART II CUMULATIVE EXERCISES

1. Conduct the appropriate test of significance for each research situation. Problems are stated in no particular order and include research situations from each chapter in Part II.

a. Is there a gender gap in use of the Internet? Random samples of men and women have been questioned about the average number of minutes they spend each day on the Internet for any purpose. Is the difference significant?

Women	Men
$\bar{X}_1 = 55$	$\bar{X}_2 = 60$
$s_1 = 2.5$	$s_1 = 2.0$
$N_1 = 520$	$N_2 = 515$

b. For high school students, is there a relationship between social class and involvement in activities such as clubs and sports? Data have been gathered for a random sample of students. Is the relationship significant?

	Social Class		
Involvement	*Middle*	*Lower*	TOTALS
High	11	19	30
Moderate	19	21	40
Low	8	12	20
TOTALS	38	52	90

c. Some respondents have been asked how many sex partners they have had over their lives. Does the number vary significantly by educational level?

Less Than High School	High School	At Least Some College
1	1	2
1	2	1
2	3	5
1	8	9
9	10	11
2	5	2
9	4	1
	3	1

d. A random sample of the U.S. population was asked how many times they had moved since they were 18 years of age. The results are presented here. On the average, how many times do adult Americans move?

$$\overline{X} = 3.5$$
$$s = 0.5$$
$$N = 1450$$

e. On the average, school districts receive $1623 per student in budget support from the state. A random sample of 107 rural schools received an average of $1605 per pupil. Is the difference significant?

$$\mu = 1623 \qquad \overline{X} = 1605$$
$$s = 701$$
$$N = 107$$

Statistical Analysis Using SPSS
The following are a number of research questions that can be answered by applying the techniques presented in Part II to the General Social Survey data set supplied with this text (*GSS2018.sav*). For each question, conduct the most appropriate test and state your conclusions. Use an alpha level of 0.05 throughout.

The questions are presented in random order and there is not a problem for every chapter or for every statistical test presented in Part II of the text. In selecting tests and procedures, you need to consider the question, the number of categories being compared, and the level of measurement of the variables.

1. Is there a statistically significant difference in the "number of hours worked last week" (*hrs1*) by sex?

2. Is there a statistically significant difference in the "number of hours worked last week" (*hrs1*) by social class (*class*) or religion (*relig*)?

3. Estimate the average number of hours worked last week (*hrs1*) and the average number of hours per week spent on the Internet (*wwwhr*) for the population of all adult Americans.

4. Is there a statistically significant relationship between support for allowing a "Muslim clergyman speaking hatred of the United States" (*spkmslm*) and education (*degree*)? (*HINT: Education is the independent variable.*)

5. Is there a statistically significant relationship between support for allowing a "Muslim clergyman speaking hatred of the United States" (*spkmslm*) to speak and how a person voted in the 2016 presidential election (*pres16*)? (*HINT: Presidential preference is the independent variable.*)

6. Is there a significant difference between Protestants and Catholics in the frequency of sex during the past year (*sexfreq*) or total number of sex partners over the past 5 years (*partnrs5*)? Treat *sexfreq* and *partnrs5* as interval-ratio variables.

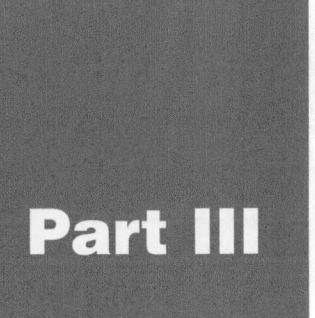

Part III

Bivariate Measures of Association

Part III covers *measures of association*, statistics that are extremely useful in scientific research and commonly reported in the professional literature. These statistics provide, in a single number, an indication of the strength and (if appropriate) the direction of a bivariate association.

It is important to remember the difference between statistical significance, covered in Part II, and association, the topic of this part. Tests for statistical significance answer a specific question: Were the differences or relationships observed in the sample caused by mere random chance? Measures of association address a different set of questions: How strong is the relationship between the variables? What is the direction or pattern of the relationship?

Thus, the information supplied by measures of association complements that supplied by tests of significance, and association and significance are two different things. It is most satisfying to find results that are *both* statistically significant and strong, but it is common to find mixed or ambiguous results: relationships that are statistically significant but weak, not statistically significant but strong, and so forth.

Chapter 12 introduces the basic ideas behind the analysis of association in terms of bivariate tables and column percentages and presents measures of association for nominal- and ordinal-level variables. Chapter 13 presents Pearson's *r*, the most important measure of association and the only one designed for interval-ratio–level variables.

12

Bivariate Association for Nominal- and Ordinal-Level Variables

LEARNING OBJECTIVES By the end of this chapter, you will be able to:

1. Analyze the *importance* (vs. statistical significance) of a bivariate relationship using measures of association.
2. Explain the concept of association in the context of bivariate tables.
3. List and explain the three characteristics of a bivariate relationship.
4. Calculate column percentages and measures of association for bivariate tables.
5. Analyze a bivariate relationship using column percentages and a measure of association.
6. Compute and interpret Phi, Cramer's V, and lambda for nominal-level variables.
7. Explain the concept of proportional reduction in error.
8. Compute and interpret Gamma and Spearman's rho, for ordinal-level variables.
9. Use SPSS to calculate column percentages for a bivariate table and produce nominal and ordinal measures of association.

USING STATISTICS

The statistical techniques presented in this chapter are used to measure the association between nominal- and ordinal-level variables. Examples of situations in which these techniques are useful include

1. Is there a relationship between age and concern for the environment? **See page 347**
2. What factors can be used to predict anxiety about statistics? **See page 348**
3. Are more educated people happier than less educated people? **See page 342**
4. Do burnout and loss of morale in teachers increase as years of experience increases? **See page 338**

As we have seen over the past several chapters, tests of statistical significance are extremely important in social science research. When we work with random samples rather than populations, tests of significance are indispensable for dealing with the possibility that our research results are the products of mere random chance.

Often, however, significance testing is only the first step in the analysis of research results. These tests do have limitations, and statistical significance is not

necessarily the same thing as relevance or importance. Furthermore, all tests of significance are affected by sample size: Tests performed on large samples may result in decisions to reject the null hypothesis when, in fact, the differences or relationships are quite minor.

Beginning with this chapter, we will be working with **measures of association**, which can help us assess the importance or strength of relationships. These statistics can also help us directly test the power and validity of our theories. Scientific theories are almost always stated in cause-and-effect terms (for example: "variable *X* causes variable *Y*"). As an example, recall our discussion of the contact hypothesis in Chapter 1. In that theory, the causal (or independent) variable was equal status contacts between groups, and the effect (or dependent) variable was prejudice. The theory asserts that involvement in equal-status-contact situations *causes* prejudice to decline. Measures of association help us trace causal relationships among variables, and they are our most important and powerful statistical tools for documenting, measuring, and analyzing cause-and-effect relationships.

We must point out that measures of association, as useful as they are, cannot *prove* that two variables are causally related. Even if there is a strong (and significant) statistical association between two variables, we cannot necessarily conclude that one variable is a cause of the other. We will explore causation in more detail in Part IV, but for now you should keep in mind that causation and association are two different things. We can use a statistical association between variables as evidence for a causal relationship, but association by itself is not proof that a causal relationship exists.

Another important use for measures of association is prediction. If two variables are associated, we can predict the score of a case on one variable from its score on the other variable. For example, if equal-status contacts and prejudice are associated, we can predict that people who have experienced many such contacts will be less prejudiced than those who have had few or no contacts.

In this chapter you will be introduced to the concept of **association** between variables in the context of bivariate tables. You will also learn how to use percentages and several different measures of association to analyze a bivariate relationship. By the end of this chapter, you will have an array of statistical tools that you can use to analyze the strength and pattern of an association between two variables.

Association and Bivariate Tables

Most generally, two variables are said to be associated if the distribution of one of them changes under the various categories or scores of the other. For example, suppose that an industrial sociologist was concerned with the relationship between job satisfaction and productivity for assembly-line workers. If these two variables are associated, then scores on productivity will change under the different conditions of satisfaction. Highly satisfied workers will have different productivity scores than workers who are low on satisfaction, and levels of productivity will vary by levels of satisfaction.

We can clarify this relationship with a bivariate table. As you recall from Chapter 11, bivariate tables display the scores of cases on two different variables.

TABLE 12.1 **Productivity by Job Satisfaction (fictitious data)**

Productivity (Y)	Job Satisfaction (X)			TOTALS
	Low	Moderate	High	
Low	30	21	7	58
Moderate	20	25	18	63
High	10	15	27	52
TOTALS	60	61	52	173

By convention, the **independent**, or **X**, **variable** (that is, the variable that is assumed to be the cause) is arrayed in the columns, and the **dependent**, or **Y**, **variable** in the rows.[1] That is, each column of the table (the vertical dimension) represents a score or category of the independent variable (X), and each row (the horizontal dimension) represents a score or category of the dependent variable (Y).

Table 12.1 displays a relationship between productivity and job satisfaction for a fictitious sample of 173 factory workers. We focus on the columns to see whether there is an association between the variables. The "within-column" frequency distributions are called the **conditional distributions of Y**, because they display the distribution of scores on the dependent variable for each condition (or score) of the independent variable.

We can inspect the conditional distributions of Y in Table 12.1 by going from column to column. If the conditional distributions change, the variables are associated. The left-hand column shows the conditional distribution of productivity (Y) for workers who were low on job satisfaction (X): 30 of the 60 were low on productivity, 20 were moderately productive, and 10 were high on productivity. The middle column shows the distribution of Y for moderately satisfied workers (21 of the 61 were low on productivity, 25 were moderate, and 15 were high) and the right-hand column displays the distribution of Y for highly satisfied workers (only 7 were low on productivity, 18 were moderate, and 27 were high).

So far, we know that the variables in Table 12.1 are associated (because the conditional distributions of Y change across the scores of X). In the next section, we extend the analysis and show how to develop additional useful information about the relationship.

Three Characteristics of Bivariate Associations

To fully investigate a bivariate association, we seek answers to three questions:

1. Does an association exist?

2. How strong is the association?

3. What is the pattern or direction of the association?

We will consider each question separately.

[1]In the material that follows, we will often, for the sake of brevity, refer to the independent variable as X and the dependent variable as Y.

DOES AN ASSOCIATION EXIST?

We have already seen that we can detect an association by observing the conditional distributions of Y in a table. In Table 12.1, we know that the two variables are associated to some extent because the conditional distributions of productivity (Y) are different across the various categories of satisfaction (X).

Comparisons from column to column in Table 12.1 are relatively easy to make because the column totals are roughly equal. This will not usually be the case, and it is helpful to compute column percentages (see Chapter 11) to control for varying column totals and to make the association more visible.

The general procedure for detecting association with bivariate tables is to compute percentages within the columns (vertically, or down each column) and then to compare column to column across the table (horizontally, or across the rows). Table 12.2 presents column percentages calculated from the data in Table 12.1. Note that this table reports the row and column marginals in parentheses. Tables with percentages are usually easier to read because changes in the conditional distributions of Y are easier to detect.

In Table 12.2, we can see that the largest cell changes position from column to column. For workers who are low on satisfaction, the single largest cell is in the top row (low on productivity). For the middle column (moderate on satisfaction), the largest cell is in the middle row (moderate on productivity), and, for the right-hand column (high on satisfaction), it is in the bottom row (high on productivity). Even a cursory glance at Table 12.2 reinforces our conclusion that an association does exist between these two variables.

If two variables are not associated, then the conditional distributions of Y will not change across the columns. The distribution of Y would be the same for each condition of X. Table 12.3 illustrates a "perfect non-association" between height and productivity. Table 12.3 is only one of many patterns that indicate "no association." The important point is that the conditional distributions of Y are the same. Levels of productivity do not change at all for the various heights; therefore, no association exists between these variables.

HOW STRONG IS THE ASSOCIATION?

Once we know there is an association between two variables, we need to know how strong it is. This is essentially a matter of determining the amount of change in the conditional distributions of Y. At one extreme is the case of "no association," where the conditional distributions of Y do not change at all

TABLE 12.2 Productivity by Job Satisfaction (percentages)

Productivity (Y)	Job Satisfaction (X)			TOTALS
	Low	*Moderate*	*High*	
Low	50.0%	34.4%	13.5%	33.5% (58)
Moderate	33.3%	41.0%	34.6%	36.4% (63)
High	16.7%	24.6%	51.9%	30.1% (52)
TOTALS	100.0% (60)	100.0% (61)	100.0% (52)	100.0% (173)

TABLE 12.3 Productivity by Height (an illustration of no association)

Productivity (Y)	Height (X)		
	Short	*Medium*	*Tall*
Low	33.3%	33.3%	33.3%
Moderate	33.3%	33.3%	33.3%
High	33.3%	33.3%	33.3%
TOTALS	99.9%	99.9%	99.9%

TABLE 12.4 Productivity by Height (an illustration of perfect association)

Productivity (Y)	Height (X)		
	Short	*Medium*	*Tall*
Low	0.0%	0.0%	100.0%
Moderate	0.0%	100.0%	0.0%
High	100.0%	0.0%	0.0%
TOTALS	100.0%	100.0%	100.0%

(see Table 12.3). At the other extreme is a perfect association, the strongest possible relationship.

A perfect association exists between two variables if each value of the dependent variable is associated with one and only one value of the independent variable.[2] In a bivariate table, the variables would have a perfect relationship if all cases in each column are located in a single cell and there is no variation in Y for a given value of X (see Table 12.4).

A perfect relationship would be taken as very strong evidence of a causal relationship between the variables, at least for the sample at hand. In fact, the results presented in Table 12.4 would indicate that, for this sample, height is the sole cause of productivity. Also, in the case of a perfect relationship, predictions from one variable to the other could be made without error. If we know that a particular worker is short, for example, we could be sure that he or she is highly productive.

Virtually all relationships will fall somewhere between the extremes of no association and perfect association, and we need to develop some way of describing these intermediate relationships consistently and meaningfully. For example, Tables 12.1 and 12.2 show an association between productivity and job satisfaction. How could this relationship be described in terms of strength? How close is the relationship to perfect? How far away is it from no association?

We rely on measures of association to give us precise, objective indicators of the strength of a relationship. Virtually all of these statistics are designed so

[2]Each measure of association to be introduced in the following chapters incorporates its own definition of a "perfect association," and these definitions vary somewhat, depending on the specific logic and mathematics of the statistic. That is, for different measures computed from the same table, some measures will possibly indicate perfect relationships when others will not. We will note these variations in the mathematical definitions of a perfect association at the appropriate times.

that they have a lower limit of 0.00 and an upper limit of 1.00 (± 1.00 for ordinal and interval-ratio measures of association). A measure of 0.00 indicates no association between the variables (the conditional distributions of Y do not vary), and a measure of 1.00 (± 1.00 in the case of ordinal and interval-ratio measures) indicates a perfect relationship. The exact meaning of values between 0.00 and 1.00 varies from measure to measure, but, for all measures, the closer the value is to 1.00, the stronger the relationship (the greater the change in the conditional distributions of Y).

A Simple Measure of the Strength of an Association. We begin to consider the many measures of association later in this chapter. At this point, we consider the **maximum difference**, a more informal way of assessing the strength of a relationship based on comparing column percentages across the rows. To find the maximum difference, compute the column percentages as usual and then skim the table across each of the rows to find the largest difference—in any row—between column percentages. For example, the largest difference in column percentages in Table 12.2 is in the top row between the "Low" column and the "High" column: $50.0\% - 13.5\% = 36.5\%$. The maximum difference in the middle row is between "moderates" and "lows" ($41.0\% - 33.3\% = 7.7\%$), and in the bottom row it is between "highs" and "lows" ($51.9\% - 16.7\% = 35.2\%$). Both of these values are less than the maximum difference in the top row.

Once you have found the maximum difference in the table, you can use the scale presented in Table 12.5 to describe the strength of the relationship. For instance, we could describe the relationship between productivity and job satisfaction in Table 12.2 as strong.

Be aware that the relationships between the maximum difference and the descriptive terms (weak, moderate, and strong) in Table 12.5 are arbitrary and approximate. We will get more precise and useful information when we compute and analyze a measure of association.

Also, the maximum difference is easiest to find and most useful for small tables. In large tables, with many (say, more than three) columns and rows, it can be cumbersome to find the high and low percentages, and it is advisable to consider only measures of association as indicators of the strength for these tables. Finally, note that the maximum difference is based on only two values (the high and low column percentages within any row). Like the range (see Chapter 4), this statistic can give a misleading impression of the overall strength of the relationship. *(For practice in computing percentages, see problems 12.1 to 12.9.)*

TABLE 12.5 The Relationship Between the Maximum Difference and the Strength of the Relationship

Maximum Difference	Strength
If the maximum difference is:	*The strength of the relationship is:*
Between 0 and 10 percentage points	Weak
Between 10 and 30 percentage points	Moderate
More than 30 percentage points	Strong

WHAT IS THE PATTERN AND/OR THE DIRECTION OF THE ASSOCIATION?

To find the pattern of the association, we need to ascertain which values or categories of one variable are associated with which values or categories of the other. We have already remarked on the pattern of the relationship between productivity and satisfaction. Table 12.2 indicates that low scores on satisfaction are associated with low scores on productivity, moderate satisfaction with moderate productivity, and high satisfaction with high productivity.

When one or both variables in a bivariate table are measured at the nominal level, we can discuss the pattern of the relationship *only*. Since the scores of nominal-level variables have no numerical order (by definition), associations including nominal-level variables cannot have a direction (that is, they cannot be positive or negative), although they may have a pattern. However, when both variables are at least ordinal in level of measurement, the association may also be described in terms of direction. The direction of the association can be either positive or negative.

In a **positive association**, the variables vary in the same direction: They will increase or decrease together. High scores on one variable are associated with high scores on the other variable, and low scores on one variable are associated with low scores on the other. The cases will tend to fall along a diagonal from the upper left of a bivariate table to the lower right. Table 12.6 displays, with fictitious data, a positive relationship between education and use of public libraries. As education increases (as you move from left to right across the table), library use also increases (the percentage of "high" users increases). The association between job satisfaction and productivity, as displayed in Tables 12.1 and 12.2, is also a positive association.

In a **negative association**, the variables vary in opposite directions. High scores on one variable are associated with low scores on the other and increases in one variable are accompanied by decreases in the other. Table 12.7 displays a negative relationship, again with fictitious data, between education and television viewing. The amount of television viewing decreases as education increases. In other words, as you move from left to right across the top of the table (as education increases), the percentage of heavy viewers decreases.

Measures of association for ordinal and interval-ratio variables are designed so that they will be positive for positive associations and negative for negative associations. Thus, a measure of association preceded by a plus sign indicates a positive relationship between the two variables, with the value $+1.00$ indicating a perfect positive relationship. A negative sign indicates a negative relationship, with -1.00 indicating a

TABLE 12.6 **Library Use by Education (an illustration of positive relationship)**

Library Use (Y)	Education (X)		
	Low	*Moderate*	*High*
Low	60%	20%	10%
Moderate	30%	60%	30%
High	10%	20%	60%
TOTALS	100%	100%	100%

TABLE 12.7 **Amount of Television Viewing by Education (an illustration of negative relationship)**

TV Viewing (Y)	Education (X)		
	Low	Moderate	High
Low	10%	20%	60%
Moderate	30%	60%	30%
High	60%	20%	10%
TOTALS	100%	100%	100%

perfect negative relationship. *(For practice in determining the pattern of an association, see problems 12.1 to 12.4, 12.5a and c, and 12.13. For practice in determining the direction of a relationship, see problems 12.5b, 12.6 to 12.9, and 12.14.)*

ASSOCIATION AND STATISTICAL SIGNIFICANCE

Tables 12.1 and 12.2 show that there is a strong, positive relationship between job satisfaction and productivity. If the 173 cases in the table were a random sample, it would be necessary to also test the relationship for statistical significance. In fact, the obtained chi square for the table is 24.2, which is significant at the 0.05 level. This means that job satisfaction and productivity are (probably) related in the population.

Usually, we want to argue for a causal relationship between our variables and we are in the strongest position to do so when we find an association that is both strong and statistically significant. Remember, however, that association and significance are two different things and it is possible for a relationship to be strong but insignificant, significant but weak, and so forth.

ASSOCIATION AND CAUSATION

The association between productivity and satisfaction is strong, positive, and significant, and it might be tempting to conclude that satisfaction *causes* productivity ("happy workers are busy workers"). While this conclusion is consistent with the evidence, remember that association does not prove causation. In fact, although we have assumed that satisfaction is the independent variable in this relationship, we could have argued the reverse causal relationship ("busy workers are happy workers"). Tables 12.1 and 12.2 are consistent with both causal arguments.

Social Research and Data Analysis 12.1 | **The Importance of Percentages and Avoiding Errors of Interpretation**

Percentages are among the humblest of statistics, but they can be misread and misused. To avoid problems, always:

- array tables with the independent variable in the columns
- calculate percentages within the columns, so that the percentage in each column adds up to 100

- look for patterns and direction (if applicable) by comparing the column percentages

Remember: "calculate percentages down and compare across."

To illustrate, consider the relationship between support for affirmative action and racial or ethnic group,

(continued)

Social Research and Data Analysis 12.1 *(Continued)*

shown in the table below for a nationally representative sample. Note that race or ethnicity is the independent variable and is placed in the columns. A person's group membership could be a cause of attitudes and opinions. The reverse cannot be true: A person's opinion cannot cause them to change their race or ethnicity.

Support for Affirmative Action by Racial or Ethnic Group (percentages)

| Affirmative Action | Racial or Ethnic Group | | | TOTALS |
	White	*Black*	*Hispanic*	
Support	202 (19.5%)	108 (47.0%)	24 (28.2%)	334 (24.7%)
Oppose	836 (80.5%)	122 (53.0%)	61 (71.8%)	1019 (75.3%)
TOTALS	1038 (100.0%)	230 (100.0%)	85 (100.0%)	1353 (100.0%)

Chi square = 77.18, $df = 2$, $p < 0.01$

Source: 2018 General Social Survey.

These results support the view that there is a relationship between these variables. The relationship is statistically significant at less than the 0.01 level and the conditional distributions change across the table. Blacks are the most supportive of affirmative action, followed by Hispanics and whites. The maximum difference between the columns is about 27 percentage points, which means that the relationship is moderate to strong.

This analysis should seem straightforward—it seems reasonable that support for affirmative action should be linked to group membership. But what if we had mistakenly switched the position of the variables and placed support for affirmative action in the columns? If we computed percentages down and compared across, we could make some serious errors of analysis. Consider the incorrectly percentaged table:

Racial or Ethnic Group by Support of Affirmative Action (with percentages calculated incorrectly)

| Racial or Ethnic Group | Affirmative Action | | TOTALS |
	Support	*Oppose*	
White	60.5%	82.0%	76.7%
Black	32.3%	12.0%	17.0%
Hispanic	7.2%	6.0%	6.3%
TOTALS	100.0%	100.0%	100.0%

A casual glance seems to indicate a relationship. If we looked only at the "Support" column of the table (as people sometimes do), we might conclude that whites are more supportive of affirmative action than the other groups. But the "Oppose" column shows that whites are also the huge majority (82.0%) of those who oppose the policy. How can this be?

The column percentages in this table reflect the fact that whites vastly outnumber the other groups:

There are almost 5 times as many whites as blacks in the sample and 12 times as many whites as Hispanics. This table could easily lead to false conclusions about this relationship.

Professional researchers sometimes compute percentages in the wrong direction or ask a question about the relationship incorrectly, so you should always check bivariate tables to make sure that the analysis agrees with the patterns in the table.

STATISTICS **In Everyday Life**

Discrimination in Football?

Even after American professional football was integrated in the middle of the 20th century, racial discrimination continued on a number of levels. Most obvious, perhaps, was the perfect relationship between race and the quarterback position. This central leadership role was reserved for whites, a practice that reflected the widespread stereotype that blacks lacked the intelligence and decision-making ability required to play the position.

Today, racial discrimination in professional sports is much weaker than in the past and minority-group athletes face less overt discrimination in their competition for any position, including quarterback. In fact, African Americans are vastly overrepresented in professional football: They are 69% of all NFL players but only 13% of the general population. This makes the NFL one of the most integrated areas of U.S. society, but African Americans still make up only about 30% of starting quarterbacks,* a pattern that may reflect lingering—if more subtle—prejudices and stereotypes. What data would you need to account for these patterns?

*The percentage is not exact because starters change from week to week and the definition of "race" is ambiguous. For example, Patrick Mahomes, starting quarterback of the Kansas City Chiefs, is biracial; should he be counted as a "black" quarterback?

Source: Lapchik, R. 2019. *The 2018 Racial and Gender Report Card: National Football League.* Available at https://www.tidesport.org/copy-of-nba

Measures of Association: Introduction

Column percentages provide useful and detailed information about bivariate associations and should always be computed and analyzed. However, they can be awkward and cumbersome to use, particularly for larger tables. Measures of association, on the other hand, summarize the overall strength (and direction for ordinal-level variables) of a bivariate association in a single number and provide a more compact and convenient format for interpretation and discussion.

There are many measures of association, but we will focus on a few of the most widely used. We will cover these statistics by the level of measurement for which they are most appropriate, first covering measures appropriate for nominal variables and then dealing with measures for ordinal-level variables. Finally, in Chapter 13, we will consider Pearson's r, a measure of association or correlation for interval-ratio–level variables. To analyze relationships for variables that are at different levels (for example, one nominal-level variable and one ordinal-level variable), we generally use the measures of association appropriate for the lower of the two levels of measurement.

Measures of Association for Nominal Variables

There are two types of measures of association for nominal-level variables, one based on chi square (see Chapter 11) and the other on a logic called "proportional reduction in error" or PRE. We consider these two classes of measures separately.

CHI SQUARE–BASED MEASURES OF ASSOCIATION: PHI AND CRAMER'S V

There are several chi square–based measures of association, but we will consider only two. The first, called **phi** (ϕ), is appropriate for 2×2 tables (that is, tables with two rows and two columns) and is easy to calculate. As shown in Formula 12.1, phi is simply the square root of chi square divided by N.

FORMULA 12.1
$$\phi = \sqrt{\frac{\chi^2}{N}}$$

For tables larger than 2×2, the upper limit of phi can exceed 1.00 and this can make the statistic difficult to interpret. For larger tables, a more general form of the statistic called Cramer's V is commonly used.

FORMULA 12.2
$$V = \sqrt{\frac{\chi^2}{N(\min r - 1, c - 1)}}$$

where $(\min r - 1, c - 1)$ = the minimum value of number of rows (r) minus 1 or number of columns (c) minus 1

To calculate V, follow the instructions in the "One Step at a Time" box. Cramer's V has an upper limit of 1.00 for a table of any size and will be the same value as phi if the table has either two rows or two columns.

Both phi and Cramer's V can be interpreted as indices that measure the strength of the association between two variables. Table 12.8 provides guidelines for the interpretation of the strength of nominal-level measures of association. As with Table 12.5, the descriptive terms are intended as general guides only.

ONE STEP AT A TIME Computing Cramer's V

Step	Operation
1.	Find the number of rows (r) and number of columns (c) in the table. Subtract 1 from the lesser of these two numbers to find $(\min r - 1, c - 1)$.
2.	Multiply the value you found in step 1 by N.
3.	Divide the value of chi square by the value you found in step 2.
4.	Take the square root of the quantity you found in step 3. The resulting value is V.

ONE STEP AT A TIME Interpreting Cramer's V

Step	Operation
1.	See Table 12.8 for general guidelines.
2.	Use the column percentages to identify the pattern.

TABLE 12.8 Guidelines for Interpreting the Strength of the Relationship for Nominal Level Measures of Association

Measure of Association	Strength
If the value is:	*The strength of the relationship is:*
Between 0.00 and 0.10	Weak
Between 0.11 and 0.30	Moderate
Greater than 0.30	Strong

TABLE 12.9 Academic Achievement by Club Membership

	Membership			
Achievement	*Greek*	*Other*	*None*	TOTALS
Low	4 (17.4%)	4 (15.4%)	17 (65.4%)	25 (33.3%)
Moderate	15 (65.2%)	6 (23.1%)	4 (15.4%)	25 (33.3%)
High	4 (17.4%)	16 (61.5%)	5 (19.2%)	25 (33.3%)
TOTALS	23 (100.0%)	26 (100.0%)	26 (100.0%)	75 (99.9%)

To illustrate the computation of V, consider Table 12.9, which shows the relationship between membership in various types of student organizations and academic achievement for a sample of college students. The obtained chi square for this table is 31.5, which is significant at the 0.05 level. The table has the same number of rows and columns, so we may use either $(r-1)$ or $(c-1)$ in the denominator. In either case, the value of the denominator is N multiplied by $(3-1)$, or 2. Cramer's V is

$$V = \sqrt{\frac{\chi^2}{N(\min r - 1, c - 1)}}$$

$$V = \sqrt{\frac{31.50}{(75)(2)}}$$

$$V = \sqrt{\frac{31.50}{150}}$$

$$V = \sqrt{0.21}$$

$$V = 0.46$$

A Cramer's V of 0.46 indicates a strong association between club membership and academic achievement. From the column percentages, we can see that, for this sample, members of fraternities and sororities tend to be moderate, members of other organizations tend to be high, and nonmembers tend to be low in academic achievement.

THE LIMITATIONS OF PHI AND CRAMER'S V

One limitation of these measures is that they are only general indicators of the strength of the relationship. The closer the measure is to 0.00, the weaker the relationship and the closer to 1.00, the stronger the relationship. Values between 0.00 and 1.00 can be described as weak, moderate, or strong, according to the

terms displayed in Table 12.8, but they have no direct or meaningful interpretation. On the other hand, phi and V are easy to calculate (once the value of chi square has been obtained) and have been commonly used indicators of the importance of an association.[3] *(For practice in computing Cramer's V, see problems 12.1 to 12.4, 12.5a and c, and 12.13. Remember that for tables with either two rows or two columns, phi and Cramer's V will have the same value.)*

A MEASURE OF ASSOCIATION BASED ON PROPORTIONAL REDUCTION IN ERROR

There are several nominal-level measures of association based on the logic of **proportional reduction in error (PRE)**. We will focus on **lambda** (symbolized by the Greek letter λ).

The logic of these measures is based on two different predictions about the scores of cases on the dependent variable. In one prediction, we ignore information about the cases given by the independent variable and, in the second, we take the independent variable into account. Lambda tells us how much knowledge of the independent variable improves our predictions.

For lambda, we first predict the category into which each case will fall on the dependent variable (Y) while ignoring the independent variable (X). Since we would be predicting blindly in this case, we would often predict the value of a case on the dependent variable incorrectly.

For the second prediction, we take information about the cases provided by the independent variable into account. If the two variables are associated, this additional information will reduce our errors of prediction—that is, we should misclassify fewer cases. Furthermore, the stronger the association between the variables, the greater the reduction in errors. In the case of a perfect association, we would make no errors at all when predicting the score on Y from the score on X. When there is no association between the variables, on the other hand, knowledge of the independent variable will not improve the accuracy of our predictions.

To calculate lambda, we find two quantities: first, the number of prediction errors made while ignoring the independent variable (E_1) and, second, the number of prediction errors made while taking the independent variable into account (E_2). These two sums are then compared to derive the statistic. The "One Step at a Time" box states the computational procedures for tables of any size.

An example will clarify the computation of lambda. We will illustrate with Table 12.10, a 2×2 table that displays a relationship between height and gender.

First, we find E_1, the number of prediction errors made while ignoring the independent variable (gender), by working with the row marginals and making a prediction for *all* cases. We could predict that all subjects are tall, a prediction that will result in 48 errors. That is, for this prediction, all 100 cases would be placed in the top row. Only 52 of the cases actually belong in this row, so this prediction would result in $(100 - 52)$, or 48, errors. If we had predicted that all subjects were

[3]Two other chi square–based measures of association, T^2 and C (the contingency coefficient), are sometimes reported in the literature. Both of these measures have serious limitations. T^2 has an upper limit of 1.00 only for tables with an equal number of rows and columns, and the upper limit of C varies, depending on the dimensions of the table. These characteristics make these measures more difficult to interpret and thus less useful than phi or Cramer's V.

TABLE 12.10 Height by Gender

Height	Gender Male	Female	TOTALS
Tall	44	8	52
Short	6	42	48
TOTALS	50	50	100

short, on the other hand, we would have made 52 errors ($100 - 48 = 52$). We will take the *lesser* of these two numbers and refer to this quantity as E_1, for the number of errors made while ignoring the independent variable. So $E_1 = 48$.

Next, we predict the score on Y (height) again, but this time we take X (gender) into account by moving from column to column. Because each column is a category of X, we thus use X in making our predictions. For each column, we predict that all cases will fall in the cell with the largest frequency. For males, we predict that all 50 cases will be tall and make 6 errors ($50 - 44 = 6$). For females, our prediction is that all are short, and 8 errors will be made. By moving from column to column, we have made a total of 14 errors of prediction, a quantity we will label E_2 ($E_2 = 6 + 8 = 14$).

If the variables are associated, we will make fewer errors under the second procedure than under the first and E_2 will be smaller than E_1. In our example, E_2 is indeed less than E_1, so gender and height are associated. Our errors were reduced from 48 to only 14. To find the *proportional* reduction in error, use Formula 12.3:

FORMULA 12.3

$$\lambda = \frac{E_1 - E_2}{E_1}$$

For the sample problem, the value of lambda would be

$$\lambda = \frac{E_1 - E_2}{E_1}$$

$$\lambda = \frac{48 - 14}{48}$$

$$\lambda = \frac{34}{48}$$

$$\lambda = 0.71$$

The value of lambda ranges from 0.00 to 1.00. A value of 0.00 means that the variables are not associated at all (E_1 is the same as E_2), and a value of 1.00 means that the association is perfect (E_2 is 0, and scores on the dependent variable can be predicted without error from the independent variable).

Unlike phi or V, the value of lambda between the extremes of 0.00 and 1.00 has a precise meaning: It tells us how much the independent variable (X) helps us to predict (or, more loosely, understand) the dependent variable (Y). When multiplied by 100, the value of lambda indicates the strength of the association in terms of the percentage reduction in error. Thus, the lambda calculated for

ONE STEP AT A TIME | **Computing Lambda**

These instructions are stated in general terms and will work with tables of any size.

Step	**Operation**
1.	To find E_1, subtract the largest row subtotal (marginal) from N.
2.	To find E_2, start with the far left-hand column and subtract the largest cell frequency in the column from the column total. Repeat this step for all columns in the table.
3.	Add up all the values you found in step 2. The result is E_2.
4.	Subtract E_2 from E_1.
5.	Divide the quantity you found in step 4 by E_1. The result is lambda.

ONE STEP AT A TIME | **Interpreting Lambda**

Step	**Operation**
1.	See Table 12.8 for general guidelines.
2.	For a PRE interpretation, multiply the value of lambda by 100. This percentage expresses the extent to which predictions of the dependent variable are improved by taking the independent variable into account.
3.	Use the column percentages to identify the pattern of the relationship.

Table 12.10 could be interpreted by saying that knowledge of gender improves our ability to predict height by 71%. That is, we are 71% better off knowing gender when attempting to predict height.

THE LIMITATIONS OF LAMBDA

Lambda has two characteristics that should be stressed. First, lambda is asymmetric. This means that the value of the statistic will vary, depending on which variable is taken as independent. Thus, you should exercise some caution in the designation of an independent variable. If you consistently follow the convention of arraying the independent variable in the columns, the asymmetry of the statistic should not be confusing.

Second, when one of the row totals is much larger than the others, lambda can be misleading. It can be 0.00 even when other measures of association are greater than 0.00 and the conditional distributions for the table indicate that there is an association between the variables. This means that great caution should be exercised in the interpretation of lambda when the row marginals are very unequal. In fact, in the case of very unequal row marginals, a chi square–based measure of association would be the preferred measure of association. *(For practice in computing lambda, see problems 12.1 to 12.4, 12.5a and c, and 12.13.)*

Social Research and Data Analysis 12.2 Using Cramer's *V* and Lambda

In Social Research and Data Analysis 12.1, we used the maximum difference to determine that the relationship between racial or ethnic group and support for affirmative action was moderate to strong in strength. For convenience, the table is reproduced here. Does Cramer's *V* confirm that description?

Support for Affirmative Action by Racial or Ethnic Group (percentages)

Affirmative Action	Racial or Ethnic Group			TOTALS
	White	Black	Hispanic	
Support	202	108	24	334
Oppose	836	122	61	1019
TOTALS	1038	230	85	1353

Chi square = 77.18, *df* = 2, *p* < 0.01

$$V = \sqrt{\frac{\chi^2}{N(\min r - 1, c - 1)}}$$

$$V = \sqrt{\frac{77.18}{(1353)(1)}}$$

$$V = \sqrt{\frac{77.18}{1,353}}$$

$$V = \sqrt{0.06}$$

$$V = 0.24$$

Based on Table 12.8, *V* is consistent with the maximum difference and the relationship is moderate to strong.

Would we get the same result with lambda? To calculate the statistics, we first find E_1 and E_2:

$$E_1 = 1353 - 1019 = 334$$
$$\text{For Whites, } E_2 = 1038 - 836 = 202$$
$$\text{For Blacks, } E_2 = 230 - 122 = 108$$
$$\text{For Hispanics, } E_2 = 85 - 61 = 24$$
$$E_2 = 334$$

Lambda is

$$\lambda = \frac{E_1 - E_2}{E_1} = \frac{334 - 334}{334} = 0.00$$

Contrary to the maximum difference and *V*, lambda indicates no relationship between the variables. How can this be? Remember that lambda can be misleading when row totals are very different, as is the case here. The other measures of association convey a more reasonable summary of the relationship and would be the preferred statistics in this case.

STATISTICS In Everyday Life

Confidence in Higher Education

As in so many other areas of public life, support for colleges and universities has a political dimension. Generally, Republicans are significantly more suspicious of higher education, particularly when it comes to issues of free speech and the political views of college professors.* Below is a general example of this relationship based on the results of a survey administered to a representative sample of Americans.

Views of Higher Education	Political Party Affiliation		TOTALS
	Republican or Leaning Republican	Democrat or Leaning Democrat	
Mostly Positive	271 (30.9%)	600 (55.8%)	871 (44.6%)
Neither Positive nor Negative	264 (30.1%)	327 (30.4%)	591 (30.3%)
Mostly Negative	342 (39.0%)	149 (13.8%)	491 (25.1%)
TOTALS	877 (100.0%)	1076 (100.0%)	1953 (100.0%)

Chi square = 188.53, *df* = 2, *p* < 0.05 Cramer's *V* = 0.31 Lambda = 0.07

(*continued*)

STATISTICS **In Everyday Life (*Continued*)**

Democrats are much more supportive of higher education and the relationship is statistically significant. We have mixed results in the two measures of association, with Cramer's *V* indicating a stronger relationship than lambda. Given the difference in row totals, we should probably choose *V* and call this relationship moderate to strong.

What social or political forces might account for this relationship? What additional questions might you ask about the relationship between these variables?

Source: Pew Research Center, 2018. *American Trends Panel, Wave 30.* https://www.pewforum.org/dataset/american-trends -panel-wave-30/

*Parker, Kim. 2019. "The Growing Partisan Divide in Views of Higher Education." Pew Research Center. https://www .pewsocialtrends.org/essay/the-growing-partisan-divide-in-views-of-higher-education/

Measures of Association for Collapsed Ordinal Variables

There are two types of ordinal-level variables. Some have only a few (no more than five or six) scores, and we will refer to these as *collapsed ordinal variables*. A number of measures of association have been invented for use with these variables but, rather than attempt to cover them all, we will focus on a statistic called **gamma (*G*)**.

The other type of ordinal variable has many possible scores and resembles interval-ratio–level variables. We will call these *continuous ordinal variables*. A scale that measures an attitude with many different items could produce this type of variable. **Spearman's rho (*r_s*)** is designed to measure the association between variables of this type.

Ordinal-level statistics are more sophisticated than lambda and Cramer's *V* but it is important to remember that they still address the same three basic questions about a bivariate association:

- Are the variables associated?
- How strong is the association?
- What is the direction of the association?

GAMMA

Gamma can be interpreted using the proportional reduction in error (PRE) logic introduced for lambda. The value of gamma, when expressed as a percentage, tells us the extent to which knowledge of one variable improves our ability to predict the other variable.

There are two crucial differences between gamma and lambda. First, the scores of ordinal-level variables have order and can be arrayed from high to low. Thus, gamma tells us the direction of the relationship—positive or negative—in addition to its strength.

Second, with gamma, the PRE logic is based on comparing the rank order of pairs of cases on the two variables. If a case ranks higher than another case on one variable, does it also rank higher on the other variable? Gamma compares the

TABLE 12.11 **Burnout by Length of Service (fictitious data)**

| | Years of Service (X) | | |
Burnout (Y)	Low	High	TOTALS
Low	25 (52.1%)	20 (38.5%)	45 (45.0%)
High	23 (47.9%)	32 (61.5%)	55 (55.0%)
TOTALS	48 (100.0%)	52 (100.0%)	100 (100.0%)

number of pairs of cases ranked in the same order on both variables (N_s) with the number of cases ranked in different order (N_d).

Computing Gamma. See the "One Step at a Time" box for instructions on computing gamma for tables of any size. To illustrate the computational routine, we will use a 2×2 table. Larger tables require more computational steps but follow the same logic.

Assume that a researcher is concerned about the causes of "burnout" (that is, demoralization and loss of commitment) among elementary school teachers and wonders whether there is a relationship between burnout (Y) and years of service (X). Table 12.11 presents a relationship between "length of service" and "burnout" for a fictitious sample of 100 teachers.

How many pairs of cases are ranked in the same order in this table? Suppose Mr. Jones is a new teacher (scores low on experience, X) who has a lot of enthusiasm (scores low on burnout, Y). He would be placed in the upper-left cell of Table 12.11.

Now suppose that Ms. Kelly is nearing the end of her career (scores high on X) and is feeling discouraged (scores high on Y). She would find herself in the lower-right cell.

If we formed a pair of cases with these two individuals, they would rank in the same order on both variables: On X, Ms. Kelly would rank higher than Mr. Jones (she has taught for more years), and on Y, she would also rank higher (she is more burned out). In fact, any pair of cases formed between the upper-left cell and the lower-right cell would contribute to N_s. So, to find the total number of pairs of cases ranked the same on both variables in a 2×2 table, multiply the number of cases in the upper-left cell by the number of cases in the lower-right cell:

$$N_s = (25)(32) = 800$$

What about cases ranked in different order? Imagine Mr. Gordon, a teacher who has many years of experience (ranks high on X) and retains his enthusiasm for the work (ranks low on Y). He would be placed in the upper-right cell of Table 12.11.

Finally, imagine Ms. Griffin, a new teacher (who ranks low on X) who has already lost her interest in teaching (ranks high on Y). She would be sorted into the lower-left cell of the table and, if we formed a pair of cases with these two, they would rank differently on the variables. On X, Mr. Gordon ranks higher than Ms. Griffin because he has taught longer but, on Y, Ms. Griffin ranks higher because she is more burned out.

Any pair of cases formed between the upper-right cell and the lower-left cell would contribute to N_d, the total number of pairs of cases ranked differently. To

find total N_d for a 2 × 2 table, multiply the number of cases in the upper-right cell by the number of cases in the lower-left cell:

$$N_d = (20)(23) = 460$$

The formula for gamma is:

FORMULA 12.4

$$G = \frac{N_s - N_d}{N_s + N_d}$$

where N_s = the total number of pairs of cases ranked in the same order on both variables

N_d = the total number of pairs of cases ranked in different order on the variables

For Table 12.11, gamma would be

$$G = \frac{N_s - N_d}{N_s + N_d}$$

$$G = \frac{800 - 460}{800 + 460}$$

$$G = \frac{340}{1260}$$

$$G = +0.27$$

Interpreting Gamma. Remember that gamma is an ordinal measure and tells us two different things: the strength of the association and its direction. We will deal with these separately.

In terms of the strength of the association, we can use the logic of PRE and say that a gamma of +0.27 means that we would reduce our errors by 27% (0.27 × 100) when predicting the order of pairs of cases on one variable from the order of pairs of cases on the other (as opposed to predicting order while ignoring the other variable). Also, we can use Table 12.12 to interpret the strength of gamma in a format similar to Tables 12.5 and 12.8. As before, the descriptive terms are arbitrary and intended as general guidelines only. A gamma of +0.27 is weak to moderate in strength.

In terms of direction, this is a positive relationship: Burnout increases as length of service increases. Knowing the rankings of two teachers on length of service

TABLE 12.12 Guidelines for Interpreting the Strength of the Relationship for Ordinal-Level Measures of Association

Measure of Association	Strength
If the value is:	*The strength of the relationship is:*
Between 0.00 and 0.30	Weak
Between 0.31 and 0.60	Moderate
Greater than 0.60	Strong

(Kelly is higher on length of service than Jones) will help us predict their ranking on burnout (we would predict that Kelly will also be higher than Jones on burnout).

We should also use the column percentages to help us analyze the direction of the relationship. In a positive relationship, high scores would cluster with high scores and low scores with low. Hence, the cases would tend to fall in a diagonal from upper left to lower right, as they do in Table 12.11. In a negative relationship, high scores on one variable would be associated with low scores on the other, and cases would tend to fall in a diagonal from lower left to upper right. See Tables 12.6 and 12.7 for examples of positive and negative relationships.

Finally, note that strength and direction are two different things. That is, a gamma of −0.35 is exactly as strong as a gamma of +0.35 but is opposite in direction.

Unlike lambda, gamma is a symmetrical measure of association. Its value will be the same regardless of which variable is taken as independent. *(To practice computing and interpreting gamma, see problems 12.5b, 12.6 to 12.9, and 12.14.)*

ONE STEP AT A TIME **Computing Gamma**

Make sure the table is arranged so that columns increase in value from left to right and rows increase from top to bottom.

1. To compute N_s, start with the upper-left-hand cell. Multiply the number of cases in the cell by the *total* number of cases in all cells below and to the right. Repeat this process for *all* cells in the table. Remember that cells that have no cells below and to the right cannot contribute to N_s.
2. To compute N_d, start with the upper-right-hand cell. Multiply the number of cases in the cell by the *total* number of cases in all cells below and to the left. Repeat this process for *all* cells in the table. Remember that cells that have no cells below and to the left cannot contribute to N_s.

To Solve Formula 12.4:

3. Subtract N_d from N_s.
4. Add N_s and N_d.
5. Divide the value you found in step 3 by the value you found in step 4. This value is gamma.

ONE STEP AT A TIME **Interpreting Gamma**

To Interpret Strength:

1. Use Table 12.12 to describe strength in general terms.
2. Multiply gamma by 100. This value represents the percentage improvement in predicting the dependent variable by taking the independent variable into account.

To Interpret Direction:

1. The sign of gamma indicates direction. However, *be careful when interpreting direction with ordinal variables*. Coding for these variables is arbitrary and a positive gamma may actually indicate a negative relationship and vice versa. See the box "Interpreting Direction with Ordinal-Level Variables."
2. Always consider the column percentages. If the percentages tend to fall in the diagonal from upper-left to lower-right, the relationship is positive. If they tend to fall in the diagonal from upper-right to lower-left, the relationship is negative.

Social Research and Data Analysis 12.3 | Interpreting Direction with Ordinal-Level Variables

When determining the direction of a relationship, you might be tempted to look only at the sign of the measure of association (as in gamma $= -0.30$). This is certainly understandable, but it can lead to confusion and mistakes because the coding schemes for ordinal-level variables are arbitrary and a higher score may mean "more" or "less" of the variable being measured.

For example, in the 2018 General Social Survey, a positive gamma of 0.17 can be found for the association between *Respondent's Highest Degree* (DEGREE) and *Confidence in U.S. Congress* (CONCONG). At first glance, you might conclude that a higher level of education is associated with a greater degree of confidence in U.S. Congress, but a closer inspection should reveal the opposite to be true. Take a look at how the two variables are coded below. What do you notice?

Respondent's Highest Degree	Confidence in U.S. Congress
DEGREE	CONCONG
1. LT HIGH SCHOOL	1. COMPLETE CONFIDENCE
2. HIGH SCHOOL	2. A GREAT DEAL OF CONFIDENCE
3. JUNIOR COLLEGE	3. SOME CONFIDENCE
4. BACHELOR	4. VERY LITTLE CONFIDENCE
5. GRADUATE	5. NO CONFIDENCE AT ALL

Respondent's Highest Degree goes up as the score on the variable gets higher (1–5). For *Confidence in U.S. Congress*, however, the opposite is true. As the score gets higher (1–5), confidence goes down. So what does this mean for our positive gamma statistic? It means that a higher degree is actually associated with a *lower* level of confidence!

One lesson to learn from this finding is that caution is always needed when interpreting the direction of an association between ordinal-level variables. A tip that we recommend is that when setting up your own questionnaire, try coding variables in a consistently positive direction, so that a higher score on *Confidence in U.S. Congress*, for example, means that the respondent actually has more confidence, not less.

USING SPSS | Crosstabs and Gamma

In this installment of "Using SPSS," we will once again use the Crosstabs procedure, but this time to produce measures of association. We will examine the relationship between *degree* and happiness (*happy*). Are more educated people happier? Both variables are ordinal in level of measurement, but we will request nominal measures of association in addition to gamma.

To begin:

1. Click the SPSS icon on your desktop.
2. Load the *GSS2018* database.
 a. Find the **File** command on the far left of the menu bar and then click **File →
 Open → Data**.
 b. Find the *GSS2018* database and click the file name.
3. From the main menu bar, click **Analyze, Descriptive Statistics**, and **Crosstabs**.
4. In the **Crosstabs** dialog box:
 a. Find *happy* and move it to the **Rows** box.
 b. Find *degree* and move it into the **Columns** box.
5. Click the **Statistics** button and select Chi-square, Phi and Cramer's *V*, Lambda, and Gamma.

(continued)

USING SPSS

(Continued)

6. Return to the **Crosstabs** window and click the **Cells** button and select "**column**" in the **Percentages** box.
7. Return to the **Crosstabs** window and click **OK**.
8. *NOTE: The output table has been edited to fit in this space. The statistics we requested are reported under the table, as you would see in the professional research literature.*

Happiness by Degree

Happy:	LT HS	HS	Jr. Coll.	BA	Grad.	TOTALS
	\multicolumn{5}{c}{R's Highest Degree}					
Very	30.4%	26.4%	22.4%	37.8%	36.8%	29.9%
Pretty	46.9%	58.4%	64.3%	51.2%	54.3%	55.8%
Not Too	22.7%	15.1%	13.3%	11.0%	8.6%	14.3%
TOTALS	100.0%	100.0%	100.0%	100.0%	100.0%	100.0%
	(260)	(1176)	(196)	(465)	(247)	(2344)

$\chi^2 = 53.88$, $df = 8$, $p < 0.00$. Lambda = 0.00; Cramer's $V = 0.11$; Gamma = -0.15.

As usual, SPSS produces a lot of information, which we will examine step by step.

- Chi square is significant ($p < 0.00$). This is important since we are working with a random sample. We can be confident that these variables are also associated in the population.
- The column percentages change, so these variables are associated. Looking at the top row ("very happy"), happiness tends to increase as education increases and the most educated respondents are most likely to report being very happy.
- The measures of association also indicate that the variables are related. Cramer's V and Gamma suggest that the strength of this association is weak to moderate. (We can disregard the lambda of 0.00 since it is contradicted by the other statistics and the pattern of column percentages.)
- Gamma is negative but the column percentages show that this relationship is positive ("happiness increases as education increases"). How can this be? The scores on *happy* are "reversed" and lower scores mean greater happiness. (See Appendix G or click **Utilities → Variables** on the SPSS menu bar to see the coding scheme.) See the box "Interpreting Direction with Ordinal-Level Variables" and be careful when interpreting direction.

What can we conclude? There is a statistically significant, weak to moderate, positive relationship between happiness and education. Happiness tends to increase with education and the more educated tend to be the happiest.

A Measure of Association for Continuous Ordinal Variables: Spearman's Rho (r_s)

Some ordinal-level variables have a broad range of scores. The scores of these variables may be collapsed into a few broad categories (such as high, moderate, and low), organized into a bivariate table, and analyzed with gamma.

TABLE 12.13 **Scores on Involvement in Jogging and Self-Esteem**

Jogger	Involvement in Jogging (X)	Self-Esteem (Y)
Wendy	18	15
Debbie	17	18
Phyllis	15	12
Stacey	12	16
Evelyn	10	6
Tricia	9	10
Christy	8	8
Patsy	8	7
Marsha	5	5
Lynn	1	2

However, collapsing scores may obscure or lose some important distinctions between cases.

For example, suppose a researcher wished to test the claim that jogging is beneficial, not only physically but also psychologically. Do joggers have an enhanced sense of self-esteem? To deal with this issue, 10 female joggers are evaluated on two scales, both of which range from 0 to 20. The first scale measures involvement in jogging and the second measures self-esteem. Scores are reported in Table 12.13.

The scores could be collapsed and, perhaps, split into high and low categories, but this raises several potential difficulties. First, the scores seem continuous, and there are no obvious or natural division points in the distribution that would allow us to clearly distinguish between high scores and low ones. Second, and more important, grouping these cases into broader categories loses information. That is, if both Wendy and Debbie were placed in the category "high" on involvement, the fact that they had different scores would be lost. If these differences are important and meaningful, then we should choose a measure of association that allows us to retain as much detail in the scores as possible.

Spearman's rho (r_s) is a measure of association for ordinal-level variables that have a broad range of many different scores and few ties between cases on either variable. As is appropriate for ordinal-level variables, Spearman's rho is computed from the ranks of the scores, not the scores themselves. First, the cases are ranked from high to low on each variable, and then the ranks are used to produce the final measure. Table 12.14 displays the original scores and the rankings of the cases on both variables.

Computing Spearman's Rho. To rank the cases, first find the highest score on each variable and assign it rank 1. Wendy has the high score on X (18) and is thus ranked number 1. Debbie, on the other hand, is highest on Y and is ranked first on that variable. All other cases are then ranked in descending order.

If cases have the same score, assign them the average of the ranks they would have used up had they not been tied. Christy and Patsy have identical scores of 8 on involvement. Had they not been tied, they would have used up ranks 7 and 8. The average of these two ranks is 7.5, and this average is assigned to all tied cases. (For example, if Marsha had also had a score of 8, three ranks—7, 8, and 9—would have been used, and all three tied cases would have been assigned rank 8.)

TABLE 12.14 Computing Spearman's Rho

	Involvement (X)	Rank	Self-Image (Y)	Rank	D	D²
Wendy	18	1	15	3	−2	4
Debbie	17	2	18	1	1	1
Phyllis	15	3	12	4	−1	1
Stacey	12	4	16	2	2	4
Evelyn	10	5	6	8	−3	9
Tricia	9	6	10	5	1	1
Christy	8	7.5	8	6	1.5	2.25
Patsy	8	7.5	7	7	0.5	0.25
Marsha	5	9	5	9	0	0
Lynn	1	10	2	10	0	0
					$\Sigma D = 0$	$\Sigma D^2 = 22.5$

The formula for Spearman's rho is

FORMULA 12.5

$$r_s = 1 - \frac{6\Sigma D^2}{N(N^2 - 1)}$$

where ΣD^2 = the sum of the squares of the differences in ranks
See the "One Step at a Time" box for detailed instructions on computing Spearman's rho. For our sample problem:

$$r_s = 1 - \frac{6\Sigma D^2}{N(N^2 - 1)}$$

$$r_s = 1 - \frac{6(22.5)}{10(100 - 1)}$$

$$r_s = 1 - \frac{135}{990}$$

$$r_s = 1 - 0.14$$

$$r_s = 0.86$$

Interpreting Spearman's Rho. Spearman's rho is an index of the strength of association between the variables and can be interpreted using the guidelines presented in Table 12.12. The direction of the relationship is indicated by the sign of the statistic. A positive value means that the cases tend to be ranked in the same order on the two variables, and a negative value means that they tend to be ranked in reverse order.

The Spearman's rho of 0.86 in our example problem indicates a strong, positive relationship between these two variables. The respondents who were highly involved in jogging also ranked high on self-image. These results are supportive of claims regarding the psychological benefits of jogging.

If the value of rho is squared, the statistic can be interpreted with the logic of PRE. Rho squared (r_s^2) represents the proportional reduction in errors of

prediction when predicting rank on one variable from rank on the other variable, as compared to predicting rank while ignoring the other variable. In our example, $r_s = 0.86$ and $r_s^2 = 0.74$. Thus, our errors of prediction would be reduced by 74% if we used the rank on jogging to predict rank on self-image. *(For practice in computing and interpreting Spearman's rho, see problems 12.10 to 12.12.)*

ONE STEP AT A TIME **Computing Spearman's Rho**

1. Set up a computing table like Table 12.16 to organize the computations.
2. In the far left-hand column, list the cases in order, with the case with the highest score on the independent variable (X) stated first.
3. In the next column, list the rank of each case on X, beginning with rank 1 for the highest score. If any cases have the same score, assign them the average of the ranks they would have used up had they not been tied.
4. In the next column, list the scores on Y and then, in a new column, rank the cases from high to low on Y. Assign the rank of 1 to the case with the highest score on Y, and assign any tied cases the average of the ranks they would have used up had they not been tied.
5. For each case, subtract the rank on Y from the rank on X and write the difference (D) in a new column. The sum of this column must be zero. If it is not, you have made a mistake and need to re-compute.
6. Square the value of each D and record the result in a new column. Add this column up to find ΣD^2, and substitute this value into the numerator of Formula 12.5.

To solve Formula 12.5:

1. Multiply ΣD^2 by 6.
2. Square N and subtract 1 from the result.
3. Multiply the quantity you found in step 2 by N.
4. Divide the quantity you found in step 1 by the quantity you found in step 3.
5. Subtract the quantity you found in step 4 from 1. This is r_s.

ONE STEP AT A TIME **Interpreting Spearman's Rho**

To Interpret Strength:

1. Use Table 12.2 to describe strength in general terms.
2. Square the value of rho and multiply by 100. This value represents the percentage improvement in our predictions of the dependent variable by taking the independent variable into account.

To Interpret Direction:

Use the sign of r_s. Be careful when interpreting direction with ordinal-level variables. Remember that coding schemes for these variables are arbitrary and that a positive r_s may mean that the actual relationship is negative, and vice versa.

Interpreting Statistics Age and Concern for the Environment

Is concern for the environment linked to age? Are younger people more concerned about air and water pollution and the damage caused by greenhouse gases? Are older people more complacent about these issues?

To explore this possible relationship, we choose two variables from the 2018 General Social Survey: age and level of concern with the effect of greenhouse gases on the rise in world temperatures (*tempgen*). Age is an interval-ratio level variable with a broad range of scores. To make it useable for bivariate table analysis, we recoded it into four categories of roughly equal size.

Our dependent variable (*tempgen*) had five possible responses, ranging from the view that greenhouse gases are "extremely dangerous" to "not very dangerous." We first generated a bivariate table using all categories of *tempgen*, but the resultant 5 × 4 table (with 20 cells) was too large for easy analysis. To simplify matters, we used the recode command to collapse *tempgen* into two categories and generated a new table, which is presented as Table 12.5. We should note that the larger table displayed essentially the same relationship as the smaller one.

Looking first at the column percentages, we see very little change from column to column: roughly 60% of each age group regards greenhouse gases as "very dangerous." This appears to be a weak, nonsignificant relationship and this impression is reinforced by the statistics listed under the table. Chi square is not significant ($p > .05$) and the measures of association are consistent in labeling the relationship as weak (see Tables 12.8 and 12.12). Gamma is positive, which means that cases have a slight tendency to fall in the diagonal from upper left to lower right, with younger people a little more concerned than the other age groups. However, this relationship is so weak that we can disregard this information. In summary, there is a statistically nonsignificant, weak relationship between age and views about the danger of greenhouse gases. According to these results, concern for the environment is not related to age.

TABLE 12.15 **Opinion on the Danger of Greenhouse Gases by Age**

Danger of Greenhouse Gases:	Age				
	18–33	*34–48*	*49–63*	*64+*	TOTALS
Very Dangerous	64.4%	58.0%	58.2%	58.8%	59.8%
Not Very Dangerous	35.6%	42.0%	41.8%	41.2%	40.2%
TOTALS	100%	100%	100%	100%	100.0%
	(177)	(200)	(182)	(194)	(753)

$\chi^2 = 2.12$, $df = 3$, $p > 0.05$; Lambda $= -0.00$. Cramer's $V = 0.05$ Gamma $= 0.06$

Source: The General Social Survey (GSS) is a project of the independent research organization NORC at the University of Chicago, with principal funding from the National Science Foundation.

READING STATISTICS Bivariate Tables

Social science researchers prefer interval-level data over nominal or ordinal-level data because it enables them to carry out more sophisticated statistics as we will see in Chapter 13. However, we can learn a lot from ordinal-level data as well. In a recent study of 91 students in a large public college in the Midwest, several researchers set out to discover why undergraduate students in the social sciences often experience anxiety when taking courses in statistics.

The sample of students was comprised of mostly juniors and seniors with a cumulative GPA of 3.35. Their general confidence was rated on a scale 0 to 10. Their expected grade in their statistics course was scaled from 0 to 4. Their level of test anxiety varied from "very relaxed" (1) to "very anxious" (5). Their math skills were rated from "poor" (1) to "excellent" (5). Their computing skills varied from "poor" (1) to "excellent" (5). Finally, their perception that it is important to take a statistics course was rated between "strongly disagree" (1) and "strongly

agree" (5). As you can see, most of the variables were measured at the ordinal level.

In the authors' table of results, gamma statistics are used to show the bivariate associations that are strong enough to enable them to be 95% certain that they did not occur due to chance. The findings show that statistics anxiety is negatively associated with confidence, expected grade, math skills, computer skills, and importance, and positively associated with test anxiety.

Later in the paper, they used several models to predict which students will have more statistics anxiety. We will do work like this in Chapter 15. In the discussion of the paper, the authors suggest that professors should not assume that all students have statistics anxiety. Instead they should identify and help those that do, while being careful not to alienate those that do not.

Source: Condron, D., Becker, J., and Bzhetaj, L. (2018). Sources of Students' Anxiety in a Multidisciplinary Social Statistics Course, *Teaching Sociology*, 46(4), 346–355.

SUMMARY

1. Measures of association provide information that is complementary to tests of significance.
2. We have a variety of tools to analyze the association between variables, including bivariate tables and various statistics. We use these tools to develop answers to three questions regarding the existence, strength, and pattern or direction of relationships.
3. Phi, Cramer's V, and lambda are measures of association for nominal variables and each is appropriate

for a specific situation. These statistics indicate *only* the strength of the relationship. The pattern of the relationship can be determined by inspecting column percentages.
4. Gamma and Spearman's rho are measures of association for collapsed and continuous ordinal-level variables, respectively. They indicate both the strength and direction of the relationship.

SUMMARY OF FORMULAS

12.1	Phi	$\phi = \sqrt{\dfrac{\chi^2}{N}}$	12.4	Gamma	$G = \dfrac{N_s - N_d}{N_s + N_d}$
12.2	Cramer's V	$V = \sqrt{\dfrac{\chi^2}{N(\min r - 1, c - 1)}}$	12.5	Spearman's rho	$r_s = 1 - \dfrac{6\Sigma D^2}{N(N^2 - 1)}$
12.3	Lambda	$\lambda = \dfrac{E_1 - E_2}{E_1}$			

GLOSSARY

Association. The relationship between two (or more) variables. Two variables are said to be associated if the distribution of one variable changes for the various categories or scores of the other variable.

Column percentages. Percentages computed within each column of a bivariate table.

Conditional distribution of Y. The distribution of scores on the dependent variable for a specific score or category of the independent variable when the variables have been organized into table format.

Cramer's V. A chi square–based measure of association for nominally measured variables that have been organized into a bivariate table with any number of rows and columns.

Dependent variable. In a bivariate relationship, the variable that is taken as the effect.

Gamma (G). A measure of association for "collapsed" ordinal variables in bivariate table format.

Independent variable. In a bivariate relationship, the variable that is taken as the cause.

Lambda. A proportional reduction in error (PRE) measure for nominally measured variables that have been organized into a bivariate table.

Maximum difference. A way to assess the strength of an association between variables that have been organized into a bivariate table. The maximum difference is the largest difference between column percentages for any row of the table.

Measures of association. Statistics that quantify the strength of the association between variables.

N_d. In a bivariate table, the number of pairs of cases ranked in different order on two variables.

N_s. In a bivariate table, the number of pairs of cases ranked in the same order on two variables.

Negative association. A bivariate relationship where the variables vary in opposite directions. As one variable increases, the other decreases, and high scores on one variable are associated with low scores on the other.

Phi (ϕ). A chi square–based measure of association for nominally measured variables that have been organized into a bivariate table with two rows and two columns (a 2 × 2 table).

Positive association. A bivariate relationship where the variables vary in the same direction. As one variable increases, the other also increases, and high scores on one variable are associated with high scores on the other.

Proportional reduction in error (PRE). A logic used in the design of certain measures of association. A PRE measure indicates how much knowledge of the independent variable improves our predictions of the dependent variable.

Spearman's rho (r_s). A measure of association for "continuous" ordinal variables.

X. Symbol used for any independent variable.

Y. Symbol used for any dependent variable.

PROBLEMS

12.1 SOC The administration of a local college campus wants to increase the mandatory student fee to finance an upgrading of the football program. A survey of a sample of faculty regarding this proposal has been completed. Is there any association between support for raising fees and the gender, discipline, or tenured status of the faculty? Use column percentages, the maximum difference, and an appropriate measure of association to describe the strength and pattern of these associations.

a. Support for raising fees by gender:

| | Gender | | |
Support?	Males	Females	TOTALS
For	12	7	19
Against	15	13	28
TOTALS	27	20	47

b. Support for raising fees by discipline:

| | Discipline | | |
Support?	Liberal Arts	Science and Business	TOTALS
For	6	13	19
Against	14	14	28
TOTALS	20	27	47

c. Support for raising fees by tenured status:

| | Status | | |
Support?	Tenured	Nontenured	TOTALS
For	15	4	19
Against	18	10	28
TOTALS	33	14	47

12.2 SOC A researcher has conducted a survey on sexual attitudes for a sample of 317 teenagers. The respondents were asked whether they considered premarital sex "Always wrong" or "Not always wrong." The tables below summarize the relationship between responses to this item and two other variables. For each table, assess the strength and pattern of the relationship and write a paragraph interpreting these results.

a. Attitudes toward premarital sex by gender:

| Premarital Sex | Gender | | TOTALS |
	Female	Male	
Always wrong	90	105	195
Not always wrong	65	57	122
TOTALS	155	162	317

b. Attitudes toward premarital sex by courtship status:

| Premarital Sex | Ever "Gone Steady"? | | TOTALS |
	No	Yes	
Always wrong	148	47	195
Not always wrong	42	80	122
TOTALS	190	127	317

12.3 GER A survey of senior citizens who live in either a housing development specifically designed for retirees or an age-integrated neighborhood has been conducted. Is type of living arrangement related to sense of social isolation?

| Sense of Isolation | Living | | TOTALS |
	Housing Development	Integrated Neighborhood	
Low	80	30	110
High	20	120	140
TOTALS	100	150	250

12.4 SOC The tables below show four dependent variables cross-tabulated against gender. Analyze these relationships with column percentages, the maximum difference, and an appropriate measure of association. Summarize the results of your analysis in a paragraph that describes the strength and pattern of each relationship.

a. Support for the legal right to an abortion by gender:

| Right to Abortion? | Gender | | TOTALS |
	Males	Females	
Yes	310	418	728
No	432	618	1050
TOTALS	742	1036	1778

b. Support for capital punishment by gender:

| Capital Punishment? | Gender | | TOTALS |
	Males	Females	
Favor	908	998	1906
Oppose	246	447	693
TOTALS	1154	1445	2599

c. Approve of suicide for people with an incurable disease by gender:

| Right to Suicide? | Gender | | TOTALS |
	Males	Females	
Yes	524	608	1132
No	246	398	644
TOTALS	770	1006	1776

d. Support for sex education in public schools by gender:

| Sex Education? | Gender | | TOTALS |
	Males	Females	
Favor	685	900	1585
Oppose	102	134	236
TOTALS	787	1034	1821

12.5 PS Below are some tables of fictional data on politics. You already have the results of the chi square tests from problem 11.7. Now, analyze these relationships in terms of association, using an appropriate measure of association. Write a paragraph summarizing your results.

a. Presidential preference in 2016 by race/ethnicity:

| Preference | Race/Ethnicity | | | Totals |
	White	Black	Latino	
Trump	289	5	44	338
Clinton	249	95	66	410
Totals	538	100	110	748

b. Presidential preference in 2016 by education:

Preference	Less Than HS	HS Graduate	College Graduate	Postgrad. Degree	Totals
Trump	30	180	118	10	338
Clinton	35	120	218	37	410
Totals	65	300	336	47	748

Education spans the middle four columns.

c. Presidential preference in 2016 by religion:

Preference	Protestant	Catholic	Jewish	None	Other	Totals
Trump	165	110	10	28	25	338
Clinton	245	55	20	60	30	410
Totals	410	165	30	88	55	748

Religion spans the middle five columns.

12.6 SOC A small sample of non-English-speaking immigrants to the United States has been interviewed about their level of assimilation. Is the pattern of adjustment affected by length of residence in the United States? For each table, compute gamma and summarize the relationship in terms of strength and direction.

a. English fluency:

English Fluency	Less Than Five Years (Low)	More Than Five Years (High)	TOTALS
Low	20	10	30
High	5	15	20
TOTALS	25	25	50

b. Total family income:

Income	Less Than Five Years (Low)	More Than Five Years (High)	TOTALS
Low	18	8	26
High	7	17	24
TOTALS	25	25	50

c. Extent of contact with country of origin:

Contact	Less Than Five Years (Low)	More Than Five Years (High)	TOTALS
Rare	5	20	25
Frequent	20	5	25
TOTALS	25	25	50

12.7 PA Various supervisors in the city government of Shinbone, Kansas, have been rated on the extent to which they practice authoritarian styles of leadership and decision making. The efficiency of each department has also been rated, and the results are summarized below. Use column percentages, the maximum difference, and an appropriate measure of association to describe this association. What is the direction of the association?

Efficiency	Low	High	TOTALS
Low	10	12	22
High	17	5	22
TOTALS	27	17	44

Authoritarianism spans Low and High.

12.8 PA All applicants to the Police Academy are given an aptitude test, but the test has never been evaluated to see whether test scores are in any way related to job performance. Describe the strength and direction of the relationship in a sentence or two. Should the test continue to be administered? Why or why not?

Efficiency	Low	Moderate	High	TOTALS
Low	12	6	7	25
Moderate	9	11	9	29
High	5	9	10	24
TOTALS	26	26	26	78

Test Scores spans Low, Moderate, High.

12.9 SW A sample of children has been observed and rated for symptoms of depression. Their parents have been rated for level of alcohol use. Is there any relationship between these variables? Write a few sentences stating your conclusions.

Symptoms of Depression	Low	Moderate	High	TOTALS
Few	7	8	9	24
Some	15	10	18	43
Many	8	12	3	23
TOTALS	30	30	30	90

Level of Alcohol Use spans Low, Moderate, High.

12.10 SOC Eleven neighborhoods have been rated by an urban sociologist on a "quality-of-life" scale (which includes measures of affluence, availability of medical care, and recreational facilities) and a social cohesion scale. The results are presented here. Higher scores indicate higher quality of life and greater social cohesion. Are the two variables associated? What are the strength

and direction of the association? Summarize the relationship in a sentence or two. *(HINT: Don't forget to square the value of Spearman's rho for a PRE interpretation.)*

Neighborhood	Quality of Life	Social Cohesion
Beaconsdale	17	8.8
Brentwood	40	3.9
Chesapeake Shores	47	4.0
College Park	90	3.1
Denbigh Plantation	35	7.5
Kingswood	52	3.5
North End	23	6.3
Phoebus	67	1.7
Queens Lake	63	3.0
Riverview	65	9.2
Windsor Forest	100	5.3

12.11 [SOC] Following are the scores of a sample of 15 nations on a measure of religious diversity (the greater the number of different religions, the higher the score) and a measure of economic inequality (the higher the score, the greater the inequality). Are these variables related? Are religiously diverse nations more economically unequal?

Nation	Religious Diversity	Income Inequality
India	4.0	35.2
South Africa	3.6	63.4
Kenya	3.1	48.5
Canada	5.3	34.0
Malaysia	6.3	46.3
Kazakhstan	5.0	26.5
Egypt	1.1	31.8
United States	4.1	41.0
Sri Lanka	5.6	39.2
Mexico	1.1	48.2
Spain	3.9	36.0
Australia	5.6	34.7
Finland	3.5	26.8
Ireland	1.7	31.9
Poland	1.2	32.1

Sources: Religious Diversity Score: Pew Research Center, 2014, https://www.pewforum.org/2014/04/04/religious-diversity-index-scores-by-country/; Income Inequality: World Bank, 2019, Gini Index, https://www.indexmundi.com/facts/indicators/SI.POV.GINI/rankings

12.12 [SOC] In a 2013 article by Parrillo and Donoghue, thirty ethnic, racial or national groups were rated by a sample of college students on a Social Distance Scale. Lower scores represent less social distance. Below are the scores for white and Black students on 20 of the groups. How similar are these rankings?

Group	Average Scores White Students	Average Scores Black Students
Americans	1.1	1.3
British	1.3	1.7
Canadians	1.3	1.6
Italians	1.3	1.6
African-Americans	1.4	1.1
Indians (American)	1.5	1.6
Polish	1.6	2.0
Chinese	1.7	1.8
Filipinos	1.7	1.8
Jews	1.7	2.1
Puerto Ricans	1.7	1.4
Russians	1.7	2.0
Japanese	1.8	2.0
Mexicans	1.8	1.8
Other Hispanics/Latinos	1.8	1.5
Vietnamese	1.8	2.0
Indians	1.9	1.9
Koreans	1.9	2.1
Arabs	2.2	2.1
Muslims	2.2	2.3

Statistical Analysis Using SPSS

12.13 In problem 11.18, you used **Crosstabs** to analyze the relationship between gender and fear of walking alone at night (*fear*), support for capital punishment (*cappun*), and job satisfaction (*satjob*). Now, we will add measures of association to the analysis.
- Click the SPSS icon on your desktop.
- Load the *GSS2018* database.
- Click **Analyze → Descriptive → Crosstabs**.
- Find *fear, cappun,* and *satjob* in the list of variables and move them to the **Rows** box.
- Find *sex* in the list of variables and move it to the **Columns** box.
- Click the **Statistics** button and select **Chi-square, Phi and Cramer's V,** and **Lambda**.
- Return to the **Crosstabs** window, click the **Cells** button, and select **column** in the **Percentages** box.
- Return to the **Crosstabs** window and click **OK**. The results of the test will be printed to the SPSS output window.

Examine the column percentages as well as the measures of association and report these results in a few sentences.

12.14. Analyze the relationship between education (*degree*) and fear of walking alone at night (*fear*), support for capital punishment (*cappun*), and frequency of prayer (*pray*). Degree is an ordinal-level variable, so we will use gamma as our measure of association.
- Click the SPSS icon on your desktop.
- Load the *GSS2018* database.
- Click **Analyze → Descriptive → Crosstabs**.
- Find *fear, cappun,* and *pray* in the list of variables and move them to the **Rows** box.

- Find *degree* in the list of variables and move it to the **Columns** box.
- Click the **Statistics** button, and select **Chi-square** and **Gamma**.
- Return to the **Crosstabs** window, click the **Cells** button, and select **column** in the **Percentages** box.
- Return to the **Crosstabs** window and click **OK**. The results of the test will be printed to the SPSS output window.

Examine the column percentages as well as the measures of association, and report these results in a few sentences.

YOU ARE THE RESEARCHER

Understanding Political Beliefs, Part II

At the end of Chapter 11, you investigated possible causes of people's beliefs on several controversial issues. Now you will extend your analysis by using **Crosstabs** to get the statistics presented in this chapter. There are two projects and you should complete both using the GSS2018 file.

PROJECT 1: Explaining Beliefs

In this project, you will again analyze beliefs about assisted suicide (*letdie1*), gay marriage (*marhomo*), and immigration (*letin1a*). You will select an independent variable *other than the one you used in Chapter 11* and use SPSS to generate chi square, column percentages, and measures of association.

Step 1: Choose an Independent Variable

Select an independent variable that seems likely to be an important cause of people's attitudes about these issues. Be sure to select an independent variable that has *only* 2–5 categories and use the recode command if necessary. You might consider gender, level of education (use *degree*), religion, or age (the recoded version—see Chapter 10) as possibilities, but there are many others. Record the variable name and state exactly what the variable measures in the table below. Also note the level of measurement of the variable.

SPSS Name	What Exactly Does This Variable Measure?	Level of Measurement

Step 2: Stating Hypotheses

State hypotheses about the relationships you expect to find between your independent variable and each of the dependent variables. State these hypotheses in terms of which category of the independent variable you expect to be associated with which

category of the dependent variable (for example, "I expect that men will be more supportive of the legal right to an abortion for any reason").

Dependent Variable	Hypothesis
letdie1	
marhomo	
letin1a	

Step 3: Running Crosstabs

After loading the 2018 GSS, click **Analyze → Descriptives → Crosstabs** and place the dependent variables in the **Rows:** box and the independent variable you selected in the **Columns:** box. Click the **Statistics** button to get chi square and the measures of association and the **Cells** button for column percentages.

Step 4: Recording Results

These commands will generate a lot of output and it will be helpful to summarize your results in the following table. For the chi square test, write "Yes" in the right-hand column of the table if the value of "Asymp. Sig (2-sided)" is less than 0.05.

Dependent Variable	Phi or Cramer's *V*	Lambda	Gamma	Chi Square Significant at 0.05?
letdie1				
marhomo				
letin1a				

Step 5: Analyzing and Interpreting Results

Choose *one* measure of association for each relationship. Your choice should be based primarily on level of measurement but should take account of other statistical characteristics as well.

Write a short summary of results for each dependent variable in which you identify the variables being tested, the results of the chi square test (see Chapter 11), and the strength and pattern or direction of the relationship. It is probably best to characterize the relationship in general terms and then cite the statistical values in parentheses. For example, we might summarize the relationship between education (*degree*) and happiness (*happy*) by saying: "The relationship between education and happiness was significant and weak to moderate in strength (chi square = 62.46, *df* = 8, *p* < 0.00, Gamma = −0.22). Happiness tends to increase as education increases." You should also note whether or not your hypotheses were supported.

PROJECT 2: Exploring the Impact of Various Independent Variables

In this project, you will examine the relative ability of a variety of independent variables to explain or account for a single dependent variable. You will again use the

Crosstabs procedure in SPSS to generate statistics and use the alpha levels and measures of association to judge which independent variable has the most important relationship with your dependent variable.

Step 1: Choosing Variables

Select a dependent variable. You may use any of the variables from Project 1 or select a new dependent variable. *Be sure that your dependent variable has no more than five values or scores.* Good choices for dependent variables include any measure of attitudes or opinions. *Do not* select characteristics like race, sex, or religion as dependent variables.

Select three independent variables that seem likely to be important causes of the dependent variable you selected. *Be sure that your independent variable has no more than five or six categories.* You might consider gender, level of education (use degree), religion, or age (the recoded version—see Chapter 10) as possibilities, but there are many others.

Record the variable names and state exactly what each variable measures and the level of measurement in the table below.

SPSS Name	What Exactly Does This Variable Measure?	Level of Measurement
Dependent Variable		
Independent Variables		

Step 2: Stating Hypotheses

State hypotheses about the relationships you expect to find between your independent variables and the dependent variable. State these hypotheses in terms of which categories of the variables you expect to be associated with each other.

Independent Variables (SPSS Names)	Hypothesis
1. _____	
2. _____	
3. _____	

Step 3: Running Crosstabs

Place your dependent variable in the **Rows:** box and all three of your independent variables in the **Columns:** box. Click the **Statistics** button to get chi square and the measures of association. Click the **Cells** button for column percentages.

Step 4: Recording Results

Your output will consist of three tables, and it will be helpful to summarize your results in the following table.

Independent Variable	Phi or Cramer's V	Lambda	Gamma	Chi Square Significant at 0.05?
1.				
2.				
3.				

Step 5: Analyzing and Interpreting Results

Choose *one* measure of association for each relationship. Your choice should be based primarily on level of measurement but should take account of other statistical characteristics as well.

Write a short summary of results of each test using the same format as in Project 1. Remember to explain whether or not your hypotheses were supported. Finally, assess which of the independent variables had the most important relationship with your dependent variable. Use the alpha level and the value of the measures of association to make this judgment.

13

Association Between Variables Measured at the Interval-Ratio Level

LEARNING OBJECTIVES By the end of this chapter, you will be able to:

1. Interpret a scatterplot for interval-level data.
2. Calculate and interpret the slope (*b*), the *Y* intercept (*a*), and Pearson's *r* and r^2.
3. Use the least-squares regression line to predict values of *Y*.
4. Explain the concepts of total, explained, and unexplained variance.
5. Use regression and correlation techniques to analyze and describe a bivariate relationship.
6. Construct a correlation matrix and interpret the data.
7. Test Pearson's *r* for significance.

USING STATISTICS

The statistical techniques presented in this chapter are used to measure the strength and direction of the association between variables measured at the interval-ratio level. Examples of research situations in which these techniques are useful include

1. An organizational researcher is interested in the factors that lead employees to share information with others. Specifically, she wants to study the effects of organizational spirituality and trust on knowledge sharing. **See page 377**

2. A sociologist wants to study the association between marital status, household labor, and sleep. **See page 361**

3. A demographer is interested in the relationship between poverty and the teen birth rate. **See page 375**

In this chapter, we will consider statistical techniques for analyzing the association, or correlation, between variables measured at the interval-ratio level.[1] The techniques presented here are also commonly used on ordinal-level variables, especially when they are "continuous" or have many values (see Chapter 12).

[1]The term *correlation* is commonly used instead of *association* when discussing the relationship between interval-ratio variables. We will use the two terms interchangeably.

As we shall see, these statistics are rather different in their logic and computation from the measures of association covered in Chapter 12. Let us stress at the outset, therefore, that we are still asking the same three questions:

- Is there a relationship between the variables?
- How strong is the relationship?
- What is the direction of the relationship?

It will quickly become apparent that the mathematics in this chapter can be challenging and time-consuming. We use a small data set to illustrate the calculations, and you should familiarize yourself with these mathematical routines. However, researchers today use computerized statistical packages such as SPSS to calculate these statistics and, after working through our simplified example problem, we will do the same. Likewise, the end-of-chapter problems offer some problems with small data sets that you can use to be sure you understand the mathematical routines, but most problems use SPSS.

Scatterplots and Regression Analysis

The usual first step in analyzing a relationship between interval-ratio variables is to construct and examine a **scatterplot**. Like bivariate tables, these graphs allow us to identify several important features of the relationship. An example will illustrate how to construct and interpret scatterplots.

Suppose a researcher is interested in analyzing how dual–wage-earner families (families where both husband and wife have jobs outside the home) cope with housework. Specifically, the researcher wonders whether the number of children in the family is related to the amount of time the husband contributes to housekeeping chores. The relevant data for a sample of 12 families are displayed in Table 13.1.

TABLE 13.1 **Number of Children and Husband's Contribution to Housework (fictitious data)**

Family	Number of Children	Hours per Week Husband Spends on Housework
A	1	1
B	1	2
C	1	3
D	1	5
E	2	3
F	2	1
G	3	5
H	3	0
I	4	6
J	4	3
K	5	7
L	5	4

FIGURE 13.1 Husband's Housework by Number of Children

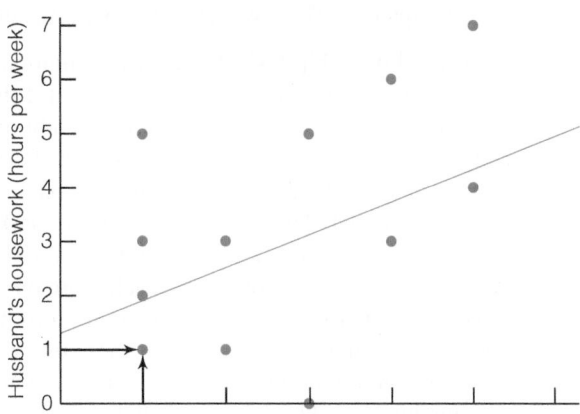

CONSTRUCTING SCATTERPLOTS

A scatterplot, like a bivariate table, has two dimensions. The scores of the independent (X) variable are arrayed along the horizontal axis, and the scores of the dependent (Y) variable are arrayed along the vertical axis. Each dot on the scatterplot represents a case in the sample, and the dot's location represents the scores of the case on both variables.

Figure 13.1 shows a scatterplot displaying the relationship between "number of children" and "husband's housework" for the sample of 12 families presented in Table 13.1. Family A has a score of 1 on the X variable (number of children) and 1 on the Y variable (husband's housework) and is represented by the dot above the score of 1 on the X axis and directly to the right of the score of 1 on the Y axis. All 12 cases are similarly represented by dots on Figure 13.1. Also note that, as always, the scatterplot is clearly titled and both axes are labeled.

ADDING A "FREEHAND" REGRESSION LINE

The overall pattern of the dots or cases summarizes the nature of the relationship between the two variables. The clarity of the pattern can be enhanced by drawing a straight line through the cluster of dots such that the line touches every dot or comes as close to doing so as possible. We will soon consider a precise technique for fitting this line to the pattern of the dots but, for now, an approximation will suffice. This summarizing line, called the **regression line**, has already been added to the scatterplot.

USING SCATTERPLOTS TO ANSWER THE THREE QUESTIONS

Scatterplots can be used for a variety of purposes. They provide at least impressionistic information about the existence, strength, and direction of the relationship and can also be used to check the relationship for linearity (that is, how well the dots approximate a straight line). Finally, the scatterplot can be used to predict the score of a case on one variable from the score of that case on the

other variable. We examine these uses first in terms of the three questions we first asked in Chapter 12. Then, we will investigate matters of linearity and prediction.

- **Does a Relationship Exist?** The basic definition of an association was stated in Chapter 12: Two variables are associated if the distributions of Y (the dependent variable) change for the various conditions of X (the independent variable). In Figure 13.1, scores on X (number of children) are arrayed along the horizontal axis. The dots above each score on X are the scores (or conditional distributions) of Y. The dots represent scores on Y for each value of X.

 There is a relationship between these variables, because the conditional distributions of Y (the dots above each score on X) change as X changes. The existence of an association is further reinforced by the fact that the regression line lies at an angle to the X axis. If these two variables had not been associated, the conditional distributions of Y would not have changed, and the regression line would have been parallel to the horizontal axis.

- **How Strong Is the Relationship?** The strength of the bivariate association can be judged by observing the spread of the dots around the regression line. In a perfect association, all dots would lie on the regression line. The more the dots are clustered around the regression line, the stronger the association.

- **What Is the Direction of the Relationship?** The direction of the relationship can be detected by observing the angle of the regression line. Figure 13.1 shows a positive relationship: As X (number of children) increases, husband's housework (Y) also increases. Husbands in families with more children tend to do more housework. If the relationship had been negative, the regression line would have sloped in the opposite direction to indicate that high scores on one variable were associated with low scores on the other.

To summarize these points, Figure 13.2 shows a perfect positive and a perfect negative relationship and a "zero relationship," or "nonrelationship," between two variables.

FIGURE 13.2 Positive (a), Negative (b), and Zero (c) Relationships

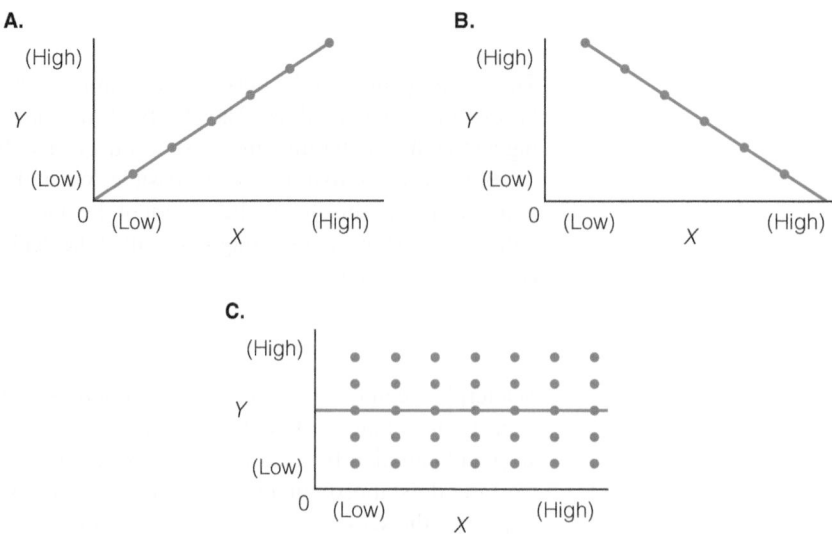

STATISTICS In Everyday Life

Who Does the Housework?
Studies of the division of labor in American households commonly find large differences between the contributions of husbands and wives, with women contributing substantially more than men. One recent study went beyond gender differences and explored the effect of women's status on their activities. Based on a large, representative sample of U.S. women, the study found that married mothers did more housework, slept less, and had less leisure time than never-married, divorced, or cohabiting mothers regardless of their involvement in the paid labor force. They also found that, consistent with traditional gender-role expectations, women with a live-in male partner spent more time on housework. Want to learn more? The citation for the article is given below.

Source: Pepin, J., Sayer, L., and Casper, L. 2018. "Marital Status and Mother's Time Use: Childcare, Housework, Leisure, and Sleep." *Demography* 55: 107–133. https://doi.org/10.1007/s13524-018-0647-x

USING SCATTERPLOTS TO CHECK FOR LINEARITY

Correlation analysis with interval-ratio–level variables assumes that the variables have an essentially **linear relationship**. In other words, the observation points or dots in the scatterplot must form a pattern that can be approximated with a straight line. Significant departures from linearity would require the use of statistical techniques beyond the scope of this text. Examples of some common curvilinear relationships are presented in Figure 13.3.

If the scatterplot shows that the variables have a nonlinear relationship, the techniques described in this chapter should be used with great caution or not at all. Checking for the linearity of the relationship is perhaps the most important

FIGURE 13.3 Some Nonlinear Relationships

FIGURE 13.4 Predicting Husband's Housework

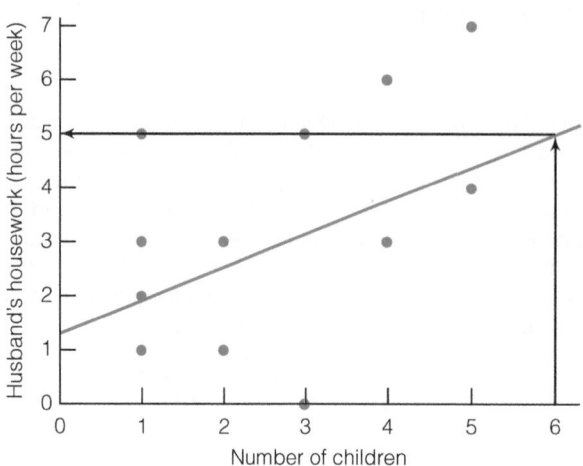

reason for examining the scatterplot before proceeding with the statistical analysis. If the relationship is nonlinear, you might need to treat the variables as if they were ordinal rather than interval-ratio in level of measurement.

USING SCATTERPLOTS FOR PREDICTION

A final use of the scatterplot is to predict scores of cases on one variable from their score on the other. Suppose that, based on the relationship between number of children and husband's housework displayed in Figure 13.1, we wish to predict the number of hours of housework a husband with a family of six children would do each week. The sample has no families with six children, but if we extend the axes and regression line in Figure 13.1 to incorporate this score, a prediction is possible. Figure 13.4 reproduces the scatterplot and illustrates how the prediction would be made.

The predicted score on Y—which is symbolized as Y' ("Y-prime") to distinguish predictions of Y from actual Y scores—is found by first locating the relevant score on X ($X = 6$ in this case) and then drawing a straight line from that point to the regression line. From the regression line, another straight line parallel to the X axis is drawn across to the Y axis. The predicted Y score (Y') is found at the point where the line crosses the Y axis. In our example, we would predict that in a dual–wage-earner family with six children the husband would devote about 5 hours per week to housework.

USING SPSS	**Scatterplots**

We can illustrate the features of a scatterplot with SPSS. We will use the *States* database and our dependent variable will be *BirthRate* (the number of births per 1000 population). The independent variable will be *BA* (the percentage of the state's population with at least a bachelor's degree). Is there a correlation between level of education and the birth rate?

(continued)

USING SPSS

(Continued)

Follow these steps to get a scatterplot with the regression line:

1. Load the *States* database
 a. Click **File → Open → Data**.
 b. Find the *States.sav* database.
2. From the menu bar across the top of the SPSS window, click **Graphs**, **Legacy Dialog**, and **Scatter/Dot**.
3. On the **Scatter/Dot** window, click "Simple Scatter" and then click **Define**.
4. The "Simple Scatterplot" window will appear.
 a. Find *BirthRate* in the list of variables and click the arrow to move the variable name to the "Y Axis:" box.
 b. Find *BA* in the list of variables and click the arrow to move the variable name to the "X Axis:" box.
 c. Click **OK** and the scatterplot will be sent to the SPSS output window.
5. To add the regression line:
 a. Double-click on the graph and the "Chart Editor" window will appear.
 b. Click the **Elements** command at the top of the window and, in the drop-down menu, click **Fit Line at Total**.
 c. Close the "Chart Editor" window to return to the Output window of SPSS.

The scatterplot will look like this:

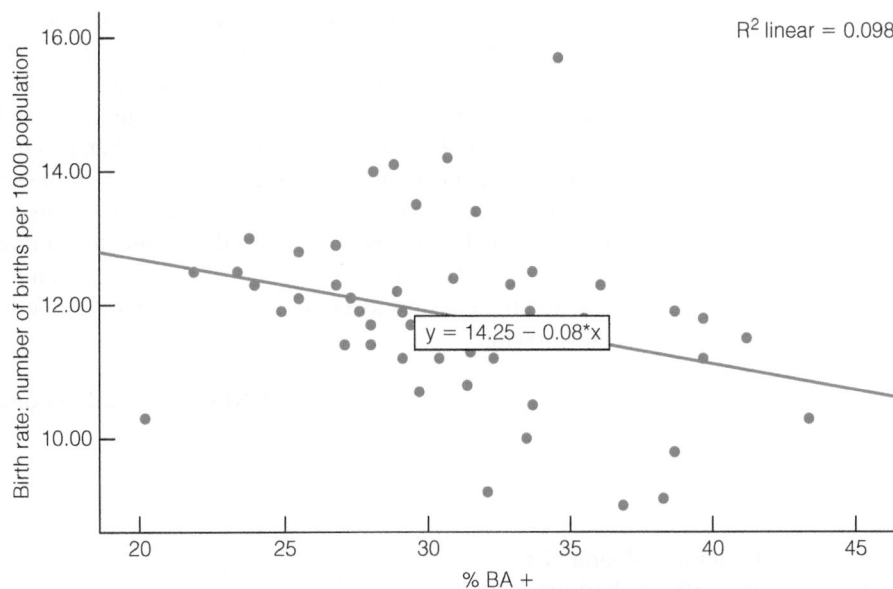

How strong is the relationship? The dots are widely spread around the regression line so this is, at best, a weak to moderate relationship. What is the direction of the relationship? Since the regression line slopes down from upper left to lower right, this is a negative relationship. As the level of education in a state increases, the birth rate tends to decrease: more educated states have lower birth rates.

We will identify and discuss the other elements in this graph (the equation in the center of the graph and the notation "R^2 Linear = 0.098" at the upper right) later in the chapter. *(For practice in constructing and interpreting scatterplots, see problem 13.6.)*

The Regression Line

Predicting the score on Y with scatterplots can be crude and impressionistic. A better way to predict would be to find the straight line that most accurately describes the relationship between the two variables. How can we find this line?

Recall that our criterion for the freehand regression line was that it touch all the dots or come as close to doing so as possible. Also, recall that the dots above each value of X can be thought of as conditional distributions of Y, the dependent variable. Within each conditional distribution of Y, the mean is the point around which the variation of the scores is minimized. In Chapter 3, we noted that the mean of any distribution of scores is the point around which the variation of the scores, as measured by squared deviations, is minimized:

$$\sum (X_i - \overline{X})^2 = \text{minimum}$$

Thus, if the regression line were drawn so that it touches each **conditional mean of** Y, it would be the straight line that comes as close as possible to all the scores.

Conditional means are found by summing all Y values for each value of X and then dividing by the number of cases. For example, four families had one child ($X = 1$), and the husbands of these four families devoted 1, 2, 3, and 5 hours to housework. The conditional mean of Y for $X = 1$ is 2.75 (11/4 = 2.75). Husbands in families with one child worked an average of 2.75 hours per week doing housekeeping chores. Conditional means of Y are computed in the same way for each value of X. They are displayed in Table 13.2 and plotted in Figure 13.5.

Let us quickly remind ourselves of the reason for these calculations. We are seeking the single best-fitting regression line for summarizing the relationship between X and Y, and we have seen that a line drawn through the conditional means of Y will minimize the spread of the observation points. It will come as close to all the scores as possible and will therefore be the single best-fitting regression line.

TABLE 13.2 Conditional Means of Y (husband's housework) for Various Values of X (number of children)

Number of Children (X)	Husband's Housework (Y)	Conditional Mean of Y
1	1, 2, 3, 5	2.75
2	3, 1	2.00
3	5, 0	2.50
4	6, 3	4.50
5	7, 4	5.50

FIGURE 13.5 Conditional Means of Y

Now, a line drawn through the points on Figure 13.5 (the conditional means of Y) will be the best-fitting line we are seeking, but you can see from the scatterplot that the line will not be straight. In fact, only rarely (when there is a perfect relationship between X and Y) will conditional means fall in a perfectly straight line.

Because we still must meet the condition of linearity, let us revise our criterion and define the regression line as the unique straight line that touches all conditional means of Y or comes as close to doing so as possible. Formula 13.1 defines this "least-squares" regression line, or the single straight regression line that best fits the pattern of the data points.

FORMULA 13.1

$$Y = a + bX$$

where: Y = score on the dependent variable

a = the Y intercept, or the point where the regression line crosses the Y axis

b = the slope of the regression line, or the amount of change produced in Y by a unit change in X

X = score on the independent variable

The formula introduces two new concepts. First, the **Y intercept (a)** is the point at which the regression line crosses the vertical, or Y, axis. Second, the **slope (b)** of the least-squares regression line is the amount of change produced in the dependent variable (Y) by a unit change in the independent variable (X). Think of the slope of the regression line as a measure of the effect of the X variable on the Y variable: The greater the effect of X on Y, the higher the value of the slope (b). If the two variables are unrelated, the least-squares regression line would be parallel to the X axis, and b would be 0.00 (the line would have no slope).

With the least-squares formula (Formula 13.1), we can predict values of Y in a much less arbitrary and impressionistic way than through mere eyeballing. This is because the least-squares regression line is the single straight line that best fits the data and comes as close as possible to all of the conditional means of Y. Before seeing how predictions of Y can be made, however, we must first calculate a and b.

Computing and Interpreting the Regression Coefficients (*a* and *b*)

In this section, we cover the computation and interpretation of the coefficients in the equation for the regression line: the slope (b) and the Y intercept (a). We begin with the computation of the slope because the value of b is needed to compute a.

COMPUTING THE SLOPE (*b*)

The formula for the slope is

FORMULA 13.2

$$b = \frac{\Sigma(X - \overline{X})(Y - \overline{Y})}{\Sigma(X - \overline{X})^2}$$

TABLE 13.3 Computation of the Slope (b)

1	2	3	4	5	6
X	$X - \bar{X}$	Y	$Y - \bar{Y}$	$(X - \bar{X})(Y - \bar{Y})$	$(X - \bar{X})^2$
1	−1.67	1	−2.33	3.89	2.79
1	−1.67	2	−1.33	2.22	2.79
1	−1.67	3	−0.33	0.55	2.79
1	−1.67	5	1.67	−2.79	2.79
2	−0.67	3	−0.33	0.22	0.45
2	−0.67	1	−2.33	1.56	0.45
3	0.33	5	1.67	0.55	0.11
3	0.33	0	−3.33	−1.10	0.11
4	1.33	6	2.67	3.55	1.77
4	1.33	3	−0.33	−0.44	1.77
5	2.33	7	3.67	8.55	5.43
5	2.33	4	0.67	1.56	5.43
32	−0.04	40	0.04	18.32	26.68

$$\bar{X} = \frac{32}{12} = 2.67$$

$$\bar{Y} = \frac{40}{12} = 3.33$$

The numerator of this formula is called the *covariation* of X and Y. It is a measure of how X and Y vary together, and its value will reflect both the direction and the strength of the relationship. The denominator is the sum of the squared deviations around the mean of X.

The calculations necessary for computing the slope should be organized into a computational table, as in Table 13.3, which has a column for each of the quantities needed to solve the formula. The data are from the dual–wage-earner family sample (see Table 13.1).

The first column of the table lists the original X scores for each case, and the second column shows the deviations of these scores around their mean. The third and fourth columns repeat this information for the Y scores and the deviations of the Y scores. Column 5 shows the covariation of the X and Y scores. The entries in this column are found by multiplying the deviation of the X score (column 2) by the deviation of the Y score (column 4) for each case. Finally, the entries in column 6 are found by squaring the value in column 2 for each case. See the "One Step at a Time" box for detailed instructions on computing the slope.

Table 13.3 gives us all the quantities we need to solve Formula 13.2. Substitute the total of column 5 in Table 13.3 in the numerator and the total of column 6 in the denominator:

$$b = \frac{\Sigma(X - \bar{X})(Y - \bar{Y})}{\Sigma(X - \bar{X})^2}$$

$$b = \frac{18.32}{26.68}$$

$$b = 0.69$$

> **ONE STEP AT A TIME** Computing the Slope (*b*)
>
> ### To Compute the Slope (b), Solve Formula 13.2
>
> 1. Use a computing table like Table 13.3 to help organize the computations. List the scores of the cases on the independent variable (*X*) in column 1.
> 2. Compute the mean of *X* by dividing the total of column 1 by the number of cases.
> 3. Subtract the mean of *X* from each *X* score and list the results in column 2. *NOTE: The sum of column 2 must be zero (except for rounding error). If this sum is not zero, you have made a mistake in computations.*
> 4. List the score of each case on *Y* in column 3. Compute the mean of *Y* by dividing the total of column 3 by the number of cases.
> 5. Subtract the mean of *Y* from each *Y* score and list the results in column 4. *NOTE: The sum of column 4 must be zero (except for rounding error). If this sum is not zero, you have made a mistake in computations.*
> 6. For each case, multiply the value in column 2 by the value in column 4. Place the result in column 5. Find the sum of this column.
> 7. Square each value in column 2 and place the result in column 6. Find the sum of this column.
> 8. Divide the sum of column 5 by the sum of column 6. The result is the slope (*b*).

INTERPRETING THE SLOPE (*b*)

The value of the slope tells us the size of the change in *Y* for every unit change in *X*. In this case, the slope means that husbands increase their contribution to housework by 0.69 hour for each additional child.

COMPUTING THE Y INTERCEPT (*a*)

Once the slope has been calculated, finding the intercept (*a*) is relatively easy. We computed the means of *X* and *Y* while calculating the slope, and we enter these figures into Formula 13.3.

FORMULA 13.3
$$a = \overline{Y} - b\overline{X}$$

For our sample problem, the value of *a* would be

$$a = \overline{Y} - b\overline{X}$$
$$a = 3.33 - (0.69)(2.67)$$
$$a = 3.33 - 1.84$$
$$a = 1.49$$

Thus, the least-squares regression line will cross the *Y* axis at the point where *Y* equals 1.49.

Now that we have values for the slope and the *Y* intercept, we can state the full least-squares regression line for our sample data:

$$Y = a + bX$$
$$Y = 1.49 + (0.69)X$$

ONE STEP AT A TIME Computing the Y Intercept (a)

To Compute the Y Intercept (a), Solve Formula 13.3

1. Multiply the slope (*b*) by the mean of *X*.
2. Subtract the value you found in step 1 from the mean of *Y*. This value is *a*, or the *Y* intercept.

Predicting Scores on *Y* with the Least-Squares Regression Line

The regression formula can be used to estimate, or predict, scores on *Y* for any value of *X*. Previously, we used the freehand regression line to predict a score on *Y* (husband's housework) for a family with six children ($X = 6$). Our prediction was that, in families of six children, husbands would contribute about 5 hours per week to housekeeping chores. By using the least-squares regression line, we can see how close our impressionistic, eyeball prediction was.

$$Y = a + bX$$
$$Y = 1.49 + (0.69)(6)$$
$$Y = 1.49 + 4.14$$
$$Y = 5.63$$

Based on the least-squares regression line, we would predict that in a dual–wage-earner family with six children, the husband would devote 5.63 hours a week to housework. What would our prediction of the husband's housework be for a family of seven children ($X = 7$)?

Note that our predictions of *Y* scores are "educated guesses." We will be unlikely to predict values of *Y* exactly except in the (rare) case where the bivariate relationship is perfect and perfectly linear. However, the accuracy of our predictions will increase as relationships become stronger. This is because the dots are more clustered around the least-squares regression line in stronger relationships. *(For practice in computing* b *and* a *and using the regression line to predict scores on* Y *from scores on* X, *see problems 13.1 to 13.3.)*

ONE STEP AT A TIME Using the Regression Line to Predict Scores on Y

1. Choose a value for *X*. Multiply this value by the value of the slope (*b*).
2. Add the value you found in step 1 to the value of *a*, the *Y* intercept. The resulting value is the predicted score on *Y*.

The Correlation Coefficient (Pearson's *r*)

The slope of the least-squares regression line (b) is a measure of the effect of X on Y and it will increase as the effect of X increases. However, b does not vary between 0 and 1.00 and is therefore awkward to use as a measure of association. Instead, researchers rely heavily (almost exclusively) on a statistic called **Pearson's *r***, or the correlation coefficient, to measure association between interval-ratio variables. Like the ordinal measures of association discussed in Chapter 12, Pearson's *r* varies from 0.00 to ± 1.00, with 0.00 indicating no association and $+1.00$ and -1.00 indicating perfect positive and perfect negative relationships, respectively. The formula for Pearson's *r* is:

FORMULA 13.4
$$r = \frac{\Sigma(X - \bar{X})(Y - \bar{Y})}{\sqrt{[\Sigma(X - \bar{X})^2][\Sigma(Y - \bar{Y})^2]}}$$

Note that the numerator of this formula is the covariation of X and Y, as was the case with Formula 13.2. A computing table like Table 13.4, with a column added for the sum of $(Y - \bar{Y})^2$, is strongly recommended as a way of organizing the quantities needed to solve this equation (see Table 13.4).

For our sample problem involving dual–wage-earner families, the quantities displayed in Table 13.4 can be substituted directly into Formula 13.4:

$$r = \frac{\Sigma(X - \bar{X})(Y - \bar{Y})}{\sqrt{[\Sigma(X - \bar{X})^2][\Sigma(Y - \bar{Y})^2]}}$$

$$r = \frac{18.32}{\sqrt{(26.68)(50.68)}}$$

$$r = \frac{18.32}{\sqrt{1352.14}}$$

$$r = \frac{18.32}{36.77}$$

$$r = 0.50$$

TABLE 13.4 Computation of Pearson's *r*

1	2	3	4	5	6	7
X	$X - \bar{X}$	Y	$Y - \bar{Y}$	$(X - \bar{X})(Y - \bar{Y})$	$(X - \bar{X})^2$	$(Y - \bar{Y})^2$
1	−1.67	1	−2.33	3.89	2.79	5.43
1	−1.67	2	−1.33	2.22	2.79	1.77
1	−1.67	3	−0.33	0.55	2.79	0.11
1	−1.67	5	1.67	−2.79	2.79	2.79
2	−0.67	3	−0.33	0.22	0.45	0.11
2	−0.67	1	−2.33	1.56	0.45	5.43
3	0.33	5	1.67	0.55	0.11	2.79
3	0.33	0	−3.33	−1.10	0.11	11.09
4	1.33	6	2.67	3.55	1.77	7.13
4	1.33	3	−0.33	−0.44	1.77	0.11
5	2.33	7	3.67	8.55	5.43	13.47
5	2.33	4	0.67	1.56	5.43	0.45
32	−0.04	40	0.04	18.32	26.68	50.68

ONE STEP AT A TIME	Computing Pearson's *r*

1. Add a column to the computing table you used to compute the slope (*b*). Square the value of $(Y - \overline{Y})$ and record the result in column 7.
2. Find the sum of column 7.
3. Multiply the sum of column 6 by the sum of column 7.
4. Take the square root of the value you found in step 3.
5. Divide the quantity you found in step 4 into the sum of column 5, or the sum of the cross products. The result is Pearson's *r*.

An *r* value of 0.50 indicates a moderately strong, positive linear relationship between the variables. As the number of children in the family increases, the hourly contribution of husbands to housekeeping duties also increases. *(For practice in computing and interpreting Pearson's* r, *see problems 13.1 to 13.5. For practice in using SPSS to produce and interpret Pearson's* r, *see problems 13.7 to 13.12.)*

Interpreting the Correlation Coefficient r^2

Pearson's *r* is an index of the strength of the linear relationship between two variables. While a value of 0.00 indicates no linear relationship and a value of ±1.00 indicates a perfect linear relationship, values between these extremes have no direct interpretation. But we can describe relationships in terms of how closely they approach the extremes (for example, coefficients approaching 0.00 can be described as "weak" and those approaching ±1.00 as "strong"); however, this description is subjective. Also, we can use the guidelines stated in Table 12.12 for ordinal-level measures of association to describe strength in terms of weak, moderate, and strong. Remember that these labels are just guidelines and will not be appropriate or useful in all possible research situations.

Fortunately, we can use a more direct interpretation of *r* that involves calculating an additional statistic called the **coefficient of determination**. This statistic, which is simply the square of Pearson's *r* (r^2), can be interpreted using a logic similar to that of proportional reduction in error (PRE). As you recall, the logic of PRE measures of association is to predict the value of the dependent variable under two different conditions. First, *Y* is predicted while ignoring the information supplied by *X*; second, the independent variable is taken into account. With r^2, both the method of prediction and the construction of the final statistic require the introduction of some new concepts.

When working with variables measured at the interval-ratio level, the predictions of the *Y* scores under the first condition (while ignoring *X*) will be the mean of the *Y* scores. Given no information on *X*, this prediction strategy will be optimal because we know that the mean of any distribution is closer to all the

FIGURE 13.6 Predicting *Y* Without *X* (dual-career families)

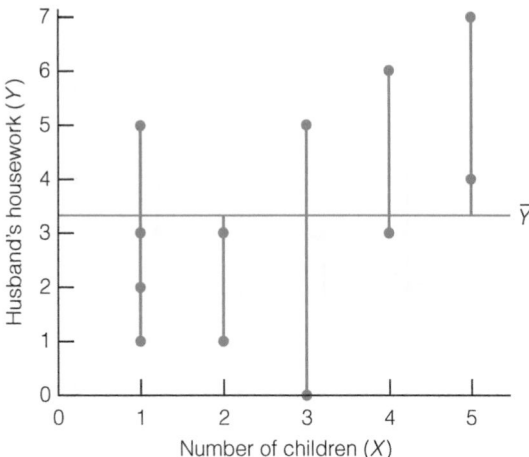

scores than any other point in the distribution. Recall the principle of minimized variation introduced in Chapter 3 and expressed as

$$\sum (X - \overline{X})^2 = \text{minimum}$$

The scores of any variable vary less around the mean than around any other point. If we predict the mean of *Y* for every case, we will make fewer errors of prediction than if we predict any other value for *Y*.

We will still make many errors in predicting *Y* even if we faithfully follow this strategy. The amount of error is represented in Figure 13.6, which displays the relationship between number of children and husband's housework, with the mean of *Y* noted. The vertical lines from the actual scores to the predicted scores represent the amount of error we would make when predicting *Y* while ignoring *X*.

We can define the extent of our prediction error under the first condition (while ignoring *X*) by subtracting the mean of *Y* from each actual *Y* score and squaring and summing these deviations. The resultant sum, which can be noted as $\Sigma(Y - \overline{Y})^2$, is called the **total variation** in *Y*. We now have a visual representation (Figure 13.6) and a method for calculating the error we incur by predicting *Y* without knowledge of *X*. As we shall see, we do not actually need to calculate the total variation to find the value of the coefficient of determination, r^2.

Our next step will be to determine the extent to which knowledge of *X* improves our ability to predict *Y*. If the two variables have a linear relationship, then predicting scores on *Y* from the least-squares regression equation will incorporate knowledge of *X* and reduce our errors of prediction. So, under the second condition, our predicted *Y* score for each value of *X* will be

$$Y' = a + bX$$

Figure 13.7 displays the data from the dual-career families with the regression line drawn in. The vertical lines from each data point to the regression line

FIGURE 13.7 Predicting Y with X (dual-career families)

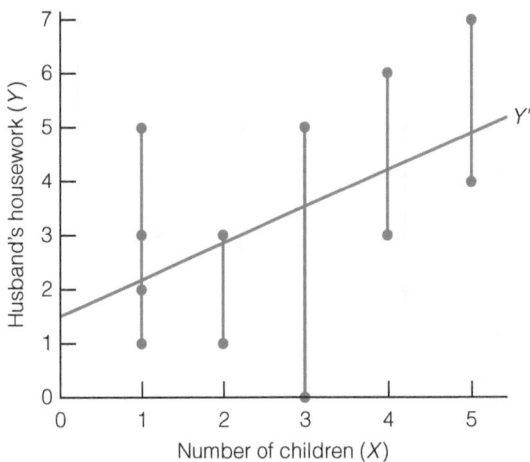

represent the amount of error in predicting Y that remains even after X has been taken into account.

We can precisely define the reduction in error that results from taking X into account. Specifically, two different sums can be found and then compared with the total variation of Y to construct a statistic that will indicate the improvement in prediction.

The first sum, called the **explained variation**, represents the improvement in our ability to predict Y when taking X into account. This sum is found by subtracting \overline{Y} (our predicted Y score without X) from the score predicted by the regression equation (Y', or the Y score predicted with knowledge of X) for each case and then squaring and summing these differences. These operations can be summarized as $\Sigma(Y' - \overline{Y})^2$ and the resultant figure could then be compared with the total variation in Y to ascertain the extent to which our knowledge of X improves our ability to predict Y. Specifically, it can be shown mathematically that

FORMULA 13.5

$$r^2 = \frac{\Sigma(Y' - \overline{Y})^2}{\Sigma(Y - \overline{Y})^2} = \frac{\text{Explained variation}}{\text{Total variation}}$$

Thus, the coefficient of determination, or r^2, is the proportion of the total variation in Y attributable to, or explained by, X. Like other PRE measures, r^2 indicates precisely the extent to which X helps us predict, understand, or explain Y.

Earlier, we referred to the improvement in predicting Y with X as the explained variation. The use of this term suggests that some of the variation in Y will be "unexplained," or not attributable to the influence of X. In fact, the vertical lines in Figure 13.7 represent the **unexplained variation**, or the difference between our best prediction of Y with X and the actual scores. The unexplained variation is thus the scattering of the actual scores around the regression line; it can be found by subtracting the predicted Y score from the actual Y score for each case and then squaring and summing the differences. These operations can be summarized as $\Sigma(Y - Y')^2$, and the resultant sum would measure the amount

of error in predicting Y that remains even after X has been taken into account. The proportion of the total variation in Y unexplained by X can be found by subtracting the value of r^2 from 1.00. Unexplained variation is usually attributed to the influence of some combination of other variables, measurement error, and random chance.

As you may have recognized by this time, the explained and unexplained variations bear a reciprocal relationship with one another. As one increases in value, the other decreases. Also, the stronger the linear relationship between X and Y, the greater the value of the explained variation and the lower the unexplained variation. In the case of a perfect relationship ($r = \pm 1.00$), the unexplained variation would be 0 and r^2 would be 1.00. This would indicate that X explains, or accounts for, all the variation in Y and that we could predict Y from X without error.

On the other hand, when X and Y are not linearly related ($r = 0.00$), the explained variation would be 0 and r^2 would be 0.00. In such a case, we would conclude that X explains none of the variation in Y and does not improve our ability to predict Y.

Relationships intermediate between these two extremes can be interpreted in terms of how much X increases our ability to predict, or explain, Y. For the dual-career families, we calculated an r of 0.50. Squaring this value yields a coefficient of determination of 0.25 ($r^2 = 0.25$), which indicates that number of children (X) explains 25% of the total variation in husband's hours of housework (Y). When predicting the number of hours per week that husbands in such families would devote to housework, we will make 25% fewer errors by basing the predictions on number of children and predicting from the regression line, as opposed to ignoring the X variable and predicting the mean of Y for every case. Also, 75% of the variation in Y is unexplained by X and presumably due to some combination of the influence of other variables, measurement error, and random chance. *(For practice in the interpretation of* r^2, *see any of the problems at the end of this chapter.)*

ONE STEP AT A TIME **Interpreting the Strength and Direction of Pearson's *r***

To Interpret the Strength of Pearson's *r*:

You can use Table 12.12 to describe strength in general terms but the preferred method would be to square the value of r and multiply by 100. This value represents the percentage of variation in Y that is explained by X. In our example problem, the number of children explained 25% ($r^2 = 0.25$) of the variation in husband's housework.

To Interpret the Direction of the Relationship:

If r has a plus sign (or if there is no sign), the relationship is positive and the variables change in the same direction. If r has a minus sign, the relationship is negative and the variables change in opposite directions. The relationship in our example problem was positive: as the number of children increased, the number of hours husbands contributed to housework also increased.

Social Research and Data Analysis 13.1	Computing the Regression Coefficients and Pearson's *r*

Professional sports have become increasingly driven by data and analytics. In the National Basketball Association (NBA), this has led teams to adopt new strategies for winning, such as focusing on the 3-point shot. In the 2017–18 season, NBA players attempted 29 3-pointers per game; that's the most in the history of the league.

In this edition of Social Research and Data Analysis, we will examine a small sample of NBA teams to see if we can identify an association between average 3-point attempts per game and victories during the regular season. A small sample of teams will not be perfectly representative, so we have chosen two teams (one from each of the two conferences) that were in the lowest 25th percentile of wins in their conference, two between the 25th and 50th percentiles, two between the 50th and 75th percentiles, and two between the 75th and 100th percentiles.

The data are presented in the following table. The average number of 3-point attempts is represented by X in the table, and the number of wins is represented by Y. Columns have been added for all sums necessary for the computation of the slope (b) and Pearson's r.

Team	(X)	$(X - \overline{X})$	(Y)	$(Y - \overline{Y})$	$(X - \overline{X})(Y - \overline{Y})$	$(X - \overline{X})^2$	$(Y - \overline{Y})^2$
MW	38.00	5.25	60.00	21.00	110.25	27.56	441.00
BN	36.00	3.25	42.00	3.00	9.75	10.56	9.00
WS	33.00	0.25	32.00	−7.00	−1.75	0.06	49.00
NY	30.00	−2.75	17.00	−22.00	60.50	7.56	484.00
GS	34.00	1.25	57.00	18.00	22.50	1.56	324.00
OC	33.00	0.25	49.00	10.00	2.50	0.06	100.00
MS	29.00	−3.75	36.00	−3.00	11.25	14.06	9.00
PH	29.00	−3.75	19.00	−20.00	75.00	14.06	400.00
Total	262.00		312.00		290.00	75.50	1816.00

$$\overline{X} = 32.75$$
$$\overline{Y} = 39$$

The slope (b) is

$$b = \frac{\Sigma(X - \overline{X})(Y - \overline{Y})}{\Sigma(X - \overline{X})^2}$$

$$b = \frac{290}{75.5}$$

$$b = 3.84$$

A slope of 3.84 means that for every unit change in X (for every additional unit change in the 3-point shot average per game), there was an increase of 3.84 units in Y (wins).

The Y intercept (a) is

$$a = \overline{Y} - b\overline{X}$$
$$a = 39 - (3.84)(32.75)$$
$$a = 39 - 125.76$$
$$a = -86.76$$

The least-squares regression equation is

$$Y = a + bX = -86.76 + (3.84)X$$

The correlation coefficient is

$$r = \frac{\Sigma(X - \overline{X})(Y - \overline{Y})}{\sqrt{[\Sigma(X - \overline{X})^2][\Sigma(Y - \overline{Y})^2]}}$$

$$r = \frac{290}{\sqrt{(75.5)(1816)}}$$

$$r = \frac{290}{\sqrt{137,108}}$$

$$r = \frac{290}{370.28}$$

$$r = 0.78$$

These variables have a strong positive association. The number of wins increases as the average 3-point attempts per game increases. The coefficient of determination, r^2, is $(0.78)^2$, or 0.61. This indicates that 61% of the variance in wins is explained by the average number of 3-point shot attempts.

So we have learned that teams that attempt more 3-pointers tend to win more games, but more research would be needed to see what effects other variables have, such as 3-pointers made or total points scored. Multivariate models are needed for this kind of analysis like those we will discuss in Chapters 14 and 15.

The Correlation Matrix

Social science research projects include many variables, and the data analysis phase of a project often begins with the examination of a **correlation matrix**, a table that shows the relationships between all possible pairs of variables. The correlation matrix gives a quick, easy-to-read overview of the interrelationships in the data set and may suggest strategies or "leads" for further analysis. These tables are commonly included in the professional research literature, and it will be useful to have some experience reading them. The "Using SPSS" section shows how to produce and interpret a correlation matrix, using the *States* database supplied with this text.

USING SPSS

Producing a Correlation Matrix

In this installment of "Using SPSS," we will produce and interpret a correlation matrix, using the *States* database supplied with this text. We will use a procedure called **Correlate**, which is somewhat easier to use than the **Regression** procedure but does not produce the regression coefficients (*a* and *b*).

To begin:

1. Click the SPSS icon on your desktop.
2. Load the *States* database.
 a. Find the **File** command on the far left of the menu bar and then click **File → Open → Data**.
 b. Find the *States* database file supplied with this text. You can download this file from the website for this text if you haven't already.
3. From the main menu bar, click **Analyze**, **Correlate**, and **Bivariate**.
4. In the **Bivariate Correlations** dialog box, find *BirthRate*, *BA*, *FamPoor17*, and *TeenBR* and move them to the **Variables** window.
5. Click **OK** and the output will be printed to the SPSS output screen.
6. The matrix presents three pieces of information for each bivariate relationship: Pearson's *r*, the statistical significance or "Sig. (2-tailed)" of the relationship, and *N* or the number of cases. Since the 50 states are not a random sample and the number of cases is always 50, we can focus only on the correlations. An abbreviated form of the matrix is presented in Table 13.5.

TABLE 13.5 A Correlation Matrix Showing the Relationship Between Four Variables

	1	2	3	4
	Birth Rate	*BA*	*FamPoor17*	*TeenBR*
1. Birth Rate	1.00	−0.31	0.13	0.45
2. BA	−0.31	1.00	−0.72	−0.84
3. FamPoor17	0.13	−0.72	1.00	0.79
4. TeenBR	0.45	−0.84	0.79	1.00

KEY: "Birth Rate" is number of births per 1000 population.
 "BA" is percentage of the population with a college degree or more.
 "FamPoor17" is percentage of families below the poverty line in 2017.
 "TeenBR" is the birth rate per 1000 females age 15–19.

(continued)

USING SPSS

(Continued)

Analyzing a Correlation Matrix

The matrix uses variable names as rows and columns, and the cells in the table show the bivariate correlation (Pearson's *r*) for each combination of variables. Note that the row headings duplicate the column headings. To read the table, begin with birth rate, the first variable in the far left-hand column (column 1) and top row (row 1). Read down column 1 or across row 1 to see the correlations of this variable with all other variables, including the correlation of birth rate with itself (1.00) in the top left-hand cell. To see the relationships between other variables, move from column to column or row to row.

Note that the diagonal from upper left to lower right of the matrix contains the correlation of each variable with itself. Values along this diagonal will always be 1.00 and, since this information is not useful, it could easily be deleted from the output.

Also note that the cells below and to the left of the diagonal duplicate the cells above and to the right. For example, look at the second cell down (row 2) in column 1. This cell displays the correlation between birth rate and percentage of the population that is college educated, as does the cell in the top row (row 1) of column 2. In other words, the cells below and to the left of the diagonal are mirror images of the cells above and to the right of the diagonal. Commonly, research articles in the professional literature will delete the redundant cells in order to make the table more readable.

What does this matrix tell us? Starting at the upper left of the table (column 1), we can see that birth rate has a weak to moderate negative relationship with the level of college education. This means that birth rate decreases as education increases: states with a higher percentage of college-educated populations have lower birth rates. Continuing down column 1, we see that birth rate has a weak positive relationship with poverty (birth rates increase as poverty increases) and a moderate positive relationship with teen births (states with higher birth rates tend to have higher rates of teen births).

To assess the other relationships in the data set, move from column to column and row to row, one variable at a time. For each subsequent variable, there will be one fewer cell of new information. For example, consider education (the percentage of the state's population that has at least a college degree), the variable in column 2 and row 2. We have already noted its weak to moderate negative relationship with birth rate, and we can ignore the correlation of the variable with itself. This leaves only two new relationships, which can be read by moving down column 2 or across row 2. Education has a strong negative relationship with poverty (the higher the percentage of the population that is college educated, the lower the rate of poverty), and a strong negative relationship with teen births (more educated states tend to have fewer teen births).

For poverty, the variable in column 3, there is only one new relationship: a strong positive correlation with teen births (the higher the poverty, the higher the rate of teen births). For column 4 (teen births), there are no new relationships to consider: We have already examined the correlations between this variable and all other variables.

In closing, remember that correlation and causation are two different things. We can see some strong relationships in Table 13.5, and it is tempting to make inferences about causation. For example, it is probably not surprising to see a strong relationship between education and poverty, but the question of causation is not so clear. Do low levels of education cause poverty or does poverty cause low levels of education? Which variable is cause and which is effect? What other information would you need to disentangle the dynamics of these relationships?

| READING STATISTICS | Predicting Knowledge Sharing in Organizations, Part 1 |

Most academic and professional research involves the study of multiple variables and the relationships between them. In many cases, a researcher will begin with a dependent variable that they would like to explain and then build out a list of independent variables with which it might be associated. Ideally, these variables will be measured at the interval-ratio level since that enables the researchers to perform the most sophisticated types of analysis for correlations and regression. But even when they are forced to work with nominal or ordinal-level variables, they can use techniques such as treating an ordinal-level variable as if it was interval level or converting a nominal variable into a "dummy" variable (discussed later in this chapter) with the values of 1 and 0 so that it becomes interval in form. Once all of the variables have been cleaned and modified, the next step is to examine the bivariate associations between them in a correlation matrix. Let's take a look at how this was done in a recent study on knowledge sharing in organizations.

In this study, the authors wanted to explain several factors (or independent variables) that

lead to higher levels of knowledge sharing among a sample of full-time workers in three private telecom organizations in India. The independent variables were organizational spirituality (e.g., the belief that one's work for an organization serves a higher purpose), psychological flourishing (e.g., feelings of competence and purpose in life), and organizational trust (e.g., belief that one will be treated fairly by supervisors and co-workers). In the Pearson correlation matrix (reproduced below), we see that significant positive associations were found between all four variables (knowledge sharing attitude, organizational spirituality, psychological flourishing, and organizational trust). For example, the association between knowledge sharing attitude and organizational spirituality was .393*** indicating an association that is significant at the 99% confidence level (or $p < 0.01$).

These positive associations are interesting but perhaps not surprising. We will revisit this article in the Reading Statistics box in Chapter 14 once we have covered some information about multivariate analyses.

Inter-Correlation, Mean and SD

	Mean	SD	KSA	OS	PF	OT
Knowledge sharing attitude (KSA)	5.152	1.029	1	0.350***	0.393***	0.427***
Organizational spirituality (OS)	4.726	1.079		1	0.348***	0.544***
Psychological flourishing (PF)	5.543	0.854			1	0.429***
Organizational trust (OT)	4.885	0.997				1

***Significance at $p < 0.01$

Source: Khari, Chitra, and Shuchi Sinha. (2018). "Organizational Spirituality and Knowledge Sharing: A Model of Multiple Mediation," *Global Journal of Flexible Systems Management* 19, 337–348.

Correlation, Regression, Level of Measurement, and Dummy Variables

Correlation and regression are powerful and useful techniques, so much so that they are often used to analyze relationships between ordinal-level variables. This practice is generally not a problem, especially with ordinal variables that are "continuous" (see Chapter 12). However, this flexibility does not

STATISTICS	In Everyday Life

Teen Pregnancies

Table 13.5 shows a strong relationship between poverty and teen births: the poorer the state, the higher the rate of births to teens. Besides the association with poverty, the rate of teen pregnancy tends to be strongly correlated with a number of other variables. For example, teens are less likely to get adequate prenatal care and babies born to teen moms are more likely to be underweight and have other health problems. Hence, it is good news that the rates of teen pregnancy have been declining since 1990 and, in 2017, hit an historic low. Furthermore, rates declined not only for the total population but also for black, Hispanic, and white teens. Does this trend reflect increased abstinence, more access to birth control, or some other factors? How could you find out?

Source: U.S. Department of Human and Health Services, 2019. "Trends in Teen Pregnancy and Childbearing." https://www .hhs.gov/ash/oah/adolescent-development/reproductive-health-and-teen-pregnancy/teen-pregnancy-and-childbearing /trends/index.html

extend to nominal-level variables such as religious denomination or gender. The scores of nominal variables are not numbers and have no mathematical quality. We might represent a Protestant with a score of "2" and a Catholic with a score of "1," but the former score is not "twice as much" as the latter. The scores of nominal-level variables are labels, not numbers, and it makes no sense to compute a slope or to discuss positive or negative relationships for these variables.

This is unfortunate because many of the variables that are most important in everyday social life—gender, marital status, race, and ethnicity—are nominal in level of measurement and cannot be included in a regression equation or a correlational analysis.

Fortunately, researchers have developed a way to change the scoring on nominal-level variables by creating **dummy variables**. Dummy variables can be of any level of measurement, including nominal, and have exactly two categories, one coded as 0 and the other coded as 1. Treated this way, nominal-level variables, such as gender (with males coded as 0 and females coded as 1), race (with whites coded as 0 and non-whites as 1), and religious denomination (with Catholics coded as 0 and non-Catholics coded as 1), are commonly included in regression equations.

To illustrate, imagine that we were concerned with the teen pregnancy rate and suspect that southern states have a higher percentage of births to teenagers. In Table 13.5, we saw that "Teen Births" had a strong negative relationship with education and a strong positive relationship with poverty. What is the relationship with region? We can find out by using region as a dummy variable, coding the South as 0 and all non-southern states as 1.

To investigate this relationship, we use a new SPSS procedure called **Regression**. Unlike **Correlate**, the new procedure can produce the regression coefficients (*a* and *b*). See the "Using SPSS" box for an example of using and interpreting a dummy variable in a regression equation.

USING SPSS

Conducting a Regression Analysis with a Dummy Variable

In this installment of "Using SPSS," we will use the *States* database to investigate the relationship between region (South/Non-South) and teen pregnancy.

To begin:

1. Click the SPSS icon on your desktop.
2. Load the *States* database.
 a. Find the **File** command on the far left of the menu bar and then click
 File → Open → Data.
 b. Find the *States* database file supplied with this text.
3. From the main menu bar, click **Analyze**, **Regression**, and **Linear**.
4. On the **Linear Regression** screen, place *TeenBR* in the **Dependent** window and *South* in the **Independent** window. *NOTE: South is coded so that southern states are scored as 1 and non-southern states as 0.*
5. Click **Statistics** and check **Descriptives**. Click **Continue** to return to the **Linear Regression** screen.
6. Click **OK** and results will be printed to the output screen in several separate boxes, only some of which are relevant for our purposes. To conserve space, the output boxes will not be reproduced here.
7. The "Correlations" box displays the correlation between the variables ($r = 0.625$).
8. The "Coefficients" box displays the value of *a* and *b* under "Unstandardized Coefficients." Look in the column labeled B. The value in the first row is *a* and the entry in the second row is *b*.

The output shows a strong positive relationship ($r = 0.625$) between the region dummy variable and teen births. The r^2 is 0.39, which means that region explains 39% of the variation in teen births.

Turning to the regression equation, we can use the values for *a* and *b* in the "Coefficients" output box to write the regression equation:

$$Y = a + bX$$
$$Y = 16.769 + (8.72)X$$

Teen birth rate is the dependent (or *Y*) variable and region (*X*) is the independent variable. The regression line crosses the vertical axis of the scatterplot at the point where $Y = 16.769$. The value for the slope ($b = 8.72$) indicates a positive relationship: As region "increases" (or moves toward the higher score associated with southern states), the percentage of teen births tends to increase. In other words, southern states averaged higher rates of births to teens. Note that the *sign* of the slope (*b*) would have been negative had we reversed the coding scheme and labeled "South" as 0 and non-South as 1, but the *value* of *b* would have stayed exactly the same. The coding scheme for dummy variables is arbitrary and, as with ordinal-level variables, the researcher needs to be clear about what the values of a dummy variable indicate. (For experience in working with dummy variables, see problem 13.8.)

Testing Pearson's *r* for Significance

When the relationship measured by Pearson's *r* is based on data from a random sample, it will usually be necessary to test *r* for its statistical significance. That is, we will need to know whether a relationship between the variables can be assumed to exist in the population from which the sample was drawn. To illustrate

Social Research and Data Analysis 13.2 **Fertility and Women's Education**

What happens to the birth rate when women become more educated? Do more-educated women choose adult roles other than wife and mother and reduce their number of children? We can get an idea about these relationships with the International Population database (*International.sav*) supplied with this text.

The variables are:

- Total fertility rate (*TotalFertRate*), the average number of children born to a woman during her lifetime, and
- Gender equality in education (*GESecondary*), the ratio of girls to boys at the secondary level of schooling.

Start SPSS, load the *International* database, and use the **Correlate** command (see "Using SPSS: Producing a Correlation Matrix"). Enter the variable names in the **Correlate** window and click **OK**.

The moderate to strong negative relationship ($r = -0.59$) between the variables shows that nations with more-educated women have lower fertility rates. This is consistent with the idea that more-educated women make different choices about family life and work and tend to have fewer children.

this test, the dual–wage-earner family sample introduced at the beginning of this chapter will be used. As you recall, the correlation between number of children and husband's hours of housework was moderate and positive ($r = 0.50$).

The null hypothesis will state that there is no linear association between the two variables in the population from which the sample was drawn. The population parameter is symbolized as ρ (rho), and the appropriate sampling distribution is the t distribution.

To conduct this test, we need to make a number of assumptions in step 1. Most are familiar, but several are new. First, we must assume that both variables are normal in distribution (**bivariate normal distributions**). Second, we must assume that the relationship between the two variables is roughly linear in form.

The third assumption involves a new concept: **homoscedasticity**. Basically, a homoscedastic relationship is one in which the variance of the Y scores is uniform for all values of X: if the Y scores are evenly spread above and below the regression line for the entire length of the line, the relationship is homoscedastic.

A visual inspection of the scatterplot will usually be sufficient to appraise the extent to which the relationship conforms to the assumptions of linearity and homoscedasticity. As a rule of thumb, if the data points fall in a roughly symmetrical, cigar-shaped, linear pattern, then it is appropriate to proceed with this test of significance. Any significant evidence of nonlinearity or marked departures from homoscedasticity may indicate the need for an alternative measure of association and thus a different test of significance.

Step 1. Making Assumptions and Meeting Test Requirements.

Model: Random sampling
Level of measurement is interval-ratio
Bivariate normal distributions
Linear relationship
Homoscedasticity
Sampling distribution is normal

Step 2. Stating the Null Hypothesis.

$$H_0: \rho = 0.00$$

$$H_1: \rho \neq 0.00$$

Step 3. Selecting the Sampling Distribution and Establishing the Critical Region. With the null hypothesis of "no relationship" in the population, the sampling distribution of all possible sample rs is approximated by the t distribution. Degrees of freedom are equal to $(N - 2)$.

$$\text{Sampling distribution} = t \text{ distribution}$$
$$\text{Alpha} = 0.05$$
$$\text{Degrees of freedom} = N - 2 = 12 - 2 = 10$$
$$t(\text{critical}) = \pm 2.228$$

Step 4. Computing the Test Statistic. The equation for computing the test statistic is given in Formula 13.6.

FORMULA 13.6
$$t(\text{obtained}) = r\sqrt{\frac{N - 2}{1 - r^2}}$$

Substituting the values into the formula, we would have:

$$t(\text{obtained}) = r\sqrt{\frac{N - 2}{1 - r^2}}$$

$$t(\text{obtained}) = (0.50)\sqrt{\frac{12 - 2}{1 - (0.50)^2}}$$

$$t(\text{obtained}) = (0.50)\sqrt{\frac{10}{1 - 0.25}}$$

$$t(\text{obtained}) = (0.50)\sqrt{\frac{10}{0.75}}$$

$$t(\text{obtained}) = (0.50)\sqrt{13.33}$$

$$t(\text{obtained}) = (0.50)(3.65)$$

$$t(\text{obtained}) = 1.83$$

Step 5. Making a Decision and Interpreting the Results of the Test. The test statistic does not fall in the critical region as marked by $t(\text{critical})$, so we fail to reject the null hypothesis. Even though the variables are substantially related in the sample, we cannot conclude that they are also related in the population. The test indicates that the sample value of $r = 0.50$ could have occurred by chance alone if the null hypothesis is true and the variables are unrelated in the population. *(For practice in conducting and interpreting tests of significance with Pearson's* r, *see problems 13.1, 13.2, and 13.4.)*

STATISTICS In Everyday Life

Correlation, Causation, and Cancer

Measures of association like Pearson's r played an important role in building the case for a causal relationship between smoking tobacco and cancer. Many studies have demonstrated the association between smoking and cancer, and the best of these followed large groups of individuals over long periods of time. This design allowed researchers to refute the objection that "correlation is not causation." The studies started with a large number of smokers in good health. If the smokers developed cancer later in the study, cancer must be the dependent variable. That is, given the time order, smoking could cause cancer but the reverse could not be true: getting cancer could not have caused people to start smoking.

Researchers were able to refute other objections and counterarguments with their growing stock of evidence and statistics. The effect of these studies has been dramatic: As recently as the 1950s, about half of all men smoked and smoking was equated with sophistication and mature adulthood, not illness and disease. Now, the percentage of adult smokers has fallen to a record low of 14%.

Source: Centers for Disease Control. 2018. "CDC Says Smoking Rates Fall to Record Low in the US." https://www.cnbc.com/2018/11/08/cdc-says-smoking-rates-fall-to-record-low-in-us.html

Interpreting Statistics: The Correlates of Crime

What causes crime? Sociologists have been researching this question since the discipline was first founded, and they have conducted extensive research and have expended an enormous amount of energy on theory construction. While we cannot contribute to this voluminous body of work in a text devoted to statistical analysis, we can investigate some of the relationships and correlations that are of continuing interest to criminologists.

One prominent school of criminological thought argues that crime is related to poverty. A central proposition of this approach links crime rates to disadvantage, impoverishment, and unemployment. In this installment of "Interpreting Statistics," we will look at some empirical relationships between crime and poverty using the 50 states as our sample.

We will measure crime with the homicide rate (the number of homicides per 100,000 population). The homicide rate is one of the more trustworthy measures of crime. The rates for many other types of crimes, such as rape and assault, are gross underestimates because victims of these crimes are less likely to report the incident to the police and, thus, not all incidents are included in the official counts of criminal activity. Homicides are much more likely to come to the attention of the authorities, if for no other reason than that there is a dead body that must be dealt with. For the independent variable, poverty, we will use the percentage of the families in the state that are living below the poverty line.

The basic descriptive statistics for the variables are reported in Table 13.6. Homicide rates in 2018 ranged from a high of 12.4 murders per 100,000 population in Louisiana to a low of 1.0 murder per 100,000 population in New Hampshire. Poverty rates ranged from a high of 15.0% in three states to a low of 5.0%, also in New Hampshire.

TABLE 13.6 Homicide Rates and Family Poverty, Basic Statistics

Variable	Mean	Standard Deviation	Range
Homicide rate	5.01	2.63	11.4
Family poverty	9.09	2.53	10.0

The first step in assessing an association between interval-ratio–level variables is to produce a scatterplot. Figure 13.8 plots the homicide rate on the vertical, or Y, axis and the poverty rate along the horizontal, or X, axis.

What can we say about this relationship? The regression line is not horizontal, so there is a relationship between these variables. The dots (states) are well scattered around the regression line, and we can see that this is a moderately strong relationship. The regression line slopes up from left to right, so the relationship is positive: States that have higher poverty rates tend to have higher homicide rates.

The next step in assessing this relationship is to specify the regression line. We'll skip the mechanics of computation here and simply report the values for the regression coefficients (a and b).

$$Y = a + bX$$
$$Y = -0.90 + (0.65)X$$

The regression line will cross the Y axis at the point where $Y = -0.90$, and every unit increase in X (percentage poor) produces an increase of 0.65 in Y (the homicide rate).

FIGURE 13.8 Homicide Rate by Family Poverty

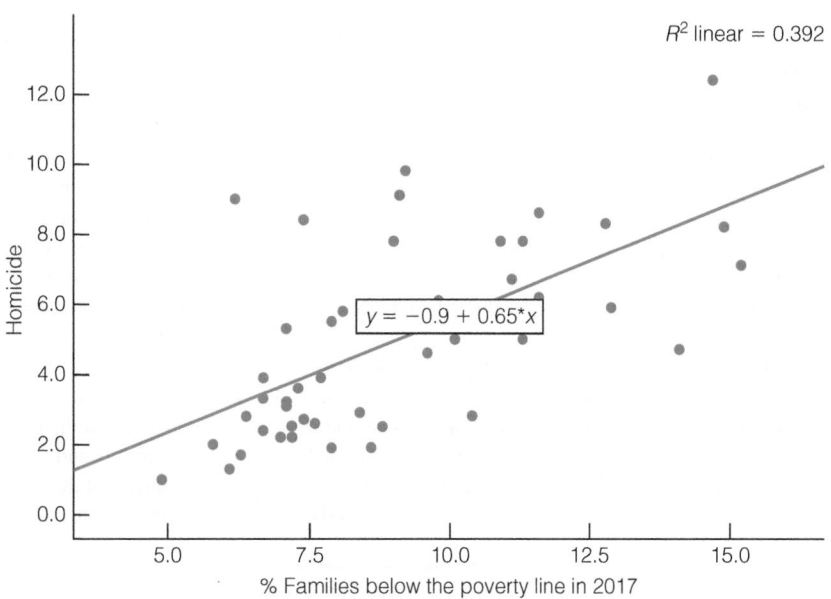

Next, we will calculate and interpret Pearson's r. Again, we will skip the details of the computation and report that $r = 0.63$, which reinforces the impression that there is a moderately strong relationship between homicide and poverty. The coefficient of determination (r^2) is 0.39, which means that poverty rate, by itself, explains 39% of the variation in the homicide rate.

In sum, the linear regression equation, the correlation coefficient, and the coefficient of determination suggest that there is a moderately strong, positive relationship between poverty and crime. The amount of unexplained variation (61%) suggests that many other variables besides poverty have an important influence on the homicide rate.

We should note two limitations of this simple test. First, correlation is not the same thing as causation. Just because two variables are correlated does not mean that they have a causal relationship. These results could be taken as support for the proposition that higher rates of poverty lead to higher rates of violence, but the mere existence of a relationship—even a strong relationship in the direction predicted by the theory—does not prove that one variable causes the other.

Second, consider a related point: The crime of homicide is, by definition, an individual act of behavior, but the data we used in this test were collected from states. Just because there is an association between the variables at the macro (state) level does not necessarily mean that the variables are related in the same way at the micro (individual) level.[2] All we know from our analysis is that states with higher rates of poverty tend to have higher rates of homicide. Our theory would lead us to assume that it is the victims of inequality (poor people) who are the offenders, but this conclusion is not proven by this analysis. In fact, it may be the wealthier residents of the poorer states who are committing the murders. We would need much more information—on both the micro and macro levels—before we could come to any firm conclusions about the theory that poverty and crime are related.

SUMMARY

This summary is based on the example used throughout the chapter.

1. We began with a question: Is the number of children in dual–wage-earner families related to the number of hours per week husbands devote to housework? We presented the observations in a scatterplot (Figure 13.1), and our visual impression was that the variables were associated in a positive direction. The pattern formed by the observation points in the scatterplot could be approximated with a straight line; thus, the relationship was roughly linear.

2. The least-squares regression line ($Y = a + bX$) is the single straight line that best fits the data because it minimizes the variation in Y. In our example, we found a slope (b) of 0.69, which indicates that each additional child (a unit change in X) is accompanied by an increase of 0.69 hour of housework per week for the husbands.

We also predicted, based on this formula, that in a dual–wage-earner family with six children ($X = 6$), husbands would contribute 5.63 hours of housework a week ($Y = 5.63$ for $X = 6$).

3. Pearson's r is a statistic that measures the overall linear association between X and Y. Our impression from the scatterplot of a substantial positive relationship was confirmed by the computed r of 0.50. We also saw that this relationship yields an r^2 of 0.25, which indicates that 25% of the total variation in Y (husband's housework) is accounted for, or explained, by X (number of children).

4. Assuming that the 12 families were a random sample, we tested the Pearson's r for its statistical significance and found that, at the 0.05 level, we could not conclude that these two variables were also related in the population.

[2]This problem is called the *ecological fallacy*.

5. We acquired a great deal of information about this bivariate relationship. We know the strength and direction of the relationship and have also identified the regression line that best summarizes the effect of X on Y. We know the amount of change we can expect in Y for a unit change in X. In short, we have a greater volume of more precise information about this association between interval-ratio variables than we ever did about associations between ordinal or nominal variables. This is possible because the data generated by interval-ratio measurement are more precise and flexible than those produced by ordinal or nominal measurement techniques.

SUMMARY OF FORMULAS

Least-squares regression line:

13.1 $$Y = a + bX$$

Slope (b):

13.2. $$b = \frac{\Sigma(X - \overline{X})(Y - \overline{Y})}{\Sigma(X - \overline{X})^2}$$

Y intercept (a):

13.3 $$a = \overline{Y} - b\overline{X}$$

Pearson's r:

13.4 $$r = \frac{\Sigma(X - \overline{X})(Y - \overline{Y})}{\sqrt{[\Sigma(X - \overline{X})^2][\Sigma(Y - \overline{Y})^2]}}$$

Coefficient of determination:

13.5 $$r^2 = \frac{\Sigma(Y' - \overline{Y})^2}{\Sigma(Y - \overline{Y})^2} = \frac{\text{Explained variation}}{\text{Total variation}}$$

Test statistic:

13.6 $$t(\text{obtained}) = r\sqrt{\frac{N - 2}{1 - r^2}}$$

GLOSSARY

Bivariate normal distributions. The model assumption in the test of significance for Pearson's r that both variables are normally distributed.

Coefficient of determination (r^2). The proportion of all variation in Y that is explained by X. Found by squaring the value of Pearson's r.

Conditional means of Y. The mean of all scores on Y for each value of X.

Correlation matrix. A table that shows the correlation coefficients between all possible pairs of variables.

Dummy variable. A nominal-level variable dichotomized so that it can be used in regression analysis. A dummy variable has two scores, one coded as 0 and the other as 1.

Explained variation. The proportion of all variation in Y that is attributed to the effect of X. Equal to $\Sigma(Y' - \overline{Y})^2$.

Homoscedasticity. The model assumption in the test of significance for Pearson's r that the variance of the Y scores is uniform across all values of X.

Linear relationship. A relationship between two variables in which the observation points (dots) in the scatterplot can be approximated with a straight line.

Pearson's r. A measure of association for variables that have been measured at the interval-ratio level.

Regression line. The single best-fitting straight line that summarizes the relationship between two variables. Regression lines are fitted to the data points by the least-squares criterion, whereby the line touches all conditional means of Y or comes as close to doing so as possible.

Scatterplot. Graphic display device that depicts the relationship between two variables.

Slope (b). The amount of change in one variable per unit change in the other; b is the symbol for the slope of a regression line.

Total variation. The spread of the Y scores around the mean of Y. Equal to $\Sigma(Y - \overline{Y})^2$.

Unexplained variation. The proportion of the total variation in Y that is not accounted for by X. Equal to $\Sigma(Y - Y')^2$.

Y intercept (a). The point where the regression line crosses the Y axis.

Y'. Symbol for predicted score on Y.

PROBLEMS

The data sets in the first five problems are unrealistically small to allow you to practice the mathematical routines required to conduct correlational and regression analysis. Most problems use SPSS.

13.1 [PS] Why does voter turnout vary from election to election? For municipal elections in five different cities, information has been gathered on the percentage of citizens who voted (the dependent variable) and three different independent variables: unemployment rate, average years of education for the city, and the percentage of all political ads that used "negative campaigning" (personal attacks, negative portrayals of the opponent's record, etc.). For each relationship between turnout (Y) and the independent variables:

 a. Compute the slope (b) and find the Y intercept (a). *(HINT: Remember to compute b before computing a. A computing table such as Table 13.3 is highly recommended.)*

 b. State the least-squares regression line and predict the voter turnout for a city in which the unemployment rate was 12%, a city in which the average years of schooling was 11, and an election in which 90% of the ads were negative.

 c. Compute r and r^2. *(HINT: A computing table such as Table 13.3 is highly recommended. If you constructed one for computing b, you already have most of the quantities you will need to solve for r.)*

 d. Assume these cities are a random sample and conduct a test of significance for each relationship.

 e. Describe the strength and direction of each relationship in a sentence or two. Which (if any) relationships were significant? Which of the independent variables had the strongest effect on turnout?

City	Turnout (Y)	Unemployment Rate (X_1)	Average Years of School (X_2)	Percentage of Negative Ads (X_3)
A	55	5	11.9	60
B	60	8	12.1	63
C	65	9	12.7	55
D	68	9	12.8	53
E	70	10	13.0	48

13.2 [SOC] Occupational prestige scores for a sample of fathers and their oldest son and oldest daughter are presented below.

Family	Father's Prestige	Son's Prestige	Daughter's Prestige
A	80	85	82
B	78	80	77
C	75	70	68
D	70	75	77
E	69	72	60
F	66	60	52
G	64	48	48
H	52	55	57

Analyze the relationship between father's and son's prestige and the relationship between father's and daughter's prestige. For each relationship:

 a. Compute the slope (b) and find the Y intercept (a).

 b. State the least-squares regression line. What prestige score would you predict for a son whose father had a prestige score of 72? What prestige score would you predict for a daughter whose father had a prestige score of 72?

 c. Compute r and r^2.

 d. Assume these families are a random sample and conduct a test of significance for both relationships.

 e. Describe the strength and direction of the relationships in a sentence or two. Does the occupational prestige of the father have an impact on his children? Does it have the same impact for daughters as it does for sons?

13.3 [GFR] The residents of a housing development for senior citizens have completed a survey in which they indicated how physically active they are (on a scale that ranges from 0 for "inactive" to 20 for "extremely active") and how many visitors they receive each week. Are these two variables related for the 10 cases reported here? Compute r and r^2. Find the least-squares regression line. What would be the predicted number of visitors for a person whose level of activity was 5? How about a person who scored 18 on level of activity?

Case	Level of Activity	Number of Visitors
A	10	14
B	11	12
C	12	10
D	10	9
E	15	8
F	9	7

(continued)

(*Continued*)

Case	Level of Activity	Number of Visitors
G	7	10
H	3	15
I	10	12
J	9	2

13.4 MR In Dave Eggers' novel *The Circle,* a mysterious technology company has developed a strangely data-driven work environment in which employees are assigned a performance score that seems to be the product of factors in their social activities and environment, such as the extent of their engagement in work community and the number of times they post on the company's social media site. Using the fictional data below, compute r and r^2 for each combination of two variables from this novel and test the correlations for their significance. Write a paragraph interpreting the relationships between these variables, and what they might mean for the company.

Employee	Community Engagement	Number of Social Media Posts	Performance Score
A	33	9	36
B	78	15	60
C	62	18	64
D	50	10	56
E	45	12	55
F	56	8	52
G	85	20	25
H	25	0	42
I	13	5	89
J	33	9	88

13.5 SP A health researcher is interested in the association between peer pressure and electronic cigarette usage among adolescents. In her survey of adolescent electronic cigarette users, she has found the data below on the percentage of friends that are vaping and the number of electronic cigarette pods consumed per month. Determine the size and the direction of the correlation.

Person Number	Percent of Vaping Friends	Number of Pods per Month
1	50	10
2	65	10
3	63	12
4	45	12

(*continued*)

(*Continued*)

Person Number	Percent of Vaping Friends	Number of Pods per Month
5	50	8
6	21	10
7	20	8
8	6	9

Statistical Analysis Using SPSS

13.6 SOC Load the *International* data set and get scatterplots for the independent variable *GNIcap* (Gross National Income per capita, a measure of affluence) and each of these dependent variables: *BirthRate* (birth rate or number of births per 1000 population), *DeathRate* (death rate or number of deaths per 1000 population), *LifeExpMales* (life expectancy or the average number of years of life a newborn male infant can expect to live), and *LifeExpFem* (the average number of years of life a newborn female infant can expect to live).

- Click **Graphs, Legacy Dialog,** and **Scatter/Dot.**
- On the **Scatter/Dot** window, click **Simple Scatter** and then click **Define.**
- On the **Simple Scatterplot** window
 - Find *BirthRate* in the list of variables and move the variable name to the **Y Axis:** box.
 - Find *GNIcap* and click the arrow to move the variable name to the **X Axis:** box.
 - Click **OK** and the scatterplot will be sent to the SPSS output window.
- To add the regression line to the graph:
 - Click anywhere on the graph and the **Chart Editor** window will appear.
 - Click the **Elements** command at the top of the window and, in the drop-down menu, click **Fit Line at Total.**
 - Close the **Chart Editor** window to return to the output window of SPSS.
- Repeat the **Graphs, Legacy Dialog,** and **Scatter/Dot** commands and, when you get to the **Simple Scatterplot** window, substitute *DeathRate* for *BirthRate.* Repeat the instructions above to get a scatterplot for *DeathRate* and *GNIcap.*
- Repeat the **Graphs, Legacy Dialog,** and **Scatter/Dot** commands once more and substitute *LifeExpMales* for *DeathRate* in the **Simple Scatterplot** window. Repeat the instructions above to get a scatterplot for *LifeExpMales* and *GNIcap.*
- Repeat the **Graphs, Legacy Dialog,** and **Scatter/Dot** commands once more and substitute

LifeExpFem for *DeathRate* in the **Simple Scatterplot** window. Repeat the instructions above to get a scatterplot for *LifeExpFem* and *GNIcap*.

a. Describe each of these graphs in terms of the direction and strength of the relationship they display. Use the regression line and the spread of the dots around the line to make these determinations.

b. Do any of these scatterplots violate the assumption of linearity? How?

13.7 [SOC] Load the *GSS2018* data set and use the **Correlate** command to get correlations between *age* (the independent or X variable) and four dependent (or Y) variables: church attendance (*attend*), number of children (*childs*), hours spent on the Internet per week (*wwwhr*), and hours of TV watching each day (*tvhours*).

- Click the SPSS icon on your desktop.
- Load the *GSS2018* database.
- From the main menu bar, click **Analyze, Correlate**, and **Bivariate**.
- In the **Bivariate Correlations** dialog box, find *age, attend, childs, tvhours*, and *wwwhr* in the variable list on the left, and move them to the **Variables** window on the right.
- Click **OK** and the output will be printed to the SPSS output screen.

Focus on the relationships with *age*, and record the value of *r* for each relationship below. Calculate r^2 and record those values as well.

	attend		childs		wwwhr		tvhours	
	r	r^2	r	r^2	r	r^2	r	r^2
age								

a. Write a summary of these relationships in which you note the strength and direction of each relationship with age. What percentage of the variance in each dependent variable was explained by age?

b. Which relationships were statistically significant at the 0.05 level? To find out, look at the second line in each cell of the correlation matrix (**Sig. 2-tailed**). Values below 0.05 are significant.

13.8 [SOC] Load the *States* data set and use the **Correlate** command to get correlations between *South* (South/Non-South dummy variable), *BA* (percentage of the state's population with at least a college degree), *TrafDths17* (the rate of traffic fatalities in 2017),

Internet (the percentage of households using the Internet), and *ChildPoor* (the percentage of children living in poverty).

- Click the SPSS icon on your desktop.
- Load the *States* database.
- From the main menu bar, click **Analyze, Correlate**, and **Bivariate**.
- In the **Bivariate Correlations** dialog box, find *South, BA, TrafDths17, Internet*, and *ChildPoor* in the variable list on the left and move them to the **Variables** window on the right.
- Click **OK** and the output will be printed to the SPSS output screen.

Take region (*South*) and education (*BA*) as independent variables and the other three variables as dependent. Record the value of *r* for each relationship below. Calculate r^2 and record those values as well.

	TrafDths17		Internet		ChildPoor	
	r	r^2	r	r^2	r	r^2
South						
BA						

a. Write a summary of these relationships in which you note the strength and direction of each relationship. What percentage of the variance in each dependent variable was explained by region? By education?

b. Do you think these relationships are causal? How? Do one of the following:
- Pick one of the six relationships in the table and explain why you think it is causal. Or,
- Pick one relationship and explain why you think it is a case of correlation but not causation.

13.9 [SOC/CJ] Use the *States* data set and the **Correlate** command to explore some of the correlates of crime. Use two measures of crime as dependent variables: the homicide rate (*Homicide*) and the rate of auto theft (*CarTheft*). Examine the relationships of these variables with three possible independent variables: population growth rate (*PopGrow*), expenditures on police (*CopSpend*), and unemployment rate (*Unemp*).

- Click the SPSS icon on your desktop.
- Load the *States* database.
- From the main menu bar, click **Analyze, Correlate**, and **Bivariate**.
- In the **Bivariate Correlations** dialog box, find *Homicide, CarTheft, PopGrow, CopSpend,*

and *Unemp*, and move them to the **Variables** window.

• Click **OK** and the output will be printed to the SPSS output screen.

Record the value of *r* for each relationship. Calculate r^2 and record those values in the table as well.

	Homicide		CarTheft	
	r	r^2	*r*	r^2
PopGrow				
CopSpend				
Unemp				

a. Write a summary of these relationships in which you note the strength and direction of each relationship. What percentage of the variance in each measure of crime variable was explained by population growth rate? By population density? By unemployment?

b. Which of these relationships seems causal? Why?

13.10 SOC/CJ Follow up on problem 13.9 by finding the regression coefficients for the relationship between homicide rate and unemployment. Use the *States* data set and the **Regression** command.

• Load the *States* database.
• From the main menu bar, click **Analyze**, **Regression**, and **Linear**.
• On the **Linear Regression** screen, place *Homicide* in the **Dependent** window and *Unemp* in the **Independent** window.
• Click **OK** and results will be printed to the output screen.
• The **Coefficients** box displays the values of *a* and *b* under **Unstandardized Coefficients**. Look in the column labeled **B**. The value in the first row is *a* and the entry in the second row is *b*.

a. State the regression equation with the values for *a* and *b*.

b. Use the equation to predict homicide rates for states with unemployment rates of 2% ($X = 2$), 5% ($X = 5$), 10% ($X = 10$), and 15% ($X = 15$). What happens to homicide rate as unemployment increases? Does this mean that unemployment *causes* murder? Why or why not?

13.11 SOC Use the *International* data set and the **Correlate** command to explore some of the correlates of religiosity around the globe. Use

Religiosity (the percentage of the population that says the religion is very important in everyday life), *GNIcap* (Gross National Income per capita—this is a measure of affluence), and *Urban* (percentage of population living in cities).

• Click the SPSS icon on your desktop.
• Load the *International* database.
• From the main menu bar, click **Analyze**, **Correlate**, and **Bivariate**.
• In the **Bivariate Correlations** dialog box, find *Religiosity*, *GNIcap*, and *Urban*, and move them to the **Variables** window.
• Click **OK** and the output will be printed to the SPSS output screen.

Record the value of *r* for each relationship. Calculate r^2 and record those values in the table as well.

	GNIcap		Urban	
	r	r^2	*r*	r^2
Religiosity				

Write a summary of these relationships in which you note the strength and direction of each relationship. What percentage of the variance in *Religiosity* was explained by *GNIcap*? By *Urban*?

13.12 SOC Follow up on problem 13.11 by finding the regression coefficients for the relationship between religiosity and urbanization. Use the *International* data set and the **Regression** command.

• Load the *International* database.
• From the main menu bar, click **Analyze**, **Regression**, and **Linear**.
• On the **Linear Regression** screen, place *Religiosity* in the **Dependent** window and *Urban* in the **Independent** window.
• Click **OK** and results will be printed to the output screen.
• The **Coefficients** box displays the values of *a* and *b* under **Unstandardized Coefficients**. Look in the column labeled **B**. The value in the first row is *a* and the entry in the second row is *b*.

a. State the regression equation with the values for *a* and *b*.

b. Use the equation to predict religiosity for nations that are 25% urbanized ($X = 25$), 50% urbanized ($X = 50$), 75% urbanized ($X = 75$), and 99% urbanized ($X = 99$). What happens to cell phone use as urbanization increases?

Who Watches TV? Who Succeeds in Life?

Two projects are presented to help you apply the statistical skills developed in this chapter. In the first, you will analyze the correlates of time spent watching TV. You will select four independent variables and assess their impact on *tvhours* (number of hours per day spent watching TV). In the second project, you will choose either *rincome* or *perstg10* as your dependent variable. Both variables measure social class standing and you will select independent variables you believe might be associated with an individual's level of success.

PROJECT 1: Who Watches TV?

In this exercise, your dependent variable will be *tvhours*. Scores on this variable range from 0 to a high of 24 hours (all day!), with most cases clustered around 1 to 4 hours of TV watching per day.

Step 1: Choosing Independent Variables

Select four variables from the 2018 General Social Survey (GSS) that you think might be important causes of *tvhours*. Your independent variables *cannot* be nominal in level of measurement unless you recode the variable into a dummy variable (see below). You may use any interval-ratio or ordinal-level variable.

Dummy Variables. To include a dummy variable in your analysis, recode the variable so that it has only two values: 0 and 1. Some possibilities include recoding *sex* so that males = 0 and females = 1, *racecen1* so that whites = 0 and nonwhites = 1, or *relig* so that Protestants = 0 and non-Protestants = 1. See Chapter 10 for instructions on recoding.

Once you have selected your variables, list them in the table below and describe exactly what they measure.

SPSS Variable Name	What Exactly Does This Variable Measure?

Step 2: Stating Hypotheses

State hypotheses about the relationships you expect to find between your independent variables and *tvhours*. State these hypotheses in terms of the direction of the relationship you expect to find. For example, you might hypothesize that time spent watching TV will increase as age increases.

SPSS Variable Name	Hypothesis

Step 3: Running Bivariate Correlations

Click **Analyze** → **Bivariate** → **Correlate** and place all variables in the **Variables:** box. Click **OK** to get your results.

Step 4: Recording Results

Use the table below to summarize your results. Enter the r for each independent variable in each cell. Calculate r^2 and record those values as well. As you read the correlation matrix, ignore correlations of variables with themselves and any redundant information.

	Independent Variables							
	1._____		2._____		3._____		4._____	
	r	r^2	r	r^2	r	r^2	r	r^2
tvhours								

Step 5: Analyzing and Interpreting Results

Write a short summary of results for each independent variable. Your summary needs to identify the variables being tested and the strength and direction of the relationship. It is probably best to characterize the relationship in general terms and then cite the statistical values in parentheses. Be sure to note whether your hypotheses were supported. *Be careful when interpreting direction* and refer back to the coding scheme to make sure you understand the relationship.

PROJECT 2: Who Succeeds?

For this exercise, choose either income (*rincome*) or the respondent's occupational prestige (*prestg10*) as your dependent variable. Most Americans would regard these variables as measures of success in life. What are the correlates and antecedents of affluence and prestige?

Step 1: Choosing Independent Variables

Select four variables that you think might be important causes of your dependent variable. An obvious choice is education (use *educ*, the interval-ratio version, not *degree*, the ordinal version). Remember that independent variables *cannot* be nominal in level of measurement unless you recode the variable into a dummy variable (see the next page). You may use any interval-ratio or ordinal-level variable.

Dummy Variables. To include a dummy variable in your analysis, recode the variable so that it has only two values: 0 and 1. Some possibilities include recoding *sex* so that males = 0 and females = 1, *racecen1* so that whites = 0 and nonwhites = 1, or *relig* so that Protestants = 0 and non-Protestants = 1. See Chapter 10 for instructions on recoding.

Once you have selected your variables, list them in the table below and describe exactly what they measure.

SPSS Variable Name	What Exactly Does This Variable Measure?

Step 2: Stating Hypotheses

State hypotheses about the relationships you expect to find between your independent variables and the dependent variable you selected. State these hypotheses in terms of the direction of the relationship you expect to find. For example, you might hypothesize that income will increase as age increases.

SPSS Variable Name	Hypothesis

Step 3: Running Bivariate Correlations

Click **Analyze** → **Bivariate** → **Correlate** and place all variables in the **Variables:** box. Click **OK** to get your results.

Step 4: Recording Results

Use the table below to summarize your results. Enter the *r* for each independent variable in each cell. Calculate r^2 and record those values as well. As you read the correlation matrix, ignore correlations of variables with themselves and any redundant information.

	Independent Variables							
	1._____		2._____		3._____		4._____	
	r	r^2	r	r^2	r	r^2	r	r^2
rincome or *prestg10*								

Step 5: Analyzing and Interpreting Results

Write a short summary of results for each independent variable. Your summary needs to identify the variables being tested and the strength and direction of the relationship. It is probably best to characterize the relationship in general terms and then cite the statistical values in parentheses. Be sure to note whether your hypotheses were supported. *Be careful when interpreting direction*, and refer back to the coding scheme to make sure you understand the relationship.

PART III CUMULATIVE EXERCISES

For each situation, compute and interpret the appropriate measure of association. Also, compute and interpret percentages for bivariate tables. Describe relationships in terms of the strength and pattern or direction.

1. For 10 cities, data have been gathered on total crime rate (major felonies per 100,000 population) and the percentage of people who are new immigrants (arrived in the United States within the past 5 years). Are the variables related?

City	Total Crime Rate	Percentage Immigrants
A	1500	10
B	1200	18
C	2000	9
D	1700	11
E	1600	15
F	1000	20
G	1700	9
H	1300	22
I	900	10
J	700	15

2. There is some evidence that people's involvement in their communities (membership in voluntary organizations, participation in local politics, and so forth) has been declining, and television has been cited as the cause. Do the following data support the idea that TV is responsible for the decline?

Hours of Community Service	Television Viewing			TOTALS
	Low	*Moderate*	*High*	
Low	5	10	18	33
Moderate	10	12	10	32
High	15	8	7	30
TOTALS	30	30	35	95

3. A national magazine has rated 10 states in terms of "quality of life" (a scale that includes health care, availability of leisure facilities, unemployment rate, and a number of other variables) and the quality of the system of higher education. Both scales range from 1 (low) to 20 (high). Is there a correlation between these two variables?

State	Quality of Life	Quality of Higher Education
A	10	10
B	12	13
C	15	18
D	18	20
E	10	15
F	9	11
G	11	12
H	8	6
I	13	9
J	6	8

4. Racial intermarriages are increasing in the United States, and the number of mixed-race people is growing. The following table shows the relationship between race of parents and the racial category, if any, with which a sample of people of mixed race identifies. Is there a relationship between these variables?

Do You Consider Yourself to Be	Parents' Race			
	Black/White	White/Asian	Black/Asian	TOTALS
Black	2	0	3	5
White	8	4	0	12
Asian	0	3	3	6
None of the above	10	8	4	22
TOTALS	20	15	10	45

Statistical Analysis Using SPSS

Following are a number of research questions that can be answered by applying the techniques presented in Part III. For each situation, choose the most appropriate measure of association for the level of measurement of the variables and state your conclusions. The questions are presented in random order.

5. **Using the General Social Survey Data Set** (*GSS2018.sav*)
 a. Are scores on "frequency of sex during the last year" (*sexfreq*) or total number of sex partners over the past 5 years (*partnrs5*) associated with income (*rincome*) or *age*? Assume that all variables are interval-ratio in level of measurement.
 b. Is *fear* associated with educational level (*degree*) or marital status (*marital*)?
 c. Is support for spanking (*spanking*) associated with religious denomination or *sex*? To measure religious denomination, use *religR* (recoded *relig*). The former variable (*religR*) has the "usual" five categories for measuring religious denomination in North America. The latter variable (*relig*) has 13 categories, including Muslim and Buddhist.
 d. In the Part II Cumulative Exercises, you tested the relationship between *Spkmslm* and *degree* for significance. Now, choose an appropriate measure of association and ascertain the strength and direction of the relationship.
 e. Is support for capital punishment (*cappun*) associated with presidential preference in the 2016 election (*pres16*)? Is there a relationship between *cappun* and support for using spanking (*spanking*) to discipline children? Take *cappun* as the independent variable.

6. **Using the States Data Set** (*States.sav*)
 a. For the 50 states, is there a relationship between the amount of time it takes to drive to work (*WrkTrav*) and the traffic fatality rate (*TrafDths17*)?
 b. What are the strength and direction of the relationship between affluence (*MdHHinc*) and the percentage of the population that is divorced (*PopDiv*)?
 c. What are the strength and direction of the relationship between percentage of the population that doesn't speak English at home (*NotEngHome*) and poverty (*FamPoor17*)? What about the relationship between percentage not speaking English (*NotEngHome*) and crime rate (*Homicide*)?

Part IV Multivariate Techniques

The two chapters in Part IV introduce multivariate techniques—statistics that allow us to analyze the relationships among more than two variables at a time. These statistics are extremely useful for probing possible causal relationships between variables and are commonly reported in the literature.

Chapter 14 covers multivariate analysis for nominal- and ordinal-level variables that have been organized into table format. The chapter discusses the logic and procedures for analyzing the effect of a third (or control) variable on the relationship between the independent and dependent variables. The possible outcomes of controlling for a third variable are introduced and analyzed one at a time, and the discussion is summarized in Table 14.5.

Chapter 15 presents partial correlation, a statistical technique that is analogous to controlling for a third variable with bivariate tables, as presented in Chapter 14. We make several references to the concepts presented in Chapter 14 and to Table 14.5; however, it is not necessary to read Chapter 14 to understand the material presented in Chapter 15.

This final chapter also presents multiple regression and correlation, some of the most powerful and useful tools available to social science researchers. The mathematics underlying these techniques can become quite complicated, so the chapter focuses on the simplest possible applications.

14

Elaborating Bivariate Tables

LEARNING OBJECTIVES By the end of this chapter, you will be able to:

1. Explain the purpose of multivariate analysis and the purpose of a control variable.
2. Construct and interpret partial tables.
3. Compute and interpret measures of association for partial tables.
4. Recognize and interpret direct, spurious or intervening, and interactive relationships.
5. Compute and interpret partial gamma.
6. Explain the limitations of elaborating bivariate tables.
7. Use SPSS to conduct multivariate analysis with partial tables.

USING STATISTICS

The statistical techniques presented in this chapter are used to analyze how bivariate relationships behave when a third variable is introduced. Examples of situations in which these techniques are useful include:

1. Is the association between education and the use of one's skills at work different between males and females? **See page 410**

2. Does organizational spirituality affect knowledge sharing in an organization? Is this relationship affected by psychological flourishing or organizational trust? **See page 415**

3. What factors influence people's opinions on abortion? What variables might interfere with those associations? Find out for yourself on page 421

Few questions can be answered completely using only the bivariate measures of association we covered in Chapters 12 and 13. In this chapter and the next, we see how these techniques can be extended to probe the relationships among three or more variables. This chapter presents multivariate techniques for nominal and ordinal variables that are in table format. Chapter 15 presents some techniques that can be used when the variables have been measured at the interval-ratio level.

Before considering the techniques themselves, we should consider why they are important and what they might be able to tell us. There are two general reasons for using multivariate techniques. First, and most fundamental, is the goal of simply gathering additional information about a specific bivariate relationship by observing how that relationship is affected (if at all) by the presence of a third

variable (or a fourth or a fifth). Multivariate techniques will increase our information on the bivariate relationship and (we hope) enhance our understanding of that relationship.

A second and related rationale for multivariate statistics involves the issue of causation. Although multivariate statistical techniques cannot prove that causal connections exist between variables, they can provide valuable evidence in support of causal arguments and are important tools for testing and revising theory.

Controlling for a Third Variable

For variables arrayed in tables, multivariate analysis proceeds by observing the effects of other variables on the bivariate relationship. That is, we observe the relationship between the independent (X) and dependent (Y) variables after a third variable (which we will call Z) has been controlled. We do this by reconstructing the relationship between X and Y for each value or score of Z. If the control variable has an effect, the relationship between X and Y will change under the various conditions of Z. On the simplest application of multivariate analysis and observe how the relationship between an independent variable (X) and a dependent variable (Y) is affected by a single control variable (Z). Once the basic techniques are grasped, they can be readily expanded to situations involving more than one control variable.

THE BIVARIATE RELATIONSHIP

To illustrate, suppose that a researcher wishes to analyze the relationship between how well an individual is integrated into a group or organization and that individual's level of satisfaction with the group. The researcher has decided to focus on college students and their level of satisfaction with the college as a whole. The necessary information on satisfaction (Y) is gathered from a sample of 270 students, and all are asked to list the student clubs or organizations to which they belong. Integration (X) is measured by dividing the students into two categories: The first includes students who do not belong to any organizations, and the second includes students who are members of at least one organization. The researcher suspects that students who are members will report higher levels of satisfaction than nonmembers. The relationship between these two variables is displayed in Table 14.1.

TABLE 14.1 **Satisfaction with College by Number of Memberships in Student Organizations**

| | Memberships (X) | | |
Satisfaction (Y)	*None*	*At Least One*	TOTALS
Low	57 *(54.3%)*	56 *(33.9%)*	113
High	48 *(45.7%)*	109 *(66.1%)*	157
TOTALS	105 *(100.0%)*	165 *(100.0%)*	270
	Gamma = 0.40		

FIGURE 14.1 A Direct Relationship Between *X* and *Y*

$X \longrightarrow Y$

The conditional distributions of satisfaction (*Y*) change across the two conditions of membership (*X*), so the variables are associated. The column percentages also show that this is a positive relationship: Members tend to be highly satisfied and nonmembers tend to be low in satisfaction. In other words, satisfaction increases as membership increases. The existence and direction of the relationship are confirmed by the gamma of +0.40.[1]

In short, Table 14.1 suggests that integration and satisfaction are related, and these results can be taken as evidence for a causal relationship between the two variables. The causal relationship is summarized symbolically in Figure 14.1, where the arrow represents the effect of *X* on *Y.*

PARTIAL TABLES

Membership is not the sole cause of satisfaction (if it were, the gamma would be +1.00). Other variables affect this relationship, and the researcher will need to consider these third variables (*Z*s) in a systematic and logical way. By doing so, the researcher will gather more information about the bivariate relationship and can further probe the possible causal connection between *X* and *Y.*

For example, perhaps both satisfaction (*Y*) and membership (*X*) are affected by grades (*Z*) and students with higher grade-point averages (GPAs) are more satisfied with the university and are also more likely to participate actively in the life of the campus by joining organizations. How can the effects of this third variable on the bivariate relationship be investigated?

Consider Table 14.1 again. This table provides information about the distribution of 270 cases on two variables. We can see that 165 students are members, that 48 students are both nonmembers and highly satisfied, that the majority of the students are highly satisfied, and so forth. What the table cannot show us is the distribution of these students on GPA. For all we know at this point, the 109 highly satisfied members could all have high GPAs, low GPAs, or any combination of scores on this third variable. GPA is "free to vary" in this table, because the distribution of the cases on this variable is not accounted for.

We can control the effect (if any) of a third variable by fixing its distribution so it is no longer free to vary. We do this by sorting the cases according to their score on the third variable (*Z*) and then observing the relationship between *X* and *Y* for each value (or score) of *Z*. In our example, we construct separate tables displaying the relationship between membership and satisfaction for each category of GPA.

Tables that display the relationship between *X* and *Y* for each value of *Z* are called **partial tables**. In Table 14.2, the student sample has been subdivided into two groups based on GPA (*Z*): 135 students had "low" GPAs and the same number had "high" GPAs. The two partial tables in Table 14.2 show the relationship between membership (*X*) and satisfaction (*Y*) for each value of the control variable.

This type of multivariate analysis is called **elaboration** because the partial tables present the original bivariate relationship in a more detailed or elaborate form. The cell frequencies in the partial tables are subdivisions of the cell frequencies reported in Table 14.1. For example, if the frequencies of the cells in

[1]See Chapter 12 for a review of this measure of association.

TABLE 14.2 Satisfaction by Membership, Controlling for GPA

A. High GPA	Memberships (X)		
Satisfaction (Y)	*None*	*At Least One*	TOTALS
Low	29 (*54.7%*)	28 (*34.1%*)	57
High	24 (*45.3%*)	54 (*65.9%*)	78
TOTALS	53 (*100.0%*)	82 (*100.0%*)	135

Gamma = 0.40

B. Low GPA	Memberships (X)		
Satisfaction (Y)	*None*	*At Least One*	TOTALS
Low	28 (*53.8%*)	28 (*33.7%*)	56
High	24 (*46.2%*)	55 (*66.3%*)	79
TOTALS	52 (*100.0%*)	83 (*100.0%*)	135

Gamma = 0.39

the partial tables are added together, the original frequencies of Table 14.1 will be reproduced.[2]

Also note how this method of controlling for other variables can be extended. In our example, the control variable (GPA) had two categories, and, thus, there were two partial tables, one for each value of the control variable. If the control variable had had more than two categories, we would have had additional partial tables. Also, we can control for more than one variable at a time by sorting the cases on all scores of all control variables and producing partial tables for each combination of scores on the control variables. Thus, if we had controlled for both GPA and gender, we would have had four partial tables to consider. There would have been one partial table for males with low GPAs, a second for males with high GPAs, and two partial tables for females with high and low GPAs.

Our next step is to trace the effect of Z by comparing the partial tables with each other and with the original bivariate table. By analyzing the partial tables, we can observe the effects (if any) of the control variable on the original bivariate relationship.

Interpreting Partial Tables

The cell frequencies in the partial tables can follow a variety of forms, but we will concentrate on three basic patterns, determined by comparing the partial tables with each other and with the original bivariate table:

1. **Direct relationships:** The relationship between X and Y is the same in all partial tables and in the bivariate table.

[2]The total of the cell frequencies in the partial tables will always equal the corresponding cell frequencies in the bivariate table, except when, as often happens in "real-life" research situations, the researcher is missing scores on the third variable for some cases. These cases must be deleted from the analysis; as a consequence, the partial tables will have fewer cases than the bivariate table.

2. Spurious relationships or intervening relationships: The relationship between X and Y is the same in all partial tables but much weaker than in the bivariate table.

3. Interaction: The partial tables and the bivariate table all show different relationships between X and Y.

Each pattern has different implications for the relationships among the variables and for the subsequent course of the statistical analysis. We will consider each pattern in detail and then summarize the discussion in Table 14.5.

DIRECT RELATIONSHIPS

This pattern is often called **replication**, because the partial tables reproduce (or replicate) the bivariate table: The column percentages and the measures of association are the same for all tables. This outcome indicates that the control variable (Z) has no effect on the relationship between X and Y.

Table 14.2 provides an example of replication. The relationship between membership (X) and satisfaction (Y) was investigated with GPA (Z) controlled. Table 14.2A shows the relationship for high-GPA students, and Table 14.2B shows the relationship for low-GPA students. The partial tables show the same conditional distributions of Y: For both high- and low-GPA students, about 45% of the nonmembers are highly satisfied, versus about 66% of the members. This same pattern was observed in the bivariate table (Table 14.1). Thus, the conditional distributions of Y are the same in all partial tables and the same as they were in the bivariate table.

Patterns will be easier to identify if we calculate measures of association. For Table 14.2, we see that the gamma for high-GPA students is 0.40 and that the gamma for low-GPA students is 0.39. The bivariate gamma (from Table 14.1) is 0.40. The fact that the partial and bivariate gammas are essentially the same value reinforces our conclusion that this is a case of replication. As a rough guideline, we can say that a relationship is direct if the measures of association computed on the partial tables are within ± 0.10 of the bivariate measure of association.

This pattern means that the control variable has no important impact on the bivariate relationship (if it did, the pattern in the partial tables would be different from that in the bivariate table) and may be ignored in any further analysis. In terms of the original research problem, the researcher may conclude that students who are members are more likely to express high satisfaction with the university regardless of their GPA. In other words, GPA has no effect on the relationship between membership and satisfaction. Low-GPA students who are members of at least one organization are just as likely to report high satisfaction as high-GPA students who are members of at least one organization. (*See any of the end-of-chapter problems for practice in detecting possible direct relationships.*)

SPURIOUS OR INTERVENING RELATIONSHIPS

In this pattern, the relationship between X and Y is much weaker in the partial tables than in the bivariate table but the same across all partial tables. Measures of association for the partial tables are much lower in value (perhaps even dropping to 0.00) than the measure computed for the bivariate table.

This outcome is consistent with two different causal relationships among the three variables. The first is called a spurious relationship or **explanation**; in this situation, Z is conceptualized as being antecedent to both X and Y (that is, Z is thought to occur earlier in time than the other two variables). Z causes both X and Y, and the original bivariate relationship is said to be spurious. The apparent bivariate relationship between X and Y is explained by, or due to, the effect of Z. Once Z is controlled, the association between X and Y disappears and the measures of association for the partial tables are dramatically lower than the bivariate measure of association.

To illustrate, suppose that the researcher in our example had also controlled for class standing by dividing the sample into upperclass (seniors and juniors) and underclass (sophomores and freshmen) students. The reasoning of the researcher might be that dissatisfied students would have transferred or dropped out before they achieved upperclass standing and upperclass students are more satisfied because of simple self-selection. Also, students who have been on campus longer will be more likely to have located an organization of sufficient appeal or interest to join. This might especially be the case for organizations based on major field (such as the Accounting Club), which underclass students are less likely to join, or honorary organizations, for which underclass students are unlikely to qualify.

FIGURE 14.2 A Spurious Relationship

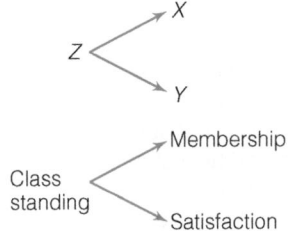

These thoughts about the possible relationships among these variables are expressed in diagram form in Figure 14.2. The absence of an arrow from membership (X) to satisfaction (Y) indicates that these variables are not truly associated. Rather, both are caused by class standing (Z). If this causal diagram is a correct description of the relationship among these three variables (if the association between X and Y is spurious), then the association between membership and satisfaction should disappear once class standing has been controlled. That is, even though the bivariate gamma was 0.40 (Table 14.1), the gammas computed on the partial tables will approach 0. Table 14.3 displays the partial tables generated by controlling for class standing.

TABLE 14.3 Satisfaction by Membership, Controlling for Class

A. Upperclass Students	Memberships (X)		
Satisfaction (Y)	*None*	*At Least One*	TOTALS
Low	8 (*25.0%*)	32 (*24.8%*)	40
High	24 (*75.0%*)	97 (*75.2%*)	121
TOTALS	32 (*100.0%*)	129 (*100.0%*)	161
	Gamma = 0.01		

B. Underclass Students	Memberships (X)		
Satisfaction (Y)	*None*	*At Least One*	TOTALS
Low	49 (*67.1%*)	24 (*66.7%*)	73
High	24 (*32.9%*)	12 (*33.3%*)	36
TOTALS	73 (*100.0%*)	36 (*100.0%*)	109
	Gamma = 0.01		

It may be helpful to have a rough guideline for identifying this type of relationship. We can say that a relationship is spurious if the measures of association for the partial tables are each more than 0.10 less than the bivariate measure of association.

The partial tables indicate that membership is no longer related to satisfaction, once class standing is controlled. Regardless of their number of memberships, upperclass students are more likely to be high on satisfaction and underclass students are more likely to be low on satisfaction. In the partial tables, the distributions of *Y* no longer vary by the conditions of *X*, and the gammas are virtually 0. These results indicate that *X* and *Y* have no direct relationship: Their apparent relationship is spurious. Class standing (*Z*) is a key factor in accounting for satisfaction, and the analysis must be reoriented with class standing as an independent variable.

FIGURE 14.3 An Intervening Relationship

This outcome (partial measures much weaker than the original measure but equal to each other) is consistent with a second causal model in which *Z* connects *X* with *Y*. This pattern is called **interpretation** and is illustrated in Figure 14.3: *X* causes *Z*, which in turn causes *Y*. This pattern indicates that *X* and *Y* are associated primarily through the control variable *Z*.

STATISTICS **In Everyday Life**

Spuriousness

Spurious relationships are part of everyday life, not just scientific research. We sometimes leap to incorrect conclusions about "what causes what" as we consider the news or chat with friends, and being aware of spurious relationships can enhance your appreciation of the complexity of social life and the subtlety of casual relationships. As you think about spuriousness, it may be helpful to refer to some simplified examples. Here are four statements, each of which asserts that one variable (*X*) causes another (*Y*). Each relationship is spurious. Can you identify the common cause (*Z*) of both *X* and *Y* for each statement? (The answers appear below.)

1. People in hospitals are more likely to die than those not in a hospital. Therefore, being in a hospital (*X*) causes death (*Y*).

2. The greater the number of firefighters at a fire, the greater the damage. Therefore, firefighters (*X*) cause fire damage (*Y*).

3. Men who listen to a lot of 1950s rock and roll are more likely to suffer from sexual impotence. Therefore, rock and roll music (*X*) causes sexual impotence (*Y*).

4. The more bathrooms you have in your house, the higher your SAT scores will be. Therefore, bathrooms (*X*) cause higher SAT scores (*Y*).

Answers:

1. *Z* = illness. People who are very sick are more likely to be in a hospital and more likely to die.

2. *Z* = extent of the fire. More serious fires attract more firefighters and result in more damage.

3. *Z* = age. Older men are both more likely to listen to this type of music and more likely to experience sexual impotence.

4. *Z* = income. People with higher incomes have larger homes with more bathrooms. They are also more likely to be able to afford better education and SAT prep classes to help their children attain higher scores on the SAT.

The pattern displayed in Table 14.3 does not allow the researcher to distinguish between spurious relationships (Figure 14.2) and intervening relationships (Figure 14.3). The differentiation between these two types of causal patterns may be made on temporal or theoretical grounds but not on statistical grounds. *(See any of the end-of-chapter problems for practice in detecting possible spurious or intervening relationships.)*

INTERACTION

FIGURE 14.4 An Interactive Relationship

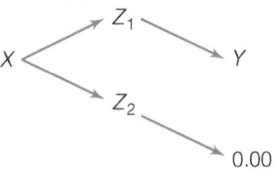

In this pattern, also called **specification**, the relationship between X and Y changes markedly for the different values of the control variable. The partial tables differ from each other and from the bivariate table. Interaction can appear in a variety of ways in the partial tables. One possible pattern, for example, is for one partial table to display a stronger relationship between X and Y than that displayed in the bivariate table, while the relationship between X and Y drops to 0 in a second partial table. Symbolically, this outcome could be represented as in Figure 14.4, which would indicate that X and the first category of Z (Z_1) have strong effects on Y, but, for the second category of Z (Z_2), there is no association between X and Y.

This type of interaction might be found, for example, in a situation in which *all* employees of a large corporation were required to attend a program (X) designed to reduce prejudice against African Americans (Y). Such a program would be likely to have stronger effects on white employees (Z_1) than on African American employees (Z_2). That is, the program would have an effect on prejudice only for certain types of subjects. The partial table for white employees (Z_1) might show a strong relationship between program attendance and reduction in prejudice, whereas the partial table for African American employees might show no relationship. African American employees, being unprejudiced against themselves in the first place, would be less affected by the program.

FIGURE 14.5 An Interactive Relationship

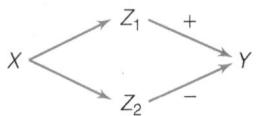

Interaction can take other forms. For example, the relationship between X and Y can vary not only in strength but also in direction between the partial tables. This causal relationship is symbolically represented in Figure 14.5, where X and Y are positively related for the first category of Z (Z_1) and negatively related for the second category of Z (Z_2).

The researcher investigating the relationship between club membership and satisfaction with college life establishes a final control for race and divides the sample into white and black students. The partial tables are displayed in Table 14.4 and show an interactive relationship. The relationship between membership and satisfaction is different for white students (Z_1) and for black students (Z_2), and each partial table is different from the bivariate table. For white students, the relationship is positive and stronger than in the bivariate table. White students who are also members are much more likely to express high overall satisfaction with the university, as indicated by both the percentage distribution (83% of the white students who are members are highly satisfied) and the measure of association (gamma = 0.67 for white students).

For black students, the relationship between membership and satisfaction is quite different, nearly the reverse of the pattern shown by white students. The great majority (80%) of the black students who are members report low satisfaction, and the gamma for this partial table is negative (-0.31). Thus, satisfaction

TABLE 14.4 Satisfaction by Membership, Controlling for Race

A. White Students	Memberships (X)		
Satisfaction (Y)	None	At Least One	TOTALS
Low	40 (50.0%)	20 (16.7%)	60
High	40 (50.0%)	100 (83.3%)	140
TOTALS	80 (100.0%)	120 (100.0%)	200

Gamma = 0.67

B. Black Students	Memberships (X)		
Satisfaction (Y)	None	At Least One	TOTALS
Low	17 (68.0%)	36 (80.0%)	53
High	8 (32.0%)	9 (20.0%)	17
TOTALS	25 (100.0%)	45 (100.0%)	70

Gamma = −0.31

increases with membership for white students but decreases with membership for black students.

Once again, a rough guideline might be helpful for identifying this type of relationship. We can say that a relationship is interactive if the measures of association for the partial tables are more than 0.10 different from each other and more than 0.10 different from the bivariate measure of association.

Based on these results, one might conclude that the implications of joining student organizations vary by race and have different effects for black students. It may be, for example, that black students join different kinds of organizations than do white students. If the black students belonged primarily to a black student association that had an antagonistic relationship with the university, then belonging to such an organization could increase dissatisfaction with the university as it increased awareness of racial problems. *(See any of the end-of-chapter problems for practice in detecting possible interactive relationships.)*

STATISTICS In Everyday Life

Interaction

It might be helpful to think about interaction as an unexpectedly intense reaction when variables are combined. For example, suppose a study of the causes of street crime found moderately strong associations with poverty ("areas with high poverty have higher levels of street crimes") and age ("areas with high percentages of teenagers and young adults have higher levels of street crime") but very strong relationships when the variables were combined ("areas that have *both* high poverty *and* younger populations have *much* higher levels of street crime"). This may be a case of interaction: Each independent variable has its own relationship with crime, but their combined effects have surprisingly strong impacts on the dependent variable.

TABLE 14.5 A Summary of Possible Results When Controlling for Third Variables

Compared with Bivariate Table, Partial Tables Show	Pattern	Implications for Further Analysis	Likely Next Step	Theoretical Implications
Same relationship between X and Y (e.g., gammas for partial tables are no more than ±0.10 different from bivariate gamma	Direct relationship (replication)	Disregard Z	Analyze another Z variable	Theory that X causes Y is supported
Weaker relationship between X and Y (e.g., gammas for partial tables are all at least 0.10 lower than bivariate gamma)	Spurious relationship (explanation)	Incorporate Z	Focus on relationship between Z and Y	Theory that X causes Y is not supported
	Intervening relationship (interpretation)	Incorporate Z	Focus on relationships between X, Y, and Z	Theory that X causes Y is partially supported but must be revised to take Z into account
Mixed (e.g., there is a difference of at least ±0.10 between gammas for partial tables and between gammas for partial tables and bivariate gamma)	Interaction (specification)	Incorporate Z	Analyze sub-groups (categories of Z) separately	Theory that X causes Y is partially supported but must be revised to take Z into account

In closing this section, let us stress that, for the sake of clarity, we have presented examples that have been unusually "clean." In an actual research project, controlling for third variables will probably have results that are considerably more ambiguous and open to interpretation than the examples presented here. In the case of spurious relationships, for example, the measures of association computed for the partial tables will probably not actually drop to 0, even though they may be dramatically lower than the bivariate measure. (The pattern where the partial measures are roughly equivalent and much lower than the bivariate measure but not 0 is sometimes called *attenuation*.) It is probably best to consider the foregoing examples as ideal types against which any given empirical result can be compared.

Table 14.5 summarizes this discussion by outlining guidelines for decision making for each of the three outcomes discussed in this section. Because your own results will probably be more ambiguous than the ideal types presented here, you should regard this table as a set of suggestions and not as a substitute for your own creativity and sensitivity to the problem under consideration.

Partial Gamma (G_p)

When the results of controlling for a third variable indicate a direct, spurious, or intervening relationship, it is often useful to compute an additional measure that indicates the overall strength of the association between X and Y after the effects of the control variable (Z) have been removed. This statistic, called **partial gamma** (G_p), is somewhat easier to compare with the bivariate gamma than are

the gammas computed on the partial tables separately. G_p is computed across all partial tables by Formula 14.1:

FORMULA 14.1

$$G_p = \frac{\Sigma N_s - \Sigma N_d}{\Sigma N_s + \Sigma N_d}$$

where: N_s = the number of pairs of cases ranked the same across all partial tables

N_d = the number of pairs of cases ranked differently across all partial tables

In words, ΣN_s is the total of the N_ss from all partial tables, and ΣN_d is the total of the N_ds from all partial tables. (See Chapter 12 to review the computation of N_s and N_d.)

To illustrate the computation of G_p, let us return to Table 14.2, which showed the relationship between satisfaction and membership while controlling for GPA. The gammas computed on the partial tables were essentially equal to each other and to the bivariate gamma. Our conclusion was that the control variable had no effect on the relationship, and this conclusion can be confirmed by also computing partial gamma.

From Table 14.2A High GPA	From Table 14.2B Low GPA
$N_s = (29)(54) = 1566$	$N_s = (28)(55) = 1540$
$N_d = (28)(24) = 672$	$N_d = (28)(24) = 672$

$$\Sigma N_s = 1566 + 1540 = 3106$$

$$\Sigma N_d = 672 + 672 = 1344$$

$$G_p = \frac{\Sigma N_s - \Sigma N_d}{\Sigma N_s + \Sigma N_d}$$

$$G_p = \frac{3106 - 1344}{3106 + 1344}$$

$$G_p = \frac{1762}{4450}$$

$$G_p = 0.40$$

The partial gamma measures the strength of the association between X and Y once the effects of Z have been removed. In this instance, the partial gamma is the same value as the bivariate gamma ($G_p = G = 0.40$) and indicates that GPA has no effect on the relationship between satisfaction and membership.

When class standing was controlled (Table 14.3), clear evidence of a spurious relationship was found, because the gammas computed on the partial tables

dropped almost to 0. Let's compute the value of partial gamma (G_p) for this second control:

From Table 14.3A Upperclassmen	From Table 14.3B Underclassmen
$N_s = (8)(97) = 776$	$N_s = (49)(12) = 588$
$N_d = (32)(24) = 768$	$N_d = (24)(24) = 576$

$$\Sigma N_s = 776 + 588 = 1364$$

$$\Sigma N_d = 768 + 576 = 1344$$

$$G_p = \frac{\Sigma N_s - \Sigma N_d}{\Sigma N_s + \Sigma N_d}$$

$$G_p = \frac{1364 - 1344}{1364 + 1344}$$

$$G_p = \frac{20}{2708}$$

$$G_p = 0.01$$

Once the effects of the control variable are removed, there is no relationship between X and Y. The low value of G_p confirms our previous conclusion that the bivariate relationship between membership and satisfaction is spurious and actually due to the effects of class standing.

In a sense, G_p tells us no more about the relationships than we can see for ourselves from a careful analysis of the percentage distributions of Y in the partial tables or by a comparison of the measures of association computed on the partial tables. The advantage of G_p is that it tells us, in a single number, the precise effects of Z on the relationship between X and Y. While G_p is no substitute for the

ONE STEP AT A TIME **Computing Partial Gamma (G_p)**

To Compute N_s and N_d for All Partial Tables

1. Add up all values for N_s from all partial tables to find ΣN_s.
2. Add up all values for N_d from all partial tables to find ΣN_d.

To Find Partial Gamma (G_p) by Solving Formula 14.1

1. Subtract ΣN_d from ΣN_s.
2. Add ΣN_d to ΣN_s.
3. Divide the quantity you found in step 1 by the quantity you found in step 2. The result is G_p.

Social Research and Data Analysis 14.1 | **Elaborating Bivariate Tables**

Using data from 2018 General Social Survey, we found that there is a positive correlation between the highest degree that a person has earned and the extent to which people say that their job allows them to use their skills. Let's look at the degree and direction of that association more closely and then elaborate by controlling for gender.

Our first step in this task was to reverse the coding of "Job allows Respondent use of skills" (MYSKILLS) as we have advised previously so that an increase in the score indicates a greater use of one's skills at work. Next, we examined the zero-order correlation between the highest degree the respondent has earned (DEGREE) and the use of

one's skills at work. The gamma for this association came to 0.19. That's pretty weak, but it's not uncommon. In your research you might find that it is difficult to find even moderate associations between most ordinal-level variables.

Now let's control for gender. Upon adding gender as a control variable, we find that the association between highest degree and use of job skills for males is 0.19 and for females it is 0.19. So what should that lead you to conclude? The zero-order correlation was 0.19, and the partial correlations were both 0.19 as well. This is a direct association between education and one's use of skills at work.

analysis of the partial tables per se, it is a convenient way of stating our results when working with direct or spurious relationships.

Although G_p can be calculated in cases of interactive relationships (see Table 14.4), it is rather difficult to interpret in these instances. If substantial interaction is found in the partial tables, this indicates that the control variable has a profound effect on the bivariate relationship. Thus, we should not attempt to separate the effects of Z from the bivariate relationship, and, because G_p involves exactly this kind of separation, it should not be computed.

Where Do Control Variables Come From?

In one sense, this question is easy to answer: Control variables come mainly from theory. Social research proceeds in many different ways and is begun in response to a variety of problems. However, virtually all quantitative research projects are guided by an explicit theory or by some question about the relationship between two or more variables. The ultimate goal of social research is to develop defensible generalizations that improve our understanding of the variables under consideration and link back to theory at some level (see Figure 1.1).

Thus, research projects are anchored in theory; the concepts of interest, which will later be operationalized as variables, may be first probed at the theoretical level. The social world is exceedingly complex, and any attempt to explain it with a simple bivariate relationship will fail. Theories are, almost by definition, multivariate, even when they focus on a particular bivariate relationship. Thus, theory may suggest the control variables that need to be incorporated into the analysis. In the example used throughout this chapter, we tried to suggest that any researcher probing the relationship between involvement in an organization and satisfaction would, while thinking over the possibilities, identify a number of additional variables that should be explicitly incorporated into the analysis.

Textbook descriptions of the research process are simplified and suggest that research flows smoothly from conceptualization to operationalization to quantification to generalization. In reality, research is full of surprises, unexpected outcomes, and unanticipated results. Research in "real life" is more loosely structured and requires more imagination and creativity than textbooks can fully convey. Our point is that the control variables that might be appropriate to incorporate in the data-analysis phase will be suggested or implied in the theoretical backdrop of the research project, along with the researcher's imagination and sensitivity to the problem being addressed.

These considerations have taken us well beyond the narrow realm of statistics and back to the planning stages of the research project. At this early time, the researcher must make decisions about which variables to measure during the data-gathering phase and, thus, which variables might be incorporated as potential controls. Careful thinking and an extended consideration of possible outcomes at the planning stage will pay significant dividends during the data-analysis phase. Ideally, all relevant control variables will be incorporated and readily available for statistical analysis.

USING SPSS

Crosstabs and Measures of Association

In this installment of "Using SPSS," we will use the **Crosstabs** procedure to further examine the relationship between *degree* and happiness (*happy*). We examined the bivariate relationship in Chapter 12 and found a weak to moderate relationship. The sign of gamma was negative but *happy* is reverse scored and the relationship is actually positive. In this chapter, we will see if the relationship is affected by sex.

To begin:

1. Click the SPSS icon on your desktop.
2. Load the *GSS2018* database.
3. From the main menu bar, click **Analyze**, **Descriptive Statistics**, and **Crosstabs**.
4. In the **Crosstabs** dialog box:
 a. Find *happy* and move it to the **Rows** box.
 b. Find *degree* and move it into the **Columns** box.
 c. Find *sex* and move it to the bottom box ("Layer 1 of 1"). SPSS will produce partial tables for each value of the variable entered in this box.
5. Click the **Statistics** button and select Gamma.
6. Return to the **Crosstabs** window and click the **Cells** button and select "column" in the **Percentages** box.
7. Return to the **Crosstabs** window and click **OK**.

SPSS produces a single large table, with the partial table for men in the top panel, the partial table for women in the middle panel, and the bivariate table in the bottom panel. You should inspect this output carefully, comparing the column percentages for the partial tables with each other and with the bivariate table. Gammas are reported in the second box.

To conserve space, we reproduce only the top rows ("Very happy") of the partial tables and the bivariate table.

(continued)

USING SPSS

(Continued)

Happiness by Degree, Controlling for Sex

	LT HS	HS	Jr. Coll.	BA	Grad.	TOTALS
A. Men (N = 1050)						
Very Happy	31.4%	26.8%	16.2%	35.1%	39.6%	29.6%
			Gamma = −0.13			
B. Women (N = 1294)						
Very Happy	29.5%	26.1%	25.8%	40.0%	34.6%	30.1%
			Gamma = −0.16			
C. Total Sample (N = 2344)						
Very Happy	30.4%	26.4%	22.4%	37.8%	36.8%	29.9%
			Gamma = −0.15			

For both men and women, happiness increases as education increases, as it does for the total sample. The gammas indicate a weak to moderate, positive relationship (remember that *happy* is reverse scored). The gammas for men and women are very similar, although women show a slightly stronger relationship between education and happiness. SPSS reports the partial gamma (G_p) as a "first-order partial" just below the "Total" gamma. This statistic is −0.15, similar to the gammas for the partial and bivariate tables.

We would probably conclude that this is a direct relationship: Happiness tends to increase with education regardless of gender. Our next step might be to select another control variable (e.g., age, race, marital status, or religion) and observe its effects (if any) on the relationship between happiness and education.

The Limitations of Elaborating Bivariate Tables

The most important limitation of elaborating bivariate tables is sample size. Elaboration is relatively inefficient because it requires that the researcher divide the sample across a series of partial tables. If the control variable has more than two or three possible values or if we attempt to control for more than one variable at a time, many partial tables will be produced. The greater the number of partial tables, the more likely we are to run out of cases to fill all of the cells. Empty or small cells, in turn, can create serious problems in terms of generalizability and confidence in our findings.

To illustrate, the example used throughout this chapter began with two dichotomized variables and a four-cell table (Table 14.1). Each of the control variables was also dichotomized, and we never confronted more than two partial tables with four cells each, or a total of eight cells for each control variable. If

we had used a control variable with three values, we would have had 12 cells to fill up, and if we had attempted to control for two dichotomized variables at the same time, we would have had 16 cells in four different partial tables. Clearly, as control variables become more elaborate and/or as the process of controlling becomes more complex, empty or small cells will increasingly become a problem.

Two potential solutions to this problem immediately suggest themselves. The easy solution is to reduce the number of cells in the partial tables by collapsing categories within variables. If all variables are dichotomized, for example, the number of cells will be kept to a minimum. A better solution is to work with only large samples so that there will be plenty of cases to fill up the cells. Unfortunately, the easy solution will often violate common sense (collapsed categories are more likely to group dissimilar elements together); the better solution is not always feasible (mundane matters of time and money rear their ugly heads).

A third solution to the problem of empty cells requires the (sometimes risky) assumption that the variables of interest are measured at the interval-ratio level. At that level, the techniques of partial and multiple correlation and regression, to be introduced in Chapter 15, are available. These multivariate techniques are more efficient than elaboration because they utilize all cases simultaneously and do not require that the sample be divided among the various partial tables.

Interpreting Statistics: Another Look at Concern for the Environment

In Chapter 12, we looked at the relationship between age and concern for the dangers to the environment posed by greenhouse gases (*tempgen*) and found a nonsignificant, weak relationship. In this installment of "Interpreting Statistics," we will introduce a different independent variable—political ideology (*polviews*)—and examine its effects on *tempgen*. For ease of analysis, we have recoded *polviews* into three categories: liberal, moderate, and conservative. Is there a political dimension to the environmental movement? Recall that *tempgen* is also recoded into just two categories (see Chapter 12). Which political ideology is linked with the belief that greenhouse gases are dangerous? The bivariate relationship is presented in Table 14.6.

TABLE 14.6 Opinion on the Danger of Greenhouse Gases by Political Ideology

Danger of Greenhouse Gases	Political Ideology			TOTALS
	Liberal	*Moderate*	*Conservative*	
Very dangerous	81.0%	63.0%	36.9%	59.6%
Not very dangerous	19.0%	37.0%	63.1%	40.4%
TOTALS	100% (205)	100% (284)	100.0% (236)	100.0% (725)

Chi Square = 90.94, *df* = 2, *p* < 0.05. Gamma = 0.56

TABLE 14.7 **Opinion on the Danger of Greenhouse Gases by Political Ideology, Controlling for Gender**

A. Males

Danger of Greenhouse Gases	Political Ideology			TOTALS
	Liberal	*Moderate*	*Conservative*	
Very dangerous	77.8%	63.9%	31.1%	56.8%
Not very dangerous	22.2%	36.1%	68.9%	43.2%
TOTALS	100%	100%	100.0%	100.0%
	(90)	(119)	(106)	(315)

Gamma = 0.58

B. Females

Danger of Greenhouse Gases	Political Ideology			TOTALS
	Liberal	*Moderate*	*Conservative*	
Very dangerous	83.5%	62.4%	41.5%	61.7%
Not very dangerous	16.5%	37.6%	58.5%	38.3%
TOTALS	100%	100%	100.0%	100.0%
	(115)	(165)	(130)	(410)

Gamma = 0.54

$G_p = 0.56$

The relationship is statistically significant and moderate to strong. The positive sign of gamma indicates that the cases tend to fall in the diagonal from upper left to lower right, a pattern made clear by the column percentages. A large majority of liberals see greenhouse gases as "very dangerous" while the majority of conservatives have the opposite opinion, with moderates intermediate. This issue has a strong partisan dimension.

Is the relationship affected by sex? The results of controlling for sex are presented in Table 14.7.

The control for gender has little effect on the bivariate relationship. For both males and females, concern for the danger of greenhouse gases is greatest for liberals and lowest for conservatives. The gammas for the two partial tables are rather close in value to each other and to the bivariate gamma, as is the partial gamma ($G_p = 0.56$). There is a direct relationship between concern for greenhouse gases and political ideology and liberals are more likely to be alarmed regardless of gender.

What other variables might affect this relationship? We might investigate age (are older liberals just as concerned as younger liberals?), education (is the bivariate relationship the same for all levels of education?), race or ethnicity (are greenhouse gases an important issue for nonwhite liberals?), and a host of other possibilities. The greater the extent to which the relationship maintained itself across the series of control variables, the greater our confidence that political ideology is directly related to concern for greenhouse gases. On the other hand, the finding that some control variables displayed spurious or interactive relationships might point to additional, useful avenues of investigation.

READING STATISTICS Predicting Knowledge Sharing Attitudes, Part 2

In Reading Statistics in the previous chapter, we learned that the authors of the study on knowledge sharing found a positive correlation in their sample of telecom workers between organizational spirituality, psychological flourishing, organizational trust, and knowledge sharing attitude. But as we mentioned in that discussion, this information did not tell us that much on its own. Later in the study, the authors used some of the techniques we covered in this chapter to more fully develop their findings.

In professional research, intervening variables are often called "mediating variables." Mediating variables are those that work in between an independent variable and a dependent variable to make an effect come to fruition. In other words, X might not have a direct effect on Y, but when Z gets involved, Z can cause that effect (Y) to actually happen. That's exactly what occurred in this study. After the authors examined the correlation matrix (or the zero-order correlations) between all of those variables, they took the analysis a step further to see if organizational trust (a Z variable) and/or psychological flourishing (another Z variable) would act as mediators (or intervening variables) between organizational spirituality (X) and knowledge sharing attitudes (Y).

They started by examining the effect of organizational spirituality on knowledge sharing while controlling for psychological flourishing and organizational spirituality. This yielded no direct effect. Then they controlled for the other two variables, and they found that there were two indirect effects present, one that showed that organizational spirituality affected knowledge sharing attitude *through* psychological flourishing, and the other that showed that organizational spirituality affected knowledge sharing attitude *through* organizational trust. These are intervening associations like those discussed in this chapter, but as we mentioned above, they are often called mediating variables in academic literature.

Source: Khari, Chitra, and Shuchi Sinha. (2018). "Organizational Spirituality and Knowledge Sharing: A Model of Multiple Mediation," *Global Journal of Flexible Systems Management*, 19, 337–348.

SUMMARY

1. Most research questions require the analysis of the interrelationships among many variables, even when the researcher is primarily concerned with a specific bivariate relationship. With multivariate statistical techniques, researchers can gather additional information about the variables of interest and probe possible causal interrelationships.

2. When variables have been organized in bivariate tables, multivariate analysis proceeds by controlling for a third variable. Column percentages and measures of association are used to compare partial tables with each other and with the original bivariate table.

3. A direct relationship exists between the independent (X) and dependent (Y) variables if, after controlling for the third variable (Z), the relationship between X and Y is the same across all partial tables and the same as in the bivariate table. This pattern suggests a causal relationship between X and Y.

4. If the relationship between X and Y is the same across all partial tables but much weaker than in the bivariate table, the relationship is either spurious (Z causes both X and Y) or intervening (X causes Z, which causes Y). Either pattern suggests that Z must be incorporated into the analysis.

5. Interaction exists if the relationship between X and Y varies across the partial tables and between each partial table and the bivariate table. This pattern suggests that no simple or direct causal relationship exists between X and Y and that Z must be explicitly incorporated into the causal scheme.

6. Partial gamma (G_p) measures the strength of the association between X and Y after the effects of the control variable (Z) have been removed. Partial gamma should not be computed when the partial tables show substantial interaction.

7. Potential control variables must be identified before the data-gathering phase of the research project. The theory that guides the research project, along with creative thinking and some imagination, will suggest the variables that should be controlled for and measured.

8. Controlling for third variables by constructing partial tables is inefficient, in that the cases must be spread out across many cells. If the variables have many categories and/or the researcher attempts to control for more than one variable simultaneously, "empty cells" may become a problem. It may be possible to deal with this problem by either collapsing categories or gathering larger samples. If interval-ratio level of measurement can be assumed, the multivariate techniques presented in the next chapter will be preferred, because they do not require the partitioning of the sample.

SUMMARY OF FORMULAS

14.1	Partial gamma:	$G_p = \dfrac{\Sigma N_s - \Sigma N_d}{\Sigma N_s + \Sigma N_d}$

GLOSSARY

Direct relationship. A multivariate relationship in which the control variable has no effect on the bivariate relationship.

Elaboration. The basic multivariate technique for analyzing variables arrayed in tables. Partial tables are constructed to observe the bivariate relationship in a more detailed or elaborated format.

Explanation. See **Spurious relationship**.

Interaction. A multivariate relationship wherein a bivariate relationship changes across the categories of the control variable.

Interpretation. See **Intervening relationship**.

Intervening relationship. A multivariate relationship wherein a bivariate relationship becomes substantially weaker after a third variable is controlled for. The inde-pendent and dependent variables are linked primarily through the control variable.

Partial gamma (G_p). A statistic that indicates the strength of the association between two variables after the effects of a third variable have been removed.

Partial tables. Tables produced when controlling for a third variable.

Replication. See **Direct relationship**.

Specification. See **Interaction**.

Spurious relationship. A multivariate relationship in which a bivariate relationship becomes substantially weaker after a third variable is controlled for. The independent and dependent variables are not causally linked. Rather, both are caused by the control variable.

Z. Symbol for any control variable.

PROBLEMS

14.1 [SOC] Problem 12.6 concerned the assimilation of a small sample of immigrants. One table showed the relationship between length of residence in the United States and fluency with the English language:

	Length of Residence		
English Fluency	Less Than Five Years (Low)	More Than Five Years (High)	TOTALS
Low	20	10	30
High	5	15	20
TOTALS	25	25	50

a. If necessary, find the column percentages and compute gamma for this table. Describe the bivariate relationship in terms of strength and direction. Is the relationship between residence and language affected by the gender of the immigrant? Here are the partial tables showing the bivariate relationships for males and females:

A. Males	Length of Residence		
English Fluency	Less Than Five Years (Low)	More Than Five Years (High)	TOTALS
Low	10	5	15
High	2	8	10
TOTALS	12	13	25

B. Females

English Fluency	Less Than Five Years (Low)	More Than Five Years (High)	TOTALS
Low	10	5	15
High	3	7	10
TOTALS	13	12	25

b. Find the column percentages and compute gamma for each of the partial tables. Compare these percentages and gammas with each other and with the bivariate table. Compute partial gamma (G_p). Does gender change the bivariate relationship?

Is the relationship between residence and language affected by the origin of the immigrant? Here are the partial tables showing the bivariate relationships for Asian and Hispanic immigrants:

A. Asian Immigrants

Length of Residence

English Fluency	Less Than Five Years (Low)	More Than Five Years (High)	TOTALS
Low	8	4	12
High	2	5	7
TOTALS	10	9	19

B. Hispanic Immigrants

Length of Residence

English Fluency	Less Than Five Years (Low)	More Than Five Years (High)	TOTALS
Low	12	6	18
High	3	10	13
TOTALS	15	16	31

c. Find the column percentages and compute gamma for each of the partial tables. Compare these percentages and gammas with each other and with the bivariate table. Compute partial gamma (G_p). Does the origin of the immigrants change the bivariate relationship?

14.2 SOC Data on suicide rates, age structure, and unemployment rates have been gathered for 100 census tracts. Suicide and unemployment rates have been dichotomized at the median so that each tract could be rated as high or low. Age structure is measured in terms of the percentage of the population age 65 and older. This variable has also been dichotomized, and tracts have been rated as high or low. The following tables display the bivariate relationship between suicide rate and age structure and the same relationship controlling for unemployment:

Suicide Rate	Population 65 and Older Low	High	TOTALS
Low	45	20	65
High	10	25	35
TOTALS	55	45	100

a. Calculate column percentages and gamma for the bivariate table. Describe the bivariate relationship in terms of strength and direction.

Is the relationship between age structure and suicide rate affected by the rate of unemployment?

A. High Unemployment

Population 65 and Older (%)

Suicide Rate	Low	High	TOTALS
Low	23	10	33
High	5	12	17
TOTALS	28	22	50

B. Low Unemployment

Population 65 and Older (%)

Suicide Rate	Low	High	TOTALS
Low	22	10	32
High	5	13	18
TOTALS	27	23	50

b. Calculate percentages and gamma for each partial table. Compare the partial tables with each other and with the bivariate table. Compute partial gamma (G_p).

c. Summarize the results of the control. Does unemployment rate have any effect on the relationship between age and suicide rate? Describe the effect of the control variable in terms of the patterns of percentages, the values of the gammas, and the possible causal relationships among these variables.

14.3 [SW] Do long-term patients in mental health facilities become more withdrawn and reclusive over time? A sample of 608 institutionalized patients was rated on a scale that measures "reality orientation." Is there a relationship with length of institutionalization?

	Length of Institutionalization		
Reality Orientation	*Less Than Five Years*	*More Than Five Years*	TOTALS
Low	200	213	413
High	117	78	195
TOTALS	317	291	608

a. Calculate percentages and gamma for the bivariate table. Describe the bivariate relationship in terms of strength and direction.

 Is the relationship affected by the gender of the patient?

A. Females

	Length of Institutionalization		
Reality Orientation	*Less Than Five Years*	*More Than Five Years*	TOTALS
Low	95	120	215
High	60	37	97
TOTALS	155	157	312

B. Males

	Length of Institutionalization		
Reality Orientation	*Less Than Five Years*	*More Than Five Years*	TOTALS
Low	105	93	198
High	57	41	98
TOTALS	162	134	296

b. Calculate percentages and gamma for each partial table. Compare the partial tables with each other and with the bivariate table. Compute partial gamma (G_p).

c. Summarize the results of the control. Does gender have an effect on the relationship between length of institutionalization and reality orientation? Describe the effect of the control variable in terms of the patterns of percentages, the values of the gammas, and the possible causal relationships among these variables.

14.4 [SOC] A job-training center is trying to justify its existence to its funding agency. To this end, data on four variables have been collected for each of the 403 trainees served over the past three years: (X) whether the trainee completed the program, (Y) whether the trainee got and held a job for at least a year after training, (Z_1) the gender of the trainee, and (Z_2) the race or ethnicity of the trainee. Is employment related to completion of the program? Is the relationship between completion and employment affected by gender or race and ethnicity?

Held Job for at Least One Year	Training Completed?		
	No	*Yes*	TOTALS
No	126	72	198
Yes	60	145	205
TOTALS	186	217	403

a. Calculate percentages and gamma for the bivariate table. Describe the bivariate relationship in terms of strength and direction.

 Is the relationship affected by the race of the trainee?

A. Whites

Held Job for at Least One Year	Training Completed?		
	No	*Yes*	TOTALS
No	47	38	85
Yes	33	85	118
TOTALS	80	123	203

B. Nonwhites

Held Job for at Least One Year	Training Completed?		
	No	*Yes*	TOTALS
No	79	34	113
Yes	27	60	87
TOTALS	106	94	200

Is the relationship affected by the gender of the trainee?

A. Males

Held Job for at Least One Year	Training Completed?		
	No	*Yes*	TOTALS
No	62	36	98
Yes	30	73	103
TOTALS	92	109	201

B. Females Held Job for at Least One Year	Training Completed?		
	No	*Yes*	TOTALS
No	64	36	100
Yes	30	72	102
TOTALS	94	108	202

b. Calculate percentages and gamma for each partial table. Compare the partial tables with each other and with the bivariate table. Compute partial gamma (G_p).

c. Summarize the results of the controls. Does gender or race and ethnicity have an effect on the relationship? Describe the effect of the control variables in terms of the patterns of percentages, the values of the gammas, and the possible causal relationships among these variables.

14.5 |SOC| What are the social sources of support for the environmental movement? A recent survey gathered information on level of concern for such issues as global warming and acid rain. Is concern for the environment related to level of education? What effects do the control variables have?

Concern	Level of Education		
	Low	*High*	TOTALS
Low	27	35	62
High	22	48	70
TOTALS	49	83	132

a. Calculate percentages and gamma for the bivariate table. Describe the bivariate relationship in terms of strength and direction.

Is the relationship affected by gender or race and ethnicity?

A. Males Concern	Level of Education		
	Low	*High*	TOTALS
Low	14	17	31
High	11	22	33
TOTALS	25	39	64

B. Females Concern	Level of Education		
	Low	*High*	TOTALS
Low	13	18	31
High	11	26	37
TOTALS	24	44	68

A. Whites Concern	Level of Education		
	Low	*High*	TOTALS
Low	19	11	30
High	18	44	62
TOTALS	37	55	92

B. Nonwhites Concern	Level of Education		
	Low	*High*	TOTALS
Low	8	24	32
High	4	4	8
TOTALS	12	28	40

b. Calculate percentages and gamma for each partial table. Compare the partial tables with each other and with the bivariate table. Compute partial gamma (G_p).

c. Summarize the results of the controls. Does gender or race and ethnicity have an effect on the relationship? Describe the effect of the control variables in terms of the patterns of percentages, the values of the gammas, and the possible causal relationships among these variables.

Statistical Analysis Using SPSS

14.6 |SOC| Do people's views on moral and political questions change as they grow older? Are these changes affected by gender? In this problem, we will use recoded age (see Chapter 10) as the independent variable, and support for capital punishment (*cappun*) and support for assisted suicide (*letdie1*) as the dependent variables. We will analyze the bivariate relationships and then look at the effect of controlling for gender.

- Click the SPSS icon on your desktop.
- Load the GSS2018 database.

- Follow the instructions in the "You Are the Researcher" feature in Chapter 10 to recode *age*.
- Click **Analyze → Descriptive → Crosstabs**.
- Find *cappun* and *letdie1* in the list of variables and move them to the **Rows** box.
- Find recoded *age* in the list of variables and move it to the **Columns** box. *Make sure you use the recoded version!*
- Find *sex* and move it to the bottom box (**Layer 1 of 1**).
- SPSS will produce partial and bivariate tables for each combination of variables entered in these boxes.
- Click the Statistics button; select **chi square** and **Gamma**.
- Return to the **Crosstabs** window, click the **Cells** button, and select **column** in the **Percentages** box.
- Return to the **Crosstabs** window and click **OK**.

a. Start by looking at the bivariate results for recoded *age* and *cappun*. Look at the pattern of column percentages and then at the gammas. Characterize the strength (e.g., weak, moderate, or strong?) and direction (positive or negative?) of the bivariate relationship. Does support for the death penalty change by age? Which age group is the most supportive?

b. Next, look at the partial tables and the associated gammas. Are the column percentages different from the bivariate table? Are they different from each other? Are the gammas for the partial tables different from the bivariate gamma or different from each other? Based on these comparisons, is the relationship between recoded age and *cappun* direct, spurious or intervening, or interactive?

c. If the relationship is direct, spurious, or intervening, note the partial gamma (G_p), which is reported as the "first-order partial" in the box reporting gamma.

d. Write a brief summary of your analysis.

e. Repeat steps a to d for the relationship between recoded *age* and *letdie1*.

14.7 [SOC] Who is more afraid? Is fear related to social class? Is this relationship affected by gender? In this problem, we will use *class* as the independent variable and *fear* as the dependent variable. We will analyze the bivariate relationships and then look at the effect of controlling for gender.

- Load the *GSS2018* database.
- Click **Analyze → Descriptive → Crosstabs**.
- Find *fear* in the list of variables and move it to the **Rows** box.

- Find *class* and move it to the **Columns** box.
- Find *sex* and move it to the bottom box (**Layer 1 of 1**).
- Click the **Statistics** button; select **chi square** and **Gamma**.
- Return to the **Crosstabs** window, click the **Cells** button, and select **column** in the **Percentages** box.
- Return to the **Crosstabs** window and click **OK**.

a. Start with the bivariate relationship. Look at the pattern of column percentages and then at the gammas. Characterize the strength (e.g., weak, moderate, or strong?) and direction (positive or negative?) of the bivariate relationship. Does fear change by class? Which class is the most fearful?

b. Next, look at the partial tables and the associated gammas. Are the column percentages different from the bivariate table? Are they different from each other? Are the gammas for the partial tables different from the bivariate gamma or different from each other? Based on these comparisons, is the relationship direct, spurious or intervening, or interactive?

c. If the relationship is direct, spurious, or intervening, note the partial gamma (G_p), which is reported as the "first-order partial" in the box reporting gamma.

d. Write a brief summary of your analysis.

14.8 [SOC] What is the relationship between level of education (*degree*) and support for the theory of evolution (*evolved*)? Is this relationship affected by age? In this problem, *degree* is the independent variable and *evolved* is the dependent variable. We will analyze the bivariate relationships and then look at the effect of controlling for age group.

- Load the *GSS2018* database.
- Click **Analyze → Descriptive → Crosstabs**.
- Move *evolved* to the **Rows** box.
- Move *degree* to the **Columns** box.
- Move the recoded version of age to the bottom box (**Layer 1 of 1**). If necessary, see the instructions in the "You Are the Researcher" feature of Chapter 10 to recode *age*.
- Click the **Statistics** button; select **chi square** and **Gamma**.
- Return to the **Crosstabs** window, click the **Cells** button, and select **column** in the **Percentages** box.
- Return to the **Crosstabs** window and click **OK**.

a. Start with the bivariate relationship. Look at the pattern of column percentages and then at the gammas. Characterize the strength (e.g., weak,

moderate, or strong?) and direction (positive or negative?) of the bivariate relationship. Does *evolved* change by *degree*? Which level of education is the most supportive of the theory of evolution?

b. Next, look at the partial tables and the associated gammas. Are the column percentages different from the bivariate table? Are they different from each other? Are the gammas for the partial tables different from the bivariate gamma or different from each other? Based on these comparisons, is the relationship direct, spurious or intervening, or interactive?

c. If the relationship is direct, spurious, or intervening, note the partial gamma (G_p), which is reported as the "first-order partial" in the box reporting gamma.

d. Write a brief summary of your analysis.

14.9 ⬜SOC Are attitudes about sexual morality related to age? Is the relationship affected by gender? In this problem, recoded *age* is the independent variable and *premarsx* (attitudes about premarital sex) is the dependent variable. If necessary, see the instructions in the "You Are the Researcher" section of Chapter 10 to recode *age*. Gender will once again serve as the control variable.

- Load the *GSS2018* database.
- Click **Analyze → Descriptive → Crosstabs**.
- Move *premarsx* to the **Rows** box.
- Move recoded *age* to the **Columns** box.
- Move *sex* to the bottom box (**Layer 1 of 1**).
- Click the **Statistics** button; select **chi square** and **Gamma**.

- Return to the **Crosstabs** window, click the **Cells** button, and select **column** in the **Percentages** box.
- Return to the **Crosstabs** window and click **OK**.

a. Start with the bivariate relationship. Look at the pattern of column percentages and then at the gammas. Characterize the strength (e.g., weak, moderate, or strong?) and direction (positive or negative?) of the bivariate relationship. Does *premarsx* change by age group? Which age group is most opposed to premarital sex?

b. Next, look at the partial tables and the associated gammas. Are the column percentages different from the bivariate table? Are they different from each other? Are the gammas for the partial tables different from the bivariate gamma or different from each other? Based on these comparisons, is the relationship direct, spurious or intervening, or interactive?

c. If the relationship is direct, spurious, or intervening, note the partial gamma (G_p), which is reported as the "first-order partial" in the box reporting gamma.

d. Write a brief summary of your analysis.

(HINT: Be careful in interpreting direction for this table.)

14.10 ⬜SOC Conduct the analysis in problem 14.9 again, but substitute *marhomo* for *premarsx*. Does support of gay marriage vary by age group? Is the relationship affected by gender?

YOU ARE THE RESEARCHER

A Multivariate Analysis of Attitudes on Abortion

In this project, you will investigate the possible causes of attitudes about abortion, using the *GSS2018* data set and *abany* and *abpoor* as dependent variables.

Step 1: Choosing Independent Variables

Choose two ordinal-level variables that might affect attitudes about this controversial subject. We are using bivariate tables, so select independent variables with relatively few categories, say six or fewer. List your independent variables here:

SPSS Name	What Exactly Does This Variable Measure?	Level of Measurement

Step 2: Choosing a Control Variable

Select a variable that you think might affect the relationships between attitudes toward abortion and your independent variables. Your control variables can be any level of measurement but should have *only* two or three scores. If necessary, you can use the **recode** command to reduce the number of categories. List your control variable here.

SPSS Name	What Exactly Does This Variable Measure?	Level of Measurement

Step 3: Stating Hypotheses

State hypotheses about the relationships you expect to find between your independent variables and the two dependent variables. State these hypotheses in terms of the direction of the relationship you expect to find. For example, you might hypothesize that approval of abortion will decline as age increases (a negative relationship).

Dependent Variables	Hypotheses	
	Independent Variable #1	Independent Variable #2
abany		
abpoor		

Step 4: Running Crosstabs

Follow the instructions from problems 14.6 to 14.10. Place *abany* and *abpoor* in the **Rows** box, your independent variables in the **Columns** box, and your control variable in the bottom box (**Layer 1 of 1**). Be sure to select chi square and a measure of association appropriate for the level of measurement of your variables with the **Statistics** button and column percentages with the **Cells** button.

Step 5: Recording Results

These commands will generate a lot of output, and it will be helpful to summarize your results in a table such as the one on the next page. Fill in the table with the value and, if applicable, the sign of your measure of association. If you use gamma, note the value of partial gamma for each relationship as well.

First Independent Variable (SPSS name: _____)	Control Variable (SPSS name: _____)		
Bivariate Meas. of Assoc. = _____	Name of 1st category of Control Variable: _____	Name of 2nd category of Control Variable: _____	Name of 3rd category of Control Variable: _____ (if necessary)
	Meas. of Assoc. for Partial Table = _____	Meas. of Assoc. for Partial Table = _____	Meas. of Assoc. for Partial Table = _____ (if necessary)
Partial Gamma (G_p)* =			
Second Independent Variable (SPSS name: _____)	Control Variable (SPSS name: _____)		
Bivariate Meas. of Assoc. = _____	Name of 1st category of Control Variable: _____	Name of 2nd category of Control Variable: _____	Name of 3rd category of Control Variable: _____ (if necessary)
	Meas. of Assoc. for Partial Table = _____	Meas. of Assoc. for Partial Table = _____	Meas. of Assoc. for Partial Table = _____ (if necessary)
Partial Gamma (G_p)* =			

* Partial Gamma is reported as "first-order partial."

Step 6: Analyzing and Interpreting Results

For each combination of dependent and independent variables, interpret the impact, if any, of the control variable. Look for evidence of direct, spurious, or interactive relationships using Table 14.5 as a guide. Remember that your results may not fit any of these outcomes perfectly: Real-life research often generates ambiguous, difficult-to-understand results. Write a paragraph of interpretation for each set of partial tables. Were your expectations confirmed? How?

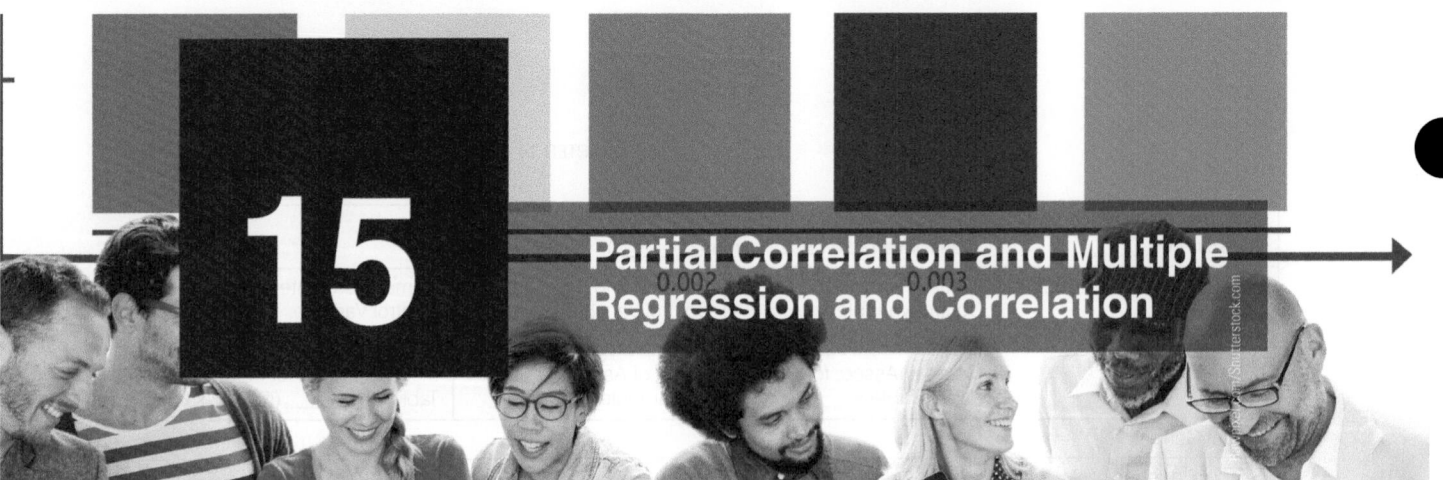

15

Partial Correlation and Multiple Regression and Correlation

LEARNING OBJECTIVES By the end of this chapter, you will be able to:

1. Compute and interpret partial correlation coefficients.
2. Find and interpret the least-squares multiple regression equation with partial slopes.
3. Find and interpret standardized partial slopes or beta-weights (β).
4. Calculate and interpret the coefficient of multiple determination (R^2).
5. Explain the limitations of partial and multiple regression analysis.
6. Use SPSS to generate partial correlations and conduct multiple regression analysis.

USING STATISTICS

The statistical techniques presented in this chapter are used to analyze the association between more than two interval-ratio–level variables. Examples of research situations in which these techniques are useful include:

1. A graduate student is interested in measuring the interrelationships between education, income, and age. **See page 431**

2. A sociology professor wants to know if students in sociology classes gain more empathy at the end of the term compared to non-sociology students, while controlling for other variables. **See page 448**

3. A researcher is interested in sorting out the effects of a mother's education and a father's education on their child's education. **See page 441**

4. A demographer is investigating the relationship between fertility rates and the educational levels of females in a group of nations. Her data show a moderate, negative relationship between the variables. But is there a difference in this relationship between affluent and less affluent nations? **See page 431**

As we mentioned at the beginning of Chapter 14, social science research is multivariate by nature and involves the simultaneous analysis of scores of variables. Some of the most powerful and widely used statistical tools for multivariate analysis are introduced in this chapter. We cover techniques that are used to analyze causal relationships between variables and to make predictions, both crucial endeavors in any science.

These techniques are based on Pearson's r (see Chapter 13) and are most appropriately used with high-quality, precisely measured interval-ratio variables. As we have noted on many occasions, such data are relatively rare in social science research, and the techniques presented in this chapter are commonly used on variables measured at the ordinal level and with nominal-level variables in the form of dummy variables (see Chapter 13).

The techniques presented in this chapter are generally more flexible than those presented in Chapter 14. They produce more information and provide a wider variety of ways of disentangling the interrelationships among the variables.

We first consider partial correlation analysis, a technique analogous to controlling for a third variable by constructing partial tables (see Chapter 14). The second technique, multiple regression and correlation, allows researchers to assess the effects, separately and in combination, of more than one independent variable on the dependent variable.

Throughout this chapter, we focus on research situations involving three variables and very small samples to illustrate the mathematical routines and basic logic. Extensions to situations involving four or more variables are relatively straightforward, but the computations become complex. Thus, after working through a simplified example problem, we will use SPSS to execute the computations, as we did with Pearson's r (Chapter 13). Likewise, there are a few end-of-chapter problems with small data sets so that you can be sure you understand the mathematical routines, but most problems require the use of SPSS.

Partial Correlation

In Chapter 13, we used Pearson's r to measure the strength and direction of bivariate relationships. To provide an example, we looked at the relationship between husband's contribution to housework (the dependent, or Y, variable) and number of children (the independent, or X, variable) for a sample of 12 families. We found a positive relationship of moderate strength ($r = 0.50$) and concluded that husbands increase their contribution to housework as the number of children increases.

You might wonder whether this relationship holds true for *all* types of families. For example, might husbands from different religious orientations respond differently? Would politically conservative husbands behave differently than husbands who are liberal? Would more-educated husbands respond differently than less-educated husbands? We can address issues like these by means of a technique called **partial correlation**, in which we observe how the bivariate relationship changes when a third variable, such as religion, political ideology, or education, is introduced. Third variables are often referred to as Z variables or **control variables**.

Partial correlation proceeds by first computing Pearson's r for the bivariate relationship (sometimes called the **zero-order correlation**) and then computing the partial (or first-order) correlation coefficient. If the partial correlation coefficient differs from the zero-order correlation coefficient, we conclude that the third variable has an effect on the bivariate relationship and might need to be incorporated into the analysis. Before considering matters of computation, we'll

consider the possible relationships between the partial and bivariate correlation coefficients and what they might mean.

TYPES OF RELATIONSHIPS

Three types of relationships were introduced in Chapter 14 and summarized in Table 14.5. We repeat the information here for the convenience of those readers who did not cover Chapter 14 and to place these patterns in the context of partial correlation.

Direct Relationship. One possibility is that the partial correlation coefficient is essentially the same value as the bivariate coefficient. Imagine, for example, that we found a partial correlation coefficient of $+0.49$ after controlling for husband's education, compared to the zero-order Pearson's r of $+0.50$. This would mean that the third variable (husband's education) has no effect on the relationship between number of children and husband's hours of housework. In other words, husband's level of education has no effect on the relationship between number of children and their contribution to housework. This outcome is consistent with the conclusion that there is a *direct* or causal relationship (see Figure 15.1) between X and Y and that the third variable (Z) is irrelevant and should be discarded from further consideration. The next step in the research project might be to run additional tests with other likely control variables (e.g., the researcher might control for the religion or ethnicity of the family, perhaps with the use of dummy variables).

FIGURE 15.1 A Direct Relationship Between *X* and *Y*

$X \longrightarrow Y$

Spurious and Intervening Relationships. A second possible outcome occurs when the partial correlation coefficient is much weaker than the bivariate correlation, perhaps even dropping to zero. This outcome is consistent with two different relationships. The first is called a *spurious* relationship: The control variable (Z) is a cause of both the independent (X) and the dependent (Y) variable (see Figure 15.2). This outcome would mean that X and Y are not actually related. They appear to be related only because both are dependent on a common cause (Z). Once Z is taken into account, the apparent relationship between X and Y disappears.

FIGURE 15.2 A Spurious Relationship Between X and Y

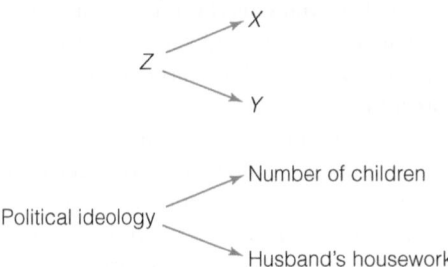

What would a spurious relationship look like? Imagine that we controlled for the political ideology of parents and found that the partial correlation coefficient was much weaker than the bivariate Pearson's r. This might indicate that the number of children does not actually change the husband's contribution to housework (that is, the relationship between X and Y is not direct). Rather, political ideology is the mutual cause of both of the other variables: Perhaps more conservative families are more likely to follow traditional gender-role patterns (in which husbands contribute less to housework) *and* have more children. For more examples of spurious relationships, refer back to the "Statistics in Everyday Life" box in Chapter 14.

FIGURE 15.3 An Intervening Relationship Between X and Y

The second explanation of this kind of pattern (partial correlation much weaker than the bivariate correlation) is called an *intervening* relationship between the variables (see Figure 15.3). In this situation, X and Y are not linked directly but are connected through the control variable. Again, once Z is controlled, the apparent relationship between X and Y disappears.

How can we tell the difference between spurious and intervening relationships? This distinction cannot be made on statistical grounds since spurious and intervening relationships look exactly the same in terms of statistics. The researcher may be able to distinguish between these two relationships in terms of the time order of the variables (i.e., which came first) or on theoretical grounds, but not on statistical grounds.

FIGURE 15.4 An Interactive Relationship Among X, Y, and Z

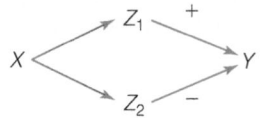

Interaction. A third possible relationship between variables should be mentioned, even though it *cannot* be detected by partial correlation analysis. This relationship, called interaction, occurs when the relationship between X and Y changes markedly under the various values of Z. For example, if we controlled for social class and found that husbands in middle-class families increased their contribution to housework as the number of children increased while husbands in working-class families did just the reverse, we would conclude that there was interaction among these three variables. In other words, there would be a positive relationship between X and Y for one category of Z and a negative relationship for the other category, as illustrated in Figure 15.4.

Interactive relationships were explored in Chapter 14. We will not examine them further here, however, because they cannot be detected by partial correlation.

COMPUTING AND INTERPRETING THE PARTIAL CORRELATION COEFFICIENT

The logic and computation of the partial correlation coefficient requires some new concepts and terminology. We'll introduce the terminology first and then move on to the formula.

Terminology. Partial correlation requires that we deal with more than one bivariate relationship, and we need to differentiate between them with subscripts. Thus, the symbol r_{yx} will refer to the correlation coefficient between variable Y and variable X; r_{yz} will refer to the correlation coefficient between Y and Z; and

r_{xz} will refer to the correlation coefficient between X and Z. Recall that correlation coefficients calculated for bivariate relationships are often referred to as zero-order correlations.

Partial correlation coefficients, or first-order partials, are symbolized as $r_{yx.z}$. The variable to the right of the dot in the subscript is the control variable. Thus, $r_{yx.z}$ refers to the partial correlation coefficient that measures the relationship between variables X and Y while controlling for variable Z. The formula for the first-order partial is

FORMULA 15.1

$$r_{yx.z} = \frac{r_{yx} - (r_{yz})(r_{xz})}{\sqrt{1 - r_{yz}^2}\sqrt{1 - r_{xz}^2}}$$

Note that you must first calculate the zero-order coefficients between all possible pairs of variables (variables X and Y, X and Z, and Y and X) before solving this formula.

Computation. To illustrate the computation of a first-order partial, we will return to the relationship between number of children (X) and husband's contribution to housework (Y) for 12 dual-career families. The zero-order r between these two variables ($r_{yx} = 0.50$) indicated a moderate, positive relationship (as number of children increased, husbands tended to contribute more to housework). Suppose the researcher wished to investigate the possible effects of husband's education on the bivariate relationship. The original data (from Table 13.1) and the scores of the 12 families on the new variable are presented in Table 15.1.

The zero-order correlations, as presented in the correlation matrix in Table 15.2, indicate that the husband's contribution to housework is positively related to number of children ($r_{yx} = 0.50$), that better-educated husbands tend to

TABLE 15.1 Scores on Three Variables for 12 Dual–Wage-Earner Families

Family	Husband's Housework (Y)	Number of Children (X)	Husband's Years of Education (Z)
A	1	1	12
B	2	1	14
C	3	1	16
D	5	1	16
E	3	2	18
F	1	2	16
G	5	3	12
H	0	3	12
I	6	4	10
J	3	4	12
K	7	5	10
L	4	5	16

TABLE 15.2 Zero-Order Correlations

	Husband's Housework (Y)	Number of Children (X)	Husband's Years of Education (Z)
Husband's Housework (Y)	1.00	0.50	−0.30
Number of Children (X)		1.00	−0.47
Husband's Years of Education (Z)			1.00

do less housework ($r_{yz} = -0.30$), and that families with better-educated husbands have fewer children ($r_{xz} = -0.47$).

Is the relationship between husband's housework (Y) and number of children (X) affected by husband's years of education (Z)? Substituting the zero-order correlations into Formula 15.1, we would have

$$r_{yx.z} = \frac{r_{yx} - (r_{yz})(r_{xz})}{\sqrt{1 - r_{yz}^2}\sqrt{1 - r_{xz}^2}}$$

$$r_{yx.z} = \frac{(0.50) - (-0.30)(-0.47)}{\sqrt{1 - (-0.30)^2}\sqrt{1 - (-0.47)^2}}$$

$$r_{yx.z} = \frac{0.36}{\sqrt{0.91}\sqrt{0.78}}$$

$$r_{yx.z} = \frac{0.36}{(0.95)(0.88)}$$

$$r_{yx.z} = \frac{0.36}{0.84}$$

$$r_{yx.z} = 0.43$$

Interpretation. The first-order partial ($r_{yx.z} = 0.43$) measures the strength of the relationship between husband's housework (Y) and number of children (X) while controlling for husband's education (Z). It is lower in value than the zero-order coefficient ($r_{yx} = 0.50$), but the difference in the two values is not great. This result suggests a direct relationship between variables X and Y. That is, when controlling for husband's education, the statistical relationship between husband's housework and number of children is essentially unchanged. Regardless of education, husband's hours of housework increase with the number of children.

Our next step in statistical analysis would probably be to select another control variable. The more the bivariate relationship retains its strength across a series of controls for third variables (Zs), the stronger the evidence for a direct relationship between X and Y.

In closing, we should mention an additional possible outcome, in which the partial correlation coefficient is greater in value than the zero-order coefficient.

FIGURE 15.5 A Possible Causal Relationship Among Three Variables

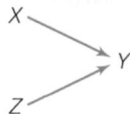

This outcome would be consistent with a causal model in which the variable taken as independent and the control variable each had a separate effect on the dependent variable and were uncorrelated with each other. This relationship is depicted in Figure 15.5. The absence of an arrow between X and Z indicates that they have no mutual relationship.

This pattern means that both X and Z should be treated as independent variables, and the next step in the statistical analysis would probably involve multiple correlation and regression, techniques that will be presented in the remainder of this chapter. *(For practice in computing and interpreting partial correlation coefficients, see problems 15.1a and b, 15.4a, and 15.5.)*

ONE STEP AT A TIME **Computing the Partial Correlation Coefficient**

Compute Pearson's r for all pairs of variables. Be clear about which variable is independent (X), which is dependent (Y), and which is the control (Z).

To Solve Formula 15.1

1. Multiply r_{yz} by r_{xz}.
2. Subtract the value you found in step 1 from r_{yx}. *This value is the numerator of Formula 15.1.*
3. Square the value of r_{yz}.
4. Subtract the quantity you found in step 3 from 1.
5. Take the square root of the quantity you found in step 4.
6. Square the value of r_{xz}.
7. Subtract the quantity you found in step 6 from 1.
8. Take the square root of the quantity you found in step 7.
9. Multiply the quantity you found in step 8 by the value you found in step 5. *This value is the denominator of Formula 15.1.*
10. Divide the quantity you found in step 2 by the quantity you found in step 9. *This is the partial correlation coefficient.*

ONE STEP AT A TIME **Interpreting the Partial Correlation Coefficient**

Choose the description below that comes closest to matching the relationship between the zero-order correlation (r_{yx}) and the partial correlation ($r_{yx.z}$).

1. The partial correlation coefficient ($r_{yx.z}$) is roughly the same value as the zero-order correlation (r_{yx}). A good rule of thumb for "roughly the same" is a difference of less than 0.10. This outcome is evidence that the control variable (Z) has no effect and that the relationship between X and Y is direct.
2. The partial correlation coefficient is much less (say, more than 0.10 less) than the bivariate correlation. This is evidence that the control variable (Z) changes the relationship between X and Y. The relationship between X and Y is either spurious (Z causes both X and Y) or intervening (X and Y are linked by Z).
3. Be aware that X, Y, and Z may have an interactive relationship in which the relationship between X and Y changes for each category of Z. Partial correlation analysis cannot detect interactive relationships.

Social Research and Data Analysis 15.1 **Education and Income**

An analysis of a representative sample of Americans ages 25 to 65 found a correlation of 0.24 between years of schooling and income, a weak to moderate, positive relationship. Income increases as education increases and education explains about 6% of the variation in income.

Is this relationship affected by age? The partial correlation between education and income, controlling for age, is 0.23, essentially the same value as the zero-order correlation. The relationship between education and income is direct: regardless of age, income increases as education increases. You may find this analysis reassuring as you move to the completion of your degree.

Source: 2018 General Social Survey

Social Research and Data Analysis 15.2 **Using SPSS to Determine If the Relationship Between Fertility and Women's Education Is Affected by Affluence**

In Social Research and Data Analysis 13.2, we found a moderate to strong negative relationship ($r = -0.58$) between women's education (X) and fertility rates (Y) for the nations in the *International* database. Now, we ask if the bivariate relationship is spurious and caused by the affluence (Z) of the nations. Do these variables appear to be related only because women in more affluent nations have more education *and* fewer children?

Start SPSS, load the *International* database, and click **Analyze** and **Correlate** and choose **Partial** (not Bivariate) from the sub-menu. Enter *TotalFertRate* (Total Fertility Rate) and *GESecondary* (the ratio of boys to girls at the secondary level of education) in the top ("**Variables:**")

window. In the bottom ("**Controlling for:**") window, enter *GNIcap* (Gross National Income per capita). This variable measures national affluence: The higher the score on this variable, the richer the nation. Click **OK**.

The output box will display a correlation matrix with the control variable (*GNIcap*) listed on the left. Find the partial correlation between "GESecondary" (gender parity in education) and "Total Fertility Rate" in the matrix. The value of -0.57 is virtually identical to the bivariate correlation (-0.58). This suggests that the relationship between education and fertility is direct. Regardless of the level of affluence of the nation, women's education is associated with fertility and more-educated women tend to have fewer children.

Multiple Regression: Predicting the Dependent Variable

In Chapter 13, the least-squares regression line was introduced as a way of describing the overall linear relationship between two interval-ratio variables and of predicting scores on Y from scores on X. This line was the best-fitting line to summarize the bivariate relationship and was defined by this formula:

FORMULA 15.2
$$Y = a + bX$$

The least-squares regression line can be modified to include (theoretically) any number of independent variables. This technique is called **multiple regression**. For ease of explication, we will confine our attention to the case involving two independent variables. The least-squares multiple regression equation is

FORMULA 15.3
$$Y = a + b_1 X_1 + b_2 X_2$$

where b_1 = the partial slope of the linear relationship between the first independent variable (X_1) and Y

b_2 = the partial slope of the linear relationship between the second independent variable (X_2) and Y

Some new notation and some new concepts are introduced in this formula. First, while the dependent variable is still symbolized as Y, the independent variables are differentiated by subscripts. Thus, X_1 identifies the first independent variable, and X_2, the second. The symbol for the slope (b) is also subscripted to identify the independent variable with which it is associated.

PARTIAL SLOPES

A major difference between the multiple and bivariate regression equations concerns the slopes (bs). In the case of multiple regression, the bs are called **partial slopes** (or unstandardized regression coefficients), and they show the amount of change in Y for a unit change in the independent variable while controlling for the effects of the other independent variables in the equation. The partial slopes are thus analogous to partial correlation coefficients and represent the direct effect of the associated independent variable on Y.

Computing Partial Slopes. The partial slopes for the independent variables are determined by Formulas 15.4 and 15.5:[1]

FORMULA 15.4
$$b_1 = \left(\frac{s_y}{s_1}\right)\left(\frac{r_{y1} - r_{y2}r_{12}}{1 - r_{12}^2}\right)$$

FORMULA 15.5
$$b_2 = \left(\frac{s_y}{s_2}\right)\left(\frac{r_{y2} - r_{y1}r_{12}}{1 - r_{12}^2}\right)$$

where b_1 = the partial slope of X_1 on Y
b_2 = the partial slope of X_2 on Y
s_y = the standard deviation of Y
s_1 = the standard deviation of the first independent variable (X_1)
s_2 = the standard deviation of the second independent variable (X_2)
r_{y1} = the bivariate correlation between Y and X_1
r_{y2} = the bivariate correlation between Y and X_2
r_{12} = the bivariate correlation between X_1 and X_2

To illustrate the computation of the partial slopes, we will assess the combined effects of number of children (X_1) and husband's education (X_2) on

[1]Partial slopes can be computed from zero-order slopes, but Formulas 15.4 and 15.5 are easier to use.

husband's contribution to housework. All the relevant information can be calculated from Table 15.1 and is noted here:

Husband's Housework	Number of Children	Husband's Education
$\overline{Y} = 3.3$	$\overline{X}_1 = 2.7$	$\overline{X}_2 = 13.7$
$s_y = 2.1$	$s_1 = 1.5$	$s_2 = 2.6$
	Zero-Order Correlations	
	$r_{y1} = 0.50$	
	$r_{y2} = -0.30$	
	$r_{12} = -0.47$	

The partial slope for the first independent variable (X_1, or number of children) is

$$b_1 = \left(\frac{s_y}{s_1}\right)\left(\frac{r_{y1} - r_{y2}r_{12}}{1 - r_{12}^2}\right)$$

$$b_1 = \left(\frac{2.1}{1.5}\right)\left(\frac{0.50 - (-0.30)(-0.47)}{1 - (-0.47)^2}\right)$$

$$b_1 = 1.4\left(\frac{0.50 - 0.14}{1 - 0.22}\right)$$

$$b_1 = 1.4\left(\frac{0.36}{0.78}\right)$$

$$b_1 = (1.4)(0.46)$$

$$b_1 = 0.65$$

For the second independent variable (X_2, or husband's education), the partial slope is

$$b_2 = \left(\frac{s_y}{s_2}\right)\left(\frac{r_{y2} - r_{y1}r_{12}}{1 - r_{12}^2}\right)$$

$$b_2 = \left(\frac{2.1}{2.6}\right)\left(\frac{-0.30 - (+0.50)(-0.47)}{1 - (-0.47)^2}\right)$$

$$b_2 = 0.81\left(\frac{-0.30 - (-0.24)}{1 - 0.22}\right)$$

$$b_2 = 0.81\left(\frac{-0.06}{0.78}\right)$$

$$b_2 = (0.81)(-0.08)$$

$$b_2 = -0.07$$

ONE STEP AT A TIME | **Computing Partial Slopes**

These procedures apply when there are two independent variables and one dependent variable. For more complex situations, use a computerized statistical package such as SPSS to do the calculations.

Compute the Partial Slope Associated with the First Independent Variable by Using Formula 15.4

1. Divide s_y by s_1.
2. Multiply r_{y2} by r_{12}.
3. Subtract the value you computed in step 2 from r_{y1}.
4. Square the value of r_{12}.
5. Subtract the value you computed in step 4 (r_{12}^2) from 1.
6. Divide the value you computed in step 3 by the value you computed in step 5.
7. Multiply the value you computed in step 6 by the value you computed in step 1. *This value is the partial slope associated with the first independent variable.*

Compute the Partial Slope Associated with the Second Independent Variable by Using Formula 15.5

1. Divide s_y by s_2.
2. Multiply r_{y1} by r_{12}.
3. Subtract the value you computed in step 2 from r_{y2}.
4. Square the value of r_{12}.
5. Subtract the value you computed in step 4 from 1.
6. Divide the value you computed in step 3 by the value you computed in step 5.
7. Multiply the value you computed in step 6 by the value you computed in step 1. *This value is the partial slope associated with the second independent variable.*

ONE STEP AT A TIME | **Interpreting Partial Slopes**

The value of a partial slope is the increase in the value of Y for a unit increase in the value of the associated independent variable while controlling for the effects of the other independent variable.

Finding the Y Intercept. Now that partial slopes have been determined for both independent variables, the Y intercept (a) (or the Constant) can be found. Note that a is calculated from the mean of the dependent variable (symbolized as \overline{Y}) and the means of the two independent variables (\overline{X}_1 and \overline{X}_2).

FORMULA 15.6

$$a = \overline{Y} - b_1\overline{X}_1 - b_2\overline{X}_2$$

Substituting the proper values for the example problem at hand, we would have

$$a = \overline{Y} - b_1\overline{X}_1 - b_2\overline{X}_2$$
$$a = 3.3 - (0.65)(2.7) - (-0.07)(13.7)$$
$$a = 3.3 - 1.8 - (-1.0)$$
$$a = 2.5$$

ONE STEP AT A TIME Computing the Y Intercept

Find the Y Intercept Using Formula 15.6

1. Multiply the mean of X_2 by b_2.
2. Multiply the mean of X_1 by b_1.
3. Subtract the quantity you found in step 1 from the quantity you found in step 2.
4. Subtract the quantity you found in step 3 from the mean of Y. *The result is the value of* a, *the* Y *intercept.*

Predicting Y′ with the Least-Squares Multiple Regression Line. For our example problem, the full least-squares multiple regression equation is

$$Y = a + b_1X_1 + b_2X_2$$
$$Y = 2.5 + (0.65)X_1 + (-0.07)X_2$$

As was the case with the bivariate regression line, this formula can be used to predict scores on the dependent variable from scores on the independent variables. For example, what would be our best prediction of husband's housework (Y′) for a family of four children ($X_1 = 4$) where the husband had completed 11 years of schooling ($X_2 = 11$)? Substituting these values into the least-squares equation, we would have

$$Y' = 2.5 + (0.65)(4) + (-0.07)(11)$$
$$Y' = 2.5 + 2.6 - 0.8$$
$$Y' = 4.3$$

Our prediction would be that this husband would contribute 4.3 hours per week to housework. This prediction is a kind of "educated guess," which is unlikely to be perfectly accurate. However, we will make fewer errors of prediction using the least-squares line (and, thus, incorporating information from the independent variables) than we would using any other method of prediction (assuming that there is a linear association between the independent and the dependent variables). *(All end-of-chapter problems offer opportunities to find unstandardized slopes and the* Y *intercept and to predict* Y *scores. The first three problems stress hand calculations and all others require the use of SPSS.)*

ONE STEP AT A TIME Using the Multiple Regression Line to Predict Scores on Y

1. Choose a value for X_1. Multiply this value by the value of b_1.
2. Choose a value for X_2. Multiply this value by the value of b_2.
3. Add the values you found in steps 1 and 2 to the value of a, the Y intercept. *The resulting value is the predicted score on* Y.

Multiple Regression: Assessing the Effects of the Independent Variables

The least-squares multiple regression equation (Formula 15.3) is used to isolate the separate effects of the independent variables and to predict scores on the dependent variable. However, in many situations, using this formula to determine the relative importance of the various independent variables will be awkward—especially when the independent variables differ in terms of units of measurement (e.g., number of children vs. years of education). When the units of measurement differ, a comparison of the partial slopes will not necessarily tell us which independent variable has the strongest effect and is thus the most important. Comparing the partial slopes of variables that differ in units of measurement is a little like comparing apples and oranges.

We can make it easier to compare the effects of the independent variables by converting all variables in the equation to a common scale and thereby eliminate variations in the values of the partial slopes that are solely a function of differences in units of measurement. We can, for example, standardize all distributions by changing the scores of all variables to Z scores. Each distribution of scores would then have a mean of 0 and a standard deviation of 1 (see Chapter 5), and comparisons between the independent variables would be much more meaningful.

COMPUTING THE STANDARDIZED REGRESSION COEFFICIENTS

To standardize the variables to the normal curve, we could actually convert all scores into the equivalent Z scores and then re-compute the slopes and the Y intercept. This would require a good deal of work; fortunately, a shortcut is available for computing the slopes of the standardized scores directly. These **standardized partial slopes**, also called **beta-weights**, are symbolized β upper case Greek letter beta.

Beta-Weights. The beta-weights (or the standardized regression coefficients) show the amount of change in the standardized scores of Y for a one-unit change in the standardized scores of each independent variable while controlling for the effects of all other independent variables.

Formulas and Computation for Beta-Weights. When we have two independent variables, the beta-weight for each is found by Formulas 15.7 and 15.8:

FORMULA 15.7
$$\beta_1 = b_1 \left(\frac{s_1}{s_y} \right)$$

FORMULA 15.8
$$\beta_2 = b_2 \left(\frac{s_2}{s_y} \right)$$

We can now compute the beta-weights for our sample problem to see which of the two independent variables has the stronger effect on the dependent variable. For the first independent variable, number of children (X_1):

$$\beta_1 = b_1 \left(\frac{s_1}{s_y} \right)$$

$$\beta_1 = (0.65) \left(\frac{1.5}{2.1} \right)$$

$$\beta_1 = (0.65)(0.71)$$

$$\beta_1 = 0.46$$

For the second independent variable, husband's education (X_2):

$$\beta_2 = b_2 \left(\frac{s_2}{s_y} \right)$$

$$\beta_2 = (-0.07) \left(\frac{2.6}{2.1} \right)$$

$$\beta_2 = (-0.07)(1.24)$$

$$\beta_2 = -0.09$$

Comparing the values of the beta-weights, we see that number of children (X_1) has a stronger effect than husband's education (X_2) on husband's housework (Y). Furthermore, the net effect (after controlling for the effect of husband's education) of the first independent variable is positive, while the net effect of the second independent variable is negative.

ONE STEP AT A TIME Computing Beta-Weights (β)

These procedures apply when there are two independent variables and one dependent variable. For more complex situations, use a computerized statistical package such as SPSS to do the calculations.

To Compute the Beta-Weight Associated with the First Independent Variable Using Formula 15.7

1. Divide s_1 by s_y.
2. Multiply the value you found in step 1 by the partial slope of the first independent variable (b_1). *This value is the beta-weight associated with the first independent variable.*

To Compute the Beta-Weight Associated with the Second Independent Variable Using Formula 15.8

1. Divide s_2 by s_y.
2. Multiply the value you found in step 1 by the partial slope of the second independent variable (b_2). *This value is the beta-weight associated with the second independent variable.*

ONE STEP AT A TIME Interpreting Beta-Weights (β)

A beta-weight (or standardized partial slope) shows the increase in the value of Y for a unit increase in the value of the associated independent variable while controlling for the effects of the other independent variable after all variables have been standardized (or transformed to Z scores).

THE STANDARDIZED LEAST-SQUARES REGRESSION LINE

Using standardized scores, the least-squares regression equation can be written as

FORMULA 15.9

$$Z_y = a_z + \beta_1 Z_1 + \beta_2 Z_2$$

where Z indicates that all scores have been standardized to the normal curve

The standardized regression equation can be further simplified by dropping the term for the Y intercept, because this term will always be 0 when scores have been standardized. This value is the point where the regression line crosses the Y axis and is equal to the mean of Y when all independent variables equal 0. This relationship can be seen by substituting 0 for all independent variables in Formula 15.6:

$$a = \overline{Y} - b_1\overline{X}_1 - b_2\overline{X}_2$$
$$a = \overline{Y} - b_1(0) - b_2(0)$$
$$a = \overline{Y}$$

Because the mean of any standardized distribution of scores is 0, the mean of the standardized Y scores will be 0 and the Y intercept will also be 0 ($a = \overline{Y} = 0$).

Thus, Formula 15.9 simplifies to

FORMULA 15.10

$$Z_y = \beta_1 Z_1 + \beta_2 Z_2$$

The standardized regression equation for our sample problem, with beta-weights noted, would be

$$Z_y = (0.45)Z_1 + (-0.09)Z_2$$

and it is immediately obvious that the first independent variable has a much stronger direct effect on Y than does the second independent variable.

As we have seen, multiple regression analysis permits the researcher to summarize the linear relationship among two or more independent variables and a dependent variable. The unstandardized regression equation (Formula 15.2) permits values of Y to be predicted from the independent variables in the original units of the variables. The standardized regression equation (Formula 15.10) allows the researcher to assess easily the relative importance of the various independent variables by comparing the beta-weights. *(All end-of-chapter problems offer opportunities to compute and interpret beta-weights.)*

Multiple Correlation

We use the multiple regression equations to disentangle the separate direct effects of each independent variable on the dependent variable. Using **multiple correlation** techniques, we can also ascertain the combined effects of all independent variables on the dependent variable. We do so by computing the **multiple**

correlation coefficient (R) and the **coefficient of multiple determination** (R^2). The value of the latter statistic represents the proportion of the variance in Y that is explained by all the independent variables combined.

We have seen that number of children (X_1) explains 25% of the variance in Y ($r_{y1}^2 = (0.50)^2 = 0.25$, which, when multiplied by 100, is 25%) by itself and that husband's education explains 9% of the variance in Y ($r_{y2}^2 = (-0.30)^2 = 0.09$, which, when multiplied by 100, is 9%). The two zero-order correlations cannot simply be added together to ascertain their combined effect on Y, because the two independent variables are also correlated with each other and, therefore, they will "overlap" in their effects on Y and explain some of the same variance. This overlap is eliminated in Formula 15.11:

FORMULA 15.11
$$R^2 = r_{y1}^2 + r_{y2.1}^2(1 - r_{y1}^2)$$

where R^2 = the coefficient of multiple determination
r_{y1}^2 = the zero-order correlation between Y and X_1, the quantity squared
$r_{y2.1}^2$ = the partial correlation of Y and X_2, while controlling for X_1, the quantity squared

The first term in this formula (r_{y1}^2) is the coefficient of determination for Y and X_1, or the amount of variation in Y explained by X_1 by itself. To this quantity we add the amount of the variation remaining in Y (given by $1 - r_{y1}^2$) that can be explained by X_2 after the effect of X_1 is controlled ($r_{y2.1}^2$). Basically, Formula 15.11 allows X_1 to explain as much of Y as it can and then adds in the effect of X_2 after X_1 is controlled (thus eliminating the "overlap" in the variance of Y that X_1 and X_2 have in common).

COMPUTING AND INTERPRETING R^2

To observe the combined effects of number of children (X_1) and husband's education (X_2) on husband's housework (Y), we need two quantities. The correlation between X_1 and Y ($r_{y1} = 0.50$) has already been found. Before we can solve Formula 15.11, we must first calculate the partial correlation of Y and X_2 while controlling for X_1 ($r_{y2.1}$). This quantity can be found by Formula 15.1 but we will simply report that $r_{y2.1} = -0.08$.

Formula 15.11 can now be solved for our example problem:

$$R^2 = r_{y1}^2 + r_{y2.1}^2(1 - r_{y1}^2)$$

$$R^2 = (0.50)^2 + (-0.08)^2(1 - 0.50^2)$$

$$R^2 = 0.25 + (0.006)(1 - 0.25)$$

$$R^2 = 0.25 + 0.005$$

$$R^2 = 0.255$$

The first independent variable (X_1), number of children, explains 25% of the variance in Y by itself. To this total, the second independent variable (X_2), husband's

education, adds only half a percent, for a total explained variance of 25.5%. In combination, the two independent variables explain a total of 25.5% of the variation in the dependent variable. *(All end-of-chapter problems offer opportunities to find and interpret* R^2. *The first three problems stress hand calculations and all others require the use of SPSS.)*

ONE STEP AT A TIME — Computing the Coefficient of Multiple Determination (R^2)

These procedures apply when there are two independent variables and one dependent variable. For more complex situations, use a computerized statistical package such as SPSS to do the calculations.

To Compute the Coefficient of Multiple Determination by Using Formula 15.11

Step Operation

1. Find the value of the partial correlation coefficient $r_{y2.1}$ using Formula 15.1.
2. Square the value you found in step 1.
3. Square the value of r_{y1}.
4. Subtract the value you found in step 3 from 1.00.
5. Multiply the value you found in step 4 by the value you found in step 2.
6. Add the value you found in step 5 to r_{y1}^2 (the value you found in step 3). *The result is the coefficient of multiple determination* (R^2).

ONE STEP AT A TIME — Interpreting R^2

The coefficient of multiple determination (R^2) is the total amount of the variation in Y explained by all independent variables combined.

USING SPSS — **Regression Analysis**

For obvious reasons, researchers use computerized statistical packages like SPSS rather than hand calculations to conduct regression analysis. In this installment of "Using SPSS," we will demonstrate regression analysis using the *States* data set. We'll take the rate of traffic fatalities (*TrafDths17*) as the dependent variable. Our independent variables will be the median age of the state's population (*MdAge*) and median household income (*MdHHinc*). Will the age of the state's population be related to the rate of traffic deaths? Will more affluent states be more or less dangerous for drivers?

To begin:

1. Load the *States* database.
2. From the main menu bar, click **Analyze, Regression,** and **Linear**.
3. On the **Linear Regression** screen, place *TrafDths17* in the **Dependent** window and *MdAge* and *MdHHinc* in the **Independent** window.

(continued)

USING SPSS

(Continued)

4. Click **Statistics** and check **Descriptives**. Click **Continue** to return to the **Linear Regression** screen.
5. Click **OK** and results will be printed to the Output screen in several separate boxes, only some of which are relevant for our purposes.
6. Find the zero-order correlations between the three variables in the "Correlations" box.
7. Find the values for R^2 in the "Model Summary" box.
8. The "Coefficients" box displays the value of a and b under "Unstandardized Coefficients." Look in the column labeled B. The value in the first row is a and the entry in the second row is b.
9. In the same box, find the beta-weights listed under "Standardized Coefficients" and "Beta."

We can begin the analysis by noting the zero-order correlations. The rate of traffic fatalities has a weak negative relationship ($r = -0.16$) with median age (the older the population, the lower the rate of traffic deaths) and a strong negative relationship ($r = -0.70$) with median income (more affluent states have lower rates of traffic deaths).

The multiple regression line, which could be used to predict scores on Y for various combinations of scores on the independent variables, would be:

$$TradDths17 = 3.25 + (-0.02)MdAge + (-0.00)MdHHinc*$$

Together, the two independent variables explain 54% ($R^2 = 0.54$) of the variance in the traffic fatality rates across the states. The beta-weights show that the net effects of both independent variables are negative and that $MdHHinc$ has a much stronger effect.

$$TradDths17 = (-0.20)MdAge + (-0.72)MdHHinc$$

In summary, age and affluence account for 54% of the variance in the rate of traffic deaths. Median household income has the stronger effect.

*The b for $MdHHinc$ is reported in scientific notation: $-1.975E-5$. To find the actual value, move the decimal point five (E-5) places to the left: -0.0000197.

Social Research and Data Analysis 15.3 **Multiple Regression and Correlation**

Who has more influence on a person's educational achievements: Mom or Dad? We can investigate this question with the 2018 General Social Survey. We'll use *educ* (highest year of schooling) as the dependent variable and *paeduc* (father's highest year of schooling) and *maeduc* (mother's highest year of schooling) as the independent variables.

The descriptive statistics are:

	Respondent's Years of Education (Y)	Father's Years of Education (X_1)	Mother's Years of Education (X_2)
Mean =	14.22	11.90	11.91
Standard deviation =	2.93	4.17	3.80

(continued)

Social Research and Data Analysis 15.3 (*Continued*)

The zero-order correlations among these three variables are

	Respondent's education (Y)	Father's education (X_1)	Mother's education(X_2)
Respondent's education (Y)	1.00	0.42	0.39
Father's education (X_1)		1.00	0.69
Mother's education (X_2)			1.00

These results suggest moderate relationships between each independent variable and educational achievement. All relationships are positive: educational achievement increases as father's and mother's years of education increase.

To find the multiple regression equation, we must find the partial slopes.

For father's education (X_1):

$$b_1 = \left(\frac{s_y}{s_1}\right)\left(\frac{r_{y1} - r_{y2}r_{12}}{1 - r_{12}^2}\right)$$

$$b_1 = \left(\frac{2.93}{4.17}\right)\left(\frac{(0.42) - (0.39)(0.69)}{1 - (0.69)^2}\right)$$

$$b_1 = (0.70)\left(\frac{0.42 - 0.27}{1 - 0.48}\right)$$

$$b_1 = (0.70)\left(\frac{0.15}{0.52}\right)$$

$$b_1 = (0.70)(0.29)$$

$$b_1 = 0.20$$

For mother's education (X_2):

$$b_2 = \left(\frac{s_y}{s_2}\right)\left(\frac{r_{y2} - r_{y1}r_{12}}{1 - r_{12}^2}\right)$$

$$b_2 = \left(\frac{2.93}{3.80}\right)\left(\frac{(0.39) - (0.42)(0.69)}{1 - (0.69)^2}\right)$$

$$b_2 = (0.77)\left(\frac{0.39 - 0.29}{1 - 0.48}\right)$$

$$b_2 = (0.77)\left(\frac{0.10}{0.52}\right)$$

$$b_2 = (0.77)(0.19)$$

$$b_2 = 0.15$$

The Y intercept would be

$$a = \overline{Y} - b_1\overline{X}_1 - b_2\overline{X}_2$$
$$a = 14.22 - (0.20)(11.90) - (0.15)(11.91)$$
$$a = 14.22 - 2.38 - 1.79$$
$$a = 10.05$$

The multiple regression equation is

$$Y = a + b_1X_1 + b_2X_2$$
$$Y = 10.05 + (0.20)X_1 + (0.15)X_2$$

What level of education would we predict for someone whose father had completed 10 years of schooling ($X_1 = 10$) and whose mother had completed 14 ($X_2 = 14$)?

$$Y' = 10.05 + (0.20)(10) + (0.15)(14)$$
$$Y' = 10.05 + 2.00 + 2.10$$
$$Y' = 14.15$$

To assess which of the two independent variables has the stronger effect on education, the standardized partial slopes must be computed.

For father's education (X_1):

$$\beta_1 = b_1\left(\frac{s_1}{s_y}\right)$$

$$\beta_1 = (0.20)\left(\frac{4.17}{2.93}\right)$$

$$\beta_1 = (0.20)(1.42)$$

$$\beta_1 = 0.28$$

For mother's education (X_2):

$$\beta_2 = b_2\left(\frac{s_2}{s_y}\right)$$

$$\beta_2 = (0.15)\left(\frac{3.80}{2.93}\right)$$

$$\beta_2 = (0.15)(1.30)$$

$$\beta_2 = 0.20$$

(*continued*)

Social Research and Data Analysis 15.3 | *(Continued)*

The standardized regression equation is

$$Z_y = \beta_1 Z_1 + \beta_2 Z_2$$
$$Z_y = (0.28)Z_1 + (0.20)Z_2$$

The independent variables have roughly equal effects on educational attainment, with father's education slightly more powerful.

To assess the combined effects of the two independent variables on education, we will compute the coefficient of multiple determination using Formula 15.11. To solve this formula, we must first compute $r_{y2.1}$ or the partial correlation between education (*educ*) and mother's education (*maeduc*) while controlling for father's education (*paeduc*). We find this quantity by solving Formula 15.1 but will save time and simply report that the value for $r_{y2.1}$ is 0.16.

$$R^2 = r_{y1}^2 + r_{y2.1}^2(1 - r_{y1}^2)$$

$$R^2 = (0.42)^2 + (0.16)^2(1 - (0.42)^2)$$

$$R^2 = 0.18 + (0.03)(1 - 0.18)$$

$$R^2 = 0.18 + (0.03)(0.82)$$

$$R^2 = 0.18 + 0.03$$

$$R^2 = 0.21$$

The first independent variable, father's education, explains 18% of the variation in educational achievement by itself. Mother's education explains an additional 3% of the variation in educational achievement. Taken together, the two independent variables explain a total of 21% of the variation in educational achievement.

Interpreting Statistics: Another Look at the Correlates of Crime

In Chapter 13, we assessed the association between poverty and crime. We found a moderate to strong, positive relationship ($r = 0.63$) between a measure of poverty and the homicide rate for the 50 states, an indication that there may be an important relationship between these two variables. In this installment of "Interpreting Statistics," we return to this relationship and add several independent variables to the analysis.[2]

The second independent variable is median age (*MdAge*). Research on street crimes and serious felonies like homicide commonly find that rates of violent crime are highest for people in their teens and twenties but decline dramatically as age rises. Thus, we can expect a substantial negative relationship between median age and homicide rate (the lower the median age in a population, the higher the homicide rate).

The third independent variable comes from a body of research that argues that there is a subculture of violence, a stronger tendency to use force and aggression, in the South. In other words, the everyday use of force is endorsed more strongly in the norms and values of the southern regional subculture than in other

[2]This analysis uses three independent variables, one more than earlier examples in this chapter. The additional independent variable will not complicate interpretation unduly but the computations for three (or more) independent variables are complex and will not be shown in this analysis. Use a computerized statistical package—like SPSS—for situations involving three or more independent variables.

TABLE 15.3 Zero-Order Correlations

	Homicide	Poverty	Age	Region
Homicide	1.00	0.63	−0.17	0.47
Poverty		1.00	−0.08	0.67
Age			1.00	−0.01

parts of the nation. Thus, we should expect to find higher rates of homicide in southern states.

Although there are compelling reasons for including region in this analysis, we must first confront an important problem. Region (South, North, Midwest, etc.) is a nominal-level variable, but regression analysis requires that all variables be measured at the interval-ratio level. We can resolve this problem by treating region as a dummy variable (see Chapter 13). As you recall, dummy variables have exactly two categories, one coded as a score of 0 and the other as a score of 1. Treated this way, nominal-level variables such as gender and race are commonly included in regression equations. In this case, we will create a variable that is coded so that southern states are scored as "1" and non-southern states as "0." Because "South" is the higher score, we expect that region will have a positive relationship with homicide.

Before beginning the multivariate analysis, we should examine the correlation coefficients between all possible pairs of variables. Table 15.3 reports the zero-order correlations.

As we saw in Chapter 13, there is a moderate to strong, positive relationship between poverty and homicide: States with higher rates of poverty have higher rates of homicide ($r = 0.63$). Turning to the other variables, there is, as expected, a negative relationship between age and homicide rate ($r = -0.17$) but the relationship is only weak to moderate. Finally, there is—as expected—a moderate positive relationship ($r = 0.47$) between region and homicide rate. Southern states have higher homicide rates.

There is a strong positive relationship ($r = 0.67$) between poverty and region (southern states have more poverty), and a very weak negative relationship ($r = -0.01$) between region and age. Finally, there is a weak negative relationship ($r = -0.08$) between age and poverty: States with a lower median age tend to be more impoverished.

We begin the multivariate analysis by asking whether the original relation between homicide and poverty persists after controlling for the effects of age and region, controlling for these variables one at a time. Recall that a partial correlation coefficient indicates the strength and direction of a bivariate relation after controlling for the effects of a third variable. Table 15.4 reports the results of controlling for age and then for region.

The partial correlation with age as the control variable is almost exactly equal to the zero-order correlation. Thus, we can conclude that age has no effect, an outcome that would increase our confidence that there is a direct, causal relationship between poverty and the homicide rate.

However, the partial correlation when controlling for region (0.47) is noticeably weaker than the zero-order correlation (0.63). On one hand, this outcome weakens the argument that there is a direct causal relationship between poverty

TABLE 15.4 **Partial Correlation Coefficients for the Relationship Between Homicide Rate and Poverty**

Zero-Order Correlation Between Homicide and Poverty	Partial Correlation After Controlling for	
	Age	*Region*
0.63	0.62	0.47

and violence. Region has a substantial effect on the bivariate relationship and should be taken into account in any further analysis. On the other hand, the bivariate relationship between homicide and poverty is positive and moderate in strength even after controlling for region. Thus, we would probably conclude that there is a relationship between poverty and homicide rate, even though the relationship is affected by region.

Next, we can assess the relative importance of the three independent variables on homicide. Table 15.5 presents the essential information—beta-weights and R^2—that will help us sort out the relative importance and total impact of these variables on the homicide rate.

By itself, poverty explained 40% of the variance in homicide rates among the states ($r^2 = (0.63)^2 = 0.40$). To this, age and region add another 1%, for a total of 41% ($R^2 = 0.41$). These three variables together account for 41% of the variation in homicide rates from state to state. This is quite a substantial percentage, even though about 59% of the variation remains unaccounted for or unexplained.

We might be able to raise the value of R^2 by adding more independent variables to the equation. However, it is common to find that additional independent variables explain smaller and smaller proportions of the remaining variance and have a diminishing effect on R^2. This can be caused by many factors, including the fact that the independent variables will usually be correlated with each other (e.g., see the relationship between region and poverty in Table 15.3) and will overlap in their effects on the dependent variable.

The beta-weights in Table 15.5 show the effect of each variable while controlling for all other variables in the equation (as opposed to controlling for other variables one at a time as we did in Table 15.4). Poverty has the strongest effect when controlling for the other variables in the equation, and this reinforces the argument that there is an important relationship between poverty and crime: Regardless of age and region, crime tends to increase as poverty increases. Age

TABLE 15.5 **Regression Analysis with Homicide Rate as the Dependent Variable**

Variable	Beta-Weights (β)
Poverty	0.55
Age	−0.12
Region	0.10
	$R^2 = 0.41$

STATISTICS In Everyday Life

Statistics and Baseball

"What's that pitcher's WHIP?" "That's pretty good, but what is his WAR?" These are the kinds of questions that are on baseball fans' minds these days. A pitcher's WHIP is the pitcher's Walks and Hits Per Innings Pitched. That's a simple ratio between the combined total of walks and hits for every full inning pitched. A player's WAR is the player's Wins Above Replacement. It considers many aspects of performance to determine how many more wins the player brings their team compared to a typical replacement player at the same position. The WAR can actually be computed in several different ways. Most fans probably do not even know what goes into the statistic.

Baseball has always been a paradise for statistics junkies and the possibilities have increased exponentially since the advent of big data analytics and advanced video technologies. Today, virtually every movement in every game is captured and recorded for analysis. Each pitch generates 20 separate pieces of data (speed, arm angle of the pitcher, movement of the ball, etc.) and each game can generate a terabyte of data (that's enough storage for 500 movies). This mountain of data is used to evaluate player performance in minute detail, help prepare for games, and can dictate in-game strategies (like where to position the shortstop for a specific batter in a specific situation).

As we noted at the beginning of this text, big data and modern analytics have found their way into business, finance, marketing, and numerous other sectors of modern society. It is worth noting that these trends rely on, among many other factors, supercomputers and advanced statistical analysis, much of which is based on the techniques of multiple regression and correlation covered in this chapter.

and region have roughly equal effects on crime, although the direction of the relationship varies. With other variables controlled, homicide rates decrease as age increases. Finally, even with the effects of poverty and age controlled, crime is associated with "southernness."

In conclusion, this multivariate analysis indicates that poverty, age, and region all have important effects on homicide rate. The beta-weights show that poverty has the strongest effect and, taken together, the three variables account for 41% of the variance in homicide rate from state to state.

The Limitations of Multiple Regression and Correlation

Multiple regression and correlation are powerful tools for analyzing the interrelationships among three or more variables. The techniques presented in this chapter permit the researcher to predict scores on one variable from two or more other variables, to distinguish among independent variables in terms of the importance of their direct effects on a dependent variable, and to ascertain the total effect of a set of independent variables on a dependent variable. In terms of the flexibility of the techniques and the volume of information they can supply, multiple regression and correlation represent some of the most powerful statistical techniques available to social science researchers.

Powerful tools are not cheap. They require high-quality data, and measurement at the interval-ratio level can be difficult to attain in social science research. Furthermore, these techniques assume that the interrelationships among the

variables follow a particular form. First, they assume that each independent variable has a linear relationship with the dependent variable. How well a given set of variables meets this assumption can be quickly checked with scatterplots.

Second, these techniques assume that there is no interaction among the variables in the equation. If there is interaction among the variables, it will not be possible to estimate or predict the dependent variable accurately by simply adding the effects of the independent variables. There are techniques for handling interaction among the variables in the set, but these techniques are beyond the scope of this text.

Third, multiple regression and correlation assume that the independent variables are uncorrelated with each other. Strictly speaking, this condition means that the zero-order correlation among all pairs of independent variables should be zero; however, practically, we act as if this assumption has been met if the intercorrelations among the independent variables are low.

To the extent that these assumptions are violated, the regression coefficients (especially partial and standardized slopes) and the coefficient of multiple determination (R^2) become less and less trustworthy and the techniques less and less useful. If the assumptions of the model cannot be met, the alternative might be to turn to the multivariate techniques described in Chapter 14. Unfortunately, those techniques supply a lower volume of less precise information about the interrelationships among the variables.

We should note that we have covered only the simplest applications of partial correlation and multiple regression and correlation. In terms of logic and interpretation, the extensions to situations involving more variables are relatively straightforward. However, the computations for these situations are extremely complex. If you are faced with a situation involving more than three variables, turn to one of the computerized statistical packages commonly available on college campuses (e.g., SPSS, R, or SAS). These programs require minimal computer literacy and can handle complex calculations in, literally, the blink of an eye. Efficient use of these packages will enable you to avoid drudgery and will free you to do what social scientists everywhere enjoy doing most: pondering the meaning of your results and, by extension, the nature of social life.

READING STATISTICS Do Sociology Courses Increase Empathy in Students?

Most professional quantitative studies analyze interrelationships among many variables, and many use the regression techniques described in this chapter. The results of these projects often feature tables of numbers referred to as ordinary least-squares regression tables. Since there are multiple names for some of the statistics you will see in the literature, we summarize them below:

Symbol	Statistic	Alternate Name
b	Partial Slope	Unstandardized Regression Coefficient
β	Beta-weight	Standardized Regression Coefficient
R^2	Coefficient of Multiple Determination	R^2
a	y-intercept	Constant

(continued)

READING STATISTICS *(Continued)*

An example of this kind of summary table might look like this:

TABLE 1 Ordinary Least-Squares Regression Coefficients for the Effects of X_1, X_2, and X_3 on Y

Independent Variables	b	β
X_1	0.17	0.47
X_2	0.23	0.32
X_3	0.27	0.16
Constant	0.031	
R^2	0.31	

This table reports that the first independent variable, X_1, has a partial slope of 0.17 in its association with the dependent variable, Y, when controlling for all of the other independent variables (X_2 and X_3) in the equation. The Constant tells us that the value of Y is 0.031 when the values of all of the X variables are zero. Finally, the R^2 statistic indicates that the three independent variables explain a combined amount of 0.031 in the variation in Y. In other words, the full model explains 3.1% of the variation in the dependent variable.

MULTIPLE REGRESSION IN THE PROFESSIONAL LITERATURE

Do students in sociology courses gain more empathy over the course of their term than students in non-sociology courses? Since sociologists are often interested in inequalities and the well-being of minorities and others that are disadvantaged, it would make sense for their students to develop some of the same values. In a recent study featured in *Teaching Sociology*, several researchers sought to explore this association by studying hundreds of undergraduate students in sociology and non-sociology courses to see if those in the sociology courses developed more empathy.

In the results section of the study, the authors tested multiple models with different combinations of independent variables to predict the dependent variable (change in empathy). Below is an abridged version of the final model that explained just over 17% of the variation in empathy change.

From this abridged table, which includes just a few of the independent variables, we can see that

TABLE 2 Empathy Change Regressed on Student and Instructor Variables

	Final Model	
Variable	b	β
Intercept	0.726	
Introduction to Sociology (reference = comparison group)	0.113***	0.163
Social Problems (reference = comparison group)	0.075*	0.112
Student predictor variables		
Conservative (reference = liberal)	−0.187**	−0.174
Instructor predictor variables		
Perceived LGBTQ	0.043	0.050
Model Statistics		
R^2		0.172

*$p < .05$, **$p < .01$, ***$p < .001$

students in an Introduction to Sociology class showed a change in their empathy score relative to those in non-sociology courses by 0.113 when controlling for all of the other variables in the equation, and the *** indicates that this was significant at the 0.001 level. So we can be 99.9% confident that the effect did not occur due to chance. Social Problems students also showed more empathy change, but the beta-weight for this variable is smaller than that for Introduction to Sociology, indicating a smaller effect.

Students that identified as conservative in their political views showed less empathy change than liberal students (−0.187) but since we do not see an asterisk (*) we can conclude that this was not significant. Similarly, the effect for the perception that the instructor was LGBTQ was not significant. Moving back to the beginning of the table, we see that the Intercept is 0.726. This is the value of empathy change when all other variables are held constant (at zero). The R^2 is at 0.172 so the full model explained 17.2% of the variation in empathy change.

Source: Rockwell, A., Vidmar, C., Harvey, P., and Greenwood, L. (2019) "Do Sociology Courses Make More Empathetic Students? A Mixed-Methods Study of Empathy Change in Undergraduates." *Teaching Sociology*, 47 (4): 284–302.

SUMMARY

1. In partial correlation, we observe what happens to a bivariate relationship when a third variable is controlled. With this technique, we can detect direct and spurious or intervening relationships between X and Y.

2. We can predict scores on a dependent variable from scores on more than one independent variable by using the multiple regression equation with partial slopes. We can disentangle the relative importance of the independent variables with standardized partial slopes or beta-weights.

3. The multiple correlation coefficient (R^2) shows the combined effects of all independent variables on the dependent variable in terms of the proportion of the total variation in Y that is explained.

4. Partial correlation and multiple regression and correlation are some of the most powerful tools available to the researcher and demand high-quality measurement and relationships among the variables that are linear and noninteractive. Further, correlations among the independent variables must be low (preferably zero).

SUMMARY OF FORMULAS

Partial correlation coefficient:

$$15.1 \qquad r_{yx.z} = \frac{r_{xy} - (r_{yz})(r_{xz})}{\sqrt{1 - r_{yz}^2}\sqrt{1 - r_{xz}^2}}$$

Least-squares regression line (bivariate):

$$15.2 \qquad Y = a + bX$$

Least-squares multiple regression line:

$$15.3 \qquad Y = a + b_1 X_1 + b_2 X_2$$

Partial slope for X_1:

$$15.4 \qquad b_1 = \left(\frac{s_y}{s_1}\right)\left(\frac{r_{y1} - r_{y2}r_{12}}{1 - r_{12}^2}\right)$$

Partial slope for X_2:

$$15.5 \qquad b_2 = \left(\frac{s_y}{s_2}\right)\left(\frac{r_{y2} - r_{y1}r_{12}}{1 - r_{12}^2}\right)$$

Y intercept:

$$15.6 \qquad a = \bar{Y} - b_1\bar{X}_1 - b_2\bar{X}_2$$

Standardized partial slope (beta-weight) for X_1:

$$15.7 \qquad \beta_1 = b_1\left(\frac{s_1}{s_y}\right)$$

Standardized partial slope (beta-weight) for X_2:

$$15.8 \qquad \beta_2 = b_2\left(\frac{s_2}{s_y}\right)$$

Standardized least-squares regression line:

$$15.9 \qquad Z_y = a_z + \beta_1 Z_1 + \beta_2 Z_2$$

Standardized least-squares regression line (simplified):

$$15.10 \qquad Z_y = \beta_1 Z_1 + \beta_2 Z_2$$

Coefficient of multiple determination:

$$15.11 \qquad R^2 = r_{y1}^2 + r_{y2.1}^2(1 - r_{y1}^2)$$

GLOSSARY

Beta-weight (β). See **Standardized partial slope.**

Coefficient of multiple determination (R^2). A statistic that equals the proportion of the total variation in the dependent variable that is explained by all independent variables combined.

Control variable. In multivariate statistical analysis, a third variable (Z) whose effect on a bivariate relationship is held constant.

Multiple correlation. A multivariate technique for examining the combined effects of more than one independent variable on a dependent variable.

Multiple correlation coefficient (R). A statistic that indicates the strength of the correlation between a dependent variable and two or more independent variables.

Multiple regression. A multivariate technique that breaks down the separate effects of the independent variables on the dependent variable; used to make predictions of the dependent variable.

Partial correlation. A multivariate technique for examining a bivariate relationship while controlling for other variables.

Partial correlation coefficient. A statistic that shows the relationship between two variables while control-

ling for other variables; $r_{yx.z}$ is the symbol for the partial correlation coefficient when controlling for one variable.

Partial slope. In a multiple regression equation, the slope of the relationship between a particular independent variable and the dependent variable while controlling for all other independent variables in the equation.

Standardized partial slope (beta-weight). The slope of the relationship between a particular independent variable and the dependent variable when all scores have been normalized and all other variables are controlled.

Zero-order correlation. Correlation coefficient for bivariate relationships.

PROBLEMS

The data sets in the first three problems are unrealistically small to allow you to practice the computational routines presented in this chapter. Most problems use SPSS.

15.1 $\boxed{\text{SOC}}$ In problem 13.1, data regarding voter turnout in five cities were presented. For the sake of convenience, the data for three of the variables are presented again here, along with descriptive statistics and zero-order correlations.

City	Turnout	Unemployment Rate	% Negative Ads
A	55	5	60
B	60	8	63
C	65	9	55
D	68	9	53
E	70	10	48
Mean =	63.6	8.2	55.8
s =	5.5	1.7	5.3

	Unemployment Rate	Negative Ads
Turnout	0.95	−0.87
Unemployment Rate		−0.70

a. Compute the partial correlation coefficient for the relationship between turnout (Y) and unemployment rate (X) while controlling for the effect of negative advertising (Z). What effect does this control variable have on the bivariate relationship? Is the relationship between turnout and unemployment direct? *(HINT: Use Formula 15.1.)*

b. Compute the partial correlation coefficient for the relationship between turnout (Y) and negative advertising (X) while controlling for the effect of unemployment rate (Z). What effect does this control variable have on the bivariate relationship? Is the relationship between turnout and negative advertising direct? *(HINT: Use Formula 15.1. You will need this partial correlation to compute the multiple correlation coefficient.)*

c. Find the unstandardized multiple regression equation with turnout as the dependent variable and unemployment (X_1) and negative ads (X_2) as the independent variables. What turnout would be expected in a city in which the unemployment rate was 10% and 75% of the campaign ads were negative? *(HINT: Use Formulas 15.4 and 15.5 to compute the partial slopes and then use Formula 15.6 to find the Y intercept, a. The regression line is stated in Formula 15.3. Substitute 10 for X_1 and 75 for X_2 to compute predicted Y.)*

d. Compute beta-weights for each independent variable. Which has the stronger impact on turnout? *(HINT: Use Formulas 15.7 and 15.8 to calculate the beta-weights.)*

e. Compute the multiple correlation coefficient (R) and the coefficient of multiple determination (R^2). How much of the variance in voter turnout is explained by the two independent variables? *(HINT: Use Formula 15.11. You calculated $r^2_{y2.1}$ in part b of this problem.)*

f. Write a paragraph summarizing your conclusions about the relationships among these three variables.

15.2 [PS] Problem 13.4 presented data on 10 precincts. The information is reproduced here.

Precinct	Percentage Democrat (X_1)	Percentage Minority (X_2)	Percentage Voter Turnout (Y)
A	33	9	36
B	78	15	60
C	62	18	64
D	50	10	56
E	45	12	55
F	56	8	52
G	85	20	25
H	25	0	42
I	13	5	89
J	33	9	88

Take voter turnout as the dependent variable and:
a. Find the multiple regression equation (unstandardized).
b. What turnout would you expect for a precinct in which 0% of the voters were Democrats and 5% were minorities?
c. Compute beta-weights for each independent variable and compare their relative effect on turnout. Which was the more important factor?
d. Compute R and R^2.
e. Write a paragraph summarizing your findings.

15.3 [SW] Twelve families have been referred to a counselor, and she has rated each of them on a scale that measures family cohesiveness. Also, she has information on family income and number of children currently living at home. Take family cohesion as the dependent variable.

Family	Cohesion Score (Y)	Income (X_1)	Number of Children (X_2)
A	10	30,000	5
B	10	70,000	4
C	9	35,000	4
D	5	25,000	0
E	1	55,000	3
F	7	40,000	0
G	2	60,000	2
H	5	30,000	3
I	8	50,000	5
J	3	25,000	4
K	2	45,000	3
L	4	50,000	0

a. Find the multiple regression equation (unstandardized).
b. What level of cohesion would be expected in a family with an income of $20,000 and 6 children?
c. Compute beta-weights for each independent variable and compare their relative effect on cohesion. Which was the more important factor?
d. Compute R and R^2.
e. Write a paragraph summarizing your findings.

Statistical Analysis Using SPSS

15.4 [SOC] Load the *GSS2018* data set and analyze the relationships between church attendance (the dependent or Y variable) and two independent variables: age and number of children. You analyzed some of these relationships in problem 13.7. Now, we will use both partial correlation and multiple regression and correlation to further examine these relationships.
- To get the zero-order correlations:
 - Click **Analyze → Correlate → Bivariate**.
 - Enter *attend, age,* and *childs* in the **Variables:** window.
 - Click **OK**.
- Partial correlation analysis: Is the relationship between *attend* and *age* affected by *childs*?
 - Click **Analyze → Correlate → Partial**.
 - Enter *attend* and *age* in the **Variables:** window and *childs* in the **Controlling for:** window.
 - Click **OK**.
- Partial correlation analysis: Is the relationship between *attend* and *childs* affected by *age*?
 - Click **Analyze → Correlate → Partial**.
 - Enter *attend* and *childs* in the **Variables:** window and *age* in the **Controlling for:** window.
 - Click **OK**.
- Multiple regression and correlation: What are the effects of *childs* and *age* on *attend*?
 - Click **Analyze → Regression → Linear**.
 - Move *attend* to the **Dependent** window and *age* and *childs* to the **Independent** window.
 - Click **Statistics** and check **Descriptives**. Click **Continue** to return to the **Linear Regression** screen.
 - Click **OK**.
a. Summarize the results for the partial correlation analysis in a paragraph in which you report the value of the zero-order and partial correlations for all relationships. Does the relationship between *attend* and *age* seem to be direct? How about the relationship between *attend* and *childs*?

b. State the unstandardized multiple regression equation. *(HINT: The values for a and b are in the "Coefficients" box of the output, under the column labeled B. The value in the first row is a, and the values below are the b's.)*

$$Y = \underline{\quad} + \underline{\quad} X_1 + \underline{\quad} X_2$$

c. State the standardized multiple regression equation. What is the direction of each relationship? Which independent variable had the stronger effect on *attend*? *(HINT: The beta-weights are in the "Coefficients" box, under "Standardized Coefficients" and "Beta.")*

$$Z_y = \underline{\quad} Z_1 + \underline{\quad} Z_2$$

d. Report the value of R^2. What percentage of the variance in *attend* is explained by the two independent variables combined? How does this compare to the amount of the variance explained by each independent variable alone? *(HINTS: R^2 is in the "Model Summary" box. You can compute r^2 values from the rs in the "Correlations" window of the output.)*

Bivariate:	r	r^2
attend and		
childs		
age		
Multiple:		$R^2 =$

15.5 SOC Load the *GSS2018* data set and analyze the relationships between *wwwhr*, number of hours spent on the Internet each week (the dependent or *Y* variable), and two independent variables: age (*age*) and years of education (*educ*). We will use both partial correlation and multiple regression and correlation to examine these relationships.

- Partial correlation analysis: Is the relationship between *wwwhr* and *age* affected by *educ*?
 - Get the zero-order correlations first:
 - Click **Analyze → Correlate → Bivariate**.
 - Enter *wwwhr*, *educ*, and *age* in the **Variables:** window.
 - Click **Analyze → Correlate → Partial**.
 - Enter *wwwhr* and *age* in the **Variables:** window and *educ* in the **Controlling for:** window.
 - Click **OK**.
- Multiple regression and correlation: What are the effects of *educ* and *age* on *wwwhr*?
 - Click **Analyze → Regression → Linear**.
 - Move *wwwhr* to the **Dependent** window and *age* and *educ* to the **Independent** window.

- Click **Statistics** and check **Descriptives**. Click **Continue** to return to the **Linear Regression** screen.
- Click **OK**.

a. Summarize the results for the partial correlation analysis in a paragraph in which you report the value of the zero-order and partial correlations. Does level of education have an effect on the relationship between age and time spent on the Internet? Does the relationship between *wwwhr* and *age* seem to be direct?

b. State the unstandardized multiple regression equation. *(HINT: The values for a and b are in the "Coefficients" box of the output, under the column labeled B. The value in the first row is a, and the values below are the b's.)*

$$Y = \underline{\quad} + \underline{\quad} X_1 + \underline{\quad} X_2$$

c. State the standardized multiple regression equation. What is the direction of each relationship? Which independent variable had the stronger effect on *wwwhr*? *(HINT: The beta-weights are in the "Coefficients" box, under "Standardized Coefficients" and "Beta.")*

$$Z_y = \underline{\quad} Z_1 + \underline{\quad} Z_2$$

d. Report the value of R^2. What percentage of the variance in *wwwhr* is explained by the two independent variables combined? How does this compare to the amount of the variance explained by each independent variable alone? *(HINTS: R^2 is in the "Model Summary" box, and you can compute r^2 values from the rs in the "Correlations" window of the output.)*

Bivariate:	r	r^2
wwwhr and		
educ		
age		
Multiple:		$R^2 =$

15.6 SOC/CJ Use the *States* data set to analyze the homicide rate (*Homicide*). The independent variables will be state expenditures per pupil on education (*EdSpend*) and unemployment rate (*Unempl*).
- Click **Analyze → Regression → Linear**.
- Move *Homicide* to the **Dependent** window and *EdSpend* and *Unempl* to the **Independent** window.
- Click **Statistics** and check **Descriptives**. Click **Continue** to return to the **Linear Regression** screen.
- Click **OK**.

a. State the unstandardized multiple regression equation. *(HINT: The values for* a *and* b *are in the "Coefficients" box of the output, under the column labeled B. The value in the first row is* a, *and the values below are the* b's.)

$$Y = \underline{\quad} + \underline{\quad} X_1 + \underline{\quad} X_2$$

b. State the standardized multiple regression equation. What is the direction of each relationship? Which independent variable had the stronger effect on *Homicide*? *(HINT: The beta-weights are in the "Coefficients" box, under "Standardized Coefficients" and "Beta.")*

$$Z_y = \underline{\quad} Z_1 + \underline{\quad} Z_2$$

c. Report the value of R^2. What percentage of the variance in *Homicide* is explained by the two independent variables combined? How does this compare to the amount of the variance explained by each independent variable alone? *(HINTS: R^2 is in the "Model Summary" box, and you can compute r^2 values from the rs in the "Correlations" window of the output.)*

Bivariate:	r	r^2
Homicide and		
EdSpend		
Unempl		
Multiple:	$R^2 =$	

15.7 SOC/CJ Use the *States* data set to analyze the rate of car theft (*Carthft*). The independent variables will be per pupil spending on education (*EdSpend*) and unemployment rate (*Unemp*).
- Click **Analyze → Regression → Linear**.
- Move *Carthft* to the **Dependent** window and *EdSpend* and *Unempl* to the **Independent** window.
- Click **Statistics** and check **Descriptives**. Click **Continue** to return to the **Linear Regression** screen.
- Click **OK**.

a. State the unstandardized multiple regression equation. *(HINT: The values for* a *and* b *are in the "Coefficients" box of the output, under the column labeled B. The value in the first row is* a, *and the values below are the* b's.)

$$Y = \underline{\quad} + \underline{\quad} X_1 + \underline{\quad} X_2$$

b. State the standardized multiple regression equation. What is the direction of each relationship? Which independent variable had the stronger effect on *Carthft*? *(HINT: The beta-weights are in the "Coefficients" box, under "Standardized Coefficients" and "Beta.")*

$$Z_y = \underline{\quad} Z_1 + \underline{\quad} Z_2$$

c. Report the value of R^2. What percentage of the variance in *Carthft* is explained by the two independent variables combined? How does this compare to the amount of the variance explained by each independent variable alone? *(HINTS: R^2 is in the "Model Summary" box, and you can compute r^2 values from the rs in the "Correlations" window of the output.)*

Bivariate:	r	r^2
Carthft and		
EdSpend		
Unemployment		
Multiple:	$R^2 =$	

15.8 SOC In problem 13.8, you used the *States* database to analyze the bivariate relationships between region (South/Non-South), education (percentage of the population with a college degree), and rate of traffic deaths. Now, you will take education and region as the independent variables and look at their combined effects on traffic death rate.
- Click **Analyze → Regression → Linear**.
- Move *TrafDths17* to the **Dependent** window and *BA* and *South* to the **Independent** window.
- Click **Statistics** and check **Descriptives**. Click **Continue** to return to the **Linear Regression** screen.
- Click **OK**.

a. State the unstandardized multiple regression equation.

$$Y = \underline{\quad} + \underline{\quad} X_1 + \underline{\quad} X_2$$

b. State the standardized multiple regression equation. What is the direction of each relationship? Which independent variable had the stronger effect on *TrafDths17*?

$$Z_y = \underline{\quad} Z_1 + \underline{\quad} Z_2$$

c. Report the value of R^2. What percentage of the variance in *TrafDths17* is explained by the two independent variables combined? How does this compare to the amount of the variance explained by each independent variable alone? *(HINTS: R^2 is in the "Model Summary" box,*

and you can compute r^2 values from the rs in the "Correlations" window of the output.)

Bivariate:	r	r^2
TrafDths17 and		
BA		
South		
Multiple:	$R^2 =$	

15.9 SOC In problem 13.11, you used the *International* database to analyze the bivariate relationships between religiosity, affluence (gross national income per capita), and urbanization. Now, you will take affluence and urbanization as the independent variables and look at their combined effects on religiosity.

- Click **Analyze → Regression → Linear**.
- Move *Religiosity* to the **Dependent** window and *GNIcap* and *Urban* to the **Independent** window.
- Click **Statistics** and check **Descriptives**. Click **Continue** to return to the **Linear Regression** screen.
- Click **OK**.

a. State the unstandardized multiple regression equation.

$$Y = \underline{\quad} + \underline{\quad} X_1 + \underline{\quad} X_2$$

b. State the standardized multiple regression equation. What is the direction of each relationship? Which independent variable had the stronger effect on *Religiosity*?

$$Z_y = \underline{\quad} Z_1 + \underline{\quad} Z_2$$

c. Report the value of R^2. What percentage of the variance in *Religiosity* is explained by the two independent variables combined? How does this compare to the amount of the variance explained by each independent variable alone?

Bivariate:	r	r^2
Religiosity and		
GNIcap		
Urban		
Multiple:	$R^2 =$	

15.10 Use the *International* database to analyze the correlates of political and civil rights (*Rights*). Take affluence and income inequality as the independent variables and look at their combined effects on *Rights*.

- Click **Analyze → Regression → Linear**.
- Move *Rights* to the **Dependent** window and *GNIcap* and *IncInequal* to the **Independent** window.
- Click **Statistics** and check **Descriptives**. Click **Continue** to return to the **Linear Regression** screen.
- Click **OK**.

a. State the unstandardized multiple regression equation.

$$Y = \underline{\quad} + \underline{\quad} X_1 + \underline{\quad} X_2$$

b. State the standardized multiple regression equation. What is the direction of each relationship? Which independent variable had the stronger effect on *Rights*?

$$Z_y = \underline{\quad} Z_1 + \underline{\quad} Z_2$$

c. Report the value of R^2. What percentage of the variance in *Rights* is explained by the two independent variables combined? How does this compare to the amount of the variance explained by each independent variable alone?

Bivariate:	r	r^2
Rights and		
GNIcap		
IncInequal		
Multiple:	$R^2 =$	

A Multivariate Analysis of TV Viewing and Success

The two projects below continue the investigations begun in Chapter 13. You will use the **Regression** command to analyze the combined effects of your independent variables (including dummy variables if you used any) on the dependent variables from Chapter 13.

PROJECT 1: Who Watches TV?

This project follows up on Project 1 from Chapter 13. The dependent variable is *tvhours*, which, as you recall, measures the number of hours per day the respondent watches television.

Step 1: Choosing Independent Variables

Choose two of the four independent variables you selected in Project 1 from Chapter 13. All other things being equal, you might choose the variables that had the strongest relationship with *tvhours*. Remember that independent variables *cannot* be nominal in level of measurement unless you recode the variable into a dummy variable. You may use any interval-ratio or ordinal-level variables with more than three categories or scores. Once you have selected your variables, list them in the table below and describe exactly what they measure.

SPSS Variable Name	What Exactly Does This Variable Measure?

Step 2: Stating Hypotheses

Restate your hypotheses from Chapter 13 and add some ideas about the relative importance of your two independent variables. Which do you expect to have the stronger effect? Why?

SPSS Variable Name	Hypotheses

Step 3: Running the Regression Procedure

Click **Analyze** → **Regression** → **Linear**, and place *tvhours* in the **Dependent:** box and your selected independent variables in the **Independent(s):** box. Click **Statistics** and **Descriptives**, and then click **OK** to get your results.

Step 4: Recording Results

Summarize your results by filling in the blanks below with information from the **Coefficients** box. The *Y* intercept (*a*) is reported in the top row of the box as a **constant**. The slopes (*b*) are reported in the **Unstandardized Coefficients** column under **B**, and the beta-weights (*β*) are reported in the **Standardized Coefficients** column under **Beta**.

1. $a =$ _____

2. **a.** Slope for your first independent variable $(b_1) =$ _____

 b. Slope for your second independent variable $(b_2) =$ _____

3. State the least-squares multiple regression equation (see Formula 15.3).

$$Y = \underline{\quad} + \underline{\quad} X_1 + \underline{\quad} X_2$$

4. **a.** Beta-weight of your first independent variable $(\beta_1) =$ _____

 b. Beta-weight of your second independent variable $(\beta_2) =$ _____

5. State the standardized least-squares regression line.

$$Z_y = \underline{\quad} Z_1 + \underline{\quad} Z_2$$

6. $R^2 =$ _____

Step 5: Analyzing and Interpreting Results

Write a short summary of your results. Your summary should identify your variables and distinguish between independent and dependent variables. You must also report the strength and direction of relationships as indicated by the slopes and beta-weights and the total explained variance (R^2). How much of the variance does your first independent variable explain? How much more is explained by your second independent variable?

PROJECT 2: What Are the Correlates of Success?

In this second project, we will continue the investigation into the correlates of success, which we began in Chapter 13.

Step 1: Choosing Independent Variables

Use two of the four independent variables you selected in Project 2 from Chapter 13. All other things being equal, you might choose the variables that had the strongest relationship with the variable you selected to be the dependent variable (either *rincome* or *prestg10*). Remember that independent variables *cannot* be nominal in level of measurement unless you recode the variable into a dummy variable. You may use any interval-ratio or ordinal-level variables with more than three categories or scores. Once you have selected your variables, list them in the table below and describe exactly what they measure.

SPSS Variable Name	What Exactly Does This Variable Measure?

Step 2: Stating Hypotheses

Restate your hypotheses from Chapter 13 and add some ideas about the relative importance of your two independent variables. Which do you expect to have the stronger effect? Why?

SPSS Variable Name	Hypotheses

Step 3: Running the Regression Procedure

Click **Analyze** → **Regression** → **Linear**, and place your dependent variable in the **Dependent:** box and your selected independent variables in the **Independent(s):** box. Click **Statistics** and **Descriptives**, and then click **OK** to get your results.

Step 4: Recording Results

Summarize your results by filling in the blanks below with information from the **Coefficients** box. The Y intercept (a) is reported in the top row of the box as a **constant**. The slopes are reported in the **Unstandardized Coefficients** column under **B**, and the beta-weights are reported in the **Standardized Coefficients** column under **Beta**.

1. $a =$ _____

2. **a.** Slope for your first independent variable (b_1) = _____

 b. Slope for your second independent variable (b_2) = _____

3. State the least-squares multiple regression equation (see Formula 15.3).

$$Y = \underline{\ \ } + \underline{\ \ } X_1 + \underline{\ \ } X_2$$

4. **a.** Beta-weight of your first independent variable (β_1) = _____

 b. Beta-weight of your second independent variable (β_2) = _____

5. State the standardized least-squares regression line.

$$Z_y = \underline{\ \ } Z_1 + \underline{\ \ } Z_2$$

6. $R^2 =$ _____

Step 5: Analyzing and Interpreting Results

Write a short summary of your results. Your summary should identify your variables and distinguish between independent and dependent variables. You must also report the strength and direction of relationships as indicated by the slopes and beta-weights and the total explained variance (R^2). How much of the variance does your first independent variable explain? How much more is explained by your second independent variable?

PART IV CUMULATIVE EXERCISES

1. Two research questions that can be answered by one of the techniques presented in chapters in Part IV are stated here. For each research situation, choose either the elaboration technique or regression analysis. The level of measurement of the variables should have a strong influence on your decision.

 a. For a sample of college graduates, did having a job during the school year interfere with academic success? For 20 students, data have been gathered on college grade-point average (GPA), the average number of hours each student worked each week, and College Board scores (a measure of preparedness for college-level work). Take GPA as the dependent variable.

 b. Only about half of this sample graduated within four years (1 = yes, 2 = no). Was their progress affected by their level of social activity (1 = low, 2 = high)? Is the relationship between these variables the same for both males (1) and females (2)?

GPA	Hours Worked per Week	College Board Score	Graduated in Four Years?	Social Activity	Gender
3.14	13	550	1	1	1
2.00	20	375	1	2	2
2.11	22	450	1	1	2
3.00	10	575	1	2	1
3.75	0	600	1	1	1
3.11	21	650	1	1	2
3.22	7	605	1	2	1
2.75	25	630	1	1	2
2.50	30	680	1	1	1
2.10	32	610	1	1	2
2.45	20	580	2	2	2
2.01	40	590	2	1	1
3.90	0	675	2	2	2
3.45	0	650	2	2	1
2.30	25	550	2	2	1
2.20	18	470	2	1	2
2.60	25	600	2	2	1
3.10	15	525	2	2	1
2.60	27	480	2	1	2
2.20	20	500	2	2	1

Statistical Analysis Using SPSS

Following are a number of research questions that can be answered by applying the techniques presented in Part IV. (*NOTE: Many of these problems follow up on Part III Cumulative Exercises, and you may need information from the earlier exercises to complete these problems.*)

Using the General Social Survey (*GSS2018.sav*) Data Set

2. In the Part III Cumulative Exercises, you examined the relationships between "frequency of sex during the last year" (*sexfreq*), total number of sex partners over the past five years (*partnrs5*), and income (*rincome*) or age.

 a. Does the relationship between *sexfreq* and *age* change when you control for *rincome*?

 b. Does the relationship between *partnrs5* and *age* change when you control for *rincome*?

 c. Take *sexfreq* as the dependent variable and conduct a regression analysis with *rincome* and *age* as the independent variables.

 State the unstandardized multiple regression equation.

 $$Y = \underline{} + \underline{} X_1 + \underline{} X_2$$

 State the standardized multiple regression equation.

 $$Z_y = \underline{} Z_1 + \underline{} Z_2$$

 Report and explain the value of R^2.

3. In the Part III Cumulative Exercises, you examined the relationships between *fear* and educational level (*degree*). Does this relationship change when you control for gender? Is the relationship between *fear* and *degree* direct? How do you know?

4. In the Part III Cumulative Exercises, you examined the relationships between support for spanking (*spanking*) and religious denomination (*religR*). Does this relationship change when you control for gender? Is the relationship between *spanking* and *religR* direct? How do you know?

5. Previously, you examined the relationships between support for capital punishment (*cappun*) and presidential preference in the 2016 election (*pres16*). Does this relationship change when you control for race or ethnicity (*race*)?

Using the States (*States.Sav*) Data Set

6. In Part III, you examined the relationship between the amount of time it takes to drive to work (*WrkTrav*) and the traffic fatality rate (*TrafDths17*).

 a. Does the relationship between *WrkTrav* and *TrafDths17* change when you control for the age of the population (*MdAge*)?

 b. Does the relationship between *WrkTrav* and *TrafDths17* change when you control for income (*MdHHinc*)?

 c. Take *TrafDths17* as the dependent variable and conduct a regression analysis with *WrkTrav* and *MdAge* as the independent variables.

 State the unstandardized multiple regression equation.

 $$Y = \underline{} + \underline{} X_1 + \underline{} X_2$$

State the standardized multiple regression equation.

$$Z_y = __ Z_1 + __ Z_2$$

Report and explain the value of R^2.

7. Take *Internet* as the dependent variable and conduct a regression analysis with education (*BA*) and affluence (*MdHHinc*) as independent variables.

State the unstandardized multiple regression equation.

$$Y = __ + __ X_1 + __ X_2$$

State the standardized multiple regression equation.

$$Z_y = __ Z_1 + __ Z_2$$

Report and explain the value of R^2.

Area Under the Normal Curve

Column (a) lists Z scores from 0.00 to 4.00. Only positive scores are displayed, but, since the normal curve is symmetrical, the areas for negative scores will be exactly the same as areas for positive scores. Column (b) lists the proportion of the total area between the Z score and the mean. Figure A.1 displays areas of this type. Column (c) lists the proportion of the area beyond the Z score, and Figure A.2 displays this type of area.

FIGURE A.1 Area Between Mean and Z

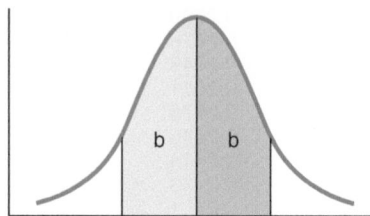

FIGURE A.2 Area Beyond Z

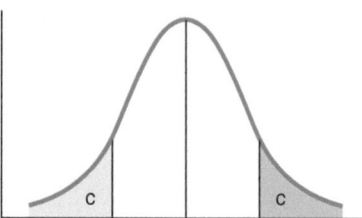

(a) Z	(b) Area Between Mean and Z	(c) Area Beyond Z	(a) Z	(b) Area Between Mean and Z	(c) Area Beyond Z
0.00	0.0000	0.5000	0.21	0.0832	0.4168
0.01	0.0040	0.4960	0.22	0.0871	0.4129
0.02	0.0080	0.4920	0.23	0.0910	0.4090
0.03	0.0120	0.4880	0.24	0.0948	0.4052
0.04	0.0160	0.4840	0.25	0.0987	0.4013
0.05	0.0199	0.4801	0.26	0.1026	0.3974
0.06	0.0239	0.4761	0.27	0.1064	0.3936
0.07	0.0279	0.4721	0.28	0.1103	0.3897
0.08	0.0319	0.4681	0.29	0.1141	0.3859
0.09	0.0359	0.4641	0.30	0.1179	0.3821
0.10	0.0398	0.4602	0.31	0.1217	0.3783
0.11	0.0438	0.4562	0.32	0.1255	0.3745
0.12	0.0478	0.4522	0.33	0.1293	0.3707
0.13	0.0517	0.4483	0.34	0.1331	0.3669
0.14	0.0557	0.4443	0.35	0.1368	0.3632
0.15	0.0596	0.4404	0.36	0.1406	0.3594
0.16	0.0636	0.4364	0.37	0.1443	0.3557
0.17	0.0675	0.4325	0.38	0.1480	0.3520
0.18	0.0714	0.4286	0.39	0.1517	0.3483
0.19	0.0753	0.4247	0.40	0.1554	0.3446
0.20	0.0793	0.4207			

(a) Z	(b) Area Between Mean and Z	(c) Area Beyond Z	(a) Z	(b) Area Between Mean and Z	(c) Area Beyond Z
0.41	0.1591	0.3409	0.98	0.3365	0.1635
0.42	0.1628	0.3372	0.99	0.3389	0.1611
0.43	0.1664	0.3336	1.00	0.3413	0.1587
0.44	0.1700	0.3300	1.01	0.3438	0.1562
0.45	0.1736	0.3264	1.02	0.3461	0.1539
0.46	0.1772	0.3228	1.03	0.3485	0.1515
0.47	0.1808	0.3192	1.04	0.3508	0.1492
0.48	0.1844	0.3156	1.05	0.3531	0.1469
0.49	0.1879	0.3121	1.06	0.3554	0.1446
0.50	0.1915	0.3085	1.07	0.3577	0.1423
0.51	0.1950	0.3050	1.08	0.3599	0.1401
0.52	0.1985	0.3015	1.09	0.3621	0.1379
0.53	0.2019	0.2981	1.10	0.3643	0.1357
0.54	0.2054	0.2946	1.11	0.3665	0.1335
0.55	0.2088	0.2912	1.12	0.3686	0.1314
0.56	0.2123	0.2877	1.13	0.3708	0.1292
0.57	0.2157	0.2843	1.14	0.3729	0.1271
0.58	0.2190	0.2810	1.15	0.3749	0.1251
0.59	0.2224	0.2776	1.16	0.3770	0.1230
0.60	0.2257	0.2743	1.17	0.3790	0.1210
0.61	0.2291	0.2709	1.18	0.3810	0.1190
0.62	0.2324	0.2676	1.19	0.3830	0.1170
0.63	0.2357	0.2643	1.20	0.3849	0.1151
0.64	0.2389	0.2611	1.21	0.3869	0.1131
0.65	0.2422	0.2578	1.22	0.3888	0.1112
0.66	0.2454	0.2546	1.23	0.3907	0.1093
0.67	0.2486	0.2514	1.24	0.3925	0.1075
0.68	0.2517	0.2483	1.25	0.3944	0.1056
0.69	0.2549	0.2451	1.26	0.3962	0.1038
0.70	0.2580	0.2420	1.27	0.3980	0.1020
0.71	0.2611	0.2389	1.28	0.3997	0.1003
0.72	0.2642	0.2358	1.29	0.4015	0.0985
0.73	0.2673	0.2327	1.30	0.4032	0.0968
0.74	0.2703	0.2297	1.31	0.4049	0.0951
0.75	0.2734	0.2266	1.32	0.4066	0.0934
0.76	0.2764	0.2236	1.33	0.4082	0.0918
0.77	0.2794	0.2206	1.34	0.4099	0.0901
0.78	0.2823	0.2177	1.35	0.4115	0.0885
0.79	0.2852	0.2148	1.36	0.4131	0.0869
0.80	0.2881	0.2119	1.37	0.4147	0.0853
0.81	0.2910	0.2090	1.38	0.4162	0.0838
0.82	0.2939	0.2061	1.39	0.4177	0.0823
0.83	0.2967	0.2033	1.40	0.4192	0.0808
0.84	0.2995	0.2005	1.41	0.4207	0.0793
0.85	0.3023	0.1977	1.42	0.4222	0.0778
0.86	0.3051	0.1949	1.43	0.4236	0.0764
0.87	0.3078	0.1922	1.44	0.4251	0.0749
0.88	0.3106	0.1894	1.45	0.4265	0.0735
0.89	0.3133	0.1867	1.46	0.4279	0.0721
0.90	0.3159	0.1841	1.47	0.4292	0.0708
0.91	0.3186	0.1814	1.48	0.4306	0.0694
0.92	0.3212	0.1788	1.49	0.4319	0.0681
0.93	0.3238	0.1762	1.50	0.4332	0.0668
0.94	0.3264	0.1736	1.51	0.4345	0.0655
0.95	0.3289	0.1711	1.52	0.4357	0.0643
0.96	0.3315	0.1685	1.53	0.4370	0.0630
0.97	0.3340	0.1660	1.54	0.4382	0.0618

(a) Z	(b) Area Between Mean and Z	(c) Area Beyond Z	(a) Z	(b) Area Between Mean and Z	(c) Area Beyond Z
1.55	0.4394	0.0606	2.11	0.4826	0.0174
1.56	0.4406	0.0594	2.12	0.4830	0.0170
1.57	0.4418	0.0582	2.13	0.4834	0.0166
1.58	0.4429	0.0571	2.14	0.4838	0.0162
1.59	0.4441	0.0559	2.15	0.4842	0.0158
1.60	0.4452	0.0548	2.16	0.4846	0.0154
1.61	0.4463	0.0537	2.17	0.4850	0.0150
1.62	0.4474	0.0526	2.18	0.4854	0.0146
1.63	0.4484	0.0516	2.19	0.4857	0.0143
1.64	0.4495	0.0505	2.20	0.4861	0.0139
1.65	0.4505	0.0495	2.21	0.4864	0.0136
1.66	0.4515	0.0485	2.22	0.4868	0.0132
1.67	0.4525	0.0475	2.23	0.4871	0.0129
1.68	0.4535	0.0465	2.24	0.4875	0.0125
1.69	0.4545	0.0455	2.25	0.4878	0.0122
1.70	0.4554	0.0446	2.26	0.4881	0.0119
1.71	0.4564	0.0436	2.27	0.4884	0.0116
1.72	0.4573	0.0427	2.28	0.4887	0.0113
1.73	0.4582	0.0418	2.29	0.4890	0.0110
1.74	0.4591	0.0409	2.30	0.4893	0.0107
1.75	0.4599	0.0401	2.31	0.4896	0.0104
1.76	0.4608	0.0392	2.32	0.4898	0.0102
1.77	0.4616	0.0384	2.33	0.4901	0.0099
1.78	0.4625	0.0375	2.34	0.4904	0.0096
1.79	0.4633	0.0367	2.35	0.4906	0.0094
1.80	0.4641	0.0359	2.36	0.4909	0.0091
1.81	0.4649	0.0351	2.37	0.4911	0.0089
1.82	0.4656	0.0344	2.38	0.4913	0.0087
1.83	0.4664	0.0336	2.39	0.4916	0.0084
1.84	0.4671	0.0329	2.40	0.4918	0.0082
1.85	0.4678	0.0322	2.41	0.4920	0.0080
1.86	0.4686	0.0314	2.42	0.4922	0.0078
1.87	0.4693	0.0307	2.43	0.4925	0.0075
1.88	0.4699	0.0301	2.44	0.4927	0.0073
1.89	0.4706	0.0294	2.45	0.4929	0.0071
1.90	0.4713	0.0287	2.46	0.4931	0.0069
1.91	0.4719	0.0281	2.47	0.4932	0.0068
1.92	0.4726	0.0274	2.48	0.4934	0.0066
1.93	0.4732	0.0268	2.49	0.4936	0.0064
1.94	0.4738	0.0262	2.50	0.4938	0.0062
1.95	0.4744	0.0256	2.51	0.4940	0.0060
1.96	0.4750	0.0250	2.52	0.4941	0.0059
1.97	0.4756	0.0244	2.53	0.4943	0.0057
1.98	0.4761	0.0239	2.54	0.4945	0.0055
1.99	0.4767	0.0233	2.55	0.4946	0.0054
2.00	0.4772	0.0228	2.56	0.4948	0.0052
2.01	0.4778	0.0222	2.57	0.4949	0.0051
2.02	0.4783	0.0217	2.58	0.4951	0.0049
2.03	0.4788	0.0212	2.59	0.4952	0.0048
2.04	0.4793	0.0207	2.60	0.4953	0.0047
2.05	0.4798	0.0202	2.61	0.4955	0.0045
2.06	0.4803	0.0197	2.62	0.4956	0.0044
2.07	0.4808	0.0192	2.63	0.4957	0.0043
2.08	0.4812	0.0188	2.64	0.4959	0.0041
2.09	0.4817	0.0183	2.65	0.4960	0.0040
2.10	0.4821	0.0179	2.66	0.4961	0.0039

(a) Z	(b) Area Between Mean and Z	(c) Area Beyond Z	(a) Z	(b) Area Between Mean and Z	(c) Area Beyond Z
2.67	0.4962	0.0038	3.13	0.4991	0.0009
2.68	0.4963	0.0037	3.14	0.4992	0.0008
2.69	0.4964	0.0036	3.15	0.4992	0.0008
2.70	0.4965	0.0035	3.16	0.4992	0.0008
2.71	0.4966	0.0034	3.17	0.4992	0.0008
2.72	0.4967	0.0033	3.18	0.4993	0.0007
2.73	0.4968	0.0032	3.19	0.4993	0.0007
2.74	0.4969	0.0031	3.20	0.4993	0.0007
2.75	0.4970	0.0030	3.21	0.4993	0.0007
2.76	0.4971	0.0029	3.22	0.4994	0.0006
2.77	0.4972	0.0028	3.23	0.4994	0.0006
2.78	0.4973	0.0027	3.24	0.4994	0.0006
2.79	0.4974	0.0026	3.25	0.4994	0.0006
2.80	0.4974	0.0026	3.26	0.4994	0.0006
2.81	0.4975	0.0025	3.27	0.4995	0.0005
2.82	0.4976	0.0024	3.28	0.4995	0.0005
2.83	0.4977	0.0023	3.29	0.4995	0.0005
2.84	0.4977	0.0023	3.30	0.4995	0.0005
2.85	0.4978	0.0022	3.31	0.4995	0.0005
2.86	0.4979	0.0021	3.32	0.4995	0.0005
2.87	0.4979	0.0021	3.33	0.4996	0.0004
2.88	0.4980	0.0020	3.34	0.4996	0.0004
2.89	0.4981	0.0019	3.35	0.4996	0.0004
2.90	0.4981	0.0019	3.36	0.4996	0.0004
2.91	0.4982	0.0018	3.37	0.4996	0.0004
2.92	0.4982	0.0018	3.38	0.4996	0.0004
2.93	0.4983	0.0017	3.39	0.4997	0.0003
2.94	0.4984	0.0016	3.40	0.4997	0.0003
2.95	0.4984	0.0016	3.41	0.4997	0.0003
2.96	0.4985	0.0015	3.42	0.4997	0.0003
2.97	0.4985	0.0015	3.43	0.4997	0.0003
2.98	0.4986	0.0014	3.44	0.4997	0.0003
2.99	0.4986	0.0014	3.45	0.4997	0.0003
3.00	0.4986	0.0014	3.46	0.4997	0.0003
3.01	0.4987	0.0013	3.47	0.4997	0.0003
3.02	0.4987	0.0013	3.48	0.4997	0.0003
3.03	0.4988	0.0012	3.49	0.4998	0.0002
3.04	0.4988	0.0012	3.50	0.4998	0.0002
3.05	0.4989	0.0011	3.60	0.4998	0.0002
3.06	0.4989	0.0011			
3.07	0.4989	0.0011	3.70	0.4999	0.0001
3.08	0.4990	0.0010	3.80	0.4999	0.0001
3.09	0.4990	0.0010	3.90	0.4999	<0.0001
3.10	0.4990	0.0010	4.00	0.4999	<0.0001
3.11	0.4991	0.0009			
3.12	0.4991	0.0009			

Distribution of *t*

Use this table to find the critical region (step 3 of the five-step model) for tests of significance with sample means when sample size (*N*) is small. First, choose an alpha level and a one- or two-tailed test. Then find degrees of freedom to find the *t* score that marks the beginning of the critical region.

Degrees of Freedom (*df*)	Level of Significance for One-Tailed Test					
	0.10	0.05	0.025	0.01	0.005	0.0005
	Level of Significance for Two-Tailed Test					
	0.20	0.10	0.05	0.02	0.01	0.001
1	3.078	6.314	12.706	31.821	63.657	636.619
2	1.886	2.920	4.303	6.965	9.925	31.598
3	1.638	2.353	3.182	4.541	5.841	12.941
4	1.533	2.132	2.776	3.747	4.604	8.610
5	1.476	2.015	2.571	3.365	4.032	6.859
6	1.440	1.943	2.447	3.143	3.707	5.959
7	1.415	1.895	2.365	2.998	3.499	5.405
8	1.397	1.860	2.306	2.896	3.355	5.041
9	1.383	1.833	2.262	2.821	3.250	4.781
10	1.372	1.812	2.228	2.764	3.169	4.587
11	1.363	1.796	2.201	2.718	3.106	4.437
12	1.356	1.782	2.179	2.681	3.055	4.318
13	1.350	1.771	2.160	2.650	3.012	4.221
14	1.345	1.761	2.145	2.624	2.977	4.140
15	1.341	1.753	2.131	2.602	2.947	4.073
16	1.337	1.746	2.120	2.583	2.921	4.015
17	1.333	1.740	2.110	2.567	2.898	3.965
18	1.330	1.734	2.101	2.552	2.878	3.922
19	1.328	1.729	2.093	2.539	2.861	3.883
20	1.325	1.725	2.086	2.528	2.845	3.850
21	1.323	1.721	2.080	2.518	2.831	3.819
22	1.321	1.717	2.074	2.508	2.819	3.792
23	1.319	1.714	2.069	2.500	2.807	3.767
24	1.318	1.711	2.064	2.492	2.797	3.745
25	1.316	1.708	2.060	2.485	2.787	3.725
26	1.315	1.706	2.056	2.479	2.779	3.707
27	1.314	1.703	2.052	2.473	2.771	3.690
28	1.313	1.701	2.048	2.467	2.763	3.674
29	1.311	1.699	2.045	2.462	2.756	3.659
30	1.310	1.697	2.042	2.457	2.750	3.646
40	1.303	1.684	2.021	2.423	2.704	3.551
60	1.296	1.671	2.000	2.390	2.660	3.460
120	1.289	1.658	1.980	2.358	2.617	3.373
∞	1.282	1.645	1.960	2.326	2.576	3.291

Source: Table III of Fisher and Yates: *Statistical Tables for Biological, Agricultural and Medical Research*, published by Longman Group Ltd., London (1974), 6th edition (previously published by Oliver & Boyd Ltd., Edinburgh).

Appendix C

Distribution of Chi Square

Use this table to find the critical region (step 3 of the five-step model) for tests of significance with chi square. Choose an alpha level and find the degrees of freedom to find the chi square score that marks the beginning of the critical region.

df	0.99	0.98	0.95	0.90	0.80	0.70	0.50	0.30	0.20	0.10	0.05	0.02	0.01	0.001
1	$.0^3157$	$.0^3628$	0.00393	0.0158	0.0642	0.148	0.455	1.074	1.642	2.706	3.841	5.412	6.635	10.827
2	0.0201	0.0404	0.103	0.211	0.446	0.713	1.386	2.408	3.219	4.605	5.991	7.824	9.210	13.815
3	0.115	0.185	0.352	0.584	1.005	1.424	2.366	3.665	4.642	6.251	7.815	9.837	11.341	16.268
4	0.297	0.429	0.711	1.064	1.649	2.195	3.357	4.878	5.989	7.779	9.488	11.668	13.277	18.465
5	0.554	0.752	1.145	1.610	2.343	3.000	4.351	6.064	7.289	9.236	11.070	13.388	15.086	20.517
6	0.872	1.134	1.635	2.204	3.070	3.828	5.348	7.231	8.558	10.645	12.592	15.033	16.812	22.457
7	1.239	1.564	2.167	2.833	3.822	4.671	6.346	8.383	9.803	12.017	14.067	16.622	18.475	24.322
8	1.646	2.032	2.733	3.490	4.594	5.527	7.344	9.524	11.030	13.362	15.507	18.168	20.090	26.125
9	2.088	2.532	3.325	4.168	5.380	6.393	8.343	10.656	12.242	14.684	16.919	19.679	21.666	27.877
10	2.558	3.059	3.940	4.865	6.179	7.267	9.342	11.781	13.442	15.987	18.307	21.161	23.209	29.588
11	3.053	3.609	4.575	5.578	6.989	8.148	10.341	12.899	14.631	17.275	19.675	22.618	24.725	31.264
12	3.571	4.178	5.226	6.304	7.807	9.034	11.340	14.011	15.812	18.549	21.026	24.054	26.217	32.909
13	4.107	4.765	5.892	7.042	8.634	9.926	12.340	15.119	16.985	19.812	22.362	25.472	27.688	34.528
14	4.660	5.368	6.571	7.790	9.467	10.821	13.339	16.222	18.151	21.064	23.685	26.873	29.141	36.123
15	5.229	5.985	7.261	8.547	10.307	11.721	14.339	17.322	19.311	22.307	24.996	28.259	30.578	37.697
16	5.812	6.614	7.962	9.312	11.152	12.624	15.338	18.418	20.465	23.542	26.296	29.633	32.000	39.252
17	6.408	7.255	8.672	10.085	12.002	13.531	16.338	19.511	21.615	24.769	27.587	30.995	33.409	40.790
18	7.015	7.906	9.390	10.865	12.857	14.440	17.338	20.601	22.760	25.989	28.869	32.346	34.805	42.312
19	7.633	8.567	10.117	11.651	13.716	15.352	18.338	21.689	23.900	27.204	30.144	33.687	36.191	43.820
20	8.260	9.237	10.851	12.443	14.578	16.266	19.337	22.775	25.038	28.412	31.410	35.020	37.566	45.315
21	8.897	9.915	11.591	13.240	15.445	17.182	20.337	23.858	26.171	29.615	32.671	36.343	38.932	46.797
22	9.542	10.600	12.338	14.041	16.314	18.101	21.337	24.939	27.301	30.813	33.924	37.659	40.289	48.268
23	10.196	11.293	13.091	14.848	17.187	19.021	22.337	26.018	28.429	32.007	35.172	38.968	41.638	49.728
24	10.856	11.992	13.848	15.659	18.062	19.943	23.337	27.096	29.553	33.196	36.415	40.270	42.980	51.179
25	11.524	12.697	14.611	16.473	18.940	20.867	24.337	28.172	30.675	34.382	37.652	41.566	44.314	52.620
26	12.198	13.409	15.379	17.292	19.820	21.792	25.336	29.246	31.795	35.563	38.885	42.856	45.642	54.052
27	12.879	14.125	16.151	18.114	20.703	22.719	26.336	30.319	32.912	36.741	40.113	44.140	46.963	55.476
28	13.565	14.847	16.928	18.939	21.588	23.647	27.336	31.391	34.027	37.916	41.337	45.419	48.278	56.893
29	14.256	15.574	17.708	19.768	22.475	24.577	28.336	32.461	35.139	39.087	42.557	46.693	49.588	58.302
30	14.953	16.306	18.493	20.599	23.364	25.508	29.336	33.530	36.250	40.256	43.773	47.962	50.892	59.703

Source: Table IV of Fisher and Yates: *Statistical Tables for Biological, Agricultural and Medical Research*, published by Longman Group Ltd. London (1974), 6th edition (previously published by Oliver & Boyd Ltd., Edinburgh). Reprinted by permission of Addison Wesley Longman Ltd.

Distribution of *F*

Use this table to find the critical region (step 3 of the five-step model) for analysis of variance tests. Choose an alpha level of either 0.05 or 0.01 and then find the degrees of freedom to find the *F* ratio that marks the beginning of the critical region.

$p = 0.05$

$n_1 \rightarrow$ $n_2 \downarrow$	1	2	3	4	5	6	8	12	24	∞
1	161.4	199.5	215.7	224.6	230.2	234.0	238.9	243.9	249.0	254.3
2	18.51	19.00	19.16	19.25	19.30	19.33	19.37	19.41	19.45	19.50
3	10.13	9.55	9.28	9.12	9.01	8.94	8.84	8.74	8.64	8.53
4	7.71	6.94	6.59	6.39	6.26	6.16	6.04	5.91	5.77	5.63
5	6.61	5.79	5.41	5.19	5.05	4.95	4.82	4.68	4.53	4.36
6	5.99	5.14	4.76	4.53	4.39	4.28	4.15	4.00	3.84	3.67
7	5.59	4.74	4.35	4.12	3.97	3.87	3.73	3.57	3.41	3.23
8	5.32	4.46	4.07	3.84	3.69	3.58	3.44	3.28	3.12	2.93
9	5.12	4.26	3.86	3.63	3.48	3.37	3.23	3.07	2.90	2.71
10	4.96	4.10	3.71	3.48	3.33	3.22	3.07	2.91	2.74	2.54
11	4.84	3.98	3.59	3.36	3.20	3.09	2.95	2.79	2.61	2.40
12	4.75	3.88	3.49	3.26	3.11	3.00	2.85	2.69	2.50	2.30
13	4.67	3.80	3.41	3.18	3.02	2.92	2.77	2.60	2.42	2.21
14	4.60	3.74	3.34	3.11	2.96	2.85	2.70	2.53	2.35	2.13
15	4.54	3.68	3.29	3.06	2.90	2.79	2.64	2.48	2.29	2.07
16	4.49	3.63	3.24	3.01	2.85	2.74	2.59	2.42	2.24	2.01
17	4.45	3.59	3.20	2.96	2.81	2.70	2.55	2.38	2.19	1.96
18	4.41	3.55	3.16	2.93	2.77	2.66	2.51	2.34	2.15	1.92
19	4.38	3.52	3.13	2.90	2.74	2.63	2.48	2.31	2.11	1.88
20	4.35	3.49	3.10	2.87	2.71	2.60	2.45	2.28	2.08	1.84
21	4.32	3.47	3.07	2.84	2.68	2.57	2.42	2.25	2.05	1.81
22	4.30	3.44	3.05	2.82	2.66	2.55	2.40	2.23	2.03	1.78
23	4.28	3.42	3.03	2.80	2.64	2.53	2.38	2.20	2.00	1.76
24	4.26	3.40	3.01	2.78	2.62	2.51	2.36	2.18	1.98	1.73
25	4.24	3.38	2.99	2.76	2.60	2.49	2.34	2.16	1.96	1.71
26	4.22	3.37	2.98	2.74	2.59	2.47	2.32	2.15	1.95	1.69
27	4.21	3.35	2.96	2.73	2.57	2.46	2.30	2.13	1.93	1.67
28	4.20	3.34	2.95	2.71	2.56	2.44	2.29	2.12	1.91	1.65
29	4.18	3.33	2.93	2.70	2.54	2.43	2.28	2.10	1.90	1.64
30	4.17	3.32	2.92	2.69	2.53	2.42	2.27	2.09	1.89	1.62
40	4.08	3.23	2.84	2.61	2.45	2.34	2.18	2.00	1.79	1.51
60	4.00	3.15	2.76	2.52	2.37	2.25	2.10	1.92	1.70	1.39
120	3.92	3.07	2.68	2.45	2.29	2.17	2.02	1.83	1.61	1.25
∞	3.84	2.99	2.60	2.37	2.21	2.09	1.94	1.75	1.52	1.00

Values of n_1 and n_2 represent the degrees of freedom associated with the between and within estimates of variance, respectively.

Source: Table V of Fisher and Yates: *Statistical Tables for Biological, Agricultural and Medical Research*, published by Longman Group Ltd., London (1974), 6th edition (previously published by Oliver and Boyd Ltd., Edinburgh). Reprinted by permission of Addison Wesley Longman Ltd.

$p = 0.01$

$n_1 \rightarrow$ $n_2 \downarrow$	1	2	3	4	5	6	8	12	24	∞
1	4052	4999	5403	5625	5764	5859	5981	6106	6234	6366
2	98.49	99.01	99.17	99.25	99.30	99.33	99.36	99.42	99.46	99.50
3	34.12	30.81	29.46	28.71	28.24	27.91	27.49	27.05	26.60	26.12
4	21.20	18.00	16.69	15.98	15.52	15.21	14.80	14.37	13.93	13.46
5	16.26	13.27	12.06	11.39	10.97	10.67	10.27	9.89	9.47	9.02
6	13.74	10.92	9.78	9.15	8.75	8.47	8.10	7.72	7.31	6.88
7	12.25	9.55	8.45	7.85	7.46	7.19	6.84	6.47	6.07	5.65
8	11.26	8.65	7.59	7.01	6.63	6.37	6.03	5.67	5.28	4.86
9	10.56	8.02	6.99	6.42	6.06	5.80	5.47	5.11	4.73	4.31
10	10.04	7.56	6.55	5.99	5.64	5.39	5.06	4.71	4.33	3.91
11	9.65	7.20	6.22	5.67	5.32	5.07	4.74	4.40	4.02	3.60
12	9.33	6.93	5.95	5.41	5.06	4.82	4.50	4.16	3.78	3.36
13	9.07	6.70	5.74	5.20	4.86	4.62	4.30	3.96	3.59	3.16
14	8.86	6.51	5.56	5.03	4.69	4.46	4.14	3.80	3.43	3.00
15	8.68	6.36	5.42	4.89	4.56	4.32	4.00	3.67	3.29	2.87
16	8.53	6.23	5.29	4.77	4.44	4.20	3.89	3.55	3.18	2.75
17	8.40	6.11	5.18	4.67	4.34	4.10	3.79	3.45	3.08	2.65
18	8.28	6.01	5.09	4.58	4.25	4.01	3.71	3.37	3.00	2.57
19	8.18	5.93	5.01	4.50	4.17	3.94	3.63	3.30	2.92	2.49
20	8.10	5.85	4.94	4.43	4.10	3.87	3.56	3.23	2.86	2.42
21	8.02	5.78	4.87	4.37	4.04	3.81	3.51	3.17	2.80	2.36
22	7.94	5.72	4.82	4.31	3.99	3.76	3.45	3.12	2.75	2.31
23	7.88	5.66	4.76	4.26	3.94	3.71	3.41	3.07	2.70	2.26
24	7.82	5.61	4.72	4.22	3.90	3.67	3.36	3.03	2.66	2.21
25	7.77	5.57	4.68	4.18	3.86	3.63	3.32	2.99	2.62	2.17
26	7.72	5.53	4.64	4.14	3.82	3.59	3.29	2.96	2.58	2.13
27	7.68	5.49	4.60	4.11	3.78	3.56	3.26	2.93	2.55	2.10
28	7.64	5.45	4.57	4.07	3.75	3.53	3.23	2.90	2.52	2.06
29	7.60	5.42	4.54	4.04	3.73	3.50	3.20	2.87	2.49	2.03
30	7.56	5.39	4.51	4.02	3.70	3.47	3.17	2.84	2.47	2.01
40	7.31	5.18	4.31	3.83	3.51	3.29	2.99	2.66	2.29	1.80
60	7.08	4.98	4.13	3.65	3.34	3.12	2.82	2.50	2.12	1.60
120	6.85	4.79	3.95	3.48	3.17	2.96	2.66	2.34	1.95	1.38
∞	6.64	4.60	3.78	3.32	3.02	2.80	2.51	2.18	1.79	1.00

Values of n_1 and n_2 represent the degrees of freedom associated with the between and within estimates of variance, respectively.

Using Statistics:
Ideas for Research Projects

This appendix presents outlines for four research projects, each of which requires you to use SPSS to analyze the 2018 General Social Survey (GSS), the database that has been used throughout this text. The research projects should be completed at various intervals during the course, and each project permits a great deal of choice on the part of the student.

The first project stresses description and should be done after completing Chapters 2–4. The second involves estimation and should be completed in conjunction with Chapter 7. The third project uses inferential statistics and should be done after completing Part II, and the fourth project combines inferential statistics with measures of association (with an option for multivariate analysis) and should be done after Part III or Part IV.

Project 1: Descriptive Statistics

For use after Chapters 2–4

1. Select five variables from the 2018 General Social Survey *other than* those you have used for end-of-chapter exercises *(NOTE: Your instructor may specify a different number of variables)*, and use the **Frequencies** command to get frequency distributions and summary statistics for each variable. Click the **Statistics** button on the **Frequencies** command window and request the mean, median, mode, standard deviation, and range. Make a note of all relevant information when it appears on screen, or print a hard copy. Appendix G lists the variables in the 2018 GSS, which we have used throughout the text and which are included with GSS2018-Sample. The full 2018 GSS includes many more variables than these 50 and there are several ways to browse through them. One is by loading the 2018 GSS into SPSS and clicking on **Utilities** from the main menu bar and then clicking **Variables**. An alphabetical list of all variables will appear with a description of each variable next to "Label."

2. For each variable, get bar or line charts to summarize the overall shape of the distribution of the variable.

3. Inspect the frequency distributions and graphs, and choose *appropriate* measures of central tendency and, for ordinal- and interval-ratio–level variables, variation. Also, for interval-ratio and ordinal variables with many scores, check for skew both by using the line chart and by comparing the mean and median. Write a sentence or two of description for each variable, being careful to include a description of the overall shape of the distribution (see Chapter 2), the

central tendency (Chapter 3), and the variation (Chapter 4). For nominal- and ordinal-level variables, be sure to explain any arbitrary numerical codes. For example, on the variable *class* in the 2018 GSS (see Appendix G), a 1 is coded as "lower class," a 2 indicates "working class," and so forth. This is an ordinal-level variable, so you might choose to report the median as a measure of central tendency. If the median score on *class* were 2.45, for example, you might place that value in context by reporting that "the median is 2.45, about halfway between 'working class' and 'middle class.'"

Here are examples of minimal summary sentences for your report, using *fictitious* data:

- For a nominal-level variable (e.g., gender), report the mode and some detail about the overall distribution. For example: "Most respondents were married (57.5%), but divorced (17.4%) and single (21.3%) individuals were also common."

- For an ordinal-level variable (e.g., occupational prestige), use the median (and perhaps the mean and mode) and the range. For example: "The median number of times that respondents read the newspaper was about once a week (a score of 3) and scores ranged from 1 ('Every day') to 5 ('Never'). The most common score was 2 ('A few times a week') and the average score was 2.13."

- For an interval-ratio–level variable (e.g., age), use the mean (and perhaps the median or mode) and the standard deviation (and perhaps the range). For example: "Average age for this sample was 42.3. Respondents ranged from 18 to 94 years of age with a standard deviation of 15.37."

Project 2: Estimation

For use after Chapter 7

In this exercise, you will use the 2018 GSS sample to estimate characteristics of the U.S. population. You will use SPSS to generate the sample statistics and then use either Formula 7.2 or Formula 7.3 to find the confidence interval and state each interval in words.

A. Estimating Means

1. There are relatively few interval-ratio variables in the 2018 GSS, and for this part of the project you may use ordinal variables that have at least three categories or scores. Choose a total of three variables that fit this description *other than* the variables you used in Chapter 7. *(NOTE: Your instructor may specify a different number of variables.)*

2. Use the **Descriptives** command to get means, standard deviations, and sample size (*N*), and use this information to construct 95% confidence intervals for the mean of each of your variables. Make a note of the mean, standard deviation, and sample size, or print a hard copy. Use Formula 7.2 to compute the confidence intervals.

3. For each variable, write a summary sentence reporting the variable, the interval itself, the confidence level, and the sample size. Write in plain English, as if you were reporting results in a newspaper. Most importantly, you should make it clear that you are estimating characteristics of the population of the entire United States. For example, a summary sentence might look like this: "Based on a random sample of 1231 respondents, I estimate at the 95% level that U.S. drivers average between 64.46 and 68.22 miles per hour when driving on interstate highways."

B. Estimating Proportions

1. Choose three nominal- or ordinal-level variables *other than* the variables you used in Chapter 7. *(NOTE: Your instructor may specify a different number of variables.)*

2. Use the **Frequencies** command to get the percentage of the sample in the various categories of each variable. Change the percentages (remember to use the **valid percents** column) to proportions and construct confidence intervals for one category of each variable (e.g., the percent female for *sex*) using Formula 7.3.

3. For each variable, write a summary sentence reporting the variable, the interval, the confidence level, and the sample size. Write in plain English, as if you were reporting results in a newspaper. Remember to make it clear that you are estimating a characteristic of the U.S. population.

4. For any one of the intervals you constructed in either Project A or Project B, identify each of the following concepts and terms, and briefly explain their role in estimation: sample, population, statistic, parameter, EPSEM, representative, confidence level.

Project 3: Significance Testing

For use after Part II

In this exercise, you will use the 2018 GSS data base to conduct tests of significance. You will use the five-step model to conduct the tests.

A. Two-Sample *t* Test (Chapter 9)

1. Choose two different dependent variables from the interval-ratio or ordinal-level variables that have three or more scores. *(NOTE: Your instructor may specify a different number of variables.)* Choose independent variables that might logically be a cause of your dependent variables. Remember that, for a *t* test, independent variables can have *only* two categories, but you can still use independent variables with more than two categories by (a) using the **Grouping Variable** box to specify the exact categories (e.g., select scores of 1 and 5 on *marital* to compare married with never-married respondents), or (b) collapsing the scores of variables with more than two categories by using the

Recode command. Independent variables can be any level of measurement, and you may use the same independent variable for both tests. (See "You Are the Researcher" in Chapter 10 for instructions on how to recode variables.)

2. Start SPSS, load the 2018 GSS data set, and click **Analyze**, **Compare Means**, and then **Independent Samples T Test**. Name your dependent variable(s) in the **Test Variable** window and your independent variable in the **Grouping Variable** window. You will also need to specify the scores used to define the groups on the independent variable. See Chapter 9 for examples. Make a note of the test results (group means, obtained *t* score, significance, sample size) or print a hard copy. Repeat the procedure for the second dependent variable.

3. Write up the results of the test. At a minimum, your report should clearly identify the independent and dependent variables, the sample statistics, the value of the test statistic (step 4), the results of the test (step 5), and the alpha level you used.

B. Analysis of Variance (Chapter 10)

1. Choose two different dependent variables from the interval-ratio or ordinal-level variables that have three or more scores. *(NOTE: Your instructor may specify a different number of variables.)* Choose independent variables that might logically be a cause of your dependent variables and that have between three and five categories. You may use the same independent variables for both tests.

2. Start SPSS, load the 2018 GSS data set, and click **Analyze**, **Compare Means**, and then **One-way Anova**. The **One-way Anova** window will appear. Find your dependent variable in the variable list on the left and click the arrow to move the variable name into the **Dependent List:** box. Note that you can request more than one dependent variable at a time. Next, find the name of your independent variable and move it to the **Factor:** box. Click **Options** and then click the box next to **Descriptive** in the **Statistics** box to request means and standard deviations. Click **Continue** and **OK**. Make a note of the test results or print a hard copy. Repeat, if necessary, for your second dependent variable.

3. Write up the results of the test. At a minimum, your report should clearly identify the independent and dependent variables, the sample statistics (category means), the value of the test statistic (step 4), the results of the test (step 5), the degrees of freedom, and the alpha level you used.

C. Chi Square (Chapter 11)

1. Choose two different dependent variables of any level of measurement that have five or fewer (preferably two to three) scores. *(NOTE: Your instructor may specify a different number of variables.)* For each dependent variable, choose an independent variable that might logically be a cause. Independent variables can be any level of measurement as long as they have five or fewer (preferably two to three) categories. Output will be easier to analyze if you use variables with few categories. You may use the same independent variable for both tests.

2. Click **Analyze, Descriptive Statistics**, and then **Crosstabs**. The **Crosstabs** dialog box will appear. Highlight your first dependent variable and move it to the **Rows** box. Next, highlight your independent variable and move it to the **Columns** box. Click the **Statistics** button and click the box next to **chi square**. Click the **Cells** button and click the box next to **column** under **Percentages**. Click **Continue** and **OK**. Make a note of the results or print a hard copy. Repeat for your second dependent variable.

3. Write up the results of the test. At a minimum, your report should clearly identify the independent and dependent variables, the value of the test statistic (step 4), the results of the test (step 5), the degrees of freedom, and the alpha level you used. It is almost always desirable to report the column percentages as well.

Project 4: Analyzing the Strength and Significance of Relationships

For use after Part III or IV

A. Using Bivariate Tables

1. From the 2018 GSS data set, select either

 • One dependent variable and three independent variables (possible causes)

 OR

 • One independent variable and three possible dependent variables (possible effects)

 Variables can be from any level of measurement but must have only a few (two to five) categories or scores. *(NOTE: Your instructor may specify a different number of variables.)*

2. Develop research questions or hypotheses about the relationships between variables. Make sure the causal links you suggest are sensible and logical.

3. Use **Crosstabs** to generate bivariate tables. See Chapters 11 and 12 for examples. Place your dependent variable(s) in the rows and independent variable(s) in the columns of the crosstabs tables. In the **Crosstabs** dialog box, click the **Statistics** button and choose **chi square**, **phi** or **Cramer's *V***, and **gamma** for every table you request. Also, click the **Cells** button and get column percentages for every table you request. Make a note of results as they appear on the screen or print hard copies.

4. Write a report that presents and analyzes these relationships. Be clear about which variables are dependent and which are independent. For each combination of variables, report the test of significance and an *appropriate* measure of association. In addition, for each relationship, report and discuss column percentages, pattern or direction of the relationship, and strength of the relationship.

Optional Multivariate Analysis. Pick one of the bivariate relationships you produced in step 3 and find a logical control variable for this relationship. Run **Crosstabs** for the bivariate relationship again while controlling for the third variable. Compare the partial tables with each other and with the bivariate table. Is the original bivariate relationship direct? Is there evidence of a spurious or intervening relationship? Do the variables have an interactive relationship? Write up the results of this analysis and include them in your summary paper for this project.

B. Using Interval-Ratio Variables

1. From the 2018 GSS, select either

 • One dependent variable and three independent variables (possible causes)

 OR

 • One independent variable and three dependent variables (possible effects)

 Variables should be interval-ratio in level of measurement, but you may use ordinal-level variables as long as they have more than three scores. *(NOTE: Your instructor may specify a different number of variables.)*

2. Develop research questions or hypotheses about the relationships between variables. Make sure the causal links you suggest are sensible and logical.

3. Use **Regression** and **Scatterplot** (click **Graphs**, **Legacy Dialogs**, and then **Scatter**) to analyze the bivariate relationships. Make a note of results (including r, r^2, slopes, beta-weights, and a) as they appear on the screen, or print hard copies.

4. Write a report that presents and analyzes these relationships. Be clear about which variables are dependent and which are independent. For each combination of variables, report the strength and direction of the relationship. Include r, r^2, and the beta-weights in your report.

Optional Multivariate Analysis. Pick one of the bivariate relationships you produced in step 3 and find another logical independent variable. Run **Regression** again with both independent variables and analyze the results. How much improvement is there in the explained variance after the second independent variable is included? Write up the results of this analysis and include them in your summary paper for this project.

An Introduction to SPSS

Computers have affected virtually every aspect of human society, and, as you would expect, their impact on the conduct of social research has been profound. Researchers routinely use computers to organize data and compute statistics and the skills needed to use computers successfully are quite accessible. This appendix will prepare you to use a statistics program called SPSS—the Statistical Package for the Social Sciences. After you finish this appendix, you will be ready to complete the exercises found at the end of most chapters of this text.

This appendix is intended only as an *introduction* to SPSS. It will give you enough information so that you can complete the assignments in the text. It is unlikely, however, that this appendix will answer all your questions. So this is a good place to tell you that SPSS has an extensive and easy-to-use **Help** facility that will provide assistance as you request it. You should familiarize yourself with this feature and use it as needed. To get help, simply click on the **Help** command on the toolbar across the top of the SPSS screen.

Once you have entered your data or gathered some from the Internet, you can easily produce an abundance of statistical information without doing any computations or writing any programs yourself. With SPSS, you can save hours of time, devote more attention to analysis and interpretation, and conduct research projects with large data sets. Mastery of this technology might be very handy in your senior-level courses, in a wide variety of jobs, or in graduate school.

Getting Started—Databases and Computer Files

Before statistics can be calculated, SPSS must have some data to process. A **database** is an organized collection of related information, such as the responses to a survey. For purposes of computer analysis, a database is organized into a **file**—a collection of information that is stored under the same name in the memory of the computer, on a disk or flash drive, or on some other medium. Words as well as numbers can be saved in files. Data can be stored in files indefinitely. Because it can take months to conduct a thorough data analysis, the ability to save a database is another advantage of using computers.

For the SPSS exercises in this text, we will rely primarily on the General Social Survey (GSS) for 2018. This database includes about 1000 items of information (age, gender, opinion about capital punishment, and so forth) for thousands of respondents. Some of you will be using a student version of SPSS, which is limited in the number of cases and variables it can process. To accommodate those limits, we also supply a shortened version of the GSS that includes 50 items of information and fewer than 1500 respondents.

The GSS is an especially valuable resource for (at least) two reasons. First, the respondents to the survey are chosen so that the sample is representative of the entire U.S. population. A representative sample reproduces, in miniature form, the characteristics of the population from which it was taken (see Chapters 6 and 7). So when you analyze the *GSS*, you are in effect analyzing U.S. society as of 2018. The data are real, and the relationships you will analyze reflect some of the most important and sensitive issues in American life.

Second, the GSS has been administered regularly since 1972 and provides a continuous record of how American society has evolved and changed over the last four decades on issues ranging from abortion to euthanasia. The GSS has been the basis for hundreds of research projects by professional social researchers.

Although we will rely primarily on the 2018 *GSS*, we will also use several other databases.

- The *States* database includes a variety of information for the 50 states.

- The *International* database presents mostly population data (e.g., birth rates and growth rates) for a group of about 100 nations.

- The *CrimeTrends84-17* database includes five measures of criminal activity in the 50 states. We will use this database to construct graphs in Chapter 2.

- The *Pew Core Trends Survey* includes about 70 variables—with an emphasis on Internet use patterns—for a representative sample of adult Americans and features.

Appendix G presents codebooks for all databases. A codebook lists all the variables in the database, explains what each variable measures, and, where necessary, lists the meaning of the scores for each variable.

It is important that you understand the difference between a statistics package like SPSS and a database like the *GSS2018* and what we are ultimately after here. A database consists of information. A statistics package processes the information in the database and produces statistics. Our goal is to apply the statistics package to the database to produce output (for example, statistics and graphs) that we can analyze and use to answer questions. The process might be diagrammed as in Figure F.1.

Statistics packages like SPSS are general research tools that can be used to analyze databases of all sorts; they are not limited just to the databases supplied with this text. In the same way, our databases could be analyzed with statpaks other than SPSS, including SAS and Stata, that may be available on your campus.

Starting SPSS and Loading a Database

You are probably using SPSS in a computer lab, and you can begin running the program immediately. However, if you are using the student version of the program on your personal computer, the first thing you need to do is install the software. Follow the instructions that came with the program and return to this appendix when the installation is complete.

FIGURE F.1 The Data Analysis Process

Database	→	Statpak	→	Output	→	Analysis
(raw information)		(computer programs)		(statistics and graphs)		(interpretation)

To start SPSS, find the icon (or picture) on the screen of your monitor that has an "IBM SPSS Statistics" label attached to it. Move the cursor on the monitor screen over this icon and then double-click the left button on the mouse. This will start SPSS.

After a few seconds, the SPSS screen will appear. You will probably see a Welcome Screen that lists recently used data files. The 2018 GSS may be listed here but, for now, close this screen by clicking the "x" in the upper right corner of the screen.

The next screen you will see is the SPSS Data Editor screen. Note that there is a list of commands across the very top of the screen. These commands begin with **File** at the far left and end with **Help** at the far right. This is the main menu bar for SPSS. When you click any of these words, a menu of commands and choices will drop down. Basically, you tell SPSS what to do by clicking on your desired choices from these menus. Sometimes submenus will appear, and you will need to specify your choices further.

The first thing we need to do is open a data set. Click the **File** command on the left of the menu bar and a submenu will drop down. Click **Open** and then **Data** from the submenu. You need to find the *GSS2018* data file, which may already be stored on the computer, probably under the name *GSS2018.sav* or something similar. If not, you can download the data set from the Web site for this text. Check with your instructor to make sure you know how to access the data set. Once you have located the data, click on the file name and SPSS will load it. After the data set is loaded, your computer screen will look like this:

SPSS provides the user with a variety of options for displaying information about the data file and output on the screen. We recommend that you tell the

TABLE F.1 **Summary of Commands**

To start SPSS	Click the SPSS icon.
To open a data file or database	Click **File**, **Open**, and **Data**. Find the file name of the data set you want to open and double-click it.
To set the display options for lists of variables	Click **Edit** and **Options**. On the "General" tab, make sure "Display names" and "Alphabetical" are selected and then click **OK**.

program to display lists of variables by name (e.g., *age, abany*) rather than by labels (e.g., AGE OF RESPONDENT, or ABORTION IF WOMAN WANTS FOR ANY REASON). Lists displayed this way will be easier to read and compare to Appendix G. To do this, click **Edit** on the main menu bar and then click **Options** from the drop-down submenu. A dialog box labeled "Options" will appear with a series of tabs along the top. The "General" options should be displayed; if not, click on this tab. On the "General" screen, find the box labeled "Variable Lists" and, if they are not already selected, click **Display names** and **alphabetical** and then click **OK**. If you make changes, a message may appear on the screen that tells you that changes will take effect the next time a data file is opened.

In this section, you learned how to start up SPSS, load a data file, and set some of the display options for this program. These procedures are summarized in Table F.1.

Working with Databases

Note that the SPSS Data Editor window is organized into a two-dimensional grid, with columns running up and down (vertically) and rows running across (horizontally). Each column is a variable or item of information from the database. The names of the variables are listed at the tops of the columns. Remember that you can find the meanings of some of these variable names in the codebooks in Appendix G.

Another way to decipher the meanings of variable names is to click **Utilities** on the menu bar and then click **Variables**. The **Variables** window will open. This window has two parts. On the left is a list of all variables in the database, arranged in alphabetical order and with the first variable highlighted. On the right is the "Variable Information" window, with information about the highlighted variable. The first variable in the *GSS2018* database is listed as *abany*. The "Variable Information" window displays a fragment of the question that was actually asked during the survey ("ABORTION IF WOMAN WANTS FOR ANY REASON") and shows the possible scores on this variable (a score of 1 = yes and a score of 2 = no), along with some other information.

The same information can be displayed for any variable in the data set. For example, find the variable *marital*. You can do this by using the arrow keys on your keyboard or the slider bar on the right of the variable list window. You can also move through the list by typing the first letter of the variable name you are interested in. For example, type "m" and you will be moved to the first variable name in the list that begins with that letter. Now you can see that *marital* measures marital status and that a score of "1" indicates that the respondent was married, and so forth. What do *relig* and *marhomo* measure? Close this window by clicking the **Cancel** button at the bottom of the window.

TABLE F.2 Summary of Commands

To move around the Data Editor window	Use any of these options: 1. Click the cell you want to highlight. 2. Use the arrow keys on your keyboard. 3. Move the slider buttons. 4. Click the arrows on the right-hand or bottom margins.
To get information about a variable	1. See Appendix G for the variables used in the text. 2. Click **Utilities** and then **Variables**. Scroll through the list until you find the variable in which you are interested. Information about the variable will appear in the window on the right.

Examine the Data Editor window a little more. Each row of the window (reading across, from left to right) contains the scores of a particular respondent on all the variables in the database. Note that the upper-left cell is highlighted (outlined in a darker border than the other cells). This cell contains the score of respondent 1 on the first variable. The second row contains the scores of respondent 2, and so forth. You can move around in this window with the arrow keys on your keyboard.

In this section, you learned to read information in the data display window and to decipher the meanings of variable names and scores. These commands are summarized in Table F.2. We are now prepared to perform some statistical operations with the 2018 GSS database.

Putting SPSS to Work: Producing Statistics

At this point, the database on the screen is just a mass of numbers with little meaning for you. That's okay because you will not actually have to read any information from this screen. All of the statistical operations you will conduct will begin by clicking the **Analyze** command from the menu bar, selecting a procedure and statistics, and then naming the variable or variables you would like to process.

To illustrate, let's have SPSS produce a frequency distribution for the variable *sex*. Frequency distributions are tables that display the number of times each score of a variable occurred (see Chapter 2). So when we complete this procedure, we will know the number of males and females in the 2018 GSS sample.

With the *GSS2018* data set loaded, click **Analyze** on the menu bar. From the menu that drops down, click **Descriptive Statistics** and then **Frequencies**. The **Frequencies** window appears, with the variables listed in alphabetical order in the box on the left. The first variable (*abany*) will be highlighted. Use the slider button or the arrow keys on the right-hand margin of this box to scroll through the variable list until you highlight the variable *sex*, or type "s" to move to the approximate location.

Once the variable name has been highlighted, click the arrow button in the middle of the screen to move the variable name to the box on the right-hand side of the screen. SPSS will produce frequency distributions for all variables listed in this box, but for now we will confine our attention to *sex*. Click the **OK** button in the Frequencies window and, in seconds, a frequency distribution will be produced.

SPSS sends all tables and statistics to the Output window, or SPSS viewer. This window is now "closest" to you, and the Data Editor window is "behind" the Output window. If you want to return to the Data Editor, click on any part of it if it is visible; it will move to the "front" and the Output window will be "behind"

TABLE F.3 An Example of SPSS Output

(RESPONDENTS SEX)

		Frequency	Percent	Valid Percent	Cumulative Percent
Valid	MALE	1052	44.8	44.8	44.8
	FEMALE	1296	55.2	55.2	100.0
	Total	2348	100.0	100.0	

it. To display the Data Editor window if it is not visible, minimize the Output window by clicking the "_" box in the upper-right corner.

Frequencies. The output from SPSS is reproduced as Table F.3. What can we tell from this table? The score labels (male and female) are printed at the left, with the number of cases (Frequency) in each category of the variable one column to the right. As you can see, there are 1052 males and 1296 females in the sample. The next two columns give information about percentages, and the last column to the right displays cumulative percentages. We will defer a discussion of this last column until Chapter 2.

One of the percentage columns is labelled "Percent" and the other is labelled "Valid Percent." The difference between these two columns lies in the handling of missing values. The "Percent" column is based on all cases, including people who did not respond to the item (NA) and people who said they did not have the requested information (DK). The "Valid Percent" column excludes all missing scores. Because we will almost always want to ignore missing scores, we will pay attention only to the "Valid Percent" column. Note that for *sex*, there are no missing scores (gender was determined by the interviewer), and the two columns are identical.

Printing and Saving Output

Once you have gone to the trouble of producing statistics, a table, or a graph, you will probably want to keep a permanent record. There are two ways to do this. First, you can print a hard copy of the contents of the Output window to take with you. To do this, click on **File** and then click **Print**. Alternatively, find the icon of a printer (third from the left) in the row of icons just below the menu bar and click on it.

The other way to create a permanent record is to save the output to the computer's memory, to a flash drive, or to some other device. To do this, click **Save** from the **File** menu. The **Save** dialog box opens. Give the output a name (some abbreviation such as "frequency-sex" might do) and, if necessary, specify the location you are using. Click **OK**, and the table will be permanently saved.

Ending Your SPSS Session

Once you have saved or printed your work, you may end your SPSS session. Click on **File** from the menu bar and then click **Exit**. If you haven't already done so, you will be asked whether you want to save the contents of the output window. You may save the frequency distribution at this point if you wish. Otherwise, click **No**. The program will close, and you will be returned to the screen from which you began.

Codebooks for All Data Sets

This appendix presents the codebooks for all data sets used in this text. A codebook lists all of the variables in the data set along with explanations for what each variable measures and, if necessary, the meaning of the codes or scores for that variable. All data sets can be downloaded from the Web site for this text.

A. Codebook for the General Social Survey, 2018

The General Social Survey (GSS) is a public opinion poll that has been regularly conducted by the National Opinion Research Council since 1972. This database is used in many of the SPSS demonstrations and in end-of-chapter exercises.

There are two versions of the 2018 GSS available at the Web site for this text.

The full 2018 GSS (*GSS2018*) is quite large and includes more than 1000 variables and about 2400 cases. We also supply a smaller version (*GSS2018-Sample*) that can be used with the student version of SPSS. This database has 50 variables and about 1500 cases.

We cannot list all of the variables in the full 2018 GSS so this codebook lists the variables in the smaller GSS 2018 (*GSS2018-Sample*) as well as the variables used in demonstrations and exercises. If you would like to explore all of the variables in the full 2018 GSS, use the General Social Survey Data Explorer at https://gssdataexplorer.norc.org/variables/vfilter.

The codebook lists each item in the data set. The variable names are those used in the data files. The questions have been reproduced exactly as they were asked (with a few exceptions to conserve space), and the numbers beside each response are the scores recorded in the data file.

The data set includes variables that measure demographic or background characteristics of the respondents, including gender, age, race, religion, and several indicators of socioeconomic status. Also included are items that measure opinion on current and controversial topics such as abortion, capital punishment, and homosexuality.

Note that the variable names in the left column are abbreviations of the related survey item. In some cases (e.g., *age*), the variable names are easy to figure out. In other cases, the variable names (like *abany* and *fefam*) are not so obvious. Early versions of SPSS required that variable names be no more than eight characters and this sometimes necessitated extreme abbreviation. Although SPSS no longer restricts variable names to eight characters (as reflected in the other databases), we retain these short names for comparability with older versions of the GSS.

Appendix G also shows the wording of the item that generated the variable. For example, *abany* consists of responses to a question about legal abortion and the variable name is formed from the question: "Should an *ab*ortion be possible

for *any* reason?" The *fefam* variable consists of responses to the statement "It is much better for everyone involved if the man is the achiever outside the home and the woman takes care of the home and family."

Where necessary, Appendix G lists the codes (or scores) of a variable along with their meanings. Thus, a score of 1 on *abany* means that the respondent approves of a legal abortion "for any reason" and a score of 0 on *attend* means that the respondent never attends religious services.

Notice that some of the possible responses are labeled IAP ("Not Applicable"), NA, or DK. The first of these responses means that the item in question was not given to a respondent. The full GSS is very long, and, to keep the time frame for completing the survey reasonable, not all respondents are asked every question. NA stands for "No Answer" and means that the respondent was asked the question but refused to answer. DK stands for "Don't Know," which means that the respondent did not have the requested information. All three of these scores are "missing values," and, as "non-information," they should be eliminated from statistical analysis. Missing values are common on surveys, and as long as they are not too numerous, they are not a particular problem.

Please tell me if you think it should be possible for a woman to get a legal abortion if ...

abany	She wants it for any reason.
	1. Yes
	2. No
	0. IAP, 8. DK, 9. NA
abpoor	The family has very low income and can't afford any more children
	(Same scoring as *abany*)
age	Age of respondent
	18–89. Actual years
	99. NA
attend	How often do you attend religious services?

0. Never	1. Less than once per year
2. Once or twice a year	3. Several times per year
4. About once a month	5. 2–3 times a month
6. Nearly every week	7. Every week
8. More than once a week	
9. DK or NA	

cantrust — Generally speaking, would you say that people can be trusted or that you can't be too careful in dealing with people?

1. Always trusted	2. Usually trusted
3. Usually not trusted	4. Always not trusted
0. IAP, 8. can't choose, 9. NA	

cappun — Do you favor or oppose the death penalty for persons convicted of murder?
3. Favor
4. Oppose
0. IAP, 8. DK, 9. NA

childs | How many children have you ever had?
1–7. Actual number
 8. Eight or more
 9. DK, NA

class | Subjective class identification
 1. Lower class
 2. Working class
 3. Middle class
 4. Upper class
 0. IAP, 8. DK, 9. NA

concong | How much confidence do you have in the U.S. Congress?
 1. Complete confidence
 2. A great deal of confidence
 3. Some confidence
 4. Very little confidence
 5. No confidence at all
 0. IAP, 8, Can't choose, 9, NA

degree | Respondent's highest degree
 0. Less than high school
 1. High school
 2. Assoc./junior college
 3. Bachelor's
 4. Graduate
 7. IAP, 8. DK, 9. NA

educ | Highest year of school completed
 0–20. Actual number of years
 97. IAP, 98. DK, 99. NA

fear | Is there any area right around here—that is, within a mile—where you would be afraid to walk alone at night?
 1. Yes
 2. No
 0. IAP, 8, DK, 9. NA

fefam | It is much better for everyone involved if the man is the achiever outside the home and the woman takes care of the home and family.
 1. Strongly agree
 2. Agree
 3. Disagree
 4. Strongly disagree
 0. IAP, 8. DK, 9. NA

fepol | Do you agree or disagree with this statement: Most men are better suited emotionally for politics than most women.
 1. Agree
 2. Disagree
 0. IAP, 8. Not sure, 9. NA

grass	Should marijuana be made legal? 1. Legal 2. Not legal *0. IAP, 8. DK, 9. NA*
gunlaw	Would you favor or oppose a law which would require a person to obtain a police permit before he or she could buy a gun? 1. Favor 2. Oppose *0. IAP, 8. DK, 9. NA*
happy	Taken all together, how would you say things are these days—would you say that you are very happy, pretty happy, or not too happy? 1. Very happy 2. Pretty happy 3. Not too happy *0. IAP, 8. DK, 9. NA*
helpblk	Some people think that Blacks have been discriminated against for so long that the government has a special obligation to help improve their living standards. Others believe that the government should not be giving help. Where do you place yourself on this scale? 1. Should help 2. 3. Agree with both 4. 5. No special treatment *0. IAP, 8. DK, 9. IAP*
homosex	What about sexual relations between two adults of the same sex? Is it 1. Always wrong 2. Almost always wrong 3. Wrong only sometimes 4. Not wrong at all *0. IAP, 8. Can't choose, 9. NA*
hrs1	How many hours did you work last week? 1–89. Actual hours *−1. IAP, 98. DK, 99. NA*
incgap	Differences in income in America are too large. 1. Strongly agree 2. Agree 3. Neither agree nor disagree 4. Disagree 5. Strongly disagree *0. IAP, 8. Can't choose, 9. NA*

income16 Respondent's total family income from all sources

1. Less than 1000	2. 1000 to 2999
3. 3000 to 3999	4. 4000 to 4999
5. 5000 to 5999	6. 6000 to 6999
7. 7000 to 7999	8. 8000 to 9999
9. 10,000 to 12,499	10. 12,500 to 14,999
11. 15,000 to 17,499	12. 17,500 to 19,999
13. 20,000 to 22,499	14. 22,500 to 24,999
15. 25,000 to 29,999	16. 30,000 to 34,999
17. 35,000 to 39,999	18. 40,000 to 49,999
19. 50,000 to 59,999	20. 60,000 to 74,999
21. 75,000 to 89,999	22. 90,000 to 109,999
23. 110,000 to 129,999	24. 130,000 to 149,999
25. 150,000 to 169,999	26. 170,000 and over

27. Refused, 98. DK, 99 NA

intcntct How much of your contact with family or close friends is through text messages, mobile phones, or other devices that use the Internet?
1. All or almost all
2. Most of it
3. About half of it
4. Some of it
5. None or almost none of it
0. IAP, 8. Can't choose, 9. NA

letdie1 When a person has a disease that cannot be cured, do you think doctors should be allowed by law to end the patient's life by some painless means if the patient and his family request it?
1. Yes
2. No
0. IAP, 8. DK, 9. NA

letin1a Do you think the number of immigrants to America nowadays should be:
1. Increased a lot
2. Increased a little
3. Remain the same as it is
4. Reduced a little
5. Reduced a lot
0. IAP, 8. DK, 9. NA

marhomo Homosexual couples should have the right to marry one another.
1. Strongly agree
2. Agree
3. Neither agree nor disagree
4. Disagree
5. Strongly disagree
0. IAP, 8. DK, 9. NA

marital Are you currently married, widowed, divorced, separated, or have you never been married?

 1. Married
 2. Widowed
 3. Divorced
 4. Separated
 5. Never married
 9. NA

muslims What is your personal attitude toward Muslims?

 1. Very positive
 2. Somewhat positive
 3. Neither positive nor negative
 4. Somewhat negative
 5. Very negative
 0. IAP, 8. DK, 9. NA

obey How important is it for a child to learn obedience to prepare him or her for life?

 1. Most important
 2. 2nd most important
 3. 3rd most important
 4. 4th most important
 5. Least important
 0. IAP, 8. DK, 9. NA

owngun Do you have any guns in your home?

 1. Yes
 2. No
 0, IAP, 3. Refused, 8. DK, 9, NA

paeduc Father's highest year of school completed

 0–20. Actual number of years
 97. IAP, 98. DK, 99. NA

parsol Compared to your parents when they were the age you are now, do you think your own standard of living now is much better, somewhat better, about the same, somewhat worse, or much worse than theirs was?

 1. Much better
 2. Somewhat better
 3. About the same
 4. Somewhat worse
 5. Much worse
 0. IAP, 8. DK, 9. NA

partnrs5 How many sex partners have you had over the past five years?

 0. No partners 1. 1 partner
 2. 2 partners 3. 3 partners
 4. 4 partners 5. 5–10 partners
 6. 11–20 partners 7. 21–100 partners
 8. More than 100 partners
 9. 1 or more, don't know the number, 95. Several, 98. DK, 99. NA,
 −1. IAP

partyid Do you think of yourself as a Democrat, Republican, Independent, or what?

 0. Strong Democrat
 1. Not very strong Democrat
 2. Independent, close to Democrat
 3. Independent
 4. Independent, close to Republican
 5. Not very strong Republican
 6. Strong Republican
 7. Other party, 8. DK, 9. NA

poleff11 People like me don't have any say about what the government does.

 1. Strongly agree
 2. Agree
 3. Neither agree nor disagree
 4. Disagree
 5. Strongly disagree
 0. IAP, 8. Can't choose, 9. NA

polhitOK Are there any situations you can imagine in which you would approve of a policeman striking an adult male citizen?

 1. Yes
 2. No
 0. IAP, 8. DK, 9. NA

polviews I'm going to show you a seven-point scale on which the political views that people might hold are arranged from extremely liberal to extremely conservative. Where would you place yourself on this scale?

 1. Extremely liberal
 2. Liberal
 3. Slightly liberal
 4. Moderate
 5. Slightly conservative
 6. Conservative
 7. Extremely conservative
 0. IAP, 8. DK, 9. NA

premarsx There's been a lot of discussion about the way morals and attitudes about sex are changing in this country. If a man and a woman have sex relations before marriage, do you think it is always wrong, almost always wrong, wrong only sometimes, or not wrong at all?

 1. Always wrong
 2. Almost always wrong
 3. Wrong only sometimes
 4. Not wrong at all
 0. IAP, 8. DK, 9. NA

pres16 In 2016, did you vote for Clinton (the Democratic candidate) or Trump (the Republican candidate)? (Includes only those who said they voted in this election.)

 1. Clinton
 2. Trump
 3. Other, 4. No presidential vote, 0. IAP, 8. DK, 9. NA

prestg10 Prestige of respondent's occupation (*The higher the score, the greater the prestige.*)
 17–87 Prestige score
 0, IAP, NA or DK

racecen Race or ethnicity of Respondent (NOTE: This variable is a recode of *racecen1*. It is not available in the *GSS2018 Sample* data set.)
1. White
2. Black
3. American Indian and Alaska Native
4. Asian American and Pacific Islander
5. Hispanic

racecen1 Race or ethnicity of respondent (based on the categories used in the U.S. Census)

1. White	2. Black
3. American Indian or Alaska Native	4. Asian Indian
5. Chinese	6. Filipino
7. Japanese	8. Korean
9. Vietnamese	10. Other Asian
11. Native Hawaian	12. Guamanina or Chamorro
13. Samoan	14. Other Pacific Islander
15. Some other race	16. Hispanic
	0. IAP, 98. DK, 99. NA

(NOTE: For the *GSS2018 Sample* data file, this variable can be recoded to match the categories in *racecen*. See the "You Are the Researcher" feature in Chapter 10 for instructions on using the **recode** command.)

region Region of interview
1. New England (ME, VT, NH, MA, CT, RI)
2. Mid-Atlantic (NY, NJ, PA)
3. East North Central (WI, IL, IN, MI, OH)
4. West North Central (MN, IA, MO, ND, SD, NE, KS)
5. South Atlantic (DE, MD, WV, VA, NC, SC, GA, FL, DC)
6. East South Central (KY, TN, AL, MS)
7. West South Central (AK, OK, LA, TX)
8. Mountain (MT, ID, WY, NV, UT, CO, AR, NM)
9. Pacific (WA, OR, CA, AL, HI)

relig What is your religious preference? Is it Protestant, Catholic, Jewish, some other religion, or no religion?
1. Protestant
2. Catholic
3. Jewish
4. None
5. Other
8. DK, 9. NA

reliten Would you call yourself a strong (preference named in *relig*) or not very strong (preference named in *relig*)?
1. Strong
2. Not very strong
3. Somewhat strong
0. IAP, 4. No religion, 8. DK, 9. NA

sex Respondent's gender
1. Male
2. Female

sexfreq About how many times did you have sex during the last 12 months?
0. Not at all
1. Once or twice
2. About once a month
3. 2 or 3 times a month
4. About once a week
5. 2 or 3 times a week
6. More than 3 times a week
−1. IAP, 8. DK, 9. NA

spanking Do you strongly agree, agree, disagree, or strongly disagree that it is sometimes necessary to discipline a child with a good, hard spanking?
1. Strongly agree
2. Agree
3. Disagree
4. Strongly disagree
0. IAP, 8. DK, 9. NA

spkrac There are always some people whose ideas are considered bad or dangerous by other people. Consider someone who believed that Blacks are genetically inferior. . . . should he be allowed to make a speech in your community or not?
1. Yes, allowed
2. Not allowed
0. IAP, 8. DK, 9. NA

suicide1 Do you think a person has the right to end his or her life if the person has an incurable disease?
1. Yes
2. No
0. IAP, 8. DK, 9. NA

tempgen In general, do you think that a rise in the world's temperature caused by the "greenhouse" effect is
1. Extremely dangerous for the environment
2. Very dangerous
3. Somewhat dangerous
4. Not very dangerous
5. Not dangerous at all for the environment
0. IAP, 8. DK, 9. NA

thnkself How important is it for a child to learn to think for himself or herself to prepare him or her for life?
1. Most important
2. 2nd most important
3. 3rd most important
4. 4th most important
5. Least important
0. IAP, 8. DK, 9. NA

tvhours On the average day, about how many hours do you personally watch television?
00–24. Actual hours
−1. IAP, 98. DK, 99. NA

wwwhr Not counting e-mail, about how many hours per week do you use the Web?
0–120. Actual hours
−1. IAP, 998. DK, 999. NA

B. Codebook for *States* Database

AggAss	Number of aggravated assaults per 100,000 population, 2017
Asian	Percentage of the population Asian, 2017
AvgHHInc	Average household income, 2017
BA	Percentage of the population with a BA degree or more, 2017
BirthRate	Number of births per 1000 population, 2017
Black	Percentage of the population black, 2017
Burglary	Number of burglaries per 100,000 population, 2017
Cartheft	Number of car thefts per 100,000 population, 2017
ChildPoor	Percentage of children age 18 or younger living in poverty, 2017
CopSpend	Spending on police per capita, 2017
DeathRate	Number of deaths per 1000 population, 2017
EdSpend	Spending on education per pupil, 2017
FamPoor00	Percentage of families below the poverty line in 2000
FamPoor17	Percentage of families below the poverty line in 2017
FemHead	Percentage of households female-headed, 2017
Hispanic	Percentage of the population Hispanic, 2017
Homicide	Number of homicides per 100,000 population, 2017
Homicide95	Number of homicides per 100,000 population, 1995
HS	Percentage of the population with a high school degree or more 2017
Internet	Percentage of households connected to the Internet, 2017
MdAge	Median age, 2017

MdHHInc	Median household income
MdIncFemale	Median income for females working full time, year round, 2017
MdIncMale	Median income for males working full time, year round, 2017
MeanHHsize	Average household size, 2017
NotEngHome	Percentage of the population that does not speak English at home, 2017
PopDiv	Percentage of the population divorced, 2017
PopGrow	Population growth rate, 2010–2017
PopMarr	Percentage of the population married, 2017
PopSingle	Percentage of the population single, 2017
Rape	Number of rapes per 100,000 population, 2017
Robbery	Number of robberies per 100,000 population, 2017
South	South, non-south dummy variable
	0 = Non-South
	1 = South
State	Name of state
TeenBR	Births per 1000 females age 15–19
Theft	Thefts per 100,000 population, 2017
TrafDths17	Traffic fatalities per 100 million miles driven, 2017
TrafDths90	Traffic fatalities per 100 million miles driven, 1990
TrafDthsAlch	Percentage of fatal crashes with alcohol in driver
Unemplymnt	Unemployment rate, 2017
Vote16	Percentage of the population voting for president, 2016
White	Percentage of the population white, 2017
WrkTrav	Average minutes of travel to work, 2017

C. Codebook for International Population (*Intl-POP*) Database

Agric	Percentage of work force in agriculture
BirthRate	Number of births per 1000 population, 2017
Contracep	Percentage of married women using contraception, 2017
DeathRate	Number of deaths per 1000 population, 2017
FemaleWork	Labor force participation rate, females age 15 and older
Gengapind	Gender gap index. A measure of gender parity across health, education, economy, and politics. The closer to 1.00, the greater the gender equality.

GESecondary	Gender equality at the secondary level of education: Ratio of females to males
GNIcap	Gross national income per capita
IncInequal	Gini index of income inequality. The higher the score, the greater the income inequality.
IncLevel	Income level (World Bank classification), 2017 1. Low income 2. Lower middle income 3. Upper middle income 4. High income
Industry	Percentage of work force in industry
InfMortRate	Infant mortality rate: Death rate for ages 0–1, 2017
LifeExpFem	Life expectancy at birth for women, 2017
LifeExpMale	Life expectancy at birth for men, 2017
Male Work	Labor force participation rate, males age 15 and older
Nation	Name of nation
NaturalInc	Rate of Natural Increase in population
NetMigRate	Annual migration (people leaving minus people entering) per 1000 population, 2017
Pop15	Percentage of population younger than 15
Pop2018	Population, 2018
Pop2050	Projected population, 2050
Pop65	Percentage of population older than 65
Popgrow	Projected 2050 population as a multiple of 2017 population. For example, a score of 2 on this variable means that the population is projected to double by 2050.
Region	Region of the world 1. East Asia and the Pacific 2. Eastern Europe and Central Asia 3. Latin America and the Caribbean 4. Middle East and North Africa 5. South Asia 6. Sub-Saharan Africa 7. Western Europe 8. North America 9. Oceania
Religiosity	Percentage of the population who agree that "religion is an important part of everyday life."

Rights	Civil and political rights
	1. Free
	2.
	3. Partly free
	4.
	5.
	6. Not free
Services	Percentage of work force in service sector
TotalFertRate	Total fertility rate: Average number of children a woman has in her lifetime
Urban	Percentage of the population living in urban areas
WmnPar	Percentage of seats in National Parliament held by women

D. Codebook for State Crime Trends (*CrimeTrends84-10*) Database

AggAssault	Number of aggravated assaults per 100,000 population
Burglary	Number of burglaries per 100,000 population
CarTheft	Number of auto thefts per 100,000 population
Homicide	Number of homicides per 100,000 population
Robbery	Number of robberies per 100,000 population
Year	Years from 1984 to 2017

Answers to Odd-Numbered End-of-Chapter Problems and Cumulative Exercises

In addition to answers, this section suggests problem-solving strategies and provides examples of how to phrase or interpret conclusions and answers. You should try to solve and interpret the problems on your own before consulting this section.

In solving these problems, we let our calculator or computer do most of the work. We worked with whatever level of precision these devices permitted and didn't round off until the end or until we had to record an intermediate sum. We always rounded off to two places of accuracy (that is, to two places beyond the decimal point, or to hundredths). If you follow these same conventions, your answers will match ours. However, you should realize that small, generally trivial discrepancies between your answer and ours still might occur. If the difference doesn't seem trivial, you should double-check to make sure you haven't made an error or solve the problem again using a greater degree of precision.

Finally, please allow us a brief disclaimer about mathematical or typographical errors in this section. Let us assure you, first of all, that we know how important this section is for most students and that we worked hard to be certain that these answers are correct. Human fallibility being what it is, however, we know that we cannot make absolute guarantees. Should you find any errors, please let us know so that we can make corrections in the future.

Chapter 1

1.5
a. Nominal, discrete
b. Ordinal, discrete (The categories can be ranked in terms of degree of honesty with "Returned the wallet with money" the "most honest.")
c. Ordinal, discrete
d. Interval-ratio, continuous ("Years" has equal intervals and a true zero point.)
e. Interval-ratio, discrete

f. Interval-ratio, discrete
g. Nominal, discrete (The various patterns are different from each other but cannot be ranked from high to low.)
h. Interval-ratio, continuous
i. Ordinal, discrete
j. Number of accidents: interval-ratio discrete; Severity of accident: ordinal, discrete

1.7

	Variable	Level of Measurement	Type	Application
a.	Opinion	Ordinal	Discrete	Inferential
b.	Letter grade	Ordinal	Discrete	Descriptive (two variables)
	Percent correct	Interval-ratio	Continuous	
c.	Party	Nominal	Discrete	Inferential
	Gender	Nominal	Discrete	
	% in favor	Interval-ratio	Continuous	
d.	Homicides	Interval-ratio	Discrete	Descriptive (two variables)
e.	Satisfaction	Ordinal	Discrete	Descriptive (one variable)

1.9

	Variable	Level of Measurement	Type
a.	Own racial ID	Nominal	Independent
	Desirability of categories	Ordinal	Dependent
b.	# of relationships	Interval-ratio	Dependent
	# of times having intercourse	Interval-ratio	Dependent
	GPA	Interval-ratio	Independent
c.	% minority voters	Interval-ratio	Independent
	Turnout	Interval-ratio	Dependent
	% minority officials	Interval-ratio	Independent
d.	Per capita income	Interval-ratio	Independent
	School enrollment (Total)	Interval-ratio	Dependent
	School enrollment (Boys)	Interval-ratio	Dependent
	School enrollment (Girls)	Interval-ratio	Dependent
e.	Support for gun control	Ordinal	Dependent
	Years of schooling	Interval-ratio	Independent
	Gender	Nominal	Independent
	Region	Nominal	Independent
	Party preference	Nominal	Independent

Chapter 2

2.1 **a.** Complex A: $(5/20) \times 100 = (0.25) \times 100 = 25.00\%$

Complex B: $(10/20) \times 100 = 50.00\%$

b. Complex A: $4:5 = 0.80$

Complex B: $6:10 = 0.60$

c. Complex A: $(0/20) = 0.00$

Complex B: $(1/20) = 0.05$

d. $(6/(4 + 6)) \times 100 = (6/10) \times 100 = 60.0\%$

e. Complex A: $8:5 = 1.60$

Complex B: $2:10 = 0.20$

2.3 Drug overdose rate = $(5088/20,278,447) \times 100,000 = (.0002509) \times 100,000 = 25.09$

Firearm death rate = $(2724/20,278,447) \times 100,000 = (.0001343) \times 100,000 = 13.43$

Homicide rate = $(1269/20,278,447) \times 100,000 = (.0000626) \times 100,000 = 6.26$

2.5 For gender:

Sex	Frequency
Male	9
Female	6
TOTAL	15

For age, set $k = 10$. $R = 77 - 23$, or 54, so we can round off interval size to 5 ($i = 5$). The first interval will be 20–24, to include the low score of 23, and the highest interval will be 75–79.

Age	Frequency
20–24	1
25–29	2
30–34	3
35–39	2

(Continued on next column)

(Continued)

Age	Frequency
40–44	1
45–49	3
50–54	1
55–59	1
60–64	0
65–69	0
70–74	0
75–79	1
TOTAL	15

2.7

Males

Value	Frequency	Percentage	Cumulative Percentage
1	2	13.33	13.33
2	2	13.33	26.66
3	2	13.33	39.99
4	1	6.67	46.66
5	4	26.67	73.33
6	4	26.67	100
	15		

Females

Value	Frequency	Percentage	Cumulative Percentage
1	1	6.25	6.25
2	4	25.00	31.25
3	2	12.50	43.75
4	2	12.50	56.25
5	1	6.25	62.50
6	0	0.00	62.50
7	3	18.75	81.25
8	3	18.75	100
	16		

2.9 Set $k = 10$. $R = 92 - 5$, or 87, so set i at 10.

Score	Frequency	Midpoint	Percentage	Cumulative f	Cumulative %
0–9	3	4.5	12.0%	3	12.0%
10–19	7	14.5	28.0%	10	40.0%
20–29	6	24.5	24.0%	16	64.0%
30–39	0	34.5	0.0%	16	64.0%
40–49	2	44.5	8.0%	18	72.0%
50–59	2	54.5	8.0%	20	80.0%
60–69	3	64.5	12.0%	23	92.0%
70–79	0	74.5	0.0%	23	92.0%
80–89	0	84.5	0.0%	23	92.0%
90–99	2	94.5	8.0%	25	100.0%
TOTAL	25		100.0		

2.11 Answers should describe the rise in all crime rates except burglary, up to a peak in the early 1990s, followed by a decline. Rates for aggravated assault and homicide have increased in recent years. Burglary rates declined throughout the period, but the rate of decline has steadied after 2000 and then declined steeply since 2013.

2.13 One possible description of *marital* might be "The most common marital status was 'married' followed by 'never married.'" The smallest categories were "'widowed' and 'separated.'" A description of hours worked should note the peak in the middle of the graph at 40 hours.

Chapter 3

3.1 "Region of birth" and "religion" are nominal-level variables, "support for legalization" and "opinion of food" are ordinal, and "expenses" and "number of movies" are interval-ratio. The mode, the most common score, is the only measure of central tendency available for nominal-level variables. For the two ordinal-level variables, *don't forget to array the scores from high to low* before locating the median. There are 10 freshmen (N is even), so the median for freshmen will be the score halfway between the scores of the two middle cases. There are 11 seniors (N is odd), so the median for seniors will be the score of the middle case. To find the mean for the interval-ratio variables, add the scores and divide by the number of cases.

Variable	Measure	Freshmen	Seniors
Region of birth	Mode	North	North
Legalization	Median	3	5
Expenses	Mean	58.50	72.55
Movies	Mean	5.80	5.18
Food	Median	6	4
Religion	Mode	Protestant	Protestant, None

3.3

Variable	Level of Measurement	Measure of Central Tendency
Gender	Nominal	Mode = Male
Class	Ordinal	Median = Middle
Years	Interval-ratio	Mean = 26.15
Education	Ordinal	Median = High school
Marital status	Nominal	Mode = Married
No. of children	Interval-ratio	Mean = 2.39

3.5

Variable	Level of Measurement	Measure of Central Tendency
Marital status	Nominal	Mode = Married
Race/Ethnic group	Nominal	Mode = White
Age	Interval-ratio	Mean = 27.53
Attitude	Ordinal	Median = 7

3.7

Variable	Level of Measurement	Measure of Central Tendency
Gender	Nominal	Mode = Male
Support	Ordinal	Median = 1
Education	Ordinal	Median = 1
Age	Interval-ratio	Mean = 40.93

3.9 Attitude scales almost always generate ordinal-level data, so the appropriate measure of central tendency would be the median. The median is 9 for the students and 2 for the neighbors. Incidentally, the means are 7.80 for the students and 4.00 for the neighbors.

3.11 Mean = 1.31, median = 0.72. The higher value for the mean indicates a positive skew. Note the very high score of Denmark.

3.13 To find the median, you must first rank the scores from high to low. Both groups have 25 cases (N is odd), so the median is the score of the 13th case. For freshmen the median is 35, and for seniors the median is 30. The mean score for freshmen is 31.72, and for seniors the mean is 28.60.

3.15 The owners are using the mean while the players are citing the median. This distribution has a positive skew (the mean is greater than the median), as is typical for income data. Note the difference in wording in the two reports: The owners cite the "average" (mean) and the players cite the "typical player" (the score of the middle case or the median).

Chapter 4

4.1 The high score is 50 and the low score is 10, so the range is 50 − 10, or 40. The standard deviation is 12.28.

3.17

Variable	Level of Measurement	Measure of Central Tendency
helpblk	Ordinal	Median = 3.0
hrs1	Interval-ratio	Mean = 41.28
partnrs5	Ordinal	Median = 1.00
region	Nominal	Mode = 5 (South Atlantic)

NOTE: *partnrs5* has an unusual coding scheme. Scores 0–4 are actual numbers (a score of "1" means one partner) but higher values represent broad categories. A score of 6, for example, means "between 11 and 20 partners." The variable is interval-ratio from scores 0 to 4 and then ordinal for the remaining scores. Given this "hybrid" scoring scheme, for purposes of description, it is best to be conservative and treat the variable as ordinal.

4.3

Variable	Males		Females	
	R	s	R	s
Age	32	9.21	50	15.19
Political Stories	14	5.06	14	4.04
Support for the President	10	3.35	8	2.50

4.5

	Canadian Provinces		U.S. States	
	2000	*2016*	*1999*	*2017*
Mean	48,161.54	76,130.31	42,235.15	60,152.92
Median	47,300.00	69,995	40,058.00	57,051
Range	23,400	58,341	22,548	35,103
Std Dev.	6,753.64	16,999.07	6,635.91	10,697.71

For the Canadian provinces, the mean and median increase across the time period. For both years, the distributions show a positive skew (the mean is greater than the median). The standard deviation and the range increase, indicating that there is greater variability in 2016. The patterns are similar for the U.S. states.

4.7

State	Labor Force Participation Rate		Percentage HS Graduates		Mean Income	
	Male	*Female*	*Male*	*Female*	*Male*	*Female*
Mean	72.03	60.75	85.94	87.31	54,271.30	50,760.40
s	3.67	4.14	2.95	2.95	4,245.89	4,076.04

Males and females are very similar in terms of percentage of high school graduates, but females are less involved in the labor force and, on average, earn about $3500 less per year than males. The females in these 10 states are more variable in their labor force participation but are similar to males in variation on the other two variables.

4.9

Mean	1.17
Median	1.06
R	1.48
Q	0.55
s	0.36

4.11

	2018	2019
Mean	98.65	160.85
s	31.67	45.90

4.13

	1973	1975
Mean	12.21	19.06
s	12.20	9.63

4.17

	Age 55–65	Age 65–75
Mean	31.72	28.60
s	12.53	13.35

Those between 55 and 65 have slightly higher support for marijuana legalization than those between 65 and 75. The variation in support is higher among those between 65 and 75.

4.19

	FamPoor00	FamPoor17
Mean	8.82	9.09
s	2.80	2.53
R	12.50	10.00

Chapter 5

5.1

X_i	Z Score	% Area Above	% Area Below
5	−1.67	95.25	4.75
6	−1.33	90.82	9.18
7	−1.00	84.13	15.87
8	−0.67	74.86	25.14
9	−0.33	62.93	37.07
11	0.33	37.07	62.93
12	0.67	25.14	74.86
14	1.33	9.18	90.82
15	1.67	4.75	95.25
16	2.00	2.28	97.72
18	2.67	0.38	99.62

5.3

	Z Scores	Area
a.	0.60	27.43%
b.	0.90	18.41%
c.	0.10	53.98%
d.	0.30	61.79%
e.	0.60	72.57%
f.	1.10	86.43%
g.	0.10 & 1.10	32.45%
h.	0.60 & 1.10	13.86%
i.	0.60 & −0.40	38.11%
j.	0.10 & −0.40	19.52%

5.5

X_i	Z Score	Number of Students Above	Number of Students Below
60	−2.00	195	5
57	−2.50	199	1
55	−2.83	200	0
67	−0.83	159	41
70	−0.33	126	74
72	0.00	100	100
78	1.00	32	168
82	1.67	10	190
90	3.00	1	199
95	3.83	0	200

Note: Number of students (a discrete variable) has been rounded off to the nearest whole number.

5.7

	Z Scores	Area
a.	−2.20	1.39%
b.	1.80	96.41%
c.	−0.20 & 1.80	54.34%
d.	0.80 & 2.80	20.93%
e.	−1.20	88.49%
f.	0.80	21.19%

5.9

	Z Scores	Area
a.	−1.00 & 1.50	0.7745
b.	0.25 & 1.50	0.3345
c.	1.50	0.0668
d.	0.25 & −2.25	0.5865
e.	−1.00 & −2.25	0.1465
f.	−1.00	0.1587

5.11 Yes. The raw score of 110 translates into a Z score of 2.88. 99.80% of the area lies below this score, so this individual was in the top 1% on this test.

5.13 For the first event, the probability is 0.0919; for the second, the probability is 0.0655. The first event is more likely.

5.15 a.

	Z Score—Freshman	Z Score—Senior	Student Did Better as a
A	0.57	1.25	Senior
B	−0.29	0.50	Senior
C	−1.14	−2.50	Freshman (−1.14 is closer to the mean than −2.50)
D	2.86	2.25	Freshman
E	1.29	1.50	Senior

b.

Freshman Year

Score	Z Score	Probability
Less than 52	−0.14	0.4443
Less than 57	0.57	0.7157
Between 40 and 50	−1.86 & −0.43	0.3022
More than 51	−0.29	0.6141
More than 62	1.29	0.0985

Senior Year

Score	Z Score	Probability
Less than 88	−1.00	0.1587
Less than 98	1.50	0.9332
Between 80 and 100	−3.00 & 2.00	0.9758
More than 97	1.25	0.1056
More than 85	−1.75	0.9599

Part I Cumulative Exercises

1. The level of measurement is the most important criterion for selecting descriptive statistics. The following table presents the values for the statistics that are relevant for each variable.

Religion is nominal, so the only statistic available is the mode. For the two ordinal-level variables ("strength" and "comfort"), the median and the range are the most appropriate choices but it is common for social science researchers to also report the standard deviation and mean. For interval-ratio variables like "pray" and "age," the mean and standard deviation are the preferred summary statistics.

	Variables				
	Religion	Strength	Pray	Comfort	Age
	Level of Measurement				
	Nominal	Ordinal	Interval-Ratio	Ordinal	Interval-Ratio
Mode	Protestant	9 & 2	0	1	30
Md		6	0	2	37
\bar{X}		5.27	1.13	2.13	38.07
R		9	6	4	37
s		2.95	1.63	1.31	10.90

3. The output for this job will begin with a summary table with all statistics. A modified version of the table, with inappropriate statistics omitted, is reproduced below.

	Variables					
	childs	degree	racecen1	spanking	tvhours	relig
	Level of Measurement					
	I-R	Ord.	Nom.	Ord.	I-R	Nom.
Mode	0	1	1	2	2	1
Md	2	1		2	2	
\bar{X}	1.86	1.6		2.18	2.94	
R	8	4		3	24	
s	1.67	1.21		0.87	2.84	

Chapter 7

7.1 **a.** 5.2 ± 0.11
 b. 100 ± 0.71
 c. 20 ± 0.40
 d. 1020 ± 5.41
 e. 7.3 ± 0.23
 f. 33 ± 0.80

7.3

Confidence Level	Alpha	Area Beyond Z	Z Score
95%	0.05	0.0250	± 1.96
94%	0.06	0.0300	± 1.88 or ± 1.89
92%	0.08	0.0400	± 1.75 or ± 1.76
97%	0.03	0.0150	± 2.17
98%	0.02	0.0100	± 2.33 or ± 2.32

7.5 **a.** 2.30 ± 0.04
 b. 2.10 ± 0.01, $0.78 \pm .07$
 c. 6.00 ± 0.37

7.7 **a.** 478.23 ± 1.97. The students spent between $476.26 and $480.20 on books.
 b. 1.5 ± 0.04. The students visited the clinic between 1.46 and 1.54 times on the average.
 c. 2.8 ± 0.13
 d. 3.5 ± 0.19

7.9 0.14 ± 0.07. The estimate is that between 7% and 21% of the population consists of unmarried couples living together.

7.11 **a.** $P_s = 823/1496 = 0.55$. The confidence interval is 0.55 ± 0.03. Between 52% and 58% of the population agrees with the statement.
 b. $P_s = 650/1496 = 0.43$. The confidence interval is 0.43 ± 0.03.
 c. $P_s = 375/1496 = 0.25$. The confidence interval is 0.25 ± 0.03.
 d. $P_s = 1023/1496 = 0.68$. The confidence interval is 0.68 ± 0.03.
 e. $P_s = 800/1496 = 0.53$. The confidence interval is 0.53 ± 0.03.

7.13

Alpha (α)	Confidence Level	Z Score	Confidence Interval
0.10	90%	± 1.65	100 ± 0.74
0.05	95%	± 1.96	100 ± 0.88
0.01	99%	± 2.58	100 ± 1.15
0.001	99.9%	± 3.32	100 ± 1.49
0.0001	99.99%	± 3.90	100 ± 1.76

7.15 The confidence interval is 0.51 ± 0.05 and the estimate would be that between 46% and 56% of the population prefer candidate A. The population parameter (P_u) is equally likely to be anywhere in the interval (that is, it's just as likely to be 46% as it is to be 56%), so a winner cannot be predicted.

7.17 The sample proportion is 0.23 (35/150) and the confidence interval is 0.23 ± 0.08. At the 95% confidence level, the estimate would be that between 240 (15%) and 496 (31%) of the 1600 freshmen would be extremely interested. The estimated numbers are found by multiplying N (1600) by the upper (0.31) and lower (0.15) limits of the interval.

7.19 **a.** The confidence interval for *age* is 48.97 ± 0.73 and the confidence interval for *hrs1* is 41.28 ± 0.76.
 b. The confidence interval for the proportion of Catholics in the population is 0.21 ± 0.02 and the confidence interval for the proportion of supporters of the death penalty in the population is 0.63 ± 0.02.

Chapter 8

8.1 **a.**

Z(Critical)
$+1.65$ or -1.65
± 1.65
± 1.89
$+2.33$ or -2.33
± 2.33

 b.

df	t(Critical)
30	± 1.697
23	± 2.500
120	± 2.617
30	$+2.457$ or -2.457
60	$+1.671$ or -1.671

 c.

1	Z(obtained) $= -3.76$
2	t(obtained) $= -2.21$
3	Z(obtained) $= -5.74$
4	Z(obtained) $= 0.66$
5	Z(obtained) $= -0.77$

8.3 **a.** Z(obtained) $= -41.00$
 b. Z(obtained) $= 29.09$

8.5 Z(obtained) $= 6.04$

8.7 **a.** Z(obtained) $= -13.66$
 b. Z(obtained) $= 25.50$

8.9 t(obtained) $= 4.50$

8.11 Z(obtained) $= 3.06$

8.13 Z(obtained) $= -1.48$

8.15 **a.** Z(obtained) $= -1.49$
 b. Z(obtained) $= 2.19$
 c. Z(obtained) $= -8.55$
 d. Z(obtained) $= -18.07$
 e. Z(obtained) $= 2.09$
 f. Z(obtained) $= -53.33$

8.17 t(obtained) $= -1.14$

8.19 **a.** Z(obtained) $= -1.28$
 b. Z(obtained) $= 3.57$
 c. Z(obtained) $= 34.96$
 d. Z(obtained) $= -19.09$
 e. Z(obtained) $= 35.00$
 f. Z(obtained) $= -140.00$

Chapter 9

9.1 **a.** $\sigma_{\bar{X} - \bar{X}} = 1.39$, Z(obtained) $= -2.52$
 b. $\sigma_{\bar{X} - \bar{X}} = 1.61$, Z(obtained) $= 2.48$

9.3 **a.** $\sigma_{\bar{X} - \bar{X}} = 10.57$, Z(obtained) $= 1.70$
 b. $\sigma_{\bar{X} - \bar{X}} = 11.28$, Z(obtained) $= -2.48$

9.5 **a.** $\sigma_{\bar{X} - \bar{X}} = 0.08$, Z(obtained) $= -11.25$
 b. $\sigma_{\bar{X} - \bar{X}} = 0.12$, Z(obtained) $= -3.33$
 $\sigma_{\bar{X} - \bar{X}} = 0.15$, Z(obtained) $= 20.00$

9.7 These are small samples (combined N's of less than 100), so be sure to use Formulas 9.5 and 9.6 in step 4.
 a. $\sigma_{\bar{X} - \bar{X}} = 0.12$, t(obtained) $= -1.33$
 b. $\sigma_{\bar{X} - \bar{X}} = 0.13$, t(obtained) $= 14.85$

9.9 **a.** (Brazil) $\sigma_{\bar{X} - \bar{X}} = 0.11$, Z(obtained) $= 0.91$
 b. (Ukraine) $\sigma_{\bar{X} - \bar{X}} = 0.15$, Z(obtained) $= 2.00$
 c. (Vietnam) $\sigma_{\bar{X} - \bar{X}} = 0.10$, Z(obtained) $= 2.00$
 d. (South Africa) $\sigma_{\bar{X} - \bar{X}} = 0.09$, Z(obtained) $= -4.44$
 e. (Egypt) $\sigma_{\bar{X} - \bar{X}} = 0.10$, Z(obtained) $= -3.00$

9.11 $P_u = 0.45$, $\sigma_{p-p} = 0.06$, Z(obtained) $= 0.67$

9.13 **a.** $P_u = 0.46$, $\sigma_{p-p} = 0.06$, Z(obtained) $= 2.17$
 b. $P_u = 0.80$, $\sigma_{p-p} = 0.07$, Z(obtained) $= -1.57$
 c. $P_u = 0.73$, $\sigma_{p-p} = 0.08$, Z(obtained) $= 0.63$

9.15 **a.** $P_u = 0.34$, $\sigma_{p-p} = 0.04$, Z(obtained) $= 1.50$
 b. $P_u = 0.53$, $\sigma_{p-p} = 0.04$, Z(obtained) $= -2.75$
 c. $P_u = 0.43$, $\sigma_{p-p} = 0.03$, Z(obtained) $= 4.67$
 d. $\sigma_{\bar{X} - \bar{X}} = 0.43$, Z(obtained) $= 1.86$
 e. $\sigma_{\bar{X} - \bar{X}} = 0.14$, Z(obtained) $= -5.71$
 f. $\sigma_{\bar{X} - \bar{X}} = 0.08$, Z(obtained) $= -5.50$

9.17

Variable	Categories	N	Mean	s	t Score	df	Sig.
educ	Protestant	1139	13.73	2.82	2.91	1629	0.004
	Catholic	492	13.26	3.33			
prestg10	Protestant	1100	44.40	13.42	−0.38	1567	0.70
	Catholic	469	44.68	12.92			

The difference between Protestants and Catholics on years of education (*educ*) is significant at the 0.05 level. Protestants average significantly more years of education.

The difference between Protestants and Catholics on occupational prestige (*prestg10*) is not significant at the 0.05 level.

Chapter 10

10.1

	Grand Mean	SST	SSB	SSW	F Ratio
a.	12.17	231.67	173.17	58.5	13.32
b.	6.87	455.73	78.53	377.20	1.25
c.	31.65	8362.55	5053.35	3309.2	8.14

10.3

	Grand Mean	SST	SSB	SSW	F Ratio
a.	4.39	86.28	45.78	40.50	8.48
b.	16.44	332.44	65.44	267.00	1.84

For problem 10.3a, with alpha $= 0.05$ and $df = 2, 15$, the critical F ratio would be 3.68. We would reject the null hypothesis and conclude that decision making *does* vary significantly by type of relationship. By inspection of the group means, it seems that the "cohabitational" category accounts for most of the differences.

10.5

Grand Mean	SST	SSB	SSW	F Ratio
9.28	213.61	2.11	211.50	0.08

10.7

Grand Mean	SST	SSB	SSW	F Ratio
5.40	427.32	124.06	303.26	6.00

10.9

	Grand Mean	SST	SSB	SSW	F Ratio
Mexico	3.78	300.98	154.08	146.90	12.59
Canada	6.88	156.38	20.08	136.30	1.77
U.S.	5.13	286.38	135.28	151.10	10.74

At the 0.05 level of significance and $df = 3, 36$, the critical F ratio is 2.92. There is a significant difference in support for suicide by class in Mexico and the United States but not in Canada. The category means for Mexico suggest that the upper class accounts for most of the differences. For the United States, there is more variation across the category means, and the working class seems to account for most of the differences. Going beyond the ANOVA test and comparing the grand means, we see that support is highest in Canada and lowest in Mexico.

10.11 For this exercise, we recoded *relig* and grouped all of the religions scored 6 ("Buddhism") through 13 ("Inter-Nondenominational") into the "Other" category (a score of 5). We called the recoded variable *religR* (*relig*-recoded) and saved the new variable to the data file.

a. *educ*

	Group Means	df	Sig.	F Ratio
Protestant	13.73			
Catholic	13.26			
Jewish	16.36	2324	0.000	11.943
None	14.26			
Other	13.73			

b. *prestg10*

	Group Means	df	Sig.	F Ratio
Protestant	44.40			
Catholic	44.68			
Jewish	53.00	2229	0.001	4.582
None	44.18			
Other	46.96			

Chapter 11

11.1 a. 1.11 **b.** 0.00 **c.** 1.52 **d.** 1.46

11.3 a. A computing table is highly recommended as a way of organizing the computations for chi square:

Computational Table for Problem 11.3

(1)	(2)	(3)	(4)	(5)
f_o	f_e	$f_o - f_e$	$(f_o - f_e)^2$	$(f_o - f_e)^2/f_e$
6	5	1	1	0.20
7	8	−1	1	0.13
4	5	−1	1	0.20
9	8	1	1	0.13
26	26	0		$\chi^2 = 0.66$

There is 1 degree of freedom in a 2×2 table. With alpha set at 0.05, the critical value for the chi square is 3.841. The obtained chi square is 0.66, so we fail to reject the null hypothesis of independence between the variables. There is no statistically significant relationship between race and services received.

b.

	Race	
Received Services?	*Black*	*White*
Yes	60.00%	43.75%
No	40.00%	56.25%
TOTALS	100.00%	100.00%

The relationship is not significant, but the pattern is that blacks are more likely to receive services than whites.

11.5 a.

Computational Table for Problem 11.5

(1)	(2)	(3)	(4)	(5)
f_o	f_e	$f_o - f_e$	$(f_o - f_e)^2$	$(f_o - f_e)^2/f_e$
21	17.5	3.5	12.25	0.70
29	32.5	−3.5	12.25	0.38
14	17.5	−3.5	12.25	0.70
36	32.5	3.5	12.25	0.38
100	100.00	0.0		$\chi^2 = 2.16$

With $df = 1$ and alpha $= 0.05$, the critical value for the chi square is 3.841. The obtained chi square of 2.16 does not fall within this area, so the null hypothesis cannot be rejected. There is no statistically significant relationship between unionization and salary.

Column Percentages

	Status	
Salary	*Union*	*Nonunion*
High	60.00%	44.62%
Low	40.00%	55.38%
TOTALS	100.00%	100.00%

The relationship is not significant but unionized fire departments tend to have higher salary levels.

11.7 The obtained chi square is 5.13, which is significant ($df = 1$, alpha $= 0.05$). The column percentages show that more affluent communities have higher-quality schools.

11.9 The obtained chi square is 6.67, which is significant ($df = 2$, alpha $= 0.05$). The column percentages show that shorter marriages have higher satisfaction.

11.11 The obtained chi square is 12.58, which is significant ($df = 4$, alpha $= 0.05$). The column percentages show that proportionally more of the students living "off campus with roommates" are in the high-GPA category.

11.13 The obtained chi square is 19.34, which is significant ($df = 3$, alpha $= 0.05$). Legalization was not favored by a majority of any region, but the column percentages show that the West was most in favor.

11.15 The obtained chi square is 33.79, which is significant ($df = 3$, alpha $= 0.05$).

11.17 a.

Preference	Gender	
	Male	Female
Trump	46.67%	43.70%
Clinton	53.33%	56.30%
TOTALS	100.00%	100.00%

$\chi^2 = 0.67$, $df = 1$, $p > 0.05$

b.

Preference	Race/Ethnicity		
	White	Black	Latino
Trump	53.71%	5.00%	40.00%
Clinton	46.28%	95.00%	60.00%
TOTALS	100.00%	100.00%	100.00%

$\chi^2 = 82.21$, $df = 1$, $p < 0.05$

c.

Preference	Education			
	Less Than HS	HS Graduate	College Graduate	Postgrad. Degree
Trump	46.15%	60.00%	35.12%	21.28%
Clinton	53.85%	40.00%	64.88%	78.72%
TOTALS	100.00%	100.00%	100.00%	100.00%

$\chi^2 = 51.20$, $df = 3$, $p < 0.05$

d.

Preference	Religion				
	Protestant	Catholic	Jewish	None	Other
Trump	40.24%	66.67%	33.33%	31.82%	45.45%
Clinton	59.76%	33.33%	66.67%	68.18%	54.55%
TOTALS	100.00%	100.00%	100.00%	100.00%	100.00%

$\chi^2 = 42.83$, $df = 4$, $p < 0.05$

Part II Cumulative Exercises

1. One of the challenges of statistical analysis is to make reasonable decisions about which test to use in which situation. You can minimize the confusion by approaching the decision systematically. We'll use the first problem of this exercise to consider some ways in which reasonable decisions can be made.

The situation calls for a test of hypotheses ("Is the difference significant?"), so our choice of procedures is limited to Chapters 8–11. Next, note that "number of minutes" is interval-ratio in level of measurement and the research question asks us to compare two groups or samples: men and women. Which test should we use?

Chapter 8 (one-sample tests) and Chapter 10 (tests involving more than two samples or categories) are not relevant. Chi square (Chapter 11) won't work unless we collapse the scores on Internet use into a few categories. This leaves Chapter 9. A significance test with sample means fits the situation. We have large samples (combined Ns greater than 100), so Formulas 9.2 and 9.4 will be relevant.

a. $\sigma_{\bar{X}-\bar{X}} = 0.14$, Z(obtained) $= 35.71$

The difference in Internet minutes is significant. Men, on the average, use this technology more frequently.

b. The table format is a sure tip-off that the chi square test is appropriate. The obtained chi square is 0.88, which is not significant at the 0.05 level. There is no statistically significant relationship between involvement and social class.

c. "Number of partners" sounds like an interval-ratio–level variable, and education has three categories. Analysis of variance is an appropriate test for this situation. The F ratio is 0.13—not at all significant—so we must conclude that this dimension of sexuality does not vary by level of education.

Grand Mean	SST	SSB	SSW	F Ratio
4.04	262.96	3.24	259.71	0.13

d. The problem asks for a characteristic of a population ("How many times do adult Americans move?") but gives information only for a random sample. The estimation procedures presented in Chapter 7 fit the situation, and, because the information is presented in the form of a mean, you should use Formula 7.2 to form the estimate. At an alpha level of 0.05, the confidence interval would be 3.5 ± 0.02.

e. The research question focuses on the difference between a single sample and a population, so Chapter 8 is relevant. The population standard deviation is unknown and we have a large sample, so Formula 8.2 will be applicable. t(obtained) is -0.26, which is not significant at alpha $= 0.05$. The sample is not significantly different from the population—rural school districts are the same as the universe of all school districts in this state with regard to budget support from the state.

Chapter 12

12.1 a.

Support?	Males	Females
	Gender	
For	44.44%	35.00%
Against	55.56%	65.00%
TOTALS	100.00%	100.00%

Phi $= 0.10$; Lambda $= 0.00$; Max. diff. $= 9.44$. This relationship is weak to moderate.

b.

Support?	Liberal Arts	Science and Business
	Discipline	
For	30.00%	48.15%
Against	70.00%	51.85%
TOTALS	100.00%	100.00%

Phi $= 0.18$; Lambda $= 0.00$; Max. diff. $= 18.15$. This relationship is moderate to strong.

c.

Support?	Tenured	Nontenured
	Status	
For	45.45%	28.57%
Against	54.55%	71.43%
TOTALS	100.00%	100.00%

Phi $= 0.16$; Lambda $= 0.00$; Max. diff. $= 16.88$. This relationship is moderate to strong.

12.3 Phi $= 0.59$; Lambda $= 0.54$; Max. diff. $= 60.00$. This is a strong relationship.

12.5 a. Cramer's $V = 0.33$; Lambda $= 0.12$
b. Cramer's $V = 0.26$; Lambda $= 0.18$
c. Cramer's $V = 0.24$; Lambda $= 0.16$

12.7 Gamma $= -0.61$; Max. diff. $= 33.56$. This is a strong, negative relationship. As authoritarianism increases, efficiency decreases. *NOTE:* These variables are associated, but which causes which? The problem assumes that authoritarianism is the independent variable ("mean bosses cause lazy workers") but could the causal relationship go in the other direction ("lazy workers make bosses mean")? Remember that correlation and causation are two different things.

12.9 $G = -0.17$. This is a weak, negative relationship. The higher the level of alcohol use among parents, the fewer the symptoms of depression.

12.11 $r_s = 0.10$

12.13

sex and	Phi or V	Lambda	
fear	Phi = 0.22*	λ = 0.04	This is a moderate to weak relationship. The column percentages indicate that females (42.7%) are more likely to say "yes" than males (21.6%).
cappun	Phi = 0.08	λ = 0.00	This is a weak relationship. Males are slightly more in favor (67.3%) than females (59.6%).
satjob	V = 0.05	λ = 0.00	This is a very weak relationship. Females are slightly more "very satisfied" (49.4%) than males (47.4%).

*Ignore the negative sign for Phi.

Chapter 13

13.1 *(HINT: When finding the slope, remember that "Turnout" is the dependent, or Y, variable.)*

	For Turnout (Y) and		
	Unemployment	Education	Neg. Campaigning
a	39.00	−94.73	114.01
b	3.00	12.67	−0.90
Reg. equation	Y = 39.0 + 3.0X	Y = −94.73 + 12.67X	Y = 114.01 + (−0.90)X
r	0.95	0.98	−0.87
r^2	0.90	0.96	0.76
t(obtained)	5.21	8.49	−3.08

13.3 *(HINT: When finding the slope, remember that "Number of visitors" is the dependent, or Y, variable.)*

Slope (b) = −0.37
Y intercept (a) = 13.42
r = −0.31
r^2 = 0.10

13.5 The correlation between the variables (r = 0.47) is positive and moderate to weak (r^2 = 0.22) in strength.

13.7

	attend		childs		wwwhr		tvhours	
	r	r^2	r	r^2	r	r^2	r	r^2
age	0.18*	0.03	0.38*	0.14	−0.26*	0.0	0.21*	0.04

*Significant at 0.05.

13.9

	Homicide		CarTheft	
	r	r^2	r	r^2
PopGrow	−0.11	0.01	0.33	0.11
CopSpend	0.17	0.03	0.30	0.09
Unemplymnt	0.63	0.40	0.30	0.09

13.11

	GNIcap		Urban	
	r	r^2	r	r^2
Religiosity	−0.79	0.62	−0.56	0.31

Part III Cumulative Exercises

1. Assuming that crime rates and "Percentage immigrants" are both measured at the interval-ratio level, Pearson's r would be the appropriate measure of association. This is a moderate negative relationship (r = −0.47), and percentage of immigrants explains 22% of the variation in crime rate for these 10 cities. As the percentage of immigrants increases, the crime rate decreases.

3. The appropriate measure for these variables is Spearman's rho, which is 0.74. This is a strong positive relationship, and states with higher quality of life tend to have better systems of higher education.

5. **a.** Among other results, you should find moderate, negative relationships between *age* and *partnrs5* (−0.32) and *sexfreq* (−0.41). In other words, sexual activity tends to decline with age. There is a weak positive relationship between *partnrs5* and *sexfreq* (0.13): Frequency of sexual activity tends to increase as the number of partners increases (r = .33). *(In your opinion, which of the variables in this relationship is the cause and which is the effect? Why?)* The relationships between *rincome* and the other variables are weak.

b. The relationship between *fear* and *degree* is weak to moderate ($G = 0.13$), as is the relationship between *fear* and *marital* (Cramer's $V = 0.09$).

c. The relationship between *spanking* and *religR* is moderate (Cramer's $V = 0.13$), as is the relationship between *spanking* and *sex* (Cramer's $V = 0.06$). Use column percentages to describe the pattern of the relationship.

d. These are ordinal-level variables, so gamma is appropriate. This is a moderate, "negative" relationship ($G = -0.35$). As education increases, support increases.

e. There is a moderate to strong relationship between *pres16* and *cappun* (phi $= 0.37$) and a moderate relationship between *spanking* and *cappun* ($G = 0.32$). *(NOTE: Ignore the negative sign for phi.)*

Chapter 14

14.1

Bivariate gamma: 0.71

Partial Gammas		
Controlling for gender	Male	0.78
	Female	0.65
	$G_p =$	0.71
Controlling for origin	Asian	0.67
	Hispanic	0.74
	$G_p =$	0.72

The bivariate relationship is strong and positive: the longer the residence, the greater the fluency in English. The bivariate relationship is not affected by the gender or origin of the immigrants. These results would be taken as strong evidence of a direct (causal) relationship between length of residence and fluency in English.

14.3

Bivariate gamma: -0.23

Partial Gammas, Controlling for Gender	
Female	-0.34
Male	-0.10
$G_p =$	-0.25

This is an interactive relationship. Length of institutionalization has a greater effect on the reality orientation of females than of males. Partial gamma should be disregarded for interactive relationships.

14.5

Bivariate gamma: 0.25

Partial Gammas		
Controlling for gender	Males	0.24
	Females	0.26
	$G_p =$	0.25
Controlling for race	Whites	0.62
	Nonwhites	-0.50
	$G_p =$	0.49

The bivariate relationship is moderate to weak and positive. Concern increases as education increases. Controlling for gender has no effect, but controlling for race shows an interactive relationship. Concern increases with education for whites but decreases with education for nonwhites.

14.7

Bivariate gamma: 0.13

Partial Gammas		
Controlling for gender	Males	0.11
	Females	0.12
	$G_p =$	0.11

The bivariate relationship is weak. Fear is highest in the lower class and tends to decrease as class increases. The relationship is not affected by gender but lower class females are especially likely to report that they are afraid, so there might be some interaction between the variables.

14.9

Bivariate gamma: -0.24

Partial Gammas		
Controlling for gender		
	Males	-0.24
	Females	-0.25
	$G_p =$	-0.25

Overall, disapproval of premarital sex increases as age increases. The same pattern holds for males and females. This appears to be a direct relationship.

Chapter 15

15.1 a. For turnout (Y) and unemployment (X) while controlling for negative advertising (Z), $r_{yx.z} = 0.95$. The relationship between X and Y is not affected by the control variable Z.

b. For turnout (Y) and negative advertising (X) while controlling for unemployment (Z), $r_{yx.z} = -0.89$. The bivariate relationship is not affected by the control variable.

c. Turnout (Y) = 70.25 + (2.09) unemployment (X_1) + (−0.43) negative advertising (X_2). For unemployment (X_1) = 10 and negative advertising (X_2) = 75, turnout (Y) = 58.90.

d. For unemployment (X_1): $\beta_1^* = 0.66$. For negative advertising (X_2): $\beta_2^* = -0.41$. Unemployment has a stronger effect on turnout than negative advertising. Note that the independent variables' effects on turnout are in opposite directions.

e. $R^2 = 0.98$

15.3 a. Cohesion (Y) = 4.31 + (0.65) children (X_1) + (−0.00001) income (X_2).

b. For children (X_1) = 6 and income (X_2) = 20,000, cohesion (Y) = 8.41.

c. For children (X_1): $\beta_1^* = 0.37$. For income (X_2): $\beta_2^* = -0.06$. Number of children has a stronger effect than income. Note that the independent variables' effects on cohesion are in opposite directions.

d. $R^2 = 0.14$

15.5 a. The zero-order correlation between *wwwhr* and *age* is −0.26. The partial correlation coefficient when controlling for *educ* is −0.27. The relationship seems to be direct: Internet use decreases with age regardless of level of education.

b. *wwwhr* (Y) = 16.15 + (−0.27) *age* (X_1) + (0.77) *educ* (X_2).

c. *wwwhr* (Z_y) = (−0.27) *age* (Z_1) + (0.13) *educ* (Z_2).

d. $R^2 = 0.08$

15.7 a. *CarTheft* (Y) = 203.52 + (−0.01) *EdSpend* (X_1) + (35.15) *Unemply* (X_2).

b. *CarTheft* (Z_y) = (−0.36) *EdSpend* (Z_1) + (0.31) *Unemply* (Z_2).

c.

Bivariate:	r	r^2
Carthft10 and		
EdSpend	−0.34	0.12
Unemplymnt	0.30	0.09
Multiple:	$R^2 = 0.22$	

15.9 a. *Religiosity* (Y) = 93.95+ (−0.001) *GNIcap* (X_1) + (−0.16) *Urban* (X_2).

b. *Religiosity* (Z_y) = (−0.72) *GNIcap* (Z_1) + (−0.11) *Urban* (Z_2).

c.

Bivariate:	r	r_2
Religiosity and		
GNIcap	−0.79	0.62
Urban	−0.56	0.31
Multiple:	$R^2 = 0.63$	

Part IV Cumulative Exercises

1. a. The choice of multivariate procedures will depend on the level of measurement and the number of possible scores for each variable. Regression analysis is appropriate for interval-ratio, continuous variables. In this situation, we have three variables: GPA, hours worked, and College Board scores. College Board scores are arguably only ordinal in level of measurement, but the scores have a wide range and seem continuous. GPA is the dependent variable, and the zero-order correlation with hours worked is −0.83, a strong and negative relationship, which indicates that having a job did interfere with academic success for this sample. The zero-order correlation between GPA and College Board scores is 0.55, a relationship that is consistent with the idea that College Board scores predict success in college.

The results of the regression analysis with hours worked (X_1) and College Board scores (X_2) as independent variables:

$$Y = 1.71 + (-0.04)X_1 + (0.003)X_2$$
$$Z_y = (-0.75)Z_1 + (0.43)Z_2$$
$$R^2 = 0.86$$

The beta-weights indicate that hours worked has a stronger direct effect on GPA than College Board scores. Even for students who were well prepared for college and had high College Board scores, having a job had a negative impact on GPA. The high value for R^2 means that only a small percentage of the variance in GPA (14%) is left unexplained by these two independent variables.

b. In this situation, we have three dichotomous variables, two ordinal level and one nominal. The elaboration technique is appropriate to analyze these relationships. The bivariate table shows that there is a relationship between graduating and level of social activity for this sample: 70% of the students with low levels of social activity graduated in four

years versus only 30% of the students with high levels. Bivariate gamma is 0.69, indicating a strong relationship.

Controlling for gender reveals some interaction. The gamma for males (0.77) is stronger than the bivariate gamma, and the gamma for females (0.60) is weaker. In other words, while level of social activity has an effect on graduation rates, the effect is stronger for males than for females. This suggests that gender should be incorporated into the analysis, with a view to developing an understanding of why it would have different effects for males and females.

3. Bivariate gamma: 0.11

Partial Gammas, Controlling for Gender	
Female	0.11
Male	0.16
$G_p = 0.15$	

5. Bivariate phi: 0.37*

Partial Phi, Controlling for Race/Ethnicity	
Whites	0.38
Blacks	0.07
American Indian	0.75 (*NOTE: N* = 16)
Asian American	0.09 (*NOTE: N* = 28)
Hispanic American	0.32 (*NOTE: N* = 31)

*Disregard the negative sign for phi.

7. *Internet* $(Y) = 62.54 + (0.15)BA(X_1) + (0.00)$
 MdHHInc (X_2)

$$Z_y = (0.21)Z_1 + (0.68)Z_2$$
$$R^2 = 0.73$$

Glossary

Each entry includes a brief definition and the chapter number where the term was introduced.

Alpha (α). The probability of error, or the probability that a confidence interval does not contain the population value. Alpha levels are usually set at 0.10, 0.05, 0.01, or 0.001. Chapter 7

Alpha level (α). The proportion of area under the sampling distribution that contains unlikely sample outcomes, given that the null hypothesis is true. Also, the probability of Type I error. Chapter 8

Analysis of variance. A test of significance appropriate for situations in which we are concerned with the differences between more than two sample means. Chapter 10

ANOVA. See **Analysis of variance**. Chapter 10

Association. The relationship between two (or more) variables. Two variables are said to be associated if the distribution of one variable changes for the various categories or scores of the other variable. Chapter 12

Bar chart. A graph used for discrete variables. Categories are represented by bars of equal width, the height of each corresponding to the number (or percentage) of cases in the category. Chapter 2

Beta-weight (β). See **Standardized partial slope**. Chapter 15

Bias. A criterion used to select sample statistics as estimators. A statistic is unbiased if the mean of its sampling distribution is equal to the population value of interest. Chapter 7

Bivariate normal distributions. The model assumption in the test of significance for Pearson's r that both variables are normally distributed. Chapter 13

Bivariate table. A table that displays the joint frequency distributions of two variables. Chapter 11

Boxplot. A graph that presents information about the central tendency and variation of a variable. Chapter 4

Cells. The cross-classification categories of the variables in a bivariate table. Chapter 11

Central Limit Theorem. A theorem that specifies the mean, standard deviation, and shape of the sampling distribution, given that the sample is large. Chapter 6

χ^2(critical). The score on the sampling distribution of all possible sample chi squares that marks the beginning of the critical region. Chapter 11

χ^2(obtained). The test statistic computed from sample results. Chapter 11

Chi square (χ^2) test. A nonparametric test of hypothesis for variables that have been organized into a bivariate table. Chapter 11

Class intervals. The categories used in the frequency distributions for interval-ratio variables. Chapter 2

Cluster sampling. A method of sampling by which geographical units are randomly selected and all cases within each selected unit are tested. Chapter 6

Coefficient of determination (r^2). The proportion of all variation in Y that is explained by X. Found by squaring the value of Pearson's r. Chapter 13

Coefficient of multiple determination (R^2). A statistic that equals the proportion of the total variation in the dependent variable that is explained by all independent variables combined. Chapter 15

Column. The vertical dimension of a bivariate table. By convention, each column represents a score on the independent variable. Chapter 11

Column percentages. Percentages computed within each column of a bivariate table. Chapter 12

Conditional distribution of Y. The distribution of scores on the dependent variable for a specific score or category of the independent variable when the variables have been organized into table format. Chapter 12

Conditional means of Y. The mean of all scores on Y for each value of X. Chapter 13

Confidence interval. An estimate of a population value in which a range of values is specified. Chapter 7

Confidence level. A frequently used alternate way of expressing alpha, the probability that an interval estimate will not contain the population value. Confidence levels of 90%, 95%, 99%, 99.9%, and 99.99% correspond to alphas of 0.10, 0.05, 0.01, 0.001, and 0.0001, respectively. Chapter 7

Continuous variable. A variable with a unit of measurement that can be subdivided infinitely. Chapter 1

Control variable. In multivariate statistical analysis, a third variable (Z) whose effect on a bivariate relationship is held constant. Chapter 14

Correlation matrix. A table that shows the correlation coefficients between all possible pairs of variables. Chapter 13

Cramer's V. A chi square–based measure of association for nominally measured variables that have been organized into a bivariate table with any number of rows and columns. Chapter 12

Critical region (region of rejection). The area under the sampling distribution that, in advance of the test itself, is defined as including unlikely sample outcomes, given that the null hypothesis is true. Chapter 8

Cumulative frequency. An optional column in a frequency distribution that displays the number of cases within an interval and all preceding intervals. Chapter 2

Cumulative percentage. An optional column in a frequency distribution that displays the percentage of cases within an interval and all preceding intervals. Chapter 2

Data. Information expressed as numbers. Chapter 1

Data reduction. Summarizing many scores with a few statistics. Chapter 1

Deciles. The points that divide a distribution of scores into tenths. Chapter 3

Dependent variable. A variable that is identified as an effect or outcome. The dependent variable is thought to be caused by the independent variable. Chapters 1 and 12

Descriptive statistics. The branch of statistics concerned with (1) summarizing the distribution of a single variable or (2) measuring the relationship between two or more variables. Chapter 1

Deviation. The distance between the score and the mean. Chapter 4

Direct relationship. A multivariate relationship in which the control variable has no effect on the bivariate relationship. Chapter 14

Discrete variable. A variable with a basic unit of measurement that cannot be subdivided. Chapter 1

Dummy variable. A nominal-level variable dichotomized so that it can be used in regression analysis. A dummy variable has two scores, one coded as 0 and the other as 1. Chapter 13

Efficiency. The extent to which the sample outcomes are clustered around the mean of the sampling distribution. Chapter 7

Elaboration. The basic multivariate technique for analyzing variables arrayed in tables. Partial tables are constructed to observe the bivariate relationship in a more detailed or elaborated format. Chapter 14

EPSEM. The Equal Probability of SElection Method for selecting samples. Every element or case in the population must have an equal probability of selection for the sample. Chapter 6

Expected frequency (f_e). The cell frequencies that would be expected in a bivariate table if the variables were independent. Chapter 11

Explained variation. The proportion of all variation in Y that is attributed to the effect of X. Equal to $\Sigma(Y' - \overline{Y})^2$. Chapter 13

Explanation. See **Spurious relationship**. Chapter 14

F ratio. The test statistic computed in step 4 of the ANOVA test. Chapter 10

Five-step model. A step-by-step guideline for conducting tests of hypotheses. A framework that organizes decisions and computations for all tests of significance. Chapter 8

Frequency distribution. A table that displays the number of cases in each category of a variable. Chapter 2

Frequency polygon. See **Line chart**. Chapter 2

Gamma (G). A measure of association for "collapsed" ordinal variables in bivariate table format. Chapter 12

Goodness-of-fit test. An additional use for chi square that tests the significance of the distribution of a single variable. Chapter 11

Histogram. A graph used for interval-ratio variables. Class intervals are represented by contiguous bars of equal width (equal to the class limits), the height of each corresponding to the number (or percentage) of cases in the interval. Chapter 2

Homoscedasticity. The model assumption in the test of significance for Pearson's r that the variance of the Y scores is uniform across all values of X. Chapter 13

Hypothesis. A specific statement, derived from a theory, about the relationship between variables. Chapter 1

Hypothesis testing. Statistical tests that estimate the probability of sample outcomes if assumptions about the population (the null hypothesis) are true. Chapter 8

Independence. The null hypothesis in the chi square test. Two variables are independent if, for all cases, the classification of a case on one variable has no effect on the probability that the case will be classified in any particular category of the second variable. Chapter 11

Independent random samples. Random samples gathered in such a way that the selection of a particular case for one sample has no effect on the probability that any other particular case will be selected for the other samples. Chapter 9

Independent variable. A variable that is identified as a cause. The independent variable is thought to cause the dependent variable. Chapters 1 and 12

Inferential statistics. The branch of statistics concerned with making generalizations from samples to populations. Chapter 1

Interaction. A multivariate relationship wherein a bivariate relationship changes across the categories of the control variable. Chapter 14

Interpretation. See **Intervening relationship**. Chapter 14

Interquartile range (Q). The distance from the third quartile to the first quartile. Chapter 4

Intervening relationship. A multivariate relationship wherein a bivariate relationship becomes substantially weaker after a third variable is controlled for. The independent and dependent variables are linked primarily through the control variable. Chapter 14

Lambda (λ). A proportional reduction in error (PRE) measure for nominally measured variables that have been organized into a bivariate table. Chapter 12

Level of measurement. The mathematical characteristic of a variable and the major criterion for selecting statistical techniques. Variables can be measured at any of three levels, each permitting certain mathematical operations and statistical techniques. The characteristics of the three levels are summarized in Table 1.2. Chapter 1

Line chart. A graph used for interval-ratio variables. Class intervals are represented by dots placed over the midpoints, the height of each corresponding to the number (or percentage) of cases in the interval. All dots are connected by straight lines. Chapter 2

Linear relationship. A relationship between two variables in which the observation points (dots) in the scatterplot can be approximated with a straight line. Chapter 13

Marginals. The row and column subtotals in a bivariate table. Chapter 11

Maximum difference. A way to assess the strength of an association between variables that have been organized into a bivariate table. The maximum difference is the largest difference between column percentages for any row of the table. Chapter 12

Mean. The arithmetic average of the scores. \overline{X} represents the mean of a sample, and μ, the mean of a population. Chapter 3

Mean square estimate. An estimate of the variance calculated by dividing the sum of squares within (SSW) or the sum of squares between (SSB) by the proper degrees of freedom. Chapter 10

Measures of association. Statistics that summarize the strength and direction of the relationship between variables. Chapters 1 and 12

Measures of central tendency. Statistics that summarize a distribution of scores by reporting the most typical or representative value of the distribution. Chapter 3

Measures of variation. Statistics that indicate the amount of variety, or heterogeneity, in a distribution of scores. Chapter 4

Median (Md). The point in a distribution of scores above and below which exactly half of the cases fall. Chapter 3

Midpoint. The point exactly halfway between the upper and lower limits of a class interval. Chapter 2

Mode. The most common value in a distribution or the largest category of a variable. Chapter 3

μ. The mean of a population. Chapter 6

μ_p. The mean of a sampling distribution of sample proportions. Chapter 6

$\mu_{\overline{X}}$. The mean of a sampling distribution of sample means. Chapter 6

Multiple correlation. A multivariate technique for examining the combined effects of more than one independent variable on a dependent variable. Chapter 15

Multiple correlation coefficient (R). A statistic that indicates the strength of the correlation between a dependent variable and two or more independent variables. Chapter 15

Multiple regression. A multivariate technique that breaks down the separate effects of the independent variables on the dependent variable; used to make predictions of the dependent variable. Chapter 15

N_d. In a bivariate table, the number of pairs of cases ranked in different order on two variables. Chapter 12

N_s. In a bivariate table, the number of pairs of cases ranked in the same order on two variables. Chapter 12

Negative association. A bivariate relationship where the variables vary in opposite directions. As one variable increases, the other decreases, and high scores on one variable are associated with low scores on the other. Chapter 12

Nonparametric test. A "distribution-free" test. These tests do not assume a normal sampling distribution. Chapter 11

Nonprobability sample. Any sample that does not meet the EPSEM criterion. Such samples have a variety of uses in social science research but cannot be used to generalize to a population. Chapter 6

Normal curve. A theoretical distribution of scores that is symmetrical, unimodal, and bell-shaped. The standard normal curve always has a mean of 0 and a standard deviation of 1. Chapter 5

Normal curve table. Appendix A; a detailed description of the area between a Z score and the mean of any standardized normal distribution. Chapter 5

Null hypothesis (H_0). A statement of "no difference." In the context of single-sample tests of significance, the null hypothesis states that the population from which the sample was drawn has a certain characteristic or value. Chapter 8

Observed frequency (f_o). The cell frequencies actually observed in a bivariate table. Chapter 11

One-tailed test. A type of hypothesis test used when (1) the direction of the difference can be predicted or (2) concern focuses on outcomes in only one tail of the sampling distribution. Chapter 8

One-way analysis of variance. Applications of ANOVA in which the effect of a single independent variable on a dependent variable is observed. Chapter 10

Outliers. Scores that are very high or very low compared to most scores. Chapter 3

P_s. Any sample proportion. Chapter 6

P_u. Any population proportion. Chapter 6

Parameter. A characteristic of a population. Chapter 6

Partial correlation. A multivariate technique for examining a bivariate relationship while controlling for other variables. Chapter 15

Partial correlation coefficient. A statistic that shows the relationship between two variables while controlling for other variables; $r_{yx.z}$ is the symbol for the partial correlation coefficient when controlling for one variable. Chapter 15

Partial gamma (G_p). A statistic that indicates the strength of the association between two variables after the effects of a third variable have been removed. Chapter 14

Partial slope. In a multiple regression equation, the slope of the relationship between a particular independent variable and the dependent variable while controlling for all other independent variables in the equation. Chapter 15

Partial tables. Tables produced when controlling for a third variable. Chapter 14

Pearson's r. A measure of association for variables that have been measured at the interval-ratio level. Chapter 13

Percent change. A statistic that expresses the magnitude of change in a variable from time 1 to time 2. Chapter 2

Percentage. The number of cases in a category of a variable divided by the number of cases in all categories of the variable, the entire quantity multiplied by 100. Chapter 2

Percentile. A point below which a specific percentage of the cases fall. Chapter 3

Phi. A chi square–based measure of association for nominally measured variables that have been organized into a bivariate table with two rows and two columns (a 2 × 2 table). Chapter 12

Pie chart. A graph used for discrete variables with only a few categories. A circle (the pie) is divided into segments proportional in size to the percentage of cases in each category of the variable. Chapter 2

Pooled estimate. An estimate of the standard deviation of the sampling distribution of the difference in sample means based on the standard deviations of both samples. Chapter 9

Population. The total collection of all cases in which the researcher is interested. Chapter 1

Positive association. A bivariate relationship where the variables vary in the same direction. As one variable increases, the other also increases, and high scores on one variable are associated with high scores on the other. Chapter 12

Post hoc test. A technique for determining which pairs of means are significantly different. Chapter 10

Proportion. The number of cases in one category of a variable divided by the number of cases in all categories of the variable. Chapter 2

Proportional reduction in error (PRE). A logic used in the design of certain measures of association. A PRE measure indicates how much knowledge of the independent variable improves our predictions of the dependent variable. Chapter 12

Quantitative research. Research projects that collect data or information in the form of numbers. Chapter 1

Quartiles. The points that divide a distribution into quarters. Chapter 3

Range (R). The highest score minus the lowest score. Chapter 4

Rate. The number of actual occurrences of some phenomenon or trait divided by the number of possible occurrences per some unit of time. Rates are usually multiplied by some power of 10. Chapter 2

Ratio. The number of cases in one category divided by the number of cases in some other category. Chapter 2

Real class limits. The class intervals of a frequency distribution when stated as continuous categories. Chapter 2

Regression line. The single best-fitting straight line that summarizes the relationship between two variables. Regression lines are fitted to the data points by the least-squares criterion, whereby the line touches all conditional means of Y or comes as close to doing so as possible. Chapter 13

Replication. See **Direct relationship**. Chapter 14

Representative sample. A sample that reproduces the characteristics of the population from which it was drawn. Chapter 6

Research. Any process of gathering information systematically and carefully to answer questions or test

theories. Statistics are useful for research projects that collect numerical information or data. Chapter 1

Research hypothesis (H_1). A statement that contradicts the null hypothesis. In the context of single-sample tests of significance, the research hypothesis says that the population from which the sample was drawn does not have a certain characteristic or value. Chapter 8

Row. The horizontal dimension of a bivariate table, conventionally representing a score on the dependent variable. Chapter 11

Sample. A carefully chosen subset of a population. In inferential statistics, information is gathered from a sample and then generalized to a population. Chapter 1

Sampling distribution. The distribution of a statistic for all possible sample outcomes of a certain size. Under conditions specified in two theorems, the sampling distribution will be normal in shape, with a mean equal to the population value and a standard deviation equal to the population standard deviation divided by the square root of N. Chapter 6

Sampling frame. A list of all the cases in a population. The list can be numbered and used to draw a simple random sample, a systematic random sample or a stratified random sample. Chapter 6

Scatterplot. Graphic display device that depicts the relationship between two variables. Chapter 13

Σ (uppercase Greek letter sigma). "The summation of." Chapter 3

Significance testing. See **Hypothesis testing**. Chapter 8

Simple random sample. A method for choosing cases from a population by which every case and every combination of cases has an equal chance of being included. Chapter 6

Skew. The extent to which a distribution of scores has a few scores that are extremely high (positive skew) or extremely low (negative skew). Chapter 3

Slope (b). The amount of change in one variable per unit change in the other; b is the symbol for the slope of a regression line. Chapter 13

Spearman's rho (r_s). A measure of association for "continuous" ordinal variables. Chapter 12

Specification. See **Interaction**. Chapter 14

Spurious relationship. A multivariate relationship in which a bivariate relationship becomes substantially weaker after a third variable is controlled for. The independent and dependent variables are not causally linked. Rather, both are caused by the control variable. Chapter 14

Standard deviation. The statistic computed by summing the squared deviations of the scores around the mean, dividing by N, and, finally, taking the square root of the result. The most important and useful descriptive measure of variation; s represents the standard deviation of a sample; σ represents the standard deviation of a population. Chapter 4

σ_{p-p}. Symbol for the standard deviation of the sampling distribution of the differences in sample proportions. Chapter 9

$\sigma_{\bar{x}-\bar{x}}$. Symbol for the standard deviation of the sampling distribution of the differences in sample means. Chapter 9

Standard error of the mean. The standard deviation of a sampling distribution of sample means. Chapter 6

Standardized partial slope (beta-weight or β). The slope of the relationship between a particular independent variable and the dependent variable when all scores have been normalized and all other variables are controlled. Chapter 15

Stated class limits. The class intervals of a frequency distribution when stated as discrete categories. Chapter 2

Statistics. A set of mathematical techniques for organizing and analyzing data. Chapter 1

Stratified sample. A method of sampling by which cases are selected from sublists of the population. Chapter 6

Student's t distribution. A distribution used to find the critical region for tests of sample means when s is unknown and sample size is small. Chapter 8

Sum of squares between (SSB). The sum of the squared deviations of the sample means from the overall mean, weighted by sample size. Chapter 10

Sum of squares within (SSW). The sum of the squared deviations of scores from the category means. Chapter 10

Systematic sampling. A method of sampling by which the first case from a list of the population is randomly selected. Thereafter, every kth case is selected. Chapter 6

t(critical). The t score that marks the beginning of the critical region of a t distribution. Chapter 8

t(obtained). The test statistic computed in step 4 of the five-step model. The sample outcome expressed as a t score. Chapter 8

Test statistic. The value computed in step 4 of the five-step model that converts the sample outcome into either a t score or a Z score. Chapter 8

Theory. A generalized explanation of the relationship between two or more variables. Chapter 1

Total sum of squares (SST). The sum of the squared deviations of the scores from the overall mean. Chapter 10

Total variation. The spread of the Y scores around the mean of Y. Equal to $\Sigma(Y - \bar{Y})^2$. Chapter 13

Two-tailed test. A type of hypothesis test used when (1) the direction of the difference cannot be predicted or (2) concern focuses on outcomes in both tails of the sampling distribution. Chapter 8

Type I error (alpha error). The probability of rejecting a null hypothesis that is, in fact, true. Chapter 8

Type II error (beta error). The probability of failing to reject a null hypothesis that is, in fact, false. Chapter 8

Unexplained variation. The proportion of the total variation in Y that is not accounted for by X. Equal to $\Sigma(Y - Y')^2$. Chapter 13

Variable. Any trait that can change values from case to case. Chapter 1

Variance. The sum of the squared deviations of the scores around the mean, divided by N. A measure of variation used primarily in inferential statistics and also in correlation and regression techniques; s^2 represents the variance of a sample; σ^2 represents the variance of a population. Chapter 4

Variation. The amount of variety, or heterogeneity, in a distribution of scores. Chapter 4

X. Symbol used for any independent variable. Chapter 12

X_i. Any score in a distribution. Chapter 3

Y. Symbol used for any dependent variable. Chapter 12

Y intercept (a). The point where the regression line crosses the Y axis. Chapter 13

Y'. Symbol for predicted score on Y. Chapter 13

Z. Symbol for any control variable. Chapter 14

Z scores. Standard scores; the way scores are expressed after they have been standardized to the theoretical normal curve. Chapter 5

Z(critical). The Z score that marks the beginning of the critical region on a Z distribution. Chapter 8

Z(obtained). The test statistic computed in step 4 of the five-step model. The sample outcomes expressed as a Z score. Chapter 8

Zero-order correlation. Correlation coefficient for bivariate relationships. Chapter 15

Index

Note: Page numbers followed by "n" refer to footnotes.

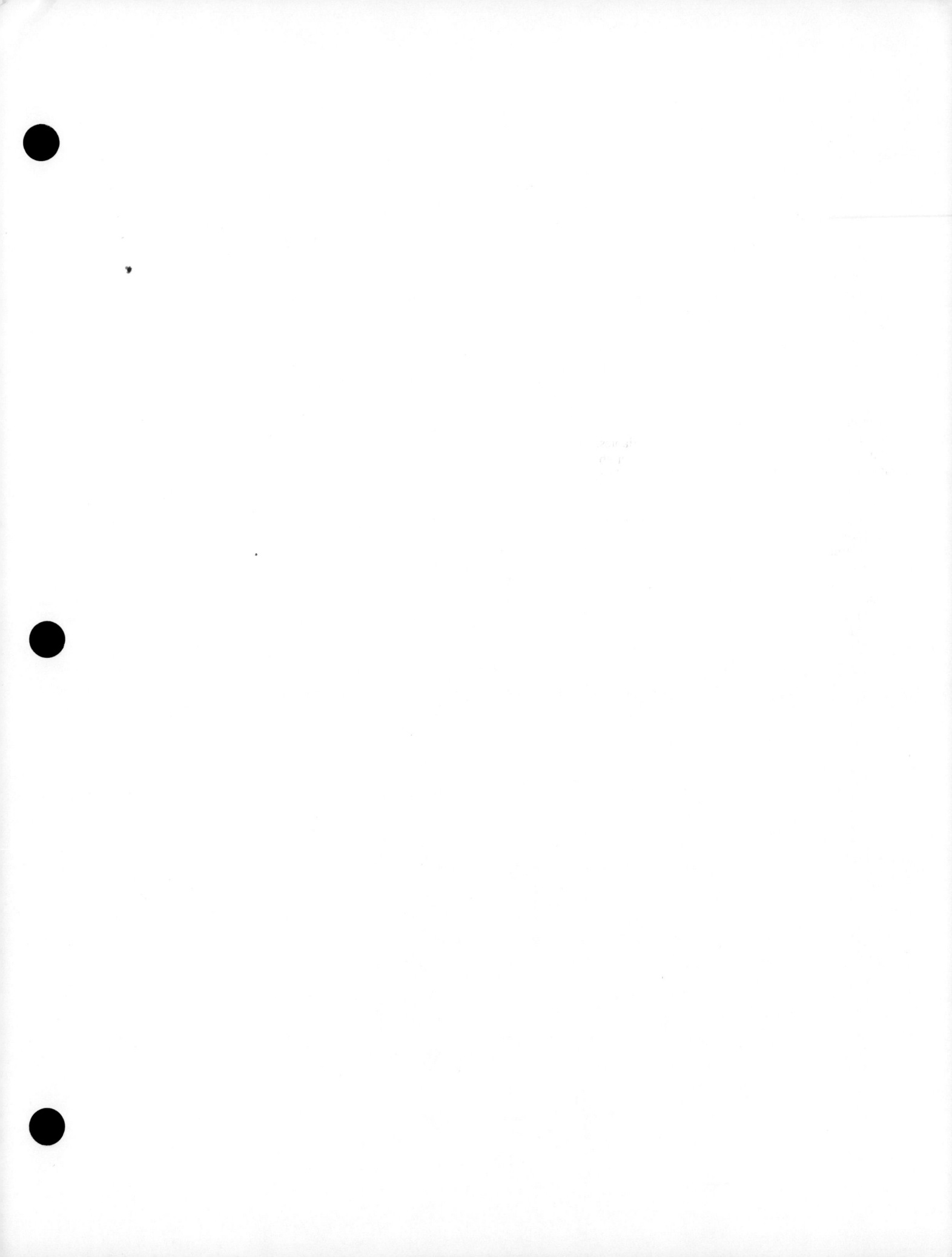

FREQUENTLY USED FORMULAS

CHAPTER 2

Proportion

$$p = \frac{f}{N}$$

Percentage

$$\% = \left(\frac{f}{N}\right) \times 100$$

CHAPTER 3

Mean

$$\overline{X} = \frac{\Sigma(X_i)}{N}$$

CHAPTER 4

Standard deviation

$$s = \sqrt{\frac{\Sigma(X_i - \overline{X})^2}{N}}$$

CHAPTER 5

Z scores

$$Z_i = \frac{X_i - \overline{X}}{s}$$

CHAPTER 7

Confidence interval for a sample mean

$$\text{c.i.} = \overline{X} \pm Z\left(\frac{s}{\sqrt{N-1}}\right)$$

Confidence interval for a sample proportion

$$\text{c.i.} = P_s \pm Z\sqrt{\frac{P_u(1 - P_u)}{N}}$$

CHAPTER 8

Means

$$Z(\text{obtained}) = \frac{\overline{X} - \mu}{s/\sqrt{N-1}}$$

Proportions

$$Z(\text{obtained}) = \frac{P_s - P_u}{\sqrt{P_u(1 - P_u)/N}}$$

CHAPTER 9

Means

$$Z(\text{obtained}) = \frac{(\overline{X}_1 - \overline{X}_2)}{\sigma_{\overline{x}-\overline{x}}}$$

Standard deviation of the sampling distribution for sample means

$$\sigma_{\overline{x}-\overline{x}} = \sqrt{\frac{s_1^2}{N_1 - 1} + \frac{s_2^2}{N_2 - 1}}$$

Pooled estimate of population proportion

$$P_u = \frac{N_1 P_{s1} + N_2 P_{s2}}{N_1 + N_2}$$

Standard deviation of the sampling distribution for sample proportions

$$\sigma_{p-p} = \sqrt{P_u(1 - P_u)}\,\sqrt{(N_1 + N_2)/N_1 N_2}$$

Proportions

$$Z(\text{obtained}) = \frac{(P_{s1} - P_{s2})}{\sigma_{p-p}}$$

CHAPTER 10

Total sum of squares

$$\text{SST} = \Sigma X^2 - N\overline{X}^2$$

Sum of squares between

$$\text{SSB} = \Sigma N_k(\overline{X}_k - \overline{X})^2$$

Sum of squares within

$$\text{SSW} = \text{SST} - \text{SSB}$$

Degrees of freedom for SSW

$$\text{dfw} = N - k$$

Degrees of freedom for SSB

$$\text{dfb} = k - 1$$

Mean square within

$$\text{MSW} = \frac{\text{SSW}}{\text{dfw}}$$

Mean square between

$$\text{MSB} = \frac{\text{SSB}}{\text{dfb}}$$

F ratio

$$F = \frac{\text{MSB}}{\text{MSW}}$$

CHAPTER 11

Chi square

$$\chi^2(\text{obtained}) = \sum \frac{(f_o - f_e)^2}{f_e}$$

CHAPTER 12

Phi

$$\phi = \sqrt{\frac{\chi^2}{N}}$$

Cramer's V

$$V = \sqrt{\frac{\chi^2}{(N)(\text{Minimum of } r - 1, c - 1)}}$$

Lambda

$$\lambda = \frac{E_1 - E_2}{E_1}$$

CHAPTER 13

Gamma

$$G = \frac{N_s - N_d}{N_s + N_d}$$

Spearman's rho

$$r_s = 1 - \frac{6\sum D^2}{N(N^2 - 1)}$$

CHAPTER 14

Least-squares regression line

$$Y = a + bX$$

Slope

$$b = \frac{\sum(X - \overline{X})(Y - \overline{Y})}{\sum(X - \overline{X})^2}$$

Y intercept

$$a = \overline{Y} - b\overline{X}$$

Pearson's r

$$r = \frac{\sum(X - \overline{X})(Y - \overline{Y})}{\sqrt{[\sum(X - \overline{X})^2][\sum(Y - \overline{Y})^2]}}$$

CHAPTER 16

Partial correlation coefficient

$$r_{yx.z} = \frac{r_{yx} - (r_{yz})(r_{xz})}{\sqrt{1 - r_{yz}^2}\sqrt{1 - r_{xz}^2}}$$

Least-squares multiple regression line

$$Y = a + b_1X_1 + b_2X_2$$

Partial slope for X_1

$$b_1 = \left(\frac{s_y}{s_1}\right)\left(\frac{r_{y1} - r_{y2}r_{12}}{1 - r_{12}^2}\right)$$

Partial slope for X_2

$$b_2 = \left(\frac{s_y}{s_2}\right)\left(\frac{r_{y2} - r_{y1}r_{12}}{1 - r_{12}^2}\right)$$

Y intercept

$$a = \overline{Y} - b_1\overline{X}_1 - b_2\overline{X}_2$$

Beta-weight for X_1

$$b_1^* = b_1\left(\frac{s_1}{s_y}\right)$$

Beta-weight for X_2

$$b_2^* = b_2\left(\frac{s_2}{s_y}\right)$$

Standardized least-squares regression line

$$Z_y = b_1^*Z_1 + b_2^*Z_2$$

Coefficient of multiple determination

$$R^2 = r_{y1}^2 + r_{y2.1}^2(1 - r_{y1}^2)$$

GLOSSARY OF SYMBOLS

The number in parentheses is the chapter in which the symbol is introduced.

a	Point at which the regression line crosses the Y axis (13)
ANOVA	Analysis of variance (10)
b	Slope of the regression line (13)
b_i	Partial slope of linear relationship between the ith independent variable and the dependent variable (15)
b_i^*	Standardized partial slope of linear relationship between the ith independent variable and the dependent variable (15)
df	Degrees of freedom (8)
f	Frequency (2)
F	F ratio (10)
f_e	Expected frequency (11)
f_o	Observed frequency (11)
G	Gamma (12)
G_p	Partial gamma (14)
H_0	Null hypothesis (8)
H_1	Research or alternate hypothesis (8)
Md	Median (3)
Mo	Mode (3)
N	Number of cases (2)
N_d	Number of pairs of cases ranked in different order on two variables (12)
N_s	Number of pairs of cases ranked in the same order on two variables (12)
%	Percentage (2)
p	Proportion (2)
p_s	Sample proportion (7)
p_u	Population proportion (7)
PRE	Proportional reduction in error (12)
Q	Interquartile range (4)
r	Pearson's correlation coefficient (13)
r^2	Coefficient of determination (13)
R	Range (4)
r_s	Spearman's rho (12)
$r_{xy.z}$	Partial correlation coefficient (15)
R^2	Coefficient of multiple determination (15)

s	Sample standard deviation (4)
s^2	Sample variance (4)
SSB	Sum of squares between (10)
SST	Total sum of squares (10)
SSW	Sum of squares within (10)
t	Student's t score (8)
V	Cramer's V (12)
X	An independent variable (12)
\bar{X}	Mean of a sample (3)
X_i	Any score in a distribution (3)
Y	A dependent variable (12)
Y'	A predicted score on Y (13)
Z score	Standard scores (5)
Z	A control variable (14)

GREEK LETTERS

α	Probability of Type I error (8)
β	Probability of Type II error (8)
λ	Lambda (12)
μ	Mean of a population (3)
μ_p	Mean of a sampling distribution of sample proportions (6)
$\mu_{\bar{x}}$	Mean of a sampling distribution of sample means (6)
ρ	Pearson's correlation coefficient for a population (13)
σ	Population standard deviation (4)
σ_p	Standard deviation of a sampling distribution of sample proportions (6)
σ_{p-p}	Standard deviation of a sampling distribution of differences in sample proportions (9)
$\sigma_{\bar{x}}$	Standard deviation of a sampling distribution of sample means (6)
$\sigma_{\bar{x}-\bar{x}}$	Standard deviation of a sampling distribution of differences in sample means (9)
σ^2	Population variance (4)
Σ	The summation of (3)
ϕ	Phi (12)
χ^2	Chi square (11)
χ_c^2	Chi square corrected by Yates' correction (11)